WORLD-CLASS DEFENSE SYSTEMS

BMY

25.00

JANE'S
ARMOUR AND ARTILLERY

TWELFTH EDITION

EDITED BY
CHRISTOPHER F FOSS

1991-92

ISBN 0 7106 0964 7
JANE'S DEFENCE DATA
"Jane's" is a registered trade mark

In the USA and its dependencies
Jane's Information Group Inc, 1340 Braddock Place, Suite 300
Alexandria, Virginia 22314-1651, USA

Printed and bound in the United Kingdom by Biddles Ltd, Guildford and King's Lynn

'M109L' Self-Propelled Howitzer

'105/14' Pack Howitzer

155/41 'PALMARIA' Self-Propelled Howitzer

SGR 1169 - PHOTOS: SG RESTANI

Contents

Alphabetical list of advertisers

Classified list of advertisers

The following companies have informed us that they are involved in the field of manufacture listed below.

Aerospace systems
Armscor
SFIM

Air defence systems
Raufoss

Air bombs
Defex

Ammunition
Armscor
BPD
Empresa Nacional Santa Barbara
Raufoss

Ammunition (armoured cars)
Empresa Nacional Santa Barbara
Raufoss

Ammunition loading
Creusot-Loire Industrie
Empresa Nacional Santa Barbara
Pakistan Ordnance Factories

Ammunition loading plants
Empresa Nacional Santa Barbara

Ammunition (tank and artillery)
BPD
Defex
Empresa Nacional Santa Barbara
Pakistan Ordnance Factories
Raufoss

Anti-aircraft guns
Empresa Nacional Santa Barbara
OTO Melara

Anti-aircraft guns (self-propelled)
Creusot-Loire Industrie
Empresa Nacional Santa Barbara
OTO Melara

Anti-aircraft guns (towed)
Creusot-Loire Industrie
Empresa Nacional Santa Barbara

Anti-aircraft sights
SFIM

Anti-aircraft sights (gunner sights)
NANOQUEST
SFIM

Anti-aircraft systems
OTO Melara

Armament systems
Armscor
Creusot-Loire Industrie
Defex
Empresa Nacional Santa Barbara
SFIM

Armoured cars & armoured personnel carriers
Creusot-Loire Industrie
Defex
Empresa Nacional Santa Barbara
OTO Melara
Steyr-Daimler-Puch

Armoured fighting vehicles
Armscor
Blohm + Voss
BMY
Creusot-Loire Industrie
Empresa Nacional Santa Barbara
OTO Melara
Steyr-Daimler-Puch

Armoured turrets and cupolas
Blohm + Voss
Creusot-Loire Industrie
Defex
Empresa Nacional Santa Barbara
OTO Melara
Steyr-Daimler-Puch

Artillery accuracy equipment
NANOQUEST

Artillery equipment
Empresa Nacional Santa Barbara

Artillery training equipment
BMY

Artillery systems
Armscor

Auxiliary military equipment
Empresa Nacional Santa Barbara

Bombs
Defex
Empresa Nacional Santa Barbara
Pakistan Ordnance Factories

Brakes and braking systems
Steyr-Daimler-Puch

Camouflage nets/concealment sets
Defex

Carbide cutting tools
Pakistan Ordnance Factories

Coastal defence guns
Creusot-Loire Industrie
Empresa Nacional Santa Barbara
OTO Melara

Combat vehicles
BMY
Creusot-Loire Industrie
Empresa Nacional Santa Barbara
Steyr-Daimler-Puch

Commander's day/night sights
SFIM

Conversion & modernisation of armoured combat vehicles
BMY
Creusot-Loire Industrie
Defex
Empresa Nacional Santa Barbara
SFIM
Steyr-Daimler-Puch

Cupolas
Blohm + Voss
Creusot-Loire Industrie
Empresa Nacional Santa Barbara

Defence vehicles
Armscor
Blohm + Voss
Empresa Nacional Santa Barbara

Design service military vehicles components
Steyr-Daimler-Puch

Diesel engines
Steyr-Daimler-Puch

Electro-optics
Armscor
Defex
NANOQUEST
SFIM

Explosives military & commercial
Defex
Empresa Nacional Santa Barbara

Field artillery
Armscor
BMY
Empresa Nacional Santa Barbara
OTO Melara

Fire-control equipment
Defex
SFIM

Fire-control systems
Zengrange

Fuzes
Defex
Empresa Nacional Santa Barbara
Pakistan Ordnance Factories

Gun control systems
SFIM

Gun systems
Armscor
BMY
Creusot-Loire Industrie
Empresa Nacional Santa Barbara

Gyro-stabilised sights
SFIM

Gyro-stabilised for combat vehicles
NANOQUEST

Hand-held computers
Zengrange

Howitzers
BMY
Creusot-Loire Industrie
Defex
Empresa Nacional Santa Barbara
OTO Melara

Hulls and turrets
Blohm + Voss
BMY
Creusot-Loire Industrie
Empresa Nacional Santa Barbara
Steyr-Daimler-Puch

Hydraulic controls/systems
Steyr-Daimler-Puch

Infantry combat vehicles
Blohm + Voss
Creusot-Loire Industrie
Empresa Nacional Santa Barbara
OTO Melara
Steyr-Daimler-Puch

Infra-red equipment
NANOQUEST
SFIM

Infra-red/panoramic sighting devices
NANOQUEST
SFIM

Internal security vehicles
Empresa Nacional Santa Barbara
Steyr-Daimler-Puch

Laser rangefinders
Defex
NANOQUEST
Pakistan Ordnance Factories

Light ammunition
Defex
Empresa Nacional Santa Barbara
Pakistan Ordnance Factories

Mechanical time fuzes for artillery shells
Pakistan Ordnance Factories

Mechanised infantry combat vehicles
Creusot-Loire Industrie
Empresa Nacional Santa Barbara

Medium battle tank engines
Empresa Nacional Santa Barbara

Medium & light battle tanks
Blohm + Voss
Creusot-Loire Industrie
Empresa Nacional Santa Barbara
OTO Melara

Medium & light battle tank turrets
Blohm + Voss
Creusot-Loire Industrie
Empresa Nacional Santa Barbara
OTO Melara

Military computers & microprocessors
Zengrange

Military electronics
SFIM

Military vehicles
Blohm + Voss
BMY
Creusot-Loire Industrie
Defex
Empresa Nacional Santa Barbara

Mine resistant vehicles
Armscor

Missiles
Armscor
OTO Melara
Pakistan Ordnance Factories

Missile components
BPD
Pakistan Ordnance Factories
Raufoss
SFIM

Mortar cannons
Defex
Empresa Nacional Santa Barbara

Multi-rocket launchers (towed & self-propelled)
BPD
Creusot-Loire Industrie
Defex
Empresa Nacional Santa Barbara

Multi-role artillery fuzes
Empresa Nacional Santa Barbara

Munitions & ordnance
BPD
Defex
Empresa Nacional Santa Barbara
Pakistan Ordnance Factories

Muzzle boresights
NANOQUEST

NBC systems
Zengrange

Night vision equipment
Armscor
Defex
NANOQUEST
Pakistan Ordnance Factories
SFIM

Packaging (wooden boxes and fibre containers) for ammunition fuzes, howitzers, mortars & hand grenades
Pakistan Ordnance Factories

Passive night driving vehicles
Defex

Personnel transport vehicles
Creusot-Loire Industrie
Defex
Empresa Nacional Santa Barbara

Portable anti-tank weapons
Defex
Pakistan Ordnance Factories

Production & marketing of complete integrated defence systems
BMY
Empresa Nacional Santa Barbara

Propellants
Armscor
Empresa Nacional Santa Barbara
Pakistan Ordnance Factories

Pyrotechnics
Armscor
Defex
Pakistan Ordnance Factories
Raufoss

Rangefinders
NANOQUEST

Reconnaissance vehicles
Creusot-Loire Industrie
Empresa Nacional Santa Barbara
Steyr-Daimler-Puch

Research and development
Armscor
Blohm + Voss
BMY
Creusot-Loire Industrie
Empresa Nacional Santa Barbara
Pakistan Ordnance Factories
Steyr-Daimler-Puch

Rocket launchers
Armscor
Creusot-Loire Industrie
Empresa Nacional Santa Barbara
Pakistan Ordnance Factories

Rockets, rocket systems
BPD
Pakistan Ordnance Factories
Raufoss

Scout cars
OTO Melara

Security vehicles
Empresa Nacional Santa Barbara
OTO Melara

Self-propelled anti-aircraft guns
Creusot-Loire Industrie
Empresa Nacional Santa Barbara
OTO Melara
Pakistan Ordnance Factories

Self-propelled artillery
Armscor
BMY
Empresa Nacional Santa Barbara
OTO Melara

Self-propelled guns & howitzers
BMY
Creusot-Loire Industrie
Empresa Nacional Santa Barbara
OTO Melara

Self-propelled tank destroyers
Creusot-Loire Industrie
Empresa Nacional Santa Barbara
OTO Melara
Steyr-Daimler-Puch

Sights
NANOQUEST
SFIM

Simulators
Defex

Small arms, anti-aircraft gun
Empresa Nacional Santa Barbara

Systems integration
Empresa Nacional Santa Barbara
SFIM

Tank guns
Empresa Nacional Santa Barbara
OTO Melara
SFIM

Tank sights
NANOQUEST

Tank tracks (for all types of military vehicles)
Empresa Nacional Santa Barbara
George Blair

Tanks (MBTs heavy)
Blohm + Voss

Tanks (MBTs, medium & light)
Blohm + Voss
Creusot-Loire Industrie
Empresa Nacional Santa Barbara
OTO Melara
Steyr-Daimler-Puch

Tanks (MBTs, medium & light turrets)
Blohm + Voss
Creusot-Loire Industrie
Empresa Nacional Santa Barbara
OTO Melara
Steyr-Daimler-Puch

Target designations
SFIM

Technical assistance (operational and maintenance: tank and artillery)
Creusot-Loire Industrie
Steyr-Daimler-Puch

Technical support services
BPD
Creusot-Loire Industrie
Steyr-Daimler-Puch

Thermal imaging systems
NANOQUEST
SFIM

Thermal imaging systems for use on armoured vehicles
NANOQUEST
SFIM

Towed anti-aircraft guns
Creusot-Loire Industrie
Empresa Nacional Santa Barbara

Towed anti-tank guns
Creusot-Loire Industrie
Empresa Nacional Santa Barbara

Towed guns and howitzers
Armscor
Creusot-Loire Industrie
Empresa Nacional Santa Barbara

Tracked armoured personnel carriers
Creusot-Loire Industrie
Empresa Nacional Santa Barbara
Steyr-Daimler-Puch

Training aids
Steyr-Daimler-Puch

Transmissions
Defex

Turrets for armoured fighting vehicles
Blohm + Voss
BMY
Creusot-Loire Industrie
Defex
Empresa Nacional Santa Barbara
OTO Melara

Vehicles, military
Creusot-Loire Industrie
Defex
Empresa Nacional Santa Barbara

Vehicles, military (track/wheel)
BMY
Creusot-Loire Industrie
Empresa Nacional Santa Barbara
Steyr-Daimler-Puch

Weapon fuzing & arming
Defex
Empresa Nacional Santa Barbara

Wheeled armoured vehicles
Armscor
Creusot-Loire Industrie
Defex
Empresa Nacional Santa Barbara
OTO Melara

340

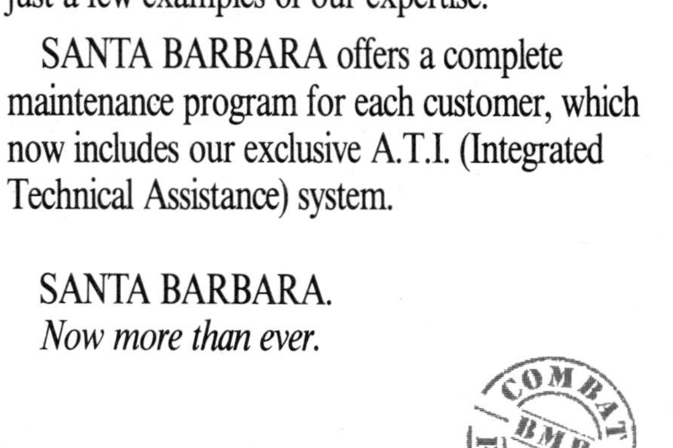

One of the prototypes of the Swedish Army's new Combat Vehicle 90 armoured vehicles being put through its paces during winter trials. This vehicle has been developed by Bofors and Hägglund Vehicle to meet the Swedish Army's unique operational requirements, with the first production order being placed in 1991. This vehicle is fitted with a two-man turret armed with a Bofors 40mm L/70 gun which can fire a variety of ammunition types including APFSDS-T

JANE'S ARMOUR AND ARTILLERY 1991-92

Jane's Information Group, 163 Brighton Road, Coulsdon, Surrey CR5 2NH, UK
Jane's Information Group Inc, 1340 Braddock Place, Suite 300
Alexandria, Virginia 22314-1615, USA

Foreword

Since the last edition of *Jane's Armour and Artillery* was published in October 1990, the world has once again seen dramatic events which have, and will continue to have, a major impact on defence forces and their supporting industries.

Early in 1991 the Allied Coalition forces successfully evicted Iraqi forces from Kuwait following a month long aerial bombardment of key targets in Iraq and Kuwait.

The aerial bombardment effectively paralysed much of the Iraqi command, control and communications facilities, supply lines, air defence weapons and their associated radars, the defence production base, civil infrastructure and the ability of Iraq to see exactly what the Allied Coalition forces future intentions were.

Although the air campaign was highly successful, a ground campaign was still required to retake Kuwait as only armour and infantry units, backed up by artillery, engineers and close air support, can take and hold ground.

The ground war lasted just 100 hours with little loss of life or equipment on the Allied Coalition side, although the Iraqi forces suffered massive losses in both men and equipment.

Although many Iraqi units quickly surrendered when faced by the massive Allied Coalition firepower, a number of Iraqi units, especially the Republican Guard, put up fierce resistance and in some cases even counter-attacked.

Even though detailed analysis of the conflict has yet to be released, there are already a number of lessons to be learned, or in some cases relearned.

The success of the Allied Coalition forces in the Middle East has, in some quarters, prompted the questioning of the whole rationale of developing new weapon systems. If the current systems proved to be so highly effective, why continue to develop new ones?

It should be remembered, however, that many of the weapons that proved so effective were developed a long time ago. For example, the first prototype of the US M1 series Abrams MBT was completed as far back as 1978 and the M2 Bradley IFV the early 1970s.

Much of the equipment used by the Iraqi Army was well over 20 years old and their latest MBTs, the Soviet T-72 series, first entered service with the Soviet Army in 1972/73. Even when the Soviet Union has exported its latest equipment, it is often of inferior quality to that being supplied to the Soviet Army. In the case of MBTs, in the past there have been broadly three types of armour packages: Soviet Army, Warsaw Pact and export.

While Operation Desert Storm was very much a high technology war, especially as far as the United States forces were concerned, training and motivation played a key part and it was demonstrated that highly professional forces were much more effective than the Iraqi Army, many of whom had been conscripted.

Much of the combat took place at night or in poor visibility, and the effectivness of thermal night vision equipment fitted to vehicles such as the General Dynamics, Land Systems Division, M1A1 Abrams and the British Vickers Defence Systems Challenger 2 MBTs gave them a decisive edge over Iraqi vehicles, most of which were fitted with image intensification or infra-red night vision equipment with limited capabilities.

At present, thermal night vision devices are normally found only on MBTs, but in the future they will be much more widely deployed on the battlefield, especially on reconnaissance vehicles and infantry fighting vehicles that form a key part of the combined arms team.

In a number of cases reconnaissance units, which even in the days of satellite and electronic surveillance are still the eyes and ears of the army at brigade and divisional level, only had image intensification night vision devices and MBTs were often carrying out the reconnaissance roles, as not only did they have better thermal night vision devices but also improved battlefield survivability and agility.

Friendly fire was a major cause of casualties in the British and United States ground forces, even though everything possible was done to reduce this.

The speed of the Allied Coalition forces advance, the lack of any clearly identifiable terrain features and poor vehicle recognition, especially as both sides used the same vehicles in some cases, coupled with the long ranges of some target engagements, all contributed to an excessive loss of life on the Allied Coalition side.

In the future, more emphasis will be placed on developing identification friend or foe (IFF) devices to reduce friendly fire casualties. Indeed, an IFF system could well form part of a complete integrated vehicle protection system that would also include threat displays, sensors and decoy launchers as well as a variety of obscurants.

The British and French armies both fitted many of their armoured vehicles with additional armour protection for Operation Desert Storm. In the case of the former this included explosive reactive armour and passive armour for the Challenger 1 MBT and passive armour for the Warrior mechanised combat vehicle. These armour packages were designed, developed and dispatched to the Middle East by Royal Ordnance and Vickers Defence Systems at very short notice.

The question must be asked, why were these armour packages not previously developed for vehicles deployed by the British Army in Germany as the weapons they would have faced if the Soviet Western Group of Forces had ever attacked NATO were far superior to those fielded by Iraq?

While most MBTs and some armoured personnel carriers/infantry fighting vehicles have an NBC system, many of the key supporting vehicles, such as self-propelled artillery systems often lack such a system. It is essential that future self-propelled artillery systems, and their supporting ammunition resupply vehicles, will have to be fitted with an NBC system.

The high ambient temperatures encountered in the Middle East showed that in addition to having an NBC system, armoured vehicles also had to have an effective air-conditioning system capable of functioning for long periods without breaking down.

The British and United States armies spend a considerable amount of time training, but much of this training takes place within a confined space under very strict controls. Often units have to use the same training ground and therefore become very familiar with the terrain and its features. In such circumstances there is not a pressing need for a land navigation system, but in the featureless terrain of the Middle East the urgent operational requirement for such a system soon became apparent and a number of vehicles were fitted with a Global Position System or other land navigation systems.

In times of peace, there is no pressing operational requirement for such items as additional armour protection for improved battlefield survivability, NBC system, air-conditioning system, land navigation system or an effective IFF system. Moreover, the installation of these systems into a modern AFV can increase the unit cost by a considerable amount, thus reducing the total number of systems that can be procured for a given amount.

During peacetime there can be a great temptation to reduce the cost of weapon systems by reducing some of its key subsystems. In time of actual conflict the existence of some of these key subsystems will decide the fate and survival of the weapon system and its crew.

The same applies to the adequate funding of ammunition and spare parts. When defence cuts are made, or budgets are overspent, the resources devoted to ammunition and consumable spare parts is often the first to be reduced, with often a major impact on the production base.

The Warsaw Pact has now ceased to exist and Soviet troops and equipment have started to withdraw from Eastern Europe, and in some cases have already left.

The failure of the coup in the Soviet Union in August 1991 is likely to hasten this process even further and it is now possible that we will see even more dramatic cuts in the Soviet Armed Forces and their extensive industrial base.

The decision by central European countries to reduce their armed forces has meant that there are considerable quantities of equipment now surplus to requirements.

In the case of some of the NATO countries, this equipment, such as M48, M60 and Leopard 1 MBTs, armoured personnel carriers, artillery systems and rocket launchers, is being passed to other NATO countries such as Greece, Norway, Portugal, Spain and Turkey. There may well be a potential for some upgrading of this equipment if its effectiveness is to be maintained.

The already dramatic reduction in defence expenditure has forced the closure of a number of leading defence equipment manufacturers, while others have already decided to pull out of the defence market or reduce their dependence on defence contracts.

In the future, there are likely to be more mergers and takeovers with the consequent loss of both the research and development and production bases. For the customer, there will be less choice in the market place.

While some defence contracts are on a cost plus fee basis, many of the more recent ones are on a fixed price basis and some companies firmly believe that it is not in their interest to enter into such contracts as the financial penalties for failure are far too great for many companies to bear.

In some countries the defence industrial base has already been eroded to such a degree that the country is no longer self-sufficient, even in ammunition.

Some of the recent developments in the fields of MBTs, light tanks and artillery systems are highlighted.

Main Battle Tanks

The Brazilian defence industry has been hard hit by world events and

ɔp left: MINI-RPV Surveillance System. Top right: ROOIKAT Integrated Weapons System AFV for High Mobility Warfare.
ottom left: ROOIVALK Combat Support Helicopter. Bottom right: VALKIRI Multiple Artillery Rocket System

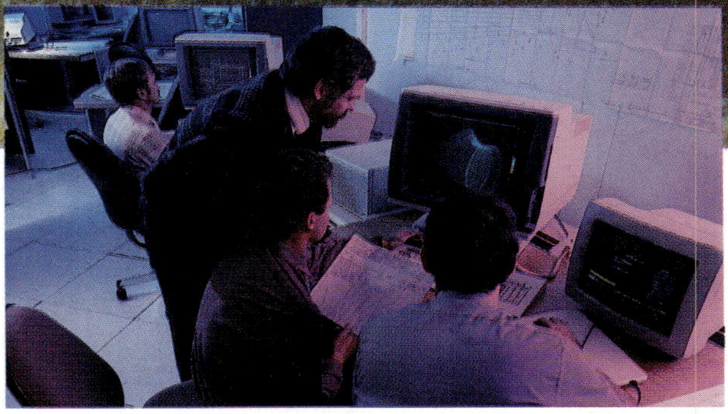

BPD
for ARTILLERY
the entire range of cartriges between 27 and 203 mm
IN ANY CONFIGURATION
special types
• extended range
• cargo rounds
• APFSDS
• illuminating
• smoke
• subammunition-carrier mortar bombs
BPD DIFESA E SPAZIO
00034 COLLEFERRO (ROME) ITALY
TEL. (6) 97891
FAX (6) 97892205
TLX 611434 BPD CF I

ENGESA has still to find a launch customer for its private venture EE-T1 Osorio MBT. In recent years Iraq has been a major customer of ENGESA and the loss of this market has caused severe problems, not only for them, but also for AVIBRAS who supplied large numbers of multiple rocket launchers to Iraq.

While some manufacturers have sold significant numbers of systems that have been developed as a private venture, many of the more recent private venture vehicles have fallen by the wayside as export customers do not like to place orders for equipment that has not been ordered by the home army.

General Dynamics, Land Systems Division, has now shipped the first M1A1 MBTs to the Egyptian Army. Most of the 555 vehicles due to be delivered will be assembled in Egypt. In addition 700 M60A1 MBTs from the United States are being provided free of charge, with Egypt just paying the cost of delivery. In order for the total number of its MBT fleet to remain constant, Egypt must destroy an equivalent number of its older Soviet-supplied vehicles in return.

GIAT Industries will deliver the first production Leclerc MBT to the French Army in December 1991, but at present firm orders have only been placed for 26 vehicles. Although the original requirement was for some 1400 Leclerc MBTs, figures of well under 1000 vehicles are now being mentioned, and any further significant reduction in the number of units to be built is bound to have a major impact on unit costs.

GIAT Industries are already marketing the Leclerc in Europe and the Middle East and as it has been selected by the French Army they have high hopes that it will capture a considerable part of the export market for MBTs in the coming years.

Production of the Leopard 2 continues in Germany but under present plans production will be completed in 1992 and there are no plans for a vehicle to succeed it. There is a possibility that earlier Leopard 2s could be brought up to the Leopard 2 (Improved) standard which has a significant increase in battlefield survivability.

Due to German Government regulations, the Leopard 2 can only be exported to a small number of countries and at present the only potential customer is Sweden which has a requirement for a new MBT to replace the ageing S-tanks.

The new Italian MBT, the C1 Ariete, is now moving towards the production phase after many delays, but it remains to be seen as to how many vehicles will be built.

The Type 90 MBT is now in production for the Japanese Ground Self-Defence Force and it has a significant increase in armour, mobility and firepower over the older Type 74 MBT. Japan has also developed a new ARV based on components of the Type 90 MBT so both vehicles should enter service at roughly the same time.

Sweden has decided not to proceed with the development of the locally designed MBT 2000 but will purchase an existing MBT off the shelf. The contenders for this include the French Leclerc, German Leopard 2 (already trialled), the UK Challenger 2 and the US M1A1 (two vehicles already trialled). As the Swedish contract is one of the few remaining MBT contests, competition is expected to be fierce.

The latest version of the Soviet T-80 MBT, the T-80U, was first seen in 1989. This has a new engine, increased armour protection and a new turret, all of which contribute to a significant increase in its capabilities. So far this does not appear to have been exported.

There is still no sign of a successor to the T-80 and given the current economic and political problems of the Soviet Union, there must now be considerable doubt as to whether a brand new Soviet MBT will enter production in the immediate future. However, it is probable that prototypes of new Soviet MBTs are currently being tested.

In the UK, the British Army's Chieftain Replacement Programme is finally over as in mid-1991 the Challenger 2 was selected and Vickers Defence Systems were awarded a contract for about 130 vehicles. Overseas marketing of the Challenger 2 is now building up and one of the prototypes will be going to the Middle East for trials next year. At the same time it was announced that the existing Challenger 1 fleet was to be upgraded in a number of key areas. It is perhaps a lost opportunity that the Challenger 1s are not to be upgraded to the full Challenger 2 standard as this would have provided much more work for the British tank industry and also the British Army would have a common MBT fleet with the obvious training and logistical advantages.

Vickers Defence Systems are in full production of the Challenger Armoured Repair and Recovery Vehicle, some of which were deployed to Saudi Arabia and took part in Operation Desert Storm. Vickers Defence Systems has also been awarded another export contract for its well established Mk 3 MBT.

In the United States, production of the General Dynamics, Land Systems Division, M1A1 MBT has been completed at the Detroit Arsenal Tank Plant, although this facility continues to supply components to the only remaining US MBT production facility, the Lima Army Tank Plant.

Although production of the M1A1 is now running down, production is expected to continue for some years, both for the US Army and for export. Egypt is taking the aforementioned 555 in-country assembled M1A1s with Saudi Arabia taking 315 vehicles which will be delivered from 1993 through to 1995. The US Army has ordered 62 M1A2s and some of the older M1s may well be upgraded in the future which could help to keep the production base alive until the next generation tank is ready for production.

Under the US Army's Armored Systems Modernisation programme, a number of Advanced Technology Transition Demonstrators (ATTD) are being built to prove key subsystems before the full scale engineering development starts.

Contracts have already been awarded for the Block III MBT (two teams), the Advanced Field Artillery System (one team) and the Combat Mobility Vehicle (one team).

No further improvements are planned for the M60 series of MBTs and substantial numbers of these are now being supplied to a number of countries, mostly NATO, at virtually no cost.

Yugoslav-supplied M-84 MBTs were used by the Kuwaiti Army in the recapture of Kuwait early in 1991, but the future of the whole Yugoslav defence industry must now be in some doubt until the current problems within that country are solved.

Late in 1990 the Yugoslav Federal Directorate of Supply and Procurement stated that a new MBT was being developed, but it is no longer clear as to whether this project is still moving ahead.

Light Tanks

Shortly before this book went to press, the United States Army finally released its request for proposals for an Armored Gun System (AGS). This will replace the M551 Sheridan light tanks used by the 82nd Airborne Division as well as being issued to the new light divisions who at present do not have any airportable tanks.

The AGS is a non-developmental item (NDI) and competition, both from within the United States and abroad, will be intense and some of the foreign contenders are already trying to tie up with United States defence equipment manufacturers so that they are better placed in the competition.

The US Marine Corps have funded the Light Armored Vehicle - Assault Gun with the prototypes to be delivered for US Marine Corps trials next year, but as this issue of *Jane's Armour and Artillery* was going to press there was considerable doubt as to whether this vehicle would move into the production phase owing to a shortage of funding.

The Light Armored Vehicle - Assault Gun is fitted with a new Cadillac Gage Textron two man turret armed with a 105 mm EX-35 rifled tank gun fed by an automatic loader.

A number of competitors for the US Army's Armored Gun System (AGS) were intending to use the turret of the Light Armored Vehicle - Assault Gun, in their entry to the AGS competition.

Artillery Systems

GIAT Industries of France has won its first export order for the private venture 105 mm LG1 light gun and this is now in full production at Bourges, with first production guns being completed in late 1991.

Building on their first export success, GIAT Industries are now marketing this weapon to a number of other countries including Canada and Thailand, with the British Royal Ordnance 105 mm Light Gun being the main competition.

Vickers Shipbuilding and Engineering have now started production of the 155 mm AS90 self-propelled artillery system for the Royal Artillery with first production systems to be delivered next year. This will enter service with a 39-calibre barrel, but it could have entered service with the new 52-calibre barrel which will be the new NATO standard 155 mm ordnance.

Vickers Shipbuilding and Engineering and Royal Ordnance have both completed prototypes of a lightweight 155 mm towed howitzer weighing under 4000 kg for a potential US Army/US Marine Corps requirement for a substantial number of weapons to replace the existing 155 mm M198 towed systems.

As private venture systems cost a substantial amount of money to design, develop and test, it would perhaps have been more beneficial if the two companies had tried to work together rather than compete in the same market place.

In the United States, a consortium lead by FMC Corporation was awarded the contract for the Advanced Field Artillery System - Advanced Technology Transition Demonstrator. This will have a number of advanced features and be armed with a 155 mm ordnance that may well use liquid propellant rather than the conventional bagged charges.

Production of the 155 mm AS90 self-propelled artillery system for the British Army is now underway with first production systems to be completed next year. At present a total of 179 systems has been ordered for the Royal Artillery with an option on another 50. It remains to be seen as to whether this option is exercised.

Land-based Air Defence

Details of self-propelled and towed anti-aircraft guns are now given in *Jane's Land-Based Air Defence*. This publication also includes details of manportable surface-to-air missile systems, static and towed surface-to-air missile systems, towed anti-aircraft gun sights, national

and international air defence systems and a comprehensive inventory listing.

Where an armoured vehicle chassis is used as a basis for an air defence system, for example the German Gepard twin 35 mm self-propelled anti-aircraft gun system based on a modified Leopard 1 MBT chassis, the Roland surface-to-air missile system based on a French AMX-30 MBT chassis and the German Marder mechanised combat vehicle chassis, brief details and usually a photograph will be found in *Jane's Armour and Artillery* under their respective chassis.

Subsystems

Details of ammunition, armoured fighting vehicle armament (including anti-tank guided weapons), turrets and cupolas, engines, transmissions and powerpacks are now given in *Jane's Armoured Fighting Vehicle Retrofit Systems.*

In addition to containing the above four sections from *Jane's Armour and Artillery,* this volume contains detailed sections covering vehicle-mounted anti-tank guided weapons, automatic loaders, land navigation systems, suspensions and tracks, armour systems, fire detection and suppression systems, laser detectors, smoke dischargers and decoys, weapon control systems, fire control systems and commander's, gunner's and driver's day and night vision equipment that is currently in production or under development.

Many of the more recent armoured fighting vehicles are now offered with a variety of different subsystems such as powerpacks, fire control systems, armament and vision equipment.

This enables a manufacturer to tailor a vehicle to meet the specific operational requirements of the customer.

As the cost of new armoured vehicles increases it is expected that more countries will turn to upgrading their existing vehicles rather than purchasing new vehicles.

Many countries already plan to upgrade their vehicles, especially MBTs, mid-way through their service lives to extend their combat effectiveness.

The publication of *Jane's Armoured Fighting Vehicle Retrofit Systems* has enabled these and many other essential subsystems that go into a modern AFV to be covered in much more detail than in the past.

Military Logistics

Jane's Armour and Artillery does contain brief details and photographs, where appropriate, of specialised variants of AFVs such as armoured recovery vehicles, engineering vehicles and bridgelayers.

Full details of these and the countless other vehicles, tracked and wheeled, armoured and unarmoured, that support the front line AFVs will be found in our sister yearbook, *Jane's Military Vehicles and Logistics.*

The success of the Allied Coalition forces during Operation Desert Storm would not have been possible without massive logistical backup. Not only did this include manoeuvring massive quantities of ammunition, fuel, food and other stores, but also the movement of large numbers of armoured fighting vehicles by tank transporter and trucks over long distances. The crews for these armoured vehicles were then flown forward by C-130 transport aircraft or helicopter so that both vehicle and crew arrived ready for action.

It was apparent that while the more recent MBTs such as the US General Dynamics, Land Systems Division, M1A1, had excellent cross-country mobility, its support vehicles were based on older chassis which simply lacked the mobility and reliability of the vehicles that they were required to operate with.

When a major new weapon system is developed, for example an MBT, wherever possible the key supporting vehicles such as the armoured recovery vehicle, bridgelayer and armoured engineer vehicle should be developed more or less concurrently.

If possible these support vehicles should share the same MBT chassis with the obvious logistical and training advantages.

If this does not occur, then the full potential of the new weapon system will not be realised.

Addenda

The reader is recommended to consult the addenda for the latest information and photographs on a number of new systems that were disclosed after the main sections of *Jane's Armour and Artillery* were compiled. The addenda is correct to 1 September 1991 and includes brief details on the Alvis Storm 105 light tank which is being developed in response to a US Army requirement for an Armored Gun System.

Acknowledgements

One of the last tasks of the editor is to thank everyone who has been involved, however small their part, in the preparation and production of this yearbook; it is very much a team effort.

The editor would like to take this opportunity to thank the many manufacturers, defence forces and individuals who have contributed to the new edition of *Jane's Armour and Artillery.* Special thanks are due to Mainly Military, Raymond Cheung, Henry Dodds, Kensuke Ebata, Adrian English, Robin Fletcher, Terry Gander, Michael Green, Murray Hammick, Paul Handle, David Isby, Franz Kosar, Gunter Lippert, Robert Lockie, John W Loop, Barry Marriot, Julio Montes, Richard M Ogorkiewicz, Rupert Pengelly, Donald Spaulding, Dr Arthur G Volz, Steven Zaloga and C R Zwart for their most valuable assistance.

The drawings used in *Jane's Armour and Artillery* are to a scale of 1/76th and except where specified have been prepared by Paul Hocking. Steve Zaloga has allowed us to reproduce some of his excellent drawings of Soviet equipment and a number of additional superb illustrations have been provided by Vasco Barbic.

Tony Cullen has been responsible for the multiple rocket launchers, coastal defence systems, and armour and artillery equipment in service.

My wife, Sheila, has been of invaluable assistance in the preparation of this book.

The staff of *Jane's Defence Weekly,* especially Ian Kemp, have also provided material for this edition.

A special word of thanks is also due to the advertising staff of Jane's Information Group for their assistance in collecting information and material from exhibitions which the editor has been unable to visit.

Keith Faulkner, Managing Editor of Jane's Defence Data, has given the editor maximum possible support throughout the preparation cycle of *Jane's Armour and Artillery.*

Kevin Borras has sub-edited the many entries that make up this volume while Ruth Simmance and her team of Sarah Erskine, Christine Varndell and Abigail Anfam played a vital role in the preparation of this edition.

Special thanks are due to the production team headed by David Moyes.

This edition of *Jane's Armour and Artillery* has over 400 new photographs and 50 new entries, with many of the existing entries having been rewritten or expanded to take into account new developments. In particular this applies to a number of Eastern systems, as a considerable amount of information is now becoming available.

The editor would also like to thank the typesetters and binders for the most valuable part they have played in the production of this book.

Additional information and photographs for the next edition should be sent to the editor as soon as possible. The address for all correspondence is Jane's Defence Data, Sentinel House, 163 Brighton Road, Coulsdon, Surrey, CR5 2NH, UK. Telephone (+44) 081-763 1030 and Fax (+44) 081-763 1005.

September 1991 — Christopher F Foss

Abbreviations

AA	anti-aircraft
AAAV	Advanced Amphibious Assault Vehicle
AAG	anti-aircraft gun
AAK	Appliqué Armor Kit (US)
AAV	Assault Amphibian Vehicle
ACCV	Armored Cavalry Cannon Vehicle (US)
ACE	Armored Combat Earthmover (US)
ACEC	*Ateliers de Constructions Électriques de Charleroi*
ACLOS	Automatic-to-Command Line-of-Sight
ACRV	armoured command and reconnaissance vehicle
ACV	Armored Cannon Vehicle (US)
ACV	Armoured Combat Vehicle
ACVT	Armored Combat Vehicle Technology Programme (US)
ADAM	Area Denial Artillery Munition
ADAMS	Air Defense Advanced Mobile System
ADATS	Air Defence Anti-Tank System
ADEA	Army Development and Employment Agency
AEV	armoured engineer vehicle
AF	air force
AFAP	artillery-fired atomic projectile
AFARV	Armored, Forward Area, Re-arm Vehicle (US)
AFAS	Advanced Field Artillery System
AFD	automatic feeding device
AFF	Ammunition Factory Footscray
AFSV	Armoured Fire Support Vehicle
AFV	Armored Family of Vehicles (US)
AFV	armoured fighting vehicle
AGF	Army Ground Forces
AGL	above ground level
AGLS	Automatic Gun Laying System
AGS	Armored Gun System (US)
AGV	Assault Gun Vehicle
AGVT	Advanced Ground Vehicle Technology
Ah	ampère hour
AIFS	Advanced Indirect Fire System
AIFV	Armored Infantry Fighting Vehicle (US)
AIFV	armoured infantry fighting vehicle
AIPS	Advanced Integrated Propulsion System (US)
ALAAVS	Advanced Light Armored/ Amphibious Vehicle System (US)
ALT	Armoured Launching Turret
AMCCOM	Armament Munitions and Chemical Command
AMDS	Anti-missile Discarding Sabot
AMF	*Amphibische Mehrzweck-Fahrzeuge* (multi-purpose amphibious vehicle)
AML	*Automitrailleuse Légère* (light armoured car)
AMR	*Automitrailleuse de Reconnaissance*
AMX	*Atelier de Construction d'Issy-les-Moulineaux*
AOI	Arab Organisation for Industrialisation
AOS	add-on stabilisation
AP	armour-piercing
APAM	anti-personnel, anti-materiel
APC	armoured personnel carrier
APC	armour-piercing capped
APC-T	armour-piercing capped tracer
APCT-BF	armour-piercing capped tracer - base fuze
APDS	armour-piercing discarding sabot
APDS-T	armour-piercing discarding sabot tracer
APE	*Amphibisches Pionier-Erkundungsfahrzeug* (amphibious front line reconnaissance vehicle)
APERS	anti-personnel
APERS-T	anti-personnel tracer
APFIDS	armour-piercing fragmentation incendiary discarding sabot
APFSDS	armour-piercing fin-stabilised discarding sabot
APFSDS(P)	armour-piercing fin-stabilised discarding sabot (practice)
APFSDS-T	armour-piercing fin-stabilised discarding sabot tracer
APG	Aberdeen Proving Grounds
APGM	Autonomous Precision-Guided Munition
APHE	armour-piercing high explosive
API	armour-piercing incendiary
APIT	armour-piercing incendiary tracer
APM	anti-personnel mine
APSE	armour-piercing secondary effect
APSE-T	armour-piercing secondary effect-tracer
APT	armour-piercing tracer
APTE	Abrams Power Train Evolution (US)
APU	auxiliary power unit
APV	armoured patrol vehicle
AR/AAV	Armored Reconnaissance/Airborne Assault Vehicle (US)
ARDEC	Armament Research Development and Engineering Centre
ARDNOT	Automatic Day/Night Optical Tracker
ARE	*Atelier de Construction Roanne*
ARETS	Armour Remoted Target System
ARMAD	Armoured and Mechanised Unit Air Defence
ARMSCOR	Armament Manufacturing Corporation (South Africa)
ARMVAL	Anti-Armor Vehicle Evaluation (US)
ARP	anti-radiation projectile
ARRADCOM	Armament Research and Development Command
ARRV	Armoured Repair and Recovery Vehicle
ARSV	Armored Reconnaissance Scout Vehicle (US)
ARV	armoured recovery vehicle
ASA	Advanced Security Agency
ASARC	Army Systems Acquisition Review Council
ASM	Armored Systems Modernization (US)
ASTROS	Artillery Saturation Rocket System
ASV	ammunition supply vehicle
AT	anti-tank
ATACS	Advanced Tank Cannon System (US)
ATDU	Armoured Trials & Development Unit
ATG	anti-tank gun
ATGW	anti-tank guided weapon
ATLV	Artillery Target Location Vehicle
ATM	anti-tank mine
ATR	Automotive Test Rig
ATS	*Atelier de Construction de Tarbes*
ATTS	Air-Transportable Towed System
AVGP	Armoured Vehicle General Purpose
AVH	Armoured Vehicle Heavy
AVL	Armoured Vehicle Light
AVLB	armoured vehicle-launched bridge
AVM	Armoured Vehicle Medium
AVR	armoured vehicle reconnaissance
AVRE	Assault Vehicle Royal Engineers
BAOR	British Army of the Rhine
BARV	Beach Armoured Recovery Vehicle
BATES	Battlefield Artillery Target Engagement System
BB	base bleed
BCC	battery control centre
BD	base detonating
BE	base ejection
bhp	brake horsepower
BITE	built-in test equipment
BL	blank
BLR	*Blindado Ligero de Ruedas*
BL-T	blank tracer
BMARC	British Manufacture and Research Company
BMF	Belgian Mechanical Fabrication
BMR	*Blindado Medio de Ruedas*
BMS	Battlefield Management Systems
BNS	Bill Night Sight
BOCV	Battery Operations Centre Vehicle
BOFI	Bofors Optronic Fire control Instrument
BTA	best technical approach
CAD	computer assisted design
CAF	Canadian Armed Forces
CAL	Canadian Arsenals Limited
CAP	Combustible Augmented Plasma
CARRV	Challenger Armoured Repair and Recovery Vehicle
CAT/FCS	Command Adjusted Trajectory/ Fire Control System (US)
CAT/LCV	Combined Arms Team/Lightweight Combat Vehicle (US)
CATT-B	Component Advanced Technology Test-Bed (US)
CAWS	Cannon Artillery Weapons Systems (US)
CBC	*Companhia Brasileira de Cartuchos*
CCC	combustible cartridge case
CCTV	closed-circuit TV
CCV	Close Combat Vehicle
CCV-L	Close Combat Vehicle – Light
CDM	Coastal Defence Missile
CDP	Canadian dry pin
CENTCOM	Central Command
CET	Combat Engineer Tractor
CEV	combat engineer vehicle
CF	controlled fragmentation
CFE	Conventional Forces Europe
CFV	Cavalry Fighting Vehicle (US)
CGS	Crew Gunnery Simulator
CHARM	Challenger Chieftain Armament
CHIP	Challenger Improvement Programme
CILAS	*Compagnie Industrielle des Lasers*
CIS	Chartered Industries of Singapore
CITV	Commander's Independent Thermal Viewer (US)
CIWS	Close-In Weapons System
CLAMS	Clear Lane Marking System (US)
CLAWS	Close Combat Light Armor Weapon System (US)
CLGP	Cannon-Launched Guided Projectile (US)
CMV	Combat Mobility Vehicle (US)
Comp B	Composition B
COMVAT	Combat Vehicles Armament Technology (US)
CORAD	Co-ordinated Roland Air Defence
COTAC	*Conduite de Tir Automatique pour Char* (tank automatic fire)
COV	Counter Obstacle Vehicle (US)
CP	concrete-piercing
CPV	command post vehicle
CRR	*Carro de Reconhecimento Sobre Rodas* (reconnaissance tracking scout car)
CRT	cathode ray tube
CSB	Combat Support Boat
CSF	Combined Service Forces
CSI	computer synthesised image
CSS	Computer Sighting System
CTI	Central Tyre Inflation
CTRA	*Carro de Transporte Sobre Rodas Anfibo* (amphibious tracking scout car)
CTT	Challenger Training Tank
CV90	Combat Vehicle 90
CVAST	Combat Vehicle Armament System Technology (US)
CVRDE	Combat Vehicle Research and Development Establishment
CVR(T)	Combat Vehicle Reconnaissance (Tracked)
CVR(W)	Combat Vehicle Reconnaissance (Wheeled)
CVT	controlled variable time
CWR	continuous wave radar
CWS	cupola weapon station
DARCOM	US Army Materiel Development and Readiness Command
DARPA	Defense Advanced Research Projects Agency
DCA	*Defense Contre Avions* (anti-aircraft)
DDA	Detroit Diesel Allison
DDS	Department of Defense Support
DDU	digital display unit
DEFA	*Direction des Études et Fabrications d'Armement*
DESO	Defence Export Sales Organisation
DFCS	Digital Fire Control System
DFSV	Direct Fire Support Vehicle
DHSS	Data Handling SubSystem
DIA	Defense Intelligence Agency (US)

PROVEN PERFORMANCE
Simrad Laser Rangefinder Sights

Simrad LP101 fitted with Simrad KN250 Image Intensifier mounted on 106 mm reccoilless gun.

SIMRAD LP101

- Miniaturised combined weapon sight and laser rangefinder
- Built-in ballistic computer provides super-elevation aiming mark
- Replaces traditional weapon sights
- For mounting on anti-tank and other direct firing weapons
- Can be fitted with a Simrad image intensifier for night operations

SIMRAD OPTRONICS PRODUCT RANGE:

- Laser rangefinders for infantry, artillery, mechanized troops, air defence and navy
- Laser rangefinder modules for integration in sights
- Laser threat warning receivers
- IR thermal imager for navy and coastal artillery
- Night sights and night observation devices for various applications

Simrad OPTRONICS

Simrad Optronics A/S
P.O.Box 6114 Etterstad
0602 Oslo 6, NORWAY
Telephone: 47 2 67 04 90
Telex: 76136 sim n. • Telefax:47 2 19 29 91

UK subsidiary:
Simrad Optronics Ltd.
No. 3 Oakwood Industrial Park
Gatwick Road, Crawley
West Sussex RH 10 2AZ, ENGLAND
Telephone: 0293 560413/560415
Telex: 0051 87 84 98 SIM UK G. • Telefax: 44 293 560 418

ON TARGET... AGAIN, AGAIN, & AGAIN

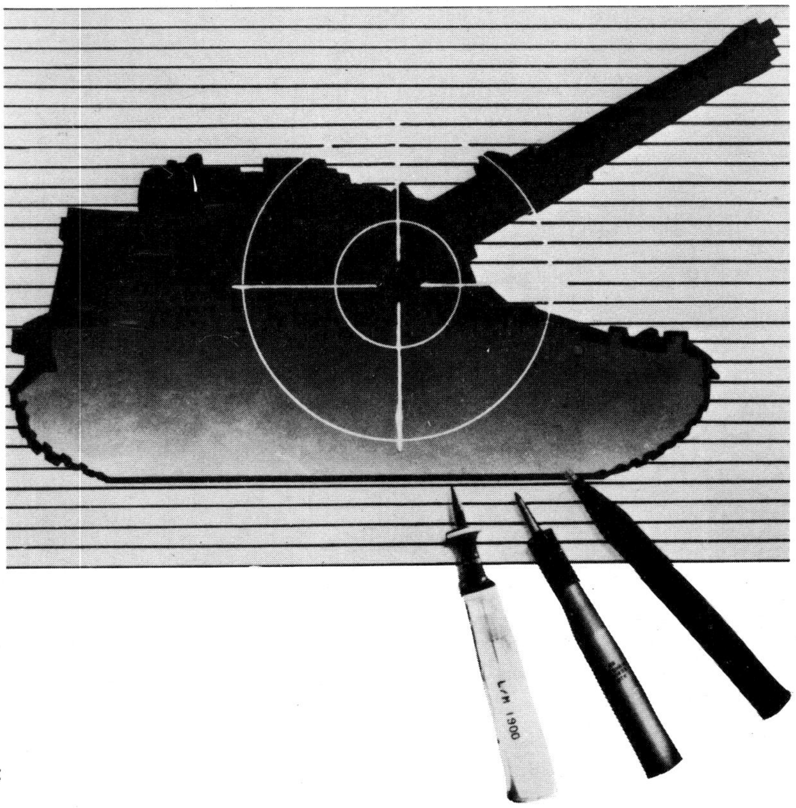

Reliable technology & the highest quality. Our weapons have been designed to be fine-tuned to your defence needs, with a stress on reliable technology & the highest quality. Dedicated to provide the most modern and sophisticated ammunition to the armies of today the Pakistan Ordnance Factories are a step ahead of the competition.

For details please contact:

Director Exports POFs

Pakistan Ordnance Factories

Phones: 0092 - 566 - 82019/83219 - 0092 - 51 - 566030 (10 LINES) SWITCH BOARD TELEX: 5840, 54178 POFAC PK FAX: 0092 - 51 - 584175

DIVADS	Division Air Defense Gun System (US)
DoD	Department of Defense (US)
DOP	Department of Productivity
DPICM	Dual Purpose Improved Conventional Munition (US)
DSACS	Direct Support Armored Cannon System (US)
DSETS	Direct Support Electrical Test System
DSO	Defence Sales Organisation
DS/T	practice discarding sabot/tracer
DSWS	Division Support Weapon System
DTAT	*Direction Technique des Armements Terrestres*
DU	depleted uranium
DWFK	Deep Water Fording Kit
EAAK	Enhanced Appliqué Armor Kit (US)
EBG	*Engin Blindé Génie (armoured combat vehicle)*
EBR	*Engin Blindé de Reconnaissance (armoured reconnaissance vehicle)*
EEC	European Economic Community
EFAB	*Établissement d'Etudes et de Fabrications d'Armement de Bourges*
EFC	equivalent full charge
EFCR	equivalent full charge rounds
EFM	Explosives Factory Maribyrnong
EFP	Expanded Feasibility Phase
EFP	Explosively Formed Penetrator
ELKE	Elevated Kinetic Energy Weapon (US)
ELS	Electro-magnetic Launcher System
ELSAP	*Elektronische Schiessanlage für Panzer* (electronic fire system for tanks)
EM	electro-magnetic
EMC	Executive Management Committee
EMDG	Euromissile Dynamics Group
EMG	externally mounted gun
EMP	electro-magnetic pulses
ENGESA	*Engesa Engenheiros Especializados*
EOD	explosive ordnance disposal
EPC	*Engin Principal de Combat* (future main battle tank)
EPC	electronic plane conversion
EPG	European Production Group
EPG	European Programme Group
ER	enhanced radiation
ER	extended range
ERA	extended range ammunition
ERC	*Engin de Reconnaissance Canon*
ERFB	extended range full bore
ERGP	extended range guided projectile
ERMIS	Extended Range Modification Integration System
ERSC	extended range sub-calibre
ERV	emergency rescue vehicle
ESD	Electronique Serge Dassault
ESPAWS	Enhanced Self-Propelled Artillery Weapons System (US)
ESRS	electro slag refined steel
EW	electronic warfare
EWK	*Eisenwerke Kaiserslautern Göppner*
EWS	external weapon station
FAAD	Forward Area Air Defense (US)
FAAR	Forward Area Alerting Radar (US)
FAASV	Field Artillery Ammunition Support Vehicle (US)
FACE	Field Artillery Computer Equipment
FARS	field artillery rocket system
FARV-A	Future Armored Resupply Vehicle – Artillery (US)
FAST	Forward Area Support Team
FAV	Fast Attack Vehicle (US)
FCC	Fire Command Centre
FCCVS	Future Close-Combat Vehicle System (US)
FCE	fire control equipment
FCS	fire control system
FDC	fire direction centre
FDCV	fire direction centre vehicle
FEBA	Forward Edge Battlefield Area
FG	field gun
FH	field howitzer
FIFV	Future Infantry Fighting Vehicle (US)
FISTV	Field Support Team Vehicle (US)
FITOW	Further Improved TOW (US)
FLIR	forward-looking infra-red
FLOT	forward line of own troops
FMC	Food Machinery Corporation
FMS	Foreign Military Sales
FN	*Fabrique Nationale*
FOO	forward observation officer
FOV	field-of-view
FRAG	fragmentation
FROG	free rocket over ground
FSCV	fire support combat vehicle
FSED	full-scale engineering development (US)
FST	Future Soviet Tank/Follow-on Soviet Tank
FSV	fire support vehicle
FTMA	Future Tank Main Armament
FTS	Future Tank Study
FUG	*Felderitö Usó Gépkosci*
FV	fighting vehicle
FV/GCE	fighting vehicle gun control equipment
FVRDE	Fighting Vehicle Research and Development Establishment
FVS	Fighting Vehicle System (US)
FVSC	Fighting Vehicle Systems Carrier (US)
FY	fiscal year
G	gendarmerie
GAO	General Accounting Office
GAP	Gun Aiming Post
GCE	gun control equipment
GCT	*Grande Cadence de Tir (high rate of fire)*
GD	General Dynamics
GD, LSD	General Dynamics, Land Systems Division
GDU	Gun Display Unit
GEMAG	General Electric Mobile Air Defense Gun
GH	gun/howitzer
GIAT	*Groupement Industriel des Armements Terrestres*
GLC	gun lay computer
GLH-H	Ground-Launched Hellfire-Heavy (US)
GLLD	Ground Laser Locator Designator (US)
GLS	*Gesellschaft für Logistischen Service*
GMC	General Motors Corporation
GM, MVO	General Motors, Military Vehicle Operations
GOCO	government-owned, contractor-operated (US)
GP	guided projectile
GPMG	General Purpose Machine Gun
GPO	gun position officer
GPS	gunner's primary sight
GSR	General Staff Requirement (US)
GSRS	General Support Rocket System
GST	General Staff Target
GST	*Gesellschaft für System-Technik*
GTCS	Gun Test and Control System
GTI	German Tank Improvement
GW	guided weapon
HAB	Heavy Assault Bridge (US)
HAWK	Homing-all-the-way-killer (US)
HB	heavy barrel
HC	high capacity
HCER	high capacity extended range
HCHE	high capacity high explosive
HCT	HOT Compact Turret
HE	high explosive
HE-APERS-FRAG	high explosive anti-personnel fragmentation
HEAP-T	high explosive anti-personnel tracer
HEAT	high explosive anti-tank
HEAT-FS	high explosive anti-tank fin-stabilised
HEAT-MP	high explosive anti-tank multi-purpose
HEAT-MP(P)	high explosive anti-tank multi-purpose (practice)
HEAT-T	high explosive anti-tank tracer
HEAT-T-HVY	high explosive anti-tank tracer heavy
HEAT-T-MP	high explosive anti-tank tracer multi-purpose
HEAT-TP-T	high explosive anti-tank target practice tracer
HEDP	high explosive dual purpose
HEER	High Explosive Extended Range (US)
HE-FRAG	high explosive fragmentation
HE-FRAG-FS	high explosive fragmentation – fin-stabilised
HE-FS	high explosive – fin-stabilised
HEI	high explosive incendiary
HEIT	high explosive incendiary tracer
HEL	High Energy Laser (US)
HEL	Human Engineering Laboratory (US)
HELP	Howitzer Extended Life Program (US)
HEMAT	Heavy Expanded Mobility Ammunition Trailer (US)
HEMTT	Heavy Expanded Mobility Tactical Truck (US)
HEP	high explosive plastic
HEPD	high explosive point detonating
HE/PR	high explosive practice
HEP-T	high explosive practice tracer
HERA	high explosive rocket-assisted
HE-S	high explosive spotting
HESH	high explosive squash head
HESH-T	high explosive squash head tracer
HE-T	high explosive tracer
HET-PF	high explosive tracer – percussion fuze
HE-T SD	high explosive tracer – self-destruct
HFCC	Howitzer Fire Control Computer (US)
HFHTB	Human Factors Howitzer Test-Bed (US)
HFM	Heavy Force Modernisation (US)
HIMAG	High-Mobility Agility Test Vehicle (US)
HIP	Howitzer Improvement Programme (US)
HMC	Howitzer Motor Carriage
HMLC	High Mobility Load Carrier
How	howitzer
hp	horsepower
HPFP	High Performance Fragmentation Projectile
HPS	Helmet Pointing System (or Sight)
HPT	high-pressure test
HSTV(L)	High Survivability Test Vehicle (Lightweight) (US)
HTTB	High Technology Test Bed (US)
HVAP	high velocity armour-piercing
HVAPDS-T	high velocity armour-piercing discarding sabot tracer
HVAPFSDS	high velocity armour-piercing fin-stabilised discarding sabot
HVAP-T	high velocity armour-piercing tracer
HVM	Hypervelocity Missile
HVMS	Hypervelocity Support Weapon
HVSD	High Value Sight Defense (US)
HVSS	horizontal volute spring suspension
HVTP-T	high velocity target practice tracer
IAFV	infantry armoured fighting vehicle
IAI	Israel Aircraft Industries
IASD	Instant Ammunition Selection Device
ICC	information co-ordination centre
ICM	improved conventional munition
ICM BB	improved conventional munition base bleed
ICV	Infantry Combat Vehicle
IDF	Israel Defence Forces
IEPG	Independent European Program Group
IFCS	Integrated Fire Control System
IFF	Identification Friend or Foe
IFV	infantry fighting vehicle
IFVwCM	Infantry Fighting Vehicle with Integrated Countermeasures (US)
II	image intensification
ILL	illuminating
IMI	Israel Military Industries
IOC	initial operational capability
IPF	Initial Production Facility
IPO	International Programme Office
IR	infra-red
IRBM	intermediate-range ballistic missile
IS	internal security
ISV	Internal Security Vehicle
ITV	Improved TOW Vehicle (US)
IVPDL	Inter-Vehicle Positioning and Data Link (US).
IWS	Improved Weapon System
JGSDF	Japanese Ground Self-Defence Force
JPO	Joint Project Office
JSC	Joint Steering Committee
JSDFA	Japanese Self-Defence Force Agency

KE kinetic energy
KEM Kinetic Energy Missile (US)
KIFV Korean Infantry Fighting Vehicle

LAAG light anti-aircraft gun
LAD light aid detachment
LADS light air defense system (US)
LAPES Low Altitude Parachute Extraction System
LATS Light Armoured Turret System
LAV Light Armored Vehicle (US)
LAV Light Assault Vehicle (US)
LAW light anti-tank weapon
LED light emitting diode
LEW Lyttleton Engineering Works
LIA Linear Induction Accelerator
LLAD low level air defence
LLLTV low-light level television
LMG light machine gun
LOAL Lock-On After Launch
LOBL Lock-On Before Launch
LOS Line-of-Sight
LOSAT Line-of-Sight Anti-tank
LP Liquid Propellant
LPC Launcher Pod Carrier
LPTS Lightweight Protected Turret System
LRAT long-range anti-tank
LRBB long-range base bleed
LRF Low Recoil Force
LRHB long-range hollow base
LRN Low Recoil Noricum
LRU line-replaceable unit
LTD laser target designator
LTFCS Laser Tank Fire Control System
LVA landing vehicle assault (US)
LVT landing vehicle tracked (US)
LVTC landing vehicle tracked command (US)
LVTE landing vehicle tracked engineer (US)
LVTH landing vehicle tracked howitzer (US)
LVTP landing vehicle tracked personnel (US)
LVTR landing vehicle tracked recovery (US)
LWML Light Weight Multiple Launcher
LWMS Light Weight Modular Sight
LWT light weapon turret

MAC Medium Armored Car (US)
MADLS Mobile Air Defence Launching System
MAOV Mobile Artillery Observation Vehicle
MAP Military Aid Programme
MAV maintenance assist vehicle
MBA main battle area
MBB *Messerschmitt-Bölkow-Blohm*
MBT main battle tank
MCCS Marconi Command and Control Systems
MCRV Mechanised Combat Repair Vehicle
MCSK Mine Clearance System Kit (US)
MCT Medium Combat Tractor (US)
MCT MILAN Compact Turret
MCWS Minor Calibre Weapons Station (US)
MENS Mission Element Need Statement (US)
MEV medical evacuation vehicle (US)
MEWSS Mobile Electronic Warfare Support System
MF multi-function
MFC mortar fire controller
MFF munition filling factory
MG machine gun
MGB Medium Girder Bridge
MICOM Missile Command (US)
MICV mechanised infantry combat vehicle
MILAN *Missile d'Infantrie Léger Antichar (light infantry anti-tank missile)*
MILES Multiple Integrated Laser Engagement System (US)
MIPS Medium Integrated Propulsion System (US)
MLC Military Load Class
MLC Modular Load Carrier
MLRS Multiple Launch Rocket System (US)
MMBF mean miles between failures
MMS mast-mounted sight
MoD Ministry of Defence
MOLF Modular Laser Fire Control
MoU memorandum of understanding
MPGS Mobile Protected Gun System (US)
MPWS Mobile Protected Weapon System (US)
MPS Maritime Prepositioning Ships (US)
MRBF mean rounds before failure
MRS multiple rocket system
MRS muzzle reference system
MRVR Mechanised Repair and Recovery Vehicle (US)
MSDS Marconi Space and Defence Systems
MSTAR Manportable Surveillance and Target Acquisition Radar
MSV Modular Support Vehicle
MT mechanical time
MTB Mobility Test-Bed
MTI moving target indication
MTL Materials Technology Laboratory (US)
MTR Mobile Test Rig (US)
MTSQ mechanical time and superquick
MTU *Motoren-und-Turbinen-Union*
MULE Modular Universal Laser Equipment (US)
MV muzzle velocity
MVEE Military Vehicles and Engineering Establishment
MWS Manned Weapon Station (US)

NATO North Atlantic Treaty Organisation
NBC nuclear, biological, chemical
NBMR NATO Basic Military Requirement
NTC National Training Center (USA)
NUGP nominal unit ground pressure

OBR Optical Beam Riding
OCC *Obus à Charge Creusê (shaped charge shell)*
ODE Ordnance Development and Engineering (Singapore)
OP observation post
OT operational test
OTA overflight top attack
OTEA Operational Test and Evaluation Agency (US)
OVT Oceonics Vehicle Technology

P^{3}I Pre-Planned Production Improvements
PAR pulse acquisition radar
PAT power-assisted traverse
PCB printed circuit board
PD point detonating
PE Procurement Executive
PFD proximity fuze disconnector
PFHE pre-fragmented high explosive
PFPX pre-fragmented proximity fuzed
P How pack howitzer
PIE pyrotechnically initiated explosive
PIVADS Product Improved Vulcan Air Defense System (US)
PLARS Position Location and Reporting System (US)
PLS Palletised Load System (US)
PM *porte mortier* (mortar carrier)
PMO Program Management Office (US)
POS *Postes Optiques de Surveillance*
PPI plan position indicator
PRAC practice
PRAC-T practice tracer
PRI projector reticle image
PTO power take-off
PWI-SR(GR) *Panser Wagen Infanterie-Standaard (Groep)*

QCB Quick Change Barrel
QE quadrant elevation

RA Royal Artillery
RAAMS Remote Anti-Armor Mine System (US)
RAC Royal Armoured Corps
RADIRS Rapid Deployment Multiple Rocket System (US)
RAM-D reliability, availability, maintainability and durability
RAO Rear Area Operations
RAP rocket-assisted projectile
RARDE Royal Armament Research and Development Establishment
RATAC Radar for Field Artillery Fire
RCAAS Remote-Controlled Anti-Armor System (US)
RCC Roland Co-ordination Centre (US)
RCDU Remote Controlled Defence Unit
RCT Royal Corps of Transport
RCV Robotic Command Vehicle
RDF/LT Rapid Deployment Force Light Tank (US)
RDJTF Rapid Deployment Joint Task Force (US)
RDT&E Research Development Test and Evaluation
REME Royal Electrical and Mechanical Engineers
RFP request for proposals
RHA rolled homogeneous armour
RHA Royal Horse Artillery
RISE Reliability Improved Selected Equipment
RMG ranging machine gun
RO Royal Ordnance
ROBAT Robotic Counter-Obstacle Vehicle (US)
ROC required operational characteristics
RoC Republic of China
ROF rate of fire
RoK Republic of Korea
ROKIT Republic of Korea Indigenous Tank
ROR range only radar
RP rocket propelled
RPV remotely piloted vehicle
RR Recoilless Rifle
RSAF Royal Small Arms Factory
RSS Rosette Scanning Seeker
RTT *Roues Transporteur de Troupes*
RUC Royal Ulster Constabulary

SABCA *Société Anonyme Belge de Constructions Aéronautiques*
SADARM Sense And Destroy Armor (US)
SADF South African Defence Force
SAE Society of Automotive Engineers
SAL semi-active laser
SAM surface-to-air missile
SAMM *Société d'Applications des Machines Motrices*
SANTAL *Système Anti-aérien Léger (light anti-aircraft system)*
SAPI semi-armour piercing incendiary
SATCP *Système Anti-aérien à Très Courte Portée (very short-range anti-aircraft system)*
SCG Self Changing Gears
SCORE Stratified Charge Omnivorous Rotary Engine
SD self-destruct
SEME School of Electrical and Mechanical Engineering
SEN shell extended range NORICUM
SFCS Simplified Fire Control System
SFIM *Société de Fabrication d'Instrument de Mesure*
SFIRR solid fuel integral rocket/ramjet
SFM Sensor Fuzed Munitions
SHORAD Short Range Air Defence System
SH/PRAC squash head practice
shp shaft horse power
SIPS Small Integrated Propulsion System
SKOT *Średni Kolowy Opancerzny Transportér*
SLAP saboted light armor penetrator
SLEP Service Life Extension Program (US)
SLR Super Low Recoil
SM smoke
SMCO Standard Manufacturing Company
SMG sub-machine gun
Smoke BE smoke base ejection
Smoke WP smoke white phosphorus
SP self-propelled
SPAAG self-propelled anti-aircraft gun
SPAAM self-propelled anti-aircraft missile
SPAG self-propelled assault gun
SPARK solid propellant advanced ramjet kinetic energy missile
SPATG self-propelled anti-tank gun
SPAW self-propelled artillery weapon
SPG self-propelled gun
SPH self-propelled howitzer
SPL self-propelled launcher
SPM self-propelled mortar
SPSM *Sensorgezundete Panzerabwehr SubMunition*
SRC Space Research Corporation
SRG shell replenishment gear
SRU slip ring unit
SRV Surrogate Research Vehicle (US)
SSG Single Shot Gun
SSK Single Shot Kill
STA shell transfer arm

ABBREVIATIONS

STAFF Small Target Activated Fire and Forget (US)
STARTLE Surveillance and Target Acquisition Radar for Tank Location and Engagement (US)
STE/ICE Simplified Test Equipment/Internal Combustion Engine (US)
STUP spinning tubular projectile

TACMS Army Tactical Missile System (US)
TACOM Tank Automotive Command (US)
TAD Trailing Arm Drive (US)
TADDS Target Alert Display Data Set (US)
TADS Target Acquisition and Designation System (US)
TAM *Tanque Argentino Mediano*
TAS tracking adjunct system
TAT-251 Tactical Armament Turret 25 mm, 1 man (US)
TBAT TOW/Bushmaster Armored Turret (US)
TCU tactical control unit
TD tank destroyer
TDCS Tank Driver Command System
TDR Target Data Receiver
TEL Transporter, Erector, Launcher
TELAR Transporter, Erector, Launcher and Radar
TES target engagement system
TGMTS Tank Gunnery Missile Tracking System
TGP Terminally Guided Projectile
TGS Tank Gun Sight
TGSM terminally guided submunition
TGTS tank gunnery training simulator
TI thermal imaging
TICM thermal imaging common modules
TIRE Tank Infra-Red Elbow
TIS thermal imaging system
TLR Tank Laser Rangefinder
TLS Tank Laser Sight
TMS Turret Modernisation System
TOE table of organisation and equipment
TOGS Thermal Observation and Gunnery System
TOPAS *Transporter Obrneny Pasovy*
TOTE Tracker, Optical Thermally Enhanced
TOW Tube-launched Optically tracked Wire-guided (US)
TP target practice
TP-FL target practice flash
TPFSDS-T target practice fin-stabilised discarding sabot tracer
TP-SM target practice smoke
TP-SP target practice spotting
TP-T target practice tracer
TRACKSTAR Tracked Search and Target Acquisition Radar System (US)
TRADOC Training and Doctrine Command (US)
TRLV Tracked Rapier Launch Vehicle
TRSV Tracked Rapier Support Vehicle
TSFCS Tank Simplified Fire Control System
TSQ time and superquick
TT *transport de troupes* (troop transporter)
TTB Tank Test-Bed (US)
TTS Tank Thermal Sight
TUA TOW Under Armor (US)
TUR *Tiefflieger-Überwachungs-radar (low level surveillance radar)*
TWMP Track Width Mine Plough

UDR Ulster Defence Regiment
UET Universal Engineer Tractor (US)
ULC Unit Load Container
USAADS US Army Air Defense School
USMC United States Marine Corps
USN United States Navy
UTS Universal Turret System

VAB *Véhicule de l'Avant Blindé* (front armoured car)
VAB Vickers Armoured Bridgelayer
VADAR *Véhicule Autonome de Défense Anti-aérienne Rapproché (autonomous close-action anti-aircraft vehicle)*
VADS Vulcan Air Defense System (US)
VAE *Vehículo Armado Exploration*
VAPE *Vehículo APoyo y Exploration*
VARRV Vickers Armoured Repair and Recovery Vehicle
VARV Vickers Armoured Recovery Vehicle
VBC *Véhicule Blindé de Combat (armoured combat vehicle)*
VBL *Véhicule Blindé Léger (light armoured vehicle)*
VCA *Véhicule Chenillé d'Accompagnement* (tracked support vehicle)
VCC *Veicolo Corazzato de Combattimento*
VCG *Véhicule de Combat du Genie (armoured engineer vehicle)*
VCI *Véhicule de Combat d'Infanterie; Vehículo combate infanteria (infantry combat vehicle)*
VCR *Véhicule de Combat à Roues (wheeled combat vehicle)*
VCR variable compression ratio
VCR/AT *Véhicule de Combat à Roues/Atelier Véhicule*
VCR/IS *Véhicule de Combat à Roues/ Intervention Sanitaire*
VCR/PC *Véhicule de Combat à Roues/Poste de Commandement*
VCR/TH *Véhicule de Combat à Roues/ Tourelle HOT*
VCR/TT *Véhicule de Combat à Roues/ Transport de Troupes*
VCTM *Vehículo de Combate Transporte de Mortero* (armoured mortar carrier)
VCTP *Vehículo de Combate Transporte de Personal* (armoured personnel carrier)
VDA *Véhicule de Defense Anti-aérienne (anti-aircraft defence vehicle)*
VDAA *Véhicule d'Auto-Défense Anti-aérienne*
VDM viscous damped mount
VDSL Vickers Defense Systems Ltd
VDU visual display unit
VEC *Vehículo de Exploración de Caballerie*
VEDES Vehicle Exhaust Dust Ejection System (US)
VGMU Vulcan Gunner Monitor Unit (US)
VHIS visual hit indicator system
VIB *Véhicule d'Intervention du Base*
VIRSS Visual and Infra-Red Smoke Screening System
VLC *Véhicule Léger de Combat* (light armoured car)
VMBT Vickers Main Battle Tank
VRL *Véhicule Reconnaissance Léger (light reconnaissance vehicle)*
VSEL Vickers Shipbuilding and Engineering Ltd
VT variable time
VTP *véhicule transport de personnel (personnel carrier)*
VTT *véhicule transport de troupe (troop transporter)*
VVSS vertical volute spring suspension
VWC Vulcan Wheeled Carrier (US)
VXB *Véhicule Blindé à Vocations Multiples* (multi-purpose armoured car)

WAM Wide Area Mines (US)
WAPC Wheeled Armoured Personnel Carrier
WARPAC Warsaw Pact
WFSV Wheeled Fire Support Vehicle
WMRV Wheeled Maintenance and Recovery Vehicle
WP white phosphorus
WP-T white phosphorus tracer

YMRS Yugoslav multiple rocket system

Tanks

MBTs and medium tanks

ARGENTINA

TAM Medium Tank

Development
The German company of Thyssen Henschel developed a medium tank called the TAM (Tanque Argentino Mediano) to meet the requirements of the Argentinian Army. The first prototype of the TAM was completed in Germany in 1976 and the second prototype in 1977. Chassis construction and final assembly took place in Buenos Aires with turret construction and integration of the gun and fire control system at Rio Tercero. Since the first TAM prototypes were delivered to Argentina development has continued by TAMSE and over a period of time more than 1450 engineering modifications were introduced into production vehicles. Additional information on the TAM vehicle is given in the entry for the Thyssen Henschel TH 301 (TAM) tank under Germany. An ICV called the VCTP has been developed to work with the TAM, details of which are given in the *Armoured personnel carriers* section.

The total Argentinian requirement was for 512 105 mm armed TAM medium tanks and VCTP infantry combat vehicles. It is understood that for financial reasons production was completed after about 350 vehicles had been built.

In mid-1983 Peru ordered 80 TAMs, but these were not delivered and nor were 60 TAMs and VCTPs ordered by Panama early in 1984. In both cases the vehicles were not delivered owing to budgetary reasons. In early 1988 a TAM was demonstrated in Ecuador alongside an Austrian SK 105, US Stingray and French AMX-13 light tank, although in the end none of these vehicles were ordered. In the Ecuador presentations the TAM tank came first gaining 950 points out of a possible 1000 with its nearest competitor gaining only 750 points.

Variants

VCA 155
Full details of the VCA 155, a 155 mm self-propelled artillery system based on a lengthened TAM chassis are given in the *Self-propelled guns and howitzers* section under Argentina. This uses the same turret as installed on the Italian OTO Melara Palmaria 155 mm self-propelled artillery system.

VCRT
In late 1986 it was confirmed that the Argentinian Army had developed a new armoured recovery vehicle based on its VCTP infantry combat vehicle and known as the VCRT. This has a new superstructure and recovery rig supplied by Krupp MaK of Germany.

The VCRT has a rear-mounted dozer blade to provide a secure anchor for the main 30-tonne winch which is located within the sloping hull front. Mounted on the right side of the superstructure rear is a crane jib which can be traversed through 213.5° and lift a maximum load of 22 tonnes.

The first prototype of the VCRT was completed in mid-1987 and it is expected that production vehicles will be conversions of chassis already built.

TAM medium tank taking part in high-altitude trials in Ecuador in early 1988

Rocket launchers
The Israel Military Industries LAR 160 multiple rocket system has had trials on an Argentinian TAM chassis with the original turret replaced by two 18-round launchers.

The multiple rocket launcher version is called the VCLC (Vehiculo de Combate Lanza Cohetes). In addition to being fitted with the LAR 160 system, also known as the CAL, it can also be fitted with a new 350 mm system called the CAM, details of which are given in the *Multiple rocket launchers* section.

Production of the TAM and VCTP was completed in 1985/1986 and according to reports in Argentina in early 1986 some 100 vehicles were still in the factory undelivered.

Status: Production complete. In service with Argentina.

Manufacturer: TAMSE
Avenida, Rolon 1445
Boulogne Sur Mer
CP 1609
Buenos Aires
Argentina
Telephone: (763) 114/4937/0714
Telex: 26208 TAMSE AR
Fax: (763) 2464

TAM medium tank with roof-mounted 7.62 mm machine gun

Scale model of VCRT armoured recovery vehicle in travelling configuration (Terry J Gander)

SPECIFICATIONS

CREW	4
COMBAT WEIGHT	30 000 kg
POWER-TO-WEIGHT RATIO	17.6 hp/tonne
GROUND PRESSURE	0.788 kg/cm²
LENGTH	6.75 m
WIDTH	3.29 m
HEIGHT OVERALL	2.66 m
HEIGHT HULL TOP	2.42 m
GROUND CLEARANCE	0.45 m
TRACK	2.62 m
MAXIMUM ROAD SPEED	75 km/h
RANGE	590 km
RANGE WITH LONG-RANGE TANKS	940 km
FUEL CAPACITY	640 + 400 l
FORDING	1.5 m
GRADIENT	60%
SIDE SLOPE	30%
VERTICAL OBSTACLE	1 m
TRENCH	2.5 m
ENGINE	MTU MB 833 Ka 500 6-cylinder diesel developing 720 hp at 2200 rpm
TRANSMISSION	Renk HSWL-204 planetary with 4 speeds forward and reverse
STEERING	double differential with hydrostatic steering control
SUSPENSION	torsion bar
ELECTRICAL SYSTEM	24 V
BATTERIES	8 × 12 V, 100 Ah
ARMAMENT	
main	1 × 105 mm gun
coaxial	1 × 7.62 mm MG
anti-aircraft	1 × 7.62 mm MG
SMOKE LAYING EQUIPMENT	2 × 4 smoke dischargers
AMMUNITION	
main	50
machine gun	6000
GUN CONTROL EQUIPMENT	electro-hydraulic/manual
by commander	yes
by gunner	yes
commander's override	yes
Turret traverse	360°
Gun elevation/depression	+18°/-7°
Gun stabiliser	
vertical	yes
horizontal	no

BRAZIL

ENGESA EE-T1 Osorio Main Battle Tank

Development

The EE-T1 Osorio main battle tank has been developed by ENGESA to meet the requirements of both the home and export markets. The first prototype was completed in October 1984; a second prototype followed in April 1985. In addition there is an automotive test rig. Vickers Defence Systems of the UK has designed and built two turrets for the Osorio. The first of these was delivered to Brazil in Spring 1985. It was fitted with a Royal Ordnance 105 mm L7A3 gun with a modified recoil system; the second turret was armed with a 120 mm GIAT smoothbore gun and was delivered to Brazil late in 1985. In mid-1985 the first prototype went to Saudi Arabia for trials. The second prototype of the Osorio (P2) went to Saudi Arabia in the Summer of 1987 for competitive trials with the Challenger 1, AMX-40 and M1A1 MBTs and in early 1988 the EE-T1 (120 mm) and M1A1 were both shortlisted to meet the requirements of Saudia Arabia, although no firm orders had been placed for the EE-T1 by April 1991. The Brazilian Army has a requirement for 500 MBTs of this type, although Bernardini is also competing for this order with a 105 mm version of its Tamoyo MBT. To reduce development time and cost, available well-proven components have been used wherever possible in the design.

The concept of the EE-T1 is that the basic design can be adapted to individual customer requirements and it is capable of modernisation in the future as the threat changes and new systems and equipment become available.

Description

The layout of the EE-T1 is conventional with the driver's compartment at the front, turret in the centre and engine and transmission at the rear. The hull and turret are of all-welded steel construction with the front and sides well-sloped to provide the maximum possible protection within the weight limits of the vehicle.

According to ENGESA the armour of the EE-T1 is a combination of the ENGESA designed bi-metallic plate and a new composite armour also developed by ENGESA.

Second prototype of Osorio MBT (P2) armed with 120 mm GIAT smoothbore gun

Prototype of EE-T1 Osorio MBT armed with 105 mm L7A3 gun and roof-mounted 7.62 mm machine gun

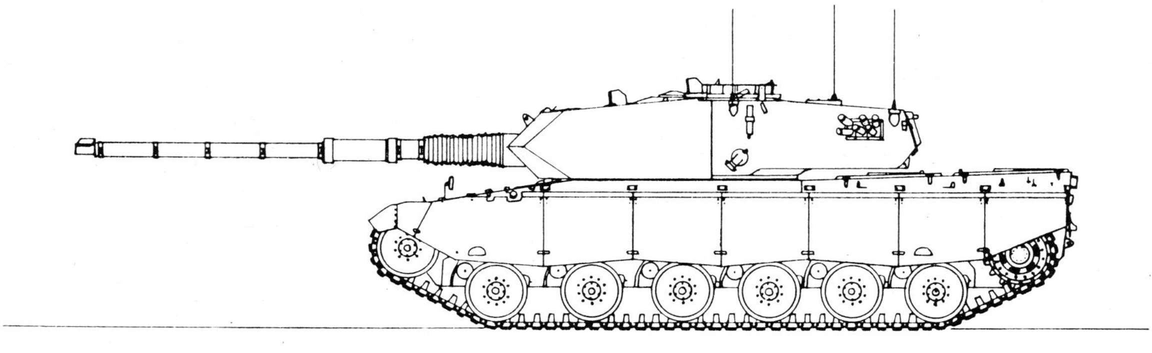

Provisional drawing of EE-T1 armed with 105 mm gun

Second prototype of Osorio MBT (P2) armed with GIAT 120 mm smoothbore gun with turret traversed right

Second prototype of Osorio MBT (P2) armed with GIAT 120 mm smoothbore gun and roof-mounted 12.7 mm M2 HB machine gun during trials in Brazil

The driver sits at the front of the hull on the left and has a single piece hatch cover and a periscope for driving closed down; the periscope can be replaced by a passive periscope for night driving.

The commander sits on the right of the turret with the gunner forward and below his position, the loader sitting on the turret left. Both the commander and loader have hatch covers in the turret roof and the commander has periscopes for all-round observation.

The MWM TBD 234 four-stroke, water-cooled, turbo-charged V-12 diesel is mounted at the rear and coupled to a ZF LSG 3000 fully-automatic transmission. The power pack can be removed in 10 minutes and re-installed in another 10 minutes. It can be operated outside the chassis for trouble shooting and tuning. Power-to-weight ratio is 26.9 hp/tonne and acceleration from 0 to 30 km/h takes 10 seconds. The fully automatic transmission has a torque converter (with lock up) and a hydraulic brake (retarder) with four forward and two reverse gears. The EE-T1 has a combined brake system in which only the retarder (installed in the transmission) works in the first stage with the disc brakes coming into operation in the second stage. A parking brake is also fitted.

Suspension is hydro-pneumatic, each side having six dual rubber-tyred road wheels, idler at the front, drive sprocket at the rear and three track return rollers. Dampers are provided on the first, second and sixth road wheel stations. In late 1983 Dunlop Aviation Division of the UK was awarded a contract for the design and development of the hydro-pneumatic suspension system for the Osorio; this has been given the company name of Dunloride and the first two sets were delivered late in 1984. The centre guide, double-pin tracks have replaceable rubber pads.

Main armament is a Royal Ordnance L7A3 rifled 105 mm gun with a modified recoil system or a GIAT 120 mm smoothbore tank gun with an elevation of +20°, a depression of −10° and turret traverse of 360°. Turret traverse and weapon elevation are electric with both the commander and gunner having controls, the former having overriding control. A 7.62 mm electrically fired machine gun is mounted coaxially to the left of the main armament and a 7.62 mm or 12.7 mm machine gun can be mounted on the roof for anti-aircraft defence. Mounted either side of the turret is a bank of six electrically operated smoke dischargers.

When fitted with a 120 mm gun, 38 rounds are carried and when fitted with a 105 mm gun, 45 rounds are carried. In each case 12 rounds are carried in the turret with the remainder in the hull. In the event of the ammunition in the turret bustle igniting, the main force of the resulting explosion will be vented upwards through blow off panels on the turret roof provided that the ammunition doors are shut.

Two main fire control options are offered for the EE-T1. The first is an integrated fire control system with a day/night gunner's periscope with laser rangefinder and a day/night commander's periscope. The gunner's fire control system takes into account the following data: ammunition type, ammunition temperature, range, target speed, cant angle and cross-wind. The commander can aim and fire at targets by overriding the gunner and using the latter's laser rangefinder and fire control system by remote control.

The advanced fire control system (the British Marconi Command and Control Systems Centaur) uses independent line of sight and coincidence window principles, which allows the main armament to be laid and fired while the tank is moving with very high first hit probability, includes a SFIM gyro-stabilised day gunner's periscope with laser rangefinder, SFIM gyro-stabilised panoramic day commander's periscope with laser rangefinder and a gyro-stabilised panoramic thermal periscope with micro monitors in both gunner and commander eyepieces. The following data can be fed into the fire control system manually or automatically: ammunition type, ammunition temperature, range, target speed, cant angle, cross-wind, ambient temperature, ambient pressure and bore wear. After calculating all of the information available an aiming mark is injected into the gunner's sight or commander's sight.

Optional equipment for both versions includes a loader's cupola with periscope with a night-vision option and a combat periscope.

The main systems of the chassis and fire computer have advanced built-in auto diagnosis.

The EE-T1 MBT has two fire-extinguishing systems. The engine bay system is operated manually or automatically; the crew bay fire-suppression system is operated automatically.

Turret options include radios, collective or individual NBC protection, air conditioning, azimuth indicator in mils, clinometer in mils, electronic compass and a laser detector. Chassis options include a passive periscope for the driver, crew heater, land navigation system and multi-fuel APU. The latter is mounted at the rear of the vehicle on the right hand side.

Variants

ENGESA is currently studying the possibility of installing a 155 mm gun turret on the chassis of the EE-T1 as well as twin 30 mm or twin 35 mm anti-aircraft turrets. Armoured recovery and bridgelayer versions of the EE-T1 will be developed.

Status: Development complete. Ready for production on receipt of orders.

Manufacturer: ENGESA Engenheiros Especializados SA,
Avenue Tucunare, 125/211
Caixa Postal/PO Box 152/154/06400 Barueri, SP
Brazil
Telephone: (55 11) 421 4711
Telex: 1171302 enes br
Fax: (55 11) 421 4445

SPECIFICATIONS (data in square brackets relates to 105 mm version where different to 120 mm model)

CREW	4
COMBAT WEIGHT	43 700 [40 900] kg
POWER-TO-WEIGHT RATIO	25 [26.9] hp/tonne
GROUND PRESSURE	0.85 [0.72] kg/cm²
LENGTH GUN	
forwards	10.10 [9.36] m
rear	9.30 [8.56] m
LENGTH HULL	7.13 m
WIDTH	3.26 m
WIDTH OVER TRACKS	3.2 m
HEIGHT TO TURRET TOP	2.37 m
HEIGHT TO TOP OF SIGHT	2.89 [2.68] m
GROUND CLEARANCE	0.46 m
TRACK	2.63 m
TRACK WIDTH	570 mm
LENGTH OF TRACK ON GROUND	4.49 m
MAX ROAD SPEED	70 km/h
MAX RANGE	550 km
FORDING	1.2 m
with preparation	2 m
GRADIENT	60%
SIDE SLOPE	30%
VERTICAL OBSTACLE	1.15 m
TRENCH	3 m
ENGINE	MWM TBD 234 4-stroke, 12-cylinder, water-cooled turbo-charged diesel developing 1100 hp at 2350 rpm
TRANSMISSION	ZF LSG 3000 fully-automatic with torque converter, four forward and two reverse gears
SUSPENSION	hydro-pneumatic
ARMAMENT	
main	1 × 120 mm gun [1 × 105 mm]
coaxial	1 × 7.62 mm MG
anti-aircraft	1 × 7.62 mm or 1 × 12.7 mm MG
smoke dischargers	2 × 6
AMMUNITION	
main	38 rounds of 120 mm (12 in turret, 26 in hull) 45 rounds of 105 mm (12 in turret, 33 in hull)
machine gun	5000 rounds of 7.62 mm or 3000 rounds of 7.62 mm and 900 rounds of 12.7 mm
GUN CONTROL EQUIPMENT	
Turret power control	electric/manual
by commander	yes
by gunner	yes
Commander's override	yes
Gun elevation/depression	+20°/−10°
Max rate of elevation	200 mils/s
Min rate of elevation	0.25 mils/s
Turret traverse	360°
Max rate of traverse	600 mils/s
Min rate of traverse	0.25 mils/s

Bernardini MB-3 Tamoyo Battle Tank

Development

In the late 1970s Bernardini designed the X-30 tank based on an M41 light tank chassis. It weighed 30 tonnes and would have been armed with a 90 mm or 105 mm gun but was not built; further work resulted in the Tamoyo tank.

Construction of the first prototype began in 1982 and the chassis ran for the first time in November 1983. Turret/hull integration took place in mid-1984 and trials began later the same year. By early 1988 three 90 mm prototypes of the MB-3 had been built. The Tamoyo was originally designated the XMB-3 but after trials with the prototype were successfully completed the X was dropped from the designation, so the vehicle became the MB-3. The Brazilian Army is understood to have a requirement for 500 tanks and both the ENGESA EE-T1 and MB-3 are competing for this order.

Description

The all-welded hull of the MB-3 Tamoyo is divided into three compartments, the driver's at the front, fighting in the centre and the engine and transmission at the rear. Spaced armour is incorporated in some areas and laminate armour is said to be under consideration for production vehicles.

The driver sits at the left front of the vehicle and has periscopes for forward observation and a single piece hatch cover opening to the right.

The all-welded turret is in the centre of the hull with the commander and gunner on the right and the loader on the left. Main armament comprises a Bernardini 90 mm gun that can fire a specially developed APFSDS-T round. Mounted coaxially with the main armament is a 12.7 mm M2 HB machine gun. A 7.62 mm machine gun is mounted on the turret roof for anti-aircraft defence and a bank of four electrically-operated smoke dischargers is on either side of the turret. The fire control system of the Tamoyo includes a computer, day/passive night sights for commander and gunner, and each sight has a laser rangefinder. The gunner has an auxiliary telescope mounted coaxially with the main armament.

The Brazilian-built Saab-Scania DSI-14 diesel will develop 500 hp at 2100 rpm, but it is hoped to increase this to 650/700 hp, so raising the power-to-weight ratio to 22.5 hp/tonne. The engine is coupled to the same Allison Transmission Division of General Motors, CD-500-3 transmission as in the M41 light tank.

Cutaway drawing of the Bernardini Tamoyo III MBT which is armed with a Royal Ordnance 105 mm L7A3 rifled tank gun, coaxial machine gun and roof-mounted 7.62 mm anti-aircraft machine gun

Prototype of Bernardini MB-3 Tamoyo MBT armed with 90 mm gun

Prototype of Bernardini Tamoyo III MBT which has a different turret and weapon system to that of the Bernardini MB-3 Tamoyo

Suspension is of the torsion bar type and consists of six dual rubber-tyred road wheels with the drive sprocket at the rear, idler at the front and three track return rollers. Hydraulic shock absorbers are fitted at the first, second and sixth road wheel stations.

Standard equipment includes an NBC system, heater, fire extinguishers, bilge pump, hull escape hatch, radios and internal communications equipment.

Variants

Tamoyo III

In mid-1987 Bernardini completed the prototype of the Tamoyo III tank. This is powered by the Detroit Diesel 8V-92TA engine developing 736 hp at 2300 rpm coupled to Allison Transmission Division of General Motors, CD-850 series automatic transmission with two forward and one reverse gears. The suspension is identical to the earlier models with six road wheels, the idler at the front, drive sprocket at the rear and three track return rollers.

The weight of the Tamoyo III has increased to 31 tonnes as both spaced and composite armour is used in the hull and turret. The redesigned turret now mounts a Royal Ordnance 105 mm L7A3 rifled tank gun for which 50 rounds of ammunition are carried, 18 in the turret and the remainder in the hull.

General Electric of the USA has delivered an HMPT-500 hydro-mechanical steering transmission to Bernardini and this will be fitted in one of the Tamoyo vehicles in the future.

Basic specifications of the Tamoyo III are as follows:

WEIGHT	31 000 kg
LENGTH	
gun forwards	8.9 m
hull	6.3 m
WIDTH	
(over side skirts)	3.2 m
HEIGHT	
(commander's cupola)	2.65 m
RANGE	600 km
ROAD SPEED	67 km/h
ACCELERATION	
(0 - 32 km/h)	7.2 s

At least two other variants are under consideration, an armoured recovery vehicle and an anti-aircraft vehicle, the latter probably armed with a Bofors 40 mm L/70 anti-aircraft gun.

Status: Development of the 90 mm model is complete and production will commence on receipt of firm orders. Prototype of 105 mm version completed in mid-1987.

Manufacturer: Bernardini S/A Indústria e Comércio
Rua Hipólito Soares No 79
04201 Sao Paulo
SP Brazil
Telephone: (011) 274 8033
Telex: (011) 21605 bsaibr
Fax: (011) 274 8567

SPECIFICATIONS		MAX ROAD SPEED	67 km/h	SUSPENSION	torsion bar
CREW	4	FUEL CAPACITY	700 l	ARMAMENT	
COMBAT WEIGHT	30 000 kg	RANGE	550 km	main	1 × 90 mm gun
UNLOADED WEIGHT	28 000 kg	FORDING	1.3 m		(or 105 mm as option)
POWER-TO-WEIGHT		GRADIENT	60%	coaxial	1 × 12.7 mm MG
RATIO	16.6 hp/tonne	SIDE SLOPE	30%	anti-aircraft	1 × 7.62 mm MG
	(or 24.5 hp/tonne with	VERTICAL OBSTACLE	0.71 m	AMMUNITION	
	8V-92TA engine)	TRENCH	2.4 m	(90 mm)	68
GROUND PRESSURE	0.72 kg/cm²	ENGINE	Saab-Scania DSI-14	GUN CONTROL	
LENGTH			diesel developing	EQUIPMENT	
hull	6.5 m		500 hp at 2100 rpm	Turret power control	electric/manual
gun forwards	8.77 m		(optional General	Gun elevation/depression	+18°/-6°
gun rear	7.4 m		Motors 8V-92TA	Turret traverse	360°
WIDTH	3.22 m		developing 736 hp at	SMOKE-LAYING	
HEIGHT			2300 rpm)	EQUIPMENT	2 × 4 smoke
overall	2.5 m	TRANSMISSION	Allison Transmission		dischargers
to hull top	2.2 m		Division of General	ELECTRICAL SYSTEM	24 V
GROUND CLEARANCE	0.5 m		Motors CD-850	BATTERIES	4 × 12 V
TRACK	2.6 m		automatic or General		
TRACK WIDTH	530 mm		Electric HMPT-500-3		
LENGTH OF TRACK			hydro-mechanical		
ON GROUND	3.9 m		with 3 forward and one		
			reverse gears		

CHINA, PEOPLE'S REPUBLIC

NORINCO Type 80 Main Battle Tank

Development

The existence of a new Chinese MBT first became apparent in 1985. It was called the Type 69 III by the Chinese but late in 1986 it was renamed the Type 80.

The Type 80 is a brand new chassis with a turret similar to that installed on the Type 69 MBT but armed with a 105 mm rifled tank gun and with a computerised fire control system.

Description

The hull of the Type 80 is of all-welded steel construction and has the driver's compartment at the front, the turret in the centre and the engine and transmission at the rear.

The driver sits at the front of the hull on the left side and has a single piece hatch cover that lifts and swings to the left. In front of this are two day periscopes, one of which can be replaced by an infra-red periscope for driving at night.

Mounted on the well-sloped glacis plate is a splash board to stop water rushing up and into the driver's compartment while the tank is fording.

To further improve the Type 80's protection a panel of composite armour can be added at the front.

The turret is in the centre of the vehicle with the commander on the left, the gunner forward and below the commander and the loader on the right. Composite armour can also be added to the turret for improved protection.

Type 80 MBT from rear showing stowage basket and mounting arms on hull rear for long range fuel tanks

The commander has a cupola that can be traversed manually through a full 360° and a single forward-opening hatch cover with two integral periscopes for observation to the sides. Forward of this hatch cover are a further three periscopes for observation over the frontal arc. The loader has a circular hatch cover that opens to the left.

Main armament comprises a 105 mm rifled gun that fires Western- or Chinese-produced ammunition. This has powered elevation from -4° to +18° with turret having powered traverse through 360°. The 105 mm gun, which is similar in appearance to the Western L7/M68, is fitted with a fume extractor and a thermal sleeve. The 105 mm gun fires the following NORINCO developed rounds; APFSDS-T (m/v 1455 m/s), HEAT-T (1173.5 m/s), HESH (731.5 m/s) and HE (850 m/s).

A 7.62 mm machine gun is mounted coaxially with the 105 mm gun and there is a 12.7 mm machine gun on the loader's hatch for anti-aircraft use.

Mounted either side of the turret are four forward-firing electrically-operated smoke dischargers. A stowage basket runs around the sides and rear of the turret.

The fire control system includes a ballistic computer, a laser rangefinder integrated into a stabilised sight, sensors and a gun stabilisation system and is believed to be the NORINCO ISFCS-212 image-stabilised fire control system that has been tested in a Type 59 MBT.

Main components of the ISFCS-212 include a stabilised gunner's sight incorporating a laser rangefinder, console, tilt sensor, ballistic computer, crosswind sensor, control unit and a two-axis gun stabiliser.

The ISFCS-212 enables the gunner to engage stationary and moving targets while his own tank is stationary or moving, under both day and night conditions with a high first-round hit probability. NORINCO claims that a typical target engagement time from detection to firing is just six seconds. The system has two operating modes, image-stabilised and graticule automatic setting.

The gunner's sight is independently stabilised and the low light level sight is interchangeable with a ×1 periscope.

Four sensors feed information into the ballistic computer automatically (target angular rate in elevation and traverse, tilt and crosswind) with a further six manual inputs, charge temperature, air temperature, muzzle velocity, manual crosswind, manual distance and synthetic corrections in both elevation and traverse, and types of 105 mm ammunition.

The system has an interface with the commander's day/night periscope to enable him to override the gunner or to slew the turret to engage a different target.

The engine is probably a turbo-charged version of the Type 12150L-7BW installed in the Type 69, but whereas the latter has a single rectangular exhaust outlet on the left side of the hull, just above the fourth road wheel, Type 80 has two circular exhaust outlets on the right

Type 80-II MBT which is armed with a 105 mm rifled gun fitted with fume extractor and light alloy thermal sleeve

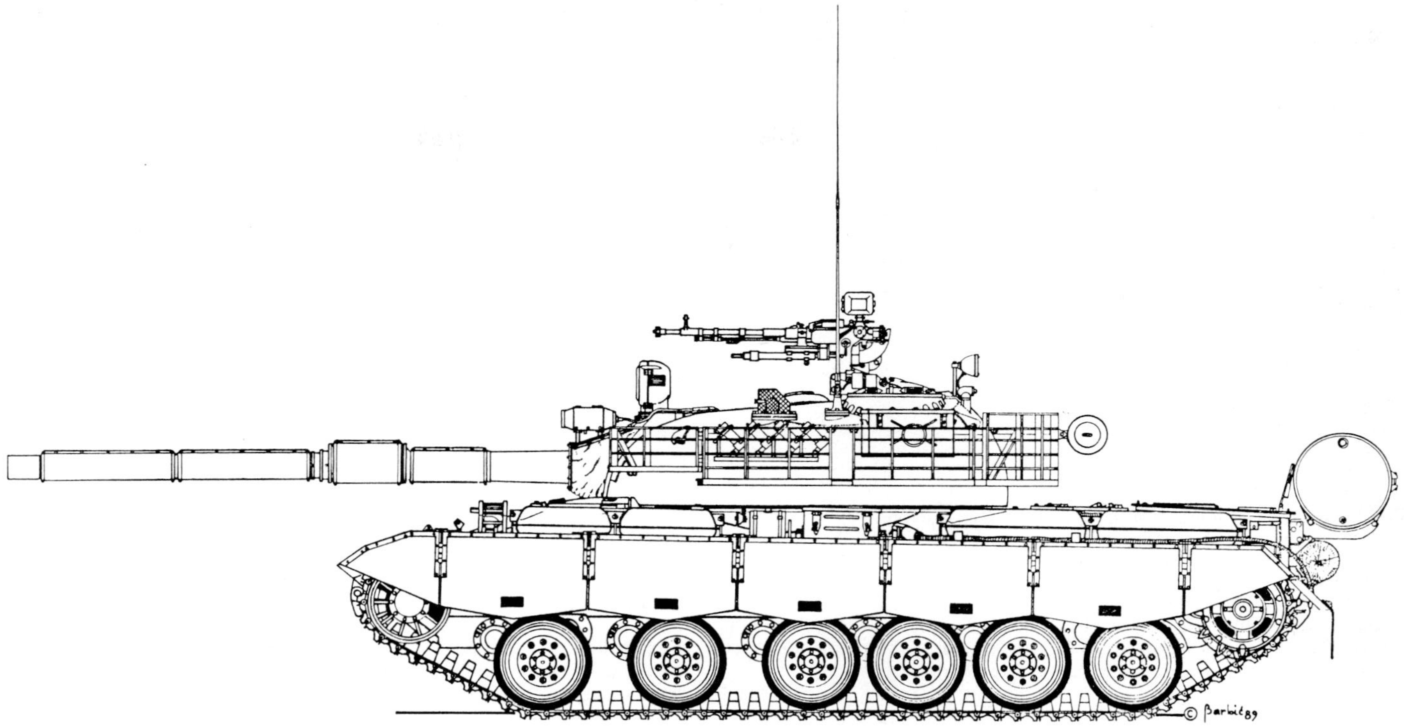

Provisional drawing of Type 80-II MBT complete with long range fuel tanks and unditching beam at rear (not to 1/76th scale) (Vasco Barbic)

Type 80 MBT armed with 105 mm rifled tank gun fitted with fume extractor and thermal sleeve

Scale model of the Type 85-II MBT which has a welded steel turret unlike earlier Chinese MBTs which had a cast steel turret

side of the hull. Diesel fuel can be injected so the tank can lay its own smoke screen while moving.

To extend the operational range of the tank two large drum fuel tanks can be mounted at the rear; these can be jettisoned if required.

The torsion bar suspension is new with six dual rubber-tyred road wheels, drive sprocket at the rear, idler at the front and three track-return rollers. The first, second, fifth and sixth road wheel stations have hydraulic shock absorbers.

The upper part of the track is covered by skirts with a wavy bottom which are hinged up to allow access to the suspension for maintenance.

Standard equipment includes an overpressure NBC system, night vision equipment for the commander, gunner and driver, and provision to install a snorkel in the roof forward of the loader's hatch; this allows the tank to submerge to a depth of 5 m and cross rivers up to 600 m wide.

Communications equipment installed includes a YRC-83 radio and VIC-8 intercom system. To further increase the tank's survivability on the battlefield an automatic explosion suppression and fire extinguishing system is fitted.

Variants

Type 80-II

This is virtually identical to the basic Type 80 except that it has a combat weight of 38.5 tonnes and a power-to-weight ratio of 18.96 hp/tonne. Specifications are identical except that it has an overall length of 9.336 m. The Basic Type 80 has a Type 889 radio and a Type VIC-8 intercom while the Type 80-II has a Type VRC-83 radio. The Type 80-II may also have a semi-automatic transmission in place of the standard manual transmission and both the Type 80 and the Type 80-II can be fitted with a different fire control system consisting of a projecting sight, laser rangefinder, ballistic computer and a bi-directional stabiliser.

Type 85-II and Type 85-IIA

Early in 1989, NORINCO stated that it had developed two further vehicles based on the Type 80 MBT chassis, these being called the Type 85-II and Type 85-IIA. These have been designed to meet the requirements of the People's Liberation Army (PLA) with firing trials being carried out in mid-1989.

Whereas all previous Chinese MBTs had a cast steel turret, both Type 85s have a welded turret which will be easier to fit with additional armour protection to improve its battlefield survivability. Composite armour can be fitted to both the hull and turret.

The exact difference between the Type 85 and Type 85-II has not been disclosed although the latter is slightly heavier at 39.5 tonnes and is also slightly shorter in length.

Armament is the same as in the Type 80 except that the Type 85-II can carry 46 rounds of 105 mm while the Type 85-IIA can carry 44 rounds. It has the same engine as the Type 80, but this may be coupled to a semi-automatic transmission.

The basic Type 80 has an individual NBC system, while the Type 80-II, Type 85-II and Type 85-IIA have a collective system, the latter tanks also have a Type 889B radio system.

Status: Development complete. Ready for production. Some sources have indicated that this vehicle is already in production.

Manufacturer: Chinese state factories
Enquiries to China North Industries Corporation
PO Box 21377A, Yuetan Nanjie
Beijing
People's Republic of China
Telephone: (86) 6898/3461/3471/7570
Telex: 22339 cnin cn

SPECIFICATIONS (TYPE 80)		LENGTH OF TRACK		AMMUNITION	
CREW	4	ON GROUND	4.064 m	main	44
COMBAT WEIGHT	38 000 kg	MAX ROAD SPEED	55 - 60 km/h	coaxial	2250
POWER-TO-WEIGHT		MAX ROAD RANGE	430 km	anti-aircraft	500
RATIO	19.2 hp/tonne	FUEL CAPACITY	1400 l	GUN ELEVATION/	
LENGTH		VERTICAL OBSTACLE	0.8 m	DEPRESSION	+18°/-4° powered/
gun forwards	9.328 m	TRENCH	2.7 m		manual
gun rear	8.611 m	FORDING	1.4 m	TURRET TRAVERSE	360°,
hull	6.325 m	FORDING WITH SNORKEL	5 m		powered/manual
WIDTH		GRADIENT	60%		
over tracks	3.354 m	SIDE SLOPE	40%		
over skirts	3.372 m	ENGINE	V-12 diesel developing		
HEIGHT			730 hp		
turret roof	2.290 m	SUSPENSION	torsion bar		
including 12.7 mm AA MG	2.874 m	ARMAMENT			
firing	1.784 m - 1.798 m	main	1 × 105 mm gun		
GROUND CLEARANCE	0.48 m	coaxial	1 × 7.62 mm MG		
TRACK	2.7 m	anti-aircraft	1 × 12.7 mm MG		

NORINCO Type 79 Main Battle Tank

Development

This made a brief appearance in the October 1984 parade in Beijing when it was referred simply as Main Battle Tank. Its chassis is similar to the Type 59/Type 69 but has the towing eyes under the nose and a headlamp arrangement similar to the Type 80 MBT.

Chinese Type 79 MBT with 105 mm gun and smoke dischargers mounted either side of main armament

Some reports have indicated that the Type 79 is a Type 69 fitted with a 105 mm gun and other minor improvements with some vehicles being built as Type 79s and some converted from existing Type 69 vehicles. As far as it is known the Type 79 has not been offered for export.

Description

Its turret is slightly different to the Type 59/Type 69 as there is a new large rectangular window located ahead of the commander's cupola which could be an internally mounted laser rangefinder. Mounted either side of the turret is a bank of four electrical smoke dischargers which are very similar to the German Wegmann type and to the immediate rear of these is a metal stowage box which is believed to house additional grenades.

Main armament comprises a 105 mm rifled tank gun which is provided with a fume extractor and a thermal sleeve. A 7.62 mm machine gun is mounted coaxially with the main armament and a 12.7 mm machine gun is mounted on the loader's cupola for local and anti-aircraft defence. There are slight differences in the rear of the hull which indicate a different powerpack.

Chinese sources state that the Type 79 weighs 37.5 tonnes compared to the 36 tonnes of the Type 59 and has a maximum road speed of 50 km/h. Standard equipment includes a simplified fire control system and night vision equipment for commander, gunner and driver.

The upper part of the suspension of the Type 79 is covered by five removable skirts and the tank has also been observed fitted with new tracks with removable rubber pads in place of the more normal all-steel track fitted to other Chinese MBTs.

Status: Production complete. In service with the Chinese Army.

Manufacturer: Chinese state factories.
Enquiries to China North Industries Corporation
PO Box 21377A, Yuetan Nanjie
Beijing
People's Republic of China
Telephone: (86) 6898/3461/3471/7570
Telex: 22339 cnin cn

NORINCO Type 69 Main Battle Tank

Development

The Type 69 MBT made its first public appearance during a parade near Zhangiakou outside Beijing in September 1982. A few sources state that production of the Type 69 began in 1969; most sources say that volume production did not start until the early 1980s. Details of recent Chinese tank production is given in the following entry on the Type 59 MBT.

Iraq placed orders with China for between 100 and 200 Type 69-I tanks armed with the 100 mm gun and first deliveries of these were made in 1983 via Saudi Arabia. Total deliveries of the Type 69 MBT to the Gulf combatants are estimated at between 1800 and 2500 over a three year period up to late 1987.

In early 1987, Thailand placed orders for substantial quantities of Chinese military equipment including Type 69-II MBTs and by late 1988, 100 of these had been delivered out of a total requirement of around 500 vehicles. They equip the 1st Cavalry Division, 2nd Armoured Division and the tank battalions of the four reserve infantry divisions. The Type 69-II is called the Type 30 MBT by the Royal Thai Army.

Pakistan is now manufacturing the 105 mm Type 69-II MBT under licence. Additional details are given under Pakistan.

Description

First production Type 69s were produced with rifled and smoothbore guns but extensive trials demonstrated that the rifled gun was more accurate and had greater armour penetration characteristics than the smoothbore gun. After 150 Type 69s had been built with smoothbore

Type 69 MBTs on parade in Zhangiakou in September 1982. Note infra-red night vision equipment, laser rangefinder over main armament and hole in glacis plate for 7.62 mm machine gun (via Conmilit)

The laser rangefinder is mounted externally above the 100 mm gun and is vulnerable to shell splinters and small arms fire. The TSFCS-L, which is an option for the Type 69-II, has the laser rangefinder combined with the gun sight in a single unit mounted inside the tank.

The Type 69 can also be fitted with the TSFCS-C, which can also be retrofitted into existing T-series tanks of Chinese and Soviet origin. This system consists of the gunner's sight, laser rangefinder, fire control computer, control panel and sensors for elevation and azimuth rates and cant, cross-wind, air temperature, and ammunition charge temperature are entered into the computer manually as required by the crew. The computer calculates target range, direction, lead angle and also interfaces with the weapon stabilisation system.

The driver's day periscope can be replaced by a night periscope with a range of 60 m and a 30° field-of-view. The tank commander's Model 69 infra-red sight has a day magnification of ×5 and a 12° field-of-view and a night magnification of ×6 and 8° field-of-view. The Model 70 gunner's infra-red sight has a magnification of ×7 with a 6° field-of-view and an effective range of 800 m.

The Type 69-II has a model 889 radio and a Model 803 intercom for the commander, gunner, loader and driver. There are two command versions of the MBT. The Type 69 II Mk B command tank has one Model 889 radio and one Model 892 radio while the Type 69 II Mk C has two Model 889 radios.

Some Type 69 MBTs have been observed fitted with an external stowage bin on the turret rear and armour protection for the anti-aircraft machine gun. Thai Army Type 69-IIs have had their 12.7 mm weapons replaced by US 12.7 mm M2 Browning MGs.

The Type 69 MBT has a complete NBC system, semi-automatic fire extinguishing system and like Soviet T-series tanks can lay a smoke screen by injecting diesel fuel into the exhaust system on the left side of the hull. The vehicle is also coated with infra-red reflecting paint.

Type 69-II MBT armed with 100 mm rifled gun with thermal sleeve (via G Jacobs)

Chinese Type 69-II MBT with laser over main armament. This vehicle was captured by Iraq during the Iran/Iraq conflict

Variants

Type 69-II

This is almost identical to the Type 69-I but has a 100 mm rifled gun firing Chinese developed HEAT, HE, APHE and APFSDS ammunition. At least three types of APFSDS have been developed including one with a semi-combustible cartridge case. The Type 69-II has a different fire control system to the Type 69-I's and the main armament is stabilised in both elevation and azimuth. The 100 mm gun has an increased hit probability as the Tank Simplified Fire Control System (TSFCS) has been installed. The Type 69-II also has side skirts.

The TSFCS consists of three main components, the TLR1A tank laser rangefinder, a ballistic computer and the TGS-A tank gun sight. The TLR1A measures the distance to the target and this information is fed automatically into the ballistic computer which supplies information to control the automatic range-setting mechanism of the sight and simultaneously the gun control system, and set gun elevation required in order to hit the target.

guns it was decided to concentrate on the rifled gun which is the current production model. The Type 69-I has the smoothbore gun; the Type 69-II has the rifled gun and a different fire control system.

The Type 69-I MBT is a further development of the Type 59 and differs from it mainly in the areas of armament, fire control and night vision equipment. The actual layout of both vehicles is virtually identical so is not repeated.

The Type 69-I is armed with a 100 mm smoothbore gun as distinct from the 100 mm rifled gun installed in the earlier Type 59 MBT. This is slightly longer than the rifled gun and has a bore evacuator near the end of the muzzle. Types of ammunition that can be fired by the

Type 69-II MBT with 100 mm rifled gun with fume extractor near muzzle, laser rangefinder over main armament and side skirts

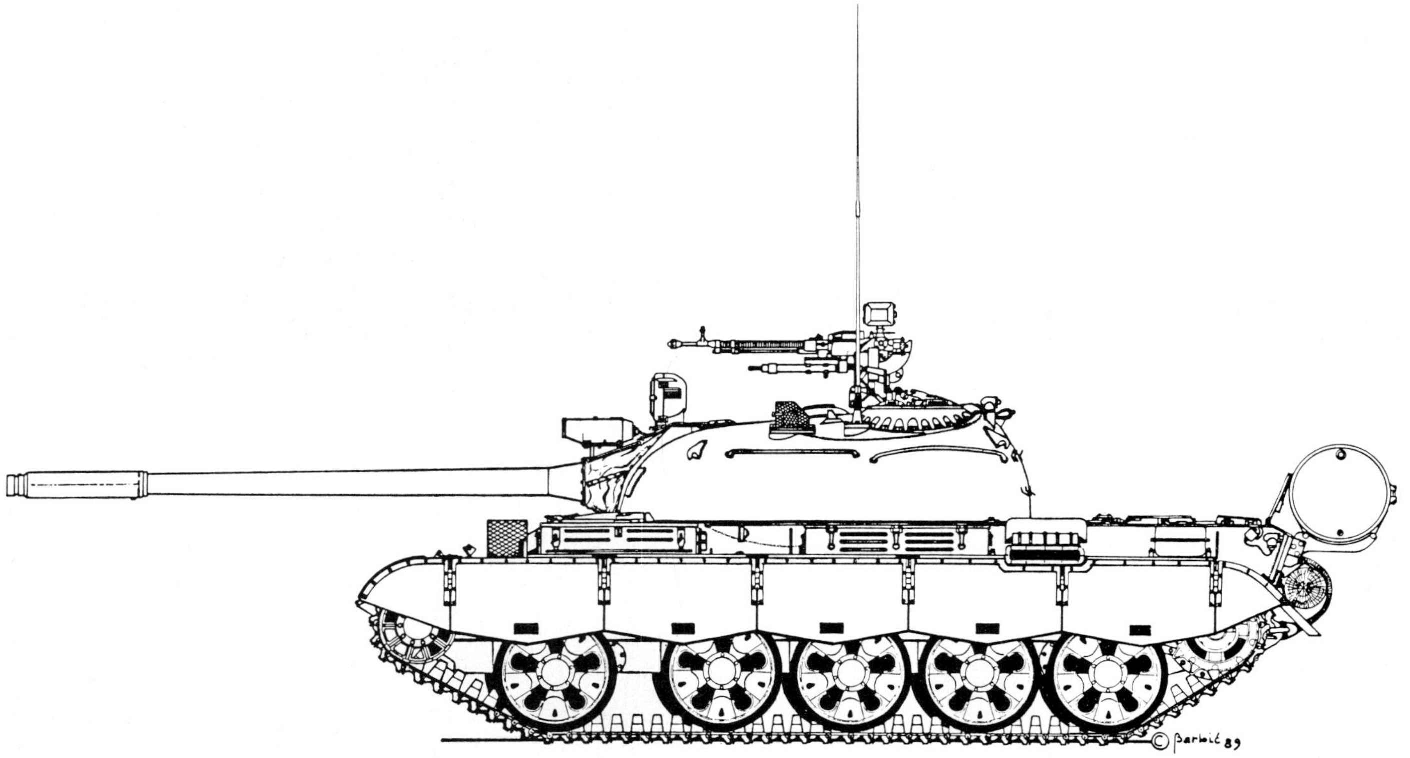

Type 69-II MBT with laser rangefinder over main armament and long range fuel tanks at hull rear (Vasco Barbic) (not to 1/76th scale)

smoothbore gun include anti-personnel, high explosive, high explosive anti-tank and high velocity armour-piercing discarding sabot.

A laser rangefinder, similar to that on the Type 59 MBT described in the following entry, is mounted externally over the main armament in front of the mantlet. This is reported to be manufactured by the Second Electronics Equipment Factory in Yángzhóu, Jiangsu Province.

Night vision equipment includes an infra-red searchlight on the commander's cupola, an infra-red sight for the gunner, an infra-red searchlight above and to the immediate right of the main armament and an infra-red driving light on each running board.

Mounted externally on the loader's cupola is a 12.7 mm Type 54 anti-aircraft machine gun. Mounted coaxially with the main armament is a 7.62 mm Type 59T machine gun and there is a similar weapon in the bow of the tank.

Twin 37 mm SPAAG

This is essentially the chassis of the Type 69 MBT fitted with a power operated two man turret armed with twin 37 mm cannon. This was first disclosed in 1986 and at that time only a prototype existed.

The commander is seated on the left and the gunner on the right with the twin 37 mm cannon between them, the latter have an elevation of +87° with a depression of -10° with turret traverse a full 360°.

No radar system is fitted, so this is a clear weather system only. NORINCO states that the twin 37 mm cannon have a maximum range of 4000 m in the air defence role with a maximum effective range of 3500 m. Three types of ammunition are fired; AP-T, HE-T and HE. Full details of this are given in *Jane's Land-Based Air Defence 1991-92*, page 58.

Twin 37 mm self-propelled anti-aircraft gun with radar scanner erected; this was first shown in 1988

Twin 37 mm SPAAG

During a defence exhibition held in Beijing late in 1988, another twin 37 mm self-propelled anti-aircraft gun system was shown for the first time.

This has a slightly different power operated turret of all-welded steel construction and is armed with the same twin 37 mm cannon fitted with flash suppressors. These weapons have a cyclic rate of fire of 360 to 380 rds/barrel/min with a total of 500 rounds of ready-use ammunition being carried. It has a crew of four and combat weight is 35 tonnes.

Mounted on the turret rear is a surveillance radar with a maximum range of 15 km and the fire control system includes a laser rangefinder with a maximum range of 8 km.

As far as is known, this system has yet to enter production. Full details of this are given in *Jane's Land-Based Air Defence 1991-92*, page 58.

Twin 57 mm Type 80 SPAAG

This is the Chinese equivalent of the old Soviet designed ZSU-57-2 twin 57 mm SPAAG based on a modified T-54 tank chassis. The Type 80 is a modified Type 69-II MBT chassis fitted with an open-topped turret armed with twin 57 mm cannon for which a total of 300 rounds of ammunition are carried.

Turret traverse is powered through a full 360° with a weapon elevation +85° and depression -5°. Maximum horizontal range is 12 000 m and maximum vertical range of 8800 m and effective anti-aircraft range is 5500 m. Types of ammunition fired are APC-T for engaging ground targets and HE-T for engaging air targets.

The Type 80 has a six man crew and a combat weight of 31 tonnes. Full details of this are given in *Jane's Land-Based Air Defence 1991-92*, page 58.

Twin 37 mm self-propelled anti-aircraft gun, clear weather version

Twin 57 mm Type 80 self-propelled anti-aircraft gun system

Type 84 AVLB laying its two part bridge over the front of the vehicle

Type 84 AVLB

The Type 84 armoured vehicle-launched bridge is essentially a Type 69 MBT with its turret removed, fitted with a bridge launching mechanism similiar to that on the German Leopard 1 Biber system. When opened out, the bridge, which is launched over the front of the vehicle, is 18 m long, can span gaps of up to 16 m and take tracked and wheeled vehicles weighing up to 40 tonnes. Full details of the Type 84 are given in *Jane's Military Vehicles and Logistics 1991-92*, page 76.

Type 653 ARV

This is based on the chassis of the Type 69 MBT and is designed not only to recover this and other vehicles on the battlefield but also to undertake major repairs such as changing complete powerpacks, clearing obstacles and preparing fire positions. Standard equipment includes a front-mounted hydraulically operated dozer blade, a hydraulic crane on the right side of the hull, a main winch with a capacity of 70 tonnes, auxiliary winch, tools, tow bars and cables. Full details are given in *Jane's Military Vehicles and Logistics 1991-92*, page 18. In addition to being used by the Chinese Army, and small quantities of these have been sold to Bangladesh, Iraq, Pakistan and Thailand as part of the Type 69 MBT sale.

Type 653 armoured recovery vehicle

SPECIFICATIONS		HEIGHT		SIDE SLOPE	40%
CREW	4	to axis of AA MG	2.807 m	ENGINE	Type 12150L-7BW V-12 diesel developing 580 hp at 2000 rpm
COMBAT WEIGHT	36 500-37 000 kg	with AA MG at max elevation	3.909 m		
POWER-TO-WEIGHT RATIO	15.89 hp/tonne	AXIS OF FIRE	1.75 m	SUSPENSION	torsion bar
GROUND PRESSURE	0.82-0.83 kg/cm²	GROUND CLEARANCE	425 mm	ARMAMENT	
LENGTH		TRACK	2.64 m	main	1 × 100 mm gun
gun forwards	8.657 m	LENGTH OF TRACK		coaxial	1 × 7.62 mm MG
gun rear	8.485 m	ON GROUND	3.845 m	bow	1 × 7.62 mm MG
hull	6.243 m	MAX ROAD SPEED	50 km/h	anti-aircraft	1 × 12.7 mm MG
WIDTH		MAX ROAD RANGE	420-440 km	AMMUNITION	
over skirts	3.298 m	VERTICAL OBSTACLE	0.8 m	main	34
over hull	3.27 m	TRENCH	2.7 m	7.62 mm	3400
		FORDING	1.4 m	12.7 mm	500
		GRADIENT	60%		

Status: In production. In service with China, Iran, Iraq, Pakistan (with local assembly/co-production to be undertaken at Taxila) and Thailand. Although most sources have indicated that Pakistan is undertaking production of a much improved Chinese Type 69-II MBT, it is possible that this could be based on the Type 80 or Type 85 MBTs or use components of some of these vehicles. Additional details are given under Pakistan later in this section.

Manufacturer: Chinese state factories
Enquiries to China North Industries Corporation
PO Box 21377A, Yuetan Nanjie
Beijing
People's Republic of China
Telephone: (86) 6898/3461/3471/7570
Telex: 22339 cnin cn

NORINCO Type 59 Main Battle Tank

Development/Description

In the early 1950s the Soviets supplied China with a quantity of T-54 MBTs, production of which was subsequently undertaken in China under the designation Type 59. The first production models were similar to the Soviet T-54 and not fitted with infra-red night vision equipment or a stabiliser for the main armament. Later models were fitted with a fume extractor as on the T-54A. More recent production Type 59s have been fitted with an infra-red searchlight for both the commander and gunner and a larger infra-red searchlight has been mounted above the main armament and moves in elevation with it. Some vehicles have been seen fitted with a laser rangefinder to the right of the infra-red searchlight mounted over the main armament.

The British company MEL supplied China with 30 sets of passive night vision equipment for the Type 59 MBT. These include the commander's image-intensified periscope Type DC 1026/00, the gunner's image-

Type 59 MBT fitted with side skirts and showing laser rangefinder over 100 mm gun

Pakistani Army Type 59 with standard 100 mm gun (via G Jacobs)

Chinese Type 59 MBT with 105 mm gun (via G Jacobs)

Type 59 MBT upgraded by Royal Ordnance for the export market and armed with a 105 mm L7 series rifled tank gun

Type 59 MBT from rear with 105 mm rifled tank gun (via G Jacobs)

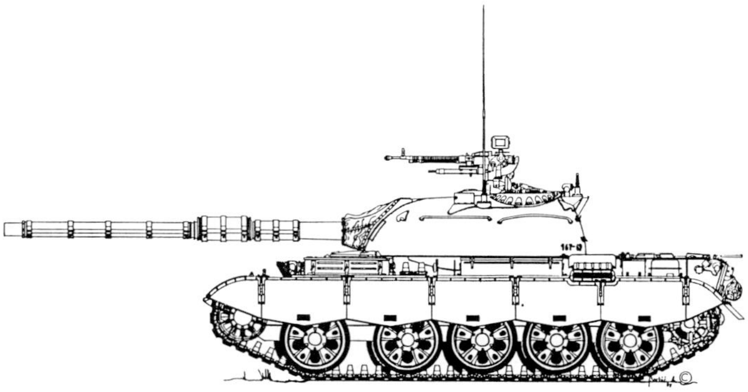

Type 59-II MBT with 105 mm rifled tank gun (not to 1/76th scale) (Vasco Barbic)

intensified periscope Type DC 1024/00 and the driver's image-intensified periscope Type DC 1028/00.

The Chinese have developed and placed in production a 100 mm APFSDS-T projectile for the Type 59 MBT. This is designated the AP100-2 and has a long rod penetrator and a semi-combustible cartridge case. The round has a saddle-type sabot and eight fins. All that remains after firing is the stub case which is ejected when the breech opens. NORINCO has also developed a 100 mm APFSDS-T round with a conventional brass cartridge case. This has a m/v of 1480 m/s and will penetrate 150 mm of armour at an angle of 65° at a range of 2400 m.

It is believed that production of the Type 59 tank began in 1957 to 1960 at Baotou in the Beijing military region. Throughout the 1970s between 500 and 700 Type 59 tanks were produced each year. In 1979 this rose to 1000, dropping to 500 in 1980 due to budget restrictions, then increasing to 600 in 1981, to 1200 in 1982 and 1500 to 1700 in 1983. The 1982 and 1983 figures probably include the Type 69 series tank as well.

Variants

NORINCO Type 59 Retrofit Package

In 1986 NORINCO announced that it was offering an upgrade package for the Type 59 MBT. The original 520 hp diesel will be replaced by a new 730 hp diesel to give a much higher power-to-weight ratio and therefore greater acceleration on the battlefield.

Other improvements include APFSDS-T ammunition for the 100 mm gun, rubber bushed tracks, NBC system, friction shock absorbers, gun stabilised in both elevation and traverse and a new fire control system incorporating spot injection in the gunner's sight. Since then, NORINCO has offered a complete retrofit package for Type 59 and T-54/T-55 tanks and details of this are given in the following entry.

Type 59 with 105 mm gun

This was first seen during a parade held in Beijing and is believed to be a Western 105 mm rifled tank gun fitted with the normal fume extractor and thermal sleeve. US sources refer to this version of the Type 59 as the M1984; it is believed that the Chinese designation for this vehicle is the Type 59-II MBT. China has developed an APFSDS round for the 105 mm rifled tank gun which is now fitted to a number of other Chinese MBTs including the Type 79 (qv), Type 80 (qv) and Type 85 (qv).

Type 59 with 105 mm L7 gun

As a private venture Royal Ordnance of the United Kingdom has fitted a Type 59 MBT with the 105 mm L7A3 rifled tank gun. This together with three other upgraded Type 59s were tested in Pakistan in 1987. Additional details are given under Pakistan in this section. Many manufacturers now offer items of equipment for installation on the Type 59 and the equivalent Soviet T-54/T-55.

Type 59 MBT with IR18 Thermal Imager

In 1985 the Royal Ordnance owned Type 59 MBT fitted with a 105 mm L7A3 rifled tank gun was fitted with a Barr and Stroud IR18 based thermal fire control system and this carried out a series of successful live firing trials at Bovington, Dorset. For these trials the IR18 based thermal sensor was mounted on a bracket on the left side of the mantlet so moving in elevation with the 105 mm gun.

Type 59 Chassis with Marksman Turret

Late in 1985 a Type 59 chassis was fitted with the private venture Marconi Command and Control Systems Marksman twin 35 mm air defence turret.

Basic ARV

This may well be a local modification and is essentially a Type 59 with its turret removed, armed with a single 12.7 mm machine gun. As this vehicle is not believed to have a winch, it is limited to towing operations.

SPECIFICATIONS	
CREW	4
COMBAT WEIGHT	36 000 kg
POWER-TO-WEIGHT RATIO	14.44 hp/tonne
GROUND PRESSURE	0.8 kg/cm²
LENGTH GUN FORWARDS	9 m
LENGTH HULL	6.04 m
WIDTH	3.27 m
HEIGHT	2.59 m
GROUND CLEARANCE	0.425 m
TRACK	2.855 m
TRACK WIDTH	580 mm
LENGTH OF TRACK ON GROUND	3.84 m
MAX ROAD SPEED	40-50 km/h
FUEL CAPACITY	
main	815 l
external	400 l
RANGE	
main fuel supply	420-440 km
main + external supply	600 km
FORDING	
without preparation	1.4 m
with preparation	5.5 m
GRADIENT	60%
VERTICAL OBSTACLE	0.79 m
TRENCH	2.7 m
ENGINE	Model 12150L V-12 liquid-cooled diesel developing 520 hp at 2000 rpm
TRANSMISSION	manual, 5 forward and 1 reverse gears
SUSPENSION	torsion bar
ELECTRICAL SYSTEM	24 V
ARMAMENT	(see text)
main	1 × 100 mm Type 59 gun
coaxial	1 × 7.62 mm Type 59T MG
bow	1 × 7.62 mm Type 59T MG
anti-aircraft	1 × 12.7 mm Type 54 MG
AMMUNITION	
main	34
coaxial/hull	3500
anti-aircraft	200
GUN ELEVATION/ DEPRESSION	+17°/-4°
TURRET TRAVERSE	360°
ARMOUR	
Hull front upper	97 mm at 58°
Hull front lower	99 mm at 55°
Hull sides upper	79 mm at 0°
Hull sides lower	20 mm at 0°
Hull rear upper	46 mm at 0°
Hull rear lower	46 mm at 0°
Hull top	33 mm
Hull floor - front	20 mm
Hull floor - rear	20 mm
Turret front	203 mm at 0°
Turret sides	150 mm at 0°
Turret rear	64 mm at 0°
Turret roof	39 mm at 79°
Mantlet	included in above figures

Status: Production probably complete. In service with:

Country	**Quantity**	**Comment**
Albania	15	
Bangladesh	70 to 100	also has ARV
China	6000	estimate
Congo	15	
Korea, North	175	
Pakistan	1300	some upgraded with 105 mm gun
Tanzania	30	
Vietnam	350	
Zimbabwe	35	

Manufacturer: Chinese state factories
Enquiries to China North Industries Corporation
PO Box 21377A, Yuetan Nanjie
Beijing
People's Republic of China
Telephone: (86) 6898/3461/3471/7570
Telex: 22339 cnin cn

NORINCO T-series Modification Packages

Development/Description
Based on its experience in the design, development and production of MBTs, NORINCO is now offering complete upgrade packages to enable existing users of Chinese Type 59 or the original Soviet T-54/T-55 MBTs, to upgrade them in their own facilities so as to extend their operational lives.

No less than eight concepts in Type 59 and T-54/T-55 upgrades are offered to enable the individual user to tailor the upgrade package to meet his own specific operational and financial requirements.

The eight concepts are as follows:-

Concept	**Composition**
I	1 + 3 + 5 + 6
II	1 + 4 + 5 + 6
III	2 + 3 + 5 + 6
IV	2 + 4 + 5 + 6
V	1 + 3 + 5 + 7
VI	1 + 4 + 5 + 7
VII	2 + 3 + 5 + 7
VIII	2 + 4 + 5 + 7

FIREPOWER

Composition 1
New 100 mm rifled gun fitted with thermal sleeve and firing three new types of 100 mm ammunition, APDS, HEAT and HE, brief specifications of these are as follows:-

Type	**APDS**	**HEAT**	**HE**
MUZZLE VELOCITY	1480 m/s	1000 m/s	900 m/s
CHAMBER PRESSURE	3300 kg/cm²	2530 kg/cm²	3000 kg/cm²
PENETRATION	150 mm RHA at 65°	120 mm RHA at 65°	not app

Composition 2
New 105 mm rifled gun fitted with thermal sleeve and fume extractor and capable of firing four new types of ammunition, APFSDS, HEAT HESH and HE. Brief specifications of these are as follows:-

Ammunition type	**APFSDS**	**HEAT**	**HESH**	**HE**
MUZZLE VELOCITY	1455 m/s	1173 m/s	731 m/s	850 m/s
LENGTH OF ROUND	887 mm	992.5 mm	941.5 mm	998.65 mm
WEIGHT OF ROUND	18.7 kg	21.8 kg	21.3 kg	27.1 kg
DIAMETER OF PROJECTILE	33 mm	105 mm	105 mm	105 mm
WEIGHT OF PROJECTILE	4.18 kg	10.16 kg	11.25 kg	n/av
PENETRATION (mm/°)	150/65	177.8/65	50-60 mm	n/app

NB: HE projectile has a lethal radius of 20 m and a maximum range of 13 000 m

INCREASED HIT PROBABILITY

Composition 3
Light-spot injection fire control system with integrated bi-directional stabilised gun laying system.

Composition 4
Image-stabilised fire control system with integrated bi-direction stabilised gun laying system.

There are two types of fire control systems, simplified and image stabilised and two types of gun laying system, bi-directional and integrated bi-directional.

GUN LAYING SYSTEM SPECIFICATIONS

Type	**Bi-directional stabilisation**	**Integrated bi-directional stabilisation**
STABILISED ACCURACY		
elevation	+-1	+-1
traverse	+-3	+-2 -1.5
MAX GUN LAYING SPEED		
elevation	6/4.5°/s	6/4.5°/s
traverse	15°/s	36/32°/s
MIN AIMING SPEED		
elevation	0.07°/s	0.06/0.04°/s
traverse	0.09°/s	0.04/0.03°/s

FIRE CONTROL SPECIFICATIONS

Type		Simplified FCS		Image-stabilised FCS
	single-directional	light-spot	bi-directional	
Performance	**data setting mode**	**injection mode**	**data setting mode**	
CALCULATED ACCURACY	0.05	0.1	0.1	0.1
CALCULATED RANGE	300-3000 m	200-4000 m	300-3000 m	200-4000 m
CORRECTION SENSOR	no	elevation, traverse, gun cant angle	elevation, traverse, gun cant angle	elevation, traverse, gun cant angle, crosswind
ARTIFICIAL INTEGRATED CORRECTION	charge temperature, climate temperature, muzzle velocity, cross-wind, distance integrated correction of impacting points			
FIRING FUNCTION	stationary/stationary	short stop to moving	short stop to moving	moving to moving

IMPROVED NIGHT FIGHTING CAPABILITY

Composition 5

The night sight and observation used in the modification package are fitted with second generation LLLTV tubes. Two types of gunner's sight are available, one independent and one connected to the centre computer. The commander has a day/night periscopic sight while the driver has an IR periscope.

NIGHT VISION EQUIPMENT SPECIFICATIONS

Type	Day/night gunner sight vision: connected with computer	independent	commander's sight	driver's sight
MAGNIFICATION	×7	×7	×5	×1
FIELD-OF-VIEW	6°	6°	8°	30°
VISUAL RANGE	1000 m	900 m	600 m	80/100 m

INCREASED SURVIVABILITY

Composition 6

Fitting of add-on composite armour plate on the glacis plate and protection for the turret front and sides, fitting of armoured skirts, bank of four electrically operated smoke dischargers either side of the turret and installation of a fire and explosion detection and suppression system inside of the vehicle.

Composition 7

Installation of explosive reactive armour on hull, turret and on each side skirt plate.

Upgraded Type 59 MBT with 105 mm rifled gun fitted with thermal shield and fume extractor

ARMOUR PROTECTION SPECIFICATIONS

Explosive reactive armour - provides protection against HEAT projectiles.

Composite armour of upper front hull - provides protection against penetration from 105 mm AP rounds at a range of 1500 m.

Lateral protection of skirts - provides protection for 40 mm HEAT anti-tank rockets.

Boom protection on the turret - provides protection against HEAT attack at certain angles of attack.

Composition 8

This includes installation of a fire extinguishing and explosion suppression system and smoke dischargers.

The fire extinguishing and explosion suppression system consists of optical detectors, control panel, manual switch and the fire extinguishers with both manual and automatic modes of operation. The system can detect automatically the penetration of the high-energy jets and output the warning electronic signals within 10 ms. The system can automatically suppress the oil/gas explosion caused by the penetration of an APDS round within 80 ms. The system can also carry out conventional fire extinguishing operations.

Mounted on either side of the turret is a bank of four forward firing electrically operated smoke dischargers which can fire a smoke grenade to a maximum range of 100 m. Inside of the vehicle is the control box allowing the commander to select two rounds, four rounds, single rounds or all eight rounds. For an eight round salvo, the smoke screen covers an angle of 120° with a width of 120 m. The smoke screen takes about 5 s to form and lasts 120 seconds.

Status: Development complete. Key parts of these retrofit packages are already in production for other Chinese vehicle programmes. It is understood that some parts of this modification package have been adopted by Pakistan, including the 105 mm rifled tank gun.

Manufacturer: Chinese state factories
Enquiries to China North Industries Corporation
PO Box 21377A
Yuetan Nanjie
Beijing
People's Republic of China
Telephone: (86) 6898/3461/3471/7570
Telex: 22339 cnin cn

Close-up of turret showing 12.7 mm anti-aircraft machine gun, smoke dischargers and boom shield

SPECIFICATIONS

CREW	4
COMBAT WEIGHT	37 000 to 37 500 kg
POWER-TO-WEIGHT RATIO	15.7/15.5 hp/tonne
GROUND PRESSURE	0.83 to 0.84 kg/cm²
LENGTH GUN FOWARD	9.224 m
WIDTH WITH SKIRT	3.307 m
HEIGHT WITH AA MG	2.807 m
GROUND CLEARANCE	0.425 m
TRACK	2.640 m
LENGTH ON TRACK ON GROUND	3.845 m
MAXIMUM ROAD SPEED	50 km/h
CRUISING RANGE	420 to 440 km
FUEL CAPACITY	927 l
FORDING	1.4 m
GRADIENT	60%
SIDE SLOPE	30%
VERTICAL OBSTACLE	0.8 m
TRENCH	2.7 m
ENGINE	580 hp diesel
ARMAMENT	
main	1 × 105 mm
coaxial	1 × 7.62 mm MG
anti-aircraft	1 × 12.7 mm MG
AMMUNITION MAIN	41 to 44

CZECHOSLOVAKIA

Czechoslovakian Tank Production

In the immediate post Second World War period Czechoslovakia produced the T-34/85 medium tank and the 100 mm SU-100 tank destroyer, mainly for export to the Middle East. From the mid-1960s Czechoslovakia produced the T-55 for both the home and export markets and, according to American Department of Defense reports, tank production in Czechoslovakia and Poland was running at 800 a year between 1976 and 1980. It has been confirmed that Czechoslovakia has produced the T-62 MBT for export (especially to the Middle East) and that these differed in minor details from their Soviet counterparts.

In 1979/80 the Martin Machine Tool Enterprise started tooling up to produce the T-72 MBT. The transmission is produced by CKD-Praga which managed T-34/85 production in Czechoslovakia. According to the 1983 edition of the US Government publication *Soviet Military Power*, only 100 T-72s were produced in Czechoslovakia and Poland during 1982. It also stated that production of the T-55 was still being undertaken outside the USSR (for example in Czechoslovakia and Poland) at the rate of 500 per year, presumably for the export market. Production of the T-55 has now been completed in Czechoslovakia. US sources have given the following tank production figures for Czechoslovakia and Poland in recent years: 550 in 1983, 450 in 1984, 700 in 1985, 700 in 1986, 750 in 1987 and 700 in 1988. No figures are available for 1989 although due to the recent events in Eastern Europe production has now been reduced.

In May 1991 Czechoslovakia confirmed that their T-72 MBT plant at Divince was still producing vehicles for both the home and export markets and was considering the sale of some 300 T-72 MBTs to Syria.

By late 1990 it is understood that the Czechoslovakian MBT fleet consisted of 50 T-34/85 vehicles used for training, 1547 T-54s, 1543 T-55s and 815 T-72s. Of these it is estimated that a total of 1578 will be disposed of under the CFE treaty with these including the T-34/85s and the T-54/T-55 MBTs that have not been upgraded.

Czechoslovakian T-72 MBTs

Modernised T-55 MBT

Several years ago the Czechoslovakian Army started to field modernised T-55 in significant numbers as insufficient T-72s were being built to replace the older tanks on a one-for-one basis.

The modifications include full length track skirts which cover the upper part of the track, thermal sleeve for 100 mm gun, additional passive armour protection covering the frontal 180° of this turret (this being very similar to that fitted to Soviet T-62 tanks), 12.7 mm ammunition boxes stowed externally on the right side of the turret and a bank of forward firing smoke dischargers installed on the right side of the turret to the immediate rear of the turret armour. Lateral protection is also provided for the fuel tanks on either side of the hull above the roadwheels.

To improve the first range hit probability of the 100 mm gun, a new fire control system has been fitted which includes a laser rangefinder mounted externally over the 100 mm gun, ballistic computer and a meteorological sensor which is mounted on the turret roof at the rear.

Additional details of this modernisation programme are given in the entry for the T-54/T-55 under the USSR.

Modernised Czechoslovakian T-55 MBT clearly showing additional armour around turret front, smoke dischargers, skirts and thermal sleeve for 100 mm gun

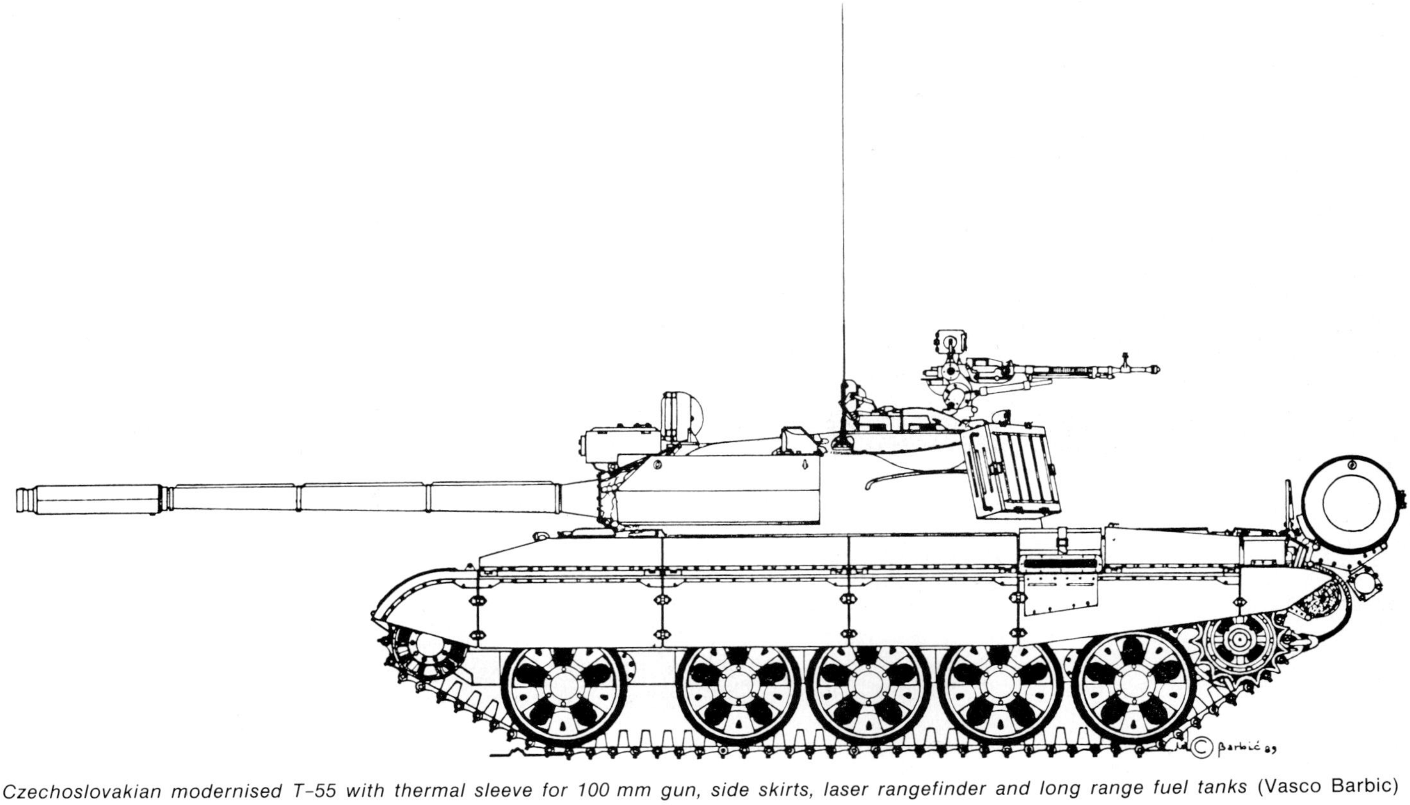

Czechoslovakian modernised T-55 with thermal sleeve for 100 mm gun, side skirts, laser rangefinder and long range fuel tanks (Vasco Barbic) (not to 1/76th scale)

DENMARK

Danish MBT Modernisation

At present the Danish MBT fleet consists of 90 Centurion MBTs armed with 105 mm guns, laser rangefinder and night vision equipment, 16 Centurion MBT armed with 105 mm guns but without laser rangefinder or night vision equipment, 111 Centurions armed with 20 pounder (83.8 mm) guns without night vision equipment and 120 Leopard 1 MBTs which were delivered new between 1976 and 1978. In addition there are 16 Centurion Mk 2 ARVs and some six Centurion AVLBs.

The Leopard 1 MBTs are concentrated in the Jutland Division while the Centurions are in Zealand.

The Danish Army has proposed a three-phased tank modernisation programme which, depending on funding, would consist of:-

1) Procurement of 110 Leopard 1A3s from German Army stocks
2) Modernisation of 120 existing Leopard 1s to Leopard 1A5 standard
3) Modernising part of the existing Centurion MBT fleet.

Late in 1989 the Danish Army Material Command received bids from some 17 companies to modernise up to 106 Centurion MBTs to extend their operational lives.

The original intention was to fit 16 Centurions with a new Swedish Ericsson night sight that incorporates a laser rangefinder and a low light-level TV, while the driver's station will be fitted with an image intensification periscope. These will then be identical to the 90 105 mm Centurions that were modernised in 1984.

All 106 Centurions were then to undergo a complete rebuild which includes the installation of a new powerpack consisting of a diesel engine coupled to a fully automatic transmission, hydro-pneumatic suspension system for improved cross-country mobility and a new electric gun control system and stabilisation system to increase first round hit probability. Some companies have bid for the complete turret upgrade and the other for the hull upgrade.

Following the signing of the Conventional Forces Europe Treaty and reduced defence expenditure, the proposal to upgrade the fleet of Centurion MBTs will not now go ahead and Germany is to provide Denmark with 110 Leopard 1A3/1A4 MBTs.

EGYPT

Egyptian Tank Fleet

Total Egyptian front-line tank strength is believed to consist of 900 T-54/T-55s (some of which have been supplied to Iraq), 600 T-62s and 759 M60A3s. Royal Ordnance has been supplying new 115 mm gun barrels to Egypt for several years. In early 1984 the American Kollsman Instrument Company was awarded a two-year contract valued at $5 million for the installation of a maintenance facility in Egypt to support the repair of the M60A3 Integrated Tank Fire Control System. This equipment was installed in the Egyptian Army's Tank Depot Number 1.

In the 1970s several hundred Egyptian T-54/T-55/T-62 MBTs were fitted with German AEG/Telefunken white/infra-red searchlights to the right of the main armament and some were also fitted with the Iskra laser rangefinder. Some Egyptian T-62s have also been fitted with two launchers either side of the turret rear for ground-to-ground smoke rockets, and a drawing of this version appears in the entry for the T-62 later in this section.

Late in 1984 General Dynamics, Land Systems Division was awarded a $150 million contract to build a new tank plant outside Cairo. The facility is called Factory 200 and will initially overhaul US-supplied M60A3 and M88A1 armoured vehicles.

Production of the M1A1 Abrams MBT is expected to be undertaken at the ETP and it will also make some spare parts for the US supplied M60A3 MBTs and M88A1s as well as overhauling the engine, transmission and other key components.

Early in 1990 Egypt announced that it had accepted an offer from the United States for 700 M60A1 MBTs free of charge, with only Egypt having to pay for the shipping costs. When these are received an equal number of Soviet supplied vehicles will have to be disposed of.

Egyptian production of M1A1 Abrams MBT

Late in 1988 the United States and Egypt signed a Memorandum of Understanding allowing Egypt to co-produce up to 555 General Dynamics, Land Systems Division, M1A1 MBTs. Under the terms of this MoU some 40 per cent of the tank's components will be manufactured in Egypt with the remainder being produced in the USA and shipped to Egypt for final assembly.

The first M1A1 MBTs will be shipped direct from the United States to Egypt and they would be followed by partly assembled tanks and then kits. In the longer term Egypt would manufacture in country such high-usage components as tracks, roadwheels, sprockets and hubs while high technology and sensitive components such as the engine, transmission, fire control system and armour would continue to come from the United States.

The M1A1 for the Egyptian Army would be identical to the US Army version and be powered by a gas turbine; there is no plan to replace this by a diesel engine. Main armament would comprise the 120 mm

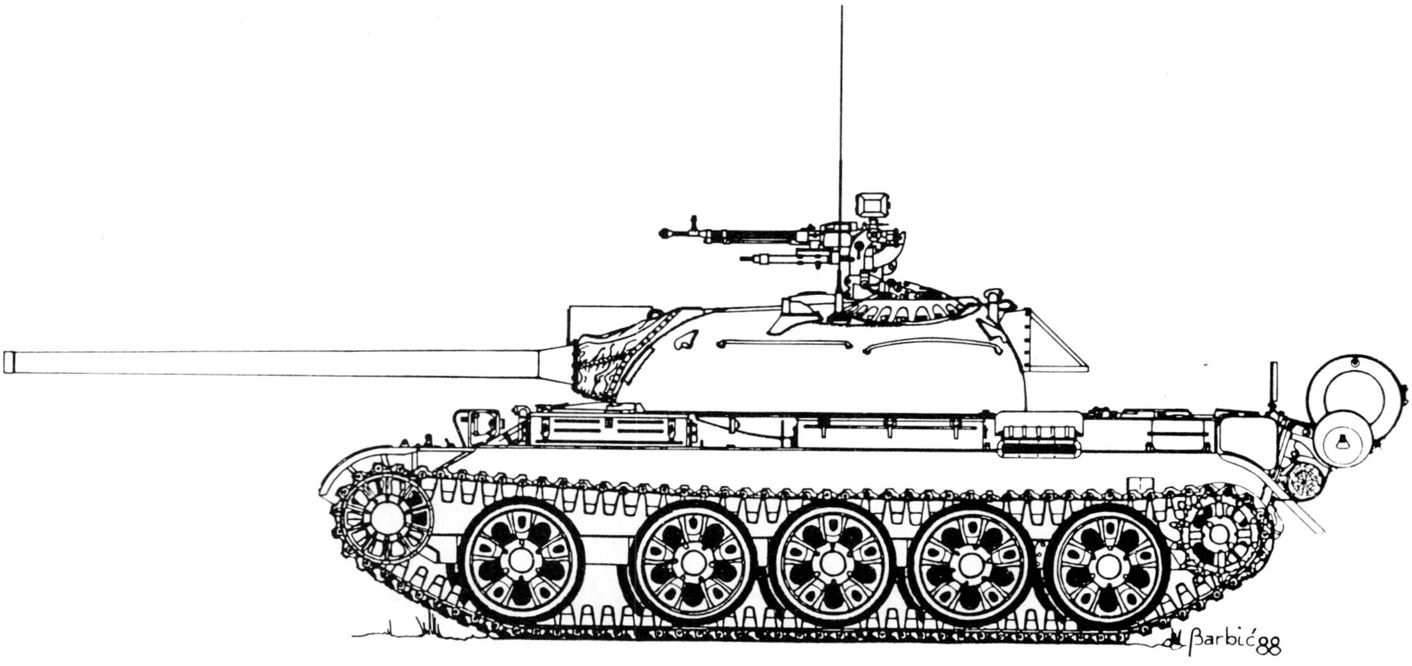

Egyptian Army T-54 upgraded in Egypt by fitting a Yugoslav Iskra laser rangefinder, German AEG-Telefunken infra-red/white-light searchlight (with stowage box on turret rear) and additional water tanks (external) on rear fenders (not to 1/76th scale) (Vasco Barbic)

Ramses II MBT undergoing mobility trials in the United States prior to shipment to Egypt

T-55 tank upgraded with 105 mm L7 series gun supplied by Royal Ordnance, infra-red/white light searchlight is mounted coaxially to 105 mm gun (Christopher F Foss)

smooth bore gun for which at present the Egyptians do not produce ammunition, so in the early stages at least this would have to be brought from the United States. Egypt will not, however, be allowed to receive the 120 mm APFSDS-T M829 round which has a DU penetrator, but Honeywell and Olin have both produced an equivalent round with a conventional penetrator specifically for the export market.

Following trials in Saudi two M1A1s went to Egypt in late 1987 and these underwent a familiarisation and performance evaluation by the Egyptian Army which was completed in December 1987.

Egypt will not be allowed to export any of the tanks without the permission of the United States and in the final stages of the project a maximum of 35 per cent of the M1A1 will be made in Egypt. It is not expected that local assembly will get underway until 1992.

In April 1989 the US DoD announced that General Dynamics, Land Systems Division, had been awarded a contract for manufacturing technical assistance (MTA) to support the co-production of M1A1 MBTs in Egypt.

Initial funding of this incrementally funded contract is $12.5 million but the co-production programme will eventually be worth approximately $1.2 billion over a 10 year period as other contracts for tank hardware components and plant facilitisation are issued.

The MTA effort will initially support the first two production increments of a planned six increment programme extending through to 1998.

The first increment involved the manufacture and assembly of M1A1 MBT components and sub-assemblies in the US for shipment to Egypt where they will be used for an Egyptian co-production tank assembly process.

General Dynamics, Land Systems Division and General Dynamics Services Company will jointly perform the MTA contract.

Ramses II MBT

In November 1984 Teledyne Continental Motors of the USA was awarded a contract for upgrading the firepower and mobility of a single T-54 MBT. This was originally called the T-54E but was subsequently renamed Ramses II. The first prototype of the Ramses II was sent to Egypt for extensive firepower and mobility trials in January 1987 and these were completed in late 1987.

Late in 1989 Egypt signed a technical assistance agreement with TCM to support continued Egyptian testing of the Ramses II, with testing commencing in the Summer of 1990.

The rear hull has been modified to accommodate the new powerpack consisting of a TCM AVDS-1790-5A turbo-charged diesel developing 908 bhp (which has 80 per cent commonality with the engine installed in the M60A3 MBT) coupled to a Renk RK-304 transmission. Two new exhaust pipes, one either side of the hull rear, replace the single exhaust outlet in the left side of the hull.

The original suspension has been replaced by new Teledyne Continental Motors Model 2880 in-arm hydro-pneumatic suspension units each fitted with a M48 type road wheel, idler at the front, large drive sprocket at the rear, two new track-return rollers and US pattern tracks replacing the original Soviet tracks.

A SABCA Titan Mk 1 laser fire control system has been installed which includes a modified Avimo TL10-T sight incorporating the laser rangefinder and an integrated in-eyepiece CRT alpha-numeric graphic display. The fire control system also includes an original SABCA double digital processor, image intensification periscopic night sight, automatic attitude and atmospheric sensors and associated control boxes.

The gun and turret stabilisation system are provided by Cadillac Gage of the US.

Other modifications include a fire detection and suppression system, new final drives, new fuel tanks, new air filtration system, British Catton tracks, new communications system and a new turret basket.

The original 100 mm gun has been replaced by the 105 mm M68 ordnance which is already fitted in Egyptian M60A3 MBTs, the original

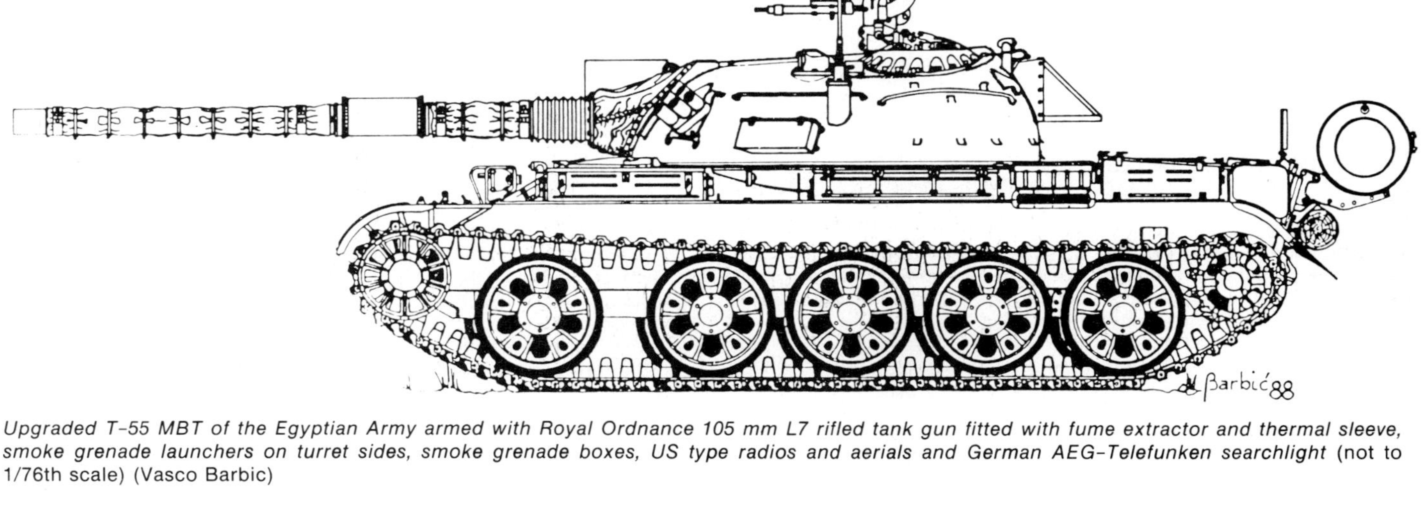

Upgraded T-55 MBT of the Egyptian Army armed with Royal Ordnance 105 mm L7 rifled tank gun fitted with fume extractor and thermal sleeve, smoke grenade launchers on turret sides, smoke grenade boxes, US type radios and aerials and German AEG-Telefunken searchlight (not to 1/76th scale) (Vasco Barbic)

Egyptian T-62 MBT fitted with 105 mm rifled tank gun fitted with thermal sleeve and fume extractor

Soviet-supplied T-34/85 tank upgraded by the Egyptian Army Number 5 Tank Workshop (Christopher F Foss)

breech of the 100 mm DT-10T gun has been retained and modified and the recoil system has also been modified. A muzzle reference system is fitted as standard and an M60 day/night searchlight is mounted over 105 mm gun.

Mounted either side of the turret is a bank of electrically operated smoke dischargers. A collective type NBC system has been installed.

German modernised T-54

Under a contract awarded early in 1988, the German company of Jung Jungenthal is modernising an Egyptian Army T-54. There is a separate entry for this MBT under Germany later in this section.

Egyptian T-62 with 105 mm gun

The Egyptian Army has evaluated a T-62 MBT with its original 115 mm gun replaced by a 105 mm L7A1 gun modified by NORICUM of Austria and designated the TG N-105. The complete kit comprises the following: breech modified to allowing opening on left, connections for automatic opening of the breech, existing recoil system is retained or new recuperator and recoil cylinders can be incorporated, existing cartridge case ejector is modified to allow it to be used for 105 mm cases, existing ammunition stowage, including the fuel tanks, is modified to accommodate the 105 mm ammunition and emergency firing unit is fitted. This consists of an impact generator and an emergency firing trigger which is foot actuated by the gunner. The elevation quadrant, traverse indicator and sight graticule have also been modified.

Egyptian modernised T-34/85 tank

The Egyptian Army Number 5 Tank Workshop has modernised one example of the T-34/85 tank which it is now offering for export, or existing T-34s could be brought up to the new configuration in the user's own facilities. The first prototype of the upgraded T-34/85 tank was shown in late 1987.

According to the Egyptian Army, the 480 hp diesel of the T-34/85 lasted only 150 km before requiring an overhaul and gave a power-to-weight ratio of 15 hp/tonne.

In the upgraded T-34/85, the original engine has been replaced by a T-54 diesel developing 520 hp which gives a power-to-weight ratio of 16.25 hp/tonne and lasts 4500 km before requiring an overhaul.

Other automotive improvements include modified fuel, air and cooling systems, rear engine deck cover now has torsion bars to assist in opening, modified final drive, control linkage and rubber dampers and finally rubber side skirts have been fitted to reduce sand thrown up by the tracks.

An explosion and fire detection and suppression system has been installed in the crew compartment that can suppress the fire in 80 milliseconds. The modified T-34/85 can also lay its own smoke screen by injecting diesel fuel into the rear exhaust pipes.

FINLAND

MBT Modernisation

Early in 1989 it was announced that Finland was to spend FIM340 million ($85 million) to upgrade its fleet of 185 Soviet supplied T-55 MBTs rather than purchase new vehicles.

It is estimated that the cost of upgrading the T-55s, which entered service in the 1960s, is about one fifth the total cost of purchasing new equipment.

By early 1989 work on upgrading the first batch of 20 vehicles was underway with the extensive modifications including the installation of a new 12.7 mm NSV anti-aircraft machine gun, installation of side skirts, four German type smoke dischargers either side of the turret, boxes of machine gun ammunition on the turret sides, white light searchlight and thermal sleeve for main armament. A new gunner's sight has been fitted which incorporates a laser rangefinder for increased hit probability. Passive night vision equipment may also be fitted.

The upgraded T-55s are known as the T-55M and these serve alongside the 50 or so T-72 MBTs purchased from the Soviet Union in the mid-1980s.

Late in 1988 the Finnish MoD placed an FIM75 million ($10 million) for an initial quantity of Marconi Command and Control Systems Marksman twin 35 mm anti-aircraft turrets which will be installed on T-55 chassis in Finland. The contract also includes an option on a further quantity in 1992.

FRANCE

GIAT Industries Leclerc MBT

Development

In December 1982, following the collapse of the French/German MBT project, the French prime minister advised Parliament that studies would begin in 1983 for a successor to the current AMX-30 and AMX-30 B2 MBTs used by the French Army.

Project definition on the Engin Principal de Combat (EPC) was completed by the Atelier de Construction d'Issy-les-Moulineaux in 1985 and by that time five test rigs were already being used in component development, one for the suspension, three for automotive trials and the fifth for the weapon system.

On January 30th 1986 it was announced that the EPC would be named the Leclerc.

Project definition for the Leclerc was completed in mid-1986 and in the same year it was decided to build six complete prototypes of the vehicle and the first of these was completed late in 1989. The Leclerc made its first public appearance at the 1990 Satory exhibition of defence equipment for the ground forces.

The first funding for the Leclerc was in the 1989 French Defence Budget when an initial 16 vehicles were ordered.

First production Leclerc's will be delivered to the French Army in 1992 with the hull being built at the Atelier de Construction Roanne (ARE) and the turret at the ATS Tarbes. The latter will send the turret to the ARE who will install it on the chassis and deliver the complete tank to the French Army.

Tactical trials of the Leclerc are due to take place in 1993 at squadron level with the first regiment being equipped in 1995 and the first division in 1996.

2nd prototype of the GIAT Industries Leclerc MBT showing additional armour protection on the sides of the hull towards the front of the vehicle

Although the original intention was to procure up to 1400 Leclerc MBTs, this figure may be reduced to 1100 or even less, depending on funding and the possible reduction in size of the French Army. If the figure is 1100 units, then 160 will be issued to each of the six armoured divisions, 70 vehicles in reserve and 70 vehicles with the schools. Production at the ARE is expected to build up to about 100 vehicles a year by 1996.

Of the total cost of the tank, 64 per cent is the turret and 36 per cent the hull and of the latter 11 per cent is the chassis, 5 per cent electrics and 20 per cent automotives.

Description

The Leclerc will have a conventional layout with the driver sitting at the front, the turret in the centre and the powerpack at the rear.

The driver is seated at the front left of the vehicle with reserve ammunition in a drum to his right. He has a single piece hatch cover that opens to the left which is provided with three periscopes, the centre one of which is a SOPELEM OB 60 day/night periscope. The driver steers the vehicle with a conventional steering wheel.

The gunner of the Leclerc is seated on the right of the turret with the commander on the left, both being provided with a single piece hatch cover. The commander has eight day periscopes and the gunner is also provided with day periscopes.

Main armament comprises a GIAT 120 mm smooth bore gun designated the CN 120-26 which has a chamber the same size as that of the 120 mm smooth bore gun fitted to the Leopard 2 and M1A1 so allowing the French Leclerc MBT to use German or United States ammunition.

The French gun however has a calibre length of 52 compared to 44 of the other weapons and therefore a higher muzzle velocity of 1750 m/s when firing APFSDS projectiles. The gun is fitted with a thermal sleeve and a compressed air system for blowing out propellant fumes. Prototypes have a muzzle reference system fitted although this may not be fitted into production vehicles.

Mounted in the turret bustle is the automatic loader developed by Creusot-Loire Industrie which contains 22 rounds of ready used ammunition that is normally loaded from outside through a small hatch in the back of the turret bustle. The autoloader can distinguish between five different types of ammunition and will automatically recognise and respond to ammunition selection commands.

The automatic loader consists of a series of linked circular troughs, each with an inner tray containing a round, which form a continuous

Close up of GIAT Industries Leclerc MBT showing 12.7 mm coaxial MG and commanders panoramic sight

Close-up of the gunner's sight on the GIAT Industries Leclerc MBT with armoured shutters in open position

Mounted in the turret bustle of the GIAT Industries Leclerc MBT is the Creusot-Loire Industrie automatic loader for the 120 mm smooth bore gun

electrically driven chain. The inner tray slides out with its round to bridge the space between the automatic loader and the gun with the latter depressed.

The 120 mm gun is automatically depressed to -1.8° for reloading and the automatic loader is separated from the turret by bulkheads with blowout panels being provided in the roof.

In addition, it can be loaded by the gunner from outside using the 18 rounds stowed in the drum located to the right of the driver.

The installation of the automatic loader enables a cyclic rate of fire of 15 rds/min to be achieved with 10 to 12 rds/min being the effective rate of fire.

The two main types of ammunition are fired, APFSDS with a m/v of 1750 m/s and HEAT with a m/v of 1100 m/s, with both having a combustible cartridge case containing a dual-base solid propellant. Its primer is also combustible. A DU 120 mm APFSDS round is under development and is expected to be fielded in 1994. GIAT Industries is also developing a 120 mm anti-helicopter round.

The APFSDS round is called the OFL 120 FA and has a tungsten penetrator with a length-to-diameter ratio of 20 : 1.

A 12.7 mm machine gun is mounted coaxially with the main armament and a 7.62 mm machine gun is mounted externally on the turret roof and used by both the commander and gunner. The 7.62 mm anti-aircraft machine gun can be aimed and fired under complete armour protection.

Mounted along either side of the turret roof, towards the rear, are nine launcher tubes which can launch various grenades, for example smoke, anti-personnel and infra-red decoy.

The fire control system includes a modular thermal imaging camera, CILAS laser rangefinder and stabilised sights for the commander and gunner. The stabilised gunner's sight is the HL60 and has been developed by SAGEM. It incorporates a daylight channel plus a SAT Athos thermal image and an integral Avimo laser rangefinder. The gunner's sight pivots in elevation with the gun but contains a mirror independently stabilised in elevation and in azimuth. The sight incorporates daylight and thermal TV channels and a Nd-YAG laser rangefinder with a maximum range of 8000 m.

The gunner's H60 sight is also linked to visual displays at the commander's station so enabling the tank commander to fire the main armament using his duplicate controls. The commander has a SFIM roof-mounted stabilised panoramic HL70 sight which also incorporates a laser rangefinder, eight periscopes and three control panels. The first panel is a master control and monitoring unit and has five systems, start up switches for the turret services, fire control system, sights, thermal imager and automatic loader.

Using automatic built-in test equipment, all of the systems are operational one minute after switch-on over the full operational temperature range.

The second panel contains the main armament controls, warning lights and controls the use of the automatic-data transmission system. This allows the burst transmission of standardised format messages which can be modified using a keyboard entry system.

The third panel contains the secondary armament controls, and the sensor inputs. These include the met sensor, ammunition selector (up to six types can be programmed) and a repeat fire selector which reloads the same ammunition type. This panel also contains the vehicle screening smoke and self-defence grenade firing buttons.

The VDU, in addition to displaying the thermal picture from the gunner's sight, is the crew interface with the VETRONICS.

Each of the commander's eight periscopes has a switch which when activated brings the gun onto its line of sight.

The gunner has two control panels, VDU, three observation periscopes and his main sight. The controls provide all of the functions provided for the tank commander with the exception of the turret services start up switches. The gunner has no backup sight available.

The fire control system will enable six different targets to be engaged in 60 seconds and the tank to engage both stationary and moving targets while moving itself across country. First round hit probability is being quoted as 95 per cent. Typical engagement cycle will be four to six seconds.

The complete powerpack of the GIAT Industries Leclerc MBT comprises the engine, automatic transmission and cooling system

A battlefield management system is installed together with a multiplexed numeric data link with two computers. The systematic use of digital electronics in all subsystems of the Leclerc and turret means that the tank uses some 30 central processing units with 8, 16 or 32 bit microprocessors. The crew will have access to all of the subsystems of the tank. For example the tank commander can find out the amount of fuel remaining in the fuel tanks just as easily as the driver, the number of rounds left in the automatic loader or where the tank is at any given time.

Positioning data is upgraded as the Leclerc moves and rendezvous points for replenishment can be anticipated. The vehicle can automatically transmit combat status and location data to field command vehicles.

The shell of the hull and turret of the Leclerc MBT will be of welded steel construction to which the composite armour will be added, this will provide a very high degree of protection against both KE and CE attack.

The armour system is of a modular design so that it can be easily replaced or upgraded so allowing new armour developments to be incorporated in the future when this becomes available. The high power-to-weight ratio of the Leclerc will enable the tank to be up-armoured to meet new threats without substantially lowering its power-to-weight ratio.

Turret traverse and gun elevation are electric and both the commander and gunner will be able to lay and fire the main and secondary armament. Manual turret controls will be provided for emergency use.

Prime contractor for the gun control equipment is CSEE and the turret can be slewed through 180° in less than six seconds.

The Leclerc will be powered by a UNI-DIESEL V8X-1500 Hyperbar diesel developing 1500 hp coupled to a SESM ESM 500 fully-automatic transmission with five forward and two reverse gears and a hydrostatic transmission.

A Turbomeca TM-307 auxiliary power unit will be installed to allow the main engine to be shut down if required to conserve fuel.

After evaluating both hydro-pneumatic and torsion bar with rotary damper suspension systems the former was selected for the Leclerc with SAMM being the prime contractor. Three different tracks are under consideration for the Leclerc, a dry-pin track similar to that installed on the current AMX-30, a steel track with connectors and a new light alloy track with connectors. The Leclerc has six road wheels, the drive sprocket at the rear, idler at the front and track return rollers.

Standard equipment on the Leclerc is a new design of NBC system, deep fording, fire detection and suppression system for both crew and engine compartments, laser detector, and a communications system from Thomson CSF and TRT.

Other contractors apart from those previously mentioned include SAMM for hydraulic motors, CGA for turret stabilisaton, SFIM commander's sight, 7.62 mm and 12.7 mm MGs will be provided by GIAT/MAS, periscopes from GIAT/MAS, Turbomeca for the turbo-charger, Paris-Rhone for the starter, CSEE provides the electric gun and turret drive, hand controls for tank commander and gunner, video monitoring unit and one of the fire control computer PCB, SAGEM for gunner's day and night sight, dynamic harmonisation system, card universal link coupler for simple bus and the instrument panel, CGA-HBS for slaving, for digital computers, and TRT and SAT for the thermal imaging system. The fuel tanks hold a total of 1300 litres of fuel but two long range fuel tanks can be fitted at the rear with each of these containing 200 litres.

Variants

It is anticipated that there will be only two variants of the Leclerc MBT, an armoured recovery vehicle and an SIR mobile command post. Feasibility studies for both of these were started in 1988.

In mid-1990 it was stated that GIAT Industries and Krupp-MaK (of Germany), were studying an ARV which would be able to recover the Leclerc MBT on the battlefield.

This new ARV would have a longer chassis with seven road wheels, a combat weight of 54 tonnes and a four man crew. It would be powered by the same 1500 hp engine, transmission and hydropneumatic suspension as the Leclerc with armament consisting of a 12.7 mm machine gun and smoke/grenade launchers.

To enable it to carry out its recovery role, equipment fitted would include a 35-tonne main winch with 160 m of cable, a 1.5-tonne auxiliary winch with 200 m of cable, crane with a lifting capacity of 36 tonnes and a traverse of 220° and a hydraulically operated dozer/stabilising blade mounted at the front of the hull.

Status: Entering production for the French Army at the Atelier de Construction Roanne (ARE).

Prime contractor: GIAT Industries
13 route de la Minière
78034 Versailles Cedex
France
Telephone: (1) 30 97 37 37
Fax: (1) 30 97 39 00

Fifth prototype of the GIAT Industries Leclerc MBT with 120 mm ammunition with semi-combustile cartridge case arranged in front of vehicle

SPECIFICATIONS	
CREW	3
COMBAT WEIGHT	53 500 kg
POWER-TO-WEIGHT RATIO	28.3 hp/tonne
GROUND PRESSURE	0.9 kg/cm²
LENGTH GUN FORWARD	9.87 m
LENGTH HULL	6.88 m
WIDTH	3.71 m
HEIGHT TURRET ROOF	2.46 m
GROUND CLEARANCE	0.5 m
LENGTH OF TRACK IN GROUND CONTACT	4.318 m
MAXIMUM ROAD SPEED	75 km/h
MAXIMUM CROSS-COUNTRY SPEED	50 km/h
MAXIMUM REVERSE SPEED	38 km/h
ACCELERATION	0/32 km/h in 5.5 s
FUEL CAPACITY	1300 l
MAXIMUM ROAD RANGE	550 km
VERTICAL OBSTACLE	
forwards	1.25 m
reverse	0.6 m
TRENCH	3 m
FORDING	1 m
FORDING WITH PREPARATION	4 m
ENGINE	UNI-DIESEL-1500 V8X 8-cylinder Hyperbar diesel developing 1500 hp
TRANSMISSION	ESM 500 automatic with 5 forward and 2 reverse gears, microprocessor controlled hydraulic multiple plate clutch
APU	Turbomeca TM-307 gas-turbine with turbo-charger
SUSPENSION	hydropneumatic
VERTICAL ROAD WHEEL TRAVEL	31.5 cm
ROAD WHEEL REBOUND	12.5 cm
ARMAMENT	
main	1 × 120 mm gun
coaxial	1 × 12.7 mm MG
anti-aircraft	1 × 7.62 mm MG
SMOKE LAYING EQUIPMENT	2 × 9 either side of turret
AMMUNITION	
main	40 (22 ready use)
GUN CONTROL EQUIPMENT	
Turret power control	electric/manual
by commander	yes
by gunner	yes
Turret traverse	360°
Max turret traverse speed	40°/s
Max gun elevation speed	30°/s
Commander's fire control override	yes
Gun elevation/depression main armament	+15°/-8°

GIAT Industries AMX-40 MBT

Development

The AMX-40 MBT, armed with a 120 mm smooth bore gun, was developed by GIAT Industries as a private venture. It is no longer being marketed and details were given in *Jane's Armour and Artillery 1990-1991*, pages 21 to 23.

GIAT Industries AMX-30 Main Battle Tank

Development

Design work was carried out on the AMX-30 at the Atelier de Construction d'Issy-les-Moulineaux under the direction of the Direction des Études et Fabrications d'Armement (now Direction Technique des Armements Terrestres), with the first two prototypes being completed in 1960. A further seven prototypes built with the Hispano-Suiza HS 110 12-cylinder multi-fuel engine which was also fitted to production tanks were completed early in 1963 and in July of that year the tank was adopted by the French Army as the replacement for its American-supplied M47 tanks. Before production began two pre-production tanks were built.

Production of the AMX-30 began in 1966 at the Atelier de Construction Roanne (ARE) which had previously been building the AMX-13 light tank, production of which was transferred to Creusot-Loire to enable the factory to concentrate on the AMX-30. It has also been built under licence in Spain (qv). In 1987 the Republic of Cyprus placed an order with France for the supply of 15 AMX-30 B2 MBTs and a single AMX-30D armoured recovery vehicle, and these were delivered in 1988. Late in 1989 the Republic of Cyprus placed an order with GIAT for the supply of a further 35 AMX-30 B2 MBTs and another AMX-30D armoured recovery vehicle.

The AMX-30 and improved AMX-30 B2 will remain in service with the French Army until replaced by the Leclerc MBT.

AMX-30 series MBTs were used by France, Saudi Arabia and Qatar during operations conducted early in 1991 to retake Kuwait following its invasion by Iraq in 1990.

Description

The hull of the AMX-30 is made of rolled steel plates welded together. It is divided into three compartments: driver's at the front, fighting in the centre and the engine at the rear.

The driver is seated at the front of the vehicle on the left with a single piece hatch cover opening to the left and three periscopes. The centre periscope, depending on the model of the tank, can be either a day periscope which can be replaced by an infra-red or image intensification night periscope (Thomson-CSF TH 9478), or a SOPELEM OB-16-A periscope. This has a binocular system for infra-red light by night and a monocular system for day driving. The infra-red system has a magnification of ×1 and a 35° field-of-view and the day system has a similar magnification and a 24° field-of-view.

The other three crew members are seated in the turret with the commander and gunner on the right and the loader, who also operates the radio, on the left. The commander's cupola is a TOP 7 with 10 periscopes for all-round vision and a single piece hatch cover opening to the rear. Mounted on the forward part of the commander's cupola is a SOPELEM M 270 prism head. This comprises an armoured housing and a thick glass behind which is a prism, swivelling in elevation and reflecting the image of the terrain towards the M 267 day sight or the OB-23-A infra-red telescope (magnification ×4, 9° field-of-view), the object lenses of which fit in its lower part. On the right of the prism is a swivelling arm which supports the 7.62 mm machine gun and its PH-9-A infra-red searchlight which has a 500 m range when being used in the infra-red mode and a 700 m range in the white light mode. The weapon can be elevated from -10° to +45° by a handwheel in the turret roof. The prism may also be used to aim the coaxial 20 mm cannon, in which case the head (of the prism) is electrically servo-controlled by the elevation swivel of the 20 mm cannon. The M 270 has a magnification of ×10 and allows the tank commander to locate and identify targets and bring the turret to bear onto the target. The tank commander also operates the SOPELEM M 208 rangefinder which can be used either as a telescope with a magnification of ×6 and a range of 600 to 3500

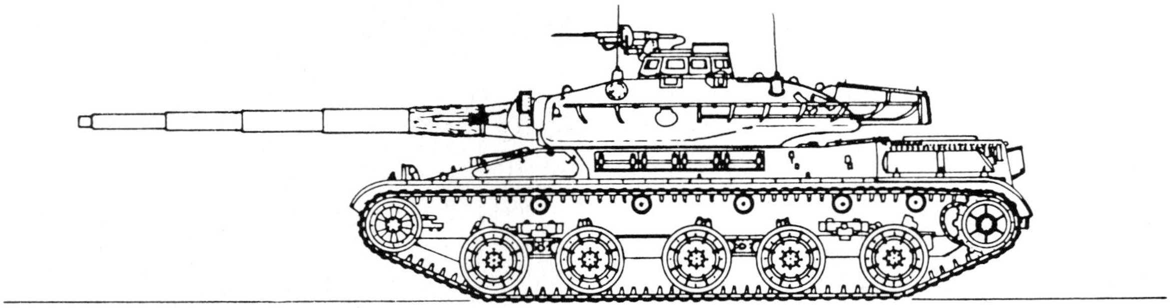

AMX-30 MBT

m or by means of a superimposed image when it has a magnification of ×12 and a similar range.

Mounted on the left side of the turret, coaxial with the main armament, is a SOPELEM PH-8-B searchlight, which has a maximum range of 2000 m when used in the white light mode and a range of 1000 m in the infra-red mode.

The gunner, who is seated below and in front of the commander, has an M 271 day sight with a magnification of ×8 which can be changed for an OB-17-A sight. This is mounted in the roof of the turret and has a luminous graticule, magnification of ×5.4 and a 7° field-of-view. When used with the PH-8-B infra-red searchlight it has a range of 800 m. The gunner also has two periscopes.

The loader is seated on the left of the turret and is provided with a single piece hatch cover that opens to the rear and two periscopes. A small circular hatch in the left side of the turret is used for replenishing ammunition and ejecting spent cartridge cases manually.

The engine is to the immediate rear of the bulkhead separating the engine compartment from the fighting compartment. The Hispano-Suiza HS 110 engine is manufactured by Renault, and operates on oil, petrol or paraffin. The engine is water-cooled and air is drawn in above the chassis in the rear part of the tank and blown down through the radiator by a fan driven by the engine. The speed of the fan is governed by the water temperature.

The complete powerpack, comprising the engine, combined gearbox and steering unit, and clutch assembly can be removed in 45 minutes by a team of three. The powerpack can also be run outside the tank before installation.

The transmission consists of an automatic clutch, combined gearbox and steering unit, brakes and two final drives. In the basic version the centrifugal type clutch is activated electrically by the gear shift lever; a non-synchronised reverser enables the same number of gears to be engaged in reverse as forwards. The combined gearbox and steering mechanism contains the mechanically operated gearbox giving five speeds both forwards and in reverse and a triple differential steering system.

The brakes are hydraulically operated and are used as both service and parking brakes. Each final drive comprises spur-type right-angle gears and an epicyclic gear train.

The torsion bar suspension consists of five rubber-tyred twin road wheels with the idler at the front and the drive sprocket at the rear. There are five track return rollers which support the inside of the track only. The first, second, fourth and fifth road wheels are mounted on bogies and the first and fifth road wheels are provided with hydraulic shock absorbers. The centrally guided steel track has removable rubber pads and each track weighs 1580 kg. When new it has 83 links.

The AMX-30 can ford to a maximum depth of 1.3 m without preparation and 2 m with preparation. When deeper rivers are encountered a snorkel is erected over the loader's hatch. Two types of snorkel are available, a wide one for training and a much thinner one for operations, similar in concept to the Soviet snorkels fitted to the T-54/T-55 and T-62 MBTs. Before entering the water a ring is inflated around the turret, mantlet and cupola using the electrically driven compressor, and two blanking plates are fitted over the engine compartment louvres; these are carried on the right side of the glacis plate when not in use.

Standard equipment on all AMX-30s includes a battery-operated electric pump for refuelling and a lubricating pump using the tank's compressed air circuit. The power receptacle in the driver's compartment can be used for recharging the tank's batteries from another tank, and also for supplying the electric pump used for refuelling. The tank is also equipped with an NBC system, heater, automatic fire alarm system, radios, crew intercom system and an infantry telephone at the rear.

Main armament of the AMX-30 is a 105 mm rifled gun designated the CN-105-F1 with a length of 56 calibres. The gun does not have a muzzle brake or a fume extractor but is fitted with a magnesium alloy thermal sleeve. A compressed air system evacuates any fumes from the barrel. The recoil system consists of two diametrically opposed hydraulic brakes and an oleo-pneumatic recuperator for counter-recoil of the barrel.

The 105 mm gun can fire APFSDS, HEAT, HE, phosphorus smoke or illuminating rounds of a French design and can also fire standard 105 mm ammunition as used with the L7 series of weapons mounted in the Leopard 1 and M60 series of MBTs. A 105 mm DU round is now under development by GIAT Industries.

A total of 47 rounds of 105 mm ammunition is carried of which 19 are in the turret (18 in the bustle) and 28 in the hull to the right of the driver.

Mounted to the left of the main armament is a 20 mm Model F2 (Type M693) cannon, which can be elevated with the main armament but can also be elevated on its own to a maximum of +40° for use against slow-flying aircraft and helicopters. The cannon has a maximum effective range of 1500 m and can be fired by the gunner or the tank commander. The 20 mm cannon can be either dual feed (with HEI rounds with a muzzle velocity of 1050 m/s and armour-piercing rounds with a muzzle velocity of 1250 m/s), or single feed firing American M56 type ammunition. A total of 500 rounds of ready use ammunition is carried with a further 550 rounds held in reserve. When originally introduced into service a 12.7 mm machine gun was mounted to the left of the main armament. This was provided with 600 rounds of ammunition.

Mounted to the right of the commander's cupola is a 7.62 mm Model F1 machine gun which can be aimed and fired from within the cupola. The weapon has an elevation of +45° and a depression of -10°. A total of 2050 rounds of 7.62 mm ammunition is carried of which 550 are ready for immediate use. The empty cartridge cases are automatically ejected outside the tank. Maximum effective range of the 7.62 mm machine gun is quoted as 700 m.

Two smoke dischargers mounted either side of the turret can lay a smoke screen which will cover the tank in eight seconds.

French Army AMX-30 MBT based in Germany showing 20 mm cannon mounted coaxially with 105 mm gun (Pierre Touzin)

AMX-30 of the French Army showing thermal sleeve of 105 mm gun (Pierre Touzin)

AMX-30 fitted with snorkel for deep fording (Pierre Touzin)

French Army AMX-30 B2 from rear (Pierre Touzin)

GIAT Industries AMX-30 B2 MBT of the French Army

The hydraulic aiming control system of the AMX-30 is a SAMM model CH 27-1S with elevation controlled through a hydraulic actuating cylinder/shock absorber and traverse controlled through a hydraulic motor. The gunner is provided with dual handle controls and the tank commander has a single handle control which can also be used to override the gunner.

A more simplified version of the AMX-30 was developed for export with no night vision or NBC system and the commander's cupola was replaced by an S 470 cupola with an externally mounted 12.7 mm machine gun rather than a 7.62 mm machine gun as installed on French Army AMX-30s, and the coaxial weapon a 7.62 mm machine gun rather than a 12.7 mm machine gun or a 20 mm cannon.

It is expected that the French Army's AMX-30 MBT fleet will be fitted with additional armour protection on turret front, sides and top.

AMX-30S

For desert operations the AMX-30S has been developed with the following modifications: the fitting of sand shields, reduction in the ratios of the gearbox which limits its speed to 60 km/h and an engine developing only 620 hp at 2400 rpm. At least one foreign army (Saudi Arabian) has adopted the AMX-30S and its tanks are fitted with a CILAS/SOPELEM M 409 sight. This has a day sight with a magnification of ×8 and an 8° field-of-view, infra-red night sight with a magnification of ×4.5 and a 10° field-of-view and a laser rangefinder with a range of between 400 and 10 000 m. This development allows the tank commander to range onto a target without having to traverse the turret. In addition the laser rangefinder increases the first-round hit probability.

AMX-30 B2

The AMX-30 B2, announced in June 1979, is essentially an AMX-30 with an integrated fire control system based on a laser rangefinder and a thermal system, new gearbox and other improvements. The French Army ordered 50 vehicles in both the 1981 and 1982 defence budgets and first deliveries were made to the 503rd Regiment at Mourmelon in January 1982. The French Army is expected to take delivery of some 271 AMX-30 B2 MBTs. In addition, 693 AMX-30s will be converted to the AMX-30 B2 configuration. Between 1984 and 1988 511 modernised AMX-30 B2s were scheduled to be delivered to the French Army with a further 182 being delivered after 1988. Orders were for 182 conversions in 1984-85 and 273 for the period 1986-1988. Final deliveries of rebuilt AMX-30 B2s to the French Army are expected to be made in 1993. The 1989 French Defence Budget provided funding for 46 AMX-30 B2 retrofits with 25 expected to be delivered from earlier contracts in same period. More recently, however, the AMX-30 B2 modernisation programme for the French Army has been cut back.

The COTAC FCS, officially designated the APX M581, has been designed by the AMX-APX in collaboration with the Ateliers de Construction de Tarbes. Its main components are the gunner's APX M544 telescopic (×10 magnification) sight which is combined with an APX M579 electronic control system and an APX M421 optical module containing a computer-controlled graticule. The laser rangefinder is the APX M550 and is made by CILAS under the designation TCV 80. It has a maximum range of 10 000 m and is accurate to ±5 m. The SOPELEM telescopic sight has a magnification of ×10 and is directly coupled to the main armament. In addition to the M581 rangefinder telescope the gunner has a rotatable M282 periscope, a fixed M223 periscope and a TV monitor associated with a thermal monitor.

The COTAC FCS also incorporates an accelerometer-type trunnion tilt sensor and a control panel. The tank commander manually inserts information via the control panel concerning the type of ammunition being fired, drift and jump angles, cross wind velocity, altitude and ambient temperature. Apart from the ammunition details, most of this information has to be inserted only once a day. To determine lead angles for the engagement of moving targets, a two-axis gyroscope is mounted on the sight to measure target tracking rates in both azimuth and elevation.

Other parameters are automatically acquired by the gunner's rangefinder telescope including target distance, its displacement speed in elevation and azimuth and turret slant. These parameters are processed by a lead computer built into the telescope which transmits the elevation and azimuth corrections to the gunner's day sight by means of an optical compensator and the graticule of the thermal camera for engaging targets at night.

In addition to the 10 M268 periscopes in his cupola the commander also has an M496 telescope with a magnification of ×8 for firing the 105 mm gun, for the secondary armament (eg 20 mm cannon and 7.62 mm machine gun) and aiming assembly comprising an M591-02 prism head and an OB 49 monocular telescope.

Mounted externally on the right side of the turret is the thermal camera with both the tank commander and gunner being provided with a TV monitor screen. The screens display an aiming mark which enables the commander or gunner to engage targets at ranges of up to 1000 m.

The loader has a rotatable M282 periscope and two fixed M223 periscopes with the driver having a central M223 periscope (or an OB 31A image intensification periscope) and two fixed M223 periscopes.

The engine of the AMX-30 B2 has been developed from that of the AMX-30 and develops 700 hp at 2600 rpm; as an option GIAT Industries can offer a super-charged version with an output of 800 hp.

The HS-110-2 is fitted with new and more powerful turbo-compressors which provide a 45 hp increase as well as improving the operational life of the engine and its torque characteristics.

A new hydrostatically controlled engine system now proposed by GIAT incorporates oil heat exchangers cooled by air instead of engine coolant. The air filters now have removable covers to facilitate field maintenance.

To allow the AMX-30 B2 to fire the newer APFSDS ammunition a new gunner's sight graticule is fitted to match the ballistics of the round while the turret and ammunition racks have been modified to accept the differently shaped APFSDS round.

Two different gearboxes are offered. First the ENC 200 which is a gearbox of the lock-up torque converter type with five forward and reverse speeds and hydrostatic steering controlled by a conventional steering wheel instead of sticks. This system allows on-the-spot turning and gear changing in bends.

The second gearbox is the 5 SD which is a further development of that installed in the AMX-30 with five speeds with interlocked pneumatic control.

The suspension uses new torsion bars which give improved cross-country mobility. As an option it is possible to fit the AMX-30 B2 with connector tracks which are quieter and offer less resistance than conventional tracks.

The AMX-30 B2 is also fitted with a new collective pressurisation system with two circuits, one with complete filtration for improved NBC protection.

In addition to these improvements, GIAT offers a number of other options for the AMX-30 B2. French Army AMX-30 B2s have the OB 49

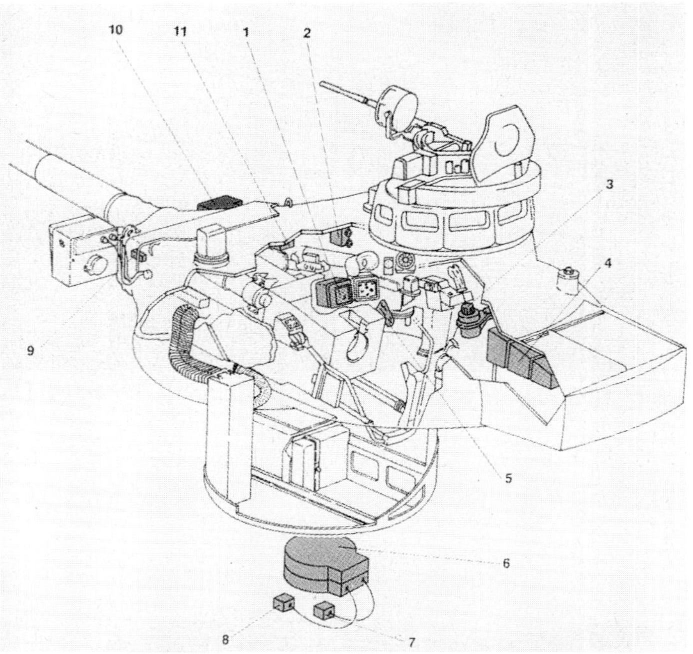

Turret of GIAT Industries Super AMX-30 MBT showing main components **(1)** *firing control console,* **(2)** *stabilised sighting caps control,* **(3)** *traverse sensor,* **(4)** *electronic box (servo control stabilisation and firing control),* **(5)** *gunner electro-hydraulic control,* **(6)** *electric rotary joint,* **(7)** *swaying gyrometer,* **(8)** *speed sensor,* **(9)** *elevation sensor,* **(10)** *stabilised head,* **(11)** *pitching gyrometer*

roof-mounted sight with two day and one night channels. This can be replaced by the M427 with day and image intensification night channels and a laser rangefinder or a roof-mounted SFIM M527 gyro-stabilised sight also fitted to AMX-32 and AMX-40. The gunner on French Army AMX-30 B2s has an M581 rangefinder telescope but this can be replaced by an OB 47. The driver's infra-red periscope can be replaced by an OB 31 or CN2-516 image intensification periscope.

Other options include new 125 Ah batteries with a longer life and needing less maintenance, new filters, DUK DUR 440 radiation counter, air-conditioning system and three different land navigation systems (Arverne, Navyx and Sydade) all of which have been developed by Crouzet SA.

French Army AMX-30 B2s have a thicker gun shield; additional add-on armour protection can be provided for the turret sides and armoured skirts could be fitted.

In place of the standard bank of two electrically-operated smoke dischargers either side of the turret, the GIAT/Lacroix AFV protection system could be fitted. This can fire a wide range of different types of grenade to meet various battlefield threats. An exhaust system smoke generator, similar to that on the AMX-40 MBT, can also be fitted to the AMX-30 B2 if required.

AMX-30 Venezuela

In 1985 SABCA of Belgium was awarded a contract to modernise Venezuelan AMX-30 MBTs with a new fire control system. Teledyne Continental Motors of the USA has also modernised at least one Venezuelan AMX-30 with a 908 hp version of the AVDS-1790 series diesel engine coupled to a fully-automatic transmission, speed is increased to 65 km/h and the operating range with increased fuel capacity is up from 535 km to 721 km. The AMX-30 was also fitted with a new fire control system incorporating a laser rangefinder, weapon stabilisation system and sensors for wind, temperature and humidity, and a modernised commander's station. The driver's station has also been modernised, a fire detection and suppression system installed and the vehicle can now lay its own smoke screen by injecting diesel fuel into the exhaust system.

Other options include appliqué armour protection and the replacement of the torsion bar suspension by the Model 2880 suspension units for improved cross-country mobility.

Variants

GIAT Industries AMX Super 30

In 1988 GIAT Industries announced that it was offering to modernise the AMX-30 MBT for the export market under the name of the AMX Super 30. The modernisation would be carried out in three key areas, mobility, firepower and logistics.

The current powerpack would be replaced by an 850 hp UNI-DIESEL engine coupled to a fully automatic German ZF LSG 3000 transmission already in production for a number of applications. A new cooling system allows the modernised AMX-30 to operate in ambient temperatures up to +50° C.

Other improvements include installation of a steering wheel rather than the tillers presently fitted, new gearbox control console, new instrument panel to right of driver, Halon fire extinguishing system and new turret electric rotary joint, new fuel tanks and improved torsion bars, bump stops and shock absorbers.

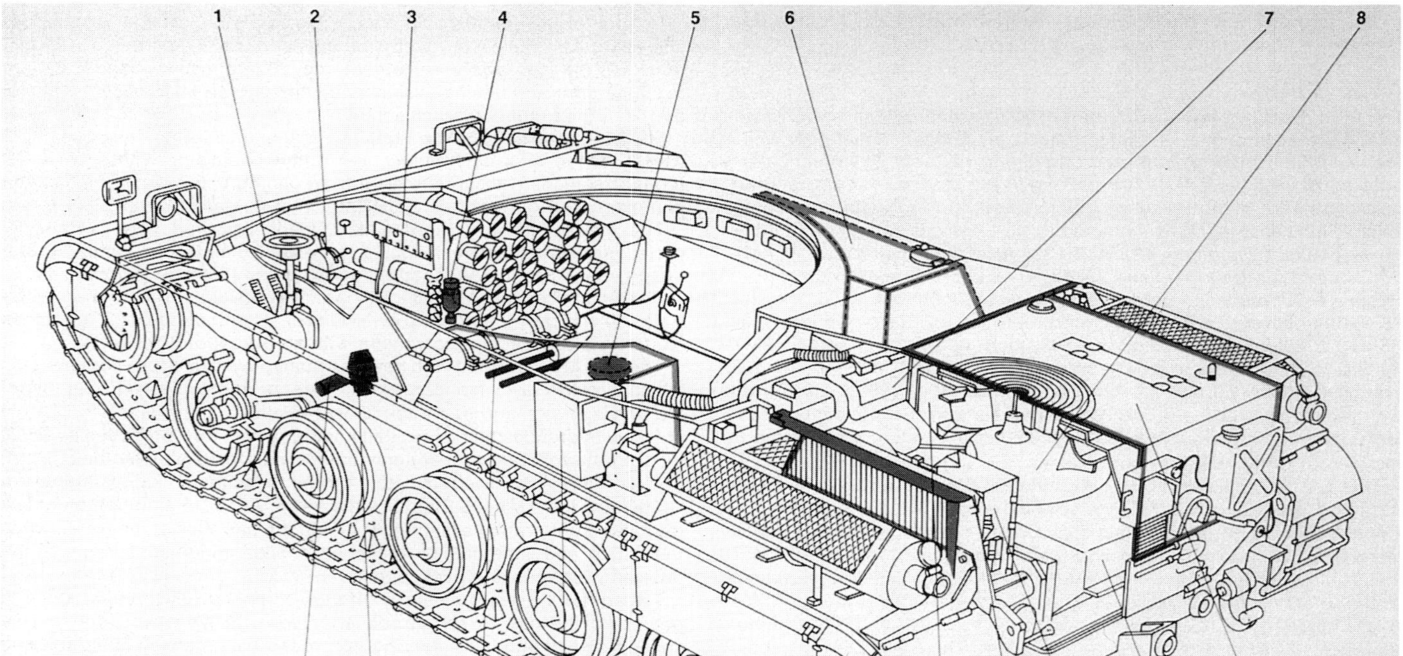

Main components of GIAT Industries Super AMX-30: **(1)** *steering wheel,* **(2)** *gearbox control console,* **(3)** *dashboard,* **(4)** *fire extinguisher (halon),* **(5)** *electric rotary joint,* **(6)** *fuel tanks,* **(7)** *engine,* **(8)** *air cooling fan,* **(9)** *engine cover,* **(10)** *gearbox,* **(11)** *radiator,* **(12)** *gearbox oil cooler,* **(13)** *fan framework,* **(14)** *air filter and pre-filter,* **(15)** *torsion bars,* **(16)** *suspension stops,* **(17)** *shock absorbers*

AMX-30D with crane traversed to front of vehicle (Pierre Touzin)

AMX-30 Combat Engineer Tractor (EBG) from front with dozer blade lowered and crane traversed to right (Christopher F Foss)

AMX-30 bridgelayer in travelling order

AMX-30 with Pluton tactical nuclear missile system (C R Zwart)

Compared to the existng AMX-30, the AMX Super 30 will have improved cross-country mobility and greater power-to-weight ratio.

A new fire control system has also been installed which includes a stabilised sighting system incorporating a laser rangefinder mounted to the right of the 105 mm gun.

Optional equipment includes a thermal camera, slant sensor and atmospheric sensor mounted on the turret roof to the rear of the commander's cupola.

GIAT Industries also refer to this upgrade package as the AMX-30 C2.

AMX-30 with GT601

For trials GIAT Industries has installed a Garrett GT601 gas turbine engine in an AMX-30 chassis, and replaced the original transmission with a new ENC 200 transmission.

Spanish AMX-30 Upgrade

In January 1987, after studying a number of proposals and trials with prototype vehicles, Spain approved a plan to upgrade 150 of its AMX-30 MBTs in two stages, first the installation of a new powerpack and second a new fire control system. Details of this plan are given under Spain.

Super AMX-30 MBT

A German consortium comprising AEG, Krupp Atlas, MTU, Wegmann, Diehl, AF and GLS has proposed a modernised version of the AMX-30 MBT. This consortium is led by GLS except for Saudi Arabia where it is led by MTU.

The Super AMX-30 has already been demonstrated in Saudi Arabia and has a new powerpack consisting of an MTU MB 833 Ka-501 diesel developing 850 hp at 2400 rpm coupled to the new ZF LSG 3000 fully-automatic transmission. The new powerpack, which can be removed from the AMX-30 in one hour, also has air filtration, cooling and electrical generating systems. The fuel capacity has been increased to 1028 litres, which increases the operational range by 20 per cent. A fire detection and suppression system has been fitted to the engine compartment and the driver's compartment has been modernised.

Each road wheel station has been fitted with a telescopic damper, new road wheels have been fitted and original tracks replaced by the new Diehl Type 234 which have double the life of the original tracks.

The Super AMX-30 has been fitted with the Krupp Atlas MOLF 30 (Modular Laser Fire Control) fire control system. The gunner has a new stabilised sight which incorporates day and night sights plus a laser rangefinder. The commander's sight allows him to observe targets via the gunner's sight.

The original hydraulic turret/weapon control system has been replaced by a new electric system.

A new NBC protection system has been installed on the turret bustle and if required additional armour protection can be provided around the turret.

AMX-30D Armoured Recovery Vehicle

This is designated the AMX-30D and is designed to carry out three major tasks: the recovery of disabled and damaged vehicles, replacement of major components such as engines, and engineer work. Equipment provided includes a hydraulically operated dozer blade at the front of the hull which is used to stabilise the vehicle while the crane is being used or during winching or for dozing. Mounted on the right side of the vehicle is a hydraulic crane which can lift 12 000 kg through 240° or 15 000 kg when being used to the front of the vehicle. The main winch has a maximum capacity of 35 000 kg and is provided with 90 m of 34 mm diameter cable while the auxiliary winch has 3500 kg capacity and is provided with 120 m of 11.2 mm diameter cable. There are two variants of the AMX-30D, the AMX-30DI which can lift a maximum load of 15 000 kg through a full 240° rather than just to the front of the vehicle and the AMX-30D(S) which is for desert operations and has modifications similar to those fitted to the AMX-30S MBT. The AMX-30D has a loaded weight of 36 000 kg and normally carries a spare engine at the rear. Details of this version are given in *Jane's Military Vehicles and Logistics 1991-92*, page 21.

AMX-30 Bridgelayer

The AMX-30 bridgelayer is not used by the French Army, although a few are used by Saudi Arabia. It is basically an AMX-30 MBT with its turret removed and replaced with a scissors type bridge which is launched over the rear of the hull in five minutes. The bridge is a class 50 and when opened out is 22 m long and will span a gap of up to 20 m. The bridge is 3.1 m wide or 3.92 m when fitted with widening panels. Weight is 42 500 kg with the bridge and 34 000 kg without. Full details of the AMX-30 armoured bridgelayer are given in *Jane's Military Vehicles and Logistics 1991-92*, page 78.

Pluton Surface-to-surface Missile System

This is the AMX-30 chassis fitted with a launcher system for the Pluton tactical nuclear missile system and is used only by the French Army.

AMX-30 DCA with twin 30 mm cannon which is in service only with Saudi Arabia

Roland 2 SAM system of French Army on modified AMX-30 chassis (Pierre Touzin)

The Pluton has a maximum range of 120 km and can be fitted with a tactical nuclear warhead of either 15 or 25 kt. The French Army has five regiments (3, 4, 15, 32 and 74) of Pluton, each of three batteries with each battery having two launchers. The Pluton will be replaced in the mid-1990s by the HADES truck-launched missile system which will have a much longer range than Pluton.

AMX-30 Combat Engineer Tractor

The AMX-30 Combat Engineer Tractor, or Engin Blindé de Génie, has been developed and will replace the AMX-13 VCG, at present used by the French Army engineers. The French Army has a total requirement for 126 EBGs of which the first 71 have been funded.

The EBG is designed for use in forward areas and is equipped with a front-mounted hydraulically operated dozer blade which is fitted with six scarifying teeth, hydraulic winch with a maximum capacity of 20 000 kg, hydraulic PTO and a working arm pivoted at the front of the hull on the right side. The arm is used to lift obstacles out of the way and can be fitted with pincers to remove tree trunks (similar to those fitted to the Soviet IMR combat engineer vehicle) or an auger which can bore 220 mm diameter holes to a depth of 3 m.

Armament consists of a 7.62 mm machine gun, four smoke dischargers, launching tube for demolition charges and four mine launching tubes. The 142 mm demolition charge has four folding fins, weighs 17 kg of which 10 kg is explosive and has a maximum range of 300 m. Each of the four mine-launching arms has a tube containing six 139 mm diameter anti-tank mines, each of which weighs 2.34 kg and contains 0.7 kg of explosive. The vehicle has a crew of three: the commander, a sapper and a driver, maximum road speed of 65 km/h and can ford to a depth of 2.5 m without preparation or 4 m with preparation. Details of this vehicle are given in *Jane's Military Vehicles and Logistics 1991-92,* page 2.

The French Army EBGs are to be provided with an MLC 65 (Military Load Class) folding assault bridge (details of which are given in *Jane's Military Vehicles and Logistics 1991-92*, page 79).

The TCP (Travure Courte Portee) folding assault bridge has been developed by CNIM who are currently supplying the French Army with the Pont Flottant Motorise (PFM) motorised pontoon bridge system.

This bridge is fitted onto the EBG from a single axle trailer using only the EBG's crane. Once in position, the scissors bridge is opened by a combination of cables and gravity and can be used to span a gap of up to 12 m wide. When opened out the bridge is 14 m long, 4 m wide and weighs a total of 4800 kg.

AMX-30 MDR Flail Tank

Development of this system has been cancelled. For trials purposes an AMX-30 MBT has been fitted with Israeli mine clearing rollers at the front of the hull.

Shahine firing unit as delivered to Saudi Arabia

Shahine acquisition unit as delivered to Saudi Arabia

Roland Anti-aircraft Missile System

This is a modified AMX-30 MBT chassis with two Euromissile Roland surface-to-air missiles in the ready to launch position and a further eight rounds in reserve in the hull, four either side, ready for automatic reloading. The surveillance radar is mounted on the turret rear with the tracking radar being mounted on the turret front. This version is in service with France (181), Iraq (13), Qatar (3), Spain (18 ordered, first deliveries late 1988) and Nigeria (16). Full details of this system are given in *Jane's Land-Based Air Defence 1991-92*, page 112.

Shahine Anti-aircraft Missile System

This was developed by Thomson-CSF specifically to meet the requirements of Saudi Arabia and consists of a firing unit with six missiles in the ready to fire position on a modified AMX-30 chassis and an acquisition unit with a large surveillance radar on a similar chassis. This is an all weather system. The firing unit carries no reserve missles, these being brought up by a cross-country truck fitted with a crane. Full details of this system are given in *Jane's Land-Based Air Defence 1991-92*, page 103.

Twin 30 mm Anti-aircraft Gun System

This is an updated version of the turret installed on the AMX-13 light tank chassis used by the French Army, but installed on an AMX-30 chassis specifically to meet the requirements of Saudi Arabia who took delivery of a total of 53 vehicles. The chassis is fitted with a power-operated turret armed with twin 30 mm cannon for which a total of 600 rounds of ready use ammunition are carried (300 for each gun) with a further 900 rounds being carried in reserve. Mounted on the turret rear is the radar system. Full details of this system are given in *Jane's Land-Based Air Defence 1991-92*, page 62.

The AMX-30 chassis was also fitted with a Thomson-CSF Sabre twin 30 mm air defence turret, a further development of the system sold to Saudi Arabia, but this never entered production.

155 mm GCT Self-propelled Howitzer

Details of this system are given in the *Self-propelled guns and howitzers* section.

Driver Training Tank

This is basically the AMX-30 MBT with its turret replaced by an observation type turret and is used for driver training.

Manufacturer: GIAT Industries
13 route de la Minière
78034 Versailles Cedex
France
Telephone: (1) 30 97 37 37
Fax: (1) 30 97 39 00

Status: In production. In service with the following countries:-

MODEL	**AMX-30**	**AMX-30S**	**AMX-30 B2**	**AMX-30**	**AMX-30**	**AMX-30**	**GCT**	**Shahine**	**DCA**	**Roland**	**Pluton**
TYPE	MBT	MBT	MBT	AVLB	ARV	CET	SPG	SPAAM	SPAAG	SPAAM	SSM
Chile	21	0	0	0	0	0	0	0	0	0	0
Cyprus	0	0	50***	0	0	2***	0	0	0	0	0
France	1084*	0	271**	0	134	126†	190	0	0	181	44
Greece	190	0	0	0	14	0	0	0	0	0	0
Iraq	0	0	0	0	0	0	85	0	0	13	0
Nigeria	0	0	0	0	0	0	0	0	0	16	0
Qatar	0	24	0	0	1	0	0	0	0	3	0
Saudi Arabia	0	290	0	12	57	0	51	96††	53	0	0
Spain	299‡	0	0	0	10	0	0	0	0	18‡‡	0
UAE (Abu Dhabi)	64	0	0	0	4	0	0	0	0	0	0
Venezuela	81	0	0	0	4	0	0	0	0	0	0

* *being upgraded to AMX-30 B2 standard*
** *new-build AMX-30 B2*
*** *Total order, not all yet delivered, see text*
† *total requirement*
†† *32 acquisition and 64 firing units*
‡ *Marines also have 36 vehicles*
‡‡ *first deliveries late 1988*
The figures for Iraq are for the total number delivered and few of these are believed to remain operational

---

SPECIFICATIONS (specifications in square brackets relate to AMX-30 B2 where this differs from AMX-30)

CREW	4
COMBAT WEIGHT	36 000 [37 000] kg
UNLOADED WEIGHT	34 000 [35 300] kg
POWER-TO-WEIGHT RATIO	20 [18.91] hp/tonne
GROUND PRESSURE	0.77 [0.90] kg/cm²
LENGTH GUN	
forwards	9.48 m
rear	8.73 m
LENGTH HULL	6.59 m
WIDTH	3.1 m
HEIGHT	
to hull top	1.5 m
to turret top	2.29 m
overall, including searchlight	2.86 m
FIRING HEIGHT	1.81 m
GROUND CLEARANCE	0.44 m
TRACK	2.53 m
TRACK WIDTH	570 mm
LENGTH OF TRACK ON GROUND	4.12 m
MAX SPEED	
1st gear	7 km/h
2nd gear	15 km/h
3rd gear	26 km/h
4th gear	43 km/h
5th gear	65 km/h
AVERAGE SPEED	
road	50 km/h
cross-country	35-40 km/h
FUEL CAPACITY	970 [900] l
MAX ROAD RANGE	500-600 [400-450] km
FORDING	1.3 m
with preparation	2.2 m
with snorkel	4 m
GRADIENT	60%
SIDE SLOPE	30%
VERTICAL OBSTACLE	0.93 m
TRENCH	2.9 m
ENGINE (see text)	Hispano-Suiza HS 110 12-cylinder, water-cooled super-charged multi-fuel developing 720 hp at 2000 rpm [700 hp at 2600 rpm]
AUXILIARY ENGINE	none
TRANSMISSION (see text)	mechanical with 5 gears in both directions
STEERING (see text)	triple differential
CLUTCH	centrifugal
SUSPENSION	torsion bar
ELECTRICAL SYSTEM	28 V
BATTERIES	8 × 12 V, 100 Ah
ARMAMENT	
main	1 × 105 mm gun
coaxial	1 × 20 mm cannon (or 1 × 12.7 mm MG)
anti-aircraft	1 × 7.62 mm MG
SMOKE-LAYING EQUIPMENT	2 smoke dischargers either side of turret
AMMUNITION	
main	47
coaxial	1050 [480]
anti-aircraft	2050 [2070]
GUN CONTROL EQUIPMENT	
Turret power control	electro-hydraulic/manual
by commander	yes
by gunner	yes
commander's override	yes
Commander's fire-control override	yes
Max rate power traverse	360° in 12-13 s
Max rate power elevation	5.5°/1 s
Main armament elevation/depression	+20°/-8°
Secondary armament elevation/depression	+40°/-8°
7.62 mm armament elevation/depression	+45°/-10°
Gun stabiliser	
vertical	no
horizontal	no
Elevation quadrant	yes
Traverse indicator	yes
ARMOUR (estimated)	
Hull front	79 mm
Hull sides front	57 mm
Hull sides rear	30 mm
Hull top	15 mm
Hull bottom	15 mm
Hull rear	30 mm
Turret front	80.8 mm
Turret sides	41.5 mm
Turret rear	50 mm
Turret top	20 mm

GERMANY

Kampfwagen 2000

Late in 1988 the tactical requirement for the German Army's future MBT Kampfpanzer 2000 was established.

It is understood that this is for a conventional layout with the low profile turret in the centre of the chassis. Main armament will comprise a large calibre 140 mm smoothbore gun fed by an automatic loader with the turret housing the driver and gunner/commander.

Four NATO countries, France, Germany, the United Kingdom and the United States; have already signed an MoU for a new 140 mm smoothbore gun with a common chamber size which will be of a conventional configuration using a solid propellant charge system.

Leopard 2 Main Battle Tank

Development

Once production of the Leopard 1 was under way Porsche was awarded a contract to carry out further development of the tank to increase its combat effectiveness. This contract expired in 1967, at which time Germany and the USA were already developing the MBT-70, and the agreement covering it did not allow for a national tank development programme.

Germany did however develop new components for the Leopard 1, some of which were to provide a basis for a new tank. In 1968 Krauss-Maffei was awarded a contract worth DM 25 million to build two prototypes of a new MBT called Experimental Entwicklung (Experimental Development). This resembled the late production Leopard 1A3/1A4 and had an improved fire control system, different stabilisation systems, new engine and transmission and was armed with a 105 mm rifled tank gun, a 7.62 mm coaxial machine gun and a remote-controlled 7.62 mm anti-aircraft machine gun.

In 1969 it became apparent that the MBT-70 would not get past the prototype stage and efforts were made to adapt certain MBT-70 components to a new German tank. This development study was known as the Eber (or Boar). There was also a study based on the previously mentioned two prototypes, for a new tank called the Keiler (Wild Boar).

In 1970 the MBT-70 project was cancelled and the USA went on to develop a more austere version of the MBT-70 called the XM803 which was soon cancelled by Congress because of its rising costs. The USA then started on a new MBT project that culminated in the XM1 MBT, first production models of which were completed early in 1980.

In 1970 Germany took the decision to proceed with a new tank called the Leopard 2 and 16 hulls and 17 turrets were completed by Krauss-Maffei between 1972 and 1974. All of the prototypes used the Renk transmission and MTU diesel from the MBT-70. A study was also made of mounting the American 152 mm Shillelagh weapon system into the Leopard 2 but this idea was soon dropped.

Ten of the prototypes were equipped with a Rheinmetall 105 mm smoothbore gun and the remainder were fitted with a Rheinmetall 120 mm smoothbore gun. Two of the chassis were fitted with a hydro-pneumatic suspension but the advanced torsion bar suspension with integrated friction dampers was subsequently adopted.

Various types of fire control systems were fitted in the prototype vehicles and one even had a remote-controlled 20 mm anti-aircraft cannon mounted on the turret roof.

In 1974 the USA and Germany signed a Memorandum of Understanding (MoU) under which both countries affirmed their intention to make all reasonable efforts to standardise their tank programmes. In 1977 the MoU was amended to include efforts to standardise some tank components between the two countries. These included the engine, transmission, gunner's telescope, night vision equipment, fire control system, tracks and main armament.

To meet the requirements of the USA, Germany built another model called the Leopard 2 AV (Austere Version). It had a different fire control system and a turret and hull which incorporated a spaced multi-layer armour, as well as many other detail improvements influenced by the conflict in the Middle East in 1973. Two Leopard 2(AV) hulls and three turrets were built. One complete Leopard 2(AV) tank and one chassis were delivered to the USA for trials in September 1976, but as expected the USA chose one of the two competing American designs, the Chrysler XM1. The USA did however adopt the Rheinmetall 120 mm smoothbore gun and first production M1A1s with this weapon were completed in August 1985.

In 1977 the German Army selected Krauss-Maffei as prime contractor for series production for the Leopard 2 and placed an order, with options, for 1800 MBTs of which 990 were built by Krauss-Maffei and the remaining 810 by Krupp MaK.

The first pre-production Leopard 2 MBT was delivered to the German Army late in 1978 for training. Krauss-Maffei and MaK delivered three pre-production Leopard 2 MBTs in early 1979 and after firing trials at Rheinmetall's range at Unterlüss, it was delivered to Troop Combat School 2 at Munster. The first production Leopard 2 was handed over by Krauss-Maffei in October 1979 in Munich. Six tanks were delivered in 1979, 100 in 1980, 220 in 1981 and by 1982 production was running at 300 a year (25 a month). In 1982 it was announced that the cost of the Leopard 2 programme for the West German Army was DM5100 million with the first production lots being for 380, 450 and 300 tanks. The last two production lots comprised 300 and 370 tanks.

In June 1987 an order for the 6th production batch of 150 Leopard 2s was placed by the German Government. Of these, 55 per cent were built by Krauss-Maffei with the remaining 45 per cent by Krupp-MaK. In 1988 the contract was placed for the 7th production batch of 100 Leopard 2s with the contract for the 8th production batch of 75 vehicles being placed in 1990. Under present contracts final deliveries will be made in 1992.

The first production lot of 380 German Army Leopard 2s are being returned to the manufacturers in batches for overhaul. At the same time the interim PZB 200 image intensification night sights are being replaced by integral thermal sights which were installed as standard from the second production batch of the Leopard 2.

Other minor changes in the second production batch of Leopard 2s include the elimination of the sidewind sensor, relocation of the tank

One of the seventh production lot of Leopard 2 MBT with modified frontal armour, all hatches closed and 7.62 mm MG 3 machine gun mounted at loader's station

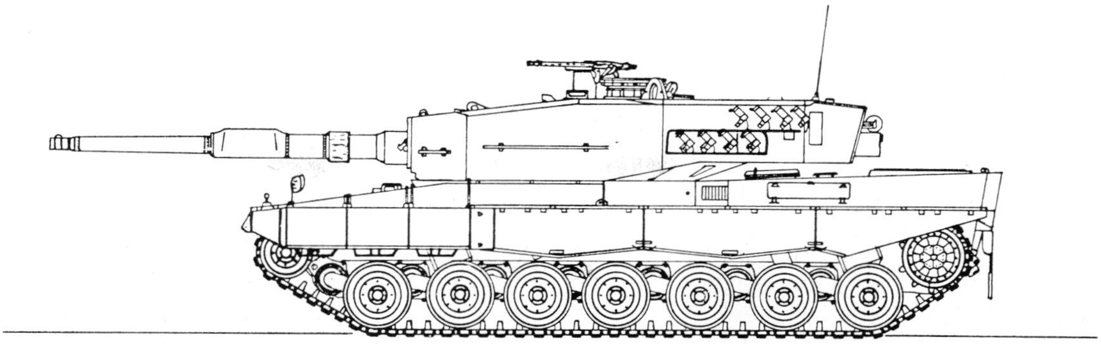

Leopard 2 MBT armed with 120 mm smoothbore gun

filler openings, improved exhaust gratings, modified tool stowage and a covering for the commander's cupola.

The final fourth production batch changes included a new paint scheme, digital core for the fire control computer, welding shut the ammunition hatch in the left side of the turret (this was eliminated on later production vehicles) and the installation of a fire suppression system.

Dutch Leopard 2 Programme

In March 1979 the Dutch placed an order for 445 Leopard 2s for delivery from 1982 to 1986 to replace 369 Centurion MBTs and 130 AMX-13 light tanks; of these Krauss-Maffei built 278 and Krupp-Mak the remaining 167. Dutch industry participated in the production of parts and components at approximately 60 per cent of the value of the order. In addition Dutch industry was provided with new export possibilities for products not directly concerned with the Leopard 2 contract. The first four Leopard 2 MBTs for the Dutch Army were completed in mid-1981. Main contract deliveries began in July 1982 and by November that year 10 vehicles were being delivered a month with final deliveries taking place in July 1986. They have different 7.62 mm machine guns (FN in place of Rheinmetall MG3s), smoke dischargers, passive night periscope for the driver, radios and intercom equipment to the German version.

In 1983 the Royal Netherlands Army's 41st Armoured Brigade stationed in West Germany became the first in the Dutch Army to be equipped with the Leopard 2. The Dutch Army also ordered 20 driver-training versions of the Leopard 2 MBT.

Swiss Leopard 2 Programme

In August 1983, the Swiss Army announced that after evaluating the Leopard 2 and M1 MBTs the Leopard 2 had been selected. A total of 380 were ordered in one lot, the first 35 coming direct from Krauss-Maffei in Germany and the remainder made under licence in Switzerland. The first 35 were delivered in 1987 and used for training. The first battalion was fully operational on the tank in January 1988.

Prime contractor for the licence production in Switzerland is Contraves with final assembly taking place at the Federal Construction Works at Thun, where the Pz 61 and Pz 68 MBTs were built and where these are overhauled today. Production of the Leopard 2 in Switzerland started in 1987 with the first Swiss-built vehicles delivered in December 1987 and production running at six per month through to March 1993. It is expected that between 60 and 70 per cent of the Leopard 2 will be built in Switzerland, with NAW building the engine and SLM the transmission. The Swiss Leopard 2 is similar to the German vehicles but with Swiss radios and intercoms, and Swiss coaxial and anti-aircraft machine guns. Switzerland has also decided to install the British Kidde-Graviner Crew Bay fire and explosion suppression system as did the German Army for the fifth production batch of Leopard 2 MBTs, beginning late in 1985.

Description

The hull of the Leopard 2 has a spaced multi-layer armour and is divided into three compartments: driver's at the front, fighting in the centre and engine at the rear.

The driver is seated at the front of the hull on the right side and is provided with a single piece hatch cover that opens to the right and three observation periscopes. The centre periscope can be replaced by a passive night periscope. Some of the ammunition supply is stowed to the left of the driver.

The turret is in the centre of the vehicle with the commander and gunner on the right and the loader on the left. The commander is provided with a circular hatch cover that opens to the rear and periscopes for all-round observation. A PERI-R 17 primary stabilised panoramic periscope with magnification of ×2 and ×8 is mounted in front of the commander's hatch and can be traversed through 360°, enabling the commander to observe the terrain and lay the main armament.

The gunner has a dual-magnification stabilised EMES 15 sight with integrated laser rangefinder, and thermal image unit (WBG) which are linked to the fire control computer. He also has an auxiliary sighting telescope FERO-Z 18 with a magnification of ×8. The tank commander uses the thermal sight, which is integrated with the gunner's EMES 15, to observe the battlefield. The picture is transmitted to the commander's PERI-R 17 so he can see the same frame as the gunner. Furthermore the commander can control the RPP 1-8 by a computer controlled testing board.

The EMES 15 is a primary stabilised binocular sighting instrument for the gunner. The mirror head of the periscope is stabilised around two axes; the day path has a magnification of ×12 and a 5° field-of-view.

The laser rangefinder has a range of 9990 m and is accurate to ±10 m with measuring distances shown to three digits together with the fire preparation and selected type of ammunition in the lower part of the gunner's sight.

The fire control computer calculates successively the angle of sight and lateral angular lead for the main armament. The following parameters are taken into account: target distance (from laser rangefinder), angle of tilt of the vehicle, direction of motion in regard to the target, lateral

German Leopard 2 MBT from rear showing turret stowage details

Dutch Leopard 2 MBT from front with commander's and loader's hatches open but without 7.62 mm machine gun mounted on roof (C R Zwart)

Leopard 2 of the German Army with Loader's 7.62 mm at maximum elevation (Pierre Touzin)

Concept for improved armour protection on Leopard 2 (Improved) clearly showing new armour array on the turret front and new gunner's sight arrangement through the roof rather than the turret frontal armour

Leopard 2 of the German Army showing its deep fording capability when fitted with a snorkel (Pierre Touzin)

wind and ballistic data of ammunition. The data calculated by the fire control computer is fed into the weapon slave system which guides the weapon to the line of sight of the EMES 15 or PERI-R 17. The thermal image unit integrated in the EMES 15 enables the armament to be aimed and fired at night or in bad weather or at camouflaged targets.

The gunner also has a roof-mounted observation periscope. The loader is seated on the left side of the turret and has a single piece hatch cover that opens to the rear and a single observation periscope. An ammunition resupply hatch is provided in the left side of the turret and there is a stowage basket at the turret rear. The latter has been eliminated on later production batches. This has been basically changed with the introduction of the seventh production batch.

The engine compartment at the rear of the hull is separated from the fighting compartment by a fireproof bulkhead. The MTU MB 873 engine is coupled to a Renk HSWL 354 hydro-kinetic planetary gear shift with an integral service brake.

The suspension with torsion bars consists each side of seven dual rubber-tyred road wheels with the idler at the front and the drive sprocket at the rear, and four track return rollers. Advanced friction dampers are provided at the first, second, third, sixth and seventh road wheel stations. The Diehl tracks have rubber-bushed pins with removable rubber pads, which can be replaced by snow grousers. The rear two-thirds of the top of the tracks are covered by steel-reinforced rubber skirts which can be folded up to allow access to the suspension for maintenance. The front third of the top of the track is covered by special armoured boxes which can be rotated upwards both for maintenance and to reduce the overall width of the vehicle for rail travel.

Standard equipment on the Leopard 2 includes an NBC system, powerpack preheating, crew compartment heater, fire-extinguishing system, electric bilge pumps and a hull escape hatch behind the driver.

The 120 mm smoothbore gun has been developed by Rheinmetall and weighs 3700 kg (barrel, thermal shroud, bore evacuator and breech) with a barrel length of 5.6 m and has a dropblock breech mechanism. Firing is electric with emergency firing induced by means of an impulse generator. When fired, the weapon recoils and is brought back into position by a recuperator. At the same time the breech block opens automatically. The 120 mm smoothbore gun fires two types of ammunition, APFSDS-T (DM-13) and HEAT-MP-T (DM-12), each of which has a practice version. The APFSDS-T has an effective range of well over 2000 m and the HEAT-MP-T has a high degree of effectiveness against both soft and hard targets. Both fin-stabilised rounds have a semi-combustible cartridge case with a metal base stub which is ejected into a bag under the gun.

In 1983 the second generation APFSDS-T round was introduced under the designation DM-23. This has a new monobloc tungsten penetrator. A third generation APFSDS-T round is now under development under the designation of the DM-33. This has a much greater length-to-diameter ratio but has yet to enter production. The ammunition of the Leopard 2 is interchangeable with that of the American M1A1 that entered production in 1985, as the M1A1's main gun is a licence version of the Leopard 2's Rheinmetall 120 mm gun.

Rheinmetall is developing two private venture projectiles which will increase the effectiveness of tank armament at longer battlefield ranges than existing 105 mm and 120 mm projectiles. The first of these is a tank cannon launched anti-helicopter round and the second a long-range anti-tank round.

Of 42 rounds of ammunition carried some are stored to the left of the driver and some in the left side of the turret bustle. If the latter is hit it will explode upwards due to blow-out panels.

A 7.62 mm Rheinmetall MG 3 machine gun is mounted coaxially to the left of the main armament and a similar weapon can be mounted on the loader's hatch. Mounted on either side of the turret are eight smoke dischargers.

Leopard 2 (Improved)

The first Leopard 2 (Improved) MBT was completed in the Autumn of 1990 by Krauss Maffei, prime contractor for the Leopard 2 MBT. Compared to the current production Leopard 2, the new Leopard 2 (Improved) has a significant increase in battlefield survivability.

Improvements in the chassis include stronger torsion bars, stronger and greater capacity bump stops and additional armour on the nose, glacis plate and the front side skirts.

Trials of a standard production Leopard 2 weighted to simulate the new improvements have already shown that cross-country mobility of the heavier version is not degraded to a significant extent.

The current production Leopard 2 weighs 55.15 tonnes with the 1500 hp MTU MB 873 Ka 501 12-cylinder diesel giving a power-to-weight ratio of 27 hp/tonne.

The new Leopard 2 (Improved) weighs 62 tonnes and will have a power-to-weight ratio of 24 hp/tonne. Because of the additional armour protection, acceleration, top speed and operational range will be slightly reduced.

If required the MTU diesel could be upgraded to maintain the original power-to-weight ratio, but at present there is no operational requirement for this. The Leopard 2 (Improved) could also be fitted with an auxiliary power unit (APU) to allow the main engine to shut down with the APU keeping the batteries charged.

The turret has been redesigned and has additional armour protection to its front, sides and roof. The turret front now has a distinctive arrowhead shape and, according to Krauss-Maffei, will defeat all known KE and CE attack. The turret roof has also been up-armoured to provide increased protection against top attack weapons such as sub-munitions.

The present electro-hydraulic gun control and stabilisation system will be replaced by a new AEG all electric system.

The Carl Zeiss commander's roof-mounted stabilised panoramic sight now incorporates a thermal night channel as well as the standard day channel with the option of a laser rangefinder.

The Carl Zeiss stabilised gunner's sight on current production Leopard 2s is routed through the frontal armour but on the Leopard 2 (Improved) it has been raised and is now routed through the roof in a similar manner to that of the British Challenger 2, so improving armour integrity.

The fire control computer is of the latest digital type which is also installed in current production Leopard 2s and has been retrofitted to earlier vehicles as they come back to Krauss-Maffei for overhaul.

The turret also has the first phase of a combat vehicle command and control system installed together with a navigation aid.

The crew compartment is fitted with a Deugra Crew Bay fire and explosion detection and suppression system which is manufactured under licence from Kidde-Graviner of the UK. This is also fitted to current production Leopard 2s and is being backfitted to earlier production vehicles.

Buffel armoured recovery vehicle of the German Army towing a Leopard 2 MBT

Leopard 2 driver training vehicle which has turret replaced by observation type turret and simulated gun barrel

Armament of the Leopard 2 (Improved) is the standard 120 mm smoothbore gun which will fire HEAT-MP and APFSDS ammunition from a variety of sources.

The Leopard 2 (Improved) is now funded by the German BWB with a further two prototypes to be completed early in 1992. It is anticipated that as the earlier Leopard 2s come back to Krauss-Maffei for overhaul they will be upgraded to the Leopard 2 (Improved) standard, although this depends on funding and the user may well elect not to go for every improvement.

Buffel Armoured Recovery Vehicle

In August 1990, Krupp MaK announced that it had been selected by the BWB as the main contractor for series production of the Buffel armoured recovery vehicle (ARV) based on automotive components of the Leopard 2 ARV.

A total of 100 Buffel ARVs have been ordered, 75 for the German Army and 25 for the Netherlands Army with first production vehicles to be completed late in 1991. Of the 100 vehicles, 55 will be built by Krupp MaK and 45 by Krauss-Maffei.

To carry out its recovery role the Buffel ARV is fitted with a hydraulic crane which is pivoted at the front of the hull on the right side, winch, dozer/stabilising blade and other specialised equipment. Additional details of the vehicle are given in *Jane's Military Vehicles and Logistics 1991-92*, page 23.

Leopard 2 Driver-training Vehicle

The German Army has now taken delivery of 22 driver-training versions of the Leopard 2 while the Dutch Army has now taken delivery of 20 vehicles. These are essentially a Leopard 2 with its turret replaced by an observation type turret and weighted to give the vehicle similar handling characteristics to the standard Leopard 2. This is also provided with a dummy 120 mm gun barrel.

Variants

The chassis of the Leopard 2 could also be used to mount the turret of the Gepard anti-aircraft tank now in service with Belgium, Germany and the Netherlands (with Dutch rather than German radars). Late in 1983 the German Government announced that it would not sell Leopard 2 MBTs to Saudi Arabia although it would allow the sale of an anti-aircraft version, but no sales had been made as of early 1991.

A Leopard 2 prototype was fitted with the American Textron Lycoming AGT 1500 gas turbine in 1978 for trials purposes by Krupp MaK at Kiel.

For trials purposes the new MTU 833 V-12 diesel developing 1475 hp at 3000 rpm has been installed in a Leopard 2 MBT. This produces the same output as the MTU 873 Ka-501 currently installed in the Leopard 2 but is lighter, more compact and more fuel efficient.

Status: In production. In service with Germany, the Netherlands and Switzerland (licenced production).

Manufacturers: Krauss-Maffei Wehrtechnik GmbH
Krauss-Maffei Strasse 2,
8000 Munich 50
Federal Republic of Germany
Telephone: (089) 88 99 0
Telex: 5 23 163 31
Fax: (089) 8 12 01 90

Krupp MaK Maschinenbau GmbH
PO Box 9009
2300 Kiel 17
Federal Republic of Germany
Telephone: (0431) 3811
Telex: 02 99 877 mak d

SPECIFICATIONS	
CREW	4
COMBAT WEIGHT	55 150 kg
POWER-TO-WEIGHT RATIO	27 hp/tonne
GROUND PRESSURE	0.83 kg/cm²
LENGTH	
gun forwards	9.668 m
gun rear	8.49 m
hull	7.722 m
WIDTH	
skirts unfolded	3.54 m
including skirts	3.7 m
HEIGHT	
to turret top	2.48 m
to commander's periscope	2.787 m
hull top	1.769 m
firing height	2.01 m
GROUND CLEARANCE	
front	0.537 m
rear	0.487 m
TRACK	2.785 m
TRACK WIDTH	635 mm
LENGTH OF TRACK ON GROUND	4.945 m

MAX ROAD SPEED	
forwards	72 km/h
reverse	31 km/h
FUEL CAPACITY	1200 l
MAX ROAD RANGE	550 km
FORDING	1 m
with preparation	2.25 m
with snorkel	4 m
GRADIENT	60%
SIDE SLOPE	30%
VERTICAL OBSTACLE	1.1 m
TRENCH	3 m
ENGINE	MTU MB 873 Ka 501 4-stroke, 12-cylinder diesel, exhaust turbo-charged, liquid-cooled developing 1500 hp at 2600 rpm
TRANSMISSION	Renk HSWL 354 hydro-kinetic planetary gear shift, 4 forward and 2 reverse gears
CLUTCH	torque converter
SUSPENSION	torsion bar, rotary friction damper and hydraulic bump stops

ELECTRICAL SYSTEM	24 V
BATTERIES	8 × 12 V, 125 Ah
ARMAMENT	
main	1 × 120 mm gun
coaxial	1 × 7.62 mm MG
anti-aircraft	1 × 7.62 mm MG
SMOKE-LAYING EQUIPMENT	8 smoke dischargers either side of turret
AMMUNITION	
main	42
MG	4750
GUN CONTROL EQUIPMENT	
Turret power control	electro-hydraulic/manual
by commander	yes
by gunner	yes
commander's override	yes
Commander's fire control override	yes
Gun elevation/depression	+20°/-9°
Turret traverse	360°
Gun stabiliser	
vertical	yes
horizontal	yes
Range setting device	yes
Elevation quadrant	yes
Traverse indicator	yes

Leopard 1 Main Battle Tank

Development

In November 1956 a Military Requirement was issued for a new MBT with the following specifications: combat weight of 30 tonnes, power-to-weight ratio of 30 hp/tonne, width of 3.15 m, height of 2.2 m, maximum road speed of 65 km/h, air-cooled multi-fuel engine, 105 mm gun and two machine guns. In June 1957 the Federal Republic of Germany and France agreed to develop a tank to meet these specifications and in July 1957 the Joint Technical Characteristics were issued, the only difference between the two countries being in the overall width of the tank. In September 1958 Italy joined the project, which resulted in supplements to the Technical Characteristics being issued in October 1961.

Germany formed two design teams, Teams A and B. In France the Direction des Etudes et Fabrication d'Armement (now the Direction Technique des Armements Terrestres) was the overall contractor with the design work being carried out at the Atelier de Construction d'Issy-les-Moulineaux at Satory.

The German Team A comprised Porsche, Jung, Luther and Jordan and MaK and Team B comprised Ingenieurbüro Warneke, Rheinstahl Hanomag and Henschel. Turrets and armament were designed by Rheinmetall and Wegmann.

Team A's first prototype was completed in June 1960 and the second in August the same year. Team B completed its first prototype in May 1960.

The four prototypes were delivered for trials early in 1961. The trials, which were carried out at Trier and later at Meppen, were completed in April 1962. It was then decided to concentrate on the Team A design. Before the trials were completed in April 1962 a further order was placed with Team A for 26 tanks with turrets designed by Wegmann, and Team B was awarded a contract for a further six tanks with turrets designed by Rheinmetall. Following the decision to concentrate on the Team A design only two of the second series from Team B were completed. The 26 tanks were known as the second series and had numerous improvements including increased armour protection, improved suspension, 10-cylinder diesel engine coupled to a new transmission, modified fire control equipment and the British 105 mm L7 series gun. Trials with the second series were completed early in 1963.

Before the second series had even completed their trials a pre-production batch of 50 tanks was ordered, which incorporated many improvements as a result of the earlier trials including the replacement of the 12.7 mm ranging machine gun with an optical rangefinder.

Comparative trials between the German design and the French AMX-30 were carried out from 1962, but in July 1963 the Defence Committee of the Federal German Parliament decided to go ahead with the production of the German design, which later became known as the Leopard, even though the main comparative trials had yet to take place.

In July 1963 Krauss-Maffei of Munich was nominated prime contractor for the Leopard 1 MBT and Krupp MaK of Kiel was selected as general contractor for production of the armoured recovery vehicle, armoured engineer vehicle and the armoured bridgelayer. Krupp MaK also built a small number of Leopard 1 MBTs under sub-contract to Krauss-Maffei.

The first production Leopard was handed over to the Federal German Army in September 1965, and since then the tank has been adopted by nine other countries including Italy, where it was manufactured under licence by OTO Melara (qv). Production of the Leopard 1 was completed by Krauss-Maffei in 1979 but resumed again by both Krauss-Maffei and Krupp-MaK early in 1981 to meet the requirements of Greece and Turkey.

Description (Leopard 1A1)

The all-welded steel hull is divided into two compartments, the crew compartment at the front and engine compartment at the rear.

The driver is seated at the front of the hull on the right and is provided with a single piece hatch cover that opens to the left. In front of the hatch are three periscopes, the centre one of which can be replaced by an infra-red or image-intensification periscope for night driving.

The all-cast turret is mounted in the centre of the hull with the commander and gunner seated on the right and the loader on the left. Both the commander and the loader have a single piece hatch cover that opens to the rear. The commander is provided with eight periscopes for all-round observation, one of which can be replaced by an infra-red image intensification periscope for night observation. The gunner's station has the sights and one periscope, the loader has two periscopes for observation of a designated battlefield sector. Mounted in the turret roof in front of the commander's hatch is a TRP 2A zoom periscope with a magnification of ×4 up to ×20. This is swivel-mounted and can also be moved in elevation by hand within the gun elevation range. When the commander actuates a switch the periscope is automatically slaved to the gun. For night observation the commander can replace the zoom periscope with an infra-red sight.

The gunner is seated in front of and below the commander and is provided with a TEM 2A rangefinder which can be used in either the coincidence or stereoscopic modes. It has magnifications of ×8 and ×16 and is mechanically linked to the gun and provided with super-elevation cams to compute the super-elevation angles for two types of ammunition. The gunner also has a TZF 1A telescope mounted coaxially with the main armament which has a magnification of ×8 and is provided with a moveable scaled graticule to set the super-elevation for the different types of ammunition. There is an ammunition resupply hatch in the left side of the turret and a stowage basket at the turret rear. Mounted over the main armament is an XSW-30-U infra-red/white searchlight, which can be removed and stowed at the rear of the turret when not required. In the infra-red mode in conjunction with an infra-red sight, which can be mounted in place of the commander's TRP 2A periscope, it has a range of up to 1200 m depending on conditions and in the white light mode a range of 1500 m. The Leopard 1A1A2 has an LLLTV observation and sighting system.

The engine compartment at the rear is separated from the fighting compartment by a fireproof bulkhead. The complete powerpack, consisting of the engine, transmission and cooling system, is provided with quick-disconnect couplings which allow the entire powerpack to be replaced in the field within 20 minutes.

The steer/shift transmission is coupled directly to the engine and has four forward and two reverse gears as well as a pivot turn capability, a torque converter and lock-up clutch. The gears can be shifted electro-hydraulically without interruption of tractive forces and the speed range can be pre-selected manually, gear changing being automatic.

The running gear includes seven dual light metal rubber-tyred road wheels with the drive sprocket at the rear, idler at the front and four track return rollers on either side. The first, second, third, sixth and seventh road wheel stations are provided with hydraulic shock absorbers. The Diehl tracks are provided with rubber-bushed track pins and the rubberised track pads can be replaced by snow grousers. Steel-reinforced rubber skirts for the top of the tracks can be folded vertically for maintenance. The track skirts perform two functions: they increase the ballistic protection, and also reduce the whirl-up of dust.

The NBC system is installed at the front of the hull and produces an over-pressure in the crew compartment. In normal use the system provides the crew with fresh air, and in the NBC mode filters contaminated particles by separate purification processes.

Standard equipment includes an automatic fire-extinguishing system, heater, hull escape hatch and infra-red reflection suppressing decontaminable paint.

For underwater operations there are two fording shafts available: one for fording to a depth of 2.25 m and an extra shaft for fording to a depth of 4 m. Either shaft is to be fitted over the commander's hatch, down which air for the engine is drawn and the exhaust leaves via the exhaust pipes in the normal manner. Before entering the water all openings are sealed, the turret ring by an inflatable rubber ring. A hydraulic system controls the engine combustion air intakes, dust ejection blower valves of the combustion air cleaners, exhaust flaps of the engine and heater, pressure ventilation in the engine compartment as well as the dust ejection valves and the air intake of the NBC system. Two bilge pumps remove any water that enters the vehicle.

Leopard 1A2 with thermal sleeve for 105 mm L7A3 gun, infra-red/white light searchlight on mantlet

German Army Leopard 1A3 showing track skirts, thermal sleeve for 105 mm gun and 7.62 mm anti-aircraft machine gun

Belgian Leopard 1 fitted with snorkel for deep fording (C R Zwart)

German Army Leopard 1A4 MBT showing welded turret and bustle

Main armament of the Leopard 1 is a British-supplied 105 mm L7A3 rifled tank gun which consists of a single piece barrel with a screwed-on breech-ring and a bore evacuator. The barrel can be changed in the field in about 20 minutes. The semi-automatic breech mechanism automatically opens after each round is fired, ejecting the empty cartridge case into the spent cartridge container under the breech.

The Leopard 1 can fire all the standard 105 mm rounds manufactured by Canada, France, Germany, Israel, the United Kingdom and the USA. Of the 55 rounds carried, 42 are in the hull and 13 in the turret.

Krauss-Maffei states that the first round hit probability of the Leopard 1 weapon system, even when on the move and firing at moving targets, was significantly increased by the incorporation of the gun stabilisation system and modern fire control systems, the latest of which has a primary stabilised line of sight, laser rangefinder and an integral thermal imaging system. From 1986-92 approximately 1300 Leopard 1A1A1/A1A2 tanks are being retrofitted and designated Leopard 1A5.

Mounted coaxially with the main armament is a 7.62 mm Rheinmetall MG 3 machine gun which is provided with 1250 rounds of ready-use ammunition. The coaxial machine gun is operated by the gunner but can also be operated by the commander in the override mode. A second MG 3 machine gun is mounted at the commander's or loader's station for anti-aircraft defence. Four electrically operated smoke dischargers are mounted on either side of the turret.

Variants

Leopard 1A1A1

The first model of the Leopard to enter service with the Federal German Army was called the Leopard 1 of which 1845 were built in the first, second, third and fourth batches. From 1971 they were fitted with a thermal sleeve for the main armament, gun stabilisation system, new tracks and lateral skirts. The stabilisation system controls the gun in elevation and traverse so that the gunner is able to observe the terrain, acquire and then engage the target while moving across country with a high degree of first-round hit probability. With these modifications the Leopard 1 became the Leopard 1A1. These tanks have since been fitted with Blohm and Voss add-on armour on the turret and gun shield and are now known as the Leopard 1A1A1. The additional add-on armour consists of flexibly mounted, screwed-on armour steel plates with two-faced rubber lining. The armour also covers the turret bustle back and the gun shield is also reinforced with armour steel plates. In addition, armour steel plates are welded on to the sloped front roof section. The extra weight caused by the fitting of this additional armour is about 760 kg, which makes the vehicle approximately the same weight as the Leopard 1A4. Another improvement on the Leopard 1A1A1 is the significantly enhanced combustion air intake filters.

Leopard 1A2

The fifth production batch of Leopards for the Federal German Army consisted of 342 tanks of which 232 were built with the modifications of the Leopard 1A1 (but without the spaced armour on turret and mantlet) and in addition have a turret of thick cast steel, an improved NBC system and passive image-intensification night vision equipment for both the commander and driver.

Leopard 1A3

A total of 110 Leopard 1A3 tanks out of the fifth production batch were delivered to the Federal German Army, incorporating all the improvements of the Leopard 1A1 and 1A2 but with a new welded turret of spaced armour with a wedge-shaped mantlet. The turret rear stowage bin has been incorporated into the contour of the turret which accommodates the searchlight when it is not mounted over the main armament. In addition, the loader's periscope is movable in both elevation and azimuth. In 1981 the Greek Government ordered 106 Leopard 1A3 MBTs plus an option for an additional 101 vehicles for delivery from February 1983 through to April 1984. Krauss-Maffei of Munich built 73 MBTs and Krupp MaK of Kiel 33 MBTs and four ARVs. The Turkish order for 81 Leopard 1s comprised four ARVs built by Krupp MaK and 77 Leopard 1A3 MBTs (54 built by Krauss-Maffei and 23 by Krupp MaK).

Leopard 1A4

This was the final production model of the Leopard 1 for the Federal German Army of which 250 were built, 215 by Krauss-Maffei and 35 by Krupp MaK. It is similar to the Leopard 1A3 but has an integrated fire control system consisting of a stabilised panoramic telescope for the tank commander, gunner's primary sight with integral stereoscopic rangefinder coupled to the fully-stabilised main armament and controlled by a ballistic computer. Late in 1985 it was stated that Turkey was to receive 150 German Army Leopard 1A4s which would be replaced by a similar number of Leopard 2s, but only 150 Leopard 2s were ordered. Before delivery to Turkey 150 tanks will be converted to the T1 configuration of the 77 Leopard 1s already in use by the Turkish Army. The first four Leopard 1A4 MBTs were handed over to the Turkish Army in mid-1990 with the final vehicles expected to be handed over late in 1991. According to the German MoD, the total value of the 150 vehicles, including spares, was DM580 million ($348 million). The final 100 vehicles will be assigned to another application.

Leopard 1A5

In 1982-83 six German Army Leopard 1 MBTs were fitted with different fire control systems for comparative trials. These were the AEG-Telefunken Lemstar M/EMES 17, Krupp-Atlas Elektronic FLP-10/EMES 18, and the Zeiss AFLS-L/EMES 12A4, all of these systems being completed by a thermal imaging system for application under poor sighting and night combat conditions. The German Army finally accepted

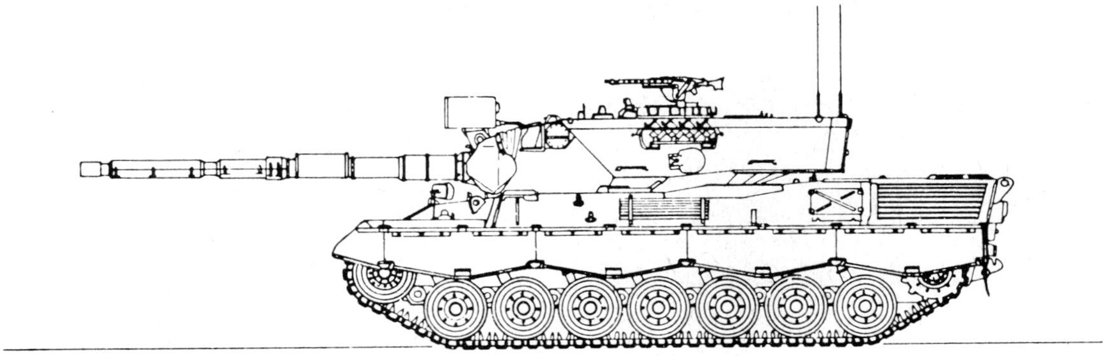

Leopard 1A3 MBT

the Krupp-Atlas EMES 18 combined with the thermal imaging system in 1984. First converted tanks were delivered to the German Army in December 1986 and a total of 1300 are being delivered through to 1992. Conversion work will be undertaken by Wegmann of Kassel under a contract worth DM829 million; the major sub-contractors will be Krupp-Atlas for the EMES 18 fire control system and Carl Zeiss for the thermal imaging system. In order to overcome future threats on the battlefield measures are also to be undertaken to enhance the ballistic protection of the Leopard 1 by add-on armour and secondary protective devices such as an explosion suppression system for the crew compartment. With these modifications the designation changes to Leopard 1A6.

Optional equipment for Leopard 1 MBT

Krauss-Maffei offers a large number of modification kits for the Leopard 1 MBT such as additional armour for the turret and mantlet, armoured skirts, automatic transmission and gear shift, the new German fire control system EMES 18 that can have an integral thermal imaging system based on the American Common Modules, passive LLLTV sighting system PZB 200, brackets on glacis plate for snow grousers, dozer blade (already adopted by Australia, Canada, Denmark), cleaning system for driver's periscopes, improved turret and trunnion bearings, modifications to FCS to allow a 105 mm gun to fire APFSDS rounds, improved combustion cleaners, external stowage boxes on hull sides (already adopted by Australia, Belgium, Denmark and the Netherlands), snorkel, passive periscope for commander and driver, track with replaceable track pads, stabilisation system for the main armament, and a tropical kit.

The PZB 200 system consists of a combined low-light-level TV camera for terrain surveillance and sighting of targets, which is mounted on the gun mantlet and so moves in elevation with the main armament, monitor that can be seen both by the commander and gunner, control panel with on/off switch, dimmer control for graticule illumination, focusing switch and selector switch for extreme illumination levels and simple built-in test equipment.

Denmark and Norway have already awarded contracts for the retrofit of the EMES 18 fire control system.

Under contract to the Canadian Department of National Defense, Krauss-Maffei has developed and integrated into the Leopard C1 a gun/turret electric drive unit. This replaces the standard hydraulic drive, TST is responsible for the electric drives while Krauss-Maffei is responsible for integrating it into the vehicle.

Norway has adapted the turret electric drive system for fitting into their existing Leopard/MBTs in conjunction with the EMES 18 for control system.

Leopard 1 Dozer Blade Unit

The complete dozer blade unit consists of the dozer blade, two pivot-mounted push arms and a hydraulic unit for lifting and lowering the blade. The complete dozer blade can be attached or removed in approximately 10 minutes without the need of a crane or an ARV and can be locked mechanically for travelling. All control elements are packaged into a watertight unit so that earth moving is possible as well as fording operations. A utility outlet for power supply is standard equipment on the vehicle. A simple control panel is attached to the driver's station and the power supply and control cable passes through the open driver's hatch or, alternatively, one of the slots normally occupied by one of the vision blocks.

To enable this dozer blade kit to be used with the Leopard C1 MBTs operated by Canada, an Engineer Equipment Interface Modification Kit has been developed. This allows the tank to operate the dozer blade or one of two mine clearing systems, permitting operation with the driver's hatch closed down. A scoop welded on the turret roof plate of the chassis supplies power and control for the three adaptable engineer equipments at two receptacles. The exterior cable routed from the armour scoop to the add-on equipment is protected by armoured covers.

The dozer blade is used by Australia, Canada and Denmark.

Leopard 1 with 120 mm Gun

For trials purposes, Rheinmetall has fitted one German Army Leopard 1 series MBT with the same 120 mm smoothbore gun as installed in the Leopard 2 and United States M1A1 MBTs. The said MBT and an additional one fitted with the Rheinmetall 120 mm gun, dummies of the latest state-of-the-art add-on armour and other improvements, such as infra-red suppressing measures, were subjected to compatibility trials in the first half of 1987.

Both prototypes were based on a Leopard 1A1A1 tank; the second prototype with the welded turret. Both prototypes were fitted with the EMES 18 fire control system.

When fitted with the 120 mm smoothbore gun 42 rounds are carried, the same as in the Leopard 2, with 13 in the turret, two ready use and 27 in the hull.

Leopard 1 with 105 mm IWS

Early in 1991, a Leopard C1 MBT of the Canadian Armed Forces was fitted with the new private venture Royal Ordnance 105 mm Improved Weapon System (IWS).

The IWS has been developed by the Ammunition Division of Royal Ordnance (the 105 mm APFSDS round) and the Guns & Vehicles Division of Royal Ordnance (the actual gun).

The objective was to design a weapon system that would increase the firepower of existing in-service MBTs armed with the 105 mm L7/M68 rifled tank gun such as the Leopard 1 and M60, that would have performance levels comparable with current in-service 120 mm gun systems but not using a depleted uranium (DU) tungsten allow penetrator.

The 105 mm IWS fits directly into the existing vehicle mounting and utilises the existing gun cradle and recoil system.

It is also capable of firing all existing NATO approved 105 mm rounds (excluding APDS) as well as existing training ammunition. Built-in safety features ensure that the new high pressure APFSDS round cannot be incorporated into standard 105 mm L7/M68 guns or into an IWS incorporating L7 components.

The new high pressure ordnance weighs 1334 kg and is made of Electro Slag Refined (ESR) steel to withstand increased pressures with the barrel being chrome plated to increase barrel life which is expected to be a minimum of 450 EFC (new ammunition).

It has a vertical sliding breech mechanism, fume extractor, concentric thermal sleeve and a screw-on muzzle brake with an efficiency of 25 per cent has been incorporated to reduce recoil loads.

The IWS has a recoil distance of 279 mm and can also be fitted with a muzzle reference system.

The new 105 mm APFSDS round has been designed to fit into existing ammunition stowage arrangements without any reduction in the number of rounds being carried by the MBT.

The new APFSDS round weighs a total of 19 kg and is 1.03 m long with a muzzle velocity of 1420 m/s. The Tungsten-Nickel-Iron long rod penetrator, manufactured by Royal Ordnance Speciality Metals, is located in a three segment sabot which has been designed to reduce parasitic weight to a minimum. The length-to-diameter weight is 23/1.

According to RO, the new penetrator has a mass approximately 40 per cent greater than that of current APFSDS rounds. No actual details of penetration have been revealed but at least 520 mm of RHA at a range of 2000 m is probable.

The propelling charge comprises a stub brass cartridge case and a combustible sleeve which together house the multi-base granular propellant and primer.

Panther Tank Destroyer

It is probable that some of the Leopard 1 MBTs withdrawn from front line service by the German Army will be converted in Panther tank destroyers by Wegmann.

The 105 mm turret is removed and replaced by a new turret fitted with a scissors type arm on top of which is mounted a pod containing long-range TRIGAT anti-tank guided missiles and their associated optics.

Leopard 1A1A1 MBT fitted with Rheinmetall 120 mm smoothbore gun

Leopard C1 MBT of the Canadian Armed Forces fitted with the Royal Ordnance 105 mm Improved Weapon System

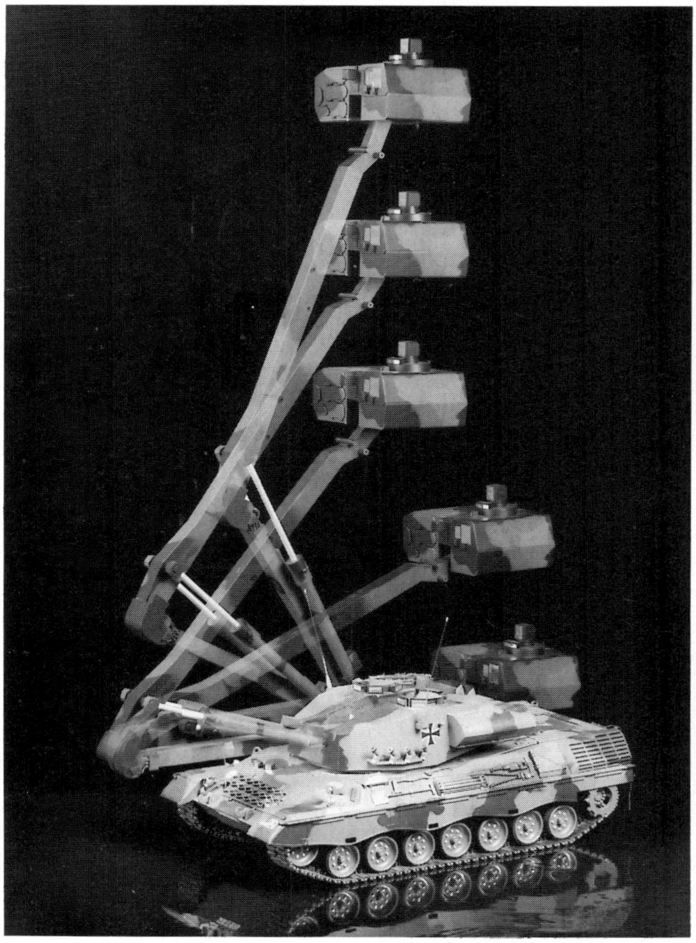

Model of the Wegmann Panther tank destroyer with arm in elevated and travelling position

When travelling the arm retracts down into the horizontal position and rests between the left and right halves of the turret with the pod containing the missiles towards the rear.

Australian Leopards

After evaluation of both the American M60 and the Leopard 1A2 between 1972 and 1973 the Australian Government placed an order for the Leopard 1A3, fitted with the Belgian SABCA fire control system and a tropical kit. These tanks, and ARVs and bridgelayers, were delivered to Australia between 1976 and 1978 and replaced the Centurions in the 1st Royal Australian Armoured Regiment. In Australian Army service the Leopard 1 is designated as follows: Leopard AS 1 MBT, Leopard AS 1 armoured recovery vehicle medium, Leopard AS 1 AVLB, Leopard AS 1 medium tank dozer and Leopard AS 1 medium tank clearing (mine-clearing equipment not yet procured).

Belgian Leopards

After evaluating a number of tanks, Belgium placed its first Leopard 1 order in 1967, the first NATO country (apart from Germany) to order the tank. The first production Leopard for Belgium was completed in February 1968. The Belgian vehicles had their MG 3 machine guns replaced by FN 7.62 mm MAG weapons plus minor stowage changes. From 1975 the tanks were fitted with stowage boxes similar to those on the Dutch Leopards, thermal sleeve for the main armament and the Cadillac Gage stabilisation system. The first third were also fitted with the SABCA FCS which has since been adopted by Australia and Canada.

The SABCA FCS basically consists of an optical sight with an integral laser rangefinder, seven sensors, and an analogue computer. The computer determines the angles between the line of sight and the gun axis from the information it receives about the range of the target and other variables. The output of the computer is transformed through a 2° of freedom gimballed mirror system, with torque DC motor drives and a compensated resolver feedback network, into a displacement of cross-hairs in the gunner's sight and to the gun via the gun stabilisation. The cross-hairs are brought back onto the target when the gun is automatically laid with the correct target elevation (or super-elevation) and azimuth. The sensors measure ambient temperature, air pressure, temperature in the ammunition stowage area, gun wear, cross wind, trunnion cant and rate of turret traverse. Early in 1980 the Belgian Government placed an order for the second and third batch of SABCA FCS for the Belgian Army Leopard 1. Following trials with a prototype Leopard 1 fitted with the German Blohm and Voss add-on armour, as already installed on German and Dutch Leopard 1 MBTs, the Belgian Army is to refit all of its tanks with this system, conversion work to be carried out at the Rocourt Arsenal in Belgium.

Australian Army Leopard AS 1 medium tank dozer with dozer blade lowered (Australian Army)

Australian Leopard bridgelayer with lower part of bridge sliding forward before being lined up with remainder of bridge (Paul Handel)

Prototype of the modernised Leopard 1 MBT of the Belgian Army with additional armour for turret and new SABCA and OIP-Instrubel day/night sight visible on the top right side of the turret

Early in 1987 the Belgian MoD awarded SABCA and OIP a BFr 177 million contract for the development of a modification kit to the fire control system incorporating a thermal imaging system which is now being installed in the 334 Leopard 1 MBTs of the Belgian Army; originally delivered by Krauss-Maffei from 1968.

OIP is responsible for the development of the infra-red telescope, including the magnification and focusing mechanisms, development of the new periscopic head which will include the development of a hard coating technique and development of an integrated day/night sight.

SABCA is responsible for the adaptive development of a complete thermal imaging system based on the TICM (thermal image common module) II module from GEC Avionics of the UK. This development includes the cooler, scanner, electronics, remote monitors and cables.

SABCA is responsible for modification of the current mirror drive included in the sight (in association with SERE-BEZU of France), development of the new sights electronics, development of a muzzle reference sensor (MRS) for automatic boresight correction (in association with Leitz of Canada), new main armament elevation resolver coaxial with the trunnion axis (in association with Wegmann of Germany), subsequent modification of the current fire control system, integration of the new complete TFCS (Tank Fire Control System) in the Leopard 1 and will also study, in association with Wegmann, the interface modification of the turret and some of its systems.

SABCA and OIP will jointly be responsible for the integration and factory test of the new sight (under the leadership of OIP) and integration and factory test of the new complete TFCS (under the leadership of SABCA).

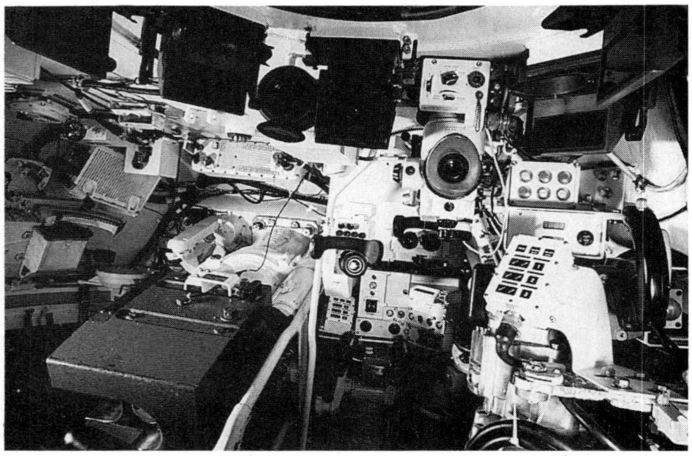

Inside view of modernised Leopard 1 MBT of the Belgian Army with commander's position on right and gunner's position forward and below this

Leopard 1 ARV of the Canadian Armed Forces being used as an AEV and laying pipeline fascine during training (Canadian Armed Forces)

There are four phases of the programme. Phase 1 covered system definition and was completed in mid-1986. Phase 2 covers development and evaluation, Phase 3 covers industrialisation and pre-series production and Phase 4 covers series production. Phases 3 and 4 will run from 1988 through to 1996.

Under Phase 2 integration and factory test of the sight was to be carried out in October 1987 with the integrated and factory tested TFCS being handed over to the Belgian MoD at the end of April 1988 and final MoD review taking place in late July 1988. The first prototype modernised Leopard 1 MBT was delivered to the Belgian authorities on October 6th 1988.

The Phase 3 industrialisation phase was completed by mid-1989 with five pre-production systems completed by the end of 1990 and production systems being delivered through to 1997 at the rate of about 50 tanks a year.

The installation of the new fire control system will provide the Leopard 1 with a high performance thermal imaging device in a day/night sight with integrated laser rangefinder.

OIP-Instrubel has been created by Oldelft of the Netherlands in order to take over the former OIP-Optics which went bankrupt in 1987.

At a later date further armour protection is expected to be added to the Belgian Leopard 1 MBTs.

Canadian Leopards

To replace their Centurion tanks the Canadians selected a modified version of the Leopard 1A3 fitted with the Belgian FCS. The total value of the Canadian order, for 114 MBTs, six bridgelayers and eight ARVs, was $187 million, of which $115 million was for the vehicles, $2.7 million for the loan of 35 Leopards until the Canadian Leopards were ready and the remaining $69.3 million for logistic supplies, training, ammunition and other equipment.

The first Leopard MBTs, called the C1 by the Canadians, were handed over in June 1978. The Canadian Leopards are based in Europe and Canada. Under the terms of the contract Krauss-Maffei took Canada's entire fleet of Centurion tanks. Some of the turrets, complete with 105 mm guns, were subsequently bought by Austria for static defence use.

The Canadian Armed Forces subsequently ordered nine of the latest version of the Leopard 1 Armoured Engineer Vehicle called Badger and the first vehicles were handed over in March 1990. In German Army service the Badger is usually referred to as the Pionierpanzer 2 Dachs Armoured Combat Engineer Vehicle.

Denmark

Denmark ordered 120 Leopard 1A3s in June 1974, the first of which were delivered in March 1976 and the last in November 1978.

Denmark is modernising the existing Leopard 1s to the 1A5 configuration and obtaining a further 110 Leopard 1A3s from German Army stocks.

Greek Leopards

The DM300 million order placed in April 1981 by the Greek Government for 110 Leopard 1s comprised 106 Leopard 1A3 MBTs and four ARVs, plus an option on a further 113 vehicles comprising 101 MBTs, eight AVLBs and four ARVs which is intended to be combined with the build up of an assembly line in Greece. Deliveries ran from February 1983 through to April 1984, with Krauss-Maffei of Munich building 73 MBTs and Krupp MaK of Kiel building 33 MBTs and four ARVs. The Greek Leopard 1s have the EMES 12A3 fire control system and the PZB 200 LLLTV system. The original option was never exercised and late in 1988 it was stated that Germany was to provide 75 Leopard 1 MBTs from its conversion programme.

Italian Leopards

Although Italy was one of the three countries originally involved in the search for a European tank it was not until 1970 that the Italians ordered the Leopard 1. The first 200, of which 92 were delivered in 1971 and the remaining 108 in 1972, were bought direct from Krauss-Maffei as it took several years for OTO Melara to tool up for production of the tank. The first Italian Leopard was completed in 1974 and the last of the original order for 600 was completed early in 1978. The Italian Army then ordered another 120 Leopard 1 MBTs which have now been delivered. Krupp MaK also delivered 69 ARVs and 12 AEVs to Italy. OTO Melara has built 160 specialised versions of the Leopard 1 for the Italian Army, 68 ARVs, 28 AEVs and 64 AVLBs. Further details are given under Italy. For trials purposes Italy has ordered four sets of the Blohm and Voss add-on armour system as installed on German and Dutch Leopard 1 MBTs.

It is expected that in the 1990s at least part of the Leopard 1 fleet will be fitted with a new fire control system incorporating a laser rangefinder.

In 1991, for trials purposes, a Leopard 1 MBT of the Italian Army was fitted with a private venture OTO Melara 76 mm OTOMATIC air defence

Canadian Armed Forces Leopard C1 fitted with front-mounted dozer blade and turret traversed left (C R Zwart)

Leopard 1 MBT of the Italian Army fitted with OTO Melara 76 mm OTOMATIC air defence turret

turret and this is to be tested by the Italian Army and Air Force for a period of 12 months through to the Spring of 1992.

The two-man power operated turret is fitted with a 76 mm gun that can fire a variety of ammunition types including a special APFSDS round. To engage aircraft or attack helicopters a five round burst would typically be used. The turret is fitted with tracking and surveillance radars.

Netherlands Leopards

After evaluating the Leopard, Chieftain and MBT-70 in 1967 and 1968, the Netherlands ordered 468 Leopard 1 MBTs between 1968 and 1970. The Leopards were delivered between 1969 and 1972 and many components were made by the DAF company and other Dutch firms.

The Dutch Leopards have different radios, Dutch type smoke dischargers and three stowage panniers. At a later date the tanks were fitted with an Honeywell gun stabilisation system, modified optical sights for British L52 APDS ammunition, passive periscope for driver and 7.62 mm FN MAG coaxial machine guns rather than German MG 3 weapons.

All 468 Leopard 1s of the Dutch Army have been modified by 574 Tankwerkplaats at Leusden and fitted with appliqué turret armour supplied by Blohm and Voss of Germany. This is the same as that fitted to the Leopard 1A1A1 of the German Army. When modified they are designated the Leopard 1-V, V standing for Verbeterd, or improved. New track skirts are also being fitted. The EMES 12A3 AFSL-Z (Dutch configuration) gunner's fire control system is installed at the same time as the new armour. The EMES 12A3 is supplied by Honeywell, Zeiss with Oldelft as the Dutch subcontractor.

Following a competition between the United Kingdom (L64), Israel (M111) and the United States (M735) for a new APFSDS round for the 105 mm L7 guns, the British L64 was selected as the replacement for the L52 APDS used.

Two Netherlands Leopard 1-Vs are to be modernised by Krauss-Maffei and Wegmann of Germany and brought up to the same standard as the Leopard 1A5 of the German Army. These will be evaluated by the Netherlands Army and a decision will be taken in 1991 as to whether or not to move ahead with the programme. During the prototype modernisation phase, the Netherlands industry will be allowed to participate to gain experience which could be used in the overall modernisation programme.

Early in 1991 however it was stated that the Netherlands Government was considering phasing out the whole Leopard 1 MBT fleet by 1995 with some vehicles to go into reserve and others passed on to NATO countries such as Greece and Portugal.

Norwegian Leopards

After testing the S-tank and the Leopard the Norwegians selected the Leopard MBT and the first order was placed in November 1968. The Leopards were delivered between 1970 and 1971. Like the Netherlands, the Norwegians use the British L52 APDS round. All Norwegian Leopard 1 MBTs are being upgraded to Leopard 1A5 standard, but retaining the cast turret. The hydraulic gun control equipment is being replaced by an all electric system.

Leopard AEV with auger being used to front of vehicle

Dutch Gepard twin 35 mm SPAAG which has different tracking and surveillance radars

Leopard 1 ARV of the Dutch Army with crane elevated and traversed to front of hull (C R Zwart)

Close-up of the turret of a Netherlands Army Leopard 1-V with part of add-on turret armour removed to show mounting points (C R Zwart)

Netherlands Army Leopard 1-V showing additional add-on armour to turret (C R Zwart)

Leopard 1 driver training tank of the German Army

German Gepard twin 35 mm SPAAG with tracking and surveillance radars retracted (US Army/Michael Green)

Turkey

Under an agreement signed in Bonn in November 1980, Turkey received 77 Leopard 1A3 MBTs (54 built by Krauss-Maffei and 23 by Krupp MaK) and four ARVs (built by Krupp MaK) which were delivered between September 1982 and December 1983. These were fitted with the EMES 12A3 fire control system and the PZB 200 LLLTV system.

Turkey is taking delivery of 150 of the 250 Leopard 1A4s currently used by the German Army. These are being modified and will all be delivered by late 1991. A further quantity of 80 Leopard 1A1A1s have already been delivered by the German Government.

Leopard Armoured Recovery Vehicle (Bergepanzer 2)

The Leopard ARV was designed by Porsche and the first production vehicle was completed in September 1966 by Krupp MaK of Kiel. The ARV is based on the chassis of the MBT but has a new hull. Standard equipment includes a dozer blade at the front of the hull, hydraulically operated crane on the right side of the hull which is pivoted at the front, main winch with a capacity of 35 tonnes (which can be increased to 70 tonnes with a guide pulley), electric wrench and a welding system. The vehicle, which weighs 39 800 kg and has a crew of four, has been designed for recovery of damaged vehicles, towing disabled vehicles, changing vehicle components and lifting complete vehicles weighing up to 20 tonnes, carrying a spare Leopard MBT powerpack, dozing, and refuelling and defuelling other vehicles. Armament consists of a bow-mounted 7.62 mm MG 3 machine gun, 7.62 mm MG 3 anti-aircraft machine gun and smoke dischargers. In 1978 the Federal German Army took delivery from Krupp MaK of 100 product-improved Leopard ARVs which have a hydraulic jack mounted at the rear of the hull on the right side and a more powerful crane (Bergepanzer 2 LS). Details of the Leopard ARV are given in *Jane's Military Vehicles and Logistics 1991-92,* page 23.

Leopard Armoured Engineer Vehicle

This is based on the Leopard ARV and entered production in 1968. The only differences between the ARV and the AEV are that a heat exchanger has been installed, explosives are carried for demolition work, an auger is carried in place of the spare powerpack on the rear of the hull and the dozer blade can be fitted with scarifiers to rip up the surface of roads. Full details of the Leopard AEV are given in *Jane's Military Vehicles and Logistics 1991-92,* page 3.

Leopard Bridgelayer (Biber)

After trials with two competing designs, model B was selected for production and the first production vehicles were completed by MaK in 1975. The aluminium bridge is launched over the front of the vehicle as follows: on arrival at the obstacle the vehicle lowers the support blade at the front of the hull, the lower half of the bridge slides forwards until its end is lined up with the end of the upper half, the two sections are locked together, extended over the gap and lowered into position, the cantilever arm is withdrawn, the support blade raised and the vehicle can then be driven away. The bridge is 22 m long when opened out and can span a gap of up to 20 m. Weight of the bridge and vehicle is 45 300 kg. Full details of the Leopard bridgelayer are given in *Jane's Military Vehicles and Logistics 1991-92,* page 82.

Leopard Driver Training Tank

This is essentially an MBT with its turret removed and replaced by an observation cabin with seats for the instructor and two trainee drivers with the third trainee driver seated in the normal position at the front of the hull. If required the instructor can take over control of the vehicle at any time. The driver training tanks used by Belgium and the Netherlands do not have the dummy gun barrel as fitted to the German Leopard driver training tanks. The Australian driver training tanks are essentially Leopard 1 MBTs with the turret replaced by an observation cabin.

Pionierpanzer 2 Dachs Armoured Combat Engineer Vehicle in travelling configuration (Pierre Touzin)

Pionierpanzer 2 Dachs Armoured Combat Engineer Vehicle

Under contract to the German MoD, Krupp MaK has developed a new Leopard Armoured Engineer Tank based on the chassis of the Leopard 1. This has a front-mounted dozer blade and an extendable hydraulic arm with a bucket on one side of the hull. In 1987 Krupp-MaK was awarded a contract for the conversion of 104 Leopard 1 ARVs and the 36 existing Leopard 1 AEV into the Pioneerpanzer 2 Armoured Combat Engineer Vehicle. At the same time the Canadian Armed Forces placed an order for nine vehicles and the first of these were handed over in March 1990.

Main tasks of this vehicle is the preparing of river crossing sites and the preparation and removal of obstacles and blockages on the battlefield. Additional tasks include the recovery of vehicles and equipment, lifting of loads, towing, recovering vehicles using the main winch, recovery of tanks and other vehicles during deep fording operations and carrying out engineer work.

Full details of this vehicle are given in *Jane's Military Vehicles and Logistics 1991-92*, page 3.

Gepard Anti-aircraft Vehicle

This is used by Belgium, Germany and the Netherlands (this version having different tracking and surveillance radars) and essentially consists of a modified Leopard 1 MBT fitted with a power-operated two man turret armed with twin 35 mm Oerlikon-Contraves 35 mm cannon, tracking radar on the turret front and surveillance radar on the turret rear. Each 35 mm cannon has a total of 310 rounds of ready use ammunition for use in the air defence role and 20 rounds of APDS-T for use in engaging ground armoured targets.

For details of the Gepard self-propelled anti-aircraft gun system see *Jane's Land-Based Air Defence 1991-92,* page 68.

Roland 2 SAM on Leopard 1 Chassis

In 1982 Krupp MaK, Blohm and Voss and Euromissile proposed that the chassis of the Leopard ARV could be modified to accept the Euromissile Roland 2 SAM system, but as of early 1991 this had not even reached the prototype stage.

76 mm Anti-aircraft Tank

Krupp MaK and OTO Melara of Italy have suggested that the Leopard 1 MBT or Palmaria chassis could be fitted with the same turret armed with the 76 mm gun as installed on the OTO 76/62 76 mm self-propelled anti-aircraft gun. This system has been proposed as an anti-helicopter system and is designated the HEFAS (Hubschrauber-Erdziel, Flugabwehr-System - Helicopter Ground Target and Air Defence System). For trials purposes, OTO Melara has fitted the 76 mm two-man power operated turret of the OTOMATIC air defence system on a modified Leopard 1 chassis of the Italian Army with trials to be carried out between 1991-92.

Leopard 155 mm Self-propelled Gun

This was developed as a private venture by Krauss-Maffei and GIAT of France and basically consists of a Leopard chassis with the turret of the French 155 mm GCT self-propelled gun which is normally mounted on the AMX-30 chassis. This has not so far been adopted by any country.

Leopard 1 orders as of 1st January 1991

	MBT	ARV	AEV	Bridgelayer	Gepard	Training	Total
Australia	90	6	nil	5	nil	nil	101
Belgium	334	36	6	nil	55	12	443
Canada	114	8	9	6	nil	nil	139
Denmark	120	nil	nil	nil	nil	nil	120
Germany	2437	544*	36	105	420	60	3602
Greece	106	4	nil	nil	nil	nil	110
Italy	920†	137	40	64	nil	nil	1161
Netherlands	468	52	25	25	95	12	677
Norway	78	6	nil	nil	nil	nil	84
Turkey	77	4	nil	nil	nil	nil	81

* 444 standard plus 100 product improved
† 200 from Germany, remainder built under licence in Italy
NB: The German Army is now reducing the quantities of both its Leopard 1 MBTs and family vehicles.
Denmark is to receive 110 Leopard 1 MBTs from Germany.
Greece is receiving a further 75 Leopard 1A5 MBTs from Germany.
Turkey is taking delivery of 150 Loepard 1A5 MBTs and 8 ARVs from Germany with a further 80 Leopard 1A1A1s supplied in 1990.

Status: Production as required. In service with Australia, Belgium, Canada, Denmark, Germany, Greece, Italy, Netherlands, Norway and Turkey.

Manufacturers: Krauss-Maffei Wehrtechnik GmbH
Krauss-Maffei Strasse 2
8000 Munich 50
Federal Republic of Germany
Telephone: (089) 88 99 0
Telex: 5 23 163 31
Fax: (089) 8 12 01 90

Krupp MaK Maschinenbau GmbH
PO Box 9009
2300 Kiel
Federal Republic of Germany
Telephone: (0431) 3811
Telex: 02 99 877 mak d

SPECIFICATIONS

Model	**Leopard 1**	**Leopard A1A1, A2, A3, A4**
CREW	4	4
COMBAT WEIGHT	40 000 kg	42 400 kg
UNLOADED WEIGHT	38 700 kg	40 400 kg
POWER-TO-WEIGHT RATIO	20.75 hp/tonne	19.57 hp/tonne
GROUND PRESSURE	0.86 kg/cm^2	0.88 kg/cm^2
LENGTH GUN FORWARDS	9.543 m	9.543 m
LENGTH HULL	7.09 m	7.09 m
WIDTH OVERALL	3.25 m (without skirts)	3.41 m (with skirts)
HEIGHT (to top of commander's periscope)	2.613 m	2.764 m*
FIRING HEIGHT	1.88 m	1.88 m
GROUND CLEARANCE	0.44 m	0.44 m
TRACK	2.7 m	2.7 m
TRACK WIDTH	550 mm	550 mm
LENGTH OF TRACK ON GROUND	4.236 m	4.236 m
MAX ROAD SPEED	65 km/h	65 km/h
FUEL CAPACITY	955 l	985 l
MAX RANGE		
road	600 km	600 km
cross-country	450 km	450 km
FORDING	2.25 m	2.25 m
with preparation	4 m	4 m
GRADIENT	60%	60%
SIDE SLOPE	30%	30%
VERTICAL OBSTACLE	1.15 m	1.15 m
TRENCH	3 m	3 m
ENGINE	MTU MB 838 Ca M500, 10-cylinder multi-fuel developing 830 hp at 2200 rpm	
TRANSMISSION	ZF 4 HP 250 planetary-gear shift with hydraulic torque converter, 4 forward and 2 reverse gears	
STEERING	double differential	
CLUTCH	torque converter with mechanical interlock	
SUSPENSION	torsion bar	torsion bar
ELECTRICAL SYSTEM	24 V	24 V
BATTERIES	8 × 12 V, 100 Ah (each)	8 × 12 V, 100 Ah (each)
ARMAMENT		
main	1 × 105 mm gun	1 × 105 mm gun
coaxial	1 × 7.62 mm MG	1 × 7.62 mm MG
anti-aircraft	1 × 7.62 mm MG	1 × 7.62 mm MG
SMOKE-LAYING EQUIPMENT	4 smoke dischargers either side of turret	
AMMUNITION		
main	55	55
7.62 mm	5500	5500
GUN CONTROL EQUIPMENT		
Turret power control	electro-hydraulic/manual	
by commander	yes	yes
by gunner	yes	yes
commander's override	yes	yes
Commander's fire-control override	yes	yes
Max rate of power traverse	360° in 15 s	360° in 15 s
Max rate of power elevation	5.4°/s	5.4°/s
Gun elevation/depression	+20°/-9°	+20°/-9°
Gun stabiliser		
vertical	yes (refitted)	yes
horizontal	yes (refitted)	yes
Range setting device	yes	yes
Elevation quadrant	yes	yes
Traverse indicator	yes	yes
ARMOUR		
Hull nose	70 mm at 55°	70 mm at 55°
Hull glacis	70 mm at 60°	70 mm at 60°
Hull glacis top	25 mm at 83°	25 mm at 83°
Hull sides upper	35 mm at 50°	35 mm at 50°
Hull sides lower	25 mm at 0°	25 mm at 0°
Hull top	10 mm	10 mm
Hull floor	15 mm	15 mm
Hull rear	25 mm at 12°	25 mm at 12°
Turret mantlet	60 mm	n/av
Turret front	52 mm	n/av
Turret sides	60 mm	n/av
Turret rear	60 mm	n/av

Thyssen Henschel TH 301 (TAM) Tank

Development

In mid-1974 the German company of Thyssen Henschel was awarded a contract for the design and development of a new medium tank for the Argentinian Army under the designation of the TAM (Tanque Argentino Mediano), as well as an infantry combat vehicle under the designation of the VCI (Véhiculo Combate Infanteria), subsequently renamed VCTP. The contract covered the design, development and construction of three prototypes of both the TAM and the VCI.

The main reason for the decision to develop a tank in the 30-tonne class rather than the normal 40- to 50-tonne class was that many bridges and roads in South America could not stand up to the heavier tanks. An added advantage was that it enabled Argentina to have a tank and an infantry combat vehicle that were both based on the same chassis with resulting logistical, training and economic advantages.

The first prototype of the TAM was completed in September 1976 with the second and third prototypes being finished the following year. On completion of company trials in Germany the TAMs were shipped to Argentina for both operational and firing trials. Chassis construction and final assembly were carried out in Buenos Aires while construction of the turret and the integration of the gun and fire control system took place at Rio Tercero. The total Argentinian requirement was for 512 TAMs and VCTP infantry fighting vehicles, but production was completed after about 350 vehicles had been built. Additional information on the TAM programme is given in this section under Argentina.

In 1981 Malaysia signed a contract with Thyssen Henschel for 102 members of this family in three versions: MBT (Lion), armoured personnel carrier (Tiger) and armoured recovery vehicle (Elephant), but no deliveries were made.

The TH 301 was tested at the Thai Cavalry Centre in 1985/86, this being the first time that this vehicle had been tested in a non-NATO country. These trials included firing 130 rounds of German ammunition from the Rheinmetall 105 mm Rh 105-30 gun against both stationary and moving targets. The vehicle was fitted with an AEG-Telefunken fire control system, FWM electro-hydraulic gun control system and with the main armament stabilised in both elevation and traverse, gunner's sight

TH 301 with 105 mm gun in travelling lock, showing PZB 200 LLLTV camera which moves in elevation with main armament mounted on right side of mantlet

with laser rangefinder, gyro-stabilised panoramic telescope for the tank commander and PZB 200 passive LLLTV system over the main armament with screens for both commander and gunner.

In early 1988 a TAM was trialled in Ecuador and a TH 301 was demonstrated in Malaysia, but as of 1991 none of these trials had resulted in firm orders being placed.

Description

The chassis of the vehicle is based on the one used for the Marder ICV used in large numbers by the German Army, but strengthened to take account of the increased weight of the vehicle and the stresses caused by firing the main armament.

The hull of the TAM is all welded with the well-sloped glacis plate providing protection against small arms fire, shell splinters and armour-piercing rounds up to 40 mm in calibre.

The driver is seated at the front of the hull on the left side and steers the vehicle with a conventional steering wheel. He is provided with a single piece hatch cover that opens to the right and three periscopes. Two escape hatches are provided, one in the floor of the vehicle and one at the rear.

The engine is mounted to the right of the driver with the glacis plate hinged on the right to allow access to the engine for maintenance. On the prototype the engine was identical to the one installed in the Marder ICV and developed 600 hp at 2200 rpm but production vehicles have a super-charged engine that develops 720 hp at 2400 rpm. The engine is coupled to a Renk HSWL-204 planetary gearbox which gives four gears in both directions. The steering system is of the stepless hydrostatic type.

The all-welded turret is mounted to the rear of the vehicle with the commander and gunner on the right and the loader on the left. The commander is provided with a single piece hatch cover and eight periscopes. Mounted in front of his hatch is a non-stabilised Steinheil

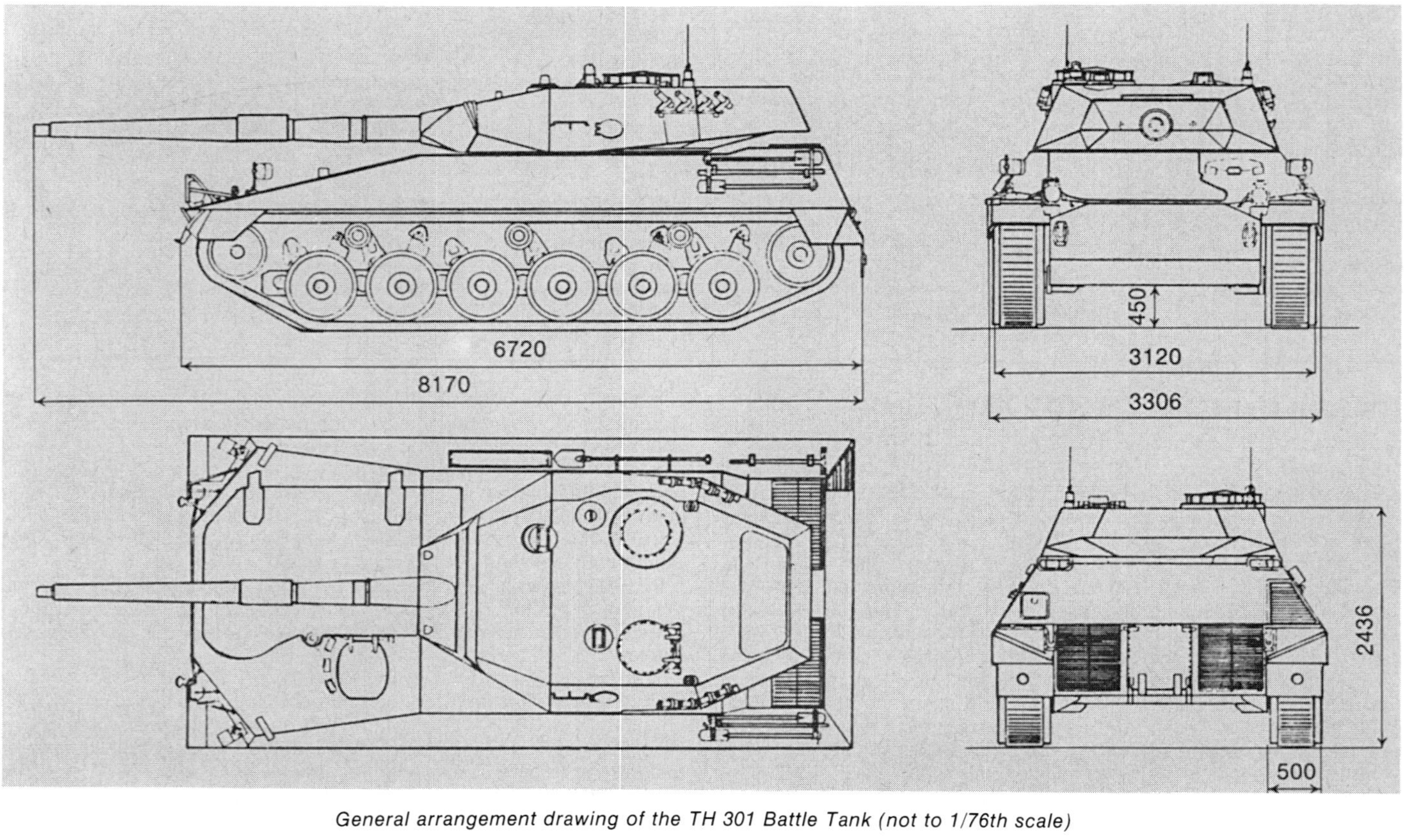

General arrangement drawing of the TH 301 Battle Tank (not to 1/76th scale)

Thyssen Henschel TH 301 from the rear showing long range fuel tanks and entry hatch in hull rear (Christopher F Foss)

Lear Siegler TRP-2A panoramic sight with a magnification of between ×6 and ×20, identical to that fitted to the Leopard 1 MBT, and which can be replaced by an infra-red sight. The commander also operates the coincidence rangefinder. The gunner is seated forward and below the commander and is provided with a Zeiss TZF sight with a magnification of ×8, combined with a swivelling and tilting periscope. The loader is provided with a single piece hatch cover that opens to the rear with a tilting periscope mounted in front of him. An ammunition-loading port is provided in the left side of the turret. Mounted in the hull rear is a hatch that can be used either as an emergency exit or to reload the vehicle rapidly with ammunition.

The torsion bar suspension system consists of six rubber-tyred road wheels with the drive sprocket at the front, idler at the rear and three track return rollers. The first, second, fifth and six road wheel stations have hydraulic shock absorbers. The tracks are Diehl with removeable rubber pads. The tops of the tracks are covered by detachable rubber skirts.

To extend the TAM's range two 200 litre long-range fuel tanks can be fitted at the rear. The basic vehicle can ford to a depth of 1.4 m but with preparation it can ford to 2.25 m, or 4 m with a snorkel. Standard equipment includes an NBC system, combustion heater for both cold engine starts and the crew compartment, electric bilge pumps, and a fire-extinguishing system that can be operated manually or automatically.

Main armament is a 105 mm gun developed in Argentina, although the Rheinmetall Rh 105-30 can also be fitted. Both weapons fire all standard 105 mm tank ammunition. It is provided with a bore evacuator, thermal sleeve and is fully stabilised in both planes. Fifty rounds of 105 mm ammunition are carried of which 20 are in the turret. Mounted coaxially with the main armament is a 7.62 mm machine gun. A similar weapon is mounted on the roof of the tank for anti-aircraft defence. Four Wegmann smoke dischargers are mounted on either side of the hull.

Optional equipment includes additional armour protection, radios and an intercom system.

Variants

TH 301

Using company funds, Thyssen Henschel has built a fourth prototype of the TAM called the TH 301 (this was originally called the TAM-4), which was completed late in 1978.

Externally it is very similar to the model designed for Argentina but internally has significant differences in its powerpack and fire control system.

In addition to eight periscopes for all-round observation, the commander has a PERI R12 sight as fitted to the final production model of the Leopard 1, the Leopard 1A4. This stabilised sight has an elevation of +60°, a depression of -10° and can be traversed through a full 360°. It has a magnification of ×2 or ×8 and is fitted with an external wiper.

The gunner has a fully stabilised periscope with a magnification of ×8 and an integral laser rangefinder. Additional vision equipment includes one periscope for the gunner and one periscope for the loader.

Its main armament is a Rheinmetall-developed 105 mm gun designated the Rh 105-30 firing all standard 105 mm tank rounds and with an elevation of +18° and depression of -7°. The turret can be traversed through a full 360° in 15 seconds and the main armament is fully stabilised in both planes.

Information on target range is fed to the computer manually by the gunner or automatically by the laser rangefinder, which has a range of 400 to 5000 m. Angle of cant and lead angle is also fed automatically and the type of ammunition to be fired is fed in manually.

For engaging targets at night there is an LLLTV camera mounted on the mantlet which moves in elevation with the main armament. Both commander and gunner have one TV monitor screen and in addition to the optical cross-hairs being displayed a second set of electronically generated cross-hairs is superimposed on the TV picture. The latter changes in accordance with the variable angle of aim (independent of ammunition type and target range), and is also corrected for vehicle cant. The two sets of cross-hairs are then brought together and the main armament fired.

The TH 301 is powered by an MB 833 Ka 500 diesel developing 750 hp coupled to a Renk HSWL-204 transmission which is fitted with a hydrodynamic retarder which is used to augment the steering effort and to brake the vehicle.

Under development is another version of the TAM which has the same fire control system as the TH 301 but with a 120 mm gun and powered by a 750 hp diesel.

57 mm Support Tank (Marder chassis)

Fitted with two-man turret armed with 57 mm Bofors gun and ATGW launcher, developed to prototype stage but not placed in production. Details given in *Jane's Armour and Artillery 1985-86*, page 28. An anti-aircraft gun version was also proposed and details of this were given in *Jane's Armour and Artillery 1983-84*, page 541.

37 mm Support Tank (Marder chassis)

Marder chassis fitted with two-man turret with externally-mounted 35 mm cannon and ATGW launcher. Proposal only.

155 mm SPG

Argentina placed an order with OTO Melara of Italy for 25 155 mm Palmaria turrets which will be installed on lengthened TAM chassis, with seven road wheels either side instead of six. The turrets were delivered in 1986/87 with the first chassis being completed in Argentina in 1987. Full details of this are given in the *Self-propelled Guns and Howitzers* section under Argentina.

30 mm SPAAG

This system, now no longer offered, was called the Dragon and consisted of a TAM/Marder type chassis fitted with a French Thomson-CSF turret with twin 30 mm cannon. Details were given in *Jane's Armour and Artillery 1984-85*, pages 504-505.

Status: Production complete in Argentina and can be undertaken in Germany or Argentina under receipt of orders.

Manufacturer: Thyssen Henschel
Postfach 102969
D-3500 Kassel
Federal Republic of Germany
Telephone: (0561) 8011
Telex: 099 750 thksr d
Fax: (0561) 801 6733

SPECIFICATIONS (TAM, data in square brackets refers to the TH 301, where different)

CREW	4
COMBAT WEIGHT	30 500 [31 000] kg
UNLOADED WEIGHT	28 000 kg
POWER-TO-WEIGHT RATIO	23.27 [24.59] bhp/tonne
GROUND PRESSURE	0.77 [0.79] kg/cm²
LENGTH GUN FORWARDS	8.23 [8.17] m
LENGTH HULL	6.775 [6.72] m
WIDTH	3.12 [3.306] m
HEIGHT TO TURRET TOP	2.42 [2.436] m
GROUND CLEARANCE	0.44 [0.45] m
TRACK	2.62 m
TRACK WIDTH	500 mm
LENGTH OF TRACK ON GROUND	3.9 m
MAX ROAD SPEED	
1st gear	16 km/h
2nd gear	31 km/h
3rd gear	47 km/h
4th gear	75 [72] km/h
FUEL CAPACITY	650 l
MAX ROAD RANGE	550 [500] km
with auxiliary tanks	900 [750] km
FORDING	1.4 m
with preparation	2.25 m
with snorkel	4 m
GRADIENT	60%
SIDE SLOPE	30%
VERTICAL OBSTACLE	0.9 m [1 m]
TRENCH	2.9 m
TURNING RADIUS	
1st gear	6.5 m
2nd gear	13.5 m
3rd gear	20 m
4th gear	30 m
ENGINE	super-charged MTU MB 833 Ka 500 6-cylinder diesel developing 720 hp at 2200 rpm [750 hp at 2400 rpm]
TRANSMISSION	Renk HSWL-204 planetary with 4 speeds both forward and reverse
STEERING	double differential with hydrostatic steering control
SUSPENSION	torsion bar
ELECTRICAL SYSTEM	24 V
BATTERIES	8 × 12 V, 100 Ah
ARMAMENT	
main	1 × 105 mm gun
coaxial	1 × 7.62 mm MG
anti-aircraft	1 × 7.62 mm MG
SMOKE-LAYING EQUIPMENT	4 smoke dischargers either side of turret
AMMUNITION	
main	50
machine gun	6000
GUN CONTROL EQUIPMENT	
Turret power control	electro-hydraulic/manual
by commander	yes
by gunner	yes
commander's override	yes
Turret traverse	360°
Commander's fire-control override	yes
Gun elevation/depression	+18°/-7°
Gun stabiliser	
vertical	yes
horizontal	yes

Super M48 MBT

Development

The Super M48 is a private venture development between Wegmann of Kassel which is responsible for the turret and GLS, a subsidiary of Krauss Maffei, which is responsible for the chassis.

Other major sub-contractors include Renk (transmission), MTU (diesel engine), Krupp Atlas Elektronik (fire control system), Blohm and Voss (armour) and FWM for weapon stabilisation.

The full Super M48 retrofit package offers improvements in the three main areas of the vehicle, armour protection, mobility and firepower, but the user can select only those items that he requires or can afford.

Prior to being involved in the Super M48 progamme, Wegmann rebuilt 650 M48 series tanks to the M48A2GA2 configuration for the German Army (details of this conversion are given in the entry for the M48 later in this section) and converted, together with GLS which was responsible for the chassis, five M48s (plus kits for 165 conversions carried out in Turkey) with many improvements including an MTU MB 837 Ea-500 diesel engine. Additional details are given in this section under Turkey.

For trials a German M48 was rebuilt by FFG with a GT601 gas turbine.

Description

The original 90 mm gun is replaced by the British Royal Ordnance 105 mm L7A3 rifled tank gun for which ammunition is available worldwide; 40 rounds of 105 mm ammunition are carried in the hull and on the turret platform.

A new electro-hydraulic gun/turret drive and weapon stabilisation system is installed, although various components of the existing turret control system can be retained.

The Super M48 has a MOLF 48 fire control system. The new roof-mounted primary gunner's sight has three optical paths for the day sight, laser rangefinder and thermal night sight all of which pass through a common head mirror.

The computer processes the input data in its ballistic core and converts this information into the appropriate lead and super-elevation angle for both the 105 mm gun and 7.62 mm coaxial machine gun. A high degree of redundancy has been maintained by retaining the optical base rangefinder mounted in the forward part of the roof and the standard mechanical computer. The Super M48 can engage both stationary and moving targets while itself stationary or moving, as a two-axis stabilisation system is fitted as standard. The stabilisation system has four modes of operation: 'stab on', in which stabilised gun and turret is slaved to the gunner's primary stabilised sight; 'stab ready', self-stabilised gun and turret, manual laying with hydraulic power and line of sight electrically slaved in elevation and azimuth; 'observation', manual laying by means of hand cranks, line of sight electrically slaved to elevation and azimuth and 'turret off', manual laying by means of handcranks, gunner aims using FERO Z19 panoranic telescope.

The turret front and sides are fitted with add-on armour which is fastened to the basic structure of the turret by elastic elements. When one of these has been damaged or hit it can be quickly replaced by another panel.

The Super M48 can also be fitted with a bank of four Wegmann electrically-operated smoke/fragmentation dischargers either side of the turret towards the rear.

The original M48 had a very short operating range of 113 km because of its petrol engine. The Super M48 is powered by the MTU MB 837 Ka-501 diesel developing 1000 hp at 2300 rpm which gives a significant increase in operating range. The MTU diesel engine has an intercooler, exhaust turbo-chargers, flat radiators and air cleaner system with a dust evacuation blower.

The old transmission has been replaced by a Renk RK 304 fully-automatic steering transmission. The brake system has a retarder and multiple disc brakes which are integrated with the transmission. Quick-disconnect couplings and mountings allow fast installation and removal of the complete powerpack.

The installation of the new engine and transmission means that the top decking of the engine compartment has to be modified. The fuel

Close-up of Super M48 turret showing commander's cupola

Close-up of new tracks and skirts on Super M48

Super M48 MBT clearly showing additional armour protection for turret front and sides and 105 mm L7A3 rifled tank gun with thermal sleeve

system has been changed and now consists of five storage and one delivery tanks with a total capacity of 1050 litres.

The final drives have been modified to cope with the increased input and double-acting hydraulic shock absorbers have been installed at roadwheel stations one, two, five and six to reduce vertical pitching. The older volute bump stops have been replaced by hydraulic bump stops which in conjunction with improved quality torsion bars give flatter spring characteristics and soften the suspension.

This suspension system improvement package gives very good pre-stabilisation for the fire control system and a much improved ride for the crew of four when the tank is moving across country.

The US pattern tracks have been replaced by a new track with removeable rubber pads and spoon-shaped end connectors which not only have an increased life but reduce track noise.

The armoured side skirts protect the sides of the vehicle and help to reduce the ingress of dust into the air cleaners.

A new electrical system has been installed that includes the engine-mounted generator, two additional batteries, new cables and harnesses, modified Leopard 1 slip ring and a deck clearance device for the 105 mm gun.

The driver has a new instrument panel and a Halon fire suppression system has been installed in the crew compartment and a fire-wire warning and extinguishing system for the engine compartment.

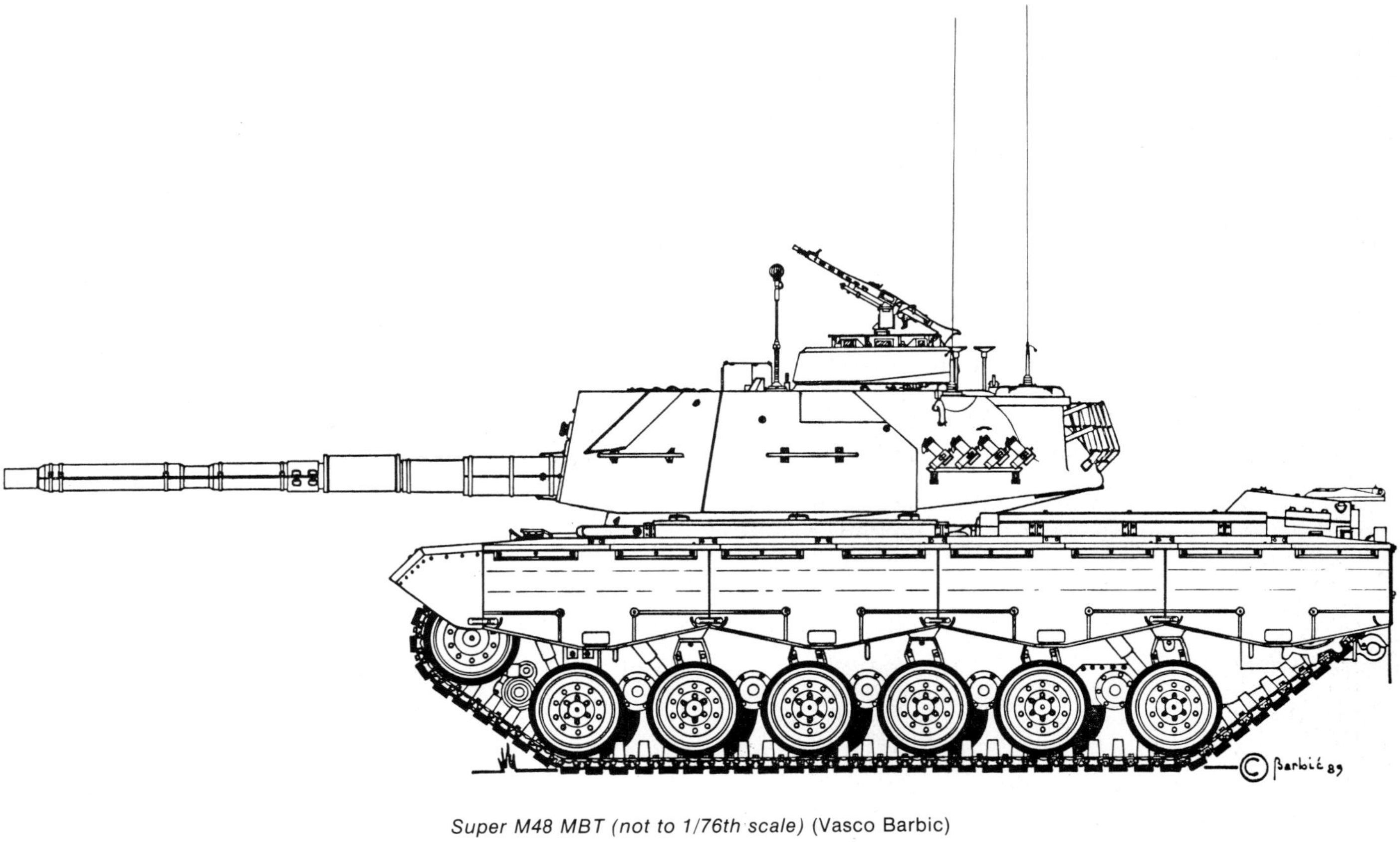

Super M48 MBT (not to 1/76th scale) (Vasco Barbic)

SPECIFICATIONS (outline)					
CREW	4	GROUND CLEARANCE	0.4 m	TRANSMISSION	Renk RK 304 fully automatic with 4 forward and 4 reverse gears
COMBAT WEIGHT	53 000 kg	TRACK WIDTH	711 mm		
POWER-TO-WEIGHT RATIO	18.86 hp/tonne	MAX ROAD SPEED	56.3 km/h		
LENGTH		FUEL CAPACITY	1050 l		
gun forwards	9.3 m	FORDING	1.219 m	ARMAMENT	
hull	6.87 m	GRADIENT	60%	main	1 × 105 mm
WIDTH OVER TRACKS	3.65 m	SIDE SLOPE	30%	coaxial	1 × 7.62 mm MG
HEIGHT		VERTICAL OBSTACLE	0.915 m	anti-aircraft	1 × 7.62 mm MG
to top of commander's cupola	2.9 m	TRENCH	2.59 m		
to hull top	1.475 m	ENGINE	MTU MB 837 Ka-501 diesel developing 1000 hp at 2300 rpm		

Status: Development complete.

Manufacturers: GLS GmbH
Allacher Strasse 230e
D-8000 Munich 50
Federal Republic of Germany
Telephone: (089) 88 99 0
Telex: 5 215 583
Fax: (089) 8 12 63 80

Wegmann and Co GmbH
August-Bode-Strasse 1
D-3500 Kassel
Federal Republic of Germany
Telephone: (0561) 1050
Telex: 99859
Fax: (0561) 105 2208

Jung Jungenthal Upgraded T-54/T-55 MBT

Development/Description
Early in 1991 Jung Jungenthal completed the first prototype of an upgraded Soviet T-54/T-55 MBT that it has developed to meet the requirements of the Egyptian Government. Although the prototype is based on the T-54/T-55 chassis it is also applicable to the Chinese equivalent of the vehicle, the Type 59 MBT.

Prior to this the company has been involved in the development of many of the German post-war armoured fighting vehicles including the Leopard 1 MBT, Leopard armoured recovery vehicle, Leopard armoured engineer vehicle, Transportpanzer 1 (6 × 6) and Spahpanzer 1 (8 × 8) reconnaissance vehicle to name but a few. In many cases the company has manufactured the complete hull of the vehicle which has then been sent to the prime contractor for assembly.

The company has also developed and placed in production conversions of the M47 and M48 tanks into armoured recovery vehicles. Details of these are given in *Jane's Military Vehicles and Logistics 1991-92*, page 24. The boom and main winches used in these conversions are the same as those used in the Leopard 1 armoured recovery vehicles.

The T-54/T-55 has been upgraded in the key areas of armour, mobility and firepower.

Battlefield survivability has been considerably improved by the fitting of additional passive armour onto the glacis plate of the vehicle while the frontal 180° of the turret has a horseshoe type armour array that is similar to that installed on some Soviet T-54/T-55/T-62 MBTs. This armour provides both chemical energy (CE) and kinetic energy (KE) protection.

The original T-54/T-55 is not fitted with side skirts but the modernised T-54/T-55 is provided with these with those either side of the driver's compartment being armoured. These hinge up vertically to allow access for maintenance purposes.

The engine and crew compartments are fitted with a fire detection/suppression system provided by Deutsche Graviner.

The original 100 mm D-10 series rifled tank gun is retained although a 105 mm L7 or M68 rifled tank gun can be installed if required. A 7.62 mm machine gun is mounted coaxial with the main armament and a 12.7 mm machine gun is mounted on the turret roof for local and air defence purposes.

A bank of four German type electrically operated smoke dischargers can be mounted either side of the turret firing to the front of the vehicle.

The original gun control equipment has been replaced by a new all electric gun control and stabilisation system developed by SIG of Switzerland which gives a higher first round hit probability. A new turret race ring has also been fitted.

Model of Jung Jungenthal upgraded T-54/T-55 MBT with additional armour on glacis plate, turret front and forward part of hull sides

The Soviet torsion bars for the T-54/T-55 have been retained but new GLS hydraulic bump stops have been installed and the original tracks and road wheels have been replaced by those fitted to the Soviet T-62. New ZF final drives have been fitted.

The hull now has a different shape with new Superflexit bag type fuel tanks installed under armour protection along the sides of the vehicle. These contain 1200 litres of diesel fuel which is an increase in capacity over those fitted in the standard T-54/T-55 MBT.

The original Soviet diesel engine has been installed but the manual transmission has been replaced by the latest German ZF LSG 3000 series fully automatic power shift transmission which has been in volume production for some years.

The powerpack is fitted with improved Behr cooling system and new combustion air filters. Like the original Soviet T-54/T-55 MBT, the upgraded vehicle can lay its own smoke screen by injecting diesel fuel into the exhaust outlet.

Status: Prototype.

Manufacturer: Jung Jungenthal GmbH
D-5242 Kirchen
Federal Republic of Germany
Telephone: (2741) 6830
Telex: 875319
Fax: (2741) 68 32 46

SPECIFICATIONS					
CREW	4	HEIGHT	2.4 m	ARMAMENT	
COMBAT WEIGHT	41 000 kg	GROUND CLEARANCE	0.5 m	main	1 × 100 mm
LENGTH		MAX ROAD SPEED	48 km/h	coaxial	1 × 7.62 mm MG
with 100 mm gun	9 m	RANGE	400 km	anti-aircraft	1 × 12.7 mm MG
hull	6.54 m	FUEL CAPACITY	1200 litres	smoke dischargers	2 × 4

GREECE

Greek MBT Modernisation

Early in 1991 the Greek tank fleet consisted of 106 Leopard 1A3s, 350 M47 and 1300 M48 (with more being delivered) and 190 AMX-30.

Of the M48s, 225 are up to M48A5 standard and 212 have been brought up to M48A3 standard. The DM300 million order placed in April 1981 by the Greek Government for 110 Leopard 1s comprised 106 Leopard 1A3 MBTs and four ARVs. Deliveries ran from February 1983 through to April 1984 with Krauss-Maffei building 73 MBTs and Krupp MaK of Kiel building 33 MBTs and four ARVs.

Greece has an option on a further 101 Leopard 1 MBTs, eight AVLBs and four ARVs which is intended to be combined with the build up of a tank assembly line in Greece, but as of April 1991 this option had not been exercised. In mid-1983 the United States Department of Defense notified Congress of a proposed letter of offer to Greece for the sale of 110 M60A3 MBTs, at a total cost, with spares and other related equipment, of $186 million. However, this option was not taken up by Greece. In Spring 1986, Vickers Defence Systems (Valiant), Krauss-Maffei, GIAT, OTO-Melara, ENGESA and Bernardini all submitted proposals to Greece regarding production of a new MBT.

In 1986 the Greek Army issued a requirement for a new MBT to which 11 manufacturers responded. The first vehicles would be supplied direct followed by a batch in kit form, then local production in Greece, but some key components such as armament and powerpack coming from abroad. But, as before none of these proposals has yet been taken up by Greece and it is unlikely that they will be in the short term.

In April 1988 the Greek Navy selected the German MEKO 200 frigate to meet its future requirements and as part of the offset Germany is to provide additional military equipment including 75 updated Leopard 1A3 MBTs.

The Greek Army is also considering the possibility of upgrading its AMX-30 MBTs which could be accomplished by GIAT, Teledyne Continental or GLS/Wegmann. The M48 fleet is expected to be modernised with a new fire control system and laser rangefinder.

Under the Southern Region Ammendment, the first of 50 US supplied 105 mm armed M48A5 tanks were delivered to Greece in August 1988. These were supplied by the United States with Greece paying the cost of transport and any upgrading work required. the matter is being carried out at the Hellenic Army Tank Plant in Velestino. It is expected that a further 250 M48A5s will be purchased from the US using long term, low interest loans.

INDIA

Vijayanta Production

In August 1961 following the evaluation of competing British and German MBT designs to meet an Indian requirement for a new MBT to be manufactured in India, an agreement was signed between Vickers Limited of the United Kingdom and the Indian Government. This agreement covered building prototypes in the United Kingdom, supplying 90 production tanks and building a new facility at Avadi, near Madras, to undertake production of the Vickers MBT. The Indians call the tank the Vijayanta, or Victorious.

The first two prototypes were completed in 1963 and one was sent to India and the other retained in the United Kingdom for development work. The first production tanks were delivered from Vickers Elswick works in 1965. The first Indian Vijayanta, which was mainly built from components supplied by the United Kingdom, rolled off the production line at Avadi in January 1965. Since then India has progressively undertaken the production of more and more of the tank and eventually almost all of the tank was produced in India. Production of the Vijayanta is now understood to have been completed after about 2200 vehicles were built. Full details of the Vijayanta tank are given in the United Kingdom section under the Vickers MBT. Early in 1981 Marconi Radar Systems was awarded a $6 million contract for its SFCS 600 tank fire control system for the Vijayanta MBT with a $6 million option for additional systems. Each contract is for 70 systems making a total of 140, but as of early 1991 the option on the additional 70 systems had not been exercised. The Tank Electronics Support Centre at Madras is now producing the Bharat Electronics Tank Fire Control System Mk 1 A (AL 4420) for the Vijayanta.

The Mk 1 A features an improved sight mount and a fire control linkage designed to minimise the play between the sight mount and linkage and between the linkage and the 105 mm rifled gun. A muzzle reference system is also provided to check for and correct misalignment between the gun and sight axes caused by thermal deformation.

Bharat has also developed the much more sophisticated Mk 1B system (AL 4421) which incorporates a British Barr and Stroud Tank Laser Sight and a computer to increase first round hit probability.

Modes of operation of the Mk 1B are autolay with the gun driven directly by the computer, stabilised in which the gunner uses the existing gun control equipment controllers to bring the aiming mark onto the target and manual where the gunner operates conventional traverse and elevation handwheels to align the aiming mark with the target.

Indian-built Vijayanta MBT armed with 105 mm gun showing infra-red/white-light searchlight mounted on left side of turret and bank of six electrically operated smoke dischargers either side of turret

In 1983 Vijayantas fitted with a British Vickers Defence Systems powerpack (using a Rolls-Royce Condor diesel) and a Cummins powerpack (using a Cummins KTA 111 diesel) were tested by the Indian Army. In mid-1983 it was announced that India was to order 250 Leyland L60 engines for its Vijayantas in two batches, the first consisting of 150 engines.

In Spring 1987 the Indian Army stated that the Vijayanta was to be provided with a new engine, new fire control system, additional armour protection, locally designed APFSDS round and a muzzle reference system. This project is called Bison.

In mid-1987 trials of three repowered Vijayantas were completed. One was powered by a Perkins Engines Condor CV-12 diesel developing 750 hp, one by an MTU 837 V-8 diesel and one by a Kieloskar (Cummins diesel) all of which were coupled to the existing TN-12 transmission. One of these engines is expected to be selected to repower at least 400 of the Vijayanta fleet in the near future.

As of April 1991 the Indian Army had still not selected any of the three engines to repower the Vijayanta MBT. Perkins Engines has signed a Memorandum of Understanding with Greaves Cotton and Company Limited (Thapar Group of Companies) under which Greaves Cotton will act as indigenous source for the Condor V-12 diesel engine and ancilliary package for the Vijayanta before delivery to the Indian Army. Initially Perkins would supply most of the parts, with the proportion of Indian sources components being gradually increased.

The armour to be fitted to the Vijayanta is believed to be the advanced Kanchan composite armour fitted to the Arjun which was designed by the Defence Metallurgical Research Laboratory at Hyderabad.

Although the Vijayanta is essentially the Vickers Mk 1 MBT, fully described under the United Kingdom, Indian specifications differ from the latter in some areas and Indian sources have provided the following information:

WEIGHT	40 000 kg
LENGTH (gun forwards)	9.788 m
WIDTH (with skirts)	3.168 m
HEIGHT (cupola doors shut)	2.711 m
MAX ROAD SPEED	48.3 km/h
GRADIENT	60%
VERTICAL OBSTACLE	0.914 m
FORDING	1.143 m
OPERATING RANGE	350 km
COMMANDER'S SIGHT	periscope binocular and sighting periscope
LOADER'S PERISCOPE	No 23 Mk 1
GUNNER'S SIGHT	sighting periscope No 30 Mk 1
DAY VISION	
commander	episcope No 7 Mk 1 (M1)
driver	periscope AFV 101A
NIGHT VISION	
commander	periscope, armoured vehicle, IR 101A
gunner	sight periscope, armoured vehicle, IR 102C
driver	binocular, armoured vehicle, IR 101A
ENGINE	Leyland L60 developing 535 hp at 2375 rpm

One of the prototypes of the Arjun MBT which is armed with a 120 mm rifled tank gun fitted with a thermal sleeve and fume extractor. The turret front and mantlet is very similar to that of the German Leopard 2 MBT

Kartik AVLB

Seen for the first time early in 1989 was an armoured vehicle launched bridge based on the lengthened Vijayanta chassis used for the 130 mm Catapult self--propelled artillery system. This has been fitted with a scissors type bridge of Eastern European design and is launched and recovered over the front of the chassis. Details of this are given in *Jane's Military Vehicles and Logistics 1991-92*, page 84.

Vijayanta 130 mm SPG

The Indian Army is believed to have about 100 Catapult self-propelled guns in service. This is a Vijayanta chassis with an additional road wheel either side. The turret has been removed and the chassis fitted with a Soviet 130 mm M-46 field gun firing over the rear. The weapon has limited traverse and about 30 projectiles and their associated charges are carried.

Vijayanta with GBT 155 Turret

Vijayanta MBT chassis fitted with the British Vickers Shipbuilding and Engineering GBT 155 turret with a Royal Ordnance Nottingham 39-calibre ordnance underwent extensive firepower and mobility trials in India. Details of this turret are given in the *Self-propelled guns and howitzers* section under UK.

T-55 MBT Upgrade

The 800 fleet of T-55 MBTs is now being modernised in batches at the Nasik ordnance facility with the installation of a 105 mm rifled gun, driver's night periscope and the Bharat Electronics Ltd Tank Fire Control System (TFCS) very similar to that being fitted in the Vijayanta.

New Arjun MBT

In 1972 the Indian Army issued a requirement for a new MBT to replace the Vijayanta in production and work on this began at the Combat Vehicle Research and Development Establishment in 1974. This vehicle was called the Chetak (originally called the MBT-80) and weighs 50 tonnes. All the major components including engine, transmission, 120 mm rifled gun, ammunition (including APFSDS), advanced armour and the fire control system are being produced in India. The prototype Arjun was shown for the first time in March 1985. In April 1985 it was stated that by 1988 20 prototypes of the Arjun MBT would be completed to enable extensive trials of key subsystems such as armament, fire control, powerpack and suspension prior to the start of full-scale production. By the time the first prototype of the Arjun was unveiled in April 1984 Rs 300 million had already been spent on the project.

The Explosives Research and Development Laboratory at Pune in Maharashtra has developed a new high-energy propellant which will be used in the 120 mm rifled gun to give a higher muzzle velocity and therefore greater penetration characteristics.

One of the main problem areas has been the engine. It was originally intended to install a 1500 hp gas turbine but this was subsequently dropped and a 12-cylinder, air-cooled, variable compression ratio diesel with a planned output of 1500 hp was to be developed instead. First models however are said to have developed only 500 hp, further development has increased this figure and with turbo-chargers a figure of between 950 and 1000 hp has been obtained. As time goes on it is expected to develop between 1200 and 1500 hp. Suspension is of the hydro-pneumatic type. Initially six prototypes of the Arjun have been built, fitted with a German MTU diesel developing 1100 hp coupled to a ZF automatic transmission. It is reported that none of these were armoured but have hulls and turrets of mild steel. Production vehicles will have a new composite armour, developed by the Indian Defence Metallurgical Laboratory, called Kanchan armour. This will be produced by Mishra Dhatu Nigam. The thermal night vision equipment has been developed by the DRDO laboratory at Dehra Dun.

Between 1983 and 1989 India is reported to have imported 42 foreign engines and transmissions for the prototypes of the Arjun at a total cost of $15 million. Consideration is being given to manufacturing a foreign engine under licence for the Arjuna with the contenders including the AGT 1500 (as in M1), V-12 Condor (as in Challenger) and MB 873 Ka 501 (as in Leopard 2).

By late 1987 10 prototypes of the Arjun, or MBT 90 as it is also sometimes referred to, had been completed under the designation of the Mark I. Of these, six have been handed over to the Indian Army for extensive trials and the remaining four have been retained for further development work and trials at the Combat Vehicle Research and Development Establishment (CVRDE).

It is understood that the vehicle does not yet meet the requirements of the Indian Army in many key areas including the design of the hull and turret, lack of system integration, small number of ready rounds, slow rate of loading, and weight.

The Arjun's fire control system is believed to include a day/thermal night sight with laser rangefinder, computer and various sensors and is a further development of the Tank Fire Control System developed by Bharat Electronics Limited for retrofitting to the Vijayanta MBT.

The 120 mm rifled gun will fire APFSDS, HE, HEAT, HESH and smoke rounds developed by the Indian Explosives Research and Development Establishment at Pune. A machine gun is mounted coaxially with the main armament and another machine gun is mounted at the loader's station. Mounted either side of the turret is a bank of electrically-operated smoke dischargers.

The Arjun has an NBC system designed and built by the Bhabha Atomic Research Centre (BARC). No detailed specifications of the Arjun have been released so far although it is now understood to weigh 52 tonnes, has a maximum speed of 70 km/h, can climb a 60 per cent slope, has a vertical obstacle climbing ability of 0.9 m and can cross a trench 3 m wide.

Indian Army T-72 MBT (via G Jacobs)

Late in 1988 the Indian Government approved $207 million to allow production of the Arjun to commence with about 1500 vehicles expected to be in service by the year 2000. By late 1988 10 prototypes had been built and production of an initial batch of 24 vehicles will start in 1990 with the first of these to be completed in 1991. Long lead items for production vehicles are believed to have been ordered and it is expected that production will eventually be around 150 vehicles a year. Due to problems with the locally designed powerpack, the first production Mk 1 vehicles will use an imported engine and transmission.

Current MBT Fleet

In 1991 the Indian tank fleet consisted of about 2200 Vijayantas, 800 T-54/T-55s and 750 T-72s. The original intention was to order only a small number of T-72s pending the production of the Arjun MBT, but as the latter is now behind schedule licence production of the T-72M is now being undertaken at Avadi and the first vehicles rolled out in 1987. The first Indian built T-72, a T-72 M1 was handed over to the Indian Army early in 1988. These are mainly comprised of components supplied from the USSR but the eventual aim is to produce as much of the tank in India as possible, the target being 95 per cent local content. The remaining Centurion tanks (some of the earlier ones have found their way to South Africa) have been withdrawn from service and disposed of. Late in 1988 it was reported that a small number of T-80 MBTs had been loaned by the Soviet Union to India for trials and that some of these may be purchased as the Arjun will not be in service in substantial numbers until the mid-1990s.

INTERNATIONAL

Jaguar Main Battle Tank

Development

Late in 1988 Cadillac Gage Textron of the United States and the China National Machinery & Equipment Import & Export Corporation announced that they were jointly developing a new MBT for the export market called the Jaguar, with a target price of $1 million for production vehicles at 1988 prices.

Cadillac Gage Textron were provided with two Type 59 MBTs by China to reduce both development time and cost.

By mid-1989 the Jaguar chassis was complete and undergoing automotive trials in Detroit while the turret was undergoing initial firing trials. The turret and hull were mated in October 1989 and the complete vehicle then underwent extensive trials in Nevada.

It was expected that the second prototype would be built in China by the China National Machinery & Equipment Import & Export Corporation with Cadillac Gage Textron providing key sub-systems.

Following the events in China in mid-1989, the Chinese technicians working with Cadillac Gage Textron in the United States returned to China and the company carried on with the project on its own.

The second Type 59 originally supplied by China is being used as spare parts.

The original intention was to supply brand new vehicles for the export market from China, but in the short term Cadillac Gage Textron is aiming at the retrofit market, especially those with access to FMS funding. Although the prototype is based on the Type 59 MBT it is also applicable to other vehicles including the Soviet T-54/T-55. These conversions would require no assistance from China as virtually all of the key components come from the United States or the United Kingdom.

Description

The layout of the Jaguar is conventional with the driver's compartment at the front, fighting compartment and turret in the centre and engine and transmission at rear.

An additional layer of armour protection has been applied to the chassis and turret. The driver is seated at the front left and is provided with a single piece hatch cover, day periscopes, one of which can be replaced by a passive night vision periscope and to his right is stowed some 105 mm ammunition.

The prototype Jaguar is based on a modified Type 59 chassis with five rubber tyred road wheels, idler at the front and drive sprocket at the rear and two track return rollers. The upper part of the running gear is covered with a rubber skirt. The prototype Jaguar has upgraded torsion bars which give increased vertical wheel travel.

Standard Chinese steel tracks have been retained on the prototype although these tend to have a shorter life than their western equivalent. If required, Cadillac Gage Textron hydro-pneumatic suspension units could be incorporated to give a significant improvement in cross-country mobility.

Main armament comprises a modified 105 mm gun fitted with a fume extractor and thermal sleeve that fires standard NATO ammunition including APFSDS. In the Type 59 the commander and gunner are on the left with the loader on the right, but in the Jaguar these positions are reversed with the commander on the right, gunner forward of the commander and the loader on the left. Both commander and loader are provided with a single piece hatch that opens to the rear.

A 7.62 mm machine gun is mounted coaxial with the 105 mm gun and a 12.7 mm M2 machine gun is mounted on the roof for anti-aircraft defence. Mounted either side of the forward part of the hull is a bank of four electrically operated smoke dischargers firing forwards.

The Jaguar is fitted with Cadillac Gage Textron electro-hydraulic gun control and stabilistation system. The main armament is fully stabilised and enables the Jaguar to engage stationary or moving targets when it is stationary or moving with a high first round hit probability. Stabilisation error is less than 0.75 mil depending on the terrain being traversed at the time.

Prototype of the Jaguar MBT showing 105 mm gun fitted with a thermal sleeve and fume extractor

Prototype of the Jaguar MBT with roof-mounted wind sensor for the fire control system

The prototype has a British Marconi Command and Control Systems Digital Fire Control System (DFCS) which has already been fitted to the Cadillac Gage Textron Stingray 105 mm armed light tanks delivered to Thailand. The fire control system incorporates image intensification night vision equipment and a laser rangefinder with a maximum range of 9990 m that is accurate to plus or minus 10 m.

The rear-mounted powerpack consists of a Detroit Diesel 8V-92TA eight-cylinder diesel developing 750 hp coupled to an Allison Transmission Division of General Motors XTG-411 fully automatic transmission The powerpack is similar to that installed in the Stingray light tank but has a hydraulically driven cooling system.

The prototype is fitted with a fire detection and suppression system developed by the Santa Barbara Research Centre (SBRC), a subsidiary of Hughes Aircraft.

A wide range of optional equipment is available including different fire control systems, vision equipment, hydropneumatic suspension, radios and an NBC system.

Status: Prototype under going trials, not yet in production or service.

Marketing: Cadillac Gage Textron
PO Box 1027
Warren
Michigan 48090
USA
Telephone: (313) 777 7100
Telex: 200707
Fax: (313) 776 9731

SPECIFICATIONS	
CREW	4
COMBAT WEIGHT	42 000 kg
POWER-TO-WEIGHT RATIO	17.9 hp/tonne
GROUND PRESSURE	0.98 kg/cm
LENGTH	
gun forwards	9.6 m
hull	6.8 m
WIDTH	3.27 m
HEIGHT	2.63 m
MAX ROAD SPEED	55 km/h
FORDING	1.1 m
GRADIENT	60%
SIDE SLOPE	30%
ENGINE	Detroit Diesel 8V-92TA 8-cylinder diesel developing 750 hp
TRANSMISSION	Allison Transmission Division of General Motors XTG-411 fully automatic with 4 forward and 2 reverse gears
SUSPENSION	torsion bar
ELECTRICAL SYSTEM	24 V
ARMAMENT	
main	1 × 105 mm
coaxial	1 × 7.62 mm MG
anti-aircraft	1 × 12.7 mm MG
smoke dischargers	2 × 4
AMMUNITION	
105 mm	34
7.62 mm	3500
12.7 mm	600
GUN CONTROL EQUIPMENT	electro-hydrdaulic/ manual
by commander	yes
by gunner	yes
TURRET TRAVERSE	360°
GUN ELEVATION/ DEPRESSION	+17°/-4°
RANGE SETTING DEVICE	yes, laser

Vickers/FMC VFM 5 Battle Tank

Development

In October 1985 at the Association of the United States Army (AUSA) Exhibition, FMC showed the Close Combat Vehicle (Light) (CCV L). This sophisticated light tank with a crew of three and an autoloader was originally designed to meet an anticipated US Army requirement for an Armored Gun System.

In 1985 Vickers Defence Systems and FMC teamed up to build a variant of the CCV L more suitable for export markets. The prototype of the VFM 5 Battle Tank was completed in May 1986 and shown for the first time at the 1986 British Army Equipment Exhibition. As of early 1991, the Vickers/FMC VFM 5 battle tank was still at the prototype stage and had yet to be demonstrated outside of the UK.

The VFM 5 battle tank is fully airportable in the US C-130 and C-141 transport aircraft.

Description

The hull of the VFM 5 is of welded aluminium armour with appliqué high-hardness steel plates on the front and sides.

The VFM 5 is protected to the same levels as many other first-generation battle tanks. Its layout is conventional with the driver's compartment at the front, the fighting compartment in the centre and the engine compartment at the rear.

The driver sits in the centre of the hull at the front. His rear-hinged hatch has five periscopes to give excellent visibility to the front and sides. The centre periscope can be replaced by a passive periscope for night driving. The driver has an adjustable sprung seat with a folding backrest to allow free access to the turret. Stowage for 11 rounds of main armament ammunition is provided on each side of the driver's position.

The turret of the VFM 5 is of welded aluminium armour wih appliqué high-hardness steel plates all around. Protection is similar to that given by the hull. The turret is mounted on a high-accuracy low-friction bearing

Prototype of VFM 5 showing 105 mm gun with muzzle brake, muzzle reference system, thermal sleeve and bore evacuator (Christopher F Foss)

Prototype of VFM 5 firing its 105 mm gun during trials in 1987

ring which is tilted down at the front by 3°. This forward tilt allows 10° of main armament depression at the front, while permitting limited depression over the rear and reducing the turret height. This means the VFM 5 can be made to meet the stringent size, weight and height requirements for the Low Altitude Parachute Extraction System (LAPES), should it be required.

Main armament is a 105 mm Low Recoil Force gun which is fitted with an integral muzzle brake to further reduce trunnion pull. The turret is designed to accept any of the available 105 mm guns manufactured by Royal Ordnance, Rheinmetall or Watervleit Arsenal.

The 105 mm gun of the VFM 5 is fitted with a Vickers rigid thermal sleeve that reduces gun barrel bend and provides a number of advantages over other types of sleeve. According to Vickers, their rigid thermal sleeve is resilient to damage, unaffected by rain, ensures uniform heat distribution around the barrel, reduces the thermal signature of the barrel and may also be transferred to the replacement barrel.

Forty-one rounds of main armament ammunition are carried, 22 (11 either side of the driver) stowed in the front of the hull and 19 in the fighting compartment. All ammunition is stowed below the turret ring.

A 7.62 mm L8 machine gun, mounted coaxially with the main armament, is fed from a box containing 200 rounds of ready use ammunition; a further 2400 rounds are carried in close proximity to the weapon.

A 12.7 mm or 7.62 mm machine gun is externally mounted over the loader's station. One hundred rounds of 12.7 mm or 200 rounds of 7.62 mm are available for ready use. If the 12.7 mm machine gun is fitted, 500 rounds for it are carried internally.

A bank of six grenade dischargers is fitted on each side of the turret. Smoke or fragmentation grenades can be fired electrically by the commander.

For the weapon control, Marconi Command and Control Systems has developed an integrated system incorporating gun stabilisation and computerised fire control.

The gun control system gives two-axis stabilisation with full duplex control for the gunner and commander. Power traverse is through 360° and power elevation from −10° to +20°.

The electro-mechanical system eliminates the risk of hydraulic oil leaking or catching fire in battle.

The computerised fire control system linked to the laser rangefinder gives a high probability of a first round hit at all battle ranges and is easy to use even after very little training.

The gunner's main combat sight is a NANOQUEST ×10 telescopic sight bolted rigidly to the gun rotor. It incorporates a laser rangefinder and muzzle reference system. This muzzle reference system allows the gunner to check his gun/sight alignment while under armour protection. The system can be used at all angles of elevation or depression.

Prototype of VFM 5 battle tank showing 105 mm gun fitted with thermal sleeve, fume extractor and muzzle reference system

Prototype of VFM 5 firing its 105 mm gun during trials in 1987

For wide-angle observation a NANOQUEST periscope is linked to main armament and fitted in front of the gunner. An optional thermal imaging system can be fitted. For this the gunner will have a monitor to the right of his GS10 sight, linked to the commander's sight, allowing him to engage targets at night and in poor visibility.

The commander has a Pilkington PE Raven II day/night sight, five periscopes for all-round observation and a rear-opening single piece hatch cover. The commander's sight has magnifications of ×1 and ×8 as well as an image intensification night channel.

The loader has a rear-opening single piece hatch cover and a single periscope AFV No 30 Mk 1 with 360° rotation for observation.

The VFM 5 has torsion bar suspension and six dual rubber-tyred road wheels on each side of the vehicle. The torsion bars give an upward suspension travel of 25.4 cm jounce and 10.2 cm rebound. A linear hydraulic shock absorber is fitted at five road wheel stations providing high cross-country mobility and stable firing platform characteristics.

The tracks are 38 cm wide and are of the double-pin type with detachable, large area, ground pads.

The VFM 5 is powered by a General Motors, Detroit Diesel Model 6V-92 TA six-cylinder diesel developing 552 bhp at 2300 rpm. This two-cycle engine is turbo-charged and fitted with an after cooler.

The engine is coupled to a General Electric HMPT-500-3 automatic transmission which is the latest upgraded model of that installed in the M2 IFV and M3 CFV. This hydro-mechanical transmission gives continuously variable gear ratios in three forward and one reverse speed ranges. Responsive hydromechanical steering and hydrostatic dynamic braking are incorporated into the gearbox. Mechanically-actuated service and parking brakes are also incorporated into the gearbox casing.

A water radiator and oil cooler provide cooling for the powerpack. Air is drawn through the radiator into the engine compartment then expelled by a fan in the hull roof. A heavy duty air cleaner protects the engine in dusty or sandy environments. A 300 A generator provides electrical power. Four batteries are fitted in the right side of the engine compartment and two more in the turret. Access to these batteries for daily checks is via a hatch in the hull or turret roof.

The rear hull ramp folds down for access to the engine compartment. The engine can be slid out onto the ramp by the crew without special tools. Access covers are provided so that maintenance checks can be easily carried out by the crew. The powerpack is protected from fire by a firewire connected to an automatic extinguisher system. Hand extinguishers are provided for the crew. The VFM 5 may also be fitted with a crew compartment fire suppression system.

Status: Development complete. Ready for production.

Manufacturers: Vickers Defence Systems
Armstrong Works
Scotswood Road
Newcastle-upon-Tyne NE99 1BX
England
Telephone: (091) 273 8888
Telex: 53104
Fax: (091) 273 2324

FMC Corporation
Defense Systems International Division
881 Martin Avenue Box 58123
Santa Clara
California 05052
USA
Telephone: (408) 289 0111
Telex: 6714210
Fax: (408) 289 2150

SPECIFICATIONS		ACCELERATION		ARMAMENT	
CREW	4	(0 to 32 km/h)	6 s	main	1 × 105 mm gun
COMBAT WEIGHT	19 750 kg	FUEL CAPACITY	570 l	coaxial	1 × 7.62 mm MG
UNLOADED WEIGHT	18 400 kg	MAX RANGE (at 40 km/h)	483 km	anti-aircraft	1 × 12.7 mm MG
POWER-TO-WEIGHT		FORDING	1.32 m	AMMUNITION	
RATIO	28 hp/tonne	GRADIENT	60%	105 mm	41
GROUND PRESSURE	0.69 kg/cm²	SIDE SLOPE	40%	7.62 mm	2600
LENGTH		VERTICAL OBSTACLE	0.76 m	12.7 mm	500
gun forwards	8.61 m	TRENCH	2.13 m	GUN CONTROL	
hull	6.2 m	ENGINE	Detroit Diesel Model	EQUIPMENT	
WIDTH	2.69 m		6V-92 TA developing	Turret power control	electro-mechanical/
HEIGHT			552 hp at 2300 rpm		manual
overall	2.62 m	TRANSMISSION	General Electric	by commander	yes
reduced	2.35 m		HMPT-500-3 automatic	by gunner	yes
GROUND CLEARANCE	0.41 m	SUSPENSION	torsion bar	Max rate of turret	
TRACK	2.16 m	ELECTRICAL SYSTEM	24 V	traverse	360° at 40°/s
TRACK WIDTH	380 mm	BATTERIES		Gun elevation/depression	+20°/-10°
LENGTH OF TRACK		hull	4 × 12 V	Gun stabilisation	yes
ON GROUND	3.654 m	turret	2 × 12 V		
MAX ROAD SPEED	70 km/h				
on 10% slope	37 km/h				
on 60% slope	8.2 km/h				

NATO Industrial Advisory Group Tank Study

Early in 1989 a total of 30 defence contractors from nine NATO countries (Belgium, Canada, France, Germany, Italy, Netherlands, Spain, United Kingdom and the United States) assembled in Brussels to organise a study aimed at ensuring that all tanks fielded by NATO nations after the year 2000 could be rearmed, refuelled and repaired during a future conventional war at any depot across Western Europe.

The NIAG tank study group was split into two subsections, one concentrating on fire control and the weapon system and the other on armour and the platform.

The aim was not to field an identical MBT throughout NATO but to draw up a common design and specification for a future MBT that should take into account the different requirements and timescales of the nine nations engaged in the study.

France, Germany, the United Kingdom and the United States are already working towards a 140 mm smooth bore tank gun which will have a common chamber size. This large calibre weapon would require an automatic loader so enabling the turret crew to be reduced to two men, commander and gunner.

NATO's common tank should be of conventional layout with driver at front, turret in centre and powerpack at the rear.

The common design and specification would allow individual countries to put in their own systems such as powerpack and suspensions, but they all have the same main armament using similar ammunition, they would also be able to communicate with each other on the battlefield using similar radios.

The final report was submitted to NATO Land Warfare officials in early 1990.

IRAQ

Modified Iraqi Main Battle Tanks

Development

As a result of combat experience during eight years of fighting against Iran, Iraq has modified many Soviet armoured fighting vehicles to meet its own specific requirements and some of these were shown during an international defence exhibition held in Baghdad in April 1989.

In 1989 Iraq also stated that it was assembling the Soviet T-72M1 MBT In Iraq under the local name of the Assad Babyle (Babylon Lion), with local industry already manufacturing the ordnance, breech mechanism, ammunition and some of the electronics.

During the Baghdad International Exhibition for Military Production held in the Spring of 1989, Iraq exhibited a number of upgraded MBTs and details of these are given below.

Although the Iraqi Army referred to these as T-54/T-55 conversions some of the chassis/turrets were in fact Chinese supplied Type 59/Type 69 chassis, large numbers of which were supplied to Iraq in the 1980s conflict with Iran.

As far as it is known, the only upgraded T-54/T-55 MBTs encountered by the Allied Forces during Operation Desert Storm were a few of the T-54/T-55 with additional armour over their frontal arc which gave a high degree of protection against HEAT projectiles.

Prior to the invasion of Kuwait by the Iraqi Army, its MBT fleet is believed to have consisted of some 5500 vehicles of all types of which about 3500 to 4000 were lost during the conflict.

T-55 with 125 mm smooth bore gun

This is a T-55 with its original 100 mm gun replaced with the complete 125 mm smoothbore gun as installed in the T-72M1 MBT and an automatic ammunition handling system. This has enabled the crew to be reduced to three men, commander, gunner and driver.

In addition the vehicle is fitted with new side skirts, passive night vision equipment and bank of four electrically operated smoke dischargers mounted either side of the turret firing forwards.

The installation of the 125 mm 2A46 gun and its automatic ammunition handling system has necessitated the raising of the turret roof complete with cupolas.

Mounted on the turret rear is a stowage basket that is similar to that installed on recent Chinese MBTs.

T-55 with 125 mm gun

This also has the complete 125 mm 2A46 gun as installed in the T-72M1 MBT as fitted to the above version but in this case the turret appears to be slightly raised and the roof has not been modified so extensively.

This version also has an automatic ammunition handling system, new gunner's roof-mounted sight as there is no optical port to the left side of the gun as in normal T-54/T-55 MBTs, new night vision equipment, bank of four smoke dischargers, turret stowage basket that extends around the turret rear and new side skirts. The example shown in Baghdad in 1989 was fitted with Soviet type mine clearing rollers at the front of the hull.

T-55 with additional armour

Some T-55s have been fitted with applique multi-layer passive armour to the glacis, hull sides and turret front, side and rear.

In addition the vehicle is fitted with new night vision equipment and a bank of four electrically operated smoke dischargers either side of the turret firing forwards.

Rear view of T-55 fitted with 125 mm 2A46 smooth bore gun and showing raised roof and door in turret rear

T-55 fitted with 125 mm smoothbore gun as installed in the T-72M1 MBT

Close up of T-55 series MBT showing additional armour fitted to turret, glacis and hull sides and bank of four smoke dischargers

T-55 fitted with 125 mm 2A46 gun and mine clearing rollers (Christopher F Foss)

T-55 with applique armour fitted to turret sides, glacis plate and forward part of hull sides. Note bank of four smoke dischargers either side of turret (Christopher F Foss)

T-55 with applique armour package from the rear and showing thickness of side armour. Note additional armour package on arms to rear of turret (Christopher F Foss)

T-55 fitted with new superstructure housing 160 mm mortar with latter in horizontal position for travelling (Christopher F Foss)

These improvements have, however, increased the weight of the vehicle so the power-to-weight ratio has been reduced from 16 hp/tonne to 14.7 hp/tonne.

T-54 with 160 mm mortar

This is essentially a T-54 MBT with its turret removed and replaced by a fixed superstructure enclosing a 160 mm mortar which is understood to be the Soviet M-160 with a maximum range of 8040 m. The mortar is breech loaded and can be elevated from +50° to +80° with a very limited traverse. Access doors are provided in the sides and rear of the superstructure and when travelling the mortar is in the horizontal position.

T-54 with observation mast

This is a T-54 with its turret removed and replaced by an hydraulically operated telescopic mast with an armoured observation cabin on top. This can be raised to a maximum height of 25 m but for travelling is laid in the horizontal position.

Status

All of these were shown in public in April 1989 but as far as it is known, only the T-55 with additional armour entered service with the Iraqi Army in quantity.

ISRAEL

Merkava Main Battle Tank

Development

Israeli experience in the 1967 campaign proved that mobility was no substitute for armour protection and they therefore decided at an early stage that the main emphasis would be placed on armour with firepower and mobility second and third priorities. Design work on the tank started as early as 1967 but detailed design work, under the direction of General Israel Tal, did not begin until August 1970. Prior to the construction of the first prototypes of the Merkava MBT a number of test rigs based on M48 and Centurion tank chassis were completed to prove the basic concept. The first prototype of the Merkava was completed in 1974. The United States has contributed over $100 million towards the development and production of the Merkava MBT. Total cost of research, development, trials and the building of the prototype vehicles was about $65 million.

In May 1977 Israel finally announced that it had developed a new MBT called the Merkava (or Chariot) to the prototype stage and that a series production run of 40 tanks was being built. First production tanks were delivered to the Israeli Army in 1979. The Merkava was first used in action during fighting in the Lebanon in the Summer of 1982. First production Merkava Mk 2s were delivered to the Israeli Army in December 1983.

No production figures of the Merkava have been released but it is estimated that by early 1991 total production amounted to about 1000 vehicles with current production version being the Mk 3 which became operational with the Israeli Armour Corps in early 1990.

As the Mk 1 and Mk 2 vehicles come back for base overhaul they are being upgraded with some of the components of the Mk 3, but this excludes the 120 mm Israel Military Industries smooth bore gun.

An outline comparison of the three Marks of the Merkava MBT is as follows:

Merkava Mk 1 MBT showing turret profile and 105 mm M68 gun fitted with fume extractor and thermal sleeve. Note exhaust port at front of hull above drive sprocket (Israeli Ministry of Defence)

Model	Mk 1	Mk 2	Mk 3
WEIGHT	63 000 kg	63 000 kg	62 000 kg
MAIN ARMAMENT	105 mm	105 mm	120 mm
MAIN GUN AMMUNITION	62	62	50
60 mm MORTAR	external	internal	internal
ENGINE HP	900	900	1200
TRANSMISSION	semi-automatic	automatic, electronic control	automatic, electronic control
SUSPENSION SYSTEM	1st generation	2nd generation	2nd generation
FIRE CONTROL SYSTEM	electronic computer, laser rangefinder, night vision equipment	new electronic computer, new laser rangefinder, new night vision equipment	new electronic computer, new laser rangefinder, new night vision equipment, stabilised vertical and horizontal line of sight
TURRET CONTROL SYSTEM	electro-hydraulic	electro-hydraulic	electric/electronic
ARMOUR	spaced armour	combined spaced and add on special armour	modular special armour
AMMUNITION SURVIVABILITY	heat resistant	heat resistant	heat resistant and resistant to chain detonation
NBC SYSTEM	over-pressure	over-pressure	over-pressure, central crew air-conditioning
ELECTRO-MAGNETIC WARNING SYSTEM	elementary	elementary	improved

Description (Merkava Mk 1)
The hull of the Merkava is made of cast and welded armour with a well-shaped glacis plate with the right side higher than the left. Behind the first layer of cast armour is a space filled with diesel fuel and then another layer of armour. This spaced armour gives the tank protection from HEAT projectiles and ATGWs.

The layout of the Merkava is unconventional, with the turret and fighting compartment at the rear of the vehicle. The driver is seated on the left side of the hull, forward of the turret, with the engine compartment to his right. The driver is provided with a one-piece hatch cover that opens to the left and three observation periscopes for driving with the hatch closed; the centre one can be replaced by a passive one for night driving. The driver can reach his compartment through the main crew compartment as the backrest of the driver's seat folds forwards.

Urdan Industries Limited is one of the major suppliers of parts for the Merkava supplying the hull front and deck, turret and turret basket, gun shield and cradle, and parts for the suspension.

The Teledyne Continental AVDS-1790-6A V-12 diesel develops 900 hp and is coupled to an Allison Transmission Division of General Motors CD-850-6BX transmission. The engine is in fact a more powerful version of the engine fitted to the M60 and M60A1 MBTs and the transmission is also similar to that installed in these tanks. Access to the engine compartment is via two flaps which are opened by springs after the locks are released. The engine can be replaced in the field in about 60 minutes. The air cooling filter vent is positioned in the upper part of the hull, forward of the driver's seat, with the outlet being located on the opposite side. The exhaust outlet is on the right side of the hull, above the skirting plates over the second road wheel.

The wedge-shaped turret, which has been designed to accept either a 105 mm (M68) or 120 mm gun, is cast with a welded front. It has a small cross-section and a large overhang at the rear. The radios and hydraulics are mounted in the turret bustle. The commander is seated on the right side of the turret with the gunner seated forward and below the commander. The commander has no cupola but is provided with a hatch cover that opens to the rear and five periscopes for all-round observation. The commander's hatch can be raised manually to give direct all-round observation while retaining full overhead protection. Mounted forward of his hatch cover in the roof of the tank is a sight that can be traversed through 360°, with a zoom magnification of from ×4 to ×20. The rotating head of the commander's periscope is linked to the turret traverse system by a counter-rotating device. The gunner's optics are in the forward part of the turret roof and right-angled ribs in front of the optics stop shell-splinters and small arms fire from damaging them. The gunner's periscope has magnifications of ×1 and ×8 and incorporates a laser rangefinder. The loader is seated on the left rear side of the turret and is provided with a single piece hatch cover opening to the rear and a single periscope.

There are three hatch covers in the rear of the hull: the left one gives access to the batteries and the right one gives access to the NBC pack. The centre one is a two-part door, the upper part opening upwards and the lower part downwards, through which ammunition or wounded can be loaded. This hatch can be opened from the outside but locked from the inside. A 60 litre water tank is provided above the rear hatch. The infantry telephone is mounted at the rear of the hull on the left side. More recently some Merkava MBTs have been observed fitted with closely spaced chains with ball ends around the lower part of the turret bustle. These detonate HEAT projectiles before they can hit the turret ring.

The Mark 1 Merkava has six 790 mm Centurion-type rubber-tyred road wheels either side with the drive sprocket at the front, idler at the rear and four track return rollers. Each road wheel is suspended by a separate helical spring with suspension arms for two road wheels, each caged

Merkava Mk 2 MBT from side clearly showing additional armour protection on turret sides and chains hanging from turret bustle to detonate HEAT projectiles (Israeli Ministry of Defence)

in a housing. The first, second and fourth track return rollers support the inside of the track only. Urdan Industries has developed a new road wheel for Merkava and Centurion MBTs which has the same weight but twice the life.

A white light and an infra-red driving light mounted on either side of the glacis plate can be folded down in action to avoid damage from shell splinters and small arms fire. The 1 kW searchlight is mounted vertically in the bustle under armour protection to the rear of the loader's position and is controlled by the tank commander. The Merkava is equipped with an NBC system and a Spectronix explosion suppression system. This was not fitted to original Merkava MBTs but is now standard on all new production vehicles and has been backfitted to older vehicles as they return for overhaul.

Main armament is a standard 105 mm M68 rifled tank gun fitted with a thermal sleeve, manufactured in Israel by Israel Military Industries and also fitted to most other Israeli tanks. It has an elevation of +20° and a depression of -8.5°. A travelling lock is provided for the 105 mm gun on the right side of the glacis plate. The Merkava carries a total of 92 rounds of 105 mm ammunition of which eight are stowed below the turret ring for ready use and the remainder in the hull rear, 12 in two-round containers and 72 in four-round containers. No ammunition is stowed above the turret ring or in the hull front, and all ammunition is in special containers. In addition to the standard 105 mm HEAT and HESH rounds the gun will fire an APFSDS-T round called the M111 developed by Israel Military Industries, which is claimed to be superior to the American M735. More recently Israel Military Industries has introduced a new 105 mm APFSDS-T round called the M413 which has a muzzle velocity of 1455 m/s when fired from an L7/M68 gun with a maximum effective range of 6000 m.

The 105 mm gun is fitted with a thermal sleeve designed and built by Vidco Industries and this is also fitted to Israel M48, M60 and Centurion series MBTs. This eliminates most of the error caused by barrel bending due to sun, wind, rain and other effects and improves first-round hit probability.

A 7.62 mm machine gun is mounted coaxially to the left of the main armament and a similar weapon is mounted at the commander's and loader's station; these weapons can be lowered to reduce the profile of the Merkava.

The coaxial machine gun is fed from a 2000-round continuous belt which is between the plates of the spaced armour. The machine guns are 7.62 mm MAGs, manufactured under licence from FN Herstal. Some Merkavas have been observed with a 12.7 mm M2 HB machine gun over the gun barrel. This was originally fitted for training but has been used in urban fighting in the Lebanon instead of the 105 mm M68 gun. The 12.7 mm machine gun is fitted with one box of ready use ammunition and fired by an electrically-operated solenoid from within the turret. A 60 mm Soltam Commando mortar is carried by the Merkava for which 30 mortar bombs are stored in a compartment in the turret rear. This mortar fires high explosive, smoke and illumination bombs and helps conserve 105 mm ammunition.

During the 1973 Middle East war many Israeli tanks ran out of ammunition and so the Merkava has been designed to carry a large supply in the rear of the hull. The Merkava can also be used as a command post with the ammunition supply containers removed. By reducing the ammunition load the Merkava can also carry troops, for example 10 infantry can be carried by reducing the ammunition load by 45 rounds or a commando squad of three together with their radios if the ammunition load is reduced by 25 rounds. It must be emphasised that the ability to carry infantry is only an option for use in special circumstances as the infantry have no vision devices at all.

Merkava has a digital fire control system designed by Elbit Computers Limited of Haifa and a laser rangefinder which can be used by the commander or gunner. This fire control system is called the Matador Mk 1. The laser rangefinder is manufacturered by EL-OP Electro Optics Industries and feeds information into the computer. The system is built around a Central Processing Unit and consists of operation units, control and feedback servo loops and sensors. The system includes three operation units, gunner's, commander's and loader's. The gunner's is the main unit and provides all manual inputs necessary for ballistic computation. It also includes a logistic panel which enables system bore sighting and system BITE, as well as the display of pre-select inputs. The unit also includes the following manual inputs: type of ammunition (six types) and recoil compensation insertion for each type of ammunition, in elevation and deflection. The commander's unit provides a readout of the system's display, range and ammunition inputs. The loader's unit provides ammunition inputs.

The control loop transfers the computer super-elevation information to the hydraulic gun elevation drive and the ballistic move drive. In addition deflection data is transferred to the moving graticule. The feedback loop ensures that the actual super-elevation and graticule deflection data are identical to the computed data and will correct the error accordingly.

The system includes automatic sensors, laser rangefinder, turret cant angle indicator and target angular velocity sensor. Optional sensors include cross wind velocity, charge temperature, barrel bend and ambient air density.

Manual operation of the electrical system if the computer fails is made possible with a handwheel drum and mounted scales, connected to the mechanical gearbox, to give elevation angle compensation and range to target for all types of projectiles. The commander can, if required, take over control of the main armament and fire the gun. The main armament is stabilised in both planes as an American Cadillac Gage stabilisation system is fitted as standard. This is manufacturered under licence in Israel by PML Precision Mechanism Ltd. It is probable that a thermal imaging system will be fitted in the future.

It has been confirmed that EL-OP Electro Optics Industries Limited supplies the commander's night vision goggles, night driver's periscope, part of the fire control system and laser rangefinder and the tank commander's turret control handle. Tadiran supplies communications equipment, Ashot Ashkelon the tank tracks, Shahal the shock absorbers.

Variants

155 mm Self-propelled Gun

Details of the Soltam Slammer 155 mm self-propelled howitzer based on a Merkava chassis are given in the *Self-propelled Guns and Howitzers* section under Israel. This system is still at the prototype stage and by early 1991 two prototypes had been completed.

Dozer/Mine Clearing Equipment

Some Merkava MBTs have attachment plates under the hull front to enable them to be fitted with Israeli designed and produced mine clearing roller equipment and dozer blades.

Rear view of Merkava Mk 3 MBT showing external stowage and chains hanging below turret bustle (Israeli Ministry of Defence)

Head-on view of Merkava Mk 2 MBT showing additional armour protection on either side of turret and thermal sleeve for 105 mm gun (Israeli Ministry of Defence)

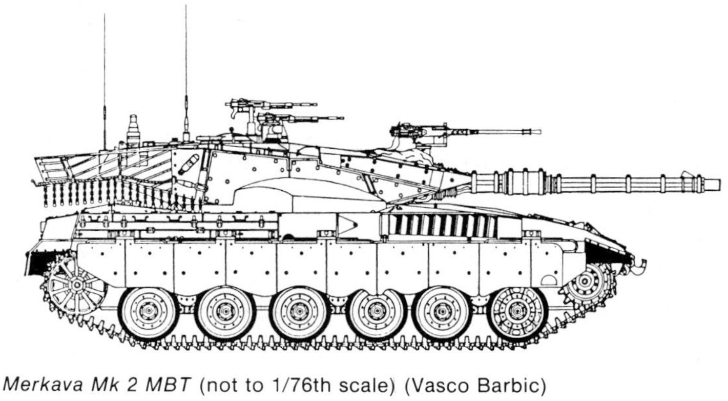

Merkava Mk 2 MBT (not to 1/76th scale) (Vasco Barbic)

Merkava Mk 3 MBT clearly showing new side skirts and 120 mm smoothbore gun which is fitted with a thermal sleeve and fume extractor (Israeli Ministry of Defence)

Merkava Mk 3 MBT fitted with more powerful engine, 120 mm smoothbore gun and new special armour (Israeli Ministry of Defence)

Platoon communications versions of the Merkava are also in service; these have additional communications equipment and additional antenna.

Merkava Mark 2

In 1983 the Merkava Mark 2 replaced the earlier Mark 1 in production. Main improvements over the original production model include a layer of special armour on the front and sides of the turret, special armour on the hull front, the 60 mm mortar is mounted in the left side of the turret roof and can be loaded and fired from within the turret, the steel skirts that protect the suspension are backed by special armour. The fire control system has a number of improvements including Nd-YaG laser rangefinder in place of the original Neodymium-glass laser rangefinder, and a more advanced computer. The new side skirts are attached by heavy spring hinges so that the skirts do not become attached while crossing rough terrain. The fire control system installed in the Merkava Mk 2 is called the Matador Mk 2. The original 900 hp diesel engine has been retained but the new Israeli designed and built transmission is much more efficient and has increased the range of the Merkava by 25 per cent to 500 km with only a slight increase in fuel capacity. Production of the Mk 2 was completed late in 1989.

Merkava Mark 3

Early in May 1989 Israel finally revealed the Merkava Mark 3 MBT which is already in production and service with the Israeli Army. This has many improvements over the earlier Mark 1 and 2 versions and these can be summarised as follows:-

1) Main armament is a 120 mm smoothbore gun fitted with a thermal sleeve. This has an elevation of +20° and a depression of -8° with a total of 50 rounds of ammunition being carried. A 7.62 mm MG is mounted coaxial with the main armament and there are also two 7.62 mm roof-mounted machine guns; a total of 10 000 rounds of 7.62 mm ammunition are carried. Some of the 50 rounds of ready use ammunition are in a drum on the floor of the turret compartment that provides the same degree of protection as the individual containers. The 60 mm mortar is aimed, loaded and fired from within the turret and in addition to firing HE bombs against anti-tank teams it can also fire illuminating rounds.

Mounted either side of the turret is the Israel Military Industries CL-3030 instantaneous self-screening system for combat vehicles.

2) Turret controls are now all electric with controls for both gunner and commander. Commander's cupola type day/night sight has a magnification of ×4 and ×14 with an optical relay to the gunner's sight which is a two axis stabilised day/night sight with a magnification of ×12. The gunner's sight incorporates a Nd-YAG laser rangefinder. The advanced fire control system has been developed by EL-OP and Elbit. The fire control system is integrated with the turret and gun control equipment and allows line-of-sight stabilisation for the tank commander and gunner. It allows the Merkava Mk 3 to engage moving targets while it is also moving.

3) An advanced threat warning system developed by Amcoram is fitted with threat warnings being displayed on a small panel at the commander's station.

4) The armour is of a modular special type and covers both turret and hull. This armour provides a much higher degree of protection over that of the Merkava Mk 1 and 2 and can also be changed in the field. The new modular armour is attached to the hull and turret of the Merkava by bolts and as new armour systems become available they can replace the earlier armour packages.

5) The suspension is new and consists of 12 road wheels independently mounted on trailing arms and sprung by pairs of concentric coil springs. The first and last road wheel stations either side have a hydraulic rotary damper developed by SHL. Ashot Ashkelon provide the final drives and suspension system for the Merkava Mk 3.

6) It is powered by a Teledyne Continental Motors AVDS-1790-9AR air cooled diesel developing 1200 hp and this is coupled to a new Ashot transmission.

7) NBC system is still of the over pressure type but with a central filter and an air conditioning capability.

Status: In production. In service with the Israeli Army.

Manufacturer: Israeli Ordnance Corps facility at Tel a Shomer, near Tel Aviv, Israel.
Enquiries to SIBAT-Foreign Defence Assistance and Defence Export
Ministry of Defence
8 David Elazar Street
Hakiryah 61909
Tel Aviv
Israel
Telex: 361203.

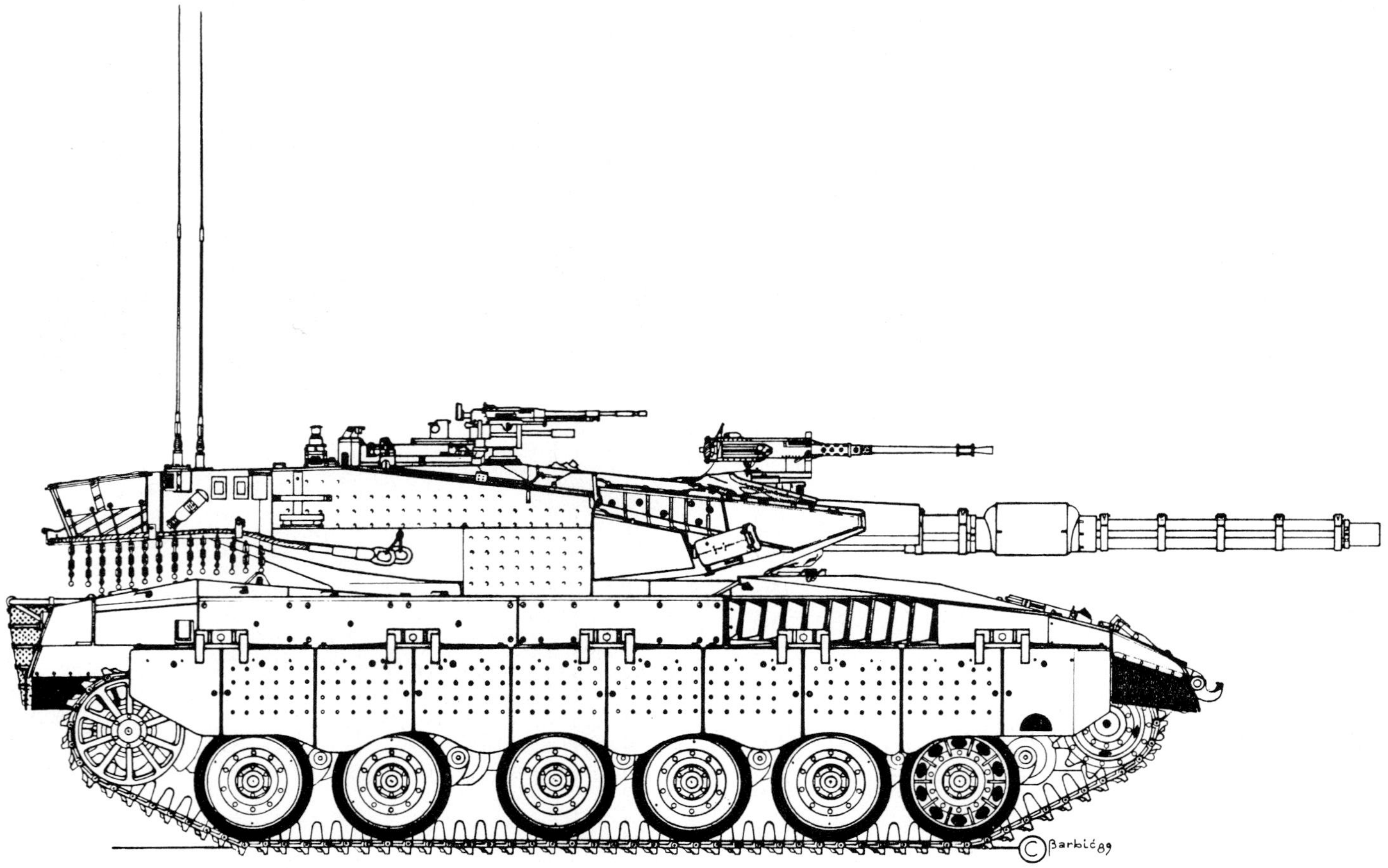

Merkava Mk 3 MBT with 12.7 mm M2 machine gun mounted over 120 mm smoothbore gun and showing chains under turret bustle (Vasco Barbic) (not to 1/76th scale)

SPECIFICATIONS (Merkava Mk 1) (Specifications in brackets relates Mk 3)

CREW	4
COMBAT WEIGHT	63 000 kg (62 000 kg)
UNLOADED WEIGHT	61 000 kg (60 000 kg)
POWER-TO-WEIGHT RATIO	14.28 hp/tonne (19.35 hp/tonne)
GROUND PRESSURE	0.9 kg/cm² (0.96 kg/cm²)
LENGTH GUN FORWARDS	8.63 m (8.78 m)
LENGTH HULL	7.45 m (7.60 m)
WIDTH	3.7 m
HEIGHT	
to commander's cupola	2.75 m (2.76 m)
to turret roof	2.64 m
FIRING HEIGHT	2.15 m
GROUND CLEARANCE	0.47 m (0.53 m)
TRACK WIDTH	640 mm (660 mm)
LENGTH OF TRACK ON GROUND	4.78 m
MAX ROAD SPEED	46 km/h (55 km/h)
ACCELERATION	
0 to 32 km/h	13 s
RANGE	400 km (500 km)

FORDING	
without preparation	1.38 m
with preparation	2 m (2.4 m)
FUEL CAPACITY	900 l
GRADIENT	70%
SIDE SLOPE	38%
VERTICAL OBSTACLE	0.95 m (1 m)
TRENCH	3 m (3.5 m)
ENGINE	Teledyne Continental AVDS-1790-6A V-12 diesel developing 900 hp (Teledyne Continental AVDS 1790-9AR V-12 turbocharged diesel developing 1200 hp at 2400 rpm)
TRANSMISSION	Allison Transmission Division of General Motors CD-850-6BX semi-automatic (Ashot fully automatic)
SUSPENSION	Independent, helical spring and volute bumper spring

ELECTRICAL SYSTEM	24 V
BATTERIES	8 × 12 V (500 Ah)
ARMAMENT	
main	1 × 105 mm gun (1 × 120 mm gun)
coaxial	1 × 7.62 mm MG
anti-aircraft	2 × 7.62 mm MG
mortar	1 × 60 mm
AMMUNITION	
main	62 (nominal, can be as high as 85) (50)
coaxial/anti-aircraft	10 000
GUN CONTROL EQUIPMENT	
Turret power control	electro-hydraulic (electric/manual)
	manual
by commander	yes
by gunner	yes
Turret traverse	360°
Gun elevation/depression	+20°/-8.5° (+20°/-8°)
Gun stabiliser	
vertical	yes
horizontal	yes

Israeli Upgraded M60 Series MBT

Development

At present, the Israeli M60 fleet is believed to consist of about 950 M60/M60A1 and 450 of the more recent M60A3 MBTs. As a result of combat experience, these have been upgraded in many areas and early in 1989 it was revealed that the IDF had started another major upgrade to its M60 series to extend the their operational lives into the 1990s.

There is insufficient money in the IDF budget to replace all existing tanks with the more recent Merkava and today about 75 per cent of the IDF's armour units still have older tanks.

Initial upgrade

From the 1970s, the M60s were upgraded in a number of areas including addition of Israeli developed thermal sleeve for 105 mm gun, addition of Israeli developed explosive reactive armour, installation of a new Urdan commander's cupola (also adopted by the US Army for its M48A5 tanks), two roof-mounted 7.62 mm machine guns and the Israeli Military Industries CL-3030 instantaneous self-screening system either side of the main armament. Details of the Blazer ERA and CL-3030 system are given in the following Centurion entry.

Latest upgrade

The key part of the latest upgrade is the new passive armour, while other changes include a new engine, tracks and fire control system.

The passive armour package has been fitted to the glacis, nose, turret front and sides and the forward part of the roof. Armoured skirts have also been fitted to provide lateral protection, something that earlier Israeli upgrade kits failed to provide. The forward part of the skirt, which covers the crew compartment, is of a different design to the rear.

The new armour package also provides some additional protection against top attack, at least on the forward part of the roof.

Openings have been left in the right side of the armour package for the gunner's sight and in the left side of the armour package for the coaxial machine gun.

M60 of the Israeli Army with Israel Aircraft Industries, Ramta Structures and Systems, Track Width Mine Plough system and mounting points for Blazer explosive reactive armour

The latest upgrade to the Israeli M60 MBT includes a new armour package which gives increased protection over the earlier Blazer ERA package. The latest upgraded M60 is called the MAGACH-7

Passive armour will give the M60s protection against kinetic energy (KE) and chemical energy (CE) attack, unlike the earlier explosive reactive armour used on M60 upgrades which only gave protection against CE like the Soviet AT-3 Sagger anti-tank guided weapon used by Arab armies.

The installation of additional armour and other improvements has led to the installation of a new engine and tracks so that the vehicle retains its cross country mobility. The new tracks, produced by Urdan Industries, are the same as those used on the Merkava.

The Israeli Armoured Corps has standardised on the US Teledyne Continental AVDS-1790 series diesel engine for many years and this has been retrofitted to all M60s and Centurions. It was already fitted to the M60s when delivered.

The upgraded MAGACH-7 MBT is fitted with the Teledyne Continental Motors AVDS-1790-5A V-12 diesel engine developing 908 hp which has also been fitted with an automatic starter.

The Matador computerised fire control system developed by Elbit and EL-OP Electro Optics Industries Limited is also being installed. The system is already in service on Merkava and includes a laser rangefinder. The M60A3 version of the M60s fielded by the IDF already have a modern fire control system including a laser rangefinder and this may point to the new upgrade being limited to earlier M60/M60A1 versions.

The 105 mm M68 gun, manufactured locally by Israel Military Industries and installed in all Israeli M60, Centurion and Merkava Mk 1 and 2 vehicles, remains unchanged.

In addition to firing the first generation Israel Military Industries M111 APFSDS round it also fires the second generation APFSDS-T M413 round which has greater armour penetration characteristics.

Pelé Robotic M60A1

Israel Aircraft Industries has developed a remote-controlled M60A1 called the Pele (Miracle). In the first application, which is being suggested for a mine clearance role, the tank is controlled by an operator who, at present, has to be within sight of the vehicle. The long term aim is to make the system autonomous.

Other applications being proposed include forward observation, surveillance and target acquisition, attraction of enemy fire, emergency control of the vehicle and a moving full-scale target simulator.

It can be installed on any MBT in less than eight hours and requires little training to operate. The Pele incorporates an operator video system with one onboard camera for driving and another for surveillance, target acquisition and arming. Maximum range of the system is at present 3000 m.

Automatic modes of operation include an automatic driver-heading control and navigation mode in which the tank follows a pre-programmed way-points plan. Part time autonomous modes are achieved by an onboard processing capability and a simple-to-operate portable operations unit that can be used from a static position or another vehicle.

According to Israel Military Industries, Pele is an intermediate solution between a combined manned/unmanned select platform operative and a robotic system.

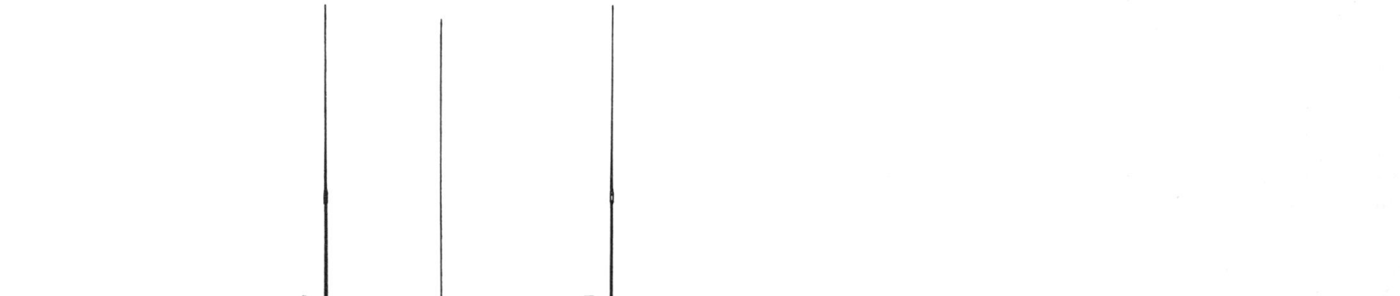

M60A1 MBT (command) of the Israeli Army fitted with 12.7 mm M2 machine gun over 105 mm M68 tank gun and Blazer explosive reactive armour system (not to 1/76th scale) (Vasco Barbic)

Israel Aircraft Industries Pele Robotic M60 MBT with operator in RAMTA light (4 x 4) armoured vehicle

Status: Upgrade programme underway. In service with Israeli Army.

Manufacturer: The upgraded M60 programme is carried out at Israeli Ordnance Corps Workshops at Tel a Shomer, near Tel Aviv, Israel.

M60A1 of Israeli Army fitted with Blazer ERA and showing position of two 7.62 mm and one 12.7 mm machine guns

Israeli Upgraded Centurion (Sho't) Main Battle Tank

Development/Description

The Centurion MBT entered service with the Israeli Armour Corps in 1959 but was not initially liked by its crews as it was more complex than the Sherman tanks used at the time.

One of the main drawbacks of the Centurion was its Meteor petrol engine which gave the tank a relatively low power-to-weight ratio and a very high fuel consumption which limited its operational range. The cooling system also gave a lot of trouble as did the air cleaners.

In 1967 a decision was taken to replace the Meteor petrol engine with a diesel and after trials with three different engines the Teledyne Continental AVDS-1790-2A diesel was selected, as it was also being installed in the Israeli Armoured Corps' M48A2 tanks which were being modernised then. The original Merritt-Brown Z51R manual gearbox was replaced by the Allison Transmission Division of General Motors Corporation CD-850-6 automatic, which greatly reduced driver fatigue as well as simplifying training.

To accommodate the new powerpack the rear of the hull had to be enlarged, but even so the engine had to be installed at an inclination of 3.5° front side up. This meant that elevated top decks had to be fitted to accommodate the cooling air outlets that vented upwards through louvres above the transmission compartment. Externally the modified Israeli Centurions, officially designated the Upgraded Centurion (not the Ben-Gurion), is recognisable by its raised engine decks and air filter boxes on the trackguards.

The Upgraded Centurion, which entered service with the Israeli Armoured Corps in May 1970 and was shown for the first time in public in February 1973, also has increased fuel capacity, more efficient air-cooled braking system, new fire extinguishing system, improved ammunition layout with 72 rounds of 105 mm ammunition carried, and a new electrical system.

Some Centurions have been observed with a 12.7 mm M2 machine gun over the gun barrel. This was originally fitted for training but was used in urban fighting in the Lebanon in lieu of the 105 mm M68 gun. The 12.7 mm machine gun is fitted with one box of ready use ammunition and fired by an electrically operated solenoid from within the turret.

The Upgraded Centurion has a maximum road speed of 43 km/h and twice the cruising range of the Centurion Mk 5 on which it is based.

Variants

Most of the Centurion, M48 and M60 series MBTs of the Israeli Army are now fitted with the Israel Military Industries Blazer explosive reactive armour package, details of which are given later in this entry.

Most Centurion, M48 and M60 series MBTs of the Israeli Army have also been fitted with the Israel Military Industries CL-3030 instantaneous self-screening system for combat vehicles.

The two basic components of the system are the two launch containers, mounted one either side of the main armament, and the IS-10 system which holds 10 smoke projectiles in two cassettes of five projectiles each. The smoke projectiles contain red phosphorous which is non-toxic and when fired these create a screen at a distance of 40 to 50 m from the tank that is initially 60 m wide and 4 to 8 m high.

Upgraded Centurion MBT of Israeli Army showing modified rear engine deck to accommodate Continental AVDS-1790-2A diesel engine and Allison Transmission of General Motors CD-850-6 transmission

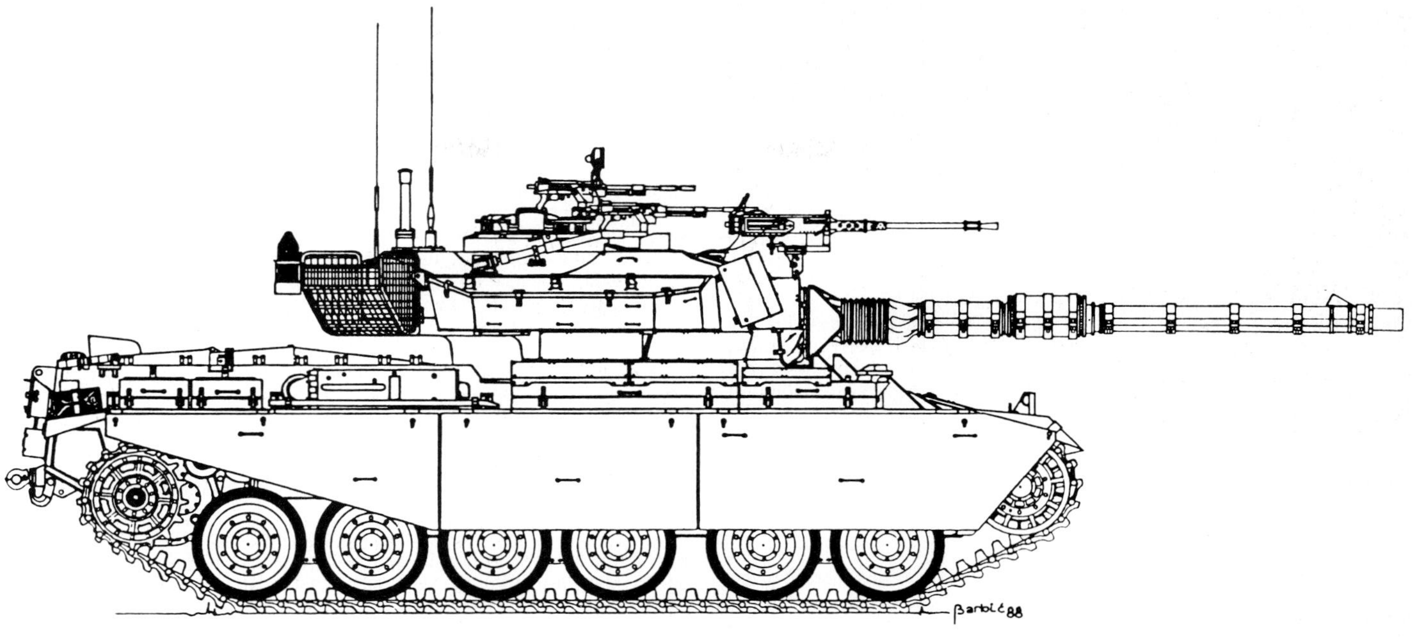

Latest version of Sho't Upgraded Israel Centurion MBT fitted with Blazer Explosive Reactive Armour, roof mounted mortar, two 7.62 mm machine guns on roof, 12.7 mm machine gun over 105 mm gun, thermal sleeve for 105 mm gun, new low visibility aerials, new exhaust ports, different stowage basket at hull rear and Matador computerised fire control system (not to 1/76th scale) (Vasco Barbic)

The Upgraded Centurion and other Israeli MBTs can also be fitted with the Ramta Structures and Systems Track Width Mine Plough (*Jane's Military Vehicles and Logistics 1991-92*, page 234) and the Urdan RKM mine clearing roller system (*Jane's Military Vehicles and Logistics 1991-92*, page 234). It can also be fitted with the RKM bulldozer attachment (*Jane's Military Vehicles and Logistics 1991-92*, page 5).

The Israeli Army also uses the Centurion ARV Mk 2 and some tanks have had their turrets removed for use as tugs. As a private venture Soltam has fitted a Centurion chassis with a turret-mounted 155 mm gun/howitzer, full details of which are given in the *Self-propelled guns and howitzers* section under Israel, although it has yet to be placed in production.

Centurion APC

Some Israeli Centurions have had their turrets removed for use as armoured personnel carriers in southern Lebanon. The hull armour around the turret ring has been built up to the front, sides and rear and 7.62 mm/12.7 mm machine guns have been mounted around the top to provide suppressive fire. In combat thick armour side skirts are fitted in place of the thin steel skirts normally fitted to the Centurion tank and Blazer explosive reactive armour is also provided. This version of the Centurion (Sho't) can carry up to eight fully equipped infantrymen. It is normally armed with 3 × 7.62 mm MAG and 3 × 52 mm or 60 mm light mortars. The Centurion APC is also used by combat engineers carrying specialised engineering equipment such as marking pennants.

290 mm Multiple Rocket System

Details of this system, which is based on a Centurion chassis, are given in the *Multiple rocket launchers* section.

Note on Israeli MBTs

During the fighting in the Lebanon in the Summer of 1982 many Israeli M48, M60 and Centurion series MBTs were observed to be fitted with appliqué armour. In late 1983 it was confirmed that this was explosive reactive armour and had been developed by Rafael Armament Development Authority with production and marketing now being undertaken by Israel Military Industries. This armour, called Blazer, is suitable for installation on all types of MBT and adds less than one tonne to the vehicle weight. It consists of small panels of armour bolted onto the hull and turret reacting to HEAT attack but not activated by small arms ammunition, fire or artillery fragments. Once a panel has been hit it no longer provides protection but the manufacturer believes that the chance of a panel being hit twice is remote. On an M60A1 type MBT about 10 m^2 are protected including the turret roof. Blazer provides a high degree of protection against attack from HEAT projectiles such as those fitted to Soviet RPG-7 and Sagger type weapons. An illustration of an M60A1 with Blazer explosive reactive armour appears in this section in the M60 MBT entry.

Blazer is not initiated by small arms ammunition up to 23 mm API at all angles of attack from zero range, artillery and mortar fragments from 2 m and arc welding and autogen cutting.

Blazer provides complete arrest of the AT-3 Sagger at all angles of attack over 70 per cent of the frontal arc of the hull. On the front of the hull Blazer provides complete arrest of RPGs at all angles of attack (0° incidence) and complete arrest of AT-3 Sagger at 30° (NATO 60°) incidence on the side of the turret.

In early 1984 it was announced that the Israeli Ordnance Corps had developed a low cost fire control system which would be fitted initially to Israeli M48 and Centurion MBTs and at a later date to the T-55s used by some reserve units. The fire control system can be installed in under a day and includes a computer developed by Elbit Computers and electronics and optics by EL-OP Electro Optics Industries Limited.

Additional details of the Israeli modifications to the T-54/T-55 tanks are given in this section under T-54/T-55.

Manufacturer: The Upgraded Centurion programme was carried out at Israeli Ordnance Corps Workshops at Tel a Shomer, near Tel Aviv, Israel.

Israeli Upgraded Centurion MBT fitted with Israel Aircraft Industries, Ramta Structures and Systems, Track Width Mine Plough (TWMP) system

Israeli Centurion APC showing external stowage and machine guns (IDF)

Israeli Sherman Tanks

Note: As far as it is known no Sherman gun tanks remain in frontline service with the Israeli Army.

M1 Super Sherman

In 1956 France delivered a batch of 100 M4A1E3 and M4A1E8 Sherman tanks fitted with 76.2 mm guns. These tanks were renamed M1 Super Sherman by Israel and were used during the 1956 and 1967 Arab/Israeli wars, after which they were transferred to border security units or rebuilt for other roles.

M50 Mark 1 and M50 Mark 2 Shermans

During 1954 Israeli technical experts went to Bourges Arsenal in France to design a turret mounting for the Sherman using the French 75 mm CN75-50 tank gun. Two prototypes were built, one on an M4A2 hull and the other on the single M10 hull. Modifications undertaken included several to the ammunition ejection system, replacement of the welded-on steel box on the turret rear by a cast counterweight filled with lead, a specially adapted telescopic sight and a new ammunition feed system.

A pre-production model was completed in 1955 and by 1956 the first M50 Mark 1 Shermans had been rebuilt from M4A4 hulls and French-supplied turret castings in Israeli workshops. One company of M50 Mark 1s was subsequently used in the 1956 war.

As a result of combat experience several modifications were carried out on the production line. Because of the modifications the last of the initial batch of 50 Sherman M50 Mark 1 conversions was not delivered until April 1959. By this time the original Continental petrol engines, narrow tracks and VVS suspensions were beginning to present considerable maintenance problems due to the increased weight of the turret and armament. Their replacement with a 460 hp diesel and HVSS suspension was started. The M4A3 hull was considered ideal for this as it was already fitted with a radiator. Several hundred M4A3 hulls were bought from a variety of sources. The first batch of Cummins diesel engines arrived in 1960 to be fitted into the M50 Mark 2, as this version was designated. The prototypes had been tested late in 1959. Later two smoke dischargers were fitted to either side of the turret.

The 75 mm CN75-50 gun fired the following types of ammunition:

Type	**AP**	**AP**	**HE**
DESIGNATION	POT-51A	PCOT-51P	n/a
WEIGHT OF COMPLETE ROUND	21 kg	21 kg	20.6 kg
WEIGHT OF PROJECTILE	6.4 kg	6.4 kg	6.2 kg
MUZZLE VELOCITY	1000 m/s	1000 m/s	1000 m/s
PENETRATION	110 mm/0°	170 mm/0°	n/app
	60 mm/60°	40 mm/60°	

None of these remain in front-line service with the Israeli Army but some may have been exported.

M51 Sherman

Although the M50 Sherman was capable of successfully engaging Soviet T-34/85 and T-54/T-55 series tanks the Egyptians began receiving increased quantities of Soviet IS-3 heavy tanks in the early 1960s and a newer and more capable gun was needed to meet the threat.

The Israelis initially looked at the French 105 mm CN-105-F1 56-calibre tank gun with a maximum muzzle velocity of 1000 m/s firing a HEAT projectile. However it was found that the gun was too long and the muzzle velocity too high to operate efficiently from a modified Sherman turret, primarily because of the lack of space during recoil.

The Israelis then returned to Bourges Arsenal to modify a CN-105-F1 to meet their requirements. This resulted in the barrel length being reduced by 1.5 m to obtain an acceptable muzzle velocity of 800 m/s. The new gun was designated the 105 mm D1504 tank gun, 44 calibre with ammunition made in Israel.

The Sherman model chosen for this conversion was the M4A1. The tanks were stripped back to their hulls, fitted with a Cummins 460 hp diesel engine, E8 HVSS suspension and modified steering, transmission and exhaust, wider tracks and a new turret mantlet, turret bustle and ammunition stowage to house the new 105 mm gun and its ammunition. Other improvements included replacing the original bulky battery charger with a small and compact dynamo, fitting a white light/infra-red searchlight over the main armament, fitting two smoke dischargers either side of the turret and replacement of the locking assembly and hydraulic controls by a new SAMM CH 23-1 system based on that fitted to the AMX-13 light tank then in service with the Israeli Army. After trials with the prototypes it was found necessary to fit the gun with a muzzle brake.

The modifications took about 25 000 man hours per tank and resulted in the combat weight being increased to 39 tonnes. The maximum road speed is 45 km/h and the operational range is 270 km.

The M51 was first used operationally in the pre-1967 war border conflict with Syria. It was subsequently used in the 1967 war, the 1968-70 war of attrition and the 1973 war. In 1973 it successfully engaged Soviet T-62 MBTs used by Egypt and Syria.

The 105 mm gun fires a non-rotating hollow charge anti-tank round (the OCC 105 F1), HE model 60, smoke and practice rounds. The first is similar to that fired by the French AMX-30 MBT and has a muzzle velocity of 905 m/s and will penetrate 360 mm of armour at an incidence of 0° or 150 mm at 60°.

Following the 1967 war the regular armoured units were re-equipped with more modern MBTs, allowing some of the M1/M50/M51 series to be transferred to the reserve mechanised brigades and border defence units. It is known that Chile has taken delivery of a number of M51 Sherman tanks from Israel. The remainder of the older Shermans were then reworked into a variety of specialised versions; descriptions follow.

Armoured Ambulance

This is a rebuilt Sherman M50 with the turret removed and engine repositioned in the front. It is similar in overall shape to the armoured command post vehicle and M50 self-propelled howitzer and first entered service in 1969 for use in frontline areas during the War of Attrition. The patients were carried on stretchers in the rear of the vehicle which was provided with extensive medical equipment. To facilitate rapid removal of wounded men from the battlefield, a trapdoor was provided in the floor of the vehicle.

Artillery Observation Post Vehicle

This is a Sherman M1 with its turret removed and replaced by a scissors type hydraulically operated arm, on top of which is mounted a platform for artillery observation personnel. It is known as the cherrypicker and was used for the first time during the War of Attrition.

Armoured Command Vehicle

This is a rebuilt Sherman M50 with turret removed, engine repositioned to the front and superstructure added to the rear, on top of which is mounted the commander's cupola with a pintle-mounted 7.62 mm machine gun. Its overall shape is similar to the armoured ambulance and it entered service in 1967.

Medical evacuation tank based on rebuilt Sherman chassis with HVSS suspension (Israeli Army)

Trail Blazer tank based on Sherman chassis with front and rear blades raised and hydraulic crane in travelling position on right side of hull (Israeli Army)

M51 Sherman tank in service with Chile and clearly showing numerous external modifications carried out by Israel

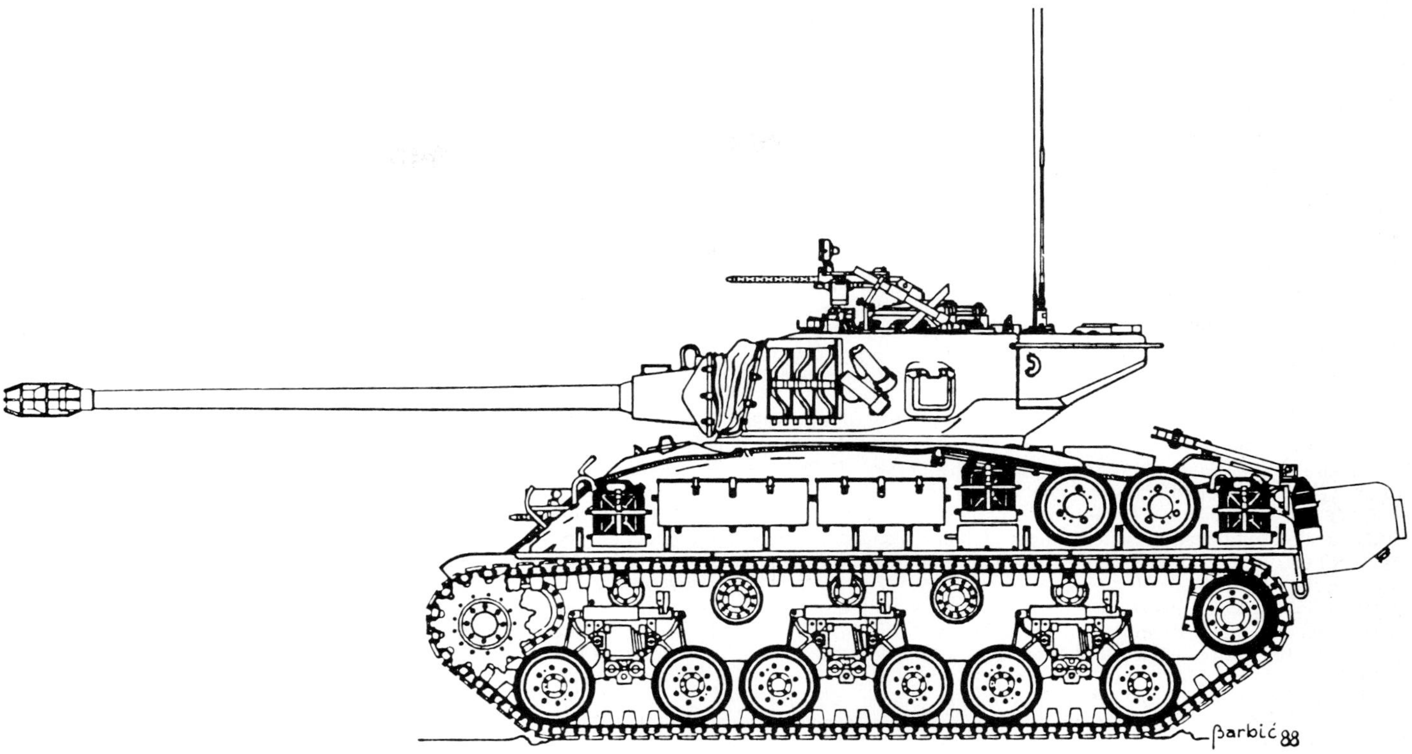

Israel M51 HV Sherman which has the same engine as the M109 series of self-propelled howitzers, mortar mounted on turret roof, 7.62 mm machine gun on a scissors mount as found on the Israeli Urdan cupola. Extra stowage bins have been fitted as well as a large stowage basket at the rear of the hull. The exhaust outlet is now on the rear engine deck and the radio antennas are similar to those fitted on the M60A1 and M60A3 tanks used by the Israeli Army (not to 1/76th scale) (Vasco Barbic)

Engineer and Mine clearing Versions

The Israeli Army is known to have used M4 flail type mine clearing tanks and a small quantity of Sherman bridgelayers but it is not known if any of them are still in use. As roller- and plough-type mine clearing devices have been developed in Israel in recent years it is probable that no mine clearing versions of the Sherman now remain in front line service.

The Israeli Army uses two modified Sherman tank chassis in the armoured engineer role. The first of these is a Sherman with a cast hull and horizontal volute suspension system, turret traversed to rear and armament removed, searchlights fitted, and a hydraulically operated dozer blade fitted at the front of the hull. The bow machine gun has also been removed on this model.

The second vehicle is called the Trail Blazer by the Israeli Army and uses a Sherman chassis with a hull and horizontal volute suspension system and the turret removed. Mounted at the front of the hull on the right side is a large hydraulically operated crane, similar to that installed on the Leopard 1 and AMX-30 ARVs, which is used for removing obstacles and changing powerpacks in the field. When not required this lies along the right side of the hull. Mounted at the front and rear of the Trail Blazer are hydraulically operated blades; the rear one is normally used when the Trail Blazer is in the recovery mode, while the front one can be used to clear obstacles as well. The winch compartment is in the centre of the hull where the turret was previously situated and stowage boxes are mounted along each side of the hull.

L-33 155 mm Gun/Howitzer

Details of this conversion will be found in the *Self-propelled guns and howitzers* section under Israel.

155 mm Self-propelled Howitzer M50

Details of this conversion will be found in the *Self-propelled guns and howitzers* section under Israel.

160 mm Self-propelled Mortar

This conversion was undertaken by the Soltam company of Haifa from 1969 and consists of a Sherman (or M7 Priest) chassis with the turret and superstructure removed and replaced by vertical sides, front and rear, with the front being capable of being lowered to the horizontal. The 160 mm breech-loaded mortar mounted in the centre of the vehicle fires forwards, has a limited traverse and can be elevated from +43° to +70°. The mortar fires an HE bomb weighing 40 kg to a maximum range of 9600 m, HE content being 5 kg, or two types of smoke bomb. Average rate of fire is five to eight bombs a minute. A total of 56 mortar bombs are stowed along the sides of the vehicle and below the floor on which the mortar is mounted. Access to the latter bombs is by a hatch to the right of the driver. A 7.62 mm or a 12.7 mm machine gun is mounted on each side of the vehicle for anti-aircraft use. The vehicle has a crew of eight, loaded weight of 36 654 kg, and is powered by a 430 hp Cummins diesel, which gives the vehicle a maximum road speed of 43 km/h and a range of 300 km.

Multiple Rocket Launcher

The Sherman chassis has also been used as the launcher for the 290 mm MRL developed by Israel Military Industries, available details of which are given in the *Multiple rocket launchers* section.

Status: Sherman gun tanks are no longer in front line service with the Israeli Army but some have been passed to the Lebanese Christian Militia and Chile is known to have taken delivery of a number of M51 Sherman tanks.

160 mm self-propelled mortar with front lowered almost to horizontal and showing on-board equipment carried. Some of the mortar projectiles are stowed down either side of the hull

ITALY

C1 Ariete Main Battle Tank

Development
In 1982 the Italian Army issued a broad specification for a new main battle tank to be designed and built in Italy. By 1984 the specifications had been agreed with industry and design work on key subsystems such as powerpack and armament was well under way.

In the same year OTO Melara and IVECO FIAT formed a consortium with headquarters in Rome to handle the Italian Army's new generation of tracked and wheeled armoured vehicles. For the C1 Ariete MBT OTO Melara has overall responsibility with IVECO-FIAT being responsible for the powerpack and suspension system. For the B1 Centaurs 105 mm (8 × 8) tank destroyer the positions are reversed.

The first prototype of the C1 was completed in 1986 and by 1988 all six prototypes had been completed and trials by the Italian Army and the manufacturer were well under way. By the Spring of 1989 the six prototypes had clocked up over a total of over 16,000 km.

OTO Melara hopes the C1 will be accepted for the Italian Army in 1991 to allow long-lead items to be ordered. It is envisaged that between 250 and 350 tanks will be ordered and initial production vehicles could be completed in 1993.

The ongoing CFE talks and the future size and structure of the Italian Army may well have an impact on the C1 Ariete MBT programme and some sources have indicated that as few as 150 vehicles may eventually be ordered.

The Italian Army has not issued any formal requirements for ARV, armoured vehicle launched bridge or armoured engineer vehicle versions of the C1 Ariete.

Description
The C1's hull and turret are of all-welded steel construction with an additional layer of advanced armour over the frontal arc, including the nose, glacis plate and turret front and sides.

Exact protection levels are classified, but the 54 tonne weight limit allows for protection against 105 mm APFSDS attack at least and against the HEAT warheads of most light anti-tank weapons. At one time it was proposed that the forward skirts be armoured in a similar manner to those of the Leopard 2, but this idea has now been dropped. Nor are there any plans to fit the vehicle with explosive reactive armour.

Layout is similar to that of most other current-generation tanks, with the driver's compartment front right, the turret in the centre and the powerpack at the rear. The driver's single piece hatch cover lifts and swings to the left. It has three integral forward-facing periscopes, the central one of which can be replaced by an MES VG/DIL 100 passive periscope for night driving. The driver's seat is hydraulically adjustable up or down and he directs the vehicle using a steering wheel rather than tillers. To his left are stowed 27 rounds of 120 mm ammunition and in the floor immediately behind him is the hull escape hatch. The fighting compartment is fitted with a fire and explosion detection and suppression system.

The turret is in the centre of the hull with the commander on the right, the gunner ahead of and below him and the loader on the left. The sides and rear of the turret are vertical but the front is well sloped like those of the M1A1 and Challenger to provide the best possible protection. The commander has a rear-opening single piece hatch cover and eight periscopes for all round observation.

The loader has a rear-opening single hatch cover and two roof-mounted periscopes looking forward and to the left of the vehicle. Forward of his hatch is a laser warning sensor with the display at the commander's station.

There are blow-out panels in the turret roof to vent any ammunition explosion upward and away from the turret crew, plus an ammunition resupply hatch in the left side and a large stowage basket at the rear of the bustle.

The C1 Officine Galileo Tank Universal Reconfigurable Modular System (TURMS) is common to the B1 Centauro 8 × 8 105 mm armed tank destroyer and a slightly modified version with increased sight elevation will be installed on the VCC-80 infantry fighting vehicle.

The system's main components are the commander's primary stabilised panoramic day sight, the gunner's primary stabilised periscope laser sight, a ballistic computer, sensors, a muzzle reference system, and commander's, gunner's and loader's control panel. The commander's sight has a magnification of ×2.5 and ×10 and for night vision a thermal image from the gunner's sight is displayed on a separate TV monitor at the commander's station. Developed jointly with SFIM of France, the roof-mounted sight can be traversed through 360° and elevated from -10° to +60°. The commander's head remains stationary while the sight head moves.

The gunner's roof-mounted sight integrates four main modules - primary stabilised head mirror, visual unit, laser and thermal image unit

One of the six prototypes of the C1 Ariete MBT showing 120 mm smoothbore gun fitted with fume extractor, thermal sleeve and muzzle reference system

C1 Ariete MBT with shutters on gunner's roof-mounted sight open and showing on-board stowage

in a single housing protected by armoured shutters which open left and right. The daylight ×5 magnification and thermal night vision with wide and narrow fields-of-view are all routed via the common head mirror.

The ballistic computer performs all fire computations and controls and manages the optical sight, laser rangefinder and servos as well as the sensors, built in test equipment and training. It also enables the system to be reconfigured from the normal operating level to back-up modes in the event of partial failure. The key TURMS sensors provide meteorological, tank attitude and barrel wear data. The meteorological and wind sensors are mounted on the turret roof.

The C1 can engage both stationary and moving targets while either stationary or moving itself. The tank commander would normally acquire the first target and hand it over to the gunner for engagement while he searches for additional targets. The gunner's back-up sight is an Officine Galileo C-102 coaxial telescope with a magnification of ×8 with and three manually selectable aiming graticules.

Main armament is an OTO Melara 120 mm 44 calibre auto-frettaged smooth bore gun fitted with a muzzle reference system, thermal sleeve and fume extractor. The coaxial recoil system consists of a recoil buffer and a nitrogen-actuated recuperator, with an internal reservoir installed on the gun cradle. The vertical sliding wedge breechblock remains open after the weapon has recoiled and the stub of the cartridge case has been ejected.

The gun chamber is identical in size to that of the 120 mm smooth bore gun installed in the Leopard 2 and the M1A1 so their ammunition is interoperable. SNIA is prime contractor for the C1's APFSDS and HEAT-MP ammunition. A total of 42 rounds of 120 mm ammunition are carried, 15 in the turret bustle separated from the crew compartment by access doors, and the remaining 27 in the hull.

One of prototypes of C1 MBT showing 120 mm gun fitted with thermal sleeve, fume extractor and muzzle reference system

The fixed mantlet has vertical slots for the auxiliary gunner's telescope, the 120 mm gun and the 7.62 mm coaxial machine gun on its left. The latter is fired electrically by the gunner, who also has a mechanical back-up. Another 7.62 mm machine gun is mounted on the turret roof for local and anti-aircraft defence. Operated by the tank commander and mounted on a spring-balanced cradle, it can be traversed rapidly through 360° and elevated from -9° to +65°.

Turret traverse and weapon elevation is electro-hydraulic with manual backup. The commander and gunner have dual control handles, and the complete system, including gun control and stabilisation, is supplied by Officine Galileo.

A bank of four electrically operated, forward-firing smoke dischargers, identical to those fitted to the B1 Centauro and VCC-80 infantry fighting vehicle is mounted on each side of the turret firing towards the front.

The complete powerpack of engine, transmission and cooling system will be delivered to OTO Melara by IVECO FIAT ready for installation in the vehicle, and can be replaced in less than an hour. The new FIAT V-12 MTCA engine, a 12 cylinder, four-stroke, direct injected, water-cooled, supercharged, after-cooled diesel, develops 1200 hp and gives a power-to-weight of 25 hp/tonne. Supercharging is by two turbochargers, one for each bank, arranged at the rear of the engine. Their inlets are cooled by air-water heat exchangers. A V-6 version of the MTCA powers both the B1 Centauro and the VCC-80. The air intake is above the engine compartment and there are exhaust outlets either side of the hull at the rear.

The engine is coupled directly to a German ZF LSG 3000 fully automatic transmission which provides four forward and two reverse speeds, three turning radii and pivot. Forward and reverse gears are selected through a power-operated shift control, with emergency mechanical shifting of the second forward and second reverse speeds available as a back-up. The LSG 3000 will eventually be built under licence in Italy by IVECO FIAT.

The transmission incorporates an auxiliary retarder to absorb about 75 per cent of the complete braking energy according to an integration logic with the mechanical brakes. The hydraulically operated main and emergency brakes act in conjunction with the automatic transmission retarder and the mechanical disc brakes on the final drive. The parking brake is of the mechanical type and acts on both disc brakes.

The two GRP main fuel tanks, located either side of the rear of the fighting compartment are selectable through a two-way valve actuated by the driver. An auxiliary tank filled from the main tank pipes ensures that fuel is available even when the tank is going up slopes and the tanks are partially empty. The fuel is forced into the engine by an electrically operated pump.

The torsion bar suspension either side consists of seven dual rubber-tyred road wheels, idler at the front, drive sprocket at the rear and four track return rollers. The first, second, third, sixth and seventh road wheels have hydraulic shock absorbers and all seven suspension arms either side have hydraulic bumpers to limit excessive road wheel travel. The Italian-designed track has rubber-bushed connectors, the individual track shoes being joined by end connectors and centre guiding teeth.

The upper part of the suspension is covered by skirts that can be quickly hinged upward to allow access to the suspension for maintenance.

The 24 V electrical system incorporates a total of eight 12 V, 100 Ah batteries to give a total of 400 Ah at 24 V. Four batteries are located on each side of the hull for ease of access via a panel above the fifth and sixth road wheel stations.

The C1's SP-180 NBC pack, designed and developed by Sekur, has unprotected and protected modes. Under normal unprotected conditions it provides forced air ventilation sucking in air from outside through a cyclone filter. In the protected mode the forced air from the cyclone filter enters a battery of composite filters containing the final filter and the carbon filter. The outlet is in the left side of the turret toward the rear.

SPECIFICATIONS		MAX RANGE (cruising)	over 550 km	ARMAMENT	
CREW	4	FORDING		main	1 × 120 mm gun
COMBAT WEIGHT	54 000 kg	(deep fording with		coaxial	1 × 7.62 mm MG
POWER-TO-WEIGHT		preparation)	4 m	anti-aircraft	1 × 7.62 mm MG
RATIO	23.14 hp/tonne	(with preparation)	2.1 m	SMOKE DISCHARGERS	2 × 4
GROUND PRESSURE	0.85 kg/cm^2	(without preparation)	1.2 m	AMMUNITION	
LENGTH		GRADIENT	60%	main	40 rounds (approx)
including main		side slope	30%	MG	2500 rounds (approx)
armament	9.669 m	VERTICAL OBSTACLE	2.1 m	GUN CONTROL	
hull	7.59 m	TRENCH	3 m	EQUIPMENT	
WIDTH		ENGINE	IVECO FIAT V-12	Turret power control	electro-hydraulic/
overall	3.601 m		MTCA, turbo-charged,		manual
over tracks	3.45 m		intercooled, 12-cylinder	by commander	yes
HEIGHT			diesel developing	by gunner	yes
turret roof	2.5 m		1250 hp at 2300 rpm	Turret traverse	360°
hull top	1.87 m	TRANSMISSION	ZF LSG 3000 automatic	Commander's fire-	
FIRING HEIGHT	1.98 m		with 4 forward and 2	control override	yes
GROUND CLEARANCE	0.44 m		reverse gears	Gun elevation/depression	+20°/-9°
TRACK	2.802 m	BRAKES	hydraulic actuated with	Gun stabiliser	
TRACK WIDTH	618 mm		transmission retarder	(2-axis stabilisation)	yes
MAX SPEED (cruising)	over 65 km/h	SUSPENSION	torsion bar	Night vision device	yes
ACCELERATION		ELECTRICAL SYSTEM	24 V		
(0 to 32 km/h)	6 s	BATTERIES	8 × 12 V, 100 Ah		

Status: Development complete. Ready for production.

Prime contractors: Consortium IVECO FIAT/OTO Melara
22 Via Vincenzo Bellini
00198 Rome
Italy
Telephone: (06) 859004
Fax: (06) 853701

OF-40 Main Battle Tank

Development

The OF-40 MBT has been designed from 1977 by OTO Melara and FIAT specifically for the export market, with the former responsible for overall design and production and the latter for the powerpack. In appearance the OF-40 resembles the late production Leopard 1A4 and uses some components of the Leopard 1 MBT. The first prototype of the OF-40 (O for OTO Melara, F for FIAT and 40 for the approximate weight in tons) was completed in 1980.

The United Arab Emirates placed an order for 18 OF-40 Mk 1s and the first of these were delivered in 1981. Since then the United Arab Emirates has taken delivery of a further 18 OF-40 Mk 2 MBTs and three armoured recovery vehicles while the original 18 OF-40 Mk 1 MBTs have been brought up to Mk 2 standard.

The OF-40 MBT has also been tested in Thailand, demonstrated in Egypt, and offered for local production in Spain and Greece. As of early 1991 the only customer for the OF-40 has been the United Arab Emirates.

Description

The all-welded hull is divided into three compartments: driver's at the front, fighting in the centre and engine and transmission at the rear.

The driver is seated at the front on the right side and has a single piece hatch cover that lifts and swings to the left. There are three unity periscopes in front of him, the centre one of which can be replaced by an Aeritalia image intensification periscope for night driving.

To the left of the driver is the NBC pack and 42 rounds of 105 mm ammunition and behind him there is an oval emergency hull escape hatch.

The all-welded turret is in the centre of the hull with the commander and gunner on the right and the loader on the left. The commander has a circular hatch cover that opens to the rear and eight unity periscopes for all-round observation, one of which can be replaced by an Aeritalia image intensification periscope. The commander's seat can be elevated by a manually operated hydraulic system and a foot operated valve enables him to lower the seat quickly in an emergency.

The commander does not have a cupola but mounted in the roof forward of his hatch is a day/night sight from Officine Galileo, developed in a joint venture with SFIM of France. This is stabilised and fitted with image intensification night vision equipment. This enables the tank commander to detect and identify the target and he can then either hand over to the gunner or engage the target himself.

The gunner is seated in front of and below the commander and has one forward facing roof-mounted periscope and an optical sight, the Aeritalia C215 articulated telescope with a 7.5° field-of-view and a

OF-40 Mk 2 MBT with loader manning roof-mounted 12.7 mm M2 machine gun and showing LLLTV mounted on mantlet coaxially with 105 mm gun

OF-40 Mk 2 MBT with turret traversed to rear and 105 mm gun in travel lock. The mount for the 12.7 mm machine gun is mounted on the roof although the actual weapon is not installed

OF-40 Mk 2 MBT from rear showing exhaust outlet above last road wheel

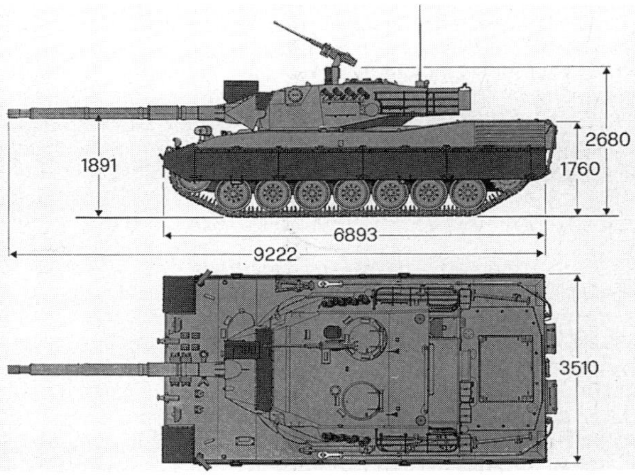

Side and top drawings of OF-40 Mk 2 MBT (not to 1/76th scale)

OF-40 Mk 2 MBT without anti-aircraft machine gun

magnification of ×8. It is mounted coaxially with the 105 mm gun in an M114 telescope mount.

The loader sits on the left of the turret and has a circular hatch cover that opens to the rear, in front of which are two roof mounted periscopes that give observation to the front and left side.

The main armament of the OF-40 is an OTO Melara designed and manufactured 105 mm 52-calibre rifled tank gun with falling wedge breech-block, concentric buffer and spring recuperator. The gun is fired electrically but can also be fired by a manually actuated impulse generator. During counter-recoil the automatic breech-block opens and ejects the spent cartridge case into a bag under the breech.

The barrel is provided with a thermal sleeve and a bore evacuator. The gun fires all standard NATO 105 mm ammunition including APDS, canister, HEAT, HESH and smoke, and a well-trained crew can achieve a rate of fire of 9 rds/min. As an option the main armament can be stabilised in both elevation and traverse.

Fifty-seven rounds of 105 mm ammunition are carried, 42 to the left of the driver and 15 in the turret for ready use.

A 7.62 mm or 12.7 mm machine gun is mounted coaxially to the left of the main armament and a similar weapon is mounted on the turret roof for anti-aircraft use.

Mounted externally on either side of the turret are four electrically operated smoke dischargers.

The fire control system has been designed by Officine Galileo and consists of the gunner's fire control system, gunner's telescope which moves in elevation with the main armament, commander's stabilised day/night roof mounted sight and the turret/gun electro-hydraulic drive system.

The gunner's fire control system consists of an Aeritalia C215 optical sight with a magnification of ×8 and range scale for APDS, HEAT and HESH 105 mm projectiles and 7.62 mm machine gun ammunition; Alenia VAQ-33 laser rangefinder with a range of between 400 and 9995 m; laser feeding and control equipment; fire control computer; control device and readout for the tank commander, gunner's controls and elevation sensor.

The Alenia VAQ-33 laser rangefinder consists of two main components, the laser optical head (transmitter/receiver) which weighs seven kg, and the Laser Electronic Unit (LEU) which weighs 15 kg. The VAQ-33 conveys both the range validation and the range of two targets digitally. It has a neodymium glass laser transmitting short coherent pulses of infra-red radiation at a wavelength of 1.06 μm. Its pulse repetition rate is 12 pulses per minute or bursts of three pulses in four seconds, with an interval of 10 seconds between bursts.

The commander's roof-mounted VS 580-B sight can be traversed through a full 360°, has a magnification of ×8, stadiametric rangefinding capability and range scales for APDS, HEAT and HESH ammunition.

Turret traverse and gun elevation are electro-hydraulic, with turret traverse at a rate of 360° in 17 seconds and gun elevation from -9° to +20° at 7°/s. The turret/gun drive system is controlled from the gunner's handle but an override system allows the commander to take over the laying and firing of the main armament if required and manual controls are provided for emergency use.

The engine, transmission and cooling system are assembled to form the powerpack which is supplied by FIAT ready for installation in the tank and can be removed by four men with a crane in under 45 minutes. The cables and wires are provided with quick disconnect couplings and if required the powerpack can be run outside the vehicle for testing.

The engine is an MTU 10-cylinder, four-stroke, precombustion chamber, super-charged multi-fuel developing 830 hp at 2200 rpm and is fitted with a tropicalisation kit to control the fuel supply and prevent the engine from overheating.

Two fuel tanks, with a total capacity of 1000 litres, are mounted one each side of the engine compartment and fuel can be drawn from one or both tanks.

Combustion air is taken in by two super-charged blowers through the two air cleaners. The two dry air cleaners are fitted in the side recess in front of the engine compartment and air enters through the intake scoops on the hull deck. Screens mounted in front of the scoops prevent large particles such as leaves entering the vehicle.

A cyclone filter battery purifies the air from coarse dust and large surfaced micro-top filters retain the fine dust. The coarse dust ejected by the cyclone filters is continuously blown outside by a dust blower installed in each air cleaner. The exhaust gases are taken from each cylinder liner by flexible corrugated tube compensators through an exhaust manifold into the exhaust mufflers in the side recesses of the engine compartment.

The engine of the OF-40 is liquid cooled through a self-contained cooling system with internal pressure controlled by a pressure relief and under pressure valve. On either side of the transmission is a cooler with a control fan on the vertical axis between them to suck the necessary cooling air from the top through the cooling air intake grille and force it through the two coolers. When deep fording, this duct is filled with water and the fan and coolers are therefore completely flooded.

Engine torque is transmitted to a torque converter which is followed by a four-speed planetary gearbox with two reverse speeds. The individual speeds are shifted electro hydraulically by a shift lever alongside the driver.

There is a separate clutch between the engine and ZF transmission which can be operated manually from the crew compartment. This provides easy starting at low temperatures as the starter has to crank the engine only and not the transmission.

The steering works through a longitudinally superimposed gearbox so steering movements do not cause essential power losses in the gearbox. Steering movements are mechanically transmitted from the control handle at the driver's station to the control valve.

The final drives are on either side at the rear of the vehicle and are of the planetary type. The transmission is connected to the final drives by an internally toothed sleeve which can be loosened from the outside when the powerpack has to be removed.

The suspension is of the torsion bar type with seven dual rubber tyred road wheels on each side. The first three and last two road wheels have hydraulic shock absorbers and bump stops to prevent excessive wheel travel. The adjustable idler is at the front, drive sprocket at the rear and there are five track support return rollers on each side.

The track is the rubber-bushed connector type with the individual track shoes joined by end connectors and centre guiding teeth. The upper part of the track and suspension is covered by a steel reinforced rubber skirt which keeps dust down and provides a degree of protection against attack by HEAT projectiles.

The OF-40 has three independent braking systems; service, parking and emergency systems.

The tank can ford to a depth of 1.2 m without preparation and an optional semi-deep fording hydraulic system enables it to cross streams to a maximum depth of 2.25 m. A snorkel can be fitted which enables the OF-40 to ford to a depth of four m. An electrically operated bilge pump is fitted in both the engine and crew compartment, each with the capacity of 120 litres a minute.

The overpressure NBC system is mounted at the front of the hull to the left of the driver. Fresh air is drawn into the vehicle by a blower and is then separated from the coarse dust in a cyclone filter. The NBC protection filter is down stream of the coarse dust filter and retains all NBC agents. The dust collection compartment of the cyclone filter is continuously kept free of dust by a dust ejector blower. Pressure in the tank is monitored by a pressure gauge in the crew compartment.

An automatic fire extinguishing system in the engine compartment has spray tubes and nozzles connected to a battery of bottles in the crew compartment containing the fire extinguishing agent. This system is activated at 180°C through a fire wire in the engine compartment.

Variants

OF-40 Mark 2 MBT

The OF-40 Mark 2 MBT is essentially the Mark 1 fitted with the Galileo OG14L2A fire control system which includes a stabilisation system for the 105 mm gun and sensors for wind velocity, powder temperature, ambient temperature and type of ammunition. The OG14L2B is similar but has a stabilised line of sight.

In the OF-40 Mark 2 the tank commander has a day/night periscope that is self-stabilised during panoramic search and slaved to the gun stabilisation when used for firing.

OF-40 ARV with dozer/stabiliser blade raised and crane traversed to rear to lift new OF-40 powerpack

The gunner has a sight with a magnification of ×7 and ×14 that incorporates a laser rangefinder as well as a telescopic sight, with a magnification of ×8, coaxial with the 105 mm gun.

Mounted externally on the right side of the mantlet is an LLLTV camera with the monitor in the turret.

Deliveries of the second batch of 18 OF-40 Mk 2 tanks to the United Arab Emirates began in 1984 and were completed in 1985. The Mk 1 tanks have now been brought up to Mk 2 standard in the United Arab Emirates.

OF-40 ARV

Three OF-40 armoured recovery vehicles have been delivered to the United Arab Emirates by OTO Melara. These have a front mounted winch with a tractive effort of 35 000 kg in first gear; this can be increased with the aid of a tackle. Mounted on the left side of the hull at the front is a hydraulic jib with a maximum lifting capacity of 18 000 kg with the front mounted stabiliser blade in position. The blade can also be used as a dozer blade to clear away obstacles and fill in trenches. The OF-40 ARV has a crew of four and basic specifications are length 7.68 m, height 2.35 m, ground clearance 0.44 m and combat weight 45 000 kg. If required a replacement OF-40 powerpack can be carried on the rear decking. Full details of this vehicle are given in *Jane's Military Vehicles and Logistics 1991-92,* page 25.

Systems on OF-40 Chassis

The hull of the OF-40 can also be used to mount other weapon systems such as a twin 35 mm anti-aircraft turret similar to the one on the Gepard anti-aircraft gun system. The OF-40 chassis, with a different powerpack, is also used as the basis for the OTO Melara Palmaria 155 mm self-propelled howitzer described later in this book. The Palmaria chassis also forms the basis for the 76 mm OTOMATIC self-propelled air defence system which is illustrated under the Palmaria entry.

SPECIFICATIONS (Mk 1)		FORDING		AMMUNITION	
CREW	4	without preparation	1.2 m	main	57
COMBAT WEIGHT	45 500 kg	with preparation	2.25 m	MG	5700
UNLOADED WEIGHT	43 100 kg	with snorkel	4 m	GUN CONTROL	
POWER-TO-WEIGHT		GRADIENT	60%	EQUIPMENT	
RATIO	18.24 hp/tonne	SIDE SLOPE	30%	Turret power control	electro-hydraulic/
GROUND PRESSURE	0.92 kg/cm^2	VERTICAL OBSTACLE	1.1 m		manual
LENGTH		TRENCH	3 m	by commander	yes
gun forward	9.222 m	TURNING RADIUS	pivot turns	by gunner	yes
gun rear	8.114 m	ENGINE	MTU 90° 10-cylinder, 4-	commander's override	yes
hull	6.893 m		stroke, pre-combustion	Commander's fire-	
WIDTH			chamber, super-charged	control override	yes
without skirts	3.35 m		multi-fuel developing	Max rate power	
with skirts	3.51 m		830 hp at 2200 rpm*	traverse	360°/17 s
HEIGHT		TRANSMISSION	power shifting 4 forward	Max rate power	
turret roof	2.45 m		and 2 reverse planetary	elevation	7° s
top of SFIM sight	2.68 m		type with hydraulic	Gun elevation/depression	+20°/-9°
top rear hull deck	1.76 m		torque converter	Gun stabiliser	
FIRING HEIGHT	1.89 m	SUSPENSION	torsion bar	vertical	no (optional)
GROUND CLEARANCE	0.44 m	ELECTRICAL SYSTEM	24 V	horizontal	no (optional)
TRACK WIDTH	584 mm	BATTERIES	8 × 12 V, 400 Ah	Range setting	
LENGTH OF TRACK ON		ARMAMENT		device	yes
GROUND	4.25 m	main	1 × 105 mm gun	Elevation quadrant	yes
MAX ROAD SPEED		coaxial	1 × 7.62 mm MG	Traverse indicator	yes
forwards	60 km/h	AA	1 × 7.62 mm MG		
reverse	25 km/h		(or optional 12.7 mm	* (950 hp option)	
FUEL CAPACITY	1000 l		MG)		
MAX ROAD RANGE	600 km	SMOKE DISCHARGERS	4 either side of turret		

Status: Production complete but can be resumed if further orders are received. In service with United Arab Emirates.

Manufacturer: OTO Melara
via Valdilocchi 15
19100 La Spezia
Italy
Telephone: (39 187) 530 111
Telex: 270368-211101 oto i
Fax: (39 187) 430 669.

Italian Tank Production

In 1956 Italy, France and Germany drew up a requirement for a European MBT which would be well armed, lighter and more mobile than other tanks then in service. In the end France placed the AMX-30 in production, Germany placed the Leopard 1 in production and Italy obtained a licence to manufacture the American M60A1. Two hundred were built in Italy by OTO Melara for the Italian Army, and a further 100 were supplied direct from the USA. These M60A1s replaced only a small part of the obsolete Italian tank fleet of M47s.

In the Summer of 1964 Italy evaluated the Leopard 1 in Sardinia and after many delays in 1970 placed an order for 800 Leopard 1 MBTs, the first 200 of which were to be supplied direct from Germany by Krauss-Maffei and the remaining 600 built under licence in Italy by a consortium headed by OTO Melara. The first Italian built Leopard 1 was completed in 1974 and the last of the original order for 600 was completed in 1978. Since then a further order was placed for an additional 120 which were completed by 1983. OTO Melara subsequently received an order from the Italian Army for 160 specialised versions of the Leopard 1 including bridgelayers (64), ARVs (68) and armoured engineer vehicles (28), production of which has now been completed. Before the delivery of these specialised versions Krupp MaK of Germany delivered 69 Leopard 1 ARVs and 12 AEVs to Italy.

To enable the Leopard 1 to remain effective into the 1990s the Italian Army is considering upgrading the tank in a number of areas, although for cost reasons it is unlikely that all of these will be carried out. The turret will be fitted with similar add-on armour to that supplied by Blohm and Voss for German, Belgian and Dutch Leopard 1s.

There is a competition to install a fire control system in at least part of the Italian Army's Leopard 1 MBT fleet.

Barr and Stroud of the UK has already entered into a joint venture agreement with Officine di Citadella, a member of the Firma Consortium, to install a computerised fire control system called the Sirio.

Another possibility is the installation of the Officine Galileo computerised fire control system installed in the C1 Ariete MBT, B1 Centauro 8 × 8 tank destroyer (in service with the Italian Army) and the VCC-80 infantry fighting vehicle.

Early in 1991, the Italian Army and Air Force started an evaluation programme of the OTOMATIC 76 mm air defence system turret installed on a modified Leopard 1 MBT of the Italian Army.

Status: Production complete. In service with the Italian Army.

Manufacturer: OTO Melara
via Valdilocchi 15
19100 La Spezia
Italy
Telephone: (39 187) 530 111
Telex: 270368-211101 oto i
Fax: (39 187) 530 669

JAPAN

Future Japanese Main Battle Tank

In FY84 the Technical Research and Development Institute of the Japanese Self-Defence Agency started studies of key components for a future MBT, even though the TK-X (under the designation Type 90) will not enter service until the early 1990s.

Initial component development will cost 200 million yen. It is expected that the new MBT, which could enter service in the early part of the next century as the replacement for the Type 74, will have a diesel engine or a gas turbine engine, and have semi-active suspension.

Type 90 Main Battle Tank

Development

Late in August 1987 the Japanese Ground Self-Defence Force finally unveiled their new TK-X MBT which was to have been type classified as the Type 89, but it was subsequently type classified as the Type 90 MBT.

Research and development of a new MBT began in 1976 under the designation of the STC, subsequently called the TK-X, by the Technical Research Headquarters of the Japanese Self-Defence Agency.

Funding for the engine, gun, ammunition and fire control system was started in FY77 with funding for the suspension and new armour in FY78. It was originally intended to have a Japanese 120 mm gun but a decision was finally taken to use the German Rheinmetall 120 mm smoothbore gun which is also installed in the Leopard 2 and M1A1 MBTs. The 1979 budget provided additional funds for the gun and ammunition as well as hull and turret with increased funding being made available in the following years.

Prime contractor for the Type 90 is Mitsubishi Heavy Industries, with the Japan Steel Works being responsible for the armament and Mitsubishi Electric for the fire control system.

The first two prototypes were built with funding from FY84 and FY85 budgets and a further four prototypes were completed by late 1988.

It is anticipated that when type classified as the Type 90 MBT, the first production batch will comprise 30 vehicles.

The prototype shown in late 1987 was the first of the two second phase prototypes and production vehicles will probably differ from these in minor details as a result of user trials.

Under the concept study, components of the TK-X were built and tested between FY77 and FY81. The first prototype was built between FY82 and FY84 and technical tests were carried out between FY85 and FY86. The second prototype was built between FY86 and FY87 and tested from FY87 to FY88. Second prototype operational tests were carried out from FY88 to FY89 with standardisation taking place in FY90.

It is expected that the Type 90 will enter service with the Japanese Ground Self Defence Force in 1992 with units on the island of Hokkaido being first equipped.

The FY90 Japanese defence budget requested 36.3 billion yen for 30 Type 90 MBTs. Late in 1990 it was announced that under the five year defence plan from FY91 to FY95, a total of 132 Type 90 MBTs would be purchased (down from the 150 originally envisaged). By the end of FY96 a total of 1136 MBTs will be in service with a further 201 Type 61s being phased out of service by then.

Description

No firm details of the armour type used in the construction of the Type 90 MBT have been released, but it is understood that extensive use of composite armour is made in both hull and turret over the frontal area of the tank. The Mitsubishi Steel Works and Kyoto Ceramic Company have been responsible for much of the armour development.

The layout of the Type 90 MBT is conventional with the driver's compartment at the front, turret in the centre and engine and transmission at the rear. The driver is seated at the front left and has a single piece hatch cover with three periscopes being provided for forward observation, the centre one can be replaced by a passive periscope for night driving.

Type 90 MBT fitted with front-mounted dozer blade

Type 90 MBT with front mounted dozer blade (Kensuke Ebata)

Type 90 MBT from rear showing exhaust outlets either side of hull (Kensuke Ebata)

The turret is similar to that of the German Leopard 2 MBT with vertical front, sides and rear with a bustle that extends well over the rear engine compartment.

The commander is seated on the right of the turret with the gunner on the left and the automatic loader being mounted in the turret bustle. Blowout panels are provided in the turret bustle roof. The commander is provided with vision blocks for all round observation and both commander and gunner have roof hatches.

Main armament comprises a 120 mm Rheinmetall smoothbore gun made under licence from Germany, which is fitted with a thermal sleeve, fume extractor and muzzle reference system. The 120 mm gun will fire APFSDS and HEAT-MP rounds with a semi-combustible cartridge case.

A 7.62 mm machine gun is mounted coaxial to left of the main armament and a 12.7 mm machine gun is mounted on the roof for anti-aircraft defence. This is mounted on the forward part of the turret roof so that it can be used by the commander or gunner. Mounted either side of the turret is a bank of three electrically operated smoke dischargers firing forwards.

The gunner has a roof mounted day/thermal night sight while the commander has a roof mounted sight that can be traversed through 360°. The computerised fire control system includes a laser rangefinder which is incorporated into the gunner's sight, digital ballistic computer, wind and cant sensors.

Either the commander or the gunner can lay the main armament with the commander having overriding control.

Suspension is of the hybrid type with torsion bars for the centre four road wheels and hydro-pneumatic units for the first and last road wheel stations, with six dual rubber tyred road wheels, drive sprocket at the rear, idler at the front and return rollers. The upper part of the track is covered by a lightweight skirt. The suspension enables the tank to change attitude in longitudinal direction only, for example on reverse slopes. Unlike the current Type 74, it will not be able to compensate for transverse changes.

Standard equipment includes an NBC system and a SIMRAD laser warning device mounted on the forward part of the turret roof.

Type 90 ARV

The Type 90 armoured recovery vehicle is based on the chassis of the Type 90 MBT and has a new superstructure with crane mounted at front right, dozer/stabiliser blade at front of hull and hydraulically operated winch. Additional details are given in *Jane's Military Vehicles and Logistics 1991-92*, page 25.

Status: In production for Japanese Ground Self-Defence Force with first production vehicles to be completed in 1992.

Manufacturer: Mitsubishi Heavy Industries, Sagamihara Limited
Mitsubishi Heavy Industries
5-1, Marunouchi 2-chome
Chiyoda-ku
Tokyo
Japan
Telephone: (212) 3111
Telex: J22282/J22443

SPECIFICATIONS		MAX ROAD SPEED	70 km/h	ARMAMENT	
CREW	3	MAX RANGE	300 km	main	1 × 120 mm gun
WEIGHT	50 000 kg	FORDING	2 m	coaxial	1 × 7.62 mm MG
POWER-TO-WEIGHT		GRADIENT	60%	anti-aircraft	1 × 12.7 mm MG
RATIO	30 hp/tonne	SIDE SLOPE	40%	SMOKE-LAYING	
GROUND PRESSURE	0.94 kg/cm^2	VERTICAL OBSTACLE	1 m	EQUIPMENT	2 × 3 smoke
LENGTH OVERALL	9.7 m	TRENCH	2.7 m		dischargers
LENGTH HULL	7.5 m	ENGINE	Mitsubishi M-12ZG	GUN CONTROL	
WIDTH OVERALL	3.4 m		10-cylinder water-	EQUIPMENT	
HEIGHT OVERALL	2.3 m		cooled diesel	Turret power control	electric/manual
GROUND CLEARANCE	0.45 m (variable		developing 1500 hp at	by commander	yes
	between 0.2 and 0.6 m		2400 rpm	by gunner	yes
	front and rear)	TRANSMISSION	automatic	Gun stabiliser	
LENGTH OF TRACK ON		SUSPENSION	torsion bar/hydro-	vertical	yes
GROUND	4.6 m		pneumatic	horizontal	yes

Type 74 Main Battle Tank

Development

Mitsubishi Heavy Industries and the Japanese Ground Self-Defence Force began project definition studies for a new MBT to succeed the Type 61 in 1962. To prove the basic concept, a number of test rigs were built and tested between 1964 and 1967. Construction of the first two prototypes began at the Maruko works of Mitsubishi Heavy Industries late in 1968 and was completed in September 1969. They were called the STB-1, and contained many features of other tanks under development at that time. The STB-1 had an automatic loader for the British 105 mm L7A1 gun which was subsequently manufactured in Japan by the Japan Iron Works Limited, with a 12.7 mm (0.50) Browning M2 HB machine gun mounted towards the rear of the turret which could be aimed and fired from within the tank.

The STB-1 was followed in 1971 by the STB-3, which had the automatic loader removed as it proved both too complex and too expensive. The remote controlled 12.7 mm (0.50) machine gun was also replaced by a more simple mount and the turret was slightly different in shape with a much longer bustle.

The first production contract was placed before the final model, the STB-6, appeared in 1973. The first Type 74 MBT was completed in September 1975 and by January 1980 some 225 had been built.

Under the Fourth Defence Plan, which ran from FY72 to 1976, 177 tanks were ordered.

Under the FY81 Mid-term Defence Programme Estimate (FY81 MDPE) which began in FY83, 246 Type 74 MBTs were ordered giving a total of 873 Type 74s in service by the end of 1991.

Recent procurement of the Type 74 MBT has been as follows:-

1980	60	1985	60
1981	72	1986	56
1982	72	1987	56
1983	60	1988	56
1984	60	1989	56

The Type 74 MBT has now been replaced in production by the Type 90 MBT.

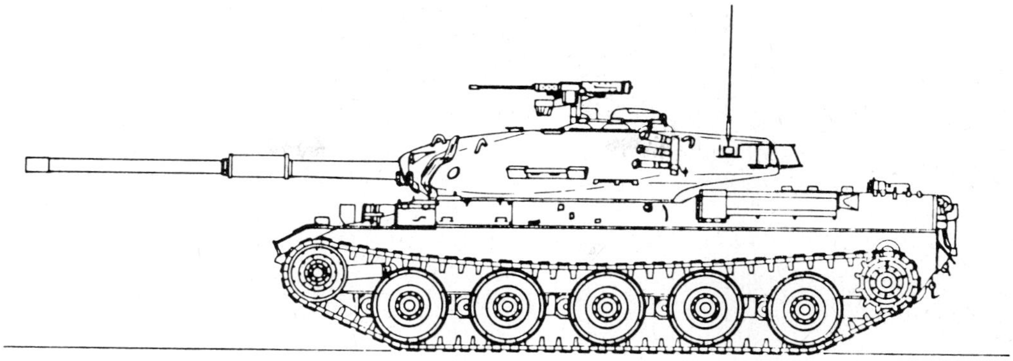

Type 74 MBT

Description

The hull of the Type 74 MBT is all welded and is divided into three compartments: driver's at the front, fighting in the centre and engine at the rear.

The driver is seated at the front of the vehicle on the left side and is provided with a single piece hatch cover that opens to the left. Three JM 17 Mod 2 periscopes are mounted forward of this hatch cover and an infra-red periscope can be mounted in the centre of the hatch cover and used in conjunction with the infra-red driving lights.

The turret is all cast with the commander and gunner seated on the right and the loader on the left. The commander's cupola can be traversed through 360° and has a single piece hatch cover that opens to the rear. There is a J3 infra-red periscope sight with an integral laser rangefinder and a magnification of ×1 and ×8 in the forward part of the commander's cupola and he also has five periscopes, two on either side and one to the rear. The gunner is seated forward of the commander and is provided with a J2 IR periscopic sight with a magnification of ×1 and ×8 in the turret roof and a telescope linked to the main armament. The loader has a single piece hatch cover that opens to the rear and a periscope which can be traversed through a full 360° mounted forward of his hatch. There is an external stowage basket at the rear of the turret.

The engine and transmission are mounted at the rear of the tank with the exhaust pipes and silencers mounted on the running boards to the rear of the turret. The maximum quoted road speed of the Type 74 is 53 km/h but reliable sources have indicated that its top speed is at least 60 km/h.

The suspension is hydro-pneumatic and can be adjusted to suit the type of terrain being crossed. There are five dual rubber tyred road wheels with the drive sprocket at the rear and the idler at the front. There are no track return rollers. The hydro-pneumatic suspension can be operated either by the commander or the driver and enables the tank to be inclined 6° forwards or backwards, 9° left or right and raised or lowered giving a minimum ground clearance of 200 mm and a maximum ground clearance of 650 mm.

Standard equipment includes infra-red driving lights, infra-red searchlight to the left of the main armament and an NBC system. In the white light mode the searchlight has a maximum range of 3000 m and in the infra-red mode a range of 1000 m. One Type 74 MBT in each company is fitted with a dozer blade.

Main armament consists of a 105 mm rifled tank gun (the British Royal Ordnance L7 series made under licence in Japan) which fires the standard range of ammunition including APDS-T and HESH-T. Early in 1984 it was announced that the Japanese Ground Self-Defence Force was to introduce the American 105 mm M735 APFSDS-T round for the Type 74 MBT with prime contractor in Japan to be Daikin Industry. Fourteen rounds of ready use ammunition are carried in the turret. Many Type 74 MBTs are now being fitted with a thermal sleeve for their 105 mm rifled guns. The gun has a drop block-breech mechanism and a new concentric recoil mechanism to reduce the volume of the upper part of the gun as well as reducing the frontal area of the turret. The fire control system includes a Nippon Electric laser rangefinder mounted in the commander's sight, which also provides inputs to the Mitsubishi Electric ballistic computer connected to the gunner's sight. Target range is fed to the computer automatically but trunnion tilt, barrel wear, ammunition type and temperature are fed in manually. The main armament is stabilised in both the vertical and horizontal planes.

A 7.62 mm Type 74 machine gun is mounted coaxially with the main armament and a 12.7 mm (0.50) Browning M2 HB machine gun is pintle-mounted in the centre of the turret forward of the commander's and loader's positions. This has an elevation of +60° and a depression of −10°. Three smoke dischargers are mounted either side of the turret.

Late in 1988 it was observed that a number of Type 74 MBTs had been fitted with a thermal sleeve for their 105 mm guns, passive thermal night vision equipment and armour on the top and forward part of the turret.

Variants

Type 78 Armoured Recovery Vehicle

The prototype of an ARV based on the chassis of the Type 74 MBT was completed in 1974 and standardised as the Type 78 ARV in 1978. In appearance it is very similar to the French AMX-30D and German

Type 74 MBT with infra-red/white light searchlight mounted to left of main armament removed. This Type 74 is also fitted with a thermal sleeve for its 105 mm rifled tank gun (Kensuke Ebata)

Type 74 MBT with 12.7 mm anti-aircraft machine gun (Ryuta Watanabe)

Type 87 twin 35 mm self-propelled anti-aircraft gun system with tracking and surveillance radars erected

Leopard 1 ARVs with the crew compartment at the front, hydraulically operated dozer blade at the front of the hull and a hydraulic crane on the right side of the hull. Details of the Type 78 ARV are given in *Jane's Military Vehicles and Logistics 1991-92*, page 26.

Type 87 Twin 35 mm SPAAG

This is essentially a Type 74 MBT chassis fitted with a new two-man power operated turret armed with twin 35 mm Oerlikon-Contraves 35 mm KDA cannon, tracking and surveillance radars. Prototype systems were designated the AWX but production systems are known as the Type 87 twin 35 mm self-propelled anti-aircraft gun system. The first production batch comprises a total of 12 vehicles.

Status: Production complete. In service with the Japanese Ground Self-Defence Force.

Manufacturer: Mitsubishi Heavy Industries, Sagamihara Limited
Mitsubishi Heavy Industries
5-1, Marunouchi 2-chome
Chiyoda-ku
Tokyo
Japan
Telephone: (212) 3111
Telex: J22282/J22443

Type 78 ARV with hydraulic crane in operating position (Kensuke Ebata)

SPECIFICATIONS		FUEL CAPACITY	950 l	SMOKE-LAYING	
CREW	4	MAX RANGE	300 km	EQUIPMENT	3 smoke dischargers
COMBAT WEIGHT	38 000 kg	FORDING	1 m		either side of turret
UNLOADED WEIGHT	36 300 kg	with preparation	2 m	AMMUNITION	
POWER-TO-WEIGHT		GRADIENT	60%	main	55
RATIO	18.94 hp/tonne	SIDE SLOPE	40%	coaxial	4500
GROUND PRESSURE	0.85 kg/cm^2	VERTICAL OBSTACLE	1 m	anti-aircraft	660
LENGTH GUN FORWARDS	9.41 m	TRENCH	2.7 m	GUN CONTROL	
LENGTH GUN REAR	7.84 m	ENGINE	Mitsubishi 10ZF Type	EQUIPMENT	
LENGTH HULL	6.7 m		22 WT 10-cylinder	Turret power control	electric/manual
WIDTH	3.18 m		air-cooled diesel	by commander	yes
HEIGHT			developing 720 hp at	by gunner	yes
including A/A MG	2.67 m (with 0.65 m		2200 rpm	Turret traverse	360°
	ground clearance)	TRANSMISSION	Mitsubishi MT75A	Gun elevation	+9° (+15°
to turret top	2.48 m (with 0.65 m		manual with 6 forward		using suspension)
	ground clearance)		and 1 reverse gears	Gun depression	-6° (-12°
to turret top	2.03 m (with 0.2 m	STEERING	double differential		using suspension)
	ground clearance)	SUSPENSION	hydro-pneumatic,	Gun stabiliser	
GROUND CLEARANCE	adjustable from		variable	vertical	yes
	0.2 - 0.65 m	ELECTRICAL SYSTEM	24 V	horizontal	yes
TRACK WIDTH	550 mm	ARMAMENT		Elevation quadrant	yes
LENGTH OF TRACK		main	1 × 105 mm gun	Traverse indicator	yes
ON GROUND	4 m	coaxial	1 × 7.62 mm MG		
MAX ROAD SPEED	53 km/h	anti-aircraft	1 × 12.7 mm MG		

Type 61 Main Battle Tank

Development

Design work on the first Japanese post-war tank began under the direction of the Ground Armaments Directorate at the Technical Research and Development Headquarters of the Japanese Self-Defence Force in 1954. The first four prototypes were completed in 1957 and comprised two model ST-A1s and two model ST-A2s.

The four prototypes were followed by two ST-A3 and 10 ST-A4 tanks which were almost identical to production tanks. In April 1961 the tank was standardised as the Type 61 MBT and first production tanks were completed in 1962. Production was very slow initially: 10 tanks were produced in 1962 with a similar number the following year, 20 in 1964, 33 in 1965 and 27 in 1966. The 100th Type 61 was completed in November 1966. By late 1970 250 tanks had been completed and it is believed that total production amounted to about 560 units. The Type 61 is still used by the Japanese Ground Self-Defence Force but has been supplemented by the Type 74 MBT, also manufactured by Mitsubishi.

As of 1983 the Japanese Ground Self-Defence Force had 559 Type 61 MBTs on strength. The first two Type 61s were retired in December 1984, 36 Type 61s were retired in 1985, a further 25 in 1986 and since then additional vehicles have been phased out of front line service. Late in 1990 it was announced that by the end of FY96 the MBT force will be 1136 vehicles, 69 less than the current strength with a total of 201 Type 61 MBTs to be retired in this period.

Description

The all-welded steel hull of the Type 61 is divided into three compartments: driver's at the front, fighting in the centre and the engine at the rear.

Type 61 MBT with commander's hatch open (Keiichi Nogi)

The driver is seated at the front of the hull on the right side and is provided with a single piece hatch cover that opens to the right. Three periscopes are mounted forward of the hatch.

The turret is cast with an overhanging bustle similar to that of the M47 tank and a light sheet steel stowage box at the very rear.

The commander and gunner are seated on the right of the turret with the loader seated on the left. The commander's domed shaped cupola, which can be traversed through 360°, has a single piece hatch cover in its rear, which opens backwards. The cupola has four vision blocks and mounted in the lower part of the cupola is a coincidence rangefinder with a magnification of ×12. The commander is also provided with a periscopic sight with a magnification of ×7 mounted in the forward part of the cupola roof. The gunner is seated forward of the commander and is provided with a telescopic sight with a magnification of ×6 and a periscope with a magnification of ×4. The loader is seated on the left side of the turret and has a single piece hatch cover opening to the rear. A single periscope is mounted in the turret roof in front of the hatch cover.

The engine is mounted at the rear of the hull and is cooled by two axial-flow cooling fans mounted over the top of the engine.

The torsion bar suspension consists of six dual rubber tyred road wheels with the drive sprocket at the front and the idler at the rear. There are three track return rollers and the first, second, fifth and sixth road wheel stations are equipped with hydraulic shock absorbers.

Main armament is a Type 61 90 mm rifled tank gun which is manufactured by the Japan Steel Works. This is fitted with a fume extractor and a T-type muzzle brake. Types of ammunition fired include APC with the projectile weighing 10.95 kg and HE with the projectile weighing 10.62 kg. Mounted coaxially with the main armament is a 7.62 mm (0.30) Browning M1919A4 machine gun while a 12.7 mm (0.50) Browning M2 HB machine gun is mounted on the commander's cupola for anti-aircraft use. This can be aimed and fired from within the tank and some vehicles have had a shield provided for this weapon. More recently many Type 61s have been fitted with a bank of three electrically operated smoke dischargers either side of the turret rear.

The Type 61 has no NBC system or deep wading equipment. Some models have been fitted with an infra-red searchlight to the left of the main armament and infra-red driving lights.

Type 61 MBTs clearly showing T-type muzzle brake on 90 mm gun (Ryuta Watanabe)

Variants

Armoured Vehicle-launched Type 67 Bridge

This is the basic MBT with its turret removed and replaced by a scissors bridge laid to the front of the vehicle. When opened out the bridge is 12 m long and will span a gap of up to 10 m. It has a maximum capacity of 40 000 kg. Launching time is between three and five minutes and the bridge can be picked up from either end. The Type 67 weighs 36 700 kg with the bridge and has a crew of three. Full details of the Type 67 AVLB are given in *Jane's Military Vehicles and Logistics 1991-92*, page 85.

Armoured Engineer Vehicle Type 67

This vehicle has been designed for engineer roles on the battlefield and is equipped with a dozer blade at the front and a light crane. The Type 67 weighs 35 000 kg and has a crew of four. Available details are given in *Jane's Military Vehicles and Logistics 1991-92*, page 7.

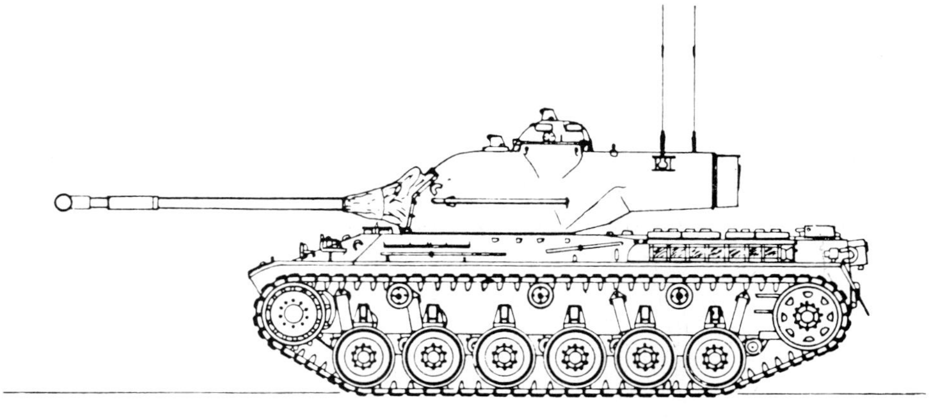

Type 61 MBT without 12.7 mm machine gun on commander's cupola

Type 67 armoured bridgelayer laying its scissors bridge over front of vehicle (K Nogi)

Type 70 armoured recovery vehicle with A-frame to rear (Kensuke Ebata)

Armoured Recovery Vehicle Type 70

The Type 70 ARV is the basic MBT with its turret removed and replaced by a new superstructure with a winch mounted in the bustle at the rear. An A-frame is pivoted towards the front of the hull; when not in use it lies back over the rear of the hull. A hydraulically operated dozer blade mounted at the front of the hull is used to stabilise the vehicle during recovery operations, or when the A-frame is in use. The Type 70 weighs 35 000 kg and has a crew of four. Armament consists of an 81 mm mortar which can be mounted on the glacis plate and a 12.7 mm and a 7.62 mm machine gun. Details of the Type 70 ARV are given in *Jane's Military Vehicles and Logistics 1991-92*, page 26.

Type 61 Training Tank

For training purposes some Type 61 MBTs have been fitted with wooden mine sweeping equipment at the front of the hull to resemble Soviet T-54/T-55 mine clearing tanks.

Status: Production complete. In service only with the Japanese Ground Self-Defence Force.

Manufacturer: Mitsubishi Heavy Industries Limited
Maruko
Tokyo
Mitsubishi Heavy Industries
5-1, Marunouchi 2-chome
Chiyoda-ku
Tokyo
Japan
Telephone: (212) 3111
Telex: J22282/J22443

SPECIFICATIONS	
CREW	4
COMBAT WEIGHT	35 000 kg
UNLOADED WEIGHT	33 500 kg
POWER-TO-WEIGHT RATIO	17.14 hp/tonne
GROUND PRESSURE	0.95 kg/cm^2
LENGTH GUN FORWARDS	8.19 m
LENGTH HULL	6.3 m
WIDTH	2.95 m
HEIGHT	
including AA MG	3.16 m
hull turret roof	2.49 m
GROUND CLEARANCE	0.4 m
TRACK WIDTH	500 mm
LENGTH OF TRACK ON GROUND	3.7 m
MAX ROAD SPEED	45 km/h
MAX RANGE	200 km
FORDING	0.99 m
GRADIENT	60%
VERTICAL OBSTACLE	0.685 m

TRENCH	2.489 m
TURNING RADIUS	10 m
ENGINE	Mitsubishi Type 12 HM 21 WT, V-12, 4-cycle direct-injection, turbo-charged diesel developing 600 hp at 2100 rpm
AUXILIARY ENGINE	none
TRANSMISSION	mechanical, 5 forward and 1 reverse speeds with a 2-speed auxiliary reduction unit
STEERING	controlled differential
SUSPENSION	torsion bar
ELECTRICAL SYSTEM	24 V
BATTERIES	4 × 12 V, 200 Ah
ARMAMENT	
main	1 × 90 mm gun
coaxial	1 × 7.62 mm MG
anti-aircraft	1 × 12.7 mm MG

SMOKE-LAYING EQUIPMENT	2 × 3 smoke dischargers
GUN CONTROL EQUIPMENT	
Turret power control	hydraulic/manual
by commander	yes
by gunner	yes
Turret traverse	360°
Gun stabilisation	
vertical	no
horizontal	no
ARMOUR	
Hull glacis	46 mm
Hull sides	25 mm
Hull rear	15 mm
Turret front	64 mm

KOREA, NORTH

North Korean AFV production

By 1991 the North Korean tank fleet is believed to consist of 200 old T-34/85 tanks which are used for training, 2000 T-54/T-55 tanks and 1500 T-62 MBTs plus up to 175 Type 59 MBTs.

In addition to producing Soviet designed T-62 MBTs under licence for home and export markets, with the latter known to include Iran, it is believed that production of the BMP-1, or a modified version of this vehicle, is also being undertaken in North Korea. North Korean production of the T-62 is believed to have commenced in the late 1970s and was completed in the early to mid-1980s. It is different in a number of areas from the original Soviet built vehicles and some items were not interchangeable.

KOREA, SOUTH

Type 88 Main Battle Tank

Development

Following proposals from a number of armoured fighting vehicle manufacturers, in 1980 the South Korean Government selected the Land Systems Division of General Dynamics (then Chrysler Defense) to design and build two prototypes of a new MBT to meet its own specific requirements. The first of two prototypes of the XK-1 MBT was completed in 1983, the Automotive Test Rig (ATR). It was shipped to Aberdeen Proving Grounds in November 1983 for automotive performance, endurance and reliability testing. The ATR was a fully payloaded tank fitted with a non-operational turret.

The second prototype, called the Fire Control Test Rig (FCTR) was rolled out at a ceremony at Selfridge Air National Guard base in December 1983 and shipped to Aberdeen Proving Grounds in February 1984 to begin fire control tests.

Production of the XK-1 began in 1984 and first production vehicles were completed in 1985. The vehicle was subsequently type classified as the Type 88 MBT and the tank made its first appearance in September 1987, by which time several battalions had been equipped with it. The first production batch of 210 vehicles was completed in 1987 with the second batch to consist of 325 vehicles. The total South Korean Type 88 requirement is now understood to be for 833 vehicles.

Many key components of the Type 88, such as the CDC Ballistic Computer, MTU diesel engine, ZF Transmission and SFIM roof-mounted sight are now being manufactured in-country.

Description

The layout of the Type 88 MBT is conventional with the driver's compartment at the front, fighting compartment in the centre and engine and transmission at the rear.

Over the vulnerable frontal arc of the Type 88 MBT, including the nose, turret front and sides, advanced armour of the Chobham type is fitted which is manufactured in the United States; this provides protection from both kinetic and chemical energy attack.

The driver is seated front left and is provided with a single piece hatch cover pivoted on the left that lifts upwards to open. This has three integral day periscopes, the centre one of which can be replaced by a passive night driving periscope. The commander is seated on the right of the turret with the gunner below and forward of his position, and the loader on the left.

The commander has a French SFIM two-axis independent stabilised panoramic sight which has a magnification of ×3 and ×10, periscopes for all-round observation and a single piece hatch cover that opens to the rear. This is manufactured in South Korea by Samsung Aerospace.

The gunner's two-axis stabilised day/night sight device incorporates a laser rangefinder and thermal imaging system that is similar to that installed in the M1A1 MBT and has magnifications of ×3 and ×10. Hughes Aircraft Co of the United States delivered these at the rate of 10 systems per month. The gunner's articulated auxiliary sight is provided by the Electro-Optical Division of the Kollmorgen Corporation which was awarded a contract for 160 sights early in 1987. This has a magnification of ×8. In October 1989 Kollmorgen stated that it had been awarded a further contract worth $2.5 million for a further 315 auxiliary sights and that 536 had already been supplied under previous contracts.

The first batch of 210 vehicles were fitted with a Gunner's Primary Sight (GPS) designed by Hughes Aircraft but the second batch has a Gunner's Primary Tank Thermal Sight (GPTTS) designed by Texas Instruments which incorporates a CO^2 laser rangefinder. Texas Instruments delivers the components for the GPTTS to Samsung Aerospace for final assembly in South Korea. In April 1989 Hughes Aircraft announced that it had been awarded a $42.9 million contract to produce 156 Gunner's Primary Sight systems to Hyundai Precision and Industry Company Limited in competition with Texas Instruments. This contract calls for Hughes to deliver 156 subsystems over a 14 month period with an option on a further 250 subsystems.

Turret drive and weapon elevation is electro-hydraulic with manual controls for emergency use. Stabilisation is provided in both elevation and traverse.

The fire control system includes a digital ballistic computer with a number of sensors including cross-wind, and allows the tank to engage both stationary and moving targets while it is stationary or moving itself.

Main armament comprises a US designed 105 mm M68A1 rifled tank gun which is also fitted on many South Korean M48A5 MBTs and for which ammunition is made in South Korea, including APFSDS. The barrel has a fume extractor, thermal sleeve and a muzzle reference system.

A 7.62 mm machine gun is mounted coaxial with the main armament, the loader has a 7.62 mm roof mounted machine gun while the gunner has a 12.7 mm roof mounted machine gun. Mounted either side of the forward part of the turret is a bank of six electrically operated smoke dischargers.

An unusual feature of the Type 88 is the hybrid suspension system with improved torsion bars in the centre and hydro-pneumatic suspension units at each end. This allows the tank to kneel so that the main armament can be depressed to -10°. The remote track adjusting system has two mode settings.

The drive sprocket is at the rear, idler at the front, with six dual rubber tyred road wheels, track return rollers with the upper part of the track being covered by an armoured skirt.

The rear mounted powerpack consists of a German MTU 1200 hp diesel coupled to a German ZF LSG 3000 fully automatic transmission with acceleration from 0 to 32 km/h in 9.4 seconds.

Standard equipment includes passive image intensification periscope for the driver, hydraulic bilge pump, heater, automatic Halon fire detection and suppression system for both crew and engine compartments and an AN/VRC-12 radio and an AN/VIC-1 intercom system.

The NBC system consists of an M8A1 alarm system and a M13A1 gas particular filter.

Armoured Vehicle Launched Bridge

Early in 1989 Vickers Defence Systems of the UK was awarded a contract by Hyundai Precision and Ind Co to design and build the prototype of an Armoured Vehicle Launched Bridge (AVLB) system and bridge.

One of the prototypes of the Type 88 MBT clearly showing its adjustable suspension being used to lower front of vehicle

One of the prototypes of the Type 88 MBT with covers of the gunner's sight in the open position

Type 88 MBT showing roof-mounted 7.62 mm and 12.7 mm machine guns and commander's roof-mounted SFIM sight

Type 88 MBT of the South Korean Army

The AVLB will launch a scissors type bridge similar to the British No 8 tank bridge which is carried and launched over the front of a Chieftain. this bridge has an overall length of 22 m and can span a gap of up to 20 m depending on the firmness of the bank.

The bridge and its associated launching system was built at Vickers Defence Systems, Newcastle-upon-Tyne facility and shipped to South Korea in 1990 where it was integrated with the chassis, based on the Type 88 MBT, designed and built by Hyundai Precision and Ind Co.

Armoured Recovery Vehicle

Hyundai Precision and Ind Co has been developing an ARV based on the chassis of the Type 88 in co-operation with Krupp-MaK of Germany. The recovery system of this vehicle is similar to that of the Bergepanzer 3 of the West German Army.

Status: In production. In service with South Korean Army.

Manufacturer: Hyundai Precision & Ind Co Ltd at Changwon.
Enquiries to Hyundai Precision & Ind Co Ltd.
140-2, Gye-dong, Chongro-Ku
Seoul
South Korea
Telephone: (82-2) 746 1114
Telex: HDPIC K23238
Fax: (82-2) 741 2248

SPECIFICATIONS

CREW	4
COMBAT WEIGHT	51 000 kg
POWER-TO-WEIGHT RATIO	23.5 hp/tonne
GROUND PRESSURE	0.86 kg/cm^2
LENGTH GUN FORWARDS	9.672 m
LENGTH HULL	7.477 m
HEIGHT TO TURRET TOP	2.248 m
WIDTH OVER SKIRTS	3.594 m
GROUND CLEARANCE	0.46 m
MAX ROAD SPEED	65 km/h
AVERAGE CROSS-COUNTRY SPEED	40 km/h
SPEED ON 60% GRADIENT	8 km/h
CRUISING RANGE	500 km
FORDING WITHOUT KIT	1.2 m
FORDING WITH KIT	1.8 m
GRADIENT	60%
VERTICAL OBSTACLE	1 m
TRENCH	2.74 m

ENGINE	MTU 871 Ka-501 diesel developing 1200 hp at 2600 rpm
TRANSMISSION	ZF LSG 3000 automatic with mechanical lock up, 4 forward and 2 reverse gears
BRAKING	hydraulic and mechanical, two circuits
SUSPENSION	torsion bar/hydro-pneumatic
ARMAMENT	
main	1 × 105 mm gun
coaxial	1 × 7.62 mm MG
commander	1 × 12.7 mm MG
loader	1 × 7.62 mm MG
SMOKE-LAYING EQUIPMENT	2 × 6 smoke dischargers

AMMUNITION	
105 mm	47
12.7 mm	2000
7.62 mm coaxial	7200
7.62 mm loaders	1400
GUN CONTROL EQUIPMENT	
Turret power control	electro-hydraulic/manual
by commander	yes
by gunner	yes
Commander's fire control overide	yes
Turret traverse	360°
Gun elevation/depression	+20°/−10°
Gun stabiliser	
vertical	yes
horizontal	yes

South Korean M48 Rebuild Programme

Many of the M48s have been brought up to M48A5 standard and are now fitted with a 105 mm M68 gun, infra-red driving lights, infra-red searchlight over the main armament, smoke dischargers either side of the turret, side skirts that are hinged at the top to allow access to the suspension and tracks for maintenance and replacement of the original petrol engine by a Continental AVDS-1790-2D diesel developing 750 hp at 2400 rpm.

Prime contractor for the South Korean M48 conversion programme was Hyundai Precision and Ind Co with major subcontractors as follows:

JE-IL Precision Co Ltd
Hull and turret electrical kit.

Kangwon Industries Ltd
Top engine deck grille, elbows of air cleaner, commander's cupola, hull welding and machining, engine and transmission mounts.

Goldstar Cable Co Ltd
Track shoes.

Ssangyong Heavy Industries Co Ltd
Continental AVDS-1790-2D diesel engine.

Tong Myung Heavy Industries Co Ltd
Turret and gun control equipment.

Pyung Hwa Industrial Co Ltd
Road wheel assembly.

The USA supplied South Korea with between 500 and 600 M47 tanks but only about 300 of these are now believed to be operational. About 1050 M48 tanks have also been supplied, of which 597 at least have been brought up to M48A5 standard. This upgrading programme is now complete.

Manufacturer: Hyundai Precision & Ind Co Ltd at Changwon.
Enquiries to Hyundai Precision & Ind Co Ltd.
140-2, Gye-dong, Chongro-Ku
Seoul
South Korea
Telephone: (82-2) 746 1114
Telex: HDPIC K23238
Fax: (82-2) 741 2248

Close-up of South Korean built M48 cupola showing externally mounted 7.62 mm M60 series machine gun

M48A5 of South Korean Army fitted with searchlight over main armament, 105 mm M68 gun and side skirts (via Kensuke Ebata)

M48s of South Korean Army (via Kensuke Ebata)

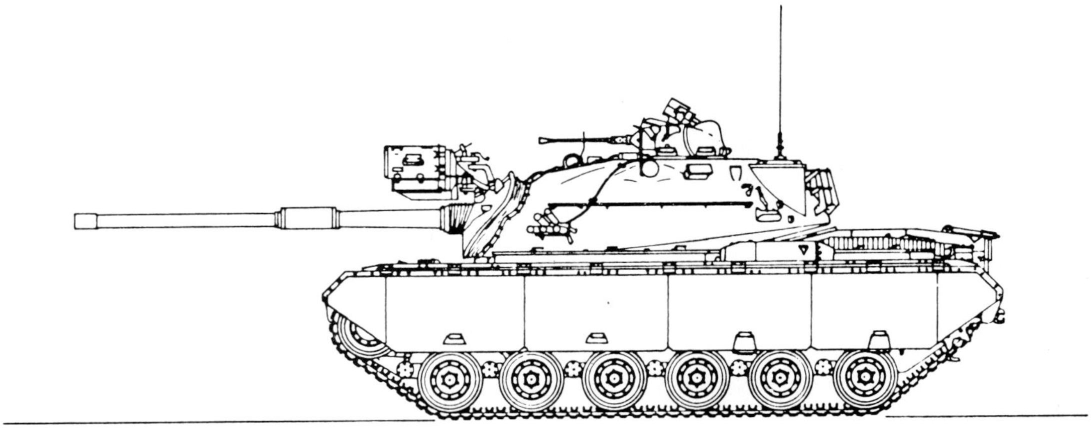

Rebuilt M48 of ROK Army with 105 mm gun and side skirts

PAKISTAN

Pakistan MBT modernisation plans

Early in 1991 the Pakistani tank fleet consisted of about 1200 Type 59s, 40/50 T-54/T-55s, 150 M47Ms and 345 M48A5s. The first 145 M48s were rebuilt to the M48A5 configuration between 1977 and 1979 and a further 100 M48A5s were ordered and delivered from the USA in 1982 at a cost of $71 million, another 100 were ordered in 1984.

With Chinese assistance, Pakistan has established the Heavy Rebuild Factory at Taxila. The facility covers 48 acres and is capable of rebuilding tanks and powerpacks as well as manufacturing individual optical and electrical components.

The Engine Rebuild Plant was completed at Taxila first, the plant for the rebuilding of the hull was completed shortly afterwards. The facility was fully operational by 1979/80 and normally rebuilds about 100 Type 59 and 250 Type 59 tank engines a year and by late 1989 had rebuilt 1000 Type 59 tanks and 2000 Type 59 engines. Of the 11 000 components used in the rebuild of the Type 59 some 8000 are now manufactured locally.

More recently the Heavy Rebuild Factory (HRF) at Taxila was renamed Heavy Defence Industries (HDI).

The first prototype of the Khalid MBT (also believed to be known as the P90) was completed in 1991. It is understood that this is based on a Chinese Type 69-II MBT but with a number of modifications to meet the requirements of the Pakistani Army including a 105 mm rifled tank gun with fume extractor and thermal sleeve, improved armour protection, new computerised fire control system incorporating a laser rangefinder and improved night vision equipment. In addition it will have a fire/explosion detection/suppression system for its fighting compartment.

Although most sources have indicated that this will be based on a Type 69-II MBT it is quite possible that it will use the Type 80 or Type 85 MBTs as these have a number of improvements over the Type 69.

The total Pakistani requirement could be as many as 2000 vehicles with production running at between 150 and 200 vehicles a year.

The Military Vehicle Research and Development Establishment has developed to the prototype stage the M47M armoured vehicle launched bridge system. This lays a scissors bridge with a maximum span of 21.4 m. Full details are given in *Jane's Military Vehicles and Logistics 1991-92*, page 86.

In August 1989 an agreement was signed between Pakistan and General Dynamics, Land Systems Division, to establish an overhaul facility in Pakistan for M47 and M48 tanks, M109 and M110 self-propelled howitzers, M113 series APCs and M88A1 armoured recovery vehicles. Pakistan is also manufacturing the M113A2 APC under licence and may in the future build the FMC Armoured Infantry Fighting Vehicle.

Iran had also been overhauling Pakistani M47 and M48 tanks before the downfall of the Shah.

In July 1987 the Pakistani Army carried out firing trials with four Chinese-built Type 59 MBTs upgraded by Royal Ordnance of the UK, CRIS (Conception, Réalisation et Implantation des Systèmes), France, NORINCO of China and Cadillac Gage Textron of the USA.

Each manufacturer upgraded the Type 59 MBT in five key areas. The existing 100 mm rifled tank gun was replaced by a 105 mm rifled tank gun which fires standard NATO ammunition including APFSDS-T rounds, now manufactured by Pakistan Ordnance Factories.

Thermal sights have been fitted for the commander and gunner in place of the first generation infra-red devices. The fighting compartment has been fitted with a fire/explosion detection and suppression system which the current Type 59 MBT lacks.

A computerised fire control system incorporating a laser rangefinder has been installed and a stabilisation system fitted to allow stationary and moving targets to be engaged with a high probability of a first round hit.

In early 1988 another series of firing trials was carried out by the three finalists, Royal Ordnance, Cadillac Gage Textron and NORINCO but as of March 1991 no company had been selected. It is expected that up to 400 Type 59s at least will be brought up to this standard, the work being carried out at the Taxila Heavy Rebuild Factory.

In 1987 Pakistan also carried out trials of three M48A5 105 mm armed tanks with a new fire control system provided by Cadillac Gage Textron, Kollsman Instrument Company and Honeywell, all of the USA.

In mid-1987, MECAR of Belgium was awarded a contract worth nearly BFr500 million ($12.9 million) for the supply of APFSDS-T ammunition for the 90 mm M47/M48 and 100 mm Type 59/T-54/T-55 tanks. It is understood that this order is for Pakistan.

In 1988 two 120 mm armed M1A1s were trialled in Pakistan which has indicated that it has a requirement for up to 400 MBTs of the M1A1 type armed with a 120 mm gun, but in the short term at least no purchase or local manufacture of the M1A1 is expected to take place in Pakistan.

M48A5 MBT of the Pakistani Army with 105 mm rifled tank gun

Type 69 MBT of the Pakistani Army with 105 mm rifled tank gun, smoke dischargers and turret stowage basket

POLAND

Polish MBT production

Poland produced the T-34/85 from 1951 to 1956 and from 1956 to 1964 produced the T-54 and T-54M at Labedy. This was followed in production by the T-55 which has been built both for the Polish Army and for export. Many of the Polish vehicles were fitted with a large stowage box mounted on the left side of the turret. Between 1976 and 1980 Czechoslovakia and Poland produced a total of 800 MBTs a year. According to the 1983 edition of *Soviet Military Power*, published by the US Department of Defense, the T-55 remained in production in 1982 in the Warsaw Pact (but not in the USSR) at the rate of 600 tanks a year. Czechoslovakia and Poland are the only two known manufacturers outside the USSR.

In 1979 the tank plant at Katowice started to tool up for production of the T-72 MBT but first production vehicles were not completed until 1981. According to the US Department of Defense Czechoslovakia and Poland produced a total of 20 T-72s in 1981 and 100 in 1982.

In early 1985 the US Department of Defense issued revised production figures for non-Soviet Warsaw Pact countries. These were 700 tanks in 1980, 500 in 1981, 600 in 1982, 550 in 1983 and 450 in 1984. In 1985 production of tanks in Czechoslovakia and Poland increased to 700 vehicles. Recent production figures are 700 in 1986, 750 in 1987 and 700 in 1988. Since then figures have declined and it is understood that while production of the T-72 continues in Czechoslovakia, mainly for the export market, production in Poland has now been completed.

Polish-built T-55A(M) fitted with 12.7 mm DShKM anti-aircraft machine gun carrying out dozer operations. Note large stowage box on left side of turret

ROMANIA

TR-77 Main Battle Tank

This was first seen in public during a parade held in Bucharest in 1977 and is based on the T-55 MBT chassis. The modifications are extensive and include an extended hull rear to accommodate the new 600 hp diesel engine, new suspension with each side now having six road wheels, idler at the front, drive sprocket at the rear, two return rollers and a skirt. Mounted on the roof is a 12.7 mm anti-aircraft machine gun with some of the ammunition boxes being stowed externally on the turret sides. The TR-77 has also been referred to as the TR-580 or the M-1977, with the latter being the US Military designation as this was the year it was first seen in public.

Romanian TR-77 MBTs on parade

SPECIFICATIONS

as per T-55 except:-

COMBAT WEIGHT	40 000 kg
LENGTH	9.25 m
WIDTH	3.3 m
HEIGHT	2.4 m
MAXIMUM ROAD SPEED	65 km/h
RANGE	
(internal tanks)	400 km
(external tanks)	600 km
GRADIENT	60%
TRENCH	2.7 m
VERTICAL OBSTACLE	0.8 m

Status: Production complete. In service with Romanian Army.

Manufacturer: Romanian state factories.

TR-77 MBT from rear showing MG ammunition stowage boxes on turret sides and side skirts

TR-80 Main Battle Tank

This is believed to be a further development of the earlier TR-77 and has also been referred to as the TR-800 or the M-1978, the latter being a US military designation based on the year that it was first seen in public.

In appearance the TR-80 is similar to the TR-77 but its 100 mm gun has a thermal sleeve, a Chinese type laser rangefinder is mounted externally over the rear of the 100 mm gun and the fire control system includes a digital ballistic computer.

Status: Production probably complete. In service with Romanian Army.

Manufacturer: Romanian state factories.

T-55A Upgrade Package

Romania is now offering a package to upgrade existing T-55A MBTs and parts of this may have come from the TR-77 and TR-80 MBTs. The modifications are extensive and include 620 hp diesel engine, new driver's seat, installation of a Ciclop fire control system, new suspension and track, thermal sleeve for 100 mm gun, installation of smoke grenade launcher system, new anti-aircraft machine gun, R10-TM regular relay assembly, VKU 330.4 collector ring assembly, OU-3G infra-red headlamp installed, L-2G infra-red weapon sight assembly installed, 7.62 mm PKT machine gun, cleaning system for vision devices, main clutch control system modernised, steering gear system modernised, anti-radiation shielding added, R 1231 radio installed, HEAT screens fitted, anti-napalm and shielding installed.

Status: Production as required. Believed to be incorporated into some Romanian MBTs and some Iraqi MBTs.

Manufacturer: Romanian state factories.

Romanian Army T-55 fitted with locally built D-5M mine clearing rollers

TR-125 Main Battle Tank

In 1989 it was disclosed that Romania was manufacturing an improved version of the Soviet T-72 series MBT under the designation of the TR-125 which has a number of significant differences to the original Soviet vehicle which entered production as far back as 1971-72.

The T-72 weighs 41 tonnes whereas the TR-125 weighs 48 tonnes. The additional weight may be due to increased armour protection over the frontal arc which is also a feature of more recent T-72s built in the USSR.

The TR-125 has seven road wheels of a different design, a new one-piece skirt and is powered by a more powerful 880 hp diesel engine which gives a better power-to-weight ratio.

Maximum road speed is quoted as 60 km/h with an operational range of 450 to 540 km and a fuel consumption of 285/325 litres/100 km.

Armament of the TR-125 is identical to the Soviet T-72 and eight forward firing smoke dischargers are mounted on the left side of the turret.

Status: In production. In service with Romanian Army.

Manufacturer: Romanian state factories.

Romanian TR-125 MBT showing new roadwheels and skirts

SPAIN

Spanish AMX-30 Tank Production

Following delivery of the first 19 vehicles from France, Spain was building the AMX-30 MBT under licence from France from 1974, initially from components supplied by France but gradually the Spanish content increased until about 85 per cent of the tank was built in Spain. A total of 19 tanks were supplied direct from France plus 10 AMX-30D ARVs. It is estimated that five were completed in 1974, 21 in 1975, 52 in 1976, 66 in 1977 and at least 80 in 1978. No funds were allocated for AMX-30 production in 1981 and final deliveries to the Spanish Army took place in 1983-84 after 280 tanks had been completed. The Spanish tanks are almost identical to the French models but have a MG3S 7.62 mm machine gun for the commander and other minor differences. Production of the AMX-30 was undertaken by the Sevilla Artillery Factory of SANTA BARBARA, Empresa Nacional de Optica (ENOSA) providing the optical equipment, and Experiencias Industriales SA (EISA) the electrical and hydraulic equipment.

In 1980 Spain purchased the entire licence from GIAT in order to continue development itself.

Modernised AMX-30

In 1984 SANTA BARBARA announced that it had successfully built and tested a modernised AMX-30. Main improvements can be summarised as follows: the installation of a new powerpack comprising a General Motors Detroit Diesel engine developing 920 hp; Allison Transmission Division of General Motors CD-850-6A automatic transmission and an FAS cooling system; new suspension and tracks; new fire control system; air-conditioning system; explosion suppression system; protective side skirts and a driving tachograph.

In 1984 Spain signed a contract with Euromissile for the supply of 18 Roland 2 SAM systems and 414 missiles. Much of this work is being carried out in Spain including manufacture of the AMX-30R launcher vehicle. The first Spanish Roland SAM system on an AMX-30 chassis was handed over to the Spanish Army late in 1988.

Early in 1987 the Spanish Army decided to rebuild 150 of its AMX-30s with a new powerpack consisting of a German MB 883 MTU 850 hp diesel engine coupled to a ZF LSG 3000 fully automatic transmission. The actual conversion work will be carried out by SANTA BARBARA,

Spanish built AMX-30 MBT from front

the MTU diesel engines will be built under licence in Spain by Empresa Nacional Bazan.

The AMX-30 will also be modernised in the future by a new fire control system including a laser rangefinder and possibly explosive reactive armour. The fire control system to be fitted will be a locally built Hughes type Mk 9A/D with laser rangefinder.

Spanish Tank Fleet

Early in 1985 the Spanish Army stated that its operational tank fleet consisted of 127 M41 light tanks, 375 M47, 164 M48 and 299 AMX-30 MBTs. In 1986 the Spanish Army stated that it had a total of 958 MBTs which would fall to 907 in 1994 and remain at this figure to the year 2000.

Talbot Main Battle Tank Conversions

In 1951 Barreiros Diesel SA was formed for the conversion of petrol engines to diesel operation. It later moved into manufacture of its own diesel engines followed by trucks and other vehicles. In 1965 it formed a partnership with Chrysler Corporation of the United States whose defence division was manufacturing M60 MBTs as well as supplying spare parts for the M48 MBT. In the 1960s the company built trucks for the Spanish Army and in the 1970s started to overhaul and modernise M47 and M48 tanks and M113 series armoured personnel carriers for the Spanish Army. Talbot also repairs engines and transmissions for AFVs as well as manufacturing wheels and tracks and other suspension items. In 1978 the company went into partnership with PSA and its name was changed to Talbot SA. The main plant in Spain is at Villaverde near Madrid employing around 14 000 people. The plant has a special military division which employs over 400 people and can convert 12 to 15 MBTs a month on a single shift basis, though this figure could easily be doubled. It has a separate line for the complete overhaul of the M113 APC and its many variants, including the mortar carrier. MBT conversions currently offered by Talbot are listed below with a résumé of army/marine conversions.

M47E1

Basic M47 tank fitted with AVDS-1790-2A 12-cylinder diesel engine coupled to CD-850-6A transmission, bow machine gun removed, Cadillac Gage Textron turret and weapon control system installed, coaxial machine gun is now an MG42 and commander's machine gun the Browning 12.7 mm M2 HB, engine exhaust smoke laying system installed and four smoke dischargers mounted either side of turret. Retains 90 mm gun.

M47E1 MBT with standard 90 mm gun and turret traversed to rear

M47E2

Fitted with AVDS-1790-2D engine, CD-850-6A transmission, 105 mm Rh-105 30 gun, TEESS engine exhaust smoke generating system and smoke dischargers. The driver has an AN/VVS-2 passive night driving periscope.

M48A3E

M48A3 with Continental AVDS-1790-2A diesel engine coupled to CD-850-6A transmission, Cadillac Gage Textron turret and weapon control system installed, TEESS engine exhaust smoke laying system installed and four smoke dischargers mounted either side of turret. Retains 90 mm gun.

M48A5E

Almost same as American M48A5, has Continental AVDS-1790-2A 12-cylinder diesel engine coupled to CD-850-6A transmission, main armament can be 105 mm M68 or Rh-105 30, standard fire control system.

M48A5E1

As M48A5E but has AVDS-1790-2D 12-cylinder diesel and a new fire control system. Gunner has M35 periscope, laser rangefinder, passive

M47E2R recovery tank in travelling configuration carrying spare engine

M48A5E1 with 105 mm M68 gun and bank of four smoke dischargers either side of turret. 12.7 mm anti-aircraft machine gun not fitted

M47E2 with new engine, 105 mm Rheinmetall Rh-105 30 gun and bank of four smoke dischargers on turret side

night vision viewer, control panel, output unit, ammunition selection unit, graticule projection unit, handgrips with switches for laser/lead/battle range, cant sensor and analogue electronic computer. Commander has M17 rangefinder and ammunition selection unit, loader has ammunition selection display unit and driver has three M27 periscopes, the centre one of which can be replaced by an AN/VVS-2 passive night driving periscope. It has a TEESS engine exhaust smoke generating system and smoke dischargers installed.

M47E2I Engineer Tank

This has a new powerpack consisting of a Continental AVDS-1790-2D 12-cylinder diesel coupled to an Allison CD-850-6A transmission and a new front superstructure. Mounted at the front of the hull is a hydraulically operated blade which can be used for clearing obstacles, preparing fire positions or to stabilise the vehicle when the winch is in use. Main winch is provided with 100 m of cable and has a maximum capacity of 70 tonnes. Mounted on the right side of the hull at the front is a hydraulically operated crane which is traversed to the rear and lowered to the horizontal for travelling. The crane can be fitted with an auger for boring operations. The M47E2I has a crew of four and is armed with one 7.62 mm bow mounted machine gun and one roof mounted 12.7 mm machine gun. Mounted on the left side of the superstructure are six smoke dischargers firing forwards. The driver has an AN/VVS-2 passive night driving periscope.

M47E2R Recovery Tank

This is almost identical to the M47E2I engineer tank but does not have the auger and is used to recover damaged and disabled vehicles. Full details of this vehicle, which has yet to enter series conversion, are given in *Jane's Military Vehicles and Logistics 1991-92*, page 28.

M47E2VLP Bridgelayer

This is an M47 (or M48) chassis fitted with a new powerpack and can lay a scissors type bridge over the front of the vehicle in a similar manner to the M48/M60 AVLBs used by the US Army. Full details of this vehicle, which has yet to enter series conversion, are given in *Jane's Military Logistics 1989*, page 115.

M47 Leguan Bridgelayer

A Spanish Army M47 chassis has been fitted with a German MAN 26 m Leguan bridgelaying system for trials purposes. This bridge is launched horizontally over the front of the vehicle. Mounted at the front of the M47 is a dozer/stabilising blade.

Spanish M48s

Eighteen M48s of the Spanish Marines have been fitted with the new AVDS-1790-2A series diesel engine but retain the 90 mm gun, these are the M48Es. Spain purchased 54 M48A2 tanks from Germany and these were all upgraded to M48A5 standard (Spanish designation M48A5E). A total of 108 American supplied M48 tanks have also been brought up to M48A5E standard. In 1983 the Spanish Army selected a Hughes fire control system for installation in the M48. Hughes Electro-Optical and Data Systems Group, Tactical Programmes Division, delivered the first of 54 systems in Summer 1983. Under the second phase of the contract a further 110 systems were delivered in Summer 1984. Many of the components of this fire control system were produced under licence from Hughes by EISA/ENOSA in Spain. About 70 per cent of the components of the M48's fire control system is the same as that of the M60A3's fire control system, as well as key modules for the laser rangefinder for the M1 Abrams MBT.

Spanish M47s

By 1983 at least 330 M47s had been rebuilt for the Spanish Army to M47E1 standard and 46 to M47E2 standard. In 1981 an additional 100 M47s were purchased from Italy and Germany and it is reported that 50 of these will be rebuilt to recovery versions, 25 to engineer tank and the remaining 25 to bridgelayers, in addition to the prototypes of these vehicles already built, but as of mid-1991 work on these had not started.

SPECIFICATIONS

Model	**M47E1**	**M47E2**	**M48A3E**	**M48A5E**
CREW	4	4	4	4
COMBAT WEIGHT	47 000 kg	46 700 kg	47 100 kg	49 000 kg
UNLOADED WEIGHT	41 000 kg	41 200 kg	44 400 kg	44 000 kg
POWER-TO-WEIGHT RATIO (hp/tonne)	16.17	16.27	16.13	15.51
GROUND PRESSURE	0.93 kg/cm²	0.93 kg/cm²	0.83 kg/cm²	0.86 kg/cm²
LENGTH GUN				
forwards	8.51 m	9.16 m	8.69 m	9.3 m
rear	7.09 m	7.74 m	7.42 m	8.04 m
LENGTH HULL	6.36 m	6.36 m	6.88 m	6.88 m
WIDTH	3.39 m	3.39 m	3.63 m	3.63 m
HEIGHT				
without AA MG	3.02 m	2.96 m	2.87 m	2.87 m
with AA MG	3.35 m	3.35 m	3.28 m	3.28 m
GROUND CLEARANCE	0.47 m	0.47 m	0.42 m	0.42 m
TRACK	2.79 m	2.79 m	2.92 m	2.92 m
TRACK WIDTH	580 mm	580 mm	710 mm	710 mm
LENGTH OF TRACK ON GROUND	3.91 m	3.91 m	4 m	4 m
MAX ROAD SPEED FORWARDS	56 km/h	56 km/h	50 km/h	50 km/h
FUEL CAPACITY	1500 l	1500 l	1500 l	1500 l
MAX ROAD RANGE	600 km	600 km	550 km	550 km
FORDING	1.2 m	1.2 m	1.2 m	1.2 m
GRADIENT	60%	60%	60%	60%
VERTICAL OBSTACLE	0.91 m	0.91 m	0.91 m	0.91 m
TRENCH	2.59 m	2.59 m	2.59 m	2.59 m
ENGINE				
Continental V-12, AVDS	1790-2A	1790-2D	1790-2A	1790-2D
output (hp/rpm)	760/2400	760/2400	760/2400	760/2400
TRANSMISSION	Allison Transmission Division of General Motors CD series 850-6A			
SUSPENSION	torsion bar	torsion bar	torsion bar	torsion bar
ELECTRICAL SYSTEM	24 V	24 V	24 V	24 V
BATTERIES	6 × 12 V	6 × 12 V	6 × 12 V	6 × 12 V
ARMAMENT				
main	1 × 90 mm	1 × 105 mm	1 × 90 mm	1 × 105 mm
coaxial	1 × 7.62 mm	1 × 7.62 mm	1 × 7.62 mm	1 × 7.62 mm
anti-aircraft	1 × 12.7 mm	1 × 12.7 mm	1 × 12.7 mm	1 × 12.7 mm
AMMUNITION				
main	77	42	62	54
coaxial	2750	2500	5900	8000
anti-aircraft	1250	2175	600	600

Status: Production complete. In service with the Spanish Army.

Manufacturer: Peugeot Talbot Espana SA
Special Products Division
Apartado No 140, 28080 Madrid
Spain
Telephone: 3472931
Telex: 27590, 42853, 43603

SWEDEN

New Swedish Main Battle Tank

Development/Description

The Swedish Army has a requirement for a new MBT to replace part of its present fleet later this decade and original contenders for this included the Swedish HB Utveckling AB Stridsvagn 2000 (MBT 2000) (*Jane's Armour and Artillery 1990-91*, page 79), German Leopard 2, US M1A1/M1A2 Abrams, British Challenger 2 and the French Leclerc.

One German Leopard 2 and two US M1A1 Abrams have already been trialled in Sweden and in the Spring of 1991 it was stated that the Swedish Stridsvagn 2000 had been dropped from the competition with the front runners being the Krauss-Maffei Leopard 2 and the General Dynamics, Land Systems Division, M1A1/M1A2.

It is expected that between 200 and 300 MBTs would be procured for use by two armoured brigades and one mechanised brigade based in Northern Sweden.

Funding for the purchase is expected to be included in the next five year budget that runs from 1992-93 through to 1996-97 with the first vehicles being delivered in 1995-96 and first vehicles operational in 1996-97.

Bofors Stridsvagn 103 Main Battle Tank

Development

In the early 1950s the head of the tank design section of the Vehicle Division of the Swedish Army Ordnance, Sven Berge, studied a number of tank designs including the French AMX-13 light tank, which features an oscillating turret with an automatic loader for its main armament. This tank suffers a number of disadvantages and Sven Berge then proposed a turretless tank with a fixed gun, the gun being elevated and traversed by altering the pitch of the hull and aimed in traverse by turning the whole vehicle.

One of the first problems to be overcome was that of laying the gun accurately in azimuth by relying only on its tracks. In the Winter of 1957/58 an Ikv-103 assault gun was fitted with an external crowbar steering system which proved the basic concept. In 1959 a similar system was fitted to the heavier M4 Sherman tank, but this time between the tracks and the hull. The Sherman installation also proved that the system would work on a heavy vehicle rather than on a light vehicle such as the Ikv-103.

In mid-1958 Bofors was awarded a contract to design the turretless tank with its fixed gun mounting, automatic loading system, adjustable hydro-pneumatic suspension system and the new steering system. The following year Bofors was awarded a contract to build two prototype tanks and the company selected two major subcontractors, Landsverk for the suspension and Volvo for the powerpack.

In mid-1960 the Swedish Army placed a pre-production order for 10 vehicles. The first two prototypes completed in mid-1961 differed in some detail from later vehicles as the suspension had no return rollers, no front support was provided for the 105 mm gun and five machine guns were fitted, one on the commander's cupola and two in boxes on each side of the hull firing forwards. They were powered by a Boeing 502/10MA gas turbine and a Rolls-Royce B81 eight cylinder petrol engine.

The pre-production vehicles had the right pair of machine guns replaced by a 12.7 mm (0.50) ranging machine gun, and the suspension incorporated two track return rollers. Development costs of the tank, which was subsequently named the Stridsvagn 103, or Strv 103, amounted to $8.4 million, including the two prototype vehicles and the 10 pre-production vehicles.

The first production Strv 103s were completed in 1966 and production continued until June 1971 by which time 300 had been built. The first production vehicles, known as the Strv 103A, were not fitted with a flotation screen or the dozer blade. Later production vehicles, known as the Strv 103B, had both the flotation screen and the dozer blade and earlier production tanks were subsequently brought up to Strv 103B standard. The Strv 103 is in service with three of Sweden's six armoured brigades, the other three having British supplied Centurion MBTs.

Description

The all-welded hull of the S-tank has the engine and transmission at the front, fighting compartment in the centre and the magazines at the rear. There is a series of horizontal ribs on the glacis plate to deflect armour-piercing rounds. The two engines are geared together to a common output and the diesel is normally used all the time with the gas turbine being used only when the tank is in action. The gas turbine can be used to assist in cold weather starting and also acts as a power source when the diesel engine is not operating. Two fire extinguishers in the engine compartment can be operated from inside or outside the vehicle. The glacis plate and the main armament have to be removed before the engine can be changed, which takes about four hours.

The transmission consists of a Bofors 3-step automatic main gearbox with forward and reverse function in bevel gear. Steering is a regenerative double differential system with a hydrostatic steering device, steering being accomplished by handle bars on the tiller columns.

The driver, who also lays and fires the main armament, is seated on the left side of the hull and is provided with a single piece hatch cover, a Jungner OP-1L combined periscope and binocular sight to his front and a single periscope to the left. The OP-1L has a field of view of 100° and a magnification of ×1, ×6, ×10 or ×18 with the sight graticule being in the right eyepiece. A Simrad LV 300 laser rangefinder is incorporated into the OPS-1L sight. The radio operator seated to the rear of the driver is provided with a single piece hatch cover and two periscopes, and is also provided with controls to drive backwards if required. The commander is seated on the right side of the tank, slightly to the rear of the radio operator and is provided with four periscopes and a Jungner OPS-1 combined periscope and binocular sight. The sight is not operated with a laser rangefinder, but line-up needles are provided in the left eye-piece. The commander's OPS-1 unit is fully stabilised in elevation from -11° to +16°, and his cupola can be traversed through 208°. The cupola is stabilised in azimuth and enables the commander to locate the target. He then uses the handle bars on the tiller columns to lay the tank onto the target, selects the type of ammunition and then fires the main armament. All the periscopes and sights on the S-tank have armoured shutters, which have two functions: they provide a measure of protection against shell splinters and also prevent the tank's position

Bofors Strv 103C S-tank showing additional fuel cans along hull and two Bofors Lyran illumunition launchers on roof

Strv 103B MBT with flotation screen erected before amphibious operation

Strv 103B S-tank showing external fuel tanks at hull rear

being given away by the sun glinting on its periscopes. There is a hull escape hatch in the fighting compartment.

The hydro-pneumatic suspension system consists of four dual rubber tyred road wheels identical to those used on the Centurion MBT, with the idler at the rear, drive sprocket at the front and two track return rollers. To provide a more stable firing platform the suspension is locked when the main armament is fired. The first and fourth road wheels are mounted on leading arms with the second and third on trailing arms. A positive displacement pump transfers fluid between the front and rear units, while a servo system compensates for the alteration in the length of track in contact with the ground, and the height of the hull above the ground. The track is of the two pin type with replaceable rubber pads.

Main armament of the S-tank is a 105 mm rifled tank gun 62 calibres long designated the L74. The gun, which has a bore evacuator but no muzzle brake, is basically a longer version of the British L7 series gun as installed in many other tanks. The gun is fed from a magazine at the rear of the hull that holds 50 rounds of ammunition in 10 racks each holding five rounds. A typical load would comprise 25 APFSDS, 20 HE and five smoke rounds, but any combination is possible. The automatic loader enables a high rate of up to 15 rds/min to be achieved. When all the ammunition has been expended the magazines are reloaded through two hatches at the rear by the crew in about 10 minutes. The empty cartridge cases are automatically ejected outside the hull and if the automatic loading system fails the radio operator can hand crank ammunition to the breech of the gun. Buttons on the main tiller box are used to select the type of ammunition to be fired. Sweden has recently adopted the Israeli M111 APFSDS-T round for the S-tank as the replacement for the APDS round currently used. The 105 mm gun can also fire HE and smoke rounds if required.

Two fixed 7.62 mm Ksp 58 machine guns mounted on the left side of the hull are laid in a similar manner to the main armament and fired alternately. Once ammunition has been expended, they have to be reloaded by one of the crew leaving the vehicle. A 7.62 mm Ksp 58 machine gun mounted on the left of the commander's cupola can be aimed and fired from within the vehicle and can also be used for anti-aircraft defence. For target illumination at night two Bofors Lyran launchers are fitted on the roof.

A dozer blade is carried folded under the nose of the tank and when required is swung forward and secured by two rods. It is operated by adjusting the hydro-pneumatic suspension. The S-tank has no NBC system but is prepared for one.

A flotation screen carried around the top of the hull takes between 15 and 20 minutes to erect and the tank is then propelled in the water by its tracks at a speed of 6 km/h. When afloat, the driver stands on top at the rear with a remote throttle control and steers by reins attached to the main tiller.

S-tank Modernisation Programme

Following successful trials of an S-tank with its Rolls-Royce K60 engine replaced by a Detroit Diesel 6V-53T developing 290 hp, all vehicles have undergone a retrofit programme with this engine together with its gearing and auxiliaries. In addition the modernised S-tank has a modified transmission consisting of a three-step automatic gearbox developed by Bofors, bevel gearing with forward/reverse function and an electronics unit, new radiators, generator, silencer and controls. The graticule in the existing sight is to be replaced and a Simrad Nd YAG laser rangefinder will be incorporated into the gunner's sight by Bofors Aerotronics. In 1983 Bofors Aerotronics delivered the prototype of a fire control computer for the S-tank using the same software as in the improved Centurion (qv Centurion entry under United Kingdom) but with modified hardware. Bofors is also studying ways of compensating for barrel droop such as a muzzle reference system. Additional fuel cans arranged along the sides of the tank will increase range and protection.

By late 1984 10 prototypes of the modernised S-tank had been completed. In late 1983 Bofors was awarded the production contract to rebuild Strv 103B tanks to the new Strv 103C configuration at the rate of about seven vehicles per month. First rebuilt vehicles were handed over in Spring 1986 and final deliveries made in 1989. The Strv 103C will also be fitted with two Bofors Lyran illumination launchers with ammunition for these kept on the gunner's/driver's hatches.

During mobilisation, the tank is fitted with 2 × 9 fuel cans which each contain 22 litres of fuel. All Strv 103C tanks are fitted with a dozer blade.

Studies have also been carried out regarding auxiliary protection, compass, NBC protection and night fighting equipment. Development of auxiliary protection, probably explosive reactive armour and night fighting equipment started in 1986.

S-tank with mine clearing rollers

Early in 1989 the first Swedish developed mine clearing rollers built to production standard were delivered to the Swedish Army. Trials have shown that these can detonate 15 to 20 heavy anti-tank mines at the same time as continuing to breach mines across the same clearing width. Mine Roller I has been designed for mounting on all types of vehicle and two types of coupling have been produced; Coupling one is for the Centurion while Coupling two is for the S-tank. The overall weight of the mine clearing roller is six tonnes while the coupling weighs a further one tonne. Further details are provided in *Jane's Military Vehicles and Logistics 1991-92*, page 239.

Bofors Strv 103C S-tank with fuel cans alsongside hull

All Strv 103 S-tanks are fitted with dozer blade that folds back under the nose of the vehicle. This is shown here in operating position

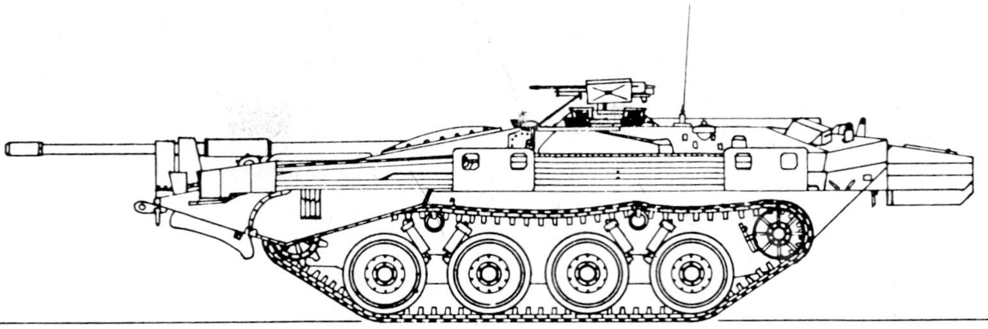

Bofors Strv 103B MBT

SPECIFICATIONS (Strv 103B)	
CREW	3
COMBAT WEIGHT	39 700 kg
UNLOADED WEIGHT	37 000 kg
POWER-TO-WEIGHT RATIO	18.4 hp/tonne
GROUND PRESSURE	1.04 kg/cm^2
LENGTH GUN FORWARDS	8.99 m
LENGTH OF HULL (excluding stowage boxes at hull rear)	7.04 m
WIDTH	3.63 m
WIDTH OVER TRACKS	3.26 m
HEIGHT	
to top of commander's cupola	2.14 m
including machine gun	2.43 m
FIRING HEIGHT	1.7 m
GROUND CLEARANCE	
hull centre	0.5 m
hull sides	0.4 m
TRACK	2.6 m
TRACK WIDTH	670 mm
LENGTH OF TRACK ON GROUND	2.85 m
MAX SPEED	
road	50 km/h
water	6 km/h
FUEL CAPACITY	960 l
MAX ROAD RANGE	390 km
FORDING	1.5 m
with preparation	amphibious
GRADIENT	60%
VERTICAL OBSTACLE	0.9 m
TRENCH	2.3 m
TURNING RADIUS	pivot in neutral
ENGINES	Rolls-Royce K60 multi-fuel developing 240 bhp at 3750 rpm and Boeing 553 gas turbine developing 490 shp at 38 000 rpm
AUXILIARY ENGINE	none
TRANSMISSION	Volvo with 2 forward and 2 reverse gears
STEERING	regenerative double-differential with hydrostatic steering drive
SUSPENSION	hydro-pneumatic
ELECTRICAL SYSTEM	24 V
BATTERIES	2 × 12 V, 114 Ah
ARMAMENT	
main	1 × 105 mm gun
coaxial	2 × 7.62 mm MG
anti-aircraft	1 × 7.62 mm MG
SMOKE-LAYING EQUIPMENT	8 smoke dischargers
AMMUNITION	
main	50
MG	2750
GUN CONTROL EQUIPMENT	
Turret power control	n/app
by commander	yes
by driver/gunner	yes
commander's override	yes
Commander's fire-control override	yes
Gun elevation/depression	+12°/-10°
Gun stabiliser	
vertical	no
horizontal	no

Note: Strv 103 C is similar except for combat weight of 42 500 kg, unloaded weight of 39 800 kg, ground pressure of 1.17 kg/cm^2, track width 635 mm, engine is Detroit Diesel 6V-53T developing 290 hp at 2800 rpm and transmission is Bofors 3-step automatic main gearbox with forward/reverse function in bevel gear. Also has two Bofors Lyran launchers and laser rangefinder.

Status: Production complete. In service with the Swedish Army.

Manufacturer: Bofors AB (now known as Swedish Ordnance)
S-691 80 Bofors
Sweden
Telephone: (46) 586 810 00
Telex: 73219
Fax: (46) 586 58145

SWITZERLAND

Leopard 2 (Pz 87 Leo) Main Battle Tank

Between August 1981 and June 1982 two West German Leopard 2 (120 mm smoothbore gun) and two American M1 (105 mm rifled gun) tanks underwent technical and user trials in Switzerland. In August 1983 Switzerland selected the Leopard 2 and late in 1984 the Swiss Parliament authorised the purchase of 380 Leopard 2s. The first 35 Leopard 2s came direct from Krauss-Maffei in Munich, the remainder are being manufactured under licence in Switzerland with Contraves as prime contractor. Final assembly takes place at the Swiss Federal Armament Works in Thun where the production and overhaul of the Pz 61 and Pz 68

Swiss Army Leopard 2 which is called the Pz 87 Leo and is very similar to those used by Germany and Netherlands (Swiss Army)

Swiss Army Leopard 2 which is called the Pz 87 Leo and is very similar to those used by Germany and Netherlands (Swiss Army)

was undertaken. The Swiss Federal Armament Works is also responsible for assembling the turrets and power units. the Rheinmetall 120 mm smoothbore gun is also built under licence.

Between 60 and 70 per cent of the Leopard 2s are built in Switzerland. Swiss Leopard 2s are similar to the German vehicles but with Swiss radios, antennas and machine guns, an improved driver's hatch, digital computer in place of the analog computer, Deugra fire/explosion detection and suppression system for the crew compartment, improved NBC protection, hydraulic track tensioning units, Baird passive night driving periscope, optical master warning for the driver when driving with the hatch open and some other minor modifications.

The first 35 Leopard 2s, or Panzer 87s as they are designated by the Swiss Army, were delivered in 1987, with the first battalion being formed in 1988. Production is expected to be at the rate of 72 tanks per year with final deliveries taking place in 1993. The first two Swiss built Pz 87s were delivered to the Federal Defence Procurement Authority (GRD) on December 17th, 1987, as originally scheduled. Late in 1984 Hughes Electro-Optical and Data Systems Group signed a manufacturing and technical assistance agreement with Wild Heerbrugg Ltd of Switzerland, under which the latter company would build the laser tank fire control system for the Swiss Panzer 87 Leopard, or Pz 87 Leo as it is usually known.

Pz 61 and Pz 68 Main Battle Tanks

Development

In 1951 the Technical Section of the Swiss General Staff carried out a study to see if an MBT could be designed in Switzerland. This study concluded that it was possible and two years later design work on the first Swiss MBT began at the Swiss Federal Armament Works in Thun.

The first prototype of the Swiss MBT, designated the KW30, was completed in 1958 and was armed with a Swiss-designed 90 mm gun. The second prototype was completed the following year. These two prototypes were followed by 10 pre-production tanks, Pz 58 (Panzer 58), which were built between 1960 and 1961 and armed with a British 20-pounder (83.4 mm) gun as installed in the Pz 55. In March 1961 a production order for 150 Pz 61s (armed with British 105 mm gun) was placed with deliveries to the Swiss Army taking place between January 1965 and December 1966. The Pz 61s are operated by the tank battalions of the Swiss 4th Mechanised Division.

Description

The hull of the Pz 61, like its turret, is a one piece casting supplied to Thun by Georg Fischer of Schaffhausen, and is divided into three compartments: driver's at the front, fighting in the centre and the engine at the rear.

The driver is seated at the front of the hull in the centre and is provided with a one piece hatch cover hinged at the rear. Three periscopes are mounted forward of the hatch.

The turret, which unlike most other contemporary tanks has no bustle, is in the centre of the hull with the commander and gunner seated on the right and the loader on the left. The commander's cupola has a single piece hatch cover that opens to the rear and eight periscopes. The commander also operates the split image coincidence rangefinder which has a magnification of ×8. The gunner is seated forward and below the commander and has a periscope with a magnification of ×8 and ×2.7. The loader's cupola is provided with six vision blocks and split hatch covers that open left and right. The loader's cupola is a little higher than the commander's whose area of observation is therefore restricted.

The engine compartment is at the rear, separated from the fighting compartment by a fireproof bulkhead. The complete powerpack, consisting of the engine, auxiliary engine, transmission, cooling and exhaust system, can be removed from the vehicle in about one hour. The engine is supplied from Germany and is coupled to a Swiss SLM transmission consisting of a semi-automatic gearbox with six forward and two reverse gears and a double differential steering system with a hydrostatic steering drive which provides progressive and continuous steering control.

The suspension of the Pz 61 consists of six dual rubber tyred road wheels, each independently located and sprung by Belleville washers. The idler is at the front and the drive sprocket at the rear. There are three track return rollers. The tracks of the Pz 61 are of cast manganese and have no rubber pads.

Main armament of the Pz 61 is a British designed 105 mm L7 series gun manufactured in Switzerland with a number of modifications, known as the Pz Kan 61 and firing a Swiss designed HE (muzzle velocity 600 m/s) round in addition to the normal APDS (muzzle velocity 1470 m/s), HESH (muzzle velocity 730 m/s) and smoke rounds. Switzerland has adopted the Israel Military Industries M111 APFSDS projectile. The gun control system for the Pz 61 is a French SAMM type CH 25 electro-hydraulic system similar to that developed for the French AMX-30 MBT except that the gun of the Pz 61 is elevated by a motor rather than a jack as in the case of the AMX-30.

Mounted on the forward part of the loader's hatch is a 7.5 mm MG 51 machine gun for anti-aircraft defence which can be elevated from −4 to +77°, and mounted on either side of the turret are three L Pz 51 (80.5 mm) smoke dischargers.

Standard equipment on the Pz 61 includes NBC system, hull escape hatch and a drinking water tank. There is no provision for deep fording. The Pz 61 AA9 is the earlier Pz 61 with an SE-412 radio and dry air filter installed, the 20 mm coaxial cannon replaced by a standard 7.5 mm machine gun and improved maintenance facilities.

Pz 68 MBT (later called the Pz 68 Mark 1)

Further development of the Pz 61 resulted in the Pz 68. The first prototype was completed in 1968 with production being authorised the same year at a cost of SFr 460 million. Between January 1971 and July 1974 170 Pz 68s were delivered to the Swiss Army. Major differences between the Pz 61 and the Pz 68 can be summarised as the installation of a stabilisation system for the main armament allowing the tank to engage targets accurately while moving across country, ammunition resupply hatch in the left side of the turret, more powerful engine and modified transmission, wider tracks with replaceable rubber pads and a greater length of track in contact with the ground. Mounted at the rear of the turret is a large stowage basket and deep fording equipment can be installed. The main armament is equipped with a fume extractor but no muzzle brake. Pz 68 and the Pz 61 tanks have been fitted with a Bofors Lyran system

Pz 68 Mk 2 MBT with driver's hatch closed and loader manning 7.5 mm machine gun (Swiss Army)

Pz 61 AA9 MBT with 7.5 mm MG replacing 20 mm cannon, thermal sleeve for 105 mm gun and driver's hatch closed (Swiss Army)

(Swiss Army designation is the 7.1 cm Le GW 74 for which 12 projectiles are carried) on the turret roof between the commander's and loader's cupolas. This system can launch an illuminating rocket to a maximum range of 1300 m. The Pz 68s are used by the 4th and 11th Mechanised Divisions. All but 13 per cent of the Pz 68 was manufactured in Switzerland, the engine, auxiliary engine and turret controls coming from abroad.

Pz 68 Mark 2

This was approved for production in 1974 at a cost of SFr 146.3 million, and is basically the Pz 68 Mark 1 with an alternator, thermal sleeve for main armament, and a system for extracting carbon monoxide. Between March and December 1977 50 Pz 68 Mk 2 tanks were delivered to the Swiss Army.

Pz 68 Mark 3

This has all the improvements of the Mks 1 and 2 but also has a larger turret. Production was authorised in 1975 at a cost of SFr 447 million and 110 tanks were delivered between 1978 and September 1979.

Pz 68 Mark 4

This is similar to the Mk 3 and 60 were ordered in 1978 for delivery between July 1983 and April 1984. The Pz 68 Mark 4 cost SFr 207 million. In general the terms Mks 1, 2, 3 and 4 are not used in Switzerland. They are normally called the 1st, 2nd, 3rd (or Pz 68/75) and 4th series.

Pz 68 Upgrade

It was announced in early 1988 that 195 Pz 68 MBTs were to be upgraded by the installation of a new fire control system supplied by Honeywell's German subsidiary and licence built in Switzerland by Contraves. The fire suppression system would also be upgraded with all work being carried out at the Swiss Federal Armament Works at Thun where the Leopard 2 (Pz 87) is built. Of the total of 195 Pz 68 MBTs to be upgraded, 25 will be from the second batch (Pz 68 Mk 2) with the remainder all of the third (Pz 68 Mk 3) and fourth (Pz 68 Mk 4) batches. When upgraded the Pz 68 will be re-designated the Pz 68/88.

Additional modifications will include a muzzle reference system, primary stabilised sighting system for the gunner with an eyepiece for the tank commander. Suspension will be upgraded with hydraulic bump stops and a improved collective NBC system will be fitted. New paint will be introduced and the steel fuel tanks replaced by GRP tanks.

Pz 68 Mk 3 MBT with commander's, loader's and driver's hatches open (Swiss Army)

120 mm gun for Pz 68

The Swiss Federal Armament Works at Thun has developed to the prototype stage a 120 mm smoothbore gun called the Compact Tank Gun which has been successfully trialled in a Pz 68 MBT of the Swiss Army.

The available 120 mm guns would not fit into the relatively small Pz 68 turret and no extensive turret modifications were allowed; also a moderate trunnion load was required. For interoperability reasons the gun was required to fire the same ammunition as the Leopard 2 MBTs used by the Swiss Army.

The barrel calibre length was increased to L/49 as compared to the L/44 of the Leopard 2 with the front end of the barrel having a dovetail attachment to enable a muzzle reference sensor to be installed. The barrrel is also fitted with a fume extractor of filament wound glass fibre composite.

The breech is the semi-automatic vertically sliding wedge type with an electrical firing system.

Variants

Anti-aircraft Tank

The prototype of this system, which had the same twin 35 mm turret as fitted to the German Gepard SPAAG, was handed over to the Swiss Ordnance Procurement Agency in April 1979. It has been decided that the system will not be procured at present.

Armoured Recovery Vehicle

Development of an ARV based on the chassis of the Pz 61 MBT began at Thun in 1961 with the first prototype being completed in 1967/68. After trials with this prototype a modified version based on the chassis of the Pz 68 was placed in production. First production vehicles, known as the Entpannungspanzer 65 (or Entp Pz 65) were completed in 1970. The vehicle is fitted with a main winch with 120 m of cable and a maximum capacity of 25 000 kg, which can be increased to 75 000 kg with the use of snatch blocks. The auxiliary winch used to pull out the main cable has 240 m of cable. Mounted at the front of the vehicle is a hydraulically operated dozer blade which is used to stabilise the vehicle or for dozing. Pivoted at the front of the hull is an A-frame with a maximum lifting capacity of 15 000 kg. A full range of tools and cutting equipment is carried. The Entp Pz 65 has a crew of five and a loaded weight of 38 000 kg. Armament consists of a single 7.5 mm machine gun and eight smoke dischargers. Full details of the Entp Pz 65 are given in *Jane's Military Vehicles and Logistics 1991-92*, page 30. Last production Entp Pz 65 was handed over in May 1985.

Armoured Bridgelayer

The prototype of the bridgelayer was based on the chassis of a Pz 61 tank but production vehicles, designated the Brückenlegepanzer 68, were based on the Pz 68 tank chassis. The bridge is launched as follows:

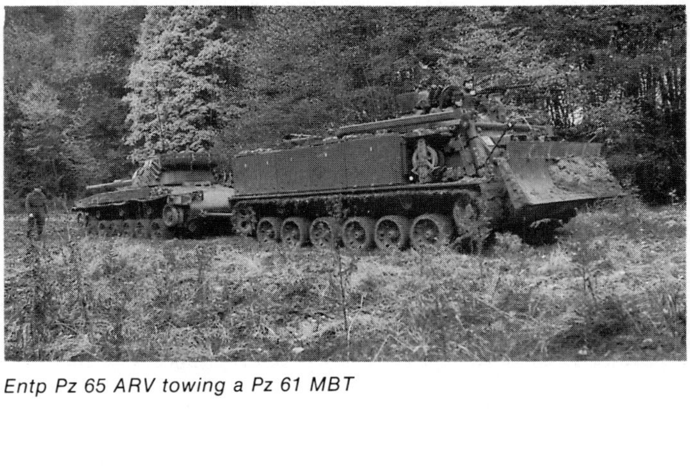

Entp Pz 65 ARV towing a Pz 61 MBT

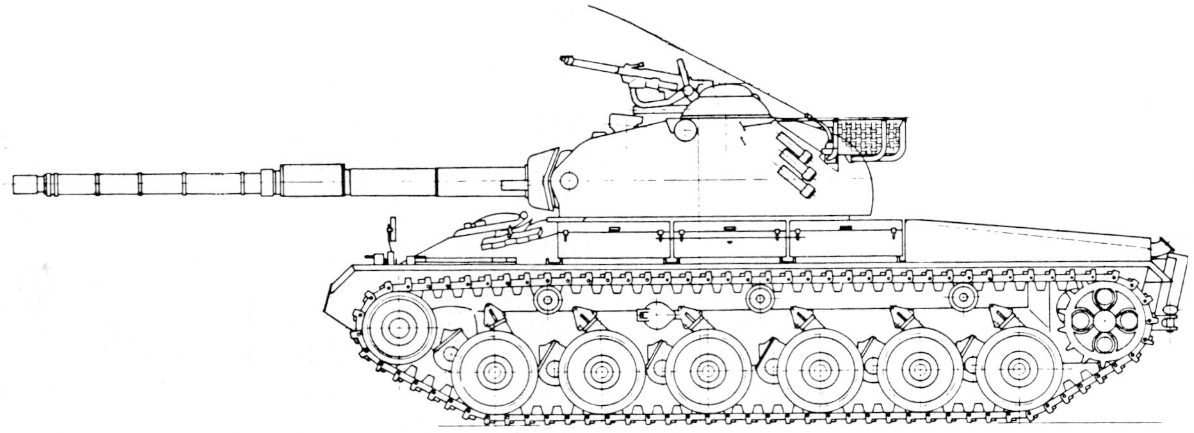

Pz 61 AA9 MBT with 20 mm coaxial cannon replaced by 7.62 mm machine gun (not to 1/76th scale) (Swiss Army)

Brü Pz 68 armoured bridgelayer in travelling configuration (Swiss Army)

Uparmoured Pz 68 MBT fitted with 120 mm smoothbore Compact Tank Gun (Swiss Army)

the vehicle stops short of the obstacle and the bridge is tilted forwards, a beam is slid over to the far bank and the bridge is then advanced along the beam. Once the bridge has reached the far bank, the beam is retracted. The bridge has an overall length of 18.23 m and a maximum capacity of 60 000 kg, its normal capacity being 50 000 kg. The bridgelayer has a crew of three and weighs 44 600 kg complete with the bridge. Full details of the Brü Pz 68 are given in *Jane's Military Vehicles and Logistics 1991-92*, page 88.

155 mm Self-propelled Gun

The prototype of a 155 mm self-propelled gun called the Panzer-Kanone 68, based on the chassis of a Pz 68 MBT, was built some time ago. It was not placed in production as the Swiss Army procured additional American 155 mm M109 SPGs.

Target Tank

A special version of the Pz 68 has been developed for training anti-tank crews, called the Panzerzielfahrzeug, or Pz Zielfz 68. It has a heavily-armed turret with a dummy gun and additional armour is provided for the hull. The tops of the tracks are covered by armoured skirting. The Pz Zielfz 68 weighs 38 000 kg and has a crew of two.

SPECIFICATIONS

Model	**Pz 61**	**Pz 68**
CREW	4	4
COMBAT WEIGHT	38 000 kg	39 700 kg
UNLOADED WEIGHT	37 000 kg	38 700 kg
POWER-TO-WEIGHT RATIO	17 hp/tonne	16.62 hp/tonne
GROUND PRESSURE	0.85 kg/cm²	0.86 kg/cm²
LENGTH GUN FORWARDS	9.43 m	9.49 m
LENGTH GUN REAR	8.28 m	8.6 m
LENGTH HULL	6.78 m	6.88 m
WIDTH	3.08 m	3.14 m
HEIGHT		
including AA MG	2.85 m	2.88 m
commander's cupola	2.72 m	2.75 m
FIRING HEIGHT	1.93 m	1.96 m
GROUND CLEARANCE	0.42 m	0.41 m
TRACK	2.59 m	2.59 m
TRACK WIDTH	500 mm	520 mm
LENGTH OF TRACK ON GROUND	4.13 m	4.43 m
MAX ROAD SPEED	55 km/h	55 km/h
FUEL CAPACITY	760 l	710 l
MAX RANGE		
road	300 km	350 km
cross-country	160 km	160 km
FORDING	1.1 m	1.1 m
GRADIENT	60%	60%
VERTICAL OBSTACLE	0.75 m	0.75 m
TRENCH	2.6 m	2.6 m
ENGINE	MTU MB 837 Ba-500 8-cylinder diesel developing 630 hp at 2200 rpm	MTU MB 837 Ba-500 8-cylinder diesel developing 660 hp at 2200 rpm
AUXILIARY ENGINE	OM 636 diesel developing 31 hp at 2500 rpm	OM 636 diesel developing 35 hp at 2600 rpm
TRANSMISSION	SLM semi-automatic 6 forward and 2 reverse gears	SLM semi-automatic 6 forward and 6 reverse gears
STEERING	double-differential with hydrostatic steering control	
SUSPENSION	Belleville washers	Belleville washers
ELECTRICAL SYSTEM	24 V	24 V
BATTERIES	4 × 6 V, 200 Ah	4 × 6 V, 200 Ah
ARMAMENT		
main	1 × 105 mm gun	1 × 105 mm gun
coaxial	1 × 7.5 mm MG	1 × 7.5 mm MG
anti-aircraft	1 × 7.5 mm MG	1 × 7.5 mm MG
SMOKE-LAYING EQUIPMENT	3 smoke dischargers either side of turret	
AMMUNITION		
main	52	56
7.5 mm	5400	5200
GUN CONTROL EQUIPMENT		
Turret power control	electro-hydraulic/manual	electro-hydraulic/manual
by commander	yes	yes
by gunner	yes	yes
commander's override	yes	yes
Commander's fire-control override	yes	yes
Max rate power traverse	360° in 9.5 s	360° in 9.5 s
Max rate power elevation	20.8°/s	20.8°/s
Gun elevation/depression	+21°/-10°	+21°/-10°
Gun stabiliser		
vertical	no	yes
horizontal	no	yes
Range setting device	yes	yes
Elevation quadrant	yes	yes
Traverse indicator	yes	yes
ARMOUR	15-120 mm	15-120 mm

Status: Production complete. In service only with the Swiss Army.

Manufacturer: Swiss Federal Armament Works, Thun, Switzerland.

TAIWAN

M48H Main Battle Tank

Development/Description

Late in 1988 the Taiwanese Army announced that it had successfully fired a home produced version of the US built 105 mm M68 rifled tank gun from a modified M48 tank.

The M68 is the US version of the British Royal Ordnance L7 weapon but with a US designed breech. It has been used for many years, not only in the US M60/M60A1/M60A3, but also the M1, M48A5 and South Korean Type 88 as well as Israeli tanks including the Merkava Mk 1 and Mk 2.

The United Services Institute of Taiwan was responsible for producing and testing the gun and the Taiwanese Army plans to arm their new MBT, the M48H, with this 105 mm rifled tank gun.

The M48H, also referred to as the Brave Tiger MBT, was shown for the first time in 1990 amid severe criticism from opposition politicians of its high costs and its seemingly out-dated appearance.

The M48H is a hybrid vehicle which uses brand new M60 hulls purchased from General Dynamics, Land Systems Division, before the production line closed down in 1987, fitted with modified M48 turrets. This has an advanced fire control system that incorporates a ballistic computer and a thermal imaging sight that is being developed by the Chung Shan Institute of Science and Technology for the vehicle.

It is understood that the roof-mounted stabilised sight, which incorporates a thermal imaging channel, is similar to that installed on the United States M1 Abrams MBT and is slaved to the 105 mm gun. This sight has dual axis stabilisation which allows improved target tracking on the move. The laser rangefinder is believed to be of the Neodium/YAG type.

Late in 1988, Urdan Industries of Southfield, Michigan, USA, was awarded a $3.8 million fixed price contract for 520 low profile commander's cupolas for installation on M48 tanks by the US Army Tank Automotive Command. This order was an FMS sale to Taiwan with the actual work being carried out at Netanya, Israel, with production completed late in 1989.

The M48H is powered by a Teledyne Continental Motors AVDS-1790-2C diesel engine which also powered the final production model of the M60 series, the M60A3. This develops 750 hp at 2400 rpm and is coupled to a CD-850 series automatic transmission.

From a firepower and mobility aspect, the M48H is on par with the US Army M60A3, but its turret armour is inferior although this could easily be improved by the addition of applique passive or explosive reactive armour.

An explosion/fire detection/suppression system is fitted to the M48H MBT, this being provided by the Santa Barbara Research Centre with the initial order being for 171 systems.

Taiwanese M48H Brave Tiger MBT which is based on modified M60 MBT chassis and M48 series MBT turret (DTM)

Until the introduction of the M48H, the most powerful tank in the Taiwanese rank fleet was the 90 mm M48 series of vehicle of which 286 are believed to be in service, together with 325 M24 Chaffe light tanks (which have the NAPCO diesel powerpack) and up to 800 M41 light tanks of which 650 are fully operational.

Exact numbers of M48H tanks to be built are not known, although early in 1985 orders were placed with General Dynamics, Land Systems Division, for 450 M60 chassis and 265 assembly ready hulls for Taiwan.

It is believed that a total of 450 M48H MBTs will be built with production running at 15 vehicles a month and final deliveries due in 1992/93. One hundred of the M48A3s are to be converted to M48A5 and in addition Taiwan has recently purchased 150 M60A3 MBTs from the US for $31.4 million.

Status: In production and in service with Taiwan. The 1989/90 budget included $151 million for the M48H programme and conversion of 90 mm armed M48A3s to the 105 mm armed M48A5.

Manufacturer: Prime contractor is Fighting Vehicles Development Centre.
ROC Army
Taichung
Taiwan 400

TURKEY

Turkish MBT Fleet

By the Spring of 1991 it is estimated that the Turkish MBT fleet consisted of 500 M47s (all of which were in reserve), 3000 M48s (of which 2120 had been upgraded to the M48A5T1 configuration) and 77 Leopard 1A4s MBTs. Of these only the latter were delivered brand new to Turkey.

Under the Conventional Forces Europe reductions, Turkey is expected to take delivery of additional armoured fighting vehicles and artillery systems. These will include additional Leopard 1 MBTs (from Germany and in addition to the 150 mentioned later) and M60A1/M60A3 (from the United States).

The Wegmann Company of Germany rebuilt five Turkish M48 tanks to a similar standard as the 650 M48A2GA2s it supplied to the German Army (see entry on M48 in this section). In addition the Turkish vehicles have been fitted wth an MTU MB 837 Ea-500 diesel engine with all peripheral equipment (for example control units and cooling system), water pre-heating system, fuel tanks with pumps and controls, exhaust system and electrical equipment including control panel and batteries. There are also modifications to the chassis which include ammunition stowage, dust shields, a heating system and a fire extinguishing system. Wegmann was then awarded a contract by the German Government to provide 165 kits, plus personnel support, to enable Turkey to carry out the work itself.

For some time Germany has been providing assistance to the Turkish Army, including the transfer of equipment such as M48A2 tanks that were being phased out of service. The 1983-84 plan included some DM 32.7 million for the Turkish tank maintenance plants at Arifiye and Kayseri, as well as the provision of spare parts for the M48 and Leopard 1.

Under an agreement signed in Bonn in November 1980, Turkey has now received 77 Leopard 1A3 MBTs (54 built by Krauss-Maffei and 23 by Krupp-MaK) and four ARVs (built by Krupp-MaK). The first of these were delivered in 1982 and final deliveries were made in 1984.

More recently, Germany has agreed to transfer 150 Leopard 1A4s used by the German Army to Turkey, with a number of modifications to their fire control systems. The first batch will comprise 75 vehicles. When modified to meet Turkish requirements these will be known as the AT3. Turkey may also take delivery of a further eight Leopard 1 ARVs from Germany. The first batch of four Leopard 1A4s were handed over in 1990 with final deliveries due in late 1991.

In late 1979 a team of Turkish officers visited the Anniston Army Depot in the USA where the M48A5 tank conversion programme was being carried out. It was determined that Turkey would be able to convert its ageing fleet of M48 series tanks into the M48A5 configuration, with a target date of 1983 for first conversion vehicles to be completed.

In addition, Turkey has already taken delivery of 80 Leopard 1A1A1 MBTs from German Army Stocks.

M48 with Wegmann retrofitting kit as supplied to Turkish Army showing thermal sleeve for 105 mm L7 gun

In 1982 the USA leased heavy industrial equipment to Turkey for installation in the depot at Kayseri which was to become the first M48 tank conversion plant. Turkey ordered its first two conversion kits from the USA in 1982 with a further 400 being ordered later the same year.

Between August 1982 and May 1983 the industrial plant arrived in Turkey and was installed and the operators trained. Conversion work on the first five M48s started early in May 1983 and by September the same year initial trials of these five vehicles, designated the M48A5T1, had been completed.

The second M48A5 conversion facility at Arifiye started conversion work in January 1984 and by the Summer of that year the first Turkish M48 battalion had converted to the M48A5T1 MBT.

It is understood that the final conversion kits were shipped by the United States late in 1990 with the final batch of M48A5T1s to be delivered in 1991/1992.

The M48A5T1 is the equivalent of the US Army M48A5 and is described in detail in the entry on the M48 tank later in this section.

Late in 1985 the US Department of Defense notified Congress of a letter of offer to Turkey, valued at $206 million, for the supply of 760 tank conversion kits, comprising tank thermal sights, ballistic computers and add-on stabilisation systems. Prime contractors for this package are Texas Instruments (thermal sights), Computer Devices Company (ballistic computers) and Cadillac Gage Textron (stabilisation system).

When fitted with the thermal sights, ballistic computer and stabilisation system, the vehicle is known as the M48A5T2.

In August 1986 Cadillac Gage Textron announced that it had been awarded a $30 million contract by the US Army Armament, Munitions and Chemical Command for 760 tank stabilisation systems. First deliveries were made in 1987, with expected follow on orders bringing the total to more than 2000 units through 1995.

Further improvements under consideration include additional armour protection and improved suspension.

For some years Turkey has been making the US 105 mm FP105 APFSDS rounds for the 105 mm armed Leopard 1 and M48A5T1 MBTs.

UNION OF SOVIET SOCIALIST REPUBLICS

Future Soviet Tanks (FSTs)

In the mid-1980s NATO intelligence agencies predicted what new Soviet MBTs would be developed in the future based on available information from a wide variety of sources.

These were given the designations of the Future Soviet Tank 1 (FST-1) and Future Soviet Tank 2 (FST-2), although they have also been referred to as the Follow-on Soviet Tank 1 and Follow-on Soviet Tank 2.

Details of the original anticipated FST-1 and FST-2 were given in *Jane's Armour and Artillery 1989-90* page 81. Since then, however, the United States Army has stated that the FST-1 is in fact a late production T-72 MBT with explosive reactive armour. This is covered in the entry on the T-72 series and has the Soviet designation of the T-72M2 or the NATO STANAG Soviet Medium Tank M1988.

The Soviet Medium Tank M1989 is a further development of the T-80 with a number of improvements including new passive armour and a diesel engine. Details of this are given in the T-80 MBT entry.

T-80 Main Battle Tank

Development

The T-80 MBT was, like the T-64, developed by the A F Kartsev OKB tank design bureau at the Ural plant at Nizhni-Tagil where the earlier T-64 was also designed and built. Production is believed to have started in 1983 with first production vehicles being completed in 1984. So far it has only been identified in front line service with the Soviet Army although it is understood that trials vehicles have been supplied to India and Syria.

No production figures for the T-80 MBT have been released although under the Conventional Forces Europe agreement the Soviets have declared the following quantities of the T-80 MBT and its variants:

Region	IV-1	IV-2	IV-3	IV-4	V.1
T-80	112	93	81	nil	19
T-80B	3518	3015	2885	2885	503
T-80BK	217	219	208	182	nil
T-80BV	594	594	569	569	nil
T-80BVK	23	23	22	22	nil
T-80U	410	410	3	nil	nil

Note: The regions are as defined under the Conventional Forces Europe agreement: Region IV is defined as Central Europe while Region V covers the flanks of the Atlantic to the Urals (ATTU).

The above variants can be basically summarised as follows:

T-80

First production model and believed to be in production from 1978 to 1982 when it was replaced by the T-80B. Combat weight is 39.7 tonnes.

T-80B

This replaced the T-80 in production from 1983 and has a similar combat weight of 39.7 tonnes.

T-80BK

Command version of the T-80B with additional communications equipment, same combat weight.

T-80BV

T-80B with explosive reactive armour (ERA) fitted to the hull and turret which increases its combat weight to 41.7 tonnes. This is referred to by NATO as the T-80 Model 1984 (M1984).

T-80BVK

Command version of the T-80BV with additional communications equipment, same combat weight.

T-80U

This was first observed by the West in 1989 and has been referred to by NATO as the SMT (Soviet Medium Tank) M1989. This has a new turret, different engine and many other improvements. Combat weight is 43 tonnes.

T-80 clearly showing details of road wheels and snorkel carried on turret rear

Soviet T-80 MBT with roof-mounted 12.7 mm machine gun covered up but clearly showing two long-range fuel drums on hull rear, exhaust outlet for gas turbine below fuel drums and snorkel carried on turret rear. This vehicle is not fitted to receive explosive reactive armour

Description

The overall layout of the T-80 is similar to that of the T-64 series with the driver's compartment at the front, two man turret in the centre and engine and transmission at the rear. There are, however, many detailed differences.

The glacis plate is of the laminate type for improved protection against kinetic energy and HEAT attack and there is a dozer blade under the nose of the vehicle.

The turret is steel with an inner layer of special armour; the gunner sits on the left and the tank commander on the right.

Whereas the T-64 is powered by a five cylinder, opposed piston, liquid-cooled diesel engine developing 750 hp coupled to a manual transmission with seven forward and one reverse gears, the T-80 has a new gas turbine engine developing 985 hp coupled to a manual transmission with five forward and one reverse gears. This is the first Soviet operational tank to be powered by a gas turbine.

The rear hull top is different to the T-64's, the T-80 also has a distinct oblong exhaust outlet in the hull rear.

The T-80 MBT uses the same 125 mm 2A46 gun and horizontal ammunition system as the T-72 whereas the T-64 uses a 125 mm 2A26 gun with vertical ammunition stowage. The fire control system is probably an improvement over that fitted to earlier Soviet tanks.

The T-64 has hydro-pneumatic suspension but the T-80 reverted back to torsion bar suspension with each side consisting of six forged steel-aluminium rubber tyred road wheels, drive sprocket at the rear, idler at the front and five return roadwheels. The rubber tyred road wheels are in two halves that are bolted together.

As the track of the T-80 is slightly wider than the T-64's and has a longer length of track in contact with the ground, the ground pressure of T-80 is improved as is its power-to-weight ratio. The new track design has end connectors and central guide horns.

The road wheel spacing is not identical and there are distinct gaps between the second and third, fourth and fifth, and fifth and sixth road wheels. The side skirt covers the return rollers.

The T-80 has the same 125 mm 2A46 smoothbore gun as the T-72 with a horizontal ammunition stowage system. This can fire either the AT-8 Songster ATGW or four types of separate loading ammunition, eg projectile and semi-combustible cartridge case. These four rounds are HE-FRAG (FS) which is designated OF-19 with a muzzle velocity of 850 m/s, HEAT-FS which is designated BK-14M, APFSDS-T designated BM-9 with a muzzle velocity of 1800 m/s and the more recent BM-12 which is also an APFSDS-T projectile with a longer penetrator. In 1991 Soviet sources stated that the T-80 MBT fired the following types of projectile from its 125 mm smoothbore gun: APFSDS Type 3BM32 with a muzzle velocity of 1700 m/s; HEAT Type 3BK18 with a muzzle velocity of 905 m/s; HE-FRAG with a muzzle velocity of 850 m/s; Flechette with a muzzle velocity of 850 m/s.

A 7.62 mm PKT machine gun is mounted coaxially to the right of the main armament and a 12.7 mm NSV machine gun is mounted on the commander's cupola. Full details of the 125 mm gun/missile launcher are given in the following entry for the T-64B.

The 125 mm gun/missile launcher tube is fitted with a thermal sleeve and a fume extractor.

Banks of electrically operated smoke dischargers are mounted either side of the 125 mm gun/missile launcher, normally five on the left and four on the right.

A total of six AT-8 Songster ATGW are carried and these are identical to those launched by the T-64B MBT deployed some years ago.

The missile guidance box is mounted on the right side of the turret roof in front of the commander's cupola and can be removed and stowed inside the vehicle if required.

Standard equipment includes snorkels for deep fording operations that are carried on the turret rear when not required, an overpressure type NBC protection system, night vision equipment for all three crew members (believed to be infra-red type), unditching beam carried across the hull rear and a laser warning device activated by laser rangefinders, laser designators or precision-guided munitions fitted with a laser guidance device. Mounted on the turret rear is a large circular container which carries two snorkels. The larger one is the snorkel for the gas turbine, with another one being fitted onto the radiator grill by means of two adaptors. This provides an air intake for the gas turbine.

To extend the operational range of the T-80, additional fuel tanks can be mounted at the hull rear. These can be quickly jettisoned if required.

Each of the two main fuel drums at the hull rear hold an additional 200 litres of fuel. An additional third fuel tank may be mounted crosswise on the engine hatch cover. This does not affect the combustion air intake which is directly behind the turret ring. The air intake is covered by two hinged rectangular grilles that open forward, so uncovering a filter that is inserted underneath.

Variants

T-80B

The original production model of the T-80 was not built in large numbers, production then switched to the T-80B.

T-80BK

This is the command version of the T-80B and has additional communications equipment and therefore additional antennas.

T-80BV

In the mid-1980s the T-80B started to be fitted with blocks of Explosive Reactive Armour (ERA). This is referred to by the US Army as the T-80 Model M1984 as this was the year that it was first identified.

When fitted with its explosive reactive armour package, the T-80 is virtually immune over their frontal arc to penetration from all current NATO ATGWs which rely on a HEAT warhead to penetrate armour.

With the explosive reactive armour fitted the smoke grenade launchers are moved from either side of the main armament back to either side of the turret and positioned between the turret side and the explosive reactive panels.

The configuration of explosive reactive armour panels on the glacis plate of the T-80 and T-64B appear to be identical, but on the turret and hull sides there are differences between the T-80 and T-64B, although explosive reactive armour arrays on the latter in Western Group of Forces are not identical.

For example, in the case of the T-80 no panels are fitted to the side skirts while the T-64B has up to four layers of explosive reactive armour.

On the turret of the T-80, the panels are joined to form a shallow chevron pointing forward, the cross-section view in the entry for the

T-80BV MBT from the rear showing explosive reactive armour on sides of turret and 12.7 mm MG boxes on rear of snorkel

T-80BV MBT showing explosive reactive armour boxes on hull and turret

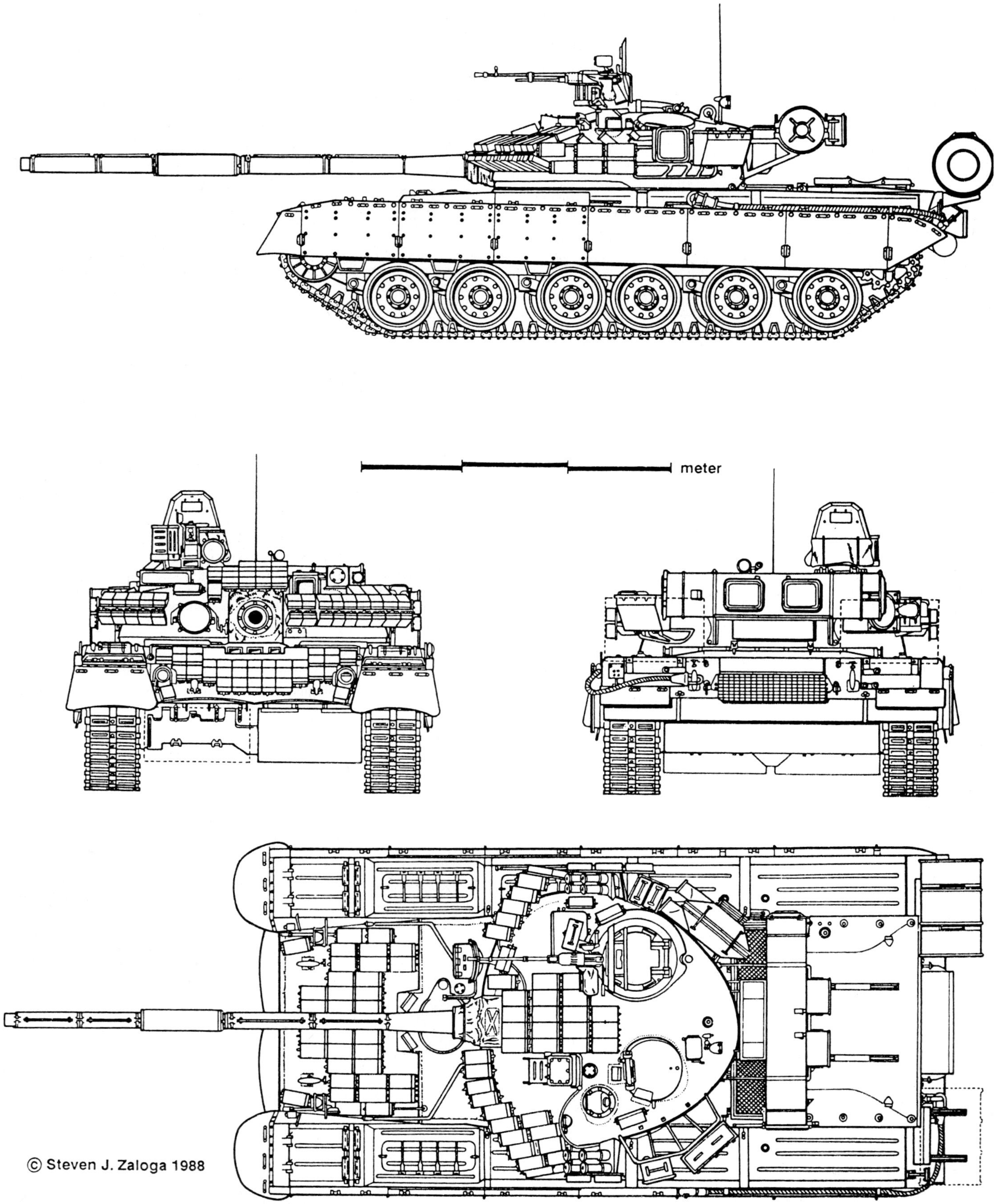

The T-80BV is the T-80B fitted with explosive reactive armour to its hull and turret, the forward part of the skirts are also reinforced on this model (not to 1/76th scale) (Steven Zaloga)

T-64/T-72 shows these differences clearly. Explosive reactive armour is also fitted to the forward part of the turret roof to provide protection against top attack weapons.

The explosive reactive armour package on the T-80 provides a high degree of protection against ATGWs such as MILAN, HOT, TOW and Swingfire over the frontal arc. It does not provide any added protection against APDS or APFSDS attack.

When the HEAT projectile reaches the MBT its high speed jet initiates the explosive between the two plates and drives the plates aside. These moving plates perturb and eat the mass of the incoming jet. The jet is then unable to achieve any significant penetration of the main tank armour.

Mounted on the roof forward of the tank commander's cupola is the laser designation system which projects a modulated beam; the system is protected by a rectangular armoured box cover.

T-80BVK

This is the command version of the T-80BV and is fitted with additional communications equipment and antennas.

T-80 with mine clearing equipment

All T-80s are built with a dozer blade mounted under the nose of the vehicle and with ribs to enable the vehicle to be fitted with mine clearing

Close-up of ERA on turret front of T-80 clearly showing the arrows to indicate to the crew the direction that the packs should be fitted to obtain maximum possible protection

T-80 MBT from the rear showing additional fuel tanks, unditching beam and snorkel on turret rear

Close-up of T-80U MBT showing commander's cupola and armoured side skirts

T-80U MBT showing new armour array on hull, turret and side skirts

equipment such as the KMT-4M mine clearing plough. Normally one tank per Soviet tank company is fitted with mine clearing equipment.

T-80 with additional passive armour

Although many Soviet T-80 MBTs have been fitted with explosive reactive armour (ERA) arrays, a number of vehicles have been observed fitted with an additional layer of armour protection to the glacis plate in a similar manner to that fitted to the older T-72.

It is not clear as to whether this is a conventional steel armour or, like that fitted to the T-72, is a multi-type armour array.

It is, however, possible that T-80s had an armour of a different type to current production vehicles and that this has been back fitted to bring these up to a common level of protection. These vehicles retain their nose-mounted dozer blades.

T-80U MBT

This was first observed in 1989 and is referred to by NATO as the SMT (Soviet Medium Tank) M1989.

Although based on the chassis of the T-80B it has many improvements including the replacement of the gas turbine engine by a more fuel efficient diesel engine, different engine decking, new commander's cupola with 12.7 mm NVS machine gun now capable of being fired from within the turret under remote-control in a similar manner to the earlier T-64, a bank of four electrically operated smoke dischargers either side of the turret firing to the frontal arc, smaller snorkel kept on rear of turret when not required, the commander's sight appears to be new and may well incorporate an image intensification channel as there is no infra-red searchlight on the commander's cupola, there is no guidance box forward of the commander's cupola for the AT-8 Songster ATGW, improved fire control system and a new armour package.

The armour package on the turret and hull is quite different than that fitted to earlier T-80/T-72 and T-64 MBTs. At first it was believed that this was a new passive type but it is now believed to be a second generation explosive reactive armour system with the headlamps being mounted over the armour.

The turret roof between the commander's and gunner's hatches has been provided with additional protection against top attack weapons and a collar of rubber skirts hangs from the turret front. The exact purpose of the latter is not certain but it could be to reduce the signature of the vehicle or deflect grenades scattered from artillery projectiles or rocket launchers away from the top of the vehicle.

The T-80U has full length side skirts but those above the first three road wheels are armoured and are provided with lifting handles.

SPECIFICATIONS (Provisional)

CREW	3
COMBAT WEIGHT	42 500 kg
POWER-TO-WEIGHT RATIO	25.9 hp/tonne
GROUND PRESSURE	0.86 kg/cm²
LENGTH	
gun forwards	9.9 m
hull	7.4 m
WIDTH	3.4 m
HEIGHT (without AA MG)	2.2 m
GROUND CLEARANCE	0.38 m
LENGTH OF TRACK ON GROUND	4.4 m
TRACK WIDTH	580 mm
MAX ROAD SPEED	70 km/h
ROAD RANGE	
without long-range fuel tanks	450 km
with long-range fuel tanks	600 km
FUEL CAPACITY	
internal	1000 l
external	400 l
FORDING	
without preparation	1.8 m
with preparation	5.0 m
GRADIENT	60%
SIDE SLOPE	40%
VERTICAL OBSTACLE	1.0 m
TRENCH	2.85 m
ENGINE	gas turbine developing 1100 hp
TRANSMISSION	manual, 5 forward and 1 reverse gears
STEERING	clutch and brake
SUSPENSION	torsion bar
ELECTRICAL SYSTEM	24 V
ARMAMENT	
main	1 × 125 mm gun/missile launcher
coaxial	1 × 7.62 mm MG
anti-aircraft	1 × 12.7 mm MG
SMOKE DISCHARGERS	8 to 12
AMMUNITION	
main	36*
coaxial	1250
anti-aircraft	300
ATGW	6
GUN CONTROL EQUIPMENT	
Turret power control	electric/manual
Turret traverse	360°
Gun elevation/depression	+14°/-5°
Gun stabiliser	
vertical	yes
horizontal	yes
Rangefinding device	yes, laser

* Excluding AT-8 Songster ATGW as mix of ammunition/Songster depends on mission requirement

Status: In production. In service with the Soviet Union.

Manufacturer: Soviet state factories.

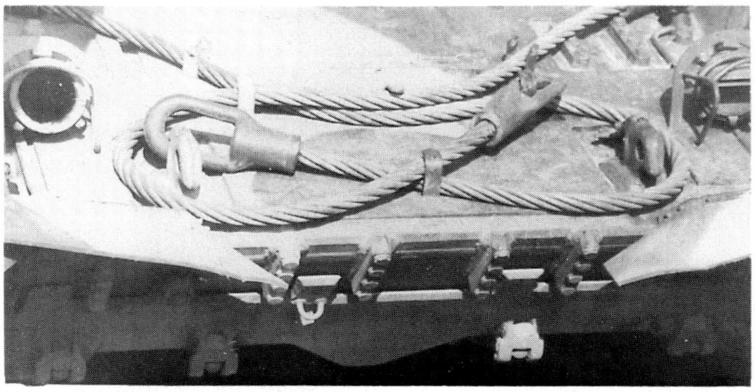

Close-up of glacis plate of T-80 fitted with additional armour protection and showing mounting points for mine clearing equipment

T-64 and T-72 Main Battle Tanks

Development

Following the commencement of full scale production of the T-62 MBT in 1961, the Soviets produced a number of prototype tanks, one of which entered production in 1966 as the T-64 and was first seen in 1970. It was deployed with the Southern Group of Soviet Forces in Hungary in 1980. As far as it is known the T-64 has not been issued to any other member of the Warsaw Pact, nor has it been exported. From all accounts the initial T-64 was not considered to be a satisfactory design and serious problems were encountered with its engine, suspension, fire control system and automatic loader.

It was originally reported that production of the T-64 series was undertaken from 1966 to 1971, but more recently the USA has stated that production continued until 1981 when 200 vehicles were delivered. Late production vehicles were the T-64B variant which can fire an AT-8 Songster ATGW through its 125 mm main armament.

The T-64 MBT was designed by the A F Kartsev OKB MBT design bureau based at the Ural HMB Plant at Nizhni-Tagil which subsequently also designed the T-80 MBT.

The Soviet Union has not released any production figures for the T-64 family of MBTs but under the Conventional Forces Europe treaty they did disclose the existence of the following T-64 MBT:

Region	IV.1	IV.2	IV.3	IV.4	V.1
T-64A	1386	1247	1109	704	139
T-64AK	220	220	202	142	nil
T-64B	1192	1247	841	612	303
T-64BV	159	159	157	nil	nil
T-64B1	420	384	367	35	36
T-64B1K	20	10	10	7	10
T-64BV1K	7	7	7	nil	nil
T-64R	578	574	574	nil	4

The T-72 MBT was designed by the A A Morozov OKB design bureau based at the Malyshev HMB Plant at Kharkov which previously designed the T-62 MBT.

The T-64 was followed by the T-72 which entered production in 1971 and was fully operational by 1973; this was first seen in October 1977 when the French Minister of Defence, Yvon Bourges, paid an official visit to the Taman Guards Division. The tank was seen in public for the first time during the November 1977 military parade in Moscow.

During fighting in Lebanon in 1982 the Israeli Army destroyed a number of T-72 MBTs with standard 105 mm guns firing APFSDS ammunition, TOW ATGWs (helicopter launched), 155 mm and 203 mm improved conventional munitions and US Rockeye cluster bombs with anti-tank submunitions.

The Soviet Union has not released any production figures for the T-72 family of MBTs but under the Conventional Forces Europe treaty they did disclose the existence of the following quantities of T-72 vehicles:

Region	IV.1	IV.2	IV.3	IV.4	V.1
T-72	1437	747	702	279	690
T-72K	64	55	51	nil	9
T-72A	906	748	689	nil	158
T-72AK	53	32	18	9	21
T-72B	1162	1109	1036	nil	53
T-72BK	49	33	21	nil	16
T-72B1	1342	1148	973	nil	194
T-72B1K	79	78	71	nil	nil

The above figures are the situation as of November 1990. For the Conventional Forces Europe treaty Region IV is defined as Central Europe while Region V covers the flanks of the Atlantic to the Urals (ATTU).

Recent Soviet MBT production, according to United States DoD has been:

Year	Quantity
1980	3100
1981	2000
1982	2500
1983	2700
1984	3200
1985	3200
1986	3300
1987	3500
1988	3500
1989	1700
1990	1300
1991	800

Within the Soviet Union, T-72 MBT production was undertaken at Kharkov, Omsk and Nizhni-Tagil, but by early 1990 at least one of these plants had stopped production. In addition, the T-72, or versions of the vehicle, have also been made under licence by Czechoslovakia (still in production, qv), India (still in production, qv), Iraq (did assemble but now stopped due to damage inflicted during Gulf Conflict), Poland (production complete, qv), Romania (probably still in production, qv) and Yugoslavia (still in production, qv).

Description of T-64

This has a similar layout to the T-72 but is armed with the 125 mm 2A26 smoothbore gun with a vertical ammunition stowage system, whereas the T-72 and T-80 are armed with a 125 mm 2A46 gun with a horizontal ammunition feed system. It has the same NBC system, but hydro-pneumatic suspension, narrower tracks, five cylinder opposed diesel engine and a slightly different turret. The suspension consists of six small dual road wheels with the drive sprocket at the rear, idler at the front and four track return rollers which support the inside of the double-pin track only. The first, second, fifth and sixth road wheel stations are

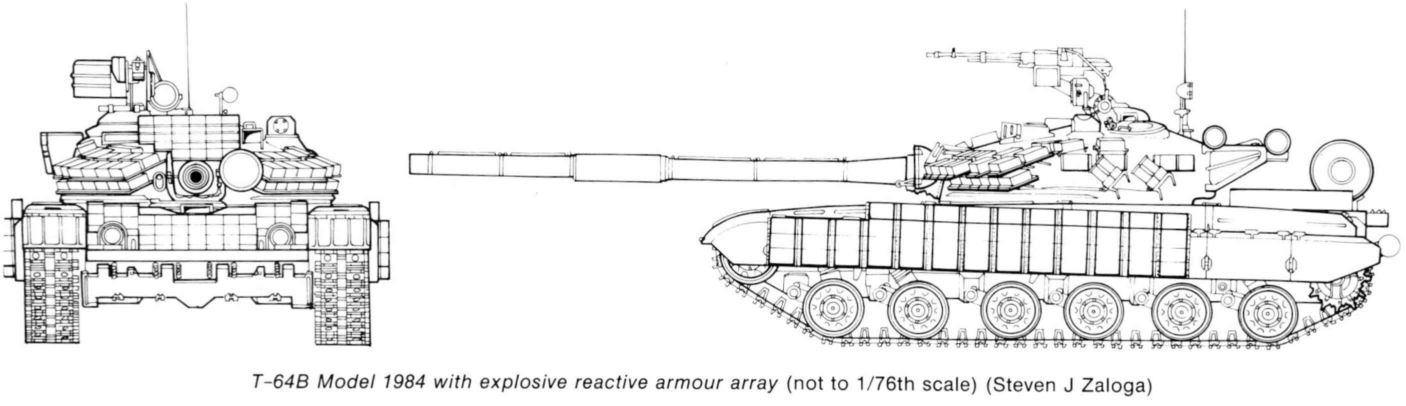

T-64B Model 1984 with explosive reactive armour array (not to 1/76th scale) (Steven J Zaloga)

T-64 series MBT from the rear showing the long range fuel drums mounted above the rear engine deck

Close up of turret of T-64B MBT showing smoke dischargers and gunner's sight

provided with a hydraulic shock absorber. Over the top of the suspension, which slopes downward towards the rear, is a rail on which panels of additional armour can be attached.

The infra-red searchlight is mounted on the left rather than the right of the main armament as in the case of the T-72M. Two or three boxes of 12.7 mm ammunition are mounted on the left side of the turret, the snorkel is carried on the top of the turret at the rear and at the very rear of the turret is a detachable stowage box. The T-64 has two snorkels for deep fording, one fitted to the turret and the other over the engine compartment.

The 125 mm gun is stabilised in both elevation and traverse with the ordnance being fitted with a thermal sleeve and fume extractor. The commander's sight is designated the TKN-3V. The 12.7 mm anti-aircraft machine gun has electric elevation from -5 to +70° with manual controls being provided for emergency use. The anti-aircraft sight is designated the PZU-5.

There is also a command version of the T-64 called the T-64K. When stationary a 10 m telescopic mast is erected over the turret and held in position by stays which are pegged to the ground. The command version does not normally have the 12.7 mm anti-aircraft gun fitted. Unlike the T-72, the 12.7 mm anti-aircraft machine gun of the T-64 can be aimed and fired from within the tank. The main armament can be laid and fired while the T-64 is moving across country and the commander can override the gunner if required. Standard equipment includes an NBC system, infra-red night vision equipment for the commander, gunner and driver, ability to be fitted with snorkels for deep fording and a laser warning device of a similar type to that fitted on the more recent T-80 MBT.

T-64 Variants

T-64 (pre-production model)

A variety of prototypes were built with different hull and turret fronts.

T-64 (basic)

This was the original production model armed with a slightly shorter 125 mm gun which was not fitted with a thermal sleeve. A handrail was provided on both sides of the forward part of the turret just above the turret ring, and stowage boxes were mounted along the right hand sponson. Some vehicles were not provided with attachment points for mine clearing equipment under the nose of the vehicle. As none of these have been referred to under the terms of the CFE Treaty they have probably been upgraded to the T-64A configuration.

T-64A

This is also referred to as the T-64 M1981/1 by the United States Army and features a modified sight for the gunner with an enlarged opening, bump stop for the fourth roadwheel, smoke grenade launchers fitted either side of the 125 mm gun and hinges for the attachment of side skirts. Each bank of six smoke launchers fire grenades to a range of 150 to 200 m. The United States Army has referred to the version with smoke grenade launchers as the T-64A Model 78. The T-64 M1980/2 is the United States Army designation for the M1981/1 but fitted with skirts.

The command version of the T-64A is referred to by the Soviet Union as the T-64AK.

The handrails on either side of the turret front were also omitted and the forward stowage boxes on the right hand sponson were replaced by additional external fuel tanks as on the left hand sponson. The forward skirts were originally spring loaded so that they faced the front, but these were subsequently replaced by skirts.

T-64B

The T-64B has the same 125 mm gun/missile launcher as the more recent T-80 MBT (qv). It also fires the same series of 125 mm separate loading ammunition as the T-72 and T-80 although its automatic loader and ammunition handling system are different.

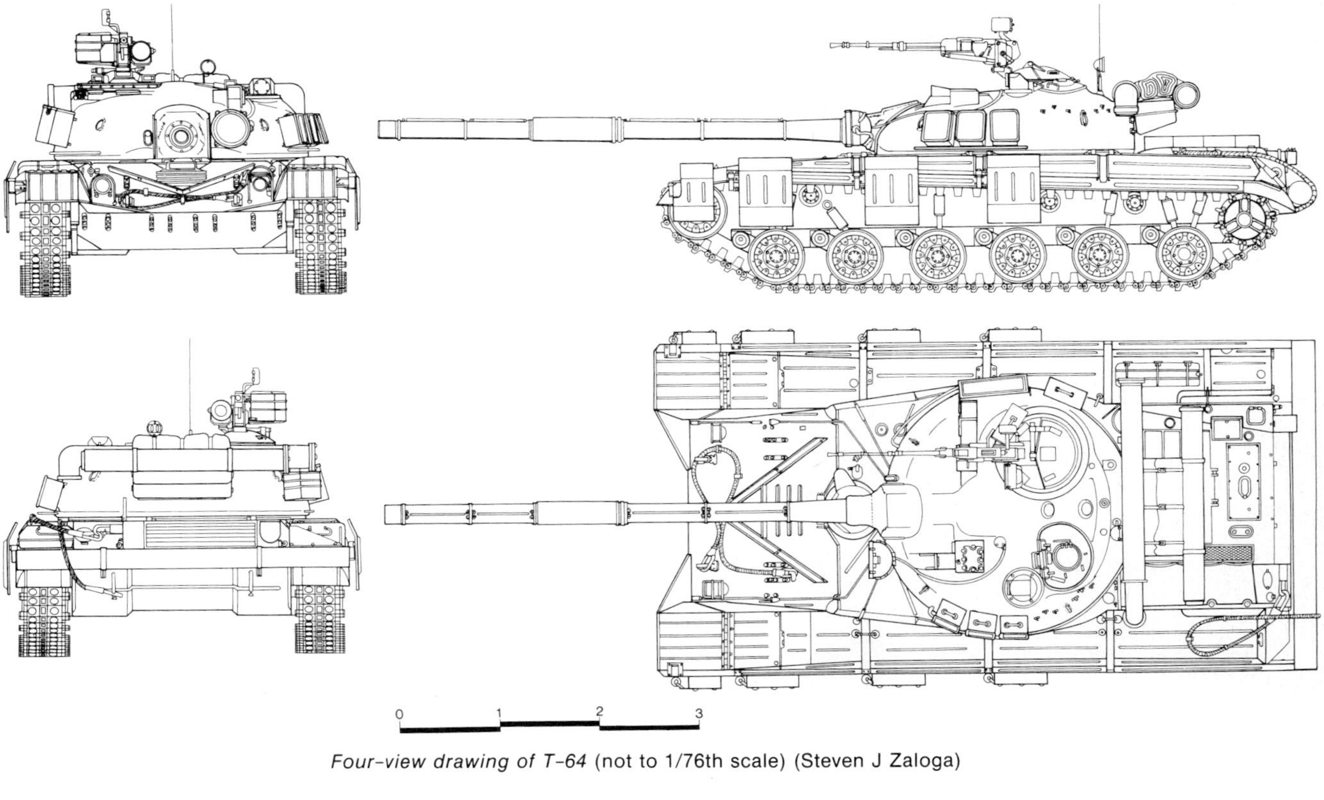

Four-view drawing of T-64 (not to 1/76th scale) (Steven J Zaloga)

T-64A MBT from above clearly showing turret detail and smoke dischargers mounted on forward part of turret

T-64BV MBT fitted with explosive reactive armour to hull and turret and also showing additional cladding around turret area

The T-64 gunner can select the type of round he wishes to fire by pushing a button and a rate of fire of six to eight rds/min can be achieved. The 125 mm gun has a sighted range out to 4000 m using the day sight and 800 m using the night sight. Maximum effective range of the APFSDS-T round is 2500 m. According to Soviet sources, the 125 mm smoothbore gun of the T-64B fires the following types of ammunition with a maximum rate of fire of eight rounds per minute: APFSDS with a muzzle velocity of 1700 m/s; HE-FRAG with a muzzle velocity of 850 m/s; HEAT with a muzzle velocity of 950 m/s; Beehive anti-personnel with a muzzle velocity of 905 m/s.

The 125 mm gun is stabilised, it does not have the same firing on the move capability as Western MBT such as the M1 or Leopard 2.

The T-64B's 125 mm ordnance also fires the AT-8 Songster ATGW, which is kept in the automatic loader in two separate parts like standard APFSDS or HEAT-FS rounds and loaded using the automatic loader.

There are at least two versions of the T-64B. New construction T-64Bs appear to lack the usual forward gunner's sight. The rebuilt T-64Bs, which are modifications of earlier T-64s.

The AT-8 Songster could be used to engage targets such as MBTs and NATO ATGW systems armed with HOT, TOW and Swingfire systems out to about 4000 m and can also engage helicopters. A boost motor launches the AT-8 Songster from the gun barrel and the main motor then cuts in and carries it all the way to the target. It takes nine or 10 seconds to reach a range of 4000 m.

The missile has a muzzle velocity of 125 m/s which increases to 500 m/s and weighs 25 kg at launch. The radio transmitter for the AT-8 is mounted in a removable steel box in front of the right commander's cupola although the missile is guided to the target by the gunner. All the gunner has to do is to keep his sight on the target to ensure a hit. The box in front of the commander's cupola can be moved and stowed elsewhere.

The AT-8 Songster has an HEAT warhead and can probably penetrate at least 600 mm of steel armour but cannot penetrate Chobham type and explosive reactive armours. It has a maximum range of 4000 m and according to Soviet sources the T-64B can fire a maximum of four AT-8 Songsters per minute. A T-64B normally carries six AT-8 Songsters plus 36 rounds (projectile and charge) of 125 mm ammunition.

The T-64B has a laser rangefinder for improved first round hit probability, explosive reactive armour and strengthened turret hatch protection.

Some US sources have indicated that the T-64B also has a sensor package capable of detecting helicopters at a range of up to 10 km which can then be engaged by the AT-8 Songster ATGW.

The T-64B has a different explosive reactive armour package to the T-80. The former has about 111 blocks, the latter has between 185 and 221 blocks.

On the T-64B the reactive armour covers the glacis plate, the forward part of the turret front, sides and roof and hull sides extending to the five road wheels. When fitted with explosive reactive armour, the smoke dischargers are moved from either side of the main armament to either side of the turret rear, roughly in line with the fourth and fifth road wheels.

T-64B fitted with explosive reactive armour firing AT-8 Songster ATGW

The drawing shows the main differences between the side turret explosive reactive armour packages of the T-64B and T-80.

A number of T-64Bs are believed to have been built from scratch, but most vehicles were conversions of earlier T-64As.

T-64BV

This is the correct Soviet designation for the T-64B fitted with explosive reactive armour.

T-64B1

The difference between the T-64B and T-64B1 is not known.

T-64B1K

This is the command tank version of the T-64B1 and has additional communications equipment installed.

T-64BV1K

This is the T-64B fitted with explosive reactive armour and fitted with additional communications equipment for use in the command role.

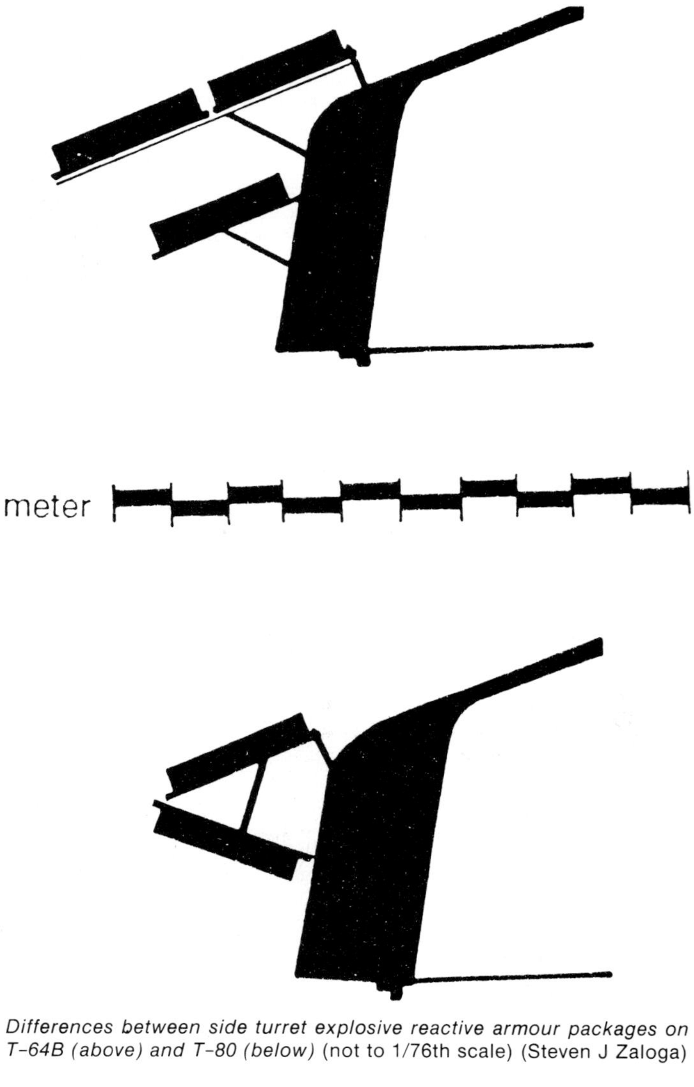

Differences between side turret explosive reactive armour packages on T-64B (above) and T-80 (below) (not to 1/76th scale) (Steven J Zaloga)

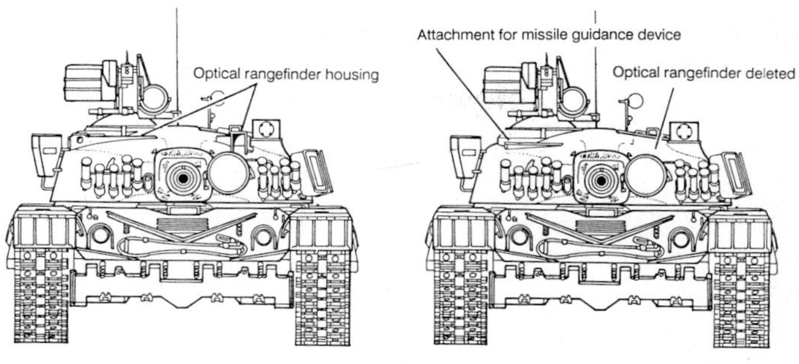

T-64A Main Battle Tank **T-64B Kobra Missile Tank**

Head-on comparative scale drawings of the T-64A and T-64B MBTs (not to 1/76th scale) (Steven J Zaloga)

T-64B (Reactive) fitted for, but not with, explosive reactive armour

T-64R

This is believed to be the T-64 retrofitted with the 125 mm 2A46 gun (as fitted to T-72 and T-80) and capable of firing the AT-8 Songster ATGW.

T-64 with mine clearing equipment

All T-64s are fitted with a dozer blade that folds back under the nose of the vehicle when not required and all have provision for the installation of mine clearing equipment such as the KMT-4 mine clearing roller.

Description of T-72

The hull of the T-72 is divided into three compartments: driver's at the front, fighting in the centre and the engine compartment at the rear. It is believed that the turret has conventional cast armour with a maximum thickness of 280 mm, the nose is about 80 mm thick and the glacis is of a new laminate armour 200 mm thick, which when inclined gives between 500/600 mm of protection. Late production T-72s do however incorporate advanced armour protection in their turrets.

The glacis is well sloped, transversely ribbed and has a deep V splashboard. The driver is seated at the front of the hull and is provided with a single piece hatch cover that opens to the right, in front of which is a single wide-angle TVNE-4E observation periscope. The driver's headlamps are designated FG 125.

The other two crew members are seated in the turret with the gunner on the left and the commander on the right. The commander's contra-rotating cupola has a single piece hatch cover that opens forward with two rear facing TNPA vision blocks. In the forward part of the cupola is a combined TKN-3 day/night sight with an infra-red searchlight, mounted over the top, and to either side of the combined day/night sight is another TNP-160 periscope. Forward and slightly below the commander's cupola is an optical stadiametric rangefinder.

The gunner's hatch opens forward and has a circular opening for mounting the snorkel for deep fording operations and in front of the gunner's hatch is a TNP-160 periscope, while a TNPA-65 vision block is fitted in the hatch cover itself. In front and to the left of the gunner's hatch is a panoramic day/night sight which is used in conjunction with the infra-red searchlight mounted to the left and in front of the sight. The gunner's sight is the TPD-2-49 while his night sight is the TPN 1-49-23. Forward and below the gunner's hatch is a laser rangefinder which may be connected to the left side of the optical rangefinder on the commander's side.

The infra-red searchlight, is mounted on the right side of the main armament rather than the left as in the case of the earlier T-64, although it is occasionally mounted on the left side (initial production vehicles).

Two light steel stowage boxes are mounted on the turret, one at the rear and the other on the right slightly behind the commander's position. The snorkel is carried on the left side of the turret to the rear. Whereas the turret on the T-64 is roughly circular, that of the T-72 has a distinct bulge and external stowage on the two tanks is also different. There are also significant differences in the engine deckings which indicate that they are powered by different engines.

Fuel cells extend along the right side of the hull top; on the left hull top are stowage boxes and a single oil cell. There are four internal oil tanks in the T-72, one to the left of the driver and two to his right and

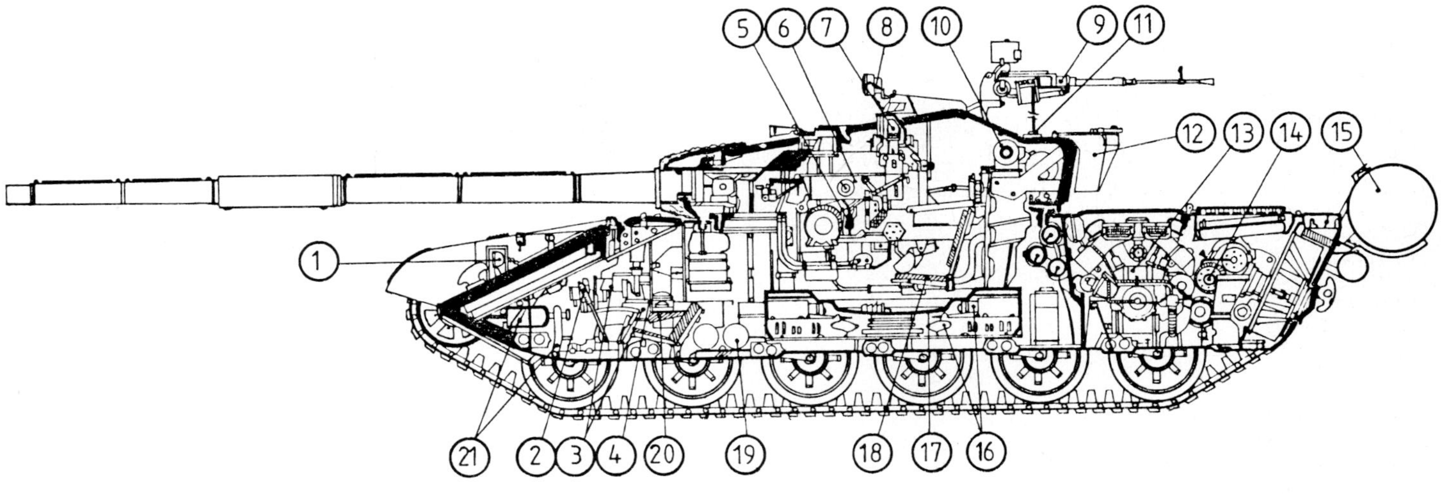

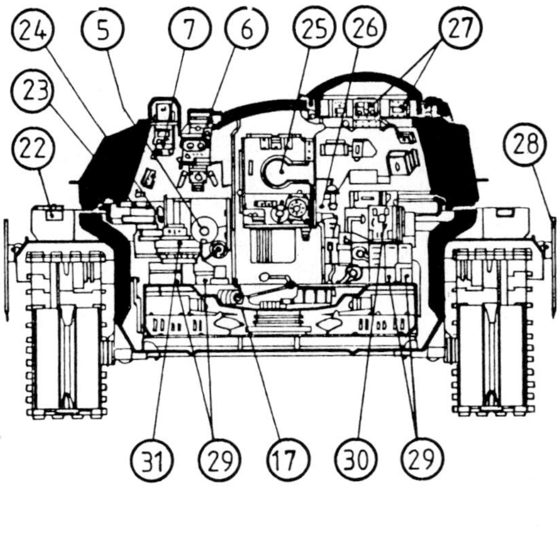

T-72 MBT showing **(1)** *driver's FG 125 headlamp* **(2)** *steering tillers* **(3)** *NBC system* **(4)** *gear lever* **(5)** *gun elevation mechanism* **(6)** *TPD-2-49 gunner's sight* **(7)** *TPN 1-49-23 gunner's night sight* **(8)** *searchlight for use with TKN 3 observation device* **(9)** *12.7 mm anti-aircraft machine gun to rear* **(10)** *hoist for ammunition containers* **(11)** *antenna base* **(12)** *turret bin for deep fording equipment and cold rations* **(13)** *780 bhp diesel engine* **(14)** *gearbox* **(15)** *long-range fuel tanks which can be jettisoned* **(16)** *charges and projectiles in containers on ammunition hoist platform* **(17)** *ammunition hoist platform* **(18)** *gunner's seat* **(19)** *NBC decontamination equipment* **(20)** *driver's adjustable seat* **(21)** *parking brake* **(22)** *external stowage for spares, tools and accessories* **(23)** *manual drive for turret traverse gear* **(24)** *turret traverse indicator* **(25)** *125 mm gun sliding wedge breech-block* **(26)** *7.62 mm PKT coaxial machine gun* **(27)** *tank commander's periscopes* **(28)** *skirting plates* **(29)** *7.62 mm PKT machine gun ammunition boxes* **(30)** *radio set* **(31)** *hydraulic turret traverse gear*

Close-up of turret of T-72M2 (STANAG Soviet Medium Tank M1986) showing additional applique roof armour and additional top armour either side of driver's position. Note also smoke mortars on left side of turret.

Close-up of glacis plate of Iraqi T-72M1 MBT clearly showing additional layer of armour on glacis plate and dozer blade retracted under nose of vehicle (Christopher F Foss)

one on the floor in the space between the rear of the ammunition carousel and the fire wall/engine bulkhead. An unditching beam is normally carried at the rear and there is also provision for carrying two 200 litre fuel drums at the rear to increase operational range. These can be quickly jettisoned if required by the tactical situation.

The T-72 B1 is powered by a V-12 piston multi-fuel air-cooled engine which develops 840 hp. The engine will run on three fuels with the driver being provided with a dial to set the engine for the type of fuel being carried. The three settings are D for Diesel, B for Benzene and K for Kerosene.

The suspension each side consists of six road wheels with the idler at the front, drive sprocket at the rear and three return rollers which support the inside of the track only. Track is of the single pin type with rubber-bushed pins.

Four removable spring loaded skirt plates fitted over the forward part of the track are unclipped in action and spring forward at an angle of 60° from the side of the vehicle to give a measure of protection against HEAT projectiles.

A dozer blade mounted under the nose of the tank is used for clearing obstacles and preparing fire positions and, like most other Soviet tanks, the T-72 can be fitted with mine clearing equipment such as the KMT-5, KMT-6 and KMT-6 M2. The dozer blade can be brought into the operating position in one or two minutes and enables the tank to prepare its own defilade position without calling on engineer support. A typical position for a T-72 MBT would be 10 m long, 4.5 m wide and 1.2 m deep with a total volume of 54 m^3. The T-72 is provided with an NBC system and can be fitted with a snorkel for deep fording. The tank takes about 20 min to prepare for amphibious use and is ready for action within two min of leaving the water.

Main armament is a 125 mm (2A46) smoothbore gun fitted with a light alloy thermal sleeve and a bore evacuator. The gun fires three main types of separate loading ammunition, APFSDS with a maximum range of 2100 m, HEAT-FS with a maximum direct fire range of 4000 m and HE-FRAG(FS) with a maximum indirect fire range of 9400 m.

Type	APFSDS-T	APFSDS-T	HE-FRAG (FS)	HEAT-FS
DESIGNATION	BM-9	BM-12	OF-19	BM-14M
WEIGHT OF PROJECTILE	5.67 kg	5.67 kg	23 kg	n/av
WEIGHT OF PENETRATOR	3.615 kg	3.615 kg	not app	not app
DIAMETER OF PENETRATOR	44 mm	44 mm	not app	not app
MUZZLE VELOCITY	1800 m/s	1800 m/s	850 m/s	950 m/s

In 1991 Soviet sources stated that the T-72B1 fired the following types of projectile from its 125 mm smoothbore gun, these being the same as those fired by the more recent T-80 MBT: APFSDS Type 3BM32 with a muzzle velocity of 1700 m/s; HEAT Type 3BK18 with a muzzle velocity of 905 m/s; HE-FRAG with a muzzle velocity of 850 m/s; Flechette with a muzzle velocity of 850 m/s.

Of 39 rounds of ammunition carried, 12 are APFSDS-T, 21 HE-FRAG(FS) and the remaining six HEAT-FS. The ammunition is of the separate loading type with a consumable cartridge case, all that remains of the latter after firing is a stub. The additional rounds of ammunition are stowed in racks behind the turret basket and in indentations in the rear floor fuel cell and second forward right cell near the driver. The carousel automatic loader is mounted on the turret floor and also on

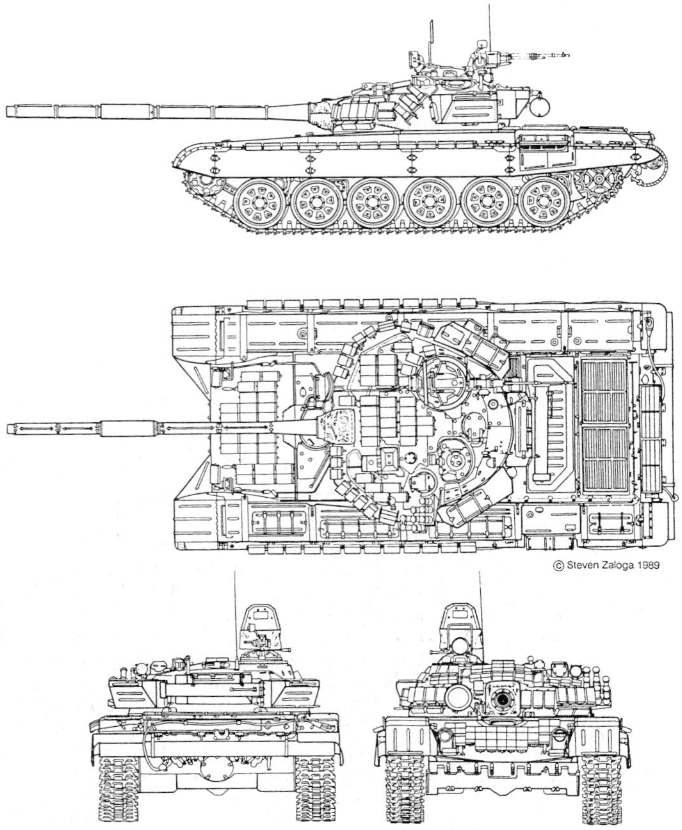

T-72M2 with explosive reactive armour (not to 1/76th scale)

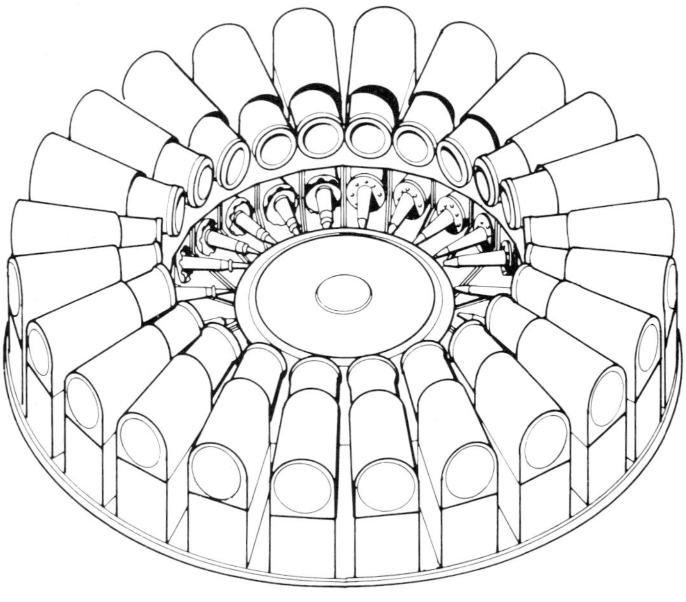

Artist's impression of T-72's ammunition carousel with charge above and projectile below (not to 1/76th scale) (Steven J Zaloga)

Late production T-72G MBT without optics in right side of turret roof and clearly showing V-type splash board on glacis plate

the rear wall of the turret. The projectile is loaded in the lower half of a carrier, the cartridge and propellant in the upper half. The carousel carries 24 ready use projectiles. When the 125 mm gun loads it must pick up the carrier and ram both the projectile and powder charge. This enables a rate of fire of 8 rds/min to be achieved but there are some doubts about the reliability of the automatic loader. The main armament is stabilised and enables the T-72 to shoot on the move with a high probability of a first-round hit. Some Soviet sources have indicated that the practical rate of fire of the 125 mm gun using the automatic loader is five to six rounds per minute.

A 7.62 mm PKT machine gun is mounted coaxially to the right of the main armament and has 250 rounds of ready use ammunition, and a new design 12.7 mm NSV machine gun is mounted on the commander's cupola. The 12.7 mm machine gun can however only be used with the commander exposing the upper part of his body. Maximum sight range in the ground-to-ground role is 2000 m, maximum sight range in the anti-aircraft role is 1500 m.

T-72s built for export have a slightly different fire control system and automatic loader and do not have the internal lining that is standard on Soviet T-72s, consisting of a layer of synthetic material that contains lead and provides some degree of protection against the effects of neutron radiation and electro-magnetic pulses.

Variants

Since the T-72 MBT entered production in 1971 a large number of variants have been produced both for the home and export markets and many earlier vehicles have been subsequently brought up to more recent production standards. Listed below is a resume of these versions based on Soviet sources:

T-72 This was the original version to enter production.

T-72K Command version based on original T-72 MBT.

T-72A This is believed to incorporate the 125 mm 2A26 gun and automatic loader of the T-64.

T-72A (rebuilt) These have plastic armour skirts as fitted to the T-72M and some may also have the TPD-2-49 laser rangefinder. It has an automatic computation system and sight angle setter which computes and corrects for a number of factors in tank gunnery including range to target (measured by the laser rangefinder), muzzle velocity of ammunition, ambient temperature, charge temperature, air pressure, barrel wear and closing distance to the target when firing on the move.

T-72AK Command version based on T-72A with additional communications equipment.

T-72B This is also referred to as the T-72B46 and was the first model built in large numbers. Has 2A46 gun with automatic loader with L-2 Luna infra-red searchlight to right of main armament. Was also exported.

T-72B Late production (modified), some of these had more advanced armour protection including two large panels of applique armour on either side of the turret front. Sometime referred to as T-72B(M).

T-72B (rebuilt) Coincidence rangefinder replaced by TPD-2-49 laser rangefinder with optical port in front of commander's cupola for older TPD-2 blanked over with armour plate.

T-72BK Command version of T-72B with R-130 radio. A special 10 m antenna is carried and fitted to a special pot attached to the turret roof to rear of gunner's hatch cover. These are normally used at battalion and regimental headquarters.

T-72B1 This is an upgraded T-72B and has also been referred to as the T-72M1 for export.

T-72B1K This is the command version of the T-72B1 with additional communications equipment installed.

Export T-72G

In the 1970s the Soviets were working on a more advanced model with new ceramic laminate armour in the front of the turret and improvements to the fire control system, including the installation of the TPD-2-49 laser rangefinder. In the late 1970s Czechoslovakia and Poland started to build a slightly less advanced version called the T-72G, although it is sometimes called the T-72M locally.

T-72G (early version) Initially produced in Czechoslovakia and Poland and fitted with TPD-2-49 laser rangefinder without the pronounced thickening of the turret front found on the T-72M series. Hull details similar to Soviet T-72B with four fold out flipper armour panels each side, not the plastic armour skirts found on its Soviet counterpart.

Close up of the commander's cupola on a T-72 series MBT showing additional applique armour on turret roof

T-72B showing additional armour on turret front and bank of smoke dischargers mounted on left side of turret

T-72G (intermediate production) Germany obtained its first T-72Bs from the Soviet Union, but from 1980 purchased them from Czechoslovakia or Poland. In 1980 they displayed a modified version of the T-72G with the same type of plastic armour skirt and hull side as the Soviet T-72M in place of the flipper type armour used on the Czechoslovakian and Polish built vehicles.

T-72G (late production) In October 1981 the German Army showed a T-72G not only with Soviet type skirts but also with the same arrangement of smoke mortars shown on Soviet T-72M series. The same arrangement also appeared on exported T-72G tanks in Iraq. The Polish Army has fitted a bank of four smoke launchers each side on a special saw-tooth mounting as well as fitting plastic armour.

M-84 Yugoslavia started to build a modified version of the T-72G as the M-84 in 1985 and there is a separate entry for this under Yugoslavia.

Improved Soviet T-72M series

(The following designations are from NATO sources. Under the Conventional Forces Europe Treaty the Soviet Union provided a list of all MBTs deployed in the ATTU. None of the designations below appeared on this list. It is however believed that the T-72M1 is the T-72B1.)

T-72M (STANAG Soviet Medium Tank M1980/1) This incorporates a new form of laminate ceramic armour that could be incorporated in a

T-72B tank of the Soviet Army fitted with Explosive Reactive Armour to its hull and turret

cavity in the cast steel front turret armour. In addition it has the new TPD-2-49 laser rangefinder in place of the older TPD-2 rangefinder found on the T-64 and T-72A/B. Two small canisters containing fire extinguishers are fitted, one to the front right of commander's cupola and one on centre rear of the turret. This was also the first version of the Soviet manufactured T-72 series to incorporate the plastic armour side skirts in place of the flipper armour panels found on earlier models. This is believed to be a plastic material moulded over a metal fabric. The skirt covers the upper part of the suspension. Separate panels protect the sides of the fuel and stowage panniers.

T-72M1 This has a number of modifications including improved engine cooling and a plate of 20 mm aplique armour added to the top of the glacis plate. When originally fielded the T-72M1 did not have smoke mortars installed.

T-72M1 (STANAG M1981/2) This was the designation given to a modified version of the T-72M1 that appeared during exercises in Byelorussia in the summer of 1981. A panel of steel armour was placed over the rear engine deck about 100 to 150 mm from the normal deck.

T-72M1 (STANAG M1981/3) T-72M1 fitted with 12 smoke mortars on front of turret.

T-72M1 (late production) (STANAG Soviet Medium Tank M1984) Snorkel stowage moved to rear of aft stowage bin, third stowage bin added on left side of turret. Later models had applique armour matting on turret roof and either side of the driver's station and was also fitted to some T-64 MBTs and BMP-2 IFVs. First seen in 1984 Moscow parade.

T-72M2 (STANAG Soviet Medium Tank M1986) This is also called the Super Dolly Parton or the pot bellied armour tank and is similar to late production T-72M1 with applique roof armour but in addition has considerably thickened turret armour.

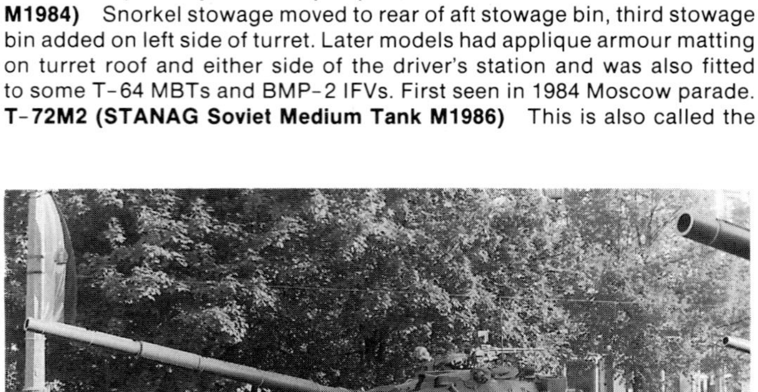

T-72M MBT of the then East German Army showing side skirts anc exhaust outlet over last road wheel

T-72M2 (STANAG Soviet Medium Tank M1988) Late versions of SMT M1986 had a new smoke mortar arrangement with eight mortars mounted on the left side near the gunner's hatch and in addition one of the ammunition boxes for the 12.7 mm NSV machine gun was deleted. This version is fitted with explosive reactive armour in addition to an enlarged gunner's night sight housing.

On the T-72M2 (SMT M1988) the glacis plate is covered with a layer of single ERA blocks while the turret is covered by one, two or three layers with one being the more normal. A single layer of ERA also covers the sides of the vehicle to provide protection to the suspension and upper sides of the hull.

There are two types of Soviet ERA blocks. First the standard shoe type box which has overall dimensions of about 25.5 cm by 13.5 cm and second the wedge type block. Both of these often have an arrow indicating the way that they should be installed to provide the maximum degree of protection.

The standard ERA block has an ERA element (two plates of steel separated by explosive) which can be positioned in one of two positions for maximum protection while the wedge type has a fixed ERA element.

In 1989 the Soviets provided the following information on a late production T-72 MBT fitted with an ERA package, although the weight quoted is probably an underestimate:-

COMBAT WEIGHT	41 500 kg
125 mm ROUNDS CARRIED	46
7.62 mm PKT MG AMMUNITION CARRIED	2000
12.7 mm NSV MG AMMUNITION CARRIED	300
RADIO	PP-173
ENGINE TYPE V-84	840 hp
TOTAL FUEL CAPACITY WITH LONG RANGE TANKS	1590 l
TOTAL FUEL CAPACITY WITHOUT LONG RANGE TANKS	1200 l
FUEL CONSUMPTION	4.5 l/km
MAXIMUM ROAD SPEED	60 km/h
TRENCH	2.8 m
VERTICAL OBSTACLE	0.8 m
FORDING	1.8 m

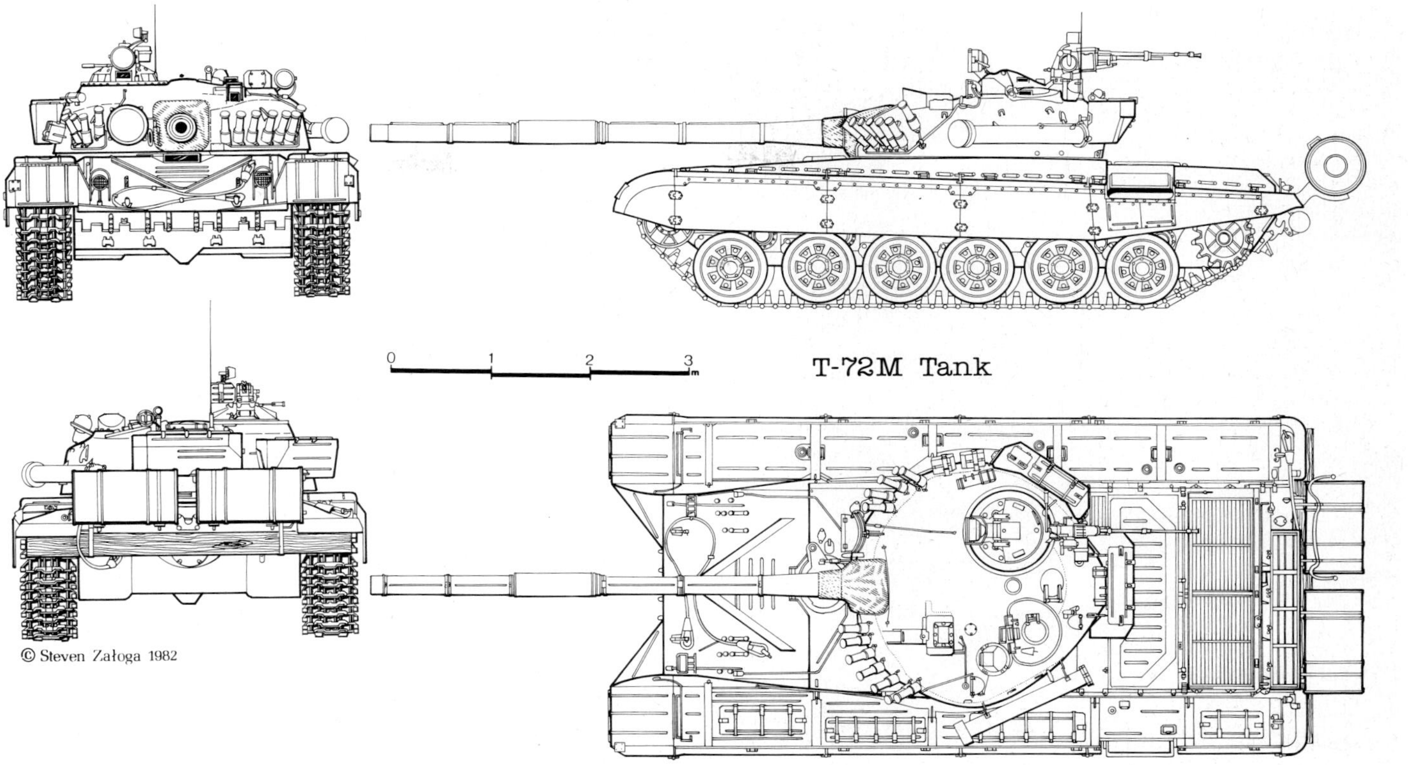

Four-view drawing of late production T-72 MBT designated T-72M1, this model fitted with 12 smoke dischargers, seven on left side of turret and five on right (not to 1/76th scale) (Steven J Zaloga)

T-72MS This is a new heavily reconfigurated version of the T-72 and first mentioned in *Soviet Military Power 1988*. It has a new suspension system, new engine and provision for mounting of explosive reactive armour. The export model may be designated the T-72S. It can also fire a laser guided projectile and some reports have indicated that some versions of T-72 have been modified to fire the AT-8 Songster (Kobra) ATGW.

TR-125 This is the Romanian version of the T-72 for which there is a separate entry.

T-72 Tank Distinguishing Features

TURRET FEATURES

		T-72A	T-72B46	T-72B (M)	T-72G (early)	T-72G (standard)	T-72G (final)	T-72M	T-72M1 (early)	T-72M1 (standard)	T-72M1 (1984 mod)	T-72M1 (1986 mod)
Rangefinder	Optical	X	X	X								
	Laser				X	X	X	X	X	X	X	X
Frontal armour	Basic	X	X									
	Thickened			X	X	X	X					
	Dolly Parton							X	X	X	X	
	Super											X
Smoke Mortars	None	X	X	X	X	X		X	X			
	Twelve						X			X	X	
	Eight											X
Search light	Right	X										
	Left		X	X	X	X	X	X	X	X	X	X
Fire extinguishers	Yes						X	X	X	X		
	No	X	X	X	X	X					X	X
Main gun	2A26	X										
	2A46		X	X	X	X	X	X	X	X	X	X
Roof liner	Yes										X	X
	No	X	X	X	X	X	X	X	X	X		
Hand-holds	Yes	X	X		X							
	No			X	X	X	X	X	X	X	X	X
Stowage bins	Two	X	X	X	X	X	X	X	X	X		
	Three										X	X

HULL FEATURES

		T-72A	T-72B46	T-72B (M)	T-72G (early)	T-72G (standard)	T-72G (final)	T-72M	T-72M1 (early)	T-72M1 (standard)	T-72M1 (1984 mod)	T-72M1 (1986 mod)
Optional skirts	Gill armour	X	X	X	X							
	Rubber skirts					X	X	X	X	X	X	X
Added glacis armour	Yes								X	X	X	X
	No	X	X	X	X	X	X	X				

T-72 with mine clearing equipment

All members of the T-72 MBT family, with the exception of command tanks, can be fitted with mine clearing equipment at the front of the hull, for example the KMT-6. All vehicles have a dozer blade which folds back under the nose when not required.

T-72 with PW-LWD mine clearance system

For some years the Polish Army has deployed a tank-mounted rocket propelled mine clearing system on a T-54/T-55 chassis called the PW-LWD.

This consists of a rocket which is attached to a 170 m explosive filled hose. The complete system is carried in a boat shaped container which slides onto a special mount on the hull top at the rear of the vehicle.

More recently the system has appeared on a Polish T-72 MBT which can also be fitted with other types of nose-mounted mine clearance systems.

The vehicle halts on arrival at the minefield, and the PW-LWD system is fired. The explosive hose falls to the ground and is detonated, setting off any mines in its path.

MTU-72 AVLB

The Soviet Army has recently started to deploy an armoured vehicle launched bridge system based on the T-72 MBT chassis which is similar in appearance and concept to the older MTU-20 which is based on the T-55 chassis (qv). It is estimated that the bridge, when opened out, is 20.6 m long.

BREM-1 Armoured Recovery and Repair Vehicle

This is based on the chassis of the T-72 and mounted at the front of the hull on the left side is a hydraulic crane which can lift 12 tonnes, a main winch with a capacity of 25 tonnes which can be increased to 100 tonnes, auxiliary winch, hydraulically operated dozer/stabilising blade at the front of the hull, towing equipment and a complete range of tools and recovery equipment. Full details are given in *Jane's Military Vehicles and Logistics 1991-92* page 30.

BREM-1 ARV in travelling configuration

Soviet IMR-2 Combat Engineer Vehicle in action (Jane's Soviet Intelligence Review)

IMR-2 Combat Engineer Vehicle

A T-72 MBT chassis is used as the basis for the new IMR-2 combat engineer vehicle, fully described in *Jane's Military Vehicles and Logistics 1991-92* page 8. This is the replacement for the IMR based on the T-55 chassis.

The crane can be fitted with a number of attachments including pincers for uprooting trees. Pivoted at the front of the vehicle is a dozer blade that can be used in a V-configuration or as a straight dozer blade. When not required it is raised clear of the ground.

SPECIFICATIONS

Model	**T-64B**	**T-72B1**
CREW	3	3
COMBAT WEIGHT	39 500 kg	44 500 kg
POWER-TO-WEIGHT RATIO	17.7 hp/tonne	18.9 hp/tonne
GROUND PRESSURE	0.86 kg/cm²	0.84 kg/cm²
LENGTH		
gun forwards	9.9 m	9.24 m
hull	7.4 m	6.95 m
WIDTH		
without skirts	3.38 m	3.6 m
with skirts	4.64 m	4.75 m
HEIGHT (without AA MG)	2.2 m	2.37 m
GROUND CLEARANCE	0.377 m	0.47 m
TRACK WIDTH	580 mm	580 mm
LENGTH OF TRACK ON GROUND	4.4 m	4.25 m
MAX ROAD SPEED	75 km/h	80 km/h
RANGE		
road, without long-range fuel tanks	400 km	480 km
road, with long-range fuel tanks	550 km	550 km
FUEL CAPACITY	1000 l	1000 l
FORDING		
without preparation	1.8 m	1.8 m
with preparation	5.0 m	5.0 m
GRADIENT	60%	60%
SIDE SLOPE	40%	40%
VERTICAL OBSTACLE	0.8 m	0.85 m
TRENCH	2.28 m	2.8 m
ENGINE	5DTF 5-cylinder opposed piston, diesel, liquid-cooled, 700 hp	V-12 multi-fuel (V-46) 840 hp at 2000 rpm
TRANSMISSION	synchromesh, hydraulically assisted with 7 forward and 1 reverse gears	
STEERING	clutch and brake	clutch and brake
SUSPENSION	hydro-pneumatic	torsion bar
ELECTRICAL SYSTEM	24 V	24 V
ARMAMENT		
type	2A26	2A46
main	1 × 125 mm gun	1 × 125 mm gun
coaxial	1 × 7.62 mm MG	1 × 7.62 mm MG
anti-aircraft	1 × 12.7 mm MG	1 × 12.7 mm MG
AMMUNITION		
main	36*	45
coaxial	1250	2000
anti-aircraft	300	300
GUN CONTROL EQUIPMENT		
Turret power control	electric/manual	electric/manual
Turret traverse	360°	360°
Gun elevation/depression	+14°/-6°	+14°/-6°
Gun stabiliser		
vertical	yes	yes
horizontal	yes	yes

Note: Late in 1985 an official Soviet publication gave the following data on the T-72: height to top of commander's cupola 2.3 m, ground clearance 0.43 m, length of hull excluding unditching beam 6.84 m, length of track on ground 4.2 m, track 2.79 m, length overall 9.53 m.

*+6 ATGW

Status: The T-72 MBT is in production. Within the USSR production is undertaken at defence plants (which have a zavod number) attached to civilian facilities managed by the Heavy and Transport Machine Building Industry. These are Nizhni-Tagil Railroad Enterprise (Zavod No 183), Kharkhov (Malyshev) Locomotive Enterprise (Zavod No 75), Omsk Railroad Enterprise (Zavod No 13) and Chelybinsk Heavy Machinery Enterprise (Zavod No 178). It is believed that only one of these plants is currently producing the T-72 MBT. Production continues in Czechoslovakia at the Divince tank plant. Production was also undertaken in Poland at Labedy but this has now been completed. Iraq assembled the T-72 as the Babylon Lion but it is understood that the facilities were destroyed in the Gulf War in 1991, although the Soviet Union stopped supplies to key components before then. The T-72 is also being built in India at the Heavy Vehicle Factory at Avadi (qv), Romania (called TR-125) and in Yugoslavia who designates it as the M-84 (qv). In both cases the late production model T-72G is built.

Country	**Quantity**	**Comment**
Algeria	100	
Bulgaria	340	first deliveries in early 1980s
Cuba	50	number unconfirmed
Czechoslovakia	815	local production
Finland	80	
Germany	549	phasing out
Hungary	138	
India	750	local production
Iraq	100	number approximate
Libya	300	
Poland	757	local production
Romania	100	local production
Syria	1100	
USSR	20 000	
Yugoslavia	300+	local production

T-64, production completed, in service only with the Soviet Union.

Manufacturer: Soviet state factories.

T-62 Main Battle Tank

Development

The T-62 MBT was developed from the earlier T-54/T-55 series in the late 1950s and entered production in 1961. It was first seen in public during a parade held in Moscow in May 1965. Production of the T-62 continued in the Soviet Union until 1975 by which time about 20 000 tanks had been completed. It was also produced for export in Czechoslovakia between 1973 and 1978 with about 1500 being built. It was also produced in North Korea for home and export markets.

No production figures for the T-62 have been released although under the Conventional Forces Europe Treaty the Soviets declared the following quantities of the T-62 and its variants:

Region	IV-1	IV-2	IV-3	IV-4	V.1
T-62	1555	1322	798	11	233
T-62K	110	57	50	nil	53
T-62M	181	126	115	nil	55
T-62MK	2	2	2	nil	nil
T-62MV	113	113	112	nil	nil
T-62M1	60	60	60	nil	nil

The above figures are the situation as of November 1990. For the Conventional Forces Europe Treaty Region IV is defined as Central Europe while Region V covers the flanks of the Atlantic to the Urals (ATTU).

Main recognition features of the T-62 compared with the earlier T-54/T-55 are a longer and wider hull, different spacing on the road wheels as the T-62 has a distinct gap between the third and fourth and fourth and fifth road wheels, shape of the turret, and the longer and fatter gun barrel with a fume extractor towards its muzzle.

Description

The hull of the T-62 is divided into three compartments, driver's at the front, fighting in the centre with the engine and transmission compartment at the rear. The driver is seated at the front of the vehicle on the left side and is provided with a single piece hatch cover that opens to the left. Two vision blocks are mounted forward of his hatch cover, each with an integral defrosting element. The left vision block can be replaced by a TVN-2 infra-red periscope which has a 30° field-of-view and a maximum range of 60 m. A hull escape hatch behind the driver's seat opens to the inside of the vehicle. Mounted on the glacis plate is a wave deflector, and mounted to the rear of this on the right side are a white light and an infra-red headlamp.

The cast turret is in the centre of the tank with the commander and gunner seated on the left and the loader on the right. Both are provided with a single piece hatch cover that opens to the rear and can be locked vertical. Rails outside the turret can be used by infantry or for stowing personal equipment.

The commander's cupola has four periscopes, two mounted in the hatch cover and two in the forward part of his cupola. The commander's sight, designated the TKN-3, is a day/night binocular periscope with an integral infra-red capability mounted in the forward part of his cupola. For day use it has a magnification of ×5 and a 10° field-of-view and for night it has a magnification of ×4.2 and an 8° field-of-view. Effective range when used in conjunction with the OU-3GK searchlight is 400 m. The handles of the sight are used to rotate the commander's cupola and operate the searchlight, target designation equipment and other systems.

The gunner has a TSh2B-41u telescope with a rotating graticule for super-elevation required for the different types of ammunition and dual magnification, ×3.5 with an 18° field-of-view and ×7 with a 9° field-of-view, filter capabilities, stadiametric rangefinder (at the bottom of the graticule) and an integral wiper. The maximum sighting ranges are 4000 m for APFSDS, 3700 m for HEAT, 4800 m for HE 18 and 3600 m for HE 11. The gunner's infra-red sight is the TPN1-41-11 periscope which is used in conjunction with the main L-2G searchlight mounted coaxially to the right of the main armament and has an effective range of 800 m. The TPN1-41-11 has a magnification of ×5.5 and a 6° field-of-view.

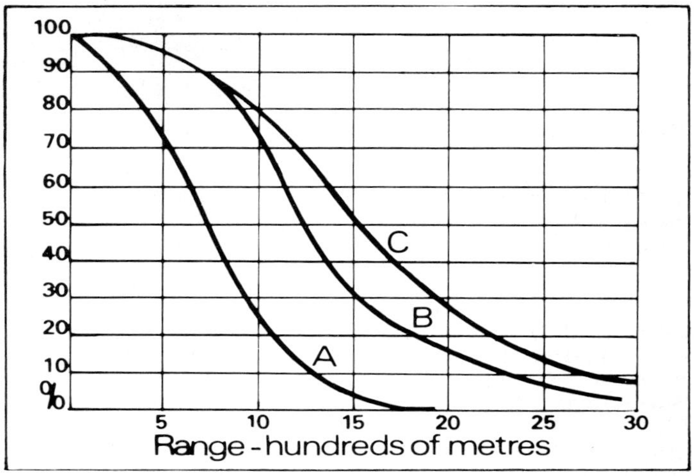

First-round hit probability of T-62 MBT with 115 mm gun showing **(A)** *firing HEAT projectile from stationary T-62 at target moving at 19 km/h* **(B)** *firing APFSDS projectile from stationary T-62 at target moving at 19 km/h* **(C)** *firing APFSDS projectile from stationary T-62 at stationary target*

T-62M MBT with 12.7 mm DShKM machine gun on loader's cupola. This vehicle is used for Opposition Forces (OPFORS) training by US Army in Germany (Michael Green/US Army)

T-62M MBT from rear showing long-range fuel tanks on hull rear (Michael Green/US Army)

The gunner is also provided with a Type TNP-165 periscope with a magnification of ×1. The loader on the right side of the turret is provided with a single TNP-165 periscope which can be used to the front or rear of the vehicle.

Mounted at the rear of the turret, to the left of the spent cartridge ejection door, is an electrically operated blower operated by the driver using the KUV-3 ventilator control box.

To the rear of the turret, over the engine compartment, is a large rectangular sheet steel plate that covers the engine louvres when the tank is snorkelling. The engine is equipped with a pre-heater and is normally started by compressed air, although an electrical auxiliary system is also provided. The engine is coupled to the manual transmission and changing up or down is accomplished by double declutching. The two-stage planetary steering system, which also serves as brakes, transmits torque to the final drives, each of which is a two step, step-down reduction gear.

The torsion bar suspension consists of five dual rubber-tyred road wheels with the drive sprocket at the rear and the idler at the front. A hydraulic shock absorber is provided at the first and last road wheel stations. The all steel track has steel pins that are not secured at the outer end and are free to travel towards the hull. A raised piece of metal welded to the hull, just forward of the sprocket, drives the track pins back into position each time they pass. Each track has 96 links and weighs 1386 kg when new. Like the T-54/T-55, the T-62 is now being fitted with an improved live track.

A centralised ethylene-bromide fire extinguishing system is activated automatically by heat sensors of which there are eight in the engine, transmission and fighting compartments, or manually by the tank commander or driver.

The T-62 MBT has a PAZ nuclear collective protection system which consists of a radiation detector/actuator (RBZ-1m), five separate explosive squib mechanisms and a blower/dust separator. The box like radiation detector/actuator (a radiation threshold detector) is mounted on the right side of the turret compartment behind the compressed air tanks.

The detector/actuator senses the initial pulse of radiation (gamma or neutron pulse) which precedes the blast wave and then activates the explosive squib mechanism. This is a system of spring loaded shutters, dampers, or louvres that are held open by a detent pin. When the squib mechanism is activated, an explosive charge detonates and forces the detent pin out of place, thus allowing the shutter, damper or louvre to close. The explosive squib mechanism's function is to close the engine louvres, sight aperture, bulkhead ventilation fan, air baffles to the transfer case and the air intake to the blower/dust separator. The blower/dust separator is in the turret below the shell ejection port and is an electric motor mounted with a set of fan blades which draws air into the vehicle and spins it at approximately 7000 rpm.

It should be noted that the blower/dust separator removes nuclear fallout only: it does not protect crew from chemical or biological contaminants as the air is not passed through a chemical filter. The tank must pass through the contaminated area as quickly as possible and must then be decontaminated before becoming fully operational again. Late production models of the T-62 are fitted with a chemical filter.

Mounted at the rear of the hull is an unditching beam and two drum fuel tanks can be installed at the rear to increase operational range. All vehicles have three external fuel cells on the right side of the vehicle for diesel fuel while the single tank on the left side is for auxiliary oil. The driver can select which fuel to use first, normally using the two rear drums, then the three external cells and finally the main fuel tank.

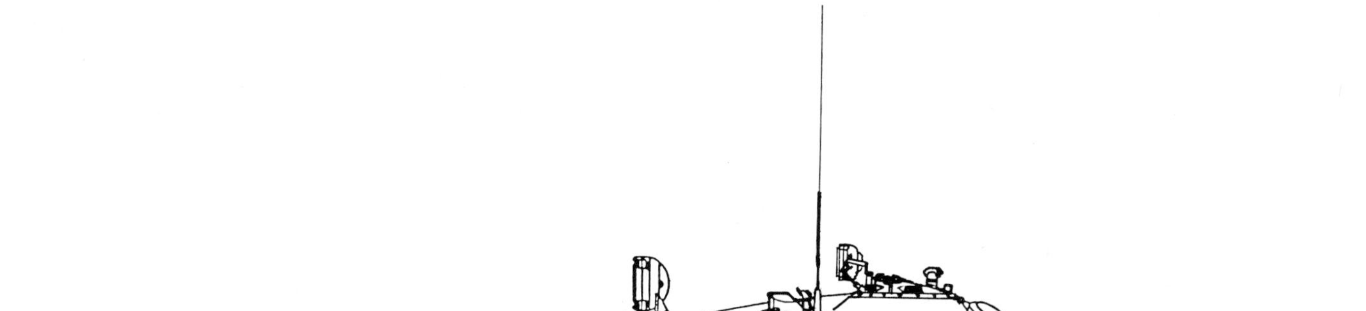

Standard T-62 MBT with long range fuel tanks at rear with unditching beam underneath (not to 1/76th scale) (Vasco Barbic)

The T-62 can, like other Soviet tanks, lay its own smoke screen by spraying diesel oil into the exhaust manifold when it is sufficiently hot, creating thick white smoke that exits from the exhaust ports on the left side of the tank. When laying the smoke screen the tank is always in second or third gear. This consumes about 10 litres of fuel per minute and produces a smoke screen 250 to 400 m long which lasts for some four minutes, but this does depend on the wind strength.

The tank can ford to a maximum depth of 1.4 m without preparation. It can also ford deep water when fitted with a snorkel. It takes up to eight hours to prepare the tank for deep fording and two types of snorkel are available; a large diameter one for training and a thinner one for operational use which is normally carried in sections on the rear of the turret. The tank is usually driven across the river bed in first gear and navigated by its GPK-59 gyro-compass and a radio link to the far bank. Once ashore, it takes only one or two minutes to prepare the tank for action again.

Main armament of the T-62 MBT is a U-5TS (2A20) 115 mm smoothbore gun fitted with a bore evacuator, with a maximum rate of fire of 4 rds/min when at a standstill. After firing, the main armament of the T-62 automatically elevates to an angle of +3° 30' for loading. The turret cannot be traversed while the weapon is being loaded. The 115 mm gun can fire HE-FRAG (FS) (OF-18), HEAT-FS (BK-4 and BK-4M) and APFSDS (BM-6) ammunition.

An integral spent shell ejection system, activated by the recoil of the gun, ejects the empty cartridge case out of the turret through a trapdoor in the turret rear. Of the 40 rounds (37 in command tanks) of 115 mm ammunition carried, two ready rounds are kept in the turret, one round by the gunner's feet, one by the loader's feet, 16 in the forward part of the tank to the right of the driver and 20 in the rear of the fighting compartment.

A 7.62 mm PKT machine gun mounted coaxially to the right of the main armament has a practical rate of fire of 200 to 250 rds/min and is fed by a belt containing 250 rounds.

The gun elevating and traversing mechanism consists of electric, hydraulic and manual controls. The gunner can elevate or depress the gun (electric/hydraulic) and both the commander and gunner can traverse the turret by electric controls through a full 360°. Only the gunner can traverse the turret and elevate the gun manually. The gun is fully stabilised in both horizontal and vertical planes.

Variants

T-62M (but called T-62A by NATO)

The main differences between the T-62 and the T-62M are the contour and the size of the turret and the removal of the fixed loader's hatch on the T-62 and its replacement by a rotating cupola mounting a 12.7 mm DShKM anti-aircraft machine gun, which can be used only from outside the tank. Some machine gun ammunition boxes are stored externally on the right of the turret. There are also differences in the design of the engine covers on the T-62A. The T-62 can also be fitted with mine clearing equipment like that fitted to the T-54/T-55, details of which are given in the entry for the T-54. According to American reports the T-62 tank is being updated in a similar fashion to the American M60 series improvement programme with laser rangefinder, ballistic computer, replacement of L-2G infra-red searchlight with new model and image intensification sights for both commander and gunner. The laser rangefinder is of the box type similar to that fitted on the T-54/T-55 and illustrated in the following entry.

T-62K Command Tank

In addition to having additional communications equipment, the T-62K command tank has the TNA-3 land navigation system. This appeared in 1973 and consists of six components; gyroscopic compass, one or two compass indicators, latitude correction device, an odometer, power converter and a calculator.

The gyroscopic compass indicates on what azimuth the vehicle is travelling and is mounted in the left rear of the turret compartment and can determine direction regardless of its position within its mounting. The compass indicator is mounted in the driver's compartment or in the tank commander's cupola; some T-62Ks have two compass indicators, one at each position.

The latitude correction device, odometer, and power converter are on the left side of the tank hull in the area of the turret. This adjusts the gyroscopic compass to conform to changes in latitude. The odometer converts the rotation of the driveshaft into distance travelled in metres. The power converter supplies the electricity for the TNA-3 land navigation system.

The calculator is in the turret to the right of the gunner and takes the signal from the other components and produces a readout which is used to navigate the vehicle.

While the system is warming up, which takes about 15 minutes, the T-62K cannot be moved without damaging the gyroscopic compass. Due to drifting of the gyroscopic compass, and other mechanical factors, the system accumulates error over a period of time and after several hours needs to be reset. The system does not automatically convert slope distance travelled into correct horizontal map distances, but the operator can manually adjust for this error by moving a correction setting. In addition, this correction allows for adjustment for varying road conditions such as dry pavement, mud or snow.

Once the crew member provides the system with his initial location and the pre-determined objective's location, the calculator gives a continuous en route numerical readout of the position of the tank, and its distance and azimuth to the objective. The readout from the calculator gives the distance in metres to the right and up, from the original location. From this readout the operator can compute his grid co-ordinates.

The command version of the T-62M is designated the T-62MK by the Soviet Army.

New track

In the late 1970s the Soviets introduced a new live track for retrofitting to the T-54, T-55 and T-62 MBTs. This is called the RMSh and requires a new drive sprocket.

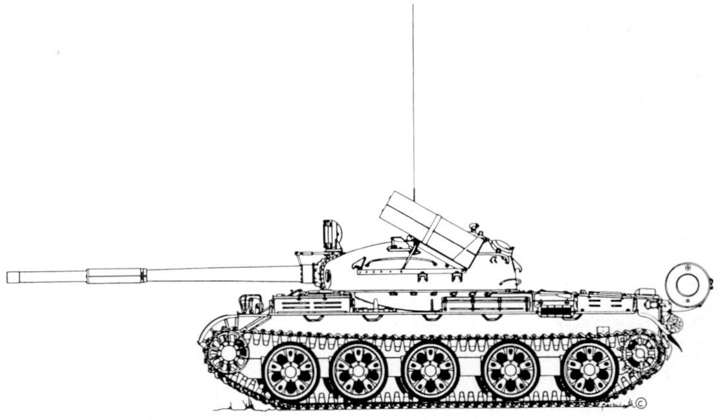

Egyptian Army T-62 MBT fitted with Sakr ground-to-ground smoke rockets (not to 1/76th scale) (Vasco Barbic)

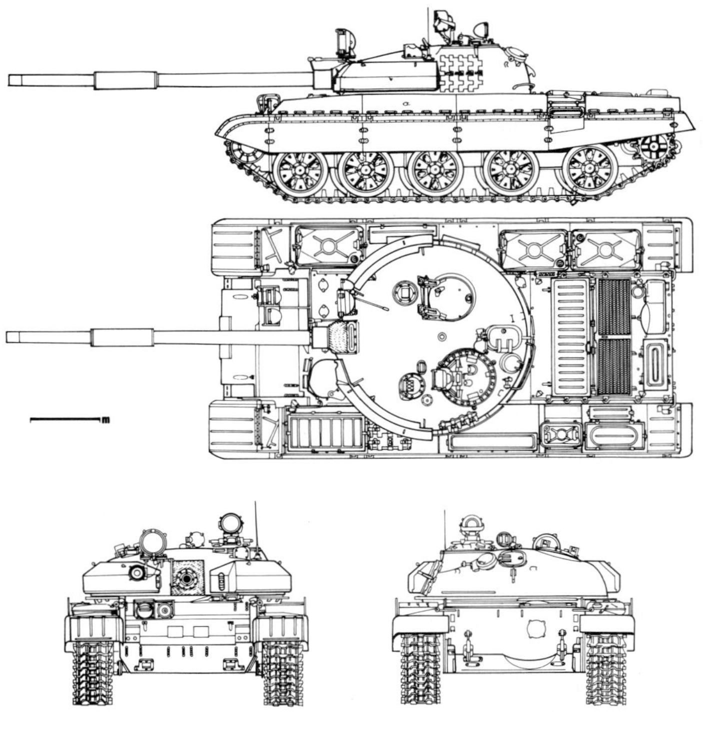

Four-view drawing of T-62 fitted with passive armour package (not to 1/76th scale) (Steven J Zaloga)

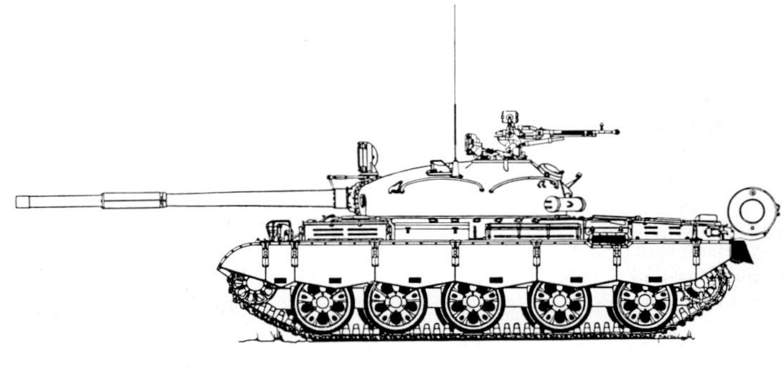

T-62M MBT as used by Iraqi Army in Middle East conflict showing side skirts (not to 1/76th scale) (Vasco Barbic)

Soviet T-62 series MBT in Afghanistan fitted with additional armour protection to hull and turret, side skirts and laser rangefinder over 115 mm gun to improve first round hit probability

Knocked out T-62 series MBT from the rear showing snorkel and unditching beam mounted on rear of hull

T-62 Flamethrower

A flamethrower version of the T-62 is in service under the Soviet designation of TO-62. This has a flame gun mounted coaxially with the 115 mm gun with an effective range of 100 m.

Modified T-62 MBTs

A number of T-62 tanks, including some which were deployed in Afghanistan, have been fitted with additional armour protection. This includes lateral protection for the fuel tanks above the road wheels, additional armour protection for the glacis plate, a horseshoe type armour array that runs from either side of the 115 mm gun halfway round the turret. This armour is similar to that fitted to the T-55AM2B upgrade (qv) and has a space to its rear which has a special filling for improved protection. These were also fitted with an improved fire control system with laser rangefinder over main armament and new night sights for commander and gunner. It is believed that this version is designated the T-62M1 while the T-62MV is the T-62M with explosive reactive armour.

Egyptian T-62s

Some Egyptian T-62s have been observed fitted with twin launchers on either side of the turret rear for indigenously developed Sakr ground-to-ground smoke rockets.

RO 115 mm Tank Barrel

British Royal Ordnance is now offering a 115 mm tank barrel as the replacement for the existing 115 mm U-5TS fitted in the T-62, and substantial quantities of these have been provided to Egypt who can no longer obtain barrels from the USSR.

T-62 with 105 mm L7 Gun

For trials in Egypt, NORICUM of Austria has fitted a Royal Ordnance 105 mm L7 rifled tank gun in place of the standard 115 mm U-5TS gun. Basically this conversion package includes the L7 105 mm rifled gun which utilises the existing recoil system, modified cartridge ejection door in the turret rear, modified ammunition stowage, new graticule for sight and emergency firing unit. Further improvements could include passive night vision equipment and a simple electronic fire control system. This has yet to enter production.

T-62 with GIAT 120 mm smoothbore gun

In mid-1986 GIAT proposed the replacement of the Soviet T-62's 115 mm U-5TS (2A20) smoothbore gun by the GIAT 120 mm smoothbore gun as fitted to the GIAT AMX-40 MBT. In this conversion the original elevation and traverse mechanism is retained and the existing 115 mm gun is replaced by the 120 mm smoothbore gun with loading from the right instead of the left, with the existing recoil system also being modified.

SPECIFICATIONS	
CREW	4
COMBAT WEIGHT	40 000 kg
UNLOADED WEIGHT	38 000 kg
POWER-TO-WEIGHT RATIO	14.5 hp/tonne
GROUND PRESSURE	0.77 kg/cm²
LENGTH GUN	
forward	9.335 m
rear	9.068 m
LENGTH HULL	6.63 m
WIDTH	3.3 m
HEIGHT	2.395 m
GROUND CLEARANCE	0.43 m
TRACK	2.64 m
TRACK WIDTH	580 mm
LENGTH OF TRACK ON GROUND	4.15 m
MAX SPEED	
1st gear	14.5 km/h
2nd gear	20 km/h
3rd gear	29 km/h
4th gear	45.5 km/h
5th gear	50 km/h
reverse	7 km/h
FUEL CAPACITY	
internal	675 l
external	285 l
supplementary external	400 l
MAX RANGE	
paved road	450 km
dirt road	320 km
paved road with additional fuel tanks	650 km
dirt road with additional fuel tanks	450 km

FUEL CONSUMPTION	
paved road	3 - 3.3 l/km
dirt road	1.9 - 2.1 l/km
FORDING	1.4 m
with preparation	5 m
GRADIENT	60%
VERTICAL OBSTACLE	0.8 m
TRENCH	2.85 m
TURNING RADIUS	skid turns
ENGINE	Model V-55-5 V-12 water-cooled diesel with injection pump fuel system, pressure lubricated, developing 580 hp at 2000 rpm
AUXILIARY ENGINE	none
TRANSMISSION	manual with 5 forward and 1 reverse gears
STEERING	2-stage planetary
CLUTCH	10 driving, 9 driven steel discs and a release mechanism
SUSPENSION	torsion bar
ELECTRICAL SYSTEM	24 V
BATTERIES	4 × 12 V, 150 Ah
ARMAMENT	
main	1 × 115 mm gun
coaxial	1 × 7.62 mm MG
anti-aircraft	1 × 12.7 mm MG (on T-62M only)
SMOKE-LAYING EQUIPMENT	diesel fuel injected into exhaust system
AMMUNITION	
main	40
coaxial	2500
anti-aircraft	300

GUN CONTROL EQUIPMENT	
Turret power control	electric-hydraulic/manual
by commander	yes (powered only)
by gunner	yes (electric-hydraulic/manual)
Turret traverse	360°
Gun elevation/depression	+16°/-6°
commander's override	yes
Gun stabiliser	
vertical	yes
horizontal	yes
Elevation quadrant	yes
Traverse indicator	yes
ARMOUR	
Hull front upper	102 mm at 60°
Hull front lower	102 mm at 54°
Hull sides upper	79 mm at 0°
Hull sides lower	15 mm at 0°
Hull rear upper	46 mm at 0°
Hull rear lower	46 mm at 0°
Hull top	31 mm
Hull floor front	20 mm
Hull floor rear	20 mm
Turret front	242 mm
Turret sides	153 mm
Turret rear	97 mm
Turret top	40 mm
Turret mantlet	included in above figures
Turret hatches	30 mm/31 mm

Status: Production complete. In service with the following countries:

Country	Quantity	Comment
Afghanistan	170	
Algeria	300	
Angola	100+	
Cuba	300	
Egypt	600	UK supplied new barrels
Ethiopia	100	status uncertain
Iran	150+	from North Korea
Iraq	400	approx number
Israel	120	
Korea, North	1500	local production
Libya	900	approx number
Mongolia	n/av	
Syria	1000	
USSR	9000	
Vietnam	200	approx number
Yemen	30	

Manufacturer: Soviet state factories. The T-62 was also produced in Czechoslovakia for export markets. Production of the T-62 was once undertaken in North Korea and some of these have been exported to Iran.

T-54 and T-55 Main Battle Tanks

Development

During 1944 the Soviet Union designed a new medium tank called the T-44. This was produced in small numbers between 1945 and 1949, but proved unreliable in service. Main improvements over the earlier T-34/85 were its torsion bar suspension, transverse-mounted engine and transmission, well-shaped hull and a turret with a similar shape to the T-34/85, except that it did not have the thick turret neck. The tank had a crew of four as the bow machine gunner was eliminated in favour of a fixed machine gun firing forwards, a feature later adopted for the T-54. The turret of the T-44 was a modified T-34/85 turret without the prominent collar at the turret base. The T-44 was used in combat towards the end of the Second World War and again during the Hungarian uprising of 1956. Rebuilt T-44s, known as T-44Ms, were still in service in the 1970s, and they have also been used for training.

The T-44 was followed by the T-54, the first prototype of which was completed in 1946 with production following in 1947 in Kharkov. No production figures for the T-54 and T-55 have been released but it is estimated that over 50 000 were built in the Soviet Union with production also being undertaken in China (as the Type 59), Czechoslovakia and Poland. Production of the T-55 is believed to have continued at the Omsk tank plant until as late as 1981, long after the T-62 had gone out of production. The chassis of the T-54, in a much modified form, was also used as the basis for the ZSU-57-2 self-propelled anti-aircraft gun system.

No production figures for the T-54/T-5 series of MBT have been released by the Soviet Union although under the Conventional Forces Europe Treaty they did declare the following quantities of T-54/T-55 vehicles:

Region	IV-1	IV-2	IV-3	IV-4	V.1
T-54	nil	250	248	nil	25
T-54K	1	1	nil	nil	nil
T-54B	901	296	244	nil	330
T-54BK	24	16	10	nil	8
T-54M	642	243	31	nil	359
T-54MK	25	8	3	nil	17
T-55	468	301	231	nil	167
T-55K	74	6	5	nil	68
T-55A	877	709	477	nil	168
T-55AD	237	177	176	nil	60
T-55AK	80	72	20	nil	8
T-55AM	nil	nil	79	nil	67
T-55AMD	21	21	21	nil	nil
T-55AMK	1	1	1	nil	nil
T-55AMV	39	39	36	nil	nil
T-55M	1006	496	342	nil	517
T-55MK	nil	18	13	nil	2
T-55MV	143	143	116	nil	nil
T-55MVK	2	2	2	nil	nil

The figures are the situation as of November 1990. For the Conventional Forces Europe Treaty Region IV is defined as Central Europe while Region V is defined as covering the flanks from the Atlantic to the Urals (ATTU).

Description

The all-welded hull of the T-54 is divided into three main compartments: driver's at the front, fighting in the centre and the engine and transmission at the rear.

The driver is seated at the front of the tank on the left and is provided with a single piece hatch cover that swings to the left. Two periscopes are provided forward of this hatch, one of which can be replaced by an infra-red periscope which is used in conjunction with the infra-red light mounted on the right side of the glacis plate. The driver's infra-red system is known as the TVN-2. A narrow board mounted at right angles to the glacis plate stops water rushing up the glacis plate when the tank is fording shallow rivers. To the right of the driver are ammunition stowage, batteries and a small fuel tank. A hull escape hatch is provided behind the driver.

The turret is a one piece casting with the top welded into position. The top consists of two D-shaped pieces of armour welded together down the centre. The commander is seated on the left of the turret, with the gunner on the same side but in front of the commander. The commander has a cupola which can be traversed through 360°, with a single piece hatch cover that opens forwards with a single periscope each side. A TPK-1 sight with a single periscope either side is mounted in the forward part of the cupola top. When in action the commander is constantly rotating his cupola for possible targets. On sighting a target he lines it up with his TPK-1 sight and then slews the turret onto the target. The gunner then lays the gun accurately onto the target and fires the main armament. The gunner has a TSh 2-22 sight with a magnification of ×3.5 or ×7. The loader is seated on the right of the turret and is provided with a single piece hatch cover that opens to the rear and a single periscope. The T-54 does not have a rotating turret floor, but the later T-55 has a partial floor, as does the T-62.

The engine is mounted transversely at the rear of the hull. The T-54 uses an electrical start-up system for the engine as its primary means, with a compressed air system for back up in cold weather. The T-55 uses a compressed air engine start-up system primarily, with an electrical back up. This is due to the fact that the T-55 has an AK-150 air compressor to refill the air pressure cylinders; the T-54 has no such air compressor.

The torsion bar suspension consists of five single rubber-tyred road wheels with a distinct gap between the first and second road wheels. The drive sprocket is at the rear and the idler at the front. There are no track return rollers. The first and fifth road wheel stations are provided with a hydraulic shock absorber. The all-steel track has steel pins that are not secured at the outer end and are free to travel towards the hull. A raised piece of metal welded to the hull just forward of the sprocket drives the track pins back into position each time they pass. Like the T-62, the T-54 and T-55 are now being fitted with a new track.

Basic T-54 MBT without anti-aircraft machine gun used by US Army for training (Michael Green/US Army)

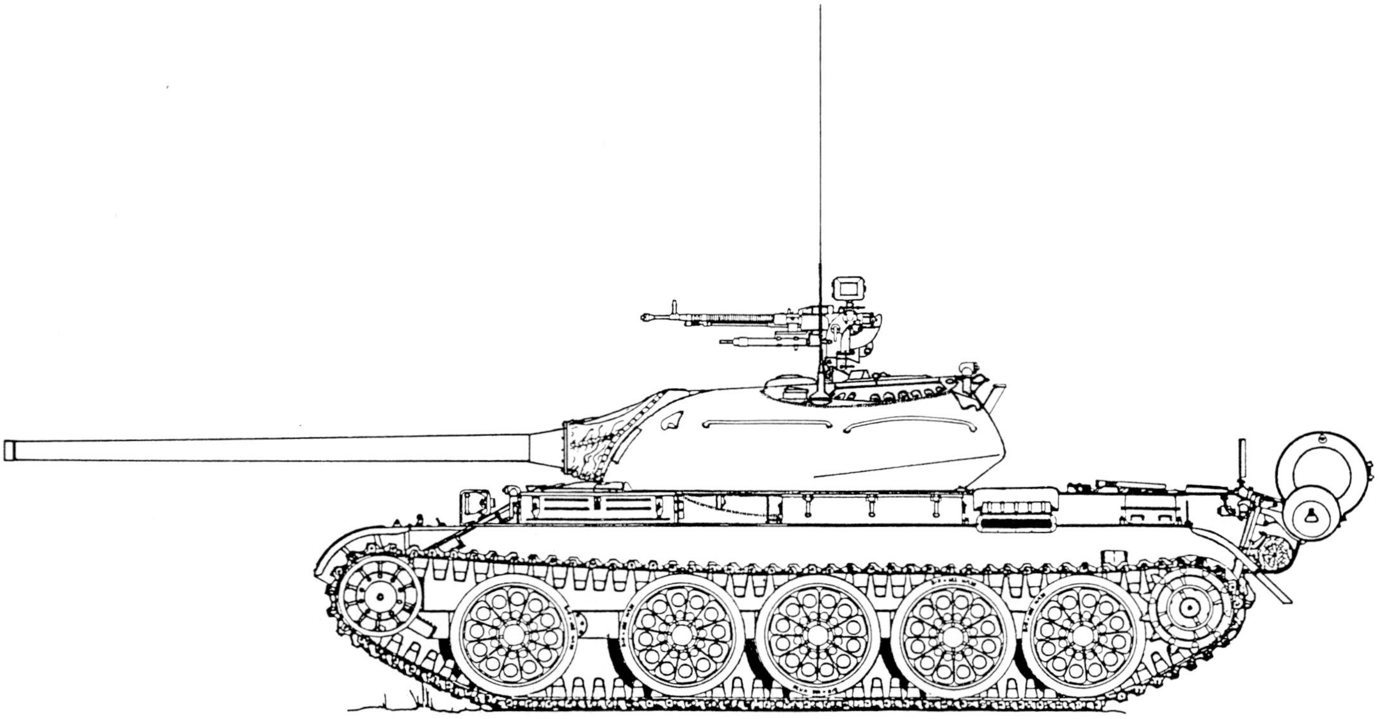

Early production T-54 Model 1949 showing distinct turret rear (not to 1/76th scale) (Vasco Barbic)

The T-54 and T-55 can be fitted with a snorkel for deep fording. Two types of snorkel are available, a thin one for operational use and a thick one for training. These snorkels take between 15 and 30 minutes to fit and on reaching the far bank are blown off. The operational snorkel is mounted over the loader's periscope and when not fitted is carried disassembled at the rear of the turret, or at the rear of the hull. The thicker snorkel is mounted over the loader's hatch cover. The tank normally crosses the river in first gear and navigates with the aid of an onboard gyrocompass.

As originally built the T-54 was not fitted with an NBC system, which was installed on later production tanks and subsequently refitted to earlier production models. The tank can lay its own smoke-screen by injecting diesel fuel into the exhaust on the left side, producing a cloud of smoke some 300 m long which lasts about two minutes.

To extend the tank's operational range two drum type fuel tanks can be fitted at the rear. On the right running board are four flat light steel tanks. The first, second and fourth tanks each contain 93 litres of fuel while the third contains lubricating oil. The driver can select which fuel tanks to use first, for example the drums at the rear, followed by the three external tanks and finally the main fuel tanks. There are three stowage boxes on the left running board, the first one for tools and other maintenance equipment, the second for gun-cleaning equipment and the third for infra-red equipment. Mounted at the very rear of the hull is an unditching beam.

When the T-54 entered service it was not fitted with night vision equipment. The first production tank to have this fitted was the T-54B and early tanks were then refitted with it. The tank is normally fitted with the following equipment: infra-red driving light which is used in conjunction with the driver's infra-red periscope and has a range of 40 to 60 m. Mounted on the forward part of the commander's cupola is an infra-red searchlight with a range of 400 m, while mounted to the right of the main armament is another infra-red searchlight with a range of 800 to 1000 m.

Main armament of the T-54 is a 100 mm D-10T (originally called the M1944) rifled tank gun which was originally developed around naval gun ammunition and also used in a modified form in the SU-100 assault gun. The D-10T has a horizontal sliding wedge breech-block with the recoil system consisting of a hydraulic buffer and a hydro-pneumatic recuperator, an overall length of 5.608 m and weight of 1948 kg. The weapon has a maximum range in the indirect fire role of 14 600 m and can fire the following types of ammunition: AP (BR-412), APC-T (BR-412D), HE (F-412), HE-FRAG (OF-412), HEAT-FS (ZBK-5M) and HVAPDS-T (BM-8).

In 1991 the Soviet Union stated that the T-54/T-55 series MBT fired the following main types of ammunition: APFSDS designated the ZBM-25 with a muzzle velocity of 1430 m/s; HEAT designated the ZBK-17M with a muzzle velocity of 1085 m/s; HE-FRAG with a muzzle velocity of 900 m/s.

MECAR of Belgium has developed a new 100 mm APFSDS round for the T-54/T-55, which is called the M1000 and is already in production for one Far East customer.

An average rate of fire for a T-54 is 4 rds/min. The D-10T gun is not stabilised but the D-10TG, first fitted to the T-54A, is stabilised in the vertical plane only while the D-10T2S, which is stabilised in both planes, is fitted in the T-54B, T-54C, T-55, T-55A and has been refitted in earlier tanks.

The turret traversing and gun elevating system was manual on early tanks but later models have full power traverse and elevation with manual controls for use in an emergency.

Mounted to the right of the main armament is a 7.62 mm SGMT machine gun and a similar weapon fixed in the centre of the glacis plate fires forward and is operated by the driver by pressing a button on the right steering lever. Mounted at the loader's position is a 12.7 mm DShKM anti-aircraft machine gun.

Variants

T-54 Model 1946 (Soviet designation)

Prototype vehicles with cast turret which was undercut at the rear, never entered volume production or service.

T-54 Model 1949 (Soviet designation)

New turret with increased protection over frontal arc but the undercut turret rear was retained, 7.62 mm MG mounted in the glacis plate and operated by driver, mounted on the right track fender were two flat fuel tanks connected to the main fuel system.

T-54 Model 1951 (Soviet designation)

This had a new turret which incorporated increased protection over the frontal arc with the undercut rear being eliminated. This turret was used for all subsequent models and modifications of the T-54/T-55.

T-54

The second production model has the normal type turret with two cupolas. On the right cupola is a 12.7 mm DShKM machine gun. When fitted with infra-red night vision equipment it is known as the T-54(M) and was also built in China under the designation Type 59 (see China entry in this section).

T-54 series MBT without 12.7 mm anti-aircraft machine gun and narrow board on glacis plate (US Army/Michael Green)

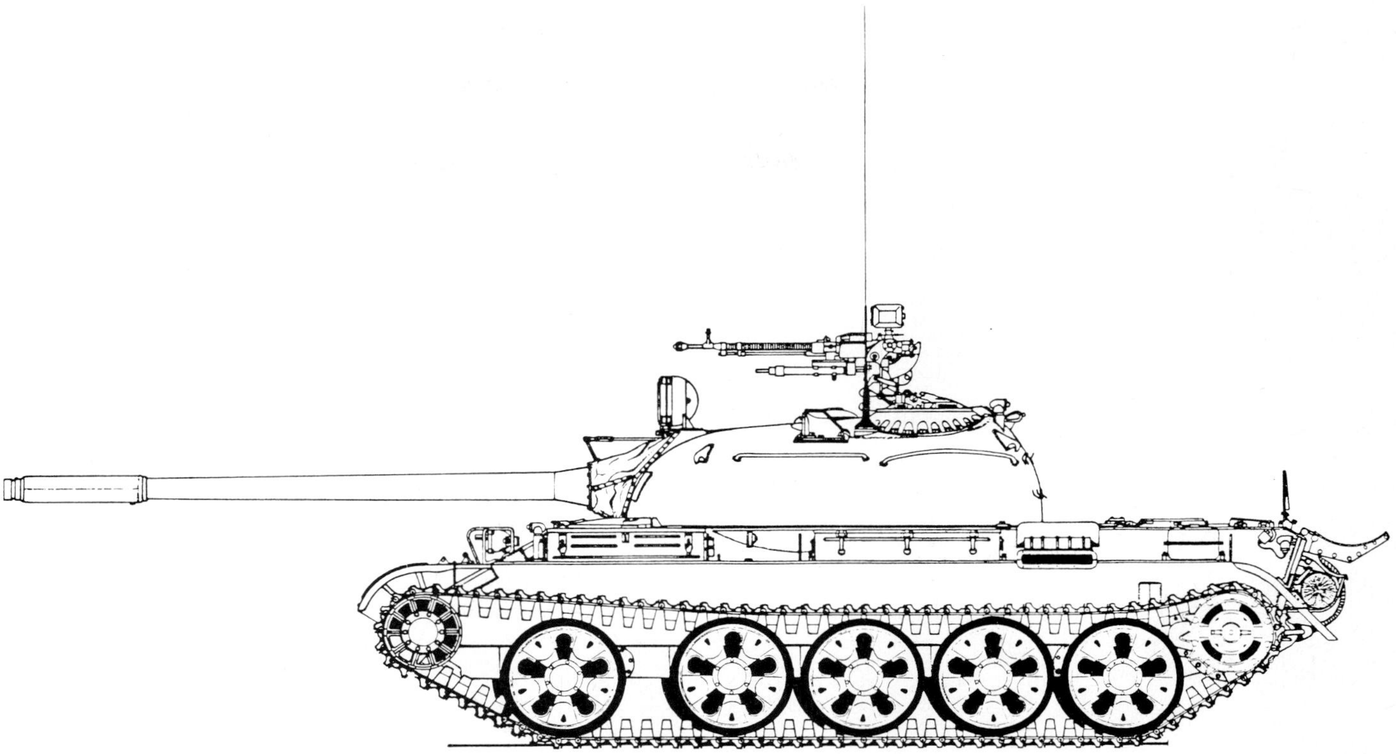

T-54B tank with 12.7 mm anti-aircraft machine gun but without long range fuel drums at rear (not to 1/76th scale) (Vasco Barbic)

T-54A

This model was first seen in the mid-1950s and its main armament has a bore evacuator which was subsequently fitted to all further T-54/T-55 tanks. Its D-10TG gun is stabilised in the vertical plane only and other improvements include power elevation for the main armament, electric oil pump, bilge pump, modified air filter and an automatic fire-extinguishing system. When originally built the tanks were not fitted with infra-red night vision equipment, but most have now been fitted, and designated T-54A(M). The gunner's sight is a model TSh 2A-22.

T-54B (Soviet designation)

This entered service in the late 1950s and was the first to be fitted with a complete infra-red night vision equipment. Mounted over the main armament is the L-2 infra-red searchlight, the commander's smaller searchlight is designated the OU-3 while the gunner's standard MK-4 observation device has been replaced by the TPN-1 night observation device. The number of external fuel tanks was increased from two to three.

T-55M (Soviet designation)

First prototype was completed in 1954 and was first seen in public during a parade held in Moscow in November 1961. Major differences between this and the T-54 can be summarised as follows: the T-55M has no loader's cupola or 12.7 mm DShKM anti-aircraft machine gun (this has since been fitted to some models), no turret dome ventilator forward of the loader's hatch, 580 hp engine, modified transmission, rotating turret floor, 100 mm D-10T2S gun stabilised in both planes, increased 100 mm ammunition stowage and simplified pattern road wheels. When fitted with a 12.7 mm anti-aircraft machine gun this version is referred to as the T-55(M) in the west.

T-55A

This model was seen for the first time in May 1963. Major differences from earlier members of this family are the replacement of the 7.62 mm SGMT by the more recent 7.62 mm PKT weapon, the elimination of the 7.62 mm bow machine gun (which allowed six more rounds of 100 mm ammunition to be carried), the base of the commander's cupola being smooth rather than bolted, the ability to lay a smoke-screen through exhaust outlet, NBC protection system, air compressor for starting, redesigned front fuel and ammunition compartments, infra-red night vision equipment for commander, gunner and driver, improved deep fording capability, improved shock absorbers and fire suppression system, and both the loader's and driver's hatch covers being raised. The T-55A has an anti-radiation lining in the inside of the hatches, which accounts for the thickened hatches, as well as the turret and fighting compartment. When fitted with a 12.7 mm anti-aircraft machine gun the T-55A is known as the T-55A(M).

T-54/T-55 Command Vehicles

Command versions of the T-54 are designated the T-54K, T-54BK and T-54MK while command models of the T-55 are designated the T-55K, T-55AK, T-55AMK, T-55MK and T-55MVK.

Soviet T-55A MBT fitted with 12.7 mm DShKM anti-aircraft machine gun at US Army's National Training Center, Fort Irwin, California (Michael Green/US Army)

T-55 MBT showing additional armour protection on hull and turret and laser rangefinder over 100 mm gun

T-55M MBTs fitted with 12.7 mm anti-aircraft machine gun (covered up) and turret traversed to rear

Peruvian Army T-55 series tank clearly showing 12.7 mm DShKM anti-aircraft machine gun. Note hole in glacis plate for 7.62 mm machine gun, snorkel on right of turret and combing around loader's hatch (US Army)

Close-up of T-55AM turret showing laser rangefinder over mantlet and mast-mounted meteorological sensor and laser warning device. This vehicle is not fitted with the horseshoe armour around the turret

Egyptian T-54/T-55

Details of the modifications carried out and proposed on the Egyptian T-54/T-55 tanks are given under Egypt.

Finnish T-55

Details of these modified vehicles are given under Finland in this section.

T-55AD (Soviet designation)

T-55 with radiation liners and additional radio set.

T-55AM (Soviet designation)

T-55A with 12.7 mm DShK cupola.

T-55AMD (Soviet designation)

T-55AM with additional radio set.

T-55AM2B

This has a number of improvements in the key areas of armour, mobility and firepower. First converted tanks were observed in service with Czechoslovakia in 1984 and since then Bulgaria, Hungary and Poland have also started to update some of their vehicles to a similar standard, although often there are differences between countries. For example Polish vehicles have the laser rangefinder integrated with the gunner's sight and a hammer-shaped crosswind sensor on the front of the turret roof.

In most countries the upgraded T-55s go to the motorised rifle divisions while the tank divisions have the T-72 series MBT.

The Kladivo fire control system fitted to the T-55AMB2 includes a laser rangefinder mounted externally above the mantlet of the 100 mm gun, ballistic fire control computer, wind sensor, mast-mounted meterological sensor that also incorporates a laser warning device, operating switch, input device, commander's periscopic sight, ammunition selection switch and power supply.

The laser may be triggered up to 10 times a minute but its performance is degraded in rain, fog and snow.

The rear mast is an armoured mast on the turret rear which contains temperature and air pressure sensors as well as a laser warning device to alert the crew if they are being lased by enemy tanks or target designators.

The operating switch allows the gunner, in case of malfunction, to switch off the complete fire control system and carry out a normal target engagement.

The input device is operated by the tank commander and in addition to the digital input keyboard for the manual input of data in the fire control computer, there is a digital display field. The computer is capable of estimating target speed and direction for the establishment of correct lead angle.

The commander's periscope is installed in the commander's cupola and allows independent observation of the target by the commander and the automatic transfer of the target to the gunner.

The loader operates the ammunition selection switch and has a choice of five different ammunition types which are selected by the tank commander, depending on the target to be engaged. These are HE, HEAT, incendiary, APFSDS and machine gun.

The fire control computer is probably of Czechoslovakian design while the power supply unit maintains a constant voltage for the system.

A typical target engagement takes six to eight seconds using the Kladivo fire control system with the sequence of events being as follows:

1. The tank commander selects a target using his command periscope and transfers the target to the gunner using the turret traverse override. The gunner then manually centres the target in the crosshairs using the guns stabiliser and at this point the crosshairs are green in colour
2. The loader selects the ordered ammunition and operates the ammunition selection switch
3. The commander operates the laser rangefinder by pushing the button on the input device and inputs the value into the fire control computer
4. After the tank commander has established the speed of his own tank and the target's speeds he uses the input device to input these values to the fire control computer. Information from the meterological sensor on wind speed, air pressure and temperature are automatically relayed to the computer
5. The system notifies the commander that it is ready to fire via a light on the input device and the commander verifies the data by pushing a button on the input device and at that point the gunner receives a light signal in his optics
6. When the system is ready to fire and if the gunner has maintained his sight picture, the gunner's optics crosshairs will turn red and the gunner can fire.

The installation of the Kladivo fire control system has enabled the T-55 AM to engage moving targets out to a range of 1600 m with an improved first round hit capability. Other improvements incorporated into the T-55AMB2 include:

1. Full track skirts which keep the dust down and provide some degree of flank protection for light anti-tank weapons

T-55A tanks fitted with laser rangefinder over main armament and two infra-red driving lights on glacis plate

Czechoslovakian T-55A MBTs crossing a bridge during exercises

2. Horse-shoe type armour either side of the turret in a similar manner to that fitted to Soviet T-62s in Afghanistan for increased protection against light anti-tank weapons. The rear part of the horseshoe armour is hollow and is filled with a substance for improved protection
3. Additional armour protection on the glacis plate for increased protection against kinetic energy projectiles, additional protection from mines under nose of vehicle and narrow vertical steel-reinforced plastic shields on the outer edges of the sponsons
4. Bank of eight smoke grenade launchers on right side of turret firing forwards. These have a range of about 250 m and may well be activated by the laser warning device
5. Some vehicles have been fitted with improved engines and suspensions which has compensated for the increase in vehicle weight by the application of new armour to hull and turret
6. Thermal sleeve for 100 mm gun, although this has not appeared on all T-55AMs
7. New tracks, with wear resistant links, are from the more recent T-72 MBT, as are the road wheels.
8. The 100 mm gun can fire a missile designated the 9M117 'Bastion' which has been given the NATO designation of the AT-10 which is believed to have a maximum range of 4000 m with a typical ammunition load being six ATGWs and 43 rounds of 100 mm ammunition. The missile is guided to the target by the gunner.

NB: It is possible that the version without the additional armour pack is the T-55AM while that with the armour package is the T-55AMB2.

T-55 with ERA

A number of T-55 MBTs used by the Soviet Naval Infantry have been observed fitted with a laser rangefinder over their main armament to improve first round hit probability while the barrel has also been fitted with a thermal sleeve.

Explosive reactive armour has been fitted to the hull and turret for improved protection against weapons with a HEAT warhead. A new commander's periscope has been installed and the standard 12.7 mm DShKM anti-aircraft machine gun has been replaced by the more recent 12.7 mm NSV machine gun which is also installed on the T-64, T-72 and T-80 MBTs.

The original tracks have been replaced by T-72 type tracks. It is not clear if the Soviet Naval Infantry T-55 incorporate all of the features.

This version of the T-55 may have the Soviet designation of the T-55MV or T-55AMV.

Vietnamese T-54/T-55

Some Vietnamese T-54/T-55 tanks have been fitted with additional bolt-on armour protection.

Soviet Naval Infantry vehicles with PT-76 light amphibious tank in foreground and modernised T-55 with explosive reactive armour and laser rangefinder over mantlet in the background

Differences between production tanks

Polish-produced tanks often have different stowage arrangements which include the mounting of a rectangular box on the left side of the turret, a smaller square stowage box on the left side of the turret rear and a slightly different rear decking. In 1978 a number of T-54s and T-55s were observed with a box type laser rangefinder mounted over the main armament.

BTS-1 Armoured Recovery Vehicle

This was the first version of the T-55 to be used as an ARV and is simply a T-54A with its turret removed. It is called the BTS-1 (Bronetankoviy Tyagach Srdniy—Medium Armoured Tractor) by the Soviet Army.

BTS-2 Armoured Recovery Vehicle

This was the second version to enter service and also has its turret removed and is fitted with a winch and a large spade at the hull rear for vehicle recovery. It has tool and specialised equipment stowage and a small tripod jib crane can be quickly erected. This version was also built in Czechoslovakia as the WZT-1.

BTS-3 Armoured Recovery Vehicle

This was fielded in the 1960s and is also known as the SPK-12G and is fitted with a hydraulic crane on the front left of the hull which can be used to the front or rear. It has a telescopic jib and can be used to lift complete tank turrets as well as powerpacks. Other equipment includes a front-mounted BTU-55 dozer blade, rear-mounted spade, winch, tools and other specialised equipment. The Czechoslovakian equivalent is the WZT-2.

BTR-4 Armoured Recovery Vehicle

This is similar to the BTS-2 but has a crane pivoted on the left side of the hull with a telescopic jib which can lift a maximum weight of 1500 kg. It also has a winch, rear-mounted spade and extensive external stowage.

Czechoslovakian MT-55 ARV

This Czechoslovakian designed and built ARV has a crane pivoted at the hull rear on the right side and when travelling the telescopic jib is traversed forwards. A cargo carrying area similar to that installed on Soviet ARVs is provided and the vehicle commander has a distinctive cupola on the forward right side of the hull. Additional details are given in *Jane's Military Vehicles and Logistics 1991-92* page 20.

KAM-1 Armoured Recovery Vehicle

To meet Finnish Army requirements VAMMAS of Finland has built the prototype of the KAM-1 ARV on a T-55 chassis. This has a crane, dozer blade and recovery winch. Details are given in *Jane's Military Vehicles and Logistics 1991-92* page 20. A more recent model is the KAM-2.

Polish T-54/T-55 ARVs

Poland has developed at least two ARVs based on T-54 or T-55 MBT chassis, designated WZT-1 and WZT-2. The WZT-1 is the Polish equivalent of the BTS-2 with some differences while the WZT-2 is a new design and equivalent to the Soviet BTS-3 ARV.

Czechoslovak MT-55 Bridgelayer

This has been developed by Czechoslovakia as the replacement for the earlier MT-34 which was based on the T-34 tank chassis. The MT-55 is fitted with a scissors bridge which is launched hydraulically over the front of the vehicle. When opened out it is 18 m long and will span a gap of 16 m, and will take a maximum load of 50 000 kg. Weight is 37 000 kg and it has a crew of two. Full details of the MT-55 armoured bridgelayer are given in *Jane's Military Vehicles and Logistics 1991-92* page 77.

Czechoslovakian MT-55 ARV in travelling configuration

T-55 MTU-20 bridgelayer in travelling configuration

BLG-60 bridgelayer in travelling configuration

TO-55, flamethrower version of T-55 with flame gun to right of 100 mm gun

ZSU-57-2 SPAAGs in travelling order

German BLG-60 Bridgelayer

This was developed by the former East German and Poland and is used by them in place of the Czechoslovak MT-55 bridgelayer. The scissors bridge is launched over the front of the vehicle; it is 21.6 m long and will span a gap of 20 m. Maximum capacity is 50 000 kg. The BLG-60 weighs 37 000 kg and has a crew of two or three. Full details of the BLG-60 are given in *Jane's Military Vehicles and Logistics 1991-92* page 81. More recently an improved version of the BLG-60 bridgelayer has been introduced under the designation of the BLG-60 M2. This features improved reliability of the hydraulic system, increased capacity with increased safety for vehicles using the bridge, use of a number of BLG-60 M2 bridges for multiple crossings of gaps and is more reliable in the field. It is understood that former East German bridgelayers have been taken into service with the German Army.

Soviet MTU-20 Bridgelayer

The MTU-20 was developed in the late 1960s as the replacement for the older MTU and is launched in a similar manner except that when travelling the ends of the bridge are folded on top to reduce the overall length of the equipment. When opened out the bridge is 20 m long and will span a gap of up to 18 m; maximum capacity is 60 000 kg. The MTU-20 weighs 37 000 kg and has a crew of two. Full details of the MTU-20 armoured bridgelayer are given in *Jane's Military Vehicles and Logistics 1991-92* page 89.

Soviet MTU-1 Bridgelayer

The MTU-1 bridgelayer entered service with the Soviet Army in the late 1950s and is basically a T-54 with its turret removed and replaced by a launching system for a bridge 12.3 m long. The bridge will span a gap of up to 11 m and has a maximum capacity of 50 000 kg. The MTU-1 weighs 34 000 kg and has a crew of two. Full details of the MTU-1 armoured bridgelayer are given in *Jane's Military Vehicles and Logistics 1991-92* page 90.

Combat Engineer Vehicle (IMR)

This is essentially a T-55 MBT with its turret removed and replaced by a hydraulically operated crane that can be traversed through a full 360°. This is fitted with a pair of pincer type grabs which are used to remove trees and other obstacles, and can be replaced by a small bucket. Mounted at the front of the hull is a hydraulically operated dozer blade which can be used in a straight or V-configuration, but cannot angle doze. Full details of the IMR are given in *Jane's Military Vehicles and Logistics 1991-92* page 9.

Dozer T-54

The T-54 and T-55 can be fitted with either the BTU bulldozer blade for clearing soil and obstacles or the STU blade for clearing snow.

Egyptian T-54 and T-55

Before the 1973 Middle East conflict many Egyptian T-54 and T-55 tanks were fitted with a German AEG infra-red/white light searchlight to the left of the main armament and a Yugoslav Iskra laser rangefinder. Additional details of Egyptian T-54/T-55 modification programmes are given in a separate entry under Egypt in this section.

Flamethrower TO-55

A flamethrower version of the T-55 is known to be in service with the Soviet Army and Naval Infantry under the designation TO-55. This has its 7.62 mm coaxial machine gun replaced by a flamethrower for which 460 litres of liquid are carried. Maximum range of the flame gun is 200 m. The 100 mm gun is retained. Due to the amount of space taken up internally by the flame gun and its associated equipment there has been a reduction in the amount of on board ammunition carried by this version.

SU-122 tank destroyer

The 122 mm SU-122 tank destroyer was introduced into the Soviet Army in 1949 and was a T-54 type chassis with a superstucture similar to that of the SU-100 100 mm assault gun. Mounted in the front of the hull was a 122 mm gun with limited elevation/depression and traverse. These were withdrawn from the late 1950s and converted into armoured recovery vehicles. This has the NATO designation of the M1977 and details are given in *Jane's Military Vehicles and Logistics 1991-92* page 31.

ZSU-57-2 SPAAG

The ZSU-57-2 twin 57 mm self-propelled anti-aircraft gun system was developed to meet the requirements of the Soviet Army and was first seen in 1957. It is a chassis based on T-54 components but with only four road wheels each side and fitted with a open topped turret armed with twin 57 mm automatic cannon.Turret traverse is a full 360° with weapon elevation from -5° to +85° and a total of 316 rounds of ready use ammunition are carried on clips of four rounds. This is a clear weather system and has been replaced in front line service by the more capable ZSU-23-4. It does however remain in service with some other countries. Full details of the ZSU-57-2 are given in *Jane's Land-Based Air Defence 1991-92* page 81.

SPECIFICATIONS
The IDF has provided detailed technical information on the T-54/T-55 modified with 105 mm gun and this is provided below, as are some areas in which it differs from the original Soviet built vehicle.

Model	T-54	T-55
CREW	4	4
COMBAT WEIGHT	35 400 kg	36 500 kg
POWER-TO-WEIGHT RATIO	14.68 hp/tonne	15.34 hp/tonne
GROUND PRESSURE	0.84 kg/cm^2	n/av
LENGTH		
gun forwards	9.20 m	9.20 m
gun in travel	8.20 m	8.20 m
without gun	6.30 m	6.30 m
HEIGHT		
with AA MG	2.85 m	2.85 m
without AA MG	2.40 m	2.40 m
to hull top	1.47 m	1.47 m
OVERALL WIDTH	3.28 m	3.28 m
GROUND CLEARANCE	0.45 m	0.45 m
TRACK	2.63 m	2.63 m
TRACK WIDTH	580 mm	580 mm
MAX ROAD SPEED		
(5th gear)	43 km/h	45.48 km/h
(4th gear)	27 km/h	28.99 km/h
(3rd gear)	18 km/h	20.21 km/h
(2nd gear)	13 km/h	14.66 km/h
(1st gear)	6 km/h	6.85 km/h
(reverse)	6 km/h	6.85 km/h
FUEL CAPACITY	817 l	938 l
FORDING	1.27 m	1.27 m
GRADIENT	60%	60%
SIDE SLOPE	30%	30%
VERTICAL OBSTACLE	0.8 m	0.8 m
TRENCH	2.70 m	2.70 m
ENGINE hp/rpm	520/2200	560/2200
TRANSMISSION	manual, 5 forward, 1 reverse gears	
STEERING	geared turn clutch and brake	
SUSPENSION	torsion bar	torsion bar
ELECTRICAL SYSTEM	24 V	24 V
BATTERIES	4	4
ARMAMENT		
main	1 × 105 mm gun	1 × 105 mm gun
coaxial	1 × 7.62 mm MG	1 × 7.62 mm MG
anti-aircraft	2 × 7.62 mm MG	2 × 7.62 mm MG
AMMUNITION		
105 mm	35	43
7.62 mm	6900	6900
TURRET POWER CONTROL	electric	electric
TIME TAKEN FOR 360° TRAVERSE	32 s	20 s
by commander	yes	yes
by gunner	yes	yes
GUN ELEVATION/ DEPRESSION	+19°/-6°	+19°/-6°

Indian T-54 and T-55
Some Indian T-54s have had their Soviet 12.7 mm DShKM anti-aircraft machine guns replaced by a standard 12.7 mm (0.50) Browning M2 HB machine gun as fitted to other Indian AFVs. India placed sheet steel tubes on the 100 mm barrels to resemble the bore evacuator of the L7 gun to serve as a means of distinguishing Indian T-54/T-55 tanks from Pakistani T-59s.

Iraqi T-54/T-55 MBTs
Details of some of the Iraqi conversions to their T-54/55 MBTs are given under Iraq.

Israeli T-54 and T-55
The Israeli Army captured large numbers of T-54 and T-55 tanks during both the 1967 and 1973 conflicts, many of which have been modified for their own use. These modifications which have not been carried out on every Israeli T-54 or T-55 include replacing the 100 mm gun by a 105 mm M68 rifled gun, ammunition racks modified to accept 105 mm ammunition, new communications equipment, commander's seat modified, gunner's seat replaced, installation of azimuth indicator, driver's hatch can now be opened from the outside, commander's traverse control installed, sighting system modified for 105 mm ammunition, replacing the coaxial machine gun by a 7.62 mm (0.30) Browning machine gun and the 12.7 mm DShKM anti-aircraft machine gun by a 12.7 mm (0.50) Browning M2 HB machine gun, new fire control and electrical system, air-conditioning system, new radio mounts on turret rear, American infantry telephone on hull rear, Browning 0.30 machine gun at loader's station, exhaust outlet angled upwards, additional track stowage, fire-extinguishing system installed and new night vision equipment. These tanks are designated the TI-67.

Israeli T-54/T-55/T-62 Model S
This was announced in 1984 and in addition to all of the previous modifications such as replacement of the 100 mm gun by the 105 mm rifled gun, it has many other improvements including the American Detroit Diesel 8V-71T engine developing 609 hp, new semi-automatic hydro-mechanical transmission equipped with a torque converter, new air cleaners as fitted to the Israeli M60 and Merkava tanks, Blazer explosive reactive armour added to the hull and turret, Cadillac Gage Textron weapon stabilisation system, installation of EL-OP computerised Matador fire control system, new low-profile commander's cupola, infra-red detectors, passive night vision equipment for commander, gunner and driver, Spectronix fire detection and suppression system, new turret basket and extensive external stowage, modernised driver's station including replacement of sticks by a steering wheel, new final drives, new fuel system with all fuel now internal and improved suspension. These as well as modernised T-62s are now being offered for export.

RO 105 mm L7 Conversion
Royal Ordnance of the United Kingdom have developed a 105 mm L7A3 conversion package for the T-54/T-55 tank and this is now being offered on the world market. It is also applicable to the Chinese Type 59 MBT.

The original 100 mm gun is replaced by the well known 105 mm L7A3 which has been turned through 180° as Soviet tanks are loaded from the right, a revised recoil system has been designed which includes new buffer components and modified recuperator, two new hull fuel tanks to incorporate 105 mm ammunition, modified main and emergency firing

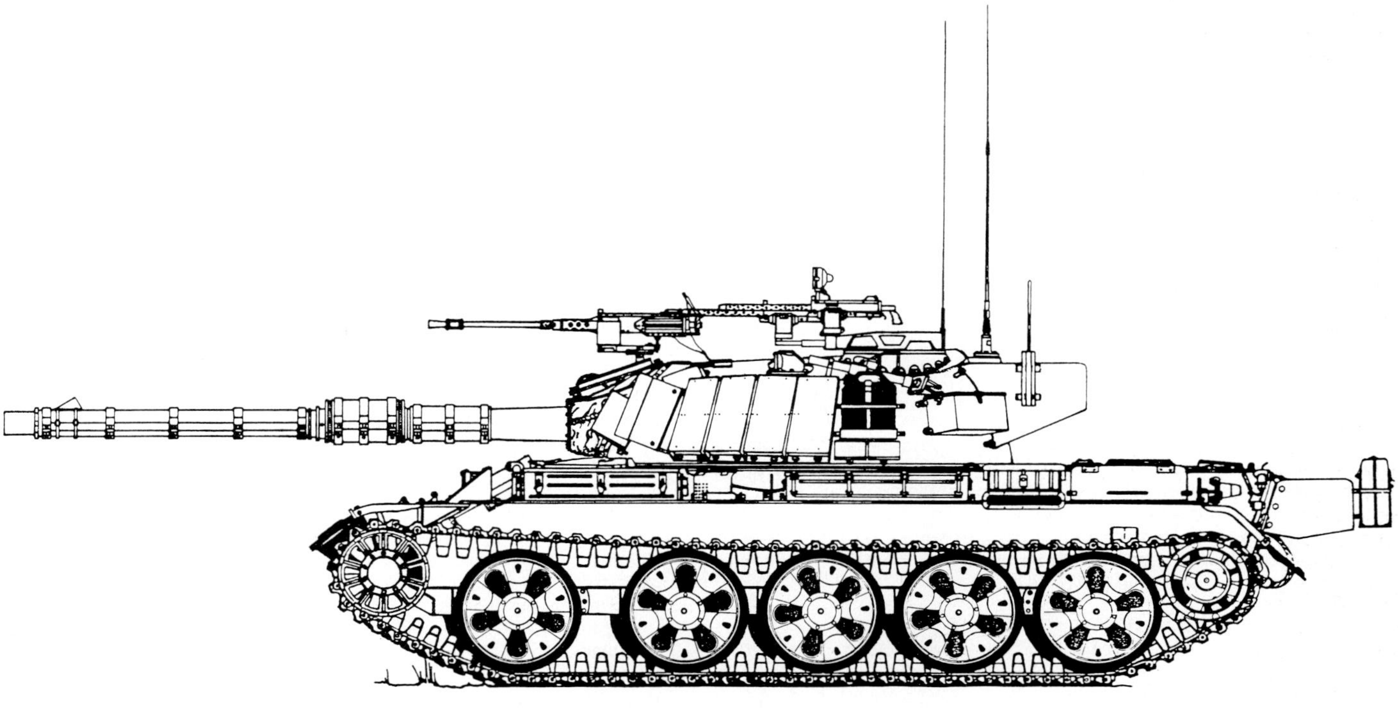

Israeli improved T-54/T-55 called the TI-67 and fitted with Blazer ERA (not to 1/76th Scale) (Vasco Barbic)

BMR mine clearing vehicle fitted with KMT M1988 mine clearing system. In the background is a T-54 fitted with the older KMT type system

In left background is T-55A with dozer blade, while in foreground is T-55 with mine clearing ploughs

circuits and a new telescopic graticule to suit new ammunition and 7.62 mm coaxial machine gun. Following trials in Egypt the Egyptian government placed a production order for this conversion.

T-54/T-55 Retrofits

Details of the Teledyne Continental Motors rebuilt T-54 for the Egyptian Army, called Ramses II, are given under Egypt. Royal Ordnance of the UK also offers a number of retrofit packages for the T-54/T-55 in the three areas of: firepower (replacement of 100 mm gun by 105 mm gun), protection (ranging from additional armour protection to NBC system and fire detection and suppression system) and mobility (including new suspension and powerpack). The Israeli company of NIMDA also offers a retrofit package for the Type 59 and T-54/T-55 MBTs. The engine used in this conversion is the United States Detroit Diesel Model 8V-92TA developing 850 hp coupled to a XTG-411 automatic transmission. This conversion kit is also applicable to the T-62 and AMX-30 MBTs.

In mid-1990 it was announced that two UK companies, A F Budge (Military) and Perkins Engines (Shrewsbury) had teamed up to offer a variety of retrofit packages for the Soviet T-54/T-55/T-62 series of MBTs used by many countries.

The aim is to enable users to select from a variety of options those which match operational and financial requirements.

Modules 1 and 2 of the retrofit package consists of the Perkins Engines (Shrewsbury) V-8 Condor engine developing 800 hp coupled to an Allison Transmission Division of General Motors XTG-411 automatic transmission.

Modules 3 and 4 cover the suspension and running gear while modules 5 to 10 cover the replacement of the existing 100 mm gun by a 105 mm weapon, laser rangefinder, improved gunner's sight, new stabilisation system and passive night vision equipment for the commander, gunner and driver.

Modules 11 to 13 include automatic fire detection/suppression system, add-on armour and the installation of smoke dischargers either side of turret.

Romanian T-54/T-55

Available details of the Romanian T-54/T-55 are given in this section under Romania.

BMR mine clearing vehicle

This was first observed in Afghanistan and uses a T-54/T-55 chassis with a new superstructure with a very well sloped glacis plate. The driver is seated at the front left with the vehicle commander to his right. The latter is provided with a cupola on which is mounted a 12.7 mm machine gun. On the right side of the hull is a bank of smoke dischargers firing forwards. This has been observed to be fitted with the KMT M1988 mine clearing rollers as well as the KMT-7 system.

Mine clearing T-54 and T-55 tanks

These tanks can be fitted with a wide range of mine clearing equipment including Czechoslovak roller and plough type systems, Soviet roller system models PT-54, PT-54M and PT-55, Soviet plough system type KMT-4, and the Soviet plough and roller type system called the KMT-5, KMT-5M, KMT-6 and KMT-6 M2. Details of these systems are given in *Jane's Military Vehicles and Logistics 1991-92* page 240. A more recent development is the mounting on the rear of the tank of rocket-propelled charges which are launched across the minefield and once on the ground are detonated. Details of this system are given in *Jane's Military Vehicles and Logistics 1991-92* page 238.

SPECIFICATIONS

Model	**T-54**	**T-55**
CREW	4	4
COMBAT WEIGHT	36 000 kg	36 000 kg
UNLOADED WEIGHT	34 000 kg	33 700 kg
POWER-TO-WEIGHT RATIO	14.44 hp/tonne	16.11 hp/tonne
GROUND PRESSURE	0.81 kg/cm^2	0.81 kg/cm^2
LENGTH GUN		
forwards	9 m	9 m
rear	8.485 m	8.485 m
LENGTH HULL	6.45 m	6.45 m
WIDTH	3.27 m	3.27 m
HEIGHT TO CUPOLA	2.4 m	2.4 m
FIRING HEIGHT	1.75 m	1.75 m
GROUND CLEARANCE	0.425 m	0.425 m
TRACK	2.64 m	2.64 m
TRACK WIDTH	580 mm	580 mm
LENGTH OF TRACK ON GROUND	3.84 m	3.84 m
MAX ROAD SPEED	48 km/h	50 km/h
TOTAL FUEL CAPACITY	812 l	960 l
MAX ROAD RANGE	400 km	500 km
MAX RANGE (with long range fuel tanks)	600 km	600 km
FUEL CONSUMPTION	1.9 l/km	1.9 l/km
FORDING	1.4 m	1.4 m
with preparation	4.546 m	4.546 m
GRADIENT	60%	60%
VERTICAL OBSTACLE	0.8 m	0.8 m
TRENCH	2.7 m	2.7 m
ENGINE MODEL	V-54	V-55
ENGINE TYPE	V-12 water-cooled diesel	
ENGINE hp/rpm	520/2000	580/2000
AUXILIARY ENGINE	none	none
TRANSMISSION	manual with 5 forward and 1 reverse gears	
STEERING	geared turn clutch and brake	
CLUTCH	multi-plate	multi-plate
SUSPENSION	torsion bar	torsion bar
ELECTRICAL SYSTEM	28 V	28 V
BATTERIES	4 × 12 V, 280 Ah	4 × 12 V, 280 Ah
ARMAMENT		
main	1 × 100 mm gun	1 × 100 mm gun
coaxial	1 × 7.62 mm MG	1 × 7.62 mm MG
bow	1 × 7.62 mm MG	none
anti-aircraft	1 × 12.7 mm MG	none
SMOKE-LAYING EQUIPMENT	diesel fuel injected into exhaust system	
AMMUNITION		
main	34	43
7.62 mm	3000	3500
12.7 mm	500	none
GUN CONTROL EQUIPMENT		
Turret power control	electro/hydraulic with manual controls for emergency use	
by commander	yes	yes
by gunner	yes	yes
Max power traverse rate	360° in 21 s	360° in 21 s
Gun elevation/depression	+17°/-4°	+17°/-4°
Gun stabiliser		
vertical	no	yes
horizontal	no	yes
Rangefinder type	stadiametric	stadiametric
Elevation quadrant	yes	yes
Traverse indicator	yes	yes
ARMOUR		
Hull front upper	97 mm at 58°	97 mm at 58°
Hull front lower	99 mm at 55°	99 mm at 55°
Hull sides upper	79 mm at 0°	79 mm at 0°
Hull sides lower	20 mm at 0°	20 mm at 0°
Hull rear upper	46 mm at 0°	46 mm at 0°
Hull rear lower	46 mm at 0°	46 mm at 0°
Hull top	33 mm	33 mm
Hull floor front	20 mm	20 mm
Hull floor rear	20 mm	20 mm
Turret front	203 mm at 0°	203 mm at 0°
Turret sides	150 mm at 0°	150 mm at 0°
Turret rear	64 mm at 0°	64 mm at 0°
Turret roof	39 mm at 79°	39 mm at 79°
Mantlet	included in above figures	

Status: Production complete. In service with the following countries:

Country	Quantity	Comment
Afghanistan	400	all Type 59
Albania	15	
Algeria	390	
Angola	300	
Bangladesh	70	inc Type 59 and ARV
Bulgaria	1600	
Cambodia	100	some Type 59
Central African Republic	4	
China	small number	plus many Type 59
Congo	35	plus 15 Type 59
Cuba	800	
Czechoslovakia	3090	most locally built some being modernised
Egypt	900	some with 105 mm gun
Ethiopia	1000	
Finland	185	some being modernised
Guinea	10	
Hungary	1300	some being modernised
India	800	
Iran	200	plus some Type 59
Iraq	500	plus some Type 59
Israel	400	all 105 mm
Korea, North	1600	estimate including 175 Type 59
Laos	30	
Lebanon	100	from Iraq in 1988
Libya	1700	
Mali	25	
Mongolia	100	
Mozambique	80	
Nicaragua	130	
Nigeria	60	
Pakistan	50	plus 1100 Type 59
Peru	300	
Poland	2093	most locally built some being modernised
Romania	1300	
Somalia	110	
Sudan	150	
Syria	2100	
Tanzania	30	all believed to be Type 59
Togo	2	
USSR	19 000	
Vietnam	1000 approx	plus 160 Type 59
Yemen	700	
Yugoslavia	850	
Zambia	10	
Zimbabwe	40	including some Type 59

In addition they are used by a number of irregular units in the Middle East and elsewhere and some countries also use them for training. Under the ongoing CFE talks some of the above figures for countries such as Poland are being considerably reduced. For example in December 1990 the Polish Army had 492 T-55AM MBTs and 1601 T-55 MBTs, the former will be retained but the latter will be reduced to 1120 vehicles.

Manufacturers: Three Soviet tank plants plus Labedy in Poland and Martin in Czechoslovakia. Poland produced the T-54A and T-54A(M) from 1956 to 1964 and the T-55 from 1964 onward. Was manufactured in China as the Type 59 (qv).

T-34/85 Medium Tank

Details of the T-34/85, armed with an 85 mm gun developed during the Second World War, were given in *Jane's Armour and Artillery 1990-91* pages 111 to 113.

UNITED KINGDOM

Chieftain Replacement Programme

Only about half of the Royal Armoured Corps Chieftain MBT fleet has been replaced by the Challenger 1 and in November 1987 the Staff Requirement (Land) 4026 was endorsed.

There were four contenders to meet this requirement, Vickers Defence Systems Challenger 2, the General Dynamics, Land Systems Division M1A2, the Krauss-Maffei Leopard 2 (Improved) and the GIAT Leclerc.

In the Autumn the contenders for the Chieftain Replacement Programme submitted their bids to the British MoD but on 1 November in a written answer in Parliament the Secretary of State for Defence said, "I had hoped to announce a decision on equipment for our future main battle tank fleet by the end of this year. It is now clear that, owing to the uncertainties created by recent events, I will be unable to do so. I expect to be able to announce a decision in the early Spring of 1991."

Each of the contenders was asked to tender for 200 MBTs together with various options for additional vehicles up to a maximum of 360, this would have allowed most of the British Army's front line Chieftains to be replaced.

Under the Conventional Forces Europe Treaty, the UK has agreed to reduce its MBTs from the current figure of 1198 down to 1015. As of 1990 the fleet consisted of:

Type	UK	Elsewhere	Total
Chieftain	244	508	752
Challenger 1	44	364	408
Centurion	11	27	38

The CRP bids submitted in late September 1990 were valid until the middle of January 1991. Subsequently the MoD asked the contenders to rebid by mid-January with the bids to be valid until the Spring. Reasons for this were cited as the British Army was not certain of the number of MBTs that it required under the options for change proposals and it wanted to see how the Challenger 1 performed in the Middle East.

On 19 February the Ministry of State for Defence Procurement stated in Parliament: "We had hoped to announce a decision on equipment for our future MBT in the early spring. Hon Members will appreciate that it would not be realistic to take such a decision at present and that it is sensible to defer it."

New and final bids were again submitted on 10 April 1991 and were for between 150 and 200 MBTs with the initial order being for around 125 vehicles. On 21 June 1991 the MoD announced that it had selected the Challenger 2 MBT to meet its CRP requirement.

Vickers Defence Systems Challenger 2 Main Battle Tank

Development

Work on the Challenger 2 started in November 1986 as a private venture and in March 1987 Vickers Defence Systems made its first presentation to the British MoD. Throughout 1987 the Challenger 2 design was further refined and in February 1988 Vickers Defence Systems submitted its formal proposal to the the British MoD following the issue of the Staff Requirement.

Early in 1988 Vickers Defence Systems took the decision to build eight turrets using company funding and the first shell was completed in the autumn of 1988.

Vickers Defence Systems and General Dynamics, Land Systems Division, both put their firm production proposals to the British MoD in mid 1988 and in December of that year it was announced that Vickers Defence Systems were to be awarded a £90 million fixed price contract to undertake a demonstration phase (also referred to as the proof-of-principle phase) which lasted to the end of September 1990.

Under the terms of the contract, which was awarded on 20th January 1989, the company was required to demonstrate within the contracted time that Challenger 2 is able to meet the staff requirement and can be successfully developed and produced to the required standard, so as to achieve the required in-service date and at a price which Vickers Defence Systems has already offered.

An important part of the demonstration phase was be to demonstrate that a new depleted uranium projectile together with a new charge system for the 120 mm CHARM gun could be successfully developed to the standard and in the timescale required.

One of the prototypes of the Vickers Defence Systems Challenger 2 MBTs showing TOGS over 120 mm L30 gun and 7.62 mm McDonnell Douglas Helicopters Chain Gun mounted coaxial with the gun

At the same time, it was stated that the current Challenger 1 was to be retrofitted with the new Royal Ordnance/RARDE 120 mm CHARM system as a replacement for the existing L11A5.

Intermediate milestones were established within the demonstration phase, at which Vickers Defence Systems was required to demonstrate satisfactory progress.

MILESTONE 1-SEPTEMBER 1989

1 Turret mock-up complete and turret simulator fully assembled and functional
2 Full ballistic performance proved by fire-at trials on four armour rigs
3 Turret fixings for armour proved by fire-at trials on armour rigs
4 Proof of mantlet integrity by fire-at trials on rigs
5 Weapon system test and integration facility (WSTIF) installed and commissioned at Leeds
6 Turret T1 and T8, shells completed and painted and all appropriate subsystems available and fitted
7 Vehicle V1-hull and turret shells complete and painted. All appropriate subsystems delivered. Vehicle on final assembly
8 Vehicles V2 to V9, all hulls and turrets on final weld or initial assembly
9 Design of CHARM firing stand complete. Drawings issued for manufacture. First CHARM gun delivered by RO Nottingham
10 All hull modifications designed and drawings issued to manufacturer
11 Six short engineering studies complete on studies including air conditioning/crew cooling and rationalisation of hull electrics

MILESTONE 2-MARCH 1990

1 Prototype build standard established for configuration control during vehicle trials
2 Prototype vehicle acceptance specification written and issued
3 Prototype weapon system equipment delivered and tested. Basic performance proved. Turret T1 commissioned and delivered to Chertsey
4 Turret T8 proof and endurance trial completed
5 Clearance for live crew firing obtained on Vehicle V1
6 Assembly of prototype Vehicles V2 to V6 complete. Fabrication of prototype vehicles V7 to V9 complete
7 CHARM gun firing stand fabricated

In January 1989 Invertron Simulated Systems Limited announced that it was to supply a Commander and Gunner Crew Station Simulator (CGCSS) for the Challenger 2 MBT.

The system is used by Vickers Defence System to demonstrate to the British Army the efficiency and ease fo use of the new sights and fire control equipment and in addition it will provide a valuable tool for analysing the ergonomics of the proposed turret layout before Challenger 2 goes into full production.

Design of the Challenger 2 was essentially completed by August 1989 with the last bought-out items being delivered in February 1990. Of the nine Challenger 2 prototypes built, seven were built at Leeds and two at Newcastle-upon-Tyne. Two additional turrets were also built at Leeds.

By early August 1990 seven Challenger 2s had been completed and the final two were being fitted out. These all differed in minor ways but No 9 incorporated all of the 150 hull improvements that are also refittable to Challenger 1 at regimental or base workshop level.

Vickers Defence Systems completed the demonstration phase of Challenger 2 on schedule at the end of September 1990. In June 1991 the British Government selected the Challenger 2 and placed an order with first vehicles to be delivered in 1994. Production is expected to run at about 35-40 vehicles per year.

Description

Turret

The turret is a new design, incorporating many of the features used in the turret of the Vickers Mk 7/2 and the turret designed and built for the ENGESA P2 Osorio MBT. It incorporates second generation Chobham armour with a significant increase in protection against both KE and CE attack and also incorporates a databus.

The turret crew positions on Challenger 2 are similar to Challenger 1 with the commander being seated on the right, gunner forward and below the commander with the loader on the left side. Each of them is provided with a lap-type seat belt.

Both commander and loader have a single piece hatch cover that opens to the rear.

The tank commander has a total of six periscopes for all round observation, although there is an area of dead ground to his front due to the position of the roof-mounted sight. A new design with a total of eight periscopes has been designed to overcome this problem.

Under each periscope is a red button, when this is pressed by the tank commander the turret slews so that it lines up with the periscope.

The loader has a single roof-mounted periscope that can be traversed and in the future this could be removed and its place taken by a commander's independent thermal viewer with full 360° traverse. At present the turret has to be traversed in order to use the thermal imaging system which could be a tactical disadvantage.

The NBC system is mounted in the turret bustle and fully meets all known threats. The crew environmental control system provides both heating and cooling for the crew compartment. This is the first time such as system has been installed in a British MBT.

Main armament of the Challenger 2 MBT consists of a Royal Ordnance Nottingham 120 mm rifled tank gun designated the L30 which is of ESR steel, autofrettaged, chrome lined and fitted with a thermal sleeve, fume extractor and muzzle reference system. It has a split-block sliding breech mechanism with an elastomeric obturating pad.

The 55 calibre L30 is the first British tank gun to be chrome lined which is claimed to give the barrel a longer life as well as giving more consistent accuracy.

In addition to firing all current 120 mm ammunition, with the exception of the obsolete APDS round, the L30 also fires a new DU round with a new Royal Ordnance stick charge propellant system.

The first DU round is part of the CHARM 1 system (gun, charge and projectile) with the actual projectile being designated the L26. The so called CHARM 3 projectile is also of the DU type but has a greater length-to-diameter ratio and greater penetration characteristics. Development

One of the nine prototypes of the Vickers Defence Systems Challenger 2 MBTs at speed during trials in 1990. The door on the TOGS system mounted over the 120 mm L30 rifled tank gun is open

of CHARM 3 round is expected to be complete to coincide with the in-service date of Challenger 2.

Ammunition is of the separate loading type, eg projectile and charge will all explosive ammunition being stowed below the turret ring for increased battlefield survivability. The charges are stowed in armoured bins.

Mounted coaxial to the left of the main armament is a McDonnell Douglas Helicopters 7.62 mm Chain Gun which is already in service with the British Army being installed in the GKN Defence Warrior mechanised combat vehicle. A similar weapon will be mounted on the turret roof and operated by the loader.

Mounted either side of the turret front is a bank of five Peak Engineering electrically operated smoke dischargers that cover the forward arc of the vehicle.

Heart of the fire control system of the Challenger 2 is the latest generation digital computer from Computing Devices Company (CDC) of Canada which is an improved version of that installed in the M1A1 Abrams. It also has growth capacity for future enhancement such as battlefield information control system (BICS) and navigation aids. It will also be capable of being used for training purposes.

The commander has a roof-mounted SFIM stabilised sight which is similar to that installed on the French GIAT Industries Leclerc MBT and gives him full 360° capability without moving his head. This has a magnification of ×3.2 and ×10.5 in the day mode and also has a laser rangefinder.

A thermal image with a magnification of ×4 and ×11.5 can be injected into the sight with a separate thermal image monitor being provided to the left of the sight for surveillance purposes.

The SFIM roof-mounted sight allows the tank commander to lay and fire the main armament or override the gunner with the laser rangefinder so allowing the commander to acquire and range on to the targets in advance of the gunner. Once the gunner has acquired the target the commander could revert back to searching for additional new targets.

A typical target engagement sequence takes less than eight seconds with a single sequence providing fast and accurate engagement of targets using optical or thermal sights.

As with all British MBTs since the Second World War, Challenger 2 has an all electric gun control and stabilisation system from Marconi Command and Control Systems. In addition it also incorporates a MIL STD 1553 databus, this being the first to be installed in a British AFV. Turret traverse is 360° with weapon elevation from -10 to +20°.

The gunner's roof-mounted stabilised sight has been developed by Barr and Stroud with SAGEM of France being the major sub-contractor and includes day optics, thermal image input and laser rangefinder. The gunner's sight has a magnification of ×3 and ×10 in the day mode and also has a traverse of 7° left and right. A thermal image with the same magnification as that of the commander can be injected at the flick of a switch.

Like the tank commander, the gunner has a separate thermal image monitor to the left of his sight for surveillance purposes.

Mounted over the 120 mm L30 gun is the Barr and Stroud thermal imaging system camera that moves in elevation with the main armament. This is the same camera as that fitted to the Challenger 1 but the actual system is much simpler.

The gunner also has a NANOQUEST L30 reversionary mode telescope mounted coaxial with the main armament with an etched graticule, this is used for muzzle reference purposes.

According to Vickers Defence Systems, Challenger 2 has plenty of growth potential including the installation of a 140 mm gun, this being possible by the width of the mantlet, roof-mounted commander's

One of the prototypes of the Vickers Defence Systems Challenger 2 MBTs showing commander's roof-mounted sight

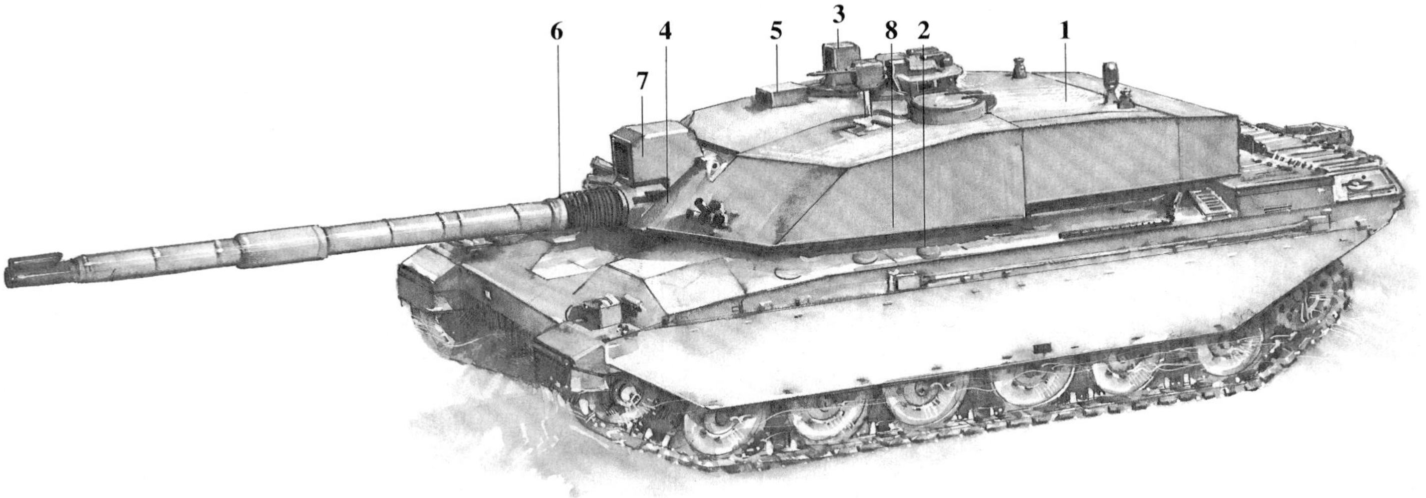

Provisional outline of Challenger 2 MBT showing main areas of improvement over current Challenger 1 MBT: **(1)** *solid state gun control equipment* **(2)** *second generation fire control equipment* **(3)** *commander's gyro-stabilised panoramic day sight* **(4)** *rotor and mantlet gun mounting* **(5)** *gunner's gyro-stabilised sight* **(6)** *gunner's telescopic sight* **(7)** *thermal imager – TOGS* **(8)** *latest Chobham armour*

independent thermal viewer which would replace the loader's periscope, improved protection, including explosive reactive armour arrays covering the sides of the vehicle, automatic muzzle reference system, automatic target detection and tracking systems and various defensive aids that could detect and counter a variety of threats.

The existence of the 1553 databus and the CDC computer make many of the electronics improvements that much easier.

Hull

The hull of the Challenger 2 is virtually identical to that of the current Challenger 1 and like the turret, incorporates stealth technology to minimise the radar signature.

Many improvements have been carried out to the chassis, for example the driver's steering controls are more rugged, fire retardant bag type fuel tanks have been fitted, new lights installed and engine decks redesigned for ease of operation.

Challenger 2 is powered by the Perkins Engines (Shrewsbury) 12 cylinder diesel developing 1200 hp at 2300 rpm coupled to the David Brown Vehicle Transmissions TN54 transmission which is already installed in the Challenger Armoured Repair and Recovery Vehicles, this was also fitted to the seventh Challenger 1 MBT regiment and may well be back fitted to the remainder of the current Challenger 1 MBT fleet.

The TN54 epicyclic transmission has six forward and two reverse gears with automatic control embodying double differential regenerative steering with hydrostatically transmitted power giving infinitely variable output. The TN54 transmission also gives increased manoeuverability in confined spaces.

Challenger 2 is also fitted with the Dowty Defence and Air Systems Digital Automotive System Control Unit (DASCU) and a new Plessey APU that is also installed in the Challenger Armoured Repair and Recovery Vehicle.

The hydropneumatic suspension system fitted to all members of the Challenger 1 family has been upgraded as the original system is somewhat dated and new materials are now available, new seals are also incorporated.

Hydraulic track tensioners have been installed on the Challenger 2 and these allow the track to be correctly tensioned from under armour protection by the driver.

Two double pin tracks have been trialled on Challenger 2s, one from Diehl of Germany and the other from Blair of the UK.

Challenger 2 has some 33 improvements to the main engine GUE (general unit electrical), 11 improvements to the gearbox assembly, 11 improvements to the running gear and no less than 37 improvements to the vehicle electrics.

Challenger 2 can also lay its own smoke screen by injecting diesel fuel into the exhaust outlets on either side of the hull top towards the rear.

All Challenger 2s are fitted to carry two standard 45 gallon (Imperial) drums at the rear of the hull with racks being provided with a quick-release mechanism. These are not connected to the vehicle's fuel supply.

The Challenger 2 is also fitted to accept the Pearson Engineering Combat Dozer Blade which was adopted by the British Army in 1990.

Status: Vickers Defence Systems completed the demonstration phase of Challenger 2 on schedule at the end of September 1990. Production order placed in June 1991 with first deliveries due in 1994.

Manufacturer: Vickers Defence Systems
Manston Lane
Crossgates
Leeds, LS15 8ST
United Kingdom
Telephone: (0532) 648123
Telex: 55134
Fax: (0253) 607674

SPECIFICATIONS		TRENCH	2.34 m	SMOKE DISCHARGERS	2 × 5
CREW	4	FORDING	1.07 m	AMMUNITION	
COMBAT WEIGHT	62 500 kg	ENGINE	Perkins Engines	main	up to 52 projectile
POWER-TO-WEIGHT			(Shrewsbury) Condor		stowage positions
RATIO	19.2 hp/tonne		V12 12-cylinder diesel	7.62 mm	4000
GROUND PRESSURE	0.9 kg/cm^2		developing 1200 bhp	GUN CONTROL	
LENGTH GUN FORWARDS	11.55 m		at 2300 rpm	EQUIPMENT	
LENGTH GUN REAR	9.86 m	AUXILIARY ENGINE	Plessey	Turret power control	electric/manual
LENGTH HULL	8.327 m	TRANSMISSION	David Brown Vehicle	by commander	yes
WIDTH OVERALL	3.52 m		Transmissions TN54	by gunner	yes
WIDTH OVER TRACKS	3.42 m		epicyclic transmission	Turret traverse	360°
HEIGHT TO TURRET			with 6 forward and 2	Gun elevation/depression	+20°/–10°
ROOF	2.49 m		reverse gears	Commander's override	yes
GROUND CLEARANCE	0.50 m	STEERING	Commercial	Commander's fire control	
TRACK	2.12 m		Hydraulics double	override	yes
LENGTH OF TRACK ON			differential, hydrostatic	Gun stabiliser	
GROUND	4.79 m		control	vertical	yes
TRACK WIDTH	650 mm	SUSPENSION	hydropneumatic	horizontal	yes
MAXIMUM ROAD SPEED	56 km/h	ELECTRICAL SYSTEM	24 V	Range setting device	laser
AVERAGE CROSS		BATTERIES	6	Elevation quadrant	yes
COUNTRY		ARMAMENT		Traverse indicator	yes
SPEED	40 km/h	main	1 × 120 mm L30 gun		
FUEL CAPACITY	1797 litres	coaxial	1 × 7.62 mm MG		
VERTICAL OBSTACLE	0.9 m	anti-aircraft	1 × 7.62 mm MG		
GRADIENT	60%				

Vickers Defence Systems Challenger 1 Main Battle Tank

Development

In September 1978 the British Ministry of Defence placed an order with the then Royal Ordnance Factory Leeds for the supply of 243 Challenger 1s, essentially a further development of the Shir 2 originally developed for Iran. This order was valued at £300 million.

At the same time it was announced that a 9th armoured regiment would be formed in BAOR by re-rolling an armoured reconnaissance regiment using Chieftains from War Reserve stocks.

The Challenger 1 was accepted by the General Staff in December 1982 by which time production was already well under way at Royal Ordnance Leeds.

The first British Army order for Challenger 1 was for 243 tanks, sufficient to equip four regiments. The first of these were handed over to the British Army by Royal Ordnance Leeds in March 1983. In June 1984 the MoD placed an order for an additional 64 Challenger 1 MBTs, sufficient to equip a fifth regiment in BAOR. Challenger 1 has already been tested in Abu Dhabi and in mid-1985 two vehicles were tested in Saudi Arabia, which had a number of modifications including replacement of IFCS by the EFCS 600 system. In mid-1987 a Challenger 1 MBT underwent trials in Saudi Arabia with two M1A1s, two AMX-40s and an EE-T1 Osorio MBT. Challenger 1 has also been demonstrated in Morocco, but by the Spring of 1991 no export sales of the vehicle had been made.

The Royal Tank Regiment received its first vehicles late in 1984 and by May 1985 the Royal Hussars were completely equipped with the Challenger 1 MBT. A further 18 Challenger 1s were ordered in June 1985.

In June 1986 the then Royal Ordnance Weapons and Fighting Vehicle Division took over the responsibility for design authority and future post design services work on Challenger 1, this was previously undertaken by the Royal Armament Research and Development Establishment at Chertsey.

In July 1986 it was announced that Vickers Defence Systems was to buy the Royal Ordnance Leeds tank factory at a price of around £11 million and that a new facility, identical to Vickers Defence Systems Armstrong Works at Newcastle-upon-Tyne, would be built at a cost of £14 million. The new Vickers Defence Systems' Leeds factory was fully operational by late 1987.

At the same time, the British MoD placed a further order for 76 Challenger 1 MBTs at a cost of £100 million. The last of 420 Challenger 1s was to have been delivered to the British Army in September 1989 but this was subsequently stretched until mid-1990.

In 1991 the MoD stated that the Challenger 1 was to be upgraded in a number of key areas including an improved computerised sighting system and improved powered cupola for the commander.

Description

Challenger 1 is essentially the FV4030/3 (Shir 2) modified to suit the requirements of the British Army. The layout of Challenger 1 is similar to that of Chieftain with the driver's compartment at the front, turret and fighting compartment in the centre, and engine and transmission at the rear. The turret and hull incorporate Chobham armour for increased battlefield survivability.

The driver has a single piece hatch cover that lifts and swings forward horizontally to allow him to drive in the head out position. To the rear of this is a single wide-angle periscope which can be replaced by a Pilkington PE Badger passive periscope for driving at night. The driver can also leave the vehicle via the fighting compartment.

The commander is seated on the right of the turret with the gunner forward and below the commander and the loader on the left side of the turret.

The commander has a modified No 15 cupola which has been designated the No 32 cupola. This has a No 37 day sight which is capable of being quickly replaced by a Rank Pullin image intensification swap sight. The Rank Pullin Control No 37 night sight (SS 120) fitted to the early production vehicle provides the commander with night surveillance, target acquisition and, if used with the PRI AV No 22 or No 24, the ability to lay the main armament. It has a magnification of ×1 in the day mode and ×5 in the night mode when filters allow it to be used as a stand-by sight in full light conditions. The cupola also has nine periscopes for all-round observation.

The No 37/swap sight was an interim solution as in July 1981 the MoD awarded a development contract to Barr and Stroud for the Thermal Imaging Surveillance and Gun Sighting Sights (also known as TOGS - Thermal Observation and Gunnery Sights). A major sub-contractor to Barr and Stroud is Pilkington PE. The turret had been designed to accept this system when it was available. A single thermal imager in an armoured box on the right of the turret provides separate outputs for both commander and gunner in the relaxed observation mode or as a gun sight. Delivery of Class II Thermal Imaging Common Modules for TOGS began in late 1983 for Challenger and Chieftain MBTs. Rank Taylor Hobson provides the infra-red scanning modules while Marconi Avionics provides the processing electronics.

In April 1987 the British MoD announced that orders had been placed for the final production batches for the Thermal Observation and Gunnery Sights worth about £35 million. This order completes the buy of TOGS for the British Army's Challenger 1 and Chieftain fleets. Half the order was placed with Barr and Stroud, the original producer of the system, with half going to Avimo. This was the third order for TOGS and Barr and Stroud supply 85 per cent of the MoD's total TOGS requirement.

The loader has a roof-mounted ×1 periscope swivel mounted forward of the two-piece hatch cover that opens front and rear.

Challenger 1s are armed with the standard 120 mm L11A5 rifled tank gun but late in 1988 it was announced that the Challenger 1 was to be backfitted with the new Royal Ordnance Nottingham/Royal Armament Research and Development Establishment 120 mm L30 smoothbore high technology gun. In addition to being installed in current British Army Challenger tanks is also to be fitted in the prototypes of the Challenger 2. First pre-production guns were completed at RO Nottingham in the Spring of 1989. The new gun forms part of the CHARM (Challenger Armament) programme which also includes a new change system and a new DU projectile with a higher muzzle velocity and greater armour penetration characteristics.

The 120 mm barrel is fitted with a thermal sleeve, fume extractor and a muzzle reference system. Types of ammunition that can be fired are APDS-T (L15A4), DS-T (L20A1), HESH (L31), HESH practice (L32A5), Smoke WP (L34) and the new APFSDS-T (L23A1). There are up to 42 charge stowage and 64 projectile stowage positions. Each charge location takes either one DS charge or two HESH/Smoke charges. A typical mix would be 20 DS and 44 HESH/Smoke with all charges stowed in special containers below the turret ring. A fire suppressant fluid in these containers reduces the risk of fire and, on later vehicles, these containers are armoured.

A 7.62 mm L8A2 machine gun is mounted coaxially with the main armament and a 7.62 mm L37A2 machine gun is mounted at the commander's cupola. Mounted either side at the front of the turret is a cluster of five electrically-operated smoke dischargers, each with a coverage of 100°.

The Challenger 1 has the Marconi Command and Control Systems Improved Fire Control System (IFCS), this is fully described in the entry for the Chieftain MBT and has a target tracking rate of 30 mils/s in traverse and 10 mils/s in elevation. Slewing rate in traverse is 0.2 to 480 mils/s and 0.2 to 200 mils/s in elevation. The computer used in the IFCS is a more modern design giving greater stretch potential for the future. The gun control system is also similar to the Chieftain's but a number of units have been redesigned to replace the thermionic valve equipment with solid-state equivalents.

The gunner has a periscope Tank Laser Sight No 10 Mark 1 with a magnification of ×1 and ×10 with an 8.5° field-of-view. The laser rangefinder is an Nd YAG with an operating range of 300 to 10 000 m and an accuracy of ±10 m for 90 per cent of shots. The gunner also has an emergency No 87 periscopic sight. This is stowed under armour but can quickly be deployed and collimated with the gun. It projects through an armoured hatch in the turret roof, rather than a conventional telescope as used in Chieftain, to avoid detracting from the frontal integrity of the armour.

Mounted at the rear of the turret is an NBC No 15 Mk 1 environmental control system.

Challenger 1 is powered by a Perkins Engines (Shrewsbury), Condor 12V 1200 diesel developing 1200 bhp at 2300 rpm, which is fitted with two Garret-AiResearch turbo-chargers. The transmission is a David Brown Vehicle Transmissions TN37. Torque converter is a Borg Warner with lock-up clutch giving four forward and three reverse gears. Steering is a Commercial Hydraulics STN37 double differential with hydrostatic, infinitely variable control. Additional details of the engine are given in the entry for the Khalid MBT.

The Challenger 1 powerpack weighs 5.49 tonnes and has been designed for quick replacement in the field and can be removed by a REME LAD in under 45 minutes. The FV434 armoured repair vehicle cannot be used

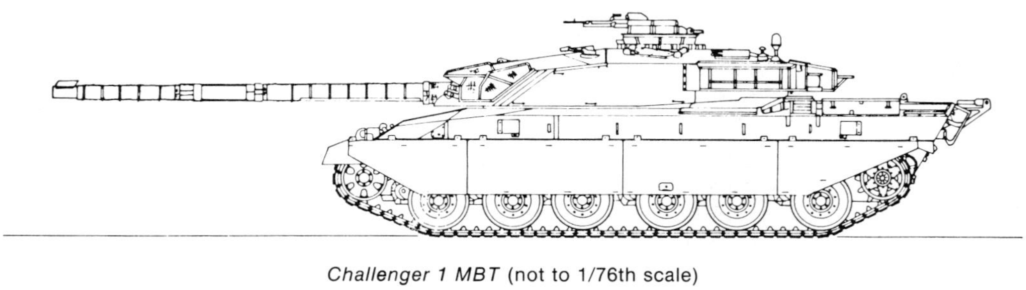

Challenger 1 MBT (not to 1/76th scale)

Challenger 1 MBT of the British Army showing 120 mm L11A5 gun fitted with thermal sleeve and TOGS doors on right side of turret open (MoD)

to change the Challenger 1 powerpack so a new version of the Chieftain ARV called the Armoured Repair and Recovery Vehicle (ARRV) has been developed (qv Chieftain). A Challenger 1 Armoured Repair and Recovery Vehicle (CR ARRV) has been designed by Vickers Defence Systems at Newcastle-upon-Tyne and is now in production for the British Army (qv).

The hydro-pneumatic suspension of Challenger 1 has been designed and developed by MVEE in conjunction with Air Log Limited (originally Laser Engineering (Development)), with production being undertaken by Vickers Defence Systems. Challenger 1 has six aluminium road wheels with the drive sprocket at the rear, idler at the front and two track return rollers either side. The upper part of the track is covered by conventional aluminium skirts similar to those fitted to Chieftain. The tracks of Challenger 1 are of a new design but are not interchangeable with those currently fitted to Chieftain.

Variants

Challenger 1 Improvement Programme

The Challenger 1 Improvement Programme (CHIP) covers both automotive and weapon system enhancements.

Most of the automotive improvements have now been tested and accepted for service by the British Army on the Challenger Armoured Repair and Recovery Vehicle (CR ARRV). Retrofit to Challenger 1 MBTs is dependent on availability of funding.

The TN54 transmission, produced by David Brown Vehicle Transmissions, is a direct replacement for the TN37 transmission. It uses similar epicyclic geartrains, but has a different clutch layout, providing six forward and two reverse gears rather than the four forward and three reverse gears of TN37. TN54 provides improved manoeuvrability in confined spaces, and superior acceleration and cross-country performance.

Dowty Defence and Air Systems has developed the Digital Automotive System Control Unit (DASCU), which replaces and carries out all the functions of the Main Engine Control Unit (MECU) and Gearbox Controller Automatic (GCA). DASCU is a twin microprocessor-based system which contains built-in test equipment (BITE) to monitor the engine, transmission and associated sensors. A key feature is DASCU's reversionary mode capability, which allows vehicle mobility to be maintained in the event of sensor failures. The DASCU can be interrogated from outside the vehicle using a Husky M208 hand-held first line test set, facilitating rapid fault diagnosis.

Plessey Aerospace has developed a new auxiliary power unit (APU) for Challenger 1, which is fitted to the 7th Regiment vehicles and also to CR ARRV. The new APU consists of a Perkins diesel engine driving a 350 A oilspray-cooled brushless generator, and has been designed for interchangeability with the H30 APU.

An oilspray-cooled main engine driven 500 A generation system, also developed by Plessey Aerospace, has been successfully trialled on Challenger 1 MBT, and is fitted to CR ARRV.

Trials of a new cast steel double pin track, produced by Blair to a RARDE Chertsey design, are ongoing. The new track, designed to replace the existing single dry-pin track, is a rubber-bushed live track joined by two end connectors and a centre connector, with replaceable rubber track pads, offers reduced rolling resistance and longer life.

Weapon system enhancements, nearing completion of development, include solid-state gun control equipment (SSGCE), improvements to the fire control system and TOGS, and a powered commander's cupola.

While developing the Challenger 2 MBT, Vickers Defence Systems found some 150 automotive areas that could be improved on Challenger 1 and a troop in one of the British Army of the Rhine's armoured regiments has been fitted with some of these improvements for user trials at no cost to the MoD.

These include some of the items already fitted to Challenger 2 such as the hydraulic track tensioner, new lights, maintenance free batteries and improved engine deck covers that can be lifted by just one man without using a crane. These improvements could be backfitted to Challenger 1s at regimental level or during base overall.

Challenger 1 in Middle East

Challenger 1 MBTs were deployed to Saudi Arabia late in 1990 and subsequently took part in Operation Desert Storm. The 7th Armoured Brigade deployed two regiments of Challenger 1 MBTs each with 57 MBTs (four squadrons) while the fourth brigade deployed one regiment of Challenger 1 MBTs with 43 MBTs (three squadrons). Challenger Armoured Repair and Recovery Vehicles were also deployed as were additional Challenger 1 MBTs as battlefield replacement vehicles.

In early 1991 these were improved in a number of key areas including the installation of Royal Ordnance Explosive Reactive Armour (ERA) on their nose and glacis plate and Vickers Defence Systems passive armour along the sides of the hull.

Royal Ordnance also improved the firepower of the Challenger 1 120 mm L11 rifled tank gun with a new projectile and charge system. The new APFSDS projectile gave greater accuracy and penetration and is believed to incorporate features of the new system developed for the L30 rifled tank gun installed in the Challenger 2 and in production for installation in the current Challenger 1 MBT fleet. The charge was designated the L14A1 and the projectile the XL26E1.

During Operation Desert Storm not a single Challenger 1 MBTs was knocked out by Iraqi Forces and they destroyed some 300 Iraqi MBTs.

Vickers Defence Systems Challenger 1 MBT deployed in Saudi Arabia in 1991 showing additional armour protection on sides of hull and additional fuel drums mounted on hull rear

Challenger 2 MBT

Details of this are given in previous entry.

Driving Training Tanks

In February 1988 the British Army placed an order with Vickers Defence Systems at Leeds for the supply of 17 Challenger Training Tanks valued at £18 million. The first Challenger Training Tank (CTT) was completed in August 1989 with the main production run being carried out between May and September 1990. The vehicles are used to provide realistic driver and maintenance training at Royal Armoured Corps and Royal Electrical and Mechanical Engineer Establishments.

The CTT is the first vehicle of its type to be used by the British Army and will replace Challenger 1 gun tanks at present used for this role.

As the CTT weighs the same as the Challenger 1 MBT so the automotive performance of the two vehicles is identical. In addition to being used for driver training, the CTT can also be used maintenance training, MBT recovery or as a dozer vehicle.

CTT is essentially a Challenger 1 MBT with its turret replaced with a new non-rotating turret that accommodates an instructor and up to four students. The new turret provides roll-over protection and the instructor has duplicate instrumentation and controls that allow him to control turret equipment, monitor and driver's performance and injects automotive faults into the driver's instrument panel. In an emergency the driver can also stop the vehicle.

Combat Dozer Blade

Early in 1990, the British Army selected the Pearson Combat Dozer Blade (UDK1) for service on its Challenger 1 MBTs and every Challenger in the British Army will be modified to accept it.

Typical roles of the UDK1 include digging hull down fighting positions, crossing anti-tank ditches and similar obstacles, construction of defensive earthworks, removing obstacles and creating roadways for follow-on vehicles through wood, bush and difficult terrain.

The dozer can be mounted or removed from the Challenger 1 MBT in under 15 minutes and it is fully interchangeable with other equipment such as the Pearson mineplough. The dozer unit contains its own powerpack with only electrical power being required from the vehicle.

The UDK1 will also be fitted to the new Chieftain AVRE currently under development by Vickers Defence Systems for the British Army.

The UDK1 meets the requirements of the British Army in a number of areas including: minimum possible weight, when in the stowed position it does not interfere with the main armament in any way and has a minimal impact on track approach, it has negligible effect on mobility or fightability, it has a continuous working cycle, earthmoving capability sufficient to fully utilise the tractive ability of a vehicle such as the Challenger, strength to withstand the forces that such a vehicle could impose, vehicle modifications would not affect performance in any way when the bulldozer was not fitted and would allow simple modification of existing vehicles, suprity at specific tasks such as digging a tank down scrape, improved reliability and simple maintenance and ease of operation.

Challenger Armoured Repair and Recovery Vehicle (CR ARRV)

In June 1985 it was announced that Vickers Defence Systems of Newcastle-upon-Tyne had been awarded a contract for the development and initial production of 30 Challenger Armoured Repair and Recovery Vehicles for the British Army. Royal Ordnance Leeds was also competing for this contract.

The first of six pre-production CR ARRVs was completed in August 1987 and all six were completed by 1988. Of these, two were used for trials by the Royal Electrical and Mechanical Engineers and the other

Close up of Vickers Defence Systems Challenger 1 deployed in Saudi Arabia with turret traversed to right and showing additional armour on hull sides

Prototype of the Vickers Defence Challenger Training tank from the rear

Challenger 1 MBT fitted with Pearson Engineering Combat Dozer Blade (UDK1) in lowered position. Typically at least one Challenger in each squadron of Challengers would be fitted with UDK1 dozer system

Challenger Armoured Repair and Recovery Vehicle deployed in Saudi Arabia early in 1991 by the British Army

four were retained by Vickers Defence Systems. Eventually the pre-production vehicles will be brought up to production standard.

Late in 1989 an additional 47 CR ARRVs were ordered from Vickers Defence Systems and in 1990 a further three were ordered bringing the total order for the British Army to 80 vehicles.

Production is undertaken at both Newcastle-upon-Tyne and Leeds, the first production vehicles were completed at Newcastle-upon-Tyne in April 1990 with the first vehicles being completed at Leeds in November 1990. A quantity of Challenger Armoured Repair and Recovery Vehicles were deployed to Saudi Arabia in late 1990 and used in Operation Desert Storm.

The Challenger Armoured Repair and Recovery Vehicle is fully described in *Jane's Military Vehicles and Logistics 1990-91* page 32.

Standard equipment on the vehicle includes a hydraulic winch, independent auxiliary winch, hydraulically operated crane that can lift a complete Challenger powerpack and a front-mounted multi-purpose blade which can be used as an earth anchor, dozer blade or crane stabiliser.

Challenger 1 SPG

For trials purposes, a Challenger 1 MBT chassis has been fitted with the private venture Vickers Shipbuilding and Engineering GBT 155 turret system. Full details of this turret are given in the *Self-propelled guns and howitzers* section.

Challenger 1 with Marksman Anti-aircraft Turret

In June 1985 the Marconi Command and Control Systems Marksman twin 35 mm anti-aircraft turret was taken to Royal Ordnance Leeds mounted on a Centurion tank chassis which had been used for firing trials.

The complete turret was then transferred to the Challenger 1 MBT hull which had been fitted with a suitable adaptor ring. This took one hour 29 minutes after which Challenger 1 Marksman was taken out and driven at speed on the test track facility. While on the test track the radar was operating and the gun stabilisation system was successfully tested.

In late summer 1985 a second Marksman turret was integrated with a Challenger 1 chassis for Proof and Endurance trials. The trials were successfully completed and demonstrated that Challenger 1 provided a suitable firing platform for the Marksman turret.

Challenger 1 MBT with TRIGAT

Vickers Defence Systems has proposed that the Challenger 1 MBT could be used as the British Army's platform for the long range TRIGAT ATGW system. The Challenger 1 MBT would have its turret removed and replaced by a new split turret with an elevatable arm on top of which is a pod with eight long range TRIGAT ATGWs in the ready to launch position.

Between the pods is a 25 mm M242 Chain Gun and mounted on the rear of the hull is a magazine containing an additional 16 TRIGAT ATGWs.

SPECIFICATIONS	
CREW	4
COMBAT WEIGHT	62 000 kg
UNLOADED WEIGHT	60 000 kg
POWER-TO-WEIGHT RATIO	19.35 bhp/tonne
GROUND PRESSURE	0.97 kg/cm^2
LENGTH	
from front mudguard to gun crutch	8.327 m
from gun muzzle to gun crutch (gun forward)	11.56 m
from front mudguard to gun muzzle (gun rearward)	9.8 m
WIDTH	
overall	3.518 m
over tracks	3.42 m
HEIGHT	
top of commander's sight	2.95 m
turret roof	2.5 m
GROUND CLEARANCE	0.5 m
TRACK	2.12 m
TRACK WIDTH	650 mm
LENGTH OF TRACK ON GROUND	4.79 m
MAX ROAD SPEED	
1st gear	13 km/h
2nd gear	22 km/h
3rd gear	34 km/h
4th gear	56 km/h
reverse 1st gear	14 km/h
reverse 2nd gear	22 km/h
reverse 3rd gear	36 km/h

MINIMUM TURNING RADIUS	
1st gear	6.8 m
2nd gear	11.1 m
3rd gear	17.4 m
4th gear	28.7 m
reverse 1st gear	7.3 m
reverse 2nd gear	11.1 m
reverse 3rd gear	18.6 m
FUEL CAPACITY	1797 l
USABLE FUEL CAPACITY	1592 l
FORDING	1.07 m
GRADIENT	58%
VERTICAL OBSTACLE	0.9 m
TRENCH	2.8 m
ENGINE	Perkins Engines (Shrewsbury) Condor 12V 1200 12-cylinder diesel, water-based cooled, developing 1200 bhp at 2300 rpm
AUXILIARY ENGINE	Coventry Climax H30 diesel, 3-cylinder, developing 37 bhp at 3000 rpm.*
TRANSMISSION	David Brown Vehicle Transmissions TN37. Borg-Warner torque converter with lock-up clutch giving 4 forward and 3 reverse gears
STEERING	Commercial Hydraulics STN 37 double differential, hydrostatic control

SUSPENSION	hydro-pneumatic, 450 mm total wheel travel
ELECTRICAL SYSTEM	24 V
BATTERIES	6 × 6TN (4 in hull, 2 in turret)
ARMAMENT	
main	1 × 120 mm L11A5 gun
coaxial	1 × 7.62 mm L8A2 MG
anti-aircraft	1 × 7.62 mm L37A2 MG
SMOKE-LAYING EQUIPMENT	2 × 5 smoke dischargers
AMMUNITION	
main	up to 64 projectile stowage positions (see text)
7.62 mm	4000
GUN CONTROL EQUIPMENT	
Turret power control	electric/manual
by commander	yes
by gunner	yes
Turret traverse	360°
Gun elevation/depression	+20°/-10°
Max rate power traverse	480 mils/s
Max rate power elevation	200 mils/s
commander's override	yes
Commander's fire control override	yes
Gun stabiliser	
vertical	yes
horizontal	yes
Range setting device	laser
Elevation quadrant	yes
Traverse indicator	yes

* On later vehicles, Perkins diesel type 4108 with a capacity of 1.76 litres giving 37 bhp at 2850 rpm

Status: Production complete, but can be resumed if further orders are received. In service with the British Army.

Manufacturer: Vickers Defence Systems
Manston Lane
Crossgates
Leeds, LS15 8ST, UK
Telephone: (0532) 648123.
Telex: 55134.
Fax: (0532) 607674.

Model of Challenger 1 MBT fitted with elevating arm on the top of which is mounted a pod containing sighting system, eight TRIGAT ATGW's and a 25 mm Chain Gun

Khalid Main Battle Tank

Development

In December 1974 Iran ordered 125 Shir 1 (FV4030/2) and 122 Shir 2 (FV4030/3) MBTs but the order was cancelled by the new Iranian Government in February 1979. The first three FV4030/2 prototypes were completed by January 1977. By this time production of the FV4030/2 was well under way at the Royal Ordnance Factory Leeds (since taken over by Vickers Defence Systems) with first production tanks scheduled for delivery in 1980.

In November 1979 Jordan placed an order with the United Kingdom for 274 Khalid MBTs worth £266 million for delivery from 1981.

Further development of the Shir 2 (FV4030/3) resulted in the Challenger 1 MBT which entered service with the British Army in 1983 and for which there is a separate entry.

Description

The Khalid is essentially the FV4030/2 with minor modifications to suit Jordanian requirements and is based on a late production Chieftain with major changes in the fire control system and the powerpack.

The powerpack consists of the Perkins Engines (Shrewsbury) Condor 12V 1200 diesel, the David Brown Vehicle Transmissions TN37 transmission and a cooling system by Airscrew Howden.

The Perkins Engines (Shrewsbury) Condor 12V 1200 is a 60° V-form, 12-cylinder, direct injection, four-cycle, compression ignition engine which develops 1200 hp at 2300 rpm. It is based on a conventional commercial design employing cast iron for major components and the majority of component parts are common to Perkins Engines (Shrewsbury) industrial and automotive CV range of diesels. The powerpacks for the Khalid (FV4030/2) and Challenger 1 are almost identical and both have twin electric starters.

The TN37 fully automatic transmission has been designed to provide four speeds forwards and three in reverse, using a three-element single-stage torque converter in conjunction with epicyclic gear trains. These are operated by multi-plate clutches, activated by an automatic electronic controller. Steering is by regenerative double differential type hydrostatic systems. The main vehicle service and parking brakes are incorporated within the transmission. The TN37 is arranged with a single input and two in-line outputs at right angles to the input. Flexible gear couplings are fitted to the outputs for connection to the final drives.

The cooling group essentially consists of two air-to-water radiators, two air-cooled charge air-coolers and three 380 mm mixed flow fans, mounted on top of the vehicle drive transmission. Cooling air enters through armoured louvres, passes through the heat exchangers to the fans and discharges through armoured louvres. A separate fan was required for cooling the auxiliary power unit when used during the vehicle 'silent watch' situation.

The Khalid has a bogie type suspension which is a further development of that fitted to Chieftain but with nearly twice the suspension travel.

The fire control system of the Khalid is the Computer Sighting System which is similar to the Chieftain Improved Fire Control System described in the entry for the Chieftain. The Barr and Stroud Tank Laser Sight is also fitted to the Khalid.

The commander's cupola is a No 15 which has been modified to accept the No 84 sight. The No 84 sight has been developed by Pilkington PE

Khalid MBT clearly showing 120 mm L11A5 rifled tank gun fitted with thermal sleeve and muzzle reference system

Khalid from rear with 120 mm L11A5 series gun in travelling lock

Limited and is a combined day/passive night sight plus projector recticle image unit, and provides the commander with a 24-hour day/night vision and firing capability. The No 84 sight incorporates two independent channels for day and night use and interfaces with the main armament via the projector reticle image unit which injects optical graticule information into the sight and also enables spot injection for the CSS. It has a fully armoured hood, can be elevated from −10 to +35° and is also provided with a wiper blade.

The main armament consists of a standard 120 mm L11A5 rifled tank gun, 7.62 mm L8A2 machine gun mounted coaxially with the main armament, 7.62 mm L37A2 machine gun which can be aimed and fired from inside the commander's cupola and six electrically operated smoke dischargers either side of the turret.

The driver can exchange his day driving periscope for a Pilkington PE image intensification night periscope.

Early in 1987 Jordan placed a multi-million pound order with Graviner of the UK for its crew bay explosion suppression system. These systems were retrofitted to all 274 Khalid MBTs, 293 Tariq modernised Centurions, M60A1/M60A3 MBTs and the new ENGESA EE-11 Urutu (6 × 6) APC ordered for police use.

Jordan has also taken delivery of a quantity of Royal Ordnance L23A1 APFSDS rounds and is considering a number of improvements for Khalid including night vision equipment and an additional armour protection package.

SPECIFICATIONS (as for Chieftain Mark 5 except)

WEIGHT	58 000 kg
POWER-TO-WEIGHT RATIO	20.68 hp/tonne
LENGTH (hull)	8.39 m
WIDTH	3.518 m
HEIGHT	3.012 m

Status: Production complete. In service with the Jordanian Army. In 1989 Royal Ordnance received several contracts from Jordan for the supply of spare parts of the Khalid MBT. Iraq has also supplied Jordan with one Chieftain Mk 3/3(P) and 89 Mk 5/5(P) MBTs plus 60 M47s and 20 CVR(T) Scorpion vehicles.

Manufacturer: Vickers Defence Systems
Manston Lane
Crossgate
Leeds, LS15 8ST, UK
Telephone: (0532) 648123.
Telex: 55134.
Fax: (0532) 607674.

Chieftain Main Battle Tank

Development

In 1956 Leyland Motors, which was the design parent for the Centurion Mk 7, built three prototypes of a vehicle called the FV4202. This was similar in some respects to the earlier Centurion but had only five road wheels and a new turret without a mantlet, and the driver was seated reclining at the front of the hull allowing the hull's height to be reduced. Both the latter characteristics were subsequently adopted for the Chieftain MBT(FV4201), the characteristics of which were issued in 1958. The first mockup was completed in early 1959 and the first prototype late in the same year. The Chieftain was first shown to the public in 1961 when it was fitted with a false covering over the front of its hull to conceal the well-shaped glacis plate.

A further six prototypes were built between July 1961 and April 1962 and in May the following year the Chieftain was accepted for service with the British Army. Two production lines for the Chieftain were established, one at Royal Ordnance Leeds (taken over by Vickers Defence Systems in 1986) and the other at Vickers' plant at Elswick.

About 900 Chieftain MBTs were built for the British Army, with production completed early in the 1970s. In 1971 Iran placed an order for about 707 Chieftain MBTs: the Mk 3/3(P) and Mk 5/3(P) plus a quantity of ARVs and bridgelayers, all of which were delivered by 1978.

Iran also took delivery of 187 improved Chieftains called the FV4030/1. These carried more fuel than the Mark 5/5P, had improved mine protection and additional shock absorbers on the rear station. They also incorporated electronic control of the Self-Changing Gears TN12 transmission to permit automatic operation and easier driver training.

In 1974 Iran ordered 125 Shir 1 and 1225 Shir 2 MBTs for delivery from 1980 but the order was cancelled in 1979 before deliveries could begin from the Royal Ordnance Leeds.

The Shir 1 (FV4030/2) was subsequently ordered by Jordan under the name Khalid and the Shir 2 (FV4030/3) is the basis for the Challenger 1 MBT for the British Army. There are separate entries for both vehicles earlier in this section.

During fighting in the Middle East Iraq captured between 200 and 300 Chieftain MBTs from Iran, many of which were undamaged. Late in 1988 Iraq gave Jordan a large amount of military equipment captured during fighting with Iran. This included 60 M47 tanks, 90 Chieftains, 19 Scorpions and 35 armoured personnel carriers.

In August 1981 Oman took delivery of 12 British Army Chieftain Mk 7/2C MBTs. These were on loan until 15 new Chieftains, called the Qayis Al Ardh by Oman, or Mk 15 by the manufacturer, were delivered between 1984 and 1985. These Chieftains, as well as the Scorpions used by Oman, are fitted with the NANOQUEST L20 sight which incorporates the Ferranti Type 520 laser rangefinder.

Chieftains of the Kuwaiti Army engaged Iraqi MBTs, including T-72s, during the Iraqi invasion of Kuwait in the summer of 1990.

Description (Mk 5)

The hull of the Chieftain is of cast and rolled steel sections welded together and is divided into three compartments: driver's at the front, fighting in the centre and engine at the rear.

The driver is seated at the front of the hull and is provided with a single piece hatch cover that lifts and opens to the right. To the rear of the hatch cover is a single wide-angle AFV No 36 Mk 1 periscope which can be replaced by an infra-red periscope for night driving, which is being superseded by a Badger passive night driving periscope developed by Pilkington PE Limited.

The turret is made of cast and rolled steel sections with the loader seated on the left and the commander and gunner on the right. The

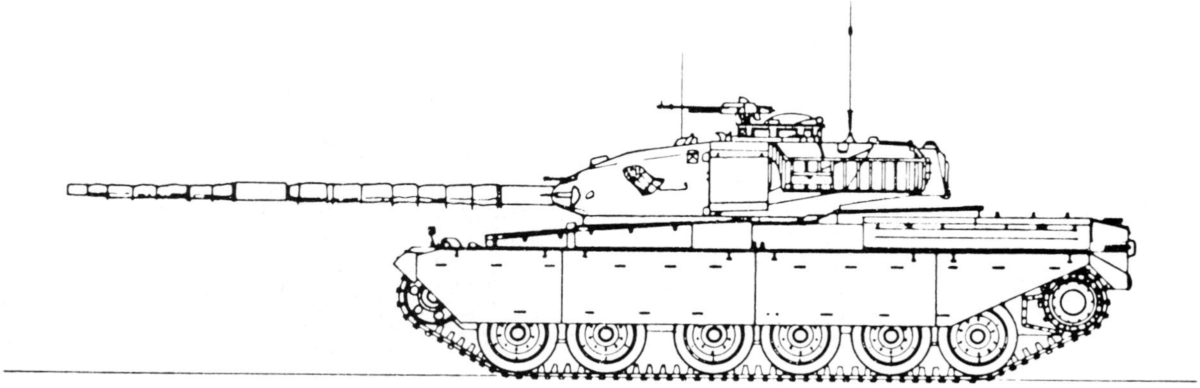

Chieftain Mk 3 MBT

British Army of the Rhine Chieftain MBT showing thermal sleeve and muzzle reference system of 120 mm L11 series rifled tank gun and white light/infra-red searchlight on left side of turret (Pierre Touzin)

commander's cupola can be traversed through 360° by hand and has a single piece hatch cover that opens to the rear, nine AFV No 40 Mk 2 observation periscopes and a single AFV No 37 Mk 4 sighting periscope. Mounted on the right side of the commander's cupola is a spotlight which is mounted coaxially with the cupola-mounted machine gun.

The gunner is seated in front of and below the commander and has a sight periscope No 59 with a magnification of ×1 and ×8 or a Barr and Stroud Tank Laser Sight Unit. The gunner also has a No 70 telescopic sight. Both the commander and gunner can replace their day sights with an infra-red sight with a magnification of ×3: the commander's night sight is the L1A1 and the gunner's an L3A1.

The loader has a two-piece hatch cover that opens to the front and rear of his position and a folding rotatable periscope AFV No 30 Mk 1.

Mounted on the left side of the turret is a Marconi Command and Control Systems infra-red/white light searchlight. The very high intensity light operates as a target illuminator and is mounted directly onto the gun turret so that it remains in line with the gun during traverse. In elevation a servo-control system on the mirror assembly ensures that the light beam remains aligned automatically onto the target. The searchlight has a range in the infra-red mode of at least 1000 m and a range in the white-light mode of at least 1500 m. The searchlight overhangs the side of the turret and has an armoured cover.

British Army Chieftains have now been fitted with the Barr and Stroud Thermal Observation and Gunnery Sight which is also installed on the Challenger MBT. This is mounted on the left side of the turret and replaced the infra-red/white light searchlight. The commander and gunner are both provided with separate outputs in the relaxed observation mode or as a gun sight.

The engine is at the rear of the tank and is coupled to a TN12 gearbox.

The TN12 epicyclic gearbox incorporates a Merritt-Wilson differential steering system and electro-hydraulic gear selector. It has six forward and two reverse gears plus an emergency mechanical selector of second gear forward and low reverse.

Of five portable fire extinguishers provided two large ones are mounted in the engine compartment. A Firewire automatic fire-detection system actuates a warning horn and internal flashing indicators and creates an audible signal in the crew's headphones.

The Horstmann suspension consists of three bogies each side, each with two dual road wheels and a set of three horizontal springs. The drive spocket is at the rear, idler at the front and there are three track return rollers. The first road wheel station has a hydraulic shock absorber. The tracks are of cast steel with dry pins and replaceable rubber pads. Each track has 96 links when new and the total weight of the pair of tracks on the Chieftain is 4719 kg. The top half of the track is covered by an armoured skirt which can be removed for maintenance.

Chieftain Mk 3 MBT from the rear with turret traversed to rear

Chieftain Mk 5/2K showing 120 mm L11 rifled tank gun fitted with thermal sleeve and muzzle reference system

Close-up of Chieftain MBT with dozer blade

Chieftain of the British Army of the Rhine fitted with Stillbrew passive armour protection and showing TOGS camera mounted on left side of turret (MoD)

An air ventilation and filtration system is mounted in the rear of the turret which provides the crew with clean air for normal ventilation and works in conjunction with an NBC filter pack which does not require the crew to wear respirators inside the tank.

Mounted on the turret roof is an infra-red detector which has three silicon photo-voltaic cells covering 360°. Any infra-red light source detected can be localised to within an arc of about 62°. A snorkel kit developed for the Chieftain was not issued for service use. The Chieftain can be fitted with a navigational aid.

Some Chieftains have a dozer blade mounted on the front of the hull consisting of an electro-hydraulic power pack which is fitted in place of the right hand front stowage bins. The blade itself is aluminium and is operated by the driver who is provided with a joystick control unit. Full details of this are given in *Jane's Military Vehicles and Logistics 1991-91* page 12.

Main armament is a 120 mm L11A5 rifled tank gun which is provided with a fume extractor, a muzzle reference system, and a thermal sleeve. It has a vertical sliding breech-block and a maximum rate of fire of eight to 10 rounds for the first minute and 6 rds/min thereafter.

The ammunition is of the separate loading type, ie projectile and charge. The projectiles are stowed alongside the driver, under the gun and in the turret and the charges are stored in pressurised bins below the turret ring. The gun can fire the following types of ammunition manufactured by Royal Ordnance: APDS-T (L15A4), DS-T (L20A1), HESH (L31), SH/Practice (L32A5), Smoke-WP (L34) and APFSDS-T (L23A1).

Mounted coaxially to the left of the main armament is a 7.62 mm L8A1 machine gun. The commander's cupola is provided with a 7.62 mm L37A1 machine gun which can be aimed and fired from within the tank. A 12.7 mm (0.50) L21A1 ranging machine gun can be mounted over the main armament. This has now been removed from all British Army Chieftain MBTs and ranging is achieved using the Tank Laser Sight. Mounted either side of the turret is a bank of six electrically operated smoke dischargers.

The gun control system is a Marconi Fighting Vehicle Gun Control Equipment (FV/GCE) No 7 Mk 4 which is a further development of the system used so successfully on the Centurion. This has the following modes of operation: stabilised power control, power control, emergency battery control and emergency hand control.

When originally introduced into service the Chieftain was provided with a 12.7 mm ranging machine gun (RMG) as the main means of aiming its 120 mm gun. For an APDS-T engagement the gunner fires short bursts from the RMG using four predetermined graticule marks. The burst which lands over the target indicates which main armament graticule mark should be used. Using this method the target can be engaged quite rapidly. For the lower velocity HESH ammunition up to 2000 m the RMG is used for obtaining the close bracket required for this round which has a higher trajectory. The main armament is not fired until a hit is virtually a certainty.

In the early 1970s the Barr and Stroud Tank Laser Sight Unit was introduced; the first model was the No 1 Mk 1 (in production from 1973/74), which was followed by the No 1 Mk 2 (1974/76), No 3 Mk 1 (1976/78) and the No 4 Mk 1 (which entered production in 1978). The Tank Laser Sight Unit houses the laser transmitter module and optical system, the receiver system and the optical sight system. The line of sight in elevation is synchronised with the axis of the gun by means of a precision parallel linkage, and in azimuth by coincident turret mounting. Final boresighting is achieved by instrument-mounted controls. Gun laying is achieved through a ballistic graticule. Laser ranging may be initiated by the gunner or remotely by the tank commander; the range is displayed in the left eyepiece and remotely displayed at the commander's station. In a 'smoke' situation, target range uncertainty is effectively diminished by selection of first and last range logic. The laser has an operating range of 500 to 10 000 m and is accurate to ±10 m for 90 per cent of shots.

The British Army's Chieftain fleet (except Mk 1s) has been retrofitted with the fully-integrated IFCS (Improved Fire Control System) developed and manufactured by Marconi Command and Control Systems Limited. IFCS provides a significantly higher first-round hit probability on stationary tank targets of at least 3000 m and on moving targets out to more than 2000 m.

Originally developed for the Chieftain, the IFCS has now been adapted for use in other tanks with different armament and sights. The IFCS has been installed in the Leopard 1 (with successful firings in 1979), and has been adopted for the Khalid and Challenger 1 as the Computer Sighting System (CSS).

The IFCS uses the Marconi 12-12P digital computer, which is programmed for the specific combination of tank, gun and ammunition carried. It automatically gathers and updates information from various sensors, calculates the ballistic solution and correct laying offsets for each target engagement, and controls the automatic laying of the gun in line and elevation (with tracking of moving targets) in readiness to fire. Both gunner and commander have full fire controls, and the commander has override control and can manually insert data whenever desired.

The system comprises four subsystems; the data handling subsystem (DHSS) with the 12-12P digital computer, commander's control and monitor unit, and the firing handles; the sighting subsystem with the tank laser sight (TLS) and aiming ellipse electronics; the sensor subsystem of various sensors located about the tank: for example, wind direction and velocity, air temperature and pressure, trunnion tilt, angle of sight, charge temperature, barrel wear, target displacement and ammunition type, and processing electronics; and the gun control equipment (GCE) which has been uprated for IFCS as the FV/GCE No 10 and No 11. The meteorological sensor has now been removed from British Army Chieftain MBTs.

For trials purposes a Chieftain has been fitted with a Barr and Stroud Tank Laser Sight with neodymium-YAG laser rangefinder, DF-series micro computer and the IR18 thermal imager. The IR18 was mounted in the space normally occupied by the Marconi infra-red/white light searchlight in the left side of the turret and moves in elevation with the main armament.

In March 1983 the British Ministry of Defence awarded Pilkington PE a contract worth over £4 million for the supply of nearly 1000 PRI (Projector Reticle Image) units for the Chieftain. This followed an earlier contract for PRI units for the Challenger 1 MBT. The PRI is incorporated in the tank's fire control system and provides an optical link between the main armament and the commander's sighting system. This allows him to lay the main armament by day and night and also enables the aiming mark derived from the IFCS to be overlaid onto the commander's view.

At least 324 Chieftain MBTs have now been fitted with the Barr and Stroud Thermal Observation and Gunnery Sight (TOGS) which is fully described in the Challenger 1 MBT entry. On Challenger 1 TOGS is mounted on the right side of the turret but on Chieftain it is on the left side in place of the current infra-red/white light searchlight. Two armoured boxes, one fitted with electrically operated fast-acting armoured doors, are fitted to the side of Chieftain's turret.

Marks of the basic Chieftain MBT

Mk 1 of which 40 were built and issued for training in 1965/66, 585 bhp engine

Mk 1/2 is Mk 1 brought up to Mk 2 standard and used for training

Mk 1/3 is Mk 1 with a new powerpack and used for training

Mk 1/4 is Mk 1 with a new powerpack and modified RMG, used for training

Mk 2 was the first model to enter service, first issued in 1967, with a 650 bhp engine

Mk 3 entered service in 1969, has improved auxiliary generator, 650 bhp engine, dry-air cleaner element, modified No 15 Mk 2 cupola with L37A1 7.62 mm machine gun, oil-filled top rollers, axle arms and track tensioner

Mk 3/G was a prototype only with turret air-breathing for engine aspiration

Mk 3/2 is Mk 3/G modified

Mk 3/S is production model of Mk 3/G with turret air-breathing and commander's firing switch

Mk 3/3 is Mk 3 with extended range RMG, fitted to accept the Barr and Stroud laser rangefinder, modified 720 bhp engine, new low-loss air cleaner system, turret aspiration and modified NBC pack
Mk 3/3P is Mk 3/3 export version for Iran
Mk 4, of which only two were built, had increased fuel capacity and other minor modifications
Mk 5 is a development of the Mk 3/3 with uprated 720 bhp engine and strengthened gearbox, gunner's telescope and commander's collimator fitted with graticule for use with extended range RMG ammunition, modified exhaust system, air cleaning filters for generator, battery heating and lagging, new type gun clamp to accept new thermal sleeve, modified charge bins, improved No 43 gunner's telescope mounting, improved stowage, 50 per cent reduction in RMG ammunition-carrying capacity, increase in APDS-T projectiles carried, No 37 Mk 4 commander's sight, stabilised infra-red detector, commander's machine gun elevation to +90° and new NBC pack
Mark 5/2K is Mk 5 for Kuwait, produced by Royal Ordnance Leeds and Vickers at Elswick with final vehicles completed late in 1979
Mk 5/3P is a Mk 5 for Iranian Army
Mk 6 is Mk 2 with a new powerpack and modified RMG
Mk 7 is Mk 3 and Mk 3/S with an improved engine and modified RMG
Mk 8 is Mk 3/3 with above modifications
Mk 9 is Mk 6 with IFCS
Mk 10 is Mk 7 with IFCS
Mk 11 is Mk 8 with IFCS, Stillbrew, TOGS and No 11 NBC system
Mk 12 is Mk 5 with IFCS, Stillbrew, TOGS and No 11 NBC system.

Variants

Chieftain with Stillbrew Armour

In 1986, 23 Base Workshop, Royal Electrical and Mechanical Engineers, Wetter, Germany, started to fit Chieftain MBTs of BAOR with Stillbrew appliqué armour. This provides a substantial increase in armour protection for the Chieftain MBT at a relatively small cost, with little degradation in automotive performance.

Stillbrew armour was originally conceived at the Royal Armament Research and Development Establishment (Chertsey) where Chobham armour was developed some years ago.

RARDE (Chertsey) made presentations of the new passive armour, subsequently code-named Stillbrew, to the British Army with the idea of improving armour protection of the British Army's Chieftain MBTs. At that time there was no formal requirement for additional armour protection for Chieftain.

Royal Ordnance Leeds (now Vickers Defence Systems) became involved in transferring the armour system from concept to a viable engineering solution for Chieftain. Firing trials against the new passive armour mounted in stands had successfully taken place.

A formal development contract was placed with Leeds by project manager Chieftain, with RARDE (Chertsey) being principal sub-contractor. This was one of the first arrangements in which RARDE (Chertsey) became involved with industry as a sub-contractor.

A short time after the contract was placed, RARDE and Leeds had designed and built the first armour system to be installed on Chieftain and this vehicle went to the ranges for full vehicle firing trials. These and subsequent trials demonstrated that the new passive armour exceeded all the requirements of the General Staff Target.

In 1985 two Chieftains with the new armour system were each subjected to mobility trials, half on roads and half cross-country. One of these trials was conducted at the Armoured Trials and Development Unit (ATDU) at Bovington Camp, Dorset, and the other at RARDE (Chertsey). These showed that there is no degradation in the vehicle's mobility as a result of the additional armour although its power-to-weight ratio is slightly reduced.

Final acceptance was given in 1986 by which time 23 Base Workshop was already involved as the armour is fitted when the Chieftains return for normal base overhaul.

No details of the actual Stillbrew armour have been released but it is probably an outer shell of steel with layers of composite armour behind. On Chieftain, Stillbrew is fitted over the front of the turret and hull top to the rear of the driver's position to give additional protection to the gap between the hull and turret, always a vulnerable part of an MBT.

Chieftain with Hydrostrut Suspension

For trials purposes a Chieftain MBT has been fitted with the Hydrostrut suspension system developed by Vickers Defence Systems Leeds and Air Log Limited. The Hydrostrut has features of the hydrogas suspension units fitted to the Challenger 1 MBT which were designed by Air Log Limited and built by Vickers Defence Systems Leeds. The Hydrostrut unit is mounted on a twin suspension bracket. No hull modifications are needed and many standard components, such as hubs and top rollers, are retained. Each road wheel is independently sprung and obviates the problem of the existing system where travel is limited if two adjacent wheels rise together. To compensate for variations in ambient temperature, terrain profile or changes in vehicle weight, the gas pressure and therefore spring stiffness and ride height can be quickly adjusted in the field.

Chieftain 900 MBT

This private venture MBT based on the Chieftain MBT is no longer marketed. Full details were given in *Jane's Armour and Artillery 1985-86*, pages 83-85.

Close-up of Chieftain 900 turret fitted with the Marconi Command and Control Systems Centaur computerised fire control system and showing new sighting system

Chieftain AVRE fitted with nose-mounted dozer blade and carrying two pipe fascines

Chieftain 900 with Centaur FCS

In 1990/91, Marconi Command and Control Systems, as a private venture, integrated their Centaur computerised fire control system into a Chieftain 900 MBT.

This was demonstrated in the UK in the Spring of 1991 to UK and foreign military personnel.

In addition to the Centaur fire control system it also featured the Marconi Command and Control Systems all-electric gun control equipment, stabilisation system and a complete set of sighting systems.

The latter included a roof-mounted SFIM VS 580 panoramic sight for the commander, SFIM stabilised sight for the gunner as well as a roof-mounted SFIM/TRT Casmir stabilised thermal sight that injects a thermal view into both the commander's and gunner's sight. Also mounted on the turret roof is a Hughes Leitz automatic muzzle reference system.

Chieftain AVRE

Although there was to be a Chieftain Armoured Vehicle Royal Engineer (FV4203) this never entered service. In 1986 the Engineer Workshops of 40 Army Engineer Support Group at Willich, Germany, converted 12 Chieftain gun tanks into Chieftain AVREs for use in BAOR.

The conversion involves the replacement of the Chieftain turret by a metal plate on which a set of rails is mounted. The rails carry up to three Maxi-fascine rolls or five rolls of Class 60 trackway.

Fascines and trackway are emplaced by raising the rear of the rails using modified M2 bridge/ferry hydraulic jacks.

The rails can be used to carry other combat engineering stores or a spare No 9 bridge for use with the Chieftain AVLB. The bridge is winched into its transport position by a winch on top of the hull and current combat engineer trailers such as those for Giant Viper can be towed. There is no provision for a demolition gun as fitted to the Centurion AVRE.

Chieftain AVRE has a crew of three and retains the dozer blade fitted to some Chieftain MBTs and can also be fitted with mine ploughs.

The Chieftain AVRE produced at Willich are however considered to be interim vehicles and late in 1989 Vickers Defence Systems at Newcastle-upon-Tyne was awarded a contract by the British MoD for the design, development and production of 48 Chieftain gun tanks into Chieftain Armoured Vehicle Royal Engineer. The 48 vehicle contract covers two prototype vehicles and 46 production vehicles (one batch of 30 and one of 16). Once trials with the prototypes have been completed these will be reconfigured to the final production configuration and delivered to the British Army.

First two prototypes were completed in 1991 with production running through to late 1994. Conversion work will be undertaken by Vickers

Chieftain MBT fitted with Pearson Engineering Track Width Mine Plough

Chieftain ARV of the British Army in travelling configuration (C R Zwart)

Full scale mock-up of Chieftain Armoured Vehicle Royal Engineer fitted with Pearson Engineering Track Width Mine Plough but without any equipment on top hampers

Chieftain Mk 6 AVLB in travelling configuration carrying No 8 Tank Bridge and fitted with Pearson Engineering Track Width Mine Plough on front of hull

Chieftain AVLB laying its No 8 Tank Bridge in position (MoD)

Defence Systems at Newcastle-upon-Tyne using Chieftain MBT chassis provided from Army stocks.

Production Chieftain AVREs have a new superstructure, winch with a capacity of 10 tonnes at hull rear, Atlas hydraulic crane and top-mounted hampers on which the engineer equipment is carried.

Main roles of the Chieftain AVRE can be summarised as:

1. Mine clearing using the Track Width Mine Plough or Giant Viper rocket propelled system
2. Laying class 60 aluminium trackway
3. Removing battlefield obstacles using front-mounted Combat Dozer Blade or crane
4. Filling in ditches using Combat Dozer Blade or plastic pipe fascines.

Armoured Repair and Recovery Vehicle

At present the British Army uses the FV434 armoured repair vehicle to change the powerpacks of the Chieftain MBT. It cannot however lift the complete powerpack of the new Challenger 1 MBT, for which the Chieftain Armoured Repair and Recovery Vehicle (ARRV) has been introduced. This is essentially a Chieftain ARV fitted with a hydraulic crane that can lift a complete Challenger 1 MBT powerpack. ARRVs were conversions from existing ARVs. Vickers Defence Systems are now supplying Hydraulic Track Tensioners for the Chieftain AVLB and ARRV.

Chieftain Armoured Recovery Vehicle (FV4204)

The Chieftain ARV was manufactured by Vickers at Elswick and was based on the chassis of the Mk 5 Chieftain MBT. The main double capstan winch with electro-hydraulic controls is provided with 122 m of 28 mm diameter cable. The similar auxiliary winch is hydraulically operated and is provided with 260 m of 11 mm diameter cable. Power for both winches is taken from a PTO on the main engine. Mounted at the front of the ARV is a hydraulically operated earth anchor which, when lowered, allows the vehicle to exert a pull of up to 90 000 kg. The Chieftains delivered to Iran had an Atlas AK 6000M crane which can lift a maximum load of 5803 kg. The Chieftain ARV has a crew of four and a loaded weight of 56 000 kg. Armament consists of a cupola-mounted 7.62 mm machine gun and smoke dischargers. Full details of the Chieftain ARV are given in *Jane's Military Vehicles and Logistics 1991-92* page 33. A total of 257 ARVs were built at Newcastle-upon-Tyne. Of this figure Iran ordered 73 (of which 41 were delivered) with Jordan taking a further 30 vehicles.

Chieftain Mk 6 AVLB

In March 1984 Vickers Defence Systems was awarded a £4.5 million contract from the British MoD to convert 11 Chieftain Mk 1/4 gun tanks into Chieftain Mk 6 AVLBs. User trials were completed in November 1985, the first three vehicles were handed over in June 1986 with final deliveries in September 1986.

The conversion work included removing the complete turret and replacing it with an armoured roof plate with a commander's hatch, fitting armoured lugs for mounting the launching mechanism to the hull and roof plate to maintain the same geometrical position as on production AVLBs, installing the hydraulic bridge-launching mechanism, stripping the vehicle and refitting the driver's/crew compartment and external parts to accommodate the AVLB components, modifying the hull to accept the HD1400 constant running pump installation, manufacturing the maraging steel launch structure and providing mounting points at the front of the hull for the Pearson Engineering Track Width Mine Plough System which is already in service on Centurion Assault Vehicle Royal Engineers.

Chieftain Armoured Vehicle-Launched Bridge (FV4205)

The Chieftain AVLB is basically a Chieftain MBT chassis without a turret and fitted with a hydraulic system for laying and recovering a bridge. The vehicle can carry and lay either a No 8 or a No 9 Tank Bridge. The No 8 Tank Bridge is carried folded and launched over the front of the vehicle. It has an overall length of 24.384 m and can span a gap of up to 22.86 m (hard banks) or 22.25 m (soft banks). The No 9 Tank Bridge is carried horizontal and is swung vertically through 180° and laid in position over the front of the vehicle. The No 9 Tank Bridge is 13.411 m long and can span a gap of up to 12.192 m (firm banks). The

Chieftain AVLB (with No 8 Tank Bridge) weighs 53 300 kg and has a crew of three. Full details of the Chieftain AVLB are given in *Jane's Military Vehicles and Logistics 1991-92* page 33. A total of 51 AVLBs were built by Leeds, 37 for the UK and 14 for Iran.

Chieftain with Marksman turret

For demonstration purposes a Chieftain has been fitted with a Marconi Command and Control Systems Marksman twin 35 mm air defence turret. Late in 1988 the company was awarded an initial contract from Finland who will install it on existing T-55 chassis.

Chieftain 155 mm SPG

This is a Chieftain chassis fitted with the Vickers Shipbuilding and Engineering Ltd GBT 155 155 mm turret, fully described in the *Self-propelled guns and howitzers* section.

SPECIFICATIONS

Model	**Mk 3**	**Mk 5**
CREW	4	4
COMBAT WEIGHT	54 100 kg	55 000 kg
UNLOADED WEIGHT	51 460 kg	53 500 kg
POWER-TO-WEIGHT RATIO	13.49 bhp/tonne	13.63 bhp/tonne
GROUND PRESSURE	0.84 kg/cm^2	0.9 kg/cm^2
LENGTH GUN		
forwards	10.79 m	10.795 m
rear	9.865 m	9.87 m
LENGTH HULL	7.52 m	7.518 m
WIDTH		
including searchlight	3.657 m	3.657 m
over skirts	3.504 m	3.504 m
over tracks	3.327 m	3.327 m
HEIGHT OVERALL	2.895 m	2.895 m
GROUND CLEARANCE	0.508 m	0.508 m
TRACK	2.718 m	2.718 m
TRACK WIDTH	610 mm	610 mm
LENGTH OF TRACK ON GROUND	4.8 m	4.8 m
MAX ROAD SPEED	48 km/h	48 km/h
FUEL CAPACITY	950 l	950 l
MAX RANGE		
road	400-500 km	400-500 km
cross-country	200-300 km	200-300 km
FORDING	1.066 m	1.066 m
GRADIENT	60%	60%
VERTICAL OBSTACLE	0.914 m	0.914 m
TRENCH	3.149 m	3.149 m
ENGINE	Leyland L60, 2-stroke, compression ignition, 6-cylinder (12 opposed pistons) multi-fuel	
ENGINE MODEL	No 4 Mk 6A	No 4 Mk 8A
OUTPUT	730 bhp at 2100 rpm	750 bhp at 2100 rpm
AUXILIARY ENGINE	Coventry Climax H30 No 4 3-cylinder, vertically opposed compression ignition developing 23 bhp at 2000 rpm	
TRANSMISSION	TN12 with 6 forward and 2 reverse gears plus emergency mechanical selection for 2nd gear forward and low reverse	
STEERING	triple differential incorporated in TN12	
CLUTCH	centrifugal	centrifugal
SUSPENSION	Horstmann	Horstmann
ELECTRICAL SYSTEM	28.5 V	28.5 V
BATTERIES	200 Ah (4 × 12 V) batteries in hull for engine starting and general electrical services. 100 Ah (2 × 12 V) in turret for radio load and emergency power supply to fighting equipment	

Model	**Mk 3**	**Mk 5**
ARMAMENT		
main	1 × 120 mm gun	1 × 120 mm gun
coaxial	1 × 7.62 mm MG	1 × 7.62 mm MG
anti-aircraft	1 × 7.62 mm MG	1 × 7.62 mm MG
RMG *	1 × 12.7 mm MG	1 × 12.7 mm MG
SMOKE-LAYING EQUIPMENT	2 × 6 smoke dischargers mounted on each side of turret	
AMMUNITION		
main	53	64
7.62 mm	6000	6000
12.7 mm*	600	300
GUN CONTROL EQUIPMENT		
Turret power control	electric/manual	electric/manual
by commander	yes	yes
by gunner	yes	yes
Max rate power traverse	360° in 16 s	360° in 16 s
Gun elevation/depression	+20°/-10°	+20°/-10°
commander's override	yes	yes
Commander's fire control override	yes	yes
Gun stabiliser		
vertical	yes	yes
horizontal	yes	yes
Range-setting device	yes**	yes**
Elevation quadrant	yes	yes
Traverse indicator	yes	yes

* When fitted with the IFCS the RMG and its ammunition is removed. IFCS is only in British Army Chieftains and not in Iranian or Kuwaiti vehicles.
** Retrofitted British Army vehicles have Tank Laser Sight fitted.

Status: Production complete. In service with:-

Country	**Quantity**	**Comment**
Iran	100	probably not all operational, also ARV and AVLB
Iraq	150	status uncertain, some 90 have been given to Jordan
Jordan	90	status uncertain, plus some Chieftain ARV
Kuwait	143	number delivered but most lost in fighting
Oman	15	delivered 1984/1985
UK	732	including reserves, 324 fitted with TOGS, also uses AVLB, AVRE, ARV and ARRV

Manufacturers: Vickers Defence Systems
Scotswood Road
Newcastle-upon-Tyne NE99 1BX
England
Telephone: (091) 273 8888
Telex: 53104
Fax: (091) 273 2324

Vickers Defence Systems
Manston Lane
Crossgates, Leeds LS15 8ST
England
Telephone: (0532) 648123
Telex: 55134
Fax: (0532) 607674

Vickers Advanced Main Battle Tank Mark 7

This vehicle, essentially a Leopard 2 chassis with a Vickers Defence Systems' turret is no longer being marketed. Full details were given in *Jane's Armour and Artillery 1990-91* pages 127 to 129.

Centurion Main Battle Tank

Development

In 1943 the Department of Tank Design was asked to start design work on a new heavy cruiser tank under the designation A41. This tank was required to have good armour protection, be armed with a 17-pounder gun and have a good cross-country performance: a high road speed was not considered essential at that time. The first mock-up of the A41 was completed in 1944 and six prototype tanks were completed in early 1945. These were sent to Germany but arrived too late to see any combat. The A41 later became known as the Centurion Mk 1 and the uparmoured A41A became the Centurion Mk 2, both armed with the 17-pounder gun.

The Centurion was first used in action by the British Army in Korea and has since seen combat in the Middle East (with Egypt, Israel, Iraq, Jordan and in the Lebanon), with the Indian Army during the Indo-Pakistan war, in Vietnam with the Australian Army and in Angola with the South African Army.

Throughout its life the Centurion has proved capable of being uparmoured and upgunned. It was originally armed with the 17-pounder which was replaced first by the 20-pounder and finally by the 105 mm L7 series gun which was subsequently adopted by many other countries and is now fitted to many tanks including the Leopard 1, Merkava, M48A5 (and other M48s), M60, Pz 61 and Pz 68, Type 74, S Tank, Vickers MBT, South Korea K-1 (or Type 88), and the American M1. Other improvements carried out during its life included increased fuel capacity, contra-rotating commander's cupola and improved stowage. All models of the tank used the same basic engine and transmission.

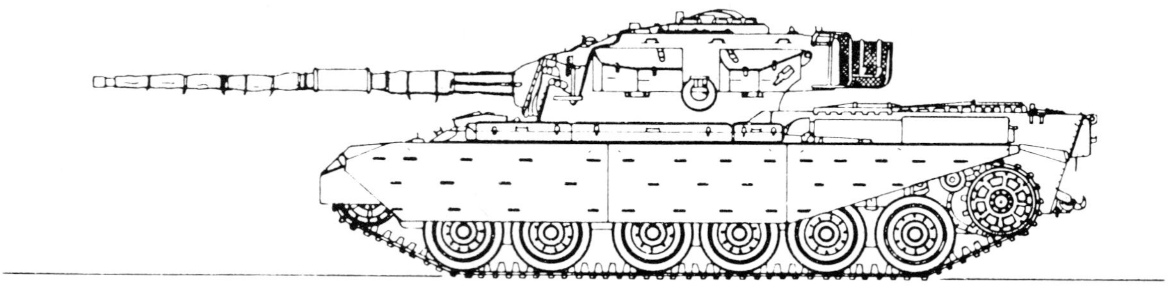

Centurion Mk 10 MBT

Production of the Centurion was undertaken by the Royal Ordnance Factories at Leeds and Woolwich, Vickers at Elswick and Leyland Motors at Leyland. The majority were built at Leeds and Elswick. Production of the Centurion was finally completed in 1962 by which time 4423 tanks had been built, of which at least 2500 were exported.

In the British Army the Centurion was replaced from the late 1960s by the Chieftain.

Description (Mk 13)
The all-welded hull of the Centurion is divided into three compartments: driver's at the front, fighting in the centre, and engine and transmission at the rear.

The driver sits at the front of the hull on the right side and has two hatch covers that open either side of his position, each with a periscope. To the left of the driver is some ammunition.

The turret in the centre of the hull is of cast construction with the roof welded into position. There is an ammunition resupply hatch in the left side of the turret and stowage boxes on either side of the turret. Most vehicles have a wire stowage rack at the rear of the turret. The loader is seated on the left of the turret, commander on the right, and the gunner in front of and below the commander.

The commander's cupola can be traversed through 360° by hand and is provided with split-hatches, periscopic sight with a ballistic pattern and seven periscopes for all-round observation. A searchlight mounted at the commander's station can be fitted with an infra-red filter. The gunner has a periscopic sight with ballistic pattern graticule, linked to a range drum for targets between 3000 and 8000 m. The commander's and gunner's sights are linked by a heat-compensated bar.

The loader is provided with twin hatch covers that open front and rear and a single observation periscope.

The engine compartment at the rear is separated from the fighting compartment by a fireproof bulkhead. Air is drawn in through the engine deck louvres by two fans. Power is transmitted from the engine through the clutch to the transversely mounted Merritt-Brown transmission. This is a combined change-speed and steering mechanism and incorporates a differential. Drive from the output epicyclic gears at each end of the gearbox is transmitted through an internally toothed coupling ring and an externally toothed driving shaft to the final drive.

The Horstmann suspension consists of three units on each side, each carrying two pairs of road wheels sprung by one set of concentric springs. The first and last road wheel units are fitted with shock absorbers. The drive sprocket is at the rear, idler at the front and there are six track return rollers, the four dual roller centre ones and the single front and rear ones which support the inside of the track only. The tops of the cast manganese steel tracks are covered by removable armour skirts which provide protection against HEAT projectiles.

As originally built the Centurion was not fitted with any infra-red night vision equipment, but in the 1960s many British Army Centurions were fitted with infra-red driving lights, infra-red searchlight to the left of the main armament and infra-red sights for both the commander and gunner. The tank has no NBC system or amphibious capability although a deep fording kit was developed. A dozer blade can be mounted at the front of the hull for clearing obstacles and preparing fire positions. One of the shortcomings of the Centurion was its short operational range, although later production tanks had increased fuel capacity and the Mk 5 could tow a monowheel trailer full of fuel to increase its operational range. In addition to being used by the British Army the monowheel trailer was also used by Sweden and the Netherlands.

Main armament of the Centurion Mk 13 is the 105 mm L7A2 rifled tank gun which is provided with a fume extractor, and many countries have

Danish Army Centurion Mk 5/2 fitted with cupola-mounted 12.7 mm M2 HB machine gun and Solartron Defence Systems tank gunnery and tactical training system

fitted a thermal sleeve. It has an effective range of 1800 m with APDS-T rounds or between 3000 and 4000 m using HESH. A well-trained crew can fire 8 rds/min.

The weapon is normally aimed using a 12.7 mm (0.50) ranging machine gun mounted coaxially with the main armament which has a maximum range of 1800 m and fires in three round bursts using tracer ammunition. The gun is fully stabilised and the gunner can select any one of the following modes of operation: manual elevation and traverse, non-stabilised power traverse, stabilised powered elevation and traverse or emergency, single-speed power traverse.

The 105 mm gun fires the following types of ammunition, all of which are of the fixed type with the projectile securely attached to a brass cartridge case which contains the propellant and initiated by an electric primer: APDS-T (L28A1), APDS-T (L52A1), APFSDS-T (L64 and later developments), DS/T (L45A1), HESH (L37), and Smoke (L39), all manufactured by Royal Ordnance. It will also fire ammunition manufactured by many other countries such as Austria, Canada, France, Germany, Israel and USA.

Mounted coaxially to the left of the main armament is a 7.62 mm (0.30) machine gun and there is a similar weapon on the commander's cupola for anti-aircraft use. Six electrically operated smoke dischargers are mounted on either side of the turret.

The earlier 20-pounder (83.8 mm) gun fires the following types of ammunition: canister with a muzzle velocity of 914 m/s, APDS-T with a muzzle velocity of 1432 m/s, HE with a muzzle velocity of 601 m/s and smoke with a muzzle velocity of 215 m/s.

There have probably been more variants of the Centurion than any other post-Second World War vehicle in the MBT class. A full list of the gun tank models and variants of the vehicle still in service follows; experimental vehicles have been excluded.

Centurion Gun Tank Models

Centurion Mk 1 armed with 17-pounder (76.2 mm) gun; none remain in service
Centurion Mk 2 armed with 17-pounder gun; none remain in service
Centurion Mk 3 armed with 20-pounder (83.8 mm) gun; none remain in service, most brought up to Mk 5 standard
Centurion Mk 4 was to have been a close support model armed with 95 mm howitzer, but was not placed in production
Centurion Mk 5 was designed by Vickers at Elswick and armed with a 20-pounder gun
Centurion Mk 5/1 is Mk 5 uparmoured
Centurion Mk 5/2 is Mk 5 with a 105 mm gun
Centurion Mk 6 is a Mk 5 uparmoured, with additional fuel capacity at rear of hull and 105 mm gun
Centurion Mk 6/1 is a Mk 6 with infra-red night vision equipment and stowage basket at rear of turret
Centurion Mk 6/2 is Mk 6 with ranging machine gun for 105 mm gun
Centurion Mk 7 was designed by Leyland and is designated the FV4007. Armed with a 20-pounder gun with fume extractor and carries 61 rounds of 20-pounder ammunition
Centurion Mk 7/1 is a Mk 7 uparmoured and designated FV4012
Centurion Mk 7/2 is a Mark 7 with a 105 mm gun
Centurion Mk 8 was developed from the Mk 7 and has a resiliently mounted gun mantlet with no canvas cover and the commander's cupola is contra-rotating. The commander can also raise his twin hatch covers in an umbrella fashion for improved visibility without exposing himself
Centurion Mk 8/1 is Mk 8 uparmoured
Centurion Mk 8/2 is Mk 8 with 105 mm gun
Centurion Mk 9 is Mk 7 uparmoured and upgunned with 105 mm gun, designated the FV4015
Centurion Mk 9/1 is Mk 9 with infra-red night vision equipment and stowage basket on rear of turret
Centurion Mk 9/2 is Mk 9 with ranging machine gun
Centurion Mk 10 is Mk 8 uparmoured, upgunned with 105 mm gun and has an ammunition capacity of 70 rounds, designated the FV4017
Centurion Mk 10/1 is Mk 10 with infra-red night vision equipment and stowage basket on rear of turret
Centurion Mk 10/2 is Mk 10 with ranging machine gun
Centurion Mk 11 is Mk 6 with ranging machine gun, infra-red night vision equipment and stowage basket on turret rear
Centurion Mk 12 is Mk 9 with infra-red night vision equipment, ranging machine gun and stowage basket on turret rear
Centurion Mk 13 is Mk 10 with ranging machine gun and infra-red night vision equipment

Centurion (Mk 5) Bridgelayer (FV4002)

The Danish army still uses 6 Centurion FV4002 bridgelayers.

Centurion (Mk 5) Bridgelayer ARK (FV4016)

This was used only by the British Army and has been withdrawn from service.

Centurion (Mk 2) ARV (FV4006)

Until the introduction of the Chieftain ARV in 1975 the Centurion was the standard ARV of the British Army, and even today is still used in small numbers. The Mk 2 was preceded by the Mk 1 ARV but none remain in service. The ARV Mk 2 is essentially a Centurion MBT with its turret removed and replaced by an all-welded superstructure behind the driver's position. The commander's cupola can be traversed through 360° and is fitted with a 7.62 mm (0.30) machine gun. Mounted at the rear of the

Centurion Beach Armoured Recovery Vehicle (BARV) (Simon Dunstan)

Royal Engineers Centurion AVRE 165 fitted with Track Width Mine Plough

Centurion AVRE 165 (Simon Dunstan)

Centurion Mk 2 ARV in travelling configuration (Terry J Gander)

Danish 105 mm armed Centurion with new gunner's sight and night vision camera mounted to right of 105 mm gun

Jordanian Army Tariq MBT

hull are spades which are used to stabilise the vehicle when the winch is being used. The winch's 31 000 kg capacity can be increased with the aid of snatch blocks to a maximum of 90 000 kg. The vehicle has a loaded weight of 50 295 kg and a crew of four. Full details of the Centurion ARV Mk 2 are given in *Jane's Military Vehicles and Logistics 1991-92* page 35. All Centurion Mk 2 ARVs were built by Vickers Defence Systems at Newcastle-upon-Tyne with 345 being built between 1954 and 1959. Export sales included Canada (9), Netherlands (28), Sweden (27) and Switzerland (12), with additional vehicles being supplied from British Army stocks.

Centurion (Mk 5) AVRE (FV4003)

This is used only by the British Army and is basically a Mk 5 with its gun replaced by a 165 mm demolition gun and a hydraulic dozer blade mounted at the front of the hull. The AVRE can also tow a trailer carrying the Giant Viper mine clearance equipment. The Centurion AVRE has a crew of five and weighs 51 810 kg loaded. Full details of the Centurion AVRE are given in *Jane's Military Vehicles and Logistics 1991-92* page 11. The Centurion AVRE and Chieftain bridgelayer can be fitted with mine clearing ploughs. Details of these are given in *Jane's Military Vehicles and Logistics 1991-92* page 243.

The Royal Engineers now also use the Centurion 105 mm gun tank to supplement the original Centurion 165 mm AVRE; the former is now referred to as the Centurion 105 AVRE and the latter the Centurion 165 AVRE. The Centurion 105 mm AVRE can also be fitted with the Track Width Mine Plough and can tow a Royal Ordnance Giant Viper mine clearance system. The Centurion AVRE was used by the Royal Engineers during Operation Desert Storm early in 1991. These were fitted with Royal Ordnance Explosive Reactive Armour to improve their battlefield survivability.

Centurion BARV (FV4018)

This is used only by the British Army and only 12 were built for use by amphibious forces. It is basically a Centurion tank with its turret replaced by a superstructure that enables it to operate in water up to 2.895 m deep. It has a crew of four, one of whom is a trained diver. Loaded weight is 40 643 kg. Full details of the Centurion BARV were given in *Jane's Military Vehicles and Ground Support Equipment 1987* page 48. By 1990 only three vehicles were left, one of which was in reserve and two with the Royal Marines for use during amphibious operations.

Teledyne Continental Upgraded Centurion

Details of this are given in the United States of America section.

Centurion with Marksman Anti-aircraft Turret

The Centurion chassis has been used for initial firing trials of the Marconi Command and Control Systems Marksman twin 35 mm anti-aircraft turret.

Swedish Army Strv 101 Centurion MBT with 105 mm L7 series rifled tank gun

Centurion with GBT turret

For trials purposes a Centurion chassis has been fitted with the Vickers Shipbuilding and Engineering GBT 155 mm turret. Details of this are given in the *Self-propelled guns and howitzers* section under UK.

Austrian Centurions

For some years Austria has been using T-34/85, Centurion 105 mm, M47 and Charioteer turrets in the static defence role, and recently purchased 120 Centurions from the Netherlands at £3800 for static defence use. There was also a plan to purchase additional Centurions from the Netherlands, some for the static defence role and some possibly to be used in the MBT role, but it has now been decided not to go ahead with this.

Danish Centurions

Denmark took delivery of 200 Centurions in the mid 1950s, funded by the United States under the Mutual Defense Assistance Program (MDAP). These remain in service and have been brought up to Mk 5 (retaining 20-pounder gun) and Mk 5/2 (with 105 mm L7A3 gun) standard. They also have a 12.7 mm M2 HB machine gun mounted forward of the commander's cupola. In 1983 it was announced that the Danish Army was to upgrade its Centurions by replacing the existing gunner's sight by a Swedish Ericsson sight with an integrated Neodymium-YAG laser rangefinder. This sight has magnifications of ×1 and ×10 and a range from 200 to 9990 m with an accuracy of 10 m. Denmark was the third country to order this sight; one of the others is known to be Sweden for its Centurion tanks (qv). Additional details of Danish Centurion modernisation plans are given in this section under Denmark

Israeli Centurions

Details of Israeli Centurion MBTs are given in this section under Israel.

Jordanian Centurions

Centurions of the Jordanian Army have been refitted with the American Teledyne AVDS-1970 series diesel engine in place of the original Meteor petrol engine. They have also been fitted with a Belgian SABCA fire control system which incorporates a laser rangefinder manufactured under licence from the Hughes company of the USA. This system is similar to the SABCA fire control system installed in the Leopard 1 MBTs used by Australia, Belgium and Canada. They have also been fitted with an American Cadillac Gage Textron turret drive and stabilisation system and hydro-pneumatic suspension. The last of 293 conversions to the Tariq standard was completed in 1985.

In early 1986 Kidde-Graviner of the UK was awarded a contract by Jordan to supply its Crew Bay explosion and suppression system for installation in all Jordanian Army Khalid, Tariq and M60A1/M60A3 MBTs plus the ENGESA EE-11 Urutu (6 × 6) APCs which were delivered in 1987.

Singaporeon Centurions

It is believed that Singapore has a minimum of 12 Centurions of which six are based in Brunei and six based in Taiwan. All of these have 105 mm guns and new diesel engines. Some reports indicate that Singapore may have as many as 63 Centurion MBTs.

South African Centurions

Details of the extensive improvements carried out by South Africa to its Centurion MBTs, now called the Olifant, are given in the *Addenda*.

Swedish Centurions

Late in 1981 the Swedish FMV awarded a contract worth some SEK 150 million to Bofors for the upgrading of the Swedish Centurion Strv 101 and 102.

The programme is being carried out in two stages covering firepower and mobility. The first retrofitted vehicle was delivered to the Swedish Army early in 1983. This contract covers the following:

A new fire control system from Bofors Aerotronics based on its export version computer. A new gun control equipment, including solid-state amplifiers, a gyro-unit and new gunner's and commander's control handles will also be in the package.

A new gunner's day sight incorporating a laser rangefinder and spot injection unit will be provided by Ericsson. Details of this sight are given in the note on Danish Centurions earlier in this entry.

A modified cupola retaining the existing commander's sight but including an armoured hood to provide protection when observing head out.

Bofors Ordnance will provide the Lyran twin-launcher for the 71 mm illuminating system with a range of 1300 m.

Late in 1982 the Swedish Army placed an order worth some SEK 180 million with Allison (CD-850-6A automatic transmission) and Teledyne Continental (AVDS-1790-2DC diesel engine) to be used in the Strv 101 and Strv 102 Centurion MBTs. Chassis conversion work will be carried out by Hägglund Vehicles. Integration of the chassis and turret will take place at a Swedish government arsenal in southern Sweden. It is believed that about one third of the Swedish Army's Centurion MBT fleet has been fitted with a new diesel powerpack.

Early in 1989 the Swedish Army took delivery of its first Swedish made mine clearing roller systems which can be fitted to both the Centurion and S-tanks. This is called the Mine Roller I with the Coupling 1 coupling the system to the Centurion and Coupling 2 coupling the system to the Strv 103B/C or S-tanks. The overall weight of the mine clearing roller is 6000 kg while the coupling weighs 1000 kg. The system can be fixed to the vehicle in 30 to 40 minutes and it is usually used at a speed of 8 to 15 km/h.

Early in 1985 the Swedish armed forces successfully completed trials of a Centurion MBT with the British Hydrostrut suspension system which has been developed jointly by Air-Log and the now Vickers Defence Systems Leeds and has also been tested on a British Army Chieftain MBT. The Hydrostrut suspension system can be retrofitted to any existing Centurion in one day and no modifications to the hull are necessary.

It is understood that Swedish Centurions are to be fitted with FFV-developed explosive reactive armour in the future.

In the longer term a number of other improvements are under consideration in addition to the above suspension improvements. These include a new elevation and traverse gearbox, new commander's cupola, new tracks, new driver's hatch, IR night sight, land navigation system, ceramic armour, NBC system, 120 mm gun and fire/explosion detection and suppression system.

SPECIFICATIONS

Model	**Mk 5**	**Mk 13**
CREW	4	4
COMBAT WEIGHT	50 728 kg	51 820 kg
POWER-TO-WEIGHT RATIO	12.81 hp/tonne	12.54 hp/tonne
GROUND PRESSURE	0.9 kg/cm²	0.95 kg/cm²
LENGTH GUN FORWARDS	9.829 m	9.854 m
LENGTH HULL	7.556 m	7.823 m
WIDTH	3.39 m	3.39 m
HEIGHT (without AA MG)	2.94 m	3.009 m
GROUND CLEARANCE	0.457 m	0.51 m
TRACK	2.641 m	2.641 m
TRACK WIDTH	610 mm	610 mm
LENGTH OF TRACK ON GROUND	4.572 m	4.572 m
MAX ROAD SPEED	34.6 km/h	34.6 km/h
FUEL CAPACITY	458 l	1037 l
MAX ROAD RANGE	102 km	190 km
FORDING	1.45 m	1.45 m
with preparation	2.74 m	2.74 m
GRADIENT	60%	60%
VERTICAL OBSTACLE	0.914 m	0.914 m
TRENCH	3.352 m	3.352 m
ENGINE	Rolls-Royce Mk IVB 12-cylinder liquid-cooled petrol developing 650 bhp at 2550 rpm	
AUXILIARY ENGINE	Morris USHNM 4-cylinder petrol, 20 bhp at 2500 rpm	
TRANSMISSION	Merritt-Brown Z51R manual with 5 forward and 2 reverse gears	
STEERING	triple differential	triple differential
CLUTCH	triple dry plate	triple dry plate
SUSPENSION	Horstmann	Horstmann
ELECTRICAL SYSTEM	24 V	24 V
BATTERIES	4 × 6 V, 115 Ah	4 × 6 V, 115 Ah
ARMAMENT		
main	1 × 83.8 mm gun	1 × 105 mm gun
coaxial	1 × 7.62 mm MG	1 × 7.62 mm MG
anti-aircraft	1 × 7.62 mm MG	1 × 7.62 mm MG
ranging MG	none	1 × 12.7 mm
SMOKE-LAYING EQUIPMENT	2 × 6 smoke dischargers either side of turret	
AMMUNITION		
main	65	64
7.62 mm	4250	4750
12.7 mm	none	600
GUN CONTROL EQUIPMENT	electric/manual	electric/manual
Max rate power traverse	360° in 26 s	360° in 26 s
Gun elevation/depression	+20°/−10°	+20°/−10°
commander's override	yes	yes
Gun stabiliser		
vertical	yes	yes
horizontal	yes	yes
Range setting device	yes	yes
Elevation quadrant	yes	yes
Traverse indicator	yes	yes
ARMOUR		
Hull glacis	76 mm*	118 mm*
Hull nose	76 mm*	76 mm*
Hull sides front	51 mm*	51 mm*
Hull sides		
rear upper	38 mm*	38 mm*
rear lower	20 mm*	20 mm*
Hull floor	17 mm*	17 mm*
Turret front	152 mm*	152 mm*

* Estimated

Status: Production complete. Centurion MBTs are still in service with the following countries:

Country	**Quantity**	**Comment**
Austria	120	turrets in static role
Denmark	226	both 20 pounder and 105 mm
Israel	1100	all 105 mm, and ARV
Jordan	293	all 105 mm, called Tariq
Singapore	12+	based in Brunei and Taiwan
Somalia	30	from Kuwait
South Africa	300	all 105 mm, called Olifant, plus ARV
Sweden	350	all 105 mm, being upgraded
United Kingdom	nil	uses ARV, AVRE 105, AVRE 165 and BARV

Manufacturers: Leyland Motors, Leyland, Royal Ordnance Factory, Leeds (taken over by Vickers Defence Systems in 1986)
Vickers Limited, Elswick
Royal Ordnance Factory, Woolwich

Vickers Defence Systems MBT Mk 3 (I)

Development

In 1985 Vickers Defence Systems made a number of improvements to the Mark 3 MBT with the principal aim of increasing the vehicle's speed and mobility. At the same time it was decided to enhance the tank's ride characteristics by fitting a suspension with a greater wheel travel. To accommodate these alterations in the powertrain the hull of the Mark 3 was redesigned, giving the vehicle a more aesthetic appearance.

The prototype of this vehicle, called the Mk 3 (I), was completed in May 1986 and shown for the first time at the 1986 British Army Equipment Exhibition.

The Vickers MBT Mk 3 (I) has been demonstrated in Morocco but as of May 1990 no production orders had been placed.

Description

The all-welded steel hull has been modified to accommodate the new powertrain and has a more streamlined appearance with a cast front, sloping panniers and stowage bins which blend into the nose of the vehicle.

The Mark 3 (I) carries a crew of four. The driver's station is at the front of the hull on the right and has been considerably redesigned for ease of operation. All instruments, including the automatic gear selector, have been regrouped on one console to the right of the driver. In addition to the usual gauges and warning lights the panel also incorporates a warning horn to alert the driver to any abnormal reading.

The driver's visibility has also been improved by the addition of two auxiliary day periscopes which increase his lateral vision when driving in the closed down position. The driver's central day periscope can be replaced by a passive night vision persicope for night driving.

Power is provided by a Perkins Engines (Shrewsbury) CV12 TCE, four-stroke, 26 litre engine. This is currently rated at 850 hp, although there are plans to increase this to 900 hp. The air cleaner, which is mounted on the powerpack, and the cooling system have been correspondingly upgraded to give improved airflow and more effective heat dissipation.

The new T1200 gearbox is closed coupled to the engine for easier powerpack removal and maintenance. Power is transmitted from the engine to the gearbox via a centrifugal clutch.

Prototype of Vickers Mk 3 (I) MBT with smoke dischargers on turret front and stowage boxes on sides of turret rear

Operation of the vehicle is further improved by power-assisted steering which also helps to increase brake pad life. In order to accommodate the additional engine power, the final drives and idler have been uprated and replaced by more rugged units.

The engine compartment has been fitted with a Kidde-Graviner fully automatic fire detection and suppression system.

The suspension of the vehicle has been improved by increasing the ground clearance and increasing active suspension movement by 10 per cent. As a result, the total wheel travel has been increased by about 10 per cent while the same spring rate has been retained. The ride characteristics of the tank are further enhanced by the fitting of rotary dampers to the first, second and sixth wheel stations. These give a much higher degree of damping and so a better fire on the move capability.

To reduce ground pressure the tank has been fitted with wider tracks (559 mm instead of 521 mm) and road wheels, and an uprated idler.

The Mark 3 (I) is fitted with the same turret as the Mark 3 but incorporating the Marconi Command and Control Systems SFCS 600 computerised fire control system and the EC 620 all-electric gun control system.

The 105 mm L7A1 rifled tank gun is retained but fitted with a muzzle reference system to allow gun/sight alignment to be carried out from within the vehicle. A thermal sleeve is also fitted as standard.

A 7.62 mm L37A2 machine gun is mounted coaxially with the main armament while a 12.7 mm M2 HB machine gun is retained as a back-up ranging machine gun. The commander's cupola has a 7.62 mm L8A2 machine gun and either side of the turret is a bank of five smoke dischargers.

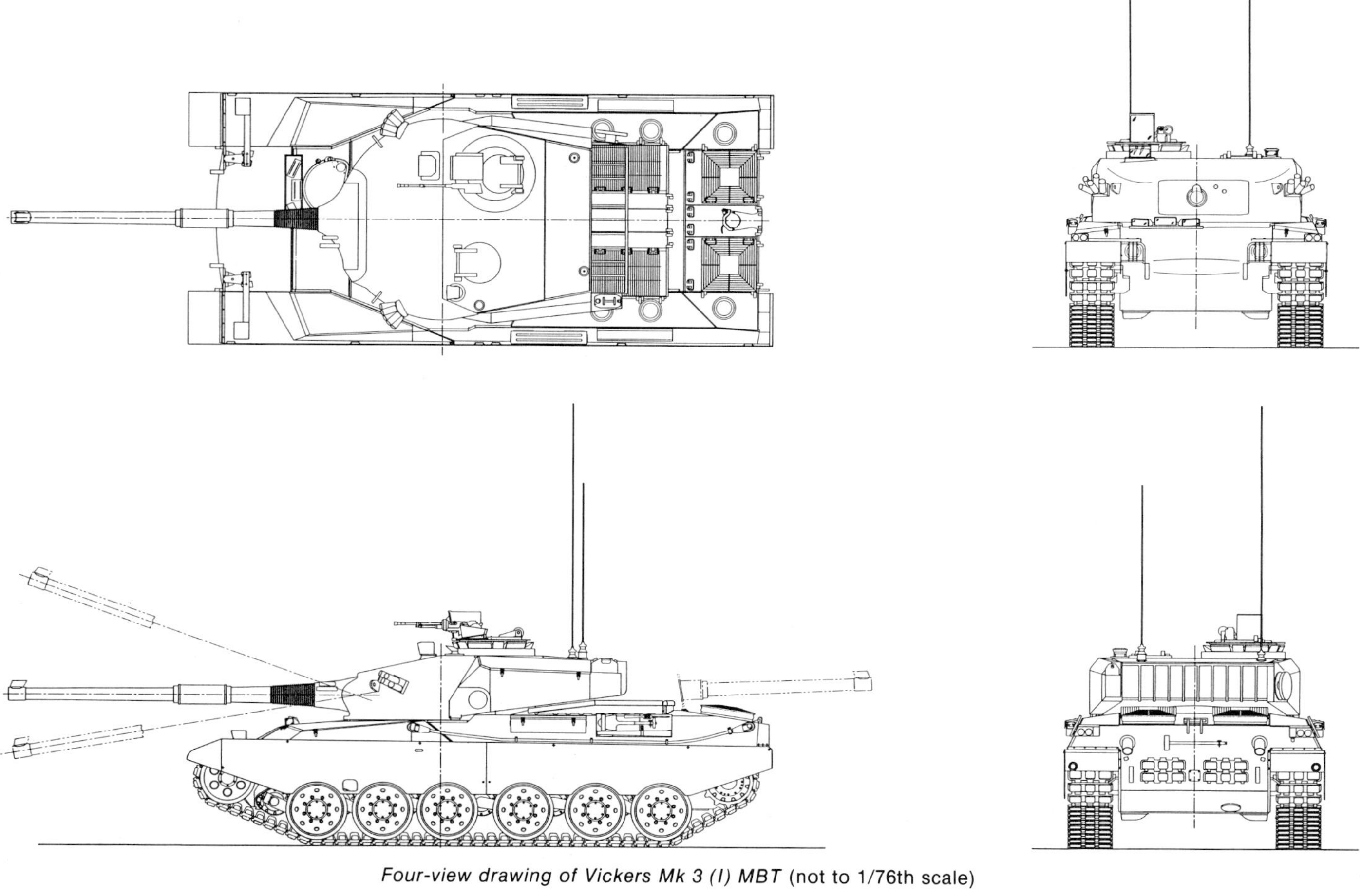

Four-view drawing of Vickers Mk 3 (I) MBT (not to 1/76th scale)

Vickers Mk 3 (I) MBT from rear firing its 105 mm L7A1 gun

Vickers Mk 3 (I) undergoing cross-country trials

SPECIFICATIONS	
CREW	4
COMBAT WEIGHT	39 600 kg
UNLOADED WEIGHT	37 100 kg
POWER-TO-WEIGHT RATIO	20.2 bhp/tonne
GROUND PRESSURE	0.82 kg/cm^2
LENGTH	
gun forwards	9.5 m
hull	7.29 m
WIDTH	3.24 m
HEIGHT	
overall without AA MG	2.97 m
to turret roof	2.44 m
GROUND CLEARANCE	0.457 m
TRACK WIDTH	559 mm
LENGTH OF TRACK ON GROUND	4.29 m
MAX ROAD SPEED	59 km/h
FUEL CAPACITY	1000 l
MAX ROAD RANGE	550 km
FORDING	1.1 m
GRADIENT	60%
SIDE SLOPE	30%
VERTICAL OBSTACLE	0.88 m
TRENCH	3 m
ENGINE	Perkins Engines (Shrewsbury) CV12 TCE diesel developing 850 hp at 2300 rpm
TRANSMISSION	Self Changing Gears T1200, 6 forward and 2 reverse gears
CLUTCH	pulsating
STEERING SYSTEM	epicyclic
BRAKES	multi-plate ball and ramp
SUSPENSION	torsion bars and rotary dampers
ELECTRICAL SYSTEM	24 V
BATTERIES	4 × 12 V, 200 Ah (maintenance free)
ARMAMENT	
main	1 × 105 mm gun
coaxial	1 × 7.62 mm L37A2 MG
anti-aircraft	1 × 7.62 mm L8A2 MG
ranging MG	1 × 12.7 mm M2
SMOKE-LAYING EQUIPMENT	5 dischargers on each side
AMMUNITION	
main	50
7.62 mm	5000
12.7 mm	1000
GUN CONTROL EQUIPMENT	
Turret power control	electric/manual
by commander	yes
by gunner	yes
Max rate of power traverse	360° in 15 s
commander's override	yes
Gun elevation/depression	+20°/-10°
Commander's fire control override	yes
Gun stabiliser	
vertical	yes
horizontal	yes
Range setting device	yes (laser rangefinder)
Elevation quadrant	yes
Traverse indicator	yes

Status: Development complete. Ready for production.

Manufacturer: Vickers Defence Systems
Armstrong Works
Scotswood Road
Newcastle-upon-Tyne NE99 1BX
England
Telephone: (091) 273 8888
Telex: 53104
Fax: (091) 273 2324

Vickers Defence Systems Main Battle Tank

Development

The Chieftain MBT was designed to meet the requirements of the British Army during the late 1950s. Realising that several countries would not buy such a heavy and expensive tank, Vickers designed a new 37-tonne MBT armed with the 105 mm gun from the Centurion and the engine, transmission, brakes, steering and fire control system of the Chieftain. The use of these Chieftain components in a tank weighing only 37 tonnes improved their reliability and durability.

In January 1961 a team of defence experts visited the United Kingdom and Germany to examine tank designs which could be produced in India. The Vickers proposal was accepted and in August 1961 an agreement was signed between Vickers and the Indian Government under which the company would establish a factory in India for the production of the Vickers MBT.

The first two prototypes of the Vickers MBT were completed in 1963; one was retained in the United Kingdom and the other sent to India. Vickers began production at its Elswick works in 1964 and the first production tank was delivered to India in 1965. The Indian factory at Avadi, near Madras, was built and the first Vijayanta (the Indian name for the Vickers tank) rolled off the production line in January 1965. Some 2200 Vijayantas have been built in India although some reports have quoted a lower figure of 1600, but production has been phased out and the Soviet T-72 produced under licence pending production of the new Indian MBT. The Vijayanta still has the original ranging machine gun but after evaluating several systems late in 1980 India selected the British Marconi Command and Control Systems SFCS 600 fire control system for installation in the tank. An initial batch of 70 systems has been supplied from the UK.

In 1968 Kuwait placed an order for 70 Vickers Mark 1 MBTs which were delivered between 1970 and 1972. In 1975 Vickers replaced the original Leyland L60 engine with the General Motors 12V-71T turbo-charged diesel developing 720 bhp at 2500 rpm, but retained the TN12 transmission. This increased the power-to-weight ratio to over 18 bhp/tonne, as well as increasing the maximum road speed to over 50 km/h.

In 1977 Kenya placed an order for 38 Vickers Mk 3 MBTs plus 3 ARVs, which were all completed by late 1980. In December 1978 Kenya placed a second order for 38 Vickers Mk 3 MBTs plus 4 ARVs which were delivered from 1981 to 1982.

In August 1981 the Nigerian Government placed an order worth some £60 million with Vickers Defence Systems for 36 Vickers Mark 3 MBTs, five ARVs and six bridgelayers. The MBTs are fitted with the General Motors 12V-71T diesel engine, Self-Changing Gears TN12 V5 transmission, NANOQUEST L23 gunner's sight which has a Simrad LV352 laser rangefinder, Pilkington PE Condor commander's day/night sight and the Marconi Command and Control Systems EFCS 600 fire control system. First production vehicles for Nigeria were completed in mid-1983. Early in 1985 Nigeria placed a repeat order for Vickers Mk 3 MBTs; first deliveries were made late in 1985 and final deliveries in late 1986. Included in this order was a further batch of five ARVs and six AVLBs. In 1989 Vickers Defence Systems supplied Tanzania with three Mk 3 ARVs to support its Chinese supplied Type 59 MBs.

In early 1984 a Vickers Mk 3 MBT was demonstrated in Thailand.

Late in 1991 it was disclosed that Vickers Defence Systems had received a contract worth £150 million for the supply of a significant number of MBTs and specialised support vehicles. It has not been disclosed which

Prototype of Vickers Mk 1 MBT without thermal sleeve for 105 mm gun and fitted with 12.7 mm RMG, 7.62 mm coaxial and 7.62 mm anti-aircraft machine guns

country ordered the vehicles or the model concerned but it is probable that this was for Nigeria with the vehicles concerned being the Mk 3 MBT as well as armoured recovery vehicles and armoured vehicle launched bridges.

Description (Mk 1)
The all-welded rolled steel hull of the Vickers MBT is divided into three compartments: driver's at the front, fighting in the centre and the engine and transmission at the rear.

The driver's compartment is on the right, with a single piece hatch cover opening to the right. Forward of the cover is a single AFV No 44 Mark 2 wide-angle periscope for closed down driving. This can be replaced by a passive periscope for driving at night. To the left of the driver are stowed 25 rounds of ammunition.

The all-welded turret has an ammunition reloading hatch in the left side and a stowage basket at the rear. The loader is seated on the left of the turret and the commander and gunner on the right.

The commander is provided with a single piece hatch cover that opens to the rear, a sight with a magnification of ×10 and six periscopes for all-round observation. The gunner is seated in front of and below the commander and is provided with a single sighting telescope with a ballistic graticule. The loader has a two-piece hatch cover that opens front and rear and is provided with an observation periscope in front of his position.

The engine, transmission, steering system and brakes are at the rear of the hull. The complete powerpack consisting of the L60 engine, transmission, radiators, fans, coolant and oil filter can be removed from the tank as a complete unit. Cooling air is drawn by way of the louvres, through the radiators and engine compartment and is discharged by fans through outlet louvres over the gearbox compartment.

The TN12 gearbox combines the Wilson epicyclic gear change principle with the Merritt steering system and provides six forward and two reverse speeds. Input to the transmission incorporates a centrifugal clutch and an input-driven pump provides oil pressure for gear engagement. Steering is controlled by hydraulically applied disc brakes with a mechanical interlock to prevent simultaneous engagement.

In addition to the belt-driven 24 V generator, an auxiliary three-cylinder engine drives a second generator to provide turret power and heat as well as charging the batteries when the main engine is not running.

The torsion bar suspension system consists of six dual rubber-tyred road wheels with the idler at the front, drive sprocket at the rear and three track return rollers. Suspension units one, two and six mount secondary torsion bars which are brought into action by stops mounted on the hull. The first, second and sixth road wheel stations are provided with a hydraulic shock absorber. The tracks are of cast manganese with removable rubber pads.

Standard equipment for the Mk 1 includes fire warning and internal fire-fighting systems. Optional equipment includes an NBC system, night vision equipment and a flotation screen carried collapsed around the top of the hull which would take between 15 and 30 minutes to erect. The tank would then be steered and propelled in the water by its tracks at a speed of 6.4 km/h. Some Indian vehicles have an infra-red/white searchlight mounted to the left of the main armament.

Main armament consists of a 105 mm L7A1 gun which is normally aimed using a 12.7 mm (0.50) ranging machine gun mounted coaxially with the main armament. This has a maximum range of 1800 m and fires three-round bursts of tracer ammunition. The 105 mm gun fires the following types of Royal Ordnance ammunition: APDS-T (L28A1), APDS-T (L52A1), APFSDS-T (L64), DS/T (L45A1), HESH (L37) and Smoke (L39), or any of the other makes of 105 mm ammunition from Austrian, French, German, Israeli or US sources.

During trials a Vickers MBT engaged 10 targets at ranges of between 600 and 1000 m in 55 seconds. Of 44 rounds of ammunition carried, five rounds are under the gun, eight on the turntable floor, three on the left side of the turret and 25 to the left of the driver.

The tank is fitted with the GEC-Marconi EC517 gun control and stabilisation system which has three modes of operation: non-stabilised, stabilised and emergency. A coaxial 7.62 mm machine gun is mounted to the left of the main armament, and a second 7.62 mm gun is mounted on the commander's cupola. On each side of the turret are six electrically operated smoke dischargers.

Variants

Vickers MBT Mk 2

This was a project only and was basically a Mk 1 with two integral missile launchers either side of the turret rear for the British Aerospace Swingfire ATGW missile.

Vickers MBT Mk 3

The all-welded rolled steel hull of the Vickers MBT is divided into three compartments: driver's at the front, fighting in the centre and the engine and transmission in the rear.

The driver's compartment is on the right, with a single piece hatch cover opening to the right. Forward of the cover is a single AFV No 44 Mark 2 wide-angle periscope for closed-down driving. This can be replaced by a passive periscope for driving at night. To the left of the driver are stowed 25 rounds of 105 mm ammunition.

The turret has a cast front welded to armour plate to give improved ballistic protection. It has an ammunition reloading hatch in the left side and a stowage basket on the rear. The loader sits on the left of the turret and the commander and gunner on the right.

The commander's cupola has 360° hand traverse and has a rear-opening single piece hatch cover. The commander has a Pilkington PE Condor combined day/night sight; this has day magnifications of ×1 and ×10 and a night magnification of ×4. Using the Condor, the commander can aim and fire the main armament at night or in poor light. The sight has an injected ballistic graticule from the collimator, a range readout from the laser rangefinder, and controls for operating the laser and for

Vickers Mk 3 MBT with thermal sleeve for main armament and Pilkington PE Condor commander's day/night sight firing main armament

laying and firing the main armament. The commander also has six periscopes for observation.

The gunner has a NANOQUEST L23 periscopic sight, with magnifications of ×1 and ×10, incorporating a Nd-YAG laser rangefinder and a ballistic graticule. The gunner's sight is linked to the gun by a temperature-compensated link bar and to a collimator in the commander's cupola. The collimator projects an illuminated ballistic graticule image into the field-of-view of the commander's sight when the cupola and the turret are lined up.

The loader has a single piece hatch cover that opens forward and an AFV No 30 Mark 1 observation periscope.

The main armament is the 105 mm L7A1 gun which fires APFSDS, APDS, HEAT, HESH, HE, Smoke and Canister rounds.

The tank is fitted with the Marconi Command and Control Systems EC620 gun control and stabilising system which has three modes of operation: non-stabilised, stabilised and emergency.

The 12.7 mm ranging machine gun is retained. It is a very effective heavy machine gun for use against lightly armoured and soft-skinned vehicles. It also provides a back-up in the event of failure of the laser rangefinder or fire control computer.

The secondary armament is a coaxially mounted 7.62 mm machine gun. A further 7.62 mm machine gun is provided on the commander's cupola, in front of and to the left of the hatch. This weapon can be elevated from -10° to the vertical, and can be mechanically cocked, aimed and electrically fired from under armour. A spotlight is fitted to the cross shaft of the machine gun mounting. Elevation is achieved by the commander's sight elevation gear; thus the machine gun and the spotlight follow the commander's line of sight through all angles of elevation.

Fifty rounds of 105 mm ammunition are carried, 18 rounds in the turret below the ring, 25 stowed horizontally in the front of the hull and seven stowed vertically in the hull centre section.

Electrically operated smoke dischargers are fitted on each side of the turret.

The engine, transmission, steering system and brakes are at the rear of the hull. The complete powerpack consisting of the engine, radiators, coolant and oil filter can be removed as a complete unit. All connections to the powerpack are by means of self-sealing couplings, plugs and sockets so that the powerpack can be readily removed from the vehicle for major overhauls. Power is provided by a Detroit Diesel 12V-71T, two-stroke, turbo-charged diesel developing 720 bhp at 2500 rpm.

The TN12 V5 automatic gearbox combines the Wilson epicyclic gear change principle with the Merritt steering system and provides six forward and two reverse gears. The driver can initiate an emergency sequential downchange to first and also engage one forward and one reverse gear

Vickers Mk 3 MBT from front and showing thermal sleeve for 105 mm gun

manually. Transmission is by a centrifugal clutch; an input-shaft driven pump provides oil pressure for gear engagement. Steering is controlled by hydraulically applied disc brakes with a mechanical interlock to prevent simultaneous engagement. The final drive gear ratio has been raised to take advantage of the extra power available.

Cooling air is drawn in through inlet louvres over the engine comapartment, through the radiators and engine compartment and is disacharged by fans through outlet louvres over the gearbox compartment.

The torsion bar suspension system consists of six dual rubber-tyred road wheels with the idler at the front, drive sprocket at the rear and three track return rollers. Suspension units one, two and six incorporate secondary torsion bars which are brought into action by stops mounted on the hull. The first, second and sixth road wheel stations have hydraulic shock absorbers. The tracks are of cast manganese steel with removable rubber pads.

Optional equipment includes added passive night vision equipment, deep wading and flotation equipment, full NBC filtration and pressurisation, heater, air-conditioning, contra-rotating gear for the comamander's cupola, and an automatic fire detection and suppression system. A 12.7 mm machine gun can also be fitted on the commander's cupola in place of the standard 7.62 mm machine gun.

Vickers Mk 3 MBT without track guards and thermal sleeve for main armament during pre-delivery trials in UK

Vickers Armoured Repair and Recovery Vehicle in travelling configuration carrying spare powerpack

Vickers Armoured Bridgelayer in travelling configuration

Vickers Mk 3 (I)
There is a separate entry for the Vickers Mk 3 (I) MBT in this section.

Vickers Armoured Bridgelayer
The Nigerian orders for Vickers Mark 3 MBTs placed in 1981 and 1985 each included six bridgelayers. The Vickers Armoured Bridgelayer (VAB) is designed to transport, launch and recover a tank bridge which is 13.41 m long and to provide a clear span of military bridge of class 60/70. The bridge is launched over the front of the vehicle with the bridge launching equipment hydraulically operated with power supplied by a pump driven from a PTO from the main engine. Full details of the vehicle are given in *Jane's Military Vehicles and Logistics 1991-92* page 93. Alternatively, the bridgelayer can incorporate a horizontally launched bridge.

Vickers Armoured Repair and Recovery Vehicle
The Vickers Armoured Repair and Recovery Vehicle is based on the Vickers MBT Mk 3 hull. Three (two with a crane and one without) were delivered to Kenya in 1980 and a further four (two with a crane and two without) in 1981-82. Nigeria has taken delivery of 10 ARRVs since 1983 in two batches of five vehicles each. In 1989 a small number, believed to be two, of Vickers ARVs were delivered to Tanzania. The vehicle has three compartments: winch compartment at the front with the driver to the left of the winch, the other three crew members in the centre of the vehicle and the engine and transmission at the rear.

The main winch, as developed for Chieftain, is of the capstan type with a nominal direct line pull of 25 000 kg working in conjunction with the anchor spade fitted to the front of the vehicle. Where necessary the line pull can be increased to a nominal 65 tons by multi-reeving of the rope using recovery equipment provided with the vehicle. The vehicle is provided with a full range of recovery equipment including pulleys, cables and tow bars and is also equipped with a crane capable of lifting 4000 kg to 3.62 m, which is primarily intended for removing/replacing the V800 power pack, or a TN12 transmission. The vehicle has a weight of 36 800 kg and is armed with a 7.62 mm machine gun and two six-barrelled smoke dischargers. Full details of the vehicle are given in *Jane's Military Vehicles and Logistics 1991-92* page 35.

Vickers Anti-aircraft Tank
This is the Vickers Mk 3 chassis fitted with the Marconi Command and Control Systems Marksman twin 35 mm turret. The Marksman turret has only been installed for trials purposes. The only customer so far is Finland who is fitting it to T-54/T-55 tank chassis.

Vickers Mk 3 with GBT turret
For trials purposes a Vickers Mk 3 chassis has been fitted with the private venture Vickers Shipbuilding and Engineering GBT 155 artillery turret. This turret, fully described in the *Self-propelled guns and howitzers* section, has also been trialled in India on a Vijayanta MBT chassis. The GBT 155 turret has yet to enter production.

SPECIFICATIONS

Model	**Mk 1**	**Mk 3**
CREW	4	4
COMBAT WEIGHT	38 600 kg	40 000 kg
UNLOADED WEIGHT	36 000 kg	37 500 kg
POWER-TO-WEIGHT RATIO	16.83 bhp/tonne	18 bhp/tonne
GROUND PRESSURE	0.87 kg/cm²	0.89 kg/cm²
LENGTH GUN FORWARDS	9.728 m	9.788 m
LENGTH HULL	7.92 m	7.561 m
WIDTH	3.168 m	3.168 m
HEIGHT		
overall without AA MG	2.64 m	3.099 m
to turret roof	2.438 m	2.476 m
GROUND CLEARANCE	0.406 m	0.432 m
TRACK	2.533 m	2.533 m
TRACK WIDTH	520 mm	520 mm
LENGTH OF TRACK ON GROUND	4.28 m	4.29 m
MAX SPEED		
road, forward	48 km/h	50 km/h
FUEL CAPACITY	1000 l	1000 l
MAX ROAD RANGE	480 km	530 km
FORDING	1.143 m	1.1 m
GRADIENT	60%	60%
SIDE SLOPE	30%	30%
VERTICAL OBSTACLE	0.914 m	0.83 m
TRENCH	2.438 m	3 m
ENGINE	Leyland L60 Mk 4B 6-cylinder water-cooled multi-fuel developing 650 bhp at 2670 rpm	Detroit Diesel 12V-71T turbo-charged diesel developing 720 bhp at 2500 rpm*
TRANSMISSION	Self-Changing Gears TN12 semi-automatic, 6 forward and 2 reverse speeds	
STEERING	triple differential	triple differential
SUSPENSION	torsion bar	torsion bar
ELECTRICAL SYSTEM	24 V	24 V
BATTERIES	4 × 6 V, 115 Ah	4 × 12 V, 200 Ah
ARMAMENT		
main	1 × 105 mm gun	1 × 105 mm gun
coaxial	1 × 7.62 mm MG	1 × 7.62 mm MG
anti-aircraft	1 × 7.62 mm MG	1 × 7.62 mm MG
ranging machine gun	1 × 12.7 mm	1 × 12.7 mm
SMOKE-LAYING EQUIPMENT	6 smoke dischargers mounted either side of turret	
AMMUNITION		
main	44	50
7.62 mm	3000	2600
12.7 mm	600	700
GUN CONTROL EQUIPMENT		
Turret power control	electric/manual	electric/manual
by commander	yes	yes
by gunner	yes	yes
Max rate of power traverse	360° in 13 s	360° in 15 s
commander's override	yes	yes
Gun elevation/depression	+20°/-7°	+20°/-10°
Commander's fire control override	no	yes
Gun stabiliser		
vertical	yes	yes
horizontal	yes	yes
Range setting device	yes	yes (laser rangefinder)
Elevation quadrant	yes	yes
Traverse indicator	yes	yes
ARMOUR		
Hull nose	80 mm	80 mm
Hull glacis	60 mm	60 mm
Hull sides front and intermediate	40 mm	40 mm
Hull sides rear	30 mm	30 mm
Hull top	25 mm	25 mm
Hull floor	17 mm	17 mm
Hull rear	20 mm	20 mm
Turret front	80 mm	n/a
Turret sides	40-60 mm	n/a
Turret rear	40 mm	40 mm
Turret top	25 mm	25 mm

* or Perkins Engines (Shrewsbury) CV12 TCA 12-cylinder turbo-charged diesel developing 800 bhp at 2300 rpm

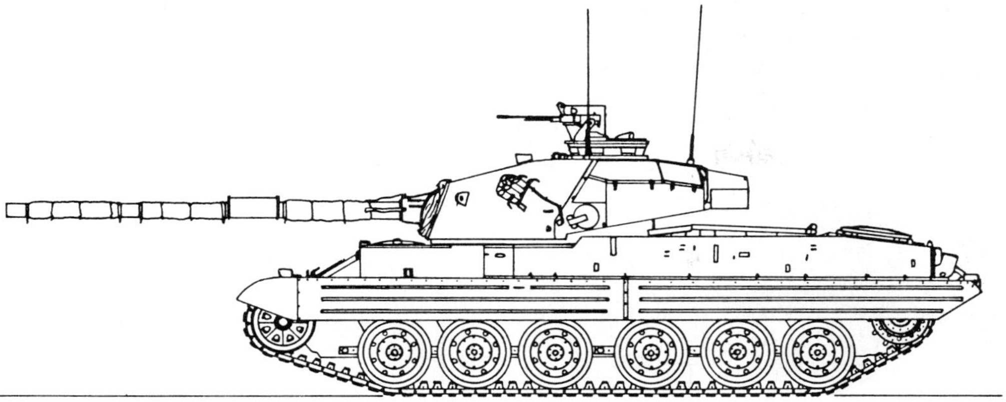

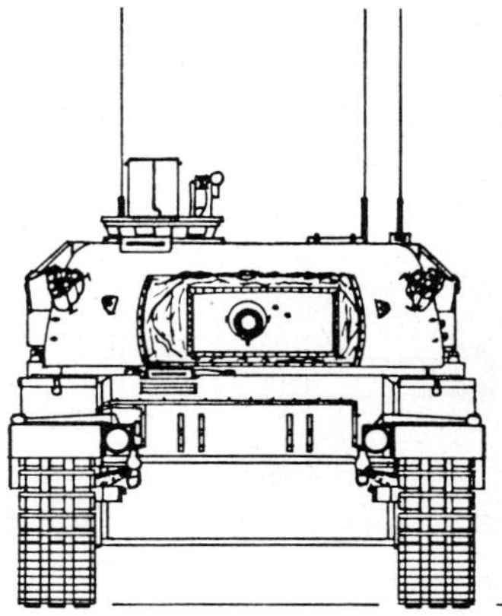

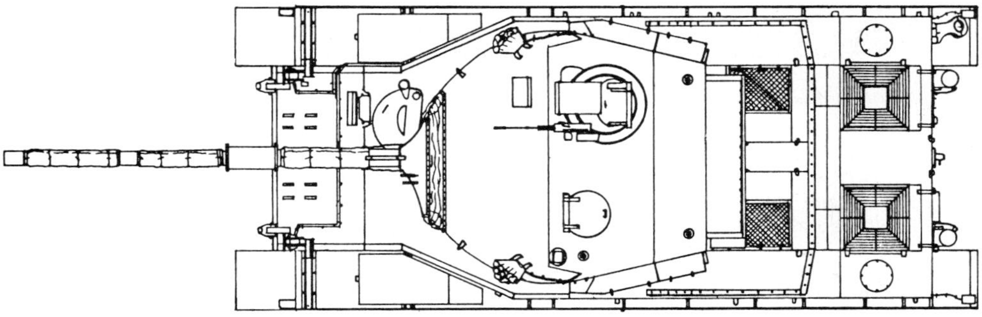

Vickers Mk 3 MBT as built for Kenyan Army

Status: Production as required. In service with the following countries:

Country	Model	Quantity	Comment
India	Mk 1	2200	Most locally built, L60 engine
Kenya	Mk 3	76	Detroit Diesel engine, plus 7 ARV
Kuwait	Mk 1	70	L60 engine. Probably lost during fighting in 1990
Nigeria	Mk 3	72	Detroit Diesel engine, plus 10 ARV and 12 AVLB. Additional order placed in 1990
Tanzania	n/app	n/app	two ARV with Detroit Diesel engine

Manufacturer: Vickers Defence Systems
Armstrong Works
Scotswood Road
Newcastle-upon-Tyne NE99 1BX
England
Telephone: (091) 273 8888
Telex: 53104
Fax: (091) 273 2324

UNITED STATES OF AMERICA

General Dynamics, Land Systems Division, M1/M1A1/M1A2 Abrams MBT

Development

The XM1 Program was established in December 1971 and in February the following year the US Army activated a task force with the user, trainer and developer participating to formulate the concept for the new battle tank. This report was published in August 1972 and the proposed characteristics were then reviewed to eliminate unnecessary features and reduce costs to the minimum. The final programme was contained in a Development Concept Paper, as amended, which was approved by the Deputy Secretary of Defense in January 1973.

In June 1973 contracts for the prototype development validation phase of the new tank, called the XM1, were awarded to the two prime contractors. Each contractor was required to develop a tank which met the material need requirements while remaining within an average 'design-to-unit hardware' cost of $507 790 in 1972 dollars for production tanks. Each contractor was required to deliver to the Army one prototype tank, one automotive test rig and one hull and turret for ballistic tests. The Defense Division of the Chrysler Corporation (now the Land Systems Division of General Dynamics Corporation), which also then manufactured the M60 series was awarded a contract worth $68.1 million, and the Detroit Diesel Allison Division of the General Motors Corporation was awarded one worth $88 million.

In February 1976 the Army accepted the prototype vehicles from both US contractors and operational and engineering testing was conducted through to April 1976.

In November 1976 the Secretary of the Army announced that the Chrysler Corporation's prototype had been selected to enter Full Scale Engineering Development (FSED). The Leopard 2 (AV) was still being tested. The original bids, submitted in June, were for $221 million from Chrysler and $208 million for General Motors but when the new bids were submitted Chrysler had dropped its bid to $196 million and the General Motors bid had increased to $232 million. Both final bids included the turbine engine, in the initial offer General Motors only bid with the diesel engine

Chrysler was awarded a three-year contract worth $196.2 million for the FSED phase, during which 11 XM1 pilot vehicles with their associated spares were produced at the then Chrysler-run Detroit Arsenal Tank Plant. The first pilot was completed in February 1978 and the last in July 1978.

In March 1982 Chrysler sold its tank building subsidiary (Chrysler Defense Incorporated) to General Dynamics for $348.5 million.

Production of the M1 (or currently M1A1) tank is in progress at the Lima Army Tank Plant in Lima, Ohio, with the first production tank completed in February 1980, and M1 assembly began at the Detroit Arsenal Tank Plant in 1982. The original intention was to procure 3312 M1s at a cost of $4900 million, but in 1978 this figure was increased to 7058 by the end of FY88; in 1984 the figure was further increased to 7467. Production increased from 30 vehicles a month in 1981 to 60 a month in 1982, which was maintained until January 1984. At that time production was increased to 70 vehicles a month.

General Dynamics, Land Systems Division, began delivering the first of 299 FY90 M1A1 MBTs in April 1991. These are all due to be delivered by September 1991, after which tank assembly at the Detroit Arsenal Tank Plant will cease and the total of 7467 will have been met. A second FY90 increment of 322 M1A1s will be manufactured at the Lima Army Tank Plant between September 1991 and June 1992. Included in this figure of 322 are 23 M1A1 MBTs for Egypt.

For FY91, the US Army has requested only 225 M1A1s.

Major improvements of the M1 over the current M60 series can be summarised as improved protection, mobility, firepower and RAM-D

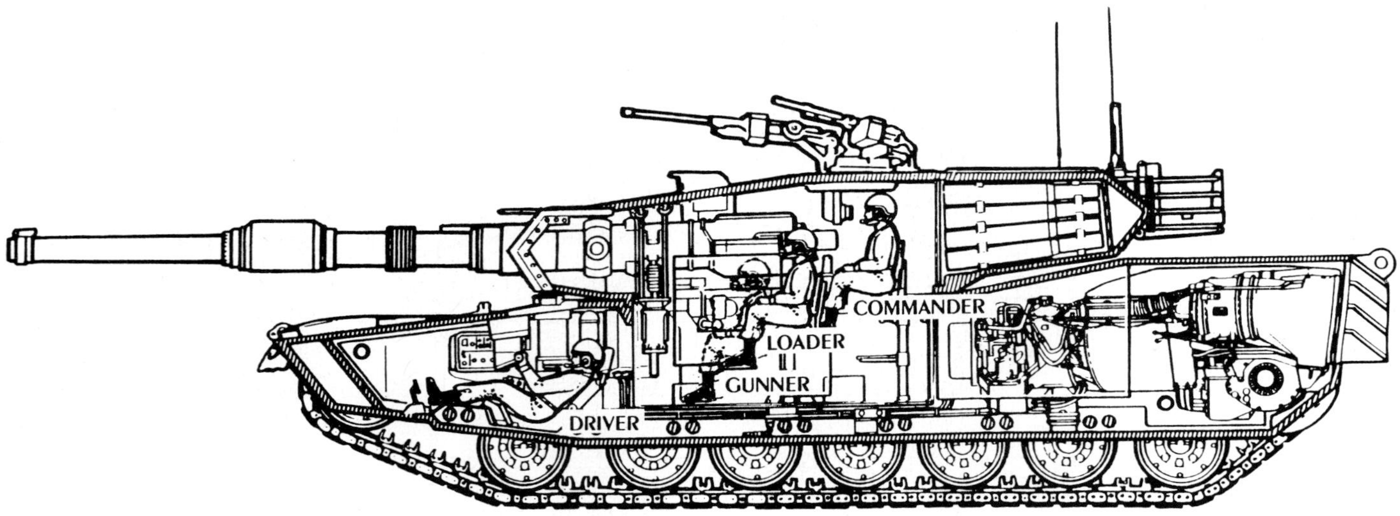

Cutaway drawing of M1A1 MBT showing position of main components. This is not representative of armour thickness (not to 1/76th scale)

(reliability, availability, maintainability and durability). Total production of the M1 amounted to 2374 tanks, finishing in February 1985 when production switched solely to the Improved M1 with improved armour protection.

There is a possibility that some earlier M1 MBTs will be upgraded with some of the improvements of the M1A1 and that this will enable the Detroit Arsenal Tank Plant to remain open.

Export sales potential, according to General Dynamics, Land Systems Division, are Saudi Arabia (total requirement is for some 700 vehicles. In July 1990 the Governement of Saudi Arabia signed an agreement for the purchase of 315 M1A2s for delivery over a three year period from 1993), Pakistan (trails already carried out), Sweden (trials with one M1A1 already carried out) and the United Arab Emirates.

The sale of 555 M1A1 MBTs to Egypt has been agreed upon and the sale will be contracted for in six increments. The first increment is already under contract. The US Government is participating with Egypt in a co-production programme that will allow Egypt to equip its army with M1A1s.

The US delivered two M1A1 MBTs in August 1990 and is scheduled to deliver a further 23 in 1991. For the remaining 530 tanks, the US, through General Dynamics, will deliver kits to the Egypt Tank Plant that will be the basis for the Egyptians to assemble their own M1A1s.

The tanks will be produced over a 10 year period. General Dynamics also has contracts to assist with in-country manufacturing and facilitation of the Egyptian Tank Plant.

M1/M1A1 in Southwest Asia

In March 1991, the US DoD released provisional information on the performance of a number of key US weapon systems deployed during Operation Desert Storm. The following is an extract on the M1/M1A1 MBT.

"After 100 hours of offensive operations, the operational readiness rates for both the VII Corps and XVIII Airborne Corps exceeded the Army's 90 per cent standard. Especially noteworthy was a night move by the 3rd Armored Division covering 200 km (120 miles). None of the more than 300 tanks in the division broke down.

Seven separate M1A1 crews reported being hit by T-72 tank rounds. These M1A1s sustained no damage, attesting to the effectiveness of our heavy armour. Other crews reported that the M1A1 thermal sight allowed them to acquire Iraqi T-72s through the smoke from oil well fires and other obscurants. The T-72 did not have the same advantage. This situation gave the Abrams a significant edge in survivability, engagement range and night manoeuvre. Additionally, tank crews reported that the M829A1 tank round was extremely effective against the T-72. In sum, the combined performance of the Abrams armour, thermal sight and ammunition attest to the systems exceptional lethality and survivability.

Of the 1955 M1A1 Abrams tanks in theatre, four were disabled and four were damaged but are repairable. No M1A1 crew members were killed in the many tank engagements. Overall M1A1 operational readiness rates exceeded 90 per cent prior to and during combat."

Description (M1A1)

The hull and turret of the M1 is of advanced armour construction similar to the Chobham armour developed in the United Kingdom and also gives protection against ATGWs and other battlefield weapons.

The driver is seated at the front of the vehicle in the centre and operates the vehicle from a semi-reclining position when driving with the hatch closed. Steering is accomplished by rotating a motorcycle type T-bar which actuates the steering lever on the transmission to produce the steering speed bias of the track. At both ends of the T-bar are twist grip controls which serve as the throttle for the electronic fuel management system. The condition of fluid levels, filters, batteries, electrical connectors and circuit breakers are displayed on the driver's

M1 MBT with 105 mm gun with driver's hatch closed (Michael Green)

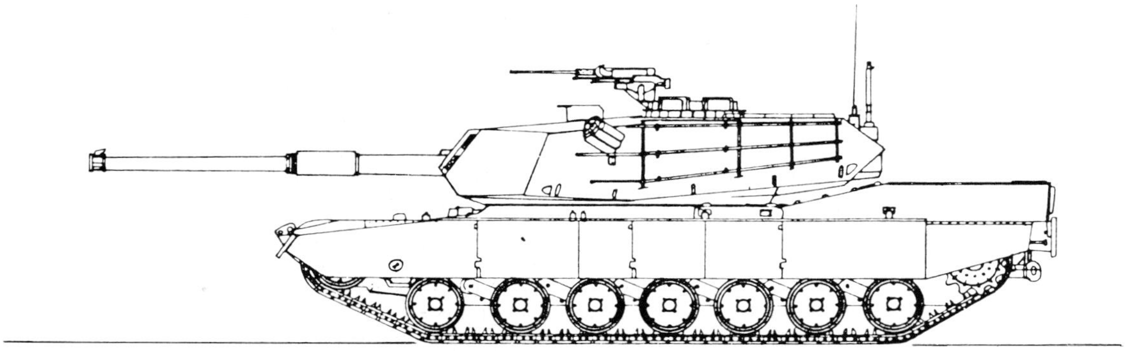

M1 MBT

M1 Abrams MBT on exercise in Germany showing extensive stowage of kit on turret and with commander's, loader's and driver's hatches open (US Army)

M1 MBT from rear with turret traversed right (Pierre Touzin)

maintenance monitoring panel. The driver is provided with a single hatch opening to the right with three integral periscopes for observation when the hatch is closed. The centre periscope can be replaced by an image intensification periscope for night driving. The driver's has 120° field-of-view and his night driving periscope will fit into the loader's periscope housing for night time surveillance.

The commander and gunner are seated on the right of the turret and the loader on the left. The commander is provided with six periscopes which cover 360°, as well as a sight with a magnification of ×3 for the 12.7 mm machine gun mounted over his position and an optical extension of the gunner's primary sight. The gunner has a primary sight (GPS) with dual day optics with a magnification of ×10 (narrow field-of-view), magnification of ×3 (wide field-of-view), close-in surveillance magnification of ×1 and an 18° field-of-view, thermal imaging night vision optics with a magnification of ×10 (narrow field-of-view), magnification of ×3 (wide field-of-view), sight stabilisation in elevation and a Hughes laser rangefinder. The turret is stabilised in azimuth with a gyro reticle compensation (gyroscope) drive to keep the aim point on target in deflection.

The gunner's auxiliary sight (a Kollmorgen model 939) has a magnification of ×8 and an 8° field-of-view. The loader is provided with a periscope with a magnification of ×1 which can be traversed through 360°.

The fire control system includes the laser rangefinder, full solution solid-state digital computer and stabilised day/thermal night sight. The stabilisation system permits accurate firing-on-the-move and the gunner merely places his reticle in GPS (a graticule is used in the gunner's auxiliary sight) on the target, uses the laser rangefinder (Neodymium-YAG) to determine the range. The computer then determines and applies the weapon sight offset angles necessary to obtain a target hit and the gunner opens fire. The main armament is equipped with a muzzle reference system to measure the bend of the gun. Information from a wind sensor mounted on the turret roof and a pendulum static cant sensor at the turret roof centre is fed automatically to the computer together with inputs from the laser rangefinder and the lead angle. The following data is manually set: battle sight range, ammunition type, barrel wear, muzzle reference compensation, barometric pressure and ammunition temperature.

The infra-red Thermal Imaging System (TIS) has been developed by the Hughes Aircraft Company and produces an image by sensing the small difference in heat radiated by the objects in view. The detected energy is converted into electrical signals which are displayed on a cathode ray tube, similar to a TV picture, and the image displayed is projected into the eyepiece of the gunner's sight. In addition, the sight displays target range information and indicates if the laser rangefinder has received more than one return. The operator can select a first or last return mode for lasing the target before firing. Ready to fire indication and confirmation that the systems are working properly are also provided.

The thermal imaging system generates a reticule pattern boresighted to the day graticule and to the laser rangefinder. This allows the gunner to operate the TIS just as he would the day sight. The infra-red sight is based on use of common modules, components standardised to specifications of the US Army Night Vision and Electro-Optics Center.

The digital fire control computer is produced in Canada by Computing Devices Company, a division of Control Data Canada Limited. The fire control computer hardware consists of an electronics unit and a separate data entry and test panel. The electronic unit contains the computing element, the power regulators and interfaces with other elements of the fire control system. The entry and test panel contains the keyboard, control switches and indicators, and a numeric display. The fire control computer carries out a continuous monitoring of its internal function and memory, and provides a visual display of any malfunction. A manually initiated self-test facility gives fault diagnosis in either unit of the system to the replaceable sub-assembly level.

Power for the electro-hydraulic gun and turret drive system is provided by an engine-driven pump through a slip ring in the turret/hull interface, to a power valve in the manifold beneath the main armament.

The crew compartment is separated from the fuel tanks by armour bulkheads. Sliding armour doors and armour protected boxes isolate the crew from on-board main armament ammunition explosion. An automatic Halon fire-extinguishing system in the tank reacts to the outbreak of a fire in two milliseconds and extinguishes fires in less than 250 milliseconds. Ready-use ammunition is stowed in the turret bustle and in the event of penetration by a HEAT projectile, the explosion would blow off the top panels with the crew being protected by the access doors which are normally kept in the closed position. The loader holds the switch closed to keep the doors open. The doors close automatically when the pressure switch is relieved. In addition to venting upwards, the turret bustle magazine vents to the rear.

The M1 is powered by a Lycoming Textron AGT 1500 gas turbine. The engine operates primarily on diesel or kerosene-based fuel, but can operate on petrol during emergencies. Approximately 70 per cent of the engine accessories and components can be removed without removing the powerpack from the tank. The complete powerpack can be removed and replaced in less than an hour compared with four hours for the current M60 series. The gas turbine delivers more horsepower to the sprocket than a comparable diesel engine because of the low cooling requirement. The exhaust for the gas turbine is at the rear of the hull with the air inlet on the hull top.

The engine is coupled to a Allison Transmission Division of General Motors X-1100-3B fully automatic transmission with four forward and two reverse speeds. The transmission also provides integral brakes, variable hydrostatic steering and pivot steering.

The improved torsion bar suspension has rotary shock absorbers at the first, second and seventh road wheel stations with 381 mm of road wheel travel compared with 162 mm in the M60 series. The top of the suspension is protected by vertical armoured skirts which hinge outwards to allow access to the suspension for maintenance. The drive sprocket is at the rear with the idler at the front and there are two return rollers. Standard equipment includes an AN/VDR-1 (RADIAC) Radiological Warning Device, chemical agent detector kit, a collective NBC protection unit and personnel heaters.

The turret has been designed to accept the standard 105 mm M68 series gun or the German Rheinmetall 120 mm smoothbore gun which has the American designation M256.

The 105 mm M68E1 gun will fire the standard ammunition as used for the M60 as well as the more recent M735 APFSDS-T round, M774 APFSDS-T and M833 APFSDS-T depleted uranium round and the M737 TPDS training round. Of 55 rounds of 105 mm ammunition carried, 52 are compartmentalised, 44 rounds stowed in the turret bustle, 22 on the left side and 22 on the right, three stowed horizontally in spallproof containers on the turret basket and the remaining eight stowed in the hull in a compartmentalised hull box.

The USA decided to adopt the 120 mm Rheinmetall smoothbore gun for the M1. The final production decision was taken in December 1984 and the first production M1 with the 120 mm gun, designated M1A1, was completed in August 1985.

Gun/turret integration was carried out by General Dynamics. The new gun mount is similar to that employed on the basic 105 mm M1 MBT. Software changes to the computer were made on completion of firing table data collection. The ammunition racks, blow-off panels and crew compartment sliding door have been redesigned.

Alliant Techsystems, the ammunition system contractor, produces the kinetic energy and HE rounds. US manufactured ammunition has been type classified to support 1985 M1A1 production. Types of ammunition fired by the 120 mm gun include:

M829 APFSDS-T, which has a DU penetrator with more advanced rounds under development
M830 HEAT-MP-T
M831 TP-T training round
M865 TPCSDS-T training round

These rounds are interoperable with the 120 mm smoothbore gun installed in the Leopard 2 with the round consisting of the projectile and semi-combustible cartridge case.

Alliant Techsystems is the system manager for the 120 mm ammunition for the M1A1 MBT with Olin Defense now being second source.

General Dynamics, Land Systems Division, M1A1 showing 120 mm smoothbore gun with fume extractor. Opening for future installation for Commander's Independent Thermal Viewer is on left side of turret roof forward of gunner's hatch

General Dynamics, Land Systems Division, M1A2 MBT showing Commanders Independent Thermal Viewer mounted in front of loaders

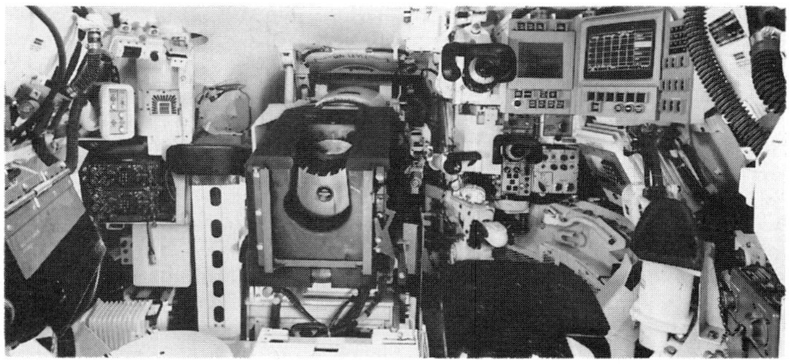

Interior improvements fo the General Dynamics, Land Systems Division, M1A2 MBT

As the M829 with its DU penetrator cannot be exported by the United States (M829E1 and M829E2) Alliant Techsystems and Olin have both developed, as a private venture, an APFSDS-T round with similar penetration characteristics using a conventional penetrator.

Mounted coaxially to the right of the main armament is a 7.62 mm M240 machine gun, and a similar weapon skate-mounted on the left side of the turret for the loader can be elevated from -30 to +65°, total traverse being 265°. A total of 11 400 rounds of 7.62 mm machine gun ammunition is carried. Mounted at the commander's station is a standard 12.7 mm (0.50) Browning M2 HB machine gun which can be elevated from -10 to +65° and can be traversed through 360°. This weapon has powered and manual controls for traverse and manual controls for elevation. The 12.7 mm machine gun has electric power traverse and can be aimed and fired from within the turret. A total of 1000 rounds of 12.7 mm machine gun ammunition is carried. Mounted on either side of the turret is a British-designed (L8A1) six-barrelled smoke discharger designated the M250.

Improved M1 MBT

The first Improved M1 MBT was completed in October 1984 and final deliveries were made in May 1986. The Improved M1 is essentially the basic M1 with improved armour protection. A total of 894 Improved M1 MBTs was built before production switched solely to the M1A1 with the 120 mm gun.

M1A1 MBT

The first production M1A1 was completed in August 1985 and of the anticipated US Army buy of 7467 tanks, 4199 are expected to be M1A1s. In addition to the improved armour of the Improved M1 MBT, the M1A1 has the 120 mm Rheinmetall smoothbore gun and an integrated NBC system. This NBC system provides the crew with conditioned air for breathing and also supplies cooling or heating for the crew as required while they are wearing their protective suits and face masks. Other improvements on the M1A1 include re-indexed torsion bars, increased damping rate of shock absorbers, strengthened compensating idler arms, modified transmission, road wheels and final drive, redesigned loader's seat, redesigned stowage under loader's seat, new loader's shoulder guard, new tank commander's panel, electrical harness rerouted, and fold-up turret bustle rack.

The M1 has three blow-off panels in the turret roof while the M1A1 has two. Of the 40 rounds of 120 mm ammunition carried by the M1A1 34 are in the turret bustle and six in a rear hull box. In addition to the smoke dischargers the M1A1 also has an engine-operated smoke-laying system.

In May 1987 General Dynamics, Land Systems Division, received a multi-year contract worth $3.5 billion from TACOM for 3299 M1A1 tanks to be delivered over a four year period. The first tank under this contract was delivered in May 1987 and final deliveries are scheduled for September 1991.

By that time, General Dynamics expects to have delivered more than 7450 Abrams MBTs to the US Army and Marine Corps. In addition, Egypt will have received 25 M1A1s as part of the co-production contract (qv).

Following the multi-year contract, General Dynamics expects the US Army and Marine Corps to purchase about 550 more M1A1s with final delivery projected for early 1993.

M1A1 with DU Armour

On March 14th 1988 the US Department of Defense announced that a new version of the M1A1 MBT (with significantly improved armour) was to enter production in the near future. This has been developed over several years of research and development and, according to the US DoD, will allow the Abrams to meet the anticipated threat well into the 1990s.

The armour design modification to the M1A1 incorporates steel-encased depleted uranium which is two and a half times the density of steel and is already used in a broad spectrum of civilian applications.

Sealed within the tank, depleted uranium has a very low level of natural radiation which is within the acceptable range established by the US Nuclear Regulatory Commission.

First production M1A1s with the new armour were completed in 1988. These tanks, which were shipped to US units in Germany, weighed about 65 US tons. Current production M1A1s with the DU armour package, US Marine Corps upgrades and the new T158 weigh 56.6 tonnes from the factory or 61.3 tonnes combat loaded. The heavy armour package deployed in Europe includes DU in the turret.

M1A2 MBT

The M1A2 MBT provides enhanced operational fightability over that of the current M1A1 MBT. The improvements of the M1A2 include Improved Commander's Weapon Station (ICWS), Commander's Independent Thermal Viewer (CITV), an Inter-Vehicular Information System (IVIS), Position/Navigation System (POS/NAV) and several survivability initiatives.

Late in 1988, the US Army awarded General Dynamics, Land Systems Division a contract worth over $350 million to carry out Full Scale Engineering Development (FSED) of the M1A2 programme.

The first production M1A2s will be completed late in 1992 with an IOC being achieved in 1993.

In 1989 General Dynamics, Land Systems Division, awarded sub-contracts for the following M1A2 subsystems:

1. Commander's Independent Thermal Viewer to Texas Instruments
2. Hull/Turret Electronics Unit (H/TEU) to Texas Instruments
3. Line of Sight/Dual Axis Head Assembly (LOS/DAHA) to Hughes Aircraft
4. Position/Navigation Unit to Smiths Industries
5. Direct Support Electrical System Test Set (DSESTS) to Chrysler Pentastar.

General Dynamics, Land Systems will manufacture the following M1A2 subsystems:

1. Commander's Integrated Display (CID)
2. Driver's Integrated Display (DID)
3. Fire Control Electronics Unit (FCEU)
4. Gunner's Control and Display Panel (GCDP)
5. Hull Power Distribution Unit (HPDU)
6. Hull/Turret Position Sensor (HTPS)
7. Improved Commander's Weapon Station
8. Radio Interface Unit (RIU)

General Dynamics, Land Systems Division, M1A1 MBT (Pierre Touzin)

Weighted M1A1 fitted with Cadillac Gage Textron 14K in-arm hydro-pneumatic suspension system undergoing trials

Under the full scale development contract, General Dynamics will build and test 10 M1A2 tanks in 1991 and 1992. In January 1990 the US government authorised the manufacturing planning Production Pilot Vehicles options. These five pilot vehicles are scheduled for delivery in 1992. Long lead low rate initial production funding was awarded for 62 tanks in April 1990. The Government of Saudi Arabia has awarded a contract to supply the Royal Saudi Arabian Land Forces with 315 M1A2 MBTs beginning in 1993 and running through 1995.

M1A1 with CVCC

In September 1988, General Dynamics, Land Systems Division, was awarded $9.3 million to modify a government owned M1A1 MBT with a new Combat Vehicle Command and Control System (CVCC).

Under the terms of the contract, the company will design, develop, integrate, test and demonstrate the CVCC system software on an improved M1A1 by 1991. CVCC will provide the tank commander with an expanded capability to communicate better with their own forces and other NATO forces. The contract, which is of the cost plus fixed fee, will run for a period of 40 months.

New laser rangefinder

The US Army has dropped the new laser rangefinder for the M1A1 and M1A2 MBTs.

Component Advanced Technology Test Bed (CATTB)

This will utilise a modified M1 chassis, additional details are given in the *Armoured fighting vehicle families* section.

M1A1 for US Marine Corps

The US Marine Corps originally intended to replace its current fleet of 716 M60A1 MBTs with 476 M1A1 MBTs at a total cost of $1357 million. More recently the US Marine Corps has decided to procure only 221 M1A1s, 66 in FY89 and 155 in FY90. The first M1A1 for the US Marine Corps was delivered in November 1990. Since then the US Army and US Marine Corps tanks have been identical. The tank has a Deep Water Fording Kit (DWFK), interface hardware to accept the Position Location Reporting System (PLRS) and additional tie down points for stowage on board ship and transportation on LCACs of the US Navy.

The DWFK was developed by General Dynamics, Land Systems Division for the US Marine Corps was procured by TACOM. Key parts of the DWFK are two vertical air intake tubes located on the left side of the hull and a larger exhaust tube pipe at the rear.

The kit can be installed by three men in less than 60 minutes and quickly removed by traversing the turret, and when fitted allows the M1A1 to ford to a depth of 1.981 m with its turret forward.

Every US Marine Corps M1A1 will be issued with a DWFK as part of its onboard equipment.

Other M1 Developments

Late in 1986 General Dynamics and others were awarded a two-year contract from TACOM to build and demonstrate a transversely mounted engine propulsion system for the M1 MBT. This will use an advanced AGT 1500 engine which achieves 15 per cent fuel savings and develops 1540 hp. The system was completed and successfully demonstrated in 1990. The engine is mounted transversely and aft against the transmission so saving 35 per cent of the engine compartment. Prime contractor is General Dynamics, Allison Transmission Division of General Motors and Lycoming Textron are sub-contractors.

FMC, General Motors and General Dynamics, Land Systems Division, each completed full scale wooden mock-ups of a 120 mm Manned Weapon Station (MWS) for the M1 in mid 1986.

In July 1987 TACOM issued an RFP for a new track for the M1/M1A1 to replace the current T156 track which has a short life.

In July 1988 it was announced that the Steel Products Division of the FMC Corporation had won the competition to produce the new T158 track for the M1/M1A1 MBT. The T158 is a joint development between the US Army Tank Automotive Command, FMC Steel Products and Goodyear.

M1A1 with full-scale mock-up of Commander's Independent Thermal Viewer (CITV) which is mounted forward of the loader's hatch

The first year of a five year contract is worth $9.6 million to FMC with the second year worth $20.9 million. Production of the T158 is undertaken by FMC at Anniston, with Goodyear being the main sub-contractor for rubber components.

The new T158 track is of the double pin type with replaceable rubber pads and is designed to last at least 3400 km before replacement, a 300 per cent increase over current T156 track. No modifications are required to install the T158 track on the M1/M1A1

An M1 with a complete in-arm hydro-pneumatic suspension system designed and built by Cadillac Gage Textron was completed in October 1985. Following company trials in 1986 it was sent to the Waterways Experimental Station (WES) where its test results over a course were compared to those of a standard M1 fitted with torsion bar suspension.

The US Army is currently testing bolt-on in-arm hydro-pneumatic suspension systems and one of these may be selected to be fitted to the Abrams M1A1 Block II or Block III tanks. Competitors for this are Cadillac Gage Textron with its 16K system and Teledyne Continental Motors.

HTT for M1

Early in 1989 Vickers Defence Systems of the UK shipped a set of hydraulic Track Tensioners (HTT) to General Dynamics, Land Systems Division, for installation on an M1 MBT.

Prototype of the private venture General Dynamics, Land Systems Division, Abrams Recovery Vehicle which is still being offered for export

M1 Abrams Bulldozer Kit

In 1978 the Defense Division of the Chrysler Corporation (now the Land Systems Division of General Dynamics Corporation) studied the possibility of adapting the M9 bulldozer kit to fit the M1 Abrams MBT. The result of this study was a recommendation that a new kit be designed with a revised mould-board geometry to improve the driver's vision and the overall system performance and yet take advantage of the low profile of the M1 Abrams. It was also recommended that improved hydraulic components be used allied to a rapid coupling and release mechanism.

Design of the new bulldozer kit was undertaken by the US Army Mobility Equipment Research and Development Command (MERADCOM), now the Belvoir Research and Development Center at Fort Belvoir, Virginia. In 1981 the Center awarded a production contract to build a prototype kit to Barnes and Reinecke, Inc, and this prototype was tested at Fort Knox during 1982.

The M1 Abrams bulldozer kit fits onto the MBT's lifting eyes and towing lugs and is powered by the tank's 24 V electrical system.

Tank Test Bed

Details of this experimental vehicle were given in *Jane's Armour and Artillery 1986-87* page 114.

M1 AVLB

General Dynamics, Land Systems Division in conjunction with MAN of Germany, is building a Heavy Assault Bridge for the Fort Belvoir Research, Development and Engineering Center. The two vehicle demonstrator programme is based on the utilisation of the MAN Leguan bridge and launcher on a modified M1 MBT chassis. The HAB design features a horizontal launch, 26 m MLC 70 (Military Load Class) bridge on a converted M1 MBT chassis which will have the same mobility as the M1 tanks it is operating with.

The Leguan bridge, which is already in service with Norway (carried on launched form a wheeled chassis) is of a commercial aluminium alloy with all hydraulic and electronic components required for launching in the launching structure itself. The bridge can be launched in less than five minutes through a computer controlled launch sequence.

Power for the Leguan hydraulics is provided by a 65 hp auxiliary power unit designed by Stewart and Stevenson which produces the pressure required during the launch/retrieval cycle.

Under the terms of the $12.46 million contract, General Dynamics, Land Systems Division, will deliver the two prototypes to the Army in October 1992.

In 1989 BMY Combat Systems was awarded a contract for the design and construction of a new 26 m MLC 70 bridge which will also be carried and launched from a M1 tank chassis. The new three part HAB will have a length of 26 m when fully opened, so enabling it to span gaps of up to 24 m.

BMY previously had built three prototypes of a 32 m HAB, two on a M1 MBT chassis and one on a M60 MBT chassis, the latter requirement has now been dropped. Details of the original BMY HAB were given in *Jane's Military Logistics 1990-91* page 124.

Combat Mobility Vehicle

In May 1991, BMY Combat Systems was awarded a $20.7 million contract by the US Army for the prototype of a single Combat Mobility Vehicle (CMV) Advanced Technology Transition Demonstrator.

Under the terms of the contract BMY will build a single vehicle based on a modified M1 Abrams MBT. The Combat Mobility Vehicle (CMV) is part of the Armored Systems Modernisation (ASM) and additional details are given in the AFV Families section later in this volume.

M1 Mine Clearing Vehicle

The mine roller kit developed by Chrysler (now General Dynamics Land Systems Division) for the M60 tank has already been Type Classified for US Army use. An adapter kit to enable the roller to be mounted on the M1 tank has also been developed. In January 1989 it was announced that Ramta Structures and Systems of Israel was to supply 400 mine clearing ploughs to the US Army for the M1A1 MBT. Details of this system, called the Track Width Mine Plough (TWMP) are given in *Jane's Military Vehicles and Logistics 1991-92* page 234. This system was used by the United States Army on its M1/M1A1 MBTs during Operation Desert Storm in early 1991.

In late 1989 the US Army started fielding the M1 Battalion Countermine Set, each of which consists of four mine clearing rollers, six mounting kits, four cleared-lane marking systems (CLAMS), four CLAMS mounting kits and 12 mine clearing blades.

M1 AVLB laying standard M48/M60 bridge in position over front of vehicle

General Dynamics, Land System Division, M1 Abrams MBT fitted with the Ramta Structures and Systems Track Width Mine Plough system

Abrams Recovery Vehicle (ARV)

A private venture, General Dynamics, Land Systems Division, has built the prototype of the Abrams Recovery Vehicle (ARV). This competed against the BMY M881E1 ARV for a US Army requirement for an Improved Recovery Vehicle and following trials, the M88A1E1 was selected, although due to cuts in the US defence budget, production of this vehicle has yet to start. General Dynamics, Land Systems Division, is however still marketing the Abrams Recovery Vehicle for the export market. The 100-hour ground war in the Persian Gulf in 1991 points out the need for a more powerful ARV capable of recovering the M1/M1A1/M1A2 MBTs, however as of May 1991 it did not appear that funding will be available for its production. The ARV is based on the chassis of the M1A1 and has a new superstructure. Pivoted at the front left side is a 35 ton (US) hydraulic crane and mounted at the front of the hull is a hydraulically operated dozer/stabilising blade. The main winch has a capacity of 70 tonnes and the auxiliary winch has a capacity of four tonnes. Details of the vehicle are given in *Jane's Military Vehicles and Logistics 1991-92* page 41.

Status: M1: Production complete, in service with US Army.
M1A1: In production, in service with US Army and US Marine Corps, will be assembled in Egypt (qv).
M1A2: On order for US Army and Saudi Arabia, under evaluation/selection by a number of other countries.

Manufacturer: Production is undertaken at the Lima Army Tank Plant in Lima, Ohio, and the Detroit Arsenal Tank Plant, Warren, Michigan.
Prime contractor is: General Dynamics
Land Systems Division
PO Box 2074
Warren
Michigan 48090-2074
USA
Telephone: (313) 825 4000

SPECIFICATIONS (Data in square brackets relates to M1A1 where different)

CREW	4
COMBAT WEIGHT	54 545 [57 154] kg
POWER-TO-WEIGHT RATIO	27 [26.24] hp/tonne
GROUND PRESSURE	0.96 kg/cm^2
LENGTH GUN	
forwards	9.766 [9.828] m
rear	8.971 [9.033] m
LENGTH HULL	7.918 m
WIDTH	3.653 [3.657] m
WIDTH REDUCED	3.479 m
HEIGHT	
to turret roof	2.375 [2.438] m
overall	2.885 [2.886] m
FIRING HEIGHT	1.89 m
GROUND CLEARANCE	
hull centre	0.482 m
hull sides	0.432 m
TRACK WIDTH	635 mm
LENGTH OF TRACK ON GROUND	4.65 m
MAX SPEED	
road	72.421 [66.77] km/h
cross-country	48.3 km/h
on 10% grade	32.2 [27.51] km/h
on 60% grade	7.2 [6.59] km/h
ACCELERATION 0 - 32 km/h	7 [6.8] s
FUEL CAPACITY	1907.6 l
MAX ROAD RANGE	498 [465] km
FORDING	1.219 m
with preparation	1.98 m
GRADIENT	60%

VERTICAL OBSTACLE	1.244 [1.066] m
TRENCH	2.743 m
TURNING RADIUS	pivot to infinitely variable
ENGINE	Lycoming Textron AGT 1500 gas turbine developing 1500 hp at 3000 rpm
TRANSMISSION	Allison Transmission Division of General Motors X-1100-3B automatic with 4 forward and 2 reverse gears
STEERING	double differential with hydrostatic control
FINAL REDUCTION RATIO	4.3:1 [4.67:1]
BRAKING SYSTEM	hydro-mechanical
SUSPENSION	advanced torsion bar
ELECTRICAL SYSTEM	24 V
BATTERIES	6 × 12 V
ARMAMENT	
main	1 × 105 mm [1 × 120 mm] gun
coaxial	1 × 7.62 mm MG gun
anti-aircraft commander	1 × 12.7 mm MG
anti-aircraft loader	1 × 7.62 mm MG
SMOKE-LAYING EQUIPMENT	6 smoke dischargers either side of turret and integral engine smoke generators

AMMUNITION	
main	55 [40]
12.7 mm	1000
7.62 mm	11 400 [12 400]
smoke grenades	24
GUN CONTROL EQUIPMENT	
Turret power control	electro-hydraulic/manual
by commander	yes
by gunner	yes
Turret traverse	360°
Max rate of power traverse	(tracking) 4.2°/s (slew rate with stabilisation) 42°/s
Max rate of power elevation	(tracking) 1.4°/s (slew rate with control handles) 22.5°/s (slew rate with stabilisation commands) 42°/s
Gun elevation/depression	+20°/-9° [+20°/-10°]
Gun stabiliser	
vertical	yes (sight)
horizontal	yes (turret)

NB: The M1A1 with the new heavy armour package weighs 61.3 tonnes

General Dynamics, Land Systems Divison, M60 Series of Main Battle Tank

Development

Early in 1956 it was decided that the M48 tank should be further developed to produce an improved tank with increased operational range and mobility which would require a minimum of refuelling and servicing as well as incorporating an improved main armament.

In November 1956 a decision was taken to install an AVDS-1790-P compression ignition engine in an M48 tank, which was subsequently tested at Yuma Test Station during the summer of 1957. As a result of a meeting in February 1958, three M48 tanks (designated the M48A2E1) were rebuilt to incorporate the new powerpack contemplated for use in the new tank which had been designated the XM60.

During October and November 1958 several main armament candidates were tested at Aberdeen Proving Ground, and based on the test results the British 105 mm L7A1 barrel with the American T254E2 breech was selected as the main armament for the XM60 and designated the M68 cannon.

In February 1957 the M60 (XM60) vehicle characteristics and production planning schedules were established and in March 1959 the M60 was classified as Standard A.

General Dynamics, Land Systems Division-built M60A3 MBT with thermal sleeve for 105 mm M68 tank gun and British-supplied smoke dischargers on turret side

Frontal view of M60 MBT (US Army)

The initial bid package for the production of M60 tanks was released in April 1959 and in June 1959 a production contract was awarded to Chrysler Corporation, Delaware Defense Plant, for the production of 180 M60 MBTs. In August 1959 an engineering release bid package was released for the second production buy of M60s to be built at Delaware. Subsequent production buys, beginning with October 1960 production, were made from the Chrysler Corporation (now operated by General Dynamics, Land Systems Division) Detroit Tank Plant, with production of the M60A3 completed in September 1987.

The M60 entered service with the United States Army in 1960 and from October 1962 the tank was succeeded in production by the M60A1 (development designation M60E1). The first M60s had M48 type turret with 105 mm gun. The M60A1 has the same basic chassis as the M60 but a redesigned turret with greater ballistic protection and other modifications. It carries 63 rounds of 105 mm ammunition rather than 60 as carried by the older M60. For most of the 1960s and early 1970s production of the M60A1 was maintained at a very low rate, the minimum necessary to sustain the production base.

As a result of the Middle East War of 1973 a major effort was undertaken to increase production of the M60A1 for two reasons: to replace M60A1 tanks supplied to Israel and to increase war reserve stocks which were then very low. It took some time to build up production of the M60A1 owing, in the main, to a shortage of hull and turret castings. By 1975 production had been boosted to 48 tanks per month, which increased to 72 a month in 1977 and 104 a month in December 1977. Peak production rate was achieved in October 1978 when 129 vehicles were completed. Production continued at a high rate until April 1979 when it started to drop and by the summer of 1980 it was running at 50 vehicles a month. The last M60A1 was completed in May 1980 and all production after this date was of the M60A3. Production of the M60A3 MBT was expected to be completed in May 1985 after over 15 000 M60 series MBTs had been built, but in May 1985 the DoD notified Congress of a letter of offer for 94 M60A3 MBTs for Egypt at a cost of $165 million. M60A3s being produced early in 1985 were for Saudi Arabia. Taiwan has also purchased M60A3 hulls from General Dynamics, Land Systems Division.

The FY80 request was for 116 M60A3s for the United States Army and a further 444 for foreign military sales, which were built between April 1981 and July 1982. There was no United States Army funding for M60A3 production after FY80 but in FY81 167 M60A3s were built for foreign military sales between July and November 1982.

In 1988 it was stated that all M60 series tanks, including those in service with the National Guard, will be replaced by 1997. M60 tank improvements will in future be limited to those related to safety and protection of the crew.

With the introduction of the M1/M1A1 and the reduction in size of the US Army, a number of M60A1 MBTs have become surplus to US Army requirements and these are to be transferred to other countries at no cost to the recipient other than the cost of transporting the vehicles. These include:

Country	**Quantity**	**Comment**
Egypt	700	M60A1
Greece	100	M60A1 (from Italy)
	110	M60A1 (from USA)
Saudi Arabia	150	M60A3
Thailand	300	M60A1

Note: In early 1991 the US stated that up to 2000 M60A1/M60A3 MBTs would be available to other NATO countries, these being Greece, Portugal, Spain and Turkey. The latter country was offered M60A1s but they wanted M60A3s with its improved fire control systems.

M60 series MBTs took part in operation Desert Storm in the Spring of 1991, they were used by Egypt (M60A1), Saudi Arabia (M60A3) and the United States Marine Corps (M60A1 with explosive reactive armour packages originally developed for the US Army).

M60A1 MBT of the US Marine Corps comes ashore during amphibious operations. Turret is traversed to rear and snorkel installed (US Marine Corps)

Description (M60A1)
The hull of the M60A1 is made of cast sections and forged floor plates welded together. It is divided into three compartments: driver's at the front, fighting in the centre and the engine and transmission at the rear.

The driver is seated at the front of the vehicle and is provided with a single piece hatch cover that opens to the right. Three M27 periscopes are mounted forward of his hatch and an M24 infra-red periscope can be installed in a mount in the centre of his hatch cover for driving at night. The M24 is now being replaced by the AN/VVS-2 night viewer, which is of the passive rather than infra-red type. A hull escape hatch is provided near the driver's position.

The all-cast turret is in the centre of the vehicle with the loader on the left and the commander and gunner on the right. There is an external stowage basket at the rear of the turret. The loader is provided with a single hatch cover that opens to the rear with an integral M37 periscope that can be traversed through 360°.

The commander has a cupola that can be traversed through 360° by hand, a single piece hatch cover that opens and swings to the rear, an M28C sight in the forward part and eight vision blocks for all-round observation. The M28C can be replaced by an M36 infra-red periscope or an M36E1 passive periscope for night vision. The gunner is seated in front of and below the commander and is provided with an M31 periscope with a magnification of ×8 and an M105D telescope with a magnification of ×8 and a 7.5° field-of-view. The M31 periscope can be replaced by an M32 infra-red periscope or an M35E1 passive periscope for night engagement of targets. The M17A1 or M17C rangefinder has a magnification of ×10, a 4° field-of-view and a range of between 500 and 4400 m.

The engine compartment at the rear of the hull is separated from the fighting compartment by a fireproof bulkhead, and is equipped with a fire-extinguishing system.

The torsion bar suspension system consists of six dual rubber-tyred road wheels with the idler at the front, drive sprocket at the rear and three track return rollers. The first, second and sixth road wheel stations are provided with a hydraulic shock absorber.

The NBC system of the M60 is of the central air filtration type which pipes fresh air to each crew member via a tube. A full range of night vision equipment is fitted as standard including an infra-red searchlight over the main armament. The latter is either the AN/VSS-1 or the more recent AN/VSS-3A. The former is a 2.2 kW Xenon unit that provides a narrow or wide beam of high-intensity visible or infra-red light. A 50 per cent increase in light intensity can be temporarily provided for 15 to 20 seconds by overriding the searchlight. It has a narrow beam width of 0.5 to 0.75° and a wide beam width of 15°. The AN/VSS-3A can be used in both the visible or infra-red modes with three types of beam, compact, spread or variable width.

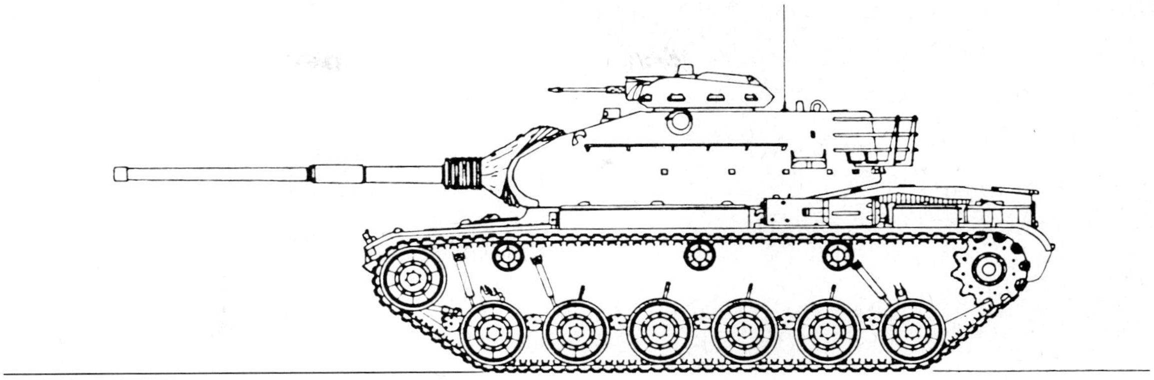

M60A1 MBT armed with 105 mm gun

The crew compartment is provided with a heater and a RADIAC NBC detector can be fitted if required. The tank can ford to a depth of 1.219 m without preparation and with preparation to a depth of 2.438 m. The tank can also be fitted with an M9 bulldozer blade kit on the front of the hull for preparing fire positions and clearing obstacles.

Main armament of the M60, M60A1 and M60A3 tanks is a 105 mm M68 rifled tank gun with a bore evacuator. A well-trained crew can fire between six and eight rds/min. Of the 63 rounds of ammunition carried, 26 are carried in the forward part of the hull, to the left and right of the driver's position, 13 in the turret for ready use, 21 in the turret bustle and the remaining three under the gun.

The 105 mm gun can fire the following types of fixed ammunition: APDS-T (M728), APFSDS-T (M735/M735A1), APFSDS-T (M774), APDS-T (M392A2), APERS-T (M494), HEAT-T (M456 series), TP-T (M467), TP-T (M490), TPDS-T (M724), TPDS (M737), HEP-T (M393A1/M393A2), Dummy (M457), TP-T (M393A1), and Smoke WP-T(M416).

Mounted in the commander's cupola is an M85 12.7 mm (0.50) machine gun with an elevation of +60° and a depression of -15°. Mounted coaxially to the left of the main armament is a 7.62 mm M73 machine gun, replaced by the M240 weapon.

A number of M60A1s have been updated with the RISE engine, main armament fully stabilised in both elevation and traverse, top-loading air cleaner fitted, new T142 tracks and improved night vision equipment. Additional details of this programme are given in the entry for the M60A3.

M60A2

This model, fully described in *Jane's Armour and Artillery 1981-82* pages 93-94, has now been phased out of service with the US Army. A total of 526 M60A2s were built and most have been sent back to the Anniston Army Depot.

M60A3

The M60A3 (development designation M60A1E3) is a product-improved M60A1 and some of the improvements in the tank, for example the add-on stabilisation system, RISE engine, and the smoke grenade launchers were first fitted to the M60A1 some years ago.

The first M60A3s completed at the Detroit Arsenal Tank Plant in February 1978 were the first of a low rate of initial production quantity of 296 M60A3s which were funded in FY76 transitional quarter and FY77. In 1978 the M60A1 monthly production rate was 116 tanks at Detroit. The first M60A3 tanks were delivered to Europe in mid 1979 and issued to the 1st Battalion of 32nd Armored. The Army has 5400 M60A3 TTS tanks of which 1686 are new production M60A3s, 114 are M60A1 passive tanks field retrofitted to M60A3 TTS configuration, and 3600 conversions carried out by the Mainz Army Depot in Germany and Anniston Army Depot in the USA. In January 1983 Saudi Arabia placed an order for 100 M60A3 MBTs at a total cost, including spares and other support equipment, of $176 million. In 1987 it ordered 150 conversion kits to convert M60A1 tanks into M60A3s at a cost of $120 million. These conversions include the installation of a Tank Thermal Sight (TTS).

M60A3 MBT from rear, clearly showing rear stowage basket and commander's cupola on right side of turret

The main improvements are in the fire control system. A Hughes laser rangefinder with a maximum range of 5000 m replaces the optical rangefinder, and a solid-state computer replaces the mechanical computer. Following successful trials with 20 prototype units Hughes was awarded its first production contract for the laser rangefinder in 1976. The contract was for 69 units at a total cost of $11 million. Range data is fed to the computer and, with other data such as cross-wind velocity, air temperature, gun trunnion tilt, air density, altitude, target tracking rate and ammunition ballistics, provides the correct azimuth and elevation firing commands to the tank gunner and commander. The improvements in the fire control system, combined with the passive night sight, improve the night-fighting capabilities of the tank.

The fire control system is composed of two major subsystems, the AN/VVG-2 laser rangefinder and the M21 ballistic computer system. The laser/sight subsystem is composed of two main units: a commander's integrated laser/sight with control unit and a laser electronics unit. Range may be fed automatically to the computer or, if more than one return is received, the commander may select the range return based on his assessment of the situation. Since the gunner's sight is boresighted with the laser rangefinder, either the commander or the gunner can fire the laser and/or the gun.

The computer subsystem includes the ammunition selector sensors. The commander or gunner may select one of the four basic types of ammunition to be used. The computer stores solutions for six ammunition types with the four basic types to accommodate ammunition stored in different locations. The switch at the bottom of the ammunition select unit is used to set up conditions according to whether the tank is moving or stationary. The computer is a solid-state hybrid and processes all the input data and commands the sight lines for laying the gun.

The top half of the gunner's control unit panel contains three manual inputs: air temperature, pressure altitude (slowly varying inputs, not required to be very accurate), and manual range in case the laser fails. The rate unit, crosswind sensor and the cant unit are automatic. The rate unit is used against moving targets from a stationary tank. When the tank is moving it is switched out and replaced by input from the gun stabilisation system gyros, since inertial tracking rates are required. The crosswind sensor can be stowed horizontally and is spring mounted to prevent damage when the tank encounters low branches. Since it is the only unit not under armour, a circuit in the computer continuously monitors its output and can switch to manual input in the event of a failure. The cant unit senses gun trunnion roll.

A row of lights on the top of the gunner's control unit is for self-test. Either a green 'System OK' light is lit, or a light comes on indicating which unit is faulty. A red light indicates failure of the system and an orange light indicates that the system will continue to function, but less accurately. This self-test system removes the trouble-shooting burden from the tank crew, and replacing system units in the field is simple. The bottom half of the panel, which is normally covered, is used for boresighting, zeroing, manual crosswind input (if desired) and the switch to select the four basic ammunition types.

In 1985 Hughes Aircraft Company received a production contract from the US Army for a further 1169 fire control systems with options for a further 585 systems.

In December 1986 Elbit Computers of Israel was awarded a contract worth $2.1 million for the manufacture of electronic units for the M60A3 fire control system.

In May 1987 Optic Electronic Corporation was awarded a contract worth $4.1 million for the supply of VVG-2 laser rangefinders for the M60A3 and $2 million for associated laser test equipment.

At the beginning of FY80 AN/VGS-2 Tank Thermal Sights (TTS) were added, which not only improved the night fighting capability of the tank but also enabled the tank to see through smoke and ground cover. The AN/VSG thermal imaging equipment replaced the gunner's passive night vision periscope. Technical details have not been released but may

US Army M60A3 with 105 mm M68 gun fitted with thermal sleeve (Pierre Touzin)

M60A3 MBT of US Army fitted with attachments for dozer blade (Pierre Touzin)

reasonably be expected to be no less than those achieved by other land battlefield IR systems based on the TI range of infra-red modules, developed by Texas Instruments Equipments Group, including the AN/TAS-4, AN/TAS-5 and AN/TAS-6.

The M60A3 has now replaced all M48A5s used by the National Guard.

Major M60A3 improvements are as follows:

Main armament fully-stabilised in both elevation and traverse*
Top-loading air cleaner installed*
AN/VSS-1 searchlight replaced by AN/VSS-3A on passive tanks
Thermal gunner's sight
T97 tracks replaced by T142 tracks with removable pads*
AVDS-1790-2C RISE (Reliability Improved Selected Equipment) engine fitted*
Thermal shroud for main armament
Laser rangefinder
650 A oil-cooled alternator
Solid-state computer
Passive night vision devices, followed by thermal sights
Coaxial machine gun replaced by M240 weapon
British-style six-barrelled smoke dischargers fitted to either side of turret
Engine smoke generator
Automatic Halon fire-extinguishing system
* also fitted to late production M60A1s.

New production M60A3s were fitted with the Teledyne developed engine exhaust smoke system and older tanks retrofitted. Fuel is sprayed into the exhaust manifold, creating smoke.

For trials purposes, three M60A1 tanks were fitted with a hydro-pneumatic suspension system developed by the National Water Lift Company. These tanks were successfully tested at both Aberdeen Proving Ground and Fort Knox and showed a significant improvement over the existing suspension system on the M60A1 and M60A3. But due to the high cost of retrofitting the Army fleet of M60s it was decided to retain the existing suspension system rather than adopt the hydro-pneumatic type.

TACOM and General Dynamics have also developed an improved air-filtration system for the M60 series called the Vehicle Exhaust Dust Ejector System (VEDES) which was being retrofitted to the fleet starting in late 1984.

The plan to fit explosive reactive armour to some US Army M60A3 MBTs was cancelled in January 1988. The plan to install a new and longer 105 mm rifled tank gun has also been cancelled.

M60A4

This was never type classified and was a proposal to further improve the M60A1/M60A3 for the National Guard.

M60A1 MBT fitted with hydraulically-operated dozer blade (US Army)

M60A1 with ERA

The US Army was to have fielded an explosive reactive armour (ERA) package for its M60A3 MBTs but although production did start and quantities were produced, it was decided not to field the system. More recently the US Marine Corps have fitted some of their M60A1s with the ERA package, and these were used during Operation Desert Storm early in 1991.

Austrian Army M60A1 MBT Upgrade

Late in 1986 an Austrian joint venture (M60 ARGE) of Steyr-Daimler-Puch and NORICUM, was awarded a contract by the Austrian Army to upgrade 118 M60A1 MBTs to the M60A3 (Passive) configuration to include new fire control system incorporating a laser rangefinder system produced by Hughes Electro-Optical and Data Systems Group and a Honeywell stabilisation system.

The new AVDS-1790-2C engine was supplied to the Austrian Army to the ARGE and purchased from the US Army's Mainz depot. The engine was upgraded to the AVDS-1790-2CA version.

In addition to upgrading the M60A1 to the M60A3 (Passive) the ARGE performed the complete overhaul of these vehicles. The ARGE purchased the conversion kits and spares for the overhaul from the FMS Corporation of Los Angeles, California, at a total cost of $36 million.

With all of these modifications the upgraded M60A1 is equivalent to the US M60A3. The first M60A3s were handed over to the Austrian Army in mid-1988 and the last ones in November 1989. The Austrian Army already has 50 brand new M60A3s.

Israeli M60 series

Outside of the United States, the Israeli Army is the largest user of the M60 series. There is a separate entry for the M60 under Israel, as this country has carried out so many improvements to the vehicle.

Saudi Arabian M60A1s

In early 1988 General Dynamics Services Company was awarded a contract by Saudi Arabia for the upgrading of 150 M60A1 MBTs to the M60A3 standard with the work to be carried out in existing Saudi Arabian facilities over a three year period. The first vehicle was handed over in December 1989 and final deliveries were made in May 1991. In addition to upgrading the M60A1s five XM30 tank turret trainers were also upgraded to the M60A3 configuration. Modifications to the M60A1 included thermal sights, new fire control system, ballistic computer and laser rangefinder and thermal sleeve for 105 mm M68 tank gun.

Super M60 MBT

This was a private venture by Teledyne Continental but is no longer being offered. Full details were given in *Jane's Armour and Artillery 1985-86*, pages 112 to 114.

Mine Roller System

General Dynamics Land Systems Division has developed a mine roller system that can be fitted to the M60A1 and M60A3 MBTs and which has already been fielded in Europe.

XM1060 ROBAT

Details of the XM1060 ROBAT (Robotic Breaching Assault Tank), based on a modified M60 series MBT chassis, were given in *Jane's Military Vehicles and Logistics 1991-92* page 247. So far the vehicle has yet to enter production.

M9 Bulldozer Kit

This was developed as a depot retrofit package and is used to give already fielded M60s bulldozing capabilities similar to those of the M728 Combat Engineer Tractor. Full details of this bulldozer are given in *Jane's Military Vehicles and Logistics 1991-92* page 15.

M60 AVLB

The M60 Armoured Vehicle Launched Bridge is an M60 fitted with a hydraulic launching mechanism and an aluminium scissors bridge which is launched over the front of the vehicle and weighs 14 470 kg. The bridge takes two minutes to lay and when open has an overall length of 19.202 m and can span a gap of up to 18.288 m. By early 1983 nearly 400 M60 AVLBs had been built. Anniston Army Depot started a programme in 1985 to convert M60A2 tanks into M60A1 AVLBs. The M60 AVLB weighs 55 746 kg with the bridge and 41 685 kg without it. Details of the M48/

US Army M728 Combat Engineer Vehicle with dozer blade raised and A-frame in travelling configuration (UK Ministry of Defence)

M60 AVLB are given in *Jane's Military Vehicles and Logistics 1991-92* page 94. In addition to being used by the US Army, M60 AVLBs are also used by Israel, Singapore (8) and Spain.

M728 Combat Engineer Vehicle

The Combat Engineer Vehicle was developed under the designation T118E1 and was standardised in 1963 as the M728. It entered production in 1965 and entered service with the US Army in 1968. By 1983 over 300 vehicles had been completed, with no subsequent production. In addition to being used by the US Army the M728 is also used by Saudi Arabia and Singapore (12).

The M728 is armed with a short-barrelled 165 mm M135 demolition gun which fires the M123A1 HEP (or HESH) round. A 7.62 mm machine gun is mounted coaxially to the main armament and a 12.7 mm machine gun is mounted in the commander's cupola for ground and anti-aircraft use.

Mounted at the front of the hull is a hydraulically operated dozer blade which is used for clearing obstacles and preparing fire positions. Pivoted towards the front of the turret is an A-frame which can be folded back over the rear of the turret when not required. The two-speed winch at the rear of the turret has a maximum capacity of 11 340 kg. An infra-red searchlight similar to that fitted to the M60A1 is mounted over the main armament of the M728. The M728 weighs 52 163 kg loaded and has a crew of four: commander, gunner, loader and driver. Full details of the M728 Combat Engineer Vehicle are given in *Jane's Military Vehicles and Logistics 1991-92* page 14.

SPECIFICATIONS

Model	**M60**	**M60A1**	**M60A3**
CREW	4	4	4
COMBAT WEIGHT	49 714 kg	52 617 kg	52 617 kg
UNLOADED WEIGHT	45 631 kg	48 684 kg	48 684 kg
POWER-TO-WEIGHT RATIO	15.08 bhp/tonne	14.24 bhp/tonne	14.24 bhp/tonne
GROUND PRESSURE	0.8 kg/cm^2	0.87 kg/cm^2	0.87 kg/cm^2
LENGTH GUN FORWARDS	9.309 m	9.436 m	9.436 m
LENGTH HULL	6.946 m	6.946 m	6.946 m
WIDTH	3.631 m	3.631 m	3.631 m
HEIGHT	3.213 m	3.27 m	3.27 m
FIRING HEIGHT	2.095 m	2.095 m	2.095 m
GROUND CLEARANCE	0.463 m	0.463 m	0.45 m
TRACK	2.921 m	2.921 m	2.921 m
TRACK WIDTH	711 mm	711 mm	711 mm
TRACK ADJUSTMENT	mechanical	mechanical	hydraulic
TRACK TYPE	T97/T142	T97/T142	T142 replaceable pads
LENGTH OF TRACK ON GROUND	4.235 m	4.235 m	4.235 m
MAX ROAD SPEED	48.28 km/h	48.28 km/h	48.28 km/h
FUEL CAPACITY	1457 l	1420 l	1420 l
MAX ROAD RANGE	500 km	500 km	480 km
FORDING	1.219 m	1.219 m	1.22 m
with preparation	2.438 m	2.438 m	2.4 m
GRADIENT	60%	60%	60%
SIDE SLOPE	30%	30%	30%
VERTICAL OBSTACLE	0.914 m	0.914 m	0.914 m
TRENCH	2.59 m	2.59 m	2.59 m
TURNING RADIUS	pivot to infinity on all models		
ENGINE	Continental AVDS-1790-2A 12-cylinder air-cooled diesel developing 750 bhp at 2400 rpm		Continental AVDS-1790-2C
TRANSMISSION	General Motors Corporation, cross-drive, single-stage with 2 forward and 1 reverse ranges		
SUSPENSION	torsion bar	torsion bar	torsion bar
ELECTRICAL SYSTEM	24 V	24 V	24 V
BATTERIES	6 × 12 V, 100 Ah	6 × 12 V, 100 Ah	6 × 12 V, 100 Ah
ARMAMENT			
main	1 × 105 mm gun	1 × 105 mm gun	1 × 105 mm gun
coaxial	1 × 7.62 mm MG	1 × 7.62 mm MG	1 × 7.62 mm MG
anti-aircraft	1 × 12.7 mm MG	1 × 12.7 mm MG	1 × 12.7 mm MG
AMMUNITION			
main	60	63	63
7.62 mm	5950	5950	5950
12.7 mm	900	900	900
GUN CONTROL EQUIPMENT			
Turret power control	electro-hydraulic/manual in all models		
by commander	yes	yes	yes
by gunner	yes	yes	yes
Max rate power traverse	360° in 15 s	360° in 15 s	360° in 15 s
Gun elevation/depression	+20°/-9°	+20°/-10°	+20°/-10°
Commander's override	yes	yes	yes
Commander's fire-control override	yes	yes	yes
Gun stabiliser			
vertical	no	yes*	yes
horizontal	no	yes*	yes
Range setting device	yes	yes	yes
Elevation quadrant	yes	yes	yes
Traverse indicator	yes	yes	yes
ARMOUR	no details released, but armour protection is considered an improvement over M48 series', especially over frontal area and turret		

* As originally built, the M60A1 was not fitted with a gun stabilisation system but these are now retrofitted to some models.

Status: Production complete. In service with the following countries:

Country	**Model**	**Quantity**	**Comment**
Austria	M60A3	168	
Bahrain	M60A3	54	delivered 1987
Egypt	M60A3	759	final deliveries 1987. Further 700 surplus M60A1s to be provided by USA at no cost
Greece	M60A1	110	to be transferred at no cost by USA
Iran	M60A1	355	only 150 approx left
Israel	M60/M60A1/M60A3	1400	many local modifications (qv)
Italy	M60A1	300	200 locally built
Jordan	M60A1	118	
	M60A3	100	
Oman	M60A1	6	delivered 1980
Saudi Arabia	M60A3	250	
Singapore	not/app	not/app	has 8 M728 CEV and 12 M60 AVLB
Sudan	M60A3	20	delivered 1982
Taiwan	M60A3	150	ordered 1990, see also M48H entry under Taiwan
Tunisia	M60A3	54	delivered 1983/1984
USA	M60/M60A1 M60A3, M728, M60 AVLB	8000	includes Marine Corps, approx number
Yemen	M60A1	64	supplied 1979

Note: See also introduction as US is expected to provide additional M60A1/M60A3 MBTs to other countries in NATO including Portugal.

Manufacturer: Detroit Arsenal Tank Plant which is now operated by General Dynamics, Land Systems Division, PO Box 2074, Warren, Michigan 48090, USA. Telephone: (313) 825 4000

M48 Series of Main Battle Tanks

Development

In October 1950 Detroit Tank Arsenal began design work on a new tank armed with a 90 mm gun and in December the same year the Chrysler Corporation was awarded a letter of intent to design the new tank under the designation T48. Design work began in late December and six prototypes were built, the first being completed in December 1951. In March 1951, before the prototypes had been completed, the Fisher Body Division of the General Motors Corporation and the Ford Motor Company at Livonia were both given a production contract for the tank. Chrysler's first production M48 rolled out of the Delaware Tank Plant and was christened by Mrs George S Patton Junior on 1 July, 1952.

The first production M48s were completed in 1952 and production of the M48 series amounted to 11 703 units, of which about 6000 were built by the Chrysler Corporation at the Delaware Defense Plant which continued production until 1959, Alco Products of Schenectady also built a number of M48A2 tanks.

Many components of the M48 are also used in the M88 ARV as well as the M53 SPG, and further development of the tank resulted in the M60 series, production of which was finally completed at the Detroit Tank Plant in September 1987. Initial production of the M60 MBT was in fact undertaken at the Delaware plant but in 1959 a decision was taken that all future production would be undertaken at the Detroit Plant.

Description (M48A3)

The cast hull of the M48 is boat-shaped with additional sections welded into position. A hull escape hatch is provided in the floor of the tank. The turret is a one-piece casting.

The driver is seated at the front of the hull in the centre and is provided with a single piece hatch cover that opens to the right, forward of which are three M27 periscopes. For driving at night an M24 infra-red periscope is placed in the turntable located in the driver's hatch.

The other three crew members are in the turret with the commander and gunner on the right and the loader on the left. The commander is provided with an M1 cupola which he can traverse by hand through 360°. This cupola is also fitted to the M48A1, M48A1C and the M67 flamethrower tanks, and is equipped with five vision blocks and an M28C sight for controlling the 12.7 mm (0.50) machine gun. The M28C has a magnification of ×1.5 and a 48° field-of-view. The coincidence rangefinder is operated by the tank commander and has a maximum range of 4400 m and a magnification of ×10. The gunner is seated forward and below the commander and has a roof-mounted periscope sight with a ×8 magnification and a telescope with a similar magnification that is linked to the main armament. The ballistic computer is an electro-mechanical device which has been designed to compute super-elevation angles for the main armament. It receives range information in the form of shaft rotation from the rangefinder. The data is then applied to the ammunition data and ballistic corrections which have been rendered manually into the computer by the gunner. The end product is the super-elevation angle which is transmitted through the ballistic drive to the periscope. The ballistic drive also transmits the same super-elevation angle to the rangefinder which is operated by the tank commander. The loader is provided with a single piece hatch cover that opens to the rear. Mounted to the top rear of the turret is a dome-shaped ventilator cover and a stowage basket is provided at the rear of the turret.

The engine is mounted to the immediate rear of the bulkhead that separates the engine and fighting compartments and the transmission is at the rear of the vehicle. The engine compartment is equipped with fire extinguishers but there is no fire warning system. Power is transmitted to the final drives through the cross-drive transmission, which is a combined transmission, differential, steering and braking unit.

The torsion bar suspension consists of six dual rubber-tyred road wheels with the idler at the front, drive sprocket at the rear and five return rollers on each side. Some earlier versions of the M48 tank were fitted with a tensioning idler between the drive sprocket and the sixth road wheel station. Hydraulic shock absorbers are provided for the first, second and sixth road wheel stations.

Infra-red driving lights are fitted as standard and most models have an infra-red/white light searchlight mounted over the main armament, which has a maximum range of 2000 m. Standard equipment includes an NBC system, heaters, external infantry phone and provision for installing a dozer blade on the front of the hull. The M48, M48C and M48A1 are fitted with the M8 blade weighing 3980 kg and the M48A2, M48A3 and M48A5 with the M8A1 blade which weighs 3810 kg. Full details of these bulldozer blades are given in *Jane's Military Logistics 1990-91* page 19.

The tank can ford to a depth of 1.219 m without preparation and to 4.438 m with a deep-fording kit. Before deep fording all the openings are sealed and an exhaust extension is fitted vertically to the right rear engine grille, and the bilge pump switched on.

Main armament of the M48A3 is an M41 (T139) 90 mm gun, which consists of the barrel, evacuator chamber, blast deflector, and breech mechanism assembly. The breech-block is of the vertical sliding type with an inertia percussion firing mechanism. The barrel has a life of 700 equivalent full charge rounds. The following types of fixed rounds can be fired, but it should be noted that not all are currently in use: APERS-T (M580), APC-T (M82), AP-T (M77), AP-T (M318), Canister (M336), Canister (M377), HE-T (M71), HEAT-T (M431), HEAT (M348/M348A1), HVAP-T (M332A1), TP-T (M353) and Smoke WP (M313).

Both MECAR of Belgium and Israel Military Industries of Israel have developed a new APFSDS round for the 90 mm gun on the M47 and M48, and both have improved armour penetration characteristics.

Nineteen rounds of 90 mm ammunition are stowed to the left of the driver with a further 11 rounds to his right, eight horizontally on the turret floor, 16 stowed vertically around the turret ring and the remaining eight for ready use in the turret.

Mounted coaxially to the left of the main armament is a 7.62 mm M73 machine gun (earlier models have the 7.62 mm (0.30) M1919A4E1 weapon) and mounted in the commander's cupola is a 12.7 mm (0.50) Browning M2 HB machine gun which can be elevated from −10 to +60°, and can be aimed and fired from within the cupola.

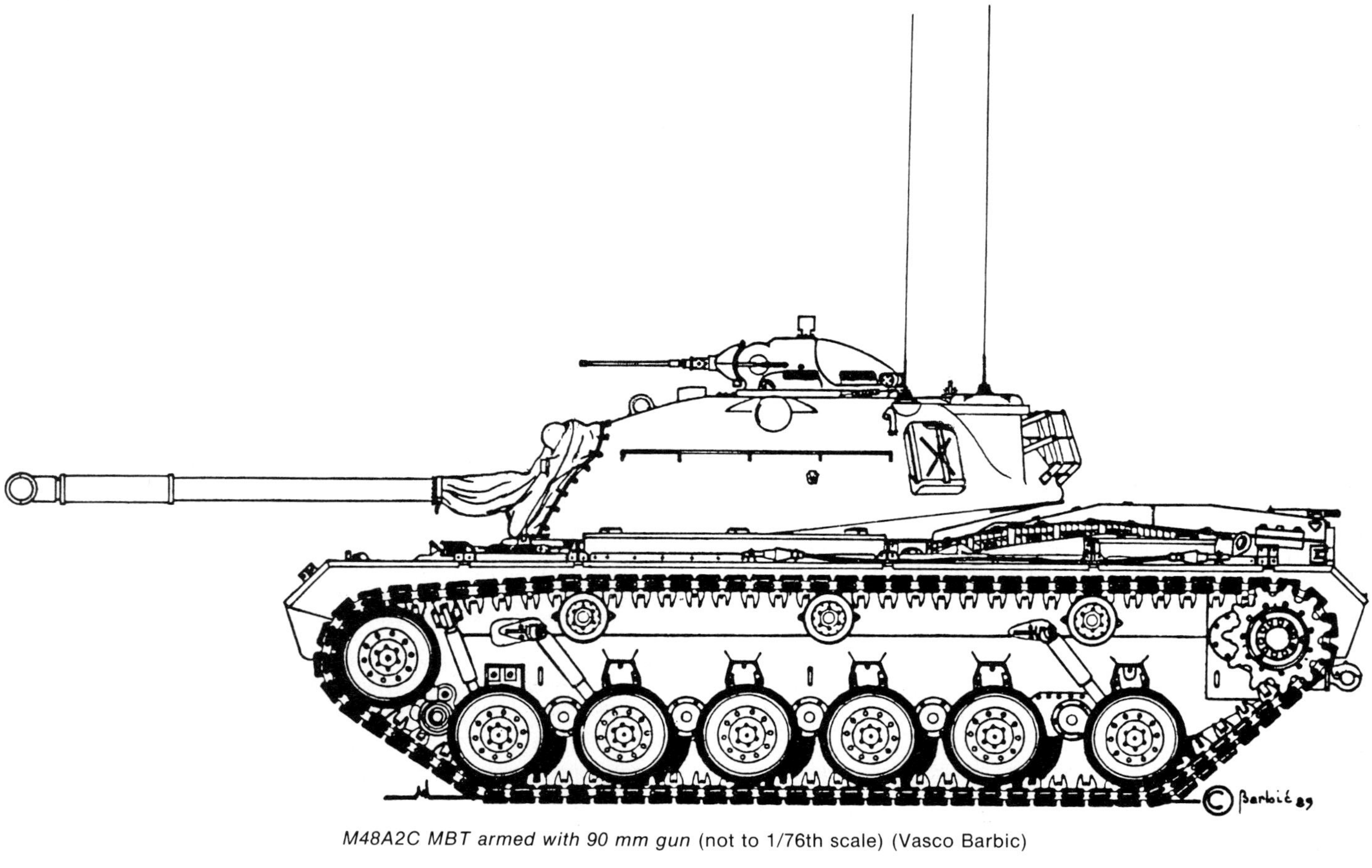

M48A2C MBT armed with 90 mm gun (not to 1/76th scale) (Vasco Barbic)

M48A2 showing 90 mm gun which is fitted with fume extractor and 'T' type muzzle brake

Variants

M48

This was the first production model of the series and has the following distinguishing features: small driver's hatch, five track return rollers, no tensioning idler, no dust shields on the fenders, 'T' or cylindrical type blast deflector on the barrel and the commander's cupola has the 12.7 mm machine gun on an open mount rather than in a fully enclosed cupola as later production tanks in this series.

M48C

This is identical to the M48 but has a hull of mild steel and is therefore unsuitable for combat. This model is used for training only and has the letter C embossed on the right front of the hull meaning that the tank is non-ballistic.

M48A1

This model has larger driver's hatch, fully enclosed commander's cupola, fender dust shields, rear track idler wheel, five track return rollers and a 'T' type blast deflector.

M48A2

Development of this model began in 1954 under the designation T48E2. The first production order was awarded to Alco Products Incorporated of Schenectady, New York in 1955, followed two years later by an order to the Chrysler Delaware Defense Plant (Lenape Ordnance Modification Center) at Newark, Delaware. Major differences between this and earlier models can be summarised as a fuel injection system for the engine, larger fuel tanks, improved engine deck to minimise infra-red detection, constant-pressure turret control system, improved fire control system, modified commander's cupola, stowage basket mounted at turret rear, and the main armament fitted with a 'T' type blast deflector. The suspension was also modified and jettisonable long-range fuel tanks could be fitted at the rear.

M48A2C

This is almost identical to the M48A2 apart from slight differences in the optical and fire control equipment. Most models do not have the track tensioner wheel.

M48A3

This had the development designation of the M48A1E2 and is basically a rebuild of the earlier M48A1 and M48A2 tanks. Major differences are the replacement of the petrol engine by the same diesel engine as installed in the M60A1 MBT, improved fire control system and commander's cupola modified by mounting a circular ring with vision blocks between the roof of the turret and the base of the commander's cupola. Most M48A3s have only three track return rollers and no rear idler. In addition they have 'T' type blast deflectors and fender dust shields.

In 1967 Bowen-McLaughlin-York was awarded a contract to remanufacture and modify 578 M48A1 tanks to M48A3 configuration. Under this programme the company rebuilt the vehicles and performed extensive modifications in accordance with 74 US Government funded, Chrysler designed kits. The principal change to this vehicle was the replacement of the petrol engine with the Continental AVDS-1790-2A diesel engine. Both remanufacture and modification were carried out in York, Pennsylvania.

The Red River Army Depot converted 400 M48A1s to M48A3 standard while the Anniston Army Depot converted 800 M48A1s to M48A3 standard.

M48A4

It was intended that turrets removed from M60s after they had been fitted with the new turret mounting the Shillelagh missile system would be fitted to older M48 chassis to be designated the M48A4. This project was cancelled after six prototypes had been completed.

M48A5

From October 1975 the Anniston Army Depot started to modernise older M48A1 and M48A3 tanks to a new standard called the M48A5. By the end of FY77, 764 had been converted, by the end of FY78 1454, and a total of 1864 at the end of FY80. The Anniston Army Depot converted 2064 M48A1/M48A3s to M48A5 configuration with final deliveries taking place in December 1979. Major modifications required to convert an M48A1 to the M48A5 standard were top-loading air cleaner, top deck grille, engine and transmission shroud, gun travel lock, exhaust grilles, powerpack (new engine fitted), final drives, tow pintle, engine and transmission mounts, hull turret seal, torsion bar knockout, bulkhead, drain valves, fuel tank and lines, track support rollers and shield, turret basket, modified turret ammunition stowage, T142 track, double bump spring and forward arm, driver's controls, driver's escape hatch, modified hull ammunition stowage, fire extinguisher, hull armour, heater, stowage boxes, driver's periscope, gun shield and cover, turret and gun control, 105 mm M68 gun (as in M60 series), composite headlamp, 2.2 kW searchlight, nylon ballistics shield, M114 mount and M105 telescope and graticule kit, M13B1 quadrant elevation, M10A6 ballistics drive, M32 and M118 periscope mount, M13B1C computer cam kit, M17B1C rangefinder and M28C periscope, M104A1 periscope mount, turret electrical kit, cupola adapter ring and retaining ring, turret manual drive and traverse gear box, turret stowage, searchlight stowage and cargo rack screen. Major modifications to convert the M48A3 to the M48A5 are the turret electrical kit, turret stowage, top-loading air cleaner, gun travel lock, solid-state regulator, turret basket, T142 track, ammunition stowage, 105 mm gun, gun shield and cover, M87 gun mount, graticule kit and computer cam kit. The fully enclosed commander's cupola was replaced by an Israeli-designed cupola which was manufactured by Associated Steel Foundries in Israel. First American-produced cupolas were delivered in mid 1976. A 7.62 mm M60D machine gun is mounted

M48A5 with 105 mm gun and two 7.62 mm M60D machine guns externally mounted on turret roof, one for commander and one for loader (Donald Spaulding)

externally at the commander's station and there are two mounting positions provided at the loader's hatch for his 7.62 mm M60D machine gun, to provide suppressive fire against enemy troop and ATGW positions. These have now been phased out of service with the National Guard and some passed onto other countries.

German M48s

The German company of Wegmann (Kassel) has rebuilt 650 M48 series tanks to a new configuration known as the M48A2GA2, first deliveries of which were made in mid 1978. The modifications include the replacement of the 90 mm gun by a 105 mm L7A3 rifled gun as installed in the Leopard 1, thermal sleeve for the main armament, new gun travelling lock, new commander's cupola, modified ammunition stowage, passive night vision equipment for the driver, commander and gunner and modifications to the fire control system. Basic data is as follows: loaded weight 47 800 kg, length gun forwards 9.35 m, length gun rear 8.1 m, height 2.9 m, ammunition stowage 46 rounds of 105 mm and 4750 rounds of 7.62 mm (2250 rounds in turret and 2500 rounds in the hull). German M48s also have four German smoke dischargers either side of the turret. The driver has the Elektro image-intensifier periscope, while mounted over the 105 mm gun is the PZB 200 passive TV aiming and observation unit from AEG-Telefunken with a screen that can be seen by both the commander and gunner. The first M48A2GA2 was handed over in June 1978 with final deliveries being made in November 1980. GLS, a subsidiary company of Krauss-Maffei, has supplied 170 of these kits to Turkey (qv Turkey in this section). In addition to the modifications carried out for the German and Turkish M48s Wegmann has suggested a number of further improvements including add-on armour for the turret, stabilised periscopic sight for gunner, laser rangefinder, thermal imager, fire control computer, fitting of hydraulic shock absorbers to improve ride, new torsion bars and an NBC system.

GLS Suspension Improvements

The German company of GLS, a subsidiary of Krauss-Maffei, has

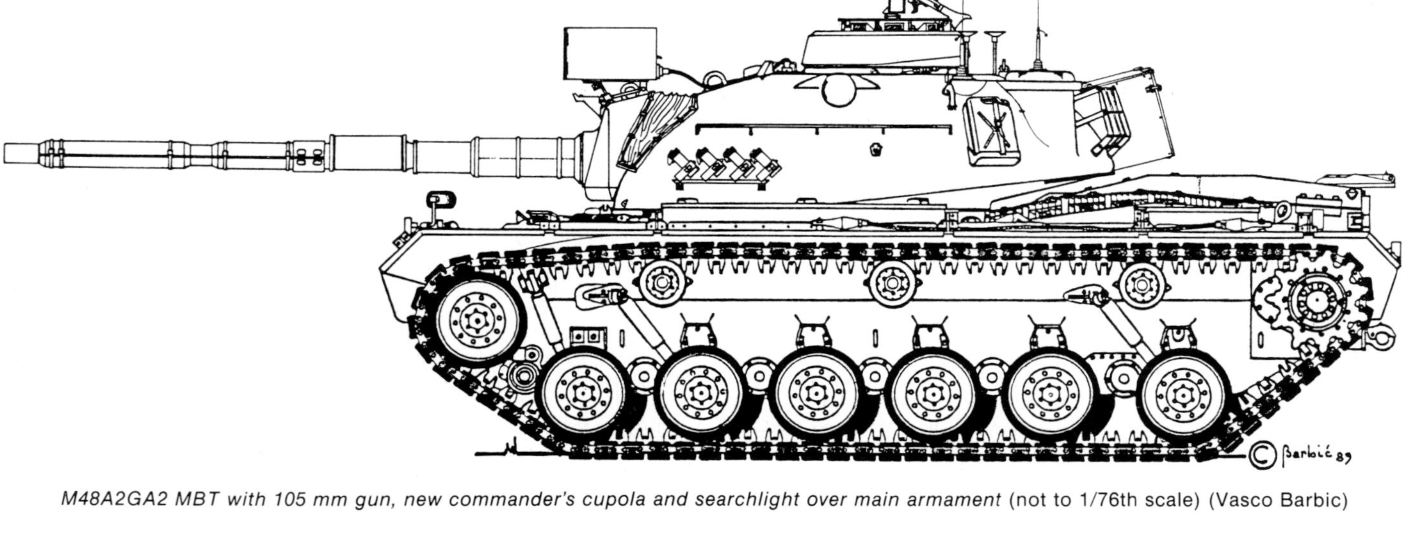

M48A2GA2 MBT with 105 mm gun, new commander's cupola and searchlight over main armament (not to 1/76th scale) (Vasco Barbic)

M48 AVLB in travelling configuration

German Army M48A2GA2 complete with infra-red/white light searchlight over main armament (Pierre Touzin)

Mock-up of M48 mine clearing vehicle

produced a suspension system retrofit package to improve the cross-country mobility of the M48 series of MBTs. All the components envisaged for retrofitting are already used on German armoured vehicles such as the Leopard 1, Leopard 2 or Marder, or have been derived from such components.

The retrofit package consists of hydraulic pump stops at all road wheel positions, hydraulic telescopic shock absorbers at first, second and sixth road wheel stations and torsion bars of high strength material.

Super M48 Main Battle Tank

The Super M48 is a private venture development by GLS of Germany and is fully covered under Germany earlier in this section.

Spanish M48s

Details of these vehicles are given in the Talbot MBT conversions entry under Spain.

Taiwanese M48H

This is fully covered under Taiwan and is an M60A1 chassis with an M48 series turret and new computerised fire control system incorporating a laser rangefinder.

Turkish M48s

Details of these vehicles and the modernisation programmes underway, are given under Turkey.

Iranian M48s

M48s of the Iranian Army have been brought up to M48A5 standard at a vehicle manufacturing plant built in Iran by Bowen-McLaughlin-York between November 1970 and February 1972.

Israeli M48s

Many Israeli M48s have been brought up to M60 standard by the installation of a Continental 12-cylinder diesel engine, 105 mm gun and the Israeli-designed and manufactured tank commander's cupola. Israel has fitted mine clearing ploughs to both the M48 and M60 tanks and provision has also been made for adding appliqué armour to the hull and turret. Details of this Israeli-developed Blazer explosive reactive armour and other modifications carried out to the M60s of the Israeli are given in the M60 entry under Israel.

M67/M67A1/M67A2 Flamethrower Tanks

None of these are now in service.

M48 Armoured Vehicle Launched Bridge

This is an M48 with its turret removed and fitted with a scissors bridge that is launched hydraulically over the front of the vehicle in three minutes. The bridge weighs 14 470 kg and has an overall length when opened out of 19.202 m, can span a gap of up to 18.288 m and has a maximum capacity of 60 000 kg. The M48 AVLB has a crew of two and weighs 55 746 kg with the bridge and 41 685 kg without it. Details of the M48/M60 AVLB are given in *Jane's Military Vehicles and Logistics 1991-92* page 94.

M48 ARV

The German companies of GLS and Jung Jungenthal offer an armoured recovery vehicle based on the M47 or M48 chassis. Details are given in *Jane's Military Vehicles and Logistics 1991-92* page 24.

New M48 Powerpacks

A number of companies offer retrofit packages for the M48, including NAPCO of the USA (still at prototype stage), Teledyne Continental Motors of the USA (AVDS-1790-C for export) and GLS of Germany (MTU MB 837 Ea-500 developing 750 hp, which has been fitted to some Turkish vehicles).

M48 Mine Clearing Vehicle

Krupp-MaK, Thyssen Henschel and Kälble have used the M48 chassis as the basis for a flail-type mine clearing tank, the first prototype of which was completed in 1985.

Status: Production complete. In service with the following countries:

Country	Model	Quantity	Comment
Germany	see comment	655	415 are 105 mm M48A2GA2
Greece	see comment	1300	600 are M48A3, some being upgraded to M48A5. In 1988 the US agreed to transfer 300 M48A5 MBTs to Greece free of charge with first 50 delivered in August 1988
Iran	M48A5	not known	all 105 mm, some non-operational
Israel	M48A5	500	all 105 mm and locally modified
Jordan	various	200 approx	not in front line use
Korea, South	various	950	597 to M48A5 standard
Lebanon	M48A5	104	delivered in 1983
Morocco	M48A5	173	M48A3 upgraded
Norway	M48A2	38	upgraded, plus 36 M48A5
Pakistan	M48A5	345	all 105 mm, further upgrade possible
Portugal	M48A5	86	some from US, some from FGR
Spain	M48E1	18	Marine Corps, 90 mm
	M48A5E	54	Army 105 mm
	M48A5	108	Army 105 mm
Taiwan	M48A1	286	upgunned to 105 mm
Thailand	M48A5	175	
Tunisia	M48A3/M48A5	28	14 of each
Turkey	M48A2	3000	2120 upgraded to 105 mm M48A5T1/M48A5T2 in Turkey, includes 170 upgraded tanks from FGR
USA	M48A5	1013	not in front line service, in store
Vietnam	M48A5	not known	called RS48

Note: Additional details of the Greek, South Korean, Spanish and Turkish M48 modernisation programmes are given under their respective countries' entries earlier in this section.

Manufacturers: Chrysler Corporation, Delaware, Ford Motor Company, Michigan, Fisher Body Division of General Motors Corporation, Michigan and Alco Products of Schenectady. Although no longer in production, the Systems Technical Support (engineering) work for the M48A5 is still being done by General Dynamics, Land Systems Division (formerly Chrysler Defense Incorporated).

SPECIFICATIONS

Model	**M48**	**M48A1**	**M48A2**	**M48A3**	**M48A5**
CREW	4	4	4	4	4
COMBAT WEIGHT	44 906 kg	47 173 kg	47 173 kg	47 173 kg	48 987 kg
UNLOADED WEIGHT	42 240 kg	43 999 kg	43 999 kg	44 452 kg	46 287 kg
POWER-TO-WEIGHT RATIO	18.03 hp/tonne	17.17 hp/tonne	17.39 hp/tonne	15.89 hp/tonne	15.89 hp/tonne
GROUND PRESSURE	0.78 kg/cm²	0.83 kg/cm²	0.83 kg/cm²	0.83 kg/cm²	0.88 kg/cm²
LENGTH GUN FORWARDS	8.444 m	8.729 m	8.686 m	8.686 m	9.306 m
LENGTH HULL	6.705 m	6.87 m	6.87 m	6.882 m	6.419 m
WIDTH	3.631 m	3.631 m	3.631 m	3.631 m	3.631 m
HEIGHT OVERALL	3.241 m*	3.13 m	3.089 m	3.124 m	3.086 m
GROUND CLEARANCE	0.393 m	0.387 m	0.385 m	0.406 m	0.419 m
TRACK	2.921 m	2.921 m	2.921 m	2.921 m	2.921 m
TRACK WIDTH	711 mm	711 mm	711 mm	711 mm	711 mm
LENGTH OF TRACK ON GROUND	4 m	4 m	4 m	4 m	4 m
MAX ROAD SPEED FORWARDS	41.8 km/h	41.8 km/h	48.2 km/h	48.2 km/h	48.2 km/h
FUEL CAPACITY	757 l	757 l	1268 l	1420 l	1420 l
MAX ROAD RANGE (without external tanks which could be fitted to M48A1 and M48A2)	113 km	113 km	258 km	463 km	499 km
FORDING	1.219 m	1.219 m	1.219 m	1.219 m	1.219 m
with preparation	2.438 m	2.438 m	2.438 m	2.438 m	2.438 m
GRADIENT	60%	60%	60%	60%	60%
VERTICAL OBSTACLE	0.915 m	0.915 m	0.915 m	0.915 m	0.915 m
TRENCH	2.59 m	2.59 m	2.59 m	2.59 m	2.59 m
ENGINE (all Continental 12-cylinder air-cooled)	AV-1790-5B/7/7B/7C	AV-1790-7C	AV-1790-8	AVDS-1790-2 A	AVDS-1790-2D*
type	petrol	petrol	petrol	diesel	diesel
output	810/2800 hp/rpm	810/2800 hp/rpm	825/2800 hp/rpm	750/2400 hp/rpm	750/2400 hp/rpm
AUXILIARY ENGINE	General Motors Corporation	Model A-41		none	none
TRANSMISSION (model)	CD-850-4/4A/4B	CD-850-4B	CD-850-5	CD-850-6	CD-850-6A
type	planetary gear shift with hydraulic torque converter with 2 forward and 1 reverse (low, high and reverse)				
STEERING	cross-drive differential in all models				
ELECTRICAL SYSTEM	24 V	24 V	24 V	24 V	24 V
BATTERIES	4 × 12 V	4 × 12 V	4 × 12 V	6 × 12 V	6 × 12 V
ARMAMENT					
main	1 × 90 mm gun	1 × 90 mm gun	1 × 90 mm gun	1 × 90 mm gun	1 × 105 mm gun
coaxial‡	1 × 7.62 mm MG	1 × 7.62 mm MG	1 × 7.62 mm MG	1 × 7.62 mm MG	1 × 7.62 mm MG
anti-aircraft§	1 × 12.7 mm MG	1 × 12.7 mm MG	1 × 12.7 mm MG	1 × 12.7 mm MG	2 × 7.62 mm MG
SMOKE-LAYING EQUIPMENT	none	none	none	none	M239 smoke grenade launchers and engine smoke laying system
AMMUNITION					
main	60	60	64	62	54
7.62 mm	5900	5900	5590	6000	10 000
12.7 mm	180	500	1365	630	none
GUN CONTROL EQUIPMENT					
Turret power control	electro-hydraulic with manual for emergency use				
by commander	yes	yes	yes	yes	yes
by gunner	yes	yes	yes	yes	yes
Rate of power traverse	Max/360° in 15 s for all models				
Gun elevation/depression	+19°/-9°	+19°/ -9°	+19°/-9°	+19°/-9°	+19°/-9°
commander's override	yes	yes	yes	yes	yes
Commander's fire control override	yes	yes	yes	yes	yes
Gun stabiliser					
vertical	no	no	no	no	no
horizontal	no	no	no	no	no
Rangefinder	M13	M13	M13A1	M17B1C	M17A1 or M17B1C
Gunner's telescope	M97C	M97C	M97C	M105	M105D
Elevation quadrant	M13A1	M13A1	M13A1	M13A3	M13B1
Traverse indicator	yes	yes	yes	yes	yes
ARMOUR					
Hull front	101/120 mm	101/120 mm	101/120 mm	101/120 mm	101/120 mm
Hull sides front	76 mm	76 mm	76 mm	76 mm	76 mm
Hull sides rear	51 mm	51 mm	51 mm	51 mm	51 mm
Hull top	57 mm	57 mm	57 mm	57 mm	57 mm
Hull floor	12.7/63 mm	12.7/63 mm	12.7/63 mm	12.7/63 mm	12.7/63 mm
Hull rear	44 mm	44 mm	44 mm	44 mm	44 mm
Turret front	110 mm	110 mm	110 mm	110 mm	110 mm
Turret sides	76 mm	76 mm	76 mm	76 mm	76 mm
Turret rear	50 mm	50 mm	50 mm	50 mm	50 mm
Turret top	25 mm	25 mm	25 mm	25 mm	25 mm

* including AA MG.
† M48A3 conversion uses AVDS-1790-2A engine while M48A1 conversion uses AVDS-1790-2D engine.
‡ 4As originally built coaxial machine gun was 7.62 mm (0.30) M1919A4E1 but most now have 7.62 mm M73 machine gun.
§ Browning 12.7 mm (0.50) M2 HB machine gun, except M48A5.

M47 Medium Tank

Development

In 1949 a decision was taken to design three new tanks, the T41 light, which became the M41 Bulldog, T42 medium and the T43 heavy, which became the M103. When the Korean War broke out the following year the T42 was not ready for production so the turret of the T42 tank (with gun T119) was mounted on the chassis of the M46 tank. This combination was known as the M46E1 and accepted for limited service as the M46A1. The front hull armour of the M46 was improved in the M46A1 by eliminating the ventilator housing between the driver's and bow machine gunner's position and increasing the front slope angle of the armour.

Production of the M47 was undertaken by the Detroit Tank Plant (which built 3440 by November 1953) and the American Locomotive Company (which built over 5000). Total production amounted to 8676 units. The M47 did not see combat in Korea and was soon replaced in the regular Army by the M48, development of which started late in 1950 by Chrysler.

Description

The hull of the M47 is made of armoured plate and cast armour sections welded together and reinforced, and has three compartments, driver and bow machine gunner at the front, fighting in the centre and the engine and transmission at the rear.

The driver is seated at the front of the vehicle on the left side with the bow machine gunner to his right. Both men are provided with a single piece hatch cover that opens to the outside of the vehicle with an integral M13 periscope. Both the driver and bow-gunner are provided with a hull escape hatch.

The cast turret is mounted in the centre of the hull with the commander and gunner on the right and the loader on the left. The commander is provided with a single piece hatch cover that opens to the rear and five vision blocks are arranged around the sides and behind his position. Mounted to his front is an M20 periscope and a similar periscope is provided forward and slightly below the commander's position for the gunner. The M20 periscope has two built-in optical systems, a ×1 system for wide-angle observation and a ×6 for sighting distant objects and ranging on targets. The M12 stereoscopic ballistic computing rangefinder is mounted in the forward part of the turret and is used by the gunner who, through the use of various adjustment knobs, indexes the proper ammunition, sets the ballistic correction, ranges and tracks the target, and boresights the gun for firing. The loader is provided with a single piece hatch cover that opens to the right front and an M13 periscope mounted forward of his hatch. The turret has an overhanging bustle with the dome-shaped turret ventilator mounted in the roof at the rear. A light steel stowage box is provided at the rear of the turret.

The engine compartment is at the rear of the vehicle and is separated from the fighting compartment by a bulkhead. The engine transmits power via the cross-drive transmission to the final drives. The torsion bar suspension consists of six dual rubber-tyred road wheels with the idler at the front and the drive sprocket at the rear. There are three track return rollers and an adjustable tensioning wheel between the sixth road wheel and the drive sprocket. Hydraulic shock absorbers are provided for the first, second, fifth and sixth road wheel stations. The steel tracks have rubber blocks and when new each track has 86 links.

The main armament of the M47 consists of a 90 mm gun M36 (T119E1) in a mount M78. The main components of the gun are the blast deflector (depending on the type of tank this can be of the 'T', cylindrical or as fitted to the M46 tank), evacuator chamber and the barrel mechanism. The mount consists of the shield group and the recoil mechanism assembly. The recoil mechanism is of the concentric hydrospring type and the breech-block is of the vertical sliding type. Once a round is loaded into the breech, a closing mechanism closes the breech which is re-opened automatically during counter-recoil of the gun. The empty cartridge case is ejected into the spent bag under the gun. The barrel has a life of 700 rounds and can fire the following types of ammunition (not all of these rounds are currently in use and the ranges quoted are with the weapon at maximum elevation; effective anti-tank range is 2000 m): APC-T (M82), APERS-T M580), AP-T (M77), AP-T (M318), Blank (M394), Canister (M336), Canister (M377), Dummy (M12), HE-T (M71), HEAT (M348A1) (fin-stabilised), HEAT (M431), HVAP-T (M332A1), TP-T (M353) (the M71 TP round has no tracer) and Smoke WP (M313). Both MECAR of Belgium and Israel Military Industries of Israel have developed new APFSDS rounds for the 90 mm gun on the M47 and M48, and these have a significant increase in armour penetration over existing rounds.

Of 71 rounds of 90 mm ammunition carried 11 are in the turret bustle for ready use. A 7.62 mm (0.30) M1919A4E1 machine gun is mounted coaxially to the left of the main armament and a similar weapon is mounted in the bow of the tank on the right side. Some countries have taken out the bow machine gun enabling the number of 90 mm rounds to be increased from 71 to 105. A 12.7 mm (0.50) Browning M2 HB machine gun is mounted on the roof of the tank for anti-aircraft use.

The M47 is not fitted with an NBC system and as built was not provided with any infra-red night vision equipment, although some countries have since fitted it. The tank can be fitted with the M6 dozer blade. Some 31 pre-production dozer blades were built followed by 468 production models.

Variants

Iranian M47M: redesign of the M47 by Bowen-McLaughlin-York to incorporate automotive and fire control components of the M60A1 MBT resulted in the M47M. The company designed and built a vehicle manufacturing plant in Iran between November 1970 and February 1972, which included heat treating, plating, X-ray capability and furnishing the operating equipment and machinery. Prototypes of the M47M were built in the United States but conversion work was carried out in Iran. In addition the company replaced the rangefinder on two M47M pilot vehicles with a laser fire control system provided by Hughes Aircraft Corporation.

South Korean M47 with white/infra-red searchlight over 90 mm gun with a "T" type blast fume extractor (US DoD)

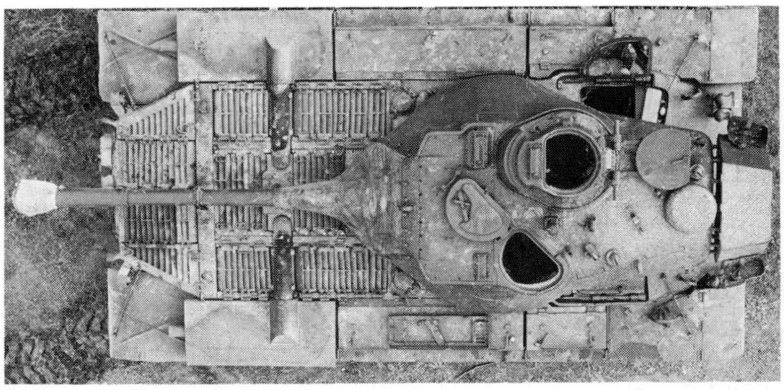

M47 from above with turret traversed to rear and commander's and loader's hatches open. This particular M47 does not have 12.7 mm anti-aircraft machine gun installed on turret roof (US Army)

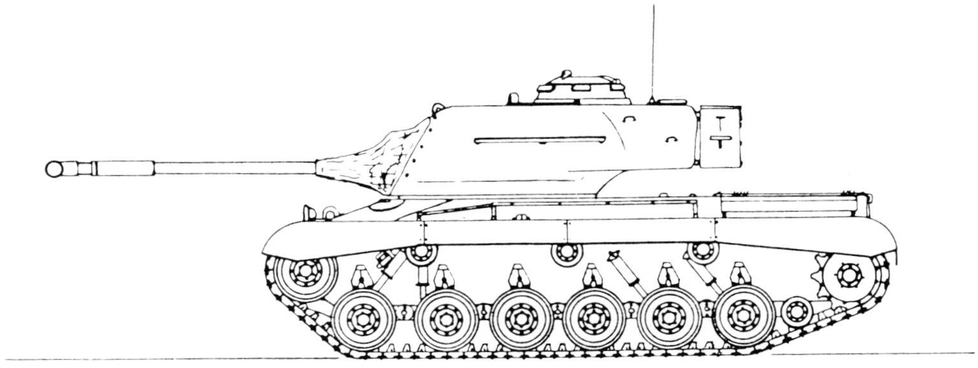

M47 medium tank

The M47M has the following components which are also interchangeable with the M48A3 and M60A1 MBTs: engine, transmission, gun elevating and traversing components, personnel heater, engine top deck and hull rear doors, air cleaners, machine gun ammunition racks, tail and warning lights, gauge instrument panel, driver's master control panel, tachometer with hour meter, speedometer with odometer, turret accessories control box, fire extinguisher control, drain valves and shock absorbers.

The bow machine gun has been removed and the gunner's position used to carry 22 rounds of ammunition. The M47M carries 79 rounds of 90 mm ammunition compared with the 71 rounds of the basic M47. The M47M can be fitted with the T80E6 steel track or the T84E1 rubber track. The auxiliary compensating idler wheels, support arms and torsion bars which are between the number six road wheel and the drive sprocket have been removed. New M60 series shock absorbers have been fitted permitting greater vehicle speed and mobility.

The larger capacity fuel tanks, coupled with the improved fuel economy of the diesel engine, have increased the operational range of the M47M to 600 km.

South Korea: has a number of M47s converted into the ARV with the 90 mm gun removed and a winch installed in the turret. Mounted at the front of the hull is an A-frame with the main cable taken out through the front of the turret. This vehicle is used primarily for changing vehicle components in the field.

Somalia: Several years ago the US Army Tank Automotive Command awarded a contract to the Automotive Systems Group of NAPCO International to overhaul three M47 medium tanks for an African country, believed to be Somalia. If these trials are successful and further funding is available, it is expected that a much larger number will be overhauled.

In addition to overhauling the M47s, including their gun control equipment, NAPCO is replacing the existing Continental AV-1790 series petrol engine developing 810 hp at 2800 rpm with a new Detroit Diesel 12V-71 QTAB engine with a similar power output but giving a substantial increase in range, around 500 km compared to the 130 km of the present petrol engine.

Spain: Details of Spanish M47 rebuilds are given under Talbot in the Spanish section.

GLS M47 ARV

Full details of the M47 ARV designed and built by GLS in conjunction with Jung Jungenthal are given in *Jane's Military Vehicles and Logistics 1991-91* page 24. Krupp MaK of Kiel, Germany, is also building an ARV based on the M47 chassis using some components already used in the Leopard 1 ARVs it has built for the home and export markets. Talbot of Spain also has an M47 ARV and an AEV; details are given under Spain.

M47M as used by Iran

Status: Production complete. In service with the following countries:

Country	**Quantity**	**Comment**
Greece	350	unmodified
Iran	100 est	out of 400 modified to M47M standard
Italy	300	unmodified, reserves
Korea, South	300	unmodified
Pakistan	150	rebuilt to M47M standard
Somalia	100	modification programme possible
Spain	375	upgraded by Talbot of Spain
Turkey	500	unmodified, reserve
Yugoslavia	60	unmodified, reserve

Manufacturers: Detroit Tank Plant and American Locomotive Company.

SPECIFICATIONS (data in square brackets relates to M47M where different from M47)

CREW	5 [4]
COMBAT WEIGHT	46 170 [46 814] kg
UNLOADED WEIGHT	42 130 kg
POWER-TO-WEIGHT RATIO	17.54 [16.11] bhp/tonne
GROUND PRESSURE	0.935 kg/cm^2
LENGTH GUN	
forwards	8.508 [8.553] m
rear	7.09 m
LENGTH HULL	6.307 [6.267] m
WIDTH	3.51 [3.39] m
WIDTH OVER TRACKS	3.377 m
HEIGHT	
to turret top	2.954 m
overall including AA MG	3.352 [3.397] m
commander's cupola	3.016 m
GROUND CLEARANCE	0.469 m
TRACK	2.794 m
TRACK WIDTH	584 mm
LENGTH OF TRACK ON GROUND	3.911 m
MAX ROAD SPEED	48 [56.3] km/h
early model	58 km/h
FUEL CAPACITY	875 [1476] l
MAX RANGE	130 [600] km
FORDING	1.219 m
GRADIENT	60%
VERTICAL OBSTACLE	0.914 m
TRENCH	2.59 m
TURNING RADIUS	skid turns
ENGINE	Continental AV-1790-5B [AVDS-1790-2A, 750 bhp at 2400 rpm], 7 or 7B, V-12, 4-cycle, air-cooled petrol developing 810 bhp at 2800 rpm
AUXILIARY ENGINE	Wisconsin TFT 2-cylinder petrol
TRANSMISSION	Allison model CD-850-4 [CD-850-6A], 4A or 4B cross drive with 2 forward and 1 reverse ranges
SUSPENSION	torsion bar
ELECTRICAL SYSTEM	24 V
BATTERIES	4 × 12 V
ARMAMENT	
main	1 × 90 mm gun
coaxial	1 × 7.62 mm (0.30) MG
bow	1 × 7.62 mm (0.30) MG [none]
anti-aircraft	1 × 12.7 mm (0.50) MG
SMOKE-LAYING EQUIPMENT	none
AMMUNITION	
main	71 [79]
7.62 mm	4125
12.7 mm	440
GUN CONTROL EQUIPMENT	
Turret power control	hydraulic and manual for emergency
Max rate power traverse	360° in 10 s
Max rate power elevation	4°/s
Gun stabiliser	
vertical	none
horizontal	none
Elevation quadrant	T21
Traverse indicator	T24
ARMOUR	
Hull front	101 mm at 60°
Hull front lower	76/89 mm at 53°
Hull sides front	76 mm
Hull sides rear	50.8 mm
Hull top	22 mm
Hull floor front	25.4 mm
Hull floor rear	12.7 mm
Turret mantlet	115 mm
Turret front	101 mm at 40°
Turret sides	63.5 mm at 30°
Turret rear	76.2 mm at 30°
Turret top	12.7 mm

Teledyne Continental Motors Centurion Modernisation Package

Development

The United States company of Teledyne Continental Motors General Products has been engaged in the development of complete retrofit packages for a wide range of tracked armoured vehicles including the Centurion, M47, M48, M60, AMX-30, M41, M42 and the Soviet T-54/T-55/T-62.

These packages are tailored to meet the specific requirements of the user and can cover the three key areas of an MBT; armour, mobility and firepower.

Teledyne Continental Motors General Products were the prime contractor for the complete upgrading of 293 Jordanian Centurion tanks to the Tariq standard (details of which are given in the Centurion entry) and more recently the company has upgraded a single T-54 MBT for the Egyptian Government under the name of Ramses II (details of this are given under Egypt).

Typically, these retrofit packages include a new powerpack incorporating a Teledyne Continental AVDS-1790 series diesel engine, Teledyne Continental hydro-pneumatic suspension system, new armament, new fire control system and, if required, additional armour to increase battlefield survivability.

As a private venture the company has also built the prototype of the Direct Fire Support Vehicle which has as externally mounted 105 mm gun that fires standard NATO ammunition. Details of this are given in the *Light tanks* section.

Teledyne Continental Motors General Products, envisions a great deal of interest, in the near term, in the upgrading of M48 and M60 assets. This can be accomplished relatively cheaply by replacing the -2D or -2C engine with an AVDS-1790-5A rated at 908 hp. This requires an upgrade of the transmission to the CD-850-6B configuration. With this the vehicle has sufficient power for the addition of armour. Depending on the mobility requirements the user can also fit the External Suspension System.

In 1989 Teledyne Continental Motors General Products made proposals of MBT modernisations:

1 M47 dieselisation (one proposal)
2 M48 dieselisation (two proposals)
3 Centurion upgrades (two proposals)
4 AMX-30 upgrade (one proposal)
5 M47 ARV conversion (one proposal)

Description

The Centurion MBT modernisation package is typical of the numerous tank modernisation packages now being offered by Teledyne Continental Motors General Products.

The main object of a complete modernisation package is to improve the Centurion's firepower, mobility and protection, with this being achieved at a fraction of the cost of the procurement of a fleet of new MBTs.

If the Centurion retains its 20-pounder gun, then this would be replaced by a 105 mm weapon which can fire the latest types of APFSDS ammunition including those with tungsten and DU penetrators with additional rounds under development. If required, however, a 120 mm upgunning kit could be developed for the Centurion, although this would have an impact on the number of rounds carried, while the additional weight would have an impact on the turret drives and stabilisation system.

If required, a new weapon/turret control stabilisation system could be fitted similar to that already fitted as standard to production M1/Leopard 2 series of MBTs. Fire control options include modified versions of existing production systems, modernising the existing system installed on the vehicle or fitting a standard production fire control system.

The new mobility kit comprises the following main components:-

a) AVDS-1790 air-cooled diesel engine developing 908 or 1050 hp
b) TCM-304 automatic hydro-kinetic transmission with four forward and two reverse gears, two stage torque converter with lockup
c) Improved final drives for increased speed and reliability
d) Installation of new Teledyne Continental Motors Model 2869 hydro-pneumatic suspension units. These provide for the independent suspension for each wheel, are externally mounted and improves ride, speed and gun stability. The existing track can be installed or a new track fitted.

The main advantages of the new mobility kit can be summarised as:-

a) 100 per cent improvement on the speed of Centurion
b) 19 per cent faster than Leopard 1
c) 140 per cent faster cross-country speed than Leopard 1
d) 500 km cruising range
e) automatic shifting transmission.

Survivability improvements include automatic fire detection and suppression system for both engine and crew compartments, installation of an NBC system (the US M13A1 gas particulate unit) and additional armour protection.

Centurion chassis upgraded by Teledyne Continental Motors General Products with Low Profile Turret 105 armed with externally mounted 105 mm gun

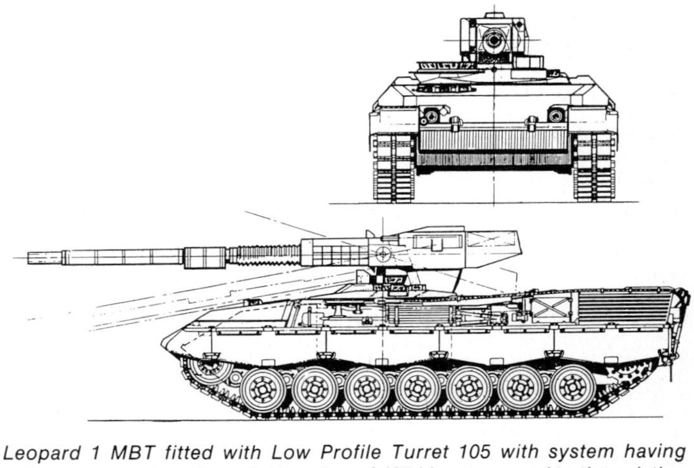

Leopard 1 MBT fitted with Low Profile Turret 105 with system having a three man crew and a weight saving of 4654 kg compared to the existing MBT which can be reapplied as hull uparmour. Fire control system includes laser rangefinder and thermal sight (not to 1/76th scale)

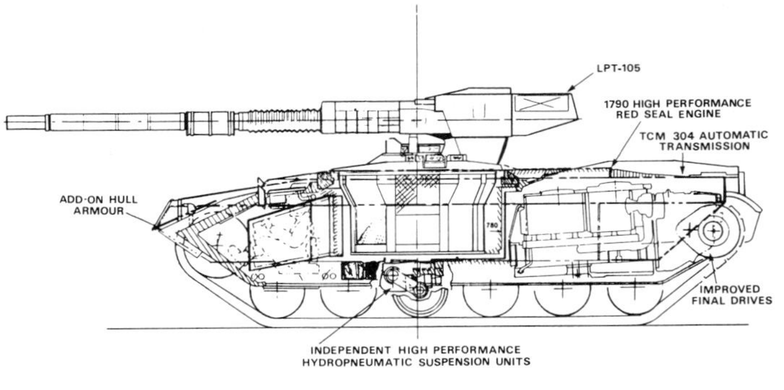

Interior arrangement of M60 series MBT modernised by Teledyne Continental Motors and fitted with Low Profile Turret 105 (not to 1/76th scale)

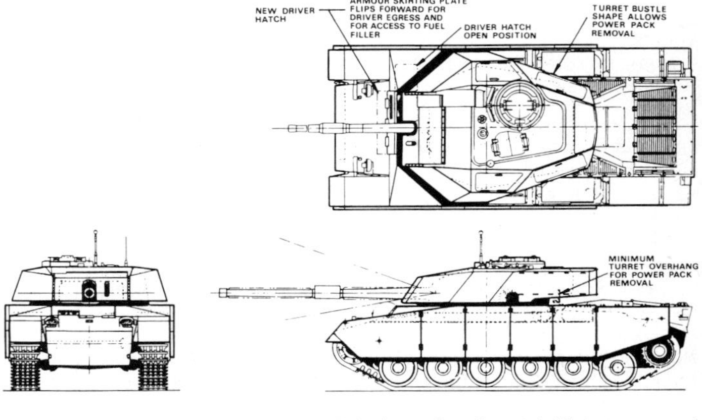

Example of an MBT fitted with Teledyne Continental Motors spaced appliqué armour kit showing main components (not to 1/76th scale)

The spaced appliqué armour package weighs 1600 kg (hull), 690 kg (skirts) and 1800 kg (turret) and provides the following protection level:-

Type	Arc	RHA equivalent	Type of weapon
APFSDS	front 60°	250 to 300 mm	Soviet 115 mm
HEAT	front 60°	300 to 350 mm	Soviet 115 mm/RPG-7

With the addition of explosive reactive armour, the protection against HEAT attack can be more than doubled.

Variants
Teledyne Continental Motors also offer the Modular Low Profile turret 105 (LPT 105), originally designed for the Direct Fire Support Vehicle (DFSV) and fully described in this entry, for installation on the Centurion, as well as other vehicles such as Leopard 1, M48, M60, M1, T-54/T-55/T-62 and AAV7A1.

The LPT, which is armed with a 105 mm gun fireing standard NATO ammunition, offers a number of advantages as follows: it reduces the profile of the tank and threrefore makes it harder to hit, reduces the overall weight of the vehicle (so allowing additional armour to be applied to the hull), reduces the crew from four down to three, provides increased protection against top attack and has further growth potential to 120 mm and a mix of guns and missiles.

Upgraded Centurion performance specifications:

Vehicle	Centurion	Leopard 1 908	Upgraded 1050 hp	Centurion
MAX SPEED				
paved road	34 km/h	62 km/h	66 km/h	67 km/h
5% gradient	22 km/h	36 km/h	38 km/h	43 km/h
10% gradient	16 km/h	24 km/h	25 km/h	28 km/h
30% gradient	4 km/h	11 km/h	10 km/h	13 km/h
60% gradient	n/a	6 km/h	5 km/h	6 km/h

Status: Development complete. Ready for production. Some parts of this modernisation package have already been used in the Jordanian Centurion upgrade programme.

Manufacturer: Teledyne Continental Motors General Products
76 Getty Street
Muskegon
Michigan 49442
USA
Telephone: (616) 724 2151
Telex: 228426
Fax: (616) 724 2796

YUGOSLAVIA

Yugoslav Tank Fleet

Shortly after the end of the Second World War Yugoslavia redesigned the Soviet T-34/85 medium tank to incorporate an improved frontal hull design and a new turret. These were not however produced in any quantity.

For some years there has been an agreement between the Swedish company of L M Ericsson and the Yugoslav company of Iskra concerning the development and production of lasers for a number of applications, including their use in MBTs such as the T-54/T-55 and the more recent M-84.

In early 1991 the Yugoslav tank fleet was believed to consist of 800 T-54/T-55 tanks, 400 M-84 MBTs and perhaps 60 M47 medium tanks out of 319 supplied by the USA many years ago. Reserve tanks are believed to include 250 T-34/85s supplied by the Soviet Union and 300 Sherman tanks supplied by the USA, although most of the latter as well as some of the M47 tanks have been expended as range targets.

M-84 Main Battle Tank

Development
In 1977 Yugoslavia decided to build the Soviet T-72 MBT under licence and by 1979 all of the documents and drawings required to build the tank had been received.

The next two years were spent carrying out detailed changes to the drawings to suit local manufacture and incorporating some Yugoslav equipment, especially in the turret. There are some 240 main contractors involved in producing the M-84 in Yugoslavia.

Yugoslav-built prototypes were completed in 1982-83 and production got underway in 1983-84 with first production vehicles being completed late in 1984. By early 1991 it is estimated that some 400 had been built in Yugoslavia under the designation of the M-84. Prior to the commencement of local production, Yugoslavia took delivery of about 50 T-72s from the Soviet Union to enable troop training to get underway.

The Yugoslav M-84 MBT is virtually identical to a late production Soviet T-72 MBT with two banks of smoke dischargers mounted on the turret front, seven on the left and five on the right. The detailed specifications of the M-84 given are from Yugoslav sources. The main areas of difference is in the computerised fire control system and the vehicle optics. The M-84 has a distinctive mast containing sensors fitted on the forward part of the turret to the immediate rear of the 125 mm gun.

The latest production model is designated the M-84A. This has a number of improvements including a more powerful 1000 hp diesel engine which gives a higher road speed, greater acceleration and improved power-to-weight ratio, and improved armour protection.

Variants
A command version is known to be in service and an armoured recovery version is believed to be in service.

Status: In production. In service with Yugoslavia. In mid 1989 Kuwait placed an order for some 200 M-84 MBTs and included in this figure are 15 command tanks, 15 armoured recovery vehicles plus ammunition, spares and training.

A small number of M-84As, believed to be 12, had already been delivered at the time of the Iraqi invasion of Kuwait in the summer of 1990. Yugoslavia continued delivery of the remainder of the order to Kuwaiti forces based in Saudi Arabia and these subsequently took part in the liberation of Kuwait in the Spring of 1991.

Note: The latest version of the M-84, designated the M-84A has a number of improvements including the installation of a V-12 water-cooled diesel engine with two turbo-chargers and is air-cooled. This develops 1000 hp at 2000 rpm and gives the vehicle a maximum road speed of 64 km/h.

Yugoslav M-84 showing 125 mm gun fitted with fume extractor and thermal sleeve

Yugoslav M-84 from the front, clearly showing mast-mounted sensor on forward part of turret roof

Contractor:
Enquiries to Federal Directorate of Supply and Procurement (SDPR)
PO Box 308
9 Nemanjina Street
Belgrade
Yugoslavia
Telephone: 621 522
Telex: 11360

SPECIFICATIONS

CREW	3
LENGTH (gun forward but without unditching beam or long range fuel drums)	9.53 m
WIDTH OVERALL	3.57 m
WIDTH OVER TRACKS	3.37 m
HEIGHT (turret roof)	2.19 m
HEIGHT TO CENTRE OF FRONT IDLER	0.835 m
HEIGHT TO CENTRE OF REAR DRIVE SPROCKET	0.731 m
FIRING HEIGHT	1.651 m
GROUND CLEARANCE	0.428 m
TRACK	2.79 m
LENGTH OF TRACK ON GROUND	4.27 m
MAXIMUM ROAD SPEED	60 km/h
RANGE	450 to 700 km
FORDING WITHOUT PREPARATION	1.2 m
FORDING WITH 5 MINUTES PREPARATION	1.8 m
FORDING WITH SPECIAL EQUIPMENT (takes 20 minutes to install)	5 m (max water speed 1.5 m/s)
GRADIENT	60%
SIDE SLOPE	40%
VERTICAL OBSTACLE	0.85 m
TRENCH	2.8 m
ENGINE	V-12 water-cooled, supercharged diesel developing 780 hp at 2000 rpm
TRANSMISSION	7 forward and 1 reverse gears with mechanical assistance

ARMAMENT	
main	1 × 125 mm gun
coaxial	1 × 7.62 mm MG
anti-aircraft	1 × 12.7 mm MG
AMMUNITION	42 (total) 22 (ready use) (in automatic loader)
TYPES	3
RATE OF FIRE	6 to 9 rpm (automatic) 2 rpm (manual)
MAXIMUM TRAVERSE SPEED	70°s
GUN CONTROL EQUIPMENT	
Turret power control	power/manual
Turret traverse	360°
Gun elevation/depression	+13° 47 min/-6° 13 min
Stabilisation system	yes, two planes
Stabilisation modes	stabilised, semi-stabilised and mechanical (manual)
TARGET TRACKING	acquisition day and night with independent line of sight
FIRING DATA	computed automatically from following sensors, target range, target speed, turret turn rate, tank tilt, air temperature, air pressure, wind speed
GUNNER'S DNNS-2 DAY/NIGHT SIGHT TYPE	monocular
measuring range	200 to 9995 m
number per minute	12
MTBF	100 000 shots

FIELD-OF-VISION	
day	9° and 20°
night	5°
MAGNIFICATION	
day	× 3 and × 7
night	× 8.5
COMPUTING RANGE	
for HEAT	200 to 6000 m
APFSDS	200 to 4000 m
HE-FRAG	200 to 4000 m
MG	200 to 1500 m
MUZZLE VELOCITY VARIATION	±50 m
SIDE WIND	±40 km/h
TANK LATERAL TILT	15°
COMMANDER'S DNKS- 2 DAY/NIGHT SIGHT TYPE	day/night, passive, binocular
DAY MAGNIFICATION	× 4.9
NIGHT MAGNIFICATION	× 4.3
FIELD-OF-VIEW	
day	10°
night	10.9°
STANDARD EQUIPMENT	intercom, radio (20 km range), NBC system, automatic fire extinguishing system, front-mounted dozer blade, camouflage, mounting points for mine clearing kit

New Yugoslav Main Battle Tank

Development/Description

Late in 1990, the Yugoslav Federal Directorate of Supply and Procurement announced that it was building the prototype of a new MBT and that production could start as early as 1995. It was not made clear as to whether this new MBT was being developed for the home or export markets.

The new MBT will be similar in appearance to the M-84 but has a combat weight of 44 tonnes, maximum road speed of 74 km/h and will be powered by a 1200 hp diesel coupled to a fully automatic transmission. The exhaust outlet will be on the left side of the vehicle as on the Soviet T-72 series of MBT.

Suspension will be of the torsion bar type with hydraulic shock absorbers being provided on the 1st, 2nd, 3rd and 6th road wheel stations. The upper part of the suspension will be covered by a light steel skirting.

Although similar in appearance to the current M-84 (qv), the new MBT will have new hull and turret armour, new NBC system and a fire detection and suppression system.

The 125 mm smoothbore gun of the M-84 will be retained but new ammunition is being developed which is planned to include a new APFSDS round with a muzzle velocity of 1800 m/s. The 125 mm gun will be fitted with a thermal sleeve, fume extractor and a muzzle reference system.

A 7.62 mm MG will be mounted coaxial with the main armament and a 12.7 mm MG will be mounted on the commander's cupola. A total of 12 smoke dischargers in three banks of four are mounted to the right of the main armament with two banks of four to the left.

A computerised fire control system will be fitted with both day and night sights provided for the commander and gunner. On the current M-84 these are of the image intensification type but on the new MBT they will be of the thermal type.

On the current M-84, the mast-mounted sensor package is mounted on the forward part of the turret roof but on the new MBT it is mounted on the turret rear.

Standard equipment will include an NBC system and the provision for deep fording, the snorkel for the latter operation being carried on the left side of the turret when not required.

Status: Under development. Not yet in production or service.

Contractor:

Enquiries to Federal Directorate of Supply and Procurement (SDPR)
PO Box 308
9 Nemanjina Street
Belgrade
Yugoslavia
Telephone: 621 522
Telex: 11360

Scale model of the new Yugoslav MBT from front showing smoke dischargers either side of 125 mm gun (Christopher F Foss)

Scale model of the new Yugoslav MBT from rear showing external stowage (Christopher F Foss)

Light tanks (under 25 000 kg)

AUSTRIA

Steyr SK 105 Light Tank

Development
The Steyr SK 105, nicknamed the Kürassier, was developed from 1965 by Saurer-Werke which was taken over in 1970 by the Steyr-Daimler-Puch company, and is based on a redesigned chassis of the Saurer APC. The SK 105 was developed to provide the Austrian Army with an independent anti-tank weapon for difficult terrain.

The first prototype was completed in 1967 with the second following in 1969. Five pre-production vehicles were completed in 1971 and by early 1990 some 600 vehicles had been built for home and export markets.

Early in 1988 the SK 105 was demonstrated in Ecuador alongside AMX-13, TAM and Stingray, but no orders were eventually placed for any of these vehicles for financial reasons. Upgraded models are known as SK 105/A1.

Description
The hull of the SK 105 is welded steel and is divided into three compartments: driver's at the front, fighting in the centre and the engine at the rear. The SK 105 is immune to 20 mm armour-piercing rounds over its frontal arc.

Add-on armour on the front part of the chassis and turret is available as an option and this provides protection against 35 mm APDS projectiles at a flank arc of ± 20°.

The driver sits at the front of the vehicle on the left and is provided with a single piece hatch cover that opens to the left. There are three periscopes forward of his hatch cover and in wet weather a small windscreen with a wiper can be fitted. The driver's centre day driving periscope can be replaced by a passive night vision periscope if required. Ammunition and the vehicle's batteries are stowed to the right of the driver.

The Steyr turret JT 1 mounted in the centre of the vehicle is derived from the FL-12 built by Fives-Cail Babcock of France and assembled in Austria with a number of improvements. The oscillating turret is similar to that fitted to some members of the AMX-13 light tank family. The commander is seated on the left of the turret and the gunner on the right. The commander is provided with seven periscopes, a periscopic sight with alternatively ×1.6 magnification and a 28° field-of-view or a ×7.5 magnification and a 9° field-of-view, and a single piece hatch cover that swivels to open. The commander's infra-red night sight has a magnification of ×6. The gunner has two observation periscopes, a telescopic sight with a magnification of ×8 and an 8.5° field-of-view and a one-piece lifting and swivelling hatch cover. Due to the design of the oscillating turret all sights are always linked to the main and secondary armament. For engaging targets at night an infra-red periscopic sight with a magnification of ×6 and a 7° field-of-view is provided for the commander. A TCV 29 laser rangefinder (range of 400 to 9995 m) is mounted on the roof of the turret. The XSW-30-U 950 W infra-red/white light searchlight is fitted on the left-front plate of the oscillating part of the turret. A fixed fan in the turret draws out fumes when the main or secondary armament is fired.

The engine and transmission are at the rear of the vehicle and the engine compartment is fitted with a fire extinguisher which can be operated by hand or automatically. When flames enter the suction part the intake pipe is closed and air is then drawn in from the crew compartment.

The engine permanently drives an oil pump with a variable output linked to a hydraulic motor acting on the differential which controls the tracks. By acting on the steering gear and the speed of the engine the driver can continuously adjust the speed ratio of both tracks. All turning radii can thus be obtained until pivoting on the spot occurs with both tracks turning in opposite directions at the same speed. Since this system needs no action from the brakes, the whole power produced by the engine is constantly available at the tracks. The main brakes of the SK 105 are hydraulic, foot-operated.

Steyr SK 105 light tank with laser rangefinder, infra-red/white light searchlight and single-stage muzzle brake for firing APFSDS projectiles

Prototype of Kürassier SK 105/A2 showing redesigned turret front and gunner's day/night sight

Steyr SK 105/A1 light tank prototype with new gunner's sight and without infra-red/white light searchlight

Prototype of SK 105/A3 light tank with new 105 mm gun fitted with muzzle brake and fume extractor and showing extensive modifications to turret (Franz Kosar)

Greif armoured recovery vehicle in travelling configuration and fitted with 12.7 mm M2 HB machine gun

The torsion bar suspension consists of five dual rubber-tyred road wheels with the drive sprocket at the rear and the idler at the front. There are three track return rollers and the first and last road wheel stations have hydraulic shock absorbers. The vehicle has tracks with rubber blocks, each track having 78 links. Steel spikes can be fitted for operations in ice and snow.

The SK 105 has individual NBC protection by means of face masks and a diesel oil heater is fitted as standard on all models. The SK 105 is not amphibious but can ford to a depth of 1 m.

Current production models have a 105 mm gun designated the 105G1 which fires the following types of spin-stabilised fixed ammunition: HE with complete round weighing 18.5 kg with a muzzle velocity of 700 m/s; HEAT with complete round weighing 17.3 kg and a muzzle velocity of 800 m/s and a maximum effective range of 2700 m which will penetrate 360 mm of armour at an incidence of 0° or 150 mm of armour at an incidence of 65°; and smoke with with complete round weighing 19.1 kg and a muzzle velocity of 695 m/s.

The 105 mm 105G1 enables APFSDS rounds to be fired which have a much greater armour-piercing capability than the HEAT projectile which is at present the main tank-killing round.

The modifications to the SK 105 include a single-stage muzzle brake, lower extractor, ammunition loading chute, six-round revolving magazines in the rear of the turret and the ammunition stowage racks in the hull and turret basket as APFSDS projectiles are longer. As the projectiles are identical to those of the AMX-30 MBT, the SK 105/A1 therefore has the same kill probability.

The APFSDS round has been developed by GIAT of France and is designated the OFL 105 G1.

The OFL 105 G1 has a muzzle velocity of 1460 m/s and will penetrate a NATO triple target at a range of 1000 m and a NATO heavy single target at a range of 1200 m. The projectile itself weighs 3.14 kg of which the penetrator accounts for 1.84 kg.

The 105 mm gun is fed from two revolver type magazines in the turret bustle, one on either side and each holding six rounds of ammunition. Once the gun is fired the empty cartridge cases are ejected out of the rear of the turret through a trapdoor hinged on the left. A rate of fire of one round every five seconds can be achieved until the two magazines are empty. Once empty the magazines have to be refilled by hand from outside the tank. A gun barrel travelling lock is fitted at the front of the hull and this folds back onto the glacis plate when not in use.

Mounted coaxially to the right of the main armament is a 7.62 mm MG 74 (Steyr) machine gun and mounted on either side of the turret are three smoke dischargers. A cupola-mounted 7.62 mm machine gun for close-in and anti-aircraft defence, positioned in front of the commander's cupola, can be fitted as an option.

Steyr SK 105/A1 Light Tank

The SK 105/A1 is a further development of the SK 105 and differs from them in modifications of the power train and the fire control equipment.

The manual shift gear is replaced by an automatic six-speed transmission with torque converter and lock-up clutch.

The fire control equipment consists of an electronic fire control system including a digital fire control computer and the respective observation and sighting system. The fire control system enables the gunner of the SK 105/A1 to independently observe, range and hit the target by an automated simple firing process. The system allows for immediate growth capabilities, such as introducing and storing parameters for additional types of ammunition.

The major components are: digital fire control computer unit, including commander's operating panel and cant angle sensor; single reticle gunner's sight, including laser rangefinder, day elbow with magnification ×8, passive night elbow with magnification ×7.2 and interchangeable with a daylight unity prism, moving-view rotating wedges module in elevation and deflection and periscopic mirror head, and gunner's control unit. In order to compute the compensation angle in elevation and deflection, the computer is fed with data from several sources, according to the type of firing: range data from the laser rangefinder, side cant of the tank from the built-in cant angle sensor, target velocity from angular velocity, and displacement sensor for tracking moving targets. The computer enables automatic computed firing corrections for firing in conformance with the type of ammunition and target range, in order to improve second-hit probability.

The commander is provided with a passive night sight, interchangeable with the daysight. In addition, the semi-automatic ammunition loading system for the gun can be upgraded to a fully automatic system, operated by means of selection switches by both commander and gunner.

Steyr SK 105/A2 Light Tank

In 1981 Steyr-Daimler-Puch completed the prototype of the Kürassier SK 105/A2, an improved version of the basic SK 105, which has a fully stabilised turret oscillating part including the turret crew seats, an electrical laying system, digital fire control computer (DFCC) and a fully automatic ammunition loading system.

The DFCC processes the following automatic input data: cant (from sensor), range (from laser rangefinder) as well as the manual input data for cross wind, atmospheric density (altitude above sea level) and barrel wear. The ballistic data of four types of ammunition can be stored and calculated. The gunner's sight with integrated laser rangefinder together with the DFCC and armament stabilisation system provides high first hit probability.

An integrated passive night sight for the gunner and a changeable passive night sight for the commander ensure full night-fighting capability.

The redesigned turret front gives upgraded ballistic protection. The SK 105/A1 and SK 105A/2 both retain the French 105 mm 105G1 guns. The SK 105A2 has an all-electric gun control system.

Steyr SK 105/A3 Light Tank

Steyr-Daimler-Puch has completed the prototype of the SK 105/A3 light tank which has a number of improvements including replacement of the 105 mm gun by a 105 mm M68 tank gun fitted with a new muzzle brake, thermal sleeve, fume extractor and modified recoil system giving a total recoil length of 650 mm. This is fitted into a new turret of the oscillating type which is claimed to have the same level of protection over its front as the US M60 MBT. In the turret bustle are two revolver-type magazines, each of which hold six rounds of ready-use ammunition. The drums are now driven electrically with the original manual system being retained for emergency use. An electrically driven rammer is also provided. Turret traverse and weapon elevation is electric with manual controls for emergency use and the commander can override the gunner. The upper part of the oscillating turret is stabilised.

The DFCC is the same as that installed in the SK 105/A2 while the engine is the more powerful 9FA developing 360 hp at 2300 rpm coupled to a six-speed transmission from ZF.

This version has a combat weight of 20 700 kg and is 8.729 m long, 2.50 m wide and 2.76 m high to gunner's sight. This was a trials vehicle to test installation of 105 mm M68 gun in a light tank. The SK 105/A3 has yet to enter production and is essentially a trials vehicle.

Upgrade Package

Steyr-Daimler-Puch are now marketing an upgrade package to enable existing SK 105 users to upgrade their vehicles to the SK 105/A1 configuration. These modifications include moving the position of the searchlight from the turret top to the left side of the 105 mm gun, replacement of coaxial machine gun by a 7.62 mm FM MAG weapon, additional armour protection over the frontal arc (20° to the longitudinal axis) to provide protection against APDS rounds up to 35 mm, modification of weapon to fire 105 mm APFSDS rounds, fire control computer and gunner's sight with integrated laser rangefinder, hydraulic bump stops for improved cross country mobility, fully automatic loading system, 250 Ah dry batteries, fire extinguishing system using Halon 1211, fully automatic ZF 6HP600 transmission with torque converter and lock up clutch. For desert operations improvements include sand shields, improved cooling system and cyclone type air cleaner. The night fighting capability can be improved by image intensification devices for commander, gunner and driver.

Variants

Greif Armoured Recovery Vehicle

The first prototype of the Greif ARV was completed in 1974 with production beginning in 1976/77. The chassis is identical to the SK 105 but a new superstructure has been built at the front of the vehicle to the right of which is mounted a hydraulic crane with a maximum lifting capacity of 6000 kg. The crane is provided with 42 m of cable and its jib can be extended from its normal length of 3 m to 3.9 m. The main winch, with a capacity of 20 000 kg, is mounted in the lower part of the hull and is led out through the front. Mounted at the front of the hull is a hydraulically operated blade which can be used for dozing or stabilising the vehicle when the winch is being used. Full designation of this vehicle is the Armoured Recovery Vehicle 4K-7FA, SB 20 (Bergepanzer Greif). Loaded weight is 19 800 kg and it has a crew of four: driver, commander and two mechanics. Full details, specifications and photographs of the Greif ARV are given in *Jane's Military Vehicles and Logistics 1991-92*, page 17.

Pionier Vehicle/Engineer Vehicle

The Pionier vehicle is designated the Engineer Tank 4KH7FA - AVE and has the same hull and chassis as the Greif ARV. It has a new winch with a capacity of 8000 kg and is fitted with a larger dozer blade at the front of the hull. On the right side of the hull, where the crane on the

ARV is situated, is a hydraulically operated excavator. This vehicle is in production for the Austrian Army. Loaded weight is 19 000 kg and it has a crew of four. Full details are given in *Jane's Military Vehicles and Logistics 1991-92*, page 1.

Driver Training Vehicle

This is called the 4KH7FA - FA (Fahrschulpanzer) and is based on the chassis of the SK 105. Each SK 105 is convertible into a driver training vehicle within two hours.

Engineer Tank 4KH7FA - AVE in operating configuration with dozer blade lowered and hydraulic arm with bucket traversed to front of vehicle

SPECIFICATIONS (SK 105/A1)

CREW	3
COMBAT WEIGHT	17 700 kg
POWER-TO-WEIGHT RATIO	18.1 hp/tonne
GROUND PRESSURE	0.67 kg/cm^2
LENGTH GUN FORWARDS	7.735 m
LENGTH HULL	5.582 m
WIDTH	2.5 m
WIDTH OVER TRACKS	2.5 m
HEIGHT	
to top of commander's cupola	2.529 m
to hull top	1.413 m
FIRING HEIGHT	1.965 m
GROUND CLEARANCE	0.4 m
TRACK	2.12 m
TRACK WIDTH	380 mm
LENGTH OF TRACK ON GROUND	3.037 m
MAX SPEED	
forward	70 km/h
reverse	7.36 km/h
FUEL CAPACITY	400 litres
MAX RANGE	520 km
FORDING	1 m
GRADIENT	75%
SIDE SLOPE	40%
VERTICAL OBSTACLE	0.8 m
TRENCH	2.41 m

TURNING RADIUS	8.5 m in 1st gear
ENGINE	Steyr 7FA 6-cylinder liquid-cooled 4-stroke turbo-charged diesel developing 320 hp at 2300 rpm
AUXILIARY ENGINE	none
TRANSMISSION	ZF automatic with 6 forward and 1 reverse gears
STEERING	hydrostatic split torque type
SUSPENSION	torsion bar
ELECTRICAL SYSTEM	24 V
BATTERIES	2 × 12 V, 180 Ah
ARMAMENT	
main	1 × 105 mm gun
coaxial	1 × 7.62 mm MG
cupola (optional)	1 × 7.62 mm MG
SMOKE-LAYING EQUIPMENT	3 smoke dischargers either side of turret
AMMUNITION	
main	42
coaxial	2000

GUN CONTROL EQUIPMENT	
Turret power control	hydraulic/manual
by commander	yes
by gunner	yes
Max rate power traverse	360° in 12-15 s
Gun elevation/depression	+12°/-8°
Max power elevation depression	4.5°-5°/s
commander's override	yes
Commander's fire control override	yes
Gun stabiliser	
vertical	no
horizontal	no
ARMOUR	
Hull front	20 mm
Hull sides	14 mm
Hull top	8 mm
Hull rear	12 mm
Turret front	40 mm
Turret sides	20 mm
Turret top	10 mm

Note: SK 105/A2 is similar except for combat weight 18 600 kg, fuel capacity 420 litres, cruising range 500 km, length 7.698 m, height to top of sight 2.561 m, ammunition is 38 × 105 mm and 1600 × 7.62 mm.

Manufacturer: Steyr-Daimler-Puch Spezialfahrzeug GmbH
P.O. Box 100
A-1111 Vienna
Austria
Telephone: (222) 76 45 11
Telex: 613221299

Status: Production as required. In service with:

Country	**Quantity**	**Comment**
Argentina	150	plus 10 ARV
Austria	250	plus 50 ARV and small number of AEV
Bolivia	34	plus 2 ARV
Morocco	109	plus 11 ARV, some SK 105 have been lost in fighting
Tunisia	54	

BRAZIL

Bernardini X1A1 Light Tank

Development

In the early 1970s the Bernardini company of Sao Paulo rebuilt two M3A1 light tanks for the Brazilian Army. Technical control of the project was under the direction of the Brazilian Army Research and Development Centre. The modifications included replacing the armour above the tracks by new sloped armour provided by the Biselli company, replacement of the original American petrol engine by a six-cylinder Saab-Scania diesel developing 280 bhp, new volute suspension designed by the Novatracao company, new turret with a French DEFA D-921A 90 F1 90 mm gun as fitted to the Panhard AML (4 × 4) and other light AFVs, and a fire control system designed by the D F Vasconcelos company.

Following trials with the two prototypes the Brazilian Army placed an order for 100 M3A1 Stuart light tanks to be rebuilt to the new standard, which were subsequently delivered to the Cavalry Regiments under the designation X1; final deliveries were made in 1978. The vehicle weighs 15 000 kg, has a road range of 450 km and can ford to a depth of 1 m without preparation.

The X1 was followed by the Carcara, or X1A1. This is essentially a stretched X1 with an additional volute spring suspension group, additional return wheel and a new tension wheel each side. Engine and turret are the same as fitted to the earlier X1 tank.

Trials with the X1A1 were completed in 1977/78 but the Brazilian Army did not place an order for the vehicle and further development resulted in the X1A2 which is based on a new chassis rather than a rebuild of an M3A1 chassis. The X1A1 is being offered for export.

Paraguayan M3/M3A1

In 1979-1980 the company rebuilt a batch of M3/M3A1s for the Paraguayan Army. These vehicles retained their 37 mm guns and turrets but had uprated powerpacks and suspensions which raised their weight by 2000 kg and range to 600 km.

XLP-10 AVLB

Following trials with prototype vehicles based on the X1A chassis, the Brazilian Army ordered 10 production models based on the X1A1 chassis. The aluminium and steel bridge is launched over the front of the vehicle, weighs 2750 kg, can span a gap of up to 10 m and take a maximum load of 20 000 kg. Full details are given in *Jane's Military Vehicles and Logistics 1991-92*, page 76.

XLF-40 Rocket Launcher

This is based on the X1A1 chassis and carries three ready-to-launch XLF-40 surface-to-surface rockets. Details are given in the *Multiple rocket launchers* section. It has not yet entered production.

Anti-aircraft Vehicle

A prototype of a self-propelled anti-aircraft gun system based on the light tank was built and first tested early in 1984. This has the Saab-Scania diesel engine developing 215 hp but retains the original M3A1 suspension. Mounted on top of the hull is the Brazilian upgraded M55 quad 12.7 mm anti-aircraft gun system.

120 mm Mortar Carrier

A prototype of a 120 mm mortar carrier was built and first tested early in 1984. This has a new hull with a six-cylinder Mercedes-Benz diesel engine developing 180 hp located in the forward part. This vehicle has a crew of three who enter the vehicle through a ramp in the rear. Brief specifications are: weight 9000 kg, length 4.6 m, width 2.24 m, height 2.22 m, ground clearance 0.45 m, maximum road speed 55 km/h, gradient 60 per cent and ground pressure 0.94 kg/cm². A pedestal-mounted 12.7 mm M2 HB machine gun is fitted for local protection.

Recovery Vehicle

A prototype of an armoured recovery vehicle was built and first tested early in 1984. This also has a forward-mounted 180 hp Mercedes-Benz diesel engine but uses redesigned volute suspension assemblies and wider tracks, as on the X1A2 light tank. Mounted at the rear of the hull, one either side of the twin opening doors, is a spade. When lowered these provide a more stable recovery platform. Mounted internally is a hydraulic winch for recovery operations and on top of the roof on the left side is a hydraulic crane. Brief specifications are: weight 10 000 kg, length 5.22 m, width 2.48 m, height 2.52 m, maximum road speed 55 km/h, gradient 60 per cent, ground pressure 0.68 kg/cm² and ground clearance 0.45 m.

SPECIFICATIONS (X1A1)

CREW	4
WEIGHT	17 000 kg
GROUND PRESSURE	0.55 kg/cm²
LENGTH GUN FORWARDS	6.36 m
LENGTH HULL	5.3 m
WIDTH	2.4 m
HEIGHT	2.45 m
to hull top	1.63 m
GROUND CLEARANCE	0.5 m
MAX SPEED	60 km/h
FUEL CONSUMPTION	1.7 km/litre
RANGE ROAD	520 km
FORDING	1.3 m
GRADIENT	60%
VERTICAL OBSTACLE	0.8 m
ENGINE	Saab-Scania diesel developing 280 hp
ARMAMENT	
main	1 × 90 mm gun
coaxial	1 × 7.62 mm MG
anti-aircraft	1 × 12.7 mm MG

Status: X1 in service with the Brazilian Army. X1A1 production as required for export only.

Manufacturer: Bernardini S/A Indústria e Comércio
Rua Hipólito Soares No 79
04201 Sao Paulo
SP, Brazil
Telephone: (011) 274 8033
Telex: (011) 21605 bsaibr

X1 light tank, essentially rebuilt M3A1 Stuart

Prototype of XLP-10 AVLB in travelling configuration (Ronaldo S Olive)

Bernardini X1A2 Light Tank

Development

Following development of the X1 light tank, essentially a modernised M3A1 Stuart light tank, Bernardini rebuilt two vehicles to a new standard known as the X1A1 (military designation XMB-1) which is basically an X1 with an additional volute spring suspension group, additional return wheel and a new tension wheel either side. Engine and turret are the same as the X1A's.

The Brazilian Army did not adopt this model but development continued and resulted in the X1A2 which is similar to the X1A1 but is based on a new chassis instead of a rebuilt M3A1 Stuart.

Between 1979 and 1983 50 X1A2s (military designation MB-2) were built for the Brazilian Army. If further orders are placed the X1A2 can be put back into production. A version with an automatic transmission was proposed called the X1A3, but this did not reach the prototype stage.

Description

The hull of the X1A2 is all-welded steel with the driver's compartment at the front, fighting compartment in the centre and the engine at the rear.

The driver is seated at the front of the hull on the left and is provided with a single piece hatch cover that opens to the left. Mounted in front of the hatch cover are three periscopes for observation when the hatch is closed.

The all-welded steel turret is in the centre of the hull with the commander, who also acts as the loader, on the left and the gunner on the right. Both are provided with a single piece hatch that opens to the rear. The commander's cupola has five observation periscopes and the gunner has three and a telescope linked to the main armament.

The vertical volute suspension has three bogies each side each with two road wheels, with the drive sprocket at the front, idler at the rear and two track return rollers.

Optional equipment includes a laser rangefinder for the main armament, full range of night vision equipment and an air-conditioning system.

Prototype X1A2s had a French D921 gun as fitted to the Panhard AML-90 (4 × 4) armoured car and early production ENGESA (6 × 6) Cascavel armoured cars but production vehicles have a 90 mm Cockerill gun made in Brazil by ENGESA which is also now fitted to current production Cascavel armoured cars. A 7.62 mm machine gun is mounted coaxially with the main armament and a 12.7 mm machine gun is mounted externally at the commander's position for anti-aircraft use. Three electrically operated smoke dischargers are mounted either side of the turret at the front.

Status: Production complete but can be resumed if additional orders are received. In service with the Brazilian Army.

Manufacturer: Bernardini S/A Indústria e Comércio
Rua Hipólito Soares No 79
04201 Sao Paulo
SP, Brazil
Telephone: (011) 274 8033
Telex: (011) 21605 bsaibr

X1A2 light tank fitted with ENGESA-built 90 mm Cockerill gun and roof-mounted 12.7 mm M2 HB anti-aircraft machine gun

SPECIFICATIONS	
CREW	3
COMBAT WEIGHT	19 000 kg
POWER-TO-WEIGHT RATIO	15.78 hp/tonne
GROUND PRESSURE	0.63 kg/cm²
LENGTH GUN FORWARDS	7.1 m
LENGTH HULL	6.5 m
WIDTH	2.6 m
HEIGHT (to turret top)	2.45 m
GROUND CLEARANCE	0.5 m
TRACK WIDTH	460 mm
MAX SPEED	
road, forwards	55 km/h
road, reverse	15 km/h
cross-country	30 km/h
FUEL CAPACITY	600 litres
MAX RANGE	
road	600 km
cross-country	350 km
FUEL CONSUMPTION	1 litre/km
FORDING	1.3 m
GRADIENT	70%
SIDE SLOPE	30%
VERTICAL OBSTACLE	0.7 m
TRENCH	2.1 m
TURNING RADIUS	7 m
ENGINE	Saab-Scania model DS-11 6-cylinder water-cooled turbo-charged diesel developing 300 hp at 2200 rpm
TRANSMISSION	manual with 2 forward and 1 reverse gears
SUSPENSION	controlled differential, manual operated by hydraulic system
SUSPENSION	vertical volute
ELECTRICAL SYSTEM	24 V
BATTERIES	2 × 12 V
ARMAMENT	
main	1 × 90 mm gun
coaxial	1 × 7.62 mm MG
anti-aircraft	1 × 12.7 mm MG
SMOKE-LAYING EQUIPMENT	3 smoke dischargers either side of turret
AMMUNITION	
main	66
coaxial	2500
anti-aircraft	750
GUN CONTROL EQUIPMENT	
Turret power control	hydraulic/manual
by commander	manual
by gunner	hydraulic
Max rate power traverse	360° in 12 s
Gun elevation/depression	+17°/−8°
Gun stabiliser	
vertical	no
horizontal	no

CHINA, PEOPLE'S REPUBLIC

NORINCO Type 63 Light Amphibious Tank

Development

In the late 1950s China received several batches of Soviet PT-76 light amphibious tanks. There is a possibility that this tank was built in China in a slightly modified version under the designation Type 60, as a number of Chinese PT-76 tanks have been observed with provision for mounting a 12.7 mm anti-aircraft machine gun on one of two pintle type mounts on the turret roof.

This was followed by the redesigned version of the Type 60 known as the Type 63 which used automotive components of the Type 77 series APC which is the Chinese version of the Soviet BTR-50PK vehicle. The Type 63 was used in combat in Vietnam by the North Vietnamese Army and during the 1971 Indo-Pakistan war by Pakistan.

The Type 63 light amphibious tank is heavier than the Soviet PT-76 but has a greater road speed and power-to-weight ratio because of its more powerful engine.

In addition to being used by the Chinese Army the Type 63 is also used by the Chinese Marines who come under naval command.

Description

The hull of the Type 63 is welded rolled steel and is divided into three compartments: driver's at the front, fighting in the centre and engine at the rear. The glacis plate of the Type 63 is much shallower than on the Type 60 or PT-76. The driver is seated at the front of the hull on the left and is provided with a single piece hatch cover that swings to the left and three periscopes forward. There is a single infra-red driving light on the right side of the hull front.

The turret is in cast steel sections welded together, with the commander and gunner seated on the left of the turret and the loader on the right. The commander's hatch opens forward and the loader's to the rear. A dome-shaped ventilator is mounted in the turret roof to the rear of the commander's and gunner's hatches. The engine and transmission are at the rear of the hull and additional fuel tanks can be fitted on top of the hull to increase operational range.

The torsion bar suspension consists of six large rubber-tyred road wheels similar to those on the Soviet PT-76, with the drive sprocket at the rear and the idler at the front. There are no track return rollers.

Main armament of the Type 63 is an 85 mm gun, probably identical to that installed in the Type 62 light tank, which fires ammunition of the fixed type including AP, APHE, HE, HEAT and smoke.

Some Type 63 light amphibious have been fitted with a Chinese-designed laser rangefinder over the rear part of the 85 mm gun. This is identical to that on some Chinese built Type 59 MBTs.

A 7.62 mm machine gun is mounted coaxially to the right of the main armament and a Type 54 heavy machine gun is mounted at the loader's station for anti-aircraft defence. The Type 54 is the Soviet M1938/46 DShKM manufactured in China.

The tank is fully amphibious being propelled in the water by two water jets of Soviet design. Before entering the water a trim board is erected at the front of the hull; when travelling this folds back onto the glacis plate. As far as it is known the Type 63 has no NBC system or night vision equipment for the commander and gunner.

NORINCO Type 63 light amphibious tank with laser rangefinder over 85 mm gun and crewman aiming type HN-5 surface-to-air missile

NORINCO Type 63 light amphibious tank with trim vane folded back onto glacis plate

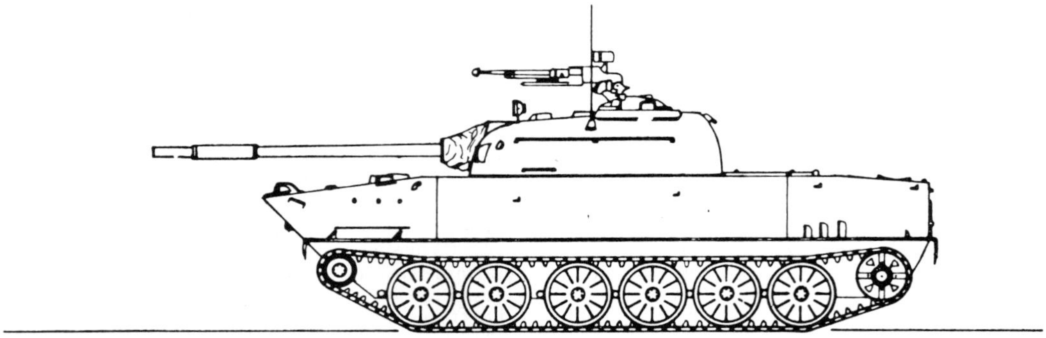

NORINCO Type 63 light amphibious tank (provisional drawing)

Status: Probably still in production. In service with:

Country	Comment	Quantity
China	1200	estimate
Korea, North	600	estimate
Vietnam	150	estimate

Manufacturer: Chinese state factories.
Enquiries to China North Industries Corporation
7A, Yuetan Nanjie
PO Box 2137
Beijing
People's Republic of China
Telephone: (86) 6898/3461/3471/7570
Telex: 22339 cnin cn

NORINCO Type 63 light amphibious tank showing amphibious capabilities

SPECIFICATIONS					
CREW	4	MAX SPEED		ARMAMENT	
COMBAT WEIGHT	18 700 kg	road	64 km/h	main	1 × 85 mm gun
POWER-TO-WEIGHT RATIO	21.39 hp/tonne	water	12 km/h	coaxial	1 × 7.62 mm MG
LENGTH GUN FORWARDS	8.437 m	FUEL CAPACITY	545 litres	anti-aircraft	1 × 12.7 mm MG
LENGTH HULL	7.15 m	MAX RANGE	370 km	AMMUNITION	
WIDTH	3.2 m	FORDING	amphibious	main	47
HEIGHT TO TURRET TOP		GRADIENT	60%	12.7 mm	500
excluding AA MG	2.522 m	VERTICAL OBSTACLE	0.87 m	7.62 mm	1000
including AA MG	3.122 m	TRENCH	2.9 m	TURRET TRAVERSE	360°
GROUND CLEARANCE	0.4 m	ENGINE	Model 12150-L 12-cylinder water-cooled diesel developing 400 hp at 2000 rpm	GUN ELEVATION/ DEPRESSION	+18°/-5°
TRACK	2.82 m	TRANSMISSION	manual, 5 forward and 1 reverse gears	ARMOUR	
LENGTH OF TRACK ON GROUND	4.44 m	ELECTRICAL SYSTEM	24 V	Hull front (max)	11 mm at 80°
				Hull sides upper	14 mm
				Hull top	10 mm
				Hull floor	10 mm
				Hull rear	10 mm
				Turret mantlet	11 mm

NORINCO Type 62 Light Tank

Description

The Type 62 light tank is virtually a scaled down Type 59 MBT. It is believed that the tank is used in place of the Type 59 in rugged terrain such as that encountered in southern China. The layout of the tank is identical to the Type 59 with the driver seated at the front of the hull on the left and the other three crew members in the turret. The commander and gunner are seated on the left of the turret with the loader on the right. The engine and transmission are at the rear of the hull. The torsion bar suspension consists of five road wheels with a distinct gap between the first and second wheels. The drive sprocket is at the rear and the idler at the front. Main armament is an 85 mm gun, probably identical to that on the Type 63 light amphibious tank, which fires fixed ammunition including AP, APHE, HE, HEAT and smoke.

A 7.62 mm machine gun is mounted coaxially to the right of the main armament and a 12.7 mm Type 54 heavy machine gun is mounted at the loader's position for anti-aircraft use. The Type 54 is the Soviet M1938/46 DShKM manufactured in China.

The Type 62 is not amphibious, has no NBC system and has not been observed with any infra-red night vision equipment.

The Type 62 was used in combat in Angola by Zaïrean forces and by Tanzania against Uganda during their 1978 October-November border war.

In addition to being used by the Chinese Army the Type 62 is also used by the Chinese Marines who come under naval command.

NORINCO Type 62 light tank

SPECIFICATIONS

CREW	4	LENGTH OF TRACK ON GROUND	3.53 m	SUSPENSION	torsion bar
COMBAT WEIGHT	21 000 kg	MAX ROAD SPEED	60 km/h	ELECTRICAL SYSTEM	24 V
POWER-TO-WEIGHT RATIO	20.47 hp/tonne	FUEL CAPACITY	730 litres	ARMAMENT	
GROUND PRESSURE	0.71 kg/cm²	MAX ROAD RANGE	500 km	main	1 × 85 mm gun
LENGTH		FORDING	1.3 m	coaxial	1 × 7.62 mm MG
gun forwards	7.9 m	GRADIENT	60%	bow	1 × 7.62 mm MG
hull	5.55 m	VERTICAL OBSTACLE	0.7 m	anti-aircraft	1 × 12.7 mm MG
WIDTH	2.86 m	TRENCH	2.55 m	AMMUNITION	
HEIGHT	2.25 m	ENGINE	liquid-cooled diesel developing 430 hp at 1800 rpm	main	47
GROUND CLEARANCE	0.42 m			coaxial	1750
TRACK	2.39 m	TRANSMISSION	manual, 4 forward and 1 reverse gears	anti-aircraft	1250
TRACK WIDTH	380 mm			TURRET TRAVERSE	360°
				GUN ELEVATION/ DEPRESSION	+20°/−4°

Status: Probably still in production. In service with:

Country	Quantity	Comment
Albania	not known	
China	800	estimate
Congo	14	
Korea, North	50	
Mali	18	
Sudan	70	
Tanzania	30	
Zaïre	50	

Manufacturer: Chinese state factories.
Enquiries to China North Industries Corporation
7A, Yuetan Nanjie
PO Box 2137
Beijing
People's Republic of China
Telephone: (86) 6898/3461/3471/7570
Telex: 22339 cnin cn

DENMARK

M41 DK-1 Light Tank

Development

The Danish Army took delivery of a quantity of M41 light tanks from the USA in the 1950s and in the mid 1980s decided to upgrade these to extend their lives into the 1990s.

After trials with prototype vehicles repowered by Falck Schmidt of Denmark, NAPCO International of the USA and FFG of Germany, the former powerpack was selected by the Danish Army.

In addition to the installation of a new diesel powerpack which increases the range of action as well as being more reliable, many other modifications were incorporated to improve its fighting capability and battlefield survivability.

A total of 53 M41 light tanks were upgraded to the new standard, called the M41 DK-1, at a cost of DKr 340 million (£31 million) with final deliveries being made in 1988.

The M41 DK-1 package is now being offered for export by the DISA Systems Group.

Description

The overall layout of the M41 vehicle has not been changed but some 70 improvements have been carried out. These include the replacement of the original petrol engine by a more fuel effecient Cummins VTA-903T diesel developing 465 hp (similar to that in the Bradley), installation of four forward firing smoke dischargers either side of the turret (as already fitted to the Danish Leopard 1), installation of an NBC system in the turret bustle, fire detection and suppression system, new battery installation, improved internal layout, redesigned stowage rack and bins, heater, new DISA 7.62 mm anti-aircraft machine gun mount on turret roof in place of the 12.7 mm M2 MG, installation of side skirts similar to those on the Leopard 1, new AEG electric gun controls, thermal night vision system for gunner with integrated laser rangefinder, Halogen searchlight mounted on gun mantlet, night vision periscope for the driver and night vision goggles for the tank commander. A scavenger system has been fitted for the 76 mm gun breech and 7.62 mm coaxial machine gun.

Prior to the completion of the modernisation programme, the Danish Army had already taken delivery of a new AAI developed 76 mm APFSDS-T round which has a significant increase in armour penetration characteristics over the existing ammunition.

SPECIFICATIONS
(as per US M41 with following exceptions)

WEIGHT	25 000 kg
POWER-TO-WEIGHT RATIO	18.6 hp/tonne
GROUND PRESSURE	0.8 kg/cm²
GROUND CLEARANCE	
front	0.46 m
rear	0.43 m
MAX ROAD SPEED	74 km/h
FUEL CAPACITY	930 litres
CRUISING RANGE	750 km
VERTICAL OBSTACLE	0.60 m
FORDING	1 m
ENGINE	Cummins VTA-903T diesel developing 465 hp at 2800 rpm

M41 DK-1 from side clearly showing new skirts, roof-mounted 7.62 mm DISA machine gun mount and extra stowage bins with bank of four electrically operated smoke dischargers on turret side

The M41 DK-1 retains the 76 mm gun of the basic M41 but fires a new AAI developed APFSDS-T round with increased armour penetration characteristics

Status: Conversion work complete. In service with Danish Army. Now being offered for export by the DISA Systems Group.

Prime contractor for exports: DISA Systems Group
9, Stationsvej
DK-3550 Slangerup
Denmark
Telephone: (45) 42337565
Telex: 42527 DISSPD
Fax: (45) 42333796

FRANCE

Creusot-Loire Industrie AMX-13 Light Tank

Development

Design work on the AMX-13 began in 1946 at the Atelier de Construction d'Issy-les-Moulineaux and the first prototype was completed two years later. Production was undertaken at the Atelier de Construction Roanne (ARE) from 1952 with the first production tanks being completed the following year. At one time production of the tank was running at 45 units per month but in the early 1960s production of the whole family was transferred to the civilian company of Creusot-Loire (now Creusot-Loire Industrie) at Châlon sur Saône as the ARE was tooling up for production of the AMX-30 MBT. The basic chassis has been used for a wide range of vehicles including the AMX VCI mechanised infantry combat vehicle, 105 mm self-propelled howitzer, 155 mm self-propelled gun Mk F3 and twin 30 mm DCA self-propelled anti-aircraft gun system.

Without changing the basic design, many improvements were announced at the 1985 Satory Exhibition of Military Equipment. New production AMX-13 vehicles are produced as standard with the new hull front, new diesel engine, automatic transmission and hydro-pneumatic suspension. Creusot-Loire Industrie can provide kits to upgrade existing in service vehicles to this new standard. Total production of the AMX-13 family of light tracked vehicles, including the light tanks, amounted to 7700 of which about 3400 were exported. In March 1990 Creusot-Loire Industrie unveiled their new MARS 15 series of light full tracked vehicles which includes a light tank. Details of this family, seen as the succesor to the AMX-13 family, are given in the *Armoured Fighting Vehicles Families* section later in this book.

Description (Basic Vehicle)

The hull of the AMX-13 is all-welded steel and is divided into three compartments with the driver's and engine compartment at the front and the turret mounted at the rear. The driver is seated at the front of the vehicle on the left side with the engine compartment to his right and the differential in front of him. The driver is provided with a single piece hatch cover that opens to the left and three periscopes, the centre one of which can be replaced by an infra-red or image-intensification periscope for night driving.

The oscillating turret, manufactured by Fives-Cail Babcock, is mounted at the rear of the hull and the type of turret depends on the model of the tank and its armament. All turrets are oscillating. In all models the commander is seated on the left of the turret and the gunner on the right. The tank commander is provided with a domed hatch cover that opens to the rear and eight periscopes. The gunner has a single piece hatch cover that opens to the rear and two periscopes.

The torsion bar suspension consists of five rubber-tyred road wheels with the drive sprocket at the front and the idler at the rear. Trials have been completed with the torsion bars replaced by Dunlop or Messier hydro-pneumatic suspension units. There are three (or in some cases two) track return rollers which support the inside of the track only. The first and last road wheel stations have hydraulic shock absorbers. The steel tracks have 85 links per side and can be fitted with rubber pads if required.

The AMX-13 does not have an NBC system, cannot be fitted for deep wading and as built was not fitted with any night fighting equipment although a number of armies have fitted their vehicles with some: for example, a Fives-Cail Babcock infra-red searchlight to the rear of the gunner's position and an infra-red sight for the gunner.

More recently, the AMX-13 has been offered fitted with passive or thermal night firing and night driving equipment, laser rangefinder and automatic display of the battle sight. Current production AMX-13s are produced with these improvements.

Variants

Model 51 with 75 mm Gun

This was the first model of the AMX-13 to enter service and is fitted with the FL-10 turret armed with a 75 mm gun with a single baffle muzzle brake. The weapon is fed from two revolver type magazines in the turret bustle, one either side, each magazine holding six rounds. Once the gun has fired, the empty cartridge case is ejected out of the rear of the turret through a trapdoor hinged on the left. A rate of fire of one round every five seconds can be achieved until the two magazines are empty. Once empty, the magazines have to be refilled by hand from outside the tank. The gun fires fixed APC and HE rounds.

A 7.5 mm or 7.62 mm machine gun is mounted coaxially to the right of the main armament and a similar weapon is often mounted externally at the commander's position. Thirty-seven rounds of 75 mm and 3600 rounds (in belts of 200) of machine gun ammunition are carried. In addition there are two electrically operated smoke dischargers either side of the turret.

AMX-13 with 90 mm Gun

This entered service in the early 1960s and is armed with a 90 mm gun in an FL-10 turret and the barrel is fitted with a thermal sleeve and a

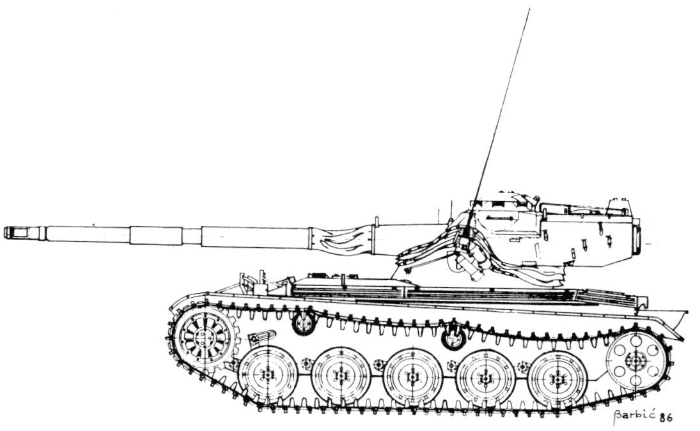

Early hull-type AMX-13 armed with 90 mm gun (not to 1/76th scale) (Vasco Barbic)

AMX-13 Model 51 with 75 mm gun showing cartridge case ejection door in turret bustle (ECP Armées)

AMX-13 armed with 90 mm gun (T J Gander)

single baffle muzzle brake. The gun fires fixed ammunition including APFSDS, canister, HE, HEAT and smoke.

The turret also has a coaxial machine gun and a similar weapon can be mounted at the commander's position. Thirty-two rounds of 90 mm ammunition are carried of which 21 are in the turret (12 in the magazines) and 11 in the hull, plus 3600 rounds of machine gun ammunition.

Existing 75 mm armed AMX-13s can be retrofitted with the 90 mm gun and this was carried out by the French Army from 1970.

AMX-13 with 105 mm Gun

This was developed specifically for export and has an FL-12 turret armed with a 105 mm gun which fires the same non-rotating rounds as the AMX-30 MBT but with a smaller and therefore lighter propellant. This turret is also fitted to the Austrian SK 105 used by Argentina, Austria, Bolivia, Morocco and Tunisia.

AMX-13 with FL-15 Turret

The AMX-13 is now manufactured by Creusot-Loire Industrie with the FL-15 turret from Fives-Cail Babcock. This is very similar to the older FL-12 but observation equipment includes seven M554 periscopes and a pair of OB44 night observation binoculars for tank commander and two M556 periscopes for the gunner.

Sighting equipment consists of M212 periscope/telescope with magnifications of ×1.6 and ×6.5 for the tank commander and a telescope with magnifications of ×8 (day) and ×6 (night) for the gunner. A fire control system, for example the SOPELEM SOPTAC 18-03 or SOPTAC 18-04, which incorporates a laser rangefinder is proposed for this turret. This will facilitate target acquisition, reduce target engagement time and increase first round hit probability.

Turret traverse is 360° and the top part of the turret elevates from −8° to +12° by the commander or gunner with the former having overriding control.

Types of ammunition fired by the 105 mm gun include GIAT Industrie APFSDS, HEAT-T, HE, smoke, illuminating, HEAT (training) and HE (training).

75 mm Gun in FL-11 Turret

This model of the AMX-13 was developed in the 1950s for use in North Africa and has an FL-11 turret, also fitted to some models of the Panhard EBR armoured car, equipped with a manually loaded 75 mm gun. This model can be distinguished from the basic AMX-13 as its turret has no overhang at the rear.

Model 51 with 75 mm Gun and SS-11 Missiles

This is the standard model with two Aerospatiale SS-11 wire-guided anti-tank missiles mounted each side of the main armament. This missile weighs 30 kg, has a minimum range of 350 m and a maximum range of 3000 m and its warhead will penetrate 600 mm of armour. As far as it is known this model is no longer in service.

Model 51 with 75 mm Gun and HOT Missiles

In the 1960s a number of AMX-13s were fitted with three launcher boxes either side of the turret for three Euromissile HOT ATGW systems. This model has not been adopted by the French Army but was offered for export.

Creusot-Loire Industrie Add-on Armour

Creusot-Loire Industrie has developed a range of add-on armour packages that are suitable for its AMX-13 range of light tanks as well as the US M113 APC series.

The add-on package for the AMX-13 light tank weighs under 650 kg with no single unit weighing over 50 kg, which enables it to be installed or removed by the crew using onboard tools.

In the case of the AMX-13 light tank the add-on armour package is installed on the turret front and sides and on the nose and glacis plate of the hull. When fitted it provides protection for both areas against penetration by 20 mm armour-piercing projectiles fired from a range of 100 m over an arc of about 180°.

The AMX-13 armour is made of high hardness steel.

In addition, Creusot-Loire Industrie offers a wide range of kits to enable users to upgrade their vehicles to the latest production standard including improvements in armour, mobility (eg new engine, transmission and suspension) and firepower (eg new armament and fire control system).

Cockerill 90 mm Regunning Packaging

Cockerill Mechanical Industrie is offering a 90 mm Mk IVA3 regunning package for the AMX-13 light tank as a replacement for the existing 75 mm or 105 mm ordnance. The new ordnance incorporates a bore excavator, muzzle brake and breech ring accessories with turret modifications including changes to the two six-round ammunition drums, turret basket ammunition racks and the existing automatic loading system. By early 1991 a total of 22 of these 90 mm re-gunning packages had been produced for export.

Netherlands AMX-13s

The Royal Netherlands Army was one of the largest users of the AMX-13 family of vehicles but all of these have now been phased out of front line service. Some have been overhauled by RDM Defense Engineering of the Netherlands, including AMX-13 light tanks, AMX VCI infantry fighting vehicles and AMX 105 mm Mk 61 self-propelled howitzers, and passed on to Indonesia.

AMX-13 GTI

GLS of Munich, Germany, a subsidiary of Krauss-Maffei, has proposed improving the mobility of the AMX-13 light tank by improving its suspension system.

The original suspension system has been replaced by a new connector type track with centreguide and replaceable track pads. The original road wheels will be replaced by M113 road wheels and the original idler will be replaced by an M113 idler. The existing sprockets and final drive will be modified as will be the existing suspension arms. The existing track return rollers will be modified and telescopic shock absorbers replaced by dual action hydraulic telescopic shock absorbers on the 1st, 2nd, 4th and 5th road wheel stations.

GLS also offers new torsion bars for the AMX-13 and also installs hydraulic bump stops.

Argentinian AMX-13s

In addition to purchasing vehicles from Creusot-Loire Industrie in France, Argentina assembled a number of vehicles from components supplied directly from France.

It is believed that most of these have had their original SOFAM petrol engines replaced by KHD V-8 diesels developing 260 hp. Peru is understood to have refitted its AMX-13 light tanks with this engine and Venezuela is refitting its vehicles in batches of 10.

Creusot-Loire Industrie AMX-13 Version 1987 with 105 mm Gun

Current production model AMX-13 light tanks from Creusot-Loire Industrie have many improvements over earlier vehicles and significant improvements in armour, mobility and fire power.

Externally the appearance of the vehicle has changed and a new hull front with improved ballistic protection is installed, the old torsion bars have been replaced by new hydro-pneumatic units, and sand guards help to keep down the dust while a 105 mm gun is fitted as standard.

The petrol-engined AMX-13 had a range of 350 to 400 km on roads but when fitted with a diesel engine this is increased to 550 to 600 km. Two diesel engines are currently offered, both with an output of 280 hp. These are the Detroit Diesel Model 6V-53T or the French Baudouin 6F 11 SRY. They give a road speed of 65 km/h and a new cooling system enables the engine to operate in high ambient temperatures.

An automatic transmission from Rockford Powertrain is proposed.

The installation of the new engine, transmission and suspension system gives a much improved cross-country mobility and provides a better ride for the crew as well as a more stable firing platform.

AMX-13 rearmed with Cockerill 90 mm Mk IVA3 gun with ordnance fitted with muzzle brake and fume extractor

Creusot-Loire Industrie AMX-13 light tank fitted with Fives-Cail Babcock FL-15 two-man turret armed with 105 mm gun

AMX-13 light tank upgraded by Creusot-Loire Industrie

AMX-13 Model 51 light tank in Singapore

AMX-13 armoured recovery vehicle in travelling configuration with A-frame over rear of hull

AMX-13 bridgelayer laying its scissors bridge in position

The automatic Fives Cail Babcock FL-15 turret has the 105 mm GIAT Industrie gun which can fire many types of ammunition including APFSDS. An extensive range of modular firing controls with integrated rangefinder SOPELEM SOPTAC 18 can be fitted without any modification.

Creusot-Loire Industrie AMX-13 Retrofit

All of the improvements and modifications incorporated into the AMX-13 Model 1987 can be applied to older vehicles. Creusot-Loire Industrie has developed modification kits which can be fitted in the user's own workshops.

Ecuadorian AMX-13

In 1988 Ecuador awarded a contract for the upgrading of its fleet of 108 AMX-13 light tanks armed with 105 mm guns. SOPELEM is to supply the basic day version of its SOPTAC 18 fire control system which comprises a laser rangefinder and its control unit, digital computer that calculates the firing elevation and a module that injects an aiming graticule into the gunner's original sight. The graticule is positioned automatically to correspond to the gun elevation calculated by the computer, which takes into account the range obtained by the laser rangefinder, atmospheric pressure, gun chamber pressure and the tank's horizontal attitude, which are obtained by integrated sensors while the side wind, type of ammunition, and barrel wear information, is fed in manually.

Fives-Cail Babcock provided kits to modify the 105 mm gun and its ammunition system to allow APFSDS rounds to be fired.

Work on upgrading the AMX-13s started in Ecuador late in 1988 and was completed in 1990. Total value of the contract to France is about $23 million

In the future Ecuador may replace the current petrol engines with a diesel engine and also give the vehicles a night fighting capability.

Singapore AMX-13 SM1

Details of this version of the AMX-13 are given later in this section under Singapore.

Venezuelan AMX-13s

In addition to the 36 AMX-13 light tanks delivered by France in the mid 1950s, late in 1988 Creusot-Loire Industrie won a contract from Venezuela for the supply of 31 90 mm armed, overhauled and modernised AMX-13 light tanks. These were ex-French Army vehicles stripped down by the company and refurbished. They now have the US Detroit Diesel Model 6V-53T engine developing 280 hp coupled to a Borg Warner three-speed fully automatic transmission with torque converter, new Chausson air-water/oil cooler with a thermostatically controlled centrifugal fan, alternator or 200 Amp generator and NATO 6TN batteries. The original torsion bar suspension has been removed and replaced by Dunlop hydro-pneumatic units for increased cross country mobility.

A SOPELEM 18-02 SOPTAC fire-control system incorporating a M213 day sight and a TCV-107 laser rangefinder has been fitted.

Main armament comprises a 90 mm gun with a 7.62 mm NF1 coaxial machine gun. A Israeli Tadiran communications system is fitted.

Final deliveries of the up-graded AMX-13 light tanks to Venezuela were made in late 1990.

Venezuelan AMX-13 Rocket Launchers

Venezuela has about 25 AMX-13 light tank chassis fitted with the Detroit Diesel 6V-53T diesel engine, with the turret replaced by the Israel Military Industries 160 mm LAR multiple rocket launcher system, details of which are given in the *Multiple rocket launchers* section.

NIMDA Upgrade Package for AMX-13

The Israeli company of NIMDA offers a complete retrofit package for the AMX-13 light tank including new Detroit Diesel Model 6V-53T developing 275 hp at 2800 rpm coupled to a new NIMDA N303 automatic transmission, new armament, computerised fire control system, additional armour protection and a fire/explosion detection and suppression system.

AMX-13 Armoured Recovery Vehicle

The AMX-13 ARV (Char de Dépannage Model 55) is used to recover other members of the AMX-13 family as well as changing major components such as turrets and engines. Equipment fitted includes a front-mounted A-frame, 15 000 kg capacity winch with 50 m of 25 mm diameter cable, secondary winch with 120 m of 6 mm diameter cable, four spades at the rear of the hull and tools and other equipment. The vehicle has a crew of three consisting of commander, driver and winch operator and weighs 15 300 kg. Armament consists of an externally mounted 7.5 mm or 7.62 mm machine gun and smoke dischargers. Details of this version are given in *Jane's Military Vehicles and Logistics 1991-92*, page 21.

AMX-13 Bridgelayer

The AMX-13 bridgelayer (Char Poseur de Pont AMX-13) is fitted with a folding class 25 bridge which has an unfolded length of 14.01 m. The bridge is launched over the rear of the vehicle and two stabilisers steady the vehicle when the bridge is being positioned. The vehicle has a loaded weight of 19 700 kg and without the bridge weighs 15 000 kg. Full details of the AMX-13 armoured bridgelayer are given in *Jane's Military Vehicles and Logistics 1991-92*, page 78.

SPECIFICATIONS (90 mm version)

CREW	3
COMBAT WEIGHT	15 000 kg
UNLOADED WEIGHT	13 000 kg
POWER-TO-WEIGHT RATIO	16.66 bhp/tonne
GROUND PRESSURE	0.76 kg/cm^2
LENGTH GUN FORWARDS	6.36 m
LENGTH HULL	4.88 m
WIDTH	2.51 m
HEIGHT (to top of commander's hatch)	2.3 m
GROUND CLEARANCE	0.37 m
TRACK	2.159 m
TRACK WIDTH	350 mm
LENGTH OF TRACK ON GROUND	2.997 m
MAX ROAD SPEED	60 km/h
FUEL CAPACITY	480 litres
MAX ROAD RANGE	350-400 km
FORDING	0.6 m
GRADIENT	60%
SIDE SLOPE	60%
VERTICAL OBSTACLE	
forwards	0.65 m
reverse	0.45 m
TRENCH	1.6 m
TURNING RADIUS	skid turns
ENGINE	SOFAM Model 8Gxb 8-cylinder water-cooled petrol developing 250 hp at 3200 rpm
AUXILIARY ENGINE	none
TRANSMISSION	manual with 5 forward and 1 reverse gears
STEERING	Cleveland (Cletrac) controlled differential
CLUTCH	single-disc
SUSPENSION	torsion bar
ELECTRICAL SYSTEM	24 V
BATTERIES	4 × 12 V, 190 Ah
ARMAMENT	
main	1 × 90 mm gun
coaxial	1 × 7.5 mm or 7.62 mm MG
anti-aircraft	1 × 7.5 mm or 7.62 mm MG (optional)
SMOKE-LAYING EQUIPMENT	2 smoke dischargers either side of turret
AMMUNITION	
main	32
MG	3600
SMOKE CANISTERS	12
GUN CONTROL EQUIPMENT	
Turret power control	hydraulic/manual
by commander	yes
by gunner	yes
Max rate of power traverse	360°/12 s
Max rate of power elevation	5° in 1s
Gun elevation/depression	+12.5°/-5°
Commander's override	yes
Gun stabiliser	
vertical	no
horizontal	no
Commander's sight	telescope L961 with ×1.5 and ×6 magnification
Gunner's sight	telescope L862 with ×7.5 magnification
ARMOUR	
Hull front	15 mm at 55° (equal to 40 mm)
Hull sides	20 mm
Hull top	10 mm
Hull rear	15 mm
Turret front	25 mm at 45° (equal to 40 mm)
Turret sides	25 mm
Turret top	10 mm

SPECIFICATIONS (105 mm version)

As 90 mm version apart from the following

LENGTH GUN FORWARDS	6.5 m
GROUND CLEARANCE	0.47 m
LENGTH OF TRACK ON GROUND	2.8 m
MAX ROAD SPEED	
diesel engine	64 km/h
MAX ROAD RANGE	
diesel engine	550-600 km
ENGINE	
diesel	Detroit Diesel 6V-53T turbo-charged, 6-cylinder water-cooled diesel developing 280 hp at 2800 rpm or Baudouin 6F 11 SRY turbo-charged 6-cylinder water-cooled developing 280 hp at 3200 rpm
TRANSMISSION	automatic
GUN CONTROL EQUIPMENT	
Turret power control	elevating and traversing gear controls (1) either by means of control grips and hydraulic motor (gunner and commander, the latter overriding the former) (2) or by means of a crank handle (gunner)
Commander's sight	M212 telescope with ×1.5 and ×6 magnification
Gunner's sight	M213 telescope with ×7.5 magnification, SOPELEM SOPTAC 18 FCS with integrated laser rangefinder SOPTAC 18-03 or 18-04
NIGHT VISION	image intensification or thermal as an option

Status: In production. In service with:

Country	Quantity	Comment
Argentina	70	was also assembled in Argentina
Chile	47	unconfirmed operator
Dominican Republic	2	75 mm model, may have 10
Ecuador	108	105 mm, plus ARV upgraded (qv)
El Salvador	12	75 mm model
Indonesia	100+	75 mm and 105 mm, plus ARV and AVLB
Ivory Coast	5	75 mm model
Lebanon	50	105 mm, out of 100+
Morocco	50	Out of 130 delivered
Nepal	5	may be non-operational
Peru	110	80 105 mm model, 30 75 mm model
Singapore	350	75 mm model, many modified
Tunisia	40	75 mm and 90 mm models
Venezuela	36	armed with 75 mm guns, 31 90 mm upgraded delivered, MRS

Note: They are no longer used by Algeria, Cambodia, Djibouti, Egypt, France, India, Israel or the Netherlands.

Manufacturer: Creusot-Loire Industrie
Immeuble Ile de France
Cedex 33,
92070 Paris la Défense
France
Telephone: 33 (1) 49 00 60 50
Telex: Montoy 615 638 F
Fax: 33 (1) 49 00 57 30

GERMANY

Thyssen Henschel TH 495 Light Tank

Development/Description

Late in 1990 Thyssen Henschel stated that it was developing, as a private venture, a 105 mm armed light tank under the designation of the TH 495.

This could meet the US Army's requirement for an Armored Gun System. No specific details of the vehicle have been released, although its 105 mm rifled tank gun will be capable of firing standard NATO ammunition types including APFSDS. It is believed that the vehicle will have the engine at the front and the turret towards the rear.

Status: Development.

Manufacturer: Thyssen Henschel
Postfach 102969
D-3500 Kassel
Federal Republic of Germany
Telephone: (0561) 8011
Telex: 99 750 thksr d
Fax: (0561) 801 6733

INDIA

New Indian Light Tank

The Indian Army at present operates about 100 Soviet-supplied PT-76 light amphibious tanks but is understood to have a requirement for up to 250 new light amphibious tanks. Trials have been carried out with the 105 mm version of the Swedish Hägglunds Ikv-91 light tank.

For trial purposes a Soviet-supplied BMP-1 infantry combat vehicle has had its one-man turret removed and replaced by a French GIAT Industrie two-man TS 90 turret armed with a 90 mm gun that can fire APFSDS rounds and a 7.62 mm coaxial machine gun.

India now makes the BMP-2 infantry combat vehicle under licence so may possibly adopt a BMP-2 chassis/GIAT Industrie TS 90 turret to meet its light amphibious tank requirements. The Indonesian Marines use a modified AMX-10P chassis fitted with the TS 90 turret to meet the requirement for an amphibious fire support vehicle.

KOREA, NORTH

M1985 Light Tank

During a parade held in the capital of North Korea several years ago a new amphibious light tank made its first public appearance. This is believed to have been designed and built in North Korea although some technology and perhaps some key components could have come from abroad.

M1985 85 mm armed North Korean light tanks on parade

The hull of the new light amphibious tank is of all-welded steel armour construction with the driver seated front left, ammunition stowage to his right, turret and fighting compartment in the centre and engine and transmission at the rear.

The turret is similar to that fitted to the Soviet PT-76 light amphibious tank with grab rails on the outside. The glacis plate is well sloped and the hull sides above the suspension slope inwards.

Suspension consists of six large rubber-tyred road wheels, drive sprocket at the rear and idler at the front. There are no track return rollers.

Main armament is believed to comprise an 85 mm gun which is fitted with a fume extractor towards its muzzle. A 7.62 mm machine gun is mounted coaxial with the main armament, and mounted over the 73 mm gun is a ATGW launcher, possibly for a Sagger AT-3. A 12.7 mm machine gun can be mounted on the turret roof for anti-aircraft defence.

In the absence of any official North Korean designation the US Army has given this vehicle the designation M1985 as this was the year it was first seen in public.

SPECIFICATIONS (provisional)

CREW	3-4
WEIGHT	19 000 kg
LENGTH	7 m
WIDTH	3.2 m
HEIGHT	2.6 m

Status: In production. In service with the North Korean Army.

Manufacturer: North Korean state factories.

NORWAY

Modernised M24 Light Tank

Development

To meet the requirements of the Norwegian Army Thune-Eureka A/S (now the Thune Myren Division of Kvaerner Eureka A/S) developed a modernised version of the M24 Chaffee light tank. The first prototype was completed in January 1973 and after extensive trials the Norwegian Army decided to refit its fleet of M24s to this new standard for anti-tank use. The Norwegian Army took delivery of 54 of these tanks, designated the NM-116, between January 1975 and October 1976. NAPCO International Incorporated of the USA has obtained a licence to market this vehicle world-wide. A number of unmodernised M24s remain in Norwegian Home Guard service.

In the future it is probable that all of the Norwegian M24 Chaffee light tanks will be phased out of service and be replaced by other AFVs, for example Leopard 1 MBTs that are now becoming surplus as a result of the Conventional Forces Europe treaty.

Description

The modernisation of the tank has been extensive and includes the installation of a new powerpack, new transmission, new gun and numerous other improvements.

The original twin Cadillac petrol engines have been replaced by a single super-charged Detroit Diesel Model 6V-53T which develops 250 hp at 2800 rpm. This is coupled to an Allison Transmission Division of General Motors pre-selector gearbox model MT 650 with a torque converter giving five forward speeds and one in reverse. In addition to the main transmission another gearbox provides the correct speed to the differential at the front of the tank.

There are two radiators with hydraulically driven and thermostatically governed fans and four heat exchangers for cooling the oil. The two heat exchangers for the transmission and differential are on the engine, one exchanger for the additional gearbox is in the air flow of one of the radiators and one for the oil of the hydraulic system is built into pipes for the water cooling of the engine. The air from the radiators blows through the engine compartment and is ejected with the exhaust.

The fighting compartment is equipped with a manual fire extinguisher and the engine compartment is fitted with a Freon fire-extinguishing system, consisting of two bottles of Freon 1301 with a tripping device hand-operated from the driver's seat and a single bottle of Freon 1301 which is automatically released when the temperature in the engine compartment reaches a predetermined level. Each bottle has sufficient capacity to extinguish any possible fire in the engine compartment.

To increase the comfort of the crew and improve the gunner's and loader's ability to operate the 90 mm gun the modified turret is fitted with a basket which both eliminates the need for the loader to work kneeling and enables a number of ready rounds to be carried in the basket.

The bow machine gunner's position in the front right side of the hull has been eliminated and the associated ball mounting has been welded shut. The space saved is used to stow 20 rounds of 90 mm ammunition. To suit the Norwegian climate a new heating system has been installed and the batteries have been moved to the floor of the fighting compartment and have hot air piped to them to increase their capacity in Winter.

The original 75 mm gun has been replaced by a French D/925 90 mm low pressure gun. The modifications include the replacement of the barrel, removal of the gyrostabiliser, modifying the breech assembly (breech ring, breech block and extractor), replacement of the telescope graticule and the adjustment of the ammunition racks. The recoil mechanism has not been modified. The gun fires the same range of GIAT Industrie ammunition as the Panhard AML light armoured car including canister, HE-FS, HEAT-FS and smoke FS.

M24 light tank as used by Norwegian Army with new French 90 mm gun, laser rangefinder over main armament, bank of German smoke dischargers either side of turret and bow machine gun removed (Norwegian Army)

The 7.62 mm (0.30) coaxial machine gun has been replaced by a 12.7 mm (0.50) Browning M2 HB machine gun and a similar anti-aircraft weapon is mounted on the roof of the turret. Four German type smoke dischargers are mounted on either side of the turret. Forty-one rounds of 90 mm and 500 rounds of 12.7 mm ammunition are carried.

Vehicles used by the Norwegian Army have a Simrad LV3 laser rangefinder mounted over the main armament and other optional equipment includes two different types of tracks (German Diehl or Swedish Hägglund), new shock absorbers, new communications equipment and either infra-red or passive night vision equipment for the driver, commander and gunner.

Dimensions of the modernised M24 are almost identical to those of the standard M24 Chaffee but the modernised version has the following performance figures:

SPECIFICATIONS

MAX ROAD SPEED	57 km/h
AVERAGE ROAD SPEED	45 km/h
AVERAGE CROSS-COUNTRY SPEED	26-30 km/h
ROAD RANGE	400 km

Variants

BK 710 Recovery Vehicle

This has been designed by A/S Moëlven Brug to meet the requirements of the Norwegian Army and is essentially an M24 Chaffee light tank with its turret removed and replaced by a hydraulic crane with a maximum winching capacity of 19 tonnes, maximum lifting capacity of 11 tonnes at 1.8 m reach, a maximum outreach of 5 m and a lifting capacity of five tonnes. When travelling the crane is normally traversed to the rear. This is in service with the Norwegian Army under the designation NM-130.

Status: In service with Norway. The modernisation programme is still available from NAPCO International and has been sold to an Asian country, probably Taiwan.

Modernised M24 light tank of Norwegian Army without stowage bins and showing modified rear engine decking

Manufacturer: Original manufacturer for the modernised M24 was Thune-Eureka A/S, Norway.
Licence is now held by NAPCO International Incorporated
1600 Second Street South
Hopkins
Minnesota 55343
USA
Telephone: (612) 931 2400
Telex: 290436
Fax: (612) 931 2402

SINGAPORE

Singapore Automotive Engineering AMX-13 SM1 Light Tank

Development

Singapore has the largest fleet of AMX-13 light tanks in the world, and for many years Singapore Automotive Engineering (SAE) has been overhauling these and other tracked and wheeled armoured vehicles for the Singapore Armed Forces.

More recently, Singapore Automotive Engineering has developed a complete chassis refurbishment programme for the AMX-13 and the first production vehicles were handed over to the Singapore Armed Forces in June 1988 under the designation of the AMX-13 SM1.

The main area of improvement is in the chassis which has been given a total automotive refit with new Detroit Diesel 6V-53T engine, new ZF 5WG-180 fully automatic transmission, new electrical system and SAE Dunlostrut hydro-pneumatic suspension system.

These improvements not only make the AMX-13 SM1 more reliable but also increase the operational range of the vehicle, give a higher speed, better acceleration and improve its cross country mobility.

Singapore Automotive Engineering are now offering this retrofit kit for other users of the AMX-13.

Description

The original petrol engine has been replaced by a more fuel efficient water-cooled, turbocharged Detroit Diesel Model 6V-53T which develops 290 bhp at 2800 rpm. This can be replaced in 50 minutes.

The engine cooling system for the upgraded AMX-13 SM1 light tank has been designed, developed and produced by the UK company Gallay and by early 1991 the company had supplied over 200 systems to Singapore Automotive Engineering.

The AMX-13 SM1 cooling system consists of three air-cooled heat exchangers which are mounted in series. Air passes from front to back through the assembly, cooling in turn the oil from the steering brakes and differential, the engine cooling water and automatic transmission fluid.

The engine is coupled to a ZF 5WG-180 fully automatic transmission with five forward and two reverse gears. This is controlled by a Electrical Control Unit (ECU) which is programmed to protect against unsafe shifts. In case of an electrical failure there is a mechanical fallback.

A new electrical system has been fitted which is more reliable and easier to maintain and repair. This is based on four 12 V batteries connected both in series and in parallel. If a single battery fails, the others can still supply power. In addition there is also a new 28 V 160 Amp fully suppressed alternator. This is more reliable than the older dynamo and is easier to maintain.

A new electrical slip ring has been fitted, as have new instruments for the driver. A new radio harness is sealed in shrinkable polyolesin tubing which provides protection against oil, fuel and moisture. It is also heat retardant and in the event of fire does not give off toxic fumes. All sections can be quickly disconnected and replaced for repairs in the field.

The original torsion bar suspension has been replaced by a new SAE Dunlostrut hydro-pneumatic suspension system which increases total wheel travel, absorbs greater shock loads and permits greater mobility over rough terrain.

The danger of shredding tracks is reduced as the suspension system maintains a more consistent tension on the track. The pressure in each unit strut can be individually adjusted.

AMX-13 SM1 light tank modernised by Singapore Automotive Engineering

PERFORMANCE COMPARISON

MODEL	**AMX-13**	**AMX-13 SM1**
POWER-TO-WEIGHT RATIO	16.6 hp/tonne	19 hp/tonne
GRADIENT	60%	60%
SIDE SLOPE	40%	40%
MAXIMUM ROAD SPEED	61 km/h	64 km/h
FUEL CAPACITY	480 litres	480 litres
FUEL CONSUMPTION		
road (litres/100 km)	110/130	100
cross-country (litres/hour)	50/60	25/30
RANGE		
road	350/400 km	450/500 km
cross country	8/9 hours	16/19 hours
WHEEL TRAVEL		
static to bump stop	115 mm	275 mm
CROSS COUNTRY OBSTACLE CLEARANCE	115 mm	275 mm
SUSPENSION WHEEL LOAD CAPACITY (G)	2.25	4.5

Status: Production. In service with the Singapore Armed Forces.

Manufacturer: Singapore Automotive Engineering Pte Ltd
5 Portsdown Road
Republic of Singapore 0513
Telephone: 4736311
Telex: SINENG RS 25755
Fax: 4710662

TAIWAN

Type 64 Light Tank

Development/Description
In the early 1970s Taiwan began design work on a vehicle based on the American M41 light tank. The first prototype was a copy of the M41 but with further modifications it was standardised in 1975 as the Type 64 Light Tank. Recent information has indicated that it was not produced on a large scale and that many Type 64s were in fact conversions of the original American M41.

The Type 64 is similar to the M41A2/M41A3 in automotive and mechanical details and has the same dimensions, chassis, engine and transmission but an improved electrical system. The hull is of all-welded construction as in the M41 but incorporates advanced welding techniques in its manufacture and uses a new high-hardness alloy steel. The all-cast turret has the same layout as the original M41 although slight alterations have been made to suit the medium stature of the Chinese tank crews. An additional layer of laminated high-hardness alloy steel armour plate is bolted/welded onto the hull and turret sides to provide greater protection against HEAT and APDS projectiles.

The Type 64 is armed with a 76 mm gun as in the M41 but made in Taiwan and claimed to have a performance superior to the M32 gun. A 7.62 mm machine gun is mounted coaxially with the main armament and a 12.7 mm machine gun is mounted at the commander's station for use in the anti-aircraft role. Both of these weapons are made in Taiwan and the coaxial weapon probably resembles a modified version of the American M60.

Chinese sources state that the Type 64 has an improved fire control system. Ammunition stowage is probably identical to that of the M41A2/M41A3.

Later Type 64 tanks may incorporate laser rangefinders, ballistic computers and passive night vision equipment. All of these have been developed by the Sun Yat-Sen Scientific Research Institute of Taiwan. It is reported that some consideration is being given to replacing the 76 mm gun.

The Type 64 has the same mobility, agility and range characteristics as the M41 but in view of its additional armour is probably heavier. Like the M41 it can be fitted for deep fording, but it is not amphibious and has no NBC system. It is fitted with IR driving lights and improved Taiwan-built communications equipment.

Status: In service with Taiwan. Production complete. Improvement and modernisation programmes may be continuing.

Manufacturer: Fighting Vehicles Command, ROC Army, Taichung, Taiwan. Development by Fighting Vehicles Development Center, ROC Army, Taichung, Taiwan 400.

UNION OF SOVIET SOCIALIST REPUBLICS

PT-76 Light Amphibious Tank

Development
The PT-76 light amphibious tank was developed shortly after the end of the Second World War by the I.V.Gavalov OKB design bureau under the designation K-90 and was accepted for service in 1950 as the PT-76. Total production is believed to have amounted to 7000 vehicles with final deliveries being made in the late 1960s. Production of the PT-76 is believed to have been undertaken in the Kirov facility in Leningrad although the Volgograd Tractor Factory may also have undertaken production of this vehicle.

At one time each Soviet tank and motor rifle regiment had a reconnaissance company whose equipment included five PT-76 light amphibious tanks and three armoured personnel carriers with each division having a separate reconnaissance company with a further five PT-76s.

Today Soviet units have replaced their PT-76s by T-54, T-55, T-62 or T-72 MBTs or even BMP infantry combat vehicles.

Many automotive and sub-components of the PT-76 light amphibious tank are also used in the BTR-50 armoured personnel carrier and other vehicles such as SA-6 Gainful SAM system and ZSU-23-4 self-propelled anti-aircraft gun system.

PT-76 light tank upgraded by NIMDA and clearly showing new turret optics and new 90 mm gun

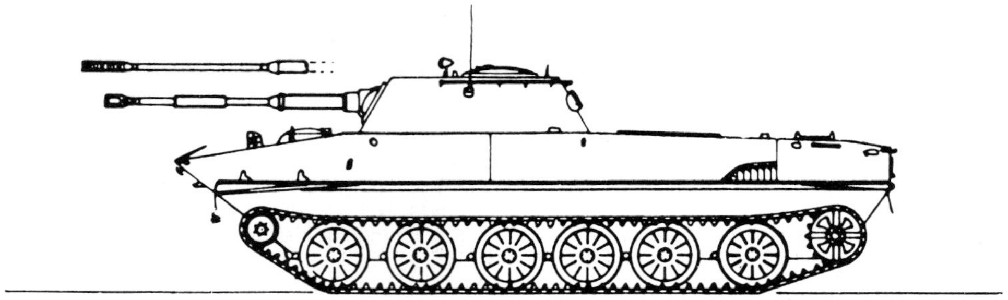

PT-76 light amphibious tank with D-56T gun with double-baffle muzzle brake. Inset is early D-56T gun with multi-slotted muzzle brake

The tank has been used in combat in Africa, the Middle East, during the Indo-Pakistan conflict and widely used in Vietnam by the North Vietnamese Army. The main drawbacks of the PT-76 are its large size necessitated by the requirement to be amphibious, lack of NBC or night fighting equipment and very thin armour that can be penetrated even by heavy machine gun fire. It does have an excellent amphibious capability and is also used by the Soviet Naval Infantry.

Under the Conventional Forces Europe (CFE) treaty, the PT-76 is classed as a combat vehicle equipped with a heavy weapon, and the Soviets have declared the following vehicles:

Region	**IV.1**	**IV.2**	**IV.3**	**IV.4**	**V.1**
Quantity	602	174	168	nil	428

Note: Regions are as defined under the CFE Treaty; Region IV is defined as Central Europe while Region V covers the flanks of the Atlantic to the Urals (ATTU) zone.

Description

The hull of the PT-76 is welded steel and is divided into three compartments with the driver at the front, fighting compartment in the centre and the engine at the rear. The driver is seated in the centre of the hull and is provided with a single piece hatch cover that swings to the right. Three periscopes are mounted forward of his hatch cover; the centre one can be raised to enable the driver to see over the front of the vehicle when the trim vane is erected. The centre periscope can be replaced by a TVN-28 infra-red periscope with a 60 m range.

The turret is all-welded steel, with the commander, who also acts as the gunner, seated on the left and the loader on the right. The turret has an oval-shaped hatch cover that hinges forwards and can be locked vertical. Mounted in the left side of the hatch cover is a circular cupola for the commander which houses three integral periscopes and can be traversed through 360° by hand. The commander is provided with an optical TSh-66 sight with a magnification of ×4 mounted to the left of the main armament and the loader has a periscope mounted in the turret roof, forward of the hatch cover.

The engine compartment is at the rear and the Model V-6 engine used in the PT-76 is one bank of that fitted to the T-54 MBT. As the tank operates in a very cold climate an engine pre-heater is fitted as standard.

The manual gearbox has five forward and one reverse gears and steering is of the clutch and brake type.

The torsion bar suspension consists of six road wheels with the idler at the front and the drive sprocket at the rear. The first and sixth road wheel stations have hydraulic shock absorbers and the steel tracks each have 96 links when new. Tracks are steel with a single dry pin.

The PT-76 is fully amphibious, being propelled in the water by two water jets mounted at the rear of the hull. The only preparation required before entering the water is to erect the trim vane at the front of the hull and switch on the two electric bilge pumps. A manual bilge pump is provided for emergency use. Steering is accomplished by opening and closing the two hatches over the rear water jets: to go left the left water jet is covered up and to turn left through 180° the left water jet sucks in water while the right water jet pushes it out.

The PT-76 is not fitted with an NBC system although there is a ventilator mounted in the turret rear and many vehicles used by marines have been fitted with a snorkel extension piece to this ventilator which often sucked exhaust gases into the fighting compartment. To increase the PT-76's operational range additional fuel tanks can be fitted; these can be of the drum type or flat type as fitted to one side of the T-54/T-55 MBT. A white light searchlight is mounted on the right side of the turret (some models have an infra-red searchlight on the turret) and one of the driving lights may be infra-red. When being used by the Soviet Naval Infantry the vehicle is often fitted with a complete set of navigation lights.

The main armament of the PT-76 is the D-56T gun which is a development of the gun used in the T-34/76 and KV-1 tanks during the Second World War. The gun has a semi-automatic vertical sliding wedge breech-block, hydraulic buffer and a hydro-pneumatic recuperator. The gun is 42 calibres long and has an overall length of 3.455 m and weighs 1150 kg. The gun has a maximum rate of fire of between 6 and 8 rds/min, maximum range in the indirect fire role of between 12 000 and 13 290 m, and can fire the following types of fixed ammunition: AP-T BR-350 series), API-T (BZR-350B), HE-FRAG (0-350A), HEAT-FS (BK-354M) and HVAP-T (BR-354P)

Other types of projectiles are also available, some with cartridges with reduced propellant charges, for HE-FRAG and AP-T projectiles.

A 7.62 mm SGMT machine gun is mounted coaxially to the right of the main armament. Many PT-76s have been fitted with 12.7 mm DShKM anti-aircraft machine gun.

Variants

The first production version of the PT-76 light tank was armed with the D-56T gun and fitted with a multi-slotted muzzle brake. This version is rarely seen today. The most common version of the PT-76 is fitted

PT-76 light amphibious tank from rear (US Army/Michael Green)

First production model of the PT-76 fitted with the D-56T gun with multi-slotted muzzle brake and roof hatch open (US Army)

PT-76 Model 2 light amphibious tank

with the D-56T gun with a double-baffle muzzle brake and a bore evacuator towards the muzzle. For training the muzzle brake is sometimes removed. The PT-76B is fitted with the D-56TM fully stabilised gun and also has an NBC system.

Chinese Type 63

This has a similar hull to PT-76 but has a different turret; details are given in this section under China.

NIMDA Upgrade Package for PT-76

The Israeli company NIMDA offers a complete retrofit package for the PT-76 light amphibious tank including a new powerpack incorporating a Detroit Diesel 6V-92T diesel developing 300 hp at 2100 rpm coupled to the original transmission with a new clutch assembly. The kit also includes a new alternator (24 V, 200 Ah), new cooling and electrical system and fuel, exhaust, air inlet, hull and top deck modifications. Another option is the replacement of the original transmission by a fully automatic transmission.

The PT-76 prototype has had its original 76.2 mm gun replaced by a new 90 mm gun, believed to be a Belgium Cockerill Mk III that can fire a wide range of ammunition including APFSDS.

The Soviet 7.62 mm machine gun has been replaced by a Western 7.62 mm machine gun and a similar weapon can be mounted on the turret roof for air defence purposes. Electrically operated smoke dischargers can be mounted either side of the turret.

To improve the first round hit probability, a new fire control system has been installed together with a new day/night sight for the gunner which incorporates a laser rangefinder.

A new solid state all-electric gun/turret stabilisation and power control system has replaced the original Soviet system and a new fire extinguishing system has been installed.

It is believed that this conversion package is aimed at Indonesia who has a quantity of PT-76 light tanks still in service.

Polish PT-76

Under the Conventional Forces Europe treaty, Poland declared that it had a total of 60 Soviet supplied PT-76 light amphibious tanks. It is understood that these are mainly held in reserve or used by the Polish 7th Coastal Defence (formerly Sea Landing) Brigade whose organisation includes three coastal defence battalions each of which has a company of 10 PT-76 light amphibious tanks. At one time the brigade also had a complete PT-76 battalion but this has now been disbanded.

The PT-76 Model 2 is used but, with the turret modified, the commander now has his own hatch that opens forwards while the gunner also has his own hatch and mans the roof-mounted 12.7 mm DShKM anti-aircraft machine gun.

SPECIFICATIONS	
CREW	3
COMBAT WEIGHT	14 000 kg
POWER-TO-WEIGHT RATIO	17.1 hp/tonne
GROUND PRESSURE	0.479 kg/cm^2
LENGTH GUN FORWARDS	7.625 m
LENGTH HULL	6.91 m
WIDTH	3.14 m
HEIGHT OVERALL	
late model	2.255 m
early model	2.195 m
FIRING HEIGHT	1.82 m
GROUND CLEARANCE	0.37 m
TRACK	2.74 m
TRACK WIDTH	360 mm
LENGTH OF TRACK ON GROUND	4.08 m
MAX SPEED	
road	44 km/h
water	10 km/h
FUEL CAPACITY	
main	250 litres
auxiliary	180 litres
MAX RANGE	
road	260 km
road with auxiliary tanks	450 km
water	65 km
FUEL CONSUMPTION	0.96 litre/km
FORDING	amphibious
GRADIENT	70%
VERTICAL OBSTACLE	1.1 m
TRENCH	2.8 m
TURNING RADIUS	skid turns
ENGINE	Model V-6 6-cylinder in-line water-cooled diesel developing 240 hp at 1800 rpm
AUXILIARY ENGINE	none
TRANSMISSION	manual with 4 forward and 1 reverse gears
STEERING	clutch and brake
SUSPENSION	torsion bar
ELECTRICAL SYSTEM	24 V
BATTERIES	2 × 12 V, 100 Ah
ARMAMENT	
main	1 × 76.2 mm gun
coaxial	1 × 7.62 mm MG
SMOKE-LAYING EQUIPMENT	diesel fuel injected into exhaust system
AMMUNITION	
main	40
MG	1000
GUN CONTROL EQUIPMENT	
Turret power control	electric/manual
by commander	yes
Max rate of power traverse	360°/21 s
Gun elevation/depression	+30°/-4°
Gun stabiliser	
vertical	only on PT-76B
horizontal	only on PT-76B
ARMOUR	
Hull front upper	11 mm at 80°
Hull front lower	14 mm at 45°
Hull side upper	14 mm at 0°
Hull side lower	14 mm at 0°
Hull rear upper	7 mm at 0°
Hull rear lower	7 mm at 45°
Hull top	7 mm
Hull belly front	5 mm
Hull belly rear	5 mm
Turret front	17 mm at 35°
Turret sides	16 mm at 35°
Turret rear	11 mm at 35°
Turret top	8 mm
Mantlet	11 mm at 33°

Status: Production complete. In service with:

Country	**Quantity**	**Comment**
Afghanistan	60	
Angola	50	
Benin	20	
Cambodia	20	
China	small number	Type 63 is local version
Congo	3	
Cuba	60	
Finland	15	
Guinea	20	
Guinea-Bissau	20	
India	100	
Indonesia	41	
Iraq	100	status uncertain
Korea, North	not known	
Laos	25	
Madagascar	12	
Nicaragua	22	
Poland	60	
USSR	1000	army and naval infantry
Vietnam	150	
Yugoslavia	13	
Zambia	30	

Manufacturer: Soviet state factories.

UNITED KINGDOM

Alvis Scorpion Light Tank

Details of the Alvis Scorpion Combat Vehicle Reconnaissance (Tracked), which is also referred to as a light tank, are given in the *Armoured Fighting Vehicle Families* section.

UNITED STATES OF AMERICA

M41 Light Tank

Development

Production of the M41 was undertaken by the Cadillac Motor Car Division of General Motors Corporation at Cleveland Tank Arsenal.

First production M41s were completed in mid 1951 and 1802 were built before production was switched to the later M41A1, which was followed by the M41A2 and finally the M41A3 which differ only in minor details. Total production of the vehicle amounted to about 5500 units. Its full United States Army designation is Tank, Combat, Full-Tracked: 76 mm Gun, M41. It has been replaced in the US Army by the M551 Sheridan but is still used by some armies. Many of the components of the M41 are also used in the M42 twin 40 mm self-propelled anti-aircraft gun system (also built at Cleveland), M44 155 mm and M52 105 mm self-propelled howitzers.

Description

The all-welded steel hull of the M41 is divided into three with the driver's compartment at the front, fighting compartment in the centre and the engine at the rear.

The driver is seated at the front of the vehicle on the left side and is provided with a single piece hatch cover that opens to the right. Three M17 periscopes are mounted forward of his position and a single one to his left. A hull escape hatch is provided beneath the driver's seat.

The turret, which is provided with a basket, is of cast and welded construction with the commander and gunner seated on the right and the loader on the left. The commander's cupola has a single piece hatch cover that opens forwards and five vision blocks, and an M20A1 periscope that can be traversed through 360°. The gunner has an M97A1 telescopic sight for aiming the 76 mm gun and an M20A1 periscope that can be traversed through 360°. The loader is provided with a single piece hatch cover that opens forwards and a single M13 periscope. Mounted at the rear of the turret is a light sheet metal stowage box and there is a dome-shaped ventilator in the turret roof towards the rear. The ventilator blower is operated from the driver's position.

The engine compartment is at the rear of the hull and is separated from the fighting compartment by a fireproof bulkhead. It is equipped with a fire extinguisher that is operated by the driver. The engine is mounted towards the front of the engine compartment with the transmission at the rear.

The torsion bar suspension consists of five dual road wheels each side with the drive sprocket at the rear, idler at the front and three track return rollers. The first, second and fifth road wheel stations are provided with a hydraulic shock absorber. The steel tracks, which have 75 or 76 links when new, have detachable rubber pads.

Standard equipment on all tanks includes a heater, deep fording equipment and electric bilge pumps. It is not fitted with an NBC system. The basic model was not fitted with night vision equipment although the final production model did have provision for an infra-red searchlight over the main armament.

Main armament of the M41 consists of a 76 mm gun M32 (T91E3) in a mount M76 (T138E1), the M41A1 has the M32A1 gun in mount M76A1. The gun has a vertical sliding breech-block, spring actuator and an inertia percussion firing mechanism. The recoil system is of the concentric hydro-spring type.

A 7.62 mm (0.30) Browning M1919A4E1 machine gun is mounted coaxially to the right of the main armament and an anti-aircraft 12.7 mm (0.50) Browning M2 HB machine gun is mounted at the commander's position. This has a traverse of 360°, elevation limits being from -10° to +65°.

The 76 mm gun fires the following types of fixed ammunition: AP-T (M339), Blank (M355A2), Canister (M363), HE (M352), HEAT-T (M496), HVAP-DS-T (M331A1/M331A2), HVAP-T (M319), TP-T (M340/M340A1) and WP Smoke (M361/M361A1).

In 1982 it was revealed that AAI Corporation had developed a new 76 mm APFSDS-T round for the M32 cannon that would defeat Main Battle Tank frontal armour.

The M41A2 is almost identical to the M41A1 except that it has a simplified turret and gun control system developed by Cadillac. This consists of manual and hydraulic power traverse for the gunner with direct mechanically linked control of oil gear pump in lieu of the electrical control in the M41, dual power traverse by commander, and manual mechanical, rack and pinion type, elevation for the gunner, with slewing elevation control for the gun. This more compact system enabled the tank to carry 65 rounds of 76 mm ammunition compared with 57 on the earlier model. These modifications were also incorporated in the M41A2 and the M41A3 both of which also have the fuel injection engine fitted.

Brazilian Army M41C light tank fitted with additional armour protection to turret, glacis plate, nose and hull sides and 90 mm gun

M41 light tank fitted with new powerpack by NAPCO of the USA

M41 light tank upgraded by Bernardini and showing new hull rear

M41E light tank fitted with General Motors 8V-71T diesel engine by Talbot of Spain

Variants

Cadillac Gage TMS

The Cadillac Gage Company, which supplied the weapon control systems for the M1/M1A1, M48, M60A1, M60A2, M60A3, Leopard 1, Leopard 2, M1, Marder, LVTP7, V-150 and V-200 Commando, M109, AIFV and special versions of the Centurion, has now developed a Turret Modernisation System for the M41. This can be fitted by the user and gives the following advantages over the existing system: improved low speed tracking and target acquisition, high precision gun positioning capability, improved first round hit capability, reduced system weight and volume, simplified system operation and maintenance, simplified overall logistical support, readily available spare parts, increased reliability and high system growth potential. This system is also available fully stabilised to allow a fire on the move capability.

QM41

This is a remote controlled tank used by the United States Navy for testing air-to-ground missiles. The tank has also been used for a variety of other trials purposes including much of the development work with the turret of the M551 Sheridan light tank/assault vehicle.

Brazilian M41 Programme

The Bernardini company of Sao Paulo, prime contractor for the X1, X1A1 and X1A2 light tanks described earlier in this section, has now converted Brazilian Army M41 light tanks into the M41B/M41C configuration.

The original 500 hp petrol engine has been replaced by a Brazilian-built Saab-Scania DS-14A 04 eight-cylinder diesel developing 405 hp which is coupled to the standard CD-500-3 transmission. To install this diesel engine the rear hull has been enlarged and a new cooling system consisting of a radiator and two 12-blade fans has been installed. Other improvements include a modified electrical system that has two 12 V batteries and four 60 Ah alternators (eliminating the need for the generator previously fitted), new instrument panel and redesigned fuel tanks.

The first 20 M41Bs retain the original 76 mm gun M32. For trials purposes one prototype was fitted with the standard Cockerill 90 mm Mk III gun. The following 120 to 200 M41Bs were fitted with the M32 gun bored out to fire the same 90 mm ammunition as the Cockerill 90 mm Mk III gun used in the Brazilian Army's ENGESA EE-9 armoured cars. This gun is designated the Ca 76/90 M32 BR1. The ordnance is shorter than the original and has a counter balance at the forward end of the barrel that gives the impression of being a bore evacuator, the original muzzle brake is retained and a torsion bar compensator is fitted at the breech.

From 1984 all rebuilt M41s were fitted with the new Ca 76/90 M32 BR2 gun which retains the basic length of the M32 and has a bore evacuator and muzzle brake and fires an APFSDS round. This model is known as the Carro Padrao (Standard Tank) and has a thermal sleeve for the 90 mm gun, side skirts and additional spaced armour for the forward part of the hull, glacis plate and turret. Either side of the turret are four smoke dischargers. The optronic fit can include night vision equipment and a laser rangefinder coupled to the gun sight. The transmission has been upgraded which improves acceleration and gives a maximum road speed of 70 km/h.

By March 1985, 386 M41 tanks had been upgraded by Bernadini for the Brazilian Army. The Brazilian Marine Corps has ordered 35 M41C tanks, these being M41Bs with an up-to-date fire control system and the 90 mm Ca 76/90 M32 BR3 gun. The Army conversion programme is now complete.

Danish M41s

Details of the M41 DK-1, in service with the Danish Army and being offered for export, are given under Denmark.

Spanish M41 Programme

The Spanish Army has 127 M41 light tanks in service with perhaps another 120 in reserve. Of these 100 may be converted for use in the anti-tank role and the remainder used in their original light tank role but with a new powerpack and possibly a 90 mm gun. Talbot of Spain has fitted an M41 with the General Motors 8V-71T diesel engine developing 400 hp at 2500 rpm, coupled to the original CD-500-3 transmission. The engine gives the M41 an operating range of 500 km and a maximum road speed of 72 km/h. The 76 mm gun is retained although a Cadillac Gage weapon control system has been installed. The Talbot designation for this vehicle is the M41E, with the E standing for Espana (Spain).

To meet the anti-tank requirement Talbot has rebuilt the M41 with the same powerpack as the M41E but with either the Euromissile HCT turret with four HOT ATGWs in the ready-to-launch position or the American Emerson ITV system as fitted to the Improved TOW Vehicle M901 used by the United States and other armies. As of early 1991 none of these upgraded vehicles had been ordered in quantity by the Spanish Army. Some $18 million has, however, been earmarked for M41 light tank improvements in the future.

Taiwanese M41s

In 1990 it was reported that Taiwan had earmarked $2.11 million for upgrades to the M41 light tank and M42 twin 40 mm self-propelled anti-aircraft gun system, both of which have similar chassis. It is understood that this upgrade will be concentrated on the powerpack, probably the replacement of the current petrol engine by a diesel. A total of $1.38 million was also allocated for the prototype of an upgraded M41 with new fire control system and gun control equipment.

M41 with New Powerpack

In addition to the Spanish and Brazilian repowered M41s previously mentioned there are at least three other repower packages being offered, although by June 1991 none of these had been sold.

NAPCO of the USA has designed a powerpack for the M41 using the Detroit Diesel 8V-71T engine developing 450 hp at 2500 rpm coupled to the original transmission, so giving an operating range of over 450 km. This vehicle is known to have been tested in Denmark, Spain and Thailand.

TECMOTIV of Canada has designed a powerpack for the M41 using the Cummins VTA-903T engine, as installed in the FMC Bradley Infantry Fighting Vehicle, coupled to the original transmission. This vehicle has a maximum operating range of 563 km.

FFG of Germany has designed a powerpack for the M41 that uses the British Perkins Engines (Shrewsbury) Condor CV-8 diesel engine which powers the Warrior, now in volume production for the British Army. This also retains the original transmission.

M41 GTI

In early 1986 the M41 GTI (German Tank Improvement) was tested in Thailand. Prime contractor for the M41 GTI is GLS of Munich which is also responsible for the chassis and Wegmann of Kassel which is responsible for the complete turret.

The original petrol engine has been replaced by an MTU MB 833 Aa-501 diesel developing 450 hp which via a new intermediate gearbox is coupled to the original transmission. Quick disconnect couplings allow the rapid removal of the complete powerpack. The installation of the new diesel engine has increased its range of action from 161 km to over

M41 fitted with Belgian Cockerill 90 mm gun Mk IV

GLS M41 German Tank Improvement

M41 light tank fitted with Cockerill LCTS 90 turret

M41 light tank chassis fitted with 105 mm armed Cadillac Gage Low Recoil Force Turret which is also on Stingray light tank

600 km on roads although total fuel capacity has also been increased from 530 litres to 800 litres.

The electrical components have been updated and the driver has a new instrument panel. A Halon fire suppression system is installed in the hull to protect the fighting compartment. A Halon firing warning and extinguishing system also protects the engine compartment.

Modifications to the suspension include new torsion bars, dual-action telescopic hydraulic shock absorbers and hydraulic bump stops for the road wheel arms. New connector type Diehl tracks with replaceable rubber pads replace the older tracks.

To reduce dust, skirts are installed on both sides of the vehicle and if required ballistic skirts can also be fitted.

The 76 mm gun is retained on the prototype although a 90 mm or 105 mm gun could be installed. The turret drive system has been replaced by a new electric system called the ELAN 41 T which is also stabilised. The fire control system is the MOLF 41 which includes a digital computer, laser rangefinder, cant and gun elevation sensors and an air sensor on the turret roof.

The periscopic gunner's sight has both day and night (thermal) capability and incorporates a laser rangefinder. The tank commander or gunner can engage stationary or moving targets while the tank itself is moving.

Mounted either side of the turret is a bank of six electrically operated smoke dischargers which can also fire fragmentation grenades.

Attached to the turret bustle is a collective NBC protection system. A new vehicle intercom system has been installed as has a new frequency-hopping radio system.

As of early 1991 this system was still at the prototype stage and had yet to enter production.

M41 with Cockerill LCTS 90 Turret

In addition to offering a gun retrofit package for the M41 light tank (see below), Cockerill Mechanical Industries have designed a two-man power-operated turret of all-welded steel construction called the LCTS 90 turret. Main armament comprises a Cockerill 90 mm Mk 7 gun, a 7.62 mm coaxial machine gun and, as an option, a 7.62 mm anti-aircraft machine gun and two banks of electrically operated smoke dischargers.

As of early 1991 this version had not entered production and was still at the prototype stage.

M41 with Cockerill 90 mm Gun

The Belgian company Cockerill has replaced the 76 mm gun of the M41 with its 90 mm gun Mk IV which fires the following types of ammunition: HEAT-T, HESH-T, HE-T, Smoke-WP-T and Canister. This model has already been adopted by Uruguay.

NIMDA Upgrade Package for M41

The Israeli company NIMDA offers a complete retrofit package for the M41 light tank which includes some or all of the following: replacement of the 76 mm gun by an Israeli 60 mm HVWS, new powerpack incorporating a Detroit Diesel engine, advanced fire control systems and optics, new suspension and improved protection and survivability. Two powerpack options are available, the 8V-71T developing 470 hp coupled to the existing CD-500-3 or an 6V-92T also developing 470 hp but coupled to X-300 or the CD-500-3 transmission.

M41 with 105 mm Turret

For trials purposes the M41 light tank chassis has been fitted with the private venture Cadillac Gage 105 mm Low Recoil Force Turret, which is also fitted to the Cadillac Gage Stingray light tank and V-600 (6×6) armoured car.

Status: Production complete. In service with:

Country	Quantity	Comment
Brazil	430	army and marines, upgraded
Chile	60	
Denmark	53	upgraded to DK-1 standard
Dominican Republic	12	
Guatemala	10	delivered 1982
Somalia	10	
Spain	247	training and reserve
Taiwan	650	could be more, some upgraded to Type 64
Thailand	150	most now in store
Tunisia	10	
Uruguay	22	some upgraded with Cockerill 90 mm

Note: M41 light tanks are no longer used by Argentina, Austria, Ethiopia, Greece, Japan, Philippines, Sudan or the United States.

Manufacturer: Cadillac Motor Car Division of General Motors Corporation
Cleveland Tank Arsenal
Ohio
USA

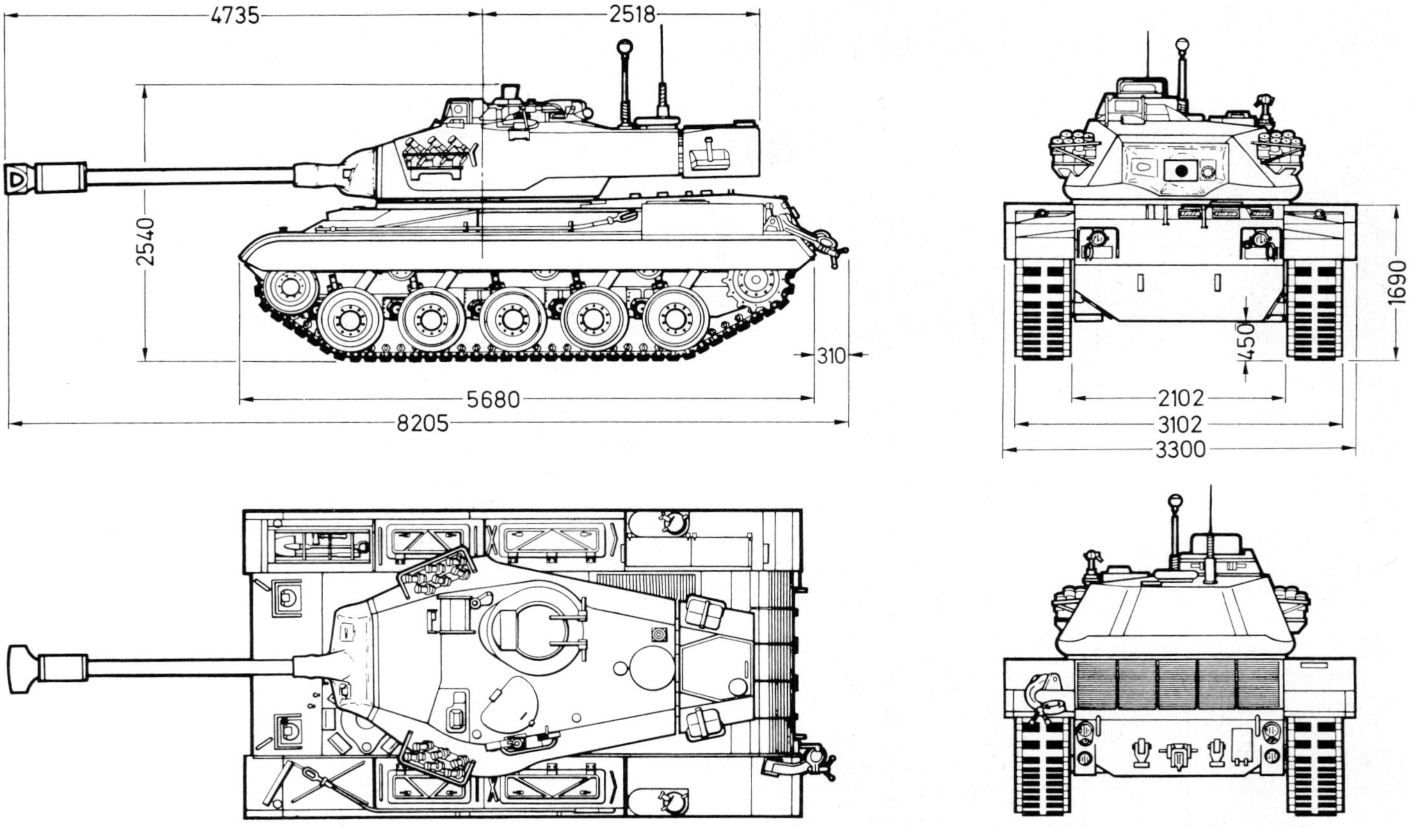

Four-view drawing of the GLS M41 German Tank Improvement showing modifications to rear of hull (not to 1/76th scale)

SPECIFICATIONS

CREW	4
COMBAT WEIGHT	23 495 kg
UNLOADED WEIGHT	18 457 kg
POWER-TO-WEIGHT RATIO	21.26 bhp/tonne
GROUND PRESSURE	0.72 kg/cm²
LENGTH GUN FORWARDS	8.212 m
LENGTH HULL	5.819 m
WIDTH	3.198 m
HEIGHT	
to cupola	2.726 m
including MG	3.075 m
GROUND CLEARANCE	0.45 m
TRACK	2.602 m
TRACK WIDTH	533 mm
LENGTH OF TRACK ON GROUND	3.251 m
MAX ROAD SPEED	72 km/h
FUEL CAPACITY	530 litres
MAX RANGE	161 km
FUEL CONSUMPTION	3.29 litres/km
FORDING	1.016 m
with preparation	2.44 m
GRADIENT	60%
SIDE SLOPE	30%
VERTICAL OBSTACLE	0.711 m
TRENCH	1.828 m
TURNING RADIUS	skid turns
ENGINE	M41 and M41A1, Continental AOS-895-3, 6-cylinder air-cooled super-charged petrol developing 500 bhp at 2800 rpm. M41A2 and M41A3 Continental or Lycoming AOSI-895-5 with fuel injection also developing 500 bhp at 2800 rpm
AUXILIARY ENGINE	GMC model A41-1
TRANSMISSION	GMC Allison Division cross-drive model CD-500-3 with 1 forward and 1 reverse ranges
FINAL REDUCTION RATIO	4.25 : 1
SUSPENSION	torsion bar
ELECTRICAL SYSTEM	24 V
BATTERIES	4 × 6TN, 100 Ah
ARMAMENT	
main	1 × 76 mm gun
coaxial	1 × 7.62 mm MG
anti-aircraft	1 × 12.7 mm MG
AMMUNITION	
main	57 rounds M41, 65 rounds M41A1, M41A2 and M41A3
coaxial	5000
anti-aircraft	2175
GUN CONTROL EQUIPMENT	
Turret power control	hydraulic/electric with manual controls
by commander	yes (only on M41A1 and later models)
by gunner	yes
Max rate power traverse	360° in 10 s
Gun elevation/depression	+19° 45'/−9° 45'
Gun stabiliser	
vertical	no
horizontal	no
Rangefinder	none
Elevation quadrant	Type M9
Gunner's sight	M97 telescope with ×8 magnification and 7° 24' field of view

Both commander and gunner have M20A1 periscope with 2 optical systems: ×1 system for observation and ×6 for ranging on targets. Illumination of graticule patterns provided for by Light Instrument M36

ARMOUR	
Hull front glacis	25.4 mm at 30°
Hull front nose	31.75 mm at 45°
Hull top	12/15 mm
Hull floor	9.25/31.75 mm
Hull rear	19 mm
Turret mantlet	38 mm
Turret front	25.4 mm
Turret sides	25 mm
Turret rear	25 mm
Turret roof	12.7 mm

Cadillac Gage Commando Stingray Light Tank

Development

The Commando Stingray light tank has been developed by Cadillac Gage Textron as a private venture. In the late 1970s/early 1980s the company noted that a number of armies required a vehicle with the firepower of a main battle tank but with much greater tactical and strategic mobility. Before design work began a number of priorities were established. These included high firepower with a 105 mm gun that would fire standard NATO ammunition, high mobility, large operational range, low profile for increased survivability, use of proven components wherever possible, light weight and ability to be carried in a C-130 Hercules transport aircraft.

The concept phase of Stingray development started in January 1983 with actual design work beginning in September the same year. Prototype construction started in February 1984 and the vehicle, without its turret, ran for the first time in August 1984. Design work on the turret began in mid 1983, the prototype turret was completed in June 1984 and installed on an M551 Sheridan light tank chassis for trials.

The Stingray light tank was shown in public for the first time in October 1984 at the Association of the United States Army (AUSA) meeting in Washington DC. By late 1985 the prototype of the Stingray light tank had completed over 5000 km of automotive trials, most of them at the Hodges Automotive Test Centre at Carson City, Nevada.

As a result of these trials only minor modifications were required including improved final drives, new shock absorbers and a modified cooling system to allow Stingray to operate in ambient temperatures of 120° F.

Three production Cadillac Gage Textron Stingray light tanks with all hatches closed and showing roof-mounted anti-aircraft machine guns

Future improvements could include a thermal imager and a telescope mounted coaxially with the main armament as a back-up sight for the gunner's M36E1 roof-mounted day/night sight.

The prototype Stingray light tank went to Thailand for trials early in 1986 and the second prototype Stingray, which was built to production drawings, was completed in mid 1986. Early in 1987 the first prototype of the Stingray was demonstrated in Malaysia.

In October 1987 Cadillac Gage Textron signed an agreement with Thailand for the supply of 106 Commando Stingray light tanks to the Royal Thai Army at a total cost of $150 million. First production, Stingray light tanks were completed in December 1988 and shipped direct to Thailand and final deliveries early in 1990.

Early in 1988 the Stingray light tank was evaluated in Ecuador alongside the SK 105, TAM and AMX-13.

Description

The hull of the Stingray is of all-welded Cadloy steel armour with protection over the frontal arc against penetration by 14.5 mm armour-piercing rounds fired from the Soviet KPVT heavy machine gun and protection against 7.62 mm armour-piercing rounds over the remainder of the vehicle. Some potential customers have indicated that they require increased armour protection and Cadillac Gage is now studying the possibility of adding explosive reactive or ceramic armour to provide increased protection.

The layout of the Stingray is conventional with the driver's compartment at the front, fighting compartment in the centre and engine and transmission at the rear.

The driver sits in the centre of the hull at the front and has a single piece hatch cover that is hinged at the rear and can be locked open. The hatch cover has three integral periscopes that give good visibility through 120° of the vehicle's front. The centre periscope can be replaced by a passive periscope for night driving. The driver has an adjustable seat and would normally enter his position via the turret. With the turret traversed fully left or right or with the 105 mm gun in its travel lock, the driver can easily enter via his hatch. The driver steers the Stingray with an oval steering wheel rather than the more usual sticks. Either side of the driver's position are 14 rounds of 105 mm ammunition. When in a combat area an optional spall blanket is hung to the sides and rear of the driver's position for added protection. The Stingray has additional internal stowage space above the tracks with an access hatch in the glacis plate either side of the driver.

The Stingray suspension is of the torsion bar type and based on that fitted to the M109 155 mm self-propelled howitzer which has been in service all over the world for over 30 years.

The suspension either side consists of six dual rubber-tyred road wheels, drive sprocket at the rear, idler at the front and three track return rollers, the last from the M41 light tank. The Cadillac Gage designed torsion bars are longer than those of the M109 and are of 166 000 psi steel with a 203.2 mm jounce, 50.8 mm rebound and bottom at 2.5 g loading. The first and last torsion bars are stiffer than the remainder and the first, second and sixth road wheel stations have a hydraulic shock absorber.

The road arms were originally designed to carry 1814 kg but in the Stingray carry only 1588 kg, therefore leaving plenty of scope for additional weight, for example appliqué armour, should this be required, without overloading the suspension.

The tracks are 38 cm wide and are of the double-pin type with detachable pads. Under development are new tracks which will be much lighter and longer lasting with a weight saving of 1000 kg per vehicle.

The Stingray is powered by the Detroit Diesel, Model 8V-92 TA eight-cylinder diesel developing 535 bhp at 2300 rpm. The engine is mounted crossways to make maximum use of the available space. The engine has twin turbo-chargers, Roots blower and an after cooler.

The engine is coupled to a Allison Transmission Division of General Motors, XTG-411-2A automatic transmission which has been installed in the much heavier M107, M108, M109 and M110 self-propelled artillery weapons for many years as well as the M992 Field Artillery Ammunition Support Vehicle (FAASV) and the M578 armoured recovery vehicle. This transmission has an automatic lock-up torque converter with a torque multiplication of 2.4, geared steer and clutch and brake steering. The four to one planetary final drive gearboxes have been designed and built by Cadillac Gage Textron specifically for the Stingray.

Cooling is by three-vane axial fans each of which move 9000 cfm, the radiators and oil coolers are sized to provide adequate cooling to 120° F. The air inlets are above the engine compartment with the air outlet at the left rear of the hull and exhaust outlet on the right side the hull. Special attention has been given to make the infra-red signature of the vehicle as small as possible. An extra heavy duty air cleaner scavenges heavier dirt into the engine exhaust.

Electrical power is provided by a 650 A oil-cooled generator and stored in six batteries in the left side of the hull. Access to these batteries for daily checks is via a door in the side of the hull. The batteries are capable of furnishing start-up current in temperatures down to -25° F.

The rear engine compartment panels fold down for access to the engine compartment while the ones on the top fold rearwards. Daily checks can easily be carried out on the engine.

The 757-litre fuel tank is integral and located between the crew and engine compartments.

Cadillac Gage Commando Stingray light tank from rear showing external turret stowage

Prototype of Cadillac Gage Commando Stingray light tank with roof-mounted 12.7 mm MG

The Royal Ordnance Nottingham 105 mm Low Recoil Force (LRF) gun has a muzzle brake added, redesigned fume extractor, cradle design and a new recoil system. The new cradle design is a monocoque structure that combines high rigidity with light weight.

The standard 105 mm L7A3/M68 gun has a recoil of 300 mm which takes 70 milliseconds and achieves a maximum trunnion force of 75 000 pounds. The LRF gun has a longer recoil and achieves a maximum trunnion force of 30 000 pounds. In addition to the new recoil system with its longer recoil, the muzzle brake also helps to reduce both trunnion pull and recoil energy.

Thirty-two rounds of 105 mm ammunition are carried, all of which are stowed below the turret ring. Eight rounds are stowed in the turret of which three are for immediate use, stowed vertically to the left of the breech, and the remainder under the spent case ejection bag. The three ready rounds are kept in special individual racks which are configured to accept all types of 105 mm ammunition. Once the loader releases the catch the first ready round swings forward as it is pivoted at the base. The spent case ejection bag holds five cases which can be disposed of through the ammunition resupply hatch in the left side of the turret.

Following initial trials with the prototype Stingray a muzzle reference system was fitted to the 105 mm LRF gun and a collapsing breech guard to allow for improved crew movement in the turret.

A 7.62 mm M240 machine gun is mounted coaxially to the left of the main armament and is provided with 400 rounds of ready-use ammunition. A further 2000 rounds of 7.62 mm ammunition are stowed in the turret area in 200-round boxes. It takes the loader seven seconds to replenish the ready-use ammunition for the coaxial machine gun. Some thought is already being given to offering a 12.7 mm machine gun in place of the 7.62 mm coaxial weapon.

A 7.62 mm (with 200 rounds of ready-use ammunition) or a 12.7 mm machine gun (with 100 rounds of ready-use ammunition) is mounted externally at the commander's station for anti-aircraft use. A total of 1000 rounds of machine gun ammunition is carried for the anti-aircraft weapon.

A bank of four Peak electrically operated smoke dischargers is mounted either side of the turret, for which 16 grenades are carried.

The three-man turret of the Stingray has been designed by Cadillac Gage Textron for installation on other chassis as well as on Stingray, including the Cadillac Gage V-600 (6 × 6), M41 and M551 tanks and heavier chassis such as the M47, M48 and Soviet T-series, or any vehicle with a chassis weight exceeding 13 tonnes.

The turret is of all-welded Cadloy steel armour construction which provides the same degree of protection as the hull. The use of flat plates in the construction of the turret enables the user to up-armour the turret at a later date without extensive rework.

The commander sits on the right of the turret with the gunner below and to his front, the loader on the left. The gunner has a roof-mounted Optic-Electronic Corporation M36E1 day/night sight. As an option this can be replaced by an M36E1 SIRE day/night sight incorporating a laser rangefinder or a thermal sight. The commander has an NV-52 day/night sight, seven periscopes for all-round observation and a single piece hatch cover that opens to the rear.

The loader has a single piece hatch cover that opens to the rear and a single forward-facing periscope for observation.

Turret traverse and weapon elevation are via a Cadillac Gage, Control Systems Operations Division, electro-hydraulic system with a manual back-up. This is based on the company's experience in the production of weapon controls for the M1 and M60 series MBTs. Controls are provided for both commander and gunner and turret traverse is a full 360° at 40°/s with weapon elevation from -7.5° to +20° at a similar speed. An automatic deck clearance system is fitted as standard and as an option a two-axis electro-hydraulic stabilisation system can be fitted to enable the 105 mm gun to be laid and fired while Stingray is moving across country.

The prototype of the Stingray light tank is fitted with the British Marconi Command and Control Systems Digital Fire Control System (DFCS) which undertook a highly successful series of trials in 1983 installed in a Cadillac Gage V-150 (4 × 4) vehicle fitted with a Cockerill 90 mm Mk 3 gun and DFCS was also selected for production vehicles.

The Stingray can be fitted with the standard M13A1 ventilated face mask system for NBC protection and the vehicle painted in chemical-resistant paint.

As an option Stingray can be fitted with a land navigation system and engine smoke generator which injects diesel oil into the exhaust to create a smoke screen behind the tank. The prototype is fitted with an HTL engine fire-warning and suppression system as well as a fire detection/suppression system for the crew compartment. Radios and crew intercom depend on the user's requirements.

Variants

Cadillac Gage Textron is now giving consideration to developing a complete family of vehicles all using the same common components, including weapons carriers.

SPECIFICATIONS

CREW	4
COMBAT WEIGHT	21 205 kg
UNLOADED WEIGHT	19 387 kg
POWER-TO-WEIGHT RATIO	25.9 hp/tonne
GROUND PRESSURE	0.72 kg/cm²
LENGTH GUN FORWARDS	9.30 m
LENGTH HULL	6.448 m
WIDTH	2.71 m
HEIGHT	
overall	2.55 m
reduced	2.4 m
GROUND CLEARANCE	0.46 m
TRACK	2.31 m
TRACK WIDTH	380 mm
LENGTH OF TRACK ON GROUND	3.632 m
MAX SPEED	67 km/h
road on 10% slope	29 km/h
road on 60% slope	6 km/h

ACCELERATION (0 to 32 km/h)	7 s
FUEL CAPACITY	757 litres
MAX RANGE (at 40 km/h)	483 km
FORDING	1.07 m
GRADIENT	60%
SIDE SLOPE	40%
VERTICAL OBSTACLE	0.76 m
TRENCH	2.13 m
ENGINE	Detroit Diesel Model 8V-92 TA developing 535 hp at 2300 rpm
TRANSMISSION	Allison Transmission Division of General Motors XTG-411-2A automatic
SUSPENSION	torsion bar
ELECTRICAL SYSTEM	24 V
BATTERIES	4 × 6TN

ARMAMENT	
main	1 × 105 mm gun
coaxial	1 × 7.62 mm MG
anti-aircraft	1 × 12.7 mm MG
AMMUNITION	
main	32
7.62 mm	2400
12.7 mm	1100
GUN CONTROL EQUIPMENT	
Turret power control	electro-hydraulic/manual
by commander	yes
by gunner	yes
Max rate of turret traverse	360° at 30°/s (stabilised)
Elevation speed	8°/s 40°/s (stabilised)
Gun elevation/ depression	+20°/-7.5°
Gun stabiliser	optional

Status: Production complete. In service with Thailand (106). Will be placed back in production on receipt of additional orders.

Manufacturer: Cadillac Gage Textron
PO Box 1027
Warren
Michigan 48090
USA
Telephone: (313) 777 7100
Telex: 200707
Fax: (313) 776 9731

Note: Production is undertaken at the Cadillac Gage Textron facility in Cocoa, Florida.

FMC Close Combat Vehicle Light

Development

The Close Combat Vehicle Light (CCV L) has been developed by the Defense Systems Group of the FMC Corporation as a private venture in response to a US Army requirement for an Armored Gun System (AGS).

The Defense Systems Group started work on the CCV L in October 1983.

At a very early stage a number of key design goals were drawn up for the CCV L, which included mobility comparable with the M1 MBT and M2 Bradley IFV, lethal main armament that would defeat threat tanks, BMPs and other battlefield targets. It should need the minimum of logistic support, be airportable by C-130 and C-141 transport aircraft and be able to use LAPES (Low Altitude Parachute Extraction System) from a C-130 aircraft, be survivable and have an early operational capability by using proven and in-production components.

The cross-section size of the C-130 fuselage and the C-141 floor loading limitations were constraints in the design of the CCV L. To have the LAPES capability the CCV L must have a maximum weight of 16 783 kg, which will be achieved by carrying less on-vehicle equipment such as fuel and ammunition, by removing track pads and carrying it without the crew. A C-130 transport will be able to carry one CCV L, the C-141 two and the future C-17 three.

The US Army has recently added a requirement for Low Velocity Air Drop (LVAD). This will further reduce the maximum weight to 16 136 kg. To meet this new objective design modifications to the FMC CCV L are currently in progress.

The first prototype of the CCV L was completed in August 1985 and following initial mobility and firing trials was airlifted in a civilian Lockheed C-130 Hercules transport aircraft from Oakland Airfield and flown non-stop to National Airport, Washington DC. It was subsequently shown in public for the first time at the Association of the United States Army meeting in October 1985.

Description

The CCV L has a hull of all-welded aluminium construction with bolt-on steel to enhance ballistic protection level. If the user required additional protection this could be provided with the aid of appliqué armour, including possible reactive armour.

The driver is seated at the front of the vehicle in the centre and has a single-piece hatch cover that opens to the rear. This has five integral wide-angle periscopes for driving when closed down and in addition, provisions are made for mounting a driver's passive night driving periscope.

Suspension is of the torsion bar type with six dual rubber-tyred road wheels, idler at the front and drive sprocket at the rear, and no track return rollers. High temperature shock absorbers, are provided at the first, second, third, fourth and sixth road wheel positions. The suspension spring system provides a 250 mm jounce.

The front-mounted idler assembly is composed of standard M113A3 and M2 parts already in the US Army inventory. The track is of the steel double-pin type with detachable rubber pads, 152 mm pitch and 381 mm wide, and has been developed by FMC's Steel Products Division.

The CCV L is powered by a six-cylinder Detroit Diesel Model 6V-92 TA developing 552 hp at 2400 rpm. This uses 65 per cent common components with the 8V-92 TA engine installed in the Oshkosh Heavy Expanded Mobility Tactical Truck (HEMTT) already in service with the US Army in significant numbers.

The engine is coupled to a General Electric HMPT 500-3 hydromechanical transmission, which is already installed in the FMC M2 Bradley IFV, with a final drive ratio of 4.4 to 1.

The rear engine compartment roof is raised well above the turret ring so the 105 mm gun cannot be depressed at all over the rear arc. The air inlet/outlets are in the roof of the engine compartment with the exhaust at the left.

At the rear of the hull is a hydraulically actuated access ramp which, when horizontal, provides a work platform for ease of maintenance. Trained personnel can remove the complete powerpack to the ramp platform, where it can be ground hopped in less than five minutes. The engine, transmission and cooling system are mounted on rails and can easily slide out onto this ramp. The M2 Bradley fan is belt driven from a transmission-mounted PTO.

The turret is of all-welded aluminium with an additional layer of steel armour like the hull, with the automatic loader on the left and commander and gunner on the right.

The gunner is seated forward of the commander and has a single-piece hatch cover that opens forwards. Well forward of this is the Hughes Aircraft Company's two-axis stabilised day/thermal night sight with laser rangefinder. Both commander and gunner have thermal displays switchable to each other's sights, so enabling the fire-control system to be used as a hunter/killer. As a back-up, the gunner also has a fibre-optic auxiliary sight mounted coaxially to the right of the main armament.

The commander is seated to the rear of the gunner and has a cupola with eight periscopes identical to the driver's for all-round observation (unity vision) and a single-piece hatch cover that opens to the rear.

In front of the commander's position is the optional commander's independent thermal viewer (CITV) developed by Hughes Aircraft Company. This is two-axis stabilised with a day/night capability and a 360° panoramic capability. When not being used by the commander the CITV retracts within the turret like a periscope.

Turret and gun drives have been developed by the Control Systems Division of Cadillac Gage Textron and are based on those installed in the M1A1 MBT currently in production and service with the US Army. The gunner has twin control handles while the commander has a single control handle.

The 105 mm gun is fully stabilised in both axis with an elevation of +20° and a depression of -10° through a frontal 270° arc. Turret traverse

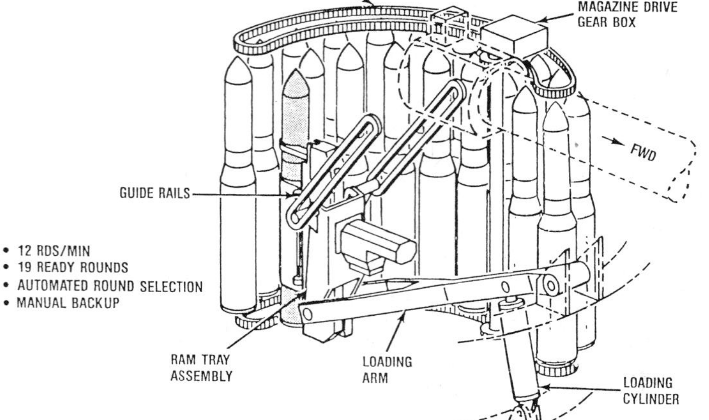

Autoloader for FMC CCV L designed by FMC's Naval Systems Division enables rate of fire of 12 rds/min to be achieved

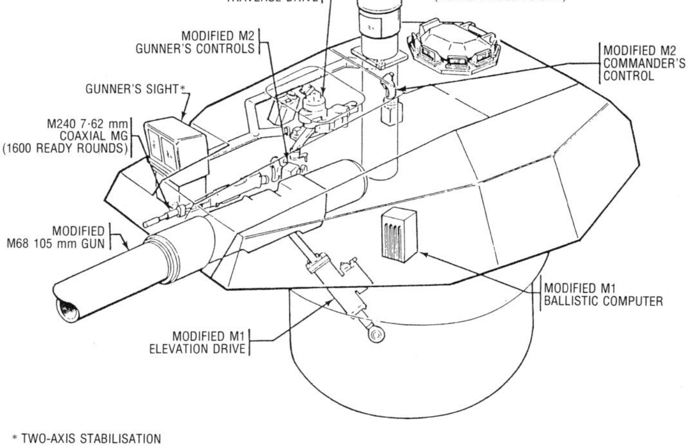

Commander and gunner sit on right side of power-operated turret with automatic loader on left

FMC CCV L with turret traversed partly left and CITV in raised position in front of commander's cupola

FMC CCV L with rear ramp lowered to allow access to the complete vehicle powerpack

is a full 360°. Turret traverse and weapon elevation are hydraulic with manual controls for emergency use.

The Computing Devices Corporation of Canada digital fire control computer is a modified version of that installed in the M1 MBT. The 105 mm gun is fitted with a muzzle reference system, and mounted on the turret roof at the rear is a wind sensor.

Main armament comprises a 105 mm M68A1 tank gun with an inverted breech fitted with a low recoil system, gun mount and muzzle brake developed by Rheinmetall of Germany.

Mounted in the left side of the turret is an automatic loader designed by FMC's Naval Systems Division which enables a rate of fire of 12 rounds a minute to be achieved.

The automatic loader holds 19 rounds of 105 mm ammunition stowed vertically base downward, a chain drive moves the rounds along a crescent-shaped pathway to the loading position. A loading arm, with a ram tray assembly, then takes the round and inserts it in the breech of the gun in one upward and forward movement.

Once the gun has fired it recoils 533 mm to the rear, and returns to battery. The 105 mm gun returns to zero degrees for case ejection and reloading. The breech is hydraulically opened and the empty case is ejected through the loading port in the turret bustle. When the gun is depressed below zero degrees, a hydraulically operated gun cover in the roof lifts to allow the weapon to be fully depressed.

The automatic loader is separated from the crew compartment by an NBC proof bulkhead as once the breech opens or a spent cartridge is ejected, the ammunition compartment would be contaminated if the vehicle were operating in an NBC enviroment. Blow-out panels are provided in the turret roof.

The 105 mm gun fires all standard US tank ammunition including APFSDS-T, HEAT and HESH and their respective training rounds.

A standard 7.62 mm M240 machine gun is mounted coaxially to the right of the main armament, provided with 1600 rounds of ready use ammunition with a further 3400 rounds of ammunition being carried in reserve.

The prototype of the CCV L is fitted with the Advanced Smoke Launcher System developed as a private venture by Tracor MBA. Either side of the turret is a bank of 16 launcher tubes which can fire L8 visual or M76 infra-red obscurants. Inside the turret is a flat panel display allowing the operator to select left or right smoke launchers and the type of round to be fired.

At present the prototype of the CCV L does not have an NBC system although it is envisaged that this would be fitted to production vehicles.

Although the prototype is fitted with the Detroit Diesel 6V-92 TA engine coupled to the General Electric HMPT 500-3 transmission, other engine/transmission options are being considered including the Cummins VT-903 diesel coupled to HMPT 500 transmission and Detroit Diesel 6V-53T coupled to X-200-5 series automatic transmission.

For the future additional armour and other systems are being considered when they have been fully developed, for example a battlefield management system, vetronics, rapid rearm capability, improvements to the fire-control system and hydro-pneumatic suspension systems (eg Cadillac Gage Textron, Teledyne or FMC Corporation).

Variants

In 1985 the International Division of FMC Corporation's Defense Systems Group entered into an agreement with Vickers Defence Systems of the United Kingdom under which Vickers was to design a new and less sophisticated three-man turret armed with a 105 mm gun. This vehicle is called the VFM 5 tank and is described in the *MBTs and medium tanks* section.

SPECIFICATIONS

CREW	3
COMBAT WEIGHT	19 414 kg
EMPTY WEIGHT	17 509 kg
POWER-TO-WEIGHT RATIO	28.4 hp/tonne
GROUND PRESSURE	0.69 kg/cm²
LENGTH	
gun forward	9.37 m
hull	6.197 m
WIDTH	
overall	2.692 m
over tracks	2.54 m
HEIGHT	
CITV raised	2.781 m
turret roof	2.349 m
rear engine	1.701 m
GROUND CLEARANCE	0.406 m
TRACK WIDTH	381 mm
LENGTH OF TRACK ON GROUND	3.657 m
MAX ROAD SPEED	70 km/h
ACCELERATION (0 to 32 km/h)	6 s
RANGE (at 40 km/h)	483 km
FUEL CAPACITY	568 litres
FORDING	1.32 m
GRADIENT	60%
SIDE SLOPE	40%
VERTICAL OBSTACLE	0.762 m
TRENCH	2.133 m
ENGINE	Detroit Diesel Model 6V-92 TA developing 552 hp at 2400 rpm
TRANSMISSION	General Electric HMPT 500-3 hydromechanical
STEERING	hydrostatic
BRAKES	multi-disc, oil cooled
FINAL DRIVE RATIO	4.4:1
SUSPENSION	torsion bar
ELECTRICAL SYSTEM	24 V
BATTERIES	4 × 6TN
GENERATOR	300 Ah
ARMAMENT	
main	1 × 105 mm gun
coaxial	1 × 7.62 mm MG
smoke dischargers	2 × 16
AMMUNITION	
main	19-43
coaxial	5000
smoke grenades	32 (ready use)
GUN CONTROL EQUIPMENT	
Turret power control	hydraulic/manual
by commander	yes
by gunner	yes
Gun elevation/depression	+20°/-10°
Turret traverse	360°
Gun stabilised	yes
Sights stabilised	yes

Status: Prototype.

Manufacturer: FMC Corporation, Ground Systems Division
2830 De La Cruz Boulevard
Santa Clara
California 95052
USA
Telephone: (408) 289 0111
Telex: 6714210
Fax: (408) 289 2150

Prototype of FMC CCV L with turret traversed to front

Teledyne Continental Direct Fire Support Vehicle

Development

In the late 1970s Teledyne Continental Motors (TCM) believed that there was possibly a market for a highly mobile tracked lightweight combat chassis for light forces that could be used for a variety of missions on the future battlefield, and various ideas were studied in 1980-81.

By 1982 the concept was far enough advanced to start detailed design work as a company funded project and soon afterwards it was decided to build a complete prototype of the vehicle, which was then referred to as the TCM Armored Gun System (AGS) or TCM-20 chassis. More recently it has been renamed the Direct Fire Support Vehicle (DFSV).

The first chassis was completed in December 1983 with the turret with its externally mounted 105 mm gun being completed in mid 1984. In October 1984 the chassis went to the Nevada test centre for durability trials and in April 1985 the chassis and turret were mated for the first time. In May 1985 the complete system was shown for the first time at the US Army Armor Conference at Fort Knox.

In 1986 the vehicle underwent both unmanned and manned firings, including fire-on-the-move. These were accomplished with full autoloading under all conditions.

In designing the DFSV, TCM decided on a number of key requirements at a very early stage. These include air transportability in a C-130 or C-141 transport aircraft, wherever possible proven and off-the-shelf components would be used, it would have a 105 mm gun fed by an automatic loader, a fire control system with the same accuracy as current M60A3 MBTs, high cross-country mobility and a low profile, both of which would help its survivability on the battlefield.

It would also be capable of being air dropped from C-130 Hercules transport aircraft so meeting the LAPES (Low Altitude Parachute Extraction System) requirement.

It was also decided that a common chassis would be designed which could be used for a wide range of other missions with the minimum of modifications.

One of the key design features of the Direct Fire Support Vehicle was a high degree of survivability and the vehicle is designed with a low silhouette positioning the crew low down in the hull. The low profile turret with its 105 mm gun allows the use of more armour to protect the crew space.

If the vehicle were placed in production, TCM would provide the final drives and hydro-pneumatic suspension and carry out vehicle integration, assembly and trials prior to delivery.

The Direct Fire Support Vehicle will be offered to the US Army to meet its requirement for an Armored Gun System.

TCM Direct Fire Support Vehicle showing 105 mm gun with fume extractor and muzzle brake

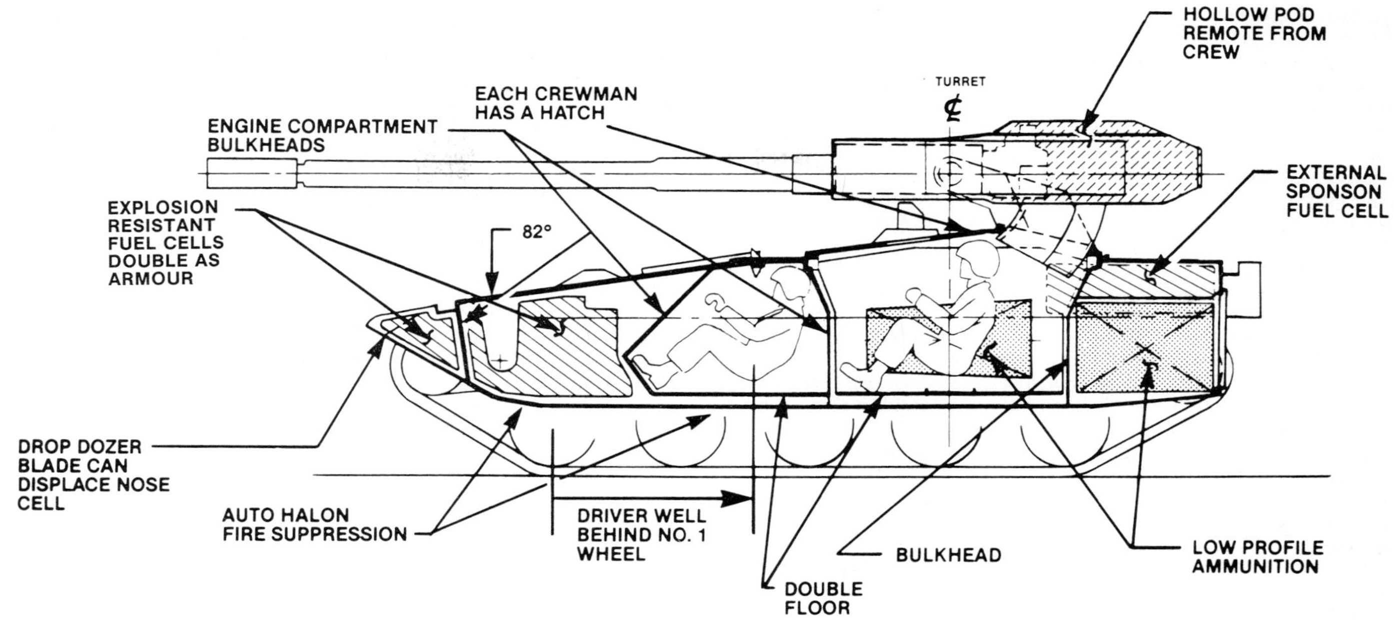

Cross-section of TCM Direct Fire Support Vehicle showing position of main components and externally mounted gun

The weight of the system will depend on its mission. For example, for applications other than that for the Armored Gun System, a weight of 21 to 25 US tons is being proposed while for the Armored Gun System a weight of 17.5 US tons is being proposed, so allowing it to meet the LAPES/LVAD requirement.

Description

The chassis has the fuel and powerpack at the front of the hull to augment the front glacis armour, with the overhead gun at the rear.

The powerpack consists of the Cummins VTA-903T eight-cylinder turbocharged diesel developing 500 hp at 2600 rpm coupled to a General Electric HMPT-500 hydromechanical transmission, air cleaner and cooling system. This can be removed as a complete unit to facilitate replacement in the field; if required it can also be ground-run outside. The engine and transmission are already in the US Army logistic system as they are used in the Bradley IFV/CFV/FVS.

The chassis incorporates RHA steel, steel and ceramic composites, ceramic appliqué, ballistic aluminium and Kevlar for armour protection. Exact levels are classified but include some RPG protection and protection from 23 mm penetration over the frontal arc.

The hull floor is double spaced for improved protection against mines and at the nose is an explosion resistant fuel cell which can, if required, be replaced by a dozer blade similar to those installed on Soviet tanks to enable the DFSV to prepare its own fire position or to clear battlefield obstacles. To the rear of this is another bulkhead, another fuel cell, and then the powerpack.

The driver is seated on the left side of the hull with the transmission to his front and engine to his right with another bulkhead to his rear. The driver's position is well to the rear of the first road wheel station so increasing his survivability in the event of the vehicle running over a mine.

The driver has a single piece hatch cover opening to the rear and three periscopes for forward observation.

The chassis has the TCM Model 2885 hydro-pneumatic suspension system with five units each side, drive sprocket at the front, idler at the rear and two track return rollers. The road wheels, wheel hubs and idler wheels are identical to those used on the 155 mm M109 self-propelled howitzer. Track is the T136 as used on the M109 or the new XT154 lightweight track.

Each Model 2885 unit weighs 72.57 kg and has a total of 482 mm of wheel travel, 381 mm jounce and 101 mm of rebound.

Main armament comprises the standard US Army 105 mm M68A1 rifled tank gun which is already installed in US Army M48A5, M60, M60A1, M60A3 and M1 MBTs, for which ammunition is available from many sources all over the world.

The 105 mm gun is mounted externally so presenting a very small target to enemy weapons and increasing the survivability of the vehicle. As the turret is to the rear of the vehicle there is much less gun overhang than on most MBTs.

The ordnance has been inverted and fitted with a muzzle brake and a long recoil mechanism (760 mm) developed by ARES from its 75 mm automatic cannon system.

The 105 mm M68A1 gun is fed from an autoloader designed by ARES and consists of a nine-round magazine which in turn is fed from two 10-round transfer drums located at the rear, so giving a total of 29 rounds of ready use ammunition. A further 11 to 15 rounds of 105 mm ammunition are stowed in the hull rear and loaded into the transfer drums manually as required.

As an alternative to the 105 mm M68A1 gun, the more recent 105 mm EX-35 could be fitted. This gun is fitted to the prototypes of the US Marine Corps Light Armored Vehicle - Assault Gun, developed by Cadillac Gage Textron.

Ammunition is transferred from the nine-round magazine to an automatic loader and then via cams to the breech.

The automatic loader enables a rate of fire of 10 rounds a minute to be achieved with up to five different types of ammunition being stowed.

The empty cartridge cases are automatically ejected externally and the system has a download capability enabling the type of ammunition to be changed if a different target presents itself. Misfired rounds can also be ejected externally.

Mounted coaxially to the left of the main armament is a 7.62 mm M240 machine gun for dealing with soft targets and a similar weapon can be

TCM Direct Fire Support Vehicle showing low profile

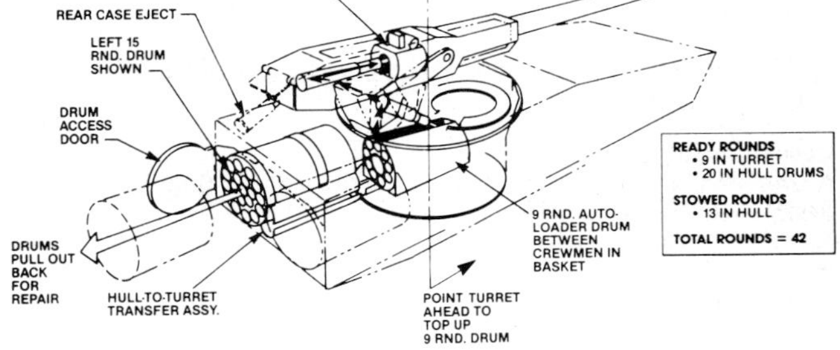

Front-mounted powerpack of TCM Direct Fire Support Vehicle allows rear of vehicle to be kept free for turret and ammunition supply

mounted externally on the commander's cupola. As an alternative a 7.62 mm M60D, 40 mm grenade launcher or 12.7 mm M2 HB machine gun can be mounted in place of the M240 weapon.

The gunner is seated on the left of the low-profile turret with the commander on the right, both crew members have a single piece hatch cover. The commander has seven periscopes with an improved field of view which provide 330° of visibility, there being a blind spot to his immediate left because of the externally mounted weapon station.

The gunner has four M27 periscopes, which are already used in the M60 series of MBT, for observation to the sides and rear through about 225°.

Turret traverse is through a full 360° and gun elevation is from -10° to +18°; the turret drive and weapon stabilisation system is based on that installed in the M2 Bradley IFV and has been designed by the General Electric Company. This is of the electric type which is considered to be safer than hydraulic gun control equipment.

The full solution digital fire control system includes a Texas Instruments low cost modular thermal sight with director head and laser rangefinder with a remote video link for the vehicle commander. The central control panel is located between the commander and gunner so both crew members can see and operate it.

The gunner's sight uses many components that are already in production for the US Army in other applications including the Tank Thermal Sight (TTS) of the M60A3 and the AN/TAS-4 FLIR night sight. The gunner uses a single eyepiece for both day (×1 and ×8) and thermal channels. Both commander and gunner can aim and fire the main armament.

A ×7 periscopic auxiliary sight developed by Kollmorgen provides eyepieces for both commander and gunner and is mechanically linked to the trunnion.

As an option a Commander's Independent Thermal Viewer (CITV) can be located on one side of the externally mounted 105 mm gun with the commander being provided with a video monitor. When not required the CITV could retract to the rear under armour protection.

An HTL Halon fire suppression system is fitted as standard and this is already in the US Army inventory. This covers both crew and engine compartments and can be operated manually or automatically.

All ammunition is stowed low down so further increasing survivability and there is another external fuel cell over the ammunition at the rear.

The prototype TCM Direct Fire Support Vehicle has the same intercom system, radios and antennas as used in the M1/M60 MBTs and the same crew compartment heater. The field telephone and slave connector are also the same as in the M60/M1.

The prototype is not fitted with an NBC system but this would be installed on production vehicles as would a flotation screen, if required by the user. With the screen erected the vehicle would be fully amphibious being propelled in the water by its tracks.

The TCM Direct Fire Support Vehicle has been designed with pre-planned product improvements (P^3I) in mind such as active height control which would enable the 105 mm gun to be depressed even further and give the vehicle unique tactical advantages.

Additional weapons in pods can be mounted either side of the 105 mm gun, for example two TOW or Hellfire ATGWs, Vought hyper-velocity missiles, anti-helicopter weapons or various countermeasures.

Variants

The chassis has also been designed for a wide range of other applications including an infantry fighting vehicle and a long-range anti-tank vehicle (LRAT) with an extendable mast-mounted sensor and vertically launched missiles. In 1985 TCM completed TACOM study contracts for both an infantry fighting vehicle with integrated countermeasures (IFVwCM) and the LRAT as part of the then US Army's Future Close Combat Vehicle System (FCCVS), with the AGS chassis forming the basis of its concepts.

Other variants could include gun/missile air defence vehicle, mortar carrier, ammunition resupply vehicle, maintenance and recovery vehicle, fuel resupply vehicle, and command post vehicle.

The Direct Fire Support Vehicle as built at present has a combat weight of 21 US tons with a width of 2.54 m, but with width increased to 2.794 m combat weight could be increased to around 25 US tonnes so enabling a 120 mm smoothbore gun to be installed as already fitted on the M1A1 MBT.

Low Profile Turret

In September 1987 Teledyne Continental Motors offered the complete weapon system of the TCM Direct Fire Support Vehicle for installation on other vehicles as the Low Profile Turret (LPT). The company envisage that it could be fitted to such chassis as the M60 MBT and the Marine Corps AAV7A1 armoured amphibious assault vehicle. In the latter case it could still carry between 10 and 12 fully equipped Marines.

Since then Teledyne Continental Motors General Products have proposed both light and heavy versions of the Low Profile Turret and brief details of these are given below:

MODEL	**Light**	**Heavy**
MAXIMUM DIAMETER	2.082 m	2.286 m
BARE WEIGHT	4354 kg	5806 kg
MAIN BEARING DIAMETER	1.854 m	1.93 m
BASKET DIAMETER	1.593 m	1.696 m
MAXIMUM SLEW RATE	40°/s	35°/s

Status: Prototype.

Manufacturer: Teledyne Continental Motors General Products
76 Getty Street
Muskegon
Michigan 49442
USA
Telephone: (616) 724 2151
Telex: 228426
Fax: (616) 724 2796

SPECIFICATIONS

CREW	3
COMBAT WEIGHT	19 051 kg
EMPTY WEIGHT	16 057 kg
POWER-TO-WEIGHT RATIO	26 hp/tonne
GROUND PRESSURE (with 381 mm track)	0.696 kg/cm²
LENGTH	
gun forward	7.49 m
hull	6.273 m
WIDTH	
over skirts	2.692 m
over tracks	2.54 m
HEIGHT	
to top of gun mount	2.54 m
to hull top	1.574 m
GROUND CLEARANCE	0.482 m
TRACK WIDTH	381 mm
LENGTH OF TRACK ON GROUND	3.555 m
MAX ROAD SPEED	72 to 80 km/h
FUEL CAPACITY	643 litres
RANGE CRUISING	482 km
FORDING	1.219 m
GRADIENT	60%
SIDE SLOPE	40%
VERTICAL OBSTACLE	0.838 m
TRENCH	2.565 m
TURNING RADIUS	pivot
ENGINE	Cummins VTA-903T V-8 turbo-charged diesel developing 500 hp [or 600 hp]
TRANSMISSION	General Electric HMPT-500 fully automatic
SUSPENSION	hydro-pneumatic
ELECTRICAL SYSTEM	24 V
BATTERIES	4 × 6 TN
ARMAMENT	
main	1 × 105 mm gun
coaxial	1 × 7.62 mm MG
anti-aircraft	1 × 7.62 mm MG
SMOKE-LAYING EQUIPMENT	optional, dischargers and/or engine exhaust
AMMUNITION	
105 mm	28-42
7.62 mm	5000
GUN CONTROL EQUIPMENT	
Turret power control	electric/manual
by commander	yes
by gunner	yes
Gun elevation/depression	+18°/-10°
Turret traverse	360°
Gun stabiliser	
vertical	yes
horizontal	yes

M551 Light Tank/Reconnaissance Vehicle

Development

A detailed development history of the M551 Sheridan, originally designated the Armored Reconnaissance/Airborne Assault Vehicle (AR/AAV) was given in *Jane's Armour and Artillery 1984-85* page 146. A total of 1700 M551s was built for the US Army between 1966 and 1970.

The role of the M551, as originally conceived, is to function as the main reconnaissance weapon for armour, infantry and airborne operations and arms teams not employing the MBT. Late in 1978 it was announced that the M551 would be phased out of service and replaced by the M60A1 MBT, apart from those vehicles allocated to the 82nd Airborne Division (57). Some 330 have been assigned to the National Training Center at Fort Irwin, California. These are essentially basic M551s but with visual modifications to the outside to disguise them as 'aggressor force' vehicles such as ZSU-23-4 SPAAG, T-72 MBT, BMP IFV, BMD airborne combat vehicle, ASU-85 airborne self-propelled anti-tank gun, and M-1974 (2S1) self-propelled howitzer.

As of October 1986, 24 National Training Center vehicles were permanently transferred to Fort Knox, Kentucky, to support 240/360 range complex study/training exercises.

A number of M551 Sheridan light tanks used during the invasion of Panama (Operation Just Cause) late in 1989. In the future the M551s

M551 Sheridan with infra-red searchlight mounted to left of main armament (US Army)

used by the 82nd Airborne Division are likely to be replaced by the projected Armored Gun System.

M551 Sheridan light tanks were among the first US Army vehicles to arrive in Saudi Arabia in the Summer of 1990 following the Iraqi invasion of Kuwait. These vehicles of the 3rd Battalion, 73rd Armored Regiment, subsequently took part in Operation Desert Storm early in 1991.

Description

The all-welded aluminium armour hull of the M551 is divided into three compartments with the driver at the front, fighting compartment in the centre and the engine at the rear.

The driver is seated at the front of the vehicle and is provided with a single piece hatch cover which, when he is driving with his head out, is swung inside the vehicle to his rear. The hatch cover has three integral M47 periscopes for day driving, each with a washer and wiper, while the centre periscope can be replaced by a passive periscope for night driving.

The turret is of all welded steel armour with the commander and gunner seated on the right and the loader on the left. The commander's cupola is provided with a split hatch cover that opens left and right and 10 vision blocks for all-round vision. The commander is also provided with a portable night-vision device with a magnification of ×4.

The gunner is seated to the front and below the commander and has an M129 telescope linked to the main armament with a magnification of ×8 and an 8° field of view. For engaging targets at night the gunner has a thermal sight as used in the M60A3 MBT (qv). The loader has a single piece hatch cover that opens to the left with an M37 periscope which can be traversed through 360° mounted in the roof forward of his hatch cover.

A large wire stowage basket is provided at the rear of the turret and some machine gun ammunition is stowed in ammunition boxes strapped to the sides of the turret, nine boxes on the left and six on the right.

The engine and transmission are at the rear of the vehicle with the aluminium cross-flow radiator mounted in the forward part of the engine compartment from the fighting compartment. The TG-250 transmission consists of a hydraulic torque converter with lock-up clutch. Planetary range gearing is combined with steering and output planetary sets to give four forward and two reverse gears. Pivot turns are possible in first and reverse gears and the brakes are integral with the transmission.

The engine compartment is equipped with a fire suppression system which can be operated by the driver, or from outside the vehicle. No fire warning system is fitted.

The torsion bar suspension consists of five dual road wheels with the idler to the front and the drive sprocket at the rear. There are no track return rollers. The first and last road wheel stations are fitted with a hydraulic shock absorber. Tracks are cast steel with rubber bushes and steel pins. Each track has 102 links when new.

It was originally intended that the Sheridan would be fully amphibious without preparation but in the end a flotation screen was designed and fitted around the top of the hull. This takes about five minutes to erect and the vehicle is then propelled in the water by its tracks.

Night vision equipment is provided for the commander, gunner and driver and most vehicles have an infra-red searchlight fitted to the left of the main armament. Standard equipment includes a heater, extraction fan in the turret roof, diesel cooker and an NBC system that pipes fresh air to the crew's face masks from a central unit.

Main armament comprises a 152 mm M81 gun/missile launcher which was designed to fire a conventional round or the Shillelagh anti-tank guided missile.

The 152 mm gun/launcher can also fire a number of conventional rounds which have a combustible cartridge case consisting of two parts, a base and a body, which are made of felted nitro-cellulose. The body is attached to the projectile base with epoxy resin and an aluminium locking ring. Types of ammunition that can be fired include HEAT-T-MP, WP, TP-T and canister.

A well-trained crew can fire a maximum of four rounds of conventional ammunition a minute.

Mounted coaxially to the main armament is a 7.62 mm M73 machine gun and mounted on the forward part of the commander's cupola is a 12.7 mm (0.50) Browning M2 HB machine gun for anti-aircraft use with an elevation of +70° and a depression of -15°. Many M551s have been provided with a shield to give the commander some protection when he is using the anti-aircraft armament.

The commander and gunner of the M551 originally had infra-red devices for night use which have a number of tactical limitations. In January 1989, US Army Tank Automotive Command commenced a two-year programme to design, field test and upgrade an undisclosed number of M551A1s for the 82nd Airborne Division.

The driver has had his infra-red night device replaced by the same device used in the M2/M3 armoured vehicles while the gunner now has the Tank Thermal Sight (TTS) used in the M60A3 MBT.

The modifications included cutting away a section of the turret armour roof, welding of a plate to support the new gun sight, installing a redesigned connecting arm which links the vehicle's 152 mm gun to the thermal sight to maintain correct alignment and mounting the driver's new viewer.

A number of modifications had to be carried out inside the vehicle, including stowage boxes, various electrical harnesses, turret manual gearbox and the azimuth indicator.

Two vehicles were converted by Anniston Army Depot, Alabama, late in 1989, and following extensive tests all M551s were upgraded by Anniston with final deliveries being completed in October 1990.

M551A1

In 1971 Frankford Arsenal awarded the Hughes Aircraft Corporation a contract worth $8.3 million for initial production of the AN/VVG-1 laser rangefinder for installation in the forward part of the commander's cupola of the M551. The rangefinder consists of a ruby laser, optics and

M551 modified to resemble Soviet ZSU-23-4 self-propelled anti-aircraft gun system (US Army National Training Center)

M551 modified to resemble Soviet BMP-1 infantry combat vehicle complete with mock-up of Sagger ATGW over 73 mm gun (US Army/Michael Green)

associated panels and electronics, and enables the commander to have accurate range information on the target within seconds. When fitted with the AN/VVG-1 the vehicle is known as the M551A1.

M551 for ARMVAL

Ten modified M551s were used for the joint US Army/Marine Corps Advanced Anti-armor Vehicle Evaluation (ARMVAL) between September 1978 and December 1980. The programme was sponsored by the Under Secretary of Defense for Research and Engineering with the Marine Corps being the Executive Service. Details of this were given in *Jane's Armour and Artillery 1989/90* page 180.

M551 with 75 mm ARES cannon

Under contract to the US Army Tank Automotive Reseach and Development Command, the Pacific Car and Foundry Company built an Elevated Kinetic-Energy Weapon Test Bed (ELKE). Details of this version, which is now stored, were given in *Jane's Armour and Artillery 1989/90* page 180.

M551 with 105 mm gun

In 1984 the Cadillac Gage company fitted an M551 Sheridan chassis with a 105 mm Low Recoil Force turret which is also installed on the Cadillac Gage Textron Stingray light tank (see earlier entry in this section).

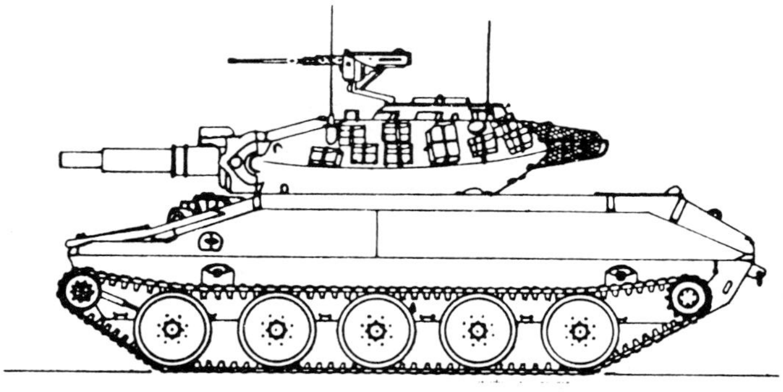

M551 Armored Reconnaissance/Airborne Assault Vehicle

SPECIFICATIONS	
CREW	4
COMBAT WEIGHT	15 830 kg
UNLOADED WEIGHT	13 589 kg
POWER-TO-WEIGHT RATIO	18.95 bhp/tonne
GROUND PRESSURE	0.49 kg/cm²
LENGTH	6.299 m
WIDTH	2.819 m
HEIGHT	
to turret top	2.272 m
overall	2.946 m
GROUND CLEARANCE	0.48 m
TRACK	2.348 m
TRACK WIDTH	444 mm
LENGTH OF TRACK ON GROUND	3.66 m
MAX SPEED	
road	70 km/h
water	5.8 km/h
FUEL CAPACITY	598 litres
MAX ROAD RANGE	600 km
FORDING	amphibious
GRADIENT	60%
SIDE SLOPE	40%
VERTICAL OBSTACLE	0.838 m
TRENCH	2.54 m
ENGINE	Detroit Diesel Model 6V-53T 6-cylinder water-cooled turbo-charged diesel developing 300 bhp at 2800 rpm
AUXILIARY ENGINE	none
TRANSMISSION	XTG-250 cross-drive with 4 forward and 2 reverse speeds
FINAL REDUCTION RATIO	2.22 : 1
SUSPENSION	torsion bar
ELECTRICAL SYSTEM	24 V
BATTERIES	2 × 6TN (or 4 in cold climate)
ARMAMENT	
main	1 × 152 mm gun/launcher
coaxial	1 × 7.62 mm MG
anti-aircraft	1 × 12.7 mm MG
SMOKE-LAYING EQUIPMENT	4 smoke dischargers either side of turret
AMMUNITION	
main	30
coaxial	3080
anti-aircraft	1000
GUN CONTROL EQUIPMENT	
Turret power control	electric/manual
by commander	yes
by gunner	yes
Max rate power traverse	(non stabilised) 360° in 10 seconds (stabilised) 360° in 7.5 seconds
Gun elevation/depression	+19.5°/-8°
Gun stabiliser	
vertical	yes
horizontal	yes
Elevation quadrant	M13A1C
Traverse indicator	M31 MOD

Status: Production complete. In service in limited numbers only with the US Army.

Manufacturer: Allison Division of General Motors Corporation
Cleveland Tank-Automotive Plant
Cleveland
Ohio
USA

AAI Rapid Deployment Force Light Tank (RDF/LT)

Development

The Rapid Deployment Force Light Tank (RDF/LT) has been designed as a private venture by AAI Corporation which has already built the prototype of the High Survivability Test Vehicle (Lightweight) under contract to the United States Army Tank-Automotive Command. Details of the HSTV(L) were given in *Jane's Armour and Artillery 1985-86* pages 154-155. The prototype was shown for the first time in October 1980.

The vehicle is airportable: the Lockheed C-5B transport aircraft can carry eight RDF/LTs, the C-141 could carry two, C-130 one and the Navy/Marine Corps CH-53E helicopter can carry one slung under its fuselage. The US Army's CH-47D helicopter can also carry the AAI RDF/LT.

Description

The hull of the RDF/LT is made of all-welded aluminium armour with the driver sitting at the front of the hull on the left and the commander/gunner to his right. Both crew members have a single piece hatch cover that opens outwards and has three integrated periscopes. Between the driver and commander/gunner, in the upper part of the glacis plate, is the hull-mounted auxiliary sight.

The main armament consists of a 75 mm ARES cannon mounted in the centre of the hull behind the crew. The 75 mm ARES cannon is fed from an automatic magazine holding 60 rounds of APFSDS and multi-purpose ammunition and when used for indirect fire has a maximum range of 12 000 m. The 75 mm APFSDS-T round is designated the XM885 and has armour penetration characteristics equivalent to the 105 mm APFSDS M774. The HE round is designated the XM884. To the right of the main armament there is a coaxial 7.62 mm machine gun.

Mounted above and behind the main armament is the stabilised rotary head which is the primary sight. The main armament is fully stabilised and the fire control system includes a digital computer. The fire control system is similar to that of the HSTV(L).

The engine and transmission are mounted at the rear of the hull and the complete powerpack is on extensible rails to facilitate maintenance in the field.

The torsion bar suspension consists of five dual rubber-tyred road wheels with a drive sprocket at the rear, idler at the front and one return roller.

Appliqué steel armour can be fitted to the RDF/LT for increased protection.

Variants

75 mm RDF/LT

In 1982 AAI announced a new version of this vehicle fitted with a new one-man turret also armed with the ARES 75 mm automatic cannon. This has a single piece hatch cover opening to the rear, six periscopes for all-round observation and forward and to the right of the hatch is a SFIM stabilised sight for target acquisition/firing.

The AAI RDF/LT shown at the 1984 AUSA Exhibition in Washington DC had a one-man welded aluminium armour plate turret with the 75 mm ARES cannon, 7.62 mm coaxial and 7.62 mm anti-aircraft machine guns on a DISA mount. In addition it had appliqué armour on its hull sides which also covered the upper part of its tracks and suspension.

AAI Combined Arms Team/Lightweight Combat Vehicle with two pods of Stinger SAMs elevated

CAT/LCV

AAI Corporation has developed an air defence variant of the Rapid Deployment Force Light Tank which is called the Combined Arms Team/

Prototype of AAI Rapid Deployment Force Light Tank with 75 mm ARES gun at 0° elevation, appliqué side armour on chassis, SFIM roof-mounted sight and 7.62 mm DISA machine gun mount

Lightweight Combat Vehicle (CAT/LCV). The turret features the XM274 75 mm gun, which is modified to fire the US Navy's Mark 199 76 mm HE projectile with Mark 404 IR proximity fuzes. In late 1987 it was stated that AAI was speeding up development of another infra-red proximity-fuzed round which was based on the existing US Army M353 projectile and Navy Mk 404 fuze. The turret features a 60-round carousel feeder (36 rounds in the outer ring and 24 rounds in the inner ring). Anti-aircraft missile pods can also be added, for example, two pods of four Stinger missiles. Another version mounts a pod of six Bofors RBS 70 missiles. The fire control system includes the Texas Instruments Tiger air defence sight which features a US 120 element common module FLIR and a CO_2 laser rangefinder which provides range and range rate information. The sight also contains a TV camera which operates in the visual spectrum.

The weapon system can also be used in the anti-armour role using APFSDS ammunition which penetrates about the same thickness of armour as a typical 105 mm APFSDS round, such as the M774. The advantage of the 75 mm automatic cannon over a 105 mm single shot gun is the much higher rate of fire and smaller silhouette in defensive scenarios.

The commander/gunner sits in the right front of the hull with the driver on the left. Both crewmen have dual controls and both can drive the vehicle and fire the guns. When stationary the commander would be free to handle command functions and would delegate gunner responsibilities to the driver. It is envisioned that radar cueing would be provided by forward area alerting radars, such as Trackstar, which shows both enemy and friendly aircraft on the gunner's/commander's video displays.

The CAT/LCV chassis is identical to the RDF/LT's shown on pages 191 and 193.

Status: Prototype. This vehicle was originally designed to meet the US Army requirement for a Mobile Protected Gun System which was never funded. The future US Army Armored Gun System (AGS) will have a 105 mm gun which rules out the AAI Rapid Deployment Light Tank in its present configuration.

Manufacturer: AAI Corporation
PO Box 126
Hunt Valley
Maryland 21030-0126
USA
Telephone: (301) 666 1400
Telex: 710 232 1800

SPECIFICATIONS		MAX ROAD SPEED	64 km/h	ARMAMENT	
(RDF/LT with crew of three and turret mentioned earlier)		FUEL CAPACITY	378 litres	main	1 × 75 mm gun
		MAX CRUISING RANGE	500 km	coaxial	1 × 7.62 mm MG
CREW	3	FORDING	1 m	anti-aircraft	1 × 7.62 mm MG (optional)
WEIGHT		GRADIENT	60%		
combat	13 426 kg	ENGINE	Detroit Diesel 6V-53T, turbo-charged, 6-cylinder diesel developing 350 hp	AMMUNITION	
unloaded	12 247 kg			main	60
POWER-TO-WEIGHT RATIO	26.07 hp/tonne			coaxial	2600
GROUND PRESSURE	0.49 kg/cm²	TRANSMISSION	Allison Transmission Division of General Motors X-200 cross drive, automatic	GUN CONTROL EQUIPMENT	
LENGTH GUN FORWARDS	8.235 m			Turret power control	hydraulic/manual
LENGTH HULL	5.569 m			by commander	yes
				by gunner	yes
WIDTH	2.54 m	SUSPENSION	torsion bar	Gun elevation/depression	+40°/-15°
HEIGHT (top of sight)	2.286 m	ELECTRICAL SYSTEM	24 V	Turret traverse	360°
AXIS OF FIRE	1.562 m	BATTERIES	6 × 12 V, 190 Ah	Turret slew rate	60°/s
GROUND CLEARANCE	0.5 m			Gun elevation rate	60°/s

AAI 13.2 Tonne Rapid Deployment Force Light Tank (RDF/LT)

Development

The original AAI Rapid Deployment Force Light Tank (RDF/LT), described in the preceding entry, can only be exported to NATO and several other key countries, therefore AAI has developed a 76 mm gun turret for export to other countries.

In 1982 AAI announced that it had built the prototype of a 13.2 tonne RDF/LT fitted with a new two-man turret armed with the same 76 mm M32 gun as the M41 light tank's. This was phased out of US Army service many years ago, although it remains in service with other countries. AAI has developed a new armour-piercing fin-stabilised discarding sabot tracer (APFSDS-T) round which has been demonstrated against NATO heavy tank targets. The prototype uses the chassis of the original RDF/LT but production vehicles would have a slightly different hull and the description relates to this. A Lockheed C-130H aircraft can carry two 13.2 tonne RDF/LTs.

In 1984 AAI announced that the RDF/LT with the 76 mm M32 gun was successfully tested in Venezuela where it travelled some 800 km over various terrain including swamp, and operated at a maximum altitude of 4000 m. During gunnery trials in Venezuela the 76 mm M32 gun fired the AAI-developed APFSDS-T round as well as a new round fitted with a US Navy IR/proximity fuze for use against aircraft and helicopters.

Description

The hull of the RDF/LT is of all-welded aluminium construction. The driver is seated at the front of the hull on the left with 30 rounds of 76 mm ammunition stowed horizontally to his right. The driver has a single piece hatch cover that opens to the left. In the forward part of this are three periscopes, the centre one of which can be replaced by a passive periscope for night driving.

The all-welded turret is in the centre of the vehicle with the commander seated on the left and the gunner on the right. Both have a single piece hatch cover, six periscopes for all-round observation and an M32 periscope for aiming the armament. The gunner's M32 periscope incorporates a laser rangefinder.

Main armament consists of a 76 mm high velocity M32 gun which is installed in the M41 tank. In addition to the range of ammunition originally developed for this weapon, AAI has developed a new round of APFSDS-T ammunition based on its experience in developing ammunition for the ARES 75 mm automatic cannon. This has already been adopted by Denmark for its M41 light tanks. According to AAI, this projectile has three times the probability of killing a T-62 tank at a normal combat range of 1500 m than the 105 mm M456 HEAT-T round and only slightly less kill probability at a similar range to the 105 mm M735 APFSDS-T round. A 7.62 mm M240 machine gun is mounted coaxially with the main armament.

A Cadillac Gage Textron stabilisation and weapons control system is fitted as standard enabling the vehicle to fire on the move with a high probability of a first round hit. Optional fire control systems include an M32 sight with an AN/VSG-2 Tank Thermal Sight or a digital tank fire control system with the AN/VSG-2 Tank Thermal Sight, AN/GVS-5 laser rangefinder and a digital ballistic computer.

The engine and transmission are mounted at the rear of the hull and the complete powerpack is on extensible rails to facilitate maintenance in the field.

The torsion bar suspension consists of five dual rubber-tyred road wheels with the drive sprocket at the rear, idler at the front and one return roller which is positioned above the second and third roadwheel stations. Track is the M113.

AAI 13.2 tonne Rapid Deployment Force Light Tank prototype showing new turret armed with 76 mm M32 gun which can fire a APFSDS-T projectile developed by AAI

SPECIFICATIONS

CREW	3
WEIGHT	
combat	13 200 kg
unloaded	11 800 kg
POWER-TO-WEIGHT RATIO	26.5 hp/tonne
GROUND PRESSURE	0.48 kg/cm^2
LENGTH GUN FORWARDS	7.34 m
LENGTH HULL	5.569 m
WIDTH	2.54 m
HEIGHT	
overall, M32 sight	2.235 m
hull top	1.562 m
GROUND CLEARANCE	0.5 m
MAX ROAD SPEED	64 km/h
FUEL CAPACITY	378 litres
MAX CRUISING RANGE	500 km
FORDING	1 m
GRADIENT	60%
ENGINE	Detroit Diesel 6V-53T, turbo-charged, 6-cylinder diesel developing 350 hp
TRANSMISSION	Allison Transmission Division of General Motors X-200 crossdrive, automatic
SUSPENSION	torsion bar
ELECTRICAL SYSTEM	24 V
BATTERIES	6 × 12 V, 190 Ah
ARMAMENT	
main	1 × 76 mm gun
coaxial	1 × 7.62 mm MG
AMMUNITION	
main	50
coaxial	2600
GUN CONTROL EQUIPMENT	
Turret power control	hydraulic/manual
by commander	yes
by gunner	yes
Gun elevation/depression	+22°/-10°
Turret traverse	360°

Status: Prototype.

Manufacturer: AAI Corporation
PO Box 126
Hunt Valley
Maryland 21030-0126
USA
Telephone: (301) 666 1400
Telex: 710 232 1800

Reconnaissance vehicles

BELGIUM

FN 4RM/62F AB Light Armoured Car

Development

The FN 4RM/62F AB light armoured car was developed by the now FN HERSTAL SA to meet the requirements of the Belgian Gendarmerie and is based on the chassis of the FN 4RM/62 Ardennes (4 × 4) 1500 kg truck. The first prototype was completed in 1962 with the second following in 1965. Between 1971 and 1972 62 production vehicles were built. FN HERSTAL SA is no longer involved in the development or production of armoured vehicles or trucks.

Description

The hull of the vehicle is all-welded steel with the driver seated at the front in the centre. The driver is provided with a single piece hatch cover that opens upwards, in the top of which are three periscopes and in the lower part a bullet-proof window with an armoured cover. There is a single door in each side of the hull; the door on the right is fitted with a rectangular observation port.

The all-welded turret is in the centre of the hull with the gunner on the left and the commander, who also acts as loader, on the right. Both the commander and gunner are provided with a single piece hatch cover that opens to the rear. The commander has eight periscopes and the gunner has three, plus an optical sight linked to the main armament. Turret traverse is electric, two-speed (fast and slow) with manual controls for emergency use.

The engine compartment at the rear of the hull is equipped with an automatic fire-extinguishing system. Standard equipment includes an NBC system, searchlights for both the commander and gunner, and channels for crossing ditches carried on the rear of the turret (on 90 mm gun model only). The vehicle has no night vision equipment and has no amphibious capability.

Variants

90 mm Gun Model

This variant is armed with a 90 mm CATI gun which fires two types of ammunition, HEAT and shrapnel. The HEAT round weighs 3.5 kg and has a muzzle velocity of 640 m/s and an effective range of 1000 m. The shrapnel round weighs 5.1 kg, has a muzzle velocity of 330 m/s and a maximum effective range of 1800 m. A 7.62 mm MAG machine gun is mounted coaxially to the right of the main armament and a similar weapon pintle-mounted in front of the commander's cupola is used for anti-aircraft defence. Mounted on either side of the turret at the rear are six electrically operated smoke dischargers.

Machine Gun and Mortar Model

This model is armed with twin 7.62 mm MAG machine guns and a 60 mm breech-loaded mortar which can be elevated separately or in parallel. It is not fitted with an anti-aircraft machine gun and the smoke dischargers are towards the front rather than the rear of the turret as in the case of the 90 mm gun model.

Armoured Personnel Carrier

This was a project only and was never built.

Machine gun and mortar model of FN 4RM/62F AB light armoured car

SPECIFICATIONS (Specifications in square brackets relate to the 90 mm gun model where this differs from the machine gun model)

CREW	3
CONFIGURATION	4 × 4
LOADED WEIGHT	8800 [8000] kg
WEIGHT ON FRONT AXLE (loaded)	3800 [3350] kg
WEIGHT ON REAR AXLE (loaded)	5000 [4650] kg
POWER-TO-WEIGHT RATIO	14.77 [16.25] hp/tonne
LENGTH GUN FORWARDS	n/app [5.42 m]
LENGTH HULL	4.5 m
WIDTH	2.26 m
HEIGHT (overall)	2.37 [2.52] m
GROUND CLEARANCE	0.324 m
TRACK	1.62 m
WHEELBASE	2.45 m
ANGLE OF APPROACH/ DEPARTURE	50°/37°
MAX ROAD SPEED	110 km/h
ACCELERATION (0 - 80 km/h)	65 s
FUEL CAPACITY	180 l
MAX ROAD RANGE (at 80 km/h)	600 km
FORDING	1.1 m
GRADIENT	60%
TURNING RADIUS	6 m
ENGINE	FN 652 6-cylinder in-line OHV petrol developing 130 hp at 3500 rpm
TRANSMISSION	manual with 4 forward and 1 reverse gears
TRANSFER CASE	2-speed
STEERING	worm and twin finger
CLUTCH	single dry plate
SUSPENSION	longitudinal leaf springs with double-action hydraulic shock absorbers front and rear
TYRES	9.00 × 20
BRAKES	
main	drum on all wheels, dual circuit, servo-assisted
parking	handbrake acting on propeller shaft
ELECTRICAL SYSTEM	24 V
BATTERIES	4 × 12 V, 50 Ah
ARMAMENT	
main	1 × 60 mm mortar [1 × 90 mm gun]
coaxial	2 × 7.62 mm MG [1 × 7.62 mm MG]
anti-aircraft	none [1 × 7.62 mm MG]
SMOKE-LAYING EQUIPMENT	2 × 6 smoke dischargers
AMMUNITION	
main	46 [40]
machine gun	4830 [3680]
smoke grenades	36
GUN CONTROL EQUIPMENT	
Turret power control	electric/manual
by commander	yes (fast only)
by gunner	yes
Gun elevation/ depression	+55° (MGs)/-10° +75° (mortar)/-10° [+27°/-12°]
Gun stabiliser	
vertical	no
horizontal	no
ARMOUR	6.5-13 mm

Status: Production complete. In service only with the Belgian Gendarmerie. It is reported that Uruguay has taken delivery of a small number of FN 4RM/62 AB light armoured cars.

Manufacturer: FN HERSTAL SA
Fabrique Nationale Nouvelle Herstal SA
B-4040 Herstal-Belgium
Telephone: (32) 41 40 81 11
Telex: 041.223
Fax: (32) 41 40 86 79

90 mm gun model of FN 4RM/62F AB light armoured car with driver's hatch closed and 7.62 mm anti-aircraft machine gun fitted

BRAZIL

ENGESA EE-18 Sucuri Tank Destroyer

Development

The EE-18 Sucuri (6 × 6) tank destroyer has been developed as a private venture by ENGESA with the first prototype being completed late in 1987.

Its main role is to engage enemy tanks using hit and run tactics but ENGESA also suggest that its speed and mobility allow it to cover the flanks of fast-moving columns and act as a highly mobile reserve to rapidly counter attack any sudden breakthrough.

The earlier EE-17 (6 × 6) Sucuri tank destroyer, which consisted of an ENGESA chassis fitted with a French Fives-Cail-Babcock FL-12 oscillating turret armed with 105 mm gun, never entered production (*Jane's Armour and Artillery 1979-80* page 107).

As of May 1991, no firm orders had been placed for the ENGESA EE-18 Sucuri and no overseas demonstrations had taken place.

Description

The hull of the EE-18 is of dual-hardness armour which consists of an outer layer of hard steel and an inner layer of softer steel roll-bonded and heat-treated to give maximum ballistic protection from small arms fire and shell splinters.

The driver is seated at the front left with the engine compartment on the right so leaving the whole of the rear clear for the turret which is of a similar construction to the hull. He is provided with a single piece hatch cover, three day periscopes, one of which can be replaced by a passive periscope for driving at night and an adjustable steering wheel.

The air intake is on the glacis plate with the exhaust and air outlet on the right side of the hull.

Steering is power-assisted on the front wheels and a central tyre pressure regulation system is fitted as standard to allow the driver to adjust the tyre pressure to suit the type of ground being crossed.

The suspension is of the independent hydro-pneumatic type with a total wheel travel of 308 mm. The axles are provided with planetary reduction and a locking system in the differentials.

The vehicle commander sits on the right with the gunner forward and below his position with the loader on the left. The commander has a single piece hatch cover and periscopes for all-round observation while the loader has a single piece hatch cover that opens to the rear and a rotating observation periscope.

The prototype is armed with an OTO Melara 105 mm soft recoil gun which is fitted with a thermal sleeve and a muzzle brake, although other types of 105 mm weapon can also be installed. The 105 mm gun can fire all standard types of ammunition including APFSDS-T, HEAT, APDS-T, HESH, smoke and training. A total of 30 rounds of 105 mm ammunition are carried, of which eight are in the turret and remaining 22 in the hull. A total of 3200 rounds of 7.62 mm machine gun ammunition are carried of which 400 are in the turret and remainder in the hull.

The fire control system of the vehicle enables the EE-18 to engage both stationary or moving targets when it is stationary or moving.

If required, the commander can override the gunner and both commander and gunner can traverse the turret at 26°/s and elevate the main armament at 35°/s (max). The fire control system includes a digital ballistic computer, moving-target tracking system, computer driven ballistic aiming mark, cant angle correction with automatic input, crosswind correction with manual range input as first emergency mode and a ballistic graticule as the second emergency mode.

Both the commander and gunner are provided with a roof-mounted sight with × 8 day and × 8 night magnification and a built-in laser rangefinder. In addition the gunner has an emergency combat sight with a magnification of × 8.

The 105 mm gun is stabilised in both planes with turret traverse and weapon elevation electric with manual control for emergency use.

The 24 V electric system uses one set of batteries for vehicle operations and the second set for starting so avoiding lack of power after long still radio operations.

Optional equipment includes passive night driving periscope for the driver, laser detector and various communications systems.

Variants

ENGESA have suggested that the chassis of the EE-18 can be fitted with a wide range of other turrets and weapon systems.

Status: Prototype. Not yet in production or service.

Manufacturer: ENGESA Engenheiros Especializados SA
Avenue Tucunaré, 125/211
Caixa Postal/PO Box 152/154
06400 Barueri, SP
Brazil
Telephone: (55 11) 421 4711
Telex: 1171302
Fax: (55 11) 421 4445

Prototype of ENGESA EE-18 Sucuri (6 × 6) tank destroyer from rear

Prototype of ENGESA EE-18 Sucuri (6 × 6) tank destroyer

SPECIFICATIONS	
CREW	4
CONFIGURATION	6 × 6
COMBAT WEIGHT	18 500 kg
POWER-TO-WEIGHT RATIO	20.50 hp/tonne
LENGTH INCLUDING GUN	7.75 m
LENGTH HULL	5.84 m
WIDTH	2.80 m
HEIGHT	2.59 m (roof of turret)
GROUND CLEARANCE	0.40 m
TRACK	2.33 m
WHEELBASE	4.04 m
ANGLE OF APPROACH/ DEPARTURE	76°/52°
MAX ROAD SPEED	105 km/h
FUEL CAPACITY	480 l
MAXIMUM ROAD RANGE	700 km
FORDING	1.3 m
GRADIENT	60%
SIDE SLOPE	30%
VERTICAL OBSTACLE	0.6 m
TRENCH	2 m
ENGINE	Scania DSI-11, 6-cylinder in-line direct injection, four-stroke, turbo-charged inter-cooled diesel developing 380 hp at 2100 rpm
TRANSMISSION	ZF 6 HP 600 fully automatic, with lock-up, retarder and torque converter, 6 speeds forward and 1 reverse
TRANSFER CASE	mechanical, 1 speed with constant mesh and split torque
SECONDARY TRANSFER CASE	mechanical, directly linked with automatic transmission
STEERING	hydraulic
SUSPENSION	independent hydro-pneumatic MacPherson with 308 mm of wheel travel
TYRES	14.00 × 22.50 run flat radial
BRAKES	disc on all wheels actuated by air over hydraulic system. Double calipers on front wheels and single calipers on rear wheels. Mechanical drum tyre parking brakes, incorporated with the differentials in intermediary and rear axles
ELECTRICAL SYSTEM	24 V
BATTERIES	4 × 12 V, 100 Ah
ALTERNATOR	28 V, 200 A
ARMAMENT	
main	1 × 105 mm gun
coaxial	1 × 7.62 mm MG
SMOKE-LAYING EQUIPMENT	6 smoke dischargers either side of turret
AMMUNITION	
main	30
coaxial	3200
smoke grenades	24 (12 inside vehicle)
GUN CONTROL EQUIPMENT	
Turret power control	electric/mechanical
by commander	yes
Turret traverse	360° at 26°/s
Gun elevation/depression	+15°/-6° at 35°/s
Gun stabiliser	
vertical	yes
horizontal	yes

ENGESA EE-9 Cascavel Armoured Car

Development

The EE-9 Cascavel was designed by ENGESA (Engenheiros Especializados SA) to meet the requirements of the Brazilian Army. Design work began in July 1970 after the company had completed the designing of the EE-11 armoured personnel carrier, with the first prototype being completed in November of that year. Following trials with prototype vehicles a pre-production order was placed by the Brazilian Army, which named the vehicle the CRR (Carro de Reconhecimento Sobre Rodas), for 10 vehicles which were delivered between 1972 and 1973.

Production of the EE-9 began at a new plant at Sao José dos Campos in 1974 with the first production vehicles being delivered the same year. The production vehicles were slightly longer and wider than the prototypes and had a different wheelbase. First production vehicles delivered to the Brazilian Army were fitted with a turret armed with 37 mm guns removed from old American-supplied M3 Stuart light tanks, but first production vehicles for export were fitted with a French CNMP-Berthiez (now Hispano-Suiza) H 90 turret. Current production vehicles have a turret of Brazilian design mounting a Brazilian 90 mm gun.

The EE-9 Cascavel shares many components with the EE-11 Urutu (6 × 6) armoured personnel carrier and many automotive components used in both vehicles are standard commercial parts available world-wide.

The EE-9 Cascavel (6 × 6) armoured car has been used in combat by the Iraqi Army during the fighting with Iran where it has carried out the traditional role of armoured reconnaissance as well as direct and indirect fire support and flank guard for tank units. During the invasion of Chad by Libyan-backed guerrillas in 1983, numbers of the early model of the EE-9 Cascavel with the French H 90 turret armed with a 90 mm gun were captured by the Chadian Army. These were supplied to the guerrillas by Libya. It has also been used by the Libyan Army in its border clashes with Egypt.

By January 1984 2550 EE-9 Cascavel armoured cars had been produced. Since then production has continued but current production figures have not been released.

Description

The all-welded hull of the EE-9 has dual hardness armour which was developed by ENGESA's Research Department with the co-operation of the University of Sao Paulo and consists of an outer layer of hard steel and an inner layer of softer steel roll-bonded and heat-treated to give maximum ballistic protection. Increased protection is given over the frontal arc and special emphasis has been given to protect the vehicle against attack from booby traps, grenades and Molotov cocktails.

The driver is seated at the front of the hull on the left side and is provided with a hatch cover opening to the right. There are three periscopes in the top part of the glacis plate giving the driver a 120° view. The driver is also provided with a small windscreen and wiper, which folds forward onto the glacis plate when not in use. The driver's seat and steering wheel are both adjustable.

Initial production vehicles were fitted with a French H 90 turret armed with a 90 mm D 921 gun but current production vehicles have an ENGESA ET-90 turret armed with an ENGESA EC-90 gun. The commander is seated on the left side of the turret and the gunner on the right, both with a single piece hatch cover that opens to the rear and four periscopes. The gunner has a telescopic sight linked to the main armament with a magnification of ×5.9. EE-9s have electric powered traverse and elevation as an option for the turret. It is believed that the EE-9s supplied to Tunisia were fitted with an OIP fire control system, probably the LRS-5.

The engine compartment is at the rear of the vehicle and access to the engine is via two hatches in the hull roof. Power is transmitted from the engine to the gearbox (a Allison Transmission Division of General Motors MT-643 automatic with four forward and one reverse gears) to the two-speed ENGESA transfer box, which splits the drive between the front and rear differentials. The front differential is bolted to the hull with the power being transmitted to the front axle by shafts, and the rear differential forms part of the ENGESA Boomerang walking beam suspension. The latter, which is also used on ENGESA 6 × 6 trucks, consists of a rigid axle connected to the hull by double leaf springs and telescopic dampers which holds two lateral walking beams. Power is taken from the drive shaft to the four rear wheels by gears inside the walking beam, which can rotate freely around the central hub. The design of the rear suspension allows up to a maximum of 0.87 m of vertical wheel travel and ensures that all four rear wheels are in contact with the ground at all times. The run-flat tyres enable the vehicle to travel for at least 40 km at a speed of 56 km/h in consistent ground after being completely deflated. The Cascavel has a double 24 V electrical system; a battery set is used for vehicle and turret operations, while a second set is used purely for engine start to avoid lack of power after long still radio operations. The Cascavel is also equipped with disc brakes and an automatic tyre inflation system.

Optional equipment includes an air-conditioning system, heater, laser rangefinder, automatic fire extinguisher and passive or active night vision equipment. The Cascavel can be fitted with various types of fire control system, for example from Marconi Command and Control systems of the UK.

The gun is called the EC-90-III and is mounted in an ET-90 turret. The rifled-bore gun fires the following types of ammunition: HE, HEAT, HESH, Smoke White Phosphorus (Smoke WP-T) and HEAT Target Practice. More recently an APFSDS round has been developed for the 90 mm gun. Of 44 rounds of 90 mm ammunition carried, 24 are in the turret and 20 in the hull. ENGESA has also developed a commander's cupola for the EE-9 which can be traversed through 360° and has vision blocks for all-round observation and a 12.7 mm (0.50) machine gun mounted externally on a bracket forward of the commander's single piece hatch cover that opens to the rear.

A 7.62 mm machine gun is mounted to the left of the main armament and a similar weapon, or a 12.7 mm (0.50) Browning M2 HB machine

ENGESA EE-9 Cascavel Mark III armoured car

ENGESA EE-9 Cascavel armoured car clearly showing large vertical wheel travel of rear walking beam suspension

gun, can be mounted on the turret roof for anti-aircraft defence. Three smoke dischargers are mounted either side of the turret and are electrically fired from within the vehicle. Recent production vehicles for Iraq have a traverse indicator, clinometer, improved seating, laser rangefinder integrated into the sight which is protected by a shutter when the main armament is fired, and a 12.7 mm FN M2 HB machine gun on an anti-aircraft mount similar in design to that used by the Soviet Union with its 12.7 mm DShKM machine guns.

A summary of the various models of the ENGESA Cascavel armoured car produced follows:

Mark I

Powered by a Perkins 6357 V engine coupled to a manual Clark transmission, no central tyre pressure regulation system. Did not enter production.

Mark II

First production model armed with a 37 mm gun, Mercedes-Benz OM 352A 170 hp diesel engine, Clark manual transmission, 12 V (hull) and 24 V (turret) electrical system, no central tyre-pressure regulation system. These were built only for the Brazilian Army and most have now been rebuilt with the new ENGESA ET-90 turret.

Mark III

First production model for export, Mercedes-Benz OM 352A 170 hp diesel, Allison Transmission Division of General Motors AT-540 automatic transmission, 12 V (hull) and 24 V (turret) electrical system, central tyre-pressure regulation system and fitted with French Hispano-Suiza H 90 turret with 90 mm gun. No longer in production.

Mark IV

First production models completed in 1977 and fitted with ENGESA ET-90 turret with ENGESA EC-90 gun, Detroit Diesel Model 6V-53 developing 212 hp coupled to an Allison Transmission Division of General Motors fully automatic transmission Model MT-643, 24 V (turret electrical system, central tyre-pressure regulation system and hull).

Mark V

Powered by Mercedes-Benz OM 352A diesel developing 190 hp coupled to a fully automatic Allison Transmission Division of General Motors AT-540 or AT-545 automatic transmission, ENGESA ET-90 turret 24 V (hull and turret) electrical system, disc brakes and a central tyre-pressure regulation system. This system enables the driver to change the tyre-pressure from his seat to suit the type of ground being crossed. For example, when travelling on roads the tyre pressure is 4 kg/cm^2, cross-country 3 kg/cm^2 and for mud, sand and snow 2 kg/cm^2.

Mark VI

As Mark V but powered by a Mercedes-Benz OM 352A diesel developing 190 hp at 2800 rpm.

Mark VII

As Mark VI but fitted with Allison Transmission Division of General Motors MT-643 fully automatic transmission.

Variants

There are no variants of the EE-9 apart from the different armament installations described above, and variations between various production batches.

ENGESA EE-9 Cascavel armoured car with ENGESA ET-90 turret with ENGESA EC-90 90 mm gun, without laser rangefinder over main armament, with commander's cupola with externally mounted 12.7 mm M2 HB machine gun and six (2 × 3) electrically operated smoke grenade launchers

Latest production ENGESA EE-9 with mount for external machine gun on commander's cupola, front-mounted winch and laser rangefinder integral with gunner's sight (Christopher F Foss)

Close-up of turret of recent production ENGESA EE-9 Cascavel armoured car showing externally mounted 12.7 mm M2 machine gun

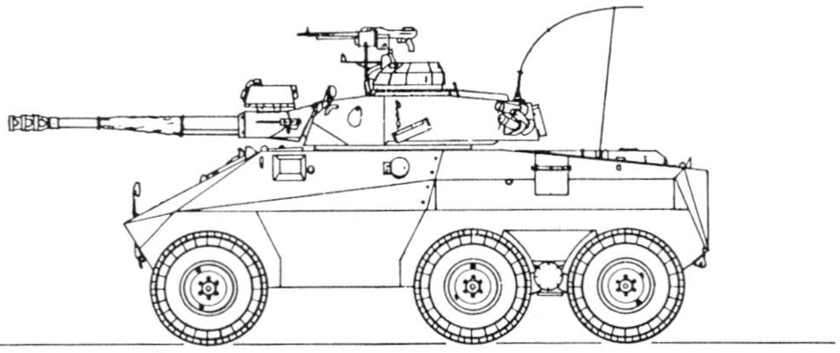

ENGESA EE-9 Cascavel (6 × 6) armoured car with ENGESA ET-90 turret armed with ENGESA EC-90 gun, commander's internally operated 7.62 mm machine gun cupola and laser rangefinder over main armament

Status: In production. In service with Bolivia, Brazil (Army and Marines), Burkina Faso, Chile, Colombia, Cyprus, Ecuador, Gabon, Iran, Iraq, Libya, Togo, Tunisia, Uruguay, Zimbabwe and other undisclosed countries.

Manufacturer: ENGESA Engenheiros Especializados SA
Avenue Tucunaré, 125/211
Caixa Postal/PO Box 152/154
06400 Barueri, SP
Brazil
Telephone: (55 11) 421 4711
Telex: 1171302
Fax: (55 11) 421 4445

SPECIFICATIONS (Mark IV)	
CREW	3
CONFIGURATION	6 × 6
COMBAT WEIGHT	13 400 kg
UNLOADED WEIGHT	10 900 kg
POWER-TO-WEIGHT RATIO	15.82 hp/tonne
LENGTH GUN FORWARDS	6.2 m
LENGTH HULL	5.2 m
WIDTH	2.64 m
HEIGHT	
top of commander's cupola	2.68 m
to turret roof	2.28 m
to hull top	1.75 m
GROUND CLEARANCE	
rear axle	0.34 m
hull centre	0.5 m
TRACK	2.1 m
WHEELBASE	2.343 m + 1.414 m
ANGLE OF APPROACH/ DEPARTURE	72°/80°
MAX ROAD SPEED	100 km/h
FUEL CAPACITY	390 l
MAX CRUISING RANGE	880 km
FORDING	1 m
GRADIENT	60%
SIDE SLOPE	30%
VERTICAL OBSTACLE	0.6 m

ENGINE	Detroit Diesel 6V-53N 6-cylinder water-cooled diesel developing 212 hp at 2800 rpm (other engines can also be installed)
TRANSMISSION	Allison Transmission Division of General Motors MT-643 automatic with 4 forward, 1 reverse gears
TRANSFER CASE	ENGESA 2-speed
STEERING	ZF integral hydraulic
SUSPENSION	
front	independent with helical springs and double action shock absorbers
rear	ENGESA Boomerang type with walking beams, leaf springs and double-action shock absorbers. Differential locking system
TYRES	12.00 × 20 (with run-flat capability)

BRAKES	
main	air over hydraulic, dual circuit, disc brakes
parking	mechanical on transfer case output
ELECTRICAL SYSTEM	24 V (4 batteries)
ARMAMENT	
main	1 × 90 mm gun
coaxial	1 × 7.62 mm MG
anti-aircraft	1 × 7.62 mm or 12.7 mm MG (optional)
SMOKE-LAYING EQUIPMENT	3 smoke dischargers either side of turret
AMMUNITION	
main	44
coaxial	2200
GUN CONTROL EQUIPMENT	
Turret power control	electric/manual
by gunner	yes
Max rate of traverse	360° in 15 s (with power traverse)
Gun elevation/depression	+15°/-8°
Gun stabiliser	
vertical	no
horizontal	no
ARMOUR	8.5-16 mm

ENGESA EE-3 Jararaca Scout Car

Development

The EE-3 Jararaca scout car was originally designed specifically for the export market and complements the other vehicles in the ENGESA range of armoured vehicles which include the EE-11 Urutu (6 × 6) armoured personnel carrier and the EE-9 Cascavel (6 × 6) armoured car. All vehicles in the ENGESA wheeled range use standard automotive components. By early 1984 360 Jararaca scout cars had been built by ENGESA. Since then production has continued for export although current production quantities have not been released. The EE-T4 Ogum light tracked vehicle is also being offered for local production overseas and was shown in Iraq in May 1989.

Description

The all-welded hull of the EE-3 consists of an outer layer of hard steel and an inner layer of softer steel roll-bonded and heat-treated to give the maximum ballistic protection. The armour is similar to that used on other members of the ENGESA family of armoured vehicles.

The driver is seated towards the front of the hull, slightly to the left of the vehicle's centre line and has a single piece hatch cover that lifts and swings to the right to open, although on recent production models it opens to the left. There are three periscopes for forward observation immediately in front of the driver. A windscreen and wiper which can be fitted in front of the driver's position for use in wet weather folds forward onto the glacis plate when not in use. The machine gunner is seated behind the driver on the right side of the hull and has a circular mounting which can be traversed through 360° and a single piece hatch cover that opens to the rear. A standard 12.7 mm (0.50) M2 HB machine gun is pintle-mounted on the forward part of the circular mounting, but there is no provision for aiming the weapon from inside the vehicle. There are vision blocks in both sides of the hull.

ENGESA EE-3 Jararaca (4 × 4) scout car fitted with one-man turret armed with twin 7.62 mm machine guns and electrically operated smoke dischargers

Late production ENGESA EE-3 Jararaca (4 × 4) scout car with one-man turret with smoke dischargers fitted but without armament installed

ENGESA EE-3 Jararaca (4 × 4) scout car with driver's hatch open and cupola with externally mounted 12.7 mm machine gun

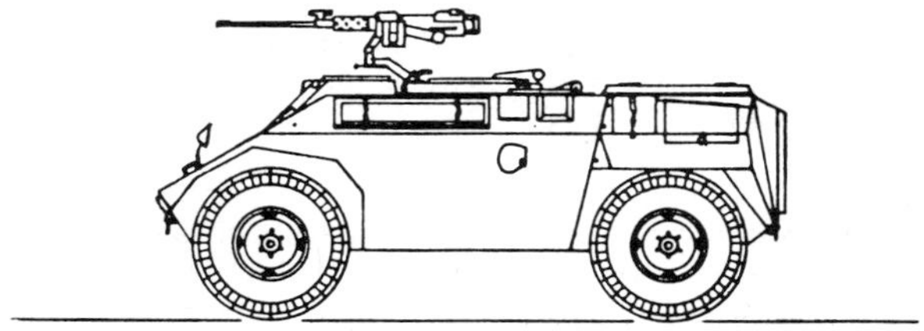

ENGESA EE-3 Jararaca scout car with 12.7 mm machine gun

The crew of three enters and leaves the EE-3 by a door in the right side of the hull that opens forward. Immediately behind the door is a single vision block under which is a firing port. The commander is seated on the left side of the hull to the rear and has a hatch cover that opens to the rear. On recent production EE-3s the commander has a hatch that opens to the left with an integral roof-mounted, forward-facing periscope. There is an observation block in the left side of the hull.

The engine is at the rear of the hull and access to it is by two large hatches in the roof. The suspension, front and rear, consists of semi-elliptical springs and hydraulic shock absorbers. The axles have hypoid gears. The wheels of the EE-3 are not within the armoured part of the vehicle as the light sheet steel covering over the top of each wheel will blow away if the vehicle hits a mine. The Jararaca is fitted with an automatic tyre inflation system as standard. The central tyre pressure regulation system enables the driver to adjust the tyre pressure to suit the type of ground being crossed without leaving his seat. For example, when travelling on roads the tyre pressure is 2.5 kg/cm^2, cross-country 2.1 kg/cm^2 and for mud, sand and snow 1.4 kg/cm^2.

The tyres of the EE-3, like those of the EE-9 Cascavel and EE-11 Urutu, have the ENGESA run-flat system which consists of an alveolar core inside the pneumatic tyre. When the tyre deflates, it contacts the inner core so ensuring continued mobility.

The standard 12.7 mm (0.50) M2 HB pintle-mounted machine gun can be replaced by a 7.62 mm machine gun or a 20 mm cannon. Other armament installations currently available include a 60 mm breech-loaded mortar, a Euromissile MILAN launcher and a cupola with internally operated 7.62 mm machine guns. The Republic of Cyprus uses the EE-3 Jararaca with the MILAN ATGW system mounted externally on the turret.

More recently ENGESA has developed two new turrets that can be fitted to the EE-3 Jararaca: the ENGESA ET-MD one-man turret with 7.62 mm NATO machine gun and 12.7 mm M2 HB machine gun, both internally operated; and the ENGESA ET-20 one-man turret with 20 mm cannon and 7.62 mm NATO machine gun.

The EE-3 has no amphibious capability. Optional equipment includes passive night vision equipment and various radios with an intercom system. The EE-3 can be fitted with an NBC system.

Variants

Apart from the NBC reconnaissance version and the different armament installations there are no variants of the EE-3 Jararaca.

Status: In production. In service with Cyprus, Gabon, Tunisia and Uruguay.

Manufacturer: ENGESA Engenheiros Especializados SA
Avenue Tucunaré 125/211
Caixa Postal/PO Box 152/154 CEP 06400
06400 Barueri, SP
Brazil
Telephone: (55 11) 421 4711
Telex: 1171302
Fax: (55 11) 421 4445

SPECIFICATIONS

CREW	3
CONFIGURATION	4 × 4
COMBAT WEIGHT	5800 kg
POWER-TO-WEIGHT RATIO	20.7 hp/tonne
LENGTH	4.163 m
WIDTH	2.235 m
HEIGHT	
without armament	1.56 m
top of 12.7 mm MG mount	2.3 m
GROUND CLEARANCE	0.335 m
TRACK	1.71 m
WHEELBASE	2.6 m
ANGLE OF APPROACH/DEPARTURE	68°/54°
MAX ROAD SPEED	100 km/h

FUEL CAPACITY	140 l
MAX ROAD RANGE	700 km
FORDING	0.6 m
GRADIENT	60%
SIDE SLOPE	30%
VERTICAL OBSTACLE	0.4 m
TURNING RADIUS	7.73 m
ENGINE	Mercedes-Benz OM 314A turbo-charged, 4-cylinder water-cooled diesel developing 120 hp at 2800 rpm
TRANSMISSION	Clark 240 V manual with 5 forward and 1 reverse gears
TRANSFER CASE	ENGESA 2-speed
STEERING	ZF integral hydraulic

CLUTCH	single dry plate hydraulically-operated
SUSPENSION	semi-elliptical springs and hydraulic shock absorbers, both have hypoid gears and tapered gear, single reduction differential
TYRES	11.00 × 20
BRAKES	
main	drum, air-assisted
parking	drum, acting on transfer case
ELECTRICAL SYSTEM	double 24 V system
BATTERIES	4 × 12 V, 95 Ah
ARMAMENT	1 × 12.7 mm MG

ENGESA EE-T4 Light Tracked Vehicle

Development

While the EE-9 Cascavel armoured car and EE-11 Urutu armoured personnel carrier were initially developed to meet Brazilian Army requirements, the EE-T1 Osorio MBT and the more recent EE-T4 Ogum have been developed with an eye on the export market as well.

In the early 1980s a number of countries expressed an interest in a small light tracked armoured vehicle which could be used for a wide range of battlefield roles ranging from anti-tank armed with missiles, reconnaissance to command and control.

A Middle Eastern country stated in late 1985 that it had a firm requirement for a highly mobile light armoured vehicle weighing around four tonnes. ENGESA started detailed design work on this vehicle, subsequently called the EE-T4 Ogum (the name of an African god of war), in November 1985. The first prototype, built with company funding as a private venture, ran for the first time in May 1986. This was a considerable achievement as virtually all the automotive components of the EE-T4 were new to ENGESA and were not used in any of its existing vehicles.

The prototype of the EE-T4 Ogum was then air lifted to the Middle East for extensive trials and demonstrations under a wide range of conditions. As a result a second prototype was built with a number of improvements. This was completed in November 1986 and early in 1987 it was sent to the Middle East for trials. The EE-T4 Ogum light tracked vehicle is also being offered for local production overseas and was shown in Iraq in May 1989.

Description

The hull of the EE-T4 is made of the famous ENGESA dual-hardness armour which consists of an outer layer of hard steel and an inner layer of softer steel roll-bonded and heat treated to give maximum ballistic protection. This armour gives protection against 7.62 mm armour-piercing attack all round. If required the vehicle could be provided with heavier armour protection but there would, of course, be a drop in the power-to-weight ratio. The small size and very low profile of the EE-T4 Ogum also help to improve its battlefield survivability.

The glacis plate is well sloped, the driver sits at the front left with the engine compartment to his right. Forward observation is by three periscopes and over his position is a single piece hatch cover that swivels to the right.

On the second prototype the driver's hatch opens to the left and is level with the hull roof, so providing improved forward visibility with observation through a single wide-angle periscope that can be replaced by a passive night driving periscope.

ENGESA EE-T4 Ogum light tracked armoured vehicle

ENGESA EE-T4 Ogum light tracked armoured vehicle fitted with ENGESA one-man turret armed with twin 7.62 mm MG

On the first prototype the driver had normal clutch, brake and accelerator pedals with the manual gear change to the right; steering was via a small steering wheel rather than sticks. As a result of user trials the manual Clark 240 V transmission with five forward and one reverse gears has been replaced by an Allison Transmission Division of General Motors AT-545 automatic with four forward and one reverse gears.

The powerpack, consisting of the engine, transmission and cooling system, is mounted to the right of the driver and can be removed as a complete unit to facilitate field replacement. Cooling air enters via louvres in the glacis plate and is expelled through louvres on the hull top to the right of the driver; the exhaust outlet is in the right side of the hull.

The EE-T4 Ogum is powered by the proven Perkins QT 20 B4 four-cylinder turbo-charged diesel developing 125 hp at 2650 rpm, giving an exceptional power-to-weight ratio, at a combat weight of 4.9 tonnes, of 25.5 hp/tonne. The cooling system comprises a water-to-air radiator, air-to-air charge cooler and an air cooler for the engine and transmission. Air circulation is forced through the engine compartment by a hydraulically driven fan.

Production vehicles would be fitted with the BMW M21D24WA, four-cylinder, four-stroke turbo-charged and inter-cooled diesel developing 129.2 hp at 4800 rpm coupled to a ZF 4 HP 22 automatic transmission with four forward and one reverse gears.

Power is taken from the transmission to the front drive sprockets via an ENGESA designed and built Cletrac steering system. The steering differential is coupled directly to the transmission and to the final drives. It is of the regenerative system and steering is achieved by the actuation of two brake discs on both left and right exits of the differential, controlled by the driver through the steering wheel. The steering discs have forced cooling which is provided by an electrically driven fan and is controlled by the discs' temperature.

Suspension is of the torsion bar type with four dual rubber-tyred road wheels, drive sprocket at the front, idler at the rear (in line with the drive sprocket and the same as the road wheel) and two track-return rollers.

ENGESA EE-T4 Ogum light tracked armoured vehicle fitted with ENGESA turret with external mount for 12.7 mm M2 HB machine gun

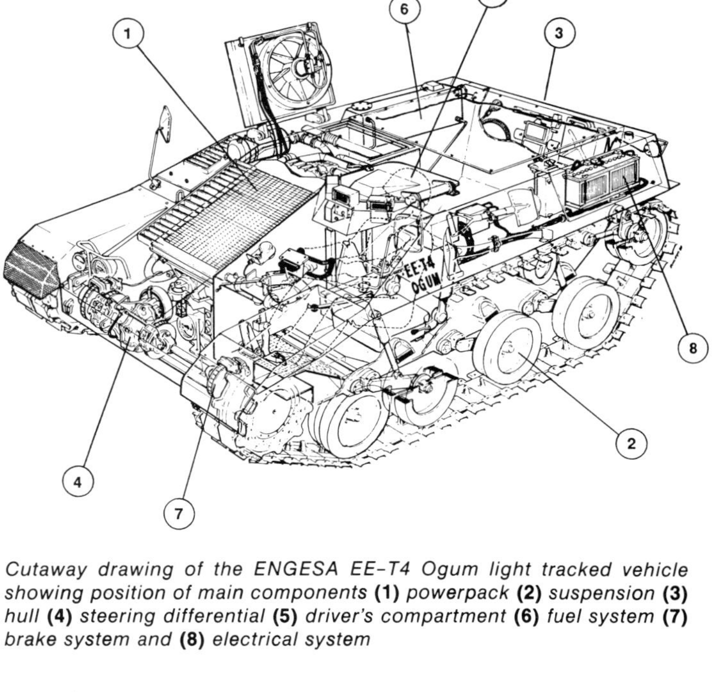

Cutaway drawing of the ENGESA EE-T4 Ogum light tracked vehicle showing position of main components **(1)** *powerpack* **(2)** *suspension* **(3)** *hull* **(4)** *steering differential* **(5)** *driver's compartment* **(6)** *fuel system* **(7)** *brake system and* **(8)** *electrical system*

Hydraulic double-action shock absorbers are provided at the first, second and third road wheels. The Diehl double-pin tracks have optional removable rubber pads to reduce noise and damage to roads during peace time deployment and are fitted with grousers for improved traction in soft soil. The brake system is a servo-hydraulic brake cylinder with two caliper discs mounted directly on both left and right final drives. The brake pedal is located inside the driver's compartment with parking brake being of the mechanical type and located at the differential inlet.

The Ogum has two fuel tanks, one either side of the rear personnel compartment. On the first prototype the batteries were on the floor but on the second prototype they have been moved to the left sponson where they are much easier to check as well as leaving the floor clear for troops or cargo.

The area to the rear of the engine and the driver's compartment is clear so that the vehicle can be adopted for a wide range of roles. The prototype has an armoured hull top on which is mounted an ENGESA cupola with an externally mounted 12.7 mm M2 HB machine gun, this can be aimed and fired from within the vehicle. Two electrically operated smoke dischargers are mounted on either side. The second prototype has been fitted with an ENGESA one-man turret armed with twin 7.62 mm machine guns.

Variants

Using the same basic chassis the following versions of the EE-T4 Ogum have been proposed:

Armoured Personnel Carrier

This would carry four troops in addition to the driver and be armed with a roof-mounted 7.62 mm MG, with 200 rounds of ready use ammunition with a further 800 rounds being carried inside the vehicle. The superstructure to the rear of the driver's position has been raised and in the roof is a circular hatch cover on the left side, where the 7.62 mm machine gun is fitted. In addition, there are two roof hatches that open on either side of the vehicle; the troops would normally enter the vehicle via a door in the hull rear.

ET-7.62 mm twin

Mounted over the rear of the vehicle is an ENGESA one man turret armed with twin 7.62 mm machine guns with an elevation of +55° and a depression of -7° with turret traverse a full 360°. In addition to the roof-mounted periscopic sight for aiming the weapons, four observation periscopes can be installed, as can two smoke dischargers either side of the turret.

Anti-tank Missile Carrier

This would be fitted with a turret that can be traversed through a full 360° and mounted either side would be an ATGW.

ET-50

This would be fitted with a one man turret which has an externally mounted 12.7 mm machine gun with 100 rounds of ready use ammunition and two smoke dischargers mounted either side of the turret. In addition to the roof-mounted periscopic sight, the gunner also has three observation periscopes.

ET-20

This would be fitted with a one man turret armed with an externally mounted 20 mm cannon with 100 rounds of ammunition for ready use and a further 200 rounds in reserve in the turret. Weapon elevation is from -10° to +45° with turret traverse a full 360°. In addition to the day/night periscope for aiming the weapons, the gunner also has four observation periscopes.

Stretched Version

ENGESA has proposed that a stretched version could be built with a longer hull and that this could be used for a wide range of roles including ambulance, command post vehicle and ammunition resupply vehicle.

120 mm Mortar Vehicle

The 120 mm mortar carrier would be similar in concept to the Soviet 240 mm self-propelled mortar M-1975 (SM-240) (2S4), with the mortar in the horizontal position for travelling and then swung through about 140° into the firing position, its baseplate placed on the ground to provide a more stable firing platform. This version would only be able to carry a few 120 mm mortar bombs so would normally be supported by another Ogum carrying additional ammunition.

ENGESA would offer various optional equipment on production vehicles such as night vision equipment, engine and/or crew compartment fire-extinguishing system, air-conditioning system and so on.

SPECIFICATIONS

CREW	2-3	ROAD WHEEL TRAVEL		TRANSMISSION	ZF 4 HP 22 automatic with 4 forward and 1 reverse gears
COMBAT WEIGHT	4400 kg	bump	150 mm	SUSPENSION	torsion bar
POWER-TO-WEIGHT RATIO	29.5 hp/tonne	rebound	90 mm	ELECTRICAL SYSTEM	24 V
GROUND PRESSURE	0.29 kg/cm²	MAX ROAD SPEED	75 km/h	BATTERIES	2 × 12 V, 105 Ah
LENGTH	3.75 m	FUEL CAPACITY	120 l		
WIDTH	2.145 m	RANGE	360 km		
HEIGHT TO HULL TOP	1.355 m	FORDING	0.67 m		
HEIGHT TO TOP OF TWIN 7.62 mm MG TURRET	1.825 m	MAX GRADIENT	60%		
GROUND CLEARANCE	0.34 m	MAX SIDE SLOPE	30%		
TRACK	1.77 m	VERTICAL OBSTACLE	0.6 m		
TRACK WIDTH	379 mm	TRENCH	1.0 m		
LENGTH OF TRACK ON GROUND	2.02 m	ENGINE	BMW M21D24WA, 4-cylinder, 4-stroke diesel, turbo-charged and inter-cooled developing 129.2 hp at 4800 rpm		

(Above relates to production vehicle with twin 7.62 mm ENGESA MG turret)

Status: Development complete. Ready for production.

Manufacturer: ENGESA Engenheiros Especializados SA
Avenue Tucunaré 125/211
Caixa Postal/PO Box 152/154
06400 Barueri, SP
Brazil
Telephone: (55 11) 421 4711
Telex: 1171302
Fax: (55 11) 421 4445

FRANCE

GIAT Industries AMX-10RC Reconnaissance Vehicle

Development

The AMX-10RC reconnaissance vehicle was designed from September 1970 at the Atelier de Construction d'Issy-les-Molineaux to meet a French Army requirement for a vehicle to replace the Panhard EBR heavy armoured car. The first of three chassis was completed in June 1971. Four TK turrets were built. Many of the automotive components of the AMX-10RC are identical to those used in the tracked AMX-10P IFV.

The AMX-10RC equips the reconnaissance regiments of the corps of the French Army as well as the cavalry regiments of the infantry divisions. The first unit to be equipped with the vehicle was the 2eme Régiment de Hussards based at Sourdun, which received its complement of vehicles late in 1979.

The original French Army requirement was for 525 AMX-10RC vehicles with the first production order being placed in 1976, but for a number of reasons only 284 vehicles were ordered with final deliveries to the French Army taking place in 1987.

By early 1991 total production of the AMX-10RC amounted to over 400 vehicles and in addition to Morocco it is now in service with one other country, believed to be Singapore who took delivery of its first vehicles early in 1991.

Deployment of the AMX-10RC in the French Army is as follows:

1st Army, three Army Corps each with one regiment of 36 vehicles.

Force d'Action Rapide (FAR), 6e DLB, with two Regiments (REC and 1st Spahis) each with 36 vehicles.

Two Infantry Divisions with one Regiment of 36 vehicles each.

Training Light Divisions, 24 vehicles.

Morocco placed an order for 108 AMX-10RC vehicles modified to meet local operating conditions, for example with no amphibious capability, in 1978 and first vehicles were delivered in 1981 for training. Deliveries were slow for financial reasons, but all vehicles have now been delivered.

The AMX-10RC has been proposed to the US Army by GIAT for the new American light divisions.

The AMX-10RC was deployed by the French Army in Chad during Operation Manta between August 1983 and November 1984.

The AMX-10RC was also deployed by the French Army in the Persian Gulf during Operation Desert Storm. Two AMX-10RC Regiments of the 6th Light Armoured Division made it possible for the US Army XVIIIth Corps to entrust the French Daguet Division with the very important west coverage mission in Iraq for the entire Allied Forces.

Since it was first introduced into service with the French Army in 1979 the AMX-10RC has been constantly improved, for example new engine, improved ammunition and additional armour protection.

Description

The all-welded aluminium hull and turret provides the crew with protection from small arms fire, flash burns and shell splinters. The driver is seated at the front of the vehicle on the left side and is provided with an adjustable seat and a large single piece hatch cover that opens to his right with three periscopes, the centre one of which can be replaced by a passive OB-31-A periscope for night driving.

The turret is in the centre of the vehicle with the commander and gunner seated on the right and the loader, who also acts as the radio operator, seated on the left. Both the commander and the loader are provided with a single piece hatch cover that opens to the rear and the loader has three periscopes, one to his front, one to his left and one to his rear.

The vehicle commander has six periscopes for all-round observation and a periscope and panoramic M389 telescope with magnifications of ×2 and ×8 which has 360° full contra-rotation and enables the commander to observe the target regardless of the position of the turret. The M389 can be elevated from −12° to +24° and has automatic projection of graticules, ×2 for machine gun and ×8 for 105 mm gun. The commander also has overriding controls enabling him to lay the gun onto the target. The gunner has three periscopes and an M504 telescope manufactured by SOPELEM and Compagnie Industrielle des Lasers (CILAS) which is the main part of the fire control system. The M504 has a magnification of ×10 and is combined with an optical compensator with control electronics for automatic input of fire corrections, M550 laser rangefinder and an M553 bore-sight unit which corrects for any abnormal droop or accidental misalignment of the main armament.

The complete COTAC fire control system comprises various sensors which provide the computer with the following data: target distance as measured by the laser rangefinder with an accuracy of ±7 mils between 400 and 10 000 m; horizontal and vertical speed of the target; angle of cant of the vehicle and windspeed, altitude and outside temperature, which is fed into the computer manually.

GIAT Industries AMX-10RC (6 × 6) reconnaissance vehicle showing cross-country mobility with commander's, loader's and driver's hatches closed

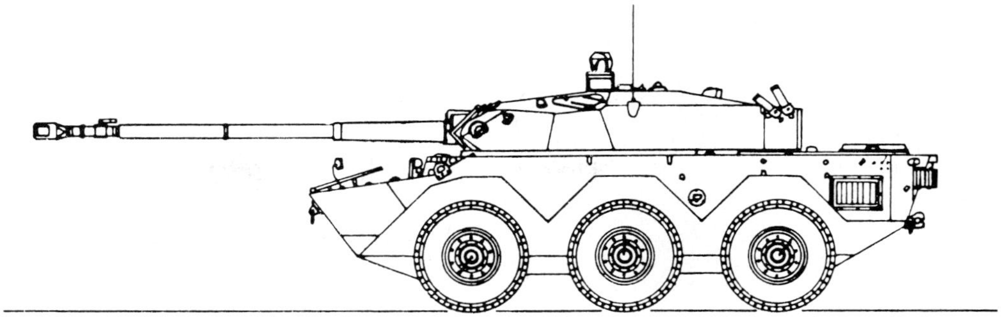

AMX-10RC (6 × 6) reconnaissance vehicle

In a typical target engagement the gunner aligns his sight with the target, tracks the target and depresses the measuring button for two or three seconds, the time taken for the target information to be fed into the computer. The introduction of fire correction data takes only 1.5 seconds and the gunner opens fire.

There are three modes of engagement, COTAC, tank laser sight and graticule judgement. No stabilisation system is fitted so moving targets cannot be successfully engaged, nor can the vehicle engage targets whilst it is moving.

For night engagements the AMX-10RC is fitted with a Thomson-CSF model DIVT 13 LLLTV. The system consists of a television camera mounted in the mantlet to the left of the main armament, electronics unit, control unit and two television screens, one for the commander and the other for the gunner. An aiming graticule is generated electronically and superimposes itself on the monitor screen. The deviation computer of the COTAC fire control system then introduces fire correction data and shifts this aiming graticule by the required value under the same conditions as for the day graticule.

For Operation Desert Storm, GIAT Industries supplied the French Army with a Thomson-CSF DIVT 16 thermal camera as the replacement for the DIVT 13. This option enabled observation and identification to take place at ranges of up to 4000 m.

The integrated test system of the modular fire control system enables any fault to be quickly located and the defective unit replaced. A test kit makes it possible to locate within a unit the defective sub-assembly which is then replaced.

The engine at the rear of the hull is identical to that of the tracked AMX-10P vehicle with which it also has many automotive components in common. The engine is coupled to the gearbox through a hydraulic torque converter with a cut-out clutch. The gearbox has two functions: pre-selection through four gear ratios in both directions with an auxiliary drive for the two water jets, and steering, with outlet epicyclic gears and hydraulically actuated disc brakes. The AMX-10RC is skid-steered in a similar manner to the tracked AMX-10P.

In 1983 it was confirmed that the last two production batches of AMX-10RC vehicles for the French Army would be fitted with the Baudouin model 6 F 11 SRX diesel engine, which develops 280 hp at 3300 rpm, in place of the standard Renault HS 115 diesel that develops 280 hp at 3000 rpm. In addition to being a more powerful engine the Baudouin is more fuel efficient which will give the vehicle a larger operating range of 1000 km. Sufficient engines have now been ordered to replace all of the Hispano-Suiza HS-115 engines installed in the AMX-10RC vehicles of the French Army.

The hydro-pneumatic suspension system, designed and manufactured by Messier Auto-Industrie, can be quickly adjusted by the driver to suit the type of ground being crossed, this being accomplished without the driver leaving his seat.

The minimum ground clearance is 210 mm, but for road travel 350 mm is normal, for cross-country travel 470 mm and for amphibious operations 600 mm.

The suspension at each wheel station consists of a road wheel arm, suspension assembly (link rod, level and cylinder) with the cylinder acting as spring and shock absorber and allowing the height to be adjusted. A centralised lubrication unit facilitates maintenance of the suspension system. The tyre pressure can be adjusted to suit the type of ground being crossed, this being accomplished without the driver leaving his seat.

The AMX-10RC is fully amphibious being propelled in the water by two water jets either side of the rear of the hull. Before entering the water a trim vane is erected at the front of the vehicle, with a transparent window in the left side to allow the driver to see ahead when afloat. The vehicles delivered to Morocco were not fitted with water jets.

GIAT Industries AMX-10RC (6 × 6) reconnaissance vehicle with suspension lowered for maximum ground clearance

GIAT Industries AMX-10RC turret fitted with German Rheinmetall 105 mm low recoil gun (Christopher F Foss)

GIAT Industries AMX-10 RAC (6 × 6) reconnaissance vehicle with TS 90 turret armed with 90 mm gun, 7.62 mm coaxial machine gun

Prototype of GIAT Industries AMX-10 RTT from rear with ramp lowered. This model has firing ports in sides and rear (Christopher F Foss)

The NBC system is mounted in the rear of the turret and maintains a slight overpressure in the fighting compartment. For use in cold climates an auxiliary heater is installed.

Main armament of the AMX-10RC is a 105 mm semi-automatic gun with a vertical wedge breech-block and a barrel with a thermal jacket and a two-stage muzzle brake. The gun itself is 48 calibres long without the muzzle brake and has a recoil length of 600 mm and a recoil force of about 13 tonnes. The recoil mechanism comprises a brake on the left side and a counter-recoil mechanism on the right side of the gun. The gun fires three types of ammunition, HEAT, HE and practice. An APFSDS round entered production in 1987. The APFSDS round uses the penetrator of the 90 mm GIAT Industries APFSDS round and has a muzzle velocity of 1400 m/s. This will penetrate a NATO triple heavy tank target at a range of 2000 m. The new APFSDS round weighs 13.1 kg complete, the projectile weighs 3.8 kg. The only modification required to the ordnance is a new muzzle brake.

Of 38 rounds of 105 mm ammunition carried 12 are for ready use in the turret. A 7.62 mm machine gun is mounted coaxially to the left of the main armament. Two smoke dischargers mounted either side of the turret towards the rear are electrically fired from within the vehicle.

The turret is equipped with a SAMM CH49 electro-hydraulic control system consisting of a servo-controlled cylinder, flow dividing unit, servo-motor, hydraulic power source, electronics box, twin handles for the gunner and a single handle for the commander.

If required the AMX-10RC can be delivered without the air-conditioning system, heater, NBC system, night fire control equipment, navigation system and amphibious capability. The vehicle can also be fitted with a thermal camera to extend observation and identification out to 4000 m in any weather condition and a front-mounted winch for self-recovery operations.

To improve the protection of AMX-10RC (6 × 6) reconnaissance vehicles deployed to Saudi Arabia for Operation Desert Storm, GIAT Industries supplied add-on armour which was designed and manufactured in France and fitted in Saudi Arabia.

Variants

AMX-10RC Driver Training Vehicle

This is simply an AMX-10RC (6 × 6) vehicle with its turret removed and fitted with a fully enclosed cabin for the instructor and several other students.

AMX-10RC with New Gun

In 1987 a new version of the AMX-10RC was unveiled for the export market. This has a German Rheinmetall 105 mm low recoil gun fitted with a thermal sleeve and muzzle brake, and fires standard NATO tank ammunition including APFSDS-T. It also has a self-recovery capability as a heavy duty winch has been installed at the front of the vehicle and the driver can adjust the tyre pressures without leaving his seat.

AMX-10 RAC

This is essentially the AMX-10RC with its original turret replaced by the GIAT Industries TS 90 turret. The vehicle has a loaded weight of 14 800 kg with the crew of three consisting of commander/loader, gunner and driver. The turret has manual traverse but can be fitted with power traverse if required. The prototype of the AMX-10 RAC has been demonstrated to a number of countries in the Middle East (Egypt and Iraq), powered by the Baudouin 6 F 11 SRX diesel engine.

Basic specifications of the AMX-10 RAC, where they differ from the AMX-10RCs are:

GIAT Industries AMX-10RC driver training vehicle (Terry J Gander)

CREW	3
COMBAT WEIGHT	15 000 kg
LENGTH GUN FORWARDS	8.57 m
LENGTH CHASSIS	6.35 m
OVERALL WIDTH	2.95 m
TOTAL HEIGHT	2.48 m
AMMUNITION	
90 mm	20 (turret), 10 (hull)
7.62 mm	3200

AMX-10 RAA

The prototype of this model was shown in 1984 and is essentially the AMX-10RC with its original turret replaced by the Thomson-CSF/SAMM SABRE turret armed with twin 30 mm cannon. This model never progressed beyond the prototype stage.

AMX-10 RTT

In 1983 GIAT Industries announced that a wheeled armoured personnel carrier version of the AMX-10RC had been designed under the designation of AMX-10 RTT (Roues Transporteur de Troupes). This has a similar suspension to the AMX-10RC but with the driver at the front of the hull on the left and the engine compartment to his right. The troop compartment is at the rear of the hull with periscopes giving observation to the sides. Over the top of the troop compartment is a GIAT Industries CIBI 12.7 mm weapon station armed with a 12.7 mm M2 HB machine gun. The crew of three consists of commander, gunner and driver and the vehicle can carry 10 fully equipped infantrymen. The first prototype was shown in June 1987 fitted with GIAT Industries 25 mm Dragar turret.

The AMX-10 RTT weighs 15 tonnes in combat order and has almost identical performance characteristics to the AMX-10RC.

Other variants

Using the front engined chassis of the AMX-10 RTT first shown in 1987, GIAT have proposed that this could be the basis for a number of other vehicles including a light tank/reconnaissance vehicle with the 105 mm turret (as fitted to the AMX-10RC) mounted at the rear of the hull and an engineer/recovery vehicle which would have a front-mounted dozer blade and rear-mounted hydraulic crane.

SPECIFICATIONS (with 0.3 m ground clearance)	
CREW	4
CONFIGURATION	6 × 6
COMBAT WEIGHT	15 880 kg
UNLOADED WEIGHT	14 900 kg
POWER-TO-WEIGHT RATIO	16.45 hp/tonne
LENGTH GUN FORWARDS	9.15 m
LENGTH HULL	6.357 m
WIDTH	2.95 m
HEIGHT	
overall	2.66 m
to turret top	2.29 m
to hull top	1.565 m
GROUND CLEARANCE (normal)	0.35 m (adjustable from 0.2-0.6 m)
TRACK	2.425 m
WHEELBASE	1.55 m + 1.55 m
MAX SPEED	
road	85 km/h
water	7.2 km/h
MAX ROAD RANGE	1000 km
FORDING	amphibious
GRADIENT	50%
SIDE SLOPE	30%
VERTICAL OBSTACLE	0.8 m
TRENCH	1.65 m
ENGINE	Hispano-Suiza HS-115 super-charged water-cooled 8-cylinder diesel developing 260 hp at 3000 rpm*
(see text)	
TRANSMISSION	preselective with 4 forward and 4 reverse gears
SUSPENSION	hydro-pneumatic
TYRES	14.00 × 20 XL
ELECTRICAL SYSTEM	24 V
BATTERIES	6 × 12 V, 100 Ah
ARMAMENT	
main	1 × 105 mm gun
coaxial	1 × 7.62 mm MG
SMOKE-LAYING EQUIPMENT	2 smoke dischargers either side of turret
AMMUNITION	
main	38
coaxial	4000
smoke grenades	16
GUN CONTROL EQUIPMENT	
Turret power control	electro-hydraulic/manual
by commander	yes
by gunner	yes
Gun elevation/depression	+20°/-8° (elevation can be increased to +40° using suspension)
Gun stabiliser	
vertical	no
horizontal	no

*See text. Final production models had a Baudouin Model 6 F 11 SRX engine developing 300 hp at 3300 rpm.

Status: Production as required. In service with France, Morocco and Singapore.

Manufacturer: Atelier de Construction Roanne (ARE)
Enquiries to GIAT Industries
13 route de la Minière
78034 Versailles Cedex
France
Telephone: (331) 30 97 37 37
Fax: (331) 30 97 39 00

GIAT Industries AMX-10 RTT (6 × 6) APC fitted with GIAT Industries 25 mm one-man Dragar power operated turret and firing ports in hull sides

Renault VBC 90 Armoured Car

Development

The VBC 90 (Véhicule Blindé de Combat) was designed by Renault Véhicules Industriels specifically for the export market and was shown for the first time at the 1979 Satory Exhibition of Military Equipment. The first production VBC 90 armoured car was completed in September 1981. It uses many automotive components of the VAB (6 × 6) APC which is in production for the export market and the VAB (4 × 4) which is in service with the French Army. The first of 28 VBC 90 armoured cars ordered by the French Gendarmerie were delivered in 1983 and are fitted with a SOPTAC 11 fire control system incorporating a laser rangefinder. Oman placed an initial order for four VBC-90 armoured cars with SOPTAC 11 fire control systems, a further two vehicles were delivered in early 1985.

Description

Its all-welded steel hull protects the crew from small arms fire and shell splinters and is divided into three compartments: driver's at the front, fighting in the centre and engine at the rear. The driver is seated at the front of the vehicle on the left side and has a single piece hatch cover that opens to the right with three bullet-proof windows for greater all-round observation, the centre one fitted with a wiper.

The GIAT Industries TS 90 turret is all-welded with the commander on the left and the gunner on the right, both with a single piece hatch cover that opens to the rear and an adjustable seat. An extractor fan is mounted in the forward part of the turret roof. Main armament consists of a long-barrelled 90 mm gun with a 35° oblique wedge breech-block, hydro-pneumatic recoil system, thermal sleeve and a muzzle brake. Twenty rounds of ready-to-use ammunition are carried, four in the turret basket and 16 in the turret bustle. The empty cartridge cases are ejected into a bag under the breech and the commander has an ammunition resupply hatch in the left side of the turret. A further 25 rounds of 90 mm ammunition are carried in the hull. Types of GIAT Industries ammunition that can be fired include canister, HE, HE long range, HEAT, smoke and APFSDS.

Mounted coaxially to the left of the main armament is a 7.62 mm machine gun and there are two electrically operated smoke dischargers mounted either side of the turret towards the rear.

A PH 9A white light searchlight is mounted coaxially to the left of the main armament and a second white light searchlight is mounted in the forward part of the turret and operated by the commander. The commander has seven (three M556s, four M554s) periscopes and the gunner five (three M556s, two M554s) and the gunner also has a GIAT Industries M563 telescopic sight with a magnification of ×5.9 for laying the main armament.

The GIAT Industries M563 sight can be replaced by the SOPELEM TJ 90A sight. Fire control options are numerous and include integration of a CILAS TCV 107 laser rangefinder in GIAT Industries M563 sight, installation of SOPELEM TJ 90A sight and also integrating CILAS TCV 107 laser rangefinder; SOPTAC 33, an adaptation of TCV 107 laser rangefinder with automatic computing and display of range and elevation of target in sight graticule; SOPTAC 34, an adaptation of TCV 107 laser rangefinder with automatic computing and display of range and elevation of target in sight graticule and lateral deflection to allow for engagement of moving targets.

The gunner can also have a SOPELEM TJ N2 90A sight with a magnification of × 6 (day) and × 6 (night). The SOPTAC 10A is the TJ N2 90A with TCV 107 laser rangefinder with range displayed to vehicle commander with automatic display of target elevation in the gunner's sight graticule. SOPTAC 11A is the TJ N2 90A with TCV 107 laser rangefinder with automatic computing and display of range to vehicle commander and automatic display of target elevation in the gunner's sight graticule and lateral deflection (target movement speed) for firing at moving targets. All SOPTAC sights can be fitted so that the gunner's field of view is also supplied to the commander's sight.

Optional equipment includes a 7.62 mm anti-aircraft machine gun. Future developments planned for the turret include independent elevation of the coaxial 7.62 mm machine gun, option of the 7.62 mm coaxial machine gun being replaced by a 12.7 mm machine gun, and electric traverse (360° in 14 seconds).

The engine, transmission and fuel tank are at the rear of the hull, with a large door, opening to the left, for easy access for maintenance.

Power is transmitted from the engine to the wheels by a torque converter and a gearbox with five forward and one reverse gears. Gears are shifted by a small electrically operated lever which also operates the clutch. The axles have differential reduction gears with double reduction and differential locking. The wheels are independently suspended by torsion bars and hydraulic shock absorbers. Steering is hydraulically assisted on the two front axles and the tyre pressures can be adjusted to suit the type of ground being crossed. Unlike the VAB family of APCs, the VBC 90 does not have any amphibious capability.

Standard equipment on all vehicles includes a heater and a power-operated winch mounted internally at the front of the hull on the right side. This is provided with 60 m of cable and has a capacity of 6000 kg. Optional equipment includes an NBC system and independent heating as well as the various fire control systems/turrets previously mentioned.

Renault VBC 90 armoured car fitted with SAMM TTB 190 turret armed with GIAT Industries 90 CS F4 high-pressure 90 mm gun, driver's hatch is open

VBC 90 (6 × 6) fitted with MECAR 90 mm KEnerga 90/46 Weapons Station

Renault VBC 90 (6 × 6) armoured car as delivered to French Gendarmerie, with driver's hatch closed and showing integral bullet-proof windows

Variants

The VBC 90 can be fitted with a number of other turrets including the following:

Hispano-Suiza Lynx 90 (already tested)

GIAT Industries turret armed with 81 mm smoothbore gun/mortar.

VBC with 90 mm KEnerga gun

For trials purposes a VBC 90 has been fitted with the Belgian 90 mm MECAR KEnerga 90/46 Weapons Station and 50 rounds of ammunition, including APFSDS, have been fired successfully.

The KEnerga 90/46 Weapons Station was also fitted with the Texas Instruments Lightweight Modular Sight (LWMS) which has a dual sensor system that enables targets to be engaged by day and night. The visible sensor has two fields-of-view, a unity window and a ×8 telescope.

The thermal sensor includes a wide field-of-view with ×4 magnification and a narrow field-of-view with ×12 or optional ×8 magnifications. The Ferranti 520 laser rangefinder has a range of 9990 m and in addition to displaying the range in the gunner's eyepiece it also displays it on the commander's display.

In the Summer of 1987, a VBC 90 (6 × 6) armoured car with the MECAR KEnerga 90/46 Weapon Station performed successfully during an intensive four-month trial in the United Arab Emirates.

VBC with TTB 190 Turret

The VBC (6 × 6) armoured car with the SAMM TTB 190 turret has now been qualified and is offered as an alternative to the previously mentioned variants.

The SAMM TTB 190 turret uses the same CS 90 F4 gun as the GIAT Industries TS 90 turret and its secondary armament consists of one 7.62 mm or one 12.7 mm coaxial machine gun and, as an option, a roof-mounted 12.7 mm machine gun. The cradle of the turret provides protection against attack from 14.5 mm rounds.

The commander sits on the left and the gunner on the right, both are provided with a combined day/night sight plus observation periscopes. The 90 mm gun has an elevation of +15° and a depression of −8° with turret traverse a full 360°. Turret traverse is electric at 35°/s.

The turret can also be fitted with a feeding barrel to facilitate rapid reloading. A wide range of options is available including various types of fire control system and a laser rangefinder.

Status: In service with France (Gendarmerie, 28 delivered from 1983) and Oman (6). Production complete.

Manufacturer: Production of the VBC 90 is undertaken by Renault Véhicules Industriels with Creusot-Loire Industrie as co-contractor under the leadership of Renault Véhicules Industriels.

Marketing: SMS Renault VI - Creusot-Loire Industrie
316 Bureaux de la Colline
92213 Saint-Cloud Cedex
France
Telephone: (1) 46 02 80 33
Telex: 250 662

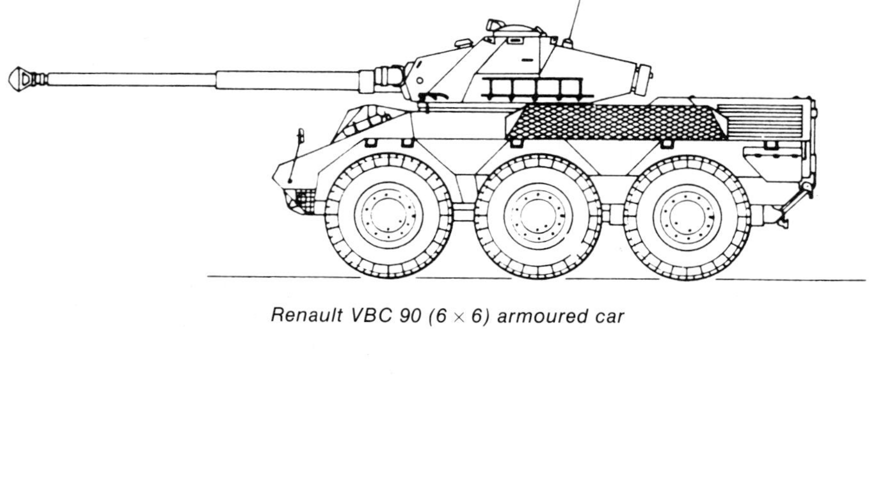

Renault VBC 90 (6 × 6) armoured car

SPECIFICATIONS
Note: This relates to VBC 90 with GIAT Industries 90 mm TS 90 turret

CREW	3
CONFIGURATION	6 × 6
POWER-TO-WEIGHT RATIO	16 hp/tonne
COMBAT WEIGHT	13 500 kg
WEIGHT UNLADEN	12 250 kg
LENGTH GUN FORWARD	8.085 m
LENGTH HULL	5.63 m
WIDTH	2.5 m
HEIGHT	
to turret top	2.552 m
to hull top	1.737 m
TRACK	2.035 m
WHEELBASE	1.5 m + 1.5 m
ANGLE OF APPROACH/ DEPARTURE	45°/45°
MAX ROAD SPEED	
5th gear	92 km/h
4th gear	58 km/h
3rd gear	34 km/h
2nd gear	20 km/h
1st gear	12 km/h
reverse	13 km/h
MAX ROAD RANGE (at 60 km/h)	1000 km
FORDING	1.2 m
GRADIENT	50%
SIDE SLOPE	30%
VERTICAL OBSTACLE	0.5 m
TRENCH	1 m
ENGINE	Renault MIDS 06.20.45 in-line water-cooled turbo-charged 6-cylinder diesel developing 220 bhp at 2200 rpm. Previous engine MAN D.2356 HM 72 in-line water-cooled 6-cylinder diesel developing 220 hp at 2200 rpm (standard on VBC 90 until 1983 but no longer available)
TRANSMISSION	Renault Transfluid 376 with 5 forward and 1 reverse gears
STEERING	recirculating ball, hydraulically assisted on first and second axles
TYRES	14.00 × 20 (run flat)
BRAKES	disc
SUSPENSION	independent (torsion bar and telescopic shock absorbers)
ELECTRICAL SYSTEM	24 V
BATTERIES	4 × 12 V, 6TN
ARMAMENT	
main	1 × 90 mm gun
coaxial	1 × 7.62 mm MG
anti-aircraft	1 × 7.62 mm MG (optional)
SMOKE-LAYING EQUIPMENT	2 smoke dischargers either side of turret
AMMUNITION	
main	45
coaxial	4000
GUN CONTROL EQUIPMENT	
Turret power control	electrically assisted/ manual (traverse only)
Max rate of power traverse	360°/20 s
Max rate of manual elevation	23°/15 s
Gun elevation/depression (except over rear)	+15°/-8°

Panhard ERC Armoured Car

Development

The ERC (Engin de Reconnaissance Cannon) range of armoured cars was developed by Panhard as a private venture from 1975 and was aimed specifically at the export market. The range was first shown in 1977 at the Satory Exhibition of Military Equipment and entered production the following year at Panhard's new factory at Marolles. First production vehicles were completed in 1979. The ERC range of armoured cars shares many automotive components with the Panhard VCR range of armoured personnel carriers which were also shown for the first time in 1977.

The French Army evaluated the ERC 90 F4 Sagaie between 1978-80 and accepted the vehicle in December 1980 although further trials, including hot weather trials in Djibouti in 1982, were subsequently carried out. First production vehicles were delivered to the French Army in 1984 and the total French Army order was for 192 ERC Sagaies with final deliveries in 1990.

The French Rapid Deployment Force uses this vehicle as its standard armoured car. The 1st Regiment Infanterie de Marine of the 9th Marine Division has converted to a reconnaissance regiment with ERC 90 vehicles. The Regiment d'Infantrie de Chars de Marine of the 9th Marine Division received all of its vehicles in 1984, the 1st Regiment of the Parachute Hussards of the 11th Parachute Division received their vehicles in 1985-86 while the 4th Chasseurs Regiment of the 27th Alpine Division received their vehicles in 1988-89. In each case they replaced AML 60 and AML 90 4 × 4 armoured cars. Each regiment has three squadrons with each of these having four platoons each with three ERC 90 F4 Sagaies and a MILAN ATGW squad.

A total of 45 ERC 90 F4 Sagaie armoured cars are forward deployed overseas, 24 with the 13th Foreign Legion in Djibouti, six with the 6th Marine Corps Battalion in Gabon, six with the 43rd BIMa in the Ivory Coast and six with the 23rd BIMa at Dakar in Senegal. By June 1991 firm orders for the Panhard ERC amounted to 380 vehicles.

Description

The all-welded steel hull provides the crew with protection from small arms fire and shell splinters. The floor of the hull consists of two plates welded together to form a flat V which helps the vehicle to slide off obstacles and stiffens the floor against damage from mines.

The driver is seated at the front of the hull, slightly to the left of the vehicle's centre line. His hatch is in two parts: one part, which contains the forward-facing periscope, folds forward onto the glacis plate and the second part folds upwards. The forward periscope can be replaced by an infra-red or image-intensification periscope for night driving, and there is an additional periscope either side of the driver's hatch cover.

The turret is mounted in the centre of the vehicle and as each version has a different turret they are described in detail under variants. All vehicles in the range are powered by a militarised version of a Peugeot V-6 petrol engine which develops 145 hp at 5500 rpm. Power is taken from the engine through a single-plate clutch to a transversely mounted Panhard gearbox with six forward and one reverse gears and incorporating a Panhard cam-type limited slip differential from which the drive is taken to either side of the vehicle. Drive is then taken by half-shafts to bevel gears at the centre wheel stations and then shafts take power to the bevel gears at the front wheel stations. The rear wheel bevel gears are driven by shafts directly from the gearbox. The drive from the centre to the front wheel on each side incorporates another cam-type limited slip differential. Power is transmitted from the bevel boxes to the wheels by a train of gears housed within trailing arms which independently locate the wheels. The front and rear wheels are sprung by single coil springs with concentric telescopic hydraulic dampers. The centre wheels, which are raised off the ground when on roads, are sprung by hydro-pneumatic units. The centre wheels are powered even when raised clear of the road surface.

Only the front wheels are steered, and steering is hydraulically assisted. The low-pressure tyres are fitted with Hutchinson cellular inner tubes, which enable the vehicle to travel for at least 100 km at 30 km/h when the tyres have been punctured by bullets.

The basic vehicle can ford to a depth of 1.2 m without preparation but two different amphibious models of the ERC have already been developed. Both models have a trim vane mounted at the front of the hull which is folded back onto the glacis plate when not in use and has a transparent panel in its centre to enable the driver to have a clear

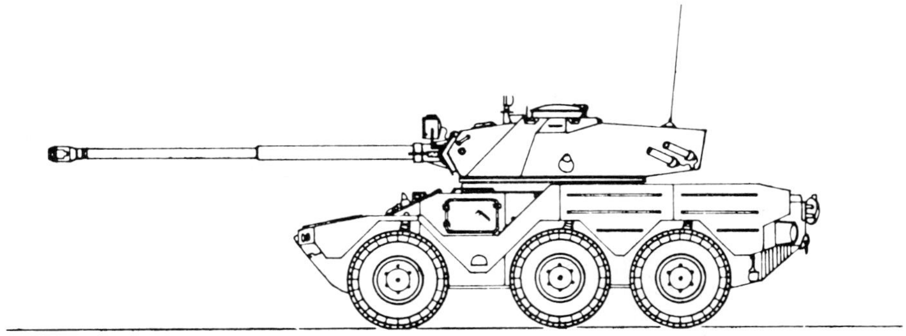

Panhard ERC 90 F4 Sagaie fitted with GIAT Industries TS 90 turret with 90 mm gun

Panhard ERC 90 F4 Sagaie 2 armoured car with SAMM TTB 190 two-man turret and roof-mounted machine gun from rear

Panhard ERC 90 F4 Sagaie of the Regiment d'Infanterie de Chars de Marine (RICM) which belongs to the FAR

Panhard ERC 90 F1 Lynx (ERC-1 Lynx) production line

Panhard ERC 90 F4 Sagaie 2 in service in Gabon

view ahead, and sheet metal floats over the wheel stations which are filled with closed-cell foam. A rear opening door is provided in the left side of the hull. Before entering the water the trim vane is erected and the two tubular ducts at the sides of the engine compartment are hydraulically swivelled into the vertical position to prevent water entering the engine compartment. The radiator compartment is allowed to flood, as when the vehicle is afloat cooling is by water-to-water heat transfer. The first amphibious model is propelled in the water by its wheels and the second by two Dowty-Messier hydrojets.

All models have a turret-mounted ventilator and internal lights as standard. Optional equipment includes an NBC system, ground navigation compass, air-conditioning equipment and a front-mounted winch with 60 m of cable and a capacity of 3500 kg.

Variants

ERC 90 F4 Sagaie (ERC-1 Sagaie)

This was the first model to enter production and is fitted with a GIAT Industries TS 90 turret armed with a 90 mm gun which can fire canister, HE, HE long range, HEAT, smoke and APFSDS ammunition.

A 7.62 mm machine gun is mounted coaxially to the left of the main armament and two smoke dischargers are mounted either side of the turret at the rear. The commander is seated on the left side of the turret and the gunner on the right, both with a single piece hatch cover that opens to the rear. The commander is provided with seven periscopes and the gunner five and the gunner also has an M563 telescopic sight with a magnification of ×5.9. Of 20 rounds of 90 mm ammunition carried in the turret, 12 are HEAT and 8 HE.

Panhard offers six options for the vehicle, each with a weight index of 1 or 2, and the user can select options up to an index of four. The options are amphibious kit propelled in the water by its wheels (1), amphibious kit propelled in the water by hydrojets (2), NBC system (1), air-conditioning system/heater (1), an additional 10 rounds of 90 mm and 1000 rounds of 7.62 mm ammunition (1) and a ground navigation system (1).

ERC 90 F4 Sagaie TTB 190

This was shown for the first time at the 1983 Satory Exhibition of Military Equipment and is the basic chassis fitted with the new SAMM TTB 190 two-man turret which is armed with the same GIAT Industries 90 mm F4 gun as the original ERC 90 F4 Sagaie with the GIAT Industries TS 90 turret.

The commander is on the left and the gunner on the right, and both have a combined day/night sight plus periscopes for observation. The 90 mm gun has an elevation of +15° and a depression of -8° with turret traverse a full 360°. Turret traverse can be electric, hydraulic or electro-hydraulic.

A 7.62 mm or 12.7 mm machine gun can be mounted coaxially to the left and a 7.62 mm anti-aircraft machine gun can be mounted externally on the commander's cupola. Total of 23 rounds of 90 mm and 2000 rounds of 7.62 mm machine gun ammunition are carried. A wide range of options is available for this turret including a stabilisation system and various types of fire control system including a laser rangefinder.

ERC 90 F4 Sagaie 2 (ERC-2 Sagaie)

This model was announced at the 1985 Satory Exhibition of Military Equipment, has a longer, wider hull and is powered by two Peugeot XD 3T four-cylinder turbo-charged diesel engines each developing 98 hp at 4150 rpm. These engine are also used in the Panhard VBL light scout car, VCR TT 2 APC (two engines) and the AML/Buffalo 4 × 4 vehicles. The engines are coupled to a ZF 4 HP 22 automatic transmission with four forward and one reverse gears and a Panhard two-speed transfer box, one for road travel and the other for cross-country.

As an alternative to the two Peugeot XD 3T engines the Sagaie 2 can also be powered by two Peugeot PVR six-cylinder water-cooled petrol engines developing 145 hp each.

The Sagaie 2 has larger 13.00 × 16.00 tyres compared to the Sagaie 1 which has 11.00 × 16 tyres, and the central pair of road wheels has been moved further back.

By 1985 two vehicles had been built, one with two petrol engines and one with two diesel engines. The vehicle can be made amphibious by filling the side bins with foam; this also provides some degree of protection against HEAT attack.

The prototypes have been fitted with the new SAMM TTB 190 turret which is the model selected by Gabon, but other turrets can be fitted including TS 90.

The desert version has no water-jets for amphibious propulsion and the buoyancy boxes are replaced by boxes for increased stowage for ammunition, fuel and water. Optional equipment includes an NBC system and air-conditioning unit.

The only known customer for this vehicle, which is also referred to as the Panhard ERC-2, is Gabon.

ERC 90 F1 Lynx (ERC-1 Lynx)

This is fitted with a Lynx 90 turret designed and manufactured by Hispano-Suiza and armed with a 90 mm F1 gun, 7.62 mm coaxial machine gun and two smoke dischargers on either side of the turret at the rear. The gun fires HEAT, HE, smoke and canister rounds.

French Army Panhard ERC 90 F4 entering the water with snorkels erected ready for amphibious operations

Panhard ERC 90 F1 Lynx armoured cars in service with Mexican Army

Panhard ERC 60-20 Serval armoured car

Panhard ERC 20 Kriss with SAMM TAB 220 20 mm turret

The commander is seated on the left of the turret with the gunner on the right. The commander is provided with a cupola with a single piece hatch cover that opens to the rear and eight periscopes for all-round observation. The gunner is provided with a single piece hatch cover that opens to the rear. Vision equipment consists of a combined day and night passive periscope model TJN 2.90 (SOPELEM), TCV 107 laser rangefinder (CILAS/SOPELEM) and 12 periscopes (nine type L 794 B and three type L 794 D). A total of 21 rounds of 90 mm ammunition is carried in the turret and if required a 7.62 mm machine gun can be fitted for anti-aircraft defence.

Panhard offers six options for the vehicle, identical to those offered for the ERC 90 F4 Sagaie except that it can carry an additional 20 rounds of 90 mm ammunition rather than the 10 on the ERC 90 F4 Sagaie.

ERC TG 120 Guepard

This model is no longer offered.

ERC 60-20 Serval

This is fitted with a 60-20 Serval turret designed and manufactured by Hispano-Suiza and is armed with a 60 mm breech-loaded mortar which can be elevated from -8° to +80° and a 20 mm cannon mounted at the right rear of the turret which can be elevated from -8° to +50°. The 60 mm mortar can be either a Thomson-Brandt HB 60 with a maximum range of 2600 m or a Thomson-Brandt HB 60 LP with a maximum range of 5000 m. The 20 mm cannon can either be a French M693 dual-feed cannon with 300 rounds of HE and 37 rounds of AP ammunition for ready use or an HS 820 (Oerlikon KAD-B16) single-feed cannon with 300 rounds of ready-use ammunition. If required, a 7.62 mm machine gun can be mounted coaxially to the right of the 20 mm cannon.

Vision equipment includes one sighting periscope, nine observation periscopes and an optional night periscope.

Panhard offers seven options for this vehicle, identical to the ERC 90 F4 Sagaie except that an additional 100 rounds of 20 mm and 1000 rounds of 7.62 mm ammunition are carried (1) and/or 20 60 mm mortar projectiles (1).

Other Versions of Panhard ERC

The following versions of this vehicle are no longer being marketed by Panhard:-

ERC 60/12 Mangouste
EMC 81 mm Mortar Gun Carrier
ERC with SAMM TTB 125 turret
ERC with SAMM S 530 A 20 mm turret
ERC with TTB 140 40 mm turret
ERC with ESD TA-20 turret
ERC with SAMM TAB 220 20 mm turret (this has a two-man turret with twin 20 mm anti-aircraft cannon and a small number have been sold to Gabon)

SANTAL SAM System

After studying the Panhard ERC (6 × 6) chassis and the SMS VAB (6 × 6) chassis, the former was selected by the French Army as the platform for the turret-mounted version of the MATRA Mistral surface-to-air missile system in 1985. This version is usually referred to as the SANTAL.

The chassis is fitted with a two-man power operated turret with a total of six fire and forget MATRA Mistral surface-to-air missiles in the ready to launch position with additional missiles being carried in reserve in panniers on either side of the hull.

Mounted on the turret rear is a Dassault Electronique surveillance radar with Thomson-TRT Defense Castor thermal camera being mounted under the left bank of four missiles.

The French Army had a requirement for up to 90 systems on the Panhard ERC (6 × 6) chassis but this programme was cancelled for financial reasons in 1989. Trials with the two prototypes were however completed and the system is still being offered for export.

Full details of the SANTAL system are given in *Jane's Land-Based Air Defence 1991-92*, page 110. The turret has also been fitted, for trials purposes, on the VAB (6 × 6) chassis.

Panhard ERC (6 × 6) armoured car chassis fitted with two man turret armed with six MATRA Mistral SAM in ready to launch position with one missile being launched

SPECIFICATIONS

Model	**Sagaie 1**	**Lynx**	**Serval**	**Sagaie 2**
CREW	3	3	3	3
CONFIGURATION	6 × 6	6 × 6	6 × 6	6 × 6
COMBAT WEIGHT	8100 kg	7600 kg	7000 kg	10 000 kg
POWER-TO-WEIGHT RATIO	17.90 hp/tonne	19.07 hp/tonne	22.14 hp/tonne	19.6 hp/tonne
LENGTH GUN FORWARDS	7.693 m	6.115 m	5.26 m	7.97 m
LENGTH HULL	5.098 m	5.235 m	5.26 m	5.57 m
WIDTH	2.495 m	2.495 m	2.495 m	2.76 m
HEIGHT				
overall	2.254 m	2.147 m	2.333 m	2.337 m
to hull top	1.502 m	1.502 m	1.502 m	1.545 m
GROUND CLEARANCE				
road	0.294 m	0.294 m	0.294 m	0.38 m
cross-country	0.344 m	0.344 m	0.344 m	n/a
TRACK	2.135 m	2.135 m	2.135 m	2.346 m
WHEELBASE	1.63 m + 1.22 m	1.63 m + 1.22 m	1.63 m + 1.22 m	2.2 m + 1.2 m
ANGLE OF APPROACH/ DEPARTURE	48°/45°	48°/45°	45°/45°	48°/45°
MAX SPEED				
road	95 km/h	95 km/h	100 km/h	110 km/h
water, wheels	4.5 km/h	4.5 km/h	4.5 km/h	n/a
water, hydrojets	9.5 km/h	9.5 km/h	9.5 km/h	7.2 km/h
MAX ROAD RANGE	700 km	700 km	800 km	600 km
FUEL CAPACITY	242 l	242 l	242 l	350 l
FORDING	1.2 m	1.2 m	1.2 m	1.2 m
GRADIENT	60%	60%	60%	50%
SIDE SLOPE	30%	30%	30%	30%
VERTICAL OBSTACLE	0.8 m	0.8 m	0.8 m	0.8 m
TRENCH	1.1 m	1.1 m	1.1 m	0.8 m
ENGINE	Peugeot V-6 petrol developing 155 hp at 5250 rpm			2 Peugeot XD 3T 4-cylinder turbo-charged diesels developing 98 hp at 2250 rpm (each)
TRANSMISSION	Panhard with 6 forward and 1 reverse gears		Panhard with 6 forward and 1 reverse gears	2 × ZF 4 HP 22 automatic
STEERING	power-assisted	power-assisted	power-assisted	hydraulic
CLUTCH	hydraulic-controlled disc	hydraulic-controlled disc	hydraulic-controlled disc	hydraulic-controlled disc
TYRES	11.00 × 16	11.00 × 16	11.00 × 16	13.00 × 16
ELECTRICAL SYSTEM	24 V	24 V	24 V	24 V
ARMAMENT				
main	1 × 90 mm gun	1 × 90 mm gun	1 × 60 mm mortar	1 × 90 mm gun
coaxial	1 × 7.62 mm MG	1 × 7.62 mm MG	1 × 20 mm (independent elevation)	1 × 7.62 mm MG
anti-aircraft	1 × 7.62 mm MG optional	1 × 7.62 mm MG optional	1 × 7.62 mm MG optional	1 × 7.62 mm MG optional
AMMUNITION				
main	20	30	50	33
coaxial	2000	3000	250	2100-2500
anti-aircraft	included in above	included in above	1000	included in above
TURRET POWER TRAVERSE	manual	manual	manual	electric/manual
GUN ELEVATION/ DEPRESSION	+15°/-8°	+35°/-8°	+80° mortar + 50° cannon/-8°	+15°/-8°

The ERC 1 armoured car is now being offered with the Mercedes-Benz OM 617 A super-charged diesel engine developing 115 hp at 4200 rpm. This gives a road range of 1000 km compared to the 800 km of the current production version. Overall height depends on the ground clearance for roads it is normally 294 mm and for cross-country operations it is 344 mm.

Status: In production. In service with Argentina (12 with Lynx 90 turret), Chad (four with Lynx 90 turret), France (192 ERC 90 F4 Sagaies), Gabon (Sagaie 2 with TTB 190 turrets and ERC 20 Kriss anti-aircraft vehicles), Ivory Coast (seven with TS 90 turret) and Mexico (40 with Lynx 90 turret).

Manufacturer: Société de Constructions Mécaniques Panhard et Levassor
18 avenue d'Ivry
75621 Paris
France
Telephone: (1) 40 77 40 00
Telex: 270 276

Panhard AML Light Armoured Car

Development

In 1956, following the successful use of the British Ferret scout car by the French Army in North Africa, the French Army issued a requirement for a similar vehicle but with a more powerful armament as the Ferret was armed only with a single 7.62 mm machine gun which, in the case of the Mk 1, was pintle mounted. At one time it was thought that the Ferret was to be manufactured under licence in France but the French armoured vehicle manufacturers resisted this suggestion.

Panhard, which was building the EBR (8 × 8) heavy armoured car at the time, was awarded a contract to build the prototype which was completed in 1959 under the Panhard designation Model 245. Saviem and DEFA-AMX also built prototype vehicles, but after trials the Panhard model was accepted for service with the French Army as the AML (Automitrailleuse Légère) and first production vehicles were delivered to the French Army in 1961. The total French Army buy was 210 AML 90s and 425 AML 60/7s; the French Gendarmerie bought 70 AML 60/7s and 45 AML 90s. The Panhard AML light armoured car was phased out of front line service with the French Army early in 1990.

Since then the AML has been built in very large numbers, although now it remains in production for export only. By 1987, 4812 AMLs had been built and the range has been continuously expanded to enable the vehicle to undertake additional roles and since then production has continued at a low rate. The AML has also been manufactured under licence in South Africa by Sandock-Austral Limited as the Eland. There is a separate entry for this vehicle under South Africa. The Panhard M3 armoured personnel carrier, for which there is a separate entry, uses 95 per cent of the automotive components of the AML, and many armies have bought complete fleets of AML armoured cars and M3 armoured personnel carriers.

Current production models of the AML have had the original Panhard petrol engine replaced by the Peugeot XD 3T turbo-charged diesel developing 98 hp and this is also offered as a retrofit package.

AML 90 with Hispano-Suiza Lynx 90 turret as delivered to Gabon

Panhard AML with H 90 turret of the French Army

Panhard AML of the French Army fitted with H 90 turret armed with GIAT Industries 90 mm gun (Pierre Touzin)

Panhard AML with HE 60-7 turret in service with French Gendarmerie in 1989

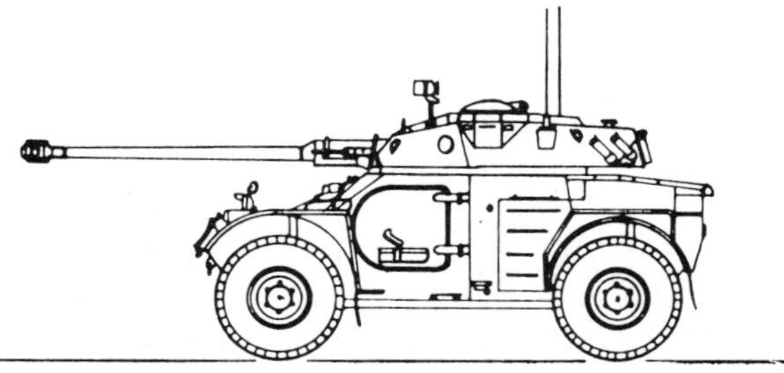

Panhard AML with H 90 turret armed with 90 mm gun

Description (AML 90)
The all-welded steel hull of the AML is divided into three compartments: driver's at the front, fighting in the centre and the engine at the rear.

The driver is seated at the front of the hull and is provided with a single piece hatch cover that opens to the right with three integral periscopes, the centre one of which can be replaced by an infra-red or image-intensification periscope for night driving.

The all-welded turret is manufactured by Hispano-Suiza. The commander is seated on the left of the turret and the gunner on the right, both with a single piece hatch cover that opens to the rear. The commander and gunner each have four L794B periscopes and the gunner also has a type M262 or M37 sighting periscope. The M37 has a magnification of ×6. A searchlight is mounted forward of the commander's hatch. Either side of the hull below the turret ring is a door, the left door, on which the spare wheel and tyre are mounted, opens to the rear, the right door opens to the front.

The engine is at the rear of the hull and two access panels are provided over the engine compartment. The gearbox is crosswise and consists of two gearboxes in one (high and low), coupled on both sides of the bevel pinion. The low-range box is for cross-country use and comprises two low gears, a top gear and one reverse gear. When the low-range box is in normal drive the four ratios of the high-range box command the four upper gears of the range: sixth, fifth, fourth and third. The high-range box is for normal road use and has three low gears and one overdrive. There are Panhard type ball differentials in the gearbox and in each rear transfer box which automatically prevent gear slip.

Drive is transmitted from the gearbox to two lateral transfer boxes via pinions to the rear wheels and via drive shafts that run along the inside of the hull to the front wheels.

The independent suspension at each wheel station consists of coil springs and hydropneumatic shock absorbers acting on the trailing arms of the wheel mechanism. The tyres are fitted with unpuncturable Hutchinson inner tubes.

The basic vehicle has no NBC system or night fighting aids, but they can be fitted as optional extras as can an air-conditioning system and an amphibious kit.

Variants

AML with H 90 Turret
This model is fitted with an H 90 turret manufactured by Hispano-Suiza and is armed with a 90 mm D 921 F1 gun which fires HEAT, HE, smoke and canister rounds.

A 7.62 mm coaxial machine gun is mounted to the left of the main armament and a similar weapon, or a 12.7 mm (0.50) machine gun, can be mounted on the roof for anti-aircraft defence. Two smoke dischargers mounted on either side of the turret are electrically fired from within the vehicle. This model was originally offered with two SS-11 or ENTAC ATGWs mounted either side of the main armament. This model is no longer in production.

AML with Lynx 90 Turret
This has now replaced the AML with H 90 turret in production and has the same armament as the H 90 turret but can be fitted with passive night vision equipment, laser rangefinder and powered traverse.

AML with HE 60-7 Turret
This model is fitted with an H 60 turret armed with twin 7.62 mm machine guns on the left and a single 60 mm mortar on the right. The machine guns can be elevated from −15° to +60° and the mortar from −15° to +80°. The Thomson-Brandt Model HB 60 can be used in the direct role up to a range of 300 m or in the indirect role up to a range of 2600 m. In the basic model 53 mortar bombs and 3800 rounds of 7.62 mm machine gun ammunition are carried; if additional radios are installed for the command role ammunition is reduced to 32 mortar bombs and 3200 rounds of machine gun ammunition. Types of mortar bomb include HE, canister, smoke and illuminating.

The turret is provided with a large two-piece hatch cover that opens to the front and rear of the vehicle, seven periscopes (model L794B) and an M112/M113 periscope with a magnification of ×5 for aiming. This model was originally offered with four ENTAC ATGWs mounted at the rear of the turret, which slid out on rails, two either side of the turret, before being launched.

AML with HE 60-12 Turret
This is almost identical to the HE 60-7 turret but is armed with a 12.7 mm (0.50) machine gun and a 60 mm HB 60 Thomson-Brandt mortar. The

machine gun can be elevated from -15° to +60° and the mortar from -15° to +80°. A total of 43 mortar bombs and 1300 rounds of 12.7 mm machine gun ammunition are carried. If additional radios are carried ammunition load is reduced to 31 rounds of 60 mm and 900 rounds of 12.7 mm machine gun ammunition. Loaded weight of an AML with the HE 60-12 turret is 4800 kg.

AML with HE 60-20 Turret

This is almost identical to the HE 60-7 turret but is armed with a 20 mm cannon and a 60 mm DTAT Model CS or an HB 60 Thomson-Brandt mortar. The cannon can be elevated from -8° to +50° and the mortar from -15° to +80°. When fitted with the DTAT mortar 53 mortar bombs and 500 rounds of 20 mm ammunition are carried, and if the Thomson-Brandt mortar is installed 43 mortar bombs are carried and 20 mm ammunition supply remains the same. If additional radios are installed for command vehicle use the DTAT mortar is provided with 32 mortar bombs and the Thomson-Brandt with 31, and in both cases 500 rounds of 20 mm ammunition are carried. If required a 7.62 mm machine gun can be mounted on the roof of the turret for anti-aircraft defence (this applies to all HE 60 turrets). This turret is no longer in production. Loaded weight of an AML with the HE 60-20 turret is 4800 kg.

AML with HE 60-20 Serval Turret

This version of the AML is fitted with an HE 60-20 Serval turret designed and manufactured by Hispano-Suiza and is armed with a 60 mm breech-loaded mortar which can be elevated from -8° to +80°, and a 20 mm cannon mounted on the right side of the turret which can be elevated from -8° to +50°. The 60 mm mortar can be either a 60 mm HB 60 Thomson-Brandt (maximum range 2600 m with M72 bomb) or a 60 mm HB 60 LP Thomson-Brandt long mortar (maximum range with LP bomb being 5000 m). The 20 mm cannon can be either a French M693 dual-feed cannon with 300 rounds of HE and 37 rounds of AP ammunition for ready use or an HS-820 Oerlikon KAD-B16 single-feed cannon with 300 rounds of ready to use ammunition. If required a 7.62 mm machine gun can be mounted coaxially to the right of the 20 mm cannon. Vision equipment includes a sighting periscope, nine observation periscopes with the option of fitting a night periscope.

AML with S 530 Turret

This was developed specifically for export and is a basic AML fitted with a turret manufactured by SAMM (Société d'Applications des Machines Motrices), armed with twin 20 mm M621 cannon with an elevation of +75°, depression of -10°, and turret traverse of 360°. Elevation and traverse are both powered, maximum elevation speed being 40°/s and maximum traverse speed being 80°/s. The all-welded steel turret's front is 15 mm thick and the sides and rear are 7 mm thick.

A total of 600 rounds of 20 mm ammunition is carried and the gunner can select either single shots, bursts or full automatic. Cyclic rate of fire is 740 rds/min per barrel. Automotive characteristics are similar to the basic AML 90 but the overall length of the vehicle with turret traversed forward is 3.95 m, length of hull is 3.79 m and height to top of turret is 2.24 m. The only known user of this version is Venezuela which has 12 vehicles in service. Full details of this are given in *Jane's Land-Based Air Defence 1991-92*, pages 67-68.

AML with CLI T.25 Turret

For trials purposes an AML chassis has been fitted with the Creusot-Loire Industrie T.25 two man turret armed with a 25 mm cannon and 7.62 mm coaxial machine gun.

Other variants of the AML

The following versions of the AML, covered in *Jane's Armour and Artillery 1989/90* are no longer offered by Panhard:- EPR scout car, AML with TG 120 turret, AML 20 scout car, AML Jungle and AML scout car.

AML Re-engining

As the 4HD petrol engine installed in the Panhard AML and the M3 is now longer produced, Panhard are offering a conversion kit to enable users of these vehicles to retrofit the new XD3T liquid cooled diesel engine in their own facilities with the minimum of specialised equipment.

The modifications can be summarised as:

HULL	new rear
cutting, drilling and welding	engine bonnet
CIRCUITS	
fuel (diesel)	modified
dashboard	modified
main electrical control box	modified
alternator	new
POWER PLANT	
XD 3T turbo-charged diesel	new
hydraulic control	
mechanical clutch	new
reinforced gearbox	new
2nd gear	synchronised
hydraulic control cooling system	new
3 pedals unit	new
rear suspension	reinforced
4 shock absorbers	changed
rear mudguards adapted to body	modified
exhaust system	new

In addition a more powerful air compressor can be fitted, as can Citröen brakes with disc brakes on front wheels.

Panhard AML with HE 60-12 turret armed with 60 mm mortar and 12.7 mm machine gun

Panhard AML with HE 60-20 Serval turret

Panhard AML fitted with Creusot-Loire Industrie T.25 turret

AML with SAMM S 530 turret armed with twin 20 mm cannon

Panhard AML 90 retrofitted with Peugeot XD 3T diesel engine

Close-up of AML 90 turret showing passive night vision equipment for gunner and externally mounted laser rangefinder

AML Weapon Modernisation

Panhard can upgrade the AML 90 in three key areas to improve its combat effectiveness.

1) replacement of current infra-red night vision equipment used by the commander, gunner and driver by new image intensification equipment
2) the installation of a laser rangefinder to increase first round hit capability
3) use of GIAT Industries APFSDS ammunition which will penetrate 50 mm of armour at an incidence of 0° at a range of 1300 m. In order to fire this ammunition the muzzle brake and recoil brake have to be modified but the same number of rounds of 90 mm ammunition can be carried in the same ready racks.

AML with 40 mm Grenade Launcher (AML 40)

This was first shown in 1990 and is essentially a Panhard AML 60-7 with the turret modified to accept a 7.62 mm machine gun (right) and the Spanish SANTA BARBARA 40 mm automatic grenade launcher which has a maximum range of 2200 m, a cyclic rate of fire of 200 rds/min and a muzzle velocity of 240 m/s. A total of 225 rounds of 40 mm and 1250 rounds of 7.62 mm ammunition are carried. As of early 1991 this model had not yet entered production.

Panhard AML with SANTA BARBARA 40 mm grenade launcher and 7.62 mm machine gun. This version is also referred to as the AML 40

SPECIFICATIONS (AML with H 90 turret)

CREW	3
CONFIGURATION	4 × 4
COMBAT WEIGHT	5500 kg
POWER-TO-WEIGHT RATIO	16.36 hp/tonne
LENGTH GUN FORWARDS	5.11 m
LENGTH HULL	3.79 m
WIDTH	1.97 m
HEIGHT	
overall	2.07 m
to hull top	1.385 m
GROUND CLEARANCE	0.33 m
TRACK	1.62 m
WHEELBASE	2.5 m
MAX ROAD SPEED	
6th gear, high range	90 km/h
5th gear, high range	61 km/h
4th gear, high range	35 km/h
3rd gear, high range	18.8 km/h
3rd gear, low range	18.8 km/h
2nd gear, low range	9.3 km/h
1st gear, low range	4.5 km/h
reverse	5.5 km/h
FUEL CAPACITY	156 l
MAX ROAD RANGE	600 km
FORDING	1.1 m
with kit	amphibious
GRADIENT	60%
SIDE SLOPE	30%
VERTICAL OBSTACLE	0.3 m
TRENCH	
with 1 channel	0.8 m
with 4 channels	3.1 m
TURNING RADIUS	6 m
ENGINE	Panhard Model 4 HD 4-cylinder air-cooled petrol developing 90 hp at 4700 rpm
TRANSMISSION	manual with 6 forward and 1 reverse gears
CLUTCH	centrifugal with electro-magnetic automatic control
SUSPENSION	independent, coil spring and hydro-pneumatic shock absorbers acting on suspension trailing arms of wheel mechanism
TYRES	11.00 × 16
BRAKES	
main	hydraulic, dual circuit
parking	handbrake operating on gearbox output shaft
ELECTRICAL SYSTEM	24 V
BATTERIES	2 × 12 V
ARMAMENT	
main	1 × 90 mm gun
coaxial	1 × 7.62 mm MG
SMOKE-LAYING EQUIPMENT	2 × 2 smoke dischargers
AMMUNITION	
main	20
coaxial	2000
smoke grenades	12
GUN CONTROL EQUIPMENT	
turret power control	manual
max rate of traverse	360° in 20 s
gun elevation/depression	+15°/-8°
gun stabiliser	
vertical	no
horizontal	no
ARMOUR	8-12 mm

Note: All current production models of the AML are powered by the Peugeot XD 3T diesel engine developing 98 hp coupled to a standard Panhard transmission which gives an increase in operating range of 700 to 800 km depending on the weight of the vehicle.

The Panhard AML 90 F1 diesel weighs 5900 kg combat loaded and has similar specifications to the H 90 version except length overall is 5.48 m, length hull 4.15 m, height to top of cupola 2.19 m, height to turret roof 2.15 m.

Status: In production. In service with Algeria (AML 60/7), Argentina (AML 90), Bahrain (AML 90), Benin (6 AML 90 diesels, 6 AML 60 diesels), Burkina Faso (AML 90), Burundi (AML 90 Lynx, AML 60), Chad (AML 90 and AML 60), Djibouti (AML 60/7 and AML 90), Ecuador (AML 90, AML 60/7), El Salvador (AML 90), France (Gendarmerie, AML 90 and AML 60/7), Gabon (AML 90, AML 60/7), Iraq (AML 90 and AML 60/7), Ireland (AML 90 and AML 60/7), Ivory Coast (AML 90 and AML 60/7), Kenya (AML 90, AML 60/7), Malaysia (AML 60, AML 90), Mauritania (AML 90 and AML 60/7), Morocco (AML 90 and AML 60/7), Niger (AML 60/7, AML 90, Serval 60/20, 4 AML 90 diesels sold in 1985-86), Nigeria (AML 90, AML 60/7), Polisario guerrillas, Portugal (AML 60/7), Rwanda (AML 60/7 and AML 90), Saudi Arabia (AML 90 and AML 60), Senegambia (AML 60/7, AML 90), Somalia (AML 60/7, AML 90), South Africa (see entry under South Africa in this section), Spain (Army AML 60/7 and AML 90, (in reserve) Marines AML 60/7), Sudan, Togo (AML 60/7, AML 90), Tunisia (AML 90), United Arab Emirates (AML 90), Venezuela (twin 20 mm self-propelled anti-aircraft gun, AML 60, AML 90), Yemen, Zaïre (AML 90, AML 60/7) and Zimbabwe (from South Africa). As far as it is known, it is no longer used by Angola, Cambodia, Ethiopia, Israel, Lebanon or Pakistan.

Manufacturer: Société de Constructions Mécaniques Panhard et Levassor
18 avenue d'Ivry
75621 Paris
France
Telephone: (1) 40 77 40 00
Telex: 270 276

Panhard VBL

Development

In 1978 the French Army issued a requirement for a vehicle weighing under 3500 kg for two basic roles, anti-tank armed with the Euromissile MILAN ATGW, and intelligence/scout armed with machine guns. After studying proposals from five companies Panhard and Renault were each awarded contracts for the design and construction of three vehicles which were delivered to the French Army for trials in 1983. The official French Army designation for the vehicle is the Véhicule Blindé Léger (VBL).

In February 1985 the French Army accepted the Panhard vehicle to meet its requirements for the VBL.

Between June and October 1988 Panhard delivered 15 pre-production vehicles to the French Army for a wide range of troop and specialist trials under a wide range of environmental conditions. Prior to this Panhard delivered three VBLs to the French Army for use in the Lebanon.

Under the French Army's first programme for the VBL, a total of 1585 vehicles will be ordered in three batches. The first of these was confirmed at the end of 1989 for a total of 569 vehicles, with first production vehicles completed late in 1990. Once production builds up it is expected that between 35 to 40 vehicles will be built for the French Army per month, with capacity to build additional vehicles for the export market.

The exact number of Panhard VBLs to be procured by the French Army is uncertain. The original requirement was for 3000 vehicles, 1000 for use in the anti-tank role and 2000 for use in the intelligence/scout role. This was subsequently increased to 3600, 3000 in standard wheelbase vehicles with the remainder in the long wheelbase version. As the French Army is to be reduced in size it is probable that the requirement is now for less than 3000 vehicles.

By early 1991 Panhard had built over 120 VBL light armoured vehicles for the export market and the vehicle had been trialled or demonstrated in many parts of the world, including the Middle East, Far East and North America.

The hulls for French Army VBLs are built by Creusot-Loire Industrie at Saint Chamond while hulls for export vehicles are built by AMEFO.

Prior to being accepted by the French Army Panhard embarked on a major sales drive for its VBL and in 1984 Mexico placed an order for 40 and the first of these were completed early in 1985. Of this order eight were for anti-tank use with the MILAN ATGW system. For export Panhard calls the VBL the ULTRAV-M11.

Wherever possible standard and proven commercial components have been used in the design, for example the diesel engine is the same as that installed in the Peugeot 505 and 605 cars, the Peugeot P4 (4 × 4) light vehicle now in service with the French Army and the Talbot Tagora car. The German ZF automatic transmission is also used in many European cars. The Peugeot P4 (4 × 4) light vehicle is now assembled at Panhard's facility at Marolles near Paris.

In September 1987 it was announced that Alvis and Panhard had reached an agreement in principle to co-operate on the development and promotion of the VBL light armoured vehicle. The two companies will work closely together to promote and develop the vehicle to meet the future potential requirements in respect of the British Army's Future Family of Light Armoured Vehicles.

Panhard has teamed with the German company of Krupp MaK to market the VBL to the German Army.

Description

The hull of the Panhard VBL, manufactured by Creusot-Loire Industrie, is of all-welded THD steel that varies in thickness from 5 to 11 mm with the engine at the front and the crew compartment at the rear. The driver sits at the front of the crew compartment on the left with the vehicle commander to the right. Both have a door in the side that opens to the front, with a bullet-proof window in the upper part. To the front of each crew member is a bullet-proof window with an electric wiper and the driver has an emergency periscope. As an option the normal single piece bullet-proof windscreen for the commander and driver can be replaced by an oblong windscreen that opens upwards for improved ventilation. When in a combat area this would be covered by an armoured shutter. Over the top of the driver's position is a single piece hatch cover that opens to the front, the commander has a single piece circular hatch cover that opens to the rear. To the upper left of the driver's rear and upper right of the commander's rear is a small bullet-proof window that gives observation to the rear, although this is not a feature of French Army production vehicles.

The upper sides of the rear crew compartment slope inward. There is a large door in the rear that opens to the left and has a bullet-proof window. In the roof at the rear is a single piece hatch cover that opens to the left side. All of the bullet-proof windows have a thickness of 33.5 to 49 mm.

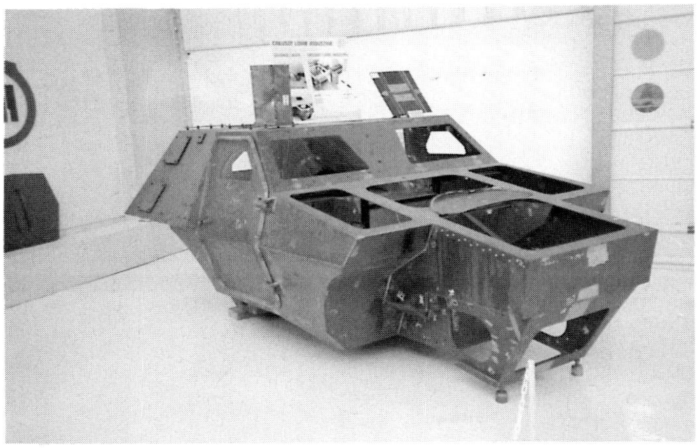

Creusot-Loire Industrie-built hull for the Panhard VBL light armoured vehicle (Christopher F Foss)

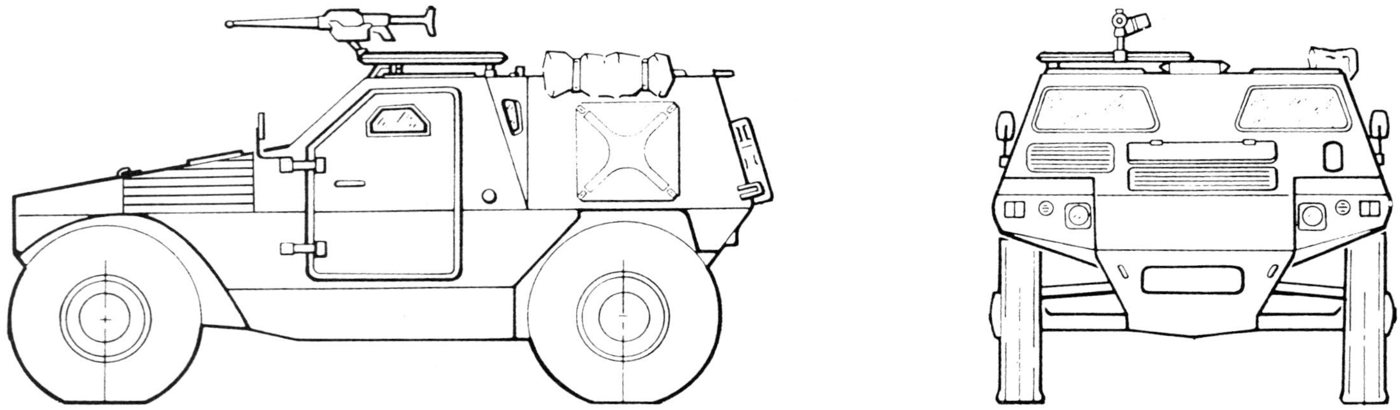

Outline drawing of production Panhard VBL (4 × 4) light vehicle with roof-mounted 7.62 mm machine gun (not to 1/76th scale)

Two VBLs with side and rear doors open. Vehicle in foreground is fitted with Euromissile MILAN ATGW launcher

Panhard VBL as delivered to Rwanda showing external stowage

Close-up of Panhard VBL showing armoured window in side door

VBL assembly line at Panhard's facility at Marolles

The Michelin combat tyres enable the vehicle to cover 50 km at a speed of 30 km/h after a puncture or being hit by bullets. A tyre pressure regulation system is fitted as standard.

The VBL is fully amphibious after two minutes of preparation and is propelled in the water by a single propeller mounted under the hull rear. Before entering the water a trim vane is erected manually at the front of the vehicle.

For export vehicles the amphibious kit will be optional, other options include an NBC system (standard on French Army vehicles), air-conditioning system, passive night driving periscope, heater and power-assisted steering.

Variants

The French Army will use the VBL in two basic roles, combat/anti-tank and intelligence.

The combat version will have a crew of three: commander, driver and MILAN operator and be armed with a 7.62 mm machine gun (with 3000 rounds of ammunition) and a MILAN launcher (with six missiles). The MILAN launcher can be fitted on the front or rear hatches of the VBL or deployed away from the vehicle. This version will also have a radio, three 5.56 mm FAMAS rifles and nine hand grenades.

The intelligence version will have a crew of two, commander/radio operator and a driver. Armament will consist of one 7.62 mm machine gun (with 3000 rounds) or one 12.7 mm machine gun (with 1200 rounds). The weapon would be fitted to a ring mount on the roof of the vehicle to the right of the driver and have an elevation from −10° to +50° and a traverse of 180° left and right. It will also have two radios, which can be used by either crew member, two 5.56 mm FAMAS rifles and six hand grenades. An LRAC anti-tank launcher and 12 rockets can also be carried.

Export variants

In addition to the two basic models required by the French Army, Panhard has proposed at least 22 other versions including command vehicle, armoured personnel carrier, internal security, anti-aircraft vehicle with two MATRA Mistral missiles, anti-tank vehicle with HOT or TOW ATGWs, radio vehicle, radar vehicle with RASIT system, electronic warfare vehicle, ammunition carrier and internal security vehicle. Other armament options are also available including 20 mm cannon. However, some of these are only suggested for the LWB version of the VBL.

Panhard offer two basic chassis, the standard model as used by the French Army, this being the model described and the ULTRAV M11 which has a stretched chassis. These can be fitted with a wide range of equipment and weapons depending on their operational role, and some of these are listed below. The Panhard ULTRAV M11 Jungle/Scout Light Vehicle

Panhard VBL in service with Gabon

(*Jane's Armour and Artillery 1989/90* page 207) is no longer being marketed.

LWB (4 × 4) VBL

Details of the long wheelbase version of the Panhard VBL are given in the *Armoured personnel carriers* section.

VBL with Surveillance Radar

For the export market a Panhard VBL has been fitted with an air defence surveillance radar system which would be used to designate targets for the VCR (6 × 6) vehicle fitted with the RBS 70 surface-to-air missile system and the ERC Kriss twin 20 mm (6 × 6) self-propelled anti-aircraft gun system. This version is used by Gabon.

The version has a crew of three and is fitted with a radar system with a range of 16 to 20 km for aircraft and 8 to 10 km for helicopters. The antenna is mounted on a telescopic mast and is raised above the vehicle when operating.

Inside the rear of the VBL are the radar transceiver display console, land navigation system and a radio communications system. This system is used by Gabon.

Panhard VBL undergoing trials with the German Army

Panhard VBL fitted with Creusot-Loire Industrie TOI one man turret armed with grenade launcher

ULTRAV M11 (US)

For the US Army Panhard has developed a special model of the VBL with a roof-mounted 12.7 mm machine gun for which 600 rounds of ammunition are carried and grenade launchers. Standard equipment would include a DUK-DUR dosimeter, DETALAC NBC alert and control system, SOFILTRA NBC protection system, NAVYX position determining system, communications equipment, winch, air-conditioning system, heater, intercom, night-vision periscope TRT OB 31, night-vision binoculars TRT OB 41 and binoculars TRT OB 42.

Basic specifications are combat weight 3930 kg, unloaded weight 3491 kg, length overall 3.82 m, height with 12.7 mm MG 2.25 m, ground clearance 0.35 m and wheelbase 2.66 m.

Status: In production. In service with:

Country	Quantity	Comment
Benin	10	
Cameroon	5	
Djibouti	7	
France	150	
Gabon	14	
Mexico	40	8 have MILAN ATGW
Niger	7	
Portugal	14	
Rwanda	16	
Togo	2	

Manufacturer: Société de Constructions Mécaniques Panhard et Levassor
18 avenue d'Ivry
75621 Paris
France
Telephone: (1) 40 77 40 00
Telex: 270 276

Panhard VBL of Gabon fitted with surveillance radar for Panhard ERC twin 20 mm SPAAG's

One of the prototypes of the VBL showing its amphibious capabilities

SPECIFICATIONS	
CREW	2 or 3
CONFIGURATION	4 × 4
COMBAT WEIGHT	
without NBC system	3550 kg
with NBC system	3590 kg
UNLOADED WEIGHT	
without NBC system	2850 kg
with NBC system	2890 kg
POWER-TO-WEIGHT RATIO	26.76 hp/tonne
LENGTH	
with extra fuel can	3.87 m
without extra fuel can	3.7 m
WIDTH	2.02 m
HEIGHT	
to hull top	1.7 m
with 7.62 mm MG	2.14 m
GROUND CLEARANCE	0.37 m
TRACK	1.69 m
WHEELBASE	2.45 m
MAX SPEED	
road	95 km/h
water	4 km/h
MAX ROAD RANGE	600 km
with on-board fuel cans	800 km
FORDING	0.9 m
GRADIENT	50%
SIDE SLOPE	30%
TURNING RADIUS	5.77 m
ENGINE	Peugeot XD 3T, 4-cylinder turbo-charged diesel developing 95 hp at 4150 rpm
TRANSMISSION	ZF automatic, 3 forward and 1 reverse gears
TRANSFER CASE	2-speed
SUSPENSION	
front	double triangle, helical springs, double effect shock absorbers
rear	torsion bars, double effect shock absorbers
TYRES	9.00 × 16
BRAKES (main)	disc on all wheels
ARMAMENT	see text
ARMOUR	5 mm to 11.5 mm

Renault VBL

This was designed to meet the requirements of the French Army in competition with the Panhard VBL which subsequently won the competition. The Renault VBL was being offered for licence manufacture overseas but as of early 1991 this offer had not been taken up. Full details of the Renault VBL were given in *Jane's Armour and Artillery 1990-91*, pages 217/218.

LOHR RPX 6000 Scout Car

Development
The RPX 6000 (4 × 4) Véhicule Blindé Léger à Roues has been developed as a private venture by the LOHR company and was shown for the first time at the 1981 Satory Exhibition of Military Equipment. As of early 1991 five prototypes of the RPX 6000 had been completed, but volume production had yet to commence.

Description
The vehicle has an all-welded steel hull with the three-man crew compartment at the front and the engine compartment at the rear. The driver is seated at the front of the hull and has a bullet-proof windscreen to his front and sides which in action are covered by hinged flaps raised by the driver from inside the vehicle. There is a single piece hatch cover over the driver's position.

In either side of the hull is a two-part entry hatch, the upper part folding to the rear and the lower part folding downwards to form a step. There is a circular roof hatch in the centre of the hull roof on which a variety of armament installations are mounted including a Euromissile MILAN MCT turret with two MILAN ATGWs ready to launch, SAMM BTM 208 one-man turret (former designation S 365-1) armed with a 12.7 mm and a 7.62 mm machine gun, Helio turret with 12.7 mm machine gun and external grenade launchers, turret-mounted 20 mm cannon, 60 mm breech loaded mortar and other light turrets not exceeding a diameter of 1.28 m. It can also be fitted with the GIAT Industries 20 mm Capre turret armed with a 20 mm cannon (500 rounds) and a 7.62 mm machine gun (2000 rounds). Optional equipment includes an NBC system, night vision equipment and amphibious capability.

Variants
Variants of the RPX 6000 suggested by the manufacturer include a command vehicle, personnel carrier, radar vehicle, command/radio vehicle and a recovery vehicle. The RPX 6000 D is powered by a BMW turbo-charged diesel developing 125 hp at 4800 rpm giving a maximum road speed of 90 km/h and a maximum range of 800 km.

In 1991 LOHR announced that a new version of the RPX 6000 (4 × 4) armoured vehicle was under development for an unidentified export customer.

This model will not have a turret and will have the commander seated on the left side of the hull and provided with a Hispano-Suiza Puma cupola provided with periscopes and a day/night observation sight.

Seated to the right of the commander is the gunner/radio operator who will be provided with a single piece hatch cover with periscopes mounted around the roof of the vehicle. Mounted on the forward part of the roof will be a 7.62 mm machine gun.

Mounted on the roof towards the rear will be forward firing smoke dischargers and mounted in the crew compartment is a rack for three Dragon ATGWs.

This version will also be powered by a new diesel engine developing 200 hp coupled to a fully automatic transmission.

LOHR RPX 90 (4 × 4 armoured car).

A single prototype of this was first shown in 1983 but it never entered production and is no longer being marketed. Details were given in *Jane's Armour and Artillery 1990-91*, page 219.

LOHR RPX 6000 scout car with GIAT Industries 25 mm DRAGAR turret

LOHR RPX 6000 scout car with Hispano-Suiza Serval 60/20 two man turret

Status: Development complete. Ready for production.

Manufacturer: LOHR Industrie
67980 Hangenbieten
France
Telephone: (1) 88 38 98 00
Telex: 870082
Fax: (1) 88 96 06 36

SPECIFICATIONS

CREW	3 (or 2 to 4)
CONFIGURATION	4 × 4
WEIGHT	
loaded	6800 kg
empty	5200 kg
LENGTH	4.585 m
WIDTH	2.164 m
HEIGHT (without armament)	1.65 m
GROUND CLEARANCE	0.4 m
TRACK	1.828 m
WHEELBASE	2.7 m
MAX ROAD SPEED	110 km/h
RANGE	600 km or 10 hours
FORDING	0.95 m
GRADIENT	60%
SIDE SLOPE	30%
VERTICAL OBSTACLE	0.4 m
ENGINE	BMW 6-cylinder petrol developing 180 hp
TRANSMISSION	automatic
TRANSFER CASE	two-speed manual with differential locks
FRONT SUSPENSION	transversal parallelogram, coil spring, hydraulic shock absorber and stabilisation bar
REAR SUSPENSION	trailing arm type, coil spring, hydraulic shock absorber and stabilisation device
STEERING	hydraulically assisted
TYRES	12.5 × 20 XS (run flat)
BRAKES	hydraulic drum on all wheels

LOHR VPX 5000 Light Armoured Tracked Vehicle

LOHR VPX 5000 light tracked armoured vehicle with GIAT Industries Mascot 7.62 mm remote mount forward and Euromissile MILAN ATGW at rear

Development

Shown for the first time at the 1981 Satory Exhibition of Military Equipment was the LOHR VPX 5000 light armoured tracked vehicle, which can undertake a wide range of roles including anti-tank and reconnaissance. This series was preceded by the VPX 110 family of light tracked vehicles which is no longer offered by the company. As of early 1991 10 prototypes of the VPX 5000 had been completed, but volume production had yet to commence.

Early in 1991 LOHR stated that development of the VPX 5000 was complete and that the company was ready to sell the licence with a technology transfer package.

Description

The VPX 5000's hull is made of all-welded steel with the engine at the right front, driver to the left of the engine and the crew compartment at the rear. There are two openings in the roof on which various armament systems can be mounted and a door in the rear of the hull on the left side.

The suspension consists of six road wheels with the idler at the front, drive sprocket at the rear and two track return rollers. The track consists of an endless rubber band internally reinforced.

Variants

The LOHR VPX 5000 can be fitted with a wide range of armament installations including the following:

Euromissile MILAN MCT compact turret with two MILAN ATGWs in ready to launch position, one either side of turret.

GIAT Industries Mascot 7.62 mm remote-controlled machine gun mount with infantry MILAN over second hatch at rear of hull.

SAMM BTM 208 turret (former designation S 365-1) armed with 12.7 mm and 7.62 mm machine guns.

Other turrets armed with 20 mm cannon and various combinations of 7.62 mm and 12.7 mm machine guns with a turret diameter of no more than 1.18 m.

In addition to being fitted with the various armament installations described the manufacturer has suggested the following roles for the vehicle: ambulance, logistic support, radio/command and recovery. At the 1983 Satory Exhibition of Military Equipment one prototype was shown fitted with the RASIT battlefield surveillance radar on a hydraulic arm that could be raised above the chassis. It has also been tested with the Dassault Electronique RODEO 3 air surveillance radar on a hydraulic arm. The Commando VPX 5000 has two 7.62 mm and one 12.7 mm machine guns and a Thomson-Brandt 81 mm mortar.

LOHR VPX 40 M 120 mm Mortar Carrier

Details of this 120 mm tracked self-propelled mortar vehicle, still at the prototype stage, are given in the *Self-propelled guns and howitzers* section.

LOHR RPX 40 M 120 mm Mortar Carrier

Details of this 120 mm wheeled self-propelled mortar vehicle, still at the prototype stage, are also given in the *Self-propelled guns and howitzers* section.

Status: Development complete, ready for production.

Manufacturer: LOHR Industrie
67980 Hangenbieten
France
Telephone: (1) 88 38 98 00
Telex: 870082
Fax: (1) 88 96 06 36

SPECIFICATIONS					
CREW	2 to 4	GROUND CLEARANCE	0.35 m	ENGINE	BMW 6-cylinder petrol developing 180 hp at 5500 rpm
COMBAT WEIGHT	4600 to 5800 kg	MAX ROAD SPEED	80 km/h		
LENGTH	4.2 m	RANGE	350 km		
WIDTH	2 m	GRADIENT	60%	TRANSMISSION	automatic
HEIGHT (without armament)	1.6 m	SIDE SLOPE	30%	TRANSFER CASE	2-speed
		VERTICAL OBSTACLE	0.5 m		
		FORDING	0.95 m		

LOHR RPX 3000 Light Armoured Vehicle

Development

The RPX 3000 light armoured vehicle was developed as a private venture by LOHR and the first prototype was completed in April 1985. The vehicle was shown for the first time at the 1985 Satory Exhibition of Military Equipment. Like the LOHR RPX 90 armoured car and the LOHR RPX 6000 scout car, the RPX 3000 is still at the prototype stage.

Prototype of LOHR RPX 3000 light armoured vehicle with swivel-mounted 7.62 mm machine gun at front and Euromissile MILAN ATGW at rear

Description

The hull of the RPX 3000 is of all-welded steel construction that provides protection from small arms fire and shell splinters.

The commander and driver sit at the front of the vehicle, each with a windscreen to their front which is covered in action by an armoured shutter. The shutter is hinged at the bottom and folds down on the front of the vehicle when not required. Each shutter has vision slits.

Both commander and driver have a rear-opening single piece hatch cover and a side door, the upper part of the door opens to the rear, the lower part folds down to form a step. There is an additional observation window between the side and front windscreen.

On the forward part of the hull roof on the right side is a swivel mount for a 7.62 mm machine gun, this can be operated by the commander or driver.

The engine compartment is in the centre of the vehicle with inlet/outlet louvres in the sides of the hull and the exhaust pipe along the right side of the hull.

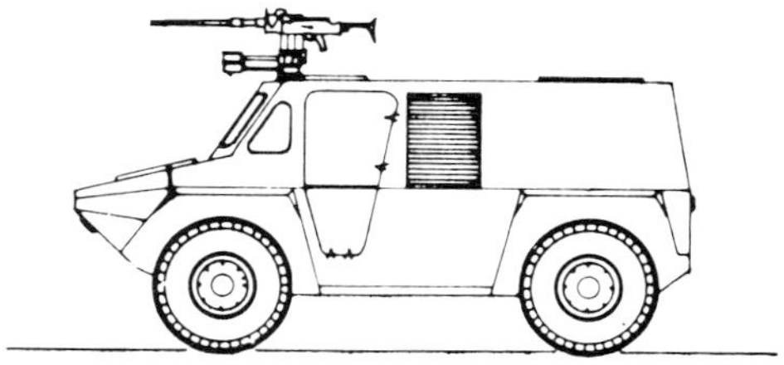

LOHR RPX 3000 light armoured vehicle

Thyssen Henschel Spähpanzer Luchs Armoured Amphibious Reconnaissance Vehicle

Development

In 1964 the German Ministry of Defence examined its requirements for a new generation of military vehicles for the 1970s. This generation was to have included an 8 × 8 armoured amphibious reconnaissance vehicle, 4 × 4 and 6 × 6 armoured amphibious load carriers and a complete range of 4 × 4, 6 × 6 and 8 × 8 tactical cargo trucks, some of which were to be amphibious. All were to share many common components which would, where possible, be from commercial sources.

In 1965 a Joint Project Office (JPO) was formed to undertake development of the complete range of vehicles. Companies in the Joint Project Office were Büssing, Klöckner-Humbolt-Deutz, Friedrich Krupp, MAN and Rheinstahl-Henschel. Daimler-Benz did not join the Joint Project Office but went ahead on its own to develop a similar range of vehicles.

The JPO and Daimler-Benz each built nine prototypes of the 8 × 8 armoured amphibious reconnaissance vehicle. The JPO delivered prototypes to the Trier testing ground in April 1968 and Daimler-Benz delivered in December the same year. Each differed in some detail as both manufacturers tested various engines and transmissions to find the right combination for the vehicle.

Extensive trials were carried out over the following two years and in January 1971 it was decided that work on the JPO vehicle would stop and that the Daimler-Benz model would enter production. In December 1973 Rheinstahl Wehrtechnik (now Thyssen Henschel) was awarded a contract worth DM300 million for 408 vehicles.

First production vehicles were completed in May 1975 and the first vehicle was officially handed over to the German Army in September 1975 and production continued until early 1978.

The 4 × 4 armoured amphibious load carrier became the Transportpanzer 2 and the 6 × 6 armoured amphibious load carrier became the Transportpanzer 1. The Transportpanzer 2 was not placed in production but further development was undertaken by EWK and the eventual result was the APE Amphibious Engineer Reconnaissance Vehicle which was fully described in *Jane's Armour and Artillery 1985-86* page 186. The 6 × 6 armoured amphibious load carrier was placed in production as the Transportpanzer 1 with Rheinstahl Wehrtechnik. Full details of the Transportpanzer 1 are given in the *Armoured personnel carriers* section.

Description

The all-welded steel hull of the Luchs provides the crew with protection from small arms fire and shell splinters. The hull and turret are immune over their frontal area to penetration from 20 mm projectiles.

The driver is seated at the front of the hull on the left side and is provided with a single piece hatch cover that opens to his right. There are three periscopes in front of the driver's hatch cover, the centre one of which can be replaced by a passive periscope for night operations.

The Rheinmetall TS-7 turret is located in the centre of the hull and has spaced armour for improved protection. The commander is seated on the left and the gunner on the right, both with a single piece hatch cover that opens to the rear. Both the commander and gunner are provided with a PERI Z-11 A-1 periscope for aiming the 20 mm cannon, and a total of 12 periscopes. Turret traverse and weapon elevation/depression are electro-hydraulic and can be operated by the commander or gunner. The fire control system includes an azimuth indicator.

An entry door is provided in the left side of the hull. The fourth crew member, the radio operator/rear driver, is seated to the rear of the turret on the left side facing the rear and is provided with a single piece hatch cover that opens to the right. There are three periscopes in front of the cover, the centre one of which can be replaced by a passive periscope for night operations.

The engine is behind the turret on the right side and access is by a large oblong hatch that opens to the right. The engine compartment, cooling and exhaust systems are separated from the fighting compartment by gastight welded bulkheads. The engine compartment is fitted with an automatic fire-extinguishing system. The complete powerpack, consisting of the engine, transmission, air filter, oil cooler and parking brake, can be removed as a complete unit and run outside the vehicle. The cooling system is at the back of the hull and the exhaust outlet is at the very rear of the hull.

Power is transmitted from the engine to a four-speed hydraulic torque converter and then via a distributor gear with differential locks to the wheels. The Luchs has the same speed in both directions.

The suspension consists of four Daimler-Benz rigid axles with differential locks which are supported by longitudinal bars, with bogies for the first and second and third and fourth axles. Each wheel station has a vertical coil spring and a hydraulic shock absorber. All eight wheels are steered but for road use the front four wheels only are normally used.

The Luchs is fully amphibious, propelled in the water by two Schottel steerable propellers at the rear of the hull, one on each side. These are powered by a PTO on the main transmission via a bevel-type distributor gear. Before entering the water a trim vane is hydraulically erected at the front of the hull and the bilge pumps, two in the fighting compartment and one in the engine compartment, are switched on.

In 1982 it was announced that a total of 423 sets of thermal night vision equipment would be purchased for the Luchs, 63 sets in 1984 and 120 a year in 1985, 1986 and 1987.

Standard equipment includes a pre-heater for the cooling liquid, engine oil and transmission oil and pre-heater for the batteries. The NBC system can also be used for ventilating the vehicle when all the hatches are closed.

The Luchs is armed with a 20 mm Rheinmetall MK 20 Rh 202 cannon which is also installed in the Marder ICV. This cannon has dual feed and the empty cartridge cases and beltlinks are ejected externally to the right of the turret. A Rheinmetall 7.62 mm MG3 machine gun mounted on a skate mount over the commander's hatch can be used for both ground and anti-aircraft defence. Four smoke dischargers are mounted either side of the turret.

Spähpanzer Luchs armoured amphibious reconnaissance vehicle from rear showing smoke dischargers on right side of turret and propellers under hull

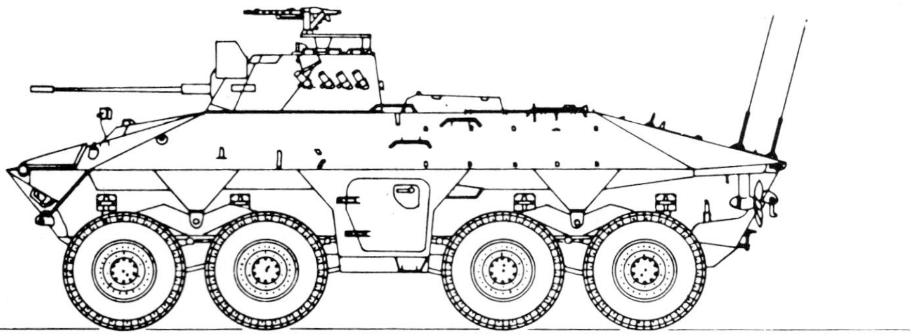

Spähpanzer Luchs armoured amphibious reconnaissance vehicle

Status: Production complete. In service with the German Army.

Manufacturer: Thyssen Henschel
Postfach 102969
D-3500 Kassel
Federal Republic of Germany
Telephone: (0561) 8011
Telex: 099 750
Fax: (0561) 801 6733

Spähpanzer Luchs with roof-mounted 7.62 mm MG3 machine gun (Pierre Touzin)

SPECIFICATIONS	
CREW	4
CONFIGURATION	8 × 8
COMBAT WEIGHT	19 500 kg
POWER-TO-WEIGHT RATIO	20 hp/tonne
LENGTH	7.743 m
WIDTH	2.98 m
HEIGHT	
to top of machine gun rail	2.905 m
to top of hull top	2.125 m
FIRING HEIGHT	2.342 m
GROUND CLEARANCE	
hull	0.44 m
axles	0.58 m
TRACK	2.54 m
WHEELBASE	1.4 m + 2.365 m + 1.4 m
ANGLE OF APPROACH/ DEPARTURE	66°/53°
MAX SPEED	
road forwards	90 km/h
reverse	90 km/h
water	9 km/h
ACCELERATION	0 - 80 km/h: 65 s
FUEL CAPACITY	500 l
MAX RANGE	800 km

FORDING	amphibious
GRADIENT	60%
SIDE SLOPE	30%
VERTICAL OBSTACLE	0.6 m
TRENCH	1.9 m
TURNING RADIUS	
8 wheels	11.5 m
4 wheels	19.4 m
ENGINE	Daimler-Benz OM 403 A 10-cylinder 90° V 4-stroke multi-fuel with 2 exhaust driven turbo-chargers developing 390 hp at 2500 rpm (diesel fuel) or 320 hp at 2500 rpm (petrol)
TRANSMISSION	ZF 4 PW 95 H 1 4-speed with torque converter
SUSPENSION	coil springs and hydraulic shock absorbers
STEERING	re-circulating ball power-assisted

TURNING RADIUS	
all wheel steering	5.75 m
front wheel steering	9.7 m
TYRES	14.00 × 20
BRAKES	
main	hydraulic, dual circuit
parking	mechanical operating on transmission
ELECTRICAL SYSTEM	24 V
BATTERIES	6 × 12 V, 100 Ah
ARMAMENT	
main	1 × 20 mm cannon
anti-aircraft	1 × 7.62 mm MG
SMOKE-LAYING EQUIPMENT	2 × 4 smoke dischargers
AMMUNITION IN TURRET	
main	375 (300 HE + 75 AP)
anti-aircraft	100-round box
GUN CONTROL EQUIPMENT	
Turret power control	electro-hydraulic/ manual
Gun elevation/depression	+69°/-15°
Gun stabiliser	
vertical	no
horizontal	no

Krupp MaK Wiesel Airportable Vehicle

Development

Porsche has developed an airportable vehicle to meet the requirements of German airborne units, called the Wiesel, or Waffentrager LL-TOW/ MK. Early in 1982 the then West Germany loaned the United States four Wiesel airportable vehicles for trials at the Yakima firing centre/proving ground between April and May 1982.

The Wiesel can be carried slung underneath helicopters such as the Puma and is airportable in the C-160 (four), C-130 (three) and CH-53 (two) aircraft.

Under a recent agreement Krupp MaK will be responsible for any series production of the Wiesel for customers from abroad and Porsche for the development of specialised versions.

In 1984 the German Army decided that it would purchase 312 Wiesel vehicles for its airborne units, armed with the Hughes Aircraft TOW ATGW and the Rheinmetall 20 mm cannon. Late in 1985 its total requirement was raised from 312 vehicles to 343.

The original prototypes of the Wiesel were powered by an Audi petrol engine coupled to a ZF automatic transmission but four of the prototypes have been rebuilt with a number of improvements including a new double sprocket and a VW five-cylinder turbo-charged diesel engine developing 86 hp.

In 1986 the German Army tested the four modified prototypes at the Trier Proving Grounds, two were armed with TOW ATGWs and two with a KUKA turret armed with a Mauser 25 mm cannon.

In April 1987 Krupp MaK completed two company demonstrator Wiesel vehicles, one of which can be armed with a 20 mm/25 mm cannon and the other with missiles such as TOW/Stinger. Both of these company demonstrator vehicles are to production standard.

In addition to being demonstrated in the USA, Wiesel has also been demonstrated in Greece, Norway and Indonesia. Indonesia has a requirement for a Wiesel type vehicle fitted with the French SAMM BTM 208 turret armed with 12.7 mm and 7.62 mm machine guns. Indonesia has a requirement for up to 700 Wiesel type vehicles, most of these being built under licence.

In December 1988 a contract was signed between the German MoD and Krupp MaK for the supply of 343 Wiesel airportable vehicles to the German Army for delivery between September 1989 and late 1992.

Of the 343 vehicles, 210 will be armed with the TOW ATGW system and the remainder with the 20 mm Rheinmetall cannon. The guns, sights, missile launchers, missiles, radios and tools for the vehicles will be provided by the German MoD.

A total of 10 vehicles, five of each version, were delivered late in 1989, with 80 following in 1990, 160 in 1991 and 93 in 1992. The production run however could be extended by further orders placed by Germany or from export customers.

The first of 343 vehicles were officially handed over to the German Army on 1 August 1990. Total value of the contract is DM208 million. The vehicles are being issued to the 1st German Airborne Brigade where they are replacing the unarmoured Kraka (4 × 2) all-terrain vehicles.

Late in 1989 the US Army placed an order for seven Wiesel vehicles, of which four will be used for robotic vehicle trials with the remaining three for Tank Automotive Command trials. All of these vehicles were delivered by late 1990.

Description

The hull is of all-welded steel construction providing the crew with protection from 7.62 mm small arms fire and shell splinters. The driver sits at the front of the vehicle on the right and has three periscopes for observation to the front and sides, the centre one can be replaced by a passive periscope for night driving.

The engine, which is a standard VW production model, is to the left of the driver with an access hatch in the glacis plate. The engine drives via a three-speed ZF automatic transmission and a Porsche two-speed reduction gear with integrated Cletrac steering gear. The complete powerpack can be removed and replaced in 10 minutes. The steering gear is operated by the driver via a steering wheel and allows a variable turning radius from infinite to 2.3 m. Pivoting on one track is possible by engaging the parking brake on one side.

The engine exhaust system is located in the cooling air exhaust flow on the left side of the vehicle. The exhaust gases are passed to the rear and cooled in the process. The closed crew compartment is at the rear of the hull and entry is by climbing over the sides of the vehicle. The flexible self-sealing rubber fuel tank has an explosion-retarding polyurethane liner and is mounted externally on the rear of the hull.

Weisel MK 20 A1 being lifted by a German Army CH-53 helicopter

Suspension either side consists of three dual road wheels, drive sprocket at the front, idler at the rear and one track return roller. A Belleville spring assembly automatically provides track tension when the vehicle is travelling. The track consists of an endless rubber band with wire reinforcements. An amphibious kit is available.

Airlift capability

According to Krupp MaK the Wiesel airportable vehicle is air-transportable in the following aircraft:

Type	**Quantity**
CH-53 helicopter	2 internally
CH-47 helicopter	2 internally
CH-130 aircraft	3 internally
C-160 aircraft	4 internally
C-141A aircraft	6 internally
C-5A Galaxy	30 internally
UH-60 helicopter	1 slung externally
Super Puma helicopter	1 slung externally

Variants

Wiesel with 20 mm Cannon (MK 20 A1)

This was built to meet the requirements of the Germany Army and is fitted with a one-man KUKA turret armed with a 20 mm Rheinmetall dual-feed cannon with 160 rounds of ready use ammunition, 60 rounds on the left side and 100 rounds on the right side, so allowing the gunner to select the type of ammunition required to engage different targets. The weapon can be elevated from -10° to +45° and traversed 50° left and right. Elevation and traverse are manual and the gunner aims the weapon via a Zeiss PERI Z-16 periscopic sight; a passive night sight is also available.

Wiesel with TOW 2 (TOW A1)

This was built to meet the requirements of the German Army and is fitted with a Hughes TOW ATGW launcher on an elevating pedestal with a traverse of 45° left and right and an elevation and depression of 10°. Seven missiles are carried of which two are for ready use. This model has a crew of three, driver, gunner and loader.

Wiesel Command and Control Vehicle

This proposed vehicle would have a fully enclosed hull, have a crew of three and be fitted with extensive communications equipment. A 7.62 mm MG3 machine gun with 100 rounds of ready use ammunition and another 400 in reserve would be fitted.

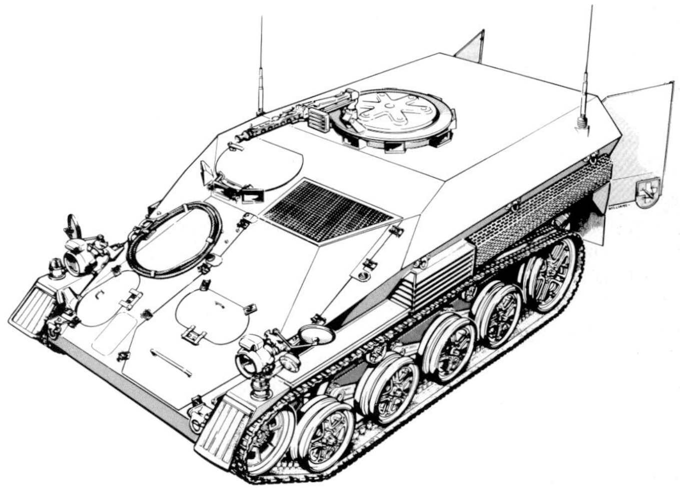

Artist's impression of Wiesel APC with a longer hull and additional roadwheel

Wiesel MK 20 A1 armed with Rheinmetall 20 mm MK 20 Rh 202 cannon in KUKA gun mount (Terry J Gander)

Wiesel TOW A1 armed with Hughes TOW 2 ATGW system with roof hatches open

German Army CH-53 helicopter unloading both versions of Wiesel airportable armoured vehicles

Wiesel Battlefield Surveillance Vehicle

This proposed vehicle would have a crew of two, radar operator and driver, and be fitted with a hydraulic mast with a RASIT battlefield surveillance radar on top. A 7.62 mm MG3 machine gun with 100 rounds of ready use ammunition and another 400 in reserve would be fitted.

Wiesel Stinger Post Vehicle

This proposed vehicle would have a crew of two and a one-man turret armed with four missiles and a 7.62 mm MG3 machine gun. The weapon can be elevated from -8° to +50° and traversed through 360°. Elevation and traverse are electrically powered and the gunner aims the weapon via a reflex sight. Ten missiles could be carried of which four are for ready use. The 7.62 mm MG3 machine gun has 100 rounds of ready use ammunition and a further 400 in reserve.

Wiesel HOT Vehicle

This would have a crew of two and a Euromissile HCT turret with two HOT missiles and a 7.62 mm MG3 machine gun. The turret can be traversed 45° left and right and the missiles elevated from -10° to +20°, elevation and traverse being electric/hydraulic with the gunner aiming via a periscopic sight. A passive night sight can also be fitted. Eight HOT missiles are carried of which two are for ready use. The 7.62 mm MG3 machine gun has 100 rounds of ready ammunition and a further 400 in reserve.

Reconnaissance Vehicle BTM 208

This is fitted with the French SAMM BTM 208 turret armed with one 7.62 mm machine gun (with 200 rounds of ready use ammunition and 300 rounds in reserve) and one 12.7 mm M2 HB machine gun (with 100 rounds of ready use ammunition and 400 rounds in reserve). As an alternative a BTM 263 turret can be fitted. It can also be fitted with the SAMM BTM 263 one-man mortar turret.

Wiesel Resupply Vehicle

This would have a combat weight of 2580 kg and act as a resupply vehicle for Wiesels fitted with the TOW ATGW system. It would have a crew of two, carry 14 TOW ATGWs and be armed with a 7.62 mm machine gun with 500 rounds of ammunition.

Wiesel Recovery Vehicle

This proposed vehicle would have a crew of two and be fitted with a slewing crane with a telescopic boom that could also be used as a winch, stabilisers at the hull rear, electric screwdriver, electric welding set, tow cable, tow bars, block and tackle, hydraulic jack, bolt cutters, tool kit and a radio.

Wiesel SAM Vehicle

This proposed vehicle would have a longer hull with an additional road wheel either side, be fitted with the Bofors RBS 70 SAM system and carry 10 missiles. A 7.62 mm MG3 machine gun with 100 rounds of ready use ammunition and a further 400 rounds in reserve would be fitted.

Wiesel Reconnaissance Vehicle

This proposed vehicle would have a longer hull with an additional road wheel either side and be fitted with the Rheinmetall turret TF 20.15 armed with a Rheinmetall 20 mm cannon with an elevation of +60°, depression of -7°, and a traverse of 360°. A total of 420 rounds of ammunition would be carried of which 220 would be for ready use.

Wiesel Ambulance

This proposed vehicle would have a longer hull with an additional road wheel either side, have a crew of two and be capable of carrying two stretcher and one seated patients or four seated patients. When being used in the cargo role a maximum of 600 kg can be carried.

Wiesel Armoured Personnel Carrier

This proposed vehicle would have a longer hull with an additional road wheel either side and carry a half-squad of infantry (six) in addition to its crew of two. The roof would be fully enclosed and fitted with a

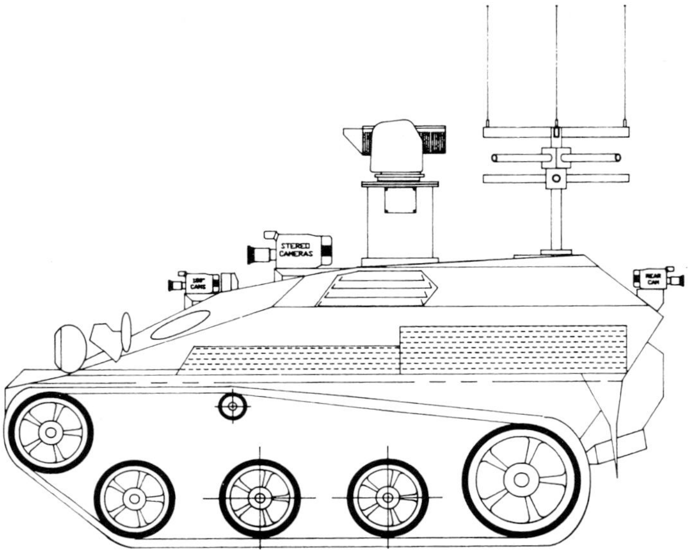

Robotic Wiesel vehicle as used by TACOM for trials purposes (US Army) (not to 1/76th scale)

commander's cupola with an externally mounted 7.62 mm MG3 machine gun with 100 rounds of ready use ammunition and a further 400 rounds in reserve. The infantry would enter and leave via a door in the hull rear.

Basic specifications of the stretched Wiesel APC are:

CREW	7
WEIGHT COMBAT	3600 kg
GROUND PRESSURE	0.34 kg/cm^2
LENGTH	3.769 m
WIDTH	1.82 m
HEIGHT	
with MG mount	2.227 m
with MG mount removed	1.9 m
deck	1.56 m
AXIS OF FIRE	2.115 m
GROUND CLEARANCE	0.3 m
MAX CRUISING SPEED	72 km/h
CRUISING RANGE	600 km
ARMAMENT	1 × 12.7 mm MG in Wegmann mount Model 2650
TRAVERSE	360°
ELEVATION/DEPRESSION	+75°/-15°
AMMUNITION	500 × 12.7 mm

Wiesel with Integrated RATAC-S

For trials, a Wiesel has been fitted with a German SEL integrated RATAC-S which combines the roles previously carried out by the RATAC (target acquisition) and RASIT (surveillance) radars. RATAC-S carries out the roles of surveillance, target acquisition, classification, precision location, automatic tracking and plotting.

Robotic Wiesel Vehicles

A prototype Wiesel vehicle was delivered to the United States late in 1988 and was used by engineers to determine what vehicle modifications will be necessary to integrate the cameras, computers and communications equipment. The seven production Wiesels were delivered in 1990.

Four of the vehicles will be modified and used at Fort Knox with a Robotic Command Center (RCC) being designed and built by FMC and due to be completed late in 1990. This RCC will be based on a modified M109 chassis and will be capable of controlling up to four robot vehicles simultaneously.

Modifications to the Wiesels will include installing special sensors and stereoscopic and conventional cameras that provide forward and peripheral views ahead of the vehicles to facilitate remote driving.

The vehicles will also be fitted with two-way radio equipment which will permit the operators to control the robots from the RCC, a north seeking gyro for navigation and a microprocessor that will manage all command and data signals travelling between the robot vehicles and the RCC.

Each robot will also be equipped with a panoramic reconnaissance platform which will include a thermal image viewer and colour camera to perform reconnaissance missions and make night driving possible.

The Robotic Wiesel vehicles will be used with the RCC to demonstrate a multi-vehicle command and control capability through the use of computer-aided remote driving (CARD), which is being developed under a TACOM contract by the Jet Propulsion Laboratory.

In CARD, a stereoscopic image of the scene ahead of a robot vehicle will be sent to an operator's display in the RCC. The operator will then electronically designate where he wants the vehicle to go and the CARD computer in the robot vehicle will generate the control signals required to direct it.

SPECIFICATIONS

Model	**TOW**	**20 mm**
CREW	3	2
COMBAT WEIGHT	2750 kg	2750 kg
UNLOADED WEIGHT	2030 kg	2030 kg
POWER-TO-WEIGHT RATIO	30.72 hp/tonne	30.72 hp/tonne
GROUND PRESSURE	0.36 kg/cm^2	0.36 kg/cm^2
LENGTH GUN FORWARDS	n/app	3.545 m
LENGTH HULL	3.31 m	3.31 m
WIDTH	1.82 m	1.82 m
HEIGHT	1.825 m	1.897 m
to hull top	1.352 m	1.352 m
FIRING HEIGHT	1.68 m	1.615 m
GROUND CLEARANCE	0.302 m	0.302 m
TRACK	1.62 m	1.62 m
TRACK WIDTH	200 mm	200 mm
LENGTH OF TRACK ON GROUND	1.82 m	1.82 m
MAX ROAD SPEED	75 km/h	75 km/h
FUEL CAPACITY	80 l	80 l
MAX RANGE*	200 km	200 km
GRADIENT	60%	60%
SIDE SLOPE	30%	30%
VERTICAL OBSTACLE	0.4 m	0.4 m
TRENCH	1.2 m	1.2 m
TURNING RADIUS		
Cletrac	8.9 m	8.9 m
pivot	5.25 m	5.25 m

Model	**TOW**	**20 mm**
ENGINE	VW 5-cylinder turbo-charged diesel developing 86 hp at 4500 rpm	
TRANSMISSION	ZF automatic type 3 HP 22 with two gear splinter group giving a total of six forward and six reverse gears	
ELECTRICAL SYSTEM	24 V	24 V
BATTERIES	2 × 12 V, 45 Ah	2 × 12 V, 45 Ah
ARMAMENT	1 × TOW launcher	1 × 20 mm cannon
AMMUNITION	7 missiles	400 rounds of 20 mm

*Range of 200 km is 60% cross-country and 40% road. Road range is 300 km.

Status: In production. In service with German Army and being used by US Army for robotic trials purposes.

Manufacturer: Krupp MaK Maschinendau GmbH
P O Box 9009
2300 Kiel
Federal Republic of Germany
Telephone: (0431) 3811
Telex: 02 99 877 mak d

Design authority:
Dr Ing h c F Porsche Aktiengesellschaft
D-7000 Stuttgart 40
Federal Republic of Germany
Telephone: (0711) 8203
Telex: 721871

HUNGARY

FUG (or OT-65) Amphibious Scout Car

Development

The FUG (Felderito Úszó Gépkocsi) fulfils a similar role in the Hungarian Army to the Soviet BRDM-1 amphibious scout car but has its engine at the rear rather than the front and two instead of one water jets for amphibious operations. The vehicle entered service with the Hungarian Army in 1964 and with the Polish and Czechoslovak armies in 1966. The Czechoslovaks call the vehicle the OT-65.

Description

The hull of the FUG is all-welded steel with the driver seated at the front on the left and the vehicle commander on the right. Both are provided with a hatch cover to their front that is hinged at the top and opens upwards, and a windscreen with a wiper. The hatch is equipped with an integral vision block for when the hatch is closed.

The crew compartment is behind the driver and the only means of entry is by the hatch in the roof that opens either side of the vehicle. The roof hatches can be locked vertical to give a measure of protection when the machine gun is being used. A ventilator is mounted in the roof of the vehicle over the commander's position. Six circular firing ports are provided, two in each side and two in the rear. A vision slit is provided to the left of the driver and the right of the commander.

The engine compartment is at the rear of the hull with the air-inlet immediately behind the crew compartment, the air-outlets in either side and the exhaust pipe on the right side.

Czechoslovak Army FUG (4 × 4) amphibious scout car with roof hatches in open position

On either side of the vehicle, between the front and rear wheels, are two belly wheels, which are lowered by the driver and driven from a PTO and give the FUG improved cross-country ability and allow it to cross trenches. When the belly wheels are lowered the vehicle is driven in first gear.

The FUG is fully amphibious, being propelled in the water by two water jets at the rear of the hull. Before entering the water a trim board is erected at the front of the vehicle which is stowed under the nose when not in use, and the bilge pumps are switched on. Standard equipment includes infra-red driving lights and some models are also fitted with an infra-red searchlight, and a central tyre-pressure system that allows the driver to adjust the tyre pressure to suit the type of ground being crossed. It has no NBC system.

Main armament comprises a pintle-mounted 7.62 mm SGMB machine gun with limited traverse, which is aimed and fired by one of the crew members with the two hatches open.

Variants

Ambulance

The FUG is known to be used as an ambulance although it is clearly unsuited to this role.

Radiological-Chemical Reconnaissance Vehicle

This performs a similar role to the Soviet BRDM-1 RKh vehicle and is used to mark lanes through contaminated areas. Mounted on either side of the hull rear is a rack that carries lane marking poles. When required the racks are swung vertically through 90° so that they are over the rear of the hull and lane marking poles can then be dispensed into the ground from inside the vehicle.

OT-65A

This is a Czechoslovak modification and is the basic vehicle with the turret of the OT-62B tracked armoured personnel carrier mounted on the roof. The turret is armed with a 7.62 mm M59T machine gun with an elevation of +20° and a depression of -10°. The turret can be traversed manually through 360°. Mounted externally on the right side of the turret is an 82 mm T-21 Tarasnice recoilless gun which can be aimed and fired from inside the turret but can be reloaded only from the outside. The T-21 fires a HEAT projectile weighing 2.13 kg with a muzzle velocity of 250 m/s which will penetrate 230 mm of armour. Maximum range is 2500 m, effective range 450 m and about 300 m against moving targets. The crew of the OT-65A enters and leaves the vehicle by a small roof hatch that opens either side of the roof immediately behind the turret. This has also been referred to as the FUGa in Eastern literature.

PSZH-IV APC

There is a separate entry in the *Armoured personnel carriers* section for the Hungarian PSZH-IV (4 × 4) armoured personnel carrier. This was originally called the FUG-66 and later the FUG-70 until it was established that it was an armoured personnel carrier and not an amphibious scout car. This has also been referred to as the FUG-2 in Eastern literature.

Status: Production complete. In service with Czechoslovakia, Hungary and Poland (limited use).

Manufacturer: Hungarian state factories.

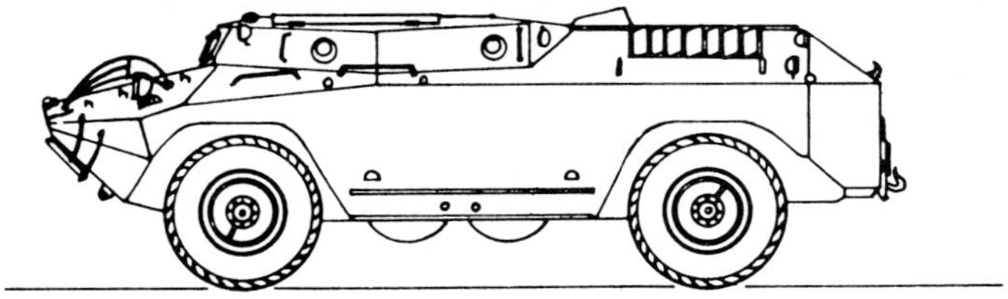

FUG (or OT-65) amphibious scout car without fitted armament

SPECIFICATIONS					
CREW	2 + 4	MAX SPEED		ENGINE	Csepel D.414.44 4-cylinder in-line water-cooled diesel developing 100 hp at 2300 rpm
CONFIGURATION	4 × 4	road	87 km/h		
COMBAT WEIGHT	7000 kg	water	9 km/h		
POWER-TO-WEIGHT RATIO	15.87 hp/tonne	FUEL CAPACITY	200 l		
LENGTH	5.79 m	MAX CRUISING RANGE (road)	600 km	TRANSMISSION	manual with 5 forward and 1 reverse gears
WIDTH	2.5 m	FUEL CONSUMPTION	0.28 l/km	TYRES	12.00 × 18
HEIGHT		FORDING	amphibious	SUSPENSION	
to hull top	1.91 m	GRADIENT	60%	front	helical springs
with turret OT-65A	2.25 m	SIDE SLOPE	30%	rear	semi-elliptical springs
GROUND CLEARANCE	0.34 m	VERTICAL OBSTACLE	0.4 m	MAIN ARMAMENT	1 × 7.62 mm MG
TRACK	1.9 m	TRENCH	1.2 m	AMMUNITION (main)	1250
WHEELBASE	3.3 m			ARMOUR (max)	13 mm

ISRAEL

RAMTA Light RAM Vehicle

Development

The light RAM vehicle has been developed as a private venture by RAMTA Structures and Systems, a division of Israel Aircraft Industries and was announced for the first time in 1987. The vehicle builds on experience with the RAM and RBY family of light 4 × 4 armoured vehicles described in the following entries.

Compared with the RAM and RBY the light RAM vehicle is shorter, incorporates advanced design elements and a new turbo-charged diesel engine.

According to the manufacturer, typical roles of the light RAM vehicle include surprise assault in rear areas, reconnaissance and fighting patrols and the rapid deployment of anti-tank and anti-guerrilla forces.

A Lockheed C-130 Hercules transport aircraft can carry three light RAM vehicles internally while the CH-53 helicopter can carry one vehicle internally or one vehicle suspended underneath.

Description

The hull of the vehicle is of all-welded steel up to 7 mm thick with the bottom of the vehicle having been designed to provide maximum protection against mines. Its very low centre of gravity prevents the RAM being overturned by mine blasts. The hull provides protection against shell splinters, shrapnel and small arms fire up to 7.62 mm armour piercing.

The front of the vehicle is normally used for kit stowage with the open-topped crew compartment in the centre and the engine compartment at the rear. The commander and driver sit at the front of the crew compartment with the remaining four crew members sitting back-to-back in pairs.

The basic model is armed with up to three 7.62 mm machine guns or one 12.7 mm and two 7.62 mm machine guns, although other armament options are possible such as various VHF and UHF communication systems.

Battlefield survivability has been enhanced by the separation of the crew compartment from the engine and fuel tanks. This has reduced the chance of a catastrophic explosion and fire.

All critical on-board systems such as fuel, brake and steering have redundant back-up.

Status: Development complete. Ready for production.

Manufacturer: RAMTA Structures and Systems, Israel Aircraft Industries
POB 323
Beer Sheba 84102
Israel
Telephone: (057) 72231; (057) 38083
Telex: 5298
Fax: 972 57 79300

SPECIFICATIONS					
CREW	4 + 2	MAX SPEED	93 km/h	TRANSMISSION	Allison Transmission Division of General Motors AT-545 automatic
CONFIGURATION	4 × 4	FUEL CAPACITY	160 l (two tanks)		
COMBAT WEIGHT	4500 kg	ROAD RANGE	800 km		
AIR TRANSPORT WEIGHT	4300 kg	FORDING	0.9 m		
UNLOADED WEIGHT	3300 kg	GRADIENT	60% at 5 km/h	TYRES	10.5 R 20 XL (8 ply)
PAYLOAD	1975 kg	SLIDE SLOPE	30%	STEERING	ZF hydraulic assisted
LENGTH	4.80 m	VERTICAL OBSTACLE	0.8 m	TURNING RADIUS	7.5 m
WIDTH	2.02 m	ENGINE	Deutz (KHD) BF4 L913 in-line turbo-charged air-cooled diesel developing 115 hp at 2550 rpm	SUSPENSION	rigid axles with semi-elliptic progressive springs and hydraulic shock absorbers
HEIGHT	1.68 m				
GROUND CLEARANCE	0.50 m				
TRACK	1.726 m			BRAKES	power, hydraulic with power booster
ANGLE OF APPROACH/ DEPARTURE	60°/42°			ELECTRICAL SYSTEM	24 V

RAMTA RAM Family of Light AFVs

Development

The RAM family of light armoured fighting vehicles has been developed by RAMTA Structures and Systems, a unit of Israel Aircraft Industries, as a successor to the company's earlier RBY Mk 1 light armoured reconnaissance vehicle (fully described in the following entry) and was announced at the Paris Air Show in June 1979.

The RAM family has now been expanded into two basic groups, RAM V-1 and RAM V-2. The former are open topped and are produced in two wheelbases, V-1 short (3.4 m wheelbase) and V-1L (3.8 m wheelbase). The V-2 series has an enclosed troop compartment and comprises the V-2 short (3.4 m wheelbase) and V-2L long (3.8 m wheelbase).

Description (RAM V-1)

The layout of the RAM V-1 is almost identical to the earlier RBY Mk 1 light armoured reconnaissance vehicle and the short model has the same wheelbase. The most significant improvements are the increased ground clearance achieved by the fitting of larger tyres, increased radius of action as the original petrol engine has been replaced by a diesel engine, and the replacement of the manual gearbox by an automatic one to reduce driver fatigue. Crew comfort has also been improved. Mounted at the front of the vehicle is a Koenig EC-100 winch with a capacity of 3629 kg.

Variants

Infantry Fighting Vehicle

This model is equipped with three 7.62 mm machine guns or one 12.7 mm and two 7.62 mm MGs, 5000 rounds of ammunition, one anti-tank launcher, six rockets, fragmentation grenades, night vision equipment and a multi-channel transceiver.

Infantry Combat Vehicle

This model carries up to three 7.62 mm machine guns or one 12.7 mm and two 7.62 mm MGs, 5000 rounds of ammunition, 52 mm mortar, 36 mortar bombs, two anti-tank rockets, fragmentation grenades, night vision equipment and a multi-channel transceiver.

TCM-20 AA

This is the basic vehicle fitted with the RAMTA Structures and Systems' TCM-20 twin 20 mm LAAG system. A total of 720 rounds of 20 mm ammunition is carried for the cannons, in addition one 7.62 mm machine gun, night vision equipment and a multi-channel transceiver are fitted. The model, which is only available based on the chassis of the RAM V-1L, has a crew of four consisting of driver/loader, commander, gunner and loader. When in the firing position mechanical jacks are lowered to the ground to provide a more stable firing platform, although it is possible to fire against both ground targets and aerial targets while the vehicle is travelling. The current TCM-20 can be replaced by the TCM Mk 3 system.

Close Range Anti-tank

This carries a 106 mm M40 series recoilless rifle built in Israel by Israel Military Industries which can be traversed through a full 360°. Eighteen rounds of ammunition are carried in the vehicle as well as night vision equipment and a multi-channel transceiver.

Long Range Anti-tank

This carries a Hughes TOW or MILAN on a telescopic pedestal, 16 TOW ATGWs, two 7.62 mm machine guns and 2500 rounds of ammunition, night vision equipment and a multi-channel transceiver.

RAM V-2 AFV

This has a fully enclosed troop compartment that carries a crew of six to nine and their personal equipment. It has a unique system of folding armour plate hatches which can be fully or partially opened so providing all-round observation while affording protection against AP small arms fire and shell fragments. The personnel, seated or standing, can fire in all directions.

Various types of armament can be fitted according to mission requirements including: one 12.7 mm machine gun or 40 mm grenade launcher, 1500 rounds of 12.7 mm ammunition or six ammunition boxes for grenade launcher, three 7.62 mm machine guns plus 5000 rounds of ammunition, one rocket launcher plus six rounds, 24 fragmentation grenades, nine smoke grenades and two smoke flares. In addition the vehicle carries night vision equipment and a multi-channel transceiver.

Status: Production as required. Morocco is known to have taken delivery of a number of these vehicles including some with a 106 mm recoilless rifle.

Manufacturer: RAMTA Structures and Systems, Israel Aircraft Industries
POB 323
Beer Sheba 84102
Israel
Telephone: (057) 72231
Telex: 5298
Fax: 972 57 79300

RAM V-1 (long) light armoured reconnaissance vehicle fitted for one 12.7 mm and two 7.62 mm MGs

RAM V-1 (long) fitted with TCM-20 twin 20 mm anti-aircraft gun system in firing position with two mechanical jacks lowered in position

RAMTA RAM (4 × 4) vehicle fitted with a retractable TOW ATGW launcher

RAM V-2L (long) AFV with 12.7 mm and 7.62 mm machine guns

SPECIFICATIONS

Model	RAM V-1 (short)	RAM V-1 (long)	RAM V-2 (short)	RAM V-2 (long)
MAX CREW	2 + 4	2 + 7	2 + 4	2 + 7
CONFIGURATION	4 × 4	4 × 4	4 × 4	4 × 4
COMBAT WEIGHT	5400 kg	5750 kg	5400 kg	5750 kg
UNLOADED WEIGHT	4100 kg	4300 kg	4300 kg	4550 kg
PAYLOAD	1300 kg	1450 kg	1100 kg	1200 kg
LENGTH	5.02 m	5.52 m	5.02 m	5.52 m
WIDTH	2.03 m	2.03 m	2.03 m	2.03 m
HEIGHT OVERALL	1.72 m	1.72 m	2.2 m	2.2 m
GROUND CLEARANCE				
hull	0.575 m	0.575 m	0.575 m	0.575 m
axles	0.31 m	0.31 m	0.31 m	0.31 m
TRACK	1.726 m	1.726 m	1.726 m	1.726 m
WHEELBASE	3.4 m	3.8 m	3.4 m	3.8 m
ANGLE OF APPROACH/ DEPARTURE	60°/43°	60°/43°	60°/43°	60°/43°
MAX ROAD SPEED	96 km/h	96 km/h	96 km/h	96 km/h
FUEL CAPACITY	120 l	160 l	120 l	160 l
ROAD RANGE	600 km	800 km	600 km	800 km
FORDING	0.75 m	0.75 m	0.75 m	0.75 m
GRADIENT	70%	60%	65%	60%
SIDE SLOPE	35%	35%	35%	35%
VERTICAL OBSTACLE	0.8 m	0.8 m	0.8 m	0.8 m
ENGINE	Deutz air-cooled, 6-cylinder diesel developing 132 bhp		Deutz air-cooled, 6-cylinder diesel developing 132 bhp	
TRANSMISSION	Allison Transmission Division of General Motors AT-540 automatic with 4 forward and 1 reverse gears		Allison Transmission Division of General Motors AT-540 automatic with 4 forward and 1 reverse gears	
TRANSFER CASE	Harwaythorn 2-speed	Harwaythorn 2-speed	Harwaythorn 2-speed	Harwaythorn 2-speed
STEERING	power-assisted ZF	power-assisted ZF	power-assisted ZF	power-assisted ZF
TURNING RADIUS	7.2 m	7.5 m	7.2 m	7.5 m
SUSPENSION	semi-elliptical springs and hydraulic shock absorbers		semi-elliptical springs and hydraulic shock absorbers	
TYRES	9.00 × 20	9.00 × 20	9.00 × 20	9.00 × 20
BRAKES				
main	hydraulic	hydraulic	hydraulic	hydraulic
parking	mechanical	mechanical	mechanical	mechanical
ELECTRICAL SYSTEM	24 V	24 V	24 V	24 V
BATTERIES	2 × 12 V	2 × 12 V	2 × 12 V	2 × 12 V
ARMOUR	8 mm	8 mm	8 mm	8 mm

Note: As an option, combat tyres can be fitted, or 12.50 × 50 tyres for improved performance in sand or mud

RAMTA RBY Mk 1 Light Armoured Reconnaissance Vehicle

Development

The RBY Mk 1 light armoured reconnaissance vehicle was developed by RAMTA Structures and Systems, a subsidiary of Israel Aircraft Industries, and was first announced at the Paris Air Show in May 1975.

The design of the vehicle is based on an analysis by RAMTA of experience obtained during two decades of field operations in the Middle East. This showed that a reconnaissance vehicle required a very low profile, good all-round visibility for all crew members, to be airportable, be able to carry different armament installations and provide the crew with the maximum protection against mines.

The RBY Mk 1 is suitable for a wide range of roles including reconnaissance, long-range patrols and commando type operations, internal security operations and the evacuation of wounded in forward battlefield areas. Its low weight enables it to be carried by light aircraft and helicopters such as the CH-53s used by the Israeli Air Force.

Description

The all-welded hull of the RBY Mk 1 has 8 mm thick armour steel (MIL-A-46100) sides inclined at an angle, and a floor of 10 mm thick plain carbon steel (SAE-1020). The shape of the hull has been designed to give the crew maximum possible protection from mine explosions. The axles and wheels are at the extreme ends of the vehicle and the effect of a mine explosion when touched off by one of the wheels is that the respective axle is damaged or cut, the glass fibre bumpers disintegrate, but the hull stays intact. The ability of the vehicle to survive mine explosions was proved in a test programme using different types of mine.

The driver is seated at the front of the vehicle on the left with the commander to his right. The front part of the crew compartment can be folded down into the horizontal for increased visibility and is provided with two integral observation hatches that are hinged at the top and open outwards. To the left of the driver and the right of the commander is another observation hatch, also hinged at the top and opening outwards.

The six passengers are seated three each side, facing outwards, back-to-back for all-round observation. There are no doors in the hull as this would weaken the structure and the crew enter and leave the vehicle by climbing over the sides.

The engine is at the rear of the hull and has a glass fibre cover and side doors for ease of access. The automotive components of the RBY Mk 1 are of a standard commercial design that meet current military requirements. Two spare wheels are carried, one at the front of the vehicle and one at the rear. Infra-red driving lights and a front-mounted winch with a capacity of 2722 kg can be fitted if required. Up to five 7.62 mm or 12.7 mm machine guns can be pintle-mounted around the top of the hull to provide suppressive fire.

RBY Mk 1 light armoured reconnaissance vehicle without front winch and armed with 7.62 mm (0.30) machine guns

RBY Mk 1 light armoured reconnaissance vehicle with commander's and driver's hatch open and showing extensive stowage at front of vehicle

Variants

Anti-tank armed with 106 mm M40 Recoilless Rifle

This model has the 106 mm M40 recoilless rifle manufactured by Israel Military Industries, with 16 anti-tank rounds. Secondary armament consists of two 7.62 mm machine guns with 5000 rounds and anti-tank missiles. A long range anti-tank model is also offered.

Communications

The RBY Mk 1 can be fitted with a variety of communications equipment.

Status: Production complete. In service with Guatemala, Honduras, Israel and one other country.

Manufacturer: RAMTA Structures and Systems, Israel Aircraft Industries
POB 323
Beer Sheba 84102
Israel
Telephone: (057) 72231
Telex: 5298
Fax: 972 57 7930

SPECIFICATIONS	
CREW	2 + 6
CONFIGURATION	4 × 4
UNLOADED WEIGHT	3600 kg
LENGTH	5.023 m
WIDTH	2.03 m
HEIGHT	
empty without armament	1.66 m
loaded without armament	1.54 m
GROUND CLEARANCE	
hull	0.48 m
axles	0.27 m
transfer case	0.375 m
TRACK	1.665 m
WHEELBASE	3.4 m
ANGLE OF APPROACH/ DEPARTURE	60°/50°
MAX SPEED	
road	100 km/h
cross-country	50 km/h
FUEL CAPACITY	140 l
MAX RANGE	
road	550 km
cross-country	400 km
FORDING	0.75 m
GRADIENT	60%
SIDE SLOPE	30%
ENGINE	Chrysler 225-2 6-cylinder water-cooled petrol developing 120 hp
TRANSMISSION	Model 435 manual with 4 forward and 1 reverse gears
TRANSFER CASE	2-speed
STEERING	worm and sector, power assisted
CLUTCH	single dry plate
SUSPENSION	semi-elliptical springs and hydraulic shock absorbers
BRAKES	
main	hydraulic
parking	mechanical mounted on drive shaft
ELECTRICAL SYSTEM	24 V
BATTERIES	2 × 12 V
ARMOUR	8 mm

Israel Military Industries Multi-Mission Combat Vehicle

Development/Description

The Multi-Mission Combat Vehicle (MACV) has been developed by Israel Military Industries as a private venture and can be adopted for a wide range of battlefield missions including reconnaissance, border patrol and anti-tank.

The driver is seated at the very front of the vehicle on the centre line with the crew and weapons space in the centre and powerpack at the very rear.

The powerpack consists of a 160 hp diesel coupled to a fully automatic transmission with hydrostatic drive to all four wheels. With a combat weight of 5.5 tonnes the MACV has a very high power-to-weight ratio of 29 hp/tonne. The tyres are of the run flat type with power to each wheel being automatically adjusted to meet changing road or terrain requirements. The long suspension travel gives a good crew ride even when travelling at speed across rough terrain.

Its very low weight enables it to be transported by aircraft and helicopters. For example a C-130 Hercules can carry three MACVs internally and ready for air drop and it can also be driven inside transport helicopters such as the CH-47 and CH-53.

The basic vehicle has a curb weight of 2700 kg but it can carry a maximum payload of 2800 kg which can be in the form of a wide range of weapons or a combination of weapons and additional armour protection according to mission requirements.

A typical crew consists of commander, driver and four infantry men with four 7.62 mm machine guns being installed to give suppressive fire cover all round the vehicle.

Various armour packages are available according to mission and threat requirements which can be installed at field or division level depending on the kit. These provide protection from small arms fire, shrapnel and even 14.5 mm armour piercing rounds.

The wheels are mounted at the extreme ends of the vehicle not only to give a good angle of approach and departure but also if the vehicle runs over a mine it will be detonated as far away from the crew as possible.

A wide range of weapons can be installed including cannon up to 60 mm in calibre, mortars, 40 mm grenade launcher, 7.62 mm and 12.7 mm machine guns and anti-tank systems such as the Hughes TOW. The latter version has one missile in the ready to launch position and a further 24 missiles being carried in reserve.

Israel Military Industries MACV fitted with pintle-mounted 7.62 mm MGs

Israel Military Industries MACV fitted with Hughes TOW ATGW system and pintle-mounted 7.62 mm MG

Israel Military Industries MACV fitted with cannon and pintle-mounted 7.62 mm MG

SPECIFICATIONS		WIDTH	2.05 m	FORDING	0.75 m
CREW	2 + 6	HEIGHT		GRADIENT	65%
CONFIGURATION	4 × 4	without armament	1.6 m	SIDE SLOPE	45%
COMBAT WEIGHT	5500 kg	GROUND CLEARANCE	0.45 m	VERTICAL OBSTACLE	0.5 m
CURB WEIGHT	2700 kg	ANGLE OF APPROACH/		ENGINE	diesel developing 160
PAYLOAD	2800 kg	DEPARTURE	60°/90°		hp at 3600 rpm
POWER-TO-WEIGHT		MAX ROAD SPEED	105 km/h	TRANSMISSION	automatic
RATIO	29 hp/tonne	SPEED ON 60% GRADE	10 km/h	TYRES	12.50 × 20 (run flat)
LENGTH	4.25 m	RANGE	850 km	TURNING RADIUS	7.5 m

Status: Development complete. Ready for production.

Manufacturer: Israel Military Industries
PO Box 1044
Ramat Hasharon 47100
Israel
Telephone: (03) 5485222
Telex: 33719
Fax: (03) 5406908

ITALY

New IVECO FIAT Wheeled Armoured Fighting Vehicles

In the early 1980s the IVECO FIAT company designed and built prototypes of a new range of 4 × 4 and 6 × 6 armoured vehicles as a private venture. This series comprised the 6634 AVL (4 × 4), 6633 AVM (4 × 4) and the 6636 AVH (6 × 6). Details of these vehicles, which have yet to enter production, are given in the *Armoured personnel carriers* section. Further development of the 6634 AVL has resulted in the Puma 4 × 4 multi-role vehicle currently being developed to meet the requirements of the Italian Army. Details of the Puma are given in the *Armoured personnel carriers* section although it can be adopted for a wide range of roles on the battlefield. Details of the Centauro B1 (8 × 8) tank destroyer are given in the following entry.

IVECO FIAT Centauro B1 (8 × 8) Tank Destroyer

Development

In the early 1990s the Italian Army will start to receive four new AFVs that have been designed and built in Italy to meet their specific requirements. These are the C1 Ariete MBT, The B1 Centauro (8 × 8) tank destroyer, VCC-80 infantry fighting vehicle and the Puma (4 × 4) multi-role light vehicle. Full details of the Puma family, which now includes a 6 × 6 vehicle, are given in the *Armoured personnel carriers* section.

These four vehicles are being developed by the Consorzio IVECO FIAT-OTO Melara based in Rome with IVECO FIAT having overall responsibility for the wheeled vehicles and OTO Melara having responsibility for the tracked vehicles.

IVECO FIAT is also responsible for the complete powerpacks of all four vehicles while OTO Melara has complete responsibility for the turrets and armament.

The B1 Centauro (8 × 8) tank destroyer (which has the IVECO FIAT designation of the Type 6638 G), will be used by six new Italian Army brigades that will be used as a rapid deployment force within Italy.

Early in 1984 the Italian Army formulated its requirements for a new highly mobile wheeled tank destroyer that was required to have a standard NATO 105 mm rifled tank gun.

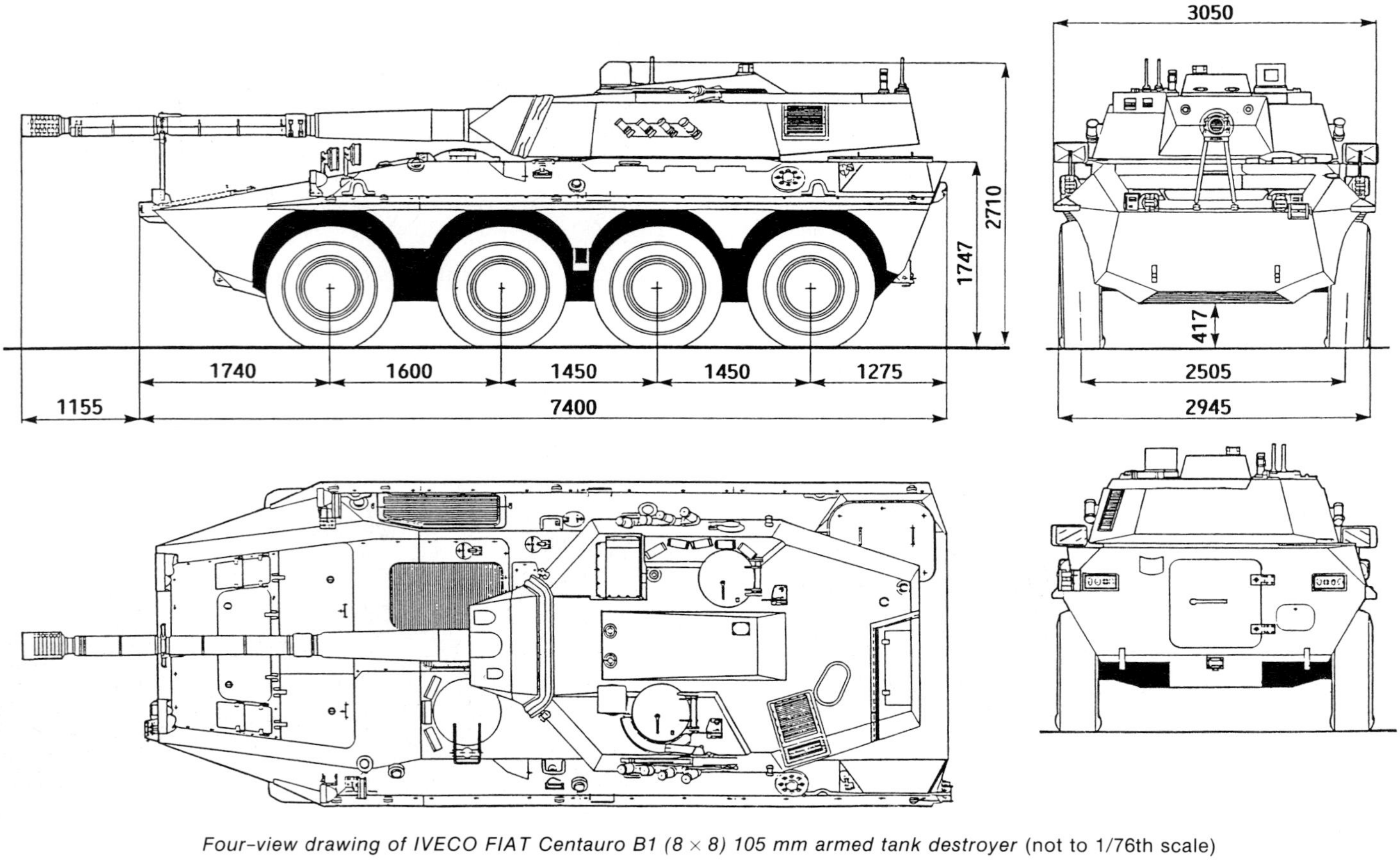

Four-view drawing of IVECO FIAT Centauro B1 (8 × 8) 105 mm armed tank destroyer (not to 1/76th scale)

Turrets for the IVECO FIAT Centauro B1 (8 × 8) tank destroyer in production at OTO Melara's La Spezia facility

Close-up of Centauro B1 turret showing commander's and gunner's sights, the former is fully stabilised and mounted in front of loader's hatch

It was also required to have high road speed for strategic mobility, long range, good cross country mobility and a computerised fire control system which would share many common components with those being developed by Officine Galileo for the C1 Ariete MBT and VCC-80 IFV.

By late 1984 the specification had been agreed, although by that time some design work had already been carried out by IVECO FIAT and OTO Melara as development work on key subsystems such as powerpack and suspension had already been accomplished.

Prior to the construction of the first complete prototype of the B1, an unarmoured test rig was built to prove the powerpack and new hydropneumatic suspension elements. This ran for the first time in April 1985 and provided a considerable amount of valuable information. Later a test bed turret was installed on the chassis to test the man/machine interface as well as firing the 105 mm gun under the worst possible operational conditions.

The first complete prototype of the B1 was completed in January 1987 with the second following in mid-1987 and by December four complete vehicles were involved in the trials programme. Prototypes 5,6, 7,8 and 9 were completed in 1988 and in addition a ballistic hull has been built.

The prototypes were following by a pre-production batch of 10 vehicles which were completed from late 1989.

Production of the B1 Centauro commenced late in 1990 with first production vehicles being completed in 1991. The original Italian Army requirement was for a total of 450 B1 Centauro vehicles but this has now been trimmed back to 400.

Production B1 Centauro vehicles differ from the prototypes in a number of key areas including improved stowage, modified turret rear, door in hull rear rather than a ramp, improved layout for the crew positions as well as improvements to the weapon system.

Within IVECO FIAT, the Defence Vehicles Division in Bolzano in Northern Italy will be responsible for the Centauro and a new assembly facility is now being constructed which will have four production lines. One for the Centauro, one for Puma and two for heavy military trucks. IVECO FIAT will assemble the complete powerpack (engine, transmission and cooling system) not only for the Centauro, but also for the Ariete, VCC-80 and Puma.

Description

The hull and turret of the Centauro are of all-welded steel armour construction which provides the crew with protection from small arms fire and shell splinters. Over the frontal arc, protection is provided against penetration from 20 mm projectiles and against 12.7 mm attack all round. The hull will be supplied already welded by another IVECO FIAT company.

The driver is seated at the front of the hull offset to the left with the powerpack to his right and separated from the remainder of the vehicle by a fireproof bulkhead. He is provided with an adjustable seat, single piece hatch cover that opens to the left and three periscopes for forward observation, the centre one of which can be replaced by a MES VG/DIL passive periscope for night driving.

The turret is mounted on the roof of the hull slightly to the rear of the vehicle's centreline with the commander on the left, gunner on the right and loader forward and below the gunner.

The commander is provided with four periscopes to give him observation to the rear. The stabilised day sight is mounted forward of the commander's hatch and gives him full 360° observation without moving his head.

The gunner leaves the vehicle via the loader's single piece hatch that opens to the rear with five periscopes being provided in the right side of the turret roof for use by the gunner and loader.

The turret of the Centauro is the complete responsibility of OTO Melara at La Spezia and deliver the turret fully integrated and tested and ready for installation on the chassis.

Main armament comprises an OTO Melara designed 105 mm 52 calibre rifled gun with a long recoil system that fires standard NATO tank ammunition as used in the L7 and M68 guns, including APFSDS. The weapon has a vertical sliding breech mechanism that remains open after the empty cartridge case has been ejected and is fitted with a multi-slotted muzzle brake, thermal sleeve with integral fume extractor and a muzzle reference system. The ordnance is autofrettaged and has a maximum recoil of 750 mm.

A total of 40 rounds of 105 mm ammunition are carried, of which 14 are in the turret and remainder in the hull. A 7.62 mm M42/59 machine gun is mounted coaxially to the left of the main armament with a similar weapon being mounted on the turret roof for anti-aircraft defence. Mounted either side of the turret is a bank of four electrically operated smoke dischargers which fire over the frontal arc of the vehicle.

Turret traverse and weapon elevation are electro-hydraulic with manual controls for emergency use with elevation being from -6° to +15°, slightly

IVECO FIAT B1 Centauro (8 × 8) tank destroyer from rear. Production vehicles have ramp replaced by door

IVECO FIAT B1 Centauro (8 × 8) tank destroyer showing commander's roof-mounted stablised sight forward of loader's position

IVECO FIAT B1 Centauro (8 × 8) tank destroyer firing 105 mm gun which is fitted with thermal sleeve, fume extractor and muzzle brake

One of the prototypes of the B1 (Centauro) (8 × 8) tank destroyer showing all wheel steering and 105 mm gun with thermal sleeve and a muzzle brake

less than MBTs because of the low turret profile. The centre of the turret roof is raised along the axis of the gun to allow the weapon to have a depression of -6°.

The B1 Centauro is fitted with the same Officine Galileo TURMS fire control system that is installed in the C1 Ariete MBT and in a slightly modified form in the VCC-80 infantry fighting vehicle which is at an earlier stage of development than the C1 or B1.

Main components of the TURMS (Tank Universal Reconfigurable Modular fire control System) are the commander's stabilised panoramic day sight, stabilised periscopic day/night laser sight for the gunner, digital ballistic computer, sensors, muzzle reference system and commander's, gunner's and loader's display panels.

The commander's stabilised panoramic sight has magnifications of ×2.5 and ×10 and for night vision the thermal image from the gunner's sight is displayed at the commander's station on a separate TV monitor. The commander's roof-mounted sight has been developed within the framework of a joint venture with the French SFIM company. Traverse is a full 360° with elevation from -10° to +60°

The gunner's roof-mounted sight integrates the four main modules (the primary stabilised head mirror, visual unit, laser transceiver and thermal image unit) in a single housing. The daylight (×5 magnification) and thermal night vision with two fields-of-view, wide and narrow, are all routed via the common head mirror.

The ballistic digital computer performs all the fire computations, controls and manages the operation of all ballistic units (optical sight, laser rangefinder and servoes) as well as the sensors, BITE (built in test equipment) and training. It also permits the reconfiguration of system operation from the normal level to backup modes in case of partial failures.

The three key sensors are meteorological, vehicle attitude and gun wear.

The gunner also has an Officine Galileo OG C-102 coaxial telescope with a magnification of ×8 and three aiming graticules which are selectable manually.

Although the main armament is stabilised, the Italian Army would normally come to the halt to open fire.

The Centauro is powered by an IVECO FIAT MTCA V-6 turbocharged diesel developing 520 hp at 2300 rpm which, with a combat weight of 24 tonnes, gives a good power-to-weight ratio of 21.66 hp tonne. A similar engine is used in the VCC-80 while some parts are also common to that installed in the C1 Ariete MBT.

Engine is coupled to a German ZF 5 HP-1500 fully automatic transmission with five forward and one reverse gears which in turn provides power to a ZF transfer case. The ZF transmission and transfer case are manufactured under licence in Bolzano. The complete Centauro power-pack can be replaced in 20 minutes.

From the transmission, power goes to the differential which distributes power to either side of the vehicle via a propeller shaft. Each of the latter then engage a series of shafts and conic couplings (one for each wheel).

Suspension is of the hydro-pneumatic type with powered steering on the front four and last two road wheels, with the latter only operating at speeds of under 20 km/h. A central tyre pressure regulation system is fitted as standard and this allows the driver to adjust the tyre pressure to suit the ground being crossed while the vehicle is still moving. The tyres are of the run flat type.

If a tracked vehicle runs over a mine it is usually immobilised and a repair team has to be brought forward to replace the damaged suspension components, or the vehicle is returned to the rear. If a wheeled vehicle such as the (8 × 8) Centauro runs over a mine a road wheel is normally blown off but it still retains some mobility. According to IVECO FIAT the Centauro can still keep mobile with two of its eight wheels blown off.

Mounted in the turret bustle is the Sekur NBC system which is of a similar design to that installed in the C1 Ariete MBT and is of the over pressure type. An air conditioning system is installed that allows the crew to operate the vehicle with the outside temperature ranging from -30° to +44°C.

Standard equipment includes a front-mounted winch, fire detection and suppression system for the engine compartment and a fire and explosion detection and suppression system for the crew compartment. A winch is located in the forward part of the hull.

Variants

As yet there are no variants of the B1 although its chassis could be adopted for a wide variety of weapon systems.

Status: In production for the Italian Army with first of 400 vehicles delivered in 1991.

Manufacturer: IVECO-FIAT SpA
Defence Vehicles Division
Via A Volta, 6
39100 Bolzano
Italy
Telephone: (0471) 905111
Telex: 400541 LANCIA 1
Fax: (0471) 905444

SPECIFICATIONS

CREW	4	ANGLE APPROACH/DEPARTURE	45°/60°	TYRES	14.00 × 20 (run flat)
CONFIGURATION	8 × 8	MAX ROAD SPEED	100+ km/h	ELECTRICAL SYSTEM	24 V
POWER-TO-WEIGHT RATIO	21.66 hp/tonne	RANGE	800 km	ARMAMENT	
COMBAT WEIGHT	24 000 kg	GRADIENT	60%	main	1 × 105 mm gun
LENGTH HULL	7.4 m	SIDE SLOPE	30%	coaxial	1 × 7.62 mm MG
LENGTH INCLUDING 105 mm GUN	8.555 m	ENGINE	IVECO FIAT MTCA V-6 turbo-charged diesel developing 520 hp at 2300 rpm	anti-aircraft	1 × 7.62 mm MG
WIDTH OVER TYRES	2.945 m	TRANSMISSION	ZF 5 HP-1500 automatic with 5 forward and 1 reverse gears	SMOKE LAYING EQUIPMENT	2 × 4 smoke dischargers
WIDTH OVER HULL	3.05 m	STEERING	power assisted	AMMUNITION	
HEIGHT TO HULL TOP	1.747 m	TURNING RADIUS	9 m	105 mm	40
HEIGHT OVERALL	2.71 m	SUSPENSION	hydropneumatic	GUN CONTROL EQUIPMENT	
GROUND CLEARANCE	0.417 m			turret power control	electro-hydraulic/manual
WHEELBASE	1.6 m + 1.45 m + 1.45 m			turret traverse	360°
TRACK	2.505 m			gun elevation/depression	+15°/-6°

FIAT-OTO Melara Type 6616 Armoured Car

Development

The Type 6616 armoured car is a joint development between FIAT and OTO Melara, with the former responsible for the hull and automotive components and the latter for the turret and armament installation. The vehicle shares many common components with the Type 6614 armoured personnel carrier which has been manufactured under licence in South Korea for the Korean Army.

The first prototype of the Type 6616 was completed in 1972 and after trials a production order was placed by the Italian Government for 50 vehicles for the Carabinieri. Production vehicles differ from the prototypes in that the turret front has been redesigned. The smoke discharger on the commander's hatch has been removed and the engine compartment grilles are now on either side of the hull, towards the rear, rather than on the top of the hull.

Production of the FIAT-OTO Melara type 6616 armoured car has been completed with 300 vehicles built, although production could start for the export market if further orders were placed. Production of the vehicle was undertaken in the FIAT facilities at Bolzano with the complete turret, ready for installation on the vehicle, being provided by OTO Melara.

Description

The all-welded steel armour hull has a hull escape hatch in either side.

The driver is seated at the front of the vehicle on the right side and is provided with a single piece hatch cover that opens to the right with an opening in the roof for a passive night driving periscope. To the front and side of the driver are five vision blocks which cover an arc of 200°, the centre three with a wiper blade.

The all-welded steel turret is in the centre of the hull with the commander on the left and the gunner on the right. Their seats, which can be adjusted for height, are mounted on the turret basket. The commander has nine periscopes for all-round observation and a single piece hatch cover that opens to the rear. The gunner has a single piece hatch cover that opens to the rear and a periscope sight mounted in the turret roof forward of the hatch cover. The Aeritalia P204 gunner's periscope has a magnification of ×1 for general observation and ×8 for aiming purposes in daylight (with a 9° field-of-view). It can be replaced by a P194 image intensifier with a magnification of ×8 and a 7.9° field-of-view for engaging targets at night. The periscopic head has an elevation of +60° and a depression of -10°. The radio is mounted in the turret bustle.

The engine compartment at the rear of the hull is separated from the fighting compartment by a fireproof bulkhead. The engine is at the very rear of the hull and is coupled to a manual five-speed transmission from which power is transmitted to the two-speed transfer case and then to the front and rear axles by propeller shafts. Both front and rear differentials are mechanically lockable. The planetary final drives, front and rear, consist of epicyclic gear trains in the wheel hubs.

The suspension, front and rear, is of the independent strut and link type with a helical spring and coaxial rubber bump stop. Each wheel station is provided with hydraulic shock absorbers and anti-roll bars are fitted as standard. The tyres are of the run-flat type, and the steering is power-assisted.

Optional equipment includes an NBC system, air-conditioning system, fire-extinguishing system and a front-mounted winch powered by a PTO with a maximum capacity of 4500 kg and provided with 40 m of 12 mm diameter cable.

The Type 6616 armoured car is fully amphibious, propelled in the water by its wheels. Four bilge pumps, each with a capacity of 180 litres per minute are fitted and it is possible to pressurise the submerged mechanical components.

The main armament consists of a Rheinmetall 20 mm MK 20 Rh 202 cannon. The empty cartridge cases are automatically ejected outside the vehicle. A total of 250 rounds of ready use 20 mm ammunition are carried with a further 150 rounds in reserve. Mounted coaxially over the main armament is a 7.62 mm machine gun with 300 rounds of ready use ammunition, and further 700 rounds carried in reserve. Mounted on either side of the turret are three smoke dischargers which are electrically fired from within the vehicle. If required a 106 mm M40 type recoilless rifle can be mounted forward of the commander's position on the turret roof.

Variants

For trials purposes the Type 6616 armoured car has been fitted with the OTO Melara OTO T 90 CKL turret which is armed with a 90 mm Cockerill Mk III gun. The vehicle can be fitted with various other turrets and cupolas as required by the user. More recently the FIAT-OTO Melara Type 6616 armoured car has been fitted with the new OTO Melara T60/70 M turret armed with a 60 mm OTO Melara High Velocity Gun System (HVGS), 7.62 mm machine gun mounted coaxially with the main armament, 7.62 mm machine gun on turret roof and two banks of three electrically operated smoke dischargers mounted either side of the turret.

The APFSDS-T round for the OTO Melara 60 mm HVGS installed in the OTO Melara T60/70 series turret was developed by the Belgian company of MECAR with the penetrator being provided by K HERTEL. The complete APFSDS-T round has a sinterised tungsten penetrator with an aluminium alloy sabot. The penetrator weighs 0.87 kg and will penetrate 130 mm of armoured steel at an angle of 60° at a range of 1500 m.

However, as of early 1991 none of these versions had entered production and the Type 6616 production line had been closed although it could be re-opened if further orders were placed.

FIAT-OTO Melara Type 6616 armoured car fitted with two-man OTO Melara turret armed with 90 mm Cockerill Mk III gun

FIAT-OTO Melara Type 6616 armoured car with two-man turret armed with 20 mm cannon and 7.62 mm coaxial machine gun

FIAT-OTO Melara Type 6616 armoured car fitted with OTO Melara T60/70 M turret

Status: Production complete. In service with Italian Carabinieri, Peru, Somalia (30 ordered in 1977 and delivered in 1978/79) and other undisclosed countries.

Manufacturer: IVECO-FIAT SpA
Defence Vehicles Division
Via A Volta 6
Bolzano
Italy
Telephone: (0471) 905111
Telex: 400541 LANCIA I
Fax: (0471) 905444

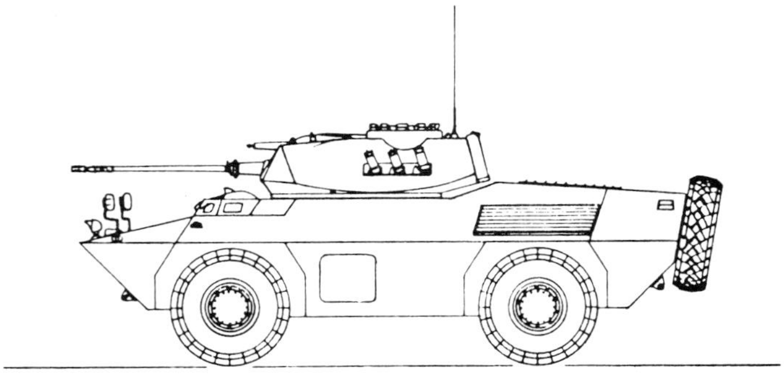

FIAT-OTO Melara Type 6616 armoured car

SPECIFICATIONS
(those in square brackets relate to the 90 mm version where this differs from the standard 20 mm version)

CREW	3
CONFIGURATION	4 × 4
COMBAT WEIGHT	8000 [8200] kg
POWER-TO-WEIGHT RATIO	20.20 [19.51] hp/tonne
LENGTH GUN FORWARDS	n/app [6.64 m]
LENGTH HULL	5.37 m
WIDTH	2.5 m
HEIGHT (to turret roof)	2.035 [2.06] m
FIRING HEIGHT	1.76 [1.785] m
GROUND CLEARANCE	
hull front	0.485 m
hull centre	0.44 m
hull rear	0.506 m
axle	0.37 m
TRACK	1.96 m
WHEELBASE	2.75 m
ANGLE OF APPROACH/ DEPARTURE	45°/45°
MAX SPEED	
road	100 km/h
water	5 km/h
FUEL CAPACITY	150 l
MAX RANGE	
road at 70 km/h	700 km
water at 5 km/h	20 km
FORDING	amphibious
GRADIENT	60%
SIDE SLOPE	30%
VERTICAL OBSTACLE	0.45 m
TURNING RADIUS	
land	7 m
water	7.5 m
ENGINE	Model 8062.4 super-charged liquid-cooled in-line diesel developing 160 hp at 3200 rpm
TRANSMISSION	manual with 5 forward and 1 reverse gears
TRANSFER CASE	2-speed
STEERING	assisted
CLUTCH	single dry plate, hydraulically operated
SUSPENSION	independent
TYRES	14.50 × 20
BRAKES	
main	drum, air over hydraulic (dual circuit)
parking	drum, hand-operated, on transfer front output shaft
ELECTRICAL SYSTEM	24 V
BATTERIES	4 × 12 V, 100 Ah
ARMAMENT	
main	1 × 20 mm cannon [1 × 90 mm]
coaxial	1 × 7.62 mm MG
SMOKE-LAYING EQUIPMENT	3 smoke dischargers either side of turret
AMMUNITION	
main	400 [16 + 20]
coaxial	1000 [1000 + 1250]
smoke grenades	39 [6 + 10]
GUN CONTROL EQUIPMENT	
Turret power control	electric/manual
Maximum elevation speed	10°/s
Maximum elevation acceleration speed	46°/s²
Maximum turret traverse speed	10°/s [30°/s]
Maximum turret traverse acceleration speed	46°/s² [45°/s²]
Traverse speed (commander controlled)	30°/s
Gun elevation/depression	+35°/-5° [+27°/-9°]
Gun stabiliser	
vertical	no
horizontal	no
ARMOUR	
Hull and turret	6 mm-8 mm

OTO Melara OTO R 2.5 Gorgona Reconnaissance Vehicle

Development

The OTO R 2.5 Gorgona vehicle has been developed as a private venture by OTO Melara and was announced in 1982. Two versions are currently being offered; an unarmed police model and a combat model which can be fitted with various types of armament. Many of the automotive components were also used in the OTO R3 reconnaissance vehicle which is no longer being marketed. Details were given in *Jane's Armour and Artillery 1988-89,* page 198.

Description

The body of the Gorgona is of welded aluminium armour 15 mm thick with the bullet-proof glass being 32 mm thick which gives the same degree of protection to the crew as the aluminium armour. This provides complete protection from 7.62 mm (NATO) small arms fire.

There is a stowage area at the front, commander, driver and two men in the centre, and the engine and transmission at the rear. Each crew member enters and leaves the vehicle via an individual side door which has an electrically operated window in its upper part. Mounted in the forward part of the vehicle is an electrically operated winch with a capacity of 2000 kg.

OTO Melara OTO R 2.5 Gorgona (4 × 4) reconnaissance vehicle from rear with front, side doors and engine cover open

The power plant is installed in the rear of the vehicle and includes the engine, automatic transmission and the transfer case. The engine oil is cooled by a heat exchanger that transfers heat to the water of the main circuit. The automatic transmission is coupled to the two-speed transfer case which permanently drives the front and rear wheels through a controlled differential. The cooling system consists of two radiators installed in the rear of the vehicle cooled by two fan-air conveyor units driven through V-belts and thermostatic devices. Electric power is supplied at 24 V by a 1200 W generator to two 90 Ah batteries. The vehicle is fully amphibious, propelled in the water by its wheels and steered in a manner similar to that of land operations.

Standard equipment includes run-flat tyres, air conditioning, heating and ventilation system, power-assisted steering and an adjustable steering wheel.

Optional equipment includes air pressurisation and filtering system, infra-red night driving equipment and two waterjets for amphibious capability.

The following armament installations can be fitted on the roof of the vehicle:

TPT shield for gunner on right side of roof

T 12.7 FA turret with 12.7 mm MG, elevation +50°, depression -10°, traverse 360°

T 7.62 FA turret with 7.62 mm MG, elevation +25°, depression -10°, traverse 360°.

OTO Melara OTO R 2.5 Gorgona (4 × 4) reconnaissance vehicle from front without armament installed

OTO Melara R 2.5 Gorgona reconnaissance vehicle showing position of automotive components (not to 1/76th scale)

Close-up of OTO Melara R 2.5 Gorgona (4 × 4) reconnaissance vehicle showing externally mounted T 7.62 FA turret

Variants
None apart from those mentioned in the text.

Status: Production as required. In service with Oman and Saudi Arabia.

Manufacturer: OTO Melara
via Valdilocchi 15
19100 La Spezia
Italy
Telephone: (0187) 530 111
Telex: 270 368
Fax: (0187) 530 669

SPECIFICATIONS

CREW	4
CONFIGURATION	4 × 4
COMBAT WEIGHT	2850 kg
MAX PERMISSIBLE WEIGHT	3150 kg
POWER-TO-WEIGHT RATIO	33.33 bhp/tonne
LENGTH	4.5 m
WIDTH	1.7 m
HEIGHT	1.62 m
GROUND CLEARANCE	0.35 m
TRACK	
front	1.418 m
rear	1.45 m
WHEELBASE	2.5 m
ANGLE OF APPROACH/ DEPARTURE	40°/40°
MAX SPEED	
road	120 km/h
water	5 km/h
RANGE	500 km
FORDING	amphibious
GRADIENT	over 60%
SIDE SLOPE	40%
TURNING RADIUS	5.5 m
ENGINE	FIAT model 8144.81.200 4-cylinder water-cooled super-charged diesel developing 95 bhp at 4200 rpm
TRANSMISSION	OTR 6, automatic, 4 forward and 1 reverse gears, hydraulic torque converter with lock up clutch (or S5/18 manual with 5 forward and 1 reverse gears)
STEERING	servo-assisted
TRANSFER CASE	2-speed
SUSPENSION	independent, MacPherson type with longitudinal torsion bars
TYRES	7.50 × 16
BRAKES	service brakes are assisted by a vacuum pump through a dual hydraulic circuit, acting on front discs and rear brake drums. Parking brake acts on rear brakes
ELECTRICAL SYSTEM	24 V
BATTERIES	2 × 12 V, 90 Ah

JAPAN

Type 87 Reconnaissance/Patrol Vehicle
Type 82 Command and Communications Vehicle

Development
In 1974 two wheeled reconnaissance vehicles were designed to meet the requirements of the Japanese Ground Self-Defence Forces. The 4 × 4 vehicle was designed and built by Komatsu Seisaku-Jyo while the 6 × 6 vehicle was designed and built by Mitsubishi Heavy Industries. After trials it was decided to concentrate on the 6 × 6 model.

Two models of the 6 × 6 vehicle have been developed, a reconnaissance patrol vehicle and a command and communications vehicle. The reconnaissance patrol vehicle is being built by Komatsu Limited and 342 million yen was allocated in the 1978 budget for the construction of two prototype vehicles. The vehicle is armed with a Swiss Oerlikon-Contraves 25 mm KBA cannon that is manufactured under licence by Nippon Seiko Jyo (Japan Steel Works). It is expected that at least 50 vehicles will be procured by the Japanese Ground Self Defence Force.

This vehicle was standardised as the Type 87 reconnaissance and patrol vehicle. The command and communications vehicle was developed by Komatsu Limited. A total of 456 million yen was allocated in the 1978 budget for the construction of four prototypes. The first prototypes were completed in 1980. A total of 250 vehicles will be produced at a unit cost of 58 million yen. Eighty per cent of its components will be identical to those used in the 6 × 6 reconnaissance patrol vehicle. This was subsequently standardised as the Type 82 command and communications vehicle with production funding being as follows:

Year	Quantity
1982	10
1983	10
1984	15
1985	18
1986	22
1987	18
1988	20

By the end of 1988 a total of 137 Type 82 command and communications vehicles were in service with the Japanese Ground Self Defence Force.

Description (Type 82)
The hull of the Type 82 command and communications vehicle is of all-welded construction with the driver sitting at the front of the hull on the right side. Above his position is a single piece hatch cover with an integral periscope that can be traversed left and right. A second crew member sits to the left of the driver and over his position is a single piece hatch cover with a rotatable periscope, forward of this is a pintle-

Type 87 reconnaissance and patrol vehicle from left side with turret traversed to right (Kensuke Ebata)

Type 82 command and communications vehicle fitted with 12.7 mm machine gun (Kensuke Ebata)

Type 87 reconnaissance and patrol vehicle with all roof hatches open (Kensuke Ebata)

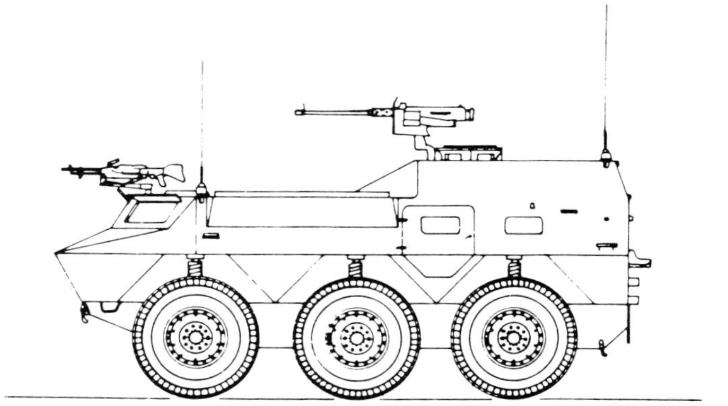

Type 82 command and communications vehicle showing externally mounted 12.7 mm machine gun (Kensuke Ebata)

Type 82 (6 × 6) command and communications vehicle from rear (Kensuke Ebata)

mounted 7.62 mm machine gun. The windscreens to the front and either side of this position are covered in combat areas by armoured shutters hinged at the top, the forward shutter being provided with two vision slits.

The engine compartment is in the forward part of the hull on the left side, to the rear of the front two crew members, with the air inlet and air outlet louvres in the roof. There is a gangway in the right side of the hull which connects the front and rear crew compartments.

The rear crew compartment, which also houses most of the communications equipment, is raised above the roof of the vehicle and has two circular hatch covers, the left one being provided with six observation periscopes and the right with an externally pintle-mounted 12.7 mm machine gun with a shield.

In the hull rear is a large door opening to the right on which is a single firing/observation port that opens upwards. In the left side of the hull, between the second and third axle, is a single forward opening door again with a single firing/observation hatch opening upwards. To the rear of this door is another single firing/observation port opening upwards. In the right side of the hull, between the first and second axle, is a forward opening door with an integral firing/observation port that opens upwards and to the rear of this are a further two firing/observation ports.

TYPE 82 SPECIFICATIONS

CREW	8
CONFIGURATION	6 × 6
COMBAT WEIGHT	13 600 kg
LENGTH	5.72 m
WIDTH	2.48 m
HEIGHT	2.38 m
GROUND CLEARANCE	0.45 m
WHEELBASE	1.5 m + 1.5 m
MAX ROAD SPEED	100 km/h
RANGE	500 km
FORDING	1 m
GRADIENT	60%
VERTICAL OBSTACLE	0.6 m
TRENCH	1.5 m
ENGINE	Isuzu 10PBI 10 cylinder water-cooled diesel developing 308 hp at 2700 rpm
TRANSMISSION	6 forward and 1 reverse gears
TURNING RADIUS	8.4 m (with 1st and 3rd axles)
TYRES	14.00 × 20

Description (Type 87)

This vehicle uses the same automotive components as the Type 82 but has a redesigned all-steel hull. The crew of five consists of the driver, assistant driver/radio operator, commander, gunner and observer.

The driver sits at the left front of the vehicle on the right and has a single piece hatch cover with an integral periscope and three fixed forward facing periscopes. The radio operator is to the left of the driver and has a single piece hatch cover which opens to the left with a single forward facing periscope.

The two-man power-operated turret is in the centre of the hull slightly forward of the second axle. The gunner sits on the left with a single piece hatch cover and two periscopes, the commander on the right with a single piece hatch cover and six periscopes.

Main armament comprises an Oerlikon-Contraves 25 mm KBA cannon with a 7.62 mm machine gun mounted coaxially. Either side of the turret, which has been designed and built by Mitsubishi Heavy Industries, is a bank of three forward firing electrically-operated smoke dischargers.

The engine compartment is at the rear of the hull on the right with an aisle for the observer on the left. In the hull roof to the rear of the turret is a single rearward facing periscope for the observer.

In either side of the hull between the first and second axles is an observation block and in the left side of the hull between the second and third axles is a crew door with an integral window and observation flap that opens towards the front.

The Type 87 reconnaissance and patrol vehicle is not fitted with an NBC system and has no amphibious capability. Specifications are similar to the Type 82 command and communications vehicle with the exception of the following:

CREW	5
COMBAT WEIGHT	14 000 kg
LENGTH	5.99 m
WIDTH	2.48 m
HEIGHT	2.8 m

Status: Both the Type 82 and Type 87 are in production and in service with the Japanese Ground Self-Defence Force.

Manufacturers: (Type 87) Komatsu Limited
2-3-6, Akasaka
Minato-ku
Japan

(Type 82) Mitsubishi Heavy Industries
5-1, Marunouchi 2-chome
Chiyoda-ku
Tokyo
Japan

MEXICO

Reconnaissance Vehicles

Mexico has completed 17 DN-3 Caballo (4 × 4) reconnaissance vehicles, some of which were fitted with a French SAMM S 365 series of one-man turrets (recently redesignated the BTM series) while others had a cupola armed with a 7.62 mm machine gun similar to that on the MOWAG Roland (4 × 4) armoured personnel carrier. The Mexican Army also uses a reconnaissance vehicle called the SEDENA 1000.

PAKISTAN

Pakistani 4 × 2 Light Armoured Vehicles

Development/Description

In the late 1980s Pakistan modified a number of standard commercial 4 × 2 tractors for military applications.

The modifications included fitting a lightly armoured body to the chassis which covers the engine at the front and extends right to the rear.

The driver and commander are seated to the immediate rear of the engine and to their front is a large windscreen. To the rear of the driver and commander is space for a further two or three men with the weapon installation being mounted on the roof.

Normal means of entry and exit is believed to be via doors in the rear of the hull. There are minor differences between vehicles with some being provided with small hatches in the sides of the hull either side of the driver's compartment.

Mounted on either side of the hull, just forward of the driver's compartment, is a bank of three electrically operated smoke dischargers that fire to the front of the vehicle. Most vehicles appear to be fitted with a single radio antenna mounted at the rear of the hull.

Variants

Anti-Tank with TOW

This version is believed to be called Cheeta and is fitted with a roof-mounted US Hughes TOW ATGW launcher with additional missiles being carried internally for manual reloading.

RBS 70 SAM Carrier

This is armed with the Swedish Bofors RBS 70 surface-to-air missile system with additional missiles being carried internally for manual reloading.

Anza SAM Carrier

This has a pedestal type mount on which are mounted four Anza (Lance) surface-to-air missiles in the ready to launch position. The Anza has been developed in Pakistan and is based on the Chinese HN-5 surface-to-air missile system which is of the infra-red fire and forget type.

MG Carrier

This has a simple open hatch on which is mounted the Chinese version of the Soviet 14.5 mm KPVT heavy machine gun. Due to the limited depression of the weapon the main role of this version is probably air defence.

Status: Production probably complete. In service with Pakistani Army.

Manufacturer: Not known.

Pakistani Cheeta 4 × 2 vehicle with TOW ATGW in operating position

Pakistani 4 × 2 vehicle with 14.5 mm heavy machine gun

ROMANIA

ARO Light Armoured Car

Development/Description

During the unrest in Romania late in 1989, a new 4 × 4 light armoured car for use in internal security situations was observed for the first time.

This is essentially an ARO 4 × 4 chassis fitted with an armoured body that provides the crew with protection from small arms fire.

The layout of the vehicle is conventional with the engine at the front, commander and driver in the centre and a small troop compartment at the rear.

In either side of the vehicle is a door that opens forwards for the commander and driver, and in its upper part is a firing port on the left side and a small window on the right side that can be covered by an armoured shutter hinged at the top.

The driver is seated at the front left with a windscreen to his front and above this is a hinged shutter that can be quickly lowered over the windscreen when in a combat situation. The commander seated to his right has a firing port which is to the right of the driver's windscreen and he also controls a white light searchlight mounted on the front right side of the vehicle.

Vision blocks or periscopes are arranged around the forward part of the vehicle's roof to allow the commander and driver observation over the frontal arc of the vehicle when the shutters are lowered. A ventilator is provided for the crew compartment.

In each side of the hull to the rear of the side door, is another firing port and in the rear of the hull are two doors that open outward. Each of these is provided with a firing port. Mounted on the roof is a small flat topped manually operated turret armed with a 7.62 mm machine gun and fitted with vision devices. This turret does not appear to have a roof hatch.

Status: Production probably complete. In service with Romania in internal security role.

Manufacturer: Romanian state factories.

SOUTH AFRICA

Sandock Austral Rooikat Armoured Car

Development

Development of the Rooikat started in 1976 and three different vehicle designs were built and evaluated by the SADF. The first vehicle had a 6 × 6 configuration, weighed 22 tonnes and was armed with a 76 mm gun mounted in a three man power operated turret in the centre of the vehicle. The driver was seated at the front with the 400/450 hp diesel at the rear.

The second vehicle, which subsequently formed the basis of the Rooikat, has an 8 × 8 configuration, weighed 27 tonnes and was also armed with the 76 mm gun mounted in a three man power operated turret in the centre of the vehicle. The driver was seated at the front and the 550 hp diesel engine mounted at the rear.

In addition there was also an 8 × 8 vehicle armed with a 105 mm high velocity gun, similar to that fitted to the Olifant MBT. It weighed over 39 tonnes. There was also an 8 × 8 APC variant with a front-mounted engine which could carry 12 troops plus the commander. The prototype of this APC still exists but no work is being carried out on it at present.

Eventually it was decided to concentrate on the 8 × 8 configuration armed with a 76 mm gun which was finally unveiled late in 1988 as the Rooikat (or Lynx).

In addition to the single prototype, five advanced development models were completed in 1986/1987 for extensive operational and technical evaluation trials. In 1987 work started on building four engineering development models for user evaluation trials. These were really pre-production models and confirmed modifications carried out as a result of the troop trials with the advanced development models.

Rooikat 8 × 8 armoured car showing suspension movement and 76 mm gun fitted with thermal sleeve

Sandock Austral, the manufacturer of the Eland and Ratel vehicles, builds the Rooikat chassis and fits the turret which is supplied by Lyttleton Engineering Works. There are, however, many key subcontractors both in the government and private sectors: for example Eloptro, part of the ARMSCOR Group, provides the optical equipment for the four man crew while Kentron provides the gyros for the stabilisation system.

For cost reasons Rooikat will not replace the Eland 90 on a one-for-one basis, but substantial numbers are expected to be procured for the Armoured Corps.

First production Rooikat armoured cars were completed in 1989 with the vehicle becoming operational with the SADF in 1990.

The main role of Rooikat will be to carry out combat reconnaissance during high-mobility operations using its speed to out-manoeuvre the enemy and strike at his flanks and deep in his rear areas. Its secondary role will be that of hunter/killer against a variety of battlefield targets including AFVs with its third role being interdiction.

Description

The hull of the Rooikat is of all-welded steel construction with the driver seated at the front, three man power-operated turret in the centre and the powerpack at the rear.

Rooikat with commander's, loader's and driver's hatches in open position going down a slope

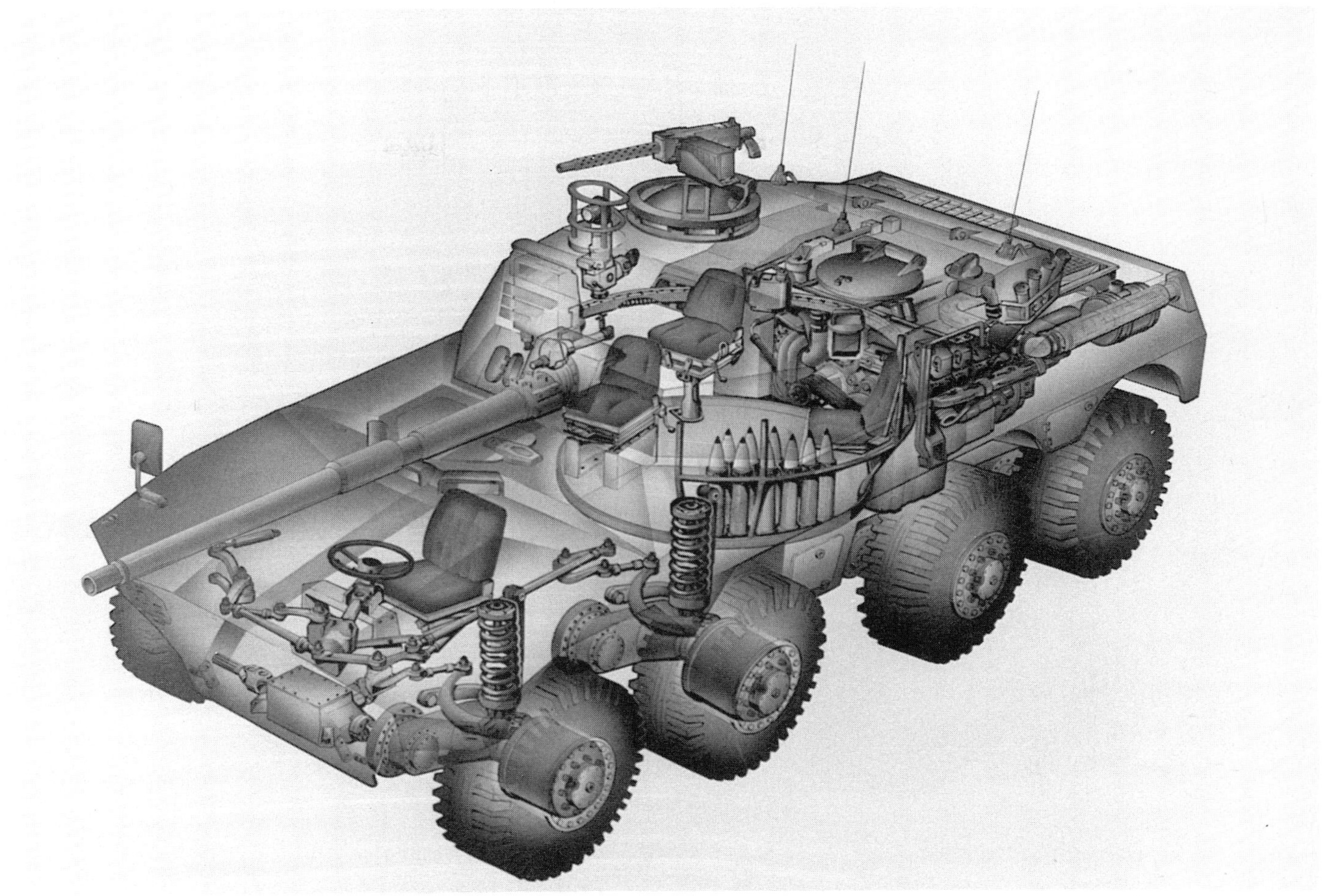

Cutaway drawing of Rooikat (8 × 8) armoured car showing position of main components

Over the frontal arc protection is provided against penetration from the 23 mm AP (armour piercing) rounds fired from the Soviet ZU-23 LAAG, widely deployed in both ground and air defence roles in Southern Africa. Trials have demonstrated that the hull design of the Rooikat provides a high degree of protection against anti-tank mines.

The driver, seated in the centre of the vehicle can enter through the fighting compartment or through the single-piece hatch cover that opens to the left. The seat is fully adjustable and when driving in the closed down position forward observation is via three periscopes, the centre one of which can be replaced by a passive night periscope developed by Eloptro. When closed down the driver's periscopes can be cleaned by a compressed air cleaning system. The headlamps are under armour protection at the front of the vehicle.

A hull escape hatch is provided in each side of the hull, between the second and third axles, thus avoiding the crew being exposed to small arms fire if they had to leave via the turret hatches.

The commander is seated on the right of the turret with the gunner forward and below his position, and the loader on the left. On the prototypes the commander is provided with periscopes to give observation through 360°, but on cost grounds these are replaced by vision blocks on production vehicles. Mounted in the roof forward of the commander's station is his day panoramic sight with a magnification of × 12. This enables him to observe the terrain through a full 360° without moving his head. This sight can also be slaved to the 76 mm gun or uncoupled. The commander is also provided with a turret-position indicator.

The loader has a single-piece hatch that opens to the rear and is provided with two periscopes to his left side of the roof. These can be traversed to allow him to observe the terrain on his side of the vehicle.

The gunner has an Eloptro roof-mounted periscope which has ×8 day and night (image intensification) channels and an integral laser rangefinder. He also has an auxiliary sight as used in the Ratel 90 vehicle. If required the commander of the Rooikat can override the gunner.

Turret traverse and weapon elevation is solid state electric with controls for both gunner and commander, with manual controls for emergency use. Traverse is a full 360°, elevation limits being from -10° to +20°.

The digital Integrated Fire Control System for the Rooikat has been designed by E.S.D. and receives information from a laser rangefinder and a number of sensors including cant and windspeed. Ammunition type is set manually. Rooikat can engage and hit targets while it is moving across rough terrain with a high probability of first round hit.

Ballistic offsets are automatically computed and implemented when the laser rangefinder is triggered, to compensate for various factors affecting the ammunition ballistics. The offsets are implemented both in the sight (electronics sight slaving) and the main weapon (auto lay) and a ready to fire indication appears in the gunner's field advising him to fire. Total reaction time from lasing to firing is typically less than two seconds.

The fire control system provides automatic fire control by computing and implementing the ballistic offsets for the ammunition type selection by taking into account target range (automatic from laser rangefinder, battle range for close battle conditions or manually entered), target speed (automatically derived by tracking the target), manually entered environmental data (air pressure, ambient temperature, ammunition temperature, headwind), cross wind speed (automatically sensed), weapon tilt (cant automatically sensed), vehicle slope (manually entered) and gun jump characteristics of main weapon (manually entered).

The IFCS has a BITE (built in test equipment) which enables any faults to be immediately traced and rectified through the substitution of line-replaceable units.

Main armament comprises a 76 mm/62 calibre gun which was developed at LEW and retains the chamber volume as the Italian OTO Melara 76 mm Compact naval gun but fires different ammunition in keeping with its role. The ammunition of the naval gun can also be fired by replacing the mechanical primer by an electric primer. The 76 mm gun was chosen for the Rooikat for a number of other reasons including the larger number of rounds capable of being carried when compared to a 105 mm gun, the ease of handling ammunition when moving across country and the optimum selection of recoil forces and firing impulses for maximum crew operational effectiveness.

The 76 mm stabilised gun has a vertical sliding semi-automatic breech block and is fitted with a thermal sleeve and concentric fibre glass fume extractor. Recoil system consists of a concentric hydrospring with external replenisher with maximum recoil being 370 mm.

Two types of 76 mm ammunition have been developed, HE-T (high explosive tracer) and APFSDS-T (armour piercing fin stabilised discarding sabot-tracer) with a total of 48 rounds of ammunition being carried, nine of which are for ready use, stowed vertically below the turret ring on the left side. For safety reasons no main gun ammunition is stowed above the turret ring.

Rate of fire is stated to be 6 rds/min, with the APFSDS-T projectile having an effective range of 2000 to 3000 m. It will penetrate the T-54/T-55 and T-62 from all angles of attack. The HE-T round has a maximum range of 12 000 m in the indirect fire role and 3000 m in the direct fire role. A range drum is fitted as standard.

Basic details of the 76 mm ammunition is:

Type	**APFSDS-T**	**HE-T**
Weight complete round	9.2 kg	12.8 kg
Length	864 mm	908.5 mm
Muzzle velocity	1610 m/s	915 m/s

Rooikat showing 76 mm gun which is fitted with a thermal sleeve and a fume extractor. A 7.62 mm machine gun is mounted coaxial with the main armament and a similar weapon is mounted on the commander's cupola for local and air defence

The APFSDS-T round has a tungsten alloy penetrator with the sabot, fins and nose shield being of aluminium and the rotating obturator of nylon 66.

The HE-T round contains 0.6 kg of RDX/TNT on a 60/40 basis with the tracer being visible for 8 to 11 seconds. The projectile is fitted with a point detonating (PD) and selectable delay fuze which provides a super-quick characteristic of 170 µs or a delay action of between 30 and 70 µs, depending on the tactical situation. Other rounds available include canister and practice.

A 7.62 mm coaxial machine gun is mounted to the left of the main armament, while a similar weapon is mounted at the commander's position for use in the ground/ground and air defence roles. This is skate-mounted so that it can be moved around the commander's hatch. An ammunition resupply hatch is provided in the left side of the turret.

Mounted either side of the turret rear is a bank of four forward firing electrically operated 81 mm smoke dischargers. The Rooikat can also lay its own smoke screen by injecting diesel fuel into the exhaust outlet at the rear of the hull on the right side.

The powerpack consists of the V-10 cylinder water-cooled diesel developing 563 hp which gives a power-to-weight ratio of 21.07 hp/tonne. This is coupled to an automatic transmission with six forward and one reverse gears.

Power is transmitted from the engine to the transmission, then to the bevel box and on to the transfer box and then to the front and rear axles.

The complete powerpack, consisting of engine, transmission and cooling system has been designed as an integral unit and can be quickly removed for replacement in the field within 45 minutes.

The engine compartment is fitted with a fire detection and suppression system which operates automatically. Maximum road speed in 120 km/h, with cross-country speed being up to 60 km/h, acceleration from 0 to 30 km/h is six seconds and the 550 litre fuel tank gives a range of 1000 km depending on terrain.

The driver can select full 8 × 8 drive or 8 × 4 drive, depending on the tactical situation. Steering is power assisted on the front four wheels. Tyres are 14.00 × 20 radials of the run flat type. Rooikat is still mobile with two same-side wheels blown off.

Suspension at each wheel station consists of internally driven trailing arms, coil springs and hydraulic shock absorbers. Total wheel travel is 327 mm and the suspension has been designed for high speed cross-country operations. At present the vehicle is not fitted with a central tyre pressure regulation system or a winch for self-recovery.

Dual circuit drum-type brakes are fitted on all eight wheels. In addition there is a retarder on the transmission, exhaust brake, crawl brake and a parking brake.

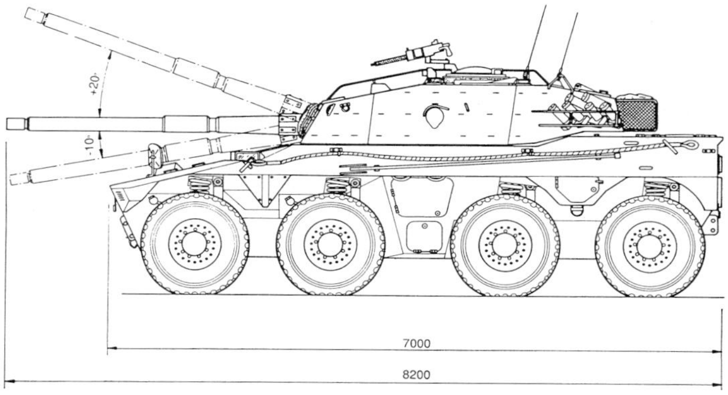

Side drawing of Rooikat (8 × 8) armoured car showing elevation/depression of 76 mm gun (not to 1/76th scale)

Standard equipment includes vehicle intercom, radios (which are mounted in the turret bustle) and a chemical and biological overpressure collective system.

Variants

It is very likely that the well chassis of the Rooikat will be used for a number of other applications including a self-propelled air defence system, surface-to-surface and surface-to-air missile systems and an armoured recovery vehicle.

Status: Production. In service with South African Armoured Corps.

Manufacturer: See text.

Enquiries to Armscor
Private Bag X337
Pretoria 0001
South Africa
Telephone: (012) 428-1911
Telex: 320217

SPECIFICATIONS	
CREW	4
CONFIGURATION	8 × 8
COMBAT WEIGHT	28 000 kg
POWER-TO-WEIGHT RATIO	21.07 hp/tonne
LENGTH INCLUDING GUN	8.2 m
LENGTH HULL	7 m
WIDTH	
overall	2.9 m
hull	2.8 m
HEIGHT	
overall	2.8 m
turret roof	2.5 m
hull top	1.76 m
GROUND CLEARANCE	0.4 m
TRACK	2.5 m
WHEELBASE	1.55 m + 2.032 m + 1.625 m
ANGLE OF APPROACH/ DEPARTURE	45°/60°
MAX ROAD SPEED	120 km/h
CROSS-COUNTRY SPEED	50 km/h
ACCELERATION	0 to 30 km/h in 6 s
FUEL CAPACITY	500 l
MAX ROAD RANGE	1000 km
FORDING	1.5 m
GRADIENT	70% (forward and reverse)
SIDE SLOPE	30%
VERTICAL OBSTACLE	1 m
TRENCH	
crawl speed	2 m
60 km/h	1 m
ENGINE	V-10 water-cooled diesel developing 563 hp
TRANSMISSION	automatic, 6 forward and 1 reverse gears
STEERING	power assisted
TURNING RADIUS	11.25 m
SUSPENSION	internally driven trailing arms, coil springs and shock absorbers
TYRES	14.00 × 20, radial, run flat
BRAKES	dual circuit drum on all 8 wheels, retarder on transmission, exhaust brake, crawl brake, parking brake
ELECTRICAL SYSTEM	24 V
BATTERIES	6
ARMAMENT	
main	1 × 76 mm gun
coaxial	1 × 7.62 mm MG
anti-aircraft	1 × 7.62 mm MG
SMOKE LAYING EQUIPMENT	8 × 81 mm smoke grenade launchers, exhaust smoke screen generator
AMMUNITION	
main	48
7.62 mm MG	3600
smoke grenades	16
GUN CONTROL EQUIPMENT	
Turret power control	electric/manual
by commander	yes
Turret traverse	360°
GUN ELEVATION/ DEPRESSION	+20°/-10°
GUN STABILISER	yes
LASER RANGEFINDER	yes
COMMANDER'S SIGHT MAGNIFICATION	×12
GUNNER'S SIGHT MAGNIFICATION	
day	×8
night	×8

Sandock-Austral Eland Light Armoured Car

Development/Description

In November 1961 South Africa concluded a contract with the French company, Panhard, for the supply of 100 AML (4 × 4) armoured cars armed with a 60 mm mortar and AA52 machine guns. An order was also placed for 150 GIAT 90 mm DEFA gun turrets and components for a further 500 units.

In January 1962 tenders were requested from 12 local South African companies for the manufacture of 350 Panhard AML armoured cars complete with turrets and 150 without turrets. Only four companies responded to the tender which was subsequently awarded to Sandock-Austral.

The original South African Defence Force order was for 600 AMLs of which 100 came direct from France and the remaining 500 were locally manufactured. Of the latter 350 were armed with 60 mm mortars and single or twin machine guns and the remaining 150 armed with a 90 mm gun.

The first locally manufactured armoured cars, called the Eland Mk 2, were delivered in July 1963 and had a local content of 40 per cent. Between eight and twelve units were delivered per month for the following 16 months. The Eland Mk 1 was the version imported from France.

The Mk 2, 3 and 4 versions of the Eland all had an imported AML engine which was air-cooled. The Mk 2 had an improved steering system, the Mk 3 had an improved brake system while the Mk 4 had an updated fuel system and electrical clutch. Although the Mk 2, 3 and 4 differed externally in minor details, they were different internally.

The Mk 5 was the first version with a locally built engine, this being a 2.5 litre water-cooled petrol engine which could be changed in the field in 90 minutes. In addition the Mk 5 also had locally developed hydraulic clutches, communications equipment, springs, shock absorbers, wheels and run flat tyres.

During 1972/1973 the assembly of 356 Eland Mk 5 armoured cars was completed and delivered at the rate of 24 vehicles per month. A programme was then completed during 1974/75 for the repair and conversion of earlier Elands. This programme included the complete rebuild of the vehicle, general repair work and overhauls which converted the vehicle to the Mk 6. This involved a total of 1016 conversions.

The latest version of the Eland is the Mk 7 which was introduced from 1979. This has new power brakes, increased storage capacity resulting from the slightly longer chassis, inclusion of a freshwater tank and a more comfortable driver's compartment. External modifications include tow bar at front, halogen lights on the mudguards and the commander's cupola with armoured glass windows which allows the commander to see through a full 360°. The Mk 7 is however slower than the Mk 1 and had a reduced range of action.

South Africa also made the complete 90 mm gun for the vehicle as well as 60 mm mortar, machine guns, optics and ammunition.

Production models of the Eland were armed with a 90 mm gun/7.62 mm coaxial and 7.62 mm anti-aircraft machine guns or a 60 mm mortar 7.62 mm machine gun. A model with a two-man turret armed with a 20 mm cannon and a coaxial 7.62 mm machine gun has been built but this has not entered production.

Ninety-five per cent of the Eland was produced in South Africa by the time production was completed. Total South African production of the Eland amounted to about 1300 vehicles. From 1990 the Eland started to be replaced by the Rooikat 76 mm (8 × 8) armoured car.

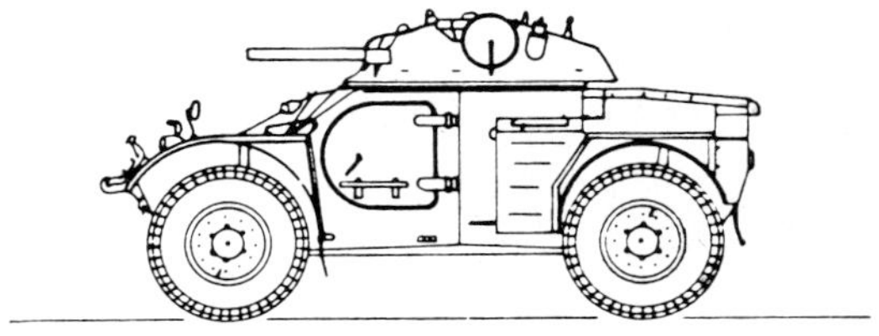

Eland light armoured car with 60 mm mortar turret

Eland 90 light armoured car from front showing extensive stowage

Status: Production completed, but can be resumed if further orders are received. In service with South Africa and Zimbabwe.

Manufacturer: Sandock-Austral Beperk Limited
West Street Industrial Sites
Boksburg
Transvaal
South Africa

Enquiries to:
Armscor
Private Bag X337
Pretoria 0001
South Africa

Telephone: (012) 292 9111
Telex: 320217

Eland 60 light armoured car with 60 mm mortar and 7.62 mm coaxial machine gun. Note mount on turret for 7.62 mm anti-aircraft machine gun (Christopher F Foss)

SPAIN

ENASA VEC Cavalry Scout Vehicle

Development

The VEC (Vehículo de Exploración de Caballeria), or cavalry scout vehicle, has been developed by ENASA to meet the requirements of the Spanish Army and shares many common components with the BMR-600 infantry fighting vehicle which is fully described in the *Armoured personnel carriers* section. The VEC is also known as the Pegaso VEC 3562. The first of five prototypes was completed in 1977/78. The total Spanish Army requirement is for 235 vehicles.

Description

The hull of the VEC is of all-welded aluminium construction providing protection over its frontal arc against 7.62 mm × 51 armour-piercing ammunition, and standard 7.62 mm × 51 rounds over the remainder of the vehicle. Some areas of the VEC have spaced armour protection.

The driver is seated at the front of the vehicle on the left side and has a single piece hatch cover that opens to the right, in front of which are three periscopes for forward observation. The centre periscope can be replaced by an infra-red or image intensification periscope for night driving.

The turret is in the centre of the vehicle, between the first and second axles. A variety of turret installations can be fitted including the same turret as fitted to the Italian FIAT-OTO Melara type 6616 (4 × 4) armoured car which is fully described in this section. In addition to these two weapon stations a variety of others ranging from 7.62 mm, 12.7 mm, 20 mm up to 90 mm in calibre can be installed. Spanish Army VECs are fitted with an Italian designed OTO Melara T 25 two-man turret manufactured under licence by SANTA BARBARA. It was expected that main armament would be an Oerlikon-Contraves 25 mm KBA cannon but early in 1987 it was announced that Spain had selected the McDonnell Douglas Helicopters M242 25 mm Bushmaster cannon as installed in the M2 Bradley; 208 weapons were ordered at a total cost of $7 million and all of these have now been delivered to Spain.

The engine compartment is to the rear of the vehicle on the left side with air inlet and outlet louvres in the roof, silencer on the left side of the hull and a hatch in the rear of the hull on the left side.

VEC Cavalry Scout Vehicle fitted with OTO Melara T 25 turret armed with 25 mm McDonnell Douglas Helicopters M242 Bushmaster cannon, 7.62 mm coaxial machine gun and electrically operated smoke dischargers

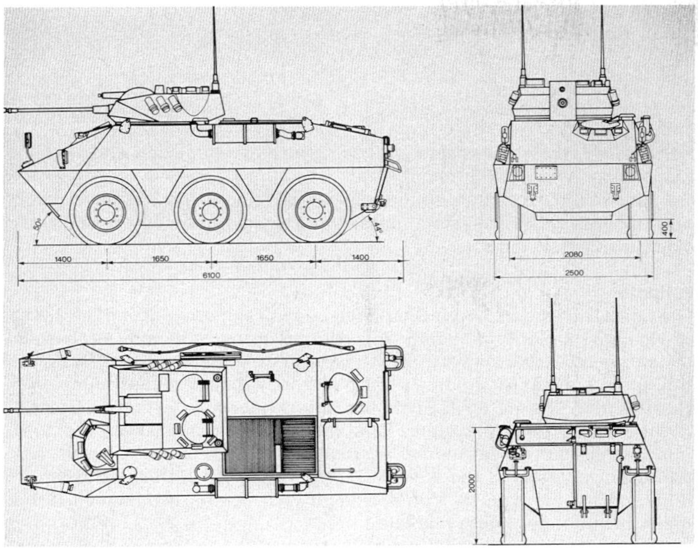

Four-view general arrangement of VEC Cavalry Scout Vehicle fitted with SANTA BARBARA built OTO Melara T 25 turret (not to 1/76th scale)

VEC Cavalry Scout Vehicle from left rear with 25 mm cannon and 7.62 mm machine gun elevated

In the hull to the rear of the turret on the right side is a seat for the fourth crew member and above this is a single piece hatch cover opening to the right with a periscope in the hull roof. The fifth crew member is seated at the very rear on the right side and has a single piece hatch cover that opens to the rear and three periscopes for observation. There is also a downward opening in the hull rear on the right side.

All three axles are of Pegaso design and manufacture and are of the double reduction type with self-locking differentials. Suspension is independent for each wheel with variable flexibility and 275 mm suspension travel achieved by an oleo-pneumatic system with MacPherson arrangement. Steering is power assisted on the front and rear axles.

Standard equipment includes tool set, 4500 kg winch, semi-automatic fire-extinguishing system protecting the engine and transmission and two 5 kg portable fire extinguishers.

Optional equipment includes an NBC system, water-jet propulsion system (one unit either side of the hull at the rear and a trim vane that folds back onto the glacis plate), 4500 kg winch, communications system, land navigation system, 14.00 × 20 tyres, run-flat tyres and an air-conditioning system.

Variants

The VEC can also be used in the anti-aircraft role (guns or missiles) and the anti-tank role with ATGWs. For trials it has also been fitted with a French Hispano-Suiza 90 mm two-man turret.

Status: Production. In service with Spanish Army.

VEC Cavalry Scout Vehicle with two-man T 25 turret showing amphibious characteristics with trim vane erected at front of hull and commander's, gunner's and driver's hatches open

Manufacturer: ENASA, Military Division
Av Aragón
402 Madrid 22
Spain
Telephone: (1) 750 10 00
Telex: 42 372
Fax: (1) 747 47 32

SPECIFICATIONS	
CREW	3 + 2
CONFIGURATION	6 × 6
COMBAT WEIGHT	13 750 kg
LOADED WEIGHT	
front axle	5000 kg
centre axle	4450 kg
rear axle	4300 kg
POWER-TO-WEIGHT RATIO	22.25 hp/tonne
LENGTH	6.10 m
WIDTH	2.5 m
HEIGHT TO HULL TOP	2 m
GROUND CLEARANCE	0.4 m
TRACK	2.08 m
WHEELBASE	1.65 m + 1.65 m
ANGLE OF APPROACH/ DEPARTURE	50°/45°
MAX ROAD SPEED	
1st gear position, 1 gear	17 km/h
2nd gear position, 1, 2, 3, 4 gear	56 km/h
3rd gear position, 1, 2, 3, 4, 5 gear	69 km/h
4th gear position, 2, 3, 4, 5, 6 gear	103 km/h
reverse	12 km/h
MAX WATER SPEED	
wheels	3 km/h
waterjets	9 km/h
FUEL CAPACITY	400 l
MAX ROAD RANGE	800 km
GRADIENT	60%
SIDE SLOPE	30%
VERTICAL OBSTACLE	0.6 m
TRENCH	1.5 m
TURNING RADIUS	7.5 m
ENGINE	Pegaso model 9157/8 6-cylinder in-line turbo-charged water cooled diesel developing 310 hp at 2200 rpm
TRANSMISSION	ZF 6 HP 500, fully automatic, 6 forward and 1 reverse gears
TORQUE CONVERTER AND HYDRAULIC RETARDER	ZF
STEERING	hydraulic, power-assisted on first and third axles
TYRES	13.00 × 20 (or 14.00 × 20)
SUSPENSION	independent each wheel, MacPherson type, hydro-pneumatic
BRAKES	
main	disc, hydraulic, dual circuit, air assisted on all wheels
parking	disc, operating on propeller transmission
emergency	hydraulic, air operated MGM chambers
ELECTRICAL SYSTEM	24 V
BATTERIES	2 × 12 V, 140 Ah
ARMAMENT	
main	1 × 25 mm cannon
coaxial	1 × 7.62 mm MG
SMOKE-LAYING EQUIPMENT	2 × 3 smoke dischargers
AMMUNITION (ready use)	
main	170 (135 HE + 35 APDS)
coaxial	250
smoke grenades	6
GUN CONTROL EQUIPMENT	
Turret power control	360° powered/manual
Elevation/depression	+50°/-10°

(NB: OTO Melara turret designation is the T 25 with Spanish designation being the CETME TC-15/M242 turret)

Note: The maximum road speed of 103 km/h is with 14.00 × 20 tyres. When fitted with 13.00 × 20 tyres maximum road speed is 96 km/h

SWITZERLAND

MOWAG Shark Weapons Carrier

Development

The Shark has been developed by the MOWAG company as a private venture specifically for the export market. It was first shown in public at the 1981 Paris Air Show fitted with the Oerlikon-Contraves Type GDD-BOE two-man turret armed with a 35 mm cannon and a coaxial 7.62 mm machine gun. The second prototype was completed in 1983 and the third in 1984.

In 1982 the Shark was successfully tested in Switzerland fitted with a French Fives-Cail Babcock FL-12 turret armed with a 105 mm gun and fitted with a SOPTAC fire control system. In Germany it was successfully tested with an experimental Rheinmetall turret armed with the Rheinmetall 105 mm Rh 105-11 super low recoil gun which fires all standard NATO tank ammunition, including APFSDS-T. It has also been fitted with the Fives-Cail Babcock 105 mm armed FL-20 two man turret, Bofors prototype 40 mm Trinity L/70 anti-aircraft gun system and air defence missile systems such as the Thomson-CSF Crotale and Oerlikon-Contraves ADATS. All of these have been used for trials purposes.

Wherever possible proven assemblies, such as engine and transmission, have been used in the design of the vehicle to simplify both training and logistics.

Description

The hull of the MOWAG Shark is of all-welded steel construction which provides complete protection up to and including the Soviet 14.5 mm KPV armour-piercing round. A 155 mm HE projectile landing 10 m away from the vehicle will cause no damage according to the manufacturer.

The driver is seated at the front of the hull on the left side and has a single piece hatch cover that opens to the left. Forward of this are three periscopes, the centre one of which can be replaced by an image intensification periscope for driving at night. A second crew member can be seated to the right of the driver or this space can be used for special equipment such as an NBC pack or for additional ammunition stowage.

The fighting compartment is in the centre of the vehicle and a wide range of armament installations can be fitted up to and including an FL-20 turret with a 105 mm gun.

The engine, transmission and fuel tanks are at the rear of the hull and engine access panels are provided in the top of the engine compartment. The air inlet louvres are in the top of the hull, one either side, the air outlets at the hull rear.

Steering is power assisted on the front and rear axles and vertical suspension travel is 420 mm.

A wide range of optional equipment can be fitted to the Shark including various types of NBC system, passive night vision equipment and fire-extinguishing systems.

Variants

The chassis of the MOWAG Shark has been designed to accept a variety of weapon stations including anti-aircraft guns, anti-aircraft missile systems (such as the Oerlikon-Contraves Air Defence Anti-Tank System), anti-tank systems (including ATGWs) and multiple rocket launchers. More recently the MOWAG Shark has been fitted with the French Crotale surface-to-air missile system. One vehicle has four missiles ready to launch and another carries the acquisition unit. The Shark chassis was also used for trials of the 40 mm Bofors Trinity Air Defence System in 1985 and for the Krauss-Maffei Wildcat twin 30 mm air defence turret which is no longer being marketed.

MOWAG (8 × 8) Shark weapons carrier fitted with Bofors Trinity 40 mm L/70 air defence module during trials in 1985

MOWAG Shark (8 × 8) with 105 mm FL-20 turret

MOWAG (8 × 8) Shark weapons carrier fitted with trials turret armed with 105 mm Rheinmetall Rh 105-11 super low recoil gun fitted with muzzle brake

SPECIFICATIONS

CREW	3 or 4
CONFIGURATION	8 × 8
WEIGHT	
loaded	22 000 kg*
empty	16 000 kg*
POWER-TO-WEIGHT RATIO	24 hp/tonne
LENGTH	7.52 m
WIDTH	3 m
HULL HEIGHT	1.9 m
AXIS OF FIRE (105 mm gun)	2.16 m
GROUND CLEARANCE	0.46 m
TRACK	2.62 m
WHEELBASE	1.51 m + 1.4 m + 1.49 m
ANGLE OF APPROACH/DEPARTURE	40°/45°
MAX ROAD SPEED	100 km/h

FUEL CAPACITY	400 l
ROAD RANGE	500 km
FORDING	1.3 m
GRADIENT	60%
SIDE SLOPE	35%
VERTICAL OBSTACLE	0.46 m
TRENCH	2.3 m
TURNING CIRCLE	12.5 m
ENGINE	Detroit Diesel 8V-71T turbo-charged diesel developing 530 hp at 2500 rpm
TRANSMISSION	Allison Transmission Division of General Motors HT 750 DRD automatic with 5 forward and 1 reverse gears
TYRES	13.00 × 20 run-flat (insert)

SUSPENSION	
1st and 4th axles	coil springs with wishbone
2nd and 3rd axles	torsion bars, all wheels have hydraulic shock absorbers
BRAKES	
main	dual circuit, hydraulic, air assisted
parking	spring loaded brake cylinder
engine	Jacobs brake system
ELECTRICAL SYSTEM	24 V

* depends on armament installation

Status: Development of the first generation MOWAG Shark was completed in the early 1980s but production never commenced. A second generation MOWAG Shark is now under development with a significantly improved performance.

Manufacturer: MOWAG Motorwagenfabrik AG
8280 Kreuzlingen
Switzerland
Telephone: (72) 71 15 15
Telex: 882 211
Fax: (72) 72 28 86

MOWAG SPY Reconnaissance Vehicle

Development

The SPY reconnaissance vehicle was developed as a private venture by the MOWAG company specifically for the export market; the first prototypes were completed in 1980. Drive train components, pressure, hydraulic and electrical components of the SPY are also used in the MOWAG Piranha range of 4 × 4, 6 × 6 and 8 × 8 armoured vehicles, which are fully described in the *Armoured personnel carriers* section.

In 1982 it was announced that an undisclosed Far Eastern country had placed an order for between 20 and 50 SPYs fitted with a MOWAG designed one-man turret armed with a 12.7 mm and a 7.62 mm machine gun.

Description

The hull of the SPY is made of all-welded steel and gives complete protection against small arms fire up to 7.62 mm AP calibre over the front, sides and rear, against anti-personnel mines exploding under the hull and against shell splinters above the hull.

The driver is seated at the front of the hull on the left with the commander immediately behind him. The driver has a single piece hatch cover and three periscopes for forward observation, the centre one of which can be replaced by an infra-red or image intensification periscope for night driving. The commander has a cupola mounted on a plinth above the roof to give observation over the driver. The cupola has a single piece hatch cover and a periscope. In the side of the hull behind the commander is a small door that opens forwards.

The engine compartment is to the right of the driver with the air inlet and air outlets in the roof and the exhaust pipe alongside the right side of the hull.

A variety of armament installations can be mounted at the rear of the hull including the MOWAG Type V 041 and Type V 042 turrets. The V 042 is armed with a remote-controlled 12.7 mm M2 HB machine gun with an elevation of +45°, depression of -8° and turret traverse of 360°. Turret traverse and weapon elevation are manual, and 100 rounds of ready use ammunition are provided. The gunner aims the machine gun with a roof-mounted PERI Z 12 sight with a magnification of ×1 and either side of this is a periscope for lateral observation.

The Type V 041 turret is armed with a 12.7 mm and a 7.62 mm machine gun with an elevation of +50°, depression of -10° and turret traverse of 360°. Turret traverse and weapon elevation are manual and 200 rounds of 12.7 mm and 500 rounds of 7.62 mm ready use ammunition are carried. The gunner aims the weapons via a PERI Z 11 sight with a magnification of ×2 and ×6 and also has seven periscopes for all-round vision.

All four wheels have independent suspension and the front axle can be disengaged for road driving. Automatic locking differentials and run-flat core tyres are fitted as standard.

Optional equipment includes various radio installations and an air-conditioning system.

Status: Production as required. In service with an undisclosed country.

Manufacturer: MOWAG Motorwagenfabrik AG,
8280 Kreuzlingen,
Switzerland
Telephone: (72) 71 15 15
Telex: 882 211
Fax: (72) 72 28 86

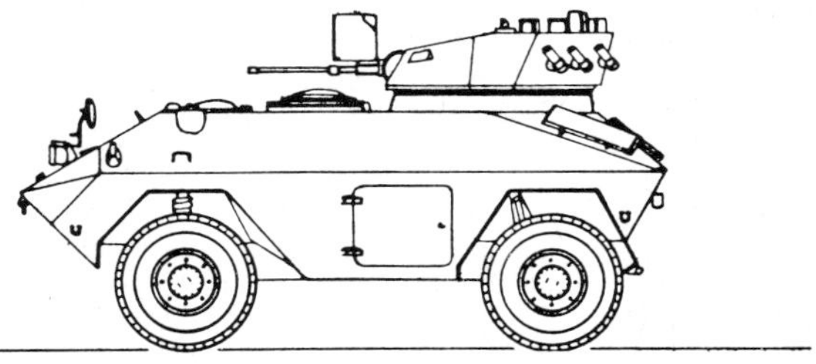

MOWAG SPY reconnaissance vehicle fitted with Type V 041 one-man turret armed with 12.7 mm and 7.62 mm machine guns

MOWAG SPY reconnaissance vehicle fitted with MOWAG-designed turret Type V 042 with remote-controlled 12.7 mm machine gun and three smoke dischargers either side

SPECIFICATIONS	
CREW	3
CONFIGURATION	4 × 4
COMBAT WEIGHT	7500 kg
UNLOADED WEIGHT	5900 kg
POWER-TO-WEIGHT RATIO	27 hp/tonne
LENGTH	4.52 m
WIDTH	2.5 m
HULL HEIGHT	1.66 m
GROUND CLEARANCE	
hull	0.5 m
axles	0.39 m
TRACK	
front	2.18 m
rear	2.205 m
WHEELBASE	2.5 m
ANGLE OF APPROACH/ DEPARTURE	50°/70°
MAX ROAD SPEED	110 km/h
FUEL CAPACITY	200 l
ROAD RANGE	700 km
FORDING	1.3 m
GRADIENT	60%
SIDE SLOPE	35%
VERTICAL OBSTACLE	0.5 m
TURNING RADIUS	6.3 m
ENGINE	Detroit Diesel V-8 developing 205 hp at 3000 rpm
TRANSMISSION	Allison Transmission Division of General Motors AT-540 automatic, 4 forward and 1 reverse gears
TRANSFER CASE	2-speed
SUSPENSION	
front	coil spring and wishbone
rear	torsion bars all wheels have hydraulic shock absorbers
TYRES	11.00 × 16
BRAKES	dual circuit, hydraulic, air-assisted
ELECTRICAL SYSTEM	24 V
ARMAMENT	
main	1 × 12.7 mm MG
coaxial	1 × 7.62 mm MG
smoke dischargers	2 × 3
AMMUNITION	
12.7 mm	3000
7.62 mm	2000
smoke grenades	20
GUN CONTROL EQUIPMENT	
Turret traverse	360°/manual
Gun elevation/ depression	+50°/-10°
GUN STABILISER	
vertical	no
horizontal	no

UNION OF SOVIET SOCIALIST REPUBLICS

BRDM-2 Amphibious Scout Car

Development

The BRDM-2 was designed by V K Rubtsov of the Dedkov OKB in the early 1960s as the successor to the earlier BRDM-1 and was first seen in public during 1966. Main improvements over the earlier vehicles are its fully enclosed armament installation, more powerful rear-mounted engine, which gives it a higher road and cross-country performance, and better water performance.

Each Soviet motorised rifle division has a total of 28 BRDM-2s, 12 in the reconnaissance battalion and four each in the tank regiment, BMP-equipped motorised rifle regiment and in each of the two BTR-60BTR-70-equipped motorised rifle regiments. Each tank division has a total of 28 BRDM-2s, 12 in the reconnaissance battalion, four in the motorised rifle regiment and four in each of the three tank regiments.

Description

The all-welded steel armour hull of the BRDM-2 provides the crew with protection from small arms fire and shell splinters.

The driver is seated at the front of the hull on the left with the vehicle commander to his right. Both are provided with a bullet-proof windscreen to their front which is covered by an armoured shutter, hinged at the top, when the vehicle is in combat areas. When the shutters are in position the driver and commander observe the terrain through periscopes around the front and sides of their position, mounted level with the roof of the vehicle. The only means of entry to the BRDM-2 is by two circular roof hatches immediately behind the commander and driver, hinged towards the centre of the vehicle and opening only as far as the vertical.

The turret is mounted in the centre of the vehicle and is the same as that mounted on the Soviet BTR-60PB and Czechoslovak OT-64 SKOT-2A armoured personnel carriers.

There is a single firing port in each side of the hull and immediately behind it are three vision blocks which protrude from the outside of the hull to give some vision to the front and rear of the vehicle.

The engine compartment is at the rear of the hull and there are two air-inlet louvres in the forward part of the engine compartment roof and four smaller air-inlet louvres to the rear. The exhaust pipes are on either side of the hull. The engine compartment is separated from the crew compartment by a bulkhead. The complete drive train of the BRDM-2 comprises the engine, single plate clutch, manual gearbox, reduction gearbox and transfer case which in turn provides power to the front and rear wheels, each of which have a self-locking differential. The manual transmission is hydraulically assisted with the 3rd and 4th gears being synchronised with lockouts in 1st, 2nd and reverse gears.

The suspension consists of four semi-elliptical springs with telescopic dual-action shock absorbers mounted two per axle. Steering is hydraulically assisted on the front two wheels with the sealed brakes having air-assisted hydraulic actuators.

On each side of the vehicle, between the front and rear wheels are two chain driven belly wheels, which are lowered by the driver and give the BRDM-2 improved cross-country performance and also allow it to cross ditches. A central tyre pressure regulation system allows the driver to adjust the tyre pressure to suit the type of ground being crossed. The driver can adjust the pressure on all four tyres or on individual tyres while the vehicle is still moving. If one of the tyres is damaged, the air compressor can maintain the tyre pressure until the vehicle is out of the combat area.

The BRDM-2 is fully amphibious being propelled in the water by a single water jet at the rear of the hull. Before entering the water a trim vane which is stowed under the nose of the hull when travelling is erected at the front of the hull. When not in use the water jet outlet is covered

BRDM-2 with commander's and driver's hatches open and trim vane folded back under nose

BRDM-2 reconnaissance vehicle with commander's hatch open

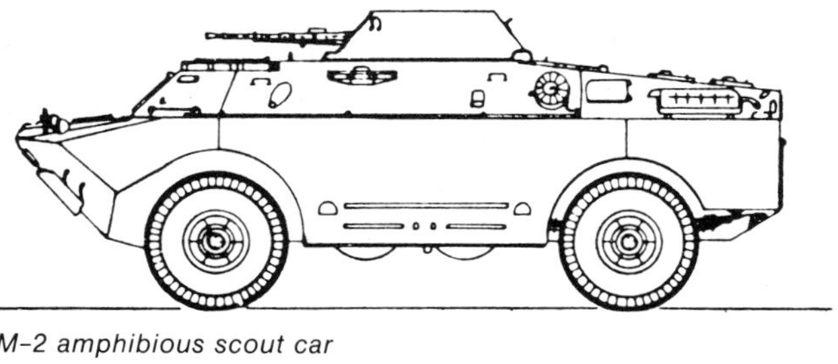

BRDM-2 amphibious scout car

by a triangular plate pivoted at the top. The hydraulic control that activates both the hydrojet drive and the trim vane is located to the front of the driver. When afloat the driver uses the steering wheel as on land with the water rudders being connected to the steering wheel. At a speed of 6 to 7 km/h, its turning radius is about 10 m.

The vehicle is fitted with an over pressure NBC system and its air inlet is on the top of the hull to the left rear of the turret. Standard equipment includes infra-red driving lights, an infra-red searchlight mounted over the commander's position which can be operated from within the vehicle, a winch mounted internally at the front of the hull, a radio and a TNA-2 land navigation system. It also has a DP-3B roentgenometer chemical detector, VPKhR troop reconnaissance device that warns the crew of chemical agents and a heater that provides hot air to the driver's windows so that they do not become frozen in very cold weather. The winch is provided with 30 m of cable and has a maximum load of 4000 kg.

Armament consists of a 14.5 mm KPVT machine gun with a 7.62 mm PKT machine gun mounted coaxially to the right. A telescopic sight is mounted to the left of the main armament. The turret has manual traverse and is provided with an adjustable seat for the gunner.

Variants

BRDM-2-RKhb (Radiobiological-Chemical-Biological Reconnaissance Vehicle)

This is easily distinguishable from the basic vehicle by two rectangular racks which contain lane-marking poles with pennants on either side of the hull at the rear. The racks are normally covered with a canvas and when required are swung through 90° so that they are vertical over the rear of the vehicle. The poles are used to mark lanes through contaminated areas. The BRDM-2-RKhb is normally issued on the scale of 29 per motorised rifle division, nine in the chemical defence battalion, four in the reconnaissance battalion, four in the single tank regiment, four in the BMD-equipped motorised rifle regiment and four in each of the two motorised rifle regiments equipped with BTR-60BTR-70 APCs. Each tank division has a similar number, but four in the BMP-equipped motorised rifle regiment and four in each of the tank regiments. There are two versions of this vehicle. The original RKha has the standard 14.5 mm KPVT heavy machine gun while the more usual RKhb has twin 7.62 mm machine guns and improved sensors.

BRDM-1U Command Vehicle

Basic BRDM-2 that retains its turret but is fitted with additional communications equipment. Has a total of four radio antennas, one on each side of the hull and two at the rear.

BRDM-2U Command Vehicle

This has its turret removed and replaced by a hatch that opens forwards. Mounted to the rear of this hatch is a generator used to power the additional communications equipment carried inside the vehicle. Two radio antennas are mounted one each side of the hull.

The BRDM-2U command vehicle is used by artillery units as well as the battalion and regimental HQs of manoeuvre and reconnaissance units.

Iraqi Upgraded BRDM-2

Shown for the first time in Baghdad in 1989 was an upgraded BRDM-2 armoured car. The original 14.5 mm KPVT heavy machine gun was replaced by a 23 mm cannon which has a significant increase in range and armour penetration characteristics over the original weapon. In addition, the elevation of the mount, originally limited to +30° has been increased to +60° to allow the vehicle to engage slow moving aircraft and helicopters with depression of the weapons being -5°. A total of 500 rounds of 23 mm and 2000 rounds of 7.62 mm ammunition are carried. As far as it is known, this upgraded BRDM-2 vehicle did not enter service with the Iraqi Army.

BRDM-2 with Sagger ATGWs

This was first used in combat by the Egyptian and Syrian forces during the 1973 Middle East campaigns. It is a BRDM-2 with its turret removed and fitted with an arm on top of which are mounted six Sagger AT-3 ATGWs. When travelling they are within the hull but in action the arm is raised above the top of the hull complete with overhead armour protection. The missiles can be launched from within the vehicle or up to 80 m away with the aid of a separation sight. When launched from within the vehicle they are guided by the missile operator/vehicle commander who is seated at the front of the vehicle on the right side and is provided with a sight in the forward part of the hull. A total of eight missiles is carried in reserve and the radio antenna is at the rear on the right side rather than at the front. Full details of the missile are given in the following entry on the BRDM-1 vehicle. In 1977 it was reported that the Sagger was being converted from its normal wire command guidance system to a semi-automatic infra-red wire guidance system, which would increase the effectiveness of the missile at its maximum range of 3000 m.

BRDM-2-RKhb (Radiobiological-chemical-biological) reconnaissance vehicle in operating configuration

BRDM-2-RKhb (Radiobiological-chemical-biological) reconnaissance vehicle in travelling configuration

Iraqi BRDM-2 with two AT-4 (left) and three AT-5 Spandrel (right) ATGWs in the ready to launch position (Christopher F Foss)

Iraqi upgraded BRDM-2 with 23 mm cannon replacing 14.5 mm KPVT heavy machine gun (Christopher F Foss)

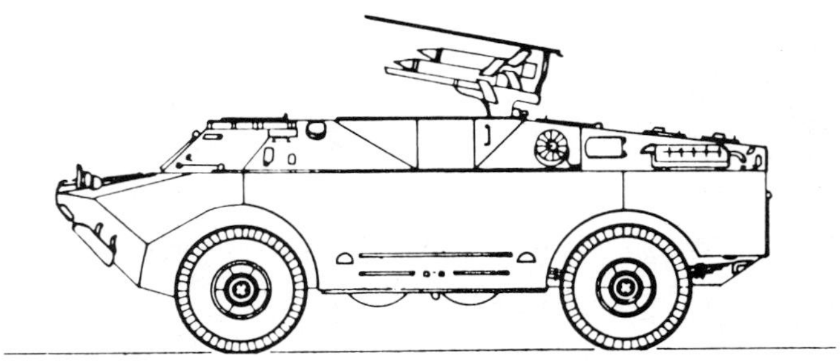

BRDM-2 with Sagger ATGWs with missiles raised ready to launch

BRDM-2 with launcher for Sagger ATGWs raised (Mainly Military)

BRDM-2 with four ready to launch Swatter-C ATGWs

Each Soviet Motorised Rifle Division has 36 BRDM-2 vehicles armed with ATGWs, nine in the anti-tank battalion, nine in the BMP-equipped motorised rifle regiment and nine in each of the two motorised rifle regiments equipped with the BTR-60/BTR-70 series APCs. Each Soviet tank division has nine BRDM-2s armed with ATGWs in the BMP-equipped motorised rifle regiments. This version is probably no longer in front line service with the Soviet Army.

BRDM-2 with Swatter-C ATGWs

This was first seen in 1973 and is similar in appearance to the BRDM-2 with the Sagger ATGW but has a quadruple launcher for the AT-2 ATGW converted from its original radio command-to-line-of-sight guidance to semi-active infra-red/command guidance. This missile weighs 29.48 kg and has a range of 3500 m compared with the 3000 m of the original Swatter. A total of eight Swatters is carried including the three in the ready to launch position. As far as is known the improved Swatter has not been fitted to the original BRDM-1 installation. This version is probably no longer in front line service with the Soviet Army.

BRDM-2 with Spandrel ATGWs

Shown for the first time at the military parade held in Moscow in November 1977 was a BRDM-2 fitted with a turret on top of which were mounted five new ready to launch missiles which have been given the NATO code name of Spandrel, or AT-5. This vehicle is sometimes referred to as the BRDM-3. To the rear of the turret in the top of the hull is a hatch which enables the launcher to be reloaded without the crew leaving the vehicle. It is estimated that in addition to the five visible missiles a further 10 are carried inside the hull. It is the second-generation type (semi-automatic, command-to-line-of-sight) and all that the operator has to do is to keep the cross-hairs of his sight on the target to ensure a hit. The sight is in the roof of the vehicle on the right side and can be traversed through 180°.

The HEAT warhead weighs about 7 kg and will penetrate between 500 mm and 600 mm of armour at an incidence of 0°. The launch tube has a blow-out cap at the front and is flared at the rear. It is thought that a bowed hatch in the roof of the vehicle to the rear of the launcher is probably used to allow the mount to be folded into the hull for reloading. Basic specifications of the AT-5 Spandrel ATGW are as follows:

LENGTH	1.3 m
DIAMETER	155 mm
MAX RANGE	4000 m
MIN RANGE	100 m
MAX FLIGHT TIME	about 20 s

BRDM-2 with five Spandrel ATGWs in ready to launch position

SA-9 (Gaskin) surface-to-air missile system with missiles elevated

Some BRDM-2 Spandrel ATGW vehicles, for example those used by Iraq, the outer launch position has been fitted with the AT-4 Spigot ATGW. This arrangement allows the carriage of an increased missile load, believed to be six AT-5 Spandrels plus eight AT-4 Spigots rather than 10 AT-5s.

SA-9 (Gaskin) System

The SA-9 (Gaskin) mobile surface-to-air missile system is based on a modified BRDM-2 (4 × 4) amphibious chassis with its belly wheels removed. The original turret has been replaced by a new one man turret with an elevating arm on either side of which are mounted two box-type launchers for the SA-9 fire-and-forget missiles.

The SA-9 first entered service with the Soviet Army in 1968 and has since been widely deployed. In the Soviet Army and some other countries it is being replaced by the SA-13 Gopher SAM system on a modified MT-LB chassis.

The SA-9 system has a maximum range of 7000 m with altitude limits being 900 to 5000 m. It is normally deployed in batteries of four systems with one of the vehicles being fitted with the Flat Box A passive radar detection system.

To reduce the overall height of the system for travelling, the missiles are normally lowered into the horizontal position either side of the vehicle. No reserve missiles are carried on the SA-9 system.

Full details of the SA-9 Gaskin are given in *Jane's Land-Based Air Defence 1991-92* page 131.

Status: Production complete. In service with Afghanistan, Afghanistan guerrillas, Algeria (including Sagger and SA-9), Angola (including SA-9), Benin (including SA-9), Botswana, Bulgaria, Cape Verde Islands, Central African Republic, Chad (status uncertain), Congo, Cuba (including SA-9), Czechoslovakia (S-9 and ATGW versions only), Djibouti, Egypt (including Sagger and SA-9), Equatorial Guinea, Ethiopia (including Sagger and Spandrel), Guinea (including SA-9), Guinea-Bissau (including SA-9), Hungary (Sagger and SA-9 only), India (including SA-9), Iraq (including Sagger and SA-9), Israel, Libya (including Sagger and SA-9), Madagascar (including SA-9), Malawi, Mali (including SA-9), Mauritania (including SA-9), Mongolia, Morocco (including Sagger, status uncertain), Mozambique (including SA-9), Nicaragua (including Sagger and SA-9), Peru, Poland (including Sagger and SA-9), Romania (including Sagger and SA-9), Sao Tomé & Principe, Seychelles, Somalia, Sudan, Syria (including Sagger and SA-9), Tanzania (including SA-9), USSR (including all variants), Vietnam (including SA-9), Yemen (including SA-9), Yugoslavia (Sagger and SA-9), Zambia and Zimbabwe.

Manufacturer: Molotov GAZ plant
Gorky
USSR

SPECIFICATIONS

Model	BRDM-2	BRDM-2 (Sagger)
CREW	4	2-3
CONFIGURATION	4 × 4	4 × 4
COMBAT WEIGHT	7000 kg	7000 kg
POWER-TO-WEIGHT RATIO	20 hp/tonne	20 hp/tonne
LENGTH	5.75 m	5.75 m
WIDTH	2.35 m	2.35 m
HEIGHT OVERALL	2.31 m	2.01 m
FIRING HEIGHT	2.13 m	n/app
GROUND CLEARANCE	0.43 m	0.43 m
TRACK	1.84 m	1.84 m
WHEELBASE	3.1 m	3.1 m
MAX SPEED		
road	100 km/h	100 km/h
water	10 km/h	10 km/h
FUEL CAPACITY	290 l	290 l
MAX ROAD RANGE	750 km	750 km
ROAD FUEL CONSUMPTION	0.35 - 0.45 l/km	0.35 - 0.45 l/km
FORDING	amphibious	amphibious
GRADIENT	60%	60%
VERTICAL OBSTACLE	0.4 m	0.4 m
TRENCH	1.25 m	1.25 m
ENGINE	GAZ-41 V-8 water-cooled petrol developing 140 hp at 3400 rpm	
TRANSMISSION	both have manual with 4 forward and 1 reverse gears	
SUSPENSION	semi-elliptical springs with hydraulic shock absorbers	

Model	BRDM-2	BRDM-2 (Sagger)
TYRES	13.00 × 18	13.00 × 18
ELECTRICAL SYSTEM	24 V	24 V
TURNING RADIUS	9 m	9 m
ARMAMENT		
main	1 × 14.5 mm MG	6 Sagger ATGWs
coaxial	1 × 7.62 mm MG	n/app
AMMUNITION		
main	500	14 Saggers (total)
coaxial	2000	n/app
GUN CONTROL EQUIPMENT		
Turret power control	manual	n/app
Gun elevation/depression	+30°/-5°	n/app
ARMOUR (max)		
Hull front upper	5 mm at 80°	5 mm at 80°
Hull front lower	7 mm at 45°	7 mm at 45°
Hull nose plate	14 mm at 14°	14 mm at 14°
Hull sides upper	7 mm at 18°	7 mm at 18°
Hull sides lower	7 mm at 0°	7 mm at 0°
Hull rear upper	7 mm at 0°	7 mm at 0°
Hull rear lower	7 mm at 0°	7 mm at 0°
Hull top	7 mm at 0°	7 mm at 0°
Belly front	2 mm	2 mm
Belly rear	3 mm	3 mm
Turret front	7 mm at 43°	n/app
Turret sides	7 mm at 36°	n/app
Turret rear	7 mm at 36°	n/app
Turret top	7 mm	n/app

BRDM-1 Amphibious Scout Car

Development

The BRDM-1 amphibious scout car first appeared in the Spring of 1959 and was designed by VK Rubtsov of the Dedkov OKB who later designed the BRDM-2. The first prototype was completed in February 1956 and went to the Black Sea area for official government trials. It was accepted for service with the Soviet Army the following year and production began late in 1957. Czechoslovakia and Poland also use the Hungarian-designed FUG which is designed to fulfil a similar role to the BRDM-1 but has its engine at the rear rather than the front as in the case of the BRDM-1. In Soviet front-line units the BRDM-1 has now been replaced by the BRDM-2 which has a higher road speed and is fitted with a turret armed with a 14.5 mm and a 7.62 mm machine gun rather than weapons on unprotected mounts as fitted to the BRDM-1.

The BRDM Model 1957 has an open roof, the BRDM-1 Model 1958 has a closed roof with twin hatches, this being the standard production model.

Total production of the BRDM-1 is believed to have amounted to around 10 000 vehicles.

Description

The hull of the BRDM-1 is made of all-welded steel with the engine at the front and the crew compartment at the rear. The driver is seated at the front of the vehicle on the left with the commander to his right. Both are provided with a hatch to their front that is hinged at the top and opens forwards, and a windscreen with a wiper blade. The hatch cover has a vision block for when the hatch is closed. A vision slit protected by a vision block is provided to the left of the driver and right of the commander. There are two firing ports in each side of the hull and two large hatches in the forward part of the roof that open to the rear. The rear of the crew compartment slopes at an angle of about 30° and is fitted with a two-piece hatch that opens either side of the superstructure. There is a firing port in each hatch.

On either side of the vehicle, between the front and rear wheels, are two belly wheels, which are lowered by the driver and give the BRDM-1 improved cross-country performance and also allow it to cross ditches. The BRDM-1 is fully amphibious, being propelled in the water by a single water jet at the rear of the hull. Before entering the water a trim board, which is stowed folded under the nose of the vehicle when not in use, is erected at the front of the hull. Standard equipment includes a system that allows the driver to adjust the tyre pressures to suit the type of ground being crossed. The vehicle is not fitted with an NBC system and when introduced was not fitted with any night vision equipment. A white light searchlight mounted on the left side of the hull is operated by the driver. A communications reel and wire is mounted on the left side of the superstructure towards the rear.

The vehicle is normally armed with a 7.62 mm SGMB machine gun pintle-mounted on the forward part of the roof, which has a limited traverse of 45° left and right and can be elevated from -6 to +23.5°. Some vehicles are fitted with a 12.7 mm DShKM heavy machine gun mounted at the forward part of the roof with a 7.62 mm SGMB machine gun being mounted at the rear.

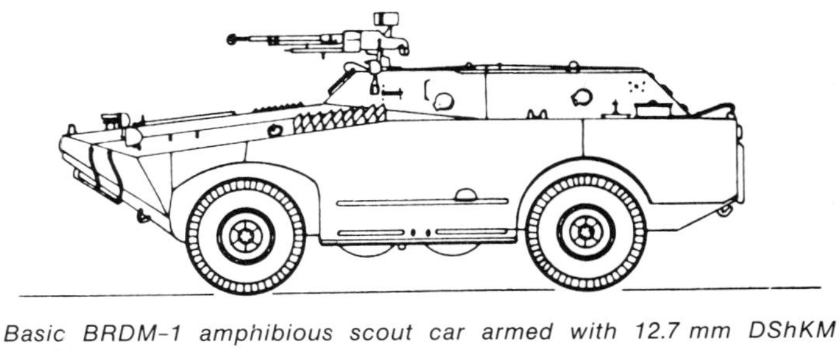

Basic BRDM-1 amphibious scout car armed with 12.7 mm DShKM machine gun

BRDM-1 command vehicle with antenna erected

BRDM-1 amphibious scout car armed with 12.7 mm DShKM heavy machine gun forwards and 7.62 mm SGMB machine gun at left rear

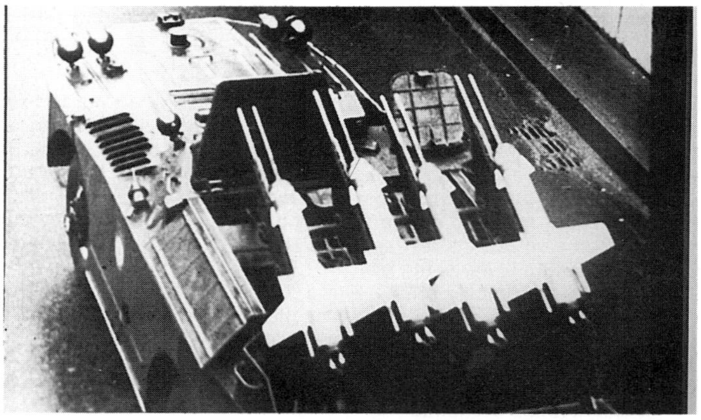

Close-up of Swatter ATGW launcher on BRDM-1 vehicle (US Department of Defence)

Variants

Command Vehicle (BRDM-U)

This can be distinguished from the basic vehicle as it has a radio antenna on each side of the hull and two radio antennas at the rear whereas the basic vehicle has only a single radio antenna on the right side of the hull.

BRDM-1-RKhb (Radiobiological-Chemical-Biological Reconnaissance Vehicle)

This is easily distinguishable from the basic vehicle as on either side of the hull at the rear of the vehicle are two rectangular racks which contain lane marking poles with pennants. The racks are normally covered with a canvas and when required are swung through 90° so that they are vertical over the rear of the vehicle. The poles are used to mark lanes through contaminated areas.

BRDM-1 with ATGWs

As far as it is known the following versions of the BRDM-1 with ATGWs are no longer in service, full details were given in *Jane's Armour and Artillery 1990-91* page 252:
BRDM-1 with three Snapper ATGW
BRDM-1 with four Swatter ATGW
BRDM-1 with six Sagger ATGW.

SPECIFICATIONS

CREW	5
CONFIGURATION	4 × 4
UNLOADED WEIGHT	5600 kg
LENGTH	5.7 m
WIDTH	2.25 m
HEIGHT	1.9 m
GROUND CLEARANCE	0.315 m
TRACK	1.6 m
WHEELBASE	2.8 m
MAX SPEED	
road	80 km/h
water	9 km/h
FUEL CAPACITY	150 l
MAX ROAD CRUISING RANGE	500 km
FUEL CONSUMPTION	0.3 l/km
FORDING	amphibious
GRADIENT	60%
VERTICAL OBSTACLE	0.4 m
TRENCH	1.22 m
ENGINE	GAZ-40P 6-cylinder in-line petrol, water-cooled developing 90 hp at 3400 rpm
TRANSMISSION	manual with 4 forward and 1 reverse gears
TRANSFER CASE	2-speed
CLUTCH	single dry plate for all models
SUSPENSION	semi-elliptical springs and hydraulic shock absorbers
TYRES	12.00 × 18
ELECTRICAL SYSTEM	24 V
MAIN ARMAMENT	1 × 7.62 mm MG
AMMUNITION	1250
READY TO LAUNCH MISSILES	n/app
ARMOUR (max)	10 mm

Status: Production complete. In service with Albania, Algeria, Angola, Bulgaria, Congo, Cuba, Czechoslovakia, Ethiopia, Guinea, Mozambique, Romania, Sudan, USSR and Zambia. In many of these countries it is now in second line units or held in reserve.

Manufacturer: Molotov GAZ plant
Gorky
USSR

UNITED KINGDOM

Alvis Saladin Armoured Car

Development

In January 1946 the British Army issued a requirement for a new armoured car as a replacement for the Daimler Mk II and AEC Mk III armoured cars which were developed during the Second World War. Design work on this new vehicle, called the FV601(A), began in the same year at the Fighting Vehicles Design Department (later to become the FVRDE and now the Royal Armament Research and Development Establishment Chertsey).

As originally conceived the FV601(A) was to have had a crew of four and be armed with a 2-pounder gun fitted with a Littlejohn Adaptor to increase the muzzle velocity. But in February 1948 it was decided that the 2-pounder was not powerful enough so a new weapon, the 76 mm gun L5, was designed by the Armament Design Establishment at Fort Halstead (now RARDE), but was not ready until 1953. In 1947 a contract was awarded to Alvis Limited of Coventry to build two prototypes of the FV601 which were completed in 1953. The first mock-up, fitted with a 2-pounder gun, was completed in 1948.

At the same time as Alvis was building the prototypes of the FV601 it was also building the FV603 armoured personnel carrier which eventually became known as the Saracen. As the Saracen was urgently required for operations in Malaya it was decided to let Alvis concentrate on this vehicle and that the six pre-production vehicles, the FV601(B), would be built by Crossley Motors at Stockport in Cheshire.

The six pre-production vehicles were completed in 1955 and had a slightly different shaped turret from those of prototype vehicles as well as different vision equipment and different hatches for the commander and gunner.

Production was authorised in 1956 and the vehicle entered production at Coventry in 1958 with first vehicles being delivered to the Army the following year. Production continued until 1972 by which time 1177 vehicles had been built. The vehicle was succeeded by the Alvis Scorpion CVR(T) which is armed with a new and lighter version of the 76 mm gun of the Saladin called the L23. The Saladin is officially known as the FV601(C) Armoured Car 76 mm Gun (Alvis Saladin Mk 2 6 × 6).

The Saladin has been replaced in all front-line British units by the Alvis Scorpion CVR(T).

Kuwaiti Army Saladin armoured cars were used against Iraqi units during the invasion of Kuwait City in the summer of 1990. Under the Conventional Treaty Europe (CFE) agreement, the British Army stated that it had a total of 13 Saladin armoured cars of which one was in the UK and 12 overseas, believed to be Cyprus.

Description

The all-welded steel hull of the Saladin is divided into three compartments: driver's at the front, fighting in the centre and engine at the rear.

The driver is seated at the front of the vehicle and is provided with a single piece hatch cover that folds forwards onto the glacis plate for improved visibility. The driver has three No 17 periscopes, one in the hatch cover and one on either side of his position.

The all-welded turret is in the centre of the hull with the gunner seated on the left and the commander, who also acts as the loader, on the right, both with a single piece hatch cover that opens to the rear. The commander has four No 17 periscopes arranged around the forward part of his hatch and a single swivelling periscope to the rear. The gunner has a periscope forward of his hatch cover, the lower part with a magnification of ×6 for fire control and the upper part with a ×1 magnification for observation. On the left side of the turret, just to the rear of the mantlet, is an extractor fan. A small rectangular observation hatch is provided in the rear of the fighting compartment, just below the turret ring. Either side of the turret below the turret ring is a crew escape hatch.

The engine compartment at the rear of the hull is separated from the crew compartment by a fireproof bulkhead and is fitted with a fire warning system and a fire-extinguishing system, the latter being operated by the driver.

Air for the engine compartment is drawn in via six louvred engine covers by two fans and then passes through the radiator and is expelled through grilles at the very rear of the hull.

The engine is coupled to a Daimler fluid coupling which in turn is coupled to a Daimler epicyclic pre-selective gearbox which transmits power to the transfer box consisting of a bevel and helical gear, incorporating reverse and a differential and enabling the five gears to be used in both directions. This transfers the drive direct to each centre bevel box and thence via transmission shafts to the front and rear axles. Each wheel houses epicyclic reduction gears and is connected to the bevel box by a transmission shaft and two universal joints.

The suspension at each wheel station consists of an upper and lower link, with the upper attached to a longitudinal torsion bar which is splined to a tube which in turn is secured to the hull. Steering is hydraulically power-assisted on the front and centre road wheels.

The Saladin has no amphibious capability although a kit is available which enables the vehicle to ford to a depth of 2.13 m. For trials purposes a Saladin was fitted with a collapsible flotation screen similar to that fitted to the CVR(T) Scorpion, but this was never adopted by the British Army.

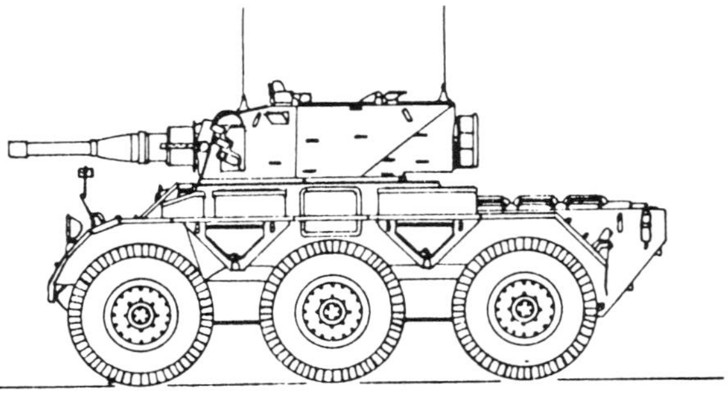

Alvis Saladin armoured car

Alvis Saladin armoured car with driver's, commander's and gunner's hatches open

Alvis Saladin armoured car (Ministry of Defence)

Alvis Saladin armoured car from rear showing external stowage (Ministry of Defence)

The Saladin does not have an NBC system or night vision equipment.

Main armament of the Saladin consists of a 76 mm L5A1 gun with a vertical sliding breech-block and a hydro-spring recoil mechanism. A 7.62 mm (0.30) machine gun model M1919A4(L3A3/L3A4) is mounted coaxially to the left of the main armament and a similar anti-aircraft weapon is mounted on the right of the commander's hatch. Six smoke dischargers fitted on either side of the turret are electrically fired from within the vehicle. The 76 mm gun fires the following types of ammunition: canister, HESH, HE, HE/PRAC, SH/P, SMK/BE and illuminating.

The 42 rounds of ammunition are stowed vertically: 11 between the commander and the gunner, 12 on the left and 11 on the right behind the driver and eight at the left rear of the hull.

Variants

The only variant of the Saladin in service is designated the FV601(D). It was built for the Federal German Border Police (Bundesgrenzschutz), who call it the Geschutzer Sonderwagen III(SW III)(Kfz 93), has no coaxial machine gun and is fitted with German smoke dischargers and driving lights. It is reported that 97 Saladins originally delivered to the Federal German Border Police were subsequently supplied to Sudan in the early 1960s.

The British Aircraft Corporation Guided Weapon Division (now British Aerospace (Dynamics) Limited) fitted a Saladin with a Swingfire ATGW either side of the turret but this was not adopted for service. There was also a project to fit the turret of the Fox armoured car to the Saladin but this did not reach even the mock-up stage.

Saladins were also used by the Australian Army and when they were withdrawn from service their turrets were removed and fitted to M113A1 APCs, which were then called M113A1 Fire Support Vehicles. Details will be found in the M113 entry in the *Armoured personnel carriers* section.

Saladin Repower

In 1991 it was announced that A F Budge Limited and Alvis had developed a repower package for the Saladin armoured car. The existing petrol engine is replaced by a Perkins 180 MTi diesel developing 180 hp.

Additional improvements have been made to the chassis including a modern cooling pack, new electrical system, upgraded brakes, fire supression system and the replacement of the original transmission by a more modern automatic transmission.

The diesel installation requires only minor changes to the engine compartment and offers the advantages of improved reliability and simplified maintenance, lower fuel consumption, increased operational range and reduced risk of fire.

This conversion package has already been trialled on an Alvis Stalwart 6 × 6 high mobility wheeled vehicle and is also applicable to the Alvis Saracen 6 × 6 armoured personnel carrier.

GKL Repower

Details of this are given in the following entry.

Status: Production complete. In service with Bahrain, Ghana, Honduras, Indonesia, Jordan (police), Kenya, Lebanon, Libya (status uncertain), Nigeria, Portugal, Sierra Leone, Sri Lanka, Sudan, Tunisia, United Arab Emirates (reserve) and Yemen. They were also used by Uganda but these are now non-operational.

Manufacturer: Alvis Limited (A member of United Scientific Group)
The Triangle
Walsgrave
Coventry
West Midlands CV2 2SP
UK

SPECIFICATIONS		VERTICAL OBSTACLE	0.46 m	SMOKE-LAYING	
CREW	3	TRENCH	1.52 m	EQUIPMENT	2 × 6 smoke
CONFIGURATION	6 × 6	TURNING RADIUS	7.31 m		dischargers
COMBAT WEIGHT	11 590 kg	ENGINE	Rolls-Royce B80	AMMUNITION	
UNLOADED WEIGHT	10 500 kg		Mk 6A	76 mm	42
POWER-TO-WEIGHT			8-cylinder petrol	7.62 mm	2750
RATIO	14.66 hp/tonne		developing 170 bhp at	GUN CONTROL	
GROUND PRESSURE	1.21 kg/cm²		3750 rpm	EQUIPMENT	
LENGTH GUN FORWARD	5.284 m	TRANSMISSION	pre-selector 5-speed	Turret power control	electric/manual
LENGTH HULL	4.93 m	STEERING	recirculating ball,	by commander	yes
WIDTH	2.54 m		hydraulic assisted	by gunner	yes
HEIGHT		CLUTCH	fluid coupling	Gun elevation/depression	+20°/-10°
to top of gunner's periscope	2.93 m	SUSPENSION	independent, double	Gun stabiliser	
to turret roof	2.19 m		wishbone and torsion	vertical	no
FIRING HEIGHT	1.854 m		bar, hydraulic shock	horizontal	no
GROUND CLEARANCE	0.426 m		absorbers	Range setting device	yes
TRACK	2.038 m	TYRES	12.00 × 20	Traverse indicator	yes
WHEELBASE	1.524 m + 1.524 m	BRAKES		ARMOUR	
ANGLE OF APPROACH/		main	hydraulic disc all	Hull glacis	12 mm at 45°
DEPARTURE	60°/50°		wheels	Hull nose	14 mm at 42°
MAX ROAD SPEED		parking	mechanical all wheels	Hull sides	16 mm at 20°
FORWARDS	72 km/h	ELECTRICAL SYSTEM	24 V	Hull top	10-12 mm
FUEL CAPACITY	241 l	BATTERIES	2 × 12 V, 60 Ah	Hull floor	8-12 mm
MAX RANGE	400 km	ARMAMENT		Hull rear	16 mm
FORDING	1.07 m	main	1 × 76 mm gun	Turret front	32 mm at 15°
with kit	2.13 m	coaxial	1 × 7.62 mm MG	Hull sides	16 mm
GRADIENT	46%	anti-aircraft	1 × 7.62 mm MG	Hull rear	16 mm
				Hull top	10 mm

GKL Saladin 90 mm

Development/Description

In 1988, the GKL Group announced that it was offering an upgraded Saladin (6 × 6) armoured car called the Saladin 90 mm.

The original 76 mm L5A1 gun has been replaced by a Belgian MECAR 90 mm KEnerga gun fitted with a muzzle brake that can fire a wide range of ammunition types including APFSDS and HEAT. This weapon is capable of defeating a NATO medium triple target at a range of 2000 m and a NATO heavy target at a range of 1000 m. The existing turret bustle has been extended to allow for an additional 20 rounds of 90 mm ammunition to be carried with a further 20 rounds being carried in the hull.

The original Rolls-Royce B80 petrol engine has been replaced by a more fuel efficient diesel engine coupled to an MT 600 fully automatic transmission with four forward and one reverse gears.

When fitted with a 250 bhp diesel engine the upgraded Saladin 90 mm has an operational range of 500 km and a power-to-weight ratio of 22 bhp/tonne.

The original 0.30 machine gun mounted coaxial to the left of the main armament could be replaced by a 7.62 mm Chain Gun while a 12.7 mm machine gun will be mounted on the turret roof. The current smoke

GKL Group Upgraded Saladin armoured car with turret traversed to rear to show new MECAR 90 mm KEnerga gun

dischargers mounted either side of the turret firing forwards could be replaced by various types of more modern smoke discharger system.

Other improvements could include a new vehicle electrical system and new instrument panel for the driver.

The GKL Group have been overhauling armoured vehicles for many years including 100 Saladin armoured cars for the Jordanian Police. There is a separate entry in the Armoured Personnel Carriers section for the GKL Upgraded Saracen (6 × 6) armoured personnel carrier and details of the Harimau scout car project are provided later in this section.

Status: Prototype. Not yet in production or service.

Manufacturer: GKL Equipment PTE Limited
Coworth Park House
Coworth Park
Ascot
Berkshire SL5 7SF
UK
Telephone: (0344) 874202
Telex: 846062 GKLLDN G

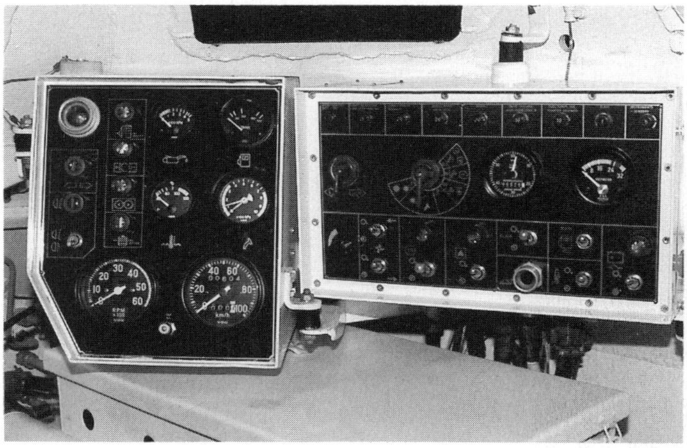

New driver's instrument panel for the GKL Group Upgraded Saladin armoured car

Alvis Scorpion Reconnaissance Vehicle

Details of this extensive range of tracked armoured vehicles are given in the *AFV families* section; details of the Alvis Stormer APC are given in the *Armoured personnel carriers* section.

Fox Light Armoured Car

Development

In the 1960s the Fighting Vehicles Research and Development Establishment, now the Royal Armament Research and Development Establishment Chertsey, designed two reconnaissance vehicles, the Combat Vehicle Reconnaissance (Tracked) Scorpion (FV101) and the Combat Vehicle Reconnaissance (Wheeled) Fox (FV721), both of which use the same Jaguar engine.

Development of the Fox began in 1965 and the following year the Daimler company of Coventry, which was building the Ferret scout car at the time, was awarded a contract to build 15 prototype vehicles. The first was completed in November 1967 and the last in April 1969. User trials began in 1968 and the first official announcement concerning the Fox was made in October 1969. The following year the Fox was accepted for service with the British Army and a production order was placed with Royal Ordnance Leeds. Production began in 1972 and the first vehicle was completed in May 1973. Production of the Fox has been completed at Royal Ordnance Leeds, which was purchased by Vickers Defence Systems in 1986. Production can be resumed if further orders are placed and marketing of the vehicle continues by Vickers Defence Systems.

Daimler completed production of the Ferret in 1971 and then closed down its armoured vehicle production facility.

The Fox is used by both regular and reserve (TA) units of the British Army, but has not replaced the Ferret which was to have been replaced by a development of the Fox known as the Vixen (FV722). This was however cancelled in 1974 owing to a shortage of funds. Alvis of Coventry supplied the welded aluminium turret for the CVR(W) Fox armoured car.

Description

The Fox is a further development of the late production Ferret light scout car and has an all-welded aluminium armour hull and turret which gives protection against medium and heavy machine gun fire and field artillery splinters.

The driver is seated at the front of the vehicle and is provided with a hatch cover that lifts and opens to the right with an integral wide-angle periscope that can be quickly replaced by a passive night periscope.

The turret is in the centre of the vehicle with the commander/loader on the left and the gunner on the right, both with a single piece hatch cover that opens to the rear. The commander has a periscopic binocular surveillance instrument in a rotating mount with a magnification of ×1 and ×10 and seven observation periscopes. The gunner has a periscopic binocular daylight sight linked to the main armament, with a magnification of ×1 and ×10, and two observation periscopes.

Mounted to the right of the main armament is a GEC Sensors SPAV L2A1 passive night sight with a high magnification of ×5.8 (8° field-of-view) and a low magnification of ×1.6 (28° field-of-view), the first for gunnery and the second for surveillance. There can be no confusion between the two as when the high magnification is being used a shutter isolates the low magnification, and when the other is being used an iris diaphragm isolates the high magnification objective. The image intensifier tube is protected from the effect of gun muzzle flash by a flash shutter that is operated electrically from the gun firing circuit. When high magnification is selected an illuminated ballistic graticule with brightness control is automatically injected into the optical system. The exposed objective window is cleared by a wiper and washer and the sight is protected by an armoured cowl with a door which is kept closed when the sight is not in use. The L2A1 night sight will also detect infra-red devices.

The radios are in the rear of the turret and the batteries are stowed below the false floor in the crew compartment.

The engine compartment is at the rear of the hull and the engine and its auxiliaries, including the main and transfer gear boxes, are mounted as a powerpack which can be removed through the rear of the hull. The engine is a militarised 4.2 litre Jaguar XK with a reduced compression ratio (7.75 to 1) in order to use military petrol. British Army vehicles are fitted with electronic ignition. The valve gear is lead-proofed and a single two-choke Marcus carburettor is fitted, as is a Ki-gas cold starting system.

The cooling system has twin radiators horizontally disposed across reset top and to the rear of the engine, and, sealed to the radiator, a pair of ducted centrifugal fans. An oil/water heat exchanger unit serves both the engine and the transmission.

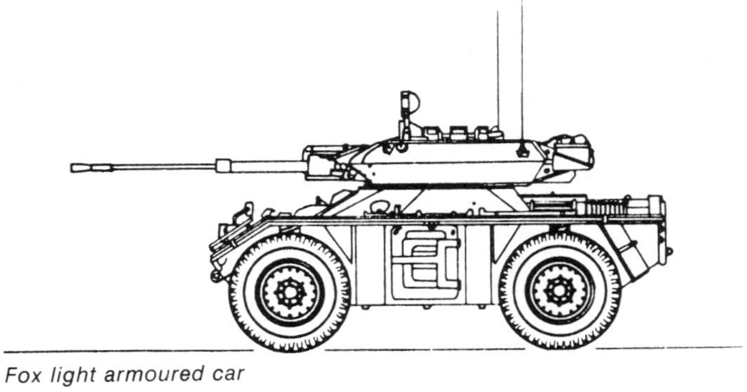

Fox light armoured car

CVR(W) Fox from rear with all hatches closed

Fox CVR(W) showing roof-mounted, hand-operated white light searchlight

British Army CVR(W) Fox with flotation screen removed and commander's, gunner's and driver's hatches open

Power is transmitted from the engine to the wheels through a fluid coupling, a five-speed pre-selecting epicyclic gearbox, and a transfer box providing reversal of all five speeds.

The independent suspension comprises in each case an upper and lower wishbone, coil spring and a hydraulic telescopic damper. The upper and lower wishbone linkages incorporate lubrication reservoirs and the dampers are enclosed in the recoil springs. The independent suspension allows wheel movements of 0.279 m and the tyres are of the run-flat type.

The Fox can ford to a depth of 1 m without preparation. A flotation screen carried collapsed around the top of the hull can be erected by the crew in two minutes and the vehicle is then propelled and steered across the river or stream by its wheels. The front of the flotation screen contains transparent panels for viewing and a bilge pump with a capacity of 205 l/min is fitted as standard on all vehicles. The Fox can travel down or up river banks with a gradient of up to 46 per cent. British Army CVR(W) Foxes have had their flotation screens removed as have the CVR(T) Scorpions.

Fox has been designed to operate in temperatures from -40 to +50°C and is air transportable. A Lockheed C-130 Hercules can carry three Fox vehicles or two for parachute dropping.

Standard equipment on British Army vehicles includes infra-red filters for the white light headlamps, spotlight for the commander, external stowage boxes either side of the hull between the wheels, drinking water tank containing 13.6 litres and a hull electric distribution box with a cooking vessel socket, inspection socket, inter-vehicle starting socket and a master switch. Optional equipment includes navigation aids, powered traverse, nuclear and chemical detection equipment and a ZB 298 surveillance radar mounted on the left side of the turret.

Main armament of the Fox is a 30 mm RARDEN cannon designed by the Royal Armament Research and Development Establishment at Fort Halstead and the Royal Small Arms Factory at Enfield. This fires a new APDS-T round developed jointly by Royal Ordnance and PATEC of the United States. In addition to the new APDS-T round the RARDEN cannon fires standard British APSE-T, HEI-T and PRAC-T rounds as well as standard Oerlikon-Contraves HE, AP and Practice rounds.

The gun is fired in rapid single shots but bursts of up to six rounds can be fired. The empty cartridge cases are automatically ejected outside the turret.

A 7.62 mm machine gun is mounted coaxially to the left of the main armament and four smoke dischargers are mounted either side of the front of the turret. The main armament, coaxial machine gun and smoke dischargers are electrically operated with the main armament and coaxial machine gun having manual overriding control.

Variants

Prior to the take over of Royal Ordnance Leeds by Vickers Defence Systems, a number of variants of the Fox light armoured car were developed to the prototype stage specifically for the export market. These included:-

1. Panga armoured reconnaissance vehicle with various machine gun turrets
2. Fox with one man power operated turret armed with 25 mm McDonnell Douglas Helicopters Chain Gun and 7.62 mm coaxial machine gun
3. Fox/MILAN with one man Euromissile MILAN Compact Turret (MCT) with two MILAN ATGW's in ready to launch position
4. Fox/Scout with turret mounted 7.62 mm machine gun

None of these vehicles have been placed in production and as the Fox production line has been closed for some years it is considered unlikely to re-open unless substantial export orders are placed. Details and photographs of the above variants of the Fox were given in *Jane's Armour and Artillery 1989/90* pages 242/243.

Status: Production as required. In service with Malawi, Nigeria and the United Kingdom.

Manufacturer: Vickers Defence Systems Limited
Manston Lane
Leeds LS15 8TN
UK
Telephone: (0532) 648123
Telex: 55134
Fax: (0532) 607674

SPECIFICATIONS

CREW	3
CONFIGURATION	4 × 4
COMBAT WEIGHT	6120 kg
UNLOADED WEIGHT	5733 kg
POWER-TO-WEIGHT RATIO	30.04 bhp/tonne
LENGTH	
gun forwards	5.08 m
gun rear	4.242 m
hull only	4.166 m
WIDTH	2.134 m
HEIGHT	
to turret top	1.981 m
overall	2.2 m
GROUND CLEARANCE	0.3 m
TRACK	1.753 m
WHEELBASE	2.464 m
MAX SPEED	
road	104 km/h
water	5.23 km/h
ACCELERATION (0-48 km/h)	9 s
FUEL CAPACITY	145.47 l
MAX ROAD RANGE	434 km

FORDING (amphibious with preparation)	1 m
GRADIENT	46%
VERTICAL OBSTACLE	0.5 m
TRENCH WITH CHANNELS	1.22 m
TURNING RADIUS	6.1 m
ENGINE	Jaguar XK 4.2 l 6-cylinder petrol developing 190 bhp at 4500 rpm
TRANSMISSION	Daimler pre-selective 5-speed unit
TRANSFER CASE	Daimler spiral bevel directional control gears and double helical drop down gear to differential provide, in combination with gear box, 5 forward and 5 reverse speeds
STEERING	power-assisted
CLUTCH	fluid flywheel
SUSPENSION	independent
TYRES	11.00 × 20, run flat

BRAKES	
main	hydraulic disc all wheels
parking	contracting band on front bevel box input
ELECTRICAL SYSTEM	24 V
BATTERIES	2 × 6TN, 100 Ah
ARMAMENT	
main	1 × 30 mm cannon
coaxial	1 × 7.62 mm MG
SMOKE-LAYING EQUIPMENT	2 × 4 smoke dischargers
AMMUNITION	
main	99
coaxial	2600
smoke grenades	16
GUN CONTROL EQUIPMENT	
Turret power control	manual, 2-speed
by commander	yes
by gunner	yes
Gun elevation/depression	+40°/-14°
Gun stabiliser	
vertical	no
horizontal	no

Alvis Ferret 80 Scout Car

Development

At the 1982 British Army Equipment Exhibition Alvis announced that it was designing a new version of the Ferret (of which 4409 were produced by the Daimler company between 1952 and 1971) under the name of Ferret 80. It uses basic Ferret components but incorporates up-to-date refinements for improved protection, performance, stability and capability. The first prototype was completed in June 1984, but as of early 1991, no orders had been placed.

Description

The hull of the Ferret 80 is of all-welded aluminium armour and well sloped to give maximum possible protection within the weight limits of the vehicle. Complete protection against 7.62 mm ball at point blank range and all angles of attack is given. Through the front 60° arc protection is provided against attack from 14.5 mm armour-piercing rounds at a distance of 800 m and against 7.62 mm armour-piercing attack at a distance of 250 m.

The driver is seated at the front of the vehicle with the commander/gunner in the centre and the engine at the rear.

The driver has three splinter-proof windscreens to his front and sides. Armoured visors are provided and a separate driver's access hatch is provided above his position.

The Ferret 80 can be fitted with a wide range of armament installations including the following according to mission requirements.

The Ferret 80 is powered by a Perkins T6/3544 six-cylinder turbo-charged diesel developing 156 bhp which has already been proved in the Alvis Scorpion and Stormer vehicles rated at 200 and 250 bhp respectively.

Power is transmitted through a five-speed automatic gearbox and transfer box connected to each wheel hub by propeller shafts through axle units. Limited slip differentials are incorporated in the transfer box and axle units. A reduction gear is located in each wheel hub. The cooling system is an adaptation of the standard system already proven in the Scorpion series, and uses a mixed flow fan mounted below ballistic louvres and above the engine which draws air through radiators.

Steering is dual circuit, power-assisted, eliminating the awkward wheel location and size in the original Ferret. Independent suspension, wheels and braking system are identical to those of the big-wheeled Ferret Mk 4. Power-assisted hydraulic brakes are fitted to all wheels and a parking brake operates on the transmission. Flat divided disc road wheels with run-flat cross-country tyres are fitted as standard.

The electric system is 24 V with the two 12 V batteries contained in the baseline hull.

Variants

In addition to the basic two-man Ferret 80, Alvis has proposed a three-man model. This has a slightly different hull top with the commander and radio operator seated in the centre, each provided with a single piece hatch cover, with the commander having a cupola with fixed periscopes that can be traversed through a full 360°. A 7.62 mm machine gun can be externally mounted on this position. Vision blocks mounted in the upper part of the hull give all-round observation.

SPECIFICATIONS

Model	Two-man reconnaissance	20 mm turret	Three-man reconnaissance
CREW	2	2	3
CONFIGURATION	4 × 4	4 × 4	4 × 4
COMBAT WEIGHT	6000 kg	7100 kg	6500 kg
WEIGHT DISTRIBUTION (front/rear)	42%/58%	43%/57%	43%/57%
POWER-TO-WEIGHT RATIO	26 hp/tonne	22 hp/tonne	24 hp/tonne
LENGTH	4.23 m	4.23 m	4.23 m
WIDTH	2.34 m	2.34 m	2.34 m
HEIGHT	2.35 m	2.44 m	2.52 m
GROUND CLEARANCE	0.33 m	0.33 m	0.33 m
TRACK (front and rear)	1.93 m	1.93 m	1.93 m
WHEELBASE	2.72 m	2.72 m	2.72 m
MAX ROAD SPEED	96 km/h	96 km/h	96 km/h
FUEL CAPACITY	159 l	159 l	159 l
MAX ROAD RANGE	560 km	560 km	560 km
FORDING	1.27 m	1.27 m	1.27 m
GRADIENT	60%	60%	60%
VERTICAL OBSTACLE	0.38 m	0.38 m	0.38 m
ANGLE OF APPROACH/DEPARTURE	65°/55°	65°/55°	65°/55°
TURNING RADIUS	7.3 m	7.3 m	7.3 m
ENGINE	Perkins T6/3544 6-cylinder in-line diesel developing 156 hp at 2600 rpm		
TRANSMISSION	Allison Transmission Division of General Motors Corporation MT 653 DR 5-speed automatic with manual selection of 1st gear		
STEERING	Cam Gears, power assisted (hydraulic) re-circulating ball		
SUSPENSION	fully independent, single coil springs, hydraulic double acting shock absorbers		
TYRES	279.4 × 508 mm on all versions		
BRAKES			
main	hydraulic disc on all wheels		
parking	mechanical, drum acting on transfer box output		
ELECTRICAL SYSTEM	24 V	24 V	24 V
BATTERIES	2 × 100 Ah	2 × 100 Ah	2 × 100 Ah

Status: A prototype fitted with the FVT 800 turret was completed in the Spring of 1984.

Manufacturer: Alvis Limited (A member of United Scientific Group)
The Triangle
Walsgrave
Coventry CV2 2SP
West Midlands
UK

Prototype of Alvis Ferret 80 (4 × 4) scout car with FVT 800 one-man turret armed with 12.7 mm and 7.62 mm machine guns

Rear view of Alvis Ferret 80 (4 × 4) scout car prototype showing stowage basket on rear of FVT 800 turret

GKL Harimau Scout Car

Development

The Harimau (4 × 4) scout car is essentially the Daimler Ferret vehicle, originally designed 40 years ago, rebuilt to incorporate new automotive and other components to extend its life for another 15 to 20 years.

The manufacturer of Harimau, The GKL Group of Companies, can either modify existing Ferret vehicles to the Harimau standard in the users own country or provide vehicles rebuilt to the new standard from stock.

The overall layout is identical to that of the Ferret and the Harimau is available without the turret and can be fitted with a variety of pintle-mounted weapons, the turret can be adapted from .30 inch MG to 7.62 mm GPMG and improved sighting/vision fitted.

The hull is of all-welded steel armour construction which provides the crew with protection from small arms fire and shell splinters.

The original Rolls-Royce B60 Mk 6A six-cylinder petrol engine developing 129 hp coupled to a Daimler pre-selective transmission has been replaced by the more powerful and fuel-efficient VM HR 692 diesel developing 140 bhp at 3700 rpm coupled to a modified Jaguar 700 series pre-selector transmission with five forward and five reverse gears. The Harimau has permanent four-wheel drive with hub reduction units at each wheel station. A non-slip differential can be provided.

The new electrical system has been installed and new lights fitted. The spare tyre and wheel normally carried on the left side of the Ferret have been replaced by a large stowage box and similar stowage facilities are provided on the right side of the hull, if run-flat tyres are fitted.

According to the manufacturers, the Harimau has the following improvements in performance over the existing Ferret vehicle:
Top speed increased by 10 per cent
Maximum gradient increased by 60 per cent
Power increased by 30 per cent
Power-to-weight ratio increased by 17 per cent
Road range increased by 30 per cent
Increased fire power
Torque increased by 28 per cent
Alternator increased by 140 per cent.

Various optional equipment includes night vision equipment, run-flat tyres and power steering.

The Harimau has been designed to operate in ambient temperatures of from -15°C, to +50°C or down to as low as -32°C with suitable lubricants and fuel.

Status: Development complete. Ready for production.

Manufacturers: GKL Equipment PTE Ltd
Coworth Park House
Coworth Park
Ascot
Berkshire SL5 7SF
UK
Telephone: (0344) 874202
Telex: 846062 GKLLDN G

GKL Harimau Scout Car fitted with 7.62 mm Chain Gun

GKL Harimau Scout Car showing new smoke dischargers

SPECIFICATIONS

CREW	2
CONFIGURATION	4 × 4
COMBAT WEIGHT	4400 kg
POWER-TO-WEIGHT RATIO	30 hp/tonne
LENGTH	3.895 m
WIDTH	1.905 m
HEIGHT	
to turret top	1.88 m
GROUND CLEARANCE	0.33 m
TRACK	
front	1.549 m
rear	1.67 m
WHEELBASE	2.86 m
ANGLE OF APPROACH/ DEPARTURE	60°/50°

MAX ROAD SPEED	100 km/h
FUEL CAPACITY	90 l
RANGE	400 km
GRADIENT	60%
VERTICAL OBSTACLE	0.406 m
TURNING RADIUS	5.8 m
ENGINE	VM HR 692 HT diesel developing 150 bhp at 3700 rpm
TORQUE	350 Nm at 2200 rpm
TRANSMISSION	Jaguar 700 series, modified, pre-select, 5 forward and 5 reverse gears with no-spin differential
STEERING	re-circulating ball

SUSPENSION	Independent, coil spring and double-acting shock absorber at each wheel station
TYRES	9.00 × 16
BRAKES	servo-assisted hydraulic, drum at each wheel station
ELECTRICAL SYSTEM	24 V
BATTERIES	2 × 12 V
ARMAMENT	See text

Daimler Ferret Scout Car

Development

In 1947 the British Army issued a requirement for a scout car to replace the Daimler Dingo scout car used during the Second World War. Late in 1948 Daimler was awarded a contract to design and build prototypes of a new scout car under the designation FV701. The first prototype was completed in 1949 and delivered to the Army for trials the following year. After trials it was adopted for service with the British Army and was named the Ferret.

The first production Ferret, the Mk 2, was completed in mid 1952 and the first production Mk 1 was completed late the same year. Throughout its production life the basic hull of the Ferret remained unchanged although slightly more powerful engines were installed as well as different turrets fitted to meet different requirements. Production of the Ferret was finally completed in 1971 by which time 4409 vehicles of all types had been built.

In 1966 the Daimler company was awarded a contract to build 15 prototypes of a new vehicle called the FV721 (or Fox as it later became known), which were completed between 1967 and 1969. The Fox was accepted for service with the British Army in 1970 but the production order was placed with the Royal Ordnance Factory at Leeds (now owned by Vickers Defence Systems) rather than Daimler at Coventry. When production of the Ferret was completed Daimler closed down its production line for wheeled armoured vehicles. The Fox is based on experience obtained with the rebuilt Ferret Mk 4 and Mk 5.

The Ferret was to have been replaced in the British Army by a development of the Fox called the FV722 Vixen, but this was cancelled in 1974 as a defence economy and Ferrets will therefore remain in service until the 1990s.

Alvis is now the design authority for the Ferret and supplies the spare parts for it.

Description (Mk 2)

The all-welded steel hull of the Ferret is divided into three compartments: driver's at the front, fighting in the centre and the engine at the rear.

The driver has three hatches, one to his front and one to each side, each of which contains an integral No 17 observation periscope. The front hatch can be folded down onto the glacis plate for increased visibility and can then be fitted with a splinter-proof windscreen with wiper blade and electric motor. The two side hatches can be opened upwards on the outside for increased visibility when not in a combat area.

The manually operated turret in the centre of the hull has a single piece hatch cover in the rear part of the roof which can be locked in three different positions, and the rear part of the turret folds down to the horizontal to enable it to be used as a seat. Mounted in the forward part of the turret roof is a sight periscope AFV No 3 Mk 1 which is used for aiming the turret-mounted machine gun.

Mounted in either side of the hull below the turret ring is a vision slit protected by a glass block. At the rear of the fighting compartment are two hatches which can be opened for increased observation. On each side of the hull, between the front and rear road wheels, is a hull escape hatch. The left one is covered by the spare wheel and the right by a stowage box.

The engine is at the rear of the hull and is fully waterproofed and will operate when completely submerged without any preparation other than venting the crankcase breather pipe. Drive is transmitted to all four road wheels through a fluid coupling, five-speed pre-selecting epicyclic gearbox and a transfer box, incorporating a forward and reverse mechanism and a differential unit to give five speeds in each direction.

The suspension at each wheel station consists of a single coil spring which encloses a double-acting shock absorber and is mounted on a stabiliser bracket at the bottom and carried in a bracket at the top attached to the hull plates. The tyres are of the run-flat type.

The Mk 2 is armed with a turret-mounted 7.62 mm (0.30) Browning machine gun which can be elevated manually from -15 to +45°, turret traverse being a full 360°. A total of 2500 rounds of machine gun ammunition is carried. Mounted either side of the hull at the front of the vehicle are three smoke dischargers which are electrically fired from within the vehicle.

The basic Ferret has no NBC system, no night vision equipment and no amphibious capability. A deep fording kit available consists of a collapsible bellows type collar fitted around the top of the commander's hatch and another fitted at the very rear of the hull. Ferrets Mks 3 and 4 are fitted with a collapsible screen around the top of their hulls which, when erected, enables them to propel themselves across lakes and rivers with their wheels. To enable the vehicle to cross trenches Ferrets Mks 1, 2 and 3 have provision for carrying steel channels on the front of their hulls.

Early production vehicles had BSF threads, different gear ratios and were powered by a Rolls-Royce B60 Mk IIIA six-cylinder petrol engine developing 120 bhp at 3750 rpm. Later production vehicles had UNF threads and the B60 Mk VIA six-cylinder engine which develops 129 bhp at 3750 rpm.

Variants

Ferret Mk 1

This is called Car Scout 4 × 4 Liaison (Ferret Mk 1) FV701(C) and has an open top which can be covered by a canvas. It is armed with a 7.62 mm Bren LMG or a 7.62 mm (0.30) Browning machine gun, with 450 rounds of machine gun ammunition being carried. Later production models are known as the Ferret Mk 1/1, FV701(J). Some Mk1/1s have been fitted with a flotation screen similar to that fitted to the Ferret Mk 4.

Ferret Mk 1/2 (FV704)

This is identical to the Mk 1 but has a crew of three and is used by infantry units as a light reconnaissance vehicle in forward areas. Its official designation is Car Scout Liaison (Ferret) Mk 1/2. It has an armoured roof with a hatch and is armed with a 7.62 mm pintle-mounted Bren light machine gun. The single piece hatch cover opens to the rear and periscopes and vision blocks are provided for observation by the commander.

Ferret Mk 2

This is basically a Mk 1 fitted with a turret and is officially known as the Car Scout 4 × 4, Reconnaissance (Ferret) Mk 2 FV701(E).

Ferret Mk 2/2

This was a local modification carried out in the Far East and is basically the Ferret Mk 2 with an extension collar fitted between the top of the hull and the machine gun turret and enables the commander/gunner to have a better field of fire.

Ferret Mk 2/3

This is a later production model of the basic Mk 2 and is officially known as the Scout Car Reconnaissance Mk 2/3 (Daimler Ferret 4 × 4) FV701(H). The description in this entry relates to this model.

Daimler Ferret Mk 1 scout car with 0.30 (7.62 mm) MG (Christopher F Foss)

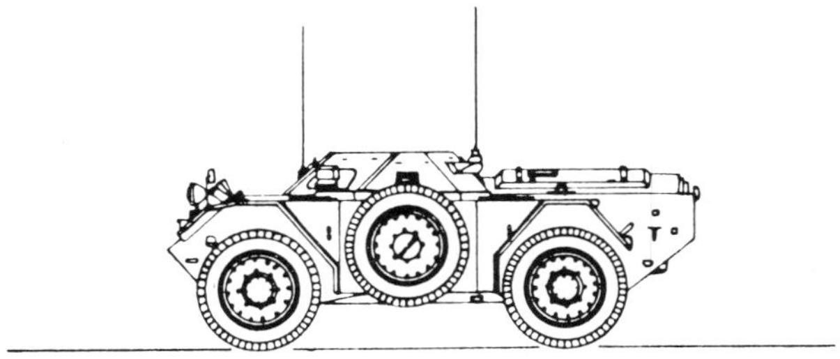

Daimler Ferret Mk 1 scout car

Ferret Mk 2/3 with turret traversed left and showing additional stowage over rear engine compartment (Peter Siebert)

Ferret Mk 1/2 Car Scout Liaison with extensive stowage but without armament (Terry J Gander)

Repowered Daimler Ferret under production test at Alvis, Coventry

Daimler Ferret Mk 2/3 scout car; turret is not fitted with machine gun

Daimler Ferret scout car upgraded by Alvis showing additional smoke dischargers either side of turret (Terry J Gander)

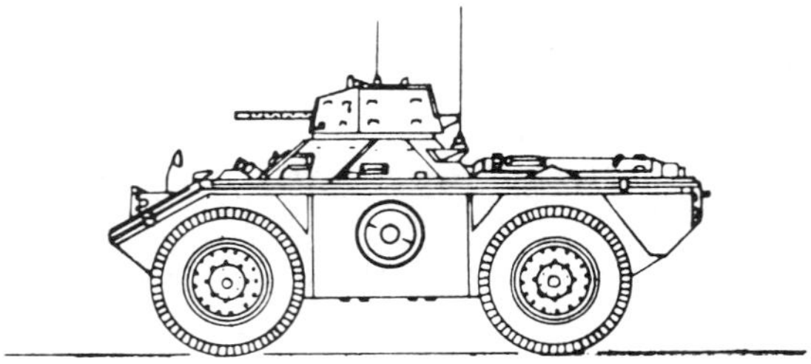

Daimler Ferret Mk 4 scout car

Ferret Mk 4 from front without armament (Ministry of Defence)

Ferret Mk 2/4

This is a Ferret Mk 2/3 with additional armour.

Ferret Mk 2/5

This is a Ferret Mk 2 brought up to Ferret Mk 2/4 standard.

Ferret Mk 2/6 (FV703)

The full designation of this model is the Scout Car Reconnaissance/ Guided Weapon Mk 2/6. It is basically a Mk 2/3 with a single Vigilant ATGW in its launcher box mounted on either side of the turret. A further two missiles are carried in their launcher boxes on the left side of the hull in place of the spare wheel.

The two ready to launch missiles have a common elevating mechanism and can be launched from inside the vehicle or away from it with the aid of a combined sight/controller and separation cable. The 7.62 mm (0.30) machine gun is retained. The Vigilant ATGW is wire guided and has a minimum range of 200 m and a maximum range of 1375 m, weighs 14 kg and has a HEAT warhead. As far as it is known, there are no known Ferret Mk 2/6 now operational. In addition to being used by the UK it was also sold to Libya and the United Arab Emirates.

Ferret Mk 2/7

This is a Ferret Mk 2/6 with the missile system removed and is therefore basically a Ferret Mk 2/3.

Ferret Mk 3

This is the Ferret Mk 1/1 brought up to the same standards as the Mk 4 but with the machine gun turret.

Ferret Mk 4 (FV711)

This model is basically an early Ferret rebuilt with stronger suspension units, original 330 mm diameter drum brakes replaced by vacuum-assisted 381 mm diameter disc brakes, and large wheels and tyres. A flotation screen is carried collapsed around the top of the hull and when erected the vehicle can propel itself across lakes and rivers by its wheels. There are watertight stowage containers either side of the hull, and the number of servicing points has been reduced.

Ferret Mk 5 (FV712)

This model is no longer in service and was used only by the British Army. It was essentially a Mk 4 with a turret mounting four British Aerospace Swingfire ATGWs.

Ferret Upgrade

In March 1988 Alvis announced that it had developed a repower package for the Daimler Ferret scout car. This package, which has been developed to the prototype stage, involves virtually no hull cutting and comprises a Perkins Phaser engine (110MT) developing 109 hp coupled to a Chrysler A727 fully automatic transmission, these are interfaced with the existing transfer box and drive lines while the cooling system has been modified.

Other modifications could include the replacement of the existing periscopes by new Helio periscopes, new smoke/fragmentation launchers, new machine gun mount and a new machine gun sight.

According to Alvis, the weight, road speed and acceleration of the Ferret are unchanged while the range, reliability and life-cycle costs have been much improved. In 1988 Alvis signed a formal joint-venture agreement with the Malaysian Mining Corporation to use the Alvis upgrade package. The Malaysian Army requirement is to be met with the Alvis Repower Package. Production contracts were signed during early 1991. According to Alvis Limited, other Ferret users are now actively considering repowering their Ferret vehicles which continue to fulfil a valuable liaison role.

Status: Production complete. In service with Bahrain, Burkina Faso, Burma, Cameroon, Central African Republic, India, Indonesia, Madagascar (status uncertain), Malaysia, Nepal, New Zealand, Portugal (including Mk 4), Qatar, Senegal, South Africa, Sri Lanka, Sudan, United Arab Emirates (reserve), United Kingdom, Yemen and Zimbabwe.

Manufacturer:
Daimler Company Limited
Coventry

Spare parts and post-design work are now the responsibility of Alvis Limited, (A member of United Scientific Group)
The Triangle
Walsgrave
Coventry
West Midlands CV2 2SP
UK

SPECIFICATIONS

Model	**Mk 1/1**	**Mk 1/2**	**Mk 2/3**	**Mk 4**
CREW	2-3	3	2	2
CONFIGURATION	4 × 4	4 × 4	4 × 4	4 × 4
COMBAT WEIGHT	4210 kg	4370 kg	4400 kg	5400 kg
UNLOADED WEIGHT	3510 kg	3660 kg	3640 kg	4725 kg
POWER-TO-WEIGHT RATIO	30.64 bhp/tonne	29.51 bhp/tonne	29.35 bhp/tonne	23.88 bhp/tonne
LENGTH	3.835 m	3.835 m	3.835 m	3.96 m
WIDTH	1.905 m	1.905 m	1.905 m	2.134 m
HEIGHT OVERALL	1.448 m	1.651 m	1.879 m	2.03 m
GROUND CLEARANCE	0.33 m	0.33 m	0.33 m	0.432 m
TRACK	1.539 m	1.539 m	1.539 m	1.75 m
WHEELBASE	2.286 m	2.286 m	2.286 m	2.286 m
MAX ROAD SPEED	93 km/h	93 km/h	93 km/h	80 km/h
FUEL CAPACITY	96 l	96 l	96 l	96 l
MAX RANGE				
road	306 km	306 km	306 km	306 km
cross-country	160 km	160 km	160 km	160 km
FUEL CONSUMPTION				
road	0.31 l/km	0.31 l/km	0.31 l/km	0.31 l/km
cross-country	0.6 l/km	0.6 l/km	0.6 l/km	0.6 l/km
FORDING	0.914 m	0.914 m	0.914 m	0.914 m
with preparation	1.524 m	1.524 m	1.524 m	amphibious
GRADIENT	46%	46%	46%	46%
VERTICAL OBSTACLE	0.406 m	0.406 m	0.406 m	0.406 m
TRENCH (with channels)	1.22 m	1.22 m	1.22 m	1.22 m
TURNING RADIUS	5.795 m	5.795 m	5.795 m	5.795 m
ENGINE	Rolls-Royce B60 Mk 6A 6-cylinder in-line water-cooled petrol developing 129 bhp at 3750 rpm			
TRANSMISSION	Daimler pre-selective epicyclic, 5-speed			
STEERING	re-circulating ball (not power-assisted)			
CLUTCH	Daimler fluid coupling			
SUSPENSION	independent, coil spring and double-acting shock absorber at each wheel station			
TYRES	9.00 × 16	9.00 × 16	9.00 × 16	11.00 × 20
BRAKES				
main	hydraulic on all wheels			
parking	hand-operated, mechanical on transmission			
ELECTRICAL SYSTEM	24 V	24 V	24 V	24 V
BATTERIES	2 × 12 V, 60 Ah	2 × 12 V, 60 Ah	2 × 12 V, 60 Ah	2 × 12 V, 60 Ah
ARMAMENT				
machine gun	1 × 7.62 mm	1 × 7.62 mm	1 × 7.62 mm	1 × 7.62 mm
missile	n/app	n/app	n/app	n/app
ready use missiles	n/app	n/app	n/app	n/app
reserve missiles	n/app	n/app	n/app	n/app
SMOKE-LAYING EQUIPMENT	all Ferret models have 2 × 3 smoke dischargers			
ARMOUR				
Hull front	12 mm at 50° and 16 mm at 35° all models			
Hull sides	12 mm at 10°, 15°, 40°; 16 mm at 20° all models			
Hull floor	10 mm driver's position, 6 mm remainder of vehicle all models			
Hull rear	12 mm and 6 mm at 45° all models			
Turret front	n/app	16 mm	16 mm at 15°	16 mm at 15°
Turret sides	n/app	16 mm	16 mm at 15°	16 mm at 15°
Turret rear	n/app	16 mm	16 mm at 15°	16 mm at 15°
Turret top	n/app	8 mm	8 mm	8 mm

Shorland S52 Armoured Patrol Car

Development

The Shorland armoured patrol car was originally designed in 1965 specifically to meet the requirements of the Royal Ulster Constabulary for use in Northern Ireland. First production vehicles were completed at Short Brothers and Harland's facility at Newtownards in 1965. By the end of 1982 was over 1000 vehicles, since when production has continued for the export market and it is now used by some 45 countries.

The first production models were powered by a four-cylinder Rover petrol engine which developed 67 hp at 4000 rpm. The Mk 2 was based on the Series 2 Land Rover chassis and was powered by a four-cylinder petrol engine which developed 77 bhp at 4000 rpm. The Mk 3 was introduced in 1972 and was fitted with a 2.6 litre six-cylinder engine which developed 91 bhp at 4500 rpm. The Mk 4, employing the new V-8 engine, went into production in 1980.

The latest Series 5 version was designed to take advantage of the Land Rover Defender 110 inch chassis with its improved coil spring suspension and wider wheel track than previously. Other improvements introduced with the new chassis were improved performance, tighter turning circle and power steering to reduce driver fatigue.

Today, the complete Series 5 family consists of the S52 Armoured Patrol Car, S53 Air Defence Vehicle, S54 Anti-hijack Vehicle and the S55 Armoured Personnel Carrier. Full details of the latter are given in the *Armoured personnel carriers* section.

Description

The Shorland S52 is based on the 110 inch heavy duty Land Rover Defender chassis which has permanent four wheel drive and a lockable central differential. Normal power is the 3.5 litre Rover V8 petrol engine with the Rover 2.5 litre Turbo diesel available as an alternative.

The chassis is strengthened to take the armoured body and is fitted with stronger axles and suspension and different final drive ratio. Nevertheless some 85 per cent of the automotive and mechanical components are stock Land Rover parts also reducing the spares and maintenance problem.

The vehicle hull provides protection against right angle fire by NATO 7.62 × 51 mm ball ammunition down to 25 m range while the bullet resistant glass provides the same degree of protection as the steel and meets the test requirements of BS 5051 part 1 Class R2.

The engine and radiator are fully armour protected and the floor is made from reinforced glass fibre to give protection against mines and improvised explosive devices. The fuel filters are under armoured hatches which can also be released from inside the vehicle. Vehicle interiors are finished in foam backed washable plastic to provide thermal installation.

The manually operated turret has full 360° rotation and is fitted with an adjustable seat for the gunner. A mechanical lock prevents rotation when not in use. An optical periscope sight is fitted, this being linked to the gun mounting. A searchlight is provided and is boresighted to the armament. The most common armament is the FN HERSTAL SA 7.62 mm GPMG but most other light machine guns can be fitted. Two four-barrelled grenade launchers can be fitted, one to each side of the turret. These fire 66 mm electrically initiated smoke/illumination/CS gas grenades. In the observation role the turret is operated with the rear part folding forwards and upwards and the turret back folding downwards.

The vehicle has a crew of three, driver and commander seated side by side in the front with the gunner in the turret. Alternative seating for the gunner is fitted behind the commander. The hull also contains the spare wheel and ammunition and there is ample room for other equipment such as radios. An escape hatch is provided in the rear panel.

Anti-hijack version of Shorland armoured patrol car to latest production standard with additional windows for improved all-round visibility

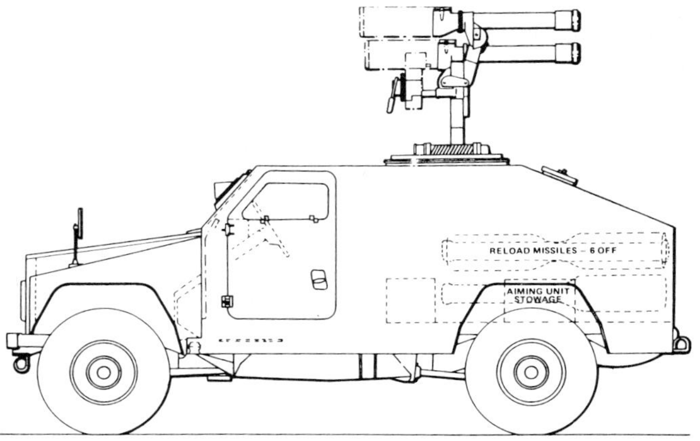

Shorland S53 mobile air defence vehicle showing position of main components (not to 1/76th scale)

Shorland S52 armoured patrol car fitted with turret armed with 7.62 mm GPMG and 2 × 4 electrically operated smoke dischargers

Shorland S52 armoured patrol car

The large bullet resistant windscreen and side windows provide excellent visibility for the driver and commander. Two extraction fans are fitted, one mounted on the sloping rear roof is controlled from the dashboard while the second one in the turret and controlled by the gunner, is specifically intended for fume extraction when the armament is being fired.

Standard equipment normally includes seat belts and air conditioning, optional equipment includes drop down visors with vision blocks in place of the bullet resistant windscreen, run flat tyres, a loud hailer, automatic fire protection and various radio installations. Right or left hand steering can be specified.

Shorland S53 Mobile Air Defence Vehicle

This has been designed by Shorts specifically for the defence of airfields and other key installations and supply routes against low level air attack.

Based on the S52 Armoured Patrol Car, the S53 is fitted with a variant of the Blowpipe/Javelin lightweight multiple launcher in place of the turret.

The system uses the standard shoulder launch aiming unit as clip-on equipment and mounts three canistered missiles. The turret ring has an integral hatch cover and periscope view block for surveillance when the hatch is closed. The ring carries a pintle mounting for the lightweight multiple launcher. This has a traverse head assembly with a sealed bearing unit which fits over the pintle. The traverse freedom relative to the turret ring is plus/minus 40° and the ring has hand grips so that the aimer can slew it to the approximate target bearing and then track with the lightweight multiple launcher. The system is manually operated.

A total of six missiles in their ammunition containers are stowed in racks on either side at the rear of the vehicle. The transport for the aiming unit is between the missile storage racks.

Shorland S54 Anti-Hijack Vehicle

This has been developed specifically for anti-hijack patrols at airports and other high risk areas and in addition to the driver's and commander's bullet resistant windscreens and side windows, additional vision blocks can be fitted in the hull and turret.

The specially designed turret has a vision block and mount which permits an HK33 rifle to be sighted discreetly on a target while both the rifle and the marksman remain concealed.

SPECIFICATIONS	
CREW	3
CONFIGURATION	4 × 4
COMBAT WEIGHT	3600 kg
POWER-TO-WEIGHT RATIO	31.2 bhp/tonne (petrol) 30.5 hp/tonne (diesel)
LENGTH	4.49 m
WIDTH	1.8 m
HEIGHT (roof)	1.8 m
GROUND CLEARANCE	324 mm
TRACK	1.5 m
WHEELBASE	2.795 m
MAX ROAD SPEED	120 km/h
FUEL CAPACITY	136 l
MAX RANGE (ROAD)	over 630 km
VERTICAL OBSTACLE	0.23 m
TURNING RADIUS	6.6 m
ENGINE	3.5 litre Land Rover V-8 water-cooled developing 114 bhp at 4000 rpm or 2.5 litre Land Rover TDI water-cooled 4-stroke turbo-charged diesel developing 107 bhp at 3800 rpm
TRANSMISSION	manual with 5 forward and 1 reverse gears
TRANSFER CASE	2-speed
STEERING	power assisted
CLUTCH	single dry plate
SUSPENSION	coil springs, anti-roll bars fitted front and rear
TYRES	9.00 × 16 (run-flat)
BRAKES	
main	hydraulic, dual circuit front and rear, disc front, drum rear
parking	mechanical drum operation on transmission
ELECTRICAL SYSTEM	12 V (24 V optional)
BATTERIES	1 × 12 V, 58 Ah
ARMAMENT	1 × 7.62 mm MG
AMMUNITION	1500
SMOKE-LAYING EQUIPMENT	4 smoke dischargers either side of turret (optional)

Status: In production. In service with over 40 countries including Argentina, Botswana, Brunei, Burundi, Cyprus, Guyana, Kenya, Lebanon, Libya, Malaysia, Portugal, Seychelles, Syria, Thailand, United Arab Emirates and United Kingdom.

Manufacturer: Short Brothers PLC
Montgomery Road
Belfast BT6 9HN
Northern Ireland
Telephone: 0232 458444
Telex: 747087
Fax: 0232 705293

Shorts S53 mobile air defence vehicle with three round Javelin lightweight multiple launcher ready to engage targets

Glover Hornet Armoured Car

Development
The Hornet armoured car, developed as a private venture by Glover Webb Limited, was shown for the first time at the 1986 British Army Equipment Exhibition.

Description
The Glover Hornet armoured car is based on the chassis of the heavy duty military Land Rover One Ten (4 × 4) vehicle which is already used by a number of armed forces.

The chassis has been fitted with an armoured steel body which provides complete protection from penetration by 7.62 mm NATO ball ammunition (L2A2) at a range of 40 m. All apertures are fitted with overlaps and splash returns and glass and vision blocks have been tested to withstand the same level of attack as the remainder of the hull. The floor area of the vehicle is protected against grenade attack by a high yield steel plate.

The layout of the vehicle is similar to the Land Rover's with the engine at the front, commander and driver in the centre (both left- and right-hand drive models are available), and a machine gun turret on the hull roof.

The commander and driver each have a forward-opening side door in the upper part of which is an observation hatch that opens to the outside. To the immediate front of both the commander and gunner is a bullet-proof windscreen with two electrically operated windscreen wipers. If required, each of the windscreens can be covered by a flap hinged at its upper part which has an observation slit.

The low-profile machine gun turret has also been designed by Glover Webb and is armed with a 7.62 mm machine gun with elevation from -10 to +55°, this is aimed via a periscopic sight to the left of the weapon in the turret roof. In addition, the turret has a rear-opening, single piece circular hatch cover and three rectangular vision blocks, one to each side and one in the rear.

The rear of the hull slopes downwards at about 45° and has an outward-opening rectangular hatch hinged at the bottom in its lower part.

Standard equipment includes a spare wheel carried internally, air conditioning, searchlight, interior light, ammunition box holders, fire extinguishers, driving mirrors, external light guards, run-flat bands and Explofoil in the fuel tank.

Optional equipment includes gun ports and additional vision blocks, public address system, electrically operated smoke dischargers, moon glow interior lights, crew seat belts, integrated air conditioning system and various radio installations.

Glover Hornet armoured car (Terry J Gander)

Glover Hornet armoured car from rear

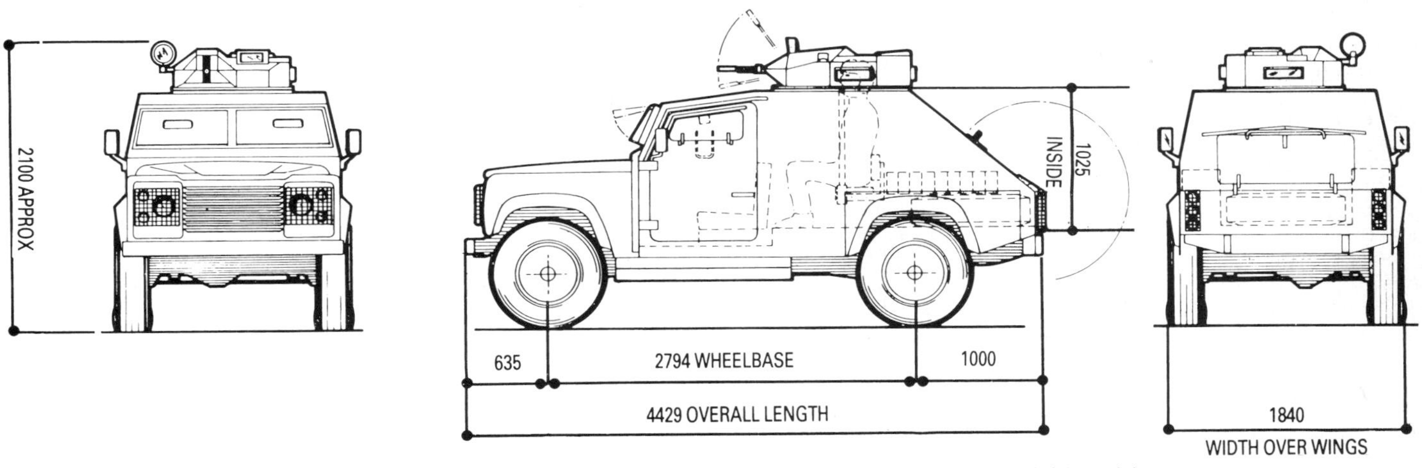

Front, side and rear drawings of Glover Hornet (4 × 4) armoured car (not to 1/76th scale)

SPECIFICATIONS		MAX ROAD SPEED	120 km/h	CLUTCH	single dry plate
CREW	3	FUEL CAPACITY	55 l	SUSPENSION	heavy duty long travel
CONFIGURATION	4 × 4	MAX ROAD RANGE	700 km		coil springs, anti-roll
COMBAT WEIGHT	3600 kg	VERTICAL OBSTACLE	0.23 m		bars fitted front and
POWER-TO-WEIGHT		TURNING RADIUS	8.75 m		rear
RATIO	31.66 bhp/tonne	ENGINE	Rover 4-stroke V-8,	TYRES	9.00 × 16
LENGTH	4.429 m		3528 cc petrol	BRAKES	
WIDTH OVER WINGS	1.84 m		developing 114 bhp at	main	dual line hydraulic
HEIGHT			4000 rpm		with disc front, drum
over turret	2.1 m	TRANSMISSION	manual with 4 forward		rear, servo-assisted
hull roof	1.85 m		and 1 reverse gears	handbrake	mechanical, on
GROUND CLEARANCE	0.32 m	TRANSFER CASE	2-speed		transfer box
TRACK	1.5 m	STEERING	recirculating ball,	ELECTRICAL SYSTEM	12 V
WHEELBASE	2.794 m		worm and nut, power-	ARMAMENT	1 × 7.62 mm MG
			assisted		

Status: Development complete. Ready for production.

Manufacturer: Glover Webb Limited
Hamble Lane
Hamble
Hampshire SO3 5NY
UK
Telephone: (0703) 452811
Telex: 477527
Fax: (0703) 452964

Hotspur Borderer Scout Car

Development

The Borderer 4284 scout car has been developed as a private venture by Hotspur Armoured Products and was shown for the first time at the 1984 British Army Equipment Exhibition.

Description

The Borderer scout car is based on the chassis of the Land Rover One Ten which has been modified and strengthened to accommodate the weight of the armoured hull.

The Hotspur steel armoured hull which fully encloses the engine and crew compartments has been designed to provide ballistic immunity at 25 m at an attack angle of 90° from ball ammunition fired from 7.62 mm type weapons and from point blank attack by lighter weapons such as the 9 mm Sterling SMG, M1 carbine and .44 Magnum. The floor armour protects the crew from mine fragments and grenades.

The engine and transmission are at the front of the vehicle with the personnel compartment at the rear. The Borderer would normally have a crew of three, commander, gunner and driver, but up to five can be carried if required.

The commander and driver sit to the rear of the engine and each has a windscreen to his front and side made of laminated armoured glass with anti-splinter screens of polycarbonate. The visors can be activated from inside the vehicle either manually or electrically. In each side of the hull is a single door that opens to the front, in the upper part of this is a bullet-proof vision block with a firing port underneath. There is also an emergency/loading hatch in the hull rear.

The petrol tank is under armour protection at the rear of the hull.

Mounted on the roof is a manually operated turret that can be traversed through 360°; this can be armed with a 7.62 mm general purpose machine gun. The machine gun is aimed via a periscope in the forward part of the turret roof, the turret also has a single piece hatch cover that opens to the rear and vision blocks for observation to the sides and rear. As an option a bank of four electrically operated smoke dischargers can be mounted on either side of the turret. The Borderer can also be supplied with a fully rotating machine gun hatch or with a solid roof plate.

Optional equipment includes a run-flat tyre system, night vision facility for the driver, 24 V electrical system, winch, tow hook and front shackles, self-sealing petrol tank, full air-conditioning system, stop lamps, intercom/siren, public address system, powered steering, and fuel tank with an additional 55 litre capacity.

Prototype of Hotspur Borderer scout car

Prototype of Hotspur Borderer scout car from rear

SPECIFICATIONS		ANGLE OF APPROACH/		STEERING	recirculating ball,
CREW	1 + 2 (or 1 + 4)	DEPARTURE	49°/30°		worm and nut
COMBAT WEIGHT	3300 kg	MAX ROAD SPEED	110 km/h	CLUTCH	single dry plate
POWER-TO-WEIGHT		FUEL CAPACITY	98 l	TRANSFER BOX	2-speed
RATIO	40.6 hp/tonne	MAX RANGE	300 km	SUSPENSION	beam axles front and
LENGTH	4.66 m	TURNING RADIUS	6.4 m		rear with coil springs
WIDTH	1.87 m	ENGINE	3.528-l, V-8, water-		controlled by hydraulic
HEIGHT			cooled, petrol		telescopic dampers
with turret	2.57 m		developing 134 hp at	BRAKES	dual, servo-assisted
hull roof	2 m		4000 rpm		discs front, drums rear
GROUND CLEARANCE	0.215 m	TRANSMISSION	manual, 4 forward and	TYRES	7.50 × 16
TRACK	1.486 m		1 reverse gears	ELECTRICAL SYSTEM	12 V
WHEELBASE	2.794 m			BATTERY	1 × 12 V, 58 Ah

Status: Production as required.

Manufacturer: Hotspur Armoured Products (a Division of Penman Engineering)
Heathall
Dumfries DG1 3NY
Scotland
Telephone: 0387 52784
Telex: 779771
Fax: 0387 67332

UNITED STATES OF AMERICA

M3 Cavalry Fighting Vehicle

Details of the M3 Cavalry Fighting Vehicle are given in the entry for Fighting Vehicle Systems in the *Armoured personnel carriers* section under United States of America.

Cadillac Gage V-600 Armoured Car

Development

The V-600 (6 × 6) armoured car has been developed as a private venture by Cadillac Gage Textron and was announced in 1985. The V-600 was previously known as the V-300 A1 but was renamed in mid 1986.

So far one prototype of the V-600 has been built and fitted with the private-venture Cadillac Gage 105 mm Low Recoil Force Turret also installed on the Cadillac Gage Textron Stingray light tank, full details of which are given in the *Light tanks* section.

By late 1985 the prototype of the V-600 had completed 5000 km of automotive trials. Gunnery trials with the 105 mm Low Recoil Force Turret, including firing the 105 mm gun at all angles of elevation and with the turret traversed 90° to the vehicle's centre line, were very successful.

The V-600 armoured car uses many automotive components of the Commando V-300 Armoured Vehicle Range (6 × 6) which is fully described in the *Armoured personnel carriers* section. Late in 1988 the Cadillac Gage V-600 (6 × 6) armoured car carried out extensive firepower and mobility trials in Egypt.

Description

The hull and turret of the V-600 are of all-welded Cadloy armoured steel that provides protection from small arms fire and shell splinters.

The driver is seated at the front left of the vehicle with the engine compartment to his right. Above his position is a single-piece hatch cover that opens to the rear and in front of this are three periscopes that give observation to the front of the vehicle. The centre periscope can be replaced by a passive night vision periscope for driving at night. In the

Cadillac Gage V-600 (6 × 6) armoured car fitted with Cadillac Gage Textron 105 mm Low Recoil Force Turret

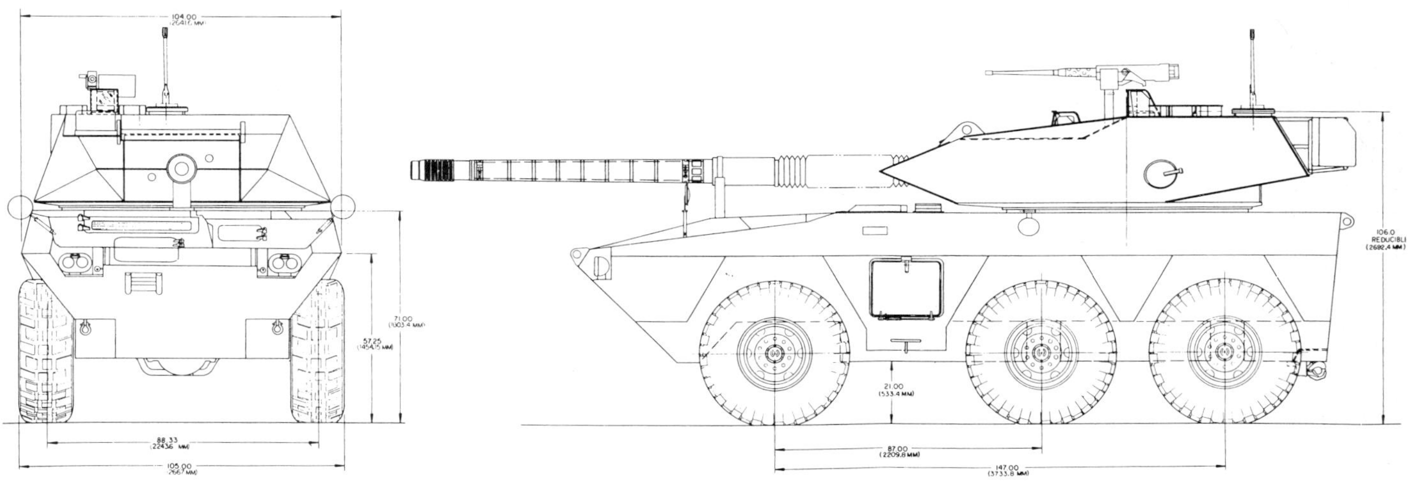

Front and side drawings of Cadillac Gage V-600 (6 × 6) armoured car with 105 mm gun in travel lock (not to 1/76th scale)

left side of the hull, to the rear of the driver and between the first and second road wheel stations, is a square hatch that opens downward to provide access to the vehicle with a bullet proof observation periscope above. Mounted on the glacis plate is a travelling lock for the 105 mm gun.

The VT-504 turbo-charged V-8 engine is to the right of the driver and develops 270 hp at 3000 rpm. It is coupled to a fully automatic Allison Transmission Division of General Motors Corporation transmission. The single-speed transfer case has a spline engagement clutch for the front axle.

All axles are of the single reduction hypoid type with silent locking differentials. Suspension of the front axle is of the coil spring type while the rear axles have torsion bars. Steering is power assisted on the front axle only. The rear two axles of the Commando V-600 are fitted with no-spin differentials that practically eliminate wheel slippage.

The turret is identical to that installed on the Cadillac Gage Stingray light tank and full details are given in the *Light tanks* section.

At the rear of the hull are two large outward-opening doors, each of which has a firing port with a vision block above. The prototype has a single door with an integral vision block. This door is used for ammunition resupply as the turret holds eight rounds, the remaining 28 rounds being carried in the hull.

Standard equipment on production vehicles would include radios, intercom for the crew of four and a manually operated fire detection and suppression system. Mounted internally at the front of the hull is a power-operated winch for self-recovery and recovering other vehicles.

Optional equipment includes NBC protection with M13A1 ventilated face masks for all crew members, chemical resistant paint, night driving lights and an automatic Halon fire detection and suppression system as well as two portable Halon bottles. If required the roof-mounted 7.62 mm machine gun can be replaced by a 12.7 mm M2 HB machine gun with 100 rounds of ready use ammunition.

Cadillac Gage V-600 (6 × 6) armoured car with all hatches closed

Status: Prototype. Not yet in production or service.

Manufacturer: Cadillac Gage Textron
PO Box 1027
Warren
Michigan 48090
USA
Telephone: (313) 777 7100
Telex: 200707
Fax: (313) 776 9731

SPECIFICATIONS		ENGINE	Cummins VT-504 V-8	ARMAMENT	
CREW	4		turbo-charged diesel	main	1 × 105 mm gun
CONFIGURATION	6 × 6		developing 270 bhp at	coaxial	1 × 7.62 mm MG
COMBAT WEIGHT	18 500 kg		3000 rpm	anti-aircraft	1 × 7.62 mm MG
POWER-TO-WEIGHT		TRANSMISSION	Allison Transmission	SMOKE DISCHARGERS	2 × 4 (optional)
RATIO	14.59 bhp/tonne		Division of General	AMMUNITION	
HULL LENGTH	6.3 m		Motors automatic 5	105 mm	8 + 26
WIDTH	2.68 m		speeds forwards and 1	7.62 mm coaxial	400 + 2000
HEIGHT			reverse	7.62 mm anti-aircraft	200 + 1000
with turret	2.74 m	TRANSFER CASE	single speed with	smoke grenades	8 + 8
reduced	2.69 m		spline engagement	GUN CONTROL	
to hull top	1.8 m		clutch for front axle	EQUIPMENT	
GROUND CLEARANCE	533 mm (hull)	STEERING	hydraulically assisted,	Turret power control	electro-hydraulic/
TRACK (front and rear)	2.24 m		front wheels only		manual
WHEELBASE		SUSPENSION	coil springs front,	by commander	yes
1st to 2nd axle	2.209 m		torsion bar rear	by gunner	yes
1st to 3rd axle	3.733 m	TYRES	14R20-XL tubeless	Max rate of power traverse	360° at 30°/s
MAX SPEED	90 km/h		radial (run flat)	Gun elevation/depression	+20°/-7.5°
RANGE	600 km	BRAKES	dual hydraulic	Gun stabiliser	
GRADIENT	60%	ELECTRICAL SYSTEM	24 V	vertical	optional
SIDE SLOPE	30%	BATTERIES	4 × 6TN	horizontal	optional
VERTICAL OBSTACLE	0.61 m				

Cadillac Gage Light Armored Vehicle Assault Gun

Development

When the Light Armored Vehicle - Assault Gun (LAV-AG) programme was started by the US Army and Marine Corps, prototypes of an Assault Gun version on the Light Armored Vehicle chassis were tested with a two man turret armed with a 90 mm Cockerill Mk III gun, but this was not adopted for service.

In February 1985 an RFP was issued for an assault gun that could be installed on the current LAV (8 × 8) vehicle with the requirement that the weapon be able to fire a kinetic energy (KE) projectile that would penetrate 200 mm of rolled homogenous armour at a range of 2000 m at 0° obliquity. A chemical energy (CE) round of the HEAT type was also required that would penetrate and blow a hole in a standard reinforced bunker, that is 200 mm of reinforced concrete.

Five companies responded to the RFP and in September 1985 the Naval Surface Weapons Centre announced that it had placed two contracts, one with Israel Military Industries for its 60 mm HVWS and another with MECAR for its 90 mm/46 KEnerga gun system with a Western Design Corporation automatic loader.

Early in 1988 an LAV was demonstrated to the US Marine Corps fitted with a 105 mm gun designated the EX-35 developed by Benet Laboratory and fitted with a muzzle brake.

In December 1988 the US Marine Corps approved a requirement for a LAV-AG and selected the Benet Laboratory 105 mm EX-35 gun as the Government Furnished Equipment (GFE) weapon.

Following competitive bidding, in mid-1990 Cadillac Gage Textron was selected to develop the Assault Gun Version of the LAV to meet the requirements of the US Marine Corps.

Under the terms of the 40 month contract, Cadillac Gage Textron will build three prototypes of the LAV-AG for extensive trials.

The turret of the LAV-AG, modified in a number of key areas, including additional armour protection, is being proposed by a number of companies for installation on a tracked chassis to meet the US Army's requirement for an Armored Gun System.

Description

Cadillac Gage Textron are designing a new two man power operated turret for the LAV-AG which will be installed on a modified chassis supplied by the Diesel Division, General Motors of Canada. The chassis will be very similar to that used for the standard LAV described in the Armoured Personnel Carriers section under Canada but will probably have additional buoyancy aids fitted either side of the turret towards the rear and a different roof arrangement as there will be no hatches in the roof of the vehicle at the rear.

The new welded steel turret will be armed with the 105 mm EX-35 gun which will be provided as government furnished equipment, as will the chassis.

The gun will be fed by a bustle-mounted automatic loader developed by Fairey Hydraulics of the UK with Hughes Aircraft supplying the sighting system, Computer Devices Corporation the computerised fire control system while the Controls Systems Division of Cadillac Gage Textron will supply the electric gun control equipment and stabilisation system.

The 105 mm EX-35 gun will be able to fire the full range of NATO ammunition types including APFSDS types. A total of eight rounds of ready use ammunition will be carried in the turret bustle with additional rounds being carried inside the hull.

Mounted coaxial with the main armament will be a 7.62 mm M240 machine gun with provision for a 7.62 mm or 12.7 mm machine gun on the turret roof. Mounted either side of the turret is a bank of four M257 electrically operated grenade launchers.

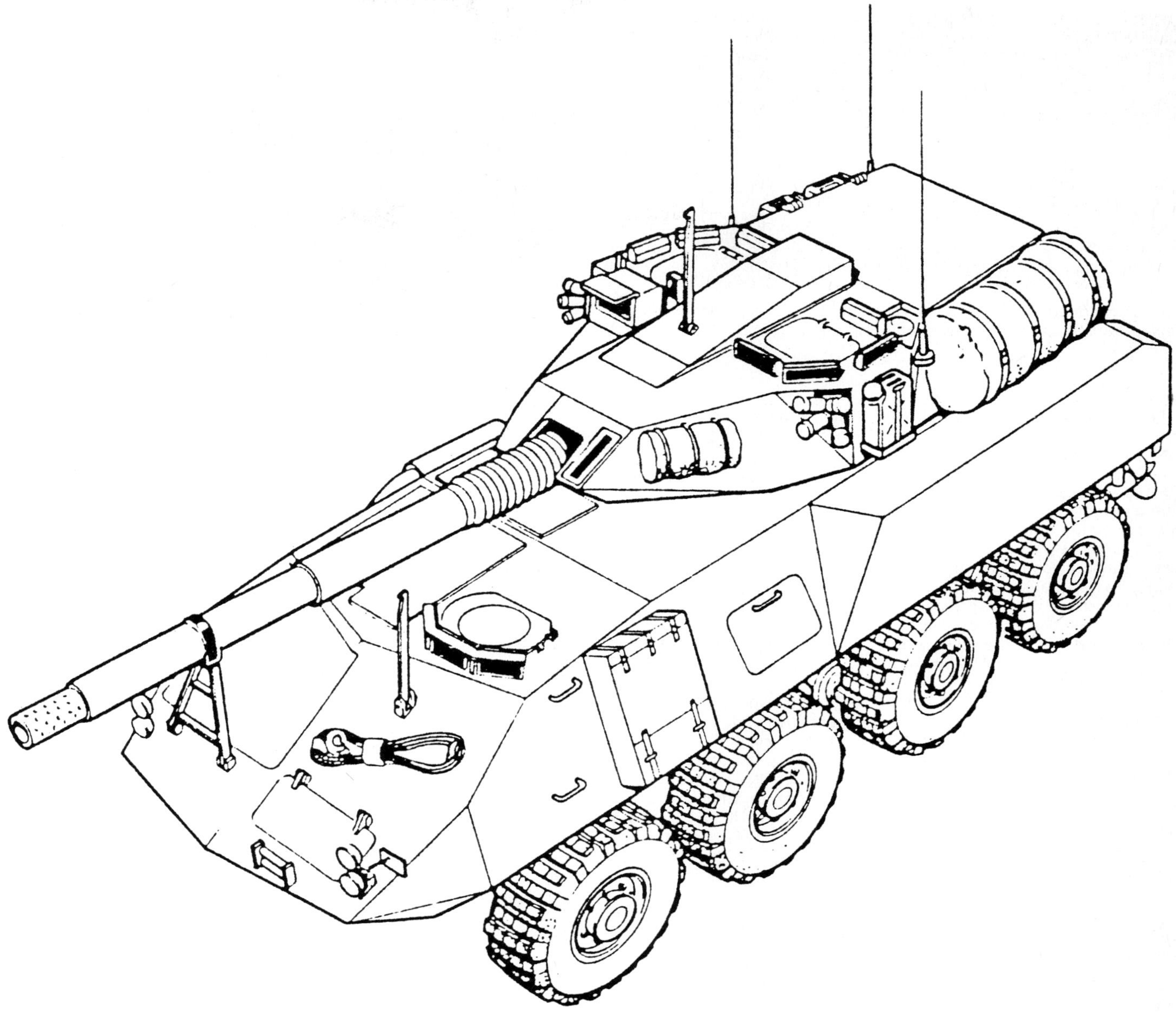

Artist's impression of Cadillac Gage Textron Light Armored Vehicle — Assault Gun with two man turret fitted with 105 mm EX-35 rifled tank gun

The commander will have a Hughes stabilised thermal day/night remote monitor and four day periscopes for observation purposes. The gunner will have a stabilised thermal day/night sight with laser rangefinder, four periscopes and a backup periscope. Both the commander and gunner will be provided with a single piece hatch cover that opens to the rear.

The research and development contract consists of 20 months to design and build the prototypes with a further 20 months of US Marine trials. In addition there is provision for an initial batch of 44 LAV-AG with a final batch of 110 vehicles.

The production decision is due in January 1994 with first deliveries being made in July 1995 and continuing for a period of two to three years.

Status: Under development for US Marine Corps.

Manufacturer: Cadillac Gage Textron
PO Box 1027
Warren
Michigan 48090
USA
Telephone: (313) 777 7100
Telex: 200707
Fax: (313) 776 9731

SPECIFICATIONS	
CREW	3
TURRET WEIGHT	3697 kg (LAV)
	4468 kg (AGS)
ARMAMENT	
main	1 × 105 mm gun
coaxial	1 × 7.62 mm MG
anti-aircraft	1 × 7.62 mm MG (or 1 × 12.7 mm)
AMMUNITION	
105 mm	8 ready use
7.62 mm	400 rounds (turret)
GUN CONTROL EQUIPMENT	electric/manual
GUN ELEVATION DEPRESSION	+15°/−8°
TRAVERSE	360°
STABILISER	
vertical	yes
horizontal	yes
RANGEFINDER	yes, laser

FMC Lynx (or M113) Command and Reconnaissance Vehicle

Development

This Command and Reconnaissance Vehicle was developed as a private venture by the FMC Corporation with the first prototype being completed in 1963. Although tested by the United States Army it was not adopted as it had already selected the M114 Command and Reconnaissance Carrier which is itself very similar in appearance to the FMC vehicle. The M114 has now been phased out of service with the United States Army but was fully described in *Jane's Armour and Artillery 1979-80* page 159.

The vehicle shares many common automotive components with the M113A1 armoured personnel carrier but is much lower, has its engine at the back instead of the front and has one fewer road wheel each side. The vehicle has been bought by two countries, Canada and the Netherlands. Canada calls the vehicle the Lynx and the first of 174 vehicles was completed in May 1968. The Dutch Army ordered 250 vehicles, the first of which was completed in September 1966 and they refer to the vehicle as the M113 C & R.

Canadian Armed Forces Lynx command and reconnaissance vehicle in United Nations markings (Canadian Forces photograph)

Late in March 1985 FMC submitted a proposal to Canada for the upgrading of Lynx vehicles, but as of early 1991 this option had not been exercised.

Description (Lynx)

The all-welded aluminium armour hull of the Lynx provides the crew with protection from small arms fire, flash burns and shell fragments.

The driver is seated at the front of the vehicle on the left side and has a single piece hatch cover that opens to the rear with an integral M19 infra-red periscope mounted in its roof and five M17 periscopes arranged around the forward part in the roof.

The commander, who also operates the main armament, is seated to the right and rear of the driver and is provided with an M26 turret that can be traversed manually through 360°, and has eight vision blocks for all-round observation.

The third crew member, the radio operator/observer, is seated to the left rear of the commander/gunner and is provided with a single piece hatch cover that opens to the rear. Three M17 and one M17C periscopes are mounted in the roof of the vehicle around the hatch cover.

Access to the engine at the rear of the hull on the right side is by a large hatch in the roof that opens to the right. In addition there is an access door in the hull rear which opens to the left. The Detroit Diesel Model 6V-53 diesel is coupled to an Allison TX100 transmission, which transmits power via a propeller shaft to the FMC DS200 controlled differential at the front of the hull, access to which is by an opening in the glacis plate.

The torsion bar suspension consists of four dual road wheels each side with the drive sprocket at the front and the idler wheel at the rear. There are no track return rollers. The tops of the tracks have a rubber cover which can be removed to reduce the overall width of the vehicle.

The Lynx is fully amphibious, being propelled in the water by its tracks. Before entering the water a trim vane is erected at the front of the hull and rectangular covers are erected around the air inlet and exhaust louvres on the top of the hull.

Standard equipment includes infra-red driving lights and a fire extinguisher mounted in the engine compartment that can be operated by the driver or from outside the vehicle.

Optional equipment includes an NBC detection and alarm system, heater, windscreen and a capstan drum which, when attached to the drive sprockets and used in conjunction with an anchor and cables, can be used for self-recovery.

The M26 turret is armed with a 12.7 mm (0.50) M2 HB machine gun that can be aimed and fired from within the vehicle. Turret traverse and elevation of the main armament are manual. A 7.62 mm (0.30) machine gun is mounted at the radio-operator/observer's station at the rear of the hull and three smoke dischargers are mounted on either side of the hull at the front.

Dutch Army command and reconnaissance vehicle fitted with Oerlikon GBD-AOA turret armed with 25 mm cannon and showing door in right side of hull in open position

Variants

Dutch Vehicle

The Dutch vehicle has a layout slightly different from that of the Canadian Lynx. The driver is seated at the front of the vehicle on the left and has a single piece hatch cover that opens to the rear with an integral M19 infra-red periscope mounted in its roof and four M17 periscopes arranged around the forward part in the roof. The radio-operator/gunner is seated to the right of the driver and has a hatch cover that opens to the rear and four M17 periscopes arranged around the forward part of the hatch cover. A 7.62 mm machine gun can be mounted forward of this hatch cover if required. There is an entry door in the right side of the hull, to the rear of the radio-operator/gunner.

When originally delivered the Dutch vehicles were armed with a 12.7 mm (0.50) machine gun but in May 1974 the Dutch Army ordered 266 Oerlikon GBD-AOA turrets armed with a 25 mm Oerlikon KBA-B cannon for installation on the vehicles.

Optional Armament Installations

FMC has also offered the following armament installations: Model 100-E cupola with a 7.62 mm (0.30) M73 machine gun, Model 74 cupola with twin 7.62 mm M73 or twin 0.30 (7.62 mm) M37 machine guns, XM27 cupola (modified M26) with 20 mm Oerlikon cannon, M113 cupola with 12.7 mm (0.50), 7.62 mm (0.30) or 7.62 mm machine gun pintle-mounted, FMC Pedestal Mount with 7.62 mm (0.30), 12.7 mm (0.50), 20 mm or 25 mm cannon that can be loaded, aimed and fired from within the vehicle, ATGWs mounted in launcher boxes on the roof of the vehicle, 106 mm M40 recoilless rifle mounted on the roof, and other types of turret as requested by the user.

Product Improved Vehicle

For trials purposes FMC has fitted a vehicle with a new tube-over-bar suspension system, similar to that fitted to its AIFV and a more powerful turbo-charged diesel engine. These modifications give the vehicle improved cross-country performance and a higher road speed.

Canadian Armed Forces Lynx command and reconnaissance vehicle with 12.7 mm M2 HB machine gun installed (Canadian Forces photograph)

Status: Production has been completed for the Canadian and Dutch armies but can be resumed if further orders are received.

Manufacturer: FMC Corporation, Ground Systems Division
2830 De La Cruz Boulevard
Santa Clara
California 95052
USA
Telephone: (408) 289 0111
Telex: 6714210
Fax: (408) 289 2150

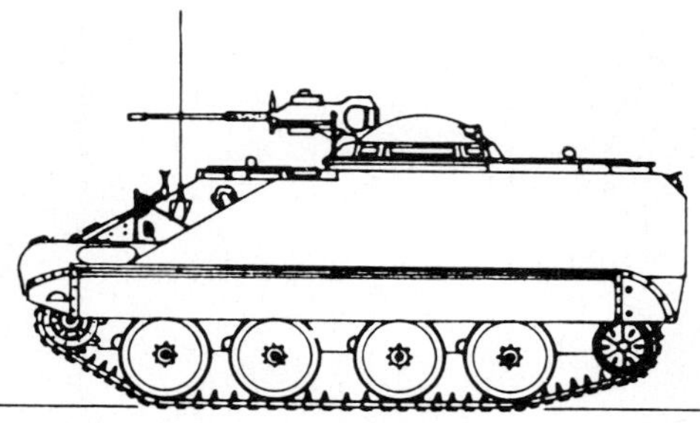

Lynx command and reconnaissance vehicle

SPECIFICATIONS (Canadian Lynx)	
CREW	3
COMBAT WEIGHT	8775 kg
UNLOADED WEIGHT	7725 kg
POWER-TO-WEIGHT RATIO	24.5 hp/tonne
GROUND PRESSURE	0.48 kg/cm^2
LENGTH	4.597 m
WIDTH	2.413 m
reduced	2.266 m
HEIGHT	
to hull top	1.651 m
with armament	2.18 m
GROUND CLEARANCE	0.41 m
TRACK	1.885 m
TRACK WIDTH	381 mm
LENGTH OF TRACK ON GROUND	2.39 m
MAX SPEED	
road	70.8 km/h
water	5.6 km/h
ACCELERATION 0-48 km/h	18.5 s
FUEL CAPACITY	303 l
MAX CRUISING RANGE	523 km
FORDING	amphibious
GRADIENT	60%
SIDE SLOPE	30%
VERTICAL OBSTACLE	0.609 m
TRENCH	1.524 m
ENGINE	Detroit Diesel 6V-53, 6-cylinder, 2-stroke, water-cooled diesel developing 215 hp at 2800 rpm
TRANSMISSION	Allison Transmission Division of General Motors Corporation TX100 with torque converter, 3 forward and 1 reverse gears
SUSPENSION	torsion bar
ELECTRICAL SYSTEM	24 V
BATTERIES	2 × 12 V
ARMAMENT	
main	1 × 12.7 mm MG
secondary	1 × 7.62 mm MG
SMOKE-LAYING EQUIPMENT	2 × 3 smoke dischargers
AMMUNITION	
main	1155
secondary	2000

Chrysler MAC-1 Armoured Car

Between 1963 and 1964 The Defense Group of the Chrysler Corporation (now the Land Systems Division of General Dynamics) delivered 15 MAC-1 (4 × 4) armoured cars to Mexico. These still remain in service. Full details of this vehicle were given in *Jane's Armour and Artillery 1979-80*, page 166.

Cadillac Gage Commando Scout

Development

The Commando Scout (4 × 4) has been designed as a private venture by Cadillac Gage Textron, manufacturer of the well-known Commando range of multi-mission vehicles. The vehicle was first announced in October 1977 at the Association of the United States Army meeting in Washington DC. The basic role of the Commando Scout is reconnaissance and security but it can also be fitted as a command or anti-tank vehicle. In 1983 it was confirmed that Indonesia had placed an order for 28 Commando Scout vehicles as well as Commando Ranger armoured personnel carriers (22). In August 1986 it was announced that a foreign country had placed a $22.8 million contract with Cadillac Gage for an undisclosed number of Commando Scout reconnaissance vehicles with half the vehicles delivered in 1986 and the remainder in 1987. Middle East sources confirm that the country is Egypt and 112 vehicles were purchased. By early 1991 a total of 140 Commando Scout (4 × 4) vehicles had been built by Cadillac Gage.

Description

The welded hull of the vehicle is made of special hardness Cadloy armour plate which will defeat at a minimum 7.62 mm armour-piercing rounds. The front, sides and rear of the hull are well sloped to afford maximum protection within the weight of the vehicle.

The front of the Commando Scout is sloped at an angle of 76° from the top of the driver's hatch to the nose. In addition to providing ballistic protection it allows the vehicle to push its way through underbrush.

The driver is seated on the left side of the hull with the engine to his right and is provided with an adjustable seat and a single piece hatch cover that slides to the front of the vehicle when he is driving with his head out. Driver vision is provided by three periscopes mounted forward of the hatch area.

The fuel tank is at the front of the hull between the wheels. The power plant is coupled to an Allison Transmission Division of General Motors Corporation four-speed automatic transmission via a Cadillac Gage power transfer unit. The complete powerpack, consisting of the engine,

Cadillac Gage Commando Scout armed with 12.7 mm and 7.62 mm machine guns in 1 m turret

Cadillac Gage Commando Scout armed with twin 7.62 mm machine guns in 1 m turret

Cadillac Gage Commando Scout fitted with command pod armed with 12.7 mm machine gun with night observation sight mounted coaxial with weapon

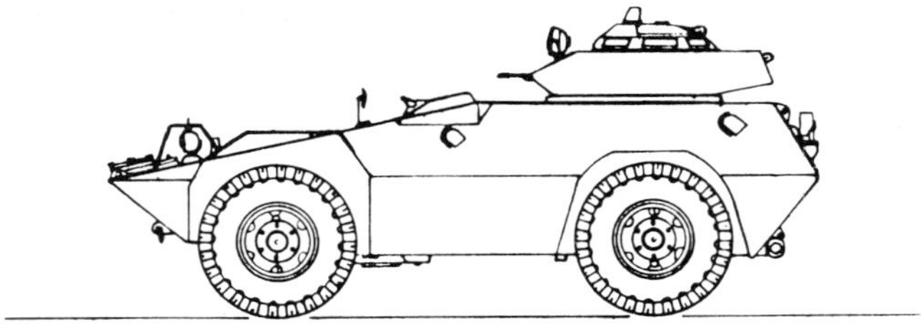

Commando Scout with 1 m turret

transfer unit, transmission and cooling system, can be removed and replaced in two hours. Access to the engine is by a large hatch in the right side of the hull through which all fluid levels (coolant, lubricant, brake fluids etc) can be checked.

The turret or pod is at the rear of the vehicle and access is by a two-part hatch in the rear of the hull with the bottom part folding downwards and the top part opening to the right.

The front suspension features coil springs over a solid axle with trailing arms. The split rear axle is tied to a heavy duty coil spring suspension. Cone-shaped passages in the rear of the hull allow for high individual vertical road wheel travel. Both axles are fitted with positive locking differentials which provide improved traction by preventing one wheel spin-out.

The integral hydraulic power steering is powered by a gear-driven pump working directly off the engine. This approach has eliminated the requirement for belt drives which slip or break and steering cylinders that require maintenance. The independent front and rear brake systems function through a split master cylinder and a back-up system supports them in the event of an engine failure. The back-up system consists of an electric motor which supplies pressure to the master cylinder. The tyres are of the run-flat type. Standard equipment includes two hand-held fire extinguishers stowed inside the vehicle, pioneer tool set, breaker bar and lug wrench, first aid kit, vehicle tool kit, and an air compressor with 15.24 m hose.

The Commando Scout is not fitted with an NBC system, has no night vision equipment and is not amphibious. Optional equipment includes a siren/public address system, radio installations, weapon stowage, water and petrol can, slave cable, 15.24 m auxiliary cable, camouflage net, smoke grenade system and fragmentation grenade system.

Variants

Twin/Combination Machine Gun (1 m) Turret

This turret is manually/electrically rotated by a Power Assist Traverse (PAT) system and can be armed with twin 7.62 mm or twin 0.50 calibre machine guns, or a combination of 7.62 mm/0.50 calibre machine guns, which can be elevated from −10 to +55°, manually traversed rate with PAT is a maximum of 45°/s. The amount of ammunition depends on the armament installed but if twin 7.62 mm calibre weapons are fitted each gun is provided with 200 rounds of ready use ammunition and 2200 rounds stowed in the vehicle. If twin 0.50 calibre weapons are fitted, each weapon has 100 rounds of ready use ammunition and 1000 rounds stowed in the hull. It can also mount a 7.62 mm MG and a 40 mm Mark 19 grenade launcher with 100 ready rounds and 200 rounds stowed in the vehicle. The turret is also provided with eight vision blocks, an M28C sight with a magnification of ×1.5 (×5 magnification optional) and a 500 000 candlepower spotlight mounted coaxially with the weapons.

40 mm/12.7 mm Turret

One-man turret armed with same turret as on US Marine Corps AAV7A1 armoured amphibious assault vehicles. This is armed with 40 mm Mk 19 grenade launcher and 12.7 mm M2 machine gun with manual traverse and elevation.

Twin Machine Gun (1 m) Turret

This is armed with twin 7.62 mm machine guns which can be fired singly or together and 400 rounds of ready use ammunition are carried with a further 2200 rounds stowed in the vehicle. It has manual traverse and elevation and the weapons can be elevated from −10 to +55°. The turret is provided with eight vision blocks, and an M28C sight with a magnification of ×1.5 (×5 magnification optional); a 500 000 candlepower spotlight is mounted coaxially with the weapons.

Command Pod

This model has a crew of three, commander, radio operator and driver and has a pod mounted on top of the hull in each side of which is a firing port and a vision block. A circular hatch cover is provided on which a 7.62 mm machine gun and shield can be mounted. 3000 rounds of 7.62 mm ammunition are carried. When fitted with radios the antennas are mounted at the rear of the hull.

Anti-tank armed with TOW

This model has a crew of two and is fitted with a Hughes TOW ATGW launcher which has a traverse of 120° and can be elevated from −10 to +30°. Two TOW ATGWs are carried inside the hull with provision for carrying a further four missiles outside. A standard TOW ground mount is carried inside the vehicle which enables the TOW system to be deployed away from the vehicle. Two positions for mounting one 7.62 mm machine gun are also provided with 2000 rounds of ammunition being carried for this weapon.

Status: Production as required. In service with Egypt and Indonesia.

Manufacturer: Cadillac Gage Textron
Post Office Box 1027
Warren
Michigan 48090
USA
Telephone: (313) 777 7100
Telex: 200707
Fax: (313) 776 9731

SPECIFICATIONS

CREW (depending on role)	1 + 1 or 1 + 2
CONFIGURATION	4 × 4
COMBAT WEIGHT	7240 kg
POWER-TO-WEIGHT RATIO	20.58 hp/tonne
LENGTH	5.003 m
WIDTH	2.057 m
HEIGHT	2.159 m
TRACK	1.66 m
WHEELBASE	2.743 m
MAX ROAD SPEED	96 km/h
FUEL CAPACITY	378 l
MAX ROAD RANGE	1287 km
FORDING	1.168 m
GRADIENT	60%
SIDE SLOPE	30%
VERTICAL OBSTACLE	0.609 m
TURNING RADIUS	7.925 m
ENGINE	Cummins V-6 diesel developing 149 hp at 3300 rpm
TRANSMISSION	Allison Transmission Division of General Motors Corporation automatic with 4 forward and 1 reverse speeds
TRANSFER CASE	2-speed with spline engagement clutch for front-axle drive
STEERING	integral hydraulic
SUSPENSION	
front	coil springs over solid axle with trailing arms
rear	independent rear axle with coil springs for each wheel
TYRES	15.50 × 21 run-flat tubeless
BRAKES	
main	dual power 4-wheel hydraulic, pump driven emergency electrical back up
parking	drum and shoe, hand operated
ELECTRICAL SYSTEM	24 V
BATTERIES	2 × 12 V, 100 Ah

NOTE: Height and weight are for vehicle with 1 m turret and twin 7.62 mm MGs.

Armoured personnel carriers

ARGENTINA

TAMSE VCTP Infantry Combat Vehicle

Development
In mid-1974 the German company of Thyssen Henschel was awarded a contract for the design and development of a new medium tank for the Argentine Army under the designation TAM (Tanque Argentino Mediano), as well as an infantry combat vehicle under the designation VCTP (Véhiculo de Combate Transporte de Personal). The contract covered the design, development and construction of three prototypes of both the tank and the VCTP. The total requirement was for just over 500 vehicles, 200 TAMs and 300 VCTPs, but production was completed after about 350 had been built owing to the then current financial situation.

Description
The VCTP is similar to the Marder ICV but has a more powerful 720 hp engine, less sophisticated two-man turret armed with a 20 mm cannon and a 7.62 mm roof-mounted machine gun, three gun ports in each side of the troop compartment at the rear, two rectangular roof hatches over the troop compartment that are hinged in the centre, and a remote-controlled 7.62 mm machine gun installation which is fitted over the troop compartment at the rear of the hull. Mounted either side of the hull, slightly forward of the turret is a bank of four electrically operated smoke dischargers firing forwards. The vehicle is fitted with an NBC system which pressurises the crew compartment for three hours. A crew compartment heating system also pre-heats the engine for cold starts.

The two man power operated turret is mounted in the centre of the hull with the commander being seated on the left and the gunner on the right, both have a single piece hatch cover that opens to the rear with the commander being slightly raised over the gunner's position for increased all round observation.

The commander has a sight and seven observation periscopes while the gunner has a sight and three observation periscopes. Turret traverse and weapon elevation is electro-hydraulic. The 20 mm cannon can be elevated from -11° 30° to +60° at a maximum speed of 60°/s with traverse through a full 360° at a maximum speed of 52°/s.

The 20 mm MK 20 Rh 202 cannon has 400 rounds of ready use ammunition in the turret (325 HE and 75 APDS) with a further 1000 rounds in reserve in the hull.

The remote-controlled turret is called the TPA-1 and is armed with a 7.62 mm machine gun and a similar weapon is mounted at the commander's station for anti-aircraft defence.

To increase the operational range of the vehicle long range fuel tanks can be mounted either side of the hull at the rear.

Variants

VCTM 120 mm mortar carrier

The mortar version of the VCTP is called the VCTM (Véhiculo de Combate Transporte de Mortero) and is fitted with a 120 mm Brandt AM-50 mortar for which a total of 61 rounds are carried. Of these 12 are of the PEPA-LA long range type with a maximum range of 9500 m with the remaining 49 being standard mortar bombs with a range of 6650 m.

The driver is seated front left with vehicle commander to his rear. The mortar is mounted in the centre of the vehicle firing forwards and when not in action a three part roof cover is positioned over the mortar, the centre part opens forward onto the glacis plate and the other parts opening left and right. There is an additional hatch to the right rear of the vehicle and mounted on the roof at the rear of a remote-controlled 7.62 mm machine gun as fitted to the basic VCTP.

VCPC command post vehicle

This is the command post member of the family and is called the Vehiculo de Combate Puesto de Comando (VCPC) and has extensive communications equipment installed. Armament consists of a pintle-mounted 7.62 mm machine gun on the cupola in the centre of the vehicle on the right side. The two oblong hatches over the rear troop compartments are retained as are the firing hatches and observation periscopes.

VCRT armoured recovery vehicle

Details of this are given in the TAM entry.

VCLC multiple rocket launcher

Details of this are given in the *Multiple rocket launchers* section.

VCA 155 self-propelled gun

Details of the Vehiculo de Combate de Artilleria based on a modified TAM chassis are given in the *Self-propelled guns and howitzers* section under Argentina.

SPECIFICATIONS

Model	**VCTP**	**VCTM**	**VCPC**
TYPE	ICV	mortar	command
CREW	2 + 10	5	3 + 6
WEIGHT COMBAT	28 tonnes	26 tonnes	24.5 tonnes
GROUND PRESSURE	0.69 kg/cm^2	0.67 kg/cm^2	0.63 kg/cm^2
LENGTH	6.83 m	6.83 m	6.83 m
WIDTH	3.32 m	3.32 m	3.30 m
HEIGHT OVERALL	2.68 m	1.86 m	2.35 m
GROUND CLEARANCE	0.45 m	0.45 m	0.45 m
TRACK	2.62 m	2.62 m	2.62 m
MAXIMUM ROAD SPEED			
forwards	75 km/h	75 km/h	75 km/h
reverse	75 km/h	75 km/h	75 km/h
FUEL CAPACITY			
main	640 l	640 l	640 l
external	400 l	400 l	400 l
RANGE			
main	590 km	590 km	590 km
external	350 km	350 km	350 km
FORDING			
without preparation	1.5 m	1.5 m	1.5 m
GRADIENT	60%	60%	60%
SIDE SLOPE	30%	30%	30%
VERTICAL OBSTACLE	1 m	1 m	1 m
TRENCH	2.5 m	2.5 m	2.5 m
ENGINE	MTU MB 833 Ka - 500 V-6 diesel developing 720 hp at 2400 rpm		
TRANSMISSION	Renk HSWL 204 with 4 forward and 2 reverse gears		
SUSPENSION	torsion bar in all versions		
ELECTRICAL SYSTEM	24 V	24 V	24 V
BATTERIES	8 × 12 V	8 × 12 V	8 × 12 V
ARMAMENT			
main	1 × 25 mm	1 × 120 mm	1 × 7.62 mm
anti-aircraft	1 × 7.62 mm	nil	nil
remote-controlled	1 × 7.62 mm	1 × 7.62 mm	nil
SMOKE DISCHARGERS	2 × 4	2 × 4	2 × 4
AMMUNITION			
25 mm	1400	nil	nil

VCTP infantry combat vehicle showing bank of four smoke dischargers on forward part of hull

VCTP 120 mm mortar carrier with remote-controlled 7.62 mm machine gun and roof hatches open to allow 120 mm mortar to be fired

VCTP command post vehicle with cupola-mounted 7.62 mm machine gun, smoke dischargers and two radio antennas

Status: Production complete but can be resumed if further orders are received. In service in Argentina.

VCTM 120 mm mortar carrier with roof hatches open to show mortar ready to fire

Manufacturer: Tanque Argentino Mediano Sociedad Del Estado
Avenida Rolon 1445
Boulogne Sur Mer
CP 1609
Buenos Aires
Argentina
Telephone: 763 114/4937/0714
Telex: 26208 TAMSE AR
Fax: 763 2464

AUSTRIA

Steyr-Daimler-Puch Kampfschützenpanzer 90 Combat Vehicle (MICV 90)

Development

The Kampfschützenpanzer 90 (KSPz 90) is being jointly developed by Steyr-Daimler-Puch of Austria and SANTA BARBARA of Spain to meet the battlefield requirements of the 1990s with the first prototype being completed in 1990.

The Austrian Army has a requirement for around 400 vehicles while the Spanish Army's requirement is for some 600 vehicles.

The MICV 90 is one of the five vehicles to be evaluated by the Norwegian Army from October 1991, the others are the Krauss-Maffei Puma, FMC M2A2 Bradley, GKN Defence Operations Warrior and the Bofors/Hägglunds Vehicle Combat Vehicle 90.

This vehicle was entered to meet 17 of the roles outlined in the British Army's FFLAV programme covered in the *AFV Families* section.

Description

The hull and turret bustle of the vehicle is of all-welded steel armour construction which provides protection against attack from small arms fire and shell splinters. The frontal arc, from 12° to both sides of the longitudinal axis of the vehicle and an angle of attack from -5 to +8° from the horizontal plane, is proof against 30 mm APDS projectiles fired from a range of 1000 m.

The driver sits at the front left and is provided with a single piece hatch cover forward of which are three periscopes. The centre periscope can be replaced by a passive night vision periscope.

The engine compartment is to the right of the driver with the air inlet louvres in the roof and the exhaust outlet and air outlet louvres on the right side of the hull.

The vehicle is powered by a six cylinder diesel developing 500 hp which is coupled to a six-speed HSWL 106 hydro-mechanical transmission with six forward gears and three reverse. The hydrostatically controlled steering assembly and reverse gears are integrated into the transmission.

The two-man power-operated turret is mounted in the centre of the vehicle offset to the right with the commander seated on the left and the gunner on the right, both with a single piece hatch cover that opens to the rear. In addition to their roof-mounted periscopes for aiming the main and secondary armament, the commander has five roof-mounted observation periscopes while the gunner has two.

Main armament comprises a 30 mm Mauser Model F cannon with a 7.62 mm machine gun mounted coaxially to the left, weapon elevation is from -10 to +50° with turret traverse being 360°. Mounted either side of the turret is a bank of three electrically operated smoke dischargers firing forwards. Turret traverse is electro-mechanic at a speed of 45°/s with elevation also being electro-mechanic at a speed of 30°/s. The commander can override the gunner and a turret and gun stabilisation (two axis gyro) system is fitted as standard.

The turret is fitted with a modular thermal and visible target acquisition system (ground/ground and limited ground/anti-aircraft) for the gunner. This system consists of the following key components:

1. Optical sight with thermal channel and gunner's display
2. Integrated laser rangefinder
3. Cant angle sensor
4. Fire control computer
5. Fire control computer control panel (gunner's station)
6. Ammunition selection unit
7. Commander's remote display (TV monitor).

In addition the commander is provided with a periscope day sight.

Other types of 20 mm, 25 mm or 30 mm cannon can also be installed.

The eight infantry sit in the rear of the vehicle with two rectangular roof hatches. The infantry normally enter and leave the vehicle via the twin doors in the hull rear.

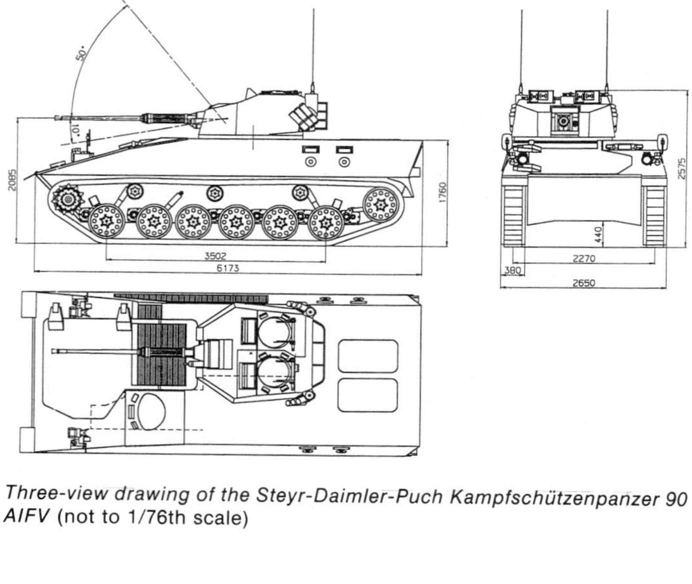

Three-view drawing of the Steyr-Daimler-Puch Kampfschützenpanzer 90 AIFV (not to 1/76th scale)

Artist's impression of Steyr-Daimler-Puch MICV 90 fitted with two man turret armed with 30 mm cannon

Mounted in either side of the rear troop compartment are two observation/aiming firing ports with an episcope above so that some of the troops can fire their infantry weapons from within the vehicle in complete safety.

Suspension is of the torsion bar type with six dual rubber-tyred road wheels, drive sprocket at the front, idler at the rear and three track return rollers. Hydraulic shock absorbers and hydraulic bump stops are provided for the first, second and last road wheel stations while the other road wheels have mechanical bump stops only. The tracks are Diehl with detachable rubber pads.

Standard equipment includes passive night driving equipment for the driver, NBC protection system, communications system, vehicle intercom with loudspeaker in the crew compartment and a heater that runs on diesel fuel. To the rear of the driver is a hull escape hatch. There are also two roof-mounted ventilation fans to remove fumes from the crew compartment when the small arms are being fired. The engine compartment can be fitted with a fire detection and suppression system which can be operated either manually by the driver or automatically.

Optional equipment could include air-conditioning system, additional armour protection and a fighting compartment fire suppression system.

Variants

There are no variants of this vehicle as yet, although these are expected to be developed in the future. Other types of armament system can also be fitted to the hull including a turret-mounted 105 mm gun.

Status: Prototypes undergoing company trials.

Manufacturer: Steyr-Daimler-Puch Spezialfahrzeug GmbH
PO Box 100
A-1111 Vienna
Austria
Telephone: 0222 76 45 11
Telex: 613221299

SPECIFICATIONS		GROUND PRESSURE	0.681 kg/cm²	ARMAMENT	
CREW	3 + 8	MAX ROAD SPEED		main	1 × 30 mm cannon
COMBAT WEIGHT	19 800 kg	forwards	70 km/h	coaxial	1 × 7.62 mm MG
POWER-TO-WEIGHT		reverse	23.4 km/h	AMMUNITION (main)	200 × 30 mm
RATIO	25.25 hp/tonne	FUEL CAPACITY	450 l	(coax)	600 × 7.62 mm
LENGTH	6.173 m	CRUISING RANGE (road)	500 km	SMOKE DISCHARGERS	2 × 3
WIDTH	2.65 m	FORDING	1.2 m	GUN CONTROL	
HEIGHT		GRADIENT	75%	EQUIPMENT	electro-mechanic/
to hull roof	1.77 m	SIDE SLOPE	40%		manual
overall	2.55 m	VERTICAL OBSTACLE	0.8 m	by commander	yes
axis of fire	2.085 m	TRENCH	2.3 m	by gunner	yes
GROUND CLEARANCE	0.45 m	TURNING RADIUS	pivot	GUN ELEVATION/	
TRACK	2.27 m	ENGINE	6-cylinder diesel	DEPRESSION	+50°/-10°
TRACK WIDTH	380 mm		developing 500 hp	TURRET TRAVERSE	360°
LENGTH OF TRACK ON		SUSPENSION	torsion bar	GUN STABILISER	
GROUND	3.517 m	ELECTRICAL SYSTEM	24 V	vertical	yes
				horizontal	yes

Steyr 4K 7FA G 127 Armoured Personnel Carrier

Development

The Steyr 4K 7FA G 127 is an uparmoured version of the earlier Saurer 4K 4FA armoured personnel carrier fitted with the engine and transmission of the tank destroyer SK 105 which is also manufactured by Steyr. The first prototype of the 4K 7FA G 127 was completed in 1976 with first production vehicles following in 1977. The Steyr 4K 7FA G 127 has been exported to Bolivia (6), Greece (qv) and Nigeria (170).

Description

The basic model is designated the 4K 7FA G 127 and is armed with a 12.7 mm (0.50) M2 HB Browning machine gun. The hull is made of all-welded steel and is immune to penetration from 20 mm armour-piercing projectiles over its frontal arc.

The driver sits at the front of the hull on the left side with the engine to his right. He has a two-piece hatch cover that opens to the left and right, in front of which are three periscopes.

The gunner sits behind the driver and has a cupola with a two-piece hatch cover opening left and right which can be locked vertical to give some protection. There are two observation periscopes in the front part of the cupola. A standard 12.7 mm (0.50) Browning machine gun is mounted on the forward part of the cupola and mounted on the rear of the cupola are four smoke dischargers which point to the rear of the vehicle. When unlocked the 12.7 mm machine gun has an elevation from -20 to +35° and when locked of -10 to +7.5°.

The personnel compartment is at the rear of the hull with the eight infantrymen seated on bench seats, four down each side of the hull. The section leader, the first man on the right bench, is provided with a periscope in the forward part of the personnel compartment, on the right side. Two hatches over the roof of the vehicle open outwards. Mounting sockets for the pivot trunnion of an adapter mount for the MG 74 machine guns are provided on the front, right, left and rear side of the hatch, above the personnel compartment. Normal means of entry and exit for the crew is by the twin doors at the rear of the hull.

The torsion bar suspension has five dual road wheels with the drive sprocket at the front, idler at the rear and three track return rollers. The first and last road wheel stations are provided with a hydraulic shock absorber. The end-connector tracks are provided with rubber pads and can be fitted with spikes for operation on ice-covered surfaces.

Standard equipment includes a ventilation system for the fighting compartment and a vehicle heater while optional equipment includes passive night vision device for driver, automatic fire extinguishing system for the engine compartment, automatic fire suppression system for the fighting compartment, collective NBC system and air-conditioning system.

Variants

4K 7FA - KSPz MICV

This model carries a total of 10 men, section leader/commander, driver, machine gunner and seven infantrymen. The driver is front left as normal

4K 7FA-KSPz infantry fighting vehicle with all hatches closed

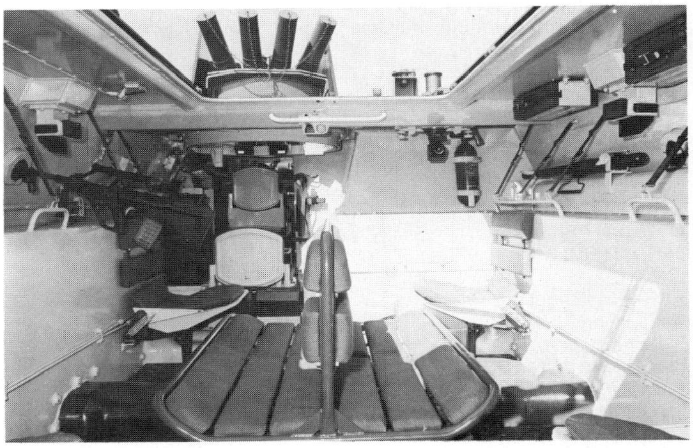

4K 7FA-KSPz mechanised infantry fighting vehicle showing Steyr 5.56 mm AUG assault rifle mounted in left firing port

4K 7FA MICV 127 showing 12.7 mm and 7.62 mm MGs in one-man turret

4K 7FA AMC 81 with roof hatches open showing 81 mm mortar

Steyr 4K 7FA G 127 armoured personnel carrier from rear with back doors open and MG 74 mounted on roof

with the section leader seated at the front right corner of the fighting compartment on a hinged single seat. He is provided with a rotatable periscope that can be traversed through 360°.

The vehicle is fitted with four ball mountings with periscopes above and the former allow for insertion of a Steyr calibre 5.56 mm AUG 508 assault rifle or a Steyr MP81 sub-machine gun, two in either side of the rear fighting compartment. In addition there are two periscopes in the hull rear.

Four of the seven infantrymen sit back to back down the centre of the vehicle so that they can use the ball-mounted weapons. The remaining members of the infantry section are seated on hinged single seats, one at each remaining corner of the fighting compartment.

Two hatches over the hull roof open outwards. Mounting sockets for the pivot trunnion of an adapter mount for the calibre 7.62 mm MG of the infantry section are provided on the front, right, left and rear side of the hatch.

4K 7FA MICV 127

This is a further development of the 4K 7FA-KSPz mechanised infantry combat vehicle. Armament consists of a 12.7 mm M2 machine gun and a coaxial 7.62 mm machine gun mounted in a Steyr one-man turret SP 1/127 with an electro-hydraulic weapon remote control system and manual/hydraulic traverse system. Mounted either side of the turret is a bank of three electrically operated smoke dischargers.

Options include a turret armed with a 20 mm, 25 mm or 30 mm cannon with an electro-hydraulic traverse system, NBC system, an automatic fire suppression system for the personnel compartment and an automatic fire extinguishing system for the engine compartment.

4K 7FA FSCV 90 Fire Support Combat Vehicle

This is the basic G 127 with a modified hull that can be fitted with various 90 mm turrets including the French GIAT Industries TS 90 (FSCV 1/90), SAMM TTB 190 (KUPz 4/90), Cadillac Gage 90 mm (FSCV 2/90), Cockerill CS 90 (FSCV 3/90).

The crew of five consists of the driver, commander, layer and two infantrymen who sit in the hull rear which retains the two doors.

Total ammunition capacity and technical specifications depend on the type of turret installed. This model is not yet in production.

4K 7FA MICV 30/1

The prototype was completed in 1984 and is essentially the 4KA 7FA-K SPz fitted with a one-man turret Steyr SPz/200 armed with a British 30 mm RARDEN cannon and a 7.62 mm coaxial machine gun. The turret is fitted with an electro-hydraulic laying system and a combined day/night sight. Combat weight of the vehicle is 15 900 kg and power-to-weight ratio is 20.1 hp/tonne.

4K 7FA-Fü Command Vehicle

This model, which is in production, has a crew of seven commander, gunner, two radio operators, two messengers and the driver. Communications equipment includes two radios with an automatic tuning unit, loud speaker and a crew intercom.

4K 7FA-San Armoured Ambulance

This unarmed model is in production. Its crew of two consists of the driver and a medical orderly. It can carry up to four sitting and two stretcher patients and carries extensive medical equipment.

4K 7FA AMC 81

Mortar carrier with a crew of five (driver, commander, gunner and two mortar men), armed with 81 mm mortar and carrying 78 mortar bombs, in production.

4K 7FA SPAAG 1/2.20 Twin 20 mm Self-propelled Anti-aircraft System

Uses a modified chassis and is fitted with a French Dassault Electronique TA 20/RA 20 turret armed with twin 20 mm cannon. The prototype of this system has been completed.

4K 7FA SPAAG 3/2.30 Twin 30 mm Self-propelled Anti-aircraft Gun System

With Thomson CSF twin 30 mm turret, has been developed to the prototype stage. The turret, however, is no longer being marketed by Thomson-CSF.

SPECIFICATIONS (4K 7FA KSpZ)		MAX ROAD SPEED		TRANSMISSION	ZF 6-S80 manual with 6
CREW	2 + 8	forward	70 km/h		forward and 1 reverse
COMBAT WEIGHT	14 800 kg	reverse	7.4 km/h		gears
POWER-TO-WEIGHT		FUEL CAPACITY	360 l	STEERING	hydrostatic split torque
RATIO	21.62 hp/tonne	MAX ROAD RANGE	520 km		type
GROUND PRESSURE	0.55 kg/cm²	FORDING	1 m	CLUTCH	single dry plate
LENGTH HULL	5.87 m	GRADIENT	75%	SUSPENSION	torsion bar
WIDTH	2.5 m	SIDE SLOPE	40%	ELECTRICAL SYSTEM	24 V
HEIGHT (without armament		VERTICAL OBSTACLE	0.8 m	BATTERIES	2 × 12 V, 180 Ah
but including MG		TRENCH	2.1 m	ARMAMENT (main)	1 × 12.7 mm MG
support)	1.611 m	ENGINE	Steyr 7FA 6-cylinder	SMOKE-LAYING	
GROUND CLEARANCE	0.42 m		liquid-cooled 4-stroke	EQUIPMENT	4 smoke dischargers on
TRACK	2.12 m		turbo-charged diesel		rear of gunner's cupola
TRACK WIDTH	380 mm		developing 320 hp at	ARMOUR	
LENGTH OF TRACK			2300 rpm	Hull (max)	25 mm
ON GROUND	3.192 m				

Status: Production complete in Austria but can recommence if additional orders are placed. In service with Bolivia, Cyprus (supplied from Greek production line), Greece and Nigeria.

Manufacturer: Steyr-Daimler-Puch Spezialfahrzeug GmbH
PO Box 100
A-1111 Vienna
Austria
Telephone: 0222 76 45 11
Telex: 613221299

NB: It is also being manufactured under licence in Greece where it is known as the Leonidas. Photographs and additional details of Greek-built vehicles appear in this section under Greece.

Saurer 4K 4FA Armoured Personnel Carrier

Development

In 1956 Österreichische Saurer-Werke (which was taken over by Steyr-Daimler-Puch AG in 1970) of Vienna started development of a full tracked armoured personnel carrier. The first prototype, the 3K 3H, was completed in 1958 and the second, the 4K 3H, the following year, both powered by a Saurer 3H 200 hp diesel engine.

The next model, the 4K 2P, had a different hull layout, new suspension and was powered by a 250 hp Saurer 2P diesel engine. In appearance it was similar to the first production vehicles which were completed in 1961 and were designated the 4K 4F. Later production vehicles were the 4K 3FA (230 hp) and the 4K 4FA (250 hp), which differed only in minor details such as their engines.

A total of 445 production vehicles were delivered to the Austrian Army in the following versions: 30 4K 4F, 211 4K 3FA, 38 4K 3FA-FS, 12 4K 4FA-Fü/FIA, 15 4K 4FA-San, 76 4K 4FA-G2, 60 4K 4FA FüA 1 to 4 series, three 4K 3FA Fül and a number of 4K 3FA GrW1 81 mm mortar carriers which were conversions of earlier vehicles. Production was finally completed in 1969. The vehicle has now been succeeded in production by a new model, described in the previous entry, called the 4K 7FA which uses the same automotive components as the SK 105 light tank also manufactured by Steyr.

These are likely to be upgraded in the future to extend their operational lives into the 1990s.

Description

The hull of the Saurer armoured personnel carrier is made of all-welded steel armour and is immune to penetration from 20 mm armour-piercing projectiles over its frontal arc.

The driver sits at the front of the hull on the left side with the engine to his right. He has a two-piece hatch cover that opens to the left and right in front of which are three periscopes.

The gunner sits behind the driver and on the basic model has a cupola with a two-piece hatch cover opening left and right and a single observation periscope. A standard 12.7 mm (0.50) Browning machine gun is mounted on the forward part of the cupola and most vehicles are fitted with a shield to give a measure of protection to the gunner. This model is known as the 4K 4FA-G1.

Steyr 4K 7FA-G1 with 12.7 mm M2 HB machine gun and cupola hatch covers open (Austrian Army)

Saurer 4K series command vehicle with additional communications equipment

The personnel compartment is at the rear of the hull with the eight infantrymen seated on bench seats, four down each side of the hull. The section leader is the first man on the right seat and is provided with a periscope in the forward part of the personnel compartment roof, on the right side. Access to the personnel compartment is by two doors in the rear of the hull. Two hatches over the roof of the vehicle open to the outside. Mounting sockets for the pivot trunnion of an adapter mount for the MG 42 machine gun are provided on the front, right, left and rear side of the hatch above the personnel compartment.

The torsion bar suspension consists of five dual road wheels with the drive sprocket at the front and the idler at the rear and two track return rollers. The first and last road wheel stations are provided with a hydraulic shock absorber.

The 4K 4FA has no NBC system, night vision equipment or amphibious capability.

Variants

4K 4FA-G2 (Grenadier-Schützenpanzer)

This is similar to the basic 4K 4FA-G1 but is fitted with an Oerlikon GAD-AOA turret which is armed with a 20 mm Oerlikon cannon model 204 GK. This is belt-fed and can be used in both the ground and anti-aircraft roles. The weapon is provided with 100 rounds of ready-use ammunition and the empty cartridge cases and links are ejected outside the turret. The 20 mm cannon has an elevation of +70° and a depression of -12°, turret traverse being 360°.

4K 3FA-Fü1 (Command Vehicle) (Führungs-SPz, for Brigade Commander)

4K 3FA-FüA (Artillery Command Vehicle) (Artillerie-Führungs-Schützenpanzer)

4K 3FA-Fü/FIA (Anti-aircraft Command Vehicle) (FIA Führungs-Schützenpanzer)

4K 3FA-FS (Wireless Teleprinter Vehicle) (Funkfernschreiber Panzer)

4K 4FA-San (Ambulance)

Full designation is 4K 4FA-San (Sanitatspänzer). This model is unarmed and has a crew of two consisting of the driver and medical orderly, and can carry four sitting and two stretcher patients. The stretchers are carried in the left side of the hull with the four sitting patients on the bench seat down the right side of the hull, strapped in with waist and shoulder straps. There are four stowage containers for medical supplies under the bench seat and other equipment carried includes a heater, oxygen and respirating equipment.

Saurer 4K 4FA-G1 of Austrian Army with all roof hatches closed and showing position of rear-mounted 7.62 mm MG (Austrian Army)

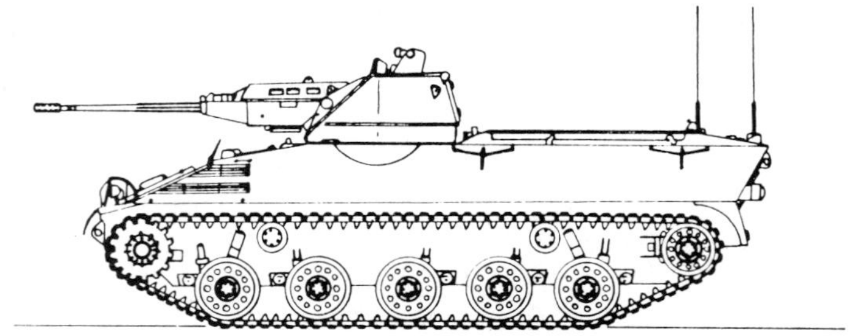

Saurer 4K 4FA-G2 with turret-mounted 20 mm cannon

4K 4FA-G2 fitted with Oerlikon-Contraves 20 mm turret type GAD-AOA

4K 4F GrW1 81 mm Mortar Carrier (Granatwerfer-SPz)

Mounted in the rear of the troop compartment, firing forwards, is an 81 mm mortar. A 120 mm self-propelled mortar was developed to the prototype stage (GrW2) but never entered service.

81 mm Multiple Rocket Launcher

This is the basic APC with an Oerlikon-Contraves 81 mm rocket launcher mounted on a turntable over the rear troop compartment. The launcher consists of two single-barrel launchers each of which has a nine-rocket magazine, mounted on a turntable. The launcher can be elevated from -10 to +50°.

Status: Production complete. In service only with the Austrian Army.

Manufacturer: Steyr-Daimler-Puch, Spezialfahrzeug GmbH
PO Box 100
A-1111 Vienna
Austria
Telephone: 0222 76 45 11
Telex: 613221299

SPECIFICATIONS					
CREW	2 + 8	GROUND CLEARANCE	0.42 m	ENGINE	Saurer model 4FA 4-stroke 6-cylinder diesel developing 250 hp at 2400 rpm
COMBAT WEIGHT		TRACK	2.12 m		
with 20 mm turret	15 000 kg	TRACK WIDTH	375 mm		
with 12.7 mm MG	12 500 kg	LENGTH OF TRACK ON GROUND	2.9 m	TRANSMISSION	manual with 5 forward and 1 reverse gears
UNLOADED WEIGHT (without turret or armament)	11 100 kg	MAX ROAD SPEED			
		1st gear	11.3 km/h	SUSPENSION	torsion bar
		2nd gear	18 km/h	ELECTRICAL SYSTEM	24 V
POWER-TO-WEIGHT RATIO		3rd gear	28.3 km/h	BATTERIES	2 × 12 V, 180 Ah
		4th gear	43.8 km/h	ARMAMENT	1 × 12.7 mm MG or 1 × 20 mm cannon
20 mm model	16.66 hp/tonne	5th gear	65 km/h		
12.7 mm model	20 hp/tonne	FUEL CAPACITY	184 l	ARMOUR	
GROUND PRESSURE	0.52 kg/cm²	MAX ROAD RANGE	370 km	Hull nose	20 mm at 45°
LENGTH	5.4 m	FORDING	1 m	Hull glacis	20 mm at 30°
WIDTH	2.5 m	GRADIENT	75%	Hull sides	14 mm
HEIGHT		SIDE SLOPE	50%	Hull top	8 mm
including 12.7 mm MG	2.1 m	VERTICAL OBSTACLE	0.8 m	Hull rear	12 mm
to hull top	1.65 m	TRENCH	2.2 m	Turret front	20-35 mm
FIRING HEIGHT	1.95 m			Turret, other surfaces	15-20 mm

Steyr-Daimler-Puch Pandur Armoured Personnel Carrier

Development

The Pandur armoured personnel carrier has been developed from 1979 as a private venture by Steyr-Daimler-Puch and was unveiled late in 1985, by which time two prototypes had been built.

The first of six pre-production vehicles was completed in June 1986 and the remaining five were completed by December 1986.

The Austrian Army has already evaluated the Pandur for possible use as an armed reconnaissance vehicle.

Wherever possible, for example with the engine and transmission, standard and in-production commercial components have been incorporated into the vehicle. The Swedish Army has also evaluated the vehicle for one year, but as of early 1991 no production orders had been placed for the Pandur (6 × 6) vehicle.

According to Steyr-Daimler-Puch, the Pandur could form the basis for a complete family of 6 × 6 and 8 × 8 vehicles all of which use common automotive components.

During the 1986 British Army Equipment Exhibition the Pandur was shown on the Royal Ordnance stand fitted with the turret of the Fox light armoured car armed with a 30 mm RARDEN cannon and a 7.62 mm coaxial machine gun.

The basic Pandur 6 × 6 vehicle has the so-called Type 330 hull and has an overall length of 5.697 m, but there is also a longer hull called the Type 332 with an overall length of 6.097 m. The latter version has a longer wheelbase between the second and third axles.

The Pandur was entered to meet six of the roles outlined in the British Army's FFLAV programme covered in the *AFV Families* section.

During the Iraqi invasion of Kuwait in August 1990, two prototypes of the Steyr-Daimler-Puch Pandur (6 × 6) APC, one of which was fitted with the MECAR 90 mm Kinetic Energy Weapons Station, were captured by the Iraqi Army. Following the loss of these two prototype vehicles, Steyr-Daimler-Puch took the decision to build two replacement vehicles.

Description

The hull of the Pandur is of all-welded steel armour construction that provides complete protection against 7.62 mm armour-piercing rounds through a full 360° and angle of fire from ±10° vertically. If required the vehicle can be built with thicker armour which provides protection over the frontal arc from 12.7 mm attack. The underside of the hull is smooth so it does not catch on any rough objects, and this also improves its speed in water.

Steyr-Daimler-Puch Pandur 6 × 6 APC fitted with firing ports and vision blocks and showing commander's cupola

Steyr-Daimler-Puch Pandur 6 × 6 with Steyr SP3/300 turret armed with 30 mm cannon and 7.62 mm coaxial machine gun (Pandur - ARSV 30)

Steyr-Daimler-Puch Pandur 6 × 6 Armoured Reconnaissance Fire Support Vehicle (ARFSV) fitted with MECAR KEnerga Weapons Station (KEWS) armed with 90 mm MECAR KEnerga 90/46 light gun system

Steyr-Daimler-Puch Pandur 6 × 6 APC from right side showing troop compartment doors, firing ports and roof-mounted recovery winch

The driver sits at the front of the vehicle on the left side and has a left-opening single piece hatch cover. In front of this are three periscopes, the centre one of which can be replaced by a night vision device of the passive type.

The vehicle commander sits to the rear of the driver and has a similar hatch cover and three periscopes.

The engine is to the right of the driver and is coupled to a fully automatic transmission with torque converter and lock-up clutch. The complete powerpack, consisting of the engine, transmission and cooling/ventilation system, can be removed as a complete unit in about 30 minutes. The air intake/outlets are in the roof of the vehicle with the exhaust outlet on the right.

From the transmission, a drive shaft leads to a central transfer box interlocked with the axial drive of the second axle. Two further shafts lead to the differential gears of the first and third axles and from these, power is transmitted to the planetary gears in the wheels. All parts of the power train are either situated within the hull or protected by tunnels or depressions of the hull.

When being driven on roads the vehicle is normally in 6 × 4 configuration, for example first and third axles driven via a central differential. For improved traction the second axle is engaged in rough terrain. The axle differentials as well as the central differential can be locked.

The troop compartment is at the rear of the hull and the troops enter and leave via twin doors in the hull rear; the eight troops sit four back to back down each side of the hull.

If required two spherical firing ports with a vision block can be provided in each side of the troop compartment to enable some of the troops to aim and fire their weapons from inside the vehicle.

The basic armoured personnel carrier has four roof hatches over the troop compartment, two each side. These are hinged in the centre and can be locked vertical.

A variety of weapons stations can be mounted on the roof of the Pandur, but this means that two of the roof hatches are not fitted.

All three axles have independent suspension with power steering on the first and second axles. The suspension of the first and second wheel stations consists of an upper longitudinal control arm with a lower transverse control arm and coil springs. The rear wheel station consists of a trailing arm and torsion bar. The identical spacing of the axles gives an even load distribution. The maximum vertical wheel travel is 300 mm. All wheels have disc brakes with the parking brake being of the drum type and manually operated.

The Pandur has a central tyre-pressure regulation system which allows the driver to adjust the tyre pressure to the type of ground being crossed at any given time. The tyre pressure can be changed while the vehicle is moving.

The amphibious model is under development and is powered in the water by two 430 mm diameter propellers mounted at the rear of the hull.

A wide range of armament installations can be mounted on the Pandur up to a maximum weight of 2100 kg. Maximum slewing ring diameter is 1600 mm, maximum turret basket diameter is 1300 mm and maximum turret basket depth is 1000 mm.

Standard equipment on production vehicles would include an NBC system, a fire detection and suppression system for both engine and fighting compartments and a heater.

Variants

Mechanised Infantry Combat Vehicle (MICV 127)

This would be fitted with a one man turret armed with a 12.7 mm MG with a 7.62 mm MG mounted coaxial to the left and a bank of three electrically operated smoke dischargers mounted either side of the turret rear firing forwards. The commander would be seated to the rear of the driver and have a raised cupola with periscopes to give observation over the frontal arc of the Pandur.

Armoured Reconnaissance Fire Support Vehicle

Fitted with a two-man turret armed with a 90 mm gun MECAR KEnerga 90/46 light gun system. This combination of Pandur and the KEWS was developed specifically to meet the requirements of a Middle East export customer.

Steyr-Daimler-Puch Pandur MICV fitted with one man turret armed with 12.7 mm MG and 7.62 mm MG

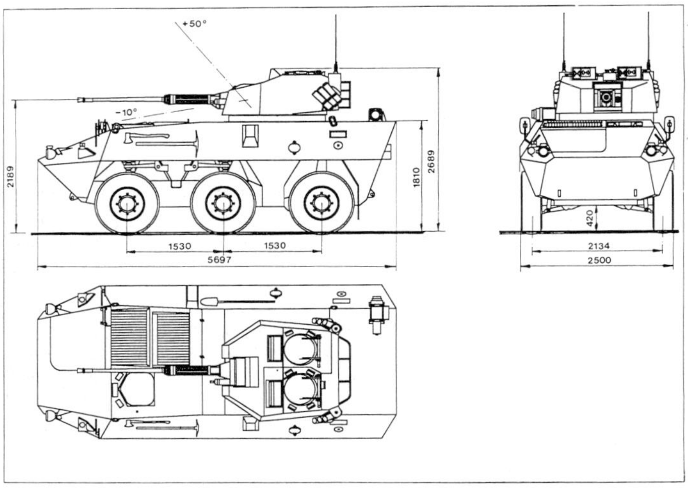

Three-view drawing of Pandur ARSV 30 (not to 1/76th scale)

The latest version of the KEWS has a number of improvements including a second generation day/night sight fire control system with integrated laser rangefinder and ballistic computer for improved first round hit probability. A 7.62 mm machine gun is mounted coaxial with the main armament and a 12.7 mm machine gun can be mounted on the turret roof for local and air defence purposes.

81 mm Mortar Carrier

Fitted with an 81 mm turntable-mounted mortar (AMC 81) or, as an alternative, a breech loaded mortar could be fitted.

Anti-Tank Guided Missile Carrier

(ATGMC/HCT) fitted with Euromissile HOT HCT launcher or a Norwegian Kvaerner-Eureka Armoured Launching Turret for TOW missile system which has two TOW ATGW in the ready to launch position. This turret is already in service with Canada, Norway and Switzerland.

Ambulance (AAMB)

This model would be unarmed.

Armoured Reconnaissance Scout Vehicle (ARSV)

This would have a two-man Steyr SP/3 turret armed with a 25 mm cannon (ARSV 25), or a two-man turret (SP/300) armed with a 30 mm cannon (ARSV 30).

The ARSV 25 is armed with a 25 mm Oerlikon-Contraves KBA cannon with a 7.62 mm MG3 coaxial machine gun with a bank of three smoke dischargers being mounted either side of the turret. Turret traverse and weapon elevation is powered with manual controls for emergency use. The gunner has an Aeritalia P204 sight with a magnification of ×8 while the commander has 9 P137 periscopes.

The ARSV 25 and ARSV 30 have a two man electro-mechanically operated turret with manual controls for emergency use.

Anti-aircraft

Fitted with power-operated turret, armed with twin 20 mm cannon.

Command

A wide range of communications equipment could be installed for use in the C³I (command, control, communications and intelligence) roles.

SPECIFICATIONS					
CREW	2 + 8/10	FORDING	1.2 m	STEERING	ZF power-assisted
CONFIGURATION	6 × 6	GRADIENT	60%	TYRES	12.5 R 20, central tyre pressure regulation system
MAX COMBAT WEIGHT	13 000 kg	SIDE SLOPE	35%	BRAKES	
UNLOADED WEIGHT	8600 kg	TRENCH	1.6 m	main	pneumatically-controlled disc, dual-circuit, all wheels
POWER-TO-WEIGHT RATIO (max weight)	18.84 hp/tonne	VERTICAL OBSTACLE	0.5 m	parking	drum brake acting on transmission output shaft
LENGTH	5.697 m	TURNING RADIUS	8.5 m	SUSPENSION	independent
WIDTH	2.5 m	ENGINE	Steyr WD 612.78 6-cylinder turbo-charged water-cooled diesel with air-intake cooling, developing 245 hp at 2400 rpm	VERTICAL WHEEL TRAVEL	150 mm - 150 mm
HEIGHT TO HULL TOP	1.81 m	TRANSMISSION	Allison Transmission Division of General Motors Corporation MT-653 DR 5-speed with torque converter and lock-up clutch	ELECTRICAL SYSTEM	24 V
GROUND CLEARANCE	0.42 m			ARMAMENT	see text
TRACK	2.134 m				
WHEELBASE	1.53 m + 1.53 m				
ANGLE OF APPROACH/ DEPARTURE	45°/45°				
MAX SPEED					
road	100 km/h				
ACCELERATION					
0 to 32 km/h	6.5 s				
0 to 48 km/h	11.5 s				
RANGE CRUISING	600 km				

Note: As an option tanks can be fitted to increase operational range to 1000 km

Status: Development was completed in 1986. Production can commence on receipt of orders.

Manufacturer: Steyr-Daimler-Puch Spezialfahrzeug GmbH
PO Box 100
A-1111 Vienna
Austria
Telephone: 0222 76 45 11
Telex: 613221299

BELGIUM

SIBMAS Armoured Personnel Carrier

Development

The SIBMAS (6 × 6) wheeled armoured vehicle range was originally designed as a private venture from 1975 by B N Constructions Ferroviaires et Métalliques. A prototype was completed in 1976 and was extensively tested both in Belgium and Malaysia. A demonstration vehicle, with improved vision for the driver and a more powerful engine, was completed in mid 1979.

The vehicle has been designed to undertake a wide range of roles including use as an anti-tank vehicle (armed with HOT, TOW and MILAN type ATGWs), APC (carrying a maximum of 16 men depending on the armament installed), internal security vehicle, mortar carrier, reconnaissance vehicle, anti-aircraft vehicle and recovery vehicle. Wherever possible standard MAN automotive components such as the engine, transmission and axles have been used.

Late in 1981 Malaysia placed an order valued at £50 million with the company for 186 vehicles with first deliveries made in June 1983. Two versions have been delivered:

AFSV-90: Armoured Fire Support Vehicle with Cockerill CM 90 turret and 90 mm Mk III gun and an OIP LRS-5 fire control system; 162 have been supplied and production was completed in late 1984.

ARV: Armoured Recovery Vehicle with front and rear spades, weight of 20 000 kg, winch with a capacity of 20 000 kg and a crane with a capacity of 10 500 kg; 24 have been supplied and production was completed in mid-1985.

CMI now has three of the original SIBMAS prototype vehicles which were overhauled early in 1986 for overseas customer demonstrations as and when required.

It is anticipated that CMI will now undertake further development of the SIBMAS vehicle, especially to try and reduce its production and operating costs.

Description

The hull of the SIBMAS is made of all-welded steel and provides the crew with complete protection against 7.62 and 5.56 mm armour-piercing rounds.

The driver sits at the very front of the vehicle and has a single piece hatch cover over his position that opens to his right, adjustable seat and an adjustable steering wheel. In front and at both sides of the driver are large bullet-proof windscreens that provide excellent vision at all times. If required these can be quickly covered by armoured shutters which are hinged at the lower part.

The turret is mounted immediately behind the driver and in each side of the hull is a large door that opens forwards.

The troop compartment is to the rear of the turret and the engine at the very rear on the left side. Six of the infantrymen sit back-to-back down the centre of the vehicle and the remaining three in the aisle between the troop compartment and the rear door. One small and three large hatches are provided over the top of the troop compartment and if required a remote-controlled machine gun can be pintle-mounted over the aisle. There are spherical firing ports and vision blocks in the troop compartment.

The engine can be removed through the roof of the vehicle in 30 minutes by two men. Quick disconnect couplings are provided for both the power and hydraulic circuits. The cooling system is behind and separated from the engine compartment by a waterproof bulkhead. Power is taken from the engine to the transmission, which is under the floor between the first and second axles, by a propeller shaft. The differentials of the front and rear two axles are lockable and a lockable differential between the two rear axles allows them to be longitudinally locked.

The SIBMAS is a fully amphibious vehicle propelled in the water by its wheels at a speed of 4 km/h. There is also a version with two propellers at the rear of the hull which can be traversed through 360° and give the vehicle a maximum water speed of 11 km/h. The propellers are driven from the PTO on the engine. An advantage of the SIBMAS is that when approaching the river bank the driver can engage first gear and as soon as the front wheels touch the bank maximum tractive effort is available as the propellers are still pushing the vehicle forward. Before entering the water a trim vane is automatically erected at the front of the hull. Two bilge pumps, each with a maximum capacity of 180 litres per minute, are fitted as standard.

The tyres are of the run-flat type. Optional equipment includes infra-red or passive night vision equipment, hydraulic winch mounted at the front of the hull with 60 m of cable and a maximum capacity of 8000 kg (which can be increased to 16 000 kg with snatch blocks), heater, air-conditioning system and an NBC system.

The SIBMAS can be fitted with a wide variety of armament installations and the following have already been fitted for trials purposes:

Cockerill CM 90 turret with 90 mm gun, 7.62 mm coaxial and 7.62 mm anti-aircraft machine gun (adopted by Malaysia)
Cockerill C25 two man power-operated turret armed with Oerlikon-Contraves 25 mm KBB cannon and 7.62 mm coaxial MG
Anti-aircraft turret with twin 20 mm cannon
Hispano-Suiza Lynx 90 turret with 90 mm gun, 7.62 mm coaxial and 7.62 mm anti-aircraft machine guns
Hispano-Suiza Serval 60/20 turret with 60 mm breech-loaded mortar, 20 mm cannon and 7.62 mm machine gun
Vickers turret with twin 7.62 mm machine guns
SAMM TG 120 (recently redesignated the TTB 120) turret with 20 mm cannon and 7.62 mm coaxial machine gun
SAMM S 365 (recently redesignated the BTM 208) turret with 12.7 mm and 7.62 mm machine guns.

Variants

Mortar carrier with higher roof and 60 mortar bombs;
Cargo carrier with payload of more than 4000 kg, higher roof and roof-mounted crane for handling cargo;
Command post vehicle with additional communications equipment and a tent which can be erected at the side of the hull;
Ambulance vehicle carrying four stretcher and three seated patients and two medical orderlies;
Recovery vehicle with front and rear stabilising blades, 20 000 kg winch and a 10 500 kg crane.

SIBMAS (6 × 6) AFSV fitted with Cockerill CM 90 turret armed with Cockerill 90 mm gun showing smoke dischargers pointing to front and rear of turret

SIBMAS (6 × 6) vehicle with Cockerill CM 90 two-man turret as delivered to Malaysia, showing amphibious capabilities

SIBMAS APC fitted with twin 7.62 mm machine gun turret

Status: Production complete. In service with Malaysia.

Manufacturer: CMI, Cockerill Mechanical Industries
Defence Division
Avenue Greiner
1 B 4100 Seraing
Belgium
Telephone: 32 + 41 - 302111
Telex: 41225 CKL SAMB
Fax: 32 + 41 - 302526

SIBMAS (6 × 6) vehicle with complete powerpack being removed through roof of vehicle by Leopard 1 ARV

SPECIFICATIONS	
CREW	3 + 11
CONFIGURATION	6 × 6
COMBAT WEIGHT	14 500-18 500 kg depending on role
UNLOADED WEIGHT	13 200 kg
MAX AXLE LOADS	
first axle	7500 kg
second axle	8500 kg
third axle	8500 kg
POWER-TO-WEIGHT RATIO	20 hp/tonne*
GROUND PRESSURE	2.4 kg/cm²
LENGTH HULL	7.32 m
WIDTH	2.5 m
HEIGHT	
to turret top	2.77 m
to hull top	2.24 m
GROUND CLEARANCE	0.4 m
TRACK	2.07 m
WHEELBASE	2.8 m + 1.4 m
ANGLE OF APPROACH/ DEPARTURE	50°/46°
MAX ROAD SPEED	100 km/h (approx)
MAX WATER SPEED	
propelled by wheels	up to 4 km/h
propelled by propellers	up to 11 km/h
ACCELERATION	
0-60 km/h	22 s
FUEL CAPACITY	400 l
MAX ROAD RANGE	1000 km (approx)
FUEL CONSUMPTION	0.36 l/km
FORDING	amphibious
GRADIENT	70%
SIDE SLOPE	40%
VERTICAL OBSTACLE	0.7 m
TRENCH	1.5 m
TURNING RADIUS	8 m
ENGINE	MAN D 2566 MK 6-cylinder in-line water-cooled turbo-charged diesel with intercooler developing 320 hp at 1900 rpm
MAX TORQUE	1270 Nm (130 kgm) between 1300 and 1600 rpm
TRANSMISSION	fully automatic ZF 6-HP 500, 6 forward and 1 reverse gears, with hydro-dynamic torque converter
STEERING	power-assisted
SUSPENSION	longitudinal suspension arms (lower) and wishbone above axle, helical springs and hydraulic shock absorbers
TYRES	14.00 × 20 or 16.00 × 20 with run-flat elements
BRAKES	
main	hydraulic air-assisted
parking	pneumatic on gearbox
TRANSMISSION	hydraulic retarder
ELECTRICAL SYSTEM	24 V
BATTERIES	4 × 12 V, 100 Ah
ARMAMENT	see text

* Based on a combat weight of 16 000 kg

CMI, Cockerill Mechanical Industries Defence Division, BMF production of M113A1/AIFV

Development

In April 1980 the Belgian Government placed an order for 514 FMC Armoured Infantry Fighting Vehicles and 525 M113A1 armoured personnel carriers. These were being built under licence by Belgian Mechanical Fabrication (BMF). This is a division of Cockerill Mechanical Industries (CMI). The Aubange factory of BMF was responsible for the hulls and hull assembly while the subcontractor NV Henschel in Antwerp carried out final assembly.

The first two vehicles were handed over in 1982 and production continued at the rate of 20 to 25 vehicles a month until April 1988. In 1983 the Netherlands ordered 26 M113A1 hulls from BMF which were converted to recovery vehicles by the Dutch company RDM.

In the Belgian Army they have now replaced the old American-supplied M75s and French-supplied AMX VCI armoured personnel carriers.

Early in 1985 Cockerill Mechanical Industries of Belgium purchased an M113A1-B APC from BMF and mounted CM 90 two-man 90 mm turret for use as a Fire Support Vehicle (FSV).

A 4.2-inch mortar transport carrier (M113A1-B) is now under consideration, with the mortar being fired from the ground rather than from the vehicle.

Late in 1989 CMI was awarded a contract valued at about $100 million to build Armoured Infantry Fighting Vehicles (AIFV) hulls for Turkey (qv).

Cockerill Mechanical Industries will build and deliver 285 hulls and then provide equipment to build the remaining 1415 vehicles in Turkey.

Delivery of the hulls is spread over three years with the first being delivered late in 1989. DAF in the Netherlands has received a $50 million sub-contract to provide weapon stations and vehicle integration. The first 20 AIFVs will be assembled by DAF after which assembly will commence in Turkey.

M113A1-B

The Belgian variants of the M113A1 are very similar to the standard US Army M113A1 but with the following modifications: suspension system similar to that of the M113A2, 200 A alternator, exterior air intake/exhaust personnel heater, an NBC system supplying air to each individual, Halon fixed fire-extinguishing system for the engine compartment and vehicle lights for Belgian and German roads.

Two basic models are produced. The first is the standard two-holed vehicle with a driver's hatch and a commander's cupola, the second is a new three-holed vehicle with a driver's hatch and two transversely adjacent cupolas. The M113A1-B vehicles are all armed with the standard 12.7 mm M2 HB machine gun on the commander's cupola, except for the unarmed ambulance vehicle. The manual machine gun has an elevation of +53°, depression of -21° and 360° traverse. Additional weapons other than personal equipment are detailed in the variants below.

Variants

M113A1-B-ATK

This is a three-holed vehicle with the 12.7 mm machine gun on the right and an unarmed cupola on the left. The vehicle has a crew of 10, carries a 7.62 mm machine gun, short range anti-tank weapon and ammunition.

M113A1-B-MIL MILAN

This is a three-holed vehicle which has the 12.7 mm machine gun on the left commander's cupola and a quick disconnect mount on the right cupola for the MILAN ATGW. The missile is fired from its tripod either attached to the cupola mount or in the field. The mount allows for the rapid escape of the vehicle over a limited distance with the launcher still attached to the cupola. Two Bofors Lyran mortar launchers are mounted on the top of the vehicle at the left rear. This model has a crew of five.

M113A1-B-TRG

This is the basic MILAN vehicle with hatches, cupolas, stowage provision and so on and is equipped with two plexiglass clear canopies, one for the observer and the other for the driver/instructor. Equipment is provided for all weather manoeuvres and the transport of observers. This vehicle can be converted into the M113A1-B-MIL.

Engineer Vehicle M113A1-B-ENG

This is a two-holed vehicle for carrying a demolition team and its equipment; it has a self-recovery winch and a crew of eight.

Command Post Vehicle M113A1-B-CP

This is a two-holed vehicle based on the M113A1, not the M577A1, and is fitted with communications equipment, map boards and other equipment for a mobile command post. It includes an optional side awning or canvas covered extension. The extension is of lightproof construction and may be connected to extensions of the same design on other M113A1-B-CPs. This has a crew of seven.

AIFV built by BMF

Belgian Army M113A1-B-MIL MILAN with Euromissile MILAN ATGW and 12.7 mm M2 HB machine gun in position (C R Zwart)

Recovery Vehicle M113A1-B-REC

This is a two-holed vehicle fitted with a hydraulic winch and fairlead assembly for recovering disabled vehicles. It also has a manual crane mounted on the top left. This model has a crew of four.

Maintenance Vehicle with crane M113A1-B-MTC

This is a two-holed vehicle with an HIAB hydraulic crane on the mid-left top, a self-recovery winch is installed in the engineer vehicle and a portable work bench, special lighting and storage for tools. It has a crew of five.

Ambulance Vehicle M113A1-B-AMB

This is a two-holed vehicle, unarmed and with a crew of three. It can carry four stretcher, two stretcher and three sitting, or six sitting patients.

AIFV-B

The Belgian Armoured Infantry Fighting Vehicle is very similar to the Dutch AIFV fully described in the AIFV entry later in this section except that the following major differences are incorporated: suspension is similar to the M113A2's (and identical to the M113A1-B's), external air intake/exhaust personnel heater, NBC system supplying air to each individual, and Halon fixed fire-extinguishing system for the engine compartment and lights for Belgian and German roads. There are three versions of this vehicle. The first Belgian production AIFVs were handed over to infantry units in September 1985.

AIFV-B with 25 mm Oerlikon KBA-BO2 Cannon, Enclosed Weapon Station (EWS), AIFV-B-25 mm

This is similar to the Dutch AIFV squad vehicle and has a crew of 10.

AIFV with 12.7 mm Cupola Weapon Station (CWS), AIFV-B-.50

This vehicle is similar to the AIFV-B-25 mm except that the 12.7 mm CWS is used in place of the EWS. This CWS permits operation of the gun in both closed hatch, remote mode and open hatch, manual flex mode. Remote operation permits an elevation range of -10 to +50° and complete 360° traverse. The flex mode has an elevation range of from -12 to +53° and 360° traverse.

The left side commander's cupola is fitted with a quick-disconnect mount for the MILAN missile launcher. As with the standard M113A1-B-MIL, the launcher remains fixed to its tripod for use on the quick-disconnect mount or in the field. With the launcher attached to the mount, limited distance operations are possible. The vehicle also has two Bofors Lyran launchers.

AIFV-B-CP Command Post

This vehicle is similar to the Dutch AIFV command post with an M113A1 type of cupola in place of a weapon station on the right. A pintle-mounted 12.7 mm M2 HB machine gun is affixed to this cupola. Special communications equipment, map boards and other equipment for operating a mobile command post are provided and a single side awning is included. This version has a crew of seven.

Status: Production of these vehicles was completed in April 1988. In service with Belgian Army.

In 1989 Cockerill Mechanical Industries was selected as the sub-contractor for the hull and main components for the Turkish Armoured Infantry Fighting Vehicle programme (see Development).

Manufacturer: CMI, Cockerill Mechanical Industries
Avenue Greiner
1 B 4100 Seraing
Belgium
Telephone: 32 + 41 - 302111
Telex: 41225 CKL SAMB
Fax: 32 + 41 - 302526

BDX Armoured Personnel Carrier

Development

In 1976 the Engineering Division of Beherman Demoen negotiated successfully with the Irish company of Technology Investments Limited for a licence to produce an improved version of the Timoney (4 × 4) armoured personnel carrier.

The first production order for the BDX was awarded by the Belgian Government in 1977 and called for the delivery of 43 vehicles to the Belgian Air Force for airfield protection duties and 80 vehicles to the Belgian Gendarmerie. The vehicles for the Air Force are armed with a ring-mounted 7.62 mm light machine gun. The Gendarmerie order consisted of 41 APCs without a turret, 26 vehicles with a dozer blade for clearing road obstacles and 13 vehicles fitted with an 81 mm mortar in the rear of the hull. First production vehicles were completed late in 1978 and production was completed for the Belgian Government in February 1981.

BDX delivered five vehicles to Argentina and the vehicle has also been demonstrated to a number of other countries including Indonesia and Malaysia. Further development of the BDX APC by Vickers Defence Systems of the UK has resulted in the Valkyr, described later under the United Kingdom.

Description

The hull of the BDX is made of all-welded steel which varies in thickness from 9.5 to 12.7 mm and provides complete protection against 5.56 and 7.62 mm armour-piercing projectiles fired at point-blank range and splinters from 105 mm artillery projectiles exploding on the ground 10 m from the vehicle. The BDX will withstand the blast from a 9 kg mine exploding 10 m from the vehicle.

The driver sits at the front of the hull in a central position and has three windscreens, one in front and one on either side, made of multiplex glass 81 mm thick which provides the same degree of protection as the rest of the hull. The large front windscreen has a wiper and can be hinged forward for improved ventilation in a non-combat area.

The crew enters and leaves the BDX by three doors, one in each side of the hull and one at the rear. All the doors open through a full 180° and can be locked half open if required. Ball mounts with vision blocks are provided in each side of the hull to the rear of the side entrance door and in the rear door.

The 10 infantrymen have individual seats with waist seatbelts. The interior of the hull is padded to reduce both noise and the internal temperature of the vehicle.

The engine is mounted to the immediate rear of the driver and is coupled to an automatic transmission which in turn provides power to the transfer box at the back of the vehicle. This transmits power to the front and rear axles by a propeller shaft. Both front and rear axles are equipped with a self-locking differential of the 'no-spin' type. The air-inlet and air-outlet louvres are in the roof over the driver's position with the air-outlet louvres located either side of the engine compartment in the roof. Protection against Molotov cocktails can be fitted if required. The exhaust pipes run along the outside of the hull.

The machine gun turret is fitted on the roof of the BDX to the rear of the engine compartment and a further circular hatch cover is provided in the roof at the rear. The vehicle, when fitted with the 81 mm mortar, has two rectangular hatches in the roof that open either side of the hull.

The BDX is fully amphibious and propelled in the water by its wheels. Optionally two water-jets can be fitted to increase its water speed.

Optional equipment includes an air-conditioning system, front-mounted dozer blade for clearing obstacles, heater, NBC system and hull-mounted smoke dischargers.

Variants

In addition to being used as an armoured personnel carrier, the BDX can be used for a wide range of other roles such as an ambulance,

BDX fitted with 81 mm mortar showing open two-part roof hatch

BDX APC used by Belgian Air Force for airfield defence armed with 7.62 mm machine gun and three smoke dischargers either side of hull (C R Zwart)

command vehicle and reconnaissance vehicle. The BDX can be fitted with a turret-mounted 20 mm cannon, turret-mounted 90 mm Cockerill gun, turret with single or twin 7.62 mm machine guns, Euromissile MCT turret with two MILAN ATGWs in ready to launch position, SNIA BPD FIROS 6 48-round 51 mm multiple rocket system, turret with 81 mm breech-loaded mortar and a coaxial 7.62 mm machine gun or an 81 mm mortar mounted in the hull firing forwards.

Late in 1980 the company completed testing of a BDX APC powered by a Detroit Diesel Model 4V-53T engine developing 180 hp at 2800 rpm.

User trials and service reports indicated that certain areas of the vehicle could incorporate new components and methods of construction which would increase the service life of the BDX and improve the ride and accessibility for the crew.

A design study was completed with a modified front to give the driver more room, a more streamlined ballistic shape to the front of the vehicle, a strengthened suspension system to take a wide variety of weapons and turrets and a redesigned high efficiency air-conditioning system and various other detailed improvements. The new design of the APC, incorporating all the improvements, was named Valkyr. This has been developed and produced by Vickers Defence Systems of the United Kingdom in association with Beherman Demoen Engineering Division. For a complete description and specifications refer to the entry under the United Kingdom in this section.

Status: Production complete. In service with Argentina, Belgian Air Force and Gendarmerie.

Note: There is a separate entry in this section for the Timoney armoured personnel carrier.

Manufacturer: Beherman Demoen Engineering (BDX)
Motstraat 54
B-2800 Mechelen
Belgium
Telephone: 00 32(01) 44 26 55
Telex: 22538
Fax: 00 32 (01) 15 42 11 34

SPECIFICATIONS		FUEL CAPACITY	248 l	CLUTCH	hydraulic, torque
CREW	2 + 10	MAX ROAD RANGE	500-900 km		converter
CONFIGURATION	4 × 4	FORDING	amphibious	SUSPENSION	independent wishbone
COMBAT WEIGHT	10 700 kg	GRADIENT	60%		with helical springs
UNLOADED WEIGHT	9750 kg	SIDE SLOPE	40%		and telescopic
POWER-TO-WEIGHT		VERTICAL OBSTACLE	0.4 m		dampers
RATIO	16.82 bhp/tonne	TURNING RADIUS	less than 7 m	TYRES	12.00 × 20
LENGTH OVERALL	5.05 m	ENGINE	Chrysler V-8	BRAKES	disc (dual circuit)
WIDTH	2.5 m		water-cooled petrol	ELECTRICAL SYSTEM	24 V
HEIGHT			developing 180 bhp at	BATTERIES	4 × 12 V
overall with twin			4000 rpm*	ARMAMENT	see text
7.62 mm MG turret	2.84 m	TRANSMISSION	Allison Transmission	ARMOUR	
to hull top	2.06 m		Division of General	Hull front	12.7 mm
GROUND CLEARANCE	0.4 m		Motors AT-540	Hull sides	12.7 mm
TRACK	1.93 m		automatic with 4	Hull top	9.5 mm
WHEELBASE	3.003 m		forward and 1	Hull floor	9.5 mm
MAX ROAD SPEED			reverse gears	Hull rear	12.7 mm
forwards	100 km/h	TRANSFER CASE	2-speed		
reverse	18 km/h	STEERING	hydraulically assisted	* Diesel engine available if required	

BRAZIL

Charrua Armoured Personnel Carrier

Development

The Charrua armoured personnel carrier is the basic member of a complete family of tracked vehicles that has been developed by Moto Pecas Transmissoes SA as a private venture.

This company has had considerable experience in the overhaul and modernisation of tracked armoured vehicles for the Brazilian Army. These include the modernisation and dieselisation of some 586 M113 series APCs, modernisation of the 105 mm M108 self-propelled howitzer and converting Sherman tanks into armoured recovery vehicles complete with A-frame and front-mounted dozer/stabilising blade.

Prior to the development of the Charrua APC, the company developed the Charrua XMP-1 APC, but this did not enter production and is no longer marketed. Details of this APC were given in *Jane's Armour and Artillery 1987-88* page 259.

Description

The hull of the Charrua is of all-welded steel armour construction that provides the crew with protection from small arms fire and shell splinters. If required, add-on armour can be fitted for increased battlefield survivability; this would provide protection against penetration from 20 mm attack.

The driver is seated at the front of the vehicle on the left side with the machine gunner to his immediate rear. The driver has a single piece hatch cover that opens to the left and three periscopes for forward observation, one of which can be replaced by a passive night vision device.

The gunner has a 12.7 mm M2 machine gun which is provided with side and rear armour protection, elevation and traverse is manual. The gunner's cupola is fitted with five periscopes.

The engine compartment is to the right of the driver and separated from the crew compartment by a bulkhead. The air inlet and outlet louvres are in the top of the hull, as is the exhaust pipe. Access panels are provided in the glacis plate for routine servicing.

There are two engine choices, a 394 hp diesel or 470 hp diesel, this being coupled to a fully automatic transmission with five forward and one reverse gears.

The troop compartment is at the rear of the hull with the nine troops being seated back to back down the center of the vehicles, four on the left and five on the right.

Infantry dismounting from the Charrua armoured personnel carrier

Charrua armoured personnel carrier showing amphibious capabilities

The troops enter and leave via a power operated ramp in the hull rear. This is also provided with two doors that open outwards, each of which has a vision block and a firing port.

Over the top of the troop compartment are two rectangular roof hatches that can be swung through 180° so that they lay back on the hull top. The troop compartment is provided with two roof-mounted ventilators.

The upper part of the hull roof slopes slightly inwards and is provided with two firing ports with a bullet proof vision block above.

Mounted either side of the troop compartment roof, at the very rear, is a bank of two electrically operated smoke dischargers firing forwards.

The Charrua is fully amphibious, being propelled in the water by two water jets mounted one either side of the hull at the rear, with maximum water speed being 8 km/h.

Before entering the water the bilge pumps are switched on and the trim vane is erected on the glacis plate.

Suspension is of the torsion bar type with each side having five dual rubber tyred road wheels, drive sprocket at the front, idler at the rear and three track return rollers. The upper part of the tracks are covered by a rubber skirt.

The vehicle is fitted with two independent fire extinguishing systems, one for the engine compartment and one for the troop compartment, with manual or automatic modes of operation.

Variants

The manufacturer has suggested that the Charrua would be adopted for a wide range of missions including:

Infantry Fighting Vehicle

Crew of two (driver and gunner), nine man squad and fitted with a turret armed with 25 mm or 20 mm cannon with a 7.62 mm coaxial machine gun.

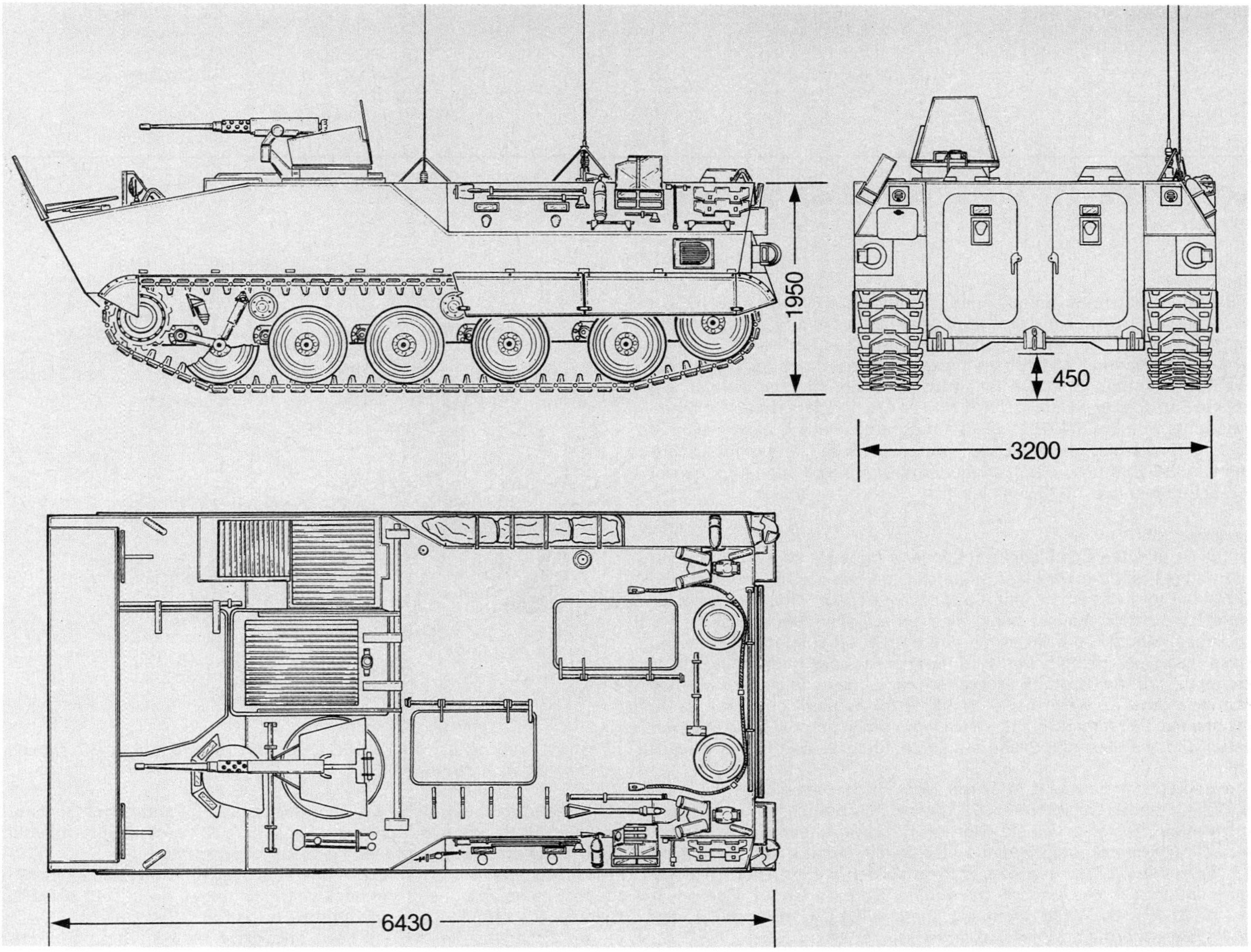

Three-view drawing of Charrua armoured personnel carrier with 12.7 mm M2 machine gun fitted (not to 1/76th scale)

Light Combat Vehicle
Three man crew consisting of driver, commander and gunner and fitted with a two man turret armed with 90 mm or 105 mm gun.

Ambulance
This would have a three man crew and carry three stretcher patients plus medical equipment or six seated wounded.

Anti-aircraft
Fitted with Italian OTO Melara SIDAM 25 turret with four 25 mm cannon, or a turret with a 40 mm L/70 gun, three man crew.

Command Post Vehicles
Similar to APC but a tent could be erected at the rear of the hull. Two man crew plus four command staff and would also have additional communications equipment.

Engineer Vehicle
This would have a similar crane to the recovery vehicle but would not have a winch and would be fitted with a front-mounted dozer blade.

Mortar Carrier
Fitted with 120 mm mortar, six man crew (commander, machine gunner and four man mortar team), 12.7 mm machine gun and four smoke dischargers.

Recovery/Crane
Three man crew and fitted with a crane, winch, cutting equipment, electric welding equipment, 12.7 mm machine gun and grenade launchers.

Status: Development complete. Ready for production.

Charrua armoured personnel carrier with trim vane folded back on glacis plate

Manufacturer: Moto Peças
Avenida Hollingsworth, 719, PO Box 622
CEP 18.100 Sorocaba SP,
Brazil.
Telephone: 0152 32 9444
Telex: 0152 145/MPTE-BR
Fax: (0152) 32 3675

SPECIFICATIONS

CREW	2 + 9
WEIGHT LOADED	18 000 kg
GROUND PRESSURE	0.46 kg/cm^2
POWER-TO-WEIGHT RATIO	
394 hp engine	21.9 hp/tonne
470 hp engine	26.1 hp/tonne
LENGTH	6.43 m
WIDTH	3.20 m
HEIGHT (hull top)	1.95 m
GROUND CLEARANCE	0.45 m
TRACK WIDTH	532 mm
LENGTH OF TRACK ON GROUND	3.63 m
MAXIMUM ROAD SPEED	70 km/h
MAXIMUM WATER SPEED	8 km/h
RANGE	500 km
FORDING	amphibious
GRADIENT	60%
SIDE SLOPE	30%
VERTICAL OBSTACLE	0.75 m
TRENCH	1.85 m
ENGINE	6 cylinder in-line turbocharged diesel developing 394 hp at 2100 rpm or 6 cylinder V turbocharged diesel developing 470 hp at 2300 rpm
TRANSMISSION	automatic with lock-up clutch with two forward and one reverse gears or manual transmission with 5 forward and 1 reverse gears
SUSPENSION	torsion bar
ARMAMENT	1 × 12.7 mm MG
SMOKE DISCHARGERS	2 × 2

ENGESA EE-11 Urutu Armoured Personnel Carrier

Development
The EE-11 Urutu armoured personnel carrier was designed by ENGESA from January 1970 with the first prototype being completed in July the same year. After trials a production order was placed for the EE-11 Urutu by the Brazilian Army which calls the vehicle the Carro de Transporte Sobre Rodas Anfibio, or CTRA for short, and production began at a new factory at Sao José dos Campos in 1974. The EE-11 shares many common components with the ENGESA EE-9 armoured car which was developed from July 1970 and is also in service with the Brazilian and other armies. By early 1984 total production of the Urutu armoured personnel carrier had reached just over 1500 vehicles.

Description (Mk V)
The hull is of all-welded construction and consists of an outer layer of hard steel and an inner layer of softer steel that are roll-bonded and heat treated to give the maximum ballistic protection. The armour gives complete protection against small arms fire at point-blank range.

The driver sits at the front of the hull on the left side and is provided with a single piece hatch cover, with three periscopes fitted in the glacis plate forward of the hatch. If required a windscreen can be erected over the driver's position when he is driving with his head out. The centre periscope can be replaced by a passive night periscope. Steering is power assisted and the steering wheel is adjustable according to the driver's position.

The engine compartment is at the front of the vehicle on the right side. The Detroit Diesel 6V-53T engine is coupled to an Allison Transmission Division of General Motors MT-643 automatic transmission. The cooling system comprises a water-to-air radiator for the engine water, oil coolers for the engine and transmission and transfer case oils. Air circulation is forced through the engine compartment by a propeller shaft and driven fan. While swimming, the cooling is performed through a keel cooler with the air inlet and outlet louvres being closed.

The air inlet louvres are on the hull top to the right of the driver while the exhaust outlets are on the right side of the hull.

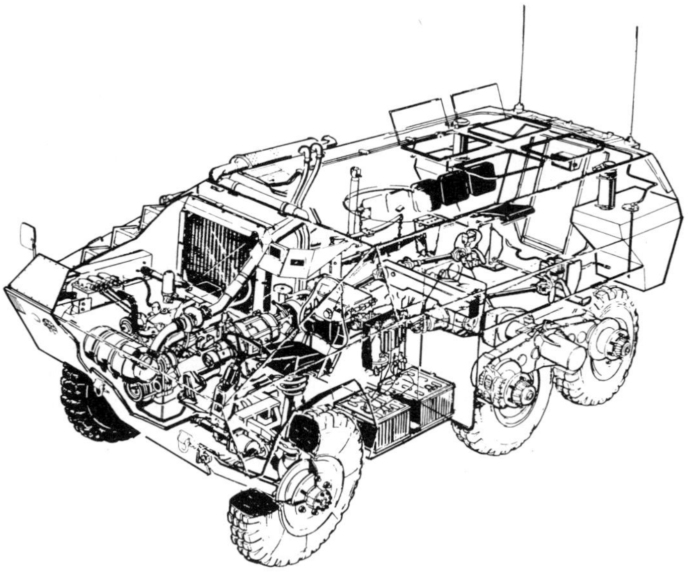

Cutaway drawing of the ENGESA EE-11 Urutu (6 × 6) APC showing position of main components

The troop compartment is at the rear of the hull and bench seats are provided down each side of the vehicle with an additional bench seat behind the driver's seat across the troop compartment.

There are two doors in the hull, one on the left side immediately behind the front axle and one at the back which is hinged on the left side. The rear door is automatically operated by the driver.

Two arrangements of firing ports and vision blocks are possible: five ball mount firing ports, two in each side of the hull and one in the rear door, and two glass bullet- and splinter-proof vision blocks in the side

ENGESA EE-11 Urutu armoured personnel carrier fitted with ET-MD one-man turret armed with 7.62 mm and 12.7 mm machine guns, trim vane erected at front of hull and driver's hatch open (Christopher F Foss)

ENGESA EE-11 Urutu AFSV fitted with ENGESA ET-90 turret armed with ENGESA EC-90 90 mm gun and coaxial 7.62 mm machine gun, and with trim vane erected at front of hull

ENGESA EE-11 Urutu fitted with raised roof in ambulance version

Recovery model of ENGESA EE-11 Urutu APC in travelling configuration with crane and jacks stowed

door, or 11 firing ports, five on each side of the hull and one in the rear door, and five bullet- and splinter-proof vision blocks, one in the left door, one on the left side, two on the right side of the hull and one in the rear door. Four roof hatches over the troop compartment open towards the outside of the vehicle.

The main armament station is behind the driver, slightly offset to the left of the vehicle's centre line. The basic model EE-11 Urutu APC is armed with a ring-mounted 12.7 mm (0.50) M2 HB machine gun, but the vehicle can be fitted with many other types of armament installation including a ring-mounted 7.62 mm machine gun, turret-mounted 60 mm Thomson-Brandt mortar, turret as fitted to the British CVR(T) Scorpion, French Dassault Electronique TA 20 twin 20 mm anti-aircraft gun system and ENGESA ET-90 turret armed with ENGESA EC-90 90 mm gun. ENGESA has developed three new turrets that can be fitted to the EE-11 Urutu armoured personnel carrier or the EE-3 Jararaca scout car. Brief details of these are:

ENGESA ET-MD one-man turret armed with a 7.62 mm machine gun and a 12.7 mm M2 HB machine gun, both internally operated

ENGESA ET-20 one-man turret armed with a 20 mm Oerlikon cannon and a 7.62 mm machine gun

ENGESA ET-25 two-man turret armed with 25 mm Oerlikon cannon and 7.62 mm coaxial machine gun.

The front suspension is independent and consists of double wishbone arms, coil springs and double action hydraulic shock absorbers. The front wheels are driven by hypoid angular transmission and bevel differential gears. For amphibious operations, the front and rear axles can be pressurised. The rear suspension consists of semi-elliptical springs and double-action telescopic shock absorbers. The articulated rear axle is the ENGESA 'Boomerang' type with walking beams and a differential locking system. The 'Boomerang' axles allow the wheels to travel 0.87 m upwards and downwards and ensure that all wheels are in constant contact with the ground. All tyres are the run-flat type with automatic inflation system. Disc brakes and planetary reduction gears on the wheel hubs are provided for all wheels. The two fuel tanks are situated on either side of the hull rear and are filled from the top. The electrical system has a double 24 V system, a battery set is used for vehicle operations while a second set is just for engine start to avoid lack of power after long, still radio operations. The EE-11 Urutu has been developed for the Brazilian Marines, which required a vehicle capable of operating in heavy surf. This vehicle is propelled in the water by two propellers with kort-nozzles and is steered by two rudders at the rear of the hull which are connected to the steering wheel. Before entering the water, a trim vane is erected pneumatically at the front of the hull by the driver. An air compressor integrated to the engine feeds three air reservoirs and all operating systems and accessories; brakes, transfer case, gear shifting, tyre inflation, rear door and amphibious elements.

It has a maximum water speed of 8 km/h and in addition to the electric bilge pumps a hand-operated emergency bilge pump is provided. The Urutu is also equipped with disc brakes and automatic tyre inflation system.

Optional equipment includes night vision equipment, various radio installations, winch with a capacity of 5000 kg and a Mercedes-Benz OM-352A (190 hp-SAE) engine with an automatic transmission Allison Transmission Division of General Motors MT-643 in place of the standard Detroit Diesel 6V-53N diesel developing 212 hp. The EE-11 Urutu can be fitted with an NBC system.

Variants

ENGESA has so far produced seven distinct models of the EE-11 Urutu, which can be summarised as follows:

Model	**I**	**II**	**III**	**IV**	**V**	**VI**	**VII**
ENGINE MANUFACTURER	MB	MB	DD	MB	MB	MB	DD
ENGINE MODEL	OM-352A	OM-352A	6V-53	OM-352A	OM-352A	OM-352A	6V-53T
OUTPUT	170 hp	170 hp	212 hp	170 hp	190 hp	190 hp	260 hp
TRANSMISSION	manual	manual MB 63/40	automatic Allison MT-643	automatic Allison AT-540	automatic Allison AT-545	automatic Allison MT-643	automatic Allison MT-643
TYRE PRESSURE REGULATION SYSTEM	no	yes	yes	yes	yes	yes	yes

Note: All are diesel engines, MB being Mercedes-Benz and DD being Detroit Diesel; MK VII also has epicycloidal reduction on wheels

ENGESA EE-11 Urutu (6 × 6) APC fitted with ENGESA ET-20 one-man turret

ENGESA EE-11 Urutu riot control vehicle fitted with one-man turret with mount for externally mounted 7.62 mm gun

ENGESA EE-11 APC (6 × 6) entering the water (Scott R Gourley)

ENGESA EE-11 Urutu APC fitted with Swedish Hägglunds turret with 20 mm cannon and pintle-mounted 7.62 mm machine gun

81 mm Mortar Carrier

Mounted in the hull rear is an 81 mm mortar firing through the roof of the vehicle. A 7.62 mm machine gun is fitted for local protection. The mortar can be elevated from +40° to +80° and if required can be dismounted from the vehicle and used in ground rôle.

Ambulance

This model is unarmed and has a higher roof and can carry between six and eight walking wounded, or a smaller number of stretcher cases. Standard equipment includes an air-conditioning system, breathing equipment and a refrigerator for storing medicines. It can also be equipped with four stretchers.

Cargo

When being used for cargo the EE-11 can carry 2000 kg.

AFSV

The ENGESA EE-11 Urutu AFSV is an amphibious Armoured Fire Support Vehicle fitted with the ENGESA ET-90 turret armed with the ENGESA EC-90 90 mm gun. This model has a crew of eight: driver, commander, gunner and five infantrymen who are seated at the very rear of the hull. A total of 48 rounds of 90 mm ammunition are carried, 12 rounds ready use in the turret and 36 rounds in the hull. The turret can be fitted with a power traverse system and a gunner's sight with an integral laser rangefinder. The first customer for this was Tunisia. In mid-1988 Venezuela placed an order for the supply of 100 EE-11 Urutu APCs including some in the Armoured Fire Support Vehicle configuration.

Command Vehicle

Basic vehicle fitted with a special radio set and all the modern requirements for a command post.

Recovery

This model has different roof hatches to permit personnel to work standing upright and is fitted with a front-mounted winch, hydraulic crane, oxy-acetylene welding machine, portable generator, electric welding transformer, electric grinder, complete range of tools and specialised equipment and a tent. Full details of this version are given in *Jane's Military Vehicles and Logistics 1991-92* page 54.

Riot Control

The ENGESA EE-11 Urutu in the riot control and internal security operation version has the same basic features as the EE-11 Urutu armoured personnel carrier and retains the same mobility and ballistic protection. Mounted at the front of the vehicle is a blade for clearing

ENGESA EE-11 APC (6 × 6) APC Carrier fitted with ENGESA one-man turret

obstacles. Typical missions of this model would include riot control, internal security, escorting convoys, transport of troops and patrolling airports and other high-risk areas.

Status: In production. In service with Bolivia, the Brazilian Army and Marines, Chile, Colombia, Cyprus, Gabon (including version with 20 mm Hägglunds turret), Guyana, Iraq (including armoured repair vehicle), Jordan (delivered in 1987, police/IS role with Kidde-Graviner Crewbay system fitted), Libya, Morocco, Tunisia (including version with turret-mounted 90 mm gun), United Arab Emirates, Uruguay (including version with 60 mm breech-loaded mortar), Venezuela (100 ordered in 1988 including some in AFSV configuration) and other undisclosed countries.

Manufacturer: ENGESA Engenheiros Especializados SA
Avenue Tucunare 125/211
Caixa Postale, PO Box 152/154
06400 Barueri, SP
Brazil
Telephone: (55 11) 421 4711
Telex: 1171302
Fax: (55 11) 421 4445

ENGESA EE-11 Mk VII Urutu with ENGESA turret armed with externally mounted 12.7 mm M2 HB machine gun and four smoke dischargers. This model has 260 hp 6V-53T diesel engine which requires larger engine compartment, hence lack of door in right side of hull

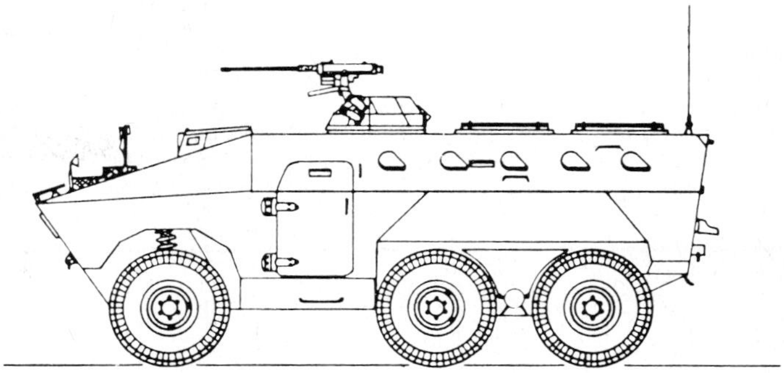

ENGESA EE-11 APC with 12.7 mm M2 MG

SPECIFICATIONS (Mk VII)

CREW	13 (12 + 1)
CONFIGURATION	6 × 6
COMBAT WEIGHT	14 000 kg
UNLOADED WEIGHT	11 000 kg
POWER-TO-WEIGHT RATIO	18.6 hp/tonne
LENGTH	6.1 m
WIDTH	2.65 m
HEIGHT	
to top of 12.7 mm MG mount	2.9 m
to hull top	2.125 m
GROUND CLEARANCE	0.38 m
TRACK	2.2 m
WHEELBASE	3.05 m
ANGLE OF APPROACH/DEPARTURE	60°/58°
MAX ROAD SPEED	105 km/h
MAX WATER SPEED	8 km/h
FUEL CAPACITY	380 l
MAX RANGE	
road	850 km
water	60 km
FORDING	amphibious
GRADIENT	60%
SIDE SLOPE	30%
VERTICAL OBSTACLE	0.6 m
TURNING RADIUS	9.8 m
ENGINE	Detroit Diesel 6V-53T 6-cylinder water-cooled diesel developing 260 hp at 2800 rpm
TRANSMISSION	Allison Transmission Division of General Motors MT-643 automatic with 4 forward and 1 reverse gears
TRANSFER BOX	2-speed
STEERING	ZF integral hydraulic
SUSPENSION	
front	independent with helical springs and double-action telescopic shock absorbers
rear	ENGESA 'Boomerang' with walking beams, leaf springs and double action shock absorbers
TYRES	14.5 × 20
BRAKES	
main	air over hydraulic, dual circuit
parking	mechanical on transfer case output
ELECTRICAL SYSTEM	24 V
BATTERIES	4 × 12 V
ARMAMENT	depends on role of vehicle
SMOKE-LAYING EQUIPMENT	2 or 3 smoke dischargers either side of turret (optional)

Bernardini AM-IV Armoured Security Vehicle

Development

The Model AM-IV armoured security vehicle has been built as a private venture by Bernardini and has been designed mainly for use in the internal security role although its 4 × 4 drive gives it good cross-country mobility. In addition to being used as an internal security vehicle, the Model AM-IV can also be used as a reconnaissance vehicle. The first prototype was completed in 1988 and by early 1990 a total of 25 production vehicles had been completed with a further 25 vehicles on order.

Description

The layout of the vehicle is conventional with the engine at the front, commander and driver in the centre and four men in the rear of the vehicle. The commander and driver are both provided with a forward opening side door while the rear door is electrically operated by the driver. The side and rear windows are provided with wire mesh protection while the driver's and commander's windows can be covered by a shutter that is hinged at the top. In addition there is a roof hatch to enable heavier weapons to be used from within the vehicle.

The all-welded steel armoured hull provides protection against 7.62 mm small arms fire and 9 mm attack with the windows providing the same

Bernardini AM-IV armoured security vehicle with all doors closed

SPECIFICATIONS

CREW	2 + 4
CONFIGURATION	4 × 4
WEIGHT LOADED	3370 kg
PAYLOAD	600 kg
POWER-TO-WEIGHT RATIO	27.89 hp/tonne
LENGTH	3.8 m
WIDTH	1.8 m
HEIGHT	2.4 m
WHEELBASE	2.285 m
MAX ROAD SPEED	100 km/h
FUEL CAPACITY	102 l (2 tanks)
MAXIMUM RANGE	860 km
GRADIENT	60%
TURNING RADIUS	5.3 m
ENGINE	Mercedes-Benz OM 314 4-cylinder diesel developing 94 hp at 2800 rpm
TRANSMISSION	manual, 4 forward and 1 reverse gears
TRANSFER CASE	two speed (high and low)
CLUTCH	hydraulically assisted, single dry disc
AXLES	floating type front and rear
STEERING	power assisted with recirculating balls
BRAKES	
main	front, ventilated disc, rear, drum
parking	drum
SUSPENSION	front and rear, leaf springs, rolling bars and hydraulic shock absorbers
TYRES	7.50 × 16 (bullet proof radial)
ELECTRICAL SYSTEM	12 V
BATTERIES	2 × 12 V, 60 Ah
ALTERNATOR	55 A

degree of protection as the hull. The interior of the vehicle is insulated.

A total of nine firing ports are provided in the front, side doors, hull sides and rear hull door that allow the occupants to use their small arms in safety from within the vehicle.

Standard equipment includes an air-conditioning system, fixed Halon fire suppression system in the engine compartment and wheel arches, three point internal seat belts for all six crew members, flashing lights on the roof and a searchlight that is mounted well above the roof that can be controlled from inside of the vehicle by the commander or driver.

Status: In production.

Manufacture: Bernardini S/A Indústria e Comércio
Rua Hipòlito Soares No 79
04201 Sao Paulo
SP Brazil
Telephone: (011) 274 8033
Telex: (011) 21605 bsaibr
Fax: (011) 274 8567

BULGARIA

BMP-23 Infantry Fighting Vehicle

This is the chassis of the Soviet 122 mm 2S1 self-propelled howitzer, fully covered in the *Self-propelled guns and howitzers* section, fitted with a locally designed two man turret armed with a 23 mm 2A14 cannon and 7.62 mm coaxial machine gun. A 9M111 (AT-4) ATGW system can be mounted on the roof and mounted either side of the turret is a bank of three electrically operated smoke dischargers firing forwards. In the turret rear is a small rectangular door that opens to the right.

The troop compartment towards the rear of the vehicle is provided with two roof hatches that open to the front of the vehicle, either side of each hatch are two roof-mounted periscopes that provide observation to the front of the vehicle.

The troops enter and leave via two doors at the rear which open outwards, each of these being provided with a firing port and a vision block. In addition there are firing ports and vision devices in the sides of the troop compartment.

The BMP-23 retains the amphibious characteristics of the 122 mm 2S1 self-propelled howitzer and before entering the water the trim vane is erected at the front of the vehicle, this being stowed flat on the glacis plate when not required.

BMP-23D Infantry Fighting Vehicle

This is a further development of the above and has a similar armament but a slightly different chassis with different roadwheels to the original BMP-23.

Bulgarian BMP-30 infantry fighting vehicle which has same turret as Soviet BMP-2 infantry fighting vehicle

BMP-30 Infantry Fighting Vehicle

This is the BMP-23 fitted with the turret of the Soviet BMP-2 and is armed with the 30 mm 2A42 cannon, 7.62 mm coaxial MG, roof-mounted ATGW and a bank of three electrically operated smoke dischargers either side of the turret firing forwards.

Status: Production. In service in Bulgaria.

Manufacturer: Bulgarian state factories.

CANADA

Future Canadian Multi-purpose Combat Vehicle

The Canadian Armed Forces are working on a new force structure as a result of the budget reductions of 1989 and 1990. It is now expected that the Army requirements will be for a 'multi-purpose combat vehicle' for Defence of Canada operations and for contingency operations. It is expected that approximately 800 vehicles will be procured from the 1990s.

Canadian M113 Series

The Canadian Armed Forces has over 900 M113A1 series armoured personnel carriers including the M577A1 command post vehicle, M548 tracked cargo carrier, M113A1 engineer vehicle with front-mounted dozer blade and the Lynx command and control vehicle which is fully covered in the *Reconnaissance vehicles* section.

M113A1 with 25 mm Cannon

Under contract to the Canadian Armed Forces, General Motors Canada has fitted an M113A1 with the two-man Delco Systems Operations power-operated turret as installed on the LAV-25 (8 × 8) vehicles as built for the US Marine Corps. The turret is armed with a 25 mm Chain Gun, 7.62 mm coaxial and 7.62 mm anti-aircraft machine guns. But as of April 1991, no decision had been taken to retrofit any Canadian vehicles with this turret.

M113A2 Engineer Vehicle

Under contract to the Canadian Armed Forces, General Motors Canada has modified an M113A2 engineer vehicle to a new configuration known as the M113A2 Engineering Variant Specially Equipped Vehicle. This retains the front-mounted hydraulic dozer blade but is also fitted with a hydraulic earth auger, hydraulic power tools, ramp restraining system, and external fuel tanks on either side of the hull. Full details of this model are given in *Jane's Military Vehicles and Logistics 1991-92* page 1. Early in 1987 the Canadian DoD awarded the company a contract to convert 55 vehicles to the Engineering Variant Specially Equipped Vehicle with a further 15 kits to be supplied to enable vehicles to be converted by the Canadian Armed Forces in Germany.

Canadian M113A2 Engineering Variant Specially Equipped Vehicle with dozer blade lowered, auger ready to use and ramp lowered

M113A1 TUA

The Canadian Armed Forces have fitted 72 of their M113A1 series APCs with the TOW Under Armour (TUA) weapon system under the Phase I programme. Under Phase II the purchase of an additional 30 systems was planned but this was cancelled in May 1991 as part of the defence review. The TUA weapon systems were manufactured under licence in Canada by Invar Manufacturing Limited under licence from Kvaerner Eureka A/S of Norway. This turret is also fitted to Norwegian M113s and Swiss MOWAG Piranha (6 × 6) APCs.

M113A1 fitted with two-man Delco Systems turret as fitted to US Marine Corps LAV-25s

Canadian Armed Forces TOW Under Armour Vehicle showing roof-mounted turret with two TOW ATGWs in ready to launch position and fuel tanks at hull rear (Canadian Armed Forces)

Manufacturer: Diesel Division, General Motors of Canada Limited
PO Box 5160
London
Ontario N6A 4N5
Canada
Telephone: (519) 452 5184
Telex: 064 5643

Diesel Division, General Motors of Canada, Armoured Vehicle General Purpose

Development

In June 1974 the Canadian Armed Forces issued a requirement for an Armoured Vehicle General Purpose (AVGP) for use in Canada. After an initial study of 14 possible vehicles three were selected for field and engineering tests, as a result of which the Swiss MOWAG Piranha (6 × 6) vehicle was chosen.

In February 1977 the Canadian Minister of Defence signed a contract worth $171 million with Diesel Division, General Motors of Canada Limited, for the supply of 350 vehicles manufactured under licence from MOWAG. The first was handed over in January 1979. The original order for 350 vehicles was subsequently increased to 441 units owing to lower production costs, and early in 1981 was again extended by 35 units bringing the total up to 476 units. A further extension of 15 units brought the total order to 491 vehicles. Delivery of this contract was completed by September 1982.

The Canadian vehicles are used by both regular and militia units; have heaters for the engine and crew compartment and are fitted with cold-starting aids. All vehicles have an AN/VVS-501 passive night driving periscope for the driver. The Canadians have three versions of the Piranha:

Cougar 76 mm Gun Wheeled Fire Support Vehicle (WFSV)

The Cougar, of which 195 were built, is the basic vehicle fitted with the turret mounted on the British CVR (T) Scorpion and the Australian M113A1 Fire Support Vehicle. The turret is armed with a 76 mm L23A1 gun with a 7.62 mm machine gun mounted coaxially with the main armament and four smoke dischargers mounted on either side of the turret. The weapons have an elevation of +35° and depression of -8°, with 360° powered turret traverse. A total of 10 rounds of 76 mm and 220 rounds of 7.62 mm machine gun ammunition are carried in the turret with a further 30 rounds of 76 mm and 15 boxes of 7.62 mm ammunition in reserve in the hull. The Cougar has a crew of three: commander, gunner and driver. It carries an AN-GVS-5 hand-held laser rangefinder.

Grizzly Wheeled Armoured Personnel Carrier (WAPC)

The Grizzly APC, of which 269 were built, is the basic vehicle fitted with a Cadillac Gage 1 m turret armed with a 12.7 mm machine gun and a 7.62 mm machine gun, with four smoke dischargers mounted on either side of the turret. The weapons have an elevation of +55° and a depression of -8°, turret traverse being a full 360°. Ten boxes of 12.7 mm and 18 boxes of 7.62 mm machine gun ammunition are carried. The Grizzly WAPC has a crew of three and carries eight fully equipped infantrymen. The vehicle has firing ports and vision blocks in the hull sides and rear. The Light Armored Vehicle and the Armored Vehicle General Purpose have been undergoing trials with a new 12.5R 75 × 20 tyre developed by Michelin. Trials to date have shown a significant improvement in mobility. In 1991 the Canadian Armed Forces awarded a contract to the Diesel Division, General Motors of Canada, to upgrade the suspension of their current 6 × 6 fleet to the heavier 8 × 8 suspension. This upgrade is expected to be completed by 1994.

Husky Wheeled Maintenance and Recovery Vehicle (WMRV)

A total of 27 of these were built for field repair and recovery work. The Husky has a crew of three and is fitted with an HIAB 650A crane with a capacity of 4536 kg. Full details of this version are given in *Jane's Military Vehicles and Logistics 1991-92*, page 43.

Bison (8 × 8) APC

There is a separate entry in this section for the new Bison 8 × 8 APC family currently in production for the Canadian Armed Forces.

LAV with Blazer 25 turret

The LAV (8 × 8) chassis has also been fitted with the private venture General Electric Blazer 25 air defence turret armed with a 25 mm GUA-12 Gatling gun, pod of four missiles (for example Stinger, Mistral), computerised fire control system with FLIR/TV options and a Dassault Electronique RA-20 or Thomson-CSF Gerfault radar system. These missiles would be used to engage targets at longer ranges (up to 6 km) with the 25 mm gun being used to engage closer targets.

Details of the United States Marine Corps Light Armored Vehicle - Air Defence, for which both General Electric and FMC are competing, are given later in this section.

LAV NBCRS

Following trials of a Nuclear, Biological, Chemical Reconnaissance System version of the German Transportpanzer 1 (6 × 6) and a modified Light Armored Vehicle - Logistics (8 × 8), it was announced in 1990 that the team of General Dynamics/Thyssen Henschel with the Transportpanzer 1 had been selected to produce the Transportpanzer (6 × 6) to meet this requirement. Details of this vehicle are given in the *Armoured Personnel Carriers* section under Germany.

LAV Programme

In September 1981 Diesel Division, General Motors of Canada, was awarded a $3.1 million contract by the United States Marine Corps for the supply of four 8 × 8 models of the Piranha vehicle, and supporting

Wheeled Armoured Personnel Carrier Grizzly with Cadillac Gage 1 m turret armed with 12.7 mm and 7.62 mm machine guns from the rear showing firing ports

Wheeled Fire Support Vehicle Cougar which has the same turret as British Alvis Scorpion CVR(T) (Canadian Armed Forces)

services, for use in competitive evaluation in the United States Marines Light Armored Vehicle (LAV) programme.

By October 1981 the company had supplied two Light Assault Vehicles fitted with a two-man Arrowpointe turret armed with a 25 mm Chain Gun, 7.62 mm M240 coaxial machine gun and eight smoke grenade dischargers, and one Assault Gun Vehicle fitted with an Arrowpointe two-man turret armed with a 90 mm Cockerill Mark III gun, 7.62 mm M240 coaxial machine gun and eight smoke dischargers. The second Assault Gun Vehicle was delivered early in 1982.

In September 1982 it was announced that the Diesel Division, General Motors of Canada, had won the LAV competition. The original five-year contract was to have been for 969 LAVs, 680 for the US Army and 289 for the Marine Corps with a total value of $477.8 million, with an option for a further 598 vehicles. The first buy, with FY82 funding, was for 60 vehicles all of which were for the Marine Corps. The first of these was handed over in October 1983. The second year buy, with FY83 funding, was for 170 vehicles of which 134 were for the Marine Corps and the remaining 36 for the Army. Early in 1984 the Army withdrew from the programme and the latter vehicles were taken over by the Marine Corps.

At the same time the original order was placed the company was awarded a $21.6 millon contract for the development and testing of five specialised versions: 81 mm mortar carrier, logistic support, command and control, maintenance/recovery and anti-tank fitted with an Emerson twin launcher as on the M901 Improved TOW Vehicle (ITV). All of these vehicles had been delivered by early 1984.

The quantities of LAV-25s procured by the US Marine Corps between FY82 and FY85 are as follows:

	FY82	FY83	FY84	FY85	Total
LAV-25	60	170	123	69	422
LAV(R)	-	-	33	13	46
LAV(L)	-	-	80	14	94
LAV(M)	-	-	-	50	50
LAV(C)	-	-	-	50	50
LAV(AT)	-	-	-	96	96
Total	60	170	236	292	758

Final deliveries of the LAV were made to the US Marine Corps in 1988.

Late in 1988 it was announced that LAV battalions would in future be called Light Armored Infantry Battalions with each battalion having five companies. Three companies will operate 14 LAV-25 and two LAV command vehicles. The weapons company will have 16 LAV anti-tank, eight LAV 81 mm mortar and eight LAV-25 for reconnaissance. The HQ and service company will have one LAV recovery vehicle, three LAV logistic vehicles and two LAV command vehicles. In the future, a fourth infantry company may be added, as well as integrating the future LAV air defence and assault gun vehicles when these have been developed and accepted for service.

US Marine Corps LAVs were used during the United States invasion of Panama in 1989 by 2nd Light Armored Infantry Battalion. The US Marines also used them during Operation Desert Storm, the recapture of Kuwait in early 1991, and the US Marines have loaned 15 LAV-25s and a recovery vehicle to the US Army's 82nd Airborne Division. Two of these vehicles have been modified for paradrop certification.

LAV Air Deployment Capability

The LAV is air-deployable in current US Air Force cargo transports. The C-5A can carry eight, the C-141 two and the C-130 Hercules one. The US Marine Corps CH-53E transport helicopter can transport one in the sling position.

In February 1987, a LAV-25 was successfully LAPESed (Low Altitude Parachute Extraction System) from a C-130 Hercules transport aircraft and was operationally ready within 12½ minutes of landing. The vehicle was also air-dropped from 1500 feet and was again operationally ready within 12½ minutes.

LAV-25(MC)

This is for the US Marine Corps as are all the following variants. The US Army designation for this vehicle was the M1047: this was very similar to the Marine Corps version but had increased ammunition stowage as it was not intended to carry troops.

The LAV-25(MC) has a Delco two-man turret armed with a 25 mm McDonnell Douglas Helicopter Company Chain Gun, 7.62 mm M240 coaxial machine gun and a pintle mount for an M60 7.62 mm machine gun. Mounted either side of the turret is a bank of M257 smoke dischargers. A stabilisation system allows the 25 mm cannon to be aimed while the vehicle is moving across country.

Up-armoured LAV

In April 1991, the US Defense Advanced Research Projects Agency (DARPA) announced that it had successfully completed a three month project to take a technology from proof-of-principle concept to fieldable prototype system within three months.

The objective of this $2 million effort was to provide 75 prototype kits of Light Applique System Technique (LAST) armour to the Marine Corps for application to LAVs deployed to Saudi Arabia by the US Army and Marine Corps.

The LAST is an attachment system for applique armour developed by Foster-Miller Inc under contract to DARPA. It consists of a hook-and-loop attachment system (from Velcro USA), armour modules (from Lanxide Armor Products and Coors Ceramics) and a protective cover (Bell Avon).

The key features of the LAST system are its light weight and ease of installation and repair with the hook-and-loop attachment system being attached to the LAV with pressure sensitive adhesive. Consequently

US Marine Corps LAV logistics vehicle with roof hatches open

LAV mortar carrier vehicle with roof hatches closed

Husky Wheeled Maintenance and Recovery Vehicle (WMRV) with crane in travelling position and stabilisers raised

US Marine Corps Light Armoured Recovery Vehicle in travelling configuration

no modification of the vehicle structure, such as drilling or welding, is required to install it.

Prototype kits consisted of the LAST system, ceramic armour tiles, a battle damage repair kit and a manual with videotape providing instructions on kit installation.

In the three month effort, an extensive list of technical and operational issues were addressed and overcome, the most challenging was the creation of an industrial base that produced over 60 000 ceramic armour tiles in less than 90 days. In addition the armour tiles also had to undergo ballistic validation testing to verify tile performance. The LAST system underwent development testing and an LAV equipped with LAST armour kit underwent live fire testing and was tested against chemical agents, all during the 90 day effort.

While the timing of the ground war in the Gulf made it impossible to field prototype LAST systems in support of Operation Desert Storm as planned, the Marine Corps has decided to equip Marine units on both the US East and West coasts with LAST kits for continued test and evaluation.

Future Possible LAV Improvements

The 6V-53T engine installed in the LAV develops 275 hp but this could be increased to 300 hp by charged injection.

In February 1986 C Company of the Marine Corps 2nd LAV Battalion, based at Camp Lejuene, North Carolina, was deployed to Fort McCoy, Wisconsin, in Winter mobility tests. The temperature during these tests varied from -20° to +20°F and to enhance operation of the LAVs (16 LAV-25s plus one LAV logistic) the vehicles were fitted with tyre chains developed by the Rudd company of Germany. It took an average of 15 minutes to install the chains on each vehicle and in addition to improving the mobility of the LAV in snow the chains were also used on hard surface roads with no damage to the vehicles or roads.

LAV with FloTrak

Vehicle Systems Development Corporation of Upland, California, has completed a preliminary design/feasibility study to adapt FloTrak to the USMC LAV family of 8 × 8 vehicles and the Canadian Armed Forces Grizzly and Cougar 6 × 6 vehicles.

This study shows that the installation of FloTrak will give a significant improvement in mobility in soft soil and in deep snow. In terms of Nominal Unit Ground Pressure (NUGP), over a 4:1 improvement is projected for the single-wheel FloTrak installation and approaching 5.5:1 for the rear tandem (half-track) wheel arrangement.

LAV HYDRA Rocket Launcher

In 1986 a LAV-25 was fitted with a 19-round HYDRA 70 70 mm (2.75-inch) rocket launcher on the right side of the turret with powered elevation and depression (-200 mils to +600 mils). About 100 rockets were successfully fired during trials in July 1986. The rockets are identical to those fired from aircraft and helicopters and would be used in both the ground-to-ground and anti-helicopter roles. The rockets are the same as those used in the 70 mm Rapid Deployment Multiple Rocket Weapon System covered in the *Multiple Rocket Launchers* section later in this book.

Logistics Vehicle

This has a higher roof with twin hatches for the rapid loading and unloading of cargo, crew of three consisting of the commander, driver and loadmaster, and a crane for loading and unloading cargo at the left hull rear. Armament consists of a pintle mount with a 7.62 mm M60 machine gun and two four-barrelled M257 smoke grenade launchers. Production of this version began in November 1985 and was completed in August 1986.

81 mm Mortar Carrier

This has a crew of five consisting of commander, driver and three mortar men and can be fitted with an 81 mm mortar (94 rounds carried). The mortar is mounted in the centre of the vehicle and fires through the three-part roof hatch. It is also armed with a pintle-mounted 7.62 mm M60 machine gun and two four-barrelled M257 smoke grenade launchers. Production of this version began in September 1985 and was completed in December 1986. A 120 mm mortar has had trials in the LAV.

Maintenance/Recovery

This has a crew of four consisting of commander, driver and two fitters and is armed with a 7.62 mm M60 pintle-mounted machine gun and two M257 four-barrelled smoke grenade launchers. Equipment fitted includes a portable auxiliary power unit, crane with a capacity of 1814 kg and a rear-mounted winch with a capacity of 13 608 kg. Four stabilisers are fitted to give a more stable platform when the crane is being used. Production of this version began in May 1986 and was completed in July 1987.

Anti-tank

This has an Emerson twin TOW launcher mounted on top of the hull above the third wheel station carrying two ready-to-launch missiles and a further 14 missiles in the hull. Secondary armament consists of a 7.62 mm or 12.7 mm machine gun on a pintle and two four-barrelled M257 smoke dischargers. A ground mount for TOW is also carried as is a hand-held laser rangefinder. This model has a crew of four: driver, commander, gunner and loader. Production of this version started in January 1987 and finished in June 1987.

Command and Control

This has a similar hull to the logistics vehicle and has extensive communications equipment installed. The crew consists of the vehicle commander and driver and there is space for a unit commander and four staff members or radio operators. Armament consists of one pintle-mounted 7.62 mm machine gun and two four-barrelled smoke dischargers. Production started in January 1987 and finished in May 1987.

Anti-aircraft

To meet the US Air Force requirement for a Mobile Weapon System (MWS) a General Motors Canada (8 × 8) chassis was fitted with a General Electric turret armed with a 30 mm four-barrelled Gatling gun and Stinger surface-to-air missiles. The MWS was successfully tested but the project was cancelled when the Army agreed to take over the role of defending US Air Force bases.

In May 1987, the US Army Tank Automotive Command solicited bids from 75 companies for an air defence version of the Light Armored Vehicle, but only two bids were received, one from the FMC Corporation and the other from the General Electric Company.

In December 1987 the FMC Corporation was awarded an initial contract worth $8.916 million while General Electric was awarded a contract worth $6.718 million. Total cost of building the four prototypes is expected to be $31.32 million. Each company has built two prototypes based on the LAV (8 × 8) chassis which has been provided by the Marine Corps.

Both entries in the competition are fitted with a two-man power-operated turret armed with a 25 mm cannon, four Stinger fire-and-forget surface-to-air missiles and a launcher with seven HYDRA-70 unguided rockets. A 7.62 mm machine gun is fitted for local protection and the turret also has two banks of electrically operated smoke dischargers. A computerised fire control system is fitted that includes a laser rangefinder.

US Marine Corps LAV-25 fitted with Delco two-man turret and armed with McDonnell Douglas Helicopter Company 25 mm M242 Chain Gun, 7.62 mm M240 coaxial machine gun and two banks of smoke dischargers

US Marine Corps LAV anti-tank vehicle with TOW launcher

The two competing systems are currently being evaluated, with one contractor expected to be selected. The Marine Corps acquisition objective is 157 systems.

Full details of the Light Armored Vehicle - Air Defense are given in *Jane's Land-Based Air Defence 1991-92* page 94.

Mobile Electronic Warfare Support System (MEWSS)

In August 1985, the US Marine Corps ordered two prototypes of the Mobile Electronic Warfare Support System (MEWSS) from General Motors of Canada, which were delivered in 1986. After extensive testing a production order for 12 vehicles was placed in September 1987. Delivery was made in early 1989.

The MEWSS has a crew of four with EW equipment including the WJ-8618B radio receiver, AN/PRD-10 radio direction finding systems, Racal AN/VLQ-19 communications jammer and radios. It also has a radio communications system, provision for the Position Location Reporting System, Land Navigation System, vehicle and shore power capabilities and two air-conditioning systems.

This has been type classified as the AN/MLQ-36 by the US Marine Corps.

Mobile Armored Reconnaissance Vehicle/Standoff Munition Disruptor

After studying the M113A2 and the LAV, the latter was selected by the US Air Force to meet its requirements for a Mobile Armored Reconnaissance Vehicle/Standoff Munition Disruptor and prototypes of the basic chassis have been ordered from General Motors of Canada. This vehicle will carry out explosive ordnance disposal duties at airfields and other security roles. Mounted on the roof will be a 12.7 mm machine gun in a simple turret (this is not part of the General Motors contract) to shoot at unexploded bombs and other ordnance.

In the Spring of 1988, the Armament Division, Eglin Air Force Base, awarded Saco Defense Incorporated a $6.275 million firm fixed-price contract for full scale development of the Mobile Armored Reconnaissance Vehicle and Standoff Munitions Disruptor. Major sub-contractors to Saco Defense are Diesel Division, General Motors of Canada for modifications to the 8 × 8 chassis, Hägglunds of Sweden for turret technology, ITI for the fire control system and XMCO for LSA provision. Two prototypes were delivered in early 1989 and will be evaluated through the year.

This system was subsequently renamed the Base Recovery Vehicle (BRV) but in the budget review for 1990 the United States Air Force lost funding for the vehicle when funds were re-programmed for use in Nicaragua and Panama.

Assault Gun Vehicle

Under contract to Tank Automotive Command, which is acting on behalf of the US Marine Corps, Cadillac Gage Textron is building prototypes of a 105 mm Light Armored Vehicle - Assault Gun (LAV - AG) on the 8 × 8 chassis. Details of this are given in the *Reconnaissance vehicles* section under the United States.

Prototype of Light Armored Vehicle (LAV) (8 × 8) Mobile Electronic Warfare System with antenna erected

One of the two prototypes of the General Electric entry in US Marine Corps competition for a Light Armored Vehicle - Air Defense

One of the two prototypes of the FMC entry in the US Marine Corps competition for a Light Armored Vehicle - Air Defense

LAV with MCWS

For trials purposes, an LAV (8 × 8) vehicle of the Marine Corps was in 1987 fitted with the AAI developed Minor Calibre Weapons Station armed with a 25 mm McDonnell Douglas Helicopter Chain Gun and a 7.62 mm coaxial MG, with the gunner being mounted below the turret ring. The Minor Calibre Weapons Station has been developed by AAI under contract to the Naval Surface Weapons Center/David Taylor Naval Ship Research and Development Center/US Marine Corps and is for trials and developmental purposes.

120 mm Mortar Carrier

For trials purposes a 120 mm mortar was fired from inside of an LAV (8 × 8) some years ago. Within the Saudi Arabian National Guard requirement is a 120 mm mortar carrier. This will have a turret-mounted mortar with a direct fire range of 1000 m, minimum unassisted indirect fire range of 8000 m, 7.62 mm coaxial machine gun, 12.7 mm anti-aircraft machine gun, self-contained mortar fire control system for both direct and indirect fire. Turret capable of 360° traverse, onboard stowage for 60 120 mm mortar bombs, minimum first minute firing rate of 8 rounds required (15 rounds desired) and a sustained firing rate of 4 rds/min.

At present there is no 120 mm turret-mounted mortar system in production that could meet this requirement, although Royal Ordnance of the UK has developed such a system to the prototype stage and this has been tested on an FMC M113 series APC.

Australian LAV

In mid-1989 the Australian Government purchased 14 LAV-25 and one LAV maintenance/recovery vehicle from United States Marine Corps stocks for $25 million. These have been issued to the Australian Army's 2nd Cavalry Regiment based at Holsworthy, New South Wales, which will move to Darwin, Northern Territory, in 1992.

The 15 LAVs are sufficient to equip one squadron of the 2nd Cavalry Regiment which at present is equipped with the M113 series of tracked vehicles. If these trials are successful then it is expected that an additional 100 vehicles will be required to bring the regiment up to full strength. Contract for Phase 2 of this programme is expected in the fourth quarter of 1991. The Canadian Armed Forces have also loaned four Bison (8 × 8) APCs to Australia for trials purposes alongside the LAV-25.

US Army LAV

Late in 1989 the United States Marine Corps loaned the 82nd Airborne Division of the United States Army 16 LAV-25, and in return the Army will loan the Marine Corps 11 Multiple Launch Rocket Systems. Two of these vehicles are being modified by the Diesel Division, General Motors of Canada, for certification trials for heavy paradrop. If successful, the US Army intends to purchase these vehicles from the US Marine Corps and modify all vehicles for paradrop.

LAV - Saudi Arabian National Guard (SANG)

In June 1990 solicitation was made to the US Congress for the sale of 1117 LAV (8 × 8) to Saudi Arabia through Foreign Military Sales (FMS) programme. These vehicles would be operated by the Saudi Arabian National Guard and replace Cadillac Gage Textron V-150 (4 × 4) vehicles currently in service.

The Letter of Agreement (LoA) was signed in September 1990 and negotiations continue towards contract.

Saudi Arabia requires 10 different vehicle variants: LAV-25, anti-tank, recovery, mortar, command, armoured personnel carrier, ammunition carrier, ambulance, engineer vehicle and assault gun.

Status: Production of the 6 × 6 model has been completed for the Canadian Armed Forces. Production of the 8 × 8 model for the USA began in 1983 with first production vehicles completed in October 1983; production for the US Marine Corps was completed in mid-1988. Production of 199 8 × 8 APCs for the Canadian Armed Forces commenced in mid-1989 with first deliveries in 1990. A total of 15 LAV (8 × 8) have been delivered to Australia for trials, 14 LAV-25 and one LAV maintenance/recovery, plus four Bison.

Manufacturer: Diesel Division, General Motors of Canada Limited
PO Box 5160
London
Ontario N6A 4N5
Canada
Telephone: (519) 452 5184
Telex: 064 5643

SPECIFICATIONS (Grizzly WAPC)	
CREW	3 + 6
COMBAT WEIGHT	10 500 kg
LENGTH HULL	5.968 m
WIDTH	2.53 m
HEIGHT OVERALL	2.53 m
HEIGHT HULL ROOF	1.85 m
GROUND CLEARANCE	0.392 m
MAX ROAD SPEED	
forward	101.5 km/h
reverse	17.9 km/h
MAX WATER SPEED	7 km/h
FUEL CAPACITY	204 l
MAX ROAD RANGE	603 km
GRADIENT	60%
SIDE SLOPE	30%
VERTICAL OBSTACLE	0.381-0.508 m
TRENCH	0.406 m
TURNING RADIUS	6.55 m

ENGINE	Detroit Diesel 6V-53T, 6-cylinder diesel developing 215 hp
TRANSMISSION	Allison Transmission Division of General Motors MT-650 automatic, 5 forward and 1 reverse gears
STEERING	power-assisted, front wheels
SUSPENSION	independent, all wheels
TYRES	11.00 × 16 with Hutchinson run-flat inserts
ARMAMENT	
main	1 × 12.7 mm M2 HB MG
coaxial	1 × 7.62 mm C5A1 GPMG
smoke dischargers	2 × 4 No. 12 Mk 1

AMMUNITION	
main	1000
coaxial	4400
GUN CONTROL EQUIPMENT	
Turret power control	electric/manual
by commander	no
by gunner	yes
Gun elevation/depression	+55°/-8°
Turret traverse	360°
Turret vision devices	1 × M28C sight 8 vision blocks
ARMOUR	
Hull front upper	8 mm at 23°
Hull front lower	8 mm at 45°
Hull sides	10 mm at 55°
Turret	9.5 mm

SPECIFICATIONS (LAV-25)

CREW	3 + 6
CONFIGURATION	8 × 8
COMBAT WEIGHT	12 882 kg
UNLOADED WEIGHT	11 099 kg
LENGTH	6.393 m
WIDTH	2.499 m
HEIGHT OVERALL	2.692 m
HEIGHT REDUCED	2.565 m
MAX ROAD SPEED	100 km/h
MAX WATER SPEED	10.46 km/h
MAX RANGE	668 km
GRADIENT	70%
VERTICAL OBSTACLE	0.5 m
TRENCH	2.057 m
ENGINE	GM Detroit Diesel 6V-53T, 6-cylinder diesel developing 275 hp at 2800 rpm
TRANSMISSION	Allison Transmission Division of General Motors MT-653 DR automatic, 5 forward and 1 reverse gears
TRANSFER CASE	Rockwell AG-VST (modified)
STEERING	power-assisted on front 2 axles
SUSPENSION	
front four wheels	independent coil springs and shock absorbers
rear four wheels	independent torsion bars and shock absorbers
BRAKES	
main	8-wheel dual air hydraulic drum brake
parking	transmission brake and transfer case lock
TYRES	11.00 × 16 with Hutchinson run-flat inserts
ARMAMENT	
main	1 × 25 mm M242 cannon
coaxial	1 × 7.62 mm M240 MG
anti-aircraft, optional	1 × 7.62 mm M60 or 1 × 12.7 mm M2 HB MG
smoke dischargers	2 × 4 (M257)
AMMUNITION	
25 mm	210 + 420 (stowed)
7.62 mm	420 + 1200 (stowed)
GUN CONTROL EQUIPMENT	
Turret power control	hydraulic/manual
Elevation/depression	+60°/-10°
Turret traverse	360°
Gun stabiliser	
vertical	yes
horizontal	yes
Periscopes	7 × M27
Sights	2 × M36E1

Diesel Division, General Motors of Canada, Bison Armoured Personnel Carrier

Development

In 1988, the Diesel Division, General Motors of Canada, designed and built within seven days, as a private venture, a new 8 × 8 armoured personnel carrier based on the chassis of the LAV (8 × 8) vehicle described in the previous entry.

The turret of the LAV was removed and a raised commander's cupola with five M17 periscopes and an externally mounted 7.62 mm machine gun was fitted to the immediate rear of the driver's station at the front left side of the hull.

The troop compartment had a raised roof that ran to the rear of the vehicle with roof hatches and a power operated ramp rather than doors as on the US Marine Corps vehicles. No firing ports were provided although two vision blocks were mounted in the rear ramp.

In addition to the commander and driver the vehicle carried eight fully equipped troops that were seated four down each side facing each other.

Standard equipment included front-mounted winch, while optional equipment included fire detection and suppression system, wire cutter, NBC defence system and night vision devices.

Following trials with the prototype vehicle, in July 1989 the Canadian Department of National Defence awarded the Diesel Division, General Motors of Canada, a contract worth $100 million to build a total of 199 vehicles to a slightly modified design for the Canadian Militia.

First production vehicles were delivered in the Summer of 1990 with deliveries running through to 1993. The Canadian Armed Forces call the vehicle the Bison with the 199 vehicles consisting of the following versions:

149 APCs

18 command post vehicles

16 81 mm mortar carriers

16 maintenance repair vehicles with same crane as Husky 6 × 6 vehicle already in service with Canadian Armed Forces. This will be able to carry a complete powerpack internally.

The Bison can be quickly adapted for specialised roles as the baseline vehicle has a rail system fitted as standard.

In addition, four vehicles have been loaned to the Royal Australian Army for trials and the manufacturer has proposed the vehicle to meet the requirements of the US Army National Guard.

Description

The basic automotive layout of the Bison APC is similar to the LAV (8 × 8) with the hull being of all-welded steel armour construction that provides protection from small arms fire and shell splinters.

The driver is seated at the front left with the engine compartment to his right. The driver has a single piece hatch cover that opens to the left and three M17 periscopes, the centre one of which can be replaced by an AN/VVS-2(V)4 passive night vision device.

The commander is seated to the rear of the driver and has a single piece hatch cover that opens to the rear and five M17 periscopes, the centre one of which can be replaced by an AN/VVS-2(V)4 passive night periscope. A 7.62 mm C6 MG is mounted on a skate type mount at the commander's station.

The troop compartment is at the rear of the hull with the troops entering and leaving via a large power-operated ramp that also has a door and two vision blocks.

The infantrymen sit on bench type seats down either side of the hull with stowage being provided under their seats and to their immediate rear. There is an additional seat to the immediate rear of the engine compartment bulkhead.

Over the forward part of the roof compartment is a three part roof hatch which opens left and right of the vehicle centre line. On the roof at the rear is a large stowage box with stowage racks being provided externally on either side of the vehicle.

Mounted on the top of either side of the hull rear is a bank of M257 four-barrelled smoke dischargers that cover the frontal arc of the vehicle.

Standard equipment includes a Halon 1301 fire suppression system in the engine and crew compartments. Front-mounted hydraulic winch with a capacity of 6800 kg, M8A1 ventilated face mask NBC system, wire cutter forward of driver's position and complete communications installation.

The Bison is fully amphibious with two minutes preparation being propelled in the water by two propellers mounted under the rear of the hull, to the rear of the propellers are two rudders.

Bison (8 × 8) APCs on the production line

Production Bison (8 × 8) APC showing external stowage

Production Bison (8 × 8) APC from rear with ramp deployed and roof hatches in open position

Command post version of Bison (8 × 8) showing additional antennas

SPECIFICATIONS	
CREW	2 + 9
CONFIGURATION	8 × 8
COMBAT WEIGHT	12 800 kg
CURB WEIGHT	10 750 kg
PAYLOAD	2045 kg
LENGTH	6.452 m
WIDTH	2.5 m
HEIGHT	2.21 m
MAX ROAD SPEED	100 km/h
MAX WATER SPEED	9.7 km/h
MAX RANGE	665 km
GRADIENT	60%
SIDE SLOPE	30%
VERTICAL OBSTACLE	0.381-0.508 m
TRENCH	2.06 m
ENGINE	GM Detroit Diesel, 6V-53T, 6-cylinder diesel developing 275 hp at 2800 rpm
TRANSMISSION	Allison Transmission Division of General Motors MT-653 DR automatic, 5 forward and 1 reverse gears
TRANSFER CASE	Rockwell AG-VST (modified)
STEERING	power-assisted on front 2 axles
TURNING RADIUS	7.75 m
SUSPENSION	
front four wheels	independent coil springs and shock absorbers
rear four wheels	independent torsion bars and shock absorbers
BRAKES	
main	8-wheel dual air hydraulic drum brake
parking	transmission brake and transfer case lock
TYRES	11.00 × 16 with Hutchinson run-flat inserts
ELECTRICAL SYSTEM	24 V
BATTERIES	2
ARMAMENT	1 × 7.62 mm MG
SMOKE DISCHARGERS	2 × 4

Status: In production. In service with Canadian Armed Forces (199 on order). Four vehicles are being trialled by Australia.

Manufacturer: Diesel Division, General Motors of Canada Limited
PO Box 5160
London
Ontario N6A 4N5
Canada
Telephone: (519) 452 5184
Telex: 064 5643

CHILE

FAMAE Piranha 8 × 8D Armoured Personnel Carrier

Development/Description

FAMAE of Chile has built the prototype of a MOWAG Piranha 8 × 8 armoured personnel carrier which is fully described under Switzerland. FAMAE calls the vehicle the Piranha 8 × 8D with the D standing for diesel.

The single prototype is identical to the Swiss vehicles except that it has a locally designed cupola armed with a 12.7 mm M2 machine gun and smoke dischargers, the propellers under either side of the hull are replaced by stowage boxes and the twin doors at the hull rear have been replaced by a power-operated ramp with an emergency door in the left side.

Status: Prototype. Not yet in production in Chile.

Manufacturer: FAMAE
Adva Pedro Montt 1568
PO Box 4.100
Santiago
Chile

Prototype of FAMAE Piranha 8 × 8D armoured personnel carrier fitted with locally produced cupola armed with 12.7 mm MG (Terry J Gander)

SPECIFICATIONS					
CREW	1 + 12	HEIGHT		MAX ROAD SPEED	100 km/h
CONFIGURATION	8 × 8	hull top	1.85 m	MAX WATER SPEED	10.5 km/h
COMBAT WEIGHT	13 300 kg	top of MG mount	2.37 m	FUEL CAPACITY	300 l
EMPTY WEIGHT	8800 kg	GROUND CLEARANCE	0.5 m (hull centre)	CRUISING RANGE	700 km
POWER-TO-WEIGHT RATIO	24.4 hp/tonne	TRACK	2.18 m	GRADIENT	70%
LENGTH OVERALL	6.365 m	WHEELBASE	1.1 m + 1.335 m + 1.04 m	SIDE SLOPE	35%
WIDTH	2.50 m	ANGLE OF APPROACH/DEPARTURE	45°/40°	VERTICAL OBSTACLE	0.5 m
				TURNING RADIUS	15.4 m

Note: remaining specifications are identical for Piranha 8 × 8 described under Switzerland.

Cardoen/MOWAG Armoured Personnel Carriers

Development/Description

By early 1991 Cardoen Industries is believed to have built over 225 MOWAG Piranha 6 × 6 vehicles under licence for the Chilean Army and a single prototype of the basic 4 × 4 model.

The Chilean-built Piranha is identical to the original Swiss model except that its fuel capacity has been increased to 400 litres compared to the 250 litres of the original model. This has increased its operational range from 600 km to about 1000 to 1200 km.

Cardoen builds the complete hull of the 6 × 6 vehicle as well as some of the subsystems with some components, such as the engine, still being imported.

So far the Chilean Army has taken delivery of 150 vehicles in the APC role, 50 120 mm mortar carriers, five command post vehicles and 20 with two-man turrets armed with French 90 mm F1 guns. The command post vehicle has a power-operated ramp in the rear in place of the normal two doors.

Variants built to the prototype stage or converted from production vehicles for trials purposes include an aerial surveillance radar carrier, anti-aircraft with the Israeli TCM-20 twin 20 mm light anti-aircraft gun system, fitted with Oerlikon-Contraves one-man 20 mm GAD-AOA turret, French Hispano-Suiza 90 mm turret, Brazilian ENGESA ET-90 90 mm turret and more recently a Cardoen designed and built 90 mm turret armed with a Cockerill 90 mm Mk III gun. Prototypes have also been built of the vehicle fitted with a 70 mm and 102 mm multiple rocket launcher. Ambulance, anti-tank, workshop and repair models have also been proposed.

In 1988 a Cardoen built MOWAG Piranha (6 × 6) vehicle was fitted, for trials purposes, with a Chinese NORINCO one-man turret with four Red Arrow 8 ATGWs in the ready to launch position.

Specifications of the Chilean-built vehicles are identical to those built by MOWAG in Switzerland and are fully covered in that entry.

Cardoen/MOWAG (6 × 6) command post vehicle with power-operated ramp at rear (T J Gander)

Cardoen/MOWAG (6 × 6) APC fitted with Israeli TCM-20 twin 20 mm anti-aircraft system

Cardoen/MOWAG (6 × 6) APC fitted with one-man NORINCO turret armed with four Red Arrow 8 ATGWs in the ready to launch position (T J Gander)

Cardoen/MOWAG (6 × 6) APC fitted with multiple rocket launcher (T J Gander)

Status: In production. In service with Chilean Army in 6 × 6 configuration only.

Manufacturer: Industrias Cardoen Chile Ltda
Los Conquistadores 1700 — Piso 28
Santiago
Chile
Telephone: 2313420; 2314219
Telex: 340997; 341630

Cardoen VTP-1 Orca Multi-purpose Armoured Vehicle

Development

The VTP-1 Orca (Killer Whale) multi-purpose armoured vehicle, developed as a private venture by Industrias Cardoen SA, was announced in 1983. The prototype is different in appearance to that in the original drawings released in 1982 (*Jane's Armour and Artillery 1983-84*, page 265). Production of the Cardoen VTP-1 Orca began in January 1985 when it is understood that the Chilean Army ordered 100 VTP-1 vehicles.

Description

The all-welded steel hull of the Orca provides complete protection from small arms fire up to 7.62 mm in calibre and shell splinters. The engine compartment is at the front of the hull with air-inlet louvres in the top to allow access to the engine for maintenance. The driver sits to the rear of the engine on the left with the commander to his right, each having a single piece hatch cover in front of which are three fixed observation periscopes.

The forward part of the troop compartment has overhead armour protection and a single-piece hatch cover each side that opens left and right with the rear part of the troop compartment roof being open. In each side of the forward part of the troop compartment are two firing ports with a vision block above. Normal means of entry for the troops is via the rear of the vehicle which has a two-part hatch, of which the lower part folds downwards and the upper part opens to the right. The infantry sit on bench seats that run down either side of the hull. A number of 7.62 mm and/or 12.7 mm machine guns are mounted around the top of the troop compartment.

Cardoen VTP-1 Orca multi-purpose armoured vehicle without armament installed (T J Gander)

Variants

In addition to its primary role of troop transport, the manufacturer has proposed the vehicle for a number of other roles including towing a 105 mm or 155 mm artillery weapon as well as carrying its crew and ammunition, 20 to 30 mm air-defence weapon, 60 to 90 mm direct fire weapon, anti-tank with ATGWs, mortar carrier with 120 mm mortar that can be traversed through 360° together with ammunition and crew, ammunition resupply vehicle, radar carrier to provide air defence weapons with an early warning capability, command vehicle with a crew of 12 + 2, vehicle-mounted communications relay station, mobile workshop, ambulance carrying up to 12 injured men, mobile hospital carrying an inflatable unit and a mobile emergency surgery.

Cardoen VTP-1 Orca multi-purpose armoured vehicle from the rear, showing two-part entry door (T J Gander)

SPECIFICATIONS		SIDE SLOPE	30%	BRAKES	
CREW	2 + 16	ENGINE	Detroit Diesel 6V-53T 6-cylinder water-cooled diesel developing 260 hp at 2800 rpm	main	air
CONFIGURATION	6 × 6			parking	mechanical
COMBAT WEIGHT	18 000 kg			ARMAMENT	see text
UNLOADED WEIGHT	13 000 kg			ARMOUR	6 mm - 16 mm
LENGTH	7.84 m				
WIDTH	2.5 m	TRANSMISSION	Allison Transmission Division of General Motors MT-653-R automatic, 5 forward and 1 reverse gears		
HEIGHT	2.5 m				
MAX ROAD SPEED	100 km/h				
FUEL CAPACITY	400 l				
MAX RANGE	1000 km				
GRADIENT	60%	TYRES	14.00 × 20		

Status: Production as required. Believed to be in service with the Chilean Army.

Manufacturer: Industrias Cardoen Chile Ltda
Los Conquistadores 1700 — Piso 28
Santiago
Chile
Telephone: 2313420; 2314219
Telex: 340997; 341630

Cardoen BMS-1 Alacran Multi-purpose Armoured Vehicle

As far as is known this half-track armoured vehicle never entered production. Full details were given in *Jane's Armour and Artillery 1987-88*, pages 269/270.

Cardoen VTP-2 Armoured Personnel Carrier

Development
The VTP-2, or Escarabajo as it is more commonly known, has been designed as a private venture by Cardoen with the first prototype completed in 1981-82.

Description
The vehicle is available with a fully enclosed body (as shown in the photographs) or with an open truck type body. The all-welded steel body can be built with one of two thicknesses of armour, 6 mm which gives a body weight of 2500 kg or 8 mm which gives a body weight of 3300 kg. The vehicle is based on the chassis of the Mercedes-Benz Unimog cross-country vehicle so spare parts are available worldwide.

The engine compartment is at the front of the vehicle with an access hatch in the bonnet; the commander and driver are in the centre with the troop compartment at the rear.

The commander and driver have a bullet-proof windscreen to their front and either side of their position is a shallow door that opens to the rear. In the upper part of this is a bullet-proof window that can be covered by an armoured shutter hinged at the bottom when in a combat area.

The troop compartment is at the rear of the hull and entry to this is via a large door in the hull rear that opens to the right. Either side of this is a vision block with a firing port underneath and in either side of the hull are a further three firing ports and a similar number of vision blocks. The troops are seated on bench seats that run down either side of the hull, with individual seats for the commander and driver. In the forward part of the roof is a circular mounting on which a 7.62 mm or 12.7 mm machine gun can be installed or a variety of other weapons systems such as an Oerlikon-Contraves 20 mm GAD-AOA one-man turret. In the rear part of the hull roof are two oblong roof hatches that are hinged to the outside and can be locked in the vertical position.

The VTP-2 can also be delivered with two doors in the hull rear, each with a vision block and firing port. In this model the troops are seated on bench seats that run down the centre of the vehicle facing outwards and the commander and driver have single piece hatch covers. The manufacturer has also proposed that a version could be built with doors in either side of the hull and periscopes in the roof for use by the commander and driver if the vehicle were fitted with an armoured shutter for the front windscreen. One model has also been built with the rear entrance consisting of two doors that open either side, each having a lower part that folds down to form a step.

Standard equipment includes a forced ventilation system. The interior of the body is covered with a thermal and acoustic insulating material. Optional equipment includes a heater and air-conditioning system.

In 1984 the company announced that the VTP-2 was being offered with a Detroit Diesel 4-53 turbo-charged diesel developing 179 hp at 2800 rpm coupled to an Allison Transmission Division of General Motors MT-653 automatic transmission. This would also have reinforced axles, hydraulic steering and an improved braking system.

Variants
According to the manufacturer the VTP-2 can be adapted for a wide range of roles including ambulance, 81 mm mortar vehicle, fire control vehicle, internal security vehicle, anti-tank with HOT, MILAN or TOW ATGWs, anti-tank with 106 mm M40 recoilless rifle, command post with extensive communications equipment and anti-aircraft with Israeli TCM-20 twin 20 mm system.

Status: Development complete. Ready for production.

Manufacturer: Industrias Cardoen Chile Ltda
Los Conquistadores 1700 — Piso 28
Santiago
Chile
Telephone: 2313420; 2314219
Telex: 340997; 341630

VTP-2 APC without armament

VTP-2 armoured personnel carrier with single rear door open

SPECIFICATIONS					
CREW	12 (2 + 10)	FUEL CAPACITY	150 l	CLUTCH	single dry plate
CONFIGURATION	4 × 4	MAX RANGE	600 km	TRANSFER CASE	2-speed
WEIGHT		GRADIENT	70%	STEERING	hydraulic
loaded, 8 mm armour	8600 kg	SIDE SLOPE	35%	TYRES	12.50 × 20 (bullet-proof)
loaded, 6 mm armour	7200 kg	VERTICAL OBSTACLE	0.5 m	NUMBER OF TYRES	4
empty, 8 mm armour	6600 kg	TURNING RADIUS	6.45 m	BRAKES	
empty, 6 mm armour	5800 kg	ENGINE	Mercedes-Benz OM-352 6-cylinder water-cooled diesel developing 120 hp at 2800 rpm	main	disc on all 4 wheels
LENGTH	5.37 m			parking	mechanical on rear 2 wheels
WIDTH	2.32 m			ELECTRICAL SYSTEM	24 V
HEIGHT	2.22 m			BATTERIES	2 × 12 V, 110 Ah
GROUND CLEARANCE	0.44 m	TRANSMISSION	manual, 4 forward and 1 reverse gears	ARMAMENT	see text
MAX ROAD SPEED	100 km/h				

Cardoen Light Armoured Vehicle

Development
The Light Armoured Vehicle (LAV) has been developed as a private venture by Industrias Cardoen to enable five or six men to be transported at high speed. The manufacturer claims that the vehicle is suitable for both military and police applications including surveillance and internal security operations.

Description
The Cardoen LAV is based on the chassis of a Toyota Land Cruiser (4 × 4) which has been fitted with a body of high resistance steel and Kevlar. Other types of chassis are already under consideration, including a Land Rover.

The prototype provides complete protection from small arms fire up to 7.62 mm in calibre. The layout of the vehicle is similar to the basic Land Cruiser with the engine at the front, commander and driver in the centre and troop compartment at the rear.

The commander and driver are both provided with a forward-opening side door in the upper part of which is a bullet proof window covered by a wire mesh, underneath of which is a circular firing port. The commander and driver also have a bullet proof window to their front between which is a circular firing port.

The troop compartment at the rear has roof hatches and doors and in either side are two circular firing ports with rectangular observation windows, which are provided with wire mesh protection.

The Cardoen LAV has a maximum road speed of 120 km/h and a range of 350 km. The tyres have elastomeric-polyurethane injected into them which make them puncture proof. As an option these can be replaced by Hutchinson run-flat tyres.

The vehicle is powered by a 4 litre engine which develops 75 hp at 4200 rpm, this being coupled to a manual transmission with four forward and one reverse gears and a two-speed transfer case. Disc brakes are fitted on all four wheels and steering is servo-assisted.

Standard equipment includes an air-conditioning system and optional equipment includes various armament installations and communications systems.

Prototype of Cardoen Light Armoured Vehicle fitted with roof-mounted flashing lights

Status: Development complete. Ready for production.

Manufacturer: Industrias Cardoen Chile Ltda
Los Conquistadores 1700 — Piso 28
Santiago
Chile
Telephone: 2313420; 2314219
Telex: 340997; 341630

Makina Carancho 180 Light Armoured Vehicle

Development
The Carancho 180 (4 × 4) light armoured vehicle was developed in the late 1970s by Makina Ltda and it is believed that a quantity of vehicles were delivered to the Chilean Armed Forces.

Wherever possible standard and proven automotive components have been used in the design of the vehicle to reduce both procurement and life cycle costs.

Description
The layout of the Carancho 180 is conventional with the fully enclosed engine compartment at the front, commander and driver to the rear and the troop compartment extending right to the rear of the hull.

The body is of all-welded armour steel which provides complete protection against penetration by 7.62 mm NATO AP rounds. The engine compartment hood and front mudguards are of sheet steel and if the vehicle runs over a mine they will be blown away, leaving the hull undamaged.

Carancho 180 (4 × 4) light armoured vehicle

Both the commander and driver have a hatch cover to the front that is hinged at the top and a forward-opening side door with an upper part that hinges downwards on the outside. This has an integral firing port.

The troop compartment has an open roof and the troops enter and leave via two doors in the hull rear; the left door has a spare wheel and tyre. The basic model has four seats, two each side facing outwards that occupy only half of the troop compartment. There is no provision for the troops to use their small arms without exposing their heads above the roof line. The four troop seats can be folded up to allow the whole floor to carry cargo.

Provision is made for installing three 7.62 mm machine guns on pintle mounts around the top of the hull, one either side and one over the commander's and driver's position.

Standard equipment includes bullet-proof windows for the commander and driver and a tow hook at the rear. Optional equipment includes run-flat tyres.

Variants
The manufacturer suggested that the Carancho 180 can be adapted for a wide range of roles such as ATGW carrier, multiple rocket launcher, fitted with 106 mm recoilless rifle, fitted with 20 mm anti-aircraft cannon, fully enclosed body fitted with a turret-mounted 7.62 mm machine gun, or an armoured ambulance, but as far as it is known none of these has entered production.

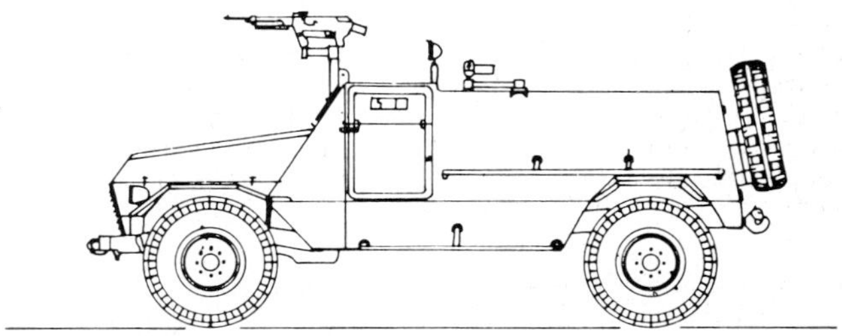

Carancho 180 LAV

SPECIFICATIONS					
CREW	2 + 4	WIDTH	2.2 m	TRANSFER CASE	2-speed
CONFIGURATION	4 × 4	HEIGHT	1.9 m	SUSPENSION	elliptical springs
WEIGHT		GROUND CLEARANCE	0.35 m	TYRES	10.5 × 20
combat	4300 kg	MAX ROAD SPEED	120 km/h	NUMBER OF TYRES	4 + 1 spare
empty	2800 kg	RANGE	450 km	BRAKES (main)	hydraulic
POWER-TO-WEIGHT RATIO	41.86 hp/tonne	ENGINE	Chrysler V-8 developing 180 hp at 4000 rpm	ELECTRICAL SYSTEM	12 V
PAYLOAD	1500 kg	TRANSMISSION	Chrysler Type A-727 automatic, 3 forward and 1 reverse gears	BATTERIES	2
LENGTH	4.9 m				

Status: Production complete. In service with Chilean Armed Forces.

Manufacturer: Makina Ltda
Ingenieros Mecanicos Industriales
Avenue Marathon 2707
Santiago 11
Chile

Makina Multi 163 Armoured Personnel Carrier

Development/Description
The Multi 163 has been designed by the Makina Company primarily for internal security operations and for patrolling airfields and other high risk areas. It is based on a modified commercial chassis and fitted with a fully armoured body. The commander and driver are seated at the front of the vehicle, each with a windscreen which can be covered by an upward opening hatch cover. The troop compartment is at the rear with doors, firing ports, two rectangular roof hatches and one circular roof hatch. The circular hatch is towards the front of the vehicle and can be fitted with a 7.62 mm machine gun or similar weapon.

Variants
Suggested variants include an ambulance, anti-aircraft vehicle, command post and an 81 mm or 105 mm mortar carrier.

Makina Multi 163 APC fitted with roof-mounted 7.62 mm machine gun with commander's and driver's hatches open

SPECIFICATIONS					
CREW	2 + 8	WIDTH	2.23 m	ENGINE	Chrysler 8-cylinder petrol developing 180 hp
CONFIGURATION	4 × 2	HEIGHT	2.1 m	TRANSMISSION	manual, 4 forward and 1 reverse gears
WEIGHT		WHEELBASE	3.327 m	TRANSFER CASE	none
loaded	5000 kg	GROUND CLEARANCE	0.325 m	TYRES	9.50 × 16.50
empty	3300 kg	MAX ROAD SPEED	110 km/h	ELECTRICAL SYSTEM	12 V
LENGTH	5.057 m	FUEL CAPACITY	114 l	ARMAMENT	see text
		RANGE	450 km		

Status: Production completed. Reported to be in service with Chilean Air Force.

Manufacturer: Makina Ltda
Ingenieros Mecanicos Industriales
Avenue Marathon 2707
Santiago 11
Chile

CHINA, PEOPLE'S REPUBLIC

NORINCO Type WZ 501 Infantry Fighting Vehicle

Development/Description
This is almost a direct copy of the Soviet BMP-1 infantry fighting vehicle, the only major differences being slight reductions in weight and maximum road speed. It is understood that the prototype of the WZ 501 IFV was copied from a single BMP-1 supplied by Egypt.

Main armament of the Type WZ 501 is a 73 mm smoothbore gun with 40 rounds of ammunition and a 7.62 mm coaxial machine gun with 2000 rounds. The original Soviet vehicle has a Sagger wire-guided ATGW mounted above the 73 mm gun with an additional four missiles carried internally for manual reloading. The Chinese WZ 501 carries four Red Arrow ATGWs which are copies of the Sagger which China is known to have produced for some years. In addition to seven 7.62 mm SMGs and two 7.62 mm LMGs, a new 40 mm manportable rocket launcher is carried, as is a man-portable surface-to-air missile called Red Tassel, a copy of the Soviet SA-7 Grail. The WZ 501 is fully amphibious and is fitted with an NBC system and infra-red night vision equipment for the commander, driver and gunner.

The Type WZ 501 infantry fighting vehicle also forms the chassis for the NFV-1. This is essentially the WZ 501 fitted with an FMC power-operated turret armed with a 25 mm cannon and 7.62 mm machine gun. Details are given under International in this section.

Variants

Type WZ 503 Infantry Fighting Vehicle

This was first shown in 1988 and is based on the chassis of the WZ 501 IFV except that its hull is 100 mm higher, now the two-man turret with 73 mm gun is installed. The two doors in the rear troop compartment are replaced by a single door that opens to the right and the troops sit on individual seats down either side of the hull rather than back to back down the centre of the vehicle.

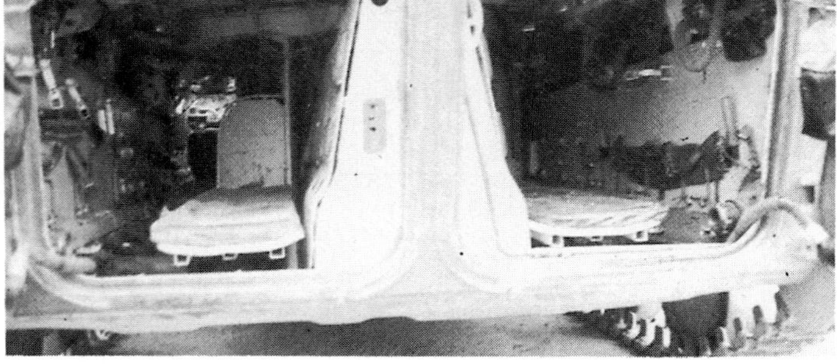

Type WZ 501 Infantry Fighting Vehicle from the rear with doors open to show interior seating arrangements

Type WZ 503 Infantry Fighting Vehicle (Eric Ditchfield)

Type WZ 501 Infantry Fighting Vehicle

Type WZ 504 Missile Launcher Vehicle with launcher in elevated position

Typically the WZ 503 is armed with a manually operated cupola armed with a 12.7 mm machine gun with no overhead protection being provided for the gunner.

Standard equipment includes five spherical firing ports, four rotatable periscopes, NBC system, fire extinguishing system and navigation system for the driver. The vehicle is fully amphibious, being propelled in the water by its tracks.

Type WZ 504 ATGW Launcher Vehicle

This was also first shown in 1988 and is based on the WZ 501 chassis. Mounted on the roof is a pod that contains four Red Arrow-73B ATGWs that are raised when required for action with the sighting system being mounted between the twin launchers. In addition to the four missiles in the ready to launch position, an additional 12 missiles are carried inside of the vehicle for manual reloading. Two ground launchers are also carried to enable missiles to be launched away from the vehicle if required by the tactical situation.

Type WZ 501A Infantry Fighting Vehicle

This is essentially the WZ 501 fitted with a new turret armed with a 25 mm cannon, 7.62 mm machine gun and eight smoke dischargers. It is the same turret that is fitted to the 25 mm IFV version of the WZ 551 (6 × 6) APC, covered in detail later in this section.

Type WZ 505 Armoured Ambulance

This is essentially a WZ 501 with a raised rear superstructure which can carry four stretcher or two stretcher and four seated wounded, or eight seated wounded. Specialised equipment includes an air-conditioning system, oxygen system and other medical equipment.

Type WZ 503 IFV from rear showing single entry door

Type WZ 506 Command Post Vehicle

This is based on the WZ 503 and is recognisable externally by its four radio antennas. Communications equipment installed includes two sets of Type 889 or VRC83 radio sets, one Type 892 or VRC84 radio set, one Type 70-zB or SR119 receiver. A generator and additional batteries are also provided.

Outline specifications of WZ 501 variants

Type	**WZ 501A**	**WZ 503**	**WZ 504**	**WZ 505**	**WZ 506**
CREW	3 + 8	2 + 13	3 + 4	2 + 6/10	2 + 6/8
COMBAT WEIGHT	13 640 kg	12 800 kg	13 000 kg	13 000 kg	12 800 kg
LENGTH	6.74 m	6.74 m	6.74 m	6.7 m	6.7 m
WIDTH	2.97 m	2.97 m	2.97 m	2.92 m	2.92 m
HEIGHT	2.522 m	2.415 m	2.085 m	2.16 m	2.417 m

SPECIFICATIONS

CREW	3 + 8
WEIGHT	
loaded	13 300 kg
empty	12 100 kg
POWER-TO-WEIGHT RATIO	22 hp/tonne
GROUND PRESSURE	0.621 kg/cm²
LENGTH OVERALL	6.74 m
WIDTH OVERALL	2.97 m
HEIGHT OVERALL	2.158 m
GROUND CLEARANCE	0.38-0.40 m
MAX SPEED	
road	65 km/h
water	7-8 km/h
MAX RANGE	
road	460-510 km
water	100 km
GRADIENT	60%
SIDE SLOPE	47%
ANGLE OF APPROACH DEPARTURE	15°/25°
VERTICAL OBSTACLE	0.6-0.8 m
TRENCH	2 m
ENGINE	Type 6v150 water-cooled diesel developing 298 hp
TRANSMISSION	manual
ARMAMENT	
main	1 × 73 mm gun
coaxial	1 × 7.62 mm MG
ATGW launcher	1 × Sagger
AMMUNITION	
73 mm	40
7.62 mm	2000
Sagger	4

Status: In production. In service with the Chinese Army.

Manufacturer: Chinese state factories
Enquiries to China North Industries Corporation
7A, Yuetan Nanjie
Beijing
People's Republic of China
Telephone: (86) 6898/3461/3471/7570
Telex: 22339

NORINCO Type 77 Armoured Personnel Carrier

Development

The Type 77 armoured personnel carrier is very similar to the Soviet BTR-50PK APC but has a much more powerful engine giving improved road and water speeds as well as higher power-to-weight ratio. Many of the automotive components of the Type 77 APC are also used in the Type 63 light amphibious tank. This has the US Army designation of the M1974, this indicating the year it was first observed.

Description

The hull of the Type 77 is of all-welded steel construction with the crew compartment at the very front, troop compartment in the centre and the engine compartment at the rear.

The driver sits at the front on the left side and has a single piece hatch cover opening to the right and three forward-facing periscopes. The commander sits in a projecting bay at the front on the right side and has three periscopes for forward observation and a single circular hatch cover opening to the rear. At the commander's station is the 12.7 mm heavy machine gun, a copy of the Soviet DShKM weapon; no protection is provided for the commander while firing this weapon.

The troop compartment has hatches in the roof and a single forward-opening door in the right side. There are three firing ports, two in the left and one in the right side. The troop compartment has a ventilating system and a white light searchlight, which can be operated from within the vehicle, is mounted in the roof at the front left.

The torsion bar type suspension consists of six rubber-tyred road wheels with the idler at the front and drive sprocket at the rear; there are no track return rollers.

The Type 77 is fully amphibious and is propelled in the water by two water jets at the rear of the hull. The only preparation required is to erect the trim vane at the front of the hull and switch on the two electric bilge pumps. There is a manual bilge pump for emergency use.

Variants

In addition to being used as an APC, the Type 77 can also be used as an ambulance, command post, fuel resupply vehicle and load carrier. The Type 77-1 has three loading ramps that can be positioned at the rear to enable an 85 mm Type 56 gun (with a crew of eight and 30 rounds of ammunition) or a 122 mm Type 54 howitzer (with a crew of eight and 20 rounds of ammunition) to be carried; the Type 77-2 has no loading ramps.

HQ-2J SAM Carrier

A much modified Type 77 APC chassis with seven, rather than six, road wheels either side has recently been developed to act as the transporter/launcher platform for the HQ-2J SAM, the Chinese equivalent of the Soviet SA-2 Guideline.

This system is not amphibious. Full details are given in *Jane's Land-Based Air Defence 1991-92*, page 236.

Status: Production. In service with Chinese Army.

Manufacturer: Chinese state factories.
Enquiries to China North Industries Corporation
7A, Yuetan Nanjie
Beijing
People's Republic of China
Telephone: (86) 6898/3461/3471/7570
Telex: 22339

Type 77-1 APC from front with 12.7 mm anti-aircraft gun

Modified Type 77 APC chassis used to transport and launch HQ-2J SAM

Type 77-1 APC from rear with ramps lowered for loading of anti-tank gun

Type 77-2 APC from side with door open and 12.7 mm anti-aircraft machine gun installed

SPECIFICATIONS

CREW	2 + 16
COMBAT WEIGHT	15 500 kg
POWER-TO-WEIGHT RATIO	25.8 hp/tonne
LENGTH	7.4 m
WIDTH	3.2 m
HEIGHT OVERALL	2.436 m
TRACK	2.8 m
TRACK WIDTH	400 mm
LENGTH OF TRACK ON GROUND	4.44 m
MAX SPEED	
road	60 km/h
water	11-12 km/h
FUEL CAPACITY	416 l
MAX RANGE	
road	370 km
water	120 km
FORDING	amphibious
GRADIENT	60%
SIDE SLOPE	40%
VERTICAL OBSTACLE	0.87 m
TRENCH	2.9 m
ENGINE	Type 12150L-2A, 4-cycle, compression ignition, direct injection, water-cooled diesel developing 400 hp at 2000 rpm
TRANSMISSION	manual
STEERING	clutch and brake
SUSPENSION	torsion bar
ELECTRICAL SYSTEM	24 V
ARMAMENT	1 × 12.7 mm MG
AMMUNITION	500 rounds

NORINCO Type YW 534 Armoured Personnel Carrier

Development

The Type YW 534 armoured personnel carrier is believed to be the successor to the older Type YW 531 series of full-tracked vehicles which has been in service with the Chinese Army for many years. The Type YW 534 is almost identical to the Type YW 531 H APC (qv) but the latter has a slightly shorter and narrower hull.

Description

The hull of the Type YW 534 is of all-welded steel armour construction that protects the crew from small arms fire and shell splinters.

The driver sits at the front of the vehicle with a left-opening single piece hatch cover and three forward-facing periscopes, one of which can be replaced by an infra-red device for night driving. The commander sits to the rear of the driver and also has a single piece hatch cover with an integral vision device.

The engine compartment is to the right of the driver with the air-inlet and air-outlet louvres in the roof and the exhaust pipe on the right side.

The troop compartment is at the rear of the vehicle with the troops entering via a large door that is hinged on the right; this has a vision block and firing port.

In the left side of the vehicle are two firing ports and periscopes. In the right side there are three or four firing ports and a similar number of periscopes, but other vehicles have been observed with three firing ports and periscopes in the left side and four in the right; these may well be differences between production vehicles. The firing ports are provided with ball and socket type mountings for rifles.

In the forward part of the troop compartment roof is a single circular roof hatch either side. To the rear of this is the 12.7 mm machine gun installation, to the rear of which are two oblong roof hatches.

The 12.7 mm machine gun can be used to engage both air and ground targets but has no overhead protection for the gunner. Mounted either side of the forward part of the hull is a bank of four electrically operated smoke dischargers.

Suspension is of the torsion bar type with each side having five dual rubber-tyred road wheels, drive sprocket at the front, idler at the rear and track return rollers. Tracks are of the rubber-bushed double-pin type with removable rubber pads.

The Type YW 534 is fully amphibious, propelled in the water by its tracks. Before entering the water a trim vane is erected at the front of the vehicle and the bilge pumps are switched on. Standard equipment includes an NBC system. All vehicles have a Type 803 intercom and a Type 889 radio set.

Variants

Type YW 307 IFV

This is essentially the Type YW 534 APC fitted with a one-man power-operated turret armed with an externally mounted 25 mm cannon, 7.62 mm coaxial machine gun and smoke dischargers. The turret can be traversed through a full 360° and the weapon elevated from -12° to +52°. Totals of 400 rounds of 20 mm and 1000 rounds of 7.62 mm ammunition are carried.

Physical characteristics are similar to the Type YW 534's except that it has a combat weight of 15 400 kg and carries seven troops plus its crew of three.

ATGW Carrier

Shown for the first time in late 1988 was an anti-tank guided weapons carrier which is believed to be based on the Type YW 534 series of armoured personnel carrier, although it could well be based on the YW 531 H/Type 85 series.

The hull top at the rear has been raised and on top of this is a turret with four Red Arrow 8 ATGWs in the ready to launch position with additional missiles being carried in reserve in the hull. The same turret has also been fitted to the YW 531 series of vehicles (qv).

Status: In production. In service with Chinese Army.

Manufacturer: Chinese state factories
Enquiries to China North Industries Corporation
7A, Yuetan Nanjie
Beijing
People's Republic of China
Telephone: (86) 6898/3461/3471/7570
Telex: 22339 CNIN CN

Type YW 534 APC with raised rear superstructure fitted with four Red Arrow 8 ATGWs in the ready to launch position (Eric Ditchfield)

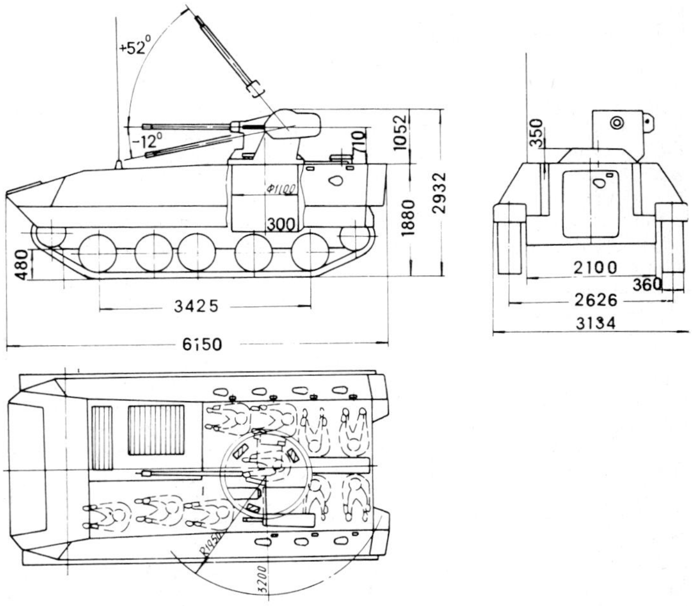

General arrangement of YW 307 IFV (not to 1/76th scale)

Type YW 534 APC

Type YW 307 IFV showing externally mounted 25 mm cannon with 7.62 mm coaxial machine gun to left

SPECIFICATIONS					
CREW	2 + 13	TRACK WIDTH	360 mm	TRENCH	2.5 m
COMBAT WEIGHT	14 300 kg	TRACK	2.626 m	ENGINE	Deutz BF8L 413F 4-cycle air-cooled, diesel developing 320 hp
POWER-TO-WEIGHT RATIO	22.06 hp/tonne	LENGTH OF TRACK ON GROUND	3.425 m		
GROUND PRESSURE	0.59 kg/cm^2	MAX SPEED			
LENGTH	6.15 m	road	65 km/h	TRANSMISSION	manual, 4 forward and 1 reverse gears
WIDTH OVERALL	3.134 m	water	6 km/h		
HEIGHT		MAX RANGE	500 km	STEERING	clutch and brake
hull top	1.88 m	FORDING	amphibious	SUSPENSION	torsion bar
over turret	2.556 m	GRADIENT	60%	ELECTRICAL SYSTEM	24 V
GROUND CLEARANCE	0.48 m	SIDE SLOPE	40%	ARMAMENT	1 × 12.7 mm MG
		VERTICAL OBSTACLE	0.7 m	SMOKE DISCHARGERS	2 × 4
				AMMUNITION	1100

NORINCO WZ 551 Armoured Personnel Carrier

Development

The existence of a new Chinese 6 × 6 armoured personnel carrier was first revealed in early 1986 and one of the 16 prototype vehicles made its first public appearance during an exhibition that took place in Beijing in November 1986. At that time it was confirmed that the vehicle had been designed and built by the NORINCO Vehicle Research Institute outside Beijing which had also built the prototype vehicles.

Using automotive components of the 6 × 6 WZ 551, NORINCO is also proposing 4 × 4 and 8 × 8 variants, but as far as it is known no actual prototypes of these have been completed.

This family of vehicles will be capable of being used for a wide range of roles, including 82 mm and 120 mm mortar carriers, and be armed with a wide range of weapons including 12.7 mm machine gun, 25 mm cannon, 90 mm or 105 mm guns.

Description

There are minor differences between prototype vehicles and the following description should be taken as provisional.

Although the 6 × 6 version of the WZ 551 is very similar in appearance to the French VAB (6 × 6) vehicle it has a number of important differences as well as being longer, wider and heavier.

The hull of the WZ 551 is of all-welded steel construction that protects the crew from small arms fire and shell splinters.

The driver sits at the front on the left with the vehicle commander to his right and the 12.7 mm machine gunner to his rear, their compartment is separated from the troop compartment by an airtight door.

The commander and driver each have a windscreen to their front which is covered when in a combat area by an armoured shutter hinged at the top. When these are lowered forward observation for each is via three roof-mounted periscopes, the centre one of which can be quickly replaced by an infra-red night vision device.

Both the commander and driver have a rear-opening circular roof hatch and a forward-opening side door in the upper part of which is an armoured shutter.

On at least one of the prototypes there is a door in the lower half of the hull between the first and second road wheels, but this is for maintenance purposes.

The engine compartment is to the rear of the driver's compartment and separated from the remainder of the vehicle by airtight bulkheads. The air inlet and outlet louvres are in the roof of the engine compartment with the exhaust pipe running along the upper left side of the hull.

The troop compartment is at the rear of the hull and the troops enter via a large door that is hinged on the right and has a firing port and a vision block. The infantry sit on bench type seats down either side of the hull facing each other. However, on the 25 mm version the troop seating layout is different.

Over the top of the troop compartment are four oblong roof hatches, two either side, that are hinged on the outside and can be locked vertical if required.

In each side of the troop compartment are four firing ports with a periscope mounted above in the roof of the vehicle. The first firing port in each side is for a 7.62 mm machine gun, with the remaining three being for assault rifles. Each firing port has a gas extractor device. The fuel tanks are located one either side of the troop compartment.

The WZ 551 is fully amphibious, propelled in the water by two shrouded propellers mounted one either side under the rear of the vehicle. These are swung backwards through 180° when in the water for improved propulsion. Before entering the water a trim vane is erected at the front of the vehicle and the bilge pumps, located in the floor at the rear, are switched on.

Standard equipment includes an NBC system, infra-red night vision equipment for the driver, automatic fire extinguishing system and run-flat tyres. If the tyres are punctured by bullets or splinters it can still travel a distance of 100 km at a speed of 40 km/h.

WZ 551 (6 × 6) APC with 25 mm cannon from rear showing propellers under hull rear

Variants

12.7 mm APC (6 × 6)

The basic vehicle which carries 11 fully equipped troops plus a crew of two. It has a circular roof hatch and is armed with an externally mounted 12.7 mm machine gun for which 500 rounds are carried. Its empty weight is 12 500 kg, maximum payload on land 7000 kg and maximum payload on water 3500 kg.

Ambulance (6 × 6)

This has a raised rear troop compartment, a crew of three and can carry four stretchers. Its empty weight is 12 600 kg. Mounted on the forward part of the roof is a cupola with an externally mounted 12.7 mm machine gun for which 500 rounds are carried.

WZ 551 (6 × 6) APC armed with 12.7 mm roof-mounted MG showing exhaust pipe alongside left side of hull

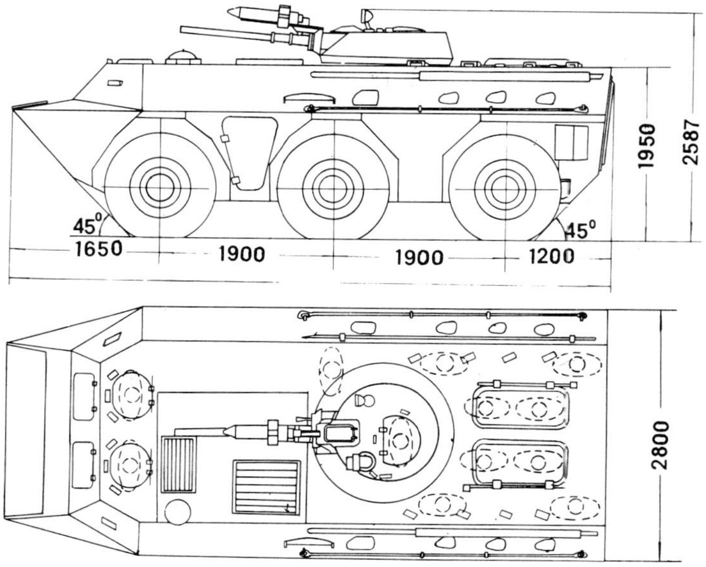

WZ 551 (6 × 6) APC with 73 mm turret (not to 1/76th scale)

WZ 551 (6 × 6) APC with one-man turret armed with 25 mm cannon, 7.62 mm coaxial machine gun and smoke dischargers

Type NGV-1 (6 × 6) IFV fitted with a GIAT Industries one-man power-operated turret armed with 25 mm cannon and 7.62 mm machine gun

73 mm IFV (6 × 6)

This has the same turret as fitted to the WZ 501 IFV described earlier in this section which in turn is a copy of the turret installed on the Soviet BMP-1 vehicle. This is armed with a 73 mm gun (40 rounds carried) with a coaxial 7.62 mm machine gun (2000 rounds carried) and a 73 mm Red Arrow 73 ATGW mounted over the main armament for which four missiles are carried. Combat weight is 14.2 tonnes.

25 mm IFV (6 × 6)

This has a crew of two and is fitted with a one-man power-operated turret on the roof to the immediate rear of the engine compartment. To the rear of the turret are two rectangular roof hatches. The turret is also installed on the YW 307 IFV.

The turret can traverse through a full 360° and the 25 mm externally mounted cannon has an elevation from -8° to +55°, a 7.62 mm machine gun is mounted coaxially to the left of the 25 mm cannon and either side of the lower part of the turret are electrically operated smoke dischargers. The gunner has a single piece hatch cover that opens to the right, two periscopes and a day/low light level night sight which is mounted in the forward part of the turret roof.

The 25 mm cannon fires HE and API rounds and the gunner can select single shots or rates of fire of 100, 300 or 500 rds/min. Two hundred rounds of ready use 20 mm ammunition are carried plus a further 200 rounds in reserve. One thousand rounds of 7.62 mm ammunition are carried.

Type NGV-1 Infantry Fighting Vehicle (6 × 6)

This was shown for the first time in late 1988 and is essentially the NORINCO WZ 551 (6 × 6) vehicle fitted with the French GIAT Dragar one-man power-operated turret. The Dragar weighs 1250 kg and in addition to being armed with the 25 mm GIAT Industries M811 cannon is also armed with a 7.62 mm machine gun and two banks of three electrically operated smoke dischargers.

Anti-tank vehicle (4 × 4)

This is fitted with a one-man turret with four Red Arrow 8 ATGWs ready to launch; an additional eight missiles are carried in reserve. The missiles have a range of 100 to 3000 m. The turret can be traversed through 360° and the launcher elevated from -7° to +13°. This version has a crew of four and the basic specifications are as follows:

WEIGHT	12 500 kg
LENGTH	6.05 m
WIDTH	2.8 m
HEIGHT	
hull top	1.95 m
launcher	2.6 m
WHEELBASE	3.2 m
MAX ROAD SPEED	85 km/h
RANGE	600 km

As an alternative the 4 × 4 version can be fitted with the one-man turret armed with the 73 mm gun, 7.62 mm coaxial machine gun and Red Arrow 73 ATGW.

122 mm SPH (8 × 8)

This is based on a lengthened 6 × 6 chassis and mounted on the hull top at the rear is a turret armed with a 122 mm howitzer which has a maximum range of 15 000 to 21 000 m depending on the type of ammunition used. The turret has a traverse of 25° left and right and the ordnance can be elevated from -3° to +70°. Forty rounds of ammunition are carried and before firing commences two spades can be lowered at the rear of the hull to provide a more stable firing platform. Basic specifications are:

CREW	4
WEIGHT	20 000 kg
LENGTH	7.9 m
WIDTH	2.8 m
HEIGHT	
overall	2.7 m
hull top	1.75 m
WHEELBASE	1.75 m + 2 m + 1.3 m
MAX ROAD SPEED	85 km/h
ROAD RANGE	500 km
GRADIENT	60%
TRENCH	2 m
ENGINE	320 hp diesel

NORINCO is also suggesting that this vehicle could be fitted with the 73 mm turret or one with four 25 mm cannon for air defence use.

SPECIFICATIONS (fitted with one-man 25 mm turret)

CREW	2 + 9
CONFIGURATION	6 × 6
COMBAT WEIGHT	15 000 kg
POWER-TO-WEIGHT RATIO	17.06 hp/tonne
LENGTH	6.65 m
WIDTH	2.8 m
HEIGHT	
hull top	1.95 m
over 25 mm turret	2.89 m
GROUND CLEARANCE	0.41 m
TRACK	2.44 m
WHEELBASE	1.9 m + 1.9 m
ANGLE OF APPROACH/DEPARTURE	45°/45°
MAX SPEED	
road	85 km/h
water	7 km/h
MAX ROAD RANGE	600 km
FORDING	amphibious
GRADIENT	60%
SIDE SLOPE	30%
TRENCH	1.2 m
VERTICAL OBSTACLE	0.5 m
TURNING RADIUS	9 m
ENGINE	Deutz F8L 413F, V-8 4-cycle air-cooled diesel developing 256 hp at 2500 rpm
TRANSMISSION	manual, 9 forward and 1 reverse gears
STEERING	hydraulic, first and second axles
SUSPENSION	coil springs and hydraulic shock absorbers
TYRES	14.00 × 20 (run-flat)
BRAKES	dual circuit, air/hydraulic
ARMAMENT	
main	1 × 25 mm cannon
coaxial	1 × 7.62 mm MG
AMMUNITION	
25 mm	400
7.62 mm	1000

Status: Prototypes undergoing trials. Not yet in production or service.

Manufacturer: Chinese state factories
Enquiries to China North Industries Corporation
7A, Yuetan Nanjie
Beijing
People's Republic of China
Telephone: (86) 6898/3461/3471/7570
Telex: 22339

NORINCO Type YW 531 H Armoured Personnel Carrier

(This has recently been renamed the Type 85 series of tracked armoured combat vehicles.)

Development

The Type YW 531 H armoured personnel carrier was developed at the same time as the Type YW 534 and is believed to be the successor to the older Type 531 which has been in service with the Chinese Army for many years. The Type YW 531 H is almost identical to the YW 534 but is slightly lighter, shorter and narrower and does not have the four electrically operated smoke dischargers either side of the forward part of the hull.

Description

Identical to the Type YW 534 fully described in an earlier entry, but has some minor differences such as three firing ports and vision devices in left hull side.

Variants

Type YW 309 Infantry Combat Vehicle

This is essentially the Type YW 531 H fitted with the turret of the WZ 501 IFV which in turn is a copy of the Soviet BMP-1. In addition to the single firing port in the rear hull door there are three firing ports and periscopes in the left side of the hull and four firing ports and periscopes in the right side of the hull.

The YW 309 has a crew of three and carries eight fully equipped infantrymen. The turret can be traversed through 360° and the 73 mm gun can be elevated from -3.5° to +30.5°. A 7.62 mm machine gun is mounted coaxially with the main armament and there is a Red Arrow 73 ATGW launcher over the main armament. This is also called the Infantry Fighting Vehicle Type 85.

Type 85 Armoured Command Post

This is similar to the YW 531 H but has a crew of two and carries six staff officers plus extensive communications equipment for which three radio antenna are provided. The driver sits front left, and the commander, who also mans the 12.7 mm anti-aircraft machine gun, is seated to his rear. The command area is at the very rear and has four circular roof hatches, one in each corner. This version weighs 13 800 kg loaded and a four-barrelled smoke discharger is mounted either side of the forward part of the hull.

Type WZ 751 Armoured Ambulance

This is also based on the YW 531 H but has a new and raised hull rear and a new larger commander's cupola to the rear of the driver which is fitted with an externally mounted 12.7 mm anti-aircraft machine gun. This has a crew of two and also carries two medical personnel plus four stretcher patients or eight seated patients or two stretcher patients and four seated patients. In addition to the stretchers, a medical kit, splint box and oxygen supply are carried and standard equipment includes an air-conditioning system. Loaded weight is 13 800 kg. This is now called the Tracked Armoured Ambulance Type 85.

Armoured Command Vehicle Type 85

This is the basic vehicle with its turret removed for use in the command role and has a crew of two plus six command staff. Externally it is recognisable by its pintle-mounted 12.7 mm machine gun and three radio antenna. Communications equipment installed includes two Type 889 radio sets or VRC-83 radio sets, one 892 radio set or VRC-84 radio set, one 70-2B receiver or one SR119 receiver installed. An auxiliary generator is also fitted as standard as are ventilation fans.

120 mm Self-propelled Mortar Type 85

This is similar to the above and has a 120 mm mortar mounted in the rear which fires to the rear with a traverse of 14° left and right with elevation from +45° to +80°. A total of 50 rounds of mortar ammunition are carried with a maximum range of 7700 m. This model has a six-man crew, including the commander and driver, and weighs 14 100 kg.

82 mm Self-propelled Mortar Type 85

This is similar to the above but with a slightly different hatch arrangement and the 82 mm mortar can be traversed through 360° with elevation being from +45° to +85°. The 82 mm Type 87 mortar has a maximum range of 3040 m with a total of 120 rounds of ammunition being carried. This model has a seven-man crew and weighs 13 800 kg. The mortar is also capable of being deployed away from the vehicle if required by the tactical situation.

120 mm Turret-mounted Mortar

It has been proposed that the Type 85 APC be fitted with the Royal Ordnance turret-mounted 120 mm breech loaded mortar system which has so far been tested on a US FMC series M113 APC. This version would weigh 16 tonnes and carry a total of 66 rounds of 120 mm mortar ammunition.

122 mm Self-propelled Howitzer

This is essentially the basic vehicle with its turret removed and the Chinese-built version of the 122 mm D-30 towed howitzer removed from its standard carriage and mounted on the hull rear top. Additional details are given in the *Self-propelled guns and howitzers* section.

Type YW 309 ICV from rear

Type YW 531 H APC showing amphibious capabilities

Type WZ 751 armoured ambulance (Christopher F Foss)

Type YW 309 ICV with Red Arrow 73 ATGW over 73 mm gun (Christopher F Foss)

Recovery Vehicle Type 85

This is basic Type 85 retaining its cupola-mounted 12.7 mm MG but fitted with a hydraulic crane on the left side of the hull with a lifting capacity of 1 tonne, a generator, welding equipment, special mounting and dismounting tools, tool cabinet and a five-man crew. Combat weight is 14 tonnes.

Maintenance Engineering Vehicle Type 85

This has a similar hull to the Type 85 ambulance with a raised roof at the rear but is fitted with a cupola-mounted 12.7 mm MG on the roof rather than the forward part of the hull. Equipment includes generator, boom, inertia dynamometer, air filter cleaner, oil filter cleaner and other special tools.

Thai purchase of YW 531 H

In 1987 Thailand took delivery of the first of some 410 Chinese Type YW 531 APCs. These have their Chinese 12.7 mm machine guns removed and replaced by US 12.7 mm machine guns which are standard on Thai tanks such as the M48A5 and on the existing M113A1 series of APC in service with Thailand in significant numbers.

Thailand has also taken delivery of a quantity of specialised versions of the Type 85 series including a 30 round multiple rocket launcher. It is believed that the launcher is the same as that used on the 130 mm (30 round) Type 82 truck-mounted multiple rocket launcher system covered later in this book.

In December 1988 it was stated that a further 360 YW 531 H series APCs had been ordered by Thailand and in January 1989 an agreement was signed between NORINCO of China and Suco Engineering Limited under which a new facility would be erected to overhaul and repair Chinese vehicles, especially the YW 531 H. At a later date the facility may assemble Chinese designed vehicles for the home and export markets.

NVH-1 MICV

This vehicle, a joint development between NORINCO of China and Vickers Defence Systems of the UK, uses a chassis almost identical to the Type YW 531 H APC's and is fully described under International in this section.

Status: In production. In service with China and Thailand.

Manufacturer: Chinese state factories.
Enquiries to China North Industries Corporation
7A, Yuetan Nanjie
Beijing
People's Republic of China
Telephone: (86) 6898/3461/3471/7570
Telex: 22339 CNIN CN

SPECIFICATIONS

Model	**YW 531 H**	**YW 309**
CREW	2 + 13	3 + 8
COMBAT WEIGHT	13 600 kg	14 700 kg
POWER-TO-WEIGHT RATIO	23.5 hp/tonne	21.8 hp/tonne
GROUND PRESSURE	0.546 kg/cm²	0.591 kg/cm²
LENGTH	6.125 m	6.27 m
WIDTH OVERALL	3.06 m	3.06 m
HEIGHT		
hull top	1.914 m	1.91 m
including MG	2.586 m	2.475 m
GROUND CLEARANCE	0.46 m	0.467 m
TRACK	2.526 m	2.526 m
TRACK WIDTH	360 mm	360 mm
LENGTH OF TRACK ON GROUND	3.275 m	3.275 m
MAX SPEED		
road	65 km/h	65 km/h
water	6 km/h	6 km/h
MAX RANGE	500 km	500 km
FORDING	amphibious	amphibious
GRADIENT	60%	60%
SIDE SLOPE	40%	40%
VERTICAL OBSTACLE	0.6 m	0.6 m
TRENCH	2.2 m	2.2 m
ENGINE	BF8L 413F 4-cycle air-cooled diesel developing 320 hp	
TRANSMISSION	manual, 4 forward and 1 and 1 reverse gears	manual, 4 forward and 1 reverse gears
STEERING	clutch and brake	clutch and brake
SUSPENSION	torsion bar	torsion bar
ELECTRICAL SYSTEM	24 V	24 V
ARMAMENT		
main	1 × 12.7 mm MG	1 × 73 mm gun
coaxial	nil	1 × 7.62 mm MG
ATGW	nil	1 × Red Arrow 73
AMMUNITION (main)	1120	40

NORINCO YW 531 Armoured Personnel Carrier

Development

The YW 531 APC was developed in the late 1960s and at various times has been called the K-63, M1967 and M1970. The official Chinese designation for this vehicle is the Type 63 but within the Chinese Army it is referred to as the YW 531. It has seen combat in Vietnam (used by both China and Vietnam), in Angola by Zaïrean forces and by Tanzania against Uganda in the 1978 October-November border war. In 1983 a quantity of these vehicles was delivered to Iraq by China.

The vehicle has a crew of four which consists of the commander, gunner, loader and driver, and can carry 10 fully equipped infantrymen. The Australian Army has tested one example of the YW 531 and reports have indicated that it has excellent cross-country mobility.

Initial production vehicles were powered by a Type 6150L water-cooled diesel engine. Current production vehicles have BF8L 413F air-cooled diesel engines and are called the YW 531C.

Description

The hull of the YW 531 is made of all-welded steel. The driver sits at the front of the hull on the left side and has a single piece domed hatch cover that opens to the left, in front of which are two periscopes. The commander sits at the front of the hull on the right side and has a single piece hatch cover that opens to the right with an integral periscope which can be traversed through 360°. The engine is behind the commander with the air-inlet and air-outlet louvres in the roof and the exhaust outlet in the right side of the roof. The third crew member is seated behind the driver and has a single domed hatch cover that opens to the left, with an integral periscope which can be traversed through 360°.

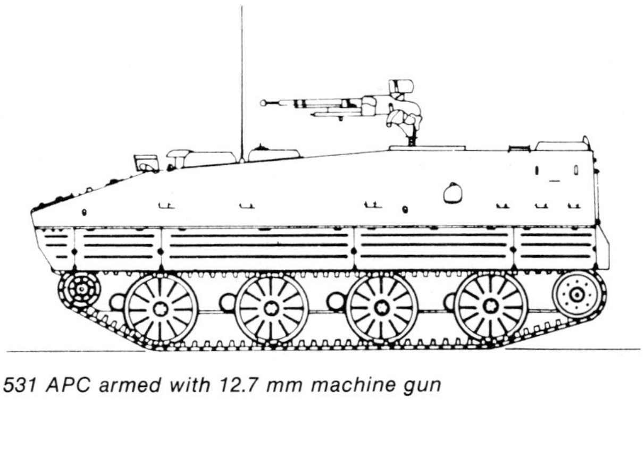

YW 531 APC armed with 12.7 mm machine gun

The gunner's position is to the rear of the vehicle in the roof. Mounted on the forward part of the gunner's hatch is a 12.7 mm heavy machine gun Type 54 (the Soviet DShKM weapon manufactured in China). The gunner also has a single piece hatch cover that opens to the rear.

The infantrymen enter and leave the vehicle by a door in the rear of the hull which is hinged on the right side. A single firing port is provided in each side of the hull. The torsion bar suspension consists of four single rubber-tyred road wheels with the drive sprocket at the front and the idler at the rear. There are no track return rollers, as the top of the track rests on the top of the road wheels.

The YW 531 is fully amphibious being propelled in the water by its tracks. Before entering the water a trim board, which is stowed on the glacis plate when travelling, is erected at the front of the hull. As far as is known the YW 531 has no NBC system or night vision equipment, although the driver can have a TVN-2 night vision device.

There are at least three sub-variants of the YW 531 APC which differ mainly in their communications equipment and details of these are given below:

Radio set		**Intercom set**		**Firing ports**			
Type	**Model**	**Number**	**Model**	**Number**	**left**	**right**	**rear**
YW 531C	889	1	803	1	2	1	1
YW 531D	889	2	803	1	1	1	1
YW 531E	889	1	803	1	1	1	1
	892	1					

Variants

YW 531C (Modified)

This has a number of improvements including a shield for the 12.7 mm machine gun, new vision ports and improved ventilation.

Type YW 701 Command Post Vehicle

This has a similar chassis and hull to the YW 750 ambulance and is powered by the 6150 L diesel developing 260 hp at 2000 rpm, this gives the vehicle a maximum road speed of 60 km/h. It has a crew of three

YW 750 armoured ambulance showing roof-mounted 12.7 mm machine gun traversed to rear

Type WY 701 command post vehicle showing higher roof and 12.7 mm machine gun

Type YW 304 82 mm mortar carrier showing open two-part roof hatch for mortar

YW 531 series APC with 12.7 mm machine gun showing front hatches open

Type 531 APC fitted with turret with four Red Arrow 8 ATGWs ready to launch

120 mm self-propelled mortar

and can carry five passengers when being used in the command role, and would normally be used by regiment and division commanders.

Communications equipment installed includes a vehicle intercom, five radio sets and a remote-controlled intercom for the vehicle commander when he has dismounted from the vehicle.

The YW 701 also has an auxiliary power unit to charge vehicle batteries, a radio for receiving meteorological reports and air raid alarms when the vehicle is stationary, a battery commander's telescope and map board, an infra-red night driving device for the commander, 12 observation blocks, air heater, forced ventilation and fans.

Armament comprises a 7.62 mm Type 56-1 light machine gun with 1000 rounds of ammunition. Four manually operated CO_2 fire extinguishers are carried.

The Type YW 701A has better communications and mobility than the YW 701. Communications equipment includes two Type 889 FM radios and one Type 892 radio. In addition there are two (Type 889 and Type 892) reserve radio sets. The vehicle is powered by the BF8L 413F engine developing 320 hp. Armament consists of one 12.7 mm anti-aircraft machine gun with 560 rounds of ammunition. Three manually operated Type 1211 fire extinguishers are provided. Another variant is the YW 701B which has one M889 and one M892 radio sets.

Ambulance

This has the Chinese designation YW 750 and has a crew of two plus two doctors and can carry four stretchered or eight seated patients. Like the YW 701 Command Post Vehicle, the YW 750 Ambulance has a higher roof line.

122 mm SPH

This is called the 122 mm Type 54-1 self-propelled howitzer by the Chinese and is a longer and slightly wider YW 531 fitted with a 122 mm howitzer in an open-topped troop compartment. The engine is a model 6150 L 6-cylinder, 4-cycle, water-cooled diesel developing 260 hp at 2000 rpm. Full details of this version are given in the *Self-propelled guns and howitzers* section.

Rocket Launcher

This is called the 130 mm Type 70 19-barrel multiple rocket launcher and has the same engine as the 122 mm self-propelled howitzer. Full details of this version are given in the *Multiple rocket launchers* section.

Type YW 304 82 mm Self-propelled Mortar

This is fitted with a turntable-mounted 82 mm mortar which can be traversed through a full 360° and elevated from +45° to +85°. It has a minimum range of 85 m and a maximum range of 3040 m. A total of 120 rounds of 120 mm mortar ammunition is carried. A 12.7 mm machine gun is mounted on the left side of the turret roof for anti-aircraft and ground defence and 560 rounds of ammunition are carried for this.

The Type YW 304 has a crew of six or eight and weighs 12 800 kg fully loaded. The physical characteristics are identical to those of the basic YW 531 APC. The mortar can be dismounted for use in the ground role.

Type YW 381 120 mm Self-propelled Mortar

This has a W86 120 mm mortar with a traverse of 14° left and right with elevation from +45° to +80°. The mortar has a maximum range of 7700 m and 50 mortar bombs are carried. A baseplate and tripod are carried externally to allow the mortar to be dismounted for use in the ground role. A 12.7 mm machine gun is fitted for anti-aircraft and local defence. The vehicle has a crew of six and weighs 14 100 kg.

Although similar in appearance to the YW 304 82 mm mortar, the 120 mm self-propelled mortar has a longer and wider chassis. Basic specifications are:

LENGTH	6.125 m
WIDTH	3.06 m
HEIGHT	2.593 m
LENGTH OF TRACK ON GROUND	3.275 m
TRACK	2.526 m
GROUND CLEARANCE	0.467 m
GROUND PRESSURE	0.566 kg/cm²

Anti-tank Vehicle

It is known that a Type 531 APC has been fitted with a turret on the hull top towards the rear with four Red Arrow 8 ATGWs ready to launch, additional missiles are carried inside for manual reloading. This missile has a range of 3000 m. It is believed that this system on Type 531 APC has yet to enter service.

Psychological Warfare

Fitted with roof-mounted loudspeakers.

Status: Production probably complete but can be resumed if further export orders are placed. In service with Albania, China, Iraq, North Korea, Sudan, Tanzania, Vietnam and Zaïre.

Manufacturer: Chinese state factories.
Enquiries to China North Industries Corporation
7A, Yuetan Nanjie
Beijing
People's Republic of China
Telephone: (86) 6898/3461/3471/7570
Telex: 22339

SPECIFICATIONS (Type YW 531C)

CREW	2 + 13
COMBAT WEIGHT	12 600 kg
POWER-TO-WEIGHT RATIO	25.39 hp/tonne
GROUND PRESSURE	0.57 kg/cm²
LENGTH	5.476 m
WIDTH OVERALL	2.978 m
HEIGHT	
overall	2.58 m
to hull top	1.887 m
GROUND CLEARANCE	0.45 m
TRACK	2.464 m
TRACK WIDTH	433 mm
LENGTH OF TRACK ON GROUND	3.095 m
MAX SPEED	
road	65 km/h
water	6 km/h
FUEL CAPACITY	450 l
MAX ROAD RANGE	500 km
GRADIENT	60%
SIDE SLOPE	40%
VERTICAL OBSTACLE	0.6 m
TRENCH	2 m
ENGINE	BF8L 413F 4-cycle turbo-charged, intercooled, V-8 diesel developing 320 hp at 2500 rpm
TRANSMISSION	manual, 5 forward and 1 reverse gears
SUSPENSION	torsion bar
ELECTRICAL SYSTEM	24 V
MAIN ARMAMENT	1 × 12.7 mm MG
elevation	0° to +90°
traverse	360°
AMMUNITION	1120 rounds

NORINCO Type WZ 523 Armoured Personnel Carrier

Development/Description

During a parade held in Beijing in October 1984 a new 6 × 6 armoured personnel carrier was seen in public for the first time. In appearance it is very similar to the Belgian SIBMAS and South African Ratel (6 × 6) APCs developed a few years ago. Some sources have speculated that the new Chinese vehicle is based on a Hanyang Motor Vehicle HY-472 (6 × 6) or Luoyang Tractor Factory LT-665 (6 × 6) chassis, or uses these vehicles' automotive components. The US Army refers to this as the M1984 armoured personnel carrier. The Chinese designation of this vehicle is the Type WZ 523.

The hull of the APC is probably of all-welded steel armour construction providing protection from small arms fire and shell splinters. The driver sits at the front of the hull with the commander to his side, probably on the right. Forward vision is through two large bullet-proof windows; in a combat area these would be covered by an armoured shutter hinged at the top. Above the commander's and driver's positions is a single piece hatch cover, hinged in the centre, which can be locked vertically. There is a door in the right side of the hull, to the rear of the first road wheel.

The engine compartment is in the centre with drive to all six wheels, transmission is probably manual. The exhaust pipe is on the right side of the hull and the engine is probably offset to the left (as on some western vehicles such as the VAB) to allow individual crew members to move from the front of the vehicle to the rear. Steering is on the front wheels only and probably power assisted to reduce driver fatigue.

The troop compartment is at the rear of the hull and the troops enter and leave via a large door in the hull rear hinged on the right with a single circular firing port. There are rectangular roof hatches above the troop compartment, opening outwards, and two observation/firing hatches in either side that open upwards.

On the hull roof, slightly forward of the troop compartment, is a 12.7 mm Type 54 machine gun with lateral protection from small arms fire.

The vehicle is fully amphibious and propelled in the water by two water jets at the rear of the hull; before entering the water a trim vane is erected at the front of the hull.

The driver has an infra-red periscope for night driving and an A 220 radio is fitted as standard. The engine for the WZ 523 is produced by the Second Automotive Plant who may also produce the vehicle.

Variants

Other variants of this 6 × 6 APC are probably under development and could include a command post, ambulance, mortar carrier and an anti-tank vehicle with missiles.

Status: Production. In service with the Chinese Army.

Manufacturer: Chinese state factories.
Enquiries to China North Industries Corporation
7A, Yuetan Nanjie
Beijing
People's Republic of China
Telephone: (86) 6898/3461/3471/7570
Telex: 22339

Type WZ 523 (6 × 6) armoured personnel carrier fitted with roof-mounted 12.7 mm machine gun

Chinese WZ 523 (6 × 6) APC showing roof-mounted 12.7 mm machine gun and roof hatches locked open (via G Jacobs)

SPECIFICATIONS		HEIGHT INCLUDING MG	2.73 m	SIDE SLOPE	30%
CREW	2 + 10	MAX ROAD SPEED	80 km/h	ENGINE	EQ 6105 water-cooled petrol developing 165 hp
CONFIGURATION	6 × 6	MAX WATER SPEED	7 km/h		
COMBAT WEIGHT	11 200 kg	FUEL CAPACITY	255 l	ARMAMENT	1 × 12.7 mm MG
LENGTH	6.02 m	MAX RANGE	600 km	AMMUNITION	600 × 12.7 mm
WIDTH	2.55 m	GRADIENT	60%		

NORINCO Chinese Internal Security Vehicle

Development/Description
Shown for the first time at a defence equipment exhibition held in Beijing late in 1988 was a new 4 × 4 internal security vehicle. It was not observed during the troubles in Beijing in June 1989.

It is believed to be based on a commercial chassis and has a combat weight of about 3800 kg.

The engine compartment is at the front with the crew compartment extending right to the rear. The commander and driver are seated towards the front, each provided with a side door with a window in its upper part. To their immediate front is a windscreen which can be covered by an armoured shutter hinged at its lower part.

The rear troop compartment is provided with a single door in each side and one at the rear. There are a total of eight vision ports in the troop compartment, three each side and one either side of the rear door.

In the forward part of the roof hatch is a single-piece hatch cover that opens to the rear and a 7.62 mm machine gun can be mounted at this position. To the immediate rear of this are flashing warning lights.

Status: Believed to be still at the prototype stage.

Manufacturer: Not known.

Chinese 4 × 4 internal security vehicle with windscreen visors lowered

Chinese 4 × 4 internal security vehicle from rear showing entry doors

CZECHOSLOVAKIA

OT-90 Armoured Personnel Carrier

Development/Description
The OT-90 is basically the Soviet designed BMP-1 Infantry Fighting Vehicle, which was also built under licence in Czechoslovakia, with its complete two-man turret removed and replaced by the turret of the OT-64C (1) (8 × 8) armoured personnel carrier armed with a 14.5 mm and a 7.62 mm machine gun. Full details of this turret are given in the following entry.

It is understood that the Czechoslovakian Army had a total of 2252 BMP-1 (which have the local designation of BVP-1) and 242 BMP-2 (which have the local designation of BVP-2). Of the former, it is believed that some 600 have been converted to the OT-90 configuration.

Czechoslovakian BMP-1 variants
BPzV is Soviet BVP-1 with modified hatches and massive sight over commander's hatch cover to rear of driver's position.
BRM-1K is the Soviet BRM-1 reconnaissance vehicle.
BVP-1K is Czechoslovakian designation for commander's version of Soviet BMP-1K vehicle.
Boure III is Czechoslovakian psychological warfare version of BMP-1 with loudspeaker system installed.
BMP-1 with loudspeaker system installed.
DP-90 is maintenance version of OT-90 APC.
DTP-90 is repair version of OT-90 APC.
MP-31 is air defence command version of Soviet BMP-1KShM command post vehicle.
MU-90 is mine laying version of OT-90 APC but has no turret with the space being covered by a steel sheet.
SVD is mine clearing version of BMP-1 with turret removed and fitted

Czechoslovakian OT-90 armoured personnel carrier

Czechoslovakian SVO mine clearance vehicle from rear

Czechoslovakian VP-90 reconnaissance vehicle

with Hedgehog type launcher in rear troop compartment that fires to front of vehicle.
VPV is crane equipped recovery variant of BMP-1. The turret and troop compartment roof hatches have been removed and mounted on the troop compartment roof. At the rear is a powered crane.
VP-90 is reconnaissance version of OT-90 with OT-64 one-man turret armed with 14.5 mm and 7.62 mm MGs.
ZT-90 is recovery version of OT-90 APC.

VPV recovery vehicle showing roof mounted crane

Status
Conversions complete. In service with the Czechoslovakian Army.

Manufacturer
Czechoslovakian state factories.

OT-64 Armoured Personnel Carrier

Development
The OT-64 (Obrneny Transporter) was jointly developed from 1959 by Czechoslovakia and Poland and is used by them in place of the similar Soviet BTR-60 (8 × 8) APC. The vehicle, which uses many automotive components of the Tatra 813 series of 8 × 8 high-mobility trucks, entered service in 1964 and has since been widely exported outside the Warsaw Pact. The Polish designation for the vehicle is the Sredni Kolowy Opancerzny Transporter. Tatra of Czechoslovakia provided the chassis and automotive components and FSC/Lubin in Poland the armoured body and weapon system. Production of the OT-64 family is believed to have continued until at least the mid-1980s.

Description (OT-64C (1))
The hull of the OT-64 is made of all-welded steel with the crew compartment at the very front, engine immediately behind and the troop compartment at the very rear.

The driver is seated at the front of the vehicle on the left side with the commander to his right, both with a single piece rear-opening door beside them. The driver has a single piece hatch cover that opens to the rear and three periscopes for vision to his front and sides. The commander has a single piece hatch cover that opens forwards with a periscope in its forward part which can be traversed through 360°. In the roof, between the driver's and commander's positions, is a searchlight which can be operated from inside the vehicle.

The engine is behind the commander and driver with the air inlet and air outlets on the top of the hull and the exhaust pipes on either side of the hull.

The turret is in the centre of the vehicle and is mounted on an eight-sided plinth to enable the weapon to have some depression.

The troop compartment is at the rear of the vehicle and the troops enter and leave by two doors in the rear of the hull, each of which has a firing port. Over the top of the troop compartment are four roof hatches, each with a firing port, hinged on the outside. If required, the roof hatches can be locked vertical. There are two firing ports in each side of the troop compartment. The infantrymen are seated down each side of the vehicle on seats which can be folded up to allow the vehicle to be used for cargo.

Steering is powered on the front four wheels and all eight tyres are provided with a tyre pressure regulation system which allows the driver to adjust the tyre pressure to suit the type of ground being crossed. The vehicle is fully amphibious being propelled in the water by two propellers mounted at the rear of the hull. Before entering the water a trim vane is erected at the front of the vehicle; when travelling it is stowed on the glacis plate. Standard equipment on all vehicles includes an over-pressure NBC system, winch mounted at the front of the hull and bilge pumps.

The one-man manually operated turret is similar to those fitted to the Soviet BRDM-2 (4 × 4) reconnaissance vehicle and BTR-60PB (8 × 8) APC and is armed with a 14.5 mm KPVT machine gun with a 7.62 mm PKT machine gun mounted to the right of the main armament and the sight to the left. The weapons have a maximum elevation of +29°, depression of -4° and 360° turret traverse. Some models have been observed with a single Sagger ATGW mounted on each side of the turret with lateral armour protection.

Variants
Note: In both Czechoslovak and Polish service these vehicles are officially designated SKOT, SKOT-2, SKOT-2A and SKOT-2AP. The designations OT-64A, OT-64B, OT-64C(1) and OT-64C(2) are often used in the West, although they are in fact bogus.

OT-64A (SKOT)
This is the basic member of the series. Those used by Czechoslovakia are unarmed while those used by Poland are fitted with a pintle-mounted 7.62 mm machine gun. There are five hatches over the troop compartment: two open to the left and right of the vehicle, the centre one opens forwards and the last two open to the left and right. All of them can be locked vertical. Some OT-64As have been observed with two Sagger ATGWs mounted over the top of the troop compartment. Four Sagger ATGWs are carried.

OT-64C(2) which has same armament as OT-64C(1) but mounted in new turret

OT-64 being used in command role

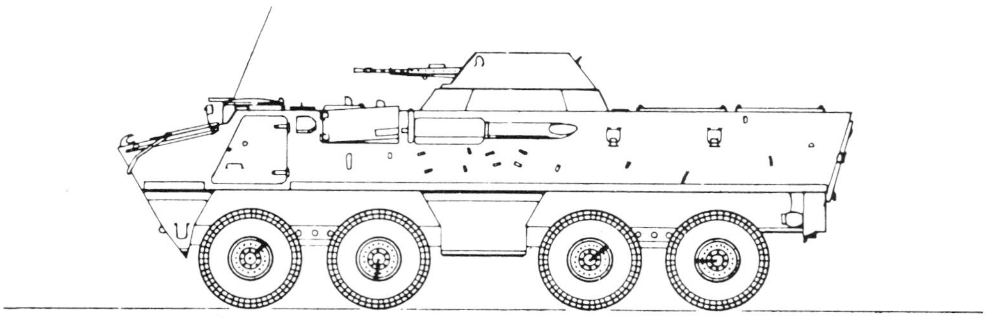

OT-64C(1) APC

OT-64B (SKOT-2)
This model has been used only by Poland and has a square plinth with a pintle-mounted 7.62 mm or 12.7 mm machine gun which has a shield which gives front and side protection to the gunner, but has an open top and is open at the rear. The OT-64B has three hatches over the troop compartment: the first one opens forwards and the other two open either side. This was probably a temporary expedient pending the delivery of turrets.

OT-64C(1) (SKOT-2A)
This is the model described in the detailed description earlier.

OT-64C(2) (SKOT-2AP)
This is used only by Poland and is similar to the OT-64C(1) except that it has a new turret with a curved top. It is armed with a 14.5 mm KPVT machine gun with a 7.62 mm PKT machine gun mounted to the right of the main armament and the sight to the left. The weapons have a depression of -4° and an elevation of +89.5° which allows them to be used against helicopters and aircraft. Some vehicles have been observed with a single Sagger ATGW mounted either side of the turret. This turret is also installed on the OT-62C tracked armoured personnel carrier.

OT-64C(1) with OT-62B Turret
Morocco is known to have some OT-64C(1) (SKOT-2A) APCs with the original turret replaced by the same turret as installed on the OT-62B full-tracked APC described in the following entry.

DPT-64 Repair
This is used by the Czechoslovakian Army and is an OT-64 modified for use in the repair role.

OT-64 R-2M Command
Czechoslovakian command variant of the OT-64. Polish Army designation for this model is the SKOT R-2AM with another variant being the SKOT R-2M.

OT-64 R-3MT Radio
Czechoslovakian version of OT-64 with radio equipment. Polish Army designation for this model is the SKOT R-3M.

OT-64 R-4MT Radio
Czechoslovakian version of OT-64 with radio equipment.

WPT Armoured Repair and Maintenance Vehicle
Repair and maintenance version of OT-64 which is fitted with light crane for changing components in the field, armament consists of a 7.62 mm PKT machine gun.

Other Polish versions of the OT-64
SKOT R-3Z is Polish OT-64 R-3Z
SKOT R-4 is Polish OT-64 R-4
SKOT R-6 is Polish OT-64 R-6
SKOT S-260 is Polish engineer vehicle for laying mines
Unidentified Polish OT-64 is an artillery resupply vehicle
SKOT WPT is Polish repair and maintenance vehicle and is equivalent of Czechoslovakian WPT.

Status: Production complete. In service with Czechoslovakia, Hungary, India, Iraq, Libya, Morocco (status uncertain), Poland, Sudan, Syria and Uganda (status uncertain).

Manufacturer: Czechoslovak state factories.

OT-64C(1) with infantry dismounting through twin doors in rear of vehicle

OT-64C(1) with roof hatches open

SPECIFICATIONS

Model	**OT-64A SKOT**	**OT-64C(1) SKOT-2A**
CREW	2 + 18	2 + 15
CONFIGURATION	8 × 8	8 × 8
COMBAT WEIGHT	14 300 kg	14 500 kg
UNLOADED WEIGHT	12 300 kg	12 800 kg
POWER-TO-WEIGHT RATIO	12.58 hp/tonne	12.41 hp/tonne
LENGTH	7.44 m	7.44 m
WIDTH	2.55 m	2.55 m
HEIGHT		
to turret top	n/app	2.71 m
to hull top	2.06 m	2.06 m
FIRING HEIGHT	n/app	2.5 m
GROUND CLEARANCE	0.46 m	0.46 m
TRACK	1.86 m	1.86 m
WHEELBASE	1.3 m + 2.15 m + 1.3 m	1.3 m + 2.15 m + 1.3 m
ANGLE OF APPROACH/ DEPARTURE	42°/43.5°	42°/43.5°
MAX SPEED		
road	94.4 km/h	94.4 km/h
water	9 km/h	9 km/h
FUEL CAPACITY	320 l	320 l
MAX ROAD RANGE	710 km	710 km
FUEL CONSUMPTION (road)	0.45 l/km	0.45 l/km
FORDING	amphibious	amphibious
GRADIENT	60%	60%
SIDE SLOPE	30%	30%

Model	**OT-64A SKOT**	**OT-64C(1) SKOT-2A**
VERTICAL OBSTACLE	0.5 m	0.5 m
TRENCH	2 m	2 m
ENGINE		
model	Tatra 928-14	Tatra 928-18
type	V-8 air-cooled diesel developing 180 hp at 2000 rpm	
TRANSMISSION	semi-automatic dual-range 5-speed plus over-drive providing 20 forward and 4 reverse gears	
STEERING	power-assisted	power-assisted
SUSPENSION	telescopic springs and hydraulic shock absorbers	
TYRES	13.00 × 18	13.00 × 18
ELECTRICAL SYSTEM	24 V	24 V
BATTERIES	4 × 12 V	4 × 12 V
ARMAMENT		
main	1 × 7.62 mm MG	1 × 14.5 mm MG
coaxial	n/app	1 × 7.62 mm MG
AMMUNITION		
main	1250	500
coaxial	n/app	2000
GUN CONTROL EQUIPMENT		
Turret power control	n/app	manual
Gun elevation/depression	+23.5°/-6°	+29°/-4°
Weapon traverse	90° (pintle)	360° (turret)
ARMOUR		
Hull	10 mm (max)	10 mm (max)
Turret	n/app	14 mm

OT-62 Armoured Personnel Carrier

Development

The OT-62 was developed jointly by Czechoslovakia and Poland and is the equivalent of the Soviet BTR-50PK APC and in appearance is almost identical to the BTR-50PU model 2 command vehicle, but the Soviet vehicles have a distinct chamfer between the top and side of their hulls. The main improvements over the Soviet vehicle are a more powerful engine which gives the vehicle higher road and water speeds, a fully enclosed troop compartment with an NBC system, and, in the case of some members of the OT-62 family, fully enclosed armament installations rather than a simple pintle-mounted machine gun.

The OT-62 entered service with Czechoslovakia in 1964 and with Poland in 1966, and production was completed in 1971. The Polish Army calls the vehicle the Transporter Opancerny Sredni, the Czechoslovak Army calls it the Transporter Obrneny Pasovy, or TOPAS. In both countries the vehicle is normally used by the tank divisions while the motorised rifle divisions use the OT-64 (8 × 8) armoured personnel carrier.

The OT-62C is also used by the 7th Polish Coastal Defence (formerly Sea Landing) Brigade which includes three battalions each of which has three infantry companies each with 10 OT-62Cs, a light tank company with 10 PT-76 plus detachments of ATGWs and combat swimmers plus support elements.

Description (OT-62A)

The hull of the OT-62 is all welded with the crew compartment at the front and the engine compartment at the rear.

The commander is seated at the front of the vehicle in the left projecting bay and is provided with a cupola that can be traversed through 360° by hand. Mounted in the forward part of his cupola is a periscope. His single piece hatch cover opens forwards and can be locked vertical. The commander is also provided with three observation periscopes in the front part of the projecting bay.

The driver is seated to the right of the commander and is provided with a single piece hatch cover in the front of the vehicle that opens upwards on the outside and has an integral vision block. The driver is provided with three periscopes below his hatch cover.

OT-62A APCs of Polish Marines with snorkel on top of hull

The right projecting bay is equipped with three observation periscopes but does not normally have a hatch cover. In the case of the OT-62B, the main armament installation is on the right projecting bay.

The troops enter and leave by the large door in either side of the vehicle which also has a circular observation/firing port. There is an additional observation/firing port in front of the side door in each side of the hull. Over the troop compartment are two large rectangular doors that open to the outside of the vehicle.

The engine is immediately behind the troop compartment with the cooling system on the left, the fuel and oil tanks on the right and the transmission at the rear. There are access panels in the top of the hull for maintenance.

The power train consists of the engine, master clutch, transmission, steering clutch, inner reduction gear, neutral steering clutch, final drive, water-jet propeller and mechanical water drainage pump drive. Power is transmitted through the transmission to the left and right inner reduction gears to the final drives which connect directly to the track drive sprockets.

The OT-62 is fitted with a fire-extinguishing system which automatically or manually extinguishes fires in the engine and transmission compartment. When activated the extinguisher is operated, engine stopped automatically, forced blowers are started and, if required, the second bottle of fire-extinguishing agent is also used.

The torsion bar suspension consists of six road wheels with the idler at the front and the drive sprocket at the rear. There are no track return rollers. The first and last road wheel stations are provided with a hydraulic shock absorber. Track tension can be adjusted from the driver's compartment.

The OT-62 is fully amphibious being propelled in the water by two water jets at the rear of the hull. Before entering the water a trim vane is erected at the front of the hull and the bilge pumps for which the outlets are on the top of the hull at the rear are switched on. The vehicle is fitted with an over-pressure NBC system, consisting of an electrically operated blower, two filter elements, control box, inlet port and hoses. It is on the right forward side of the troop compartment and in case of electrical failure can be hand operated. An infra-red driving light is mounted at the front of the vehicle on the right side.

Variants

OT-62A (or Model 1)

This was the first model to enter service and is unarmed. It is also known as the TOPAS. If required an M59A 82 mm recoilless gun can be carried and fired from the rear deck.

OT-62B (or Model 2)

The OT-62B is used by the Czechoslovak Army and is fitted with a small turret on the right projecting bay. This manually operated turret is identical to that fitted to the OT-65A (4 × 4) amphibious scout car and is armed with a 7.62 mm M59T machine gun with an elevation of +20°, depression of -10°, and turret traverse of 360°. Mounted externally on the right side of the turret is an 82 mm T-21 Tarasnice recoilless gun, which can be aimed and fired from inside the turret but can be reloaded only from the outside. The T-21 fires a HEAT projectile weighing 2.13 kg with a

OT-62B clearly showing 82 mm T-21 Tarasnice recoilless gun mounted externally on turret (US Army)

muzzle velocity of 250 m/s, which will penetrate 230 mm of armour. Maximum range is quoted as 2500 m but its range against a moving target is 300 m. When being fired when the vehicle is afloat, the water-jet propulsion system must be in operation to stop the vehicle from drifting backwards. A total of 1250 rounds of 7.62 mm and 12 rounds of 82 mm recoilless ammunition are carried. Mounted to the right of the T-21 recoilless gun is an infra-red search light. This model is also known as the TOPAS-2A.

OT-62C (or Model 3)

This is a Polish version and is officially called the TOPAS-2AP. Mounted in the centre of the hull is a manually operated turret identical to that fitted to the OT-64C(2) (8 × 8) armoured personnel carrier. This is armed with a 14.5 mm KPVT machine gun with a 7.62 mm PKT machine gun mounted coaxially to the right. The weapons have an elevation of +89.5°, depression of -4°, and turret traverse of 360°. The turret is called the WAT after the Wojskowa Akademia Techniczna (Army Technical Academy) where it was developed. The OT-62C is also used to carry two 82 mm mortar squads, each squad with a crew of four and an 82 mm mortar.

WPT-TOPAS Recovery Vehicle

This is a Polish development and is used for recovery of damaged and disabled vehicles during amphibious operations. In appearance it resembles an OT-62A but is fitted with a winch with a capacity of 2500 kg, small hand-operated crane with a capacity of 1000 kg and other specialised equipment. Armament comprises a 7.62 mm PK machine gun, RPG-7 anti-tank weapon and 12 F-1 fragmentation grenades. Details of this model are given in *Jane's Military Vehicles and Logistics 1991-92* page 19. This version is also used by Poland.

DPT-62 Repair

Czechoslovakian version of OT-62 modified for use in the repair role.

OT-62 R-3MT Radio

Czechoslovakian version of OT-62 with radio equipment. Polish version is called TOPAS R-3M.

OT-62 R-4MT Radio

Czechoslovakian version of OT-62 with radio equipment.

OT-62 R-2 Radio

Czechoslovakian version of OT-62 with radio equipment.

OT-62 R-2M Radio

Czechoslovakian version of OT-62 with radio equipment. Polish version is called TOPAS R-2M.

OT-62C which has same turret as OT-64C(2) (8 × 8) APC

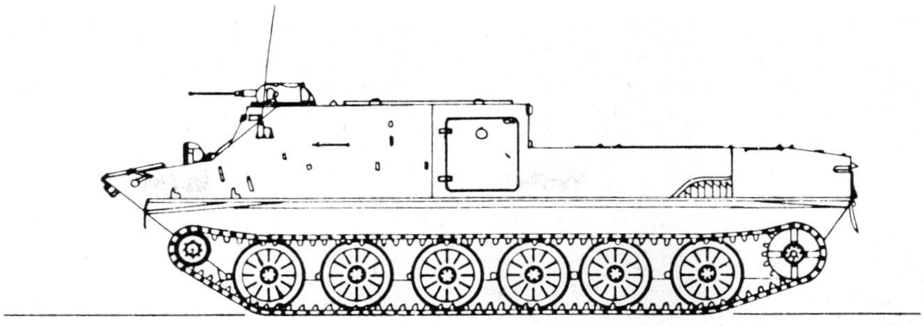

OT-62B armoured personnel carrier

SPECIFICATIONS

Model	**OT-62B**	**OT-62C**
CREW	2 + 18	3 + 12
COMBAT WEIGHT	15 000 kg	16 390 kg
POWER-TO-WEIGHT RATIO	20 hp/tonne	18.3 hp/tonne
GROUND PRESSURE	0.53 kg/cm²	0.55 kg/cm²
LENGTH	7.08 m	7 m
WIDTH	3.14 m	3.225 m
HEIGHT (including turret)	2.23 m	2.725 m
GROUND CLEARANCE	0.37 m	0.425 m
TRACK	2.74 m	2.74 m
TRACK WIDTH	360 mm	360 mm
LENGTH OF TRACK ON GROUND	4.08 m	4.08 m
MAX SPEED		
road	58.4 km/h	60 km/h
water	10.8 km/h	10.8 km/h
FUEL CAPACITY	417 l	520 l
MAX ROAD RANGE	460 km	570 km
FUEL CONSUMPTION	0.9 l/km	0.9 l/km
GRADIENT	65%	65%
SIDE SLOPE	40%	40%
VERTICAL OBSTACLE	1.1 m	1.1 m
TRENCH	2.8 m	2.8 m
ENGINE	PV-6 6-cylinder super-charged water-cooled diesel developing 300 hp at 1800 rpm	
TRANSMISSION	manual with 5 forward and 1 reverse gears	
SUSPENSION	torsion bar	torsion bar

Model	**OT-62B**	**OT-62C**
ELECTRICAL SYSTEM	24 V	24 V
BATTERIES	2 × 12 V	2 × 12 V
ARMAMENT		
main	1 × 7.62 mm MG	1 × 14.5 mm MG
coaxial	none	1 × 7.62 mm MG
anti-tank	1 × 82 mm	none
SMOKE-LAYING EQUIPMENT	diesel fuel injected into exhaust system	
AMMUNITION		
14.5 mm	none	500
7.62 mm	1250	2000
82 mm	12	none
GUN CONTROL EQUIPMENT		
Turret power control	manual	manual
Gun elevation/depression	+20°/−10°	+89.5°/−4°
ARMOUR		
Hull front upper	8 mm at 83°	8 mm at 83°
Hull front lower	10 mm at 53°	10 mm at 53°
Hull side upper	10 mm at 0°	10 mm at 0°
Hull side lower	9 mm at 0°	9 mm at 0°
Hull rear upper	7 mm at 0°	7 mm at 0°
Hull rear lower	6 mm at 42°	6 mm at 42°
Hull top	7 mm at 86°	7 mm at 86°
Belly front	6 mm	6 mm
Belly rear	6 mm	6 mm

Status: Production complete. In service with Angola, Bulgaria, Czechoslovakia, Egypt, India, Iraq, Israel, Libya, Morocco (status uncertain), Poland (Army and Marines) and Sudan.

Manufacturer: Czechoslovak state factories.

Note: Czechoslovakia is now producing the BMP-2 under licence.

EGYPT

Kader Fahd Armoured Personnel Carrier

Development

The Fahd (4 × 4) APC was developed to meet the requirements of the Egyptian Army, based on its experiences in Middle East conflicts. The Fahd was designed by Thyssen Henschel of Germany under the designation TH 390, Thyssen Henschel also built the first prototypes of the vehicle for Egypt.

Production of the Fahd is undertaken by the Kader Factory for Developed Industries, part of the Arab Organisation for Industrialisation. In the Egyptian Army it is now rapidly replacing old Soviet-supplied armoured personnel carriers such as the BTR-40 as well as the Egyptian-built Walid (or Waleed as it is called in Egypt) (qv) armoured personnel carrier.

The Fahd essentially consists of a Daimler-Benz LAP 1117/32 truck chassis fitted with an armoured body. Wherever possible standard and proven commercial components are used in the construction of the vehicle. Production of the Fahd commenced in 1985 with the first production vehicles being completed in 1986 for the Egyptian Army which is believed to have placed an initial order for 300 vehicles with an option on a further 110 vehicles. By early 1991 it is estimated that about 350 Fahd vehicles had been built for the home and export markets with production continuing.

Description

The hull of the Fahd is of all-welded steel armour construction giving complete protection from attack by 7.62 × 54 mm armour-piercing projectiles and shell splinters.

The driver sits at the front of the vehicle on the left side with the commander to his right. Both have a large bullet-proof windscreen to their front which can be covered by a shutter hinged at the top, and a side door that opens to the front. This has a window in its upper part which can be covered by a shutter if required. Above the commander's position is a single piece, rear-opening hatch cover. The driver has a forward-facing, roof-mounted periscope which can be replaced by an infra-red or passive periscope for night driving.

The troop compartment is at the rear of the hull with the infantry entering via a door in the rear, the upper part of which folds upwards and the lower part downwards to form a step. Over the top of the troop compartment are two rectangular roof hatches hinged in the centre which can be locked vertical. The infantry sit on individual bucket seats down the centre of the vehicle facing outwards. In either side of the troop compartment are four firing ports with a vision block above so the troops can fire their weapons from within the vehicle. Either side of the rear door is another vision block with a firing port underneath.

The engine is mounted towards the front of the vehicle with ventilation for the engine and cooling system provided through an opening with armoured shutters at the front of the vehicle. The air inlet for the engine is in the roof of the vehicle.

The Fahd is fitted with a central tyre-pressure regulation system allowing the driver to adjust the tyre pressure to suit the terrain being crossed. The driver can select front or rear tyres, or both together. The central tyre-pressure system stops automatically after reaching the maximum or minimum tyre pressure selected by the driver. Low-pressure, run-flat tyres are fitted as standard and allow the vehicle to travel a distance of 50 km at 30 km/h in the event of a puncture. Power-assisted steering is fitted as standard and there is a differential lock between the front and rear axles and between the rear wheels.

Up to three machine guns can be mounted on the roof of the Fahd for suppressive fire.

Fahd (4 × 4) APC from front showing winch in lower part of hull (Christopher F Foss)

Fahd (4 × 4) APC fitted with French SAMM turret

Fahd (4 × 4) APC from rear showing two-part rear door open (Christopher F Foss)

Artist's impression of Fahd (4 × 4) mine laying vehicle with six modules of anti-tank mines

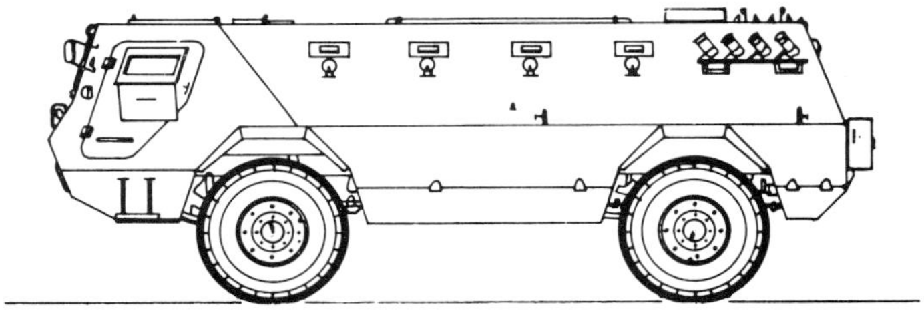

Unarmed Fahd (4 × 4) armoured personnel carrier

Fahd (4 × 4) APC fitted with French Creusot-Loire Industrie turret type T20 armed with 20 mm cannon

Optional equipment includes a front-mounted hydraulic winch with a capacity of 5000 kg, NBC and ventilating system, passive night vision equipment and smoke dischargers mounted either side of the hull.

In March 1989 it was announced that the Kader Factory for Industrialisation, a member of the Arab Organisation for Industrialisation, had placed an order with a British consortium for the supply of 110 air cooling systems for the Fahd (4 × 4) APC being built for export. Galley Limited supplied the air cooling systems to Mantech Limited who shipped them to Egypt with first deliveries made in early 1989.

The Galley designed air cooling system is a totally self-contained unit mounted at the rear of the vehicle and driven by its own three cylinder diesel engine. The unit takes warm air from within the crew compartment, cools it in a compact vapour-cycle refrigeration system and then returns it through ducts which direct the air around the vehicle.

Cooling air for the radiator and condenser is drawn into the system from the surrounding atmosphere and is discharged, together with the engine exhaust cases, through louvred panels in the hull of the vehicle. The unit is sealed for use under NBC conditions.

The air cooling unit is capable of delivering more than 7 kW of cooling and weighs 260 kg. It has been designed to operate in ambient temperatures of up to +52° C and relative humidities of up to 90%, delivering dehumidified air to the interior of the vehicle at around +20° C.

Variants

The design of the Fahd is such that a wide range of armament installations can be mounted on the hull top such as a one-man turret with single or twin 7.62 mm machine guns, 12.7 mm machine gun or a 20 mm cannon. Other variants suggested by the manufacturer include ambulance, command post vehicle, missile carrier, multiple rocket launcher, mine laying, internal security and recovery.

Specialised versions of the Fahd already announced include:

Combat vehicle with 20 mm turret, which weighs 11 tonnes fully loaded or 9800 kg empty, has a three-man crew and can carry 10 fully equipped troops.

Police security vehicle with turret armed with 7.62 mm and/or 12.7 mm MG, same crew as above, weight 11 tonnes fully loaded.

Police security vehicle with turret armed with 12.7 mm and/or 7.62 mm MG and smoke bomb launchers, combat weight 11.3 tonnes, same crew as before. Options include an air-conditioning system, night periscope for driver and commander, wireless and intercom system, navigation compass, hydraulic towing winch with capacity of 5 tonnes, obstacle removal blade at front of vehicle, loudspeaker, siren, search light and flashing light.

Mine laying vehicle. It has been proposed that the Fahd could be used as a mobile mine laying vehicle. This would have the fully armoured crew compartment at the front but the area to the immediate rear of the front axle would be a flatbed on which would be carried six modules each with 20 tubes. Each of the latter contains five anti-tank mines with an electro-magnetic sensor, anti-handling and self-destruction features.

AFV Fahd 30

This has been developed as a private venture and was announced for the first time late in 1990. It essentially consists of the standard Fahd (4 × 4) vehicle with the roof fitted with a plinth on which is mounted the complete turret of the Soviet BMP-2 infantry fighting vehicle (IFV) which is fully described under the USSR.

The commander is seated on the right of the turret with gunner on left. The commander's binocular sight has a day system with a magnification of ×5 and a 10° field-of-view, night system with a magnification of ×4.2 and an 8° field-of-view. In addition he has a day monocular periscope with a magnification of ×1.2 or ×4 and two periscopes, one to the front and one to the rear of the vehicle.

The gunner has a day/night binocular periscopic sight with a day magnification of ×6 and 10° field-of-view and a night magnification of

AFV Fahd 30 which has the turret of the BMP-2 IFV

×5.5 and a 7° field-of-view and four periscopes, three to the front and one to the rear of the vehicle.

The two-man power-operated turret is armed with a 30 mm dual feed cannon, 7.62 mm machine gun mounted coaxial with the main armament and roof-mounted AT-5 Spandrel ATGW launcher.

The latter can also be deployed away from the vehicle if required by the tactical situation and has a maximum range of 4000 m (with time of flight to this range being 20 seconds). The launcher has a traverse of 360°, elevation +15° and depression -5°. The AT-5 has a HEAT warhead and will penetrate 550 mm of RHA.

In addition to the one missile in the ready to launch position a further four missiles are carried internally, two either side of the hull.

Mounted either side of the turret is a bank of three 81 mm electrically operated smoke dischargers that fire to the front of the vehicle and mounted either side of the hull firing forwards is a bank of four 76 mm electrically operated smoke dischargers. Basic specifications where different from the standard vehicle:

CREW	10	
COMBAT WEIGHT	12 500 kg	
UNLOADED WEIGHT	10 700 kg	
HEIGHT	2.71 m	
MAXIMUM ROAD SPEED	85 km/h	
ROAD RANGE	750 km	
CROSS COUNTRY RANGE	400 km	
TRENCH	0.8 m	
ARMAMENT		
main	1 × 30 mm cannon	
coaxial	1 × 7.62 mm MG	
ATGW	1 × AT-5 Spandrel	
SMOKE DISCHARGERS		
position	hull	turret
number	2 × 4	2 × 3
calibre	76 mm	81 mm
range	50-60 m	200-300 m
smoke screen	60 × 15 m	80 × 5 m
duration	120-180 s	170-240 s
AMMUNITION (ready)		
main	500 (160 AP + 340 HE-I)	
coaxial	2000	
ATGW	1	
GUN CONTROL EQUIPMENT		
TURRET POWER CONTROL	powered/manual	
MAX TRAVERSE SPEED	36°/s (rapid) 9°/s (slow)	
MAX ELEVATION SPEED	30°/s	
GUN ELEVATION/ DEPRESSION	+80°/-5°	
TURRET TRAVERSE	360°	
STABILISATION SYSTEM	yes	
COMMANDER'S OVERRIDE	yes	

SPECIFICATIONS

CREW	2 + 10
CONFIGURATION	4 × 4
COMBAT WEIGHT	10 900 kg
UNLOADED WEIGHT	9100 kg
PAYLOAD	1800 kg
POWER-TO-WEIGHT RATIO	15.4 hp/tonne
LENGTH	6 m
WIDTH	2.45 m
HEIGHT TO HULL TOP	2.1 m
GROUND CLEARANCE	0.31-0.37 m
ANGLE OF APPROACH/ DEPARTURE	35°/35°
WHEELBASE	3.2 m
MAX ROAD SPEED	90 km/h
MAX ROAD RANGE	800 km
FORDING	0.7 m
GRADIENT	70%
SIDE SLOPE	30%
TRENCH	0.9 m
VERTICAL OBSTACLE	0.5 m
ENGINE	Mercedes-Benz OM-352 A 6-cylinder direct-injection, turbo-charged, water-cooled diesel developing 168 hp at 2800 rpm
TRANSMISSION	manual, 5 forward and one reverse gears
TRANSFER CASE	2-speed
STEERING	power-assisted
TURNING RADIUS	7 m
BRAKES	hydropneumatic
SUSPENSION	leaf springs and shock absorbers
ELECTRICAL SYSTEM	24 V
ARMAMENT	see text

Status: In production. In service with Egypt, Kuwait (IS role), Oman, Sudan and Zaïre (IS role with BTM 105 turret). The exact status of the Kuwaiti vehicles is uncertain, although it is believed that they were captured during the Iraqi invasion of 1990.

Manufacturer: Kader Factory for Developed Industries
Orouba Street
Heliopolis
POB 287
Arab Republic of Egypt
Telephone: 611142; 604324
Telex: 92136
Fax: 2608718

Kader Walid Armoured Personnel Carrier

Development

The Walid 1 armoured personnel carrier was designed and developed by the Kader Factory, Cairo with first production vehicles being completed in 1960. It is also referred to as the Waleed and was used in combat for the first time in the 1967 Middle East campaign. It basically consists of a German Magirus Deutz chassis, which was manufactured under licence in Egypt, fitted with an armoured body. The vehicle is very similar in appearance and performs a similar role to the Soviet BTR-40 armoured personnel carrier which is also used by the Egyptian Army. The BTR-40 can be distinguished from the Walid by the vertical rather than sloped sides to the troop compartment. The Walid 2 was introduced in 1981 and is based on Mercedes-Benz automotive components.

Description

The hull of the Walid is made of all-welded steel with the engine at the front, driver and commander in the centre and the troop compartment at the rear. The driver sits on the left with the commander to his right, both with a door in the side of the hull, the upper part of which folds down on the outside for increased visibility. Both the commander and driver have an individual windscreen to their front which when in action can be covered by an armoured shutter hinged at the top.

The open-topped troop compartment is at the rear of the hull and entry to it is by a door on which the spare wheel and tyre are carried. Three observation/firing ports are provided in each side of the hull with a further two at the rear of the hull, one either side of the spare wheel.

The Walid has no NBC system, no night vision equipment and no amphibious capability. A central tyre-pressure regulation system is fitted as standard.

The basic model is armed with a 7.62 mm pintle-mounted machine gun but additional weapons can be mounted around the top of the hull.

Variants

Minelayer

This is a basic Walid fitted with a mine laying chute attached to the rear door. Anti-tank mines slide down the ramp and are laid on the ground where they can subsequently be buried by hand.

Smoke launcher

The Walid has been modified by the SAKR Factory for Developed Industries to lay smoke rockets. Details of this are given in the *Multiple rocket launchers* section under Egypt.

Walid (4 × 4) APC used in minelaying role with rear door open (Christopher F Foss)

Walid (4 × 4) armoured personnel carrier (J G Figa)

SPECIFICATIONS

CREW	2 + 8-10
CONFIGURATION	4 × 4
MAX PAYLOAD	2800 kg
LENGTH	6.12 m
WIDTH	2.57 m
HEIGHT	2.3 m
MAX ROAD SPEED	86 km/h
MAX RANGE	800 km
ENGINE	168 hp diesel

Status: Production complete. In service with Algeria, Burundi, Egypt, Israel, Sudan (including 24 delivered in 1986) and Yemen.

Manufacturer: Kader Factory for Developed Industries
Orouba Street
Heliopolis
POB 287
Arab Republic of Egypt
Telephone: 611142; 604324
Telex: 92136
Fax: 2608718

EL SALVADOR

VAL (Light Assault Vehicle) Cashuat

Development

To meet its requirements for an armoured personnel carrier El Salvador built a mild steel prototype based on the chassis of the old US M37B1 (4 × 4) ¾ ton truck (*Jane's Military Vehicles and Logistics 1991-92* page 299). Trials with this prototype showed that it lacked ballistic protection.

The government of El Salvador then turned to the USA for assistance and the US Army Tank Automotive Command's RDE Center's Design and Manufacturing Technology Directorate became involved.

After studying the first prototype in El Salvador in February 1985 it was apparent that the country lacked the facilities to build the required numbers of vehicles.

Tank Automotive Command then offered a three-part plan to El Salvador. First-to build the prototype of the APC in its own facilities in Detroit and then take it to El Salvador to be used as a reference model. Secondly, to pre-cut dual hardness armour and carry out critical welding on the side panels, side rails and windshield before shipment to El Salvador, with a total of 66 kits to be provided. Finally, train a team of El Salvadorians in the USA in welding techniques.

This package was accepted by El Salvador and after carrying out some redesign work, including changing the driver's compartment doors and rear crew compartment doors from a flush design to a surface mount design, modifying the rear fenders and cutting the hood configuration straight across in front of the windscreen, the first prototype was completed in Detroit in March 1985.

Trials in the USA showed that the armour could not be penetrated by 5.56 mm ball ammunition fired from a range of 25 m. This was followed by two days of mobility trials during which the prototype was driven over 500 km of various terrain.

In April 1985 the prototype was shipped from Tank Automotive Command for its flight to El Salvador, 74 days after receiving the go-ahead. In May 1986 TACOM shipped the last of 66 kits to El Salvador which assembled them at the rate of two a week.

Description

This vehicle is essentially an M37B1 (4 × 4) truck chassis fitted with an armoured body that protects the occupants from small arms fire. Dual-hardness armour protects the crew compartment, high hardness armour protecting less crucial areas. Kevlar is fitted inside the troop compartment and on the outside of the 12.7 mm turret. At present belly armour is not fitted.

The layout is conventional with the armour-plated engine compartment at the front, driver and vehicle commander to its immediate rear, the troop compartment at the very rear of the hull. Both commander and driver have a forward opening door with a bullet-proof window in the upper part with a firing port underneath. To their front is a bullet-proof window with a firing port underneath and armour protection is provided above their position. Provision has been made for installing a radio.

The open-topped troop compartment is at the rear and the troops normally enter and leave via two outward-opening doors in the rear that

Unarmed VAL (Light Assault Vehicle) (US Army)

Unarmed El Salvadorian Cashuat weapons carrier showing two shielded mounts for 7.62 mm M60D machine guns and turret which normally mounts a 12.7 mm M2 HB machine gun (Julio A Montes)

have a firing port on either side. An additional three firing ports are provided in the upper part of each side of the troop compartment.

Before being fitted with the new armoured body, the chassis was stripped down, cleaned and reinforced, the suspension was uprated, and new brakes, shock absorbers and springs were fitted.

The tyres were filled with a Dupont component to allow the vehicle to travel up to 90 km with a perforated tyre.

Variants

There are two versions of this vehicle in service with the El Salvadorian Army. One is the armoured personnel carrier which has a crew of two and carries eight fully equipped troops. The second version is a gun carrier and is fitted with an armoured turret to the immediate rear of the commander and gunner which is armed with a 12.7 mm M2 HB machine gun with a 7.62 mm M60 machine gun and shield mounted either side towards the rear.

Status: Conversions complete. In service with El Salvadorian Army.

Manufacturer: Chassis and armour package, USA. Final assembly and testing, National Armoury, San Salvador, El Salvador.

SPECIFICATIONS

CREW	2 + 8	LENGTH	5.38 m	FUEL CAPACITY	90 l
CONFIGURATION	4 × 4	WIDTH	1.83 m	ENGINE	Detroit Diesel Model 353 developing 110 hp
WEIGHT		HEIGHT	2.63 m		
combat gun carrier	4223 kg	GROUND CLEARANCE	0.35 m	TRANSMISSION	Spicer 3053, 5 forward and 1 reverse gears
loaded APC	4454 kg	MAX ROAD SPEED	80 km/h		
unloaded gun carrier	3769 kg	MAX RANGE	480 km	ELECTRICAL SYSTEM	24 V
unloaded APC	3547 kg				

Other El Salvadorian APCs

In addition to the VAL (Light Assault Vehicle) Cashuat (4 × 4) vehicles described in the previous entry, the El Salvadorian Army also uses the following locally produced vehicles.

Thyssen Henschel UR-416 (4 × 4) APC

A total of 10 of these vehicles were purchased from Germany of which four were for the Cavalry, two for the National Guard, two for the National Police, one for the 1st Infantry Brigade and one in reserve. As of late 1990 eight vehicles remained, seven with the Cavalry and one with the National Police.

Modified M114 Carrier

The M114 Command and Reconnaissance Carrier (*Jane's Armour and Artillery 1979-80* page 159) was developed for the US Army from the late 1950s with production being undertaken by the Cadillac Division of General Motors Corporation. First production vehicles were completed in 1962 but all vehicles were phased out of service by 1982.

El Salvadorian APC based on M114 chassis and armed with 12.7 mm MG (D Spencer/Julio A Montes Collection)

El Salvadorian stretched APC on M114 chassis (Julio A Montes)

A total of 27 stripped down chassis found their way to El Salvador where some 23 vehicles were rebuilt into the following versions, with the other four being used for spare components:
4 Armoured Personnel Carriers (crew of 2 + 8)
4 Improved Armoured Personnel Carriers
4 Stretched Armoured Personnel Carriers
4 open topped Cargo/Personnel Carriers
3 or 4 Self-Propelled Anti-Aircraft Gun Systems fitted with the Yugoslav 3 × 20 mm anti-aircraft system
3 Miscellaneous (one of which was amphibious).

Armoured Truck (6 × 6)

A total of 14 of these were built locally based on a German Magirus-Deutz Jupiter 'Mazinger' truck chassis. These can carry a total of 18 troops plus a two-man crew and can be armed with a single 12.7 mm M2 MG.

Boxer Armoured Pick-up Truck (4 × 4)

About 20 of these vehicles, called the 'Astroboys' and based on a commercial Ford F250 pick-up chassis are in service and these carry a total of six men plus two-man crew.

El Salvadorian M114 vehicle fitted with turrent armed with Yugoslav 3 × 20 mm anti-aricraft gun system (Julio A Montes)

Other Vehicles

Two additional 3 × 20 mm self-propelled gun systems, (one on an International Harvester dump truck chassis (dismantled in 1988) and one on a German MAN 630 truck chassis), one armoured Jeep (4 × 4) fitted with turreted 12.7 mm M2 HB machine gun, and other types including armoured 2.5 ton (US) (6 × 6), 5 ton (US) (6 × 6) and Mercedes-Benz Unimog (4× 4) vehicles.

Status

In service with El Salvadorian Army.

El Salvadorian 'Astroboy' (4 × 4) APC (Julio A Montes)

FINLAND

SISU XA-180 Armoured Personnel Carrier

Development

The Finnish Army operates about 60 Soviet-supplied BTR-60PB (8 × 8) APCs but has a requirement for between 300 and 400 new wheeled APCs. To meet this requirement SISU built the XA-180 (6 × 6) and Valmet the Model 1912-6 (6 × 6), with both prototypes completed in 1982.

Following trials with these vehicles in December 1983 the Finnish government placed an initial order for 59 SISU XA-180 APCs at a cost of FIM 42 million, with deliveries taking place over a three-year period. Of the initial order, 50 were for the Finnish Army and the remaining nine for the Finnish United Nations Battalion. By early 1991 it is estimated that over 130 vehicles had been produced. Between February and April 1988 the Swedish UN contingent took delivery of four vehicles at a cost of FM 5.3 m.

A major reason for the decision to select the SISU XA-180 was that this vehicle uses many automotive components of the new SISU SA-150 VK (4 × 4) 6500 kg truck already in service with the Finnish Army.

Description

The hull of the XA-180 is of all-welded steel armour construction that provides the crew with protection from small arms fire and shell splinters. The commander sits at the front of the hull on the right with the driver on the left, both enter the vehicle via a forward-opening door in the left side of the hull. In the upper part of the door is a bullet-proof window which can be partly covered by a shutter. To the front of both commander and driver is a large bullet-proof windscreen that can be covered by a shutter hinged at the top when in a combat area. Above the driver's position is provision for installing a night observation device while above the commander's position is a cupola with a single piece, rear-opening hatch cover. Mounted on the forward part of the cupola is an infra-red searchlight that can be operated from within the vehicle while on the left is a pintle mount for a light machine gun.

The engine compartment is to the rear of the driver on the left side with the air inlet and air outlets in the roof. An aisle on the right side of the hull allows passage from the front crew compartment to the troop compartment at the rear of the hull.

The infantry sit on bench seats in the rear troop compartment, five down either side, and enter via two doors in the rear, the left one of which has a single firing port and vision block. In either side of the troop compartment are three vision blocks with a firing port underneath allowing the troops to fire their weapons from inside the vehicle. Over the top of the troop compartment are two roof hatches that are hinged in the centre and can be locked vertical if required.

The suspension of the XA-180 consists of leaf springs and hydraulic shock absorbers, steering is power-assisted on the front two axles. The vehicle is fully amphibious being propelled in the water by two propellers mounted under the hull rear, one either side. Before entering the water a trim vane is erected at the front of the vehicle. Standard equipment includes a front-mounted winch with a capacity of 10 tonnes, provided with 50 m of 16 mm cable, and an automatic fire suppression system for the engine compartment. If required the XA-180 can be fitted with an NBC system and its tyres fitted with snow chains.

When originally delivered to the Finnish Army the SISU XA-180 (6 × 6) vehicles were unarmed, but many vehicles have now been fitted with a cupola-mounted 12.7 mm NSV machine gun which is also installed on the more recent Soviet AFVs such as the T-72 which is also used by the Finnish Army. The Finnish United Nations vehicles have an all-welded turret armed with a Soviet 12.7 mm DShKM machine gun with a bank of five smoke dischargers mounted either side of the turret. Some Finnish Army SISU XA-180 vehicles are also fitted with a roof-mounted AT-4 Sprigot ATGW system.

SISU XA-180 (6 × 6) in United Nations markings and fitted with turret but without 12.7 mm machine gun installed

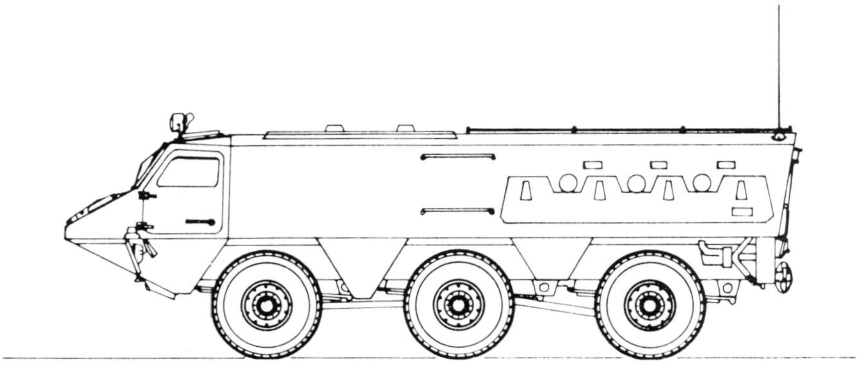

SISU XA-180 (6 × 6) APC

SISU XA-180 (6 × 6) with all doors and roof hatches locked open

Finnish Crotale New Generation on SISU XA-180 (6 × 6) vehicle in travelling configuration

SISU XA-180 (6 × 6) showing firing ports in rear troop compartment and propellers under hull rear

SISU XA-180 (6 × 6) APC fitted with Swedish Ericsson Giraffe radar system which was fitted in Finland by Jantronic Oy. When travelling, the radar folds down on top of the hull

Variants

Air defence radar

The Finnish Army has taken delivery of a number of SISU XA-180 (6 × 6) vehicles fitted with the Swedish Ericsson Giraffe air surveillance radar system on a hydraulic arm. This version of the vehicle has been developed by the Finnish company of Jantronic Oy which has manufactured the hydraulic mast, electrics, heating, raising and armouring system. To provide a more stable platform, stabilisers are lowered to the ground before the roof-mounted scissors-type mast is raised into the operating position.

The system is used to provide key target information to anti-aircraft guns and surface-to-air missile systems in the forward battlefield area.

Crotale New Generation on XA-180 chassis

In mid-1988 Finland placed a contract with Thomson-CSF of France for the Crotale New Generation (Crotale NG) air defence system mounted on the XA-180 (6 × 6) chassis. This was the first order for Crotale New Generation which has been developed as a private venture by Thomson-CSF with the VT-1 missiles being developed by LTV of the United States.

The system is mounted on the roof of the XA-180 and has a total of eight missiles in the ready to fire position plus tracking and surveillance radar.

Finland sends the XA-180 chassis to Thomson-CSF complete with NBC system, air-conditioning system, auxiliary power unit and stabilisers. Thomson-CSF then integrate the Crotale New Generation SAM system, complete trials and deliver the complete system to Finland.

It is believed that Finland has ordered a total of 20 systems with deliveries running from late 1991 through to 1993.

Full details of the Crotale New Generation are given in *Jane's Land-Based Air Defence 1991-92* page 108.

Other variants

It is believed that the Finnish Army also has a command post version and a mobile battalion radar centre, although the latter could relate to the aforementioned Giraffe system.

The manufacturer has also proposed other versions of the SISU XA-180 including mortar carrier, anti-aircraft gun, fire support/anti-tank with two-man turret armed with 90 mm gun, anti-tank with the Euromissile Mephisto ATGW system which launches 4000 m HOT missiles, ambulance and repair/recovery.

Status: In production. In service with the Finnish Army and Finnish and Swedish United Nations units in the Middle East and Ireland (2 for UN roles).

Manufacturer: Oy SISU-Auto Ab
Sisu Defence
PO Box 189
SF-13101 Hämeenlinna
Finland
Telephone: 358 17 5851
Telex: 2315
Fax: 358 17 197130

SPECIFICATIONS					
CREW	2 + 10	HEIGHT	2.3 m	STEERING	power-assisted, front two axles
CONFIGURATION	6 × 6	GROUND CLEARANCE	0.4 m	TYRES	14.00 × 20 or 14.00 × R20
COMBAT WEIGHT	15 000 kg	MAX SPEED		SUSPENSION	leaf springs, hydraulic shock absorbers
UNLOADED WEIGHT	12 000 kg	road	100 km/h	ELECTRICAL SYSTEM	24 V
POWER-TO-WEIGHT RATIO	16 hp/tonne	water	10 km/h	ARMAMENT	1 × 12.7 mm MG
PAYLOAD		MAX ROAD RANGE	800 km	SMOKE LAYING EQUIPMENT	1 × 4 smoke dischargers
road	6500 kg	ENGINE	Valmet 6-cylinder, water-cooled, turbo-charged diesel developing 240 hp		
cross-country	3000 kg	TRANSMISSION	semi-automatic, 4 forward and 1 reverse gears, 2-speed power divider		
swimming	3000 kg				
LENGTH	7.35 m				
WIDTH	2.9 m				

FRANCE

GIAT Industries AMX-10P Infantry Combat Vehicle

Development

The AMX-10P was developed from 1965 by the Atelier de Construction d'Issy-les-Moulineaux to meet a French Army requirement for a vehicle to replace the older AMX VCI based on the chassis of the AMX-13 light tank. The first prototypes were completed in 1968 and were powered by a Hispano-Suiza multi-fuel engine which developed 250 hp and gave the vehicle a maximum road speed of 65 km/h.

Production began at the Atelier de Construction Roanne (where the AMX-30 MBT is produced) in 1972 and first vehicles were delivered to the 7th Mechanised Brigade at Rheims in 1973. The largest export order to date is for over 300 vehicles for Saudi Arabia. By 1991 total production of the AMX-10P family for both home and export markets amounted to about 1750 vehicles and production was continuing at a low rate.

Description

The hull of the AMX-10P is made of all-welded aluminium with the driver's compartment at the front of the vehicle on the left, engine compartment to his right and the troop compartment at the rear of the hull. The driver has a single piece hatch cover that opens to the rear, in front of which are three periscopes, the centre one of which can be replaced by a passive periscope for driving at night.

The engine compartment is provided with a fire-extinguishing system and the complete powerpack can be replaced in two hours. The air-inlet and air-outlet louvres are in the top of the hull and the exhaust outlet is in the right side of the hull. The transmission consists of a torque converter with an electro-magnetically operated clutch and a steer/drive unit comprising a gearbox, steering mechanism and a PTO for the water jets.

The two-man Toucan II turret is mounted in the centre of the vehicle, offset slightly to the left of the vehicle's centre line, with the gunner seated on the left and the commander on the right. Both the commander and the gunner have a single piece hatch cover that opens either side of their position. The commander has an M371 sight with a magnification of ×1 and ×6, a direct sight for anti-aircraft use and an external sight for direct fire. The gunner has an OB 40 day/night periscope, which has a day magnification of ×6 (10° field-of-view) and a night magnification of ×5 (7° field-of-view). The vehicle can also be delivered with the gunner's OB 40 sight replaced by an M406 day sight with a magnification of ×2 and ×6, which is replaceable by an OB 37 image intensification sight with a magnification of ×6. The commander and gunner also have seven periscopes for all-round observation.

The GIAT Industries M693 20 mm cannon is mounted externally on the turret and has a dual-feed system allowing the gunner to select either HE or AP projectiles, and a burst selector. The cyclic rate of fire of the 20 mm cannon is 700 rpm and the maximum effective range is 1500 m.

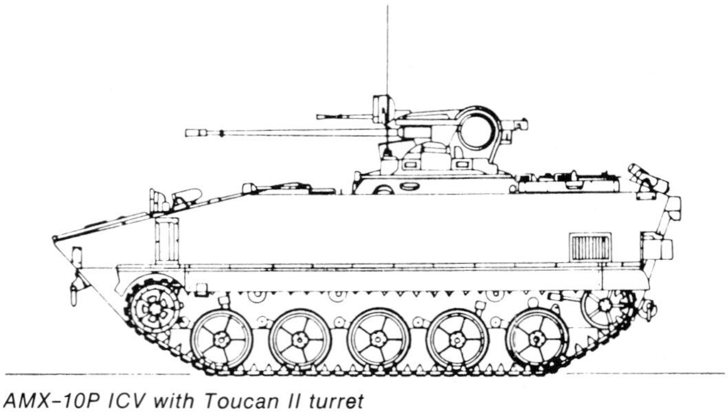

AMX-10P ICV with Toucan II turret

Of 325 rounds of 20 mm ready ammunition carried in the turret, 260 are HE and 65 AP.

Mounted coaxially above and to the right of the 20 mm M693 cannon is a 7.62 mm machine gun, which has a cyclic rate of fire of 900 rpm and a maximum effective range of 1000 m. Of 900 rounds of ammunition for the machine gun carried in the turret, 200 are for ready use. Mounted coaxially with the 20 mm cannon and 7.62 mm machine gun is a PH 9A searchlight and mounted on either side of the forward part of the turret are two electrically operated smoke dischargers.

If required the vehicle can be delivered with the 20 mm M693 replaced by a similar weapon such as the German Rheinmetall Rh 202, fitted with other turrets such as the one-man Toucan I or GIAT Capre 20 or various turrets armed with a 20 mm cannon, 12.7 mm or 7.62 mm machine guns.

The eight infantrymen are seated on individual bucket seats in the troop compartment at the rear. One man is seated forward and right of the turret facing the rear, one man to the right of the turret facing the turret, four men to the rear of the turret (two each side facing outwards), and one man on each side of the rear facing inwards.

They enter and leave by the large electrically operated ramp at the rear, which is hinged at the bottom and has two doors, each with a firing port. Over the top of the troop compartment are two roof hatches which

French Army AMX-10P ICV fitted with Euromissile MILAN ATGW from rear (Pierre Touzin)

French Army AMX-10P ICV (Pierre Touzin)

are hinged in the centre and can be locked vertical. Seven periscopes are fitted in the roof of the troop compartment, one at the front on the right, two in either side and two at the rear. The basic AMX-10P is also in service carrying two Euromissile MILAN launchers with 10 missiles carried in the hull. If required the MILAN launchers can be mounted on the sides of the hull top, one each side.

The torsion bar suspension consists of five single rubber-tyred road wheels with the drive sprocket at the front, idler at the rear and three track return rollers. The first and fifth road wheel stations are fitted with hydraulic shock absorbers. The tracks have replaceable rubber pads.

The AMX-10P is fully amphibious being propelled in the water by two water jets at the rear of the hull, one either side of the ramp. Before entering the water the two bilge pumps are switched on (one in the engine compartment and one in the troop compartment) and the trim vane is erected at the front of the hull. When not in use the trim vane is stowed on the glacis plate. The AMX-10P can be delivered without the amphibious capability.

Standard equipment includes an NBC system in the right side of the hull and a heater. The vehicle can be delivered without the NBC system or the heater and optional equipment includes an NBC detector and decontamination equipment.

In the future it is possible that the present 280 hp engine will be replaced by a more powerful version developing 350 hp, a new transmission fitted and that the turret will be fitted with a stabilisation system. An AMX-10P fitted with a hydrostatic transmission has already been tested.

Variants

AMX-10P 25 ICV

The prototype of the AMX-10P 25 was shown for the first time in June 1983. It is based on the standard AMX-10P hull without the original two-man turret. The commander now sits to the left of the turret and has a single piece hatch cover that opens to the front, five day periscopes and a day/night periscope.

Mounted in the centre of the hull roof is the new GIAT Industries Dragar one-man turret armed with a 25 mm Model 811 GIAT Industries dual-feed cannon which has 175 rounds of HE and 45 rounds of armour-piercing ammunition for ready use. A 7.62 mm machine gun is mounted coaxially with the 20 mm cannon on the right side of the turret and has 200 rounds of ready use ammunition. Mounted either side of the turret is a bank of two electrically operated smoke dischargers firing forwards.

Turret traverse and weapon elevation are electric at a speed of 45°/s and the weapons can be elevated from -10 to +45° with turret traverse being a full 360°. The gunner has four M336 periscopes for forward observation, two fixed rear vision blocks and a day/night sight for aiming the weapons. This has a night channel with a magnification of ×4.5 and a day channel with a magnification of ×1 and ×7. As an option a thermal camera can be fitted and the turret stabilised.

This model weighs 14 300 kg loaded and 11 700 kg empty and has a crew of three: commander, gunner and driver, and can carry eight fully equipped troops.

The AMX-10P 25 ICV has been developed by GIAT Industries as a private venture and has not yet been adopted by the French Army. Early in 1991 it was revealed that an AMX-10P 25 ICV had been demonstrated in Singapore.

AMX-10P Marine

The AMX-10P Marine has been developed by GIAT to meet the requirements of the Indonesian Marines. The batch of 50 AMX-10P Marine and AMX-10 PAC 90 Marine Fire Support Vehicles delivered from 1982 to Indonesia were fitted with the Baudouin 6 F11 SRX diesel developing 276 hp in place of the standard HS-115 diesel engine.

All of the vehicles were specially modified for naval operations and have 305 mm water jets to give a maximum water speed of 10 km/h, reinforced waterproofing, special protection against salt corrosion, strong transparent wave breaker on the front of the hull, four bilge pumps, streamlined track links, new air intakes, side rings and an emergency pneumatic engine starter.

The wave breaker mounted on the glacis plate can be raised and lowered by the driver hydraulically without him leaving his seat. When afloat air enters the vehicle through the large circular roof cowling and the hot air is expelled through side vents.

The original two-man turret of the AMX-10P has been removed and over the rear troop compartment of the AMX-10P Marine is the GIAT Industries CIBI 50 turret armed with an externally mounted 12.7 mm machine gun with 120 rounds of ready use ammunition. This weapon can be elevated from -10 to +45° and turret traverse is a full 360°. The AMX-10P has a crew of two and can carry a landing party of 13. The AMX-10P Marine can also be fitted with other turrets including the 25 mm Dragar.

AMX-10 Ambulance

This model is unarmed and is fitted with a fixed commander's cupola with three forward-facing vision blocks, and has a crew of three: the driver and two orderlies. The vehicle can carry three stretcher patients or one stretcher and four seated patients. The AMX-10 ambulance has an air-conditioning system, searchlight, washing facilities, oxygen system and blood transfusion equipment.

AMX-10P Driver Training Vehicle

This is the basic vehicle with its turret removed and replaced by an observation position for the instructor and drivers not under instruction.

AMX-10 ECH Repair Vehicle

This has the Toucan I one-man turret and a crew of five which consists of the commander/chief mechanic, three mechanics and the driver. On the right side of the roof at the rear is a crane with a capacity of 6000 kg. When it is being used two jacks are positioned under the rear of the hull between the ground and the floor of the vehicle. The AMX-10 ECH is used to carry out engine changes on other members of the AMX-10P family as well as changing torsion bars on both the AMX-10P and the AMX-30. This model is in production.

Full details of the AMX-10 ECH are given in *Jane's Military Vehicles and Logistics 1991-92* page 45.

AMX-10 HOT (Anti-tank)

This is the basic AMX-10P with the Toucan II turret replaced by a new two-man Lancelot turret based on TH 20 turret with four ready to launch HOT missiles and a further 14 missiles carried in the rear of the hull. The crew of five consists of the commander and gunner in the turret, two missile loaders and the driver. The turret has powered traverse through

AMX-10P 25 ICV with one-man GIAT Industries Dragar turret armed with GIAT Industries 25 mm cannon and 7.62 mm coaxial machine gun (Christopher F Foss)

AMX-10 ECH repair vehicle changing powerpack of AMX-10P (ECP Armées)

AMX-10P Marine as delivered to Indonesian Marines with GIAT Industries CIBI 50 turret

Rear view of AMX-10P HOT with turret with four Euromissile HOT ATGWs ready to launch (Christopher F Foss)

AMX-10 SAO artillery observation vehicle (ECP Armées)

AMX-10 TM towing 120 mm Thomson-Brandt mortar

360° at a speed of 50°/s and manual traverse at a speed of 10°/s. The missiles can be elevated from -12 to +18° at a speed of 18°/s (powered) and 22°/s (manual). The missiles are launched by the gunner who is seated on the left side of the turret and has an M509 sight with a magnification of ×3 and ×12, while the commander operates the M427 laser rangefinder which has a maximum range of 8000 m and a magnification of ×8 (day), and ×3.5 (night). In addition the turret has six periscopes and two vision blocks. The AMX-10 HOT has a combat weight of 14 100 kg and an unloaded weight of 12 400 kg. This model has been produced for export, the only known user being Saudi Arabia. As an option the turret can be fitted with a CASTOR thermal camera with two magnifications.

AMX-10 PC (Command Vehicle)

This is the command member of the AMX-10P and is identical to the basic vehicle except that it has additional communications equipment and a crew of six (two officers, one NCO, two radio operators and the driver). Mounted on the top of the hull at the rear is a portable generator which is positioned on the ground when the vehicle is being used in the static role. When two AMX-10 PCs are back to back a canopy can be erected between them, and in addition an awning can be erected at the side of the hull. This model is in service with the French and several other armies.

AMX-10P with RATAC

This has no turret and is fitted with a roof-mounted RATAC (Radar For Field Artillery Fire) pulse Doppler radar scanner. The vehicle has a crew of five: commander, radar operator, assistant radar operator, radio operator and driver. The radar has four operating modes: ground surveillance, acquisition and identification, angular deviation measurement and automatic tracking. The radar is in the forward part of the vehicle on the right side, with the radar operating console behind it and the plotting table at the rear on the right. Height including radar is 2.84 m. This model also has a navigation system and an identification system. It is in production. At least 24 of this version have been sold to Saudi Arabia.

AMX-10 RAV

GIAT Industries have proposed the development of a ammunition resupply vehicle for use with the 155 mm GCT. This will have a hydraulic crane to handle the pallets of ammunition.

AMX-10RC (Reconnaissance Vehicle)

This 6 × 6 reconnaissance vehicle armed with a turret-mounted 105 mm gun uses many automotive components of the AMX-10P series. Full details are given in the entry in the *Reconnaissance vehicles* section. A tracked version of the AMX-10RC, called the AMX-10C, was developed to the prototype stage, but has not been placed in production.

AMX-10 SAO (Artillery Observation Vehicle)

The AMX-10 SAO is used for battlefield observation and artillery fire control and is an AMX-10P with its Toucan II turret replaced by a new two-man turret armed with an externally mounted 7.62 mm machine gun on the right side. Mounted on the top of the turret are four electrically operated smoke dischargers firing forwards. Turret traverse is electric, two-speed, with a maximum speed of 50°/s. The turret has a laser rangefinder with a maximum range of 8000 m coupled to a day sight with a magnification of ×8 and a night sight with a magnification of ×4.5. It also has a binocular telescope with a magnification of ×2.5 and ×10, which has adjustable cross-wires, retractable sun filter and can be elevated from -10 to +45°. The crew of five consists of the commander (who normally operates the binocular telescope), second in command (usually the laser rangefinder operator), two radio operators and a driver. The AMX-10 SAO is fitted with a vehicle attitude corrector which enables it to establish its exact position. The ATILA system message transmitter unit (CMS) can be used in the SAO vehicle and if required the SICLAMEN IFF system can be fitted. Production of this model began in 1980.

AMX-10 SAT (Artillery Survey Vehicle)

This is based on the AMX-10 PC command vehicle but has been modified internally to carry gyro-stabilised theodolite, topographic survey theodolite, distance measuring equipment and a navigation system.

AMX-10 TM (Mortar Tractor)

This model tows a 120 mm Thomson-Brandt rifled mortar model MO-120-RT-61 as well as carrying 60 rounds of ammunition, of which 56 are stored horizontal and four vertical. The AMX-10 TM has a crew of six consisting of commander, aimer, artificer, loader, ammunition carrier and driver, and is fitted with a Toucan I turret armed with a 20 mm cannon and a 7.62 mm machine gun, which have elevations of +50° and depressions of -8°, turret traverse being a full 360°. The turret is manually operated: maximum traverse speed is 12°/s and maximum elevation speed 16°/s. The turret has an optical sight with magnifications of ×1 and ×6 plus periscopes. The AMX-10 TM has been produced in quantity for both the home and export markets.

AMX-10P with 120 mm Mortar

For trials purposes an AMX-10P ICV has had its turret removed and

AMX-10 VOA artillery observer vehicle with commander's hatch open (Pierre Touzin)

AMX-10 PAC 90 special marine model with trim vane erected on glacis plate for amphibious operations

a Thomson-Brandt 120 mm 120-RT-61 rifled mortar firing to the rear of the vehicle installed. Thomson-Brandt is now designing loading aids to enable rates of fire to be achieved comparable to those of the standard towed 120 mm 120-RT-61 mortar which is in service with some 13 countries.

AMX-10 VOA (Artillery Observer Vehicle)

This is based on the AMX-10PC command vehicle chassis and has been designed for use by artillery observers. Its crew of four consists of commander, observer, radio operator and driver. The turret, which is manned by one or two men, is provided with equipment for day and night observation and local defence. The equipment fitted in the turret enables the acquisition of target co-ordinates and the manual or automatic transmission of artillery messages. Armament consists of a 7.62 mm machine gun and four smoke dischargers. The onboard laser rangefinder has a magnification of ×8 and can be dismounted if required. The turret optical sight has magnifications of ×2.5 and ×10, thermal camera and periscopes.

AMX-10Ps used in the ATILA Control System

Two versions of the AMX-10P have been developed to mount parts of the ATILA automatic artillery fire control system. The AMX-10 VFA is used at regimental and battery level and the AMX-10 VLA, which does not have a computer, is used as an artillery liaison vehicle.

AMX-10 SAF

This is also known as the AMX-10PC SAF and is used with the ATILA artillery fire control system. It has a crew of five consisting of the radio operator, driver, two warrant officers and one commanding officer, and is fitted with an auxiliary power unit in the forward part of the vehicle, two air-conditioning systems for the electronics and crew, multi-purpose computer, data processing equipment centre, man-machine interface (screen/keyboard) and three radio sets.

AMX-10P with TMR 81 81 mm turret

For trials purposes an AMX-10P has been fitted with the new GIAT Industries TMR 81 81 mm mortar turret which has been designed to meet the requirements of the French Army. The vehicle has a three-man crew consisting of the driver, gunner seated in the turret and a loader in the rear of the vehicle. Turret traverse is electric through 360° and the GIAT Industries mortar can be elevated from +38° to +83°. A SOPTAM fire control system from Sopelem has been fitted and the mortar has a sustained rate of fire of one round every two to four seconds.

AMX-10 TMC-81 Fire Support Vehicle

Shown for the first time at the 1981 Satory Exhibition of Military Equipment was the AMX-10 TMC-81 fire support vehicle which has been developed specifically for the export market. It has a modified AMX-10 PAC 90 turret fitted with a Thomson-Brandt 81 mm CL 81 smoothbore gun with a semi-automatic breech and can be used for direct and indirect fire. The weapon has an elevation of +66°, depression of -7° and turret traverse of 360°. Elevation/depression and turret traverse are both manual.

A total of 108 HE projectiles with a muzzle velocity of 400 m/s and a maximum range of 7200 m are carried. In addition 10 APFSDS projectiles, which have a muzzle velocity of 1000 m/s and will penetrate 50 mm of armour at an incidence of 45° at a range of 1000 m, are carried. If required a 7.62 mm or 12.7 mm machine gun can be mounted on the turret roof for anti-aircraft defence.

The crew of four consists of the commander and gunner in the turret and driver and loader/mechanic in the hull. Mobility and amphibious characteristics remain the same as the basic AMX-10P. This version is still at the prototype stage.

AMX-10 PAC 90 Fire Support Vehicle

The AMX-10 PAC 90 has been developed primarily for the export market by the Atelier de Construction d'Issy-les-Moulineaux and was first announced in 1978. It basically consists of an AMX-10P ICV hull fitted with the GIAT Industries TS 90 turret which is also mounted on the Panhard ERC 90 F4 Sagaie (6 × 6) armoured car as well as the Renault VBC 90 (6 × 6) armoured car. The AMX-10 PAC 90 has been evaluated by several countries and is in service with Indonesia. The Indonesian Marine models have automotive characteristics and modifications similar to the AMX-10P Marines described earlier in this entry.

AMX-10P fitted with a two-man SAMM TTB 140 40 mm turret for trials purposes

The primary role of the AMX-10 PAC 90 is that of anti-tank and fire support on the battlefield at ranges of between 1000 and 2500 m, but it can also be used as a reconnaissance vehicle or as an APC as well as transporting MILAN anti-tank or 81 mm mortar teams.

The GIAT Industries 90 mm gun fires the following types of ammunition: canister, high explosive, high explosive long range, HEAT, smoke and APFSDS.

Twenty rounds of ammunition, 12 HEAT and 8 HE, are carried in the turret, 16 in the turret bustle to the rear of the commander, who also loads the gun, and four behind the commander's seat. Another 10 rounds are carried in the hull. A 7.62 mm machine gun is mounted coaxially to the left of the main armament, and 2000 rounds of ammunition for it are carried in the turret with an additional 1200 rounds carried in the hull. Two smoke dischargers are mounted either side of the turret towards the rear.

The four infantrymen sit on individual seats at the rear of the hull and enter and leave by the large ramp in the hull rear, which has two doors, each with a firing port. There are four periscopes in the rear of the troop compartment, one in each side and two at the rear.

AMX-10P with TTB 140 turret

For trials purposes an AMX-10P has been fitted with the private venture SAMM TTB 140 40 mm multi-purpose turret. This two-man turret is armed with the new Bofors 40 mm 4070B gun with an elevation of +50° and a depression of -10°, with turret traverse a full 360°. A 7.62 mm MG is mounted with the main armament.

Status: In production. In service with France, Greece, Indonesia (AMX-10 PAC 90 and AMX-10P Marine), Qatar, Saudi Arabia and the United Arab Emirates.

Manufacturer: Atelier de Construction Roanne (ARE)
GIAT Industries
13 route de la Minière
78034 Versailles
France
Telephone: (1) 30 97 37 37
Fax: (1) 30 97 39 00

SPECIFICATIONS

(data in square brackets relates to AMX-10 PAC 90 where different)

CREW	3 + 8 [3 + 4]	FUEL CAPACITY	528 l
COMBAT WEIGHT	14 500 [14 800] kg	MAX ROAD RANGE	600 km
UNLOADED WEIGHT	12 700 [13 800] kg	FORDING	amphibious
POWER-TO-WEIGHT		GRADIENT	60%
RATIO	20.68 [20.27] hp/tonne	SIDE SLOPE	30%
GROUND PRESSURE	0.53 [0.56] kg/cm²	VERTICAL OBSTACLE	
LENGTH GUN FORWARD	n/app [7.22 m]	forwards	0.7 m
LENGTH	5.778 [5.87] m	reverse	0.6 m
WIDTH	2.78 [2.83] m	TRENCH	2.1 m
HEIGHT		ENGINE	Hispano-Suiza HS-115 V-8 water-cooled super-charged diesel developing 300 hp at 3000 rpm (manufactured by Renault Véhicules Industriels)
overall	2.57 [2.73] m		
to hull top	1.92 m		
GROUND CLEARANCE	0.45 m		
TRACK WIDTH	420 mm		
LENGTH OF TRACK ON GROUND	3 m		
MAX ROAD SPEED			
4th gear	65 km/h	TRANSMISSION	pre-selective with 4 forward and 1 reverse gears
3rd gear	34.4 km/h		
2nd gear	15.4 km/h		
1st gear	9.4 km/h	SUSPENSION	torsion bar
MAX WATER SPEED		ELECTRICAL SYSTEM	24 V
(water jets)	7 [10] km/h	BATTERIES	6 × 12 V

ARMAMENT	
main	1 × 20 mm cannon [1 × 90 mm gun]
coaxial	1 × 7.62 mm MG
anti-aircraft	n/app [1 × 7.62 mm or 1×12.7 mm, optional]
SMOKE-LAYING EQUIPMENT	2 × 2 smoke dischargers
AMMUNITION	
main	760 [30]
coaxial	2000 [7.62 mm: 3200, 12.7 mm: 1000]
GUN CONTROL EQUIPMENT	
Turret power control	electric [manual]
by commander	yes [no]
by gunner	yes
Max rate power traverse	50°/s (10°/s manual) [360°/20 s]
Max rate power elevation	30°/s (26°/s manual)
Gun elevation/depression	+50°/-8° [+15°/-8°]
Turret traverse	360°
Commander's override	yes

AMX VCI Infantry Combat Vehicle

Development

The AMX VCI (Véhicule de Combat d'Infanterie) was developed in the early 1950s to meet the requirements of the French Army following the cancellation of the Hotchkiss TT 6 and TT 9 APCs. The first prototype was completed in 1955 and the first production vehicles in 1957 at the Atelier de Construction Roanne (ARE). When the ARE started production of the AMX-30 MBT, production of the complete AMX-13 tank family, including the AMX VCI, was transferred to the Creusot-Loire facility at Châlon-sur-Saône.

When originally introduced into service with the French Army the vehicle was called the Transport de Troupe Chenillé Model 56 (or TT 12 CH Mle 56 for short), which was later changed to the Véhicule Transport de Personnel (or AMX VTP) and more recently to the Véhicule de Combat d'Infanterie. It was used in large numbers by the French Army but is now being replaced by the AMX-10P amphibious ICV.

It is believed that total production of the vehicle amounted to about 3400 vehicles.

In mid-1990 Creusot-Loire Industrie unveiled their MARS 15 family of light armoured vehicles which includes an APC. Details of this are given in the *Armoured fighting vehicle families* section under France.

Description

The chassis of the AMX VCI is similar to that of the AMX-13 light tank. The hull of the vehicle is divided into three compartments with the driver and engine compartments at the front and the troop compartment at the rear.

The driver is seated at the front of the vehicle on the left with the engine compartment to his right. The driver has a single piece hatch cover that opens to the left in front of which are three periscopes. The centre periscope can be replaced by an infra-red or image intensification periscope for night driving.

The petrol engine is mounted facing the rear and can be removed through the roof in 40 minutes. The engine transmits power via the clutch to the gearbox at the front of the hull, to the right of which is the Cleveland type steering differential.

Mounted on the glacis plate is a splash board to prevent water rushing up the front of the vehicle when fording and a spare wheel is normally carried on the glacis plate.

The troop compartment is behind and above the driver with the gunner seated on the left and the vehicle commander to his right. When the VCI was originally introduced into the French Army the gunner had a single-piece hatch cover with eight vision devices which swivelled to open, forward of which was a pintle-mounted 7.5 mm machine gun. This was later replaced by a ring-mounted 12.7 mm machine gun or a Creusot-Loire CAFL 38 turret armed with a 7.5 mm or a 7.62 mm machine gun.

The vehicle can be fitted with most other modern turrets such as the Creusot-Loire Industrie T 25 (25 mm), T 20 (20 mm), TL 20 S (20 mm) and shield guns racers such as CB 20 (20 mm), CB 12.7 (12.7 mm) and TL 21 (twin 7.62 mm) to name but a few.

The vehicle commander is seated to the right of the gunner and has a single piece hatch cover that opens forwards, to the front and right side of which are three periscopes.

The troop compartment is at the rear: the 10 infantrymen sit back to back down the centre of the hull and enter and leave the vehicle by two doors in the hull rear that open outward. Each door has a single firing port. In each side of the troop compartment are two two-piece hatch covers. The lower part of each has two firing ports and folds forwards into the horizontal and the upper part folds upwards through 180° to rest on the troop compartment roof.

The torsion bar suspension consists of five rubber-tyred road wheels with the drive sprocket at the front, idler at the rear and four (in some cases three) return rollers that support the inside of the track only. The first and last road wheel stations have hydraulic shock absorbers. The steel tracks have 85 links each side when new and can be fitted with rubber pads.

When originally introduced into service the VCI was not fitted with an NBC system but one was subsequently fitted to later production vehicles for the French Army. Infra-red driving lights were standard on most vehicles but they do not have any amphibious capability.

Variants

Listed below are the current variants of the AMX VCI family in service or production, excluding trials vehicles such as the Roland or minelayer. There are separate entries for the AMX-13 light tank, AMX 105 mm Mk 61 self-propelled howitzer, and AMX 155 mm self-propelled gun Mk F3.

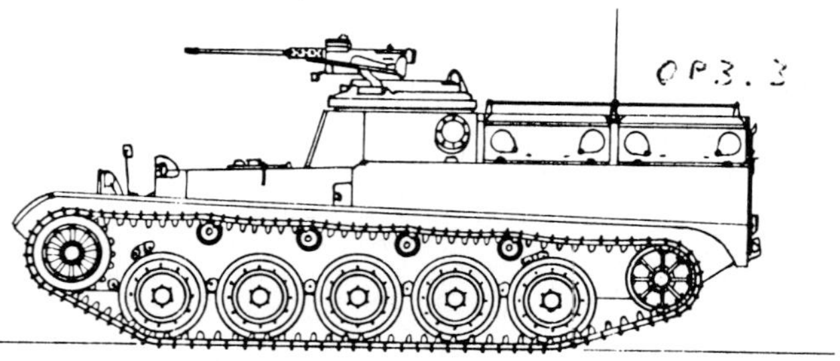

AMX VCI ICV armed with 12.7 mm (0.50) Browning M2 HB machine gun

AMX Combat Engineer Vehicle, VCG, in travelling configuration (Pierre Touzin)

French Army AMX VCI fitted with one-man turret armed with machine gun

AMX VCI fitted with Creusot-Loire Industrie CB127 shield gun racer but without 12.7 mm M2 MG fitted

AMX VCI fitted with two-man power-operated Creusot-Loire Industrie T 25 turret

French Army AMX VCI with Toucan 1 turret (Pierre Touzin)

AMX VTT/RATAC with radar lowered, fitted with Creusot-Loire Industrie's CAFL 38 turret armed with 7.62 mm machine gun (Pierre Touzin)

Ambulance (VTT/TB)

The VTT/TB (Transport de Blessés) is the ambulance member of the family, is unarmed and has a crew of four, driver, surgeon and two orderlies/stretcher bearers. The vehicle can transport three stretcher and four seated patients at once.

Anti-aircraft Vehicle

This private venture vehicle was first shown in 1983 and is essentially an AMX VCI with the machine gun turret removed and commander's roof hatch welded shut. Over this has been fitted a Dassault-Electronique TA 20/RA 20 turret armed with two 20 mm cannon, but this version of the AMX VCI has yet to enter production.

Command Post (VTT/PC)

The VTT/PC (Poste de Commandement) is the basic VCI adapted for use as a command vehicle and has a crew of six to eight depending on its use. Internally the vehicle has two map tables and additional communications equipment.

Cargo (VTT/Cargo)

This version can carry up to 3000 kg of cargo.

Combat Engineer Vehicle (VCG)

The VCG (Véhicule de Combat du Génie) is based on the VCI and is designed to carry out engineer work in the forward battle area. Mounted at the front of the hull is a hydraulically operated dozer blade for clearing obstacles and preparing fire positions. An A-frame is pivoted at the front of the hull. When not in use it is split at its upper end and lies back along either side of the hull. Other equipment includes a hydraulic winch with 63 m of cable and a maximum capacity of 4500 kg, tools and cutting equipment. The VCG has a crew of 10, commander, gunner, driver and seven men and weighs 17 800 kg.

As far as it is known this version is only used by the French Army where it is in the process of being replaced by the GIAT Industries EBG armoured engineer vehicle based on the AMX-30 MBT chassis. Full details are given in *Jane's Military Vehicles and Logistics 1991-92* page 2.

ENTAC Anti-tank Vehicle

As far as it is known this model, which was used only by the Belgian Army, is no longer in front line service. The Belgian Army also had a number of vehicles fitted with the MILAN ATGW system, these are no longer in front line service.

Fire control Vehicle (VTT/LT)

The VTT/LT (Véhicule de commandement du tir) is a development of the basic VTT/PC command vehicle, has a crew of seven and is fitted with additional communications equipment, plotting table, fire computing table, four-channel switchboard for battery communications and an external loudspeaker.

RATAC Radar Vehicle (VTT/RATAC)

This is the basic vehicle fitted with the RATAC (Radar de Tir pour Artillerie de Campagne) radar which detects, identifies, locates and automatically tracks ground targets such as tanks, vehicles, personnel, light aircraft and helicopters or shell bursts. The radar itself is mounted on the roof of the vehicle on the right side. The French Army took delivery of 115 systems on the AMX VCI chassis but these are now being transferred to the VAB (4 × 4). Export sales of RATAC on the AMX VCI chassis were made to Ecuador (4) and Morocco (12).

155 mm Support Vehicle (VTT/VCA)

The VTT/VCA (Véhicule Chenillé d'Accompagnement) is designed to accompany the 155 mm Mk F3 self-propelled gun and carries the remainder of the gun crew and ammunition. The crew of eight consists of the driver, layer, loader, gunner, and four ammunition members. A total of 25 155 mm projectiles is carried plus 25 charges and 49 fuzes. The vehicle can also tow an ARE 2T F1 ammunition trailer containing 30 projectiles plus charges and fuzes. The vehicle is normally fitted with the CAFL 38 turret armed with a 7.5 mm or a 7.62 mm machine gun or, more recently, a Creusot-Loire Industrie CB 127 shield gun racer for a 12.7 mm M2 machine gun.

81 mm Mortar Carrier (VTT/PM)

The VTT/PM (Porte Mortier) has an 81 mm mortar firing forward through the roof of the personnel compartment. It has an elevation from +43 to +80° and a total traverse of 40°. A total of 88 long mortar bombs can be carried plus 40 short mortar bombs. The crew of six consists of the driver, commander, observer and three mortarmen. A baseplate is carried enabling the mortar to be deployed away from the vehicle if required.

120 mm Mortar Carrier (VTT/PM)

The VTT/PM (Porte Mortier) has a 120 mm Thomson-Brandt type MO 120 LT mortar firing forwards through the roof of the personnel compartment. The mortar has a maximum range with PEPA/LT projectiles of 8300 m. When fired away from the vehicle with the aid of the baseplate carried the range is increased to 8950 m. The mortar has an elevation of from +45 to +77° and a total traverse of 46°. A total of 60 mortar bombs is carried and the crew of six consists of the driver, commander, layer, gunner and two mortarmen.

AMX VCI converted to flatbed configuration and fitted with GIAT Industries Minotaur mine scattering system

AMX VCI with Minotaur Mine System

GIAT Industries have modified an AMX VCI infantry combat vehicle chassis into a flatbed type chassis and installed on the rear is the GIAT Industries Minotaur scatterable minelaying system. This was developed as a private venture by GIAT Industries but entered production late in 1990 to meet the requirements of the British Army deployed to Saudi Arabia. These were installed on new build Stormer flatbed chassis which were deployed to Saudi Arabia early in 1991.

Creusot-Loire Industrie AMX-13 VTT Version 1987

This incorporates all of the automotive improvements of the modernised 105 mm AMX-13 light tank (qv) and has a Detroit Diesel engine Model 6V-53T developing 280 hp or a Baudouin 6F 11 SRY diesel engine, which is much more fuel efficient, in place of a petrol engine for increased fuel efficiency, automatic transmission and hydro-pneumatic suspension. Creusot-Loire Industrie can either deliver new vehicles to this standard or upgrade existing vehicles to this configuration.

SPECIFICATIONS

(1989 production model with diesel engine replacing original SOFAM Model 8Gxb 8-cylinder water-cooled petrol developing 250 hp at 3200 rpm)

CREW	3 + 10
COMBAT WEIGHT	15 000 kg
UNLOADED WEIGHT	12 500 kg
POWER-TO-WEIGHT RATIO	18.67 hp/tonne
GROUND PRESSURE	0.7 kg/cm²
LENGTH	5.7 m
WIDTH	2.67 m
HEIGHT	
to turret top	2.41 m
to hull top	2.1 m
GROUND CLEARANCE	0.48 m
TRACK	2.159 m
TRACK WIDTH	350 mm
LENGTH OF TRACK ON GROUND	3.012 m
MAX ROAD SPEED	64 km/h
FUEL CAPACITY	410 l
ROAD RANGE	500-550 km
FORDING	1 m
GRADIENT	60%
VERTICAL OBSTACLE	
forwards	0.65 m
reverse	0.45 m
TRENCH	1.6 m
TURNING RADIUS	skid turns
ENGINE	Detroit Diesel 6V-53T, turbo-charged 6-cylinder water-cooled diesel developing 280 hp at 2800 rpm, or Baudouin 6F 11 SRY turbo-charged 6-cylinder water-cooled diesel developing 280 hp at 3200 rpm
TRANSMISSION	manual with 5 forward and 1 reverse* *gears*
STEERING	Cleveland type differential
CLUTCH	single-disc
SUSPENSION	torsion bar**
ELECTRICAL SYSTEM	24 V
BATTERIES	4 × 12 V, 190 Ah
ARMAMENT	1 × 20 mm cannon, or 1 × 12.7 mm MG, or 1 × 7.62 mm/7.5 mm MG
SMOKE-LAYING-EQUIPMENT	none
ARMOUR	
Hull front	15 mm at 55° (corresponding to 40 mm)
Hull sides	20 mm
Troop compartment front	30 mm
Troop compartment sides	20 mm
Troop compartment hatches	15 mm
Hull top	15 mm
Hull floor front	20 mm
Hull floor rear	10 mm
Hull rear	15 mm

* or an automatic transmission with 3 forward and 1 reverse gears as an option

** or hydropneumatic suspension as an option

Status: In service with Argentina (some of which were assembled there), Belgium (transferred to the reserves), Ecuador, France, Indonesia, Italy (now in reserve), Kuwait (34 vehicles, including command vehicles, fitted with the Detroit Diesel Model 6V-53T were delivered in 1982-83), Lebanon, Morocco, Sudan (two AMX PCs and three AMX VCAs), the United Arab Emirates and Venezuela. The Netherlands has phased its AMX VCIs out of service and some of these have been passed on to Indonesia. Late in 1989 Cyprus ordered 12 VTT vehicles in the battery command post and ammunition supply versions to support the 12 155 mm self-propelled howitzers ordered at the same time. The exact status of the AMX VCI series used by Kuwait is uncertain, most were probably captured by Iraqi forces during the invasion of Kuwait in August 1990.

Manufacturer: Creusot-Loire Industrie
Immeuble Ile France
Cedex 33
92070 Paris la Defense
France
Telephone: 33 (1) 49 00 60 50
Telex: MOTOY 615 638F
Fax: 33 (1) 49 00 57 30

Renault VAB Armoured Personnel Carrier

Development

In the late 1960s the French Army decided to equip its infantry units with both tracked and wheeled vehicles. The mechanised units would be issued with the tracked AMX-10P ICV, then already under development, and the remaining units with wheeled APCs as the tracked AMX-10P was considered too expensive and sophisticated for many of the roles it was expected to undertake.

In 1970 the French Army issued a requirement for a Forward Area Armoured Vehicle (VAB or Véhicule de l'Avant Blindé) which would meet the requirements of the remaining infantry units and would be capable of undertaking a wide range of roles including use as an APC, cargo carrier and mortar towing vehicle.

To meet this requirement Panhard and Saviem/Renault Group built prototypes of both 4 × 4 and 6 × 6 vehicles between 1972 and 1973. These were tested by the Section Technique de l'Armée de Terre between May 1973 and early 1974 and in May 1974 the 4 × 4 version of the Saviem/Renault Group entry was selected for service.

No pre-production vehicles were built and first production vehicles were delivered to the French Army in 1976. The French Army has a requirement for 4430 VAB vehicles for delivery by the early 1990s. At present the French Army is ordering the 4 × 4 version only but it is expected that the 6 × 6 version, which is already in production for export, may be ordered in the future.

Prime contractor for the VAB is Renault Véhicules Industriels which supplies automotive components to Creusot-Loire Industrie at Usine de Saint-Chamond for final assembly.

VAB (6 × 6) from above without armament installed. The driver's and commander's armoured windscreen shutters are in the raised position, as are the troop compartment shutters (T J Gander)

VAB New Generation from front showing additional armour

VAB (4 × 4) fitted with RATAC radar system, and fitted with Creusot-Loire Industrie CB 52 rotating shield for 7.62 mm MG

VAB New Generation from rear showing bullet proof windows in rear troop compartment

In July 1986 Renault announced that since production of the VAB had begun late in 1976 over 3333 vehicles had been built for the home and export markets with 4300 ordered by the French Army and 900 delivered to at least 11 countries. Some 50 VABs in both 4 × 4 and 6 × 6 configurations can be built every month (for the French Army and for export). The largest export order so far is for 394 units from Morocco, including APCs command vehicles. By early 1991, total production of the VAB vehicle amounted to over 4700 units in both 4 × 4 and 6 × 6 configurations for the home and export markets.

At the Satory defence equipment exhibition held in June 1990, prototypes of the VAB New Generation (VAB NG) were shown for the first time. This has been developed and funded by Renault VI and Creusot-Loire Industrie in co-operation with the French Army and will become the updated successor of the VAB for the next two decades and also allow the rebuilding of older first generation VABs to the VAB NG standard.

The VAB NG was officially qualified by the French Army early in 1991 and production could commence as early as 1992. All existing VAB variants will be available on the VAB NG basic chassis.

The main improvements of the VAB NG cover the areas of armour protection, mobility and ergonomics with the full list of improvements being as follows:

1. Driver and commander have high protection bullet-proof windscreen
2. Front of vehicle hull now has protection against armour piercing ammunition
3. Headlamps and direction indicators at the front of the hull are now recessed low in the hull
4. Hydraulically powered front-mounted winch
5. Car type driver's dashboard and steering wheel
6. Integrated air-conditioning system with controls on driver's dashboard (option)
7. Pinion drive down transmission
8. Five speed automatic transmission with lock up device
9. Integrated NBC system
10. 250 hp twin turbocharged 6-cylinder diesel engine
11. Reinforced suspension
12. Remote-controlled tyre inflation/deflation system (option)
13. 2 × 6 DREC grenade launchers mounted either side of turret firing forwards
14. Troop compartment floor is now flat and washable
15. Troops now have individual fold up seats
16. Waterjet drive system is more compact
17. New engine cooling system
18. Optional add on armour to sides of troop compartment
19. Splinter absorbent liner for crew compartment (option)
20. Troop compartment windows are bullet-proof with armoured shutters being an option
21. The engine compartment roof has a Molotov cocktail protection system
22. Ballistic protection of air intake
23. Driver's and commander's compartment has more room
24. Driver's and commander's shutters are of composite armour (option)

In October 1986 two agreements were concluded with the Hughes Aircraft Corporation of the USA. One by Renault Véhicules Industriels for technical co-operation in TOW Under Armour systems on the VAB, and one with SMS for co-operation in the promotion and sale of these systems.

The VAB was used by France and Qatar during Operation Desert Storm, the recapture of Kuwait, early in 1991. The VAB with the HOT ATGWs successfully engaged Iraqi AFVs during this conflict.

Description

The basic model used by the French Army is the 4 × 4 VAB VTT (Véhicule Transport de Troupe) which has a crew of two (commander/machine gunner and driver) and carries 10 fully equipped infantrymen. The following description relates to this vehicle.

The all-welded steel hull of the VAB provides the crew with protection from small arms fire and shell splinters. The driver sits at the front of the vehicle on the left with the commander/machine gunner to his right. Both crew members have a side door that opens to the front, with a bullet-proof window in its upper part that is hinged at the top and opens outwards and can be covered by a shutter. In front of them is a heated bullet-proof windscreen that can be covered by a flap hinged at the top. Over the driver's position is a single piece hatch cover that opens to the front. A similar hatch is mounted over the commander's position but vehicles issued to the French Army have a Creusot-Loire Industrie rotating gun mount type CB.52 armed with a 7.62 mm machine gun with an elevation of from −15 to +45° with the shield in the normal position and from −20 to +80° in the anti-aircraft position; in both cases traverse is a full 360°. Other armament installations available include the Creusot-Loire Industrie TLi 52 A. Current production French Army VABs are fitted with the Creusot-Loire Industrie CB 127 gun ring shield for a 12.7 mm M2 HB machine gun.

The engine compartment, which is fitted with a fire-extinguishing system, is immediately behind the driver with the air-inlet and air-outlet louvres in the roof and the exhaust pipe running along the top of the hull on the right side. Power is transmitted from the engine to the wheels by a hydraulic torque converter and a gearbox with five forward and one reverse gears. Gears are shifted using a small electrically operated

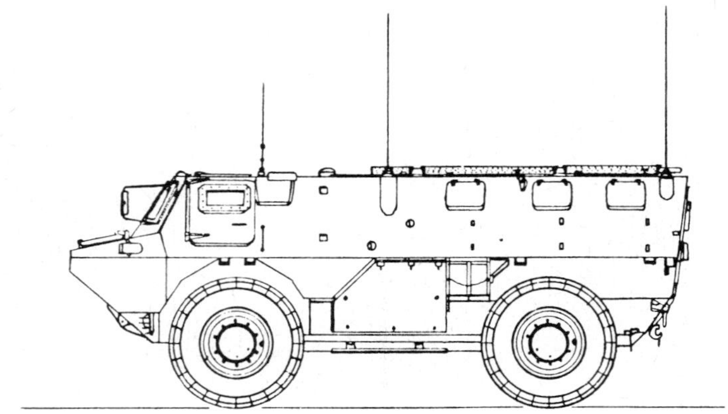

Basic (4 × 4) version of VAB as used by French Army

VCAC HOT with Euromissile UTM 800 turret with four ready-to-launch HOT ATGWs (Christopher F Foss)

lever which also operates the clutch. On the VAB New Generation, the transmission is fully automatic. The axles have differential reduction gears with double reduction and differential locking. The wheels are independently suspended by torsion bars and hydraulic shock absorbers. Steering is hydraulically assisted on the front wheels, or, in the case of the 6 × 6 model, on the front four wheels. The tyres are Michelin run-flat radials, the pressure of which can be adjusted to suit the type of ground being crossed. A remotely controlled tyre inflation/deflation device is now available as an option on the VAB family.

There is a passageway on the right side of the hull which connects the crew compartment at the front with the personnel compartment at the rear. The infantrymen enter and leave the VAB via a double door without a central pillar in the rear of the hull which open outwards. Each door has a window which can be opened to the outside and is covered by an armoured shutter.

The infantrymen sit on bench seats that run along either side of the hull. In the VAB New Generation, the infantry now have individual seats that fold up to allow cargo to be carried. In each side of the hull are three windows which can be opened to the outside and can be covered by an armoured shutter. They can be locked open on the outside to allow the infantrymen to use their small arms from inside the vehicle.

On the basic VAB there are three hatches over the troop compartment, one in the centre and two at the rear, all of which open towards the front. The troop compartment is 2.46 m long, 1.375 m wide and 1.35 m high and has a total floor area of 3.38 m^2. The seats can be quickly folded up to enable cargo to be carried. The lack of a central pillar at the rear enables cargo, up to 2000 kg, to be quickly loaded with the aid of a fork lift truck.

The VAB is fully amphibious being propelled in the water by two Dowty water jets (manufactured under licence in France by Messier) at the rear of the hull, each fitted with a deflector for steering and reverse thrust, which is hydraulically controlled by a joystick mounted on the dashboard. Before entering the water the bilge pumps are switched on and the trim vane, which is folded back onto the glacis plate when not required, is erected at the front of the hull. Early production VAB (4 × 4) vehicles for the French Army were not fitted with water jets but current production vehicles are. Not all export customers have water jets fitted.

In the French Army, the VAB is also used to transport MILAN anti-tank teams. Optional equipment for the VAB includes an air-conditioning system, armoured grenade thrower hood, firing gun port with incorporated observation port, gas dispenser, NBC system (standard for French Army vehicles), infra-red or passive night vision equipment, public address system and a front-mounted winch with a capacity of 7000 kg (which can be doubled with the aid of a pulley) and supplied with 60 m of cable.

Variants

VCI (Infantry Combat Vehicle)

The VCI (Véhicule de Combat de l'Infanterie) is currently offered in four versions:

VAB VCI TL 20S (IFV) has a crew of 10 to 12 and is fitted with a Creusot-Loire Industrie TL 20S manually operated turret armed with one 20 mm cannon. This turret has a foldable seat and no basket in the vehicle. Nine hundred rounds of 20 mm ammunition are carried.

VAB VCI T 20 has a crew of 9 to 11 and is fitted with a Creusot-Loire Industrie T 20 electrically operated turret armed with a 20 mm cannon and a 7.62 mm coaxial machine gun. Totals of 720 rounds of 20 mm and 2200 rounds of 7.62 mm ammunition are carried. In French Army service it is also organic to Pluton SSM regiments for local defence.

VAB VCI T 25 has a crew of eight, including two in the turret. The turret is an electrically powered Creusot-Loire Industrie T 25 armed with a 25 mm cannon and a 7.62 mm coaxial machine gun. Totals of 530 rounds of 25 mm and 1200 rounds of 7.62 mm ammunition are carried.

VAB VCI Dragar has a crew of eight and is fitted with the GIAT Industries Dragar turret armed with one 25 mm cannon and one 7.62 mm coaxial machine gun. A total of 620 rounds of 25 mm and 1400 rounds of 7.62 mm ammunition are carried.

VAB VCI Toucan with crew of 11 and fitted with a Toucan 1 turret armed with a 20 mm cannon and a 7.62 mm coaxial machine gun. Totals of 700 rounds of 20 mm and 200 rounds of 7.62 mm ammunition are carried.

VAB/VCI (6 × 6) fitted with Creusot-Loire Industrie T 25 two-man turret armed with 25 mm McDonnell Douglas Helicopter Chain Gun and 7.62 mm coaxial machine gun

This vehicle is in service with the French Air Force as the VIB (Véhicule d'Intervention sur Base). Production complete.

VAB for French Gendarmerie

In late 1988 it was stated that the French Gendarmerie was to purchase an initial batch of 33 VABs which would be ordered at the rate of 11 per year from 1989 to 1991.

VAB Echelon (Repair Vehicle)

This is fitted with welding equipment, work bench and tools for carrying out vehicle repairs in the field and has a crew of three. Armament consists of a Creusot-Loire Industrie TLi 52A turret over the commander's position. This model has been adopted by the French and other armies. Full details of the Renault VAB ECH repair vehicle are given in *Jane's Military Vehicles and Logistics 1991-92* page 45.

VCAC HOT (Anti-tank) Mephisto

This model is now in service with the French Army in its 4 × 4 configuration and is basically the VAB with the Euromissile Mephisto system with four ready to launch HOT missiles. When travelling the launcher is retracted into the hull making the vehicle very difficult to distinguish from the APC version. A further eight missiles are carried in reserve in the hull. This model has a crew of four.

VCAC HOT (Anti-tank) UTM 800

This has been developed for the export market and is a VAB fitted with the Euromissile UTM 800 turret with four ready to launch HOT missiles and a further 16 missiles in reserve in the hull. This model is known to be in service with Qatar, and 18 systems were purchased by Cyprus in 1987 and delivered in 1988. This model has a four-man crew and retains its amphibious capabilities.

VCAC MILAN (Anti-tank) MCT

For trials purposes a VAB (4 × 4) vehicle has been fitted with the Euromissile MILAN Compact Turret (MCT) with two 2000 m range ready-to-launch MILAN ATGWs. This model has a crew of eight.

VAB RATAC and VAB ATILA

Developed to meet the requirements of the French Army, these two variants of the VAB belong to the inventory of the '155 AUF 1' artillery regiments of the armoured divisions.

The VAB RATAC is a mobile firing radar system foı field artillery and is fitted with a roof-mounted 20 km range Doppler radar with the generating set installed on the right side of the hull. It has a five-man crew of whom two are radar operators and one a radio operator. In 1988 it was announced that the French Army would remove its RATAC radars from its AMX VCI vehicles, modernise them and install them on VAB (4 × 4) vehicles.

VAB VMO (security vehicle) of the Abu Dhabi police with Creusot-Loire Industrie CB 80 MG mount, dozer blade and Creusot-Loire Industrie firing ports

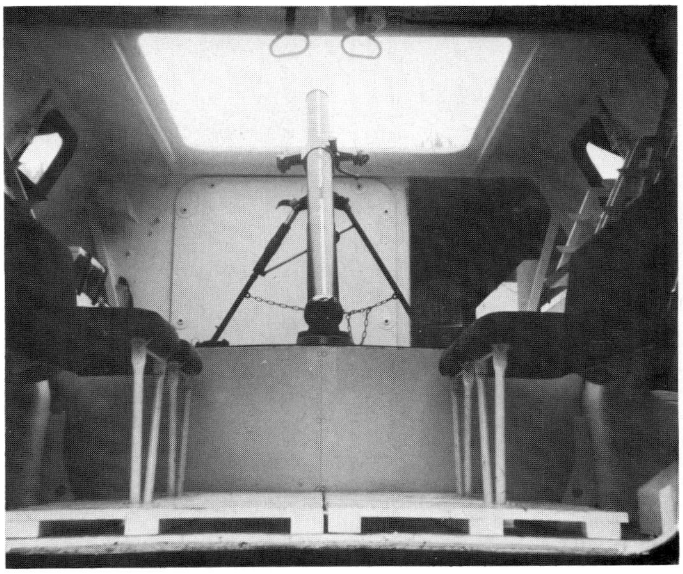

Interior view of VAB fitted with 81 mm mortar firing forwards

The VAB ATILA is a mobile automatic data transmission system in use at various command post levels in artillery regiments and is fitted out as a mobile data teleprocessing centre.

The French Army also uses the VAB RASIT ground surveillance radar system in its reconnaissance squadrons.

VAB PC (Command Vehicle)

This has been developed to meet the requirements of the French Army and is the basic VAB fitted with mapboards and additional communications equipment, with a crew of six. The VAB is also used as a fire direction centre (VAB FDC) and forward observation officer vehicle (VAB FOO) acting as part of the ATILA system and to mount parts of the RITA communications system.

VAB Transmission

This version is fitted with communications equipment and has seats for the operators. It is also used for the RITA Automatic Integrated Transmission Network and normally has a crew of four.

Engineer Vehicle

The French Army is now taking delivery of a version of the VAB modified for use by engineers. This is called the VAB Genie and carries engineer equipment and a rubber dingy on the roof. This model has a crew of nine.

Electronic Warfare Vehicle

The French Army is now taking delivery of an electronic warfare version of the VAB called the Bromure. This has the electronic warfare equipment at the very rear and the operators in the centre. This model has a crew of three.

VAB Sanitaire (Ambulance)

This is unarmed and can carry four stretcher or 10 seated patients, or two stretcher and five seated patients. This version, which has been adopted by the French Army, is fitted with an air-conditioning system and carries first aid equipment.

VMO (Internal Security Vehicle)

This is called the Véhicule de Sécurité and has been developed for export. It can be fitted with a variety of light armament installations over the

VAB (6 × 6) chassis fitted with Dassault Electronique twin 20 mm anti-aircraft turret

VAB VMO (6 × 6) with roof mounted TOI cupola fitted with externally mounted video camera

commander's position including the Creusot-Loire Industrie TLi 52 A armed with a 7.62 mm machine gun or the Creusot-Loire Industrie TLi 52 G turret. Mounted at the front of the hull is a hydraulically operated obstacle-clearing blade, and special protection is fitted against petrol bombs. Mounted in each side of the hull and at the rear is a special firing port with a vision block over it, which allows three of the crew members to use their small arms from inside the vehicle in complete safety. Equipment installed includes a loudspeaker system while a wide range of optional equipment is available including a front-mounted winch, heater, air-conditioning system, radios, and various communications systems. Abu Dhabi has a quantity of these in service in the 6 × 6 configuration with a 7.62 mm GPMG in a CB 80 mount.

VMO (Internal Security Vehicle) with TOI

In 1990 the prototype of a new VMO internal security vehicle was revealed, fitted with a Creusot-Loire Industrie TOI observation and interventional cupola. Thirteen production vehicles have been built for an export customer.

The TOI cupola has manual traverse through a full 360° and can be fitted with a wide range of optional equipment according to mission requirements, including a TV camera with power zoom, white or active spotlight, washers, reclining seat, electrical console, monitors and various grenade dischargers, 7.62 mm machine gun and DREC launchers.

VTM 120 (Mortar Towing Vehicle)

This is called the Véhicule Tracteur de Mortier, is already in service with the French Army and tows a Thomson-Brandt 120 mm mortar. The VTM 120 has a crew of six, a Creusot-Loire Industrie TLi 52 A turret at the front of the hull and carries 70 mortar bombs.

VPM 81 (mortar carrier)

This model is fitted with an 81 mm mortar that is fired through a two-part opening in the roof. Qatar uses this version in an 6 × 6 configuration and armed with a Royal Ordnance 81 mm mortar with a 7.62 mm machine gun turret installed at the front.

VAB NBC Reconnaissance Vehicle

In early 1988 the French Army unveiled a NBC reconnaissance version of the VAB called the VAB Reco. This has been developed at the ARE and is provided with nuclear fallout sensors and a series of devices to collect and analyse chemical agents from inside of the vehicle, a land navigation system is also provided. Two prototypes of the VAB Reco have been built and following successful trials this should enter production in 1992.

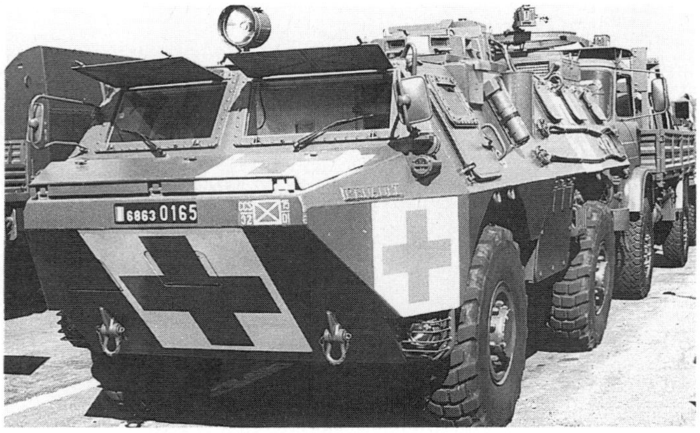

VAB Sanitaire (ambulance) of the French Army (Pierre Touzin)

VCAC HOT (anti-tank) Mephisto of French Army with launcher retracted and armed with 7.62 mm machine gun (Pierre Touzin)

VTM 120 (4 × 4) 120 mm mortar towing vehicle of the French Army (Pierre Touzin)

VAB VDAA Albi SAM system with two missiles in the ready to launch position. This vehicle is also fitted with a machine gun turret from Creusot-Loire Industrie although the machine gun is not fitted

VAB VDAA Santal SAM system with six missiles in the ready to launch position

VAB Recovery

This has been developed as a private venture and is available in both 4 × 4 and 6 × 6 configurations and have a four-man crew. Equipment fitted includes a hydraulic crane with a maximum lifting capacity of 6 tonnes/m, four hydraulic stabilisers which will be lowered to the ground to provide a more stable platform, front-mounted winch with 60 m of cable, flashing light, warning triangle, block and tackle, tow hooks, ladder, while rear compartment is fitted with a work bench, tools, welding equipment and a generator. Armament consists of a 7.62 mm or a 12.7 mm machine gun.

VAB TOW ALT

This is the VAB fitted with the Norwegian Kvaerner Eureka armoured launching turret for TOW ATGWs. A total of two missiles would be carried in the ready to launch position with a further 16 missiles being carried internally, plus a standard ground mount allowing for dismounted operations. This model has yet to enter production.

VAB TOW PIVOT

This is the basic VAB with a pivot-mounted TOW launcher with one missile in ready to launch position and a further 16 being carried internally. This model has yet to enter production.

VAB VDAA TA 20 twin 20 mm SPAAG

Omani Royal Guard has taken delivery of nine VAB (6 × 6) vehicles fitted with the Dassault Electronique one man TA 20 turret armed with twin 20 mm cannon for use in the air defence rôle. Three of the vehicles are also fitted with radar. Full details of this are given in *Jane's Land-Based Air Defence 1991-92* page 65. Production of this system is now complete.

VAB VDAA Santal SAM System

This has been developed specifically for the export market and consists of a VAB (6 × 6) fitted with a two-man power-operated turret with six MATRA SATCP surface-to-air missiles in the firing position with a Dassault Electronique Rodeo 2 radar on the turret rear. This turret was originally developed to meet the requirements of the French Army which was to install it on the Panhard ERC (6 × 6) armoured car chassis, but this system was subsequently cancelled by the French Army.

The VDAA Santal has a three-man crew consisting of the driver, commander/radio operator and gunner. In addition to the six missiles in the ready to launch position a further six missiles are carried internally for manual reloading.

A prototype of the VAB (6 × 6) Santal system was completed in 1990. Although this programme was cancelled by the French Army, the trials programme using one VAB and one ERC chassis was completed and production can start when firm orders are placed.

VAB VDAA Albi SAM System

This has been developed as a private venture and consists of a 6 × 6 or 4 × 4 VAB with a manually operated turret on the roof with two MATRA SATCP surface-to-air missiles in the ready to launch position, with an additional six missiles being carried internally for manual reloading. Turret traverse is 360° with elevation from −10 to +45°. A standard SATCP tripod is carried inside the vehicle to allow missiles to be launched away from the vehicle if required by the tactical situation.

The first prototype of the VAB VDAA Albi SAM system was completed in 1988 and it is under consideration by a number of countries.

VBC 90

Full details of this armoured car which uses components of the 6 × 6 VAB are given in the *Reconnaissance vehicles* section.

SPECIFICATIONS

(4 × 4 version; data in square brackets relate to 6 × 6 model where different)

CREW	2 + 10
CONFIGURATION	4 × 4 [6 × 6]
COMBAT WEIGHT	13 000 [14 200] kg
UNLOADED WEIGHT	10 200 [11 400] kg
POWER-TO-WEIGHT RATIO	16.92 [15.49] hp/tonne
LENGTH	5.98 m
WIDTH	2.49 m
HEIGHT (to hull top)	2.06 m
GROUND CLEARANCE	
axles	0.4 m
hull	0.5 m
TRACK	2.035 m
WHEELBASE	3 m [1.5 + 1.5]
ANGLE OF APPROACH/ DEPARTURE	45°/45°
MAX ROAD SPEED	
5th gear	92 km/h
4th gear	58 km/h
3rd gear	34 km/h
2nd gear	20 km/h
1st gear	12 km/h
reverse	13 km/h
MAX WATER SPEED	7 km/h
FUEL CAPACITY	300 l
MAX ROAD RANGE	1000 km
FORDING	amphibious
GRADIENT (starting from static)	50%
SIDE SLOPE	35%
TRENCH	n/app [1 m]
VERTICAL OBSTACLE	0.5 m
TURNING RADIUS	9 m
ENGINE	Renault MIDS 06.20.45 in-line water-cooled turbo-charged 6-cylinder diesel developing 220 bhp at 2200 rpm. Original engine was a MAN D.2356 HM 72 in-line water-cooled 6-cylinder diesel developing 220 hp at 2200 rpm (standard on VAB until 1983)
TRANSMISSION	Transfluide with 5 forward and 1 reverse gears
STEERING	recirculating ball, hydraulically assisted
TYRES	14.00 × 20 (run-flat)
BRAKES	disc
SUSPENSION	independent (torsion bar and telescopic shock absorbers)
ELECTRICAL SYSTEM	24 V
BATTERIES	4 × 12 V 6TN
ARMAMENT	see text

SPECIFICATIONS

VAB New Generation

(4 × 4 version; data in square brackets relate to 6 × 6 model where different)

CREW	2 + 10
CONFIGURATION	4 × 4 [6 × 6]
COMBAT WEIGHT	13 600 [14 800] kg
POWER-TO-WEIGHT RATIO	18.5 [17] hp/tonne
LENGTH	6.102 m
WIDTH	2.5 m
HEIGHT (to hull top)	2.1 m
GROUND CLEARANCE	
axles	0.4 m
hull	0.5 m
WHEELBASE	3 [1.5 + 1.5] m
ANGLE OF APROACH/ DEPARTURE	45°/45°
MAX ROAD SPEED	110 km/h
MIN ROAD SPEED	4 km/h
FUEL CAPACITY	280 l + four 20 l feeders
ROAD RANGE AT 60 km/h	1000 km
GRADIENT	60%
SIDE SLOPE	40%
TRENCH	n/app [1 m]
VERTICAL OBSTACLE	0. 5 m
ENGINE	Renault MIDR 06.02.26 in-line water-cooled turbocharged diesel developing 250 hp at 2500 rpm
TRANSMISSION	Allison Transmission Division of General Motors Model MT 653 DR automatic with 5 forward and 1 reverse gears, torque converter with lock up clutch
STEERING	hydraulic assisted
TYRES	14.00 × R20 run flat
BRAKES	disc all wheels, dual circuit
SUSPENSION	independent (torsion bar and telescopic shock absorbers)
ELECTRICAL SYSTEM	24 V
BATTERIES	4 × 12 V 6TN

Status: In production. In service with Brunei (26 in IS version ordered in 1988 with first deliveries made in 1990), Central African Republic, Cyprus (127 ordered in 1984/85/86 of which 27 are fitted with Creusot-Loire Industrie T 20 series turrets with a 20 mm cannon and the remaining vehicles fitted with a CB 127 gun ring mount armed with a 12.7 mm M2 HB machine gun, a further 20 additional vehicles were ordered in 1987 of which 18 were fitted with the UTM 800 ATGW turret), France, Ivory Coast (13 6 × 6), Lebanon (5 plus a further batch of 90 delivered in 1983-84), Mauritius (11 6 × 6), Morocco (394 6 × 6 including 12 mortar towing, 68 command vehicles, 20 ECH repair vehicles, 49 VCI Toucan, remainder being APCs or internal security vehicles), Oman (23 VAB 6 × 6 and 9 VBC 90), Qatar (162 including 24 UTM-800 and 81 mm mortar), United Arab Emirates (20 6 × 6 inc VMO) and other undisclosed countries, including some in the Middle East.

Manufacturers: Renault Véhicules Industriels and Creusot-Loire Industrie under the leadership of Renault VI
Marketing by SMS Renault VI-Creusot-Loire Industrie
316 Bureaux de la Colline
92213 Saint-Cloud, Cedex
France
Telephone: (1) 46 02 80 33
Telex: 250 662

Berliet VXB-170 Armoured Personnel Carrier

Development

In the mid-1960s, as a private venture, Berliet started the development of a 4 × 4 armoured personnel carrier which would use automotive components from its wide range of tactical military vehicles then in production. The first prototype of this vehicle, called the BL-12, was completed in March 1968 and shown for the first time at the 1969 Satory Exhibition of Military Equipment. A second vehicle was built for the French Army but it did not place a production order for the BL-12.

The French Gendarmerie placed an order for five slightly modified vehicles which were delivered from 1971. Following successful trials with these vehicles the French Gendarmerie placed an initial order for 50 vehicles for internal security use, the first of which were completed at Berliet's Bourge facility in 1973. The total Gendarmerie requirement was for about 400 vehicles but only 155 were ordered and delivered.

The VXB-170 (Véhicule Blindé à Vocations Multiples, or Multi-Purpose Armoured Vehicle) was one of the three entries in the competition for the VAB vehicle for the French Army. This competition was however won by the Renault (4 × 4 and 6 × 6) range of vehicles. In 1975 Berliet became part of the Renault Group which also controlled Saviem so after completion of the VXB order for the French Gendarmerie, Gabon and Senegal, production stopped.

Description

The hull of the VXB is made of all-welded steel. The driver is seated at the front of the hull on the left side with the commander to his right. In front of the driver and commander is a large windscreen which, when in action, is covered by a two-piece armoured screen hinged at the top. To the left of the driver and right of the commander is a smaller window, covered by a flap hinged at the bottom when in action. Over the top of the driver's position are three observation periscopes for use when the windscreen is covered and over the commander's position is a single piece hatch cover that opens to the right with a swivel-mounted periscope for all-round observation.

The main armament is installed in the centre of the vehicle behind the driver's and commander's position. There is an entry door in each side of the hull and a third door at the rear of the hull on the right side. In addition to the roof hatch on which the main armament is installed there are three other roof hatches: two in line with the doors in the sides of the hull, one of which opens forwards and the other to the rear and a third in the aisle which connects the personnel compartment with the door in the rear of the hull.

In the original model there were seven firing ports, two in the left side (one of which was in the door), four in the right side (of which one was in the door) and one in the door in the rear of the vehicle. Vehicles delivered to the French Gendarmerie have five bullet-proof windows,

Berliet VXB-170 APC with obstacle clearing blade (Pierre Touzin)

Berliet VXB-170 APC with one-man turret (Pierre Touzin)

two in each side of the hull (of which one is in the door) and one in the door in the rear of the vehicle. These can be opened to allow the troops to fire their weapons from inside the vehicle. The Gendarmerie vehicles also have a single piece hatch cover over the driver's rather than the commander's position and there is no hatch cover over the aisle.

The engine is at the rear of the hull on the left side and access panels are provided in the top and rear of the hull for maintenance.

Power is transmitted from the engine to the gearbox and then by a propeller shaft to the two-speed transfer box in the centre of the vehicle under the floor. Power is then taken from the transfer box to the front and rear axles via propeller shafts. The front and rear wheels have planetary reduction gears and the rear axle has a pneumatically controlled interwheel differential lock. The suspension at each wheel station consists of a concentric helical spring and a hydraulic shock absorber.

The basic vehicle is fully amphibious being propelled in the water by its wheels. Optional equipment included a heater, hydraulic obstacle-clearing blade at the front of the hull (as fitted to French Gendarmerie vehicles), infra-red or passive night vision equipment, NBC system, winch with a capacity of 3500 to 4500 kg (and 60 m of cable) and bullet-proof tyres.

The vehicles delivered to the French Gendarmerie were originally fitted with a simple pintle-mounted machine gun but some were subsequently fitted with the SAMM S365.3 (now designated the BTM 103) light turret which is armed with a 7.62 mm machine gun and a 40 mm grenade launcher. Elevation limits of the weapons are from -15 to +60°.

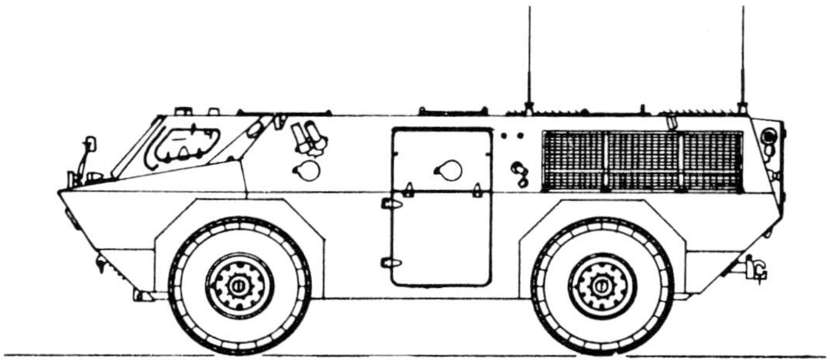

Berliet VXB-170 APC without armament

Status: Production complete. In service with France (Gendarmerie), Gabon and Senegal.

Manufacturer: Berliet, Bourge
Société des Automobiles Berliet
160 Boulevard de Verdun
92400 Courbevoie
France

SPECIFICATIONS		MAX SPEED		TRANSFER BOX	2-speed
CREW	1 + 11	road	85 km/h	STEERING	hydraulic power-assisted
CONFIGURATION	4 × 4	water	4 km/h	SUSPENSION	helical springs and hydraulic shock absorbers
COMBAT WEIGHT	12 700 kg	FUEL CAPACITY	220 l	TYRES	14.00 × 20
UNLOADED WEIGHT	9800 kg	MAX ROAD RANGE (at 50 km/h)	750 km	ELECTRICAL SYSTEM	24 V
POWER-TO-WEIGHT RATIO	13.38 hp/tonne	FORDING	amphibious	BATTERIES	4 × 12 V
LENGTH	5.99 m	GRADIENT	60%	ARMAMENT	1 × 7.62 mm MG
WIDTH	2.5 m	SIDE SLOPE	30%	SMOKE-LAYING EQUIPMENT	2 smoke dischargers either side of hull (optional)
HEIGHT (without armament)	2.05 m	VERTICAL OBSTACLE	0.3 m	ARMOUR	7 mm
GROUND CLEARANCE (transfer case)	0.45 m	TURNING RADIUS	7.8 m		
TRACK	2.04 m	ENGINE	Berliet model V 800 M V-8 diesel developing 170 hp at 3000 rpm		
WHEELBASE	3 m	TRANSMISSION	pre-selective with 6 forward and 1 reverse gears		
ANGLE OF APPROACH/ DEPARTURE	45°/45°				

Panhard VCR TT 2 Armoured Personnel Carrier

Development

The VCR (Véhicule de Combat à Roues) TT (Transport de Troupes) 2 has been designed as a private venture by Panhard as a successor to the original VCR/TT which entered production in 1978.

The VCR TT 2 shares many automotive components, for example the powerpack and suspension, with the Panhard ERC 90 F4 Sagaie 2 armoured car which was unveiled together with the VCR TT 2 for the first time in June 1985. At that time three prototypes of the VCR TT 2 had been built, but by May 1991 production of this vehicle had yet to commence.

Description

The hull of the Panhard VCR TT 2 is of all-welded steel construction that provides protection from small arms fire and shell splinters. The driver sits at the front of the vehicle in the centre and has three periscopes for forward observation, the centre one of which is mounted in a forward-opening hatch for driving in the head-out position.

The engine compartment is to the immediate rear of the driver. The VCR TT 2 is available powered by two Peugeot XD 3T diesel engines developing a total of 196 hp at 4150 rpm or two Peugeot PRV V6 petrol engines developing a total of 290 hp at 5500 rpm. These are each coupled to a ZF 4 HP 22 fully automatic transmission with four forward and one reverse gears which are in turn connected to a Panhard two-speed, road and cross-country, transfer box. Power is transmitted to all road wheels via drive shafts in the armoured hull.

Steering is power assisted on the front two road wheels and when driving on roads the centre two road wheels are normally raised clear of the ground.

Each road wheel is independent and mounted on trailing arms with helicoidal spring suspension. The front and rear wheels have hydro-pneumatic suspension units while the centre ones have hydraulic jacks.

Panhard VCR TT 2 APC fitted with SAMM BTE 105 turret armed with 12.7 mm machine gun

Panhard VCR TT 2 from rear without armament and showing propellers under hull rear

There is a single door in each side of the hull, to the rear of the first road wheel station, and a large door in the hull rear that opens to the left. This has two Creusot-Loire Industrie firing ports each with an observation block above.

Six troops sit down either side of the hull facing each other with four upward-opening hatches in the upper part of each side of the troop compartment. Over the top of the troop compartment are four roof hatches, two either side, that are hinged on the outside.

The main armament installation is mounted on the roof to the rear of the engine compartment. If for example a SAMM BTE 105 cupola is fitted armed with a 12.7 mm M2 HB machine gun, then 12 troops can be carried, but when heavier two-man turrets are mounted, for example the SAMM TTB 125 armed with a 25 mm cannon and a 7.62 mm machine gun, then the number of troops is reduced.

Optional equipment includes an air-conditioning system, NBC equipment, front-mounted winch and various night vision devices.

The VCR TT 2 is fully amphibious, propelled in the water by two steerable propellers mounted under the hull rear.

Variants
In addition to being fitted with a wide range of armament installations, Panhard has suggested a number of specialised versions of the VCR TT 2, these include command post, ambulance and workshop.

Status: Development complete. Ready for production.

Manufacturer: Société de Constructions Mécaniques Panhard et Levassor
18 avenue d'Ivry
75621 Paris
France
Telephone: 33 (1) 40 77 40 00
Telex: 270 276

SPECIFICATIONS	
CREW	2 + 10 - 12
CONFIGURATION	6 × 6
COMBAT WEIGHT	9600 kg
POWER-TO-WEIGHT RATIO	30.2 hp/tonne (petrol engines)
LENGTH	6.14 m
WIDTH	2.7 m
HEIGHT	2.27 m
GROUND CLEARANCE	0.37 m
TRACK	2.35 m
WHEELBASE	2.6 m + 1.09 m
MAX SPEED	
road	110 km/h
water	7.2 km/h
ROAD RANGE	
petrol engines	600 km
diesel engines	1000 km
GRADIENT	over 50%
SIDE SLOPE	30%
VERTICAL OBSTACLE	0.8 m
TRENCH	0.8 m
ENGINES	2 × Peugeot V6 PRV petrol engines developing a total of 290 hp or two Peugeot XD 3T diesels developing a total of 196 hp
TRANSMISSIONS	2 × ZF 4 HP 22 automatic, 4 forward and 1 reverse gears
TRANSFER BOX	Panhard, 2-speed
BRAKES	
main	hydraulic, disc
parking	mechanical
TYRES	13.00 × 16
ELECTRICAL SYSTEM	24 V

Panhard VCR Armoured Personnel Carrier

Development
The VCR (Véhicule de Combat à Roues) range of armoured personnel carriers was developed by Panhard as a private venture from 1975 and is aimed specifically at the export market. The range was first shown in 1977 at the Satory Exhibition of Military Equipment and entered production at Panhard's Marolles factory the following year. First production vehicles, equipped with the Euromissile UTM 800 turret, were completed in 1979. The VCR range of armoured personnel carriers uses 95 per cent of the automotive components of the Panhard ERC range of 6 × 6 armoured cars which was also shown for the first time in 1977. Latest production vehicles have a redesigned hull, troop compartment roof 400 mm above the commander's position, external stowage boxes, more internal hull space, twin doors at the hull rear and fuel tanks repositioned from inside the troop compartment to under the floor.

In June 1983 Panhard issued new specifications for the VCR range of 6 × 6 APCs. These showed that there had been slight changes to the vehicle with the wheelbase increased from 1.425 + 1.425 m to 1.66 + 1.425 m. This has made the hull slightly longer and increased the overall weight of the vehicle. The specifications for the VCR/TH anti-tank version remain unchanged except for an increase in loaded and empty weights.

By January 1990 total production of the Panhard VCR range of 6 × 6 APCs amounted to 262 vehicles, all for export.

Description
The all-welded steel hull of the VCR/TT (Transport de Troupes) protects the crew against small arms fire and shell splinters. The floor of the hull consists of two plates welded together to form a flat V which helps the vehicle to slide off obstacles and also offers resistance to damage from mines.

Panhard VCR/TT fitted with electronic warfare equipment (front) and VCR/IS (rear)

The driver sits at the front of the hull in the centre and has a single piece hatch cover that opens to his right with three integral periscopes for forward observation, the centre one replaceable by an infra-red or image intensification periscope for driving at night.

The commander sits to the rear of the driver on the left side at a higher level and has a single piece hatch cover that opens forwards and seven periscopes around his position.

The engine is to the rear of the driver on the right side and power is transmitted through the single clutch plate to a transversely mounted Panhard gearbox (as fitted to the AML armoured car and M3 APC) with

Panhard VCR/TT for UAE with wider hull and twin rear doors (Christopher F Foss)

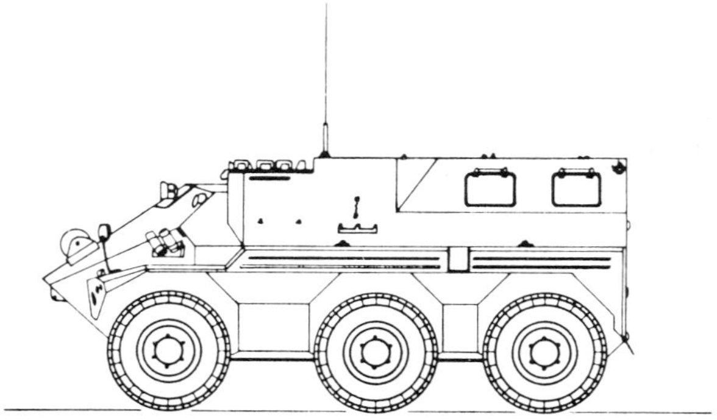

Panhard VCR/TT built to latest production standard without armament

six forward and one reverse gears and incorporating a limited slip differential. Drive from the bevel gears to the road wheels is by a train of gears housed within trailing arms which independently locate the wheels.

The front and rear wheels are sprung by single coil springs with telescopic hydraulic dampers. The centre wheels, which are raised off the ground when on roads, are sprung by hydro-pneumatic units. The centre wheels are powered even when raised clear of the road surface. All wheels have disc brakes. Only the front wheels are steered, and steering is hydraulically assisted. The low-pressure tyres are fitted with Hutchinson cellular inner tubes which enable the vehicle to travel for at least 100 km at 30 km/h when the tyres have been punctured by bullets.

The 10 infantrymen are seated on bench seats: five along each side of the hull facing each other. They enter and leave the vehicle by a large door in the rear of the hull which is hinged on the right side and has a single firing port. Two observation/firing hatches in each side of the troop compartment open upwards and can be locked open. The VCR/TTs built for the United Arab Emirates have a wider hull with double doors at the rear to give increased leg room.

The armament is mounted to the rear of the commander's position in the roof. When armed with a 7.62 mm machine gun the VCR/TT can carry nine men plus its crew of three (commander, driver and gunner, who sits on one of the bench seats when not manning the machine gun) but when heavy armament is installed, eg a 20 mm cannon, only six infantrymen can be carried plus a crew of three. The VCR/TT can be fitted with a variety of armament installations including a SAMM BTM 263 turret with 60 mm mortar and 7.62 mm machine gun, Creusot-Loire Industrie STB ring mount with a shield (which forms the hatch cover when not in use) and a 7.62 mm machine gun, Creusot-Loire Industrie CB rotating gun ring with a 7.62 mm machine gun, Creusot-Loire Industrie turret with twin 7.62 mm machine guns, Creusot-Loire Industrie gun ring Model CB 20 with 20 mm cannon, Creusot-Loire Industrie gun ring with Thomson-Brandt 60 mm breech-loaded mortar, Euromissile MCT MILAN turret with two ready to launch MILAN ATGWs, Creusot-Loire Industrie one-man turret Model TL 20 armed with a 20 mm cannon and a 7.62 mm machine gun, or a GIAT Industries Toucan I turret armed with a 20 mm cannon, 7.62 mm machine gun and four smoke dischargers. There is also a single piece hatch cover in the roof of the troop compartment at the rear which opens towards the front. A 7.62 mm machine gun on a Creusot-Loire Industrie STR rail mount can be mounted behind it. Two smoke dischargers can be mounted on either side of the hull if required.

Panhard VCR/TT with front-mounted winch and pintle-mounted Euromissile MILAN ATGW

Panhard VCR/TT chassis with GIAT Toucan I turret armed with 20 mm cannon, 7.62 mm machine gun and smoke dischargers

The VCR is fully amphibious being propelled in the water by its wheels, with steering by turning the front road wheels as for land operations. Optional equipment includes a front-mounted winch with 60 m of cable and a capacity of 3000 kg, NBC system, air-conditioning system and passive night vision equipment.

Variants

VCR/AT Repair Vehicle

The VCR/AT (Atelier Vehicle) is the basic vehicle modified to undertake repairs in the field. It has no recovery capability although it can tow other vehicles.

Equipment installed includes a generator, tools, work benches, inspection lamp, tow bars and tow cables. A penthouse can be erected at the rear to enable repairs to be carried out in bad weather. A block and tackle can be erected at the rear allowing the VCR/AT to carry out engine and other component changes. This model is used by the United Arab Emirates.

VCR/AA Anti-aircraft Vehicle

This was unveiled in 1984 and is fitted with the Bofors RBS 70 surface-to-air missile system. The VCR/AA weighs 7500 kg and has a crew of three. Mounted in the rear troop compartment is the RBS 70 missile launcher with one missile ready to launch and a further eight in reserve. For travelling the launcher is safely retracted within the hull making the vehicle almost indistinguishable from other members of the VCR family. This model has yet to enter production. Additional details of this system are given in *Jane's Land-Based Air Defence 1991-92* page 35.

VCR/TH (Tourelle Hot) Anti-tank Vehicle

This is the basic VCR/TT fitted with the Euromissile UTM 800 HOT turret mounted on a raised plinth to the rear of the commander's position. The turret has four ready to launch HOT missiles and another 10 missiles carried in reserve in the hull are loaded via a hatch in the right rear of the roof, which opens to the right. The turret has an optical sight with a magnification of ×3 (18° field-of-view) and ×12 (5° field-of-view), elevation being +10°, depression -6° and turret traverse 360°.

Mounted on the roof at the rear on the left is a GIAT Industries Mascot remote-controlled 7.62 mm machine gun, which can be elevated from -10 to +50° and traversed through a full 360°. Two hundred rounds of ready-use ammunition are provided for the weapon. Two smoke dischargers are mounted on the front of the hull at either side. Panhard has delivered 100 of these vehicles to Iraq.

VCR/IS (Intervention Sanitaire) Ambulance

The VCR/IS (Véhicule Blindé d'Intervention Sanitaire) is the ambulance member of the family and has a much higher hull behind the commander than the basic VCR. The vehicle has a crew of three (commander, driver and orderly) and can carry four stretcher or six sitting and two stretcher patients.

The casualties enter the vehicle through a door at the rear of the hull that opens to the left and has a single vision port. Unlike the basic APC, there are no observation flaps in the sides of the hull.

Equipment carried includes folding stretchers, cupboards and instrument storage lockers, water tank, refrigerator and breathing and blood transfusion equipment. If required a tent, which is stowed in boxes on either side of the hull, can be erected at the rear of the hull. This model is used by the United Arab Emirates.

VCR/PC (Poste de Commandement) Command Vehicle

This is basically the VCR/TT adapted for use as a command vehicle with mapboards, tables and additional communications equipment. It is probable that a command version based on the hull of the VCR/IS ambulance will be developed as this has more headroom than the basic vehicle. This model is used by the United Arab Emirates.

Status: Production as required. In service with Iraq (100 VCR/TH), Mexico (2 VCR/PC) and the United Arab Emirates (Abu Dhabi) (82 VCR/TT, VCR/AT, VCR/IS and VCR/PC). Some of the Iraq Panhard VCR/TH vehicles were destroyed during the 1991 Middle East conflict.

Manufacturer: Société de Constructions Mécaniques Panhard et Levassor
18 avenue d'Ivry
75621 Paris
France
Telephone: 33(1) 40 77 40 00
Telex: 270 276

Panhard VCR/TH with Euromissile UTM 800 turret with four ready to launch HOT ATGWs and Mascot 7.62 mm remote-controlled machine gun mount at rear

Panhard VCR/IS with tent erected at rear of vehicle

SPECIFICATIONS

Model	**VCR/TT** (APC)	**VCR/PC** (command)	**VCR/TH** (anti-tank)	**VCR/AT** (repair)	**VCR/IS** (ambulance)
CREW	3 + 9	6	4	4	3 + 4
CONFIGURATION	6 × 6	6 × 6	6 × 6	6 × 6	6 × 6
COMBAT WEIGHT	7900 kg	7900 kg	7800 kg	8200 kg	8100 kg
UNLOADED WEIGHT	6800 kg	7300 kg	7000 kg	7200 kg	7500 kg
POWER-TO-WEIGHT RATIO	18.35 hp/tonne	18.35 hp/tonne	18.58 hp/tonne	17.68 hp/tonne	17.90 hp/tonne
LENGTH	4.875 m	4.875 m	4.565 m	4.875 m	4.875 m
WIDTH	2.50 m	2.50 m	2.50 m	2.50 m	2.50 m
HEIGHT					
including armament	2.56 m (7.62 mm MG)	2.56 m	2.968 m	2.71 m	n/app
to hull top	2.13 m	2.13 m	2.03 m	2.13 m	2.53 m
GROUND CLEARANCE					
4 wheels	0.315 m	0.315 m	0.315 m	0.315 m	0.315 m
6 wheels	0.375 m	0.375 m	0.375 m	0.375 m	0.375 m
TRACK	2.115 m	2.115 m	2.115 m	2.115 m	2.115 m
WHEELBASE	1.66 m + 1.425 m	1.66 m + 1.425 m	1.425 m + 1.425 m	1.66 m + 1.425 m	1.66 m + 1.425 m
ANGLE OF APPROACH/ DEPARTURE	53°/45°	53°/45°	53°/45°	53°/45°	53°/45°
MAX SPEED					
road	90 km/h	90 km/h	90 km/h	90 km/h	90 km/h
water	4 km/h	4 km/h	4 km/h	4 km/h	4 km/h
MAX ROAD RANGE	700 km	700 km	700 km	700 km	700 km
FUEL CAPACITY	242 l	242 l	242 l	242 l	242 l
FORDING	amphibious	amphibious	amphibious	amphibious	amphibious
GRADIENT	60%	60%	60%	60%	60%
SIDE SLOPE	30%	30%	30%	30%	30%
VERTICAL OBSTACLE	0.8 m	0.8 m	0.8 m	0.8 m	0.8 m
TRENCH	1.1 m	1.1 m	1.1 m	1.1 m	1.1 m
ENGINE	Peugeot PRV V-6 petrol developing 145 hp at 5500 rpm				
TRANSMISSION	Panhard with 6 forward and 1 reverse gears				
STEERING	power-assisted	power-assisted	power-assisted	power-assisted	power-assisted
CLUTCH	hydraulic control	hydraulic control	hydraulic control	hydraulic control	hydraulic control
TYRES	11.00 × 16	11.00 × 16	11.00 × 16	11.00 × 16	11.00 × 16
ELECTRICAL SYSTEM	24 V	24 V	24 V	24 V	24 V
ARMAMENT	1 × 7.62 mm MG, or 1 × 20 mm cannon (see text)	optional	launcher with 4 HOT ATGWs, plus remote-controlled 7.62 mm MG	optional	none
SMOKE DISCHARGERS	optional	2 × 2	2 × 2	2 × 2	2 × 2
ARMOUR (estimated)	8-12 mm	8-12 mm	8-12 mm	8-12 mm	8-12 mm

Panhard VCR (4 × 4) Armoured Personnel Carrier

Development

This vehicle has been developed as a private venture and was announced in June 1979. It is essentially a 4 × 4 version of the Panhard VCR (6 × 6) APC which is already in production and is fully described in this section. Ninety-five per cent of the automotive components of the 4 × 4 VCR are identical to those of the 6 × 6 VCR (Véhicule de Combat à Roues).

Description

The only major difference from the 6 × 6 model is that the 4 × 4 has a slightly different roof arrangement. The main armament installation is behind the commander's position and a variety of armaments can be mounted there including a Creusot-Loire Industrie STB ring mount with a shield (which forms the hatch cover when not in use) and a 7.62 mm machine gun, Creusot-Loire Industrie CB rotating gun ring with a 7.62 mm machine gun, Creusot-Loire Industrie turret with twin 7.62 mm machine guns, Creusot-Loire Industrie gun ring model CB 20 with a 20 mm cannon, Creusot-Loire Industrie gun ring with a Thomson-Brandt 60 mm breech-loaded mortar and the Euromissile MCT MILAN turret with two MILAN ATGWs ready to launch. There is also a single piece front-opening hatch cover in the roof of the troop compartment at the rear behind which a 7.62 mm machine gun on a Creusot-Loire Industrie STR rail mount can be mounted.

The VCR is fully amphibious being propelled in the water by two water jets, one either side of the hull, which have deflectors for steering when afloat.

Variants

In addition to being used as an armoured personnel carrier, the VCR (4 × 4) can be used for other roles such as command, ambulance and cargo carrier. A VCR (4 × 4) with a redesigned hull fitted with a Euromissile Mephisto launcher system with four HOT ATGWs and another eight missiles carried in reserve was shown at the 1979 Paris Air Show. This model weighs 7800 kg and has a crew of three consisting of commander, gunner/loader and driver.

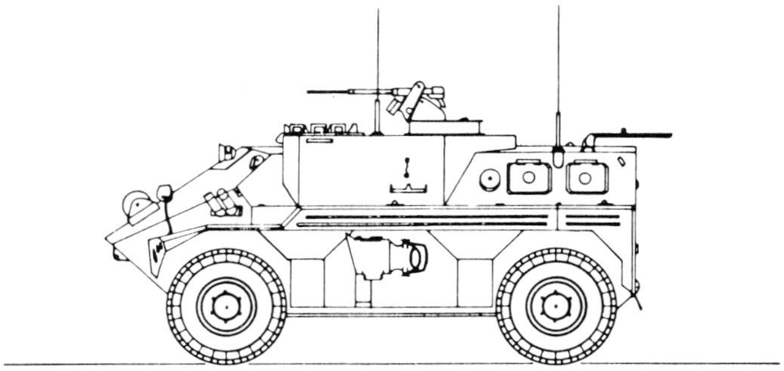

Panhard VCR (4 × 4) APC armed with Creusot-Loire Industrie CB rotating gun ring with 7.62 mm machine gun

Panhard VCR/TT with trim boards erected at front of hull after amphibious operations

Panhard VCR (4 × 4) fitted with water jets showing amphibious capabilities

SPECIFICATIONS	
CREW	3 + 9
CONFIGURATION	4 × 4
COMBAT WEIGHT	7800 kg
UNLOADED WEIGHT	6800 kg
POWER-TO-WEIGHT RATIO	19.87 hp/tonne
LENGTH	4.875 m
WIDTH	2.478 m
HEIGHT	
with 7.62 mm turret	2.694 m
to hull top	2.13 m

GROUND CLEARANCE	
max	0.37 m
min	0.315 m
TRACK	2.16 m
WHEELBASE	3.085 m
MAX SPEED	
road	100 km/h
water	7.2 km/h
FUEL CAPACITY	242 l
MAX ROAD RANGE	800 km
FORDING	amphibious
GRADIENT	60%
SIDE SLOPE	30%

VERTICAL OBSTACLE	0.4 m
ENGINE	Peugeot V-6 petrol developing 155 hp at 5500 rpm
TRANSMISSION	Panhard manual with 6 forward and 1 reverse gears
STEERING	hydraulically assisted
TYRES	11.00 × 16
ELECTRICAL SYSTEM	24 V
ARMAMENT	see text
ARMOUR (estimated)	8-12 mm

Status: Production as required. Argentina has taken delivery of 24 Panhard VCR (4 × 4) APCs.

Manufacturer: Société de Constructions Mécaniques Panhard et Levassor
18 avenue d'Ivry
75621 Paris
France
Telephone: 33 (1) 40 77 40 00
Telex: 270 276

Panhard Buffalo Armoured Personnel Carrier

Development

The Buffalo (4 × 4) APC has been developed as a private venture by Société de Constructions Mécaniques Panhard and Levassor and was shown for the first time in mid-1985.

Early in 1986 it was announced that the Buffalo would replace the M3 in production. The Buffalo is essentially a much improved version of the original M3 described in the following entry but with additional external stowage, slightly longer wheelbase and a choice of a V-6 petrol engine developing 145 hp or a diesel engine developing 95 hp. Both these engines are also used in other members of the Panhard range of 4 × 4 and 6 × 6 armoured vehicles.

By January 1991 total production of the Buffalo and M3 APC amounted to 1222 vehicles, all of which were for export.

Description

The hull of the Buffalo is of all-welded steel armour construction with the driver seated at the front of the hull, the engine to his immediate rear and the troop compartment at the rear of the hull. The troop compartment has a single large door in each side of the hull and twin doors in the hull rear. The spare wheel and tyre are carried on the right door.

Extensive external stowage is provided above the wheels and in the event of the vehicle running over a mine this will blow off so maintaining hull integrity.

A variety of light weapon stations can be mounted on the roof of the Buffalo to the rear of the engine, for example a Hispano-Suiza F135 Puma cupola with an externally mounted 7.62 mm machine gun. There

Panhard Buffalo armoured personnel carrier from above without armament installed

Panhard Buffalo APC with front-mounted obstacle clearing blade for use in IS role

is an additional hatch in the roof of the vehicle at the rear and three hatches along each side of the troop compartment which allow some of the troops to fire their weapons from within the vehicle.

Optional equipment includes a front-mounted winch and an air-conditioning system. The basic vehicle is fully amphibious propelled in the water by its wheels.

Variants

In addition to being used as an APC, the Buffalo can also be used as an internal security vehicle with a hydraulically operated obstacle clearing blade at the front of the hull, as a command vehicle with extensive communications equipment installed, an 81 mm mortar carrier, workshop and ambulance.

Panhard Buffalo armoured personnel carrier from rear without armament installed

SPECIFICATIONS	
CREW	2 + 10
CONFIGURATION	4 × 4
COMBAT WEIGHT	6600 kg
POWER-TO-WEIGHT RATIO	21.96 hp/tonne (PRV engine)
LENGTH	4.585 m
WIDTH	2.4 m
HEIGHT (without armament)	2 m
GROUND CLEARANCE	0.33 m
TRACK	
front	2.06 m
rear	2.08 m
WHEELBASE	2.835 m
MAX SPEED	
road	90 km/h
water	3.6 km/h
ROAD RANGE	600 km
FORDING	amphibious
GRADIENT	60%
SIDE SLOPE	30%
VERTICAL OBSTACLE	0.4 m
ENGINE	Peugeot V-6 PRV water-cooled petrol developing 145 hp at 5500 rpm or Peugeot XD 3T turbo-charged water-cooled diesel developing 95 hp at 4150 rpm
TRANSMISSION	manual with 6 forward and 1 reverse gears
SUSPENSION	independent, coil and spring with hydro-pneumatic shock absorbers acting on suspension trailing arm of wheel mechanism
TYRES	11.00 × 16
BRAKES	
main	hydraulic, dual circuit
parking	handbrake operating on gearbox output shaft
ELECTRICAL SYSTEM	24 V

Status: In production. In service with Benin (one in command post version), Columbia and Rwanda.

Manufacturer: Société de Constructions Mécaniques Panhard et Levassor
18 avenue d'Ivry
75621 Paris
France
Telephone: 33 (1) 40 77 40 00
Telex: 270 276

Panhard M3 Armoured Personnel Carrier

Development

The M3 armoured personnel carrier was designed as a private venture by Panhard with the first prototype being completed in 1969. This had square-cut sides with a single door in each side of the hull and twin doors in the hull rear. It was intended that each door would have a single circular firing port. Mounted on the top of the hull was a Creusot-Loire CAFL 38 turret armed with a 7.5 mm machine gun.

The first production vehicle, with a redesigned hull incorporating three hatches either side of the troop compartment, was completed in 1971. Ninety-five per cent of the Panhard M3's automotive components are identical to those of the Panhard AML series of light armoured vehicles of which over 4800 have been built. The Panhard M3 APC has now been replaced in production by the Buffalo fully described in the previous entry.

By January 1991 total production of the Buffalo and M3 APC amounted to 1222 vehicles, all of which were for export.

Description

The hull of the M3 is made of all-welded steel with the driver seated at the front of the vehicle. He has a single piece hatch cover that opens to the right in which there are three periscopes, the centre one replaceable by an infra-red or image intensification periscope for night driving.

The engine is immediately behind the driver with the air-inlet louvres above and to the rear of the driver, air-outlet louvres on either side of the roof and the two exhaust pipes running along the top of the hull, one on each side.

The gear casing is crosswise and consists of two gearboxes in one coupled on both sides of the bevel pinion. The low-range gearbox comprises two low gears, top and one reverse for use in rough country and the high-range gearbox has three low ranges and one overdrive. When the low-range gearbox is in direct drive, the four ratios of the high-range box command the four upper gears of the range (6th, 5th, 4th and 3rd).

Panhard M3 APC fitted with GIAT Industries Toucan I turret armed with 20 mm cannon and 7.62 mm machine gun (ECP Armées)

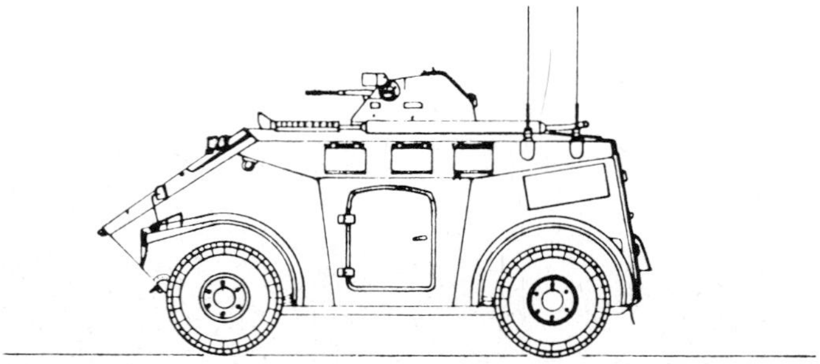

Panhard M3 APC fitted with Creusot-Loire Industrie TL 2i series turret armed with twin 7.62 mm machine guns

Three Panhard M3/VDA twin 20 mm self-propelled anti-aircraft gun systems deployed in the firing position, the vehicle in the centre being fitted with a radar system to provide target information to the other two vehicles.

Drive is transmitted from the main gearbox to the two lateral boxes, which transmit power to the front wheels by pinions and to the rear wheels by drive shafts that run along the inside of the hull.

Ball type differentials in the gearbox and in each of the rear lateral transfer boxes prevent wheel slip. The suspension at each wheel station consists of coil springs and hydro-pneumatic shock absorbers acting on the suspension arms of the wheel mechanism. The tyres have puncture-proof Hutchinson inner tubes.

There are four doors in the M3, one in each side of the hull and two in the rear. The rear doors both have a firing port. Along the upper part of each side of the hull are three hatches hinged at the top that can be locked open. When being used as an APC a maximum of 10 men can be carried in addition to the crew of two, or 1360 kg of cargo.

There are two circular hatches in the roof, one behind the engine compartment and a second one at the rear with a single piece hatch cover that opens forwards and is normally fitted with a Creusot-Loire Industrie STR rail armed with a 7.62 mm machine gun. A wide range of armament installations can be mounted on the forward hatch, brief details of which follow.

Creusot-Loire Industrie STB rotary support shield armed with 7.62 mm machine gun, elevation +45°, depression -15°, total traverse 360°.
Creusot-Loire Industrie CB rotating gun ring with 7.62 mm machine gun, elevation +45°, depression -15°, total traverse 360°.
Creusot-Loire Industrie CAFL 38 turret with 7.62 mm machine gun, elevation +45°, depression -15°, total traverse 360°.
Creusot-Loire Industrie TLi G turret armed with 7.62 mm machine gun and 40 mm smoke discharger, elevation +55°, depression -13°, total traverse 360°.
Creusot-Loire Industrie TL 2i turret armed with twin 7.62 mm machine guns with an elevation of +55°, depression -13°, total traverse 360°.

Panhard M3 APC fitted with one-man Creusot-Loire Industrie TL 2i series turret but without 7.62 mm machine guns installed (T J Gander)

Panhard M3/VLA engineer vehicle with obstacle-clearing blade lowered at front of hull

Creusot-Loire Industrie CB 127 gun ring shield armed with 12.7 mm (0.50) machine gun, elevation +65°, depression -15°, traverse 360°.
Creusot-Loire Industrie CB 20 gun ring shield armed with 20 mm cannon, elevation +65°, depression -15°, total traverse 360°.
Creusot-Loire Industrie gun ring shield with 60 mm Thomson Brandt breech-loaded mortar with an elevation of +75°, depression of -15° and a total traverse of 360°.
Euromissile MCT turret with two ready-to-launch MILAN ATGWs.
GIAT Toucan I turret armed with a 20 mm cannon and a 7.62 mm machine gun.

The basic Panhard M3 is fully amphibious without preparation, being propelled in the water by its wheels, steering when afloat by turning the front wheels as on land. Optional equipment includes an air-conditioning system and smoke dischargers.

Variants

M3/VDA Anti-aircraft Vehicle

The M3/VDA (Véhicule de Défense Antiaérienne) is a joint development for the export market by Panhard (responsible for the chassis) and Dassault Electronique (responsible for the turret). It essentially consists of a modified Panhard M3 (4×4) chassis fitted with the Dassault Electronique turret armed with twin 20 mm cannon and a Galileo sight with the option of an Electronique Serge Dassault RA-20 series radar on the turret rear. One M3/VDA with the radar can control two other systems without the radar.

The first prototype of the Panhard M3/VDA was completed in December 1973 and following trials, production commenced in April 1975. It is known to be in service with the Ivory Coast (6), United Arab Emirates (Abu Dhabi 42) and one unidentified country.

When in the firing position, four out-riggers are lowered to the ground to provide a more stable firing platform.

Full details of this system are given in *Jane's Land-Based Air Defence 1991-92* page 65.

M3/VAT Repair Vehicle

This model has a crew of five (commander, driver and three mechanics) and equipment fitted includes a pulley block with struts and tackle, cutting equipment, generator, benches, vice, tow bars, tow cables and a complete range of tools. Full details of this model are given in *Jane's Military Vehicles and Logistics 1991-92* page 46.

M3/VPC Command Vehicle

This model has additional communications equipment, two additional batteries, and map tables. Its basic crew consists of commander, assistant commander, driver and two radio operators plus command staff. Armament normally comprises a 7.62 mm machine gun on a Creusot-Loire Industrie STB mount.

M3/VLA Engineer Vehicle

This model is essentially the M3 APC fitted with a removable hydraulically operated dozer blade at the front of the hull and has a crew of six consisting of commander/gunner, driver, pioneer section commander and three pioneers.

M3/VTS Ambulance

This model has a crew of three (driver and two orderlies) and can carry four stretcher patients, or six sitting wounded, or two stretcher patients and three sitting wounded. It is unarmed and has a large single piece door in the rear of the hull that is hinged on the left. Medical equipment carried includes oxygen, blood transfusion and a complete first aid outfit.

M3/Radar

The Panhard M3 can also be used to carry electronic warfare equipment, this model is called the M3/VSB, as well as being fitted with a variety of radar systems such as the RASIT battlefield surveillance radar or the Dassault Electronique RA 20S surveillance radar. The latter is used with the Panhard M3/VDA twin 20 mm self-propelled anti-aircraft gun system when this is not equipped with its own individual radar.

Status: Production complete. Replaced in production by Panhard Buffalo APC. In service with Algeria (in 1982 ordered 44 APCs with SAMM MG turret, five VPC, two VAT, two VLA, two VTS, all for Gendarmerie use), Bahrain, Burkina Faso, Burundi (9 ordered in 1982), Chad (including some with Toucan I turret), Gabon, Iraq, Ireland, Ivory Coast, Kenya, Lebanon, Malaysia, Mauritania, Morocco, Niger, Nigeria (including four with RASIT radars, 4 VTS, 4 VAT and 6 VPC), Portugal, Rwanda, Saudi Arabia, Senegal, Spain (army reserve and marines), Sudan (including some with RASIT surveillance radar, one M3/VAT repair vehicle, two M3 with RA-20 radars and 13 M3 with Toucan I 20 mm turrets), Togo, United Arab Emirates and Zaïre.

Manufacturer: Société de Constructions Mécaniques Panhard et Levassor
18 avenue d'Ivry
75621 Paris
France
Telephone: 33 (1) 40 77 40 00
Telex: 270 276

SPECIFICATIONS	
CREW	2 + 10
CONFIGURATION	4 × 4
COMBAT WEIGHT	6100 kg
UNLOADED WEIGHT	5300 kg
POWER-TO-WEIGHT RATIO	14.75 hp/tonne
LENGTH	4.45 m
WIDTH	2.4 m
HEIGHT	
with Creusot-Loire TL 2i turret armed with twin 7.62 mm machine guns	2.48 m
to hull top	2 m
GROUND CLEARANCE	0.35 m
TRACK	2.05 m
WHEELBASE	2.7 m
ANGLE OF APPROACH/ DEPARTURE	68°/50°

MAX ROAD SPEED	
6th gear, high range	90 km/h
5th gear, high range	61 km/h
4th gear, high range	35 km/h
3rd gear, high range	18.8 km/h
3rd gear, low range	18.8 km/h
2nd gear, low range	9.3 km/h
1st gear, low range	4.5 km/h
reverse	5.5 km/h
MAX WATER SPEED	4 km/h
FUEL CAPACITY	165 l
MAX ROAD RANGE	600 km
FUEL CONSUMPTION	0.26 l/km
FORDING	amphibious
GRADIENT	60%
SIDE SLOPE	30%
VERTICAL OBSTACLE	0.3 m
TRENCH	
1 channel	0.8 m
5 channels	3.1 m
TURNING RADIUS	6.55 m
ENGINE	Panhard Model 4 HD 4-cylinder air-cooled petrol developing 90 hp at 4700 rpm

TRANSMISSION	manual with 6 forward and 1 reverse gears
CLUTCH	centrifugal with electro-magnetic automatic control
SUSPENSION	independent, coil spring and hydro-pneumatic shock absorbers acting on suspension trailing arms of wheel mechanism
TYRES	11.00 × 16
BRAKES	
main	hydraulic, dual circuit
parking	handbrake operating on gearbox output shaft
ELECTRICAL SYSTEM	24 V
BATTERIES	2 × 12 V
ARMAMENT	see text
ARMOUR	8-12 mm

ACMAT Armoured Personnel Carrier (TPK 4.20 VBL)

Development

The TPK 4.20 VBL armoured personnel carrier has been developed as a private venture by Ateliers de Construction Mécanique de l'Atlantique (ACMAT) and is essentially the well known VLRA (4 × 4) 2500 kg long-range reconnaissance vehicle fitted with an armoured body. The VLRA is used by some 37 countries in Europe, Africa, the Middle East and the Far East and is fully described in *Jane's Military Vehicles and Logistics 1991-92* page 362.

The first prototype of the ACMAT TPK 4.20 VBL was completed in 1980 and production began the same year. The company is now offering the TPK 6.40 VBL armoured personnel carrier which has a 6 × 6 configuration.

Description

The hull of the TPK 4.20 VBL is made of all-welded nickel chrome molybdenum steel 5.8 mm thick. The vehicle has a similar layout to the basic VLRA with the engine at the front, behind which are seats for the commander and driver, and the troop compartment at the rear.

Immediately in front of the driver and commander is a bullet-proof windscreen which is covered when in a combat area by a hatch hinged at the top. The commander and driver enter and leave the vehicle by a side door that opens to the rear. This has a bullet-proof window covered by a flap that opens upwards below which is a circular firing port if required. Above the commander's and driver's positions is a single piece hatch cover opening to the rear.

The troops are carried on bench seats down either side of the hull. The sides of the rear troop compartment are carried upright to give the maximum possible protection to the occupants. If fire is to be returned or if rapid dismounting is required, these sides can be hinged outward and downwards at their mid-point as shown in the accompanying photographs. The vehicle can also be delivered with a fully enclosed, removable rear compartment to give improved protection from roof-mounted snipers.

ACMAT TPK 6.40 VBL (6 × 6) APC with troops dismounting over side

ACMAT VBL light armoured car with unarmed Creusot-Loire Industrie one-man turret

ACMAT VBL light armoured car fitted with Creusot-Loire Industrie one-man turret armed with 12.7 mm M2 machine gun

ACMAT VBL with sides and rear of body folded down

If required an 81 mm Thomson-Brandt mortar can be mounted in the rear. Various 12.7 mm and 7.62 mm machine guns can be pintle-mounted in the rear troop compartment.

VBL Light Armoured Car

This has the same chassis and hull front as the VBL armoured personnel carrier but has a fully enclosed rear troop compartment. The troops are seated on bench seats down either side of the hull. In each side of the hull are four bullet-proof sliding vision blocks.

In the rear of the VBL are two outward opening doors, in each of which is a bullet-proof vision block covered by an armoured flap.

Mounted in the centre of the roof is the one-man turret armed with a 7.62 mm machine gun which is aimed using a roof-mounted periscope, or a Creusot-Loire Industrie one-man turret model TLi armed with a 12.7 mm (0.50) M2 HB machine gun. A Euromissile MILAN ATGW system can also be installed.

The VBL can be fitted with optional equipment including different radios and an air-conditioning system as well as being modified for more specific roles such as ambulance or command post.

Status: Production as required. In service with a number of countries including Gabon and Ivory Coast.

Manufacturer: ACMAT, Ateliers de Construction Mécanique de l'Atlantique
Le Point du Jour
44600 Saint-Nazaire
France
Telephone: 40 22 33 71; 40 22 29 40; 40 22 25 12
Telex: 700913
Fax: 40 66 30 96

SPECIFICATIONS (TPK 4.20 VBL; data in square brackets relates to TPK 6.40 VBL where different)

CREW	2 + 8-10 [2 + 16]
CONFIGURATION	4 × 4 [6 × 6]
COMBAT WEIGHT	7300 [8600] kg
UNLOADED WEIGHT	5880 [7180] kg
WEIGHT ON FRONT AXLE	
empty	3300 [4000] kg
loaded	3550 [4200] kg
WEIGHT ON REAR AXLE	
empty	2580 [3180] kg
loaded	3750 [4400] kg
LENGTH	5.98 [7] m
WIDTH	2.1 m
HEIGHT (hull top)	2.205 m
GROUND CLEARANCE	
axle	0.273 m
hull	0.5 m
TRACK	1.66 m
WHEELBASE	3.6 [4.01] m
ANGLE OF APPROACH/ DEPARTURE	45°/35°
MAX SPEED	95 km/h
FUEL CAPACITY	360 l
MAX RANGE	1600 km
FORDING	0.8 m
GRADIENT	60%
TURNING RADIUS	8.5 [10.35] m
ENGINE	Perkins model 6.354.4 6-cylinder diesel developing 135 hp at 2800 rpm
GEARBOX	Type 435 manual with 4 forward and 1 reverse gears [Type 540 with 5 forward and 1 reverse gears]
CLUTCH	single dry plate
TRANSFER BOX	2-speed
STEERING	worm and nut
SUSPENSION	leaf springs and double action hydraulic shock absorbers
TYRES	12.50 × 20
NUMBER OF TYRES	4 + 1 spare (run-flat) [6 + 1 spare]
BRAKES	
main	air/hydraulic
parking	mechanical
ELECTRICAL SYSTEM	24 V
BATTERIES	2 × 6 TN AS 188
ARMAMENT	see text
ARMOUR	5.8 mm

Panhard VBL Armoured Personnel Carrier

Development/Description

The standard French Army VBL (4 × 4) light armoured vehicle, fully described and illustrated in the *Reconnaissance vehicles* section, has a wheelbase of 2.45 m and a crew of two or three, depending on mission and equipment fit.

As a private venture, Panhard have built a stretched version with a wheelbase of 2.7 m which can have a crew of up to four men depending on equipment fit. This version has a total of four roof hatches, each of which swings back through 180° towards the rear with driver's at front left, commander's at front right and two rectangular hatches at the very rear.

The layout of these versions is virtually identical to that of the standard VBL with engine compartment at front, driver front left and commander front right, both with a forward opening door and the other two crew members at the very rear and can enter and leave the stretched VBL via a large door hinged on the left.

Like the standard VBL, the stretched VBL is fully amphibious being propelled in the water by a propeller mounted under the hull rear.

Panhard have already suggested the following roles for the stretched VBL:

Stretched Panhard VBL (4 × 4) with all doors and roof hatches open

Stretched Panhard VBL (4 × 4) with roof-mounted Hughes TOW ATGW system installed

Stretched Panhard VBL (4 × 4) with roof-mounted RASIT radar system

1 Patrol vehicle with four man crew and 7.62 mm roof-mounted machine gun
2 Squadron command vehicle with 7.62 mm roof-mounted machine gun and additional communications equipment
3 Scout vehicle with various armament installations including SAMM TMA 105 one man turret armed with 12.7 mm MG
4 Surveillance vehicle with RASIT radar system mounted on the roof and operator at rear
5 Air defence surveillance vehicle with roof-mounted radar system and operator at rear
6 Air-defence vehicle with MATRA Albi system with two SATCP surface-to-air missiles in the ready to launch position.

Variants
Panhard are also working on a cargo version of the VBL which would not have the roof hatches over the rear of the vehicle as well as the Super VBL which would be a 6 × 6 vehicle suitable for a wide range of roles including armoured personnel carrier. This version has a wheelbase 2.49 m (1st to 2nd axle) and 1 m (2nd to 3rd axle).

Specifications
Apart from a slight increase in weight and a wheelbase of 2.705 m, the stretched VBL has the same specification as the VBL covered in the *Reconnaissance vehicles* section.

Status: Development completed. Production as required.

Stretched Panhard VBL (4 × 4) with SAMM turret armed with 12.7 mm MG

Manufacturer: Société de Constructions Mecaniques Panhard and Levassor
18 avenue d'Ivry
75621 Paris
France
Telephone: (1) 40 77 40 00
Telex: 20 276

CBH P4 Armoured Personnel Carrier

Development
This vehicle is the Peugeot P4 (4 × 4) 750 kg light vehicle, already in volume production for the French Army (*Jane's Military Vehicles and Logistics 1991-92* page 264) fitted with an armoured body by Constructions Blindées d'Hardricourt. The vehicle is also known as the CBHcp.

The CBH was announced in 1985 by which time two prototypes had been built, one powered by a petrol engine and the other by a diesel.

Description
The welded steel armour provides the crew with complete protection against penetration from 7.62 mm NATO and 7.62 × 39 mm rounds fired at 0° from a 5 m range and 7.62 mm NATO armour-piercing rounds fired at an angle of 30° from a 30 m range. The armour is provided by Creusot-Loire Industrie.

The layout of the vehicle is conventional with the engine at the front and the driver and commander in the centre. They each have a bullet-proof window to their front and a forward-opening side door with a

CBH (4 × 4) APC from rear

CBH (4 × 4) APC

bullet-proof window in the upper part and a circular firing port below. There is also a single firing port in between the lower part of the driver's and commander's windscreens.

The four troops sit two either side at the back and enter via a large door hinged on the right which has a bullet-proof window in its upper part with a firing port underneath.

In each side of the rear troop compartment is a square bullet-proof window with a circular firing port underneath. The sides and front of the roof are sloped so that any devices thrown on the roof roll off. There is a tow hook at the rear and the vehicle can tow a trailer or weapon weighing up to 1750 kg.

The vehicle is offered with a petrol or diesel engine; optional equipment includes an air-conditioning system, smoke dischargers and various communications systems.

Variants

The CBH P4 can be fitted with a wide range of light-weight stations on the roof such as a Creusot-Loire Industrie CB 127 armed with an externally mounted 12.7 mm M2 HB machine gun. Other weapons proposed by the manufacturer include Euromissile MILAN ATGWs or MATRA Mistral surface-to-air missiles. More specialised versions include command post vehicles with extensive communications equipment fitted with various types of radar systems such as the Dassault Electronique aerial surveillance system.

Status: Development complete. Ready for production.

Manufacturers: Chassis: Automobiles Peugeot
75 Avenue de la Grande-Armée
75016 Paris
France
Telephone: (1) 40 66 55 11
Telex: 610700

Constructions Blindées d'Hardricourt
Avenue des Etangs Prés
78250 Hardricourt
France

SPECIFICATIONS		SIDE SLOPE	30%	STEERING	rack and pinion with damper
CREW	2 + 4	MAX ROAD SPEED		TURNING RADIUS	5.5 m
CONFIGURATION	4 × 4	petrol engine	118 km/h	SUSPENSION	
COMBAT WEIGHT		diesel engine	103 km/h	front	coil springs, anti-sway bar and double-acting telescopic hydraulic shock absorbers
petrol engine	3300 kg	MAX RANGE	1500 km	rear	coil springs and double-acting telescopic hydraulic shock absorbers
diesel engine	3380 kg	FUEL CAPACITY	150 l	TYRES	700 RC 16 × C Type L
MAX LOAD	1100 kg	ENGINE	Peugeot 1.971-l 4-cylinder petrol developing 79 hp at 4750 rpm or Peugeot 2.498-l 4-cylinder diesel developing 70.5 hp at 4500 rpm	BRAKES	hydraulic dual circuit, disc front and drum rear
LENGTH	4.2 m	GEARBOX	manual, 4 forward and 1 reverse gears	ELECTRICAL SYSTEM	24 V
WIDTH	1.7 m	CLUTCH	single dry disc with ball-bearing thrust mechanism	GENERATOR	1200 W
HEIGHT	2.06 m	TRANSFER BOX	2-speed	ARMOUR	6 mm
GROUND CLEARANCE	0.24 m				
TRACK	1.4 m				
WHEELBASE	2.4 m				
ANGLE OF APPROACH/DEPARTURE	42°/37°				
FORDING	0.5 m				
GRADIENT					
petrol engine	73.5%				
diesel engine	55%				

LOHR VMO Armoured Personnel Carrier

Development

The LOHR company is developing as a private venture a 4 × 4 vehicle called the VMO for the internal security role. This is based on the chassis of the Renault Vehicules Industriels JE 13 (4 × 4), although it could be built on other chassis if required by the customer.

Description

The all-welded steel armoured hull provides the occupants with protection from small arms fire. The driver is seated at the front of the vehicle on the left side with the vehicle commander to his right; both are provided with an individual seat, side opening door and windscreen to their front which can be covered by an armoured shutter if required by the tactical situation.

The 12 passengers are seated in the rear, four down the left side on individual seats and eight down the right side on twin seats, all of which face the front.

On the right side of the hull to the immediate rear of the commander, is a door that opens to the front with a window in its upper half, below which is a firing port and in addition there is a door in the hull rear.

There are a total of nine windows in the rear passenger area, five on the left and four on the right, each of which has a firing port underneath and an armoured shutter.

An emergency hatch is provided in the roof and standard equipment includes a remote controlled searchlight mounted on the right side of the hull, four flashing lights on the roof, roof-mounted periscopes for commander and driver and air-conditioning system.

Artist's impression of LOHR VMO (4 × 4) internal security vehicle

Status: Proposal.

Manufacturer: LOHR Defense
67980 Hangenbieten
France
Telephone: 88 38 98 00
Telex: 870082
Fax: 88 96 06 36

SPECIFICATIONS		GROUND CLEARANCE	0.28 m	TRANSMISSION	5 forward, 1 reverse gears
CREW	2 + 12	TRACK	1.836 m	TRANSFER BOX	single speed
CONFIGURATION	4 × 4	WHEELBASE	3.85 m	SUSPENSION	leaf springs
COMBAT WEIGHT	11 000 kg	MAX ROAD SPEED	80 km/h	STEERING	power-assisted
EMPTY WEIGHT	8500 kg	MAX RANGE	600 km	TYRES	11 R 22.5 with limited deflation
LENGTH	6.375 m	GRADIENT	60%		
WIDTH	2.5 m	SIDE SLOPE	30%		
HEIGHT		ENGINE	6 cylinder diesel developing 154 hp at 2900 rpm		
hull top	2.55 m				
overall	2.8 m				

GERMANY

Marder 2 Infantry Combat Vehicle

Development
The German Army has a two part plan to equip its armoured infantry battalions with new or upgraded infantry fighting vehicles for the 1990s and beyond. The first phase covers the upgrading of 2100 existing Marder 1 vehicles to the Marder 1 1A3 IFV standard (qv) which will be used by the 12 armoured infantry brigades.

The second phase is the fielding of the new Marder 2 Infantry Fighting Vehicle. Four companies, Krauss-Maffei, Krupp-MaK, Thyssen-Henschel and Daimler-Benz submitted concept studies for the Marder 2 and in late 1987 Krauss-Maffei was selected to proceed with concept definition. It is expected to enter service in the mid-1990s with around 1000 vehicles being required.

Description
No detailed specification of the Marder 2 has been released but it is expected to weigh around 43 tonnes and show a significant increase in armour, mobility and firepower over the existing Marder 1.

The engine will be mounted at the front of the vehicle, so leaving the whole rear of the vehicle free for the fighting compartment which is unlikely to be fitted with firing ports.

Marder 2 will be fitted with a two-man power-operated turret armed with the new Rheinmetall MK 35/50 mm Rh 503. This is an externally powered weapon which initially fires 35 × 228 mm ammunition, although Rheinmetall are also developing new HE and KE rounds for it. By merely changing the barrel, however, it will be possible to fire more powerful 50 × 300 mm ammunition.

Rheinmetall were awarded the full scale development contract for the MK 35/50 mm Rh 503 in October 1988. The contract also covers development of ammunition and linkless feed system. In the longer term, the 35 mm version would be used for training while the 50 mm version would be for combat.

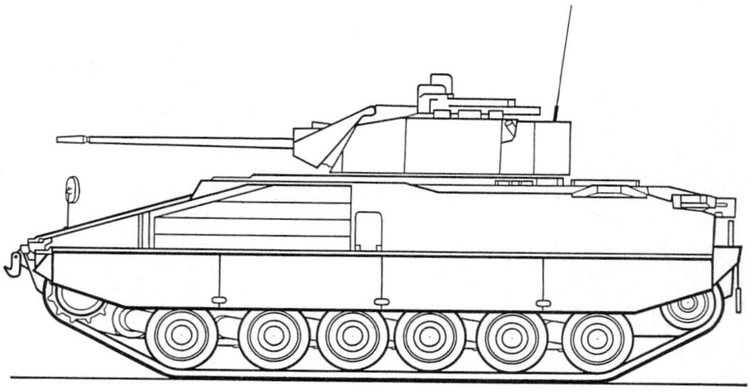

Provisional outline drawing of concept study for Marder 2 IFV (not to 1/76th scale)

Suspension of the Marder 2 will be of the torsion bar type, with six road wheels either side, drive sprocket at the front, idler at the rear and track return rollers. The upper part of the track will be covered by an armoured skirt.

Status: Concept definition.

Prime contractor: Krauss-Maffei Wehrtechnik GmbH
Krauss-Maffei Strasse 2
8000 Munich 50
Federal Republic of Germany
Telephone: (089) 88 99 0
Telex: 5 23 163 31
Fax: (089) 8 12 01 90

Marder 1 Infantry Combat Vehicle

Development
In the late 1950s a chassis was developed which could be used for a number of basic vehicles including the Jagdpanzer Kanone and Jagdpanzer Rakete (both of which have separate entries in the *Tank destroyers* section), reconnaissance tank (developed to the prototype stage but not placed in production), an ICV (which eventually became the Marder) and various other supporting vehicles, most of which did not enter service.

In January 1960 contracts for the design and construction of prototypes of the ICV were awarded to two groups of companies: the Rheinstahl group comprising Rheinstahl-Hanomag, Ruhrstahl, Witten-Annen and Büro Warnecke and the partnership Henschel Werke of Kassel and MOWAG of Kreuzlingen, Switzerland. Three vehicles were built by Hanomag, two by Henschel and two by MOWAG, known as the first series prototypes.

Between 1961 and 1963 the second series of prototypes was built, four by Hanomag, one by Henschel and three by MOWAG. There was then a pause in the development of the vehicle as priority was given to the Jagdpanzer Kanone (which entered production at the works of both Hanomag and Henschel in 1965) and the Jagdpanzer Rakete (which entered production at the works of both Hanomag and Henschel in 1967 when production of the Kanone had been completed).

In 1966 the military requirements were completed and in 1967 the construction of the third and final series of prototypes began. Ten prototypes were built, three by Hanomag, four by Henschel and three by MOWAG.

Following the takeover of Henschel Werke in 1964 by the Rheinstahl Group most of the final development work was completed by Rheinstahl. The 10 pre-production vehicles were completed in 1967/68 and in October 1968 the first vehicles were delivered to the German Army for troop trials which ran from October 1968 to March 1969. In April 1969 companies were asked to tender for series production of the vehicle which was officially named the Marder 1 in May 1969. In October of that year Rheinstahl was nominated as prime contractor with MaK of Kiel as a sub-contractor. The first production vehicle was delivered to the Army in December 1970 but the first vehicles were not officially handed over until May 1971. Two production lines were established, one at Kassel and one at MaK of Kiel. The original contract covered the construction of 1926 vehicles by Rheinstahl and 875 vehicles by MaK but in 1974 Rheinstahl received a further order for 210 vehicles and total production by MaK (now Krupp MaK) amounted to 975 vehicles. Production of the Marder 1 was finally completed in 1975.

The chassis of the Marder 1 ICV remained in production at Kassel for the Roland 2 surface-to-air missile system until 1983.

Description
The all-welded hull of the Marder 1 provides the crew with protection from small arms fire and shell splinters, with the front of the vehicle giving complete protection from 20 mm projectiles.

The driver sits at the front of the hull on the left side and has a single piece hatch cover that opens to the right, in front of which are three periscopes, the centre one of which can be replaced by a passive night driving device. One of the infantrymen sits to the rear of the driver and has a single piece hatch cover that opens to the right, with a periscope that can be traversed through a full 360°.

The engine compartment is to the right of the driver and is coupled to a Renk four-speed HSWL-194 planetary gearbox and a stepless hydrostatic steering unit which transmits power to the tracks via two final drive assemblies mounted at the front of the hull. The radiators are mounted at the rear of the hull, one either side of the ramp, with cooling air being sucked in through grilles in the top of the hull by an axial cooling fan.

The two-man turret, designed by Keller and Knappich, is mounted in the forward part of the roof with the commander on the right and the gunner on the left. The commander has a single piece hatch cover that opens to the right while the gunner has a similar hatch cover that opens to the rear.

The 20 mm Rheinmetall MK 20 Rh 202 cannon is mounted externally and has an elevation of +65°, depression of -17° and 360° turret traverse. It is fed from three different belts which allows the gunner to select one type of ammunition at a time, for example AP when engaging light AFVs and HE when engaging softer targets. The empty cartridge cases are automatically ejected outside the turret.

Both the commander and gunner have a PERI Z 11 sight for use in the ground and anti-aircraft role with a magnification of ×2 and ×6 which

Marder 1 ICV complete with side skirts but with infra-red searchlight removed from left side of turret (Pierre Touzin)

Marder 1 ICV with Euromissile MILAN mounted on right side of turret (Pierre Touzin)

Marder 1 with rear ramp lowered and 7.62 mm remote-controlled machine gun turret removed from roof (Pierre Touzin)

can be replaced by an infra-red night periscope. In addition the commander has eight periscopes for all-round observation and the gunner has three.

Mounted coaxially above and to the right of the 20 mm cannon is a 7.62 mm MG 3 machine gun and mounted to the left of the 20 mm cannon are six electrically operated smoke dischargers. Mounted externally to the left of the gun mount is an infra-red/white light searchlight that moves in elevation with the main armament. With the introduction of a thermal sight, this is being removed.

The six infantrymen are seated in the troop compartment at the rear of the hull, three down each side facing outwards. Their seating unit was designed by MOWAG and can be adjusted so that four men can sleep. Entry and exit is by a power-operated ramp at the rear of the hull that opens downwards. In each side of the troop compartment are two MOWAG-designed spherical firing ports and in the roof either side are two circular roof hatches and three periscopes. The firing ports, used in conjunction with the periscopes, allow four of the infantrymen to use their weapons from inside the vehicle in complete safety. All Marders have an NBC system.

Mounted over the top of the troop compartment at the rear of the hull is a MOWAG-designed remote-controlled 7.62 mm MG 3 machine gun, which has an elevation of +60°, a depression of -15° and a total traverse of 180°. The turrets are now being removed.

The torsion bar suspension each side consists of six dual rubber-tyred road wheels with the drive sprocket at the front and the idler at the rear, and three track return rollers. The first, second, fifth and sixth road wheel stations have a hydraulic shock absorber. The Diehl tracks have replaceable rubber pads.

The Marder 1 can ford to a depth of 1.5 m without preparation or to 2.5 m with the aid of a kit. An amphibious kit has been developed for the Marder 1 but has not been adopted.

All Marder 1s used by the German Army except command vehicles have been fitted with a Euromissile MILAN ATGW launcher.

Variants

There have been many variants of the Marder 1 ICV, especially in the early days of the project, including an anti-aircraft gun vehicle (called the Flak Zwilling but cancelled in favour of the Gepard), ambulance (2HK 2/2), cargo carrier (RU121), artillery observation vehicle (called the Beobachtungs Panzer Artillerie, but cancelled in favour of a cheaper model based on an M113A1 chassis), 120 mm mortar carrier (245 were ordered in 1969 but subsequently cancelled, instead the mortars were mounted in M113A1 APCs, the conversion work being carried out by Thyssen Henschel), multiple rocket launcher and Rapier SAM vehicle. More recently Thyssen Henschel has suggested that a 160 mm mortar be carried in place of the 120 mm mortar.

The TH 397 is the Marder 1 ICV with the rear external 7.62 mm remote-controlled machine gun removed. For trials purposes Krupp MaK has fitted a Marder with a gas turbine auxiliary power unit.

Upgraded Marder 1

From 1982 most Marder 1s were upgraded to one of two new standards, 1 A1 or A1A, at the rate of about 80 vehicles per month.

1 A1(+)

This is the complete upgrade and 674 vehicles have been brought up to this standard. Main improvements are better firepower due to double belt feed for 20 mm cannon, improved night capabilities with installation of PERI Z 59 night sight with sight head PERI Z 16 which uses image intensification technique with thermal pointer, remaining active infra-red searchlight and sight are retained, holder for commander's NBC mask on upper mounting for MILAN ATGW, new racks for water cans inside, stowage for image intensification equipment, flaps for periscopes, storage for helmets and two 5 m tow ropes. This model weighs 30 000 kg, has a crew of four and carries five instead of six infantrymen.

1 A1(-)

A total of 350 vehicles were upgraded to the A1(-) configuration, but unlike the A1(+) standard (see before) they were prepared for, but not fitted with, the thermal pointer. The latter can, however, be transferred from an A1(+) turret that is not operational.

1 A1A

A total of 1112 vehicles were modified to the previous standards but were not fitted with passive vision equipment. All Marder 1s of the German Army, except command vehicles, now carry a MILAN ATGW system with five missiles and this has meant that the number of infantrymen carried has been reduced from six to five.

1 A2

Since 1983, all the German Marder 1s have been upgraded to the A2 standard. The 674 A1(+) are being upgraded to standard A2 which excludes thermal imaging sight equipment. The remaining 1462 ICVs are being fitted with thermal imaging sight equipment instead of the image intensifier. At the same time the infra-red searchlight on the left side of the turret is being removed and the chassis (fuel tanks, cooling system and water jet cleaning system) and suspension are being modified.

1 A3

In 1988 Thyssen Henschel was awarded a contract covering the conversion of 2100 Marder 1 A1/A2 series vehicles to the A3 standard over a 10 year period at the rate of 220 vehicles a year, with the project starting in 1989. The first vehicles were handed over on 17 November 1989 to the German Army. The modifications include the following: Armour package - a new armour package weighing 1600 kg is being fitted to provide protection against penetration from the Soviet BMP-2 30 mm cannon. This package consists of: frontal armour on the hull, add-on armour on the glacis plate, conformal add-on armour on both sides of the turret, three box-type armour components on both sides of the hull (so blocking off the hull firing ports), spaced armour plates on the roof to cause premature detonation of top attack rounds, add-on armour with stowage compartment at rear door and improvements in human engineering.

The infantryman's hatch to the immediate rear of the driver's position has been eliminated and over the top of the troop compartment there are now three roof hatches arranged in a triangular pattern, two towards the front and one towards the rear.

Marder 1 with Radar System

First prototypes of the Radarpanzer TÜR (Tiefflieger-Überwachungs-Radar) based on the chassis of the Marder 1 ICV were delivered to the German Army late in 1981. This is an extensively modified raised Marder 1 hull with the turret replaced by a hydraulically operated arm on top of which is mounted a Siemens radar with a range of some 30 km. When raised into the operating position, the radar is approximately 10 m above the ground. The hull of the RadarPz TÜR has been raised to provide increased space for the crew and equipment. The interior of the hull houses the radar unit, display and control units, radio and data processing system and power supply, cooling systems and hydraulic equipment.

Basic specifications are: crew 4, length 7.2 m, width overall 3.27 m, total height travelling 3.58 m, combat weight 35 tonnes, maximum road speed 60 km/h, range 570 km on roads. Armament consists of two banks of four electrically operated smoke dischargers mounted either side of the hull rear and a 7.62 mm machine gun on each cupola. In July 1988 Siemens handed over the matured MPDR 3002-S for the TÜR to the Federal German Office for Military Procurement and Technology for extensive technical trials at the Army Military Technology Department 81. The system is expected to enter service with the German Army in the 1990s and provide target information to Roland and Gepard air defence systems.

Marder 1 Driver Training Tank

Krauss-Maffei has delivered 36 Driver Training Tank versions of the Marder 1 infantry combat vehicle to the German Army. These have all had their two-man power-operated turrets replaced by a fixed cupola with seats for the instructor and additional pupils under instruction.

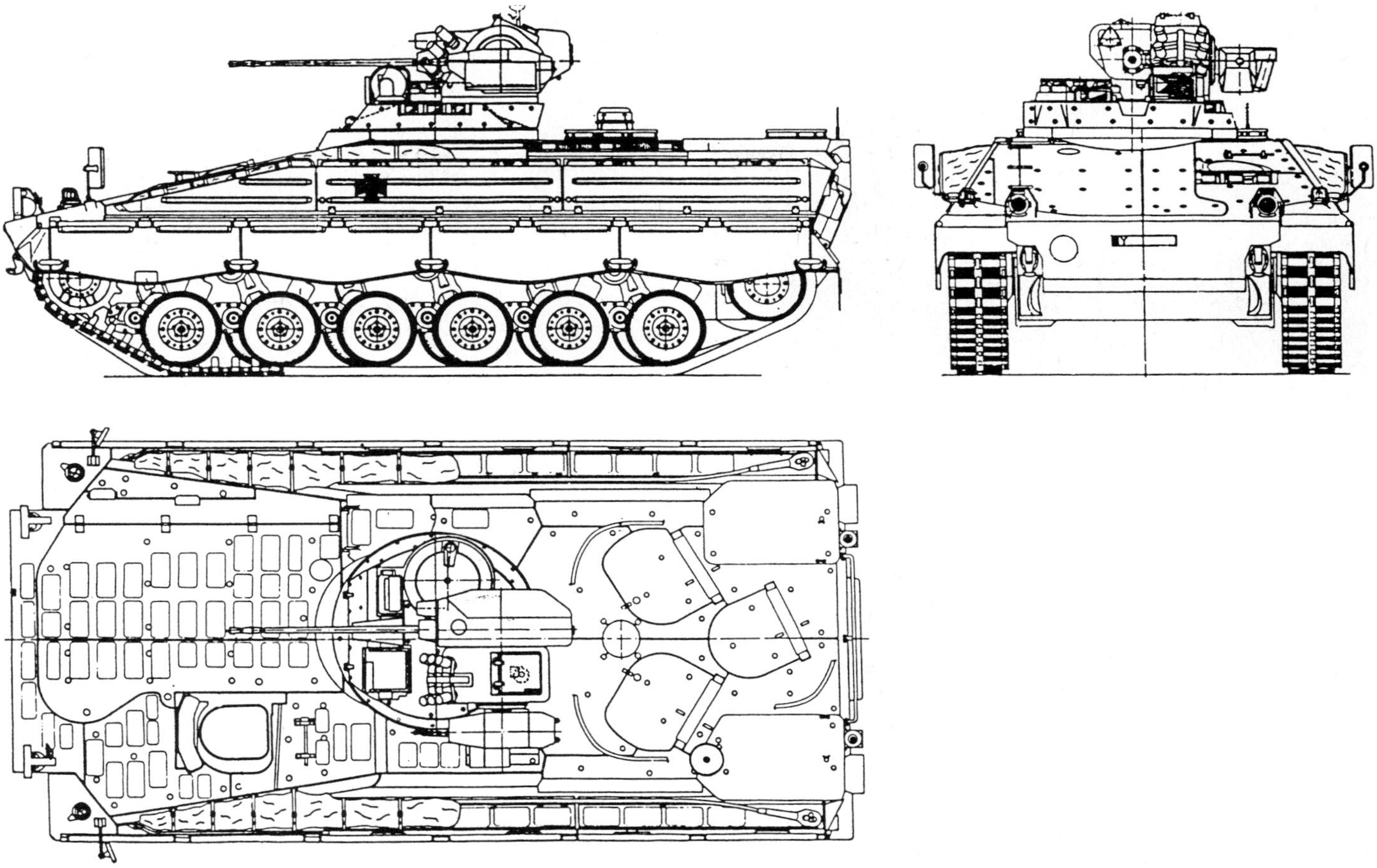

Three-view general arrangement drawing of Marder 1 A3 showing additional armour and revised roof hatches (not to 1/76th scale)

Marder 1 with LWT-3 Turret

This has completed its trials and is a Marder 1 fitted with a new turret armed with a 20 mm cannon which is fully stabilised. LWT-3 stands for Light Weapon Turret with three-axis stabilisation.

Support Tank 35 mm

Thyssen Henschel has proposed that the Marder 1 chassis be fitted with a new two-man turret armed with an externally mounted 35 mm cannon, 7.62 mm coaxial machine gun and an ATGW launcher. This would be called the 35 mm support tank and have a combat weight of 30 000 kg. The 35 mm cannon is the same as that installed in the Gepard twin 35 mm self-propelled anti-aircraft gun system already used by the German Army and has dual feed. The larger diameter turret would mean that two of the circular roof hatches over the troop compartment would have to be removed. This version never entered production.

Marder/Dragon Chassis with FL-15 Turret

In mid 1983, for trials purposes, the chassis of the Marder/Dragon was fitted with the new French Fives-Cail Babcock FL-15 turret armed with a 105 mm gun. In 1985 a Marder chassis was fitted with the new FL-20 turret. This version never entered production.

Marder 1 with EMG

Rheinstahl Henschel and Krupp MaK have also used a Marder 1 chassis for testing the Externally Mounted Gun (EMG) concept. A prototype, called the VTS-1, with a 105 mm gun mounted externally on the hull top has been built for trials and also evaluated by Sweden. The companies carried out two studies, study B with the commander and gunner turning with the gun and study D with the commander and gunner remaining stationary in the hull. An illustration of this vehicle appeared in *Jane's Armour and Artillery 1983-84*, page 15.

Clovis Combat Vehicle

Details of this vehicle, which consists of a Marder 1 type chassis fitted with a 105 mm FL-20 turret, were given in *Jane's Armour and Artillery 1987-88*, pages 37-38. It has not so far entered production or service.

Roland Surface-to-air Missile System

A total of 144 Roland 2 surface-to-air missile systems on the Marder 1 chassis has been built for the German Army with a further four systems delivered to Brazil. Two Roland missiles are carried in the ready to launch position with another eight missiles being carried in the hull, four either side of the turret, ready for automatic loading.

Full details of this version are given in *Jane's Land-Based Air Defence 1991-92*, page 112.

TAM Medium Tank Family

To meet the requirements of the Argentinian Army Thyssen Henschel has developed the TAM Medium Tank (Tanque Argentino Mediano) full details of which are given in the *Tanks* section. To operate with the TAM Thyssen Henschel has developed the VCTP ICV of which 300 were required by Argentina. The VCTP is similar to the Marder 1 but has a

Krauss-Maffei Marder 1 Driver Training Tank

Marder 1 A2 which has had the rear 7.62 mm MG turret and infra-red searchlight on the left side of the turret removed

Radarpanzer TÜR with surveillance radar lowered for travelling

Marder 1 A3 which has many improvements including additional armour protection for the hull and turret, redesigned roof hatches and stowage boxes alongside of the hull

Close-up of the KUKA turret installed on the Marder 1 A3 showing 7.62 mm machine gun mounted externally on the left side, and additional armour protection

more powerful 720 hp engine, less sophisticated two-man turret armed with a 20 mm cannon and a 7.62 mm anti-aircraft machine gun, three gun ports in each side of the troop compartment and two rectangular roof hatches that are hinged on the outside. A remote-controlled 7.62 mm machine gun is fitted at the rear of the hull. The VCTP can carry 12 men including its crew and has a maximum road speed of 75 km/h. To increase the operating range of the vehicle to 800 km two long-range drum-type fuel tanks are mounted at the rear on the left and right sides.

Using the chassis of the TAM Thyssen Henschel has developed a whole range of vehicles including a 155 mm self-propelled gun (using the turret of the French 155 mm GCT), 57 mm Support Tank, or Begleitpanzer (details of which are given in the *Tanks* section in *Jane's Armour and Artillery 1985-86*), 57 mm anti-aircraft vehicle (details of which were given in *Jane's Armour and Artillery 1983-84*, page 541) and the Dragon twin-30 mm self-propelled anti-aircraft gun system (details of which were given in *Jane's Armour and Artillery 1984-85*, page 504).

Note: The Thyssen Henschel VCTP (previously known as the VCI) designation relates only to vehicles being used by the Argentinian Army. For other markets, the company calls the vehicle the TH 302, this has slightly different specifications from the VCTP including a combat weight of 27 tonnes, length of 6.83 m, width of 3.306 m, height of 2.683 m, ground clearance of 450 mm, maximum road speed of 72 km/h and a road range of 460 km. Armament comprises a two-man turret armed with a 20 mm MK 20 Rh 202 Rheinmetall cannon, 7.62 mm machine gun mounted coaxially with the main armament, 7.62 mm anti-aircraft machine gun and a 7.62 mm machine gun in a remote mount over the rear troop compartment. In each side of the rear troop compartment are three upward-opening firing hatches with a periscope above in the roof.

SPECIFICATIONS (Key specifications of the upgraded Marder A3 are given in square brackets where applicable)

CREW	9
COMBAT WEIGHT	29 207 [35 000] kg
UNLOADED WEIGHT	28 200 [33 500] kg
POWER-TO-WEIGHT RATIO	20.54 [18] hp/tonne
GROUND PRESSURE	0.83 [0.94] kg/cm²
LENGTH	6.79 [6.88] m
WIDTH	3.24 [3.38] m
HEIGHT	
over turret top	2.985 [3.015] m
over hull top	1.9 m
FIRING HEIGHT	
20 mm cannon	2.56 [2.7] m
7.62 mm coaxial MG	2.735 m
GROUND CLEARANCE	0.44 [0.455] m
TRACK	2.62 m
TRACK WIDTH	450 mm
LENGTH OF TRACK ON GROUND	3.9 m
MAX ROAD SPEED (forward and reverse)	
4th gear	75 [65] km/h
3rd gear	47 km/h
2nd gear	31 km/h
1st gear	16 km/h

FUEL CAPACITY	652 l
MAX ROAD RANGE	520 [500] km
FORDING	1.5 m
with preparation	2.5 [2.0] m
MAX GRADIENT	60%
SIDE SLOPE	30%
VERTICAL OBSTACLE	1 m
TRENCH	2.5 m
TURNING RADIUS	
4th gear	30 m
3rd gear	20 m
2nd gear	13.5 m
1st gear	6.5 m
ENGINE	MTU MB 833 Ea-500 6-cylinder liquid-cooled diesel developing 600 hp at 2200 rpm
TRANSMISSION	4-speed HSWL-194 planetary, 4 forward and 2 reverse gears, integral steering and braking system
SUSPENSION	torsion bar
ELECTRICAL SYSTEM	24 V
BATTERIES	6 × 12 V
ARMAMENT	
main	1 × 20 mm cannon
coaxial	1 × 7.62 mm MG

SMOKE-LAYING EQUIPMENT	6 smoke dischargers
AMMUNITION	
main	1250
7.62 mm	5000
GUN CONTROL EQUIPMENT	
Commander's fire control override	yes
Turret power control	electro-hydraulic/manual
by commander	yes
by gunner	yes
Max rate power traverse	60°/s
Max rate power elevation	40°/s
Gun elevation/depression	+65°/-17°
Turret traverse	360°
Gun stabiliser	
vertical	no
horizontal	no

Note: When originally introduced into service with the German Army the Marder had a crew of 4 + 6 and was fitted with a remote-controlled machine gun over the rear of the troop compartment. Most vehicles also carry one Euromissile ATGW for which six missiles are carried inside the vehicle.

Status: Production of the Marder has been completed in both Argentina and Germany. Brazil has four Roland SAM systems based on the Marder chassis in service. Production of the Marder can be resumed if further orders are received.

Manufacturers: Production of the Marder was undertaken by Rheinstahl (now Thyssen Henschel) of Kassel and Krupp MaK of Kiel.
Thyssen Henschel
Postfach 102969
D 3500 Kassel
Federal Republic of Germany
Telephone: (0561) 8011
Telex: 99750

Roland 2 surface-to-air missile system of the German Army (Pierre Touzin)

Armoured Combat Vehicle Family

Development

Details of this range of armoured vehicles, developed as a private venture by Krauss-Maffei and Diehl and shown for the first time in mid-1986, are given in the *Armoured fighting vehicles families* section.

Thyssen Henschel Transportpanzer 1 (Fuchs) Armoured Personnel Carrier

Development

In 1964 the German Ministry of Defence examined its requirements for a new generation of military vehicles for the 1970s. This generation was to have included an 8 × 8 armoured amphibious reconnaissance vehicle, 4 × 4 and 6 × 6 armoured amphibious load carriers and a complete range of 4 × 4, 6 × 6 and 8 × 8 tactical cargo trucks, some of which were to be amphibious. All were to share many common components which would, where possible, be from commercial sources both to reduce costs and simplify procurement of spare parts.

In 1964 a Joint Project Office was formed to undertake development of the complete range of vehicles. Companies in the Joint Project Office were Büssing, Klöckner-Humbolt-Deutz, Friedrich Krupp, MAN and Rheinstahl-Henschel. Daimler-Benz did not join the Joint Project Office but went ahead on its own to develop a similar range of vehicles which were finally chosen.

After severe and extensive trials the 8 × 8 armoured amphibious reconnaissance vehicle became the Spähpanzer Luchs and 408 vehicles were built by Rheinstahl Wehrtechnik between 1975 and 1978. The 4 × 4, 6 × 6 and 8 × 8 tactical trucks were built by MAN and all were delivered to the German Army by 1980. The 4 × 4 armoured amphibious load carrier became the Transportpanzer 2 and the 6 × 6 armoured amphibious load carrier became the Transportpanzer 1. The Transportpanzer 2 was not placed in production but further development was undertaken by EWK and the eventual result was the APE Amphibious Engineer Reconnaissance Vehicle described in *Jane's Armour and Artillery 1985-86*, pages 186 to 187, but this never entered production or service.

In 1977 Rheinstahl Wehrtechnik (now Thyssen Henschel), under licence from Daimler-Benz, was awarded a production contract for 996 Transportpanzer 1 vehicles for delivery from 1979 at the rate of 160 vehicles per year. The first production vehicle was handed over to the German Army, which also calls the vehicle the Fuchs, in December 1979. Deliveries of the Transportpanzer were completed in late 1986. Major subcontractors to the company were Daimler-Benz (axles, engine and steering), Jung (hull) and Zahnradfabrik Friedrichshafen (transmission).

More recently the Thyssen Henschel Transportpanzer has been placed back into production. These new vehicles are for the German Army, replacing Transportpanzer NBC reconnaissance vehicles supplied to forces operating in Saudi Arabia. These included Israel (8), Turkey (4), United Kingdom (11) and the United States of America (60).

The Transportpanzer is used by the German Army in the following roles:

RASIT Radar

This model is fitted with the RASIT battlefield surveillance radar (radar unit, target tracking and locating unit and radio) manufactured by LMT of France. The radar is mounted in the forward part of the hull and when operating is raised hydraulically to a maximum height of 1.8 m above the vehicle's roof. A 30 m cable allows remote operation of the system. This model is known in the German Army as the Panzeraufklärungsradargerät, or PARA for short, and 110 were built.

Command and Communications Vehicle

A total of 134 of these were built under the designation of FüFu. This model has a 5 kW generator in the left wing of the rear door, fuel being supplied from the main tank, communications equipment, map board, two folding tables with personnel lamps, three folding seats and a box with a padded seat. Up to four antennas can be fitted, including a very high one at the right rear of the hull.

NBC Reconnaissance Vehicle

Some 140 of these were built. The NBC detection equipment includes radiation and chemical agent detection units, sampling instruments and markers. All operations are carried out inside the vehicle which gives protection against NBC agents. The German Army designation for this vehicle is the ABC Erkundgruppe (ABC-ErkdGrp). The first vehicles were handed over to the German Army in February 1988.

This version carries radiation and chemical agent detectors, including the MM-1 mobile mass spectrometer, sampling instruments and markers.

All of the detection operations are carried out from within the safety of the vehicle which is equipped with an NBC protection ventilation system.

In early 1990 a modified version of this vehicle, called the Nuclear Biological Chemical M93 Reconnaissance System was selected by the United States Army. Additional details of the M93 are given later in this entry.

Transportpanzer 1 with RASIT battlefield surveillance radar raised (C R Zwart)

Command and communications version of Transportpanzer 1 from front with commander's and driver's doors open (Pierre Touzin)

Command and communications variant of Transportpanzer 1 as used by German Army

Transportpanzer 1 in basic German Army configuration and showing bank of smoke dischargers on side of hull

Engineer Vehicle

Two hundred and twenty of these are used by combat engineers and carry mines, demolition equipment and other specialised equipment.

Electronic Warfare

This model, known as the TPz 1 Eloka, is fitted with the EK 33 jammer kit. Unlike other models of the Transportpanzer it has no amphibious capability, only one roof hatch and a 15 kW generator to provide power for the extensive electronic warfare system installed.

Supply Carrier

Two hundred and twenty vehicles have been issued to supply units for use in the forward area. It can also be used as an ambulance, carrying four stretcher patients or two stretcher and four sitting patients.

HELAS EW System

The Radio and Systems Division of Siemens has developed to the prototype stage a passive electronic radar reconnaissance system called the HELAS. By 1988 this had successfully completed its technical tests and after overhaul the subsystem was delivered to the German Army for tactical field trials.

HELAS consists of two reconnaissance sets and a central processing set which are mounted in a Transportpanzer 1 (6 × 6) vehicle. The sensor package is carried on a mast pivoted at the rear and when required for action is raised into the vertical position.

HELAS electronic warfare system deployed in the field and with mast in elevated position

HELAS can carry out a number of roles on the battlefield including searching for a reception of all electromagnetic radiation from radar and fire control systems in the surveillance area and localising and signal processing. The latter includes identification, classification (especially assignment to weapon platforms), threat detection, contributing towards definition of the situation, detection of new radars, operating modes and strategies and information for central evaluation.

Description

The all-welded steel hull of the Transportpanzer 1 protects the crew from small arms fire and shell splinters. The hull has a rhomboid cross section and incorporates spaced armour in certain areas.

The driver sits at the front of the hull on the left with the vehicle commander to his right. A large bullet-proof windscreen to their front can be covered by an armoured shutter hinged at the top when in action. Four periscopes are fitted in the roof of the vehicle to the front of the driver's hatch for use when the windscreen is covered by the armoured shutter, the centre one of which can be replaced by a passive periscope for driving at night. The commander has a circular roof hatch, which opens to the left rear. Both the driver and commander have a door in the side of the hull with an integral window that can be covered by a shutter in a combat area.

The engine compartment behind the driver and commander is provided with an automatic fire-extinguishing system. There is a small passage on the right side between the front of the vehicle and the passenger/cargo area at the rear. The air-inlet and air-outlet louvres are in the roof of the vehicle witn the exhaust pipe running along the left side of the hull.

The powerpack consists of an eight-cylinder Mercedes-Benz OM 402A exhaust turbo-charged diesel developing 320 hp and a six-speed planetary gear torque converter transmission on the side of the engine and connected to it by spur gearing. Above the engine are the air cleaner system and the cooling system, with a hydraulically driven fan above the transmission. Attached to the powerpack are the vehicle parking brake, 5 kW alternator, brake supply system and hydraulics, including the oil supply reservoir.

The powerpack can be removed in about 10 minutes as all disconnecting points on the powerpack for mountings, functionally connected shafts for automotive and amphibious devices, pneumatic, hydraulic and electric supply or control lines are designed as quick-disconnect points. The engine of the Fuchs can be run outside the vehicle for test purposes.

The troop/cargo compartment is at the rear of the vehicle and is 3.2 m long, 1.25 m high and 2.5 m wide at its widest point. In each side of the hull is a small vision port protected by an armoured shutter. There are three hatches in the roof of the troop compartment: a large 1 m circular one in the forward part on which the main armament is normally mounted, a single piece hatch cover to the rear of the first one on the right side which opens to the left, and to the rear and left of the second hatch is the third one which opens to the right.

Transportpanzer 1 as delivered to Venezuela in 1983 armed with 12.7 mm and 7.62 mm machine guns

German Transportpanzer 1 with 7.62 mm MG on forward part of roof (Michael C Klaver)

The 10 infantrymen are seated on individual bucket seats, five down each side of the hull. The seats can be folded up to enable cargo to be carried. Normal means of entry for the infantrymen is through two large doors in the rear of the hull. The right door has a vision port covered by an armoured shutter when not in use.

Nominal amphibious payload is 2000 kg but in the non-amphibious mode it can carry up to 4000 kg of cargo.

Steering is hydraulically assisted on the front four wheels. An electrically driven back-up pump automatically maintains the supply of hydraulic fluid to the steering box and propellers in case of hydraulic failure. The rigid axles incorporate hub-mounted planetary gears and are positioned by guide rods supported by progressively acting coil springs. All axles have differential lock ups and hydraulic shock absorbers. The tyres are of the run-flat type.

The Transportpanzer 1 is fully amphibious being propelled in the water by two Schottel four-bladed propellers beneath the floor level of the vehicle at the rear of the hull. For steering when afloat the propellers can be traversed through 360°. Before entering the water a trim vane, which is stowed on the glacis plate when travelling, is hydraulically erected at the front of the hull. The three bilge pumps each have a capacity of 180 litres per minute.

An NBC system, fitted as standard, can also be used to ventilate the crew and personnel compartments. Armament varies according to mission requirements but can consist of a 7.62 mm MG 3 machine gun mounted over the commander's position or a 20 mm Rheinmetall cannon mounted on a Keller and Knappich E-6 ring mount over the first circular roof hatch. The 7.62 mm machine gun has an elevation of +40° and a depression of -15°. The 20 mm cannon has an elevation of +75° with depression being determined by a contour limiter, and 150 rounds of ready-use ammunition are provided. All vehicles have six smoke dischargers mounted on the left side of the hull firing forwards.

Electronic Trials Vehicles

In 1987 a number of Transportpanzer 1s were being used for various electronics trials. One has been fitted with the Siemens HELAS electronic warfare system (qv) with the sensors mounted on a hydraulic arm that is raised above the vehicle when operating. It has also been used for trials of the SEL RATAC-S target acquisition and surveillance system.

Variants

In 1985 a contract was signed between Thyssen Henschel and GLS outlining the responsibilities of both firms. Thyssen Henschel, being the manufacturer of the TPz-1, would act also as prime contractor whereas GLS as partner would perform the after sales service and spare parts support.

In 1983 Venezuela placed an order for 10 Transportpanzer 1s which were delivered in September that year. These vehicles did not have the standard NBC system installed but had smoke dischargers, an 8000 kg capacity winch, an air-conditioning system and two roof-mounted weapon systems. To the rear of the engine compartment is a KUKA turret with an externally mounted 12.7 mm machine gun while behind this and facing the rear is a Wegmann 11/1 mount with a shield and a 7.62 mm machine gun. These vehicles have full amphibious capability.

The Rheinmetall 105 mm Super Low Recoil gun has been successfully tested on the MC 601 chassis, a derivative of the Transportpanzer 1 (6 × 6) vehicle.

Export vehicles will not have the extensive range of equipment fitted as standard to vehicles of the German Army and GLS is therefore offering the following equipment as options: 12 000 Kcal/hour heater, induction warming device, NBC system with an output of 3 m^3/min, Teldix FNA vehicle navigation system, six electrically operated launcher cups for smoke or fragmentation grenades on both sides of the hull, 5000 Kcal/hour air-conditioning system with two evaporators, automatic fire-extinguishing system, intercom system with external and internal telephones, passive night vision equipment for the driver, trailer coupling, non-skid chains, self-recovery winch and a maximum of six ball mounts to enable the crew to use their small arms from inside the vehicle.

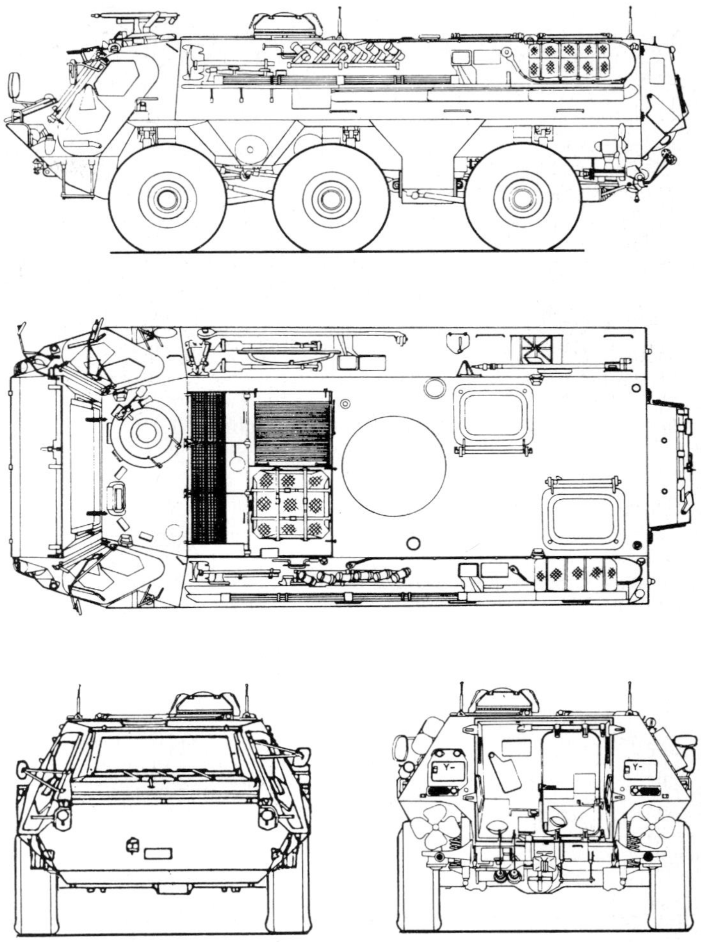

General arrangement drawing of the US Army M93 Fox NBC reconnaissance system (not to 1/76th scale)

US Army M93 Nuclear, Biological, Chemical Reconnaissance System

The United States Army purchased five Thyssen Henschel Fox (6 × 6) NBC reconnaissance vehicles from Germany which were tested alongside two 8 × 8 vehicles from the Diesel Division, General Motors of Canada with TRW being the prime contractor.

In the Spring of 1990 the US Army selected the Fox to meet its requirement for the M93 Nuclear, Biological, Chemical Reconnaissance System.

In March 1990 General Dynamics, Land Systems Division was awarded a $13.878 million contract to improve the M93 system with 66 per cent of the work being carried out in the US and the remaining 34 per cent in Germany with Thyssen Henschel being the main sub-contractor.

At the same time General Dynamics, Land Systems Division, was awarded a $15.11 million contract for the supply of eight interim M93 systems.

The original US Army requirement was for a total of 576 vehicles at a total programme cost of $790 million with the initial requirement being for 48 vehicles with a further 231 to an upgraded specification to follow from 1993, after which the original 48 will be upgraded to the latest configuration. Following the decision to cut back the US Army, the total US Army requirement has yet to be redefined.

Following the Iraqi invasion of Kuwait in the Summer of 1990, the German Government offered a number of standard production German Army NBC reconnaissance vehicles to the Allied Coalition Forces who had no vehicle of this type. Israel (8), Turkey (4), UK (11) and USA (60) were supplied with these vehicles.

The German Army vehicles were delivered back to Thyssen Henschel at Kassel for 'Americanisation' and then delivered to Saudi Arabia.

The upgrading of the vehicles included integrating US communications equipment, weapons, smoke grenade launchers, air-conditioning and other engineering changes.

US Army crews were trained to perform these functions by the Bundeswehr at their NBC School at Sonthofen, Germany.

To replace the vehicles supplied to the Allied Coalition Forces for use in Saudi Arabia, the German MoD placed an order with Thyssen Henschel for the supply of a further 45 replacement vehicles.

TPz 1 Fox EOD Vehicle

Thyssen Henschel has completed the prototype of the Fox EOD (Explosive Ordnance Disposal) vehicle based on the Transportpanzer 1 (6 × 6) chassis to meet the German Air Force requirement for an airfield EOD vehicle (EOD - EinsFz).

This is essentially a modified Transportpanzer chassis fitted with a one-man TS-15 turret and carries specialised EOD equipment.

The modifications to the chassis include:

(1) Partially reinforced armour plating, mainly in the front part, to increase protection to the crew
(2) Modified roof to allow for the installation of the turret
(3) Omission of the vehicle's smoke launcher unit and lateral vision blocks in the crew compartment
(4) Omission of the water propulsion unit
(5) Internal modifications to accept the EOD operational kit.

The one-man Rheinmetall turret system TS-15 is armed with a 20 mm cannon with a 7.62 mm machine gun being mounted coaxial to the right with three smoke dischargers being mounted either side of the turret firing to the front.

The targets to be engaged are selected by the commander seated in the front of the vehicle next to the driver. EODs are engaged using the 20 mm cannon or 7.62 mm machine gun with the turret being provided with a Steinheil LZL 2000 sight with integrated laser rangefinder. A coaxial Xenon searchlight enables targets to be engaged at night.

A total of 220 rounds of 20 mm and 500 rounds of 7.62 mm ammunition are carried for ready use.

Export Variants

Armoured mortar carrier with hull-mounted 81 mm mortar or towing 120 mm mortar.

Armoured missile launcher with turret-mounted TOW, HOT or MILAN ATGWs.

Armoured support vehicle for use in a variety of roles such as ambulance, cargo, recovery and maintenance.

Armoured infantry fighting vehicle with various armament options including 20 or 25 mm cannon with the option of a 7.62 mm remote-controlled machine gun over the rear part of the troop compartment.

Components of the Transportpanzer 1 (6 × 6) vehicle were also used in the Wildcat twin 30 mm air defence system developed under the leadership of Krauss-Maffei. This was developed to the prototype stage and tested under a wide range of operational conditions but is no longer being marketed.

The chassis of the Wildcat could also be fitted with other armament installations. For trials purposes the chassis has already been fitted with a two-man turret fitted with a Cockerill 90 mm Mark III gun and coaxial 7.62 mm machine gun.

A fully amphibious 8 × 8 version of the vehicle with a 25 mm turret was developed to the prototype stage.

Status: In production. In service with Germany, Israel (NBC version only), United Kingdom (NBC version only), United States (NBC version only) and Venezuela.

Manufacturer and prime contractor: Thyssen Henschel
Postfach 102969
D-3500 Kassel
Federal Republic of Germany
Telephone: (0561) 8011
Telex: 99750

After sales service support: GLS
Postfach 500231
D-8000 Munich 50
Federal Republic of Germany

SPECIFICATIONS					
CREW	2 + 10*	ANGLE OF APPROACH/ DEPARTURE	45°/45°	TYRES	14.00 × 20
CONFIGURATION	6 × 6	MAX SPEED		BRAKES	
COMBAT WEIGHT	17 000 kg	road	105 km/h	main	air-assisted, dual circuit
UNLOADED WEIGHT	14 400 kg	water	10.5 km/h	parking	spring-loaded acting on gearbox output shaft
POWER-TO-WEIGHT RATIO	18.82 hp/tonne	FUEL CAPACITY	390 l	ELECTRICAL SYSTEM	24 V, 250 Ah
LENGTH	6.8 m	MAX ROAD RANGE	800 km	BATTERIES	4 × 12 V, 125 Ah
WIDTH	2.98 m	FORDING	amphibious	GENERATOR	Bosch T3, 28 V, 180 A
HEIGHT (to hull top)	2.48 m	GRADIENT	70%	ARMAMENT	1 × 20 mm cannon or 1 × 7.62 mm MG
INTERNAL DIMENSIONS		TURNING RADIUS	8.5 m	SMOKE-LAYING EQUIPMENT	6 smoke dischargers
length	3.2 m	ENGINE	Mercedes-Benz model OM 402A V-8 liquid-cooled diesel developing 320 hp at 2500 rpm		
width	1.42 m	TRANSMISSION	ZF model 6 HP 500 6-speed automatic with torque converter		
height	1.25 m	STEERING	hydraulic, recirculating ball, operating on 1st and 2nd axles		
GROUND CLEARANCE					
axles	0.406 m				
hull	0.506 m				
TRACK					
front	2.54 m				
rear	2.56 m				
WHEELBASE	1.75 m + 2.05 m				

* Maximum is 14 for export vehicles
All versions are fully amphibious using propellers except EW version which is propelled in water by its wheels

HWK 11 Armoured Personnel Carrier

Forty of these vehicles were built by Henschel-Werke between 1964 and 1965 and these remain in service with the Mexican Army. Full details were given in *Jane's Armour and Artillery 1983-84*, pages 302-303.

The NIMDA company of Israel has modernised a single example of the HWK 11 APC with the original powerpack being replaced by a new one consisting of a Model 6V-53 Detroit Diesel developing 210 hp coupled to a TX-100 automatic transmission. This gives the HWK 11 a maximum speed of 60 km/h and a maximum range of 400 km.

Thyssen Henschel Condor Armoured Personnel Carrier

Development

The Condor has been developed as a private venture by Thyssen Henschel as a successor to its UR-416 armoured personnel carrier. Main improvements over the earlier vehicle can be summarised as: increased road speed and load carrying capacity, improved ballistic protection, and its being fully amphibious. Although designed primarily as an armoured personnel carrier, the vehicle can also be adopted for a wide range of other roles including anti-tank, cargo carrier, command vehicle, ambulance, fitters' vehicle and reconnaissance vehicle.

The first prototype was completed in 1978 and the vehicle is now in production. Wherever possible, standard automotive components have been used in the design of the Condor, both to reduce costs and to enable spare parts to be obtained from a commercial source.

Late in 1981 Thyssen Henschel was awarded a contract by the Malaysian Government for 459 Condor armoured personnel carriers, the last of these was completed in March 1984. Some of the Malaysian Condors were fitted with a British Helio Mirror FVT 900 one-man turret armed with an Oerlikon 20 mm cannon and variants included an ambulance, twin 7.62 mm machine guns, fitters' vehicle with a crane and a command post vehicle.

Description

The hull of the Condor is made of all-welded armoured steel that protects the crew from 7.62 mm armour-piercing rounds and 5.56 mm ball-type rounds, shell splinters and anti-personnel mines.

Bullet-proof windows are provided to the driver's front and sides. When in combat, the windows are covered by armoured flaps. A periscope and slit between the armoured flaps provide sufficient vision for the driver in combat situations.

The two rigid axles are supported by Panhard-type bars and torque tubes with axle motion damped by progressively acting coil springs. Each axle has double-acting hydraulic shock absorbers and a differential lock.

The engine, which is the same as that used in the well-known Daimler-Benz Unimog range of cross-country vehicles, is mounted to the right of the driver with the air-inlet and air-outlet louvres in the top of the hull to the right and rear of the driver and the exhaust pipe on the right side of the hull. The engine is mounted on three shock-absorbing suspension points and power is transmitted to the gearbox by a short drive shaft. The gearbox is a Daimler-Benz model UG3/40-8/13.85 GPA and has eight gears which can be used in both directions. Power is transmitted to the front and rear axles by Cardan shafts mounted inside torque tubes.

Condor APC fitted with Thyssen Henschel turret armed with twin 7.62 mm machine guns

Condor APC from rear with KUKA Type 605 12.7 mm machine gun turret, the winch is behind the cover at left rear of the vehicle

Two Thyssen Henschel Condors of the Malaysian Army fitted with a Helio Mirror FVT 900 turret armed with a 20 mm cannon

Condor armoured recovery vehicle in travelling configuration

The two rigid axles are located by Panhard-type bars and torque tubes with axle motion being damped by progressively acting coil springs. Each axle has two double-acting hydraulic shock absorbers and a differential lock. The tyres allow a maximum road speed of 50 km/h to be maintained, even after they have been punctured and deflated, to a maximum range of 30 km. If the vehicle is driven at a lower speed the range is longer.

The commander sits behind the driver facing the centre of the vehicle with a single piece hatch cover. The Condor has doors on both sides and the rear that allow rapid embarkation of the crew. Firing ports in the hull sides and rear door allow some of the troops to fire their weapons from within the vehicle.

The Condor is fully amphibious, propelled in the water by a propeller mounted under the rear of the hull. This can be fully traversed through 360° by the driver who is provided with an indicator to show the position of the propeller.

Before entering the water a trim vane is erected at the front of the vehicle. Optional equipment includes air-conditioning equipment, radios, intercom, night vision equipment, NBC system, propeller kit and a winch with 50 m of cable which can be operated to the front or rear of the Condor by remote control.

Variants

Other armament installations can be fitted including a Thyssen turret armed with twin 7.62 mm machine guns and a turret armed with ATGWs such as HOT, and 12.7 mm and 7.62 mm machine guns.

Status: Production as required. In service with Indonesia (50), Malaysia, Portugal (four ordered with German Government funding in mid-1984), Turkey, Uruguay and other countries.

Manufacturer: Thyssen Henschel
Postfach 102969
D-3500 Kassel
Federal Republic of Germany
Telephone: (0561) 8011
Telex: 99750

SPECIFICATIONS (standard version)	
CREW	2 + 12
CONFIGURATION	4 × 4
COMBAT WEIGHT	12 400 kg
UNLOADED WEIGHT	9200 kg
POWER-TO-WEIGHT RATIO	13.54 hp/tonne
LENGTH	6.05 m
WIDTH	2.47 m
HEIGHT	
to turret top	2.79 m
to hull top	2.08 m
GROUND CLEARANCE	0.475 m
TRACK	1.85 m
WHEELBASE	3.275 m
ANGLE OF APPROACH/DEPARTURE	41°/51°
MAX SPEED	
road	100 km/h
water	10 km/h
FUEL CAPACITY	280 l
MAX ROAD RANGE	900 km
FORDING	amphibious (with propeller)
GRADIENT	60%
SIDE SLOPE	30%
VERTICAL OBSTACLE	0.55 m
ENGINE	Daimler-Benz OM 352A 6-cylinder super-charged water-cooled diesel developing 168 hp at 2800 rpm
TRANSMISSION	manual giving 8 speeds in both directions
CLUTCH	single dry plate
SUSPENSION	coil springs with double-acting hydraulic shock absorbers
TYRES	13.20 × 20
BRAKES	dual circuit hydraulic disc brakes
ELECTRICAL SYSTEM	24 V
ARMAMENT	
main	1 × 20 mm cannon
coaxial	1 × 7.62 mm MG
SMOKE-LAYING EQUIPMENT	2 × 3 smoke dischargers
AMMUNITION (in turret)	
20 mm	220 rounds
7.62 mm	500 rounds
GUN CONTROL EQUIPMENT	
Turret power control	manual
by commander	no
by gunner	yes
Gun elevation/depression	+60°/−7°
Turret traverse	360°

Thyssen TM 170 Armoured Personnel Carrier

Development

The TM 170 is based on the chassis of a Daimler-Benz cross-country vehicle and spare parts are therefore easily obtainable world wide.

The TM 170 was announced for the first time in 1978 and entered production in 1979. It has been designed primarily for use as an armoured personnel carrier or an internal security vehicle, but can be adapted for a wide range of other roles such as an ambulance, anti-tank vehicle, command post or reconnaissance vehicle.

The TM 170 has been selected by the German Border Guard and State police to replace the old MOWAG MR 8 series of APCs, designated the SW1 and SW2. The TM 170 has been designated the SW4, the SW3 being the armoured version of Mercedes-Benz (4 × 4) light vehicle. The first order for 87 vehicles was placed in July 1982 with deliveries running from 1983 onwards. By early 1991 269 had been built for home and export markets.

The TM 125 and TM 90 (4 × 4) armoured personnel carriers are no longer offered.

Description

The hull of the TM 170 is all welded which provides protection from small arms fire and shell splinters. The engine is at the front of the hull, commander and driver immediately behind the engine and the troop compartment at the rear. The body is attached to the chassis at five points by gimbal mounts and can quickly be removed and lifted for repair or maintenance and overhaul such as changing the engine. The engine is coupled to a manual gearbox and the driver can select either 4 × 4 drive for cross-country use or 4 × 2 drive for roads, with the differential lock. All vehicles have run-flat tyres as standard.

The driver sits on the left with the commander to his right. There is a large windscreen which can be covered in action by two individually operated armoured shutters hinged at the bottom which fold forwards onto the bonnet when not in use. The commander and driver each have three roof-mounted periscopes for observation to the front and side of the vehicle when the armoured windscreen shutters are raised. To their side is a window which is covered in action by a shutter hinged at the bottom, while over each of their positions is a single piece hatch cover that opens to the rear of the vehicle.

Apart from the driver and co-driver a maximum of 10 fully equipped troops can be carried seated on individual seats down each side of the hull. There is a door in each side of the hull which opens parallel to the vehicle's body, immediately behind the commander and driver, with a single firing port in each. There is also an entrance hatch at the rear of the hull which folds downwards on the outside. In the centre of it is a vision block with a spherical firing port underneath and a small firing port either side. In each side of the hull, to the rear of the side entrance door, is one firing port and one vision block with a spherical firing port underneath. The spherical firing ports enable the crew to use their small arms from within the vehicle in complete safety. A rifle or SMG fitted in one of the spherical firing ports has a traverse of 16° left and right, an elevation of +23° and a depression of -8°.

In the centre of the roof is a large circular single piece hatch where a variety of armament installations can be mounted.

All vehicle types have a fording depth of 1.3 m. There is also a fully amphibious version with a trim vane which moves upwards at the front of the bonnet where it is stowed when travelling on land. When afloat the TM 170 has a maximum speed of 10 km/h being propelled and steered by two hydraulically operated propellers which can be turned through a full 360°.

The vehicle can be fitted with the following optional equipment: auxiliary heater, fire warning and extinguishing system, hydraulically operated 5000 kg capacity winch with 40 m of 13 mm diameter cable, public address system, police flashing lights, radio, NBC filtering system and run-flat tyres. A single roof-mounted tear gas/smoke discharger with aiming device can be traversed through a full 360°. This can be reloaded from within the vehicle and triggered by remote control. As an alternative banks of four dischargers each with a fixed direction can be mounted on the roof. Another option is a hydraulically operated obstacle-removal blade at the front of the hull. When not in use this can be carried at the rear of the vehicle.

The following armament installations are available for the TM 170: pintle- or ring-mounted 7.62 mm machine gun, turret armed with twin 7.62 mm machine guns, Rheinmetall, or KUKA turret armed with 20 mm cannon which can be aimed and fired from within the vehicle, turret armed with four Euromissile HOT or MILAN ATGWs in the ready-to-launch position. A surveillance cupola is also available.

Variants

A low-priced version of the TM 170 is known as the Armoured Special Vehicle TM 170 Hardliner which has a number of features comparable to those of the original TM 170 BGS version.

The two front and two side windows for the driver and commander are fitted with bullet-proof glass. When in a combat area the front windows are covered by armoured shutters hinged at the top which are lowered manually. The ones either side of the commander and driver are in the upper parts of low doors which open forwards, the windscreens on these are covered by swivel-type armoured plates. With the windows covered by bullet-proof shutters, a set of three periscopes each for the commander and driver provide forward observation and above their seats are two rear-opening circular roof hatch covers.

Either side of the vehicle are four firing ports and the crew enter via a large ramp in the hull rear which is provided with one firing port.

On the roof of the troop compartment can be mounted a turret armed with 7.62 mm or 12.7 mm machine guns or a 20 mm or 25 mm cannon. To the rear of this is a single piece circular hatch cover that opens towards the front. The TM 170 Hardliner has a fording depth of 1.3 m.

The TM 170 Hardliner Patrol is a pick-up type vehicle but with a fully enclosed cab for the commander and driver and an open troop compartment at the rear with bench seats down either side.

For riot control the TM 170 can be fitted with a 1000 litre water tank plus a CN/CS gas tank. The single CN/water cannon (monitor) ejects intermittent water/gas mixture shots up to 50 m through 360°.

NBC Scout Car

In mid-1988 an NBC scout car version of the TM 170 (4 × 4) APC was shown for the first time. This is a private venture and is a joint development between two German companies, DECO-DELTA of Stuttgart and Thyssen Maschinenbau of Witten-Annen.

TM 170 BGS APC in its standard configuration, multi-purpose mount on hull front and single grenade discharger on roof to rear of commander's cupola

TM 170 APC fitted with one-man turret armed with twin 7.62 mm machine guns and with visors lowered

TM 170 NBC scout car with windscreens covered by armoured shutters and mast retracted into vehicle

The main roles of the vehicle are the exact determination of an area exposed to nuclear or chemical contamination, carrying out measurements in contaminated areas, marking such areas from within the safety of the vehicle, detection analysis of all chemical warfare agents in the air, ground sampling without leaving the safety of the vehicle and the quick reporting of measurements to the NBC operations centre so that the appropriate action can be taken.

Status: In production. In service with German Border Guard and Police Forces and in undisclosed Near and Middle East countries.

Manufacturer: Thyssen Maschinenbau
Stockumerstrasse 28
5810 Witten-Annen
Federal Republic of Germany
Telephone: (02302) 6611
Telex: 8 229 117
Fax: (02302) 661666

SPECIFICATIONS					
CREW	2 + 10	GROUND CLEARANCE	0.48 m	VERTICAL OBSTACLE	0.6 m
CONFIGURATION	4 × 4	TRACK	1.84 m	GRADIENT	80%
COMBAT WEIGHT	11 650 kg	WHEELBASE	3.25 m	ENGINE	Daimler-Benz OM 352 turbo-charged diesel developing 168 hp at 2800 rpm
UNLOADED WEIGHT	9500 kg	ANGLE OF APPROACH/ DEPARTURE	45°/48°		
POWER-TO-WEIGHT RATIO	14.5 hp/tonne	MAX SPEED road	100 km/h	STEERING	hydraulic, power assisted
LENGTH	6.275 m	water	9 km/h		
WIDTH	2.5 m	FUEL CAPACITY	170 l	ELECTRICAL SYSTEM	24 V
HEIGHT		MAX ROAD RANGE	720 km	BATTERIES	2 × 12 V, 125 Ah
(without armament)	2.41 m	FORDING	1.2 m	ARMAMENT	optional

Thyssen UR-416 M Armoured Personnel Carrier

Development

The UR-416 was developed by Rheinstahl Maschinenbau (now Thyssen Maschinenbau) as a private venture, the first prototype being completed in 1965. Production began in 1969. By early 1991 1030 UR-416 had been built, the vast majority of which have been exported, mainly to the Near East, South America and Africa.

The vehicle has been designed primarily for internal security operations but can also be used for a wide variety of other roles such as command and communications, reconnaissance and field workshop. The UR-416 is essentially the chassis of a Daimler-Benz cross-country vehicle fitted with an armoured body: spare automotive parts are identical to those used in the truck and are therefore available from commercial sources.

Description

The current production model of the UR-416 is called the UR-416 M. The modifications and improvements have been included in the following detailed description.

The hull of the UR-416 M is of all-welded steel armour construction which protects the crew against small arms fire, shell splinters and anti-personnel mines.

The driver is seated at the front of the vehicle on the left, just behind the engine, with the vehicle commander to his right. Both have a large bullet-proof front and side window for improved visibility. Armoured flaps covering the commander's and driver's front windows are lowered by gas pressure, and side windows are protected by swivelling armoured flaps. Forward observation is maintained by two single periscopes in the forward part of the roof.

The engine is coupled to a manual gearbox which has six forward and two reverse gears. For normal road use only the rear axle is engaged but when travelling across country the front axle is also engaged and when travelling across very rough country the front and rear axle differential locks are engaged.

The eight fully equipped men are seated to the rear of the commander and driver, three down each side of the hull facing outwards and two at the back facing the rear. Each man has an individual seat which can be folded upwards. There are three doors, one in each side of the hull and one at the rear. The side doors are in two parts: the lower part folds downwards to form a step and the upper part is hinged on one side and folds flat onto the side of the hull. The rear door is a wide gas spring supported ramp providing fast mounting and dismounting of the crew with the spare wheel and tyre mounted externally.

There are altogether six firing ports, two in the rear door, one in each of the side doors and a further two ports in the hull side. In addition each side wall has an observation block fitted with a spherical ball mount underneath.

In the roof of the UR-416 M, behind the commander and driver, is a circular single piece hatch that opens to the rear and is arrested on the roof. The basic model is normally armed with a 7.62 mm machine gun with a shield, which has an elevation of +75°, a depression of -10° and a total traverse of 360°. It can also be delivered with a horseshoe-shaped rail on which a 7.62 mm machine gun and shield with a traverse of 235° can be mounted. To the rear of this armament installation is a forward-opening single piece hatch cover, normally with a single firing port which can be used when the hatch is locked vertical.

Forward of the commander's and driver's position are two side-mounted ventilators which draw in fresh air from the outside and pipe it to two channels which run down either side of the crew compartment. Each channel has six controlled air-outlets which can be adjusted by the crew. The crew compartment also has handrails, lights and mountings for personnel equipment.

A feature of the UR-416 M is that the armoured body can be removed and lifted from the chassis for major repairs and overhaul such as changing the engine. The lifting equipment consists of three jacks, one of which is fitted under each side of the hull and the third at the rear which acts as a pivot. The chassis can be driven away once the superstructure has been lifted.

Optional equipment includes an automatic fire warning and extinguishing system, independent heater, night vision periscopes, run-flat tyres, smoke dischargers (either turret- or hull-mounted), and an air-conditioning system. A 5000 kg capacity winch with 40 m of 12 mm diameter cable can also be fitted. This can either be driven by the PTO of the engine or an independent unit mounted onto one half of the exit door.

Variants

Ambulance

Carrying eight sitting or four sitting and two stretcher patients plus a crew of two.

Command and Communications

Fitted with additional communications equipment and mapboards.

Internal Security

The internal security model can be fitted with an obstacle-clearing blade at the front of the hull. The lower half of the blade is made of welded steel with web stiffening, and the upper half consists of a pipe framework with a robust wire grille which gives the driver and commander forward observation. The height of the blade can be adjusted hydraulically from the driver's seat and it can also be removed for transport when it is usually stowed at the rear of the hull.

The vehicle can be fitted with the same turrets as the reconnaissance model but one cupola has been specifically developed for the internal security role. This cupola can be rotated through 360° and is infinitely lockable in any required position. At the front of the cupola above each other are two vision blocks with bullet-proof glass and ball mounts. The cupola hatch opens to the rear and can be locked at 180°. Two vision blocks with ball mounts and a further 10 vision blocks (in double rows)

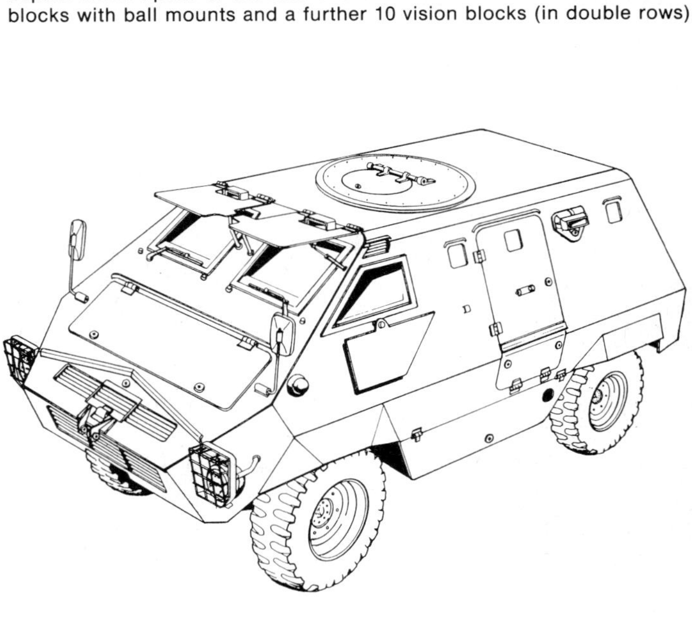

UR-416 M APC with ring mount for armament (not to 1/76th scale)

give improved observation. To the right of the two ball mounts is a flap which is opened to hold a tear gas nozzle. Tear gas mixture is provided from tanks which hold a maximum of 500 litres.

Reconnaissance

This can be fitted with most types of turret on the market today including the German Rheinmetall turret models TF 20 15 (Tur 1/1006) and the Tur 2/1007 (twin MG turret), both of which have been developed in co-operation with Thyssen Maschinenbau (formerly Rheinstahl Maschinenbau).

Workshop

This has a full range of tools, workbenches, vice and cutting equipment, and an A-frame can be erected at the front of the hull to enable the vehicle to change engines and other components. When the A-frame is in use two stabilisers are lowered at the front of the hull.

UR-416 M APC without armament showing driver's and commander's armoured shutters in locked open position

SPECIFICATIONS		MAX ROAD SPEED	81 km/h	STEERING	hydraulic, power assisted
CREW	2 + 8	FUEL CAPACITY	150 l		
CONFIGURATION	4 × 4	MAX ROAD RANGE	600-700 km	CLUTCH	single dry plate
COMBAT WEIGHT	7600 kg	FUEL CONSUMPTION		TYRES	12.50 × 20
UNLOADED WEIGHT	5400 kg	(road)	0.25 l/km	BRAKES	
POWER-TO-WEIGHT		FORDING	1.3 m	main	hydraulic, air assisted
RATIO	16.5 hp/tonne	GRADIENT	70%	parking	mechanical on rear wheels
LENGTH	5.1 m	VERTICAL OBSTACLE	0.55 m		
WIDTH	2.25 m	TURNING RADIUS	6.45 m	ELECTRICAL SYSTEM	24 V
HEIGHT	2.25 m	ENGINE	Daimler-Benz OM 352 6-cylinder water-cooled diesel developing 120 hp at 2800 rpm	BATTERIES	2 × 12 V, 110 Ah
GROUND CLEARANCE	0.44 m			ARMAMENT	
TRACK	1.78 m			basic model	1 × 7.62 mm MG
WHEELBASE	2.9 m			SMOKE-LAYING EQUIPMENT	optional
ANGLE OF APPROACH/ DEPARTURE	47°/51°	TRANSMISSION	manual, 6 forward and 2 reverse gears		

Status: Production as required. Known users of the UR-416 include Ecuador, El Salvador, Germany, Kenya, South Korea, Morocco, Netherlands, Nigeria, Pakistan, Peru, Philippines, Qatar, Saudi Arabia, Spain (National Police), Togo, Turkey and Venezuela. In many of these countries the vehicle is used by the police rather than by the army.

Manufacturer: Thyssen Maschinenbau
Stockumerstrasse 28
5810 Witten-Annen
Federal Republic of Germany
Telephone: (02302) 6611
Telex: 8 229 117
Fax: (02302) 661666

Thyssen Henschel TH 444 APC

Development

The TH 444 APC has been developed as a private venture by Thyssen Henschel, with the first prototype being completed in 1987. The vehicle has been demonstrated in Germany, Turkey and the United States.

The manufacturer has proposed that the TH 444 could be used for a wide range of military roles such as APC, ambulance, command and control vehicle and weapons carrier.

It is based on the long wheelbase version (2.85 m) of the Mercedes-Benz 250 GD cross-country chassis which is already in service with some 20 armed forces, so making training and logistics easier. To improve its power-to-weight ratio to 30.3 hp/tonne it has been fitted with a more powerful 112 hp diesel engine.

Prototype of the Thyssen Henschel TH 444 (4 × 4) APC (Christopher F Foss)

Description

The layout of the TH 444 is conventional with the engine at the front, commander and driver in the centre with the remainder of the crew at the rear.

The welded steel armour provides protection against 7.62 mm attack and its frame body design is such that the armour package (eg sides, rear and roof) can be manufactured to meet specific user requirements.

The prototype has a single piece bullet-proof windscreen and two forward opening doors in each side, each of the latter has a bullet-proof window in its upper part. The rear area also has a bullet-proof window in its upper part and at the back a rear door with a bullet-proof window in its upper part is provided with the spare wheel being mounted below. Various roof hatch arrangements can be provided including a machine gun mount above the commander's position and forward opening roof hatch.

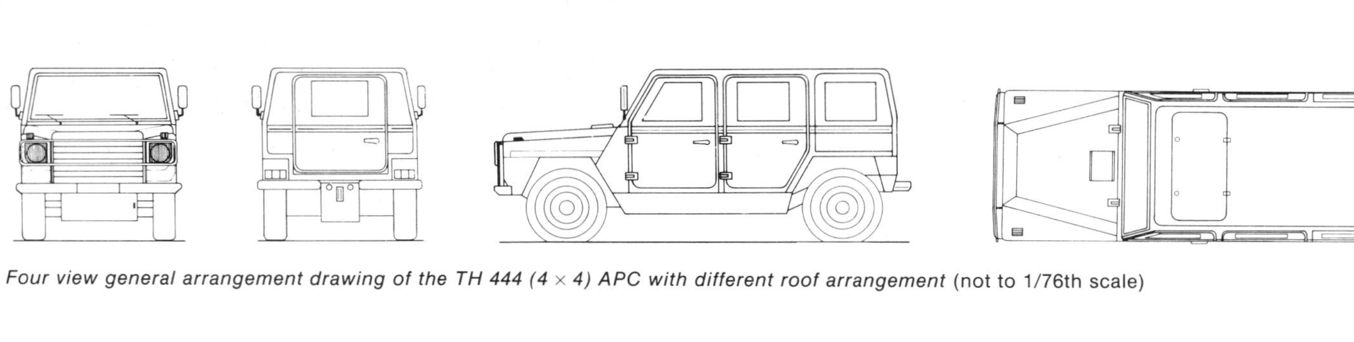

Four view general arrangement drawing of the TH 444 (4 × 4) APC with different roof arrangement (not to 1/76th scale)

SPECIFICATIONS		TRENCH	0.6 m	SUSPENSION	front and rear, coil springs and telescopic shock absorbers. Both front and rear axles are located by 1 transverse and 2 longitudinal control links
CREW	5 - 8	FORDING	0.6 m		
CONFIGURATION	4 × 4	ENGINE	112 bhp diesel		
WEIGHT LOADED	3700 kg	TRANSMISSION	manual, 4 forward and 1 reverse gears		
WEIGHT EMPTY	2900 kg				
POWER-TO-WEIGHT RATIO	30.3 hp/tonne	CLUTCH	single dry plate		
LENGTH	4.35 m	TRANSFER BOX	for mechanical front axle engagement and synchronised cross country gear		
WIDTH	1.73 m			TYRES	7.50 × 16
HEIGHT	1.85 m			NUMBER OF TYRES	4 + 1 spare
GROUND CLEARANCE	0.24 m	STEERING	re-circulating ball, power assisted	BRAKES (MAIN)	hydraulic, dual circuit with vacuum booster, disc front, drum rear
MAXIMUM ROAD SPEED	140 km/h				
ROAD RANGE	600 km	TURNING RADIUS	6.3 m		
GRADIENT	60%			ELECTRICAL SYSTEM	24 V
VERTICAL OBSTACLE	0.3 m			BATTERIES	2 × 12 V, 45 Ah

Status: Development complete. Ready for production.

Manufacturer: Thyssen Henschel
Postfach 102969
D-3500 Kassel
Federal Republic of Germany
Telephone: (0561) 8011
Telex: 99750 thksr d
Fax: (0561) 801 6733

Thyssen Henschel Breitkettenfahrzeug TH 439 APC

Development

The TH 439 full tracked APC has been developed as a private venture by Thyssen Henschel to transport men and equipment over marginal terrain especially swamp, sand, mud and snow.

It is basically the German Kassbohrer Flexmobile FM 23.150K all-terrain chassis, which is already produced for a number of civilian applications (*Jane's Military Vehicles and Logistics 1991-92*, page 464), fitted with a fully armoured body that provides its occupants with protection from 7.62 mm small arms fire and shell splinters. A higher level of ballistic protection can be provided by using a compound armour.

Description

The engine is at the front of the vehicle with the personnel compartment to the rear. The commander and driver are seated at the front of the vehicle with a large bullet-proof window to their front. When required these can be covered by a shutter hinged at the top. A forward opening door is provided in each side of the hull and in its upper part is a bullet-proof window.

In either side of the rear troop compartment are two bullet-proof windows and in the hull rear is a door that opens to the right with a window above. The troop compartment is provided with seats, extractor fans and a heater.

The vehicle uses rubber tracks that can be fitted with bolt-on double steel web plates for difficult terrain; one width of summer track and two widths of winter track, all steel, are available. The main road wheels use a double-track guided system and are self-cleaning. Either side of the vehicle are five dual road wheels with drive sprocket at the front, idler at the rear and three track return rollers.

Variants

The interior and interchangeable lateral wall and roof segments enable the vehicle to undertake a wide range of roles including APC, weapons carrier, ambulance, command post vehicle, tractor and specialised reconnaissance vehicle.

Thyssen Henschel Breitkettenfahrzeug TH 439 APC

Options include a KHD air-cooled diesel engine developing 152 hp in place of the Mercedes-Benz OM 352A diesel engine, hatches for crew in the rear part of the roofing, firing ports with vision blocks within the sidewalls and rear door and a hatch for a machine gun above the commander's/co-driver's seat.

Status: Development complete. Ready for production.

Manufacturer: Thyssen Henschel
Postfach 102969
D-3500 Kassel
Federal Republic of Germany
Telephone: (0561) 8011
Telex: 99750 thksr d
Fax: (0561) 801 6733

SPECIFICATIONS		TRACK WIDTH		FORDING	0.75 m
CREW	1 + 9	summer	700 mm	ENGINE	Mercedes-Benz OM 352A water-cooled 6-cylinder turbo-charged diesel developing 150 hp
WEIGHT LADEN	6500 kg	winter track	825 mm or 925 mm		
WEIGHT UNLADEN	5450 kg	MAX ROAD SPEED	50 km/h		
GROUND PRESSURE	0.16 kg/cm²	FUEL CAPACITY	116 l		
LENGTH OVERALL	4.2 m	ROAD RANGE	250 km		
WIDTH	2.27 m	GRADIENT	100%	TRANSMISSION	hydrostatic
HEIGHT	2.36 m	SIDE SLOPE	40%	STEERING	hydrostatic
GROUND CLEARANCE	0.33 m	VERTICAL OBSTACLE	0.45 m	TURNING RADIUS	pivot
		TRENCH	1.50 m		

GREECE

ELBO Leonidas 1 Armoured Personnel Carrier

Leonidas 1 APC

This is essentially the Austrian Steyr 4K 7FA APC manufactured under licence in Greece for the Greek Army. Its physical characteristics are identical to the original Austrian version. The first Greek Leonidas 1 armoured personnel carrier was completed in 1982 and by early 1989 about 200 vehicles had been built for the Greek Army as well as a quantity of vehicles for the Republic of Cyprus.

Leonidas 2 APC

Early in 1987 an order for an additional 300 Leonidas 2 APCs was placed by the Greek Army with deliveries to be completed by 1993.

By January 1991 a total of 110 vehicles had been completed with production running at five vehicles per month.

The 100 vehicles of the new order will be similar to the existing Leonidas 1 APC but the remainder, to be called Leonidas 2, will have a number of improvements including the incorporation of firing ports and their associate vision devices in the rear troop compartment. This version will be almost identical to the Steyr-Daimler-Puch 4K 7FA-KSPz infantry fighting vehicle described and illustrated under Austria.

In the longer term, funding permitting, the Greek Army is expected to fit its Leonidas 2 APCs with a one-man turret armed with a 25 mm cannon, 7.62 mm coaxial machine gun and smoke dischargers.

At the Defendory Exhibition held in Athens in late 1990, three Leonidas 2 APCs were displayed, each with a different turret, to meet the possible future requirements of the Greek Army. These were:

1. Spanish SANTA BARBARA TC-13 turret armed with McDonnell Douglas Helicopter M242 Chain Gun and 7.62 mm MG. This turret is still at the prototype stage.
2. DAF/FMC Enclosed Weapon Station armed with Oerlikon-Contraves 25 mm KBA-BO2 cannon and 7.62 mm machine gun. This turret has already been produced in quantity for Belgian and Dutch Army Armored Infantry Fighting Vehicles.
3. Norwegian NFT T25 turret armed with 25 mm Mauser Model E cannon and 7.62 mm machine gun. This turret is also still at the prototype stage.

The chassis of the Leonidas has already been fitted with the Brazilian ENGESA ET-90 turret armed with the EC-90 90 mm gun and the French GIAT Industries TS 90 turret armed with a long-barrelled 90 mm gun.

Status: In production. In service with Cyprus and Greece.

Manufacturer: ELBO, Hellenic Vehicle Industry SA
Industrial Area
Sindos Thessaloniki
PO Box 102 39
Greece
Telephone: (031) 798502/798602
Telex: 412586/412811
Fax: (031) 798426

ELBO Leonidas 2 APC with NFT T25 one man turret

Leonidas 1 armoured personnel carrier without 12.7 mm M2 HB machine gun but with bank of four electrically operated smoke dischargers at rear

ELBO Leonidas 2 APC with SANTA BARBARA TC-13 one man turret

Leonidas 2 APC showing firing ports in rear troop compartment but without armament installed

ELBO Pandur Armoured Personnel Carrier

Development/Description

Early in 1991, ELBO, Hellenic Vehicle Industry SA, announced that it was working on a new project, namely the development of the Austrian Steyr-Daimler-Puch Pandur (6 × 6) armoured personnel carrier. This vehicle, which has been developed as a private venture, is fully described under Austria earlier in this section.

The ELBO version will be identical to the Austrian Pandur but will be armed with a 12.7 mm FN QCB (quick change barrel) machine gun, 7.62 mm M63 coaxial machine gun and two banks of eight electrically operated smoke dischargers.

Status

Development. Not yet in production or service with the Greek Army.

Manufacturer: ELBO, Hellenic Vehicle Industry SA
Industrial Area
Sindos Thessaloniki
PO Box 102 39
Greece
Telephone: (031) 798502/798602
Telex: 412586/412811
Fax: (031) 798426

HUNGARY

PSZH-IV Armoured Personnel Carrier

Development

After developing and producing the FUG amphibious scout car the Hungarian defence industry went on to produce a wheeled APC which is used in the Hungarian Army in place of the Soviet or Czechoslovakian/Polish models, in spite of the fact that it carries a smaller landing party and has less cross-country capability.

A certain amount of confusion has arisen over this vehicle and its predecessors. The original Hungarian scout car was officially known as the FUG, and because of the small size and 4 × 4 configuration observers in the West promptly dubbed the 1966 prototype and the PSZH-IV as FUG-66 and FUG-70 as well. The so-called FUG-66, which mounted an egg-shaped turret and dummy automatic cannon, appeared only on a single manoeuvre parade in Bratislava. PSZH in Hungarian is Páncélos Száľitó Harcjármű (or armoured personnel carrier). It is also referred to as the FUG-2, although its correct Hungarian designation, under the Conventional Forces Europe (CFE) Treaty, is the PSZH D-944.

Description

The all-welded hull of the PSZH-IV has the driver and commander seated at the front of the vehicle, both with a hatch cover in front of them hinged at the top and opening forwards, and a windscreen with a wiper blade. The hatch cover is equipped with an integral vision block for observation when the hatch is closed.

A door, four vision blocks and two firing ports are provided in each side of the hull. The ports are positioned one either side of the hull door, next to a vision block. The six infantrymen are seated around the sides and rear of the fighting compartment in what must be very cramped positions.

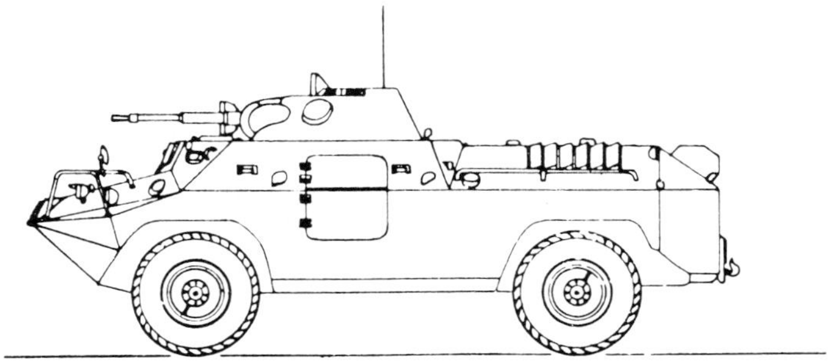

PSZH-IV armoured personnel carrier

PSZH-IV APC showing its amphibious characteristics

Hungarian Army PSZH-IVs clearly showing arrangement of commander's and driver's hatches (Truppendienst (Dr F Wiener))

The turret is in the centre of the vehicle and is a different shape from the one on the BRDM-2, with a stepped top with the forward-facing periscopes in the top part. Like the turret fitted to the BRDM-2 it has no roof hatch. The sight for aiming the armament is in the roof of the turret, just in front of the periscope. A ventilator fan is provided in the left side of the turret.

The engine is behind the turret and is much more powerful than that installed in the earlier FUG. The PSZH-IV is fully amphibious, propelled in the water by two waterjets at the rear of the hull. Before entering the water a trim board, which is stowed folded on the glacis plate when not in use, is erected at the front of the hull. Standard equipment includes infra-red driving lights, an NBC system, power-operated winch and a central tyre-pressure system that allows the driver to adjust the tyre pressure to suit the type of ground being crossed.

Main armament comprises a Soviet 14.5mm KPV machine gun and there is a 7.62mm PKT machine gun mounted coaxially to its left. An infra-red searchlight is mounted coaxially to the right of the main armament.

Variants

A turretless version of the PSZH-IV is used as a command vehicle and other variants of the vehicle exist such as an ambulance and a radiological-chemical reconnaissance vehicle. There is also a turreted command vehicle which does not carry the landing party of six but is armed with a 14.5mm machine gun.

There is also known to be an NBC reconnaissance version of the PSZH-IV (4 × 4) in service which has similar equipment to that fitted to the NBC reconnaissance version of the Soviet BRDM-2 (4 × 4) vehicle (qv *Reconnaissance vehicles* section).

NBC reconnaissance version of PSZH-IV APC

Status: Production complete. In service with Bulgaria, Czechoslovakia (OT-66), Hungary and Iraq.

Manufacturer: Hungarian state factories.

SPECIFICATIONS					
CREW	3 + 6	FUEL CAPACITY	200 l	BRAKES	dual-circuit, air/hydraulic, parking on rear axle
CONFIGURATION	4 × 4	MAX CRUISING RANGE (road)	500 km	ELECTRICAL SYSTEM	24 V
COMBAT WEIGHT	7600 kg	FORDING	amphibious	TYRES	30.00 × 18
POWER-TO-WEIGHT RATIO	13.15 hp/tonne	GRADIENT	60%	ARMAMENT	
LENGTH	5.695 m	VERTICAL OBSTACLE	0.4 m	main	1×14.5 mm MG
WIDTH	2.5 m	TRENCH (with channels)	0.6 m	coaxial	1×7.62 mm MG
HEIGHT	2.308 m	ENGINE	Csepel D.414.44 4-cylinder in-line water-cooled diesel developing 100hp at 2300rpm	AMMUNITION	
FIRING HEIGHT	2.05 m			main	500
GROUND CLEARANCE				coaxial	2000
hull	0.42 m			GUN CONTROL EQUIPMENT	
rear differential	0.3m	TRANSMISSION	manual, 5 forward and 1 reverse gears	Turret traverse	360°, manual
TRACK	1.9 m	CLUTCH	single dry plate	Gun elevation/depression	+30°/-5°
WHEELBASE	3.3 m	STEERING	hour-glass worm with pneumatic assistance	ARMOUR (max)	14 mm
MAX SPEED					
road	80 km/h				
water	9 km/h				

INDIA

Indian Armoured Personnel Carrier Development

The Indian Army APC fleet consists of some 500 BTR-50, BTR-60 (8×8), BTR-152 (6×6), OT-62, OT-64 (8×8) APCs, plus up to 800 Soviet supplied BMP-1 IFV. The Combat Vehicle Research and Development Establishment is known to be developing a new APC or IFV, but progress has been slow as main emphasis has been on the development of the new Arjun Indian MBT to succeed the Vijayanta.

In mid-1983 it was announced that India was to manufacture the Soviet BMP-2 infantry fighting vehicle under licence at the Ordnance Factory at Shankarpally in the Medak district of Andhra Pradesh. The first vehicle, assembled from components supplied in kit form from the Soviet Union, was handed over to the Indian Army in August 1987 and called the Sarath. Full local production was under way by 1991 when key components such as the 30 mm gun and AT-4 Spigot ATGW were being made in India. The Spigot is made by Dharat Dynamics Limited at Bhanoor also in the Medak district.

INTERNATIONAL

Armoured Vehicle AV 90

Development

The Armoured Vehicle AV 90 was announced in 1988 and is a joint private venture development between OTO Melara of Italy and Krupp MaK of Germany, with development commencing in 1986.

It essentially consists of the private venture OTO Melara C13 APC, fully described under Italy, fitted with a new German powerpack consisting of an MTU 8 V 183 TA 22 (Daimler Benz OM 442 A) diesel engine developing 368 hp at 2100 rpm coupled to the ZF LSG 1500 fully automatic transmission with four forward and two reverse gears and a lockable torque converter.

The cooling system has been designed by Behr and is provided with two hydrostatic axial fans while Mann and Hummel have designed the air filter. The design of the powerpack is such that it can operate in high ambient temperatures of up to +55°C (+131°F) such as that encountered in the Middle East.

The cooling system comprises a flat radiator installed above the engine plus two heat exchangers for the gear lubricant and the hydraulic fluid. After cooling the engine and ventilating the engine compartment, air is extracted by means of the two axial fans. This creates a negative pressure in the engine compartment and prevents exhaust fumes from leaking into the crew compartment.

The air filter is attached to the engine and the two dry-filter cartridges can be replaced without removing the power pack. The two fuel tanks are located in the engine compartment. The latter is fitted with a Deugra fire detection and suppression system.

The engine can be uprated to the 600 hp 8 V 183 TE 22 version without any major modification and a further increase to 670 hp is also possible, albeit at the cost of a shorter service life. This allows the vehicle to accept additional armour protection while still retaining a high power-to-weight ratio.

The road wheels of the AV 90 are identical to those used on the M113 series while the tracks are made by Diehl.

The AV 90 is being offered as a potential replacement for the FMC M113 series of vehicles, of which over 75 000 have been built since 1960.

In addition to the sale of complete vehicles, Krupp MaK is also suggesting co-production as some of the key subsystems could be made under licence. For example, the MTU diesel engine is already manufactured under licence in a number of countries as the Daimler Benz OM 442A and used to power heavy trucks.

Variants

A wide range of variants, similar to those proposed for the C13, will be offered.

Status: Prototype. Not yet in production or service.

Manufacturer: Krupp MaK Maschinenbau GmbH
PO Box 9009 Kiel
Federal Republic of Germany
Telephone: (0431) 3811
Telex: 02 99 877 mak d

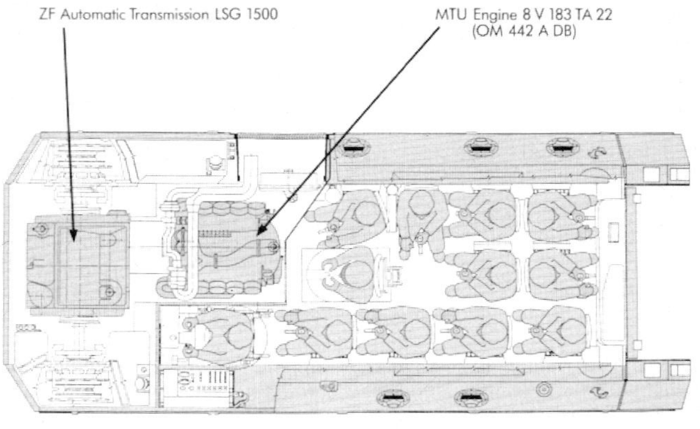

Hull layout showing position of main components and crew seating arrangements

Prototype of the AV 90 clearly showing the layer of appliqué armour on the hull front and sides

SPECIFICATIONS		RANGE AT 45 km/h	565 km	TRANSMISSION	ZF LSG 1500 fully automatic with 4 forward and 2 reverse gears
CREW	12	FUEL CAPACITY	480 l		
WEIGHT	15 000 to 20 000 kg	GRADIENT	60%		
GROUND PRESSURE	0.58 to 0.7 kg/cm²	SIDE SLOPE	30%		
LENGTH	6 m	VERTICAL OBSTACLE	0.7 m	STEERING	hydrostatic cross drive
WIDTH	2.71 m	TRENCH	2 m	BRAKES	air-cooled disc brakes and primary retarder
WIDTH REDUCED	2.57 m	ENGINE	MTU 8 V 183 TA 22 (OM 442A) 8 cylinder, 4-stroke turbocharged, direct-injected liquid-cooled diesel developing 368 hp at 2100 rpm	SUSPENSION	torsion bar
HEIGHT roof	1.72 m			ELECTRICAL SYSTEM	24 V
HEIGHT overall	2.59 m				
GROUND CLEARANCE	0.4 m				
LENGTH OF TRACK ON GROUND	3.334 m				
MAX ROAD SPEED	64 km/h				

NVH-1 Mechanised Infantry Combat Vehicle

Development

In late 1984 Norinco (China North Industries Corporation) and Vickers Defence Systems of the United Kingdom signed an agreement covering a new mechanised infantry combat vehicle (MICV) called the NVH-1.

The prototype of the NVH-1 was completed in mid-1986 and shown for the first time during the Asiandex Exhibition held in Beijing in November 1986. This included a firing and mobility demonstration.

The NVH-1, the N standing for Norinco and V for Vickers, consists of the Chinese H-1 APC with a new two-man turret designed by Vickers Defence Systems.

In early 1988 it was announced that the NVH-1 was a member of the Tracked Armoured Combat Vehicle Type 85 series (qv China in this section).

The NVH-1 has a slightly longer hull than the original Chinese vehicle and is powered by a German Deutz V-8 air-cooled diesel engine made under licence in China.

The Vickers Defence Systems two-man, all-welded steel turret is fitted with a McDonnell Douglas Helicopters 25 mm M242 Chain Gun with a 7.62 mm Chain Gun mounted coaxially to the left of the main armament.

Description

The Norinco-built hull is of all-welded armoured steel with a low profile for increased survivability. The NVH-1 has a crew of three, the commander, gunner and driver and can carry an infantry section.

The driver sits at the front of the hull on the left side with a rear-opening, single piece hatch cover and two periscopes for observation to his front, one of which can be replaced by an infra-red or image intensification periscope for night driving. To the right of the driver are the engine and manual transmission.

The two-man turret is on the roof of the vehicle to the rear of the engine compartment with the gunner on the left and commander on the right, both have individual rear-opening roof hatches.

NVH-1 from rear showing single door with firing port and vision block (Christopher F Foss)

NVH-1 with trim vane folded back on glacis plate (Christopher F Foss)

Mounted coaxially with the main armament is a 7.62mm McDonnell Douglas Helicopters Chain Gun for which 2000 rounds of ammunition are carried.

On either side of the turret are four 66mm grenade launchers which can be loaded with either smoke or anti-personnel rounds.

Both commander and gunner have linked Avimo NVL53 sights with magnifications of ×1 and ×8. Both sights are laser protected, have etched ballistic graticules and 8 and 25° fields-of-view, and are designed to accept a second generation passive night vision module which allows effective undetected night fighting. An Nd-YAG laser rangefinder can also be fitted to the gunner's sight. Target ranges from 300 to 9995 m are projected into the sight eyepiece and can operate in either day or night channels.

As an alternative, the OIP LRS5 day/night system, incorporating a sophisticated fire control computer as well as a laser rangefinder may be fitted.

The turret has seven periscopes for all-round observation and a centrally-mounted rotating periscope which can be used by the commander or gunner.

The infantry sit at the rear of the vehicle and enter via a large door in the hull rear that opens to the right; this has an integral firing port and vision block. Over the top rear of the troop compartment are two rectangular, rear-opening roof hatches. There are also six firing ports and vision blocks in the hull sides of the NVH-1.

The suspension is of the torsion bar type with five dual rubber-tyred road wheels, the idler at the rear, drive sprocket at the front and three track-return rollers. The tracks are rubber bushed.

An air-conditioning unit is also available up to full NBC specification. With the aid of optional pontoons the NVH-1 can be prepared for flotation in 20 minutes.

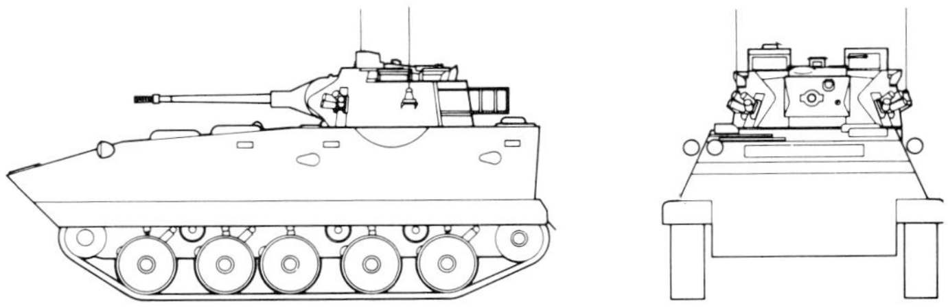

Front and side views of NVH-1 MICV with 25 mm M242 Chain Gun (not to 1/76th scale)

Variants

The NVH-4 is fully amphibious without preparation and can carry 11 men. This version has a 700mm longer chassis with an additional road wheel each side, a combat weight of 17000kg and a maximum road speed of 60km/h. When afloat that vehicle is propelled by its tracks.

As an alternative to the McDonnell Douglas Helicopters 25 mm M242 cannon, the British Royal Ordnance 30 mm RARDEN or the Swiss Oerlikon-Contraves 25 mm KBB cannon is also offered with this vehicle, although all trials to date have been with the M242 cannon.

SPECIFICATIONS

Model	**NVH-1**	**NVH-4**
CREW	2+9	2+11
COMBAT WEIGHT	16000kg	17000kg
POWER-TO-WEIGHT RATIO	20 hp/tonne	18.8 hp/tonne
GROUND PRESSURE	0.64 kg/cm²	0.59 kg/cm²
LENGTH	6.125 m	6.825 m
WIDTH	3.06 m	3.06 m
HEIGHT		
(including turret)	2.77 m	2.77 m
(hull top)	1.91 m	1.91 m
GROUND CLEARANCE	0.46 m	0.46 m
MAX ROAD SPEED	65 km/h	60 km/h
RANGE	500+ km	500+ km
FORDING	amphibious with preparation	amphibious without preparation
GRADIENT	60%	60%
SIDE SLOPE	40%	40%
VERTICAL OBSTACLE	0.6 m	0.6 m
TRENCH	2.2 m	2.5 m
ENGINE	Deutz model BF8L 413F air-cooled, turbo-charged, and inter-cooled diesel developing 320 hp at 2500 rpm	
Model	**NVH-1**	**NVH-4**
TRANSMISSION	manual, 5 forward and 1 reverse gears	
SUSPENSION	torsion bar with linear dampers	
ARMAMENT		
main	1 × 25 mm Chain Gun	1 × 25 mm Chain Gun
coaxial	1 × 7.62 mm MG	1 × 7.62 mm MG
SMOKE LAYING EQUIPMENT	4 × 66 mm smoke dischargers either side of turret	
AMMUNITION		
main	520	520
coaxial	2000	2000
GUN CONTROL EQUIPMENT		
Turret power control	electric/manual	electric/manual
by commander	yes	yes
by gunner	yes	yes
Turret traverse	360°	360°
Max turret speed		
electric	500 mil/s	500 mil/s
manual	200 mil/s	200 mil/s
Max smooth traverse speed		
electric	0.3 mil/s	0.3 mil/s
manual	0.1 mil/s	0.1 mil/s
Max elevation speed		
(manual)	100 mil/s	100 mil/s
Max smooth elevation speed		
(manual)	0.1 mil/s	0.1 mil/s

Status: Development complete. Ready for production.

Manufacturers: China North Industries Corporation
7A Yuetan Nanjie
PO Box 2137
Beijing
People's Republic of China
Telephone: (86) 6898/3461/3471/7570
Telex: 22339

Vickers Defence Systems
Armstrong Works
Scotswood Road
Newcastle-upon-Tyne NE99 1BX
UK
Telephone: (091) 273 8888
Telex: 53104
Fax: (091) 273 2324

NFV-1 Infantry Fighting Vehicle

Development/Description

The NFV-1 Infantry Fighting Vehicle is a joint development by NORINCO of China and FMC of the USA and was shown for the first time in public in November 1986.

The NFV-1 essentially consists of the NORINCO WZ501 chassis, this being a copy of the Soviet BMP-1 Infantry Fighting Vehicle with its original 73 mm turret replaced by the FMC one-man 25 mm electric drive enclosed weapon station. The WZ501 is fully described earlier in this section.

The FMC 25 mm electric drive one-man station is armed with a 25 mm McDonnell Douglas Helicopters M242 Chain gun with a 7.62 mm M240 (FN MAG) mounted coaxial to the left. Turret traverse is electric through a full 360° while elevation, also electric, is from -7° to +44°.

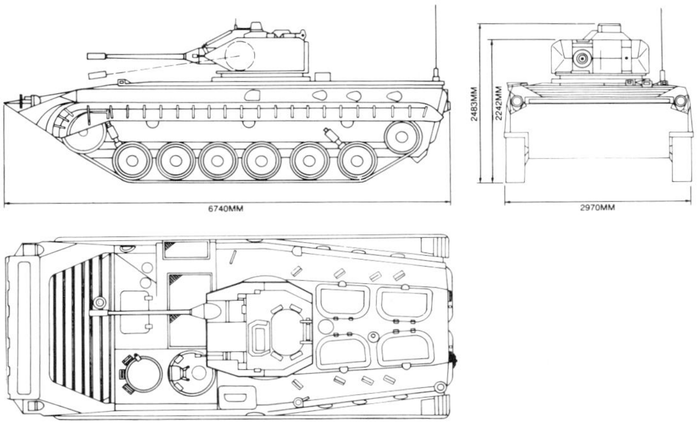

NFV-1 IFV (not to 1/76th scale)

The gunner is provided with a M36E3 sight with a day/night capability, four M27 unity periscopes for frontal and side vision with a rear periscope being an optional extra. The turret is fully stabilised in both azimuth and elevation permitting accurate firing on the move.

NORINCO and FMC had their first talks on this joint venture in February 1986 and the written agreement was signed in June 1986. NORINCO also provided the slip ring and communications system for the vehicle as well as carrying out integration of the hull with the turret.

By the time the vehicle was unveiled about 200 km of automotive trials had been conducted with no problems encountered. First firings took place in mid-1987 although the turret had already been tested on other vehicles including an M113 APC and the British Fox armoured car.

In late 1987 the NFV-1 was evaluated by the Egyptian Army following its appearance at the Cairo defence exhibition.

Status: Development complete. Ready for production.

Manufacturers: FMC Corporation
Ground System Division
2830 De La Cruz Boulevard
Santa Clara
California 95052
USA
Telephone: (408) 289 0111
Telex: 6714210
Fax: (408) 289 2150

China North Industries Corporation
7A Yuetan Nanjie
PO Box 2137
Beijing
People's Republic of China
Telephone: (86) 6898/3461/3471/7570
Telex: 22339

Prototype of NORINCO/FMC NFV-1 Infantry Fighting Vehicle

SPECIFICATIONS		FUEL CAPACITY	460 l	ELECTRICAL SYSTEM	24 V
CREW	3 + 8	GRADIENT	60%	BATTERIES	2 × 12 V
COMBAT WEIGHT	13 600 kg	SIDE SLOPE	47%	ARMAMENT	
POWER-TO-WEIGHT RATIO	21.5 hp/tonne	VERTICAL OBSTACLE	0.6 m	main	1 × 25 mm cannon
GROUND PRESSURE	0.63 kg/cm^2	TRENCH	2 m	coaxial	1 × 7.62 mm MG
LENGTH	6.74 m	ENGINE	Model 6V 150 4-cycle diesel developing 293 hp at 2600 rpm	AMMUNITION	
WIDTH	2.97 m			25 mm, ready use	200
HEIGHT				7.62 mm, ready use	230
to turret top	2.242 m	TRANSMISSION	manual	GUN CONTROL EQUIPMENT	
over turret periscope	2.483 m	CLUTCH	hydraulic	Turret power control	electro-mechanical/manual
TRACK WIDTH	300 mm	STEERING	mechanical, 2-stage planetary	Turret traverse	360°
MAX SPEED		SUSPENSION	torsion bar with hydraulic shock absorbers at first and last road wheel stations	Max traverse rate	60°/s
road	65 km/h			Gun elevation/depression	+47°/-7°
water	7 km/h			Gun stabiliser	optional
MAX CRUISING RANGE	460 km				

IRAN

Iranian Armoured Personnel Carriers

Development/Description
In recent years Iran has developed a number of armoured personnel carriers and brief details of these are given below.

Half-track APC

This has a hull of all-welded steel armour construction with the driver seated front left, engine compartment to his right and fully enclosed troop compartment extending to the rear of the vehicle. The exhaust pipe runs along the right side of the hull.

The troop compartment has a single door in either side of the hull to the rear of the front road wheels and in the upper part of each door is a firing port.

In either side of the troop compartment are two circular firing ports and in the hull rear are two doors each with a firing port.

In the forward part of the roof is a circular device that could be a roof hatch or an air inlet for the engine. Towards the rear of the roof are two or three rectangular roof hatches.

The vehicle is fully amphibious, being propelled in the water by a single water jet mounted under the rear of the hull. This is very similar to that fitted to the Soviet BTR-60 (8×8) series of amphibious armoured personnel carriers.

The rear half-track suspension appears to be of a new design with three rubber-tyred road wheels, drive sprocket at the front, idler at the rear and two track return rollers

The tracks are fitted with grousers for improved traction in soft ground. Steering is on the front wheels but it is not known if this is powered.

This APC is claimed to have a maximum road speed of 70 km/h, maximum water speed of 10 km/h and be able to carry 24 fully equipped troops plus its two-man crew.

Wheeled APC

This is believed to be a 4×4 vehicle with the engine at the front and the fully enclosed troop compartment at the rear. Commander and driver are seated to the immediate rear of the engine compartment and are provided with a windscreen to their front, roof hatches that open to the rear and entry hatches in the sides.

The troop compartment is provided with a firing port and mounted over the troop compartment is a manually operated machine gun turret. This APC can carry a total of 10 troops.

In 1988 the Islamic Revolutionary Guards Corps Industry announced that it had developed and tested a new APC called the Boraq but it is not known to which of the above vehicles, if any, that this relates to.

Status: Not known.

Manufacturer: Not known.

Iranian half-track armoured personnel carrier from rear

Iranian half-track armoured personnel carrier showing its amphibiouis capabilities

Iranian wheeled APC with roof hatches open

IRELAND

Timoney Technology Armoured Personnel Carrier

Development
In January 1972 Timoney Technology began design work on a 4×4 armoured personnel carrier which would meet the requirements of the Irish Army for a vehicle capable of world-wide deployment when operating under United Nations control.

The first prototype, called the Mk 1, was completed in mid-1973 and was followed by a further two prototypes called the Mk 3. In 1976 the Engineering Division of Beherman Demoen of Belgium negotiated a licence with Timoney Technology to undertake production of the Timoney APC. The first Belgian vehicles were completed in 1978 and Beherman Domoen was awarded a contract from the Belgian Government for 123 vehicles (43 for the Air Force and 80 for the Gendarmerie) and from the Argentine Government for five vehicles all of which were delivered by early 1981.

In September 1981 Beherman Demoen and Vickers Defence Systems of the United Kingdom signed a 10-year agreement for Vickers to develop and build an improved version of the BDX in the UK. This was subsequently called the Valkyr and full details are given in this section under the United Kingdom.

Timoney Technology built five pre-production Mk4 vehicles for the Irish Army.

There are only minor differences between the Timoney and BDX vehicles, mainly concerned with internal seating arrangements and the position of the firing ports and/or vision blocks. The Irish vehicles have a twin 7.62mm machine gun turret designed by Timoney Technology. The weapons have an elevation of +55°, a depression of −11° and 360° turret traverse. Gun elevation and depression and turret traverse are manual. The all-welded turret has a two-piece hatch cover that opens to the front and rear, periscopic sight for aiming the weapons, vision blocks, two smoke dischargers either side and a searchlight that moves in elevation with the main armament.

Description
A full description of the Timoney/BDX armoured personnel carrier together with full technical specifications will be found in this section in the BDX armoured personnel carrier entry under Belgium.

Irish Army Timoney Technology Mk 6 APC from rear with all doors closed (Denis McCarthy)

Timoney Technology Mk 6 APC fitted with Creusot-Loire Industrie TLi 127 turret armed with 12.7 mm M2 HB and coaxial 7.62 mm machine guns and smoke dischargers

Variants

Technology Investments has also suggested that the Timoney APC could be extended into a 6×6 APC which would be able to carry 17 men including its crew of two. Other projected variants include the Mk 2 amphibious armoured reconnaissance vehicle with a crew of three and six infantrymen which would have a maximum loaded weight of 6803 kg, maximum road speed of 98 km/h and would be fully amphibious. The powered turret could be fitted with a wide range of armament installations up to a 90 mm gun or a 106 mm recoilless rifle. An amphibious load carrier with a capacity of 5080 kg was also projected.

Mark 6 APC

Further development of the Mk 5 APC, which was the model produced under licence in Belgium, has resulted in the Mk 6, of which five pre-production vehicles were delivered to the Irish Army early in 1982.

The Mk 6 has many improvements over the Mk 5 including redesign of the glacis plate and of the driver's vision ports which are lower, more horizontal and slightly forward of the slope of the glacis plate. The vision ports have a pronounced reveal on the top and sides as have the three firing ports which are arranged symmetrically along each side. The sides of the vehicle are continued up vertically for some 75 per cent of the height before being angled inwards. The air intakes, which are located over the driver's compartment are also lower than in the earlier models.

The wheelbase has been increased from 2.87 m to 3 m while the track has been widened from 1.93 m to 2.2 m; these and other changes to the hull have increased weight to around 8000 kg unladen.

The original Chrysler petrol engine has been replaced by a four-cylinder Detroit Diesel 4-53T, this together with the increased fuel capacity has increased the vehicle's operating range to around 1000 km. The Allison Transmission Division of General Motors AT-540 automatic transmission has been retained.

This, like the earlier Timoney vehicles, has the patent Timoney two-speed transfer case and reduction hubs. The former takes the drive from the shaft at the rear of the engine to the rear axles and, when engaged, via a secondary shaft to the front steering axle. The reduction hubs contain single stage epicyclic reduction gears which supply the additional reduction ratios required. The main differential incorporates no-spin locking units to minimise loss of traction in off-road operations.

The suspension system has also been designed and patented by Timoney. Each of the four wheels is an independent unit suspended by two unequal transverse wishbones and sprung by a single helical spring with a concentric telescopic damper. The drive from the axle units is transmitted by means of splined shafts with universal joints at each end in the case of both the permanently driven rear axles and of the front axles which are normally only powered when operating over rough terrain.

The vehicle can be fitted with a wide range of armament installations including the Creusot-Loire Industrie TLi 127 turret armed with a 12.7 mm M2HB machine gun.

Mark 6 Fire Support Vehicle

In March 1982 Timoney Technology unveiled the Mk 6 fire support vehicle which is essentially the Mk 6 APC with the French Hispano-Suiza Lynx 90 turret armed with a 90 mm gun, 7.62 mm coaxial and 7.62 mm anti-aircraft machine guns. In addition to the crew of three, which consists of the commander, gunner and driver, the vehicle can carry nine fully equipped infantrymen.

Status: To date, 10 vehicles have been delivered to the Irish Army. Production as required.

Manufacturer: Timoney Technology Limited
Trim
Co Meath
Ireland
Telephone: (046) 31622
Telex: 43588 TIL EI
Fax: (046) 36120

SPECIFICATIONS

CREW	3 + 9
CONFIGURATION	4 × 4
COMBAT WEIGHT	9980 kg
UNLOADED WEIGHT	8000 kg
LENGTH	4.95 m
WIDTH	2.5 m
HEIGHT OVERALL	2.746 m
GROUND CLEARANCE	0.4 m
WHEELBASE	3 m
TRACK	2.2 m
MAX ROAD SPEED	95 km/h
RANGE	1000 km
FUEL CAPACITY	300 l
FORDING	amphibious
GRADIENT	70%
SIDE SLOPE	50%
VERTICAL OBSTACLE	0.4 m

ENGINE	Detroit Diesel Model 4-53T
TRANSMISSION	Allison Transmission Division of General Motors AT-540 fully automatic, 4 forward and 1 reverse gears
CLUTCH	hydraulic, torque converter
STEERING	hydraulically assisted, manual in case of hydraulic failure
BRAKES	disc, dual hydraulic circuit
SUSPENSION	independent wishbones with helical springs and telescopic dampers

TYRES	12.00 × 20 or 14.00 × 20
ARMAMENT	
main	1 × 90 mm gun
coaxial	1 × 7.62 mm MG
anti-aircraft	1 × 7.62 mm MG
SMOKE LAYING EQUIPMENT	2 × 2 smoke dischargers
GUN CONTROL EQUIPMENT	
Turret traverse	360°, manual
ARMOUR	
Floor	10 mm
Roof	12 mm
Hull front, back and sides	14 mm

ITALY

IVECO/OTO Melara VCC-80 Infantry Fighting Vehicle

Development

After the completion of feasibility studies followed by the project definition stage, a 60-month development contract was awarded to OTO Melara and IVECO FIAT by the Italian Army in February 1982. This covered the construction of three prototype vehicles, all of which were completed by 1988 and are now undergoing evaluation trials with the Italian Army. As of July 1991 the VCC-80 infantry fighting vehicle had not entered production and was not expected to do so in the near future owing to a shortage of funding.

Description

The hull and turret of the VCC-80 are of all-welded aluminium construction with an additional layer of ballistic steel armour bolted to the hull for increased protection.

The driver sits at the front left side of the vehicle and has a right-opening single piece circular hatch cover. Forward and either side of this are three periscopes, the centre one of which can be replaced by an image intensification periscope for driving at night.

The powerpack is to the right of the driver with the air-inlet and air-outlet louvres in the hull top and the exhaust outlet on the right side of the hull.

The IVECO FIAT Model 8260 four-stroke turbo-charged water-cooled diesel engine develops 520 bhp so giving the VCC-80 a power-to-weight ratio of about 24 bhp/tonne. The engine is coupled to a hydro-mechanical shift reverse and steering system with a built-in retarder giving a total of four forward and two reverse gears.

The ZF LSG 1500 automatic transmission includes a torque converter, a retarder and three epicycloidal gear trains. The steering system is integral with the transmission with hydrostatic control of the steering radii and pivot being continuous.

The control of the forward and reverse ranges is electronic and automatic, including the converter lock-up clutch. The speed ranges can be pre-selected manually among four programmes after which the gears are selected automatically.

The service braking system is of the retarder brake disc type. The retarder is included in the transmission and generates a torque resistance capable of reducing the braking energy up to 70 per cent, so slowing down the vehicle without using the brake pads.

The brake discs are of the self-ventilated type and are splined to the shaft of the final drives. The braking system is of the full power dual circuit type and is fed by a single pump driven directly by the engine.

The cooling system has been designed to enable the vehicle to operate in an ambient temperature of up to +44°C with a high percentage of dust in the atmosphere. The air fan is hydraulically operated with speed control depending on the coolant temperature. The air intake grille is located on top of the front plate of the hull.

Exhaust gases expelled from the engine turbocharger enter a manifold, provided with a silencer, from which they are forced outward through a grid on the right side of the vehicle after mixing with the expelled

One of the prototypes of the VCC-80 from the rear showing power-operated ramp, firing ports and vision blocks

Artist's impression of VCC-80 IFV fitted with OTO Melara two man turret Model T 60/70 A armed with a 60 mm rapid fire cannon

VCC-80 during troop trials with bank of four smoke dischargers on turret side

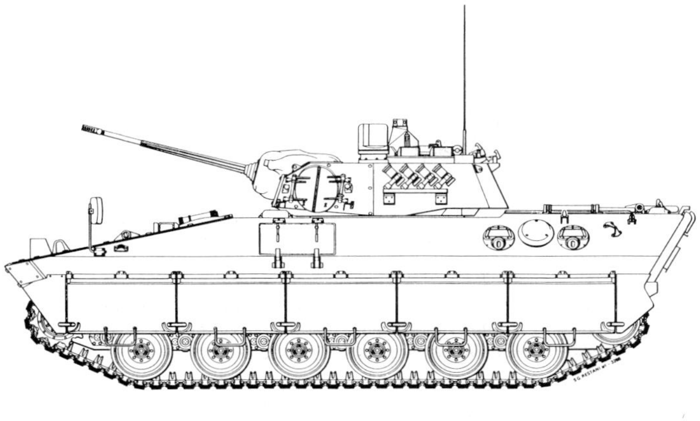

Left side elevation of VCC-80 infantry fighting vehicle (not to 1/76th scale)

air of the cooling system to reduce thermal signature. A hydraulically operated valve, installed on the exhaust gas system, is provided to allow the vehicle to ford streams and rivers.

The electrically operated turret is in the centre of the vehicle with the commander on the left and the gunner on the right, each with an individual rear-opening circular hatch cover.

The commander is also the squad leader when the infantry squad dismount from the vehicle.

The commander has five periscopes for all-round observation and in front of his hatch is a periscope that can be traversed through a full 360°. In front of the gunner is a day/night sight for aiming the 25mm cannon and 7.62mm coaxial machine gun.

The commander's panoramic sight has full 360°, elevation limits from -10° to +60° with magnification of ×2.5 and ×10. The gunner's fixed roof-mounted monocular sight has elevation limits from -10° to +60° with day magnifications of ×1 and ×10 with night magnifications being ×4 and ×12. The gunner's sight also incorporates a laser rangefinder with a maximum range of 10000 m.

The main armament of the VCC-80 is an Oerlikon-Contraves 25mm KBA-B02 cannon with a very advanced fire control system which includes a laser rangefinder and thermal image night vision equipment.

The fire control system for the VCC-80 has been designed by Officine Galileo and is virtually identical to that installed in the C1 MBT and the B1 8×8 tank destroyer.

The dual-feed 25mm cannon has a maximum range of 2500m with 200 rounds of ready-use ammunition being provided, of which 125 rounds are HEI-T and 75 APDS-T.

If required a Euromissile MILAN ATGW launcher can be fitted to the turret roof with four missiles being carried inside the troop compartment for manual reloading.

A 7.62 mm MG 42/59 machine gun is mounted coaxially to the left of the 25 mm cannon and both weapons have an elevation of +60° and a depression of -10°, the former allows the 25 mm cannon to engage slow-flying aircraft and helicopters. Turret traverse is through a full 360° in 8 seconds while elevation is at a maximum speed of 45° a second. On either side of the turret are four electrically operated smoke dischargers.

Typically the commander carries out observation of the battlefield, target search, detection and identification; he then designates the target to the gunner who carries out the engagement. The commander can operate the armament if required.

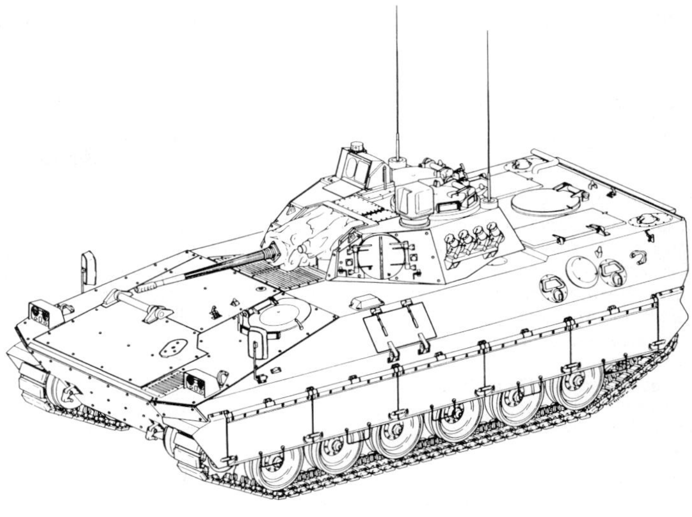

Detailed drawing of VCC-80 showing position of roof hatches in rear troop compartment (not to 1/76th scale)

The six infantrymen sit at the rear of the vehicle, two on each side facing the front and two in the centre, one of which sits facing the left side of the vehicle and the other facing the rear.

The troop compartment has five firing ports each with a vision block, to allow five of the infantrymen to fire their weapons from inside the vehicle. Two of these are in either side of the hull and one is in the rear.

The infantry enter and leave the vehicle via a hydraulically operated ramp at the rear of the hull, which has an integral door for emergency use should the ramp fail to open. Over the top of the troop compartment are two oblong and one circular hatches all of them hinged.

The torsion bar suspension system consists of six road wheels with the drive sprocket at the front, idler at the rear and four track-return rollers. The upper part of the suspension system is protected by a skirt.

Standard equipment on the VCC-80 includes a full range of passive night vision equipment, an air-conditioning system, a complete NBC system, automatic fire extinguishing system for the crew and engine compartment, and equipment to enable the vehicle to ford to a depth of 1.5m. OTO Melara also offer a hydropneumatic suspension system in place of the torsion bar system fitted to the prototype vehicles.

Variants

OTO Melara has already proposed that the VCC-80 be fitted with its 60 mm two-man T 60/70 A turret.

Status: Three prototypes being evaluated by the Italian Army.

Manufacturer: OTO Melara and IVECO FIAT
via Valdilocchi 15
19100 La Spezia
Italy
Telephone: (0187) 530 111
Telex: 270 368
Fax: (0187) 530 669

SPECIFICATIONS

CREW	3 + 6
COMBAT WEIGHT	21700 kg (approx)
POWER-TO-WEIGHT RATIO	24 hp/tonne
GROUND PRESSURE	0.66 kg/cm²
LENGTH	6.705 m
WIDTH	3.0 m
HEIGHT OVERALL	2.7 m
HEIGHT HULL TOP	1.75 m
GROUND CLEARANCE	0.4 m
MAX SPEED	70 km/h
CRUISING RANGE	500 km
FORDING (without preparation)	1.5 m
GRADIENT	60%
SIDE SLOPE	40%
VERTICAL OBSTACLE	0.85 m
TRENCH	2.5 m

ENGINE	FIAT 8260 V-6 turbo-charged, inter-cooled diesel developing 520bhp at 2300rpm
TRANSMISSION	ZF LSG 1500 fully automatic with 4 forward and 2 reverse gears
STEERING	hydrostatic
SUSPENSION	torsion bar
ELECTRICAL SYSTEM	24 V
BATTERIES	4 × 12 V, 180 Ah at 24 V
ARMAMENT	
main	1 × 25 mm cannon
coaxial	1 × 7.62 mm MG
anti-aircraft	1 × 7.62 mm MG
smoke dischargers	2 × 4

AMMUNITION	
25 mm belted	400
7.62 mm belted	1200
MILAN ATGW	4
GUN CONTROL EQUIPMENT	
Turret traverse	360° electric/manual
by commander	yes
by gunner	yes
Gun elevation/depression	+60°/-10°
Gun stabiliser	yes

OTO Melara OTO C13 Infantry Fighting Vehicle

Development

The OTO C13 infantry fighting vehicle has been designed by OTO Melara specifically for export and compared with earlier armoured personnel carriers has superior armour protection, greater mobility and a lower profile.

The C13 was announced in 1982. The first prototype, with no firing ports, was completed early in 1983 and the second prototype, which has firing ports, completed later the same year. By early 1991 five prototypes of the C13 family of tracked armoured vehicles had been completed in different configurations but volume production had yet to start. Further development by Krupp MaK has resulted in the AV90. Details of this are given under International.

Description (APC version)

The hull of the C13 is of all-welded aluminium construction with sloped front and upper hull sides to give maximum possible protection within the weight limits of the vehicle. Armour protection is increased by a steel armour plate system which bolts to the hull front and sides. This gives the C13 protection against armour-piercing machine gun projectiles, artillery and shell fragments.

OTO Melara has a new concept of additional armoured structures with light weight and very high effectiveness under advanced development.

The driver sits at the front of the vehicle on the left side and has a single piece circular hatch cover over his position which lifts and swings to his left rear to open. To his front and left side are four periscopes, one of which can be replaced by an infra-red or passive periscope for night driving.

The powerpack and all mechanical components are to the right of the driver with a large dual access hatch opening both from the outside and from the crew compartment. The air-inlet and air-outlet louvres are in the hull top and left side. The engine cooling radiator and fan are mounted above the engine with the latter being coupled to the engine and the steering control differential which is at the very front of the hull.

The machine gunner/commander is seated in the centre of the hull and has a cupola with five periscopes for frontal and side observation which can be manually traversed through 360°. Mounted on the forward part of the cupola is a 12.7 mm M2 HB machine gun. Either side of the cupola is a bank of four electrically operated smoke dischargers firing forwards. The machine gunner/commander is provided with lateral and rear armour protection.

The troop compartment is at the rear of the hull and the troops enter via a large power-operated ramp at the rear which is larger than that of the M113 and has a door in the left side. Over the top of the troop compartment are three roof hatches, one oblong hatch each side that opens upwards towards the centre of the vehicle and a circular one at the rear which has a single piece hatch cover that opens forwards and can be locked on top of the hull. Mounted in the roof of the troop compartment, one either side at the rear, are two electric fans to remove fumes from the vehicle when the weapons are fired. The troops have individual bucket-type seats that can be folded upwards.

Five vision blocks each with a firing port underneath allow most of the troops to fire their weapons from within the vehicle, three of these are in the right side of the hull and two in the left. There is ample stowage space for personnel equipment for the 12 men and the five short-barrelled 5.56 mm firing port weapons.

The suspension is of the torsion bar type and consists of six dual rubber-tyred road wheels with the drive sprocket at the front, idler at the rear and three track return rollers. Hydraulic shock absorbers are fitted at the first, second, fifth and sixth road wheel stations. The upper part of the track is covered by a ballistic shroud which can be lifted to give access to the upper part of the suspension.

Standard equipment includes two bilge pumps (one front and one rear) and an automatic fire-extinguishing system in the engine compartment. Optional equipment includes an air-conditioning system, NBC system and an amphibious kit.

Variants

OTO C13/20 Infantry Fighting Vehicle

This version has a crew of nine: commander, gunner, driver and six infantrymen, and is fitted with the OTO Melara T20 two-man power-operated turret. This is armed with a 20 mm Rheinmetall MK20 Rh202 cannon, 7.62 mm coaxial machine gun and two banks of three electrically operated smoke dischargers. This version has a combat weight of 15500 kg.

OTO C13/25 Infantry Fighting Vehicle

This version has the same crew as the C13/20 and is fitted with the OTO Melara T25 two-man power-operated turret. This is armed with a 25 mm Oerlikon-Contraves KBA cannon, 7.62 mm coaxial machine gun and two banks of three electrically operated smoke dischargers. This version has a combat weight of 15700 kg.

OTO C13/60 A and M Infantry Fighting Vehicles

These versions have a crew of seven: commander, gunner, driver and four infantrymen and a combat weight of 18000 kg. The two-man power-operated T60/70 turret, is armed with a 60 mm 60/70 hyper-velocity gun, 7.62 mm coaxial machine gun and a bank of four electrically operated smoke dischargers either side of the turret.

The OTO T60/70A turret has an automatically loaded 60 mm gun, the T60/70M has a manually loaded weapon. The former is available with two different fire control systems.

OTO C13/90 Infantry Fighting Vehicle

This version has a crew of nine: commander, gunner, driver and six infantrymen, and is fitted with the OTO T90 CKL two-man turret armed with a 90 mm Cockerill Mk III gun, 7.62 mm coaxial machine gun and two banks of three electrically operated smoke dischargers. Combat weight is 16000 kg.

OTO C13/4 × 25 Anti-aircraft Mobile System

This is armed with an OTO Melara turret with four 25 mm Oerlikon-Contraves KBA automatic cannon designed to meet the requirements of the Italian Army, in which application it is known as the SIDAM 25 and is mounted on the M113A2 APC chassis. This model would weigh 16500 kg and have a crew of three consisting of the commander, gunner and driver. The SIDAM 25 turret system is already in production for the Italian Army on a modified M113 series APC chassis.

OTO C13 TUA

The TOW Under Armor (TUA) is the basic C13 chassis fitted with the US Emerson TOW Under Armor turret as installed on the M901 Improved TOW Vehicle (ITV) used by the US Army and a number of other countries. In addition to the two ready to launch missiles a further 10 missiles are stowed internally. This model has a combat weight of 15000 kg and a crew of four: commander, gunner, driver and loader.

OTO C13 Ambulance

This version is unarmed and has an operational weight of 14650 kg and a crew of three consisting of a driver and two orderlies. It can carry three stretcher cases, two seated wounded and one stretcher case, or six seated wounded.

OTO Melara C13/20 Infantry Fighting Vehicle with T20 two-man power-operated turret

Second prototype of OTO Melara C13 APC armed with 12.7 mm M2 HB machine gun and four electrically operated smoke dischargers either side of cupola

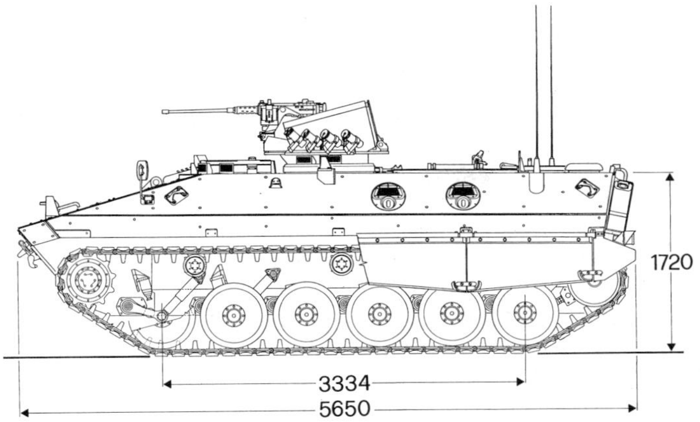

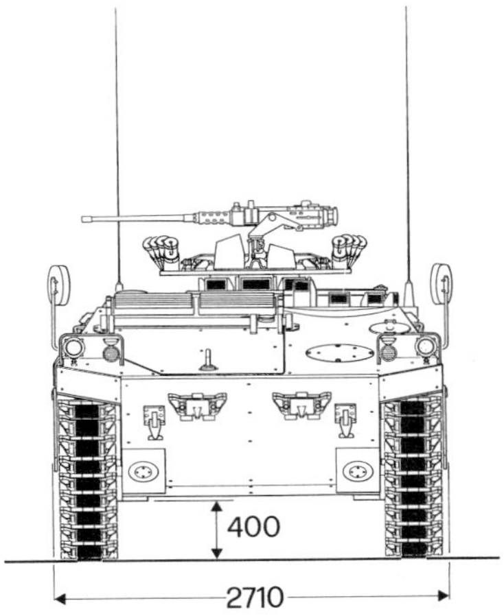

Front and side drawing of OTO Melara C13 APC in basic version with 12.7 mm MG on commanders cupola

OTO C14 carrier with fully enclosed armoured cab fitted with mine laying system and hydraulic crane at rear

Equipment fitted includes wash basin, oxygen equipment for three people, refrigerator, surgical instruments, steriliser, respirator, folding board and miscellaneous medical equipment.

OTO C14 Cargo Carrier

This has a redesigned hull but uses the same track, engine and transmission. The suspension has, however, been lengthened to enable the vehicle to carry a payload of 6500 kg.

The fully-enclosed cab is at the front of the vehicle and is of welded aluminium armour which provides protection against small arms fire and shell splinters. If required, a 12.7 mm machine gun can be mounted on the cab roof for anti-aircraft use. Fully loaded the C13 cargo carrier has a weight of 19000 kg.

The engine has a transfer to allow the transmission to be installed parallel to the engine axis. The transfer has two PTOs including one for a 10000 kg capacity winch. Full details of this version, which can also be used as a minelayer, are given in *Jane's Military Vehicles and Logistics 1991-92* page 465.

Status: Development complete. Ready for production.

Manufacturer: OTO Melara
via Valdilocchi 15
19100 La Spezia
Italy
Telephone: (0187) 530 111
Telex: 270 368
Fax: (0187) 530 669

SPECIFICATIONS (basic C13 IFV with 12.7 mm MG)

CREW	3+9
COMBAT WEIGHT	14 650 kg
UNLOADED WEIGHT	13 350 kg
POWER-TO-WEIGHT RATIO	24.5 hp/tonne
GROUND PRESSURE	0.59 kg/cm²
LENGTH	5.65 m
WIDTH	2.71 m
over tracks	2.54 m
HEIGHT	
hull top	1.72 m
inc MG	2.475 m
GROUND CLEARANCE	0.4 m
LENGTH OF TRACK ON GROUND	3.334 m
MAX ROAD SPEED	70 km/h
FUEL CAPACITY	400 l
CRUISING RANGE	500 km
GRADIENT	60%
SIDE SLOPE	30%
VERTICAL OBSTACLE	0.7 m
TRENCH	1.8 m
ENGINE	Isotta Fraschini model ID 38 SS V6 90° V-6 super-charged diesel developing 360 hp at 2800 rpm
TRANSMISSION	automatic, 5 forward and 1 reverse gears with hydro-dynamic torque converter
STEERING	hydrostatic with indirect power-assisted control operated by steering wheel
BRAKES	service brakes hydraulically actuated, ventilated disc brakes. Parking brake mechanically actuated
SUSPENSION	torsion bar
ELECTRICAL SYSTEM	24 V
ARMAMENT	1 × 12.7 mm MG
SMOKE LAYING EQUIPMENT	2 × 4 smoke dischargers
AMMUNITION	800 (12.7 mm)
GUN CONTROL EQUIPMENT	
Cupola traverse	manual 360°
ARMOUR	
aluminium	50 mm (max)
steel appliqué	6 mm to 13 mm

OTO Melara Infantry Armoured Fighting Vehicle

Development

The Infantry Armoured Fighting Vehicle (IAFV) is the result of a development programme by the Automotive Technical Service (Servizio Tecnico della Motorizzazione) of the Italian Army to increase the combat effectiveness of the M113A1 APC. The Infantry Armoured Fighting Vehicle is based on the M113A1 APC which OTO Melara manufactured under licence from the FMC Corporation of the USA from the 1960s. Main improvements over the M113A1 can be summarised as increased firepower, improved armour protection, the infantry's ability to aim and fire their weapons from inside the hull and improved seating arrangements. The IAFV is called the VCC-1, or Camillino, by the Italian Army. Final deliveries were made to the Italian Army in 1982 when production of the TOW version of the VCC-1 for Saudi Arabia began.

Note on Italian M113 production

Under an agreement between the Italian and United States Governments, OTO Melara obtained a licence to undertake production of the petrol engined FMC M113 series armoured personnel carrier in Italy.

Main sub-contractors were FIAT for the engine and suspension and Lancia for the transmission. The first batch of 40 M113 series APCs was delivered to the Italian Army in May 1964. The first order comprised a total of 3000 vehicles for the Italian Army with the second order being for 697. A total of 581 vehicles were also built for Turkey. The initial petrol engine versions were subsequently replaced in production by the diesel engined M113A1.

Total production of the M113, M113A1, variants of these vehicles, the VCC-1 and VCC-2 amounted to a total of 5486 units.

IAFV with firing ports open and armour protection for 12.7 mm machine gun raised

Description

The hull of the IAFV is made of all-welded aluminium with an additional layer of steel armour on the front and sides of the hull to improve ballistic protection.

The driver sits at the front of the hull on the left and has a single piece hatch cover that opens to the right. There are four M17 periscopes to the front and left of his position and a single M19 periscope, which can be traversed through a full 360°, mounted in the hatch cover. The M19 can be replaced by an infra-red periscope for night driving.

The engine is mounted to the right of the driver with the air-inlet and air-outlet louvres in the roof of the vehicle. The engine compartment is fitted with a fire warning and extinguishing system and the engine is coupled to an Allison Transmission Division of General Motors TX-100 transmission which consists of a three-speed gearbox and a two-stage torque converter which gives six forward and two reverse gears. Access to the engine for maintenance is by a panel in the glacis plate and panels inside the crew compartment.

The vehicle commander sits immediately behind the driver and has a single piece hatch cover that opens to the rear and five M17 periscopes. The gunner sits to the right of the commander and has a three-piece hatch cover that provides protection to his sides and rear when raised, and five M17 periscopes.

The six infantrymen sit at the rear of the vehicle, three down each side of the hull on individual seats. The first four face outwards and the last two inwards. The upper part of the hull sides of the personnel compartment are sloped to improve ballistic protection and in each side of the hull are two firing ports each with a rectangular vision block over it. The infantry enter and leave the vehicle by a power-operated ramp in the rear of the hull. There is a firing port in the right side of the ramp and an emergency door in the left side should the ramp fail to open. In the roof of the personnel compartment, at the rear, is a rotating hatch which is covered by a three-part triangular hatch cover when not in action. The forward part of the cover folds forwards and is locked vertical and the two side covers open to the outside of the vehicle and are also locked vertical to provide protection for the machine gunner. A 7.62 mm machine gun can be mounted at this position to provide covering fire to the rear of the vehicle.

Two fans in the top of the hull, one either side of the rear gunner's position, expel fumes from the personnel compartment. There are two aluminium fuel tanks each side of the ramp at the rear.

The torsion bar suspension consists of five dual rubber-tyred road wheels with the drive sprocket at the front and the idler at the rear. There are no track return rollers. The first and last road wheel stations are provided with a hydraulic shock absorber.

The basic model of the IAFV is called the Mk 1 and is armed with a pintle-mounted 12.7 mm M2 HB machine gun. The Mk 2 version, which is not in service with the Italian Army, has a remote-controlled 12.7 mm machine gun. The Mk 3 version, still at the prototype stage, has an externally mounted 20 mm cannon.

The IAFV is fully amphibious being propelled in the water by its tracks. Before entering the water a trim board is erected at the front of the hull and the two electric bilge pumps are switched on. The vehicle is fitted with infra-red driving lights but has no NBC system.

VCC-1 with TOW

Saudi Arabia placed an order worth LIt350000 million with OTO Melara for 224 VCC-1 vehicles fitted with the same Emerson TOW ATGW system as the M901 Improved TOW Vehicle. The first 10 production vehicles were delivered early in 1983 and final deliveries were made in 1984.

VCC-2

This is the Italian built M113 with appliqué armour fitted to the front and sides. There are two firing ports with a bullet-proof vision block above in either side of the rear troop compartment. The VCC-2 has the standard circular hatch for the driver, to the rear of this is a circular hatch for the commander, while the gunner has an externally mounted 12.7 mm M2 HB machine gun with lateral protection. Some of these vehicles were deployed in Lebanon with Italian Army units operating as part of the United Nations peacekeeping forces.

Other variants

OTO Melara has proposed that the vehicle could be fitted with similar turrets to the C13 including 4 × 25 mm anti-aircraft, 20 mm, 25 mm, 60 mm and 90 mm, all of which have been tested on other OTO Melara designed and built vehicles.

Saudi Arabian VCC-1 with Emerson Improved TOW System installed on roof

IAFV with all roof hatches closed and armour plates for 12.7 mm M2 machine gun lowered

Status: Production complete, but can be resumed if further orders are received. In service with Italy and Saudi Arabia (TOW version only).

Manufacturer: OTO Melara
via Valdilocchi 15
19100 La Spezia
Italy
Telephone: (0187) 530 111
Telex: 270 368
Fax: (0187) 530 669

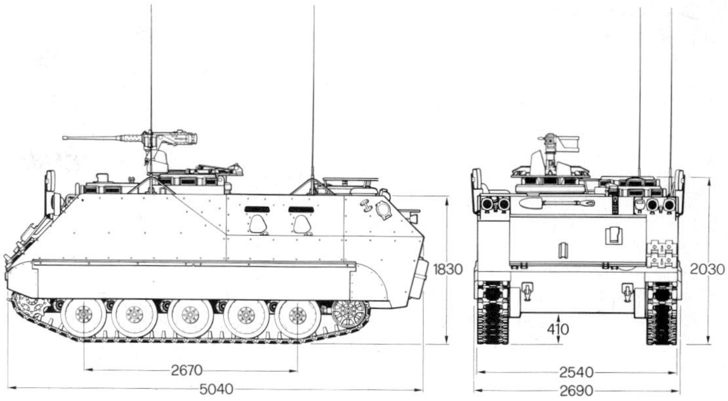

Side and front drawings of the OTO Melara IAFV armed with roof-mounted 12.7 mm MG (not to 1/76th scale)

SPECIFICATIONS		FUEL CAPACITY	360 l	SUSPENSION	torsion bar
CREW	2 + 7	RANGE	550 km	ELECTRICAL SYSTEM	24 V
COMBAT WEIGHT	11 600 kg	FORDING	amphibious	ARMAMENT	
POWER-TO-WEIGHT		GRADIENT	60%	main	1 × 12.7 mm MG
RATIO	18.53 bhp/tonne	SIDE SLOPE	30%	secondary	1 × 7.62 mm MG
GROUND PRESSURE	0.57 kg/cm²	VERTICAL OBSTACLE	0.61 m	SMOKE-LAYING	
LENGTH	5.041 m	TRENCH	1.68 m	EQUIPMENT	none
WIDTH	2.686 m	ENGINE	GMC model 6V-53	AMMUNITION	
over tracks	2.54 m		6-cylinder water-	main	1050
HEIGHT			cooled diesel	secondary	1000
over MG	2.552 m		developing 215 bhp	GUN CONTROL	
over commander's cover	2.032 m		at 2800 rpm	EQUIPMENT	
to hull top	1.828 m	TRANSMISSION	Allison Transmission	Cupola traverse	manual
GROUND CLEARANCE	0.406 m		Division of General		
TRACK	2.159 m		Motors TX-100 with 3		
TRACK WIDTH	381 mm		forward and 1 reverse		
LENGTH OF TRACK			gears coupled to		
ON GROUND	2.667 m		2-stage torque		
MAX SPEED			converter which gives		
road	64.4 km/h		6 forward and 2		
water	5 km/h		reverse gears		

New IVECO-FIAT Wheeled Armoured Personnel Carriers

Development

In early 1984 IVECO-FIAT announced that it was to develop, as a private venture, a new family of wheeled armoured vehicles which would all use proven commercial components wherever possible. The family consists of three basic vehicles, the FIAT 6634 (4×4 and 6×6) Armoured Vehicle Light (AVL), the FIAT 6633 (4×4) Armoured Vehicle Medium (AVM) and the FIAT 6636 (6×6) Armoured Vehicle Heavy (AVH).

Description/Variants FIAT 6634 AVL

The FIAT 6634 has an all-welded hull giving protection from small arms fire and shell splinters. The driver sits at the front of the hull on the left side and has a single piece hatch cover that opens to the right. To his front and sides is a bullet-proof observation block and in the roof to the front of the hatch cover is a single forward-facing periscope which can be replaced by a passive periscope for night driving. The engine compartment is to the right of the driver. In either side of the hull is a door that opens to the front of the vehicle, in the upper part of which is a vision block with a firing port underneath.

The troop compartment is at the rear of the hull and in the centre of the roof is a cupola with periscopes for all-round observation, a rear-opening single piece hatch cover and a pintle-mounted 12.7 mm M2 HB machine gun. There are six (in the 4×4 version) and eight (in the 6×6 version) individual seats in the rear, one under the cupola facing the front, one at front right side, one near each side door also facing the front and two (four in 6×6 version) at the rear facing each other. In the top of the hull to the rear of the commander's cupola are two rectangular roof hatches that open to either side. Mounted either side of the hull above each rear wheel station is a bank of three electrically-operated smoke dischargers firing forwards.

FIAT suggests that in addition to being used as a troop carrier (driver, commander/gunner and four troops), the vehicle can also be used as an anti-tank vehicle with the Euromissile MILAN MCT turret, anti-aircraft with Bofors RBS 70, Shorts Blowpipe/Javelin or MATRA Mistral surface-to-air missiles, or as an artillery observation vehicle which would be fitted with a land navigation system.

Two prototypes of the FIAT 6634 AVL were built by FIAT as a private venture and further development of this has resulted in the Puma (4 × 4) vehicle for the Italian Army. Puma will operate with the new FIAT/OTO Melara Centauro B1 8 × 8 105 mm armed tank destroyer. There is a separate entry for the Puma later in this section.

IVECO-FIAT 6633 APC without armament installed

Prototype of the FIAT 6636 AVH fitted with 90 mm/105 mm turret mock-up

Description/Variants FIAT 6633 AVM

This has a similar layout to the FIAT 6634 AVL but is longer, wider and higher. Main differences are that in either side of the rear troop compartment is a single vision block/firing port, the air inlet/outlets are in the hull roof and right side of the hull and a power-operated ramp is at the rear. When carrying 12 men seating arrangements are: eight men at rear, four down either side on two bench seats, machine gunner under cupola, one man to left front of machine gunner, driver at front left and one man to his rear.

The FIAT 6633 AVM is fully amphibious, propelled in the water by two water jets mounted one either side at the rear.

FIAT suggests that in addition to being used in the basic troop carrying rôle, eg commander/gunner, driver and 10 men, it can be fitted with other weapon stations of up to two tonnes in weight. For example, when fitted with a 25 mm turret up to six men would be carried. More specialised versions include HOT/TOW anti-tank vehicle and 81 mm mortar carrier. Two prototypes of the FIAT 6633 AVM were built, but for the time being no further work is being carried out on the vehicle.

Description/Variants FIAT 6636 AVH

This is the largest member of the family and has a layout similar to the others with the driver at the front left, engine to his right and troop compartment at the rear. It can be fitted with various weapon systems up to a weight of 3500 to 4000 kg. When being used as an APC its crew would consist of driver and commander/machine gunner with the 12 troops sitting six down each side of the rear troop compartment facing each other. Entry to the troop compartment would be via a power-operated ramp in the hull rear. Over the troop compartment at the rear are three hatches, a rectangular one either side hinged on the outside and a circular one which has a two-part cover. In the upper part of the hull between the second and third road wheels is a small door that opens to the front; this has a vision block with a firing port underneath. In either side of the troop compartment is a single vision block and firing port. This model is fully amphibious, propelled in the water by two water jets mounted at the rear of the hull. The basic model has a cupola with a 12.7 mm M2 HB machine gun but other armament stations are possible including a turret with a 25 mm cannon, anti-tank vehicle with TOW or HOT missiles and an 81 mm, 107 mm or 120 mm mortar carrier. FIAT built one prototype of the 6636 AVH which was fitted with a mock up of a three-man turret with a 90 mm/105 mm gun.

Type 6638 G

This is the FIAT designation for the Centauro B1 8 × 8 105 mm tank destroyer that it has built in co-operation with OTO Melara to meet the requirements of the Italian Army. Full details are given in *Reconnaissance vehicles* section.

SPECIFICATIONS

Model	**6633**	**6634**	**6636**
CREW	1+12	1+5	1+13
CONFIGURATION	4 × 4	4 × 4	6 × 6
COMBAT WEIGHT	9800 kg	4200 kg	16 000 kg
LENGTH	6.098 m	4.75 m	6.7 m
WIDTH	2.50 m	2.5 m	1.9 m
HEIGHT (hull top)	1.9 m	1.67 m	1.9 m
GROUND CLEARANCE	0.43 m	0.33 m	0.40 m
TRACK	2.06 m	1.745 m	2.47 m
WHEELBASE	3 m	2.8 m	1.6 m + 2 m
ANGLE OF APPROACH/ DEPARTURE	40°/45°	45°/45°	45°/45°
MAX ROAD SPEED	100 km/h	100 km/h	100 km/h
FUEL CAPACITY	270 l	150 l	400 l
RANGE	n/a	800 km	800 km
GRADIENT	60%	60%	60%
SIDE SLOPE	30%	30%	30%
ENGINE	FIAT diesel	FIAT diesel	FIAT diesel
TYPE	8060 TCA	8140 TCA	8260 TCA
NUMBER OF CYLINDERS	4	4	6
OUTPUT	200 hp/3000 rpm	120 hp/3800 rpm	400 hp/2300 rpm
TRANSMISSION	Allison AT 545	ZF 4 HP 22	ZF 6 HT 500/600
	4 F, 1 R	4 F, 1 R	6 F, 1 R
TRANSFER CASE	2-speed	nil	2-speed
SUSPENSION	independent	independent	independent
STEERING	power-assisted	power-assisted	power-assisted
TYRES	14.00 × 20	11.00 × 20	14.00 × 20
ELECTRICAL SYSTEM	24 V	24 V	24 V
BATTERIES	2×12 V, 100 Ah	2×12 V, 100 Ah	2×12 V, 100 Ah

Status: Prototypes/development. Not yet in production or service.

Manufacturer: IVECO-FIAT SpA
Defence Vehicle Division
Via A Volta, 6
39100 Bolzano
Italy
Telephone: (0471) 905 111
Telex: 400 541
Fax: (0471) 905 444

IVECO-FIAT Puma Armoured Personnel Carrier

Development

In the early 1980s IVECO-FIAT built a 4×4 armoured vehicle called the 6634 Armoured Vehicle Light (AVL), the first of a family of projected 4×4 and 6×6 armoured vehicles which is covered in the previous entry.

The Italian Army however wanted a larger vehicle that could undertake a wider range of roles on the battlefield and operate with its IVECO-FIAT B1 Centauro (8×8) tank destroyers (qv *Reconnaissance vehicles* section) already under development.

The end result was the Puma, the first prototype of which was completed in 1988 with the second prototype being completed 1989. These two prototypes were followed by a further three prototypes of the definitive armoured personnel carrier version in mid-1990.

The Italian Army also awarded the company a contract for the design of six specialised versions of the Puma; three are weapon platforms, one each for the TOW and MILAN ATGW systems and one for the Minstral SAM (all of these would have a three or four-man crew), the fourth is an 81 mm mortar carrier (four-man crew including the driver), the fifth is an ambulance (carrying two stretcher cases plus crew) while the sixth is a command post vehicle (crew of four or five).

The Puma has the company designation of the Model 6634 G and the Italian Army has a requirement for a substantial number of vehicles in both 4 × 4 and 6 × 6 configurations and first deliveries of the 4 × 4 version could take place within two years of the production order being placed.

Description

The hull of the Puma is of all welded steel armour construction that provides protection from small arms fire and shell splinters with the upper part of the hull sides sloping inwards. The hull design is unusual in that the sides are in one piece rather than two pieces welded together.

The driver sits towards the front of the vehicle on the left side and has a single piece domed hatch cover that opens to the right. To his front are three periscopes, the centre one of which can be replaced by a passive night periscope.

Mounted at the rear of the IVECO-FIAT Puma (4 × 4) armoured personnel carrier is a winch, to the immediate rear of the bank of three smoke dischargers

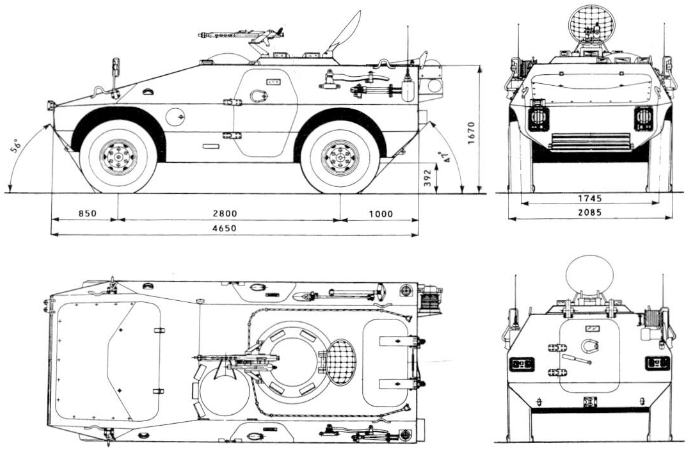

General arrangement drawing of the IVECO-FIAT Puma (4 × 4) armoured personnel carrier (not to 1/76th scale)

One of the prototypes of the IVECO-FIAT Puma (4 × 4) armoured personnel carrier without armament installed

Prototype of the 6 × 6 version of the IVECO-FIAT Puma (6 × 6) armoured personnel carrier armed with 12.7 mm M2 HB machine gun

IVECO-FIAT Puma (4 × 4) armoured vehicle fitted with Hughes TOW ATGW launcher

Prototype of the 6 × 6 version of the IVECO-FIAT Puma (6 × 6) armoured personnel carrier armed with 12.7 mm M2 HB machine gun

The engine compartment is to the front of the driver and is separated from the remainder of the vehicle by a bulkhead. Access panels are provided in the glacis plate for maintenance purposes, and the infantry are provided with individual seats that fold up.

The prototypes were powered by a FIAT 8141.47 engine developing 125hp coupled to a ZF transmission but later vehicles are powered by a FIAT 8042.45 four-cylinder turbo-charged diesel developing 180hp at 3000rpm coupled to a Reco automatic transmission with five forward and one reverse gears.

From the automatic transmission, which incorporates a torque converter, power is taken to the transfer box and then by propeller shafts to the front and rear differentials. It then goes via short universally or constant-velocity jointed shafts to the front and rear wheels.

Each of the four road wheels is independently located by means of an upper telescopic strut and a lower arrangement of transverse links, which is made up of anti-roll torsion bars and radial links. The hydropneumatic struts are made by Messier.

Steering is power assisted on the front wheels and the Michelin 11.00×16 tyres are of the run-flat type.

The engine air-inlets are under the forward part of the hull and the exhaust outlet is on the right side of the hull.

In either sie of the hull is a door that opens forwards and in its upper part is a bullet-proof vision block with a firing port underneath. In the rear of the hull is a large door that opens to the left, with a vision block in its upper part and a firing port underneath. On the right side of the hull at the rear is a winch.

The commander's cupola is mounted in the centre of the roof and has a single piece hatch cover that opens to the rear and five periscopes that give all round observation. A 7.62mm or 12.7mm machine gun can be mounted on the cupola.

To the immediate rear of the commander's cupola is a single piece hatch that opens to the rear.

Mounted on either side of the hull towards the rear is a bank of three electrically operated smoke dischargers firing forwards.

Standard equipment includes NBC system, fire detection and suppression system and integrated air-conditioning system.

Variants

In addition to being used as an armoured personnel carrier carrying a total of seven men, the Puma is suitable for a wide range of other roles including reconnaissance, anti-tank with ATGWs such as MILAN, air defence with SAM, 81mm mortar carrier, ambulance, NBC reconnaissance and internal security.

Late in 1990 the first prototype of the 6×6 version of the Puma armoured personnel carrier was built, sharing a number of common components with the 4 × 4 version including the same engine, transmission and suspension.

In addition to being used as an armoured personnel carrier, a number of specialised versions of the 6 × 6 Puma have been proposed including an 81mm mortar carrier and a command post vehicle.

The prototype of the 6×6 version of the Puma is armed with an externally mounted 12.7mm M2 HB machine gun with a bank of three electrically operated smoke dischargers mounted either side of the turret firing forwards, although a wide range of other armament systems are possible including ATGWs and surface-to-air missiles such as the MATRA Mistral.

SPECIFICATIONS

CONFIGURATION	**4 × 4**	**6 × 6**
CREW (typical)	6 + 1	8 + 1
COMBAT WEIGHT	5500 kg	7500 kg
LENGTH	4.65 m	5 m
WIDTH	2.085 m	2.3 m
HEIGHT		
to hull top	1.67 m	1.7 m
GROUND CLEARANCE	0.392 m	0.38 m
TRACK	1.745 m	1.97 m
WHEELBASE	2.8 m	1.45 m + 1.75 m
ANGLE OF APPROACH/ DEPARTURE	56°/47°	45°/45°
MAX SPEED	105 km/h	n/av
FUEL CAPACITY	150 l	270 l
RANGE	800 km	n/av
GRADIENT	60%	60%
SIDE SLOPE	30%	30%
ENGINE	FIAT Type 8042 TCA 4-cylinder diesel developing 180 hp at 3000 rpm	
TRANSMISSION	Reco automatic with 5 forward and 1 reverse gears	
SUSPENSION	independent	independent
BRAKES	disc, all wheels	disc, all wheels
STEERING	power assisted	power assisted
TURNING RADIUS	6 m	7.5 m
TYRES	11.00 × 16 (run flat)	11.00 × 16 (run flat)
ELECTRICAL SYSTEM	24 V	24 V
BATTERIES	2 × 12 V, 100 Ah	2 × 12 V, 100

Status: Under development. Not yet in production or service.

Manufacturer: IVECO-FIAT SpA
Defence Vehicle Division
Via A Volta, 6
39100 Bolzano
Italy
Telephone: (0471) 905 111
Telex: 400 541
Fax: (0471) 905 444

FIAT-OTO Melara Type 6614 Armoured Personnel Carrier

Development

The Type 6614 APC is a joint development between FIAT and OTO Melara, the former responsible for the hull and the automotive components and the latter for the armament installation. The vehicle shares many common components with the FIAT-OTO Melara Type 6616 armoured car for which there is a seperate entry in the *Reconnaissance vehicles* section.

The Type 6614 is has been manufactured under licence in South Korea (qv this section).

The FIAT Type 6614 was also used by the Italian contingent serving in the Lebanon and was tested by the Italian Army in Denmark and Norway while operating as part of the Allied Command Europe Mobile Force.

The Italian Air Force have 110 vehicles for patrolling airfields while 60 were delivered to the Italian Police. More recently the Italian Army has taken delivery of 14 vehicles for use by its alpine units. Total production of the Type 6614 APC for home and export markets amounted to 1160 vehicles.

Description

The hull of the Type 6614 is all welded steel which provides the crew with protection from small arms fire and shell splinters.

The driver sits at the front of the hull on the left side and has a single piece domed hatch cover that opens to the right. To the front and side of the driver are five vision blocks which cover an arc of 200°, the centre three with wiper blades.

The engine is mounted at the front of the vehicle to the right of the driver and is separated from the personnel compartment by a fireproof bulkhead. The engine is coupled to a manual five-speed transmission from which power is transmitted to the two-speed transfer case and then to the front and rear axles by propeller shafts. Both front and rear differentials are mechanically lockable. The planetary final drives, front and rear, consist of epicyclic gear trains in the wheel hubs.

Entry doors in each side of the hull both open forward and have a vision block and a firing port. There are another three vision blocks and firing ports in each side of the hull. At the rear of the hull is a large ramp hinged at the bottom, either side of which is a single vision block and firing port. The 10 infantrymen, one of whom acts as gunner, have individual seats which can be folded up. In the centre of the roof is mounted the main armament installation. On the basic vehicle this consists of an M113 type cupola which has five periscopes and a single piece hatch cover that opens to the rear. Mounted on the forward part of this is a 12.7mm (0.50) M2 HB machine gun. Other types of armament can be mounted if required, for example a fully enclosed turret armed with single or twin 7.62mm machine guns. Two roof hatches over the top of the troop compartment open either side of the vehicle. A ventilator is mounted in the roof at the rear.

The suspension, front and rear, is of the independent strut and link type with a helical spring and coaxial rubber bump stop. Each wheel station has two hydraulic shock absorbers, and anti-roll bars are fitted as standard. The tyres are of the run-flat type.

The vehicle is fully amphibious being propelled in the water by its wheels. Standard equipment includes four electrically operated bilge pumps with a capacity of 180 litres/min each. Optional equipment includes passive or active night vision equipment, air-conditioning system, fire-extinguishing system in the wheel arches, spare wheel and holder, two three-barrelled 76 mm smoke dischargers and a front-mounted 4500 kg capacity winch with 40 m of 12 mm diameter cable.

Variants

Although designed primarily for use as an APC the Type 6614 can also be used for other roles such as reconnaissance and internal security. Projected variants include an ambulance, cargo, command and mortar carrier. For trials purposes a Type 6614 APC has been fitted with an SNIA BPD FIROS 6 multiple rocket system.

FIAT-OTO Melara Type 6614 APC. Commander's cupola not fitted with 12.7 mm machine gun (Paul Handel)

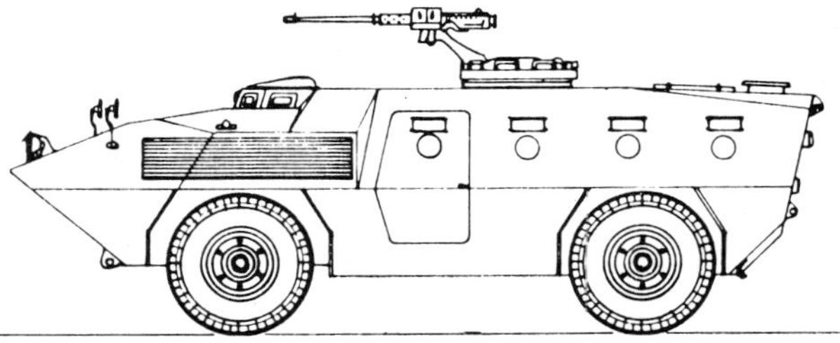

FIAT-OTO Melara Type 6614 APC with 12.7 mm machine gun

SPECIFICATIONS

CREW	1+10
CONFIGURATION	4×4
COMBAT WEIGHT	8500 kg
LOADED WEIGHT	
front axle	4500 kg
rear axle	4000 kg
POWER-TO-WEIGHT RATIO	18.82 hp/tonne
LENGTH OF HULL	5.86 m
WIDTH	2.5 m
HEIGHT	
to hull top	1.78 m
inc MG mount	2.18 m
GROUND CLEARANCE	
axles	0.37 m
hull front	0.485 m
hull centre	0.385 m
hull rear	0.385 m
TRACK	1.96 m

WHEELBASE	2.9 m
ANGLE OF APPROACH/DEPARTURE	40°/45°
MAX SPEED	
road	100 km/h
water	4.5 km/h
FUEL CAPACITY	142 l
MAX ROAD RANGE	700 km
FUEL CONSUMPTION (road)	0.2 l/km
FORDING	amphibious
GRADIENT	60%
SIDE SLOPE	30%
VERTICAL OBSTACLE	0.4 m
TURNING RADIUS	7.9 m
ENGINE	Model 8062.24 super-charged liquid-cooled in-line diesel developing 160 hp at 3200 rpm
TRANSMISSION	manual with 5 forward and 1 reverse gears

TRANSFER CASE	2-speed
STEERING	hydraulic
CLUTCH	single dry plate, hydraulically-operated
SUSPENSION	independent
TYRES	14.50 × 20 (run-flat)
BRAKES	
main	drum, air-over-hydraulic (dual circuit)
parking	drum, hand-operated, on transfer front output shaft
ELECTRICAL SYSTEM	24 V
BATTERIES	2 × 12 V, 100 Ah
ARMAMENT	1 × 12.7 mm MG
GUN CONTROL EQUIPMENT	
Cupola traverse	360° manual
ARMOUR	6-8 mm

Status: Production complete in Italy. In service with Italy (Police and Army, Air Force), South Korea, Libya, Peru (including 10 in mortar carrying model), Somalia (270 ordered in 1977 and delivered by 1979) and Tunisia (120 ordered in 1979).

Manufacturer: IVECO-FIAT SpA
Defence Vehicle Division
Via A Volta, 6
39100 Bolzana
Italy
Telephone: (0471) 905 111
Telex: 400 541
Fax: (0471) 905 444

FIAT-OTO Melara Type 6614 APC armed with 12.7 mm M2 HB machine gun, with spare wheel on hull top, driver's hatch open and weapons being used in side firing ports

Boneschi MAV 4 Multi-role Armoured Vehicle

Development

The MAV 4 multi-role armoured vehicle has been developed as a private venture by Carrozzeria Boneschi srl and has been designed to carry 14 fully equipped men (commander, driver and 12 men) and their equipment, on both road and cross country.

Prior to designing the MAV 4 mutli-role armoured vehicle, the company had built over 600 armour protected vehicles for police use and these had been sold both within Italy and abroad.

Description

The MAV 4 multi-role armoured vehicle is essentially an IVECO FIAT 80.16 (160hp engine) or 80.17 (170hp engine) chassis, fitted with a new steel armoured body 6mm thick that provides the occupants with complete protection from small arms fire. The floor and roof are also of ballistic steel armour with a thickness of 3 to 6mm.

All windows and vision blocks are of bullet proof glass with a total thickness of 38mm which provide the same level of protection as the hull.

The driver is seated at the front of the vehicle on the left side with the vehicle commander to his right, both being provided with a bullet proof window to their front and sides. In addition the vehicle commander has a sliding firing port below the window.

The 12 troops are seated on individual seats facing the centre of the vehicle, seven on the left and five on the right. They can enter the vehicle via three doors, one in each side to the rear of the driver and commander and one at the rear. The latter opens to the right and has an integral vision block with a sliding firing port underneath. The doors are provided with locks and anti-intruder devices.

In the left side of the troop compartment are four vision blocks, each with a firing port underneath while in the right side there are three vision blocks each with a firing port underneath.

There are two hatches in the roof, the forward one allows the engine to be removed while the one in the centre of the roof is for emergency exit.

The engine is located at the front of the vehicle and is protected by means of strips of ballistic steel which can be moved electrically.

Optional equipment includes air conditioning system, an anti-tear gas filtering system and a fire fighting system that protects the engine and front wheels, the fuel tank and rear wheels and can also put out a fire ahead of the vehicle. Axles with different ratios can be provided as can different tyres (including run flat), exhaust brake, air electrical connection for towed vehicle, heating system, 5000kg winch driven by PTO on transfer case, 4000kg towing hook, lockable front differential and various communications systems.

Standard equipment includes 12 arms for rifles, ceiling lights for internal lighting, special protection for electrical cables, two tone siren and special fire resistant paint.

Variants

The manufacturer has also suggested that the vehicle can undertake a wide range of other roles including riot control vehicle, mobile command post, HF radio and communications system, ambulance, mobile first aid unit, mobile repair unit and recovery unit.

Boneschi MAV 4 (4 × 4) multi-role armoured vehicle from rear

Boneschi MAV 4 (4 × 4) multi-role armoured vehicle from front

SPECIFICATIONS

CREW	2 + 12
MAXIMUM WEIGHT	9500 kg
UNLOADED WEIGHT	7450 kg
MAXIMUM TOWED LOAD	4000 kg
POWER-TO-WEIGHT RATIO	16.84 hp/tonne
LENGTH	5.75 m
INTERNAL LENGTH	3.25 m
WIDTH	2.28 m
INTERNAL WIDTH	2.12 m
HEIGHT	2.85 m
WHEELBASE	3 m
MAXIMUM ROAD SPEED	90 km/h
FUEL CAPACITY	130 l
RANGE	600 km
FORDING	0.7 m
GRADIENT	60%
SIDE SLOPE	30%
TURNING RADIUS	6.67 m
ENGINE	FIAT water-cooled diesel developing 160 hp at 3200 rpm
GEARBOX	manual, 5 forward and 1 reverse gears
CLUTCH	single plate, hydraulically operated
TRANSFER BOX	2-speed
STEERING	recirculatory ball type with integral power assistance
SUSPENSION	leaf springs with hydraulic telescopic shock absorbers
AXLES	Portal type with lockable differential at rear
WHEELS	disc type, rim size 11-20 SDC
TYRES	12.5-20 PR, single at front, dual at rear
BRAKES	
SERVICE	drum type, air over hydraulic, dual circuit
PARKING	on rear wheels
ELECTRICAL SYSTEM	24 V
BATTERIES	2 × 12 V, 110 Ah each
GENERATOR	800 W alternator
ARMOUR	6 mm (max)

Status: Development complete. Ready for production.

Manufacturer: Carrozzeria Boneschi srl
Viale Delle Industrie
61-20040 Cambiago
Milan
Italy
Telephone: (02) 9506445
Telex: 334529 BONCAR I
Fax: (02) 95067078

JAPAN

Type 89 Mechanised Infantry Combat Vehicle

Development

The 1984 Japanese defence budget included a 600 million yen request for the construction of four prototypes of a new Mechanised Infantry Combat Vehicle. In FY81 funding was provided for prototypes of the turret and hull. Following trials with the prototype vehicles it was type classified as the Type 89 Mechanised Infantry Combat Vehicle and it is now in production for the Japanese Ground Self-Defence Force with the initial requirement being for 300 vehicles.

Under the next five year defence plan, approved by the Japanese Government late in 1990, a total of 218 medium armoured vehicles, including the Type 89 MICV and the Type 87 Reconnaissance Patrol Vehicle will be purchased.

Description

The hull of the Type 89 MICV is of-all welded construction which provides the crew with protection from small arms fire and shell splinters. Mounted under the nose of one of the prototype Type 89 MICVs were attachment points for a mine clearance system.

The powerpack is mounted at the front of the vehicle on the left side with the driver being seated on the right and one of the seven infantrymen to his immediate rear.

The driver has a single piece hatch cover that opens to the right and three day periscopes, the centre one of which can be replaced by a night vision periscope.

The infantryman to the rear of the driver has a single piece hatch cover, periscopes that give observation to the front and sides and a spherical firing port to right side of the hull.

The two-man power-operated turret is in the centre of the hull with the commander being seated on the right and gunner on the left, both with single piece hatch covers that open to the rear.

Both commander and gunner have a roof-mounted periscopic sight mounted in the forward part of the turret, the gunner has two fixed periscopes to give observation to his front and left side while the commander has six periscopes that give observation to the front, sides and rear.

Main armament is an Oerlikon-Contraves 35mm cannon which has a cyclic rate of fire of 200 rds/min. First weapons were supplied direct from Switzerland with licensed production being undertaken in Japan.

A 7.62mm machine gun is mounted coaxial with the 35mm cannon and either side of forward part of the turret is a bank of three electrically operated smoke dischargers fireing forwards.

Mounted either side of the turret are a Jyu-MAT medium range wire-guided anti-tank missile developed by Kawasaki Heavy Industries. The first prototype of the basic infantry version was completed early in 1984. Once the missiles have been launched new missiles have to be reloaded manually.

The raised troop compartment is at the rear of the hull with the troops entering and leaving via two doors in the rear that open outwards, with the right door having a spherical firing port. There are a further two

Type 89 MICV with launcher for ATGW either side of turret (Kensuke Ebata)

Type 89 MICV from front showing attachment points under nose of vehicle for mine clearance equipment (Kensuke Ebata)

Type 89 MICV showing three spherical firing ports in side of hull (Kensuke Ebata)

Close-up of turret of Type 89 MICV showing launcher for ATGW either side of turret (Kensuke Ebata)

firing ports in the right side of the troop compartment and three in the left side, each of the firing ports has a roof-mounted periscopic sight. Over the top of the troop compartment is a two-part roof hatch that opens left and right of the vehicle's centreline.

Suspension is of the torsion bar type with seven dual rubber-tyred road wheels, drive sprocket at the front, idler at the rear and three track return rollers.

The vehicle does not have any amphibious capability, but is fitted with NBC system and a full range of passive night vision equipment.

Status: In production. In service with Japanese Ground Self-Defence Force.

Close-up of three spherical firing ports in left side of rear troop compartment (Kensuke Ebata)

SPECIFICATIONS		HEIGHT	2.5 m	SMOKE DISCHARGERS	2×3
CREW	3+7	MAXIMUM ROAD SPEED	70 km/h	GUN CONTROL	
COMBAT WEIGHT	26 500 kg	ENGINE	600 hp diesel	EQUIPMENT	
POWER-TO-WEIGHT-		ARMAMENT		turret traverse	powered/manual
RATIO	22.64 hp/tonne	main	1 × 35 mm cannon		
LENGTH	6.8 m	co-axial	1 × 7.62 mm MG		
WIDTH	3.2 m	ATGW	2 × Jyu-MAT		

Type 73 Armoured Personnel Carrier

Development

The development of a new armoured personnel carrier to supplement the SU 60 began in FY67 under the direction of the Technical Research and Development Institution of the Japanese Defence Agency although before this, in 1964, Mitsubishi Heavy Industries had built an amphibious all-welded aluminium APC under the designation XU as a private venture.

In January 1967 the following requirements were issued for the new APC: a maximum speed of over 60 km/h, ability to carry 12 men including the crew, to be fully amphibious, have all-welded aluminium armour, have provision for the infantry to use their small arms from inside the vehicle and be armed with a 20 mm cannon, 12.7 mm machine gun and 7.62 mm machine gun.

In 1967 the following manufacturers started on the manufacture of specific components: Hitachi Manufacturing Company (transmission), Kobe Iron Works (aluminium alloy), Komatsu Manufacturing Company (tracks), Mitsubishi Heavy Industries (engine, steering and suspension), and Nihon Electric Corporation (night vision equipment).

In 1968 an automotive test rig, called the SUT, was completed and tested. The following year Mitsubishi Heavy Industries and the Komatsu Manufacturing Company each built two prototype vehicles, one steel and one aluminium.

After trials, in December 1972, the second prototype of the Mitsubishi model, with aluminium armour, was selected for use by the Japanese Ground Self-Defence Force and after minor modifications and improvements was standardised as the Type 73 APC. Due to budgetary restrictions production has been very slow with the following funding:

FY73	34	FY80	9
FY74	18	FY81	9
FY75	17	FY82	9
FY76	7	FY83	99
FY77	6	FY84	15
FY78	6	FY85	16
FY79	6	FY86	23

Type 73 APCs clearly showing position of armament, side skirts and buoyancy aids attached to outside of road wheels (Ryuta Watanabe)

Type 73 APC showing position of two banks of three electrically operated smoke dischargers on rear of hull roof (Kensuke Ebata)

Type 75 self-propelled ground wind measuring system in travelling configuration (Kensuke Ebata)

Type 73 APC with trim vane erected at front for amphibious operations (Panzer Magazine)

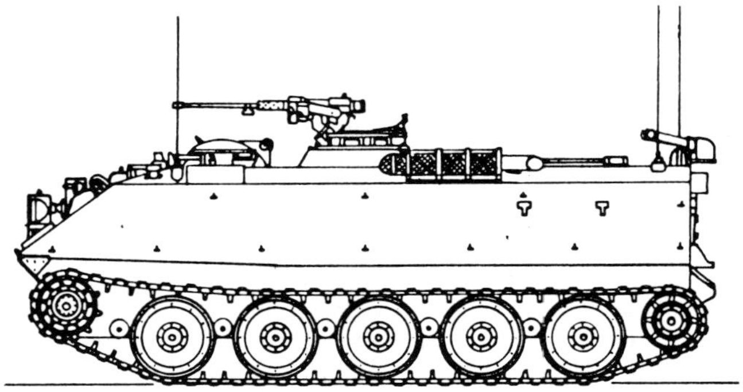

Type 73 armoured personnel carrier

Since then production has continued with a total of 225 vehicles in service in 1988/89. The Type 73 and Type SU 60 will be supplemented by the Type 89 MICV from the early 1990s.

Description

The hull of the Type 73 APC is made of all-welded aluminium armour with the bow machine gunner seated at the front of the hull on the left side. The bow machine gun has an elevation, depression and traverse of 30°. The gunner has a single piece hatch cover that opens to the left and has a single periscope which can be traversed through 360°.

The driver sits at the front of the hull on the right side and has a single piece hatch cover that lifts and swings horizontally to the rear to open, and has an integral periscope. To the front of the driver's position are three periscopes.

The commander sits slightly behind the bow machine-gunner and driver and has a single piece hatch cover that opens to the rear. Six vision blocks give the commander all-round observation.

The engine is mounted to the rear of the bow machine gunner on the left side of the hull with the air-inlet, air-outlet and exhaust pipe in the roof. The engine and transmission are mounted as a powerpack and can be removed from the vehicle as a complete unit in half an hour. The machine gunner sits behind the driver and has a cupola that can be traversed through a full 360°. The cupola has a single piece hatch cover that opens to the rear and six periscopes for all-round observation. The 12.7 mm (0.50) M2 HB machine gun is not provided with a shield as it can be aimed and fired from inside the cupola. The 12.7 mm machine gun has an elevation of +60°, depression of -10° and can be traversed through 360°.

The Type 73 carries nine infantrymen of whom one normally mans the 12.7 mm (0.50) machine gun position. The eight infantrymen sit in the troop compartment at the rear of the hull down either side on seats that can be folded up, and enter the vehicle via two doors at the rear of the hull that open outwards. There is a T-shaped firing port in each door and two in each side of the troop compartment. Over the top of the troop compartment are two single piece hatch covers that open either side of the hull top. Mounted at the very rear of the hull top, one either side, is a three-barrelled smoke discharger that fires forwards over the front of the vehicle.

The torsion bar suspension consists of five dual rubber-tyred road wheels with the drive sprocket at the front and the idler at the rear. There are no track return rollers. A hydraulic shock absorber is provided for the first road wheel station.

The basic model of the Type 73 APC is not amphibious although an amphibious kit has been developed and fitted to some vehicles. The kit consists of buoyancy aids attached to the outside of each of the road wheels, skirts fitted to cover the upper part of the tracks to improve water flow, box type structures fitted on top of the hull over the air inlet, air outlet and exhaust to stop water entering when afloat and a trim vane fitted on the front of the hull. The trim vane is in two parts and is hinged horizontally. A transparent panel in the right side provides forward vision for the driver. When afloat the vehicle is propelled in the water by its tracks.

The Type 73 is fitted with passive infra-red night vision equipment and an NBC system.

Variants

The Type 75 self-propelled ground-wind measuring unit is used in conjunction with the Type 75 self-propelled 130 mm multiple rocket launcher (full details of this are given in the *Multiple rocket launchers* section) which also uses some components of the Type 73. The measuring unit has a mast which when extended is 12.5 m high and can measure wind velocity in two directions up to a maximum of 30 km/h. The wind velocity measuring equipment was developed by Meisei Electric.

In 1978 some Type 73 APCs were fitted with three firing ports and three periscopes in either side of the hull for trials by the 7th Division. Once these trials had been completed the vehicles were converted back to their original configuration. The Type 74 105 mm self-propelled howitzer and Type 75 multiple rocket launcher both use the same engine as the Type 73. There is also a command post version of the Type 73 which has a higher roof with hatches to the rear of the commander's cupola.

Status: Production complete. In service only with the Japanese Ground Self-Defence Force.

Manufacturer: Mitsubishi Heavy Industries
5-1, Marunouchi 2-chome
Chiyoda-ku
Tokyo
Japan

SPECIFICATIONS		MAX SPEED		ELECTRICAL SYSTEM	24 V
CREW	3 + 9	road	70 km/h	ARMAMENT	
COMBAT WEIGHT	13 300 kg	water	7 km/h	main	1 × 12.7 mm MG
POWER-TO-WEIGHT		MAX RANGE	300 km	bow	1 × 7.62 mm MG
RATIO	22.6 hp/tonne	FUEL CAPACITY	450 l	SMOKE-LAYING	
LENGTH	5.8 m	FORDING (with kit)	amphibious	EQUIPMENT	2 × 3 smoke dischargers
WIDTH	2.8 m	GRADIENT	60%		
HEIGHT		VERTICAL OBSTACLE	0.7 m	GUN CONTROL	
including MG	2.2 m	TRENCH	2 m	EQUIPMENT	
to hull top	1.7 m	TURNING RADIUS	7 m	Cupola traverse	360°, manual
GROUND CLEARANCE	0.4 m	ENGINE	Mitsubishi 4ZF air-cooled 2-stroke V4 diesel developing 300 hp at 2200 rpm	Gun elevation/depression	+60°/-10° (manual)
TRACK	2.56 m				
TRACK WIDTH	380 mm				
LENGTH OF TRACK ON GROUND	3.67 m	SUSPENSION	torsion bar		

Type SU 60 Armoured Personnel Carrier

Development

In the mid-1950s the Japanese Ground Self-Defence Force issued a requirement for a full tracked armoured personnel carrier. Development began in 1956 with overall control of the project managed by the Technical Research and Development Headquarters of the Japanese Self-Defence Agency. In 1957 prototypes of the APC were completed by Komatsu and Mitsubishi. The Komatsu prototype was called the SU-I and the Mitsubishi model the SU-II.

Both prototypes weighed 10.6 tonnes, carried nine men and had a maximum road speed of 45 km/h. Between 1957 and 1958 the SU-I and SU-II were tested alongside an American-supplied M59 APC. Eleven second series prototypes were manufactured by Mitsubishi Heavy Industries and the Komatsu Manufacturing Corporation. Mitsubishi built five SU-II (Kai) Ms, one SV (Kai) M and one SX (Kai) M, M standing for Mitsubishi and Kai for modified. Komatsu built two SU-II (Kai) Ks, one SV (Kai) K and one SX (Kai) K, K standing for Komatsu. The SV was the 81 mm mortar carrier and the SX was the 107 mm (4.2-inch) mortar carrier.

In both the mortar was mounted in the rear compartment but on the SX the top of the hull was sloped to the rear to avoid muzzle blast from the 107 mm mortar, but this also decreased the weight of the vehicle.

In July 1959 four prototypes of the SU-II (Kai) were completed and tested through to October 1959. In 1960 the vehicle was standardised as the Type 60 APC with the mortar carriers being standardised as the Type 60 self-propelled 81 mm mortar and the Type 60 107 mm (4.2-inch) mortar.

The first production contract for 24 vehicles was provided for in the 1959 budget and production was undertaken by Komatsu and Mitsubishi. Final orders were placed with funding from the 1972 budget.

Description

The hull of the SU 60 is made of all-welded steel with the driver seated at the front of the hull on the right. The driver has a single piece hatch cover over his position that opens to the right, in front of which are three M17 periscopes. The bow machine gunner sits at the front of the hull on the left and has a single piece hatch cover that opens to the left rear, with an integral M6 periscope for aiming the 7.62 mm (0.30) M1919A4 bow machine gun.

The commander sits immediately behind the driver's and bow machine gunner's position and has a single piece hatch cover that opens to the left rear. Eight vision blocks provide all-round observation for the commander.

The engine is on the left side of the hull to the rear of the bow-machine gunner and power is transmitted to the differential at the front of the hull by a short propeller shaft.

The gunner sits to the rear of the commander on the right side of the hull and has a two-piece hatch cover that opens either side of his

Type SU 60 APC clearly showing roof-mounted 12.7 mm M2 HB and bow-mounted 7.62 mm machine guns (Ryuta Watanabe)

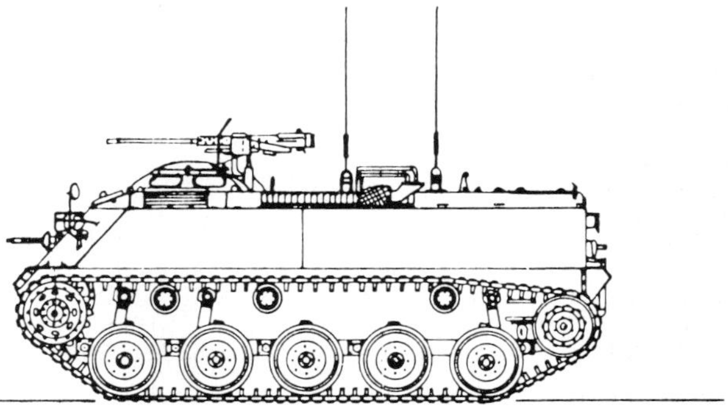

Type SU 60 armoured personnel carrier

Type SX 4.2-inch (107 mm) mortar carrier in travelling configuration (Kensuke Ebata)

Type SU 60 APC fitted with two roof-mounted Type 64 (KAM-3D) ATGW launcher on roof at rear (Kensuke Ebata)

NBC detection vehicle based on Type SU 60 APC chassis (Kensuke Ebata)

position. The 12.7 mm (0.50) M2 HB machine gun can be traversed through a full 360° and is provided with a shield.

The troop compartment is at the rear of the hull with the six infantrymen seated on individual seats three down each side of the hull. They enter and leave the vehicle by two doors in the rear of the hull that open outwards. The right door has a small observation slit. Over the top of the troop compartment is a single piece hatch cover that opens forwards. To the rear of this is a two-part hatch cover that opens either side of the hull. Each part of the hatch cover is in two halves hinged in the centre. There is no provision for the infantry to use their small arms from inside the vehicle.

The torsion bar suspension consists of five dual rubber-tyred road wheels with the drive sprocket at the front, idler at the rear and three track return rollers. The first, second and fifth road wheel stations are provided with a hydraulic shock absorber. The SU 60 APC has no NBC system, no night vision equipment and no amphibious capability.

Variants

Type SV 60 81 mm Mortar Carrier

This is essentially a Type 60 APC with an 81 mm mortar mounted in the back of the hull firing to the rear with two doors in the rear of the hull. A bipod and baseplate are carried on the glacis plate to enable the mortar to be deployed away from the vehicle. The mortar has an elevation from +40° to +85°, traverse of 40° left and right and a maximum range of 3000 m. Twenty-four mortar bombs are carried. The SV 60 has a crew of five and weighs 12 100 kg fully loaded. Armament consists of one roof-mounted 12.7 mm (0.50) anti-aircraft machine gun. Only 18 of these vehicles were built.

Type SX 60 4.2-inch (107 mm) Mortar Carrier

This has a 4.2-inch (107 mm) mortar mounted in the rear of the hull firing to the rear. A baseplate and stand are carried on the glacis plate enabling the mortar to be deployed away from the vehicle. The mortar has an elevation from +37° to +65°, traverse of 40° left and right and a maximum range of 4000 m. Eight mortar bombs are carried. The SX 60 has a crew of five, weighs 12900kg fully loaded but is fitted with the 12.7 mm anti-aircraft machine gun only. It can be distinguished from the 81 mm mortar carrier by the lack of a bow machine gun and the distinct chamfer to the rear of the top of the hull of the Type SX 60. The SX 60 has a single piece ramp at the rear of the hull hinged at its lower part. Only 18 of these vehicles were built.

Type SU 60 Anti-tank Vehicle

A small number of Type SU 60 APCs have been fitted with KAM-3D anti-tank wire-guided missiles in their launcher boxes, one on either side of the hull rear.

Other variants

There are at least three other variants of the Type SU 60 APC, an NBC detection vehicle, a dozer and a model modified to resemble the Soviet BMD airborne combat vehicle used by the Japanese Ground Self-Defence Force for training purposes. Only two dozers are in service, used by the 7th Division for clearing snow. A prototype of a 105 mm self-propelled howitzer was built under the designation SY 60 but was never placed in production.

Status: Production complete. In service only with Japanese Ground Self-Defence Force.

Manufacturers: Mitsubishi Heavy Industries, Maruko, Tokyo.
Komatsu Manufacturing Corporation.
Mitsubishi Heavy Industries
5-1, Marunouchi 2-chome
Chiyoda-ku
Tokyo
Japan

SPECIFICATIONS		MAX ROAD SPEED	45 km/h	ELECTRICAL SYSTEM	24 V
CREW	4 + 6	MAX RANGE	230 km	BATTERIES	2 × 12 V
COMBAT WEIGHT	11 800 kg	FORDING	1 m	ARMAMENT	
UNLOADED WEIGHT	10 600 kg	GRADIENT	60%	main	1 × 12.7 mm MG
POWER-TO-WEIGHT		VERTICAL OBSTACLE	0.6 m	bow	1 × 7.62 mm MG
RATIO	18.64 hp/tonne	TRENCH	1.82 m	SMOKE-LAYING	
GROUND PRESSURE	0.57 kg/cm²	TURNING RADIUS	7 m	EQUIPMENT	none
LENGTH	4.85 m	ENGINE	Mitsubishi Model 8 HA	GUN CONTROL	
WIDTH	2.4 m		21 WT, V-8 air-cooled	EQUIPMENT	
HEIGHT			4-cycle turbo-charged	Cupola traverse	360°, manual
including MG	2.31 m		diesel developing		
to hull top	1.7 m		220 hp at 2400 rpm		
GROUND CLEARANCE	0.4 m	TRANSMISSION	manual with 3 forward		
TRACK	2.05 m		and 1 reverse gears		
TRACK WIDTH	350 mm		and torque converter		
LENGTH OF TRACK		STEERING	controlled differential		
ON GROUND	3.18 m	SUSPENSION	torsion bar		

KOREA, NORTH

M-1973 Armoured Personnel Carrier

This is a licenced copy of the Chinese YW 531 full tracked APC covered on page 303. Variants include a Sagger ATGW carrier, a self-propelled anti-aircraft gun system with a 4 × 14.5 mm heavy machine gun mount, a 120 mm M1943 self-propelled mortar and a 12-tube 107 mm multiple rocket launcher.

Status: Production. In service with North Korea.

Manufacturer: North Korean state factories

KOREA, SOUTH

Korean Infantry Fighting Vehicle

Development

The Korean Infantry Fighting Vehicle has been developed to meet the requirements of the South Korean Army and is understood to have entered service in 1985.

The vehicle is based on the private venture FMC Armoured Infantry Fighting Vehicle with a number of improvements, especially in the automotive area, although the FMC Corporation was not involved in the design of the KIFV in any way.

The KIFV uses a number of foreign components including British Alcan aluminium armour, a German designed MAN diesel engine and a British Self-Changing Gears T-300 fully automatic transmission.

AP Precision Hydraulics designed and built the complete power braking and steering system using caliper brakes and supplied other key components including the engine-driven pump and the ancillary ramp door control circuit.

Description

The hull of the KIFV is of welded aluminium armour with an additional layer of spaced laminate steel armour bolted to it. This composite armour provides better protection for less weight and the closed polyurethane foam in the armour system gives increased buoyancy for amphibious operations.

The driver sits at the front of the hull on the left and has a single piece hatch cover that opens to the right. To the front and side of his position are four M27 day periscopes, the centre one of which can be replaced by a passive periscope for driving at night.

Mounted on the front of the hull above the trim vane is a bank of six forward-firing, electrically operated smoke dischargers. Day and night driving lights are fitted as standard.

The commander sits to the rear of the driver and has a rear-opening, single piece hatch cover and five periscopes, four standard M17 day periscopes and an M20A1 periscope with magnifications of ×1 and ×6, this can be replaced by a passive night periscope if required. The commander has a pintle-mounted 7.62 mm M60 series machine gun.

The MAN D-2848M diesel engine is in the front of the hull to the right of the driver and is coupled to a Self-Changing Gears T-300 automatic transmission with seven forward and seven reverse gears.

To the right of the commander is a cupola that can be traversed through a full 360°. It is fitted with a pintle-mounted 12.7 mm M2 HB machine gun that has a shield, lateral and rear armour protection and periscopes for all-round observation.

The troop compartment is at the rear of the hull and the nine infantry have individual seats. They enter and leave the vehicle via a power-operated ramp at the rear of the hull which has an integral door in the left side.

There is a roof hatch over the troop compartment and in either side are two firing ports with a vision block above, there are two firing ports in the rear ramp.

The suspension is of the torsion bar type with five dual rubber-tyred road wheels, drive sprocket at the front and idler at the rear, there are no track return rollers. The first, second and last road wheel stations have a hydraulic shock absorber. The track is believed to be T130E1 steel block, rubber-bushed type with removable rubber pads.

The KIFV is fully amphibious, propelled in the water by its tracks. Before entering the water the bilge pumps are switched on and a trim vane, which is stowed on the glacis plate when not required, is erected at the front of the hull. An NBC system is fitted as standard.

Variants

25 mm Gun Carrier

This is the basic KIFV with a one-man turret armed with a 25 mm cannon. Other weapon stations are possible including a 75 mm or 90 mm turret and a TOW anti-tank launcher.

Vulcan Weapon Carrier

This is the basic KIFV with the 20 mm Vulcan air defence system, which is made under licence in South Korea, installed on the turret roof. This version is in service with the South Korean Army.

Mortar Carrier

This has revised hatch covers with a mortar firing through the roof of the vehicle.

Command Post

This has a higher roof and extensive communications equipment and is armed with a roof-mounted 12.7 mm M2 HB machine gun.

Ammunition Carrier

This is similar to the command post and also has a roof-mounted 12.7 mm M2 HB machine gun.

Recovery Vehicle

This has a similar hull to the KIFV but is fitted with a hydraulically operated crane at the rear for changing powerpacks in the field.

Korean Infantry Fighting Vehicle showing positions of 12.7 mm and 7.62 mm machine guns

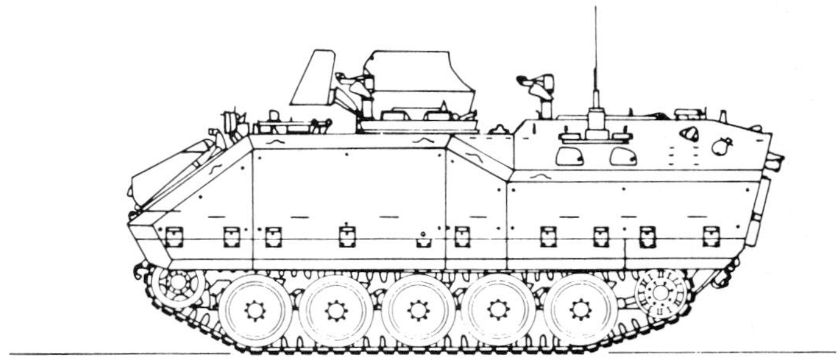

Unarmed Korean Infantry Fighting Vehicle

Korean Infantry Fighting Vehicle from rear with roof hatches open

Air defence version of Korean Infantry Fighting Vehicle armed with 20 mm Vulcan air defence system

The KIFV is also known as the Type 200, or K-200 series and it is estimated that by early 1991 about 1000 vehicles had been built for the South Korean Army.

Next Generation Vehicle

Early in 1990 it was revealed that Daewoo Heavy Industries was developing a 'Next Generation Vehicle' with trials expected to be carried out in 1992/93 with production starting in 1994/95.

The new design is understood to weigh about 20 tonnes and is powered by a 350 hp engine coupled to a fully automatic transmission. It is reported that the first prototype of this vehicle started undergoing trials in the Spring of 1991 and if these prove to be successful earlier production vehicles may be brought up to this configuration.

Design and engineering efforts focus on developing a basic platform with increased mobility, improved weapons and increased armour protection following in later stages.

The new vehicle has six road wheels either side compared to the KIFV's five.

It is believed that the prototype of the Flying Tiger (Bibo) twin 30 mm self-propelled anti-aircraft gun system is based on the lengthened chassis and that trials are already under way.

To support the South Korean Army's 155 mm M109A2 self-propelled howitzers an ammunition resupply vehicle is also being developed on this chassis. Some sources have indicated that this is called the New Generation Vehicle and is powered by a 550 hp MAN diesel engine coupled to a fully automatic transmission. The idea is to have a family of vehicles which will be powered by diesel engines with outputs from 500 up to 700 hp to give a power-to-weight ratio from 25 to 40 hp/tonne.

Basic applications are expected to include combat support, logistic support and combat services support.

Combat support would include ground and air defence weapons with calibres of 25 to 30 mm, anti-tank weapons of 90 mm to 105 mm or a range of anti-tank, air defence or tactical missiles.

Some sources have also indicated that the first versions of the new series will include an anti-tank variant fitted with a main gun no smaller than 90 mm.

Status: In production. In service with South Korean Army.

Manufacturer: Special Project Division
Daewoo Heavy Industries Limited
23rd Floor, Daewoo Center Building
541 Namdaemonno 5-ga, Chung-gu Seoul
South Korea
Telephone: 777 1856
Telex: 23301/CPO Box 7955

SPECIFICATIONS	
CREW	3 + 7
COMBAT WEIGHT	12 900 kg
UNLOADED WEIGHT	11 500 kg
POWER-TO-WEIGHT RATIO	21.7 hp/tonne
GROUND PRESSURE	0.67 kg/cm^2
LENGTH	
overall	5.486 m
hull	5.345 m
WIDTH	
overall	2.846 m
over tracks	2.545 m
HEIGHT	
hull top front	1.829 m
hull top rear	1.93 m
MG shield	2.518 m
GROUND CLEARANCE	0.41 m
ANGLE OF APPROACH/DEPARTURE	68°/45°
MAX SPEED	
road	74 km/h
water	6 km/h
FUEL CAPACITY	400 l
CRUISING RANGE	480 km
FORDING	amphibious
GRADIENT	60%
SIDE SLOPE	30%
VERTICAL OBSTACLE	0.63 m
TRENCH	1.7 m
ENGINE	MAN D-2848M V-8 diesel developing 280 hp at 2300 rpm
TRANSMISSION	Self-Changing Gears T-300 fully-automatic with 7 forward and 7 reverse gears (heavier versions of KIFV will have T320 transmission)
SUSPENSION	torsion bar in tube
ELECTRICAL SYSTEM	28 V
ARMAMENT	
main	1 × 12.7 mm M2 HB MG
secondary	1 × 7.62 mm M60 MG
SMOKE LAYING EQUIPMENT	6 smoke dischargers

KM900 Armoured Personnel Carrier

Development/Description

This is essentially the Italian FIAT-OTO Melara Type 6614 armoured personnel carrier manufactured under licence in South Korea. The basic model is designated KM900 while more specialised models are designated KM901. It is assumed that the latter include ambulance, command and mortar carriers. As far as it is known the specifications of the South Korean model are identical to those vehicles built in Italy.

Status: Production complete. In service with the South Korean Army.

Manufacturer: Asia Motors Co, Inc
1-60, Yoido-Dong
Yungdeugpo-ku, Seoul
South Korea
CPO Box 5022

NETHERLANDS

YP-408 Armoured Personnel Carrier

A total of 750 YP-408 (8×6) armoured personnel carriers were built by DAF for the Royal Netherlands Army between 1964 and 1968. These were finally phased out of service early in 1990 with their replacement being the YPR 765, a Netherlands-built version of the United States FMC Armoured Infantry Fighting Vehicle. A few YP-408s are still used for airport security, border patrol and by the Royal Military Police. Full details of the YP-408 were given in *Jane's Armour and Artillery 1989-90*, pages 365 and 366.

NORWAY

Norwegian Infantry Fighting Vehicle Competition

The Norwegian Army has a requirement for a new infantry fighting vehicle with increased armour, mobility and firepower over the existing M113 series vehicles currently in service and from 1991/92 will evaluate five different vehicles under a wide range of operational conditions in Norway:

1. FMC M2A2 Bradley infantry fighting vehicle
2. GKN Defence Warrior mechanised combat vehicle
3. Krauss-Maffei Puma armoured combat vehicle
4. Swedish Ordnance/Hägglunds Vehicle Combat Vehicle 90
5. Steyr-Daimler-Puch Kampfschutzenpanzer 90 armoured combat vehicle.

Following trials with these five vehicles, two will be selected for further modifications and these will then carry out further trials in the Winter of 1992/93.

Key Norwegian Army requirements include a turret-mounted 30 mm cannon, mobility equivalent to the Leopard 1 MBT currently in service with the Norwegian Army, have a three-man crew consisting of commander, gunner and driver and carry seven to eight fully equipped infantrymen with all of their combat equipment.

The contract is expected to be signed in 1993 with first deliveries in 1995.

Norwegian Army M113A1 vehicles

At present the Norwegian Army operates a fleet of about 250 FMC M113 series armoured personnel carriers of which about 100 have been upgraded to the Norwegian NM 135 configuration.

The NM 135 is an M113A1 fitted with a Swedish Hägglunds Vehicle one-man turret armed with a German Rheinmetall 20 mm Rh 202 cannon with a 7.62 mm machine gun mounted coaxially with the main armament to the right of the turret. Mounted on either side of the turret front is a bank of three electrically operated smoke dischargers firing forwards.

The turret is mounted behind the engine compartment and the vehicle commander's hatch moved to the immediate rear of the driver's position. The rearward opening hatch in the troop compartment roof is retained.

The Norwegian Army has also fitted some of its M113A1 APCs with the Norwegian Kvaerner-Eureka Armoured Launching Turret (ALT) which has two Hughes TOW ATGWs in the ready to launch position with additional TOW missiles being carried inside the hull for manual loading. Mounted on the turret rear are forward firing some dischargers.

The Norwegian M113A1 with the ALT is designated the NM 142 with the first batch comprising 36 vehicles, these were conversions of existing M113A1 vehicles, the second batch consisted of 44 vehicles which were purchased specifically for this purpose.

In addition to being used by Norway, the Armoured Launching Turret is also used by Canada (on M113A1s) and Switzerland (on MOWAG Piranha 6 × 6 light armoured vehicles).

For trials purposes, a Norwegian M113A1 APC has been fitted with a new Norwegian NFT turret armed with a German Mauser 25 mm cannon.

Norway also operates a total of 86 M548 tracked cargo carriers which are based on automotive components of the M113A1 armoured personnel carrier, these are used to support self-propelled artillery units equipped with 155 mm M109 series self-propelled howitzers. A locally developed recovery version of the M548 is in service called the NM 84.

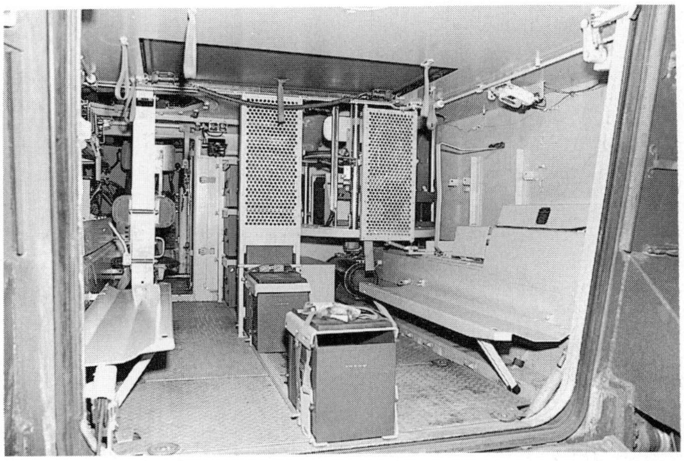

Interior view of Norwegian Army NM 135 with turret installation on right side

Norwegian Army NM 135 from rear showing redesigned roof with turret on right side

Norwegian Army M113 series APC fitted with Armoured Launching Turret with two TOW ATGW in ready to launch position

PAKISTAN

Local production of M113A2

Under an agreement signed early in 1989, Pakistan is to build 775 FMC Corporation, M113A2 APCs over a 4½ year period. The first 20 vehicles were assembled in Pakistan with the complete hulls being provided by FMC's San Jose facility.

From vehicle number 21, the hull was supplied in kit form with welding taking place at a new facility in Pakistan from late 1990.

The Pakistani Army already operates a fleet of about 900 M113 series APCs including 24 Improved TOW Vehicles. The M113s were supplied in 1963/64 in two batches, one of 73 and one of 36 vehicles. In the late 1970s two batches of the diesel powered M113A1 were supplied with the first batch comprising 300 vehicles and the second 230 vehicles and since then additional orders have been placed.

Some have recently been fitted with the Swedish Bofors RBS 70 surface-to-air missile system, this is carried internally and raised into the operating position when required for action.

Early in 1990 it was revealed that a new locally built air defence version of the M113 was in service with the Pakistani Army. Mounted on top of the hull is a pedestal armed with four locally built Chinese HN-5 (which in turn is based on the Soviet SA-7) series surface-to-air missiles and two 14.5mm KPV machine guns that can all be elevated together.

Pakistan Ordnance Factories have also developed a twin 106mm recoilless rifle mount that for trials purposes has been fitted onto a M113A1 APC of the Pakistani Army.

The system also integrates a laser rangefinder into the existing sight of the 106mm recoilless rifle to replace the elbow telescope. This allows the range to be displayed for the gunner in the sighting optic which means that he can apply the super elevation and fire without having to look at one more instrument. The effective range of the recoilless rifles is 1500m.

Local production of Armored Infantry Fighting Vehicles

In early 1991, the Pakistan Government ordered a single Armored Infantry Fighting Vehicle from the FMC Corporation for extensive trials. If these trials are successful it is expected that a substantial order will be placed with production taking place in Pakistan.

The first Armored Infantry Fighting Vehicle was shipped to Pakistan in kit form consisting of the hull and other essential parts. The initial Pakistani AIFV hull was originally sold to the Canadian Government in 1975 and had been in storage at a Canadian test area since then.

Due to the urgency of the Pakistani request, FMC negotiated to purchase the vehicle back from the Canadians and modify the hull to the current Armored Infantry Fighting Vehicle configuration prior to shipment to Pakistan.

Status
M113A1 in production in Pakistan and in service with Pakistani Army.

M113A1 APC with twin 106 mm recoilless rifle mount developed by Pakistan Ordnance Factories

M113 APC with RBS 70 surface-to-air missile system in raised position

PORTUGAL

BRAVIA Armoured Personnel Carriers

Development
The Chaimite range of 4 × 4 armoured vehicles was developed by BRAVIA (Sociedade Luso - Brasileira de Viaturas Equipamentos SARL) in the 1960s to meet the requirements of the Portuguese armed forces. The first prototype of the Chaimite was completed in 1966 and by early 1991 production had amounted to over 600 vehicles. It is reported that the Portuguese Army has 90 Chaimite armoured personnel carriers. In appearance the Chaimite is very similar to the American Cadillac Gage Commando range of 4 × 4 multi-mission vehicles which entered production in 1964, and for this reason there is often confusion about the original source of these vehicles. The basic APC is called the V-200, but BRAVIA has now developed a whole range of vehicles which use the same basic hull.

The original Chaimite vehicle has now been replaced in production by the upgraded BRAVIA Mk 1 4×4 vehicle which is almost identical in appearance.

Description (V-200)
The hull of the BRAVIA APC is made of all-welded steel. The driver sits towards the front of the vehicle with the second crew member, normally the commander, to his right. Over their positions is a two-piece hatch cover that opens either side of the vehicle. The driver and commander each have a vision block to their front between which is a single firing port. To the left of the driver and the right of the commander are a vision block and firing port.

There is a two-part door in each side of the hull, the upper part opening to the rear and the lower part folding downwards to form a step. There are two vision blocks and two firing ports in each side of the hull between the front of the vehicle and the side door.

The main armament is installed immediately behind the driver's and commander's position.

The engine is at the rear of the vehicle on the left side and is separated from the personnel compartment by a fireproof bulkhead. Access to the engine for maintenance is through two hatches in the roof and a single hatch in the left side of the hull. The engine compartment is equipped with a fire extinguisher operated by the driver.

A door at the rear of the hull on the right side opens to the left and has a single vision block and a single firing port. There is an additional forward-opening hatch in the roof of the vehicle to the right of the engine compartment and there is also a firing port and vision block below it in the right side of the hull.

Power is transmitted from the engine to the transmission then to the transfer case and then to the front and rear axles. The axles are of the fully floating double reduction type with locking differentials. The tyres are of the run-flat type.

Mounted in the front of the Chaimite is a hydraulic winch with a maximum capacity of 4530kg, provided with 38.1 m of 11mm diameter cable with a maximum breaking strength of 8090kg. Optional equipment includes infra-red night vision equipment.

The BRAVIA is fully amphibious and propelled in the water by its wheels.

The V-200 is fitted with a BRAVIA-designed turret armed with twin 7.62mm (0.30) machine guns, twin 5.56mm machine guns or one 7.62mm (0.30) and one 12.7mm (0.50) machine gun. The weapons can be elevated manually from -15 to +50°, traverse is manual through a full 360°. The

BRAVIA Mk I with four 3.5-inch rocket launchers either side of turret

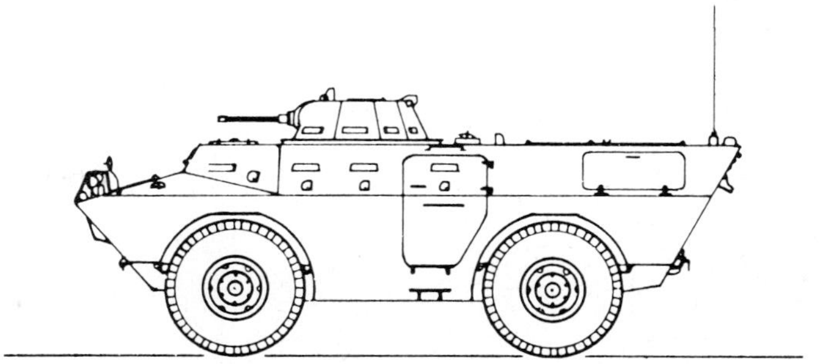

BRAVIA Mk I with one-man turret armed with twin 7.62 mm machine guns

turret with twin 7.62 mm (0.30) machine guns is provided with 500 rounds of ready-use ammunition (250 rounds per gun) plus a further 9000 rounds carried in the vehicle in reserve. The turret has a single piece hatch cover that opens to the rear, vision blocks and an M28C periscopic sight mounted in the left side of the turret for aiming the machine guns. If required the turret can be fitted with a device for launching five 60 mm anti-personnel, anti-tank, smoke, illuminating or incendiary grenades. The turret can also be fitted with four 3.5-inch rocket launchers either side of the turret, which can be fired from inside the vehicle but are reloaded from outside.

A wide range of optional equipment can be fitted, including an air-conditioning system, fire extinguishing system and explosion suppression system.

Variants

V-300

This model has a crew of four or five and can be fitted with a one- or two-man turret. The former is the Oerlikon GAD-AOA turret armed with a 20 mm KAA-001 cannon.

The following two-man turrets can be fitted to the V-300; BRAVIA turret armed with a 20 mm cannon (with 100 ready rounds and 300 rounds in reserve), 7.62 mm coaxial machine gun (500 ready rounds) and a 7.62 mm anti-aircraft machine gun (500 ready rounds) and 3000 rounds of 7.62 mm ammunition in reserve; French SAMM S 530 (new version is designated TAB 220) turret armed with twin 20 mm M621 cannon or Dassault Electronique TA-20 turret armed with twin 20 mm cannon.

V-400 armoured car (anti-tank)

The following two-man turrets can be fitted to the V-400: turret armed with MECAR 90 mm gun with an elevation of +25°, depression of -8° and 360° turret traverse, a 7.62 mm machine gun mounted coaxially with the main armament and a 7.62 mm machine gun mounted on the turret roof for anti-aircraft use. More recently the company has offered this version fitted with the 90 mm MECAR KEnerga gun. Can also be fitted with 60 mm grenade launchers as fitted to the V-200. Ammunition capacity is 61 rounds of 90 mm (of which 21 are for ready use) and 3500 rounds of 7.62 mm (of which 500 are for ready use). In April 1983 a BRAVIA Chaimite V-400 fitted with a French Hispano-Suiza Lynx 90 turret was successfully tested in Portugal.

V-500 armoured car

This is a command and communications vehicle.

V-600 armoured car

This is a mortar carrier and has an 81 mm or 120 mm mortar.

V-700 armoured car

ATGW vehicle armed with HOT, TOW or MILAN Compact Turret (MCT).

V-800 armoured car

This is the ambulance version and is unarmed.

V-900 armoured car

Crash rescue vehicle.

V-1000 armoured car

This is a riot control vehicle and can be fitted with various armament installations including a water cannon.

BRAVIA Mk I with SS-11 ATGW

Portugal has fitted a number of its BRAVIA Mk I (4×4) vehicles with a single SS-11 wire-guided ATGW either side of the turret.

BRAVIA Mk II

This is a 6 × 6 version of the BRAVIA Mk I and is powered by a 280 hp turbocharged diesel engine coupled to a transmission with five forward and one reverse gears. It can be fitted with heavier weapon systems up to a 105 mm rifled tank gun firing standard NATO ammunition. According to the manufacturer, development of this is complete and production can start on receipt of orders.

BRAVIA Mk III

This is an 8 × 8 version of the BRAVIA Mk II and is powered by a 300 hp turbocharged diesel engine coupled to a transmission with five forward and one reverse gears and can also be fitted with larger weapon systems. According to the manufacturer development of this version is complete and production can start on receipt of orders.

BRAVIA Mk I armoured car model V-400 with turret armed with MECAR 90 mm gun, 7.62 mm coaxial machine gun and 7.62 mm anti-aircraft machine gun

BRAVIA Mk I V-200 with turret-mounted 7.62 mm and 12.7 mm machine guns

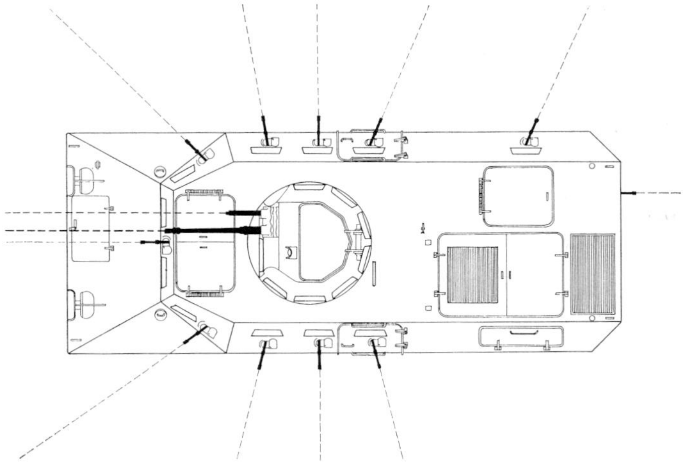

Top view of BRAVIA Mk 1 V-200 (4 × 4) vehicle showing position of weapons (not to 1/76th scale)

SPECIFICATIONS

MECAR BRAVIA Mk 1 (V-200)

CREW	11
CONFIGURATION	4 × 4
COMBAT WEIGHT	7300 kg
POWER-TO-WEIGHT RATIO	32 hp/tonne
LENGTH HULL	5.606 m
WIDTH	2.26 m
HEIGHT	
including turret	2.26 m
hull top	1.84 m
GROUND CLEARANCE	
hull	0.61 m
differential	0.41 m
TRACK	1.87 m
WHEELBASE	2.667 m
ANGLE OF APPROACH/DEPARTURE	52°/46°
MAX ROAD SPEED	
1st gear	13.7 km/h
2nd gear	22.6 km/h
3rd gear	37.9 km/h
4th gear	62 km/h
5th gear	99 km/h
reverse	13.7 km/h
MAX WATER SPEED	7 km/h
FUEL CAPACITY	300 l
MAX ROAD RANGE	
petrol engine	804-965 km
diesel engine	1367-1529 km
MAX CROSS-COUNTRY RANGE	
petrol engine	563-724 km
diesel engine	1046-1207 km
FORDING	amphibious
GRADIENT	65%
SIDE SLOPE	40%
VERTICAL OBSTACLE	0.9 m
ENGINE	V-8 turbocharged water-cooled Detroit Diesel developing 210 hp at 3000 rpm or V-6 turbo-charged water-cooled Detroit Diesel Model 6V-53T developing 280 hp at 2800 rpm or M75 V-8 water-cooled petrol developing 210 hp at 4000 rpm
TRANSMISSION	Allison Transmission Division of General Motors Corporation AT-545 automatic with 4 forward and 1 reverse gears or manual with 5 forward and 1 reverse gears
TRANSFER CASE	1 or 2-speed
STEERING	power-assisted
CLUTCH	single dry plate, hydraulically operated
SUSPENSION	semi-elliptical springs and hydraulic single-action shock absorbers
TYRES	14.00 × 20
BRAKES	
main	air/hydraulic, power assisted
parking	on output shaft of transmission
ELECTRICAL SYSTEM	24 V
ARMAMENT	see text
ARMOUR	
Hull front upper	6.35 mm at 75°
Hull front lower	7.94 mm at 40°
Hull sides upper	6.35 mm at 30°
Hull sides lower	7.94 mm at 10°
Hull top	6.35 mm
Hull floor	9.35 mm
Hull rear	6.35 mm at 30°
Turret front	7.94 mm
Turret sides	6.35 mm
Turret top	6.35 mm

Status: Production as required. The Chaimite and BRAVIA Mk I 4 × 4 APC is known to be in service with Lebanon (internal security role), Libya (status uncertain), Peru, Philippines, Portugal (Army and Navy) and other undisclosed countries.

Manufacturer: BRAVIA SARL
Porto Alto - 2135 Samora Correia
Portugal
Telephone: (063) 62256/7
Telex: 12715 BRAVIA P
Fax: (1) 284 6293

BRAVIA Tigre Mk II (6×6) APC

Development

The BRAVIA Tigre Mk II (6×6) armoured personnel carrier is essentially a BRAVIA Leopardo (6×6) 2268 kg or a BRAVIA Pantera (6×6) 4535 kg truck, fully described in *Jane's Military Vehicles and Logistics 1991-92* page 374, fitted with an armoured body for use in the armoured personnel carrier role.

Description

The layout of the BRAVIA Tigre Mk II (6×6) armoured personnel carrier is similar to the basic truck with the armour-plated engine compartment at the front, driver and commander to the rear of the engine, with the troop compartment extending right to the rear. The armoured shutters at the front of the engine compartment are controlled by the driver.

The troop compartment has an open roof with circular firing ports being provided in the sides and rear.

Optional equipment includes a winch with a capacity of 4536 kg and bullet-proof/run-flat tyres.

BRAVIA Tigre Mk II (6 × 6) armoured personnel carrier with roof-mounted 7.62 mm machine guns

Variants
V-200, basic APC carrying a maximum of 20 men and armed with 1×7.62 mm and 1×12.7 mm MG or 2×7.62 mm or 2×12.7 mm MGs.
V-300, armed with 2×20 mm or 2×25 mm cannon, 4×12.7 mm MG or 20 mm Vulcan air defence system.
V-400, armed with a 106 mm M40 series recoilless rifle.
V-500, armoured command post vehicle.
V-600, mortar carrier with 81 mm, 82 mm, 120 mm or 160 mm mortar.
V-700, ATGW carrier with HOT, TOW or MILAN ATGW.
V-800, ambulance.
V-900, crane.

Status: Production as required.

Manufacturer: BRAVIA SARL
Porto Alto - 2135 Samora Correia
Portugal
Telephone: (063) 62256/7
Telex: 12715 BRAVIA P
Fax: (1) 284 6293

SPECIFICATIONS		FORDING	0.76 m	TRANSMISSION	manual, 5 forward and
CREW	20	with preparation	1.98 m		1 reverse gears
CONFIGURATION	6×6	GRADIENT	60%	TRANSFER CASE	2-speed
COMBAT WEIGHT	9500 kg	SIDE SLOPE	30%	CLUTCH	single dry plate
TOWED LOAD	4500 kg	ENGINE	V-8 petrol developing	STEERING	cam and gear type
LENGTH	6.8 m		210hp at 4000rpm, or		with option of power
WIDTH	2.4 m		Detroit Diesel 4-		steering
HEIGHT	2.3 m		cylinder diesel	BRAKES	
GROUND CLEARANCE	0.28 m/0.32 m		developing 180hp at	parking	internal/external
TRACK			2800rpm or Perkins	service	air over hydraulic
front	1.721 m		turbocharged in-line	ELECTRICAL SYSTEM	24 V
rear	1.721 m (single tyres)		diesel developing	BATTERIES	2×12 V
	1.778 m (dual tyres)		159hp at 2800rpm or		
FUEL CAPACITY	380 l		V-8 turbocharged		
RANGE	960 km (diesel engine)		diesel developing		
MAX ROAD SPEED	100 km/h		210hp at 3000rpm		

BRAVIA Commando Mk III Armoured Personnel Carrier and Scout Car

Development
The Commando Mk III APC was developed by BRAVIA in 1977 to meet the requirements of the Portuguese National Guard and has since been exported to at least two countries for internal security use. The first prototype of the Commando Mk III APC was completed in 1977 with production vehicles following the same year. The vehicle is basically a short wheelbase version of the BRAVIA Gazela (4 × 4) one-ton truck chassis fitted with an armoured body.

In appearance the Commando Mk III is very similar to the British Shorland armoured patrol car, but the Portuguese vehicle has a longer wheelbase, is much heavier and can also be used as an APC.

By early 1991, total production of the BRAVIA Commando Mk III armoured personnel carrier amounted to about 200 vehicles.

Description
The engine is at the front of the vehicle, driver and commander in the centre and the personnel compartment at the rear. The driver sits on the left of the vehicle with the commander to his right. They each have a windscreen to their front which can be covered quickly by an internally controlled drop-down armoured visor containing a laminated glass observation block. The commander and the driver have a side door, the top part of which folds down on the outside for improved visibility and has a laminated glass observation block.

The five troops sit to the rear of the driver's and commander's position and are provided with four firing ports with a laminated observation window over each, one in each side of the hull and two in the rear. The interior of the vehicle is lined with non-inflammable pvc foam and ventilation is provided by a fresh air system with electric fan boost.

The BRAVIA-designed turret is mounted in the centre of the roof and is armed with twin 7.62 mm machine guns or one 7.62 mm and one 12.7 mm machine gun. The weapons can be elevated from -15 to +60° manually and traverse is manual through a full 360°. The turret has a single piece hatch cover that opens to the rear, vision blocks and a periscopic sight in its forward part for aiming the machine guns. The turret can be fitted with a device for launching five 60 mm anti-personnel, anti-tank, smoke, illuminating or incendiary grenades.

Optional equipment includes sand tyres, run-flat tyres, an air-conditioning system, fire extinguishing and explosion suppression system.

Variants
The Commando Mk III is also available in a 4 × 2 configuration when there is a limited off-road requirement. The vehicle is also available with an additional door to the immediate rear of the commander's and driver's doors. This is similar to the latter door.

If required the Commando Mk III can also be supplied without the manually operated machine gun turret, for use in the armoured personnel carrier role.

For use in the dedicated armoured personnel carrier role a version of the Commando Mk III is available with the hull roof extended to the rear. The troop compartment has two doors in the rear, each of which has a vision block with a firing port underneath. In addition there are two vision blocks and two firing ports in either side of the rear troop compartment. This version can be delivered with or without the machine gun turret.

SPECIFICATIONS		FUEL CAPACITY	160 l	STEERING SYSTEM	integral power with
CREW	3 + 5	MAX ROAD RANGE			pump assitance
CONFIGURATION	4 × 4	diesel engine	800 km	CLUTCH	single dry plate
COMBAT WEIGHT	4855 kg	petrol engine	600 km	SUSPENSION	heavy duty semi-
UNLOADED WEIGHT	4330 kg	GRADIENT	70%		elliptical springs and
WEIGHT ON FRONT		TURNING RADIUS	7.65 m		telescopic shock
AXLE LOADED	2236 kg	ENGINE	V-8 petrol developing		absorbers at each
WEIGHT ON REAR			180 hp at 3600 rpm or		wheel station
AXLE LOADED	2619 kg		6-cylinder in-line	TYRES	9.00 × 16 or 10.50 × 16
LENGTH	4.975 m		diesel developing	BRAKES	
WIDTH	1.93 m		120 hp at 2800 rpm or	main	hydraulic on all wheels
HEIGHT			V-8 turbocharged	parking	mechanical operating
to turret top	2.42 m		diesel developing		on rear wheels
to hull top	2.05 m		155 bhp at 3600 rpm	ELECTRICAL SYSTEM	24 V
GROUND CLEARANCE	0.21 m		or	BATTERIES	2 × 12 V, 105 Ah
TRACK			V-8 air-cooled diesel	ARMAMENT	see text
front	1.72 m		developing 168 bhp	ARMOUR	
rear	1.665 m	TRANSMISSION	manual with 4 forward	Hull	6.35-7.94 mm
WHEELBASE	3.03 m		and 1 reverse gears or	Turret front	7.94 mm
ANGLE OF APPROACH/			automatic with 3	Turret sides	6.35 mm
DEPARTURE	49°/47°		forward and 1 reverse	Turret top	6.35 mm
MAX ROAD SPEED			gears		
diesel engine	90 km/h	TRANSFER CASE	2-speed		
petrol engine	110 km/h				

BRAVIA Commando Mk III APC and scout car with driver's and commander's hatches open

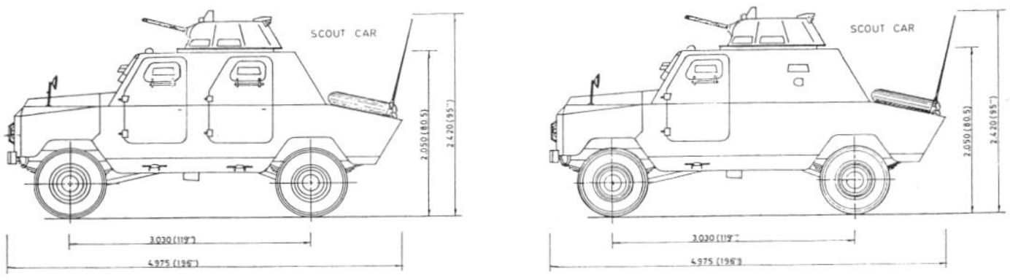

BRAVIA Commando Mk III fitted with turret mounted machine guns and showing different hull door arrangement (not to 1/76th scale)

Status: Production as required. In service with Portuguese National Guard and at least two other countries.

Manufacturer: BRAVIA SARL
Porto Alto - 2135 Samora Correia
Portugal
Telephone: (063) 62256/7
Telex: 12715 BRAVIA P
Fax: (1) 284 6293

ROMANIA

MLI-84 Infantry Combat Vehicle

Development/Description

In appearance and layout the MLI-84 is virtually identical to the Soviet BMP-1 infantry fighting vehicle but is slightly larger and heavier and has a 12.7 mm anti-aircraft machine gun mounted over the left side of the rear troop compartment. The MLI-84 is fully amphibious being propelled in the water by its tracks and is fitted with an NBC system and night vision equipment for commander, gunner and driver.

Status: Production. In service with Romanian Army.

Manufacturer: Romanian state factories.

Romanian MLI-84 IFV showing roof-mounted 12.7 mm anti-aircraft machine gun to rear to turret

SPECIFICATIONS		ENGINE	8V-1240 DT-S supercharged diesel developing 360 hp	ATGW launcher	1 × Sagger
CREW	2 + 9			anti-aircraft	1 × 12.7 mm MG
COMBAT WEIGHT	16 600 kg			AMMUNITION	
POWER-TO-WEIGHT RATIO	21.68 hp/tonne	FUEL CONSUMPTION		main	40
		road	110/150 l/100 km	coaxial	2000
LENGTH	7.32 m	cross-country	190/220 l/100 km	anti-aircraft	500
WIDTH	3.15 m	ARMAMENT		Sagger	4
HEIGHT	1.97 m	main	1 × 73 mm gun		
GROUND CLEARANCE	0.4 m	coaxial	1 × 7.62 mm MG		

MLVM Mountaineers Combat Vehicle

Development/Description

This vehicle has been designed specifically for use in mountainous terrain and can carry an assault group of seven men in addition to its two-man crew which consists of commander and driver. It can also be used to carry stores and ammunition in the forward battlefield area.

The hull and turret is of all-welded steel armour construction that provides the crew with protection from small arms fire and shell splinters.

The driver is seated at the front of the vehicle on the left side with three periscopes for forward observation, one of which can be replaced by a night vision device and a single piece hatch cover. To the rear of the driver is the vehicle commander who has a single piece hatch cover and three periscopes, one of which can be raised for forward observation. The commander also has an externally mounted infra-red searchlight.

The turret is in the centre of the hull and is armed with one 14.5 mm and one 7.62 mm machine gun with the sight on the left side of the turret. The turret is the same as that installed on the TAB-71M (8 × 8) APC.

The troop compartment is at the rear and is provided with roof hatches, six firing ports (three in each side of the hull) each with an associated periscope and a door in the hull rear which has two firing ports and two periscopes as well as containing some diesel fuel.

Suspension is probably of the torsion bar type with six rubber-tyred spoked road wheels, drive sprocket at the front, idler at the rear and track return rollers.

Romanian MLVM mountaineers combat vehicle

In addition to the small arms of the infantry an AG-7 anti-tank grenade launcher is also carried.

Variants

120 mm mortar carrier

This has a higher superstructure than the MLVM and is fitted with a 120 mm mortar that fires to the rear with a baseplate being carried on the rear of the hull to allow for dismounted use. This has the Romanian designation of the MLVM AR.

ABAL armoured vehicle for combat supply

This is very similar to the MLI-84 but has been designed to carry ammunition in the forward area, for example 64 rounds of 100 mm or 104 rounds of 76 mm ammunition with total payload being 2100 kg. Some of the ammunition is in ready-use lockers in either side of the hull with external access doors being provided while others are stowed inside. This version does not have turret-mounted machine guns although a 7.62 mm machine gun is mounted to the rear of the driver's position with a traverse of 360° and elevation from -5° to +82°. Standard equipment includes NBC system, fire fighting equipment and communications equipment.

Specifications are similar to the MLI-84 except for combat weight of 10 500 kg, unladen weight of 8400 kg, ground pressure 0.49 kg/cm², width 2.714 m, height 1.90 m, maximum road speed 32 km/h, road range 666 km, cross-country range 383 km and fuel capacity 450 litres.

Status: In production. In service with Romanian Army.

Manufacturer: Romanian state factories

Romanian ABAL armoured vehicle for combat supply

SPECIFICATIONS		MAX ROAD SPEED	48 km/h	TRENCH	1.5 m
CREW	2 + 7	RANGE		ENGINE	4-stroke supercharged diesel
WEIGHT COMBAT	9800 kg	cross-country	370 to 400 km		
POWER-TO-WEIGHT RATIO	15.7 hp/tonne	road	680 to 740 km	ARMAMENT	
GROUND PRESSURE	0.47 kg/cm²	FUEL CAPACITY	480 litres of which 360 litres are in main fuel tank and 120 litres in door tank	main	1 × 14.5 mm MG
LENGTH	5.85 m			coaxial	1 × 7.62 mm MG
WIDTH	2.87 m			GUN CONTROL EQUIPMENT	manual
HEIGHT		FORDING	0.6 m	ELEVATION/DEPRESSION	+84°/-3°
turret roof	1.95 m	GRADIENT	60%	TRAVERSE	360°
hull top	1.55 m	SIDE SLOPE	30%		
GROUND CLEARANCE	0.38 m	VERTICAL OBSTACLE	0.6 m		

TAB-77 Armoured Personnel Carrier

Development/Description

For some years Romania has been building a modified version of the Soviet BTR-70 (8 × 8) armoured personnel carrier under licence. This has the Romanian designation of the TAB-77 and differs from the Soviet BTR-70 in a number of areas.

The most significant difference is that the Romanian TAB-77 is powered by a single diesel engine whereas the BTR-70 has two petrol engines.

The TAB-77 is fully amphibious being propelled in the water by a single waterjet located in the rear of the hull and before entering the water the trim vane, which is stowed on the glacis plate when not required, is erected and the bilge pumps switched on.

The one-man manually operated turret is armed with one 12.7 mm machine gun with a 7.62 mm machine gun mounted coaxial to the right.

Variants

TAB-77 PCOMA - Armoured artillery command post vehicle.
TAB-77A 451/M - Command vehicle.
TERA-77L - TAB-77 maintenance and recovery vehicle.

Status: Production. In service with the Romanian Army and available for export.

Manufacturer: Romanian state factories.

TAB-77 (8 × 8) armoured personnel carrier

TAB-71 Armoured Personel Carrier

Development/Description

In the late 1960s the Romanian Socialist Army (ASR) approached the Military Industrial Committee of COMECON with the request to licence produce the Soviet BTR-60PB (8 × 8) APC. This request was granted and the first model, called TAB-71 entered production in 1970 and was first seen in public during a military parade in Bucharest held in August 1972.

The TAB-71 is very similar to the Soviet BTR-60PB (8 × 8) armoured personnel carrier but the two 90 hp petrol engines have been replaced in the TAB-71 by two 140 hp engines.

The TAB-71 retains the central tyre pressure regulation system, amphibious capability, front-mounted winch and NBC system of the Soviet vehicle.

Variants

TAB-71M -TAB-77 fitted with a locally designed one-man manually operated turret armed with one 12.7 mm and one 7.62 mm machine gun

TAB-71M (8 × 8) APC without machine gun and showing distinctive sight on left side of turret (Michael Green)

with a distinctive gun sight mounted on the left side of the turret. This turret is also fitted to the TAB-77 (8 × 8) and TABC-79 (4 × 4) armoured personnel carriers, and the MLVM Mountaineers Combat Vehicle.
TAB-71AR - TAB-71 modified to carrier 82 mm mortar. The turret has been removed and modifications made to the troop compartment to allow the 82 mm mortar to be raised when the roof hatches are swung open through 180° either side. It is estimated that between 50 and 100 82 mm mortar bombs are carried.
TERA-71 - TAB-71 modified for the maintenance and recovery role.
TERA-71L - TAB-71 modified for the maintenance and recovery role.
TAB-71A R1 450 - TAB-71 modified for armoured command vehicle.
TAB-71A R1 451 - TAB-71 modified for armoured command vehicle.
TAB-71A R1 452 - TAB-71 modified for armoured command vehicle.

Status: Production complete. In service with Romania and Yugoslavia.

Manufacturer: Romanian state factories

TAB-71AR mortar carrier with mortar raised and roof hatches open

SPECIFICATIONS		FUEL CAPACITY	290 l	ARMAMENT	
CREW	3 + 8	MAX ROAD RANGE	500 km	main	1 × 14.5 mm MG
CONFIGURATION	8 × 8	FORDING	amphibious	coaxial	1 × 7.62 mm MG
COMBAT WEIGHT	11 000 kg	GRADIENT	60%	AMMUNITION	
POWER-TO-WEIGHT		VERTICAL OBSTACLE	0.4 m	main	500 rounds
RATIO	25.4 hp/tonne	TRENCH	2 m	coaxial	2000 rounds
LENGTH	7.22 m	ENGINES	2 × V-6 liquid-cooled	GUN CONTROL	
WIDTH	2.83 m		petrol developing	EQUIPMENT	
HEIGHT (overall)	2.7 m		140 hp each	Turret power control	manual
GROUND CLEARANCE	0.47 m	TRANSMISSION	manual, 5 forward and	Gun elevation/depression	+85°/-5°
WHEELBASE	4.21 m		1 reverse gears	Turret traverse	360°
MAX SPEED		TYRES	13.00 × 18	ARMOUR	
road	95 km/h			Hull (max)	9 mm
water	10 km/h			Turret (max)	7 mm

TABC-79 Armoured Personnel Carrier

Development/Description

TABC-79 (4 × 4) armoured personnel carrier has been developed in Romania and uses some automotive components of the TAB-77 (8 × 8) armoured personnel carrier.

The commander and driver are seated at the front of the vehicle with the driver on the left and the vehicle commander on the right, both of these have a single piece hatch cover that opens forward and periscopes for observation to the front and sides, while the commander has an infra-red searchlight that is mounted on the vehicle's roof and can be operated from inside. To the immediate front of the commander and driver is a windscreen that can be quickly covered by an armoured shutter when in a combat area.

Romanian TABC-79 (4 × 4) APC with commander's and driver's hatches in open position

The one-man manually operated turret is in the centre of the roof and is the same as that installed on the TAB-71M (8 × 8) and TAB-77 (8 × 8) armoured personnel carriers as well as the MLVM Mountaineers Combat Vehicle.

The engine compartment is located at the rear of the TABC-79 with an access passage either side for the troops, in addition there is a small entrance door that opens forward between the first and second road wheels and the troop compartment is also provided with firing ports.

The TABC-79 is fully amphibious being propelled in the water by a waterjet mounted at the hull rear, before entering the water the bilge pumps are activated and the trim vane erected at the front of the hull, when not required the latter is stowed on the glacis plate.

Standard equipment on the TABC-79 includes a front-mounted winch, NBC system and infra-red night vision equipment.

Variants

TAB-79A POMA - TABC-79 armoured artillery observation post vehicle.
TAB-79AR -TABC-79 self-propelled 82 mm mortar system.
TABRCH-84 - TABC-79 armoured chemical and radiological reconnaissance vehicle.
TCG-80 - unidentified vehicle which may not be based on TABC-79 chassis.

Status

Production probably complete. In service with Romanian Army.

SOUTH AFRICA

Ratel Infantry Fighting Vehicle

Development

In the 1950s the United Kingdom supplied South Africa with 250 Alvis Saracen (6 × 6) armoured personnel carriers, all of which are now held in reserve after being reconditioned. In 1968 Sandock-Austral, which was then building an improved version of the French Panhard AML (4 × 4) armoured car called the Eland, started design work on a new infantry fighting vehicle, the Ratel, to meet the requirements of the South African Infantry Corps and, at a later date, the South African Armoured Corps.

The first prototype was completed in July 1974, followed by four pre-production vehicles, the first completed in 1976. The first production models were designated the Mark 1 and were followed in 1979 by the Mark 2, and final deliveries of these were made in early 1987. The differences are mainly the result of operational experience in Namibia and long-range penetration raids into Angola. Most of the modifications have subsequently been retrofitted to earlier vehicles when they return to Sandock-Austral for overhaul.

Description (FSV 90)

The hull of the Ratel is made of all-welded steel that provides complete protection from 7.62 mm small arms fire and shell splinters, with protection over the frontal arc against 12.7 mm armour-piercing rounds. Emphasis has also been placed on providing protection against mines.

The driver sits at the front of the vehicle with excellent forward visibility through three large bullet-proof windows to his front and sides. In combat areas they can be quickly covered by armoured shutters hinged at the bottom and activated by the driver with a single handle from within the vehicle. With the shutters in position the driver observes the terrain through three periscopes, one to the front and one either side. Over the driver's position is a single piece hatch cover that opens to the left. The driver's seat and steering wheel are adjustable and he can also enter his compartment from the rear so enabling drivers to be changed without one of them leaving the vehicle.

Mounted on the roof of the vehicle immediately behind the driver's position is the two-man all-welded steel turret, identical to the Eland (4 × 4) armoured car's. The commander sits on the left and the gunner on the right. The commander's cupola has vision blocks for all-round observation and a single piece hatch cover opening to the rear, plus an orientation vane sight. The gunner has four periscopes for observation plus an M494 optical sight for aiming the main and secondary armament, and a single piece hatch cover that opens to the rear.

The main armament is a 90 mm semi-automatic quick-firing gun which is a further development of the weapon fitted in the 90 mm version of the Eland (4 × 4) armoured car. This has an elevation of +15° and a depression of -8° with turret traverse being a full 360°. Depression can be obtained through 285° only because of the troop compartment to the turret rear. Gun elevation and turret traverse are manual, operated by the gunner. When travelling the turret is often traversed to the rear. The 90 mm gun fires HEAT (effective range 1200 m), HE (effective range 2200 m), and practice (effective range 1200 m) projectiles, all manufactured by Armscor. Twenty-nine rounds of 90 mm ammunition are carried in the turret and 42 rounds in the hull. A 12.7 mm M2 HB machine gun can be mounted over the 90 mm gun for training.

Mounted to the left of the 90 mm gun is a 7.62 mm coaxial machine gun for which 2000 rounds of 7.62 mm ammunition are carried in the turret with a further 4000 rounds on the hull. Both the main and coaxial weapons are fired electrically and above the coaxial machine gun is an extractor fan to remove fumes from the turret.

Mounted either side of the turret towards the rear is a bank of two 81 mm electrically operated smoke dischargers which can be operated by the commander or gunner. A 7.62 mm anti-aircraft machine gun is mounted externally at the gunner's station and on top of the turret is a hand-operated searchlight. In the left side of the turret there is an ammunition resupply hatch.

The Ratel 90 carries a crew of 10, one fewer than the Ratel IFV, to enable the 42 rounds of ammunition to be carried inside the hull. The crew consists of the vehicle commander, driver, main gunner (90 mm), rear anti-aircraft gunner, section commander and five infantrymen.

In either side of the hull behind the driver's position is a large door opened pneumatically by the driver. It opens forwards and has a vision block with a firing port underneath.

The section commander sits near the left door and the five infantrymen on bench seats down the middle of the vehicle behind the turret, two on the left and three on the right. In both sides of the troop compartment there are three bullet-proof vision blocks with a firing port underneath and over the top of the troop compartment are four roof hatches that are hinged on the outside and can be locked vertical. There is a passageway from the troop compartment to the rear of the vehicle on the right side, above which are two roof hatches that open outwards and a circular mount with a single piece hatch cover on which the 7.62 mm machine gun can be mounted. In the rear of the hull on the right side is a door. Its lower part folds down to form a step and the upper part opens to the left.

The engine compartment is at the rear of the Ratel on the left side with access panels in the roof for maintenance.

The six-cylinder direct injection turbo-charged diesel is coupled to a fully automatic powershift gearbox with a hydro-dynamic torque converter. The gearbox can also be operated manually and has a mechanical emergency gearshift. Power is transmitted to the three axles in two stages with final reduction by means of planetary gearing housed in the wheel hubs. The three axles are each provided with their own lockable differential and longitudinal differential locks. Suspension consists of progressively acting coil springs and large hydraulic shock absorbers. The hydro-pneumatic braking system can be assisted by an exhaust brake provided through the engine.

The powerpack can be removed from the vehicle by two men with a crane in less than 30 minutes, quick disconnect couplings and connectors are provided in all power, hydraulic, electric and pneumatic circuits.

Standard equipment includes a complete set of tools, spades, picks, shovels and axes, tow bar and towing cables, first aid kit, two petrol stoves, fire extinguishers, emergency spares, two 50-litre drinking water tanks, two external stowage boxes, radios, intercom and remote handset with 1000 m of cable.

Optional equipment includes night vision equipment, air-conditioning and ventilation system and the replacement of the standard D3256 BTXF six-cylinder diesel developing 282 hp by an ADE 407 TI diesel developing 315 hp.

Variants

Ratel 60 IFV

This model has a crew of 11 consisting of a vehicle commander, driver, main gunner, anti-aircraft gunner, section commander and six infantrymen. It is fitted with a two-man turret armed with a 60 mm mortar, 7.62 mm coaxial and 7.62 mm anti-aircraft machine gun with an additional

Ratel FSV 90 with 7.62 mm anti-aircraft machine gun on commander's cupola and armoured visors over driver's windows lowered (Christopher F Foss)

Ratel FSV 90 being put through its paces during trials at the ARMSCOR proving ground

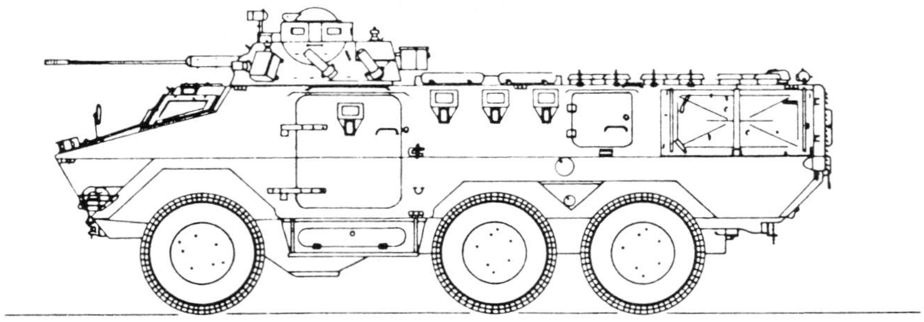

Ratel 20 IFV with two-man turret armed with 20 mm cannon and 7.62 mm machine gun

7.62 mm anti-aircraft machine gun at the rear of the vehicle on the right side. The 60 mm mortar is breech loaded and is capable of firing HE, smoke and illuminating bombs to a maximum range of 1500 m. A typical ammunition load would consist of 38 to 45 HE bombs, three smoke and three illuminating bombs.

Ratel 20 IFV
This model has the same crew as the Ratel 60 IFV but is fitted with a two-man turret armed with a 20 mm F2 cannon (M693) with a 7.62 mm machine gun mounted coaxially to the left and a similar weapon can be fitted on the turret roof for anti-aircraft defence. There is also a 7.62 mm anti-aircraft machine gun at the rear on the right side. Turret traverse and weapon elevation are manual. The weapons have an elevation of +38° and a depression of -8°. The 20 mm cannon fires armour-piercing (effective range 1000 m), high explosive (effective range 2000 m) and practice (effective range 1000 m) projectiles, all manufactured by Armscor.

Ratel 12.7 mm Command
This command vehicle has a crew of nine: the vehicle commander, driver, main gunner and six command post personnel. The two-man turret is armed with a 12.7 mm M2 HB machine gun and a 7.62 mm machine gun mounted at the hull rear as on other members of the Ratel family. This command vehicle is fitted with three radios, tape recorder with time injection, civilian type combined receiver and cassette recorder in the hull, intercom for all key personnel, internal loudspeakers, public address system, pneumatic mast and map boards.

Ratel Armoured Repair Vehicle
This is simply a Ratel 20 with a fixed lifting jib made up of four tubular steel posts mounted on the rear hull with a block and tackle arrangement to raise and carry damaged vehicles such as the Ratel. A wide range of spare parts and repair equipment is carried inside the vehicle and in a wire mesh stowage box on the front of the hull. An air compressor is carried as are spare wheels. Additional details are given in *Jane's Military Vehicles and Logistics 1991-92* page 47.

Ratel 81 mm Mortar Carrier
Production of the Ratel was to have been completed late in 1985 but this was extended to January 1987 as the South African Army wanted an 81 mm mortar version of this vehicle.

The 81 mm mortar is mounted on a turntable which can be quickly traversed through a full 360°. It fires through a two-part circular roof hatch which opens front and rear.

The 81 mm mortar carrier has a crew of six including the driver. One hundred and fifty rounds of 81 mm mortar ammunition are carried. The commander sits on a drop-down seat to the rear of the driver and has a raised cupola armed with a 7.62 mm machine gun. This version has only two firing ports and associated vision blocks, one of which is in either side.

Ratel with Swift ATGW System
This was first used in Angola in 1988 and is a Ratel (6 × 6) hull fitted with a new two-man turret on the top of which is mounted a bank of three Swift (ZT-3) laser guided ATGWs with an additional 12 missiles being carried inside for manual loading. The turret also has a 7.62 mm machine gun and two banks of smoke grenade launchers, one on either side, that fire to the front of the vehicle.

The four-man crew consists of commander, gunner, driver and loader.

Ratel 20 IFV with all roof hatches closed

Ratel command vehicle armed with 12.7 mm and 7.62 mm machine guns in two-man turret with raised cupolas for commander and gunner

Ratel 60 infantry fighting vehicle with all doors closed

Ratel fitted with two man turret with three Swift ATGW's in ready-to-launch position

The Swift (ZT-3) laser guided ATGW has a maximum range of over 4000 m and its HEAT warhead will penetrate over 650 mm of RHA.

Ratel 8 × 8 Logistic Support Vehicle

A prototype of the Ratel 8 × 8 logistic support vehicle was built and tested but it was not placed in production. Details were given in *Jane's Armour and Artillery 1986-87*, page 369.

Ratel 81 mm mortar carrier with two part roof hatch open to allow mortar to be fired (Terry J Gander)

Status: Production complete but can be resumed if further orders are placed. In service with Morocco and South Africa.

Manufacturer: Sandock-Austral Pty Limited
PO Box 6390
West Street Industrial Sites
Boksburg
Transvaal
South Africa
Telephone: (011) 52 8459
Fax: (011) 917 2100

SPECIFICATIONS

Model	**Ratel 20**	**Ratel 90**	**Ratel command**
CREW	11	10	9
CONFIGURATION	6 × 6	6 × 6	6 × 6
COMBAT WEIGHT	18 500 kg	19 000 kg	18 000 kg
UNLOADED WEIGHT	16 500 kg	17 000 kg	16 000 kg
POWER-TO-WEIGHT RATIO	15.24 hp/tonne	14.84 hp/tonne	15.67 hp/tonne
LENGTH (hull)	7.212 m	7.212 m	7.212 m
WIDTH (hull)	2.516 m	2.516 m	2.516 m
HEIGHT			
overall	2.915 m	2.915 m	2.755 m
to hull top	2.105 m	2.105 m	2.105 m
GROUND CLEARANCE	0.34 m	0.34 m	0.34 m
TRACK	2.08 m	2.08 m	2.08 m
WHEELBASE	2.809 m + 1.4 m	2.809 m + 1.4 m	2.809 m + 1.4 m
ANGLE OF APPROACH/ DEPARTURE	44°/45°	44°/45°	44°/45°
MAX ROAD SPEED	105 km/h	105 km/h	105 km/h
FUEL CAPACITY	430 l	430 l	430 l
MAX RANGE			
road	1000 km	1000 km	1000 km
cross-country	14 h	14 h	14 h
FUEL CONSUMPTION			
road	0.4 l/km	0.4 l/km	0.4 l/km
cross-country	1 l/km	1 l/km	1 l/km
FORDING	1.2 m	1.2 m	1.2 m
GRADIENT	60%	60%	60%
SIDE SLOPE	30%	30%	30%
VERTICAL OBSTACLE	0.6 m	0.6 m	0.6 m
TRENCH	1.15 m	1.15 m	1.15 m
TURNING RADIUS	8.5 m	8.5 m	8.5 m
ENGINE	OD 3256 BTXF 6-cylinder in-line turbo-charged diesel developing 282 hp at 2200 rpm		
TRANSMISSION	automatic, 6 forward and 2 reverse gears with hydro-dynamic torque converter		
STEERING	mechanical, recirculating ball with hydraulic assistance		
TYRES	14.00 × 20	14.00 × 20	14.00 × 20
BRAKES			
main	hydro-pneumatic on front wheels and air/mechanical on rear 4 wheels		
crawl	pneumatic/mechanical on rear 4 wheels		
parking	mechanical on rear 4 wheels		
AXLES			
front	steering axle with steering damper and intermediate steering arm, lockable differential		
middle	non-steering with transfer train, lockable differential (transfer and longitudinal)		
rear	non-steering and lockable differential		
SUSPENSION	solid axle with single coil springs and double acting hydraulic shock absorbers supported by wishbones and longitudinal arms		
ELECTRICAL SYSTEM	24 V	24 V	24 V
BATTERIES	2 × 12 V, 100 Ah	2 × 12 V, 100 Ah	2 × 12 V, 100 Ah
ARMAMENT			
main	1 × 20 mm cannon	1 × 90 mm gun	1 × 12.7 mm MG
coaxial	1 × 7.62 mm MG	1 × 7.62 mm MG	none
anti-aircraft, turret	1 × 7.62 mm MG	1 × 7.62 mm MG	none
anti-aircraft, rear	1 × 7.62 mm MG	1 × 7.62 mm MG	1 × 7.62 mm MG

Model	Ratel 20	Ratel 90	Ratel command
AMMUNITION			
90 mm	none	69	none
20 mm	1200	none	none
7.62 mm	6000	6000	3600
12.7 mm	none	none	300
GUN CONTROL EQUIPMENT			
Turret power control	manual, 360°	manual, 360°	manual, 360°
by commander	yes	yes	yes
Gun elevation/depression			
(main)	+38°/−8°	+15°/−8°	+38°/−8°
ARMOUR			
Hull front	20 mm at 60°	20 mm at 60°	20 mm at 60°
Hull sides upper	8 mm at 65°	8 mm at 65°	8 mm at 65°
Hull sides lower	10 mm at 90°	10 mm at 90°	10 mm at 90°
Hull top	6 mm	6 mm	6 mm
Hull floor	8 mm and 10 mm	8 mm and 10 mm	8 mm and 10 mm
Hull rear	10 mm	10 mm	10 mm

Ingwe Range of 4 × 4, 6 × 6 and 8 × 8 AFVs

Development

The Ingwe range of 4 × 4, 6 × 6 and 8 × 8 high-mobility armoured vehicles has been developed as a private venture by Sandock-Austral Pty Limited for a wide range of roles including use as an armoured personnel carrier by the army, internal security vehicle or as a vehicle for protecting high risk areas such as power stations and oil refineries.

Wherever possible standard and in-production commercial components have been used in the design and construction of the vehicle, including the engine, transmission and axles. This has kept both procurement and operating costs as low as possible. Design work on the 4 × 4 Ingwe started in November 1985 and finished early in 1986 after which several prototypes were built. A small batch of 4 × 4 vehicles was delivered to the South African Army for trials in late 1986 and these have been evaluated under a wide range of operational conditions.

Description 4 × 4

The hull of the Ingwe is of all-welded steel armour which provides the crew with complete protection from 7.62 mm small arms fire and shell splinters. The high ground clearance and well-shaped hull design provide excellent protection from mines.

The Ingwe has a crew of 12 fully equipped men. The driver sits at the front right with the vehicle commander to his left, each having a bullet-proof windscreen to his front and forward-opening side door with a bullet-proof windscreen in its upper part. Above the commander's position is a circular hatch cover that opens to the rear and on the forward part of this is a mount for a 7.62 mm MG. The driver has a square hatch that opens to the right.

The six-cylinder diesel engine is mounted under the forward part of the hull and is coupled to an automatic transmission. The complete powerpack slides out of the hull on rails for ease of repair or replacement.

The suspension consists of solid axles with coil springs and progressive telescopic shock absorbers. Tyres are 14.00 × 20 with run-flat inserts. The high ground clearance with good angles of approach and departure give good cross-country capability. The axles have differential locks.

The troop compartment is at the rear and in the side of the hull there are five bullet-proof windows each with a downward-opening firing port underneath. There are double doors at the rear, each of which is provided with a bullet-proof observation window with a firing port underneath. The spare wheel and tyre is carried on the right door.

In the roof of the Ingwe are a number of rectangular roof hatches that open at the rear and at the rear right of the roof is a cupola that can be traversed through 360° which is provided with a single piece hatch cover and a mount for a 7.62 mm MG.

A wide range of different seating arrangements are available to suit different user requirements and optional equipment includes various communications systems, searchlights, weapon mounts, light machine gun turrets and improved armour protection. A modular water cannon has been built with a capacity of 4500 litres (4 × 4 version) or 10 000 litres (6 × 6) version, this will have the turret at the front and the water module at the rear.

Variants

Using the same basic common components a complete family of 4 × 4, 6 × 6 and 8 × 8 vehicles can be provided to meet a wide range of civil and military requirements. These are summarised as follows:

Ingwe 4 × 4 paramilitary vehicle

Cargo carrier with fully enclosed armoured cab and flat load carrying area at rear, fully enclosed APC, fully enclosed armoured ambulance, fitted with water cannon and a recovery vehicle with a fully enclosed forward control cab.

Ingwe (4 × 4) vehicle showing high ground clearance

Ingwe (4 × 4) vehicle showing arrangement of roof hatches

Ingwe (4 × 4) vehicle from front

Ingwe 4 × 4 military vehicle

These would all have a slightly longer wheelbase than the above model and variants would include cargo carrier with fully enclosed armoured cab and flat load carrying area at rear, fully enclosed APC, command and control vehicle, technical support vehicle, armoured ambulance, infantry combat vehicle with 20 mm turret-mounted cannon, 60 mm or 81 mm mortar carrier, air-defence, anti-tank and armoured recovery vehicle with fully enclosed forward control cab.

Ingwe 6 × 6 military vehicle

These are essentially the AC 200 mine-protected vehicle covered in *Jane's Armour and Artillery 1987-88*, pages 357 and 358. In addition to a basic cargo carrier with a fully enclosed cab other variants could include fully enclosed APC, command and control vehicle, technical support vehicle, armoured ambulance, 20 mm or 90 mm infantry combat vehicle, 81 mm or 120 mm mortar carrier, anti-aircraft and anti-tank, water cannon and recovery vehicle.

Ingwe 8 × 8 military vehicle

Cargo carrier with fully enclosed forward control cab and flatbed cargo area at rear, logistic support with a hydraulic crane, technical support vehicle, recovery and a long wheelbase cargo carrier.

Note: All vehicles are available with various options, for example the 4 × 4 can have a manual or automatic transmission, all models have 14.00 × 20 tyres with run-flat inserts as standard although 16.00 × 20 can be fitted as an alternative.

Status: Production complete. Small numbers delivered to South African Army for trials purposes.

Manufacturer: Sandock-Austral Pty Limited
PO Box 6390
West Street Industrial Sites
Boksburg
Transvaal
South Africa

Enquiries to Armscor
Private Bag X337
Pretoria 0001
South Africa
Telephone: (012) 428 1911
Telex: 320217

SPECIFICATIONS

CONFIGURATION	**4 × 4**	**6 × 6**	**8 × 8**
POWER-TO-WEIGHT RATIO	14.5 hp/tonne	22.5 hp/tonne	18.46 hp/tonne
COMBAT WEIGHT	13 000 kg	19 000 kg	28 000 kg
WEIGHT UNLADEN	11 500 kg	17 000 kg	18 000 kg
LENGTH	5.885 m	7.125 m	8.675 m
WIDTH	2.5 m	2.5 m	2.5 m
HEIGHT	2.99 m	2.99 m	2.99 m
GROUND CLEARANCE			
hull	0.49 m	0.48 m	0.48 m
axle	0.42 m	0.42 m	0.42 m
TRACK	2.08 m	2.07 m	2.07 m
WHEELBASE	3.45 m	2.05 m + 2.64 m	2.05 m + 2.64 m + 1.55 m
ANGLE APPROACH/DEPARTURE	45°/43°	42°/51°	42°/51°
OVERHANG			
front	1.47 m	1.47 m	1.47 m
rear	0.965 m	0.965 m	0.965 m
MAX ROAD SPEED (manual/ automatic)	93 km/h/100 km/h	110 km/h	100 km/h
RANGE	950 km	1100 km	900 km
FUEL CAPACITY	310 l	645 l	645 l
GRADIENT	70%	70%	70%
SIDE SLOPE	30%	30%	30%
VERTICAL OBSTACLE	0.5 m	0.5 m	0.6 m
TRENCH	1.06 m	1.26 m	1.55 m
FORDING	1.2 m	1.2 m	1.2 m
TURNING RADIUS			
wall to wall	16.3 m	22 m	27 m
curb to curb	14.7 m		
ENGINE	ADE 366 Ti water-cooled diesel 6 cylinder developing 189 hp at 2600 rpm	Daimler-Benz OM 442LA V-8 diesel developing 429 hp at 2100 rpm	Daimler-Benz OM 443LA V-10 diesel developing 517 hp at 2100 rpm
TRANSMISSION			
automatic	ZF 6 WG 200	automatic Renk HS 106	automatic Renk HS 106
manual	ASTAS S9-65		
STEERING BOX	ZF 8046 955115	ZF 8046	ZF 8046
SUSPENSION	all have MAN rigid axles with hub-reductions and pneumatic differential locks, coil springs and telescopic shock absorbers		
TYRES	14.00 × 20	14.00 × 20	14.00 × 20

TFM Casspir Mk III Armoured Personnel Carrier

Development

The original mine protected vehicle was based on a Bedford (4 × 4) truck chassis. This was followed by the Casspir Mk I vehicle of which about 200 were built between 1979 and 1980.

In 1981 production of the Casspir was taken over by TFM Limited which developed the improved Mk II vehicle. Externally there is not a lot of difference between the Mk I and Mk II vehicles. The Mk I had a kerb-side, side escape hatch which was ineffective, a longer rear overhang and early models had a 120 hp engine.

TFM has overhauled a large number of the original Casspir Mk I vehicles since 1981.

The actual name Casspir refers both to the complete family of vehicles and to the armoured personnel carrier model of which some 700 in both Mk I and Mk II versions had been built by late 1986.

The main difference between the Mk II and Mk III vehicles is a change in axle type and ratio, as well as the equipment that it is required to carry, hence the overall increase in weight of the vehicle.

Using the same unitary hull, dimensions and driveline TFM has developed a range of supporting vehicles which have also been produced in quantity.

In designing the Casspir range of vehicles special attention has been paid to protect the vehicle against land-mines. According to the manufacturer, field repair times after a single anti-tank mine, the Soviet TM-57 or equivalent, is between one and two hours and after a multiple configuration of mines 8 to 12 hours. The practical limit to this mine protection is mine or mines equivalent of 18 to 20 kg of TNT/RDX.

The Casspir series of armoured vehicles was originally developed for use in the dense bush and desert terrain of southern Africa but they have also been used by the Army for pre-emptive strikes in the sub-continent and more recently for internal security roles.

By early 1991 total production of the Casspir Mk III APC amounted to some 2200 vehicles, including variants.

Description

The hull of the Casspir is of all-welded steel and will withstand attack by 5.56 mm or 7.62 mm NATO ball ammunition as well as grenade and mortar splinters and shrapnel.

The running components of the Casspir are derived from a commercial 15-ton truck and the vehicle is powered by a locally produced six-cylinder diesel developing 170 hp coupled to a five-speed manual transmission and a two-speed transfer case; the rear axle has a differential lock.

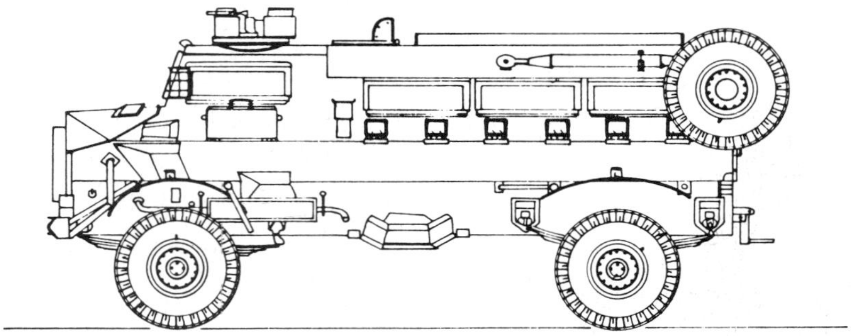

TFM Casspir Mk II APC

The engine, transmission and fuel tank are all contained within the armoured hull of the Casspir for maximum protection.

The protected engine compartment is at the front with the armoured crew compartment extending right to the rear.

The driver has a roof hatch over his position, the commander/gunner sits on the left. To their front and sides are bullet-proof windows 54 mm thick which provide the same degree of protection as the hull.

On the left side is a turret which can be armed with single or twin 7.62 mm machine guns and a 7.62 mm forward firing machine gun can be fitted to the left front windscreen.

The troop compartment extends right to the rear of the vehicle and the 12 troops sit six each side facing each other on individual seats with four-point safety harnesses.

In each side of the hull are six rectangular firing ports and three oblong bullet-proof windows.

The troops enter and leave via two doors in the rear both of which have a bullet-proof vision block with a firing port underneath. The rear doors are power-assisted and remote-controlled. There are six hatches in the roof and the crew compartment is force ventilated by two impeller fans.

The original versions of the Casspir had an open roof, but steel roofs are fitted to vehicles used in the IS role for protection against petrol bombs and grenades.

All versions of armoured personnel carriers have a 200-litre drinking water tank and two spare wheels and tyres and are designed for an operational range of between 600 and 800 km lasting for five to seven days.

Standard equipment also includes two fire extinguishers, a tow hook and D-eyes at the rear and a towing pin at the front.

Optional equipment includes an automatic rubber bullet launcher for use in the internal security role. This has a rate of fire of 160 rds/min and an effective range of 150 m. This is called the Slingshot bullet machine and works on the principal of counter-rotating discs gripping the projectile and slinging it forward at a velocity of up to 80 m/s with ammunition being fed into the machine automatically. The range of Slingshot is 150 m with a secondary bounce potential up to 170 m. Cyclic rate of fire is 170 rounds in the fully automatic mode with a traverse of 180° and elevation from -10 to +24°. Some vehicles have been fitted with 20 mm cannon removed from aircraft.

Other optional equipment includes floodlights, searchlights and a pneumatically controlled drop-down front bumper for clearing obstacles.

Variants

Police Version

This has larger windows to increase visibility for use in urban areas and the windows have grilles. The grille for the commander's and driver's windscreen can be pneumatically raised for improved visibility. A wire cutter can be mounted on the roof and the front buffer can be lowered from inside the vehicle to bulldoze barricades and other obstructions.

Blesbok Freighter

This has a fully armoured two-man cab at the front and has a dropside rear cargo area that can carry five tons. One hundred and sixty units

TFM Blesbok cargo carrier

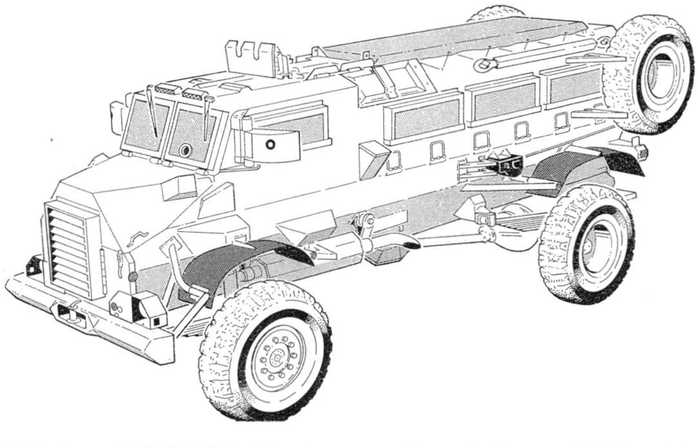

TFM Casspir Mk III APC without armament installed (not to 1/76th scale)

TFM Duiker 5000-litre tanker

TFM Casspir Mk IIIC APC without armament

Sandock-Austral RCV 9 light armoured vehicle fitted with twin side doors and wire mesh protection for all windows

each side of the troop compartment, to the immediate rear of the twin side doors is another bullet-proof window with a firing port in its upper part.

The engine compartment is at the rear of the vehicle with the spare wheel and tyre being carried on the roof. In the roof of the vehicle are three roof hatches, one above the commander's and driver's position which opens to the rear and one above either side of the troop compartment which opens to the outside.

Optional equipment includes 4 × 4 drive, communications equipment, floodlights and spotlights, weapons storage fittings, roof-mounted wire cutter, wire mesh protection for all windows and lights, ball- or pin-type towbar, drinking water containers, barricade bumper with pneumatic control, fire extinguishers, air-conditioning, front-mounted electric winch and run-flat tyres.

Variants

In addition to being used as an armoured personnel carrier, Sandock-Austral have also proposed that the RCV 9 could also be used as a bullion transport vehicle, ambulance, reconnaissance, command post vehicle and a weapons carrier. The RCV 9 is also available with a single forward-opening door in each side of the hull rather than the twin doors.

Status: Production as required. In service in internal security role in South Africa.

Manufacturer: Sandock-Austral Pty Limited
PO Box 6390
West Street Industrial Sites
Boksburg
Transvaal
South Africa

Enquiries to Armscor
Private Bag X337,
Pretoria 0001,
South Africa.
Telephone: (012) 428 1911
Telex: 320217

SPECIFICATIONS

CREW	1 + 8
CONFIGURATION	4 × 2
POWER-TO-WEIGHT RATIO	19.99 hp/tonne
WEIGHT	
combat	6900 kg
empty	6020 kg
LENGTH	
overall	5.38 m
hull	5.125 m
WIDTH	
over hull	2.10 m
doors open	3.16 m
HEIGHT	
over spare wheel	2.28 m
hull top	2.04 m
GROUND CLEARANCE	0.28 m
TRACK	1.8 m
WHEELBASE	2.995 m
ANGLE OF APPROACH/ DEPARTURE	37.5°/42°
MAX ROAD SPEED	100 km/h
RANGE (road)	500 km
GRADIENT	45%
SIDE SLOPE	45%
VERTICAL OBSTACLE	0.25 m
TRENCH	0.54 m
ENGINE	turbo-charged diesel developing 120 hp at 2800 rpm
TRANSMISSION	ZF S5-35 manual, 5 forward and 1 reverse gears
STEERING	hydraulic
TURNING RADIUS	13 m
SUSPENSION	independent on all wheels with double wishbones and coil springs, hydraulic double-acting shock absorbers
TYRES	9.00 × 16 tubeless
BRAKES	
main	dual-line, air over hydraulic disc front and rear
handbrake	mechanical on transfer case
ELECTRICAL SYSTEM	12 V

Other South African Armoured Personnel Carriers

Development/Description

In addition to the British supplied Alvis Saracen (6 × 6) APC, the South African Ratel series, the Ingwe (used only in small numbers) and the Casspir series, the South African Defence Force and police use a wide range of mine-protected vehicles and brief details of some of these are given below.

In addition, South West Africa/Namibia developed two ranges of mine-protected vehicles called the Wolf and the Albatross series and brief details of these are also provided.

Bulldog (4 × 4) mine-protected APC

Bulldog APC

The Bulldog APC has not been adopted by the South African Army although a quantity of vehicles have been supplied to the South African Air Force for air base defence. It is based on the chassis and automotive components of the SAMIL 50 range of (4 × 4) 2000 kg truck which is the standard vehicle in its class in the South African Army (*Jane's Military Vehicles and Logistics 1991-92* page 377). This differs from the current in-service Buffel (see following entry) in that it incorporates the driver's cab into the main troop compartment which has back-to-back seating for 10 men who are seated five down either side. The only means of entry and exit is via climbing over the sides of the vehicle.

SPECIFICATIONS

CREW	1 + 10
CONFIGURATION	4 × 4
WEIGHT	7700 kg
LENGTH	5.34 m
WIDTH	2.3 m
HEIGHT	3.1 m
GROUND CLEARANCE	0.46 m
MAX ROAD SPEED	90 km/h
CRUISING SPEED	80 km/h
RANGE ROAD	900 km
GRADIENT	60%
VERTICAL OBSTACLE	0.60 m
TRENCH	0.865 m
FORDING	1.2 m
TURNING RADIUS	6.5 m
ENGINE	4-cylinder air-cooled diesel
TRANSMISSION	manual, 6 forward and 1 reverse gears
SUSPENSION	hydraulic telescopic shock absorbers
STEERING	power assisted
BRAKES	air/hydraulic, dual circuit
TYRES	14.50 × 20 PR 12

Buffel (4 × 4) mine-protected APC

Rhino-protected (4 × 4) troop carrier

SAMIL 50 (4 × 4) mine-protected armoured ambulance

SAMIL 20 (4 × 4) truck with mine-proof cab (Terry J Gander)

Buffel (4 × 4) series

This has been the standard mine protected APC of the SADF since 1978 and it is estimated that around 1400 were built before production was completed. The engine is at the front of the vehicle with the driver being seated above on the left side in a fully armoured cab that is provided with bullet-proof windows to the front and sides. A spare wheel and tyre is carried on the right side of the engine compartment.

The hull floor is V-shaped to provide maximum possible protection against mines. The rear troop compartment has an open top with the 10 infantrymen being seated five down either side back to back. Each seat is provided with a full seat belt harness and the only means of entry and exit is by climbing over the sides of the vehicle. The latter can be hinged downwards if required. Two machine guns can be pintle-mounted on top of the hull. Specialised variants include a mortar carrier. Prime contractor for the Buffel family of vehicles was ARMSCOR with a Mercedes-Benz UNIMOG chassis being used as the basis.

SPECIFICATIONS

CREW	1 + 10
CONFIGURATION	4 × 4
WEIGHT LOADED	6140 kg
LENGTH	5.1 m
WIDTH	2.05 m
HEIGHT	2.995 m
GROUND CLEARANCE	0.42 m
MAX ROAD SPEED	96 km/h
RANGE ROAD	1000 km
FUEL CAPACITY	200 l
ENGINE	6-cylinder water-cooled diesel developing 125 hp at 2800 rpm
TRANSMISSION	manual, 8 forward and 4 reverse gears (driver can select (4 × 4) or (4 × 2) drive)
TYRES	12.50 × 20

There is also the Buffel Mk I tractor which has the fully armoured cab at the front but is fitted with a cargo area at the rear with a drop tailgate. This can carry 2500 kg and has also been used as a prime mover, for example towing a 20 mm light anti-aircraft gun.

Anti-Aircraft Command and Control Vehicle

This was unveiled for the first time in 1990, by which time it was already in production and service with the South African Army air defence units.

It is a mobile air defence command and control vehicle for mobile and towed anti-aircraft defence systems that provides information on the position of both friendly and hostile aircraft.

It is based on an armoured SAMIL 20 (4 × 4) truck chassis with a Bulldog braking system and a Rhino nine-man troop carrier hull.

Prime contractors for the system were ESD and ARMSCOR. A pneumatic lifting system elevates the antenna to a height of 6 m to provide for an improved communications capability.

Communications equipment installed includes one HF radio and one VHF ground-to-air radio and a 1.4 kVA diesel generator is fitted to power the onboard electrical equipment independently from the vehicle's battery system, allowing silent running for a period of up to three hours.

Armament comprises a 7.62 mm machine gun which is pintle-mounted on the roof.

Standard equipment includes 50 litre drinking water tank, camouflage nets and a compressor system.

Ystervark 20 mm Self-propelled Anti-aircraft Gun System

This entered service with the South African Army some years ago and was used operationally in Angola.

It is based on a SAMIL 20 truck chassis with a fully enclosed armoured cab. Mounted on the rear is an Oerlikon-Contraves 20 mm GAI-CO1 manually operated anti-aircraft cannon. Additional details of this system are given in *Jane's Land-Based Air Defence 1991-92* page 79.

Rinkhals Mk I

This is the standard mine protected ambulance of the South African Medical Service's Medical Battalion Corps. This was the first monocoque mine-protected vehicle to be taken into service with the South African Army with first vehicles entering service in the early 1980s. The engine is at the front with the crew compartment to its immediate rear. The vehicle can carry both sitting and walking wounded who enter the vehicle via the rear. Bullet-proof windows are provided all round and both commander and driver are provided with a side door. Provision is made for a roof-mounted 7.62 mm machine gun.

Ystervark 20 mm self-propelled anit-aircraft gun system (Christopher F Foss)

SAMIL 100 (6 × 6) mine-protected truck (Terry J Gander)

SPECIFICATIONS

CONFIGURATION	4 × 4
WEIGHT	11 970 kg
LENGTH	7.21 m
WIDTH	2.55 m
HEIGHT	3.15 m
GROUND CLEARANCE	0.355 m
FUEL CAPACITY	360 l
MAX ROAD RANGE	900 kg
FORDING	1 m
ENGINE	V-6 fuel injected diesel developing 146 hp at 1600 rpm
TRANSMISSION	manual, 6 forward and 1 reverse gears with high and low ratios
STEERING	power assisted
BRAKES	air

SAMIL 100 Kwevoel Mk II A

This the SAMIL 100 (6 × 6) 10 tonne truck (*Jane's Military Vehicles and Logistics 1991-92* page 378) fitted with a mine protected two-man cab. Other versions include the Mk I horse transporter and the Mk I tanker. There is also a mine protected recovery vehicle (*ibid* page 60). This version also forms the basis for the new 127 mm (40-round) Valkiri multiple artillery rocket system described later in this book.

SAMIL 20

This is the SAMIL 20 (4 × 4) 2 tonne truck (*Jane's Military Vehicles and Logistics 1991-92* page 376) fitted with a mine-protected two man cab. There is also another version called the Rhino-protected troop carrier which has a fully armoured rear compartment for 10 men and this is also provided with bullet-proof windows, firing ports, side and rear doors and roof hatches.

SAMIL 50

As far as it is known, the only mine-protected versions of the SAMIL 50 (4 × 4) 4800 kg truck (*Jane's Military Vehicles and Logistics 1991-92* page 377) is the SAMIL 50 recovery vehicle (*ibid* page 60).

Wolf (4 × 4) series

These were developed to meet the requirements of the South West Africa COIN unit in 1983 with first production vehicles being completed the following year. The Wolf II entered production in 1985.

The engine is at the front and fully enclosed within the armour envelope of the vehicle with the whole floor of the vehicle being in the usual V-configuration to provide maximum protection against mines.

The hull is of all-welded steel armour construction with the driver and co-driver at the front and the troop compartment extending right to the rear of the vehicle. Bullet-proof windows are provided in the front, sides and rear as are firing ports. The Wolf I had spherical firing ports while the Wolf II had simple firing ports.

Armament comprises one or two 7.62 mm MGs mounted above the co-driver's position in a remote-controlled mounting with a traverse of 180° left and right and elevation from -20° to +30°. In addition there is a roof-mounted 1 × 12.7 mm MG or twin 7.62 mm MGs, or 1 × 7.62 mm MG to the rear of the driver's and co-driver's position. In a spherical mounting in the armour glass in front of the co-driver is a 7.62 mm MG.

SPECIFICATIONS

CREW	1 + 10
CONFIGURATION	4 × 4
WEIGHT LOADED	14 200 kg
MAX ROAD SPEED	120 km/h
CRUISING SPEED	90 km/h
OFF-ROAD SPEED	90 km/h
RANGE	
road	2000 km
off-road	1600 km
FUEL CAPACITY	610 l
ENGINE	diesel developing 260 hp
TRANSMISSION	manual, 16-speed
TRANSFER CASE	2-speed
SUSPENSION	leaf springs and hollow spring shock absorbers
TYRES	16.00 × 20 18-ply

The supply truck version of the Wolf is called the Strandwolf. This retains the fully enclosed and armoured cab at the front but has the cargo area at the rear. Armament comprises a single 7.62 mm machine gun on the roof which is remote-controlled. There is also (6 × 6) recovery vehicle called the Sterk Hans which has a mine-protected cab capable of carrying four passengers.

Albatross series

These were designed and built by TFM for South West Africa. The basic vehicle is a 10 tonne (6 × 6) cargo carrier with a fully armour protected two-man cab at the front and the cargo area at the rear. The latter is provided with drop down sides and a rear tailgate. Variants include a recovery vehicle and a tractor truck for towing semi-trailers. All versions have a 7.62 mm remote-controlled machine gun above the co-driver's position.

SPAIN

Spanish Infantry Fighting Vehicle

This is based on the Austrian Steyr-Daimler-Puch KSPz 90 armoured combat vehicle and is fully described in this section under Austria. Spanish contractors expected to be included in production of the vehicle include SANTA BARBARA, ENASA, INSEL and ENOSA.

ENASA BMR-600 Infantry Fighting Vehicle

Development

In 1972 the Comisión de Desarrollo de Vehículos Blindados of the Dirección de Investigación of the Spanish Army Ministry and the Dirección de Proyectos of ENASA started development work on a wheeled IFV to meet the requirements of the Spanish Army. ENASA, in which Pegaso is integrated, was responsible for the automotive side of the vehicle while the Army concentrated on the armour and the armament installation. Between 1975 and 1976 the prototype, called the Pegaso 3.500, was tested under a wide range of operational conditions. The engine was behind the driver on the left side of the hull and the armament installation to the right of the engine compartment. Redesigned prototypes had the engine on the right side of the hull and the armament installation to the rear of the driver. The company designation for this vehicle is the BMR 3560, the specific variants offered by Pegaso are designated:

3560.50	armoured personnel carrier
3560.51	command vehicle
3560.56	radio communications vehicle
3562.03	VEC cavalry scout vehicle (see entry in *Reconnaissance vehicles* section)
3560.57	missile launcher vehicle (prototype only)
3560.53	81 mm is 3560/53E while 120 mm model is 3560/59E
3564	combat vehicle with 90 mm cannon
3560.54	ambulance
3560.55	recovery and maintenance vehicle

After comparative trials between the BMR-600 (Blindado Medio de Ruedas), the French VAB and the Swiss MOWAG Piranha vehicles, the BMR-600 was selected for the Spanish Army and an initial production order for 15 vehicles was placed in 1979. First production vehicles were handed over to the Spanish Army in 1979 and by 1988 a total of 687 vehicles had been ordered by the Spanish Army who has a total requirement for 741 vehicles of this type.

In 1982 Egypt signed a contract for 600 BMRs of which the first 250 were delivered by late 1984. These comprised 217 APCs, 13 command post vehicles, 10 recovery and maintenance vehicles and 10 ambulances.

In 1983 Saudi Arabia ordered 140 BMRs for delivery from 1985, these are used by the Navy. Four hundred BMR-600s were built in 1984 for the home and export markets and a similar number was built in 1985. Since then production has been at a lower level due to the completion of export orders for Egypt and Saudi Arabia.

Description

The hull of the BMR-600 is made of all-welded aluminium armour and incorporates spaced armour at the front giving protection against 7.62 mm armour-piercing rounds. Protection is provided over the remainder of the vehicle against 7.62 mm ball attack.

ENASA BMR-600 IFV of the Egyptian Army with firing ports in rear troop compartment and fitted with CETME TC-3 cupola armed with 12.7 mm M2 MG (Christopher F Foss)

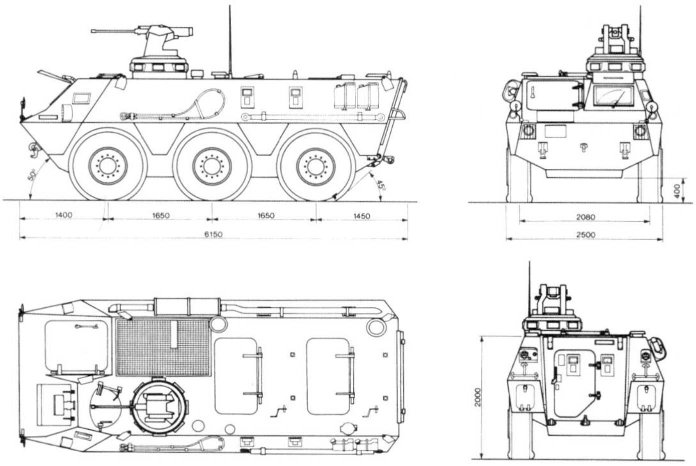

Four view scale drawing of the ENASA BMR-600 IFV with CETME TC-3 cupola armed with 12.7 mm HB MG (not to 1/76th scale)

ENASA BMR 3560.55 armoured recovery and repair vehicle with stabilisers lowered to ground and hydraulic crane traversed to right

ENASA BMR-600 fitted with CETME TC-3 cupola with externally mounted 12.7 mm M2 HB machine gun and driver's hatch locked open

The BMR-600 has a crew of two consisting of the machine gunner/radio operator and driver and can carry 10 fully equipped infantrymen and a commander who dismounts with them. The driver sits at the front of the vehicle on the left side and has a bullet-proof windscreen in front of him and a smaller bullet-proof windscreen either side (or a periscope). In action his front windscreen can be covered by a flap and forward vision is then through a periscope mounted in the roof. There is a single piece hatch cover over his position that lifts and swings to the right. The engine is to the right of the driver with the air-inlet and air-outlet louvres in the roof and the exhaust pipe on the right side of the hull. There is an engine access plate in the front of the hull. The engine compartment has a semi-automatic fire-detection and suppression system.

The machine gunner/radio operator sits behind the driver and has a CETME TC-3 cupola that can be traversed through a full 360° and is provided with a single piece hatch cover that opens to the rear and eight periscopes. A 12.7 mm M2 machine gun is mounted externally.

Over the top of the troop compartment are two rectangular roof hatches that open to the rear. The infantrymen enter and leave the vehicle by a ramp in the rear of the hull that opens downwards and has a door in the left side in case the ramp fails to open. Depending on the model, six firing ports with a vision block can be fitted to the sides and rear of the vehicle.

The BMR-600 has an independent suspension system for each wheel station that is a combination of the hydro-pneumatic and McPherson type. This allows each wheel station to be raised or lowered through 275 mm, according to the terrain being crossed. Steering is power-assisted on both the front and rear axles. All three Pegaso designed and built axles are of the double reduction type with self-locking differentials. The tyres are of the run-flat type.

The vehicle is fully amphibious being propelled in the water by water jets either side of the hull at the rear, which have deflectors for steering when afloat. When propelled in the water by the two water jets the BMR-600 has a maximum speed of 10 km/h. It is also available without the water jets when it is powered in the water by its wheels at a speed of 4.5 km/h. Before entering the water a trim vane, which is folded back onto the glacis plate when not in use, is erected at the front of the hull and the bilge pumps are switched on. Mounted at the rear of the BMR-600 is a winch with a capacity of 4500 kg. Optional equipment includes various radio systems, 14.00 × 20 tyres, run-flat tyres, 300-litre fuel tank, night vision equipment, NBC system, communications equipment, land navigation system, smoke dischargers and an air-conditioning system. It can also be delivered without the amphibious capability.

Variants

The vehicle has already been fitted with other armament installations such as the French GIAT Industries Toucan I turret and the MOWAG 7.62 mm remote-controlled machine gun mounting. Other turrets can also be fitted, for example the CETME TC-7 turret armed with two 106 mm recoilless rifles and a 12.7 mm ranging machine gun.

The BMR-600 is suitable for a wide range of other roles including an ambulance carrying four stretcher patients and two attendants, battalion, brigade or division command vehicle, fire support vehicle (for trial purposes it has already been fitted with the French GIAT Industries TS 90 turret armed with a 90 mm gun), section or company command vehicle, radio vehicle, anti-tank vehicle (it has already been successfully tested fitted with the Euromissile HCT turret with four HOT ATGWs ready-to-launch), and anti-aircraft vehicle with 20 mm Meroka or similar system.

Details of the scout and reconnaissance version of the BMR-600, the VEC, are given in the *Reconnaissance vehicles* section.

Mortar Carriers

The BMR-600 is offered in two mortar carrier configurations, one armed with a turntable mounted 81 mm mortar Model 3560/53E mortar and the other with a turntable mounted 120 mm mortar Model 3560/59E mortar. Both versions have a total crew of five men, including commander and driver with the mortar firing through a two-part roof hatch that opens left and right. A baseplate and mount are also carried to allow the mortar to be deployed away from the vehicle if required by the tactical situation.

BMR 3560.55 Armoured Recovery and Repair Vehicle

This model is in service with the Egyptian Army but has not yet been adopted by Spain. Equipment installed includes a roof-mounted hydraulic crane with a 360° traverse and a lifting capacity of 18 000 kg, four hydraulic stabilisers that are lowered to the ground two either side of the hull when the crane is in use, a winch in the hull rear with a capacity of 7000 kg and 100 m of cable, tools, lighting, welding equipment and a penthouse. Full details of this model are given in *Jane's Military Vehicles and Logistics 1991-92* page 28.

BMR 3560.54 Ambulance

This model is in service with the Egyptian Army but has not yet been adopted by Spain. This has a crew of four (driver, commander and two orderlies) and can carry four stretcher patients or two stretcher and four seated patients. A full range of medical equipment is carried.

BMR 90 mm

It is believed that the Spanish Army has taken some of the 90 mm turrets of its AML (4 × 4) armoured cars and installed them on its BMR-600s.

Status: Production. In service with Egypt, Saudi Arabia and Spain. A modified BMR-600 chassis has been proposed by Alvis to meet part of the British Army's requirement for a Future Family of Light Armoured Vehicles (FFLAV) additional details of which are given in the *AFV Families* section later in this book.

Manufacturer: Empresa Nacional de Autocamiones SA, Military Division
Avda de Aragón
402 Madrid-22
Spain
Telephone: (1) 750 10 00
Telex: 42 372
Fax: (1) 747 47 32

Production is undertaken at the company's facilities at Valladolid.

SPECIFICATIONS

CREW	2 + 11
CONFIGURATION	6 × 6
COMBAT WEIGHT	14 000 kg
LOADED WEIGHT	
front axle	5000 kg
middle axle	4450 kg
rear axle	4300 kg
POWER-TO-WEIGHT RATIO	22 hp/tonne
LENGTH	6.15 m
WIDTH	2.5 m
HEIGHT	
including armament	2.36 m
to hull top	2 m
GROUND CLEARANCE (adjustable)	0.4 m
TRACK	2.08 m
WHEELBASE	1.65 m + 1.65 m
ANGLE OF APPROACH/DEPARTURE	50°/45°
MAX SPEED	
1st gear	17 km/h
2nd gear	56 km/h
3rd gear	69 km/h
4th gear	103 km/h
reverse	12 km/h
water (water jets)	9 km/h
(wheels)	3 km/h

FUEL CAPACITY	400 l
MAX RANGE	1000 km
FORDING	amphibious
GRADIENT	60%
SIDE SLOPE	30%
VERTICAL OBSTACLE	0.6 m
TRENCH	1.6 m
TURNING RADIUS	7.5 m
ENGINE	Pegaso 9157/8 6-cylinder in-line diesel developing 310 hp at 2200 rpm
TRANSMISSION	ZF 6 HP 500 automatic with 6 forward and 1 reverse gears. Torque converter and hydraulic retarder
STEERING	recirculating ball, power-assisted on first and last axles
SUSPENSION	Messier hydro-pneumatic independent each wheel, McPherson type
TYRES	13.00 × 20

BRAKES	
main	dual hydraulic disc circuit, air-assisted on all wheels,
parking	disc operating on propeller transmission hydraulic,
emergency	air-operated
ELECTRICAL SYSTEM	24 V
BATTERIES	2 × 12 V, 140 Ah
ARMAMENT (main)	1 × 12.7 mm MG
AMMUNITION	2500 rounds
GUN CONTROL EQUIPMENT	
Turret power control	manual
Gun elevation/depression	+60°/-15°
Turret traverse	360°

Note: with 13.00 × 20 tyres the vehicle has a maximum speed of 96 km/h and with 14.00 × 20 tyres a speed of 103 km/h

ENASA BLR Armoured Personnel Carrier

Development

The BLR (Blindado Ligero de Ruedas) APC has been designed by Empresa Nacional de Autocamiones to meet the requirements of the Spanish Army and police for an APC with good cross-country mobility, provision for the crew to observe the terrain and capable of being used for a wide range of operational roles such as airport security, border patrol and IS operations. Its company designation is BLR 3545.

Description

The all-welded steel hull of the BLR gives the crew complete protection against small arms fire (up to 7.62 mm × 51 armour-piercing ammunition) over its frontal arc and against 7.62 mm × 51 ball ammunition over the remainder of the vehicle. The glass vision blocks give the same degree of protection as the vehicle's frontal arc. The driver sits at the front of the hull on the left with the vehicle commander to his right. In front of each is a large bullet-proof windscreen which in action is covered by a steel shutter hinged at the top with an integral vision slit. The windscreens have a double wiper facility, one the conventional type and the other using a solvent to deal with paints and spray. To the left of the driver and to the commander's right is a smaller bullet-proof window and above each of their positions is a single hatch cover that opens to the rear. In front of the commander and below the windscreen are two firing ports to give covering fire over the frontal arc.

There are four doors in the BLR, one in each side of the vehicle opening forwards and two at the rear, one either side of the engine compartment. The rear two open outwards and both have a vision block.

In each side of the hull there are five bullet-proof windows. In the centre of the hull roof, behind the commander's and driver's positions, is a cupola with eight vision blocks and a single piece hatch cover. This cupola can be traversed through a full 360° and has a 7.62 mm machine gun and shield for the gunner.

In the roof, over each of the side doors is a single piece hatch cover that opens to the centre of the vehicle and similar hatches are provided in the roof either side of the engine compartment.

The personnel compartment has a volume of 14 m^3 and all personnel can enter the vehicle in 16 seconds and leave in 10. The BLR is smoke- and gas-proof, ventilated by two rear suction units and an interior intake equipped with a filter.

The engine compartment is at the rear of the vehicle in the centre with the air inlet and air outlet in the top and rear of the vehicle. It can be delivered with a 170 hp engine coupled to a manual transmission or a 224 hp engine coupled to an automatic transmission.

An unusual feature of the BLR is that in addition to the engine compartment being fitted with a semi-automatic fire-extinguishing system, each of four road wheels has outlets for extinguishing fires. The tyres of the BLR are fitted with puncture-proof Hutchinson O-rings.

Optional equipment includes smoke or CS gas dischargers which can be activated by the vehicle commander. It can also be fitted with various types of communications equipment as well as a siren and loudspeakers. Optional equipment includes PTO, run-flat tyres, front-mounted winch with a capacity of 4500 kg, front-mounted obstacle-removing blade, night vision devices, radios and specialised (anti-riot equipment). According to the manufacturer the BLR can be fitted with a wide range of other armament systems including 12.7 mm machine guns, 20 and 25 mm cannon and up to 90 mm guns.

Variants

There are no variants of the BLR but it can be adopted for other roles such as command/communications vehicle or an ambulance.

ENASA BLR (4 × 4) APC showing engine compartment in centre of vehicle at rear and all side doors open

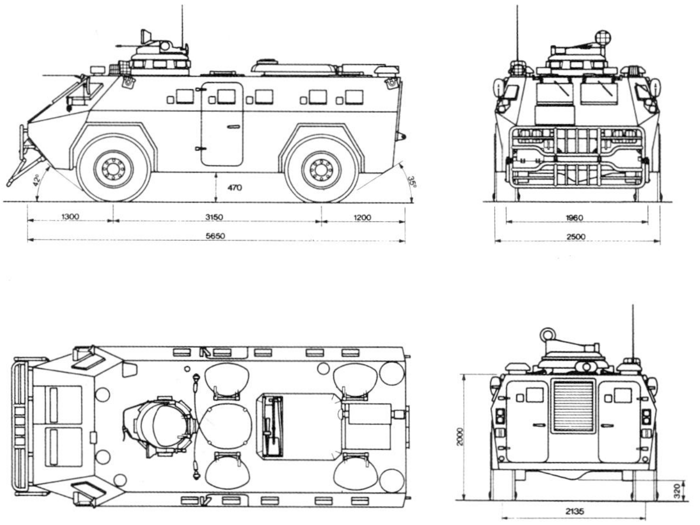

Four view drawing of ENASA BLR (4 × 4) APC fitted with turret-mounted MG and front-mounted dozer blade in retracted position (not to 1/76th scale)

SPECIFICATIONS

CREW	1 + 12
CONFIGURATION	4 × 4
COMBAT WEIGHT	12 000 kg
UNLOADED WEIGHT	9600 kg
POWER-TO-WEIGHT RATIO	17.5 hp/tonne
LENGTH	5.65 m
WIDTH	2.5 m
HEIGHT (hull top)	2 m
GROUND CLEARANCE	
axles	0.32 m
hull sides	0.47 m
TRACK	
front	1.96 m
rear	2.135 m
WHEELBASE	3.15 m
ANGLE OF APPROACH/ DEPARTURE	42°/35°
MAX SPEED	93 km/h
FUEL CAPACITY	200 l
ROAD RANGE	570 km
FORDING	1.1 m
GRADIENT	75%
SIDE SLOPE	30%
TURNING RADIUS	9 m
ENGINE	Pegaso 4-stroke, turbo-charged, direct-injection 6-cylinder in-line diesel developing 210 hp at 2100 rpm
TRANSMISSION	ZF manual, 6 forward and 1 reverse gears
TRANSFER CASE	ZF, single speed
CLUTCH	single dry plate
STEERING	hydraulic, power-assisted
TRANSFER CASE	ZF, 2-speed
SUSPENSION	semi-elliptical springs and double-action hydraulic shock absorbers at each wheel station
TYRES	13.00 × 20
BRAKES	
main	dual circuit, air
parking	mechanical
ELECTRICAL SYSTEM	24 V
BATTERIES	2 × 12 V, 150 Ah
ARMAMENT	see text

Status: Production as required. In service with Spanish Marines and Guardia Civil (rural paramilitary police).

Manufacturer: Empresa Nacional de Autocamiones SA, Military Division
Avda de Aragón
402 Madrid-22
Spain
Telephone: (1) 750 10 00
Telex: 42 372
Fax: (1) 747 47 32

ENASA BLR (4 × 4) APC with all doors and roof hatches closed

SWEDEN

Combat Vehicle 90 (Stridsfordon 90)

Development

The Swedish Army Materiel Command first looked at the light tracked vehicles on the world market which could possibly meet its future requirements. In 1984 a British Alvis Scorpion armed with a standard 76 mm gun and an Alvis Stormer were tested in northern Sweden.

The Alvis vehicles demonstrated excellent mobility, especially in deep snow, but in the end a decision was taken to design a new and heavier vehicle from scratch to meet Sweden's unique requirements.

This family was subsequently called the Stridsfordon 90, or Combat Vehicle 90, and development was entrusted to HB Utveckling AB. This company is only a holding company with a staff of less than 10 and in turn issues contracts to its two major contractors, Bofors which is responsible for the turrets and weapon systems, and Hägglunds Vehicle which is responsible for the chassis and automotive aspects. In their turn, Bofors and Hägglunds Vehicle issue contracts to their sub-contractors and suppliers.

In July 1985 HB Utveckling was awarded a contract for the design, development and construction of five prototypes of the Stridsfordon 90 family. Prior to this contracts had already been awarded for the development and construction of an automotive test rig and a new two-man turret armed with a 40 mm Bofors gun. In December 1986, HB Utveckling AB was awarded a SEK 140 million contract for further development and construction of four complete turrets, three armed with the 40 mm Bofors gun and one with a 25 mm cannon.

These four prototypes are:-

Pbv G MICV fitted with two-man Bofors turret armed with 40 mm 40/70B gun, 7.62 mm M/39 coaxial machine gun, commander's cupola, powered traverse and manual elevation, and a basic Bofors Aerotronics AB Universal Tank and Anti-Aircraft System fire control system. Mounted on either side of the turret is a bank of six electrically operated smoke dischargers and mounted on the turret rear are two Bofors Lyran 71 mm mortars which can fire an illuminating bomb out to a range of 1600 m

Pbv L MICV fitted to same turret as above and armed with 40 mm 40/70B gun and 7.62 mm M/39 machine gun, Pilkington Sabre gunner's sight, powered traverse and manual elevation

Lvkv A2 air defence vehicle with Bofors 40 mm 40/70B gun without radar for which there are two competitors, the Swedish Ericsson HARD and the French Thomson CSF TRS 2620 Gerfault

Pbv L 25 APC fitted with two-man turret designed by Hägglunds Vehicle armed with 25 mm cannon.

E-40, a fifth turret was completed in 1990 and is armed with the 40 mm 40/70B gun but with electric rather than electro-hydraulic turret controls as fitted to the other turrets.

To reduce development time and costs it was decided to use proven commercial automotive components wherever possible, such as engine, transmission and steering system.

In addition to designing the vehicle to cost targets, the Stridsfordon 90 family will be designed with a growth potential to enable the fighting members of the family to be upgraded in the future without having to be rebuilt.

It is envisaged that there will be seven basic members of the Stridsfordon 90 family which will all use the same basic hull and automotive assemblies.

The automotive test rig of Combat Vehicle 90 was completed on schedule in December 1985 and started trials in early 1986, The construction of five prototypes began in Summer 1986 and by early 1990 all of these had been delivered to the Swedish Army for extensive trials.

Early in 1991, the Swedish Army Material Command placed the first production order, worth SEK2.9 billion, for the Combat Vehicle 90.

The first version to enter production is the MICV (Pbv G) which will be fitted with a two-man turret armed with a Bofors 40 mm 40/70B cannon and a 7.62 mm M/39 coaxial machine gun.

Hägglunds Vehicle will send the completed chassis Bofors who will build the turret and integrate this with the chassis before delivering the complete system to the Swedish Army.

First production Combat Vehicle 90s will be delivered to the Swedish Army in October 1993.

First prototype of the Combat Vehicle 90 which is armed with a Bofors 40 mm L/70 gun and a 7.62 mm coaxial machine gun

The turret of the first prototype of Combat Vehicle 90 is offset slightly to the left of the vehicles centre line

Description

The hull of the vehicle is of all-welded steel armour construction although it is possible that production vehicles could also incorporate additional armour protection. No details of the armour protection level have been disclosed, but over the frontal arc, protection is probably provided for at least penetration against 20 mm attack, and possibly against 30 mm rounds fired by the BMP-2.

The engine compartment is at the front of the hull on the right side with an access hatch in the glacis plate. The driver sits on the left with a rear-opening single piece hatch cover and three periscopes for forward observation, the centre one of which can be replaced by a passive night-driving periscope.

Combat Vehicle 90 is powered by a Scania DS 14 diesel engine developing 550 hp coupled to a modified X-300-4B transmission, made under licence by Perkins Engines.

To keep the profile of the Combat Vehicle 90 as low as possible, the radiators are positioned in the right side of the hull rear, in a similar manner to that of the German Marder infantry fighting vehicle.

The Combat Vehicle 90 has a torsion bar suspension system and either side it will have seven rubber-tyred road wheels, drive sprocket at the front, idler at the rear, but no track-return rollers, although the need for return rollers will be considered during trials. The door in the hull rear opens to the right and there is a hull escape hatch in the floor.

Standard equipment on all vehicles will include a full NBC system which is being designed and built by MDH Limited of the United Kingdom, and an engine compartment fire suppression system.

Members of the CV 90 family

Armoured Personnel Carrier (CV90 25)

This is called the Pbv L and will have a crew of three: commander, gunner and driver, and be able to carry eight fully-equipped infantry (compared with 10 for the current Pbv 302) in the rear troop compartment.

The infantry will enter and leave via a door in the hull rear and they will also be able to fire their 5.56 mm rifles with the roof hatches open. The vehicle will not be fitted with firing ports and vision devices in the hull sides as this weakens the armour protection and makes stowage and seating arrangements more difficult. Vision devices and firing ports will be provided in the rear door.

It is envisaged that it will be possible to position the Bofors RBS 56 BILL anti-tank guided missile, which entered production for the Swedish Army in 1986, above the rear troop compartment to engage enemy MBTs out to 2000 m.

Mechanised Infantry Combat Vehicle (CV90 40)

This is called the Pbv G and will be fitted with a two-man power-operated turret armed with a Bofors 40 mm cannon and a 7.62 mm machine gun mounted coaxially to the right. In addition to its crew of three it will also be able to carry eight fully equipped infantry who will sit four down each side of the vehicle facing each other.

Either side of the turret will be a bank of six electrically operated smoke dischargers and on top of the roof a twin Bofors Lyran 71 mm mortar firing illuminating bombs out to a range of 1600 m. These are already standard equipment on Swedish Army S-tanks, Centurions and Ikv-91s.

After studying the new Bofors 40 mm 40/70B gun and the Israel Military Industries/OTO Melara 60 mm Hyper-velocity Medium Support Weapon, the former was selected.

The turret will have the commander on the left and gunner on the right with the former having a cupola with periscopes for all-round observation and the latter periscopes for observation to the front and right side.

The 40 mm gun will be laid and fired by the commander or gunner. The gunner will have a sight with a laser rangefinder and a thermal intensifier. Turret traverse and weapon elevation will be electric with manual controls for emergency use. Bofors Aerotronics AB have developed the U Type universal fire control system for this version.

The Bofors 40 mm gun is called the 40/70B and is based on the ordnance of the well known towed anti-aircraft gun. When mounted in the two-man turret of the Stridsfordon 90, the weapon has been inverted with the ammunition being fed from the bottom and the cartridge cases ejected upwards through a hatch on the turret roof.

Under the breech case of the 40/70B is the magazine which is divided into three compartments with eight rounds in each. Depending on the tactical situation, each magazine could contain a different type of ammunition with the change from one compartment to another accomplished by a hydraulic device.

The gun can fire single shots, bursts or full automatic fire and has an elevation of +35°, depression of -8° and turret traverse of 360°.

Ammunition is loaded into each of the three magazines manually, one by one, with eight rounds being loaded in about 20 seconds. Reserve ammunition would be stowed in the turret floor. Total ammunition storage is 240 rounds.

Types of ammunition fired by the 40/70B gun include the well known Bofors PFHE for use against aircraft and helicopters, MPT for engaging ground targets and TP-T. For engaging combat vehicles at ranges of 1500 to 2000 m Bofors has developed an APFSDS-T round. This has a tungsten rod penetrator with a length-to-diameter ratio of at least 15:1. The APFSDS-T round has a muzzle velocity of 1470 m/s, while the MP-T and PFHE Mk 2 will have a muzzle velocity of 1025 m/s.

Anti-aircraft (CV90 AA)

This version is called the Lvkv A2 and is almost identical to the Pbv G having the same two-man turret and 40 mm gun, but does not carry the eight fully equipped troops. It does, however, carry additional ammunition for the 40 mm gun.

Mounted on the turret rear is the search radar which can detect aircraft and hovering helicopters. This version will also be equipped with night vision equipment.

The radar provides the commander and gunner with small displays for target information. When travelling the radar normally is retracted to the rear of the turret bustle although it can be erected and operating while the vehicle is moving.

The prototype of the air defence version was handed over to the Swedish Army in December 1991.

The two radar contenders are the Swedish Ericsson HARD (Helicopter and Airplane Radio Detection) and the French Thomson-CSF Gerfaut systems. By late 1990 both of these had already been evaluated under a wide range of operational conditions and targets in Sweden.

This version will have a seven man crew consisting of driver in hull front, vehicle commander and gunner in turret (on the left and right respectively) with the remaining four crew members (radar operator, fire control commander, external co-ordinator and loader) in the back of the vehicle.

Armoured Command Post (Stripbv) (CV90 CP)

This would be fitted with extensive communications equipment, seats and equipment for command staff. At present it is envisaged that it would have a two-man turret. Contract was awarded for this version in April 1990.

Armoured Observation Post Vehicle (Epbv) (CV90 OPV)

This is designated the Epbv and would be fitted with an extensive array of sensors and communications equipment. It will have the two-man turret. Contract for this version was awarded in April 1990.

Armoured Recovery Vehicle (Bgbv) (CV90 APV)

This can be fitted with a rear-mounted hydraulic winch for recovery operations, hydraulic crane for hanging powerpacks in the field and a front-mounted blade which could be used as a dozer blade or to stabilise the vehicle during lifting or recovery operations. It would be fitted with a 7.62 mm machine gun turret.

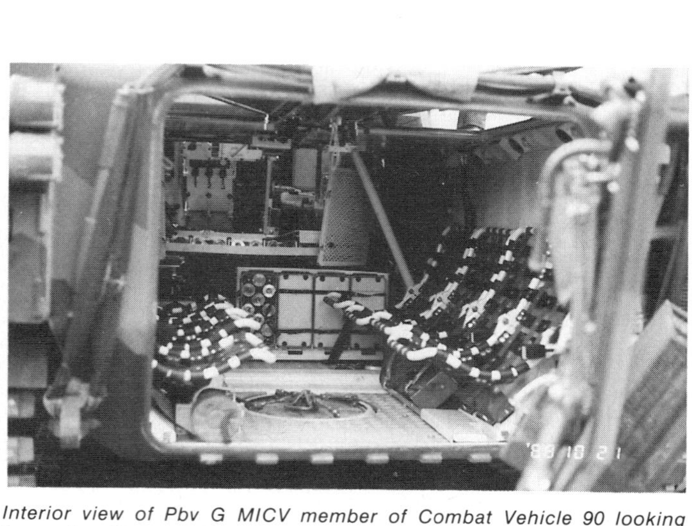

Interior view of Pbv G MICV member of Combat Vehicle 90 looking forwards with some 40 mm ammunition stowed below turret

Pbv G MICV member of Combat Vehicle 90 family firing its 40 mm Bofors gun

Mortar Carrier (Grkpbv)

This would be fitted with a 120 mm mortar firing to the rear, or the Austrian 4 × 120 mm Salvo system described later in this book. At present there is no mortar version of the Pbv 302 as today the standard 120 mm m/41C (1941) mortar of the Swedish Army is carried disassembled in a wheeled vehicle. At present there is no funding for this version.

The mortar carrier will be fitted with a 7.62 mm machine gun turret.

Once trials with the APC and MICV have been completed, these will be converted to the armoured command post and armoured observation post versions respectively.

Hägglunds Vehicle will build the chassis and Bofors the turrets and weapon systems with final integration taking place at Bofors or Hägglunds Vehicle depending on the type of vehicle.

Status: On order for the Swedish Army with first production vehicles to be completed in 1993.

The chassis of the Combat Vehicle 90 has been proposed by Alvis Defence of the UK to meet part of the British Army's FFLAV requirement, additional details of which are given in the *AFV families* section under UK.

Manufacturers: Swedish Ordnance (previously Bofors AB)
Box 500
S-690 20 Bofors
Sweden
Telephone: (46) 586 810 00
Telex: 73210
Fax: (46) 58681150

Hägglunds Vehicle AB
S-891 82
Örnsköldsvik
Sweden
Telephone: (46) 0660 800 00
Telex: 6051
Fax: (46) 0660 826 49

Prime contractor: HB Utveckling AB, Box 100 31, S-103 21 Stockholm, Sweden

SPECIFICATIONS
(basic Pbv G 40 mm - provisional)

CREW	3 + 8
COMBAT WEIGHT	22 000 kg
POWER-TO-WEIGHT RATIO	25 hp/tonne
GROUND PRESSURE	0.49 kg/cm^2
TRACK WIDTH	550 mm
LENGTH OF CHASSIS	6.4 m
WIDTH	3.1 m
HEIGHT	2.5 m
GROUND CLEARANCE	0.45 m
LENGTH OF TRACK ON GROUND	3.9 m
MAX ROAD SPEED	70 km/h
FUEL CAPACITY	525 l
MAX RANGE	300 km
ENGINE	Scania DS 14 diesel developing 550 hp
TRANSMISSION	Allison Transmission Division of General Motors X-300-4B fully automatic
STEERING	cross drive
ELECTRICAL SYSTEM	24 V
ARMAMENT	
main	1 × 40 mm cannon
coaxial	1 × 7.62 mm MG
smoke dischargers	2 × 3
Lyran launchers	2
GUN CONTROL EQUIPMENT	
Turret power control	electric/manual
Gun elevation/depression	+35°/−8°
Turret traverse	360°
GUN STABILISER	
Vertical	no
Horizontal	no

Hägglunds Vehicle Pansarbandvagn 302 Armoured Personnel Carrier

Development

Late in 1961 Hägglund and Söner was awarded a contract to develop a new full tracked amphibious APC. The first of two prototypes was completed in December 1962 and in October the following year the company was awarded an initial production contract for 700 Pansarbandvagn 302 APCs (or Pbv 302 for short). The first production Pbv 302s were completed in February 1966 and production continued until December 1971.

Since developing the Pbv 302, Hägglund and Söner has developed a complete family of light full tracked armoured vehicles which now includes the Ikv-91 light tank/tank destroyer, Brobv 941 bridgelayer (*Jane's Military Vehicles and Logistics 1991-92*, page 88) and the Bgbv 82 armoured recovery vehicle (*Jane's Military Vehicles and Logistics 1991-92*, page 29).

Description The hull of the Pbv 302 is made of all-welded rolled steel with the front of the hull providing complete protection against projectiles up to 20 mm in calibre. Above the tracks the hull sides are double skinned which gives both increased buoyancy and a measure of protection against attack from HEAT projectiles.

The driver sits at the front of the vehicle in the centre and has a single piece hatch cover that opens to the rear, in front of which are three periscopes.

The Hägglunds-designed turret to the left and slightly to the rear of the driver is armed with a 20 mm Hispano Suiza cannon with an elevation of +50° and a depression of −10°, turret traverse being a full 360°. The 20 mm cannon has a cyclic rate of fire of 500 rds/min and fires both HE and AP projectiles. Three belts each holding 135 rounds of HE ammunition are carried plus 10 magazines of AP ammunition, each magazine holding 10 rounds. The gunner has a monocular sight with a magnification of × 8 for use against ground targets. For aerial targets the turret hatch is opened to the rear and the gunner uses the open sights mounted on the weapon itself. There are three periscopes for forward and one for rear observation.

The commander sits to the right, and slightly to the rear, of the driver and has a single piece hatch cover that opens to the rear, plus five periscopes for all-round observation.

The powerpack, which consists of the engine, clutch and transmission, is mounted under the floor of the vehicle. From the transmission power is taken forward to a cross shaft with clutches and disc brakes at either end which provide a clutch and brake steering system. The air inlet, air outlet and exhaust pipe are immediately behind the driver.

The nine infantrymen sit at the rear of the vehicle, three on each side facing inwards, one at the front facing forwards and two at the rear back to back. The personnel enter and leave the Pbv 302 by two large doors in the rear of the hull.

Over the top of each side of the troop compartment is a hydraulically operated hatch which, when opened, allows the infantry to use their small arms from inside the vehicle. The hatches are operated by the commander of the vehicle and can be opened half way if required. A pressure-sensitive bar is fitted along the top of the hull under the hatch and if a weapon or hand is in contact with the strip the hatch remains open.

The torsion bar suspension consists of five dual rubber-tyred road wheels with the drive sprocket at the front and the idler at the rear. Hydraulic shock absorbers are fitted at the first and last road wheel stations. The tracks are the Hägglunds M70 type which have a life of over 10 000 km with one recondition, and have 20 per cent better traction on soft soil than earlier tracks.

The Pbv 302 is fully amphibious being propelled in the water by its tracks. Before entering the water a trim vane is erected at the front of the hull and the bilge pumps switched on. The Pbv 302 has no NBC system, although one could be fitted.

The basic Pbv 302 can be used, with minor modifications, as a cargo carrier (carrying 2000 kg of cargo) and ambulance (carrying four stretchers in its basic form and up to six with a special kit).

Variants

Pbv 302 Mk 2 APC

Details of this version, which never entered production, were given in *Jane's Armour and Artillery 1989-90*, page 385.

Product-improved Pbv 302 APC

Details of this version, which never entered production, were given in *Jane's Armour and Artillery 1989-90*, page 386.

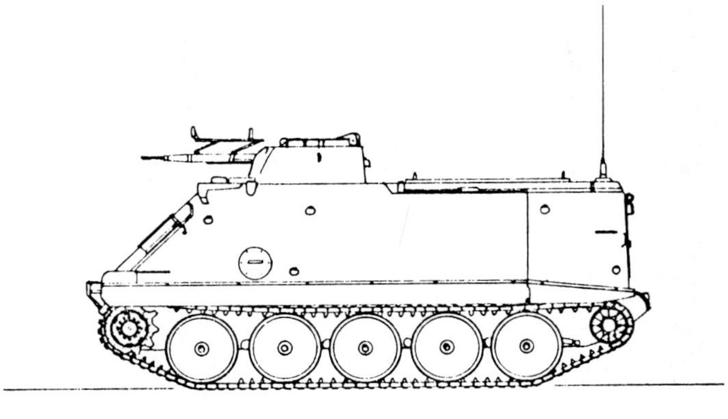

Pbv 302 amphibious APC

Bplpbv 3023 armoured fire direction post vehicle in travelling configuration showing access panels in glacis

Pbv 302 APC from rear (Christopher F Foss)

Epbv 3022 armoured observation post vehicle

Pbv 302 APC with trim vane retracted (Christopher F Foss)

Armoured Command Vehicle

This is designated the Stripbv 3021 and can be distinguished from the Pbv 302 by its four rather than two radio antennas. The radios are mounted on the right side of the troop compartment and four radio operators are normally carried plus command staff who are provided with mapboards.

Armoured Observation Post Vehicle

This is designated the Epbv 3022 and is issued on the scale of 10 per armoured brigade. The major difference is that the commander's hatch has been replaced by a new cupola manned by the fire control officer. The cupola has a combined binocular and optical rangefinder which can be changed between ×4 binocular and ×10 coincidence rangefinder using the same eyepieces. The cupola is hand-cranked in traverse and the line of sight is manually elevated, an azimuth sensor driven by the turret ring gives the accurate azimuth to the target. In the personnel compartment are three radios, two wire links and a navigation system which displays information at one of the operator's positions. Externally the Epbv 3022 is recognisable by its three radio antennas, additional cupola and additional buoyancy aids on the two trim vanes at the front of the hull.

Armoured Fire Direction Post Vehicle

This is designated the Bplpbv 3023 and has a crew of three (commander, gunner and driver) and also carries a ranging section of seven. Externally it is recognised by its four radio antennas and additional equipment carried on the roof of the troop compartment. Internally it has four radios and a fire direction computer.

Status: Production complete. In service only with the Swedish Army.

Manufacturer: Hägglunds Vehicle AB
S-891 82 Ornsköldsvik
Sweden
Telephone: (46) 0660 800 00
Telex: 6051
Fax: (46) 0660 826 49

SPECIFICATIONS	
CREW	2 + 10
COMBAT WEIGHT	13 500 kg
POWER-TO-WEIGHT RATIO	20.74 hp/tonne
GROUND PRESSURE	0.6 kg/cm²
LENGTH	5.35 m
WIDTH	2.86 m
HEIGHT	
to turret top	2.5 m
to hull top	1.9 m
GROUND CLEARANCE	0.4 m
TRACK	2.42 m
TRACK WIDTH	380 mm
LENGTH OF TRACK ON GROUND	2.98 m
MAX SPEED	
road	66 km/h
water	8 km/h
FUEL CAPACITY	285 l
MAX ROAD RANGE	300 km
FORDING	amphibious
GRADIENT	60%
VERTICAL OBSTACLE	0.61 m
TRENCH	1.8 m
ENGINE	Volvo-Penta Model THD 100B horizontal 4-stroke turbo-charged 6-cylinder in-line diesel developing 280 hp at 2200 rpm
TRANSMISSION	Volvo-Penta T60 manual with 8 forward and 2 reverse gears
STEERING	clutch and brake
CLUTCH	double-disc clutch plate
SUSPENSION	torsion bar
ELECTRICAL SYSTEM	24 V
BATTERIES	2 × 12 V
ARMAMENT (main)	1 × 20 mm cannon
SMOKE-LAYING EQUIPMENT	1 × 6 smoke dischargers
AMMUNITION (main)	405 HE plus 100 AP
GUN CONTROL EQUIPMENT	
Turret power control	manual
Gun elevation/depression	+50°/-10°
Turret traverse	360°

SKPF m/42 Armoured Personnel Carrier

SKPF m/42 modernised in 1983 with new tyres, improved armour protection and two roof-mounted 7.62 mm machine guns (FMV)

Development

The m/42 armoured personnel carrier was developed in Sweden during the Second World War. The SKPF and the VKPF were identical in appearance, the only difference being that the SKPF is based on a Scania-Vabis chassis and powered by a Scania-Vabis engine whereas the KPF was based on a Volvo chassis and powered by a Volvo engine.

Until the introduction of the Pbv 301 full tracked APC (which was essentially the Strv m/41 light tank converted into an APC configuration by Hägglund and Söner) the m/42 was the only APC available to the Swedish Army. The Pbv 301 has been phased out of front-line service and replaced by the more recent Pbv 302 APC but the m/42 remains in service with some units, especially on some of the Swedish islands. About 200 vehicles were built between 1943 and 1946 and in 1991 around 130 of these were still in service with the Swedish Army. Only the SKPF is in service, the VKPF having been phased out.

The m/42 has served overseas with the Swedish Army when it has operated as part of United Nations forces in the Congo, Cyprus and the Gaza Strip.

Description

The all-steel hull of the m/42 has the engine at the front, commander and driver in the centre and the troop compartment at the rear. The hull is diamond-shaped to provide maximum possible protection against small arms fire and mines.

The commander and driver each have a side door that opens forward and has a double vision port protected by a flap. To their front is an armoured visor hinged at the top, which is lowered when the vehicle is in a combat area. Many m/42s were fitted with a ring mount over the top of the commander's and driver's positions on which are mounted twin 8 mm machine guns.

Personnel are seated on bench seats which run down the middle of the hull and enter and leave the vehicle by a door in the rear. The upper sides and roof of the personnel compartment are open but can be covered by a tarpaulin which is normally carried on the vehicle.

A winch fitted on the left side of the hull can be used to the front or rear of the vehicle. The m/42 has no NBC system, no night vision equipment and is not amphibious.

Variants

In the 1950s these vehicles were fitted with more modern machine guns and in 1971 with new brakes, modern headlamps and interior fittings for weapons. In 1983 one of the vehicles used by a unit on Gotland Island was further modernised and fitted with additional armour protection for the troop compartment, the hull shortened at the rear and fitted with a door, the original dual wheels on the rear axle replaced by single low-profile tyres which have also been fitted to the front axle. On the roof of the vehicle, at either end of the troop compartment, a cupola has been mounted armed with an externally mounted machine gun and with a two-part roof hatch opening either side to provide some lateral protection for the machine gunner. The engine and transmission remain unchanged. The Defence Material Administration (FMV) has now decided to upgrade other vehicles to this configuration to extend their lives further.

Status: Production complete. In service only with the Swedish Army.

Manufacturer: Scania-Vabis

SPECIFICATIONS					
CREW	2 + 13	WIDTH	2.3 m	ENGINE	4-cylinder petrol developing 115 hp
CONFIGURATION	4 × 4	HEIGHT	2.9 m	TRANSMISSION	manual with 4 forward and 1 reverse gears
WEIGHT	8500 kg	TRACK	1.66 m	TRANSFER CASE	2-speed
POWER-TO-WEIGHT RATIO	13.53 hp/tonne	WHEELBASE	3.8 m	ARMAMENT	2 × 8 mm MG
LENGTH	6.9 m	MAX ROAD SPEED	70 km/h	ARMOUR	5-10 mm
		FORDING	0.736 m		

Hägglunds Vehicle Bv 206 S Armoured Personnel Carrier

Development

The Bv 206 S has been developed by Hägglunds Vehicle and the Swedish Army Material Command to meet the increased need for all-terrain mobility under armour protection.

The Bv 206 S is essentially an armoured version of the Bv 206 all-terrain vehicle originally developed to meet the requirements of the Swedish Army and is fully described in *Jane's Military Vehicles and Logistics 1991-92*, page 468.

In 1985, three test vehicles were built and after extensive technical and tactical trials, the decision was taken to build prototypes which were shown in public for the first time in mid-1990.

Description

The chassis, drive train and steering system are identical to the Bv 206 with the all-welded steel bodies and the cooling system being new.

The Bv 206 S consists of two tracked units which are linked together with a steering unit. The front unit has seats for the driver and three passengers while the rear unit has eight seats.

The track assemblies are mounted to the central beam by two transversal leaf springs. Each track assembly is built up around a tubular bar which carries the sprocket assembly, road wheels and idler.

In each track assembly there are four pairs of road wheels on trailing arms sprung by rubber tension springs. The idler at the rear with a tensioning device is also supported by a rubber spring. All four track assemblies of the vehicle are identical and interchangeable. The tracks are rubber with longitudinal textile cord and integral steel profile reinforcements.

Steering is accomplished by changing the direction between the front and rear unit by two hydraulic cylinders, servo-controlled from a conventional steering wheel. The hydraulic system is built up of commercially available components and the steering system is designed to permit a large freedom of movement between the two bodies.

Bv 206 S showing bullet-proof windows in front and rear units

Bv 206 S with roof hatch in forward unit in open position

The engine and transmission are mounted in the front unit. A shaft connects the gearbox with the drop-down gearbox. A disc brake is mounted in front of the drop-down gearbox inside of the body. Cardan shafts transmit power to the final drives on the front end of both chassis.

The welded steel bodies provide protection from small arms like 7.62 mm ball ammunition and shell fragments. All windows have the same protection level as the body. The front unit also has a circular roof hatch that opens to the rear. The rear unit has a large door in the rear that opens to the right, which has a single window in its upper part.

The Bv 206 S is fully amphibious, being propelled in the water by its tracks and has been designed to operate in temperatures from -46° to +49° C.

Variants
In addition to the basic armoured personnel carrier, other versions suggested by the manufacturer include mortar carrier, command post vehicle and recovery vehicle.

Status: Prototypes, not yet in production or service.

Manufacturer: Hägglunds Vehicle AB
S 891 82 Ornskoldsvik
Sweden
Telephone: (46) 0660 800 00
Telex: 6051
Fax: (46) 0600 826 49

SPECIFICATIONS		GROUND CLEARANCE	0.35 m	TRANSMISSION	Daimler-Benz W4A 040 automatic with 4 forward and 1 reverse gears
CREW	4 + 8	TRACK WIDTH	620 mm		
COMBAT WEIGHT	7000 kg	MAX ROAD SPEED	50 km/h		
EMPTY WEIGHT	5600 kg	MAX WATER SPEED	2.2 km/h		
PAYLOAD	1400 kg	RANGE		STEERING	hydrostatic, articulated
LENGTH	6.75 m	road	300 km	SWEPT DIAMETER	8 m
WIDTH		GRADIENT	60% (hard surface)	ELECTRICAL SYSTEM	24 V
over bodies	1.96 m		30% (snow)	BATTERIES	2 × 12 V, 85 Ah
over tracks	2 m	ENGINE	Mercedes-Benz 603A 6-cylinder in-line diesel devleoping 136 hp at 4600 rpm		
HEIGHT OVERALL					
rear body	2.03 m				
front body	2.00 m				

SWITZERLAND

New Swiss Infantry Combat Vehicle

The Swiss Army currently operates a fleet of about 1475 M113 series full-tracked armoured personnel carriers supplied by FMC of the United States over 20 years ago. Additional details of these are given in the entry for the M113 (Switzerland).

To supplement this fleet of vehicles the Swiss Army is understood to have a requirement for a new Infantry Combat Vehicle with contenders for this including a new vehicle designed by MOWAG, Krauss-Maffei of Germany with a version of the Puma armoured combat vehicle (described in the *AFV families* section) and FMC of the United States with the M2 Bradley.

The new MOWAG vehicle is understood to be called the NSPZ and will feature a hydropneumatic suspension system, two-man power-operated turret developed by Oerlikon-Contraves of Switzerland and armed with an automatic cannon and all electric gun control equipment from SIG of Switzerland.

MOWAG Improved Tornado Infantry Combat Vehicle

Development
The Improved Tornado ICV has been developed as a private venture by MOWAG with the first prototype completed in the Spring of 1980.

In addition to being employed as an ICV with a one- or two-man main weapon station, the basic vehicle can be adapted to a variety of other uses including ambulance, ATGW carrier, command vehicle, fire support combat vehicle, maintenance and recovery vehicle, mortar carrier, multiple rocket launcher carrier and a reconnaissance vehicle. As of early 1991 production of the MOWAG Improved Tornado IFV had yet to start.

Description
The hull of the MOWAG Tornado is made of all-welded steel and provides the crew with complete protection from small arms fire and shell splinters. To give increased protection against larger calibre weapons spaced armour is provided over the frontal arc of the vehicle.

The driver sits at the front of the hull on the left and has a single piece hatch cover that opens to the right in front of which are three periscopes, the centre one of which can be replaced by a passive one for night driving. The driver steers the Tornado with a steering wheel. The commander sits behind the driver and has a single piece hatch cover that opens to the right and five periscopes, three to his front and two to his left. When the Tornado is fitted with a two-man turret, the commander sits in the turret, not to the rear of the driver.

The engine compartment is to the immediate right of the driver and separated from the remainder of the vehicle by an NBC-proof bulkhead. The engine compartment has an automatic fire-extinguishing system which can also be operated manually. The air-inlet and air-outlet louvres are in the roof with the exhaust outlet in the right side of the hull. The engine, steer-shift transmission, air intake and exhaust system are rubber-mounted as a complete unit and can easily be removed from the vehicle for field replacement or maintenance.

The electric/hydraulically operated four-speed transmission with torque converter reduces driver fatigue. The four speeds can be used in either direction and high speed reverse capability is provided. The torque converter has an automatic lock-up clutch and a retarder integrated in the transmission relieves the vehicle brakes during prolonged downhill driving. The hydrostatic steering transmission allows infinitely variable steering up to a fixed radius, turning about either track or pivot steer about vertical axis in both directions.

The torsion bar suspension consists of six dual rubber-tyred road wheels with the drive sprocket at the front, idler at the rear and three track return rollers. The first, second and sixth road wheel stations are fitted with heavy duty hydraulic shock absorbers. The tracks are fitted with quickly replaceable rubber pads which can be replaced by rubber pads with spikes for driving on icy surfaces. The upper part of the track is covered by a skirt which is hinged at its upper part to allow access to the track and suspension for maintenance.

The armament station, such as an one- or two-man turret armed with a 20 mm or 25 mm cannon, is mounted in the centre of the vehicle. In 1987, for trials purposes, the Tornado was fitted with the Belgian CMI C-25 two-man power-operated turret armed with a 25 mm cannon and 7.62 mm coaxial MG.

On the top of the personnel compartment, one either side, is a remote-controlled machine gun mount designed by MOWAG, armed with a 7.62 mm machine gun with an elevation of +60°, a depression of -15° and a total traverse of 230°. The mounts are each provided with one periscope for aiming the machine gun and two vision blocks.

In each side of the hull are two spherical firing ports which enable the crew to fire an SMG from inside the hull in complete safety.

The six infantrymen are seated three down each side of the hull facing outward and enter and leave the vehicle by the large power-operated ramp in the rear of the hull which opens downwards. In an emergency this can be opened manually. There is an emergency escape hatch in the floor of the personnel compartment.

The Tornado can ford to a depth of 1.3 m without preparation. An over-pressure ventilation system is fitted as standard and has an air flow of 4 m^3/min when being used for ventilation and 3 m^3/min when used for NBC protection.

Variants
See development.

Status: Development complete. Ready for production.

Manufacturer: MOWAG Motorwagenfabrik AG
8280 Kreuzlingen
Switzerland
Telephone: (72) 71 15 15
Telex: 882 211
Fax: (72) 72 28 86

MOWAG Improved Tornado ICV fitted with one-man Oerlikon-Contraves GBD-COA turret armed with 25 mm cannon, 7.62 mm machine gun and electrically operated smoke dischargers and MOWAG remote-controlled 7.62 mm machine gun mount above either side of troop compartment

SPECIFICATIONS		MAX SPEED		ENGINE	Detroit Diesel 8V-71T, V-8 2-cycle turbo-charged diesel developing 390 hp at 2500 rpm
CREW	3 + 7	1st gear	15.3 km/h		
COMBAT WEIGHT	22 300 kg	2nd gear	29.9 km/h		
UNLOADED WEIGHT	18 800 kg	3rd gear	44.5 km/h		
POWER-TO-WEIGHT RATIO	17.7 hp/tonne	4th gear	66.1 km/h		
GROUND PRESSURE	0.63 kg/cm^2	FUEL CAPACITY	500 l	TRANSMISSION	Renk HSWL 194 automatic hydro-mechanical steer/shift, 4 speeds in both directions
LENGTH	6.7 m	MAX ROAD RANGE	400 km		
WIDTH	3.15 m	FORDING	1.3 m		
HEIGHT (hull top)	1.75 m	with preparation	1.7 m		
GROUND CLEARANCE	0.45 m	GRADIENT	60%	SUSPENSION	torsion bar
TRACK	2.7 m	SIDE SLOPE	40%	ELECTRICAL SYSTEM	24 V
TRACK WIDTH	450 mm	VERTICAL OBSTACLE	0.85 m	BATTERIES	6 × 12 V, 300 Ah
LENGTH OF TRACK ON GROUND	3.786 m	TRENCH	2.2 m	ARMAMENT	see text

MOWAG Piranha Armoured Personnel Carrier

Development

The Piranha range of 4 × 4, 6 × 6 and 8 × 8 armoured vehicles was developed by MOWAG as a private venture in the early 1970s and is designed for the home and export markets. The first prototype was completed in 1972 with first production vehicles following in 1976. In February 1977 Canada placed an initial order for 350 6 × 6 Piranhas (subsequently increased to 491) which were built under licence in Canada by the Diesel Division of General Motors Canada. Additional information on the Canadian vehicles will be found in this section under Canada. General Motors of Canada has built 758 vehicles in the 8 × 8 configuration for the US Marine Corps under the designation Light Armored Vehicle. Final deliveries were made in mid-1988.

More recently the Canadian Armed Forces have ordered an 8 × 8 APC version called the Bison, there is a separate entry for this under Canada. Additional details of other recent development in the Canadian built Light Armored Vehicle, both for the United States Armed Forces and for export, are given under Canada.

Cardoen Industries manufactures the 6 × 6 MOWAG Piranha under licence while FAMAE has built the prototype of the 8 × 8 version of the MOWAG Piranha under the designation 8 × 8D. Full details of these are given under Chile.

In mid-1990 GKN Defence of the United Kingdom announced that it had teamed with MOWAG to enter the Piranha in both its 6 × 6 and 8 × 8 configurations to meet some of the requirements of the British Army's Future Family of Light Armoured Vehicles programme.

The Piranha range has been designed to undertake a wide range of roles such as ambulance, anti-tank (armed with ATGWs), cargo, command, internal security, mortar carrier, recovery and reconnaissance. Many components of the vehicle are identical to all members of the family such as the front and rear hull sections, doors, hatches, wheel drives, wheels, differentials, suspension, steering and propellers.

GKN Defence built Piranha (8 × 8) AFVs

Under an agreement signed in 1990, MOWAG awarded GKN Defence of the UK a contract to build a significant number of Piranha (8 × 8) armoured vehicles for an undisclosed export customer.

GKN is now laying down a new production line at its Telford facility for the Piranha (8 × 8) vehicle with first production vehicles being completed in mid-1992.

GKN completed a Piranha (8 × 8) demonstrator using components supplied by MOWAG of Switzerland and this commenced trials in the UK in 1991, this being for a potential Middle East customer.

MOWAG also licenced GKN defence as System Prime Contractor for the design, integration, production and marketing for agreed customers.

In addition to existing variants, GKN Defence will design, integrate and manufacture new variants based on their extensive experience in this area of armoured vehicle technology.

Prior to this, in mid-1990, GKN Defence announced that they had come to a teaming arrangement for the Piranha range of 4 × 4, 6 × 6 and 8 × 8 armoured vehicles to meet a number of requirements of the British Army's Future Family of Light Armoured Vehicle (FFLAV) programme.

Description

The all-welded steel hull of the Piranha protects the crew from small arms fire and shell splinters. The driver sits at the front of the vehicle on the left side and has a single piece hatch cover that opens to the

MOWAG Piranha (8 × 8) APC with GKN Defence 12.7 mm turret

MOWAG Piranha (4 × 4) APC with MOWAG-designed remote-controlled 7.62 mm machine gun mount

MOWAG Piranha (6 × 6) APC with RASIT battlefield surveillance radar and remote-controlled 7.62 mm machine gun

MOWAG Piranha (6 × 6) vehicle fitted with Norwegian Kvaerner Eureka turret for launching TOW ATGWs

MOWAG Piranha (6 × 6) vehicle for use in internal security role with hydraulically operated obstacle-clearing blade on front of hull

MOWAG Piranha (8 × 8) fitted with GIAT Industries TS-90 90 mm turret

rear. In front of his hatch cover are three periscopes, the centre one of which can be replaced by a night vision periscope. The engine is to the right of the driver with the air-inlet and air-outlet louvres in the top of the hull and the exhaust outlet on the right side of the hull.

The main armament is normally installed in the centre of the hull. The troop compartment is at the rear and the troops enter and leave the vehicle by two doors in the rear of the hull that open outwards. There are two hatch covers over the roof of the troop compartment, which open outwards. MOWAG-designed spherical firing ports, with a vision block over each, can be fitted into each of the rear doors and two in either side of the hull to enable the crew to fire their weapons from inside the vehicle.

The automatic locking differentials are flexibly mounted within the hull. The suspension is of the independent type with coil springs on the front axle and torsion bars on the rear. All wheel stations have hydraulic shock absorbers. On the 6 × 6 vehicle the first axle has coil springs and the rear two axles have torsion bars and on the 8 × 8 vehicle the front two axles have coil springs and the rear two axles have torsion bars. The suspension allows the roadwheel a maximum travel of 0.32 m. The tyres are of the run-flat type.

All members of the family are fully amphibious being propelled in the water by two propellers mounted at the rear of the hull. Before entering the water a trim vane, which is stowed under the nose of the vehicle when not in use, is erected at the front of the hull. Standard equipment includes an NBC and air-conditioning system and optional equipment includes night vision devices of the infra-red or passive type.

Variants

Piranha (4 × 4)

A variety of armament installations can be mounted up to and including a 20 mm turret-mounted cannon, for example one or two 7.62 mm remote-controlled machine guns, turret armed with an externally mounted 12.7 mm (0.50) M2 HB machine gun or an Oerlikon 20 mm turret.

Piranha (6 × 6)

A variety of armament installations can be fitted including a 12.7 mm MG turret, 20 mm turret, 25 mm turret, 30 mm turret, turret armed with a 90 mm Cockerill or Mecar gun, 81 mm mortar mounted in the rear. More specialised versions include an unarmed ambulance, fitted with a roof-mounted air-conditioning system, and a radar carrier. This has a RASIT battlefield surveillance radar mounted on the roof which can be lowered inside the vehicle under armour protection. Full details of the repair version of the MOWAG Piranha (8 × 8) are given in *Jane's Military Vehicles and Logistics 1991-92*, page 47. In 1987, for trials purposes, a 4 × 4 and an 8 × 8 MOWAG Piranha were fitted with the Belgian CMI C-25 two-man power-operated turret armed with a 25 mm cannon and 7.62 mm MG.

Piranha (6 × 6) Anti-tank Vehicle

After extensive trials of three MOWAG Piranha (6 × 6) AFVs fitted with the Norwegian Kvaerner Eureka A/S (previously Thune-Eureka) Armoured Launching Turret and three vehicles fitted with a pedestal-mounted TOW launcher which retracted under armour protection when not required, the former was selected by the Swiss Army.

The Swiss 1986 Armament Programme proposed SFr941 million for the procurement of 310 MOWAG Piranha (6 × 6) anti-tank vehicles fitted with the Armoured Launching Turret (ALT) to be issued to 31 anti-tank companies at present equipped with the 106 mm 58 (US designation M40) recoilless anti-tank gun. Each company will consist of a command platoon and three anti-tank platoons. Companies operating in the mountain divisions will also be allocated a repair platoon.

In addition to the two missiles in the ready-to-launch position a further eight are carried internally for manual reloading. A standard TOW launcher is also carried.

The 310 vehicles will be delivered between 1989 and 1992 with MOWAG building the Piranha (6 × 6) vehicle and installing the ALT which will be manufactured under partial licence by the Ateliers Federaux, Thoune, the total Swiss share being 70 per cent.

The Federal Aircraft Factory at Emmen will build about 49 per cent of the aiming system, including day and night sights, and almost 55 per cent of the TOW 2 guided missile, and simulators.

Of the total of SFr 941 million, some 400 million will be for the vehicles, 310 million for the missiles and the remainder for spare parts and maintenance equipment. Of the grand total of SFr 941 million, the Swiss element will represent 617 million, or 66 per cent.

Piranha (8 × 8) Anti-tank Vehicle

MOWAG has completed the prototype of an anti-tank version of the Piranha (8 × 8) vehicle fitted with the Euromissile Mephisto system which

MOWAG Piranha (8 × 8) vehicle fitted with Euromissile Mephisto system for launching HOT ATGWs with launcher raised

MOWAG Piranha (8 × 8) fitted with two-man turret armed with 25 mm cannon and 7.62 mm machine gun

consists of a retractable launcher with four Euromissile 4000 m range HOT ATGWs with additional missiles being carried in reserve. This system is also used by the French Army on a SMS VAB (4 × 4) APC.

This version has a combat weight of 12 600 kg and is 6.4 m long, 2.5 m wide and 2.8 m high with the Mephisto sight raised. Four HOT missiles are ready to launch and a further eight rounds are in reserve.

Piranha (8 × 8)

A variety of armament installations can be fitted including all those of the 6 × 6 model plus twin 20 or 30 mm anti-aircraft guns, multiple rocket launcher with two banks of 15 81 mm launcher tubes, and mortar tractor towing a 120 mm mortar. Late in 1980 an 8 × 8 Piranha was shown in the United States fitted with an AAI developed turret armed with a 75 mm ARES cannon.

AFV-90

Late in 1985 MOWAG demonstrated the AFV-90 in Switzerland. This is essentially an improved MOWAG 8 × 8 Piranha fitted with the MECAR KEnerga Weapons Station. This is armed with the 90 mm MECAR KEnerga gun with a coaxial 7.62 mm machine gun.

The 90 mm KEnerga gun fires APFSDS-T, HEAT-T (two types), HE, SMK-WP-T and canister rounds with 10 ready rounds carried in the turret; a further 34 rounds are in the hull.

Optional equipment includes a Lear Siegler land navigation system. Other improvements include a fuel tank with increased fuel capacity which increases operational range to 1000 km, a small access hatch in the left side of the hull, a Rotzler type winch with direct pulling capacity of 8 tonnes and new sand channels.

Combat weight of the AFV-90 is 13 000 kg, length with gun forwards 6.79 m, width 2.5 m and height 2.7 m. The MOWAG Piranha (8 × 8) vehicle has also been tested fitted with French GIAT Industries TS-90 90 mm turret.

SPECIFICATIONS

Model	**4 × 4**	**6 × 6**	**8 × 8**
CREW (max)	10	14	15
COMBAT WEIGHT*	7800 kg	10 500 kg	12 300 kg
UNLOADED WEIGHT	6700 kg	8000 kg	8800 kg
POWER-TO-WEIGHT RATIO (diesel engine)	27.7 hp/tonne	28.6 hp/tonne	24.4 hp/tonne
LENGTH	5.32 m	5.97 m	6.365 m
WIDTH	2.5 m	2.5 m	2.5 m
HEIGHT (without armament)	1.85 m	1.85 m	1.85 m
GROUND CLEARANCE	0.5 m	0.5 m	0.5 m
TRACK			
front	2.18 m	2.18 m	2.18 m
rear	2.2 m	2.2 m	2.2 m
WHEELBASE	2.42 m	2.04 m + 1.04 m	1.1 m + 1.135 m + 1.04 m
ANGLE OF APPROACH/ DEPARTURE	40°/45°	40°/45°	40°/45°
MAX SPEED			
road	100 km/h	100 km/h	100 km/h
water	9.5 km/h	10.5 km/h	10.5 km/h

MOWAG Piranha (8 × 8) recovery vehicle with stabilisers lowered removing engine from MOWAG Piranha (6 × 6) APC fitted with remote-controlled 7.62 mm machine gun mount

Model	**4 × 4**	**6 × 6**	**8 × 8**
FUEL CAPACITY	200 l	200 l	300 l
MAX ROAD RANGE	700 km	600 km	780 km
FORDING	amphibious	amphibious	amphibious
GRADIENT	70%	70%	70%
SIDE SLOPE	35%	35%	35%
VERTICAL OBSTACLE	0.5 m	0.5 m	0.5 m
TURNING RADIUS	6.3 m	7.3 m	7.7 m
ENGINE (Detroit Diesel)	6V-53 developing 216 hp at 2800 rpm	6V-53T developing 300 hp at 2800 rpm	6V-53T developing 300 hp at 2800 rpm
TRANSMISSION (Allison Transmission Division of General Motors)	MT-653 automatic, 5 forward/1 reverse gears	MT-653 automatic, 5 forward/1 reverse gears	MT-653 automatic, 5 forward/1 reverse gears
SUSPENSION	independent	independent	independent
TYRES	11.00 × 16	11.00 × 16	11.00 × 16
BRAKES (main)	air hydraulic	(dual circuit)	on all vehicles
ELECTRICAL SYSTEM	24 V	24 V	24 V
ARMAMENT	depends on role	depends on role	depends on role

* Dependent on role.

Status: In production. In service with Australia (8 × 8), Canada (6 × 6 and 8 × 8), Chile (6 × 6), Ghana (4 × 4, 6 × 6 and 8 × 8), Liberia (4 × 4), Nigeria (6 × 6), Sierra Leone, Switzerland (6 × 6) and USA (8 × 8). Unconfirmed reports have indicated that Taiwan ordered 50 MOWAG Piranha (6 × 6) vehicles in 1984.

Manufacturer: MOWAG Motorwagenfabrik AG
8280 Kreuzlingen
Switzerland
Telephone: (72) 71 15 15
Telex: 882 211
Fax: (72) 72 28 86

Also manufactured under licence in Canada (qv), Chile (qv) and the United Kingdom.

MOWAG MR 8 Series of Armoured Personnel Carriers

Development
In the 1950s the MOWAG Company designed a series of 4 × 4 armoured vehicles as a private venture under the designation of the MR 8 series. The MOWAG MR 8 series of armoured personnel carriers is sometimes referred to as the WOTAN. The armoured personnel carrier version, called the MR 8-01, was subsequently tested by the Federal German Border Police (the Bundesgrenzschutz, or BGS for short). The BGS placed an order with MOWAG for 20 production vehicles which were delivered to West Germany in 1959/60 and production was then undertaken in Germany by Büssing and Henschel who together built some 600 vehicles, some of which have since been transferred from the BGS to the West German Police.

The BGS uses two basic models, the SW1 and the SW2. The SW1, or Geschützter Sonderwagen 1 (Kfz 91) is unarmed and used as an armoured personnel carrier, but some vehicles have been fitted with an obstacle-clearing blade at the front of the hull. The SW2 is fitted with a turret-mounted 20 mm cannon. In German service this is being replaced by the Thyssen Henschel TM 170 APC which is designated the SW4 by the BGS.

Description (SW1)
The all-welded hull of the MR 8-01 protects the crew from small arms fire and shell splinters. The driver sits at the front of the hull on the left with the commander to his right, both with a windscreen to their front which is covered in action by a steel shutter hinged at the top. Each shutter has a vision block for observation when it is lowered. Over the top of the driver's position is a single piece hatch cover that opens to the right and when the driver is driving with his head out a windscreen and wiper can be erected in front of him. A vision block is fitted in each side of the hull to the left of the driver and the right of the commander.

Front view of SW2 clearly showing different shape of hull and turret-mounted 20 mm cannon (BGS)

The hull top immediately behind the commander and driver has a distinct step upwards with a vision block fitted in the face of the personnel compartment for forward observation. In the roof of the vehicle, to the rear of the driver, is a six-sided cupola which is split down its centre line to open either side. Each half of the cupola has three vision blocks and can be locked vertical. In the left side of the hull is a two-part door, the left part with a firing port and the right part with a vision block. There is a similar door in the right side of the hull and a door in the rear of the hull on the right side, which has two parts with a vision block in the right side. There are also two hatches in the roof of the MR 8-01.

The engine, air-inlet and air-outlet louvres are in the left side of the hull and the exhaust outlet at the rear. Panels to allow easy access to the engine for maintenance are provided in hull side, rear and roof.

The MR 8-01 has no amphibious capability, no NBC system and no night vision equipment, although the last two could have been fitted if specified.

Variants
The SW2 has a different shaped hull top with no step and is fitted with a turret armed with a 20 mm Hispano-Suiza cannon with four smoke dischargers mounted either side of the turret. The SW2 normally has a crew of four and has a single door in each side of the hull, each with a firing port.

Other variants developed by MOWAG to the prototype stage included the MR 8-09 with a turret-mounted 20 mm cannon, MR 8-23 armed with a turret-mounted 90 mm Mecar gun, MR 8-30 fitted with an 81 mm multiple rocket system and the MR 9-32 with a 120 mm mortar.

Status: Production complete. In service only with the Federal German Border Police and Federal German Police.

Manufacturer: MOWAG Motorwagenfabrik AG
8280 Kreuzlingen
Switzerland
Telephone: (72) 71 15 15
Telex: 882 211
Fax: (72) 72 28 86

SW1 of the BGS with commander's cupola hatches in open position (BGS)

SPECIFICATIONS					
CREW	2 + 5	GROUND CLEARANCE		ENGINE	Chrysler type R 361 6-cylinder petrol developing 161 hp
CONFIGURATION	4 × 4	hull	0.5 m	TRANSMISSION	manual with 5 forward and 1 reverse gears
COMBAT WEIGHT	8200 kg	axles	0.3 m	TRANSFER CASE	2-speed
UNLOADED WEIGHT	7000 kg	TRACK	1.95 m	TYRES	10.00 × 20
POWER-TO-WEIGHT RATIO	19.63 hp/tonne	WHEELBASE	2.6 m	ELECTRICAL SYSTEM	24 V
LENGTH	5.31 m	ANGLE OF APPROACH/ DEPARTURE	45°/44°	ARMAMENT	see text
WIDTH	2.2 m	MAX ROAD SPEED	80 km/h		
HEIGHT		RANGE	400 km		
to turret top	2.2 m	FORDING	1.1 m		
to hull top	1.88 m	GRADIENT	60%		
		SIDE SLOPE	40%		

MOWAG Grenadier Armoured Personnel Carrier

Details of this vehicle, which is no longer in production, were given in *Jane's Armour and Artillery 1990-91*, pages 409-410.

MOWAG Roland Armoured Personnel Carrier

Development
The Roland was developed as a private venture by MOWAG in the early 1960s and is suitable for a wide range of roles including use as an ambulance, armoured personnel carrier, cargo/ammunition carrier, command and radio vehicle, internal security vehicle and as a reconnaissance vehicle. The prototypes of the Roland were completed in 1963 with series production following in 1964. It is most widely used as an internal security and reconnaissance vehicle, especially in Africa and South America, and its crew normally consists of commander, gunner, driver and three or four infantry.

Description
The hull of the Roland is all-welded steel and protects the crew from 7.62 mm small arms fire. The driver sits at the front of the hull on the left side and has a single piece hatch cover that opens to the right, in front of which are three periscopes. A windscreen and wiper can be fitted in front of the driver's position in wet weather. To the left of the driver is a vision block protected by a grille.

In either side of the hull, to the rear of the driver's position, is an entry hatch hinged at the top. The left hatch has a vision block protected by a grille beneath which is a firing port. A vision block and firing port are provided to the immediate rear of the right side hatch. The gunner sits on an adjustable seat in the centre of the vehicle with one man to his left facing the front and another to his right facing the rear. Over the gunner's position is a turret which can be manually traversed through a full 360°, and has a single piece hatch cover that opens to the rear and a vision block, to the right of which is a small circular hatch that opens to the right.

To the rear of the turret, on the right side, is a circular hatch cover that opens to the rear, to the left of which is the ventilator.

Another door at the rear of the vehicle on the right side opens to the right and has a vision block and a firing port. Two men are seated in the passageway that connects the main crew compartment and the rear door, one facing forward and one facing the rear.

The engine is at the rear of the hull on the left side and is separated from the crew compartment by a fireproof bulkhead. The suspension system, front and rear, consists of semi-elliptical springs and hydraulic shock absorbers. The rigid axles have hypoid gears and self-locking differentials.

The latest model of the MOWAG Roland has a slightly longer wheelbase and is fitted with an automatic instead of a manual transmission.

Standard equipment for the internal security version includes blue flashing lights, electric roof fan, heater, two-tone intermittent siren and wire mesh protection for the headlamps.

The basic vehicle can be fitted with a number of different types of light armament installation including a MOWAG-designed remote-controlled 7.62 mm machine gun mount on top of the low turret with the gunner being provided with a sight with a magnification of ×4 for aiming the weapon. Smoke dischargers can be fitted on either side of the turret if required.

Optional equipment includes an air-conditioning system, infra-red or passive night vision equipment, obstacle-clearing blade mounted at the front of the hull, searchlights, MOWAG-designed firing ports which allow the infantrymen to aim and fire their weapons from within the vehicle in safety, and MOWAG bullet-proof cross-country wheels consisting of metal discs either side of the tyre which support the tyre when it has been punctured and also provide additional traction when crossing rough country.

Variants
There are no variants of the Roland apart from the various roles that the vehicle has been designed to undertake.

MOWAG Roland armed with pintle mounted 12.7 mm M2 HB MG

MOWAG Roland fitted with obstacle-clearing blade at front of hull

MOWAG Roland APC with side and rear doors/hatches closed

SPECIFICATIONS (vehicle armed with remote-controlled 7.62 mm MG. Specifications in brackets relate to the model with automatic transmission where it differs from the manual version)

CREW	3 + 3	MAX ROAD SPEED	110 km/h	CLUTCH	single dry plate
CONFIGURATION	4 × 4	with bullet-proof wheels	80 km/h	SUSPENSION	semi-elliptical springs and hydraulic shock absorbers
COMBAT WEIGHT	4700 [4900] kg	FUEL CAPACITY	154 [170] l	TYRES	9.00 × 16
UNLOADED WEIGHT	3900 [4000] kg	MAX ROAD RANGE	550 [570] km	BRAKES	
POWER-TO-WEIGHT RATIO	42.9 [33] hp/tonne	FUEL CONSUMPTION	0.28 [0.3] l/km	main	hydraulic on all wheels
LENGTH	4.44 [4.73] m	FORDING	1 m	parking	mechanical on rear wheels
WIDTH	2.01 [2.05] m	GRADIENT	60%	ELECTRICAL SYSTEM	12 V
HEIGHT		SIDE SLOPE	30%	BATTERY	1 × 12 V, 125 Ah
to turret top, without armament	2.03 [2.1] m	VERTICAL OBSTACLE	0.4 m	ARMAMENT	1 × 7.62 mm MG
to hull top	1.62 [1.67] m	TURNING RADIUS	6.45 [6.85] m	GUN STABILISER	
GROUND CLEARANCE (hull)	0.4 [0.42] m	ENGINE	V-8 4-stroke water-cooled petrol developing 202 hp at 3900 rpm	vertical	no
TRACK		TRANSMISSION	manual with 4 forward and 1 reverse gears [automatic with 3 forward and 1 reverse gears]	horizontal	no
front	1.71 [1.66] m	TRANSFER CASE	2-speed		
rear	1.655 [1.66] m				
WHEELBASE	2.5 [2.65] m				
ANGLE OF APPROACH/ DEPARTURE	40°/36° [46°/40°]				

Status: Production complete. No longer available. In service with Argentina, Bolivia, Chile, Greece, Mexico, Peru and other countries.

Manufacturer: MOWAG Motorwagenfabrik AG
8280 Kreuzlingen
Switzerland
Telephone: (72) 71 15 15
Telex: 882 211
Fax: (72) 72 28 86

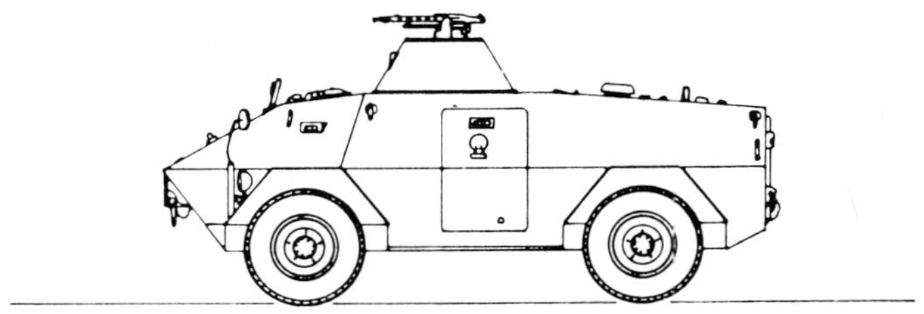

MOWAG Roland APC armed with NOWAG remote-controlled 7.62 mm GPMG on turret roof

TAIWAN

Armoured Infantry Fighting Vehicle

Development
The Armoured Infantry Fighting Vehicle (AIFV) was developed by the Republic of China Armoured Fighting Vehicle Development Centre in the 1970s and completed in 1979. It is manufactured by the Republic of China Fighting Machines Command which is responsible for final assembly and testing. About 40 firms are involved in supplying major components for the AIFV including the Taiwan Machinery Manufacturing Corporation (suspension and power plant) and the Taiwan Aluminium Corporation (hull and tracks).

Description
The AIFV is based on the American M113 APC which has been in service in Taiwan for many years but incorporates features of the M113A2, M113A3 and the FMC AIFV.

The layout of the AIFV is similar to the M113 series with the driver at the front left, engine compartment to his right and the troop compartment at the hull rear.

The hull is of all-welded aluminium armour construction with an additional layer of laminate steel armour bolted to the hull front, sides and rear. Closed cell polyurethane foam is used to fill the space between the two layers of armour and this gives increased buoyancy for amphibious operations.

The driver has a single piece hatch cover that opens to the rear and periscopes for observation. The engine compartment has a fire fighting system. The vehicle is powered by a British Perkins TV8.640 diesel developing about 215 bhp, coupled to a transmission developed by the Institute of Industrial Technology in Taiwan, and is similar to the M113A1. The air-inlet, air-outlet louvres and exhaust pipe outlet are in the roof of the vehicle.

The commander sits to the rear of the driver who has a cupola that can be traversed through a full 360°. This is provided with a single piece hatch cover, periscopes for all-round observation, 12.7 mm M2 HB machine gun and a shield that gives front and side protection against small arms fire. Either side of the shield is a bank of three electrically operated smoke dischargers firing forwards.

The troop compartment is at the rear of the vehicle and the infantry enter and leave via a power-operated ramp in the hull rear. This has an emergency door in the left side. Five firing ports, each with a periscope or a vision block are provided, two in each side of the hull and one at the rear. These allow the infantry to use their 5.56 mm weapons from within the vehicle in safety. It is believed that between six and eight infantrymen are carried plus the crew of two (commander/gunner and driver). In addition there is a hatch in the troop compartment roof.

The suspension is of the torsion bar type and consists of five rubber-tyred road wheels with the drive sprocket at the front and idler at the rear, there are no track return rollers.

The electrical system is similar to that of the M113. Standard equipment includes bilge pumps and infra-red night vision equipment. The AIFV is fully amphibious, being propelled in the water by its tracks. Before entering the water the bilge pumps are switched on and the trim vane erected at the front of the hull. An NBC system is not installed at present.

An armoured fuel tank, similar to that offered by FMC for the M113A2, is fitted either side of the hull ramp at the rear.

Physical and performance characteristics of the AIFV are similar to those of the M113A2.

Variants
The following variants are believed to be already in service or under development:
Mortar carrier (81 mm, 107 mm and 120 mm)
Command posts (command, communications and fire direction)
AIFV with 20 mm or 30 mm cannon
AIFV with Kung Feng IV 126 mm multiple rocket system, details of this version are given in the *Multiple rocket launchers* section.
AIFV with Hughes TOW ATGW system
AIFV with Kuen Wu ATGW
Others could include fire support vehicle, reconnaissance vehicle, flamethrower, cargo carrier and ambulance.

Status: Production. In service with Taiwanese Army.

Manufacturer: Fighting Machines Command
ROC Army
Taichung
Taiwan 400.

Armoured Infantry Fighting Vehicle (DTM)

Armoured Infantry Fighting Vehicle (DTM)

BMP-2 of the Soviet Army with trim vane folded back onto glacis plate, ATGW launcher on roof, applique armour on turret front and sides and bank of smoke dischargers towards turret rear

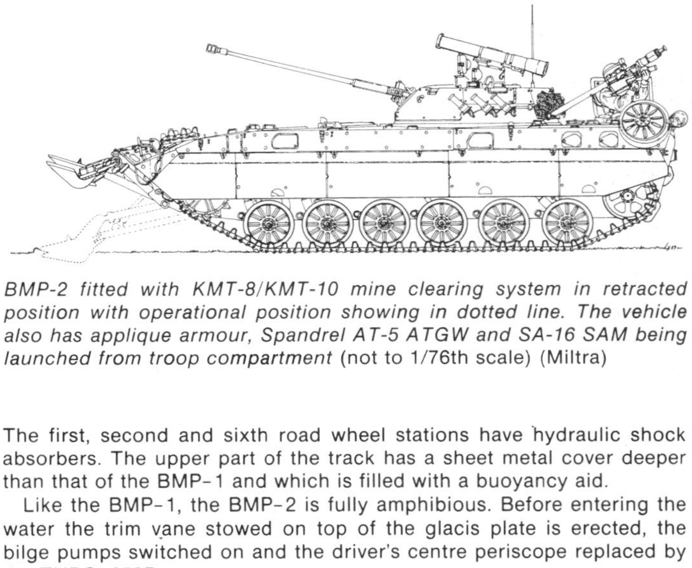

BMP-2 fitted with KMT-8/KMT-10 mine clearing system in retracted position with operational position showing in dotted line. The vehicle also has applique armour, Spandrel AT-5 ATGW and SA-16 SAM being launched from troop compartment (not to 1/76th scale) (Miltra)

BMP-2 from rear with troop compartment roof hatches locked vertical

The first, second and sixth road wheel stations have hydraulic shock absorbers. The upper part of the track has a sheet metal cover deeper than that of the BMP-1 and which is filled with a buoyancy aid.

Like the BMP-1, the BMP-2 is fully amphibious. Before entering the water the trim vane stowed on top of the glacis plate is erected, the bilge pumps switched on and the driver's centre periscope replaced by the TNPO-350B.

Standard equipment on the BMP-2 includes a full range of night vision equipment for commander, gunner and driver, fire extinguishing system, GPK-59 gyro-compass system, PAZ overpressure NBC system, engine pre-heater and turret extractor fan.

Variants

Improved BMP-2

From the late 1980s a number of improvements were carried out to production BMP-2 and most of these can be backfitted to the earlier BMP-2 and some are also applicable to the older BMP-1.

These improvements include modifications to the stabilisation system of the 30 mm cannon, BPK-1-42 gunner's sight has the warm-up switch moved to the control panel, improved internal communications, the tracks rubber bushes have conductive elements installed to ensure that the electricity flows through the track pins and links and into the ground, the tracks also have new end connectors and the track now has a chevroned cleat assembly, the spare track links have been moved from the rear exit doors to the left and right upper side walls of the troop compartment and a sensor now indicates the position of the louvre in the exhaust ejector port.

The vehicle now also includes a special mat for the transport of the seriously wounded and six slings for the lightly wounded.

BMP-2D

This is the late production BMP-2 and has a number of improvements including applique armour on the turret, provision for mounting mine clearing system under the nose of the vehicle and applique armour of the spaced type fitted along either side of the hull.

BMP-2K

This is the command version of the BMP-2 IFV and has additional communications equipment.

BVP-2

This is the Czechoslovakian Army designation for the Soviet BMP-2 AFV.

BMP-2 with Mine Plough System

Both the BMP-1 and the more recent BMP-2 can be fitted with the KMT-8 or KMT-10 plough type mine clearing system which is installed on the front of the hull in line with the tracks. The mine clearing ploughs are very narrow with just two digging tines which probably limits them to clearing smaller surface-laid mines, in particular scatterable anti-tank mines. Tilt-rod fuzes mounted between the right-hand and left-hand ploughs can detonate mines between them. To reduce the travelling width, the outer section of each plough can be folded inwards and held in place by a spring lock.

To enable the vehicle to retain its amphibious capabilities, an extra flotation device has been secured to the lower front of the hull to counteract the weight of the KMT-8 or KMT-10 system when afloat, although this could also be an additional layer of composite armour to improve protection to the forward part of the vehicle.

SPECIFICATIONS

CREW	3 + 7
COMBAT WEIGHT	14 300 kg
POWER-TO-WEIGHT RATIO	20.30 hp/tonne
GROUND PRESSURE	0.64 kg/cm^2
LENGTH	
gun forwards	6.735 m
hull	6.735 m
gun rear	7.295 m
WIDTH	
overall	3.15 m
over tracks	2.85 m
HEIGHT	
to commander's sight	2.45 m
for air transport	2.25 m
GROUND CLEARANCE	0.42 m
TRACK	2.55 m
TRACK WIDTH	300 mm
LENGTH OF TRACK ON GROUND	3.60 m
MAX ROAD SPEED	
5th gear	65 km/h
4th gear	43.3 km/h
3rd gear	29.1 km/h
2nd gear	19.6 km/h
1st gear	10.6 km/h
reverse	10.6 km/h
MAX WATER SPEED	7 km/h
FUEL CAPACITY	462 l
MAX ROAD RANGE	550-600 km
GRADIENT	60%
SIDE SLOPE	30%
VERTICAL OBSTACLE	0.7 m
TRENCH	2.5 m
ENGINE	Model UTD-20 6-cylinder, 4-stroke direct injection diesel developing 285/300 hp at 2600 rpm
TRANSMISSION	manual, 5 forward and 1 reverse gears
CLUTCH	multi-plate, dry, constant engagement
SUSPENSION	torsion bar, hydraulic shock absorbers on 1st, 2nd and 6th road wheels
ELECTRICAL SYSTEM	22-29 V
BATTERIES	2, 140 Ah
ARMAMENT	
main	1 × 30 mm 2A42 cannon
coaxial	1 × 7.62 mm PKT MG
other	1 launcher for AT-5 Spandrel or AT-4 Spigot ATGW
SMOKE LAYING EQUIPMENT	2 × 3 81 mm smoke dischargers, diesel fuel injected into exhaust
AMMUNITION	
main	500
coaxial	2000
ATGW	4
GUN CONTROL EQUIPMENT	
Turret power control	electric/manual
by commander	yes
by gunner	yes
Gun elevation/depression	+74°/-5°
Turret traverse	360°
Gun stabiliser	
vertical	yes
horizontal	yes

Status: In production. In service with Algeria, Czechoslovakia, Finland, Germany, India, Iraq, Kuwait (245 ordered in 1988) and USSR. India is undertaking licensed production of the BMP-2, and additional details of this are given under India in this section.

Manufacturers: Czechoslovakian, Indian and Soviet state factories.

BMP-1 Infantry Fighting Vehicle

Development

The BMP-1 (Boevaya Mashina Pekhota) was developed in the early 1960s as the replacement for the BTR-50P series of tracked APCs and was seen in public for the first time during the 1967 November parade in Red Square, Moscow. For a short period it was called the M-1967 and then the BMP-76PB in the West. Its correct Soviet designation is now known to be BMP-1. The vehicle uses some suspension components of the PT-76 light amphibious tank. The BMP-1 has now been replaced in production by the BMP-2, described in the previous entry.

Under the Conventional Forces Europe (CFE) Treaty signed in late 1990, the Soviet Union admitted holding the following quantities of BMP-1 IFVs in the following regions:

Region	IV.1	IV.2	IV.3	IV.4	V.1
BMP-1	3304	1791	1187	431	1513
BMP-1K	423	338	226	34	85
BMP-1P	1794	1700	1308	402	94
BMP-1PK	2625	2625	2601	2568	nil

Description

The hull of the BMP-1 is made of all-welded steel which provides the crew with protection from small arms fire and shell splinters. According to American reports the armour provides complete protection from 12.7 mm AP/API rounds over its frontal arc. The well-sloped glacis plate is ribbed, with an infra-red driving light at the front on the right and a white light on the left.

The driver sits at the front of the hull on the left side and has a single piece hatch cover that opens to the right. In front of his position are three periscopes, the centre one (TNPO-170) of which can be replaced by a vertically extensible one (TNPO-350B) which allows the driver to see ahead when the trim vane is erected for amphibious operations.

The commander is seated behind the driver and has a cupola that can be traversed through a full 360°. In the forward part of the cupola

BMP-1 chassis used by Polish Army for training purposes (Henry Dodds/ Jane's Intelligence Review)

BMP-1 from rear with both doors locked in open position (Michael Green)

BMP-1 with trim vane erected at front of vehicle (Michael Green)

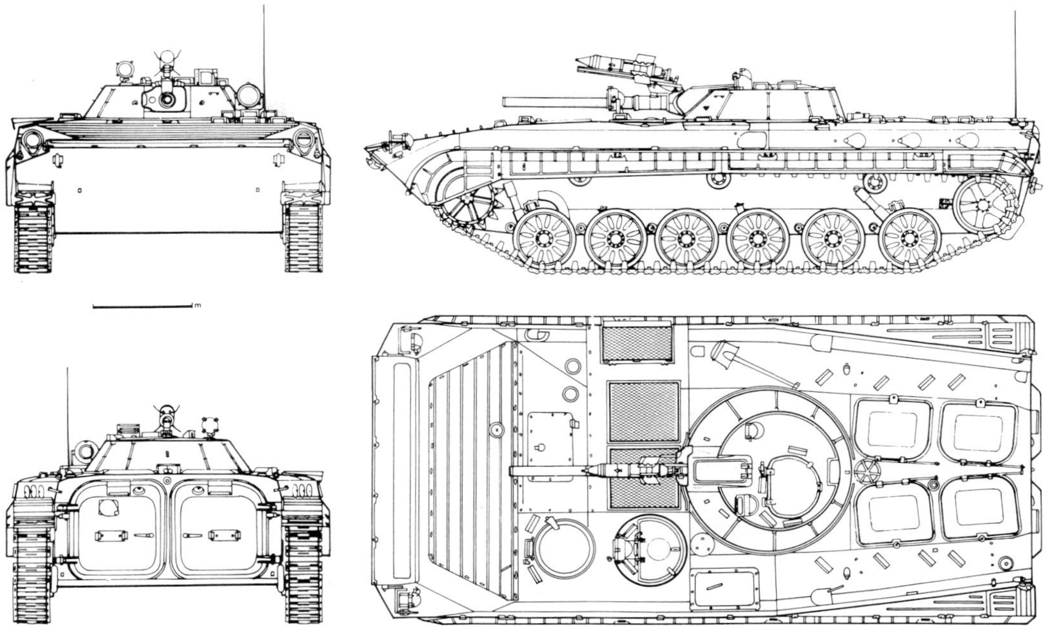

BMP-1 Model 1970 Infantry Fighting Vehicle (not to 1/76th scale) (Steven Zaloga)

are three periscopes, the centre one of which can be replaced by a TKH-3 binocular sight or a sight with a variable magnification. An infra-red searchlight is mounted on the commander's cupola and controlled from inside the vehicle.

The engine and transmission are to the right of the driver's and commander's positions with the steering system at the front of the hull. The air-inlet and outlet louvres are in the top of the hull. A pneumatic system is provided for air starting of the engine, compressed air and liquid clearing of the driver's and commander's periscopes, actuation of the control linkage operating the valve guard, air intake pipe and dust removal valve, the duplicating pneumatic stopping brake linkage and for disengaging the master clutch.

The one-man turret of the BMP-1 is identical to that installed on the BMD Airborne Combat Vehicle. The gunner has a single piece hatch cover that opens forwards, in front of which on the left side is a dual-mode 1PN22M1 monocular periscopic sight. In the day mode it has a magnification of ×6 and a 15° field-of-view, and in the image intensification mode it has a magnification of ×6.7 and a 6° field-of-view. Its maximum range at night is between 400 and 900 m. Graticule lines coincide with the different ranges in the gunner's sight and the gunner also has a stadiametric rangefinder to determine range to the target based on a target height of 2.7 m. The sight is heated and provided with a wiper arm. The gunner has an additional four observation periscopes and a white light or infra-red searchlight is mounted on the right side of the turret.

Main armament of the BMP-1 is a 73 mm model 2A28 smoothbore, low pressure, short-recoil gun which weighs 115 kg. This is fed from a 40-round magazine to the right rear of the gunner. After each round is fired the gun is returned to an elevation of +3° 30' for reloading. The weapon fires a fixed fin-stabilised HEAT round with an initial velocity of 440 m/s which increases to 700 m/s once the projectile has left the barrel and a rocket motor cuts in. The projectile is the same as that used in the SPG-9 infantry weapon and has a maximum effective range of 1300 m. The PG-9 HEAT projectile fired by the 73 mm 2A28 low pressure gun uses a small PG-15P stub casing to boost the projectile out of the gun tube at which point the PG-9V rocket motor cuts in to provide the main source of propulsion to the target. The HE-FRAG projectile is called the OG-9, but this is not believed to be used with the 73 mm 2A28 low pressure gun of the BMP-1. The HEAT projectile will penetrate 300 mm of armour at an incidence of 0°. According to American reports the weapon is not effective in high winds. Maximum rate of fire is 8 rds/min. The gun elevation and turret traverse mechanism is electric with mechanical controls for emergency use.

BMP-1s with applique side armour with Hip-C tactical transport helicopters flying overhead

Mounted coaxially to the right of the main armament is a 7.62 mm PKT machine gun which is fed from a continuous belt of 2000 rounds honeycombed in an ammunition box mounted below the weapon. There is a turret ventilator system for extracting fumes.

Mounted over the main armament is a launcher for a Sagger wire-guided ATGW. One missile is carried in the ready to launch position with a further two missiles in the turret which are loaded via a loading rail through a hatch in the forward part of the turret roof. A further two missiles are carried in the hull. The controls for the Sagger are normally kept under the gunner's seat and when required are released by pulling a handle. The controls are then locked in position between the legs of the gunner who controls the missile by using the joystick in the conventional manner. The Sagger has a minimum range of 500 m and a maximum range of 3000 m. Full details of the missile are given in the entry for the BRDM-1 in the *Reconnaissance vehicles* section. The Sagger can be used only in daylight and once launched a new missile cannot be loaded until the missile has hit the target. Missile loading time is about 50 seconds. Some modernised vehicles are to have a pintle-mounted Spigot AT-4 ATGW on the turret roof in place of the Sagger AT-3.

The troop compartment is at the rear with the eight infantrymen seated back-to-back four down each side of the hull. They enter and leave by two doors in the rear of the hull, which have integral fuel tanks (60 litres in the left door and 70 litres in the right) and are hinged on the outside. Each door has a vision device and the left door has a firing port. Over the top of the hull are four roof hatches, two each side, which are hinged in the centre and can be locked vertically. There are four firing ports in each side of the hull and four heated periscopes in the roof each side for aiming. Each weapon station has a vacuum exhaust system attached to the weapon to remove fumes.

The forward firing port on each side is usually used for the 7.62 mm PKM machine gun and the remaining three either side for the AKMS 7.62 mm assault rifle. Each PKM weapon is provided with 950 rounds of ammunition and each AKMS with 120 rounds. In addition five rounds are carried for the single RPG-7 anti-tank grenade launcher carried in the vehicle and one BMP-1 in each platoon carries an SA-7 Grail SAM, which is launched by one of the infantrymen standing up in the troop compartment with one hatch open.

The torsion bar suspension either side consists of six rubber-tyred road wheels with the drive sprocket at the front, idler at the rear and three track return rollers. The first and last road wheel stations have a hydraulic shock absorber and the top of the track has a light sheet steel cover, which is normally removed when operating in snow. The track links are of the double pin type with water scoops between the housings.

The BMP-1 is fully amphibious being propelled in the water by its tracks. Before entering the water the trim vane, which is stowed on the glacis plate of the vehicle when not in use, is erected at the front of the hull, bilge pumps switched on and driver's centre periscope replaced by an extensible one to enable him to see ahead over the trim vane. When afloat and fully loaded, the vehicle is driven in third gear but when empty second gear is used.

Should a nuclear explosion take place, the protection system ensures automatic shutdown of the engine, closing of the engine louvres, stopping the ejector and valves of the turret and troop compartment fans, stopping the fans and supercharger, turning off the electric drive of the turret, and switching on the absorbent filter air delivery system. When the shock wave has passed the driver/mechanic turns on the supercharger, which provides decontaminated air at overpressure for the inhabited compartment of the BMP-1.

The overpressure system of the BMP-1 consists of the NBC filter element and the blower/dust separator which is mounted in the roof to the rear of the commander's position. There is also a related scavenger system which removes gases from inside the vehicle when weapons are fired.

The blower/dust separator is similar to that used in the T-62 MBT and BTR-60PB/BTR-60PK APCs except that an extra set of fan blades has been added to force the air through the NBC filter. When the hatches are closed the blower/dust separator draws contaminated air into the system through an air intake immediately behind the turret. The contaminated air is then drawn through ducts around the turret ring to the blower/dust separator where fallout is separated from the air by the turning action of the propeller.

The air is then forced into a heating chamber (in cold weather) or, if necessary, directed through the NBC filter where the chemical and biological contaminants are removed. The air is then forced through the heating chamber and into the troop and driver's compartments through 11 outlet vents.

The forced air creates a build-up of air pressure inside the vehicle and so prevents NBC agents from leaking into the BMP through areas such as the firing ports which do not have air-tight seals; if the scavenger system is used when the NBC system is in operation, the effectiveness of the overpressure is reduced.

Variants

Differences between Production Vehicles

The BMP-1 first shown in Moscow in 1967 is referred to as the BMP Model 1966 and is believed to have been the pre-production vehicle; it was soon followed by the improved version, the BMP Model 1966, initial production batch. This had a curved shock absorber behind the first road wheel and the rear fender and swim vanes were modified by cutting down the number of swimming vanes from seven to four and raising the angle of the fender profile nearer to the horizontal. The fume extraction ports on the hull roof to the rear of the turret were moved outwards and were re-oriented to expel toward the rear of the BMP. The single torsion bar used to spring the rear roof troop hatches was dropped in favour of a twin torsion bar system.

This was followed by the BMP Model 1966 standard production vehicle on which the two fender tool stowage boxes were removed. The BMP Model 1966 final production type had a modified port for the PKM section machine gun on the centre of the hull side.

In 1970 the BMP-1 Model 1970 was introduced, this was probably the main production variant and has a 200 mm extended nose to improve its amphibious characteristics. The triangular air intake to the turret rear was replaced by a circular telescopic snorkel which is raised when the vehicle is afloat. The roof hatches were also slightly rearranged, the air intake located to the front left of the driver was removed and the NBC sensor cover moved inwards closer to the turret. To the left of the turret is a new prominent NBC filter cover and the PKM port has been redesigned.

Some sources have also mentioned an improved BMP which could have a stabilised 73 mm gun, semi-automatic guidance for the Sagger ATGW, improved NBC system, more powerful engine, improved automatic loader and improved night vision devices.

Soviet Designations

Under the Conventional Forces Europe (CFE) Treaty, the Soviets declared the following versions of the BMP-1 ICV and additional details of most of these are given later:

BMP, basic BMP-1 Model 1966, now obsolete
BMP-1, basic BMP-1 Model 1970
BMP-1F, BMP-1 reconnaissance model used by Hungary
BMP-1K, BMP-1 commander's model
BMP-1K3, BMP-1 commander's model
BMP-1P, BMP-1 with AT-4 9M114 ATGW and smoke grenade launchers
BMP-1PK, BMP-1P commander's model
BRM-1K, BRM-1 basic reconnaissance version - armoured cavalry (AKA M-1976)
BREM-1, recovery vehicle
BREM-4, recovery vehicle
BMP-1KShM, BMP-1 unarmed command variant
BWP, Polish designation for basic Soviet BMP-1
1V31, unidentified command variant (MP-31) based on BMP-1KShM
9S743, Bulgarian version of Soviet BMP-1KShM with unidentified radio sets installed (MP-31)
MLI-84, Romanian built BMP-1 (qv Romania).

Czechoslovakian BMP-1 designations (qv Czechoslovakia)

OT-90, BMP-1 with turret replaced by same turret as fitted to Czechoslovakian OT-64C (1) (8 × 8) APC armed with 14.5 mm and 7.62 mm machine guns
BVP-1, Czechoslovakian designation for Soviet BMP-1
BPzV, Soviet BVP-1 with modified hatches and massive sight over

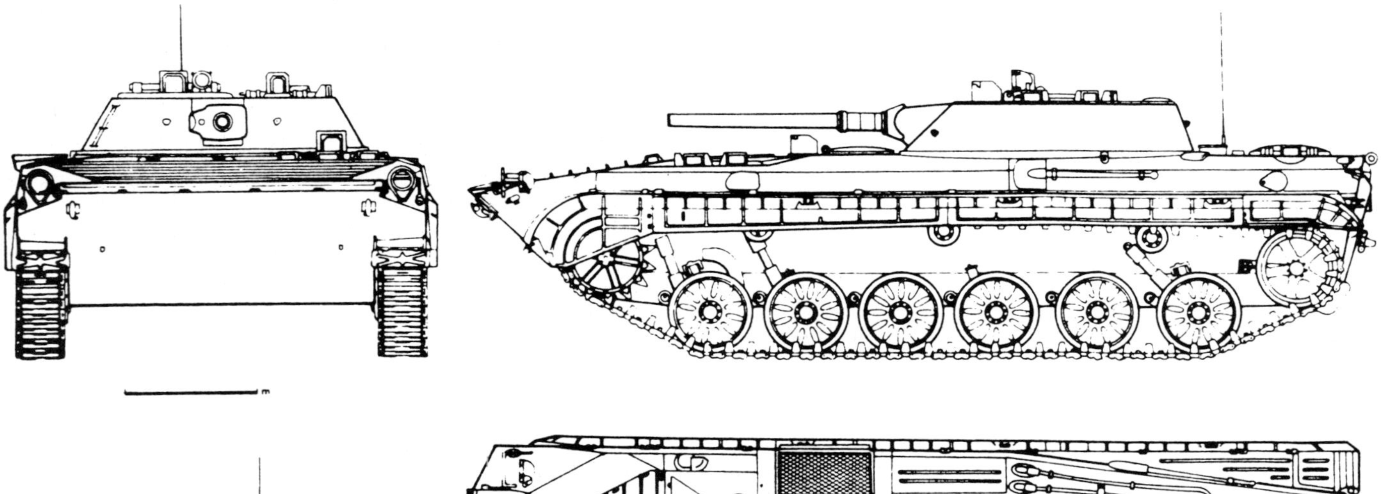

BRM Model 1976 reconnaissance vehicle (Soviet designation BRM) (not to 1/76th scale) (Steven Zaloga)

commander's hatch to rear of driver's position
BRM-1K, Soviet BRM-1 reconnaissance vehicle
BVP-1K, Czechoslovakian designation for commander's version of Soviet BMP-1K vehicle
Boure III, Czechoslovakian psychological warfare version of BMP-1 with loudspeaker system fitted
DP-90, maintenance version of OT-90
DTP-90, repair version of OT-90 APC
MP-31, air defence command version of Soviet BMP-1KShM command post vehicle (also called AKA 9S743)
MU-90, mine laying version of OT-90, no turret with space being covered by steel sheet
SVO, mine clearing version of BMP-1 with turret removed and fitted with Hedgehog type launcher in rear troop compartment
VPV, crane equipped recovery version of BMP-1. Turret and troop compartment roof hatches have been removed and mounted on troop compartment roof at the rear is a powered crane (Hungarian designation is BMP-1VPV)
VP-90, reconnaissance version of OT-90 with OT-64 turret
ZT-90, recovery version of OT-90.

Egyptian BMP-1s

It has been confirmed that from 1979 between 140 and 150 Egyptian Army BMP-1s were refitted with a French Poyaud 520 6L CS2 diesel engine developing 310 hp at 2800 rpm.

Modified BMP-1s

Also for use in Afghanistan some BMP-1s have been fitted with a 30 mm AGS-17 automatic grenade launcher on their turrets; as far as is known this is not aimed and loaded from within the turret.

Some BMPs used by the Soviet Army in Afghanistan were fitted with applique 6 mm thick steel armour on the upper side of the hull and covering the upper track and return rollers. It is believed that this modification was incorporated for protection against 12.7 mm rounds fired by DShKM heavy machine guns used by the guerrillas which can penetrate the 16/18 mm side armour of the BMP-1. The vehicle has an additional stowage box on the hull roof towards the rear. There is also a spike-type mount on the right side of the roof which is believed to be for the AT-4 Spigot ATGW manportable missile launcher.

BMP-1K (Command)

A battalion level command version of the BMP-1 is known to exist and is called the BMP-1K by the Soviet Army. It has additional communications equipment, an additional radio aerial on the right rear of the hull, a redesigned troop compartment to incorporate tables and a mapboard and the firing ports on the right side of the hull have been either eliminated or welded up. This model is also known in the US Army as the BMP M1974 Command Vehicle. Other command vehicles include the BMP-1KSh regimental and divisional level (command and staff vehicle), previously called the BMP M1978 which mounts a larger telescopic antenna and more radio equipment than the BMP M1974 but is unarmed.

BMP-1 with mine plough system

Both the BMP-1 and the more recent BMP-2 can be fitted with the KMT-8 or KMT-10 plough-type mine clearing system which is installed on the front of the hull in line with the tracks. The mine clearing ploughs are very narrow with just two digging tines which probably limits them to clearing smaller surface-laid mines, in particular scatterable anti-tank mines. Tilt rod fuzes mounted between the right-hand and left-hand ploughs can detonate mines between them. To reduce the travelling width, the outer section of each plough can be folded inwards and held in place by a spring lock.

To enable the vehicle to retain its amphibious capabilities an extra flotation device has been secured to the lower front of the hull to counteract the weight of the KMT-8 or KMT-10 system when afloat, although this could also be an additional layer of composite armour to improve protection to the forward part of the vehicle. An illustration of the KMT-8/KMT-10 appears in the previous entry on the BMP-2.

The BRM reconnaissance vehicle is based on the BMP-1 chassis but is fitted with a new two-man turret armed with the same 73 mm gun

BRM and BRM-1 (Reconnaissance)

This is known in NATO as the BMP-R or the BMP M1976 Reconnaissance Vehicle, or BRM by the Soviet Army, and has the same hull as the BMP-1 but is fitted with a larger two-man turret armed with a 73 mm gun to the rear of which are two roof hatches instead of four rectangular ones as in the case of the BMP-1. Some vehicles have been observed fitted with a bank of electrically operated smoke dischargers to the turret rear firing forwards.

BRM-1K (Reconnaissance)

The Soviet Army designation for this vehicle is the BMP M1976/2 and is similar to the above but the Sagger ATGW launcher over the 73 mm gun has been removed and a Tall Mike (NATO designation) surveillance radar is mounted in the rear part of the turret.

When not required the radar is retracted into the turret, but when required a hatch opens towards the rear and the radar is elevated above the turret roof.

The PSNR-5K radar has a maximum range of 10 000 m and other equipment carried by the vehicle includes a DKRM-1 laser rangefinder, ARRS-1 location device, IMP mine detector and 1PN33B night binoculars.

Navigation equipment carried includes TNA-1, IG11N gyro-compass and 1T25 survey device.

This version has a six man crew and armament consists of one 73 mm 2A28 gun, one 7.62 mm PKT coaxial machine gun and six smoke dischargers. As it has a different battlefield mission, it has a smaller number of firing ports and vision devices than the standard BMP-1 IFV.

PRP-3 (Radar)

This model appeared in 1975 and has been observed attached to 152 mm 2S3 and 122 mm 2S1 self-propelled units and is believed to be used as an artillery fire adjustment and/or artillery/mortar locating vehicle, as these battalions already use the ACRV-2 Model 2 for the command post/observation role. The front of the vehicle is identical to the BMP-1 but the vehicle has a new two-man turret which has two single piece hatch covers that open forwards. They both have periscopes for observation plus a large optical device in front of the hatch. Armament consists of a 7.62 mm machine gun. Mounted on the rear of the turret is a radar with a flat antenna that folds forwards when not operating. To the rear of the turret on the left side is a further circular hatch cover and a telescopic antenna. The Small Fred radar operates in the J-band,

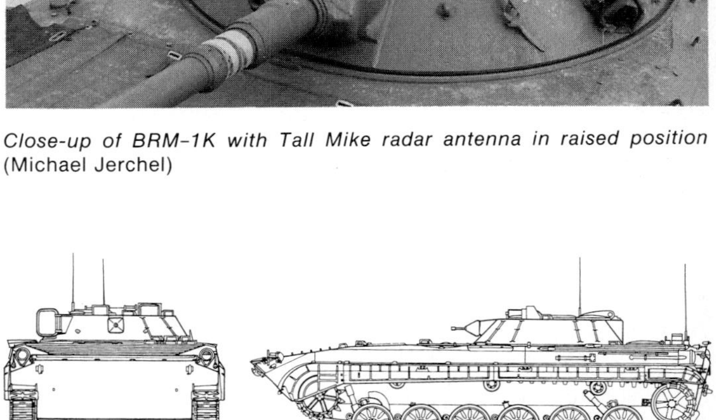

Close-up of BRM-1K with Tall Mike radar antenna in raised position (Michael Jerchel)

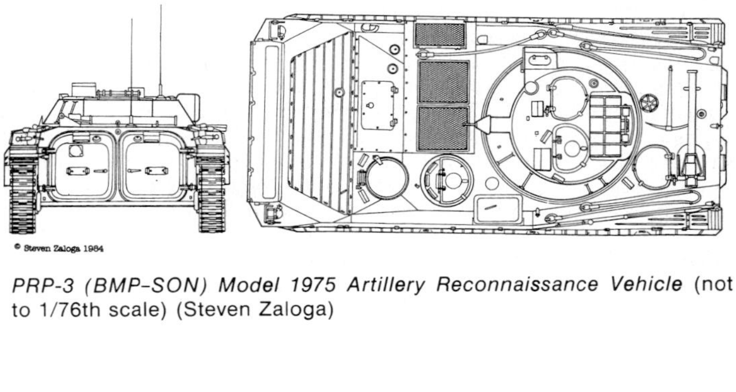

PRP-3 (BMP-SON) Model 1975 Artillery Reconnaissance Vehicle (not to 1/76th scale) (Steven Zaloga)

Model of IRM amphibious engineer reconnaissance vehicle with trim vane erected at front of hull

PRP-3 radar vehicle showing folded Small Fred radar at rear of turret and 7.62 mm PKT machine gun at front

with a detection range of 20 km and a tracking range of 7 km. This version is believed to have a crew of five. This model is known as the PRP-3 in the USSR or the BMP M1975 surveillance vehicle, by the US Army. It has a five-man crew and is fitted with extensive communications equipment and optical devices.

Scale of issue is one per towed or self-propelled howitzer battalion with another vehicle attached to the target acquisition battery of the artillery regiment.

IRM Amphibious Engineer Reconnaissance Vehicle

It was originally believed that this vehicle was based on the chassis of the 122 mm 2S1 self-propelled howitzer, but in 1986 it was disclosed that it was based on the BMP chassis. It has a new hull and uses the engine and suspension of the BMP vehicle. It has been designed to undertake a variety of specialised engineer reconnaissance roles including mine detection with two devices mounted at the front of the vehicle which can be retracted flush with the front of the hull when not required.

The IRM is fully amphibious, propelled in the water by two shrouded propellers at the rear of the vehicle. When submerged a snorkel is erected on the top of the hull, this is kept horizontal when not required.

Basic specifications of the IRM are:

WEIGHT	17 000 kg
LENGTH	8.22 m
WIDTH	3.15 m
HEIGHT	2.4 m
MAX ROAD SPEED	52 km/h
RANGE	500 km

Additional details of this vehicle, which so far has been identified only to be in service with the Soviet Army, is given in *Jane's Military Vehicles and Logistics 1991-92*, pages 8/9.

BMP-PPO Mobile Training Centre

The BMP-PPO (Podvizhniiy Punkt Obucheniya) mobile training centre is essentially a BMP-1 with its turret removed and fitted with eight roof-mounted cupolas for trainees under instruction, plus seats for the vehicle commander and driver. The commander and driver sit at the front left of the vehicle with the eight trainees at the rear, three down either side and two down the centre. Each trainee has two TNPO-170 and one Type MK-4 observation devices mounted in the forward part of the cupola and an A-2 unit of the R-124 intercom set (used via the two-way radio). The instructor has a console with three cassette recorders, AGU-10-3 amplifier, three switches, external loudspeakers and a microphone. There is also an R-123m two-way radio, A-1 and A-2 units from the tank intercom R-124 and 10 chest-mounted switches with cables.

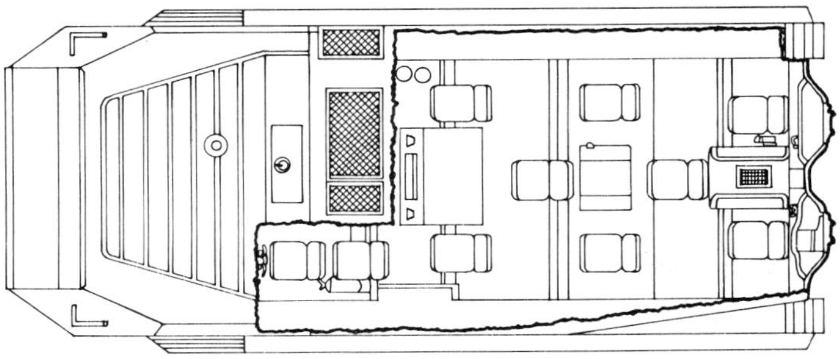

BMP-PPO Mobile Training Centre showing seating layout (not to 1/76th scale)

Iraqi BMP-1 infantry fighting vehicle with additional armour protection (Christopher F Foss)

For training each platoon would have two standard BMP-1s and one BMP-PPO, with the former being manned as normal practice. The instructor and one trainee take turns to send information while the remainder hear via loudspeakers.

BMP-1 Command Vehicle

The BMP-1 command vehicle, called the BMP KShM Command and Staff Vehicle by the Soviet Army, has the same set of communications equipment as the BMD KShM (BMD M1979/3) and like the latter, saw operational use in Afghanistan.

Finnish BMP-1

It is understood that Finland has removed the automatic loaders from its BMP-1 as these have proved to be unreliable in service.

Iraqi BMP-1 with applique armour

In 1989, a BMP-1 which had applique armour fitted to the sides of the hull to provide protection against 12.7 mm and 14.5 mm armour piercing projectiles fired from a range of 200 m was first observed in Iraq. The total package weighs 1250 kg and spaces are cut in the sides of the hull armour package to allow the infantry to use their small arms from within the vehicle. No applique armour was observed fitted to the turret or glacis plate.

Iraqi BMP-1 ambulance

To evacuate wounded personnel from forward battlefield areas, Iraq has developed an armoured ambulance based on the BMP-1 vehicle. The turret has been removed and the rear of the vehicle extended upwards to allow for stretcher and walking wounded to be more easily transported. The original doors of the BMP-1 have been removed and replaced by two new higher doors that do not contain the integral fuel tanks of the basic BMP-1 vehicle.

Status: Production complete. In service with Afghanistan, Algeria, Angola, Bulgaria, Cuba, Czechoslovakia, Egypt, Ethiopia, Finland, Germany, Hungary, India, Iran, Iraq, Jordan (BMP-1 command vehicles), North Korea, Libya, Mongolia, Poland, Romania, Syria, USSR, Yemen and Yugoslavia.

Manufacturers: Czechoslovakian and Soviet state factories. Unconfirmed reports have stated that the BMP-1 and BMP-2 are both built at Izhevsk, USSR.

Note: The Chinese build a similar version of the BMP-1 called the WZ 501, full details of which are given under China. Romania builds a modified version of the BMP-1 under the designation of the MLI-84. Full details are given under Romania.

BMP-1 converted by Iraq into a forward-armoured ambulance (Christopher F Foss)

SPECIFICATIONS	
CREW	3 + 8
COMBAT WEIGHT	13 500 kg
UNLOADED WEIGHT	12 500 kg
POWER-TO-WEIGHT RATIO	22.22 hp/tonne
GROUND PRESSURE	0.6 kg/cm^2
LENGTH	6.74 m
WIDTH	2.94 m
HEIGHT (over searchlight)	2.15 m
GROUND CLEARANCE	0.39 m
TRACK	2.75 m
TRACK WIDTH	300 mm
LENGTH OF TRACK ON GROUND	3.53 m
MAX SPEED	
road	80 km/h
water	6-8 km/h
FUEL CAPACITY	460 l
MAX RANGE	500 km
FUEL CONSUMPTION	1 l/km
FORDING	amphibious
GRADIENT	60%
SIDE SLOPE	30%
VERTICAL OBSTACLE	0.8 m

TRENCH	2.2 m
ENGINE	Type UTD-20 6-cylinder in-line water-cooled diesel developing 300 hp at 2000 rpm
TRANSMISSION	manual, 5 forward and 1 reverse gears
STEERING	clutch and brake
SUSPENSION	torsion bar
ELECTRICAL SYSTEM	24 V
ARMAMENT	
main	1 × 73 mm gun
coaxial	1 × 7.62 mm MG
other	1 launcher rail for Sagger ATGW
SMOKE-LAYING EQUIPMENT	diesel fuel injected into exhaust
AMMUNITION	
main	40
coaxial	2000
other	4 + 1 Sagger

GUN CONTROL EQUIPMENT	
Turret power control	electric/manual
by commander	no
by gunner	yes
Gun elevation/depression	+33°/-4°
Turret traverse	360°
Gun stabiliser	
vertical	no
horizontal	no
ARMOUR	
Hull front upper	7 mm at 80°*
Hull front lower	19 mm at 57°
Hull side upper	16 mm at 14°
Hull side lower	18 mm at 0°
Hull rear upper	16 mm at 19°†
Hull rear lower	16 mm at 19°†
Hull top	6 mm
Hull belly front	5 mm
Hull belly rear	7 mm
Turret front	23 mm at 42°
Turret sides	19 mm at 36°
Turret rear	13 mm at 30°
Turret top	6 mm
Turret mantlet	26-33 mm

* Does not include wave deflector, does not represent aluminium engine cover
† Does not include fuel or inner armour of door

BMD-1 Airborne Combat Vehicle

Development

The BMD-1 (Boevaya Mashina Desantnaya, or Airborne Combat Vehicle) entered service with Soviet airborne units in 1970 and was first seen in public during a parade held in Moscow in November 1973. The vehicle was called the M1970 before its Soviet designation became known. It spearheaded the Soviet invasion of Afghanistan in December 1979. It is issued to Soviet Air Assault Divisions (or Air Landing Divisions) on the scale of 330 per division, three command versions at divisional headquarters and three airborne regiments with 109 BMDs each: 10 command vehicles, nine BMDs without turrets and 90 basic BMDs. It is airportable and can be dropped by parachute.

Unconfirmed reports have indicated that India has taken delivery of a quantity of BMD-1 airborne combat vehicles from the Soviet Union and a quantity of BMD-1 vehicles were captured in Kuwait by United States forces early in 1991 during Operation Desert Storm.

Under the Conventional Forces Europe (CFE) Agreement signed late in 1990, the Soviet Union released details of some of its holdings of armoured vehicles and artillery systems, including the BMD-1 airborne combat vehicle and the BTR-D armoured personnel carrier and these are as follows:

Region	**IV.1**	**IV.2**	**IV.3**	**IV.4**
BTR-D	769	462	353	nil
BMD-1	2149	1213	765	nil
BMD-1K	5	5	5	nil

Note: Regions are as defined under the CFE agreement; Region IV is defined as Central Europe while Region V covers the flanks of the Atlantic to the Urals (ATTU) zone.

Description

The hull of the BMD-1 is welded. The driver sits at the front of the vehicle in the centre just forward of the turret and has a single piece hatch cover that opens to the right. Three periscopes are mounted forward of his hatch and the centre one can be replaced by an infra-red periscope for night driving. The vehicle commander sits to the left of the driver/mechanic and alongside the commander's seat is the radio and gyro-compass. The bow machine gunner sits to the driver's right and aims the two bow-mounted 7.62 mm PKT machine guns using a TNPP-220 periscopic sight to which the machine guns are connected by parallelogram drives. The two machine guns are mounted one either side of the vehicle's front firing forwards. A single semi-circular hatch cover is positioned either side of the forward part of the turret. There is a single fixed forward-looking periscope to the left and slightly to the rear of the left turntable-mounted periscope.

The turret is similar to the one fitted to the BMP-1 ICV and has a single piece forward-opening hatch cover to the left of the turret. The gunner has four periscopes, one mounted to each side and two forward of the hatch.

The personnel compartment at the rear of the vehicle has a concertina type cover that opens towards the front. The only means of entry and exit to this compartment is via this hatch. In here are seated the senior gunner, grenade launcher and his assistant, with a roof-mounted TNPO-170 periscope for observation which can be traversed. In the rear part of this hatch is a normal BMP-1 type firing port with observation device to enable one of the crew to use a weapon from within the vehicle.

The independent suspension consists of five small road wheels with the idler at the front and the drive sprocket at the rear. There are four track return rollers. The independent suspension combines a hydraulic system for altering the ground clearance and maintaining track tension with pneumatic springs, which enables the ground clearance to be altered from 100 to 450 mm.

The BMD is fully amphibious, being propelled in the water by two water jets at the rear of the hull. Before entering the water a trim vane which is stowed on the glacis plate when not in use is erected at the front of the hull. The vehicle has electric and manual bilge pumps, a gyro-compass, engine pre-heater, smoke-generating equipment, NBC system and a centralised ethylene-bromide fire-extinguishing system as fitted to other Soviet armoured vehicles.

The gunner is provided with a dual mode 1PN22M1 monocular periscopic sight mounted on the left side of the turret roof. In the day

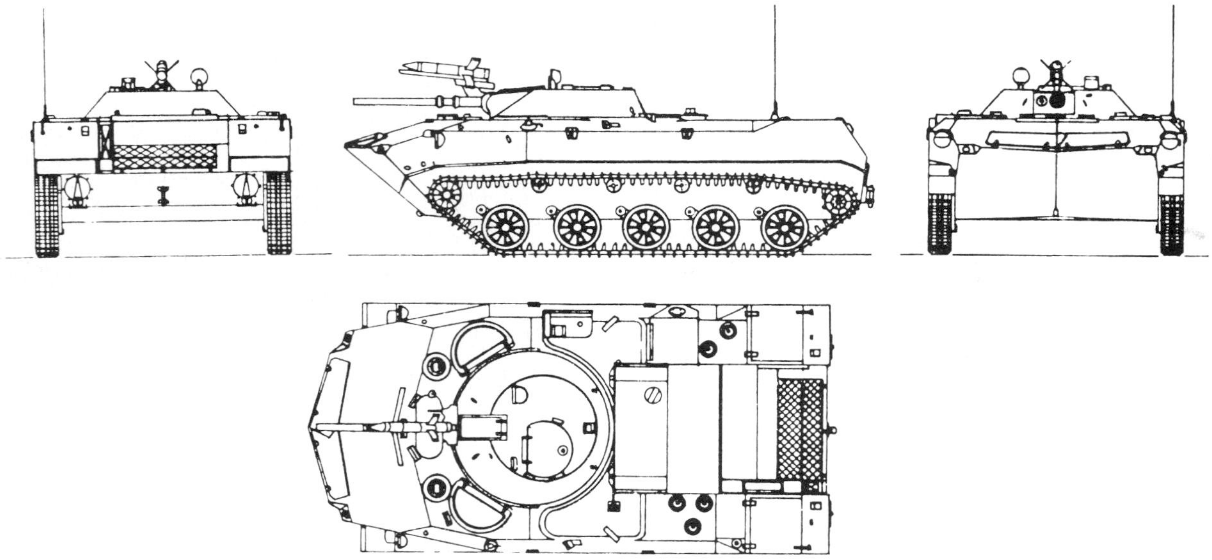

BMD-1 airborne combat vehicle with Sagger AT-3 ATGW mounted over 73 mm gun

mode it has a magnification of ×6 and a 15° field-of-view while in the image intensification night mode it has a magnification of ×6.7 and a 6° field-of-view. Its maximum range at night is between 400 and 900 m. Graticule lines coincide with the different ranges in the gunner's sight and he also has a stadiametric rangefinder to determine range to the target, based on a target height of 2.7 m. The sight is heated and has a wiper arm. A white light searchlight is mounted on the right side of the turret although some vehicles have been observed with an infra-red searchlight at this position.

Main armament of the BMD is a 73 mm model 2A28 smoothbore, low pressure, short-recoil gun which weighs 115 kg. This is fed from an automatic 40-round magazine to the right rear of the gunner. After each round is fired the gun is returned to an elevation of 3° 30' for reloading. The weapon fires a fixed fin-stabilised HEAT round with an initial muzzle velocity of 400 m/s, which increases to 665 m/s once the projectile has left the barrel and a rocket motor cuts in. The projectile is the same as that used in the SPG-9 infantry weapon and has a maximum effective range of 1300 m. According to a number of reports the weapon is not accurate in high wind.

The gun elevation and traverse mechanism is electric with mechanical controls for emergency use.

A 7.62 mm PKT machine gun is mounted coaxially to the right of the main armament and is fed from a continuous belt of 2000 rounds, honeycombed in an ammunition box mounted below the weapon. A cartridge case and link collector are mounted in the turret basket. A ventilator system is provided in the turret to extract fumes.

Mounted over the main armament is a launcher rail for a Sagger ATGW. Two missiles are carried inside the turret and are loaded via a loading rail through a hatch in the forward part of the turret roof. The controls for the Sagger are normally kept under the gunner's seat and when required for the Sagger are released by pulling a handle. The controls are then locked in position between the legs of the gunner who controls the missile by using the joystick in the conventional manner. The Sagger has a minimum range of 500 m and a maximum range of 3000 m. Full details of the Sagger ATGW are given in the entry for the BRDM-1 in the *Reconnaissance vehicles* section.

Soviet BMD airborne infantry combat vehicles are reported to be equipped with a radio 'bleeper' device that allows its crew, who are equipped with direction-finding equipment, to locate their own individual BMD when they are parachuted onto the drop zone.

Variants

Differences between BMDs

The pre-production models of the BMD first appeared in 1970 and had a squarish NBC filter cover on the right hull roof. The first production models, the BMD-1s, appeared in 1973 and had the NBC hatch replaced by a more prominent circular vent cover. To increase the operational range of the vehicle it can be fitted with additional fuel cells externally in the rear well.

Due to poor ventilation in early models of the BMD-1 which could lead to crew asphyxiation when the weapons were fired, a modernised version with new internal ventilation was developed. This was shown in 1980 and is called the BMD-1M. First models had the same wheels as the initial production BMD-1 but were distinguishable by a vent grille on the bow. Current production models have the new pattern road wheels.

BMD-1P Airborne Combat Vehicle

This is BMP-1 with AT-3 Sagger ATGW mounted over 73 mm gun barrel removed and mounted on the turret roof is an AT-4 Spigot ATGW launcher.

BMD-2 Airborne Combat Vehicle

The BMD-2 was first observed in 1988 and was designated the BMD M1988/1 until its correct Soviet designation became known.

It is believed that the scale of issue of the BMD-2 was originally nine in each of the three airborne regiments of a Soviet air assault division, but it is probable that it is now replacing the older BMD-1 on a one-for-one basis.

The one-man turret of the BMD-2 is of a new design with the gunner being seated on the left and provided with a single piece circular hatch cover that opens to the front.

In front of this is the gunner's day/night sight that appears to be identical to that fitted on the BMP-2. Additional periscopes give observation to the sides and a white light searchlight is mounted on the forward part of the turret roof.

BMD-1 airborne combat vehicle with all roof hatches open and Sagger AT-3 ATGW over 73 mm gun

BMD-2 airborne combat vehicle which has new turret armed with 30 mm cannon

BMD-1 airborne combat vehicle deployed in a static position with suspension adjusted to suit terrain

Iraqi BMD-1 airborne combat vehicle captured in Kuwait early in 1991

BMD-1 of Soviet Army clearly showing one of two 7.62 mm bow machine guns

On the left side of the turret is another sight that moves in elevation with the 30 mm 2A42 cannon and this is believed to be the high angle of fire sight which is used to aim the weapon when it is being used in the anti-aircraft/anti-helicopter role. This has an elevation of +60° and a depression of -5°. A total of 300 rounds of 30 mm and 2940 rounds of 7.62 mm ammunition are carried.

Mounted on the right side of the turret roof is a searchlight that moves in elevation with the main armament, this appears to be a new searchlight that is also fitted to the BMP-3.

Main armament comprises a 30 mm 2A42 stabilised cannon with a 7.62 mm PKT machine gun mounted coaxial to the right with a launcher for an AT-4 Spigot (maximum range 2000 m) or an AT-5 Spandrel (maximum range 4000 m) ATGW mounted on the right side of the turret roof.

In addition there is also a circular hatch cover in the turret roof towards the rear which is believed to be the missile resupply hatch as there is insufficient room to pass new missiles to the gunner via his small hatch.

It is possible that a ground launcher is also carried to allow the ATGW system to be deployed away from the vehicle if required by the tactical situation.

While the BMD-1 has two bow-mounted 7.62 mm PKT machine guns, one in either side, the BMD-2 has only one 7.62 mm bow-mounted machine gun on the right side with the machine gun port in the left side being eliminated. The front right hatch cover has also been eliminated on the BMD-2 which may indicate that the bow machine gunner is not carried in the BMD-2 and that this space is used for additional stowage.

Other features of the BMD-1 are retained on the BMD-2 including a total of three firing ports for the infantrymen carried, one in either side of the hull and one in the rear troop compartment.

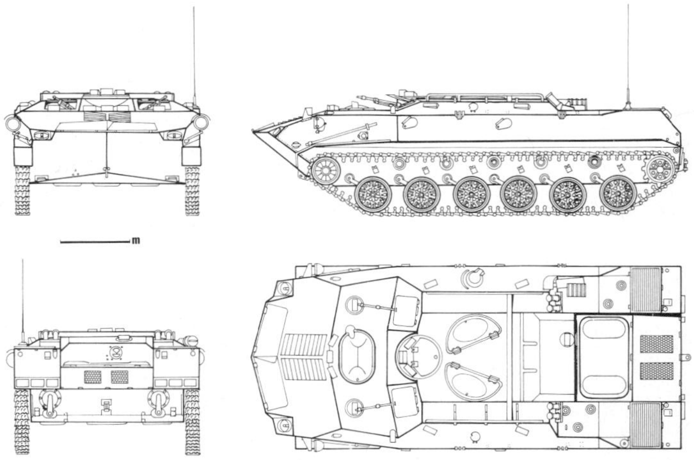

BTR-D Air Assault Transporter (not to 1/76th scale) (Steven Zaloga)

BMD-1 (left) compared to BMD-2 (right)

BTR-D armoured personnel carrier which has a longer chassis with six road wheels

82 mm Mortar Carrier

For the use in Afghanistan, some BMDs were fitted with a standard 82 mm mortar in the rear troop compartment firing to the rear. It would be more logical to remove the existing turret and mount the 82 mm mortar in this position rather than in the very small troop compartment at the hull rear.

BTR-D APC

The BTR-D was first seen during the Soviet invasion of Afghanistan and is often referred to as the BMD M1979 by NATO. It is distinguishable from the standard BMD by its longer chassis with six rather than five road wheels, different hull top and lack of a turret.

The basic BTR-D has the STANAG designation BMD M1979/1 and is a multi-purpose armoured transporter used by the Soviet VDV (Air Assault Forces) air assault divisions, for a variety of roles including troop transporter, towing support weapons such as the 23 mm ZU-23 light anti-aircraft gun and maintenance support. This version retains the hull front and side firing ports of the BMD-1 and has two forward firing smoke dischargers on each side of the hull roof in line with the fifth road wheel

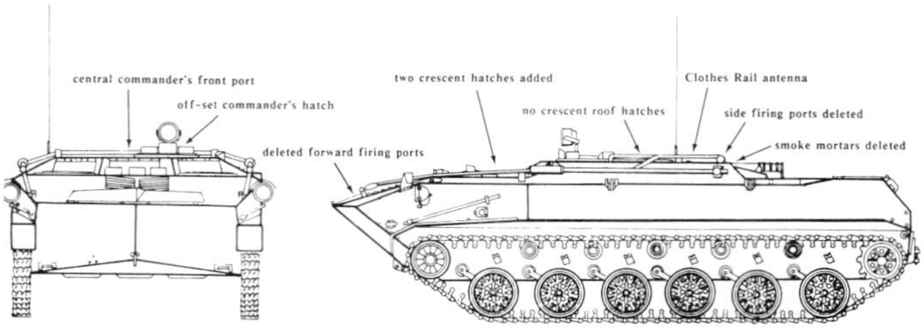

BTR-D Air Assault Command Vehicle (not to 1/76th scale) (Steven Zaloga)

BMD-KShM command vehicle in Afghanistan with Clothes Rail antenna raised (US Army)

station. There are also two crescent shaped hatches on the roof to the rear of the commander's position which projects forward onto the glacis plate. There are nine firing ports, two in the hull front each with a 7.62 mm machine gun, two in each side of the hull, one in the rear and the two front hatches below and either side of the commander's cupola can be used.

The second version is called the BMD-KShM by the Soviets and is a command post vehicle. This has a folding Clothes Rail antenna around the superstructure but the front firing ports with their 7.62 mm machine guns are faired over and there are no firing ports in the hull sides, the forward firing smoke dischargers have also been removed. The commander's hatch is offset to the left and does not project forward.

SPECIFICATIONS (Provisional)

CREW	1 + 9
WEIGHT	7500 kg
LENGTH	5.95 m
WIDTH	2.7 m
HEIGHT	1.8 m
ENGINE	5D 20 diesel developing 300 hp
MAX ROAD SPEED	60 km/h
MAX RANGE	500 km

Scale model of the BMD-1KShM chassis used with the Shmel-1 RPV (Christopher F Foss)

BMD-3

This was declared under the Conventional Forces Europe Treaty, but no firm details are available as yet.

1V118

This is a BMD used in artillery observation post role.

1V119

This is a BMD used in artillery fire direction centre role.

BMD with Shmel-1 RPV

A BMD-1KShM chassis is used as the launcher and flying control station for the Shmel-1 remotely piloted vehicle.

BREM-D

BMD-1 recovery variant.

BTR-D

BTR-D with Zu-23 (2 × 23 mm) gun towed or mounted.

SO-120 (Anona) (2S9) 120 mm Self-propelled Howitzer/Mortar

Details of this are given in *Self-propelled guns and howitzers* section.

Status: In production. In service with India (unconfirmed), Iraq and the Soviet Union.

Manufacturer: Soviet state factories. Unconfirmed reports have stated that the BMD is built at Izhevsk, USSR.

SPECIFICATIONS

CREW	3 + 4
COMBAT WEIGHT	7500 kg
POWER-TO-WEIGHT RATIO	32 hp/tonne
GROUND PRESSURE	0.57 kg/cm²
LENGTH	5.4 m
WIDTH	2.63 m
HEIGHT	1.62-1.97 m
GROUND CLEARANCE	0.1-0.45 m
TRACK WIDTH	230 mm
LENGTH OF TRACK ON GROUND	2.84 m
MAX SPEED	
road	70 km/h
water	10 km/h
FUEL CAPACITY	300 l
RANGE	320 km
FORDING	amphibious
GRADIENT	60%
VERTICAL OBSTACLE	0.8 m
TRENCH	1.6 m
ENGINE	Type 5D-20 V-6 liquid-cooled diesel developing 240 hp
TRANSMISSION	manual, 5 forward and 1 reverse gears
SUSPENSION	hydro-pneumatic
ELECTRICAL SYSTEM	24 V
ARMAMENT	
main	1 × 73 mm gun
coaxial	1 × 7.62 mm MG
forward firing	2 single 7.62 mm MGs
other	1 launcher rail for Sagger ATGW (or roof-mounted AT-4 Spigot on turret)
AMMUNITION	
main	40
coaxial	2000
Sagger	3
GUN CONTROL EQUIPMENT	
Turret power control	electrical and mechanical
by gunner	yes
Turret traverse	360°
Gun elevation/depression	+33°/-4°
Gun stabiliser	
vertical	no
horizontal	no
ARMOUR	
Hull (max)	15 mm
Turret (max)	23 mm

BTR-50P Armoured Personnel Carrier

Development

The BTR-50P entered operational service in 1954 and was seen in public for the first time during the 1957 November parade in Moscow. For many years the BTR-50P was the standard APC of the Motor Rifle Regiments of Soviet Tank Divisions and was issued on the scale of 91 (one at headquarters and three battalions each with 30 vehicles) per regiment, but the BTR-50P has now been replaced in front-line units by the BMP-1 or BMP-2 infantry combat vehicles.

Under the Conventional Forces Europe Treaty signed in 1990, the Soviets declared a total of 21 BTR-50 APCs as being in service with the Soviet Army and all of these were in Central Europe.

The Czechoslovaks have developed a much improved version of the BTR-50P under the designation OT-62, which has its own entry in this section.

Description

The BTR-50P is based on the chassis of the PT-76 light amphibious tank with a new superstructure added to the front of the vehicle. The hull of the BTR-50P is made of all-welded steel with the crew compartment at the front, open-topped troop compartment in the centre and the engine compartment at the rear.

The commander sits in a projecting bay at the front of the hull on the left side, which has three vision devices and a single piece hatch cover that opens to the front. The driver sits to the right of the commander

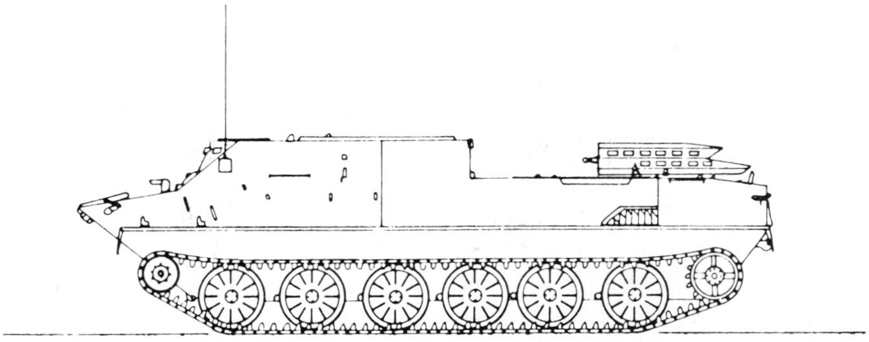

BTR-50P APC complete with ramps on rear decking

Finnish Army BTR-50PK with driver's hatch open and fitted with windscreen and wiper (Finnish Army)

BTR-50PK with trim vane erected and roof hatches open. This model has two circular firing ports in troop compartment (Israeli Ministry of Defence)

in the centre and has a single piece hatch cover immediately in front of him which opens upwards on the outside and has a single vision device. A further three vision devices are mounted in the front of the hull below his hatch cover. There is a vision block mounted in the front of the hull to the right of the driver, over the top of which is a white light or infra-red searchlight which can be operated from inside the vehicle. A single infra-red driving light is normally mounted on the glacis plate.

The 20 infantrymen sit on bench seats which run across the full width of the vehicle and enter and leave by climbing over the side of the hull. Armament consists of a pintle-mounted 7.62 mm SGMB machine gun which has an elevation of +23.5°, a depression of -6° and a total traverse of 90°.

When the BTR-50P was originally introduced there were ramps at the rear of the hull to enable a 57 mm anti-tank gun M1943 (ZIS-2), 76 mm divisional gun M1943 (ZIS-3) or an 85 mm division gun D-44 to be carried and fired from the rear decking. The weapon could also be fired when the vehicle was afloat, but only when the water jets were in operation.

The engine used in the BTR-50P is one half of that fitted to the T-54 MBT. As the vehicle operates in a very cold climate an engine pre-heater is fitted as standard. The power train consists of the engine, master clutch, transmission, steering clutch, inner reduction gear, final drives, water jet propeller and mechanical water drainage pump drive. Power is transmitted through the transmission to the left and right inner reduction gears to the final drives which are connected directly to the track drive sprockets. The vehicle has three fuel tanks, two in the right forward side of the engine compartment and one at the rear. A centralised fire-extinguishing system is provided for the engine compartment: once the sensors have detected a fire, the bottle operates, the engine is switched off, forced blowers are switched on, and, if required, the second extinguisher also operates.

The torsion bar suspension consists of six rubber-tyred road wheels with the drive sprocket at the rear and the idler at the front. There are no track return rollers. The first and last road wheel stations have a hydraulic shock absorber and the steel tracks each have 96 links when new.

The BTR-50P is fully amphibious and propelled in the water by two water jets at the rear of the hull. The only preparation required before entering the water is to erect the trim vane at the front of the hull and switch on the two electric bilge pumps. There is a manual bilge pump for emergency use. Steering is accomplished by opening and closing the two doors over the rear water jets: to go left the left water jet is covered and to turn through 180° the left water jet sucks in water and the right water jet pushes it out. The basic BTR-50P has no NBC system.

During exercises held in Egypt in 1981/82, United States Army troops used BTR-50PK APCs in extensive swimming operations, but found that the vehicle had a number of drawbacks. The engine could only be run at full power while swimming for eight minutes, due to cooling limitations. The use of reverse thrust steering with the BTR-50PK's water jet propulsion system meant that although it had a high water speed it was uncontrollable in a current faster than around 8 km/h. The engine had to be reduced to idle when track drive was engaged or disengaged, making entering or leaving water difficult, especially in a current. In conclusion, the US Army felt that the BTR-50PK was only slightly superior in its amphibious capabilities to the M113A1.

Variants

BTR-50PA

This model is almost identical to the BTR-50P but does not have the loading ramps at the rear of the hull. It is sometimes armed with a 14.5 mm KPVT heavy machine gun mounted over the commander's position.

BTR-50PK

This model has full overhead armour protection and NBC system with the troops entering and leaving the vehicle by two rectangular roof hatches that open either side. In each side of the hull there are two firing ports and in the roof at the front of the hull a single piece hatch cover that opens forwards. Two ventilators are fitted, one at the front of the troop compartment on the right side and one at the rear of the troop compartment on the right side. The BTR-50PK is normally armed with a 7.62 mm SGMB machine gun in an unprotected mount.

BTR-50PU (Command)

There are at least two models of the BTR-50PU (Command), designated the models 1 and 2. The model 1 (early model, very rarely seen today) has one projecting bay whereas the model 2 has two and is similar in appearance to the Czechoslovakian OT-62A APC but the Soviet vehicles have a distinct chamfer between the top and the sides of their hulls. The right bay on the model 2 has three vision devices and no cover.

Behind the driver is a single piece hatch cover fitted with a periscope, to the rear of which are two oval roof hatches that are hinged in the centre and can be locked vertically.

The BTR-50PU has three compartments: engine, commander's and driver's. The last two are fitted with special equipment and a total of 10 seats, four for the commander and his staff, four for the radio operators and two for the vehicle commander and driver.

In addition, the staff compartment accommodates a collapsible table for working the maps and documents, a small table for the commander, two hammocks for resting and three ladders. An emergency escape hatch

BTR-50PK(B) amphibious armoured recovery vehicle with drums on rear engine decking

BTR-50PU Model 2 Command Vehicle with all hatches open. This has two projecting bays at front, Model 1 Command Vehicle has one

Soviet BTR-50PK re-powered by NIMDA being put through its paces with trim vane erected and roof hatches open

is provided in the floor of the vehicle and the interior of the vehicle has thermal insulation.

External equipment includes infra-red searchlight, four whip antennas, an 11 m telescopic mast, battery charger, armoured box for a fuel drum and three stowage boxes.

Extensive communications equipment is provided including radio, radio relay and wire equipment. This includes an R-112 radio, R-113 radio telephone, R-105U radio telephone, R-403BM two-channel radio relay telephone half set, 10-line field telephone switchboard, four telephone sets and four reels each with 600 m of two-wire cable.

A gyro course indicator and course plotter are the basis of the navigation system with the former indicating the vehicle's course and the latter plotting it in a rectangular co-ordinate system.

The BTR-50PU can lay its own smoke screen by injecting diesel fuel into the vehicle's exhaust system. This is referred to as the TDA thermal smoke apparatus and enables the vehicle to lay a smoke screen 300-400 m long in about two minutes. In addition to the BTR-50PU Command Vehicle, the following additional command post vehicles were declared under the Conventional Forces Europe Treaty signed in 1999: BTR-50PU R-82, BTR-50PU-11, BTR-50PUM and BTR-50PUR.

BTR-50PK(B) Amphibious Armoured Recovery Vehicle

This is a specialised version of the BTR-50PK developed for the recovery of other vehicles at water obstacles. It has a combat weight of 14 000 kg and normally has a crew of two: commander and driver, with seats for four auxiliary personnel, and during rescue operations the vehicle can accommodate up to eight rescued personnel.

The BTR-50PK(B) is fitted with R 123M and R 124 radio sets, a rear-mounted towing coupling, towing gear and hook and two extra towing cables, two special quick-release shackles, standard shackles and snap hooks, searchlight, two lifebelts, life jackets and four fenders. A set of tools and fire extinguishers are also carried.

MTP Technical Support Vehicle

This is based on a BTR-50PK APC and is used for recovery and repair of armoured personnel carriers and the BMP ICV. In addition it is used to deliver POL supplies to forward units which are difficult to reach with normal truck-mounted bowsers. A distinctive feature of the MTP is the raised workshop compartment which is high enough to allow the crew to work while standing as well as providing sleeping room for the crew of three. Details of the MTP are given in *Jane's Military Vehicles and Logistics 1991-92*, page 48.

MTK Mine clearing Vehicle

This is a BTR-50PK APC with a special launcher mounted on top of the hull to the rear of the troop compartment. It fires rockets to which are attached flexible tubes containing high explosives which fall to the ground on to the minefield and are then detonated from the vehicle. The Soviet designation for the rocket system is UR-67.

MTP technical support vehicle with its higher roof

MTR-1

Repair version of BTR-50.

NIMDA Upgraded BTR-50PK

In 1989 NIMDA of Israel announced that it had completed trials of a modernised BTR-50PK armoured personnel carrier.

This is powered by a Detroit Diesel Model 6V-92T diesel coupled to the original transmission and a new clutch assembly. The kit also includes a new 24 V 200 Ah alternator, new cooling and electrical system and fuel, exhaust, air inlet, hull and top deck modifications.

As an option to the original transmission could be replaced by a new fully automatic transmission. The new engine develops 300 hp at 2100 rpm and, with a combat weight of 14.50 tonnes, gives a power-to-weight ratio of 20.68 hp/tonne.

Additional improvements include a bow-mounted machine gun to provide suppressive fire over the frontal arc, two roof-mounted 7.62 mm machine guns and the option of passive night vision equipment.

Status: Production complete. In service with Afghanistan, Albania, Algeria, Angola (and OT-62), Bulgaria (and OT-62), Congo, Cyprus, Czechoslovakia (OT-62 only), Egypt (and OT-62), Finland, Guinea, India (and OT-62), Iran, Iraq (and OT-62), Israel (and OT-62), North Korea, Libya (and OT-62), Morocco (OT-62 only, and status uncertain), Nicaragua, Poland (OT-62 only), Romania, Somalia, Sudan (and OT-62), Syria, USSR, Vietnam and Yugoslavia.
Manufacturer: Soviet state factories.

Note: The Chinese build a similar version of the BTR-50PK called the Type 77, details are given under China.

SPECIFICATIONS (BTR-50PK)

CREW	2 + 20
COMBAT WEIGHT	14 200 kg
POWER-TO-WEIGHT RATIO	16.9 hp/tonne
GROUND PRESSURE	0.51 kg/cm²
LENGTH	7.08 m
WIDTH	3.14 m
HEIGHT (to hull top)	1.97 m
GROUND CLEARANCE	0.37 m
TRACK	2.74 m
TRACK WIDTH	360 mm
LENGTH OF TRACK ON GROUND	4.08 m
MAX SPEED	
road	44 km/h
water	11 km/h
FUEL CAPACITY	400 l
MAX RANGE	400 km
FUEL CONSUMPTION	1 l/km
FORDING	amphibious
GRADIENT	70%
SIDE SLOPE	40%
VERTICAL OBSTACLE	1.1 m
TRENCH	2.8 m
ENGINE	model V-6 6-cylinder in-line water-cooled diesel developing 240 hp at 1800 rpm
TRANSMISSION	manual with 5 forward and 1 reverse gears
STEERING	clutch and brake
SUSPENSION	torsion bar
ELECTRICAL SYSTEM	24 V
ARMAMENT	1 × 7.62 mm MG
AMMUNITION	1250
GUN CONTROL EQUIPMENT	
Gun traverse	manual
Gun elevation/depression	+23.5°/-6°
Gun traverse (total)	90°
ARMOUR	
Hull front upper	8 mm at 83°
Hull front lower	10 mm at 53°
Hull sides upper	10 mm at 0°
Hull sides lower	9 mm at 0°
Hull rear upper	7 mm at 0°
Hull rear lower	6 mm at 42°
Hull top	7 mm at 86°
Hull belly front	6 mm
Hull belly rear	6 mm

MT-LB Multi-purpose Tracked Vehicle

Development
The MT-LB was developed in the late 1960s as the replacement for the older AT-P armoured tracked artillery tractor and was initially known as the M-1970 multi-purpose tracked vehicle until its correct Soviet designation became known. It is based on the MT-L unarmoured tracked amphibious carrier described in *Jane's Military Vehicles and Logistics 1991-92*, page 480.

Typical roles of the MT-LB include use as an artillery prime mover towing 100 mm anti-tank guns and 122 mm howitzers, command and radio vehicle (which usually has at least two whip antennas on the roof), artillery fire control vehicle and cargo carrier. It is also widely used as an armoured personnel carrier, especially in difficult terrain such as snow or swamp.

Description
The hull of the MT-LB is all-welded steel with the crew compartment at the front, engine immediately behind the crew compartment on the left side and the troop compartment at the rear of the hull.

The driver sits at the front of the hull on the left side and has a single piece hatch cover, in front of which are three periscopes. The commander sits to the right of the driver in the centre and has a single piece rear-opening hatch cover and two periscopes. In action, the commander also mans the turret. The machine gun turret is mounted to the right of the commander's position and is armed with a 7.62 mm PKT machine gun. Like the turrets fitted to the BRDM-2 (4×4) and BTR-60PB (8×8) vehicle, it does not have a hatch cover.

Both the driver and machine gunner have a windscreen in front of their positions which when in action is covered by a flap hinged at the top. There is a vision block in each side of the hull, to the left of the driver's and the right of the machine gunner's position.

An aisle provides access from the crew compartment at the front of the vehicle to the personnel compartment at the rear which has inward-facing folding canvas seats for the 10 infantrymen. Two hatches over the top of the troop compartment open forwards. The infantry enter and leave the vehicle by two doors in the rear of the hull, both of which are provided with a firing port. There is an additional firing port and vision block in each side of the troop compartment. An unditching beam is often carried on the roof or side of the vehicle.

The torsion bar suspension consists of six road wheels with the drive sprocket at the front and the idler at the rear with hydraulic shock absorbers on swing arms on the first and last road wheel stations. There are no track return rollers as the top of the track rests on the top of the road wheels. The first and last road wheel stations either side have a hydraulic shock absorber.

MT-LB engineer vehicle from rear with plough blade raised for travelling (Donald C Spaulding)

Yugoslav Army MT-LB with SNAR-10 artillery/mortar-locating radar mounted at rear. When travelling, antenna is folded forward on top of vehicle

The MT-LB is fully amphibious being propelled in the water by its tracks. Before entering the water a trim vane is erected at the front of the vehicle and the bilge pumps, which have a capacity of 450 litres a minute, are switched on. Standard equipment on all vehicles includes an NBC system. The MT-LB has air-actuated brakes which can be connected to a trailer. Night vision equipment includes an OU-3GK white/infra-red searchlight with a range of 400 m for the commander and a TVN-2 infra-red periscope for the driver with a range of 40 m.

It can also tow a trailer or weapon weighing up to 6500 kg or carry up to 2000 kg of cargo or stores.

Variants

Soviet MT-LB Declared Variants
Under the Conventional Forces Europe (CFE) Treaty signed in 1990, the Soviets declared the following variants of the MT-LB multi-purpose tracked armoured vehicle, and additional details of some of these are given later in this entry:

Hungarian MT-LB Engineer Vehicle, said to be based on Soviet MT-LBU chassis but has six road wheels

MT-LBO, Bulgarian designation for MT-LB vehicle

1V13, Soviet battery fire direction centre vehicle, called 1W13 by Poland, also used by Czechoslovakia and Hungary

1V14, Soviet battery command post vehicle, called 1W14 by Poland, also used by Czechoslovakia and Hungary

1V15, Soviet battalion command vehicle, called 1W15 by Poland, also used by Czechoslovakia and Poland

1V16, Soviet battalion fire direction centre vehicle, called 1W16 by Poland, also used by Czechoslovakia and Hungary

Czechoslovakian MT-LB with SNAR-10 artillery/mortar-locating radar in travelling position

MTP-LB repair vehicle in travelling configuration

MT-LBV: basic MT-LB multi-purpose tracked vehicle fitted with 565 mm wide tracks for use in snow and swamp covered areas

MT-LB multi-purpose armoured vehicle towing 122 mm D-30 howitzer

MT-LB used by the US Army for training purposes with commander's/ gunner's and driver's armoured visors in open position. The turret is not fitted with its 7.62 mm machine gun (Pierre Touzin)

MT-LB fitted with 2B9 Vasilek 82 mm automatic mortar on rear decking

MT-LB used at US Army National Training Center at Fort Irwin, California (US Army/Michael Green)

1V21, Soviet air defence management vehicle, called MP-21 by Czechoslovakia, also used by Poland
1V22, Soviet air defence management vehicle, called MP-22 by Czechoslovakia, also used by Poland
1V23, Soviet air defence management vehicle, called MP-23 by Czechoslovakia, also used by Poland
1V24, Soviet air defence management vehicle, called MP-24 by Czechoslovakia, also used by Poland
1V25, Soviet air defence management vehicle, called MP-25 by Czechoslovakia, also used by Poland
MT-LB R-330P, Soviet MT-LB with R-330 VHF jammer, also used by Hungary
MT-LB 9S743, Bulgarian MT-LB with radio system and generator at hull rear
MT-LB KShM R-80, Bulgarian MT-LB with table and increased headroom
MT-LB KShM R-81, Bulgarian MT-LB with radio equipment
MT-LB TRI, Polish engineer reconnaissance vehicle
MT-LB WPT, Polish recovery and maintenance vehicle
MT-LB R-137, Polish MT-LB with R-137 radio
MT-LB R-137T, Polish MT-LB with R-137 radio
MT-LB Sova, Bulgarian MT-LB with dismountable radar system
Beta EM, Polish communications vehicle.

MT-LBV

The MT-LB can also be fitted with 565 mm wide tracks for operation in snow and swampy ground; it is then called the MT-LBV. This version has a ground pressure of 0.28 kg/cm^2.

MT-LB Artillery Tractor

The MT-LBs used as artillery prime movers have been observed with a fully enclosed box mounted over the troop compartment roof containing the gun section equipment.

MT-LBU (Command)

This is the command version of the MT-LB and has additional radios, generator, land navigation system and a canvas cover that can be extended to the rear when the vehicle is being used in the static role.

MT-LB with Big Fred Radar (SNAR-10)

In 1981 it was revealed that the Yugoslav Army had acquired a number of MT-LBs from the Soviet Union fitted with a new artillery/mortar-locating radar which has been allocated the NATO reporting name of Big Fred. This version has the Soviet designation of SNAR-10.

When travelling the antenna folds forward onto the top of the large turret which is to the rear of the vehicle. The forward turret-mounted 7.62 mm machine gun is retained. The radar is believed to be of a similar type to the British THORN EMI Cymbeline in that the radar measures the slant range and bearing of two points in the mortar bomb/artillery shell trajectory. The time taken for the bomb/projectile to travel between the two points is also measured and the onboard computer uses this information together with the pre-set elevation angles to determine the position of the enemy mortar or artillery piece. This information is then relayed to the field artillery units and the target is engaged. The radar has a range of about 20 km. Specifications of the MT-LB with Big Fred are similar to those of the basic MT-LB except for a weight of 11 500 kg, height with antenna down of 2.9 m and a crew of four to six. This vehicle has the STANAG designation of the MT-LB M1975.

MTP-LB Repair Vehicle

The MTP-LB is designed for field maintenance, repair, and recovery of tanks and other AFVs and is recognisable by its lack of a machine gun turret.

Mounted at the front of the vehicle is an A-frame which can lift a maximum load of 1500 kg. Standard equipment includes tools, gas welding and cutting equipment, cable winch with 85 m of cable and a capacity of 6700 kg, jacking device, towing attachment hooks and a crane. Full details of the MTP-LB are given in *Jane's Military Vehicles and Logistics 1991-92*, page 48.

MT-LB (Ambulance)

This is an MT-LB used as an armoured evacuation vehicle (armoured ambulance) with stretchers fitted in the rear compartment.

MT-LB Engineer Vehicle

This is similar in appearance to the basic MT-LB but modified to mount a plough blade on the roof. Hydraulic devices at the rear of the vehicle allow manual mounting of the plough blade to the rear only. The United States has a number of these vehicles at Fort Irwin, California. Germany also has its own engineer vehicle. This is the basic MT-LB with a rectangular box on the rear deck for carrying engineer equipment. The vehicle retains its cupola-mounted 7.62 mm machine gun.

RKhM Chemical Reconnaissance Vehicle

This was originally believed to be based on the chassis of the MT-LB multi-purpose vehicle but is now known to be based on the chassis of the 122 mm 2S1 self-propelled howitzer. Details of this version are given in the *Self-propelled guns and howitzers* section.

MT-LB with Vasilek

For use in Afghanistan, a self-propelled version of the towed 2B9 Vasilek 82 mm automatic mortar described and illustrated in the *Towed guns*

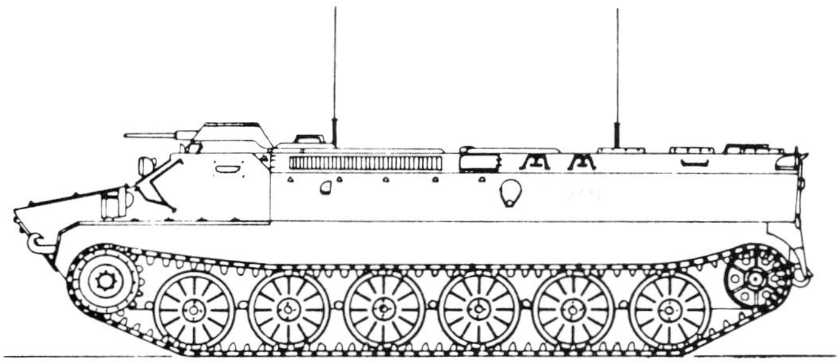

MT-LB multi-purpose tracked vehicle

SA-13 Gopher SAM system which uses the MT-LB chassis

MT-LB chassis with Polish WAT turret armed with 14.5 mm and 7.62 mm MGs.

and howitzers section was developed. The mortar has had its wheels removed and has been propped up on the upper rear deck on steel ammunition boxes.

MT-LB with AT-6 ATGW System

Details of this are given in the *Tank destroyers* section.

MT-LB 120 mm mortar

The Bulgarian Army has mounted a 120 mm mortar in the rear of the MT-LB multi-purpose armoured vehicle. Early reports that the new Soviet 2S12 Sani (Sled) self-propelled mortar system used a modified MT-LB chassis are now incorrect. The 2S12 in fact uses the GAZ-66 (4 × 4) truck with the 120 mm two wheeled mortar being carried portee fashion.

Bulgarian 120 mm mortar carrier based on MT-LB chassis

Iraqi 120 mm self-propelled mortar based on MT-LB

Other versions of MT-LB

In addition to those versions of the MT-LB previously mentioned, US Army sources list the following variants, some of which may well refer to the vehicles already mentioned: MT-LB M1979/1, MT-LB M1979/2, ATV M1979/3, ATV M1979/4 and the MT-LB M1981.

MT-LB with WAT Turret

Poland has fitted an MT-LB with the WAT turret, as installed on the OT-64C(2) (SKOT-2AP) and OT-62C APCs, which is armed with a 14.5 mm and a 7.62 mm machine gun.

SA-13 Gopher SAM system

The MT-LB multi-purpose armoured vehicle chassis is also used as the basis for the SA-13 Gopher SAM system. Mounted in the centre of the hull roof is a turret with an arm, on which is carried a total of four SA-13 missiles in the transport/launch containers, with the operator being seated below and between the missiles. The SA-13 entered service in 1977 and is replacing the SA-9 on a one for one basis. It retains the amphibious capability of the MT-LB and has a range-only radar.

Full details of the SA-13 are given in *Jane's Land-Based Air Defence 1991-92*, page 136.

Iraqi 120 mm self-propelled mortar

Early in 1989 an MT-LB was shown in Iraq for the first time, with wide tracks modified locally to carry a 120 mm mortar in the rear of the vehicle. The roof hatches have been modified and now open to either side of the hull to allow the 120 mm mortar to fire to the rear. A base plate is carried on the right side of the hull to allow the mortar to be deployed away from the vehicle if required by the tactical situation.

Iraqi 4 × 120 mm self-propelled mortar

Details of this system, shown for the first time in 1989 in Baghdad, are given in the *Self-propelled guns and howitzers* section under Iraq.

SPECIFICATIONS

CREW	2 + 11
COMBAT WEIGHT	11 900 kg
UNLOADED WEIGHT	9700 kg
MAX PAYLOAD	2000 kg
MAX TOWED LOAD	6500 kg
POWER-TO-WEIGHT RATIO	20.16 hp/tonne
GROUND PRESSURE	0.46 kg/cm²*
LENGTH	6.454 m
WIDTH	2.86 m
HEIGHT (to turret top)	1.865 m
GROUND CLEARANCE	0.4 m
TRACK WIDTH	350 mm*
LENGTH OF TRACK ON GROUND	3.7 m

MAX SPEED	
road	61.5 km/h
water	4.5 km/h
FUEL CAPACITY	450 l
MAX RANGE	500 km
FUEL CONSUMPTION	0.9-1.2 l/km
FORDING	amphibious
GRADIENT	60%
SIDE SLOPE	30%
VERTICAL OBSTACLE	0.61 m
TRENCH	2.41 m
ENGINE	YaMZ 238 V, V-8 cylinder diesel developing 240 hp at 2100 rpm
TRANSMISSION	manual with 6 forward and 1 reverse gears

STEERING	clutch and brake
SUSPENSION	torsion bar
ELECTRICAL SYSTEM	24 V
ARMAMENT	1 × 7.62 mm MG
AMMUNITION	2500
GUN CONTROL EQUIPMENT	
Turret traverse	manual, 360°
Gun elevation/depression	+30°/-5°
ARMOUR	3.19-9.87 mm (estimated)

*The MT-LB can also be fitted with 565 mm wide tracks which reduce ground pressure to 0.27 or 0.28 km/cm²

Status: In production. In service with Bulgaria, Czechoslovakia, Finland, Germany, Hungary, Iraq, Poland, USSR and Yugoslavia.

Manufacturer: Polish and Soviet state factories. Some sources have stated that the MT-LB family of tracked vehicles is produced at the Kirov Plant in Leningrad.

BTR-80 Armoured Personnel Carrier

Development

While the BTR-70 did have a number of improvements over the BTR-60 introduced in 1960-61, it still suffered from a number of significant disadvantages including its poor means of entry and exit for troops and the fact that it was powered by two petrol engines which were not fuel efficient and could catch fire.

In 1984 the Soviet Army took delivery of its first production BTR-80 (8 × 8) armoured personnel carrier which has many improvements over the earlier vehicle.

From 1988, the BTR-80 APC started to replace the older BTR-60PB APCs used by the Soviet Naval Infantry.

Under the Conventional Forces Europe (CFE) Treaty signed in 1990, the Soviet Union declared the following quantities of BTR-80 APCs in the west:

Region	**IV.1**	**IV.2**	**IV.3**	**IV.4**	**V.1**
Quantity	1130	944	696	330	186

Description

Like the BTR-70 and BTR-60, the BTR-80 has a hull of all-welded steel armour construction. Its layout is similar to that of the earlier vehicles except that the two petrol engines have been replaced by a single 260 hp V-8 diesel which gives a slightly increased road speed, and is more fuel efficient giving a greater operating range as well as reducing the risk of fire. The engine compartment has an automatic fire extinguishing system, a preheater, fuel primer, heat exchangers for the radiator and crank case, diesel fuel tanks and a bilge pump.

The one-man manually operated turret is similar to that fitted to BTR-70 and BTR-60PB but the 14.5 mm KPV heavy machine gun can be elevated to +60° compared to the +30° of earlier vehicles, increasing its effectiveness against slow flying aircraft and helicopters. The standard 7.62 mm coaxial machine gun is retained. The turret also has an improved sighting system and is being retrofitted to older vehicles such as the BRDM-2 (4 × 4) amphibious scout car and the BTR-70 APC.

Mounted on the rear of the turret is a bank of six 81 mm Model 902V forward-firing electrically operated smoke dischargers. This is operated from within the turret and each smoke grenade can create a smoke screen up to 30 m wide and 10 m high depending on the strength and direction of the wind.

A new hatch arrangement has been provided between the second and third axles, the upper part opens to the front while the lower part folds down to form a step. This allows the infantry to dismount from the BTR-80 much more quickly and with less exposure to enemy fire than with

BTR-80 (8 × 8) APC with commander's and driver's hatches open

BTR-80 (8 × 8) APC with trim vane in stowed position

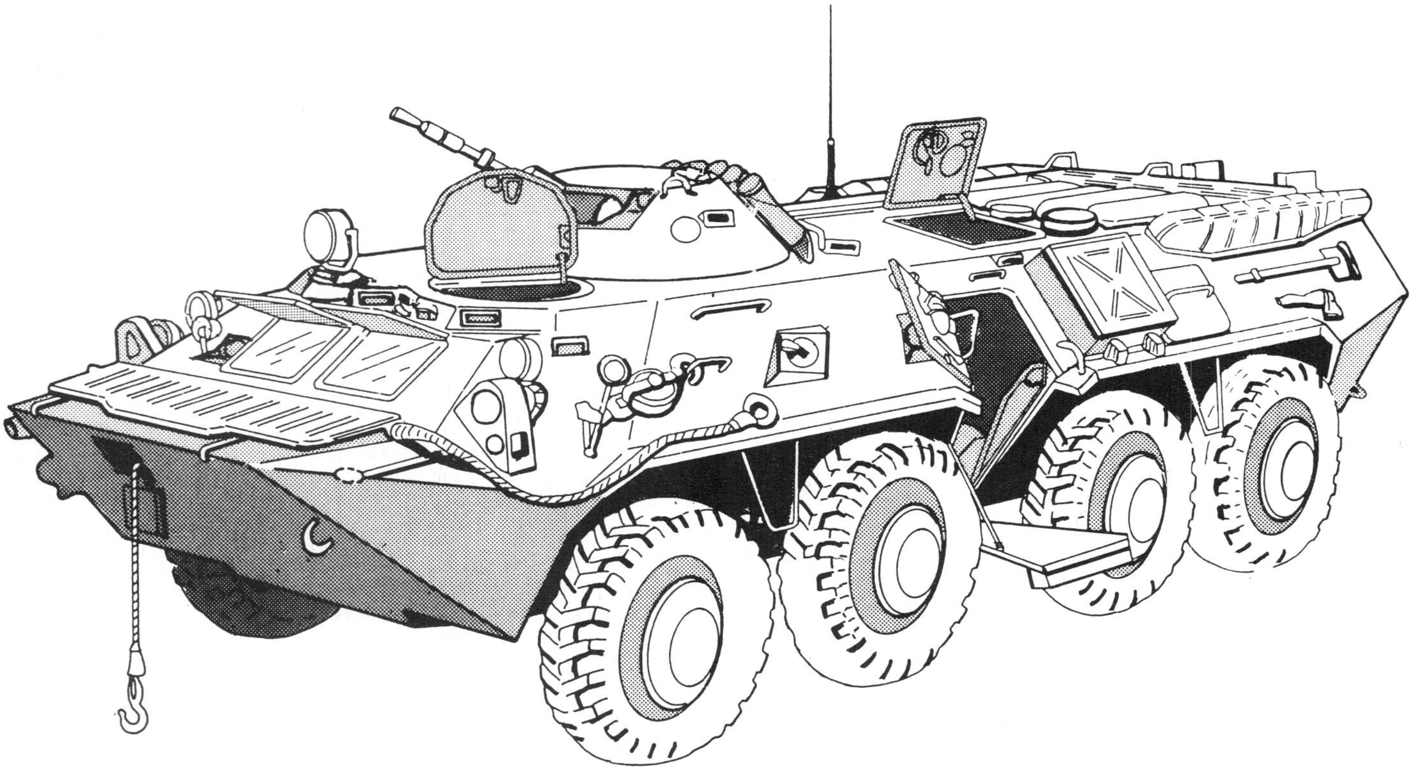

BTR-80 (8 × 8) APC with new hatch arrangement between second and third axles open and showing bank of smoke dischargers on turret rear (not to 1/76th scale)

BTR-80 (8 × 8) APC of the Soviet Army showing commander's and driver's hatches open and trim vane folded back onto glacis plate

earlier vehicles. The BTR-80 has a crew of three, commander, gunner and driver, and carries seven fully equipped infantry.

The BTR-70 has three firing ports in either side of the troop compartment whereas the BTR-80 has three firing ports in the side of the hull angled to fire obliquely forward so helping to cover the previously dead ground towards the front of the vehicle. In addition there is also a single firing port to the right of the commander's bow position which also appears on the BTR-70 and an additional firing port in each of the two roof hatches.

The two forward firing ports are designed for the 7.62 mm PK general purpose machine guns while the three firing ports either side allow the AKMS/AK-74 weapons to be fired from under cover. Small arms carried by the crew are two 7.62 mm PK general purpose machine guns (with 3000 rounds), eight 7.62 mm AKMS or 5.56 mm AK-74 assault rifles and nine type F1 hand grenades.

The commander and driver each have a total of four vision devices while the gunner has three. The BTR-80 is fully amphibious and has a front-mounted winch, overpressure NBC system, night vision equipment and a central tyre-pressure regulation system. Before entering the water, the trim vane is erected at the front of the vehicle, drain valves closed and locked, tyre pressure lowered and cover over the waterjet system opened. If the waves are over 0.5 m, air intake extension pipes are provided and stored low in the rear of the engine compartment. Low gear is then selected, transfer case locked in all-wheel drive, dust cover over the air intake filter is closed, also shut and closed are the louvres over the air intake and exhaust for the radiators.

The BTR-80 drives into the water and once afloat wheel drive is disengaged and water jet drive engaged. Steering is accomplished by means of the steering wheel which controls the rudder, which then steers the vehicle.

Variants

BTR-80 M1989/1

A command version of the BTR-80 is in service under the designation of the BTR-80 M1989/1 (Command Variant). This has additional communications equipment and antennas.

120 mm 2S23 Self-propelled Gun

Available details of this are given in the *Self-propelled guns and howitzers* section.

RKhM-4

Chemical and reconnaissance vehicle.

BTR-80 (8 × 8) APC from rear with water jet outlet closed

BTR-80 (8 × 8) APC with side hatch open

SPECIFICATIONS					
CREW	3 + 7	FUEL CAPACITY	290 l	TURNING RADIUS	7 m
CONFIGURATION	8 × 8	MAX ROAD RANGE	600 km	ARMAMENT	
COMBAT WEIGHT	13 600 kg	CROSS-COUNTRY RANGE	223 to 480 km	main	1 × 14.5 mm KPVT MG
POWER-TO-WEIGHT RATIO	19.11 hp/tonne	FORDING	amphibious	coaxial	1 × 7.62 mm PKVT MG
LENGTH	7.65 m	GRADIENT	60%	AMMUNITION	
WIDTH	2.90 m	SIDE SLOPE	30%	main	500
HEIGHT	2.35 m	VERTICAL OBSTACLE	0.5 m	coaxial	2000
GROUND CLEARANCE	0.475 m	TRENCH	2 m	GUN CONTROL EQUIPMENT	
MAX ROAD SPEED	80 km/h	ENGINE	four stroke V-8 water cooled diesel developing 260 hp	Turret traverse	manual, 360°
MAX WATER SPEED	9 km/h	STEERING	power-assisted	Gun elevation	+60°/-4°

Status: In production. In service with Afghanistan, Hungary, Soviet Union (Army and Navy Infantry).

Manufacturer: Soviet state factories, probably same as BTR-70.

BTR-70 Armoured Personnel Carrier

Development

The BTR-70 (8 × 8) amphibious armoured personnel carrier was shown in public for the first time during the November 1980 parade in Moscow.

Under the Conventional Forces Europe (CFE) Treaty signed late in 1990, the Soviet Union declared the following quantities of BTR-70 (8 × 8) series of APCs in the west:

Region	**IV.1**	**IV.2**	**IV.3**	**IV.4**	**V.1**
Quantity	3936	2690	2356	784	1246

Note: Regions are as defined under the CFE agreement: Region IV is defined as Central Europe while Region V covers the flanks of the Atlantic to the Urals (ATTU) zone.

For a short period, the US Army referred to the BTR-70 as the M1978 as this was the first time that it was observed, the former East German Army referred to the vehicle as the SPW-70.

It is believed to have been designed by the Dedkov design bureau at Gor'ky which also developed the BRDM-1, BRDM-2 and BTR-60 wheeled armoured vehicles.

According to the 1984 edition of *Soviet Military Power*, the earlier BTR-60P series remained in production for export while production of the BTR-70 was completed in 1982.

Description

The hull of the BTR-70 is of all-welded steel construction with improved protection over its frontal arc compared to the original BTR-60 series, the nose is also wider and the front of the vehicle provides added protection to the front wheels. The bow section of the BTR-70 may incorporate special layered armour for increased protection.

The commander sits at the front of the hull on the right with the driver to his left. The commander has a single piece hatch cover that opens to the front, the driver has a single piece hatch cover that opens at a 45° angle towards the centre of the vehicle. Both the commander and driver have a window to their front which is covered in combat by an armoured hatch cover hinged at the top. Each also has three forward and one side-facing periscopes, with a single firing port under the commander's last periscope only. To the rear of the commander and driver are two infantry facing the front.

Over the second axle is the turret, armed with a 14.5 mm and a 7.62 mm machine gun. This turret is identical to that fitted to the BTR-60PB (8 × 8) armoured personnel carrier and the BRDM-2 (4 × 4) armoured car. Some BTR-70 (8 × 8) APCs have been observed fitted with the complete turret of the BTR-80 (8 × 8) APC described in the previous entry.

The troop compartment is to the rear of the turret with three firing ports and one vision block in each side of the hull. Over the top of the troop compartment are two square roof hatches each with a circular

Early BTR-70 (8 × 8) APC from above showing arrangement of roof hatches

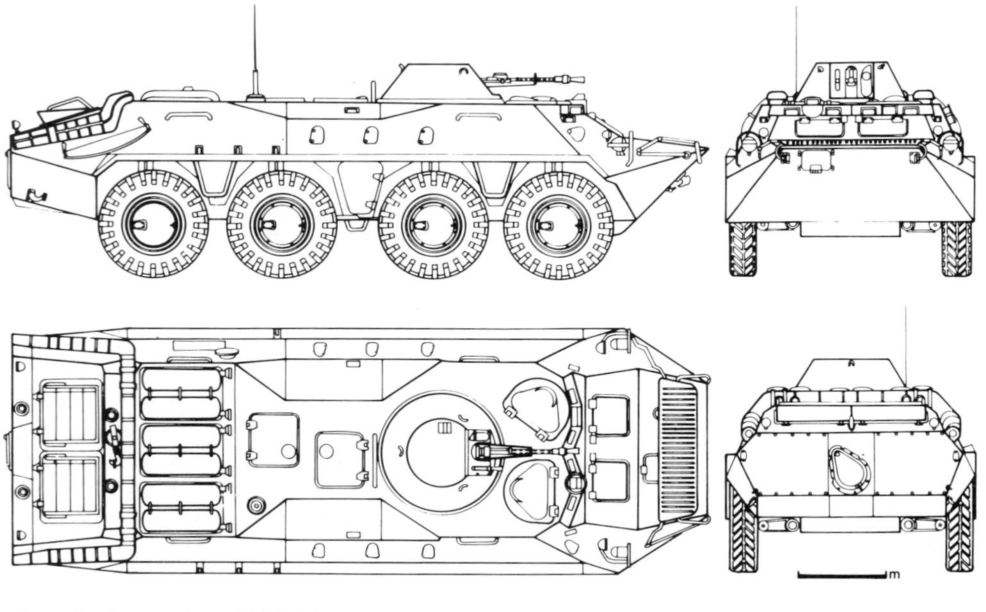

Standard production version of BTR-70 (8 × 8) armoured personnel carrier (not to 1/76th scale) (Steven Zaloga)

BTR-70 M1986/1 (8 × 8) APC of the Soviet Army with commander's and driver's roof hatches open and trim vane folded back onto the glacis. This version has the additional firing ports in the very top of each side of the troop compartment

firing port, opening to the front and rear respectively. Between the second and third axles is a small door that opens to the front. The six infantrymen sit on a bench seat down the centre, back to back, so enabling them to use the firing ports.

The engine compartment is at the rear of the hull and is fitted with a fire-extinguishing system. The air-inlets are in the top of the engine compartment and the air outlets at the rear. The exhaust pipes run from the top of the engine compartment down either side towards the rear. The original two GAZ-49B six-cylinder in-line water-cooled petrol engines have been replaced by two ZMZ-4905 petrol engines developing 120 hp each compared to the 90 hp each of the BTR-60.

Each engine has its own transmission with the right engine providing power for the first and third axles, the left for the second and fourth axles. If one engine is damaged or disabled then the vehicle can still travel on the other, although at a reduced speed.

Steering is power-assisted on the front two axles, early models of the BTR-70 had the same wheels and tyres as the BTR-60 series. Standard equipment includes an NBC system, central tyre-pressure regulation system and a front-mounted winch which is provided with 50 m of cable and has a maximum pull capacity of 6000 kg. It also has a fire detection/suppression system, DP-3B roentgenometer, a service device for chemical reconnaissance (VPKhR), special tool sets, towing mountings and lifejackets.

The wheels have split rims with pneumatic directional tread tyres fitted as standard. The driver can adjust the tyre pressure of all eight wheels from 2.8 to 0.5 kg/cm^2 without leaving his seat. If a tyre is damaged the on-board compressor compensates for the leakage of air.

The BTR-70 is fully amphibious and propelled in the water by a single water jet at the rear of the hull. On the BTR-60 this had a two-part cover but on the BTR-70 a one-part cover is fitted, hinged at the top left. Before entering the water the trim vane is erected at the front of the hull. When travelling this is stowed on top of the hull front whereas on the BTR-60 series it is under the nose of the vehicle.

Internal stowage space is provided for a light anti-tank launcher such as the RPG-7 and two AGS-17 automatic grenade launchers plus nine hand grenades.

Variants

BTR-70 with BTR-80 turret

A number of BTR-70 (8 × 8) APCs have been observed fitted with the turret of the more recent BTR-80. Details of this turret are given in the entry for the BTR-80. This version has the NATO/US designation of the BTR-70 M1986/1. In addition to being fitted with the complete BTR-80 (8 × 8) turret, it has a modified wave deflector, rear bumpers, headlight brushguards, additional armour brackets on the vehicle sides and two additional firing ports in the very top of each side of the troop compartment between the hull sides and the roof.

BTR-70Kh

Chemical reconnaissance vehicle.

SPR-2

Possible radar jamming variant.

BTR-70 with AGS-17

A number of BTR-70s were seen in Afghanistan fitted with the 30 mm AGS-17 automatic grenade launcher on the roof to the rear of the driver and commander. These vehicles retain the machine gun turret.

BTR-70MS

This is a turretless communications vehicle with the MS standing for Mashina Svyazi (signals vehicle).

BTR-70KShM

This is a command/staff vehicle with the KShM standing for Komandno-Shtabnaya Maschina.

BREM repair and recovery vehicle

This is a turretless BTR-70 fitted with a bow-mounted jib crane and other equipment, BREM stands for Bronirovannaya Remontno-Evakuatsionannaya Mashina — armoured repair and recovery vehicle.

SPECIFICATIONS		FUEL CAPACITY	350 l	TURNING RADIUS	6.3 m
CREW	2 + 9		(estimated)	SUSPENSION	torsion bar with
CONFIGURATION	8 × 8	MAX RANGE	600 km (with external		hydraulic shock
COMBAT WEIGHT	11 500 kg		fuel tanks)		absorbers
POWER-TO-WEIGHT		FORDING	amphibious	ELECTRICAL SYSTEM	12 V
RATIO	20.86 hp/tonne	GRADIENT	60%	ARMAMENT	
LENGTH	7.535 m	SIDE SLOPE	40%	main	1 × 14.5 mm MG
WIDTH	2.8 m	VERTICAL OBSTACLE	0.5 m	coaxial	1 × 7.62 mm MG
HEIGHT		TRENCH	2 m	AMMUNITION	
to top of sight	2.32 m	ENGINES	2 × ZMZ-4905 8-	main	500
to top of turret	2.235 m		cylinder petrol	coaxial	2000
GROUND CLEARANCE	0.475 m		developing 120 hp	GUN CONTROL	
TRACK	2.38 m		each	EQUIPMENT	
WHEELBASE	4.4 m	TRANSMISSIONS	manual with 4 forward	Turret power control	manual
MAX SPEED			and 1 reverse gears	Gun elevation/depression	+30°/-5°
road	80 km/h	TRANSFER CASE	2-speed	Turret traverse	360°
water	10 km/h	STEERING	power-assisted		

Status: Production complete. Known to be in service with Afghanistan, Germany and USSR (Army and Naval Infantry). Licenced production in Romania as the TAB-77 (qv).

Manufacturer: Gor'ky Automobile Plant, Gor'ky, USSR.

BTR-60P Armoured Personnel Carrier

Development

In the late 1950s a requirement was issued for a new wheeled APC to replace the non-amphibious BTR-152 (6 × 6) APC. Development of an 8 × 8 APC began in two competing design bureaus in 1957-58, this competition was called Project 49. The Dedkov OKB team won with its vehicle, the Type 62B. This was accepted for service with the Soviet Army in 1960 under the designation BTR-60P. The vehicle was seen in public for the first time during the 1961 November parade in Moscow. The BTR-60P and its variants are normally used by the Motorised Rifle Divisions, whereas the Tank Divisions have the tracked BMP-1/ BMP-2 ICV. In recent years a number of Motorised Rifle Divisions have had their BTR-60Ps replaced by the better armed BMP-1. In addition, it is used by the Soviet Naval Infantry and some BTR-60PBs are replacing BRDM-2s in some divisional reconnaissance battalions. The original BTR-60P and BTR-60PA have been virtually relegated to reserve, second-line or training use.

The BTR-60P APC is not used by Czechoslovakian or Polish motor rifle units. Regular units use the Czechoslovakian OT-64 (8 × 8) amphibious APC which has a number of advantages over the Soviet vehicle.

A small number of BTR-60PB (8 × 8) armoured personnel carriers were encountered by the US forces during the invasion of Grenada in 1983.

Under the Conventional Forces Europe (CFE) Treaty signed late in 1990, the Soviet Union declared the following quantities of BTR-60 (8 × 8) series of APCs in the west:

Region	IV.1	IV.2	IV.3	IV.4	V.1
BTR-60P	66	64	64	nil	2
BTR-60PA	9	9	4	4	nil
BTR-60PB	3229	2827	2710	1881	402
BTR-60PBK	887	830	830	742	57

Note: Regions are as defined under the CGE agreement: Region IV is defined as Central Europe while Region V covers the flanks of the Atlantic to the Urals (ATTU) zone.

BTR-60PB (8 × 8) APC with commander's and driver's forward opening armoured windscreen hatches locked in open position and turret traversed to rear (US Air Force)

Description

The hull of the BTR-60P is all-welded steel with the driver and commander seated at the front of the hull, open-topped personnel compartment behind them and the engine compartment at the very rear of the hull. The interior of the BTR-60P is open from the driver's and commander's position right through to the engine compartment at the rear.

The driver sits on the left with the vehicle commander to his right. Both have a windscreen to their front which, when in action, is covered by a flap hinged at the top. This has an integral vision device which can be replaced by an infra-red periscope for night vision. To the left of the driver and the right of the commander is a vision block. Mounted above the flap in front of the commander is an infra-red searchlight which can be operated from inside the vehicle.

The infantrymen are seated on bench seats that run across the full width of the hull. In each side of the hull are two half doors and three firing ports. If required bows and a tarpaulin cover can be fitted over the top of the troop compartment.

Main armament comprises a 7.62 mm SGMB or PKB machine gun which is pintle-mounted on the forward part of the hull. Some vehicles have been seen with the heavier 12.7 mm DShKM machine gun at this position. Mounts are also provided for 7.62 mm machine guns on either side of the hull.

The two engines are mounted at the rear of the hull: the first and third axles are powered through the transmission of the right engine and the second and fourth axles through the transmission of the left engine. Torque is transmitted from each engine to the heels through a single-plate clutch with hydraulic control, a four-speed gear box with synchromesh in third and fourth gears, a two-speed transfer case, the final drives of the two driving axles with gearless differentials and four wheel reducers. Torque from the final drives to the wheel reducers is transmitted through half-axles and gimbal drive gears.

All eight wheels are powered and the first four, which are used for steering, are power assisted. The vehicle can be driven with one wheel missing from the second axle. The suspension is of the torsion bar type with the first and second road wheels on each side being provided with two hydraulic shock absorbers and the third and fourth road wheels with a single hydraulic shock absorber. A central tyre-pressure regulation system fitted as standard on all BTR-60 series APCs enables the driver to adjust the tyre pressure to suit the ground being crossed.

The BTR-60P is fully amphibious being propelled in the water by a single water jet mounted at the rear of the hull. When not in use the exit at the rear of the hull is covered by a two-part circular plate. Before entering the water a trim vane is erected at the front of the hull, and the bilge pump is switched on. When not in use the trim vane is stowed flat under the nose of the vehicle. Power is transmitted to the water jet via a PTO and water jet reducer. Direction is changed when afloat by using a hydrodynamic rudder in the hull of the water jet propeller and by turning the front four road wheels. To reverse direction, the shutters of the water jets are closed and water is ejected through side ducts in the rear part of the vehicle that face forwards.

The basic BTR-60P has no NBC system as it has an open top. All vehicles are fitted with infra-red night vision equipment and a front-mounted winch with a maximum capacity of 4500 kg.

Variants

BTR-60PA

This model is also referred to as the BTR-60PK, which entered service in 1963, has complete overhead armour protection for the troop compartment and is fitted with an NBC system. The commander's and driver's positions are identical to those on the original BTR-60P except

BTR-60P (8 × 8) APC clearly showing forward-mounted 12.7 mm MG and lateral-mounted 7.62 mm MGs

BTR-60PA (8 × 8) APC showing arrangement of roof hatches and machine guns

BTR-60PA (8 × 8) APC modified into the command post role with telescopic mast to rear of commander's station covered up, roof-mounted generator and rail-type antenna mounted around the front, left and rear side of the hull roof

that the driver has a roof-mounted periscope and both have a single piece hatch cover that opens to the rear.

Behind the commander's and driver's hatches is a single rear-opening rectangular hatch, in front of which is a single pintle-mounted 7.62 mm SGMB or PKB machine gun. Many vehicles also have a 7.62 mm machine gun mounted either side of this hatch cover. There is another hatch cover in the roof of the personnel compartment towards the rear on the right side, which opens to the right.

Three firing ports are provided in each side of the hull but no entry doors. The infantrymen are seated on bench seats along the side and rear of the troop compartment.

BTR-60PB

The BTR-60PB is essentially the BTR-60PA fitted with a machine gun turret and other modifications. In place of the vision blocks in the hatches immediately in front of the commander and driver, there are a number of roof-mounted periscopes for observation to the front and sides of the vehicle.

The turret, which is identical to that fitted to the Soviet BRDM-2 (4 × 4) reconnaissance vehicle and the Czechoslovakian OT-64 (8 × 8) APC, is armed with a 14.5 mm KPV machine gun and a 7.62 mm PKT machine gun mounted coaxially to the right with the telescopic sight mounted coaxially to the left. The weapons have an elevation of +30°, a depression of -5° and the turret can be traversed through a full 360°. Gun elevation and depression and turret traverse are all manual.

To the rear of the turret on the right side is a single piece hatch cover that opens to the left and to the left rear of this is a single piece hatch cover that opens to the right.

In either side of the hull are a single door that opens forwards, two or three firing ports and vision blocks.

There was also a version of the BTR-60PB used as an artillery command and observation post vehicle by the former East German Army.

The BTR-60PB was preceded by the BTR-60PAI which also had a turret. Late production BTR-60PBs have the same sighting improvement as the BTR-70: a small additional periscopic sight on the turret roof.

BTR-60PBK

This is a commander's model of BTR-60PB.

BTR-60 1V18

This is an artillery observation post vehicle.

BTR-60 1V19

This is a fire direction centre vehicle.

BTR-60 R-145BM

Hungarian BTR-60 with R-145 radio equipment.

BTR-60 R-975

Hungarian BTR-60 with unidentified radio equipment.

BTR-60-E-351BR

Soviet BTR-60 with unidentified support system.

BTR-60-P-238BT

Soviet BTR-60 with P-238 field telephone system.

BTR-60-P-240BT

Soviet BTR-60 with P-240 field telephone system.

BTR-60-P-241BT

Soviet BTR-60 with P-241 field telephone system.

BTR-60-R-137B

Soviet BTR-60 with R-137B radio system.

BTR-60-R-140BM

Soviet BTR-60 with R-140 radio system.

BTR-60-R-145

Soviet BTR-60 with R-145 radio system.

BTR-60-R-156

BTR-60 with R-156 radio system.

BTR-60-R-409BM

Soviet BTR-60 with R-409 radio system.

BTR-60-R-975

BTR-60 with unidentified support system.

BTR-60AVS

Soviet BTR-60 command variant.

BTR-60PAU

Bulgarian BTR-60 with unidentified communications equipment for use by artillery units. Also used by Soviet Union.

BTR-60PB R-145

Bulgarian BTR-60, no turret, fitted with R-145 radio station.

BTR-60PU Command Vehicle

This is essentially the BTR-60P fitted with bows and a tarpaulin cover and internally with seats for the command staff, mapboards and additional communications equipment. A command version of the BTR-60PA is also in service with a roof-mounted generator, 10 m high radio antenna,

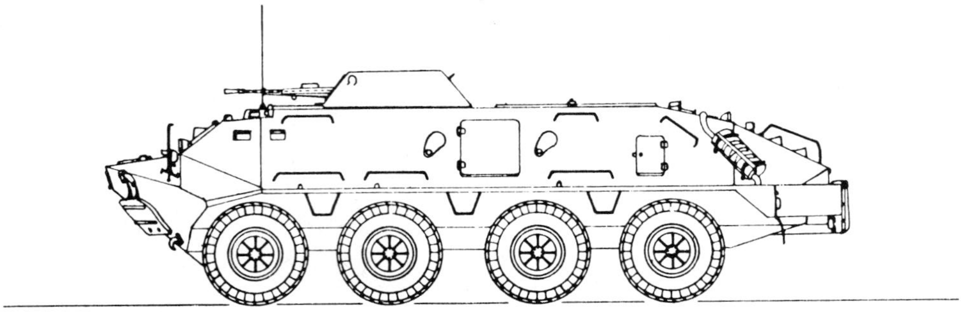

BTR-60PB (8 × 8) amphibious armoured personnel carrier

BTR-60PU series command vehicle with Hawkeye (NATO designation) antenna in raised position

which is used in the static role and a rail antenna running along the front, left side and rear of the hull.

BTR-60PU-12 Command Vehicle

The BTR-60PU-12 command and control vehicle has been in service with the Soviet Army since the late 1970s and has also been identified in service with Bulgaria, Czechoslovakia, Hungary and Yugoslavia.

The vehicle is essentially a BTR-60PA with a large stowage box on the right side of the hull to the rear of which is a water or fuel can. On the roof is a box which may house a generator to power the additional communications equipment installed. To the immediate rear of the commander's roof hatch is a telescopic mast which is raised into the operating position by turning a handle located on the right side of the mast.

The BTR-60PU-12 operates with air defence units equipped with systems such as the ZSU-23-4 self-propelled anti-aircraft gun, Gecko SA-8 and Gopher SA-13 surface-to-air missile systems.

BTR-60PU-12M

Soviet BTR-60 air defence command post vehicle.

BTR-60VVS

Soviet BTR-60 command post vehicle.

BTR-60P Maintenance Assistance

A number of older BTR-60P (8 × 8) APCs have been converted for use in the maintenance assistance role and have a raised tarpaulin cover over the troop compartment that runs almost to the rear. This may have the Soviet designation of the MTR-2.

Czechoslovakian Army BTR-60PU-12 command and control vehicle being prepared for use with mast being extended into operating position

BTR-60P (8 × 8) APC converted into maintenance assistance role with raised cover over troop compartment

BTR-60MS Radio Vehicle

This is the BTR-60P series used in the radio role and fitted with a High Ball (NATO designation) telescopic antenna.

BTR-60 twin 30 mm SPAAG

The Cuban Army uses a modified BTR-60P with the Czechoslovakian twin 30 mm towed anti-aircraft gun system mounted in an open mount on top of the vehicle.

BTR-60PB Forward Air Control Vehicle

This is essentially the BTR-60PB with the armament removed from the turret and the resulting port covered by a plexiglass observation window. To provide power for the additional communications equipment carried, a generator is mounted externally on the top of the hull at the rear.

BTR-60 ACRV M1979(2)

The BTR-60 Artillery Command and Reconnaissance Vehicle (ACRV) M1979(2) is used by towed artillery batteries and battalions as a command observation post (COP) vehicle. The designation ACRV M1979(2) was allocated by the US Army, this may well relate to one of the variants already described.

Status: Production complete. In service with Afghanistan, Algeria, Angola, Botswana, Bulgaria, Cambodia, China, Congo, Cuba, Djibouti, Ethiopia, Finland, Germany, Guinea, Guinea-Bissau, India, Iran, Iraq, North Korea, Libya, Mali, Mongolia, Mozambique, Nicaragua, Romania (built in Romania for which there is a separate entry), Somalia, Syria, USSR, Vietnam, Yemen, Yugoslavia and Zambia. Czechoslovakia and Poland use specialised versions of the BTR-60 series alongside their OT-64s.

Manufacturer: Gor'ky Automobile Plant, Gor'ky, USSR.

SPECIFICATIONS (BTR-60PA; data in square brackets relates to BTR-60PB where different)

CREW	2 + 16* [2 + 14]†
CONFIGURATION	8 × 8
COMBAT WEIGHT	9980 [10 300] kg
POWER-TO-WEIGHT RATIO	18.03 [17.47] hp/tonne
LENGTH	7.56 m
WIDTH	2.825 m
HEIGHT	
to turret top	n/app [2.31 m]
to hull top	2.055 m
FIRING HEIGHT	n/app [2.13 m]
GROUND CLEARANCE	0.475 m
TRACK	2.37 m
WHEELBASE	1.35 m + 1.525 m + 1.35 m
MAX SPEED	
road	80 km/h
water	10 km/h
FUEL CAPACITY	290 l
MAX ROAD RANGE	500 km
FUEL CONSUMPTION	0.58 l/km
FORDING	amphibious
GRADIENT	60%
SIDE SLOPE	40%
VERTICAL OBSTACLE	0.4 m
TRENCH	2 m
ENGINES	2 GAZ-49B 6-cylinder in-line water-cooled petrol developing 90 hp at 3400 rpm each
TRANSMISSION	manual with 4 forward and 1 reverse gears
TRANSFER CASE	2-speed
STEERING	power-assisted
TYRES	13.00 × 18
SUSPENSION	torsion bar with hydraulic shock absorbers
ELECTRICAL SYSTEM	12 V
ARMAMENT	
main	1 × 7.62 mm [1 × 14.5 mm] MG
coaxial	n/app [1 × 7.62 mm MG]
SMOKE-LAYING EQUIPMENT	none
AMMUNITION	
main	2000 [500]
coaxial	n/app [2000]
GUN CONTROL EQUIPMENT	
Turret power control	n/app [manual]
Gun elevation/depression	+23.5°/-6° [+30°/-5°]
Gun traverse	90° [+360° (turret)]
ARMOUR	
Hull front upper	7 mm at 86°
Hull front lower	9 mm at 47°
Hull side upper	7 mm at 35°
Hull side lower	7 mm at 0°
Hull rear upper	5 mm at 65°
Hull rear lower	7 mm at 0°
Hull top	7 mm at 86°
Hull belly front	5 mm
Hull belly rear	5 mm
Turret front	n/app [7 mm at 43°]
Turret sides	n/app [7 mm at 36°]
Turret rear	n/app [7 mm at 36°]
Turret top	n/app [7 mm]

* Max is 16, 12 are normally carried
† Max is 14, 8 are normally carried

BTR-152 Armoured Personnel Carrier

Development

In November 1946 the BM Fitterman design bureau started development work on a 6 × 6 armoured personnel carrier called the BTR-140. This was based on the chassis of the ZiS-123 truck, itself a derivative of the ZiS-151 truck but with an improved engine. The first prototypes of the BTR-140 were completed in May 1947 and Soviet Army tests were completed in 1949. The BTR-140 was accepted for service by the Soviet Army on 24th March 1950 under the designation BTR-152 and first production vehicles were completed late in 1950.

The first production models were based on the chassis of the ZIL-151 (6 × 6) truck but later vehicles (from and including the BTR-152V1) were based on the chassis of the improved ZIL-157 (6 × 6) truck. The BTR-152 has been replaced in front-line Soviet Motorised Rifle Divisions by the BTR-60P (8 × 8) amphibious APC, but large numbers remain in reserve and it is still used in North Africa and the Middle East. In addition to being used as an APC it is also used for towing artillery (eg 57 mm Ch-26), carrying cargo and mortar teams and for minelaying.

Description

The hull of the BTR-152 is made of all-welded steel with the engine at the front, commander and driver immediately behind the engine and the troop compartment at the rear.

Armoured shutters, controlled from inside the driver's compartment, protect the radiator from damage by small arms fire. Power is transmitted from the engine to the transmission and then by a propeller shaft to the transfer box in the centre of the vehicle. Power is then transmitted by a propeller shaft to the front axle and each of the rear axles.

The driver sits on the left of the vehicle with the commander to his right, both with a windscreen which can be covered by an armoured shutter hinged at its upper part. This shutter has a vision block for observation when the shutter is closed. Both the commander and driver have a door in the side of the hull and the upper part, which has a small vision slit, folds down on the outside for increased visibility. The roof over the commander's and driver's position is armoured.

The 17 infantrymen are seated in the open-topped troop compartment at the back of the vehicle, the top of which can be covered by a tarpaulin cover which is normally carried in the vehicle. The seating arrangements depend on the model: in some there are bench seats across the vehicle and others have them down either side. The infantrymen enter and leave the vehicle by twin doors in the rear of the hull which open outwards, with the left door carrying the spare wheel. There are three firing ports in each side of the hull and another two in the rear, one each side of the doors.

The BTR-152 has no NBC system, no night vision equipment and no amphibious capability. A saw is carried externally on the left side of the hull. The winch, if fitted, is mounted at the very front of the vehicle and has a maximum capacity of 5000 kg and 70 m of cable. The winch is also equipped with a special block which can be used as a pulley block to increase pulling capacity, or as a conventional block to change the direction of pull. The winch incorporates a three-speed PTO unit (low, high and reverse) controlled by a lever in the driver's compartment.

BTR-152A anti-aircraft vehicle with twin 14.5 mm machine guns

BTR-152V1 (6 × 6) APC clearly showing external air lines of central tyre pressure regulation system

BTR-152U command vehicle clearly showing higher superstructure of this model and externally mounted central tyre pressure regulation system

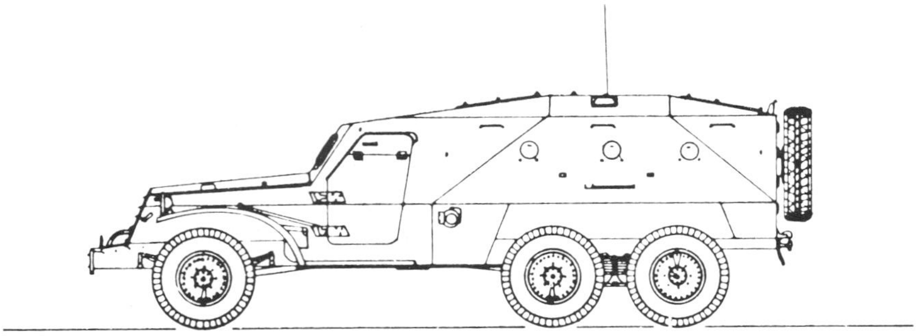

BTR-152K APC with overhead protection

Some models (see variants) have a central tyre-pressure regulation system which allows the driver to adjust the tyre pressures to suit the type of ground being crossed.

There are three sockets for mounting machine guns, one on the top of the armour cover over the commander's and driver's position and the other two on either side of the troop compartment. The forward mounting normally has a 7.62 mm SGMB machine gun, which can be elevated from -6° to +23.5° and has a traverse of 45° left and right, or a 12.7 mm DShKM heavy machine gun. The side mountings normally have 7.62 mm SGMB machine guns.

Variants

BTR-152

This was the first model to enter service and is also known in the West as the Model A. It has an open top, no winch and no central tyre-pressure regulation system.

BTR-152V

This was first seen in 1954 and was accepted for service with the Soviet Army in October 1955. This model has a central tyre-pressure regulation system for improved cross-country performance. The artillery command vehicle is called the BTR-152I.

BTR-152V1

This has been referred to as the Model B in the West and has an open top, front-mounted winch and a central tyre-pressure regulation system with external air lines. The BTR-152V1 was developed in 1959 and entered service in 1962.

BTR-152V2

Basic BTR-152 fitted with central tyre-pressure regulation system but no winch.

BTR-152V3

This is also known as the Model C in the West and has an open roof, front-mounted winch, central tyre-pressure regulation system with internal air lines and infra-red driving lights.

BTR-152K

This model, which is also known as the Model D (or 1961) in the West, has all the modifications of the BTR-152V3 plus full overhead armour protection. Over the top of the troop compartment are two roof hatches that open to the right, one with a single firing port and one with two.

BTR-152U Command Vehicle

The command vehicle is usually based on the BTR-152V1 or BTR-152V3 APC but has a much higher superstructure to allow the command staff to work upright. There are four windows in the upper part of the superstructure, two on the left, one on the right and one at the back. Access to stowage racks on the roof is by a ladder forward of the roof. The BTR-152U normally tows a trailer carrying additional equipment such as a generator. The first of these PUA (artillery command post) vehicles with a higher roof entered service in 1952 as the BTR-152B but did not have the central tyre-pressure regulation system, unlike the BTR-152I PUA which does.

BTR-152A Anti-aircraft Vehicle

This is armed with twin 14.5 mm KPV heavy machine guns in a manually operated turret (ZTPU-2) with a traverse of 360°. The guns have an elevation of +80° and a depression of -5°. The weapons have a maximum horizontal range of 8000 m and a maximum vertical range of 5000 m. Effective anti-aircraft range however is 1400 m and effective range when being used in the ground role is 2000 m. These weapons and mount are also used on the BTR-40A (4 × 4) anti-aircraft vehicle (which has the same turret as the BTR-152A), towed models of the KPV are called the ZPU-1 (single barrel), ZPU-2 (twin barrel) and ZPU-4 (quad barrel).

BTR-152K (6 × 6) APC showing roof armour, 7.62 mm machine gun and internal tyre pressure regulation system

KPV machine guns are also mounted in armoured vehicles.

The BTR-152A was designed at the same time as the BTR-152 APC and entered service in 1952. The BTR-152D also had KPV machine guns for anti-aircraft use, possibly four, plus 2000 rounds of ammunition and a crew of five. In 1955 the BTR-152E was introduced with a quad 14.5 mm KPV machine gun system, but based on the BTR-152V chassis. Full details of this version are given in *Jane's Land-Based Air Defence 1991-92*, page 87.

BTR-152 with Quad 12.7 mm M53 MGs

The Egyptians fitted a number of their vehicles with the Czechoslovak Quad 12.7 mm M53 anti-aircraft system. The M53 consists of four Soviet-designed 12.7 mm DShKM machine guns on a Czechoslovak-designed two-wheeled mount. It is believed that these are no longer in front-line service with the Egyptian Army. Some of these have, however, been encountered in Afghanistan operated by the Afghan Army.

BTR-152 with 23 mm ZU-23 AAG

During the fighting in the Lebanon in the Summer of 1982 the Israeli Army captured a number of BTR-152 APCs from the PLO which were fitted with the towed twin 23 mm automatic anti-aircraft gun ZU-23 in the rear of the troop compartment. The 23 mm ZU-23 has a maximum vertical effective anti-aircraft range of 2500 m. In addition to the version shown in the photograph there was also another with the ZU-23 mounted lower down at the hull rear which was cut away at an angle of about 45°.

SPECIFICATIONS (BTR-152V1)					
CREW	2 + 17	VERTICAL OBSTACLE	0.6 m	TYRES	12.00 × 18
CONFIGURATION	6 × 6	TRENCH	0.69 m	ELECTRICAL SYSTEM	12 V
COMBAT WEIGHT	8950 kg	ENGINE	ZIL-123 6-cylinder in-line water-cooled petrol developing 110 hp at 3000 rpm	BATTERIES	1 × 12 V
POWER-TO-WEIGHT RATIO	12.29 hp/tonne	TRANSMISSION	manual with 5 forward and 1 reverse gears (5th is overdrive)	ARMAMENT	1 × 7.62 mm MG
LENGTH	6.83 m	TRANSFER BOX	2-speed	AMMUNITION	1250 rounds
WIDTH	2.32 m	CLUTCH	dry, 2-plate	GUN ELEVATION/DEPRESSION	+23.5°/-6°
HEIGHT (hull top)	2.05 m	SUSPENSION		GUN TRAVERSE	90°
GROUND CLEARANCE	0.295 m	front	2 leaf springs and 2 dual-action hydraulic shock absorbers	ARMOUR	
TRACK (front/rear)	1.742 m/1.72 m	rear	equalising type with 2 leaf springs working in conjunction with torsion bars	Hull front	13.5 mm at 35°
WHEELBASE	3.3 m + 1.13 m			Hull sides	9 mm at 7°
MAX ROAD SPEED	75 km/h			Hull top	6 mm
FUEL CAPACITY	300 l			Hull rear	9 mm at 0°
MAX ROAD RANGE	780 km			Belly	4 mm
FUEL CONSUMPTION (road)	0.46 l/km				
FORDING	0.8 m				
GRADIENT	55%				

Status: Production complete. In service with Afghanistan, Albania, Algeria, Angola, Bulgaria (reserve), Cambodia, Central African Republic, China (Type 56), Congo, Cuba, Egypt, Ethiopia, Guinea, Guinea-Bissau, Hungary (reserve), Indonesia, Iran, Iraq, North Korea, Laos, Mali, Mongolia, Mozambique, Nicaragua, Poland (reserve), Romania (reserve), Somalia, Sri Lanka, Sudan, Syria, Tanzania, Uganda, USSR, Vietnam, Yemen, Yugoslavia, and Zimbabwe.

Manufacturer: Soviet state factories.

Note: In many of these countries the BTR-152 is in non-front line units or held in reserve.

BTR-40 Armoured Personnel Carrier

Development

Development work on a new 4 × 4 armoured personnel carrier (under the designation BTR-141) began in 1944 by the VA Dedkov OKB design bureau under the chief designer VK Rubtsov. This was based on a lengthened GAZ-63 (4 × 4) light truck chassis with a slightly more powerful engine (increased by 10 hp). The first prototype was completed in 1947 and had angled sides that tended to restrict its carrying capacity. The BTR-141 was then redesigned with vertical sides and rear. Trials with this model were carried out by the Soviet Army from 1949 and in 1950 it was accepted for service under the designation BTR-40. It entered service with the Soviet Army in 1950. The vehicle was originally used both as an armoured personnel carrier and as a command and reconnaissance vehicle, and performed a similar function to the US M3A1 scout car which was supplied to the Soviet Union by the USA during the Second World War.

BTR-40 series vehicles produced from 1956 had a central tyre-pressure regulation system and are known as the BTR-40Vs. BTR-40 series vehicles are often called the Sorokovke in the Soviet Union.

Description

The hull of the BTR-40 is made of all-welded steel with the engine at the front, commander and driver in the centre and the personnel compartment at the rear.

The driver sits on the left of the vehicle with the vehicle commander to his right, with a windscreen which can be covered by an armoured shutter hinged at its upper part. The shutter has a vision block for observation when it is lowered. Both the commander and driver have a door in the side of the hull, in the upper part of which is a vision flap that hinges open on the outside and has a vision slit.

The eight infantrymen sit in the back of the vehicle, three on a bench seat at the very rear of the hull and the other five on individual seats. The troops enter and leave the vehicle by twin doors in the rear of the hull. The troop compartment has an open top which can be covered by a tarpaulin which is normally carried in the vehicle.

There are three pintle mounts for machine guns, one on the rail that runs across the top of the troop compartment immediately behind the commander's and driver's position, and the other two one either side of the troop compartment. The basic vehicle is armed with a single 7.62 mm SGMB machine gun which is normally mounted on the front pintle. This weapon has an elevation of +23.5°, a depression of -6° and a traverse of 45° left and right of the vehicle's centre line.

Early production vehicles had no firing ports but later production vehicles have two or three in each side of the hull and two at the rear of the hull, one either side of the twin doors.

A spare wheel is carried on the rear of the hull in the centre and a saw is carried on the left side of the hull. Some vehicles have a winch with a capacity of 4500 kg mounted at the front of the hull. The BTR-40 has no NBC system, no night vision equipment, no amphibious capability and, unlike most other Soviet wheeled vehicles, has no central tyre-pressure regulation system.

Variants

BTR-40A Anti-aircraft Vehicle

This is armed with twin 14.5 mm KPV heavy machine guns in a turret (ZTPU-2) with a traverse of 360°. The guns have an elevation of +80° and a depression of -5°, both elevation and traverse being manual. The BTR-40A has a crew of five and carries 2400 rounds of 14.5 mm ammunition. The guns have a maximum horizontal range of 8000 m and a maximum vertical range of 5000 m but effective anti-aircraft range is 1400 m, and effective ground range is 2000 m. These weapons are

BTR-40 (4 × 4) APC with pintle-mounted 7.62 mm MG and front-mounted winch

also mounted on the BTR-152A (6 × 6) vehicle and towed models of the KPV are called the ZPU-1 (single barrel), ZPU-2 (twin barrel) and the ZPU-4 (quad barrel).

This vehicle was developed at the same time as the BTR-40 armoured personnel carrier. Although normally referred to as an anti-aircraft gun system, it was also employed by the Soviet Army as a long range support weapon against ground targets.

Full details of this version are given in *Jane's Land-Based Air Defence 1991-92*, page 87.

BTR-40B Armoured Personnel Carrier

This is similar to the basic model but has overhead armour protection and four roof hatches, two per side, that open to the outside of the vehicle, with a circular firing port. This entered service in 1957 and could carry six troops plus its crew of two.

BTR-40Kh Chemical Reconnaissance Vehicle

This is fitted with equipment at the rear of the hull which enables the crew to set marking pennants into the ground while the vehicle is crossing contaminated ground.

BTR-40 zhd

This rail scout vehicle was introduced in 1969 and is a BTR-40 with a metal wheel system for running on railway lines. It was derived from the earlier BA-64G rail scout vehicle based on the BA-64 (4 × 4) light armoured car.

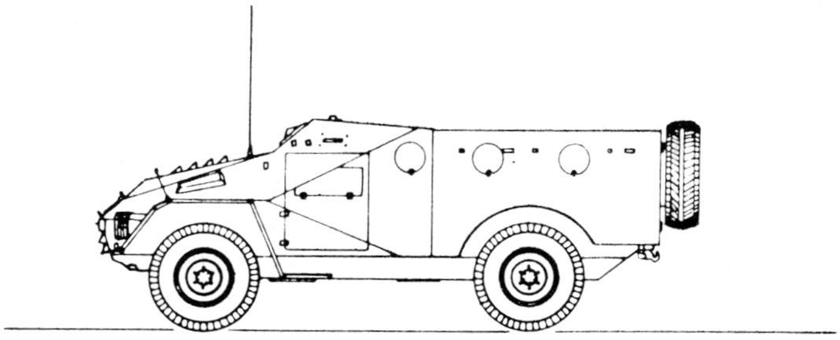

Unarmed BTR-40 (4 × 4) armoured personnel carrier

Status: Production complete. In service with Afghanistan, Albania, Bulgaria (reserve), Burundi, Cambodia, China (Type 55), Cuba, Egypt, Equatorial Guinea, Ethiopia, Guinea, Guinea-Bissau, Indonesia, Iran, North Korea, Laos, Mozambique, Nicaragua, Somalia, Syria, Tanzania, Uganda, USSR (reserve), Vietnam, Yemen, Yugoslavia and Zaïre.

Manufacturer: Soviet state factories.

SPECIFICATIONS		MAX ROAD SPEED	80 km/h	STEERING	cone worm and dual ridge roller
CREW	2 + 8	FUEL CAPACITY	120 l	CLUTCH	single dry plate
CONFIGURATION	4 × 4	MAX ROAD RANGE	285 km	TYRES	9.75 × 18
COMBAT WEIGHT	5300 kg	FUEL CONSUMPTION	0.42 l/km	BRAKES	hydraulic drum on all wheels
POWER-TO-WEIGHT RATIO	15 hp/tonne	FORDING	0.8 m	SUSPENSION	
LENGTH	5 m	GRADIENT	60%	front	longitudinal semi-elliptical springs with hydraulic shock absorbers
WIDTH	1.9 m	SIDE SLOPE	30%	rear	longitudinal semi-elliptical springs with check springs
HEIGHT		VERTICAL OBSTACLE	0.47 m	ELECTRICAL SYSTEM	12 V
without armament	1.75 m	TRENCH (with channels)	0.7 m	MAIN ARMAMENT	1 × 7.62 mm MG
BTR-40A	2.5 m	TURNING RADIUS	7 m	AMMUNITION	1250
FIRING HEIGHT (BTR-40A)	2.05 m	ENGINE	GAZ-40 6-cylinder water-cooled in-line petrol developing 80 hp at 3400 rpm	ARMOUR	8 mm max
GROUND CLEARANCE	0.275 m	TRANSMISSION	manual with 4 forward and 1 reverse gears		
TRACK		TRANSFER BOX	2-speed		
front	1.588 m				
rear	1.6 m				
WHEELBASE	2.7 m				

UNITED KINGDOM

GKN Defence Warrior Mechanised Combat Vehicle

Development

The initial proposal to consider future armoured personnel carrier requirements for the British Army in 1967 was followed by in-house Ministry of Defence feasibility studies from 1968 and 1971. Initial industrial work (Project Definition 1) and Ministry of Defence option studies to define an affordable programme took place between 1972 and 1976. Full project definition (Project Definition 2) was undertaken in 1977-78 by GKN Sankey after its competitive selection as prime contractor for the MCV-80.

In 1978 a parallel evaluation of the American XM2 (now known as the Bradley Infantry Fighting Vehicle) took place. The following year full development of the MCV-80 started in parallel with continued study of the XM2.

In June 1980 the following Parliamentary Statement was made on the British Army's APC replacement:

'The Army's present armoured personnel carrier, the FV432 series of vehicles, has been in service since the 1960s and will need to be replaced from the mid-1980s. Two vehicles have been considered for this requirement, the Mechanised Combat Vehicle (MCV-80), designed by the British firm GKN Sankey, and the American Infantry Fighting Vehicle, which would be manufactured under licence in this country.

'After careful assessment of the relevant operational, financial and industrial factors I have decided to select MCV-80 to meet this requirement. The total estimated cost of the replacement programme is about £1000 million and full development will be launched shortly.'

Detailed design work commenced in 1977 and by 1980 three prototypes of the MCV-80 were running and the company was also using test rigs, for example cooling, damper and suspension, to study specific characteristics of components for MCV-80.

In January 1984 the British Ministry of Defence announced that negotiations with GKN Sankey for an initial production order were well advanced but that future orders will be open to competitive tendering by both GKN Defence and other qualified manufacturers. At the same time it was announced that the MoD would be contracting with GKN Defence for full development of variants and derivative vehicles, subject to satisfactory completion of preliminary studies. MCV-80 (now known as Warrior) was accepted for service with the British Army in November 1984 with a reliability rating well in excess of the trials targets.

By 1984 10 prototypes had been built; one of these had a highly successful demonstration in the Middle East in 1983 and again in another Gulf country in Summer 1984, when a Warrior Platoon Vehicle covered 850 km, 650 km of which was completed in two days' running. This has led to the development of a Desert Fighting Vehicle configuration with 25 mm, 30 mm or 90 mm cannon and a crew of 10.

The MCV-80 was named Warrior by the British Army in 1985. Warrior has also been demonstrated or evaluated in Jordan, Kuwait, Saudi Arabia, Turkey and the United Arab Emirates but as of early 1991 no export contracts had been placed. It has also been evaluated in France. In June 1985 the British MoD announced that GKN Defence was to be awarded all three production contracts for the Warrior, totalling 1053 vehicles.

Total programme cost is about £725 million, which includes the provision of spares and training aids. Prime contractor for the spares and training aids is Alvis. Rediffusion Simulation and Invertron Simulated Systems are the major sub-contractors.

Production of Warrior began at Telford in January 1986 with first production vehicles being completed in the December. The first production batch comprised 290 vehicles of which 170 were section vehicles with two-man RARDEN turrets and the remaining 120 vehicles are specialised variants. The first production Warrior was officially handed over to the British Army in May 1987. The first Warrior battalion in the British army of the Rhine was fully operational in mid-1988.

The second and third batches total 763 vehicles and include platoon vehicles and variants. Of the total order for 1053 vehicles 70 per cent will be section vehicles and the remainder variants.

Vickers Defence Systems deliver the two-man power-operated turret to GKN Defence ready for installation in the section and command

GKN Defence Warrior mechanised combat vehicles with additional armour protection for Operation Granby

vehicles and Perkins Engines (Shrewsbury) deliver the complete powerpack, comprising the engine, transmission and cooling system, ready for installation in the Warrior. GKN produce the 7.62 mm Chain Gun turrets for the Recovery and Repair versions of the Warrior.

By mid-1991, over 500 Warrior vehicles had been delivered to the British Army with production continuing. It was expected that final deliveries would be made to the British Army in 1994, but late in 1990 the British MoD asked that production deliveries for the Warrior could be extended until 1996.

Each BAOR Warrior armoured infantry battalion is equipped with 53 Warrior vehicles, 36 section vehicles, nine command vehicles, four repair and recovery vehicles and four combat repair vehicles. In addition the battalion still retains 23 FV432 vehicles for a number of support roles including 81 mm mortar, MILAN team carrier and ambulance roles. It also has four Alvis Spartans fitted with a Euromissile MILAN Compact Turret (MCT) with two MILAN ATGWs in the ready to launch position and a further four Spartans for mortar fire controllers.

During Operation Desert Storm, the recapture of Kuwait early in 1991, all six variants were deployed including the Artillery Observation and Battery Command vehicles which entered service for the first time.

A number of Warriors were converted for use in the MILAN ATGW role. The order was placed on 22 October 1990 and the first kits were designed, produced and delivered in one week. This allowed MILAN to be launched from the vehicle with provision being made for internal missile stowage.

As an interim measure prior to the issue of Artillery Observation Post vehicles to units (and afterwards), some Infantry Command Vehicle Warriors were converted for the Forward Observation Role.

Warrior Command Vehicles were also deployed in the forward observation role for armoured regiments, a new task for the vehicle. An 81 mm mortar carrier was developed to a stage to allow early production when required in the future.

To improve battlefield survivability, the Warriors deployed to Saudi Arabia were fitted with passive armour developed by Vickers Defence Systems, this was fitted in Saudi Arabia.

Many minor modifications were also carried out, for example installation of maintenance free batteries, installation of Mini Pipe Fascines and satellite navigation aids for section, infantry command and REME variants.

An order for air-conditioning for deployed Warriors was placed in January 1991 and design and development commenced in conjunction with Galley and NGL. Full hability trials were completed and 1500 km of trundling trials were undertaken but no deliveries were made due to the speedy end of the conflict.

Description

The hull of Warrior is of all-welded aluminium construction which provides a significant increase in protection over that of the current FV432 APC. The driver sits at the front of the vehicle on the left side and has a single piece hatch cover hinged at the rear with a single wide-angle periscope that can be replaced by an image intensification periscope for driving at night.

The powerpack is to the right of the driver and drives, through spur final drive units, the front sprockets. The engine is a Perkins Engines Condor CV8 TCA diesel developing 550 hp which is coupled to an American Allison Transmission Division of General Motors X-300-4B transmission which is manufactured under licence in the UK by Perkins Engines. The automatic transmission has four forward and two reverse gears, and steering is through a hydrostatically controlled differential steer assembly.

The power-assisted vehicle brakes are integral with the transmission. The cooling system covers the engine, generator, transmission and vehicle brakes. The Lucas oil spray cooled 300 A generator is also driven from the gearbox by a PTO which also drives the hydraulic pump for the Airscrew-Howden fan.

Warrior mechanised combat vehicle with infantry dismounting from rear (British Army)

Warrior mechanised combat vehicle with driver's, commander's and gunner's hatches open (British Army)

GKN Defence Warrior vehicles deployed in the Middle East in early 1991 with Warrior Mechanised Combat Vehicle on the left and Warrior Repair and Recovery Vehicle on the right

Warrior Mechanised Artillery Observation Vehicle with its four aerials and dummy 30 mm cannon

Warrior Combat Repair Vehicle towing GKN Defence T4 high-mobility trailer carrying a complete Challenger MBT powerpack

Warrior Combat Repair Vehicle in travelling configuration

The two-man steel turret is in the centre of the vehicle, offset slightly to the left of the vehicle's centre line. The turret design, which was sub-contracted by GKN Defence to Vickers Defence Systems of Newcastle, has seats for the commander on the right, who dismounts with the troops, and the gunner on the left, with both crew members having individual hatch covers that open to the rear. The turret has power traverse with manual controls for emergency use. Both the commander and gunner have a Pilkington PE Raven day/night sight. In the day mode this has magnifications of ×1 and ×8, while in the image intensification night mode it has magnifications of ×2 and ×6. The commander's Raven sight has additional traverse. Late in 1986 it was announced that, as a result of competitive bidding, Avimo was awarded a contract worth over £17 million by the British MoD for the second production batch of Raven day/night sights totalling some 1500. Additional periscopes give vision to the sides and rear of the turret.

After evaluating the British RARDEN 30 mm cannon and the American 25 mm M242 Chain Gun, the former was selected for the Warrior. Following competitive bidding by Royal Ordnance Enfield, which had previously built all production RARDEN weapons, and the British Manufacture and Research Company, the latter was awarded the Warrior RARDEN production contract in 1985. Mounted coaxially with the main armament is a 7.62 mm EX-34 Chain Gun (British designation L94A1) with the expended cartridge cases being ejected outside of the turret.

Mounted either side of the turret are four electrically operated smoke dischargers.

The troop compartment is at the rear of the Warrior with the troops entering and leaving via a single power-operated door in the rear of the hull which has a vision block. In the roof of the troop compartment are two rotating periscopes, one either side, for use by the infantry. The sides of the troop compartment are sloped inward but there is no provision for the infantry to fire their small arms from within the vehicle. Firing ports/vision devices could be provided on export vehicles if required. Over the top of the troop compartment are double roof hatches. The Warrior carries eight infantrymen, one of whom also commands the vehicle and therefore dismounts with the infantry. The remaining seven sit in the troop compartment, four on the right and three on the left, in separate seats with seat belts. Personal kit can be stowed under the seats or in the sponsons which have interior grilles to hold the kit in position. Warrior carries all of the equipment required to fight on the battlefield for a period of 48 hours.

The suspension is of the torsion bar type and consists of six rubber-tyred aluminium road wheels with the drive sprocket at the front, idler at the rear and three track return rollers. Dampers, built into the suspension pivot housing, are fitted on the first, second and sixth road wheel stations. The tracks are conventional rubber-bushed cast steel with rubber pads.

Standard equipment includes an NBC system which is mounted in the side of the hull to the left rear of the driver and a full range of passive night vision equipment. Either side of the rear troop doors is a stowage box. The Warrior has no amphibious capability as this was not required by the British Army.

In 1986 Chubb Fire Security was awarded a contract by GKN Defence for the supply of engine fire protection systems for Warrior. This system uses Halon gas as the extinguisher. Situated in the front of the vehicle, the system is carried on two lightweight, rechargeable Halon canisters which vent into sparge pipes running around the engine compartment. In the event of an engine compartment fire, the crew activates the system by manual triggers releasing a four-second burst of Halon. A second Halon canister provides a back up, with further manual triggers on the outside of the vehicle's hull providing a remote extinguishing facility. In addition Warrior carries five handheld fire extinguishers, three inside and two outside.

British Army Variants

Section Vehicle

This is the basic Warrior MCV fitted with a two-man turret armed with a 30 mm RARDEN cannon and 7.62 mm L94A1 Chain Gun as described in the Description.

Command Vehicles

GKN Defence are producing three command versions of Warrior: platoon, company and battalion. These all leave Telford to the same configuration with the Army fitting the communications systems and other equipment. These all have a two-man turret armed with 30 mm RARDEN cannon and 7.62 mm L94A1 Chain Gun as previously described.

Warrior Repair and Recovery Vehicle (MRVR)

This has been designed under contract to the British Army and two prototypes were completed by GKN Defence by mid-1986. This model is in production for the British Army and a total of 67 MRVRs have been ordered with first deliveries in 1990.

The MRVR has a crew of five consisting of the commander, driver, gunner and two fitters and mounted on the left side of the hull is a

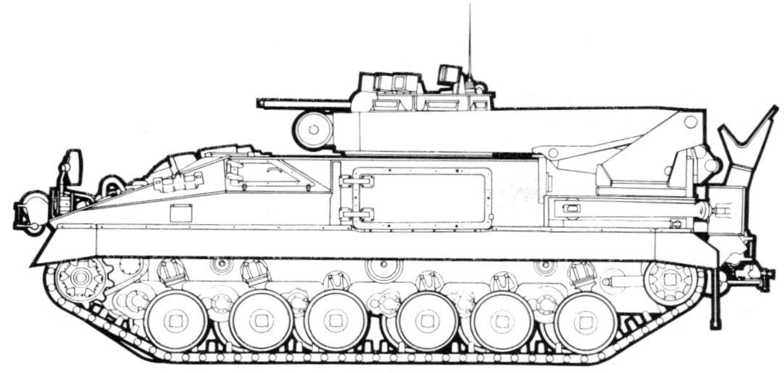

Warrior Repair and Recovery Vehicle (not to 1/76th scale)

GKN Desert Warrior fitted with two-man power-operated and stabilised Delco turret armed with McDonnell Douglas Helicopters 25 mm Chain Gun and 7.62 mm Chain Gun and a Hughes Aircraft TOW ATGW launcher either side of the turret. The vehicle has a HIRE thermal sight and high performance air conditioning

GKN Defence Desert Warrior fitted with two-man power-operated and stabilised Delco turret armed with McDonnell Douglas Helicopter 25 mm Chain Gun and 7.62 mm coaxial MG and showing firing ports

Model of Warrior turret armed with Medium Range TRIGAT ATGW system

hydraulically operated crane which can lift a maximum of 6500 kg. A hydraulically operated stabiliser leg is lowered to the ground before the crane is used. The crane can be traversed through a full 360° and can lift a complete Challenger MBT powerpack.

Mounted in the rear is a hydraulically operated winch with a capacity of 20 000 kg which can be increased to 38 000 kg when double-reeved.

The vehicle is provided with suspension lockout and mounted at the rear of the hull is an earth anchor to facilitate winching, which is provided with a pilot winch and 200 m of rope giving a capacity of 1.25 tonnes.

An air compressor is provided for the use of air tools and to assist in work in the field, and an electro-hydraulic pump allows the vehicle to change its own powerpack.

The manually operated turret is armed with a 7.62 mm L94A1 Chain Gun and smoke dischargers.

The MRVR can tow the GKN Defence T4 high mobility trailer which weighs 10 500 kg fully loaded and carries a complete Challenger MBT powerpack.

Full details of the Warrior MRVR are given in *Jane's Military Vehicles and Logistics 1991-92*, page 36. The trailer is covered in the same volume on page 517.

Warrior Combat Repair Vehicle (MCRV)

This is virtually identical to the Warrior Repair and Recovery Vehicle (MRVR) previously described but it does not have the rear-mounted earth anchor. It retains the 7.62 mm machine gun turret. The British Army has a total requirement for 67 of these vehicles with first production vehicles being delivered in 1990.

Warrior Mechanised Artillery Observation Vehicle

The MAOV is used by the forward observation officers and battery commanders in Royal Artillery field regiments. It is believed that about 100 vehicles will be procured, with the first prototype completed in 1988.

Externally the MAOV is similar in appearance to the basic Warrior vehicle but has a dummy 30 mm RARDEN cannon. The observer has a Pilkington PE Osprey combined day/thermal/laser system. This has magnifications of ×1 and ×8 in the day mode and built-in laser protection. The thermal imaging system is based on GEC Avionics Class II Thermal Imaging Common Modules already in service with the British Army. The assistant has a Pilkington PE Raven day/night sight as fitted to the standard Warrior vehicle.

Between the observer and his assistant's position is a remote display for the thermal imaging system. The vehicle is also be fitted for the new Manportable Surveillance and Target Acquisition Radar (MSTAR) which is the replacement for the ZB.298.

The vehicle is fitted with an inertial land navigation and attitude reference system, BATES (Battlefield Artillery Target Engagement System) which is being developed to replace the FACE (Field Artillery Computer System) and extensive communications equipment.

Low profile Warrior with ADATS system installed on hull roof

MAOV has a crew of six consisting of the driver, two observers in the turret, one of whom commands the vehicle, and three in the rear to operate communications equipment and THORN EMI MSTAR.

First production vehicles were delivered to the Royal Artillery in early 1990.

Battery Command Vehicle

This has been designed to meet the requirements of the Royal Artillery with first production vehicles being completed in 1990.

Warrior Export Vehicles

GKN Defence are offering a wide range of variants of the Warrior for the export market in addition to those currently in production for the British Army:

Arctic Warrior

A vehicle of similar configuration to Desert Warrior, but with a high performance integrated diesel air-to-air heater replacing the air-conditioning system, will undergo trials in Northern Europe in the Winter of 1991-92.

Desert Warrior

Shown for the first time at the 1986 British Army Equipment Exhibition was Desert Warrior, also known as the Fahris.

Desert Warrior is aimed at the export market, mainly the Middle East, and is fitted with a stabilised Delco Systems two-man power-operated turret armed with a 25 mm McDonnell Douglas Helicopters Chain Gun, 7.62 mm Chain Gun and a bank of four electrically operated smoke dischargers either side of the turret.

In 1990, Desert Warrior was shown with the Delco Systems turret with a Hughes TOW ATGW launcher mounted either side of the turret. The optional TOW system has an integrated Hughes sight that provides thermal imaging and TOW 2 ATGW tracking. All vehicles will be provided fitted for TOW to allow a TOW retrofit by field engineers if the TOW system is not fitted to all turrets initially.

Desert Warrior seats seven infantrymen facing each other, with up to seven firing ports, three in each side of the hull and one in the redesigned door.

The Desert Warrior is also fitted with a high performance combined air-conditioning and NBC filtration system.

Other variants include:

Warrior Engineer Vehicle, mechanised engineer combat vehicle with cupola-mounted 7.62 mm machine gun and THORN EMI Ranger anti-personnel mine-laying system on hull roof at rear.

Light Tank, GKN Defence has completed feasibility studies of a 105 mm light tank based on automotive components of Warrior. The 105 mm turret would be mounted at the rear of the hull as on the TH 301.

MCV/ATGW, with various ATGW systems, for example TOW, HOT or MILAN

MCV/AA, with two-man Thomson-CSF Sabre turret armed with twin 30 mm cannon

MCV/SAML, with eight British Aerospace Rapier SAMs in ready-to-launch position

MCV/MLRS, carrying Vought MLRS at rear (longer chassis)

MCV/MLC, modular load carrier (longer chassis)

Warrior 81 mm or 120 mm mortar carrier.

Warrior with CM 90 Turret

In mid-1984 Warrior was tested in Belgium fitted with the Belgian Cockerill CM 90 two-man power-operated turret with a Cockerill 90 mm Mk III gun, 7.62 mm coaxial machine gun, OIP fire control system and a Cadillac Gage Textron two-axis power drive system. On the standard Warrior the two-man 30 mm RARDEN turret is mounted slightly to the left of the vehicle's centre line, the CM 90 turret was however moved about 25 cm to the right so that it was concentric.

HOT ATGW Warrior

A Warrior chassis with its normal turret replaced by a Euromissile HOT HCT turret with four 4000 m range HOT ATGWs in the ready to launch position with a further 14 missiles carried in reserve in the hull was shown for the first time in September 1982. The one-man turret has powered traverse of 30° left and right and the launcher has powered elevation from −18 to +23°. The missile operator has a sight with a magnification of ×3 and ×12.

Warrior with TRIGAT

GKN Defence have proposed the Warrior chassis as the British Army launch platform for the long range version of the TRIGAT ATGW system to enter service in the late 1990s. The GKN Defence proposal was for a Warrior chassis fitted with an elevated arm, on the top of which is a pod containing the TRIGAT missiles and its associated sighting system. GKN Defence have also proposed that Warrior could be fitted with a turret armed with Medium Range TRIGAT ATGWs.

Low Profile Warrior

In mid-1988 GKN Defence showed the low profile Warrior fitted with the Oerlikon-Contraves/Martin Marietta ADATS air defence system with eight missiles in the ready-to-launch position. The low profile Warrior can accept a wide range of weapons, gun and missile, and a 15-tonne module alternative is also being offered.

Status: In production. In service with the British Army.

Manufacturer: GKN Defence
PO Box 106
Hadley Castle Works
Telford
Shropshire TF1 4QW
UK
Telephone: (0952) 244321
Telex: 35248/9 sanhad g
Fax: (0952) 244947

SPECIFICATIONS	
CREW	3 + 7
COMBAT WEIGHT	24 500 kg
POWER-TO-WEIGHT RATIO	22.45 bhp/tonne
GROUND PRESSURE	0.65 kg/cm^2
LENGTH	6.34 m
WIDTH	3.034 m
HEIGHT	
hull roof	1.93 m
overall	2.791 m
GROUND CLEARANCE	0.49 m
TRACK	2.54 m
TRACK WIDTH	460 mm
LENGTH OF TRACK ON GROUND	3.816 m
MAX ROAD SPEED	
forwards	75 km/h
reverse	48 km/h
ACCELERATION (0 to 48 km/h)	13.5 s
FUEL CAPACITY	770 l
MAX ROAD RANGE	660 km (at 60 km/h)
FORDING	1.3 m
GRADIENT	60%
SIDE SLOPE	40%
VERTICAL OBSTACLE	0.75 m
TRENCH	2.5 m
ENGINE	Perkins Engines CV8 TCA V-8 diesel developing 550 hp at 2300 rpm
TRANSMISSION	Allison Transmission Division of General Motors X-300-4B fully automatic with torque converter and lock-up clutch, 4 forward and 2 reverse gears
STEERING SYSTEM	differential steering system based on a variable hydrostatic drive
SUSPENSION	torsion bar
ELECTRICAL SYSTEM	24 V
ARMAMENT	
main	1 × 30 mm RARDEN cannon
coaxial	1 × 7.62 mm Chain Gun
SMOKE-LAYING EQUIPMENT	2 × 4 smoke dischargers
ARMOUR	
Hull	aluminium
Turret	steel

GKN Defence FV432 Armoured Personnel Carrier

Development

In the 1950s the Fighting Vehicles Research and Development Establishment (now the Royal Armament Research and Development Establishment Chertsey) designed the FV420 series of armoured vehicles. They never entered production but further development resulted in the FV430 series. The FV431 was an armoured load carrier which never entered production. The APC member of the family was called the FV432 and the first prototype was completed in 1961. For a short period the FV432 was known as the Trojan, but this name was subsequently dropped to avoid confusion with the car company of the same name.

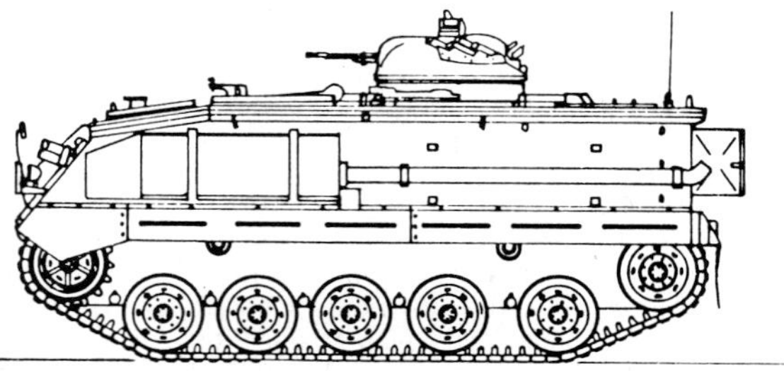

FV432 APC fitted with Peak Engineering lightweight turret armed with 7.62 mm GPMG

In 1962 GKN Sankey, now GKN Defence, was awarded a production contract for the FV432 APC and first production vehicles were completed in 1963. Production continued until 1971 by which time about 3000 FV432s had been built. The first production models were the Mk 1 and were followed by the Mk 2 and finally the Mk 2/1. Late production vehicles are distinguishable by the NBC pack in the right side of the hull which is almost flush against the hull rather than protruding as on earlier vehicles.

The FV432 replaced the Alvis Saracen (6 × 6) APC in the British Army of the Rhine many years ago and was itself supplemented from 1988 by the Warrior mechanised combat vehicle. The FV433 Abbot 105 mm self-propelled gun, manufactured by Vickers Limited at its Elswick facility, uses many automotive components of the FV432 APC.

In November 1990 the British Army stated that it had the following quantities of FV432 series vehicles:

Type	**UK**	**Elsewhere**	**Total**
FV432	180	1060	1240
FV432 RARDEN	1	12	13

Note: The above excludes specialised versions such as recovery.

FV432s were deployed to Saudi Arabia and took part in Operation Desert Storm in early 1991. A simple air-conditioning system for the FV432 was developed, but this did not enter service due to the rapid end to the conflict.

Description

The all-welded steel hull of the FV432 provides complete protection against small arms fire and shell splinters. The driver sits at the front

FV432 APC fitted with Peak Engineering turret armed with 7.62 mm GPMG

FV432 with Fox turret armed with 30 mm RARDEN cannon (Ministry of Defence)

FV432 with flotation screen removed carrying stores on roof (Pierre Touzin)

FV432 with THORN EMI Cymbeline mortar-locating radar in operating position and rear door open (Terry J Gander)

of the vehicle on the right side and has a single piece hatch cover that opens to the left. This has an AFV No 33 Mk 1 wide-angle periscope which can be replaced for night driving by an MEL L5A1 passive periscope.

The commander sits behind the driver and has a cupola that can be traversed manually through 360°. This cupola has a single piece hatch cover and three AFV No 32 Mk 1 periscopes. Mounted on the forward part of the commander's cupola is a 7.62 mm GPMG, although vehicles used in the command role and other rear area roles are normally fitted with a 7.62 mm Bren LMG. At one time it was proposed to issue many infantry units with FV432s fitted with the turret of the Fox armoured car armed with the 30 mm RARDEN cannon; this project was cancelled although a few vehicles were issued to the Berlin Brigade. Many FV432s have been fitted with a Peak Engineering lightweight turret armed with a 7.62 mm GPMG with an elevation of +50°, depression of -15° and 360° turret traverse. The turret has a single piece hatch cover that opens to the rear, three periscopes and eight smoke dischargers mounted four either side of the turret. This turret is mounted over the forward part of the circular troop compartment hatch. The original four-part hatch has been removed and replaced by a circular steel piece which contains the turret, with a hatch to the immediate rear of the turret.

The engine compartment is to the left of the driver with the air-inlet (forward) and air-outlet (rear) louvres in the roof and the exhaust pipe running along the left side of the hull. The complete engine with its oil tanks and filters is mounted on a common sub-frame which can be removed from the vehicle as a complete unit and reconnected on the ground by extended cables and fuel lines and pre-run for testing. The engine compartment is fitted with a fire wire-detection system. The engine is coupled to the General Motors Allison Division TX-200-4A semi-automatic transmission, which was built under licence in the United Kingdom by Rolls-Royce, which in turn supplies power to the steering unit at the front of the hull via a universally jointed propeller shaft. Access to the steering system for maintenance is via a forward-opening hatch in the glacis plate, behind the trim vane.

The troop compartment is at the rear of the FV432 with the 10 (five a side) infantrymen seated on bench seats that run down either side of the hull. The seats are hinged to the lower side plates on either side of the compartment and fold upwards, enabling the vehicle to carry up to 3670 kg of cargo. The infantrymen enter and leave the vehicle via a large door in the rear of the hull which opens to the right and is provided with a vision block. Over the top of the troop compartment is a circular hatch cover that opens to the left and right of the vehicle, each part being hinged in the middle, concertina fashion.

The torsion bar suspension consists of five dual rubber-tyred road wheels with the drive sprocket at the front, idler at the rear and two track return rollers. The first and last road wheel stations have a friction shock absorber and the upper part of the track is covered by a light sheet metal covering. The steel tracks are rubber-bushed and fitted with removable rubber pads.

There is an NBC system mounted in the right side of the hull which provides fresh air via ducts to both the driver's and troop compartment. Unlike the similar M113, the FV432 is not amphibious without preparation. Before entering the water a flotation screen is erected around the top of the hull, where it is held in position by 10 stays, the trim vane erected at the front of the hull and an extension piece fitted to the exhaust pipe. When afloat the vehicle is propelled in the water by its tracks. Most FV432s have now had their amphibious capability removed.

Variants

(excluding trials variants)

Ambulance

This model is unarmed and carries four stretcher patients or two stretcher and five seated patients, plus a crew of two. The stretchers are easily loaded and unloaded from the vehicle by means of sliding swivel racks.

Carl Gustaf

This is simply the standard FV432 with a bar fitted across the top of the troop compartment on which the Swedish 84 mm Carl Gustaf anti-tank weapon is mounted. The FV432 is also used to carry the Euromissile MILAN ATGW which is deployed away from the vehicle. This has replaced the 120 mm Wombat RR in the British Army.

Command

This model is fitted internally with two mapboards and additional communications equipment. To increase the available working area a penthouse measuring 3.66 × 2.74 × 1.98 m can be quickly erected at the rear of the hull. The command model has a crew of seven and weighs 15 500 kg.

81 mm Mortar Carrier

This is the basic vehicle fitted with an 81 mm L16 mortar on a turntable in the rear of the hull which can be traversed through a full 360°. The mortar has a maximum range of 5660 m and 160 mortar bombs are carried. The vehicle has a crew of six, weighs 16 400 kg and is issued on the scale of six per mechanised infantry battalion.

Maintenance Carrier

This model is designated the FV434 and is operated by the Royal Electrical and Mechanical Engineers. Its primary role is to change major components in the field such as the complete powerpack of the Chieftain

MBT. On the right side of the vehicle is an HIAB crane which has a lifting capacity of 1250 kg at 3.96 m radius to 3050 kg at 2.26 m radius. The suspension on the FV434 can be locked when the crane is being used. The FV434 has a crew of four and weighs 17 750 kg fully loaded. Full details are given in *Jane's Military Vehicles and Logistics 1991-92*, page 49.

Minelayer

The FV432 is used by the Royal Engineers to tow the Bar minelayer which can lay mines at a rate of 600 an hour. In addition a THORN EMI Ranger anti-personnel minelaying system can be mounted on the top of the FV432 to enable a combined anti-tank and anti-personnel minefield to be laid. Full details of the Bar mine system and the THORN EMI Ranger are given in *Jane's Military Vehicles and Logistics 1991-92*, pages 193 and 194.

Radar

The FV432 is used to mount the Marconi Avionics ZB 298 short-range radar (maximum range 10 000 m) and the THORN EMI Cymbeline mortar-locating radar (range 20 000 m). The FV436 fitted with the earlier THORN EMI Green Archer mortar-locating radar is no longer in service with the British Army.

Recovery

This is the basic FV432 with a winch installed in the rear troop compartment. The conversion kits were supplied by the British Rail Derby Locomotive Works from 1972. The winch is driven from the PTO on the engine transfer case and has both high- and low-speed gears. The winch has a maximum pull, using a three-part tackle, of 18 299 kg and maximum winching speeds are 18.34 m/min at 6608 kg and 122.4 m/min at 3050 kg.

Royal Artillery Vehicles

In addition to using the FV432 with the THORN EMI Cymbeline mortar-locating radar, the Royal Artillery uses the FV432 in the battery command role fitted with the Marconi Command and Control Systems Field Artillery Computer Equipment, and fitted with the Plessey sound ranging system.

Royal Engineers Vehicles

In addition to using the FV432 fitted with the THORN EMI Ranger anti-personnel minelaying system and towing the Bar minelayer the Royal Engineers use it to tow the Giant Viper mine-clearance trailer, details of which are given in *Jane's Military Vehicles and Logistics 1991-92*, page 242.

Royal Signals Vehicles

The Royal Signals use a number of specialised communication versions of the FV432 under the designation FV439, most of which have extensive antenna arrays on the roof. It is also used to house parts of the Wavell and Ptarmigan systems.

FV438 Wavell

The FV438 was originally fitted with a twin-launcher for the Swingfire ATGW but these have now been withdrawn from service and some have been used to install the Wavell Automatic Data Processing (ADP) system designed to assist formation headquarter's staff to carry out their command and control functions. The Wavell vehicle is usually referred to as the FV435 and is recognisable by the large box mounted on the roof. Basic specifications are: weight 16 tonnes, length 5.233 m, width 2.87 m and height 2.883 m.

FV432 Upgrades

The last production FV432 APC was completed by GKN Defence in 1971 but until recently very few modifications had been carried out to the basic vehicle.

Royal Signals FV439 signals vehicle with antenna being erected (Ministry of Defence)

A new fan drive motor for the FV432 has recently been developed by Commercial Hydraulics and by late 1989 over 700 had been delivered to the British Army, with production continuing. These are fitted to the FV432 to provide reliable and variable speed control to the cooling fans.

A thermostatic valve senses the water temperature and this controls the flow of oil to the fan drive, which keeps the engine at optimum operating level in all climates.

One hydraulic motor drives each of the two fan impellers while a pump, also produced by Commercial Hydraulics, is fitted to the engine to provide oil to power the motors.

GKN Defence has suggested a number of upgrades to the FV432 as part of its proposals to the British MoD to meet its requirement for a Future Family of Light Armoured Vehicles (FFLAV), additional details of which are given in the *AFV families* section.

Possible improvements include an improved NBC system and increased armour protection, upgraded K60 engine and improved steering and suspension.

SPECIFICATIONS		MAX WATER SPEED	6.6 km/h	AMMUNITION	1600 rounds of
CREW	2 + 10	FUEL CAPACITY	454 l		7.62 mm (belted when
COMBAT WEIGHT	15 280 kg	MAX ROAD RANGE			fitted with GPMG) or
UNLOADED WEIGHT	13 740 kg	diesel	480 km		50 magazines each
POWER-TO-WEIGHT		petrol	424 km		holding 28 rounds
RATIO	15.7 bhp/tonne	FORDING	1.066 m		(when fitted with Bren
GROUND PRESSURE	0.78 kg/cm²	with preparation	amphibious		LMG)
LENGTH	5.251 m	GRADIENT	60%	ARMOUR	6-12 mm
WIDTH	2.8 m	VERTICAL OBSTACLE	0.609 m		
over tracks	2.527 m	TRENCH	2.05 m		
HEIGHT		ENGINE	Rolls-Royce K60 No 4		
including machine gun	2.286 m		Mk 4F 2-stroke,		
to hull top	1.879 m		6-cylinder multi-fuel		
GROUND CLEARANCE	0.406 m		developing 240 bhp at		
TRACK	2.184 m		3750 rpm		
TRACK WIDTH	343 mm	TRANSMISSION	TX-200-4A semi-		
LENGTH OF TRACK			automatic with 6		
ON GROUND	2.819 m		forward and 1		
MAX ROAD SPEED			reverse gears		
6th gear	52.2 km/h	SUSPENSION	torsion bar		
5th gear	37.6 km/h	ELECTRICAL SYSTEM	24 V		
4th gear	27 km/h	BATTERIES	6 × 12 V		
3rd gear	19.5 km/h	ARMAMENT (main)	1 × 7.62 mm MG		
2nd gear	13.7 km/h	SMOKE-LAYING			
1st gear	9.8 km/h	EQUIPMENT	2 × 3 smoke		
reverse	8.65 km/h		dischargers		

Status: Production complete. In service with British Army and India (command post role for Abbot SPG), GKN Defence is completing post design exercises on the FV432 series. The FV432 is being supplemented in the British Army by the GKN Defence Warrior mechanised combat vehicle.

Manufacturer: GKN Defence
PO Box 106
Hadley Castle Works
Telford
Shropshire TF1 4RQ
UK
Telephone: (0952) 244321
Telex: 35248/9 sanhad g
Fax: (0952) 244947

Alvis Stormer Combat Vehicle

Development

In 1980 Alvis Limited purchased the manufacturing and marketing rights of the FV4333 from the British Ministry of Defence. The vehicle was originally designed by the Military Vehicles and Engineering Establishment, now Royal Armament Research and Development Establishment Chertsey, and the first prototype made a brief appearance at the mobility demonstration at Bovington during the 1978 British Army Equipment Exhibition. In June 1981 Alvis announced that its development of the FV4333 had been designated Stormer.

The Stormer has taken advantage of the existing technology of the Alvis Scorpion CVR(T) family and many of the proven components have been used. The transmission is the same as on the CVR(T) but with a modified input gear train (with the designation changed from TN15 to T300) and the suspension uses CVR(T) components with wheel stations increased to six per side. The hull of the Stormer is wider than the CVR(T) and its original petrol engine has been replaced by a Perkins T6/3544 turbo-charged diesel which can also be installed in the Scorpion.

In 1981 Alvis was one of three manufacturers to be awarded contracts for vehicles to meet the US Army Marine Corps requirement for a Light Armored Vehicle (LAV). Alvis was awarded a contract for one Scorpion 90, powered by a Perkins diesel and armed with a Cockerill Mark III 90 mm gun, and three Stormer APCs, each fitted with a two-man Arrowpointe turret armed with a 25 mm M242 Chain Gun and a coaxial 7.62 mm M240 machine gun. However, the production LAV contract was awarded to General Motors of Canada with the 8 × 8 version of the MOWAG Piranha.

Late in 1981 Malaysia placed an order with Alvis for 26 Scorpion 90 vehicles and 25 Stormer APCs, production of which commenced in 1982 with first deliveries made in 1983. Of the Stormer APCs, 12 vehicles are fitted with Helio FVT 900 turrets armed with a 20 mm Oerlikon cannon and a 7.62 mm machine gun and the remainder with a Thyssen Henschel TH-1 turret armed with twin 7.62 mm machine guns. In Summer 1986 the Stormer was selected by the British Army as the carrier for the Shorts Starstreak HVM air defence system.

In 1990, two Alvis Stormer full tracked vehicles each completed 16 000 km of extensive road and cross-country reliability, availability, maintainability and durability (RAM-D) trials at Alvis's Coventry facility.

Late in 1990, under a crash programme for the British Army, Alvis produced a flat bed version of the Stormer AFV carrying the French GIAT Industries Minotaur anti-tank mine scattering system. This was deployed to Saudi Arabia early in 1991 and additional details are given under variants.

The Alvis Stormer for the Shorts Starstreak High Velocity Missile system is now in production at Coventry where systems integration will also take place.

Alvis Stormer APC fitted with, No 16 cupola armed with externally mounted 7.62 mm machine gun, pintle-mounted 12.7 mm M2 machine gun, spare wheel on left hull side, flotation screen on forward part of hull, smoke dischargers and external stowage boxes on sides of hull at rear

Description

The hull of the Stormer is made of all-welded aluminium with the driver at the left front, engine compartment to his right and troop compartment at the rear.

The driver has an adjustable seat and a single piece hatch cover that opens forwards. In front of him there is a single ×1 periscope which can be replaced by a passive night driving periscope.

The Perkins T6/3544 diesel is mounted to the right of the driver with air induction through armoured louvres and Farr air cleaners. Engine cooling is by a single mixed-flow fan drawing air through the radiator, over the engine and through the outlet louvres. The exhaust is also cooled by some of this air. The engine is coupled to a T300 hot shift, foot-operated, seven-speed gearbox with centrifugal clutch. An automatic shift option for the transmission is available. The transmission contains a controlled differential steering system giving turning circle radii varying with the gear ratio, and a neutral or pivot turn when needed.

From the gearbox the drive goes to the front sprockets and then the tracks. Track adjustment is by hydraulic ram and hand pump.

The vehicle commander sits behind the driver and has a No 16 cupola with eight ×1 periscopes and a periscope with a magnification of ×1 and ×10 for aiming the 7.62 mm GPMG mounted on the right side of the cupola which can be aimed and fired from within. The cupola has a single piece hatch cover that opens to the left.

The section leader sits to the right of the vehicle commander and has four periscopes and a hatch cover.

The number of troops carried depends on the armament installation but would normally be eight, seated four down either side of the troop compartment on bench seats that can be folded up to enable stores to be carried. The infantry enter and leave the vehicle through a large single door that opens to the right and incorporates a single vision block.

Over the top of the troop compartment there is a large two-piece hatch that opens either side of the hull. On the prototype there was no provision for the infantry to aim and fire their small arms from inside the vehicle, but this capability has been provided on one of the more recent Stormer vehicles.

Alvis Stormer fitted with cupola with provision for externally mounted 7.62 mm machine gun

Alvis Stormer with Delco LAV-25 two-man turret armed with 25 mm McDonnell Douglas Helicopters Chain Gun and 7.62 mm coaxial MG

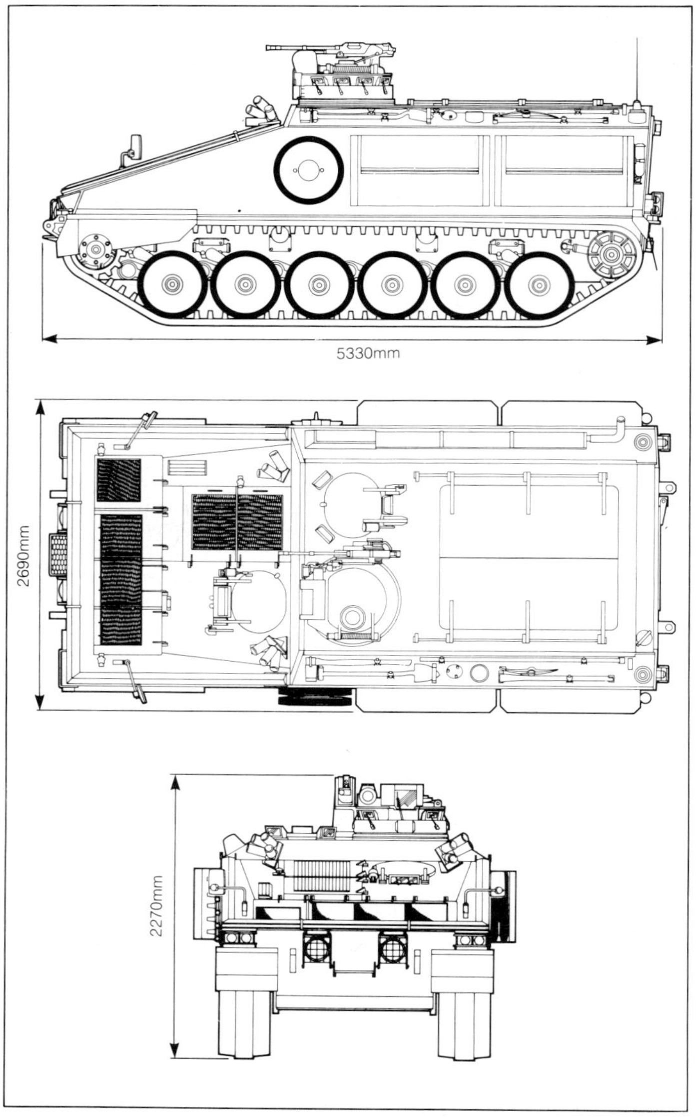

Side, top and front views of typical Alvis Stormer with No 16 cupola and externally mounted 7.62 mm machine gun (not to 1/76th scale)

The torsion bar suspension consists of six dual rubber-tyred aluminium road wheels with the drive sprocket at the front and the idler at the rear. Tracks are single pin, steel links, with rubber bushes and integrally mounted rubber pads. The first, second and sixth road wheel stations have hydraulic shock absorbers. Early Stormers did not have track return rollers but the latest versions have two per side.

The Stormer has an NBC and/or air-conditioning system in the left side of the hull, below the commander's station and can be fitted with passive night vision equipment. Two four-barrelled electrically operated smoke dischargers are mounted on the front of the hull. Appliqué armour can be added to the vehicle for increased protection.

The basic model can ford to a depth of 1 m without preparation but, with a front flotation screen it is fully amphibious, propelled in the water by its tracks at a speed of 6.5 km/h. When fitted with an appliqué propeller kit its water speed is 9.5 km/h.

Optional equipment includes: a No 27 cupola, 7.62 mm machine gun, NBC system, air-conditioning system, gun ports and vision blocks, driver's night periscope, night devices for commander, fire warning and protection equipment, automatic shift transmission, flotation screen and fittings, propeller kit and washboard, communications equipment and ground navigation system.

At the 1986 British Army Equipment Exhibition a Stormer was shown fitted with the US Delco LAV-25 turret.

Further evaluation of the Stormer design is evident as in early 1991 the Stormer VLSMS (Vehicle Launched Scatterable Mine System) vehicles were shown. These have a distinctive sharper front end and revised lighting layout. The vehicle is also longer and incorporates radar stealth technology in the outer structure of the vehicle. Further details of this version are given later.

Variants

The prototype of the Stormer was fitted with a No 16 cupola armed with a 7.62 mm GPMG but the basic hull can be adapted to mount either a gun or missile turret. This is achieved by modification of the roof section and installation of compatible turret rings. Typical variants include:

Two-man turret armed with 76 mm or 90 mm gun
Two-man turret armed with 30 mm RARDEN cannon and a 7.62 mm coaxial machine gun with both weapons having an elevation of +35°, depression of -10°, with turret traverse a full 360°. A total of 165 rounds of 30 mm ammunition would be carried, of which 60 would be for ready use in the turret with remainder in the hull
ATGW vehicle with HOT, TOW or MILAN systems (and in the future TRIGAT)
Anti-aircraft vehicle, gun or missile
Engineer vehicle
Load carrier
Minelayer, for example Royal Ordnance Bar minelaying system
120 mm or 81 mm mortar carrier
Repair/recovery vehicle
Ambulance with higher roof
Command/control vehicle.

Alvis Stormer Forward Observer Vehicle

Stormer Infantry Fighting Vehicle

The Stormer Infantry Fighting Vehicle can carry eight fully equipped infantry plus the gunner and driver. The example shown in 1984 was fitted with a Helio FVT 900 one-man turret armed with a 20 mm Oerlikon KAA cannon (with 100 ready rounds) and a 7.62 mm coaxial machine gun (200 ready rounds). It has firing ports/vision devices in the rear troop compartment.

Stormer FOV

In June 1988 Alvis announced that together with Avimo and Helio, also members of United Scientific, it was developing the Stormer Forward Observer Vehicle (FOV) as a private venture.

The Stormer FOV is fitted with a Teldix FNA 50 land navigation and orientation system, a turret-mounted laser rangefinder and an MEL BA 1402 data processing/transmission system. The Helio turret has manual traverse through a full 360°, the Avimo sight has an integral laser rangefinder with a magnification of ×8 (day) and ×6.8 (night) and the laser has a maximum range of 20 km.

The system functions as follows. The sight/cupola outputs range/ elevation/azimuth of the target when the laser rangefinder is fired and the land navigation system outputs the vehicle position/orientation. The data processing system computes the target position and the target and own vehicle position are displayed for message assembly. The operator inserts contact report codes, assembles the message and triggers the burst transmission. The message is encrypted and burst transmitted to the command vehicle. The message is de-encrypted and printed in the command post vehicle on the MEL BA M28 printer. Other artillery variants could include a battery command post and a target acquisition/ designation vehicle with a telescopic mast with a top-mounted sensor package.

Stormer VISOR

In 1989 the Vehicles and Turrets Division of Alvis formed a consortium of specialist companies to meet the future reconnaissance role. Alvis identified the following requirements:

Enhancement of passive observation capabilities
Practical reductions in the crewing demands of such systems
Cost-effective and affordable equipments, whether enhanced existing or new.

The end result was the Vehicle Integrated System for Observation and Reconnaissance (VISOR) which was first shown at the 1990 British Army Equipment Exhibition with the Alvis Stormer being used as the carrier.

For the passive observation role demonstrated in 1990, the VISOR is crewed by a commander, operator and driver. Mounted on the roof is a representative elevated sensor suite that comprises a thermal imager and colour CCD camera with pan and tilt controlled remotely from the vehicle.

The VISOR system will, however, accept other sensor combinations as required. The elevation system is a commercial item, although other systems can be fitted. A laser rangefinder is also provided.

The following facilities are provided for the three crew members:
Operator - display screens for thermal imager and TV, separate data display screen, data entry set and sensor controls
Commander - single display screen with switchable thermal imager or TV viewing and sensor controls

Stormer VLSMS with GIAT Industries Minotaur anti-tank mine scattering system on rear flatbed

Driver - single display screen showing automotive instrumentation but switchable to data display with associated data entry keypad.

Alvis is also a member of the Royal Armament Research and Development Establishment (Chertsey) VERDI (Vehicle Electronics Research Defence Initiative) project that uses an GKN Defence Warrior chassis.

Stormer VLSMS

Under Operation Granby, a quantity of Stormer Vehicle Launched Scatterable Mine Systems (VLSMS) were ordered from Alvis late in 1990 and integrated in Coventry with the Minotaur anti-tank mine scattering system from GIAT Industries of France. The UK MoD awarded Alvis the first of two contracts early in October 1990 and the first vehicle was built and delivered in just 14 weeks.

These Stormer vehicles were the first to enter service with the British Army in February 1991.

While automotive components of the new Stormer are identical to those of future British Army production Stormers, the hull is of a new design and is both longer and wider.

The driver is seated at the front left of the vehicle with the engine to his right. He has a single piece hatch cover that opens forward and a single wide-angle day periscope that can be replaced by a passive night perisope if required.

The powerpack is to the right of the driver with the vehicle commander being seated to the rear of the driver. He operates the minelaying controls and a pintle-mounted 7.62 mm machine gun. There is space to his right for a third crew member.

The driver's and commander's compartment has both air-conditioning and NBC protection allowing mines to be laid without the crew being exposed to contamination. A bank of four electrically operated, forward firing smoke dischargers is mounted on either side of the vehicle.

The vehicle hull is of all-welded aluminium armour and the hull sides have been sloped inwards at the bottom to reduce the radar signature.

Suspension is similar to the standard Stormer except that this version has longer torsion bars and greater length of track in contact with the ground.

Mounted on the flat bed is the Minotaur scatterable anti-tank mine system.

Although the Minotaur system was adopted for Operation Granby, the British Army has still to select its definitive system. The three main competitors for this are GIAT Industries of France with Minotaur, Alliant Tech Systems of the USA with Volcano and Dynamit-Nobel of Germany with Skorpion.

The Stormer flat bed can be adopted for a wide range of other roles on the battlefield including repair/recovery and towing minelaying systems.

Stormer SPAAG

The Stormer chassis can be fitted with a number of different power operated air defence turrets, both gun and gun/missile, the latter includes the US General Electric series armed with 25 mm cannon and surface-to-air missiles.

Stormer High Velocity Missile Carrier

In 1986 the Stormer chassis was selected by the British Army as the platform for its new Shorts Starstreak High Velocity Missile system which will enter service in the early 1990s. Series production of this version commenced at the new Alvis manufacturing facility at Coventry late in 1991. In this application the Stormer will be armed with eight Starstreak missiles in the ready to launch position with additional missiles being carried in reserve. Mounted on the turret roof to the right of the commander's cupola will be the Avimo acquisition sight and mounted between the two banks of four Starstreak missile will be the THORN EMI Air Defence Alert Device (ADAD) which is passive. Total British Army requirement is for about 136 Stormer Starstreak systems with an option on a further batch of 50 systems. Full details of the Starstreak air defence system are given in *Jane's Land-Based Air Defence 1991-92*, page 142.

Stormer Laserfire

The Alvis Stormer chassis has been proposed as a carrier for the British Aerospace Laserfire SAM system.

Full details of the Laserfire low-altitude surface-to-air missile system are given in *Jane's Land-Based Air Defence 1991-92*, page 141.

Family Concept

In June 1982 Alvis released outline drawings for 14 variants based on the chassis of the Stormer APC. These are anti-tank with Emerson TOW launcher, missile launcher platform, maintenance/recovery, engineer with front-mounted dozer blade, electronic warfare/battalion command and control, logistics, ambulance, light armoured squad carrier, mobile protected gun-near term/light assault, assault gun, low profile 75 mm, low profile 90 mm, air defence and mortar carrier. Since then Alvis has carried out extensive development and pre-production trials, supplementing the 32 187 km of testing by the US Marine Corps during the LAV programme. The first Stormer production vehicle with the 250 bhp engine was completed early in 1984.

SPECIFICATIONS

CREW	3 + 8
MAX COMBAT WEIGHT	12 700 kg
POWER-TO-WEIGHT RATIO	19.68 bhp/tonne
GROUND PRESSURE	0.46 kg/cm²
LENGTH	5.33 m
WIDTH	2.4 m
over stowage boxes	2.69 m
over tracks	2.31 m
HEIGHT (top of MG)	2.27 m
GROUND CLEARANCE	425 mm
TRACK	1.892 m
TRACK WIDTH	432 mm
LENGTH OF TRACK ON GROUND	3.112 m
MAX ROAD SPEED	80 km/h (forward and reverse)
MAX WATER SPEED	
tracks	5 km/h
propeller	10 km/h
ACCELERATION (0-50 km/h)	24 s
FUEL CAPACITY	405 l
MAX ROAD RANGE	650 km
FORDING	1.1 m
with preparation	amphibious
GRADIENT	60%

VERTICAL OBSTACLE	0.6 m
TRENCH	1.75 m
THEORETICAL TURNING CIRCLE RADII	
1st gear	2.04 m
2nd gear	4.64 m
3rd gear	6.58 m
4th gear	11.01 m
5th gear	20.09 m
6th gear	26.2 m
7th gear	40.89 m
ENGINE	Perkins T6/3544 water-cooled 6-cylinder turbo-charged diesel developing 250 bhp at 2600 rpm
TRANSMISSION	T300 crossdrive, semi-automatic hot-shift type, providing 7 speeds in each direction and pivot turns
STEERING	regenerative triple differential system controlled by external disc brakes

SUSPENSION	torsion bar with trailing arm geometry and lever type hydraulic dampers
WHEEL MOVEMENT	
static to bump	274 mm
static to rebound	97 mm
HAND BRAKE	hand lever connected by cable to contracting bands operating on periphery of main brake discs
FOOT BRAKE PEDAL	hydraulically connected to main brakes
ELECTRICAL SYSTEM	28.5 V
BATTERIES	4 × 6 TN, 125 Ah (2 for automotive, 2 for radio) or low-maintenance
ARMAMENT	depends on role

Status: Stormer entered production in 1982 with first deliveries made to Malaysia in 1983. Three vehicles have been delivered to the USA. Adopted by British Army for Shorts Starstreak HVM in Summer 1986. Stormer VLSMS entered service with the British Army under Operation Granby early in 1991.

Manufacturer: Alvis Limited
The Triangle
Walsgrave
Coventry
West Midlands CV2 2SP
UK
Telephone: (0203) 535455
Telex: 31459
Fax: (0203) 539280

One of the pre-production of Alvis Stormer vehicles with the Shorts Starstreak HVM installation traversed slightly left

Alvis Saracen Armoured Personnel Carrier

Development

Shortly after the end of the Second World War the Fighting Vehicles Research and Development Establishment (now the Royal Armament Research and Development Establishment Chertsey) began the design of the FV600 series of 6 × 6 vehicles. The first three models in this series were the FV601 Saladin armoured car, the FV602 command vehicle (which was cancelled in 1949 although the FV604 later took over its original role) and the FV603 Saracen APC. Owing to the situation in Malaya in the late 1940s, development of the FV603 was given precedence over that of the FV601 Saladin armoured car.

Design parent for the FV600 series was Alvis Limited of Coventry which completed the first prototype of the Saracen in 1952 and first production vehicles in December the same year. The first Mk 1s were powered by a Rolls-Royce B80 Mk 3A petrol engine which had BSF threads but later production Mk 1s were powered by a B80 Mk 6A petrol engine and had UNF threads. Early production vehicles had a slightly different turret: the rear of the turret opened either side rather than folding down horizontally as on later turrets.

Throughout the 1950s and early 1960s the Saracen was the standard APC of the British Army and was replaced from 1963 by the FV432 tracked APC. Production continued for export until 1972. Production of the Saracen APC amounted to 1838 vehicles. Saracen APCs were widely used by the British Army in Northern Ireland in the IS role but these have now been withdrawn. A few are used in Cyprus by the British Army.

British Army FV604 command vehicle with external stowage (Christopher F Foss)

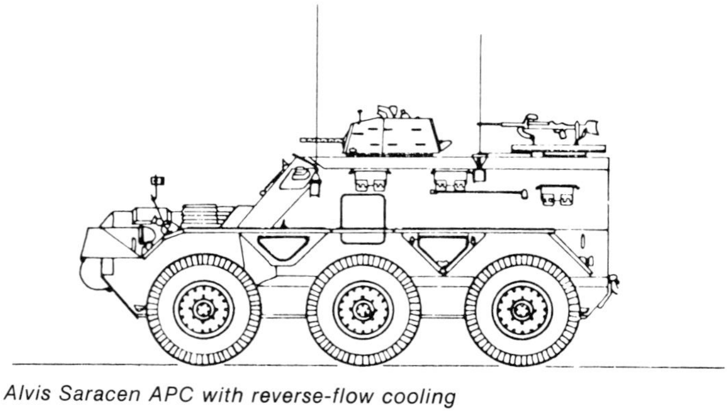

Alvis Saracen APC with reverse-flow cooling

Under the Conventional Forces Europe (CFE) agreement signed in 1990, the British MoD stated that the British Army had a total of 166 Saracen (6 × 6) APCs of which 121 were in the United Kingdom and the remaining 45 overseas.

Description

The hull of the Saracen is made of all-welded steel with the engine at the front and the personnel compartment at the rear.

Air enters the engine compartment at the front of the vehicle through armoured louvres and is expelled through the four hatch covers over the engine compartment. Power is transmitted from the engine to the road wheels via a fluid coupling, five-speed pre-selective gearbox, transfer box with forward and reverse change, and differential transverse drive direct to each centre wheel hub through universally jointed shafts. Drive to the front and rear wheel hubs is by transmission shafts and bevel boxes. Each wheel hub houses an epicyclic train which provides the final gear reduction to the wheels. Steering of the front two and centre wheels is by positive mechanical linkage, hydraulically assisted between the steering wheel and the road wheels. Hydraulic pressure for both steering and the foot brake is provided by an engine-driven hydraulic pump.

The suspension is of the double wishbone type with torsion bars and sleeves. Two telescopic double-acting hydraulic shock absorbers are fitted at each front and rear wheel station and the centre wheel stations each have a single telescopic double-acting hydraulic shock absorber. The gearbox and the transfer box are under the floor of the crew compartment in the centre and the fuel tank is under the floor of the crew compartment at the rear.

The driver sits at the front of the vehicle behind the engine compartment and has three No 17 periscopes, one in front and one either side of his position. The periscopes are mounted in hatch covers with the hatch cover to the driver's front folding forward onto the engine compartment for increased visibility and those either side folding upwards on the outside.

The section commander sits behind the driver on the left with the radio operator in a similar position to his right. The eight infantrymen sit on individual seats, four down each side of the vehicle facing each other. They enter and leave the vehicle by two doors in the rear of the hull which open outwards and have a rectangular firing port that folds downwards on the outside. There are three similar firing ports in each side of the hull and a hull escape hatch over the second axle on each side.

Alvis Saracen APC from rear (Christopher F Foss)

Alvis Saracen with reverse-flow cooling (FV603(C)) on patrol in Cyprus (Ministry of Defence)

FV610 command vehicle showing higher hull and tent erected at rear

The turret, which is similar to that fitted in the Ferret scout car Mk 2/3, has a basket-type seat and is mounted in the forward part of the roof. The rear part of the turret roof folds forwards and can be locked horizontal if required while the turret rear folds downwards on the outside to form a seat for the gunner. The turret is armed with a 7.62 mm (0.30) Browning machine gun which has an elevation of +45° and a depression of −12°, except over the rear of the vehicle, and is fitted with a No 3 Mk 1 periscopic sight which is mounted in the forward part of the turret roof.

At the rear of the roof is a ring mount which is normally fitted with a 7.62 mm Bren LMG or sometimes a 7.62 mm (0.30) Browning machine gun. Three electrically operated smoke dischargers mounted on each mudguard at the front of the vehicle launch smoke grenades to the front of the vehicle.

The Saracen APC is not fitted with an NBC system or any night vision equipment, although the latter could be fitted if required. A fresh air ventilation system is standard on all vehicles and pipes air from the outside (via two dome-shaped ventilators on the right side of the hull) to ducts that run down each side of the personnel compartment. The Saracen has no amphibious capability, although a deep fording kit was developed for the vehicle.

Variants

Saracen with Reverse-flow Cooling (FV603 (C))

This was designed by Alvis specifically for operations in the Middle East. On the basic model air is drawn in through the louvres at the front of the engine compartment and expelled through the four hatch covers over the top. On the reverse-flow cooling model air is drawn in through raised louvres at the back of the engine compartment, passed forward over the engine by the fans the blades of which have been reversed, and then expelled through the radiators and front louvres. To prevent air rushing into the radiator and impeding passage of air through it in the opposite direction, a cowl has been mounted over the front of the vehicle.

Saracen with Open Roof

Some of the Saracens delivered to Kuwait in the 1960s were supplied with an open roof and a reverse-flow cooling system.

FV604 Command Vehicle

The FV604 is essentially the FV603 APC modified for use as a command vehicle.

The inside has been fitted with mapboards and modified to accommodate a crew of six: three staff officers, two radio operators and a driver. Externally, stowage baskets have been fitted as well as an auxiliary charging plant and additional batteries to power the communications equipment installed. Early vehicles retained the turret of the standard Saracen APC but most vehicles now have a ring-mounted machine gun in place of the turret. To increase the working area for the command staff a tent can be erected at the rear of the FV604. Loaded weight of the FV604 is 10 170 kg, length 4.851 m, width 2.515 m and height 2.057 m.

FV610 Command Vehicle

The full designation of the FV610 is Armoured Carrier 6 × 6 Command GPO/CPO FV610(A). It has a crew of six which normally consists of three staff officers, two radio operators and a driver. Its main difference from the FV604 is a wider and higher hull as it is anticipated that it will be used in the static role for longer periods than the FV604. Equipment fitted to the FV610 includes additional mapboards, communications equipment, an auxiliary charging plant and additional batteries. Loaded weight of the FV610 is 10 700 kg, length 4.851 m, width 2.515 m and height 2.362 m.

FV611 Ambulance

This has the same hull as the FV610 and can carry 10 seated patients, or three stretcher and two seated patients, or two stretcher and six seated patients in addition to the crew of two which consists of the driver and the medical orderly.

Saracen Repower

In 1991 it was announced that A F Budge Limited and Alvis had developed a conversion package for the Saracen 6 × 6 APC with the backing and assistance of Perkins Engines. The conversion involves the replacement of the existing engine by a Perkins Phaser 180 MTi diesel developing 180 hp. Further modifications include enhanced drive line components, improved braking system, new cooling pack and modern electrical system.

This conversion package offers the advantages of improved fuel efficiency, improved reliability, increased operational range and a reduced fire risk.

Saracen 88

Details of this are given in the following entry.

SPECIFICATIONS (FV603 (C) with reverse-flow cooling)

CREW	2 + 10
CONFIGURATION	6 × 6
COMBAT WEIGHT	10 170 kg
UNLOADED WEIGHT	8640 kg
POWER-TO-WEIGHT RATIO	15.73 hp/tonne
GROUND PRESSURE	0.98 kg/cm²
LENGTH HULL	5.233 m
WIDTH	2.539 m
HEIGHT	
to turret top	2.463 m
to hull top	2 m
GROUND CLEARANCE	0.432 m
TRACK	2.083 m
WHEELBASE	1.524 m + 1.524 m
ANGLE OF APPROACH/DEPARTURE	53°/53°
MAX ROAD SPEED	72 km/h
FUEL CAPACITY	200 l
MAX RANGE	400 km
FORDING	1.07 m
with preparation	1.98 m

GRADIENT	42%
VERTICAL OBSTACLE	0.46 m
TRENCH	1.52 m
TURNING RADIUS	7 m
ENGINE	Rolls-Royce B80 Mk 6A 8-cylinder petrol developing 160 hp at 3750 rpm
TRANSMISSION	pre-selector 5-speed
STEERING	recirculating ball, hydraulically assisted
CLUTCH	fluid coupling
SUSPENSION	independent double wishbone and torsion bar, hydraulic shock absorbers
TYRES	12.00 × 20
BRAKES	
main	hydraulic disc on all wheels
parking	mechanical on all wheels
ELECTRICAL SYSTEM	24 V

BATTERIES	2 × 12 V
ARMAMENT	
turret	1 × 7.62 mm MG
ring-mounted	1 × 7.62 mm MG
SMOKE-LAYING EQUIPMENT	2 × 3 smoke dischargers
AMMUNITION	3000
ARMOUR	
Hull visor plate	12 mm at 40°
Hull skid plate	10 mm at 40°
Hull sides	12 mm at 20°
Hull top	8 mm
Hull floor	12 mm
Turret front	16 mm at 15°
Turret sides	16 mm at 15°
Turret rear	16 mm at 15°
Turret top	8 mm

Status: Production complete. In service with Indonesia, Jordan, Lebanon, Libya (status uncertain), Nigeria, Qatar, South Africa (reserve), Sudan (status uncertain), Thailand, United Arab Emirates (reserve) and the United Kingdom.

Manufacturer: Alvis Limited
The Triangle
Walsgrave
Coventry
West Midlands CV2 2SP
UK
Telephone: (0203) 535455
Telex: 31459
Fax: (0203) 539280

GKL Saracen 88 Armoured Personnel Carrier

Development/Description

In 1988 GKL announced that it was offering an enhancement package for the Saracen under the designation of the Saracen 88. The company can offer vehicles upgraded to the Saracen 88 configuration or supply kits to enable the user to upgrade vehicles to this configuration in his own workshops.

The improvement package has a total of 23 improvements including the replacement of the original petrol engine by a more powerful and fuel efficient diesel engine, although the user can select only those improvements that meet his own specific requirements.

The upgrade allows full combat load operation in ambient temperatures in excess of +45° with improved engine and gearbox cooling.

Other improvements include new AC vehicle electrics, new hydraulic power steering, new vehicle exterior and interior lighting, new dual braking air over hydraulic system, new vehicle electrical wiring and instrumentation, improved machine gun turret mount, new modern smoke grenade launchers, new vehicle exterior fittings and stowage and new vehicle periscopes.

A similar automotive upgrade is offered for the Alvis Saladin (6 × 6) armoured car by GKL (qv).

Status: Development complete. Conversion package ready for production on receipt of orders.

Manufacturer: GKL Equipment PTE Limited
Coworth Park House
Coworth Park
Ascot
Berkshire SL5 7SF
Telephone: (0344) 874202
Telex: 846062 GKLLDN G

Saracen 88 upgraded by GKL

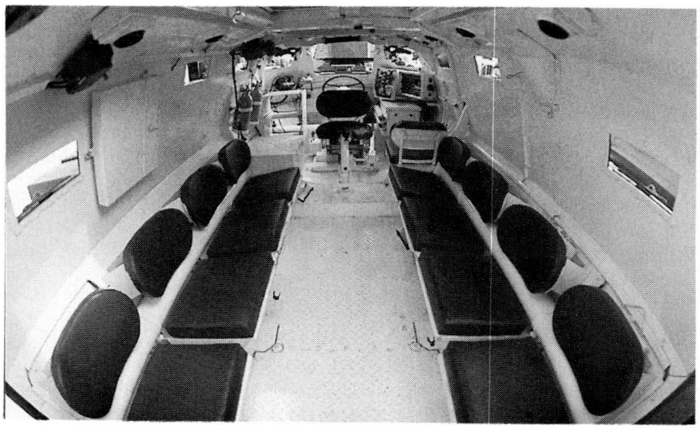

Interior of Saracen 88 upgraded by GKL which is configured for the commander plus 10 personnel

Alvis Spartan Armoured Personnel Carrier

The Alvis Spartan (FV103) is a member of the Alvis Scorpion range of light tracked vehicles and details of the complete family, including the Spartan, are given in the *Armoured fighting vehicle families* section.

Vickers Defence Systems Valkyr Family

Development

The Valkyr has been developed as a private venture by Vickers Defence Systems in association with Beherman-Demoen of Belgium which built the BDX (4 × 4) APC which is described in this section under Belgium.

Vickers was responsible for research, development and construction of the prototypes based on the initial Beherman Demoen BDX Mk II design. There are two basic variants of Valkyr which differ only in roof height, the higher roof vehicle being used for roles demanding maximum internal volume (eg APC) and the lower roof version as a Fire Support Vehicle.

A prototype low-roof FSV was shown at the 1982 British Army Equipment Exhibition. A pre-production high-roof APC was shown at the 1984 exhibition and a Fire Support Vehicle with a two-man 90 mm Cockerill turret was demonstrated at the firepower display. At the 1986 Exhibition an internal security version of the Valkyr APC was exhibited and in 1988 one was configured as a VIP transport vehicle.

Valkyr has been designed specifically to form the basis of a vehicle family fulfilling a variety of roles as a personnel carrier, fire support vehicle and internal security vehicle.

In addition to extensive trials in Belgium and the United Kingdom the Valkyr has also been tested in Brunei and Oman. Kuwait ordered two vehicles for use in the IS role and these were completed in late 1988 and has subsequently placed an order for a further batch of vehicles but work on these had not commenced at the time of the Iraqi invasion of Kuwait in August 1990.

Description

The basic monocoque hull of both versions of the Valkyr is all-welded high-hardness steel armour.

A high level of protection is available against a wide range of weapons, varying from 7.62 mm armour-piercing projectiles over the frontal arc and 7.62 mm ball over the remainder, up to M1 Garand armour-piercing at any range and any angle of attack (including the roof). This additional protection against armour-piercing rounds can be provided by a layer of appliqué armour, which in the event of penetration acts as a spall liner.

The hull is also protected from shell splinters from 155 mm howitzer projectiles. The armoured glass windscreen and vision blocks afford similar protection, while the underside of the hull gives protection against anti-personnel mines, by its special shaping and armour. In the event of an explosion, the front wings of the vehicle are designed to blow off to dissipate the force of the blast.

The driver sits centrally at the front of the vehicle. He has access to the troop compartment through a gangway on the left side of the hull.

On the APC a commander's cupola is fitted on the left side of the body between the front wheel arch and side door. A variety of cupolas are offered with various vision and weapon systems.

Access to the crew compartment is through two large rear-opening doors, and a side door on the left for the commander and driver. Stores may be loaded through a door on the right, which also acts as an emergency exit. Vision blocks are fitted above the side door and in each rear door. Weapon firing ports can be fitted to suit the user's needs.

Vickers Valkyr Fire Support Vehicle (FSV) with two-man CM 90 turret armed with Cockerill 90 mm Mk III gun, 7.62 mm coaxial and anti-aircraft machine guns and smoke dischargers

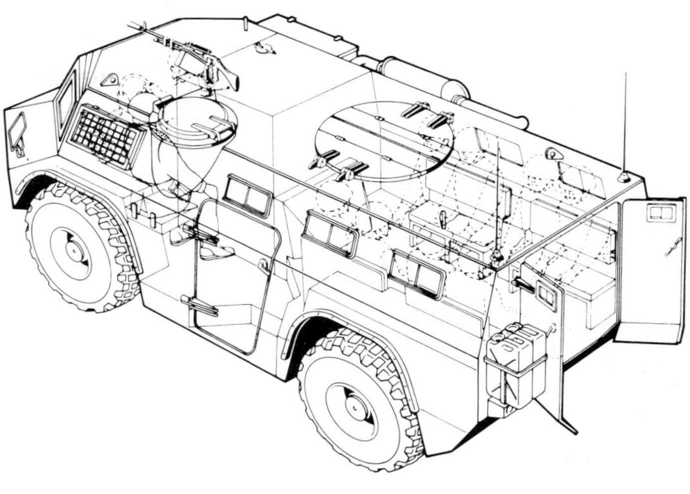

Seating arrangements of Vickers Defence Systems Valkyr (not to 1/76th scale)

Vickers Defence Systems Valkyr APC

Valkyr APC in internal security role with front-mounted obstacle-clearing blade

The fire support vehicle has a single rear door and additional doors on the left and right sides.

The engine compartment is to the rear of the driver on the right side with the air-inlet and air-outlet louvres in the roof and the exhaust pipe running alongside the upper part of the hull, on the right side. The powerpack is fully enclosed by noise absorption panels. A PTO at the front of the engine provides hydraulic power for brake and steering actuation, fan operation and brake cooling oil circulation.

A two-stage air cleaner is fitted. Large particles of dust and sand are removed by the cyclonic first stage while a dry filter element removes the remainder. An electrical restriction indicator provides a visual warning to the driver when the filter element needs replacing.

The Detroit Diesel 4-53T engine is water-cooled. The coolant flows through a gill-type radiator mounted horizontally above the engine. Air is drawn through the radiator by a hydraulically driven mixed flow fan. A second gill-type radiator is mounted below the coolant radiator and cools the axle/brake oil. The transmission is cooled by air drawn through vents in the rear over the engine transmission casings and expelled through the roof.

A Kidde-Graviner Firewire fire-detection system is fitted around the engine/transmission area, with a manually operated extinguishing system.

The engine is directly coupled to an Allison Transmission Division of General Motors AT-545 fully automatic gearbox with four forward and one reverse gears and a two-speed transfer case. For road use the rear-wheel drive is normally engaged, with the four-wheel drive being selected on soft ground as required. To improve traction an automatic locking differential is fitted to the rear axle and a manually lockable differential to the front axle.

The engine and transmission are all located within the armoured hull for maximum protection against disabling damage. Drive is provided to the hubs through the constant velocity joints on the front and universal joints on the rear. Hub reduction units are employed on all wheels.

Fully independent suspension is fitted all round and each wheel is located on double wishbones with variable rate coil springs and coaxial telescopic dampers. The steering is power-assisted rack and pinion.

The brakes are of the oil-immersed multiple disc type, which means that heat generated by braking is dispersed into the oil which is continually cycled through a remote cooler.

The footbrake operates hydraulically on all four wheels, with full power assistance. A dual hydraulic circuit, split vertically into front and rear axle systems, ensures braking capability in the event of either circuit failing. Hydraulic accumulators incorporated into the brake circuit provide a reserve of power to stop the vehicle even with the engine disabled. The mechanical handbrake operates on the rear wheels.

The large tyres of the Valkyr reduce ground pressure and improve traction on soft ground. They are also able to propel the vehicle through the water as the Valkyr is fully amphibious. Speed may be increased by the addition of a water propulsion system which can be mounted at the rear of the hull.

Valkyr as an APC can carry eight fully equipped troops plus the driver and commander/gunner. Ample storage is provided behind the seats for light anti-tank weapons, ammunition, 48-hour survival kit appropriate for NATO operations, water and spare fuel. The seats can be lifted to allow large loads to be carried and two standard NATO pallets may easily be accommodated in the hull. The Valkyr FSV has been designed to accept a complete range of turret and weapons including:

One-man turret with twin 7.62 mm machine guns
One-man turret with 7.62 mm and 12.7 mm machine guns
Two-man turret with 7.62 mm machine gun, 20 mm cannon and 60 mm gun/mortar
Two-man turret with 90 mm gun
One-man anti-aircraft turret with twin 20 mm cannon. In addition six crew members can be carried.

Optional equipment includes an air-conditioning system, NBC system, passive night vision equipment, alternative protection levels and run-flat tyres.

Variants

In addition to being used as an APC and FSV the Valkyr can be used for a wide range of other roles including anti-tank with ATGW, close support, recovery, workshop, state vehicle, command post, field ambulance, 81 mm mortar carrier and internal security vehicle. For the last role a variety of equipment can be fitted including a barricade removal device, grenade launchers, public address system, Molotov cocktail protection, roof-mounted machine gun, water cannon and armoured shutters. At the 1988 British Army Equipment Exhibition, Valkyr was shown configured as a VIP transport with interior by Rolls-Royce, Mulliner Park Ward.

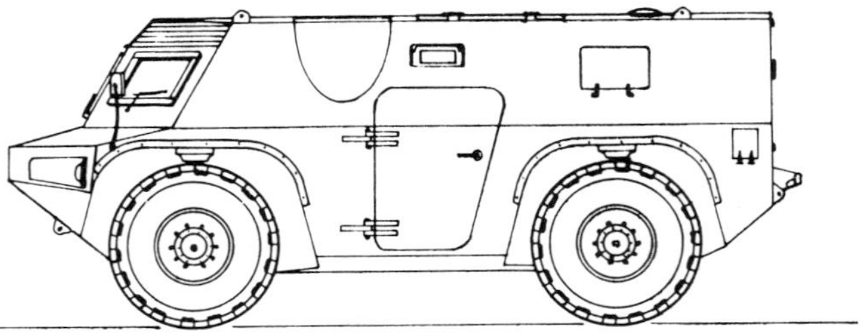

Basic APC version of Valkyr without armament

SPECIFICATIONS

Model	**APC**	**FSV**
CREW	2 + 8	2 + 6
CONFIGURATION	4 × 4	4 × 4
COMBAT WEIGHT	11 500 kg	11 500 kg
POWER-TO-WEIGHT RATIO	17.5 hp/tonne	17.5 hp/tonne
LENGTH HULL	5.6 m	5.6 m
WIDTH	2.5 m	2.5 m
HEIGHT TO HULL TOP	2.27 m	2.09 m
GROUND CLEARANCE	0.4 m	0.4 m
WHEELBASE	3.1 m	3.1 m
ANGLE OF APPROACH/ DEPARTURE	45°/50°	45°/50°
MAX SPEED	100 km/h	100 km/h
FUEL CAPACITY	200 l	200 l
MAX RANGE	700 km	700 km
FORDING	fully amphibious	fully amphibious
GRADIENT	60%	60%
SIDE SLOPE	30%	30%
VERTICAL OBSTACLE	0.4 m	0.4 m
TURNING RADIUS	8 m	8 m
ENGINE	Detroit Diesel 4-53T developing 300 hp at 2800 rpm	
TRANSMISSION	Allison Transmission Division of General Motors, AT-545 automatic with 4 forward and 1 reverse gears	
SUSPENSION	double wishbone	double wishbone
TYRES	14.00 × 20 XL	14.00 × 20 XL
BRAKES		
main	hydraulic	hydraulic
parking	mechanical	mechanical
STEERING	power-assisted rack and pinion	
ELECTRICAL SYSTEM	24 V	24 V
BATTERIES	2 × 12 V, 100 Ah	2 × 12 V, 100 Ah
ARMAMENT	see text	see text

Status: Production as required. Two vehicles were supplied to Kuwait but it is assumed that these were lost during the Iraqi invasion of 1990.

Manufacturer: Vickers Defence Systems
Armstrong Works
Scotswood Road
Newcastle-upon-Tyne NE99 1BX
UK
Telephone: (091) 273 8888
Telex: 531042
Fax: (091) 273 2324

GKN Defence Simba Mk 3 Light Combat Vehicle

Development

The Simba (4 × 4) family of Light Combat Vehicles has been designed as a private venture by GKN Defence specifically for the export market and meets the requirement for a wheeled family of vehicles capable of mounting up to 90 mm modern weapon systems. A removable roof plinth enables weapon systems to be exchanged rapidly.

The description and specification relate to the latest version of Simba which has a different hull and other improvements over the original vehicle which was described in previous editions of *Jane's Armour and Artillery*. GKN Defence is also offering the Simba as a technology transfer for local production overseas.

In mid-1991 it was expected that the Armed Forces of the Philippines (AFP) would soon sign a contract for the supply of 150 Simba (Lion) 4 × 4 APCs.

Prior to being selected by the AFP, the Simba undertook extensive in-country trials during which time some 7000 km were clocked up, both on road and cross-country.

The Philippines' Simbas will be customised to suit specific AFP requirements including a pintle-mounted 7.62 mm machine gun and larger capacity diesel fuel tanks for increased operating range.

In 1989 a joint venture company was established in the Philippines called Asian Armoured Vehicle Technologies Corporation (AAVTC), with GKN Defence having a 20 per cent interest. The other 80 per cent is held by Multi-Parts Motors International Incorporated (MMIIC) and Philippines Veterans Investment and Development Corporation (PHIVIDEC).

The AFP have a total requirement for 150 Simba vehicles, of which the first eight would be supplied ready for use from GKN Defence with the following 142 vehicles being supplied in kit form with the actual assembly being undertaken by the Philippine Corporation, (PHILCORP), who would also look after sales service of the vehicles.

GKN Defence Simba Light Combat Vehicle with 7.62 mm machine gun turret

The purchase of the first vehicles would be financed by a counter-purchase agreement and the Philippines would be allowed to export the Simba in others parts of the world including Africa and Far East.

Description

The hull of the Simba is of all-welded steel armour construction that provides the occupants with protection from 7.62 mm ball small arms fire and shell splinters. Thermal insulation on the roof and forced air circulation are fitted as standard. A heater and/or full air conditioning are available as optional extras as are various communications systems.

The driver sits at the front of the vehicle on the left and has bullet-proof windows to his front and either side and a single piece hatch cover that opens to the rear.

The engine is to the right of the driver with the air-inlet louvres at the lower front glacis plate and the outlet on the upper side of the glacis plate to the right rear of the driver's position. The Perkins TV8.540 rated at 212 bhp is coupled to a Clark combined gearbox/transfer box which transmits power to the front and rear axles.

The troop compartment is at the rear of the Simba. Armoured vision blocks are provided in each side of the troop compartment with firing ports positioned between the windows.

The troops enter and leave the Simba via a single large rear door or the door in the left hand side, each has a bullet-proof vision block with a firing port underneath.

A wide range of armament installations can be fitted to the Simba. When used as an APC the vehicle normally transports up to a crew of 12 including the commander and driver. When a one-man turret with single or twin 7.62 mm GPMG, or a 12.7 mm M2 machine gun is fitted a crew of 10 is normally carried.

Variants

Simba Low Profile APC

The basic APC version of Simba carries 10 to 12 men depending on the operational role, for example the driver front left, one man to his rear facing the rear, five men down the left side facing inwards, four men down the right side facing inwards and one man under the cupola facing the front.

Simba 90 mm Fire Support Vehicle

For trials purposes a GKN Defence Simba has been fitted with the Belgian Cockerill 90 mm CM90 turret armed with a 90 mm gun and 7.62 mm coaxial machine gun.

Simba Anti-tank

This would be fitted with the Emerson TUA (TOW Under Armour) system as fitted to the US Army M901 Improved TOW Vehicle or the Euromissile HOT HCT turret.

Simba Mortar

Fitted with a turntable-mounted 81 mm mortar in the rear of the hull.

GKN Defence Simba Light Combat Vehicle from rear without armament fitted (Christopher F Foss)

Pre-production GKN Defence Simba (left) with standard production Saxon APC (right)

Simba APC

With a GKN turret armed with a 7.62 mm or 12.7 mm machine gun, this version has a crew of 10 including the commander, gunner and driver.

Simba AIFV

This version has a removable roof plinth that can accept a wide range of 20 mm and 25 mm turrets.

Simba IS Vehicle

This version can be fitted with a wide range of equipment including side riot screens, barricade remover, public address system, siren or wailer, turret firing 38 mm CS and smoke rounds from within armour, collective CS gas filtration system and a winch.

SPECIFICATIONS

Type	**APC**	**AFV**
CONFIGURATION	4 × 4	4 × 4
CREW	12	10
COMBAT WEIGHT	9900 kg	10 600 kg
LENGTH	5.35 m	5.35 m
WIDTH	2.5 m	2.5 m
HEIGHT (hull top)	2.19 m (w/o armament)	2.5 m
GROUND CLEARANCE		
(axles)	330 mm	330 mm
(hull)	450 mm	450 mm
WHEELBASE	2.972 m	2.972 m
TRACK		
front	2.06 m	2.06 m
rear	2.12 m	2.12 m
ANGLE OF APPROACH/ DEPARTURE	45°/40°	45°/40°
FUEL CAPACITY	240 l	240 l
MAX ROAD SPEED	100 km/h	100 km/h
ROAD RANGE	650 km	650 km
FORDING	1 m	1 m
GRADIENT	60%	60%
SIDE SLOPE	40%	40%
VERTICAL OBSTACLE	0.45 m	0.45 m
TURNING RADIUS	8.25 m	8.25 m
ENGINE	Perkins TV8.540 diesel, 212 bhp at 2500 rpm	
TRANSMISSION	Clark 13.1 HR 28422 semi-automatic, 4 forward and 2 reverse gears	
ARMAMENT	pintle GPMG or GKN turret with 7.62 mm or 12.7 mm MGs	20 mm Oerlikon cannon, 7.62 mm coaxial GPMG or 90 mm Cockerill Mk III Gun, 7.62 mm coaxial GPMG
STEERING	power-assisted	power-assisted
SUSPENSION	semi-elliptical springs with hydraulic shock absorbers	
TYRES	14.75/80 × 20 (run flat)	
BRAKES		
main	drum, air/hydraulic (dual circuit)	
parking	transmission brake	
ELECTRICAL SYSTEM	24 V	24 V

Status: Development complete. Ready for production.

Manufacturer: GKN Defence
PO Box 106
Hadley Castle Works
Telford
Shropshire TF1 4RQ
UK
Telephone: (0952) 244321
Telex: 35248/9 sanhad g
Fax: (0952) 244947

GKN Defence Saxon Armoured Personnel Carrier

Development

The AT105 was developed by GKN Defence as the successor to its earlier AT104. The first prototype of the AT105 was completed in 1975 and the first production vehicles in 1976.

Main improvements over the original AT104 can be summarised as: the armoured hull has been designed to give the same degree of protection to the radiator, engine and transmission as to the crew; redesigned hull floor to give increased protection against damage from mines; more powerful engine and a shorter wheelbase for improved mobility and turning.

In mid 1982 the AT105 was named the Saxon. Early in 1983 the British Ministry of Defence placed an order with GKN Defence for an initial 47 Saxon APCs for the British Army, with first deliveries made in early 1984.

The second order for the British Army comprised 247 vehicles, deliveries of which have now been completed. Late in 1985 the British Army exercised its option for a further 200 Saxons which were all completed by early 1989. The Royal Artillery has ordered 30 Saxons for use as command post vehicles with Rapier surface-to-air missile units and final deliveries of this version were made in early 1987.

Each British Army Saxon carries a fully equipped rifle section of 10 and is being issued to 11 Type A UK-based infantry battalions that will reinforce BAOR in time of war. The British Army Saxons have a fixed observation cupola for the commander, with a socket in each corner of the cupola for a 7.62 mm GPMG DISA mount. British Army vehicles have no door in the left side of the hull, a stowage rack on the roof and external fully enclosed bins. Two variants have been supplied to the British Army, command and recovery. The latter has a side-mounted five-tonne hydraulic winch that can be used to the front or rear and is driven from a PTO from the gearbox. There are interior seats for a crew of four in addition to stowage for recovery equipment.

By March 1984 the first group of the British Army's Saxon APCs had successfully completed the first familiarisation course at the Infantry Trials and Development Unit, School of Infantry, Warminster, and first deliveries had been made to the 1st Battalion The King's Own Border Regiment for operational training.

Following successful user trials by the British Army, the vehicle is now being considered for a number of additional roles including command vehicles, ambulance, infantry support weapons carriers and for specialist Air Force requirements.

In 1988 the Hong Kong Police Force took delivery of seven Saxon vehicles for use by the Police Tactical Unit and these replaced older Saracen vehicles. Of the seven vehicles, two are command post vehicles and the remaining five are personnel carriers carrying 10 men plus commander and driver.

Saxon APC as delivered to British Army with 7.62 mm GPMG installed on DISA mount forward of commander's cupola

Saxon recovery vehicle with side-mounted 5-tonne hydraulic winch, as used by British Army

By early 1989 the British Army had taken delivery of 524 production Saxon (4 × 4) APCs and variants with a further quantity of vehicles required in the future as command post vehicles for use with the Shorts Starstreak High Velocity Missile System.

At the 1990 British Army Equipment Exhibition Saxon was relaunched with a Cummins 6BT 5.91 turbocharged six cylinder diesel developing 160 bhp coupled to a fully automatic transmission from the Allison Transmission Division of General Motors.

GKN Defence are offering Saxon to meet part of the British Army's requirement for a Future Family of Light Armoured Vehicles (FFLAV).

Late in 1990, the UK MoD placed an order with GKN Defence for an additional batch of Saxon (4 × 4) APCs worth approximately £20 million. In addition it included an order for the Saxon ambulance. This was the first production order for the Saxon ambulance which was originally developed as a private venture. It is understood that around 100 Saxon vehicles will be built under this latest contract with deliveries commencing in 1991.

The new production Saxons will have a number of improvements over current vehicles including a Cummins 6BT 5.9 litre turbo-charged six cylinder diesel developing 160 hp coupled to an Allison Transmission Division of General Motors Corporation automatic transmission.

Description (Right-hand drive model; left-hand drive also available)
The basic model has a hull of all-welded steel with the driver seated at the very front of the hull on the right side and the personnel compartment at the rear. The hull provides the crew with complete protection against small arms fire up to and including 7.62 mm armour-piercing rounds, and HE shell fragments up to 155 mm when burst at 10 m from the vehicle. The floor is V-shaped to give maximum protection against mine blast, except for the axles, which are outside the armoured section. The engine, transmission, fuel, batteries, as well as the vacuum servos and hydraulics are all contained within the armoured compartment. The engine, transmission and other automotive components are standard commercial items. British Army vehicles have a high degree of commonality in engine and axle components with other in-service vehicles.

The driver can reach his seat either from inside the vehicle or by a forward-opening hatch cover over his position. There are bullet-proof windscreens in front and on both sides.

The personnel sit four down each side of the hull on padded bench seats which are usually equipped with seat belts, and enter and leave the vehicle by two doors in the rear of the hull and a single door in each side of the hull. The British Army specification has minor differences from other in-service vehicles and these include exterior stowage bins and no left hand door. Each of the doors has a firing port and a vision block. There are additional firing ports in each side of the hull and a single firing port in the forward part of the hull on the left side. Specially designed ball mounts have also been developed which can be installed in the firing ports to allow the crew to fire their weapons from inside the vehicle in safety. The interior of the hull is lined with 26 mm thick thermal insulation and a forced air circulation system is fitted as standard on all vehicles.

The commander's cupola has four sides each with a vision block, and a single piece hatch cover that opens forwards. A 7.62 mm machine gun can be pintle-mounted at this position if required. The commander's cupola is welded to an oblong sheet of armour which in turn is bolted to the roof of the vehicle. This allows the command module to be removed and replaced by another module with a different armament installation. When the engine needs replacement the roof module is removed first and the complete engine and transmission can be lifted through the roof. Run-flat cross-country tyres are fitted as standard and in a 'shot through' condition the vehicle can travel 96 km at 48 km/h.

Optional equipment for the Saxon includes air-conditioning systems, auxiliary power unit, barricade remover, door safety mechanism, grenade launchers, hand-operated searchlight, heater, rotating/flashing beacons and siren, loud speakers and a front-mounted winch.

Variants

Saxon LHD Armoured Personnel Carrier

This is as AT105-P but fitted with a turret armed with a single 7.62 mm GPMG with an elevation of +50° and a depression of −15°, or a turret armed with twin 7.62 mm GPMGs with identical elevation and depression. The Saxon can also be used to deploy and operate a remote-controlled EOD vehicle for bomb disposal duties.

Saxon Command Vehicle

This has the commander's cupola as fitted to the Saxon but can be fitted with a GKN turret armed with single or twin 7.62 mm GPMGs or a 12.7 mm HMG. The interior is modified for use as a command vehicle: on the right side of the hull are staff positions with individual communication control modules, and on the left is a mapboard with strip lighting, and storage space behind. A teleprinter operator's position with seat and table behind are also provided. A similar configuration has been adopted for the British Army Command vehicles.

Saxon Armoured Ambulance

This is a Saxon with commander's cupola and can carry four stretcher patients, one or two orderlies, commander and driver, or two stretcher cases, four walking wounded, one orderly, commander and driver.

Saxon with ARWEN 37V

In February 1984 GKN Defence announced that a new version of the Saxon had been developed for use in IS operations. This has a one-man turret fitted with the ARWEN 37V rapid fire weapon which can fire five rounds in quick succession before it needs reloading. It is mounted in a new turret design that allows high levels of visibility and the ability to follow the flight of CS or smoke rounds.

Saxon Recovery Vehicle

This model is already in service with the British Army and is the standard Saxon modified to recover other vehicles. Mounted on the left side of the hull is a Hudson Wharton capstan 5000 kg hydraulic winch that can handle vehicles weighing up to 16 000 kg with the aid of a block and tackle.

The winch has 40 m of cable, has a slow speed of 9.1 m/min and a fast speed of 30.84 m/min. It can pull vertically at angles from 10 to 40° and horizontally from +15° to −15°. The winch can be operated by remote-control.

The vehicle is normally operated in the British Army by four members of the Royal Electrical and Mechanical Engineers (REME). If required a lightweight tent can be erected at the rear so repair work can be carried out in all weathers.

Standard equipment includes tools, spotlight, Scotch lorry skid pads and a front and rear differential lock.

Ambulance

A prototype of this was completed in 1984 and demonstrated to the British Army.

The Saxon ambulance is the standard vehicle with internal fittings to enable four general service stretchers to be carried, two either side, or two stretcher patients plus five walking wounded.

The crew would consist of the commander, driver, and one medical orderly. There is ample space inside the vehicle for medical equipment and in addition there is external stowage on the roof and sides.

Incident control vehicle

GKN Defence have completed the prototype of an incident control vehicle fitted with front-mounted obstacle clearing blade, wire mesh protection for driver's windows, turret armed with machine gun and smoke dischargers, wire mesh screens and a low light level TV surveillance system on a telescopic mount at the rear of the vehicle with a monitor/record/playback facility inside of the vehicle.

Saxon ambulance which is unarmed

Saxon internal security vehicle with roof-mounted TV system and mesh screens folded back along side of hull

Water Cannon

This version has been designed specifically for the internal security role and has a turret-mounted water cannon.

A water pump delivers 341 litres per minute at a high pressure of 44 kg/cm^2 from a 1363-litre steel or GRP tank carried inside the troop compartment. Sufficient water is carried for four minutes of continuous operation, but in normal practice only short bursts are used giving approximately 30 minutes of operation.

The water cannon is fitted with a specially developed 10 mm nozzle and trials have shown that rioters can be knocked down at a range of 15 m from the vehicle, with rioters being discouraged at a range of up to 30 m.

Full filtration is fitted to the water inlet and outlet to allow filling from any source and baffles in the water tank prevent water surging.

The water cannon can be installed or removed from a standard Saxon in four hours so enabling a rapid change in role according to the tactical situation. The water tank is fitted and removed through the rear doors.

Internal Security Turret

In 1990 GKN Defence announced that it had designed an Internal Security Turret (IST) which could be fitted to a variety of APCs including its own Saxon vehicle.

This is armed with a five shot 37 mm riot gun with 20 rounds of ready use ammunition with a further 100 rounds being carried in the turret. The turret shell is the same as that of other GKN light turrets and can be configured for alternative fits such as a light machine gun or water cannon. In addition to the main option, 66 mm smoke dischargers can be fitted to give 180° forward spread or directional employment.

Other features include a ×1 day sight or ×1.4/×6 day/night sight, 150 W coaxial spotlight, 150 cfm fume extraction, 100 Amp and 12 channel slip ring, 70 W flashing beacon and various communications fittings.

Status: In production. In service with:

Country	Quantity	Comment
Bahrain	2	
Hong Kong	7	delivered 1988
Kuwait	4	status uncertain
Malaysia	44	delivered 1978/79
Nigeria	75	
Oman	20	Royal Guard and Police
UAE	8 + 2	
UK	524	delivered 1984/1989 with a further batch of 100 to be delivered from 1991 onwards

Manufacturer: GKN Defence
PO Box 106
Hadley Castle Works
Telford
Shropshire TF1 4RQ
UK
Telephone: (0952) 244321
Telex: 35248/9 sanhad g
Fax: (0952) 244947

SPECIFICATIONS

CREW	2 + 8 (or 10)
CONFIGURATION	4 × 4
COMBAT WEIGHT	11 660 kg
UNLOADED WEIGHT	9940 kg
POWER-TO-WEIGHT RATIO (164 bhp engine)	14.06 bhp/tonne
LENGTH	5.169 m
WIDTH	2.489 m
HEIGHT	
to commander's cupola	2.628 m
top of turret, if fitted	2.86 m
GROUND CLEARANCE	
hull	0.41 m
axles	0.29 m
TRACK	
front	2.08 m
rear	1.99 m
WHEELBASE	3.073 m
ANGLE OF APPROACH/DEPARTURE	48°/40°
MAX ROAD SPEED	96 km/h
FUEL CAPACITY	153 l
MAX ROAD RANGE	480 km
FORDING	1.12 m
GRADIENT	60%
VERTICAL OBSTACLE	0.41 m
TURNING RADIUS	7.57 m
ENGINE	Bedford 500 6-cylinder diesel developing 164 bhp at 2800 rpm*
TRANSMISSION	Allison Transmission Division of General Motors AT-545 automatic with 4 forward and 1 reverse gears
TRANSFER CASE	2-speed
STEERING	Burman, power-assisted
SUSPENSION	semi-elliptical springs and hydraulic shock absorbers
TYRES	14.00/74 × 20 or 14.00 × 20 British Army vehicles are fitted with French Hutchinson run-flat cores
BRAKES	
main	drum, air/hydraulic (dual circuit)
parking	transmission brake
ELECTRICAL SYSTEM	24 V
BATTERIES	2 × 12 V, 100 Ah
ARMOUR	proof against 7.62 mm/0.30 AP rounds at point blank range

* Note: It can also be fitted with a Perkins T6.3544 diesel developing 195 bhp at 2500 rpm. From mid-1990 the Saxon has been offered with the Cummins 6BT 5.91 litre turbocharged six cylinder diesel developing 160 bhp coupled to a fully automatic transmission

Glover Transac GS 4 × 4 Armoured Personnel Carrier

Development

The Glover Transac GS 4 × 4 APC was developed as a private venture in 1986 to meet the specific requirements of Pakistan. After in-country trials with a prototype vehicle, four production vehicles were ordered and by early 1991 over 70 vehicles had been ordered by overseas customers

The vehicle is based on the chassis of the British Stonefield 4 × 4 truck which is fully described in *Jane's Military Vehicles and Logistics 1991-92*, page 404.

One of the design requirements was for the vehicle to have clean lines and a non-aggressive appearance.

Description

The layout of the vehicle is of the forward control configuration with the engine mounted centrally between the driver and co-driver/vehicle commander. An armoured engine/transmission tunnel and half bulkhead separates the driver's area from the rear crew area.

The driver and commander are seated at the front of the vehicle, each being provided with a large bullet-proof window to their immediate front and a side door that opens forwards. The troop compartment is at the rear with the five troops or police being seated down either side on bench seats. The latter can enter the vehicle via two doors in the rear or a single door in each side. Each door is provided with a bullet-proof observation block with a sliding firing port underneath and in addition there are two bullet-proof observation blocks with a firing port underneath in either side of the troop compartment giving the vehicle a total of eight.

The hull of the vehicle is of all-welded steel construction which provides the crew with complete protection from 7.62 mm and 5.56 mm ball small arms fire. The front windscreen and cab window are of bullet-proof multi-laminate glass with a polycarbonate anti-spall liner giving the same level of protection to the steel armour. Floor armour is provided to protect the occupants against anti-personnel mines, nail bombs, and pipe bombs.

Standard equipment includes air conditioning, explosion suppressed fuel tank, and a run-flat tyre system. Optional equipment includes rifle racks, public address system, roof lighting systems, machine gun hatch, patrol hatch, smoke/grenade launchers, and window grilles.

Variants

In 1988 was shown fitted with the MEL developed Setter vehicle mounted ESM system. The antenna is mounted on the roof and normally kept under clamshell doors and raised into vertical position when required for use. In 1989 the Mk II version was launched which has a number of improvements including a Renault diesel engine.

Late in 1990, the manufacturer suggested that the vehicle could undertake the following roles and missions:

(1) 14 seater riot control vehicle
(2) 4000 litre protected water cannon
(3) Command and communications unit
(4) Four stretcher armoured ambulance with four man crew
(5) Short wheelbase anti-terrorist vehicle
(6) Bomb disposal vehicle with blast container at the rear
(7) Highline escort/airport security unit
(8) Military 14 seater with 7.62 mm MG turret
(9) Border patrol and/or light reconnaissance
(10) Protected bank unit with payload of 3 tonnes
(11) Armoured patrol/escort unit, short wheelbase with body similar to French Panhard VBL.

Status: In production. In service with undisclosed foreign armies and police forces, including Pakistan and Saudi Arabia.

Manufacturer: Glover Webb Limited
Hamble Lane
Hamble
Hampshire SO3 5NY
UK
Telephone: (0703) 452811
Telex: 477527
Fax: (0703) 452964

Glover Transac GS 4 × 4 APC with wire mesh protection for windscreens. Rear troop compartment has bullet proof observation blocks and sliding firing ports

Glover Transac GS 4 × 4 APC fitted out for police role with special roof for use in hot climates

SPECIFICATIONS

CREW	2 + 10
CONFIGURATION	4 × 4
COMBAT WEIGHT	6500 kg
POWER-TO-WEIGHT RATIO	16.3 bhp/tonne
LENGTH	5 m
WIDTH	1.97 m
HEIGHT	2.3 m
GROUND CLEARANCE	0.29 m
TRACK	1.97 m
WHEELBASE	2.8 m
MAX ROAD SPEED	100 km/h
FUEL CAPACITY	159 l
MAX ROAD RANGE	650 km
GRADIENT	60%
SIDE SLOPE	40%
ENGINE	Perkins Phaser 110 T 4-cylinder in-line turbocharged diesel developing 106 bhp at 2800 rpm
TRANSMISSION	manual, 5 forward and 1 reverse gears
CLUTCH	single dry plate
TRANSFER CASE	2-speed
STEERING	re-circulating, ball and nut, power assisted
SUSPENSION	multi-leaf semi-elliptical, hydraulic dampers fitted front and rear

Humber One-ton Armoured Personnel Carrier

Development

After the end of the Second World War, an entire range of one-ton (4 × 4) vehicles was developed specifically to meet the requirements of the British Army by Rootes under the Humber name, known as the FV1600 series. As insufficient Alvis Saracen (6 × 6) APCs were available in the early 1950s it was decided to produce an APC model of the FV1600 series. About 1700 were subsequently built with the chassis supplied by Rootes at Maidstone in Kent and the armoured bodies by GKN (which later built the FV432 tracked armoured personnel carrier) and the Royal Ordnance Factory at Woolwich.

By the late 1960s most of the vehicles had been withdrawn from service and scrapped but when the situation in Northern Ireland became worse the vehicles were brought back into service as internal security vehicles, in which role they remain in service today.

Under the Conventional Forces Europe (CFE) agreement signed in 1990, the UK stated that the British Army had a total of 313 Humber (4 × 4) vehicles of which 274 were in the UK (including Northern Ireland) and 39 elsewhere.

Description

The basic APC is designated the FV1611 or Truck 1-ton Armoured 4 × 4 Humber, or more commonly the Pig. It has an all-welded hull with the

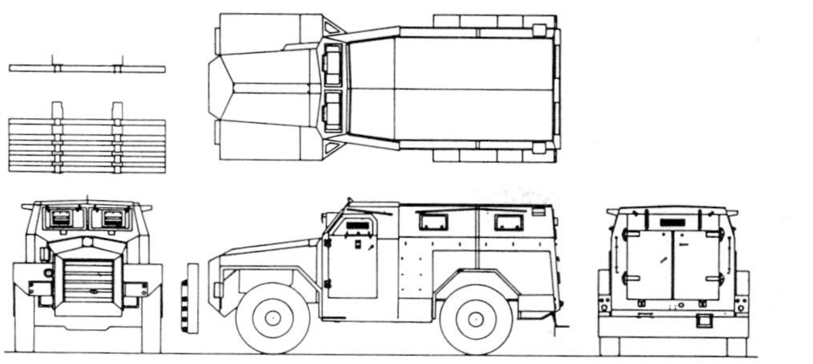

Five-view drawing of FV1611 Mk 2 1-ton armoured personnel carrier (not to 1/76th scale) (R Lockie)

FV1611 APC modified for use in Northern Ireland (Ministry of Defence)

engine at the front, commander and driver in the centre and the personnel compartment at the rear.

The driver sits on the right with the commander on the left, both with a windscreen to their front which when in action is covered by an armoured shutter hinged at the top, with an integral vision block. Most vehicles have had the windscreen removed. Both the driver and commander also have a side door which opens forwards, the top of which opens to the outside for increased visibility. The top part has an integral vision block. Over the top of the commander's and driver's position is a circular hatch that opens forwards.

The six men sit on bench seats, three down each side of the vehicle, although a maximum of eight can be carried in the rear. They enter and leave the vehicle by twin doors in the rear of the hull, each of which has a firing port. As originally built two rectangular firing ports, hinged at the bottom, were provided in each side of the troop compartment, but many vehicles have had a ventilator fitted over one of these firing ports in each side of the hull.

As a result of combat experience in Northern Ireland all vehicles have been up-armoured to enable them to withstand attack from 7.62 mm armour-piercing rounds, which has increased the vehicle's weight to about 7000 kg. Other modifications include the removal of external stowage boxes, a drop panel fitted under the rear of the hull to stop bullets going under the vehicle and striking the legs of troops standing at the rear and modifications to the rear doors: cutting off the top quarter and fitting a hinged flap to the top of the hull.

There are a number of specialised versions of the Pig in service in Northern Ireland, including vehicles with barricade-removing equipment at the front of the hull, special EOD equipment and some with a protected observation position in the roof. Some vehicles have been fitted with searchlights and tear gas launchers.

Variants

FV1609, original armoured version but the troop compartment at the rear has an open roof.

FV1612, radio vehicle with crew of three (commander, driver and radio operator).

Humber one-ton APC showing firing ports in hull side (Christopher F Foss)

FV1613, ambulance with crew of two (driver and medical orderly) and accommodation for three stretchers or eight sitting patients or one stretcher and four sitting patients.

The basic truck version, the FV1601, is no longer in service with the British Army and the FV1620 armoured Hornet vehicle, which launched the Malkara ATGW, has also been withdrawn from service.

Status: Production complete. In service with Malta and United Kingdom.

Manufacturers: Chassis, Rootes Limited, bodies GKN and ROF Woolwich.

SPECIFICATIONS		WHEELBASE	2.743 m	STEERING	re-circulating ball
CREW	2 + 6 (or 8)	MAX ROAD SPEED	64 km/h	CLUTCH	single dry plate
CONFIGURATION	4 × 4	FUEL CAPACITY	145 l	SUSPENSION	torsion bar and tele-
COMBAT WEIGHT	5790 kg	MAX ROAD RANGE	402 km at 64 km/h		scopic shock absorber
UNLOADED WEIGHT	4770 kg	ENGINE	Rolls-Royce B60 Mk		for each wheel station
POWER-TO-WEIGHT			5A 6-cylinder petrol	TYRES	11.00 × 20
RATIO	20.72 bhp/tonne		developing 120 bhp	BRAKES	
LENGTH	4.926 m		at 3750 rpm	main	hydraulic, drum
WIDTH	2.044 m	TRANSMISSION	manual with 5 forward		pattern
HEIGHT	2.12 m		and 1 reverse gears		on all wheels
TRACK	1.713 m	TRANSFER BOX	single speed	parking	hand, mechanical

Hotspur Hussar Armoured Personnel Carrier

Development

The Hussar APC and Hussar internal security vehicle were originally developed as a private venture by Hotspur Armoured Products of Wales and shown for the first time at the 1984 British Army Equipment Exhibition.

Following trials with two prototypes the first export order was placed and by early 1987 no less than four countries had placed firm orders for the vehicle.

In August 1986 Hotspur Armoured Products became a division of Penman Engineering Limited in Scotland where the design and manufacturing facility is now located.

Description

The Hussar APC is based on the chassis of the Land Rover One Ten with an additional powered axle, modified and strengthened to accommodate the weight of the armoured hull.

The Hotspur steel armoured hull, which fully encloses the engine and troop compartment, has been designed to provide ballistic immunity at 25 m at a 90° attack angle from ball ammunition fired from 7.62 mm type weapons and from point-blank attack by lighter weapons such as the 9 mm Sterling SMG, M1 carbine and .44 Magnum. The floor armour protects the crew from mine fragments and grenades.

The engine and transmission are at the front of the vehicle with the troop compartment at the rear. The engine is coupled to a manual transmission and the driver can select either 6 × 6 or 6 × 4 (first and second axles) depending on the terrain encountered.

The commander and driver sit to the rear of the engine and each has a laminated armoured glass windscreen to his front and side with anti-splinter visors of polycarbonate. The visors can be activated from inside the vehicle either manually or electrically. In each side of the hull is a single door that opens to the front, in the upper part of this is a bullet-proof vision block with a firing port underneath.

The troop compartment is at the rear of the hull and the 12 troops sit on individual flip-up seats on either side of the hull and enter via the side doors or the two doors in the hull rear. In each side of the troop compartment are two vision blocks with a firing port underneath and each of the rear doors also has a firing port with vision block underneath. The troop compartment is 3.25 m long, 1.75 m wide and 1.39 m high.

If required, a light armour turret or a fully rotating machine gun hatch mounting a 7.62 mm machine gun can be fitted to the roof. The basic turret has manual traverse but powered traverse is an option.

Standard equipment for the Hussar includes a full air-conditioning system for the troop compartment in addition to a forced draught ventilation system.

Optional equipment includes a 24 V electrical system, winch, spot lamps, intercom/siren, public address system, two four-barrel smoke/grenade launchers on the turret, self-sealing fuel tank, fuel tank with an additional 40-litre capacity, powered steering, run-flat tyres with split rims and night vison equipment for the driver.

Variants

A number of variants of the Hussar are available, which include a field ambulance, internal security vehicle and a communications/command vehicle.

Hotspur Hussar (6 × 6) APC fitted with turret

SPECIFICATIONS	
CREW	1 + 13
CONFIGURATION	6 × 6
COMBAT WEIGHT	5350 kg
POWER-TO-WEIGHT RATIO	25 bhp/tonne
LENGTH	5.994 m
WIDTH	2.489 m
HEIGHT	
with turret	2.833 m
to hull roof	2.383 m
GROUND CLEARANCE	0.21 m
TRACK	1.486 m
WHEELBASE	3.81 m
ANGLE OF APPROACH/ DEPARTURE	40°/30°
MAX ROAD SPEED	120 km/h
FUEL CAPACITY	98 l
MAX RANGE	300 km
TURNING RADIUS	8.65 m
ENGINE	3.528-l V-8, water-cooled petrol developing 134 hp at 4000 rpm
TRANSMISSION	manual, 5 forward, 1 reverse gears
STEERING	power, worm and roller
CLUTCH	single dry plate
TRANSFER BOX	2-speed
SUSPENSION	
front	dual rate coil springs, live beam axle, double acting hydraulic dampers
middle and rear	A frame, live rear axle, single rate coil springs, double acting hydraulic dampers
BRAKES	
front	hydraulic, servo-assisted disc
rear and middle	hydraulic, servo-assisted drum
TYRES	7.50 × 16
ELECTRICAL SYSTEM	12 V
BATTERY	2 × 12 V, 95 Ah

Status: Production as required. In service with Egypt (11) and three other undisclosed countries.

Manufacturer: Hotspur Armoured Products Limited
Division of Penman Engineering Limited
Heathall
Dumfries DG1 3NY
UK
Telephone: (0387) 52784
Telex: 779771
Fax: (0387) 67332

Hotspur 4984 (4 × 4) Armoured Personnel Carrier

Development

The Hotspur APC has been developed as a private venture and was shown for the first time at the British Army Equipment Exhibition held in June 1980. By early 1981 the prototype had successfully completed its trials and production began in 1982.

The 4 × 4, the Model 4984, is based on the coil spring Land Rover One Ten.

The APCs have over 90 per cent commonality of spares with the standard Land Rover and full spare parts back up is therefore available from the extensive network of Land Rover distributors all over the world. Servicing is no more difficult than with the Land Rover and all routine maintenance points are readily accessible and major items such as the engine or gearbox are easily removed and replaced.

Description

The layout of the Hotspur APCs is similar to the Land Rover One Ten with the engine at the front, driver and commander to the rear of the engine and the troop compartment at the rear. The chassis has been modified, reinforced and internally rustproofed.

The steel armour is resistant to multi-impact all-round attack from high velocity rifles at 25 m and shell splinters. The fully welded body unit, complete with seats, is bolted to the chassis and is detachable.

The V-8 engine is coupled to a manual transmission with four forward and one reverse gears, two-speed transfer box with permanent four-wheel drive, and a lockable central differential.

The radiator grille and engine compartment sides are armoured and the integral welded steel floor with angled sections gives protection against sub-surface mine fragments and grenade attack.

The commander and driver have side doors that open forwards, the upper part of the doors has transparent armour of multi-laminated glass construction with separate polycarbonate anti-spall screens resistant to attack from high velocity rifles and lesser weapons to British Standard Specification 5051 level G3. The windscreens to the front of the commander and driver are also of transparent armour with a similar level of protection. The commander and driver each have a Land Rover de-luxe seat complete with seat belt.

The model 4984 (4 × 4) can carry, including the driver, eight fully equipped infantrymen in the rear.

Locking bars are provided on all doors and rear steps, spring loaded, assist the infantry in leaving the vehicle. A grab rail runs down the roof of the vehicle and the floor of the troop compartment is non-slip.

Firing ports with vision blocks, two in each side of the hull and two in the rear doors, allow some of the infantry to fire their small arms from within the vehicle.

Hotspur 4984 (4 × 4) APC

The on/off road tyres are fitted with safety run-flat bands.

Optional equipment includes automatic engine fire-extinguishing system, auxiliary 50-litre petrol tank, barricade ram, CS gas/smoke dischargers, driving siren, external spotlamps, front-mounted twin spotlamps (internally controlled), front-mounted power winch, full air-conditioning, full harness seat belts in rear, fully adjustable Bostrum shock-absorbing driver's seat, heavy duty alternator, helicopter lifting lugs, hostile fire indicator, high-power solvent screen wash, roof-mounted lightweight armoured machine gun cupola with periscope to accept 7.62 mm GPMG or Minimi and mount, lightweight fully rotating armoured machine gun turret, public address system, radio base or slave equipment, self-sealing petrol tank, twin driving spotlamps, two-way intercom and various communications systems.

Variants

The manufacturer has other body options including command/communication vehicle, field ambulance, long distance patrol vehicle and a fuel or water tanker.

SPECIFICATIONS					
CONFIGURATION	4 × 4	WHEELBASE	2.797 m	TRANSFER BOX	2-speed
CREW	2 + 6	TRACK	1.486 m	STEERING	worm and roller power-assisted
COMBAT WEIGHT	3600 kg	TURNING RADIUS	6.4 m	SUSPENSION	coil springs, live beam axles, double-acting hydraulic dampers
LENGTH	4.438 m	FUEL CAPACITY	90 l		
WIDTH	1.79 m	ENGINE	V-8 petrol developing 134 hp at 4000 rpm		
HEIGHT	2.085 m	TRANSMISSION	manual, 5 forward and 1 reverse gears	ELECTRICAL SYSTEM	12 V, split charge
GROUND CLEARANCE	0.215 m			BATTERY	2 × 12 V, 95 Ah

Status: Production as required. In service with Sri Lanka and an undisclosed Gulf State army. Two older 6 × 6 vehicles were purchased by Finland for use with their United Nations Forces. Egypt is believed to have 63 Hotspur 4 × 4 vehicles in service.

Manufacturer: Hotspur Armoured Products Limited
Division of Penman Engineering Limited
Heathall
Dumfries DG1 3NY
UK
Telephone: (0387) 52784
Telex: 779771
Fax: (0387) 67332.

Hotspur Polisec Armoured Personnel Carrier

Development

The Polisec APC was developed as a private venture by Hotspur Armoured Products in response to a demand for a vehicle with a lower profile and less hostile appearance than other vehicles of this type and is mainly for internal security use.

Description

The Hotspur Polisec is based on the chassis of the Land Rover One Ten with an additional powered axle. The chassis has been lengthened and reinforced to accommodate the weight of the monocoque armoured hull.

The layout of the vehicle is conventional with the engine at the front, commander and driver in the centre and the troop compartment extending right to the rear.

The Hotspur steel armour hull fully encloses the crew and troop compartments and has been designed to provide ballistic immunity from 7.62 mm ball ammunition fired at a range of 25 m at an angle of 90° and from point blank attack by lighter weapons.

The engine and transmission are also protected from ballistic attack and the engine compartment also has a full fire extinguishing system.

To the front of the commander and driver is a one piece windscreen and each also has a forward-opening side door with an upper window. These doors, together with the large single door in the rear, allows the crew to dismount rapidly from the vehicle.

The seven troops are seated in the rear of the vehicle and each has an individual seat complete with seat belt.

The rear troop compartment has observation blocks and all glass is fully armoured to the same ballistic protection as the hull and is fitted with inner spall protection polycarbonate, the front and underbody guards are fitted to provide basic protection against fire bombs and other obstacles prevalent in a riot situation.

Standard equipment includes full air conditioning, riot equipment stowage, spot lamps, light bar, siren and public address system.

Optional equipment includes run-flat tyres with split rims, surveillance equipment and outer mesh grilles on all windows.

Hotspur Polisec APC with all doors open

Status: Production as required.

Manufacturer: Hotspur Armoured Products
Division of Penman Engineering Limited
Heathall
Dumfries DG1 3NY
UK
Telephone: (0387) 52784
Telex: 779771
Fax: (0387) 67332.

SPECIFICATIONS					
CREW	1 + 8	MAX ROAD SPEED	95 km/h	SUSPENSION	
CONFIGURATION	6 × 6	FUEL CAPACITY	98 l	front	dual rate coil springs, live beam axle, double-acting hydraulic dampers
GROSS WEIGHT	5350 kg	MAX RANGE	300 km	middle and rear	A frame live rear axle, dual rate coil springs, double-acting hydraulic dampers
POWER-TO-WEIGHT RATIO	25 hp/tonne	TURNING RADIUS	8.6 m	BRAKES	dual circuit
LENGTH	5.59 m	ENGINE	Rover 3.528-l V-8 water-cooled petrol developing 134 hp at 4000 rpm	front	hydraulic servo-assisted disc
WIDTH	1.8 m	TRANSMISSION	manual, 5 forward and 1 reverse gears	rear	hydraulic servo-assisted drum
HEIGHT	2.21 m	TRANSFER BOX	2-speed	TYRES	7.50 × 16
GROUND CLEARANCE	0.21 m	CLUTCH	single dry plate, hydraulic	ELECTRICAL SYSTEM	12 V
TRACK	1.486 m	STEERING	worm and roller	BATTERY	1 × 12 V, 95 Ah
WHEELBASE	3.81 m				
ANGLE OF APPROACH/DEPARTURE	35°/30°				

Hotspur Skirmisher 4585 Armoured Personnel Vehicle

Hotspur Skirmisher 4585 (4 × 4) APV fitted with roof hatch

Development

The Skirmisher 4585 armoured personnel vehicle has been developed as a private venture by Hotspur Armoured Products and was shown for the first time in 1985.

The vehicle is already in service in the Middle East and is specifically designed for high-speed patrol duties. The vehicle is based on a Land Rover One Ten chassis for ease of maintenance and availability of spare parts on a worldwide basis.

Description

The Hotspur Skirmisher is an armoured version of the Land Rover One Ten which provides seating for up to eight personnel including driver and commander within an armoured monocoque crew compartment.

The layout is conventional with the engine compartment at the front, commander and driver in the centre and the troop compartment at the rear. The vehicle has a complete monocoque hull fabricated in high-strength steel armour replacing the Land Rover-style bonnet and wings.

The all-welded steel hull gives protection against 7.62 mm NATO ball ammunition and being a total monocoque hull gives increased strength for roll-over protection. Crew survivability is further enhanced as none of the vehicles sides is flat-on to hostile fire; they have been angled to encourage ricochet and slide.

In addition to the single forward-opening door in either side of the hull, each of which has a firing port with a vision block above, there is also a large door in the rear. In each side of the rear troop compartment is a firing port with a vision block above. All windows and vision blocks are of ballistically protected glass and the commander's and driver's front and side windows can be provided with louvres.

The vehicle can be fitted with a roof-mounted rotating machine gun turret fitted with a hatch, bullet-proof observation windows and, as an option, two banks of four electrically operated smoke dischargers. The turret is armed with a 7.62 mm machine gun that is aimed via a periscope in the roof. As an alternative, the vehicle can be fitted with a fully rotating machine gun hatch or with a solid roof plate.

The petrol tank is under armour protection at the rear of the hull. Optional equipment includes a run-flat tyre system, night vision facility for the driver, 24 V electrical system, winch, tow hook and front shackles, self-sealing petrol tank, full air-conditioning system, stop lamps, intercom/siren, public address system, and fuel tank with an additional 55-litre capacity.

Status: Production as required. In service with Far and Middle East countries.

Manufacturer: Hotspur Armoured Products, a Division of Penman Engineering
Heathall
Dumfries DG1 3NY
Scotland
Telephone: (0387) 52784
Telex: 779771
Fax: (0387) 67332

SPECIFICATIONS

CREW	1 + 7	ANGLE OF APPROACH/ DEPARTURE	49°/30°	CLUTCH	single dry plate		
CONFIGURATION	4 × 4	MAX ROAD SPEED	110 km/h	TRANSFER BOX	2-speed		
COMBAT WEIGHT	3850 kg	FUEL CAPACITY	98 l	SUSPENSION	beam axles front and rear with coil spring controlled by hydraulic telescopic dampers		
POWER-TO-WEIGHT RATIO	34 hp/tonne	MAX RANGE	300 km				
LENGTH	4.66 m	TURNING RADIUS	6.4 m				
WIDTH	1.87 m	ENGINE	3.528-l, V-8 water-cooled, petrol, developing 136 hp at 4000 rpm	BRAKES	dual, servo-assisted discs front, drums rear		
HEIGHT WITH TURRET	2.57 m			TYRES	7.50 × 16		
HULL ROOF	2.0 m			ELECTRICAL SYSTEM	12 V, split charge		
GROUND CLEARANCE	0.215 m	TRANSMISSION	manual, 5 forward and 1 reverse gears	BATTERY	1 × 12 V, 95 Ah		
TRACK	1.486 m						
WHEELBASE	2.794 m	STEERING	worm and roller power steering				

Shorland S 55 Armoured Personnel Carrier

Development

The first Shorland armoured personnel carrier, the SB 301, was developed as a private venture by Shorts with the prototype completed in 1973 and the first production vehicle the following year. It has been designed specifically for internal security operations and is based on operational experience obtained with the Shorland armoured patrol car, with which it shares many common components. The Shorland SB 401, with strengthened armour and the more powerful V-8 petrol engine was introduced in 1980.

The latest version, the S 55, was designed to take advantage of the new Land Rover Defender 110 chassis with its improved coil spring suspension and wider wheel track than previously. Other improvements introduced with the new chassis were improved performance, tighter turning circle and power steering to reduce driver fatigue.

Latest production vehicles have a redesigned bonnet for easier engine access, bullet resistant glass windscreen and windows and eight firing ports, each with a vision block above. Of these, three are in either side of the hull and one in each of the rear doors.

The complete Shorland Series 5 family is comprised of the S 52 armoured car, S 53 air defence vehicle, S 54 anti-hijack vehicle and the S 55 armoured personnel carrier. Details of the first three versions are given in the entry for the Shorland armoured patrol car in the *Reconnaissance vehicles* section.

Description

The S 55 is based on the 110 inch heavy duty Land Rover (Defender) chassis which has permanent four wheel drive and a lockable central differential. Normal power is the 3.5 litre Rover V-8 petrol engine with the Rover 2.5 litre turbo-diesel available as an alternative.

The chassis has been strengthened to take the armoured body and is fitted with stronger axles and suspension and different final drive ratios. Nevertheless some 85% of the automotive and mechanical components are stock Land Rover parts so reducing spares and maintenance problems.

The vehicle body of the S 55 provides protection against right angle fire by NATO 7.62 × 51 ball ammunition down to 25 m range while the bullet resistant glass provides the same degree of protection as the steel and meets the test requirements of BS 5051 Part 1 Class R2.

The engine and radiator are fully armour protected while the floor is made from reinforced glass fibre to give protection against mines and improvised explosive devices. The fuel fillers are under armoured hatches which can only be released from inside of the vehicle.

The roof of the S 55 is ridged so that hand thrown missiles roll off even when the vehicle is stationary while the bonnet and windscreen assemblies are designed to ensure that burning fuel from petrol bombs thrown against the windscreen is directed to the ground without entering the engine compartment.

A searchlight, controlled from within the vehicle, is mounted in the forward part of the roof between the driver's and commander's seats.

The vehicle is designed to carry eight personnel including the driver and commander. The two side and two rear doors permit easy entry and exit and a grab rail suspended from the roof assists rapid movement in operational emergencies.

The large bullet resistant windscreen and side windows provide excellent visibility for the driver and commander and eight gun ports with associated vision blocks are fitted, three on each side and one in each of the rear doors.

The interior of the vehicle is finished in foam backed washable plastic to provide thermal insulation. An air-conditioning system is fitted as standard to allow for vehicle operation under all climatic conditions.

Interior of latest production Shorland looking forwards

Latest Shorland S 55 with new low profile front, bullet proof windscreen and windows and roof-mounted 7.62 mm MG mount with spare wheel and tyre on roof

Latest production Shorland S 55 with new low profile front, bullet proof windscreen and windows and public address system

Variants

Optional equipment for the S 55 includes run-flat tyres, drop down visors with bullet vision blocks in place of the bullet resistant windscreen, police/emergency lights either as individual lamps or as part of a multi-function unit, public address system, automatic fire protection and a roof hatch either with or without a machine gun mounting. Fixed grenade launchers can be fitted which can fire 66 mm electrically initiated smoke/illumination/CS gas grenades.

A special barricade remover can be installed at the front of the vehicle if required. Unlike other conventional designs this does not reduce ground clearance when not in use.

Rear view of Shorland armoured personnel carrier with back door open

Status: In production. The S 55 is in service in over 20 countries including Malaysia. In January 1988 Short Brothers announced that it had been awarded a contract to supply 40 S 55 Shorland APCs to the Turkish Ministry of the Interior, to be used by the Gendarmerie. In June 1990 Short Brothers announced that it has been awarded a contract to supply 24 Shorland S 55 APCs to the Sindh Police of Pakistan.

Manufacturer: Short Brothers PLC
Montgomery Road
Belfast BT6 9HN
Northern Ireland
Telephone: (0232) 458444
Telex: 747087
Fax: (0232) 705293

SPECIFICATIONS

CREW	2 + 6
CONFIGURATION	4 × 4
COMBAT WEIGHT	3600 kg
POWER-TO-WEIGHT RATIO	
petrol	31.2 hp/tonne
diesel	30.5 hp/tonne
LENGTH	4.25 m
WIDTH	1.8 m
HEIGHT	2.28 m
GROUND CLEARANCE	0.324 m
TRACK	1.5 m
WHEELBASE	2.795 m
MAX ROAD SPEED	120 km/h
FUEL CAPACITY	136 l
MAX RANGE	over 630 km
VERTICAL OBSTACLE	0.23 m
TURNING RADIUS	6.6 m
ENGINE	Land Rover 3.5 l V-8 water-cooled 4-stroke petrol developing 114 bhp at 4000 rpm or Land Rover TDI 2.5 l water-cooled 4-stroke diesel developing 107 bhp at 3800 rpm
TRANSMISSION	manual with 5 forward and 1 reverse gears
TRANSFER CASE	2-speed
STEERING	power-assisted worm and roller
CLUTCH	single dry plate
SUSPENSION	coil springs and telescopic hydraulic shock absorbers, anti-roll bars fitted front and rear
TYRES	9.00 × 16
BRAKES	
main	hydraulic, dual circuit front and rear, disc front, drum rear
parking	mechanical drum operating on transmission output
ELECTRICAL SYSTEM	12 V (24 V optional)
BATTERY	1 × 12 V, 58 Ah

Land Rover Armoured Patrol Vehicle

Development

The Land Rover Armoured Patrol Vehicle was developed by Glover Webb to meet the requirements of the British Army and, following extensive competitive trials against other vehicles, was selected by the British Army in 1985 for use in Northern Ireland. Over 100 vehicles were ordered and first deliveries were made in 1986.

The vehicle is based on the chassis of the Land Rover One Ten (4 × 4) vehicle which is already used by a number of armed forces. This also forms the basis of the Glover Hornet armoured car which is fully described in the *Reconnaissance vehicles* section.

The manufacturer envisages the vehicle undertaking a wide range of roles including border patrol, escort duties, anti-terrorism activities, internal security, riot control and airfield perimeter patrol.

This vehicle is commonly known as the Piglet by the British Army in Northern Ireland.

Description

The hull of the Glover Webb Armoured Patrol Vehicle is of all-welded steel armour construction which provides protection from 7.62 mm ball and 5.56 mm small arms fire. The front windscreen and cab windows are of bullet-proof multi-laminate glass with a polycarbonate anti-spall liner giving a similar level of protection to the steel armour.

The layout of the vehicle is conventional with the engine compartment at the front, driver and commander in the centre and the troop compartment at the rear. The commander and driver both have a forward-opening side door and these, as are the rear doors, are ballistically sealed with 20 mm overlaps.

The troops enter and leave via two doors in the hull rear which each have a bullet-proof vision block. There are no firing ports or vision devices in the sides of the troop compartment, although these can be fitted if specified by the user. The troops sit on bench seats that run down either side of the hull and each man has a seat belt.

In the forward part of the roof is a searchlight protected by a grille which can be traversed from within the vehicle, and over the rear troop compartment is a forward-opening hatch cover.

The light obstacle removal device is fitted at the front and a wire mesh grille protects the headlamps and radiator. Wire mesh protection is provided for the front windscreen and when this is not required it can be folded forward onto the bonnet. A vertical wire cutter is provided to prevent injury or decapitation for personnel using the roof hatch.

The floor is tough glass fibre reinforced plastic, giving protection against blast and grenade fragments.

The fuel tank is protected by Explofoil explosion suppression and also by its location behind the belly plate.

Steel well filler bands are fitted to all wheels giving a run-flat capability and a minimal loss of steering control in the event of penetration of the tyre by small arms fire.

Options include increased armour protection, integrated air conditioning system, gun ports and additional vision blocks, public address system, electrically operated smoke grenade launchers, hi-low sirens, machine gun hatch with 7.62 mm GPMG pintle, flashing lights, riot screens, front winch, internal and external fire extinguishers, radios and alternative internal layouts. The roof hatch, wire mesh protection and barricade remover fitted to British Army vehicles are optional extras for export.

Variants

A slightly different version is available for police roles and this has been supplied to the London Metropolitan Police.

Status: In production. In service with the British Army in Northern Ireland.

Manufacturer: Glover Webb Limited
Hamble Lane
Hamble
Hampshire SO3 5NY
UK
Telephone: (0703) 452811
Telex: 477527
Fax: (0703) 452964

Land Rover Armoured Patrol Vehicle as delivered to British Army for use in Northern Ireland; this is commonly known as the Piglet

Glover Webb Armoured Patrol Vehicle from rear with doors open

SPECIFICATIONS

CREW	2 + 6
CONFIGURATION	4 × 4
LENGTH	4.58 m
WIDTH	1.79 m
HEIGHT	2.05 m
GROUND CLEARANCE	0.32 m
TRACK	1.468 m
WHEELBASE	2.794 m
MAX ROAD SPEED	120 km/h
VERTICAL OBSTACLE	0.23 m
TURNING RADIUS	8.75 m
ENGINE	Rover 4-stroke V-8 3528 cc petrol developing 114 bhp at 4000 rpm
TRANSMISSION	manual, with 5 forward and 1 reverse gears
TRANSFER CASE	2-speed
STEERING	recirculating ball, worm and nut, power assisted
CLUTCH	single dry plate
SUSPENSION	heavy duty long travel coil springs, anti-roll bars fitted front and rear
TYRES	9.00 × 16
BRAKES	
main	dual line hydraulic disc front, drum rear, servo-assisted
handbrake	mechanical, on transfer case
ELECTRICAL SYSTEM	12 V

New British Army Protected Vehicles

Development/Description
In the Spring of 1991, following competitive bidding, the British MoD placed a contract with National Plastics, an operating company within Courtaulds Advanced Materials and part of Courtaulds Aerospace, a contract valued at about £50 million for the supply of about 1000 lightly armoured protected vehicles for the British Army.

The contract covers the development of prototype vehicles for validation of the light armoured concept and the engineering systems of the vehicle assembly.

The new lightly armoured protected vehicles will be based on the chassis of the Land Rover Defender (4 × 4) vehicle with National Plastics fitting the special ballistic protection system which will be an integral structure manufactured from advanced fibre reinforced composite materials.

Major sub-contractors will be Land Rover who will supply the base vehicle and Royal Ordnance Nottingham for key aspects of design and engineering.

National Plastics has an extensive range of contoured panel structures with high ballistic and mechanical performance.

No details of delivery dates or where the vehicles were deployed were released, nor was a detailed specification of the vehicle.

Status: In production for the British Army.

Manufacturer: Courtaulds Advanced Materials
PO Box 16
Foleshill Road
Coventry
West Midlands CV6 5AE
UK
Telephone: (0203) 688771
Telex: 312171 CTLDCV
Fax: (0203) 687325

Transaif Multi-role Armoured Vehicle

Development
This has been designed by Transac International of the United Kingdom with the actual vehicle being built by Thiele of Bremerhaven, Germany. Development of the Transaif began in mid 1982 in response to the company's customers who had a requirement for multi-role armoured vehicles, non-hostile in appearance, which could not be met by existing vehicles. The prototype of the Transaif multi-role armoured vehicle was completed late in 1985 and production began in 1986 for undisclosed customers.

Description
The Transaif multi-role armoured vehicle is based on an unmodified Mercedes-Benz LA 1513 (4 × 4) truck chassis for which spare parts are available from normal commercial sources worldwide.

The hull of the Transaif is of all-welded steel armour that provides complete protection from multi-shot rounds fired from 5.56 mm, 7.62 mm (Soviet), 7.62 mm (NATO) and 30.06 Garand military rifles at ranges of 0 to over 25 m. All openings have special gutters so that any flaming liquid does not run inside the vehicle.

The outside of the hull does not enable rioters to climb onto the vehicle. All door handles are recessed and in the event of great force from outside a pin allows the outside handle to shear off, so leaving the door shut.

The underfloor area has protection against fragmentation devices such as grenades. The diesel fuel tank is armoured and filled with explosafe material.

The engine is at the front of the vehicle and protected by a double louvre against petrol bombs, the armoured bonnet is raised vertically on pneumatic arms to allow access to the engine for routine maintenance.

The driver sits at the front left with the vehicle commander to the right with the troop compartment at the rear, the driver and commander have electrically operated wing mirrors.

Access is via four doors, one either side opening to the front and twin doors at the rear.

All windows in the Transaif provide the same degree of protection as the hull. The driver's and commander's front windows have paint cleaning equipment and there is also a long-range spotlight in the forward part of the roof which can be operated from inside.

The standard vehicle is fitted with a wide angle TV camera above the rear door which is linked to a monitor for the commander to assist in safe deployment. This system engages automatically when the vehicle is put into reverse gear as a driving aid.

The vehicle can carry a maximum of 14 fully equipped men who can dismount from the vehicle with their full riot equipment in just four seconds. The type of seats fitted depends on the number of men carried, but if required each man can have an individual seat with hull harness; the interior standing height is 1.85 m.

For crew comfort, the standard Transaif is fitted with a full air-conditioning system, which can be run over extended periods, driven by an independent diesel generator with its own electrical system. This auxiliary generator slides out on rails for ease of maintenance and it can be armour protected. As an option a full NBC system can be installed in the vehicle.

The side doors open forwards and each has a bullet-proof observation window with a firing port above and below, the former is for use from within the vehicle when the door is shut, while the latter is for use when the door is opened through 90° to the front. In the latter mode an armoured flap folds downwards to provide leg protection.

In each side of the vehicle there are four firing ports and three bullet-proof windows, all of the latter can be covered by sliding ballistic shutters from inside the vehicle if required.

The interior of the troop compartment is of flame-retardant material and the floor is non-slip. Two grab rails are provided in the roof. On top of the roof at the rear, on either side, is a spotlight operated from within the safety of the vehicle. There is a circular emergency escape hatch in the forward part of the roof.

Transaif multi-role armoured vehicle showing large bullet-proof windows for commander and driver

The two doors at the rear can be opened manually and as the doors open one step automatically folds out to assist the crew in leaving the vehicle. Each rear door also has a large bullet-proof window with a firing port above and below which can be used in a similar manner to the side doors. An automatic towing coupling is provided at the rear.

The tyres are of the bullet-resistant run-flat type and after penetration the vehicle can be driven safely for a distance of over 60 km only slightly reducing its top speed of 105 km/h.

For maximum crew safety an extensive fire detection and fire extinguishing system is provided by Kidde-Graviner, which is a two-shot system, the first operated automatically and the second manually. There are two sensors in the engine compartment which operate at +120°F and three under the chassis which operate at +170°F.

Many of the fittings on the Transaif depend on individual customer requirements, for example flashing lights and a cable cutter on the roof and special stowage to the rear of the commander and driver for riot shields and 12 rifles.

Communications equipment depends on user requirements and can include a crowd control loudspeaker mounted on the roof.

The vehicle can be adapted for other roles including a protected ambulance, mobile command post with extensive communications equipment, bomb disposal vehicle for two teams complete with remotely-operated bomb disposal unit, airport security, incident vehicle or, with suitable armament, border patrol. Optional equipment includes outside grilles for windows, public address system and front-mounted barricade removal equipment.

Status: Production as required. In service with undisclosed countries.

Designed and marketed: Transac International Limited
6 Magellan Terrace
Gatwick Road
Crawley
West Sussex RH10 2QY
UK
Telephone: (0293) 534639
Telex: 87326 TRNSAC G
Fax: (0293) 534326

Manufacturer: W Thiele GmbH
Bremerhaven
Germany

SPECIFICATIONS

CREW	2 + 12 (individual seats) 2 + 14 (bench seats)	GROUND CLEARANCE	295 mm	TRANSFER CASE	VG500
CONFIGURATION	4 × 4	FUEL CAPACITY	170 l	DIFFERENTIAL LOCK	yes
MAX WEIGHT	14 800 kg	MAX ROAD RANGE	680 km	STEERING	power-assisted
UNLOADED WEIGHT	10 000 kg	MAX ROAD SPEED	105 km/h	BRAKES	
POWER-TO-WEIGHT RATIO	11.35 hp/tonne	TRACK	2.028 + 1.821 m	main	dual circuit hydraulic plus single circuit air brakes
LENGTH	7.3 m	WHEELBASE	4.2 m	CLUTCH	single-disc dry clutch
WIDTH	2.5 m	ENGINE	Mercedes-Benz OM-352A 6-cylinder diesel developing 168 hp	TYRES	12/80 R20
HEIGHT	3.1 m	TRANSMISSION	manual, 5 forward and 1 reverse gears	ELECTRICAL SYSTEM	24 V

UNITED STATES OF AMERICA

FMC M2 Infantry Fighting Vehicle/M3 Cavalry Fighting Vehicle

Development

In 1967, under contract to the United States Army, the then Ordnance Division of FMC Corporation built two MICVs under the designation XM765. This was not adopted by the Army but further development by the company, as a private venture several years later, resulted in the Armored Infantry Fighting Vehicle which is currently in service with the Netherlands, Philippines and Belgium. More recently the AIFV has been selected by Turkey who will build it under licence. There is a separate entry in this section for the Armored Infantry Fighting Vehicle.

In April 1972 the United States Army issued a Request For Proposal for a new MICV. A number of companies submitted proposals to meet this requirement and three companies were short-listed: Chrysler Corporation, FMC Corporation and Pacific Car and Foundry.

In November 1972 an Engineering Development and Advanced Production Engineering contract was awarded to the Ordnance Division of FMC Corporation. The total value of this contract was $29.3 million which covered the cost of the design, development and fabrication of three prototype vehicles, a ballistic vehicle, 12 pilot vehicles and associated systems engineering, product assurance and test support.

The prototypes, called the XM723, had a crew of three (commander, gunner and driver) and carried eight fully equipped infantrymen. The driver sat at the front on the left, with the commander to his rear and the gunner in the turret on the right side of the hull. The turret was armed with a 20 mm cannon and a coaxial 7.62 mm machine gun. The troop compartment was at the rear of the hull and the infantrymen could aim and fire their weapons from inside the vehicle. All prototypes were completed by the Summer of 1975.

In August 1976 an MICV Task Force was formed by the United States Army to make an independent examination of the whole XM723 programme to determine whether the vehicle would meet the future requirements of the Army. The Task Force made a number of recommendations, of which the following were accepted by the Army in October 1976:

A common vehicle would be developed for both the infantry and scout roles as the Armored Reconnaissance Scout Vehicle (prototypes of which were built by the FMC Corporation and Lockheed Missile and Space) had been cancelled.

The vehicle would be fitted with the TOW ATGW system and 25 mm cannon in a two-man TOW/Bushmaster Armored Turret (TBAT-II).

A two-tube TOW ATGW launcher would be mounted on the left side of the turret to give the vehicle an anti-tank capability.

The firing ports would be retained.

The vehicle would be amphibious.

The vehicle would have the same level of armour protection as the XM723.

The vehicle would be issued on the scale of four per platoon, 13 per company and 41 per battalion. (The M2 would not replace the M113 one-for-one, but would replace them for specific roles such as the APC in the mechanised infantry battalions.)

The complete programme was then renamed the Fighting Vehicle System (FVS) which consisted of two vehicles, the XM2 Infantry Fighting Vehicle and the XM3 Cavalry Fighting Vehicle and was expanded in June 1977 to include a carrier for the General Support Rocket System (now

M3 Bradley CFV clearly showing covered firing ports in hull sides and with turret traversed to left (Pierre Touzin)

Sequence of three photographs showing a Bradley Infantry Fighting Vehicle launching a TOW-2B ATGW. The device above the two round launcher box is a TV camera fitted for trials purposes

known as the Multiple Launch Rocket System). The IFV and CFV appear almost identical except that it has a crew of five, no firing ports and has greater ammunition capacity.

The first two prototypes of the XM2 Infantry Fighting Vehicle were handed over to the United States Army at San Jose, California in December 1978, and the remaining six were completed by March 1979. In December 1979 the XM2 was type classified as the M2 and the XM3 as the M3. In February 1980 the US Army announced that it planned to produce 100 M2/M3s with FY80 funds. The first production contract deliveries of 100 vehicles were handed over in May 1981 and completed on schedule in July 1982. Second year deliveries of 400 vehicles were completed in May 1983.

In October 1981, during the first production contract, the M2 IFV/M3 CFV was dedicated the Bradley Fighting Vehicle after the late General of the Army, Omar N Bradley.

The first unit equipped with the M2 was a battalion at Fort Hood, Texas, in March 1983, with the seventh training centre in Vilseck, Germany, receiving deliveries in September 1983. By August 1989 FMC had delivered 4517 vehicles out of the 5531 funded, with production running at about 54 vehicles per month and by April 1991 total deliveries amounted to 5500 vehicles.

Funding for the M2/M3 series for the US Army has been as follows:

Year	**Quantity**	**Year**	**Quantity**
FY80	100	FY87	662
FY81	400	FY88	554
FY82	600	FY89	644
FY83	600	FY90	400
FY84	600	FY91	300
FY85	655	FY92	300
FY86	716	FY93	200

These production quantities do not include those to be ordered for the ADATS air defence system selected by the US Army in late 1988.

The US Army had a total requirement for 8811 vehicles of which 2300 would be M2/M3, 1371 M2A1/M3A1 and the remaining 5140 would be M2A2/M3A2.

It was expected that the US Army would place a multi-year contract with FMC Corporation, Ground Systems Division, for 3000 M2/M3 series vehicles over a five year period at 600 vehicles per year with the contract to be signed in mid-1990.

In mid-1990, it was reported that the US Army had reduced its total Bradley requirement from 8811 vehicles down to 6724 vehicles.

In the Spring of 1991 the FMC Corporation was awarded a $1.126 billion contract to supply the US Army with a further 1600 Bradley Fighting Vehicles and 250 carriers for the Multiple Launch Rocket System. The contract includes approximately 1200 Bradleys for the US Army and 400 for sale to Saudi Arabia with deliveries to run from June 1991 through to October 1994.

International Acquisition

The Bradley has been sought for procurement by other countries and in 1985 the US Congress was notified of a proposed sale of military equipment to Jordan which included 32 M3 Bradley CFVs, but this did not go ahead.

Early in 1988 Congress was advised of a Letter of Offer and Acceptance (LOA) to Saudi Arabia for 200 Bradley IFVs which was accepted. Two vehicles were delivered in FY89, 98 in FY90 and the remaining 100 in FY91.

In December 1990 a second LOA for 200 additional Bradley IFVs was approved and accepted by Saudi Arabia. The additional 400 Bradley IFV procurement by Saudi Arabia was included in the Bradley multi-year

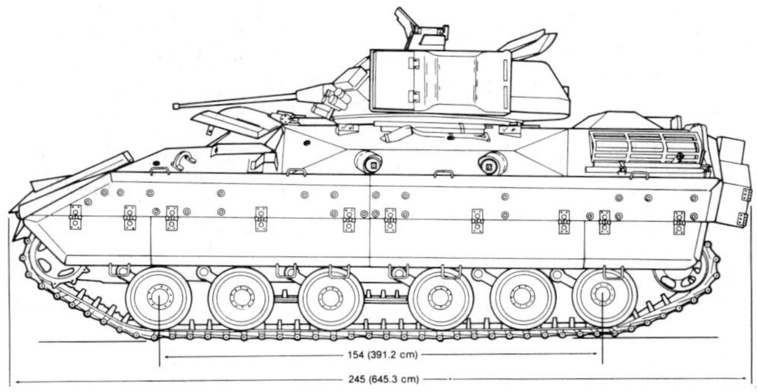

M2 Bradley Infantry Fighting Vehicle with TOW launcher retracted and hatches propped (not to 1/76th scale)

M2 Bradley Infantry Fighting Vehicle with flotation screen erected prior to amphibious operations and showing firing ports in rear hull ramp (Pierre Touzin)

contract over and above the 1200 vehicle US Army procurement. Deliveries for the second buy will be 88 in FY92 and 112 in FY93.

Bradley in the Middle East

In March 1991, the US Army issued an initial report on Army Weapons Systems Performance in Southwest Asia with the following comment on the M2/M3: "Crews reported that the sights were very effective, even during sand storms. Other crews reported that the 25 mm Bushmaster cannon was more lethal than they expected. There were no reports of transmission failure during offensive operations. Of the 2200 Bradleys in theatre, three were disabled. To date, we have no information on the number damaged. Overall Bradley operational readiness rates remained at 90 per cent or above during combat."

Description

The hull of the M2 Infantry Fighting Vehicle is made of all-welded aluminium armour with spaced laminate armour fitted to the hull, sides and rear. According to FMC the armour of the M2 can defeat 95 per cent of all types of ballistic attack encountered on the battlefield under IFV/CFV doctrine. Latest production M2A2/M3A2 Bradleys have an additional layer of applique steel armour plus explosive reactive armour for increased battlefield survivability.

The driver sits at the front of the vehicle on the left and has a single piece hatch cover that opens to the rear and four periscopes, three to the front and one to the left side. The centre front periscope can be replaced by an AN/VVS-2 passive night periscope.

The engine compartment is to the right of the driver. The Cummins VTA-903T engine is coupled to a General Electric HMPT-500 hydro-mechanical transmission. There are three speed ranges with overall transmission ratios infinitely variable in all ranges.

The vehicle is equipped with one 3.2 kg Halon fixed fire extinguisher in the engine compartment, two automatic 2.3 kg Halon fire extinguishers in the personnel compartment and one 1.2 kg portable Halon fire extinguisher.

The turret, which is of welded steel and aluminium armour construction, is mounted in the centre of the vehicle on the right side with the gunner seated on the left and the commander on the right. Each crew member is provided with a single piece hatch cover that opens to the rear. The gunner has a combined day/thermal sight with an optical relay for the commander with magnifications of ×4 and ×12, as well as periscopes for front and side observation. In addition, production vehicles are fitted with a fixed power, daylight back-up sight which will allow the gunner or commander secondary sighting capability in the event of primary sight failure.

Main armament consists of a McDonnell Douglas Helicopter Company M242 25 mm Chain Gun with a 7.62 mm M240C machine gun mounted coaxially to the right of the main armament. The 25 mm cannon has

M2A2 Bradley IFV fitted for, but not with, additional armour blocks, with twin TOW launcher in raised position and gunners day/night sight ready for action

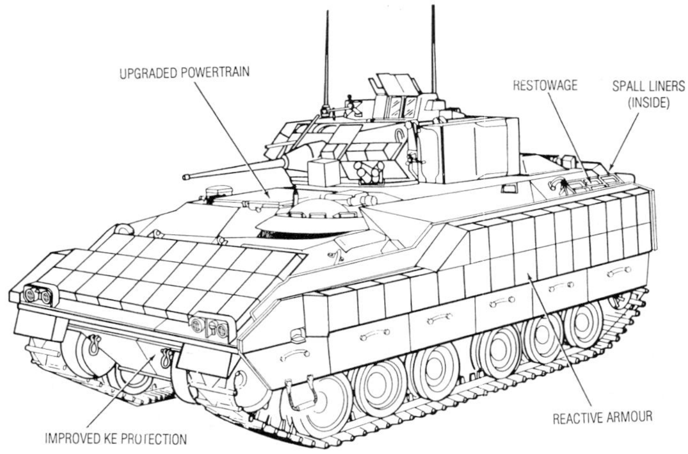

Drawing of FMC M2A2 Bradley IFV showing main improvements over earlier M2A1 vehicle (not to 1/76th scale)

M2 Bradley Infantry Fighting Vehicle with twin Hughes TOW ATGW launcher elevated into ready to fire position

M2A2 Bradley IFV showing floatation screen in retracted position around top of hull and additional armour installed on hull and turret

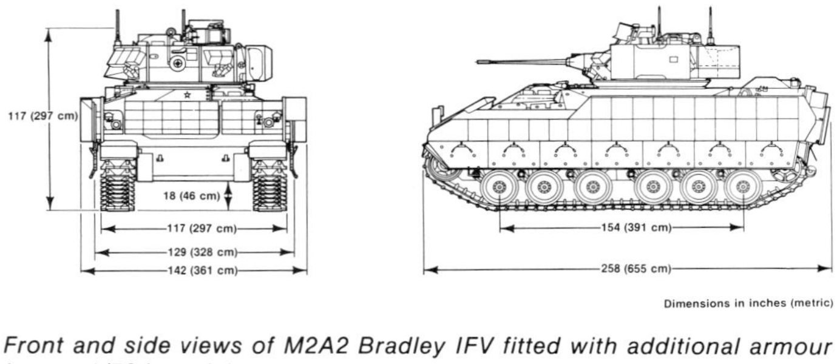

Front and side views of M2A2 Bradley IFV fitted with additional armour (not to 1/76th scale)

dual feed rate capability and the gunner can select single shots, 100 or 200 rpm rates of fire. The cannon will fire both Oerlikon 25 mm and American M790 series of ammunition including M791 APDS-T, M792 HEI-T and M793 TP-T, and the M919 APFSDS-T round which is now in production. The older M791 will defeat the Soviet BMP-1 at a range of 2500 m. The M919 has even greater penetration characteristics as compared to the current M791 APDS-T round. The empty cartridge cases for both the 25 mm cannon and 7.62 mm machine guns are automatically ejected outside the turret.

The turret has 360° electric traverse and the weapons can be elevated from -10° to +60°. The General Electric turret drive and stabilisation system allows the armament to be laid and fired while moving across rough country. The system consists of a traverse drive assembly for positioning and holding the turret, gun elevation drive assembly for positioning and holding the weapon, TOW elevation drive assembly for positioning and holding the TOW missile launcher, a TOW lift mechanism for raising and lowering the TOW launcher, electronic control assembly, three gyro blocks, gunner's handstation, commander's handstation and cabling.

The TOW weapon subsystem has been developed by the Hughes Aircraft Corporation under a contract worth $16.5 million. When travelling the twin tube TOW launcher is retracted and lies along the left side of the turret. The TOW system enables the M2/M3 vehicles to engage enemy armour out to a maximum range of 3750 m. The TOW missile launcher has an elevation of +30° and a depression of -20°.

Two M257 electrically operated smoke dischargers, with four smoke grenades in each, are mounted on the forward part of the turret, one on either side of the main armament. In addition new production vehicles are fitted with an engine smoke-generating system similar to that on most Soviet vehicles.

The M2 carries seven infantrymen: one sits forward of the turret on the left side facing the rear, one to the left of the turret facing the front, one at the left rear of the vehicle facing inwards, two sit at the right rear facing the back and two sit to the back of the turret facing the front. The M2A2 only carries a six man squad with the seat to the rear of the driver's position has been eliminated. In both vehicles, the commander also dismounts with the infantry.

The infantrymen enter and leave the vehicle via a large hydraulically operated ramp at the rear of the hull, which has an integral door in the left side. A single piece hatch cover that opens to the rear is provided over the top of the troop compartment. Six firing ports, two in each side of the hull and two at the rear, each with a periscope over it, enable the infantrymen to fire their M231 5.56 mm weapons from inside the vehicle. The M2A2 has no side firing ports but the ones in the ramp are retained.

The suspension system includes torsion bars, and on each side there are six dual rubber-tyred road wheels with the drive sprocket at the front and the idler at the rear. There are two track return rollers that support the inside of the track only, and one double roller. Hydraulic shock absorbers are fitted to the first, second, third and sixth road wheel stations. The tracks are of the single-pin type with replaceable rubber pads. In 1991, the Bradley was upgraded with the 'bigfoot' track that has a larger rubber pad area for increased track life.

In M2A1/M3A1 production vehicles a central NBC M13A1 gas particulate filter system is provided for the commander, gunner and driver but the dismounted infantry have to wear individual masks.

All M2/M3 vehicles are fully amphibious being propelled in the water by its tracks. Before entering the water a special water barrier can be erected in approximately 30 minutes by a trained crew.

By 1987 the US Army had conducted more than 10 000 successful Bradley swim exercises but 13 swim-related incidents had occurred, most of which were caused by a collapsed trim vane, usually due to incorrect erection by the crew. Since then modifications have been carried out to solve this problem.

The M2A2/M3A2 Bradley has a redesigned water barrier but there is also a programme to look at inflatable bags which would be attached to the armoured skirts on each side of the vehicle. These can be inflated using a simple on-board blower in less than four minutes without the crew leaving the vehicle. When afloat the bags will be completely submerged under water. If one of the floats is punctured by small arms fire it will be topped up with air by the blower.

The M2/M3 series of vehicle are transportable by truck, rail, cargo ship or aircraft. The Bradley can be airlifted by C-141, C-5 and C-17 aircraft. When being airlifted in the C-141 Starlifter, the head of the gunner's integral sight and the skirt plates are removed and all road wheel arm positions, six each side, are snubbed in the up position using the steel cables provided.

Variants

M2A1 and M3A1 Bradleys

Further development of the M2 and M3 has resulted in the M2A1 (development designation M2E1) and M3A1 (development designation M3E1) which entered production in 1985 with first production deliveries in May 1986.

The M2A1 and M3A1 can fire all versions of the Hughes TOW ATGW including the full diameter TOW 2 with a maximum range of 3750 m.

The NBC system is the M13A1 gas particulate filter system that has been added in both models. On the M2A1/M3A1 the commander, gunner and driver are all connected to the central NBC system while the dismounted infantrymen in the rear have their own NBC suits and masks which allow them to leave the vehicle quickly. These are generally known as Block I improvements.

Other improvements in M2A1/M3A1 include turret power indicator, padding for driver's station, redesigned weapons interlock and top deck clearance arrangements, simplified stowage strapping, new onboard rations, two camouflage nets carried and replacement of the single water tank by two smaller ones with the same total capacity.

On the M2A1 the Dragon AN/TAS-5 night sight is now carried and up to five Dragons can be carried in place of TOWs. Mine and flare stowage has been re-arranged so that more mines and fewer flares are carried. The M3/M3A1 is not equipped with the firing ports or the associated vision blocks of the M2/M2A1. The two scouts carried now have individual seats that can be raised and lowered and four periscopes in the missile loading hatch for improved observation. For increased survivability the M2A1/M3A1 also incorporates a revised fuel system and revised fire suppression system.

M2A2 and M3A2 Bradleys

In late 1984, the US Army initiated a Bradley Block II study that evaluated many enhancements to improve the survivability of the Bradley vehicle and its crew. The Block II study evolved into the High Survivability Bradley programme in 1986 which was the development period for the M2A2 and M3A2 production vehicles.

The improvements include the following: enhanced vehicle survivability by restowing ammunition; internal armour protection of key components; spall liners and mounting provisions for improved vehicle smoke screening; additional armour protection including additional armour for the hull and turret; a ballistic shroud provided for the commander's backup sight; and improved drive train and suspension.

The additional armour protection has increased its weight to around 30 000 kg. It is powered by the Cummins VTA-903 engine developing 600 hp compared to 500 hp of the original production vehicle.

The General Electric transmission has been upgraded and the intake air cleaner expanded to accommodate the additional airflow requirements of the more powerful engine. A larger exhaust pipe has been installed and modifications have also been made to the existing air muffler.

The final drives are the same as those on the MLRS system but the suspension has been upgraded. All of these improvements, except the ballistic shield of the commander's backup sight were included in the M2A2/M3A2 series vehicles.

Prototype of the Bradley with a composite hull which was completed in June 1989

The 600 hp engine was not included in the initial M2A2/M3A2 production of 662 vehicles from May 1988 but was included in all follow-up M2A2/M3A2 production from May 1989.

The cut-in dates for the various survivability modifications to the M2A2/M3A2 Bradley have been as follows:

Modification	**Date of production cut in**
Addition of steel applique armour, spall liners, relocation of ammunition and addition of attachment points for armour tiles	May 1988
Addition of armour tiles (passive or explosive reactive)	To be determined
Addition of dual shot fire extinguishing system	October 1991
Addition of 600 hp engine	May 1989
Modification of transmission	May 1989
Changes to fuel system	May 1986
Rerouting of fire extinguishing system cables	May 1986

The US Army plans to retrofit 2033 of its Bradleys to the high survivability A2 configuration.

Of the 2033 Bradleys, 662 are the new A2 model without the upgraded engine and transmission. These will be retrofitted with the new powerpack. The remaining 1371 A1 Bradleys will be fitted with all survivability enhancements. These are expected to be retrofitted as follows:-

Bradley model	**FY90**	**FY91**	**FY92**	**FY93**	**FY94**
A1 to A2 high survivability	54	239	395	443	240
A2, plus upgraded powertrain	249	292	121	0	0

Under the present plans, production of the M2A2/M3A2 will end in 1994. Plans to retrofit the first 2300 vehicles are being evaluated for other user requirements and level of upgrade. This could extend the production of the A2 Bradley and incorporate some or all of the following potential improvements: driver's thermal viewer and improved fire control, improved navigational aids like a heading reference unit, low profile antenna, improved biological/chemical protection, nuclear hardening, adoption of an advanced swim system, improved maintenance/diagnostic capabilities and higher horsepower engine and transmission.

Future Bradley designs under consideration for a potential Block III (or M2A3/M3A3 when in production) could include a composite hull, improved optics, improved 40 mm to 50 mm cannon in a new turret with an improved heavy anti-tank system, hybrid NBC system, improved electrical system featuring a vehicle databus and automated information processing, system survivability enhancements through improved mine protection and a full solution fire control system.

Bradley with Composite Hull

In September 1986 the US Army Materials Technology Laboratory (MTL) awarded a $13 million, four-year contract to FMC Corporation, Ordnance Division, for the construction of a moulded thick laminate composite (reinforced plastic) hull based on the Bradley chassis. This testbed provided materials technology for the next generation of combat vehicles for the US Army.

The composite selected for this project consists of resin-bonded Owens-Cornings S-2 glass fibre woven fabric. In fabricating the composite hull structure, three moulded sections replace 23 welded aluminium plates while still incorporating aluminium reinforcing members for strength.

The first of two prototypes was rolled out in June 1989 and then started a 9600 km durability test to verify previous laboratory predictions. This information will be used to incorporate any improvements in the second prototype.

Bradley with Composite Turret

FMC delivered to the US Army in mid 1986 a Bradley with a composite turret following a contract awarded in 1983 and this underwent extensive trials which were completed late in 1986.

Multiple Launch Rocket System

Details of this are given in the *Multiple rocket launchers* section. Full details of the Fighting Vehicle Systems Carrier, M993, on which the MLRS (M269) is carried, are given in *Jane's Military Vehicles and Logistics 1991–92*, page 474, as are other members of the family including the Armored Forward Area Rearm Vehicle (AFARV) which has been designed to rearm MBTs and other AFVs in forward battlefield areas.

The complete MLRS system is designated the M270 which comprises the MLRS carrier (M993) and the MLRS rocket itself (M269). The FVSC by itself (generic) is the XM987. It has not been approved as a stand alone type classified M987 chassis yet. When referred to as the "FVSC on which the MLRS (M269) is carried", the carrier has to carry the M993 designation. The difference between the M993 and the XM987 is the removal of the mounting provisions in the bed for the MLRS rocket system.

More recent variants on this chassis are the Armored Maintenance Vehicle and the Electronic Fighting Vehicle Systems carrier. Both of these have been extensively evaluated by the US Army.

The Electronic Fighting Vehicle Systems (EFVS) carrier is also referred to by the US Army as the XM1070 Intelligence and Electronic Warfare (IEW) system. The prototype of the EFVS carrier was completed by FMC

Fighting Vehicle Systems Carrier

Prototype of Armored, Forward-Area, Rearm Vehicle (AFARV)

Corporation in 1986 and purchased by the US Army in 1989. A further two vehicles were ordered in 1990 for delivery in 1991 and an additional four vehicles are expected to be ordered in 1992.

The EFVS is also referred to as the Ground Based Common Sensor - Heavy (GBCS-H) when fitted with its electronics and a 28 m mast for its antenna system. The Ground Based Common Sensor - Light (GBCS-L) will be on an HMMWV chassis and have a 15 m mast for use by light divisions.

The Electronic Fighting Vehicle Systems carrier has also been proposed by GKN Defence of the UK to meet part of the British Army's FFLAV programme.

In addition, it has also been proposed for air defence missile systems such as the Patriot, Chaparral and private venture Corps SAM which would have eight missiles in the ready to launch position.

COMVAT

The Combat Vehicle Armament Technology (COMVAT) system development team consists of the US Army Armament Research and Development Center, Alliant Tech Systems (ammunition development), Ares (45 mm cannon), Western Design (ammunition feed and storage system) and FMC (vehicle integration and fire control).

COMVAT has already been successfully demonstrated in 30 mm with the 45 mm COMVAT on schedule for system demonstrations in FY90 and FY91.

Potential of COMVAT include the next generation Bradley IFV, the proposed Future Infantry Fighting Vehicle (covered in the *AFV Families* section) and various external turrets. The 45 mm cannon has a cyclic rate of fire of 200 rds/min and is recoil operated. The feed system is linkless and feeds through the trunnions. Types of ammunition include APFSDS-T, HEI-T (with impact fuzing and insensitive explosives) and TP-T, all of which are of the telescoped type.

M2 Bradley with 35 mm Cannon

The Combat Vehicle Armament System Technology (CVAST) test bed underwent trials in 1984 and was an M2 Bradley chassis with a new turret designed by Ares armed with a 35 mm Ares Talon rapid firing cannon. This is only a test bed and will not enter production.

Bradley Air Defence Vehicle

The Bradley chassis has been selected by the US Army to mount its Oerlikon-Contraves ADATS air defence system selected in 1987. Mounted on the hull top is a power operated turret with the electro-optics in the centre and four ADATS missiles in the ready to launch position on either side of the turret.

Electronic Fighting Vehicle System (EFVS) with telescopic antenna erected

Fighting Vehicle Systems Carrier being used in Armored Maintenance Vehicle role

So far the US Army has purchased eight of these systems from Martin Marietta for extensive trials. Of these eight, four systems are on the modified A1 chassis and four are on the modified A2 chassis.

Details of the ADATS system, which is already in service with the Canadian Armed Forces on an M113 series chassis are given in *Jane's Land-Based Air Defence 1991-92*, page 122.

The Bradley chassis can also be used to mount a variety of gun and/or missile air defence turrets such as those designed by FMC and General Electric for the LAV air defence programme. Details of these are given in *Jane's Land-Based Air Defence 1991-92*, page 94.

Stingray

It is reported that the US Army is to install 48 Stingray laser weapons on M3 Bradley chassis. Full scale development is expected to start in 1992 although a prototype system has already been tested.

Line of Sight Anti-Tank

The Bradley chassis has been selected as the carrier for the Line of Sight Anti-Tank (LOSAT) weapon system being developed for the US Army's Armored Systems Modernisation (ASM) programme.

The LOSAT system will consist of the hypervelocity missile technology developed by LTV Aerospace mounted on a Bradley derivative chassis. The LOSAT programme is currently in development with the first unit expected to be equipped in 1998.

Ground Launched Hellfire Heavy

Details of the Ground Launched Hellfire Heavy system, which could be fitted onto a M2/M3 chassis, and currently under development by Emerson, are given in the *Tank destroyers* section.

M3A2 Bradley with explosive reactive armour blocks covered up. The M3A2 has no firing ports in the sides of the troop compartment but they are retained in the rear hull ramp (Michael Green)

The US Army has selected the Bradley chassis for the Oerlikon-Contraves/Martin Marietta ADATS system, but this is not yet in quantity production for the US Army

SPECIFICATIONS (M2, data in square brackets relate to M3 where different)

CREW	3 + 6 [3 + 2]
COMBAT WEIGHT	22 590 [22 443] kg
UNLOADED WEIGHT	19 005 [18 824] kg
AIRPORTABLE WEIGHT	20 055 [19 732] kg
POWER-TO-WEIGHT RATIO	20.38 [20.51] hp/tonne
GROUND PRESSURE	0.54 [0.53] kg/cm^2
LENGTH	6.453 m
WIDTH	3.2 m
over tracks	2.972 m
HEIGHT	
to gunner's sight	2.972 m
to turret roof	2.565 m
reduced	2.634 m
GROUND CLEARANCE	0.432 m
TRACK WIDTH	533 mm
LENGTH OF TRACK ON GROUND	3.912 m
MAX SPEED	
road	66 km/h
water	7.2 km/h
FUEL CAPACITY	662 l
CRUISING RANGE	483 km
FORDING	amphibious
GRADIENT	60%
SIDE SLOPE	40%
VERTICAL OBSTACLE	0.914 m
TRENCH	2.54 m
ENGINE	Cummins VTA-903T turbo-charged 8-cylinder diesel developing 500 hp at 2600 rpm
TRANSMISSION	General Electric HMPT-500 hydro-mechanical
STEERING	hydrostatic
SUSPENSION	torsion bar
ELECTRICAL SYSTEM	28 V
BATTERIES	4 × 12 V 6TN
ARMAMENT	
main	1 × 25 mm cannon
coaxial	1 × 7.62 mm MG
ATGW	2-tube TOW launcher
firing port guns	5.56 mm [nil]
SMOKE-LAYING EQUIPMENT	2 × 4 smoke dischargers plus exhaust generator
AMMUNITION	
25 mm	300 ready, 600 reserve [300/1200]
7.62 mm	800 ready, 1540 reserve [800/3740]
5.56 mm firing port	4200 reserve [n/app]
TOW missiles	2 in launcher, 5 in reserve [10]
LAW (M72A2)	3 stowed
GUN CONTROL EQUIPMENT	
Turret power control	electric/manual
by commander	yes
by gunner	yes
Max rate power traverse	60°/s
Max rate power elevation	60°/s
Gun elevation/depression	+60°/−10°
Turret traverse	360°
TOW launcher elevation/depression	+30°/−20°
Gun stabiliser	yes
ARMOUR	
Top and front slopes	5083 aluminium
Vertical sides and rear	spaced laminate
Bottom	5083 aluminium with anti-mine appliqué (IFV only)
Side slopes	5083 aluminium

Note:
Basic specifications of the M2A2 are as follows, the M3A3 is identical apart from the slightly lower air transportable weight in brackets.

COMBAT WEIGHT	29 940 kg
AIR TRANSPORTABLE WEIGHT	19 640 [19 960] kg
GROUND PRESSURE	0.72 kg/cm^2
LENGTH	6.55 m
WIDTH overall armour tiles	3.61 m
WIDTH over skirts	3.28 m
WIDTH over tracks	2.972 m
HEIGHT to gunner sight	2.972 m
GROUND CLEARANCE	0.46 m
MAX SPEED	
road	61 km/h
water	6.4 km/h
CRUISING RANGE	400 km
ENGINE	600 hp at 2600 rpm
AMMUNITION (7.62 mm)	1400 [3400] reserve

Status: In production. In service with the US Army and Saudi Arabia.

Manufacturer: FMC Corporation, Ground Systems Division
2830 De La Cruz Boulevard
Santa Clara
California 95052
USA
Telephone: (408) 289 0111
Telex: 6714210
Fax: (408) 289 2150

FMC Armored Infantry Fighting Vehicle

Development

In 1967, under contract to the United States Army, the then Ordnance Division of the FMC Corporation built two MICVs under the designation XM765. These were based on the proven M113 APC modified to incorporate firing ports and a fully enclosed weapon station. Further development of the XM765 by FMC as a private venture resulted in the Product Improved M113A1, the first prototype of which was completed in 1970. This had a fully enclosed weapon station in the centre of the hull, behind the driver and engine, and the commander's cupola to the rear of the turret. This arrangement gave the commander very little forward vision so the vehicle was redesigned with the commander sitting behind the driver on the left and the turret to the rear of the engine on the right. This vehicle was subsequently called the Armored Infantry Fighting Vehicle.

The first country to order the AIFV was the Netherlands which placed an order for 880 vehicles in 1975 with first deliveries of the Netherlands AIFVs, designated the YPR 765, being made in 1977. Further development was carried out on behalf of the Dutch Army by FMC. Many of the

Armored Infantry Fighting Vehicle from rear with all hatches closed

Netherlands Army Armored Infantry Fighting Vehicle with Emerson ITV turret traversed to rear and ramp lowered (C R Zwart)

components for the Dutch vehicles are made in the Netherlands under the terms of a co-production arrangement. In 1981 the Dutch Army ordered an additional 840 AIFVs of which 119 were in the Improved TOW Vehicle configuration, DAF were involved in the production/final assembly of these vehicles. The Philippines received 45 AIFVs. In July 1979 Belgium placed an order for 514 AIFVs and 525 M113A2s to be built under licence in Belgium by Belgian Mechanical Fabrication to replace M75 and AMX VCIs in service with the Belgian Army. First production vehicles were completed in 1982 and production ran at the rate of 20 to 25 vehicles (AIFV and M113A2) a month until May 1988. Additional details of the Belgian Armored Infantry Fighting Vehicles are given in this section under Belgium.

More recently the AIFV has been selected by Turkey to meet its future requirements and additional details of this programme are given under Turkey. A total of 1698 vehicles are covered, with the first 285 hulls coming from Cockerill Mechanical Industries who will then ship the equipment to enable the remainder of the order to be built in Turkey by FMC-Nurol.

In 1990 Pakistan purchased a single prototype of the Armored Infantry Fighting Vehicle in kit form and this was shipped later that year. Additional details of Pakistani M113/AIFV production are given under Pakistan.

Description

The hull of the AIFV is made of welded aluminium armour with an additional layer of FMC-developed and patented spaced laminate steel armour bolted onto the hull. The bolt-on armour provides better protection for less weight than any other armour material in current production. Closed cell polyurethane foam within the armour system gives increased buoyancy for amphibious operations.

The driver sits at the front of the hull on the left and has a single piece hatch cover that opens to the right. To the front and left side of his position are four M27 day periscopes, the centre one of which can be replaced by a passive periscope for driving at night.

The commander sits to the rear of the driver and has a single piece hatch cover that opens to the rear and five periscopes, four standard M17 periscopes and a periscope with a magnification of ×1 and ×6.

The engine is to the right of the driver and is coupled to the three-speed automatic transmission by a transfer gear case. The complete powerpack can be removed through a front hatch, which is fitted with a door that can be used for daily maintenance tasks. Normal steering and braking are by oil-cooled band-type brakes in the controlled differential. There is a set of air-cooled disc brakes on the differential output shafts for low-speed tight turns on land and for more efficient manoeuvring in the water. A large radiator and a belt-driven high-capacity fan mounted in the top of the engine compartment provide cooling to normal high ambient temperature limits.

The power train is similar to the M113A1s except for improved features such as the engine turbo-charger, larger capacity radiator, transmission modified with heavy-duty components, heavy-duty universal joints and final drives as fitted to the M548 tracked cargo carrier.

The enclosed weapon station is on the right side of the hull behind the engine. It is all-welded with an additional layer of spaced armour as fitted to the hull. The gunner has a single piece hatch cover that opens to the rear, four M27 day periscopes (two either side), a Philips sight with a magnification of ×2 or ×6 for day use and ×6 for night use.

Main armament consists of an Oerlikon 25 mm KBA-B02 cannon with dual feed, which has 180 rounds of ready-use ammunition and another 144 rounds in reserve. Mounted coaxially to the left of the main armament is an FN 7.62 mm machine gun with 230 rounds of ready-use ammunition and a further 1610 rounds in reserve in the hull. Turret traverse and gun elevation are electro-hydraulic, elevation is from -10° to +50° at a speed of 60°/s and turret traverse is through a full 360° at 60°/s. Although developed and manufactured by FMC, some weapon stations for the Dutch vehicles were manufactured and integrated by DAF. Certain interior modifications have also been carried out by DAF for a number of AIFVs. Dutch vehicles also have six smoke grenade launchers mounted on the front of the hull.

The troop compartment at the rear of the hull accommodates seven seated infantry. One man sits between the commander and the turret facing the rear and the remaining six sit three each side, back to back. Over the top of the troop compartment is a single piece hatch cover that opens to the rear and contains the troop compartment ventilator. There are five firing ports, two in each side of the hull and one in the ramp at the rear. Over each of the side firing ports is an M17 periscope and over the rear firing port an M27 periscope. To reduce accidental damage inside the vehicle, holders support the individual weapons during firing. Spent brass catching bags prevent ejected cartridge cases from hitting adjacent men.

The infantrymen enter and leave the vehicle by a power-operated ramp at the rear of the hull which opens downwards. An emergency door is provided in the left side of the ramp.

The two armoured fuel tanks at the rear of the vehicle, one either side of the ramp, are separated from the vehicle interior by armour plate.

The new torsion bar-in-tube spring arrangement, which almost doubles the effective length of the torsion spring, gives a much improved ride compared with the M113A1 APC. The track, roadwheels and most of the other suspension elements are identical to the M113A1's. The suspension either side consists of five dual rubber-tyred road wheels with the drive sprocket at the front and the idler at the rear. There are no track return rollers. The first, second and last road wheel stations either side have a hydraulic shock absorber. The track is a T130E1 steel block, rubber-bushed pin type with removable rubber pads.

The Belgian and Turkish AIFVs use torsion bar only suspension, similar to the M113A2, to achieve the same results.

The AIFV is fully amphibious and is propelled in the water by its tracks. Before entering the water a trim vane, which is stowed flat on the glacis plate when not in use, is erected at the front of the hull.

Optional equipment includes a McDonnell Douglas Helicopters 25 mm Chain Gun or a 12.7 mm (0.50) M2 HB machine gun in place of the 25 mm cannon, an M36 sight for the gunner with magnifications of ×1 and ×7, an NBC system, a heater for the personnel/cargo area and a heater for the engine coolant and battery.

Variants

Belgian Vehicles

Details of these are given in this section under Belgium.

Netherlands Models

The basic AIFV is designated the YPR 765 PRI by the Netherlands Army and variants in service are:

YPR 765 PRCO-B command vehicle, crew nine, combat weight 13 700 kg.

YPR 765 PRCO-C1 to C5, crew nine, combat weight 12 400 kg, armed with 12.7 mm (0.50) M2 machine gun. The C1 is a battalion commander's vehicle, C2 a battalion gunnery centre, C3 a mortar fire control vehicle, C4 an AA command vehicle and C5 an observation vehicle with a crew of four. All these versions have an M113 type cupola with a 12.7 mm M2 HB pintle-mounted machine gun.

YPR 765 PRRDR radar vehicle fitted with the British ZB 298 battlefield surveillance radar.

YPR 765 PRRDR-C radar/command vehicle.

YPR 765 PRGWT ambulance, crew four, weight 11 600 kg, unarmed.

YPR 765 PRI/I squad vehicle fitted with M113 type cupola with 12.7 mm (0.50) M2 machine gun, carries nine men plus driver.

YPR 765 PRMR mortar tractor (tows 120 mm Thomson Brandt mortar), crew seven, combat weight 13 000 kg, armament one 12.7 mm (0.50) M2 machine gun pintle-mounted on a standard M113 type cupola.

YPR 765 PRVR-A and PRVR-B cargo vehicles, crew two, combat weight 13 400 kg, armament one 12.7 mm (0.50) M2 machine gun pintle-mounted on a standard M113 type cupola.

YPR 765 PRAT fitted with the Emerson TOW launcher system as fitted to the M901 Improved TOW Vehicle. The PRAT (Pantser Rups Anti-tank) entered service with the Dutch Army in 1982, and has a crew of four.

Netherlands Army Armored Infantry Fighting Vehicle (YPR 765) with trim vane erected at front of hull (C R Zwart)

Netherlands YPR 765 series vehicle with cupola armed with 12.7 mm M2 HB machine gun and power operated ramp lowered at the rear (C R Zwart)

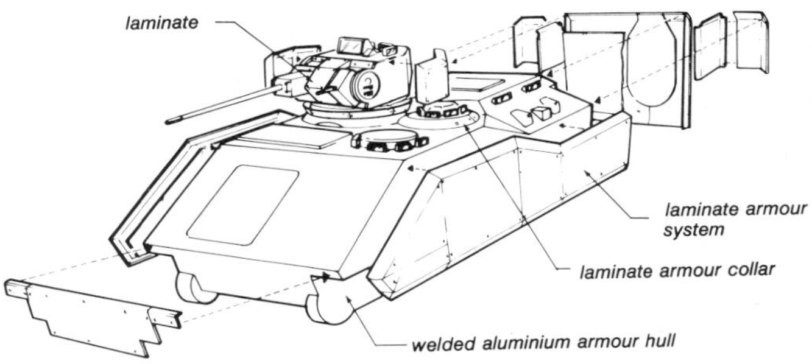

Armour arrangement on FMC Armored Infantry Fighting Vehicle (not to 1/76th scale)

YPR 806 PRBRG armoured recovery vehicle, crew four, weight 12 200 kg, armament one 12.7 mm (0.50) M2 machine gun. Details of this model are given later in this entry.

The first prototype of a mid-life upgrade vehicle for the Dutch Army is to be completed in 1993 with conversions starting in 1995.

With the addition of a variety of kits, developed by FMC, the basic AIFV Universal Vehicle Chassis can be converted to the following roles.

Command Post

This has revised seating arrangements for a command staff of six plus its crew of three (commander, gunner and driver), and is fitted with additional communications equipment, table and stowage rack for 12.7 mm (0.50) ammunition. It is normally armed with a 12.7 mm (0.50) M2 HB machine gun on a pintle mount.

Mortar Prime Mover

This tows a 120 mm Thomson Brandt or similar mortar and has racks at the rear of the troop compartment on the right side for 51 mortar bombs. A crew of seven is normally carried: three vehicle crew plus four mortar crew. It is normally armed with a 12.7 mm (0.50) M2 HB machine gun on a pintle mount, but this can be replaced by an M26 weapon station fitted with a 12.7 mm (0.50) M2 HB machine gun.

TOW ATGW Vehicle

In addition to being fitted with the Emerson ITV kit, other TOW installations are possible and the vehicle can also be fitted with other types of ATGW system such as the HOT.

Cargo Carrier

This model can carry up to 2040 kg of cargo and has a crew of two, driver and gunner. Armament consists of a 12.7 mm (0.50) M2 HB machine gun. The cargo kit includes protective grating behind the driver and gunner and tie-down devices to secure the load while travelling across country.

Ambulance

Four stretchers suspended from hanger harnesses are fitted in the troop compartment at the rear and two seats are provided for medical assistants. This model is unarmed.

Recovery Vehicle

This is similar to the XM806 which is a member of the M113A1 family but the Dutch vehicles have smoke dischargers mounted on the front of the hull, an HIAB crane mounted on the roof and additional buoyancy pods attached to the hull sides. This vehicle features the same suspension and power train as the AIFV. The trim vane at the front of the hull also has an additional buoyancy attachment.

Philippines Models

In 1979 the Philippine Government procured 45 AIFVs which have basically the same configuration as the Netherlands YPR 765 PRI vehicles. Although the enclosed weapon station was delivered ready to accept the Oerlikon 25 mm KBA-BO2 cannon, a kit was developed to convert the station to accept the 12.7 mm (0.50) M2 HB machine gun in place of the Oerlikon 25 mm cannon. Six armoured recovery vehicles were also procured (see previous variant entry).

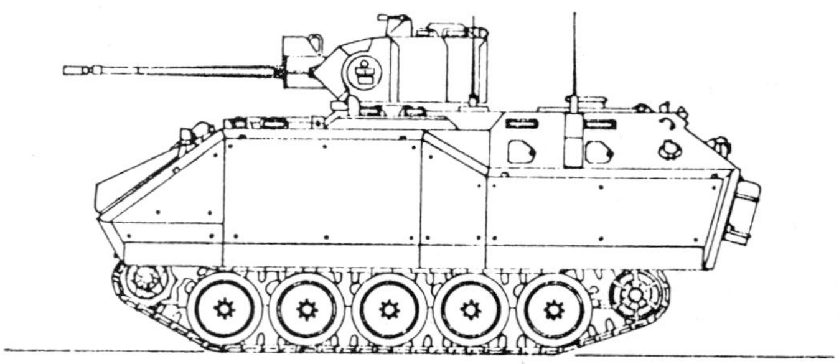

FMC Armored Infantry Fighting Vehicle

Command post version of Dutch YPR 765 showing external stowage for tent (Pierre Touzin)

SPECIFICATIONS

CREW	3 + 7
COMBAT WEIGHT	13 687 kg
UNLOADED WEIGHT	11 405 kg
POWER-TO-WEIGHT RATIO	19.29 hp/tonne
GROUND PRESSURE	0.67 kg/cm²
LENGTH	5.258 m
WIDTH	2.819 m
WIDTH OVER TRACKS	2.54 m
HEIGHT	
to top of periscopes	2.794 m
turret roof	2.619 m
to hull top, front	1.854 m
to hull top, rear	2.007 m
FIRING HEIGHT	2.33 m
GROUND CLEARANCE	0.432 m
TRACK WIDTH	381 mm
LENGTH OF TRACK ON GROUND	2.667 m
MAX SPEED	
road	61.2 km/h
on 10% grade	28.2 km/h
water	6.3 km/h
ACCELERATION	
0 to 32 km/h	10 s
0 to 48 km/h	23.1 s
FUEL CAPACITY	416 l
MAX ROAD RANGE	490 km
FORDING	amphibious
GRADIENT	60%
SIDE SLOPE	30%
VERTICAL OBSTACLE	0.635 m
TRENCH	1.625 m
TURNING RADIUS	7.62 m
ENGINE	Detroit Diesel 6V-53T V-6 liquid-cooled diesel developing 264 hp at 2800 rpm
TRANSMISSION	Allison Transmission Division of General Motors TX100-1A automatic with 3 forward and 1 reverse gears, torque converter and lock-up clutch
STEERING	FMC DS 200 mechanically controlled differential and pivot steer
SUSPENSION	torsion bar in tube
ELECTRICAL SYSTEM	28 V
BATTERIES	2 × 12 V 6TN
ARMAMENT	
main	1 × 25 mm cannon
coaxial	1 × 7.62 mm MG
SMOKE-LAYING EQUIPMENT	6 smoke grenade launchers
AMMUNITION	
25 mm	180 ready + 144 reserve
7.62 mm	230 ready + 1610 reserve
GUN CONTROL EQUIPMENT	
Turret power control	electro-hydraulic/manual
Max rate power traverse	60°/s
Max rate power elevation	60°/s
Gun elevation/depression	+50°/-10°
Gun stabiliser	optional

Status: Production as required. In service with Belgium, the Netherlands, the Philippines and Turkey. Under evaluation in Pakistan. Production is still underway in Turkey (qv) but has been completed in Belgium (qv).

Manufacturer: FMC Corporation, Ground Systems Division
2830 De La Cruz Boulevard
Santa Clara
California 95052
USA
Telephone: (408) 289 0111
Telex: 6714210
Fax: (408) 289 2150

FMC M113 Armored Personnel Carrier Family

Development

In January 1956 development began of an air-transportable, armoured, multi-purpose vehicle family, 'to provide a lightweight, armored personnel carrier for armor and infantry units capable of amphibious and air-drop operation, superior cross-country mobility and adaptation to multiple functions through applications of kits and/or modifications of its superstructure'.

Prototypes of both aluminium (T113) and steel (T117) were built and tested but development of the T117 was cancelled in favour of further development of the T113. This became the T113E1 which was standardised as the M113 in 1960 and entered production at FMC's San Jose facility in early 1960. Trials of a diesel-engined model called the T113E2 showed a substantial improvement in operating range (from 321 to 483 km) as well as a reduction in fire risk. This model was standardised as the Carrier, Personnel, Full-Tracked, Armored, M113A1, in May 1963. After standardisation 10 pre-production vehicles were built and the M113A1 replaced the M113 in production from September 1964.

Recent production figures for the M113 family of tracked vehicles are:

Year	US Army	Foreign Military Sales	Direct Sales
1977	707	2166	307
1978	1167	1734	292
1979	1302	777	213
1980	1229	905	629
1981	317	1211	518
1982	119	1186	524
1983	479	566	112
1984	587	603	23
1985	287	311	666
1986	464	130	426
1987	567	202	65
1988	706	93	17
1989*	116	0	6
Totals	8047	9884	3798

Grand Total 1977 to 3/1/89: 21 729
*to 3/1/89

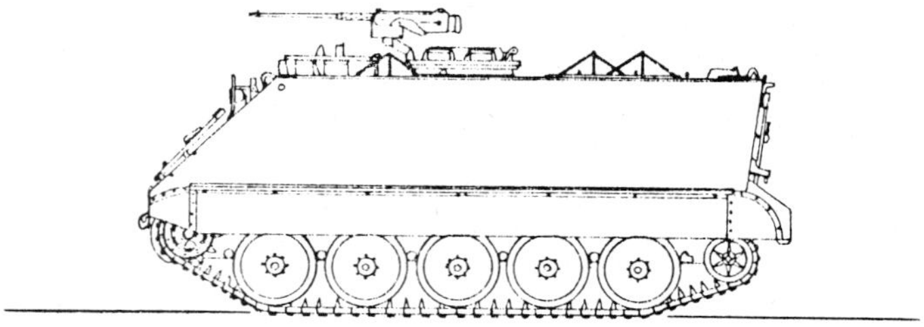

M113A1 armoured personnel carrier armed with 12.7 mm (0.50) M2 HB machine gun

Procurement history of M113 family as of 1 March 1989:

Vehicle type	**US**	**FMS**	**Direct Sales**	**Total**
M113A1 ITV	139			139
M113A1 TOW	257	211	12	480
M113A2 ITV	1403	196		1599
M113A2 TOW	5	367		372
Ambulance		11	8	19
Bulldozer		4		4
C&R			266	266
C&R2			173	173
Fitters	2	494	110	606
M106	841	20	1	861
M106A1	990	326	93	1409
M106A2	53	282	11	346
M113	10 260	4553		14 813
M113A1	10 193	13 036	2094	25 323
M113A1R			1	1
M113A2	874	3369	1898	6141
M113A2IT			6	6
M113A2R			80	80
M113A3	1048			1048
M125A1	854	1398	133	2385
M125A2	69	492	36	597
M132		201		201
M132A1	150			150
M548	2362	195	2868	5425
M548G			290	290
M548A1	256	127	22	405
M577	23	921		944
M577A1	2201	1440	222	3863
M577A2	1287	423	140	1850
M667		26		26
M730	52	64	1	117
M730A1	107	37		114
M741		220		220
M741A1		10		10
M741A2		13		13
M901	145	56		201
M901 ITV			14	14
RTL		42		42
XM474	57			57
XM474E2	240			240
XM667E1	302	3		305
XM727	124			124
XM730	424	37		461
XM741	381	46		427
XM806E1			54	54
XM806E2		27		27
Totals	35 098	28 647	8533	72 278

In addition, OTO Melara has built 4500 members of the M113 series of vehicle under licence for home and export markets.

Description

The all-welded aluminium hull of the M113 protects the crew from small arms fire and shell splinters.

The driver sits at the front of the hull on the left side and has a single piece hatch cover that opens to the rear. To his front and left side are four M17 periscopes and there is also an M19 infra-red periscope in his roof hatch.

The engine compartment to the right of the driver's position is fitted with a fire-extinguishing system that can be operated by the driver or from outside the vehicle. The air-inlet and air-outlet louvres and the exhaust pipe outlet are in the roof and there is an engine access door in the front of the hull that hinges forwards.

The power train consists of the power plant (engine, transfer gear case and transmission), steering control differential, pivot steer, final drive and associated drive shafts and universal joints.

The commander sits to the rear of the engine compartment and has a cupola that can be traversed through a full 360°, five M17 periscopes and a single piece hatch cover. Pintle-mounted on the forward part of the commander's cupola is a 12.7 mm (0.50) Browning M2 HB machine gun with an elevation of +53°, a depression of -21° and a total traverse of 360°. One hundred rounds of ready-use ammunition are carried for this weapon.

The infantrymen enter and leave the M113 via a power-operated ramp in the rear of the hull that opens downwards and has a door in the left side. Behind the commander's cupola is an oblong hatch cover that opens to the rear, behind which is a domed ventilator. The infantrymen travel on seats down each side of the hull, which can be folded up to enable the vehicle to be used as an ambulance or to carry cargo.

The torsion bar suspension either side consists of five dual rubber-tyred road wheels with the drive sprocket at the front and the idler at the rear. There are no track return rollers and the first and last road wheel stations are fitted with a hydraulic shock absorber. A rubber track shroud on each side of the hull controls the flow of water over the tracks when the vehicle is afloat. The M113 is fully amphibious being propelled in the water by its tracks. Steering when afloat is the same as on land. Before entering the water the two bilge pumps are switched on and the trim vane, which folds back onto the glacis plate when not in use, is extended at the front of the hull.

The following kits are available for members of the M113 family: anti-mine armour bolted on front half of vehicle bottom (includes buoyancy aids), anchor kit (set of two for use with capstans for self-recovery), buoyant side pods, combination bulldozer/snow plough, NBC detector and automatic alarm, full width buoyant trim vane, gun shields, heater for personnel and cargo areas, heater for engine coolant and battery, stretcher kit which provides support for four stretchers when the vehicle is being used as an ambulance, M8A3 gas-particulate unit (includes M2A2 air purifier with flexible hoses to fit M14A1 tank gas masks of driver and commander and up to two others), non-skid ramp plate kit and windscreen for driving with hatch open.

M113A2 (formerly Product Improved M113A1)

Following successful trials with prototype vehicles the United States Army in 1978 decided to carry out a product improvement programme on its fleet of M113 (5300) and M113A1 (12 700) APCs. The conversions are being carried out at the Red River Army Depot Texas, Daewoo Industries in South Korea and Mainz Army Depot, Germany. The Army plans to have 21 323 Product Improved M113A1s in service. The US Army also ordered an initial batch of 2660 M113A2s at a cost of $154 million, the first of which were completed at FMC's San Jose facility in July 1979. These improvements can be summarised as follows:

Improved engine cooling design that reverses the position of the fan and radiator and incorporates a new radiator and surge tank and a new cooling fan. This new cooling system draws in ambient air through the radiator providing increased cooling efficiency for both the engine and transmission, reduces oil film and dust build-up on the core of the radiator, provides a negative pressure engine compartment which reduces the possibility of exhaust fumes leaking into the crew compartment and finally provides a longer engine life and greater tractive effort at reduced temperatures.

The improved suspension system incorporates high-strength steel torsion bars that provide 228 mm of roadwheel travel, improved shock absorbers on the first, second and fifth road wheel stations, stronger rear idler assembly which has also been raised to reduce the chance of ground impact and finally increased ground clearance to minimise sprocket ground contact on rough terrain.

Other M113A2 variants such as the M577A2 and the M125A2 are currently available. The M548 cargo carrier with A2 improvements has been redesignated the M548A1 and first production vehicles were completed in 1982.

Rear-mounted, armoured fuel cells are a current production option and many non-United States orders are calling for their installation. This option provides more interior space and reduces the danger of fire. M113A1 vehicles have now been phased out of production. A 200 A generating system is available and is being phased into production on some vehicle models. A kit is also available.

M113A3

M113A3 production commenced in early 1987 and US Army units started receiving these vehicles in the Summer of 1987.

The M113A3 incorporates the cooling and suspension improvements of the M113A2 but also has better performance and reliability. Major improvements include the replacement of the 6V-53 (212 hp) diesel engine by the turbo-charged Detroit Diesel 6V-53T (275 hp) and the replacement of the present TX100-1 transmission, transfer gearcase, steering differential and pivot steer with the X-200-4 Allison transmission which provides four forward speeds instead of the present three, hydrostatic steering to provide smoother turning with less effort and reduced shock loading on the suspension system and greater power efficiency which results in more horsepower and fuel savings.

Basic M113 APC with 12.7 mm M2 HB machine gun (US Army)

M113A3 with external fuel cells at rear of hull

M113A3 with add-on armour kit for increased protection, smoke dischargers and armoured cupola with 12.7 mm M2 MG

The driver's controls have also been changed and the conventional sticks have been replaced by a steering wheel and brake pedal.

The US Army refers to this improvement as the RISE power train (Reliability Improvement of Selected Equipment). The M113A3 has a 200 A generating system, four batteries and improved electrical diagnostic capabilities to replace the 100 A system with two batteries in the M113A2. It also has armoured external fuel tanks and internal spall protection liners. The three improvements (RISE, AEFT and spall liners) will be applied to the M901A1 and M981 vehicles which will become the M901A2 and M981A1. The first M113 family variant to be converted to the RISE power train were M730 Chaparral carriers. The entire US fleet of 495 vehicles was converted to the M730A2 between 1987 and 1989.

Possible additional improvements

Several other improvements are being investigated for the M113 family. Each member of the M113 family is now being studied by the US Army for both active and passive NBC protection. Beginning in 1987, all M113A2/M113A3 production vehicles have been fitted with smoke grenade launchers.

In mid 1986 it was announced that the US Army was to modify its M113 series vehicles in a two-phase programme. For Block I the RISE power train is to be installed as well as internally mounted and movable spall suppression liners, external fuel tanks and mounting provisions for external armour kits. It is anticipated that between 6500 and 8000 vehicles will be fitted with this kit at a total cost of up to $923 million.

Block II includes bolt-on external kits of armour with $6.1 million being requested in FY88 for 186 kits although the total requirement is for between 4000 to 6000 kits.

Stretched M113A2

In 1976 two prototypes of a stretched M113A2 were built and tested between 1977 and 1978. The hulls are approximately 0.66 m longer than the standard vehicle's and have an additional set of road wheels each side, and are also fitted with the Detroit Diesel 6V-53T turbo-charged diesel engine, hydrostatic steering, the new cooling system and the improved suspension. The stretched M113A2 has a gross vehicle weight of 14 968 kg, a maximum payload of 4218 kg, maximum road speed of 64 km/h, range of 483 km and can carry 14 men plus the driver. The single hatch cover over the troop compartment has been replaced by two hatch covers hinged in the centre that open either side of the vehicle. Other improvements being investigated include a variable-speed fan drive, a new track, a night sight system, infra-red suppressor and laser target designator. As of early 1991 this version was still at the prototype stage.

Variants

M113/M113A1/M113A2 with Dozer Kit

The bulldozer kit adapts the M113A1 APC to carry out general bulldozing work, improvement of water entrances and grading. The installation of the kit does not change the basic vehicle or its load-carrying ability. It is fully amphibious as the front-mounted buoyant blade acts as a trim vane when raised during amphibious operations, after removal of the standard trim vane.

M106/M106A1/M106A2 107 mm Mortar Carrier

Full designation is Carrier, Mortar, 107 mm, Self-propelled. This is the basic APC with a 4.2 in (107 mm) M30 mortar mounted on a turntable in the rear of the hull that fires to the rear through an opening in the roof covered by a three-part hatch. The mortar has a traverse of 43° right and 46° left and 88 (M106A1) or 93 (M106) rounds of mortar ammunition are carried. Mounted externally on the left side of the hull is a mortar baseplate and bridge that allows the mortar to be fired away from the vehicle if required. The 12.7 mm (0.50) M2 HB machine gun is retained.

M106 107 mm (4.2 in) mortar carrier with three part roof hatch open and ramp lowered at the rear. Commander's cupola has 12.7 mm M2 HB machine gun fitted

Following extensive trials with three competing 120 mm mortar systems, the Israeli Soltam system was selected in early 1988 and this will replace the 107 mm mortar in the future.

M125/M125A1/M125A2 81 mm Mortar Carrier

Full designation is Carrier, Mortar, 81 mm, Self-propelled. This is the basic APC with an 81 mm M29 mortar mounted on a turntable with a 360° traverse in the rear of the hull. That fires through an opening in the roof, covered by a three-part hatch. The mortar can also be deployed away from the vehicle and 114 mortar bombs are carried. The 12.7 mm (0.50) M2 HB machine gun is retained.

M163 Vulcan Self-propelled Anti-aircraft Gun

This is essentially an M113A1 chassis fitted with a one-man electrically driven turret which is armed with a 20 mm M168 Vulcan gun, Navy Mk 20 Mod. A gyro lead-computing sight and a range-only radar are mounted on the right side of the turret. The chassis is designated the M741. Full details of this sytem are given in *Jane's Land-Based Air Defence 1991-92*, page 96. In addition to being used by the US Army it is also used by Israel, Jordan, Morocco, Portugal, Sudan, Thailand, United States and Yemen.

M132/M132A1 Flamethrower

This model is obsolete and no longer in service with the US Army. Details were given in *Jane's Armour and Artillery* 1982-83, page 379.

M548 Cargo Carrier

This unarmoured cargo carrier was designed from 1960 under the designation XM548 and uses many automotive components of the M113A1 APC. The XM548 was followed by a diesel-engined version called the XM548E1 which was subsequently standardised as the M548. First production vehicles were completed early in 1966 by the FMC Corporation at its Charleston, West Virginia, facility. The current M548A1 vehicles are being produced with M113A2 improvements, such as suspension and cooling system. The M730A1 includes suspension and cooling improvements. The M730A2 incorporates the 6V-53T diesel engine and the X-200-4 transmission. A full description of the vehicle is given in *Jane's Military Vehicles and Logistics 1991-92*, page 475.

MSM-Fz: This is a modified M548 with an anti-tank minelaying system for the German Army. Details are given in *Jane's Military Vehicles and Logistics 1991-92*, page 210.

Lance surface-to-surface tactical nuclear missile system uses two variants of the M548 chassis, the **M752 launch vehicle** (chassis designation M667) and the **M688 loader transporter** which carries additional missiles. **M1015 electronics carrier:** this is used to carry tactical electronic

M125A1 showing turntable-mounted 81 mm mortar (US Army)

M548 tracked cargo carrier with bows and tarpaulin cover over rear troop compartment

M163A1 20 mm Vulcan self-propelled anti-aircraft gun system of the US Army fitted with the Lockheed Electronics Company Product Improved Vulcan Air Defense System (PIVADS) kit

British Army Tracked Rapier vehicle launching a British Aerospace (Dynamics) Rapier surface-to-air missile during annual training

reconnaissance and countermeasures systems but is now being phased out of service.

The basic M548 can carry 5443 kg of cargo. FMC has built and tested as a private venture a stretched M548 which has a payload of 7258 kg. This has an additional road wheel each side and all the improvements mentioned in the stretched M113A2 described previously.

Tracked Rapier

The Tracked Rapier SAM system, which is only in service with the British Army and was originally developed to meet the requirements of Iran, is based on a rebuilt M548 tracked cargo carrier chassis. At the front of the hull is a new fully enclosed armoured compartment with the Rapier optical tracker on the roof, which is retracted for travelling, to the immediate rear of the cab is the auxiliary power unit and mounted at the rear is a turntable with eight Rapier SAMs in the ready to launch position. Tracked Rapier is supported in action by the Forward Area Support Team (FAST) vehicle which carries spare parts and the Tracked Rapier Support Vehicle (TRSV) which carries 20 Rapier SAMs. Both the latter vehicles are essentially M548 tracked vehicles.

Full details of Tracked Rapier are given in *Jane's Land-Based Air Defence 1991-92*, page 138.

M730 Chaparral SAM launcher

This is based on a modified M548 series tracked cargo carrier and mounted over the rear area is a one man power operated turret with four Chaparral SAMs in the ready to launch position with a further eight missiles being carried in reserve. In addition to being used by the US Army, the Chaparral is also used by Egypt, Israel, Morocco, Portugal, Taiwan and Tunisia.

The US Army has a total of 495 Chaparral surface-to-air missile systems and all of these have been upgraded to the M730A2 configuration. Full details of the system are given in *Jane's Land-Based Air Defence 1991-92*, page 149.

M577/M577A1/M577A2 Command Post

Full designation is Carrier, Command Post: Light Tracked, M577 or M577A1. The first four prototypes were built at Detroit Arsenal under the designation XM577 and were conversions of existing M113 APCs. The XM577 was standardised as the M577 in March 1963, after the vehicle was already in production.

The M577/M577A1 is essentially an M113/M113A1 with a higher roof to the rear of the driver's position. In the static role a tent can be erected at the rear of the hull to increase the work area. A generator is mounted externally to provide power for the additional communications equipment installed. In addition to being used as a command post the vehicle can also be used as a fire direction centre, communications vehicle and as a mobile medical treatment facility.

Stretched XM577A3 Command Post

Late in 1988 FMC completed the prototype of the Stretched XM577A3 Command Post Vehicle which has six, rather than five, roadwheels either side as well as external fuel tanks similar to those fitted to recent production M113 series vehicles. Other improvements include the complete M113A3 powerpack and the option of additional armour on the roof for increased protection against top attack and the installation of an environmental control unit. This vehicle could meet the US Army requirement for a Standard Integrated Command Post System (SICPS).

Basic specifications of the XM577A3 are:-

CREW	8
WEIGHT, COMBAT LOADED	14 424 kg
LENGTH	5.83 m
WIDTH	2.69 m
WIDTH REDUCED	2.54 m
HEIGHT	2.59 m
GROUND CLEARANCE	0.43 m
MAX ROAD SPEED	64.37 km/h
CRUISING RANGE	483 km
GRADIENT	60%
SIDE SLOPE	30%
VERTICAL OBSTACLE	0.61 m
TRENCH	2.18 m
ENGINE	Detroit Diesel Model 6V-53T diesel developing 275 hp
TRANSMISSION	Allison Transmission Division of General Motors X200-4 hydrokinetic

M730 series Chaparral surface-to-air missile launcher, based on the M548 tracked cargo carrier with four Chaparral surface-to-air missiles in ready to launch position

M577A1 Armoured Command Vehicle of the Royal Australian Armoured Corps (Barry Marriott)

XM1068 Command Post Vehicle

Early in 1990 prototype modification kits for the creation of a new command post vehicle were being field tested by the US Army. The XM1068 is based on the standard M577A2 which has been modified under the Standardised Integrated Command Post System (SICPS) programme.

The SICPS programme is designed to develop and field standard command post equipment for use at battalion through corps level and to include accessories and integrated shelters.

There are three major hardware platforms being developed under the SICPS programme, a tent command post, a rigid wall shelter for the High Mobility Multi-purpose Wheeled Vehicle and the XM1068 command post modification kit. The latter has been designed for fielding in mechanised units at operational levels from battalion through corps.

M113A2 Recovery Vehicle

Full designation is Recovery Vehicle: Full Tracked, Light Armored M113A2 (previously called the XM806) and is essentially an M113A2 fitted with a hydraulically driven winch and a fairlead assembly for recovering disabled vehicles. Mounted either side of the ramp at the rear of the hull are spades which are lowered to the ground during recovery operations. An auxiliary spade unit, carried on the roof when not required, can be used between the two spades in soft soil. The winch has a capacity of 5103 kg on a full drum and 9070 kg on a bare drum. Mounted on the top of the hull on the left side is an auxiliary crane which can lift a maximum load of 1361 kg at a radius of 1.53 m. There is also a fitter's model of the M113/M113A1/M113A2 which has an HIAB crane mounted on the left side of the roof which can lift a maximum load of 3085 kg. Full details of the ARVs and Repair Vehicles based on the M113 chassis are given in *Jane's Military Vehicles and Logistics 1991-92*, page 38.

M113 APC with TOW

The US Army has about 350 M113 series vehicles in service fitted with a pedestal-mounted TOW system that is retracted into the troop compartment when not required.

XM577A3 stretched command post from rear before ramp has been installed (Scott R Gourley)

M806A1 Armored Recovery Vehicle, Light, of the Royal Australian Armoured Corps (Barry Marriott)

M901/M901A1 Improved TOW Vehicle

Details of this version are given in the *Tank destroyers* section.

M981 Fire Support Team Vehicle

The Fire Support Team Vehicle (FISTV) M981 has been developed under contract to the US Army by Emerson Electric and is essentially the M901 Improved TOW Vehicle fitted with the AN/TVQ-2 GLLD with north-finding module and line of sight subsystem, AN/TAS-4 night sight, land navigation system and extensive communications equipment.

Three pre-production units were used in the engineering development tests with an additional three vehicles completed late in 1982.

In August 1983 Emerson was awarded a production contract for the first 70 systems with a further 459 systems being ordered in March and December 1984. The first batch of 37 production units were delivered in May 1985 with first fielding completed at Fort Stewart in August 1985 and delivery of first units to Europe in August 1986.

Major subcontractors are FMC for the M113A2 chassis, Hughes Aircraft for the AN/TVQ-2 Ground/Vehicle Laser Locator/Designator, Bendix for the North Seeking gyroscope, Emerson integrates the whole system as well as providing the targeting station, intercommunications system and the image transfer assembly.

In early 1988 Egypt took delivery of 20 Artillery Target Locating Vehicles (ATLV) which are almost identical to the FISTV used by the US Army with another five vehicles on order.

M1059 Smoke Generating System

This entered service with the US Army in 1988 and is an M113A2 with two M54 smoke generators mounted on top of the hull, which can be operated either singly or in tandem.

An electrical air compressor mounted inside the M113A2 provides air to the pulse jet engine and purges it before restarting. The air pressure is monitored by the operator at the control panel allowing hot restarts by the press of a button.

The fog oil system includes a 120 (US) gallon fog oil tank mounted inside the vehicle with two fog oil pumps. Fog oil flow can be controlled by adjusting the rheostat on the control panel.

Each smoke generator has its own remote five-gallon jerry can external fuel supply which is submersible. The integral equipment box contains a fuel pump, fuel solenoid and fuel pressure switches. Associated controls allow remote-control from inside of the vehicle. More than 200 M1059 smoke generating versions will be converted from existing M113A2 series vehicles.

M113 Technology Demonstrator

It consists of a M113 APC rebuilt by Marshall of Cambridge (Engineering) Limited with a new powerpack consisting of a Cummins Engine Company

M113A1 fitters vehicle of the Royal Australian Armoured Corps with roof-mounted crane (Barry Marriott)

FMC's Advanced Ground Vehicle Technology demonstrator showing extensive array of sensors on roof

6 CTA 8.3 diesel coupled to a Self-Changing Gears T320 fully automatic transmission. To improve cross-country mobility Horstman Defence Systems dampers have been installed on the first and last road wheel stations. This is only a trials vehicle and has yet to enter production.

Advanced Ground Vehicle Technology (AGVT) Demonstrator

In 1986 FMC demonstrated the Advanced Ground Vehicle Technology (AGVT) demonstrator in both its tele-operation and autonomous modes.

In the first of these the driver controlled the robotic vehicle from the command centre and used it to carry out a simulated route reconnaissance. Reconnaissance on the move was carried out by the commander and driver using remotely controlled colour cameras and a Forward Looking Infra-red (FLIR) camera on the robotic vehicle.

In the autonomous mode the vehicle followed a road without any assistance from the driver or commander using its onboard computers as sensors.

The prime contractor for the AGVT is FMC Corporation under contract to the US Army Tank Automotive Command and the DoD Defense Advanced Research Projects Agency (DARPA).

The AGVT demonstrator consists of two main components, a command centre and robotic vehicle. The former houses the computer and communications equipment and has been designed as a field laboratory to carry out research, development and performance evaluation of component robotic technologies.

The command centre generates and transmits mobility and sensor commands in real time to the robotic vehicle, which then decodes and executes such commands.

The robotic vehicle is a modified FMC-built M113 APC fitted with sensors and computers to carry out the missions transmitted from the command centre.

The throttle, brake, transmission, steering and pivot mechanisms are all controlled by a driver. The perception system includes colour and infra-red cameras for remote driving while other sensors provide navigation and status data.

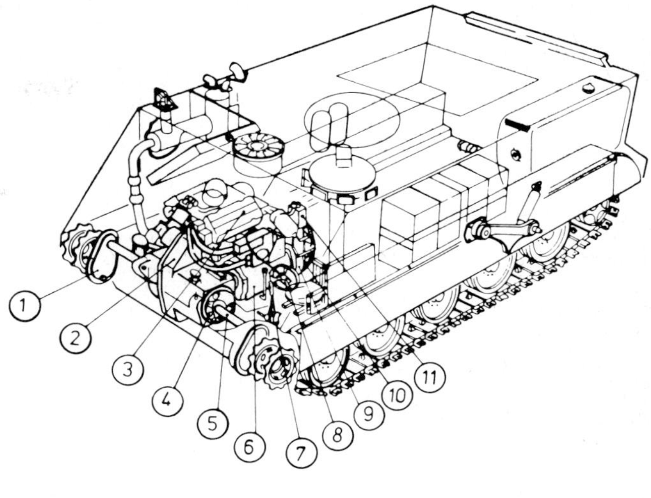

M113A1 APC fitted with Thyssen Henschel improved brake system **(1)** *final drive* **(2)** *HL-113 hydrostatic steering gear* **(3)** *diesel engine* **(4)** *dry-disc brake* **(5)** *Cardinac shaft* **(6)** *gearbox* **(7)** *parking brake* **(8)** *steering unit* **(9)** *PTO gear* **(10)** *intermediate gearing for driving axial pistons displacement pump* **(11)** *ramp hydraulic system* (not to 1/76th scale)

Armored Infantry Fighting Vehicle

This has been developed from the M113A1 as a private venture by the FMC Corporation and is fully described in its own entry earlier in this section.

Automatic Cannon Vehicle

This is an M113A2 fitted with the FMC two-man 25 mm weapon station that is a development of the weapon station installed on the M2 Bradley IFV. The ACV is a private venture by FMC and has yet to enter production.

M113 with CVAST turret

For trials an M113 series APC has been fitted with the same turret as the Combat Vehicle Armament System Technology (CVAST) test bed on the M2 Bradley chassis.

Lynx Command and Reconnaissance Vehicle

This uses many automotive components of the M113 APC and is fully described in the *Reconnaissance vehicles* section. This vehicle is only used with the Canadian and Netherlands Armies, the latter fitted with 25 mm turret.

MTU M113 Repower Package

The German company MTU is offering a M113 repower package consisting of an MTU 6V 183 TC22 diesel engine coupled to a ZF LSG 1000 automatic transmission with an integrated hydrostatic steering system. The MTU 6V 183 TC22 is based on the Daimler-Benz OM 441 LA downrated to 300 hp for use in the M113 application.

Thyssen Henschel M113

Thyssen Henschel has completed trials of an M113 series APC fitted with HL-113 hydrostatic steering gear and improved brake system. This, according to Thyssen Henschel, has a number of advantages over the torque-controlled differential steering system currently fitted to the early production models of the M113 series.

The powerpack consisting of the engine, PTO and gearbox which remain unchanged and the HL-113 hydrostatic steering gear and improved brake system replaces the standard DA 200 differential steering gear. In order to drive the hydrostatic axial piston transmission it is necessary to flange an intermediate PTO gear between the engine and the gearbox. The conversion can be undertaken by a normal repair workshop although Thyssen Henschel can undertake this if required.

M113 with Composite Armour

In 1985 FMC Corporation and Martin Marietta Aerospace, Baltimore, delivered an M113 with a lightweight composite hull to the US Marine Corps for trials as part of its amphibious vehicle development programme. The hull is made of a reinforced plastic material covered with ceramic tiles with a total thickness of about 32 mm. This is said to provide greater protection than aluminium armour plate at equal or lighter weight which would be critical in amphibious armoured vehicles.

This contract was part of the US Marine Corps Surface Mobility Exploratory Development Program to demonstrate feasibility and affordability under acceptable maintenance and repair conditions of technology developed for future amphibious vehicles.

The FMC vehicle completed field testing at Camp Pendleton, California in 1985 and is currently in full operation with its aluminium hulled counterparts. A significant feature of the FMC vehicle is a 5 to 10 dB reduction in interior crew compartment noise relative to a standard aluminium hulled M113 vehicle.

Thyssen Henschel Appliqué Armour

Thyssen Henschel has developed an appliqué armour kit for the M113 APC which weighs a total of 650 kg. When fitted the armour provides the following protection:

Hull front	14.5 mm projectiles at a range of 100 m
	20 mm (DM 43) projectiles at a range of 200 m
Hull sides	14.5 mm projectiles at a range of 100 m at an angle of 45°
	20 mm (DM 43) projectiles at a range of 200 m at an angle of 35°

Creusot-Loire Add-on Armour

Creusot-Loire Industrie has developed an add-on armour package for the M113 that can be easily installed or removed from the vehicle using onboard tools. Some panels are fixed while others are removable for access for maintenance. The amphibious characteristics of the basic M113 are not changed. Three kits have been developed. One gives complete protection against penetration from 14.5 mm API rounds at a range of 300 m through an angle of 180°, this kit weighs 445 kg.

The second kit provides protection against 12.7 mm AP rounds fired from a range of 100 m through an arc of 140° and against 20 mm AP and 23 mm AP rounds fired from a range of 200 m through an arc of 90°. Total weight of this kit is 830 kg.

The third kit, still under final development, provides protection against Soviet RPG-7 anti-tank weapons with HEAT warheads.

GIAT Industries Armour Package

GIAT Industries of France is now offering an add-on armour package for the M113 family of vehicles to provide increased protection against 14.5 mm and 20 to 23 mm attack. The package comprises panels of special hard steel that are attached to the front, sides and rear of the vehicle.

These provide complete protection through a full 360° arc from 14.5 mm rounds fired at a range of 100 m from the Soviet KPV heavy machine gun.

Over the frontal arc the GIAT Industries add-on armour provides protection against 20 mm or 23 mm armour piercing projectiles fired from a range of 200 m through an arc of 90°.

The armour package is of modular construction so that individual modules can be removed if required. The total weight of the kit depends on the version, for example the basic kit weighs 650 kg while the version that includes a false floor weighs 950 kg.

When fitted with the kit, the overall width of the M113 is increased to 2.84 m while the power-to-weight ratio is slightly reduced. The top speed of the vehicle is not effected by the installation of the package.

Other Armour Systems for M113

In addition to the Creusot-Loire and Thyssen Henschel armour packages for the M113 series already described, a number of other companies also offer such packages, these include Rafael and Urdan of Israel, LASAR of Italy and Cerachoc of France.

M113 with Bofors RBS 70 SAM

Bofors of Sweden has developed a kit which enables a standard M113 series APC to be fitted with its RBS 70 SAM system on a pedestal in the rear troop compartment. This was announced in 1988 and is now understood to be in service with the Pakistani Army with the conversions being undertaken locally.

M113 with 120 mm Mortar

In 1987 an M113A2 was fitted with a turret-mounted 120 mm breech-loaded mortar and demonstrated in the UK. The mortar is aimed, loaded and fired under complete armour protection. This system is still at the prototype stage.

The turret and 120 mm breech loaded mortar has been developed as a private venture by Royal Ordnance and can also be installed on a number of vehicles.

M113A2 APC chassis fitted with turret-mounted Royal Ordnance 120 mm breech-loaded mortar

Modified M113 series APC used for the ADATS system, this version is now in service with the Canadian Armed Forces

M113 with ADATS

The Swiss Oerlikon-Contraves ADATS (Air Defence Anti-tank System) can be installed on a wide range of tracked and wheeled armoured vehicles. The Canadian Armed Forces selected a modified M113 series APC and a total of 36 systems have been ordered. Those will be integrated in Canada, with the first vehicles already handed over. Mounted on top of the modified M113 vehicle is a turret with eight ADATS missiles, four either side, in the ready to launch position. Mounted between the missiles is the electro-optical module and mounted on the turret rear is the surveillance radar.

The US Army has also adopted the ADATS system but uses a modified M3 Bradley Cavalry Fighting Vehicle chassis. A small number have been delivered for extensive trials.

Infantry Armored Fighting Vehicle

This is a further development of the M113A1 by OTO Melara of Italy and is fully described in this section under Italy.

M113 Fire Support Vehicle

This is essentially an M113A2 APC fitted with the Cadillac Gage two-man turret armed with a 90 mm Cockerill Mark III gun, coaxial 7.62 mm machine gun, 7.62 mm anti-aircraft machine gun and, if required, smoke dischargers. Forty-two rounds of 90 mm (10 ready use and 32 stowed) and 2000 rounds of 7.62 mm ammunition would be carried. This model has yet to enter production. For trials purposes the Belgian company of Cockerill have fitted a M113A1-B APC with a Cockerill CSE 90 mm two man turret for use in the fire support role.

M113A2/TS 90 Light Tank

This is a joint development between FMC and GIAT Industries and essentially consists of an M113A2 APC with the roof modified to accept the GIAT Industries TS 90 two-man turret armed with a 90 mm gun. As of 1990 the M113A2/TS 90 system was still at the prototype stage.

Other Variants of the M113 Family

Almost every M113 user has developed variants of the M113/M113A1 family, including:

Australia: following trials with three prototypes of the M113A1 fitted with the turret of the British Scorpion CVR (T) vehicle armed with a 76 mm gun, a further 45 vehicles have been delivered to the Australian Army. The vehicle is used for fire support and retains its amphibious capability as additional buoyancy aids are fitted to the hull sides and front. This

M113A1 APC/LRV of Royal Australian Engineers (Barry Marriott)

M113A1 Medium Reconnaissance Vehicle of Royal Australian Armoured Corps (Barry Marriott)

Egyptian M113A2 APC fitted with Soviet 120 mm mortar showing baseplate stowed on rear of ramp

One of the prototypes of the Dassault Electronique ESD twin 23 mm self-propelled anti-aircraft systems designed to meet requirements of Egyptian Army

model is 4.864 m long, 2.69 m wide and 2.73 m high. The Australian Army also has in reserve 18 M113A1s fitted with the turret of the Saladin (6 × 6) armoured car which were withdrawn from service in the 1960s. Other M113 variants include a light reconnaissance vehicle with the American Cadillac Gage turret T50 armed with a 7.62 mm (0.30) and a 12.7 mm (0.50) machine gun, with 2500 rounds of 7.62 mm and 3000 rounds of 12.7 mm ammunition carried. Many Australian M113 APCs have a T50 turret armed with twin 7.62 mm (0.30) machine guns, with 5000 rounds of ammunition carried for them. Following the cancellation of project Whaler (*Jane's Armour and Artillery 1985-86*), the Australian Army is expected to upgrade at least half its fleet of 770 M113A1 vehicles to extend their lives into the coming decade. Among the Australian companies believed to be interested in this project are Goninan (teamed with FMC), Comeng (teamed with Thyssen Henschel), Ifield Engineering (which has developed a new hydrostatic transmission, steering and braking system) and Clyde Engineering (which teamed up with Krauss-Maffei for the Whaler project).

Early in 1989 Goninan proposed a special M113 improvement programme to the Australian Department of Defense under which the company would upgrade the existing M113A1 vehicles to the M113A2 or M113A3 standard, depending on the exact requirements of the Australian Army.

The Australian Army has applied the following designations to its M113 family vehicles:
M113A1 APC/LRV (armoured personnel carrier/light reconnaissance vehicle)
M113A1 APC(A) (ambulance)
M113A1 FSV (Scorpion) (fire support vehicle - Scorpion turret). This is now called the Medium Reconnaissance Vehicle
M113A1 FSV (Saladin) (fire support vehicle - Saladin turret)
M577A1 ACV (armoured command vehicle)
M125A1 APC(M) (carrier mortar)
M806A1 (armoured recovery vehicle, light)
M113A1 APC(F) (carrier, repair fitters)
M548 TLC (tracked load carrier).

Belgium: Belgian Mechanical Fabrications has built 525 M113 series vehicles under licence for the Belgian Army, additional details are given in this section under Belgium. More recently it has been announced that 21 Belgian Army M113A1-B vehicles would be converted into ground radar carriers and fitted with the Israeli EL/M-2130 radar which is being built under licence by the Belgian company of BATS (Belgian Advanced Technology Systems). CMI of Belgium is now involved in the production of AIFV hulls for Turkey. Additional details of this programme are given in this section under Turkey.

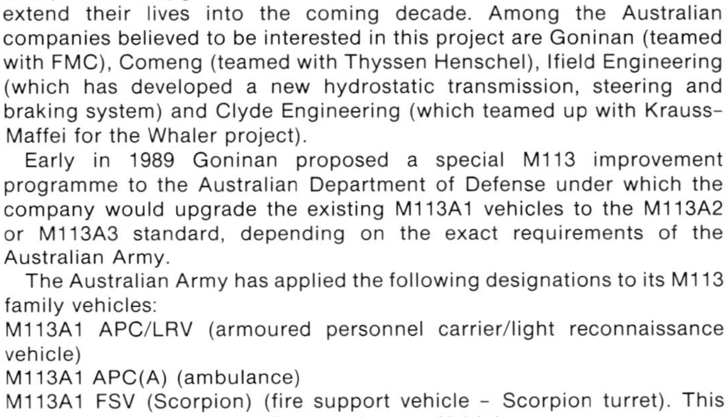

Brazilian Marines M113 APC's 12.7 mm M2 HB machine gun provided with all-round protection and shields at rear for lateral 7.62 mm machine guns (Ronaldo S Olive)

Brazil: the Brazilian Army has about 600 M113 series vehicles and Moto Pecas SA of Sorocaba, Sao Paulo State, has overhauled these and replaced the original petrol engine with a Brazilian-built Mercedes-Benz OM-352-A diesel.

Canada: details of the Canadian M113 series are given in this section under Canada.

Denmark: the Danish Army has a fleet of about 530 M113 series of APC and has recently ordered an additional 50 M113A2s from FMC which are to be fitted with the Italian OTO Melara T-25 two man power operated turret armed with a 25 mm Oerlikon-Contraves KBA cannon and 7.62 mm coaxial machine gun with the sights being provided by Zeiss of Germany. Systems integrator is E Falck Schmidt A/S who has recently upgraded the M41 light tanks of the Danish Army to the M41 DK-1 configuration.

Egypt: Egypt placed an order for 500 M113A2s in 1980 and in 1984 the Pentagon announced a letter of offer to Egypt for the supply of 354 M113A2 armoured personnel carriers, 19 mortar carriers, 13 M577A2 command post vehicles, 23 M113A2 armoured ambulances, 33 M548A1 ammunition carriers, 52 fitters vehicles and 43 M806E1 armoured recovery vehicles.

Egypt has also developed an appliqué armour kit for the M113A2. The ballistic steel plates give protection against 23 mm projectiles fired from a range of 200 m with an angle of impact of 35°. Each panel weighs 25 kg.

Following competitive trials between prototype twin 23 mm self-propelled anti-aircraft gun systems on a modified M1132A2 chassis, in 1987, Egypt selected the Dassault Electronique system. This is fitted with a one man power-operated turret armed with twin 23 mm cannon which are manufactured in Egypt and fitted with an Officine Galileo fire control system and a Dassault Electronique surveillance radar. In addition to the two 23 mm cannon, it is also armed with the locally built Sakr Eye SAM, this being an improved version of the Soviet SA-7 'Grail'. Full details of this air defence system are given in *Jane's Land-Based Air Defence 1991-92*, page 61.

The Kader Factory for Developed Industries has produced a kit to enable the M113 APC to be fitted with an Egyptian 120 mm mortar which fires to the rear of the vehicle. A total of 60 rounds of 120 mm mortar bomb ammunition are carried and if required the mortar can also be deployed away from the vehicle as a baseplate is carried on the ramp externally. A 12.7 mm machine gun and 1200 rounds of ammunition is also provided.

Germany: Germany is one of the largest users of the M113 outside the United States and has developed a number of versions to meet specific

German M113 APC fitted with Green Archer mortar-locating radar in operating position (C R Zwart)

German Army M113A1G armoured artillery observation vehicle

German M113 series APC being used in the ambulance role (C R Zwart)

German Army M557A1G armoured command post vehicle

requirements, including radar vehicles fitted with the RATAC battlefield artillery radar and the Green Archer mortar-locating radar, an artillery observation post vehicle, 120 mm mortar carrier and modifications to the basic APC.

Thyssen Henschel offers the following conversion kits for the M113 series:

KrKw	M113A1G armoured ambulance
PzM	M113A1G self-propelled mortar (120 mm)
ABRA	M113A1G artillery observation vehicle (with RATAC)
FlgltPz	M113A1G armoured air control vehicle
FltPzArt	M113A1G armoured fire control vehicle
Beob PzArt	M113A1G armoured artillery provision vehicle
RkPzArt	M113A1G artillery aiming circle vehicle
GepGefStdFzg	M113A1G command cell
GepGefStdFzg	M577A1G command post facility variant
GepGefStdFzg	M577A1G information cell
GepGefStdFzg	M577A1G air support cell

Thyssen Henschel, as prime contractor, has developed and produced an integrated fire control system for artillery batteries called the IFAB (Integrierte Feuerleitmittel für Artillerie Batterie) which consists of three main components, forward observer vehicle on M113A1 chassis, fire direction centre vehicle on M577A1G chassis and data installation kits for the actual weapons.

The artillery observation vehicle is called the Beobachtungs-panzer Artilleries and 320 vehicles were delivered to the German Army between 1982 and 1985. This vehicle has a Peri D11 double periscopic sight, incorporating a laser rangefinder, FNA615 land navigation system and a data input/output terminal linking it with the battery fire control centre.

The first IFAB command post vehicle based on a modified M577 chassis and the first armoured artillery fire control vehicle based on M113 series chassis were handed over to the German Army in April 1986. There are three versions of the M577, command cell, teleprinter cell and information cell, and 220 vehicles have been converted. There are 154 armoured artillery fire control vehicles fitted with a Falke 2 computer, three data terminals, matrix printer, teleprinter, radio and field telephone.

The mortar carrier, of which some 500 are in service, is fitted with a Tampella 120 mm mortar in the back which fires to the rear and has a traverse of 18.5° left and 25° right, elevation limits being from +45° to +80°. A total of 63 120 mm mortar bombs are carried, but when afloat only 23 can be carried. Most German M113/M113A1 vehicles are fitted with eight smoke dischargers on the front of the hull and are armed with a single 7.62 mm MG3 machine gun. M113s are now being retrofitted with a diesel engine to bring them up to M113A1 standard.

Thyssen Henschel also offer low wear track for the M113 series of armoured vehicles (the Diehl B213), M113A2 suspension, cooling, external fuel tanks at the rear, hydrostatic steering system (also covered earlier in this entry), more powerful diesel engine, supplementary armour (also covered earlier in this entry), NBC protection system and various turret systems with weapons of up to 20 mm in calibre.

The German Army has also taken delivery of 301 M548G1 tracked carriers modified by Krauss-Maffei to take the Dynamit Nobel Skorpion (MiWS) mine launching system with final deliveries taking place in 1989. Full details of this system are given in *Jane's Military Vehicles and Logistics 1991-92* page 210.

Israel: most M113s have had their trim vanes removed from the front of the hull and an extension fitted to the exhaust outlet so that exhaust does not blow back into the troop compartment when the vehicle is being driven with its roof hatches open. Other modifications include mounting up to four 7.62 mm or 12.7 mm (0.50) machine guns on the top of the hull and racks fitted on either side of the hull to enable personnel kit and other stores to be carried outside the vehicle thus giving more room inside the troop compartment. The Israelis call the M113 the Zelda. In addition to the basic APC, the Israeli Army uses many specialised versions of the M113A1 including anti-tank fitted with the Hughes TOW ATGW system, forward ambulance, forward engineer, modified M577 command vehicles, 60 mm Soltam mortar carriers and technical support.

The RAMTA Improved TOW Vehicle is a M113 fitted with a retractable telescopic mast on top of which is a Hughes TOW ATGW launcher. When not required this is retracted into the hull of the vehicle and raised hydraulically when required. The gunner stands on a special platform that rises together with the complete system. The gunner controls the lifting and lowering from the platform as well as the launching and tracking of the missile. The system has a traverse of 180° and takes 60 seconds to be raised. A total of 10 missiles are carried internally.

In 1987 Rafael Armament Development Authority announced that it has developed another armour system called the TONGA which gives protection against 14.5 mm rounds and that this had been successfully used during the Lebanon campaign.

The Israeli company of NIMDA now offer a conversion package to enable users of the M113 to convert their vehicles to the M113A1/M113A2 configuration, or from the M113A1/M113A2 to the latest M113A3 standard.

The first converstion package features a Detroit Diesel 6V-53T engine developing 216 or 265 hp coupled to an Allison Transmission Division of General Motors TX-100-1A automatic transmission, improved cooling system, new transfer case, steering/brake differential, new electrical system and so on.

M548 tracked carrier of the Israeli Army fitted with the Elta EL/M-2310 artillery fire control system with antenna in travelling configuration

M113 series APC of Israeli army fitted with Rafael TOGA passive armour system

The second conversion package includes a Detroit Diesel 6V-53T engine developing 275/300 hp, X-200-4 cross-drive transmission with hydrostatic steering.

Other options include improved protection, automatic fire detection and suppression, spall liners, improved suspension, external fuel tanks, NBC system and remote-controlled weapons ranging from 7.62 mm machine guns up to a 20 mm cannon.

The Israeli Army also uses a modified M548 tracked carrier chassis as the platform for its Elta EL/M-2310 artillery fire control system.

Italy: OTO Melara of La Spezia built some 4500 M113 series of armoured personnel carriers under licence from the United States including M577 command post vehicles and M548 tracked cargo carriers. Further development resulted in the Infantry Armored Fighting Vehicle described under Italy.

In mid 1989 OTO Melara completed the first batch of 35 SIDAM 25 (4 × 25 mm) self-propelled anti-aircraft gun systems for the Italian Army which has a total requirement for 340 systems. This consists of a modified M113 chassis with a new power operated turret armed with 4 × 25 mm Oerlikon-Contraves cannon and a day-only fire control system. The chassis are supplied by Astra to OTO Melara who then integrate the turret and deliver the complete system to the Italian Army. Key modifications to the chassis can be summarised as replacement of the existing engine by a more powerful Detroit Diesel 6V-53T developing 265 hp, two new fuel tanks mounted one either side externally at the hull rear giving a total capacity of 360 litres, modifying the top deck to incorporate an improved reverse flow cooling system, reinforcement of idler wheel mounts and repositioning the shock absorber supports and the existing fan is replaced together with a new air cleaner and silencer. The existing TX-100-1 transmission can be modified or replaced by the TX-100-1A or the new X-200-4. New and stronger torsion bars are fitted and the combat engineer vehicle has additional armour protection, front-mounted dozer blade and 3-tonne hydraulic crane. Internally there is a 10-tonne winch and stabilisers are provided at the rear. (See *Jane's Land-Based Air Defence 1991-92*, page 73.)

Italian Army SIDAM (4 × 25 mm) SPAAG. First production vehicles were completed for Italian Army in mid-1989

Spanish Army M113A1 with 120 mm mortar in rear

The Italian Army is also testing an M113 series APC fitted with a Contraves SHORAR (SHOrt-Range-Acquisition Radar) which is being used for extensive trials with both the SIDAM 25 system and manportable SAMs.

In Italian Army service the Infantry Armoured Fighting Vehicle is known as the VCC-1 or Camillino. The VCC-2 is the locally built M113 with appliqué armour fitted to the hull and sides. There are two firing ports with a bullet proof vision block above in either side of the rear troop compartment. The M113 upgraded by Astra is known as the VTC-9 in Italian Army service.

New Zealand: New Zealand purchased 72 members of this family including the M113A1 armoured personnel carrier (some of which have the T50 turret, as on Australian vehicles), M125A1 mortar carrier, M577A1 command vehicle, M579 fitters' vehicle and the M806A1 armoured recovery vehicle. The Ministry of Works and Development (MWD) has rebuilt 64 of these over a six-year period beginning in 1980 at a cost of $6 million.

Norway: details of Norwegian M113 series APCs and their variants are given in this section under Norway.

Pakistan: details of Pakistani M113 series APCs and their variants are given in this section under Pakistan.

Singapore: to meet the requirements of the Singapore Army, Chartered Industries of Singapore has designed a modification kit to convert the M113 to a 120 mm or 81 mm mortar carrier. This is now in service and offered for export. The 120 mm mortar has a traverse of 18.5° left and 25° right with elevation being from 800 mils to 1422 mils. In addition Singapore uses the M806A1 ARC, M113A1 fitter's vehicle and uses the M113A1 in the command post role.

Spain: Spain uses the M113, M113A1, M113A2, M548 tracked cargo carrier, M577 command post vehicle, M113 series recovery vehicle and the M125 81 mm mortar carrier. Talbot carries out the complete overhaul of M113s in Spain and can also bring M113 petrol-engined vehicles up to the M113A1/M113A2 standard. In place of the M106 107 mm mortar carrier the Spanish Army uses a similar vehicle with a Spanish-built 120 mm mortar which fires to the rear of the hull. A baseplate is carried externally on the left side of the hull to allow the mortar to be dismounted for use away from the vehicle.

Switzerland: the basic APC is called the Schützenpanzer 63 by the Swiss Army but many have been fitted with the Swedish Hägglund turret armed

Swiss Minenwerferpanzer Mw Pz 64 armed with 120 mm mortar (Swiss Army)

with a 20 mm cannon as fitted to the Pbv 302 APCs of the Swedish Army. When fitted with this turret the vehicle is known as the Schützenpanzer 63/73. Other versions in service include the bulldozer model (called the Geniepanzer 63), a command vehicle (Schützenpanzer Kdo Spz-63), 120 mm mortar carrier (designated the Minenwerferpanzer 64), command vehicle (designated the Kommandopanzer 63), command vehicle with turret-mounted 20 mm cannon (designated the Kommandopanzer 63/73), repair vehicle (designated the Kranpanzer 63), wireless vehicle (designated the U bermittlungspanzer 63) and an artillery command vehicle (designated the Feuerleitpanzer 63).

The Swiss Army has some 625 M113 series APCs and in 1988 it was decided that 316 Schützenpanzer 63/73 and 66 of the command post equivalent would be upgraded from 1990 to 1993 to enable them to operate with the Leopard 2 MBTs now in service with the Swiss Army.

A total of SFr123 million have been allocated for this programme. The modifications to the vehicles include additional armour protection to the hull, smoke grenade launchers, more powerful Detroit Diesel engine and external fuel tanks either side of the rear ramp. The kits are being provided by the FMC Corporation, and when upgraded the vehicles will be called the Schützenpanzer 63/89.

Vietnam: in 1985 the Vietnamese Army was observed to be still using significant numbers of American M113 series armoured personnel carriers including a version similar to the ACAV but with the 12.7 mm M2 HB machine gun replaced by a Soviet 12.7 mm DShkM machine gun. The two lateral machine guns were also retained, although it is probable that the original 7.62 mm M60 series have been replaced by Soviet weapons of the same calibre. In 1987 the Vietnamese Army was using flamethrower version of the M113 series and ACAV models with appliqué armour added.

SPECIFICATIONS

Model	**M106A1**	**M113**	**M113A1**	**M113A2**	**M113A3**	**M125A1**	**M577A1**
CREW	6	2 + 11	2 + 11	2 + 11	2 + 7	6	5
COMBAT WEIGHT	11 996 kg	10 258 kg	11 156 kg	11 341 kg	12 062 kg	11 261 kg	11 513 kg
UNLOADED WEIGHT	9005 kg	8960 kg	9702 kg	9926 kg	10 633 kg	10 539 kg	10 865 kg
POWER-TO-WEIGHT RATIO	17.92 hp/tonne	20.37 hp/tonne	19.27 hp/tonne	18.51 hp/tonne	22.8 hp/tonne	19.09 hp/tonne	18.67 hp/tonne
GROUND PRESSURE	0.59 kg/cm²	0.5 kg/cm²	0.55 kg/cm²	0.56 kg/cm²	0.56 kg/cm²	0.55 kg/cm²	0.57 kg/cm²
LENGTH	4.926 m	4.863 m	4.863 m	4.863 m	4.863 m	4.863 m	4.863 m
WIDTH	2.863 m	2.686 m	2.686 m	2.686 m	2.686 m	2.686 m	2.686 m
over tracks	2.54 m	2.54 m	2.54 m	2.54 m	2.54 m	2.54 m	2.54 m
HEIGHT							
overall	2.5 m	2.5 m	2.5 m	2.52 m	2.52 m	2.5 m	2.68 m
to hull top	1.828 m	1.828 m	1.828 m	1.85 m	1.85 m	1.828 m	2.469 m
GROUND CLEARANCE	0.41 m	0.41 m	0.41 m	0.43 m	0.43 m	0.41 m	0.41 m
TRACK	2.159 m	2.159 m	2.159 m	2.159 m	2.159 m	2.159 m	2.159 m
TRACK WIDTH	381 mm	381 mm	381 mm	381 mm	381 mm	381 mm	381 mm
LENGTH OF TRACK ON GROUND	2.667 m	2.667 m	2.667 m	2.667 m	2.667 m	2.667 m	2.667 m
MAX SPEED							
road	67.59 km/h	64.37 km/h	64.37 km/h	64.37 km/h	65.97 km/h	67.59 km/h	64.37 km/h
water	5.8 km/h	5.6 km/h	5.8 km/h	5.8 km/h	5.8 km/h	5.8 km/h	5.8 km/h
FUEL CAPACITY	360 l	302 l	360 l	360 l	360 l	360 l	454 l
MAX CRUISING RANGE	483 km	321 km	483 km	483 km	483 m	483 km	595 km
FORDING	amphibious	amphibious	amphibious	amphibious	amphibious	amphibious	amphibious
VERTICAL OBSTACLE	0.61 m	0.61 m	0.61 m	0.61 m	0.61 m	0.61 m	0.61 m
GRADIENT	60%	60%	60%	60%	60%	60%	60%
TRENCH	1.68 m	1.68 m	1.68 m	1.68 m	1.68 m	1.68 m	1.68 m
ENGINE	Detroit Diesel model 6V-53 6-cylinder water-cooled diesel developing 215 bhp at 2800 rpm	Chrysler 75M V-8 petrol developing 209 bhp at 4000 rpm	Detroit Diesel model 6V-53 6-cylinder water-cooled diesel developing 215 bhp at 2800 rpm	Detroit Diesel model 6V-53 6-cylinder water-cooled diesel developing 215 bhp at 2800 rpm	Detroit Diesel model 6V-53T 6-cylinder water-cooled diesel developing 275 bhp at 2800 rpm	Detroit Diesel model 6V-53 6-cylinder water-cooled diesel developing 215 bhp at 2800 rpm	Detroit Diesel model 6V-53 6-cylinder water-cooled diesel developing 215 bhp at 2800 rpm
TRANSMISSION	GMC, Allison TX-100-1 with 3 forward and 1 reverse ranges	GMC, Allison TX-200 with 6 forward and 1 reverse gears	GMC, Allison TX-100-1 with 3 forward and 1 reverse ranges	GMC, Allison TX-100-1 with 3 forward and 1 reverse ranges	GMC, Allison X-200-4 with 4 forward and 1 reverse ranges	GMC, Allison TX-100-1 with 3 forward and 1 reverse ranges	GMC, Allison TX-100-1 with 3 forward and 1 reverse ranges
SUSPENSION	torsion bar	torsion bar	torsion bar	torsion bar	torsion bar	torsion bar	torsion bar
ELECTRICAL SYSTEM	24 V	24 V	24 V	24 V	24 V	24 V	24 V
BATTERIES	2 × 12 V 6TN	2 × 12 V 6TN	2 × 12 V 6TN	2 × 2 V 6TN	4 × 12 V 6TN	4 × 12 V 6TN	2 × 12 V 6TN
ARMAMENT	1 × 12.7 mm MG	1 × 12.7 mm MG	1 × 12.7 mm MG	1 × 12.7 mm MG	1 × 12.7 mm MG	1 × 12.7 mm MG	1 × 7.62 mm MG
AMMUNITION	1000	2000	2000	2000	2000	1000	1000
ARMOUR	12-44 mm	12-44 mm	12-44 mm	12-44 mm	12-44 mm	12-44 mm	12-44 mm

Status: In production. In service with Argentina, Australia, Belgium, Bolivia, Brazil (army and marines, including M577 command), Cambodia, Canada, Chile, Costa Rica, Denmark, Ecuador, Egypt (including M125A2, M577A2), El Salvador, Ethiopia, Germany, Greece, Guatemala, Iran, Israel, Italy, Jordan, South Korea, Kuwait (including models with TOW ATGW), Laos, Lebanon (including M113A1, M113A2 and M577A1), Libya, Morocco, Netherlands, New Zealand, Norway, Pakistan, Peru, Philippines, (including some fitted with two 7.62mm machine guns in a one man manually operated turret), Portugal, Saudi Arabia (including M113A1, M113A2, M548 and VCC-1 from Italy), Singapore, Somalia, Spain, Sudan (including mortar carriers), Switzerland, Taiwan (more recently Taiwan has ordered 302 M113A2, 24 M113A2 ambulances and 31 M557A2 command post vehicles), Thailand (including M113A2), Tunisia (including M125A1, M557A1), Turkey, USA, Uruguay, Vietnam, Yemen and Zaïre.

Manufacturer: FMC Corporation, Ground System Division
2830 De La Cruz Boulevard
Santa Clara
California 95052
USA
Telephone: (408) 289 0111
Telex: 6714210
Fax: (408) 289 2150

M59 Armoured Personnel Carrier

Just over 4000 M59 series APCs were built by the FMC Corporation between 1954 and 1959 and some remain in service with Brazil, Greece and Turkey (reserve). Full details were given in *Jane's Armour and Artillery 1990-91*, pages 486-488.

Cadillac Gage Commando V-300 Armoured Vehicle Range

Development

The Commando V-300 (6 × 6) range of armoured vehicles has been developed as a private venture by Cadillac Gage Textron which has already designed and built the highly successful Commando V-150 (4 × 4) range of vehicles now in service with over 20 countries. The first two prototypes of the V-300 were completed in 1979 and the vehicle entered production in 1983. Late in 1982 Panama placed an order worth over $6 million for 12 V-300 Commando vehicles in four different configurations including fire support vehicle with 90 mm Cockerill Mk III gun, APC with twin 7.62 mm machine guns and recovery. All of these were delivered by the end of 1983. The company also supplied a number of vehicles for the Light Armored Vehicle competition including models fitted with the two-man turret armed with a Cockerill 90 mm Mk III gun, two-man 25 mm Chain Gun turret and two-man 25 mm Chain Gun turret fitted with a stabilisation system. In 1984 the Kuwait National Guard ordered 62 V-300 Commandos.

As of January 1991 a total of 79 Commando V-300 vehicles had been built. The status of the Kuwaiti vehicles is uncertain, although it is probable that most were lost during the Iraqi invasion.

Description

The hull of the V-300 Commando is of all-welded unitised construction of special high hardness steel ballistic plate that protects the crew from small arms fire and shell splinters.

The driver sits at the front of the hull on the left side and has a single piece square hatch cover that opens to the rear. To his front are three periscopes for forward observation and one of these can be replaced by a passive periscope for driving at night. In the side of the hull to the left of the driver is a bullet-proof vision block with a firing port underneath. To the rear of this is a small half-door that opens to the rear. This has a bullet-proof vision block underneath which is a firing port.

The engine compartment is to the right of the driver with the air inlet and air outlet in the roof and the exhaust pipe in the right side of the hull. Power from the engine is transmitted to the final drives via an Allison Transmission Division of General Motors MT-643 four-speed automatic gearbox. The transfer box contains a two-speed system with a spline engagement clutch for the axle drive.

On the glacis plate is a receptacle for a shovel and pick axe and under the nose is an internally mounted hydraulic winch which has a maximum capacity of 9072 kg. This is provided with 45 m of 12.7 mm diameter steel core cable.

The V-300 can be fitted with a number of armament installations, which are mounted in the centre of the hull over the second axle and include:
Cadillac Gage two-man turret with 90 mm Cockerill Mk III gun (described in V-150 entry).
Cadillac Gage two-man turret with 76 mm Royal Ordnance L23A1 gun.
Turret armed with 25 mm M242 McDonnell Douglas Helicopters Chain Gun, 7.62 mm coaxial and 7.62 mm anti-aircraft machine guns and smoke dischargers.
Cadillac Gage two-man turret with 20 mm Oerlikon cannon (described in V-150 entry).
Cadillac Gage one-man 1 m machine gun turret (described in V-150 entry).
Cadillac Gage one-man machine gun turret (described in V-150 entry).
Ring mount with 7.62 mm or 12.7 mm machine gun.

The troop compartment is at the rear of the V-300 and in each side of the hull are three bullet-proof vision blocks each with a firing port underneath. Over the top of the troop compartment are two rectangular roof hatches hinged on the outside. The infantry enter and leave the vehicle through two doors in the rear of the hull, each of these has a bullet-proof vision block underneath which is a firing port. As an alternative to the two-part hatch cover a circular hatch cover with provision for mounting a 7.62 mm pintle-mounted GPMG can be fitted.

The front suspension is a swing-mounted solid axle with rear suspension being independent with coil springs at each wheel station. The front solid axle is restrained longitudinally by trailing arms and laterally by stabilising rods attached to the axle and hull. The two rear axles consist of short drive shafts extending from the hull-mounted differentials to the independently sprung trailing arms-mounted wheel stations. The road wheels have a combined wheel travel of 30 cm, a jounce of 17.5 cm and a rebound of 12.5 cm. All the swing-mounted axles are of the double-reduction top-mounted type. The front and rear axles have positive No-Spin differentials to maintain traction in off road operations. The central axles contain the inter-axle differential. All wheels have hydraulic brakes and steering is power assisted on the front wheels. All wheels have run-flat tubeless tyres with a self cleaning tread design.

The V-300 is fully amphibious without preparation, being propelled in the water by its wheels at a speed of 4.8 km/h.

A wide range of optional equipment is available including night vision devices, heater kit, air-conditioning system, NBC system, wiper kit for the driver and a slave cable.

Variants

TOW Anti-tank

This would be fitted with the same mount as used in the M901 Improved TOW Vehicle based on the M113 chassis with two missiles in the ready-to-launch position and reserve missiles carried in the hull.

TOW Anti-tank

Fitted with an external top-mounted swing-up launcher with an elevation of +30°, depression of -10° and traverse of 360°. Ten TOW ATGW missiles are carried plus a 7.62 mm machine gun and 2000 rounds of ammunition.

81 mm Mortar

Fitted with turntable-mounted 81 mm mortar with elevation from +42° to +85°, total traverse of 360°, range of 150 to 4400 m with a total of 60 mortar bombs carried. Also fitted with 7.62 mm machine gun for which 2000 rounds of ammunition are carried. The 81 mm mortar can also be dismounted for use on the ground.

Ambulance

This model would be unarmed and fitted with a raised roof and ramp in place of twin doors at the rear to facilitate the loading of stretchers and walking wounded.

Others

These include cargo carrier, command vehicle (with higher roof line), recovery vehicle and anti-aircraft with a missile system such as ADATS, or a gun system such as the General Electric 20 mm Vulcan as installed on a number of V-150 Commando 4 × 4 vehicles supplied to Saudi Arabia.

Cadillac Gage V-300 Commando with two-man turret armed with 90 mm Cockerill Mark III gun

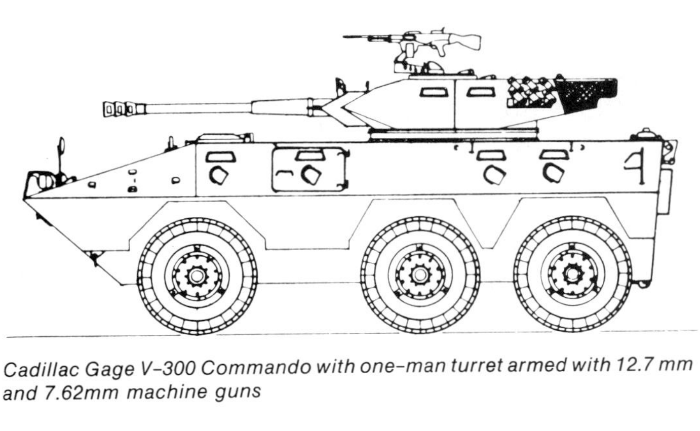

Cadillac Gage V-300 Commando with one-man turret armed with 12.7 mm and 7.62mm machine guns

Cadillac Gage V-300 command post vehicle with higher roof

SPECIFICATIONS	
CREW	3 + 9 (max)
CONFIGURATION	6 × 6
COMBAT WEIGHT	14 696 kg
POWER-TO-WEIGHT RATIO	18.94 hp/tonne
LENGTH	6.4 m
WIDTH	2.54 m
HEIGHT	
MG turret	2.59 m
20 mm turret	2.692 m
hull	1.98 m
ITV	2.756 m
hull top	1.981 m
GROUND CLEARANCE	
hull	0.533 m
axle	0.381 m
TRACK	
front	2.243 m
rear	2.243 m
WHEELBASE	2.209 m + 1.524 m
ANGLE OF APPROACH/ DEPARTURE	45°/40°
MAX SPEED	
road	95 km/h
water	5 km/h
ACCELERATION	
0 to 32 km/h	10 s
FUEL CAPACITY	265 l
MAX RANGE	700 km
FORDING	amphibious
GRADIENT	60%
SIDE SLOPE	30%
VERTICAL OBSTACLE	0.609 m
TURNING RADIUS	10 m
ENGINE	VT-504 V-8 turbo-charged diesel, 270 hp at 3000 rpm
TRANSMISSION	Allison Transmission Division of General Motors Corporation MT-643 automatic, 4 speeds forwards and 1 reverse
TRANSFER CASE	2-speed with spline engagement clutch for axle drive
STEERING	variable ratio power
SUSPENSION	solid front axle, middle and rear independent with coil springs, direct acting telescopic shock absorber acting at each wheel station
TYRES	14.5 × 21 (run-flat)
BRAKES	
main	hydraulic (all wheels) power booster with electric back up
parking	drum and shoe hand-operated
ELECTRICAL SYSTEM	24 V
BATTERIES	2 × 12 V, 100 Ah

Status: Production as required. Kuwait took delivery of 62 vehicles while Panama took delivery of 12. In both countries fighting has taken place in recent years and the exact status of the vehicles is now uncertain.

Manufacturer: Cadillac Gage Textron
PO Box 1027
Warren
Michigan 48090
USA
Telephone: (313) 777 7100
Telex: 200707
Fax: (313) 776 9731

Cadillac Gage Commando V-150 Armoured Vehicle Range

Development

In 1962 the now Cadillac Gage Textron, as a private venture, started design work on a 4 × 4 armoured vehicle which could undertake a wide range of roles such as armoured personnel carrier, convoy escort vehicle or reconnaissance vehicle. The first prototype of the vehicle, called the V-100 Commando, was completed in March 1963 and the first production vehicles the following year.

The V-100 was widely used in Vietnam for both convoy escort work and for patrolling air bases and other high value targets. The V-100 was powered by a Chrysler 361 V-8 petrol engine which developed 200 hp and gave the vehicle a maximum road speed of 100 km/h. Gross vehicle weight was 7370 kg and a maximum of 12 men including the crew could be carried. The V-100 was supplemented in production by the larger V-200, which was powered by a Chrysler 440 CID engine which developed 275 hp and gave the vehicle a maximum road speed of 96 km/h. Gross vehicle weight of the V-200 is 12 730 kg. The V-200 was sold only to Singapore and is no longer being produced. Versions of the V-200 used by Singapore include anti-aircraft with the Swedish Bofors RBS 70 surface-to-air missile system, recovery fitted with an A-frame at the front of the hull, fire support fitted with two-man turret armed with 90 mm MECAR gun with 7.62 mm coaxial and 7.62 mm anti-aircraft machine guns and APC fitted with two-man turret armed with 20 mm Oerlikon cannon, 7.62 mm coaxial and 7.62 mm anti-aircraft machine guns.

The United States Air Force still uses a number of original V-100 Commando vehicles at the Tactical Fighter Weapons Centre at Nellis Air Force Base in Nevada. Typical roles include simulating the Soviet SA-9 SAM system and command and control systems associated with the Soviet air defence systems.

In October 1971 the V-150 was introduced and replaced the V-100 and V-200 in production. Production of the V-100, V-150 and V-200 has now reached over 3195 units, of which 2328 were V-150/V-150 S 4 × 4 vehicles. The current production model is the V-150 S.

The vehicles have been exported to over 20 countries and have been used in some numbers both by the United States Army and the United States Air Force. As a result of experience in use the vehicle has been constantly updated and improved.

There is a separate entry for the Cadillac Gage V-300 (6 × 6) armoured vehicle range.

There are two other vehicles on the market today which are very similar in appearance to the Commando, the American Dragoon series and the Portuguese Chaimite, which is known to have been sold to Lebanon, Libya, Peru, Philippines and Portugal. Both these vehicles are described in this section.

Cadillac Gage delivered two V-150 S Commando vehicles for evaluation in the LAV competition. In 1984 Kuwait placed an order for 20 V-150 S Commandos and 60 V-300 Commandos.

In 1985 Cadillac Gage announced that production of the V-150 was to cease and it would be replaced in production by the V-150 S.

In August 1988 US Army Tank Automotive Command placed an order with Cadillac Gage Textron for 14 V-150 S vehicles at a cost of $7.014 million with another 10 being ordered in December 1988 at a cost of $3.113 million.

V-150 Commando fitted with turret armed with 20 mm Oerlikon cannon, 7.62 mm coaxial machine gun, 7.62 mm anti-aircraft machine gun and six smoke dischargers fitted to turret side

V-150 S armed with turret-mounted Belgian Cockerill Mk III 90 mm gun, 7.62 mm coaxial and 7.62 mm anti-aircraft machine guns and smoke dischargers

US Air Force V-100 being used by an Air Base Ground Defence team (US Air Force)

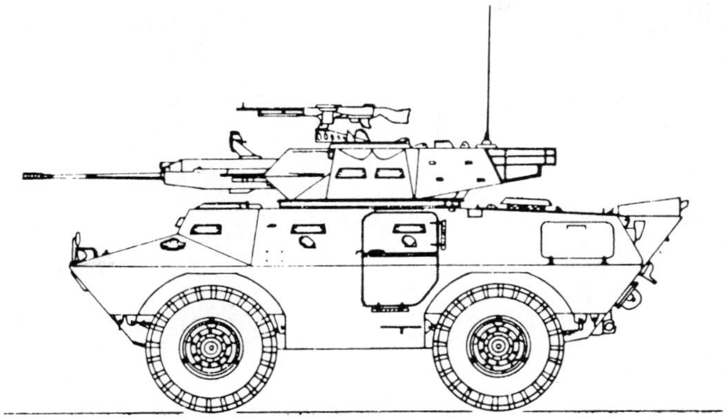

V-150 Commando with two-man turret armed with 20 mm Oerlikon cannon, 7.62 mm coaxial machine gun and 7.62 mm anti-aircraft machine gun

Description

The all-welded steel hull of the V-150 Commando armoured personnel carrier protects the crew from small arms fire up to 7.62 mm in calibre, overhead blast, shell splinters and Molotov cocktails.

The driver sits at the front of the hull on the left side and has two vision blocks in front of him and one on his left. The vehicle commander sits on the driver's right and has one vision block to his front and one on his right. Both crew members have a single piece hatch cover over their position that opens to the outside. Early production vehicles had a firing port just behind the driver's and commander's side vision block, but current vehicles do not.

The troop compartment is behind the commander's and driver's position with three infantrymen seated down each side of the hull and one to the rear of the machine gunner's position. They enter and leave the vehicle by two doors in the side of the hull, one on each side. The lower part of the door folds downwards and the upper part, which contains the vision block and firing port, opens to the rear. In each side of the hull, forward of the side door, is a firing port with a vision block over the top. All vision blocks in the V-150 have crash pads and spall shields for additional protection.

In the centre of the troop compartment roof is a cupola which can be traversed manually through 360° and has a single piece hatch cover that opens to the rear. On the forward part of the cupola, which has no observation devices, is a pintle-mounted 7.62 mm machine gun with 200 rounds of ready-use ammunition and another 3000 rounds in reserve in the hull.

There is also a door at the rear of the hull on the right side, with the lower part folding downwards and the upper part, which contains the firing port and vision block, opening upwards. Over the top of the passageway that connects the troop compartment with the rear door is a single piece circular hatch that opens forwards. A further two seats are provided in this passageway.

The engine compartment is at the rear of the hull on the left side with access panels in the top and side of the hull. The engine cooling system is a conventional automotive water-cooled system utilising a fin and tubed radiator with an additional surge tank for reserve capacity. The engine fan and venturi are designed for maximum air flow efficiency and minimum horsepower draw. The cooling system is capable of operating in ambient temperatures of 130° F. The engine compartment has a fire-extinguishing system that is operated manually by the driver.

Wherever possible, standard military automotive components are used in the V-150 Commando, such as the engine, transmission and axles. When originally introduced the Commando V-150 was powered by a Chrysler 361 V-8 petrol engine developing 200 hp at 4000 rpm coupled to a manual transmission with five forward and one reverse gears and a single-speed transfer case. Current production vehicles have a V-8 diesel that develops 202 hp at 3300 rpm coupled to an automatic transmission and a two-speed transfer case.

The solid axles have silent locking differentials and the suspension consists of semi-elliptical springs and hydraulic double-acting shock absorbers with rebound control at each wheel station. Jounce control is by hollow rubber springs at each wheel station. The run-flat tyres, which are capable of operation for a distance of 40 km, have self-cleaning treads. Mounted internally at the front of the hull is a winch with a capacity of 4536 kg, which has 48.76 m of 9.5 mm diameter cable. All members of the Commando family are fully amphibious being propelled in the water by their wheels.

Standard equipment on all vehicles includes fuel and water cans, two hand-held fire extinguishers, pioneer tool set, tow cable, breaker bar and lug wrench, first aid kit, vehicle tool kit (including 8000 kg hydraulic jack stowed inside the vehicle), pamphlet bag, oddment box, slave cable, spare vision blocks and an air compressor with hose.

Optional equipment for the Commando includes smoke or smoke/fragmentation launchers, fixed pintle socket for rear, rear ring mount with 7.62 mm pintle and cradle, ring mount and cradle for 0.50 calibre machine gun or 40 mm automatic grenade launcher, gun shield for fixed pintle socket or ring mount weapon, spotlight for fixed pintle socket or ring mount weapon with or without shield, night vision equipment, periscope for ring mount (standard on 20 mm, 76 mm and 90 mm turret ring mounts), spare periscope and stowage, weapons (where applicable), extra spare barrel stowage, hand held weapon stowage, M26 hand grenade stowage, smoke hand grenade stowage, binocular stowage, radios and associated mountings, canopy with poles (pod type vehicles only), vision blocks in pod (standard on Police ERV), gun ports in pod (standard on police ERV), sight M28C × 5.6 power (machine gun turret vehicles only), land navigation system and/or vehicle heading reference system, NBC system, public address system (standard on Police ERV), heater kit, air-conditioning system, wiper kit for driver, blackout covers, handbrake warning light, cooker kit and lifting boom.

Variants

Commando with MG Turret

Basic vehicle fitted with Cadillac Gage one-man turret which can mount twin 7.62 mm machine guns or a combination of 7.62 mm/0.50 calibre machine guns. Laying and control of the weapons are mechanical with a hand-operated gearbox that allows 360° continuous traverse, and a gunner's handle at the rear of the balanced gun cradle which allows elevation from -13° to +55°. The turret has a single piece hatch cover that opens to the rear, eight vision blocks and an M28C sight with a magnification of ×1.5. Mounted coaxially with the weapon is a 500 000 candlepower spotlight. When fitted with twin 7.62 mm machine guns 3800 rounds of ammunition are carried of which 800 rounds are in the turret for ready use and 3000 in the hull. This model has a crew of three and can carry seven infantrymen. Height to top of turret is 2.54 m and gross vehicle weight is 8981 kg.

Commando with 1 m MG Turret

Basic vehicle fitted with Cadillac Gage 1 m turret which can be armed with twin 7.62 mm or twin 0.50 calibre machine guns, or a combination of these weapons. The turret has a manual traverse through 360° and the guns have an elevation of +55° and a depression of -10°. The turret has a single piece hatch cover that opens to the rear, an M28C sight with a magnification of ×1.5, eight vision blocks and a 500 000 candlepower spotlight. Depending on the armament installed 1400 rounds of 0.50 calibre ammunition are carried, of which 400 are for ready use in the turret, or 3800 rounds of 7.62 mm ammunition of which 800 are for ready use in the turret. This model has a crew of three and can carry seven infantrymen. Height to the top of the turret is 2.59 m and gross vehicle weight is 8754 kg.

Commando with 1 m 20 mm Turret

Basic vehicle with one-man turret armed with 20 mm Oerlikon 204 GK cannon with provision for coaxial 7.62 mm machine gun. Gun can be

V-200 Commando 120 mm mortar carrier of Singapore Army with roof hatches, side and rear doors open

V-200 Commando of Singapore Army fitted with two-man Cadillac Gage turret armed with 20 mm Oerlikon cannon and two 7.62 mm machine guns

V-150 Commando fitted with command pod and externally mounted 7.62 mm machine gun

elevated from -8° to +55° at 40°/s and turret traversed through 360° at 60°/s. Traverse is electro-hydraulic with manual backup. Turret has eight vision blocks and a sight with a magnification of ×8 and ×1, a 500 000 candlepower searchlight that moves in elevation with the armament and a blower for removing turret fumes.

Four hundred rounds of 20 mm ammunition are carried, 200 in the turret ready to use and 200 in the hull, and 3200 rounds of 7.62 mm ammunition, 220 for ready use and the remainder in the hull.

This model has a crew of three consisting of commander, gunner and driver and can carry five men, and has a loaded weight of 9072 kg and a height of 2.895 m.

Commando with 20 mm Turret

Basic vehicle fitted with a two-man turret armed with a 20 mm Oerlikon 204 GK cannon, with a 7.62 mm machine gun mounted coaxially to the left of the main armament and a 7.62 mm anti-aircraft machine gun mounted on the turret roof. The cannon has an internally controlled hydraulic charger, emergency firing trigger for the commander and rate controller which allows firing of 1, 2 or 4 rds/s or fully automatic. Turret vision equipment consists of four vision blocks (two in each side of the turret), commander's periscope that can be traversed through a full 360° and a gunner's sight with a magnification of ×8 and a periscope with a projected graticule. In addition to the main sights the turret has an external anti-aircraft sight. It is also provided with a 500 000 candlepower spotlight and a turret blower. Of 400 rounds of 20 mm ammunition carried 200 are for ready use and of 3200 rounds of 7.62 mm ammunition 400 are in the turret for ready use and 200 in the anti-aircraft gun for ready use. This model has a crew of three and can carry two infantrymen. The gross vehicle weight is 9888 kg and the height to the top of the turret is 2.54 m.

Commando with 25 mm Turret

This turret was installed on the V-150 S Commando entered in the Light Armored Vehicle competition. It is armed with a 25 mm M242 McDonnell Douglas Helicopters Chain Gun, 7.62 mm M240 coaxial MG and a similar weapon on the turret roof for anti-aircraft defence. The turret carries 230 25 mm ready rounds (170 HE and 60 AP) with a further 400 in the vehicle. There are 400 rounds of 7.62 mm ready use ammunition in the turret, 200 at the anti-aircraft station and a further 1000 rounds in the vehicle.

Commando 20 mm air defence vehicle in travelling configuration

The turret has powered traverse (360° at 30°/s) and elevation (from -8° to +60° at 30°/s). Vision equipment includes an M36E1 day/night sight for the gunner and eight periscopes for the commander. The commander can also have an M36E1 day/night sight but only six periscopes are then fitted.

Commando with 40 mm/12.7 mm Turret

This is fitted with a one-man turret armed with a 40 mm Mk 19 Mod 1 grenade launcher on the left, a 12.7 mm M2 HB machine gun on the right and four smoke dischargers mounted either side of the turret at the rear. The weapon can be elevated from -8° to +45° and turret traverse is a full 360°.

Commando V-150 S

This was announced in 1981 with two vehicles being delivered for evaluation in the LAV competition late in the same year. Both vehicles had a two-man turret, one armed with a 90 mm Cockerill Mk III gun and the second with a 25 mm M242 McDonnell Douglas Helicopters Chain Gun, both having a 7.62 mm coaxial and a 7.62 mm anti-aircraft machine gun. The Commando V-150 S is 46 cm longer than the Commando V-150 which gives additional internal space so increasing the payload by 726 kg to give a total payload of 3357 kg. The wheelbase has been lengthened and the suspension strengthened so giving an improved ride across country.

Additional vehicle improvements include a new transfer case giving better grade-climbing performance, new hydraulic boost brake system with an electro-hydraulic backup and an improved cooling system with increased fan efficiency and increased cooling capacity. The V-150 S is now the standard production model of the V-150.

Commando Air Defence Vehicle

This is the basic vehicle fitted with the same turret as mounted on the M167 towed system armed with a 20 mm six-barrelled General Electric cannon and fitted with a an optical sight. To provide a more stable firing platform, three outriggers are lowered to the ground before firing commences: one of these is located at the front of the vehicle and one either side of the hull towards the rear. This version was developed to meet the requirements of the Saudi Arabian National Guard. Full details of this version are given in *Jane's Land-Based Air Defence 1991-92*, page 98.

Commando with 76 mm Turret-mounted Gun

This model is fitted with a two-man power-operated turret armed with the 76 mm L23A1 gun as installed on the British Scorpion CVR(T), with a 7.62 mm machine gun mounted coaxially to the left of the main armament and a 7.62 mm anti-aircraft machine gun on the turret roof. As of early 1991 the 76 mm version had not been placed in production.

Commando with 90 mm Turret-mounted Gun

This was originally armed with the Belgian MECAR gun but is now offered with the Cockerill Mk III gun. Mounted coaxially with the main armament is a 7.62 mm machine gun and a similar weapon is fitted on the turret roof for anti-aircraft defence. Turret traverse is a full 360° at 30°/s, and elevation is from -8° to +28° at 30°/s. Vision equipment consists of three vision blocks (two in left and one in right side of the turret), commander's periscope that can be traversed through 360°, gunner's periscope with a magnification of ×8 and a unity periscope with projected graticule. A 500 000 candlepower spotlight is mounted coaxially with the main armament. Of 39 rounds of 90 mm ammunition carried, eight are for ready use in the turret and of 3200 rounds of 7.62 mm machine gun ammunition 400 are for ready use in the turret and 200 for the anti-aircraft machine gun. This model has a crew of three and can carry four infantrymen, gross vehicle weight being 9888 kg. In 1983 the V-150 Commando with the Marconi Command and Control Systems Digital Fire control System was successfully demonstrated. This gives a high first round hit probability and is suitable for installation in a wide range of other tracked and wheeled vehicles. The DFCS is also installed in the Cadillac Gage Textron Stingray 105 mm armed light tank now in service in Thailand.

V-150 Commando fitted with 81 mm mortar with roof hatches open

Cadillac Gage V-150 upgraded in Malaysia by MMC Engineering Services Limited (Christopher F Foss)

Commando with 81 mm Mortar

This is armed with an 81 mm M29 turntable-mounted mortar that fires through the roof of the vehicle, and can be traversed through a full 360° and elevated from +40° to +85°. Its minimum range is 150 m and maximum range 4400 m. When not in action the roof is covered by concertina doors that open either side of the vehicle. Provision is made for mounting a 7.62 mm machine gun at any one of four positions around the top of the hull and 2000 rounds of ammunition are carried for it. A total of 62 81 mm mortar bombs are carried and if required the mortar can be removed from the vehicle and fired from the ground. This model has a crew of five and a gross vehicle weight of 8845 kg.

Commando with TOW ATGW System

This is almost identical to the above model but is fitted with a Hughes TOW ATGW system which has an elevation of +30°, a depression of -10° and 360° traverse. Seven TOW missiles are carried. When not in action the top of the vehicle is covered by hatches that open to the front and rear. Provision is made for mounting a 7.62 mm machine gun at any one of four positions around the top of the hull and 2000 rounds of ammunition are carried for this weapon. This model has a crew of four and can also carry two infantrymen. Gross vehicle weight is 8958 kg.

Commando Command Vehicle

This is the standard vehicle fitted with a fixed armoured pod mounted at the roof opening. In each of the four sides of the pod is a firing port and in the roof is a single piece hatch cover that opens to the rear, forward of which is a 7.62 mm pintle-mounted machine gun. Of 2000 rounds of 7.62 mm ammunition carried, 200 are for ready use. Internally the vehicle has a mapboard, table and additional communications equipment. In addition to the crew of three, seven staff members are carried. Overall height without the machine gun is 2.311 m.

Base Security Vehicle

This was originally developed for the United States Air Force and like the mortar carrier has concertina hatch covers that open either side of the vehicle. Provision is made for mounting 7.62 mm machine guns or 40 mm grenade launchers in four positions at the front, sides and rear. A total of 3000 rounds of 7.62 mm ammunition is carried. This model has a crew of three and can carry eight fully equipped infantrymen, weighs 9072 kg and is 2.311 m high.

Malaysian Upgraded V-150 Commando

In 1987 a 90 mm V-150 Commando of the Malaysian Army was modernised by MMC Engineering Services Sdn Bhd and subsequently evaluated by the Malaysian Army over 10 000 km during which trials were carried out for road endurance, cross-country, vertical obstacle and amphibious capabilities. The original petrol engine has been replaced by a more fuel-efficient diesel engine while the manual transmission has been replaced by a fully automatic transmission.

Police Emergency Rescue Vehicle (ERV)

This model has a fixed pod with six vision blocks and eight vertical gun ports. In the roof is a single piece hatch cover that opens to the rear and in front of which is a 7.62 mm or a 12.7 mm machine gun. This model has a crew of three and can carry nine additional men and weighs 9162 kg. It is used by a number of police authorities and a large number was purchased by Turkey.

Recovery Vehicle

This is fitted with a heavy-duty winch with a maximum capacity of 11 340 kg which is provided with 60.9 m of 19 mm diameter cable. The boom is pivoted at the front of the vehicle and when required hinges forward to the front of the vehicle. When it is being used two stabilisers are lowered at the front. The boom has a maximum capacity of 4536 kg with the stabilisers in position. Armament consists of a 7.62 mm or a 12.7 mm machine gun with 2200 rounds of ammunition. Full details of the Commando ARV are given in *Jane's Military Vehicles and Logistics 1991-92*, page 38.

SPECIFICATIONS (Commando V-150 with turret-mounted 20 mm cannon. Specifications in square brackets relate to V-150 S Commando)

CREW	3 + 2	TURNING RADIUS	8.382 [8.5] m	SMOKE-LAYING	
CONFIGURATION	4 × 4	ENGINE	V-504 V-8 diesel	EQUIPMENT	optional
COMBAT WEIGHT	9888 [10 090] kg		developing 202 bhp at	AMMUNITION	
POWER-TO-WEIGHT			3300 rpm	main	400
RATIO	20.42 bhp/tonne	TRANSMISSION	automatic, 4 forward, 1	7.62 mm	3200
LENGTH	5.689 [6.14] m		reverse gears	GUN CONTROL	
WIDTH	2.26 m	TRANSFER CASE	2-speed (with spline	EQUIPMENT	
HEIGHT			engagement for front	Turret power control	electro-hydraulic/
to turret roof	2.54 m		axle drive)		manual
to hull top	1.981 m	STEERING	power-assisted	by commander	yes
GROUND CLEARANCE		SUSPENSION	solid axles with semi-	by gunner	yes
axles	0.381 m		elliptical springs and	Max rate power	
hull	0.648 m		double-acting hydraulic	traverse	60°/s
TRACK			shock absorbers	Max rate power	
front	1.914 m	TYRES	14.50 × 21 run-flat	elevation	40°/s
rear	1.943 m	BRAKES		Gun elevation/depression	+60°/-8°
WHEELBASE	2.667 [3.12] m	main	hydraulic, dual	Turret traverse	360°
MAX SPEED			circuit		
road	88 km/h	parking	drum acting on		
water	5 km/h		output shaft of		
ACCELERATION			transmission		
(0-32.2 km/h)	10 s	ELECTRICAL SYSTEM	24 V		
FUEL CAPACITY	303 l	BATTERIES	2 × 12 V, 100 Ah		
MAX RANGE	643 km	ARMAMENT			
FORDING	amphibious	main	1 × 20 mm cannon		
GRADIENT	60%	coaxial	1 × 7.62 mm MG		
SIDE SLOPE	30%	anti-aircraft	1 × 7.62 mm MG		
VERTICAL OBSTACLE	0.609 m				

Note: In 1982 Cadillac Gage announced that it had developed a Cummins V-6 diesel conversion kit as a direct replacement for the V-8 diesel installed in the V-150. When fitted with this engine the V-150 has an operational range of 800 km compared to the standard V-8 model with 644 km, with an acceleration of 0 to 32 km/h in nine seconds. This kit can be installed with no special tools or other equipment at base workshop level.

Status: Production as required. In service with Bolivia (V-100), Botswana (V-150), Cameroon (V-150, additional 24 ordered in 1983 at a cost of $13.1 million), Chad (six V-150 Ss ordered early in 1986 at a cost of $3.7 million including 90 mm ammunition, spares and training), Dominican Republic (V-150), Gabon (V-150), Guatemala (V-150), Haiti (V-150), Indonesia (V-150), Jamaica (V-150), Kuwait (20 V-150 S ordered in 1984 status uncertain), Malaysia (V-100 and V-150), Panama (V-150 status uncertain), Philippines (V-150, further 25 V-150 Ss ordered mid-1987 at cost of $6.898 million), Saudi Arabia (ambulance, APC, ARV, 20 mm anti-aircraft, 81 mm mortar, 90 mm gun and Hughes TOW), Singapore (V-150 and V-200), Somalia (V-100 and V-150), Sudan (V-150, 24 V-150 Ss ordered in 1986 at a cost of $7.6 million, V-100), Taiwan (V-150), Thailand (V-150), Tunisia (V-150), Turkey (V-150) in police role, USA (V-100), Venezuela (V-150), and Vietnam (V-100).

Note: V-150 has been replaced in production by V-150 S. United States Army designation for the V-100 is the M706. They are no longer used by Ethiopia or Oman.

Manufacturer: Cadillac Gage Textron
PO Box 1027
Warren
Michigan 48090
USA
Telephone: (313) 777 7100
Telex: 200707
Fax: (313) 776 9731

AV Technology Corporation Dragoon Family of Armored Vehicles

Development

This expanded family of wheeled armoured vehicles has been developed from the field experience gained by the success of the earlier Dragoon 300 carrier design. The Dragoon 300 was originally developed by Verne Corporation and marketed by Arrowpointe Corporation.

In 1984 the two companies merged to form the privately held AV Technology Corporation, a company owned by the two founders of Verne and Arrowpointe Corporation. Details of the original Dragoon 300 armoured vehicle family were given in *Jane's Armour and Artillery 1988-89* pages 418 to 420.

The latest designed Dragoon has been purchased by the United States military and several foreign governments and is available in 10 different variants. One- and two-man turrets are available that have been easily adapted to a variety of other armoured vehicles.

The current Dragoon, like its predecessors, makes an extensive use of proven US Military, NATO and world class commercial components. There is more than seventy per cent logistics commonality with in-service tactical vehicle fleets such as the M113A2, M809 and M939. Replacement parts for the Dragoon are available in virtually every part of the world.

The Dragoon armoured vehicle family has been evaluated by a number of countries. The first production vehicles were completed for the United States Army and Navy in 1983. Since then Venezuela, Thailand and the US Navy have placed orders for Dragoon armoured vehicles which have subsequently been delivered. In 1988 AV Technology Corporation introduced the Dragoon Armored Security Vehicles (ASV) and the Patroller (ASV-XC) to police departments and other law enforcement agencies around the world. AV Technology Corporation has already delivered the standard ASV to some police departments and has delivered the newly designed Patroller ASV-XC (extended cab) to the New York and New Jersey Port Authority for use by security police at John F Kennedy Airport.

Description

The hull of the current Dragoon is constructed of high hardness steel ballistic plate which meets MIL-A-12560 and is of monocoque construction with welded seams. The standard ballistic resistance of the Dragoon is 7.62 mm NATO ball (147 gram M80 2750 ft/s) with a 0.95 probability of no penetration at 0 m and 0° obliquity on all surfaces. Additional armour protection is available on some models that upgrades the protection to 7.62 mm NATO AP (147 gram M80 2800 ft/s) with 0.95 probability of no penetration on front rear sides and top of the vehicle.

The hull has an angular surface on the front, right and left sides which project outward from the top and bottom meeting at the mid section around the perimeter. The rear surface of the hull projects outward from the bottom at an angle of 25° vertical. The basic design of the hull provides for easy changes of configuration. The Dragoon is coated with a two part epoxy primer and can be top coated with CARC. The interior of the vehicle is finished with a heavy duty epoxy coating.

The two double width side doors are 1.066 m wide and 0.889 m high. The rear door is 0.609 m wide and 0.916 m high. There are three hatches on the top of the vehicle. All doors and hatches allow for the swift passage to enter and exit the vehicle while providing security and water tightness. Power up and power down adjustment of the driver and co-driver seats allows the vehicle to be driven with the hatches open.

When the seats are lowered, visibility is through the vision blocks. The driver has four vision blocks which provide him with a 180° field-of-view, while the commander has two vision blocks giving him view of the front and right side of the vehicle. An optional weather cover kit with a windshield and wiper is available for the driver and co-driver to be used with the hatches open in adverse weather or temperatures. The vehicle is equipped with two outside rear-view mirrors. The standard vehicle comes fitted with 12 ballistic vision blocks which are located for easy viewing from all sides of the vehicle by seated personnel.

There is seating for 12 occupants including the driver and all seats are equipped with seat belts. Other optional seating arrangements are available to meet specific requirements. There are five gun ports, two each side and one in the rear that will allow the use of hand-held weapons. All have ballistic covers that can be positively secured from the inside of the vehicle.

Radio and internal communications can be easily mounted in the hull. Mounting pads and power for more common AN/VRC-12 family of radios is standard in the Dragoon, but other radio communication systems have been installed and special configurations can be tailored to meet user specifications.

The Dragoon comes standard with a 50 000 BTU heating system and a 30 000 BTU air-conditioning system. There are seven adjustable direct-flow air ducts mounted for easy access and for even distribution throughout the crew compartment. There is an additional heater in the front area of the vehicle that provides an extra level of comfort.

The 24 V water-resistant electrical system is equipped with a master power switch and two 100A/hour heavy duty batteries and a 200A heavy-duty radio suppressed military type alternator. The Dragoon is equipped with blackout headlights with high and low beam selection, stop and tail lamps, instrument lights, interior lights, horn and turn signals all mounted in protected areas. An optional 1 000 000 candle power spotlight is available and it can be mounted to the top hatch ring mount for 360° rotation.

A fixed Halon fire extinguisher system is installed with direct discharge into the engine compartment. Two hand-held Halon fire extinguishers are installed in the crew compartment.

The power-pack system is a Dragoon-exclusive feature. This system allows for the fast and easy removal of the complete engine, transmission, hydraulic, cooling and electrical system. This is removed as a single unit in less than 30 minutes without any loss of operating fluids or Freon from the air-conditioning system. Power-packs can be quickly exchanged between vehicles without any modifications.

The Dragoon has a patented cooling system including a thermostatically controlled two speed hydraulically driven fan. This allows all vehicles to operate in a temperature range of -25° F to +125° F. All routine operator fluid checks are easily accessed at the engine hood opening on the top of the vehicle. Engine oil and hydraulic filters are the spin-on type for quick service and preventative maintenance.

The Dragoon is equipped with two 170 litre armoured fuel tanks and includes an emergency fuel pump prime system. This feature is used in the event of the main fuel pump failing or to prime the engine if it should run out of fuel.

The patented Dragoon system drive line provides the most durable and rugged system with the latest number of components. The Allison Transmission Division of General Motors MT-653 heavy duty five speed forward and single speed reverse transmission has a torque converter that locks up in the fourth and fifth gears providing better fuel economy. The single speed transfer case has an easy quick disconnect four bolt universal joint to the drive shaft. This allows for rapid removal of the power-pack. The Dragoon's single speed in-out transfer box allows for selection of two or four wheel drive at any safe speed without stopping. The hydraulic system has an engine-mounted three section hydraulic pump that suppies hydraulic fluid to the winch, power steering, cooling fan, brakes, power seat and other optional accessory features like the 10 kW generator.

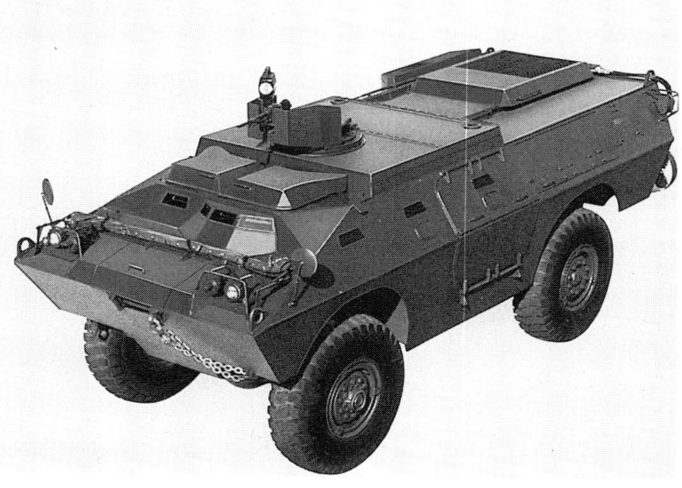

Dragoon Armored Personnel Carrier (APC) with 7.62 mm MG and searchlight

Front and rear axles are standard off-the-shelf heavy duty Rockwell military-type (the same axles used on the M939). The axles may be equipped with the optional Central Tire Inflation System (CTIS) which increases flotation and traction by driver selection and automatically adjusting tyre pressure for different terrains. All Dragoon's are equipped with power brakes and a disc-type parking brake.

Power steering is equipped with a manual override in the event of power loss. The front and rear leaf springs have a capacity of 12 700 kg and 304 mm of suspension travel. The heavy-duty shock absorbers are interchangeable. Military standard all-terrain tyres may be equipped with a run-flat device that permits continuous operation.

The Dragoon is fully amphibious and during amphibious operations it is propelled by all wheels and turning is accomplished by steering the front wheels. Three electric bilge pumps with the capacity of 190 litres/minute expel any water that may seep into the vehicle. All Dragoons come equipped with an hydraulic self recovery winch that has 53.34 m of cable.

Variants

Armored Personnel Carrier

The Dragoon Armored Personnel Carrier (also referred to as an Armored Security Vehicle) is the base vehicle configuration and is a fully equipped vehicle capable of carrying 12 occupants with equipment. It is outfitted for use as a reconnaissance, security and escort vehicle as well as a troop carrier. Each vehicle is equipped with a ring or pintle-mounted machine gun up to 12.7 mm in calibre and a variety of communications systems. This variant has a payload capacity of 3402 kg and a gross vehicle weight of 13 068 kg with a top speed of 116 km/h.

Dragoon Patroller Armored Security Vehicle

(ASV-XC Extended Cab)

The Patroller comes equipped with all the performance features of the standard Dragoon APC, but design allows for two 406 mm × 660 mm windshields and angular 304 mm × 685 mm side windows that have no spall 7.62 mm NATO AP protection. The Patrollers are equipped with windshield wiper and washers. Extra large vision blocks and gun ports are also installed. The ASV-XC has 1.995 m of interior headroom allowing the occupants sufficient area to stand upright in the vehicle if necessary for surveillance activities. This variant can also be equipped for amphibious operations. The Patroller will accommodate a crew of 12 with weapons, gear and supplies.

Dragoon Armored Command Vehicle (ACV)

The Dragoon Command Control and Communications vehicle is another variant of the APC and can be configured to the user's preferred interior layout and will accommodate a wide variety of specialised equipment to satisfy the needs of field commanders, tactical communication and electronic warfare systems. A 360° rotating machine gun ring mount with removable ballistic shield, work table, 3 kW generator, camouflage net, communication harness and extra brackets for equipment and radios are all standard equipment. The base model will accommodate a crew of five with gear, weapons, ammunition and rations.

Dragoon Armored Mortar Carrier (AMC)

The Dragoon Armored Mortar variant mounts an 81 mm mortar on a 360° rotating turntable. The mortar may be traversed and locked at a number of azimuth positions. Fine adjustment in azimuth and elevation is provided as part of the mortar-mounting system. Full width armour bi-fold doors are provided over the weapons/crew compartment. The mortar may be fired from within the vehicle. The crew is protected from small arms fire and artillery fragments by the vehicle hull. Sixty rounds of various types of mortar ammunition may be stowed in the weapons compartment and is easily accessible to the loader. Additional stowage is provided for the equipment necessary for firing off-vehicle when required. The Dragoon AMC variant will accommodate a crew of five with standard field gear, weapons, ammunition and rations.

Dragoon Light Forces Vehicle 90 mm and Two-man Turrett (LFV-90 mm)

The Dragoon Light Forces Vehicle comes equipped with the two-man turret and a 90 mm cannon and a coaxial 7.62 mm or 12.7 mm (0.50) calibre MG. The turret shell is constructed from the same high hardness steel armour plate as the vehicle hull and provides the same level of ballistic protection. The fire control power system is electro-hydraulic and is operated by a single hand grip controller in azimuth and elevation. Secondary or backup manual controls are provided for both azimuth and elevation. The fire direction system is the standard M36E1 passive day/night sight for the gunner (thermal imaging and other sight options are available). Seven M27 periscopes are provided for the commander for 360° visibility. Space is available in the turret bustle for mounting a variety of radio communications systems. Stowage is provided in the turret for 500 ready rounds of 7.62 mm ammunition and 10 ready rounds of 90 mm cannon ammunition. Additional stowage is provided for up to 50 rounds of 90 mm and 2000 rounds of 7.62 mm ammunition in the vehicle.

The turret weighs less than 1905 kg at full load and has solid state electronics and auto-deck interrupt features. NBC features and commanders control and override are optional. The LFV-90 mm will accommodate a crew of four with standard field gear, weapons, ammunition and rations. The two-man turret can be ordered separately and has been mounted to a variety of vehicles.

Dragoon Light Forces Vehicle with 90 mm two-man turret

Dragoon Armored Mortar Carrier (AMC) with 81 mm mortar

Dragoon Armored Maintenance Vehicle (AMV) changing a Dragoon powerpack under field conditions

Dragoon Light Forces Vehicle 40/.50 One-man Turret (LFV-40 mm)

The Dragoon LFV-40 mm (one-man turret) has mounts for an Mk 19 model 3 40 mm grenade machine gun and a coaxially mounted M2 HB 12.7 mm (0.50) calibre MG. The Mark 19 model 3, firing the M430 high explosive dual purpose (HEDP) round, can penetrate up to 50 mm of conventional steel armour plate at over 2000 m. Like the two-man weapon system, this turret is equipped with the standard M36E1 sight. Stowage is provided in the turret for 100 rounds of 40 mm and 200 rounds of 12.7 mm ammunition plus eight ready smoke grenades. Additional ammunition stowage is provided in the vehicle. The LFV-40 vehicle will accommodate a crew of eight with standard field gear, weapons, ammunition and rations. The one-man turret can be ordered separately and has been mounted to a variety of vehicles.

Dragoon Armored Maintenance Vehicle (AMV)

The Dragoon Armored Maintenance Vehicle and Engineer variant is equipped to provide field recovery, maintenance and logistics support for all Dragoons and many other tactical vehicles. The 2268 kg hydraulic boom crane is capable of performing a wide variety of maintenance and engineering tasks including installation and removal of vehicle power-packs, turrets, tyre changing, cargo loading and unloading and general construction.

Bi-fold top doors are provided for cargo loading and equipment access. Auxiliary power is provided for operating both air hydraulic accessories and tools. The vehicle comes standard with hydraulic levelling jacks, fuel transfer pump, portable pump, tow bar, tool set, air compressor, power cable, snatch block and spare tyre mount. The interior is equipped with built-in tool and accessory storage and a work bench with a vice. The Dragoon AMV will accommodate a crew of five with standard field gear, weapons, ammunition and rations.

Mobile Electronic Warfare System (MEWS)

The MEWS carries out advanced battlefield, direction-finding and high speed communications jamming and is fitted with an antenna that can be quickly raised well above the vehicle. The video optical surveillance vehicle has a modified AV Tech two-man turret which has been fitted with a long range day/night surveillance system that can be retracted quickly under armour protection. This provides the field commander with highly mobile, armour-protected observation capability, which, when connected to an on-board data link, gives clear front line view of the battlefield for rear commanders. The Teampack electronic detection system is normally carried in the horizontal position on the left side of the vehicle. When required for action it can be erected quickly into a vertical position. A 7.62 mm machine gun is ring mounted on top of the hatch.

Armored Logistics Support Vehicle (ALSV)

The Dragoon ALSV is configured to facilitate the transport of standard pallets and containers. It is equipped with a hydraulic boom crane capable of self loading and unloading cargo up to one metric ton. The ALSV may be fitted with a variety of self defence weapons and will accommodate a crew of three with standard field gear, weapons, ammunition and rations.

Dragoon TOW Carrier

The Dragoon TOW system mounts the TOW pedestal on a 360° rotating turntable and is protected from small arms fire and artillery fragments by the vehicle's armoured wall and top bi-fold doors. Provision is made for stowage of six TOW missiles in the vehicle. The vehicle will accommodate a crew of three with field gear, weapons and ammunition and rations.

SPECIFICATIONS (fitted with 90 mm turret)

CREW	4
CONFIGURATION	4 × 4
POWER-TO-WEIGHT RATIO	23.62 bhp/tonne
COMBAT WEIGHT	12 700 kg
UNLOADED WEIGHT	10 796 kg
LOAD	1905 kg
TOWED LOAD	12 700 kg
LENGTH	5.89 m
WIDTH	2.44 m
HEIGHT	
overall	2.819 m
hull top	2.082 m
GROUND CLEARANCE	
axles	0.381 m
hull	0.685 m
TRACK	1.981 m
WHEELBASE	3.10 m
ANGLE OF APPROACH/ DEPARTURE	60°/45°

MAX ROAD SPEED	116 km/h
MAX WATER SPEED	
wheels	5.6 km/h
propeller kit	8.8 km/h
FUEL CAPACITY	350 l
RANGE CRUISING	885 km
FORDING	amphibious
GRADIENT	60%
SIDE SLOPE	35%
VERTICAL OBSTACLE	
forwards	0.99 m
reverse	0.61 m
TURNING RADIUS	8.6 m
ENGINE	Detroit Diesel 6V-53T 6-cylinder, liquid-cooled, turbo-charged diesel developing 300 bhp at 1800 rpm
TRANSMISSION	Allison Transmission Division of General Motors MT-653 automatic, 5 forward and 1 reverse gears
TRANSFER CASE	single speed, gearless

STEERING	hydraulic, power assisted, variable ratio
SUSPENSION	semi-ellipitical springs and hydraulic shock absorbers
TYRES	14.50 × 21, 12-ply radial
BRAKES	
main	hydraulic, power boosted
parking	dual grip
ELECTRICAL SYSTEM	24 V
BATTERIES	2 × 12 V, 100 Ah
ARMAMENT	
main	1 × 90 mm gun
coaxial	1 × 7.62 mm MG
anti-aircraft	1 × 7.62 mm MG (optional)
AMMUNITION	
90 mm	60
7.62 mm	2500

Status: Production as required. In service with US forces, Thailand and Venezuela plus civil authorities.

Manufacturer: Arrowpointe Corporation
2240 Alger
Troy
Michigan 48083
USA
Telephone: (313) 528 2575
Telex: 235354 ARROW SOFD

Half-Track Armoured Personnel Carriers

A total of 42 000 half-tracks were built by US manufacturers before production was completed in 1944. It remains in service in declining numbers with a number of countries including Argentina, Cameroon, Colombia, Dominican Republic, Gambia, Greece, Israel, Madagascar, Mauritania, Mexico, Niger, Paraguay, Peru, Senegal, Taiwan, Thailand, Togo and Zaïre. In most cases the half-tracks are held in reserve or used by second line units. The largest user is Israel. Details of the half-track and most of its many variants that are still operational are given in *Jane's Armour and Artillery 1989-90* pages 477 to 479.

Cadillac Gage Commando Ranger Armoured Personnel Carrier

Development

The Commando Ranger has been developed by Cadillac Gage Textron as a private venture to meet a requirement for a low-cost vehicle capable of transporting troops at a high road speed but still retaining an off-road capability.

Late in 1977 the United States Air Force issued a requirement for a Security Police Armored Response/Convoy Truck for patrolling air bases. The Air Force decided to initiate a two-step procurement, technical and price. In April 1978 a number of American manufacturers submitted detailed technical proposals in response to the requirement and in October the same year the US Air Force selected three manufacturers' proposals as meeting the technical requirements: Cadillac Gage Textron, Oshkosh Truck Corporation and the Vehicle Systems Development Corporation.

In March 1979 the US Air Force chose the Cadillac Gage Commando Ranger and in June 1979 the company was awarded a contract worth $4.4 million for the supply of a number of vehicles plus technical and administrative data.

The first Cadillac Gage Commando Ranger, called the Peacekeeper by the United States Air Force, was handed over in April 1980. By early 1981 560 vehicles had been ordered by the US Air Force and the US Navy. By 1991 708 Cadillac Gage Commando Ranger armoured personnel carriers had been built for US forces and overseas sales.

Description

The hull of the Commando Ranger is made of all-welded 'CADLOY' high hardness armour which provides the crew with complete protection from small arms fire. The crew and vital components are protected from the effects of grenades exploding under the vehicle by bolt-on armour shields.

The Commando Ranger armoured personnel carrier is based on a standard Chrysler 4 × 4 truck chassis but with a shorter wheelbase.

The engine is at the front of the hull and is coupled to an automatic transmission with three forward and one reverse gears which is equipped with an auxiliary oil cooler. Coupled to the transmission is a two-speed transfer case. The manually selected four-wheel drive case with an

Command Post version of Cadillac Gage Commando Ranger showing antennas

Cadillac Gage Commando Ranger from rear showing firing ports (Terry J Gander)

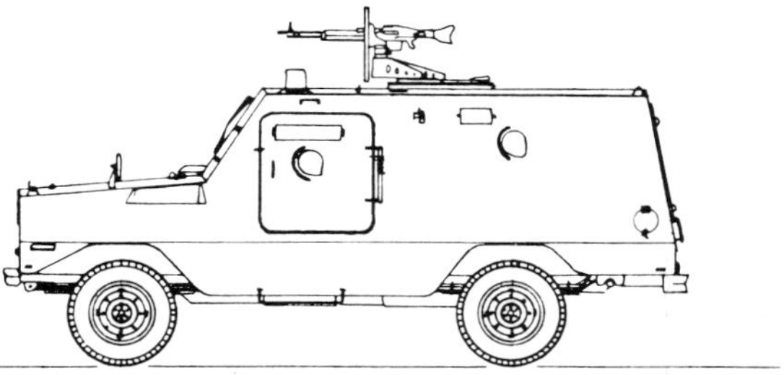

Cadillac Gage Commando Ranger fitted with 7.62 mm machine gun and shield to protect gunner

US Air Force Cadillac Gage Commando Ranger without armament installed (Terry J Gander)

interaxle differential delivers proportionalised torque to both front and rear axles. The interaxle differential enables the wheels on one axle to travel at a different speed from those on the other. Both axles are of the full floating type.

The vehicle has a dual hydraulic service brake system and a foot/lever actuated parking brake. The service brake system consists of two separate brake fluid lines and a split, front and rear, master cylinder. Power assist is supplied by means of a tandem-diaphragm brake booster that utilises engine vacuum for power. The front brakes are disc with self-adjusting calipers and the rear are drum with self-adjusting shoes.

The suspension, front and rear, consists of leaf springs with double-acting shock absorbers and jounce stops at each wheel station. The combat tyres are filled with foam and when penetrated by small arms fire allow the vehicle to travel up to 80 km at 56 km/h.

The driver sits at the front of the vehicle on the left with the commander to his right. The steering wheel can be tilted to assist the driver in entering and leaving the vehicle. Both crew members have a bullet-proof windscreen in front of them with a wiper. There is a firing port between the two windscreens. Each also has a rear-opening side door with a bullet-proof vision block in its upper half under which is a firing port.

The troop compartment is at the rear of the hull with the men seated three down each side facing each other. They enter the vehicle through two doors in the rear of the hull which open outwards. Both doors have a firing port and the left one has a vision block. In each side of the troop compartment is a vision block with a firing port underneath. The top of the hull is angled to allow hand grenades to roll off before they explode. If required the vehicle can be delivered with a roof hatch on which can be fitted a variety of light armament installations such as a pintle-mounted 7.62 mm M60 machine gun with a shield and 200 rounds of ready-use ammunition and a further 800 rounds carried in the hull.

The interior of the Commando Ranger is insulated, which reduces noise, retains cool air as the vehicle is fitted with an air-conditioning system and also aids crew safety. The air-conditioning system is fitted as standard and there is also provision for expelling air. Standard equipment includes two-speed wipers with washers, heater, windscreen defogger and internal dome light.

Optional equipment includes flashing lights, public address system, radios, spotlight, Wegmann grenade launchers, front-mounted winch and a 24 V electric system in place of the standard 12 V system.

Variants

In addition to being used as an APC the Commando Ranger can be used as a command vehicle (crew of two plus two command staff), ambulance (crew of two and carrying two stretcher plus two seated patients) and as a light reconnaissance vehicle which can be fitted with a Cadillac Gage one-man turret armed with twin 7.62 mm (or 0.30) or one 7.62 mm (0.30) and one 12.7 mm (0.50) machine guns. These have an elevation of +55°, depression of -14°, and 360° turret traverse, which is manual. The turret has an M28C roof-mounted sight with a magnification of ×1.5, eight vision blocks and a single piece hatch cover that opens to the rear.

SPECIFICATIONS

CREW	2 + 6
CONFIGURATION	4 × 4
COMBAT WEIGHT	4536 kg
POWER-TO-WEIGHT RATIO	40 hp/tonne
LENGTH	4.699 m
WIDTH	2.019 m
HEIGHT	1.981 m
GROUND CLEARANCE	
front axle	0.203 m
rear axle	0.203 m
TRACK	1.689 m
WHEELBASE	2.641 m
ANGLE OF APPROACH/DEPARTURE	25°/24°
MAX ROAD SPEED	112.65 km/h
FUEL CAPACITY	121 l
MAX RANGE (at 72 km/h)	482 km
GRADIENT	60%
SIDE SLOPE	30%
VERTICAL OBSTACLE	0.254 m
ENGINE	Dodge 360 CID V-8 liquid-cooled petrol developing 180 hp at 3600 rpm
TRANSMISSION	Dodge model A727 automatic with 3 forward and 1 reverse gears
TRANSFER CASE	2-speed
STEERING	integral power with pump assistance
SUSPENSION	leaf springs and double-acting hydraulic shock absorbers
BRAKES	
main	hydraulic, dual circuit
parking	foot/lever actuated
ELECTRICAL SYSTEM	12 V
BATTERY	1 × 12 V, 70 Ah
ARMAMENT	see text

Status: Production as required. In service with Indonesia (22, of which seven are in VIP configuration), Luxembourg (five with one-man turret armed with twin 7.62 mm MGs), US Air Force and US Navy.

Manufacturer: Cadillac Gage Textron
PO Box 1027
Warren
Michigan 48090
USA
Telephone: (313) 777 7100
Telex: 200707
Fax: (313) 776 9731

Vehicle Systems Development Corporation Lancer Armoured Patrol Vehicle

Development

Late in 1977 the United States Air Force issued a requirement for a Security Police Armored Response/Convoy Truck for patrolling air bases. The Air Force decided to initiate a two-step procurement, ie technical and price. In April 1978 a number of American manufacturers submitted detailed technical proposals in response to the requirement and in October the same year the US Air Force selected three manufacturers' proposals as meeting the technical requirements: Cadillac Gage Textron, Oshkosh Truck Corporation and the Vehicle Systems Development Corporation.

In March 1979 the Air Force awarded the initial production contract to Cadillac Gage for its Commando Ranger vehicle on the basis of a lower price per vehicle. Details of the Commando Ranger are given earlier in this section.

Discussions covering the licensed manufacture of a version of the Lancer were under way in 1991. This vehicle is being planned for the Internal Security Vehicle (ISV) mission and will incorporate provisions for several armament options including the M60 7.62 mm machine gun and the Mk 19/3 40 mm grenade launcher.

The Lancer based on the M1108 (1.25-ton) (4 × 4) tactical truck chassis is being modified for installation on the new GMT-400 series General Motors truck chassis.

Description

The Lancer Armoured Patrol Vehicle is based on an unmodified 4 × 4 commercial chassis utilising standard automotive components and suspension. The original version had four side doors, two in each side, but the current model has a single door in each side. As an Armored Patrol Vehicle it carries a crew of four, but the vehicle can also be used for other roles such as command, convoy escort, internal security vehicle, medical evacuation, reconnaissance and other paramilitary roles.

The engine is at the front of the vehicle, commander and driver in the centre and personnel compartment at the rear. Protection against small arms fire and shell splinters is provided by ArmaCore armour which combines superior ballistic qualities and low weight. The driver sits on the left with the commander to his right, both with a side door that opens forward with a spherical firing port in its lower part that enables the crew member to fire his weapon from inside the vehicle in complete safety. The doors also have a ballistic shield in their lower part which can be lowered to the ground to protect the crew members when they are dismounting from the vehicle.

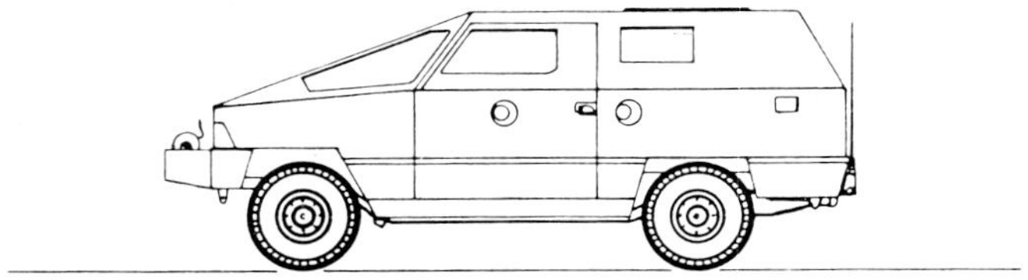

Production version of Vehicle Systems Development Corporation Lancer Armoured Patrol Vehicle

Another two crew members seated to the rear of the commander and driver have a side window under which is a spherical firing port. Over the top of the crew compartment is a circular hatch cover that opens to the left rear of the vehicle, on which various armament installations can be mounted, such as the Dragon ATGW with a Viscous Damped Mount (VDM) with four missiles, 7.62 mm machine gun with 750 rounds of ammunition, 12.7 mm M2 HB machine gun with 500 rounds of ammunition or a water cannon. It can also accommodate a protected marksman and sniper's rifle for anti-hijack operations at airports. There is a door in the rear of the hull, which has a window beneath which is a spherical firing port. If required the vehicle can be delivered with an angled roof to deflect hand grenades.

Standard equipment includes an explosion-proof fuel tank, air-conditioning system with modified fan and forced air evacuation system and 10-15 LT LR C6 Hutchinson ballistic core tyres and 9.00 JJX15 spoke steel wheels.

Mounted at the front of the hull is an electrically operated winch and optional equipment includes air-conditioning with modulated fan, run-flat tyres and night vision devices.

Variants

There are no variants of the Lancer Armored Patrol Vehicle at present, although the vehicle can carry specialised communications equipment.

SPECIFICATIONS

CREW	4
CONFIGURATION	4 × 4
COMBAT WEIGHT	2818 kg
LENGTH	4.42 m
WIDTH	1.792 m
HEIGHT (to hull top)	1.727 m
GROUND CLEARANCE	0.22 m
TRACK	1.481 m
WHEELBASE	2.54 m
ANGLE OF APPROACH/DEPARTURE	38°/30°
MAX ROAD SPEED	128.7 km/h
FUEL CAPACITY	70 l
MAX RANGE (at 72.4 km/h)	395 km
FORDING	0.457 m
GRADIENT	60%
SIDE SLOPE	30%
VERTICAL OBSTACLE	0.254 m
ENGINE	V-8 petrol developing 150 hp at 3200 rpm
TRANSMISSION	automatic 3 forward and 1 reverse gears
TRANSFER CASE	2-speed
STEERING	integral power
TYRES	9.50 × 16.5
SUSPENSION	semi-elliptical springs and hydraulic shock absorbers
BRAKES (main)	front disc, rear drum, hydraulic
TURNING RADIUS	5.18 m
ELECTRICAL SYSTEM	12 V (24 V optional)
ARMAMENT	optional

Note: The 5.7 litre V-8 petrol engine developing 150 hp can be replaced by a 5 litre V-8 petrol engine or a 3.3 litre diesel engine. Both petrol engines are available with 3- or 4-speed manual transmissions or a 3-speed automatic transmission. With the 5.7 litre petrol engine and manual transmission range is 464 km, with diesel and the same transmission range is 687 km.

Status: Development complete. Ready for production.

Manufacturer: Vehicle Systems Development Corporation
1271 W 9th Street
Upland
California 91786
USA
Telephone: (714) 981 3236
Telex: 214807 VSDC CA

Landing Vehicle Tracked (Experimental)

In March 1985 the Landing Vehicle Tracked (Experimental) was cancelled. Details of the LVT(X) programme and the earlier Landing Vehicle Assault (LVA) were given in *Jane's Armour and Artillery 1984-85*, pages 414-415 and 883. The current AAV7A1 is being further product improved until the follow-on vehicle, now called the Advanced Amphibious Assault Vehicle (AAAV) is available.

Within the Marine Corps Surface Mobility Technology Program a number of separate development programmes are under way with the aim of increasing the speed of movement from the ship to the beach. These include increased water speed, reduced vehicle weight, improved armour protection and increased fire power.

AAI has already completed the prototype of the Automotive Test Rig for the Marine Corps and this is covered later, as is the AAI developed High Water Speed Technology Demonstrator completed in 1989.

Other areas under study include a metal matrix track which will be 30 to 50 per cent lighter than the current steel tracks, a band track which will be tested on an M113, improved elastomers, retractable hydro-pneumatic suspension system, hydraulic and electric drive trains, two-speed final drive, advanced transmission control, composite components (see also entry on M113 with new armour system) and a family of weapon stations.

The 25 mm Minor Calibre Weapons Station (MCWS) has been designed and built by AAI under contract to the Naval Surface Weapons Center/David Taylor Research Center/US Marine Corps and for trials purposes has already been installed on the AAI Automotive Test Rig described later and on a Light Armored Vehicle (LAV) of the Marine Corps.

The MCWS is a lightweight, remotely controlled weapon station armed with a McDonnell Douglas Helicopters 25 mm Chain Gun with 90 rounds of AP and 200 rounds of HE ammunition for ready-use with a 7.62 mm coaxial Chain Gun, with the gunner seated in the vehicle.

AAI Propulsion System Demonstrator

Development

The Propulsion System Demonstrator (PSD) is an assault amphibious vehicle designed and built by AAI under contract to the David Taylor Research Center (DTRC), Marine Corps Program Office.

The objective of the PSD is to demonstrate the feasibility of a tracked amphibious vehicle's capability to attain a water speed in excess of 37 km/h (20 knots) and to perform the automotive functions associated with land operation of a tracked assault amphibious vehicle.

Information obtained from trials with the PSD will be used in the US Marine Corps Advanced Amphibious Assault Vehicle (AAAV programme) (qv).

Description

The PSD is a full tracked amphibious vehicle with a composite hull constructed of aluminium alloy and glass reinforced plastic. An appliqué of alumina tiles is attached to the hull exterior for armour protection.

To obtain high water speed performance the PSD is equipped with hull-mounted deployable appendages including a bow flap, chine flaps and a transom flap. Its suspension system is retractable for water operations and employs liquid spring suspension units and AAI developed 533 mm wide roller chain band track.

When on land, the PSD is powered by either a Cummins VTA 903-T700 diesel engine or a John Deere 2116R rotary engine coupled to a General Electric HMPT-500-3 full automatic transmission with final drives.

For high speed water operation, power from the diesel or rotary engine is combined with the power output of a General Electric LM120/T700/CT7 turboshaft engine to hydrostatically drive four 406 mm diameter David Taylor Research Center developed waterjets mounted in the transom flat.

AAI Corporation Minor Calibre Weapons Station (MCWS)

Low speed water operation is accomplished by two hull-mounted Dowty 305 mm diameter waterjets powered by either the diesel or rotary engine.

Operation of the PSD is accomplished by a three man crew using a redundant drive by wire control system. This control system provides vehicle control in both the land and water modes and automatically configures the vehicle for these modes of operation.

Individual stations are provided for driver, vehicle commander/gunner and the troop commander. Seating provisions are provided in the rear troop compartment for a third crew member and 14 troops. The troop compartment is provided with an overhead hatch and a rear ramp.

The AAI developed Minor Calibre Weapons Station (MCWS) is installed in the PSD. The MCWS is operated remotely by the vehicle commander and mounts a 25 mm M242 Chain Gun and a 7.62 mm EX-34 Chain Gun both of which have been developed by the McDonnell Douglas Helicopter Company, the former is already in service with the United States Marines on their LAV-25 (8 × 8) vehicles.

Other equipment installed in the PSD includes a Santa Barbara Research Center fire detection and suppression system, hydraulic/electric bilge pump system, radio system and vehicle intercom system. Navigation of the PSD is accomplished by an onboard GPS navigation system, depth finder and KVH magnetic heading system.

Status: Prototype undergoing trials.

Manufacturer: AAI Corporation
PO Box 126
Hunt Valley
Maryland 21030-0126
USA
Telephone: (301) 666 1400
Telex: 8-7849
Fax: 710 232 1800

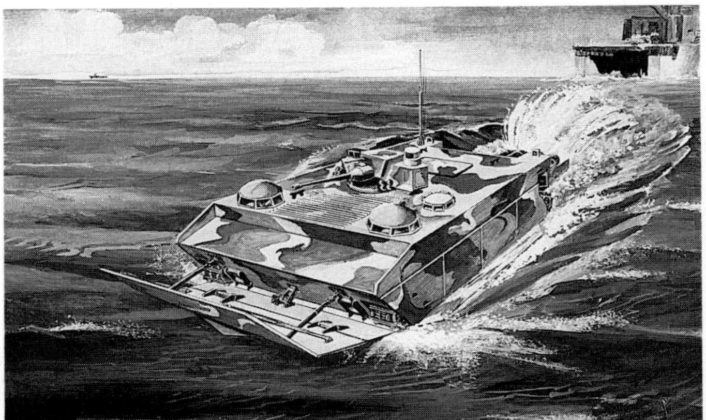

Artist's impression of the AAI Corporation Propulsion System Demonstrator in amphibious mode

SPECIFICATIONS		LENGTH (land)	8.153 m	MAX SPEED	
CREW	3 + 15	LENGTH (water)	11.582 m	land	72.421 km/h
WEIGHT	26 400 kg	WIDTH (land)	3.378 m	water	32.18+ km/h (high speed)
POWER-TO-WEIGHT RATIO		WIDTH (water)	3.835 m		7.24 km/h (low speed)
land	25.8 hp/tonne	HEIGHT	2.032 m	FUEL CAPACITY	1135 l
water	80.6 hp/tonne	GROUND CLEARANCE	0.384 m	PIVOT TURN	land/water
GROUND PRESSURE	0.52 kg/cm²				

Advanced Amphibious Assault Vehicle

Development/Description

The US Marine Corps plans to field a new amphibious vehicle called the Advanced Amphibious Assault Vehicle (AAAV) to replace the current AAV7A1 amphibious assault vehicle in the 1990s. This will use technology obtained in the AAI Automotive Test Rig (qv), AAI High Water Speed Technology Demonstrator (qv) and the AAI Propulsion System Demonstrator (qv).

The AAAV will complement the Textron Marine Systems high-speed Landing Craft, Air Cushion (LCAC).

In September 1989, Naval Sea Systems Command (NAVSEA) issued an AAAV Request for Information (RFI) to contractors and these were returned in December 1988.

Early in 1989 a request for proposals was issued and later a concept exploration (paper study) was awarded. The development teams must be composed of at least two principal team members that will offer on each phase of the AAAV programme as a single entity. Principal team members are defined as those companies that possess the production capability to eventually produce the AAAV system on their own.

The concept exploration/definition phase is divided up into two distinct periods, the Concept Design period and the Sustaining Engineering period.

Once a design team has been eliminated it will be precluded from offering on subsequent phases of the programme as a team. It does not preclude former principal team members, or their subcontractors, from becoming associated with the surviving industry development teams.

Two of the design teams will be selected to enter the demonstration and validation (demval) phase of research and development. During this phase two prototypes will be built, one by each contractor, and competitively tested. The prototypes will undergo three months of contractor tests followed by six months of government conducted DT/OT I testing.

Following the demonstration and validation phase one contractor team will be selected to further develop the AAAV with a full scale engineering development (FSED) contract. Under this contract an additional 15 test vehicles will be built and delivered over a four year period. Of these 15 vehicles, which will be used to carry out DT/OT II trials, five will be built jointly by the contractor team and five prototypes built by each principal team member.

The final production contract will be awarded to one of the full scale development team members.

The requirement is for a vehicle with a three man crew carrying 18 fully equipped troops which will have a high water-speed and have a land speed similar to that of the M1A1 MBT. It will be fitted with a power weapon station which must be capable of defeating the BMP type vehicle that will be in service in the 1990s.

Late in 1990 it was reported that the US Marine Corps was considering reducing the maximum high water speed of the AAAV from 25 to 20 knots in order to keep the cost of the programme at $2.6 million per vehicle.

The two teams competing in the project definition phase are General Dynamics, Land Systems Division teamed with AAI and FMC Corporation, Ground Systems Division.

FMC Corporation is believed to be pursuing a hydrofoil assisted planning hull design that has demonstrated more than 33 knots in a half-scale demonstrator model while the General Dynamics/AAI design is believed to utilise a more typical planning hull.

Status: Concept exploration. It was originally envisaged that some 1400 AAAVs would be procured to replace existing AAV7A1 vehicles on a one-for-one basis. Due to pressure on the US defence budget and changing operational requirements it is now expected that the total requirement will be for less than the original consort.

The original in service date was between 1998 and 1999 but this may now slip to around 2005.

Manufacturer: Not selected, so not yet in production or service.

AAI Automotive Test Rig

Development/Description

Following the decision in 1985 not to proceed with the development of the Landing Vehicle Tracked (Experimental) LVT (X), the US Navy/US Marine Corps embarked on a programme to develop advanced components, systems and technologies that could be applied to the next United States Marine Corps assault amphibious vehicle.

To demonstrate these advanced items a tracked test bed, the Automotive Test Rig (ATR) was developed by AAI for the David Taylor Research Center, Marine Corps Program Office. The objective of the ATR was to demonstrate the technical feasibility of automotive components being investigated for future Marine Corps amphibious vehicles.

The ATR has a crew of three and is fully amphibious. It is powered by a 320 hp Caterpillar Tractor Model 3208T diesel engine coupled to a hydrostatic computer-controlled drive train for land and water propulsion. The final drives are two-speed and a dual cooling system (land and water) is installed.

Suspension either side consists of five dual rubber-tyred road wheels, drive sprocket at the rear, idler at the front and track return rollers, the track is of the wire and link band type.

Water propulsion is by two Dowty two-stage 304 mm diameter waterjets with steering buckets. An unusual feature of the hydro-pneumatic suspension is that it can be fully retracted when the vehicle is afloat to reduce drag. A bilge pump and fire detection system are installed as standard.

Under contract to the Naval Surface Weapons Center, AAI has also completed the prototype of the Unmanned Minor Calibre Weapons Station (MCWS) which was installed on the ATR in the second half of 1986. A photograph of this appears in the entry for the AAI Propulsion System Demonstrator.

SPECIFICATIONS

CREW	3
GROSS VEHICLE WEIGHT	12 700 kg
GROUND PRESSURE	0.50 kg/cm^2
LENGTH	5.258 m
WIDTH	2.692 m
HEIGHT	
(without turret)	1.981 m
GROUND CLEARANCE	0.406 m
MAX SPEED	
land	64.34 km/h
water	8.04 km/h

Status: Has completed testing.

Manufacturer: AAI Corporation
PO Box 126
Hunt Valley
Maryland 21030-0126
USA
Telephone: (301) 666 1400
Telex: 8-7849
Fax: (710) 232-1800

AAI Automotive Test Rig showing amphibious characteristics

AAI Automotive Test Rig fitted with AAI developed Minor Calibre Weapons Station (MCWS)

AAI High Water Speed Technology Demonstrator (HWSTD)

Development

The High Water Speed Technology Demonstrator (HWSTD) has been designed and built by AAI under contract to the David Taylor Research Center (DTRC), Marine Corps Program Office.

The objective of the HWSTD is to demonstrate the feasibility of a tracked vehicle achieving a water speed in excess of 37 km/h (20 knots).

The HWSTD was completed by AAI in 1989 and demonstrated a water speed in excess of 51.8 km/h (28 knots) the same year.

Description

The HWSTD is a tracked amphibious test bed vehicle powered by a German MTU MT-883 diesel engine developing 1609 hp (gross) with land and water propulsion being provided by hydrostatic drive trains. Cooling system is coolant/air for the land mode and coolant/water for the amphibious mode.

In view of the planned limited land operation of the HWSTD, the land drive train is only configured for low speed operation, appropriate for short distance movement over prepared surfaces to and from water test sites.

The HWSTD is equipped with deployable appendages, a bow flap, chine flaps and transom flap and a retractable suspension system to provide a vehicle configuration suitable for high water speed operation.

It is equipped with hydro-pneumatic suspension units and AAI Center developed 406 mm diameter high performance waterjets mounted in the transom flap with steering vanes. The 431 mm wide roller chain band track has also been developed by AAI.

Operation of the HWSTD is accomplished by a two man crew using a redundant drive by wire control system. The control system provides vehicle control in both the land and water modes and automatically configures the vehicle for these modes of operation.

The HWSTD is equipped with an automatic fire detection and suppression system and a diagnostics system for monitoring vehicle subsystem status during operation. Some elements of the HWSTD are fabricated of composite materials. Hydraulic and electric bilge pumps are fitted.

Status: Test bed.

Manufacturer: AAI Corporation
PO Box 126
Hunt Valley
Maryland 21030-0126
USA
Telephone: (301) 666-1400
Telex: 8-7849
Fax: (710) 232-1800

AAI High Water Speed Technology Demonstrator in land mode configuration

AAI High Water Speed Technology Demonstrator in operation

SPECIFICATIONS

CREW	2	LENGTH		MAX WATER SPEED	51.8 km/h
WEIGHT	16 243 kg	land mode	7.137 m	MAX LAND SPEED	16.09 km/h
POWER-TO-WEIGHT RATIO		water mode	10.972 m	PIVOT TURN	land/water
land mode	30.3 hp/tonne	WIDTH		FUEL CAPACITY	397.42 l
water mode	89.9 hp/tonne	land mode	2.794 m		
GROUND PRESSURE	0.51 kg/cm^2	water mode	3.302 m		
		HEIGHT	1.981 m		
		GROUND CLEARANCE	0.363 m		

FMC AAV7A1 Amphibious Assault Vehicle

Note: This vehicle was originally called the LVTP7 and is still referred to by this designation by some users. In 1985 the US Marine Corps changed this to the Amphibious Assault Vehicle 7 (AAV7) and all new production vehicles were built to the AAV7A1 configuration and existing vehicles were upgraded to the new production standard.

Development

In March 1964 the United States Marine Corps issued a requirement for a new LVTP and, after evaluating a number of submitted proposals, a contract for the development of a new LVTP was awarded to the then Ordnance Division of the FMC Corporation. Overall project management of the programme was under the direction of the Naval Ship Systems Command.

Engineering development began in February 1966 and the first of 15 prototypes was completed in September 1967 and delivered to the Marine Corps the following month. Trials with the prototypes, designated the LVTPX12, were completed in September 1969.

In June 1970 FMC was awarded a contract worth $78.5 million for the production of 942 vehicles without weapon stations. When the government-supplied weapon stations were included the unit cost was about $129 000. The first production vehicles, which were designated the LVTP7 (Landing Vehicle, Tracked, Personnel, Model 7) were handed over to the United States Marine Corps in August 1971 and the first unit was equipped with the vehicle by March 1972. Final deliveries were made in September 1974 after which the LVTP5A1 and its variants were phased out of service.

In the United States Marine Corps, AAV7A1s are issued to Assault Amphibian Battalions which consist of a Headquarters and Service Company, and four Assault Amphibian Companies. The Headquarters and Service Company have 15 AAVP7A1s, three AAVC7A1s and one AAVR7A1. The Assault Amphibian Company has four platoons of AAV7A1s, each platoon having 10 AAVP7A1s. The company Headquarters section has one AAVR7A1, three AAVC7A1s and three AAVP7A1s. This gives the battalion a total of 187 AAVP7A1s, 15 AAVC7A1s and five AAVR7A1s. Both the 1st Marine Division (based in California) and the 2nd Marine Division (based in North Carolina) each have one Assault Amphibian Battalion. The 3rd Marine Division is based in the Far East and has two companies of AAV7A1s.

The US Marine Corps has upgraded all of its LVTP7s to the new AAV7A1 configuration and total production/conversion figures for both vehicles are as follows:

	LVT7				AAV7A1			
	Personnel	*Command*	*Recovery*	**Total**	*Personnel*	*Command*	*Recovery*	**Total**
Argentina	19	1	1	21	0	0	0	0
Brazil	0	0	0	0	10	1	1	12
Italy	24	1	0	25	0	0	0	0
Korea, South	53	5	3	61	41	1	0	42
Spain	16	2	1	19	0	0	0	0
Thailand	12	0	0	12	18	2	1	21
USA	0	0	0	0	1153	106	64	1323
Venezuela	9	1	1	11	0	0	0	0
Total	133	10	6	149	1222	110	66	1398

The AAV7A1 is expected to be replaced in United States Marine Corps service by the Advanced Amphibious Assault Vehicle (AAAV) covered in an earlier entry. To enable the current AAV7A1 to be viable through the year 2000 at least, a number of product improvement programmes are already under way and details of some of these are given later in this entry.

Description

The all-welded aluminium hull of the LVTP7 protects the crew from small arms fire, shell splinters and flash-burns.

The driver sits at the front of the hull on the left side and has a single piece hatch cover that opens to the right rear and seven vision blocks for all-round observation. An M24 infra-red periscope can be fitted in the forward part of the driver's hatch cover for driving at night.

The commander sits behind the driver and has a single piece hatch cover that opens to the rear and seven vision blocks for all-round observation. Forward of the commander's station is an M17C periscope which can be extended to enable the commander to see forward over the driver's hatch cover.

The engine is in the forward part of the hull on the centre-line and power is transmitted to the FMC HS-400 transmission via a torque converter with lock-up. The transmission combines shifting, steering and braking functions in one unit and gives four gears forward and two in reverse. The engine can be removed from the vehicle in 45 minutes and can be run outside for test purposes if required. Power is transmitted to the drive sprockets through hull-mounted final drives. A PTO mounted on the converter housing supplies power for the water jets and the cooling fan through electro-hydraulically controlled clutches.

Mounted over the powerpack is a fan and radiator. Air is drawn in and discharged through ballistic grilles in the roof. When afloat hydraulically actuated doors below the grilles seal off the air openings. Air for the engine and troop compartment then enters the vehicle through a hydro-dynamically actuated air valve mounted on the top of the hull to the right of the driver. Cooling takes place in the contact cooler which is an integral part of the hull floor.

The armament installation is beside the engine on the right side of the hull. The turret has a single piece hatch cover that opens to the rear, eight vision blocks, a sight with a magnification of ×1 and ×8 and one direct ring sight. The turret is armed with a 12.7 mm (0.50) M85 machine gun with two rates of fire, 1050 rpm (high) and 450 rpm (low). Of 1000 rounds of 12.7 mm ammunition carried 400 are for ready use. The machine gun has an elevation of +60° and a depression of -15°. Turret traverse, which is electro-hydraulic, is a full 360°.

The 25 fully equipped marines sit on three bench seats in the troop compartment at the rear of the hull, one in the centre and one on either side. The centre seat can be removed and the two side ones folded up to allow up to 4536 kg of cargo to be carried. The removable centre seat stows on the left bulkhead when not required. The marines enter and leave the vehicle through a power-operated ramp at the rear of the hull, which has a door in the left side of the ramp in case of ramp failure and a single vision block. Over the top of the troop compartment are three torsion-spring-balanced roof hatches.

The tube-over-bar suspension consists of six dual rubber-tyred road wheels with the drive sprocket at the front and the idler at the rear. There are no track return rollers. The first and last road wheel stations have a hydraulic shock absorber. The single pin, rubber bushed tracks have a moulded rubber top surface and replaceable rubber pads.

The LVTP7 is fully amphibious without preparation, being propelled in the water by two water jets which are mounted one either side of the hull at the rear. The water jets are driven through right-angled gearboxes on top of the sponsons and drive shafts that extend to the rear of the hull. The water jets, which are aluminium, are mixed flow pumps with special exhaust nozzles to maximise thrust and each pumps 52 990 litres of water per minute. At the rear of each pump is a hinged steering deflector which reverses the direction of the jet when closed. Maximum speed forwards is 13.51 km/h and maximum speed in reverse is 7.24 m/h. The tracks can also be used to propel the vehicle when afloat.

Four bilge pumps are fitted: two electric (each with a capacity of 379 litres/min) and two hydraulic (each with a capacity of 435 litres/ min).

The LVTP7 has no NBC system; it is fitted with night driving lights. A number of kits have been developed for use with the LVTP7, including one for converting the vehicle to an ambulance that can carry six stretcher patients, navigation kit, visor kit for the driver and a winterisation kit which keeps the crew and personnel compartment heated when the outside temperature is down to -54° C, the normal heater being effective down to -32° C.

In March 1988, following two years of extensive tests, the US Marine Corps selected the KVH Industries MV103 militarised digital compass for use on its AAV7A1 family of vehicles.

Modernisation

AAV7A1

In March 1977 the FMC Corporation was awarded a contract to convert 14 LVTP7s into prototypes of a new configuration known as the LVTP7A1. The entire Marine Corps LVTP7 fleet was converted between 1982 and 1986. The emphasis on the conversion was to increase the vehicle's reliability, availability, maintainability and durability while reducing life cycle costs. New items added to the LVTP7 to bring it up to the LVTP7A1, subsequently redesignated AAV7A1, standard are:

A powerpack using the Cummins VT400 engine
Off-board and on-board diagnostic instrumentation (fault isolation)
Smoke-generating capability

LVTP7 with 12.7 mm turret-mounted machine gun and with raised commander's M17C periscope to rear of driver's position

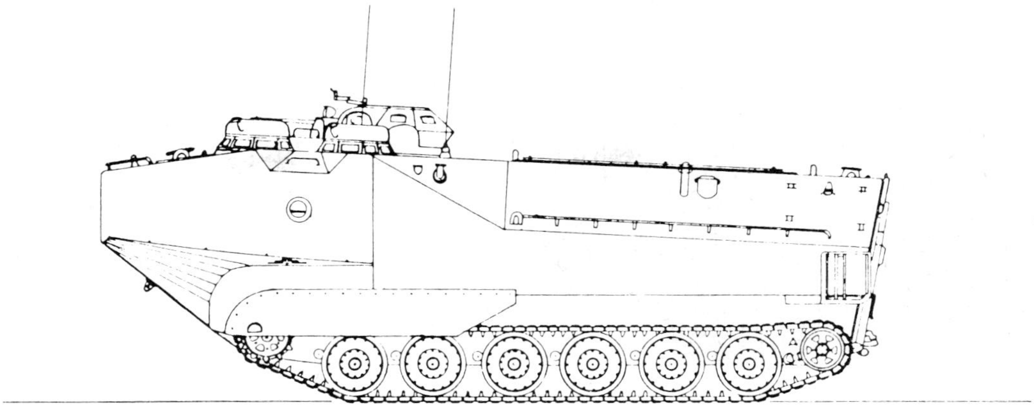

LVTP7 Amphibious Assault Vehicle

Passive night driving and night firing devices
Improved fire suppression system
Secure voice radio installation
Provision to mount Position Location and Reporting System (PLARS) installation
Improved troop/crew compartment ventilation
Improved hydraulic and electrical system
Improved electric weapon station
Improved suspension system.

In addition to previously announced improvements, the AAV7A1 has a non-integral fuel tank, watertightness improved by reworking seals which surround air intakes, exhaust plenums and cargo hatches, main wiring harness segmented for rapid repair without requiring an entirely new harness, removable headlights, troop commander's hatch and vision block raised for better visibility over the more forward driver's hatch.

The programme was known as the LVTP7 Service Life Extension Program (SLEP) with $10.5 million requested in FY80, $47.3 million in FY81 (25 conversions plus $7.2 million for RDT & E), $171.3 million in FY82 (149 conversions), $116 million in FY83 (307 conversions), $150.5 million in FY84 (263 conversions), and $115.9 million in FY85 (240 conversions).

In addition to the conversion of LVT7 to AAV7A1 configuration (this conversion programme also applies to the other two members of the family, AAVC7 and AAVR7), funding was provided to build new vehicles to AAV7A1 configuration; $12 million in FY81, $61.6 million (30 vehicles) in FY82, $116.0 million (146 vehicles) in FY83 and $115.3 million (153 vehicles) in FY84. With the later funding the additional 329 AAV7A1s for the Maritime Prepositioning Ships (MPS) was completed. In FY85 a further four vehicles were ordered at a cost of $3.5 million.

The first production AAV7A1 was handed over to the US Marine Corps in August 1983. By mid 1984 production was running at about 50 vehicles a month with final deliveries made in 1986.

Further Improvements

The US Marine Corps is considering a number of additional improvements for the AAV7A1 including:

Driver-operated extendable/retractable bow flap for increased water speed, reduced spray, improved visibility for driver during amphibious operations, and to correct negative trim problem when vehicle is lightly loaded

Improving the turret-mounted M85 machine gun feed chute to reduce stoppage rates and reload times

Universal weapon mount for TOW or Dragon ATGW, 40 mm grenade launcher, M2 12.7 mm (0.50 calibre) or 7.62 mm M60/M60D machine gun

Improved Aluminium Track which offers a 23 per cent weight reduction over existing track.

With the cancellation of the LVT(X), the AAV7A1 is now being studied for further improvements to meet the USMC Assault Amphibian Vehicle (AAV) role well into the late 1990s.

The AAV7A1 Product Improvement Program commenced with the initial fielding of the Upgunned Weapons Station during 1987 and will continue through the mid-1990s. Upgunned Weapons Stations, Bow Planes and Magnetic Heading Systems were fully fielded by 1990 as will the Automatic Fire Sensing and Suppression System. The Enhanced Appliqué Armor Kits and Improved Transmissions will be fully fielded by 1992.

Appliqué Armour

Early in 1986 the US Marine Corps Systems Acquisition Review Council considered the programme and recommended adopting as Phase I an armour kit for forward deployed AAV7A1s as well as those with the maritime pre-positioning assets.

The first contract will be for 189 kits of P900 (this is also fitted to US Army M113s) which consists of a set of punched and perforated armour plates attached to the sides and sloping parts of the vehicle.

It is expected that the remaining 1136 vehicles will be fitted with an enhanced appliqué armour kit which will protect the sides, slopes, top surfaces and all three hatches. The enhanced armour package has two layers, the first of which is about 150 mm away from the hull with an outer layer being a further 25 mm away. The enhanced appliqué armour package is known as the Phase II Program, which is also known as the Enhanced Appliqué Armor Kit (EAAK). This is required to reflect the following specifications:

14.5 mm armour piercing (AP) at 300 m, 95 per cent probability of no penetration
12.7 mm AP, 95 per cent probability of no penetration
155 mm AP a 15 m overhead, 99 per cent probability of no penetration.

The RAFAEL Armament Development Authority was selected as the sole winner of the open competition to design and fabricate three pre-production EAAKs for ballistic and field testing.

Development of the EAAK was completed in 1989 and subsequently delivered three complete pre-production prototype systems to the US Marine Corps for extensive tests. Following the successful completion of the performance and operational test the production option of the contract was exercised. Production of the EAAKs commenced in 1991.

AAV7A1 of the US Marine Corps, this can be distinguished from the earlier models by its square headlamps recessed into the nose

AAV7A1 fitted with the RAFAEL Armament Development Authority Enhanced Applique Armor Kit (EAAK)

AAV7A1 fitted with Block I Appliqué Armour Kit (AAK) (Paul Handel)

Model of AAV7A1 with roof hatches open showing CATFAE in rear troop compartment

The turret will also be fitted with appliqué armour and drums will be fitted to the upper sides of the hull to give a smoke-laying capability.

It is expected that each armour kit will weigh about 2000 kg which will not result in any performance loss while the vehicle is afloat.

In the future, the AAV7A1 and its variants are to be fitted with a bow vane to improve amphibious characteristics as well as the automatic fire detection and suppression system.

Variants

AAVC7A1 Command Vehicle

This had the development designation LVTCX2 and is similar to the basic AAV7 except that its machine gun turret has been replaced by a simple hatch cover similar to that fitted at the driver's and commander's position. Armament consists of a pintle-mounted 7.62 mm M60D machine gun with 1000 rounds of ammunition. The AAVC7A1 also has a navigation system, a loaded weight of 21 537 kg and an unloaded weight of 19 785 kg.

AAVR7A1 Recovery Vehicle

This had the development designation of the LVTRX2 and like the AAVC7A1 has had the machine gun turret replaced by a 7.62 mm M60D machine gun. Equipment carried includes benches, tools, air compressor, battery charger, generator, 14-35 VDC MIG welder, and a portable welding kit. Mounted on the roof of the AAVR7A1 is a hydraulic crane with an extending boom which can lift 2722 kg. The boom can be elevated from 0 to +65°. The vehicle also has a recovery winch with 84.73 m of 19 mm diameter cable with a breaking strength of 22 044 kg. The two-speed (high and low) winch has a maximum capacity of 13 605 kg on a bare drum at a low speed and 1878 kg on a full drum at high speed. The AAVR7A1 has a crew of five, a loaded weight of 23 601 kg and an unloaded weight of 22 563 kg. Full details of the AAVR7A1 are given in *Jane's Military Vehicles and Logistics 1991-92*, page 39.

AAVR7A1 recovery vehicle with crane stowed in travelling position

AAV7A1 with Mine Plough

The Ground Systems Division of FMC has a demonstration model of a track-width mine system which can be mounted on the front of a AAV7A1 which will proof a lane previously cleared by a M58/M59 linear demolition charge or CATFAE. Its secondary role is to provide a mine-clearing capability from surf zone exit point, across the beach and inland, in support of amphibious operations. A contract for full-scale engineering development was awarded late in 1988. A production decision is not due until trials are completed in 1990. Caterpillar Incorporated has built the prototype of a Track Width Mine Plough in response to a US Marine Corps requirement. Details of this are given in *Jane's Military Vehicles and Logistics 1991-92.*

AAVP7A1 with one-man Upgunned Weapons Station armed with 40 mm grenade launcher and 12.7 mm M2 HB machine gun

AAV7A1 with VEMASID

For trials purposes the AAV7A1 has been fitted with the Vehicle Magnetic Signature Duplicator (VEMASID) under development for the US Army. It is probable that there will be a dedicated Marine Corps engineer version of the AAV7A1 with CATFAE, plough and VEMASID.

AAV7A1 with Mine Clearance System Kit

With the cancellation of the LVTE7 many years ago, the US Marine Corps had no vehicles with a mine clearing capability.

Under development is a modular line-charge system which can be bolted into the cargo compartment of the AAV7A1. This is a modified M125 mine clearance system which uses a Mark 22 rocket motor fired from an elevated launcher rail. When launched, the rocket pulls a linear charge to a range of about 550m. Once the charge is on the ground it is detonated and this should detonate about 80 per cent of the mines for a distance of 7 m either side of the line. Production AAV7A1 vehicles have recently been adapted for mounting this Mine Clearance System Kit.

AAV7A1 with CATFAE

This is an AAV7A1 series vehicle fitted with a Catapult Launched Fuel Air Explosive (CATFAE) in the rear troop compartment. CATFAE will be capable of operating from a stationary position or on the move, afloat or ashore. The system can fire single rounds at a point target or volleys of rounds (up to 21 in total) to create a cleared lane through a minefield. Breaching a 20 m × 300 m lane through a minefield can be accomplished in 90 seconds or less with vehicle speeds up to 6 knots afloat and 24 km/h ashore. The CATFAE is still under development and full details are given in *Jane's Military Vehicles and Logistics 1991-92*, page 246.

AAV7A1 with Upgunned Weapons Station

After studying bids from Cadillac Gage and the Ordnance Division of FMC Corporation, the US Navy Sea Systems Command, acting for the US Marine Corps, placed an order with Cadillac Gage early in 1986 for the supply of 240 Upgunned Weapons Stations for installation on AAVP7A1s.

The total value of this contract was $17.236 million and the first turrets were delivered in November 1986.

The Upgunned Weapons Station is armed with a 40 mm grenade launcher and a 12.7 mm M2 HB machine gun, has electric traverse and manual elevation, and replaces the turret armed with a single 12.7 mm M85 machine gun.

The original contract for 240 Upgunned Weapons Stations included an option for a further 100 turrets, this was exercised in January 1987 at a cost of $5.823 million.

In September 1986 Optic Electronic Corporation of Dallas, Texas, was awarded a contract by Cadillac Textron to supply a modified M36E1 day/night/laser rangefinder sight for this turret.

In September 1988 AV Technology Corporation was awarded a competitive re-buy contract for the Upgunned Weapons Station for the US Marine Corps.

AAV7 with hydro-pneumatic suspension system

Undergoing trials in 1989 was an AAV7 series amphibious assault vehicle fitted with the Cadillac Gage Textron 6K hydro-pneumatic suspension system, also referred to as the in-arm suspension system, as a replacement for the current torsion bar suspension system employed on all production AAV7A1 series vehicles. The contract was awarded to Cadillac Gage Textron by the David Taylor Research Center.

Status: Production complete. In service with Argentina, Brazil, Italy, South Korea, Spain, Thailand, USA and Venezuela. Production can be resumed if further orders are placed.

Manufacturer: FMC Corporation, Ground Systems Division
2830 De La Cruz Boulevard
Santa Clara
California 95052
USA
Telephone: (408) 289 0111
Telex: 6714210
Fax: (408) 289 2150

Cadillac Gage Textron 6K in-arm suspension system fitted to an FMC AAV7 series amphibious assault vehicle

SPECIFICATIONS (data in square brackets relates to AAV7A1 where different from basic vehicle)

CREW	3 + 25
COMBAT WEIGHT	22 838 [23 991] kg
UNLOADED WEIGHT	17 441 [18 663] kg
POWER-TO-WEIGHT RATIO	17.51 [16.67] hp/tonne
GROUND PRESSURE	0.57 [0.56] kg/cm²
LENGTH	7.943 m
WIDTH	3.27 m
HEIGHT	
overall	3.263 m
to turret roof	3.12 m
GROUND CLEARANCE	0.406 m
TRACK	2.609 m
TRACK WIDTH	533 mm
LENGTH OF TRACK ON GROUND	3.94 m
MAX SPEED	
road, forwards	64 [72.42] km/h
water, waterjets	13.5 [13.2] km/h
water, tracks	7.2 km/h
FUEL CAPACITY	681 [647] l
MAX RANGE (land at 40 km/h)	482 km
MAX ENDURANCE (water)	7 hours at 2600 rpm
FORDING	amphibious
GRADIENT	60%
SIDE SLOPE	40%
VERTICAL OBSTACLE	0.914 m
TRENCH	2.438 m
TURNING RADIUS	pivot
ENGINE	Detroit Diesel model, 8V-53T, 8-cylinder, water-cooled, turbo-charged diesel developing 400 hp at 2800 rpm [Cummins VT400, 8-cylinder, water-cooled, turbo-charged diesel developing 400 hp at 2800 rpm]
TRANSMISSION	HS-400 [HS-400-3A1] with hydraulic torque converter, 4 forward and 2 reverse ratios
SUSPENSION	tube-over-bar
ELECTRICAL SYSTEM	24 V
BATTERIES	4 × 12 V 6TN
ARMAMENT (main)	1 × 12.7 mm MG
SMOKE-LAYING EQUIPMENT	none [yes]
AMMUNITION (main)	1000
GUN CONTROL EQUIPMENT	
Turret power control	electro-hydraulic/manual [electric/manual]
by commander	no
by gunner	yes
Gun elevation/depression	+60°/-15°
Turret traverse	360°
Gun stabiliser	none
ARMOUR	
Hull sides	31-45 mm
Hull top	30 mm
Hull floor	30 mm
Hull rear	35 mm
Ramp outer	25.4 mm
Ramp inner	12.7 mm

LVTP5A1 Armored Amphibious Assault Vehicle

Development

Design studies began in January 1951 and the first prototype, designated the LVTPX1, was completed in August the same year. Production of the LVTP5 (Landing Vehicle, Tracked, Personnel, Model 5) began in 1952 and continued to 1957 by which time 1124 had been built. These were built by Baldwin-Lima-Hamilton (91), FMC Corporation (313), Ingersoll (239), Pacific Car and Foundry (56) and St Louis Car Manufacturing Company (425).

In the 1960s all vehicles were fitted with a box type snorkel over the engine compartment, which together with other minor modifications, resulted in the designation being changed from the LVTP5 to the LVTP5A1. Various experiments were tried to improve the performance of the LVTP5 but in the end a new vehicle called the LVTP7 was designed by the FMC Corporation. This entered service in 1971 and by 1974 the LVTP5 family had been phased out of service. Some were transferred to the Philippines and Taiwan where they remain in service today. More recently Chile has acquired a quantity of these vehicles.

Description

The hull of the LVTP5 is of armour plate ranging from 6.35 to 15.87 mm thick welded and reinforced with internal framing members to form a rigid, watertight unit. The hull is barge shaped and has an inverted V-shaped bow and bottom for more efficient water operation. The ramp mounted in front of the LVTP5 is hydraulically operated and is composed of an inner and outer steel plate separated by equally spaced webs. The ramp is hinged to the hull and heavy live rubber seals, cemented around the hull opening, provide a watertight seal when the ramp is closed.

The driver sits at the front of the hull on the left and has four M17 periscopes, one M17C periscope and a single piece hatch cover. The commander sits at the front of the hull on the right and also has four M17 periscopes, one M17C periscope and a single piece hatch cover. The machine gun turret is between and slightly forward of the commander's and driver's positions and is armed with a 7.62 mm (0.30) M1919A4 machine gun with an elevation of +60°, a depression of -15° and a total traverse of 360°. The gunner has five vision blocks and an M25C periscope sight for aiming the machine gun. Forward and either side of the machine gun turret is a single M17 periscope facing forwards.

All ventilation openings, bilge discharge outlets and access hatches are in the top deck. In each side of the hull is an emergency escape hatch and a single vision block.

Over the top of the troop compartment is the cargo hatch, which is 2.438 m long and 2.108 m wide and is made up of two double folding sections which are hinged to the outer sides of the top deck and spring-loaded to assist in opening.

The 34 marines sit on four troop seats of eight each with the remaining two men seated on the machine gunner's platform. In an emergency 45 standing marines can be carried. When being used for cargo the LVTP5 can carry 5443 kg when afloat and 8165 kg when on land, or a 105 mm towed howitzer, its crew and 90 rounds of ammunition. The crew compartment is ventilated by three manually operated vents and an electric blower mounted in the roof.

The engine compartment is at the rear of the hull and contains the engine, cross-drive transmission, drop gears, final drives and all the accessories necessary for the operation of the engine. Limited access to the engine compartment is via a hatch in the transverse bulkhead. The main engine hatch is bolted to the top deck over the engine compartment and may be removed to permit the complete powerpack (engine and transmission) to be removed from the vehicle.

The engine is mounted on the vehicle's centreline in the engine compartment and is coupled to an Allison Transmission Division of General Motors model CD-850-4 transmission which is a combined transmission, differential and steering unit. Torque received from the engine is delivered through the cross-drive transmission, drop gears, and single-speed final drives to the track driving sprocket.

The suspension consists of nine pairs of dual steel bogie wheels suspended from rubber torsion spring suspension assemblies on each side of the vehicle. Each wheel of each pair of road wheels consists of a hub with one steel rim and one demountable solid rubber-tyred rim. Within each underwater track return channel are the adjustable front idler, five track return rollers, track return skids and rear drive sprocket.

In the water the LVTP5 is propelled by its tracks, each of which is made of 134 blocks held together by inner and outer 25 mm diameter pins. Each track block has an inverted water grouser.

Kits available for the LVTP5A1 included infra-red, stretcher (carrying 12 stretcher patients), navigation and winterisation.

LVTH6 fitted with two-man turret armed with 105 mm Howitzer (US Marine Corps)

LVTP5A1 armoured amphibious assault vehicle (US Marine Corps)

Variants

LVTH6 (Landing Vehicle, Tracked, Howitzer, Model 6)

This is fitted with a turret armed with a 105 mm M49 howitzer, a 7.62 mm (0.30) machine gun mounted coaxially with the main armament and a 12.7 mm (0.50) M2 HB machine gun mounted on the roof for anti-aircraft defence. When afloat, 100 rounds of 105 mm, 1000 rounds of 7.62 mm and 1050 rounds of 12.7 mm ammunition are carried. A total of 210 were built between 1953 and 1957, 175 by Ingersoll and 35 by Pacific Car and Foundry. They were retired from the United States Marine Corps without replacement as the 105 mm howitzer version of the LVTP7 was never built.

LVTC5 (Landing Vehicle, Tracked, Command, Model 5)

This is similar to the basic LVTP5 but has a modified cargo compartment fitted with additional communications equipment, tables and mapboards. It has a command crew of nine plus its normal crew of three, and is distinguishable from the basic vehicle by its additional radio antennas. A total of 58 LVTC5s were converted from basic LVTP5s by Ingersoll in 1955 and are also known as the LVTP5 (CMD).

LVTR1 (Landing Vehicle, Tracked, Recovery, Model 1)

In total, 65 of these were built by FMC Corporation from 1954. The vehicle is used for the recovery and maintenance of other members of the LVTP5 family and is fitted with a generator, welding equipment, air compressor, boom with a capacity of 3175 kg and a winch with a maximum capacity of 20 412 kg. This model does not have a machine gun turret.

LVTE1 (Landing Vehicle, Tracked, Engineer, Model 1)

This was a specialised model and was fitted with a special dozer blade at the front of the hull and a rocket-propelled mine-clearing system. It is no longer in service and was retired without replacement. There were many experimental models of the LVTP5 family including the LVTAAX1 anti-aircraft system armed with the turret of the M42 twin 40 mm self-propelled anti-aircraft gun.

Taiwanese LVTP5/LVTP6

In 1982 it was reported that Taiwan was to replace the LV-1790-1 V-12 petrol engine of the LVTP5/LVTP6 with a diesel engine and that initial trials with vehicles fitted with three different powerpacks had already commenced in Taiwan. These are believed to include NAPCO (GM Detroit Diesel 12V-71 with existing transmission), Summit Engineering (12V-71 diesel with CD-50-6 transmission) and Teledyne Continental (Teledyne Diesel with CD-50-6 transmission).

Status: Production complete. In service with Chile (26), Philippines (100) and Taiwan (366).

Manufacturers: Baldwin-Lima-Hamilton, FMC Corporation, Ingersoll, Pacific Car and Foundry and St Louis Car Manufacturing Company.

SPECIFICATIONS (LVTP5)	
CREW	3 + 34
COMBAT WEIGHT	
land	30 144 kg
water	37 422 kg
UNLOADED WEIGHT	29 801 kg
POWER-TO-WEIGHT RATIO	20.177 hp/tonne
GROUND PRESSURE	0.65 kg/cm²
LENGTH	9.042 m
WIDTH	3.567 m
WIDTH OVER TRACKS	3.491 m
HEIGHT	
to turret top	2.921 m
to hull top	2.565 m
TRACK WIDTH	521 mm
LENGTH OF TRACK ON GROUND	5.892 m
GROUND CLEARANCE	0.457 m
MAX SPEED	
road, forwards	48.28 km/h
road, reverse	12.87 km/h
water, forwards	10.94 km/h
water, reverse	3.219 km/h
FUEL CAPACITY	1726 l
MAX ROAD RANGE	306 km
FORDING	amphibious
GRADIENT	70%
SIDE SLOPE	60%
VERTICAL OBSTACLE	0.914 m
TRENCH	3.657 m
ENGINE	Continental LV-1790-1 V-12 water-cooled petrol developing 810 hp at 2800 rpm
TRANSMISSION	Allison Transmission Division of General Motors cross-drive model CD-850-4B or 5 hydraulic torque converter with 2 forward and 1 reverse gears
SUSPENSION	Torsilastic
ELECTRICAL SYSTEM	24 V
BATTERIES	4 × 12 V 6TN
ARMAMENT (main)	1 × 7.62 mm MG
SMOKE-LAYING EQUIPMENT	none
AMMUNITION (main)	2000
GUN CONTROL EQUIPMENT	
Turret power control	manual
by commander	no
by gunner	yes
Gun elevation/depression	+60°/-15°
Turret traverse	360°
ARMOUR	
Hull ramp outer	9.52 mm
Hull ramp inner	15.87 mm
Hull sides	12.7-15.87 mm
Hull top	6.37 mm
Hull floor	9.52-15.87 mm
Hull rear	9.52 mm
Turret	9.52 mm

YUGOSLAVIA

BVP M80A Mechanised Infantry Combat Vehicle

Development

The BVP M80A mechanised infantry combat vehicle is a further development of the M-980 (although in early 1991 the SDPR redesignated this vehicle the M-80) and was shown in public for the first time at the Cairo defence equipment exhibition held in late 1984. It is believed that the BVP M80A entered production in 1980 as the replacement for the M-980 vehicle and that about 600 vehicles had been built by early 1991.

Description

The overall layout of the BVP M80A is identical to that of the M-980 but it is slightly heavier (14 tonnes compared to 13 tonnes), slightly wider and higher, has a greater ground clearance (0.4 m compared to 0.3 m) and has a slightly higher road speed (64 km/h compared to 60 km/h). It has a more powerful engine which in addition to giving the higher road speed gives an improved power-to-weight ratio.

Standard equipment includes an NBC system, fire extinguishing equipment, system for eliminating gases in the vehicle when the weapons are fired, heater, bilge pumps and with preparation the vehicle is fully amphibious being propelled in the water by its tracks. It can also lay its own smoke screen by injecting diesel fuel into the exhaust outlet on the right side of the hull.

Variants

BVP M80AK Infantry Fighting Vehicle

This is essentially the BVP M80A previously described but fitted with a new one-man turret armed with a 30 mm M86 cannon, 7.62 mm coaxial machine gun, twin launcher for Yugoslav built Sagger ATGW and two two-round smoke dischargers that fire to the front of the vehicle.

Turret traverse is a full 360° with weapon elevation from -10° to +65°. Infantry weapons carried are four 64 mm rocket launchers, two 7.62 mm machine guns, six 7.62 mm automatic rifles and a 26 mm signal pistol.

Specifications are almost identical to the BVP M80A except for a combat weight of 14 400 kg, height to turret roof 2.398 m, height over anti-tank missiles 2.87 m, track 2.526 m, length of track on ground 3.30 m, maximum road speed 65 km/h and maximum water speed 6.8 km/h.

BVP M80A1 anti-aircraft system

This has a new two-man power-operated turret armed with twin 30 mm automatic anti-aircraft cannon with the three-man crew consisting of commander and gunner in the turret and driver in the hull.

BVP M80A KC company commander's vehicle

This is the standard vehicle modified for use by the company commander with additional communications equipment installed.

BVP M80A KB battalion commander's vehicle

This is the standard vehicle modified for use by the battalion commander with additional communications equipment installed.

BVP M80A LT anti-tank vehicle

Standard M80A with turret removed and fitted with a new turret armed with six Yugoslav built Sagger ATGWs in the ready to launch position, these have a maximum range of 3000 m. Mounted between the two banks of missiles is a 7.62 mm machine gun. This pod is probably identical to that fitted to the BOV-1 (4 × 4) ATGW carrier (qv).

BVP M80A Sn ambulance

This is the standard vehicle with its turret removed and plated over and modified for use in the ambulance role. There is a single oblong hatch in the roof that opens left and a single door in the hull rear. Its three-man crew consists of driver and three medical orderlies and it can transport four stretcher patients, four seated and two stretcher patients or eight seated patients.

Close-up of turret of BVP M80A MICV showing 20 mm cannon, 7.62 mm coaxial machine gun and twin launcher for Sagger-type ATGWs (Christopher F Foss)

BVP M80AK IFV in travelling configuration with trim vane folded back onto glacis plate

BVP M80A MICV with trim vane on glacis plate retracted, firing port covers in position and Sagger ATGWs ready to launch

Status: In production. In service with Yugoslavia.

Contractor: Federal Directorate of Supply and Procurement (SDPR)
PO Box 308
9 Nemanjina Street
Belgrade
Yugoslavia
Telephone: 621 522
Telex: 11360

BVP M80A MICV from rear with Sagger ATGW covered and showing firing ports and vision blocks in rear doors

SPECIFICATIONS		FUEL CAPACITY	510 l	ARMAMENT	
CREW	3 + 7	MAX RANGE		main	1 × 20 mm cannon
COMBAT WEIGHT	14 000 kg	road	500 km	coaxial	1 × 7.62 mm MG
POWER-TO-WEIGHT RATIO	22.5 hp/tonne	cross-country	300-350 km	other	twin Sagger ATGW launcher
		FORDING	amphibious		
GROUND PRESSURE	0.67 kg/cm²	GRADIENT	66%	AMMUNITION	
LENGTH	6.42 m	SIDE SLOPE	55%	20 mm	1400
WIDTH	2.995 m	VERTICAL OBSTACLE	0.8 m	7.62 mm	2000
HEIGHT OVERALL		TRENCH	2.4 m	Sagger	2 + 2
(turret roof)	2.20 m	ENGINE	FAMOS 10V003 4-stroke 10-cylinder direct-injection diesel developing 315 hp at 2500 rpm	TURRET TRAVERSE	
(with ATGW)	2.67 m			powered/manual	360°
GROUND CLEARANCE	0.4 m				
TRACK	2.526 m				
LENGTH OF TRACK ON GROUND	3.30 m	TRANSMISSION	manual, 5 forward and 1 reverse gears		
MAX SPEED					
road	64 km/h	CLUTCH	mechanical, dry plate		
water	7.8 km/h	SUSPENSION	torsion bar		
		ELECTRICAL SYSTEM	24 V		

M-80 Mechanised Infantry Combat Vehicle

Development

The M-80 Mechanised Infantry Combat Vehicle was seen for the first time in public during a parade in Belgrade in May 1975. It has been designed in Yugoslavia but uses a number of foreign components; for example the engine is the same as in the French AMX-10P armoured infantry carrier, the ATGW is the Soviet Sagger manufactured in Yugoslavia and the 20 mm cannon is manufactured under licence from Switzerland. In previous editions of *Jane's Armour and Artillery* this was referred to as the M-980, but in early 1991 the SDPR changed its designation to the M-80 mechanised infantry combat vehicle.

Description

The hull of the M-80 is made of all-welded steel, although some reports have indicated that aluminium is used in some non-critical areas. Over the front arc of the M-80 its armour provides complete protection against 20 mm armour-piercing projectiles at a range of 100 m while the remainder of the vehicle's armour provides protection against 7.62 mm armour-piercing projectiles fired from a range of 100 m. The M-80 carries 10 men: driver and turret gunner who remain with the vehicle at all times, vehicle commander, and seven fully equipped infantrymen. The driver sits at the front of the hull on the left and has a single piece hatch cover that lifts and swings to the right to open. In front of his hatch are three periscopes, the centre one of which can be replaced by an infra-red or image intensification periscope for driving at night.

The squad commander sits behind the driver and has a cupola that can be traversed through a full 360°, in the forward part of which is a single periscope. The cupola has a single piece hatch cover that opens forwards and can be locked vertical if required. An infra-red searchlight mounted externally on the forward part of the commander's cupola can be operated from inside the vehicle.

The engine is mounted to the right of the driver with the air-inlet and air-outlet louvres in the roof and the exhaust outlet in the right side of the hull. It is assumed that the transmission is the same as that fitted to the French AMX-10P.

The one-man turret is to the rear of the engine compartment and offset to the right of the vehicle's centreline. The turret is armed with an HS 804 20 mm cannon which has a cyclic rate of fire of 800 rds/min and is fed from a 50 round drum magazine. The cannon fires an HE projectile weighing 0.122 kg with a muzzle velocity of 880 m/s or an AP projectile weighing 0.142 kg with a muzzle velocity of 880 m/s which will penetrate 18 mm of armour at an incidence of 0° at a range of 500 m. Maximum anti-aircraft range is 1500 m and maximum range when used in the ground role is 2000 m.

Mounted coaxially to the right of the 20 mm M55 cannon is a 7.62 mm machine gun with a cyclic rate of fire of 1200 rpm, which is fed from a 50 or 300 round belt. Both weapons have a maximum elevation of +75° and a depression of -5°.

Mounted to the rear of the turret on the right side are two launcher rails for the Soviet Sagger ATGW. This missile weighs 11.29 kg and has a maximum range of 3000 m. It is estimated that a further two to four missiles are carried in reserve in the hull, although the gunner has to expose himself in order to load another missile.

The turret is provided with a day sight for engaging ground targets (magnification of ×1.1 and a 40° field-of-view), night sight (range of 800 m, magnification of ×7.3 and 7° field-of-view). There are three vision blocks in the left side and rear of the turret. In the turret roof on the left side are two forward-facing periscopes and a single piece hatch cover that opens to the rear.

The troop compartment is at the rear of the vehicle and the infantrymen enter and leave by two doors in the rear of the hull, each of which has a firing port and a vision block. In each side of the hull are three firing ports each with a periscope in the roof of the vehicle for aiming the weapon. Over the top of the troop compartment are two oval roof hatches hinged in the centre, which can be locked vertical.

The torsion bar suspension consists of five single rubber-tyred road wheels that are similar to those used on the French AMX-10P, with the drive sprocket at the front and idler at the rear. There are two track return rollers and the top of the track is covered by a skirt. Hydraulic shock absorbers are fitted at the first and last road wheel stations.

The M-80 is fully amphibious, being propelled in the water by its tracks. Before entering the water a trim vane, which is stowed on the glacis plate when not in use, is erected at the front of the hull. Standard equipment includes an NBC system which allows the vehicle to stay in a contaminated area for 10 hours, smoke-laying system, central fire-extinguishing system with automatic and manual operation and internal temperature at +15° C at ambient temperature of -15° C.

M-80 MICV from front with twin Sagger ATGWs

Status: Production complete. In service with Yugoslavia.

Contractor: Federal Directorate of Supply and Procurement (SDPR)
PO Box 308
9 Nemanjina Street
Belgrade
Yugoslavia
Telephone: 621 522
Telex: 11360

M-80 MICV from rear showing two ready to launch Sagger ATGWs mounted above turret rear

SPECIFICATIONS

CREW	3 + 7
COMBAT WEIGHT	13 700 kg
UNLOADED WEIGHT	11 700 kg
POWER-TO-WEIGHT RATIO	18.97 hp/tonne
GROUND PRESSURE	0.64 kg/cm²
LENGTH	6.4 m
WIDTH	2.59 m
HEIGHT	
over missile launcher	2.5 m
to turret roof	2.3 m
to hull roof	1.8 m
GROUND CLEARANCE	0.4 m
TRACK	2.47 m
TRACK WIDTH	300 mm
LENGTH OF TRACK ON GROUND	3.484 m
MAX SPEED	
road	60 km/h
water	7.5 km/h
MAX RANGE	
road	500 km
cross-country	300-350 km
FORDING	amphibious
GRADIENT	60%
VERTICAL OBSTACLE	0.8 m
TRENCH	2.2 m
ENGINE	HS 115-2 V-8 water-cooled turbo-charged diesel developing 260 hp at 3000 rpm
TRANSMISSION	pre-selector gearbox with hydraulic torque converter, 4 forward and 1 reverse gears
SUSPENSION	torsion bar
ELECTRICAL SYSTEM	24 V
ARMAMENT	
main	1 × 20 mm cannon
coaxial	1 × 7.62 mm MG
other	twin Sagger ATGW launcher
AMMUNITION	
20 mm	400
7.62 mm	2250
Sagger	4
GUN CONTROL EQUIPMENT	
Turret power control	n/av
by commander	no
by gunner	yes
Gun elevation/depression	+75°/-5°
ARMOUR	
Hull front	25-30 mm
Hull sides	15 mm
Hull top	8 mm
Hull floor	12 mm
Hull rear	12 mm
Turret front	20 mm

M-60P Armoured Personnel Carrier

Development

The M-60P armoured personnel carrier was first seen in public during a parade held in Yugoslavia in 1965, and for a short period was called the M-590.

Its engine is an Austrian Steyr diesel which also powers the Yugoslav GJ-800 artillery tractor (which in turn is based on the Hungarian K-800 light tracked artillery tractor) and its suspension is based on that of the Soviet SU-76 self-propelled gun's which was supplied to Yugoslavia in the immediate post-Second World War period.

Description

The hull is made of all-welded steel which provides the crew with protection from small arms fire and shell splinters.

The driver sits at the front of the hull on the left side and has a single piece rear-opening hatch cover with an integral periscope which can be replaced by an infra-red periscope for driving at night. The co-driver is seated on the right side of the hull and also has a hatch cover with an integral periscope that opens to the rear. The co-driver also operates the bow-mounted 7.92 mm machine gun which is based on the German MG42 weapon.

The commander sits to the rear of the driver and has a single piece hatch cover. The machine gunner sits to the rear of the co-driver and has a cupola which can be traversed through 360°. The cupola has a two-piece hatch cover that opens either side of the machine gunner to give him some protection when he is using the 12.7 mm (0.50) M2 HB machine gun which is mounted on the forward part of his cupola. A tripod carried on the right side of the hull enables the 12.7 mm machine gun to be removed from the vehicle and used in the ground role.

The infantrymen sit on bench seats down either side of the hull and enter and leave the vehicle by twin doors in the hull rear. There are also hatches in the roof of the troop compartment. Three firing/observation ports are provided in each side of the hull and one in each of the rear doors. A spare road wheel is carried on the left side of the hull rear and a jerrican is carried on the right side of the hull rear.

The torsion bar suspension consists of five single rubber-tyred road wheels with the drive sprocket at the front and the idler at the rear, and three track return rollers. The running gear is outside the hull armour and its top part is covered by a light sheet metal guard.

The M-60P APC is not amphibious and is not fitted with an NBC system.

Variants

M-60PB Anti-tank

Mounted on top of the hull on the left side are two 82 mm recoilless rifles on a rotating mount with a mechanical traverse and elevating mechanism. The gunner has a suspended seat which moves with the rifles and single piece hatch cover that opens forwards. To the right of the mount is a hatch for the loader. The M-60PB has also been observed with the recoilless rifles mounted on the right side of the hull top.

The recoilless rifles have a maximum range of 4500 m and a maximum rate of fire of 4 to 5 rds/min and have an elevation of +6°, depression

M-60PB armoured personnel carrier with twin recoilless 82 mm M-60 anti-tank guns

Infantry dismounting from M-60P APC

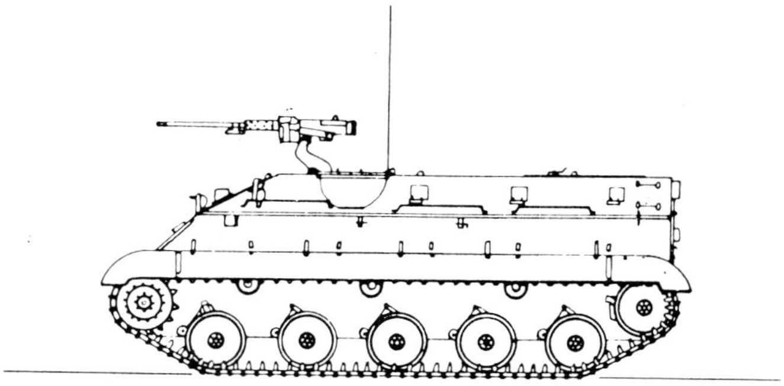

M-60P APC with 12.7 mm machine gun

of -4° 30 minutes with a prohibited zone of action of 76° in the plane of the anti-aircraft machine gun and 135° to the rear of the vehicle.

Effective range when engaging stationary targets is 1500 m, and 1000 m when engaging moving targets. The HEAT projectile will penetrate 250 mm of armour at an angle of 90° with dispersion being 2 m at 500 m and 4 m at 1000 m. The HEAT projectile itself weighs 4.2 kg, with the complete round weighing 7.2 kg; 10 rounds are carried. The M-60PB carries one fewer crew member than the basic M-60P APC.

It is also probable that there are other versions of the M-60 such as ambulance, command and radio vehicles.

Status: Production complete. In service with Yugoslavia only.

Contractor: Federal Directorate of Supply and Procurement (SDPR)
PO Box 308
9 Nemanjina Street
Belgrade
Yugoslavia
Telephone: 621 522
Telex: 11360

SPECIFICATIONS

CREW	3 + 10
COMBAT WEIGHT	11 000 kg
POWER-TO-WEIGHT RATIO	12.73 hp/tonne
GROUND PRESSURE	0.7 kg/cm²
LENGTH	5.02 m
WIDTH	2.77 m
HEIGHT	
without armament	1.86 m
including MG	2.385 m
MG at max elevation	3.51 m
GROUND CLEARANCE	0.4 m
TRACK	2.385 m
TRACK WIDTH	400 mm
LENGTH OF TRACK ON GROUND	2.94 m
MAX ROAD SPEED	45 km/h
FUEL CAPACITY	150 l
MAX RANGE	
road	400 km
cross-country	250 km
FUEL CONSUMPTION (road)	0.38 l/km
FORDING	
still water	1.35 m
running water	1.25 m
GRADIENT	60%
SIDE SLOPE	40%
VERTICAL OBSTACLE	0.6 m
TRENCH	2 m
ENGINE	FAMOS 6-cylinder in-line water-cooled diesel developing 140 hp
TRANSMISSION	manual, 5 forward, 1 reverse gears
SUSPENSION	torsion bar
ELECTRICAL SYSTEM	24 V
BATTERIES	2 × 12 V
ARMAMENT	
anti-aircraft	1 × 12.7 mm MG
bow	1 × 7.92 mm MG
ARMOUR	
max	25 mm
min	10 mm

BOV Wheeled Armoured Personnel Carrier

Development

In the early 1980s Yugoslavia developed a new family of 4 × 4 wheeled armoured fighting vehicles which share the same basic chassis, engine, transmission and suspension. This family is known as the BOV and is used both by the army and police.

Description

The basic APC, which is used by the Yugoslav Police and Militia is called the BOV-VP and carries a total of eight fully armed men plus commander and driver.

The hull of the vehicle is of all-welded steel construction that provides protection from small arms fire and shell splinters. The driver sits at the front left with the commander to his right, both with a rear-opening single piece hatch cover. The driver has a single forward facing periscope in the roof which can be replaced by a passive periscope. To the immediate front and either side of the commander and driver is a bullet proof window covered by a wire mesh screen.

The eight men are seated four either side down the centre of the vehicle and enter and leave via a single door in each side that opens towards the front. Over the top of the troop compartment is a cupola with a two part roof hatch that opens left and right and mounted on this is an externally mounted 7.62 mm M86 machine gun and a searchlight. The 7.62 mm machine gun is manned by one of the eight men. In either side of the troop compartment are three firing ports each with an associated firing port and there is an additional firing port to the right of the commanders windscreen. Windows are also provided in the rear of the troop compartment.

The engine compartment is at the rear of the hull with an air-inlet and air-out louvres on the top and an engine access door at the rear.

The vehicle is powered by a German Deutz F 6L 413 F six-cylinder diesel developing 148 hp at 2650 rpm which is coupled to a manual transmission with five forward and one reverse gears and a two speed transfer case.

Steering is power assisted to reduce driver fatigue and a central tyre pressure regulation system is fitted as standard, pressure can be adjusted from 0.7 to 0.35 bars to suit the type of ground being crossed. Main brakes are air-hydraulic with a hand operated parking brake. Suspension consists of leaf type springs with telescopic shock absorbers with 1300 - 18 PR10 cross-country tyres fitted as standard. The differential locks are controlled electro-pneumatically.

Standard equipment includes a Jugo-Webasto 7.5 kW heater, day and night vision equipment, intercom and radios.

Mounted either side of the hull is a bank of three forward firing electrically operated smoke dischargers and either side of the forward part of the hull is a wire mesh screen that lays alongside the hull when travelling and when required, for example, in a riot situation, is swung out through 90°.

BOV-1 anti-tank vehicle from front with each pod containing three ready to launch ATGWs

BOV-3 triple 20 mm self-propelled anti-aircraft gun system

BOV-M armoured personnel carrier operated by Yugoslavian Militia

Variants

BOV-1 ATGW carrier

Mounted on the roof of the vehicle is a pod containing six ATGWs based on the Soviet AT-3 Sagger but fitted with a semi-automatic guidance system. Additional missiles are carried under armour protection within the hull. Maximum rate of fire is two missiles a minute and maximum range of the missiles is 3000 m. Between the two pods of missiles is mounted a 7.62 mm machine gun. Both the BOV-1 ATGW and the two SPAAG chassis are slightly different to the APC described above as they do not have the raised troop compartment roof.

BOV-3 SPAAG

This is fitted with a power operated turret armed with three 20 mm cannon, each of which has a drum type magazine holding 60 rounds of ammunition with a total of 1500 rounds of ammunition being carried. Maximum anti-aircraft engagement altitude is quoted as 2000 m with effective anti-aircraft range being between 1000 and 1500 m. The turret is fitted with a J-171 optical sight. BOV-3 is a clear weather system only. Full details of the BOV-3 SPAAG are given in *Jane's Land-Based Air Defence 1991-92*, page 99.

BOV-30 SPAAG

This is the latest SPAAG version and is fitted with a larger enclosed turret armed with twin 30 mm cannon with a cyclic rate of fire of 600 rds/gun/min. Two three barrelled smoke dischargers are mounted either side of the turret. This model was first seen in 1985.

BOV-30 twin 30 mm self-propelled anti-aircraft gun system in travelling configuration with hatches open

Status: In production and service with the Yugoslavian Army and Militia.

Contractor: Federal Directorate of Supply and Procurement (SDPR)
PO Box 308
9 Nemanjina Street
Belgrade
Yugoslavia
Telephone: 621 522
Telex: 11360

SPECIFICATIONS

CREW	2 + 8
CONFIGURATION	4 × 4
WEIGHT	5700 kg
POWER-TO-WEIGHT RATIO	25.96 hp/tonne
LENGTH	5.70 m
WIDTH	2.534 m
WIDTH WITH SCREENS	2.92 m
HEIGHT HULL ROOF	2.335 m
GROUND CLEARANCE	0.325 m
TRACK	1.90 m
WHEELBASE	2.75 m
MAX SPEED	95 km/h
RANGE	500 to 800 km
FUEL CAPACITY	220 l
FORDING	1.1 m
GRADIENT	55%
SIDE SLOPE	30%
VERTICAL OBSTACLE	0.54 m
TRENCH	0.64 m
TURNING RADIUS	10 m
ENGINE	Deutz type F 6L 413 F 6-cylinder diesel developing 148 hp at 2650 rpm
TRANSMISSION	manual, 5 forward and 1 reverse gears
SUSPENSION	leaf springs and telescopic shock absorbers
TYRES	1300 - 18 PR10
ELECTRICAL SYSTEM	24 V
ARMAMENT	1 × 7.62 mm MG
SMOKE DISCHARGERS	2 × 3

Armoured Fighting Vehicle Families

In recent years there has been a trend towards the design of complete families of armoured fighting vehicles which have a wide range of components in common so enabling development time and costs to be dramatically reduced.

Very often these vehicles have used proven commercial components or military components that have already been used in other military vehicles. This will help to reduce life cycle costs which can be anything up to three times the original cost of the vehicle.

In addition to covering the government funded programmes such as the United Kingdom's Future Family of Light Armoured Vehicles (FFLAV) and the United States Army's Armored Systems Modernisation (ASM) programmes, some of the private venture projects are also covered.

These include the French Creusot-Loire Industrie MARS 15 series announced in 1990, the Krauss-Maffei/Diehl Puma family announced in 1986, the new Saudi Arabian 6 × 6 vehicles, the Alvis LPXV announced in 1988 and the well established Alvis Scorpion family of full tracked vehicles of which some 2700 have been built for the home and export markets since the first vehicles came off the production line in 1972.

In 1986 Royal Ordnance unveiled the RO2000 family of full tracked vehicles but this is no longer being marketed as shortly afterwards the Royal Ordnance Factory at Leeds was sold to Vickers Defence Systems. Full details of the RO2000 family of light tracked armoured vehicles were given in *Jane's Armour and Artillery 1990-91*, pages 526-528.

The private venture BMY Modular Support Vehicle (*Jane's Armour and Artillery 1989-90*, page 509) and the BMY Mobility Test Bed (*Jane's Armour and Artillery 1989-90* page 510), were never competed and for the time being at least have been abandoned.

In addition to these armoured fighting vehicle families, there are also many other vehicles which share common components and some of these are listed below. Full details are given in their respective entries. In addition most vehicles, especially MBTs and armoured personnel carriers, have many variants of the basic chassis.

A typical example is the French GIAT AMX-30 MBT whose chassis has been the basis for many vehicles including an armoured recovery vehicle, armoured vehicle-launched bridge, twin 30 mm self-propelled anti-aircraft gun, EBG armoured engineer vehicle, Roland surface-to-air missile system, driver training vehicle, GCT 155 mm self-propelled howitzer, Pluton surface-to-surface tactical nuclear missile system and the Shahine surface-to-air missile system for Saudi Arabia.

In the Soviet Union, the PT-76 light amphibious tank has also formed the basis for a complete family of light armoured fighting vehicles which are used for a wide range of battlefield roles with further development of the BTR-50P armoured personnel carrier version taking place in China and Czechoslovakia.

Country	Vehicles sharing common automotive components
Argentina	TAM medium tank and VCTP IFV, with Thyssen Henschel of Germany also offering variants, these are related to the Marder IFV used by the German Army
Austria	Steyr SK 105 and Steyr 4KA 7FA APC
Brazil	EE-9 (6 × 6) armoured car and EE-11 (6 × 6) APC
France	AMX-10RC (6 × 6) armoured car and AMX-10P tracked IFV
	Panhard ERC Sagaie (6 × 6) armoured car and Panhard VCR TT (6 × 6) APC
	AMX-13 light tank, VCI infantry fighting vehicle, 105 mm SPH, 155 mm Mk F3 SPG, DCA twin 30 mm SPAAG
	Panhard AML (4 × 4) armoured car and M3 (4 × 4) APC
	VAB (4 × 4 and 6 × 6) APC and VBC 90 (6 × 6) armoured car
Germany	Jagdpanzer Kanone and Rakete, and Marder IFV
Italy	FIAT/OTO Melara Type 6616 (4 × 4) armoured car and Type 6614 (4 × 4) APC
	FIAT Puma 4 × 4 and 6 × 6 vehicles
	OTO Melara OTO R3 Capraia reconnaissance vehicle and OTO Melara R 2.5 Gorgona reconnaissance vehicle
	Infantry Armoured Fighting Vehicle and M113A1 APC
Spain	BMR-600 (6 × 6) APC and VEC (6 × 6) cavalry scout vehicle
Sweden	Pbv 302 APC, Ikv-91 light tank, Brobv 941 bridgelayer and Bgbv 82 ARV
UK	Alvis Saladin (6 × 6) armoured car and Saracen (6 × 6) APC
	FV432 APC and Abbot 105 mm SPG
	Shorland (4 × 4) armoured patrol car and Shorland (4 × 4) APC
USA	M41 light tank, M42 twin 40 mm SPAAG, M44 155 mm SPH, M52 105 mm SPH
	Lynx command and reconnaissance vehicle and M113 APC
	M2 IFV/M3 CFV and Fighting Vehicle Systems Carrier
	Armoured Infantry Fighting Vehicle and M113A2 APC
	M110 203 mm and M107 175 mm self-propelled artillery weapons
	M109 155 mm and M108 105 mm self-propelled howitzers
Yugoslavia	BOV APC, BOV anti-tank vehicle and BOV anti-aircraft vehicles

FRANCE

Vehicule Blindé Modulaire (VBM) (Modular Armoured Vehicle)

Development/Description

To work with the GIAT Industries Leclerc MBT, currently entering production, the French Army has a requirement for a Vehicule Blindé Modulaire (VBM) (Modular Armoured Vehicle).

The French Army's VBM is not expected to enter service until later this decade and in some respects the VBM requirement is similar to that of the British Army's Future Family of Light Armoured Vehicles (FFLAV) (qv) but covers a range of vehicles from 15-35 tonnes with a power-to-weight ratio of 25-30 hp/tonne.

This would give the vehicle a similar cross-country mobility to the new GIAT Industries Leclerc MBT scheduled to replace the French Army's ageing AMX-30 B2 MBTs from 1992.

The VBM family is expected to comprise wheeled vehicles only, for example 6 × 6 and 8 × 8, and to reduce operating and procurement costs wherever possible proven and existing automotive components would be used.

According to GIAT Industries, VBM variants will include the Vehicule d'Appui Direct (VAD), the Vehicule de Transport de Troupes (VTT) and the Vehicule Poseur de Travures de Avani (PTA).

The VAD Direct Support Vehicle would be an 8 × 8 vehicle with steering on the front and rear wheels and a three man crew consisting of driver, commander and gunner with the last two seated in a power operated turret armed with an automatic weapon. The vehicle would also be able to carry some infantrymen in the rear.

The VTT armoured personnel carrier would have a two man crew and similar hull and layout to the VAD with a machine gun mounted in the centre and the troop compartment at the rear. The infantrymen would sit facing each other and enter and leave via the hull rear.

The PTA bridgelayer would be a 10-wheeled vehicle with a fully laden weight, including the bridge, of 53 tonnes, to enable it to handle the Leclerc MBT, have a two man crew and be powered by a 1100 hp engine. It will be able to lay a 24 m cantilever bridge or two 12 m long bridges. A DREC vehicle protection system would be fitted as standard.

In the Autumn of 1990, GIAT Industries, Renault Vehicules Industriels (RVI) and Creusot-Loire Industrie (CLI) signed a draft agreement to collaborate on armoured vehicles with the agreement being initially aimed at the French Army's VBM programme.

GIAT Industries is prime contractor for the Leclerc MBT and already has some experience in the design, development and production of wheeled armoured vehicles, for example the 105 mm armed AMX-10RC. RVI and CLI are already working together on the VAB family of 4 × 4 and 6 × 6 armoured personnel carriers which have been built in substantial numbers for both the home and export markets.

The other contenders for the VBM competition include Panhard who has teamed with Mercedes-Benz of Germany. Panhard have extensive experience in the design, development and production of wheeled vehicles in 4 × 4, 6 × 6 and 8 × 8 configurations. The latter include the M8 developed to the prototype stage some years ago which could pivot on its axis as the front and rear wheels could be raised off the ground.

Also teamed with Panhard is Daimler-Benz of Germany who has considerable experience in the development of high mobility off-road vehicles and has recently completed an 8 × 8 Armoured Combat Vehicle Demonstrator which is covered in the *Reconnaissance vehicles* section.

Status: Project. Not yet in production or service.

Manufacturer: Not yet selected.

Creusot-Loire Industrie MARS 15 Family of Light Armoured Vehicles

Development

The Division Mecanique Specialisée of Creusot-Loire Industrie (CLI) has developed a new range of light tracked armoured vehicles for the export market called the MARS 15. These were first shown in public at the 1990 Satory exhibition of military equipment held near Paris.

CLI see the MARS 15 family, which have combat weights of 15 to 18 tonnes depending on version, as the successor to the AMX-13 family of light armoured vehicles which were developed soon after the end of the Second World War.

Production of the AMX-13 family was originally undertaken by GIAT at the Atelier de Construction Roanne but once this facility became heavily involved in the AMX-30 MBT programme for the French Army, and later for export, production and marketing of the AMX-13 was transferred to CLI.

Since then CLI have continued to build new vehicles as required and in recent years have also offered refurbished French Army vehicles to overseas customers as well as upgrading vehicles to extend their lives into the 1990s.

In the longer term CLI believed that a new vehicle was required which would have a number of new features as well as being easier to operate and maintain, have improved protection and survivability, greater cross country mobility and lower life cycle costs.

CLI first had the idea for the MARS 15 family in 1985 with full scale development starting in 1988 and two first prototypes being completed early in 1990.

By mid-1990 CLI had invested around FFr40 million in the MARS 15 project with many other subcontractors supplying equipment and expertise. Within the Division Mecanique Specialisée there is considerable experience in AFV design and development as well as welding technology, and all of this was used in the MARS 15 programme. CLI made extensive use of CAD in the development of the MARS 15 family.

By mid-1990 two prototypes had been built, a light tank/tank destroyer armed with an up-armoured GIAT Industries TS 90 turret armed with a 90 mm gun and 7.62 mm coaxial machine gun and an infantry fighting vehicle fitted with a CLI T 25 two-man power operated turret armed with the United States McDonnell Douglas Helicopters 25 mm M242 Chain Gun and a 7.62 mm coaxial machine gun.

The third prototype of the MARS 15 is due to be completed in 1991 and will be a light tank fitted with a 105 mm rifled tank gun firing standard NATO ammunition. CLI are also considering a stretched version of the MARS 15 with six road wheels either side.

The prototypes of the MARS 15 were built at Saint-Chamond where production will be undertaken when orders are received. Creusot-Loire Industrie will also be offering the vehicle for licensed production overseas. For example, the first batch of vehicles would be supplied by CLI, the next batch in kit form then followed by local production.

The US Army has a requirement for an Armored Gun System (AGS) and to meet this requirement a MARS 15 chassis fitted with the US Cadillac Gage Textron LAV 105 turret, currently under development for the US Marine Corps, has been studied by CLI.

Description

Using the same two basis chassis a wide range of turrets can be fitted. For example, although the prototype of the armoured infantry fighting vehicle chassis has a CLI T 25 turret, it can be fitted with a wide range of other light turrets, including those manufactured by CLI. When a smaller turret is fitted more infantrymen can be carried.

Other roles envisaged include a command post vehicle, mortar carrier, artillery observation post vehicle, surveillance radar carrier, various types of anti-tank system including ATGWs, air defence systems (gun and missiles), ambulance, recovery vehicle and a self-propelled artillery system. The latter would have a GIAT Industries 155 mm artillery system in an open mount at the rear and be fitted with spades to provide a more stable firing platform.

All of these will use virtually the same basic chassis with the main difference being the area to the immediate rear of the driver's position.

The chassis is of all-welded steel armour construction which provides protection against 14.5 mm AP attack from a range of 100 m through the frontal arc of 60°, with protection against 7.62 mm AP attack through a full 360°.

CLI have a great deal of experience in the design and manufacture of high hardness armour plate so can provide additional armour protection for the MARS 15 family if required.

Considerable effort has gone into the hull design of the MARS 15 family to provide the maximum possible protection within the weight limits of the vehicle. For example, the glacis plate is well sloped and the sponsons are used to support subsystems such as the batteries, exhaust and NBC system. The latter is installed in the left side of the hull at the rear. Emphasis has also been placed on reduced thermal and radar signatures. An explosion/fire detection suppression system is fitted as standard.

The driver is seated at the front left and is provided with a single piece hatch cover and three periscopes for observation when driving closed down, the centre one of which can be replaced by a passive night vision device. He steers the vehicle with a small wheel rather than tillers and can also enter and leave the vehicle via the rear crew compartment.

MARS 15 armoured infantry fighting vehicle fitted with CLI T 25 two-man turret armed with 25 mm M242 McDonnell Douglas Helicopters Chain Gun and 7.62 mm coaxial machine gun

MARS 15 light tank fitted with GIAT Industries TS 90 two-man turret armed with a 90 mm gun that can fire a wide range of ammunition including APFSDS

The powerpack is mounted to the right of the driver and can be replaced in under 60 minutes. It consists of a French Baudouin 6F 12 SRY 400 turbocharged diesel developing 400 hp coupled to a German Renk HSWL 106 fully automatic transmission with six forward and three reverse gears and a Valeo hydrostatically operated cooling system. The air inlet is on the glacis with the exhaust outlet on the right side of the hull.

The MARS 15 has a hydropneumatic suspension system designed by SAMM which gives a total wheel travel of 350 mm. Either side of the hull are five dual rubber tyred road wheels, drive sprocket at the front, idler at the rear and three track return rollers. Hydropneumatic suspension was selected to provide better cross country mobility for the vehicle as well as a better ride for the crew.

The MARS 15 family has been designed for ease of maintenance, for example the air cleaners can be changed through the glacis plate and the four batteries are in the sponson to the left of the driver. The MARS 15 has also been designed to be used in conjunction with an interactive diagnostic system so that defects can be quickly detected.

The GIAT Industries TS 90 turret, installed on the light tank version of the MARS 15, is already in volume production and has been sold to a number of countries. For example the Panhard Sagaie 1 (6 × 6) armoured cars of the French Army have the TS 90 turret. The light tank version of the MARS 15 has a single door in the hull rear which is used for ammunition resupply or emergency means of escape.

The IAFV version of the MARS 15 has twin doors in the hull rear to allow the infantry to rapidly dismount from the vehicle. No firing ports or vision devices are provided in the rear troop compartment, although these could be fitted.

The MARS 15 is air transportable in a C-130 Hercules transport aircraft and two can be carried on a bogie railway flatcar and still stay within the loading gauge. It can also be carried by a tank transporter without a special route escort being required.

The third prototype of the MARS 15 will be completed in 1991 and will be a light tank fitted with a three-man power operated turret armed with a 105 mm tank gun firing standard NATO ammunition. This will have the same firepower as M48 and Leopard 1 MBTs plus greater mobility. CLI are also considering a stretched version of the MARS 15 with six road wheels either side.

The light tank version of the MARS 15 has a very low profile hull and is fitted with the up-armoured version of the GIAT Industries TS 90 turret armed with a 90 mm gun. The door in the hull rear can be used for emergency exit or ammunition resupply (Christopher F Foss)

Armoured Infantry Fighting Vehicle version of MARS 15 with CLI T 25 turret with commander's, gunner's and driver's roof hatches in open position

Model of MARS 15 light tank fitted with GIAT Industries three man turret armed with 105 mm gun capable of firing standard NATO ammunition (Terry J Gander)

SPECIFICATIONS

Version	**APC**	**AIFV**	**90 mm**	**105 mm**
CREW	1 + 12	3 + 8	3	4
COMBAT WEIGHT	15.3 t	16.5 t	16 t	18 t
GROUND PRESSURE	0.61 kg/cm²	0.66 kg/cm²	0.67 kg/cm²	0.68 kg/cm²
LENGTH WITH GUN	n/app	n/app	6.83 m	7.75 m
LENGTH HULL	5.99 m	5.99 m	5.6 m	5.95 m
WIDTH	2.98 m	2.98 m	2.98 m	2.98 m
HEIGHT OVERALL	1.93 m	2.64 m	2.60 m	2.75 m
POWERPACK	Baudouin diesel developing 400 hp			
MAX ROAD SPEED	75 km/h	76 km/h	75 km/h	75 km/h
MAX RANGE	600 km	600 km	600 km	600 km
VERTICAL OBSTACLE	0.8 m	0.8 m	0.8 m	0.8 m
TRENCH	1.80 m	1.80 m	1.80 m	1.80 m
GRADIENT	60%	60%	60%	60%
SIDE SLOPE	30%	30%	30%	30%
FORDING	1 m	1 m	1 m	1 m

Status: Prototypes. Not yet in production or service.

Manufacturer: Creusot-Loire Industrie
Division Mecanique Specialisée
Immeuble Ile-de-France
Cedex 33
92070 Paris La Defense
France
Telephone: (1) 49 00 60 50
Telex: MOTOY 615 638F
Fax: (1) 49 00 57 30

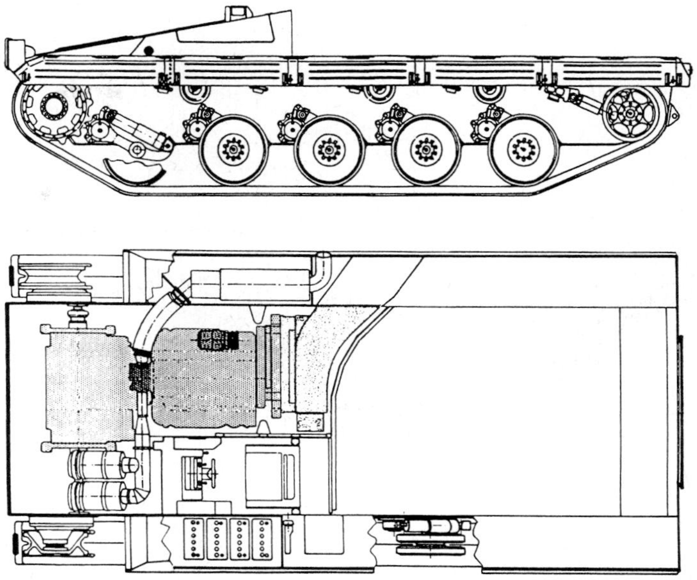

Outline drawing of the MARS 15 basic chassis which has the engine and driver's compartment at the front so leaving the whole of the rear area free for dedicated mission equipment, for example light tank or armoured personnel carrier (not to 1/76th scale)

GERMANY

Krauss-Maffei/Diehl Puma Armoured Combat Vehicle Family

Development

In 1983 Krauss-Maffei and Diehl started private-venture development of a light tracked vehicle family designed to have combat weights between 18 and 38 tonnes in three weight and performance classes, with four, five, or six pairs of roadwheels according to mission requirements.

The first prototype of this vehicle, called the Puma Armoured Combat Vehicle (ACV) with four road wheels each side was completed in Spring 1986 and fitted with a turret armed with a Diehl 120 mm mortar which can be loaded, aimed and fired under complete armour and NBC protection. This was later replaced by a 20 mm KUKA weapon station. In late 1986 and early 1987 the first prototype of the Puma ACV fitted with a 20 mm turret underwent extensive trials in Turkey alongside the British GKN Defence Warrior and the FMC Armoured Infantry Fighting Vehicle to meet an anticipated Turkish Army order. This competition was however won by FMC with its Armored Infantry Fighting Vehicle (AIFV).

The second prototype of the Puma ACV (PT-2), with five road wheels each side, was completed in mid-1988 and fitted with a Cockerill CK25B two man power-operated turret armed with an Oerlikon-Contraves 25 mm KBB automatic cannon although a wide range of other weapon systems can be fitted including the KUKA E4A1 turret with 25 mm Mauser MK 25 or the Mauser 30 mm MK 30 F cannon.

The second prototype has a redesigned and more streamlined hull and is powered by a 440 hp MAN diesel coupled to a ZF 6-speed transmission. It is configured as an infantry fighting vehicle with the troops being seated back to back down the centre of the vehicle with each man having an individual seat that folds up automatically, so making the exit of troops as quick as possible.

No firing ports are provided although the rear troop compartment is fitted with two roof hatches, six roof-mounted periscopes and a large bullet-proof observation block in each door. This version has a three man crew; commander, gunner and driver and carries eight infantrymen. If a cupola-mounted 12.7 mm MG was installed then a total of 12 infantry plus the driver could be carried. An overpressure NBC system is fitted as standard.

Early in 1989 the second Krauss-Maffei prototype was modified into an armoured personnel carrier version with four road wheels each side and accommodation for an infantry group with 10 soldiers. This vehicle was evaluated in field trials at the infantry school of the German Army.

The German Ministry of Defence has placed an order for two Puma ACVs. The first of these, PM-1, is powered by a 440 hp MAN diesel engine, has four road wheels each side and is configured as a mortar carrier. This was delivered in the third quarter of 1989. It was used for automotive trials followed by firing trials.

The second prototype Puma, designated the PM-2, is powered by a MAN diesel developing 750 hp and has six road wheels each side. This is being used as a testbed vehicle and was delivered to the German Army in the Autumn of 1990.

Prototype of Krauss-Maffei/Diehl Puma ACV with Diehl turret-mounted 120 mm mortar

Armoured command vehicle based on Puma ACV chassis

Mine scattering vehicle version of Puma ACV

Second prototype of the Puma ACV configured as an infantry fighting vehicle and fitted with a Cockerill CK25B two man power-operated turret armed with a 25 mm Oerlikon-Contraves KBB cannon and a 7.62 mm coaxial machine gun

Fourth Puma Armoured Combat Vehicle manufactured under contract to German Army and used as a test bed vehicle

There are now three weight classes of Puma; 18-25 tonnes, 26-32 tonnes and 33-40 tonnes. Depending on the weight, the suspension system can be provided with four, five or six road wheels with the option of 440 hp or 750 hp MAN diesel engines. Payloads of 7, 12, and 17 tonnes are possible depending on the ultimate mission, weapons fit, crew and type of armour required.

According to the manufacturers, the Puma family, with its more than 20 variants, is a highly mobile, versatile and cost-effective weapon system for both combat and combat support troops.

Description

The hull of the Puma ACV is of all-welded steel construction with the engine and transmission at the front left and driver towards the front on the right. The driver has a single piece circular roof hatch that lifts and swings to the right. In front of this are three periscopes, the centre one of which can be replaced by a passive periscope for night driving.

Depending on the armour used, ballistic protection ranges from 7.62 mm all round to 14.5 mm all round and 30 mm over the front 60° (30° left and right) arc. As an option additional armour protection can be added.

The all-welded steel hull of the second prototype Puma provides protection from 14.5 mm AP attack, but this could be increased to 30 mm attack. In the future it may be possible to add composite or explosive reactive armour to the Puma chassis.

The engine, transmission, cooling system and combustion air cleaners are grouped together to form a compact power unit which can be replaced as a complete unit quickly. It also permits static test runs outside the vehicle.

The Puma ACV is being offered with a choice of a MAN D 2866 KE turbocharged diesel developing 440 hp coupled to a ZF 6 HP 600/STV 600 LGH 10.000 transmission or the more powerful MAN D 2840 LE developing 750 hp coupled to a Renk HSWL 284C2 automatic transmission with integrated steering units and brake system. The choice of two engines allows the user to select the model with the power output required to meet high or low power-to-weight ratios.

Although many proven components have been used in Puma, it does incorporate new technology. For example, the steering gear is new and permits infinitely variable steering and pivot-steer of the vehicle. In the case of the 440 hp version, the steering gear delivers the correct speed for the drive shaft, there being no requirement for a final drive unit. The hydraulic track adjustment on the rear idler wheels are from the Swiss version of the Leopard 2 and are operated from the driver's station.

The suspension either side consists of four, five or six dual rubber-tyred road wheels, with the drive sprocket at the front, idler at the rear and three track-return rollers.

The road wheels, track support rollers, torsion bars, shock absorbers, road wheel arms with bearings, idler wheels, tracks and driver's seat are all from the Leopard 1 MBT. The hydraulic track adjuster, hydraulic bump stops, cooling system components and sprocket hub are all from the Leopard 2 MBT.

Five types of Diehl track are available for the Puma ACV, including that used on the Leopard 1 MBT, the Type 640A and brief details of these are given below:

Type	Weight per metre of track	Track life on Puma
840 end connector	161 kg	13 000 km
640A	154 kg	10 000 km
139E2 end connector	123 kg	5000 km
141 single pin	95 kg	4000 km
340 end connector	135 kg	10 000 km

The standard Puma has four 12 V, 100 Ah maintenance free batteries and a 5 or 9 kW generator. Another option is reinforced cabling.

Also standard on the Puma is a heating and ventilation system, NBC system and a fire extinguishing system, but options include a cooling system for use in very hot climates, explosion detection and suppression system, and deep fording equipment.

The area to the rear of the powerpack and driver's compartment depends on the variant.

To reduce both procurement and life cycle costs, proven commercial and military components have been used in the design of the Puma ACV family.

Variants

Air Defence/Anti-tank Vehicle

The articulated mast can be raised to a height of 12 m and would be either manned or unmanned and be fitted with four, six, eight or more missiles.

In the manned version, the gunner is in an armoured cabin with a traversable platform mounting a guided missile carrier with four or six missiles. In addition to having sights for guiding the missiles and vision blocks, it also has a long-range periscope which allows accurate identification of hard targets up to a range of 4000 m.

Currently HOT and TOW missiles are proposed but when fitted with air defence Stinger missiles, the vehicle can also be used for air space interdiction in the battlefield area, for example against helicopters.

Armoured Personnel Carrier/Command Vehicle

This would have a combat weight of between 20 and 22 tonnes and in addition to the driver would carry 12 fully equipped troops. The rear troop compartment would be fitted with roof hatches, a cupola with an externally mounted 7.62 mm machine gun, rear doors, and vision devices. The command post vehicle would have extensive communications equipment installed.

Infantry Fighting Vehicle

This model would weigh between 23 and 32 tonnes and have a crew of three; commander, gunner and driver. It could carry eight fully equipped infantrymen. This model would have various two-man turrets that could be armed with various types of 20 mm to 40 mm weapons.

120 mm Mortar Carrier

The 120 mm mortar carrier would weigh 25 tonnes and has a crew of four: commander, gunner, loader and driver. The 120 mm Diehl mortar is integrated in a turret which can be traversed 180° left and right of the centre line of the vehicle. The mortar can be loaded, aimed and fired under NBC and fragment protection. This version has cupolas for the commander (armed with 7.62 mm MG) and gunner. The 120 mm mortar can be elevated from +45° to +80° and fires a standard DM11 120 mm bomb to a maximum range of 6500 m or an extended range projectile to a maximum range of more than 8000 m. Sixty 120 mm mortar bombs are carried. This version is also fitted with an automatic loader to enable a high rate of fire to be achieved. To provide a standalone system, the Puma mortar turret can also be fitted with a computerised fire control system and a land navigation system.

Light Battle Tank 105 mm

The chassis of the Puma ACV has been designed to accept two- or three-man turrets armed with 105 mm weapons. This version would have a weight of between 26 and 28 tonnes.

TOW Anti-tank Vehicle

This model is fitted with a Kvaerner-Eureka launcher with two TOW missiles ready to launch or the Emerson twin TOW launcher as on the M901 Improved TOW Vehicle in service with a number of countries. This version would have two missiles ready to launch with a further 12 in reserve in the hull and would have a combat weight of 19 tonnes.

Air Defence Vehicle

This was originally offered with the twin 30 mm turret of the Wildcat self-propelled anti-aircraft gun system. Since then Krauss-Maffei has undertaken several feasibility studies concerning low-altitude surface-to-air missile systems. These have shown that the Puma in its four-wheel version is well suited for the French Thomson-CSF Crotale while the six-wheel version is well suited to the Euromissile Roland 3 and would have a weight of 32 tonnes.

Mine Scattering System

The mine scattering system was introduced into the German Army in 1986 on a modified M548 tracked carrier which has no armour protection. Krauss-Maffei suggests that the Puma ACV could be used as a mine scattering system, this version would have a crew of two and a combat weight of 21 tonnes.

Other variants

These include an armoured tactical resupply vehicle with remote-controlled arms which could resupply an MBT with ammunition and fuel while the crew of both vehicles are under complete armour protection.

The command post vehicle would have extensive communications equipment installed, a large map table and a raised hull roof so that the command staff can stand upright.

Status: Three industry prototypes of the Puma Armoured Combat Vehicle in various configurations are completed and a Puma will undergo performance trials as well as comparative trials in Norway and Switzerland in 1992-93 as both of these countries have a requirement for an infantry combat vehicle that could be met by the Puma.

Two additional prototypes have been ordered by the German Army to undergo automotive trials and both of these have now been delivered. One of these vehicles is fitted with a 120 mm mortar turret while the second vehicle has been designed to accept an elevating platform for such weapon systems as the PARS 3.

Manufacturer: Krauss-Maffei, Wehrtechnik GmbH
Krauss-Maffei Strasse 2
D-8000 Munich 50
Federal Republic of Germany
Telephone: (089) 88 99 0
Telex: 5 23 163 31
Fax: (089) 8 12 01 90

SPECIFICATIONS

MAXIMUM COMBAT WEIGHT	25 000 kg	32 000 kg	38 000 kg
CHASSIS WEIGHT IMMUNITY AGAINST 7.62 mm AP	18 000 kg	19 000 kg	21 000 kg
NUMBER OF ROAD WHEELS EITHER SIDE	4	5	6
LENGTH	5.80 m	5.80 m	6.50 m
WIDTH	3.25 m	3.25 m	3.25 m
HEIGHT	1.71/1.91 m	1.71/1.91 m	1.71/1.91 m
GROUND CLEARANCE	0.44 m	0.44 m	0.44 m
TRACK WIDTH	550 mm	550 mm	550 mm
MAX ROAD SPEED	65 km/h	65 km/h	65 km/h
ROAD CRUISING RANGE	650 km	550 km	650 km
FUEL CAPACITY	500 l	500 l	650 l
VERTICAL OBSTACLE	0.91 m	0.91 m	0.91 m
TRENCH	2.15 m	2.15 m	3.00 m
GRADIENT	60%	60%	60%
SIDE SLOPE	30%	30%	30%
FORDING WITHOUT PREPARATION	1.20 m	1.20 m	1.20 m
FORDING WITH DEEP FORDING EQUIPMENT	1.80 m	1.80 m	1.80 m
ENGINE	450 hp	750 hp	750 hp

INTERNATIONAL

NATO Multi-Role Armoured Vehicle

Development/Description

In mid-1990, the final draft document for a future 'multi-purpose base armoured vehicle' was issued for comment by the WGE 5 Working Group of the NATO Army Armaments Group (NAAG) Panel 2.

This Group first met early in 1989 and included Belgium, Canada, France, Germany, Italy, the Netherlands, Spain, United Kingdom and the United States of America.

Comments were submitted in the Summer of 1990 and the next phase is expected to the drafting of an Outline NATO Staff Target (ONST).

It is expected that the tracked vehicle would weigh around 30 tonnes and be capable of undertaking many of the roles currently being carried out by the M113 vehicles but with the new vehicle having better armour protection and mobility, with an in service date of around 2000 or later.

Where this vehicle would fit into some of the above countries armoured vehicle programmes is by no means clear. For example France is already studying the VBM programme (qv), Germany is developing the Puma Armoured Combat Vehicle (qv) while the UK has its FFLAV programme (qv).

While the US has its Future Infantry Fighting Vehicle (FIFV) as an M2 Bradley Infantry Fighting Vehicle (IFV) replacement, the new vehicle could perhaps replace or supplement the many M113 series vehicles that remain in service with the United States Army as well as many other countries.

Status: Study.

Manufacturer: Not selected.

SAUDI ARABIA

Saudi Arabian Armoured Car and Armoured Personnel Carrier

Development

The Saudi Arabian company of Abdallah Al-Faris & Co for Heavy Industries (one of the Military Industries Corporation companies), has been developing a 6 × 6 armoured fighting vehicle suitable for a wide range of operational roles on the battlefield since 1977.

The first vehicle, an armoured personnel carrier, was completed in 1981, while the second, which incorporated many improvements as a result of trials with the prototype, was completed in 1988 (*Jane's Armour and Artillery 1989-90*, page 776). This was fitted with a Cockerill LCTS two man turret armed with a Cockerill 90 mm Mk 7 gun and a 7.62 mm coaxial machine gun.

Early in 1990 the manufacturer stated that development and trials were complete prior to the pre-series production of the 6 × 6 armoured vehicle under the 'Dera Al Jazirah' (Peninsular Shield) Program of the Military Industries Corporation. The second and last prototype was subsequently named 'El Labwa' after the lioness which is known for protecting its cubs and hunting for them.

By early 1990 20 pre-series units of different specifications and capabilities were said to be in production in Saudi Arabia.

Description

The hull of the vehicle is of all-welded steel armour construction that provides protection from small arms fire and shell splinters.

The driver is seated at the front left with the vehicle commander to his right and to their front is a large one piece bullet-proof windscreen and both have a bullet-proof side window. When in a combat area these are covered by hydraulically operated window flaps hinged at the top. When these are lowered in position forward observation for the commander and driver is via roof-mounted periscopes. The driver is also provided with a circular roof hatch above his position.

In each side of the vehicle is an entry door that opens to the front and this is fitted with a firing port with an observation device above.

The troop compartment is located in the centre of the vehicle with the engine compartment located at the rear on the right side with access grilles in the roof. The troops can also leave the vehicle via a passage to the rear on the left side. An entry door is provided at the left rear and this is hinged at the bottom and folds down like a ramp.

The engine, transmission, cooling system and drive shafts are made in one unit to permit field stripping with the least effort in an emergency such as a powerpack exchange.

The independent wheel suspension with hydropneumatic spring system provide a total wheel travel of 550 mm. It is provided with a self-regulating system (electro-mechanical) that controls the pressure of the suspension and level of the vehicle while it is moving.

Steering is power assisted on the front and rear wheels and a central tyre pressure regulation system is fitted as standard, thus allowing the driver to adjust the tyre pressure whilst the vehicle is moving without leaving his position.

An NBC protection system can be provided as an optional extra since the powerpack is designed to accommodate this. The air-conditioner consists of two units of 12 000 BTU each.

An extra emergency power supply for the vehicle during standing, guarding and/or reconnaissance time is provided by a generator with an electric dynamo and hydraulic pump capable of supplying power for all equipment during the waiting period, for example air-conditioning, armoured flaps for commander and driver, battery filling, lights and turret drives.

The emergency power supply is of the diesel air-cooled type with a capacity of 13 kW and is located at the rear left side of the hull. If all of the batteries are dead, the emergency power supply unit is electrically and manually started.

To reduce life cycle costs, standard or slightly modified and proven commercial components have been used in the design of the vehicle.

The vehicle is fully amphibious being propelled in the water by two hydraulically driven propellers mounted under the rear of the hull.

The basic vehicle has been designed for use as an armoured personnel carrier carrying a total of nine men (commander, driver and seven soldiers), but when fitted with a 90 mm turret it would normally have a three man crew consisting of commander and gunner in the turret and the driver in the front of the hull.

Optional equipment includes anti-riot roof louvres and flaps for windows and roof, NBC system, crawler gear, hydraulic winch, land navigation system, automatic fire extinguishing system, fire detection and suppression system, various communications systems, passive night driving aids, various armament installations, recovery equipment, crane, elecric emergency brake, emergency steering and compass.

Saudi Arabian 6 × 6 armoured personal carrier

Variants

In addition to the basic armoured personnel carrier and light armoured fighting vehicle with 90 mm turret, projected variants could include air defence vehicle, ambulance, command post vehicle, maintenance vehicle and battlefield resupply vehicle.

Status: Development complete. Pre-production batch of 20 vehicles being completed.

Manufacturer: Abdallah Al-Faris Co for Heavy Industries
P O Box 5987
Dammam 31432
Saudi Arabia
Telephone: (03) 8500666
Telex: 897054
Fax: (03) 8562550

Saudi Arabian 6 × 6 light armoured fighting vehicle with Cockerill 90mm LCTS showing steering on front and rear axles

Saudi Arabian 6 × 6 light armoured fighting vehicle with Cockerill 90 mm LCTS showing cross country mobility

SPECIFICATIONS

CREW	9 (as APC) 3 (with 90 mm turret)
CONFIGURATION	6 × 6
COMBAT WEIGHT	16 000 kg (APC) 18 500 kg (with 90 mm turret)
POWER-TO-WEIGHT RATIO	22.8 hp/tonne
LENGTH	7.39 m
WIDTH	2.94 m
HEIGHT	2.05 m (without turret)
GROUND CLEARANCE	190 to 640 mm
WHEELBASE	2.55 m
ANGLE OF APPROACH/ DEPARTURE	45°/45°
MAX ROAD SPEED	95 km/h
CRUISING SPEED	85 km/h
MAX WATER SPEED	8/12 km/h
ACCELERATION	0/50 km/h in 18 to 22 s
FULL BRAKING SPEED AT 95 km/h	50 m
TURNING RADIUS	16 m
RANGE (CRUISING)	900 km
ENGINE	KHD air-cooled turbocharged diesel developing 400 hp
TRANSMISSION	ZF automatic with 6 forward and 1 reverse gears
APU	air-cooled diesel, 13 kW
AXLES	independent wheel suspension on all three axles and hydropneumatic spring system (McPherson) providing wheel travel of net 470 mm, gross 550 mm
STEERING	front and rear axles steerable, each with one ZF servo ball steering gear
EMERGENCY STEERING	an emergency wheel steering pump is provided on the main transmission gear to enable steering as long as wheels are rolling
BRAKES	full hydraulic brake with independent supply line for each wheel to ensure independence of each wheel from the others during emergency
EMERGENCY BRAKES	an electric driven emergency pump is supplied to ensure functioning of main brakes at all times
HAND BRAKES	spring activated parking brake on all differentials is provided and can be used as emergency brakes as well
TYRES	radial tyres 16.00 × 20 with run flat lining

UNION OF SOVIET SOCIALIST REPUBLICS

MT-SB Medium Armoured Transporter

Development/Description

The USSR has developed MT-SB Medium Armoured Transporter from the unarmoured MT-S Medium Tracked Transporter (*Jane's Military Vehicles and Logistics 1991-92*, page 470). The MT-SB chassis forms the basis for a number of new specialised vehicles including:

(1) 2S6 self-propelled anti-aircraft gun/surface-to-air missile system (*Jane's Land-Based Air Defence 1991-92*, page 83).

(2) SA-11 Gadfly self-propelled surface-to-air missile system and its associated Fire Dome and Tube Arm radar vehicles (*Jane's Land-Based Air Defence 1991-92*, page 134).

(3) SA-15 self-propelled surface-to-air missile system (*Jane's Land-Based Air Defence 1991-92*, page 138).

(4) SA-17 self-propelled surface-to-air missile system (*Jane's Land-Based Air Defence 1991-92*, page 138).

It is expected that other weapons systems/missions have been allocated to the MT-SB chassis.

Status: In production. In service with USSR.

UNITED KINGDOM

Future Family of Light Armoured Vehicles (FFLAV)

In June 1989 the British MoD briefed manufacturers from NATO and other European countries on its Future Family of Light Armoured Vehicles (FFLAV) which is expected to be the largest British AFV programme carried out since the Second World War.

There are three phases to the FFLAV. Phase 1, carried out by the MoD Design Authority contractors, covered a study of the current fleet to ascertain how closely current vehicles meet, or can be upgraded to meet the FFLAV requirement. Phase 2 covers the proposals from industry described below and Phase 3 is the assessment and analysis of all vehicles and proposals from Phases 1 and 2.

The FFLAV was first revealed in 1987 and since then requirements have been identified by the British Army and the total number of vehicles needed has increased from around 5500 to as many as 7000. The first of these are required to be in service from the mid-1990s.

When responding to the proposals, the prime contractors include budgetary costs and time scales for producing each vehicle against its role requirement in varying quantities, suggested terms and reciprocal purchase or collaborate development agreements where appropriate. Statements as to the level against each requirement with any shortfalls identified and discussed in terms of trade-off against costs and other aspects and answers to specific questions regarding the implications of increasing the performance requirements in certain areas, such as ballistic protection and mobility are also required.

The proposals were assessed on the basis of the level of achievement against the CPR, confidence level demonstrated against that achievement, value and implications of any trade-offs suggested, risk level and capital cost to the MoD. This will give the British Army a range of options from which to select to meet the FFLAV requirement.

There are 13 basic role configurations of the FFLAV, namely ambulance, armoured personnel carrier, combat engineer tractor, command, communications, liaison, load carrier, observation, reconnaissance, REME recovery, REME repair, Rover and a self-propelled weapon which will further be divided up into 50 sub-roles.

The MoD stresses, however, that some of these requirements may be filled by current British Army vehicles either re-roled or up-graded. The remainder of the requirement could be met by buying 'off-the-shelf', including possible reciprocal purchase arrangements, or by new developments, including possible collaborative ventures. Finally, a brand new fleet option could also be considered by the MoD.

The requirements in the CPR are set out as a hierarchy to allow each option to be assessed against a common format. The six main headings of the performance hierarchy are operability, firepower, survivability, movement, fightability and role specific.

The MoD is stating the requirements in nominal performance terms rather than specific data, but both tracked and wheeled vehicles will be considered. The lighter weight requirement would be met by wheeled vehicles and the heavier requirement by tracked vehicles with 5 tonnes being the lightest member of the FFLAV and 25 tonnes the heaviest. To reduce both procurement and life cycle costs, proven and in-production components would be used wherever possible.

FFLAV could replace a large part of the British Army's current AFV fleet as well as undertaking new roles at present carried out by unarmoured vehicles, especially command, communications and ambulance, and for mounting new weapon systems including TRIGAT.

Vehicles to be replaced include FV432 series of APC (although part of the fleet is being replaced by the GKN Defence Warrior), Alvis Scorpion CVR (T) series, RO/Vickers Defence Systems Fox, Daimler Ferret and the RO Combat Engineer Tractor.

Late in 1989 it was announced that a consortium led by engineer and software consultants Rex Thompson & Partners had been selected by the British MoD to help assess proposals for the FFLAV programme. Rex Thompson & Partners will assist the MoD to analyse the suitability of these proposals. The output of the assessment phase will be a series of detailed procurement options on which the MoD can base a final decision.

Phase 1 of FFLAV was completed late in 1989 with Phase 2 being completed early in 1990. Following the decision, taken in mid-1990, to reduce the size of the British Army over a four year period, it is expected that the whole future of the FFLAV will be reviewed and that the total number of vehicles will be well below the figure of 7000 envisaged several years ago.

As of June 1991 a full list of the FFLAV contenders had not been released, although the following are known to be bidding:

Alvis (UK)	existing vehicles plus Panhard VBL (4 × 4). Pegaso BMR (6 × 6) and Hägglunds Combat Vehicle 90 (chassis only)
FIAT/IVECO	Puma in 4 × 4 and 6 × 6 configurations
GKN	existing vehicles plus MOWAG Piranha (6 × 6 and 8 × 8), FMC Fighting Vehicle Systems Carrier and Krupp-MaK (specialised engineering experience)
Royal Ordnance	teamed with Krauss-Maffei and offering Puma armoured combat vehicle to meet about half of the FFLAV requirement
Steyr-Daimler-Puch	Pandur (6 × 6) to meet six of the FFLAV roles and Kampfschützenpanzer 90 ACV to meet a further 17 roles
Vickers Shipbuilding & Engineering Limited	AS90 chassis to meet combined repair and recovery vehicle requirement

Alvis LPXV Reconnaissance Vehicle

Development

The private venture Alvis LPXV (Low Profile Experimental Vehicle) was announced at the 1988 British Army Equipment Exhibition during which a non-running concept demonstrator was shown for the first time.

Alvis regard the LPXV as a testbed to explore new vehicle technologies for the 1990s which could form the basis for a family of light tanks, fire support and reconnaissance vehicles to meet the needs of world armies from the mid 1990s and well into the 21st century.

The LPXV is complementary to the Alvis Stormer APC and weapon systems carrier which has been adopted by the British Army as the platform for its Starstreak High Velocity Missile System.

The LPXV has been designed to mount a wide range of Alvis and other turrets armed with weapons up to 90 mm in calibre with maximum vehicle weight not exceeding 15 tonnes. It also has the space and payload to carry short or medium range anti-tank missiles in addition to the sophisticated surveillance equipment needed for this role.

According to Alvis, the vehicle's low silhouette and compact configuration minimise target area and allow a high degree of protection to be achieved within the vehicle's weight limit. Attention has been given to reducing signatures to a minimum and enhancing protection by every possible means. Alvis has taken advantage of the latest UK advances in stealth technology in the design of the LPXV.

In addition to offering various engines, the LPXV will also be offered with a range of different suspension options ranging from dampers to hydrogas, according to the customer's requirements.

At the 1988 British Army Equipment Exhibition the LPXV was shown fitted with the French GIAT Industries TS-90 two man turret armed with a 90 mm gun and carrying 55 rounds of ammunition, with a combat weight of 12.9 tonnes.

The upper weight limit of 15 tonnes has been selected to retain battle-ready transportability for all LPXV derivatives by truck, rail, helicopter, and fixed wing transport to enable rapid deployment to remote areas of operation.

Description

The hull of the LPXV is of all-welded aluminium armour construction which provides protection against penetration from 14.5 mm armour-piercing rounds and 155 mm artillery shell splinters.

The driver is seated at the left front with the powerpack to his right and the fighting compartment at the rear. He has an adjustable seat for driving in the head-out position and a single piece hatch cover that opens to the rear. Two periscopes are provided for forward observation, one to his front and one to he left; the former can be replaced by a passive night driving periscope.

The turbocharged and charged cooled diesel engine is coupled to a fully automatic microprocessor-controlled transmission with a full override capability.

A wide range of armament stations are being proposed including the following:

(1) GIAT Industries TS-90 two man turret with 90 mm gun and 7.62 mm coaxial machine gun, a total of 55 rounds of 90 mm and 2200 rounds of 7.62 mm ammunition are carried

(2) Bofors/Hägglunds two man turret armed with Bofors 40 mm L/70 gun and 7.62 mm coaxial machine gun, a total of 124 rounds of 40 mm and 2230 rounds of 7.62 mm ammunition are carried

(3) Alvis two man R30 turret with 30 mm RARDEN cannon and 7.62 mm coaxial machine gun, a total of 281 rounds of 30 mm and 2200 rounds of 7.62 mm ammunition are carried

(4) Alvis two man C30 turret armed with 90 mm Cockerill Mk 7 gun and 7.62 mm coaxial machine gun, a total of 43 rounds of 90 mm and 2800 rounds of 7.62 mm machine gun ammunition are carried

(5) Alvis two man M90 turret armed with 90 mm MECAR KEnerga cannon and 7.62 mm coaxial machine gun, a total of 43 rounds of 90 mm and 2800 rounds of 7.62 mm ammunition are carried

(6) Alvis two man G90 turret armed with GIAT Industries 90 mm gun and 7.62 mm machine gun, a total of 43 rounds of 90 mm and 2800 rounds of 7.62 mm machine gun ammunition are carried

(7) Fitted with Alvis OM 60 two man turret armed with 60 mm OTO Melara gun and 7.62 mm coaxial machine gun, a total of 41 rounds of 60 mm and 2400 rounds of 7.62 mm ammunition are carried

(8) Fitted with Alvis OB 35 two man turret armed with 35 mm Oerlikon-Contraves 35 mm KDE cannon and 7.62 mm machine gun, a total of 329 rounds of 35 mm and 2400 rounds of 7.62 mm ammunition are carried

On the standard vehicle, suspension is of the transverse torsion bar type with trailing arm geometry and lever type hydraulic dampers. Each side has six dual rubber-tyred aluminium alloy road wheels with the drive sprocket at the front and the idler at the rear and two track return rollers. The first two and last road wheel stations either side are fitted with high absorption double action dampers. Tracks are steel with rubber pad inserts, steel pins with rubber bushes.

A wide range of optional equipment is available including a flotation screen, communications equipment and navigational aids.

SPECIFICATIONS

CREW	3
COMBAT WEIGHT	13 000 kg
POWER-TO-WEIGHT RATIO	20.77 bhp/tonne
GROUND PRESSURE	0.53 kg/cm²
LENGTH	5.45 m
WIDTH	2.70 m
HEIGHT	
overall	2.40 m
over hull top	1.525 m
GROUND CLEARANCE	0.425 m
TRACK	2.05 m
TRACK WIDTH	432 mm
LENGTH OF TRACK ON GROUND	3.11 m
MAX ROAD SPEED	
forward	80 km/h
reverse	80 km/h
MAX WATER SPEED	
tracks	8 km/h
tracks and propulsion	12 km/h
ACCELERATION	
0 to 50 km/h	20 s
FUEL CAPACITY	450 l
MAX ROAD RANGE	650 km at 60 km/h
FORDING	1.2 m
GRADIENT	60%
VERTICAL OBSTACLE	0.46 m
TRENCH	1.75 m
ENGINE	turbocharged and charged cooled diesel developing 280 to 340 hp
TRANSMISSION	full-function microprocessor-controlled automatic with full override capability
STEERING	power-assisted control
SUSPENSION	torsion bar with trailing arm geometry and lever type hydraulic dampers
WHEEL MOVEMENT	
static to bump	274 mm
static to rebound	97 mm
BRAKES	
parking	hydraulic with power off control
main	power assisted disc
ELECTRICAL SYSTEM	28.5 V
BATTERIES	4, 110 Ah, 2 in hull, 2 in turret
ARMAMENT	see text

Status: Development.

Manufacturer: Alvis Limited
The Triangle
Walsgrave
Coventry
West Midlands CV2 2SP
UK
Telephone: (0203) 535455
Telex: 31459
Fax: (0203) 539280

Alvis LPXV fitted with GIAT Industries TS-90 turret

Alvis Scorpion Reconnaissance Vehicle

Development

In the late 1950s the British Army issued a requirement for an Armoured Vehicle Reconnaissance (AVR) to undertake the roles of reconnaissance, fire support and anti-tank. After studying a variety of proposals the Fighting Vehicles Research and Development Establishment, now the Royal Armament Research and Development Establishment Chertsey, came to the conclusion that two vehicles would be required to carry out these roles. These became the Combat Vehicle Reconnaissance (Tracked) Scorpion range and the Combat Vehicle Reconnaissance (Wheeled) Fox vehicle, both of which use the same Jaguar engine.

In September 1967 Alvis Limited of Coventry, which was at that time building the FV600 range of 6 × 6 armoured vehicles, was awarded a contract to build 17 prototypes. The first prototype was completed in January 1969 and the first official announcement of the vehicle was made in September the same year. All 17 prototypes were completed by the middle of 1970 and in May 1970 the Scorpion was accepted for service with the British Army. In October 1970 the Belgian Army placed an order for 701 Scorpions and variants, which were assembled at a British Leyland facility at Malines in Belgium. Some of the components of the British and Belgian vehicles were made in Belgium.

The first production Scorpions were delivered to the British Army in January 1972 with first deliveries being made to the Belgian Army in February 1973.

It was announced early in 1979 that the Royal Air Force Regiment had ordered 150 members of the Scorpion family (including Scorpion, Spartan, Sultan and Samson) for the defence of Royal Air Force airfields in Germany. In November 1981 the Royal Air Force Regiment at Catterick formally took delivery of its first Alvis CVR(T) Scorpions for airfield defence. This order was followed by another in 1982 for Scimitar vehicles.

Scorpion has been designed to operate in temperatures from -30 to +50°C and a Lockheed C-130 Hercules transport can carry two of them.

Many countries order Scorpions modified to meet their own particular operational requirements. For example the New Zealand vehicles, which were all delivered by late 1983, have electronic ignition systems, Dunlop power traverse and American radios, they do not have the NBC pack or the night sight to the right of the main armament. The Malaysian vehicles have the Perkins diesel engine, 90 mm gun and Wegmann smoke grenade launchers.

Standard Alvis Scorpion armed with 76 mm gun and fitted with flotation screen which is carried collapsed around hull top. A 7.62 mm machine gun is pintle-mounted on the turret roof

In June 1988 Alvis won an $85 million contract from Venezuela for the supply of an undisclosed number (believed to be 84) of Scorpion 90 light tanks, support vehicles, ammunition, communications equipment and both turret and desk top training equipment. First vehicles were completed late in 1989 and deliveries are scheduled to run through to 1992.

The Scorpion 90s are armed with the Cockerill 90 mm Mk III gun which is capable of firing APFSDS ammunition, Avimo fire control system incorporating a laser rangefinder, Helio periscopes and an external machine gun mount. The vehicles are powered by a Perkins diesel engine coupled to the TN15D transmission.

Support vehicles included in the order are the Samson recovery vehicle and the Sultan command post vehicle, this being the first time that diesel powered versions of these vehicles have been placed in production. By January 1st 1989 Alvis had completed the following quantities of Scorpion family vehicles:

Scorpion 76	1241
Scorpion 90	32
Striker	132
Spartan	967
Samaritan	101
Sultan	372
Samson	136
Scimitar	486

Current production vehicles are referred to by Alvis as the Scorpion 2 and these incorporate many improvements, some of which were incorporated in earlier production vehicles for the export market. These include the proven Perkins diesel engine which is much more fuel efficient, uprated suspension system, track and running gear, and further enhanced RAM-D characteristics with options including NBC system, air-conditioning system, fire detection and suppression, fully automatic transmission and power steering. The two man turret systems have been refined and upgraded and options now include 30 mm, 76 mm or 90 mm main armament, combined day/night sights for commander with laser rangefinder options for the gunner, advanced electro-mechanical gun and turret drive system, refined electrical system, enhanced capacity for communications equipment and improved ergonomics.

Many of these improvements are now available as retrofit options that can be undertaken by the user or the vehicles can be returned to Alvis for overhaul and modernisation.

Description

The hull of the Scorpion is made of all-welded aluminium armour and provides the crew with protection against attack over its frontal area from 14.5 mm projectiles and against 7.62 mm armour-piercing rounds over the remainder of the vehicle. The aluminium armour is also particularly effective against shell splinters. The hull is divided into three compartments: driver's at the front on the left, engine at the front on the right and the fighting compartment at the rear.

The driver has a single piece hatch cover that swings to the left, in front of which is a single wide-angle periscope which can be replaced by a Pilkington passive periscope for night driving.

The Scorpion is powered by a de-rated (from 265 to 190 bhp) and militarised Jaguar 4.2-litre engine with a Marcus carburettor and a compression ratio reduced from 9:1 to 7.75:1, which permits the use of military fuels. The transmission, which is at the front of the vehicle, has been developed on the same principles as that used for the Chieftain MBT. It is a hot-shift, foot-operated, seven-speed gearbox with a controlled differential steering system. For engine cooling a single mixed flow fan draws in air through the radiator over the gearbox, over the engine and out through the louvres.

The other two crew members are seated in the all-welded aluminium armour turret with the commander on the left and the gunner on the

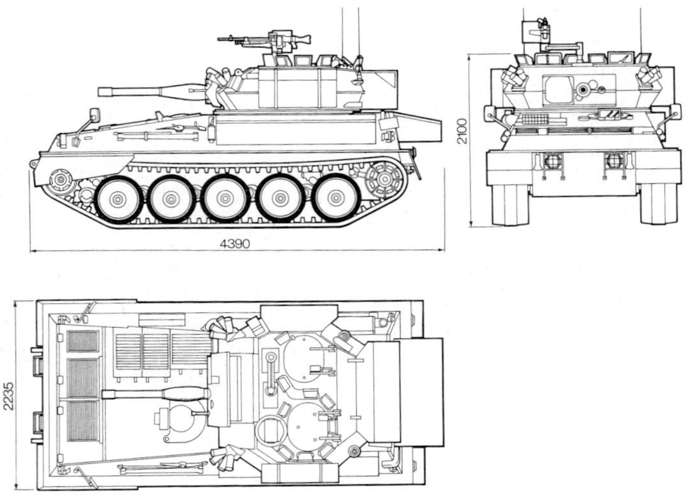

Combat Vehicle Reconnaissance (Tracked) 76 mm Scorpion (not to 1/76th scale)

right, both with a single piece hatch cover that opens to the rear. The commander has seven periscopes and a roof-mounted sight in front of his hatch cover with magnifications of ×1 and ×10. The roof-mounted sight is capable of limited rotation allowing approximately an 85° horizontal field-of-view. The gunner has two periscopes and a roof-mounted sight with magnifications of ×1 and ×10. Mounted to the right of the main armament is a GEC Sensors passive night sight with a magnification of ×5.8 (8° field-of-view) and a low magnification of ×1.6 (28° field-of-view). There can be no confusion between the two as when the high magnification is being used a shutter isolates the low magnification objective. The image intensifier diaphragm isolates the high magnification objective. The image intensifier tube is protected from the effect of gun muzzle flash by a flash shutter that is operated electrically from the gun firing circuit. When high magnification is selected an illuminated ballistic graticule with brightness control is automatically injected into the optical system. The exposed objective window is cleaned by a wiper and washer and the sight is protected by an armoured cover with a door which is kept closed when the sight is not in use. The sight will detect infra-red devices.

The radios are in the turret bustle and a light metal stowage box is mounted externally at the rear of the hull. British Army Scorpions now have a stowage box on the left side of the hull.

The torsion bar suspension consists of five rubber-tyred aluminium road wheels with the drive sprocket at the front and the idler at the rear. There are no track return rollers. Hydraulic lever type shock absorbers are provided for the first and last road wheel stations either side. The tracks are of light steel with rubber bushes and pads, and have a life of over 5000 km of mixed road and cross-country running.

British Army Scorpions have an NBC system at the rear of the hull, but if it is not installed an additional five rounds of 76 mm ammunition can be carried. Optional equipment includes an NBC detector kit, vehicle navigation system and a Galley air-conditioning system which has already been fitted on the vehicles supplied to the United Arab Emirates.

The Scorpion can ford to a depth of 1 m without preparation. A flotation screen carried collapsed around the top of the hull can be erected by the crew in five minutes and the vehicle is then propelled and steered across the stream or river by its tracks at a speed of 6.5 km/h. A propeller kit has also been developed and when fitted with it the Scorpion has a maximum water speed of 9.5 km/h.

The Scorpion is armed with a Royal Ordnance Nottingham 76 mm L23 gun which is a lighter version of the L5 used in the Saladin armoured car. This weapon has a vertical sliding breech and is loaded with fixed ammunition. It has a recoil of approximately 280 mm and is returned to the firing position by a hydro-pneumatic recuperator. During runout the breech is opened by a semi-automatic cam, the empty cartridge case is ejected and the breech then remains open ready for reloading. The gun has an elevation of +35° and a depression of −10°, turret traverse being 360°. Elevation and traverse are both manual but if required the vehicle can be delivered with powered traverse. The following types of ammunition can be fired: canister, HESH, HE, HE/PRAC, SH/P, SMK/BE and illuminating.

A 7.62 mm machine gun mounted coaxially to the left of the main armament can be used as a ranging machine gun. Mounted each side of the turret is a four-barrelled electrically operated smoke discharger.

In June 1987 the British Ministry of Defence issued a requirement for 494 radiological survey meters for Fox (101), Scorpion (105), Spartan (117) and Scimitar (171) vehicles, for delivery by February 1990. Each comprises a probe unit, radiation screen and control and display unit (probe and screen assembled to mast) fitted with interconnecting supply/signal cable.

Belgian Scorpion and Scimitar

Between 1981 and 1984 the Belgian Army's 286 Scorpion and Scimitar vehicles underwent two major repair and maintenance programmes conducted by the 65e Section of the Arsenal du Matériel Mécanique et de l'Armement at Rocourt. The first programme, called SCOREPOLE, included modifications to rectify cracks in the armour including needle peening, shot peening, adding of reinforcement ARP (Armoured Repair Plates), screw-up of the hair line cracks, sewing of the plates (metalock) and welding. The BARGEPOLE programme which was combined with a mid-life general overhaul, was a vehicle improvement programme aimed at improved safety for the crew, better operational capability, improved reliability and easier maintenance.

Irish Scorpions

Ireland ordered 14 Scorpion vehicles, the first of these being taken into service in March 1980 and operated by 1 Tank Squadron. The Irish Scorpions have a 12.7 mm M2 HB machine gun on an M63 anti-aircraft mount installed on the turret roof.

New Zealand Scorpions

New Zealand purchased 26 Scorpions as a reconnaissance vehicle and tank training expedient and these entered service in August 1982. Differences from British Army vehicles include installation of an electronic ignition system, Dunlop power traverse system, American radios and no NBC or image intensification night sight installed.

Omani Scorpions

Scorpions replaced the Alvis Saladin armoured car in Oman and between 30 and 50 vehicles were delivered between late 1982 and late 1983, plus three Samson armoured recovery vehicles. A second order for up to 30 vehicles was delivered in 1985; this included Scorpion, Sultan, Spartan and Samson vehicles. The Omani Scorpions had many modifications, including installation of a central warning system, fitting of a coolant temperature gauge connected to a fire warning horn, larger rubber mud flaps at rear, glass fibre track shrouds replaced by rubber shrouds, driver's mine plate increased to 20 mm, larger rear hull stowage bin, left tool bin replaced by larger bin with facility to stow two extra jerrycans, additional stowage bin on right side of hull, additional insulation for engine/driver bulkhead, no NBC pack so five rounds of 76 mm ammunition and three boxes of machine gun ammunition in its place, No 25 reel and bracket installed, turret fume extractor fitted, Helio MG pintle fitted adjacent to commander's hatch aperture, gun selector box lights duplicated at commander's station, stowage bins either side of turret, time switch connected to fire warning circuit to isolate horn after five to 10 minutes' operation in event of false alarm, removable sealed plates fitted to engine/driver bulkhead to permit access to distributor and oil filter, latest type fuel filler and gauze fitted, heater removed and spare driver's sight carried, Fox type silencer fitted, latest jacking strut fitted, electrical feed to elevating handle to be slit-ring type, rear turret bin catches strengthened, warning notices in Arabic, speedometer in km/h in lieu of mph and fitted with trip distance recorder as well as total distance recorder (also in km), 7.62 mm ammunition box brackets to be modified to suit pattern used in Oman, two-colour camouflage paint, air-circulation system fitted using fan normally incorporated in NBC system, mounting points to enable Clark type radio mast to be erected at front of vehicle, AV No 54 sight mounting and turnbuckle supplied, NANOQUEST L20 laser sights to complement the L22 TLS fitted in the Omani Chieftain MBTs.

Scorpion 90

For the export market Alvis has fitted a Scorpion with a 90 mm Cockerill Mk III gun. The turret of the Scorpion 90 is designated the AC 90 and is almost identical to that fitted to the basic vehicle.

The 90 mm gun has an elevation of +30° and a depression of −8° which is almost the same as that of the basic Scorpion. The first production orders are fitted with the American Cadillac Gage Textron power traverse and elevation system which has now been superseded by Marconi Command and Control Systems equipment.

The basic Scorpion carries 40 rounds of 76 mm ammunition and the Scorpion 90 carries 34 rounds. When fitted with the 90 mm Cockerill

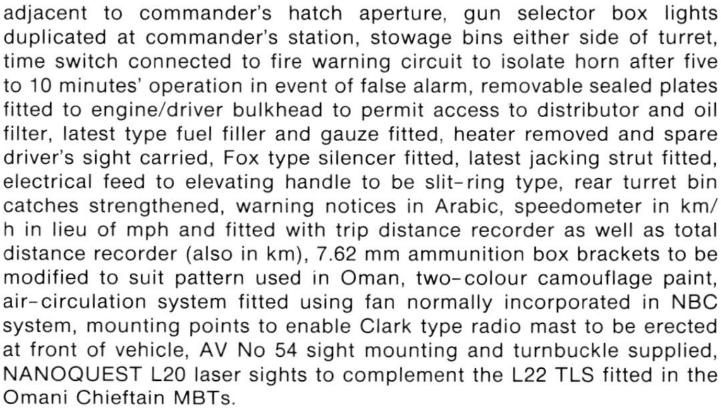

Scorpion 90 armed with 90 mm Cockerill 90 Mk III gun, roof-mounted anti-aircraft machine gun mount and without flotation screen

gun the vehicle can deal with all types of target and can inflict severe damage on MBTs. Late in 1981 Malaysia ordered 26 Scorpion 90s fitted with the Perkins diesel engine. Production commenced in 1982 and deliveries are now complete. Alvis supplied one Scorpion 90 with a Perkins diesel model T6-3544 engine, developing 200 bhp at 2700 rpm, to the USA for evaluation in the USMC Light Armored Vehicle project.

The Malaysian vehicles also have German Wegmann 76 mm smoke/fragmentation grenade dischargers fitted to give 360° coverage. Early in 1983 it was confirmed that 33 Nigerian Scorpions were fitted with the Belgian OIP-5 fire control system and a Cockerill 90 mm Mk III weapon.

In mid-1988 Venezuela placed an order with Alvis for the supply of an undisclosed number (believed to be 84) of Scorpion 90 mm vehicles (see development).

Scorpion Improvement Programme

In 1980 Alvis announced that a Scorpion with its Jaguar XK petrol engine replaced by a Perkins turbo-charged six-cylinder diesel model T6-3544 would be offered for export. This model entered production for Malaysia in 1982 with first deliveries made in 1983. The engine, for this application, develops 200 bhp at 2700 rpm and is already in production. Performance comparison between the diesel engine and the basic petrol engined versions is given below:

Model	**Diesel**	**Petrol**
WEIGHT	8260 kg	8073 kg
GROUND PRESSURE	0.384 kg/cm^2	0.376 kg/cm^2
ROAD RANGE	over 840 km	over 640 km
ACCELERATION TO 48.3 km/h	18.5 s	16 s

At least three British companies, Avimo, Barr and Stroud, and GEC Sensors, have already developed laser rangefinders for the Scorpion. The Avimo model is called the LV10 and is an integrated laser rangefinder/gunner's sight which can be fitted into the Scorpion without modification to the vehicle. Barr and Stroud has developed the LF11 laser rangefinder which has already been adopted by two countries for installation in the Scorpion and is suitable for installation in other AFVs such as the Vickers MBT. GEC Sensors has developed the SS125/126 Argus day/night sight which can be installed at either the commander's or gunner's position and can be fitted with a laser rangefinder as can the Avimo NVL53 day/night sight, if required. The Omani Scorpions have the NANOQUEST L20 sight which has a Ferranti laser rangefinder fitted.

In October 1986 Ferranti announced that NANOQUEST had ordered an additional batch of Ferranti Type 520 laser rangefinders for incorporation in the Vickers L20 gunner's sight. A total of 160 L20 sights are to be provided to a Far Eastern customer over a four year period. It is believed that the customer is Thailand which is the largest user of Scorpion in the Far East.

Some of the Scorpions used by the United Arab Emirates have been fitted with a Barr and Stroud Tank Laser Sight similar to the one fitted to the Chieftain MBT but with a neodymium-YAG laser in place of the original ruby laser and a graticule for a 76 mm instead of 120 mm gun.

Marconi Command and Control Systems has also developed a version of its SFCS 600 tank fire control system for the Scorpion as well as a powered traverse system. A system with both powered elevation and traverse has been developed by Marconi.

A complete package of improvements is now offered on export vehicles. This includes:

Avimo NV53 day/night commander's sight
Avimo NVL53 day/night gunner's sight with integral laser rangefinder
Power traverse and elevation drives
External 7.62 mm GPMG mount which can be used by commander or gunner
Revised final drive units
Revised sprocket carrier and rear idler unit
Longer life tracks
New fume extraction system
New electrics with low maintenance batteries
Improved transmission
Improved suspension and enhanced cross-country mobility
Fire suppression system
A range of thermal imaging sights is also under consideration.

Scorpion with Swingfire ATGWs

British Aerospace, (Dynamics), has proposed that a Swingfire ATGW could be mounted either side of the turret to give the vehicle a long-range anti-tank capability.

Striker Anti-tank Guided Weapon Vehicle (FV102)

The first production Striker was delivered to the British Army in June 1975 and it entered service with the BAOR in 1976. It is issued to the anti-tank units of Royal Armoured Corps. First deliveries to the Belgian Army were made in June 1976. The Belgian Army operates a total of 43 Striker vehicles.

The Striker is based on the hull of the Spartan APC with the driver at the front on the left, commander to the rear of the driver and the missile controller to the right of the commander. The commander has a No 26 cupola which has eight periscopes, monocular sight with a magnification of ×1 or ×10 with the line of sight elevating with the machine gun. The ×1 mirror assembly can be replaced by a ×1.8 image intensifier. Mounted on the right side of the cupola is a 7.62 mm machine gun which can be aimed and fired from within the vehicle and the cupola has a single piece hatch cover that opens to the left.

British Army Striker launching Swingfire ATGW

The missile controller has a split-view monocular sight with magnifications of ×1 and ×10 which can be traversed through 55° left and right, and a single piece hatch cover that opens to the right is provided to the rear of his sight.

On the roof at the rear of the vehicle is a launcher box with five British Aerospace Swingfire ATGWs. A further five missiles carried inside the vehicle are loaded manually and one of the crew has to leave the vehicle to accomplish this. The launcher box is pivoted at the rear and is elevated to an angle of about 35° before the missiles are launched.

The Swingfire missile weighs 28 kg and has a HEAT warhead. Minimum range is between 150 and 300 m depending on the distance of the vehicle and the controller when being used in the separated mode. Maximum range of the Swingfire is 4000 m. An advantage of the Swingfire ATGW system over other systems is that the missiles can be launched from behind a crest or cover so that the launcher position cannot be seen by the enemy. The missiles can be launched by the controller in the vehicle, or with the controller at a vantage point (up to 100 m from the launcher) separated from the launcher giving him protection from counter fire.

Mounted at the front of the vehicle are two four-barrel smoke dischargers which are operated from within the vehicle. The vehicle has recently been equipped with a Barr and Stroud thermal imaging night sight.

In the Spring of 1990 British Aerospace announced that it had been awarded a £35 million contract from the British MoD to upgrade the British Army Striker/Swingfire fleet over a five year period.

Much of the Swingfire's existing missile control and selection systems will be replaced and this will include the integration and installation of a new guidance and tracker systems.

The inclusion of an Automatic Command Line of Sight (ACLOS) capability leads to the improvement in guidance accuracy as well as reducing operator stress.

The ACLOS capability was fully demonstrated by British Aerospace in the 1980s under a contract from the Royal Armament Research and Development Establishment (RARDE). During these trials, targets were tracked and missiles steered onto them automatically under a variety of experimental conditions.

In addition, the contract covered revised test and training equipment for the new system together with the modification of all of the Striker vehicles used by the British Army.

Spartan Armoured Personnel Carrier (FV103)

This entered service with the BAOR in 1978 and is used for a variety of roles such as carrying Royal Artillery Javelin/Starburst manportable SAM teams, missile resupply carrier for the Striker and for carrying Royal Engineer assault teams. It is not the replacement for the FV432 APC.

The Spartan has a similar hull to that of the Striker and can carry four infantrymen in addition to the driver, vehicle commander/gunner and section commander/radio operator.

The driver is seated at the front of the hull on the left and has a single piece hatch cover that opens forwards. He is provided with a single wide-angle periscope which can be replaced with a passive periscope for night driving.

The vehicle commander/gunner is seated behind the driver and has a No 16 cupola with eight periscopes, monocular sight with a magnification of ×1 or ×10 with the line of sight elevating with the machine gun. The ×1 mirror assembly can be replaced by a ×1.8 image intensifier. Mounted on the right side of the cupola is a 7.62 mm machine gun which can be aimed and fired from inside the vehicle. The cupola has a single piece hatch cover that opens to the left.

To the right of the commander/gunner is the section commander/radio operator who is provided with three observation periscopes and a single piece hatch cover that opens to the right.

Entry to the personnel compartment at the rear of the vehicle is by a single door in the rear of the hull hinged on the right side, which has an integral vision block. Over the top of the troop compartment are two roof hatches that open either side of the vehicle. Two periscopes are provided in the left side and one in the right side of the troop compartment

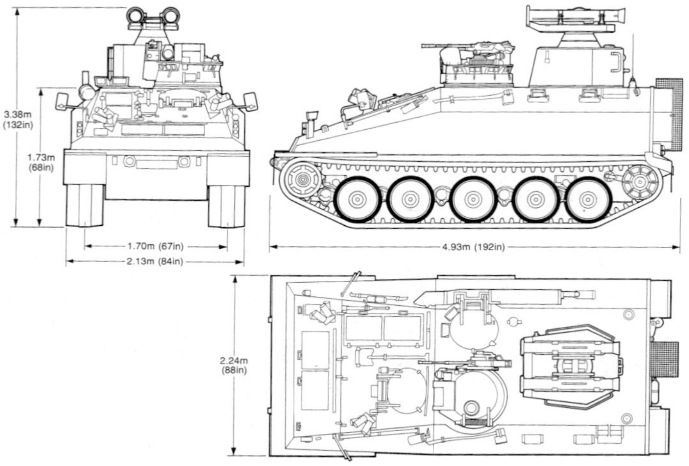

Three-view drawing of Alvis Spartan fitted with Euromissile MILAN Compact Turret (MCT) which is only used by the British Army (not to 1/76th scale)

Alvis Spartan APC used by the Belgian Army as a command post for a M109A2 self-propelled artillery battery (C R Zwart)

Spartan APC with Euromissile MILAN MCT turret with two ready to launch missiles

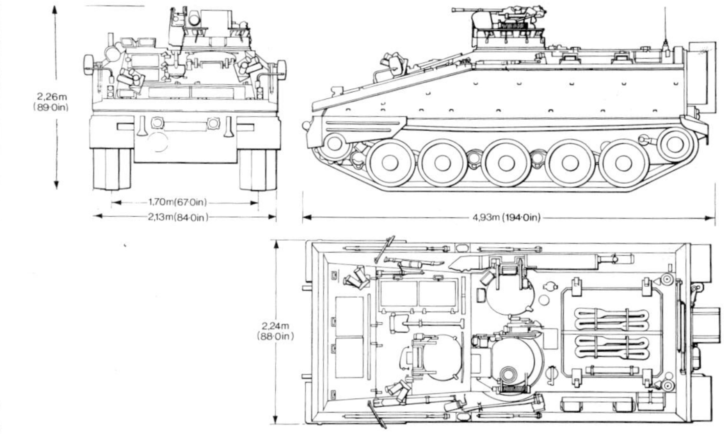

Three-view drawing of Alvis Spartan APC (not to 1/76th scale)

but there is no provision for the crew to fire their weapons from inside the vehicle. Three infantrymen are seated on the left side of the vehicle and a fourth to the rear of the vehicle commander/gunner and section commander/radio operator, facing the rear. A ZB 298 ground surveillance radar can be mounted on the roof of the vehicle if required.

Spartan Anti-aircraft Variants

It has been proposed that the Spartan could be fitted with the French Dassault Electronique TA 20 turret armed with twin 20 mm cannon, which has a depression of -5° and an elevation of +85°, turret traverse being 360°.

Spartan with Hughes TOW ATGW System

In 1973 Alvis proposed that the Spartan could be fitted with a Hughes TOW launcher on the roof and carry nine missiles in the rear of the vehicle. A tripod would also be carried enabling the missile to be used in the ground role.

Spartan with Euromissile HOT ATGW System

For trials purposes a Spartan APC has been fitted with a Euromissile HOT HCT turret with four HOT 4000 m ATGWs in the ready to launch position. This was fitted on a plinth above the troop compartment at the rear of the vehicle.

Spartan with Euromissile MILAN ATGW System

Shown for the first time in 1980 was an Alvis Spartan APC fitted with a Euromissile MILAN MCT compact turret over the roof with two missiles in the ready to launch position, can be installed on the Spartan during manufacture or retrospectively fitted to existing vehicles without major modification. In addition to the two ready to launch missiles a minimum of eight missiles would be carried internally.

In June 1985 the Spartan MCT combination was formally accepted by the British Army and orders were placed for 75 Euromissile turrets. In addition to the two ready to launch missiles a further 11 missiles are carried internally, seven on the left side and four on the right. The first Spartan MILANs for the British Army were handed over by 23 Base Workshops, REME, in Wetter, Germany, in September 1986. Each mechanised infantry battalion in BAOR will receive four Spartan MILAN vehicles. Each vehicle will also carry a standard MILAN firing post so that MILAN can be deployed away from the vehicle if required.

Stormer Armoured Personnel Carrier

Details of this vehicle, previously known as the FV4333, are given in the *Armoured personnel carriers* section, as are details of a new family of light tracked combat vehicles, including the Sagitar which can be armed with a variety of weapons including a 105 mm gun.

Samaritan Armoured Ambulance (FV104)

This model entered production early in 1978 and has the same hull as the Sultan command vehicle and has a crew of two: commander/medical orderly and driver. It is unarmed and can carry four stretcher cases, or five sitting cases, or two stretcher and three sitting cases. There is a large door at the rear of the hull, hinged on the right, with an integral vision block. The commander has five periscopes for all-round observation, plus a single wide-angle periscope with a forward-sloping window with an interchangeable passive night sight. The single piece hatch cover opens to the left. Medical supplies can be carried on the top of the hull and at the rear of the vehicle.

Samaritan armoured ambulance in travelling order (Terry J Gander)

Sultan Armoured Command Vehicle (FV105)

First production Sultans were delivered to the British Army in April 1977 and have replaced the Saracen vehicles as command vehicles (eg FV604 and FV610). The vehicle has a similar hull to the Samaritan and has a crew of five or six: commander/radio operator, radio operator, driver plus two or three additional crew members. The command area is at the rear of the vehicle and to increase the amount of working space a tent can be quickly erected at the rear of the vehicle. The vehicle is normally equipped with at least two radios (one forwards and one rearwards), and mapboards. A radio antenna can be erected at the front of the vehicle when it is being used in the static role. The commander has a Number 27 cupola with five periscopes for all-round observation,

Alvis Sultan command vehicle crossing bridge (C R Zwart)

British Army Alvis Scimitar armed with 30 mm RARDEN cannon

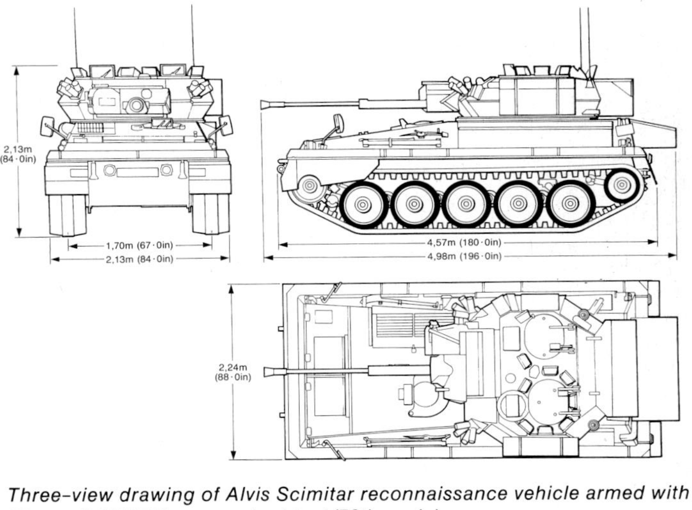

Three-view drawing of Alvis Scimitar reconnaissance vehicle armed with 30 mm RARDEN cannon (not to 1/76th scale)

Belgian Army Alvis Samson armoured recovery vehicle from rear with spade lowered and A-frame erected (C R Zwart)

plus a single wide-angle periscope with a forward-sloping window with an interchangeable passive night sight. The single piece hatch cover opens to the left. Armament consists of a pintle-mounted 7.62 mm machine gun and four smoke dischargers mounted on either side of the vehicle towards the front. This vehicle is currently being considered as a carrier for a range of electronic warfare systems.

Samson Armoured Recovery Vehicle (FV106)

This entered service with the British Army in 1978 and has a hull similar to that of the Spartan APC. The driver is seated at the front of the vehicle on the left and has a single piece hatch cover that opens forwards and a single wide-angle periscope which can be replaced by a passive night periscope. The commander's No 27 cupola has a single wide-angle periscope which can be replaced with a passive night periscope, five periscopes for all-round observation plus a single piece hatch cover that opens to the left.

The heavy-duty winch fitted inside the hull is driven from the main engine and has a variable speed of up to 122 m/min on the 229 m of wire rope. Maximum pull, with a 4:1 snatch block, is 12 000 kg. The winch leads to the rear of the vehicle and when being used for recovery operations two spades are lowered manually at the rear of the vehicle. An A-frame can be erected at the rear of the hull. Tools, a bench, tow bars and tow cables are fitted as standard. Armament comprises a pintle-mounted 7.62 mm machine gun and four smoke dischargers mounted on either side of the vehicle towards the front. Full details of the Samson ARV are given in *Jane's Military Vehicles and Logistics 1991-92,* page 34.

Scimitar Reconnaissance Vehicle (FV107)

The first prototype of the Scimitar was completed in July 1971 and accepted for service in June 1973. First deliveries were made to the British Army in March 1974 with first deliveries being made to the Belgian Army in April 1974. It has the same hull and turret as the Scorpion but is armed with a 30 mm RARDEN cannon instead of the 76 mm gun.

The 30 mm cannon was designed by the Royal Armament Research and Development Establishment at Fort Halstead and the Royal Small Arms Factory at Enfield. It fires 30 mm Hispano Suiza Oerlikon ammunition as well as a number of British rounds. In addition the 30 mm RARDEN cannon fires a new APDS-T round developed jointly by Royal Ordnance and PATEC of the United States. In addition to this APDS-T round the 30 mm RARDEN fires British APSE-T, HEI-T and PRAC-T ammunition and Oerlikon-Contraves HE, AP and PRAC rounds.

The gun is fired in rapid single shots but bursts of up to six rounds can be fired if required. The empty cartridge cases are automatically ejected outside the turret.

A 7.62 mm machine gun is mounted coaxially to the left of the main armament and four smoke dischargers are mounted either side of the front of the turret. The main armament, coaxial machine gun and smoke dischargers are electrically operated, with the main armament and coaxial machine gun having manual overriding control. The Royal Air Force uses them in the EOD role on airfields.

Streaker HMLC

Late in 1982 Alvis completed the prototype of the private venture Streaker high mobility load carrier which is based on the chassis of the Spartan APC. This is powered by the standard 4.2-litre petrol engine which could however be replaced by the Perkins T6/3544 200 bhp turbo-charged diesel coupled to an automatic transmission. Maximum road speed is 84 km/h, road range 480 km, the unloaded weight is 5465 kg and it has a payload of 3000 kg.

The baseline vehicle has a flat deck and offers two front configurations, one with a lightweight cab structure which is unarmoured, the other with the driver and engine compartment having full armour protection.

By 1984 two prototypes of the Streaker high mobility load carrier had been built. Typical roles suggested for Streaker include towing the Royal Ordnance Bar minelayer, being fitted with the Belgian SA LAU 97 40-round 70 mm multiple rocket launcher (details in the *Multiple rocket launchers* section), excavator, refueller, trench digger, missile carriage, towing 105 mm Light Gun, carrying anti-aircraft systems, general cargo carrier and Ranger anti-personnel mine layer. As of early 1991 this model had not entered production.

Full details of the Streaker are given in *Jane's Military Vehicles and Logistics 1991-92,* page 472.

Status: In production. In service, or ordered by Belgium, Brunei (16 Scorpion, 2 Sultan and 1 Samson), Honduras (12 Scorpion 76 mm and 3 Scimitar 30 mm), Iran, Ireland (12 Scorpion), Malaysia (26 Scorpion 90s with the Perkins diesel engine were ordered in late 1981 and delivered in 1983), New Zealand (in August 1980 26 Scorpions were ordered at a cost of $8.2 million, to replace M41 light tanks), Nigeria, Oman (Scorpion 76 mm, Spartan, Samson and Sultan), Philippines, Spain (17 Scorpion 76 mm with diesel engines ordered for Marines in 1984), Tanzania, Thailand, Togo (12 Scorpion 90 mm delivered late in 1986), United Arab Emirates (Abu Dhabi), United Kingdom (Army and Air Force) and Venezuela (Scorpion 90, Samson and Sultan).

Manufacturer: Alvis Limited (A member of the United Scientific Group)
The Triangle
Walsgrave
Coventry
West Midlands CV2 2SP
UK
Telephone: (0203) 535455
Telex: 31459
Fax: (0203) 539280

Alvis Streaker High Mobility Load Carrier towing Royal Ordnance Bar minelayer

SPECIFICATIONS

Model	**Scorpion**	**Scorpion 90**	**Striker**	**Spartan**	**Samaritan**	**Sultan**	**Samson**	**Scimitar**
DESIGNATION	FV101	n/app	FV102	FV103	FV104	FV105	FV106	FV107
CREW	3	3	3	3+4	2	5 - 6	3	3
COMBAT WEIGHT	8073 kg	8723 kg	8346 kg	8172 kg	8664 kg	8664 kg	8738 kg	7800 kg
POWER-TO-WEIGHT RATIO	23.54 bhp/tonne	22.92 bhp/tonne	22.77 bhp/tonne	23.25 bhp/tonne	21.93 bhp/tonne	21.93 bhp/tonne	21.74 bhp/tonne	24.5 bhp/tonne
GROUND PRESSURE	0.36 kg/cm^2	0.37 kg/cm^2	0.345 kg/cm^2	0.338 kg/cm^2	0.358 kg/cm^2	0.358 kg/cm^2	0.358 kg/cm^2	0.338 kg/cm^2
LENGTH GUN FORWARDS	n/app	5.288 m	n/app	n/app	n/app	n/app	n/app	4.985 m
LENGTH HULL (inc stowage boxes where applicable)	4.794 m	4.794 m	4.826 m	5.125 m	5.067 m	4.8 m	4.788 m	4.794 m
WIDTH	2.235 m	2.226 m	2.28 m	2.242 m	2.242 m	2.254 m	2.43 m	2.242 m
over tracks	2.134 m	2.134 m	2.134 m	2.134 m	2.134 m	2.134 m	2.134 m	2.134 m
HEIGHT OVERALL	2.102 m	2.102 m	2.28 m	2.26 m	2.416 m	2.559 m	2.254 m	2.096 m
GROUND CLEARANCE	0.356 m	0.419 m	0.356 m	0.356 m	0.356 m	0.356 m	0.356 m	0.42 m
TRACK	1.708 m	1.708 m	1.708 m	1.708 m	1.708 m	1.708 m	1.708 m	1.708 m
TRACK WIDTH	432 mm	432 mm	432 mm	432 mm	432 mm	432 mm	432 mm	432 mm
NUMBER OF LINKS PER TRACK	79	79	82/84	84	84	82/84	84	79
MAX ROAD SPEED	80.5 km/h	72.5 km/h	80.5 km/h	80.5 km/h	72.5 km/h	72.5 km/h	72.5 km/h	80.5 km/h
FUEL CAPACITY	423 l	391 l	350 l	386 l	395 l	395 l	404 l	423 l
MAX ROAD RANGE	644 km	756 km	483 km	483 km	483 km	483 km	483 km	644 km
FORDING	1.067 m	1.067 m	1.067 m	1.067 m	1.067 m	1.067 m	1.067 m	1.067 m
	all members of Scorpion family are amphibious with preparation**							
FREEBOARD								
forward with screen erected	0.965 m	0.965 m	0.89 m	0.914 m	0.832 m	0.883 m	0.965 m	0.965 m
aft with screen erected	0.711 m	0.711 m	0.813 m	1.168 m	0.559 m	0.457 m	0.815 m	0.711 m
GRADIENT	60%	60%	60%	60%	60%	60%	60%	60%
VERTICAL OBSTACLE	0.5 m	0.5 m	0.5 m	0.5 m	0.5 m	0.5 m	0.5 m	0.5 m
TRENCH	2.057 m	2.057 m	2.057 m	2.057 m	2.057 m	2.057 m	2.057 m	2.057 m
TURNING RADIUS	all members can pivot turn in neutral							
in 1st gear	1.71 m	1.71 m	1.84 m	1.84 m	1.84 m	1.84 m	1.84 m	1.71 m
in 7th gear	33.22 m	33.22 m	35.9 m	35.97 m	35.97 m	35.97 m	35.97 m	33.22 m
ENGINE	Jaguar J60 No 1 Mk 100B 4.2-l 6-cylinder petrol developing 190 hp at 4750 rpm*							
TRANSMISSION	TN15 crossdrive, semi-automatic hot-shift type, providing 7 speeds in each direction and pivot turns							
STEERING	Merritt system incorporated in transmission							
SUSPENSION	torsion bar on all members of Scorpion family							
ELECTRICAL SYSTEM	28.5 V	28.5 V	28.5 V	28.5 V	28.5 V	28.5 V	28.5 V	28.5 V
BATTERIES	4	4	4	4	4	8	4	4
MAIN ARMAMENT	76 mm	90 mm	Swingfire ATGWs	n/app	none	n/app	n/app	30 mm
MACHINE GUN	7.62 mm	7.62 mm	7.62 mm	7.62 mm	none	7.62 mm	7.62 mm	7.62 mm
SMOKE DISCHARGERS	2 × 4	2 × 4	2 × 4	2 × 4	2 × 4	2 × 4	2 × 4	2 × 4
AMMUNITION								
main	40	33	10	n/app	none	n/app	n/app	165
machine gun	3000	3000	3000	3000	none	2000	2000	3000
GUN CONTROL EQUIPMENT	manual	powered	n/app	n/app	none	n/app	n/app	manual
Gun elevation/depression	+35°/-10°	+30°/-8°	n/app	n/app	none	n/app	n/app	+35°/-10°

*All members of the Scorpion are now offered with a Perkins diesel engine developing 200 bhp at 2700 rpm and details of the Scorpion 90 relate to a vehicle with this engine.
**This capability has been removed from British vehicles.

RO2000 Family of Armoured Fighting Vehicles

This range of vehicles, first shown in 1986, is no longer being marketed by Royal Ordnance. Details were given in *Jane's Armour and Artillery 1990-91*, pages 526-528.

Tank destroyers

Other Tank Destroyers

The vehicles described in this section have mainly been developed specifically for the tank destroyer role. There are, however, many other tank destroyers in service or developed to the prototype stage that are based on other chassis, normally an armoured personnel carrier's or a reconnaissance vehicle's. Listed here is a résumé of the tracked and wheeled tank destroyers that are covered elsewhere. It should be noted that almost any vehicle can be fitted with a standard infantry MILAN or TOW system so the listing of these missiles is not exhaustive. Recoilless rifles are excluded as these are in service in declining numbers.

Missile	**Platform**	**Comment**
ADATS	M113 (for example)	This Air Defence Anti-tank System is designed for installation on a variety of tracked and wheeled chassis, with Canada using a modified M113A2 and the United States a modified M3 Bradley Cavalry Fighting Vehicle
HOT	AMX-10P	Has Lancelot turret with four missiles ready to launch, in service with Saudi Arabia
	AMX-13 light tank	Three missiles installed either side of turret, trials only
	BMR-600 (6 × 6)	Basic vehicle with HCT turret, trials only for Spanish Army
	Chaimite 700 (4 × 4)	Proposal only
	Condor (4 × 4)	Has HCT turret, trials only
	Warrior	Has HCT turret, trials only
	Panhard VCR (6 × 6)	Has UTM 800 turret, called VCR/TH, in service, photograph in *Armoured personnel carriers* section
	Panhard VCR (4 × 4)	Mephisto system, called VCR/TH, not in service
	Piranha (8 × 8)	Fitted with HOT Mephisto system for trials
	RPX 6000 (4 × 4)	Has HCT turret, trials only
	Spartan	Has HCT turret, trials only
	Simba	FS100/HOT has HCT turret, trials only
	VAB (6 × 6)	Fitted with Mephisto system, in service with French (4 × 4) Army
	VAB (6 × 6)	Has UTM 800 system, used by Qatar and Cyprus
	VPX 5000	Has HCT turret, trials only
	Wiesel	Has HCT turret, trials only
MILAN	EE-3 Jararaca (4 × 4)	Pintle-mounted, in service with Mexico
	Marder	Most German Army Marders have MILAN on turret, photograph in *Armoured personnel carriers* section
	M113	Belgium uses this vehicle with pintle-mounted MILAN
	Panhard M3 (4 × 4)	Has MCT turret, trials only
	Panhard VCR (4 × 4)	Has MCT turret, trials only
	Panhard VBL (4 × 4)	Pintle-mounted, photograph in *Reconnaissance vehicles* section (adopted by France and Mexico)
	RPX 6000 (4 × 4)	Has MCT turret, trials only
	Spartan	Has MCT turret, in service with British Army, photograph in Scorpion entry in *AFV families* section
	Transportpanzer 1 (6 × 6)	Has MCT turret, trials only
	VAB (4 × 4)	Has HCT turret, trials only
	VAB (6 × 6)	Oman uses pintle-mounted MILAN on this vehicle
	VPX 5000	Has MCT turret, trials only
Sagger	BMP-1	Launcher rail over 73 mm gun, photograph in *Armoured personnel carriers* section, standard on all vehicles
	BMD	Launcher rail over 73 mm gun, photograph in *Armoured personnel carriers* section, standard on all vehicles
	BOV-1 (4 × 4)	Yugoslav chassis with six Sagger-type ATGWs, photograph in *Armoured personnel carriers* section, standard vehicle
	BRDM-1	Sextuple launcher, photograph in *Reconnaissance vehicles* section, standard vehicle
	BRDM-2	Six missiles ready to launch, photograph in *Reconnaissance vehicles* section, standard vehicle
	M-980	Two missiles ready to launch on turret, photograph in *Armoured personnel carriers* section, standard vehicle
	OT-64A (8 × 8)	Two missiles ready to launch above troop compartment
	OT-64C (1) (8 × 8)	One missile either side of turret, photograph in *Armoured personnel carriers* section
Snapper	BRDM-1 (4 × 4)	Three missiles ready to launch, photograph in *Reconnaissance vehicles* section, standard vehicle
Spandrel	BMP-2	One missile ready to launch, photograph in *Armoured personnel carriers* section, standard vehicle
	BRDM-2 (4 × 4)	Five missiles ready to launch, photograph in *Reconnaissance vehicles* section, standard vehicle
Spigot	BMD	One missile pintle-mounted on turret, photograph in *Armoured personnel carriers* section
	BMP-1	Some vehicles have pintle-mounted Spigot in place of Sagger
	BMP-2	Some vehicles have Spigot in place of Spandrel
	BRDM-2 (4 × 4)	Some vehicles have three Spandrel and two Spigot ATGWs
	BMP-23/BMP-23D/BMP-30	One missile pintle-mounted on vehicle turret
Swatter	BRDM-1 (4 × 4)	Four missiles ready to launch, standard vehicle
	BRDM-2 (4 × 4)	Four missiles ready to launch, standard vehicle
Swingfire	Striker	Five missiles ready to launch, standard vehicle, photograph in Scorpion entry in *AFV families* section
	Scorpion	Proposal to fit Swingfire either side of turret
TOW	AIFV	Fitted with ITV launcher, used by Dutch Army
	Commando Scout (4 × 4)	Fitted with normal infantry TOW, trials only
	LAV (8 × 8)	Fitted with ITV launcher for US Marine Corps, photograph in *Armoured personnel carriers* section
	M8 (6 × 6)	Fitted with normal infantry TOW, used by Colombia
	Piranha (6 × 6)	Swiss Army has version with Kvaerner- Eureka turret
	RAM V-1 (4 × 4)	Fitted with normal infantry TOW
	Simba (4 × 4)	Fitted with ITV launcher, trials only
	OTO R3 (4 × 4)	Fitted with normal infantry TOW, trials only
	Spartan	Proposal to fit normal infantry TOW
	VCC-1	Fitted with ITV launcher, 224 supplied to Saudi Arabia, photograph in *Armoured personnel carriers* section
	V-300 (6 × 6)	Proposal to fit ITV launcher to basic vehicle
	M113	M113 with ITV launcher is known as M901 Improved Vehicle, standard in US Army and many other countries
	M113	Fitted with basic infantry TOW
	M113	Norwegian Army has Kvaerner-Eureka turret for its M113 vehicles also used by Canada (M113A1)
	Warrior	Single TOW launcher either side of turret (trials only)
Vigilant	Ferret Mk 2/6	Two missiles ready to launch, standard vehicle

AUSTRIA

Jagdpanzer SK 105 Tank Destroyer

Details of the Jagdpanzer SK 105 tank destroyer are given in the *Light tanks* section, as it has the appearance of a tank and is often used in this role.

GERMANY

Thyssen Henschel Jagdpanzer Jaguar 1 Self-propelled ATGW Vehicle

Development

The first prototype of the Jagdpanzer Rakete was completed by Hanomag in 1961-62 under the designation RU 234. This was followed by a further six prototypes which were completed from 1963, three of which were built by Hanomag (the RU 341, RU 342 and RU 343) and three by Henschel (RU 344, RU 345 and RU 346). The Jagdpanzer Rakete succeeded the Jagdpanzer Kanone in production in 1967 and 370 were built by the following year, 185 by Hanomag and 185 by Henschel.

These were fitted with two launchers for the Aérospatiale SS-11 wire-guided ATGW with a maximum range of 3000 m. Following successful trials with prototype vehicles, 316 of the 370 SS-11 armed vehicles were rebuilt with the Euromissile K3S HOT ATGW system. The first vehicles were handed over by Thyssen Henschel in 1978 with final deliveries in 1983.

Description (Jaguar 1)

The all-welded steel hull of the Jagdpanzer Jaguar 1 has the fighting compartment at the front and the engine compartment at the rear. The driver sits at the front of the hull on the left side and has a single piece hatch cover that opens to his left, in front of which are three periscopes. The bow machine gunner sits on the right side of the hull and has a single piece hatch cover that opens to the right, forward of which are three periscopes.

The engine compartment is at the rear of the hull with the engine on the left side, cooling system on the right side and the Renk HSWL transmission at the rear. The engine compartment is equipped with a fire-extinguishing system which is activated automatically in the event of fire.

The torsion bar suspension system consists of five dual rubber-tyred road wheels with the idler at the front, drive sprocket at the rear and three track return rollers. The first, second, fourth and fifth road wheel stations have a hydraulic shock absorber. The Jagdpanzer Jaguar is equipped with infra-red night vision equipment and an NBC system. The basic vehicle can ford to a depth of 1.4 m and with a kit to 2.1 m.

The system operates as follows. The operator tracks the target through his periscopic sight which is mounted in the forward part of the roof and has a two-stage magnification, the first stage being ×4 (with a 13° field-of-view) and the second ×12 (with a 4.3° field-of-view). Traverse is 140° and elevation from -20° to +20°. The missile is launched and when it has travelled between 30 and 50 m is armed. The booster stage accelerates the missile to 240 m/s in one second and this speed is then kept constant throughout the flight by the sustainer stage. Once the missile hits the target the launcher arm is depressed vertically into the magazine, which holds eight missiles, and a new missile is loaded. This takes several seconds and once accomplished the launch cycle can begin again. Average firing rate is three rds/min. Hit probability against a fixed target (2.3 × 2.3 m) or moving target (2.3 × 4.6 m) is 80 per cent at ranges of up to 500 m and almost 100 per cent at ranges between 500 and 4000 m. An Eltro thermal night sight has been developed to enable the vehicle to engage enemy armour at night and in bad weather.

The Eltro RZ 1201 thermal imaging sight cost about DM10 million to develop and uses Hughes thermal imaging modules. First company trials of the HOT with this sight, also called the NZBG (Night Target and Observation Equipment), took place in November 1984 at the Meldorf range in Germany. During these trials stationary and moving targets were engaged and hit at ranges of over 2000 m by both day and night.

The missile is carried in its launch tube and is treated like a normal round of ammunition. The total weight of missile and launch tube is 32 kg, it is 1.3 m long and 175 mm in diameter. The missile itself weighs 23 kg at launch with the warhead weighing 6 kg including fuze. The warhead is of the HEAT type, contains 3 kg of explosive and its diameter is 136 mm. Flight times are nine seconds to reach 2000 m, 13 seconds to reach 3000 m and 17.3 seconds to reach 4000 m.

In 1985 the HOT 2 entered production with delivery of first production missiles in 1986 after which time production of the HOT 1 ceased. HOT 2 has its warhead diameter increased from 136 mm to 150 mm and explosive charge increased from 3 kg to 4.1 kg, with no actual increase in missile weight. The HOT 2 warhead will penetrate 1300 mm of armour.

Other modifications to the vehicle have included the addition of spaced armour to the front and sides of the hull, consisting of a layer of steel bolted to the vehicle, with rubber mountings providing the space and acting as a buffer between the armour and the hull. The commander sits behind the bow machine gunner and has a single piece lift-and-swing hatch cover. He is provided with periscopes for all-round observation and mounted to the rear of his hatch is a periscope which can be traversed through a full 360°. A 7.62 mm MG3 machine gun is mounted at the commander's position.

Mounted in the bow of the vehicle on the right side is a 7.62 mm MG3 machine gun with an elevation of +15°, a depression of -8° and a traverse of 15° left and 15° right. A second 7.62 mm MG3 machine gun can be mounted on the roof of the vehicle at the bow gunner's position for anti-aircraft use. Eight smoke dischargers mounted on the rear decking fire forwards over the front of the vehicle.

Jagdpanzer with HOT Compact Turret HCT

In April 1981 a Jaguar 1 tank destroyer fitted with a Euromissile HCT (previously called the HAKO) turret with four HOT ATGWs and the MILAN ATGW MIRA night sight was successfully tested. The trials showed that with the MIRA sight it was possible to identify targets at ranges of over 3000 m and HOT ATGWs were subsequently fired at a 2.3 × 2.3 m target with hits being obtained. In addition aiming tests were also carried out against moving targets up to ranges of 3000 m. This was a trials installation only and never entered production or service.

Jagdpanzer Rakete with TOW (Jaguar 2)

In 1978 Thyssen Henschel completed two prototypes of the Jagdpanzer Rakete armed with the Hughes TOW system, known as the Jaguar 2. Mounted on the top of the vehicle is a Hughes TOW system which can

Jagdpanzer Jaguar 1 showing bank of electrically operated smoke dischargers on hull rear

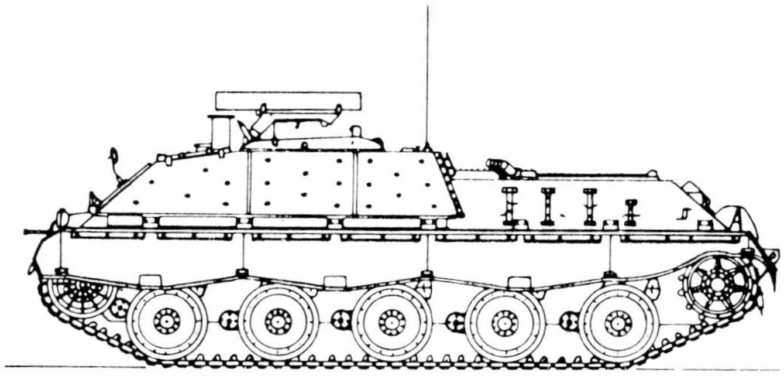

Jagdpanzer Jaguar 1 armed with HOT ATGW system

be elevated from -10° to +15° and has a traverse of 30° left and 30° right. The main advantage of this system is that the missiles are much cheaper than the HOT and the system already has the Texas Instruments AN/TAS-4 night sight. The TOW missile is faster than the HOT and has an almost identical range (3750 m compared to 4000 m with the HOT).

Early in 1983 the German BWB placed an order worth DM 25 million with Thyssen Henschel of Kassel for the conversion of 162 Jagdpanzer 90 mm self-propelled anti-tank guns to TOW ATGW carriers. The vehicles were converted between 1983 and 1985.

Status: In service only with the German Army. Production complete, but can be resumed if further orders are received.

Manufacturer: Original manufacturers were Henschel and Hanomag.
Current manufacturer would be Thyssen Henschel
Postfach 102969
D 3500 Kassel
Federal Republic of Germany
Telephone: (0561) 8011
Telex: 99750

Jaguar 2 armoured vehicle with TOW launcher in raised operating position

SPECIFICATIONS (Jaguar 1)	
CREW	4
COMBAT WEIGHT	25 500 kg
POWER-TO-WEIGHT RATIO	19.6 hp/tonne
GROUND PRESSURE	0.7 kg/cm^2
LENGTH HULL	6.61 m
WIDTH	3.12 m
HEIGHT	
including missiles	2.54 m
to hull top	1.98 m
GROUND CLEARANCE	
front	0.45 m
rear	0.44 m
TRACK	2.54 m
TRACK WIDTH	450 mm
LENGTH OF TRACK ON GROUND	3.8 m

MAX ROAD SPEED	
forward	70 km/h
reverse	70 km/h
FUEL CAPACITY	470 l
MAX ROAD RANGE	400 km
FORDING	1.2 m
with preparation	1.8 m
GRADIENT	58%
SIDE SLOPE	30%
VERTICAL OBSTACLE	0.75 m
TRENCH	2 m
ENGINE	Daimler Benz MB 837 8-cylinder water-cooled diesel developing 500 hp at 2000 rpm
TRANSMISSION	Renk HSWL 123
SUSPENSION	torsion bar
ELECTRICAL SYSTEM	24 V

ARMAMENT	
main	1 HOT ATGW launcher
bow	1 × 7.62 mm MG
anti-aircraft	1 × 7.62 mm MG
SMOKE-LAYING EQUIPMENT	8 smoke dischargers on roof
AMMUNITION	
missiles	20
machine gun	3200
ARMOUR	12- 50 mm (estimated)

Thyssen Henschel Jagdpanzer Kanone (JPZ 4-5) Self-propelled Anti-tank Gun

Development

When the German Army was formed it issued a requirement for a 90 mm self-propelled anti-tank gun, or tank destroyer as the Germans normally call this type of vehicle. The first vehicle designed to meet this requirement was basically a Swiss-designed HS-30 APC with a 90 mm gun ball-mounted in the glacis plate. Trials did not prove a success so the development of a new vehicle began, but this time a chassis was developed which could be used for a number of basic vehicles including the Jagdpanzer Kanone, Jagdpanzer Rakete (previous entry), reconnaissance tank (developed to the prototype stage but not placed in production), an MICV (which eventually became the Marder) and various other supporting vehicles.

In all five prototypes of the self-propelled anti-tank gun were built between 1960 and 1962, two by Hanomag, two by Henschel and one by the Swiss MOWAG company.

These five prototypes were followed by a second series of six which were completed between 1962 and 1963, three built by Hanomag and three by Henschel. The second series of prototypes was followed by the third and final series of six which were completed between 1963 and 1965, three built by Hanomag and three by Henschel.

The prototypes were followed by 750 production vehicles which were completed between 1965 and 1967, of which 375 were built by Hanomag and 375 by Henschel. The Jagdpanzer Rakete, which entered production after the Jagdpanzer Kanone, is almost identical to the latter except for its armament.

Description

The hull of the Jagdpanzer Kanone is all welded steel with the fighting compartment at the front and engine and transmission at the rear. The driver sits at the front of the hull on the left side and has a single piece hatch cover that opens to his right in front of which are three periscopes. The gunner sits on the right side of the hull at the front and has a telescopic sight with a magnification of ×8 linked to the main armament, and a roof-mounted periscope. The loader sits behind the driver and has a circular lift-and-swing hatch cover in front of which is a swivel-mounted periscope. The commander sits at the rear of the fighting compartment on the right side and has a circular lift-and-swing hatch cover and periscopes for all-round observation. There is a swivel-mounted periscope with a magnification of ×6 to ×20 in front of the commander's hatch.

The engine compartment at the rear on the left side is equipped with a fire-extinguishing system which is activated automatically in the event of fire. The cooling system is to the right side of the engine and the Renk HSWL 123 transmission is at the rear of the hull.

The torsion bar suspension consists of five dual rubber-tyred road wheels with the idler at the front, drive sprocket at the rear and three track return rollers. The first, second, fourth and fifth road wheel stations have a hydraulic shock absorber.

The Jagdpanzer Kanone is equipped with infra-red night vision equipment and an NBC system. The basic vehicle can ford to a depth of 1.4 m and with a kit to 2.1 m.

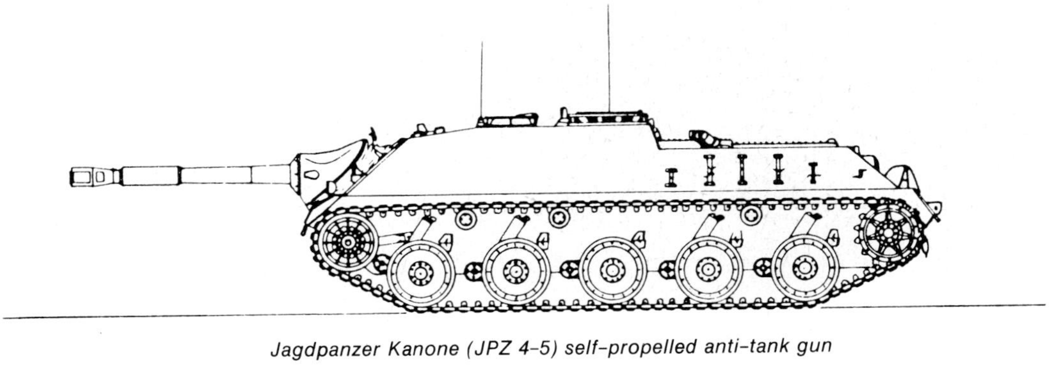

Jagdpanzer Kanone (JPZ 4-5) self-propelled anti-tank gun

Jagdpanzer Kanone with 90 mm gun removed for use in observation post role (C R Zwart)

Belgian Army Jagdpanzer Kanone from rear showing roof-mounted Bofors Lyran flare launchers (C R Zwart)

The 90 mm gun, which fires the same ammunition as the M47 and M48 tanks, is mounted in the glacis plate, slightly offset to the left of the vehicle's centreline and has an elevation of +15°, depression of -8° and a traverse of 15° left and 15° right. Elevation and traverse are both manual. The 90 mm gun fires HEAT-T and HESH-T projectiles to a maximum effective range of 2000 m. Maximum rate of fire is stated to be 12 rds/min. The barrel is fitted with a double-baffle muzzle brake and a fume extractor.

Mounted coaxially with the main armament is a 7.62 mm MG3 machine gun and a similar weapon mounted at the commander's station for anti-aircraft use can be removed and fitted at the loader's station. Mounted on the rear of the fighting compartment are eight electrically operated smoke dischargers which fire forwards over the front of the vehicle. Mounted over the main armament is an infra-red/white light searchlight, which can be removed and stowed in a box at the rear of the hull if required.

One hundred and sixty-two of the 770 Jagdpanzer Kanone self-propelled anti-tank guns have been rebuilt to Jaguar 2 configuration with the Hughes TOW ATGW system and AN/TAS-4 night sight, but some have been transferred to the territorial reserve. Details of this model are given in the previous entry.

Recently some Jagdpanzer Kanone have been used in the observation post role with the 90 mm gun removed.

Variants

Belgian Jagdpanzer Kanone

In December 1972 the Belgian Army placed an order for 80 improved versions of the Jagdpanzer Kanone, which were assembled at Anvers in Belgium, the first vehicles being completed in April 1975. Eight Belgian infantry battalions each have two platoons of four Jagdpanzer Kanones with the remaining 16 vehicles used for training or held in reserve. The Belgian Army designation for this vehicle is JPK-90.

The Belgian vehicles have the same hull, armament and engine as the original German vehicles but incorporate the complete Renk transmission, final drives, suspension and track of the Marder ICV. Two hydraulic clutches which have been fitted to the transmission allow very sharp turns to be made at all speeds.

The fire control system is a simplified version of the SABCA system developed for the Leopard 1 and already installed in Australian, Belgian and Canadian Leopard 1 MBTs. This incorporates a laser rangefinder, sensors, analogue computer and an optical sight. The vehicle also has three angular deviation indicators, one optically coupled to the commander's sight, one for the driver and one for the gunner which is coupled to his gun sight. This works as follows: if the commander sees a target to his right outside the traverse limitations of the gun, he lines up his sight onto the target, the driver then turns the vehicle until the two lines (one indicator showing the position of the gun and the other the position of the commander's sight) line up. The gunner has then only to make fine adjustments and fire the gun.

Other equipment fitted to the Belgian vehicles includes Swedish Bofors Lyran flare launchers for illuminating targets at night, passive night vision equipment and the replacement of German 7.62 mm MG3 machine guns with Belgian MAG 7.62 mm machine guns.

Basic specifications of the Belgian vehicles are similar to the German vehicles except for a loaded weight of 25 900 kg, track 2.62 m and width 3.07 m.

Jagdpanzer Kanone 105 mm

It has been proposed that the Jagdpanzer Kanone could be fitted with a 105 mm rifled gun or even a Rheinmetall 120 mm smoothbore gun.

Status: Production complete. In service with Belgian and German Armies. Can be placed back in production if further orders are received.

Manufacturers: Original manufacturers were Henschel and Hanomag.
Current manufacturer would be Thyssen Henschel
Postfach 102969
D 3500 Kassel
Federal Republic of Germany
Telephone: (0561) 8011
Telex: 99750

German Army Jagdpanzer Kanone armed with 90 mm gun with infra-red/white light searchlight over main armament

SPECIFICATIONS		FUEL CAPACITY	470 l	SMOKE-LAYING	
CREW	4	MAX ROAD RANGE	400 km	EQUIPMENT	8 smoke dischargers
COMBAT WEIGHT	27 500 kg	FORDING	1.4 m		on roof
POWER-TO-WEIGHT		with preparation	2.1 m	AMMUNITION	
RATIO	18.18 hp/tonne	GRADIENT	60%	main	51
GROUND PRESSURE	0.75 kg/cm^2	SIDE SLOPE	30%	machine gun	4000
LENGTH GUN FORWARDS	8.75 m	VERTICAL OBSTACLE	0.75 m	GUN CONTROL	
LENGTH HULL	6.238 m	TRENCH	2 m	EQUIPMENT	
WIDTH	2.98 m	ENGINE	Daimler Benz MB 837	Gun traverse	manual
HEIGHT (without AA MG)	2.085 m		8-cylinder	by commander	no
FIRING HEIGHT	1.4 m		water-cooled diesel	by gunner	yes
GROUND CLEARANCE			developing 500 hp at	Gun elevation/depression	+15°/-8°
front	0.45 m		2000 rpm	Gun traverse	15° left and right
rear	0.44 m	TRANSMISSION	Renk HSWL 123	Gun stabiliser	
TRACK	2.54 m	SUSPENSION	torsion bar	vertical	no
TRACK WIDTH	450 mm	ELECTRICAL SYSTEM	24 V	horizontal	no
LENGTH OF TRACK		ARMAMENT		ARMOUR	12 - 50 mm (estimated)
ON GROUND	3.8 m	main	1 × 90 mm gun		
MAX ROAD SPEED		coaxial	1 × 7.62 mm MG		
forward	70 km/h	anti-aircraft	1 × 7.62 mm MG		
reverse	70 km/h				

JAPAN

Type 60 Self-propelled Recoilless Gun

Development
Following the successful use by the United States Army in Korea of Jeep-mounted recoilless rifles, the Japanese Ground Self-Defence Force issued a requirement for a self-propelled tracked recoilless gun specifying that it was to be armed with two recoilless rifles which could be raised hydraulically with manual controls in emergencies, have a crew of three, a loaded weight of 5000 kg, an air-cooled diesel engine, a maximum road speed of over 55 km/h and a provision for replacing the standard tracks by special wider snow tracks.

Two prototypes of the vehicle were built under the direction of the Technical Research and Development Headquarters of the Japanese Ground Self-Defence Force, one by Komatsu (called the SS1) and the other by Mitsubishi (called the SS2), both powered by the same six-cylinder diesel which developed 110 hp. The prototypes were completed in December 1956 and were tested until December 1958.

In 1957 it was proposed to arm the vehicle with four recoilless rifles, instead of the two 105 mm weapons fitted to the prototypes, to increase the hit probability, but with the adoption of the 106 mm recoilless rifle it was decided to keep the two.

As a result of trials with the second series of prototypes it was decided to build a third and final series of prototypes with weight increased to 7500 kg, modified engine, new transmission with four forward and one reverse gears and a two-speed auxiliary transmission, and a wet multi-disc hydraulically operated clutch. Three of the third series prototypes were built under the designation SS4 and were tested from May 1959 to March 1960. As a result the clutch was changed to a wet multi-disc mechanical type. In September 1960 the vehicle was accepted for service as the Type 60 Self-propelled 106 mm Recoilless Gun.

Production of the vehicle was undertaken by Komatsu. The first production model, the model A, was replaced in 1967 by the model B which was basically a strengthened model A. In 1975 the model C, powered by a commercial model Komatsu SA4D105 four-cylinder water-cooled engine which develops 150 hp at 2800 rpm, entered production. The Type 60 was produced at the rate of 10 to 20 vehicles a year and by December 1977 223 vehicles had been built. Production was completed in FY79.

Description
The hull of the Type 60 is of all-welded steel and riveted construction. The driver is seated on the left side of the vehicle towards the front and has a single piece hatch cover that opens to the rear in front of which are three periscopes. The commander is seated in the middle of the hull to the left of the two recoilless rifles and his position is attached to the recoilless rifle mounting so when they are raised into the firing position he is at the same height as the weapons. The commander has a single domed hatch cover that opens to the rear, in front of which is his periscopic sight. There is also a vision block to his rear. The loader is seated to the left of the commander and has a single piece hatch cover that opens to the left and a retractable periscope for observation to his rear.

The engine is at the rear of the hull and power is transmitted to the transmission at the front by a propeller shaft. The torsion bar suspension consists of five dual rubber-tyred road wheels with the drive sprocket at the front, idler at the rear and three track return rollers. The first, second and fifth road wheel stations have a hydraulic shock absorber. The tracks can be fitted with rubber pads when operating on roads and wider tracks can be fitted for operating in snow. The Type 60 has no NBC system and no amphibious capability.

Main armament consists of two 106 mm Type 60 recoilless rifles which are manufactured by Nihon Seikojyo and also fitted singly on Japanese Ground Self-Defence Force Jeeps. The rifles have an overall length of 3.408 m and barrel length of 3.32 m, weigh 114 kg and have a maximum range of 7000 m and an effective range of 1100 m against tanks. A 12.7 mm spotting rifle manufactured by Howa Kogyo is mounted on the right recoilless rifle. The two recoilless rifles are mounted parallel to each other on the right side of the hull. When lowered they have an elevation of +10°, depression of -5° and a traverse of 10° left and 10° right. When raised they have an elevation of +15°, depression of -20° and a traverse of 30° left and 30° right. The weapons are raised into the firing position by a hydraulic/electric power system, but if this fails they can be raised by a hand pump. Eight rounds of HEAT or HE ammunition are carried. A 75 cm stereo rangefinder is also carried on the vehicle and more recently an infra-red night sight has been mounted on the left recoilless rifle.

Status: Production complete. In service only with the Japanese Ground Self-Defence Force.

Manufacturer: Komatsu Seisakujyo
2-3-6 Akasaka
Minato-ku
Tokyo
Japan

Type 60 self-propelled recoilless gun with dotted line showing position of weapons when raised

Type 60 self-propelled recoilless gun with wider snow tracks (Keiichi Nogi)

SPECIFICATIONS		FUEL CAPACITY	77 l	ARMAMENT	
CREW	3	MAX ROAD RANGE	130 km	main	2 × 106 mm recoilless rifle
COMBAT WEIGHT	8000 kg	FORDING	1 m		
UNLOADED WEIGHT	7600 kg	GRADIENT	60%	ranging machine gun	1 × 12.7 mm MG
POWER-TO-WEIGHT RATIO		SIDE SLOPE	30%	AMMUNITION (main)	8
120 hp engine	15 hp/tonne	VERTICAL OBSTACLE	0.55 m	GUN ELEVATION/ DEPRESSION	see text
150 hp engine	18 hp/tonne	TRENCH	1.8 m	GUN TRAVERSE	see text
GROUND PRESSURE	0.67 kg/cm²	ENGINE		ARMOUR	12 mm maximum (estimated)
LENGTH	4.3 m	models A and B	Komatsu model 6T 120-2 air-cooled, 4-cycle, 6-cylinder diesel developing 120 hp at 2400 rpm		
WIDTH	2.23 m				
HEIGHT	1.59 m				
GROUND CLEARANCE	0.35 m				
TRACK	2 m	model C	Komatsu SA4D105 air-cooled, 4 cycle, 6-cylinder diesel developing 150 hp at 2800 rpm		
TRACK WIDTH					
normal tracks	260 mm				
snow tracks	680 mm				
LENGTH OF TRACK ON GROUND	2.4 m	TRANSMISSION	manual with 4 forward and 1 reverse gears, auxiliary, 2-speed		
MAX ROAD SPEED FORWARDS					
models A and B	45 km/h	CLUTCH	dry, multi-disc		
model C	55 km/h	SUSPENSION	torsion bar		

SWEDEN

Hägglund Vehicle Infanterikanonvagn 91 Tank Destroyer

Development

The first of three prototypes was completed in December 1969 with the second two following in 1970. After extensive trials, Hägglund was awarded a production contract for the Infanterikanonvagn (Ikv)-91 in March 1972. The first two pre-production vehicles were completed in 1974 with the first production vehicle being completed in 1975. Production of the Ikv-91 was completed late in the 1978 for the Swedish Army.

Description

The hull of the Ikv-91 is all welded steel and divided into three compartments with the driver's at the front, fighting in the centre and the engine at the rear. The vehicle has complete protection against attack from 20 mm armour-piercing rounds over its frontal arc. Above the tracks the sides of the hull are double-skinned, with the main armour inside the tracks. The space is used for stowing accessories and diesel fuel, giving increased protection against HEAT and HE penetration.

The driver is seated at the front of the hull on the left side and has a single piece hatch cover opening to the left. Three periscopes are mounted forward of his hatch cover and there is a hull escape hatch to his right slightly to the rear.

The other three crew members are seated in the all-welded turret with the commander and gunner on the right and the loader on the left. The commander's cupola is fitted with M17 periscopes and a ×10 periscopic sight is mounted in the front of the cupola. The cupola has a single piece hatch cover opening to the rear which can be traversed through 240° and operated either locked to the turret or counter-rotating. Superimposed on either mode is an electric servo system which provides an accurate power traverse of the cupola. The loader's cupola has a single piece hatch cover and a periscope.

The engine compartment is at the rear, separated from the fighting compartment by a bulkhead. The complete powerpack consists of the engine, gearbox with torque converter, bevel gear and steering clutch. The powerpack is mounted diagonally in the engine compartment to shorten the length of the hull. The gearbox is fully automatic but a selector enables the driver to restrict the upshift within any desired range. Engine exhaust is used to drive the turbo-charger as well as to induce a flow of cooling air over the steering brakes by passing it through an ejector.

The Ikv-91 has the same clutch-and-brake steering system as the Pbv 302 APC, designed to permit continuous slip-steering. The steering clutch is of the double dry plate type operated by a hydraulic servo. The steering brakes at each end of the drive shafts are also used as main brakes.

The torsion bar suspension consists of six dual rubber-tyred road wheels with the drive sprocket at the rear and the idler at the front. There are no track return rollers. Shock absorbers are provided at the first and last road wheel stations.

The Ikv-91 is fitted with the Hägglunds-designed M70 track which is a single-pin, rubber-bushed link with the rubber bushings confined to one link at each joint, the pin being fixed at the other. Tests by the Swedish Army have shown this track to have a life in excess of 8500 km with one recondition. It is also quieter and gives better traction and less vibration than other comparable tracks.

The track shoes can be fitted with the same type of studs that are commercially used on trucks but for increased traction when driving cross-country in deep snow the track shoes can be removed and, if necessary, the track links can be fitted with conical spikes protruding 50 mm below the surface of the link.

The vehicle is fully amphibious, being propelled in the water by its tracks. Before entering the water a trim vane is erected at the front of the hull and low screens are raised around the air- inlets and the exhaust and air-outlets. Four bilge pumps are provided.

The Ikv-91 has been designed for operations in the northern part of Sweden and pre-heating the engine with a built-in blowtorch permits engine starts from -35°C (-31°F) and a built-in electric pre-heater for the combustion air permits immediate engine starts under less severe conditions.

An electric fan creates a slight over-pressure by sucking air from the air-inlet through a dust filtration system and blowing it through an NBC filter fitted into the fighting compartment which also has a heater. The Ikv-91 is not fitted with night vision equipment.

The gun fires the following types of fixed ammunition: fin-stabilised HEAT with the complete round weighing 10.7 kg, projectile weighing 4.5 kg and a muzzle velocity of 825 m/s; fin-stabilised HE with the complete round weighing 12.3 kg, projectile weighing 6.7 kg and a muzzle

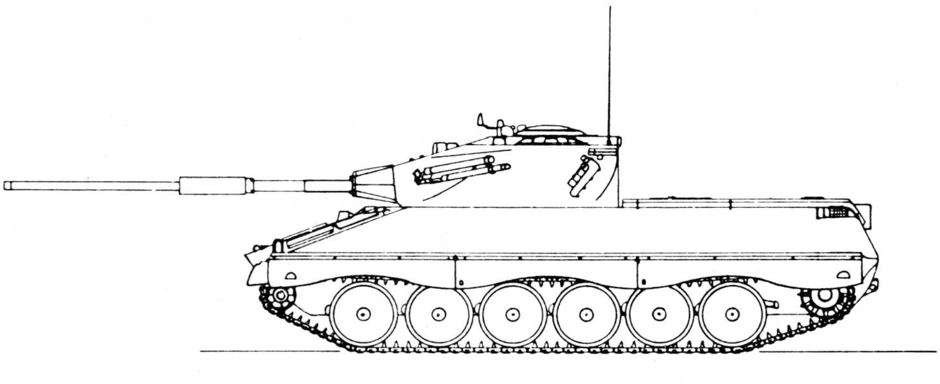

Ikv-91 tank destroyer

Ikv-91 tank destroyer used at the Swedish Infantry Combat School for training (Christopher F Foss)

Pre-production version of Ikv 91-105 light tank showing amphibious capabilities

Pre-production version of Ikv 91-105 light tank armed with Bofors 105 mm low recoil gun

velocity of 600 m/s. The HEAT round, developed by Bofors, utilises a piezo-electric fuze to reach full-calibre sensitivity and is capable of penetrating all NATO targets with substantial effects behind the plates. The HE round, which is used against personnel, has a mechanical time fuze.

Of the 59 rounds of 90 mm ammunition carried 16 are stowed in ready racks at the loader's station, 18 to the right of the driver and the remaining 25 in the chassis on the left side to the rear of the turret.

More recently the Ikv-91 has been retrofitted with a thermal sleeve for the 90 mm gun, Bofors Lyran launchers on the turret roof at the rear, external stowage for Lyran flares and an external stowage bin on the turret rear.

Fire control system

The commander has no direct access to the computer but a mechanical linkage from one of the trunnions transfers gun elevation to a collimator. When the cupola is aligned with the turret the collimator is in front of the commander's sight. The collimator then projects an image of a ballistic graticule into the commander's field of view.

The gunner has a Jungner Instrument AB monocular periscopic sight TP-1050L with a magnification of ×10 and a 6° field-of-view which also incorporates a Bofors laser rangefinder. Gun elevation is transferred from the top prism by a mechanical linkage from one of the trunnions to a differential in the gunner's sight. Electric servos deflect the line of sight in azimuth and elevation according to the actual ballistic solution determined by the computer.

The transmitter/receiver of the Neodymium-type laser rangefinder is attached to the right side of the gunner's sight and the path of the laser beam is incorporated in the optics of the sight. The range of up to two objects within the laser beam can simultaneously be displayed in the left eye-piece of the gunner's sight. The measured range is fed automatically into the computer and in the case of two echoes, the shortest range, which is usually the correct one, is fed into the computer. Ranging is triggered from the gunner's control handle as well as shifting between two simultaneously displayed echoes.

The laser beam leaves the top prism of the sight coaxially with the gunner's line of sight, which means that ranging can be made instantaneously, regardless of the super-elevation and lead angles introduced into the system. If the computer fails the electric servos of the gunner's sight are mechanically zeroed and the firing sequence can be completed by using a ballistic graticule.

The basis of the fire control system of the Ikv-91 is a hydraulic system for turret traverse and elevation, fed through a hydraulic accumulator from an intermittently working electro-hydraulic pump in the turret.

For the hydraulic system, AGA Aerotronics AB (now Bofors Aerotronics) developed an electric gun control system which includes a ballistic computer and electric servos to control the gunner's line of sight. The power traverse and power elevation velocities are normally controlled from the gunner's control handle with an overriding capability from the commander's control handle. If there is a power failure the gunner can hand crank the turret in traverse and the gun in elevation with hydraulic pumps working directly on the traverse motor and the elevation jack assembly.

The computer determines the ballistic solution for the 90 mm HE and HEAT rounds as well as the 7.62 mm coaxial machine gun as chosen by the gunner. Range is fed automatically from the laser rangefinder or, if desired, from a manual range input at the gunner's station.

The computer depends on preset values for non-standard muzzle velocities, powder temperatures and ambient conditions. These values are manually introduced by turning potentiometers on the computer panel. Trunnion cant and tracking rates are introduced from sensors, and lead angles in azimuth and elevation due to target movements are continuously calculated. When super-elevation and lead angles in azimuth and elevation are introduced into the system the gunner's line of sight is counter-rotated at exactly the same controlled rate. Thus the gunner's line of sight remains undisturbed on the target, reducing reaction time and increasing accuracy.

Variant

Ikv-91 with 105 mm gun

The Ikv 91-105 light tank is essentially the Ikv-91 tank destroyer, which has been in service with the Swedish Army since 1975, with its original Bofors 90 mm low-pressure gun replaced by a Bofors 105 mm low recoil gun fitted with a thermal sleeve, fume extractor and muzzle brake. This weapon can fire all standard 105 mm ammunition, including APFSDS rounds with their greatly increased armour-piercing capabilities.

The fire control system is a modified version of that installed in the Ikv-91 but the Ikv 91-105 can be fitted with two different types of passive night vision equipment. One is a combination of an LLLTV camera with an infra-red scanner, both of which are mounted on the gun mantlet. The other is a thermal imaging sight located on the left of the turret and with its head mirror linked mechanically to the gun. The turret traverse and gun elevation are stabilised.

The Ikv 91-105 can also be fitted with Bofors Lyran 71 mm mortar firing illuminating flares.

The amphibious characteristics of the Ikv-91 have been retained in the Ikv 91-105 by increasing the buoyancy of the vehicle. This has been achieved by extending the front of the hull and adding skirt-like floats each side. The Ikv 91-105 propels itself in the water by its tracks but propellers are fitted for increased water speed, these are powered by hydraulic motors.

Hägglund Vehicle have proposed a modified version of the 105 mm armed Ikv-91 to meet its Armored Gun System requirement. This would have a new powerpack consisting of a 550 hp diesel coupled to an Allison Transmission Division of General Motors X-300 series automatic transmission. Maximum combat weight with add-on armour would be 19 958 kg, combat weight without add-on armour would be 17 010 kg and air-transportable weight would be 15 876 kg. Main armament would comprise a 105 mm/51 calibre high pressure gun capable of firing all types of NATO ammunition.

Status: Production of the Ikv-91 for the Swedish Army is complete but can be resumed if further orders are placed. In service with Swedish Army. An early version of Ikv 91-105 has been tested in India.

Manufacturer: Hägglund Vehicle AB
S-891 02 Örnsköldsvik
Sweden
Telephone: (46) 0660 800 00
Telex: 6051
Fax: (46) 0660 826 12

SPECIFICATIONS (Ikv 91-105) (where different from 90 mm version covered below)

COMBAT WEIGHT	18 000 kg	ENGINE	Volvo TD 121 G 6-cylinder diesel developing 360 hp at 2200 rpm	ARMAMENT (main)	1 × 105 mm gun
POWER-TO-WEIGHT RATIO	20.2 hp/tonne			AMMUNITION (main)	40-50
GROUND PRESSURE	0.53 kg/cm²	TRANSMISSION	Allison Transmission Division of General Motors HT 740 D, 4 forward and 1 reverse gears	GUN CONTROL EQUIPMENT	
LENGTH				Turret power control	electro-hydraulic/manual
gun forwards	9.72 m			Turret traverse	360°
gun rear	8.67 m			Gun elevation/depression	
WIDTH	3.25 m	TORQUE CONVERTER	Allison Transmission Division of General Motors TC 497	to front of vehicle	+15°/-10°
HEIGHT (including sight)	2.495 m			over rear of vehicle	+15°/+1°
MAX WATER SPEED	12 km/h			Max elevation speed	10°/s
		STEERING	Hägglund, twin disc, dry plate	Max traverse speed	20°/s
				Gun stabiliser	
				vertical	yes
				horizontal	yes

SPECIFICATIONS (Ikv-91)

CREW	4	SIDE SLOPE	60%	SMOKE-LAYING EQUIPMENT	6 smoke dischargers mounted either side of turret
COMBAT WEIGHT	16 300 kg	VERTICAL OBSTACLE	0.8 m		
POWER-TO-WEIGHT RATIO	22.2 hp/tonne	TRENCH	2.8 m	AMMUNITION	
GROUND PRESSURE	0.49 kg/cm²	ENGINE	Volvo-Penta TD 120 A, 4-stroke, 6-cylinder turbo-charged diesel developing 330 hp at 2200 rpm	main	59
LENGTH GUN FORWARDS	8.84 m			MG	4250
LENGTH GUN REAR	7.56 m			GUN CONTROL EQUIPMENT	
LENGTH HULL	6.41 m	AUXILIARY ENGINE	none	Turret power control	electro-hydraulic/manual
WIDTH	3 m	TRANSMISSION	Allison Transmission Division of General Motors model HT 740 D with 4 forward and 1 reverse gears	by commander	yes
HEIGHT				by gunner	yes
to top of commander's cupola	2.32 m			Turret traverse	360°
to top of loader's cupola	2.155 m			Max rate power traverse	20°/s
GROUND CLEARANCE	0.37 m	STEERING	Hägglunds clutch and brake	Max rate power elevation	10°/s
TRACK WIDTH	450 mm	CLUTCH	twin dry plate	Gun elevation/depression	+15°/-10° (+1° to +15° over rear arc)
LENGTH OF TRACK ON GROUND	3.68 m	SUSPENSION	torsion bar		
MAX SPEED		ELECTRICAL SYSTEM	24 V		
road	65 km/h	BATTERIES	4 × 12 V, 228 Ah		
water	6.5 km/h	ARMAMENT			
FUEL CAPACITY	400 l	main	1 × 90 mm gun	Gun stabiliser	
MAX ROAD RANGE	500 km	coaxial	1 × 7.62 mm MG	vertical	no
FORDING	amphibious	anti-aircraft	1 × 7.62 mm MG	horizontal	no
GRADIENT	60%				

Hägglund Vehicle Pvrbv 551 Anti-tank Missile Carrier

Development

In February 1983, following successful trials with a prototype vehicle, the Swedish Defense Materiel Administration awarded a contract to Hägglund and Söner for the conversion of a number of obsolete Ikv-102 and Ikv-103 self-propelled infantry cannons, which were withdrawn from front line use some years ago, into Pvrbv 551 anti-tank missile carriers. At the same time the company was awarded a contract to convert the Ikv-102 and Ikv-103 into RBS 70 SAM vehicles. First vehicles were delivered to the Swedish Army in 1984 and production continued until mid-1986.

The modifications have been extensive and include replacement of the engine and transmission, extending the crew compartment to the rear of the hull, improving protection and fitting new communications equipment and observation devices.

The vehicles are organised into companies and supplied to both armoured and mechanised brigades.

Description

The gun has been removed from the glacis plate and the position plated over. The driver sits at the front of the hull on the left side and has a single piece hatch cover and fixed periscopes for observation to the front and sides. The commander sits to the right of the driver and has a single piece hatch cover that opens to the right rear and observation periscopes. The commander's position is raised above the driver for improved observation.

In the centre of the hull roof is a larger circular opening covered by a two-piece hatch cover opening either side. The Hughes TOW ATGW system is carried inside of the hull and raised up when required for action. Reserve missiles are stowed in the hull rear above the engine and transmission compartment and access to these is via hatches in the roof that open either side.

The suspension, which is identical to the Ikv-102's and Ikv-103's, consists of six rubber-tyred road wheels with the drive sprocket at the rear, idler at the front and two track return rollers. The original engine has been replaced by a Ford V-6 petrol engine, also installed in the

Pvrbv 551 anti-tank vehicle with Hughes TOW ATGW (FMV)

Lvrbv 701 with RBS 70 SAM launcher in elevated position ready to engage targets

Hägglund Vehicle AB Bv 206 all-terrain vehicle in service with the Swedish Army. The crew of four consists of the commander, driver, gunner and loader.

Variant

Lvrbv 701 RBS 70

This has a similar hull to the Pvrbv 551 anti-tank missile carrier but is armed with the Bofors RBS 70 surface-to-air missile system. When travelling the launcher is retracted under armour protection and when required is raised into the operating position. Reserve missiles are stowed in the hull rear above the engine and transmission compartment with access via the front edge of the magazine which folds upwards. This version has a four man crew. Full details of this version are given in *Jane's Land-Based Air Defence 1991-92*, page 119.

Status: Production complete. In service with Swedish Army.

Manufacturer: Hägglund Vehicle AB
S-891 82 Örnsköldsvik
Sweden
Telephone: (46) 0660 800 00
Telex: 6051
Fax: (46) 0660 826 12

SPECIFICATIONS	
CREW	4
COMBAT WEIGHT	9700 kg
POWER-TO-WEIGHT RATIO	14 hp/tonne
GROUND PRESSURE	0.4 kg/cm^2
LENGTH	4.81 m
WIDTH	2.54 m
WIDTH OVER TRACKS	2.23 m
HEIGHT	
hull top	1.97 m
missile and launcher up	2.46 m
GROUND CLEARANCE	0.33 m
TRACK	1.83 m
TRACK WIDTH	400 mm
LENGTH OF TRACK ON GROUND	3 m
MAX ROAD SPEED	
high range forwards	41 km/h
low range forwards	21 km/h
reverse	7 km/h
FUEL CAPACITY	240 l
MAX RANGE	300 km
FORDING	0.9 m
GRADIENT	60%
TRENCH	1.5 m
ENGINE	Ford model 2658E V-6 petrol developing 136 hp at 5200 rpm
TRANSMISSION	Mercedes W 4A-018 4 forward, 1 reverse gears
ARMAMENT	1 × TOW ATGW launcher
ELECTRICAL SYSTEM	24 V
BATTERIES	4 × 12 V, 57 Ah

UNION OF SOVIET SOCIALIST REPUBLICS

AT-6 Spiral/MT-LB Tank Destroyer

Development

In 1990 the Soviet Union revealed that it had deployed a version of the MT-LB multi-purpose armoured vehicle, fully described in the *Armoured personnel carriers* section, armed with the AT-6 Spiral (NATO designation) ATGW under the designation of the Shturm-C (Soviet industrial index number 9P149).

The radio command-guided tube-launched AT-6 Spiral (Soviet industrial number 9M114) semi-automatic command to line-of-sight (SACLOS) anti-tank missile first entered service with the Soviet Army as a helicopter-launched anti-tank weapon for the Mil Mi-24 Hind. It was subsequently fitted to other helicopters such as the Mil Mi-28 Havoc and the Kamov Ka-29 Helix-B.

MT-LB tracked vehicle with AT-6 Spiral Launcher in elevated position

Description

Externally, the MT-LB with AT-6 Spiral is virtually identical to the standard MT-LB but has a gunner's sight mounted at the front of the hull and a single rail launcher is mounted at the rear which retracts under armour protection when not required and for reloading purposes.

All the gunner has to do is to keep his day/night sight crosshairs on the target until the missile impacts. Airborne targets with crossing/approach speeds of 60 m/s and altitudes from ground level up to 3000 m can be engaged. Rate of fire of the system aboard the 9P149 is 3-4 rds/min with a total of 12 missile launcher/containers being carried.

According to the Soviet Union, crossing targets travelling at a speed of up to 60 km/h can be engaged by the system as can head-on targets travelling at a speed of up to 80 km/h. The target engagement arc is 85° left and right with elevation arc from -5° to +15°.

Like the standard MT-LB, the AT-6 Spiral MT-LB is fully amphibious being propelled in the water by its tracks. It has a two man crew and a combat weight of 12.3 tonnes.

Basic specifications of the AT-6 missile are as follows:

TYPE	SACLOS
LENGTH	1.83 m
DIAMETER	0.13 m
WING SPAN	0.25 m
LAUNCH WEIGHT	40 kg
WARHEAD	7-8 kg HEAT
RANGE	400-5000 m
ARMOUR PENETRATION	800 mm plus
SPEED	
at launch	55 m/s
in flight	350-400 m/s

Status: In service with the Soviet Army.

Manufacturer: Soviet state factories.

ASU-85 Airportable Self-propelled Anti-tank Gun

Development

The ASU-85 was developed in the late 1950s, entered service in 1960, and was first seen in public during the 1962 Moscow May Day parade. ASU is the Soviet designation for airborne assault gun and 85 is for the calibre of the gun. The vehicle is deployed only with air assault divisions and is airportable in the An-12 (Cub) aircraft. At one time 18 ASU-85s were issued per air assault division but this has been increased to a battalion of 31 vehicles. This consists of 3 companies of 10 and a command vehicle at battalion HQ. The ASU-85 was the replacement for the ASU-57 57 mm air-portable self-propelled anti-tank gun which is no longer in service with the Soviet Army. Details were given in *Jane's Armour and Artillery 1985-86*, page 444.

Description

The hull of the ASU-85 is all welded steel with the fighting compartment at the front and the engine and transmission at the rear. The glacis plate is well sloped for maximum protection and is provided with a splash board to stop water rushing up the glacis when the vehicle is fording.

The driver is seated at the front of the vehicle on the right and has a vision block to his front with a periscope above it and another vision block in the right side of the hull. The commander is seated behind the driver and has a single piece hatch cover in the roof that opens forwards and can be locked vertically. There is a periscope in front of the commander's hatch above which is an infra-red searchlight, both with a degree of traverse.

The gunner and loader are seated on the left of the main armament and enter and leave the vehicle by a single piece hatch cover in the roof that opens to the left and has a periscope for forward observation.

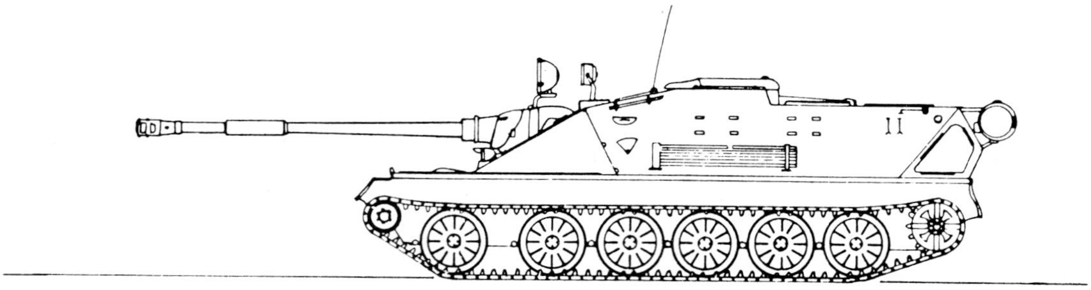

ASU-85 airportable self-propelled anti-tank gun

ASU-85 airportable self-propelled anti-tank gun with infra-red searchlight over 85 mm gun (Henry Dodds)

ASU-85 self-propelled anti-tank guns with 12.7 mm anti-aircraft machine guns at maximum elevation. Roof-mounted smoke dischargers can be seen just forward of rear external drum-type fuel tanks

There is a single vision block and firing port in the left side of the hull.

Another two hatches in the roof of the vehicle, behind the forward hatches, both open forwards. There is a distinct step down between the fighting compartment and the engine compartment and a single vision block is fitted in the rear of the fighting compartment on the right side.

The engine and transmission are at the rear of the hull and there are two large square hatches in the roof for access to the engine for maintenance. The air inlet and outlet louvres are on top of the roof at the left rear with the exhaust outlet in the rear of the hull, on the left side below the external stowage box. There is a large metal stowage box at the rear of the hull, either side of which is a long-range fuel drum.

The torsion bar suspension consists of six rubber-tyred road wheels with a distinct gap between the first and second road wheels, with the idler at the front and the drive sprocket at the rear. There are no track return rollers.

The ASU-85 has infra-red night vision equipment, is probably fitted with an NBC system but has no amphibious capability.

Main armament consists of an 85 mm gun mounted in the glacis plate slightly offset to the left of the vehicle's centreline. The weapon has a double-baffle muzzle brake and a fume extractor one third of the way from it. Mounted coaxially to the right of the main armament is a 7.62 mm SGMT machine gun and more recently some vehicles have been observed with a 12.7 mm DShKM machine gun mounted on the roof for anti-aircraft defence. Mounted over the main armament is an infra-red searchlight which moves in elevation and traverse with the gun. The gun has an elevation of +15°, depression of -4° and a total traverse of 12°. Elevation and traverse are both manual.

The main armament is the 85 mm D-70 gun and fires APHE and HVAP projectiles with a muzzle velocity of over 1000 m/s. These projectiles are more powerful than normal 85 mm field, anti-aircraft and tank guns.

Some vehicles have been seen fitted with a bank of smoke dischargers over the top of the hull towards the rear. These launch smoke grenades over the frontal arc of the ASU-85.

Variants

There are no known variants of the ASU-85.

Status: Production complete. In service with USSR only.

Manufacturer: Soviet state factories.

SPECIFICATIONS	
CREW	4
COMBAT WEIGHT	15 500 kg
POWER-TO-WEIGHT RATIO	13.5 bhp/tonne
GROUND PRESSURE	0.44 kg/cm²
LENGTH GUN FORWARDS	8.49 m
LENGTH HULL	6 m
WIDTH	2.8 m
HEIGHT (without AA MG)	2.1 m
FIRING HEIGHT	1.57 m
GROUND CLEARANCE	0.4 m
TRACK	2.66 m
TRACK WIDTH	360 mm
LENGTH OF TRACK ON GROUND	4.195 m
MAX ROAD SPEED	45 km/h
FUEL CAPACITY	250 l
MAX ROAD RANGE	260 km
FUEL CONSUMPTION	0.96 l/km

FORDING	1.1 m
GRADIENT	70%
VERTICAL OBSTACLE	1.1 m
TRENCH	2.8 m
ENGINE	Model V-6 6-cylinder in-line water-cooled diesel developing 240 hp at 1800 rpm (late production vehicles may have 280 hp engine)
TRANSMISSION	manual with 4 forward and 1 reverse gears
STEERING	clutch and brake
SUSPENSION	torsion bar
ELECTRICAL SYSTEM	2 × 12 V, 100 Ah
ARMAMENT	
main	1 × 85 mm gun
coaxial	1 × 7.62 mm MG
anti-aircraft	1 × 12.7 mm MG

AMMUNITION	
main	40
coaxial	2000 (estimated)
GUN CONTROL EQUIPMENT	
Gun elevation/depression	+15°/-4°
Total traverse	12°
Gun traverse	manual
Gun elevation	manual
Gun stabiliser	
vertical	no
horizontal	no
ARMOUR	
Hull glacis	40 mm at 60°
Hull sides upper	15 mm at 30°
Hull mantlet	40 mm at 60°
Hull top	8 mm
Hull floor	10 mm
Hull rear	10 mm

UNITED STATES OF AMERICA

Emerson M901 Improved TOW Vehicle

Development

To meet a US Army requirement for a turret-mounted and elevated TOW ATGW system, three companies built prototype systems and after trials, in December 1976, Emerson was awarded a low-rate initial-production contract for 10 vehicles.

In June 1978, the M901 Improved TOW Vehicle (ITV) was approved for full scale production and by 1991 the company had produced over 3200 systems for the US Army with additional systems being produced for the export market.

The latest production model, the M901A1, is capable of firing any of the TOW missile variants; Basic-TOW, Improved-TOW (I-TOW) or TOW-2 series produced by Hughes.

The US Army mounts the system on a modified FMC M113 series chassis, but it can be installed on a wide range of other chassis. For example the US Marine Corps uses the Light Armored Vehicle (8 × 8), the Netherlands uses the FMC Armored Infantry Fighting Vehicle while Saudi Arabia uses the Italian OTO Melara VCC-1 full tracked chassis.

Description

The M901/M901A1 Improved TOW Vehicle consists of a standard M113A1 APC with an M27 cupola mounted on the roof. On the forward part of this is an arm containing the image transfer assembly and on top of the arm is the armoured launcher assembly which contains two TOW launcher tubes, TOW sight assembly and the acquisition sight.

Improved TOW Vehicle deployed in firing position

The turret has full powered 360° traverse at 35°/s with the launcher head having an elevation of +34° and a depression of -30°. This enables the ITV to operate on gradients and side slopes with only the launcher head showing above the crest.

When travelling, the launcher head is retracted level onto the hull top so making the ITV difficult to distinguish from the standard M113A1 series APC.

Once the ITV has come to a halt, it takes only 20 seconds for the launcher to be elevated and the target to be engaged. Time from first-round impact to second-round triggering with a 12.5° target separation is 4.25 seconds.

The gunner identifies and tracks the target through the image transfer assembly, which contains the coupling optics and allows the gunner to select the standard day sight or night tracking by remote-control.

In addition, the coupling optics have a wide field-of-view so allowing the gunner to scan for targets. The guidance and sights of the ITV are identical to the normal TOW ground-based systems and if required, the components can be removed from the vehicle and fitted to a standard TOW series ground mount which can be carried inside the hull.

Once the two missiles have been launched, the launcher assembly is depressed at the rear and two new missiles are loaded, which takes only 40 seconds. The loader is provided with both side and overhead armour protection during the manual loading operation.

A further 10 missiles are carried internally and a 7.62 mm machine gun is mounted on the cupola for local defence with a total of 2000 rounds being carried for this weapon. Mounted on either side of the hull front is a bank of four electrically operated smoke dischargers.

SPECIFICATIONS (Similar to M113A1 APC except for following)

CREW	4 (deployed with mechanised infantry) 5 (deployed with cavalry)
COMBAT WEIGHT	11 794 kg
GROUND PRESSURE	0.58 kg/cm^2
HEIGHT	
launcher stowed	2.91 m
launcher erected	3.35 m

Status: In production. In service with Egypt (52), Greece (24), Jordan (50), Kuwait (58, status uncertain), Netherlands (314, see text), Pakistan (24), Saudi Arabia (224, see text) and USA (see text).

Manufacturer: Emerson Electronics and Space Corporation
8100 W Florissant Avenue
St Louis
Missouri 63136
USA
Telephone: (314) 553 3232
Telex: 44 869

Emerson Ground Launched Hellfire: Heavy (GLH-H)

Development/Description

Late in 1989, the United States Army Missile Command awarded a contract to Emerson to build a single prototype of the Ground Launched Hellfire: Heavy (GLH-H) system for trials purposes.

The prototype was completed within 15 months of contract award and is based on a government furnished M113 APC chassis, although it will also be applicable to other chassis, tracked and wheeled, including the US Army M2/M3 Bradley and the US Marine Corps Light Armored Vehicle (LAV) (8 × 8).

Emerson is prime contractor for the GLH-H with Rockwell International providing the Hellfire missile, GE Aerospace the electric turret drives and Hughes Aircraft the G/VLLD (Ground/Vehicular Laser Locator Designator).

The power operated two-man turret has an armoured box on either side with four Hellfire ATGWs in the ready to launch position along with smoke dischargers and a machine gun provided for local protection. The gunner will be seated on the left of the turret and the commander on the right, both being provided with single piece hatch covers that open to the rear.

The operator will have a choice of three operational modes: autonomous (laser designate and fire), co-operative designate (laser designate only) and co-operative fire (fire only).

Further growth is planned from fire-and-forget capabilities through to Hellfire applications of imaging infra-red or millimetre wave technologies.

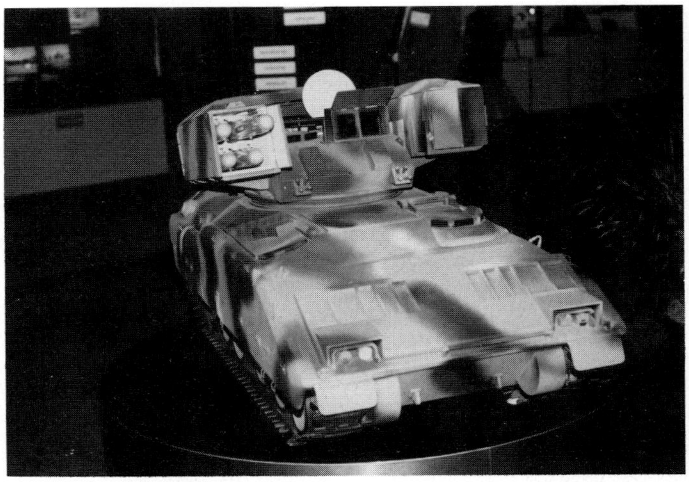

Scale model of the Emerson Ground Launched Hellfire - Heavy on a M2 Bradley IFV chassis (Ramon Lopez)

The current TOW ATGW has a maximum range of 3750 m, whereas the Hellfire has a much longer range of some 8000 m. Hellfire is operational with the US Army on the AH-64 Apache attack helicopter and the US Marine Corps' AH-1W attack helicopter and it is also entering service with Sweden in the coastal defence role.

Status: Prototype undergoing US Army trials.

Manufacturer: Emerson Electronics and Space Corporation
8100 W Florissant Avenue
St Louis
Missouri 63136
USA
Telephone: (314) 553 3232
Telex: 44 869

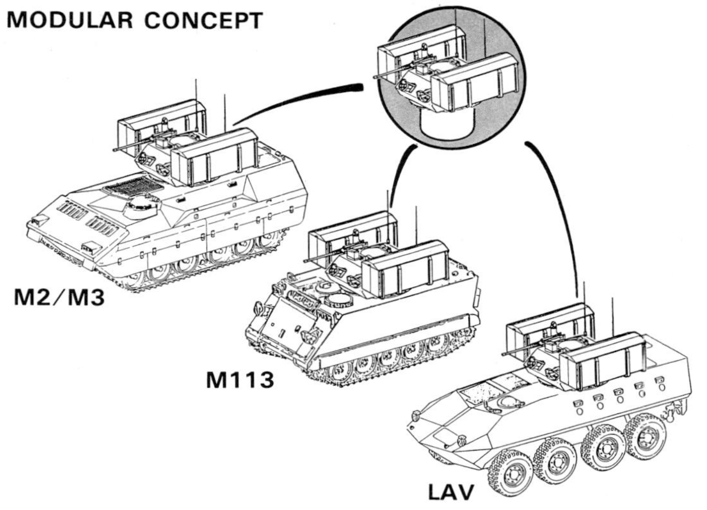

The Emerson Electric Ground Launched Hellfire - Heavy is a modular concept that can be fitted to a wide range of chassis, tracked and wheeled

Line-Of-Sight Anti-Tank System

Development

The Line-Of-Sight Anti-Tank System (LOSAT) is a member of the US Army's Armored Systems Modernisation programme (qv) and is expected to be the first member of the family of tracked armoured vehicles to enter service.

LOSAT is a member of the Common Medium Chassis family of vehicles which consists of just two vehicles, LOSAT and the Future Armored Resupply Vehicle - Artillery (FARV-A). The former is based on a modified FMC M2 Bradley chassis while the latter is based on a modified FMC Multiple Launch Rocket System chassis.

The US Army has a requirement for a total of 907 LOSAT vehicles with deployment beginning in FY99, it will be the replacement for the current M901 Improved TOW Vehicle (qv).

According to the US Army, "LOSAT fulfils the need for a dedicated, highly mobile, adverse weather, day/night, direct fire, long-range, anti-armour weapon system capable of defeating future threat armour systems. It will support the ASM Block III tank and the FIFV by assisiting in the long-range destruction of threat armoured vehicles."

Description

The chassis and hull of the LOSAT will be identical to that of the latest production version of the M2 Bradley IFV with the driver front left and the powerpack to his right.

Mounted in the rear of the hull will be the KEM (Kinetic Energy Missile) Weapon Module. This will have a weapons head assembly (WHA) with a sight mounted on top and have four or six KEM in the ready to launch position.

This will be coupled to the elevation mechanism, or Weapons Head Support Assembly (WHSA) which will raise and lower the Weapon Module. Mounted below, under armour protection in the hull, are the two auto-loaders containing additional KEMs.

The KEM is a hypervelocity munition capable of penetrating current and projected threat systems and will also have a limited self-defence capability against hovering or slow-moving attack helicopters.

The KEM is being developed by LTV Aerospace and Defense Corporation, Missiles and Electronics Group, launch weight is approximately 77 kg, missile length 2.845 m and diameter 0.162 m, range limits are from around 914-4572 m. Speed of the KEM is around 1524 m/s with armour penetration being achieved by its tungsten carbide solid nose. Although the cfficial US Army designation is KEM, LTV usually refers to the missile as the HVM, or Hypervelocity Missile.

Artist's impression of LTV LOSAT system on modified Bradley chassis launching an HVM

Guidance is by the Multifunctional Infra-red Coherent Optical Scanner (MICOS) device located in the rear of the missile near to the rocket motor. The laser sensing unit receives coded pulses containing position information from the launcher platform and transfers it to the HVMs onboard guidance package.

A Forward Looking Infra-Red sensor is used for target acquisition, target and missile tracking whilst the laser guidance link is used to transmit the necessary time-coded azimuth and elevation positional data to the KEM in flight.

An all-boost rocket motor is used to accelerate the weapon to hypervelocity speeds with attitude control motors spaced radially around the missile to provide a full three-dimensional manoeuvre capability.

Status: Development.

Manufacturer: Not selected but chassis/hull will be provided by FMC Corporation, Ground Systems Division with the HVM being developed by LTV.

Self-propelled guns and howitzers

ARGENTINA

TAMSE VCA 155 155 mm Self-propelled Artillery System

Development
The VCA 155 (Vehiculo de Combate de Artilleria) 155 mm self-propelled artillery system has been developed by TAMSE to meet the requirements of the Argentinian Army.

It is essentially a lengthened TAM MBT chassis with seven road wheels either side fitted with the same turret as installed on the Italian OTO Melara 155 mm Palmaria self-propelled howitzer fully described later in this section.

TAMSE VCA 155 self-propelled artillery system with ordnance at maximum elevation (Terry J Gander)

OTO Melara supplied a total of 25 turrets to TAMSE but by early 1991 it is understood that only two complete prototypes of the VCA 155 had been completed and that volume production had yet to start.

Description
The layout of the VCA 155 is virtually identical to the TAM medium tank fully described in the *MBTs and medium tanks* section with the driver front left, powerpack to his right and turret in the centre of the hull slightly to the rear. However, an auxiliary generator is mounted in the right side of the hull and to increase the operational range of the system two long range fuel tanks are mounted at the rear of the hull.

The all-welded steel turret is identical to that of the Palmaria covered in this section under Italy. Firing standard ammunition a maximum range of 24 000 m can be achieved while a range of 30 000 m can be achieved with special ammunition. An automatic loading system is fitted.

Status: Two prototypes completed. Not yet in production or service.

Manufacturer: TAMSE
Avenida, Rolon 1445
Boulgone Sur Mer
CP 1609
Buenos Aires
Argentina
Telephone: (763) 114/4937/0714
Telex: 26208 TAMSE AR
Fax: (763) 2464

TAMSE VCA 155 self-propelled artillery system with ordnance in travel lock

SPECIFICATIONS

CREW	5
COMBAT WEIGHT	40 000 kg
POWER-TO-WEIGHT RATIO	18 hp/tonne
GROUND PRESSURE	0.82 kg/cm²
LENGTH (overall)	10.30 m
LENGTH (chassis)	7.69 m
WIDTH (overall)	3.30 m
MAX HEIGHT	3.20 m
GROUND CLEARANCE	0.45 m
TRACK	2.62 m
MAX ROAD SPEED	60 km/h
FUEL CAPACITY	873 + 400 l
MAX ROAD RANGE	719 km (with external fuel tanks)
FORDING	1.4 m
GRADIENT	60%
SIDE SLOPE	30%
VERTICAL OBSTACLE	1 m
TRENCH	2.5 m
ENGINE	MTU MB 833 Ka 500 6-cylinder diesel developing 720 hp at 2200 rpm
TRANSMISSION	Renk HSWL-204 planetary with 4 speeds forward and reverse
STEERING	double differential with hydrostatic steering
SUSPENSION	torsion bar
ELECTRICAL SYSTEM	24 V
ARMAMENT	
main	1 × 155 mm 41 calibre howitzer
anti-aircraft	1 × 7.62 mm AA MG
SMOKE DISCHARGERS	2 × 4
MAIN ARMAMENT ELEVATION/DEPRESSION	+70°/-5°
TURRET TRAVERSE	360°
AMMUNITION CARRIED	28 × 155 mm (23 turret, 5 hull)

CHILE

Cardoen CC-SP-45 Self-propelled Howitzer

Development/Description
In mid-1989 it was stated that Cardoen Industries was to undertake local manufacture of the South African ARMSCOR 155 mm G6 self-propelled howitzer.

According to sources in Chile, a prototype had by then already been completed by Cardoen Industries and had undertaken firing trials in Northern Chile.

Assembly is to be undertaken by the company from locally made components, although major items such as the engine and transmission will be purchased overseas.

The description and specifications of the CC-SP-45 are identical to those of the South African G6, for which there is a separate entry later in this section.

Status: Prototype.

Manufacturer: Cardoen Industries
Los Conquistadores 1700
28th Floor
Santiago
Chile
Telephone 2313420/2314219
Telex: 341630 INCAR CK
Fax: 2316366

Cardoen Industries CC-SP-45 deployed in the firing position with stabilisers lowered

CHINA, PEOPLE'S REPUBLIC

NORINCO 155 mm 45 Calibre Self-propelled Gun/Howitzer

Development
Late in 1988 the prototype of a new 155 mm self-propelled gun-howitzer was shown for the first time. In many respects this is very similar to the US M109 series but is heavier and is fitted with a 45 calibre barrel based on that used in the Chinese 155 mm gun-howitzer Type WAC 21 system which is already in production and is covered in the *Towed guns & howitzers* section.

Description
The hull and turret of the system is of all-welded steel armour construction that provides protection from small arms fire and shell splinters. The layout is conventional with the driver's and engine compartments at the front with the fighting compartment and turret at the rear.

The driver is seated at the front of the hull on the left side with the engine compartment to his right. The driver has a single piece hatch that opens to the rear and in front of this are two day periscopes for use when the hatch is closed. One of these can be replaced by a passive night driving periscope.

The turret is at the rear of the hull with the commander's cupola on the right side. This has four periscopes for observation, a single piece hatch cover that opens to the rear and an externally mounted 12.7 mm machine gun for use in the ground defence and anti-aircraft roles.

In either side of the turret is a large forward opening door with an observation block in its lower part and there is also a large hatch in the left side of the turret towards the rear.

A small door in the turret rear opens to the right and a larger door in the hull rear opens to the right for ammunition resupply.

The direct fire telescope is mounted in the front of the turret to the left of the 155 mm gun/howitzer while the indirect fire sight is mounted in an armoured cupola on the left side of the turret roof.

The 45 calibre barrel is provided with a slotted muzzle brake and a fume extractor and when in the travelling position is held in place by a lock mounted on the glacis plate that is released manually.

The ordnance has a semi-automatic horizontal sliding wedge breechblock, semi-automatic loader and an electrically controlled and hydraulically operated rammer that enables projectile loading to take place at any angle of elevation with the charge being loaded manually.

The onboard fire control system includes a data display unit, computer and sensors for elevation, azimuth and tilt, with fire information coming from the battery or battalion command system. Chinese sources state that the system can have either automatic or manual operating and laying. The hydraulic equilibrator for elevation/depression is mounted on the roof of the turret to make more space for the crew. Although the turret has full 360° turret traverse, when firing it is normally 30° left and right.

When in the firing position two spades are lowered at the rear of the hull, one either side and in line with the track.

Using ERFB-BB ammunition a range of 39 000 m can be achieved with maximum rate of fire being 4 to 5 rds/min.

The torsion bar suspension either side consists of six dual rubber tyred road wheels with a distinct gap between the first and second and third road wheels with the drive sprocket at the front, idler at the rear and three track return rollers. Each road wheel station is fitted with a hydraulic shock absorber.

Standard equipment includes an NBC system, explosion detection and suppression system and a fire suppression system.

Chinese 155 mm self-propelled gun/howitzer with ordnance elevated (Tony Banks)

Chinese 155 mm self-propelled gun/howitzer with ordnance in travelling lock (Tony Banks)

The following types of ERFB ammunition can be fired:

Type	ERFB	ERFB-BB	ERFB-WP	ERFB III	ERFB BE
Length	938 mm	938 mm	938 mm	940 mm	940 mm
Weight	45.5 kg	47.9 kg	47.9 kg	45.2 kg	45.7 kg
Muzzle velocity	897 m/s	903 m/s	795.5 m/s*	789.6 m/s*	974.5 m/s*
Max range	30 km	39 km	26 km	24 km	26 km

(the system can also fire NATO 155 mm M107 projectiles)
* with M2 charge 9 system

Under development are three new carrier projectiles, ERFB-BB illuminating, ERFB-ICM and ERFB-ICM-BB.

SPECIFICATIONS	
CREW	5
COMBAT WEIGHT	32 000 kg
POWER-TO-WEIGHT RATIO	16.40 hp/tonne
LENGTH GUN FORWARDS	10.15 m
WIDTH	3.23 m
HEIGHT	
with AA MG	3.417 m
to turret roof	2.597 m
AXIS OF FIRE	2.16 m
GROUND CLEARANCE	450 mm
TURRET DIAMETER	2.60 m
MAX ROAD SPEED	56 km/h
MAX RANGE	450 km
GRADIENT	60%
SIDE SLOPE	40%
VERTICAL OBSTACLE	0.7 m
TRENCH	2.7 m
ENGINE	air-cooled diesel developing 525 hp
GEARBOX	manual with dual power take off
STEERING	hydraulic
ARMAMENT	
main	1 × 155 mm gun/howitzer
anti-aircraft	1 × 12.7 mm MG
LENGTH OF BARREL	7046 mm (without muzzle brake)
RATE OF FIRE	4-5 rpm
GUN CONTROL EQUIPMENT	
Gun elevation/depression	+72°/-3°
Turret traverse	360°
NORMAL FIRING ARC	30° left and right

Status: Prototype.

Manufacturer: Chinese state factories.

NORINCO
Chinese North Industries Corporation
7A Yuetan Nanjie
PO Box 2137
Beijing
People's Republic of China
Telephone: (86) 6898, (86) 3461, (86) 3471, (86) 7570
Telex: 22339 CNIN CN

NORINCO Type 83 152 mm Self-propelled Gun-howitzer

Description

The Type 83 152 mm self-propelled gun-howitzer was first seen publicly in October 1984 although it is thought that it entered service some years before then. The Type 83 uses an entirely new full tracked chassis that is shared by the 122 mm tracked self-propelled rocket launcher (see entry under *Multiple rocket launchers*).

The gun-howitzer carried is a derivative of the towed 152 mm gun-howitzer Type 66 fitted with a fume extractor. The howitzer is mounted in a traversing enclosed turret at the rear of the hull. The vehicle driver sits at the front left with the engine compartment to his right. The turret has a full 360° traverse and there are doors for the four-man turret crew at the hull rear and on both sides of the turret. A 12.7 mm Type 54 anti-aircraft machine gun is mounted on the roof and there is also a 7.62 mm tank machine gun fitted next to the main barrel. The turret is also equipped with a Type 40 rocket launcher. It is assumed that the Type 83 fires the same ammunition as the towed Type 66, ie HE and smoke. The maximum range is 17 230 m and the muzzle velocity at full charge is 655 m/s. An RAP known as the MP-152 has been developed and tested. Four rounds can be fired in one minute.

The fire control system is operated from a panel and all turret movements are power controlled. Time in and out of action is one minute. Panoramic and direct fire sights are provided along with infra-red night vision equipment. Communications equipment includes a Type 889D radio set and a Type 803 intercom system.

The suspension uses six road wheels with the central four wheels arranged in two pairs. There are three track return rollers, idler at the rear and drive sprocket at the front.

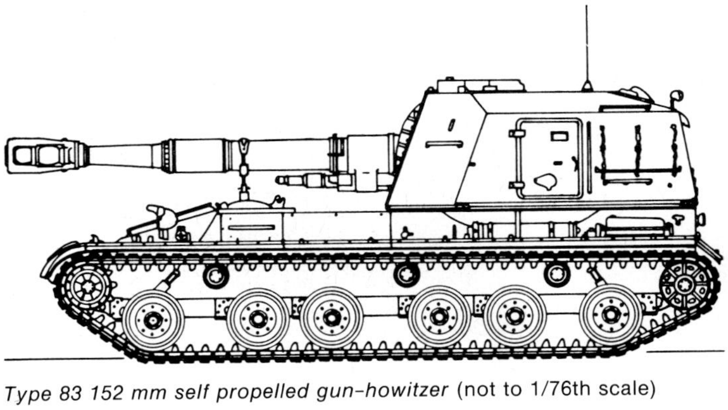

Type 83 152 mm self propelled gun-howitzer (not to 1/76th scale)

Variants

Trench Digger

The chassis of the Type 83 152 mm self-propelled gun-howitzer is used as the basis for a trench digging machine capable of digging 250 to 300 m^3 of earth per hour. The digging machine head is carried over the rear of the hull and the crew of three in a raised cab at the front. A dozer blade is also carried. Combat weight is 30 000 kg. For full details relating to this digger refer to *Jane's Military Vehicles and Logistics 1991-92*, page 556.

425 mm Mine-clearing Rocket Launcher

The Type 83 is also used as the basis of the carrier/launcher for the Type 762 425 mm mine-clearing rocket. The turret of the Type 83 152 mm self-propelled gun-howitzer is replaced by a low turret carrying launching rails for two Type 762 rockets. Each rocket weighs 600 kg at launch and has a range of 800 to 1000 m. Combat weight is 27 000 kg. For full details of the rocket and the launch vehicle refer to *Jane's Military Vehicles and Logistics 1991-92*, page 231.

Main armament of Type 83 152 mm gun-howitzer is believed to be based on towed Type 66 gun-howitzer

Type 83 152 mm self-propelled gun-howitzer

Status: Production. In service with the Chinese Army.

425 mm mine-clearing rocket launcher which uses Type 83 chassis

Manufacturer: Chinese state factories.

NORINCO
Chinese North Industries Corporation
7A Yuetan Nanjie
PO Box 2137
Beijing
People's Republic of China
Telephone: (86) 6898, (86) 3461, (86) 3471, (86) 7570
Telex: 22339 CNIN CN

SPECIFICATIONS		TRACK	2.62 m	BATTERIES	4 × Type 65
CREW	5	TRACK WIDTH	480 mm	ARMAMENT	1 × 152 mm Type 66
COMBAT WEIGHT	30 000 kg	LENGTH OF TRACK			howitzer
POWER-TO-WEIGHT		ON GROUND	4.601 m		1 × 12.7 mm Type 54
RATIO	17.33 hp/tonne	MAX SPEED			AA MG
GROUND PRESSURE	0.68 kg/cm²	road	55 km/h		1 × 7.62 mm MG
LENGTH		average on road	40-45 km/h		1 × Type 40 rocket
gun forward	7.33 m	country roads	30-35 km/h		launcher
gun to rear	9.805 m	FUEL CAPACITY	885 l	MAIN ARMAMENT	
LENGTH OF HULL	6.882 m	MAX RANGE	450 km	ELEVATION/	
WIDTH OVERALL	3.236 m	FORDING	1.3 m	DEPRESSION	+65°/-5°
HEIGHT		GRADIENT	32°	TURRET TRAVERSE	360°
overall	3.502 m	SIDE SLOPE	30°	AMMUNITION CARRIED	
centre of AA MG	3.19 m	VERTICAL OBSTACLE	0.7 m	152 mm	30 rounds
top of turret	2.682 m	TRENCH	2.7 m	12.7 mm and 7.62 mm	650 rounds
bore of gun	2.08 m	ENGINE	Type 12150L diesel	rocket launcher	4 rounds
top of hull	1.47 m		developing 520 hp		
GROUND CLEARANCE	0.45 m	ELECTRICAL SYSTEM	24 V		

NORINCO 130 mm Self-propelled Gun

Development/Description
This is essentially the Type 83 152 mm self-propelled gun/howitzer, fully described in the previous entry, with the 152 mm weapon replaced by a 130 mm ordnance taken from the towed 130 mm gun Type 59, full details of which are given in the *Towed anti-tank guns, guns & howitzers* section.

Specifications
Identical to the Type 83 except for slightly longer overall length due to the longer barrel of the 130 mm gun Type 59. The latter has an elevation from -2.5° to +45° whereas the self-propelled 130 mm on the Type 83 chassis and turret has an elevation from -5° to +65°. A total of 38 rounds of 130 mm ammunition and associated charges are carried.

Status: Prototype.

Manufacturer: Chinese state factories.

NORINCO
Chinese North Industries Corporation
7A Yuetan Nanjie
PO Box 2137
Beijing
People's Republic of China
Telephone: (86) 6898, (86) 3461, (86) 3471, (86) 7570
Telex: 22339 CNIN CN

NORINCO 122 mm Self-propelled Howitzer Type 85

Development
The 122 mm self-propelled howitzer Type 85 is the self-propelled artillery component of the range of vehicles based on the Type 85 armoured personnel carrier, formerly known as the Type YW 531 H; for details of this vehicle refer to the entry in the *Armoured personnel carriers* section.

Description
The Type 85 self-propelled howitzer uses the basic chassis of the Type 85 APC with a Chinese-produced copy of the Soviet 122 mm D-30 howitzer mounted in a semi-open superstructure at the rear. The Type 85 self-propelled howitzer has a crew of six; the driver seated in the left side of the hull and the commander and four-man gun crew housed in the fighting compartment in the superstructure. The semi-open superstructure is equipped with a canvas cover for weather protection.

The 122 mm D-30 howitzer is fitted with a multi-baffle muzzle brake and has a maximum elevation of +70°. Traverse is limited to 22.5° either side. The D-30 fires an HE shell weighing 21.76 kg to a range of 15 300 m, but an ERFB round has been developed which has a maximum range of 21 000 m. The maximum rate of fire is 6 to 8 rds/min.

122 mm self-propelled howitzer Type 85 from the front with ordnance covered up

The vehicle is amphibious but requires additional pontoons attached to both sides of the hull to achieve full flotation. No secondary armament is carried other than the crew's weapons and the vehicle is equipped with a Type 889 or VRC-83 radio set.

Status: In production. In service with the Chinese Army.

Manufacturer: Chinese state factories.

NORINCO
Chinese North Industries Corporation
7A Yuetan Nanjie
PO Box 2137
Beijing
People's Republic of China
Telephone: (86) 6898, (86) 3461, (86) 3471, (86) 7570
Telex: 22339 CNIN CN

122 mm self-propelled howitzer Type 85 from the rear with ordnance covered up

SPECIFICATIONS	
CREW	6
COMBAT WEIGHT	16 500 kg
POWER-TO-WEIGHT RATIO	19.4 hp/tonne
GROUND PRESSURE	0.67 kg/cm^2
LENGTH (overall)	6.663 m
WIDTH	3.068 m
HEIGHT	
overall	2.85 m
roof	1.914 m
GROUND CLEARANCE	0.46 m
MAX ROAD SPEED	60 km/h
SPEED IN WATER (with pontoons)	6 km/h
FUEL CAPACITY	450 l
MAX RANGE	500 km
FORDING	amphibious
GRADIENT	53.1%
VERTICAL OBSTACLE	0.6 m
TRENCH	2.2 m
ENGINE	Type BF8L413F air-cooled diesel developing 320 hp
TRANSMISSION	manual with 5 forward and 1 reverse gears
ARMAMENT (main)	1 × 122 mm D-30 howitzer
GUN ELEVATION/DEPRESSION	+70°/-5°
GUN TRAVERSE	22.5° left, 22.5° right
AMMUNITION (122 mm)	40 rounds

NORINCO 122 mm Self-propelled Howitzer Type 54-1

Description

There are two types of this self-propelled howitzer. The first version had the 122 mm howitzer mounted directly onto the chassis of the Type 531 (YW 531) APC, but the latest version uses a lengthened version of this chassis with an extra road wheel (five instead of the earlier four).

The 122 mm self-propelled howitzer Type 54-1 carries a 122 mm Type 54-1 howitzer (copy of Soviet M1938 (M-30)) in an open position on top of the chassis rear. The original howitzer shield is retained and the sides and top have a canvas cover that is stowed around the periphery of the fighting compartment when not in use. The driver sits forward to the left of the engine compartment and has an overhead hatch and two forward-facing periscopes. There are two white light headlamps, one each side, and on the right is a night-driving infra-red lamp with a night vision device for use with it in one of the driver's periscopes. The suspension uses five road wheels and there are three track return rollers. The first, second and fifth road wheel arms are fitted with shock absorber cylinders. The track has a cover along its upper length, probably for use when wading when an anti-splash board on the front hull is raised. Rubber-covered track links are used.

The Type 54-1 howitzer has a maximum range firing HE of 11 800 m. It fires the following variable-charge, case-type, separate loading ammunition:

122 mm self-propelled howitzer Type 54-1 with cover in position over main armament (Christopher F Foss)

Ammunition type	**HE**	**Smoke**	**Illuminating**[1]
DESIGNATION	Type 54	Type 54	Type 54
WEIGHT OF PROJECTILE	21.76 kg	22.55 kg	21.9 kg
WEIGHT OF FILLING	3.5 kg	3.65 kg	1.515 kg
TYPE OF FILLING	TNT	WP	-
MUZZLE VELOCITY	515 m/s	509 m/s	496 m/s
MAX RANGE	11 800 m	11 930 m	11 000 m
TYPE OF FUZE	Liu-4	Yan-2	Shi-1

[1] The illuminant burns for 25 seconds with 450 000 candela at a burst height of 450 m

Also carried on the Type 54-1 self-propelled howitzer is a 7.62 mm Type 67 machine gun (Chinese version of Soviet RP-46) with 1000 rounds of ammunition. Each vehicle has a Model A-220A radio set with a range of 16 000 m. Two manually operated CO_2 fire extinguishers are also carried.

Variants

It is known that the Chinese have built the prototype of a new self-propelled gun based on a new chassis with the engine compartment at the front and fully enclosed turret at the rear. Suspension consists of six rubber-tyred road wheels, drive sprocket at the front, idler at the rear but no return rollers. The suspension is similar to that used on the Type 77 APC which has the engine compartment at the rear. It is believed that main armament comprises a 122 mm howitzer with a 12.7 mm machine gun mounted on the roof for anti-aircraft defence.

122 mm self-propelled howitzer Type 54-1 showing shield for 122 mm howitzer

Status: Production. In service with the Chinese Army. Offered for export.

Manufacturer: Chinese state factories.

NORINCO
Chinese North Industries Corporation
7A Yuetan Nanjie
PO Box 2137
Beijing
People's Republic of China
Telephone: (86) 6898, (86) 3461, (86) 3471, (86) 7570
Telex: 22339 CNIN CN

122 mm self-propelled howitzer with new chassis

SPECIFICATIONS	
CREW	up to 7
COMBAT WEIGHT	15 300 kg
POWER-TO-WEIGHT RATIO	17 hp/tonne
GROUND PRESSURE	0.6 kg/cm²
LENGTH	5.6 m
WIDTH	3.07 m
HEIGHT	
top of gun	2.685 m
top of hull	1.976 m
GROUND CLEARANCE	0.45 m
TRACK	2.52 m
LENGTH OF TRACK ON GROUND	3.55 m
MAX ROAD SPEED	56 km/h
MAX RANGE	450 km
GRADIENT	25°
VERTICAL OBSTACLE	0.6 m
TRENCH	2 m
ENGINE	Model 6150L 6-cylinder in-line water-cooled diesel developing 260 hp at 2000 rpm
ARMAMENT	
main	1 × 122 mm howitzer
auxiliary	1 × 7.62 mm MG
SMOKE-LAYING EQUIPMENT	none
AMMUNITION	
main	40
MG	1000
GUN ELEVATION/ DEPRESSION	+63°/-2.5°
GUN TRAVERSE	22.5° left, 22.5° right
ARMOUR (Front)	12 mm

CZECHOSLOVAKIA

152 mm Self-propelled Howitzer DANA

Description

First observed during 1980, this 152 mm self-propelled howitzer is mounted on a Czechoslovakian Tatra 815 8 × 8 truck chassis. A wheeled chassis rather than the more orthodox tracked chassis for such a large calibre piece was chosen because of the good roads in the Czechoslovakian interior. This chassis is cheaper to produce and maintain than the comparable tracked chassis, and for medium-range artillery support of infantry and tank divisions, the Tatra 815's mobility is more than adequate. Known to the Czechoslovakian Army as the DANA the first examples of this vehicle were issued to the Kiev Artillery Battalion Jan Zizka of Trocnov based at Tabor, Southern Bohemia, during 1981. Its correct designation is *vzor 77* self-propelled howitzer DANA.

Four 152 mm DANA self-propelled howitzers deployed in the firing position (Jane's Intelligence Review)

The Tatra 815 chassis has been much modified to carry the howitzer but retains its centralised tyre-pressure regulation system for cross-country mobility. The engine which is normally at the front of the vehicle has been re-located to the rear. The cab is armoured sufficiently to withstand small-arms and light shell-splinter fire and has vision devices at the sides. The main driver's windscreen can be covered by steel shutters which fold downwards in action. There does not appear to be any space for the crew within the gun housing when travelling. Power-assisted steering is provided for the driver on the front four road wheels.

The main armament is a 152 mm howitzer, probably based on the standard Soviet M-1973(2S3). There is no bore evacuator as the gun housing is open along its length, ie the barrel rises and falls along an open slot. The housings on either side are probably armoured but the interior is rather cramped and there are probably only three crew. A hydraulic crane on the housing roof assists in supplying ammunition for the howitzer. It has been confirmed that an automatic loading system is installed. A hydraulic motor to supply power for turret traverse and barrel elevation is situated behind a ventilation grille on the rear left-hand side of the turret. There is also a roof access hatch for the aimer on the left-hand side of the barrel. The gun housing also mounts an external 12.7 mm DShKM machine gun for local and anti-aircraft defence, on the right-hand housing.

The range of the 152 mm howitzer is stated to be between 18 000 m using standard ammunition and 24 000 m with a rocket-assisted projectile. Rate of fire is 3 rds/min for a period of 30 minutes using ammunition from within the turret and external sources. Turret traverse is a full 360°. In action three stabilisers are lowered to the ground, one at each side between the second and third axles and the other at the rear.

If the howitzer is based on the normal Soviet conventions, the following types of ammunition can be fired:

TYPE	APHE	HE	Illuminating	Smoke
DESIGNATION	BR-540	OF-540	S-540	D-540
PROJECTILE WEIGHT	48.8 kg	43.51 kg	n/av	44 kg

Standard equipment includes an NBC system for both the turret and crew compartments.

Czechoslovakian 152 mm (8 × 8) self-propelled howitzer DANA in travelling configuration

SPECIFICATIONS (provisional)

CONFIGURATION	8 × 8
COMBAT WEIGHT	23 000 kg
LENGTH	
gun forward	10.4 m
hull	8.87 m
WIDTH	
turret	2.97 m
hull	2.722 m
HEIGHT	
to top of machine gun	3.525 m
to cab roof	2.03 m
to engine deck	1.88 m
FIRING HEIGHT	2.4 m
WHEELBASE	1.65 m + 2.97 m + 1.45 m
MAX SPEED	80 km/h
RANGE	1000 km
FORDING	1.4 m
TRENCH	2 m
GRADIENT	60%
ENGINE	Tatra T-928 V-12 multi-fuel developing 345 hp at 2200 rpm
GEARBOX	dual-range, 5-speed plus overdrive providing 20 forward and 4 reverse gears
STEERING	power-assisted
ARMAMENT	
main	1 × 152 mm howitzer
AA	1 × 12.7 mm MG
GUN ELEVATION/ DEPRESSION	+60°/-3°

Status: Production probably complete. In service with Czechoslovakia, Libya and Poland.

Manufacturer: ZTV Dubnica nad Váhom
Zavody Tazeho
Strojarstava
N P Kosice
Czechoslovakia

Chassis normally produced at Tatra, Národi Podnik, Kohprivnice, Czechoslovakia.

EGYPT

122 mm D-30 Self-propelled Howitzer Programme

Development

The Egyptian Army has a large number of Soviet-supplied 122 mm D-30 towed howitzers and in recent years a production line for this weapon has been established in Egypt by the Abu-Zaabal Engineering Industries Company, with other concerns manufacturing the full range of D-30 ammunition. The Egyptian Army issued a requirement for a self-propelled version of the D-30 following attempts to mount the howitzers on converted T-34/85 chassis and prototypes for this programme have been constructed by Royal Ordnance of the UK and BMY of the USA. Both prototypes have undergone a series of firing and other trials in Egypt using Egyptian Army personnel and the first testing phase was completed by end of May 1985. Following trials with both vehicles further modifications were required which were carried out by the respective manufacturers. Once these were completed further trials were carried out in Egypt including mobility trials totalling some 5000 km. It is anticipated that the selected design chassis will be manufactured in the country of origin with final assembly taking place in Egypt. It is possible that local manufacture may be introduced at a later stage of production.

The 122 mm D-30 howitzer involved in this programme is manufactured in Egypt and its ballistic and other details are the same as those of the Soviet original.

Trials with the competing designs were completed in late 1987 but as of June 1991 neither had been selected for production.

Royal Ordnance SP122

Description

Development of the Royal Ordnance SP122 has been carried out by Royal Ordnance in conjunction with the Ministry of Defence and Military Production of the Arab Republic of Egypt. The SP122 was designed and built by the then Royal Ordnance Weapons and Fighting Vehicles Division with its HQ at Royal Ordnance Leeds. The latter facility was however taken over by Vickers Defence Systems late in 1987, and the SP122 is now the responsibility of Guns & Vehicles Division, Royal Ordnance. SP122 is a member of the RO2000 family of vehicles which was officially launched at the 1986 British Army Equipment Exhibition with SP122 being the RO2001. Full details of the RO2000 family of full tracked vehicles were given in *Jane's Armour and Artillery 1990-91*, pages 526 to 528. The RO2000 family of vehicles is no longer being marketed by Royal Ordnance. As of August 1991 work on the RO2000 family of vehicles had been suspended and it is considered unlikely that any further work will be carried out on this series unless the SP122 project moves into the production phase.

The SP122 has been designed with ease of manufacture as a priority and makes extensive use of armoured plate steel with no curves involved.

On the SP122 the 122 mm D-30 howitzer is mounted centrally in a fully enclosed compartment at the rear of the vehicle. The engine is mounted forward on the right-hand side of the forward hull with the driver on the left. The driver has a single wide-angle periscope for use when closed down and night vision equipment is optional. The remainder of the crew of five are inside the main fighting compartment and include the commander, gunner, loader and ammunition preparation number. The commander sits under a domed cupola fitted with a periscope with a full 360° traverse and located to the right-hand side of the howitzer. The gunner sits to the left of the howitzer and uses a panoramic telescope for indirect fire and telescopic sight OD4M-45 for direct fire roles. Above the gunner is a hatch above which is mounted a 0.50 (12.7 mm) Browning M2 HB machine gun with a 360° traverse and an anti-aircraft capability. In the upper compartment wall to the left of the gunner is an upward-opening rectangular hatch for side vision and improved ventilation. The commander has an entry hatch next to his position in the side wall with an emergency exit on the opposite side and there is a further entry door in the rear wall. The fighting compartment has space for 72 projectiles and charges but up to 80 can be carried as an option.

The SP122 is powered by a Perkins turbo-charged diesel engine coupled to a Self-Changing Gears T320 fully automatic transmission.

The suspension uses five road wheels each side carried on transverse torsion bars and with telescopic dampers on the front and rear wheel positions. The track uses cast steel links of the dry-pin type with rubber bush inserts and rubber roadpads.

Status: Trials with the competing designs were completed late in 1987 but as of June 1991 neither had been selected for production.

Manufacturer: Royal Ordnance Nottingham
Commercial Director
Guns & Vehicle Division
Royal Ordnance PLC
Kings Meadow Road
Nottingham NG2 1EQ
UK
Telephone: (0602) 863341
Telex: 37531
Fax: (0602) 861436.

Royal Ordnance SP122 self-propelled 122 mm D-30 in travelling configuration with ordnance in lock (Christopher F Foss)

SPECIFICATIONS	
CREW	5
COMBAT WEIGHT	20 000 kg
GROUND PRESSURE	0.52 kg/cm^2
LENGTH	7.7 m
WIDTH	
over external stowage	3.2 m
hull	2.81 m
HEIGHT (top of roof)	2.7 m
MAX ROAD SPEED	55 km/h
MAX GRADIENT	60%
VERTICAL OBSTACLE	0.75 m
TRENCH	2.2 m
ENGINE	Perkins TV8.640 8-cylinder turbo-charged diesel developing 300 hp
TRANSMISSION	Self-Changing Gears T320 automatic 6-speed epicyclic unit with manual hold control
STEERING	regenerative, disc brake
SUSPENSION	transverse torsion bar
ELECTRICAL SYSTEM	24 V
BATTERIES	4 × 6TN, 100 Ah
ARMAMENT	
main	1 × 122 mm howitzer
secondary	1 × 12.7 mm MG
SMOKE-LAYING EQUIPMENT	grenade dischargers optional
AMMUNITION (main)	72 (80 optional)
GUN CONTROL EQUIPMENT	
Howitzer elevation limits	−5°/+70°
Howitzer traverse limits	30° left, 30° right

BMY SPH 122 mm Self-propelled Howitzer

Description

The BMY SPH 122 mm self-propelled howitzer is based on the chassis of the BMY M109 155 mm self-propelled howitzer/BMY Field Artillery Ammunition Support Vehicle. Both the M109 and M992 FAASV are already in service with the Egyptian Army.

The BMY prototype was completed in early 1984 and carried out a series of firing and other trials in the United States before beginning its Egyptian trials in April 1985. The BMY SPH uses the same basic aluminium construction as other vehicles in the M109 range and carries the 122 mm D-30 howitzer cannon in a fixed superstructure at the rear. A dual-purpose sighting system is fitted, comprising an indirect fire sighting device and a direct fire telescopic sight. The former incorporates a range drum, which directly computes the required angle elevation corresponding to the target without additional calculation. The latter is used for the engagement of both stationary and moving targets.

The howitzer breech mechanism is a mechanical type semi-automatic gear with automatic case extraction after firing. The breech block is prevented from interfering with the extraction of the fired case and the breech remains open for re-loading of the next round.

Hatches are located in the superstructure sides and rear and there is a roof hatch over the gunner's position with accommodation for a 12.7 mm Browning M2 HB machine gun. The driver's position, engine, transmission and suspension of the M109 vehicle family are retained.

BMY SPH 122 mm self-propelled howitzer with 122 mm D-30 ordnance in travelling lock (Christopher F Foss)

The main fighting compartment houses four of the crew of five and there is space for the stowage of 77 HE, four HEAT and four illumination rounds. There is also space for the crew's personal 7.62 mm rifles.

SPECIFICATIONS

CREW	5
COMBAT WEIGHT	24 948 kg
UNLOADED WEIGHT	22 000 kg
LENGTH (with armament)	6.957 m
WIDTH	3.15 m
HEIGHT (without AA MG)	3.28 m
MAX ROAD SPEED	56.3 km/h
FUEL CAPACITY	484 l
MAX RANGE (cruising)	300 km
FORDING	1.07 m
GRADIENT	60%
SIDE SLOPE	40%
VERTICAL OBSTACLE	0.61 m
TRENCH	2.0 m
ENGINE	Detroit Diesel Model 8V-71T turbo-charged, 2-stroke, liquid-cooled 8-cylinder diesel developing 405 hp at 2350 rpm
TRANSMISSION	Allison Transmission Division of General Motors, XTG-411-4 cross-drive with 4 forward and 2 reverse gears
SUSPENSION	torsion bar
ELECTRICAL SYSTEM	24 V
ARMAMENT	
main	1 × 122 mm howitzer
anti-aircraft	1 × 12.7 mm MG
SMOKE LAYING EQUIPMENT	none
AMMUNITION (main)	77 HE, 4 HEAT, 4 illuminating
MG	500
GUN CONTROL EQUIPMENT	
Gun elevation limits	-5°/+70°
Gun traverse limits	30° left, 30° right

Status: Trials with the competing designs were completed late in 1987 but as of June 1991 neither had been selected for production.

Manufacturer: BMY-Combat Systems
A Division of HARSCO
PO Box 15512
York
Pennsylvania 17405
USA
Telephone: (717) 225 4781
Fax: (717) 225 4615

FRANCE

GIAT Industries 155 mm GCT Self-propelled Gun

Development
The 155 mm GCT (Grande Cadence de Tir) was developed from 1969 to meet a French Army requirement for a self-propelled gun to replace the 105 mm and 155 mm self-propelled weapons then in service, both of which are based on the chassis of the AMX-13 light tank. The first prototype was completed in 1972 and was shown for the first time at the 1973 Satory Exhibition of Military Equipment. Between 1974 and 1975 six pre-production vehicles were completed and the vehicle entered production in 1977. First production GCTs were delivered in 1978 to Saudi Arabia. The 155 mm GCT self-propelled howitzer was officially selected by the French Army in July 1979 and is now deployed, in five-gun batteries, with each regiment having four batteries.

Production funding for the 155 mm GCT self-propelled gun, or 155 AU F1 as it is referred to by the French Army, is as follows:

Year	Quantity
1981	47
1982	30
1983	26
1984	27
1985	28
1986	nil
1987	20
1988	nil
1989	17

Saudi Arabia ordered 51 155 mm GCT self-propelled guns which were all delivered by 1982. In 1982 Iraq placed an order worth FFr4000 million for 85 GCTs, plus ammunition and spares, and Saudi Arabia loaned Iraq a quantity of GCTs until France commenced full scale deliveries.

Description
The 155 mm GCT basically consists of a modified AMX-30 MBT chassis fitted with a new turret armed with a 155 mm gun and an automatic loading system. The vehicle has a crew of four which consists of the commander, gunner, loader and driver.

The hull is almost identical to that of the AMX-30 MBT apart from the following modifications, which result in the chassis being 2000 kg lighter than the MBT's: the 105 mm ammunition racks in the hull have been removed, and a 5kVA 28 V generator and a ventilator system to supply the turret with cold air have been installed.

The driver is seated at the front of the vehicle on the left side and has a single piece hatch cover that opens to the left, in front of which are three periscopes. The centre periscope can be replaced by an infra-red or image intensification periscope for night driving or a combined SOPELEM OB-16-A day/night periscope.

The turret is in the centre of the vehicle. The engine is immediately behind the bulkhead that separates the engine compartment from the turret compartment. The engine is water-cooled and air is drawn in above the chassis in the rear part of the vehicle and blown through the radiator by a fan driven by the engine. The fan's speed is governed by the temperature of the water. The complete powerpack, consisting of the engine, combined gearbox and steering unit and clutch assembly, can be removed in 45 minutes by a team of three. The powerpack can also be run outside the vehicle before installation. The transmission consists of an automatic clutch, combined gearbox and steering unit, brakes and two final drives. The centrifugal type clutch is activated electrically by a gear shift lever. The combined gearbox and steering mechanism contains a mechanically operated gearbox giving five speeds both forward and reverse, and a triple differential steering system. The brakes are hydraulically operated and are used both as service and parking brakes. Each final drive comprises a spur-type right angle and an epicyclic gear train.

The torsion bar suspension consists of five dual rubber-tyred road wheels with the idler at the front, drive sprocket at the rear and five track return rollers that support the inside of the track only. The first, second, fourth and fifth road wheels are mounted on bogies and the first and last road wheel stations have a hydraulic shock absorber.

The turret is all-welded with the commander and gunner seated on the right and the loader on the left. The commander's cupola is equipped with periscopes for all-round observation and a single piece hatch cover that opens to the rear. The loader, who also operates the anti-aircraft machine gun, also has a single piece rear-opening hatch cover. The crew enter the turret by a door in each side.

The 155 mm 40-calibre barrel has a vertical sliding wedge breech-block hermetically gas sealed by a metal blanking plate. The breech is hydraulically opened with manual controls for emergency use. The barrel has a double-baffle muzzle brake. The 155 mm gun has an elevation of +66°, depression of -4° and the turret can be traversed through 360°. Turret traverse and gun elevation are hydraulic with manual controls for emergency use.

The weapon takes one to two minutes to bring into action and one minute to bring out of action. Average rate of fire is 8 rds/min with automatic loading and 2 to 3 rds/min with manual loading. For burst fire six rounds can be fired in 45 seconds. During French Army tests 108 projectiles were fired in one hour.

155 mm GCT self-propelled gun firing with turret traversed left

155 mm GCT self-propelled gun

155 mm GCT self-propelled gun from rear with turret doors open

Forty-two projectiles and 42 cartridge cases are carried in the turret rear arranged in seven racks of six identical shells and seven racks of six cartridge cases; in addition 40 charges are housed in a fixed container in the turret basket. A typical ammunition load would consist of 36 (6 × 6) HE shells plus six smoke (1 × 6) or 30 HE (5 × 6), six smoke (1 × 6) and six illuminating (1 × 6); but any combination of HE, illuminating or smoke shells is possible in racks of six similar shells. The various types of shell can be selected and fired with automatic loading without any preparation of the loading system.

The gunner can select single shots or bursts of six rounds. The turret is resupplied with ammunition by two large doors in the turret rear which fold down into the horizontal to provide a platform for the crew. The turret can be reloaded by a crew of four in 15 minutes or by a crew of two in 20 minutes. The gun can continue to fire while being reloaded.

The 155 mm gun can fire the following types of ammunition:
155 mm HE 56/59 with a maximum range of 23 000 m
155 mm HE (hollow base) with a maximum range of 23 000 m
155 mm HE (base bleed) with a maximum range of 29 000 m
155 mm HE US M107 with a maximum range of 18 000 m
plus other types of projectiles including illuminating, smoke and carrier.

The fire control system of the GCT consists of the following:

(1) Gimbal suspension providing for the mounting of the goniometer in the turret and permitting it to move freely by 10° relatively to the axis in all directions
(2) One hermetically sealed conventional optical goniometer. Traverse angles are read on engraved drums and elevation angles appear on an elevation scale
(3) One azimuth plate that can be manually adjusted and is provided with a spirit level to set the goniometer vertically
(4) A contra-rotation device making it possible to preserve the vertical position of the goniometer during turret rotation
(5) A direct fire sight for anti-tank firing.

A 7.62 mm or a 12.7 mm machine gun can be mounted at the loader's station for anti-aircraft defence. The 7.62 mm machine gun has an elevation of +45°, depression of -10°, total traverse of 360° and is provided with 2050 rounds of ammunition, of which 550 rounds are for ready use. The 12.7 mm machine gun has an elevation of +50°, a depression of -20°, total traverse of 360° and is provided with 800 rounds of ammunition of which 100 are for ready use. Two electrically operated smoke dischargers are mounted on the forward part of the turret, below the main armament.

The standard optical control system can be replaced by a CITA 20 system which consists of a land navigator coupled with a gyroscope, when the 155 mm GCT comes to a halt it knows immediately the co-ordinates of its position and the bearing of its ordnance. So far all three users of the 155 mm GCT have also purchased the Thomson-CSF ATILA fire control system.

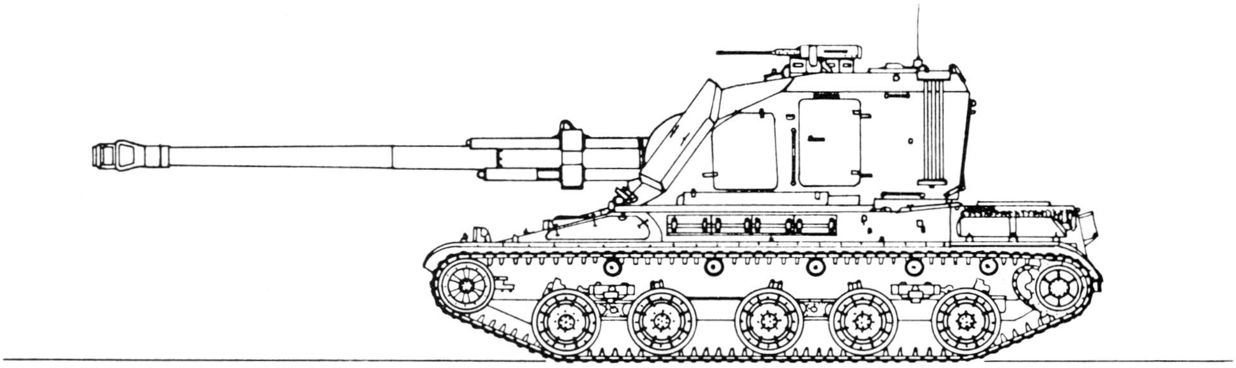

155 mm GCT self-propelled gun

SPECIFICATIONS		GRADIENT	60%	ARMAMENT	
CREW	4	SIDE SLOPE	30%	main	1 × 155 mm gun
COMBAT WEIGHT	43 500 kg	VERTICAL OBSTACLE		anti-aircraft	1 × 7.62 mm or
UNLOADED WEIGHT	39 500 kg	forwards	0.93 m		12.7 mm MG
POWER-TO-WEIGHT		reverse	0.48 m	SMOKE-LAYING	
RATIO	16.55 hp/tonne	TRENCH	1.9 m	EQUIPMENT	2 × 2 smoke
LENGTH GUN FORWARDS	10.25 m	ENGINE	Hispano-Suiza HS 110		dischargers
LENGTH GUN REAR	9.51 m		12-cylinder, water-	AMMUNITION	
LENGTH HULL	6.7 m		cooled super-charged	main	42
WIDTH	3.15 m		multi-fuel developing	anti-aircraft	2050 rounds of
over tracks	3.115 m		720 hp at 2000 rpm		7.62 mm or 800 rounds
HEIGHT (to turret top)	3.25 m	TRANSMISSION	mechanical with 5		of 12.7 mm
FIRING HEIGHT	2.16 m		gears in both	GUN CONTROL	
GROUND CLEARANCE	0.42 m		directions	EQUIPMENT	
TRACK	2.53 m	STEERING	triple differential	Turret power control	hydraulic/manual
TRACK WIDTH	570 mm	CLUTCH	centrifugal	Max rate of power traverse	10°/s
LENGTH OF TRACK		SUSPENSION	torsion bar	Max rate of power elevation	5°/s
ON GROUND	3.9 m	ELECTRICAL SYSTEM	28 V	Gun elevation/depression	+66°/-4°
MAX ROAD SPEED	60 km/h	BATTERIES	8 × 12 V, 100 Ah	Turret traverse	360°
MAX ROAD RANGE	450 km at 40 km/h	AUXILIARY POWER		Max range	29 000 m
FORDING	2.1 m	UNIT	Microturbo gas turbine	Rate of fire	6 rounds in 45 s

Status: In production with about 400 systems built. In service with France (164), Iraq (85) and Saudi Arabia (51). The number of Iraqi 155 mm GCT systems lost during the recent conflict is not known.

Manufacturer: Atelier de Construction Roanne (ARE).

GIAT Industries
13 route de la Minière
78034 Versailles
France
Telephone: (33) 30 97 37 37
Fax: (33) 30 97 39 00

Creusot-Loire Industrie 155 mm Self-propelled Gun Mk F3

Development

The 155 mm Self-propelled Gun Mk F3, or the Canon de 155 mm Mle F3 Automoteur (Cn-155-F3-Am), was developed in the early 1950s and is essentially an AMX-13 tank chassis with the rear idler removed and the hull modified to accept a 155 mm gun.

The 155 mm gun was designed by the Atelier de Construction de Tarbes (ATS), the chassis by the Atelier de Construction Roanne (ARE) while integration of the gun with the chassis and firing trials were undertaken by the Etablissement d'Etudes et de Fabrications d'Armement de Bourges (EFAB).

As the ARE was tooling up for production of the AMX-30 MBT, production of the complete AMX-13 tank family, including the 155 mm self-propelled gun, was transferred to Creusot-Loire.

Over 600 Mk F3 guns have been produced.

Description

The hull of the 155 mm self-propelled gun Mk F3 is all welded with the driver's compartment at the front on the left, engine compartment to his right and the main armament at the rear. A splash board mounted at the front of the hull stops water rushing up the glacis plate when the vehicle is fording. One spare road wheel is often carried on the glacis plate. The driver is provided with a single piece hatch cover that opens to the left, in front of which are three periscopes, the centre one replaceable by an infra-red or image intensification periscope for night driving. The commander is seated behind the driver and has a two-piece hatch cover that opens either side of his position and periscopes in front of him.

The torsion bar suspension consists of five single rubber-tyred road wheels with the drive sprocket at the front and the fifth road wheel acting as the idler, and three track return rollers. The first and last road wheel stations have hydraulic shock absorbers. The steel tracks can be fitted with rubber pads if required. There are stowage containers along each side of the hull and standard equipment includes a loudspeaker and a cable reel with 400 m of cable. Each side of the hull at the rear is a large spade which, on arrival at the gun position, is released manually; the vehicle then reverses slightly to bed them in. The vehicle does not have an NBC system and has no amphibious capability.

The other eight members of the gun crew are normally carried in a 6 × 6 truck or an AMX VCA tracked vehicle which also carries 25 projectiles, 25 charges plus fuzes. For short distances, for example when moving from one gun position to another, four of the gun crew can be carried on the self-propelled gun itself.

When travelling the gun is horizontal and locked 8° to the right of the vehicle's centreline. The gun has a 33-calibre barrel, screw breech, double baffle muzzle brake and a sustained rate of fire of 1 rd/min which can be increased to 3 rds/min for short periods. The weapon is provided with an APX L877G direction finder with a magnification of ×4 and 175 mils field-of-view and an APX M321 sighting telescope for direct fire with a magnification of ×4 and 180 mils field-of-view.

The gun has an elevation of from 0° to +67°. When elevated between 0° and +50° it has a traverse of 20° left and 30° right, and when elevated from +50° to +67° it has a traverse of 16° left and 30° right. Elevation and traverse are both manual.

The 155 mm self-propelled gun can fire the following types of separate loading ammunition:

155 mm self-propelled gun Mk F3 in firing position (P Touzin)

155 mm self-propelled gun Mk F3 in travelling configuration (Christopher F Foss)

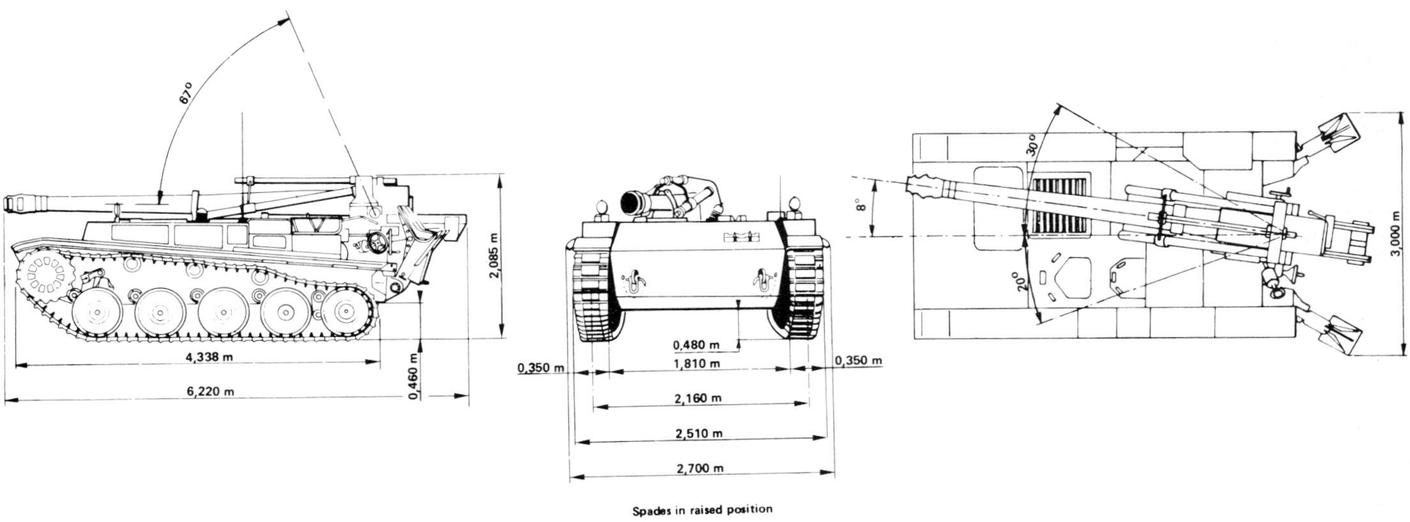

155 mm self-propelled gun Mk F3 (not to 1/76th scale)

155 mm HE projectile Mk 56 weighing 43.75 kg to a maximum range, with charge 9A, of 20 047 m, maximum muzzle velocity 725 m/s
155 mm illuminating shell weighing 44 kg to a maximum range of 17 750 m, illumination time of two minutes and intensity of 800 000 candlepower
155 mm smoke shell weighing 44.25 kg to a maximum range of 17 750 m
With base bleed system ranges are improved, for example to 23 800 m for OE ammunition weighing 42.5 kg to a maximum range of 25 300 m
155 mm HE projectile (type M107 USA) weighing 43.1 kg to a maximum range of 14 700 m (maximum muzzle velocity of 570 m/s) with American charge and 18 000 m (maximum muzzle velocity 690 m/s) with French charges.

Variants

Creusot-Loire Industrie offers the Mk F3 with a Detroit Diesel 6V-53T diesel engine and a kit is available to convert existing petrol-engined versions to this configuration. More recently the Baudouin 6 F 11 SRY diesel developing 280 hp at 3200 rpm and an automatic transmission have been offered by Creusot-Loire Industrie for this and other members of the AMX-13 family.

155 mm Mk F3 with 39-calibre ordnance

At the 1990 Satory exhibition of military equipment, Creusot-Loire Industrie announced that they had completed the prototype of a 155 mm Mk F3 self-propelled artillery system with the original ordnance replaced by the 39-calibre ordnance used in the GIAT 155 mm Towed Gun TR covered in the following section.

This has a maximum range of 24 800 to 30 000 m depending on the type of ammunition used and with the aid of the power assisted ramming system a rate of fire of up to three rounds in 20 seconds can be achieved. Elevation and traverse is hydraulic.

This version is called the AMX-13 155 AM 39 and has a combat weight of 17 400 kg. In addition to being fitted with the new 155 mm ordnance, Creusot-Loire Industrie can also modernise the chassis in other areas including new engine, transmission and electrics.

Creusot-Loire Industrie 155 mm Mk F3 with 39-calibre ordnance which was first shown in 1990

Status: In production. In service with Argentina (24 - 12 locally produced), Chile (12), Cyprus (ordered 12 systems plus associated AMX VCA tracked ammunition carriers and battery command vehicles late in 1989), Ecuador (12), France (222), Kuwait (80 - at least 8 loaned to Iraq), Morocco (64), Qatar (8), Sudan (6), United Arab Emirates (20) and Venezuela (20). Status of Kuwait vehicles is uncertain at present.

Manufacturer: Creusot-Loire Industrie
Immeuble Ile de France
Cedex 33
92070 Paris la Defense
France
Telephone: (1) 49 00 60 50
Telex: MOTOY 615638 F
Fax: (1) 49 00 57 30

SPECIFICATIONS

CREW (on weapon)	2
COMBAT WEIGHT	17 400 kg
POWER-TO-WEIGHT RATIO	
petrol engine	14 hp/tonne
diesel engine	16 hp/tonne
GROUND PRESSURE	0.84 kg/cm^2
LENGTH INCLUDING GUN	6.22 m
WIDTH	2.7 m
over tracks	2.51 m
HEIGHT TRAVELLING	2.085 m
GROUND CLEARANCE	0.48 m
TRACK	2.16 m
TRACK WIDTH	350 mm
LENGTH OF TRACK ON GROUND	2.997 m
MAX ROAD SPEED	
petrol engine	60 km/h
diesel engine	64 km/h
FUEL CAPACITY	450 l
MAX ROAD RANGE	
petrol engine	300 km
diesel engine	450 km
FUEL CONSUMPTION	1.4 l/km for petrol engine version, for diesel engine it is 1 l/km
FORDING	1 m
GRADIENT	40%
VERTICAL OBSTACLE	
forwards	0.6 m
reverse	0.4 m
TRENCH	1.5 m
TURNING RADIUS	skid turns
ENGINE	SOFAM model 8Gxb 8-cylinder water-cooled petrol developing 250 hp at 3200 rpm or General Motors Detroit Diesel 6V-53T turbo-charged, 6-cylinder water-cooled diesel developing 280 hp at 2800 rpm or Baudouin 6 F 11 SRY turbo-charged 6-cylinder water-cooled diesel developing 280 hp at 3200 rpm
TRANSMISSION	manual with 5 forward and 1 reverse gears (2nd, 3rd, 4th and 5th are synchronised) or automatic transmission with three forward and one reverse gears
STEERING	Cleveland type differential
CLUTCH	single dry plate
SUSPENSION	torsion bar
ELECTRICAL SYSTEM	24 V
BATTERIES	4 × 12 V, 190 Ah
ARMAMENT	1 × 155 mm howitzer
GUN ELEVATION/DEPRESSION	+67°/none
GUN TRAVERSE	
at 0° to +50° elevation	20° left and 30° right
at +50° to +67° elevation	16° left and 30° right
ARMOUR	
Hull front	15 mm at 55° (corresponding thickness is 40 mm)
Hull sides	20 mm
Hull top	10 mm
Hull floor front	20 mm
Hull floor rear	10 mm
Hull rear	15 mm

105 mm Mk 61 Self-propelled Howitzer

Development

The 105 mm Mk 61 self-propelled howitzer, or to give the vehicle its correct designation the Obusier de 105 Model 50 sur Affût Automoteur, was designed to meet the requirements of the French Army in the late 1940s. It basically consists of a modified AMX-13 light tank chassis with a 105 mm howitzer mounted at the rear in a lightly armoured superstructure. After trials with various prototypes with different types of 105 mm howitzer the Mk 61 was finally accepted for service in 1958 with production of the vehicle being undertaken by the Atelier de Construction Roanne. The Mk 61 has been replaced in the French Army by the 155 mm GCT self-propelled gun which is based on the chassis of the AMX-30 MBT.

Description

The hull of the 105 mm Mk 61 self-propelled howitzer is all welded with the driver's compartment at the front on the left, engine compartment to his right and main armament at the rear. A splash board is often mounted at the front of the hull to stop water rushing up the glacis plate when the vehicle is fording and a spare wheel is sometimes carried on the glacis plate. The driver has a single piece hatch cover that opens to the left, in front of which are three periscopes, the centre one replaceable by an infra-red or image intensification periscope for night driving.

The torsion bar suspension consists of five single rubber-tyred road wheels with the drive sprocket at the front, idler at the rear and three track return rollers. The first and last road wheel stations have hydraulic shock absorbers. The steel tracks can be fitted with rubber pads if required to reduce damage to roads.

The fighting compartment is at the rear of the hull and entry to it is by two doors in the rear of the hull. The commander is seated on the right side at the front of the fighting compartment and has a two-piece hatch cover that opens either side of his position. Some models have a commander's cupola with a single piece domed hatch cover plus periscopes for all-round observation. Over the rear part of the fighting compartment is a large two-piece hatch that opens either side of the vehicle. Some Mk 61s have a single vision port protected by an armoured shutter in each side of the hull.

The Mk 61 has no amphibious capability, and no NBC system.

Main armament comprises a 105 mm M1950 howitzer with a 23- or 30-calibre barrel which is provided with a double-baffle muzzle brake but no fume extractor. This weapon has an elevation of +66°, a depression

French Army 105 mm Mk 61 self-propelled howitzer (ECP Armées)

French Army 105 mm Mk 61 self-propelled howitzer (ECP Armées)

of -4° 30' and a traverse of 20° left and 20° right. Both elevation and depression are manual. It can fire the following types of ammunition: 105 mm HE projectile weighing 16 kg with a muzzle velocity of 670 m/s to a maximum range of 15 000 m, plus all American 105 mm rounds 105 mm HEAT projectile with a muzzle velocity of 700 m/s which will penetrate 350 mm of armour at an incidence of 0° or 105 mm of armour at 65°.

A total of 56 rounds of separate-loading ammunition is carried of which six are normally HEAT. Fire control equipment consists of an L841.9 goniometer with a magnification of ×4 and an L881 telescope for anti-tank use with a magnification of ×6. A 7.62 mm machine gun is mounted on the roof for anti-aircraft defence and a second machine gun is normally carried inside the hull for ground defence use.

Variants

105 mm Self-propelled Howitzer (Turret)

This is similar to the Mk 61 but has a turret-mounted 105 mm howitzer which has a traverse of 360° and can be elevated from -7° to +70°. Eighty rounds of 105 mm separate loaded ammunition are carried. This model was not adopted by the French Army although four were bought by the Swiss Army for trials. This version is no longer being offered.

The chassis of the Mk 61 was also used as the basis for a mine-laying vehicle which did not proceed past the prototype stage.

Status: Production complete. In service with France (reserve), Indonesia and Morocco.

Manufacturers: Atelier de Construction Roanne
Creusot-Loire Industrie
Enquiries to Creusot-Loire Industrie
Immeuble Ile de France
Cedex 33
92070 Paris la Defense
France
Telephone: (1) 49 00 60 50
Telex: MOTOY 615638 F
Fax: (1) 49 00 57 30.

SPECIFICATIONS	
CREW	5
COMBAT WEIGHT	16 500 kg
UNLOADED WEIGHT	13 700 kg
POWER-TO-WEIGHT RATIO	15.15 hp/tonne
GROUND PRESSURE	0.8 kg/cm²
LENGTH	5.7 m
WIDTH	2.65 m
HEIGHT (cupola)	2.7 m
GROUND CLEARANCE	
front	0.32 m
rear	0.275 m
TRACK	2.159 m
TRACK WIDTH	350 mm
MAX ROAD SPEED	60 km/h
FUEL CAPACITY	415 l
MAX ROAD RANGE	350 km
ROAD FUEL CONSUMPTION	1.3 l/km
FORDING	0.8 m
GRADIENT	60%
SIDE SLOPE	30%
VERTICAL OBSTACLE	
forwards	0.65 m
reverse	0.45 m
TRENCH	1.6 m
TURNING RADIUS	skid turns
ENGINE	SOFAM 8Gxb 8-cylinder water-cooled petrol developing 250 hp at 3200 rpm
TRANSMISSION	manual with 5 forward (2nd, 3rd, 4th and 5th are synchronised) and 1 reverse gears
STEERING	Cleveland type differential
CLUTCH	single dry plate
SUSPENSION	torsion bar
ELECTRICAL SYSTEM	24 V
BATTERIES	4 × 12 V, 190 Ah
ARMAMENT	
main	1 × 105 mm howitzer
anti-aircraft	1 × 7.5 mm or 7.62 mm MG
other	1 × 7.5 mm or 7.62 mm MG for ground use
SMOKE-LAYING EQUIPMENT	none
AMMUNITION	
main	56
MG	2000
GUN CONTROL EQUIPMENT	
Gun elevation/depression	+66°/-4° 30'
Gun traverse	manual, 20° left and right
ARMOUR	
Hull front	15 mm at 40° (corresponding to 40 mm)
Hull sides	20 mm
Hull top	10 mm
Hull floor	
front	20 mm
rear	10 mm
Hull rear	15 mm
Superstructure front and sides	20 mm
Superstructure rear	15 mm
Superstructure top	10 mm

LOHR VPX 40 M 120 mm Mortar Carrier

Development

The VPX 40 M 120 mm mortar carrier, designed as a private venture by the LOHR company, was first shown at the 1987 Satory Exhibition of Military Equipment. The vehicle shares a number of common components with the LOHR VPX 5000 family of light armoured tracked vehicles first shown in 1981.

Description

Mounted in the rear part of the VPX 40 M is a Thomson-Brandt Armements MO-120-RT-61 rifled mortar firing to the rear of the vehicle. This has an elevation from +45° to +85° and can fire a PR14 high explosive bomb to a maximum range of 8135 m, a PRPA rocket-assisted projectile to a maximum range of 13 000 m, as well as the normal smoke, illuminating and anti-armour projectiles. Between 10 and 20 rounds of 120 mm mortar projectiles are carried.

The hull of the VPX 40 M is of all-welded armour plate which provides protection from small arms fire, although the top of the hull is open.

The engine is mounted at the hull rear and is coupled to an automatic gearbox and a two-speed manual transfer box.

The suspension is of the trailing arm and oleo-pneumatic type with six dual rubber-tyred road wheels, hydraulic tension idler at the rear, drive sprocket at the front and two track return rollers. The track consists of an internally reinforced endless rubber band. To provide a more stable firing platform the suspension of the VPX 40 M is raised when the mortar is firing at the rear so that shock of the 120 mm mortar firing is absorbed directly by baseplate being in contact with the ground.

LOHR regard this as a demonstrator vehicle which can be adapted to other types of chassis.

LOHR VPX 40 M 120 mm mortar carrier deployed in firing position with ready use ammunition in centre of vehicle

LOHR VPX 40 M 120 mm mortar carrier in firing position with suspension raised to provide more stable firing platform, showing crew member loading mortar bomb into Thomson Brandt 120 mm MO-120-RT-61 rifled mortar

SPECIFICATIONS		HEIGHT (travelling)	2 m	TRANSMISSION	automatic, 4 forward and 1 reverse gears
CREW	4	GROUND CLEARANCE (travelling)	0.3 m	TRANSFER BOX	2-speed
COMBAT WEIGHT	4500 kg	MAX ROAD SPEED	80 km/h	SUSPENSION	trailing arm and oleo-pneumatic cylinder
UNLOADED WEIGHT	3300 kg	RANGE	400 km	BRAKES	disc
POWER-TO-WEIGHT RATIO	27 hp/tonne	FORDING	0.8 m	STEERING	hydrostatic
GROUND PRESSURE		GRADIENT	60%		
loaded	0.444 kg/cm²	SIDE SLOPE	45%		
empty	0.326 km/cm²	ENGINE	turbo-charged diesel developing 125 hp		
LENGTH	4.15 m				
WIDTH	2.1 m				

Status: Prototype

Manufacturer: LOHR Industrie
67980 Hangenbieten
France
Telephone: (1) 88 38 98 00
Telex: 870082
Fax: 88 96 06 36

LOHR RPX 40 M 120 mm Mortar Carrier

Development

The RPX 40 M 120 mm mortar carrier, designed as a private venture by the LOHR company, was first shown at the 1987 Satory Exhibition of Military Equipment. The vehicle shares many automotive components with the LOHR RPX 3000 light armoured vehicle.

Description

Mounted in the rear part of the hull is a Thomson-Brandt Armements MO-120-RT-61 rifled mortar firing to the rear of the vehicle. This has an elevation from +45° to +85° and can fire a PR14 high explosive bomb to a maximum range of 8135 m, a PRPA rocket-assisted projectile to a maximum range of 13 000 m, as well as the normal smoke, illuminating and anti-armour projectiles. Between 10 and 20 rounds of 120 mm mortar projectiles are carried.

The hull of the RPX 40 M is of all-welded armour plate which provides protection from small arms fire. The fully enclosed crew compartment is at the front, engine in the centre and the mortar at the rear. The crew can enter via a door in each side of the hull and there are also two circular roof hatches between which a 7.62 mm machine gun can be mounted.

The engine is coupled to a fully automatic transmission and a two-speed transfer case.

When deployed in the firing position, the rear two road wheels are raised so that the shock of the 120 mm mortar firing is absorbed directly by the baseplate being in contact with the ground.

Variants

The LOHR company is using this as a demonstrator and it can also be adapted to other types of chassis.

LOHR RPX 40 M 120 mm mortar carrier in firing position with rear wheels raise to absorb shock during firing of 120 mm mortar

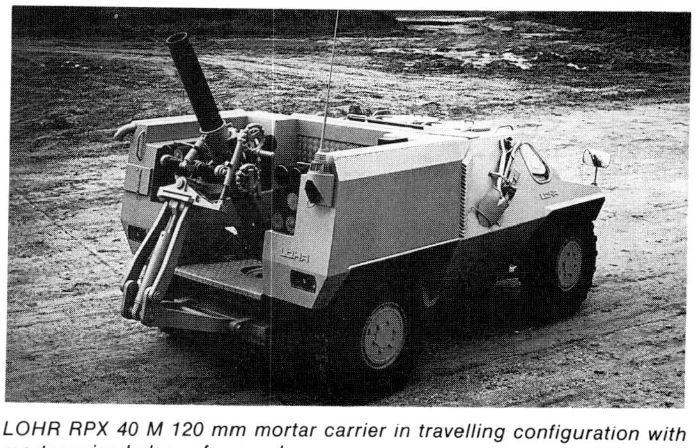

LOHR RPX 40 M 120 mm mortar carrier in travelling configuration with mortar raised clear of ground

SPECIFICATIONS					
CREW	4	WIDTH	2.2 m	ENGINE	turbo-charged diesel developing 125 hp
CONFIGURATION	4 × 4	HEIGHT	1.73 m	TRANSMISSION	automatic, 4 forward and 1 reverse gears
COMBAT WEIGHT	4700 kg	GROUND CLEARANCE	0.3 m	TRANSFER BOX	2-speed
UNLOADED WEIGHT	3500 kg	WHEELBASE	2.25 m	SUSPENSION	oleo-pneumatic
POWER-TO-WEIGHT RATIO	26.59 hp/tonne	MAX SPEED	100 km/h	STEERING	hydraulically assisted
LENGTH OVERALL	4 m	RANGE	600 km	TYRES	11.00 × 16 XS
LENGTH HULL	3.75 m	FORDING	0.8 m		
		GRADIENT	60%		
		SIDE SLOPE	30%		

Status: Prototype. Not yet in production or service.

Manufacturer: LOHR Industrie
67980 Hangenbieten
France
Telephone: (1) 88 38 98 00
Telex: 870082
Fax: (1) 88 96 06 36

LOHR 120 mm Self-propelled Mortar System

Development/Description
In addition to developing two 120 mm armoured self-propelled mortar systems mounted on tracked and wheeled chassis, which are described in the previous two entries, the LOHR company has also developed, as a private venture, a 120 mm mortar system on a Renault TRM 2000 (4 × 4) cross-country truck. Full details of this truck are given in *Jane's Military Vehicles and Logistics 1991-92*, page 324.

The prototype of the system based on a Renault TRM 2000 (4 × 4) truck chassis was first shown in June 1990 and in 1991 the LOHR company stated that a second prototype, based on a Mercedes-Benz UNIMOG (4 × 4) cross-country chassis was under development.

This would have a four door fully enclosed cab with seats for the crew of six, including the driver, 40 rounds of 120 mm mortar bombs and a quick-release system to allow the mortar to be uncovered.

The standard Thomson-Brandt Armements MO-120-RT-61 rifled mortar is mounted on the rear of the vehicle and fires over the front.

When in the firing position a large hydraulically operated spade is lowered at the rear of the vehicle to provide a more stable firing platform. The recoil force is filtered by shock absorbers fixed on the chassis.

The mortar has an elevation from +45° to +85° and can fire a PR14 high explosive bomb to a maximum range of 8135 m, a PRPA rocket-assisted projectile to a maximum range of 13 000 m, as well as normal smoke, illuminating and anti-armour projectiles.

The system takes only 30 seconds to come into action and 20 seconds to come out of action, a considerable reduction in the time taken by a normal towed mortar system. Aiming is identical to the standard mortar and rate of fire is 16 to 20 rds/min with a total of 24 mortar bombs being carried.

The system has a six man crew and weighs 6300 kg complete with mortar, it is air transportable in a C-130 and C-160 tactical transport aircraft.

LOHR 120 mm self-propelled mortar on Renault TRM 2000 (4 × 4) chassis in firing position (not to 1/76th scale)

Status: Prototype. Not yet in production or service.

Manufacturer: LOHR Industrie
67980 Hangenbieten
France
Telephone: (1) 88 38 98 00
Telex: 870082
Fax: (1) 88 96 06 36

LOHR 120 mm self-propelled mortar on TRM 2000 (4 × 4) chassis in firing position

GERMANY

Wegmann/MaK Panzerhaubitze 2000 (PzH 2000)

Development

Following the cancellation of the international 155 mm SP-70 programme in July 1986, the German Government awarded contracts totalling DM186 million to two company consortia for the development and construction of prototypes of a new 155 mm self-propelled howitzer for the German Army called the Panzerhaubitze 2000 (PzH 2000).

The German Army requirements include:

Ability to carry 60 projectiles and corresponding charges
Installation of automatic loader to enable a high rate of fire to be achieved
Maximum range of 30 km without using assisted projectiles
NBC system must be fitted
Have good mobility and reliability
Be able to operate as an independent unit
Have increased protection against top attack weapons.

The two consortia consisted of first Krauss-Maffei, KUKA and Rheinmetall with the second consisting of MaK and Wegmann.

Under Phase I, each of the two consortia built a prototype system which was tested by the German Army and in late 1990 the Krupp MaK/Wegmann consortia version was selected.

Under Phase II a development contract was subsequently awarded for an additional four vehicles for further trials with a production decision expected in the mid-1990s.

At one time it was expected that the total German Army requirement would be for up to 1254 systems but with the restructuring of the German Army and the decrease in East/West tension, this number is expected to be reduced.

Wegmann, as prime contractor, is responsible for the complete system, turret and autoloader, while MaK is responsible for the chassis and Rheinmetall for the 52 calibre ordnance.

Description

The hull and turret of the PzH 2000 is of all welded steel armour construction, with the driver being seated at front right, powerpack to his left and turret towards the rear.

The driver has a single piece hatch cover that opens to the rear for driving in the head out position and mounted forward of this hatch are three periscopes that give observation to the front of the vehicle, the centre periscope can be replaced by a passive night vision periscope.

The driver, who can also exit via his immediate rear, controls the vehicle using a steering wheel with essential information being provided by means of digital data processing with multi-function colour display and colour displays on the steering wheel.

The engine and transmission can be removed as a complete unit with the air intakes being located on the top of the hull and the exhaust outlet at the front of the hull on the left side.

The suspension is of the torsion bar type with drive sprocket at the front, idler at the rear, seven double tyred road wheels plus track return rollers. Linear shock absorbers and hydraulic bump stops using proven components from the Leopard MBT are installed. The upper part of the track is covered by a skirt.

The turret is at the rear of the vehicle and is provided with two roof hatches, one forward on the left side and one on the right side towards the rear. Mounted on the left hatch is a 7.62 mm machine gun for defence purposes.

The turret has a total of 16 smoke dischargers, four either side of the 155 mm ordnance firing forwards and eight at the rear of the turret on the left side firing to the rear.

The Rheinmetall 52 calibre 155 mm ordnance is fitted with a multi-slot muzzle brake and a quick-change bayonet fixture to secure the barrel to the breech and recoil system.

The breech mechanism is a conventional sliding block with an improved design of obturating ring. A standard percussion ignition system is used with the standard ignition tube primers being contained in a new 30 round automatic magazine attached to the rear face of the breech.

The ordnance has a range of 30 km with standard ammunition or 40 km with assisted projectiles. It can fire ammunition already in service or under development based on the Quadrilateral Ballistic Memorandum.

An automatic electrically driven projectile loading system enables loading to take place within the full traverse and elevation range. The 60-round projectile magazine has an automatic data management system and a total of 67 charges are carried.

The PzH 2000 has three ammunition flow modes:

(1) Automatic/manual ammunition loading and unloading
(2) Automatic/semi-automatic/manual loading from the magazine
(3) Automatic/manual through loading from the rear of the vehicle.

A burst rate of fire of three rounds in 10 seconds can be achieved with sustained rate of fire being 8 rds/min.

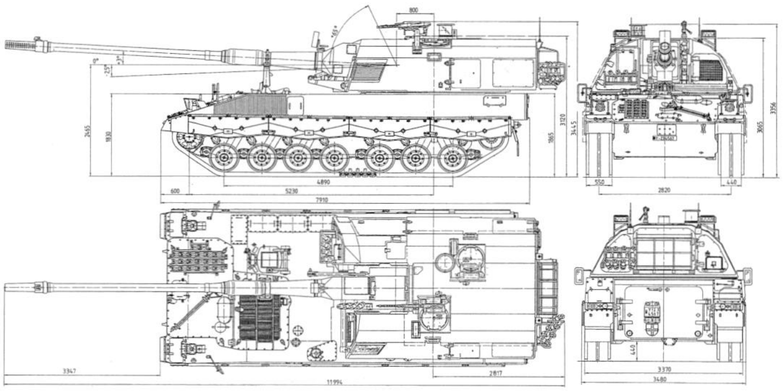

Four-view general arrangement drawing of PzH 2000 without adaptive armour (not to 1/76th scale)

PzH 2000 from the rear with hull doors open

PzH 2000 with 155 mm 52 calibre ordnance at maximum elevation

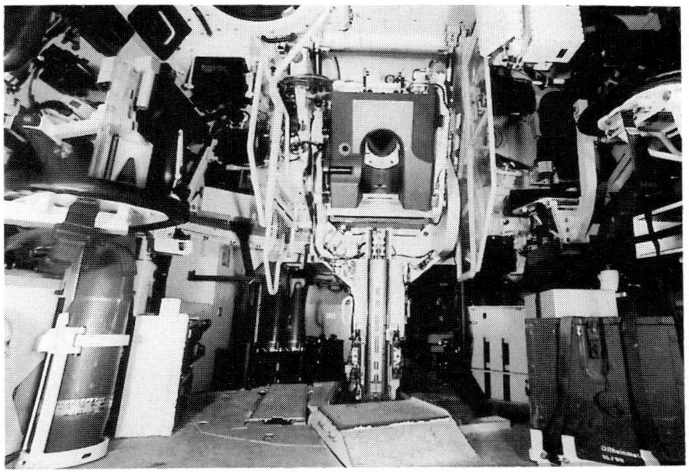

Interior view of PzH 2000 looking fowards with breech in centre

Fresh ammunition is loaded into the PzH 2000 from the rear of the vehicle in the lower part of the hull.

The weapon integrated fire control system and the all electric gun control equipment installed in the PzH 2000 enable firing with indirect laying from unprepared firing positions as well as firing with direct laying with a high first round hit probability and short reaction times.

The commander has the capability of direct control as well as the monitoring of the gunner and the two loaders during system operation and target engagement.

The PzH 2000 is operated from data and inertial reference terminal using a mission oriented menu-driven user operating system which completely automates the system operation and communication with the fire control command via data radio link or via voice radio communications.

The onboard fire control equipment integrated with the gun control system for indirect target laying consists of an integrated navigation system with appropriate digital computers for automatic operation in the primary mode and optical/mechanical equipment for the manual backup mode.

A roof-mounted panoramic day/thermal sight with laser rangefinder and digital ballistics for automatic primary mode operation in combat with direct laying, optical/mechanical equipment for manual backup.

For improved survivability, especially against top attack weapons, additional armour protection can be fitted to the roof of the vehicle. This can be quickly attached by the crew using onboard equipment. Survivability is also enhanced by compartmentalised charge storage separate from the crew compartment.

Standard equipment includes a fire warning and extinguishing system and an NBC protection and ventilation system.

Status: Under development.

Prime contractor: Wegmann and Co GmbH
August-Bode-Strasse
D-3500 Kassel
Federal Republic of Germany
Telephone: 0561 105-0
Fax: 0561 105 22 08

In co-operation with:
MaK Systemegesellschaft
Falckesteiner Str 2
D-2300 Kiel 17
Telephone: 0431 3995-02
Fax: 0431 3995-155

SPECIFICATIONS		TRACK WIDTH	550 mm	SUSPENSION	torsion bar
CREW	5	LENGTH OF TRACK ON GROUND	4.89 m	ELECTRICAL SYSTEM	24 V
COMBAT WEIGHT	5500 kg	ROAD SPEED	60 km/h	ARMAMENT	
POWER-TO-WEIGHT RATIO	18 hp/tonne	CRUSING RANGE ROAD	420 km	main	1 × 155 mm howitzer
GROUND PRESSURE	0.96 kg/cm²	GRADIENT	50%	anti-aircraft	1 × 7.62 mm
LENGTH OVERALL	11.994 m	SIDE SLOPE	25%	SMOKE-LAYING EQUIPMENT	4 × 4 smoke dischargers
LENGTH HULL	7.91 m	VERTICAL OBSTACLE	1 m	AMMUNITION (155 mm)	60 projectiles and 67 charges
WIDTH	3.48 m	TRENCH	3 m	GUN CONTROL EQUIPMENT	powered/manual
WIDTH OVER TRACKS	3.37 m	ENGINE	MTU 881 diesel developing 1000 hp	ELEVATION/DEPRESSION	+65°/-2.50°
HEIGHT		TRANSMISSION	automatic four speed forward, 2 speed reverse, integrated steering and clutch	TURRET TRAVERSE	360°
to top of MG	3.445 m				
to turret roof	3.12 m				
GROUND CLEARANCE	0.44 m				

Rheinmetall 155 mm M109A3G Product Improvement Programme

Description

In order to update the existing fleet of 155 mm M109G self-propelled howitzers of the German Bundeswehr, Rheinmetall Defence Engineering is undertaking a programme to modify the vehicles to a new standard. The main modification is the replacement of the existing 155 mm barrel with a new 39-calibre barrel based on the ordnance of the FH-70 towed howitzer but other modifications are being introduced at the same time.

The installation of the new 39-calibre barrel allows the full family of improved 155 mm ammunition to be fired. This family includes high explosive, illuminating and smoke, all of which can be fired to a maximum range of 24 700 m. Enhanced range projectiles such as base bleed can be fired to a range of 30 000 m. Various carrier projectiles such as the RB 63 and Rh 49 can also be fired. The new ordnance is mounted in the cradle and recoil system of the M109G and has a revised obturation system. An external rotation key is located in a cradle extension under the barrel and the barrel dust cover of the original M109G has been modified. Other innovations to the barrel include a new flexible bellows and a fume extractor, the same as that intended for the cancelled SP-70, is located half-way along the tube. The original M109G muzzle brake is retained but is held in place by new fasteners. Other modifications to the ordnance involve machining alterations to the breech ring and buffer.

The new ordnance weighs 2380 kg, 1420 kg being the barrel. The barrel is 6.874 m long with the length of rifling being 5.0577 m. Maximum muzzle velocity is 827 m/s.

Modifications to the turret include a new projectile magazine at the rear of the turret bustle. This can be reloaded through two outward-opening doors and holds 22 projectiles. The vehicle can carry a total of 34 projectiles. New provisions are made for propellant charge stowage. Other changes include door locks, a blast cover for the panoramic sight, and new sealing for the shields. A hand drive mechanism is provided for the traversing gear and new elevation balancing cylinders are provided.

Changes to the rest of the vehicle include a new barrel clamp, revised ventilators, a new air filter system, a new instrument layout for the driver, the location of a driver's control panel in the main compartment, a turbo-supercharger system, reinforced torsion bars on the suspension, and a driver's hatch lock. Alterations have been made to the engine powerpack to allow it to be replaced more readily.

Other modifications are that the turret hydraulics have been altered to suit the metric system and the traversing gear hydraulic motor is now the same as that used on the Leopard 1 battle tank. An electrical firing circuit has been introduced and various new fire control data panels have been incorporated.

Rheinmetall M109A3G 155 mm self-propelled howitzer with new long-barrelled ordnance

SPECIFICATIONS

Comparison between M109G and M109A3G

Model	M109G	M109A3G
CALIBRE	155 mm	155 mm
NUMBER OF GROOVES	48	48
TWIST, CLOCKWISE, CONSTANT	8° 55' 37'	8° 55' 35'
LENGTH OF BARREL	4422 mm	6874 mm
LENGTH OF RIFLED PART	2900 mm	5057.7 mm
CHAMBER VOLUME	13.489 l	18.845 l
WEIGHT, BARREL COMPLETE	1916 kg	2380 kg
WEIGHT, LINER	940 kg	1420 kg
MAX RANGE (standard round)	18 600 m	24 700 m (increased range up to 30 km)
LAYING RANGE (elevation)	-44 to +1245 mils	-44 to +1245 mils
(traverse)	360°	360°
EFFECT VALUE OF MUZZLE BRAKE	40%	40%
MAX RECOIL OVER 910 MILS	465 mm	465 mm
MAX GAS PRESSURE WITH CHARGE 8 DM 52	3200 bar	2300 bar
MAX GAS PRESSSURE WITH CHARGE FH 155-1	not app	3628 bar
MUZZLE VELOCITY WITH CHARGE 8 DM 52	685 m/s	685 m/s
MUZZLE VELOCITY WITH CHARGE FH 155-1	not app	827 m/s

Status: Modification programme in progress for the German and Norwegian Armies. The overhaul and conversion work is being carried out by a German Army depot.

Manufacturer: Rheinmetall Defense Engineering
Ulmenstrasse 125
PO Box 6609
D-4000 Düsseldorf 1
Federal Republic of Germany
Telephone: (0211) 4471
Telex: 8584963
Fax: (0211) 483290

M44T 155 mm Self-propelled Howitzer

Development

Late in 1986 a German consortium consisting of MTU, GLS and Rheinmetall completed the prototype of an upgraded 155 mm M44 self-propelled howitzer subsequently designated the M44T. This was built to meet the requirements of the Turkish Army and the prototype commenced mobility and firepower trials at the Turkish School of Artillery in 1987. Compared to the original M44, the M44T has increased mobility, firepower and reliability.

A total of 168 M44s of the Turkish Army are being upgraded in Turkey with some of the component parts being made under licence in Turkey.

The M44T conversion is of modular design so that the user can select only those parts that meet his operational requirement and financial situation.

Description

The M44T has its original petrol engine replaced by an MTU MB 883 Aa V-6 water-cooled diesel developing 450 hp at 2300 rpm coupled to the original CD-500-3 transmission via a ZF gearbox.

Increased fuel tankage of 780 litres, together with the diesel engine, provides a range of approximately 620 km compared to the 122 km of the original petrol engined M44.

Various other components and subsystems have been modernised and a Halon fire warning and extinguishing system has been installed. A new driver's instrument panel has also been fitted.

Replacement of the existing electrical system has also been carried out using new cables, fuse boxes and batteries. New actuators for steering, braking and acceleration have been fitted, and for cold starting of the engine at temperatures below -18° C a pre-heating system has been fitted. The top engine deck and exhaust system have also been modified.

Modifications to the suspension include new hydraulic shock absorbers and torsion bars and some adaptation of the hydraulic bump stops. A new track has also been installed with replaceable rubber pads, snow grips can be fitted if required. New rubber dust skirts have been fitted to help to keep the dust down.

The original ordnance has been replaced by a 39-calibre barrel and breech similar to that used on the Rheinmetall M109A3G (qv). This provides a range, firing standard projectiles of 24 700 m and enhanced range projectiles can now reach 30 000 m. Elevation is from -5° to +65° with ammunition stowage for 30 projectiles and changes being provided.

The ordnance has automatic primer feeding as well as automatic opening and closing of the breech mechanism, the recoil system, balancers and ammunition racks have also been modified. The ordnance has a life of 10 000 standard rounds.

Status: Conversions underway in Turkey for the Turkish Army.

M44T showing new 155 mm ordnance with its double baffle muzzle brake

Manufacturers: MTU GmbH
PO Box 2040
D-7990 Friedrichshafen
Federal Republic of Germany
Telephone: (07541) 29-1
Telex: 734280-0 mt d
Fax: (07541) 29 22 47/29 84 86

Rheinmetall Defense Engineering
Ulmenstrasse 125
PO Box 6609
D-4000 Düsseldorf 1
Federal Republic of Germany
Telephone: (0211) 4471
Telex: 8584963
Fax: (0211) 483290

GLS GmbH
Allacher Strasse 230e
D-8000 München 50
Federal Republic of Germany
Telephone: (089) 88 99 0
Telex: 5215 583
Fax: (089) 812 63 80

HUNGARY

82 mm Self-propelled Automatic Mortar

Development

In 1990 Hungary revealed that it had developed an 82 mm self-propelled automatic mortar with an autonomous automatic fire control system and was offering this on the export market. It is not known whether this was developed as a private venture or to meet the requirements of the Hungarian Armed Forces.

Description

This system essentially consists of the ACRV (or MT-LBu to give its correct Soviet designation) chassis modified to accept the Soviet 82 mm Vasilyek automatic mortar covered in the *Towed guns and howitzers* section.

The ACRV (Artillery Command and Reconnaissance Vehicle) is used for a wide range of roles in the Soviet Army and additional details are in the entry for the Soviet 122 mm 2S1 self-propelled howitzer in the *Self-propelled guns and howitzers* section.

Hungarian 82 mm self-propelled automatic mortar with autonomous automatic fire control system deployed in firing position

The ACRV (Artillery Command and Reconnaissance Vehicle) is used for a wide range of roles in the Soviet Army and additional details are in the entry for the Soviet 122 mm 2S1 self-propelled howitzer in the *Self-propelled guns and howitzers* section.

it is virtually indistinguishable from standard ACRV vehicles.

The 82 mm Vasilyek system can be used both in the direct and indirect fire roles and fires an 81 mm mortar bomb to a maximum range of 4300 m, minimum range of 100 m, with a maximum rate of fire of 100 rpm. It takes two minutes to come into action.

The 81 mm mortar bomb has a maximum muzzle velocity of 270 m/s and weighs 3.1 kg, they are fed into the mortar from the left side in clips of four rounds which weigh a total of 16.75 kg.

The PAN-1 sight has magnifications of ×3 (18° field-of-view) and ×2.5 (9° field-of-view).

The system is fitted with an automatic fire control system that consists of a meteorological station, cant measuring device, target distance and angle measuring system, expedient computer, processing equipment and keyboard and mortar operation equipment that also includes a servo drive. The distance to the target is measured using a laser rangefinder.

Variants

It has also been proposed that the system can be installed on a modified BMP-1 chassis. This would have its turret removed, hull raised at the rear with the 82 mm Vasilyek system being mounted in the centre of the vehicle.

Status: Development complete. Ready for production.

Manufacturer: Hungarian state factories.

Enquiries to: IDEX
Foreign Trading, Contracting and Engineering Company Limited
POB 24
H-1251 Budapest
Hungary
Telephone: (36 1) 115 9290
Fax: (36 1) 135 1393

SPECIFICATIONS

CREW	3	TRACK	2.5 m	STEERING	clutch and brake
COMBAT WEIGHT	13 000 kg	MAX ROAD SPEED	61.5 km/h	SUSPENSION	adjustable
POWER-TO-WEIGHT RATIO	22.76 hp/tonne	FUEL CAPACITY	550 l	ARMAMENT	1 × 82 mm mortar
LENGTH TRAVELLING	7.655 m	RANGE	500 km	AMMUNITION	128
LENGTH OPERATIONAL	7.21 m	GRADIENT	60%	GUN CONTROL EQUIPMENT	
WIDTH	2.85 m	SIDE SLOPE	40%	Gun power control	powered/manual
HEIGHT	2.1 m	ENGINE	JAMZ-238 V-8 diesel developing 296 hp	Elevation/depression	+76°/-4° 30'
GROUND CLEARANCE	0.4 m	TRANSMISSION	manual, 5 forward and 1 reverse gears		

INDIA

130 mm Self-propelled Gun (Catapult)

Development/Description

As a result of the effectiveness of the Soviet 130 mm Field Gun M-46 in the Indo-Pakistan conflicts, the Indian Army has mounted some on Vijayanta tank chassis. The conversion is now in Indian Army service which has an inventory of about 1500 Vijayantas and about 550 130 mm M-46 guns.

The gun has a range of 27 150 m and fires an HE projectile weighing 33.4 kg with a muzzle velocity of 930 m/s. It can also fire an APHE projectile weighing 33.6 kg, also at a muzzle velocity of 930 m/s which can penetrate 230 mm of armour at an incidence of 0° at a range of 1000 m.

It would appear that the Vijayanta/130 mm M-46 combination can carry 30 rounds of ammunition. It is reportedly known to the Indian Army as 'Catapult', with around 100 systems believed to be in service. Some have recently been observed fitted with a horizontal shield over the gun and crew compartments to provide protection against top attack weapons.

Trials of the private venture Vickers Shipbuilding and Engineering (VSEL) GBT 155 155 mm turret fitted to a Vijayanta MBT chassis commenced in 1990. If these joint Indian Army/VESL development trials are successfully completed then the system, described in this section under the United Kingdom, could well be adopted by the Indian Army in the future.

IRAQ

Majnoon 155 mm and Al Fao 210 mm Self-propelled Artillery Systems

Development

The existence of these two 6 × 6 artillery systems, which both share the same chassis and turret, was first revealed during a defence exhibition held in Baghdad in May 1989. At the time of the invasion of Kuwait in 1990 both of these artillery systems were still at the prototype stage and none were encountered by Allied Forces during the campaign to recapture Kuwait early in 1991. It is probable that all work on these two weapons has now ceased.

Description

Each system consists of two units, front and rear, connected by a pivot at the top. This allows movement left and right but not up and down and in the lower part is the propeller shaft and two hydraulic rams to provide damping between the two units.

The driver is seated at the very front of the forward unit with the powerpack to his rear. The driver has a large windscreen to his front that is provided with a wiper and a windscreen either side. On the prototypes no shutters are provided for these windows. He enters his compartment via a single piece hatch cover over his position. The Mercedes-Benz 560 hp diesel is coupled to a manual transmission.

The turret is mounted at the rear of chassis and two doors in the turret rear open left and right; in addition there is a large door in the hull rear that opens to the right for ammunition resupply. There are also small hatches in lower part of the hull either side.

The tyres are 21.00 × R25 and a central tyre pressure regulation system is fitted as standard. No spades are fitted to provide a more stable firing platform.

The only difference between the two systems is that they have different weapons, the Majnoon having a 155 mm weapon fitted with a fume extractor and a three slotted muzzle brake while the 210 mm weapon has a long barrel with a fume extractor and a double baffle muzzle brake. In both cases the direct fire sight is mounted to the left of the ordnance.

Both weapons fire ERFB and ERFB-BB ammunition. For some years Iraq has been using 155 mm ERFB and ERFB-BB ammunition for its GH N-45 and G-5 artillery systems and the 210 mm systems uses a similar type of ammunition but with a different calibre and much larger HE content. At present the Al Fao 210 mm system is the only self-propelled artillery system in this calibre in the world.

SPECIFICATIONS

MODEL	**Majnoon**	**Al Fao**
CONFIGURATION	6 × 6	6 × 6
WEIGHT	43 000 kg	48 000 kg
LENGTH OVERALL	12 m	15 m
WIDTH	3.5 m	3.5 m
HEIGHT	3.6 m	3.6 m
MAX ROAD SPEED	90 km/h	90 km/h
CROSS-COUNTRY SPEED	60-70 km/h	60-70 km/h
ARMAMENT (main)	1 × 155 mm	1 × 210 mm
LENGTH OF BARREL	8.06 m	11.13 m
NUMBER OF GROOVES	48	64
LENGTH OF RECOIL	1.041 m	1.041 m
ELEVATION	0° to +72°	0° to +55°
TRAVERSE	40° left/right	40° left/right
RANGE ERFB	30 200 m	45 000 m
MUZZLE VELOCITY ERFB	900 m/s	997 m/s
RANGE ERFB-BB	38 800 m	57 340 m
MUZZLE VELOCITY ERFB-BB	889 m/s	992 m/s
PROJECTILE WEIGHT	45.5 kg	109.4 kg
RATE OF FIRE	4 rpm	4 rpm
SUSTAINED RATE OF FIRE	1 rpm	1 rpm

Status: Prototypes only. All development work has now probably ceased.

Manufacturer: Iraq state facilities with some components being provided from Europe.

Majnoon 155 mm self-propelled gun (Christopher F Foss)

Al Fao 210 mm self-propelled artillery system (Christopher F Foss)

Iraqi MT-LB 120 mm Self-propelled Mortar System

Development/Description

Shown for the first time at a defence equipment exhibition held in Baghdad in May 1989, was an MT-LB series multi-purpose armoured vehicle modified by Iraq to carry four 120 mm mortar tubes in a similar manner to the Austrian NORICUM SM-4 120 mm Salvo Mortar described in *Jane's Armour and Artillery 1990-91*, page 544. As far as it is known none of these systems were encountered by Allied Forces during Operation Desert Storm in early 1991.

When travelling the mortars are horizontal on the roof of the vehicle but when required for action are swung back through about 150° hydraulically with the large webbed rectangular base plate providing the firing platform.

It takes 120 seconds for the system to come into action and about 60 seconds for it to be taken out of action.

The mortars can be elevated from +40° to +80° with traverse being 10° left and right. The mortar barrels are 2.85 m long with maximum muzzle velocity using charge eight being 400 m/s. There are a total of 10 charges (one to 10).

The base plate weighs 630 kg, each barrel weighs 135 kg while the total system, excluding the carrier vehicle, weighs 3000 kg.

The operator's controls are on the left side of the hull and to provide a more stable firing platform two hydraulic stabilisers are lowered to the ground before firing commences. These are positioned one either side of the hull.

A 16 kg mortar bomb can be fired to a maximum range of 8200 m while a 12 kg mortar bomb goes out to 10 000 m. Four 120 mm mortar bombs can be fired in just one second, or the operator can fire each mortar tube individually. The controls are unclipped from the side of the vehicle and the mortars are normally fired by remote-control from a distance of about 30 m.

The system is operated by a four man crew with a total of 104 rounds of ammunition being carried.

Status: Believed to be still at the prototype stage. It is probable that all work on this system has now ceased.

Manufacturer: Conversion carried out by Iraqi Army facility.

Iraqi 120 mm mortar system on MT-LB chassis from the side showing the large welded steel baseplate (Christopher F Foss)

ISRAEL

Soltam Rascal 155 mm Light Self-propelled Howitzer

Development/Description

The Rascal 155 mm light self-propelled howitzer has been developed by Soltam Limited to overcome limitations of weight and size of self-propelled artillery systems imposed by the construction of roads, railways and bridges. It is fully air transportable in a Lockheed C-130 transport aircraft.

The Rascal is also known as the Diablillo and as of early 1991 a single prototype had been completed as a private venture.

Wherever possible proven components have been used in the design of the vehicle to reduce both procurement and operating costs.

The hull of the Rascal is of all-welded steel armour that provides protection from small arms fire and shell splinters. The driver is seated at the front of the vehicle on the left side and has windows to his front and sides and a single piece hatch cover over his position that opens to the rear. When in a combat area the windows are covered by armoured shutters hinged at the top. The other two crew members are seated at the rear.

The commander's compartment is in the centre of the vehicle and is provided with a roof hatch and observation windows. The engine is mounted to the rear of the driver with the air inlet and air outlets in the roof and left side of the vehicle.

Mounted at the rear is the ordnance which is mounted on a turntable to give limited traverse to the left and right. The 155 mm weapon can have either a 39 or 45 calibre ordnance with elevation from -3° to +65°. Elevation and traverse is hydraulic with manual controls for emergency use. The main engine of the Rascal provides the hydraulic power required for elevation and traverse movements, compressed air for the pneumatic rammer and hydraulic power for the winch.

A total of 40 155 mm projectiles, 45 propellant charges plus associated fuzes are carried on the Rascal with the 155 mm projectiles being stowed upwards on either side of the weapon. The charges are stowed in an armoured container to the immediate front of the weapon.

Maximum ranges with a 39 or 45 calibre ordnance are given below:

Projectile type	**39 calibre**	**45 calibre**
M107	21 900 m	18 000 m
Regular unassisted	25 000 m	31 000 m
Special with RAP or base bleed	30 000 m	39 000 m

The suspension either side consists of six dual rubber tyred road wheels with a distinct gap between the first and second road wheel, large drive sprocket at the front, idler at the rear and two track return rollers. When in the firing position two hydraulic spades are lowered, one either side at the rear, to provide a more stable firing platform. When the fire mission has been completed, the vehicle moves off retracting the spades in the process.

The vehicle is powered by a diesel engine coupled to a fully automatic transmission, hydraulic-powered steering gear. A winch is mounted at the front of the vehicle and night vision equipment can be provided if required by the user.

Rascal 155 mm light self-propelled howitzer in firing position with spades deployed at the rear

Rascal 155 mm light self-propelled howitzer in travelling configuration

SPECIFICATIONS		HEIGHT OVERALL	2.30 m	ARMAMENT	1 × 155 mm howitzer
CREW	4	GROUND CLEARANCE	0.40 m	AMMUNITION	40 projectiles, 45 charges
COMBAT WEIGHT	19 500 kg	TRACK WIDTH	380 mm		
UNLOADED WEIGHT	16 500 kg	FUEL CAPACITY	500 l	GUN CONTROL EQUIPMENT	
POWER-TO-WEIGHT RATIO	17.95 hp/tonne	CRUISING RANGE	350 km		
		ENGINE	350 hp diesel	Gun power control	hydraulic/manual
LENGTH TRAVELLING	7.5 m	TRANSMISSION	fully automatic	Gun elevation/depression	+60°/-3°
WIDTH	2.46 m	SUSPENSION	torsion bar	Traverse	40° left and right

Status: Development complete. Ready for production.

Manufacturer: Soltam Limited
PO Box 1371
Haifa 31013
Israel
Telephone: (04) 996211
Telex: 46277
Fax: (04) 994020

Soltam Slammer 155 mm Self-propelled Howitzer

Development

The Slammer 155 mm self-propelled howitzer was developed by Soltam to meet the requirements of the Israeli Artillery Corps and was first revealed in early 1990. Assisting Soltam was Israel Aircraft Industries Limited, Electronic Division/MBT Weapons Systems.

By early 1991 a total of two prototypes of the Soltam Slammer 155 mm self-propelled howitzer had been completed and delivered to the Israeli Artillery Corps for trials.

Description

The Slammer 155 mm self-propelled howitzer is based on a modified Merkava MBT chassis fitted with a new welded steel turret designed by Soltam that has a 155 mm 45-calibre ordnance with a total of 75 projectiles and charges being carried.

The ordnance is fitted with a fume extractor and a double baffle muzzle brake. When travelling the ordnance is held in position by a travel lock that is mounted on the forward part of the glacis plate and this is operated from the crew compartment by remote-control.

Firing an ERFB-BB projectile, the 45-calibre ordnance has a maximum range of 40 000 m.

The 155 mm 45-calibre ordnance and recoil system is of Soltam's well-proven type already used in its towed weapons. The breech block assembly is of the semi-automatic wedge type that contains an automatic primer feeding system that enables manual reloading of the primer without opening the breech. Turret traverse and weapon elevation is hydraulic with manual controls for emergency use.

A maximum rate of fire of 9 rds/min can be achieved due to the automatic computerised loading system while a burst rate of fire of three rounds in 15 seconds can be achieved.

The rate of fire can be achieved using the onboard ammunition supply or from ground piled ammunition.

The loading cycle is operated by two crewmen only, with the commander operating the computer and charge loader.

The automatic loader has five main subsystems: projectile storage system; projectile transfer system; loading tray with flick rammer; charge loading tray and elevator for external charge supply and projectile elevator for reloading the external storage for directly loading the gun.

The internal projectile storage contains 60 projectiles ready for automatic loading with the remaining 15 stored in other locations. The system enables the handling of all kinds of projectiles in use without any adaptation.

Charge loading is accomplished manually using a loading tray with the ignition primer being inserted automatically.

All systems have a manual backup so that in the case of failure the loading system may be operated partly or completely manually by only three crewmen so allowing a continuous firing rate of 4 rds/min.

The computer also controls the functioning of the gun. The Loader Control System (LCS) consists of five main units.

(1) The commander's panel provides the means for the commander to control the automatic loader and has a dedicated keyboard and supporting electronic circuits
(2) The Central Control Unit (CCU) is based on the CPU-8086 and produces all of the system's logic equations. The unit transfers commands through the serial communications (RS 422) to the computerised units and controls the display on the commander's panel
(3) The Terminal Units (TUs) are based on the 8031 controller for purposes of independent control of the drive elements according to a functionally determined division. With the assistance of the terminal unit, a local mode can also be used in working with selected elements
(4) For guiding operators and making round identification and fuzing, the Operator's Panel (OP) includes a LCD display with fixed instructions and one dot matrix line
(5) The Loader Keyboard Panel (LKP) includes breech-block closing switch, fire and local activation of the trays.

The main operational roles are firing from internal storage, firing for elevator - ground piled ammunition, loading from elevator - external pile, synthesising fire programs, unloading, manual firing, identification and fuzing and checks.

Slammer 155 mm self-propelled howitzer with ordnance elevated (Peter Allen-Frost)

Slammer 155 mm self-propelled howitzer from rear (Peter Allen-Frost)

Although the prototype has a 45-calibre ordnance, this could be replaced by a 52-calibre ordnance.

Standard equipment includes an NBC system of the overpressure type and an inertial navigation and aiming system designed for autonomous operations.

According to Soltam, the ordnance and automatic loader, or parts of the system, could be installed in other self-propelled artillery systems and used to upgrade other self-propelled systems such as the M109 and M44.

OUTLINE SPECIFICATIONS

CREW	4
LENGTH	
L/45 ordnance	10.4 m
L/52 ordnance	11 m
WIDTH	3.7 m
HEIGHT	3.4 m

Status: Development complete. Ready for production.

Manufacturer: Soltam Limited
PO Box 1371
Haifa 31013
Israel
Telephone: (04) 996211
Telex: 46277
Fax: (04) 994020

Soltam 155 mm Self-propelled Gun/Howitzer M72

Details of this 155 mm turret system, developed some years ago but not placed in production, were given in *Jane's Armour and Artillery 1990-91*, pages 562 to 563.

Soltam L33 155 mm Self-propelled Gun/Howitzer

Development
The L33 (the length of the barrel in calibres) was developed in the late 1960s by Soltam Limited and is basically a Sherman M4A3E8 chassis fitted with a Cummins diesel engine, all-welded steel superstructure and a Soltam 155 mm M-68 gun/howitzer. The L33 entered service in 1973 and was used operationally for the first time during the 1973 Middle East conflict.

Description
The driver is seated at the front of the hull on the left side and has a bullet-proof windscreen to his front and sides and a single piece hatch cover that opens to the rear over his position. The commander is seated to the rear and above the driver's position and has a windscreen to his front and sides and a single piece hatch cover. The anti-aircraft gunner's position is at the front of the hull on the right side. A 7.62 mm (0.30) machine gun is pintle-mounted on the forward part of the cupola which can be traversed through a full 360° and has a two-piece hatch cover that opens either side. The crew enter and leave the vehicle by a single door in each side of the hull that opens to the rear. At the rear of the hull are two large doors for ammunition resupply.

The engine is under the gun compartment at the rear of the hull. Power is transferred to the transmission in the nose of the vehicle by a two-part propeller shaft. The controlled differential transmits engine power to the final drive unit and also contains a brake system for steering and stopping the vehicle. The suspension each side consists of three two-wheeled bogies, three return rollers with idler at the rear and the drive sprocket at the front.

The 155 mm M-68 gun/howitzer is mounted in the forward part of the superstructure and has an elevation of +52°, a depression of -3° and a total traverse of 60°. Both elevation and traverse are manual. The weapon has a monobloc barrel and incorporates a horizontal sliding semi-automatic breech-block. Mounted on the cradle are the hydro-pneumatic recuperator, a variable hydraulic damper which varies as a function of elevation (for example 800 mm at 52° elevation and 1100 mm at 0° elevation) and a pneumatic rammer which permits loading of the gun at all angles of elevation. The barrel has a single-baffle muzzle brake, bore evacuator and a travelling lock.

The L33 has a panoramic telescope for indirect fire engagements and a telescope for direct fire.

Ammunition is separate loading with obturation achieved by self-sealing obturator discs. Firing is accomplished by a cartridge primer inserted into the breech-block and fired by a trip action mechanism. The HE projectile weighs 43.7 kg and has a maximum muzzle velocity of 725 m/s and a maximum range of 20 000 m. The following propelling charges are used: M1A2 zone 3 (muzzle velocity 284 m/s), M1A2 zone 4 (muzzle velocity 326 m/s), M1A2 zone 5 (muzzle velocity 390 m/s), M9A2 zone 6 (muzzle velocity 468 m/s), M9A2 zone 7 (muzzle velocity 562 m/s), M9A2 zone 9 (muzzle velocity 647 m/s), M9A2 zone 9 (muzzle velocity 725 m/s). The M-68 fires all NATO 155 mm projectiles including illuminating and smoke. Of the 60 rounds of ammunition carried, 16 are for ready use.

This self-propelled howitzer can be supplied with a 39-calibre barrel that increases the range to 24 000 m using conventional ammunition. This is called the L39 but as far as it is known this has not entered production or service.

Soltam L33 155 mm self-propelled gun/howitzer with ordnance in travelling lock and roof-mounted 7.62 mm machine gun in position (Israeli Army)

Status: Production complete. In service with the Israeli Army reserves.

Manufacturer: Soltam Limited
PO Box 1371
Haifa 31013
Israel
Telephone: (04) 996211
Telex: 46277
Fax: (04) 994020

SPECIFICATIONS					
CREW	8	LENGTH OF TRACK ON GROUND	4.6 m	TRANSMISSION	manual with 5 forward and 1 reverse gears
COMBAT WEIGHT	41 500 kg	MAX ROAD SPEED	38 km/h	ELECTRICAL SYSTEM	24 V
POWER-TO-WEIGHT RATIO	10.08 hp/tonne	FUEL CAPACITY	820 l	ARMAMENT	
LENGTH GUN FORWARDS	8.47 m	MAX ROAD RANGE	260 km	main	1 × 155 mm gun/howitzer
with 39-calibre barrel	9.36 m	FORDING	0.9 m	anti-aircraft	1 × 7.62 mm MG
LENGTH HULL	6.47 m	GRADIENT	60%	AMMUNITION (main)	60
WIDTH	3.5 m	SIDE SLOPE	30%	GUN CONTROL EQUIPMENT	
over tracks	2.86 m	VERTICAL OBSTACLE	0.91 m	Gun elevation/depression	+52°/-4°
HEIGHT	3.45 m	TRENCH	2.3 m	Gun traverse (total)	60°
GROUND CLEARANCE	0.43 m	TURNING RADIUS	9.5 m	ARMOUR	12 - 64 mm
TRACK	2.28 m	ENGINE	Cummins VT 8-460-Bi diesel developing 460 hp at 2600 rpm		
TRACK WIDTH	580 mm				

155 mm Self-propelled Howitzer M-50

Development/Description
This self-propelled howitzer was developed in the late 1950s to meet the requirements of the Israeli Army by the French Etablissement d'Etudes et Fabrications d'Armement de Bourges and entered service in 1963. It basically consists of a Sherman chassis fitted with the French 155 mm Model 50 howitzer at the rear of the hull. The Model 50 fires an HE projectile weighing 43 kg with a maximum velocity of 650 m/s to a maximum range of 17 600 m. The ordnance is fitted with a multi-baffle muzzle brake, hydro-pneumatic recoil system and a screw breech mechanism. Maximum elevation is +69° and a travelling lock is mounted at the front of the hull.

As the Model 50 howitzer is mounted at the rear of the hull, the engine has been moved forward to the right side of the hull with the driver seated on the left. Ammunition is carried along the sides of the hull vertically and in two racks at the very rear of the hull, one on each side. In action a platform is unfolded into the horizontal at the rear of the hull. Stowage compartments are provided externally just over the tops of the tracks.

155 mm self-propelled howitzer M-50 based on Sherman chassis with Horizontal Volute Suspension System (Israeli Army)

Shermans with both Horizontal Volute Suspension Systems and Vertical Volute Suspension Systems were used for these conversions. The Israeli Army also used the towed version of the Model 50 howitzer but it has now been withdrawn from service.

There is also an armoured ambulance and an armoured artillery observation vehicle based on the chassis of the self-propelled howitzer, brief details of which are given under Israel in the *MBTs and medium tanks* section.

SPECIFICATIONS

CREW	8
WEIGHT	31 000 kg
LENGTH	6.1 m
WIDTH	2.98 m
HEIGHT	2.8 m

Status: Production complete. In service with the Israeli Army reserves.

ITALY

OTO Melara Palmaria 155 mm Self-propelled Howitzer

Development

The Palmaria was developed from 1977 by OTO Melara specifically for export and first production vehicles followed in 1982. The first customer to order the Palmaria was Libya with an order for 210. Nigeria ordered 25 during 1982 and a further 25 Palmaria turrets were ordered by Argentina to be fitted on TAM medium tank chassis. These were completed late in 1986 by which time final deliveries had been made to Libya and Nigeria. Argentina has built two lengthened chassis which will be fitted with the turret of the Palmaria, but volume production has yet to start. The Argentinian system is called the VCA 155 and is described in detail earlier in this section.

In 1990 OTO Melara completed a further batch of 155 mm Palmaria systems for an undisclosed customer.

Description

The chassis is similar to the OF-40 MBT's, apart from the engine, and is fully described in the *MBTs and medium tanks* section. The driver is seated at the front of the hull with the other four crew members, commander, gunner, charge handler and magazine operator seated in the aluminium turret in the centre of the hull.

The commander is seated in the right forward part of the turret and has eight periscopes for all-round observation and a single piece hatch cover that opens to the rear. There are two external stowage racks at the rear of the turret. In either side of the turret there is a large rectangular hatch, the left one opening to the rear and the right one opening forwards. A 7.62 or 12.7 mm anti-aircraft machine gun can be mounted on the roof for anti-aircraft defence and four smoke dischargers are mounted either side of the turret.

The vehicle has an auxiliary power unit to supply the hydraulic and electric power required for operating the turret, consisting of a diesel engine driving, through a gearbox, a generator and a hydraulic pump. Standard equipment includes an NBC system, an automatic fire-extinguishing system, bilge pumps and an emergency escape hatch in the floor.

The OTO Melara-designed barrel is 41 calibres long, is fitted with a fume extractor and a muzzle brake and is loaded automatically.

Using the automatic loading system the howitzer can fire one round every 15 seconds, one round every minute for one hour or one round every three minutes continuously. The automatic loader is at the rear of the turret and can be loaded via a hatch in the turret side. The barrel is loaded at an angle of +2° and the propellant charge is loaded into the chamber manually. Using automatic loading the turret can accommodate 23 rounds ready for use with a further seven ready for use elsewhere in the hull. The projectiles are stowed in a linked row

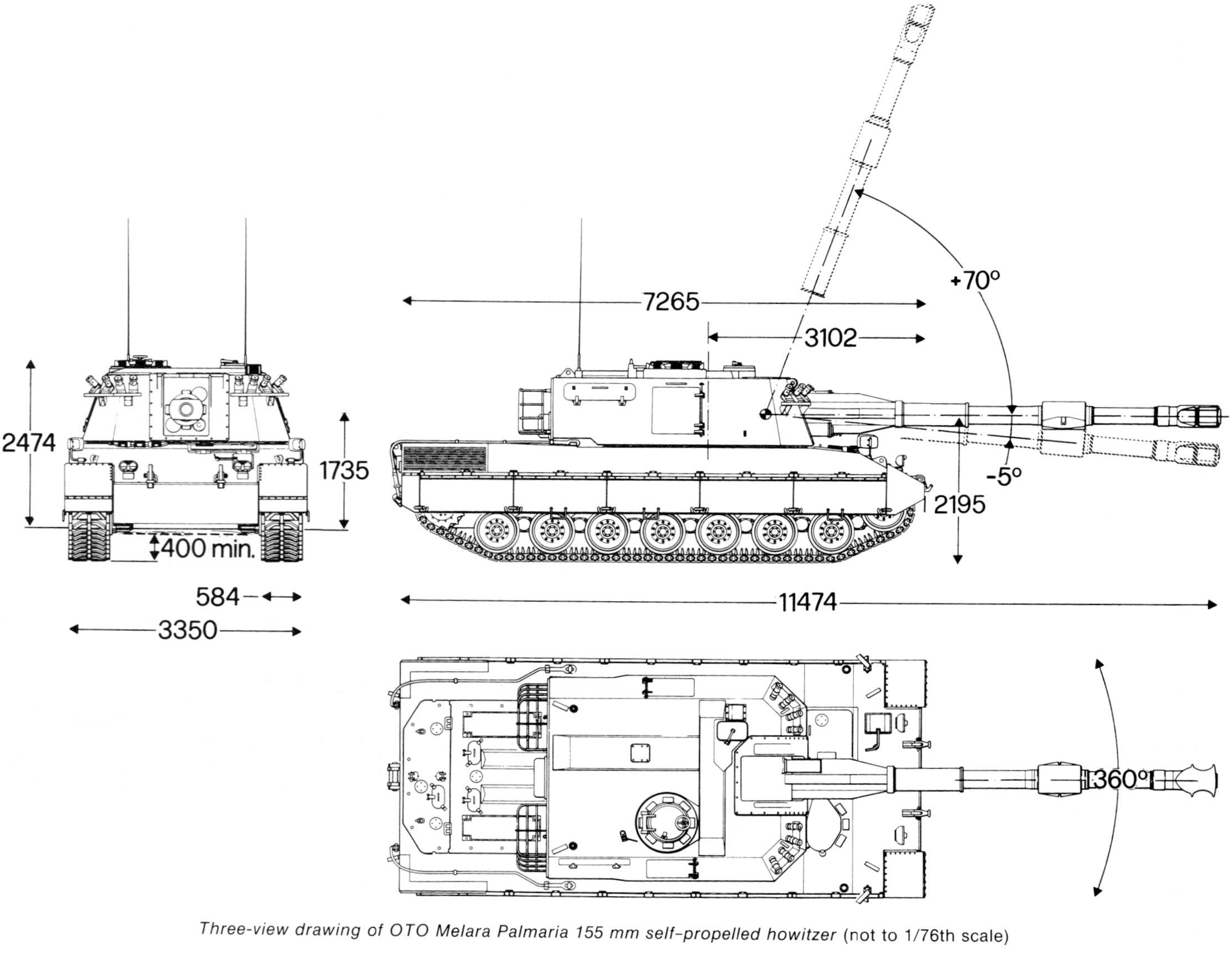

Three-view drawing of OTO Melara Palmaria 155 mm self-propelled howitzer (not to 1/76th scale)

of tubes at the rear of the turret ready for use. The firing percussion mechanism is actuated by an electro-hydraulic valve. Firing is by a push-button and using this system an initial burst rate of three rounds in 20 seconds can be achieved. The barrel elevating and balancing mechanism consists of a hydraulically actuated cylinder linked to the cradle and pivoting on the turret structure. The limits of elevation are from -4° to +70°. There are two modes of operation for the mechanism, hand and power. Using the hand mode the barrel can be moved using a wheel to operate a hand pump linked to the elevation cylinder. With the power mode, with the hydraulic power on, the gun layer can lay the weapon directly, using a control handle. After firing the barrel is returned to the loading position of +2° automatically. Traversing is by a hydraulically operated and controlled gear box; traverse is a full 360°.

The recoil system consists of two buffers and one recuperator. Firing at charge 8 at an elevation of +70° the recoil force is approximately 45 000 kg.

The Italian company of Simmel developed a new range of ammunition for the Palmaria self-propelled howitzer, brief details of which are given below.

Production version of OTO Melara Palmaria 155 mm self-propelled howitzer with turret traversed

Type	**P3 HE**	**P3 HE LT**	**P3 HE RAP**	**P4 Illuminating**	**P5 Smoke**
WEIGHT OF PROJECTILE	43.5 kg	43.5 kg	43.5 kg	43.5 kg	43.5 kg
WEIGHT OF EXPLOSIVE	11.7 kg	11.7 kg	8 kg	n/app	n/app
WEIGHT OF PROPELLANT	n/app	1 kg	2.8 kg	n/app	n/app
LENGTH	933 mm	933 mm	933 mm	933 mm	933 mm
MAX RANGE	24 700 m	27 500 m	30 000 m	24 700 m	24 700 m

It can also fire other types of ammunition including that used by the FH-70, Bofors FH-77B as well as US ammunition types including the M107, M549A1 and M483A1.

The Palmaria has three sighting systems installed, a P170 direct firing periscope, a P186 panoramic sight for indirect sighting and a dial sight.

The P170 can be used for both observation and sighting and in latter mode has a magnification of ×8. The optical unit of the sighting system can be replaced by a light intensification system that transforms the day periscope into a low light device.

The P186 panoramic sight is used for indirect firing and is provided with an illuminated reticle for day or low light level use. The reticle is fixed and pre-aligned with the optical axis.

The Palmaria can also be provided with an Electronic Plane Converter for the rapid and accurate aiming both in direct and indirect fire.

The primary fire control system of the Palmaria is the Aeritalia P186 telescope which is mounted on a Bofors telescope mount. This is used for laying the weapon in azimuth and for indirect fire control. The sight has a magnification of ×4 with a 10° field-of-view and weighs 8 kg. Internal communications are via the normal intercommunication system and an extra external telephone point can be provided. The turret has adequate space for normal radio installations.

Variants

OTO Melara has proposed that the chassis of the Palmaria should be used for other applications, for example fitted with a similar turret to the Otomatic 76 mm self-propelled anti-aircraft gun system and it has also been fitted with the turret of the ATAK 35 twin 35 mm anti-aircraft gun system.

SPECIFICATIONS	
CREW	5
COMBAT WEIGHT	46 000 kg
UNLOADED WEIGHT	43 000 kg
POWER-TO-WEIGHT RATIO	16.3 hp/tonne
LENGTH	
gun forward	11.474 m
gun rear	9.6 m
hull	7.265 m
WIDTH	3.35 m
HEIGHT	
without machine gun	2.874 m
axis of bore	2.195 m
GROUND CLEARANCE	0.4 m
TRACK WIDTH	584 mm
MAX SPEED (forwards)	60 km/h
FUEL CAPACITY	800 l
MAX CRUISING RANGE	500 km
GRADIENT	60%
SIDE SLOPE	30%
VERTICAL OBSTACLE	1 m
TRENCH	3 m
FORDING	1.2 m
with preparation	4 m
ENGINE	MTU MB 837 Ea-500 (or Ka-500 developing 1000 hp) 4-stroke, pre-combustion chamber, turbo-charged, 8-cylinder multi-fuel developing 750 hp
TRANSMISSION	Renk RK 304 power shifting 4-speed (4 forward and 2 reverse) planetary type with hydraulic torque converter and lock up clutch
SUSPENSION	torsion bar
ELECTRICAL SYSTEM	24 V
BATTERIES	6 × 12 V, 100 Ah
GENERATOR	9 kW, 3-phase with rectifier
ARMAMENT	
main	1 × 155 mm howitzer
anti-aircraft	1 × 7.62 mm MG
SMOKE LAYING EQUIPMENT	2 × 4 smoke dischargers
AMMUNITION (main)	30
GUN CONTROL EQUIPMENT	
Turret power control	hydraulic/manual
by commander	no
by gunner	yes
Gun elevation/depression	+70°/-5°
Turret traverse	360°

Status: Production as required. In service with Argentina (25 turrets for modified TAM chassis, but not yet in service), Libya (210) and Nigeria (25). OTO Melara completed another batch of 155 mm Palmaria systems for another country in 1990.

Manufacturer: OTO Melara SpA
Via Valdilocchi 15
La Spezia
Italy
Telephone: (39-187) 530 111
Telex: 270 368 OTO I
Fax: (39-187) 530 669

OTO Melara 155 mm M109L Self-propelled Howitzer

Development

Italy purchased 221 M109s from the United States without armament. The armament was manufactured and installed in Italy by OTO Melara and subsequently issued to the Italian Army.

In 1970-71 OTO Melara produced a long barrel version of the M109 capable of firing FH-70 projectiles to a maximum range of 24 000 m. At that time there was no Italian Army requirement for this increase in range but early in 1984 the Italian MoD ordered three examples of this for extensive trials. The trials were subsequently successful and a decision was taken in 1986 to upgrade a total of 280 Italian Army M109 series self-propelled weapons to the M109L configuration. In addition to taking delivery of 221 M109s without armament the Italian Army also took delivery of 62 M109A1B systems. The first batch of 32 M109Ls were completed in 1986 and final deliveries will be made in 1992.

Description

The hull of the M109 has been modified so that it is the equivalent of the M109A3 which is compatible with the new elevating mass and the aiming system of the M109G.

The main change on the M109L is the replacement of the existing 23-calibre barrel by a 39-calibre barrel. Externally this is configured to the M109 mounting but internally it is ballistically similar to the FH-70 configuration and can thus fire existing M107 ammunition or FH-70

OTO Melara M109L with 155 mm 39-calibre ordnance at maximum elevation

ammunition. Firing L15 projectiles the M109L has a maximum range of 24 000 m or 30 000 m with RAPs.

In addition to the installation of a new ordnance, the following have been modified:

Recoil buffers and run out buffer
Cradle to fit the new ordnance
Replacement of the equilibrator and elevating cylinder
Barrel clamp to fit new barrel.

The new barrel is fitted with a revised muzzle brake with the maximum recoil increase on the M109L being only 100 mm. The existing breech and breech ring of the original M109 have been retained.

The elevating mass, as modified, is interchangeable with the one on the basic M109, furthermore, the mechanical interface between the elevating mass and saddle as well as the mountings of the optical device, do not require any modification.

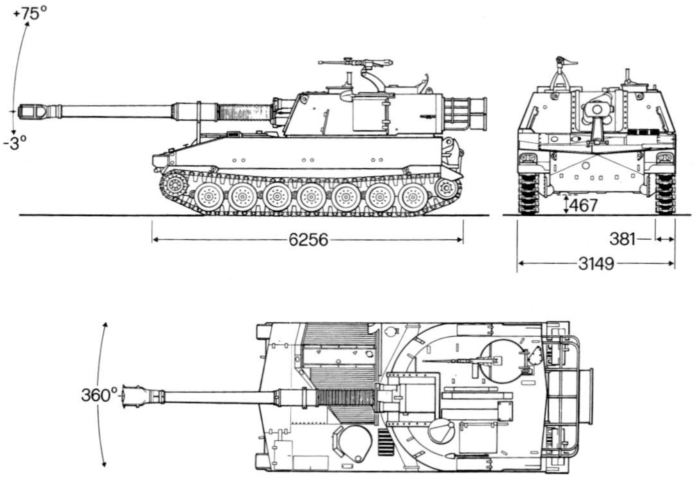

Three-view drawing of M109 upgraded to M109L configuration by OTO Melara (not to 1/76th scale)

Specifications of the OTO Melara 155 mm 39-calibre barrel are as follows:

CALIBRE	155 mm
LENGTH	39 calibres
POWDER CHAMBER VOLUME	18.84 dm^3
NUMBER OF RIFLING GROOVES	48
TWIST UNIFORM RIGHT HAND	1 turn in 20 calibres
GUN BARREL WEIGHT	1670 kg
ELEVATING MASS WEIGHT	4200 kg
MAXIMUM RECOIL LENGTH	915 mm

Status: Conversions underway for the Italian Army.

Manuafacturer: OTO Melara SpA
Via Valdilocchi 15
La Spezia
Italy
Telephone: (39-187) 530 111
Telex: 270 368 OTO I
Fax: (39-187) 530 669

SPECIFICATIONS		VERTICAL OBSTACLE	0.533 m	BATTERIES	4 × 12 V, 6TN
CREW	6	TRENCH	1.828 m	ARMAMENT	1 × 155 mm howitzer
WEIGHT COMBAT	24 800 kg	ENGINE	Detroit Diesel Model	anti-aircraft	1 × 12.7 mm MG
WEIGHT UNLOADED	20 000 kg		8V-71T, turbo-	SMOKE LAYING	
GROUND PRESSURE	0.68 kg/cm^2		charged, 2-stroke,	EQUIPMENT	2 × 3 smoke
LENGTH OVERALL	8.133 m		liquid cooled 8-		dischargers
LENGTH HULL	6.256 m		cylinder diesel	AMMUNITION	
WIDTH OVERALL	3.295 m		developing 405 bhp at	main	28
WIDTH OVER TRACKS	3.149 m		2300 rpm	12.7 mm	500
HEIGHT INC AA MG	3.28 m	TRANSMISSION	Allison Transmission	GUN CONTROL	
GROUND CLEARANCE	0.467 m		Division of General	EQUIPMENT	
TRACK	2.768 m		Motors Corporation	Turret power control	hydraulic/manual
TRACK WIDTH	381 mm		XTG-411-2A cross	by commander	no
LENGTH OF TRACK			drive with 4 forward	by gunner	yes
ON GROUND	3.962 m		and 2 reverse gears	Gun elevation/depression	+75°/-3°
MAX ROAD SPEED	56 km/h	SUSPENSION	independent torsion	Turret traverse	360°
FUEL CAPACITY	511 l		bar		
MAX ROAD RANGE	349 km	ELECTRICAL SYSTEM	24 V		

JAPAN

New Japanese 155 mm Self-propelled Howitzer

Development/Description

In 1983 the Japanese Self-Defence Force Technical Research Institute commenced development of a new self-propelled howitzer which will have a turret equipped with the ordnance of the 155 mm FH-70. This has the provisional designation Type 75 155 mm SPH Kai.

Early in 1989 Mitsubishi Heavy Industries was awarded a contract to build the prototype of the new self-propelled howitzer which will be based on the chassis of the new Type 89 Mechanised Infantry Combat Vehicle.

The turret and loading system for the new 155 mm self-propelled howitzer will be the responsibility of Nihon Seiko Jyo/Japan Steel Works which also designed and built the turret for the earlier Type 75 self-propelled howitzer system, production of which was completed in 1988.

In 1983 Komatsu began licensed production of the 203 mm (8-inch) M110A2 self-propelled howitzer with the gun mount being supplied by the Japan Steel Works. By the end of 1988 201 M110A2s were in service with the Japanese Ground Self-Defence Force.

Type 75 155 mm Self-propelled Howitzer

Development

In 1965 the Japanese Ground Self-Defence Force bought 30 105 mm M52A1 and 10 155 mm M44A1 self-propelled howitzers from the USA to meet its immediate requirements for self-propelled artillery. These have now been phased out of service. In 1967 the development of a new 105 mm self-propelled howitzer began, the first prototype of which was completed in 1969-70. This was subsequently accepted for service as the Type 74 105 mm self-propelled howitzer, but only 20 were built as by then the 105 mm howitzer was considered to be less effective than the 155 mm howitzer.

In 1969 the Technical Research and Development Headquarters of the Japanese Self-Defence Force started design work on a new 155 mm self-propelled howitzer. Mitsubishi Heavy Industries built the hull while Nihon Seiko Jyo/Japan Steel Works built the gun and turret. Two prototypes completed in 1971/72 differed only in their ammunition loading systems. Trials with the two prototypes were performed between 1973 and 1974 and in October 1975 the vehicle was standardised as the Type 75 155 mm self-propelled howitzer. By late 1978 the first 20 vehicles had been delivered to the Japanese Ground Self-Defence Force. By end of 1988 201 Type 75 self-propelled howitzers were in service with the Japanese Ground Self-Defence Force and production was complete.

Description

The hull and turret of the Type 75 are made of all welded aluminium. The crew of six consists of the commander, layer, two loaders, radio-operator and driver.

The hull is divided into three compartments: driver's at the front on the right, engine to his left and the turret at the rear. The driver has a single piece hatch cover that opens to the right, in front of which are three periscopes. The driver can also enter and leave his compartment via the turret.

The turret is fully enclosed with the commander and gunner seated on the right and the loaders on the left. The commander has a single piece hatch cover that opens to the rear and a single periscope for forward observation. One of the two loaders is also provided with a single piece hatch cover that opens to the rear, in front of which is a 12.7 mm (0.50) M2 HB anti-aircraft machine gun with a small shield. In each side of the turret is a square hatch that opens to the rear.

At the rear of the hull are two doors which open either side of the vehicle. The torsion bar suspension consists of six dual rubber-tyred aluminium road wheels with the drive sprocket at the front and the sixth road wheel acting as the idler. The first and fifth road wheel stations are provided with a hydraulic shock absorber. There are no track return rollers.

The Type 75 is provided with an NBC system which is in the left side of the fighting compartment, fire-extinguishing system, heater, and infra-red night vision equipment. An amphibious kit was developed for the vehicle but was not adopted.

Main armament consists of a 155 mm 30-calibre weapon with a double-baffle muzzle brake, fume extractor and a stepped thread interrupted screw type breech-block which opens to the right. A gun travelling lock is fitted on the glacis plate.

The weapon fires a Japanese HE projectile (maximum charge 9) to a maximum range of 19 000 m or an American M107 HE projectile (maximum charge 7) to a maximum range of 15 000 m. An RAP 155 mm projectile currently under development in Japan is expected to have a range of around 24 000 m.

The gun has an elevation of +65°, a depression of -5° and the turret can be traversed through a full 360°. Gun elevation and turret traverse are hydraulic by means of a single lever: movement in the horizontal plane controls the turret position in azimuth and movement in the vertical plane controls the gun in elevation. The traverse of the turret and the elevation of the gun are stopped whenever the layer halts the movement of the lever. Manual controls are provided for emergency use and the turret can also be traversed by the commander.

The gun has a variable recoil mechanism which consists of twin hydraulic cylinders, one on the left above the cradle and the other on the right side under the cradle. The variable recoil system is fitted to prevent the breech hitting the floor when the gun is being fired at a high angle. The recuperator is pneumatic and the recoil damper, which is on the centre line under the cradle, is hydraulic.

When the gun is fired the weapon automatically returns to an elevation of +6° for reloading and returns to its firing elevation once reloaded. Mounted in the rear part of the turret are two drum type rotating magazines, an extensible loading tray and a hydraulic rammer. Each drum holds nine projectiles, is rotated electrically or manually and can be reloaded from inside the vehicle or from outside via two small circular doors in the turret rear. The axis of the drums, two-stage extensible loading tray and the rammer are on the same axes as the gun when it is returned from its firing position for loading. Each of the two magazines holds nine projectiles which, coupled with the use of the extensible loading tray and the hydraulic rammer, enable 18 rounds to be fired in three minutes, ie 6 rds/min. A further 10 projectiles are carried inside the gun compartment plus 56 fuzes and 28 bagged charges.

Fire control equipment includes a J2 panoramic sight with a magnification of ×4 and a 10° field-of-view, a J3 direct telescopic sight with a magnification of ×4 and a 10° field-of-view, electric elevation sight which includes a cant angle sensor, angle indicator and an inverter and finally a collimator. All fire control equipment is mounted in front of the commander's position and is operated by the layer.

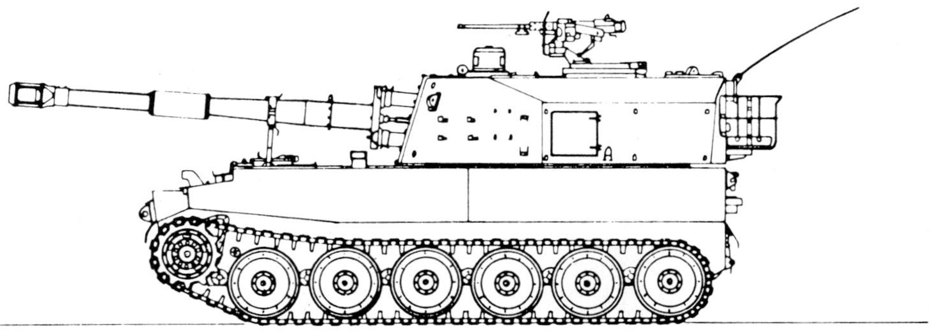

Type 75 155 mm self-propelled howitzer

SPECIFICATIONS		FUEL CONSUMPTION	2.1 l/km	SMOKE-LAYING	
CREW	65	FORDING	1.3 m	EQUIPMENT	none
COMBAT WEIGHT	25 300 kg	GRADIENT	60%	AMMUNITION	
POWER-TO-WEIGHT		SIDE SLOPE	30%	main	28
RATIO	17.78 hp/tonne	VERTICAL OBSTACLE	0.7 m	anti-aircraft	1000
GROUND PRESSURE	0.64 kg/cm^2	TRENCH	2.5 m	GUN CONTROL	
LENGTH GUN FORWARDS	7.79 m	ENGINE	Mitsubishi 6ZF 2-cycle	EQUIPMENT	
LENGTH HULL	6.64 m		6-cylinder air-cooled	Turret power control	hydraulic/manual
WIDTH	3.09 m		diesel developing	by commander	yes
HEIGHT (to turret roof)	2.545 m		450 hp at 2200 rpm	by gunner	yes
GROUND CLEARANCE	0.4 m	TRANSMISSION	manual with 4 forward	Max rate power	
TRACK	2.5 m		and 1 reverse gears	traverse	360°/30 s
TRACK WIDTH	500 mm	SUSPENSION	torsion bar	Gun elevation/depression	+65°/-5°
LENGTH OF TRACK		ELECTRICAL SYSTEM	24 V	Commander's override	yes
ON GROUND	5.2 m	BATTERIES	4 × 12 V		
MAX ROAD SPEED	47 km/h	ARMAMENT			
FUEL CAPACITY	650 l	main	1 × 155 mm howitzer		
MAX RANGE	300 km	anti-aircraft	1 × 12.7 mm MG		

Status: Production complete. In service with the Japanese Ground Self-Defence Force.

Manufacturers: Mitsubishi Heavy Industries (hull and final assembly), Japan Steel Works/Nihon Seiko Jyo (gun and turret).

Type 75 155 mm self-propelled howitzer fitted with 12.7 mm anti-aircraft machine gun on turret roof (Kensuke Ebata)

Type 74 105 mm Self-propelled Howitzer

Development

To meet the immediate requirements of the Japanese Ground Self-Defence Force for self-propelled artillery, 30 105 mm M52A1 and 10 155 mm M44A1 self-propelled howitzers were bought from the USA in 1965. These have now been phased out of service. Two years later the development of a 105 mm self-propelled howitzer using the engine and transmission of the Type 73 armoured personnel carrier began under the direction of the Technical Research and Development Headquarters of the Japanese Self-Defence Force.

The first two prototypes were completed in 1969-70 with Komatsu responsible for the hull and Japan Steel Works for the turret and main armament. In 1974 the vehicle was standardised as the Type 74 105 mm self-propelled howitzer and production began in 1975. Only 20 were built, the last being completed in 1978, as it was decided to concentrate on the more efficient 155 mm self-propelled howitzer which was eventually standardised as the Type 75. The 20 Type 74s are attached to the 2nd Artillery Regiment at Asahikawa, Hokkaido, the northern part of the Japanese Islands.

Description

In appearance the Type 74 is very similar to the British Vickers Abbot 105 mm self-propelled gun. The all-welded hull is divided into three compartments: driver's at the front on the right, engine to the left and gun compartment at the rear.

The driver has a single piece hatch cover that opens to the right, in front of which are three observation periscopes. The turret, which is all welded, is at the rear of the hull and has both roof and side hatches. The torsion bar suspension consists of five dual rubber-tyred road wheels with the drive sprocket at the front and idler at the rear. There are no track return rollers. The first and fifth road wheel stations have a hydraulic shock absorber.

A flotation screen is carried collapsed around the top of the hull and when it is erected the vehicle can propel itself across rivers by its tracks.

Type 74 105 mm self-propelled howitzer with 12.7 mm M2 HB anti-aircraft machine gun mount with shield for gunner installed on turret roof

The 105 mm gun has a muzzle brake and fume extractor. No details of the range of the 105 mm weapon have been released but it can be assumed that it has a similar range to that of the American 105 mm M2A1 towed howitzer which is used in some numbers by the Japanese Ground Self-Defence Force and fires an HE projectile weighing 19.06 kg to a maximum range of 11 270 m.

Status: Production complete. In service only with the Japanese Ground Self-Defence Force.

Manufacturers: Komatsu (hull and final assembly), Japan Steel Works (turret and main armament).

SPECIFICATIONS					
CREW	5	GROUND CLEARANCE	0.4 m	ENGINE	Mitsubishi 4ZF 2-stroke air-cooled V-4 diesel developing 300 hp at 2200 rpm
WEIGHT	16 500 kg	MAX ROAD SPEED	50 km/h		
POWER-TO-WEIGHT RATIO	18.18 hp/tonne	RANGE	300 km		
LENGTH	5.9 m	FORDING	1.2 m		
WIDTH	2.9 m	with screen erected	amphibious	SUSPENSION	torsion bar
HEIGHT	2.39 m	GRADIENT	60%	ARMAMENT	
		VERTICAL OBSTACLE	0.61 m	main	1 × 105 mm howitzer
		TRENCH	2 m	anti-aircraft	1 × 12.7 mm MG

KOREA, NORTH

North Korean Self-propelled Artillery

Development/Description

North Korea appears to be following in the footsteps of countries such as China and Syria which have mated Soviet vehicle chassis and towed artillery in the absence of suitable ready-made hardware. The US Defense Intelligence Agency has reported the existence of the following weapons:

Designation	Calibre	Type	Comment
M-1974	152 mm	SP gun/howitzer	
M-1975	130 mm	SP gun	
M-1977	122 mm	SP howitzer	may be locally built Chinese Type 54-1 (qv)
M-1978	175 mm	SP gun	separate entry
M-1981	122 mm	SP gun	
M-1985	122 mm	SP gun	

A typical North Korean mechanised infantry brigade has two artillery battalions equipped with 18/20 M-1977 self-propelled gun/howitzers. Some of these brigades have substituted 12 M-1981/M-1985 for the M-1977s.

North Korean armoured brigades have two artillery battalions equipped with 12/14 M-1981/M-1985 self-propelled guns. Within armoured and mechanised infantry divisions, the divisional self-propelled artillery regiments consist of two battalions of M-1977 self-propelled howitzers, one battalion of M-1974 self-propelled guns and one battalion of M-1981/M-1985 self-propelled guns.

Corps and Army level self-propelled artillery brigades are equipped with M-1975 or M-1978 self-propelled guns and it is believed that by the early 1980s North Korea had deployed one army level artillery brigade with 24/36 M-1978 self-propelled guns and at least three army level artillery brigades with 36/60 M-1978 self-propelled guns each.

The infantry divisions and regiments are usually still equipped with towed artillery.

175 mm M-1978 Self-propelled Gun Koksan

Development/Description
This weapon was first observed by United States sources in the city of Koksan in 1978 and is therefore referred to as the M-1978 self-propelled gun Koksan.

A quantity of these was supplied to Iran by North Korea and some of these were subsequently captured by Iraq in 1987-88.

In North Korean service, it is believed that the 175 mm Koksan is employed in regiments with a total of 36 weapons.

The system is based on a modified Chinese Type 59 MBT chassis with the existing turret removed and plated over. On the top of the chassis has been mounted a 175 mm gun of a previously unknown type in an open mount. Some sources have stated that the calibre of the weapon is 170 mm or 180 mm, and not 175 mm.

At the very rear of the chassis are two large spades that are lowered to the ground manually before firing commences; when travelling these are vertical. A large ordnance travelling lock is pivot-mounted on the glacis plate and this folds forward onto the ground when deployed in the firing position.

The long ordnance is fitted with a multi-slotted muzzle brake but has no fume extractor. The screw type breech mechanism opens to the right with a loading tray being mounted on the left side of the weapon. In addition there is probably a power rammer.

Elevation and traverse appears to be powered although traverse is very limited to avoid overstressing the chassis.

It is possible that the weapon used may be an ex-Soviet naval or coastal weapon that was supplied to North Korea in the 1950s for use in the coastal defence role. When these were replaced by surface-to-surface missiles, the coastal guns were surplus to requirements and could have then become available for mounting on the tracked chassis.

As there is insufficient space for the gun crew and ammunition on the Koksan, another vehicle is required to support the system in action and this would carry the remainder of the crew and some ready use ammunition.

Koksan self-propelled gun from rear showing two large spades

Koksan self-propelled gun with ordnance in travelling lock

Status: Production probably complete. In service with Iran and North Korea.

Manufacturer: North Korean state factories.

KOREA, SOUTH

M109A2 155 mm Self-propelled Howitzer

Development/Description
Samsung Shipbuilding and Heavy Industries (SHI), with BMY of the USA, is currently engaged in a co-production programme for the 155 mm M109A2 self-propelled howitzer.

This programme covers a total of 272 vehicles, of which two complete M109A2s came from BMY as did 20 sets of components for complete assembly and final checks in South Korea, with the final 250 being built by Samsung Shipbuilding and Heavy Industries, but with BMY supplying key sub-assemblies such as turret shell, engine, transmission and armament. The last of these 250 vehicles was completed in 1990.

Under a co-production agreement revealed early in 1989, South Korea is to get a further 242 155 mm M109A2 self-propelled howitzers in addition to the 272 supplied under the original agreement.

As part of the latest agreement, Watervliet Arsenal was in January 1989 awarded a $3 million increment as part of a $14.4 million firm-fixed-price project order for the manufacture of 242 155 mm M185 cannons for BMY by November 1990.

BMY is expected to supply about 50 per cent of the total value of the latest South Korean order with deliveries to run through to 1991.

It is understood that there is also a project to develop a new 155 mm self-propelled howitzer in South Korea which will incorporate a number of foreign automotive components which will also be used in other tracked armoured vehicles.

NETHERLANDS

RDM Defense Engineering

Development/Description
Recent production has included manufacture of parts for Dutch Army 155 mm M109A2 self-propelled howitzers, major assemblies for the Leopard 2 MBT for both Dutch and German Armies, fitting FMC command and reconnaissance carriers of the Dutch Army with Oerlikon-Contraves 25 mm GBD-A0A turrets, modifying Dutch Army FMC AIFVs to carry the Emerson Improved TOW mount, and production of a large series of AIFVs (together with DAF, Eindhoven) for the Dutch Army.

In 1988 RDM Defense Engineering delivered the first upgraded M114/39-calibre towed howitzers, details of this programme are given in the *Towed guns and howitzers* section under the Netherlands.

For the M109A2 the company manufactured the gun and mount assembly, installing the turret and completing system integration before delivering the complete vehicle.

RDM Defense Engineering M109A2 before delivery to Dutch Army

Manufacturer: RDM Defense Engineering
PO Box 913
3000 AX Rotterdam
Netherlands
Telephone: (31) (10) 487 9111
Telex: 20753
Fax: (31) (10) 487 2772

ROMANIA

Model 89 122 mm Self-propelled Howitzer

Development/Description

This is the latest artillery system to enter service with the Romanian Army and is very similar to the Soviet 122 mm M-1974 (2S1) self-propelled artillery system that entered service with the Soviet and Polish Armies in 1971.

The Model 89, or M-89 as it is also referred to, is essentially a new Romanian chassis that utilises suspension elements of the Romanian MLI-84 infantry combat vehicle (qv) which is the Romanian equivalent of the Soviet BMP-1.

The turret and hull are of all-welded steel armour construction. The driver is seated at the front of the vehicle on the left with the engine compartment to his right and the turret at the very rear.

The torsion bar suspension either side consists of seven single rubber tyred road wheels with the drive sprocket at the front, idler at the rear and three track return rollers.

Whereas the original Soviet 2S1 is propelled in the water by its tracks at a maximum speed of 4.5 km/h; the Romanian version is propelled in the water by two water jets which give a higher water speed as well as greater manoeuvrability whilst afloat.

The outlets for these water jets are in the hull rear, one either side of the large single piece door that opens to the left which is supplied with a firing port and a periscope above.

The turret of the Model 89 is manufactured under licence from the Soviet Union by Bulgaria, who in turn supply it to Romania.

Firing an OF-462 HE projectile, a maximum range of 15 300 m can be achieved, or 21 900 m using an HE rocket assisted projectile.

Like the 122 mm 2S1, there is a long stowage box on the left side of the turret and on the turret rear are slotted covers that cover the forward part of the suspension when afloat to improve water flow.

Status: It is believed that only a small quantity of these, perhaps as low as 12, have been completed.

Manufacturer: Chassis and vehicle integration, Romanian state factories, turret, Bulgarian state factories.

Romanian 122 mm Model 89 self-propelled howitzer showing water jets at rear

Romanian 122 mm Model 89 self-propelled howitzer

SOUTH AFRICA

ARMSCOR G6 155 mm Self-propelled Howitzer

Development

In the 1960s the South African Artillery deployed three main artillery weapons, the British 25-pounder Field Gun (G1), the 5.5-inch Medium Gun (G2) and the Canadian-built Sexton 25-pounder self-propelled gun.

During early South African operations in Angola the lack of range of South African artillery compared to opposing Soviet guns and multiple rocket launchers led to the fielding of the 155 mm G5 gun howitzer and the 127 mm (24-round) Valkiri multiple rocket system.

While the towed G5 is perfect for many South African Artillery requirements it does not have the cross-country mobility of the Ratel (6 × 6) infantry vehicle which has been in service since 1976.

The South Africans weighed the options of wheels or tracks for their new self-propelled artillery system and came down firmly in favour of wheels. The possible areas of operations of the G6 are likely to be some distance from their main bases, perhaps up to 1000 km and tracked vehicles are not only slower on hard roads but more liable to breakdown and are very noisy internally. Fuel consumption of a wheeled vehicle is a lot less than comparable tracked vehicles so less logistic support is required. Finally, South Africa has extensive experience of building wheeled armoured vehicles going back some 50 years. The decision was then taken to go for a wheeled artillery system.

Development work was carried out in South Africa with Lyttleton Engineering Works (LEW) being prime contractor, with considerable assistance from other Armscor subsidiaries and contractors in the private sector.

The G6 prototype was completed in 1981 and shown for the first time in public in September 1982. By late 1986 four complete G6s had been built plus one additional turret.

Of these four, two can be considered as prototypes (one of mild steel and one armoured) and two as advanced development vehicles for they incorporated many improvements as the result of user trials and changes in operational requirements.

The first advanced development vehicle was completed early in 1984 and the second during 1986. Significant changes included permanent 6 × 6 drive, four grenade launchers either side of the turret, increased armour protection, fully independent suspension all round (the prototypes had a walking beam rear suspension), a redesigned turret, modified ordnance with a new muzzle brake, powerpack (consisting of engine, transmission and cooling system) on a sub-frame for ease of maintenance and replacement, a new fire control system, improved tyres with stronger side walls and redesigned ammunition stowage.

Four engineering development models were completed during 1987 for final troop trials and confirmation of the modifications carried out on earlier vehicles.

The first production G6s were completed during 1988 with Lyttleton Engineering Works (LEW) again being the prime contractor. Sandock Austral delivers the chassis to LEW which will integrate the turret and delivering the complete system to the user.

As with the earlier G5, the G6 is a complete system and not only includes the vehicle but also the ammunition system, meteorological station, muzzle velocity analyser, fire control system and a special artillery helmet radio communication system. Some of these systems have been in production for the G5 for some years. This support equipment is continuously upgraded or replaced by improved products.

With the South African Artillery the G6 is deployed in batteries of eight with each regiment having three batteries. For many roles, batteries will probably be split into two troops, each with four G6s.

Three of the advanced development G6 155 mm self-propelled howitzers were successfully used during the final South African campaign in Angola.

In 1990, Abu Dhabi, part of the United Arab Emirates, placed an order with Armscor for the supply of 78 G6 155 mm self-propelled howitzers with the first of these being delivered early in 1991. It is believed that these will be organised into three regiments each of 24 weapons, with the remaining weapons being used for training and war reserve.

One prototype of 155 mm G6 self-propelled howitzer in travelling configuration with hatches closed and 155 mm ordnance in travelling lock at front of hull

Description

The hull of the G6 is all-welded steel armour with protection against a variety of threats including artillery splinters. The G6 now has a double armoured floor for increased protection against mines.

The chassis is essentially divided up into three compartments, driver's at the front, engine in the centre and fighting compartment at the rear.

The driver sits at the front between the front two road wheels and as there is no hull armour above the latter any blast resulting from the vehicle running over a mine will be vented upwards and away from the driver.

To the front and sides of the driver are large bullet-proof windows which afford excellent visibility through the front 180°. If required the central front window can be covered by an armoured shutter hinged at the bottom simply by activating a lever. When this is raised the driver observes terrain through a periscope. He enters and leaves the vehicle via a roof hatch that opens to the left.

The driver's position has been completely redesigned, allowing optimum visibility of and accessibility to all controls. Key vehicle automotive characteristics, for example low oil pressure or fuel running short, are constantly monitored and the driver is alerted by visual and audible alarms before any serious damage occurs.

Steering is power assisted and the G6 now has permanent 6 × 6 drive (prototypes had the option of 6 × 4 with front axle being disengaged) and the driver can select either fully automatic or semi-automatic drive.

Forward of the driver's compartment is a lateral wedge-shaped box that doubles as a container for 14 projectiles (seven each side) and as a bush-clearing device capable of cutting through trees and shrubs.

The engine compartment is to the rear of the driver's position with air-inlet louvres in the front of the engine through the roof and the air-outlet louvres in the sides, all of which can be opened to allow access to the engine for maintenance. The engine compartment is fitted with a fire detection and suppression system.

The G6 is powered by an air-cooled diesel developing 525 hp which is coupled to an automatic transmission with a torque converter giving a total of six speeds forward and one in reverse. On the prototype the transmission was to the rear of the engine and protruded into the fighting compartment, but on production vehicles it is parallel to the engine.

Drive shafts take the power from the transmission via differentials to each road wheel station.

The crew compartment with the turret above has seats for the commander, layer, breech operator, loader and an ammunition handler and there is an escape hatch in the floor.

On prototypes the turret crew entered the G6 via a door in the rear of the vehicle, but following a major re-design they can now enter via the main door in the rear right-hand side of the turret or via two roof hatches in the turret, one either side.

The turret has four firing ports, two either side, with vision blocks above for close-in defence. Standard 5.56 mm R4 rifles and grenades are carried for close defence.

Mounted either side of the forward part of the turret is a bank of four 81 mm electrically operated grenade launchers which fire smoke grenades. Late production G6s have brush guards provided for these smoke dischargers.

The layer sits to the left of the breech and has an optic-mechanical indirect sight system of the digital type mounted directly on the trunnion with a compensating system for trunnion cant and an optical telescopic sight for direct firings. As an option an optic-mechanical direct sight which is provided with a laser rangefinder, day sight, passive night sight and a ballistic fire computer can be provided. The direct fire sight has a range of up to 5000 m.

Sophisticated electronic navigation and north-finding equipment is currently under development for integration into the fire control system of the G6. This will give the gun full autonomous navigation capability, thereby significantly decreasing the reaction time and greatly simplifying the laying procedure.

The commander sits on the right and uses a cupola with vision blocks for all-round observation. Inside of the fighting compartment braking controls are provided for use should the driver be disabled. The 12.7 mm HB or a 7.62 mm AA MG was originally mounted on the commander's cupola but has now been moved to the left cupola to allow the commander to concentrate on commanding the gun.

The auxiliary power unit (APU) is mounted in the rear side of the turret bustle to charge batteries and drive the air-conditioning system. The latter can be fitted with a full BC system if required. Radios are also mounted in the turret bustle.

In addition to the previously mentioned sights, the G6 is coupled to the battery fire control computer which provides laying information and ammunition details (projectile, charge, fuze) to each G6 where it is displayed on the gun display unit (GDU), for the commander. Information is relayed to the GDU by radio or line from the battery HQ.

The ordnance can be fitted with a gun monitoring system indicating number of rounds fired, recoil length and equivalent full charges (EFC) fired. A thermal warning device is incorporated to indicate high barrel temperatures.

The 45-calibre autofrettaged ordnance of the G6 is a modified version of that used in the towed G5 and has a single-baffle muzzle brake, a semi-automatic screw breech mechanism and an electrical firing system. A fume extractor of reinforced epoxy is fitted some two thirds of the way along the ordnance. The recoil system consists of a buffer, recuperator and replenisher.

When travelling the turret is traversed to the front and the ordnance held in a travelling lock on the hull top to the rear of the driver's position which can be unclamped without anyone leaving the vehicle.

The turret can be traversed through a full 180° (90° either side) under electro-hydraulic control with manual controls provided for emergency use, although for practical purposes the firing arc of the turret is 40° left and right. The 155 mm ordnance can be elevated from -5° to +75° under electro-hydraulic control with manual back-up provided.

A semi-automatic hydraulic flick rammer is mounted on the left of the breech with the projectile being placed on the rammer tray by hand. This is then aligned with the breech and rammed. The rammer can be used at all elevation angles and is operated by the loader. Propelling charges are loaded directly into the breech by hand.

A well trained crew can fire 4 rds/min and firing with the maximum charge can be continued for a period of up to 15 minutes, after which chamber temperatures have to be monitored to prevent possible cook-offs. A thermal warning device is installed to warn the crew of high barrel temperature conditions. A burst rate of fire of three rounds in 21 seconds can be achieved. The G6 takes 60 seconds to come into action and 30 seconds to come out of action.

The G6 fires the same M1 series of extended range full bore (ERFB) projectiles as the towed G5 with a three-charge, six-zone type charge

Latest production G6 with brush guard protectors for smoke dischargers (Christopher F Foss)

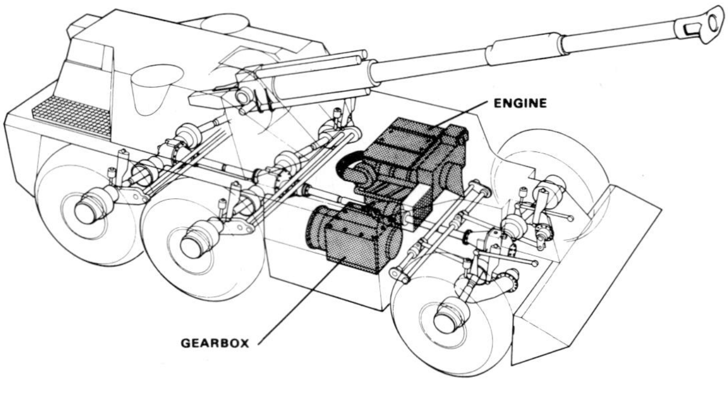

G6 155 mm self-propelled howitzer showing position of main components

Latest production G6 from rear with turret door open (Christopher F Foss)

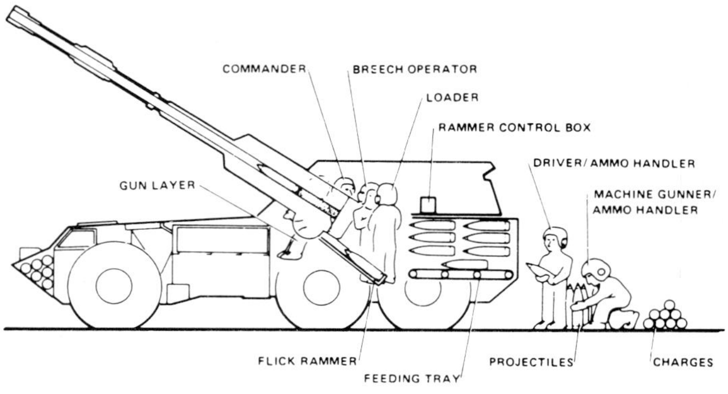

Ammunition storage and handling system of G6 155 mm self-propelled howitzer

Mounted under the hull rear of the G6 155 mm self-propelled howitzer are two stabilisers which can be quickly lowered

system packed in linen bags. Following successful trials a combustible cartridge case commenced volume production in 1987 for use with both G5 and G6 systems. The new charge system comprises five zones made up of three charges (combustible cartridge cases) and is optimised for use with 45-calibre ballistic systems.

Types of ammunition currently in production are HE, submunition, smoke, illuminating, leaflet and RP. The cargo rounds are of the base ejection type. All rounds can be of the standard or base bleed type.

M1 HE projectiles are of the high fragmentation type with welded splines. One type uses AISI 9260 steel filled with TNT which fragments into 3032 pieces, the other has a Composition B filling and fragments into 4756 pieces. Basic data for the projectiles are as follows:

Type	**HE**	**HE BB**	**Smoke**	**Ill**	**RP**	**Leaflet**
LENGTH	938 mm	958 mm	938 mm	938 mm	938 mm	938 mm
WEIGHT	45.5 kg	47 kg	43.5 kg	43.9 kg	43.5 kg	43 kg
PAYLOAD	8.7 kg	8.7 kg	13.2 kg	13 kg	13 kg	13.1 kg

The length of all projectiles includes the nose-mounted fuze. The Illuminating projectile has 1.5 million (for a minimum of 90 seconds) candlepower while the Smoke (SCR SMK) produces smoke for 90 seconds minimum from four canisters.

The cargo, or submunition round, contains 56 bomblets which cover an area of 30 × 100 m and self-destruct on the ground after three seconds if they do not impact a target. This round is sometimes referred to as the CLSTR. The leaflet round is also known as the Propaganda round (PROP).

Using Charge 3 M53A1 the HE projectile can be fired to a maximum range of 30 800 m while the base bleed round goes out to 39 000 m, or slightly more at altitude and in higher temperatures.

At 75 per cent of maximum range, the 50 per cent zone for dispersion is 0.96 per cent of range in length and less than two mils in breadth.

All 155 mm projectiles and charges are stowed below the turret ring for safety. It takes three men 15 minutes to reload the G6 with 45 projectiles and 50 charges.

The G6's latest advanced development is a major improvement in that fuzed projectiles and correct charges are carried from an ammunition support vehicle through the rear of the G6 on conveyor chutes, for example one for the projectile and one for the charge. In action, the G6 would normally use ammunition from its support vehicle so it would always be able to retain its basic load of ammunition.

The G6 can be fired with no preparation, but gun relaying is faster when the four stabiliser legs are lowered hydraulically to the ground by the driver or commander, two are at the rear and one between the first and second wheels either side.

The rear crew compartment is provided with an explosion detection and suppression system with no less than four outlets provided for the suppressant around the turret. Blow out panels are provided in the hull rear and the stowage of all ammunition below the turret also assists in the improved survivability of the G6.

The driver's compartment is provided with forced ventilation, while the rear fighting compartment has full air-conditioning.

The production G6 has permanent 6 × 6 drive and diff-locks on all differentials for improved traction and is fitted with a central tyre pressure regulation system that enables the driver to adjust the tyre pressure to suit the type of ground being crossed while moving.

The original walking beam rear suspension has now been replaced by independent suspension with torsion bars, hydraulic shock absorbers and bump stops all round. The 21.00 × 25 tyres are fitted with run flat elements.

Power assisted brakes are provided for all six wheels. There is also a crawl brake, a retarder on the transmission, an exhaust brake and a parking brake.

The G6 is supported in action by a cross-country vehicle carrying additional ammunition and support equipment.

Latest production G6s incorporate a number of improvements and these are being back fitted to earlier vehicles as they come back for overhaul. The driver has been provided with his own separate air-conditioning system which is installed in the nose of the vehicle. A new and upgraded turret air-conditioning system has been installed with increased capacity and greater reliability. The auxiliary power unit is not only more powerful but has increased RAM-D. The biological/chemical protection system is now integral in the turret bustle and is of the overpressure type with improved seals with the filters being changed from outside the vehicle. The recoil system has been upgraded and is now a more engineered system with improved RAM-D and improved mean time to repair. A new back-up laying system has been installed which is cheaper and has its own computer and sensors. A barrel cool down system has been installed, this is a blow type system run from the vehicle air-conditioning system and enables a higher rate of fire to be obtained. Finally a number of more general improvements have been incorporated including improvements to the structure, better corrosion protection and improved vision from the turret.

Status: In production, in service with Abu Dhabi (United Arab Emirates) and South Africa.

Manufacturer: Enquiries to Armscor, Private Bag X337, Pretoria 0001, South Africa. Telephone: (012) 428 1911. Telex: 320217. Fax: (012) 428 5635

SPECIFICATIONS		FUEL CAPACITY	700 l	BATTERIES	
CREW	6	ROAD RANGE	700 km	hull	2 × 24 V, 175 Ah
CONFIGURATION	6 × 6	FORDING	1 m	turret	4 × 24 V, 370 Ah
COMBAT WEIGHT	46 000 kg	GRADIENT	50%	ARMAMENT	1 × 155 mm howitzer
UNLOADED WEIGHT	42 500 kg	SIDE SLOPE	30%		1 × 12.7 mm M2 HB or
POWER-TO-WEIGHT		VERTICAL OBSTACLE	0.50 m		1 × 7.62 mm AA MG
RATIO	11.4 hp/tonne	TRENCH	1 m		2 × 4 grenade
LENGTH		TURNING RADIUS	12.5 m		launchers
gun forwards	10.335 m	ENGINE	air-cooled diesel	AMMUNITION	45 × 155 mm
hull	9.2 m		developing 525 hp		projectiles
WIDTH	3.4 m	TRANSMISSION	automatic with torque		50 × 155 mm charges
HEIGHT			converter providing 6	TURRET TRAVERSE	180° (but only 90°
including MG	3.8 m		forward and 1 reverse		used)
turret top	3.3 m		gears	WEAPON ELEVATION	−5° to +75°
fighting compartment top	1.9 m	SUSPENSION	independent with		
GROUND CLEARANCE	0.45 m		torsion bars and		
TRACK	2.8 m		hydro-pneumatic		
MAX SPEED			damping system		
road	90 km/h	TYRES	21.00 × 25 (run flat)		
cruising	85 km/h	ELECTRICAL SYSTEM	24 V		

SWEDEN

Bofors 155 mm Bandkanon 1A Self-propelled Gun

Development

The 155 mm Bandkanon 1A was developed by Bofors in the late 1950s with the first prototype being completed in 1960. The prototype had a slightly different chassis from later vehicles and its suspension consisted of six road wheels with the drive sprocket at the front, idler at the rear and three track return rollers. In 1965 Bofors was awarded a production contract for the vehicle worth SEK 60 million. The first production vehicle was completed the following year and production continued until 1968. The 155 mm Bandkanon 1A uses many components of the Bofors S-tank including the suspension, powerpack and transmission. The single prototype was followed by 26 production vehicles which are deployed in northern Sweden.

In August 1988 the Defence Material Administration of Sweden (FMV) awarded a contract to Bofors for the modernisation of the Swedish Army's 155 mm Bandkanon 1As. Under the terms of the contract Bofors will replace the existing powerpack with a new unit consisting of a Detroit Diesel 6V-53T developing 290 hp coupled to a new automatic transmission developed and produced by Bofors. (This powerpack is the same as that being installed in Strv 103B S-tanks). The contract, the first part of which is worth SEK 75 million including training and maintenance documentation, followed Swedish Artillery trials held over a 12-month period using a prototype fitted with the new powerpack.

Description

The hull of the vehicle is made of all welded steel with the engine and transmission at the front, the driver seated immediately in front of the turret and the turret at the rear.

The powerpack consists of a diesel and a gas turbine geared together to a common output. The diesel is normally used all the time, the gas turbine being engaged when driving across country. The transmission consists of a Volvo torque converter coupled to a gearbox with two forward and two reverse gears (one gear is for road use and the other for cross-country). The steering system is regenerative double-differential with a hydrostatic system and steering is by handlebars on the tiller column. The powerpack, complete with the transmission, can be removed from the vehicle as a complete unit to facilitate replacement in the field. The cooling system for the diesel is on the right side and the cooler for the steering system on the left.

Existing powerpacks are being replaced by new units consisting of a Detroit Diesel model 6V-53T developing 290 hp coupled to a new automatic transmission developed and produced by Bofors.

The driver, who sits slightly to the left of the vehicle's centre line, has a single piece hatch cover in front of which is a periscope.

The all welded turret is in two halves, left and right, both of which have a door in the front and three vision blocks, two for observation to the front and one to the side of the vehicle. In the left half of the turret are the commander, gun layer and radio operator and in the right half is the loader and anti-aircraft machine gunner. Both turret halves have a single piece hatch cover.

The hydro-pneumatic suspension can be locked in position when the gun is being fired to provide a more stable firing platform. The suspension consists of six dual rubber-tyred road wheels with the drive sprocket at the front and the last road wheel acting as the idler. There are no track return rollers.

The L/50 155 mm gun weighs 4800 kg complete with muzzle brake and replaceable liner and is 7.75 m long. The liner itself weighs 1340 kg.

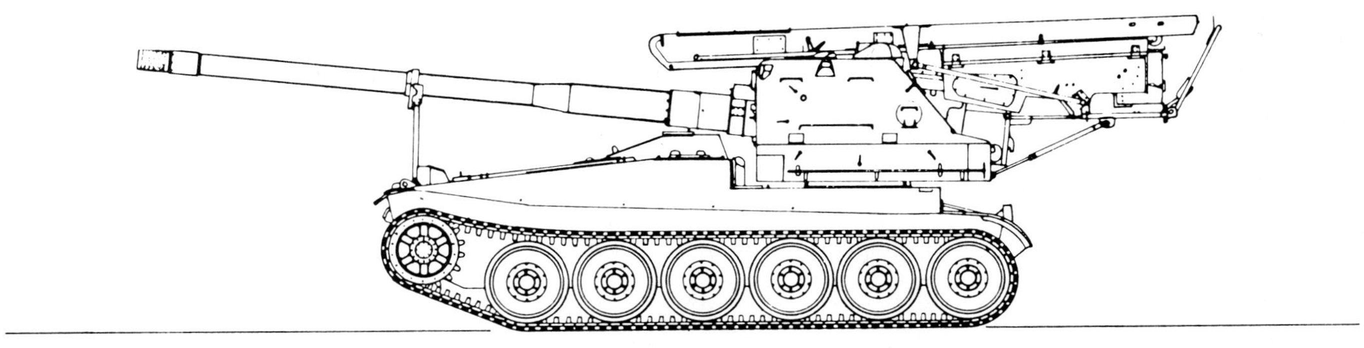

Bofors-designed 155 mm Bandkanon 1A self-propelled gun in travelling order

155 mm Bandkanon 1A self-propelled gun in travelling position

155 mm Bandkanon 1A from rear showing ammunition handling system

The gun barrel has a pepperpot muzzle brake and a semi-automatic wedge-shaped breech-block that opens downwards. When travelling the barrel is held by a lock at the front of the hull.

The elevating mechanism is counter-balanced by a spring balancing gear mounted in the pivot of the cradle. The elevating device consists of a hand wheel mounted in the left cabin, which operates a worm gear on the elevating pinion shaft. The elevating device also has an auxiliary electric motor for rapid laying between the loading and firing position. The control lever for the auxiliary motor is in the left cabin and the motor itself is in the right cabin.

The gun can be elevated by the electric motor from +2° to +38° at 5°/s. Manual elevation is from -3° to +40°. Traverse is manual, 15° left and 15° right at an elevation of above 0° and a traverse of 15° left and 4° right when the barrel is below 0°.

The gun is fed from a 14-round clip. The loading device consists of a loading tray which is suspended on rollers beneath the magazine. The loading tray transports the ammunition from the magazine to the ramming position, from where it is rammed by a rammer built into the loading tray. The loading tray, as well as the rammer, is operated by springs which are cocked by the run-out of the gun and the movement of the loading tray. The rounds are fed to the loading tray by alternating spring-operated feeding trays, one each side of the loading tray. Maximum rate of fire is 14 rounds in 48 seconds.

The magazine itself consists of seven drop compartments, holding two rounds each, from where the rounds drop down onto the feeding trays. Both the magazine and the feeding devices are protected by armour and the two armour-plated doors on the top of the magazine, which open vertically for reloading, are operated by a single electric motor. Reloading with 14 rounds takes two minutes.

When the magazine has been expended the gun is elevated to an angle of +38°, the two covers on top of the magazine are opened vertically, and the hoist on top of the turret slides along to the end of the slide bar where it picks up a new clip of ammunition. This is placed in the magazine, the doors closed and the gun is then ready to fire again. The complete loading sequence takes only two minutes. The first round is manually loaded after which loading is automatic. The gunner can select either single shots or fully automatic. The gun fires an HE projectile weighing 48 kg to a maximum range of 25 600 m. Maximum muzzle velocity with charge 3 is 865 m/s and with charge 1 is 600 m/s. The sight has a magnification of ×4 and a 10° field-of-view.

Status: Production complete. In service with Swedish Army.

Manufacturer: AB Bofors (now Swedish Ordnance)
Box 500, S-691 80 Bofors
Sweden
Telephone: (46) 586 81000
Telex: 73210
Fax: (46) 586 58145

SPECIFICATIONS

CREW	5
COMBAT WEIGHT	53 000 kg
POWER-TO-WEIGHT RATIO	10.18 hp/tonne
GROUND PRESSURE	0.9 kg/cm²
LENGTH	
including gun	11 m
without gun or loading arm	6.55 m
WIDTH	3.37 m
over tracks	3.26 m
HEIGHT	
with AA MG	3.85 m
to turret top	3.25 m
GROUND CLEARANCE (hull centre)	0.37 m
TRACK	2.59 m
TRACK WIDTH	670 mm
LENGTH OF TRACK ON GROUND	4.3 m
MAX ROAD SPEED	
direct gear	28 km/h
cross-country gear	9 km/h
FUEL CAPACITY	1445 l
MAX ROAD RANGE	230 km
direct gear	28 km/h
cross-country gear	9 km/h

FUEL CAPACITY	1445 l
MAX ROAD RANGE	230 km
FORDING	1 m
GRADIENT	60%
SIDE SLOPE	30%
VERTICAL OBSTACLE	0.95 m
TRENCH	2 m
TURNING RADIUS	
direct gear	40 m
cross-country gear	13 m
ENGINES	Rolls-Royce K60 diesel developing 240 hp at 3750 rpm or Detroit Diesel 6V-53T developing 290 hp at 2800 rpm. Boeing Model 502/10MA gas turbine developing 300 shp at 38 000 rpm
TRANSMISSION	Volvo FBTV-2 with direct and cross-country speeds or Bofors automatic
STEERING	regenerative double-differential with hydrostatic steering drive

SUSPENSION	hydro-pneumatic
ELECTRICAL SYSTEM	24 V
BATTERIES	4 × 12 V
ARMAMENT	
main	1 × 155 mm gun
anti-aircraft	1 × 7.62 mm MG
SMOKE-LAYING EQUIPMENT	none
AMMUNITION (main)	14
GUN CONTROL EQUIPMENT	
Turret power control	manual
by commander	no
by gunner	yes
Max rate of traverse	0.25°/turn of wheel
Max rate of elevation	
electric	5°/s
manual	-3° to +40°
Gun traverse	
0° elevation and above	15° left and 15° right
below 0° elevation	15° left and 4° right
ARMOUR	
Hull	10-15 mm
Turret	20 mm
Magazine sides	20 mm
Magazine covers	10 mm
Magazine bottom	15 mm

SWITZERLAND

SFAW 155 mm M109 Upgrade

Development/Description
Based on the 155 mm Bison Fortress Gun, fully covered in the *Coastal artillery guns and missiles* section, the Swiss Federal Armament Works (SFAW) is proposing an upgrade programme for the 155 mm M109 self-propelled howitzer which is used by the Swiss Army in significant quantities.

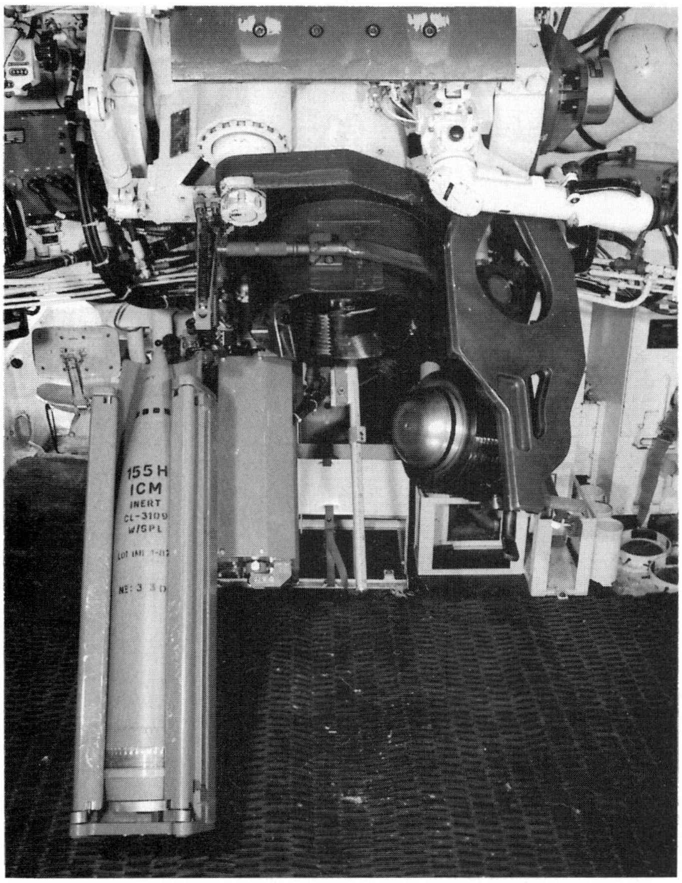

Interior of upgraded M109A1B of the Swiss Army showing installation of flick rammer (left) with breech mechanism open

By retrofitting a 47-calibre barrel, the firing range can be increased by some 40 per cent and according to SFAW this is an optional solution in terms of cost effectiveness.

The recoil force is practically unchanged and the elevating mass is only 100 kg heavier than for the standard barrel, therefore the suspension and chassis are hardly affected.

SFAW has also increased the number of projectiles and charges that can be carried to 42 as it is considered that the original quantity of 28 rounds is insufficient for the variety of ammunition types used today. The additional ammunition supply is carried in the extended turret bustle which is separated from the crew compartment by a blast-proof sliding door.

The effective engagement of fast moving targets requires a higher rate of fire which is achieved by the installation of a flick rammer which enables a burst of three rounds to be fired in 10 to 12 seconds. The semi-automatic loading rammer can operate at any quadrant angle.

Swiss Army M109s

In 1968 the Swiss Parliament voted to place an order for 146 M109s which subsequently entered service under the designation of the Panzerhaubitze 66. In 1974, an additional 120 M109A1Bs were obtained and at the same time the existing M109s were upgraded to the M109A1B configuration by the installation of a longer barrel.

The new M109A1Bs are known as the Panzerhaubitze 74 and retrofitted vehicles as the Panzerhaubitze 66/74.

The 1979 armament programme called for the purchase of a further batch of about 207 M109A1Bs to enable the Swiss Army to field a total of nine M109 series regiments. The last batch of vehicles are known as the Panzerhaubitze 79.

Swiss M109s have a bank of three electrically operated smoke dischargers mounted on either side of the turret firing forwards and German Diehl tracks for improved traction.

Future upgraded M109s are expected to be designated the Panzerhaubitze 88.

Status: Development complete. Ready for production.

Manufacturer: Swiss Federal Armament Works (SFAW)
CH-3602 Thun
Allmendstrasse 86
Switzerland
Telephone: (033) 28 11 11
Telex: 921 256
Fax: (033) 28 20 47

Swiss Army M109A1Bs fitted with new 47-calibre ordnance, extended turret bustle for increased ammunition stowage and flick rammer

TAIWAN

XT-69 155 mm Self-propelled Howitzer

Development/Description
The XT-69 155 mm self-propelled howitzer was developed by the Combined Service Forces (CSF) with the assistance of a number of civilian companies including the Taiwan Aluminium Corporation (hull and tracks) and the Taiwan Machinery Manufacturing Corporation (chassis and parts of the weapon and possibly the engine and suspension). The Hsing-Ho Arsenal was responsible for the hydraulics and the Chi-Ning Arsenal for the ordnance. Both of the arsenals come under the control of the Combined Service Forces.

The XT-69 is basically the American 155 mm M109/105 mm M108 self-propelled howitzer chassis with the fully enclosed turret removed and the 155 mm howitzer mounted in the open topped compartment at the rear of the hull. This has no overhead armour protection and therefore no NBC system.

The 155 mm howitzer is a modified version of the American M114 towed system and uses the same ordnance but fitted with a multi-baffle muzzle brake. A travelling lock is provided for the ordnance on the glacis plate and this folds down when not required. When originally introduced the ordnance had no muzzle brake.

A ramp is fitted at the rear of the vehicle and is used both as a means of exit and for ease of ammunition loading. Hydraulic spades at the rear of the hull are lowered to the ground before firing begins.

It is assumed that the weapon has the same traverse of 24° left and 25° right and elevation from -2° to +63° as the towed weapon, there is a possibility that traverse and elevation are hydraulic with manual controls for emergency use.

Maximum range of the M114 is 14 600 m although some Taiwanese sources have given the XT-69 a range of 15 000 m. It is estimated that 25 projectiles and charges are carried and that there is a crew of five.

XT-69 155 mm self-propelled howitzer with muzzle brake (L J Lamb)

Physical and automotive characteristics of the XT-69 are similar to the American M109 described later in this section except for a lower height and weight. Standard equipment includes infra-red driving lights and a 12.7 mm M2 HB anti-aircraft machine gun mounted on the right side to the rear of the howitzer.

Status: Production of this system is understood to have been completed and in time these will probably be upgraded to the XT-69 155 mm self-propelled extended range gun. In service with Taiwan.

Manufacturer: Combined Service Forces, Taiwan.

XT-69 155 mm Self-propelled Extended Range Gun

Development/Description
This was developed by the Combined Service Forces and is identical in layout and concept to the basic XT-69 described in the previous entry.

It is fitted with a 155 mm 45-calibre weapon which is said to have been developed from the South African 155 mm G5 or the SRC International GC 45, although the elevating mass is more like the Soltam 155 mm M-71 gun/howitzer than these two weapons.

The ordnance has a multi-baffle muzzle brake, fume extractor and a travelling lock that folds back onto the glacis plate when not in use. Maximum range is said to be 30 000 m with the Extended Range Full Bore projectile, and over 35 000 m with a base bleed projectile. A maximum rate of fire of 6 rds/min can be attained due to a pneumatically assisted loading device which enables the weapon to be loaded at any angle of elevation and also gives uniformity from shot to shot. Elevation is estimated to be from 0° to +70° and traverse 30° left and right. The weapon is used in conjunction with a battery fire control system developed by the Sun Yat-sen Scientific Research Institute in Taiwan.

This system has a crew of five and is fitted with infra-red night vision equipment for the driver and a 12.7 mm M2 HB anti-aircraft machine gun. No NBC system is fitted.

Status: In service with Taiwan.

Manufacturer: Combined Service Forces, Taiwan.

XT-69 155 mm self-propelled extended range gun in travelling order (L J Lamb)

UNION OF SOVIET SOCIALIST REPUBLICS

240 mm Self-propelled Mortar M-1975 (SM-240) (2S4)

Development

The 240 mm self-propelled mortar known to the West as the M-1975 is known as the SM-240 (2S4) by the Soviet Army, although its more common name is the Tyulpan, or Tulip Tree. These weapons are normally retained at front or high command control levels for special purposes, as the M-1975 can fire a nuclear projectile. This system reached a full operational capability with the Soviet Army in 1975, although it has yet to appear in public. It is believed that each 240 mm mortar regiment has between 34 and 56 systems.

It is estimated that well over 400 of these systems have been completed, although the Soviet Union has not released any production figures. Under the Conventional Forces Europe (CFE) Treaty signed in 1990, the Soviet Union admitted having the following quantities of vehicles in the West:

Region	**IV.1**	**IV.2**	**IV.3**	**IV.4**	**V.1**
Quantity	**66**	**18**	**1**	**nil**	**48**

The above figures were the situation as of November 1990. For the Conventional Forces Europe Treaty, Region IV is defined as Central Europe while Region V covers the flanks of the Atlantic to the Urals (ATTU).

In addition to being used by the USSR the 240 mm weapon has also been identified to be in service with Czechoslovakia, Iraq and the Lebanon. The latter are supplied to the Christian Army via Iraq.

Description

The M-1975 consists of a much-modified GMZ tracked minelaying vehicle (see *Jane's Military Vehicles and Logistics 1991-92*, page 214) carrying a 240 mm M-240 breech-loading mortar (see entry in the *Towed guns and howitzers* section in this edition) on the hull rear. The mortar is carried complete with a baseplate and is hydraulically lowered from its travelling position around a pivot on the hull rear. The baseplate is hinged to the hull rear so that when emplaced the mortar barrel faces away from the hull. The 240 mm mortar is lowered into the firing position under remote-control and when in position can be elevated from +45° to +80° with a traverse of 8° left and right. Some rounds are probably carried inside the vehicle hull and it is probable that some form of assisted loading is provided. The rate of fire is probably about one round a minute, with firing being by remote-control.

Firing a standard 130 kg HE round designated the F-864, a minimum range of 800 m and a maximum range of 9700 m can be achieved. In addition there is understood to be an extended range round with a maximum range of 20 000 m. Other rounds include a chemical projectile, nuclear, and a concrete-piercing round for use against urban targets has been reported.

The hull of the vehicle is of all welded steel armour with the engine compartment at the front right and driver front left. The commander is seated to the rear of the driver and has a raised cupola with an externally mounted 12.7 mm anti-aircraft machine gun. A further hatch is to the rear of the commander's cupola with a similar outward opening hatch on the opposite side of the vehicle roof.

Ammunition is stowed internally and is loaded into the mortar using a hand-operated crane mounted at the rear of the vehicle on the left side. The mortar sighting system is on the right side of the mortar.

Suspension is of the torsion bar type with each side having six road wheels with the drive sprocket at the front, idler at the rear and four track return rollers.

Status: Production complete. In service with Czechoslovakia, Iraq, Lebanon and the Soviet Union.

Manufacturer: Soviet state factories.

240 mm self-propelled mortar M-1975 (SM-240) (2S4) deployed in firing position

Czech 2S4 self-propelled 240 mm mortar

240 mm self-propelled mortar M-1975 (SM-240) (2S4) in travelling configuration

203 mm Self-propelled Gun M-1975 (SO-203) (2S7)

Description

The Soviet Ground Forces began receiving the 203 mm Model 1975 (M-1975) self-propelled gun during 1975. It is estimated that well over 1000 of these systems have been completed, although the Soviet Union has not released any production figures. Under the Conventional Forces Europe (CFE) Treaty signed in 1990, the Soviet Union admitted having the following quantities of vehicles in the West:

Region	IV.1	IV.2	IV.3	IV.4	V.1
Quantity	311	212	148	nil	99

The above figures were the situation as of November 1990. For the Conventional Forces Europe Treaty, Region IV is defined as Central Europe while Region V covers the flanks of the Atlantic to the Urals (ATTU). The gun equips heavy artillery units at Front level and is capable of firing nuclear projectiles. It is believed that a 2S7 artillery regiment has a total of 24 weapons in three batteries each with eight weapons.

The M-1975 is based on a large chassis that is probably the largest armoured vehicle currently in the Soviet inventory. The chassis has an armoured cab at the front capable of carrying two to four crew, this also has an over pressure NBC system, and immediately behind the cab is the engine compartment. One report states that the engine is derived from that used on the T-80 MBT. The transmission is forward mounted below the cab to power the forward track drive sprockets. The suspension is of a new type with seven road wheels. There are a series of stowage bins mid-way along the hull and the gun is mounted at the rear in a conventional fashion. The gun is equipped with a form of power-assisted loading device with the operator seated in a 'crow's nest' position at the left rear corner of the vehicle. At the extreme rear is a large hydraulically operated recoil spade.

The M-1975 has no cover for the four-man gun crew when in action and it is possible that a second tracked vehicle carries the ammunition and the bulk of the gun crew. It is believed that the 203 mm weapon has a maximum range of 37 500 m using standard ammunition and 50 000 m using extended range ammunition. It is understood that the 2S7 has a maximum rate of fire of 2 rds/min and a sustained rate of fire of one round every two minutes. Types of ammunition fired include concrete piercing, high explosive fragmentation and tactical nuclear, although some reports have also mentioned a chemical round.

Status: Production. In service with Czechoslovakia and the Soviet Union.

Manufacturer: Soviet state factories.

203 mm M-1975 (SO-203) (2S7) self-propelled gun in travelling configuration

Soviet 203 mm M-1975 (SO-203) (2S7) self-propelled gun with spade lowered at rear and loading system on right side of gun mount

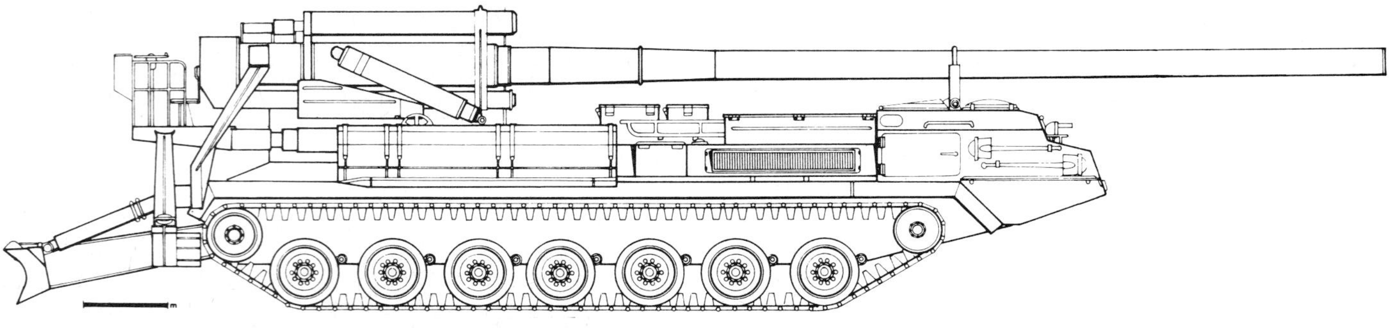

Provisional drawing of Soviet 203 mm M-1975 (SO-203) (2S7) self-propelled gun with ordnance in travelling lock and spade at rear lowered (not to 1/76th scale) (Steven Zaloga)

SPECIFICATIONS (provisional)

CREW (on vehicle)	4	LENGTH		RANGE	500 km
COMBAT WEIGHT	40 000 kg	overall	12.8 m	ENGINE	450 hp diesel
POWER-TO-WEIGHT RATIO	11.25 hp/tonne	chassis	10.5 m	GUN CONTROL EQUIPMENT	
		WIDTH	3.2 m	weapon control	manual/powered
		HEIGHT	2.8 m	elevation	0° to +55°
		GROUND CLEARANCE	0.4 m	traverse	30°
		MAX ROAD SPEED	50 km/h		

152 mm Self-propelled Gun (2S5)

Development/Description
Since 1978 the Soviet Union has deployed two new 152 mm guns, one self-propelled and one towed. The former has the Soviet industrial number of the 2S5 and is believed to have entered service with the Soviet Army in 1980. Both weapons use the same ordnance and full details of the 152 mm M-1976 (Soviet industrial number 2A36) towed weapon are given in the *Towed guns and howitzers* section.

It is estimated that well over 2000 of these systems have been completed, although the Soviet Union has not released any production figures. Under the Conventional Forces Europe (CFE) Treaty signed in 1990, the Soviet Union admitted having the following quantities of vehicles in the West:

Region	IV.1	IV.2	IV.3	IV.4	V.1
Quantity	499	497	398	253	2

The above figures were the situation as of November 1990. For the Conventional Forces Europe Treaty, Region IV is defined as Central Europe while Region V covers the flanks of the Atlantic to the Urals (ATTU).

The 152 mm ordnance is mounted at the rear of the hull with no protection being provided for the crew. Elevation and traverse limits are believed to be identical to those of the 2A36 towed weapon, ie traverse 25° left and right with elevation from -2.5° to +57°. The ordnance is fitted with a five part multi-slotted muzzle brake but no fume extractor is fitted. A large spade is lowered at the rear of the hull before firing commences. This weapon has a nuclear capability. Maximum range with a conventional HE round is 27 000 m, or 40 000 m with a rocket-assisted projectile. In addition to high explosive and tactical nuclear projectiles the 2S5 can also fire chemical, concrete piercing and improved conventional munitions. The 2S5 is commonly known as the Giatsint (Hyacinth) by the Soviet Army. The layer is seated at the rear of the hull on the left side and a projectile and charge ramming system is fitted to enable a high rate of fire to be achieved. When travelling the ordnance is held in a travel lock.

Status: Production. In service with Soviet Union.

Manufacturer: Soviet state factories.

152 mm self-propelled gun 2S5 deployed in firing position

152 mm self-propelled gun 2S5 deployed in firing position

SPECIFICATIONS (provisional)

CREW	6
COMBAT WEIGHT	30 000 kg
LENGTH GUN FORWARDS	8.8 m
LENGTH HULL	7.3 m
WIDTH	3.2 m
HEIGHT	2.9 m
GROUND CLEARANCE	0.4 m
MAX ROAD SPEED	50 km/h
FUEL CAPACITY	850 l
FORDING	1 m
GRADIENT	60%
SIDE SLOPE	30%
VERTICAL OBSTACLE	0.7 m
TRENCH	2 m
ARMAMENT	1 × 152 mm gun
GUN CONTROL EQUIPMENT	powered/manual
Gun elevation/depression	+57°/-2.5°
Gun traverse	25°, left and right
Rate of fire	6 rpm
ARMOUR	15 mm (max)

152 mm 2S19 Self-propelled Artillery System

Development
In 1990 the Soviet Union disclosed that it had developed and placed in production a new 152 mm self-propelled artillery system that has the industrial number of the 2S19, although it is also referred to as the MSTA-S.

The 2S19 uses the same ordnance as the MSTA-A 2A65 152 mm gun howitzer which is referred to by NATO as the M-1986.

The 152 mm 2S19 is believed to be the replacement for the 152 mm 2S3 which first entered service with the Soviet Union as far back as 1971.

Like many of the more recent Soviet Army systems, the 2S19 has yet to be seen in public, the description should be taken as provisional.

No production figures for the 2S19 are available, although under the Conventional Forces Europe (CFE) Treaty signed in 1990, the Soviet Union admitted it had deployed the following quantities of the 2S19 in the West:

Region	IV.1	IV.2	IV.3	IV.4	V.1
Quantity	14	11	8	nil	3

Description
The 2S19 is based on a full tracked MBT chassis with the driver at the front, fully enclosed turret in the centre and powerpack at the rear and in many respects is similar to the French 155 mm GCT self-propelled artillery system based on an AMX-30 MBT chassis.

The hull and turret is of all welded steel armour construction which provides protection from small arms fire and shell splinters, and, according to the Soviets, mines.

The suspension is of the torsion bar type with six road wheels, idler at the front, drive sprocket at the rear and track return rollers with the upper part of the track being covered by a skirt.

Main armament comprises a long barrelled 152 mm gun fitted with a fume extractor and a muzzle brake, when the 2S19 is travelling the ordnance is held in position by a travelling lock mounted on the glacis plate.

The 152 mm gun fires an HE-FRAG (high explosive fragmentation) projectile to a maximum range of 24 700 m, although using an extended range projectile its range is probably approaching 40 000 m. Other types of projectile can also be fired including illuminating, incendiary, smoke and tactical nuclear.

Turret traverse is powered through a full 360° with weapon elevation from -3° to +68°. According to the Soviets, the 2S19 has a maximum rate of fire of 8 rds/min, this being achieved as an automatic loading system is provided. This selects and then places the fuzed projectile into the breech with the charges being loaded semi-automatically.

In addition to selecting the projectile from the magazine, the automatic loading mechanism controls the number of rounds to be fired and has built-in test equipment. Laying in elevation is automatic with laying in traverse being semi-automatic.

A total of 50 projectiles and charges is carried. Spades are not required on the 2S19 as it is very stable when in the firing position.

If required the 2S19 can be reloaded through the turret rear while the system is firing in a similar manner to the French 155 mm GCT. This enables the system to move off to its next fire position with a full load of ammunition on board.

The commander's cupola is on the right side of the turret roof and is armed with a 12.7 mm machine gun which is remote-controlled from inside the turret and can be used to engage both ground and air targets. Three smoke dischargers are mounted either side of the turret.

A 16 kW autonomous gas turbine auxiliary power unit (APU) is provided in the turret and this provides power within 30 to 60 seconds of being switched on. This allows the system to be fully operational at temperatures ranging from -50° to +50° C with the main diesel engine being switched off to conserve fuel supplies.

Standard equipment includes an NBC system and night vision equipment for the driver.

Status: In production. In service with the USSR.

Manufacturer: Soviet state factories.

Scale model of the 152 mm 2S19 self-propelled gun

SPECIFICATIONS (provisional)

CREW	5	RANGE	500 km	SMOKE DISCHARGERS	6
WEIGHT	42 000 kg	FORDING	1 m	AMMUNITION	
POWER-TO-WEIGHT RATIO	20 hp/tonne	GRADIENT	60%	152 mm	50
LENGTH OVERALL	11.917 m	SIDE SLOPE	30%	12.7 mm	300
WIDTH	3.38 m	TRENCH	2.8 m	GUN CONTROL EQUIPMENT	
HEIGHT TURRET ROOF	2.985 m	ENGINE	840 hp diesel	Gun elevation/depression	+68°/-3°
MAX ROAD SPEED	60 km/h	SUSPENSION	torsion bar	Turret traverse	360°
		ARMAMENT			
		main	1 × 155 mm gun		
		anti-aircraft	1 × 12.7 mm MG		

152 mm Self-propelled Gun/Howitzer M-1973 (2S3)

Development

The correct Soviet designation for this weapon is SO-152 Akatsiya (Acacia) and it is now known that its industrial number is 2S3. It entered service with the Soviet Army in 1971 and in the absence of an official Soviet designation was called the 152 mm self-propelled gun/howitzer M-1973 by the West. The M-1973 is based on the chassis used for the SA-4 Ganef SAM system but has six instead of seven road wheels with a distinct gap between the first and second and second and third road wheels. In appearance it is very similar to the American M109 self-propelled howitzer which entered service in 1962.

It is issued on the scale of 18 per tank division and 18 per motorised rifle division and has now virtually replaced the 130 mm Field Gun M-46 in the Soviet Army Artillery Brigades and artillery divisions. The batteries originally had six guns giving the regiment a total of 18 weapons, but many batteries now have eight guns giving the regiment a total of 24 guns. A Soviet Front artillery division has five brigades: two howitzer brigades (one with 72 towed D-30 howitzers and one with 72 2S3 self-propelled howitzers), one gun brigade with 72 2S5 self-propelled guns, one multiple rocket launcher brigade with 72 BM-22 (8 × 8) rocket systems and one 2S7 self-propelled brigade with 72 guns. Late production models of the 2S3 have been referred to as the 2S3M.

It is estimated that well over 10 000 of these systems have been completed, although the Soviet Union has not released any production figures. Under the Conventional Forces Europe (CFE) Treaty signed in 1990, the Soviet Union admitted having the following quantities of vehicles in the West:

Region	IV.1	IV.2	IV.3	IV.4	V.1
Quantity	2010	1812	1598	1097	198

The above figures were the situation as of November 1990. For the Conventional Forces Europe Treaty, Region IV is defined as Central Europe while Region V covers the flanks of the Atlantic to the Urals (ATTU).

The replacement for the 152 mm self-propelled gun/howitzer M-1973 (2S3) is believed to be the 152 mm 2S19 covered in the previous entry.

Description

The all-welded steel hull of the M-1973 is divided into three compartments; driver's at the front on the left, engine to the right of the driver and the turret at the rear.

The driver has a single piece hatch cover that opens to the rear, in front of which are two periscopes. On the forward part of the glacis is a splash board to stop water rushing up the glacis plate when the vehicle is fording a stream. The air-inlet and air-outlet louvres are on the top of the hull with the exhaust outlet on the right side of the hull, just above the track guard.

The large all-welded turret has a sloped front and well-sloped sides with a vision block in each side. The commander is seated on the left of the turret with the gunner forward and below him and the loader on the right. The commander has a cupola which can be traversed through a full 360° and a single piece hatch cover that opens to the rear. Vision devices are provided around the lower part of his cupola. Mounted on the forward part of the hatch is a 7.62 mm machine gun which can probably be aimed and fired from inside the turret. There is an OU-3K infra-red/white searchlight mounted to the left of the machine gun. The gunner has a rather large periscope type sight which is mounted in the turret roof in front of the commander's cupola. An ammunition resupply hatch is provided in the right side of the turret and there is an oval hatch at the rear of the hull which opens downwards.

Main armament comprises a 152 mm weapon, designated the 2A33, which is based on the towed 152 mm D-20 gun/howitzer but with a bore evacuator added behind the muzzle brake. The barrel has a double-baffle muzzle brake, fume extractor and a gun barrel travelling lock. The normal projectile fired by the M-1973 is the HE-FRAG OF-540, which is fitted with an RGM-2 fuze, weighs 43.5 kg, contains 5.76 kg of TNT, and has a maximum muzzle velocity of 655 m/s and maximum range of 18 500 m. Other types of projectile fired include chemical, HEAT-FS (also referred to as HEAT-SS, with latter standing for spin stabilised), HE/RAP (high explosive rocket assisted projectile) with a range of 24 000 m, AP-T, illuminating, smoke, 0.2 kt tactical nuclear, incendiary, flechette,

152 mm self-propelled gun/howitzer M-1973 (2S3) captured by Iraq with turret traversed to front

152 mm self-propelled gun/howitzer M-1973 (2S3) from rear showing commander's cupola and turret details. This is a late production model and has no hatches in turret rear and different arrangement of hatches in the hull rear

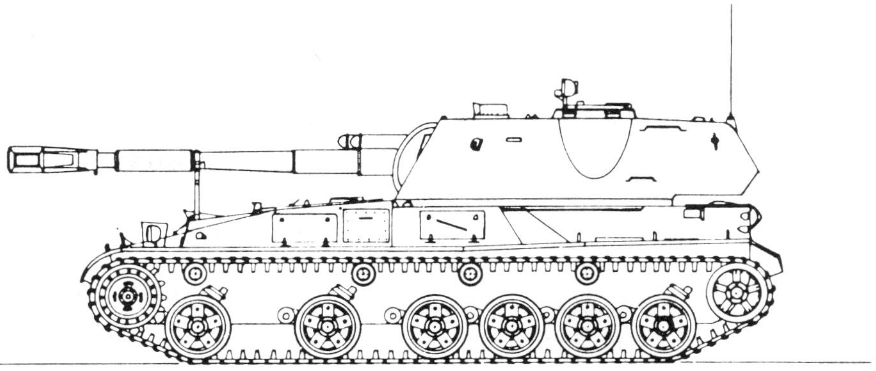

Provisional drawing of Soviet self-propelled gun/howitzer M-1973 (2S3)

scatterable mines (anti-tank and anti-personnel) and semi-active laser. The gun has an elevation of +60°, a depression of -3° and turret traverse is a full 360°. It is possible that some type of automatic loading system is fitted as there are two hatches in the rear of the turret similar to those on the Japanese Type 75 155 mm self-propelled howitzer. In the rear of the hull are two circular hatches, one either side of a hatch that opens downwards. To the rear of each of these is a short conveyor belt that takes the projectiles and charges further into the vehicle for loading into the ordnance.

The torsion bar suspension consists each side of six dual rubber-tyred road wheels with the drive sprocket at the front, idler at the rear and four return rollers. The first and last return rollers support the inside of the track only. The first, second and sixth road wheel stations have a hydraulic shock absorber and there is a distinct gap between the first and second and second and third road wheels.

The M-1973 has infra-red night vision equipment and is assumed to have an NBC system. It has no amphibious capability and normally carries an unditching beam at the rear of the hull.

M-1973 (2S3) 152 mm self-propelled gun/howitzer with ordnance in travelling lock

Variants

The chassis of the M-1973 (2S3) shares a number of common components with the SA-4 Ganef surface-to-air missile system, 240 mm self-propelled mortar M-1975 and the 152 mm self-propelled gun 2S5.

SPECIFICATIONS					
CREW	4 + 2 in ammunition carrier	MAX ROAD SPEED	62 km/h	RATE OF FIRE	
COMBAT WEIGHT	30 000 kg	FUEL CAPACITY	850 l	(max)	4 rds/min
POWER-TO-WEIGHT RATIO	17.33 hp/tonne	MAX RANGE	500 km	(sustained)	60 rds/min (first hour)
GROUND PRESSURE	0.6 kg/cm²	FORDING	1.5 m	UNIT OF FIRE	60 rounds
LENGTH GUN FORWARDS	8.4 m	GRADIENT	60%	AMMUNITION	
LENGTH HULL	7.8 m	SIDE SLOPE	30%	main	46
WIDTH	3.2 m	VERTICAL OBSTACLE	1.1 m	7.62 mm	1000
HEIGHT	2.8 m	TRENCH	2.5 m	GUN CONTROL EQUIPMENT	
GROUND CLEARANCE	0.4 m	ENGINE	V-59 V-12 water-cooled diesel developing 520 hp	Turret power control	powered/manual
LENGTH OF TRACK ON GROUND	4.94 m	SUSPENSION	torsion bar	Gun elevation/depression	+60°/-3°
		ARMAMENT		Turret traverse	360°
		main	1 × 152.4 mm gun	ARMOUR	
		anti-aircraft	1 × 7.62 mm MG	turret	15 mm (max)
				hull	20 mm (max)

Status: Production complete. In service with Hungary, Iraq, Libya, the Soviet Union (in which it is now being supplemented by a new 152 mm self-propelled howitzer designated the 2S19) and Syria.

Manufacturer: Soviet state factories.

122 mm Self-propelled Howitzer M-1974 (2S1)

Development

The correct military designation for this weapon is SO-122 but it is commonly known as the Gvozdika (Carnation). Its industrial number is 2S1 and NATO usually refers to it as the M-1974, the year it was first seen in public. It entered service with the Soviet and Polish armies in 1971 and was first seen in public during a parade held in Poland in July 1974. The M-1974 shares many automotive components with the MT-LB multi-purpose tracked armoured vehicle described in the *Armoured personnel carriers* section. In appearance it is very similar to the American M109 self-propelled howitzer which entered service in 1962.

The M-1974 is issued on the scale of 36 per artillery division, 36 per motorised rifle division and 72 per tank division.

Each 2S1 battalions originally had three batteries each with six guns, but many now have eight guns, giving the battalion a total of 24 systems.

It is estimated that well over 10 000 of these systems have been completed, although the Soviet Union has not released any production figures. Under the Conventional Forces Europe (CFE) Treaty signed in 1990, the Soviet Union admitted having the following quantities of vehicles in the West:

Region	**IV.1**	**IV.2**	**IV.3**	**IV.4**	**V.1**
Quantity	2309	2090	1910	1362	219

The above figures were the situation as of November 1990. For the Conventional Forces Europe Treaty, Region IV is defined as Central Europe while Region V covers the flanks of the Atlantic to the Urals (ATTU).

Description

The all-welded steel hull of the M-1974 is divided into three compartments: driver's at the front on the left, engine to the right of the driver and the turret at the rear.

The driver has a single piece hatch cover that opens to the rear and, in front of his position, a windscreen covered by a hatch hinged at the top. In the top of the hull, to the right of the driver's position, is a large engine access hatch that opens to the right. There is a smaller hatch on the right side of the glacis plate. The exhaust pipe is on the right side of the hull towards the front.

The welded turret, which does not overhang the side of the hull, has a well-sloped front and sides with the commander seated on the left,

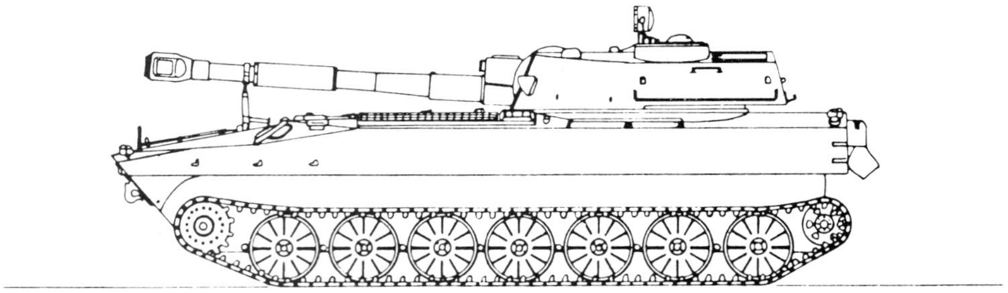

Provisional drawing of 122 mm self-propelled howitzer M-1974 (2S1)

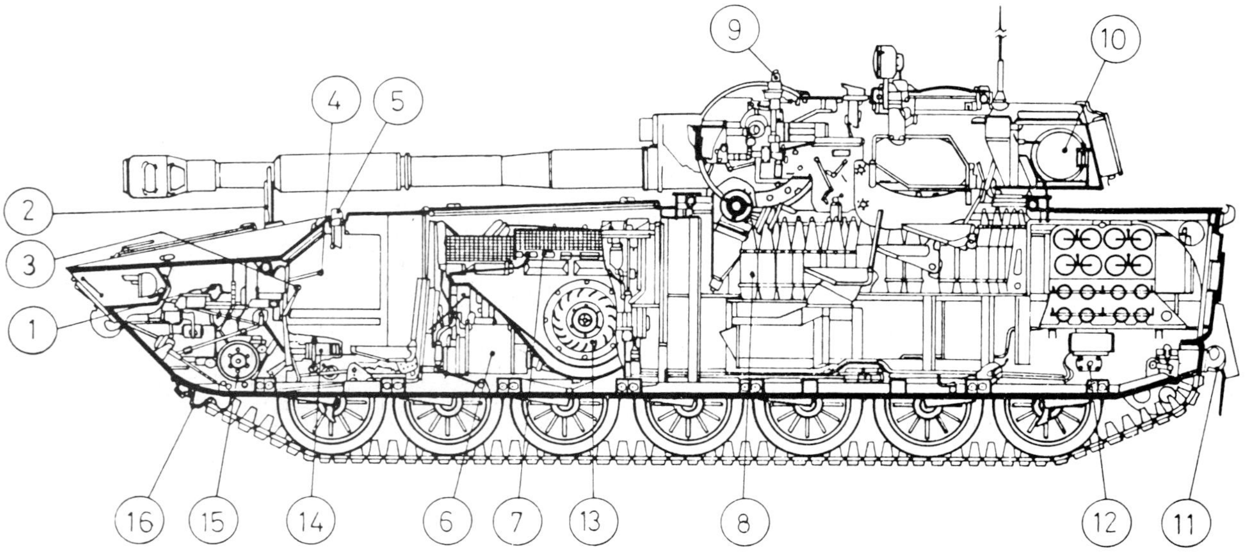

122 mm self-propelled howitzer M-1974 (2S1) **(1)** *pneumatic system* **(2)** *travel lock system* **(3)** *controls for turning clutch and brakes* **(4)** *main drive transmission control* **(5)** *viewing devices* **(6)** *engine pre-heating system* **(7)** *engine oil system and main transmission* **(8)** *stowage* **(9)** *measuring instrument mount* **(10)** *air filtering system* **(11)** *hydraulic equipment* **(12)** *hydraulic shock absorber* **(13)** *cooling system jacket* **(14)** *space reduction gear* **(15)** *final drive* **(16)** *drive sprocket*

gunner in front of and below the commander and the loader on the right. The commander has a cupola which can be traversed through a full 360° and a single piece hatch cover that opens to the front. The commander's cupola, has three periscopes for forward observation and an infra-red searchlight. The indirect sight is mounted in the roof of the turret, forward of the commander's cupola with the direct sight being mounted to the left of the main armament. The loader's hatch opens forward, and in front of it is a swivel type periscope. A long stowage box is often mounted on the left side of the turret and there is a large door in the rear of the hull, hinged on the left side.

The M-1974's main armament, designated the 2A31, is a modified version of the 122 mm D-30 towed howitzer. This has a semi-automatic vertical sliding wedge breech-block and a firing pin that can be recocked in the event of a misfire. A power rammer and extractor are fitted to the folding guard rail to enable a higher rate of fire to be achieved and to permit loading at any angle of elevation. A maximum sustained rate of fire of 5-8 rds/min can be obtained with a sustained rate of fire of 70 rounds for the first hour and of the 40 122 mm projectiles normally carried, 32 are HE, six smoke and two HEAT-FS.

Soviet Army 122 mm self-propelled howitzer M-1974 (2S1) in travelling configuration

The HE projectile has a maximum range of 15 300 m, and in addition to the ammunition listed below, leaflet, HE/RAP (range of 21 900 m), armour piercing high explosive, flechette, and chemical projectiles are also available.

Type	HE	HEAT-FS*	Illumination	Smoke
PROJECTILE DESIGNATION	OF-462	BK-6M	S-462	D-462
WEIGHT OF PROJECTILE	21.72 kg	21.63 kg	22.4 kg	22.3 kg
MAX MUZZLE VELOCITY	690 m/s	740 m/s	690 m/s	690 m/s

* will penetrate 460 mm of armour at 1000 m.

The suspension system is similar to the MT-LB multi-purpose tracked vehicle's and consists of seven road wheels with the drive sprocket at the front, idler at the rear, but no track return rollers. An unusual feature of the M-1974 is that the suspension can be adjusted to give different heights, which is of particular use when the vehicle is being transported by tactical transport aircraft. The M-1974 is normally fitted with 400 mm wide tracks but when operating in snow or swampy ground, 670 mm wide tracks are fitted to give improved traction and a lower ground pressure.

Standard equipment includes infra-red night vision lights and an NBC system. It is fully amphibious, being propelled in the water by its tracks.

Variants

Artillery Command and Reconnaissance Vehicle MT-LBu (IV12 series)

To operate with the 122 mm M-1974 self-propelled howitzer and the 152 mm M-1973 self-propelled gun/howitzer, the Soviet Union has developed the Artillery Command and Reconnaissance Vehicle (ACRV) which is based on the chassis of the M-1974 but with a much larger superstructure. The ACRV has an NBC system and like the M-1974 is fully amphibious, being propelled in the water by its tracks. Track shrouds fitted on the forward fender to improve water flow are often removed on land and stowed at the rear of the vehicle. There are at least three versions of the Artillery Command and Reconnaissance Vehicle. These ACRVs are called MT-LBus in the Soviet Army as they are based on the chassis of the MT-LB multi-purpose armoured vehicle used in large numbers by the Soviet forces.

Soviet Army 122 mm self-propelled howitzer M-1974 (2S1), also known as SO-122 or Gvozdika (Carnation), from the rear. The devices on the turret rear are attached above the forward part of the suspension during amphibious operations

The basic ACRV normally has a driver and has seating for up to seven men. Typically it has a five-man crew consisting of driver, commander, radio operator, navigator and rangefinder operator.

The Soviet Army is now known to refer to these vehicles as the 1V12 system which applies to those command and control vehicles assigned to 122 mm 2S1 and 152 mm 2S3 self-propelled artillery systems.

Designation	Role	NATO Designation
1V13	deputy battery commander	M-1974-1
1V14	battery commander	M-1974-2A
1V15	battalion commander	M-1974-2B
1V16	deputy battalion commander	M-1974-3

The Polish designations for these vehicles are 1W12, 1W13, 1W14, 1W15 and 1W16.

Each battery has two vehicles, with the 1V13 deputy battery commander's vehicle serving as the battery fire direction centre, while the 1V14 serves as the observation post for the battery commander. In addition, the battalion command platoon also includes one 1V15 for use by the battalion commander and one 1V16 for the battalion fire direction under the control of the deputy battalion commander.

ACRV M-1974-1 (IV13)

(Battery Fire Direction Centre)

This model is issued on the scale of three per self-propelled battalion. It has a crew of five or six with the driver seated at the front on the left with a single piece hatch cover that opens to the rear and the vehicle commander to his right with a single piece hatch cover that opens forwards. Both have a windscreen in front with a shutter which is lowered in a combat area. The remainder of the crew are seated at the back, with one or two probably in the circular cupola at the rear. This is often fitted with an externally mounted 12.7 mm DShKM anti-aircraft machine gun which in turn is mounted on a rotating cupola. The circular cupola is provided with vision equipment and single piece hatch cover. Onboard equipment includes a land navigation system, gyro heading indicator, day and night observation and rangefinding equipment, extensive communications facilities and possibly a digital computer.

Basic specifications are as follows:

CREW	5 (commander, driver/ mechanic, commander/ machine gunner, navigator and rangefinder operator)
WEIGHT	14 000 kg
LENGTH	7.21 m
WIDTH	2.85 m
HEIGHT	2.3 m
MAX SPEED	
road	80 km/h
water	5 km/h
ROAD RANGE	525 km
ARMOUR	
Hull	15 mm (est)
Turret	20 mm (est)

ACRV M-1974-1 with roof hatches open and mount for 12.7 mm DShKM machine gun but not weapon itself

ACRV M-1974-2 (IV14/IV15)

(Battery and Battalion Command Vehicle)

This is fitted with various optical and electro-optical equipment, including

ACRV M-1974-2 battery command vehicle

MT-LBus jamming vehicle with antenna erected

Soviet Dog Ear vehicle which is based on ACRV tracked chassis

RKhM chemical reconnaissance vehicle showing rectangular box on rear decking for marking pennants in travelling position

a battery-powered laser rangefinder with a maximum range of 20 000 m but a practical range of only 10 000 m. Its recognition features are a box on the right side of the hull, a box on the turret side and small boxes on the turret roof.

The battery commander's vehicle is designated the ACRV M1974-2A and has whip antennas while the battalion commander's vehicle, the ACRV M1974-2B, has a 10 m telescopic antenna which is stowed on the upper left hull when not required.

The ACRV M1974-2A is used both as a command and observation vehicle, is fitted with a laser rangefinder in the turret and also carries a dismountable rangefinder such as the DAK-1 (NATO Codename Sage Gloss).

ACRV M-1974-3 (IV16)

(Battalion Fire Direction Centre)

Main identification features of this version are the lack of a box on the right side of the hull, clean turret top and side and additional antenna on the hull roof.

This version contains the battalion's electronic field artillery computer and is fitted with a 10 m telescopic antenna which is stowed on the upper left hull when not deployed.

1V21 series air defence management vehicles

The Soviet Union has also deployed a whole family of ACRV based vehicles for use in the air defence command and control roles, these are designated the 1V21, 1V22, 1V23, 1V24 and 1V25. These have also been exported to Czechoslovakia and Poland with the former designating them the MP21, MP22, MP23, MP24 and MP25.

MT-LBus jamming vehicle

This was first identified in Afghanistan and is an ACRV series chassis with an auxiliary power unit mounted at the rear of the hull, while mounted on the roof is an antenna for the Type R-330P VHF jamming set. When in use the 11 element fan type antenna is in the vertical position, but it can be lowered into the horizontal position if required.

Dog Ear Radar Vehicle

A variant of the basic M-1974 chassis carrying a radar known by the NATO reporting name of Dog Ear has been observed. This radar uses a curved grid rotating aerial mounted over the rear of the hull and is used in conjunction with air defence units equipped with ZSU-23-4 or 2S6 twin 30 mm self-propelled anti-aircraft guns and SA-9 Gaskin or SA-13 Gopher missile systems, indicating that the Dog Ear is a form of target surveillance acquisition radar.

Mine Clearing Vehicle M-1979

This has a turret-like superstructure that contains three rockets on launch ramps. These, together with the upper part of the superstructure, are hydraulically elevated for firing. It is estimated that the range of the rockets is between 200 and 400 m with each rocket connected, via a towing line, to about 150 m of mine clearance hose that is stowed folded in the uncovered base of the turret on the roof of the vehicle.

The hose is connected by a cable to the vehicle which allows the crew to position the hose in the optimum breaching position after launching. The hose is then command detonated.

RKhM Chemical Reconnaissance Vehicle

It is now known that the RKhM chemical reconnaissance vehicle is based on the chassis of the M-1974 (2S1) self-propelled howitzer and not the MT-LB multi-purpose vehicle. It has a raised superstructure on which the 7.62 mm machine gun cupola is mounted and on the rear decking are boxes for the marking pennants.

Status: In production. In service with Algeria, Angola, Bulgaria, Czechoslovakia, Ethiopia, Hungary, Iraq, Libya, Poland, Syria, USSR and Yugoslavia.

Manufacturer: Bulgarian, Polish and Soviet state factories.

SPECIFICATIONS		MAX SPEED		STEERING	clutch and brake
CREW	4 + 2 in ammunition carrier vehicle	road	61.5 km/h	SUSPENSION	adjustable
COMBAT WEIGHT	15 700 kg	water	4.5 km/h	ARMAMENT (main)	1 × 122 mm howitzer
POWER-TO-WEIGHT RATIO	15.28 hp/tonne	FUEL CAPACITY	550 l	AMMUNITION (main)	40
GROUND PRESSURE	0.5 kg/cm²	MAX ROAD RANGE	500 km	UNIT OF FIRE	80 rounds
LENGTH	7.26 m	FORDING	amphibious	GUN CONTROL EQUIPMENT	
WIDTH	2.85 m	GRADIENT	60%	Turret power control	electric/manual
HEIGHT	2.715 m	VERTICAL OBSTACLE	1.1 m	Gun elevation/depression	+70°/-3°
GROUND CLEARANCE	0.46 m adjustable	ENGINE	YaMZ-238V, V-8 water-cooled diesel developing 240 hp	Turret traverse	360°
TRACK WIDTH	400 or 670 mm	TRANSMISSION	manual, 5 forward and 1 reverse gears	ARMOUR	
				hull	15 mm
				turret	20 mm

120 mm SO-120 Self-propelled Howitzer/Mortar (2S9)

Development

The 120 mm SO-120 (2S9) Anona (Anemone) self-propelled howitzer/mortar was first seen in public in May 1985 and it appears to be an airborne artillery assault vehicle that has been developed to carry out two tactical functions. One role is as a conventional artillery equipment to replace existing mortars and howitzers in the Soviet Army's air assault divisions and the second role could be as a direct fire anti-tank weapon system firing HEAT projectiles. The SO-120 is used by the Soviet Army *Vozdushno-desantnyye Voyska* (or VDV) air assault divisions. The 2S9 entered service with the Soviet Army in quantity in 1984 and was deployed in Afghanistan.

Although it was originally developed to meet the requirements of the Soviet air assault divisions, more recently it has also been identified to be in service with regular ground forces and Soviet Naval Infantry. A quantity was also left behind in Afghanistan when the Soviets withdrew.

The chassis used in the S-120 (2S9) system was developed originally by the I V Gavalov Design Bureau located at the Volgograd Tractor Plant.

It is estimated that well over 1000 of these systems have been completed, although the Soviet Union has not released any production figures. Under the Conventional Forces Europe (CFE) Treaty signed in 1990, the Soviet Union admitted having the following quantities of vehicles in the West:

Region	**IV.1**	**IV.2**	**IV.3**	**IV.4**	**V.1**
Quantity	452	236	110	nil	216

The above figures were the situation as of November 1990. For the Conventional Forces Europe Treaty, Region IV is defined as Central Europe while Region V covers the flanks of the Atlantic to the Urals (ATTU).

More recently, the Soviet Union has developed the 120 mm 2S23 self-propelled howitzer/mortar system. This is essentially the BTR-80 (8 × 8) armoured personnel carrier, fully described in the *Armoured personnel carriers* section, fitted with a modified version of the turret fitted to the 2S9. Additional details are given in the following entry.

Description

The SO-120 (2S9) has a crew of four; commander, driver/mechanic, gunner and loader. The driver/mechanic is seated in the front centre of the hull and is provided with three periscopes with the centre block replaceable by an infra-red night-driving device.

The commander is located at the front left of the hull. He is provided with three vision devices and the communication sets, a gyro compass and land navigation devices.

The gunner is on the left of the turret. His position has a panoramic sight fitted to the turret roof and a telescopic sight for direct fire. The loader is to the right of the gun. He is provided with a single vision block and a seat that folds away while he is serving the gun.

The SO-120 (2S9) hull is a stretched version of the BMD airborne combat vehicle and is divided into three compartments; the command compartment, the fighting compartment and the engine compartment. The command compartment is in the front portion of the hull, forward of the turret mount. It accommodates the commander and the driver/mechanic. The fighting compartment has stowage for 60 rounds of ammunition behind the turret and is in the middle of the hull.

The turret is located above the fighting compartment. At the rear of the hull is the engine compartment. The power plant is a 5D20 diesel developing 300 hp, coupled to a five-speed manual transmission. Fuel is stored in the engine compartment.

The two-man turret is located in the middle of the hull and is of welded steel construction with 16 mm thick frontal armour. The turret roof has two hatches, one for the gunner and the other for the loader. Turret armour is probably 7 mm thick. Weapon traverse is limited to 35° either side. Mounted in either side of the turret rear is a TNPO-170A observation device.

The track is the same as that used on the BMD airborne combat vehicle and the suspension is hydro-pneumatic with an adjustable ground clearance of between 100 and 450 mm. The ground clearance is adjustable by the driver/mechanic from within the vehicle. There are six road wheels each side and five return rollers.

The SO-120 (2S9) is amphibious and is powered in the water by two water jets at the rear. Power can be delivered to the water jets and the track at the same time. When in the water, splash vanes can be erected in front of the driver's position.

The SO-120 (2S9) is armed with a 120 mm breech-loaded mortar (2A60) with a barrel approximately 1.8 m long. The mortar is probably provided with an interrupted-screw breech mechanism and a chamber detent to retain a round in place when the barrel is elevated. Ammunition is fixed and loading is manual, although ramming is automatic. After the loader has selected a round from a ready rack it is placed in a feed tray. After an electrical button has been pressed a rammer automatically seats the round in the chamber and closes the breech. The rammer is a pneumatic device operating at a pressure of 150 kg/cm^2. As the breech opens after firing the rammer bleeds compressed air into the chamber to force firing fumes from the muzzle.

The 120 mm mortar has a rate of fire of 6 to 8 rds/min. The direct fire HEAT projectile is estimated to defeat up to 600 mm of armour. In the indirect fire role the mortar fires high explosive, white phosphorous and smoke rounds. According to Soviet sources, the maximum range of the 120 mm weapon is some 8000 m. Mounted below the turret rear is an ammunition loading hatch and mounted on top of the hull is a device for loading ammunition from the ground directly into the fighting compartment. This allows sustained fire missions to be carried out whilst still retaining onboard ammunition supply.

The SO-120 (2S9) takes 30 seconds to come into action and a similar time to come out of action. When in the firing position the suspension is raised to provide a more stable firing platform.

The SO-120 (2S9) has no coaxial machine in the turret and there are no bow or turret-mounted machine guns. Standard equipment includes an NBC system, night vision equipment and an electric bilge pump.

For air transport the SO-120 (2S9) can be carried by any Soviet Military Air Transport (VTA) medium or heavy cargo aircraft. It can be para-dropped from the An-32 'Cline', Il-76 'Candid' or An-22 'Cock' aircraft using the PRSM-915 heavy drop system, from heights between 300 and 1500 m.

Status: Production probably complete. In service with Afghanistan and USSR (airborne, ground and Naval Infantry).

Manufacturer: Soviet state factories.

120 mm SO-120 self-propelled howitzer/mortar (2S9)

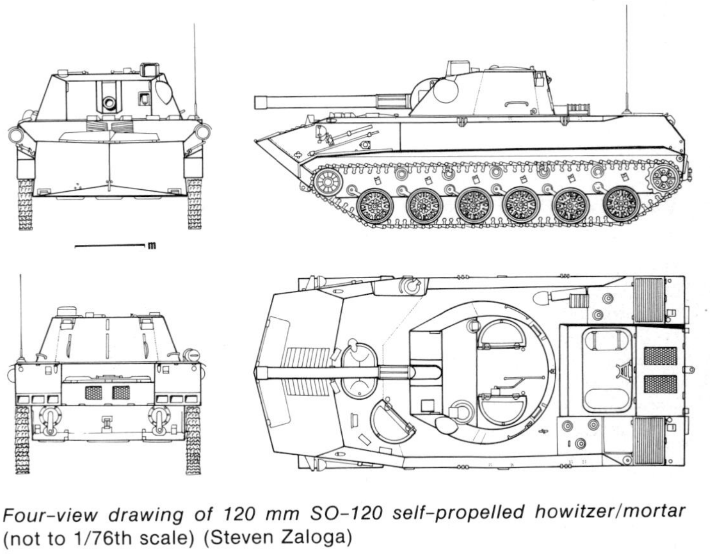

Four-view drawing of 120 mm SO-120 self-propelled howitzer/mortar (not to 1/76th scale) (Steven Zaloga)

SPECIFICATIONS					
CREW	4	TRACK WIDTH	250 mm	ENGINE	5D20 diesel developing 300 hp
COMBAT WEIGHT	8000 kg	LENGTH OF TRACK ON GROUND	3.23 m	TRANSMISSION	manual, 5 forward and 1 reverse gears
GROUND PRESSURE	0.5 kg/cm²	MAX SPEED			
LENGTH	6.02 m	road	60 km/h	SUSPENSION	hydro-pneumatic, adjustable
WIDTH	2.63 m	water	9 km/h		
HEIGHT		RANGE		ARMAMENT	1 × 120 mm mortar
at max ground clearance	2.3 m	roads	500 km	AMMUNITION	60
in firing configuration	1.9 m	water	75 to 90 km	GUN CONTROL EQUIPMENT	
GROUND CLEARANCE (variable)	100 to 450 mm	FORDING	amphibious		
		GRADIENT	60%	Gun elevation/depression	+80°/-4°
TRACK	2.38 m	MAX SIDE SLOPE	33%	Gun traverse	35° left/35° right

120 mm 2S23 Self-propelled Howitzer/Mortar System

Development/Description

Late in 1990, the Soviet Union revealed that it had developed the 120 mm 2S23 self-propelled howitzer/mortar system. This is essentially a modified version of the turret used with the 2S9 system, covered in the previous entry, mounted on a modified BTR-80 (8 × 8) armoured personnel carrier chassis.

The 2S9 (or SO-120) Anona (Anemone) chassis was developed by the I V Gavalov Design Bureau located at the Volgograd Tractor Plant and was first seen in 1985, although it entered service in 1984.

The all welded steel turret of the 2S23 does differ from that observed on the 2S9 (SO-120) in a number of areas including the installation of three electrically operated smoke dischargers either side of the turret firing forwards and a raised commander's cupola towards the rear of the turret on the left side which overhangs the side of the turret.

On the 2S9 (SO-120) the infra-red searchlight is mounted on the roof in front of the commander's hatch on the right side of the turret, but on the 2S23 it is mounted on the forward part of the commander's cupola which is also provided with vision devices. Forward of the cupola is a 7.62 mm pintle-mounted machine gun. The turret is also provided with vision blocks for direct observation.

The 120 mm main armament, designated the 2A60, has a maximum range of 8000 m with a maximum rate of fire being 6 to 8 rds/min.

Types of Soviet 120 mm mortar bomb fired include the OF-843 which weighs 16.2 kg complete with GVMZ-7 point detonating fuze (or the M-12 RD fuze) which contains 2.68 kg of TNT. The OF-843A is a similar mortar bomb and contains 1.58 kg of Amatol 80/20.

In addition to firing Soviet 120 mm mortar bombs, the 120 mm weapon of the 2S23 can also fire mortar bombs fired by the French Thomson-Brandt MO-120-RT-61 mortar system that is used by many countries around the world. This fires a wide range of mortar bombs including HE, HE extended range, illuminating and smoke.

Turret traverse of the 2S23 is 35° left and right and weapon elevation is from -1° to +85°. The direct fire sight is located to the left of the 120 mm main armament while the indirect fire sight is mounted in the turret roof on the left side. The indirect sight housing of the 2S23 appears to be slightly different to that of the earlier 2S9 (SO-120) system.

Although the chassis of the 2S23 is similar to that of the BTR-80, fully described in the *Armoured personnel carriers* section, it does have a number of differences including removal of all firing ports and their associated vision devices, removal of the side doors between the second and third road wheels, different roof hatches and removal of the commander's roof-mounted infra-red searchlight.

2S23 from right side showing 120 mm main armament and bank of three forward firing smoke dischargers

Although the doors in the upper hull sides have been removed, a door is fitted in the lower part of the hull between the second and third wheels as an emergency exit.

The 2S23 retains the amphibious characteristics of the original BTR-80 and has a single water jet at the rear of the hull with a trim vane being mounted on the front of the vehicle. This folds back onto the glacis plate when not required.

The standard BTR-80 (8 × 8) APC weighs 13.6 tonnes combat loaded compared to the 2S23 which weighs 14.5 tonnes, including 30 120 mm mortar bombs. Soviet sources give the 2S23 a maximum road speed of 80 km/h and a maximum water speed of 10 km/h.

Standard equipment on the 2S23 includes an NBC system, night vision equipment and a central tyre pressure regulation system that allows the driver to adjust the tyre pressure to suit the type of ground being crossed.

Status: Prototype first shown in 1990. Believed to be entering production.

Manufacturer: Soviet state factories.

UNITED KINGDOM

VSEL AS90 155 mm Self-Propelled Gun

Development

During the production run of their towed howitzer FH-70, Vickers Shipbuilding and Engineering Limited (VSEL) identified a world need for improved self-propelled artillery. Subsequent concept studies offered a cost-effective solution using VSEL's private venture GBT 155, a turret based on the 155 mm 39 calibre ordnance of FH-70 and designed specifically for export.

The GBT 155 turret, armed with the same 155 mm ordnance as the current AS90 self-propelled howitzer, was officially unveiled at the 1982 British Army Equipment Exhibition and is now being marketed worldwide. Full details are given in the following entry.

Whilst the conversion of existing MBT chassis to the basic SP configuration initially appeared to be an acceptable cost effective solution to some potential users, to others it presents difficulties with ammunition re-supply and retrofitting of technological advances, and limits access to the breech. All these limitations are exacerbated by the MBT engine being positioned at the rear of the vehicle's chassis. Furthermore the heavy armour of an MBT may pose logistical problems in some parts of the world in that some bridges could not accept the total weight of the chassis and turret, and where ground pressures need to be kept low.

VSEL's market research further highlighted the worldwide need for a purpose-built advanced self-propelled howitzer albeit configured on the classical SP howitzer lines with the engine frontally positioned alongside the driver.

Conceptual studies commenced in 1982 for the design of a modern artillery chassis using a variant of the GBT 155, incorporating all and more of its features but having a larger turret ring to increase crew space to a maximum and a frontally positioned powerpack. From the outset AS90 was specifically designed to meet the needs of modern artillery warfare until well into the next century. AS90 uses modern but well proven technology yet has been designed to accept further technological developments as they become available. This was considered to be a prime requirement since artillery systems tend to have a much longer life, at least 25 years more, than other weapon systems. Therefore future updates need to be accomplished without the need for extensive rebuilding.

The three principal factors of concern in the AS90 concept if it were to survive the increased firepower, improved armour, electronic and NBC warfare of the future battlefield were judged to be performance, crew comfort and survivability and a systems approach to the vehicle's equipment features to prove the efficiency of materials and equipments, for the complete AS90 system. As a result, key features in the design

AS90 undergoing firing trials with 155 mm 39-calibre barrel

AS90 with ordnance elevated and hull door open

of AS90 include, in addition to the stretch potential mentioned earlier, ease of operation and maintenance, reliability, and modular construction to suit different user requirements. The same rate of fire and accuracy as FH-70 was stipulated as was a rapid into and out of action time to avoid counter-battery fire, and total manual back up in the event of battle damage or power failure.

The AS90 has been designed to be easily manufactured and wherever possible standard and proven components have been used. For example, even the variety of bolt sizes has been kept to a minimum and the elevation and traverse motors are identical.

The project definition for AS90 was completed in March 1985 and the first prototype was shown at the British Army Equipment Exhibition in June 1986. Following the exhibition a series of firing trials, augmented by mobility trials, was undertaken leading to the award of Live Crew Clearance from the British Ordnance Board in March 1987. A series of endurance and automotive trials was then initiated, with AS90 travelling some 3000 km and firing in excess of 1500 rounds without a mission relevant failure. These trials, completed in June 1987, covered all aspects of army use. AS90 has also been tested with 47 and 52 calibre barrels built by Royal Ordnance Nottingham.

Following the withdrawal of the United Kingdom from the International 155 mm SP-70 project late in 1986, the search for a new self-propelled howitzer started.

This eventually resulted in the selection in June 1989 of the Vickers Shipbuilding and Engineering Limited 155 mm AS90 self-propelled howitzer to meet its future requirements.

The fixed price contract is worth £300 million and covers the completion of development of the AS90, the production of 179 systems and the provision of initial spares. In addition there is an option for a further batch of 50 vehicles which may be exercised by the MoD within a period of two years.

Each regiment of AS90s will have 32 weapons in four batteries each of eight guns. First production AS90s will be completed in 1991 with the first regiment, 26 Field Regiment, RA, becoming operational early in 1992.

Throughout 1989-90, AS90 has undergone a full series of reliability growth trials for the British Army as part of the trials programme. By the Spring of 1991, the two prototype vehicles had travelled 28 630 km and fired 6170 rounds (3380 EFCs).

Serious consideration is already being given by VSEL to the use of the AS90 chassis as the basis for a whole family of armoured fighting vehicles including command post vehicle, ammunition resupply vehicle, anti-tank (for example TRIGAT), air defence (gun or missile), armoured personnel carrier, and armoured recovery/repair vehicle.

VSEL has also entered a combined repair/recovery vehicle on the AS90 chassis as part of the British Army's FFLAV programme as well as to carry scatterable mine systems.

Description

The chassis of AS90, like the turret, is of all-welded steel armour construction with a maximum thickness of 17 mm. Steel rather than aluminium was chosen as it is not only easier to build, but also easier to repair in the field.

The driver sits at the front left with the powerpack to his right and normally enters via a single piece hatch over his position that hinges to the left side. This can be locked open for driving in the head-out position. Positioned in front of the driver is a single vision block with black-out blind and laser filter. The vision block can be changed for a night vision device by the driver while under armour without breaking the NBC seal.

The driver has a fully adjustable seat and can also enter and leave his position via the fighting compartment at the rear. The driver steers the vehicle using a steering wheel rather than sticks.

The powerpack, consisting of the engine, transmission and cooling system, can be removed as a complete unit in well under an hour. The engine compartment is provided with the Kidde-Graviner fire detection and suppression system which can be operated automatically or manually. Hinged louvres and access covers are provided above the engine compartment for ease of access.

The barrel clamp is located on the front of the vehicle to provide a travelling support for the gun. Operation of the barrel clamp is from within the vehicle from the driver's position.

The AS90 is powered by the Cummins VTA 903T-660 14.8-litre V-8 diesel developing 660 bhp at 2800 rpm. This is a more powerful version of the engine installed in the US Army's M2/M3 Bradley Infantry Fighting Vehicle/Cavalry Fighting Vehicle and the Multiple Launch Rocket System. The VTA 903T-660 has plenty of stretch potential, having already been run at 750 hp for trials purposes. In addition to having responsibility for the complete AS90 powerpack, the Cummins Engine Company of the UK is also responsible for the vehicle brakes.

The engine is coupled to a fully automatic German ZF model LSG 2000 automatic transmission with four forward and two reverse gears. A fully automatic rather than a semi-automatic or manual transmission was selected to reduce driver fatigue and training.

A diesel-powered auxiliary power unit is installed in the forward part of the fighting compartment so the main engine does not need to be constantly run to recharge the batteries.

The area to the rear of the engine compartment is the fighting compartment, which extends to the rear of the vehicle. There is a large door opening to the right in the rear of the chassis for access and ammunition resupply and mounted either side of this on the outside are two integral armoured containers which can be used to store supplies and other equipment.

The vehicle runs on 12 double road wheels, each having hydro-pneumatic suspension, with a drive sprocket at the front and an idler at the rear. Use of the hydro-pneumatic suspension units contributes to the increased headroom in the fighting compartment by eliminating the false floor needed with torsion bar suspension, in addition to giving AS90 a low profile. The system has been designed to give the vehicle, with its large wheel travel, progressive rate of springing and powerful damping on every wheel station, a very smooth ride enabling high average speeds to be maintained across rolling irregular terrain without impairing the safety and comfort of the crew.

The turret is of all welded steel construction with a maximum thickness of 17 mm and has a large turret ring diameter of 2.7 m. The commander has a cupola on the right side of the turret roof and there is the option of an air sentry's hatch with pintle mounting for a 12.7 mm machine gun secondary armament on the left-hand side. There is a large hatch on the left side of the turret.

The commander's and layer's seats are on the right of the turret. The two ammunition handlers are to the left. The commander controls all powered functions of the turret. Turret traverse is through a full continuous 360° and weapon elevation is from -5° to +70°. Elevation and traverse are both electric at 10°/s, and full manual controls are provided as a back up.

An electric traverse and elevation system, with silent operation from the integral power supply, was chosen rather than hydraulic since it was considered to be safer, more reliable and would require less maintenance. The only hydraulics in the turret are those applicable to the balancing gear, loading tray and rammer.

AS90 is fitted with the 155 mm 39-calibre barrel designed and built by Royal Ordnance Nottingham and already installed in the GBT 155 turret system. The recoil system has two diametrically opposed buffers and one recuperator, each with its own integral reservoir, and a maximum

AS90 undergoing mobility trials with ordnance in travelling lock

recoil length of 800 mm. The barrel can be drawn out through the front of the turret for replacement in less than one hour. The removal of the complete elevating mass is simple as the saddle is constructed in two parts and can be easily detached from the trunnions.

The ordnance has a double-baffle muzzle brake, fume extractor and a split block breech. The breech combines the speed of action associated with the sliding block principle and the robustness of the Crossley pad obturating system and has an integral 12-round primer magazine.

AS90 can fire a burst of three rounds in 10 seconds, or 6 rds/min for a short period with a sustained rate of fire of 2 rds/min. The ordnance incorporates an electrically initiated percussion firing mechanism using standard DM191A1 or US M82 igniter tubes. Forty-eight projectiles and charges are carried and the ordnance will fire all standard NATO projectiles including the HE M107 and extended range types.

Of the 48 projectiles carried, 31 are stowed in the turret bustle in four magazine modules, each of which has a motor that moves the required projectile to the correct position to enable it to be pulled forward onto a shell transfer arm. This pivots about the left trunnion and aligns with either magazine tray irrespective of gun elevation. The shell transfer arm realigns to lock with the elevating mass and the projectile is moved sideways by a motor-driven shell trolley onto the loading tray. The loading tray is in line with the breech so the projectile is held in position by clamps and then rammed by the flick rammer. The charge is then loaded, the breech closed, the loading tray and rammer retracted and the ordnance fired by the commander or layer. The hydraulic loading system has full manual backup and safety interlocks will not allow it to be operated below safe pressures.

Using standard ammunition and charges, the 39-calibre barrel enables a range of 24 700 m to be achieved. Using an Extended Range Full Bore projectile a range of over 32 000 m can be achieved.

AS90 is fitted with a full navigation and auto-lay capability using the MAP's standard DRU based on Ring Laser Gyro technology giving the gun complete autonomy. This enables the gun to be laid automatically in bearing as well as elevation.

Muzzle velocity measuring equipment, air-conditioning and NBC pack are provided as standard, the latter two being mounted on the rear of the turret bustle.

In addition to Cummins Engines, other major subcontractors to VSEL are Air-Log and Avimo.

The Air-Log contract is worth £16 million and covers the supply of the complete running gear for AS90. This includes an advanced Air-Log hydropneumatic suspension system, Diehl double pin track and sprockets, GLS top rollers and roadwheels and Air-Log designed and manufactured idler gear. The latter incorporates an advanced hydraulic track tensioner with an overload protection device that allows the idler wheel to retract if excessive track tension is built up.

Avimo was awarded a contract worth over £3 million for the design, development and series production of the day/night direct fire telescope for the AS90.

Variants

The modular design of the AS90 will allow the system to be upgraded or modified without extensive reworking. Optional equipment includes ordnances up to 52 calibres and in the future perhaps even liquid or electro-magnetic armament systems, currently under various stages of development in both the USA and UK.

Incorporation of a 52-calibre barrel has already been mentioned but in the longer term the application of liquid propellant guns has been studied. A practical laser head for integration into the AS90 cradle for laser ignition to replace percussion primers is also being considered. Studies are already underway by VSEL to investigate options for product improvement programmes including increased automation and a reduction in the crew.

Status: In production. In June 1989 the British Army selected the AS90 to meet its future self-propelled artillery requirements. Vickers Shipbuilding and Engineering was awarded a £300 million fixed price contract for the supply of 179 AS90 systems with the 39-calibre barrel. The contract also covered the completion of development and the provision of initial spare parts. Also included was an option on a further 50 systems, bringing the total to 229. First production vehicles will be completed in 1991 with the first regiment becoming operational in BAOR in 1992.

Manufacturer: Vickers Shipbuilding and Engineering Limited
Barrow-in-Furness
Cumbria
LA14 1AF
UK
Telephone: (0229) 823366
Telex: 65411 VSEL G
Fax: (0229) 23366

SPECIFICATIONS	
CREW	5
COMBAT WEIGHT	45 000 kg
POWER-TO-WEIGHT RATIO	14.66 hp/tonne
GROUND PRESSURE	0.90 kg/cm^2
LENGTH	
gun forward	9.7 m
hull	7 m
WIDTH	3.3 m
HEIGHT OVERALL	3 m
GROUND CLEARANCE	0.41 m
TRACK	double pin
TRACK WIDTH	550 mm
LENGTH OF TRACK ON GROUND	4.59 m
MAX ROAD SPEED	55 km/h
FUEL CAPACITY (approx)	640 l
ROAD RANGE (approx)	370 km
FORDING	1.5 m
GRADIENT	60%
SIDE SLOPE	25%
VERTICAL OBSTACLE	0.88 m
TRENCH	2.8 m
ENGINE	Cummins VTA 903T-660 V-8 diesel developing 660 bhp at 2800 rpm
TRANSMISSION	ZF LSG 2000 fully automatic, 4 forward and 2 reverse gears
AUXILIARY GENERATOR	4-stroke diesel
SUSPENSION	hydro-pneumatic
ELECTRICAL SYSTEM	24 V
BATTERIES	
turret	4 × 12 V, 100 Ah
hull	4 × 12 V, 100 Ah
ARMAMENT	
main	1 × 155 mm 39-calibre
anti-aircraft	1 × 12.7 mm MG
SMOKE-LAYING EQUIPMENT	2 × 5 smoke dischargers
AMMUNITION	
155 mm	48 projectiles and charges
12.7 mm AA	1000 rounds
GUN CONTROL EQUIPMENT	Turret traverse 360°, electric
gun elevation	-5° to +70, electric
ARMOUR	17 mm (steel)

VSEL GBT 155 155 mm Gun Turret

Development

The Vickers Shipbuilding and Engineering Limited (VSEL) GBT 155 155 mm Gun Turret has been produced as a result of the company's involvement in the international FH-70 and the now cancelled SP-70 howitzer projects. Research has shown that there is a market for a modular 155 mm gun turret that can be fitted into virtually any AFV heavy enough to withstand the recoil forces of a 155 mm howitzer. The prototype turret was developed and produced during 1981 and has since undergone successful static and chassis-mounted firing trials. Early in 1987 the GBT 155 155 mm gun turret was evaluated in India and mounted on the chassis of the Vijayanta MBT chassis, this being the Vickers Mk 1 built in India.

As of July 1991 the GBT 155 155 mm gun turret had yet to enter production.

Description

The GBT 155 is a modular design intended for fitting directly onto almost any modern main battle tank chassis. To date firing/mobility trials have been completed on Centurion, Chieftain, Challenger and Vickers Mk 3 chassis.

The turret structure is armoured steel, constructed to provide strength and stiffness and ensure protection against small arms fire and shell splinters.

Ground level access to the fighting compartment is provided by local modifications to the chassis. Access hatches in the turret sides provide additional entry points for the crew and there is also a hatch in the commander's cupola.

The magazine at the rear of the turret stows 31 projectiles and incorporates simple indexing mechanisms which transfer the projectiles to a central point where they are manually withdrawn and placed on the loading tray. A total of 21 propellant charges are stowed in sealed containers mounted above the shell magazine and a further 11 at the front left of the turret. Additional ammunition can be stowed within the chassis, the quantity of which depends on the chassis employed.

The turret has a crew of four; the layer at the forward right with the commander behind and two loaders on the left.

The ordnance, cradle and saddle design is such that the barrel can easily be withdrawn from the front of the turret leaving the breech in place. In addition the whole of the elevating mass complete with saddle can be quickly removed from the turret leaving all other turret-mounted equipment intact.

The recoil system has two diametrically opposed buffers and one recuperator, each with its own integral reservoir to reduce external piping and space requirements. The system is protected by an easily removed armoured cover.

The ordnance is a 155 mm 39-calibre barrel, with a vertical sliding split-block breech ensuring positive obturation, a double-baffle muzzle brake, and a fume extractor. The ordnance conforms to the Quadrilateral Ballistic Agreement governing the standardisation of 155 mm ordnance and ammunition between the UK, USA, Germany and Italy.

The new design of self-obturation breech mechanism combines the rapid action of a sliding breech-block with a Crossley Pad obturator to provide a positive seal. The breech can be operated in both manual and semi-automatic modes and in the latter runout energy is used to open the breech and load an igniter tube. Both the German DM191A1 and the American M82 percussion igniter tubes can be used. The breech is connected to the barrel by a conventional interrupted screw-thread.

The barrel is machined from an electro-slag refined (ESR) forging of monobloc construction and is autofrettaged.

The howitzer can fire the full range of ammunition intended for the FH-70 and the M198 to a range of 24 700 m with standard projectiles and to 31 500 m with base bleed projectiles.

Due to the weight of 155 mm projectiles a semi-automatic loading system has been incorporated at the rear of the elevating mass to enable a rapid burst rate of fire to be achieved and a high rate of continuous fire to be sustained. This system provides a burst rate of three rounds in 13 seconds and an intense rate of 6 rds/min. This equipment comprises a loading tray and power rammer for loading the projectiles. The propellant charges are much lighter than the shells and are manually loaded.

Ammunition can be taken from either the onboard stowage or from an external source. Suitably positioned hatches in the chassis provide access for this purpose and additional hatches in the turret provide alternative routes for use in an emergency.

Power for the loading system is provided by a small hydraulic powerpack installed in the front left corner of the turret. The electric motor driving this pack is identical to the motors used in the traverse and elevation drives.

Elevation and traverse motions are achieved using identical electric motors driving through gearboxes onto an elevation arc and a traversing rack respectively. Both motions are controlled by the layer's two-axis joystick. The commander has an override control in traverse only for designating targets in the direct fire role. Manual back up for both motions is provided by means of non-backdriving handwheels.

As the barrel is muzzle heavy, a nitrogen-charged gas cylinder is used to provide balance over the full elevation range of -5° to +70°. Compensation equipment is fitted to allow for changes in ambient temperature.

The system is operated on conventional external optical reference principles and is fully compatible with standard artillery procedures. Overall accuracy and ease of operation are improved over the traditional cross levelling systems and due to the feature of auto-lay, re-laying between rounds is very rapid. The initial 'coming into action' procedures of passing the line with director recording 'Gun Aiming Post' (GAP) positions and the initial gun lay procedure are accomplished very quickly simply by following the procedure indicated by the illuminated push-buttons on the computer control panel.

Inertial navigation and auto-orientation can be provided as an option which will afford a full vehicle autonomous operation capability.

For direct fire against targets up to 2000 m away a day/night sight is mounted on the elevating mass adjacent to the layer. The graticule is calibrated for range and crossing target velocity and, since the layer has both elevation and traverse controls, simple aiming of the gun can be achieved.

The power supply for all turret services is provided by a small diesel-driven generator and battery pack mounted on the rear of the turret. This obviates the need to run the vehicle main engine during fire missions.

An extractor fan is fitted as standard to the fighting compartment to remove any residual fumes and to provide adequate ventilation and air circulation for the crew.

The GBT 155 turret has been designed as a basic weapon system to meet the requirements of most customers and allows the choice of fire control equipment and communications by the customer. Optional extras include fire-detection and fighting equipment, smoke dischargers and air-conditioning.

Besides the chassis already trialled, possible carriers for the GBT 155 turret include the Leopard I, M48, M60 and a chassis of Soviet design. Options for the turret include inertial navigation for gun autonomy and barrel lengths of up to 52 calibres for ranges beyond 40 km.

VSEL GBT 155 155 mm gun turret on Vickers Mark 3 MBT chassis

SPECIFICATIONS

CREW	4 + 1 (driver)
LENGTH OF TURRET	
including barrel cradle	4.26 m
including barrel	8.63 m
WIDTH OF TURRET	3.1 m
HEIGHT	
to turret roof	1.325 m
to turret cupola	1.515 m
WEIGHT OF TURRET	13 500 kg
GUN ELEVATION	-5° to +70° at 10°/s
TURRET TRAVERSE	360° at 10°/s
MODE OF CONTROL	layer's joystick (commander has override)
MAIN ARMAMENT	1 × 155 mm
MAX RANGE	24 700 m
with base bleed projectile	31 500 m
MINIMUM RANGE	2500 m (high angle)
SECONDARY ARMAMENT	1 × 7.62 mm or 1 × 12.7 mm MG
EXTERNAL REPLENISHMENT SYSTEM	manually assisted
TURRET PROJECTILE STOWAGE CAPACITY	31
TURRET CHARGE STOWAGE CAPACITY	32
INTERNAL PROJECTILE HANDLING SYSTEM	manually assisted
SHELL LOADING SYSTEM	automatic
RATES OF FIRE	
Burst	3 rounds in 13 s
Intense	6 rds/min
Sustained (one hour)	2 rds/min
ELECTRICAL SYSTEM	24 V

Status: Development complete. Ready for production.

Manufacturer: Vickers Shipbuilding and Engineering Limited
Barrow-in-Furness
Cumbria LA14 1AF
UK
Telephone: (0229) 823366
Telex: 65411 VSEL G
Fax: (0229) 23366

Royal Ordnance 155 mm Ordnance Systems

Development/Description

The Guns and Vehicles Division of Royal Ordnance has developed a range of 155 mm ordnance systems that can be installed in new vehicles or retrofitted to existing vehicle platforms.

The 39 calibre 155 mm ordnance is, for example, fitted to production versions of the Vickers Shipbuilding and Engineering AS90 self-propelled artillery system ordered by the British Army in 1989, while the 47 and 52 calibre ordnances have been installed in the AS90 for trials and development work.

The M109 system has also been fitted with the Royal Ordnance 155 mm 39, 47 and 52 calibre ordnance for trials and development work.

All of these 155 mm systems have chambers that conform to the Quadrilateral Ballistics MoU and can fire all 155 mm ammunition conforming to the Quadrilateral Ballistics MoU.

They all have rapid breech operation based on the sliding block principle with Crossley pad obturation, the flick rammer is hydraulically powered and electronically controlled. Firing mechanism is electric/percussion with a 13 mm vent tube or DM191/M82. Tube loading is manual assist.

A burst rate of fire of three rounds in 10 seconds can be achieved with an intense rate of fire of 6 rds/min for a period of three minutes. Breech operation is automatic following loading of first projectile and charge.

The following reversionary modes are available:
Manual tube loading - manual insertion
Power rammer - manual rammer
Firing mechanism - lanyard firing

Optional enhancements include automatic primer tube loading, barrel temperature indicator and fall back detection.

SPECIFICATIONS

Version	**39 cal**	**47 cal**	**52 cal**
LENGTH OF ORDNANCE	7.092 m	8.338 m	9.024 m
WEIGHT OF ORDNANCE	2615 kg	2895 kg	3020 kg
MUZZLE VELOCITY WITH L15 PROJECTILE AND CHARGE 8	827 m/s	805 m/s	830 m/s
RANGE	24 700 m	24 000 m	24 900 m
MUZZLE VELOCITY WITH NEW RO CHARGE	—	945 m/s	945 m/s
RANGE	—	30 000 m	30 000 m

Status: The 39 calibre ordnance is in production for the AS90.

Manufacturer: Royal Ordnance
Guns & Vehicles Division
Kings Meadow Road
Nottingham NG2 1EQ
UK
Telephone: (0602) 863341
Telex: 377531
Fax: (0602) 861436

Vickers Defence Systems Abbot 105 mm Self-propelled Gun

Development

In the late 1950s a decision was made to design a new self-propelled gun which would be armed with a new 105 mm gun in a turret with a 360° traverse and use automotive components, including the engine, transmission, and suspension, of the FV430 series of tracked vehicles. The FV432, the standard APC of the British Army, is a member of the FV430 series.

The first of 12 prototypes of the new self-propelled gun, called the FV433, was completed by the design parent, Vickers, in 1961. Six of the prototypes were powered by a Rolls-Royce K60 multi-fuel engine and the other six by a Rolls-Royce B81 petrol engine. After trials a production contract was awarded to Vickers. The first production Abbot was completed in 1964 and production continued until 1967.

The Abbot is deployed with close support regiments of the Royal Artillery in the British Army of the Rhine. Each regiment has three batteries each with eight Abbots. The Abbot is supported by the Alvis Stalwart (6 × 6) High Mobility Load Carrier which carries additional ammunition on pallets.

A total of 146 production Abbots was built for the British Army. In addition India took delivery of 68 Value Engineered Abbots while the British Army took 20.

Abbot 105 mm self-propelled gun (UK Ministry of Defence)

With the British Army, the Abbot is employed in Canada for training at BATUS, in the United Kingdom at the Royal School of Artillery, Larkhill and with four regiments in the British Army of the Rhine. The latter are to be replaced from 1992 by the Vickers Shipbuilding and Engineering AS90 155 mm self-propelled artillery system.

In November 1990 the British Army stated that it had a total of 151 Abbot guns in service of which 18 were in the UK and 133 elsewhere (for example Canada and Germany).

Description

The all-welded steel hull of the Abbot is divided into three compartments: driver's at the front on the right, engine to the left of the driver and the gun compartment at the rear.

The driver has a two-piece hatch cover that opens either side of his position in front of which is a single wide-angle periscope that can be replaced by an infra-red periscope for driving at night.

Power from the multi-fuel engine is taken through a Rolls-Royce transfer drive to an Allison TX200 six-speed automatic transmission. A short propeller shaft transmits drive from the transmission to the Cletrac type steering unit and from there to the drive sprockets. The complete powerpack, consisting of the engine, transfer drive and transmission, can be removed as a unit to facilitate replacement in the field. The engine compartment is provided with a fire warning system.

The other three crew members are in the all-welded steel turret with the commander on the right, gunner in front of and below the commander and the loader on the left. The commander's cupola can be traversed through a full 360°, has a single piece hatch cover that opens to the rear and three periscopes in the forward part. The loader has a single piece hatch cover that opens to the rear and there is a small hatch in

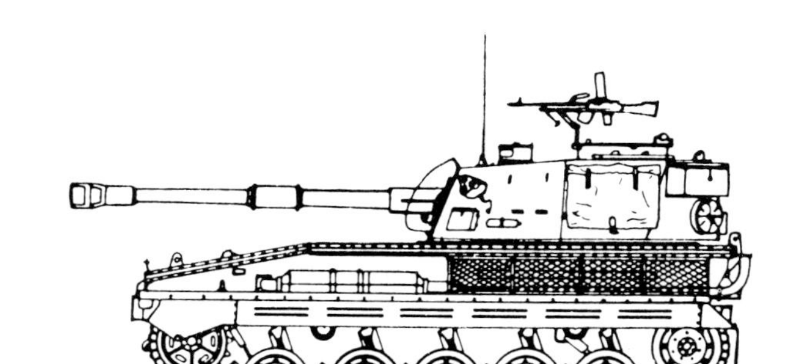

Abbot 105 mm self-propelled gun

the turret rear that opens to the left. A large door, hinged on the right at the rear of the hull, is used for ammunition resupply.

The torsion bar suspension consists of five dual rubber-tyred road wheels with the drive sprocket at the front, idler at the rear and two track return rollers. The first and last road wheel stations have a hydraulic shock absorber. The tracks are cast manganese steel with rubber pads. The tops of the tracks are covered with light steel guards to reduce dust and over the top of these are stowage racks.

The Abbot can ford to a depth of 1.219 m without preparation. A flotation screen is carried folded around the top of the hull and when erected, which takes about 15 to 30 minutes, the vehicle is propelled in the water by its tracks at 5 km/h.

Main armament consists of a 105 mm gun, the L13A1, designed by the Royal Armament Research and Development Establishment at Fort Halstead, Kent, with a double baffle muzzle brake and a fume extractor. The weapon has an elevation of +70°, depression of -5° and turret traverse of 360°. Gun elevation and depression are manual, while the turret traverse is powered, with manual controls for emergency use. The gun is mounted in a ring cradle with twin-hydraulic buffers and a single hydro-pneumatic recuperator, and has a semi-automatic vertical sliding breech-block which opens on run-out, ejecting the empty case.

The ammunition, which is the same as that used in the towed 105 mm Light Gun, is separate loading: the projectile is loaded with the aid of an electrically operated rammer and the cartridge by hand. Maximum rate of fire is 12 rds/min; 40 projectiles are carried of which six are HESH. The following types of ammunition can be fired:

HE(L31) with the projectile weighing 16.1 kg to a maximum range of 17 000 m
HESH(L42) with the projectile weighing 10.49 kg
Illuminating (L43)
SH/PRAC (L41)
Smoke (L45) with the projectile weighing 15.98 kg
Smoke, Coloured, Red (L37)
Smoke, Coloured, Orange (L38)

There are eight charges in the cartridge system. These are fired from two separate cartridges: supercharge cartridge containing supercharge only and normal cartridge containing charges one to five. In addition to these, two sub-zone charges (A and B) are provided. These are fired from a normal cartridge case, emptied of its original charge bags.

The fire control system consists of a periscopic dial sight mounted in the roof and protected by a rotating cupola and a direct fire sight. A 7.62 mm Bren light machine gun is pintle-mounted on the commander's cupola and there are three electrically operated smoke dischargers mounted either side of the turret, towards the front.

Variant

Value Engineered Abbot

The Value Engineered Abbot, the first prototype of which was completed in 1967, is the standard Abbot with the following modifications: no anti-aircraft machine gun, charging by an air-cooled alternator driven from the engine rather than two air-cooled alternators, commander's cupola does not rotate and he has a single periscope and twin hatch covers that open either side of his position, dial sight replaced by a German sight in an unarmoured hood, no fire warning system or flotation screen, loader has a split hatch cover, manganese steel tracks are not fitted with rubber pads, only one set of batteries is installed, night vision and NBC system are not provided, the Rolls-Royce K60 Mk 60G/1 runs on diesel fuel only and develops 213 bhp at 3750 rpm, smoke dischargers are not fitted and turret traverse is by a two-speed hand wheel rather than powered.

Total production of the Abbot for the British Army amounted to 146 vehicles. The British Army also purchased 20 Value Engineered Abbots for use in Canada while India purchased 68 Value Engineered Abbots, plus a small quantity of FV432 APCs for use in the command post role.

Status: Production complete. Abbot is in service with the United Kingdom. Value Engineered Abbot in service with India and the United Kingdom.

Manufacturer: Vickers Defence Systems
Armstrong Works
Scotswood Road
Newcastle-upon-Tyne NE99 1BX
UK
Telephone: (091) 273 8888
Telex: 53104
Fax: (091) 273 2324

SPECIFICATIONS (British Army Abbot. Data in square brackets relate to Value Engineered Abbot where this differs from the standard Abbot)

CREW	4
COMBAT WEIGHT	16 556 [15 900] kg
UNLOADED WEIGHT	14 878 [14 200] kg
POWER-TO-WEIGHT RATIO	14.49 [14.11] bhp/tonne
GROUND PRESSURE	0.89 [0.81] kg/cm²
LENGTH GUN FORWARDS	5.84 [5.714] m
LENGTH HULL	5.709 [5.333] m
WIDTH	2.641 m
HEIGHT (without armament)	2.489 m
GROUND CLEARANCE	0.406 m
TRACK	2.184 m
TRACK WIDTH	343 mm
LENGTH OF TRACK ON GROUND	2.844 m
MAX ROAD SPEED	
1st gear	8.9 km/h
2nd gear	12.3 km/h
3rd gear	17.5 km/h
4th gear	24.3 km/h
5th gear	33.8 km/h
6th gear	47.5 km/h
reverse	7.8 km/h
MAX WATER SPEED	5 km/h
FUEL CAPACITY	386 l
MAX ROAD RANGE	390 km
FORDING	1.219 [1.117] m
with preparation	amphibious [n/app]
GRADIENT	60%
SIDE SLOPE	30%
VERTICAL OBSTACLE	0.609 m
TRENCH	2.057 m
ENGINE	Rolls-Royce K60 Mk 4G 6-cylinder, in-line vertically opposed, multi-fuel developing 240 bhp at 3750 rpm [K60 Mk 60G/1 diesel developing 213 bhp at 3750 rpm]
TRANSMISSION	Allison Transmission Division of General Motors TX200 automatic with 6 forward and 1 reverse gears
SUSPENSION	torsion bar
ELECTRICAL SYSTEM	24 V
BATTERIES	4 × 12 V [2 × 12 V]
ARMAMENT	
main	1 × 105 mm howitzer
anti-aircraft	1 × 7.62 mm MG [none]
SMOKE-LAYING EQUIPMENT	2 × 3 smoke dischargers [none]
AMMUNITION	
main	40 [36]
anti-aircraft	1200 [none]
GUN CONTROL EQUIPMENT	
Turret power control	powered/manual [manual]
Gun elevation/depression	+70°/-5°
Turret traverse	360°
ARMOUR	
Hull front	12 mm
Hull sides	12 mm
Hull top	10 mm
Hull floor	6 mm
Hull rear	10 mm
Turret front	10 mm
Turret sides	10 mm
Turret rear	10 mm
Turret top	12 mm

UNITED STATES OF AMERICA

M110 series of 203 mm (8-inch) Self-propelled Howitzers

Development

The standard heavy self-propelled artillery pieces of the United States Army in the early 1950s were the M41 (155 mm), M43 (203 mm), M53 (155 mm) and the M55 (203 mm), which could not however be carried by transport aircraft in service at that time. Future United States Army requirements for new heavy artillery emphasised air-transportability, time into and out of action, common parts and interchangeability area of fire. A feasibility study encompassing these requirements was carried out and presented and approved at a meeting in January 1956. The Pacific Car and Foundry Company (now PCF Defense Industries) submitted a concept study for a new family of self-propelled weapons and was subsequently awarded a contract for the design, development and construction of six prototype vehicles: two 175 mm self-propelled guns designated the T235, three 8-inch (203 mm) self-propelled howitzers designated the T236 and one 155 mm self-propelled gun designated the T245.

Major design features included the interchangeability of the 175 mm gun, 8-inch howitzer and 155 mm gun in a common mount, on a common chassis; use of the 8-inch howitzer and 155 mm gun field piece weapons and portions of the M17 standard mount, drastic reduction in the size and weight over conventional equipment which was made possible by the use of a new hydraulic lock-out system.

The chassis was also considered suitable for use as a light recovery vehicle and in 1957 the programme was expanded to include both armoured (T120) and unarmoured (T119 and T121) recovery vehicles.

Ordnance service tests with the first prototype chassis began late in 1958 but in 1959 a policy was established that diesel rather than petrol engines would be used for future vehicles and three of the prototypes, the T235, T236 and the T120 were retrofitted with diesel engines and designated the T235E1, T236E1 and the T120E1. Trials with the T235E1 and T236E1 were completed early in 1961 and in March of that year both vehicles were standardised, the T235E1 as the M107 and the T236E1 as the M110. The T120E1 was subsequently standardised as the M578 ARV but the T245, T119 and the T121 were not developed further.

In June 1961 the Pacific Car and Foundry Company (now PCF Defense Industries) was awarded an initial production contract for both the M107 and the M110. First production vehicles were completed in 1962 and the first M110 battalion was formed at Fort Sill early in 1963. Production of the M110 was also undertaken at a later date by the FMC Corporation of San José and Bowen-McLaughlin-York (BMY) of York, Pennsylvania. Original production of the M110 was completed in the late 1960s by which time about 750 vehicles had been produced. In FY78 209 M110A2s were ordered at a cost of $109.3 million for delivery from 1979. No further funds for the vehicle were requested in FY79 or FY80 as with the last buy the United States Army has 93 per cent of its planned total, and its 175 mm M107 self-propelled guns have been converted to M110A1 and M110A2 configuration.

The first M110A1 was produced by Bowen-McLaughlin-York in May 1980 (it produced its first M110 during 1965). The first M110A2 was produced in February 1980.

The M110 series was issued to the United States Army on the scale of one battery of four M110s per infantry division and one battalion of 12 M110s to each mechanised infantry division and armoured division. Under the US Division 86 plan, M110A2s are being integrated with MLRSs into a combined M110A2/MLRS battalion. This will have 12 M110A2s (two batteries each with six M110A2s) and one MLRS battery with 9 MLRSs.

More recently there have been further US artillery changes, especially in VII Corps based in Germany. These have included conversion of all 155 mm battalions from three six-gun batteries to three eight-gun batteries, conversion of the 8-inch battalions from 3 × 4 and 3 × 6 to three batteries each with eight guns and conversion of divisional 8-inch and MLRS battalions to separate MLRS batteries.

Description

The M110 is operated by a team of 13, of which five (commander, driver and three gunners) are carried on the gun and the rest in the M548 tracked cargo carrier which also carries the ammunition.

The hull of the M110 is identical to that of the M107 and is made of all-welded armour and high-tensile alloy steel with the driver at the front of the hull on the left, engine to his right and the main armament at the rear.

The driver, who is the only member of the crew to be seated under armour, has a single piece hatch cover in front of which are three M17 periscopes. The engine is coupled to the Allison Transmission Division of General Motors XTG-411-2A cross-drive transmission at the front of the hull.

The torsion bar suspension consists of five dual rubber-tyred road wheels with the drive sprocket at the front and the fifth road wheel acting as the idler. There are no track return rollers. Attached to each road wheel is a hydraulic cylinder which serves as a shock absorber, hydraulic bump-stop and suspension lock-out which transmits recoil shock directly to the ground. The tracks are of the single pin type with removable rubber pads.

The M110 has infra-red night vision equipment but no NBC system or amphibious capability. Some M110s of the United States Army have been fitted with a lightweight cover around the howitzer to provide some protection for the crew in wet weather.

M110A2 self-propelled howitzer of Pakistani Army

British Army M110A2 deployed in firing position (Michel C Klaver)

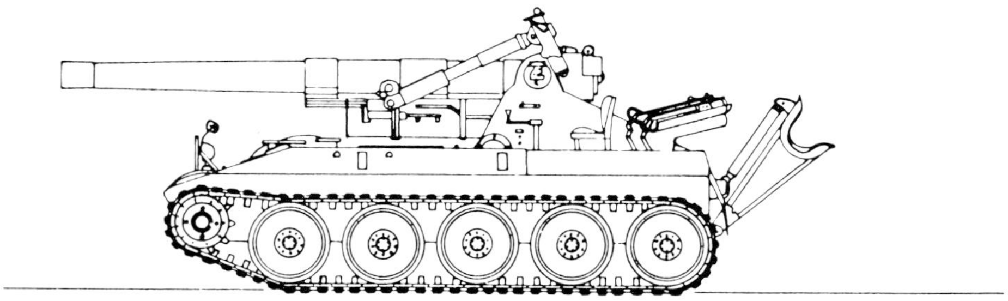

M110 203 mm self-propelled howitzer in travelling order

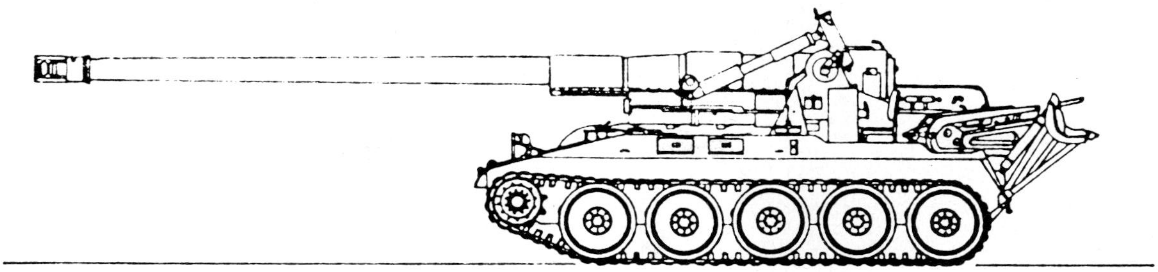

M110A2 203 mm self-propelled howitzer in travelling order

The M110 is armed with a 203 mm (8-inch) M2A2 howitzer in mount M158, and has an elevation of +65°, depression of -2° and a traverse of 30° left and 30° right. Elevation, depression and traverse are hydraulic with manual controls for emergency use. The weapon has a hydro-pneumatic recoil system, interrupted screw breech-block and a percussion firing mechanism. Mounted at the rear of the chassis on the left side is a rammer and loader assembly which lifts a projectile from the rear or left side of the vehicle, positions and rams it into the chamber. Hydraulic power for operating the system is obtained from the vehicle's system but it can also be operated by handcranks. Mounted at the rear of the chassis is a large hydraulically operated spade which is lowered into position before firing begins. The weapon fires the following types of ammunition:

HE(M106) with the projectile weighing 92.53 kg, maximum muzzle velocity of 587 m/s to a maximum range (charge 7) of 16 800 m

HE (M404) (carries 104 M43A1 grenades) with a projectile weighing 90.72 kg, maximum muzzle velocity of 587 m/s to a maximum range (charge 7) of 16 800 m

Agent GB or VX (M426) with the projectile weighing 92.53 kg, maximum muzzle velocity of 587 m/s to a maximum range (charge 7) of 16 800 m

Nuclear (M422)

The ammunition is separate loading and two rounds are carried on the vehicle and the rest in the supporting vehicle. The normal rate of fire is one round every two minutes but two rounds a minute can be fired for short periods.

Fire control equipment consists of a panoramic sight M115 (magnification of ×4 and 10° field-of-view) for indirect fire, telescope M116C (magnification of ×3 and 13° field-of-view) for direct fire, elevation quadrant M15 and gunner's quadrant M1A1.

Variants

Apart from the M110A1 and M110A2 described later, there are no variants of the M110 although the chassis of the M107 175 mm self-propelled gun and the M578 ARV are identical to the M110.

PCF Defense Industries has designed an improvement retrofit package for the M110A2. This has improvements to the hydraulics, electrical system, emplacement and recoil systems, a shelter of Kevlar/aluminium construction to provide ballistic protection for the crew, M13A1 NBC system and an automatic fire control system. Consideration is also being given to uprating the General Motors 8V-71T engine from the present 405 bhp to 500 bhp. This package has been tested at Fort Sill.

M110A1 and M110A2

In 1969 the United States Armament Command began the development of a new version of the M110 which would have a longer range and fire a new range of improved ammunition. This was standardised as the M110A1 in March 1976 and entered service in January 1977. The M110A1 replaced both the M110 and M107, which were phased out of service with the United States Army in Europe by 1980. It then cost less than $100 000 to convert the M107/M110 to the M110A1/M110A2 standard.

The M110A1 has a new and much longer barrel called the M201, a direct fire elbow telescope, the M139, and a chassis identical to the M110's. The M110A2, which was standardised in 1978, is the M110A1 fitted with a double-baffle muzzle brake and can fire charge 9 of the M118A1 propelling charge whereas the earlier M110A1 can fire only up to charge 8. The weapon can fire the following projectiles:

Agent GB or VX (M426) with the projectile weighing 92.53 kg, maximum muzzle velocity of 711 m/s to a maximum range (charge 8) of 22 900 m

HE (M404) with the projectile weighing 90.72 kg, maximum muzzle velocity of 711 m/s to a maximum range (charge 8) of 17 200 m. This is also referred to as an Improved Conventional Munition and carries 104 anti-personnel grenades

HE (M509A1) carrying 180 anti-personnel/anti-material grenades to a maximum range (charge M188A1) of 22 900 m. This is also referred to as an Improved Conventional Munition and carries dual purpose grenades maximum range (charge M188A1) of 22 900 m. This is also referred to as an Improved Conventional Munition and carries dual purpose grenades.

HE (XM711) which is being developed as the replacement for the M106

HERA (M650) to a maximum range (charge 9) of 30 000 m. This RAP is also used as a spotting round for the XM753 nuclear projectile

Nuclear (RAP) (M753) which is the replacement for the M422 nuclear projectile, maximum range (charge 9) of 30 000 m. This has a W79-1 warhead. This projectile was not deployed outside the United States

Nuclear (M422). This is the standard 203 mm nuclear projectile, maximum velocity of 711 m/s to a maximum range (charge 8) of 18 100 m. This has a W33 warhead. The M424 HE-S is the spotting projectile used with the M422.

M110A2 (Belgium)

In 1982 the Arsenal du Matériel Mécanique et de l'Armement at Rocourt near Liege started to convert Belgian M110s to the M110A2 configuration. This programme, which was combined with a general overhaul of the vehicles, was completed in 1986.

Robotic Technology Demonstration Project

The US Army Human Engineering Laboratory, lead agency in DARCOM for robotics, is undertaking a feasibility study to determine whether a robot can perform loading tasks on the M110A2 203 mm self-propelled howitzer.

To the rear of the M110A2 will be positioned a trailer carrying ammunition. When a crew member calls for a round by pushing a button indicating which type of round he wants, the robot will search for the round, pull it out of the rack, and transfer it to the fuzing machine where the noseplug will be removed and the fuze inserted.

The projectile will then be transferred to the existing hoist on the weapon where the crew will take over. They will set the fuze, ram the round into the barrel, load the propellant charge, close the breech and fire the weapon.

SPECIFICATIONS (M110)

(data in square brackets relate to M110A2)

CREW	5
COMBAT WEIGHT	26 534 [28 350] kg
UNLOADED WEIGHT	24 312 [25 492] kg
POWER-TO-WEIGHT RATIO	15.26 [14.28] bhp/tonne
GROUND PRESSURE	0.76 kg/cm^2
LENGTH GUN FORWARDS (spade up)	7.467 [10.731] m
LENGTH HULL (without gun or spade)	5.72 m
WIDTH	3.149 m
HEIGHT	
to top of barrel travelling	2.93 [3.143] m
to top of mount	2.809 [2.743] m
to top of hull	1.475 [1.47] m
GROUND CLEARANCE	0.44 [0.393] m
TRACK	2.692 m
TRACK WIDTH	457 mm
LENGTH OF TRACK ON GROUND	3.936 m
MAX ROAD SPEED	56 [54.7] km/h
FUEL CAPACITY	1137 [984] l
MAX ROAD RANGE	725 [523] km
FUEL CONSUMPTION	1.568 [1.88] l/km
FORDING	1.066 m
GRADIENT	60%
SIDE SLOPE	30%
VERTICAL OBSTACLE	1.016 m
TRENCH	2.362 [1.905] m
ENGINE	Detroit Diesel model 8V-71T, turbo-charged 2-stroke, liquid-cooled 8-cylinder diesel developing 405 bhp at 2300 rpm
TRANSMISSION	Allison Transmission Division of General Motors, XTG-411-2A cross drive with 4 forward and 2 reverse gears
SUSPENSION	torsion bar
ELECTRICAL SYSTEM	24 V
BATTERIES	4 × 12 V model 6TN
ARMAMENT	1 × 203 mm howitzer
SMOKE-LAYING EQUIPMENT	none
AMMUNITION	2 rounds
GUN CONTROL EQUIPMENT	
Turret power control	hydraulic/manual
Gun elevation/depression	+65°/-2°
Gun traverse	30° left and 30° right

M110A2 self-propelled howitzer of the Japanese Ground Self-Defence Force from the rear (Kensuke Ebata)

Status: Production of the M110 has been completed.

Manufacturer: It was manufactured by Pacific Car and Foundry Company (now PCF Defense Industries), Renton, Washington, FMC Corporation, San José, California and Bowen-McLaughlin-York (BMY) of York, Pennsylvania. Telephone: (717) 225 4781. Fax: (717) 225 4615.

User	M110	M110A1	M110A2
Belgium	-	-	11*
Germany	-	-	226*
Greece	31	-	-
Iran	38**	-	-
Israel	48	-	-
Italy	-	-	23*
Japan	-	-	201 (locally built)
Jordan	-	-	102 (29 new, rest upgraded)
Korea, South	99	-	-
Netherlands	-	-	76 (37 new, rest upgraded)
Pakistan	-	-	40
Spain	12	-	-
Taiwan	-	-	75 to 150
Turkey	-	-	16
UK	-	-	16*
USA	-	-	1046† (Army)
	-	-	25† (Marines)

* Upgraded to M110A2. to be replaced by MLRS
The British Army deployed 12 M110A2 howitzers in Operation Desert Storm.
** Total delivered, some destroyed or captured in Gulf War
† Mixture of upgraded M110/M110A1 and new M110A2

M107 175 mm Self-propelled Gun

Development

The standard heavy self-propelled artillery pieces of the United States Army in the early 1950s were the M41 (155 mm), M43 (203 mm), M53 (155 mm) and the M55 (203 mm), which could not however be carried by transport aircraft in service at that time. Future United States Army requirements for new heavy artillery emphasised air-transportability, time into and out of action, common parts and interchangeability area of fire. A feasibility study encompassing these requirements was carried out and presented and approved at a meeting held in January 1956. The Pacific Car and Foundry Company (now PCF Defense Industries) submitted a concept study for a new family of self-propelled weapons and was subsequently awarded a contract for the design, development and construction of six prototype vehicles: two 175 mm self-propelled guns designated the T235, three 8-inch (203 mm) self-propelled howitzers designated the T236 and one 155 mm self-propelled gun designated the T245.

Major design features included the interchangeability of the 175 mm gun, 8-inch howitzer and 155 mm gun in a common mount, on a common chassis; use of 8-inch howitzer and 155 mm gun field piece weapons and portions of the M17 standard mount; drastic reduction in the size and weight over conventional equipment which was made possible by the use of a new hydraulic lock-out system.

The chassis was also considered suitable for use as a light recovery vehicle and in 1957 the programme was expanded to include both armoured (T120) and unarmoured (T119 and T121) recovery vehicles.

Ordnance service tests with the first prototype chassis began late in 1958 but in 1959 a policy was established that diesel rather than petrol engines would be used for future vehicles and three of the prototypes, the T235, T236 and T120, were retrofitted with diesel engines and designated the T235E1, T236E1 and the T120E1. Trials of the T235E1 and T236E1 were completed early in 1961 and in March 1961 both vehicles were standardised, the T235E1 as the M107 and the T236E1 as the M110. The T120E1 was subsequently standardised as the M578 ARV but the T245, T119 and T120 were not developed further.

In June 1961 the Pacific Car and Foundry Company (now PCF Defense Industries) was awarded an initial production contract for both the M107 and M110. First production vehicles were completed in 1962 and the first M107 battalion was formed at Fort Sill in January 1963. Production of the M107 was also undertaken at a later date by the FMC Corporation of San José and Bowen-McLaughlin-York (BMY) of York, Pennsylvania, with the first production model coming off the York line during 1965. Production was completed in May 1980 by which time 524 units had been completed by BMY.

In the United States Army M107s were deployed in battalions of 12 guns, held at Corps level. All M107s in the United States Army and Marine Corps were converted to M110A2s by 1981 and other countries are, or already have, converted their M107s to the M110A2 configuration.

Description

The M107 is operated by a team of 13, five of which (commander, driver and three gunners) are carried on the gun and the rest in the M548 tracked cargo carrier, which also carries the ammunition.

The hull of the M107 is identical to the M110's and is made of all-welded cast armour and high tensile alloy steel, with the driver at the front of the hull on the left, engine to his right and the main armament at the rear.

The driver, who is the only member of the crew seated under armour, has a single piece hatch cover in front of which are three M17 periscopes. The engine is coupled to the Allison Transmission Division of General Motors XTG-411-2A cross-drive transmission which is at the front of the hull.

The torsion bar suspension consists of five dual rubber-tyred road wheels with the drive sprocket at the front and the fifth road wheel acting as the idler. There are no track return rollers. Attached to each road wheel arm is a hydraulic cylinder which serves as a shock absorber, hydraulic bump-stop and suspension lock-out which transmits recoil shock directly to the ground. The tracks are of the single pin type with removable rubber pads.

The M107 has infra-red night vision equipment but does not have an NBC system or any amphibious capability.

The M107 is armed with a 175 mm M113 gun (development designation T256E3) in mount M158, and has an elevation of +65°, a depression of -2° and a traverse of 30° left and right. Elevation, depression and traverse are all hydraulic, with manual controls available for emergency use. The gun has a hydropneumatic recoil system, minimum recoil being 0.711 m and maximum recoil being 1.778 m. The breech-block is of the Welin-step thread type with a percussion firing mechanism. Mounted at the rear

M107 175 mm self-propelled gun of the Israeli Army deployed in firing position (Israel Defence Forces)

Israeli Army M107 mm self-propelled gun in travelling configuration (Israeli Ministry of Defence)

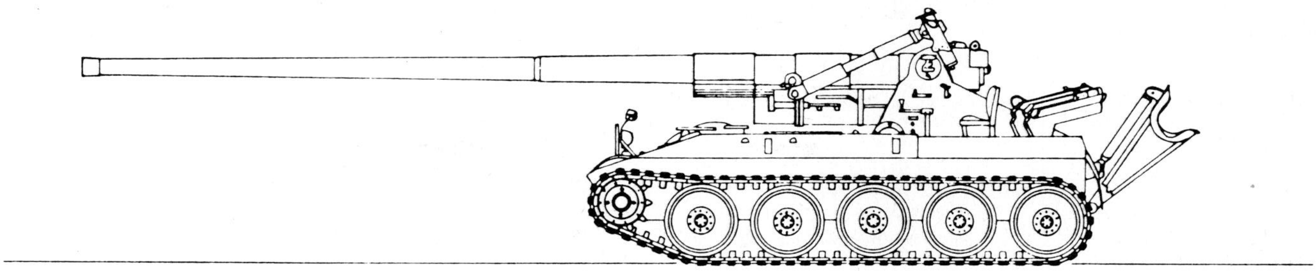

M107 mm self-propelled gun in travelling order

of the vehicle is a rammer and loader assembly which lifts a projectile from the rear or left side of the vehicle, positions and rams it into the chamber. Power for operating this system is obtained from the vehicle's hydraulic system but it can also be operated by handcranks.

The 175 mm gun fires only an HE projectile M437A2 or M437A1. The projectile weighs 66.78 kg with the M437A1 containing 13.6 kg of TNT and the M437A2 containing 14.96 kg of Composition B. Range of the 175 mm gun is as follows:

	Muzzle velocity	**Max range**
Charge 1	509 m/s	15 100 m
Charge 2	702 m/s	22 100 m
Charge 3	912 m/s	32 700 m

In addition the M107 can fire the Space Research Corporation ERSC Mk 7 MOD 7 projectile to a maximum range of some 40 000 m, but this was not developed for the US Army, although it has been used by Israel. The ammunition is of the separate loading type: two rounds are carried on the vehicle and the rest in the supporting vehicle. The normal rate of fire is one round every two minutes but 2 rds/min can be fired for short periods. When travelling the barrel is retracted slightly to the rear to reduce the overall length of the complete vehicle. Mounted at the rear of the chassis is a large hydraulically operated spade which is positioned before firing begins.

Fire control equipment includes a panoramic sight M115 on mount M137 (magnification of ×4 and a 10° field-of-view), elevation quadrant M15, gunner's quadrant M1A1 and direct fire telescope M116C on mount M138 (magnification of ×3 and 13° field-of-view).

Variants

There are no variants of the M107 although the M110 8-inch (203 mm) self-propelled howitzer and the M578 ARV use the same chassis as the M107.

Status: Production complete. In service with the following countries:

Country	**Quantity**	**Remarks**
Greece	36	
Iran	8	some lost/ destroyed in Gulf War
Israel	140	
Italy	18	reserve
Korea, South	n/av	
Spain	12	
Turkey	36	

Manufacturers: Pacific Car and Foundry Company (now PCF Defense Industries), Renton, Washington, USA; FMC Corporation, San José, California; McLaughlin-York (BMY), York, Pennsylvania, USA. Telephone: (717) 225 4781. Fax: (717) 225 4615.

SPECIFICATIONS

CREW (on gun)	5
COMBAT WEIGHT	28 168 kg
UNLOADED WEIGHT	25 915 kg
POWER-TO-WEIGHT RATIO	14.37 hp/tonne
GROUND PRESSURE	0.81 kg/cm²
LENGTH GUN FORWARD (spade up)	11.256 m
LENGTH HULL (without gun or spade)	5.72 m
WIDTH	3.149 m
HEIGHT	
to top of barrel, travelling	3.679 m
to top of mount	2.809 m
to top of hull	1.475 m
GROUND CLEARANCE	0.466 m
TRACK	2.692 m
TRACK WIDTH	457 mm
LENGTH OF TRACK ON GROUND	3.936 m
MAX ROAD SPEED	56 km/h
FUEL CAPACITY	1137 l
MAX ROAD RANGE	725 km
FUEL CONSUMPTION	1.568 l/km
FORDING	1.066 m
GRADIENT	60%
SIDE SLOPE	30%
VERTICAL OBSTACLE	1.016 m
TRENCH	2.362 m
ENGINE	Detroit Diesel model 8V-71T, turbo-charged, 2-stroke, liquid-cooled 8-cylinder diesel developing 405 bhp at 2300 rpm
TRANSMISSION	Allison Transmission Division of General Motors XTG-411-2A cross drive with 4 forward and 2 reverse gears
SUSPENSION	torsion bar
ELECTRICAL SYSTEM	24 V
BATTERIES	4 × 12 V model 6TN
ARMAMENT	1 × 175 mm gun
SMOKE-LAYING EQUIPMENT	none
AMMUNITION (main)	2 rounds
GUN CONTROL EQUIPMENT	
Turret power control	hydraulic/manual
Max rate power traverse	100 mils/s
Max rate power elevation	100 mils/s
Gun elevation/depression	+65°/−2°
Gun traverse	30° left and 30° right

Advanced Field Artillery System (AFAS)

Development

The aim of AFAS is to field a weapon system with a 52-calibre cannon, advanced gun propulsion system, automatic ammunition handing and a significant improvement in survivability over existing systems whose original design is now over 30 years old.

It is expected to have a range of 40 km and be able to fire within 30 seconds of coming to a halt with a rate of fire of 3-6 rds/min. It will have a crew of between four and six distributed between it and the FARV-A and carry up to 60 projectiles.

It is expected that the US Army will eventually order 782 AFASs plus 782 FARV-As which would be issued on a one-for-one basis. The eventual figures, however, could well be affected by a number of factors including the size and shape of the US Army in the 21st century as well as budgetary constraints.

The AFAS is one of the four members of the Common Heavy Chassis which forms part of the Armored System Modernisation (ASM) programme. In addition to the AFAS it also includes the Future Infantry Fighting Vehicle (FIFV), Block III MBT and a Combat Mobility Vehicle (CMV).

The Common Medium Chassis is also part of the ASM and consists of the FARV-A (Future Armored Resupply Vehicle - Artillery) and LOSAT (Line of Sight Anti-Tank).

Additional details on background to the AFAS were given in *Jane's Armour and Artillery 1989-90* page 572. The Human Factors Howitzer Test Bed (covered in the entry on the M109) also has an input to AFAS in a number of areas including reduced manning requirements.

It is expected that deployment of the AFAS will commence in FY 2004 with the FARV-A being deployed the same year. AFAS is being preceded by AFAS Advanced Technology Transition Demonstrator (ATTD) which is covered in the following entry.

Status: Early development.

Manufacturers: Not yet selected.

FMC Advanced Technology Transition Demonstrator (ATTD)

Development/Description

In May 1991, the US Army awarded the FMC Corporation, Ground Systems Division, a $66.8 million cost plus incentive fee contract to build the Advanced Technology Transition Demonstrator (ATTD) for the US Army's future Advanced Field Artillery System (AFAS) (qv).

Production AFAS will be based on the heavy protection level chassis of the US Army's Armored Systems Modernisation (ASM) programme described in the *AFV families* section.

Under current plans, the US Army is expected to purchase 782 AFAS systems with first equipped units entering service in 2004. According to FMC, follow-on development and production contracts for AFAS could have an estimated value of $3 to $4 billion.

Although the award was made to FMC Corporation's Ground Systems Division, who currently manufacture the Bradley family of tracked armoured vehicles, some 76 per cent of the work will be carried out by FMC's Naval Systems Division based in Minneapolis, Minnesota, who have extensive experience in gun systems and ammunition handling.

The AFAS ATTD will utilise a surrogate chassis being provided by FMC Ground Systems with the 52 calibre ordnance being provided as GFE (Government Furnished Equipment).

Key subsystems include an ammunition handling suite, fire control/command and control suite and crew station and operator controls. The US Army has still to decide on whether AFAS will use a liquid propellant or modular charge system.

Under the terms of the four year phased development contract, a single ATTD AFAS will be built and tested by FMC and delivered to the US Army for field trials by January 1995. This phase will be followed by full scale development, estimated to be worth $750 million, and then full scale production, with a competition for each of these phases.

At present there is only one manufacturer of self-propelled artillery systems in the US, BMY Combat Systems, who are currently producing the M109A6 Paladin (qv). They did not bid for the ATTD but are expected to compete for the full scale development and production phases.

Other members of the FMC ATTD team are understood to include Magnavox, Delco Electronics, Wegmann, Rheinmetall, Dynamics Research Corporation, Vector Research Inc, Armored Vehicle Technologies Associates and LB and M Associates.

Also competing for the ATTD contract were General Motors, Military Vehicles Operations and United Armaments Company, owned jointly by General Dynamics, Land Systems Division and Vickers Shipbuilding and Engineering Limited.

The ATTD is being built to reduce risk in the mission module components and system integration prior to full prototypes of the system being built. This allows testing of key components prior to full scale development.

Status: Development.

Manufacturer: FMC Corporation
Naval Systems Division
4800 East River Road
Minneapolis
Minnesota 55421-1498
USA
Telephone: (612) 571 9201
Fax: (612) 337 3001

Liquid Propellant Artillery Systems

Development/Description

The advantages of using liquid propellants (LP) for artillery in place of time-honoured solid propellants have attracted US Army artillery planning authorities since the 1950s. By using LP, considerable savings could be introduced into logistical supply systems and there are also many advantages that LP would offer in onboard storage, ease of onboard handling and operation, and potential crew safety enhancement in the event of enemy strikes or onboard fires.

By the late 1960s General Electric had begun a private venture LP technology programme based at its Burlington facility in Vermont. Early work centred on the use of bulk-loaded bi-propellants for small calibre test guns but by 1974 the emphasis had changed to regeneratively pumped loading methods and monopropellants. By then the work was centred at Pittsfield, Massachusetts.

By 1975 it had been decided that the regeneratively pumped direct injection LP gun concept held the most promise. At that time Otto Fuel II (normally used for torpedos) was the most widely used monopropellant but later a less toxic monopropellant developed by the US Naval Ordnance Station at Indianhead was introduced.

In 1978 the US Army provided General Electric with funds to carry out feasibility studies to increase the calibre of their test guns from the 25 mm used until then to 105 mm. Subsequent testing involved a 105 mm M68 tank gun tube which was selected purely for test purposes and not with the intention that a tank gun was the prime priority. By 1983 more than 2000 LP test firings had been made using test rigs that included calibres of 0.35-inch (89 mm), 25 mm, 30 mm and 105 mm. By 1988 the firing total had reached 3500, with some of the 30 mm firings involving five-round bursts at a cyclic rate of 500 rds/min.

During 1983 General Electric began the construction of a specialised three-cell test facility for LP guns at Malta, New York. By then, other companies had expressed an interest in the LP concept and the US Army provided numerous small study and investigative contracts to various commercial and academic concerns.

In 1986 the US Army's Ballistic Research Laboratory provided General Electric with a $42 million contract for a practical 155 mm LP howitzer demonstrator.

By 1988 Gun No 1 had commenced its firing trials. The test gun involved is mounted on a M115 towed carriage normally used for the 8-inch (203 mm) towed howitzer. The tube involved is a modified XM283 with a specially designed swing breech. Firing is carried out at a fixed 0° elevation. Recoil is 1.752 m.

In July 1990, General Electric stated that it was to install its Defender 155 mm liquid propellant gun on an M109 self-propelled howitzer chassis.

In addition, this will have a fire control system, elevation drives, three round automatic loading system and a liquid propellant storage and feed system.

In 1991 the company expects to demonstrate the following capabilities with Defender:

(1) Ranges from less than 6 km to more than 40 km
(2) Firing of three rounds in 15 seconds
(3) Having three rounds land simultaneously on target
(4) Velocity repeatability of 0.2 per cent with the first-generation gun already demonstrating 0.2 per cent
(5) Propellant fill accuracy of 0.032 per cent, a figure already attained by the first-generation gun.

It is anticipated that with a computerised fire control system, Defender should eventually be capable of placing six rounds simultaneously on a target at ranges from 6 km to 40 km.

Sensitivity test of the Defenders liquid propellant has already been completed. Direct hits by 81 mm HEAT rounds have failed to ignite the

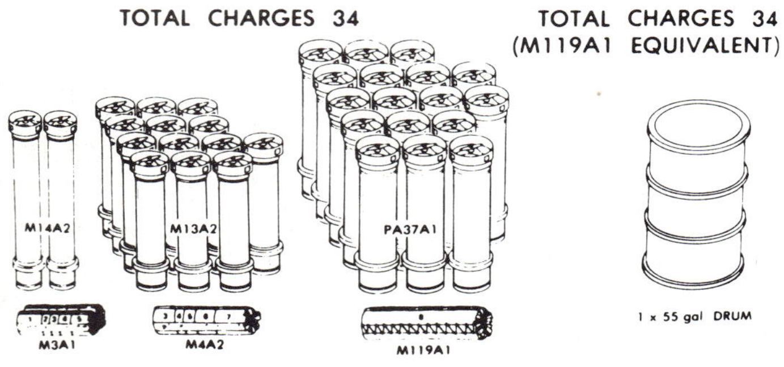

A graphic indication of the savings possible with liquid propellants (LP) showing how the equivalent of 34 propellant charges carried on a M109A2 155 mm self-propelled howitzer could be accommodated in a single 55-gallon (250-litre) drum of LP

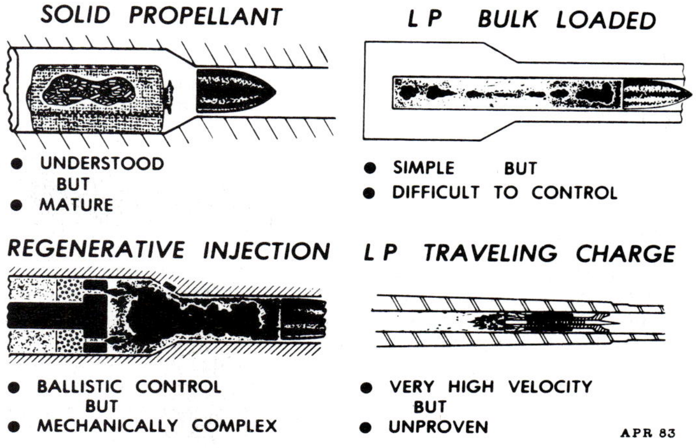

The evolution of artillery propellants from solid propellants (top left) to the futuristic liquid propellant (LP) travelling charge (bottom right)

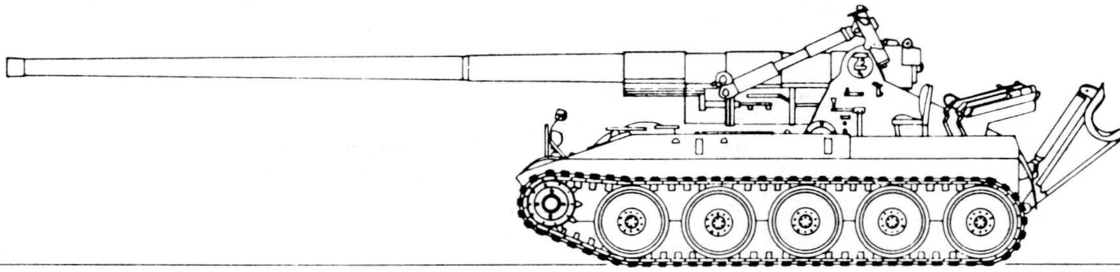

M107 mm self-propelled gun in travelling order

of the vehicle is a rammer and loader assembly which lifts a projectile from the rear or left side of the vehicle, positions and rams it into the chamber. Power for operating this system is obtained from the vehicle's hydraulic system but it can also be operated by handcranks.

The 175 mm gun fires only an HE projectile M437A2 or M437A1. The projectile weighs 66.78 kg with the M437A1 containing 13.6 kg of TNT and the M437A2 containing 14.96 kg of Composition B. Range of the 175 mm gun is as follows:

	Muzzle velocity	**Max range**
Charge 1	509 m/s	15 100 m
Charge 2	702 m/s	22 100 m
Charge 3	912 m/s	32 700 m

In addition the M107 can fire the Space Research Corporation ERSC Mk 7 MOD 7 projectile to a maximum range of some 40 000 m, but this was not developed for the US Army, although it has been used by Israel. The ammunition is of the separate loading type: two rounds are carried on the vehicle and the rest in the supporting vehicle. The normal rate of fire is one round every two minutes but 2 rds/min can be fired for short periods. When travelling the barrel is retracted slightly to the rear to reduce the overall length of the complete vehicle. Mounted at the rear of the chassis is a large hydraulically operated spade which is positioned before firing begins.

Fire control equipment includes a panoramic sight M115 on mount M137 (magnification of ×4 and a 10° field-of-view), elevation quadrant M15, gunner's quadrant M1A1 and direct fire telescope M116C on mount M138 (magnification of ×3 and 13° field-of-view).

Variants

There are no variants of the M107 although the M110 8-inch (203 mm) self-propelled howitzer and the M578 ARV use the same chassis as the M107.

Status: Production complete. In service with the following countries:

Country	**Quantity**	**Remarks**
Greece	36	
Iran	8	some lost/ destroyed in Gulf War
Israel	140	
Italy	18	reserve
Korea, South	n/av	
Spain	12	
Turkey	36	

Manufacturers: Pacific Car and Foundry Company (now PCF Defense Industries), Renton, Washington, USA; FMC Corporation, San José, California; McLaughlin-York (BMY), York, Pennsylvania, USA. Telephone: (717) 225 4781. Fax: (717) 225 4615.

SPECIFICATIONS

CREW (on gun)	5
COMBAT WEIGHT	28 168 kg
UNLOADED WEIGHT	25 915 kg
POWER-TO-WEIGHT RATIO	14.37 hp/tonne
GROUND PRESSURE	0.81 kg/cm²
LENGTH GUN FORWARD (spade up)	11.256 m
LENGTH HULL (without gun or spade)	5.72 m
WIDTH	3.149 m
HEIGHT	
to top of barrel, travelling	3.679 m
to top of mount	2.809 m
to top of hull	1.475 m
GROUND CLEARANCE	0.466 m
TRACK	2.692 m
TRACK WIDTH	457 mm
LENGTH OF TRACK ON GROUND	3.936 m
MAX ROAD SPEED	56 km/h
FUEL CAPACITY	1137 l
MAX ROAD RANGE	725 km
FUEL CONSUMPTION	1.568 l/km
FORDING	1.066 m
GRADIENT	60%
SIDE SLOPE	30%
VERTICAL OBSTACLE	1.016 m
TRENCH	2.362 m
ENGINE	Detroit Diesel model 8V-71T, turbo-charged, 2-stroke, liquid-cooled 8-cylinder diesel developing 405 bhp at 2300 rpm
TRANSMISSION	Allison Transmission Division of General Motors XTG-411-2A cross drive with 4 forward and 2 reverse gears
SUSPENSION	torsion bar
ELECTRICAL SYSTEM	24 V
BATTERIES	4 × 12 V model 6TN
ARMAMENT	1 × 175 mm gun
SMOKE-LAYING EQUIPMENT	none
AMMUNITION (main)	2 rounds
GUN CONTROL EQUIPMENT	
Turret power control	hydraulic/manual
Max rate power traverse	100 mils/s
Max rate power elevation	100 mils/s
Gun elevation/depression	+65°/-2°
Gun traverse	30° left and 30° right

Advanced Field Artillery System (AFAS)

Development

The aim of AFAS is to field a weapon system with a 52-calibre cannon, advanced gun propulsion system, automatic ammunition handing and a significant improvement in survivability over existing systems whose original design is now over 30 years old.

It is expected to have a range of 40 km and be able to fire within 30 seconds of coming to a halt with a rate of fire of 3-6 rds/min. It will have a crew of between four and six distributed between it and the FARV-A and carry up to 60 projectiles.

It is expected that the US Army will eventually order 782 AFASs plus 782 FARV-As which would be issued on a one-for-one basis. The eventual figures, however, could well be affected by a number of factors including the size and shape of the US Army in the 21st century as well as budgetary constraints.

The AFAS is one of the four members of the Common Heavy Chassis which forms part of the Armored System Modernisation (ASM) programme. In addition to the AFAS it also includes the Future Infantry Fighting Vehicle (FIFV), Block III MBT and a Combat Mobility Vehicle (CMV).

The Common Medium Chassis is also part of the ASM and consists of the FARV-A (Future Armored Resupply Vehicle - Artillery) and LOSAT (Line of Sight Anti-Tank).

Additional details on background to the AFAS were given in *Jane's Armour and Artillery 1989-90* page 572. The Human Factors Howitzer Test Bed (covered in the entry on the M109) also has an input to AFAS in a number of areas including reduced manning requirements.

It is expected that deployment of the AFAS will commence in FY 2004 with the FARV-A being deployed the same year. AFAS is being preceded by AFAS Advanced Technology Transition Demonstrator (ATTD) which is covered in the following entry.

Status: Early development.

Manufacturers: Not yet selected.

FMC Advanced Technology Transition Demonstrator (ATTD)

Development/Description

In May 1991, the US Army awarded the FMC Corporation, Ground Systems Division, a $66.8 million cost plus incentive fee contract to build the Advanced Technology Transition Demonstrator (ATTD) for the US Army's future Advanced Field Artillery System (AFAS) (qv).

Production AFAS will be based on the heavy protection level chassis of the US Army's Armored Systems Modernisation (ASM) programme described in the *AFV families* section.

Under current plans, the US Army is expected to purchase 782 AFAS systems with first equipped units entering service in 2004. According to FMC, follow-on development and production contracts for AFAS could have an estimated value of $3 to $4 billion.

Although the award was made to FMC Corporation's Ground Systems Division, who currently manufacture the Bradley family of tracked armoured vehicles, some 76 per cent of the work will be carried out by FMC's Naval Systems Division based in Minneapolis, Minnesota, who have extensive experience in gun systems and ammunition handling.

The AFAS ATTD will utilise a surrogate chassis being provided by FMC Ground Systems with the 52 calibre ordnance being provided as GFE (Government Furnished Equipment).

Key subsystems include an ammunition handling suite, fire control/ command and control suite and crew station and operator controls. The US Army has still to decide on whether AFAS will use a liquid propellant or modular charge system.

Under the terms of the four year phased development contract, a single ATTD AFAS will be built and tested by FMC and delivered to the US Army for field trials by January 1995. This phase will be followed by full scale development, estimated to be worth $750 million, and then full scale production, with a competition for each of these phases.

At present there is only one manufacturer of self-propelled artillery systems in the US, BMY Combat Systems, who are currently producing the M109A6 Paladin (qv). They did not bid for the ATTD but are expected to compete for the full scale development and production phases.

Other members of the FMC ATTD team are understood to include Magnavox, Delco Electronics, Wegmann, Rheinmetall, Dynamics Research Corporation, Vector Research Inc, Armored Vehicle Technologies Associates and LB and M Associates.

Also competing for the ATTD contract were General Motors, Military Vehicles Operations and United Armaments Company, owned jointly by General Dynamics, Land Systems Division and Vickers Shipbuilding and Engineering Limited.

The ATTD is being built to reduce risk in the mission module components and system integration prior to full prototypes of the system being built. This allows testing of key components prior to full scale development.

Status: Development.

Manufacturer: FMC Corporation
Naval Systems Division
4800 East River Road
Minneapolis
Minnesota 55421-1498
USA
Telephone: (612) 571 9201
Fax: (612) 337 3001

Liquid Propellant Artillery Systems

Development/Description

The advantages of using liquid propellants (LP) for artillery in place of time-honoured solid propellants have attracted US Army artillery planning authorities since the 1950s. By using LP, considerable savings could be introduced into logistical supply systems and there are also many advantages that LP would offer in onboard storage, ease of onboard handling and operation, and potential crew safety enhancement in the event of enemy strikes or onboard fires.

By the late 1960s General Electric had begun a private venture LP technology programme based at its Burlington facility in Vermont. Early work centred on the use of bulk-loaded bi-propellants for small calibre test guns but by 1974 the emphasis had changed to regeneratively pumped loading methods and monopropellants. By then the work was centred at Pittsfield, Massachusetts.

By 1975 it had been decided that the regeneratively pumped direct injection LP gun concept held the most promise. At that time Otto Fuel II (normally used for torpedos) was the most widely used monopropellant but later a less toxic monopropellant developed by the US Naval Ordnance Station at Indianhead was introduced.

In 1978 the US Army provided General Electric with funds to carry out feasibility studies to increase the calibre of their test guns from the 25 mm used until then to 105 mm. Subsequent testing involved a 105 mm M68 tank gun tube which was selected purely for test purposes and not with the intention that a tank gun was the prime priority. By 1983 more than 2000 LP test firings had been made using test rigs that included calibres of 0.35-inch (89 mm), 25 mm, 30 mm and 105 mm. By 1988 the firing total had reached 3500, with some of the 30 mm firings involving five-round bursts at a cyclic rate of 500 rds/min.

During 1983 General Electric began the construction of a specialised three-cell test facility for LP guns at Malta, New York. By then, other companies had expressed an interest in the LP concept and the US Army provided numerous small study and investigative contracts to various commercial and academic concerns.

In 1986 the US Army's Ballistic Research Laboratory provided General Electric with a $42 million contract for a practical 155 mm LP howitzer demonstrator.

By 1988 Gun No 1 had commenced its firing trials. The test gun involved is mounted on a M115 towed carriage normally used for the 8-inch (203 mm) towed howitzer. The tube involved is a modified XM283 with a specially designed swing breech. Firing is carried out at a fixed 0° elevation. Recoil is 1.752 m.

In July 1990, General Electric stated that it was to install its Defender 155 mm liquid propellant gun on an M109 self-propelled howitzer chassis.

In addition, this will have a fire control system, elevation drives, three round automatic loading system and a liquid propellant storage and feed system.

In 1991 the company expects to demonstrate the following capabilities with Defender:

(1) Ranges from less than 6 km to more than 40 km
(2) Firing of three rounds in 15 seconds
(3) Having three rounds land simultaneously on target
(4) Velocity repeatability of 0.2 per cent with the first-generation gun already demonstrating 0.2 per cent
(5) Propellant fill accuracy of 0.032 per cent, a figure already attained by the first-generation gun.

It is anticipated that with a computerised fire control system, Defender should eventually be capable of placing six rounds simultaneously on a target at ranges from 6 km to 40 km.

Sensitivity test of the Defenders liquid propellant has already been completed. Direct hits by 81 mm HEAT rounds have failed to ignite the

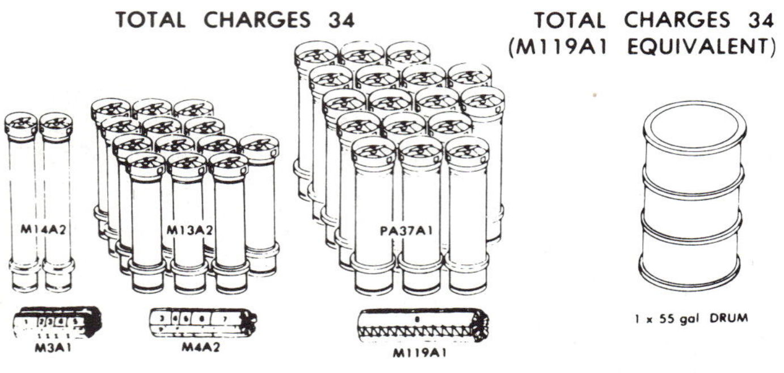

A graphic indication of the savings possible with liquid propellants (LP) showing how the equivalent of 34 propellant charges carried on a M109A2 155 mm self-propelled howitzer could be accommodated in a single 55-gallon (250-litre) drum of LP

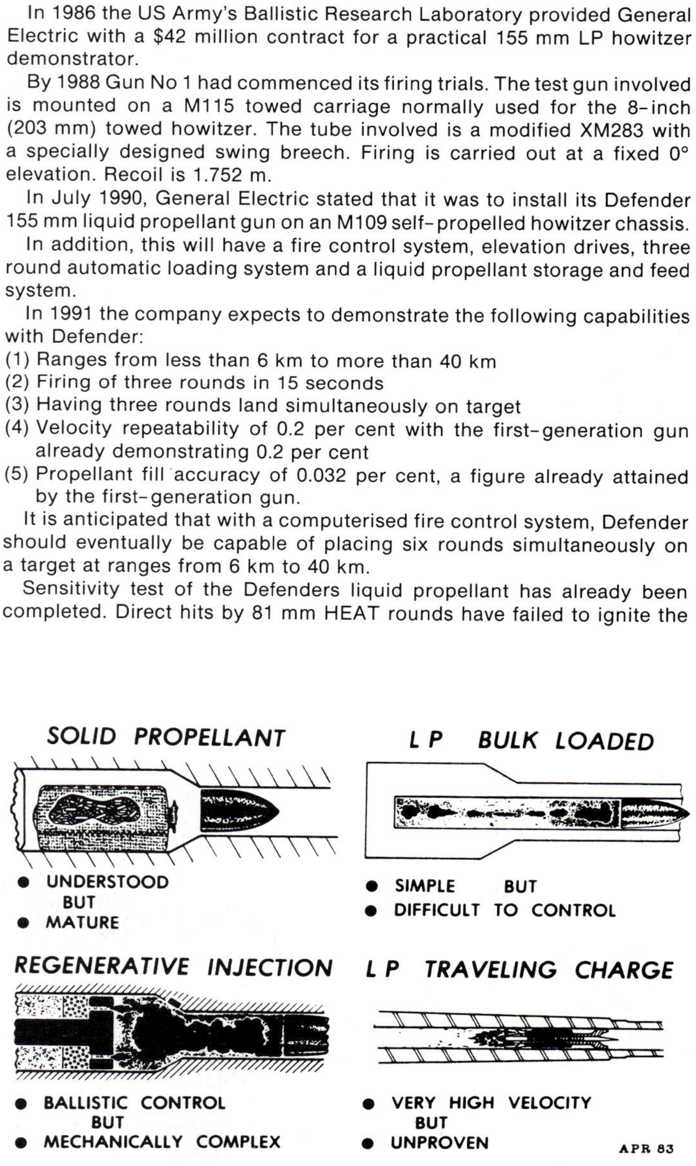

The evolution of artillery propellants from solid propellants (top left) to the futuristic liquid propellant (LP) travelling charge (bottom right)

propellant according to General Electric. In addition to enhancing gun crew survivability, liquid propellant also produces no tell-tale muzzle flash so making visual detection of the weapon system more difficult.

General Electric is designing a breech mechanism so that standard solid propellants could be used with the aid of a removable breech plug. When the latter is removed it would be supplemented by a sleeve to restore the chamber volume to a size prescribed for solid propellants.

Status: Development - see text.

Prime contractor: General Electric Aerospace
Defense Systems Division
Tactical Systems Department
Pittsfield
Massachusetts 01201
USA
Telephone: (413) 494 4255

BMY M109 Series of 155 mm Self-Propelled Howitzers

Development

In January 1952 a conference was held in Washington DC on the subject of self-propelled artillery which indicated that there was an urgent need for improved self-propelled artillery. Shortly after this conference preliminary concept studies began for a self-propelled howitzer to replace the 155 mm M44. The first studies for the design of the new vehicle, called the Howitzer, 156 mm, Self-propelled, T196, were presented to CONARC in August 1952, but were rejected, as were additional studies presented in September 1953. At a conference in May 1954 a concept was finally approved for presentation to CONARC.

In June 1954 a conference was held at Fort Monroe to review the military characteristics and to determine the whole self-propelled programme. It was then decided that future concepts of the T196 would be prepared along the design proposed for the Howitzer, 110 mm, Self-propelled, T195. Concept studies continued along these lines and in June 1956 it was finally decided to use the basic hull and turret of the T195 but with a 155 mm howitzer instead of the original 156 mm howitzer.

In October 1956 the mock-up of the T196 was reviewed and verbal authority was given to proceed with the development of the first prototype. Main differences from the T195 were power elevating and turret traverse mechanisms, different ammunition racks and two spades at the rear of the hull.

The first prototype of the T196 was completed in 1959, about six months later than the 105 mm T195. During preliminary User Evaluation at Fort Knox a number of failures occurred in the suspension. The prototype differed from later vehicles in that it had a different shaped hull and turret, the seventh road wheel acted as the idler and it was powered by a Continental petrol engine. In 1959 a policy was established that diesel rather than petrol engines would be used for future combat vehicles and the prototype of the T196 was then fitted with a diesel engine and redesignated the T196E1.

In February 1961 an order was placed for two T196E1 pre-production vehicles which were delivered within six months. After further trials the T196E1 was classified as a Limited Production Type in December 1961. Two months before this decision, in October, a letter order was given to the Cadillac Motor Car Division for one year's production of the T196E1 at the Cleveland Army Tank Plant. The first production vehicles were completed in October 1962. In January 1963 an extension was authorised to continue the classification of Limited Production and in July the same year the T196E1 was classified as standard A and designated the Howitzer, Medium, Self-propelled: 155 mm, M109. Early in 1963 a contract was awarded to Cadillac for the second year of production, but the contract for the third year of production, awarded in December 1963, went to the Chrysler Corporation, although production remained at the Cleveland Army Tank Plant. First M109s were issued to the United States Army in June 1963.

Further development of the M109 resulted in the M109A1 and the M109A2 and by early 1979 production had exceeded 4000 units. M109 production for the US Armed Forces by the Cleveland Army Tank Plant was as follows:

FY	US Army	US Marine Corps
1962	217	30
1963	150	58
1964	360	-
1965	360	-
1966	454	-
1967	420	36
1968	-	-
1969	-	26

BMY began production of the M109 in 1974 and issued the following yearly production figures in 1989:

FY	M109A1B	M109A2	Total
1974	99	-	99
1975	222	-	222
1976	164	-	164
1977	183	-	183
1978	166	12	178
1979	-	116	116
1980	74	231	305
1981	49	339	388
1982	260	276	536
1983	120	199	319

One of the prototypes of the BMY Combat Systems M109A6 Paladin firing its 155 mm ordnance at maximum elevation

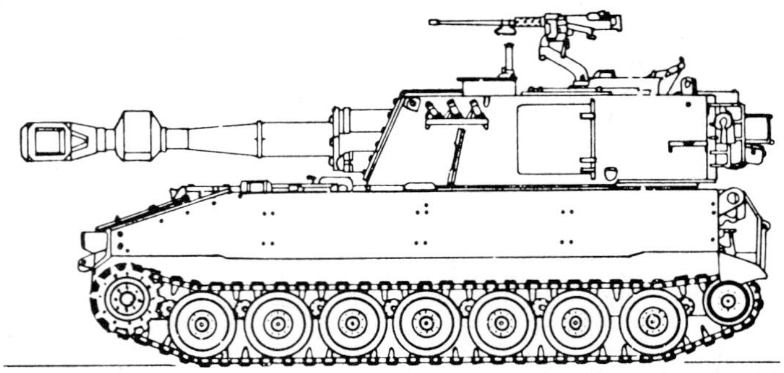

M109 155 mm Self-propelled Howitzer in travelling order

M109 HIP (M109A6)

Following the demise of the Divisional Artillery Support Weapon System (DASWS), a number of companies were involved with the M109-related project known as the Howitzer Extended Life Program (HELP). This programme eventually merged with the M109 Howitzer Improvement Program HIP to bring the M109 up to a new M109A6 standard. Approval for the programme was given in November 1984 and solicitations were released in February 1985. In October 1985 BMY was awarded a development contract to carry out the HIP programme which involved the conversion of eight prototypes (six American and two for the Israeli Defence Forces) over a three-year period. A Special In-Process review of the programme was carried out in April 1986 with a further In-Process Review being carried out during the second quarter of FY87.

The first of the American and Israeli HIP prototypes were rolled out on 30 March 1988 at BMY's York facility and subsequently undertook extensive prototype qualification tests at Aberdeen and Yuma proving grounds. From September 1988 to February 1989 the American prototypes (the M109A3E2) underwent a technical testing programme and operational tests were carried out at Fort Sill from May 1989 onwards.

The US Army made a production decision for the M109A6 on 5 September 1990 with the first production vehicles to be completed early in 1992. The first buy is for 44 M109A6s and was placed in September 1990 with the second contract for 60 vehicles valued at $59.2 million, being placed in April 1991. The US Army has named the M109A6 the Paladin.

BMY Combat Systems is assured of initial production contracts involving approximately 150 vehicles, but follow-on contracts will then be offered on a competitive bid basis.

It is planned that 853 US Army M109s (out of a total fleet of around 3500) will be upgraded to the M109A6 configuration.

BMY Combat Systems is building the new M109A6 turrets with the upgraded chassis being provided by Letterkenny Army Depot. BMY Combat Systems will integrate the chassis and turret and deliver the complete system to the US Army.

For the M109A6 programme, BMY Combat Systems teamed with Alliant Techsystems (fire control equipment), GE (onboard diagnostics/ prognostics), Sondertechnik (ammunition loading assist device), ECC (training material) and Carrier for the NBC microclimate cooling system (MCS). Letterkenny Arsenal will co-ordinate modifications to the M109 chassis. MBT supplies the Israeli fire control equipment.

The M284 cannon is a modified M185 cannon with many improvements including a reinforced muzzle brake, redesigned chamber and forcing cone, two shallow grooves for torque key replaces one deep groove, new breech housing assembly, wiring of fasteners, added XM49 firing mechanism and improved firing block, improved crank lead springs, strengthened breech handle springs and modified plunger, modification of breech ring and longer ordnance.

The M182 gun mount is a modified M178 with an added temperature sensor, redesigned recuperator cylinder, redesigned torque key, improved seals in recoil cylinders and improved seals and new bearing installed in elevation/equilibration cylinders.

With M203/M203A1 charge a range of 30 000 m can be achieved.

Features involved in the M109A6 include the following items:

New turret with improved armour and Kevlar ballistic lining
Full width turret bustle
Automatic fire control system (AFCS)
Onboard ballistic fire control computer and navigation system providing automatic gun pointing
New gun drive servos
SINCGARS radio
Built-in training software
M182 gun mount
M284 cannon which allows the M203 charge to be used
New traversing mechanism
Upgraded suspension system
Driver's night vision system
Electrical and hydraulic system improvements, including a new 650 amp alternator
Increased ammunition stowage
Micro-climate cooling system
Remotely operated barrel travel lock
Segregated hydraulic compartment and hydraulic line fuzing
Engine/transmission disconnect for cold starting
Engine starter protective circuit
Air cleaner blower relay
Engine sensor and connector covers
Crossover tube protective cover
New crew compartment water and oil drains
Slave start circuit
New sensors to monitor power pack parameters
Ventilated face piece system for NBC conditions
Automatic halon fire extinguishing system in engine compartment
Improved hatch and night vision provisions for the commander
External stowage baskets
Final drive disconnect
Propellant segregation
Power pack removal aids
Digital communications link
Four man crew.

Beginning in 1991, the US Army conducted live fire testing against two prototypes remaining from the development programme. The test verified the increase in survivability of the M109A6 resulting from the vulnerability reduction measures taken in the design.

A flick rammer loader assist was installed on the Israeli version of HIP which yielded burst rates of fire of three rounds in less than 13 seconds.

Following the receipt of the initial production howitzers, the US Army will conduct a Production Qualification Test and a Follow-on Operational Test to verify the performance demonstrated during development testing has not been adversely affected by the production process.

According to BMY Combat Systems, the M109A6 has increased artillery survivability and combat power:

Factor	**% increase**
Survivability from new tactics	150%
Survivability when hit	50%
Response times	100%
Range	30%
Operational availability	40%
Crew effectiveness	25%

Description (M109)

The M109 has a crew of six which consists of commander, gunner, three ammunition members and the driver. The hull is made of all-welded aluminium with the driver at the front of the hull on the left, engine to his right and the turret at the rear.

The driver has a single piece hatch cover that opens to the left, in front of which are three M45 periscopes which can be covered by small metal flaps to prevent damage.

The engine is coupled to the Allison Transmission Division of General Motors XTG-411-4A cross drive transmission which is at the front of the hull.

The all-welded aluminium turret at the rear of the hull has a square hatch in each side that opens to the rear, and twin doors in the turret rear. The commander is seated on the right side of the turret and has a cupola which can be traversed through 360°, a single piece hatch cover that opens to the rear and an M27 periscope. Pintle-mounted on the forward part of the commander's cupola is a 12.7 mm (0.50) M2 HB anti-aircraft machine gun. The gunner is seated on the left side of the turret and has a square single piece hatch cover that opens to the right. Twin doors are provided at the rear of the turret for ammunition resupply. Mounted at the rear of the hull, each side of the hull door, is a large spade which is manually lowered to the ground before firing.

The torsion bar suspension consists of seven dual rubber-tyred road wheels with the drive sprocket at the front and the idler at the rear. There are no track return rollers. The tracks are of the single-pin, centre guide type with replaceable rubber pads.

The M109 is fitted with infra-red night vision equipment but does not have an NBC system. The basic vehicle can ford to a depth of 1.828 m without preparation but can be fitted with an amphibious kit consisting of nine air bags, four each side of the hull and one at the front. The bags, which are not carried on the vehicle as part of its normal equipment, are inflated from the vehicle and it can then propel itself across rivers by its tracks at 6.43 km/h.

Main armament is a 155 mm howitzer M126 in mount M127, with a fume extractor and a large muzzle brake. The recoil system is hydropneumatic and the breech-block is of the Welin-step thread type. The weapon has an elevation of +75°, a depression of -5° and the turret can be traversed through a full 360°. Gun elevation and depression and turret traverse are hydraulic with manual controls for emergency use. Normal rate of fire is 1 rd/min but 3 rds/min can be fired for a short period. Fire control equipment includes an elbow telescope M118C for direct fire with a magnification of ×4 and a 10° field-of-view, panoramic telescope M117 for indirect fire with a magnification of ×4 and a 10° field-of-view, quadrant fire control M15 and gunner's quadrant M1A1. The weapon can fire the following types of ammunition:

Agent H/HD (M110) with the projectile weighing 42.91 kg, maximum muzzle velocity of 562.4 m/s to a maximum range (charge 7) of 14 600 m
Cannon Launched Guided Projectile M712 to a range of 16 400 m
HE (M107) with the projectile weighing 42.91 kg, maximum muzzle velocity of 562.4 m/s to a maximum range (charge 7) of 14 600 m
HE (M795) which is replacement for M107, projectile weighs 46.90 kg

US Army M109A2 155 mm self-propelled howitzer with commander manning 12.7 mm M2 HB machine gun (C R Zwart)

HE (M449A1) (carries 60 anti-personnel grenades) with the projectile weighing 43.09 kg, maximum muzzle velocity of 563 m/s to a maximum range of 14 600 m
HE (M483A1 series) (carries 64 M42 and 24 M46 dual-purpose grenades) with the projectile weighing 46.53 kg, maximum muzzle velocity of 535.2 m/s to a maximum range (charge 7) of 14 320 m
HE (M692) (carries 36 anti-tank mines) with the projectile weighing 46.49 kg, maximum muzzle velocity of 535.2 m/s to a maximum range (charge 7) of 14 320 m
M718/M741 (containing nine anti-tank mines) to maximum range of 14 320 m
HE (M731) (carries 36 anti-personnel mines) with the projectile weighing 46.49 kg, maximum muzzle velocity of 535.2 m/s to a maximum range (charge 7) of 14 320 m
HERA (M549) with the projectile weighing 43.54 kg, maximum range, charge 7, 19 300 m
Illuminating (M485) with the projectile weighing 41.73 kg, maximum muzzle velocity of 576.9 m/s to a maximum range (charge 7) of 13 586 m
Illuminating (M818) with projectile weighing 46.26 kg, maximum muzzle velocity 536 m/s to a maximum range of 11 600 m
M454 Nuclear
Smoke (M825) with projectile weighing 46.53 kg, muzzle velocity of 535.2 m/s and maximum range 14 320 m
Smoke BE(M116 series) with the projectile weighing 42.22 kg, maximum muzzle velocity of 562.4 m/s to a maximum range (charge 7) of 14 600 m
Smoke WP (M110 series) with the projectile weighing 44.4 kg, maximum muzzle velocity of 563.9 m/s to maximum range (charge 7) of 14 600 m
VX or GB (M121A1) projectile weighing 45.4 kg, maximum muzzle velocity of 562.4 m to a maximum range (charge 7) of 14 600 m
M804 practice
GB2 (M687) contains Agent GB2 with projectile weighing 42.18 kg.

Variants

M109A1

The M109A1 is basically the M109 fitted with a new and much longer barrel, designated the M185, improved elevation and traverse system and a strengthened suspension system. The M109A1 fires an HE projectile to a range of 18 100 m compared with the 14 600 m of the M109, while an RAP extends the range to 24 000 m. The M109A1 was type classified as Standard A in October 1970 with first conversion kits being made available early in 1972 and the first vehicles, which were converted from standard M109s, becoming operational in 1973. The M109A1 weighs 24 070 kg fully loaded and is 9.042 m long including the barrel. The weapon can fire the following projectiles:
Agent H/HD (M110) with the projectile weighing 42.91 kg, maximum muzzle velocity of 684 m/s to a maximum range (charge 7) of 18 100 m
HE (M795) which is replacement for M107, projectile weighs 46.9 kg
HE (M107) with the projectile weighing 42.91 kg, maximum muzzle velocity of 684.3 m/s to a maximum range (charge 8) of 18 100 m
HE (M483A1) (carries 64 M42 and 24 M46 dual-purpose grenades) with the projectile weighing 46.53 kg, maximum muzzle velocity of 650 m/s to a maximum range (charge 8) of 17 500 m
HE (M692) (carries 36 anti-tank mines) with the projectile weighing 46.49 kg, maximum muzzle velocity of 650 m/s to a maximum range (charge 8) of 17 740 m
HE (M731) (carries 36 anti-personnel mines) with the projectile weighing 46.49 kg, maximum muzzle velocity of 650 m/s to a maximum range (charge 8) of 17 740 m/s
HERA (M549A1) with the projectile weighing 43.54 kg to a maximum range of 23 500 m
M864 Improved Conventional Munition, base bleed, type classified in December 1987, not yet in service
Illuminating (M485) with the projectile weighing 41.73 kg, maximum muzzle velocity of 684.5 m/s to a maximum range (charge 8) of 17 500 m
Illuminating (M818) with projectile weighing 46.26 kg, maximum muzzle velocity of 684.3 m/s to a maximum range of 18 100 m
M804 Practice
Smoke BE (M116) with the projectile weighing 42.22 kg, maximum muzzle velocity of 684 m/s to a maximum range (charge 8) of 18 100 m
Smoke WP (M110 series) with the projectile weighing 44.4 kg, maximum muzzle velocity of 684.3 m/s to a maximum range of 18 100 m
Smoke (M825) with projectile weighing 43.53 kg, muzzle velocity of 660 m/s and maximum range of 17 500 m
VX or GB (M121A1) with the projectile weighing 45.4 kg, maximum muzzle velocity of 684.3 m/s to a maximum range (charge 8) of 18 100 m
Nuclear M454 with W48 warhead and weighing 54.43 kg; length is 863.6 mm. Maximum range is 14 800 m
HE ICM M449A1 carrying 60 anti-personnel grenades to a maximum range of 18 100 m; weight approximately 43.1 kg and length 899 mm
Binary M687, producing GB agent from two separate canisters. Weight 42.18 kg and length 863 mm. Maximum range is 14 400 m
ADAM M692/M731 Area Denial Artillery Munition containing 36 self-destruct anti-personnel mines that are ejected in flight. Maximum range 17 740 m, weight 46.72 kg and length 899 mm
HEAT CLGP M712 Copperhead, maximum glide range 16 400 m; M823 Trainer is used for crew training and is not fired
AT RAAMS M718/M741 containing nine anti-tank mines that are ejected during flight. Weight is 46.72 kg and length 899 mm, maximum range 17 740 m
Practice M804 used in place of M107
GB2 (M687) contains Agent GB2 with projectile weighing 42.18 kg.

M109A2

The M109A2 entered production at BMY Combat Systems, a division of HARSCO, in 1978, with first deliveries made early in 1979. The first production quantity of 103 vehicles was with FY77 funding. Major changes from the M109A1 include a redesigned rammer and improved recoil mechanism, engine operation warning devices, redesigned hatch and door latches, improved hydraulic system and a bustle designed to carry an additional 22 rounds of ammunition.

M109A2 production carried out by BMY for the US Army was as follows:

FY	US Army total
1976	12
1977	103
1978	250
1979	136
1980	96
1981-1983	-
1984	123
1985	70

In addition to the above 36 M109A2s were produced for the National Guard using FY82 funds. There has been no recent procurement of the M109A2/M109A3 by the US Army or Marine Corps.

Israeli Army M109 AL (M109A1) with ordnance in travelling lock and showing extensive external stowage (Israeli Army)

M109A3

The M109A3 is a US Army depot converted M109A1. The M109A3 conversion includes:
140 Boresight Alignment Driver
M127 to M178 Gun Mount Kit
Selected RAM and safety kits, which include: fuel system air purge, drivers instrument panel, bustle/rack, propellant stowage, torsion bar, counter recoil buffer, upper recoil cylinder.

M109A4

The M109A4 is a converted M109A2 or M109A3. The major changes are the incorporation of nuclear, biological and chemical (NBC) requirements and reliability and maintainability (RAM) improvements. Selected improvements are found in the following kits: traverse mechanism, NBC equipment, hydraulic power pack filter, starter protection, external slave start, floor drains.

M109A5

The M109A5 is an upgraded M109A4, with the incorporation of the Reserve Component/Modified Armament System (RC/MAS). The RC/MAS is a range capability improvement kit and firing the M203A1 charge ranges of 30 km can be achieved with rocket assisted projectiles. The RC/MAS kit includes:
M284 cannon assembly
M182 gun mount.

M109A3 (Austria)

In 1987 the Austrian Army fielded two versions of the M109, the M109A1 and the M109A2. Currently Austria is planning to improve their fleet to near the US Army M109A6 Paladin configuration. An additional purchase of this new configured howitzer is also being considered.

M109A2 (Belgium)

In August 1983 the Belgian Minister of Defence signed a contract with BMY for the purchase of 127 M109A2 self-propelled howitzers. The first vehicles were delivered in August 1984 and all were delivered by the end of 1985. These vehicles are used by the six field artillery battalions (108), Brasschaat Artillery School (nine) and the maintenance reserve (seven). The older M109s are to be upgraded to M109A3 standard by the Arsenal du Matériel Mécanique et de l'Armement, at Rocourt, near Liège.

M109A3 (Denmark)

Early in 1989 the Danish Army stated that the modernisation of its 76 M109 155 mm self-propelled howitzers to the M109A3 standard had been completed. The modernisation was carried out in Denmark with kits being supplied by FMS Corporation of Los Angeles, California.

M109 (Germany)

The original German Army vehicles were designated the M109G having a sliding breech developed by Rheinmetall which extends the effective range of the weapon to 18 000 m.

In 1983 the German Army purchased 586 kits to convert its M109Gs to the M109A3G from the FMS Corporation, Los Angeles, California. The newly modified M109A3G has an extended range of 24 000 m and uses a newly developed gun tube made by Rheinmetall but still uses the original M109G sliding breech mechanism. The M109A3G also has a different fire control system and three forward firing smoke dischargers either side of the turret. There is a separate entry under Germany, for the modernised Rheinmetall M109A3G.

M109 (Israel)

The M109A1s of the Israeli Army are designated the M109 AL and have additional stowage racks on the hull and turret sides to allow them to carry more ammunition inside. Israel is considering the M109A6 Paladin for future procurement.

German Army M109G showing smoke dischargers on side of turret

Urdan ARTRAIL Ammunition Resupply Trailer

This has been designed by Urdan Industries of Israel and has been tested with the M109A2 self-propelled howitzer; full details are given in *Jane's Military Vehicles and Logistics 1991-92*, page 406. It carries 44 155 mm artillery projectiles, propellant charges and fuzes. It weighs 6200 kg fully loaded and 4000 kg empty.

M109L (Italy)

There is a separate entry under Italy for the M109L developed by OTO Melara to meet the requirements of the Italian Army.

Korea, South

There is a separate entry in this section for the South Korean M109A2.

The first order was for 272 systems, but more recently a further order for 242 systems was placed with production being undertaken locally.

M109 (Netherlands)

The Netherlands has M109s with their turrets removed for training drivers. In place of the turret is a fully enclosed cabin with observation windows. The Dutch Army was the first in Europe to introduce the conversion of the M109 to the M109A3. In 1980 the Dutch Army released its contract to RDM for the conversion of 136 of its M109s to the M109A3 configuration. RDM purchased all kit equipments from FMS Corporation of Los Angeles, California, who was also its contractor for an earlier parts contract for 86 M109A2s co-produced in the Netherlands for the Dutch Army.

M109 (Norway)

In 1983, the Norwegian Army Material Command exercised an option under the German Army's contract with FMS Corporation, Los Angeles, California, for the procurement of 126 conversion kits to modify its fleet of M109Gs to the M109GN which is virtually identical to the German M109A3G described under Germany earlier in this section. The conversion work was carried out by Kongsberg Vapenfabrik, now Norsk Forsvarsteknologi AS, in Narvik. Deliveries took place from 1988 to 1990.

M109 (Spain)

It is believed that Spain has a total of 96 M109 series 155 mm self-propelled howitzers in service and 48 M108 105 mm self-propelled howitzers. The Spanish Army plans to modernise their fleet to near US Army M109A6 Paladin configuration.

M109 (Switzerland)

In 1968 the Swiss Parliament voted for an order of 140 M109s which entered service as the Panzerhaubitze 66. In 1974 a further 120 M109A1s

M109A1 of Swiss Army which calls it PzHb74 (Swiss Army)

were obtained and at the same time the existing Swiss M109s were converted to M109A1s by fitting the longer barrel. The new M109A1s are known as the Panzerhaubitze 74 and retrofitted vehicles as the Panzerhaubitze 66/74. The 1979 armament programme called for a further purchase of about 200 M109A1s to enable the Swiss Army to field a total of nine 155 mm artillery regiments. This last purchase is known as the Panzerhaubitze 79.

There is a separate entry on the latest Swiss 155 mm M109 Bison self-propelled artillery system under Switzerland.

M109 (Taiwan)

Details of the Taiwanese 155 mm self-propelled howitzers with the ordnance in an open mount are given in this section under Taiwan.

Royal Ordnance M109 Package

There is a separate entry under the United Kingdom for the Royal Ordnance M109 modernisation package.

Human Factors Howitzer Test Bed (HFHTB)

Details of this were given in *Jane's Armour and Artillery 1990-91*, page 601.

M992 Field Artillery Ammunition Support Vehicle

Originally developed as a private venture by Bowen-McLaughlin-York, the XM992 was Type Classified M992 in 1983.

The M992 Field Artillery Ammunition Support Vehicle (FAASV) is basically the chassis of a standard M109 155 mm self-propelled howitzer with the turret replaced by a high fully enclosed superstructure. Inside this superstructure, which has the same protection as the rest of the vehicle, can be stacked 93 155 mm projectiles, 99 propellant charges and 104 fuzes. These items of ammunition can be loaded into the superstructure by a front-mounted crane and are fed from the vehicle along a conveyor to the recipient self-propelled howitzer; this crane was not fitted to US Army M992 vehicles but is offered for export; it was fitted to the Egyptian vehicles. All ammunition handling within the M992 is mechanical. In use the M992 backs up to the user M109 and the conveyor delivers the 155 mm ammunition at the rate of up to 8 rds/min to the M109 bustle or lower hatches. The M992 can have a crew of two plus six passengers and the weight fully loaded is 25 900 kg. Intended primarily for use with M109 field units the M992 can also be used by units with 175 mm, 203 mm, 120 mm and 105 mm self-propelled weapons. With 203 mm units the M992 can carry 48 203 mm projectiles, 53 charges and 56 fuzes.

It has been suggested that the M992 could form the basis for a family of battlefield vehicles including a Fire Direction Centre Vehicle (see this page), Command Post Vehicle (CPV), Medical Evacuation Vehicle (MEV), Maintenance Assistance Vehicle (MAV) and Armoured Forward Area Rearm Vehicle (AFARV). American FAASVs lack the front-mounted hydraulic crane for ammunition supply.

Recent US Army testing and decisions related to the M109A6 Howitzer Improvement Program reaffirmed the need for a FAASV in direct, organic support of each self-propelled gun. Plans are underway to procure additional FAASVs (about 150) so that they are available for approximately 800 M109A6s.

By mid-1991, a total of 792 FAASVs had been ordered from BMY Combat Systems:

United States	675 (delivered)
Egypt	51 (delivered)
Saudi Arabia	60 (ordered late 1990, delivery 1991-92)
Spain	6 (delivered)

In addition a total of 119 Fire Direction Centre vehicles had been built:

Egypt	72
Greece	41
Taiwan	6

Full details of the M992 FAASV and its variants are given in *Jane's Military Vehicles and Logistics 1991-92*, page 473.

Fire Direction Centre Vehicle

The Fire Direction Centre Vehicle (FDCV), based on the chassis and hull of the M992 FAASV, has been tested at the US Army School of Artillery at Fort Sill to prove the feasibility and function of the concept under field and NBC conditions. The equipment used in the FDCV is a mix of 'off-the-shelf' or prototype equipment assembled for the concept for which the internal head-room and armour of the M992 hull is eminently suitable. If required extra bolt-on armour can be provided to protect the specialist personnel and equipment carried.

The following equipment was carried in the prototype FDCV: mast system with OE254 antenna which is 11.28 m high when elevated; mast elevation control box; mast elevation control power unit; packet digital radio; RT524/VRC receiver/transmitter radios (4); printer; liquid cooling vest connect points (9 stations); tactical graphics terminal; tactical display; ANS 2000 land navigation system display; ANS 2000 land navigation system heading reference unit; M13A1 filtered air system for NBC ventilated face piece; M43 detector; M42 alarm; M49 NBC collective protective system; APU; air-conditioner (36 000 BTU cooling/heating); ventilation filter/blower; HEL computer (PDP1134); plotter; CRT terminal; digitiser; liquid cooling vest control box; M10 protective entrance; battery boxes and batteries; liquid cooling vest chilling unit; four extra personnel seats; additional stowage compartments. A total of 119 production vehicles have been completed for the export market.

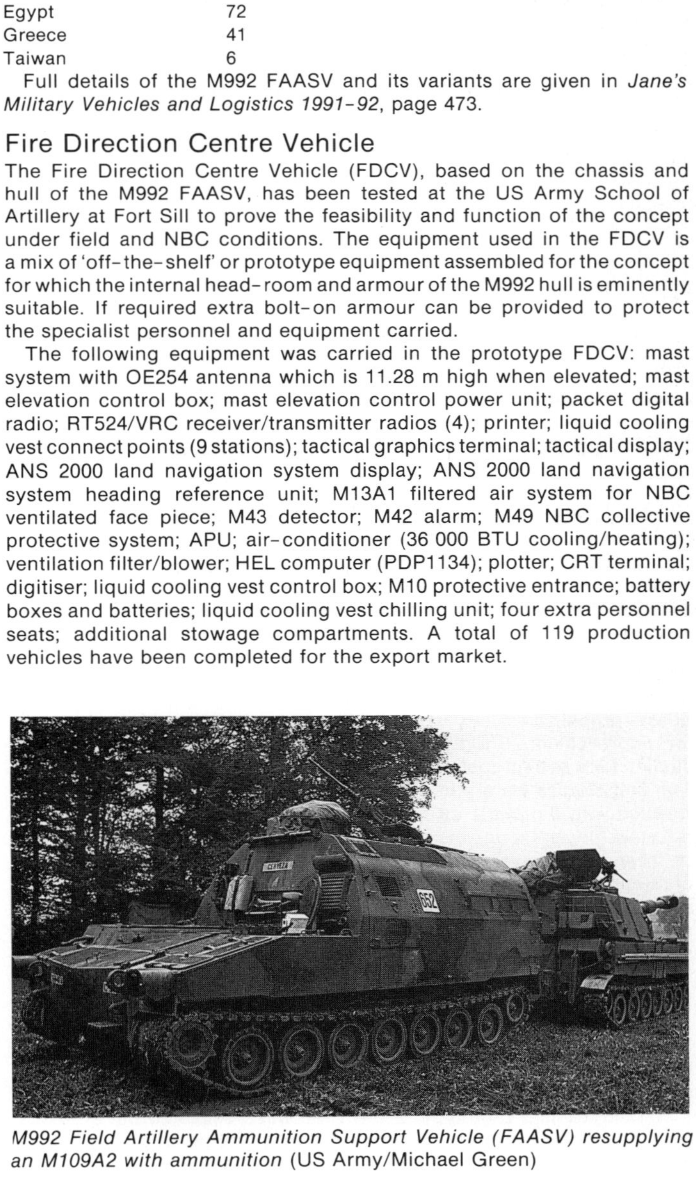

M992 Field Artillery Ammunition Support Vehicle (FAASV) resupplying an M109A2 with ammunition (US Army/Michael Green)

SPECIFICATIONS (M109A2)	
CREW	6
COMBAT WEIGHT	24 948 kg
UNLOADED WEIGHT	21 110 kg
LENGTH GUN	
FORWARDS	9.12 m
LENGTH HULL	6.19 m
WIDTH	3.15 m
HEIGHT	
including anti-aircraft MG	3.28 m
reduced	2.8 m
GROUND CLEARANCE	0.46 m
TRACK	2.788 m
TRACK WIDTH	381 mm
LENGTH OF TRACK	
ON GROUND	3.962 m
MAX ROAD SPEED	56.3 km/h
FUEL CAPACITY	511 l
MAX ROAD RANGE	349 km
FORDING	1.07 m
GRADIENT	60%
SIDE SLOPE	40%
VERTICAL OBSTACLE	0.53 m
TRENCH	1.83 m
ENGINE	Detroit Diesel Model 8V-71T, turbo-charged, 2-stroke, liquid-cooled 8-cylinder diesel developing 405 bhp at 2300 rpm
TRANSMISSION	Allison Transmission Division of General Motors XTG-411-4A cross drive with 4 forward and 2 reverse gears
SUSPENSION	independent torsion bar
ELECTRICAL SYSTEM	24 V
BATTERIES	4 × 12 V model 6TN
ARMAMENT	
main	1 × 155 mm howitzer
anti-aircraft	1 × 12.7 mm or 7.62 mm MG
SMOKE-LAYING EQUIPMENT	none
AMMUNITION	
main	34 plus 2 CLGP
12.7 mm anti-aircraft	500
GUN CONTROL EQUIPMENT	
Turret power control	hydraulic/manual
by commander	no
by gunner	yes
Gun elevation/depression	+75°/-3°
Turret traverse	360°

Status: The M109 and its derivatives are in service with the US Army, Marine Corps and National Guard. Other nations using M109-series weapons are as listed below.

During the first production phase carried out initially by the Cleveland Army Tank Plant, American-configured M109s were delivered to the following nations: Austria, Belgium, Canada, Denmark, Ethiopia, Iran, Israel, Jordan, Netherlands, Spain, Switzerland and the United Kingdom. Italy, Netherlands and Norway formed co-production agreements. Canada, Iran, Israel, Netherlands, Spain, Switzerland and the United Kingdom have converted their M109s to the M109A1 configuration. M109Gs were delivered to Germany, Italy and Norway.

BMY M109A1B FMS production since 1972 has resulted in systems for Greece, Iran, Israel, Italy, Jordan, Kuwait, Morocco, Netherlands, Pakistan, Portugal, Spain, Taiwan and the United Kingdom. BMY M109A2 production since 1980 has resulted in deliveries to Austria, Jordan, Kuwait, Netherlands, Pakistan, Portugal, Spain, Taiwan and the United Kingdom.

Since December 1983 BMY has made direct sales to Belgium, Canada, Egypt, Greece and South Korea.

By mid-1991 M109-series sales were as follows:

Country	M109	M109A1 kit	PIP kit	M109A1B	M109A2
Austria	38	-	-	-	18
Belgium	-	-	41	-	127D
Canada	-	-	50	-	26D
Denmark	76	-	-	-	-
Egypt	-	-	-	-	164D
Ethiopia	12	-	-	-	-
Germany	609*	-	2	-	-
Greece	-	-	-	51	72D
Iran	-	50	-	390	-
Israel	-	60	-	294 + 75D	-
Italy	221*	-	-	62	-
Jordan	-	-	126	-	108
Korea, South	-	-	-	-	242
Kuwait	-	-	-	6**	31**
Morocco	-	-	-	36	36
Netherlands	-	-	135	3	91
Norway	127*	-	-	-	-
Pakistan	-	-	-	-	152
Peru	-	-	-	12	-
Portugal	-	-	-	-	6
Saudi Arabia	-	-	-	87	24
Spain	18	18	-	60	6
Switzerland	-	146	-	327	-
Taiwan	-	-	-	-	125
United Kingdom	-	-	50	-	69
Totals	1101	274	404	1403	1297

D - direct sales
*M109G
**Status uncertain following Iraqi invasion and Operation Desert Storm
Thailand has recently ordered 20 M109A5 for delivery in 1992

Manufacturers: Original manufacturer was Cadillac Motor Car Division at Cleveland Army Tank Plant, later Chrysler Corporation at same plant. Current manufacturer is BMY Combat Systems, York, Pennsylvania, USA. Telephone: (717) 225 4781. Fax: (717) 225 4615.

Also involved in M109A2 production was RDM Defense Engineering, PO Box 913, 3000 AX, Rotterdam, The Netherlands, which manufactured the M109A2 mounting and the turret installation equipment, as well as carrying out the system integration for NATO armament conversions, and for US Army equipment based in Europe (qv).

M44 155 mm Self-propelled Howitzer

Details of this system, developed in the 1940s, were given in *Jane's Armour and Artillery 1990-91*, page 603.

M108 105 mm Self-propelled Howitzer

Development

In January 1952 a conference was held in Washington DC on the subject of self-propelled artillery which indicated that there was an urgent need for improvement. Shortly after this conference preliminary concept studies for a self-propelled howitzer to replace the 105 mm M52 began. The first studies were presented to CONARC in August 1952 but were rejected with a request for further studies. These were undertaken and a further presentation was made to CONARC in September 1953. One of these studies was selected for development under the designation of the 110 mm self-propelled howitzer T195. The first mock-up was built and approved in March 1954 and authority was also given to build prototype vehicles as well as the placing of orders for long lead items such as engines and transmissions. In May 1954 a decision was made to freeze the project although items already ordered were not cancelled. In June 1954 it was decided that the 156 mm (later 155 mm) T196 self-propelled howitzer would use the same chassis and turret as the T195. It was also decided at the time that the 110 mm howitzer would be replaced by a 105 mm calibre weapon as this ammunition was already available in large quantities.

In 1958 the first prototype of the T195 was completed but a number of failures occurred in the suspension. The prototype differed from later vehicles in that it had a different shaped hull and turret, the seventh road wheel acted as the idler and the 105 mm howitzer was fitted with a muzzle brake.

In 1959 a policy was established that diesel rather than petrol engines would be used for future vehicles and the prototype of the T195, which was powered by a Continental petrol engine, was refitted with a diesel engine and called the T195E1. Trials with the latter were carried out in 1960 but there were failures of critical suspension and final drive components. These deficiencies were subsequently corrected and in December 1961 the T195E1 was classified for limited production as the Howitzer, Light, Self-propelled, 105 mm, M108.

Production of the M108 was undertaken by the Cadillac Motor Car Division of the General Motors Corporation: the first production vehicle was completed in October 1962 and production continued until 1963. The vehicle was in production for only a short period, as the US Army decided to concentrate on the self-propelled 155 mm M109 rather than the 105 mm M108. The M108 is almost identical to the M109, which differs only in its armament, loading system, fire control system, and ammunition stowage racks and is fitted with spades at the rear of the hull.

Description

The M108 has a crew of five: the commander, gunner, two ammunition members and the driver. The hull is made of all-welded aluminium with the driver at the front of the hull on the left, engine to his right and the fighting compartment at the rear.

M108 105 mm self-propelled howitzer with ordnance in travelling lock (C R Zwart)

The driver has a single piece hatch cover that opens to the left, in front of which are three M45 periscopes which can be covered by small metal flaps to prevent damage.

The engine is coupled to the Allison Transmission Division of General Motors XTG-411-2A cross-drive transmission which is at the front of the hull.

The all-welded aluminium turret is at the rear of the hull and has a square hatch in each side which opens to the rear and twin doors in the rear. The commander is seated on the right and has a cupola which can be traversed manually through 360°, a single piece hatch cover that opens to the rear and an M27 periscope. Pintle-mounted on the forward part of the commander's cupola is a 12.7 mm (0.50) M2 HB anti-aircraft machine gun. The gunner is seated on the left side of the turret and has a square single piece hatch cover that opens to the right. Twin doors are provided at the rear of the hull for ammunition resupply.

The torsion bar suspension consists of seven dual rubber-tyred road wheels with the drive sprocket at the front and the idler at the rear. There are no track return rollers. The tracks are of the single pin, centre guide type with replaceable rubber pads.

The M108 is fitted with infra-red night vision equipment but no NBC equipment. The basic vehicle can ford to a depth of 1.828 m without preparation. It can also be fitted with an amphibious kit which consists of nine air bags, four each side of the hull and one at the front of the hull. The bags, which are not carried on the vehicle as part of its normal equipment, are inflated from the vehicle and when they are inflated the vehicle can propel itself across rivers by its tracks at 6.43 km/h.

Main armament is a 105 mm howitzer M103 in mount M139. The recoil system is of the hydrospring, constant retarding force type while the breech-block is of the sliding wedge drop block type. The weapon has an elevation of +74°, depression of -4° and 360° turret traverse. Gun elevation/depression and turret traverse are both manual. Normal rate of fire is 1 rd/min but 3 rds/min can be fired for short periods. Fire control equipment consists of an M15 elevation quadrant, M117 panoramic telescope with a magnification of ×4 and a 10° field-of-view and an M118 elbow telescope with a magnification of ×4 and a 10° field-of-view.

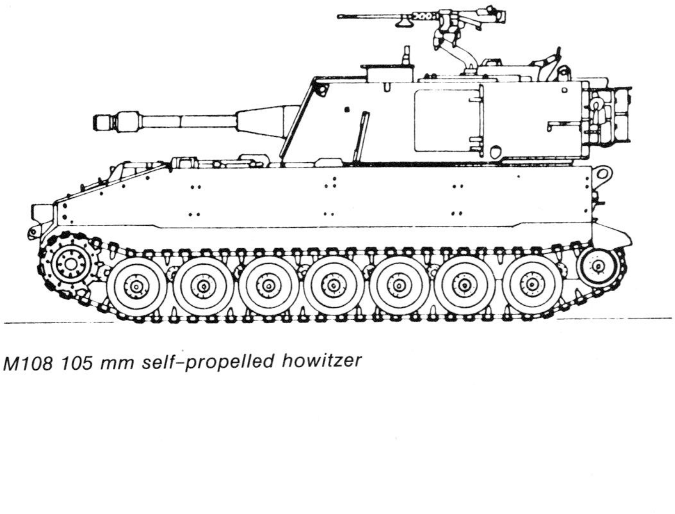

M108 105 mm self-propelled howitzer

The following types of ammunition can be fired:

Agent (M60) with the complete round weighing 19.46 kg, maximum muzzle velocity 494 m/s and maximum range (charge 7) of 11 500 m
Agent (M360) with the complete round weighing 19.89 kg, maximum muzzle velocity 494 m/s and a maximum range (charge 7) of 11 500 m
APERS-T (M546) with the complete round weighing 17.35 kg, maximum muzzle velocity of 549 m/s and maximum range (charge 7) of 12 400 m
HE (M1) with the complete round weighing 18.107 kg, maximum muzzle velocity of 494 m/s and a maximum range (charge 7) of 11 500 m
HE (M444) (carries 18 M39 grenades) with the complete round weighing 19.05 kg, maximum muzzle velocity of 494 m/s and a maximum range (charge 7) of 11 500 m
HE (M413) (carries 18 M35 grenades) with the complete round weighing 19.05 kg, maximum muzzle velocity of 494 m/s to a maximum range of 11 500 m
HEP-T (M327) with the complete round weighing 15.17 kg, maximum muzzle velocity of 559 m/s and a maximum range of 8685 m
HERA (M548) with the complete round weighing 17.46 kg, maximum muzzle velocity of 548 m/s and a maximum range of 15 000 m
Illuminating (M314 series) with the complete round weighing 21.06 kg, maximum muzzle velocity of 494 m/s and a maximum range (charge 7) of 11 500 m
Illuminating (M314A3) with complete round weighing 21.06 kg, maximum muzzle velocity of 494 m/s to a maximum range of 11 500 m
Leaflet (M84B1) with the complete round weighing 18 kg, maximum muzzle velocity of 433 m/s and a maximum range (charge 7) of 9091 m
Smoke (M60 series) with the complete round weighing 19.46 kg, maximum muzzle velocity of 494 m/s and a maximum range (charge 7) of 11 500 m
Smoke (M84) with the complete round weighing 19.03 kg, maximum muzzle velocity of 494 m/s and a maximum range (charge 7) of 11 500 m
M629 tactical CS gas with the projectile weighing 19.05 kg, maximum muzzle velocity of 494 m/s and a maximum range (charge 7) of 11 500 m.

Command Post Vehicles

Following their replacement by 155 mm M109A2 self-propelled howitzers, a number of Belgian Army M108 105 mm self-propelled howitzers have been converted into VBCLs (Véhicule Blinde de Commandement et de Liaison) by the Arsenal du Matériel Mécanique et de l'Armement at Rocourt.

Status: Production complete. In service with Brazil, Spain, Taiwan, Tunisia and Turkey.

Manufacturer: Cadillac Motor Car Division of General Motors Corporation.

SPECIFICATIONS

CREW	5
COMBAT WEIGHT	22 452 kg
UNLOADED WEIGHT	18 436 kg
POWER-TO-WEIGHT RATIO	18.038 bhp/tonne
GROUND PRESSURE	0.71 kg/cm^2
LENGTH	6.114 m
WIDTH	3.295 m
reduced	3.149 m
HEIGHT	
including AA MG	3.155 m
reduced	2.794 m
GROUND CLEARANCE	0.451 m
TRACK	2.768 m
TRACK WIDTH	381 mm
LENGTH OF TRACK ON GROUND	3.962 m
MAX SPEED	
road	56 km/h
water	6.43 km/h
FUEL CAPACITY	511 l
MAX ROAD RANGE	390 km
FORDING	1.828 m (amphibious with kit)
GRADIENT	60%
VERTICAL OBSTACLE	0.533 m
TRENCH	1.828 m
ENGINE	Detroit Diesel model 8V-71T, turbo-charged, 2-stroke, liquid-cooled 8-cylinder diesel developing 405 bhp at 2300 rpm
TRANSMISSION	Allison Transmission Division of General Motors XTG-411-2A cross drive with 4 forward and 2 reverse gears
SUSPENSION	torsion bar
ELECTRICAL SYSTEM	24 V
BATTERIES	4 × 12 V 6TN
ARMAMENT	
main	1 × 105 mm howitzer
anti-aircraft	1 × 12.7 mm MG
SMOKE-LAYING EQUIPMENT	none
AMMUNITION	
main	87
anti-aircraft	500
GUN CONTROL EQUIPMENT	
Turret power control	manual
by commander	no
by gunner	yes
Gun elevation/depression	+74°/-4°
Turret traverse	360°

Towed anti-tank guns, guns and howitzers

ARGENTINA

SALA 30 155 mm Howitzer Project

Development/Description
SALA 30 is intended to replace the current Argentine CITEFA L33 howitzer (see next entry) and the entire system was intended to be ready for presentation to army officials in March 1986. However, defence budget cuts have prevented this and the project is apparently running well behind schedule.

SALA 30 (Sistema de Artilleria de Larga Alance - long-range artillery system) has two components. They are the CALA 30 155 mm howitzer and its range of ammunition, known as PALA 30. The CALA 30 has a 45-calibre barrel with a range of 30 000 m. Elevation is +70°, traverse 40° and the barrel is traversed through 180° for travelling. A pneumatic loading system is used and the howitzer can travel on a two- or four-wheel carriage powered by an APU. Tyre pressures for the two-wheeled options have been specified as no greater than 5 kg/cm^2 and no more than 3 kg/cm^2 for the four-wheeled carriage. The target weight for the howitzer is less than 13 000 kg. Fire control is to include both optical and laser targeting systems.

The SALA 30 will be able to fire standard NATO ammunition and the PALA 30 ammunition will include the CX4 and CX5 (F1) projectiles developed at the Instituto de Investigaciones Cientificas Y Tecnicas de las Fuerzas Armadas (CITEFA). These are based on the French M56 and M56/69 explosive and fragmentation projectiles. Base bleed projectiles may be used. A hollow charge projectile known as the PACU (Projectile de Artillera de Culote Hueco) may be fired and there is a requirement to fire extended range ERFB projectiles.

Status: Development.

Manufacturer: Instituto de Investigaciones Cientificas Y Tecnicas de las Fuerzas Armadas (CITEFA)
Zufriategui y Varela
1603 Villa Martelli
Provincia de Buenos Aires
Argentina

Enquiries to Direccion General de Fabricaciones Militares Argentina
Sales Department
Avenue Cabildo 65(1462)
Buenos Aires
Argentina

---

155 mm Howitzer L33 X1415 CITEFA Models 77 and 81

Development/Description
These weapons are basically the top carriage (barrel, cradle, recoil system and equilibrators) of the French 155 mm self-propelled gun Mk F3 (which has been used by the Argentine Army for some years) mounted on a new bottom carriage designed by the Scientific and Technical Research Institute of the Armed Forces (CITEFA).

The top carriage is a welded steel structure which contains the elevating brackets and traverse mechanism, and forms the connection with the cradle trunnion attachment and the bottom carriage. The traversing mechanism is mounted inside the front and bottom of the top carriage and to obtain acceptable handwheel loads, two pneumatic equilibrators are connected to the top carriage and cradle.

The bottom carriage is a steel welded and machined unit. For firing the wheels are raised clear of the ground and the weapon rests on a firing base at the front and on the trails at the rear. The firing base is a circular steel structure mounted on a ball socket, suspended underneath the bottom carriage, which compensates for uneven ground contour. The firing base is normally carried on the bottom carriage when travelling as the ground clearance of 0.3 m makes removal unnecessary during movement. The suspension is of the spring type.

The L33 X1415 fires an HE projectile weighing 43 kg with a maximum muzzle velocity of 765 m/s, and illuminating and smoke ammunition.

The Model 81 is a slightly revised version of the Model 77, fitted with a locally produced copy of the 155 mm Mk F3 barrel. Some changes have been made to the trail legs and to the recuperator and equilibrators. The Model 81 has exactly the same specifications as the Model 77 and fires the same ammunition.

Status: Production complete. In service with Argentina.

Manufacturer: Instituto de Investigaciones Cientificas Y Tecnicas de las Fuerzas Armadas (CITEFA)
Zufriategui y Varela
1603 Villa Martelli
Provincia de Buenos Aires
Argentina

Enquiries to Direccion General de Fabricaciones Militares Argentina
Sales Department
Avenue Cabildo 65(1462)
Buenos Aires
Argentina

155 mm howitzer L33 X1415 CITEFA Model 77 in firing position with barrel at maximum elevation

155 mm howitzer L33 Model 81 in direct firing configuration

SPECIFICATIONS		LENGTH	10.15 m	RATE OF FIRE	
CALIBRE	155 mm	WIDTH	2.67 m	max	4 rds/min
BARREL LENGTH	5.115 m	HEIGHT	2.2 m	sustained	1 rd/min
MUZZLE BRAKE	double-baffle	GROUND CLEARANCE	0.3 m	MAX RANGE	
CARRIAGE	split trail	TRACK	2.31 m	normal round	22 000 m
SHIELD	no	ELEVATION/DEPRESSION	+67°/0°	RAP	25 300 m
WEIGHT (travelling order)	8000 kg	TRAVERSE (total)	70°		

AUSTRALIA

Project HAMEL

Development/Description
As far back as 1969 the Australian Army began to consider possible replacements for its 105 mm M101 (M2A2) howitzers. At that time no decision was made but subsequently the need for a replacement initiated a search for a new weapon. As a result, Material Branch (Army Office) established Project HAMEL to find the Australian Army's new field gun.

The weapons considered included the American M102 and M204 105 mm howitzers, an Indian 105 mm gun, product-improving the existing M101A1 and/or the ammunition, the American XM164 howitzer and the British L118/L119 Light Gun. In the end, the choice narrowed to the American M204 and the British L118/L119.

The Australian Army finally selected the British Royal Ordnance Light Gun as it 'provides the most advanced gun and ammunition technology available in a weapon system of this type'. Both the L118 which fires 105 mm Abbot ammunition, and the L119 have been procured, the longer-ranged L118 for operational use and the L119, which fires existing American M1 ammunition, for training. The Light Gun is being licence-produced in Australia at the Ordnance Factories at Maribyrnong and Bendigo, and the Australian Government approved an A$13.3 million sum to cover the initial establishment and the proving of local manufacturing capabilities. Once this phase had been successfully completed subsequent orders were worth A$139 million. Ammunition will also be produced in Australia.

Six Light Guns for evaluation and design model purposes have been obtained from the United Kingdom, and the first of these was fired in Australia at Puckapunyal in June 1983. Initial production of the barrel, muzzle brake, breech ring and breech-block has begun at the Ordnance Factory Maribyrnong. The barrels are rifled and finished at the Ordnance Factory Bendigo where production of the cradle, recoil system, electrical firing system, breech mechanism and the bulk of the complete equipment schedule items takes place. The sighting equipment is produced by Fairey Australasia Pty Limited while Maribyrnong manufactures the dial sight carrier. Main production of the first 59 guns began in late 1984 and the first two guns were handed over for Army acceptance trials in July 1985. These trials were carried out with two comparative Royal Ordnance Light Guns. The first production gun was accepted by the Australian Army at the School of Artillery, Sydney, in August 1988. By then the production contracts had increased in value to A$139 and involved the production of 59 guns for the Regular Army and a further 46 to partially re-equip the Army Reserve. It is expected that guns for the Reserve Army will be issued during the early 1990s. The Australian Army has three regular field regiments and six reserve batteries equipped with the M101.

One of six 105 mm Light Guns delivered from UK for Project HAMEL seen with M101 105 mm Howitzer and 105 mm Pack Howitzer (Paul Handel)

In service the Australian-produced version of the Light Gun is known as the Hamel Gun.

In Spring 1986 the New Zealand Government approved the purchase of 24 105 mm Light Guns from Australia at a cost of NZ$21 million with deliveries running from 1986 to September 1989.

Manufacturer: Australian Ordnance Factory Bendigo and Maribyrnong
Enquiries to Director, Marketing, Office of Defence Production
PO Box E 4010
Queen Victoria Terrace
Canberra ACT 2600
Australia
Telephone: (6162) 482872
Telex: AA 62063
Fax: (6161) 485413

AUSTRIA

NORICUM GH N-45 155 mm Gun Howitzer

Development
Since 1979 NORICUM GmbH has been producing an improved version of the 155 mm SRC International GC 45 155 mm Gun Howitzer (see entry under Belgium in this section) under the designation GH N-45 (Gun Howitzer, NORICUM, 45-calibre).

NORICUM (at one time using the name Voest-Alpine AG, NORICUM handles sales and marketing) assembled all but two of the 12 guns (two six-gun batteries) supplied to the Royal Thai Marines and then began deliveries of over 300 guns to various customers. Known customers include Jordan from where GH N-45s have been passed to Iraq. The Austrian Army has a few GH N-45s for evaluation (fitted with APUs) but these are not used to fire Extended Range Full Bore (ERFB) projectiles for treaty reasons. ERFB and base bleed ammunition is manufactured by NORICUM together with Hirtenberger Patronenfabrik.

In June 1987 the Thai Cabinet approved a purchase of six GH N-45 weapons at a cost of $7 million, all of which have now been delivered. These models are fitted with an auxiliary power unit. ENGESA of Brazil has a licence to undertake manufacture of the GH N-45 weapon but production will not start until firm orders are received.

NORICUM GH N-45 155 mm systems were used by both sides in the Iran/Iraq conflict and quantities of these weapons were captured during Operation Desert Storm early in 1991.

NORICUM GH N-45 155 mm gun howitzers showing towed version (left) and powered version (right)

Deliveries of GH N-45 systems to Saudi Arabia were stopped late in 1990 after only a few had been delivered and NORICUM is now ceasing production of artillery systems.

Description
The GH N-45 is similar in appearance to the GC 45 in most respects but has been redesigned in certain key areas to improve handling in

NORICUM GH N-45 155 mm gun howitzer breech and rammer (Christopher F Foss)

Towed version of NORICUM GH N-45 155 mm gun howitzer

the field and reliability, and to make the weapon easier to manufacture.

The carriage uses a walking beam suspension for good mobility and a secondary float is fitted for easy emplacement, displacement and general weapon handling. There is a quick-release in the traversing gear which brings the barrel from the travelling jack-knife position to the firing position in a few seconds and there is a steering device that allows the weapon to travel in the tracks of the towing vehicle. The GH N-45 can be air-lifted by a C-130 transport aircraft without preparation and the normal towing vehicle is a five- or 10-tonne (6 × 6) truck with the travelling speed on all types of terrain being limited only by the capabilities of the towing truck.

The GH N-45 can be delivered in a basic towed form or with an APU; an APU kit can be supplied for fitting to towed models on location. The kit takes two days to install and the main power source is a Porsche four-cylinder, four-cycle air-cooled 2.3-litre engine delivering 110 or 120 hp. The APU is mounted on the forward part of the carriage with the driver on the right. The APU enables the weapon to propel itself at a maximum speed of 30 km/h and two 30-litre fuel tanks provide sufficient fuel for ranges of up to 130 km. With the APU the weapon can climb 40 per cent gradients at a speed of 5 km/h.

The GH N-45 has a 45-calibre barrel with a large chamber volume to enable base bleed ammunition to be fired to a maximum range of 39 600 m. The barrel has a semi-automatic breech mechanism with a pneumatic rammer.

Extras available for the GH N-45 include a power elevating and traverse system, and an emergency ammunition bracket holding six stand-by projectiles, both for the APU-powered version. Also available for the APU version is an ammunition handling device. The jib and power unit for this device are mounted on the top right of the barrel, and are used to raise the ammunition tray from ground level to the loading position. Tracks can be fitted to the APU-powered version to increase traction over soft ground and enable gradients of up to 45 per cent to be climbed. Also available for the APU version is a remote control system that enables a 6 × 6 tractor to use the traction of the powered carriage over adverse terrain using a 10 × 10 configuration.

Ammunition for the GH N-45 includes all standard NATO ammunition, including the Hirtenberger-produced M107 equivalent, the SN-107. It also fires a Hirtenberger ERFB projectile known as the SEN to a range of 30 300 m and a ERFB Base Bleed (SEN-BB) to a range of 39 600 m. Accuracy claimed for the ERFB Base Bleed projectile is 0.35 per cent of range and 0.7 mils in deflection, up to maximum range. Hirtenberger also produces N-37, N-89 and N-10 charges. The normal unlimited rate of fire is 2 rds/min (charges 3 to 7) and one round a minute (charges 8 to 10), maximum is 7 rds/min for two minutes (charge 10) and a three-round burst can be fired in 16 seconds.

SPECIFICATIONS

Model	**Towed**	**APU**
CALIBRE	155 mm	155 mm
BARREL LENGTH	7.046 m	7.046 m
LENGTH OF RIFLING	5.82 m	5.82 m
MUZZLE BRAKE	multi-baffle	multi-baffle
CARRIAGE TYPE	split trail	split trail
SHIELD	no	no
TOTAL WEIGHT	10 070 kg	12 382 kg
LENGTH TRAVELLING		
barrel forward	13.97 m	13.97 m
barrel aft	9.72 m	9.725 m
LENGTH, FIRING	11.4 m	11.4 m
WIDTH		
travelling	2.5 m	2.75 m
firing	9.931 m	9.931 m
HEIGHT (travelling)	2.05 m	2.05 m
GROUND CLEARANCE	0.31 m	0.315 m
WHEEL TRACK		
main	2.47 m	2.47 m
trail	0.865 m	0.865 m
TYRES		
main	9.00 × 20R	11.00 × 20R
trail	8.25 × 15R	8.25 × 15R
ELEVATION/DEPRESSION	+72°/−5°	+72°/−4°
TRAVERSE (left/right)	30°/40°	30°/40°
RANGE		
ERFB	30 300 m	30 300 m
ERFB BB	39 600 m	39 600 m
RATE OF FIRE		
normal	2 rds/min unlimited	2 rds/min unlimited
max	7 rds/min	7 rds/min
burst	3 rounds/16 s	3 rounds/16 s
CREW	6	6

Status: Production complete although quantities of stock weapons are available for sale. In service with Austria (trials batch only), Iran, Iraq, Jordan, Libya, Saudi Arabia and Thailand. NORICUM also built the following additional artillery systems, but they never entered production:
155 mm M114/39 (*Jane's Armour and Artillery 1990-91*, page 610)
105 mm anti-tank gun ATG N 105 (*Jane's Armour and Artillery 1990-91*, page 610)

Manufacturer: NORICUM Maschinenbau und Handel GmbH
Postfach 3
A-4010 Linz
Austria
Telephone: (0732) 27 45 46
Telex: 22491 nl a

NORICUM GH N-45 155 mm gun howitzer captured by Iraq from Iran with ordnance traversed to rear

BELGIUM

SRC International GC 45 155 mm Gun Howitzer

Development

The GC 45 155 mm Gun Howitzer was developed from 1975 as a private venture by SRC International of Belgium which is a company established jointly by the now defunct Space Research Corporation of Canada and PRB of Belgium (now wholly owned by PRB). The first prototype of the GC 45 was completed in 1977 and 12 weapons were subsequently purchased by the Royal Thai Marines. Of these, two were produced and assembled in Canada and 10 were produced in Canada and assembled in Austria by Voest-Alpine. Licence for the production of a weapon based on this design has been transferred to Voest-Alpine which now produces this with a number of improvements as the GH N-45, details of which are given in an earlier entry.

GC 45 stands for Gun Calibre 45 with the latter figure indicating the length of the gun in calibres.

Description

The GC 45 is of the split-trail carriage configuration and has a 45-calibre long barrel which has been designed to fire all standard as well as new 155 mm NATO ammunition and charges, with the same muzzle velocity and range as would be achieved with a 39-calibre barrel of the FH-70 or M198 type. This performance is achieved at a lower breech pressure owing to a larger chamber, matched to the 45-calibre length. The feature of the GC 45 is that it meets the NATO requirement for a range of 30 000 m with a completely ballistic solution, without rocket booster or other external devices. When fired with the SRC 155 mm ERFB projectile and the top zone M11 charge, the maximum range of 30 000 m is achieved. At high angle, low charge (M3A1 zone 2) the GC 45 achieves a minimum range of 2800 m with the same projectile. Using an ERFB/Base Bleed projectile with the M11 charge the maximum range is 39 000 m.

All structural components are built of a high-strength alloy steel and the barrel is made of high-yield steel and autofrettaged throughout. The breech assembly incorporates an automatic breech-opening cam system: the breech opens automatically when the barrel assembly returns to battery and closes automatically when the cam is swung out of engagement.

The elevating system incorporates pneumatic equilibrators and results in small handwheel loads during elevation. Adjustment capability in the equilibrator system geometry, together with different nitrogen pressures, allows for ease of elevating at all gun system operating ambient temperatures.

A low-friction traverse bearing eases traversing. In addition, a friction device system is incorporated to minimise feed-back loads into the traversing system from any gun barrel whip.

A swing-aside loading tray mounted on the rear cradle of the recoil system moves in elevation with the barrel assembly. A pneumatically operated telescoping ramming cylinder is rotated into place behind the projectile for loading at high elevation angles, obviating the necessity of lowering the weapon between shots for ramming. The pneumatic rammer also provides a high degree of uniformity in ram from shot to shot. A pneumatic system consisting of a bottle, mount, valving, regulator, accumulator, hoses and fittings is installed to power the ramming cylinder. An added feature of this system is that the bottle can be recharged in the field.

The suspension of the GC 45 incorporates a walking beam (two wheels each side) which enables the gun to be towed at speeds of up to 90 km/h. The freely pivoting walking beams also allow for greater mobility during towing in rough terrain.

The design of the screw jack and float system, which is mounted under the forward part of the carriage, incorporates a telescoping system and ball screw jack. One crew member can easily jack the gun in 90 seconds. Disengagement of the ball jack will result in the immediate settling of the gun on its undercarriage wheels.

The trails and spades, with the built-in tow bar system, allow for efficient handling during emplacement and disemplacement. The spades are pivoted, allowing for four angular positions, so that handling of these heavy components is minimised. It is possible to spread the trails to very large angles to permit large traverse capability of the barrel assembly at all angles of elevation.

The trail lifting assemblies at the outboard rear of each trail are designed to assist the gun crew during the emplacement and disemplacement operations. They are fully swivelling and fitted with rubber tyres, and are raised and lowered by handwheels at the forward end of each assembly. The force at the handwheel is transmitted to the wheel through a ball-screw actuator connected to the wheel forks by a ball and crank arrangement. The ball-screw is reversible and the trails can be simply dropped to the ground during emplacement. An additional feature of these assemblies is that they enable the gun crew to manhandle the gun through 360° in traverse.

The GC 45 fires the following separate-loading ammunition: an SRC ERFB Mk 10 Mod 2 projectile weighing 45.4 kg and containing 8.8 kg of HE, illuminating projectile weighing 45.2 kg, smoke projectile weighing 45.7 kg and WP projectile weighing 47.7 kg. An ERFB/Base Bleed projectile weighing 47.5 kg fired with an M11 charge achieves a maximum range of 39 000 m. Firing the ERFB Mk 10 Mod 2 projectile the weapon has the following range capabilities:

Charge	Muzzle velocity	Max range
M3A1 Zone 3	266 m/s	5900 m
M3A1 Zone 4	302.1 m/s	7300 m
M3A1 Zone 5	359.9 m/s	9300 m
M4A2 Zone 6	458.1 m/s	12 400 m
M4A2 Zone 7	541.9 m/s	15 100 m
M119 Zone 8	658 m/s	19 200 m
M2 Zone 9	795.5 m/s	24 700 m
M11	897 m/s	30 000 m

Status: Production complete. In service with Royal Thai Marines (12).

Manufacturer: SRC International SA
114 boulevard Brand Whitlock
B-1200 Brussels
Belgium
(SRC International SA is no longer in existence)

SRC International GC 45 155 mm gun howitzer in firing position

SPECIFICATIONS	
CALIBRE	155 mm
BARREL LENGTH	6.975 m
MUZZLE BRAKE	multi-baffle
CARRIAGE	split trail
SHIELD	none
WEIGHT	
travelling order	8222 kg
firing position	8222 kg
at lunette, towed mode, muzzle aft	1815 kg
at lunette, towed mode, muzzle forward	1134 kg
LENGTH	
travelling, muzzle aft	13.614 m
travelling, muzzle forward over trails	9.144 m
firing	10.82 m
WIDTH	
travelling	2.692 m
firing	10.364 m
HEIGHT (travelling)	3.28 m
GROUND CLEARANCE	0.355 m
ELEVATION/DEPRESSION	+69°/-5°
TRAVERSE (total)	80°
RATE OF FIRE	
sustained	2 rds/min
max for 15 minutes	4 rds/min
RANGE	
normal	30 000 m
base bleed	39 000 m
CREW	8
TOWING VEHICLE	5-ton (6 × 6) truck

90 mm MECAR KEnerga 90/46 Anti-tank Gun

Details of this anti-tank gun, which has yet to enter production, were given in *Jane's Armour and Artillery 1990-91* page 612.

SRC Artillery Systems

Following the revelation that the Belgian SRC Group of Companies had been involved in a number of projects in Iraq, in the Spring of 1990 the company ceased operations. Details of the SRC Group of Companies' then current artillery projects were given in *Jane's Armour and Artillery 1989-90* as follows:
VSP 203 203 mm self-propelled (6 × 6) howitzer (page 528)
FGT-203 203 mm towed field gun (page 591)
FGH-155 155 mm howitzer (page 591)
M46/84 155 mm gun (page 593).

The SRC 155 mm M46/84 gun system is essentially the Soviet 130 mm M46 weapon upgraded and fitted with a 45-calibre ordnance. Yugoslavia still markets this conversion and full details are given later in this section under Yugoslavia.

In addition to its involvement in the so called Super Gun, SRC Group of Companies was also involved with the Iraq 155 mm Majnoon (6 × 6) and the 210 mm Al Fao (6 × 6) self-propelled artillery systems first shown in 1989 and covered in the previous section under Iraq. Neither of these systems ever entered production or service with the Iraqi Army.

CHILE

Cardoen CC-45 155 mm Gun-Howitzer

Development/Description
This is essentially the South African Armscor G-5 155 mm gun-howitzer manufactured under licence for the export market by Cardoen. As far as it is known, the system is identical to the G-5 covered in detail later in this section although the examples observed in Chile do not appear to be fitted with the flick rammer.

In March 1989 it was stated that Cardoen Industries had successfully trialled the first locally produced prototype of the G-5 which incorporated key components supplied by Armscor in 1988.

Associated equipment for the CC-45 including the fire control system, data links and ammunition such as ERFB and ERFB-BB will be supplied by Armscor.

Status: By early 1990 some six systems had been assembled/produced in Chile by Cardoen Industries. As of early 1991 there were no known orders for the Chilean produced CC-45 artillery system or its self-propelled counterpart.

Manufacturer: Cardoen Industries
Los Conquistadores 1700 - 28th Floor
Santiago, Chile.
Telephone: 2313420-2314219
Telex: 341630 INCAR CK
Fax: 2316366

Cardoen Industries CC-45 155 mm gun-howitzer in APU mode

CHINA, PEOPLE'S REPUBLIC

NORINCO 155 mm Gun-Howitzer Type WA 021

Development

The 155 mm gun-howitzer Type WA 021 (also referred to as the WAC 21) is very similar in many respects to the Austrian NORICUM GH N-45 gun-howitzer to the extent that the carriage of the Type WA 021 can accommodate the ordnance of the NORICUM weapon. The Type WA 021 is produced by the Heping Machinery Factory of the China North Industries Corporation (NORINCO). It would appear that an interim 155 mm weapon known as the MF 45 was produced in China, probably as part of the development programme that led to the Type WA 021.

By the end of 1986 10 Type WA 021s had been produced, two as prototypes and the remainder as pre-production examples.

The WA 021 was first seen in public during a defence equipment exhibition held in Beijing in 1986. The WA 021 system is based on a design supplied to the now defunct SRC International of Belgium headed by Dr Gerald Bull who also designed the GC 45 (qv).

In 1991 officials of NORINCO stated that the WA 021 was in service with the Chinese Army.

Description

The Type WA 021 has an auto-fretted barrel 45 calibres long fitted with a multi-baffle muzzle brake. A screw-type breech mechanism opening to the right is employed and the chamber volume is 22.95 litres. The rifling employs 48 grooves with a twist of 1 in 20 calibres, and the grooves are understood to be three times deeper than the rifling depths used on comparable Western designs.

The recoil system consists of the recuperator cylinder (top) and the recoil cylinder (below). The recoil mechanism can change the recoil distance with the variation of elevation angle so there is no need to dig a firing pit when the weapon fires at a high elevation.

The upper carriage includes the elevating mechanism, traversing mechanism, equilibrator and gear for traverse. The equilibrator is a pneumatic-hydraulic balancer. When the temperature changes and the unbalanced torque increases or decreases, the torque of the hand wheel in the elevating mechanism will change. In this case the hydraulic pressure can be adjusted to obtain suitable torque of the hand wheel in the elevating mechanism.

The carriage is of the split trail type and employs a screw jack firing platform under the carriage pivot. For travelling the barrel is swung through 180° and held in a clamp on the left-hand trail leg. A walking beam suspension is used and pivoting dolly wheels are provided on each trail leg for ease of handling. A flick rammer is located on the left-hand trail leg. Elevating and traversing systems are located to the left of the barrel. Sighting equipment includes the M115 panoramic sight and the M137 telescopic sight. An APU could be fitted.

Batteries of up to six WA 021s can be controlled by a fire control system mounted on a command vehicle. The system includes a 16-bit fire control computer, a command control console, a tape driver/printer, a company command relay display unit, gun display units, automatic observation equipment and a laser rangefinder.

The Type WA 021 fires a range of Extended Range Full Bore (ERFB) projectiles. These include ERFB HE, ERFB BB HE, ERFB BB WP, ERFB illuminating, ERFB smoke, ERFB BB illuminating, ERFB improved conventional munition and ERFB improved conventional munition BB. The improved conventional munition is understood not to be in production and carries 72 bomblets. The ERFB projectiles have welded nubs. Range with ERFB is 30 000 m and 39 000 m with ERFB BB. Dispersion at maximum range is stated to be 0.45% longitudinal and 0.7 mil in traverse. Barrel service life is 1000 EFCs.

The normal sustained rate of fire is 2 rds/min. This can be increased to between 4 and 5 rds/min for short periods.

155 mm Gun-Howitzer with APU

The 155 mm gun-howitzer Type WA 021 is also available fitted with an APU which enables the gun to go into or move out of the firing position as well as travel in the self-propelled mode for a short distance.

The APU can also be mounted on other artillery systems such as the 130 mm and 152 mm types.

The APU consists of a four cylinder diesel developing 77 hp at 3600 rpm that is mounted on the forward part of the carriage giving the system a maximum speed of 18 km/h.

The four main wheels are hydraulically powered with the front wheels being used for differential steering.

155 mm Coastal Defence System

Details of this, which is based on the WA 021, are given in the *Coastal artillery guns and missiles* section.

Status: In production and in service with the Chinese Army.

Manufacturer: Heping Machinery Factory
China North Industries Corporation
7A Yuetan Nanjie
PO Box 2137
Beijing
People's Republic of China
Telephone: (86) 6898/3461/3741/7570
Telex: 22339 cnin cn

155 mm gun-howitzer Type WA 021 in firing position

155 mm gun-howitzer Type WA 021 in firing position with breech open

SPECIFICATIONS

CALIBRE	155 mm
LENGTH OF BARREL ASSEMBLY	7.582 m
LENGTH OF BARREL	7.045 m
LENGTH OF RIFLING	5.83 m
MUZZLE BRAKE	multi-baffle
CARRIAGE TYPE	split trail
SHIELD	no
WEIGHT TRAVELLING	9500 kg
LENGTH TRAVELLING	
barrel forward	13.512 m
barrel rear	9.068 m
LENGTH FIRING	11.4 m
WIDTH	
travelling	2.67 m
firing	9.931 m
HEIGHT	
travelling, barrel forward	3.048 m
travelling, barrel rear	2.23 m
firing overall, 0° elevation	2.187 m
to bore of gun	1.659 m
GROUND CLEARANCE	0.28 m
RECOIL LENGTH	
long	1.524 m
short	1.041 m
ELEVATION/DEPRESSION	+72°/-5°
TRAVERSE (left/right)	30°/40°
RANGE	
ERFB (M11 charge)	30 000 m
ERFB BB (M11 charge)	39 000 m
min, ERFB	3500 m
RATE OF FIRE	
sustained	2 rds/min
max	4-5 rds/min
TOWING VEHICLE	6 × 6 7- to 10-ton truck
MAX TOWING SPEED	
road	90 km/h
cross-country	15-50 km/h

NORINCO 155 mm Gun-Howitzer Type GM-45

Development/Description

The 155 mm Type GM-45 is a conversion of the existing 130 mm Gun Type 59-1 (Soviet M-46) or the 152 mm gun-howitzer Type 66 (Soviet D-20) to accommodate the ordnance of the 155 mm gun-howitzer WA 021. The resultant performance is the same as that for the WA 021 and it is stated that the barrel and chamber structure used for the ordnance is the same as that for the Austrian GH N-45.

The GM-45 is fired with its travelling wheels on the ground. The firing rate is 2 to 4 rds/min. Recoil is variable and loading and breech operation are manual.

To increase the rate of fire a flick rammer has been installed on the left side and hydraulically operated wheels have been added to each trail assembly to ease the opening and closing of the trails.

Details for the GM-45 are the same as those for the WA 021 except for the following:

WEIGHT TRAVELLING	8900 kg
WEIGHT FIRING	8150 kg
LENGTH TRAVELLING	12 m
WIDTH TRAVELLING	2.5 m
HEIGHT TRAVELLING	2.2 m
AXIS OF BORE	1.38 m
ELEVATION/DEPRESSION	+55°/-2° 30'
TRAVERSE	
(at elevation of -2° 30' to +15°)	25° left and right
(at elevation of +16° to +55°)	14° left and 11° right
RECOIL LENGTH	
max	775 mm
min	125 mm
RATE OF FIRE	4 rds/min
TOWING SPEED	
(by truck)	15 to 60 km/h
(by tracked vehicle)	35 km/h

Status: Available.

Manufacturer: China North Industries Corporation
7A Yuetan Nanjie
PO Box 2137
Beijing
People's Republic of China
Telephone: (86) 6898/3461/3741/7570
Telex: 22339 cnin cn

NORINCO 152 mm Gun Type 83

Development/Description

The 152 mm Type 83 gun is stated to equip reserve artillery units. It uses a long barrel, believed to be about 45 calibres long, on a much revised carriage and recoil system based on that of the 152 mm gun-howitzer Type 66 (see following entry).

The barrel is fitted with a single-baffle muzzle brake and the right-hand pitch rifling uses a uniform twist. A shield is fitted and the split trail legs are equipped with hydraulically operated wheels to assist handling. The carriage rests on a firing platform when emplaced with the main wheels raised off the ground. When travelling the barrel is swung through 180° and clamped over the closed trail legs. The trails are then carried on a steerable limber. Short moves can be made without having to stow the barrel in the reverse position. On-carriage fire control instruments include the Type 58 panoramic telescope, the Type 58 indicator and the Type GS-43 telescope for use against direct fire targets such as tanks.

The Type 83 gun fires an HE projectile also designated the Type 83. This projectile weighs 48 kg, is 830 mm long and uses a propellant case 960 mm long that weighs 37 kg with a full charge and maximum muzzle velocity being 955 m/s. It can also fire a NORINCO ERFB-BB projectile to a maximum range of 38 000 m.

Fire control system installed on the Type 83 includes a Type 58 panoramic sight with a magnification of ×3.7, Type 58 collimator sight and a Type CS-43 direct sight with a magnification of ×5.5.

152 mm Gun Type 83 in travelling configuration

SPECIFICATIONS					
CALIBRE	152 mm	LENGTH (travelling)	9.4 m	ELEVATION/DEPRESSION	+45°/-2.5°
MUZZLE BRAKE	single-baffle	WIDTH (travelling)	2.67 m	TRAVERSE (left/right)	24°/26°
CARRIAGE TYPE	split trail	HEIGHT		RANGE	
SHIELD	yes	travelling	1.865 m	max	30 370 m
WEIGHT		to bore of gun	1.4 m	direct fire	1170 m
travelling	10 500 kg	RECOIL		RATE OF FIRE	3-4 rds/min
firing	9700 kg	normal	910 mm	TRAVELLING SPEED	15-60 km/h
		max	950 mm		

Status: In production. In service with the Chinese armed forces.

Manufacturer: China North Industries Corporation
7A Yuetan Nanjie
PO Box 2137
Beijing
People's Republic of China
Telephone: (86) 6898/3461/3741/7570
Telex: 22339 cnin cn

NORINCO 152 mm Gun-Howitzer Type 66

Development/Description

The 152 mm gun-howitzer Type 66 is the Chinese version of the Soviet 152 mm gun-howitzer D-20 and in appearance differs from the Soviet original in few details. The Type 66 may or may not be fitted with a shield. It is normally towed by a 6 × 6 truck and can be towed at speeds of up to 60 km/h. The carriage is equipped with a Type 58 panoramic sight, a Type 58 collimator and a Type 66 telescopic sight with a magnification of ×5.5.

The Type 66 is used to fire three types of projectile using a variable six-charge bag propellant system. Two types are HE and smoke. Details are as follows:

Type	**HE**	**Smoke**
DESIGNATION	Type 66	Type 66
WEIGHT OF PROJECTILE	43.56 kg	43.56 kg
LENGTH OF PROJECTILE	702 mm	689.7 mm
WEIGHT OF FILLING	5.86 kg	6.626 kg
TYPE OF FILLING	TNT	WP
MUZZLE VELOCITY	655 m/s	655 m/s
MAX RANGE	17 230 m	17 230 m
FUZE	Liu-4	Yan-2

The third type is an RAP projectile known as the Type MP-152. This weighs 43.56 kg and provides a range increase of 27 per cent to 21 880 m. Accuracy is claimed to be the same as that for a conventional projectile.

152 mm gun-howitzer Type 66 in firing position

Status: Production. In service with the Chinese Army, Albania, North Korea, Vietnam and Zaïre.

Manufacturer: China North Industries Corporation
7A Yuetan Nanjie
PO Box 2137 Beijing
People's Republic of China
Telephone: (86) 6898/3461/3741/7570
Telex: 22339 cnin cn

SPECIFICATIONS					
CALIBRE	152.4 mm	SHIELD	optional	ELEVATION/DEPRESSION	+45°/-5°
BARREL LENGTH (overall)	5.195 m	WEIGHT	5720 kg	TRAVERSE	58°
MUZZLE BRAKE	double-baffle	LENGTH (travelling)	8.69 m	RATE OF FIRE	6-8 rds/min
BREECH MECHANISM	semi-automatic vertical sliding block	WIDTH (travelling)	2.42 m	RANGE	17 230 m
		HEIGHT		CREW	10-12
		travelling	2.52 m	TOWING VEHICLE	6 × 6 truck
CARRIAGE	split trail with castor wheel on each	to axis of bore	1.22 m	MAX TOWING SPEED	60 km/h

NORINCO Auxiliary Propulsion Unit

Development/Description

NORINCO have developed an auxiliary propulsion unit (APU) for installation on towed artillery weapons such as the 130 mm Field Gun Type 59-1 which will enable it to be brought into action by a smaller crew as well as giving it independent mobility in the forward battlefield area.

The APU is mounted on the forward part of the carriage and consists of a medium power engine and a hydraulic transmission system. It can also be installed on other types of artillery systems, for example 152 mm calibre weapons, and may be related to the APU installed on the 155 mm gun-howitzer Type WA 021 described earlier in this section.

Key features of the APU have been summarised by NORINCO as high transmission efficiency, light weight, small volume, reliable operation and easy to operate and maintain.

The APU weighs 1800 kg with the 75 hp engine giving the system a maximum speed of 15 km/h with sufficient fuel being provided for between two and three hours of continuous operation.

Steering is believed to be achieved by altering the speed of the two main wheels, although each trail also has a small rubber tyred road wheel which also assists in bringing the weapon into action.

Status: Development complete. Ready for production.

Manufacturer: China North Industries Corporation
7A Yuetan Nanjie
PO Box 2137 Beijing
People's Republic of China
Telephone: (86) 6898/3461/3471/7570
Telex: 22339 cnin cn

NORINCO 130 mm Field Gun Type 59-1

Development/Description

The first 130 mm field gun produced by China was the Type 59, a direct copy of the Soviet M-46. At around the same time as producing the Type 59 China was also producing its 122 mm field gun Type 60, a copy of the Soviet D-74, and some of the features of the Type 60 were carried over to the production of the Type 59 to produce the Type 59-1. Thus the Type 59-1 uses a scaled-up version of the Type 60's muzzle brake, recoil mechanism and breech system all mounted on a carriage that features some components from the Type 60.

This amalgamation of Type 59 and Type 60 features has produced a weapon that is more manoeuvrable than the original Type 59 (Soviet M-46). The carriage no longer requires a limber for towing and weight has been saved by using a smaller shield.

The Type 59-1 fires two types of projectile, HE and illuminating, although at one stage of the gun's development an AP-T projectile was proposed capable of penetrating 250 mm of armour set at 0° at a range of 1500 m. Details of the two projectiles are as follows:

Type	**HE**	**Illum***
DESIGNATION	Type 59	Type 59
WEIGHT OF PROJECTILE	33.4 kg	29.1 kg
LENGTH OF PROJECTILE	668 mm	663 mm
WEIGHT OF FILLING	3.6 kg	1.96 kg
MUZZLE VELOCITY	930 m/s	952 m/s
MAX RANGE	27 490 m	25 000 m†
FUZE	Liu-5	Shi-1

* burns for 40 s producing 500 000 candela
† air burst

Both projectiles are fired using a two-charge full variable charge or a three-charge reduced variable charge.

130 mm field gun Type 59-1 in firing position

Two enhanced range projectiles have been developed for the Type 59-1 gun. One is an RAP known as the MP-130. This weighs 33.4 kg and provides a maximum range of 34 360 m, an increase of 25 per cent. An RAP known as the MP-132 has been developed and tested.

The second projectile is an ERFB HE that is fired using a multi-section charge system employing four bags, an increment and 2.5 kg of loose propellant. Details of this ERFB round are as follows:

WEIGHT OF COMPLETE ROUND	
with full charge	58 kg
with reduced charge	50.7 kg
packed	79 kg
WEIGHT OF PROJECTILE	32.41 kg
LENGTH OF PROJECTILE	799 mm
WEIGHT OF CHARGE	3.3 kg
TYPE OF CHARGE	TNT
MUZZLE VELOCITY	944 m/s
MAX CHAMBER PRESSURE	3150 kg/cm^2
MAX RANGE	32 000 m
FUZE	Liu-5

The following new projectiles have been developed for this weapon.

ERFB-BB, weight of projectile 33.4 kg, m/v 930 m/s, maximum range 37 000/38 000 m.

ERFB, weight of projectile 32.41 kg, m/v 994 m/s, maximum range 32 000 m.

ERFB/HB HE, weight of projectile 32.7 kg, m/v 940 m/s, maximum range 30 000 m.

Cargo carrying sub-munitions, maximum range 25 000 m.

Smoke, weight of projectile 32.47 kg, m/v 935 m/s, maximum range 26 000 m.

HE Incendiary, maximum range 27 000 m. There is also an incendiary round with a range of 21 000 m.

Variants

More recently the Chinese have developed a self-propelled version of the 130 mm Field Gun Type 59-1 and available details of this are given in the previous section.

Status: Production probably complete. In service with the Chinese Army and with Cameroon, Egypt (qv local production), North Korea, (where it is also produced), Oman, Pakistan, Somalia, Sudan, Thailand (18 delivered free of charge) and Vietnam.

Manufacturer: China North Industries Corporation
7A Yuetan Nanjie
PO Box 2137 Beijing
People's Republic of China
Telephone: (86) 6898/3461/3741/7570
Telex: 22339 cnin cn

SPECIFICATIONS	
CALIBRE	130 mm
MUZZLE BRAKE	double-baffle
BREECH MECHANISM	semi-automatic vertical sliding breech block
CARRIAGE	split trail with castor wheel on each
SHIELD	yes
WEIGHT	6300 kg
LENGTH (travelling)	10.8 m
WIDTH (travelling)	2.42 m
HEIGHT	
travelling	2.75 m
to axis of bore	1.214 m
ELEVATION/DEPRESSION	+45°/-2.5°
TRAVERSE	58°
RATE OF FIRE	8-10 rds/min
RANGE (HE)	27 490 m
CREW	8-10
TOWING VEHICLE	6 × 6 truck

NORINCO 122 mm Howitzer D-30

Development/Description

The 122 mm Howitzer D-30 produced by China North Industries Corporation is the same weapon as the Soviet D-30 and differs from the Soviet original only in detail. Combat weight is 3200 kg with rate of fire being 8 rds/min.

The Chinese D-30 fires an HE projectile weighing 21.76 kg that is identical to the Soviet original. However, the Chinese D-30 also fires a locally developed long-range HE projectile using a special charge that produces a maximum range of 18 000 m (standard range is 15 300 m). The long-range HE projectile is longer and uses a more aerodynamic body with a wide knurl at the bottom of the body that serves to increase friction between the driving band and the body surface to prevent sliding and shedding of the drive band under high chamber pressures. The knurl on the driving band edge also improves the chamber gas sealing. A 122 mm cargo round with six layers of HEAT bomblets has been developed to the prototype stage.

In 1990 NORINCO stated that it offered the following ammunition for its versions of the Soviet D-30 as well as for the original Soviet weapon:
ERFB/BB with a maximum range of 21 000 m.
ERFB HE with a maximum range of 18 000 m.
HE/HEAT cargo round with a maximum range of 15 000 m.
HE/incendiary with a maximum range of 15 000 m.
Smoke with a maximum range of 15 000 m.
Illuminating with a maximum range of 15 000 m and a minimum range of 4000 m.

On-carriage fire control instruments mounted on the Chinese D-30 include a Type PG-1M panoramic sight, a Type OP4M-45 telescopic sight and a Type K-1 indicator.

The D-30 is used as the main armament of the Type 85 self-propelled howitzer. For details refer to the *Self-propelled guns and howitzers* section.

Status: Production. In service with the Chinese armed forces.

Manufacturer: China North Industries Corporation
7A Yuetan Nanjie
PO Box 2137 Beijing
People's Republic of China
Telephone: (86) 6898/3461/3741/7570
Telex: 22339 cnin cn

Chinese copy of Soviet 122 mm howitzer D-30 in firing position

SPECIFICATIONS

CALIBRE	122 mm
MUZZLE BRAKE	multi-baffle
WEIGHT FIRING	3200 kg
LENGTH (travelling)	5.4 m
WIDTH (travelling)	1.95 m
HEIGHT (travelling)	1.66 m
ELEVATION/DEPRESSION	+70°/−7°
TRAVERSE	
(at elevation from −7° to +22°)	360°
(at elevation from +22° to +70°)	66° (between movable trails) 124° (between fixed mount and left and right trails)
DIRECT FIRE RANGE	1000 m
MUZZLE VELOCITY	
(HE projectile)	690 m/s
RATE OF FIRE	6–8 rds/min
MAX TOWED SPEED (road)	60 km/h

NORINCO 122 mm Howitzer Type 54-1

Development/Description

The 122 mm Howitzer Type 54-1 is the latest Chinese version of the Soviet 122 mm howitzer M1938 (M-30) and is virtually identical to the Soviet original apart from some small manufacturing expedients.

The 122 mm Type 54-1 Howitzer is fitted with a Type 58 panoramic sight and a collimator sight Type 58.

The Type 54-1 uses a nine-charge variable propellant charge system and fires four types of projectile: HE, smoke, illuminating and incendiary. No details are available regarding the incendiary projectile and it is thought to be little used. Details of the other three projectiles are as follows:

Type	**HE**	**Smoke**	**Illum***
WEIGHT OF PROJECTILE	21.76 kg	22.55 kg	21.9 kg
LENGTH OF PROJECTILE	559 mm	551 mm	531 mm
WEIGHT OF FILLING	3.5 kg	3.65 kg	1.515 kg
MUZZLE VELOCITY	515 m/s	509 m/s	496 m/s
MAX RANGE	11 800 m	11 930 m	11 000 m†
FUZE	Liu-4	Yan-2	Shi-1

* burns for up to 30 s producing 450 000 candela
† air burst

A leaflet shell containing 1.1 kg of leaflets is also used. A 122 mm cargo round with six layers of HEAT bomblets has been developed to the prototype stage.

This howitzer is mounted on the 122 mm self-propelled howitzer Type 54-1. For details refer to the entry in the *Self-propelled guns and howitzers* section.

122 mm howitzer Type 54-1 in firing position

Status: Production complete. In service with the Chinese and some other armies.

Manufacturer: China North Industries Corporation
7A Yuetan Nanjie
PO Box 2137 Beijing
People's Republic of China
Telephone: (86) 6898/3461/3741/7570
Telex: 22339 cnin cn

SPECIFICATIONS

CALIBRE	121.92 mm
BARREL LENGTH (overall)	2.8 m
MUZZLE BRAKE	no
RECOIL SYSTEM	hydraulic buffer and hydro-pneumatic recuperator
BREECH MECHANISM	hinged screw
CARRIAGE	split trail
SHIELD	yes
WEIGHT (total)	2500 kg
LENGTH (travelling)	5.9 m
WIDTH (travelling)	1.975 m
HEIGHT	
travelling	1.82 m
to axis of bore	1.2 m
ELEVATION/DEPRESSION	+60.5°/−3°
TRAVERSE	49°
RATE OF FIRE	5–6 rds/min
RANGE (HE)	11 800 m
CREW	8
TOWING VEHICLE	6 × 6 truck
MAX TOWING SPEED	20/25 km/h

NORINCO 100 mm Field Gun

Development/Description

This gun is essentially a copy of the Soviet BS-3 (M1944) and was introduced to replace the 85 mm Type 56 in border area artillery regiments. The gun is effective as either a heavy anti-tank weapon or counterbattery artillery piece. The gun is essentially the same as found on the SU-100 and T-54 tank. The Soviet UBR-412 AP-T round is used in the anti-tank role, while the HE-FRAG UOF-412 round is also produced in the PRC (with point detonating fuze Type 429).

The gun has a double-baffle muzzle brake, with an easily identified massive square-shaped breech system. It has a cradle-enclosed recoil system and vertical sliding-wedge breech-block operating semi-automatically. The carriage has double tyres on each wheel. The breech-ring apparently doubles for the long gun tube. The Chinese copy is intended primarily for the anti-tank role, and its effective range is about 1040 m. There are 18 guns in an Artillery Regiment but these will be found mixed with the 85 mm Type 56 at the army level.

This gun may be designated Type 59, but this is unconfirmed.

Status: Production complete. In service with the Chinese Army.

Manufacturer: China North Industries Corporation
7A Yuetan Nanjie
PO Box 2137
Beijing
People's Republic of China
Telephone: (86) 6898/3461/3741/7570
Telex: 22339 cnin cn

SPECIFICATIONS

CALIBRE	100 mm
BARREL LENGTH	9.21 m
ELEVATION/DEPRESSION	+45°/-5°
TRAVERSE	about 50° on carriage
FIRING WEIGHT	3450 kg (est)
SHELL WEIGHT	
AP-T	29.8 kg
HE	27 kg
MUZZLE VELOCITY	895 m/s
MAX RANGE	20 000 m
RATE OF FIRE	7 rds/min
ARMOUR PENETRATION	
0°/1000 m	157 mm (AP-T)
0°/500 m	170 mm (AP-T)
AMMUNITION	HE-FRAG projectile weighing 15.6 kg; AP-T round, and smoke available in Chinese service
CREW	6

NORINCO 100 mm Anti-tank Gun Type 86

Development/Description

The Type 86 Anti-tank Gun appears to be a modification of the 85 mm Field Gun Type 56 (see following entry) incorporating a new ordnance and recoil system, but it may be a development of the Type 73 anti-tank gun which is believed to be a copy of the Soviet T-12 towed anti-tank gun which has a smooth barrel.

The Type 86 ordnance is smoothbored and features a distinctive muzzle brake. The breech mechanism has a vertical sliding block and the sighting mechanism is located to the left of the gun. The carriage uses split trails with a castor wheel on the left trail leg to assist in bringing the gun into action. A shield with sloping sides is provided.

Although the Type 86 is primarily an anti-tank weapon it can be used for the indirect fire role, even though the potential maximum range is limited by the maximum gun elevation of +38°. Maximum rate of fire is quoted as 8 to 10 rds/min.

A night vision device for sighting purposes is available.

The Type 86 fires the following types of ammunition:

APFSDS Type 86 firing a tungsten penetrator at a muzzle velocity of 1610 m/s to an effective range of 1800 m; when firing this round the maximum rate of fire is 4 to 5 rds/min
APDS Type 73 firing a tungsten penetrator at a muzzle velocity of 1500 m/s to an effective range of 1700 m
HEAT Type 73 firing a fin-stabilised projectile at a muzzle velocity of 1010 m/s.
APFSDS Type 73 with tungsten alloy penetrator with a maximum direct fire range of 1700 m
APFSDS Type 73 with the projectile having a length-to-diameter ratio of 13.3/1 and a maximum range in the direct fire role of 1730 m
HE Type 73 firing a fin-stabilised projectile at a muzzle velocity of 900 m/s to a maximum range of 13 700 m.

When firing these projectiles the usual recoil length is between 1 and 1.14 m; maximum recoil length is 1.18 m.

SPECIFICATIONS

CALIBRE	100 mm
WEIGHT (travelling)	3660 kg
LENGTH (travelling)	9.52 m
WIDTH (travelling)	2.12 m
HEIGHT (travelling)	1.838 m
(centre of bore)	1.088 m
AXIS OF FIRE	1.08 m
RECOIL	
normal	1.05 m
max	1.18 m
ELEVATION/DEPRESSION	+38°/-4°
TRAVERSE	50° (left and right)
MAX RANGE	13 654 m
MAX TOWING SPEED	50 km/h
RATE OF FIRE	8-10 rds/min

Status: Production probably complete. In service with the Chinese Army.

Manufacturer: China North Industries Corporation
7A Yuetan Nanjie
PO Box 2137 Beijing
People's Republic of China
Telephone: (86) 6898/3461/3741/7570
Telex: 22339 cnin cn

100 mm anti-tank gun Type 86 in firing position

100 mm anti-tank gun Type 86 in travelling mode

NORINCO 85 mm Field Gun Type 56

85 mm field gun Type 56 in firing position

Development/Description

The Soviet 85 mm Divisional gun D-44 was first supplied to China during the Korean War. China began production of this gun during the early 1960s under the designation Type 56. Some slight changes were made to the D-44 design to suit Chinese production methods one of them being the location of a push rod firing device in the centre of the elevating handwheel. The Type 56 does not appear to use the Soviet infra-red searchlight and sight system. Some guns have a castor wheel under the closed trails.

Ammunition currently available for the Type 56 includes HE, HEAT-FS and HESH. In the past AP-T and HE-FRAG have been used. The only current projectile for which definite information is available is the Type 56 HEAT-FS. This one-piece round weighs 12.5 kg and is 990 mm long. The projectile weighs 7 kg and the TNT/RDX 50:50 shaped charge weighs 0.647 kg. The direct fire maximum range is 970 m at which it will penetrate 100 mm of armour set at an angle of 65°. A Dian-1A fuze is used and the muzzle velocity is 845 m/s.

Optical equipment fitted includes a Type 58 panoramic sight with a magnification of ×3.7, Type 85 direct fire sight with a magnification of ×5.5 and a collimator Type 58.

SPECIFICATIONS

CALIBRE	85 mm
BARREL LENGTH (overall)	4.693 m
MUZZLE BRAKE	double-baffle
RECOIL SYSTEM	hydraulic recoil buffer and hydropneumatic recuperator
BREECH MECHANISM	semi-automatic vertical sliding block
CARRIAGE	split tubular trail
SHIELD	yes
WEIGHT (travelling)	1750 kg
LENGTH (travelling)	8.34 m
WIDTH (travelling)	1.73 m
HEIGHT (travelling)	1.42 m
ELEVATION/DEPRESSION	+35°/-7°
TRAVERSE	54°
RATE OF FIRE	15-20 rds/min
RANGE	
max	15 650 m
HEAT	970 m
CREW	6-8
MAX TOWING SPEED	
road	60 km/h
cross-country	15 km/h
TOWING VEHICLE	4 × 4 truck

Status: Production probably complete. In service with the Chinese Army. Also in service with Afghan Resistance, Albania, Cameroon, Congo, North Korea, Mozambique, Pakistan (190), Sri Lanka (330), Tanzania (75), Vietnam and Zaïre (24).

Manufacturer: China North Industries Corporation
7A Yuetan Nanjie
PO Box 2137 Beijing
People's Republic of China
Telephone: (86) 6898/3461/3741/7570
Telex: 22339 cnin cn

Other Chinese Towed Artillery Systems

NORINCO 152 mm Type 54 Howitzer

This is the Chinese-built version of the Soviet D-1 and has a combat weight of 3600 kg, rate of fire of 4 rds/min and a maximum range of 12 400 m.

NORINCO 122 mm Type 83 Howitzer

This new light howitzer was first seen in public in 1984 during the National Day Parade and the following outline specifications are available:

CALIBRE	121.92 mm
LENGTH TRAVELLING	7.862 m
WIDTH	2.157 m
WEIGHT TRAVELLING	2668 kg
ELEVATION/DEPRESSION	+65°/-3°
TRAVERSE (TOTAL)	65°
RATE OF FIRE	7-8 rpm
MAX RANGE	18 000 m with extended range ammunition
MIN RANGE	3700 m
WEIGHT OF COMPLETE ROUND	27.36 kg
MUZZLE VELOCITY	618 m/s

NORINCO 122 mm Type 60 Field Gun

This is the Chinese-built version of the Soviet D-74 and has a combat weight of 5500 kg and a maximum range of 24 000 m.

NORINCO 100 mm Type 73 Anti-tank Gun

This is possibly a copy of the Soviet T-12 anti-tank gun and has a maximum range of 1700 m.

Status: In service with the Chinese Army. The 122 mm Type 83 howitzer is the only one of the above that is still in production today.

Manufacturer: China North Industries Corporation
7A Yuetan Nanjie
PO Box 2137 Beijing
People's Republic of China
Telephone: (86) 6898/3461/3741/7570
Telex: 22339 cnin cn

CZECHOSLOVAKIA

100 mm Field Gun M53

Development/Description

The 100 mm Field Gun M53 developed in Czechoslovakia in the early 1950s fulfils a similar role to the Soviet 100 mm M1944 (BS-3) field gun and fires the same ammunition. Its ammunition is not interchangeable with the more recent Soviet smoothbore 100 mm T-12 anti-tank gun. The M53 can be used in both the field and anti-tank roles and has a much longer range than the 100 mm M1944's 15 400 m range as its barrel can be elevated to +42° compared with the +20° of the M1944.

The 100 mm Field Gun M53 can be distinguished from the 100 mm M1944 (BS-3) by single road wheels and a shield with a straight top and sides that slope to the rear rather than the M1944's dual road wheels and a shield with a wavy top. The M53 can be fitted with the large diameter APN-3-5 infra-red night device over the rear part of the barrel.

The M53 fires the following ammunition (a Czechoslovak HEAT projectile is also used):

100 mm field gun M53 in firing position (via Franz Kosar)

Ammunition type	**APC-T**	**FRAG-HE**
PROJECTILE DESIGNATION	BR-412D*	OF-412†
FUZE MODEL	DBR-2	V-429
WEIGHT OF PROJECTILE	16 kg	15.59 kg
WEIGHT OF BURSTING CHARGE	0.064 kg	1.46 kg
TYPE OF BURSTING CHARGE	RDX/alum	TNT
MUZZLE VELOCITY	1000 m/s	900 m/s
ARMOUR PENETRATION AT 0° OBLIQUITY	185 mm/1000 m	n/app

Ammunition type	**HEAT-FS**	**HVAPDS-T**
PROJECTILE DESIGNATION	ZBK-5M‡	BM-8‡
FUZE MODEL	VP-9	n/app
WEIGHT OF PROJECTILE	12.36 kg	5.69 kg
WEIGHT OF BURSTING CHARGE	1.038 kg	n/app
TYPE OF BURSTING CHARGE	RDX/wax	n/app
MUZZLE VELOCITY	900 m/s	1415 m/s
ARMOUR PENETRATION AT 0° OBLIQUITY	380 mm/any range	200 mm+/1000 m

* older design projectiles BR-412 and BR-412B also used with rounds giving penetrations of 135 and 150 mm
† other projectiles available, some cartridges have reduced propelling charges
‡ employment with this weapon not confirmed. It is also reported that the 100 mm field gun M53 fires an HVAP projectile weighing 10.2 kg with a muzzle velocity of 1200 m/s

Status: Production complete. In service only with the Czechoslovak Army.

Manufacturer: Czechoslovak state factories.

SPECIFICATIONS

CALIBRE	100 mm
BARREL LENGTH (overall)	6.735 m
MUZZLE BRAKE	double-baffle
RECOIL SYSTEM	hydraulic buffer and hydro-pneumatic recuperator
BREECH MECHANISM	horizontal sliding wedge
CARRIAGE	split trail with castor wheel on each
SHIELD	yes
WEIGHT	
travelling order	4280 kg
firing position	4210 kg
LENGTH (travelling)	9.1 m
WIDTH (travelling)	2.36 m
HEIGHT (travelling)	2.606 m
AXIS OF BORE (at 0°)	1.25 m
GROUND CLEARANCE	0.35 m
TRACK	1.98 m
TYRES	11.00 × 22
ELEVATION/DEPRESSION	+42°/-6°
TRAVERSE (total)	60°
RATE OF FIRE	8-10 rds/min
MAX RANGE	21 000 m
CREW	6
TOWING VEHICLE	Tatra-138 (6 × 6) truck

85 mm Field Gun M52

Development/Description

The 85 mm Field Gun M52 is the Czechoslovak equivalent of the Soviet 85 mm Divisional Gun D-44 and fires the same ammunition. A slightly modified version of the M52 is designated the M52/55, and both can be fitted with an infra-red night device.

The 85 mm Field Gun M52 is very similar in appearance to the Czechoslovak 100 mm Field Gun M53, which also has a double-baffle muzzle brake but on which the sides of the shield slope to the rear and the top of the shield is straight rather than wavy as in the case of the M52. The trails of the M53 also have castor wheels to assist in bringing the weapon into and out of action.

The M52 fires the following fixed ammunition:

85 mm field gun M52 deployed in firing position (Austrian Army)

Ammunition type	**AP-T**	**FRAG**
PROJECTILE DESIGNATION	BR-365*	O-365K†
FUZE MODEL	MD-5	KTM-1
WEIGHT OF PROJECTILE	9.2 kg	9.5 kg
WEIGHT OF BURSTING CHARGE	0.068 kg	0.77 kg
TYPE OF BURSTING CHARGE	RDX/alum	TNT
MUZZLE VELOCITY	820 m/s	805 m/s
ARMOUR PENETRATION AT 0° OBLIQUITY	123 mm/1000 m	n/a

Ammunition type	**HEAT-FS**	**HVAP-T**
PROJECTILE DESIGNATION	BK-2M‡	BR-356P§
FUZE MODEL	n/a	n/a
WEIGHT OF PROJECTILE	7.34 kg	5.06 kg
WEIGHT OF BURSTING CHARGE	0.96 kg	n/a
TYPE OF BURSTING CHARGE	RDX/wax	n/a
MUZZLE VELOCITY	840 m/s	1070 m/s
ARMOUR PENETRATION AT 0° OBLIQUITY	n/a	130 mm/1000 m

* other projectiles available
† other projectiles available, some cartridges have reduced charges
‡ employment with this weapon not confirmed
§ spool type, possibly replaced by improved model

Austria uses a Belgian-designed HEAT-T round called the Leuchtspur-Hohlgranate 72 which has a muzzle velocity of 820 m/s and will penetrate 350 mm of armour

Status: Production complete. In service with Austria (250) and Czechoslovakia.

Manufacturer: Czechoslovak state factories.

SPECIFICATIONS
(M52; data in square brackets relates to M52/55 where different)

CALIBRE	85 mm
BARREL LENGTH (overall)	5.07 m
MUZZLE BRAKE	double-baffle
RECOIL SYSTEM	hydraulic recoil buffer and hydro-pneumatic recuperator
BREECH MECHANISM	semi-automatic vertical sliding wedge
CARRIAGE	split trail
SHIELD	yes
WEIGHT	
travelling order	2130 [2168] kg
firing position	2095 [2111] kg
LENGTH (travelling)	7.52 m
WIDTH (travelling)	1.98 m
HEIGHT (travelling)	1.515 m
AXIS OF BORE (at 0°)	1.135 m
GROUND CLEARANCE	0.35 m
TRACK	1.63 m
TYRES	8.25 × 20
ELEVATION/DEPRESSION	+38°/-6°
TRAVERSE (total)	60°
RATE OF FIRE	20 rds/min
RANGE	
max	16 160 m
effective FRAG	925 m
effective HVAP-T	1150 m
CREW	7
TOWING VEHICLE	Praga V3S (6 × 6) truck
MAX TOWED SPEED	
road	50 km/h
cross-country	10 km/h

EGYPT

Factory 100 Artillery Systems

Abu Zaabal Engineering Industries Co, also known as Factory 100, has reverse engineered the Soviet 122 mm D-30 towed howitzer, Chinese 130 mm Field Gun Type 59 and the Soviet 23 mm ZU-23 towed anti-aircraft gun. The first 122 mm D-30 was completed in Egypt early in 1984 with Royal Ordnance Nottingham providing some part-machined components which were completed in Egypt.

Factory 100 is also closely involved with Royal Ordnance Nottingham in the upgunning of the Soviet-supplied T-55 MBT with the 105 mm L7A3 rifled gun and will also be involved in whichever of the two competing 122 mm D-30 self-propelled howitzers is eventually selected for production or co-production in Egypt. Details of both competing designs are given in the *Self-propelled guns and howitzers* section under Egypt. Factory 100 will also be involved in the twin 23 mm self-propelled anti-aircraft gun system which was selected in late 1987.

SPECIFICATIONS

Original Designation	**D-30**	**Type 59**
EGYPTIAN DESIGNATION	D-30-M	M59-1M
CALIBRE	122 mm	130 mm
BARREL LENGTH	4.27 m	7.15 m
MAX PRESSURE	2500 kg/cm²	3150 kg/cm²
WEIGHT		
travelling	3210 kg	6300 kg
firing	3150 kg	6100 kg

Original Designation	**D-30**	**Type 59**
LENGTH (travelling)	5.4 m	10.8 m
WIDTH (travelling)	1.95 m	2.42 m
HEIGHT (travelling)	1.66 m	2.75 m
GROUND CLEARANCE	0.325 m	0.3 m
ELEVATION/DEPRESSION	+70°/-7°	+45°/-2.5°
TRAVERSE (total)	360°	58°
RATE OF FIRE	6-8 rds/min	6-8 rds/min
MAX RANGE	15 300 m	27 150 m
TIME INTO ACTION	90-210 s	180 s
TIME OUT OF ACTION	90-210 s	180 s
CREW	7	9
MAX TOWING SPEED	60 km/h	60 km/h

Status: Production as required. In service with Egyptian Army.

Manufacturer: Abu Zaabal Engineering Industries Co
PO Box 5888
Heliopolis West
Cairo
Arab Republic of Egypt
Telephone: 2917305/2917033
Telex: 21487 hvrt un
Fax: 2916962

Egyptian 122 mm D-30-M towed howitzer in firing position (Christopher F Foss)

Egyptian 130 mm M59-1M gun in firing position (Christopher F Foss)

FINLAND

VAMMAS 155 mm M-83 Howitzer

Development/Description

VAMMAS Ordnance (previously the Tampella Defence Division) has been engaged in the design, development and production of artillery and their associated ammunition systems for more than 50 years.

VAMMAS now has four divisions, ammunition, ordnance, military rebuilding and modernisation and engineering. The ammunition division manufactures shell bodies for artillery in calibres of 105 mm, 122 mm, 130 mm, 152 mm and 155 mm as well as base bleed and hollow base artillery projectile bodies and mortar bomb bodies.

In January 1991 the original Tampella products, the Tampella 155 mm field guns, Tampella 130 mm coastal gun and the Tampella 60, 81, 120 and 160 mm mortars assumed the name VAMMAS.

The VAMMAS 155 mm M-83 howitzer builds on experience obtained in the development and production of the Tampella 155 mm M-74 gun-howitzer and the 122 mm M-60 field gun covered in the following entries. It is believed that the Finnish Army designation for this system is the 155K83.

Description

The VAMMAS 155 mm M-83 howitzer has a split trail carriage and can be fitted with a 39 or 45 calibre ordnance which has a distinctive T-type muzzle brake. As an alternative to the normal 155 mm ordnance, the gun can be supplied with a 122 mm, 130 mm or 152 mm barrel so allowing Chinese/Soviet types of ammunition to be fired rather than 155 mm NATO type ammunition.

The semi-automatic horizontal sliding breech mechanism and a special loading mechanism enable a high rate of fire to be achieved with obturation being obtained by a steel ring.

The special loading system is believed to be operated by compressed air with the cylinder for the latter being located on the right side of the mount. The sights, elevation and traverse wheels are located on the left side of the mount.

According to VAMMAS the weapon is very easy to handle so that only a few men are required to bring the weapon into firing or travelling configuration.

When in the travelling configuration the ordnance is traversed through 180° and locked in position over the closed trail legs. When in the firing position the weapon is supported on a firing jack under the forward part of the carriage and on the two trails, each of which is fitted with a spade. Steel stakes are provided for the spades and these are carried in racks on the inside of each trail.

For both calibre lengths, the total weight in firing position is 9500 kg with elevation being from -5° to +70° and traverse 45° left and right.

The 155 mm M-83 fires the standard NATO M107 projectile as well as hollow base and two types of base bleed projectile, one type with nubs and the other without nubs.

The design of the charge system is optimised so that range areas cover each other in low and high elevation angles. Compared to the standard NATO charge system (at least four different charges), the VAMMAS system consists of three different charges, incremental charge and two types of full charges.

VAMMAS 155 mm M-83 howitzer deployed in firing position

The make up of the incremental charge is similar to NATO charges M4A2 and M3A1, including four increments, Zones 5, 6, 7 and 8. Full and top charges are similar to NATO M2 and M11, Zones 9 and 10. The latter is only used in the 45 calibre gun.

Some 155 mm M-83 systems have been observed to be fitted with a muzzle velocity measuring system over the ordnance.

SPECIFICATIONS

CALIBRE	155 mm	155 mm
CALIBRE LENGTH	39	45
BARREL LENGTH	5990 mm	7100 mm
WEIGHT IN FIRING POSITION	9500 kg	9500 kg
ELEVATION/DEPRESSION	+70°/-5°	+70°/-5°
TRAVERSE	45° (left/right)	45° (left/right)
RATE OF FIRE		
burst	3 rounds/12 s	3 rounds/12 s
normal	10 rds/min	10 rds/min
MAXIMUM RANGE	30 000 m	39 600 m
MAX TOWING SPEED	80 km/h	80 km/h

Status

Production as required.

Manufacturer: VAMMAS
POB 18
SF-38201 Vammala
Finland
Telephone: 358 32 1971
Telex: 22283 VAMAS SF
Fax: 358 32 41148

Tampella 155 mm Gun-Howitzer M-74

Development/Description

Relatively little information has been released by Tampella on its 155 mm Gun-Howitzer M-74 but it appears to be a progressive refinement of its essentially similar M-68, using the same ammunition and extending the barrel length to L/39. The main alterations appear to be with the saddle and the equilibrators. A further charge has been added to the propellant system used as the maximum range has been increased from 23 000 to 24 000 m. The prominent barrel counterweight of the M-68 is absent and the muzzle brake design has been altered. According to Tampella, the piece can be brought into action, and out again, by only two men, although it is not specified how. An HE projectile weighing 43.6 kg is fired at a maximum muzzle velocity of 850 m/s to a range of approximately 24 000 m. Using a base bleed projectile the range is 30 000 m.

Tampella 155 mm gun-howitzer M-74 being brought into firing position

Status: Production complete. In service with Finnish Army.

Manufacturer: Oy Tampella Ab (Now renamed VAMMAS qv)
Pääkonttori
Lapintie 1
SF-33100 Tampere 10
Finland
Telephone: (31) 11 44 11
Telex: 22 117 tamsec sf

SPECIFICATIONS

CALIBRE	155 mm
BARREL LENGTH (L/39)	5.99 m
MUZZLE BRAKE	single-baffle
BREECH MECHANISM	horizontal sliding
CARRIAGE	split trail
SHIELD	no
WEIGHT	9500 kg
LENGTH (travelling)	7.5 m
WIDTH (travelling)	2.58 m
HEIGHT (travelling)	2 m
ELEVATION/DEPRESSION	+52°/-5°
TRAVERSE	90°
MAX RANGE	24 000 m 30 000 m (base bleed)

Tampella 122 mm M-60 Field Gun

Development/Description

The 122 mm M-60 field gun was developed by Tampella to meet the requirements of the Finnish Army. The first prototype was produced during 1960 and the first production version was produced during 1964.

The carriage comprises the top carriage, bottom carriage and trails. In action a firing jack is lowered under the lower carriage and the two trails are opened out. At the end of each trail is a loose spade which is staked to the ground by three wedges. A further two loose spades and six wedges are carried on the carriage to enable the gun crew to change the direction of fire without taking up the original loose spades. When travelling the loose spades and wedges are stowed on the trails. The bogie has four pneumatic tyres and each road wheel is provided with an air brake.

Tampella 122 mm M-60 field gun in firing position

The recoil mechanism is under the barrel and the counter recoil system above it with pneumatic equilibrators either side of the barrel. Recoil varies in relation to elevation but maximum recoil is 1 m. The elevating wheel and the panoramic sight are on the left side of the carriage and the traversing gear is on the right.

The semi-automatic horizontal sliding breech-block remains open on recoil for loading of the next round. Ammunition is separate loading with the HE projectile weighing 25 kg and having a maximum muzzle velocity of 950 m/s.

When travelling the barrel is swung through 180° and locked in position over the closed trails. To improve the cross-country capability of the M-60 the four wheels can be powered by a hydraulic motor run from the towing vehicle.

Status: Production complete. In service only with the Finnish Army.

Manufacturer: Oy Tampella Ab (Now renamed VAMMAS qv)
Pääkonttori
Lapintie 1
SF-33100 Tampere 10
Finland
Telephone: (31) 11 44 11
Telex: 22 117 tamsec sf

Note: Details of the Tampella 155 mm M-68 Gun/Howitzer and Tampella 155 mm M-71 Gun/Howitzer are given in the entries in the Israeli section for the Soltam 155 mm M-68 Gun/Howitzer and the Soltam 155 mm M-71 Gun/Howitzer. Both weapons use a similar carriage to the M-60.

SPECIFICATIONS					
CALIBRE	122 mm	WEIGHT (travelling order)	8500 kg	TRAVERSE (total)	90°
BARREL LENGTH	6.466 m	LENGTH (travelling)	7.2 m	RATE OF FIRE	4 rds/min
MUZZLE BRAKE	single-baffle	WIDTH (travelling)	2.58 m	MAX RANGE	25 000 m
BREECH MECHANISM	horizontal sliding	HEIGHT (travelling)	2 m	CREW	8
CARRIAGE	split trail	GROUND CLEARANCE	0.38 m	TOWING VEHICLES	Sisu A-45 (4 × 4) truck
SHIELD	no	TRACK	2.2 m		
		ELEVATION/DEPRESSION	+50°/−5°		

FRANCE

GIAT Industries 155 mm Towed Gun TR

Development

This 155 mm towed gun (Le Canon de 155 mm Tracte) has been developed to meet the requirements of the French Motorised Infantry Divisions and has now started to replace the old 155 mm Model 50 and 105 mm M101 (except in the Airborne Division and overseas) towed howitzers. The weapon was shown for the first time at the 1979 Satory Exhibition of Military Equipment. Following trials with prototype systems, a pre-production batch of six weapons was delivered by GIAT for troop trials from December 1987. The main production run commenced in 1989 and is expected to peak at about 30 weapons a year. The original French Army requirement was for around 200 weapons but this has now been reduced to about 100, of which 40 were funded in the 1989 French Defence budget.

The GIAT Industries 155 mm towed gun TR was deployed to Saudi Arabia and subsequently took part in Operation Desert Storm, the recapture of Kuwait, in the Spring of 1991.

Description

A crew of eight is required for the tractor and gun: commander, layer-gunner who also drives the gun in the self-propelled mode, loader who operates the loading device during firing, controls the raising and lowering of the gun and operation of the trails when moving into and out of action, two men who are responsible for the preparation of the propelling charges and supplying the gun, two loaders who are responsible for the preparation of the projectiles and supplying the loading mechanism and finally the tractor driver. If the tasks are divided differently the crew can be reduced to seven, six of whom serve the gun. If required the gun can be brought into and out of action by a crew of three.

The 155 mm barrel has a double-baffle muzzle brake and when travelling is swung through 180° and secured in position over the trails. The barrel is almost identical to that used in the self-propelled 155 mm GCT (full details of which are given in the *Self-propelled guns and howitzers* section), except that the latter has a vertical sliding breech-block whereas the towed gun's is horizontal sliding.

Mounted on the front of the carriage is a 39 hp engine which drives three hydraulic pumps, one for each of the two road wheels and one to provide power for functions such as elevation, traverse, raising the suspension, trail wheel jacks and the shell-loading mechanism. Travelling under its own power the gun has a maximum road speed of 9 km/h, can climb a gradient of 60 per cent and ford to a depth of 1 m without preparation. With preparation it can ford to a depth of 1.2 m.

The gun will normally be towed by a Renault TRM 10 000 (6 × 6) truck which carries the crew of eight and 48 rounds of ammunition. This combination has a maximum road speed of 80 km/h and a road range of 600 km, but it can be towed by various other types of prime mover.

On arriving at the battery position the weapon is first uncoupled from the prime mover and then the barrel is swung from its travelling position through 180° into its firing position. The wheels are raised off the ground so that the weapon rests on its firing jack under the carriage. The trails are then swung apart, which is a simple operation as castor wheels are fitted on each trail. When opened out the small wheels are raised clear

GIAT Industries 155 mm TR towed gun being taken out of action

Renault TRM 10 000 (6 × 6) truck towing GIAT Industries 155 mm towed gun TR

GIAT Industries 155 mm towed gun TR in firing position

GIAT Industries 155 mm towed gun TR in self-propelled mode (Pierre Touzin)

of the ground and the trails staked into position. The maximum time taken to bring the weapon into battery is 1.5 minutes and it can be brought out of battery in the same time thanks to the hydraulic system.

The layer is seated on the left of the mount and has hydraulic aiming controls, and the sights which can be used by day and night are the GA 81 goniometer, indirect sight and the direct fire telescopic sight. After indicating the elevation angle on a graduated drum, the barrel position is checked using the level shown on the elevating mechanism's elevation meter. The ranging (GR) and firing (GT) angles can be read directly through the goniometric sight. The direct fire telescope can be used to engage targets up to a maximum range of 2000 m. In an emergency the gun can be fired without the trails being deployed.

It has similar firepower characteristics to the 155 mm GCT self-propelled gun and is inferior only in its rate of fire as the GCT has an automatic loading system which permits a higher rate of fire to be achieved. The automated hydraulic system for ramming the projectile into the ordnance operates at all angles of elevation, reduces crew fatigue and ensures that all projectiles are rammed at the same pressure, giving a stable muzzle velocity. The ramming system permits a high rate of fire of three rounds in the first 18 seconds and 6 rds/min thereafter.

Due to its total interoperability, the GIAT Industries 155 mm towed gun TR can fire projectiles with the French or NATO propelling system to the following ranges:
HE hollow base to 24 000 m
HE base bleed to 30 000 m
HE 56/59 to 23 000 m
HE M107 to 18 500 m
HE ERFB BB to 32 000 m
Smoke or illuminating projectiles can also be fired as can cargo bomblets (63) to a maximum range of 28 000 m
155 mm HE hollow base shell Type F1 (HE 155 F1) with the projectile weighing 43.25 kg (containing 8.83 kg of HT 50/50) and a muzzle velocity of 830 m/s to a maximum range of 24 000 m
155 mm phosphorus smoke shell HB Type F2A (OFUM 155 F2A) with the projectile weighing 43.15 kg (containing 8.7 kg of white phosphorus) and a muzzle velocity of 830 m/s to a maximum range of 24 000 m
155 mm training shell Type F1 (OX 155 F1) with the projectile weighing 43.5 kg (containing 1.2 kg of black powder) and a muzzle velocity of 830 m/s to a maximum range of 24 000 m
155 mm illuminating shell Type F1 (OECL 155 F1) with the projectile weighing 43.5 kg and a muzzle velocity of 760 m/s to a maximum range of 19 250 m
155 mm anti-tank mine shell Type H1 (OMI 155 H1) with the projectile weighing 46 kg; this shell contains six anti-tank mines each weighing 0.55 kg and although fired at a maximum of charge 5, no maximum range has yet been announced
155 mm HE base BB shell Type H2 (BBE 155 H2) with the projectile weighing 43.5 kg (containing 10 kg of HT 50/50) and a muzzle velocity of 830 m/s to a maximum range of 29 000 m
GIAT Industries cargo round projectile OGRE 155 G1 containing 63 bomblets to a maximum range of 28 500 m
155 mm M107 (US) projectile to a maximum range of 18 500 m plus ammunition used by the FH-70

The TR has been provided with a number of emergency systems to allow continued use in the event of damage. These are a battery for the hydraulics which makes it possible to fire a further six rounds or take the weapon out of action if the thermal engine stops, connection of the gun to the tractor's hydraulic system to power all of the gun's firing operations and the coming into and out of action (or two guns can be connected together), manual pumps for elevation and traverse and manual mechanisms for traversing, opening and closing the breech and firing.

Variants

155/45 gun

In 1990 GIAT Industries announced that it had completed the prototype of a 155 mm towed gun TR with a 45 calibre ordnance that enables a range of 28 000 m to be achieved with basic projectiles and 39 000 m with ERFB-BB projectiles. This has similar specifications to the 155/39 except that it is slightly longer when firing and weighs 10 800 kg.

155/52 gun

In 1990 GIAT Industries announced that it had completed the prototype of a 155 mm towed gun TR with a 52 calibre ordnance that enables a range of 30 000 m to be achieved with basic projectiles and 41 000 m with ERFB-BB projectiles. This conforms to the new NATO ballistic MoU and was fired for the first time early in 1990. At the present time there is no formal French Army requirement for this system although in the future it is possible that it may want to upgrade its current 155/39 calibre guns to the 155/52 configuration.

The main modifications include replacement of the barrel by the new 52 calibre ordnance, modified tow bar, modified rammer and some modifications to the recoil system.

The 155/52 has similar specifications to the 155/39 except that it is slightly longer when in the firing position and weighs 11 000 kg.

Status: In production. In service with the French Army. 155/45 and 155/52 versions still at prototype stage.

Manufacturer: Etablissement d'Etudes et de Fabrications d'Armement de Bourges.
GIAT Industries
13 Route de la Minière
F-78034 Versailles
France
Telephone: (331) 30 97 37 37
Fax: (331) 30 97 39 00

SPECIFICATIONS		WIDTH		RANGE	
CALIBRE	155 mm	travelling	3.09 m	standard projectile	24 000 m
BARREL LENGTH	6.2 m	firing	8.4 m	base bleed projectile	32 000 m
MUZZLE BRAKE	double-baffle	HEIGHT (firing)	1.65 m	TUBE LIFE	3000 rounds
RECOIL	hydro-pneumatic	GROUND CLEARANCE		CREW	7
BREECH MECHANISM	horizontal wedge	(travelling)	0.5 m	TOWING VEHICLE	Renault TRM
CARRIAGE	split trail	ELEVATION/DEPRESSION	+66°/-5°		10 000 (6×6)
SHIELD	no	TRAVERSE (total)	65° (27°L 38°R)		truck
WEIGHT (travelling order)	10 750 kg approx	RATE OF FIRE	3 rounds first 18 s	AUTONOMOUS SPEED	9 km/h
LENGTH			6 rds/min for 2	GRADIENT	60%
travelling, barrel over trails	8.75 m		minutes	FORDING	1.2 m
firing	10 m				

GIAT Industries 155 mm Howitzer M114F

GIAT Industries M114F 155 mm modernised howitzer in firing position

Development/Description

In 1987 GIAT Industries showed the prototype of an upgraded 155 mm M114 towed howitzer under the designation of the M114F, this being developed as a private venture and is aimed specifically at the export market. GIAT Industries can either upgrade existing M114s in France or supply kits to enable users to upgrade M114s in their own facilities. By 1990 development of the M114 was complete and it had successfully completed exhaustive firing trials in France.

The modification consists of the following:

(1) Replacement of the original barrel with the GIAT Industries 39 calibre NATO barrel 155 TR (eg ordnance, muzzle brake, sleeve and breech) without modification of the original gun cradle
(2) Replacement of the equilibrators by new gas compensators equilibrators
(3) Adding a loading cradle
(4) Slightly modifying the elastic couplings
(5) Advancing the firing jack by 20 cm
(6) Fitting of castor wheels to reduce time taken to come into or be taken out of action
(7) One single barrel handwheel for aiming in both elevation and traverse.

The original M114, developed over 50 years ago, has a crew of 11 whereas the M114F has a seven man crew with time into action being three minutes compared to six.

According to GIAT Industries, the M114F has the following ranges:
18 500 m with US M107 HE projectile
24 000 m with GIAT OE HB 155 F1 projectile
30 000 m with GIAT OE 155 BB G1 projectile
32 000 m with ERFB-BB projectile.

The charges involved (from one to nine) may be loaded in cartridge bags or combustible containers.

Status: Prototype.

Manufacturer: GIAT Industries
13 Route de la Minière
F-78034 Versailles
France
Telephone: (331) 30 97 37 37
Fax: (331) 30 97 39 00

SPECIFICATIONS		LENGTH (travelling)	10 m	TRAVERSE	25° right/24° left
CALIBRE	155 mm	WIDTH (travelling)	2.44 m	RATE OF FIRE	4 rds/min
BARREL LENGTH	6.2 m	WIDTH OF TRACK	2.08 m	MAX RANGE	32 000 m
MUZZLE BRAKE	double baffle	HEIGHT		CREW	7
RECOIL	hydro-pneumatic	(max, travelling with lunette			
BREECH MECHANISM	horizontal wedge	0.7 m from ground)	2.3 m		
SHIELD	yes	(trunnion)	1.42 m		
WEIGHT	7300 kg	ELEVATION/DEPRESSION	+63°/-2°		

155 mm Howitzer Model 50

155 mm howitzer Model 50 of French Army being towed by Berliet GBU 15 truck (6×6) (Pierre Touzin)

Development

This 155 mm howitzer was developed in France in the immediate post-Second World War period and was also manufactured in Sweden by Bofors for the Swedish Army which calls it the 15.5 cm Field Howitzer Fr. The French call the weapon the OB 155-50 BF. The Model 50 was widely used by the Israeli Army but has been withdrawn in favour of self-propelled artillery pieces such as the M109. The Israeli Army also uses a self-propelled version of the Model 50 which is based on a modified Sherman chassis, details of which are given in the *Self-propelled guns and howitzers* section under Israel. In the French Army the Model 50 is now being replaced by the new 155 mm towed gun TR and has been supplemented in the Swedish Army by the Bofors FH-77A. A number of these weapons were captured by the Israeli Army from the PLO during the fighting in the Lebanon in the Summer of 1982.

Description

The barrel, with the breech ring, slides in the cradle during recoil and the breech ring is secured to the piston rod of the recoil buffer and to the recuperating rod. The recoil length varies automatically with elevation. The barrel, breech ring and cradle, which together form the elevating mass, are secured to the upper mounting by cradle trunnions. Elevating mass moves vertically in relation to the upper mounting when the elevating device is turned. The upper mounting rests on the lower carriage and turns horizontally in relation to it when the traversing device is turned. On the bottom of the lower mounting there is a pivot plate. The trails are pivoted in the sides of the lower carriage. Secured to each trail is a wheel bar which in turn holds two pneumatic puncture-proof tyres.

The weapon has a split trail carriage and on the right trail there is an unlimbering eye which during transport is hooked to the corresponding hook on the towing vehicle. During transport the elevating mass is locked to the trails by a locking device in the rear part of the cradle. The weapon is fitted with a brake system which is operated with compressed air from the towing vehicle.

In the firing position the gun rests on the pivot plate and the outer end of the trails. The trails, when opened out, are secured to the ground by pickets. The pivot plate is pivoted in two directions at right angles to each other and thus can be placed so that the carriage remains horizontal regardless of the inclination of the ground.

The Model 50 fires an HE projectile weighing 43 kg with a maximum muzzle velocity of 650 m/s.

Status: Production complete. In service with France, Lebanon, Sweden, Switzerland and Tunisia.

Manufacturer: French state arsenals.

155 mm howitzer Model 50 captured in Lebanon in summer 1982 (Israeli Ministry of Defence)

SPECIFICATIONS	
CALIBRE	155 mm
BARREL LENGTH (from front of muzzle brake to rear part of breech ring)	4.41 m
LENGTH OF RIFLED BORE	2.727 m
MUZZLE BRAKE	multi-baffle
RECOIL	hydro-pneumatic
BREECH MECHANISM	screw
CARRIAGE	split trail
SHIELD	no
WEIGHT	
travelling order	9000 kg
firing position	8100 kg
LENGTH	
travelling	7.8 m
firing	7.15 m
WIDTH	
travelling	2.75 m
firing	6.8 m
HEIGHT	
travelling	2.5 m
firing	1.65 m
GROUND CLEARANCE	0.25 m
TRACK	2.31 m
ELEVATION/DEPRESSION	+69°/-4°
TRAVERSE (total)	80°
RATE OF FIRE (max)	3-4 rds/min
MAX RANGE	
standard ammunition	18 000 m
with RAP	23 300 m
CREW	11
TOWING VEHICLE	France, Berliet GBU 15 (6 × 6) truck

GIAT Industries 105 mm LG1 Light Gun

Development

The 105 mm LG1 Light Gun has been developed by GIAT Industries as a private venture specifically for the export market.

By 1987 three prototypes had been built by the Etablissement d'Etudes et de Fabrications d'Armement de Bourges for extensive trials.

These were followed by three pre-production weapons, one of which has been tested in the Far East.

Late in 1990, an undisclosed export customer, believed to be Singapore, placed an order with GIAT Industries for the 105 mm LG1 light gun with first weapons being delivered late in 1991. Also competing for this contract was Royal Ordnance with its 105 mm Light Gun.

Description

The GIAT Industries 105 mm LG1 Light Gun is similar to the earlier GIAT Industries LTR Light Gun with a split trail carriage, small shield that elevates with the ordnance and a 30 calibre ordnance that is fitted with double baffle muzzle brake.

When deployed in the firing position it is supported on its two trails and a circular firing platform that is lowered under the forward part of the carriage.

The 105 mm LG1 Light Gun can be brought into or taken out of action in just 30 seconds as a simple hydraulic system operated by a hand pump assists in opening out the trails.

According to GIAT Industries, only two types of ammunition are required to cover the range of between 1.4 and 17.5 km. The standard US 105 mm M1 ammunition covers 1.4 up to 11.5 km while the new GIAT Industries base bleed projectile covers the range from 9.5 up to 17.5 km. In addition, GIAT Industries has also proposed a hollow base projectile covering the range 2.5 up to 15 km.

The GIAT Industries 105 mm LG1 Light Gun can be towed by a wide range of light vehicles including an ACMAT (4 × 4) truck, Peugeot P4 (4 × 4), Renault TRM 2000 (4 × 4), Land Rover (4 × 4) and Toyota (4 × 4).

SPECIFICATIONS	
CALIBRE	105 mm
LENGTH OF BARREL	3.15 m
MUZZLE BRAKE	double baffle
RECOIL	hydro-pneumatic
BREECH MECHANISM	vertical sliding, semi-automatic
CARRIAGE	split trail
SHIELD	yes
WEIGHT	1485 kg
WEIGHT ON TOW HOOK	
barrel to rear	200 kg
barrel deployed	120 kg
LENGTH	
barrel to rear	5.32 m
barrel deployed	6.76 m
HEIGHT TRAVELLING	2 m
GROUND CLEARANCE	0.35 m
ELEVATION/DEPRESSION	+70°/-3°
TRAVERSE	36°
MAX RATE OF FIRE	12 rds/min
MAX RANGE	
HE M1	11 500 m
HE HB	15 000 m
HE BB	17 500 m
CREW	6

GIAT Industries 105 mm LG1 Light Gun in travelling configuration

GIAT Industries 105 mm LG1 Light Gun

Status: In production for export market with first deliveries to Singapore late in 1991.

Manufacturer: GIAT Industries
13 Route de la Minière
F-78034 Versailles
France
Telephone: (331) 30 97 37 37
Fax: (331) 30 97 39 00

GIAT Industries 105 mm LG1 Light Gun in firing position and showing jack under forward part of carriage

GIAT Industries 105 mm M101A1 Modernisation Package

Development/Description

GIAT Industries has developed and sold a modernisation package for the US 105 mm M101A1 towed howitzer to an undisclosed export customer.

The modifications can be carried out by GIAT Industries or the latter can supply kits to enable the user to upgrade them in his own facilities using his own personnel.

The modernisation includes replacing the original barrel with the 105 mm LG1 barrel with the breech tube and breech of the original M101A1 being retained, installing a new muzzle brake and balancer, and installing ballast on the outrigger.

The upgraded M101A1 is some 280 kg heavier than the standard weapon but can fire all standard NATO ammunition including the US M1 projectile to a maximum range of 11 500 m. In addition it can fire a GIAT 105 BB base bleed projectile to a range of 17 500 m or a GIAT HB HE projectile to a maximum range of 15 000 m.

Status: Production as required. In service with an undisclosed country.

Manufacturer: GIAT Industries
13 Route de la Minière
78034 Versailles
France
Telephone: (331) 30 97 37 37
Fax: (331) 30 97 39 00

105 mm M101A1 towed howitzer modernised by GIAT Industries

INDIA

Indian Ordnance Factories 105 mm Light Field Gun

Development

The Indian Ordnance Factories 105 mm Light Field Gun bears a resemblance to the British 105 mm Royal Ordnance Light Gun, partly due to the fact that both were derived from the same 105 mm L13 gun used on the FV433 Abbot self-propelled gun. The first 105 mm barrel developed in India from the British original was used on the 105 mm Field Gun Mark 1 (see next entry). This uses tubular bow trails similar to those of the British 105 mm Light Gun.

Description

The 105 mm Light Field Gun has been developed for use in a variety of types of terrain ranging from mountains to deserts. To reduce weight the weapon has been constructed using light, high-strength alloy steels to the extent that the weight in action is 2380 kg. The ordnance consists of a monobloc autofrettaged barrel with a vertical, sliding, hand-operated breech block. Firing is electrical. The cradle, saddle and under carriage are lightweight fabricated structures. Other features include spring-type balancing gears, a telescopic type of hydro-pneumatic recuperator, a suspension system and bow-shaped tubular trails. The normal towing position is with the barrel folded back across the trails. A 360° traversing firing platform is provided. The Light Field Gun can be parachute dropped and carried by helicopter.

The gun fires separate ammunition using a seven-charge system (1, 2, 3, 4, 4.5, 5 and Super). Projectiles include HE, smoke, base ejection, star and HESH. Maximum range is 17 200 m. The normal rate of fire is 4 rds/min and the intense rate of fire is 6 rds/min for up to 10 minutes (Charge 5). The sustained fire rate is one round every two minutes for up to two hours (Charge Super).

Indian Ordnance Factories 105 mm Light Field Gun (Christopher F Foss)

Status: In production. In service with the Indian Army.

Manufacturer: Gun Carriage Factory
Jabalpur - 482 011
India

Enquiries to: The Secretary
Ordnance Board Factory
10/A Aukland Road
Calcutta - 700 001
India

SPECIFICATIONS	
CALIBRE	105 mm
LENGTH OF ORDNANCE	4.4 m
BARREL LENGTH	3.8905 m
MAX RECOIL LENGTH	1.06 m
NATURE OF RIFLING	uniform 1 in 19 calibres right handed
NUMBER OF RIFLING GROOVES	28
MUZZLE BRAKE	double-baffle
BREECH MECHANISM	vertical sliding
CARRIAGE	bow
SHIELD	yes
WEIGHT (firing)	2380 kg
LENGTH	
firing	7.216 m
travelling	5.054 m
WIDTH (travelling)	1.82 m
HEIGHT	1.494 m
GROUND CLEARANCE	0.33 m
TRACK	1.465 m
ELEVATION/DEPRESSION	+73°/-5°
TRAVERSE	
on carriage	10°
on platform	360°
RATE OF FIRE	
normal	4 rds/min
intense	6 rds/min max up to 10 min (Charge 5)
sustained	1 round/2 min max up to 2 hrs
MAX RANGE	17 200 m
TOWING VEHICLE	Shaktiman model 4200 truck

Other Indian Artillery Developments

Development/Description

In addition to the Indian Ordnance Factories 105 mm Light Field Gun (see previous entry), two other Indian towed artillery equipments have been developed and placed into production.

The lightest of these is the 75 mm Pack Gun/Howitzer Mark 1 which has a 24-calibre barrel and was designed for use in mountainous terrain. The carriage is of the split trail type and if required the wheels can be removed to allow the weapon to pivot on the trails and a firing platform under the carriage. The barrel has a single- or double-baffle muzzle brake. At least two slightly different models have been developed. The Mark 1A has a two-part flat shield with the top part of each half folding forwards. The Mark 1B has a one-piece shield with a flat lower half and a top half that slopes to the rear. This 75 mm pack gun/howitzer is in service with the Indian Army and it is believed that it is no longer in production.

The second weapon developed in India is the 105 mm Field Gun, the development of which started in 1972 at the Armament Research and Development Centre in Kirkee in collaboration with the Kanpur Ordnance Factory, where the first prototype was manufactured. The carriage was built by the Ambarnath Gun Carriage Factory, the elevation and traversing mechanism by the Ambarnath Machine Tool Prototype Factory. The Indian barrel for the 105 mm Field Gun is not auto-frettaged and is based on the British L13 gun used in the FV433 Abbot self-propelled guns, many of which were delivered to the Indian Army. The carriage is of the split trail type.

The 105 mm Field Gun was placed in production to replace some of the British-supplied 25-pounder field guns. It has been reported that the gun has a maximum range of 17 400 m and a maximum rate of fire of 5 rds/min. It has also been reported that this weapon has now been replaced in production by the 105 mm Light Field Gun (see previous entry).

155 mm FH-77B

India has now taken delivery of all of the 410 155 mm FH-77B field howitzers together with its associated ammunition, fire control systems, tractors and training equipment. Under the second and third phases of the programme it was expected that local assembly and then local production would be undertaken in India, but as of mid-1991 this had not occurred.

130 mm self-propelled gun

The Indian Army has modified a number of Vijayanta (Vickers Mk 1 MBTs) to mount the Soviet 130 mm M-46 field gun. The modifications to the chassis have been extensive and include an additional road wheel either side. Details of this system are given in the *Self-propelled guns and howitzers* section under India.

Status: Both the 75 mm gun/howitzer and 105 mm Field Gun are in service with the Indian Army.

105 mm gun in firing position (Indian Army)

Indian 75 mm pack gun/howitzer Mk 1B (Indian Army)

INTERNATIONAL

155 mm Field Howitzer 70 (FH-70)

Development

In the early 1960s Germany, the United Kingdom and the USA agreed that they had a requirement for a new 155 mm field howitzer. Germany and the USA wanted to replace their 155 mm M114 howitzers and the United Kingdom wanted to replace its 5.5-inch guns. In the end the USA went on to develop the 155 mm M198 which has no APU and Germany and the United Kingdom agreed to develop a new weapon which became known as the FH-70. The FH-70 MoU was signed in 1968 and the requirement was for a 155 mm howitzer to meet the following military and tactical requirements:

Continuous high rate of fire with a burst fire capability
High mobility with minimum effort for deployment
Increased range and lethality and a new family of ammunition.

The United Kingdom was the project leader for the FH-70 and Germany for the self-propelled version called the SP-70, development of which was suspended in 1986. Details were given in *Jane's Armour and Artillery 1986-87*, page 505.

The first six prototypes were completed between 1969-70 and after accepting the Agreed Operational Characteristics, Italy joined the collaborative project in 1970 as a full partner. The second batch of eight prototypes was completed between 1971 and 1973 and the first international trials battery of six FH-70s was formed in 1975.

A total of 19 prototypes of the FH-70 were completed and after extensive trials the weapon was accepted for service in 1976 with the first production weapons being completed in 1978. Development cost of the FH-70 and its ammunition was about $30 million and unit cost in 1976 was stated to be about $300 000. Germany ordered 216, Italy 164 and the United Kingdom 71 FH-70s. In May 1982 Saudi Arabia placed an order with International Military Services of the United Kingdom for the supply of 72 FH-70 howitzers plus ammunition, muzzle velocity measuring equipment, fire control systems and training equipment. By early 1983 the first 18 weapons had been delivered. The British designation for the FH-70 is the L121 while the Germans call it the FH 155-1. Production for all three countries was completed in 1982. It has also been evaluated in Australia, Brazil, Canada, India, Norway and Oman. Malaysia took delivery of nine FH-70 systems in 1989.

The trilateral responsibilities for production are as follows:

United Kingdom: carriage, traversing gear, high explosive shell and propelling charge (cartridges Nos 1 and 2)

Germany: ordnance, loading system, auxiliary propulsion unit, suspension, sighting equipment, smoke and illuminating ammunition and propelling charges (cartridge No 2)

Italy: cradle, recoil system, sights bracket, elevating gear and arc, high-explosive shell, smoke and illuminating ammunition and propelling charges (cartridges Nos 2 and 3).

Description

The elevating mass comprises the ordnance, cradle with integrated recoil system, and the loading mechanism. The ordnance consists of a monobloc barrel fitted with a double-baffle muzzle brake, a breech mechanism that incorporates a semi-automatic vertically sliding block incorporating an automatic tube loader with a capacity of nine tubes, spent tube ejection and an emergency firing mechanism.

The cradle, which carries the ordnance on slides, includes the recoil system which consists of a buffer with high-angle cut-off gear and a recuperator. The mechanical main firing system, which is independent of elevation, is carried through one of the cradle trunnions.

The semi-automatic loading system operates at all elevations and traverses and consists of a loading tray which presents the round to the chamber. Together with an automatic tube loader, this system permits the normal rate of fire of six rds/min and a burst rate of three rounds in 15 seconds.

An automatic loading ('flick ramming') system is available as an optional extra and provides a burst fire rate of three rounds in 8 seconds.

The carriage is of lightweight construction with saddle, split trails, self-digging spades, main and trail wheels which are operated hydraulically, a soleplate and a detachable APU. When travelling the barrel is rotated through 180° and clamped between the trails.

The APU is housed in a space frame attached to the forward part of the carriage and consists of a commercial Volkswagen 1800 cc engine with a gearbox, differential, batteries and a hydraulic pump. In the APU mode the FH-70 can attain speeds of up to 16 km/h and is capable of negotiating slopes of up to 34 per cent. When being towed it can ford to a depth of 1.5 m and when using the APU mode to 0.75 m.

The functions of the APU include driving the main wheels for moving the howitzer and providing hydraulic power for steering, raising and lowering both the main and trail wheels. This facility allows for rapid unhooking, lowering the howitzer onto its soleplate, easy opening and lowering of the trails when going into action as well as for extracting the spades from the ground, lifting the equipment, rapid closing of the trails and limbering up. In an emergency, or when the APU has been removed, power is provided by a hand pump.

The main wheels are hydraulically suspended on swinging arms attached to the carriage body and are cushioned by hydraulic accumulators. Similar suspension elements are used for raising and lowering the trail wheels. Braking of both main wheels is possible from the towing vehicle or from the APU driver's position on the left side of the carriage.

The trail wheels are attached to the trail end and can be steered, raised and lowered. The hydraulic units providing these functions are controlled from the driver's position and can operate with the trails open or closed. The steering is capable of 60° of movement either side of the centre and for spreading the trails, the steering wheels can be turned through 90°.

The sighting system which is mounted on the left of the saddle consists of a support bracket, a dial sight carrier incorporating a Quadrant Elevation scale and levelling bubbles, a periscopic dial sight and a direct fire telescope. A direct fire night sight can also be fitted. All scales, graticules and bubbles are self-illuminated. The sight carrier incorporates transducers for azimuth and elevation and in the German and Italian armies these are connected to a digital display unit (DDU) which also receives firing data (azimuth and elevation) from a fire control computer. The layer lays the gun by nulling out the differences on the DDU and then laying the sight on the gun aiming point. The DDU is not used in United Kingdom service.

The FH-70 can be towed on a road at a maximum speed of 100 km/h. Germany uses the MAN (6 × 6) 7000 kg truck to tow the FH-70, Italy

FH-70 of the German Army in auxiliary powered configuration with ordnance locked over closed trails and spades on trails

155 mm projectile being placed on loading tray of Italian FH-70 (UKLF)

uses the FIAT 6605 TM (TM69) and the United Kingdom uses the Foden (6 × 6) medium mobility vehicle. The FH-70 is airportable in a Lockheed C-130 Hercules aircraft or underslung under the CH-47D helicopter.

A new family of 155 mm ammunition has been developed for the FH-70 which gives the weapon a significant increase in range and lethality for both direct and indirect firing against armoured targets. In addition the FH-70 can fire NATO standard ammunition including the American Copperhead CLGP. The new range of ammunition consists of the following projectiles:

HE shell (L15A1) which weighs 43.5 kg and is thin walled, with high fragmentation effect and HE capacity of 11.3 kg.

Base Ejection Smoke shell (DM105), four smoke canisters fall to the ground giving a dense cloud of smoke in 30 seconds.

Illuminating (DM106) which provides illumination of 2.1 million candelas for 60 seconds.

The ERP (Extended Range Projectile) which will increase the range of the FH-70 to over 30 000 m are US M549A1 Rocket Assisted Projectile, RO30 Base Bleed and PRB Base Bleed projectiles.

The eight-zone charge system uses triple-based propellant and is divided into three separate cartridges, containing zones 1 and 2, 3 to 7 and 8 respectively. The charge system is ballistically balanced, has a clean combustion and is initiated by a percussion inner tube (DM 191A1). Muzzle velocity ranges from 213 to 827 m/s. Using the Bishopton-produced BIS 14 propelling charge the FH-70 can fire a ERFB base bleed projectile to a range of 31 000 m.

Using the APU the FH-70 can be unhooked, manoeuvred into position and put into action in less than two minutes, and brought out of action in a similar period. If a target moves outside the traverse limits the trails can be raised hydraulically, thereby pulling the spades from the ground, and the gun wheeled about its soleplate to a new target. With the trails lowered the first shot will embed the spades.

Apart from the usual preliminary checking of the equipment, the gun detachment has only to open the breech by hand, load the first shell and propelling charge and place the second shell on the loading tray. The breech is closed and an ignition tube is automatically loaded from the nine-tube magazine. After this sequence, which takes less than five seconds, the FH-70 is ready for action. When the FH-70 is ready and laid correctly, all that is required for firing is for the layer to pull the firing handle. Thereafter the sequence of operation becomes semi-automatic. The recoil system and muzzle brake jointly absorb the recoil forces as the gun recoils on the cradle slides to a maximum of 1.4 m. On running out to the firing position, the following operations take place simultaneously. The breech is opened automatically by a cam, the spent tube is ejected and the next shell is lifted to the breech and aligned in the chamber before final hand ramming. The detachment then ram the shell and load the propelling charge, place the next shell in the loading tray, close the breech, if necessary re-lay the gun, and again fire. A flick ramming device is available as an optional extra giving a burst rate of three rounds in eight seconds and reducing crew fatigue.

To assist gun loading over prolonged periods of firing, VSEL has developed an ammunition hoist device which effectively reduces the handling weight of a 155 mm projectile by about half. The device is suspended from a steel joist gantry mounted to the right hand side of the main carriage, just forward of the breech. It is a sprung-loaded bobbin from which issues a steel wire cable with a projectile clamp at the end. When the clamp is secured over a projectile, the loader can lift the projectile with at least half of the weight being taken up by the sprung-wound bobbin, and the gantry allows him to swing the projectile upwards and over to the loading tray where the projectile clamp is released. The device cuts loader fatigue over long periods and was troop-tested at the Royal School of Artillery, Larkhill, during mid-1982. It is now available as an optional extra.

Other options include a direct fire image intensification night sight, deflector boxes and mainwheel mudguards.

Variants

Japanese FH-70

The Japanese Ground Self-Defence Force has selected the FH-70 to replace its 155 mm M114 towed howitzers and production is being undertaken under licence in Japan. Prime contractor in Japan is the Japan Steel Works. A total of 176 weapons are now being built, the first batch of 20 being procured with 1983 funding; these were provided direct from the European production base. Rheinmetall was responsible for the Japanese contract. The ordnance of the FH-70 is to be used on a new self-propelled howitzer for the Japanese Ground Self-Defence Force which will use the chassis of the new Japanese MICV.

Export version

A simplified sighting system is available for the world market which utilises the British Royal Ordnance 105 mm Light Gun sighting system with a modified graticule in the direct fire telescope. It can be fitted to the standard gun via an adaptor on the left hand trunnion.

FH-70 with 52 calibre ordnance

For trials purposes Rheinmetall has fitted an FH-70 with a 52 calibre ordnance that conforms with the new quadrilateral ballistic agreement. This enables a range of 30 000 m to be achieved with the L15A1 projectile or 40 000 m with base bleed projectiles.

Status: Production complete in Europe but can be resumed if further orders are placed. Production under licence continues in Japan. In service with:

Country	Quantity	Comment
Germany	216	delivered
Italy	164	delivered
Japan	176	local production
Malaysia	9	delivered 1989
Saudi Arabia	72	delivered
UK	71	delivered (68 admitted under CFE Treaty)

Manufacturers: Vickers Shipbuilding and Engineering Limited
Barrow-in-Furness
Cumbria LA14 1AF
UK
Telephone: (0229) 823366
Telex: 65411 vsel g
Fax: (0229) 23366

Rheinmetall GmbH
Ulmenstrasse 125
D-400 Düsseldorf
Federal Republic of Germany
Telephone: (0211) 4471
Telex: 8584963

OTO Melara SpA
19100 La Spezia
via Valdilocchi 15
Post Box 337
Italy
Telephone: (0187) 530 111
Telex: 270 368 oto i
Fax: (0187) 530 111

Rear view of 155 mm FH-70 deployed in firing position and showing position of spades and trail wheels

Saudi Arabian 155 mm FH-70 in firing position showing APU mounted on forward part of carriage

SPECIFICATIONS	
CALIBRE	155 mm
BARREL LENGTH	6.022 m
MUZZLE BRAKE	double-baffle
BREECH MECHANISM	semi-automatic wedge
CARRIAGE	split trail with APU
SHIELD	no
WEIGHT	
travelling	9300 kg
firing	9300 kg
LENGTH	
travelling	9.8 m
firing	12.43 m
WIDTH	
travelling	2.58 m
firing	7.5 m

HEIGHT	
travelling	2.45 m
firing	2.192 m
AXIS OF BORE (firing)	1.525 m
GROUND CLEARANCE (travelling)	0.3 m
TRACK	2.195 m
WHEELBASE	5.15 m
FORDING	
towed	1.5 m
APU in use	0.75 m
TURNING RADIUS	9.25 m
TYRES	14.00 × 20
ELEVATION/DEPRESSION	+70°/-4.5°
TRAVERSE (total)	56°
RATE OF FIRE	
normal	6 rds/min
burst	3 rounds/13 s
sustained	2 rds/min for 1 hr

MAX RANGE	
standard projectile	24 700 m
base bleed projectile	31 500 m
CREW	7 or 8
TOWING VEHICLES	
Germany	MAN (6 × 6) 7000 kg truck
Italy	FIAT 6605 TM (6 × 6) truck
United Kingdom	Foden medium mobility vehicle (6 × 6)

IRAQ

122 mm Saddam Howitzer

Development/Description

This is essentially a locally produced copy of the Yugoslav version of the Soviet 122 mm D-30 towed howitzer. The Saddam weighs 3350 kg in the firing position and 3440 kg in the travelling configuration, and traverse is a full 360° with elevation being from -7° to +70°. Rate of fire is 6 to 8 rds/min with a maximum range of 17 133 m being achieved. The HE projectile weighs 22 kg while the charge weighs 3.8 kg.

Status: Following the 1991 Gulf Conflict, it is assumed that production of the 122 mm Saddam howitzer has been suspended. In service with the Iraqi Army.

NB: The Iraqi Army had a large inventory of Soviet supplied 130 mm M-46 towed field guns and it was intended that these would be upgraded to 155 mm/45 calibre configuration using the Yugoslav conversion kit described later in this section. As far as it is known this programme had not commenced at the time of the 1991 Gulf Conflict. If it had gone ahead then Iraq would have had ammunition commonality with its South African supplied 155 mm G5 and Austrian supplied 155 mm GH N-45 towed artillery systems which it purchased in significant numbers.

122 mm Saddam howitzer deployed in firing position

ISRAEL

Soltam 155 mm Model 839P and 845P Howitzers

Development

The Model 839P is a version of the Soltam M839 howitzer fitted with an auxiliary power unit (APU) for self-propulsion and for various gun services such as opening and closing the trail legs. It was first demonstrated in 1983. It was followed in 1984 by the Model 845P which is similar but fitted with a 45-calibre barrel in place of the 39-calibre barrel of the Model 839P. In many ways the Model 845P is similar to the Model 839P.

Description (Model 839P)

The complete weapon consists of two major assemblies, plus the APU, these are the ordnance and the carriage. The driver's position is on the left of the barrel and his controls include a starter, a speed range selector, a single control joystick and a parking brake. The driver's control position is attached to the central saddle structure along with the driver's seat and a safety belt. For firing the control position is folded downwards. The howitzer can be controlled either from the seat or from the ground with the driver walking alongside.

The ordnance for the Model 839P is 39 calibres long (Model 845P 45 calibres) and is a monobloc autofrettaged component with a horizontal sliding breech-block. Obturation is by self-sealing obturator discs. Standard type primer cartridges are used and are ejected from the breech-block automatically on run out. A single-baffle muzzle brake is fitted along with a fume extractor located just over halfway along the barrel. Progressive twist rifling is employed. The hydraulic recoil buffer incorporates a variable length mechanism as well as a run out control plunger, and the recuperator is hydro-pneumatic. Elevating and traverse gears incorporate 'no back' mechanisms and pneumatic equilibrators balance the elevated mass. A quick-release device is incorporated into the elevating mechanism for loading. A pneumatic rammer is mounted on the elevating mass and can operate at all angles of elevation.

The carriage comprises two main parts, the superstructure (the upper carriage) and the basic structure (the lower carriage).

The superstructure includes the saddle and cradle which enable the gun to be laid in both elevation and azimuth. The basic structure, which is the main support when the weapon is in both firing and transport configurations, consists of its central structure with the main bearing, trails and the suspension.

The cradle carries all of those parts of the gun that take part in recoil as well as the vertical movement during elevation and depression. The central part of the cradle carries the barrel assembly. On the top are the recuperator cylinders, at the bottom the buffer cylinder and the elevation arc. At its rear end are the trunnions, left and right, and at its front end the pivot pins for the equilibrators. The recoil length cut off gear is located at the lower right front end of the cradle.

The saddle is of all-welded construction and screwed to the bearings rotating part at the bottom and carrying the cradle at its top. The equilibrators act as a self-adjusting force for the barrel allowing easy laying and adjusting in elevation and depression.

The traversing mechanism is mounted on the front part of the saddle and serves to lay the gun in azimuth. For fast deflection, or when rotating the barrel from the firing into the travelling position, or vice versa, the mechanism can be disengaged so allowing free lateral movement of the barrel.

When deployed in the firing position, the system rests on three points, the firing platform and the trail spades. Lifting the howitzer off the ground

Soltam Model 839P 155 mm howitzer in travelling configuration showing position of APU

Soltam 155 mm M-71 gun/howitzer in travelling configuration

Soltam Model 845P 155 mm howitzer in self-propelled configuration

Soltam Model 839P 155 mm howitzer deployed in firing position

so that it is supported on the firing platform, is carried out by an electric hydraulic pump, battery operated with a mechanical backup.

The suspension consists of four wheels, two of which are mounted together on a walking beam. This allows the weapon to be towed at any reasonable towing speed. Each individual wheel has a pneumatic brake cylinder operated by compressed air that is provided from the towing vehicle.

The weapon can be used for direct and indirect fire missions and is fitted with two different sighting systems, both of which are mounted on the left side of the weapon.

The carriage is fitted with an ammunition handling crane mounted on the right trail leg. On the left trail leg is the APU which is a Deutz F4L/912 air-cooled diesel unit providing 80 hp at 2600 rpm. The engine powers a hydraulic unit which in turn powers the driving wheels via hydraulic motors on the wheels. The system also powers a hydraulic ram on the left trail leg to assist in spreading and closing the trail legs and for lifting the howitzer off its wheels as well as assisting the howitzer off its towing vehicle. The firing platform is also hydraulically operated. For silent or emergency operation the hydraulic system includes an electric motor powered by two in-line connected 12 V batteries and a hand pump is also supplied.

The Model 839P is able to drive under its own power at speeds up to 17 km/h on roads and the maximum operating range is 70 km. Only one joystick control lever is required for direction or turning and when released acts as a brake.

The Model 845P differs from the Model 839P only in barrel length and slight changes in overall weight.

The Model 839P and Model 845P can be supplied without the APU. Guns without the APU are known as the Model 839 or Model 845. Their ballistic performances remain the same as guns fitted with an APU.

Both types of howitzer can use a wide range of ammunition and the following tables provide an indication of their performances.

Projectile (HE)	**Range**	**Range**
Model	M839P	M845P
M107	21 900 m	18 460 m
M107A1	21 900 m	24 000 m
M56A1	23 480 m	25 800 m
M953	25 000 m	27 500 m
M549(RAP)	30 000 m	30 000 m
ERFB	25 900 m	30 000 m
ERFB-BB	31 000 m	39 000 m

Propelling charges	
M1A2	zones 2-5
M4A2	zones 3-7 if required
M9A2	zones 6-9
M10A2	zone 10
M11A2 (M845P only)	zone 11

SPECIFICATIONS

Model	**M839P**	**M845P**
CALIBRE	155 mm	155 mm
BARREL LENGTH	6.67 m	7.7 m
LENGTH OF RIFLING	5.038 m	5.857 m
MUZZLE BRAKE	single-baffle	single-baffle
BREECH MECHANISM	horizontal sliding block	
CARRIAGE	split trail with APU	
SHIELD	no	no
WEIGHT	10 850 kg	12 400 kg
LENGTH		
at 0° elevation	10.8 m	12.8 m
at 70° elevation	6.6 m	8 m
WIDTH	8 m	9 m
HEIGHT		
at 0° elevation	2.1 m	2.1 m
at 70° elevation	7.2 m	8.1 m
AXIS OF BORE AT 0°	1.65 m	1.65 m
GROUND CLEARANCE	0.38 m	0.38 m
TRACK		
driving wheels	2.2 m	2.2 m
auxiliary wheels	1 m	1 m
TURNING CIRCLE		
outside, self-propelled	15 m	13 m
outside, towed	21.5 m	21.5 m
TYRES		
driving	11 × 20	11 × 20
support	6.50 × 10	8.50 × 10
SELF-PROPELLED SPEED		
low range	0-8 km/h	0-8 km/h
high range	0-17 km/h	0-17 km/h
GRADIENT	34%	34%
FUEL CAPACITY	50 l	85 l
DRIVING RANGE	70 km	120 km
ELEVATION/DEPRESSION	+70°/-3°	+70°/-3°
TRAVERSE (total)	78°	78°
RATE OF FIRE		
short period	4 rds/min	4 rds/min
sustained period	2 rds/min	2 rds/min
MAX RANGE (ERFB-BB)	31 000 m	39 000 m

Status: Production as required.

Manufacturer: Soltam Limited
POB 1371
Haifa
Israel
Telephone: 896211
Telex: 46277
Fax: (4) 894020

Soltam 155 mm M-71 Gun/Howitzer

Development/Description

The Soltam 155 mm M-71 Gun/Howitzer is a further development of the Soltam 155 mm M-68 Gun/Howitzer and uses the same recoil system, breech and carriage. The first prototype was completed in 1974 and first production units the following year. The main differences between the M-71 and the M-68 are that the former has a longer 39-calibre barrel and is fitted with a rammer driven by compressed air which permits rapid loading at all angles of elevation. The rechargeable cylinder for the M-71 is mounted on the right trail. The barrel has 48 grooves, length of rifling is 5.038 m, width of groove 6.6 mm, width of land 3.5 mm and depth of groove is 1.2 mm. Full details of the carriage, recoil system, breech and ammunition are given in the entry for the M-68.

Variants

Soltam has developed to the prototype stage a self-propelled model of the M-71 gun/howitzer based on a Centurion MBT chassis. Details of this system, which never entered production or service, were given in *Jane's Armour and Artillery 1990-91*, page 562.

Status: Production complete but can be resumed if additional orders are placed. In service with Israel, Singapore, South Africa (where it is known as the G4), Thailand and three other undisclosed armed forces.

Manufacturer: Soltam Limited
POB 1371
Haifa
Israel
Telephone: 896211
Telex: 46277
Fax: (4) 894020

Soltam 155 mm M-71 gun/howitzer in travelling position with barrel swung through 180° and locked over trails (Israeli Ministry of Defence)

Soltam 155 mm M-71 gun/howitzer in firing position

SPECIFICATIONS		SHIELD	no	TRAVERSE (total)	84°
CALIBRE	155 mm	WEIGHT (travelling order)	9200 kg	RATE OF FIRE	
BARREL LENGTH	6.045 m	LENGTH (travelling)	7.5 m	short period	4 rds/min
(including breech ring)	6.67 m	WIDTH (travelling)	2.58 m	sustained	2 rds/min
MUZZLE BRAKE	single-baffle	HEIGHT (travelling)	2.115 m	MAX RANGE	23 500 m
RECOIL	hydro-pneumatic	GROUND CLEARANCE (travelling)	0.38 m	CREW	8
BREECH MECHANISM	horizontal wedge	TRACK	2.2 m	TOWING VEHICLE	5-ton (6 × 6) truck
CARRIAGE	split trail	ELEVATION/DEPRESSION	+52°/-3°	MAX TOWING SPEED	100 km/h

Soltam 155 mm M-68 Gun/Howitzer

Development/Description

The Soltam M-68 Gun/Howitzer uses a carriage similar to that of the Tampella 122 mm M-60 Field Gun described in the Finnish section, and further development by Soltam has resulted in the M-71 Gun/Howitzer which is also described in this section. The first Israeli prototype was completed in 1968, with the first production example produced in 1970.

The carriage comprises a top carriage, bottom carriage and two trails. In action a screw type firing jack is lowered under the lower carriage and the two trails are opened out. A loose spade is placed at the end of each trail and staked to the ground by three wedges. A further two loose spades and six wedges are carried to enable the gun crew to change the direction of firing without taking up the original loose spades. When travelling the loose spades and wedges are stowed on the trails. The bogies have four pneumatic tyres and each wheel has a hydraulic brake.

The recoil mechanism is under the barrel and the counter recoil system is on the top of it with the pneumatic equilibrators either side of the barrel. Recoil varies in relation to elevation but maximum recoil is 1 m. The elevating wheel is on the left side of the carriage as is the panoramic sight, and the traversing gear is on the right.

The barrel is of the monobloc type with a screw-on breech mechanism and a single-baffle muzzle brake with a fume extractor fitted to the rear. The barrel has an overall length of 5.18 m and 4.185 m long rifling.

The semi-automatic horizontal sliding breech-block remains open on recoil for loading of the next round. The breech incorporates a specially developed gas check ring which effectively prevents the escape of propellant gases to the rear. Provision is made for opening the breech-block for loading the first round. Loading from the left is possible at all angles of elevation.

The M-68 fires an HE projectile weighing 43.7 kg with a maximum muzzle velocity of 725 m/s to a maximum range of 21 000 m. The weapon fires all standard NATO 155 mm projectiles including illuminating and smoke, and in addition the M-68 fires Tampella-designed projectiles using a nine-charge system (charges 1 to 9) with a muzzle velocity of 253 to 820 m/s to a maximum range of 23 500 m.

When travelling the barrel is swung through 180° to the rear and is locked in position over the closed trails.

Variants

Soltam has developed a self-propelled model of the M-68 called the L-33 based on a modified Sherman chassis, full details of which are given in the *Self-propelled guns and howitzers* section under Israel.

Soltam 155 mm M-68 gun/howitzer in firing position

SPECIFICATIONS					
CALIBRE	155 mm	LENGTH (travelling)	7.2 m	MAX RANGE	21 000 m (see text)
BARREL LENGTH	5.18 m	WIDTH (travelling)	2.58 m	CREW	8
MUZZLE BRAKE	single-baffle	HEIGHT (travelling)	2 m	TOWING VEHICLE	5-ton (6 × 6) truck
RECOIL	hydro-pneumatic	GROUND CLEARANCE (travelling)	0.38 m	MAX TOWING SPEED	100 km/h
BREECH MECHANISM	horizontal	TRACK	2.2 m		
CARRIAGE	split trail	ELEVATION/DEPRESSION	+52°/-5°		
SHIELD	no	TRAVERSE (total)	90°		
WEIGHT		RATE OF FIRE			
travelling order	9500 kg	short period	4 rds/min		
firing position	8500 kg	sustained	2 rds/min		

Status: No longer in production, but available to special order. Soltam has not released any details of sales of this weapon but *Jane's Armour and Artillery* has been informed from other sources that the M-68 is in service with Thailand (Marines).

Manufacturer: Soltam Limited
POB 1371
Haifa
Israel
Telephone: 896211
Telex: 46277
Fax: (4) 894020

Soltam 155 mm Upgraded M-46 Field Gun

Development

Based on their extensive experience in the design, development and production of towed and self-propelled artillery systems, Soltam have recently completed the prototype of a Soviet 130 mm M-46 field gun upgraded with a 155 mm 45 calibre ordnance.

Soltam can either provide the kit to enable the user to upgrade existing 130 mm M-46 weapons in his own facilities, or the company can undertake this work in its own facilities in Israel.

Description

The 155 mm conversion kit comprises the 45-calibre ordnance, breech ring assembly, breech block assembly, semi-automatic breech block operating mechanism, new muzzle brake, fume extractor and cradle rear bushing and threaded collar.

The upgraded M-46 is fitted with a monobloc autofrettaged 155 mm ordnance 45 calibres long that incorporates a horizontal sliding breech block that is semi-automatically operated. It uses standard M82 primers that are automatically ejected after firing.

According to Soltam, the upgraded weapon has a number of unique features:

(1) Firing cannot commence unless the breech block is completely closed and locked
(2) It has a unified sighting system with the indirect and direct elbow telescope being located on the same side of the weapon and operated by the same gunner
(3) High rate of accurate fire, the horizontal breech mechanism together with the pneumatic ramming device, enable a high rate of fire, uniform loading and ramming at any elevation without the need to lower the ordnance for loading
(4) The gun crew area is clear from blast over pressure effects and clean from residual post-firing gases.

The complete weapon system can be divided into two key parts, the ordnance and the carriage.

The ordnance comprises the barrel assembly, recoil mechanism, equilibrators and sight instrument.

The new barrel is fitted with a muzzle brake. Obturation is achieved between the barrel and the breech block by a set of steel rings being pressed face to face during firing by means of the gas pressure in the chamber.

The breech mechanism has three major components, breech ring assembly, breech block assembly and breech block semi-automatic operating mechanism.

The recoil mechanism consists of the recuperator, buffer and recoil length cut-off gear.

The carriage comprises two main parts, the superstructure (upper carriage) and the basic structure (the lower carriage) which are not modified.

The basic structure is the main support of the weapon in both action and travelling and consists of the central structure, trails and running gear.

The cradle carries all of those parts of the gun participating in recoil as well as in the vertical movement during elevation and depression. The central part of the cradle carries the barrel assembly and on top are the recuperator cylinders with the buffer cylinder and elevating arc at the bottom. At its rear ends are the trunnions, left and right, and at its front end the pivot pins for the equilibrators. The recoil length cut-off gear is located at the lower right front end of the cradle.

The saddle is unmodified and is an all-welded construction screwed to the bottom and carries the cradle at its top. The equilibrators act as a self-adjusting force for the barrel to allow for ease of elevation and depression. The traversing mechanism is mounted on the front part of the saddle.

The running gear consists of two wheels each of which has a pneumatic brake cylinder operated by compressed air from the towing vehicle.

Soltam 155 mm upgraded M-46 gun in travelling configuration

Both direct and indirect fire sights are fitted on the left side.

The Soltam upgraded M-46 can fire all of the standard 155 mm types of ammunition available in the world today, starting with the standard M107 family of ammunition and ending with the more recent ERFB BB long-range ammunition family.

Ranges of 30 000 m can be achieved with unassisted projectiles or up to 40 000 m with assisted projectiles.

In the Soltam conversion, only those assemblies directly related to the ordnance have been modified, all of the existing spare parts inventory concerning the actual carriage can be used for the upgraded weapon.

As an option, a pneumatic ramming system, to enable uniform loading and ramming at all elevations, so allowing a firing rate of four rounds per minute can be fitted.

SPECIFICATIONS	
CALIBRE	155 mm
BARREL LENGTH	7.7 m (with breech ring)
DIAMETER OF CHAMBER	172 mm
VOLUME OF CHAMBER	27.7 l
NUMBER OF GROOVES	48
LENGTH OF RIFLING	5851 mm
WIDTH OF GROOVE	6.6 mm
WIDTH OF LAND	3.5 mm
DEPTH OF GROOVES	1.2 mm
INCLINATION OF RIFLING	constant
BREECH BLOCK	horizontal wedge
OBTURATION	self sealing
FIRING MECHANISM	trip action
WEIGHT TRAVELLING	8850 kg
LENGTH TRAVELLING	11.73 m
WIDTH TRAVELLING	2.45 m
HEIGHT TRAVELLING	2.55 m
MAX TOWING SPEED	100 km/h
ELEVATION/DEPRESSION	+45°/-2.5°
TRAVERSE	25° (left and right)

Status: Development complete. Ready for production.

Manufacturer: Soltam Limited
PO Box 1371
Haifa
Israel
Telephone: 896211
Telex: 46277
Fax: (4) 894020

Soltam 155 mm M114S Howitzer

Development/Description

A number of manufacturers have developed modernisation packages for the old US 155 mm M114 towed howitzer that utilise a 39 calibre ordnance.

The upgraded Soltam M114S utilises a 33 calibre barrel which enables a standard 155 mm M107 projectile to be fired to a maximum range of 18 100 m compared to 14 600 m of a standard M114.

The first prototype of this was completed in 1988 and production of the conversion packages will commence on receipt of firm orders. The conversion can be carried out in the user's own facilities using a kit supplied by Soltam or the weapons can be returned to Soltam who will carry out the upgrade work.

According to Soltam, the 33 calibre version does not require so extensive modifications to the existing weapon compared to 39 calibre weapon modernisation packages.

Soltam 155 mm M114S upgraded howitzer in firing position

The new 33 calibre barrel has a high efficiency muzzle brake and is fitted with its original breech ring and breech mechanism. The new ordnance has a progressive 1:20-twist rifling with 48 lands and grooves. It provides for a projectile travel of 4.14 m with a chamber volume of 17.5 m^3 with M107 HE projectile. A new primer housing can be fitted as an option which enables the more modern M82 primer to be used.

The cradle remains unchanged although a front bushing has to be added by inserting a new bushing into the original one without the need of extensive machining. The existing spring loaded equilibrators have to be slightly modified by replacing the upper connecting joints that are connected to the original points on the saddle.

If however the user requires a 39 calibre barrel, then this can be supplied by Soltam as can a pneumatically operated rammer for higher rates of fire, loaded at all angles of elevation and reduced crew fatigue.

SPECIFICATIONS

CALIBRE	155 mm
MUZZLE BRAKE	single baffle
BREECH MECHANISM	stepped thread, interrupted screw
CARRIAGE	split trail
SHIELD	yes
WEIGHT	5900 kg
LENGTH	
travelling	9.2 m
firing	6.7 m
WIDTH	2.4 m
HEIGHT	1.8 m
ELEVATION/DEPRESSION	+63°/−2°
TRAVERSE	25° (right), 24° (left)
RANGE	
M107 projectile	18 100 m
M549 projectile	24 000 m
ERFB	20 100 m
ERFB BB	25 000 m

Status: Development completed. Ready for production on receipt of orders.

Manufacturer: Soltam Limited
POB 1371
Haifa
Israel
Telephone: 896211
Telex: 46277
Fax: (4) 894020

Israel Military Industries 60 mm Towed HVMS 60

Development/Description

The Towed HVMS 60 anti-tank gun is a towed version of the Israel Military Industries 60 mm Hyper-velocity Medium Support Weapon (HVMS 60) designed for mounting in AFV turrets.

Israel Military Industries 60 mm Towed HVMS 60 prototype

Early prototypes of the 60 mm HVMS were mounted on converted ex-British 6-pounder anti-tank gun carriages but these mounts were for trials purposes only. The proper towed version is still at an early stage of development and uses a simple carriage with a small shield and tubular trails. To date only prototype examples have been produced but it is known that production versions will have a form of upwards-folding trail to reduce the length of the towed load. An automatic loading device may be fitted over the breech.

No further details are available.

Status: Prototypes.

Manufacturer: Israel Military Industries
Marketing Division
POB 1044
Ramat Hasharon 47100
Israel
Telephone: (3) 5485222
Telex: 33719
Fax: (3) 5406908

ITALY

OTO Melara 155 mm 155/39 TM Howitzer

Development/Description

During late 1987 OTO Melara produced the first conversion of an American 155 mm M114 howitzer accommodating a new 155 mm barrel 39 calibres long. The new barrel is identical in design and ballistic terms to the ordnance fitted to the 155 mm M109L self-propelled howitzer upgraded from standard M109 self-propelled howitzers by OTO Melara from 1986 onwards (for details of the M109L refer to M109 entry under Italy in the *Self-propelled guns and howitzers* section).

The conversion of standard M114 towed howitzers to the new 155/39 TM howitzer standard mainly involves the installation of the M109L ordnance and associated components. New components include the gun barrel, muzzle brake, breech ring, breech block and an automatic breech operating mechanism; all these components are the same as those used on the M109L. The equilibrators have the same configuration as the original M114 parts but are modified to compensate for the increased elevating mass. Changes are also made to the equilibrator supports. Modifications are introduced to the recoil and recuperator systems. The original firing platform is retained but to improve firing balance and stability it is moved forward using a spacer. Other modifications are made to the cradle bushes, the front cradle and new spade bushes are fitted to improve travelling and firing balance.

The weight of the recoiling mass is increased from 1700 kg to 2450 kg. Weight when firing is 7850 kg. The introduction of the automatic breech opening device introduces some traverse limitations. At elevation angles between 0° and +45° the traverse remains unaltered (24° left, 25° right) but at elevation angles between +45° and +60° it is limited to 17° left and 25° right.

The 39 calibre barrel is ballistically identical to the FH-70 and permits the use of the existing FH-70 ammunition family. Thus a standard M107 projectile fired at a muzzle velocity of 827 m/s has a range of 24 000 m.

OTO Melara 155 mm 155/39 TM howitzer conversion

OTO Melara 155 mm 155/39 TM howitzer conversion in firing position

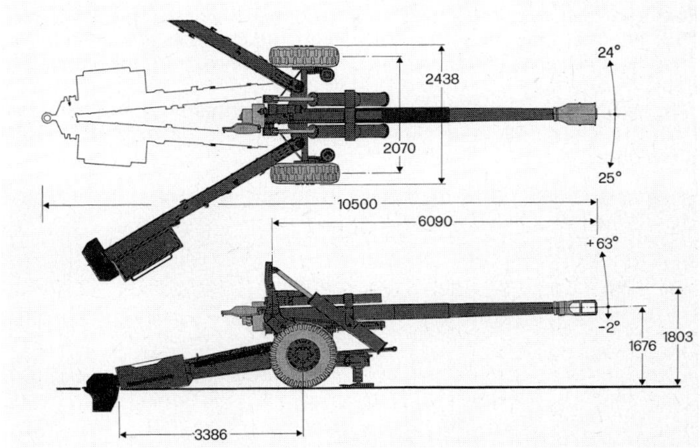

Top and side drawings of OTO Melara 155 mm 155/39 TM howitzer in firing position (not to 1/76th scale)

SPECIFICATIONS		LENGTH		MAX RANGE	
CALIBRE	155 mm	firing	10.5 m	M107 projectile	18 200 m
BARREL LENGTH	6.045 m	WIDTH		L15A1 projectile	18 500 m (charge 2, zone 7)
MUZZLE BRAKE	double baffle	travelling	2.438 m		
BREECH MECHANISM	semi-automatic, step-threaded	AXIS OF BORE	1.676 m	M483A1 projectile	16 500 m
		GROUND CLEARANCE	0.229 m	L15A1 projectile	24 000 m (charge 8)
RECOIL	hydro-pneumatic	TRACK	2.07 m	M549A1	24 000 m (charge 2, zone 7)
CARRIAGE	split-trail	TYRES	14.00 × 20		
SHIELD	no	ELEVATION/DEPRESSION	+63°/-2°	M549A1	30 000 m (charge 8)
WEIGHT		TRAVERSE	24° left, 25° right	CREW	11
travelling	7850 kg			TOWING VEHICLE	FIAT 6605 TM (6 × 6) truck
firing	7850 kg				

Status: Prototype.

Manufacturer: OTO Melara SpA
15 via Valdilocchi
19100 La Spezia
Italy Telephone: (0187) 530 111
Telex: 270 368 oto i
Fax: (0187) 530 111

OTO Melara 105 mm Model 56 Pack Howitzer

Development

The 105 mm Model 56 Pack Howitzer was developed by the Italian Army and produced by OTO Melara in the mid-1950s and entered production in 1957. Since then over 2500 have been built and sales have been made to more than 30 countries.

The Model 56 has the advantage that it can be dismantled into 11 sub-assemblies for transport across rough terrain, can be lifted by helicopter such as the Wessex or Bell UH-1, and its profile can be lowered for anti-tank use.

Description

The barrel assembly is composed of the barrel, breech ring, slipper and muzzle brake. The weapon has a multi-baffle muzzle brake but German weapons have a large single-baffle muzzle brake. The vertical sliding breech-block has a percussion firing mechanism which can be cocked automatically or by hand. Firing is either by hand lever or by lanyard.

The gun carriage comprises the undercarriage, saddle, cradle, upper and lower recoil systems, two balancing springs, two sights and the shield. The undercarriage consists of a cross head housing, the saddle centre pivot pin, two stub axles, spring equalising rocker arm and two articulated arms. Each of the stub-axle supporting arms has two bearings into which may be fitted the stub-axle which carries the wheel assembly and its handbrake. Each of the articulating brackets carries a trail leg of two or three sections which can be folded up and interlocked. The end section of the trail may be anchored to the ground by a knife or rock spade.

The saddle with its centre pivot mounting is provided with two side members with rear-mounted lugs which carry the elevating and traversing gear. The recoil system is contained in two cradles, upper and lower, which slide on top of each other. The upper cradle contains the two recuperator-buffers and the lower one contains one buffer and two recuperators. The two balancing springs have helical springs operating in sliding sleeves and the cradle is mounted on two trunnion bearings.

The sighting equipment consists of an indirect sight on the left and an anti-tank direct sight on the right. The splinter-proof shield mounted on the saddle is in two halves, both of which can be removed to lighten the weapon.

As a field weapon the 105 mm Pack Howitzer wheels are overslung and the barrel has an elevation of +65°, a depression of -5° and a traverse of 18° left and right. As an anti-tank weapon the wheels are underslung and the barrel has an elevation of +25°, a depression of -5° and a traverse of 28° left and right.

The weapon can be dismantled into 11 sub-assemblies, the heaviest of which weighs 122 kg, in three minutes and can be reassembled in four minutes.

The 105 mm Pack Howitzer fires the same range of ammunition as the American 105 mm M101 including HE(M1) with the complete round weighing 21.06 kg with a muzzle velocity of 472 m/s and the fixed HEAT (M67) round which weighs 16.7 kg, has a muzzle velocity of 387 m/s and will penetrate 102 mm of armour.

Status: In production. In service with Argentina, Australia, Bangladesh, Belgium, Brazil, Burkina Faso, Canada (designated the C5), Chile, Djibouti, Equador, France (war reserve stocks only), Germany, Greece, India, Iraq, Italy, Kuwait (status uncertain), Malaysia, Morocco, New Zealand, Nigeria, Pakistan, Peru, Philippines, Portugal, Saudi Arabia, Somalia (90), Spain (army and marines), Sudan, Thailand, United Arab Emirates, United Kingdom (training; under the CFE Treaty the UK stated that it had a total of 52), Venezuela, Zambia and Zimbabwe.

Manufacturer: OTO Melara SpA
via Valdilocchi 15
19100 La Spezia
Italy
Telephone: (0187) 530 111
Telex: 270 368 oto i
Fax: (0187) 530 111

Canadian Armed Forces 105 mm C5 Pack Howitzer firing, this particular weapon is not fitted with its shield (Canadian Armed Forces)

OTO Melara 105 mm Model 56 Pack Howitzer

OTO Melara 105 mm Pack Howitzer in travelling configuration

SPECIFICATIONS
(data in square brackets relates to anti-tank configuration)

CALIBRE	105 mm
BARREL LENGTH	1.478 m
MUZZLE BRAKE	multi-baffle
RECOIL	hydraulic buffer and helical recuperator
BREECH MECHANISM	vertical sliding wedge
CARRIAGE	split trail
SHIELD	optional
WEIGHT	
travelling order, 3 section trails	1290 kg
firing position, 3 section trails	1290 kg

LENGTH	
travelling	3.65 m
firing	4.8 [4.7] m
WIDTH	
firing	2.9 [4] m
travelling	1.5 m
HEIGHT	
travelling	1.93 m
firing	1.93 [1.55] m
AXIS OF BORE (firing)	1.07 [0.693] m
TRACK	1.14 [1.32] m
ELEVATION/DEPRESSION	+65°/-5° [+25°/-5°]
TRAVERSE (total)	36° [56°]

RATE OF FIRE	
for 30 minutes	4 rds/min
for 1 hour	3 rds/min
anti-tank	8 rds/min
MAX RANGE	
(normal HE projectile)	10 575 m
CREW	7
TOWING VEHICLES	LWB (4 × 4) Land Rover Unimog (4 × 4) truck AR-59 (4 × 4) truck

KOREA, SOUTH

105 mm M101A1/155 mm M114A1 Howitzers

Development/Description
South Korea has produced the American 105 mm M101A1 and 155 mm M114A2 howitzers for both the home and export markets. These have the same physical characteristics as the original US-built versions, but the M114A2 can fire a RAP to 19 500 m.

Status: Production complete. In service with South Korean Army and offered for export.

Manufacturer: Daewoo Corporation
541, 5-Ga
Namdaemoon-Ro
Jung-Gu
Seoul 100-095
Republic of Korea
Telephone: 771 91
Telex: K23341/4; K24295

105 mm KH178 Light Howitzer

Development
In the preliminary stages of the design of a new 105 mm light howitzer for use by the Korean Army, the Kia Machine Tool Company procured examples of the British Royal Ordnance 105 mm Light Gun and the German Rheinmetall conversion of the American 105 mm M101 howitzer. Aspects of both have been incorporated in the KH178 105 mm light howitzer, production of which began in 1984, and it is now in service with the South Korean Army.

Description
The KH178 comprises the CN78 cannon, the RM78 recoil mechanism, the CG78 carriage and a new fire control system. The barrel is 34 calibres long and is fitted with a double-baffle muzzle brake. A horizontal sliding breech-block is used with a percussion firing mechanism. The barrel is autofrettaged and has a life of 7500 EFC. Progressive twist rifling is used, changing from 1:35 to 1:18 at the muzzle.

The recoil mechanism is of the hydro-pneumatic type with a constant recoil length. The recoil and recuperator cylinders are mounted over and under the barrel. In order to reduce elevation handwheel loads the equilibrators are of a new type using a hydropneumatic and spring system.

On the split trail carriage the trail legs are reinforced and a shield may be fitted. Run-flat tyres are used. All the fire control equipment and some accessory devices are fitted with self-luminous sources.

The KH178 can fire all current standard 105 mm ammunition plus a new single increment super propellant M200. This gives the weapon a range of 14 700 m. Firing an RAP projectile the range is 18 000 m.

Status: Production as required. In service with the South Korean Army.

Manufacturer: Kia Machine Tool Company Limited
Fourth Floor
Kia Building
15 Yoido-dong
Youngdeungpo-gu
Seoul
Republic of Korea

105 mm KH178 howitzer in firing position

SPECIFICATIONS

CALIBRE	105 mm
BARREL LENGTH	4.48 m
TUBE LENGTH	3.922 m
MUZZLE BRAKE	double-baffle
CARRIAGE	split trail
SHIELD	optional
WEIGHT (total)	2650 kg
LENGTH	7.56 m
WIDTH	2.146 m
HEIGHT	1.575 m
GROUND CLEARANCE	0.386 m
ELEVATION/DEPRESSION	+65°/-5°
TRAVERSE (total)	45.5°
RATE OF FIRE	
max	15 rds/min
sustained	5 rds/min
MAX RANGE	
HE	14 700 m
RAP	18 000 m
MUZZLE VELOCITY	662 m/s
TOWING SPEED	
road	56 km/h
cross country	16 km/h
TOWING VEHICLE	2½-tonne (6 × 6) truck

155 mm KH179 Howitzer

Description
Development of the KH179 155 mm howitzer began in 1979 and was completed in 1982. It has been in service with the South Korean Army since early 1983.

The 155 mm KH179 howitzer is a conversion of the American M114A1 howitzer and carriage to mount a new 39-calibre long barrel. Certain other parts of the carriage have been altered to suit the new barrel and new fire control equipment has been fitted. The KH179 is light enough to be air-lifted by a CH-47C/D helicopter and can also be carried inside a C-130 Hercules transport aircraft.

The new barrel is of monobloc construction with an interrupted screw breech mechanism. The rifling has a constant 1 in 20 twist and there are 48 grooves. High strength alloy steel is used throughout the barrel construction and heat treatment and autofrettaging are employed to provide maximum barrel life. The maximum barrel pressure is 3208.5 kg/cm² and the chamber capacity 18.845 cm³. Separate recoil and recuperator systems are used in what is stated to be a 'variable length hydropneumatic independent design'.

On the carriage the layer is on the left of the breech to operate the traverse handwheel on the left of the top carriage. Another member of the gun crew turns the elevating handwheel on the right of the breech. The layer is provided with a dial sight and a telescope for aiming, the telescope for direct fire. All fire controls and some of the accessory equipment on the KH179 are equipped with radioactivated light sources. Pneumatic equilibrators are connected between the yoke of the recoil mechanism and the top carriage to keep the handwheel loads within acceptable limits.

For direct firing, a panoramic telescope with a × 4 magnification is fitted on the left side. For direct firing the elbow telescope with a × 3.5 magnification and a 13.5° field-of-view is mounted on the right and has an effective performance up to 1500 m.

155 mm KH179 howitzer in firing position

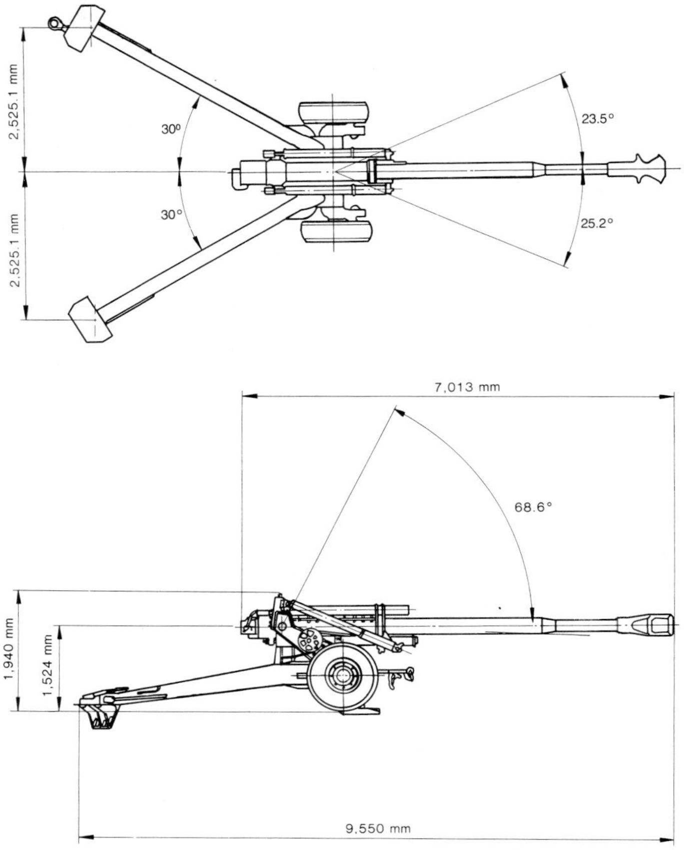

Top and side views of 155 mm KH179 howitzer in firing position (not to 1/76th scale)

The KH179 can fire existing NATO 155 mm ammunition and can also accommodate enhanced performance ammunition such as the elongated ERFB projectiles. Range with RAP is stated to be 30 000 m and for normal HE projectiles the range is 22 000 m. Maximum rate of fire is 4 rds/min dropping to 2 rds/min for sustained fire.

Status: Production as required. In service with the South Korean Army.

Manufacturer: Kia Machine Tool Company Limited
Fourth Floor
Kia Building
15 Yoido-dong
Youngdeungpo-gu
Seoul
Republic of Korea

SPECIFICATIONS

CALIBRE	155 mm
BARREL LENGTH (total)	7.013 m
LENGTH OF TUBE	6.096 m
MUZZLE BRAKE	double-baffle
BREECH MECHANISM	interrupted thread
RECOIL MERCHANISM	variable, hydro-pneumatic
RECOIL LENGTH	
(long)	1.524 m
(short)	1.041 m
CARRIAGE	split trail
SHIELD	no
WEIGHT (total)	6890 kg
LENGTH (overall, travelling)	10.389 m
WIDTH (overall)	2.438 m
HEIGHT (overall, travelling)	2.77 m
GROUND CLEARANCE	0.28 m
ELEVATION/ DEPRESSION	+68.6°/0°
TRAVERSE	23.5° left, 25.2° right
RATE OF FIRE	
max	4 rds/min
sustained	2 rds/min
MAX RANGE	
standard HE	22 000 m
RAP	30 000 m
MUZZLE VELOCITY	826 m/s
TOWING SPEED	
road	70 km/h
cross country	16 km/h
TOWING VEHICLE	5-tonne (6 × 6) truck

NETHERLANDS

RDM 155 mm M139 Howitzer and M114/39 Modified Howitzer

Development

RDM Defense Engineering has developed a conversion kit which enables the old American 155 mm M114 towed howitzer to fire new ammunition including ERFB. This kit is a further development of the SRC International 155 mm M114/39 modification kit which was fully described and illustrated in *Jane's Armour and Artillery 1982-83* page 550. By mid-1983 the M114/39 was being evaluated by the Dutch Army and has been evaluated by a number of other countries. RDM Defense Engineering have converted two Canadian M114s to the M114/39 configuration and these have completed their trials in Canada. The remaining 53 conversions will be carried out in Canada using kits provided from the Netherlands, but as of early 1991 this programme had yet to commence.

An MoU was signed in late 1986 between Denmark, the Netherlands and Norway for a total of 226 M114/39 conversion kits, including 82 for the Dutch Army. RDM Defense Engineering is the prime contractor, the major sub-contractors for offset work being DISA of Denmark and Norsk Forsvarsteknologi as of Norway. The first M114/39 was delivered to the Norwegian Army in August 1988 and the project was successfully completed early in 1990.

RDM now produces, in addition to the M114/39 modification kit, brand new guns under the designation M139. These are virtually identical to the M114/39 and their ballistic performance is the same.

RDM Defense Engineering handed over the first rebuilt systems in Spring 1988. Each of the three countries will be responsible for converting those weapons for its own army with each of the three companies feeding components to each other.

Following a competition, Bofors was selected to manufacture the barrels for the M114/39 conversion programme; the first sample barrel was delivered in 1984 for extensive tests and the first production barrels were delivered in 1987.

Although deliveries to the three countries were completed in 1990, RDM are still marketing the system to a number of countries and can still produce conversion kits to order, or new guns.

Description

The modification is accomplished using the maximum number of existing gun parts and new components only where necessary. The major change involves a new 39-calibre barrel with a larger chamber, increased twist rifling and a muzzle brake which makes the weapon, with current projectiles and charges, ballistically similiar to the M109A2 self-propelled and M198 towed howitzers.

RDM M114/39 155 mm howitzer in firing position

The fitting of a new ordnance has required a number of changes to the carriage including moving the jack forwards and modifying the trails to compensate for the forward shift of the gun's centre of gravity, fitting larger capacity equilibrators with temperature compensating devices, a new metering rod in the recoil system with new seals and the modification of the gear trains in both the elevating and traversing systems resulting in higher mechanical advantages. Full details of these modifications are as follows:

Cannon assembly

The new 39-calibre barrel has a constant 1 in 20 calibres twist rifling with 48 lands and grooves. This provides for a projectile travel of 5.08 m with a chamber volume of 19 000 cm^3. The barrel is autofrettaged to provide maximum fatigue life and is fitted with a screw block breech mechanism with a conventional type obturator pad and split ring seal obturator.

A three-port high efficiency muzzle brake is fitted to reduce recoil. In order to adapt the new barrel, the existing cradle has been replaced by a new structure which includes an anti-rotation device.

Recoil mechanism

The original recoil and metering rod, which is in the recoil assembly, has been replaced by one which is optimised with respect to increased muzzle velocity, heavier weight of the ordnance and therefore longer recoil distance. The original seals have been replaced with modern components in order to seal off the larger fluid pressure generated during recoil. This enables the weapon to be fired under all operational conditions without damage. The original front yoke has been replaced with a new structure to allow adaptation of the new ordnance and to provide attachments for the new equilibrators. The original spring equilibrators were unable to compensate for the increased unbalanced moment and have been replaced by higher capacity pneumatic equilibrators.

Other components

The original equilibrator bracket is replaced by one which allows for temperature compensation compatible with the pneumatic equilibrators. The gear trains in the elevating and traversing mechanism have been slightly modified in order to increase the mechanical advantage which has enabled the handwheel loads to be kept within acceptable operational limits.

In order to compensate for the shift in the centre of gravity of the weapon the jack and the float assembly have been moved forward. This is accomplished by securing a spacer to the bottom carriage which in turn holds the jack. This provides a stable firing platform under all conditions.

Due to the forward shift of the centre of gravity, the original trails have been modified, this also allows the crew to manhandle the weapon easily when it is standing on its wheels. All other original components, including the sight and sight mount are retained. This simplifies both training and logistic support.

A pneumatic power rammer can be incorporated as an optional extra to permit power loading at all angles of elevation. The pneumatic rammer is powered by compressed air from an air cylinder mounted on the right trail leg.

RDM Defense Engineering has developed to the prototype stage a VW air-cooled diesel auxiliary power unit that can be installed on the forward part of the carriage. This develops 45 kW and enables the M139 and M114/39 to move around under their own power at 10 km/h.

RDM Defense Engineering can either convert existing guns at its own facilities or supply kits to enable armies to undertake the conversion themselves.

BALLISTIC PERFORMANCE

Charge	Zone	M107 projectile		ERFB-HE projectile		ERFB + base bleed	
		Muzzle Velocity (m/s)	Max Range (m)	Muzzle Velocity (m/s)	Max Range (m)	Muzzle Velocity (m/s)	Max Range (m)
M3A1	1	—	—	—	—	—	—
	2	237.7	5000	*	*	*	*
	3	277.4	6500	*	*	*	*
	4	318.5	8300	*	*	*	*
	5	374.9	9800	*	*	*	*
M4A2	3	292.6	7200	*	*	*	*
	4	336.8	8500	*	*	*	*
	5	393.2	10 300	*	*	*	*
	6	475.5	12 400	465	12 700	*	*
	7	565.4	14 800	552.7	15 700	*	*
M119A1	8	684.3	18 100	671.5	20 100	665	23 600
M9	9			785	24 600	785	30 400
M203/L3	S			812	25 500	812	32 400

* The usefulness of the ERFB projectile becomes apparent only with higher charges. ERFB can be used with smaller charges but is not cost effective.

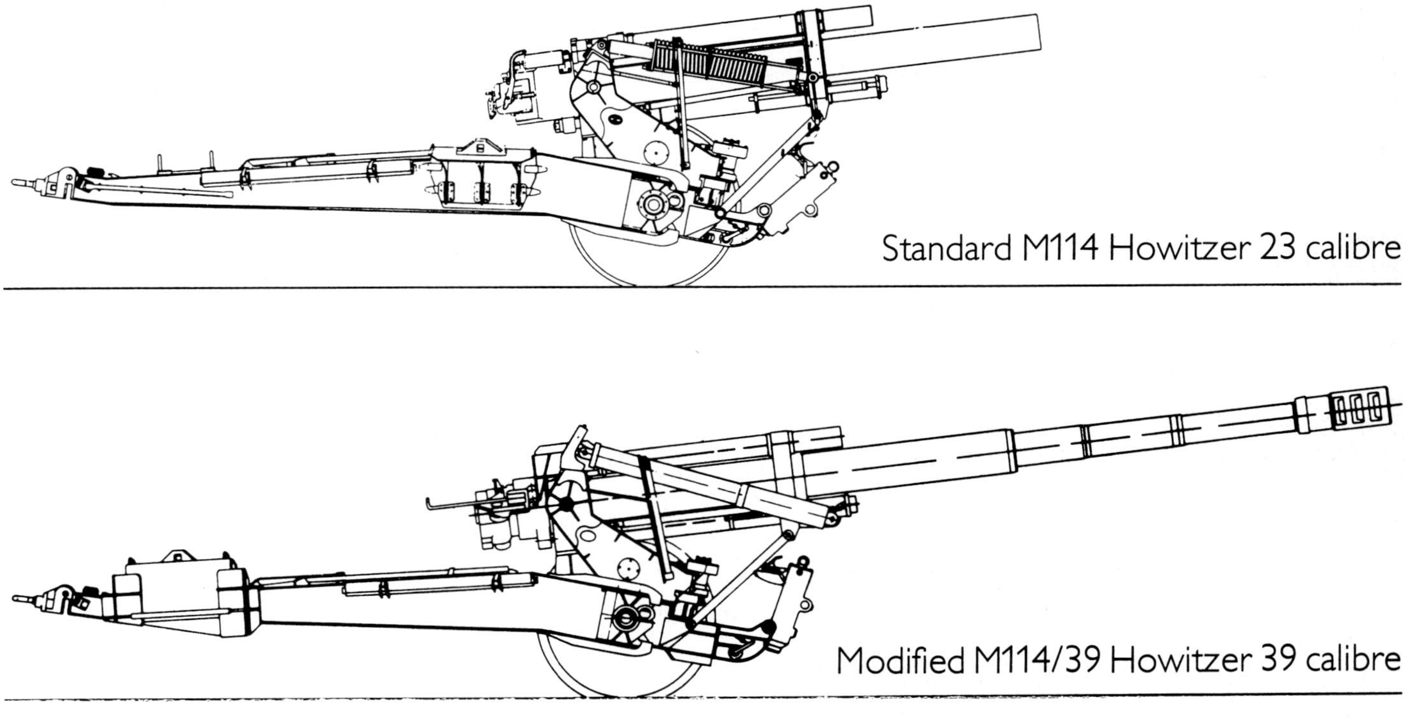

Comparative drawings of standard 155 mm M114 (top) with 23 calibre ordnance and modernised M114/39 (bottom) with 39 calibre ordnance (not to 1/76th scale)

SPECIFICATIONS

CALIBRE	155 mm
BARREL LENGTH (muzzle face to rear face not including muzzle brake)	6.016 m
MUZZLE BRAKE	multi-baffle
CARRIAGE	split trail
SHIELD	optional
WEIGHT	7600 kg
LENGTH (travelling)	10 m
WIDTH (travelling)	2.44 m
HEIGHT (travelling with lunette 0.737 m from ground)	2.31 m
TRUNNION HEIGHT	1.42 m
TRACK	2.08 m
ELEVATION/DEPRESSION	+63°/-2°
TRAVERSE	right 25° left 24°
RATE OF FIRE (max)	4 rds/min
RANGE	
M107 projectile	18 100 m
ERFB	24 600 m
ERFB BB	32 400 m
CREW	11
TOWING VEHICLE	5-ton (6 × 6) truck

RDM M114/39 155 mm howitzer in travelling configuration being towed by 6 × 6 truck

Status: Production as required. In service with Denmark (96 ordered), Netherlands (82 ordered) and Norway (48 ordered).

Manufacturer: RDM BV General Engineering Department
PO Box 913
3000 AX Rotterdam
Netherlands
Telephone: (31 10) 487 2747
Telex: 20753 rdm nl
Fax: (31 10) 487 2299

RDM 105 mm M101/33 Howitzer

Development

Building on their experience in the production of modernisation packages for the 155 mm M114 towed howitzer, fully covered in the previous entry, RDM have more recently developed an upgrade package for the widely deployed US 105 mm M101 series of towed howitzer that was originally developed over 50 years ago.

When upgraded, the M101 can fire more modern ammunition with enhanced range and lethality.

The modernisation is accomplished by using a maximum of existing gun parts and only adopting new components where required.

In late 1990, an upgraded M101/33 howitzer was demonstrated to the Canadian Armed Forces in Canada and RDM can provide conversion packages on receipt of firm orders.

Description

The major modification includes the installation of a new 33-calibre barrel with progressive rifling and a new muzzle brake. In addition, the M101 is modified in a number of other areas including the incorporation of larger capacity equilibrators, new metering and recoil rod in the recoil system and new seals, replacement of the original elevating arc and reinforcement on the drawbar.

The new 105 mm/33 calibre barrel is ballistically similar to the latest generation of 105 mm artillery ammunition and suitable for high pressure with the design being for 4500 bars.

It has progressive rifling (1.35 to 1.18) over a length of 2794 mm whereas rifling is constant (1.18) over the last 306 mm (muzzle end). The barrel has 36 lands and grooves and provides for a shot travel of 3100 mm with a chamber volume of 2507 cm², this is identical to the standard M101 so allows the upgraded weapon to fire all existing types of 105 mm ammunition.

The modified breech mechanism is still of the horizontal sliding type but due to the higher forces the original breech ring is replaced by a stronger one which allows it to use all modern ammunition types.

The remaining parts of the original breech mechanism can still be used including the firing mechanism. In order to reduce the increased load on trunnions and recoil when using new ammunition, a new three slotted muzzle brake is fitted to the threaded end of the barrel. This muzzle brake provides for a substantial reduction in the recoil forces while still maintaining blast overpressure within acceptable limits.

A modification to the recoil system was required due to the increase of the recoil momentum resulting from the use of the 33 calibre barrel and more modern ammunition. The increased recoil momentum is absorbed partly by the muzzle brake and partly by the recoil mechanism.

The original control rod with the recoil cylinder is replaced and is tailored with respect to the increased recoil momentum of the heavier cannon. To limit the recoil the diameter of the recoil cylinder is increased. The recoil distances have not been changed and are still the same as the original, in addition the original seals are replaced in order to retain the larger fluid pressure generated during the recoil.

As the original two spring equilibrator is unable to compensate for the increased tipping movement a new equilibrator has been mounted in such a way that the equilibrator now houses three nested springs instead of the original two. The new equilibrator reduces the weight moment at all angles of elevation.

The cradle system does not need to be changed, although a small modification is required in order to connect the new equilibrator. No modifications are required to the carriage of the M101. A small reinforcement on the drawbar connection of the trails is required.

RDM 105 mm M101/33 howitzer deployed in firing position

The original M101 howitzer provides for a maximum elevation of +65° but in order to increase the overlap in ranges the original elevation arc has been replaced which allows the elevating arc to be elevated to +70°.

SPECIFICATIONS

CALIBRE	105 mm
BARREL LENGTH	3.513 m
MUZZLE BRAKE	multi-baffle
CARRIAGE	split trail
SHIELD	yes
WEIGHT (M101)	2150 kg
WEIGHT (M101A1)	2380 kg
LENGTH TRAVELLING	7.4 m
WIDTH TRAVELLING	2.15 m
HEIGHT TRAVELLING	1.55 m
TRUNNION HEIGHT	0.97 m
TRACK	1.85 m
ELEVATION/DEPRESSION	+60°/-6°
TRAVERSE	46° (total)
MAXIMUM RANGE	
M1 projectile, M67 charge	11 900 m
M1 projectile, Super charge*	15 700 m
ERMI projectle, Super charge*	19 500 m
M760, M200 charge	14 400 m
M548 RAP, M176 charge	15 500 m
XM913 RAP, M200 charge	19 600 m

* Currently under development by Eurometaal of the Netherlands

Status: Development complete. Ready for production.

Manufacturer: RDM BV General Engineering Department
PO Box 913
3000 AX Rotterdam
Netherlands
Telephone: (31 10) 487 2747
Telex: 20753 rdm nl
Fax: (31 10) 487 2299

ROMANIA

100 mm Mountaineer Howitzer

Development/Description

This 100 mm weapon has been developed specifically to meet the requirements of artillery units operating in mountainous terrain and in some respects is similar to the Italian 105 mm OTO Melara Pack Howitzer.

It has a split trail carriage with the rear half of each trail folding upwards and forwards through about 180° so that it rests on the remainder of the trails. The trails are then clamped together and a single rubber tyred road wheel can be mounted under the rear of the closed trails for ease of handling in the field.

The sight is mounted on the left side and the ordnance is fitted with a multi-baffle muzzle brake.

The weapon is normally towed by a DAC 444 truck or can be disassembled into three loads and carried on small carriages, each of which is towed by a horse.

Romanian 100 mm mountaineer howitzer in travelling configuration

SPECIFICATIONS

CALIBRE	100 mm
MUZZLE BRAKE	multi-baffle
CARRIAGE	split trail
SHIELD	no
WEIGHT	1500 kg
ELEVATION/DEPRESSION	+65°/-5°
TRAVERSE	50°
RANGE (max)	10 000 m
RANGE (direct fire)	600 m
WEIGHT OF HE PROJECTILE	17.9 kg
RATE OF FIRE	6 rds/min
CREW	7
TOWING VEHICLE	DAC 444 truck horses

Status: Production as required. In service with Romanian forces.

Manufacturers: Romanian state factories.

Romanian 100 mm mountaineer howitzer in firing configuration

152 mm Model 1981 Gun

Development/Description

The 152 mm Model 1981 gun is mounted on a split trail two wheeled carriage with each of the trails being provided with a castor wheel to assist in bringing the weapon into and taking it out of action.

The carriage appears to be very similar to that used for the Soviet 152 mm D-20 gun howitzer although the Romanian weapon has a much longer barrel and therefore has a longer range. The D-20 has a maximum range, firing standard ammunition, of 17 410 m.

The 152 mm Model 1981 fires a HE projectile weighing 43.56 kg with a maximum muzzle velocity of 825 m/s, with the propelling charge weighing 11.95 kg to a maximum range of 24 000 m.

Romanian sources have also mentioned a 152 mm M1985 towed artillery system, but firm details of this system are not yet available.

Romanian 152 mm M1981 deployed in firing position

SPECIFICATIONS

CALIBRE	152.4 mm
LENGTH OF BARREL	7.105 m (without muzzle brake) 8.025 m (with muzzle brake)
LENGTH OF RIFLING	6.102 m
NUMBER OF GROOVES	48
RECOIL	910 mm
WEIGHT	
firing	7500 kg
travelling	7550 kg
LENGTH TRAVELLING	11.17 m
WIDTH TRAVELLING	2.53 m
GROUND CLEARANCE	0.35 m
ELEVATION/DEPRESSION	+60°/-2°
TRAVERSE (total)	50°
MAX RANGE	24 000 m

Status: Production probably complete. In service with the Romanian Army.

Manufacturer: Romanian state factories.

152 mm M1981 Howitzer

Development/Description

The carriage has split box type trails with a castor wheel being provided to bring the weapon into action; when not required this is swung back and rests over the trail. The carriage of this weapon is similar in some respects to that used for the Soviet 152 mm Gun-Howitzer D-20 and 122 mm Field Gun D-74, as a circular firing pedestal is fitted as standard and is inverted and secured just forward of the shield when travelling. When in the firing position the firing pedestal is lowered so allowing the weapon to be traversed fairly quickly through 360°.

The ordnance is fitted with a large double baffle muzzle brake with a distinctive circular ring to its rear.

Status: Production complete. In service with the Romanian Army.

Manufacturer: Romanian state factories.

Romanian 152 mm M1981 Howitzer in travelling configuration

152 mm Model 1947 Gun

Development/Description

This fires the same ammunition as the 152 mm M1985 Gun although the Model 1947 has a shorter ordnance. It is probable that this weapon is in fact the Soviet 152 mm gun M-47 which was developed at the same time as the Soviet 130 mm field gun M-46. While the latter was deployed by the Soviet Army in large numbers and exported to many countries, the M-47 is believed to have been built only in small numbers.

Status: Production complete. In service with the Romanian Army.

Manufacturer: Romanian state factories.

SPECIFICATIONS

CALIBRE	152.4 mm
LENGTH OF BARREL	6.65 m (without muzzle brake) 7.17 m (with muzzle brake)
LENGTH OF RIFLING	5.47 m
NUMBER OF GROOVES	48
RECOIL	1250 mm
WEIGHT	
firing	7700 kg
travelling	8450 kg
LENGTH TRAVELLING	11.32 m
WIDTH TRAVELLING	2.45 m
GROUND CLEARANCE	0.40 m
ELEVATION/DEPRESSION	+45°/-2° 30'
TRAVERSE (total)	50°
MAX RANGE	20 470 m

130 mm Gun Model 82

Development/Description

No firm details of this weapon are available although it could be a Romanian version of the Soviet 130 mm Field Gun M-46.

Status: Production probably complete. In service with the Romanian Army.

Manufacturer: Romanian state factories.

100 mm Towed Anti-Tank Gun M1977

Development/Description

The 100 mm towed anti-tank gun was developed specifically to meet the requirements of the Romanian Army for a towed weapon to engage tanks and other armoured fighting vehicles under both day and night conditions, as well as being used in the fire support role.

The ammunition fired by the M1977 is identical to that fired by the 100 mm gun D-10 series guns installed in the T-54/T-55 MBTs so the weapons may well be related.

Romanian sources give the M1977 a maximum range of 4000 m against tanks and armoured vehicles in the indirect fire mode and 1100 m in the direct fire mode firing an armour piercing projectile against a target height of 2 m. In the indirect fire mode using HE ammunition it has a maximum range of 20 700 m.

To bring the weapon into action from its travelling configuration takes 60 seconds.

Romanian sources provide the following ammunition data:

Type	HE-FRAG	APC-T	HEAT	HEAT
Designation	OF-412	BR-412D*	3BK-5M	BK-412R
M/velocity (max)	900 m/s	887 m/s	900 m/s	950 m/s
M/velocity (reduced)	600 m/s	n/app	n/app	n/app
Weight of propellant	5.5 kg**	5.5 kg	4.5 kg	4.5 kg
Weight of projectile	15.6 kg	15.88 kg	12.2 kg	10.625 kg
Max range	20 700 m	4000 m	2200 m	2200 m

* Also BR-412 and BR-412B with a muzzle velocity of 895 m/s

** Weight of reduced propellant charge is 4.5 kg.

Optical equipment installed on the M1977 includes a mechanical sight, panoramic sight with a magnification of ×4 and a 10° field-of-view, optical sight with magnifications of ×4 and ×8 and a collimator.

SPECIFICATIONS

CALIBRE	100 mm
CREW	8
LENGTH OF ORDNANCE	5350 mm
LENGTH OF RIFLING	4630 mm
NUMBER OF RIFLED GROOVES	40
PITCH OF RIFLING	30 calibres
WIDTH OF LAND	3.15 mm ± 0.3 mm
WIDTH OF GROOVE	4.7 mm ± 0.33 mm
NORMAL RECOIL	750 mm to 830 mm
MAX RECOIL	840 mm
TIME OF RECOIL	
RETURN CYCLE	1-2 s
AXIS OF FIRE AT 0°	0.9 m
LENGTH TRAVELLING	9.37 m
HEIGHT TRAVELLING	1.9 m
WIDTH	2.24 m
GROUND CLEARANCE	0.36 m
ELEVATION/DEPRESSION	+37°/-5°
ELEVATION SPEED	0.76°/revolution
TRAVERSE (total)	55°
TRAVERSE SPEED	1°/revolution
MAX RANGE	
OF-412 HE-FRAG projectile	22 700 m
APC-T projectile	4000 m
HEAT projectile	2200 m

Status: Production probably complete. In service with Romanian Army.

Manufacturer: Romanian state factories.

76.2 mm Mountain Gun

Development/Description

This 76.2 mm Mountain Gun has been developed specifically to meet the requirements of Romanian mountain units and can be quickly disassembled into eight pack animal loads.

When being towed by a vehicle or horse team the weapon can be brought into action in about 1 minute. When being carried by pack horse the weapon takes between 6 and 8 minutes to come into action.

The weapon can fire HE and HEAT projectiles with the ordnance having a ballistic life of about 15 000 rounds. The ratio for calculating the number of rounds which can be fired with the other charges, and wearing the barrel in the same degree, as the 4th charge does, is shown in the following table:

Charge	Number of rounds wearing the barrel in the same degree
4	1
3	3
2	7
1	10

Optical equipment fitted to the weapon consists of a telescope with a magnification of ×3.

Status: Production probably complete. In service with the Romanian Army.

Manufacturer: Romanian state factories.

SPECIFICATIONS					
CALIBRE	76.2 mm	WIDTH IN ACTION	2.65 m	RATE OF FIRE	
LENGTH OF BARREL	1178 mm (without breech mechanism)	HEIGHT	1.67 m	first minute	25 rounds
		GROUND CLEARANCE	0.18 m	in three minutes	40 rounds
		TRACK	1.28 m	in five minutes	50 rounds
MAX RECOIL LENGTH	830 mm	ELEVATION/DEPRESSION		in 10 minutes	60 rounds
WEIGHT		with lengthened trails	+50°/−15°	in 15 minutes	70 rounds
travelling (towed)	680 kg	with shortened trails	+45°/−10°	in 30 minutes	90 rounds
travelling (horse towed)	705 kg	on slope with shortened		in 60 minutes	120 rounds
combat	690 kg	trails	+70°/+20°	in 120 minutes	210 rounds
LENGTH TRAVELLING	2.42 m	TRAVERSE (total)	50°	in each further hour	80 rounds
LENGTH IN ACTION	3.100 m	MAX RANGE (HE)	8600 m	LIFE OF BARREL	15 000 rounds
WIDTH TRAVELLING	1.31 m	MAX RANGE (HEAT)	1000 m		

SINGAPORE

ODE 155 mm FH-88 Gun Howitzer

Development

Ordnance Development and Engineering of Singapore (ODE), a member of Chartered Industries of Singapore, has developed a 155 mm gun howitzer with a 39-calibre barrel known as the FH-88. Five prototypes were produced over a period of four years beginning in 1983, these were followed by a pre-production batch of six guns that incorporated a number of improvements as a result of trials with the prototype weapons. First production FH-88s were completed in 1987 with the weapon becoming operational with the Singapore Army the following year. The gun-howitzer is produced in Singapore with some components such as the barrel being imported as forgings with final machining carried out in ODE facilities.

Description

The monobloc barrel is 39 calibres long and is autofrettaged. A double-baffle muzzle brake is fitted. The breech is of the hinged, interrupted screw type and automatically opens during the latter half of the return stroke of the recoil movement. The camplate lifting mechanism incorporates a safety lock to prevent accidental closing of the breech screw and the firing mechanism cannot be activated if the breech screw is not fully closed. Obturation is achieved using a resilient pad and split rings. The firing mechanism is a mechanical 'trip action' type which is actuated either by pulling the lanyard or operating the firing handle at the layer's seat. Primer feeding and extraction is automatic using a 12-round magazine. A flick rammer is used with power derived from an independent powerpack.

The hydropneumatic recoil mechanism is of the independent type single recoil system. The recoil mechanism, connected to the recoiling and static parts of the gun, consists of a recoil brake and a recuperator/counter-recoil mechanism. The governor rod in the recoil brake varies the orifice for hydraulic oil to flow throughout the recoiling stroke, absorbs and ensures minimum and uniform recoil on the carriage. The pressurised nitrogen gas in the recuperator/counter-recoil mechanism is to return the barrel after recoil and also holds the barrel in its fully run out position at all elevations.

There are two independent pneumatic equilibrators which balance the mount weight of the elevated mass throughout the full range of elevation. The equilibrator is made up of a cylinder, piston rod and a receiver using nitrogen gas as the working medium.

The cradle, which is made of high strength alloy steel, supports and guides the recoiling parts. It houses the recoil mechanism and provides attachments for the equilibrators, camplate arm, elevating arc, flick rammer and the sighting system.

The saddle, which is made of high strength alloy steel, comprises two side frames welded to a circular baseplate. It supports the elevating mass through the trunnion bearings and houses the elevating and traversing mechanism and provides attachments for the equilibrators and sighting equipment.

ODE 155 mm FH-88 gun howitzer travelling under power from APU (Karen Walker)

ODE 155 mm FH-88 gun howitzer in travelling configuration (Karen Walker)

The elevating mechanism, which with the aid of the equilibrators, enables the ordnance to be elevated or depressed with accuracy and ease. The traversing mechanism, mounted in front of the saddle, enables the ordnance to be swivelled horizontally.

The flick rammer is an electronically controlled hydraulic system which in turn operates the mechanical system. The latter basically consists of the swivel arm and the ramming unit. The swivel arm, with the loading tray attached, is pivoted at the rear end of the cradle.

The chassis serves as the base frame to integrate the upper carriage to the lower carriage and during firing it transmits all of the load on the upper carriage to the ground through the firing platform and trail legs.

The chassis is an all-welded structure of high strength steel on a closed box design with provision for the mounting of the main bearing, rocker beams, firing platform, trail legs and the APU.

The firing platform is connected to the chassis assembly through jack arrangements and lifts the weapon off the ground for firing. The lowering of the platform is achieved through the tripod arrangement being operated by a telescopic cylinder which is completely isolated from all firing loads when the platform is fully deployed. An indicator is incorporated to show when the platform is fully deployed.

The trails are made of high strength steel and together with the firing platform, serve as the main support for the weapon. In the closed position they serve as a base for locking the barrel in a fixed position for travelling. Hydraulic powered operations permit the trails to be spread or retracted within 10 seconds.

The two trail wheels, attached to the trail ends, can be steered, raised or lowered by hydraulically powered operations controlled by the driver.

The spade, an integral part of the trails, transmits the firing loads to the ground. As the spades are permanently hinged at the end of the trails, no carrying and mounting of spades are required during deployment.

The carriage is of the split trail type and is fitted with an APU. The gun can be towed at speeds up to 80 km/h and is self-propelled at speeds up to 16 km/h. The APU is a 96 hp Deutz air-cooled turbo-charged diesel which powers a hydrostatic drive system on the front wheels. Back-up systems include a battery-operated pump and a manual pump. The APU can be operated in tandem with the towing vehicle for maximum traction and can be remote controlled. The APU has a 50-litre fuel tank providing an operating range of 60 km.

The diesel powered APU consists of a hydrostatic drive system, braking system, operation hydraulic system and a backup system. The basic functions of the APU can be summarised as follows:

(1) Self-propelled capability of the howitzer
(2) Mechanised hydraulic operation during deployment
(3) Ability to assist the prime mover in manoeuvring over rough terrain.

The rapid change in the zone of fire by the FH-88 to engage targets outside of the primary zone is achieved by:

(1) Raising the firing platform
(2) Lowering and steering trail wheels
(3) Executing pivot turn
(4) Steering and raising trail wheels
(5) Embedding spades
(6) Lowering firing platform

The basic sighting system of the FH-88 consists of a cant compensating mount, peri-telescopic sight (×4 magnification) and a direct aiming sight (×6 magnification). However this can be upgraded to a sophisticated electronic sight and linked to battery fire control computers.

The trail legs have wheels for ease of handling. They can be opened and closed by hydraulic power and the firing platform is raised and lowered by a double-stage hydraulic cylinder. The trail spades are self-embedding. With a crew of six the gun can be brought in or out of action in less than one minute.

The ODE 155 mm FH-88 gun howitzer can fire standard NATO ammunition such as the M107 HE projectile to a maximum range of 19 000 m. The ERFB/HB projectile can be fired to 24 000 m while an ERFB BB can be fired to 30 000 m. ODE claims a range accuracy of less than 0.22 per cent with deflection less than 0.35 per cent at two-

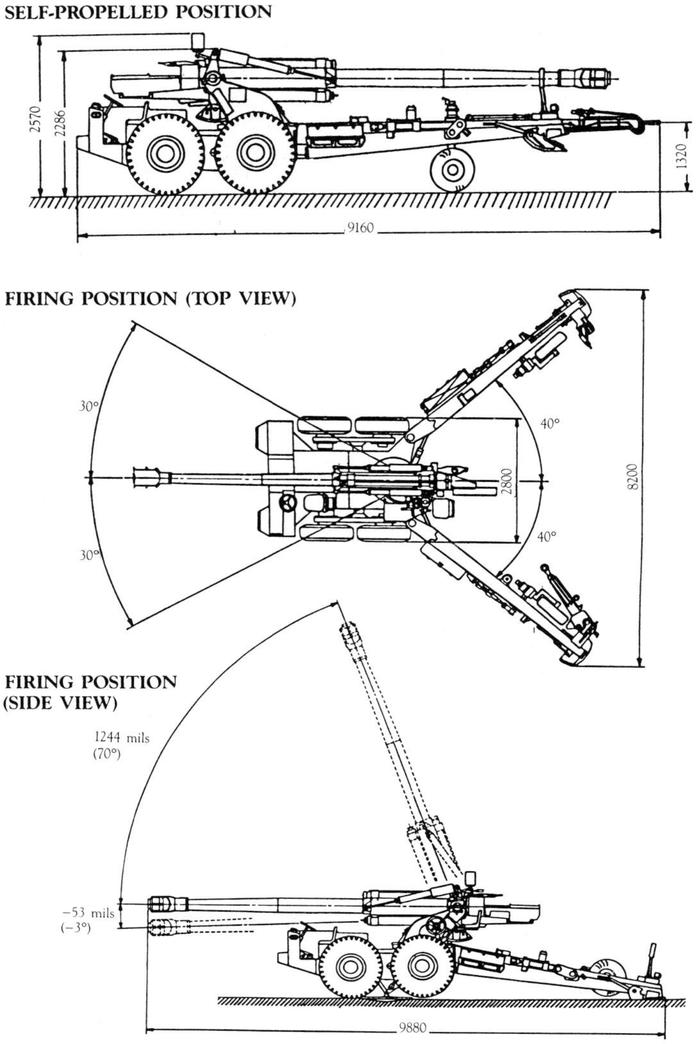

General arrangement drawings of ODE 155 mm FH-88 gun howitzer in travelling configuration (top) and firing position (bottom) (not to 1/76th scale)

thirds of maximum range. The flick rammer can be used to achieve a burst rate of three rounds in 15 seconds. The sustained fire rate is 2 rds/min for one hour.

Variants

In 1990 the prototype of an FH-88 was unveiled fitted with a 52 calibre ordnance. This has a chamber volume of 23 litres, so conforming to NATO standards, and is designed to fire all types of NATO ammunition with a maximum range in excess of 40 km being achieved with extended range ammunition.

The FH-88 has been developed as a private venture and is essentially an FH-88 with a new 52 calibre ordnance that conforms to the NATO MOU, modified carriage and modified towing mechanism to take into account the longer ordnance.

Firing trials with the prototype are expected to be completed late in 1992 and a second prototype is now being built that will incorporate improvements as a result of trials with the first prototype.

Basic specifications of the FH-88 52 calibre are:

WEIGHT TRAVELLING	13 200 kg
LENGTH TRAVELLING	10.95 m
WIDTH TRAVELLING	2.8 m
HEIGHT WITH DISPLAY UNIT	2.55 m
HEIGHT WITHOUT DISPLAY UNIT	2.286 m
MAX RANGES	
M107 projectile	24 000 m
ERFB HB	30 000 m
ERFB BB	40 000 m

Status: The FH-88 with a 39 calibre barrel is in production and in service with Singapore.

Manufacturer: Ordnance Development and Engineering of Singapore
15 Chin Bee Drive
Singapore 2261
Telephone: 2655322
Telex: 27144 ODE RS

ODE 155 mm FH-88 gun howitzers in firing position

SPECIFICATIONS		GRADIENT	45%	MAX RANGE	
CALIBRE	155 mm	MAX SPEED		M107 projectile	19 000 m
LENGTH OF BARREL	6.1 m	towed on main roads	80 km/h	ERFB HB projectile	24 000 m
MUZZLE BRAKE	double-baffle	towed on secondary roads	50 km/h	ERFB BB projectile	30 000 m
BREECH MECHANISM	hinged, interrupted screw	self-propelled, roads	16 km/h	ACCURACY	
CARRIAGE	split trail	self-propelled, cross-country	8 km/h	Probable error (range)	less than 0.22% of range at two-thirds max range
SHIELD	no	tandem drive on tow	8 km/h		
WEIGHT	12 800 kg	remote driving	4 km/h	Probable error (deflection)	less than 0.035% of range at two-thirds of max range
LENGTH		ELEVATION/DEPRESSION	+70°/-3°		
travelling	9.16 m	TRAVERSE	60°		
firing	9.88 m	RATE OF FIRE		CREW	6
WIDTH (travelling)	2.8 m	sustained	2 rds/min for 60 mins		
HEIGHT (travelling)	2.57 m	max	8 rds/min alt min for 20 min		
GROUND CLEARANCE	0.35 m				
FORDING	0.76 m	burst	3 rounds in 15 s		

SOUTH AFRICA

Armscor 155 mm G5 155 mm Gun Howitzer

Development

During operations carried out in 1975 the South African Defence Forces found their own artillery outranged and outgunned by various forms of Soviet artillery used by the Angolan forces. Following this experience the South African artillery initiated a staff requirement for a new artillery system that would encompass not only a new artillery piece but also an ammunition system, a new gun tractor, fire control equipment, a fire control computer system and other accessories. The full development of this new system began during 1976 and for the gun component a number of existing designs were examined and some were procured. Of these, a number of Soltam 155 mm M-71 gun-howitzers were placed into service and were given the designation of G4.

From this examination the GC 45 was selected as a starting point for future development and one of these development pieces based on the ex-SRC GC 45 was mounted for test purposes on an American 155 mm M59. This was one of a series of seven development models which have culminated in the G5, now in production. Throughout the development and testing cycle of the G5 a thorough systems approach has been adopted and considerable computer technology has been employed.

According to Armscor spokesmen the original design of the SRC GC 45 has now been developed to such a point that the G5 bears little resemblance to the original. Considerable changes have been made in internal ballistics, in the barrel construction, numerous carriage and cradle alterations and strengthened trails. An APU has been added and the muzzle brake altered.

The G5 became operational with the South African Army during 1983. Early in 1989 is was announced that Chile's Cardoen Industries had successfully trialled their first locally produced G5 system. Chile calls the G-5 the CC-45 and there is a separate entry for this under Chile.

The G5 155 mm gun howitzer has seen combat service with the South African Army in Southern Angola and South West Africa (now Namibia) and by the Iraqi Army during its war with Iran. Quantities of these weapons were captured by the Allied Coalition forces during Operation Desert Storm, the liberation of Kuwait, which took place early in 1991.

Description

The 155 mm G5 howitzer uses a 45-calibre barrel designed around the performance of the ERFB round originally developed by SRC. The G5 is produced and constructed at the Lyttelton Engineering Works at Verwoerdburg near Pretoria using high grade steel billets manufactured at the Union Steel Corporation mill at Vereeniging. The monobloc autofrettaged barrel has a single-baffle muzzle brake and uses a breech mechanism very similar to that on the American 155 mm M109. Situated behind the breech is a pneumatic piston-type ramming mechanism controlled manually for positioning the projectiles in the chamber at a uniform position whatever the elevation. Charges are loaded by hand.

155 mm G5 howitzer firing in direct fire role

The carriage uses a slightly angled trunnion arrangement for balancing the barrel and the dynamic forces produced on firing. When firing the carriage is raised hydraulically on a firing platform with a wishbone and piston arrangement that cannot be fouled by debris and which is raised clear for cross-country movements. Power for the firing platform hydraulics is taken from an APU situated in front of the carriage in a space frame. The APU is powered by an air-cooled 79 hp diesel engine based on a Magirus-Deutz design and supplies hydraulic power for raising and lowering the trail wheels, opening and closing the trails and raising and lowering the firing platform. The controls for the APU are situated on the right of the barrel with the barrel facing forward, and the driver/operator sits on a seat forward and to the left of the APU. The APU has a fuel tank capacity of 102 litres.

When travelling the barrel of the G5 is carried over the trails in a clamp. The usual G5 gun crew consists of five men, although in an emergency only two are required. Using the APU the trail wheels are lowered and the trail spades released to fall into position. The trail legs can then be spread using the trail wheels and the firing platform lowered. Elevation and traverse controls are manual.

The G5 features a gun display unit which displays all relevant data related to the laying of the gun.

The layer's position is to the left of the breech and he is provided with an upgraded metricated No 9 dial sight and a direct-fire telescope for use up to 5000 m. The AS80 Artillery Fire Control System is designed for use with the G5, and also used are the EMVA Mark 10B muzzle velocity analyser and the S700 meteorological ground station. A special helmet radio has been developed for the gun crew.

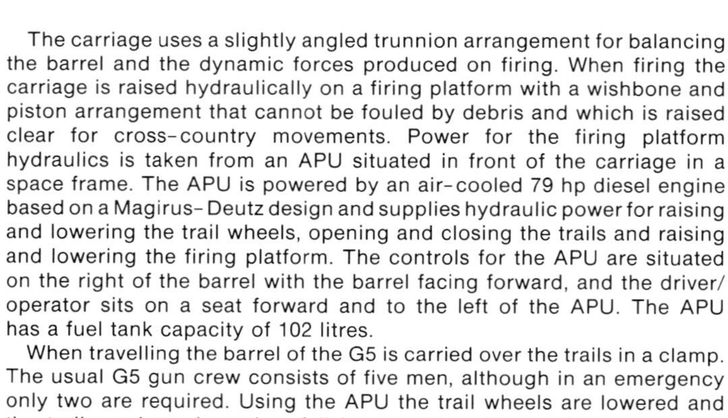

Armscor 155 mm G5 howitzer in travelling configuration (not to 1/76th scale)

The G5 is used in conjunction with a special gun tractor based on a SAMIL 100 (6 × 6) chassis. The gun tractor not only tows the G5 but has accommodation for the full crew of eight (only five normally serve the gun in action). The gun tractor is equipped to carry 15 projectile pallets each weighing 189 kg and a hydraulic crane at the rear of the vehicle is used to handle these.

The ammunition used with the G5 is of the ERFB type in a series known as the M1. Included in the ammunition available are HE, smoke, Illuminating, WP, RP submunition and leaflet. The M1 HE shells are of the high fragmentation type and have welded nubs. One type of M1 HE projectile uses AISI 9260 steel filled with TNT to fragment into 3032 pieces. Another is filled with 50/50 RDX/TNT to fragment into 3756 pieces. The cargo rounds mentioned above are of the base ejection type. All rounds can be of the standard M1 base bleed type.

Full details of the projectiles are as follow:

Type	**HE**	**HE BB**	**Smoke**	**Illum**	**WP**	**RP**	**Leaflet**
LENGTH							
(fuzed)	938 mm	958 mm	938 mm	938 mm	938 mm	938 mm	938 mm
WEIGHT							
(nominal)	45.5 kg	47 kg	45.7 kg	45.2 kg	47.7 kg	45.5 kg	45.5 kg
PAYLOAD	8.7 kg	8.7 kg	10.9 kg	11.8 kg	7.6 kg	13 kg	13 kg

The charge system used with the G5 is of the three-charge, six-zone type and is packed into linen bags or a combustible cartridge. Details are as follows:

Charge	**Muzzle velocity**	**Max range (see level)**
1 M52A1	350 m/s	5700-9200 m
2 M52A1	483 m/s	8700-14 400 m
2 + 1 increment	642 m/s	12 600-19 000 m
2 + 2 increment	795 m/s	17 500-25 400 m
2 + 2 increment (BB)	795 m/s	21 500-30 500 m
3 M53A1	897 m/s	21 700-30 200 m
3 M53A1 (BB)	897 m/s	27 000-39 000 m

The fuzes used with the M1 projectiles include the PD M841, the M8611 electronic time and the M8513 proximity. The M841 and M8513 is used with HE and WP. The M8611 is used with smoke, illuminating, RP, submunition and leaflet. The PD M841 fuze incorporates a rain/cloud desensitising device to replace the American M572 PD fuze and is compatible with base bleed projectiles.

Since the G5 first entered production, a number of improvements have been carried out to the weapon and its associated ammunition system.

The improvements include a temperature warning device that allows higher rates of fire to be achieved, new tow bar that allows the G5 system to be towed by virtually any prime mover as well as improved RAM-D characteristics. The latter include improved sealing, improved oil flow in the recoil system and improved clearances. The breech mechanism has been fitted with an added stop mechanism for when the breech is open.

Troop of 155 mm G5 howitzers in the field

The ammunition system has been improved and the linen bag charge system is no longer produced. The new charge system is of the combustible cartridge type using a moulded nitrocellulose based cartridge case.

The five zone modular charge system comprises a stand-alone Charge 1, a Charge 2 which can be fired alone or with one or two Charge 2 increments and a stand-alone Charge 3 for the top zone. Fitting a base bleed unit to the ERFB projectile increases the range by up to 30 per cent at sea level.

Status: In production. In service with Iraq, South Africa and Qatar (12 ordered in 1991).

Manufacturer: Lyttelton Engineering Works (Pty) Limited
368 Selbourne Avenue
Verwoerdburg 0140
South Africa.

Enquiries to Armscor
Private Bag X337
Pretoria 0001
South Africa
Telephone: (012) 428 1911
Telex: 320217

155 mm G5 howitzer from the rear showing flick rammer mounted on the left side of the breech mechanism

SPECIFICATIONS					
CALIBRE	155 mm	HEIGHT		TRAVERSE	
BARREL LENGTH	6.975 m	travelling	2.3 m	total, up to 15° elevation	82°
MUZZLE BRAKE	single-baffle	barrel forward, clamped	3.3 m	total above 15° elevation	65°
BREECH MECHANISM	interrupted thread	HEIGHT OF TOWING EYE (travelling)	0.95 m	RATE OF FIRE (max)	3 rds/min for 15 mins
CARRIAGE	split trail with APU	GROUND CLEARANCE	0.31 m	MAX RANGE (sea level)	
SHIELD	no	FORDING	0.6 m	standard	30 000 m
WEIGHT	13 750 kg	GRADIENT	45%	BB	39 000 m
LENGTH		AXIS OF BORE (firing)	1.77 m	MIN RANGE (+75°)	3000 m
travelling	9.5 m	TYRES		CREW	5
firing	11 m	main wheels	14.00 × 20	TOWING VEHICLE	SAMIL 100 (6 × 6) gun tractor
on tow (barrel to rear)	12.1 m	trail wheels	7.50 × 16	SPEED IN APU MODE	16 km/h
WIDTH		ELEVATION/DEPRESSION	+75°/-3°		
travelling	3.3 m				
firing	8.7 m				

SPAIN

SITECSA 203 mm Gun

Development/Description

Late in 1990, SITECSA completed the first prototype of a 203 mm (8 inch)/45 calibre gun that it has developed as a private venture in association with the China North Industries (NORINCO).

This weapon is based on a design originally prepared in conjunction with Dr Gerry Bull who developed the GC 45/45 calibre gun in the 1970s.

During its initial development trials, which took place in November 1990, the weapon fired 15 projectiles.

The 203 mm weapon uses a modified SITECSA 155 mm 155/45 ST 012 gun-howitzer carriage with a double baffle muzzle brake and a flick rammer. It is believed that this weapon has a maximum range of between 40 and 50 km firing a projectile weighing between 88 and 100 kg.

China North Industries (NORINCO) were involved in the design and development of the muzzle brake, ordnance and breech. The latter is said to have an unusual bayonet breech configuration that features a Crossley pad-type breech seal secured by means of a screw thread machined into the outer surface of the chamber surround.

Status: Prototype.

Manufacturer: SITECSA Desarrollos de Sistema y Tecnologias SA
Claudio Coello, 124
28006 Madrid
Spain
Telephone: (34) 1 563 48 85/563 58 47
Telex: 43743 SETIC-E
Fax: (34) 1 261 10 81

SITECSA 203 mm/46 calibre gun deployed in firing position

SITECSA 155 mm 155/45 ST 012 Gun-Howitzer

Development

In 1985 SITECSA began a technological collaboration programme with the Canadian Space Research Corporation (SRC) to develop a 155 mm gun-howitzer with a 45-calibre barrel. The resultant gun, sometimes known as the ST-155/45 or ST-012, may be regarded as a modernised GC 45 with a reduced number of components, a stronger barrel and deeper rifling. Some of the development work has been linked with the Chinese work that produced the NORINCO WA 021 gun-howitzer (qv) and some of the parts for the Spanish prototype were produced by NORINCO. The Spanish Army has a requirement to replace its old US supplied M114 towed howitzers and in 1988 it evaluated a single prototype of the SITECSA 155 mm 155/45 ST 012B gun howitzer and a single prototype of the Spanish Santa Barbara 155 mm SB 155/39 towed weapon covered later in this section. Although another alternative is to upgrade the M114 to the M114/39 or M114/45 configuration.

As of mid-1991, the SITECSA 155 mm 155/45 ST 012/012B gun/howitzer had not entered production.

Description

The 155/45 ST 012 is produced in two versions, the ST 012A being a straightforward towed version while the ST 012B has a 130 hp auxiliary power unit (APU). While the pre-series production examples have a 45-calibre long barrel, the design and development of a version with a 50-calibre barrel has commenced. This version will have an identical performance to that of the 45-calibre version but the barrel maximum pressures and accelerations are decreased to allow a longer barrel life. A pneumatic rammer is provided.

In appearance and design terms the 155/45 ST 012 follows the same general lines of the earlier GC 45. The barrel, which is fitted with a multi-baffle muzzle brake, can be fitted with either a screw breech or a sliding horizontal breech mechanism. For towing, the barrel is slung to the rear and clamped over the split trail carriage. When fitted, the APU provides power for self-propulsion, for opening and closing the trails and raising and lowering the firing platform located under the carriage front axles. Jockey wheels are provided to assist the handling of the trail legs and when manoeuvring the gun in and out of position. Time to emplace the towed version is four minutes.

The 155/45 ST 012 can fire all standard NATO 155 mm projectiles and also ERFB and ERFB BB projectiles. The maximum range with standard M107 projectiles fired at a muzzle velocity of 675 m/s is 17 800 m while firing ERFB at a muzzle velocity of 897 m/s it is 30 000 m. Firing ERFB BB the maximum range is 39 600 m.

ERFB projectiles include an ERFB HE weighing 45.45 kg and carrying an explosive charge weighing 8.8 kg, an ERFB BB weighing 47.6 kg with an explosive charge weighing 8.8 kg, an ERFB WP weighing 47.04 kg, an ERFB smoke weighing 46.54 kg an ERFB illuminating weighing 46.37 kg.

The rate of fire is five rounds in the first minute and two rounds per minute for continuous fire.

SITECSA 155/45 ST 012B gun-howitzer deployed in firing position

SITECSA 155/45 ST 012B gun-howitzer fitted with an APU

Polyvalent Carriage

More recently SITECSA has developed the Polyvalent carriage which can be fitted with:
(1) Barrels of 39 calibres with 18 litre chamber capacity
(2) Barrels of 45 or 52 calibres with 23 litre chamber capacity
(3) Barrels of 50 calibres with 32 litre chamber capacity
(4) Barrels of 52 calibres with 23 litre chamber capacity
(5) Automatically opening screw breech mechanism
(6) Ability to fire standard NATO ammunition as well as ERFB and ERFB BB types.

The recoil system has been designed to absorb the recoil and moderate the counter recoil of all of the above barrels simply by changing the pressure of the hydraulic accumulators. A typical barrel change can be carried out by three men in the field is less than 60 minutes. The change of pressure in the hydraulic accumulators in less than five minutes.

Brief specifications of these barrels are:

Type	**39 calibre**	**45 calibre***	**50 calibre**	**52 calibre**
TOTAL WEIGHT OF BARREL	1425 kg	1650 kg	1995 kg	3020 kg
VOLUME OF CHAMBER	18.80 l	23 l	32 l	23 l
SHOT TRAVEL OF PROJECTILE IN BORE	503.70 cm	589.00 cm	666.50 cm	n/a
LENGTH OF BARREL	604.50 cm	704.50 cm	782.10 cm	n/a

* can be extended to 52 calibres

RANGE WITH ERFB PROJECTILE:

Model	**155/39**	**155/45**	**155/52**
max velocity			
Zone No 8[1]	675 m/s	658 m/s	681.7 m/s
Zone No 9[1]	815.2 m/s	795.5 m/s	821.3 m/s
Zone No 10[2]	prohibited	897 m/s	930.7 m/s
max range			
Zone No 8[1]	20 202 m	19 600 m	20 439 m
Zone No 9[1]	25 598 m	24 800 m	25 845 m
Zone No 10[2]	not/app	30 000 m	31 759 m

[1] Charge type M2
[2] Charge type M11

Late in 1990, the third prototype of the system was completed with a Polyvalent carriage as described above. This system has a number of further improvements including:
(1) Different automatic elevation system
(2) New auxiliary power unit
(3) New firing jack
(4) New wheels to provide better cross-country mobility
(5) Deutz BF 6L913 diesel replaced by Lambardino DS 105 air-cooled diesel developing 110 hp
(6) New muzzle brake
(7) Overall weight of the gun reduced by 1500 kg.

Third prototype of the SITECSA 155/45 gun-howitzer, which is fitted with an APU, deployed in firing position

Status: Prototypes.

Manufacturer: SITECSA Desarrollos de Sistemas y Tecnologias SA
Claudio Coello, 124
28006 Madrid
Spain.
Telephone: (34) 1 563 48 85/563 58 47
Telex: 43743 SETIC-E
Fax: (34) 1 261 10 81

SPECIFICATIONS	
CALIBRE	155 mm
BARREL LENGTH	45 calibres
MUZZLE BRAKE	multi-baffle
BREECH MECHANISM	interrupted thread or sliding block
CARRIAGE	split trail with optional APU
SHIELD	no
WEIGHT	
towed version	10 000 kg
version with APU	12 000 kg
LENGTH	
firing	11.4 m
travelling	9.088 m
travelling, barrel to rear	12.512 m
WIDTH	
firing, over trails	9.931 m
travelling	2.5 m
HEIGHT	
travelling	2.23 m
travelling, barrel to rear	3.049 m
WHEELBASE	2.3 m
GROUND CLEARANCE	0.42 m
FORDING (with APU)	0.7 m
GRADIENT (with APU)	40%
TYRES	9 × 20
ELEVATION/DEPRESSION	+72°/-5°
TRAVERSE (left/right)	40°/30°
RATE OF FIRE	
burst	5 rounds in 1 min
sustained	2 rds/min
MAX RANGE	
M107	17 800 m
ERFB	30 000 m
ERFB BB	39 600 m
TOWING VEHICLE	6 × 6 10-ton truck
MAX SPEED (APU)	40 km/h
MAX CROSS COUNTRY SPEED (APU)	15 km/h
RADIUS OF OPERATION (APU)	100 km
APU POWER	130 hp

SANTA BARBARA 155 mm SB 155/39 Towed Howitzer

Development

The SB 155/39 towed howitzer has been developed by the Fabrica de Artilleria de Sevilla, part of SANTA BARBARA, to meet the requirements of the Spanish Army for a weapon to replace its American supplied 155 mm M114 towed howitzers.

In addition to the basic SB 155/39 towed howitzer, one other version is under development, one with an auxiliary power unit (REMA) (first prototype completed in 1987).

The prototype began firing trials in March 1985. The SB 155/39 will be employed at brigade and division level for defensive and offensive action. By early 1991 four prototypes of this system had been completed but volume production had yet to start.

The Spanish Army has evaluated the SITECSA 155/45 ST 012B and the latest model of the SANTA BARBARA SB 155/39 REMA as far back as 1988, but as of mid-1991, no production orders had been placed for either weapon.

Description

The weapon has a conventional split trail and to reduce its overall length for travelling or air transport the barrel can be swung back through 180° and locked in position over the closed trails. The trails are of a welded box section with the spade folding up under either end. To assist in bringing the SB 155/39 into and out of action, a rubber-tyred wheel is mounted on the inside of each trail leg, this is swung up through 90° when the weapon is in the firing or travelling position.

When in the firing position the wheels are raised clear of the ground and the carriage is supported by a circular firing jack. Each main road wheel has a hydro-pneumatic suspension system which is also used to lift the wheels clear of the ground when bringing the system into action.

The ordnance has a double-baffle muzzle brake, hydro-pneumatic recoil system and a vertical sliding breech mechanism. Elevation is from -5° to +70° with a traverse of 30° left and right. Elevation and traverse are manual with controls on the left side of the carriage. The goniometer has a magnification of ×4 and can also be used as a telescope to engage armoured vehicles at close range.

Prototype of SB 155/39 REMA in self-propelled configuration

Prototype of SB 155/39 REMA deployed in firing position and showing ordnance travelling lock on right trail leg

Interior ballistics of the ordnance conform with those of the 155 mm FH-70 and M198 systems already in service in NATO. A power-assisted rammer is fitted as standard. Using charge 8, an HE projectile weighing 43.5 kg can be fired to a maximum range of 24 000 m. With a rocket assisted projectile range is increased to 30 000 m. SANTA BARBARA produce an HE projectile known as the SB-811 which carries a 10.5 kg warhead and has a range of 24 000 m.

With a well trained crew a burst rate of fire of 4 rds/min can be achieved.

The SB 155/39 is towed by a 6 × 6 truck which carries the crew of six and ammunition. Maximum towing speed on roads is 80 km/h, 40 km/h cross-country.

Variants

SB 155/39 Remolccado con motor auxiliar (REMA)

This is essentially the standard towed howitzer with a 70 hp four-stroke diesel engine mounted on the forward part of the carriage which, via hydraulic motors, propels the road wheels. In addition to propelling the weapon at a speed of 12 km/h, the engine also enables it to negotiate gradients of up to 30 per cent, to change position quickly after a few rounds have been fired to avoid counter battery fire, and, in conjunction with the 6 × 6 truck, negotiate very rough terrain as it then has an 8 × 8 configuration.

Canon de Costa 155/39

Details of this version are given in the *Coastal artillery* section.

Rear view of 155 mm SB 155/39 towed howitzer prototype with trails open showing two small wheels on inside of trails

SPECIFICATIONS (SB 155/39 towed howitzer)

CALIBRE	155 mm	LENGTH		MAX RANGE	
LENGTH OF BARREL	39 calibres (6045 mm)	travelling	9.6 m	standard projectile	24 000 m
RIFLING		firing	12.2 m	enhanced range projectile	30 000 m
number of grooves	48	HEIGHT		RATE OF FIRE	2 rds/min for 1 hour
pitch	20 calibres (3100 mm)	travelling	2.4 m		3 rds/min for 30 minutes
RECUPERATOR	oil/pneumatic	firing	1.525 m		
BRAKE	rod and contra-rod type, hydraulic with modifications	WIDTH	2.5 m		1 rds/min sustained
		GROUND CLEARANCE	0.3 m		4 rds/min (burst)
		ELEVATION/DEPRESSION	+70°/-3°	CREW	6
RECOIL (max)	1400 mm	TRAVERSE	30° left and right		
WEIGHT (travelling)	9500 kg				

Status: Prototypes (4). Not yet in production or service. This has been evaluated by the Spanish Army along with the SITECSA 155 mm 155/45 ST 012B gun-howitzer (qv previous entry) as a possible replacement for the 96 155 mm M114s used by the Spanish Army. Although the M114s could be upgraded to the M114/39 to M114/45 configuration such as that offered by SITECSA in the following entry.

Manufacturer: Fabrica de Artilleria de Sevilla.
Enquiries to: SANTA BARBARA SA
Julian Caucinillo, 32
28037 Madrid
Spain
Telephone: 5050100
Fax: 5850268

SITECSA M114 155/45 and M114 155/39 Conversions

Development/Description

Using experience gained in the development of their 155/45 ST 012 gun-howitzer (qv) SITECSA are offering an upgrade package for the American M114 howitzer. The package has four objectives: to improve weapon performance, qualifying it to fire new projectiles and charges; improving manoeuvrability; reducing the number of men required for the crew; and making the M114 easier to use under combat conditions.

SITECSA is offering packages with either a 39-calibre or 45-calibre barrel. In both cases the modifications include a new ordnance, similar to that used on the 155/45 ST 012, which is fitted with a multi-baffle muzzle brake and an automatic screw breech mechanism. Breech obturation is obtained by using a mushroom plastic pad and bearing plate. The autofrettaged barrel has 48 rifling grooves with rifling of one in 29 calibres. New pneumatic equilibrators are fitted which do not need constant adjustment due to changes in temperature. A pneumatic rammer allows for ramming at any angle. The rammer is mounted on the left trail leg and consists of a rammer, cylinder, loading tray and stand pipe, Two components of the recoil system piston rod and extended piston rod are modified to cater for the new recoil length which varies from 1.014 to 1.524 m. The cradle has been reinforced and modifications have been made to the equilibrator side plate support. A front support for the elevation system gear ring is also fitted.

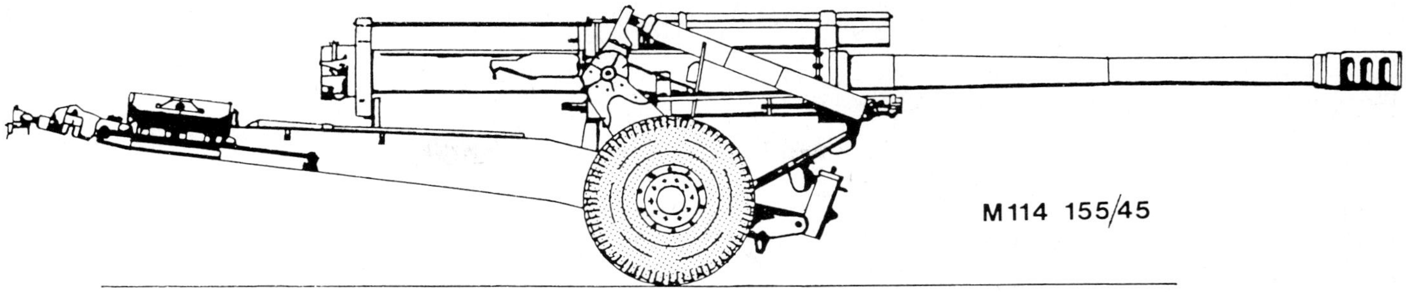

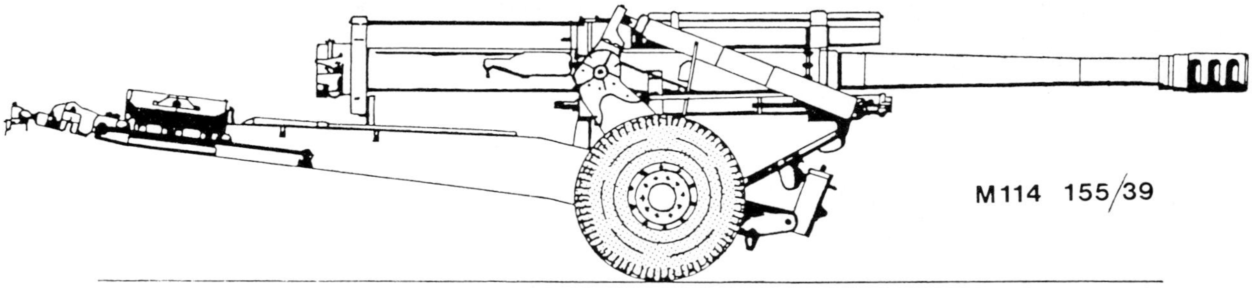

SITECSA modernised M114 howitzers with M114 155/45 (top) M114 155/39 (bottom)

The existing jack and float are hydraulically operated and the trails have been balanced, both features which improve handling and reduce the time needed to bring the weapon into action. To reduce the overall length of the weapon when travelling the ordnance can be withdrawn to the rear using a new mechanism.

According to SITECSA a modernised M114 with a 45-calibre barrel can reach a range of 30 000 m with ERFB rounds and 39 600 m using ERFB BB. With a 39-calibre the ranges are 24 000 and 30 000 m.

It is claimed that seven men are required to operate a modernised M114 compared to the 11 required for a standard M114.

It has been suggested that similar ordnance modifications could be made to existing M108 and M109 self-propelled howitzers.

Status: Available

Manufacturer: SITECSA Desarrollos de Sistema y Tecnologias SA
Claudio Coello, 124
28006 Madrid
Spain.
Telephone: (34) 1 563 48 85/563 58 47
Telex: 43743 SETIC-E
Fax: (34) 1 261 10 81

SPECIFICATIONS (45-calibre barrel)

CALIBRE	155 mm
WEIGHT	7350 kg
LENGTH TRAVELLING	9.761 m
short distance	11.825 m
WIDTH	2.438 m
OUTER WIDTH OVER TRAILS	4.7 m
HEIGHT	1.99 m
HEIGHT OF TRUNNIONS IN FIRING POSITION	1.49 m
GROUND CLEARANCE	0.279 m
ELEVATION/DEPRESSION	+60°/-2°
TRAVERSE (left/right)	24°/25°
RATE OF FIRE	
normal	2 rds/min
max	5 rds/min
MAX RANGE (ERFB BB)	39 600 m

105 mm m/26 Field Howitzer

Development/Description

It is believed that between 800 and 850 of these weapons were built by the Sociedad Espanola de Constraccion Naval, Reinosa factory under the designation of R-50 (Reinosa 1950). Later models were the models 1950-53, Reinosa 58 and NR-61.

The m/26 has been designed to fire both American and Spanish ammunition. It will fire an American HE projectile weighing 14.57 kg, Spanish HE projectile weighing 15.27 kg or a HEAT projectile which will penetrate 85 mm of armour at an incidence of 0° at a range of 1000 m.

In recent years these guns have undergone a product improvement programme carried out by the Sociedad Anonima de Placencia de las Armas of Madrid.

105 mm m/26 field howitzer being towed by M34 (6 × 6) 2½-ton truck

SPECIFICATIONS (105 mm m/26)

CALIBRE	105 mm
BARREL LENGTH	3.349 m
MUZZLE BRAKE	double-baffle
RECOIL	hydro-pneumatic
BREECH MECHANISM	horizontal sliding
CARRIAGE	split trail
WEIGHT	1950 kg
LENGTH (travelling)	6.08 m
WIDTH (travelling)	2.1 m
HEIGHT (travelling)	2.2 m
TRACK	1.732 m
ELEVATION/DEPRESSION	+45°/-5°
TRAVERSE (total)	50°
RATE OF FIRE	6 rds/min
MAX RANGE	11 450 m
CREW	6
TOWING VEHICLE	M34 (6 × 6) 2½-ton truck

Status: Production complete. In service with the Spanish Army which still has between 120 and 200 L/46 guns that are derivatives of the Soviet M1931 or M1937 (A-19) gun. Germany provided Spain with the technical designs of these weapons during the Second World War and production was then undertaken at the Fábrica Nacional de Trubia under the German designations of the 390/lr Model 1931 and 390/2r Model 1931/37. It is believed that the remaining guns have been refurbished with modified ordnance, wheels and the addition of a firing baseplate.

SWEDEN

Bofors 155 mm Field Howitzer 77A (FH-77A)

Development

Studies between 1965 and 1968 to establish the future requirements of Swedish Army artillery weapons led to the recommendation that a 155 mm howitzer with an auxiliary propulsion unit (APU) should be developed. It was required to have a long range, high rate of fire, use an improved type of ammunition and have good mobility, especially across rough country and snow.

AB Bofors was subsequently awarded a development contract by the Army Materiel Department of the Defence Materiel Administration to develop a new weapon which became the FH-77A. The first prototype was built in 1973 with the main production run being carried out from 1978 to 1984 for the Swedish Army. Further development resulted in the FH-77B covered in the following entry.

Description

The FH-77A is normally towed by a Saab-Scania SBAT 111S (6 × 6) truck at a maximum road speed of 70 km/h. When travelling across rough country the main wheels of the howitzer can be engaged thus making an 8 × 8 combination. Maximum speed in this configuration is 8 km/h and once this speed has been exceeded the wheels of the FH-77A are disengaged.

Mounted on the forward part of the carriage is the Volvo B20 engine which is connected to two variable hydraulic pumps. These are connected to hydraulic motors in the driving wheels and when being used in the self-propelled mode steering is by means of levers at the driver's position on the left side of the carriage. Both driving and steering are accomplished by the large road wheels. The two support wheels, one on each trail, pivot freely and are also hydraulically operated and assist in limbering and unlimbering the weapon.

The elevating mass, which is supported in the cradle carrier, consists of a cradle which carries the barrel with the breech mechanism, recoil mechanism, and loading tray and rammer for the loading system. The semi-automatic breech mechanism has a vertical sliding breech-block which opens downwards. The firing device is electro-mechanical. Elevation and traverse are both hydraulic at a maximum speed of 3°/s.

On the right side of the FH-77A is the loading table on which the projectiles are placed with the aid of an ammunition crane. The cartridge cases are placed manually on the loading tray. The loading device is operated by the loader who is seated on the right side of the howitzer.

Swedish Army weapons have the BEAB BAAB RIA sight which is operated by the driver/layer on the left side of the carriage.

The crew of six of the FH-77A consists of the commander, gunner who lays the gun and also acts as the driver, ammunition handler who places the cartridge case onto the loading table, loader who feeds the projectile from the loading table and operates the rammer, crane handler who lifts three projectiles at a time onto the loading table and the ammunition handler who screws the fuzes into the projectile.

On arriving at the battery position the howitzer can be unlimbered by raising the caster wheels on the carriage trails with the aid of the hydraulics. The lunette is thereby raised and unhooked from the pintle. Once the howitzer has been unhooked the carriage trails are spread by releasing the locking device between the carriage trails. The howitzer is then swung first to one side and then the other, the carriage trails locking in their outer positions. The supporting wheels are turned up, and the howitzer, oriented in the firing direction ordered, is driven in reverse to dig in the trail spades. If the howitzer needs to be re-oriented to a new firing direction beyond its traverse limits it is driven forwards to free the trail spades and the supporting wheels are turned down. Once in the new firing position the supporting wheels are turned up again, and the howitzer is again reversed to dig in the trail spades.

The FH-77A is also provided with hand pumps, one for laying and ramming and one for the caster wheels, which can be used to obtain the pressure required in the hydraulic systems if the auxiliary engine should be out of order. These hand pumps can also be used for quiet operation of the howitzer at night.

The unit loads of ammunition (both projectiles and charges) facilitate handling and can be stacked on each other. Each contains 12 projectiles or 12 cartridge cases and is kept unopened all the way to the ammunition supply at the gun. With the aid of the ammunition crane three shells at a time can be taken directly from the unit load and transferred to the loading table of the gun. The cartridge cases are kept in special plastic tubes, with covers on one end. The tubes, which are inserted horizontally into the unit loads, can be opened and the cartridge case removed without having to break the unit load. If the situation should require manual moving of the cartridge cases, the unit load is divided into separate tubes, in which the cartridge cases are carried.

The firing sequence is as follows. A cartridge case is placed on the loading tray and then the loader feeds a projectile from the loading table and actuates the ramming device when the projectile has slipped down into the cartridge case on the loading tray. The layer then fires the weapon. The cartridge case is ejected on recoil after which the loading cycle is repeated.

The HE projectile is designated the M/77 and weighs 42.4 kg complete with Bofors ZELAR fuze and contains 8 kg of TNT. The empty plastic cartridge case has a steel base and weighs 5.9 kg, cartridge case plus increments 1 to 4 weigh 11.9 kg complete and cartridge case plus increments 5 to 6 weigh 16.7 kg. The maximum muzzle velocity with the

Bofors 155 mm Field Howitzer 77A from rear showing hydraulic loading crane to right of breech

Bofors 155 mm Field Howitzer 77A in travelling position under own power

Bofors 155 mm Field Howitzer 77A in firing position

various charges is as follows: charge 1, 310 m/s; charge 2, 371 m/s, charge 3, 472 m/s, charge 4, 583 m/s, charge 5, 692 m/s and charge 6, 774 m/s. A unit load of 12 projectiles weighs 620 kg, one of 12 cartridges with increments 1 to 4, 250 kg and one with 12 cartridges with increments 5 to 6 weighs 310 kg.

Variants

The 120 mm KARIN mobile coast defence gun uses the same carriage as the FH-77A. For details of KARIN see the separate entry in the *Coastal artillery* section.

Status: Production complete. In service with the Swedish Army.

Manufacturer: Swedish Ordnance (previously AB Bofors)
S-691 80 Bofors
Sweden
Telephone: (46) 586 810 00
Telex: 73210 bofors s
Fax: (46) 586 581 45

SPECIFICATIONS

CALIBRE	155 mm
BARREL LENGTH	5.89 m
MUZZLE BRAKE	pepperpot
BREECH MECHANISM	vertical sliding
CARRIAGE	split trail with APU
SHIELD	no
WEIGHT	
travelling order	11 500 kg
on driving wheels	9500 kg
on rear support wheels	2000 kg
LENGTH	
travelling	11.6 m
firing	11.2 m
WIDTH	
travelling	2.64 m
firing	7.18 m
HEIGHT	
travelling	2.75 m
firing at 0° elevation	2.75 m
firing at 50° elevation	6.75 m
AXIS OF BORE	1.6 m
GROUND CLEARANCE	0.445 m
TRACK	
driving wheels	2.2 m
support wheels	2.23 m
ELEVATION/DEPRESSION	+50°/−3°
TOTAL TRAVERSE	
elevation of less than +5°	50°
elevation of more than +5°	60°
RATE OF FIRE	3 rounds/6-8 s 6 rounds/20-25 s 6 rounds every other minute for 20 minutes
MAX RANGE (charge 6)	22 000 m
CREW	6
TOWING VEHICLE	Saab-Scania SBAT 111S (6 × 6) truck

Bofors 155 mm Field Howitzer 77B (FH-77B)

Development

The Bofors 155 mm FH-77B (now frequently referred to just as the FH-77) was originally developed as an export model of the FH-77A (see previous entry). The main change was the ability to fire NATO standard ammunition, other improvements being made to its general handling, cross-country mobility and the introduction of a mechanised ammunition handling system. Following an FH-77B order from Nigeria, the largest export order in Swedish history came when India placed an initial order for 410 FH-77B weapons and equipment. AB Bofors acted not only as manufacturer but as the major supply contractor for items such as Saab-Scania trucks, Barracuda camouflage netting, Marconi Command and Control Systems Quickfire fire control systems, Fairey Australia muzzle velocity radars, BEAB sighting equipment and surveying equipment from Wild of Switzerland and navigation equipment from Ferranti UK. First production FH-77Bs for India left the Bofors factory in August 1986.

The first Indian order for the FH-77B is for a total of 410 weapons with final systems delivered in 1990. The company delivered FH-77Bs to India in complete batteries to enable Indian artillery units to become fully operational as complete units.

The ammunition being supplied by Bofors includes smoke (FFV), illuminating (Bofors) and new Bofors High Explosive Extended Range (HEER) projectiles for which India was the first customer. Two versions of the HEER are being supplied, one with boat tail and the other with base bleed. LIAB, a Bofors subsidiary, are providing 155 mm fuzes.

When the original order was placed by India, it was envisaged that the total Indian requirement would consist of some 1500 weapons.

The second batch of FH-77Bs would have been assembled in India from components supplied by Bofors with the third batch being made under licence in India. As of mid-1991, India had not placed any additional orders for the FH-77B beyond the original order for 410 systems.

In March 1990 Bofors announced that the Swedish Material Defence Administration had placed a $162 million contract with Bofors for the supply of FH-77B artillery systems, spares, software, ammunition and associated charges.

Bofors 155 mm FH-77B in firing position with ordnance at high elevation. Note ammunition crane with three projectiles on right side of carriage

Prototype installation of FIN 1150 land navigation system on a 155 mm FH-77B field howitzer

Indian Army 155 mm Bofors FH-77B showing new double baffle muzzle brake which was fitted to all final production systems and backfitted to earlier weapons

Bofors FH-77B deployed in the firing position

The Swedish Army FH-77Bs will be similar to those produced for India but with some modifications to enable them to be operated at lower temperatures.

The Nigerian FH-77Bs and the original Indian FH-77Bs had a parallel pepperpot muzzle brake, but following trials with the more recent ammunition types, including HEER used with charge 9, a new and more efficient muzzle brake of the double baffle type was developed. This was fitted to late production FH-77Bs and is being backfitted to earlier weapons.

Ammunition in the Swedish Army order includes the Bofors developed HEER (High Explosive Extended Range) projectile, developed to meet the requirements of the Indian Army which, when used with charge 9, has a range of 30 000 m. HEER-BB (High Explosive Extended Range - Base Bleed) is also included in the order and the FH-77B will also be able to use ammunition already stockpiled by the Swedish Army.

Description

The FH-77B differs from the earlier FH-77A in three main areas; a slightly longer barrel, the introduction of a breech screw mechanism using bag charges, and a new hydraulic loading system. The FH-77B can be towed at speeds of up to 70 km/h, normally by the Saab-Scania SBAT 111s (6 × 6). For cross-country travel the driver of the tractor vehicle can remotely engage the gun road wheels by starting the gun hydraulic motor making an 8 × 8 combination. Maximum speed in this configuration is 7 km/h and when this speed is exceeded the gun wheels are disengaged. Steering is by regulating the drive to the independent road wheels of the gun. The same hydraulic motor can also be used for several other functions. To get the gun into action the driver on the gun uses the hydraulic power to raise the gun caster wheels on the trail to disengage from the tractor drawbar. In the self-propelled mode the trails are split, again using the hydraulic power. The gun driver then reverses the gun to dig the trail spades into the ground. For making changes of fire direction the same process can be repeated. Using this system the driver on the gun can alone make rapid fire direction changes, get the gun in and out of action, and move the gun from one firing location to another.

The ammunition loading system uses a hydraulic crane driven from the hydraulic system. The crane grab is designed to lift any 155 mm projectiles in groups of three at a time, offered to the grab either horizontally or vertically, and almost any projectile length can be handled, including the longer enhanced-range projectiles. From the crane the projectiles are loaded onto a loading table and are then guided onto the loading tray, while another loader places the projectile charge onto a separate loading trough/rammer. In two operations the projectile loading tray is swung over to line up with the chamber and the charge tray swings over behind it. The shell is then hydraulically rammed forward into the chamber with the charge pushed right forward behind the projectile. Once the projectile is seated in the driving lands the loading trough/rammer is withdrawn until the charge is stripped from the tray by a retaining catch. The rammer mechanism is fully withdrawn, the breech screw is closed and a primer cartridge automatically loaded into the primer seat from a magazine holding eight tubes. Using this system, projectiles and charges can be loaded into the gun at angles of elevation of up to +70°. The rate of fire, with three men on the gun can be three rounds in 12 seconds. The subsequent rate of fire is governed only by the ammunition supply, and an added bonus of the hydraulic ramming is that uniform ramming decreases the resultant projectile dispersion.

The full crew of the FH-77B is six: the gun commander, a layer and four loaders. One loader operates the mechanism for loading the projectiles from the loading table to the loading tray, one prepares the projectiles by inserting the fuzes, setting them and loading the projectiles in groups of three into the loading crane, one places the charges onto the loading tray and another loader operates the ammunition crane. If necessary, only the layer, crane operator and projectile loader combined, and a charge loader can operate the gun in action. There is a special internal communication system built into each gun, separate from the usual fire control communication systems and with a 10 m range based on an induction loop with receivers built into the crew's ear protectors. The gun laying is fully hydraulic. Two levers, one on the right for elevation and depression and one on the left for traverse allow barrel movement. Several types of sight may be fitted but the usual model fitted is the electronic Philips BEAB BAAB RIA. The layer sits high on the left of the breech with a protective open cage over the seat and sight (the loading crane is on the right of the breech). A ×4 telescope sight is fitted for direct fire. The BAAB NK24 night sight fitted in front of the ordinary telescopic sight permits direct fire under night conditions. As the FH-77B depends on hydraulic power for many of its systems, an electric motor for all operations using an internal or external power supply or two hydraulic hand pumps are provided in case the main system fails. One, situated just forward of the layer's position and in the power compartment itself, can provide manual hydraulic power for traverse, elevation and ramming. The other pump is on the other side of the gun and provides manual hydraulic power for operating the support wheels.

Unlike the FH-77A the ammunition system of the FH-77B is based on a bagged charge system which enables a wide range of ammunition to be fired, including the following main types of projectile:
Bofors HE M/77 B which is a high capacity shell weighing 42.6 kg and containing 8 kg of TNT or Composition B. This has a maximum range of 24 000 m when fired from the FH-77B howitzer. The round is fully compatible with all Western-manufactured guns and internationally adopted fuzes.
HE M107, the standard NATO 155 mm howitzer projectile weighing 42.91 kg. It can be fired to a range of approximately 19 000 m.
ERFB Mark 10 BB produced by PRB and weighing 47.6 kg. This assisted shell can be fired to a range of up to 30 000 m.
Luchaire HE BB which can be fired to a range of 28 000 m.
Development of a base bleed round designated HEER (High Explosive Extended Range) has been completed by Bofors. HEER is compatible with all Western-manufactured 155 mm guns and international fuzes. Maximum range is 30 000 m and the weight is the same as conventional 155 mm rounds. This involves less stress on the barrel than firing other types of projectiles such as base bleed with a greater weight. Production is also made easier by omitting ERFB nubs.
Other rounds available are different types of smoke (including infra-red smoke) and the MIRA illuminating round in two versions, with maximum ranges of 17 000 m and 24 000 m respectively. For both the time of illumination is 60 seconds, descending velocity 4 m/s and the mean illuminating intensity is 2.2 Mcd.
A sensor-fuzed munition known as BONUS is in a pre-development state. BONUS series production will start in 1992.

Standard NATO and other propellant charges can be used with the FH-77B but AB Bofors is now producing its own charges especially for the FH-77B. Charges available include M3 which provides a muzzle velocity of 375 m/s to give a range of 10 300 m, and M4 which provides a maximum muzzle velocity of 563 m/s to give a range of 15 500 m.

Charges include the Bofors Charge 8 which is roughly equivalent to M119 but gives more favourable pressure and burning properties at extreme temperatures. Charge 8 provides a muzzle velocity of 685 m/s to give a range of 19 000 m. Charge 9 is a new Bofors development to provide a muzzle velocity of 827 m/s and a range with standard projectiles of 24 000 m and with base bleed projectiles out to 30 000 m.

Variant

CD 77

In order to provide a version of the 120 mm KARIN/CD 80 coast defence gun firing NATO standard ammunition AB Bofors has developed a version of the 155 mm FH-77B known as the CD 77. The CD 77 retains the same ordnance and carriage as the FH-77B and uses the same mechanical handling system but allied to the fire control system used with the 120 mm KARIN/CD 80 or a similar system. Thus the CD 77 will be deployed as an individual gun and will use remote fire control coupled to an automatic sight. Weights and dimensions are the same as those for the FH-77B.

Autonomous FH-77B

During late 1987 the first example was demonstrated of a FH-77B fitted with a combined land navigation gyro-stabilised system linked to the sight equipment and fire control to permit autonomous positioning of the gun. In place of the normal sight a land navigation system (FIN 1150) from the then Ferranti Defence Systems of Edinburgh, Scotland, was used. The system gives the gun position and the attitude of the barrel (ie traversing, elevation and trunnion angle). This alleviates the need for surveying and alignment normally associated with bringing a gun into action and permits 'shoot and scoot' tactics to be employed more readily.

During one demonstration a gun with a crew of four was deployed and four rounds were fired in 2 minutes 15 seconds. The gun was then redeployed approximately 50 m and three further rounds were fired. Total time taken was 5 minutes and 10 seconds.

Status: In production. In service with India (410), Nigeria (48) and Sweden.

Manufacturer: Swedish Ordnance (previously AB Bofors)
S-691 80 Bofors
Sweden
Telephone: (46) 586 810 00
Telex: 73210 bofors s
Fax: (46) 586 581 45

SPECIFICATIONS					
CALIBRE	155 mm	HEIGHT (travelling)	2.82 m	RATE OF FIRE	
LENGTH OF BARREL	6.045 m	AXIS OF BORE (firing)	1.7 m	max	3 rounds/12 s or 10 rds/min
MUZZLE BRAKE	pepperpot	TRACK			
BREECH MECHANISM	screw	driving wheels	2.2 m	MAX RANGE	
CARRIAGE	split trail with APU	castor wheels	2.2 m	standard	24 000 m
SHIELD	no	GROUND CLEARANCE	0.42 m	extended range	30 000 m
WEIGHT (total)	12 000 kg	FORDING DEPTH	0.75 m	CREW	6
LENGTH		MAX SPEED		ENGINE	petrol or diesel
travelling	11.6 m	towed	70 km/h		
firing	11.16 m	self-propelled	8 km/h	TOWING VEHICLE	Saab-Scania SBAT IIIS (6 × 6) truck or equivalent vehicle
WIDTH		ELEVATION/DEPRESSION	+70°/-3°		
travelling	2.65 m	TOTAL TRAVERSE	60°		
firing	7.18 m				

Bofors 150 mm m/39 Field Howitzer

Development/Description

This weapon is called the 15 cm Haub m/39 by the Swedish Army. To reduce the overall length for travelling the barrel can be withdrawn out of battery to the rear and secured between the trails. The first model, with solid tyres, was designated the m/39 and the later model with pneumatic tyres to permit higher towing speeds is designated the m/39b. Ammunition is of the separate loading type with the HE projectile weighing 41.5 kg and having a maximum muzzle velocity of 580 m/s.

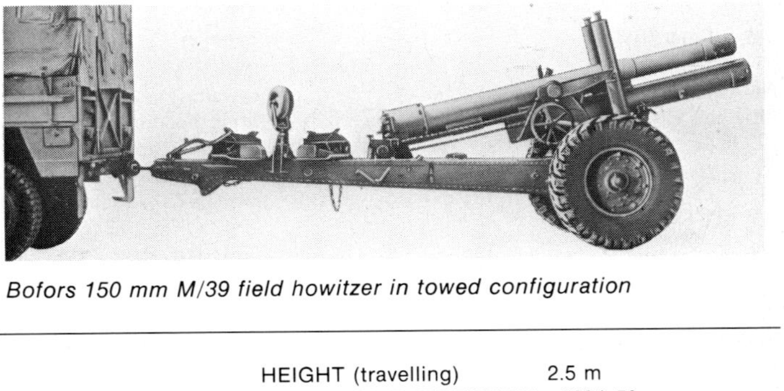

Bofors 150 mm M/39 field howitzer in towed configuration

SPECIFICATIONS					
CALIBRE	149.1 mm	LENGTH		HEIGHT (travelling)	2.5 m
BARREL LENGTH	3.6 m	travelling, barrel forwards	7.37 m	ELEVATION/DEPRESSION	+66°/-5°
MUZZLE BRAKE	pepperpot	travelling, barrel out of battery	6.55 m	TRAVERSE (total)	45°
CARRIAGE	split trail	WIDTH		RATE OF FIRE	4-6 rds/min
SHIELD	no	travelling	2.5 m	MAX RANGE	14 600 m
WEIGHT (travelling order)	5720 kg	firing	5.53 m	CREW	8
				TOWING VEHICLE	Volvo TL31 L3153 (6 × 6) truck

Status: Production complete. In service with the Swedish Army and Navy.

Manufacturer: Swedish Ordnance (previously AB Bofors)
S-691 80
Bofors
Sweden
Telephone: (46) 586 810 00
Telex: 73210 bofors s
Fax: (46) 586 581 45

Bofors 105 mm Type 4140 Field Howitzer

Development

The 105 mm Type 4140 Field Howitzer was developed in the post-Second World War period by Bofors and is called by it the 10.5 cm Field Howitzer L/32 4140 4-trailed. In concept it is similar to the 10.5 cm Leichte Feldhaubitze 43 developed between 1943 and 1945 by the Skoda company of Czechoslovakia to meet the requirements of the German Army. This weapon was not placed in production.

Description

The monobloc barrel is fitted with a single-baffle muzzle brake. The vertical sliding breech mechanism opens downwards during the last part of the recoil and ejects the empty cartridge case. The barrel is supported on two claws in a box-shaped cradle with claw guards. The buffer is hydraulic and the recuperator hydro-pneumatic. The cylinders for the buffer and the recuperator are built into the box of the cradle and can be cooled by filling the cradle with cooling liquid. The length of recoil is dependent on elevation and recoil length is set automatically.

The loading of the gun is facilitated by a loading tray mounted on parallelogram levers on the cradle. The tray is automatically lowered to the loading position as soon as the case from the previous round has been ejected and the gun returned to the run-out position. The gun fires fixed rounds to enable a high rate of fire to be achieved. The cradle carrier consists of a pivot with welded box-shaped mounting sides. The pivot is mounted on ball-bearings in the lower mounting.

The gun is hand laid with the elevating handwheel on the right side of the gun. It has two handwheels for traversing, one on the left side of the gun and the other, which is used when the gun is laid by one operator, is on the right side to the immediate rear of the elevating handwheel. The traversing movement is transmitted via a number of gears to a worm gear, the pinion of which is secured to the lower mounting by a coupling. The traversing mechanism can be disconnected by means of this coupling and the gun can then be rapidly traversed. The elevating scale is on the right side of the gun mounting with the adjustable traversing scale on the lower mounting.

The balancing device is hydro-pneumatic, of the horizontal type, with one cylinder on either side of the barrel. The fluid in the balancing cylinder is the same type as the buffer fluid. The shield, which is 4 mm thick, consists of a movable central part fixed to the cradle and two fixed parts, one on each side of the cradle carrier. The lower mounting is shaped as a box with a pivot bearing and has a fastening device for the supporting plate and the trails.

The supporting plate is collapsible and its shaft is shaped with double concentric tubes. On the outer tube are lugs which serve as supports for the front trails during the conversion from travelling to firing position. The trails are of a welded design and have fixed ground plates which are secured with pickets. All trails can be turned sideways and the front trails can also be turned upwards. In the firing position the front trails are vertically locked to the lower mounting and hydraulic buffer devices on the lower mounting automatically keep the front plates against the ground. When travelling the gun is clamped by the front trails being folded up and locked against the cradle.

The wheels, which have pneumatic tyres with inner tubes, are mounted on pendulum arms, which are locked to the cradle carrier by half-bolts in both the travelling and firing positions. The brakes on the carriage are operated by compressed air from the towing vehicle, but can be lever operated to serve as a parking brake.

The Type 4140 fires an HE projectile with the complete round weighing 21 kg, the projectile itself weighing 15.3 kg and the bursting charge weighing 2.3 kg; maximum muzzle velocity is 640 m/s.

Bofors 105 mm Type 4140 field howitzer in travelling configuration

Bofors 105 mm Type 4140 field howitzer in firing position with trails staked to ground

SPECIFICATIONS					
CALIBRE	105 mm	SHIELD	yes	TRACK	1.5 m
BARREL LENGTH	3.36 m	WEIGHT		ELEVATION/DEPRESSION	+60°/-5°
MUZZLE BRAKE	single-baffle	travelling order	3000 kg	TRAVERSE	360°
RECOIL	hydraulic buffer and hydro-pneumatic recuperator	firing position	2800 kg	RATE OF FIRE	25 rds/min
BREECH MECHANISM	vertical sliding	LENGTH (travelling)	6.8 m	MAX RANGE	15 600 m
CARRIAGE	four split trail	WIDTH (firing)	4.38 m	CREW	4
		HEIGHT (travelling)	1.81 m	TOWING VEHICLE	Volvo TL31 L3153 (6 × 6) truck
		AXIS OF BORE (firing)	1.2 m		
		GROUND CLEARANCE	0.35 m		

Status: Production complete. In service only with the Swedish Army.

Manufacturer: Swedish Ordnance (previously AB Bofors)
S-691 80 Bofors
Sweden
Telephone: (46) 586 810 00
Telex: 73210 bofors s
Fax: (46) 586 581 45

SWITZERLAND

105 mm Field Howitzer Model 46

Development/Description
This is called the 10.5 cm Haubitze 1946 L22 (10.5 cm Hb 46) by the Swiss Army and was built at the Swiss Federal Armament Works at Thun to a Swedish design. It was in production between 1943 and 1953.
It fires a separate-loading HE projectile weighing 15.1 kg with a maximum muzzle velocity of 490 m/s.

105 mm Field Howitzer Model 46 (Swiss Army)

SPECIFICATIONS

CALIBRE	105 mm
BARREL LENGTH	2.31 m
MUZZLE BRAKE	multi-baffle
CARRIAGE	split trail
SHIELD	yes
WEIGHT (firing position)	1840 kg
ELEVATION/DEPRESSION	+67°/-5°
TRAVERSE (total)	72° to 56° (depending on elevation)
RATE OF FIRE	6-10 rds/min
MAX RANGE	10 000 m
CREW	7

Status: Production complete. In service with the Swiss Army.

Manufacturer: Swiss Federal Armament Works
CH-3602 Thun
Allmendstrasse 86
Switzerland
Telephone: 033 28 11 11
Telex: 921 256
Fax: 033 28 20 47

TAIWAN

155 mm Extended Range Gun

Development/Description
In the late 1970s the Combined Service Forces developed manufacturing capability for a version of the Soltam 155 mm M-68/M-71 family of towed howitzers which are fully described in this section under Israel. It is believed that two prototypes were built in Taiwan, as from the late 1970s all effort was concentrated on the towed Extended Range Gun.
The 155 mm towed Extended Range Gun uses the same ordnance and recoil system as the XT-69 self-propelled weapon which is based on the chassis of the M108/M109 which is fully described in the *Self-propelled guns and howitzers* section under Taiwan.
It is mounted on a two-wheeled split trail carriage similar to that of the American 155 mm M114 howitzer and incorporates some of the features found on the SRC International modernised M114/39 howitzer (*Jane's Armour and Artillery 1982-83* page 550).
A handwheel is fitted at the elevation controls and the pneumatic equilibrators are of the high capacity type. Mounted over the left trail are a pneumatic rammer, pressurised air bottle, swing-aside loading tray and a telescopic rammer mounted near the rear cradle. This enables a high rate of fire to be achieved and the weapon to be loaded at all angles of elevation.
The 45-calibre ordnance is fitted with a multi-baffle muzzle brake and a fume extractor. The weapon can fire all standard US/NATO ammunition as well as ERFB (with a maximum range of 30 000 m) which is now being manufactured in Taiwan. It is believed that 155 mm base bleed ammunition is under development which will extend the range to about 39 000 m.

Status: Believed to be in production for Taiwanese Army.

Manufacturer: Combined Service Forces
Taiwan

T65 155 mm Howitzer

Development/Description
This is essentially the US 155 mm M114A1 manufactured in Taiwan for both the home and export markets. The manufacturer gives the following specifications.

TOTAL WEIGHT	5000 kg
LENGTH	7.3 m
ELEVATION/DEPRESSION	+65°/0°
TRAVERSE	
right	25.2°
left	23.5°
MAX RANGE	15 000 m
MAX MUZZLE VELOCITY	564 m/s
RATE OF FIRE	1-3 rds/min

Status: Production complete. In service with Taiwan.

Manufacturer: Hsing Hua Company Limited
P O Box 8746 Taipei
Taiwan

T65 155 mm howitzer with trails open but without spades fitted

T64 105 mm Howitzer

Development/Description
This is essentially the US 105 mm M101A1 manufactured in Taiwan for both the home and export markets. The manufacturer gives the following specifications.

TOTAL WEIGHT	2030 kg
LENGTH	6 m
WIDTH	2.15 m
HEIGHT	1.52 m
ELEVATION/DEPRESSION	+65°/-5°
TRAVERSE	23° left and right
MAX RANGE	11 600 m
MAX MUZZLE VELOCITY	472 m/s
RATE OF FIRE	2-4 rds/min

Status: Production complete. In service with Taiwan.

Manufacturer: Hsing Hua Company Limited
PO Box 8746 Taipei
Taiwan

T64 105 mm howitzer with trails open

UNION OF SOVIET SOCIALIST REPUBLICS

240 mm Mortar M-240

Development
The 240 mm mortar M-240 was first shown publicly during the military parade held in Moscow in November 1953. At that time it was known in the West as the M1953 or M53. It is still in service with Soviet Army heavy artillery brigades held at Front level. Each of these brigades has a brigade headquarters, a target acquisition battery, two heavy mortar battalions equipped with the M-240, and two heavy artillery battalions, some of which still use the 203 mm howitzer M1931 (B-4M). With these battalions the M-240 is used for the demolition of buildings and fortifications that cannot be dealt with by normal artillery units, especially in built-up areas.

Under the Conventional Forces Europe Treaty the Soviet Union stated that it had none of these systems deployed West of the Urals.

Description
The 240 mm M-240 mortar consists of a barrel with breech and breech block frame, a frame with shock absorbers, the mount with training and equilibrating mechanisms, a two-wheel travelling carriage with the suspension, a boom for changing from the firing to the travelling configuration, a baseplate and a towing bar with a lunette. The sights are carried separately and are mounted on the mortar only when firing. The shock absorbers are used to protect the sights from firing vibrations and also provide the link between the ordnance and mount. It is also used when the mounting returns to the loading position after firing. The boom provides stability when firing and also has two winches to convert the mortar from the firing to the travelling configuration.

The M-240 has a minimum range of 800 m and a maximum range of 9700 m with traverse limited to 18°. Elevation limits are from +45° to +65°. The sights, elevation and traverse gears are on the left-hand side of the barrel. The mortar is normally towed, at a maximum speed of 40 km/h, muzzle first by an AT-P, AT-L or AT-S tractor which also carries the crew of 11. Additional vehicles carry the ammunition and emplacing equipment. On arrival at the firing position, which has to be on firm ground, the mortar is uncoupled from the tractor and the towing lunette is removed. The large circular welded baseplate is lowered to the ground and packed with earth to provide a stable firing platform. The smoothbore barrel is 5.34 m long and for loading is swung into the horizontal position. At the lower end of the barrel is the breech block and projectile guide. Bringing the M-240 into action takes at least 25 minutes, slightly less to move it after firing. Weight in action is approximately 4150 kg.

The HE bomb weighs 130 kg of which 34 kg is payload. The bomb, which is almost 1.5 m long, is brought to the mortar on a two-wheeled trolley and a team of five is used for loading. Large tongs/gripping pincers are used to lift the bomb from the trolley onto the projectile guide with two men on each handle and the fifth steadying the fins. The bomb is then pushed into the barrel and the breech is closed. The barrel is then raised to the firing position. The rate of fire is about one round a minute.

The 240 mm M-240 towed mortar has been replaced in most front line Soviet Army units by the self-propelled 240 mm mortar system the M-1975 (US Army designation) (SM-240) 2S4 (Soviet industrial number) which is fully described in the *Self-propelled guns and howitzers* section.

SPECIFICATIONS

CALIBRE	240 mm
LENGTH OF BARREL OVERALL	5340 mm
LENGTH (travelling)	6.51 m
WIDTH (travelling)	2.49 m
HEIGHT (travelling)	2.21 m
ELEVATION	+45° to +65°
TRAVERSE (total)	18°
RATE OF FIRE	1 rd/min
MAX RANGE	9700 m
CREW	9

Status: Production complete. Known to be in service with Egypt, Syria and the Soviet Union.

Manufacturer: Soviet state factories.

240 mm M-240 mortar in loading position

Egyptian Army 240 mm M-240 mortar in travelling configuration

180 mm Gun S-23

Development

The 180 mm Gun S-23 was developed in the early 1950s from a naval weapon and was seen in public for the first time in the 1955 Moscow May Day parade. For many years the weapon was referred to in the West as the 203 mm gun-howitzer M1955, but weapons captured in the Middle East by Israeli forces were found to have a calibre of 180 mm rather than 203 mm.

Under the Conventional Forces Europe (CFE) Treaty signed in 1990, the Soviet Union stated that it has none of these towed artillery systems deployed West of the Urals.

Description

When travelling the ordnance is withdrawn to the rear out of battery and linked to the trails to reduce the overall length of the weapon. The carriage is of the split trail type with box section trails and large dual rubber tyres which are filled with sponge rubber. A limber with slightly smaller diameter wheels is provided. In action the S-23 is supported on a base which, when travelling, is retracted under the carriage just below the equilibrators.

It fires a bag-type variable charge, and a separate-loading FRAG-HE projectile OF-43 weighing 84.09 kg with a maximum muzzle velocity of 790 m/s, HE/RAP with a maximum muzzle velocity of 850 m/s, a concrete-piercing G-572 projectile weighing 97.7 kg and a 0.2 kt tactical nuclear projectile.

In the Soviet Army the 180 mm gun S-23 is issued on the scale of 12 weapons in the Heavy Artillery Brigade of the Artillery Division.

Status: Production complete. In service with Egypt, India, Somalia, Syria and USSR.

Manufacturer: Soviet state arsenals.

180 mm S-23 guns being towed across Red Square, Moscow, by AT-T Heavy Tracked Artillery Tractors (TASS)

SPECIFICATIONS					
CALIBRE	180 mm	WIDTH (travelling)	2.996 m	RANGE	
BARREL LENGTH	8.8 m	HEIGHT (travelling)	2.621 m	max	30 400 m
MUZZLE BRAKE	pepperpot	AXIS OF BORE (at 0°)	1.99 m	with RAP	43 800 m
BREECH MECHANISM	screw	GROUND CLEARANCE	0.4 m	CREW	16
CARRIAGE	split trail	ELEVATION/DEPRESSION	+50°/-2°	TOWING VEHICLE	AT-T Heavy Tracked Artillery Tractor
SHIELD	no	TRAVERSE (total)	44°		
WEIGHT (in action)	21 450 kg	RATE OF FIRE			
LENGTH (travelling)	10.485 m	short period	1 rd/min		
		sustained	1 round/2 minutes		

152 mm Gun 2A36 (M1976)

Development

In the 1970s, the Soviet Union developed a new 152 mm towed gun which was allocated the industrial number of the 2A36. This was first seen by NATO sources in 1976 and allocated the Western designation of the 152 mm gun M1976.

Within the Soviet Army the 2A36 is commonly referred to as the Giatsint (Hyacinth) and is believed to have entered service in 1981 as the replacement for the 130 mm field gun M-46.

The 152 mm 2A36 was first seen in public during a parade held in Moscow in May 1985 when it was towed by a KrAZ-260 (6 × 6) truck.

The 152.4 mm ordnance of the 2A36 is identical to that used by the 2S5 self-propelled artillery system which entered service with the Soviet Army as far back as 1980. The 2S5 has yet to be seen in public.

The 152 mm 2A36 is normally deployed in batteries of six or eight guns with three batteries making up a battalion.

Under the Conventional Forces Europe (CFE) Treaty signed in 1990, the Soviet Union admitted having the following quantities of 152 mm 2A36 (M1976) weapons deployed in the West:

Region	**IV.1**	**IV.2**	**IV.3**	**IV.4**	**V.1**
Quantity	488	236	208	nil	252

The above figures are the situation as of November 1990. For the Conventional Forces Europe Treaty, Region IV is defined as Central Europe while Region V covers the flanks of the Atlantic to the Urals (ATTU).

Although the 2A36 was first observed being towed by the new KrAZ-260 (6 × 6) truck, it can also be towed by the KrAZ-255B (6 × 6) truck or ATT, ATS or ATS-59 full tracked vehicles.

Close-up of the breech of the Soviet 152 mm 2A36 gun (M1976) (Christopher F Foss)

152 mm gun 2A36 (M1976)

152 mm gun 2A36 (M1976) from the rear in travelling configuration

Description

The 49 calibre barrel of the 2A36 is fitted with a multi-slotted muzzle brake weighing 141 kg with the recoil system consisting of a buffer and a recuperator which are mounted above the ordnance towards the rear.

The horizontal sliding breech mechanism opens to the right automatically.

The ordnance pivots on the cradle which is of cast and welded steel construction. Elevation and traverse is manual with the former being from -2° 30' to +57° with traverse being 25° left and right.

The direct and indirect sights are mounted on the left side of the carriage as are the elevation and traverse mechanisms.

Some of the eight man crew are provided with protection from shell splinters by the armoured shield that slopes to the rear and extends over the wheels.

Mounted below and to the rear of the breech is the load assist system which is referred to as a quick firing loading system by the Soviets. This includes a hydraulic rammer operated from the hydropneumatic accumulator with controls being provided for ramming and return.

A fuzed projectile is placed on the loading tray which is mounted to the right of the breech. When the breech is opened this slides to the left until it lines up with the breech. The projectile is then rammed into the ordnance, the loading tray then slides back to the right.

The cartridge cased charge is then located in a similar manner and when the breech is closed the rammer automatically returns to the side and the weapon is then ready to fire. If the hydraulic rammer fails, then loading takes place manually, although in this case the rate of fire is reduced.

The Soviets quote a rate of fire of six rounds a minute for the 2A36 and say that a battery can put almost one tonne of ammunition onto a target in one minute.

Types of separate loading ammunition fired include high explosive fragmentation (HE-FRAG) and armour piercing tracer (AP-T). The former weighs 46 kg and has a maximum muzzle velocity, using the top charge, of about 800 m/s. Before loading the projectile the fuzing mode is selected either for fragmentation or high explosive action.

The AP-T round is used in the direct fire mode against tanks and other armoured fighting vehicles. The propelling charge, which weighs a maximum of 11 kg, is placed in a conventional cartridge case.

The standard HE-FRAG projectile has a maximum range of 27 000 m although a rocket assisted HE-FRAG projectile, called the active-reactive projectile by the Soviets, can also be fired. This has a maximum range of 40 000 m.

Other types of projectile that can probably be fired by the 152 mm 2S36 include chemical, smoke, concrete piercing, incendiary and tactical nuclear.

The ammunition system used by the 152 mm 2A36 is of a new design and is not interoperable with that of earlier Soviet artillery systems, such as the 152 mm 2S3, and may well be of a new streamlined design.

The walking beam suspension has a total of four rubber tyred braked road wheels for high speed towing as well as for improved cross-country mobility with the tyres having a normal tyre pressure of 4.8 kg/cm^2. Total wheel travel is 150 mm. Each of the front road wheels has a hydraulic shock absorber and a manual handbrake.

When in firing position, the 152 mm 2A36 is supported on a circular jack that is located under the forward part of the carriage and on each of the box type welded trails which is provided with a spade.

The standard spade is the winter spade but when firing from soft terrain much larger summer spades are used.

When coming out of action, the jack is raised clear of the ground and secured under the front, spades removed and stowed on the trails, trails brought together and clamped with a tie mechanism. The hinged tow bar is fitted to the right trail and the weapon is then hitched up to the prime mover.

Status: In production. In service with Iraq and the Soviet Union.

Manufacturer: Soviet state factories.

SPECIFICATIONS		LENGTH		RATE OF FIRE	5-6 rds/min
CALIBRE	152.4 mm	travelling	12.92 m	RANGE	
BARREL LENGTH	8.197 m	firing	12.3 m	standard ammunition	27 000 m
MUZZLE BRAKE	5-part multi-baffle	WIDTH	2.788 m	RAP	40 000 m
BREECH MECHANISM	semi-automatic horizontal sliding	HEIGHT	2.76 m	CREW	8
		GROUND CLEARANCE	0.475 m	TOWING VEHICLE	KrAZ-260 (6 × 6) truck
CARRIAGE	split trail	TRACK	2.34 m		
SHIELD	yes	TYRES	1080 × 300 mm		
WEIGHT		ELEVATION/DEPRESSION	+57°/-2.5°		
travelling	9800 kg	TRAVERSE	25° left, 25° right		
firing	9760 kg				

152 mm Gun-Howitzer 2A65 (M1987)

Development/Description
In addition to the 152 mm 2A36 gun described in the previous entry, the Soviet Union has also deployed a new 152 mm gun-howitzer designated the 2A65 which has been allocated the NATO designation of the M1987.

152 mm Gun-Howitzer 2A65 (M1987) in firing position

The 152 mm 2A36 is also referred to as the MSTA-B by the Soviet Union with the weapon also forming the main armament of the new 152 mm self-propelled artillery system the 2S19 which is also referred to as the MSTA-S. Full details of the latter are given in the *Self-propelled guns and howitzers* section.

No firm details of the 2A65 are available at present although it may well be the 152 mm gun-howitzer D-20, fully described in the following entry, upgraded in a number of areas including the installation of a new and longer barrelled ordnance.

Under the Conventional Forces Europe Treaty signed in 1990, the Soviet Union admitted to having the following quantities of the 152 mm 2A65 weapon deployed in the West:

Region	**IV.1**	**IV.2**	**IV.3**	**IV.4**	**V.1**
Quantity	297	274	265	72	23

The above figures are the situation as of November 1990. For the Conventional Forces Europe Treaty, Region IV is defined as Central Europe while Region V covers the flanks of the Atlantic to the Urals (ATTU).

Status: Production. In service with the Soviet Army.

Manufacturer: Soviet state factories.

152 mm Gun-Howitzer D-20

Development
The 152 mm Gun-Howitzer D-20 was developed shortly after the end of the Second World War by the F F Petrov design bureau and was first seen in public during the May Day parade in Red Square, Moscow, in 1955.

Under the Conventional Forces Europe (CFE) Treaty signed in 1990, the Soviet Union admitted having the following quantities of 152 mm D-20 weapons deployed in the West:

Region	**IV.1**	**IV.2**	**IV.3**	**IV.4**	**V.1**
Quantity	937	562	537	192	98

The above figures are the situation as of November 1990. For the Conventional Forces Europe Treaty, Region IV is defined as Central Europe while Region V covers the flanks of the Atlantic to the Urals (ATTU).

Description
The D-20 uses the same carriage and recoil system as the 122 mm Field Gun D-74 which was also shown for the first time in 1955. The D-20 can be distinguished from the D-74 by a much shorter and fatter stepped barrel and a larger double-baffle muzzle brake.

152 mm Gun-Howitzer D-20 in firing position

The shield has an irregular top with a sliding centre section and the top of the shield can be folded down to reduce the overall height of the weapon. A circular firing pedestal fitted as standard on all D-20s is inverted and secured just forward of the shield when travelling. In the firing position the firing pedestal is lowered, allowing the weapon to be traversed quickly through a full 360°. On each of the split box-section trails is a castor wheel (on top) and a spade (underneath).

The ordnance of the 152 mm self-propelled gun-howitzer 2S3 is a development of that used on the towed D-20.

It fires the following case-type, variable-charge, separate-loading ammunition:

Ammunition type	**FRAG-HE**	**CP**	**AP-T**
PROJECTILE DESIGNATION	OF-540*	G-545†	BR-540
FUZE MODEL	RGM-2	KTD	MD-7
WEIGHT OF PROJECTILE	43.51 kg	56 kg	48.78 kg
WEIGHT OF BURSTING CHARGE	6.25 kg	4.22 kg	1.2 kg
TYPE OF BURSTING CHARGE	TNT	TNT	n/app
MUZZLE VELOCITY	655 m/s	670 m/s	600 m/s
ARMOUR PENETRATION AT 0°	n/app	n/app	124 mm/1000 m

* other FRAG-HE projectiles also available
† other CP projectiles available

Other types of ammunition include chemical, HE/RAP (range of 24 000 m), HEAT, illuminating (S-540), smoke (D-540) and tactical nuclear (0.2 kt). More recent ammunition types include HEAT-SS (spin-stabilised), flechette, scatterable mines (anti-tank and anti-personnel) and semi-active laser.

At least two types of chemical projectiles have been identified. One weighs 40 kg of which 2.8 kg comprises the Sarin CW agent which is dispersed by the TNT bursting charge. The 152 mm viscous Lewisite projectile weighs 42.5 kg of which 5.4 kg comprises the CW agent which is dispersed by the explosion of the TNT bursting charge.

Yugoslavia produces a long-barrelled version of the D-20 known as the 152 mm Howitzer M84. See entry under Yugoslavia for illustrations.

152 mm gun-howitzer D-20 on parade in Nicaragua

SPECIFICATIONS		LENGTH (travelling)	8.69 m	CREW	10
CALIBRE	152.4 mm	WIDTH (travelling)	2.32 m	TOWING VEHICLES	AT-S Medium Tracked Artillery Tractor
BARREL LENGTH (overall)	5.195 m	HEIGHT (travelling)	1.925 m		
MUZZLE BRAKE	double-baffle	AXIS OF BORE (at 0°)	1.22 m		Ural-375 (6 × 6) truck
BREECH MECHANISM	semi-automatic vertical sliding wedge	GROUND CLEARANCE	0.4 m		Tatra 813 (8 × 8) truck
		TRACK	2.03 m	MAX TOWING SPEED	
CARRIAGE	split trail with castor wheel on each	TYRES	12.00 × 20	road	60 km/h
		ELEVATION/DEPRESSION	+63°/-5°	cross country	15 km/h
SHIELD	yes	TRAVERSE (total)	58°	UNIT OF FIRE	65 rounds
WEIGHT		RATE OF FIRE	5-6 rds/min		
travelling order	5700 kg	RANGE	17 410 m		
firing position	5650 kg	with RAP	24 000 m		

Status: Production complete. In service with Algeria, Afghanistan, Bulgaria, Czechoslovakia, China (manufactured in China as the Type 66 for which there is a separate entry), Cuba, Egypt, Ethiopia, Hungary, India, Nicaragua, Romania, USSR, Vietnam and Yugoslavia.

Manufacturer: Soviet state factories.

152 mm Howitzer M1943 (D-1)

Development

The 152 mm Howitzer M1943 (D-1) was introduced into the Soviet Army in 1943 as the replacement for the earlier 152 mm Howitzer M1938 (M-10). It was standardised in August 1943 and designed by the F F Petrov design bureau at Artillery Plant No 9 in Sverdlovsk. The 152 mm Howitzer M1943 (D-1) is basically a strengthened carriage and recoil system of the 122 mm Howitzer M1938 (M-30) fitted with the ordnance of the 152 mm Howitzer M1938 (M-10) but with a large double-baffle muzzle brake. The 152 mm Howitzer M1943 (D-1) weighs 3640 kg in travelling order compared with the 4550 kg of the earlier 152 mm M1938 (M-10), and both weapons use the same ammunition.

Under the Conventional Forces Europe (CFE) Treaty signed in 1990, the Soviet Union admitted having the following quantities of 152 mm D-1 weapons deployed in the West:

Region	**IV.1**	**IV.2**	**IV.3**	**IV.4**	**V.1**
Quantity	15	13	nil	nil	2

The above figures are the situation as of November 1990. For the Conventional Forces Europe Treaty, Region IV is defined as Central Europe while Region V covers the flanks of the Atlantic to the Urals (ATTU).

Description

The carriage is of the riveted box-section split trail type. The top half of the shield slopes to the rear and in the centre of the shield is a section which slides upwards to enable the weapon to be elevated. The recoil system is under the barrel and the counter-recoil system is on top of the barrel. The elevating mechanism is on the right side of the carriage and the traverse mechanism is on the left side. Maximum towing speed is 48 km/h. Each trail has two spades, one fixed for use on hard surfaces and one hinged spade for use in soft ground.

The M1943 (D-1) fires the following variable-charge, case-type, separate-loading ammunition:

152 mm howitzer M1943 (D-1) in travelling order

Ammunition type	**FRAG-HE**	**FRAG-HE**	**CP**
PROJECTILE DESIGNATION	OF-530	OF-530A	G-530
WEIGHT OF PROJECTILE	40 kg	40 kg	40 kg
WEIGHT OF BURSTING CHARGE	6.86 kg	5.66 kg	5.1 kg
TYPE OF BURSTING CHARGE	TNT	TNT	TNT
MUZZLE VELOCITY	508 m/s	508 m/s	508 m/s

* also RGM and D-1 fuzes

Other projectiles include chemical, HEAT which will penetrate 300 mm of armour at 0° obliquity at 1000 m, illuminating, semi-AP weighing 51.1 kg with m/v of 432 m/s which will penetrate 82 mm of armour at 1000 m and smoke

Status: Production complete. In service with Afghanistan, Albania, China (Type 54: 6 in each artillery regiment of the artillery division), Cuba, Egypt, Ethiopia, Hungary, Iraq, Mongolia, Mozambique, Poland, Somalia, Syria, USSR and Vietnam.

Manufacturer: Soviet state factories.

SPECIFICATIONS		WEIGHT		TRAVERSE (total)	35°
CALIBRE	152.4 mm	travelling order	3640 kg	RATE OF FIRE	4 rds/min
BARREL LENGTH (overall)	4.207 m	firing position	3600 kg	SUSTAINED RATE OF FIRE	65 rds/h
MUZZLE BRAKE	double-baffle	LENGTH (travelling)	7.558 m	RANGE	
RECOIL SYSTEM	hydraulic buffer and hydro-pneumatic recuperator	WIDTH (travelling)	1.994 m	max	12 400 m
		HEIGHT (travelling)	1.854 m	effective semi-AP	510 m
		AXIS OF BORE (at 0°)	1.24 m	CREW	7
BREECH MECHANISM	hinged screw	GROUND CLEARANCE	0.37 m	TOWING VEHICLE	AT-S Medium Tracked Artillery Tractor
CARRIAGE	split trail	TRACK	1.8 m		
SHIELD	yes	TYRES	11.80 × 35	UNIT OF FIRE	60 rounds
		ELEVATION/DEPRESSION	+63.5°/-3°		

152 mm Gun-Howitzer M1937 (ML-20)

This weapon is no longer in front line service with the Soviet Union, although it may be found in service with some countries that have received Soviet aid. Full details were given in *Jane's Armour and Artillery 1990-91*, page 663.

130 mm Field Gun M-46

Development

The 130 mm Field Gun M-46 was developed in the early 1950s and was first seen in public during the 1954 May Day parade. It replaced the 122 mm M1931/37 (A-19) Field Gun and is ballistically similar to 130 mm guns used by the Soviet Navy. The M-46 has seen combat in many parts of the world but has now been replaced in most front line Soviet units.

The 130 mm field gun M-46 has been replaced in virtually all front line Soviet artillery units by more modern weapons.

Under the Conventional Forces Europe (CFE) Treaty signed in 1990, the Soviet Union admitted to having the following quantities of the 130 mm M-46 weapon deployed in the West:

Region	**IV.1**	**IV.2**	**IV.3**	**IV.4**	**V.1**
Quantity	13	1	nil	nil	12

The above figures are the situation as of November 1990. For the Conventional Forces Europe Treaty, Region IV is defined as Central Europe while Region V covers the flanks of the Atlantic to the Urals (ATTU).

Description

When travelling the barrel is withdrawn by a mechanism on the right trail from battery to the rear to reduce the overall length of the weapon. The carriage is of the split trail type and is provided with a two-wheeled limber. When travelling the spades are removed and carried on each of the two trails. The recoil system is mounted under the barrel, and in front of the shield, which has been removed on some models, is an inverted U-shaped collar. The M-46 has direct fire sights including an APN-3 active/passive night sight.

It fires case-type, variable-charge, separate-loading ammunition, including:

Ammunition type	**FRAG-HE**	**APC-T**
PROJECTILE DESIGNATION	OF-482M	BR-482B
FUZE MODEL	RGM-2	DBR
WEIGHT OF PROJECTILE	33.4 kg	33.6 kg
WEIGHT OF BURSTING CHARGE	4.63 kg	0.127 kg
TYPE OF BURSTING CHARGE	TNT	RDX/alum
MUZZLE VELOCITY	930 m/s	930 m/s
ARMOUR PENETRATION AT 0°	n/app	230 mm/1000 m

Other types of projectile include SP-46 illuminating weighing 25.8 kg with an m/v of 687 m/s and smoke target marking with an m/v of 930 m/s. An RAP is known to be in service having been used by Syria in the 1973 Middle East War.

Israel Military Industries produces a 130 mm HE projectile identical to the Soviet original but fitted with a PD M739 fuze. This projectile is used by the Israeli Army. PRB of Belgium developed a base bleed projectile that can be fired from the M-46 to a maximum range of 37 000 m.

At least two chemical projectiles have been fielded for this weapon. The 130 mm Sarin projectile weighs 33.4 kg of which 1.6 kg comprises the Sarin CW agent which is dispersed by a TNT bursting charge. The other is a VX projectile weighing 33.4 kg of which 1.4 kg is the VX CW agent.

Variants

Modified 130 mm M-46s with a longer barrel, recuperator and cradle appeared in Soviet Army service in the mid-1970s. India has deployed a self-propelled model based a lengthened Vijayanta MBT chassis, brief details of which are given in the *Self-propelled guns and howitzers* section under India. This is called the Catapult by the Indian Army.

Israeli Army 130 mm field gun M-46 in travelling configuration on modified carriage (Israeli Ministry of Defence)

130 mm field gun M-46 captured by Israeli Army during invasion of Lebanon in summer of 1982 (Israeli Ministry of Defence)

155 mm Conversions

The Chinese NORINCO concern is offering a 155 mm conversion known as the GM-45 (see entry under People's Republic of China for details) and Yugoslavia is offering a conversion (see under Yugoslavia for details). Soltam of Israel is also offering a conversion pack for the 130 mm M-46 to 155 mm/45 calibre standard and details of this are given under Israel.

Status: Production complete. In service with Algeria, Angola, Bulgaria, Cambodia, China (Type 59), Congo, Cuba, Egypt, Ethiopia, Finland, Guinea, Guinea Bissau, Guyana, India, Iran, Iraq, Israel, North Korea, Libya, Mali, Mongolia, Morocco, Mozambique, Nigeria, Oman, Pakistan (and Chinese Type 59-1, for which there is a separate entry), Peru, Poland, Somalia, Sudan, Syria, Tanzania, USSR, Vietnam, Yemen, Yugoslavia and Zaïre.

Manufacturer: Soviet state factories.

SPECIFICATIONS			
CALIBRE	130 mm	LENGTH (travelling)	11.73 m
BARREL LENGTH (overall)	7.6 m	WIDTH (travelling)	2.45 m
MUZZLE BRAKE	pepperpot	HEIGHT (travelling)	2.55 m
RECOIL SYSTEM	hydraulic buffer and hydro-pneumatic recuperator	AXIS OF BORE (at 0°)	1.38 m
BREECH MECHANISM	horizontal sliding wedge breech-block	GROUND CLEARANCE	0.4 m
CARRIAGE	split trail with limber	TRACK	2.06 m
SHIELD	optional	TYRES	13.50 × 38
WEIGHT		ELEVATION/DEPRESSION	+45°/-2.5°
travelling order	8450 kg	TRAVERSE (total)	50°
firing position	7700 kg	RATE OF FIRE	5-6 rds/min
		RANGE	27 150 m
		CREW	8
		TOWING VEHICLES	AT-S Medium Tracked Artillery Tractor; ATS-59 Medium Tracked Artillery Tractor; M1972 Medium Tracked Artillery Tractor; AT-P Armoured Tracked Artillery Tractor

122 mm Field Gun D-74

Development/Description

The 122 mm Field Gun D-74 was developed in the late 1940s by the F F Petrov design bureau and was seen for the first time in public during the 1955 May Day parade in Moscow. It appears that the 122 mm Field Gun D-74 and the 130 mm Field Gun M-46 were both designed to meet a requirement for a gun with a longer range to replace the 122 mm Corps Gun M1931/37 (A-19). After extensive troop testing the 130 mm Field Gun M-46 was selected in preference to the D-74 as it fired a heavier projectile to 27 150 m compared with the 24 000 m of the D-74. The D-74 was however produced in some numbers, most of which were subsequently exported. It is now considered obsolete by the USSR. The 122 mm Field Gun D-74 uses the same carriage as the 152 mm Gun-Howitzer D-20 but the D-20 can easily be distinguished from it by a much thicker barrel and a larger muzzle brake.

As far as it is known the 122 mm field gun D-74 is no longer in front line service with the USSR but it may be held in reserve.

The shield has an irregular top with a centre section which slides upwards so the ordnance can be elevated. A circular firing pedestal fitted as standard on all D-74s is inverted and secured just forward of the shield when travelling. In the firing position the firing pedestal is lowered to the ground, allowing the D-74 to be quickly traversed through a full 360°. On each of the split box-section trails is a castor which folds up on to the top of the trail when travelling and a spade which folds up underneath the trail when travelling.

The D-74 fires variable-charge, case-type, separate-loading ammunition as follows:

Ammunition type	**FRAG-HE**	**APC-T**
PROJECTILE DESIGNATION	OF-472	BR-472
FUZE MODEL	V-429	DBR
WEIGHT OF PROJECTILE	27.3 kg	25 kg
WEIGHT OF BURSTING CHARGE	2.95 kg	0.091 kg
TYPE OF BURSTING CHARGE	TNT	RDX/alum
MUZZLE VELOCITY	885 m/s	885 m/s
ARMOUR PENETRATION AT 0°	n/app	185 mm/1000 m

Other types of projectile include chemical, illuminating (m/v 885 m/s) and smoke (m/v 885 m/s with projectile weighing 21.7 kg)

122 mm field gun D-74 on firing pedestal with spades lowered and support wheels raised (US Army)

SPECIFICATIONS		WEIGHT		SUSTAINED RATE OF FIRE	75 rounds (first hour)
CALIBRE	121.92 mm	travelling order	5550 kg	RANGE	
BARREL LENGTH (overall)	6.45 m	firing position	5500 kg	max	24 000 m
MUZZLE BRAKE	double-baffle	LENGTH (travelling)	9.875 m	effective APHE	1070 m
RECOIL SYSTEM	hydraulic buffer and	WIDTH (travelling)	2.35 m	CREW	10
	hydro-pneumatic	HEIGHT (travelling)	2.745 m	TOWING VEHICLES	AT-L Light Tracked
	recuperator	AXIS OF BORE (at 0°)	1.668 m		Artillery Tractor
BREECH MECHANISM	semi-automatic	GROUND CLEARANCE	0.4 m		AT-S Medium Tracked
	vertical	TRACK	2.03 m		Artillery Tractor
	sliding wedge	TYRES	12.00 × 20		Ural-375D (6 × 6)
CARRIAGE	split trail with castor	ELEVATION/DEPRESSION	+45°/-5°		Truck
	wheel on each	TRAVERSE (total)	58°	UNIT OF FIRE	90 rounds
SHIELD	yes	RATE OF FIRE	6-7 rds/min		

Status: Production complete. In service with China (Type 60), Cuba, Egypt, North Korea, Nigeria, Peru and Vietnam.

Manufacturer: Soviet state factories.

122 mm Howitzer D-30

Development

The 122 mm Howitzer D-30 was introduced into service with the Soviet Army in the early 1960s as the replacement for the 122 mm Howitzer M1938 (M-30). It was designed by the F F Petrov design bureau at Artillery Plant No 9 at Sverdlovsk. The main improvements over the earlier weapon are increased range and the ability to traverse the weapon through 360° quickly.

Under the Conventional Forces Europe (CFE) Treaty signed in 1990, the Soviet Union admitted to having the following quantities of the 122 mm D-30 weapon deployed in the West:

Region	**IV.1**	**IV.2**	**IV.3**	**IV.4**	**V.1**
Quantity	1904	784	648	258	1120

The above figures are the situation as of November 1990. For the Conventional Forces Europe Treaty, Region IV is defined as Central Europe while Region V covers the flanks of the Atlantic to the Urals (ATTU).

Late production models of the D-30 are designated the D-30M, these have a number of modifications including new double baffle muzzle brake, a new central baseplate which is square rather than round and a towing lunette assembly. Ballistically it is identical to the earlier D-30.

Description

The D-30 is towed muzzle first and on arrival at the battery position the crew first unlock the barrel travelling lock which then folds back onto the central trail. The firing jack under the carriage is then lowered to the ground, raising the wheels clear of the ground. The two outer trails are each spread through 120° and the firing jack is raised until all three trail ends are on the ground when they are staked into position. The lunette is mounted either under the muzzle brake or on the barrel just to the rear of the muzzle brake. In either case it is normally swung through 180° so that it lies under the barrel before firing begins.

An unusual feature of the D-30 is that the recoil system is mounted over the barrel. Some of the projectiles fired by the D-30 are interchangeable with that fired by the 122 mm Howitzer M1938 (M-30) but the D-30 also fires a HEAT projectile of the non-rotating fin-stabilised type.

The D-30 fires the following variable-charge, case-type, separate-loading ammunition:

122 mm D-30M Howitzer captured by South African Army in Angola in travelling configuration showing breech details

Ammunition type	**FRAG-HE**	**HEAT-FS**
PROJECTILE DESIGNATION	OF-462*	BK-6M†
FUZE MODEL	RGM-2	GPV-2
WEIGHT OF PROJECTILE	21.76 kg	21.63 kg
WEIGHT OF BURSTING CHARGE	3.675 kg	n/a
TYPE OF BURSTING CHARGE	TNT	RDX
MUZZLE VELOCITY	690 m/s	740 m/s
ARMOUR PENETRATION AT 0°	n/a	460 mm/any range

* there are variants of projectile design giving different weights. Various other fuzes may also be used.
† rotating BP-463 may also be used.

Other types of projectiles include chemical, illuminating (S-462 weighing 22.4 kg), smoke (D-462 weighing 22.3 kg), leaflet, flechette, incendiary and semi-active laser (SAL). More recently an RAP has been introduced with a maximum range of 21 900 m.

There are at least two chemical projectiles for the D-30. The Sarin projectile weighs 22.2 kg of which 1.3 kg comprises the Sarin CW. The 122 mm viscous Lewisite projectile weighs 23.1 kg of which 3.3 kg comprises the viscous Lewisite CW agent. In both cases the CW agent is dispersed by the explosion of a TNT bursting charge.

In the Soviet Army the D-30 is issued on the scale of 36 per Tank Division (two battalions each with 18 weapons) in the Artillery Regiment which also has a battalion of 18 152 mm (2S3) self-propelled gun/howitzers. Each Motorised Rifle Division has 72 D-30s with each Motorised Rifle Regiment having a battalion of 18 D-30s with the remaining two battalions of 18 weapons in the Artillery Regiment.

The main armament of the 122 mm self-propelled howitzer 2S1 is based on the ordnance of the D-30.

The D-30 is also produced in China (see separate entry), Iraq (called the Saddam, qv) and Yugoslavia (122 mm Howitzer D-30J, qv).

Soviet 122 mm D-30M howitzer being loaded onto a PKP trailer which is towed by a PTS series tracked amphibious vehicle

Israeli Army 122 mm D-30 Howitzer in firing position with projectile and charge under wheel (Israeli Ministry of Defence)

SPECIFICATIONS					
CALIBRE	121.92 mm (with muzzle brake)	LENGTH (travelling)	5.4 m	RANGE	
BARREL LENGTH	4.875 m	WIDTH (travelling)	1.95 m	max	15 400 m
MUZZLE BRAKE	multi-baffle	HEIGHT (travelling)	1.66 m	with RAP	21 900 m
BREECH MECHANISM	semi-automatic vertical sliding wedge breech-block	AXIS OF BORE (at 0°)	0.9 m	CREW	7
SHIELD	yes	GROUND CLEARANCE	0.325-0.345 m	TOWING VEHICLES	MT-LB Multi-Purpose Tracked vehicle Ural-375D (6 × 6) truck
WEIGHT		TRACK	1.85 m		
travelling order	3210 kg	TYRES	9.00 × 20		
firing position	3150 kg	ELEVATION/DEPRESSION	+70°/-7°		
		TRAVERSE	360°		
		RATE OF FIRE	7-8 rds/min	UNIT OF FIRE	80 rounds
		SUSTAINED RATE OF FIRE (first hour)	75 rounds		

Status: Production. In service with Afghanistan, Algeria, Angola, Benin, Bulgaria, Chad, China, Cuba, Czechoslovakia, Egypt, Ethiopia, Finland, Guinea Bissau, Hungary, Iraq (local production), Israel, North Korea, Libya, Madagascar, Mali, Mongolia, Morocco, Mozambique, Nicaragua, Peru, Poland, Romania, Somalia, Syria (some of which have been removed from their carriages and mounted on T-34 tank chassis), USSR, Vietnam, Yugoslavia (local production), Zaïre and Zambia. Abu Zaabal Engineering Industries is now making this weapon in Egypt and Royal Ordnance, in England, under contract to Egypt has built the prototype of a self-propelled version. BMY of the United States has also built a prototype of a self-propelled version of the D-30 for Egypt based on the M109 chassis. Details of both Egyptian self-propelled models of the D-30 are given in the *Self-propelled guns and howitzers* section. China has also built a self-propelled version of the D-30 on a full tracked chassis. Romania has produced this under licence since the early 1980s.

Manufacturer: Soviet state factories.

122 mm Howitzer M1938 (M-30)

Development

The 122 mm Howitzer M1938 (M-30) entered service with the Soviet Army in 1939 after standardisation in September 1938. It was designed by the F F Petrov design bureau at Artillery Plant No 172 at Perm. Until the introduction of the 122 mm Howitzer D-30 it was the standard division howitzer of the Warsaw Pact and issued on the scale of 36 per Motorised Rifle Division (two battalions each consisting of 18 M1938s with three batteries each with six howitzers) and 54 per Tank Division (three battalions each consisting of 18 M1938s with three batteries each with six howitzers). It is no longer in front line service with the USSR.

Under the Conventional Forces Europe Treaty signed in 1990, the Soviet Union stated that it had a total of 69 of these weapons West of the Urals.

Description

The carriage is of the riveted box-section split trail type and is identical to that used for the 152 mm Howitzer M1943 (D-1), which can easily be distinguished from the 122 mm weapon by its longer and fatter barrel with a double-baffle muzzle brake. The top half of the shield slopes to the rear and in the centre of the shield is a section which slides upwards so the weapon can be elevated. The recoil system is under the barrel and the counter-recoil system on top of it. The elevating mechanism is on the right side of the carriage and the traverse mechanism is on the left side. The 122 mm howitzer can be fired without spreading its trails, in which case traverse is limited to 1.5°. Maximum towing speed is 48 km/h. Each trail has two spades, one fixed for use on hard surfaces and one hinged for softer ground.

The M1938 (M-30) fires the following variable-charge, case-type, separate-loading ammunition:

Ammunition type	**FRAG-HE**	**HEAT**
PROJECTILE DESIGNATION	OF-462*	BP-463†
FUZE MODEL	RGM-2	n/a
WEIGHT OF PROJECTILE	21.76 kg	13.3 kg
WEIGHT OF BURSTING CHARGE	3.675 kg	n/a
TYPE OF BURSTING CHARGE	TNT	n/a
MUZZLE VELOCITY	515 m/s	570 m/s
ARMOUR PENETRATION AT 0°	n/a	200 mm/1000 m

* other FRAG and FRAG-HE projectiles available
† fired at full charge. Better design than earlier BP-460A (weight 14.8 kg) which was fired at charge 4 giving only 325 m/s
Other projectiles include chemical, illuminating (S-462) weighing 21 kg, smoke (D-462) weighing 22.4 kg and leaflet (A-462) weighing 21.5 kg.

The 122 mm Howitzer M1938 (M-30) used by Bulgaria have new disc wheels with spoke-type impressions but retain the sponge rubber filled tyres of the original Soviet model.

The M1938 has also been manufactured in China as the Type 54 and Type 54-1 and is issued on the scale of 12 per infantry division and 12 per armoured division. A self-propelled model of the Type 54-1 in existence is essentially a Type 531 fully-tracked APC with the Type 54-1 mounted on the rear of the vehicle firing forwards. Details of this appear in the *Self-propelled guns and howitzers* section.

122 mm howitzer M1938 (M-30) in travelling configuration

122 mm howitzer M1938 (M-30) with standard sponge rubber filled tyres showing sliding vertical section in centre of shield

SPECIFICATIONS		LENGTH (travelling)	5.9 m	RANGE	
CALIBRE	121.92 mm	WIDTH (travelling)	1.975 m	max	11 800 m
BARREL LENGTH (overall)	2.8 m	HEIGHT (travelling)	1.82 m	effective HEAT	630 m
MUZZLE BRAKE	no	AXIS OF BORE (at 0°)	1.2 m	CREW	8
RECOIL SYSTEM	hydraulic buffer and hydro-pneumatic recuperator	GROUND CLEARANCE	0.33 m	TOWING VEHICLES	AT-S Medium Tracked Artillery Tractor
		TRACK	1.6 m		Ural-375 (6 × 6) truck
		TYRES	11.80 × 15		MT-LB Multi-Purpose Tracked Vehicle
BREECH MECHANISM	hinged screw	ELEVATION/DEPRESSION	+63.5°/-3°		
CARRIAGE	split trail	TRAVERSE (total)	49°		
SHIELD	yes	RATE OF FIRE	5-6 rds/min	UNIT OF FIRE	80 rounds
WEIGHT		SUSTAINED RATE OF FIRE			
travelling order	2450 kg	(first hour)	75 rounds		
firing position	2450 kg				

Status: Production complete. In service with Afghanistan, Albania, Algeria, Bulgaria, China (Type 54, for which there is a separate entry), Congo, Cuba, Czechoslovakia, Egypt, Ethiopia, Finland, Guinea-Bissau, Hungary, Indonesia, Iraq, Israel (training), North Korea, Mongolia, Mozambique, Nicaragua, Poland, Romania, Somalia, Sudan, Syria, Uganda, USSR, Vietnam, Yemen, Yugoslavia and Zaïre. In many of these countries this weapon is held in reserve or used for training.

Manufacturers: Artillery Plant No 9, Sverdlovsk; Artillery Plant No 172, Perm; Artillery Plant No 221, Stalingrad, USSR.

122 mm Corps Gun M1931/37 (A-19)

Details of this weapon, which is no longer in service with the Soviet Army, were given in *Jane's Armour and Artillery 1990-91*, page 668.

120 mm 2B16 Combination Gun

Development/Description

In the late 1980s the Soviet Union deployed a new small towed artillery system that has the industrial designation of the 2B16 and, according to Soviet sources, is a combination gun and also referred to as the ANRA-K (NONA-K).

Under the Conventional Forces Europe Treaty signed in 1990, the Soviet Union stated that it had a total of 76 of these weapons deployed West of the Urals.

When in the firing position the 2B16 Combination Gun is supported on a small circular baseplate located under the forward part of the carriage with the wheels raised clear of the ground, and on the two trails.

The trails are of a box section and locked together at the rear for towing, probably by a GAZ-66 (4 × 4) 2 tonne truck which also carries the crew and a quantity of ready use ammunition.

The ordnance is fitted with a large multi-baffle muzzle brake with the recoil system being mounted above the ordnance and extending about half-way along the barrel.

120 mm 2B16 Combination Gun from rear

120 mm 2B16 Combination Gun from front

When in the travelling configuration the ordnance is withdrawn out of battery to the rear to reduce the overall length of the system.

A thin shield is provided either side of the ordnance to give some degree of protection to the gun crew from shell splinters with the left side of the shield being slightly higher than the right as this also protects the sighting system. The upper part of the shield can be folded forward through 180° if required, although there is an aperture in the shield for the direct fire telescopic sight.

The elevation and traverse handwheels are located on the left side of the carriage as are the direct and indirect sights.

The breech is believed to be of the vertical sliding type and what appears to be a chamber indent device is fitted to retain the round in place when the weapon is fired at high elevation.

It is believed that a power rammer is normally fitted to assist in loading ammunition at higher elevations as well as increasing its rate of fire.

Speculation is that the 2B16 ordnance is related to that of the 120 mm S-120 self-propelled howitzer/mortar (2S9) covered in detail in the previous section. This was originally developed to meet the requirements of the Soviet Air Assault Divisions.

Recent information has indicated that the 2S9 is now also used by the Soviet Naval Infantry and some tank and motorised rifle divisions as a replacement for their 120 mm towed mortars which not only take time to come into action but are also vulnerable to counter battery fire.

The turret of the 2S9 is also fitted on the BTR-80 8 × 8 APC to form the 2S23 system, details of this are given in the *Self-propelled guns and howitzers* section. It is not believed that the 2S23 is in large scale service yet.

The ordnance of the 2S9 can be elevated from -4° to +85° with a traverse of 35° left and right with a maximum effective range of the system being 8800 m. Types of ammunition fired include HE, HEAT, white phosphorous and smoke.

SPECIFICATIONS

CALIBRE	120 mm
CREW	5
WEIGHT	1100 kg
LENGTH	5.9 m
WIDTH	1.79 m
MAX RANGE	8000 m
RATE OF FIRE	8-10 rds/min
PROJECTILE WEIGHT	20.5 kg
MAX TOWING SPEED	80 km/h

Status: In production. In service with USSR.

Manufacturer: Soviet state factories.

100 mm Anti-tank Gun T-12/MT-12

Development/Description

The 100 mm smoothbore anti-tank gun T-12 is the replacement for the 85 mm Anti-Tank Gun D-48. The fin-stabilised non-rotating APFSDS and HEAT projectiles fired by the T-12 resemble those fired by the 115 mm gun of the T-62 MBT. The T-12 reached full operational capability with the Soviet Army in 1965.

It fires the following fixed ammunition: APFSDS projectile weighing 5.5 kg, with a muzzle velocity of 1500 m/s which will penetrate 406 mm of armour at an incidence of 0° at a range of 500 m; and 225 mm at 1000 m. HEAT projectile weighing 9.5 kg with a muzzle velocity of 990 m/s which will penetrate 400 mm of armour at an incidence of 0°. Grazing range against a 2 m high target is 1880 m using APFSDS projectile.

The T-12 can be distinguished from the 85 mm Anti-Tank Gun D-48 by its larger breech-ring, slightly different shaped shield with an opening in the left side for the sight and its smaller diameter pepperpot muzzle brake which is also parallel to the muzzle. Sighting equipment includes a PG-1M panoramic sight, OP-4M-40 direct sight and an APN-5-40 infra-red nightsight which is mounted over the rear part of the barrel.

A slightly modified version of the T-12 is known in the West as the T-12A and this was referred to by the then East German Army as the MT-12. It differs from the basic T-12 in weighing 3100 kg, and uses 9.00 × 20 tyres. When travelling the T-12A is 9.64 m long and 2.31 m wide. The MT-12 variant, formerly known as the T-12A, has a winged shield angled to the rear on both sides and an additional recoil cylinder above the breech on the right side.

100 mm anti-tank gun MT-12

The weapon is issued in the Soviet Army on the scale of 12 (two batteries each with six T-12s) in the anti-tank battalion of the Motorised Rifle Division and 36 (two battalions each of 18 T-12s) in independent anti-tank regiments.

Status: Production complete. Known to be in service with Hungary and the Soviet Union and with Yugoslavia. It is probably in service with countries in the Middle East and North Africa.

Manufacturer: Soviet state factories.

SPECIFICATIONS

CALIBRE	100 mm
BARREL LENGTH (overall)	8.484 m
MUZZLE BRAKE	pepperpot
CARRIAGE	split trail
SHIELD	yes
WEIGHT	
travelling order	3000 kg
firing position	3000 kg
LENGTH (travelling)	9.162 m
WIDTH (travelling)	1.7 m
HEIGHT (travelling)	1.448 m
AXIS OF BORE (at 0°)	0.8 m
GROUND CLEARANCE	0.35 m
TRACK	1.585 m
TYRES	6.50 × 20
ELEVATION/DEPRESSION	+20°/-10°
TRAVERSE (total)	27°
MAX RATE OF FIRE	10 rds/min
MAX RATE OF FIRE (aimed)	6 rds/min
RANGE	
max effective	8200 m FRAG-HE*
max effective	2000 m APFSDS
CREW	6
TOWING VEHICLES	MT-LB Multi-Purpose Tracked Vehicle ZIL-131 (6 × 6) truck Ural-375D (6 × 6) truck
UNIT OF FIRE	60 rounds

* A maximum range of 18 000 to 21 000 m can be achieved with trails dug in to provide 45° elevation.

100 mm Field Gun M1944 (BS-3)

Development/Description

The 100 mm Field Gun M1944 (BS-3) was developed from a Soviet naval gun of this calibre and entered service with the Soviet Army in 1944. It is no longer in front-line service with the Soviet Union having been first replaced by the 85 mm anti-tank gun D-48 and then by the 100 mm anti-tank gun T-12. It is still however used by Third World countries and large numbers are probably held in reserve.

The M1944 (BS-3) fires the following ammunition:

Ammunition type	**FRAG-HE**	**APC-T**
PROJECTILE DESIGNATION	OF-214*	†
FUZE MODEL	V-429	DBR-2
WEIGHT OF PROJECTILE	15.59 kg	16 kg
WEIGHT OF BURSTING CHARGE	1.46 kg	0.064 kg
TYPE OF BURSTING CHARGE	TNT	RDX/alum
MUZZLE VELOCITY	900 m/s	1000 m/s
ARMOUR PENETRATION AT 0°	n/app	185 mm/1000 m

Ammunition type	**HVAPDS-T**	**HEAT-FS**
PROJECTILE DESIGNATION	BM-8‡	ZBK-5M‡
FUZE MODEL	n/app	VP-9
WEIGHT OF PROJECTILE	5.69 kg	12.36 kg
WEIGHT OF BURSTING CHARGE	n/app	1.038 kg
TYPE OF BURSTING CHARGE	n/app	RDX/wax
MUZZLE VELOCITY	1415 m/s	900 m/s
ARMOUR PENETRATION AT 0°	200 mm+/1000 m	380 mm/any range

* other projectiles available, some cartridges have reduced propelling charges

† older design projectiles BR-412 and BR-412B also used with rounds giving penetrations of 135 mm and 150 mm respectively

‡ employment with this weapon not confirmed; normally used by D-10 tank gun in T-54/T-55

100 mm field gun M1944 (BS-3) captured by Israel in 1973 (Tom Woltjer via David Isby)

The ammunition used by the M1944 can also be fired by the 100 mm KS-19 anti-aircraft gun, T-54/T-55 MBTs and the SU-100 assault gun. The M1944 can be distinguished from other Soviet and Czechoslovak weapons by the dual tyres on the carriage wheels and the shield with stowage boxes on the front that overhangs the road wheels.

Status: Production complete. In service with Afghanistan, Bulgaria, China (known as Type 59 - unconfirmed), Congo, Cyprus, Egypt, India, North Korea, Mali, Mongolia, Mozambique, Poland, Romania, Somalia, Sudan, Syria (some were mounted on a modified T-34 tank chassis) and Vietnam.

Manufacturer: Soviet state factories.

SPECIFICATIONS

CALIBRE	100 mm
BARREL LENGTH (overall)	6.069 m
MUZZLE BRAKE	double-baffle
RECOIL SYSTEM	hydraulic buffer and hydro-pneumatic recuperator
BREECH MECHANISM	semi-automatic vertical sliding wedge
CARRIAGE	split trail
SHIELD	yes
WEIGHT	
travelling order	3650 kg
firing position	3650 kg
LENGTH (travelling)	9.37 m
WIDTH (travelling)	2.15 m
HEIGHT (travelling)	1.5 m
AXIS OF BORE (at 0°)	1.01 m
GROUND CLEARANCE	0.33 m
TRACK	1.8 m
TYRES	6.50 × 20
ELEVATION/DEPRESSION	+45°/-5°
TRAVERSE	58°
RATE OF FIRE	8-10 rds/min
RANGE	
max	21 000 m
effective HEAT-FS	1000 m
CREW	6
TOWING VEHICLES	AT-P Armoured Tracked Artillery Tractor Ural-375D (6 × 6) truck

85 mm Auxiliary-propelled Field Gun SD-44

Development/Description

The 85 mm auxiliary-propelled field gun SD-44 was designed by the F F Petrov design bureau and is basically the 85 mm Divisional Gun D-44 fitted with an auxiliary propulsion unit: an M-72 two-cylinder petrol engine which develops 14 hp and gives a maximum road speed of 25 km/h. Fuel is carried in one of the trails. The engine is mounted on the left trail and ready-use ammunition is carried in the right one. The engine provides power to the two rubber-tyred carriage wheels via a drive shaft and a differential. The conventional steering wheel and steering column are attached to a castor just behind the spades. The lunette is retained to enable the weapon to be towed if required, maximum towing speed being 60 km/h. A two-part folded rammer is mounted vertically on either side of the front of the shield. An infra-red night vision device can be fitted if required. The SD-44 has sometimes incorrectly been called the D-48.

85 mm auxiliary-propelled field gun SD-44 in travelling configuration

The SD-44 fires the following fixed ammunition:

Ammunition type	**FRAG-HE**	**AP-T**
PROJECTILE DESIGNATION	O-365K*	BR-365†
FUZE MODEL	KTM-1	MD-5
WEIGHT OF PROJECTILE	9.6 kg	9.2 kg
WEIGHT OF BURSTING CHARGE	0.77 kg	0.068 kg
TYPE OF BURSTING CHARGE	TNT	RDX/aluminium
MUZZLE VELOCITY	792 m/s	792 m/s
ARMOUR PENETRATION AT 0°	n/app	125 mm/1000 m

Ammunition type	**HVAP-T**	**HEAT-FS**
PROJECTILE DESIGNATION	BR-365P‡	BK-2M§
FUZE MODEL	n/a	n/a
WEIGHT OF PROJECTILE	5.06 kg	7.34 kg
WEIGHT OF BURSTING CHARGE	n/a	0.96 kg
TYPE OF BURSTING CHARGE	n/a	RDX/wax
MUZZLE VELOCITY	1030 m/s	840 m/s
ARMOUR PENETRATION AT 0°	180 mm/1000 m	300 mm/ any range

* other projectiles available, some cartridges have reduced charges
† other projectiles available
‡ spool type, possibly replaced by improved model
§ employment with 85 mm field guns not confirmed

In the Soviet Army the SD-44 was issued on the scale of 18 per Airborne Division, 6 per anti-tank battery in each of the three Airborne Regiments. Soviet Airborne Divisions have now been restructured as Airborne Rifle Divisions and their provisional TOEs show that they no longer have any SD-44s.

Status: Production complete. In service with Bulgaria, Cuba and Poland.

Manufacturer: Soviet state factories.

SPECIFICATIONS	
CALIBRE	85 mm
BARREL LENGTH (overall)	4.693 m
MUZZLE BRAKE	double-baffle
RECOIL SYSTEM	hydraulic recoil buffer and hydro-pneumatic recuperator
BREECH MECHANISM	semi-automatic vertical sliding wedge
CARRIAGE	split trail
SHIELD	yes
WEIGHT	
travelling order	2250 kg
firing position	2250 kg
LENGTH (travelling)	8.22 m
WIDTH (travelling)	1.78 m
HEIGHT (travelling)	1.42 m
AXIS OF BORE (at 0°)	0.825 m
GROUND CLEARANCE	0.35 m
TRACK	1.434 m
WHEELBASE	3.8 m
TYRES	6.25 × 20
ELEVATION/DEPRESSION	+35°/-7°
TRAVERSE (total)	54°
RATE OF FIRE	10 - 15 rds/min
RANGE	
max	15 650 m*
effective HVAP	1150 m
CREW	7
TOWING VEHICLES	AT-P Armoured Tracked Artillery Tractor GAZ-66 (4 × 4) truck GAZ-63 (4 × 4) truck

* With trails dug in to provide 45° elevation a range of 18 970 m can be achieved.

85 mm Anti-tank Gun D-48

Development/Description

The 85 mm Anti-tank Gun D-48 was originally given the Western designation 100 mm Field Gun M1955 following its first public appearance in the 1955 May Day Parade in Moscow. Subsequent investigations have now determined that the gun is, in fact, a special high-performance 85 mm anti-tank gun designed by the F F Petrov design bureau to replace the 100 mm Field Gun M1944 (BS-3) in anti-tank warfare.

The ammunition for the D-48 was developed for high performance by necking down the cases for the 100 mm ammunition to accommodate a new 85 mm projectile. Two basic forms of projectile include a full calibre hard-core HVAP projectile and at least one type of high explosive round. Soviet references mention an AP projectile weighing 9.3 kg, an HE projectile weighing 9.7 kg and a muzzle velocity of over 1000 m/s. If the AP projectile is fired at 1000 m/s, the HVAP projectile would be fired at nearly 1200 m/s. Estimated penetration could then be about 190 mm for AP and 240 mm for HVAP at a range of 1000 m (0° incidence).

The D-48 was replaced in the Soviet anti-tank artillery formations during the mid-1960s by the 100 mm anti-tank gun T-12. It is a very long-barrelled gun with the entire recoil system over the breech area. A single castor wheel assists in bringing the weapon into and out of action and when travelling the breech is clamped between the twin box-section trails. The D-48 can be fitted with an infra-red night vision device.

It is most likely that the D-48 served as the basis for the 85 mm gun D-70 mounted on the ASU-85 airborne assault gun.

Status: Production complete. Confirmed as being in service with Romania and Soviet reserve units, but reported as being in service in Afghanistan, Congo, Cuba, Ethiopia, India, North Korea, Mongolia, Mozambique, Somalia, Sudan and Vietnam.

Manufacturer: Soviet state factories.

85 mm anti-tank gun D-48 towed by AT-P Armoured Tracked Artillery Tractor

SPECIFICATIONS	
CALIBRE	85 mm
BARREL LENGTH	
overall	6.49 m
without muzzle brake	6.29 m
MUZZLE BRAKE	multi-perforated
BREECH MECHANISM	semi-automatic vertical sliding wedge
CARRIAGE	split trail
SHIELD	yes
WEIGHT	2350 kg
LENGTH TRAVELLING (est)	8.717 m
WIDTH TRAVELLING (est)	1.585 m
HEIGHT TRAVELLING (est)	1.89 m
AXIS OF BORE AT 0° (est)	0.83 m
GROUND CLEARANCE (est)	0.35 m
TRACK (est)	1.2 m
TYRES	6.50 × 20
ELEVATION/DEPRESSION	+35°/-6°
TRAVERSE (total)	54°
RATE OF FIRE	
max	15 rds/min
normal	8-9 rds/min
RANGE	18 970 m
CREW	6
TOWING VEHICLES	AT-P Armoured Tracked Artillery Tractor Ural-375D (6 × 6) truck

85 mm Divisional Gun D-44

85 mm divisional gun D-44-N with infra-red searchlight over ordnance

Development

The 85 mm Divisional Gun D-44 was developed by the F F Petrov design bureau during 1943-44 as the replacement for the 76 mm Division Gun M1942 (ZIS-3) but not issued until after the war. The barrel is a development of the 85 mm gun mounted in the T-34/85 medium tank and is basically identical to both this and the 85 mm M1939 (KS-12) anti-aircraft gun. Ammunition, which is interchangeable with that used in the Czechoslovak 85 mm M52 field gun and the auxiliary-propelled version of the D-44 called the SD-44, is as given in a previous entry in this section.

As far as is known, this weapon is no longer in front-line use with the Soviet Union, but is still used for training.

Description

The shield has a wavy top and a small sliding panel in front of the main gun shield moves in elevation with the barrel. The recoil system is mounted behind the shield and above the cradle and breech-ring. A single castor mounted towards the spade on one of the trails assists in manoeuvring the weapon into or out of action. When fitted with an infra-red night device, which is mounted to the rear of the shield, the D-44 is known as the D-44-N. An auxiliary-propelled version of the D-44 is known as the SD-44, for which there is a separate entry earlier in this section.

Status: Production complete. In service with Albania, Algeria, Angola, Bulgaria, China (Type 56), Cuba, Egypt, Guinea, Guinea-Bissau, Hungary, Iran, Iraq, North Korea, Laos, Mali, Morocco, Mozambique, Poland, Romania, Somalia, Sri Lanka, Sudan, Syria, USSR (Reserve and Militia) and Vietnam.

Manufacturer: Soviet state arsenals including Artillery Plant No 9, Sverdlovsk, USSR.

SPECIFICATIONS		SHIELD	yes	ELEVATION/DEPRESSION	+35°/-7°
CALIBRE	85 mm	WEIGHT		TRAVERSE (total)	54°
BARREL LENGTH (overall)	4.693 m	travelling order	1725 kg	RATE OF FIRE	15-20 rds/min
MUZZLE BRAKE	double-baffle	firing position	1703 kg	RANGE	
RECOIL SYSTEM	hydraulic recoil buffer	LENGTH (travelling)	8.34 m	max	15 650 m*
	and hydro-pneumatic	WIDTH (travelling)	1.78 m	effective HVAP	1150 m
	recuperator	HEIGHT (travelling)	1.42 m	CREW	8
BREECH MECHANISM	semi-automatic	AXIS OF BORE (at 0°)	0.825 m	TOWING VEHICLES	BTR-152 (6 × 6) APC
	vertical	GROUND CLEARANCE	0.35 m		ZIL-157 (6 × 6) truck
	sliding wedge	TRACK	1.434 m		
CARRIAGE	split trail	TYRES	6.50 × 20		

* With trails dug in to provide 45° elevation a maximum range of 18 870 m can be achieved.

82 mm Vasilyek Automatic Mortar

Development/Description

The existence of the 82 mm Vasilyek (Cornflower) (2B9) automatic mortar has been known for some time but few details are as yet available in the West. It is a breech-loaded, recoil-operated mortar fed by four-round clips. The cyclic rate of fire is 40 to 60 rds/min but a clip of four rounds can be fired in two seconds.

The 82 mm barrel is mounted on a towed, split trail, wheeled carriage and in action the weapon rests on the trail legs and a forward extendable leg that is lowered by a hand wheel, with the main carriage raised off the ground. The ordnance then rests on a turning plate allowing 10° of traverse but the saddle is not fixed directly to the plate. Instead, it is supported by hydraulic elevators that can extend the ordnance into a high angle for indirect fire. For this the saddle, cradle, ammunition feed tray and ordnance are elevated together and moved forward so that at full elevation the trunnions are above the carriage axle. Elevation is from 0 to 80°.

Range for direct fire is 1000 m and combat ranges are between 100 and 5000 m. The weapon can be towed but is normally transported under canvas covers on the cargo area of a modified GAZ-66 (4 × 4) 2000 kg truck.

A number of tracked MT-LB multi-role armoured vehicles have been fitted with an 82 mm Vasilyek automatic mortar.

Hungarian 82 mm Vasilyek

In 1990 it was revealed that Hungary had developed a self-propelled version of this system mounted on a modified Artillery Command and Reconnaissance Vehicle (ACRV) and details of this are given in the *Self-propelled guns and howitzers* section under Hungary.

Status: In service with the Soviet Army.

Manufacturer: Soviet state factories.

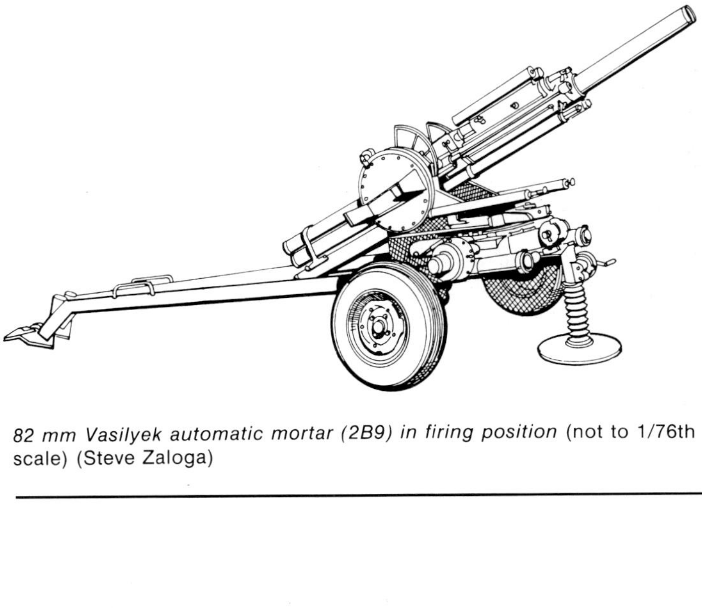

82 mm Vasilyek automatic mortar (2B9) in firing position (not to 1/76th scale) (Steve Zaloga)

82 mm Vasilyek automatic mortar (2B9) in firing position

76 mm Mountain Gun GP (M1966)

Development/Description

The 76 mm mountain gun GP is known in the West by the reporting designation of M1966 but has been referred to as the M1969. The gun was first reported in 1966 and has been used by Soviet troops operating in Afghanistan. Details of the older 76 mm Mountain Gun M1938 were given in *Jane's Armour and Artillery 1987-88*, page 682.

The 76 mm mountain gun GP has a split trail and a carriage arrangement similar to that used by the OTO Melara Model 56 Pack Howitzer, including a split shield. The short barrel has no muzzle brake and reciprocates in a square cross-section slipper. The breech is a horizontal sliding wedge opening to the right. Panoramic and direct fire optical sights are located to the left of the barrel. The split trails can fold sideways and the right-hand trail leg is provided with a metal dolly wheel for ease of handling with the trails folded. The rubber-tyred wheels are on stub axles that can be raised and lowered to alter the height of the gun in action and to level the carriage on rough ground. It is assumed that the gun can be broken down into loads for pack transport.

It has been reported that the gun can fire any type of ammunition used by the 76 mm divisional gun M1942 (ZIS-3) including HE/FRAG, HEAT-FS, HEAT, HVAP and AP-T. The HE/FRAG projectile is fired at a muzzle velocity of 600 m/s while the HEAT projectile can penetrate 300 mm of armour set at right angles at any range. Maximum rate of fire is 15 rds/min with sustained fire being 100 rounds in one hour.

Status: Production complete. In service with Afghanistan and Soviet Union.

Manufacturer: Soviet state factories.

76 mm mountain gun GP (M1966) being manhandled into position

76 mm mountain gun GP (M1966) in firing position

SPECIFICATIONS		WEIGHT		TRAVERSE (total)	50°
CALIBRE	76.2 mm	travelling	780 kg	MAX RANGE	10 500 to 11 500 m
MUZZLE BRAKE	no	firing	780 kg	RATE OF FIRE	
BREECH MECHANISM	horizontal sliding block	LENGTH (travelling)	4.8 m	max	15 rds/min
		WIDTH (travelling)	1.5 m	sustained	100 rds/hr
CARRIAGE	split trail	HEIGHT (travelling)	1.4 m	CREW	7
SHIELD	yes, probably optional	ELEVATION/DEPRESSION	+65°/-5°		

76 mm Divisional Gun M1942 (ZIS-3)

Details of this weapon, introduced into service with the Soviet Army in 1942 and still used in some Third World countries, were given in *Jane's Armour and Artillery 1990-91*, page 674.

UNITED KINGDOM

Vickers Shipbuilding and Engineering 155 mm Ultralightweight Field Howitzer (UFH)

Development

The Ultralightweight Field Howitzer (UFH) has been developed as a private venture and its origins can be traced back to the early 1980s when Vickers Shipbuilding and Engineering Limited (VSEL) originally conceived that there was a potential market for a lightweight 155 mm towed howitzer.

The decision of the US Army to adopt the Royal Ordnance 105 mm Light Gun and the award of a contract to BMY/RO for the prototype of a 155 mm Lightweight Towed Howitzer Demonstrator (LTHD), led VSEL to shelve the whole project for the time being. Another factor in this decision was that they were heavily involved in the development of their private venture GBT 155 turret system and the AS90 self-propelled artillery system which was adopted by the British Army in mid-1989.

In the Spring of 1987 the project was revived and redefined with the object of having a weapon with the same range of the US Army's M198 155 mm towed howitzer but weighing no more than 4000 kg. The current M198 weighs 7163 kg which limits its air mobility and it can only be carried by two helicopters, the US Army CH-47 or US Marine Corps CH-53.

The US Army was fully briefed on the system and agreed that if VSEL built a prototype of the system with its own money it would carry out a complete evaluation of the system.

In September 1987 the VSEL main board gave approval to build two prototypes of the system which subsequently became known as the UFH and the first of these was completed in September 1989 with the second prototype being completed late in 1989.

The decision was taken to build two prototype systems to enable one to be used for the US Army trials while the other one is retained by VSEL for further development work and firing trials in the UK.

Although the UFH was originally targetted at the US Army, it was the US Marine Corps which took the initiative and they are looking for a lightweight 155 mm system to replace all current 105 mm and 155 mm towed artillery systems in service, a total of 610 systems.

Following the successful lift of UFH and a seven man crew by a UH-60L Blackhawk in August 1990, US Army interest has increased and an Operational and Organisational Plan for an advanced lightweight 155 mm Towed Cannon System (LT WT 155) was approved for Commander TRADOC in January 1991. The US Army requirement would be for 960 systems.

In the longer term, perhaps from 2005, the British Army may have a requirement for a new lighter weight 155 mm towed system to replace both the 105 mm Light Gun and the current 155 mm FH-70s.

The complete upper part of the UFH was test fired at Eskmeals in June 1989 with a total of 50 rounds being fired at all elevations of which 12 were zone 8S (top charge).

Meanwhile interest is being shown in the UFH in Europe, the Gulf States and Asia and in late 1991 one of the two prototypes of the UFH underwent a three week evaluation in an undisclosed country in the Middle East.

Following its unveiling at the 1989 Association of the United States Army Exhibition in Washington DC, one of the two prototypes went to the United States for early evaluation. This evaluation, under the supervision of the US Army Armament Research and Development Command on behalf of the US Marine Corps, took place in three phases through to 1990. Phase 1 was at Yuma Proving Ground where the system fired 437 rounds, 50 per cent at Charge 7W or higher (25 per cent at Charge M203 8S). It also completed limited land mobility trials and airlift certification in a single and split mode. At the end of Phase 1 the system was awarded limited live crew clearance for the USA.

First prototype of the private venture VSEL 155 mm UFH deployed in the firing position clearly showing two forward trails which fold back alongside the weapon for travelling

First prototype of the private venture VSEL 155 mm UFH from the rear which has the same range as the current 155 mm M198 towed weapon but is much lighter due to the extensive use of lightweight materials

Phase 2 was conducted at the US Marine Corps Base at Camp Lejeune and at the Naval Base at Little Creek, Virginia. During Phase 2 130 rounds were fired with a live crew and the system achieved a single lift with the UH-60L Blackhawk helicopter. Amphibious trials were carried out successfully at Little Creek.

The final phase took place at Aberdeen Proving Ground where the system carried out successful climatic chamber firings at temperatures ranging from -25° to +145°C. These climatic firings were followed by air transportability (split lift) trials and 622 km of land mobility trials on test tracks ranging from trails to Belgian blocks and included wading to a depth of 1.5 m.

Before designing the UFH VSEL laid down five key requirements. It must weigh less than 4000 kg, have a maximum range of 24 000 m with zone 8 charge, be stable at all firing elevations, be capable of being carried slung under a UH-60 Block 1 helicopter and it must be capable of being split into two loads for transport by lighter helicopers.

To enable these key requirements to be met a number of design principles were laid down including extensive use of lightweight high strength materials, minimum reinforcements and the whole system would be geometrically optimised.

Description

Although the UFH uses advanced materials in its construction, it is simple to operate and maintain under field conditions. VSEL regard this as a low risk programme due to the extensive use of titanium alloy fabrication and its efficient structural design.

VSEL 155 mm UFH deployed in firing position. The trails are of welded titanium alloy with the rear trails folding vertically for transport and forward trails folding horizontally alongside the weapon

VSEL 155 mm UFH slung underneath a UH-60L Blackhawk helicopter during trials in the USA in 1990

The 39 calibre ordnance of the UFH is essentially that of the M284 barrel used by the US Army's M109 HIP and fitted with the M199 muzzle brake as used by the M198 howitzer but modified to take the towing eye. The conventional screw breech is hydraulically operated and opens vertically. The complete ordnance for the UFH was built by Watervliet Arsenal for VSEL.

A standard manual primer loading system is fitted although an automated version is under development.

VSEL stress that there is considerable flexibility in the design of the UFH and the first two systems are concept demonstrators rather than complete prototypes. Having identified areas for improvement during the US evaluation trials and in trials in the UK, VSEL are now building two pre-production prototypes. Before the design is even frozen and a Technical Data Pack (TDP) completed, it will be refined even further.

The UFH has the same elevation as the M198 (-5° to +72°) and the same traverse (22.5° left and right), although zone 8S can only be fired at elevations of above +20°. The prototype weighs well under 4082 kg although the target for production weapons would be 3736 kg.

To keep the weight down extensive use has been made for extruded titanium alloy fabrications which were designed by VSEL but subcontracted to other companies as well as efficient structural design. Some of the parts perform two functions, for example the Air Log designed hydropneumatic suspension system also operates as a hydraulic jack.

The titanium body of the UFH acts as the sole plate to which the two stabilisers and two short box type trails with spades are attached, these open front and rear and are self-levelling and the trails lock up on firing. The rear two trails fold up vertically for travelling while the front stabilisers fold back alongside the body.

The leading arm suspension is of the hydropneumatic type with standard truck wheels and tyres being fitted, these have individual control and are fitted with vehicle towing disc brakes and hand brakes.

The saddle is also of titanium with worm type elevating gear on the left and a simple traverse gearbox mounted on the left side and moving with the saddle. A universal sight mount is fitted on the left side, but there is provision made for a second sight on the right side for customers wishing a two man sighting system.

The cradle consists of four external extruded titanium tubes with the trunnions at the rear, the two aluminium nitrogen cylinders are at the rear of the cradle. The hydropneumatic recoil system is of a novel design with a maximum recoil length of 1.4 m.

A loading tray is provided to the rear of the breech which is opened hydraulically from the right side. VSEL are considering future options which could include a flick rammer and automatic breech opening, but the aim for the present is to keep the weapon as simple as possible.

The VSEL UFH can be brought into action in under two minutes and split into two loads in about 20 minutes.

When in the travelling configuration it is towed by its muzzle in a similar manner to the Soviet 122 mm D-30 howitzer by any truck with a two tonne capacity or greater fitted with air operated brakes.

It can be internally transported in a number of transport aircraft including the C-130, C-141, C-5 and C-160 Transall. It can also be transported slung under the UH-60L/UH-60M Blackhawk, CH-53E/CH-53D helicopters and the MV-22 Osprey.

SPECIFICATIONS		WIDTH		RATE OF FIRE	
CALIBRE	155 mm	towing	3.173 m	intense	4 rpm
BARREL LENGGTH	6.096 m	firing	3.72 m	sustained	2 rpm
MUZZLE BRAKE	double-baffle	HEIGHT (towing)	2.34 m	MAX RANGE	
RECOIL	hydropneumatic	TRUNNION HEIGHT		unassisted	24 000 m
MAX RECOIL	1.4 m	(firing)	0.65 m	assisted	30 000 m
BREECH MECHANISM	screw	GROUND CLEARANCE			
SHIELD	no	(towing)	0.66 m		
WEIGHT	3864 kg	WHEELBASE	2.3 m		
LENGTH		ELEVATION/DEPRESSION	+70°/-2.5°		
towing	9.527 m	TRAVERSE (on gun)	25° (left and right)		
firing	10.210 m				

Status: Prototypes only. Not yet in production or service.

Manufacturer: Vickers Shipbuilding and Engineering Limited
Barrow-in-Furness
Cumbria LA14 1AF
UK
Telephone: (0229) 823366
Telex: 65411 VSEL G
Fax: (0229) 823366

Royal Ordnance 155 mm Light Towed Howitzer

Development

Early in 1990, Royal Ordnance, Guns & Vehicles Division, completed the first of two prototypes of its private venture 155 mm Light Towed Howitzer (LTH).

The LTH was preceded by the Lightweight Towed Howitzer Demonstrator (LTH-D) in 1986. Prime contractor for this was BMY with Royal Ordnance, Phoenix Engineering and Hercules Incorporated being main subcontractors.

This project was zero funded in November 1987 but the curvilinear recoil system concept was successfully tested using the slave cradle hardware late in 1988.

LTH is being developed to meet two market requirements, first a high performance airportable weapon for rapid deployment forces and second a more mobile 155 mm howitzer for land forces. This could replace existing 105 mm and 155 mm weapons with Royal Ordnance seeing both the UK and USA as potential markets, although Vickers Shipbuilding and Engineering Limited (VSEL) have also developed, as a private venture, their 155 mm Ultralightweight Field Howitzer (qv) with a similar market in mind.

The Guns & Vehicle Division is the prime contractor for the LTH with the four major subcontractors being FMC Naval Systems Division, Hercules Aerospace Products Group, International Automotive Design and Phoenix Engineering Inc.

International Automotive Design manufactured the running gear based around production Dunlop Hydrostrut units. Hercules manufactured the composite trail, metal trail and components, RO manufactured the remaining majority of components except for the 155 mm ordnance that was procured by Phoenix Engineering from Watervliet Arsenal.

Development of titanium fabrication technology was carried out with Machine Metal Fabrications, a division of Titanium International. FMC is responsible for assisting with US prototype assembly and testing.

Description

Although the LTH is of conventional appearance with a split trail, conventional upper carriage (saddle) and vertically angled balancing/elevation mechanism, it is in many respects a unique design and incorporates many new features.

The main components of the LTH are the cannon assembly, track and track support structure, recoil system, cradle, upper carriage, combined balancing and elevating (CBE) system, lower carriage running gear, firing platform, trail legs and trail end gear.

The LTH weighs under 3850 kg in standard configuration, this being achieved by the use of composites and other low density/high strength materials with weapon stability being retained by the use of a curvilinear recoil system.

The 39 calibre ordnance has been manufactured by Watervliet Arsenal and is a modified version of the M284 ordnance installed in the US Army's M109A6 HIP (Howitzer Improvement Programme) self-propelled howitzer. The ordnance is fitted with a standard M199 muzzle brake with a (30 per cent efficiency) as fitted to the M198 towed howitzer and a modified M284 breech mechanism.

To enable it to be used in the LTH the ordnance has additional keyways and other features to enable it to connect to the new recoil structure and is also of lighter construction whilst retaining the fatigue life of over 3500 EFCs.

Full scale mock-up of the Royal Ordnance 155 mm Light Towed Howitzer

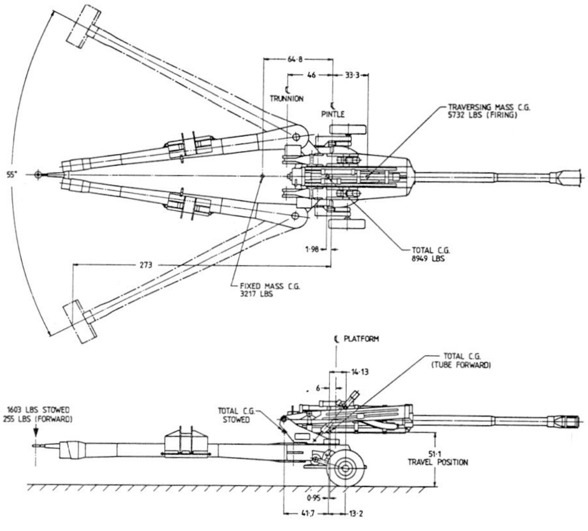

Top and side elevations of the Royal Ordnance 155 mm Light Towed Howitzer (not to 1/76th scale)

RO/BMY 155 mm proof of principle demonstrator during initial firings

The modified M284 breech mechanism has been rotated through 60° and now opens downwards with a modified spring pack to assist in breech closure. The semi-automatic cam follower is removed as manual operation is adequate to meet the rate of fire requirement of four to six rounds a minute with firing being accomplished by a lanyard.

The LTH has a maximum range of 30 000 m firing assisted ammunition and 24 000 m using unassisted ammunition.

Elevation, depression and traverse are identical to that of the current M198 towed system.

The cannon assembly is attached to the recoil system and tracks by a breech band and fabricated track/recoil support structure. A pair of curvilinear tracks are mounted on the cannon assembly by means of the support structure and run from the breech band forward to cater for a maximum recoil length of 1.828 m (top charge) with a total track lift being a maximum of 355 mm.

The track is supported by three fabricated structures mounted around the cannon tube. The track supports and tracks are also aligned and held together by the recoil system cylinders which pass above and below the cannon tube.

The recoil system consists of a single recoil brake above the ordnance with the replenisher mounted to it and a single recuperator below the ordnance with an integral reservoir. For high angle firing, the recoil length is reduced by rotating the recoil brake piston using the cut-off gear. In this case maximum recoil is 1.168 m with no recoil pit being required.

The curvilinear recoil system combines the curved recoil tracks with a specially designed recoil system. The curvilinear recoil system artificially increases the weight of the gun during firing by accelerating the recoiling parts upwards, introducing a stabilising force downwards.

As the LTH fires the ordnance recoils rearwards and pivots slightly upwards as the curvilinear action operates. The LTH cradle is of titanium alloy with a front pivoting bearing for the cannon tube to recoil within, with separate pivots being provided for the recoil brake and recuperator piston rods. The track rollers are mounted near the trunnion end of the cradle and support the rear end of the recoilling mass, taking recoil, curvilinear and rifling torque loads.

The upper carriage is an aluminium alloy fabrication of conventional saddle appearance and provides mounting points for the CBE, traversing and elevating handwheels and cradle. The CBE comprises two hydropneumatic cylinders either side of the upper carriage with separate gas and hydraulic systems. The gas system provides a balancing force to support the elevating mass and to compensate for the changing out-of-balance moment with the gun elevation.

The lower carriage is a titanium alloy fabrication incorporating the pintle, bearing pad for downloads from the upper carriage, mountings for the running gear and short legs with vertical hinges for attachment to the trail legs.

The running gear comprises a hydrogas strut trailing link suspension system with titanium swing arms, hydraulic raise and lower system. The wheels have run flat tyres and are raised and lowered by valves and handpumps. Each wheel has twin caliper disc brakes operated by air over hydraulic system powered from the towing vehicle. To provide stationary parking brakes and to assist in manoeuvring of the weapon, individual handbrake actuators are fitted to each wheel.

The firing platform is of titanium fabrication connected to the lower carriage by a gimbal mount and allows the platform to be positioned in two places, stowed position for travelling which is also used for swing fire operations and a firing position during all fire operations.

The trail legs are of tubular composite (carbon) fibre construction and taper towards the rear. They are connected by hinge joints at the front to the lower carriage and incorporate mounting features at the rear for the trail end spades. Bolted to the end of each trail is the trail leg end gear which comprises an ordnance tube clamp, tow hook, electrical and pneumatic connections.

When travelling the spades are stowed one on each trail. The 155 mm ordnance has an elevation of +72°, depression of -10° with a total traverse of 45°. If a fire mission requires greater than 400 mils traverse from the centre arc, then the spades are released. The howitzer centre of gravity is moved over the firing platform by means of a screw mechanism that pulls the gun over the firing platform. The complete weapon is then rotated about the firing platform to the new azimuth. The howitzer centre of gravity is returned to its firing position by reverse operation of the screw mechanism and the spades fixed into position.

The indirect sight is mounted on the left side of the LTH as is the elevation and traversing mechanism with the direct sights on the right.

The indirect sighting system consists of the M137 panoramic sight, M171 telescope and quadrant mount and M17 fire control quadrant. The direct sight consists of the M172 telescope and quadrant mount, M138 elbow telescope and the M18 fire control quadrant.

When deployed in the firing position the legs of the split trail carriage are splayed out and lock at 55° and the system is supported on three points, the firing platform and the two spades.

To come out of action, the wheels are pushed down by means of hand pumps either side of the lower carriage until the limit is reached. The firing platform is then wound to its stowed position and the spades released to allow the trails to be unlocked and brought together.

When in the towed configuration, the split trail legs are locked together and the traverse mechanism is disconnected by means of a quick-release pin and the traverse parts manually rotated through 180° so that they are over the locked trails. The ordnance is retained by means of a clamp attached to one of the trail legs.

The LTH can be towed by a 5000 kg truck and for airlift can be split into two loads. The design of the LTH is modular so enabling it to be tailored to meet customers' requirements with the lighter version making extensive use of composite materials.

The LTH could also be fitted with a 52 calibre ordnance and a rammer assist device to enable it to be loaded at all elevation angles. This version would have a burst rate of fire of three rounds in 15 seconds. The 52 calibre version would weigh 4355 kg, a semi-automatic vertical sliding breech would be fitted and a maximum range of 38 000 m with an assisted projectile would be achieved or 28 000 m with a standard projectile.

Status: Prototype.

Manufacturer: Royal Ordnance
Guns & Vehicles Division
Royal Ordnance Nottingham
Kings Meadow Road
Nottingham NG2 1EQ
UK
Telephone: (0602) 863341
Telex: 37531
Fax: (0602) 861436

SPECIFICATIONS

CALIBRE	155 mm
MUZZLE BRAKE	double baffle
BREECH MECHANISM	screw, semi-automatic
CARRIAGE	split trail
SHIELD	no
WEIGHT	3850 kg
RECOIL LENGTH	
long	1.828 m
short	1.193 m
LENGTH FIRING	11.938 m
LENGTH TRAVELLING	11.861 m (ordnance forwards)
WIDTH TRAVELLING	2.616 m
WIDTH FIRING	6.229 m
HEIGHT FIRING	1.892 m
HEIGHT TRAVELLING	3.632 m
TRUNNION HEIGHT	1.168 m
TYRES	Michelin 10R × 17.5 with run flat insert
WHEEL TRAVEL	
static to bump	152 mm
static to rebound	50 mm
static to firing position	483 mm
MAX RANGE	
M203A1 Zone 8 charge	30 000 m (assisted)
M203A1 Zone 8 charge	24 000 m (unassisted)
MIN RANGE	
M3A1 Zone 2 charge	3000 m
RATE OF FIRE	
max	4 rpm for 6 mins
sustained	1-2 rpm for 30 mins and 1 rpm thereafter
ELEVATION/DEPRESSION	+72°/-5°
TRAVERSE	22.5° (left and right)
SPEED SHIFT RANGE	360°
TOWING SPEEDS	
primary roads	72 km/h
secondary roads	48 km/h
cross-country	8 km/h
TOWING VEHICLES	M900 series (6 × 6) HMMWV (4 × 4) (limited)
AIR TRANSPORTABILITY	
C-130 aircraft	2 stowed
C-141 aircraft	3 palletised
helicopter	CH-47/CH-53D/CH-53E/EH-101/UH-60E

5.5-inch Medium Gun

Development

The 5.5-inch Medium Gun was developed by the Armament Research and Development Establishment in the late 1930s to meet a requirement for a 5-inch gun to fire a projectile weighing 40.824 kg (90 lb) to a maximum range of 14 630 m. The gun entered production in 1941 and was first used in action in the Middle East the following year.

The 5.5-inch Medium Gun remained in service in diminishing numbers with front-line units of the British Army until 1978 and has now been replaced by the 155 mm FH-70. It was finally phased out of service with the British Army in 1980. In 1982-83 four 5.5-inch Medium Guns were re-introduced into service with the British Army at the Royal School of Artillery, Larkhill, Wiltshire.

Under the Conventional Forces Europe Treaty signed in 1990, the UK stated that it had a total of nine 5.5-inch medium guns, all of which were in the UK.

Description

The carriage of the 5.5-inch Medium Gun is the conventional split trail type. In the firing position each trail is fitted with a spade, which is carried on the top of the trail when travelling. The ordnance has a Welin breech screw mechanism with the recoil system mounted under the rear part. The two spring-balancing cylinders are mounted vertically each side of the ordnance.

When originally introduced the 5.5-inch gun fired an HE projectile weighing 45.35 kg (100 lb) to a maximum range of 14 800 m but in 1943 an HE projectile weighing 36.28 kg (80 lb) was introduced. This had a muzzle velocity of 594 m/s and a maximum range of 16 460 m. Chemical, incendiary and smoke projectiles were developed but none are now used.

In South Africa Armscor still produces ammunition for the 5.5-inch Medium Gun which it has designated the 140 mm G2. In production are two projectiles, an HE weighing 36.3 kg and a Smoke weighing 45.9 kg. A new range of charges has been developed to propel these projectiles. There are four charges plus a Super charge which provides a muzzle velocity of 594 m/s and a maximum range of 16 500 m (Charge 2 has a muzzle velocity of 381 m/s and Charge 4 556 m/s). The range performance of each charge is as follows:

Charge 1	7000 m
Charge 2 (FNH 023 propellant)	10 100 m
Charge 3	12 600 m
Charge 4 (WM 109 or FNHP 038)	15 400 m
Super charge (WM 130)	16 500 m

Armscor have also developed a base bleed (BB) projectile for the 140 mm G2 gun. The BB projectile has a maximum range of 21 000 m and has been qualified with M572, M841 and M8513 PD fuzes. Armscor has quoted that the BB projectile has an even better accuracy performance than its specified probable error in range of 0.6 per cent. The fragmentation effectiveness of the projectile has also been increased by using a high-strength steel and a RDX/TNT filling.

Status: Production completed. In service with Burma, Greece, India, New Zealand, Pakistan, Portugal (reserve), South Africa (designated the 140 mm G2), United Kingdom (training only) and Zimbabwe.

5.5-inch Medium Gun with breech open at the Royal School of Artillery (Christopher F Foss)

SPECIFICATIONS

CALIBRE	139.7 mm
BARREL LENGTH	4.175 m
MUZZLE BRAKE	none
RECOIL	hydropneumatic
BREECH MECHANISM	interrupted screw thread
CARRIAGE	split trail
SHIELD	no
WEIGHT	
travelling order	5850 kg
firing position	5850 kg
LENGTH (travelling)	7.518 m
WIDTH (travelling)	2.54 m
HEIGHT (travelling)	2.616 m
ELEVATION/DEPRESSION	+45°/−5°
TRAVERSE (total)	60°
RATE OF FIRE	2 rds/min
MAX RANGE	
with 45.35 kg HE projectile, m/v 510 m/s	14 800 m
with 36.28 kg HE projectile, m/v of 594 m/s	16 460 m
CREW	10
TOWING VEHICLE	5-ton (6 × 6) truck

Royal Ordnance 105 mm Light Gun

Development

In 1965 the British Army issued a requirement for a new 105 mm Light Gun to replace the Italian OTO Melara 105 mm Pack Howitzer then in use with the Light Regiments of the Royal Artillery. The requirement was for a gun with a longer range, a more stable firing platform and the ability to be towed across rough country at high speed.

Design work began at the Royal Armament Research and Development Establishment at Fort Halstead the following year. Trials with the prototypes were successful and in 1973 the 105 mm Light Gun (L118) was accepted for service with the British Army. First production guns were handed over at Royal Ordnance Nottingham in October 1974. The 105 mm Light Gun, which uses the same range of 105 mm ammunition as fired by the Vickers Abbot self-propelled gun, has now replaced all 105 mm Pack Howitzers in regular units of the Royal Artillery and has been issued to TA units. Each 105 mm Light Gun Regiment has three batteries each with eight guns.

Late in 1981 Australia selected the 105 mm Light Gun to meet its future requirements where it will be known as the Hamel Gun. Details of this programme are given in this section under Australia.

In early 1986, following an extensive series of trials and demonstrations, the Light Gun was type classified by the US Army as the M119. The Light Gun version involved is the L119 with a barrel suitable for firing American 105 mm M1 ammunition. The type classification followed the evaluation of six Light Guns leased to the US Army in late 1984, followed in mid-1985 by a further 14 guns. These were evaluated by the 9th Infantry Division at Fort Lewis, Washington, as the XM119. Early in 1986 the US Army stated that its total procurement objective for the M119 Light Gun was 583 systems. Royal Ordnance Nottingham will build 147 of these with the remainder being manufactured under licence in the United States. Watervliet Arsenal, New York, manufacture the M119 cannon tubes while Rock Island Arsenal, Illinois, produce the recoil mechanism and assemble the weapons for delivery to the US Army.

The United States Army Armament, Munitions and Chemical Command completed fielding of 36 new M119 howitzers to two battalions of the 7th Infantry Division (Light) at Fort Ord, California, in December 1989. By early 1990 the Division had its full complement of 54 weapons.

By early 1991 production of the Light Gun, excluding licenced production, had passed 650 units with production continuing.

In the Spring of 1989 Royal Ordnance and Oerlikon Aerospace Inc of Canada had signed a MoU under which they would jointly manufacture the Light Gun if this weapon is selected by the Canadian Armed Forces as the replacement for its current 105 mm weapons. The gun would be assembled in Canada with a significant number of parts being manufactured locally.

In mid-1991 the Brazilian Marines placed an order for a small batch of Royal Ordnance 105 mm Light Guns in the L118 configuration.

Description

The 105 mm Light Gun consists of the following main components: elevating mass, saddle and controls, trail and spades, platform, suspension and sighting system.

Royal Ordnance L118 105 mm Light Gun deployed in the firing position on its turntable

The elevating mass consists of the ordnance, recoil system, cradle including trunnions, balancing gear and electric firing gear. The gun is fitted with a multi-baffle muzzle brake which can be easily removed for cleaning. The vertical sliding breech-block is actuated by a lever mounted on the top. The hydropneumatic recoil system has a separate recuperator and the buffer is fitted with a cut-off gear to shorten the recoil length as elevation is increased. The recoil system is mounted in a lightweight fabricated cradle which carries the elevating arc and a simple helical compression spring-balancing gear. The firing gear, which is mounted on the cradle, is fully waterproofed and is of the electric magneto type specially designed for use with the Cap Conducting Composition primer of the Abbot ammunition system.

The saddle is a lightweight fabrication on which the complete elevating mass is carried. It provides a top traverse of 11°. The elevating gear is a simple mechanical system providing constant handwheel effort throughout the range of elevation. A clutch is provided to reduce abnormal firing loads on the equipment.

The trail is fabricated in high strength corrosion-resistant steel and is bow shaped enabling the breech operator and loader to remain within the trail and maintain a high rate of fire at all elevations. The platform is carried on top of the trail during travelling. The forward end carries the saddle, traversing gear, suspension and the layer's seat. At the rear are the non-rotating tow eye, over-run brake, barrel clamp and spades. The spade system caters for varying firing and ground conditions and there are three types of spade. These are a combined rock and digger spade for use, normally with the platform, on rock or firm ground, a field spade for use on very soft ground or when firing without a platform, and a snow spade.

The platform is a circular lightweight fabrication which gives a firm base and permanent gun stability under all conditions. The tyres run on the outer edge of the platform which is connected to the underside of the gun by three wire stays.

The wheels, which have large section 9.00×16 tyres, incorporate special hydraulically operated over-run brakes to ensure safe towing at high speed with a light vehicle such as a one-tonne Land Rover (4×4). The brakes can also be operated during firing by a lever at the rear end of the trail. For manhandling purposes, an individual handbrake is fitted at each wheel. The trailing arm suspension has laminated torsion springs and shock absorbers. The suspension system remains in operation during firing and assists in maintaining the stability of the equipment and reducing carriage stresses.

The indirect and direct sight system and controls are operated by the layer while seated. The indirect laying has been simplified by the abolition of a separate setting for angle of sight. The quadrant elevation is set on the elevation scale. A direct-fire telescope is fitted which incorporates a moving illuminated graticule which is adjusted for lead to allow for target movement and range. A direct-fire night sight is also available. Trilux nuclear light sources are fitted to illuminate all scales and graticules, thus eliminating the use of electric batteries on the equipment.

The L35 charge contains the first five charge zones in a single brass cartridge case. The propellant charge is ignited by the L10 electrical primer which is more reliable than a percussion primer. The Supercharge, L36, for maximum range, is a separate brass case and also has an L10 electrical primer. To enable targets to be engaged between 2400 m and 3600 m, polycarbonate spoiler rings are clipped onto the nose of the projectile; these reduce the range by causing aerodynamic drag.

In mid-1986 it was revealed that Royal Ordnance was working on extended range (base bleed) and cargo rounds as a private venture. The former has been fired and has achieved a range of 21 000 m compared to the 17 200 m of the standard projectile. As of mid-1991 none of these had entered production and significant work was yet to be carried out on them.

The USA is working on the XM919 rocket assisted cartridge to increase range of its 105 mm systems and a Dual Purpose Improved Conventional Munition (DPICM) called the XM915 is also under development.

The 105 mm Light Gun fires the following separate loading ammunition: HE (L31) with the projectile weighing 16.1 kg; HESH (L42) with the

US Army 105 mm M119 Light Guns awaiting delivery at Royal Ordnance Nottingham

projectile weighing 10.49 kg (no longer produced by Royal Ordnance); illuminating (L43); SH/practice (L41) (no longer produced by Royal Ordnance); smoke (L45) with the projectile weighing 15.89 kg; smoke, coloured, red (L37) and smoke, coloured, orange (L38).

The charge system for the L118 Light Gun is as follows:

Charge 1	2500-5700 m (min range using shell spoilers)
Charge 2	2700-7200 m (min range using shell spoilers)
Charge 3	5900-9500 m
Charge 4	7900-12 200 m
Charge 4½	8700-13 600 m
Charge 5	15 300 m max
Charge Super	17 200 m max

It can also fire American 105 mm M1 semi-fixed ammunition by using a complete interchangeable percussion-fired ordnance (L119) which is available. The charge coverage is as follows:

Charge 1	1800-3400 m
Charge 2	2100-4100 m
Charge 3	2600-4800 m
Charge 4	4100-6300 m
Charge 5	4200-7900 m
Charge 6	5000-9500 m
Charge 7	6200-11 500 m

The 105 mm Light Gun can be carried slung under a Puma helicopter as one load. The elevating mass can be removed to give two light helicopter (eg Wessex) loads and the complete weapon can be reassembled with one simple tool in less than 30 minutes.

The normal towing attitude is achieved by jacking up the equipment with a jack stowed on the trail, removing the quick-release wheel and traverse gear pin, swinging the gun through 180°, clamping the barrel to the trail and finally replacing the wheel. The complete conversion takes 60 seconds. The weapon can also be towed with the gun in the forward (eg firing) position. The Royal Artillery use the 1-tonne Land Rover to tow the Royal Ordnance Light Gun while the US Army uses the AM General HMMWV.

Firing US M1 ammunition and using the M200 charge system originally developed for the M204, a range of 14 000 m is achieved. Apart from the different ordnance, there are a number of other differences between the US and UK versions of the 105 mm Light Gun.

These include modifications of the sight to permit use of the deflection method of laying, revision of the graticule pattern of the sight to incorporate 5 mil graduations instead of 10 mil, addition of attachment brackets for the battery computer, provision of attachment brackets for M90 muzzle velocity measurement system, labelling of all items containing radioactive elements (tritium sights), modification of tie down brackets, use of standard US paint type and US camouflage pattern, use of nitrogen in the recoil system instead of air, use of US tail light patterns, incorporating of recent UK MoD minor modifications, elevation level guidance plate for M200 charge system, modified engraving and limited US thread forms in some areas of the weapon.

Variant

To meet the requirements of the Swiss Army the L127A1 Light Gun has been developed by Royal Ordnance Nottingham, with the ordnance being designated the L27A1. This is very similar to the British Army version but can fire Swiss ammunition. Two guns were delivered to Switzerland in 1979-80 with a further four following in 1981, but no production order was placed.

Royal Ordnance L118 105 mm Light Gun on skis

Royal Ordnance L118 105 mm Light Gun in travelling configuration with turntable on top of bow trails (Royal Ordnance)

Royal Ordnance L119 105 mm Light Gun deployed in the firing position and with wheels mounted on turntable

Status: In production. In service with Australia (local production), Bahrain (8), Botswana (6), Brazil (4), Brunei, Ireland (12), Kenya (40), Malawi (12), Morocco, New Zealand, Oman (39), United Arab Emirates, United Kingdom (180) and the United States of America.

Manufacturer: Royal Ordnance
Guns and Vehicles Division
Royal Ordnance Nottingham
Kings Meadow Road
Nottingham NG2 1EQ
UK
Telephone: (0602) 863341
Telex: 37531
Fax: (0602) 861436

Differences between 105 mm Light Gun Ordnances

Weapon designation	Ordnance designation	Rifling length	Number of grooves	Twist	Firing mechanism	Muzzle brake	Max range
L118A1	L19A1	3.21 m	28	1 in 19 constant	electric	double baffle	17 200 m
L119A1	L20A1	2.779 m	36	1 in 18 to 1 in 35	percussion	single baffle	11 500 m
L127A1	L27A1	3.052 m	32	1 in 25.6	percussion	double baffle	n/a

SPECIFICATIONS

CALIBRE	105 mm
MUZZLE BRAKE	double-baffle
RECOIL SYSTEM	hydro-pneumatic
RECOIL LENGTH	
at 0° QE	1.07 m
at 70° QE	0.33 m
BREECH MECHANISM	vertical sliding
CARRIAGE	bow
SHIELD	none
WEIGHT	
travelling order	1860 kg
firing position	1860 kg
load on towing eye	168 kg
elevating mass	1066 kg
lower carriage and essential stores	794 kg
LENGTH TRAVELLING	
gun forwards	6.629 m
gun over trail	4.876 m
LENGTH FIRING (gun at 0°)	7.01 m
WIDTH	
travelling	1.778 m
FIRING	1.778 m
HEIGHT TRAVELLING	
Ordnance forwards	2.63 m
Ordnance trails	1.371 m
TRACK	1.4 m
TYRES	9.00 × 16
ELEVATION/DEPRESSION	+70°/-5.5°
TRAVERSE	
total	11°
on platform	360°
RATE OF FIRE	
for 1 min	8 rds/min
for 3 min	6 rds/min
sustained	3 rds/min
MAX RANGE	17 200 m
MIN RANGE	2500 m

25-pounder Field Gun

25-pounder on tow by ACMAT (4 × 4) truck (Raymond Maloney)

Development
The 25-pounder Field Gun was developed in the late 1930s and entered production in 1939. As an interim measure a large number of 18-pounder field carriages were fitted with the ordnance of the 25-pounder, officially called the 25-pounder Mk 1 but more commonly known as the 18/25-pounder. The first production 25-pounders with the humped box trail, called the 25-pounder Mk 2, were completed in 1940. In 1942 the ordnance was fitted with a two-port Solothurn muzzle brake to enable it to fire an armour-piercing projectile with a higher muzzle velocity, which became the 25-pounder Mk 3 and is the only version likely to be encountered today.

Description
There is a turntable mounted under the humped box trail. When lowered to the ground the wheels of the 25-pounder run around the outside edge of the turntable, allowing the gun crew to change the direction of fire rapidly. A similar feature has been adopted for the 105 mm Light Gun developed in the United Kingdom. The recoil system is mounted under the ordnance.

Ammunition is of the separate loading type with the HE projectile weighing 25 lb (11.34 kg) and having a maximum muzzle velocity of 518 m/s to a maximum range of 12 250 m. The armour-piercing projectile weighs 9.2 kg, has a muzzle velocity of 609 m/s and will penetrate 70 mm of armour at an incidence of 0° at a range of 365 m. Other types of separate-loading ammunition still in use include smoke and illuminating.

Status: Production complete. In service with Bangladesh, Burma, Cyprus, Ireland, India, Malta (saluting purposes only), Pakistan, Portugal, Qatar, Sierra Leone, South Africa (G1), Sri Lanka, Sudan, United Arab Emirates, United Kingdom (training only), Yemen and Zimbabwe.

SPECIFICATIONS					
CALIBRE	87.6 mm	LENGTH (travelling)	7.924 m	CREW	6
BARREL LENGTH	2.35 m	WIDTH (travelling)	2.12 m	TOWING VEHICLE	Bedford (4 × 4) 4000 kg truck
MUZZLE BRAKE	yes	HEIGHT (travelling)	1.65 m		
RECOIL	hydro-pneumatic	AXIS OF BORE (firing)	1.168 m		
BREECH MECHANISM	vertical sliding	GROUND CLEARANCE	0.342 m		
CARRIAGE	box	TRACK	1.778 m		
SHIELD	yes	TYRES	9.00 × 16		
WEIGHT		ELEVATION/DEPRESSION	+40°/-5°		
travelling order		TRAVERSE			
with platform	1800 kg	total	8°		
firing position		on turntable	360°		
with platform	1800 kg	RATE OF FIRE	5 rds/min		
		MAX RANGE	12 250 m		

UNITED STATES OF AMERICA

8-inch Howitzer M115

Development
The Westervelt Board recommended the development of a howitzer with a calibre of about eight inches (203 mm) mounted on a carriage with an elevation of 0 to +65° and a traverse of 360°, and a maximum range of 18 000 yards (16 460 m). It was also stated that it would be desirable for a carriage to be developed which could mount either an 8-inch howitzer or a 155 mm gun. Two 8-inch howitzers were subsequently developed, the M1920 and the M1920M1, plus a carriage which could also be used for a 155 mm gun.

Owing to a shortage of funds the project was suspended until 1927. Between 1927 and 1929 the carriage was redesigned for high-speed transport and the breech-ring of the 8-inch howitzer modified. The howitzer was known as the T2 and was to have a centrifugal cast barrel but as facilities for centrifugal casting were not then available it was decided to design a barrel of cold-worked forged steel under the designation T3. Tests of the T3 howitzer began at Aberdeen Proving Ground in 1939 and the following year the T3 howitzer was standardised as the 8-inch Howitzer M1 and the 155 mm Gun/8-inch Howitzer Carriage T2E1 with minor modifications was standardised as the M1.

In the post-Second World War period the complete weapon was redesignated the Howitzer, Heavy, Towed: 8-inch: M115, which consisted of Cannon Howitzer M2 or M2A1, Recoil Mechanisms M4 series (M4, M4A1, M4A2 or M4A3), and Carriage M1.

Description
The cannon consists of the barrel, breech-ring, and breech mechanism. The M2A1 cannon is ballistically and physically interchangeable with the cannon M2, the major difference being that the M2A1 has been designed with stronger steels. The carriage consists of equilibrator assemblies, elevating and traversing mechanisms, two single-wheel, single-axle heavy limber M5, two-axle bogie with eight tyres and two trails. Four spades, carried on the trails, are used to emplace the weapon. In recent years some armies tow the weapon without the limber.

The M115 fires the following separate-loading ammunition:
HE (M106) with the projectile weighing 92.53 kg, maximum muzzle velocity of 587 m/s to a maximum range (charge 7) of 16 800 m
HE (M404) (carries 104 M43A1 grenades) with the projectile weighing 90.72 kg, maximum muzzle velocity of 587 m/s to a maximum range (charge 7) of 16 800 m

8-inch howitzer M115 firing with ordnance recoiling to the rear (US Army)

HE (M509) (carries 195 M42 grenades) with the projectile weighing 93.66 kg, maximum muzzle velocity of 594.4 m/s to a maximum range of 16 000 m
Agent GB or VX (M426) with the projectile weighing 92.53 kg, maximum muzzle velocity of 587 m/s to a maximum range (charge 7) of 16 800 m
Nuclear (M422) which has W33 warhead in 5 to 10 kt range.

The carriage of the 8-inch Howitzer is also used to mount the 155 mm Gun M1 (Long Tom) which is still used by some armies. In many countries the 8-inch M115 has been replaced by self-propelled version called the M110, details of which are given in the *Self-propelled guns and howitzers* section.

Status: Production complete. In service with Denmark, Greece, Iran, Italy, Jordan, South Korea, Spain, Taiwan and Turkey.

SPECIFICATIONS		WEIGHT		TYRES	11.00 × 20
CALIBRE	203.2 mm	travelling order	14 515 kg	ELEVATION/DEPRESSION	+65°/−2°
BARREL LENGTH	5.142 m	firing position	13 471 kg	TRAVERSE (total)	60°
MUZZLE BRAKE	none	LENGTH (travelling)	10.972 m	RATE OF FIRE	
RECOIL SYSTEM	hydro-pneumatic	WIDTH		max	1 rd/min
BREECH MECHANISM	interrupted screw stepped thread	travelling	2.844 m	sustained	1 round per 2 min
		firing	6.857 m	MAX RANGE	16 800 m
CARRIAGE	split trail with limber	HEIGHT (travelling)	2.743 m	CREW	14
		GROUND CLEARANCE	0.318 m	TOWING VEHICLES	M125 (6 × 6) truck
SHIELD	no	TRACK	1.98 m		

155 mm Lightweight Indirect Fire Weapon System (LWIFWS)

Development/Description

The United States Marine Corps has drafted a Required Operational Capability (ROC) document for a new 155 mm lightweight towed howitzer that is also referred to as the Lightweight Indirect Fire Weapon System (LWIFWS).

Prior to this the US Army was working on a 155 mm Lightweight Towed Howitzer Demonstrator (LWTD), but this was cancelled before a complete weapon system was assembled. Additional details on the LWTD are given in the entry, under the United Kingdom, for the new Royal Ordnance 155 mm Lightweight Towed Howitzer project.

The two private venture contenders for this are the Royal Ordnance 155 mm Lightweight Towed Howitzer and the Vickers Shipbuilding and Engineering Limited (VSEL) 155 mm Ultralightweight Field Howitzer (UFH), also covered in this section under the UK.

Status: Not yet funded by the US Army or Marine Corps.

155 mm Howitzer M198

Development

Development of a 155 mm Howitzer to replace the 155 mm Howitzer M114/M114A1 began in 1968 under the direction of the Project Manager for Cannon Artillery Weapon Systems of the United States Army Armament Command at Rock Island, Illinois.

A test rig was built to demonstrate that a lightweight towed 155 mm howitzer could fire the increased range of ammunition with reasonable probability of stable dynamics and structural soundness. Design and fabrication of an advanced development prototype began in 1969 and firing trials in 1970.

Rock Island was responsible for the carriage and recoil mechanism, Frankford Arsenal for the fire control equipment, Watervliet Arsenal for the ordnance and Picatinny Arsenal and the Harry Diamond Laboratory for the ammunition.

The first two prototypes were delivered in April and May 1972 and were followed by a further eight prototypes.

After trials the XM198 was standardised as the M198 and production commenced at Rock Island Arsenal in 1978 with the first production weapons being completed in July 1978. Since then large numbers have been built for the home and export markets but there has been no US Army funding since FY82. It was planned to start funding again in FY89 through to FY92 to complete reserve component fielding but this was stopped due to budget restrictions. By 1988 it is believed that about 1800 M198s had been built of which over 1000 had been delivered to the US Army and Marine Corps.

The M198 replaces M114A1s in separate field artillery batteries assigned to Corps/Army and the general support field artillery battalions of the light divisions. It is deployed in 24 gun battalions, with each battalion having three batteries each with eight guns. The first battalion of M198s became operational at Fort Bragg in April 1979. The first Marine Corps battalion to be equipped with the M198 was the 2nd battalion, 10th Marines, at Camp LeJeune.

The M45 recoil mechanisms were produced by Rock Island Arsenal, carriages by the ConDiesel Mobile Equipment of Condec Corporation of Old Greenwich, Connecticut, fire control components by Numax Electronics Corporation, Hauppauge, Long Island, New York and the ordnance by Watervliet Arsenal.

Description

The weapon is mounted on a split trail carriage fitted with a two-position rigid suspension system which can be rotated upwards to raise the wheels, allowing the weapon to rest on a non-anchored firing platform. When travelling the ordnance remains forward but for storage it can be traversed through 180° and locked in position over the trails.

The carriage consists of the top carriage, elevating mechanism, equilibrators, traversing mechanism, bottom carriage, wheel suspension assembly, speed shift, trails and firing base. An auxiliary power unit, the MC-4000, has been developed for use with the M198 during amphibious operations, but no procurement decision has yet been made. A proposal has been made that US Marine Corps M198s should be equipped with 42-pitch FloTrak wheel surrounds in order to increase

155 mm howitzer M198 towed by AM General M939 series tactical truck

155 mm howitzer M198 in firing position

US Marine Corps 155 mm howitzer M198 in firing position with breech open (US Marine Corps)

mobility when towed across soft terrain such as beach sand or snow. The FloTraks improve surface flotation and thus reduce motion resistance produced by the M198's high pressure tyres.

The recoil mechanism is of the hydro-pneumatic type with a variable recoil length. The autofrettaged ordnance has a screw-on double-baffle muzzle brake and a screw breech mechanism with conventional type obturator pad and split ring seal obturator.

The fire control equipment includes an M137 panoramic telescope with a magnification of ×4 and a 10° field-of-view, two elevation quadrants (M17 and M18) and an M138 elbow telescope with a magnification of ×8 and a 6° field-of-view.

The internal configuration of the Cannon Assembly M199 is in accordance with the German/British/US/Italian quadrilateral agreement and will fire all current standard and developmental ammunition of the anti-material and anti-personnel types. The weapon can fire the following types of ammunition:

Agent H/HD(M110) with the projectile weighing 42.91 kg with muzzle velocity of 684.3 m/s to a maximum range of 18 100 m
Anti-tank (M718) (containing nine anti-tank mines) with the projectile weighing 46.72 kg with a muzzle velocity of 660 m/s to a maximum range of 17 740 m
Anti-tank (M741) (containing nine anti-tank mines) with the projectile weighing 46.72 kg with a muzzle velocity of 660 m/s to a maximum range of 17 740 m
Cannon Launch Guided Projectile (M712) to a range of 16 400 m
HE(M107) with the projectile weighing 42.91 kg with a muzzle velocity of 684.3 m/s to a maximum range of 18 100 m
HE (M449) (containing 60 anti-personnel grenades) with the projectile weighing 43.09 kg to a maximum range of 18 100 m
HE (M483) (containing 88 dual-purpose grenades) with the projectile weighing 46.53 kg with a muzzle velocity of 660 m/s to a maximum range of 17 500 m
HE (M692) (containing 36 anti-personnel mines) with the projectile weighing 46.5 kg with a muzzle velocity of 660 m/s to a maximum range of 17 740 m
HE (M731) (containing 36 anti-personnel mines) with the projectile weighing 46.5 kg with a muzzle velocity of 660 m/s to a maximum range of 17 740 m
HE (M795), registration projectile for M483, under development to a maximum range of 22 400 m
HERA (M549 series) (RAP) with the projectile weighing 43.54 kg, with a maximum muzzle velocity of 826 m/s to a maximum range of 30 000 m
DPICM (Dual Purpose Improved Conventional Munition) M864 (Type classified late in 1987)
Illuminating (M118 series) with the projectile weighing 46.26 kg with a muzzle velocity of 684.3 m/s to a maximum range of 18 100 m
Illuminating (M485 series) with the projectile weighing 41.73 kg with a muzzle velocity of 684 m/s to a maximum range of 17 500 m
M454, Nuclear, current projectile has W48 warhead with a yield of 0.1 kiloton
REMBASS (XM694E1), under development
Smoke WP (M110 series) with the projectile weighing 44.4 kg with a muzzle velocity of 684.3 m/s to a maximum range of 18 100 m
Smoke BE (M116 series) with the projectile weighing 42.22 kg with a muzzle velocity of 684.3 m/s to a maximum range of 18 100 m
Smoke (M825), base ejection, with a muzzle velocity of 797 m/s to a maximum range of 22 600 m
VX or GB (M121A1) with the projectile weighing 45.5 kg with a muzzle velocity of 684.3 m/s to a maximum range of 18 100 m

SPECIFICATIONS	
CALIBRE	155 mm
BARREL LENGTH	6.096 m
MUZZLE BRAKE	double-baffle
RECOIL	hydropneumatic
BREECH MECHANISM	screw
CARRIAGE	split trail
SHIELD	no
WEIGHT	
travelling order	7163 kg
firing position	7163 kg
LENGTH	
travelling	12.34 m
firing	11 m
with ordnance over trails	7.44 m
WIDTH	
travelling	2.794 m
firing	8.534 m
HEIGHT	
travelling	2.9 m
min firing	1.803 m
with ordnance over trails	2.121 m
GROUND CLEARANCE (travelling)	0.33 m
TRACK	2.362 m
TYRES	16.5 × 19.5
ELEVATION/DEPRESSION	+72°/-5°
TRAVERSE (total)	45°
RATE OF FIRE	4 rds/min
MAX RANGE	
M107 projectile	18 150 m (Z8 charge)
M483A1 projectile	22 000 m (Z8C charge)
M549A1 RAP	30 000 m (Z8S charge)
NOMINAL TOWING SPEED	
cross-country	8 km/h
improved roads	72 km/h
secondary roads	40 to 48 km/h
CREW	11
TOWING VEHICLE	M813 or M939 5-ton (6 × 6) truck

Status: Production complete. In service with US Army and Marine Corps. Overall security assistance orders include Australia (38), Bahrain (14), Ecuador (12), Honduras (4), Lebanon (36), Pakistan (95), Saudi Arabia (66), Somalia (18), Thailand (62) and Tunisia (57).

Manufacturer: Rock Island Arsenal
Rock Island
Illinois 61201
USA

155 mm Gun M59

Details of this weapon, which was developed before the Second World War, were given in *Jane's Armour and Artillery 1990-91*, page 683. It reamins in service with Austria (static role), Denmark, Jordan, South Korea, Pakistan, Taiwan, Turkey and Yugoslavia.

155 mm Howitzer M114

Development

The standard 155 mm howitzer of the US Army between 1920 and 1941 was the M1917 (a French Schneider) and the M1918 (the French weapon manufactured in the United States). Improvements were made to the weapons in the 1920s to enable them to be towed by vehicles rather than horses, and in 1934 the development of a new split-trail carriage was authorised under the designation T3. Development of the T3 was subsequently cancelled in favour of the development of a new 155 mm howitzer (the T3) and a new carriage (the T2), which were standardised as the 155 mm Howitzer M1 and Carriage M1 in April 1941. The carriage was the same as that used for the 4.5-inch Field Gun. By the time production of the M1 had been completed, just over 6000 had been built.

In the post-Second World War period the complete weapon was re-designated the Howitzer, Medium, Towed: 155 mm, M114, consisting of the Cannon (M1 or M1A1), Carriage (M1A1) and Recoil System (M6, M6A1, M6B1 or M6B2). The M114A1 is almost identical to the M114 except that it has an M1A2 carriage.

Description

The howitzer consists of the breech-ring, breech-block, breech mechanism and barrel. The carriage M1A1 or M1A2 consists of top and bottom carriages, wheels and brake mechanisms, trails, equilibrators and the firing jack. The carriage is of the unsprung two-wheel split-trail type with the wheels fitted with combat tyres for high-speed towing. The carriage is equipped with airbrakes that are operated from the prime mover and there are handbrakes on each wheel for parking.

The barrel is balanced by spring-type equilibrators. The recoil mechanism, elevating and traversing mechanism, left and right shields, telescope mount and panoramic telescope are attached to the top carriage. The bottom carriage supports the weight of the top carriage and the barrel, and transmits operating stresses to the firing jack and trails.

The counter-recoil system consists of the recuperator cylinder, counter-recoil cylinder and the counter-recoil and recuperator cylinder head stuffing box which links the two cylinders hydraulically. The recoil system consists of the recoil cylinder, replenisher and variable recoil cam assembly.

The M1 barrel is similar to the M1A1 but the M1A1 uses steels with increased physical properties. The M1A1 carriage has a rack-and-pinion-type firing jack and the M1A2 has a screw-type firing jack. The travelling lock of the M1A2 is equipped with a firing jack hanger whereas the M1A1's is not.

The M114 fires the following separate-loading ammunition:

Agent H/HD (M110) with the projectile weighing 42.91 kg, maximum muzzle velocity of 562.4 m/s to a maximum range (charge 7) of 14 600 m

HE (M107) with the projectile weighing 42.91 kg, maximum muzzle velocity of 563.9 m/s to a maximum range (charge 7) of 14 600 m

HE (M449) (carries 60 anti-personnel grenades) with the projectile weighing 43.09 kg, maximum muzzle velocity of 563.9 m/s to a maximum range (charge 7) of 14 600 m

Illuminating (M485) with the projectile weighing 41.73 kg, maximum muzzle velocity of 576 m/s to a maximum range (charge 7) of 13 600 m

Smoke (M116 series) with the projectile weighing 42.22 kg, maximum muzzle velocity of 563.9 m/s to a maximum range of 14 600 m

Smoke WP (M110 series) with the projectile weighing 44.4 kg, maximum muzzle velocity of 563.9 m/s to a maximum range (charge 7) of 14 600 m

VX or GB (M121A1) projectile weighing 45.5 kg, maximum muzzle velocity of 563.9 m/s to a maximum range of 14 600 m

M454 tactical nuclear to a maximum range of 14 900, this is fitted with an air burst fuze

M712 Copperhead Cannon Launched Guided Projectile

M804 practice, to a maximum range of 14 600 m.

Variants

M114A2

The US Army has retrofitted existing M114A1s with a new and longer barrel (as fitted to the M198), redesignating them as the M114A2, and the maximum projectile/range is given below:

Projectile	Max range
M107 (HE)	14 600 m
M549A1 (RAP)	19 300 m
M449A1 (ICM)	14 600 m
M485 (Illuminating)	13 600 m
M483A1 (DPICM)	14 500 m
M692 (ADAM-L)	14 300 m
M731 (ADAM-S)	14 300 m
M454 (nuclear)	14 900 m
M116A1 (HC)	14 600 m
M110A1 (WP)	14 600 m
M110 (H/HD)	14 600 m
M121 (chemical)	14 600 m
M718 (RAAMS-L)	14 300 m
M741 (RAAMS-S)	14 300 m
M712 (Copperhead)	n/a
M804 (practice)	14 600 m

US Army M114 155 mm howitzer in firing position (US Army)

US Marine Corps M114 155 mm howitzer in travelling configuration without shield (Ray Young)

M114 (Brazil)

Following a series of trials with various 155 mm howitzers, the Brazilian Army has decided to update its existing M114 howitzers. It has around 150 M114s of which about 90 are to be upgraded. It is expected that one updated system will be procured for trials followed by the purchase of six kits. Progressive local production will then follow.

Conerns known to have taken an interest in the programme include CBV teamed with RDM Defence Engineering, ENGESA teamed with GIAT, and Vilares teamed with OTO Melara.

Modified M114 (Netherlands)

Details of the M114/39 Modified Howitzer originally developed by SRC International and now produced by RDM Defense Engineering are given in this section under the Netherlands. Denmark, the Netherlands and Norway have now completed conversion of a total of 230 M114s to the M114/39 configuration.

Modified M114 (Austria)

Details of the Voest-Alpine conversion for the M114, which never entered service, were given in *Jane's Armour and Artillery 1990-91*, page 610.

Modified M114 (Germany)

The German Rheinmetall company developed to the prototype stage a modernised M114 under the designation of the FH 155 (L). The modifications included fitting a new shield with increased protection, new barrel, breech-ring, breech and muzzle brake. It has not so far been adopted.

M114 (Greece)

The Greek Army is expected to convert a number of its existing M114 towed howitzers with a new 39 calibre ordnance, with Hellenic Arms Industry to be prime contractor in Greece. Contenders for this contract are believed to include GIAT Industries of France, OTO Melara of Italy, RDM Defence Engineering of the Netherlands and SITESCA of Spain.

Modified M114 (Israel)

Full details of this conversion, developed as a private venture by Soltam of Israel are given under Israel.

Modified M114 (Italy)

Details of the OTO Melara conversion are given in this section under Italy.

M114F (France)

GIAT Industries of France have developed a conversion package for the M114, details are given under France.

M114 (Spain)

SITECSA of Spain are offering conversions of the M114 to accommodate either 45-calibre or 39-calibre barrels. For details see entry under Spain.

KH179

Details of the South Korean KH179, a conversion of the M114, are given under Korea, South.

SPECIFICATIONS		WEIGHT		RATE OF FIRE	
CALIBRE	155 mm	travelling order	5800 kg	first ½ min	2 rounds
BARREL LENGTH		firing position	5760 kg	first 4 min	8 rounds
muzzle to rear end		LENGTH (travelling)	7.315 m	first 10 min	16 rounds
of breech mechanism	3.778 m	WIDTH (travelling)	2.438 m	per hour, sustained	40 rounds
excluding breech	3.626 m	HEIGHT (travelling)	1.803 m	MAX RANGE	14 600 m
MUZZLE BRAKE	no	AXIS OF BORE (at 0°)	1.676 m	CREW	11
RECOIL SYSTEM	hydro-pneumatic	GROUND CLEARANCE	0.229 m	TOWING VEHICLES	5-ton (6 × 6) truck
BREECH MECHANISM	stepped thread, interrupted screw	TRACK	2.07 m		
		TYRES	14.00 × 20		
CARRIAGE	split trail	ELEVATION/DEPRESSION	+63°/-2°		
SHIELD	optional	TRAVERSE	(right) 25° (left) 24°		

Status: Production complete. In service with Argentina, Austria, Belgium, Brazil, Cambodia, Canada, Chile (including Marines), Denmark, El Salvador, Ethiopia, Germany, Greece, Iran, Iraq, Israel, Italy, Jordan, South Korea (licence production), Laos, Lebanon, Libya, Morocco, Netherlands, Norway, Pakistan, Peru, Philippines, Portugal, Saudi Arabia, Senegal, Singapore, Somalia (6), Spain, Sudan (12), Taiwan (including self-propelled model on M108/M109 chassis, details of which are given in the *Self-propelled guns and howitzers* section), Thailand, Tunisia, Turkey, USA, Uruguay, Vietnam, Yemen, and Yugoslavia (which produces a similar weapon called the M65, details of which are given under Yugoslavia). The M114s of Denmark, the Netherlands and Norway have now all been upgraded to the M114/39 configuration with first systems delivered in 1989.

105 mm Howitzer M102

Development

The requirement for a light towed 105 mm howitzer to replace the 105 mm M101 in selected units was issued in 1960. The first prototype was completed at Rock Arsenal in 1962 and was type classified as Standard A in December 1963. First production M102s were completed at Rock Island in December 1965 and the howitzer was first used in combat in South Vietnam the following month. The weapon subsequently underwent major modifications to solve design problems.

Over 1200 M102s were built by Rock Island Arsenal before production was finally completed in the early 1970s. A further 22 were built for foreign military sales in FY80. The main advantage of the M102 over the earlier M101 is that it weighs only 1496 kg compared with 2258 kg and it can also be traversed rapidly through 360° to engage targets in other sectors. In 1985 the US Army had a total of 526 M102 howitzers in its inventory. The M102 is being replaced in many units by the 105 mm Light Gun which is designated the M119 in US Army service.

Description

The 105 mm Howitzer M102 consists of the M137 cannon, M37 recoil mechanism, M31 carriage and the fire control system.

The two-wheeled box carriage is constructed of aluminium. In the firing position a circular baseplate is lowered to the ground under the forward part of the carriage, the complete carriage can be quickly swung through 360° by means of a roller located at the rear of the trail assembly. The baseplate is staked to the ground before firing begins. The elevating mechanism utilises a pair of elevating screw assemblies and permits an elevation of +75° and depression of -5°. The variable length recoil system used on this weapon eliminates the need for a recoil pit.

The fire control system consists of the M1A2 quadrant, M14A1 quadrant, M113A1 panoramic telescope and an M114A1 elbow telescope. Radioactive light sources are used to illuminate the fire control system. The tritium light sources are a heavy radioactive isotope of hydrogen gas, sealed in a pyrex glass container, coated with phosphor. Weapons Command initiated a self-illumination development programme at Frankford Arsenal in January 1969 in response to numerous reports of poor instrument light performance in Vietnam. The equipment successfully completed testing in February 1970. Production contracts for the modification kits were awarded in September 1973 and the first kits were issued in 1975.

The M102 fires the same range of ammunition as the M101 but with a slightly higher muzzle velocity and a slightly longer range:

Agent (M60) with the complete round weighing 19.46 kg, maximum muzzle velocity of 494 m/s and a maximum range (charge 7) of 11 500 m

Agent (M360) with the complete round weighing 19.89 kg, maximum muzzle velocity of 494 m/s and a maximum range (charge 7) of 11 500 m

APERS-T (M546) with the complete round weighing 17.35 kg, maximum muzzle velocity of 549 m/s and a maximum range (charge 7) of 12 400 m

HE (M1) with the complete round weighing 21.06 kg, maximum muzzle velocity of 494 m/s, to a maximum range (charge 7) of 11 500 m

HE (M413) (carries 18 M35 grenades) with the complete round weighing 19.06 kg, maximum muzzle velocity of 494 m/s, and a maximum range (charge 7) of 11 500 m

HE (M444) (carries 18 M39 grenades) with the complete round weighing 19.06 kg, maximum muzzle velocity of 494 m/s, and a maximum range (charge 7) of 11 500 m

HEP (M327) with the complete round weighing 15.27 kg, maximum muzzle velocity of 559 m/s and a maximum range of 8685 m

HERA (M548) (RAP) with the complete round weighing 17.46 kg, maximum muzzle velocity of 548 m/s and a maximum range of 15 100 m

Illuminating (M314 series) with the complete round weighing 21.06 kg, maximum muzzle velocity of 494 m/s and a maximum range (charge 7) of 11 500 m

Smoke (M60 series) with the complete round weighing 19.46 kg, maximum muzzle velocity of 494 m/s, and a maximum range (charge 7) of 11 500 m

Smoke (M84 series) with the complete round weighing 19.03 kg, maximum muzzle velocity of 494 m/s, and a maximum range (charge 7) of 11 500 m

Tactical CS (M629) with the projectile weighting 19.05 kg, muzzle velocity of 494 m/s and a maximum range of 11 500 m

M629, gas, TAC CS to a maximum range of 11 500 m

M760, HE, to a maximum range of 14 000 m.

105 mm howitzer M102 belonging to the 7th Infantry Division at Fort Ord, California (Mike Green)

SPECIFICATIONS		WEIGHT		ELEVATION/DEPRESSION	+75°/-5°
CALIBRE	105 mm	travelling order	1496 kg	TRAVERSE	360°
BARREL LENGTH	3.382 m	firing position	1496 kg	RATE OF FIRE	10 rds/min
MUZZLE BRAKE	none	LENGTH (travelling)	5.182 m	MAX RANGE	
RECOIL	hydro-pneumatic	WIDTH (travelling)	1.964 m	with HE (M1)	11 500 m
BREECH	vertical sliding wedge	HEIGHT (travelling)	1.594 m	with HERA (M548)	15 100 m
		GROUND CLEARANCE		CREW	8
CARRIAGE	box trail	(travelling)	0.33 m	TOWING VEHICLE	(6 × 6) truck
SHIELD	no	TYRES	7.00 × 16		

Status: Production complete. In service with Brazil, Cambodia, El Salvador, France, Greece, Guatemala, Honduras, Israel, Jordan, South Korea, Laos, Saudi Arabia, (National Guard), Turkey, USA, Uruguay, Vietnam and Yemen.

Manufacturer: Rock Island Arsenal
Rock Island
Illinois 61201
USA

105 mm Howitzer M101

Development

In 1919 the Westervelt Board recommended the development of a howitzer of about 105 mm calibre, firing a projectile weighing 30 to 35 lb (13.61 to 15.87 kg), with a maximum range of about 12 000 yards (10 973 m). The carriage would permit elevation of the howitzer from -5° to +65°, and a total traverse of 360°. In addition the carriage would have to accept a 75 mm gun.

Development followed of the 105 mm Howitzer M1920, 105 mm Howitzer Carriage M1920E (split trails) and the 105 mm Howitzer Carriage M1921E (box trail). The M1921E was considered the superior of the two carriages and was then developed to become the M1925E1 while the howitzer became the M1925E. At the same time a further howitzer was developed under the designation T2 and a further carriage also under the designation of the T2. In 1928 these were standardised as the 105 mm Howitzer M1 and the 105 mm Howitzer Carriage M1 but, owing to a shortage of funds, neither was placed in production.

In 1934 the M1 howitzer was modified to enable it to fire fixed ammunition and was subsequently standardised as the 105 mm Howitzer M2. This, when modified for mounting on the M2 carriage, was standardised as the 105 mm Howitzer M2A1 in March 1940.

In 1933 it was decided to redesign the existing M1 carriage but three years later it was decided to develop new carriages under the designation of the T3 and the T4. The T4 was found to be the better of the two and further development resulted in the T4E1 and finally the T5. The T5 was standardised as the 105 mm Howitzer Carriage M2 early in 1940. Shortly afterwards the 105 mm Howitzer Carriage M2A1 was standardised.

By the end of the Second World War 8536 105 mm towed howitzers had been built and post-war production continued at Rock Island Arsenal until 1953, by which time 10 202 had been built. In 1985 the US Army had six M101 and 494 M101A1 howitzers in its inventory. The M101 105 mm howitzer was replaced in US Army service by the M102 howitzer which in turn is now being replaced by the British Royal Ordnance 105 mm Light Gun.

Description

In the post-Second World War period the complete weapon was re-designated the Howitzer, Light, Towed: 105 mm, M101 and M101A1. The M101 consists of Cannon M2A1 or M2A2, Carriage M2A1 and Recoil System (M2A1, M2A2, M2A3, M2A4 or M2A5). The M101A1 is identical except that it has an M2A2 carriage. The difference between the Cannon M2A1 and the M2A2 is that the M2A1's muzzle end is straight and the M2A2's is bell-shaped. The M2A2 carriage is fitted with a main shield group composed of right and left upper and lower shields, right and left top flaps, a bottom flap, and left and right auxiliary shields; while the earlier M2A1 is equipped with a left and right main shield and a top shield.

The barrel consists of the barrel assembly, breech-ring and locking ring. The carriage is of the single axle split trail type with a drawbar for securing to the prime mover. The carriage consists of the equilibrator, shield, elevating mechanism, cradle gears, elevating arcs, traversing mechanisms, top carriage, wheels and trails. Fire control equipment consists of a Telescope Elbow M16A1D with a magnification of ×3 and a 13° field-of-view, a Telescope Panoramic M12A7S with a magnification of ×4 and a 10° field-of-view and Quadrant M4A1.

The M101 fires the following semi-fixed ammunition:
Agent (M60) with the complete round weighing 19.46 kg, maximum muzzle velocity of 472.4 m/s and a maximum range (charge 7) of 11 270 m
Agent (M360) with the complete round weighing 19.89 kg, maximum muzzle velocity of 472.4 m/s and a maximum range (charge 7) of 11 270 m

Australian Army M101A1 in firing position (Australian Army)

105 mm modified M101 howitzer of German Army with new barrel fitted with muzzle brake (German Army)

APERS-T (M546) with the complete round weighing 17.35 kg, maximum muzzle velocity of 504 m/s and a maximum range (charge 7) of 11 600 m
HE (M1) with the complete round weighing 21.06 kg, maximum muzzle velocity of 472.4 m/s and a maximum range (charge 7) of 11 270 m
HE (M444) (carries 18 M39 grenades) with the complete round weighing 19.05 kg, maximum muzzle velocity of 472.4 m/s and a maximum range (charge 7) of 11 270 m
HEP (M327) with the complete round weighing 15.17 kg, maximum muzzle velocity of 559 m/s and a maximum range of 8685 m
HERA (M548) (RAP) with the complete round weighing 17.46 kg, maximum muzzle velocity of 548 m/s and a maximum range of 14 600 m
Illuminating (M314 series) with the complete round weighing 21.06 kg, maximum muzzle velocity of 472.4 m/s and a maximum range (charge 7) of 11 270 m
Smoke (M60 series) with the complete round weighing 19.46 kg, maximum muzzle velocity of 472.4 m/s and a maximum range (charge 7) of 11 270 m
Smoke (M84 series) with the projectile weighing 19.03 kg, maximum muzzle velocity of 472.4 m/s and a maximum range of 11 270 m
Tactical CS (M629) with muzzle velocity 472.4 m/s and maximum range of 11 270 m
M546 (APERS-T) with a maximum range of 11 600 m.

Variants

M101 Modified (Germany)

The German Rheinmetall Company has fitted all M101s used by the Federal German Army with a new longer barrel with a single-baffle muzzle brake and new sights. It fires an HE projectile weighing 21.06 kg, with a maximum muzzle velocity of 600 m/s to a maximum range of 14 100 m.

Upgraded M101 (France)

Details of the GIAT Industries upgraded M101 are given under France.

105 mm M101 howitzer of Belgian Parachute Commando battery (C R Zwart)

105 mm M101 series howitzer of Brazilian Marines in firing position (Ronaldo S Olive)

Upgraded M101 (Netherlands)

Details of the RDM upgraded M101 are given under the Netherlands.

M7 105 mm Self-propelled Howitzer

This is armed with an M2 or M2A1 howitzer with an elevation of +35°, a depression of -5°, and a traverse of 30° left and 15° right. Details were given in *Jane's Armour and Artillery 1983-84*, page 521.

Status: In service with Argentina, Australia, Austria, Bangladesh, Belgium, Benin, Bolivia, Brazil, Burkina Faso, Burma, Cambodia, Cameroon, Canada (called C1, some built in Canada in 1950s), Chile, Colombia, Cyprus, Denmark, Dominican Republic, Ecuador, El Salvador, Ethiopia, France, the Gambia, Germany, Greece, Guatemala, Guinea, Guinea-Bissau, Haiti, Honduras, Indonesia (in 1981 the Defense Department notified Congress that it intended to sell Indonesia 133 M101A1s at a cost of $16 million), Iran, Iraq, Israel, Italy, Ivory Coast, Japan, Jordan, South Korea, Laos, Lebanon, Liberia, Libya, Mexico, Morocco, Mozambique, Netherlands, New Zealand, Nicaragua, Norway, Pakistan, Paraguay, Peru, Philippines, Portugal, Saudi Arabia, Senegal, Sudan, Taiwan (local production as T64), Thailand (has a modified version which is sometimes called the M425), Togo, Tunisia, Turkey, Uruguay, Venezuela, Vietnam, Yemen and Yugoslavia. Production for foreign customer requirements was completed by Rock Island Arsenal in November 1983.

Manufacturer: Rock Island Arsenal
Rock Island
Illinois 61201
USA

SPECIFICATIONS

CALIBRE	105 mm
BARREL LENGTH	
muzzle to rear face of breech-ring	2.574 m
bore	2.363 m
MUZZLE BRAKE	none
RECOIL SYSTEM	hydro-pneumatic
BREECH MECHANISM	horizontal sliding wedge
CARRIAGE	split trail
SHIELD	yes
WEIGHT (travelling order)	
M101	2030 kg
M101A1	2258 kg
WEIGHT (firing position)	
M101	2030 kg
M101A1	2258 kg
LENGTH (travelling)	5.991 m
WIDTH	
travelling	2.159 m
firing	3.657 m
HEIGHT (travelling)	
M101	1.524 m
M101A1	1.574 m
FIRING HEIGHT	3.124 m
AXIS OF BORE (at 0°)	1.3 m
GROUND CLEARANCE	0.356 m
TRACK	1.778 m
TYRES	9.00 × 20
ELEVATION/DEPRESSION	+66°/-5°
TRAVERSE (total)	46°
RATE OF FIRE	
max	10 rds/min
sustained	3 rds/min
MAX RANGE	11 270 m
CREW	8
TOWING VEHICLES	2½-ton (6 × 6) truck

YUGOSLAVIA

155 mm Gun M46/84

Development/Description

The 155 mm Gun M46/84 is a conversion of the Soviet 130 mm Gun M-46 to accommodate a new barrel 45 calibres long and capable of firing ERFB ammunition. This conversion gives a range increase of 45 per cent over the M-46 when firing ERFB-BB ammunition, while the efficiency of the 155 mm ERFB projectile is about three times that of the 130 mm OF-482 projectile. It is believed that this conversion was originally designed by the now defunct Space Research Corporation of Belgium (*Jane's Armour and Artillery 1989-90* page 593). Iraq intended to upgrade its 130 mm M-46 weapons to this standard but conversion had not commenced by the time of the invasion of Kuwait.

In addition to being used to upgrade the 130 mm M-46 it can also be used to upgrade its Chinese equivalent the Type 59-1. Iraq intended to upgrade its 130 mm M-46 weapons to this standard but conversion had not commenced by the time of the invasion of Kuwait.

The M46/84 conversion involved replacing the existing 130 mm barrel by a new barrel 45 calibres long which is designed to make use of the original 130 mm sliding breech and cradle parts.

The barrel is fitted with a high efficiency double-baffle muzzle brake of a new design. An updated equilibrator sealing system is introduced involving a nitrogen pressure adjustment and an updated sealing system for the recuperator mechanism is also required.

The existing 130 mm sights are also revised to accommodate the new 155 mm ballistics with the optical sight graticule being changed and a new range quadrant drum with respective range scales is mounted.

The conversion allows existing 130 mm M-46 and Type 59-1 users to fire standard M107 as well as ERFB and ERFB-BB ammunition to the following ranges:

	PROJECTILES					
Charge zone	**ERFB-BB**		**ERFB**		**M107**	
	m/v m/s	**range m**	**m/v m/s**	**range m**	**m/v m/s**	**range m**
3			266	5900	292	7100
4			302	7300	334	8800
M4A2 5			360	9400	390	10 200
6			458	12 500	469	12 200
7			542	15 200	555	14 500
M2 8			658	19 300	675	17 850
9			796	25 100		
M11 10	897	38 600	897	30 300		

155 mm gun M46/84 in travelling configuration

155 mm gun M46/84 in firing position

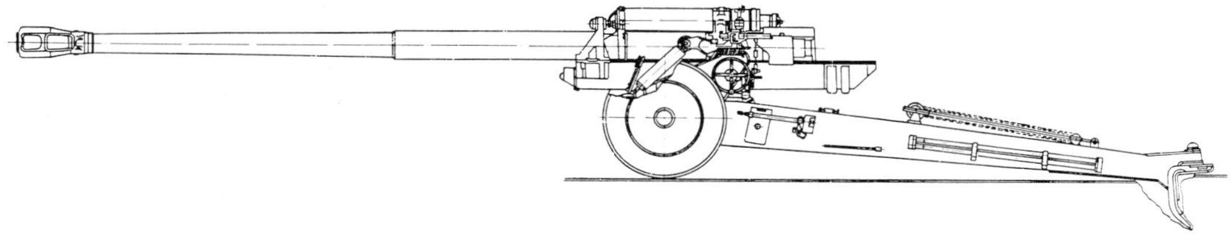

Outline drawing of 155 mm Gun M46/84 deployed in firing position (not to 1/76th scale)

SPECIFICATIONS	
CALIBRE	155 mm
BARREL LENGTH	6.975 m
MUZZLE BRAKE	double baffle
BREECH MECHANISM	horizontal sliding wedge
CARRIAGE	split trail with limber
SHIELD	yes
WEIGHT	
travelling	8428 kg
in action	7680 kg
MASS OF RECOILLING PARTS	2930 kg
MASS OF ELEVATING PARTS	3860 kg
LENGTH	
travelling	11.17 m
WIDTH	
travelling	2.4 m
HEIGHT	
travelling	2.65 m
axis of bore (at 0°)	1.38 m
GROUND CLEARANCE	0.35 m
ELEVATION/DEPRESSION	+45°/-2.5°
TRAVERSE	50°
RECOIL LENGTH	
-2.5° to +20°	1.28 m
+36° to +45°	0.775 m
RATE OF FIRE	
max	6 rds/min
normal	4 rds/min
in 5 minutes	15 rounds
in 10 minutes	25 rounds
in 30 minutes	50 rounds
in 60 minutes	90 rounds
RANGE	
ERFB	30 000 m
ERFB BB	38 900 m
M107	17 850 m
CREW	8 + 1

Status: Development complete.

Contractor: Federal Directorate of Supply and Procurement (SDPR)
9 Nemanjina Street
Belgrade
Yugoslavia
Telephone: 621 522
Telex: 11360

155 mm Howitzer M65

155 mm Howitzer M65

Development/Description

The 155 mm Howitzer M65 is a copy of the American 155 mm Howitzer M114A1 which is also in service with the Yugoslav Army. The M65 is virtually identical to the American model and differs only in detail.

The M65 fires the American M107 HE projectile using a seven-charge propellant system. The M107 HE projectile weighs 43 kg and the propellant used can weigh up to 6 kg. Maximum muzzle velocity is 564 m/s and the barrel life is quoted as approximately 15 000 rounds. Normal rate of fire is 3 rds/min but for prolonged firing this drops to one round a minute.

The M65 can be towed at a speed of 20 km/h.

For training the M65 can be supplied with a 20 mm sub-calibre barrel with the appropriate 20 mm ammunition. This training barrel is mounted in the normal 155 mm barrel. Each M65 is supplied with a kit of tools, spares and accessories, and further tools, spares and accessories can be supplied for holding at battery level.

SPECIFICATIONS					
CALIBRE	155 mm	CARRIAGE	split trail	GROUND CLEARANCE	0.28 m
BARREL LENGTH		SHIELD	yes	ELEVATION/DEPRESSION	+63°/−2°
(with breech ring)	3.778 m	WEIGHT (firing position)	5791 kg	TRAVERSE (total)	48° 42'
MUZZLE BRAKE	no	LENGTH (travelling)	8.08 m	RATE OF FIRE	3 rds/min
RECOIL SYSTEM	hydro-pneumatic	WIDTH (travelling)	2.44 m	MAX RANGE	14 955 m
BREECH MECHANISM	stepped thread/ interrupted screw	HEIGHT		MUZZLE VELOCITY	564 m/s
		top of right shield	1.804 m	CREW	10
		barrel at 0° elevation	1.423 m		

Status: Production complete. In service with the Yugoslav Army.

Contractor: Federal Directorate of Supply and Procurement (SDPR)
PO Box 308
9 Nemanjina Street
Belgrade
Yugoslavia
Telephone: 621 522
Telex: 11360

152 mm Howitzer M84

Development/Description

The 152 mm Howitzer M84 is based on a similar carriage to that of the Soviet 152 mm gun-howitzer D-20 and can fire D-20 ammunition. The barrel is much longer than that of the D-20 and has a semi-automatic vertical wedge-type breech mechanism. A counter-recoil system, consisting of a hydraulic brake, compensator and hydro-pneumatic recuperator, is located over the barrel and when combined with the high efficiency (approximately 50 per cent) multi-baffle muzzle brake the resultant recoil length is short even at high angles of elevation. The maximum rate of fire is 6 rds/min.

The welded steel carriage has a shield and split box section trails fitted with castor wheels to assist in handling. Two hydraulic pumps are provided to assist the nine man crew in bringing the weapon from the travelling to the firing position. These set the weapon on a circular firing platform mounted on the forward part of the carriage and raise the wheels and suspension clear of the ground. Time into action is from three to five minutes.

Sighting equipment includes a range quadrant, a direct fire ×5.5 optical sight, a ×3.7 panoramic sight, a gunner's quadrant, a collimator and a sight illuminating set.

152 mm howitzer M84 in firing position

The 152 mm Howitzer M84 takes between three and five minutes to be brought into action and a similar time to be taken out of action.

It is provided with a single and double line braking installation and a hand brake as well as a lighting system.

The M84 can fire the complete range of Soviet D-20 ammunition including the OF-540 FRAG-HE to a range of 17 190 m with a round muzzle velocity of 647 m/s. The M84 can also fire a M84 HE round weighing 43.56 kg of which 7.677 kg is the cast TNT:RDX (50:50) explosive payload. When firing the M84 HE at a muzzle velocity of 810 m/s the maximum range is 24 400 m (minimum range 5000 m). The propellant charge system uses one full charge and six increments.

The Yugoslav illuminating projectile is called the M88 and falls to the ground at a speed of 5 m/s and gives a maximum intensity of 1.3 million candellas for a maximum period of 60 seconds.

In addition there is a HE/ICM (High Explosive/Improved Conventional Munition) projectile which carries 63 KB-2 bomblets with a maximum range of 22 500 m. Each bomblet is 40 mm in diameter, 85 mm long and weighs 250 g, of which 35 g is FH-5 explosive.

Status: In service with the Yugoslav army.

Contractor: Federal Directorate of Supply and Procurement (SDPR)
PO Box 308
9 Nemanjina Street
Belgrade
Yugoslavia
Telephone: 621 522
Telex: 11360

152 mm howitzer M84 in travelling configuration

SPECIFICATIONS					
CALIBRE	152.4 mm	LENGTH		TRAVERSE	50°
MUZZLE BRAKE	multi-baffle	travelling	11.21 m	RATE OF FIRE	
RECOIL SYSTEM	hydro-pneumatic	firing	9.67 m	max	6 rds/min
BREECH MECHANISM	semi-automatic vertical sliding wedge	WIDTH		normal	4 rds/min
		travelling	2.415 m		15 rounds in 3 min
		firing	5.73 m		40 rounds in 20 min
CARRIAGE	split trail	HEIGHT (travelling)	2.16 m		100 rounds in 1 hour
SHIELD	yes	HEIGHT OF BORE FIRING	1.385 m	MAX RANGE (M84 HE)	24 400 m
WEIGHT		GROUND CLEARANCE	0.37 m	MUZZLE VELOCITY	810 m/s
travelling	7126 kg	TYRES	12.00 × 20	CREW	9
firing	7080 kg	ELEVATION/DEPRESSION	+63°/−5°	TOWING VEHICLE	FAP 2026 BS/AV (6 × 6) truck

122 mm Howitzer D-30J

Development/Description
The 122 mm Howitzer D-30J is a Yugoslav version of the Soviet 122 mm Howitzer D-30 and differs from it in few respects, so for full details refer to the entry under Soviet Union. Details are provided here for the few changes from the Soviet original, using data supplied by the Yugoslavs.

Illustrations of the D-30J indicate that the D-30J is fitted with a simpler double-baffle muzzle brake than the Soviet original. Travelling and combat weights are given as 3440 kg and 3350 kg respectively while the barrel mass is 1050 kg.

The maximum range when firing a standard HE projectile with a full charge is given as 17 300 m. While the D-30J can fire the full family of ammunition developed for the Soviet D-30, the Yugoslavs have under development an M82 HEAT projectile fitted with a UT PE M82 fuze and an M82 illuminating projectile fitted with a ETU 102 time fuze. The Yugoslav M76 projectile has a maximum range of 17 300 m.

Status: Available.

Contractor: Federal Directorate of Supply and Procurement (SDRP)
9 Nemanjina Street
Belgrade
Yugoslavia
Telephone: 621 522
Telex: 11360

105 mm Howitzer M56

Development/Description
Yugoslavia, in addition to using the American 105 mm M101 (M2A1) howitzer and the German 105 mm M18, M18M and M18/40, also uses a 105 mm howitzer of local design and manufacture called the M56. In many respects the M56 is similar to the M18/40 but fires semi-fixed ammunition of the US pattern, rather than the separate-loading type used by the German 105 mm howitzers.

The semi-fixed ammunition fired by the M56 consists of the following three types: HE shell M1, weighing 15 kg and filled with 2.2 kg of explosive, fired at a muzzle velocity of 570 m/s, and with a fuze that can be set for graze or 0.05 second delay; smoke shell WP M60 filled with white phosphorous and weighing 15.8 kg, this shell has the same fuze as the HE M1; the AP-T shell M67 HESH-T weighing 10 kg and fired at a muzzle velocity of 670 m/s to a maximum range of 12 000 m. In use the M67 HESH range is normally limited to 870 m. The 2.2 kg of explosive filling can penetrate an armour plate 100 mm thick set at an angle of 30°. On such a plate the HESH projectile will chip off a film of material opposite the strike of the projectile that is 25 mm thick with the fragments weighing up to 5 kg and travelling at a velocity of 300 m/s.

The M56 may be fitted with a 20 mm sub-calibre training device.

The barrel is mounted on a tubular split trail carriage with a split shield that slopes to the sides and rear. The most common model has pressed alloy wheels with solid rubber tyres similar to those on the German M18 series weapons, but some M56s are fitted with pneumatic-tyred wheels similar to those fitted to the American 105 mm M101 (M2A1). With the pneumatic wheels the M56 can be towed at speeds up to 70 km/h. Normally the M56 is supplied with a range quadrant, panoramic telescope (dial sight), direct firing anti-tank telescope and a gunner's quadrant, all equipped with lights for night firing. The direct fire telescope has a magnification of ×2.8. The panoramic telescope has a magnification of ×4 and weighs 6 kg.

While the carriage of the M56 is normally used for firing with the trail legs spread, in an emergency it can be fired direct from the travelling position. In this configuration the maximum elevation is limited to +16° and the total traverse to 16°.

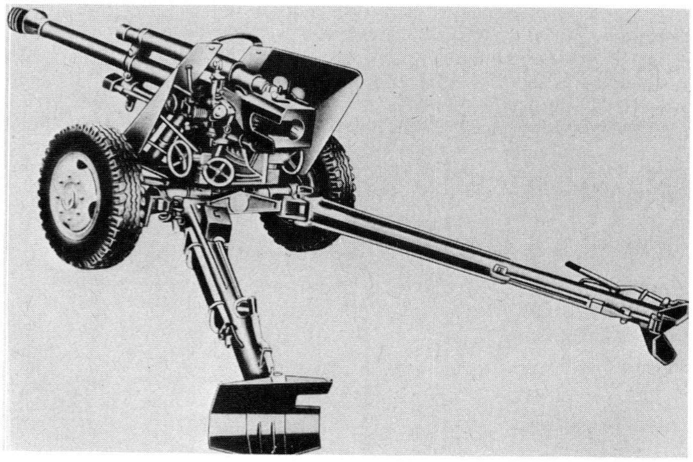

105 mm M56 howitzer in firing position. This model has American type wheels with pneumatic tyres

105 mm M56 howitzer in travelling order. This model has German type wheels with solid rubber tyres

SPECIFICATIONS					
CALIBRE	105 mm	WEIGHT		GROUND CLEARANCE	0.335 m
BARREL LENGTH (overall)	3.48 m	travelling order	2100 kg	TRACK	1.8 m
MUZZLE BRAKE	multi-baffle	firing position	2060 kg	ELEVATION/DEPRESSION	+68°/−12°
RECOIL SYSTEM	hydraulic recoil buffer and hydro-pneumatic recuperator	LENGTH		TRAVERSE (total)	52°
		travelling	6.17 m	RATE OF FIRE	16 rds/min
		firing	5.46 m	MAX RANGE	13 000 m
		WIDTH (travelling)	2.15 m	CREW	7
BREECH MECHANISM	horizontal sliding wedge	HEIGHT		TOWING VEHICLE	TAM 1500 (4 × 4) truck
		travelling	1.56 m		
CARRIAGE	split trail	at max elevation	3.8 m		
SHIELD	yes	AXIS OF BORE (at 0°)	1.13 m		

Status: Production. In service with Burma, Cyprus, El Salvador, Guatemala, Indonesia, Iran and Yugoslavia.

Contractor: Federal Directorate of Supply and Procurement (SDPR)
PO Box 308
9 Nemanjina Street
Belgrade
Yugoslavia
Telephone: 621 522
Telex: 11360

76 mm Mountain Gun M48

Development/Description

The 76 mm Mountain Gun M48, which is often called the 'Tito Gun', was developed in the post-Second World War period specifically to meet the requirements of Yugoslav mountain units, although it can also be used as a field gun. Its low weight has attracted the attention of other countries, especially in the Third World, where weight, rather than range, is the overriding factor.

There are at least four variants of the M48. The M48 (B-1) has pneumatic tyres and a maximum towing speed of 60 km/h. It can also be towed by animals in tandem or broken down into eight pack loads. The M48 (B-1A1-I) has the pneumatic tyres and wheels as fitted to the M48 (B-1), plus some of the features of the suspension of the M48 (B-1A2), but can also be used as a field piece and cannot be towed by animals or broken down for pack transport. The M48 (B-1A2) has light alloy wheels with solid rubber tyres and modified suspension, but has a maximum towing speed of 30 km/h. The latest model is called the B-2 about which little is known at present.

Ammunition is of the semi-fixed type with four charges and based on that used for the obsolete Soviet 76 mm Regimental Gun M1927 which fired fixed ammunition: HE M55 projectile weighing 6.2 kg with a muzzle velocity between 222 and 398 m/s; HEAT projectile weighing 5.1 kg which will penetrate 100 mm of armour at a range of 450 m; and a smoke shell WP M60 weighing 6.2 kg.

All models of the M48 have split trails with each trail hinged in the centre and folding forwards through 180° for towing. At the end of each trail is a slot for the stake. The shield has winged sides and the centre of the shield has to be raised to obtain maximum elevation.

76 mm Mountain Gun M48 (B1)

76 mm Mountain Gun M48 with rectangular cross-section trail legs in horse or mule-towing configuration

SPECIFICATIONS (M48 (B-1))	
CALIBRE	76.2 mm
BARREL LENGTH	1.178 m
MUZZLE BRAKE	multi-baffle
RECOIL SYSTEM	hydraulic
BREECH MECHANISM	semi-automatic sliding wedge breech-block
CARRIAGE	split trail, folding
SHIELD	yes
WEIGHT	
travelling order	720 kg
firing position	705 kg
LENGTH (travelling)	2.42 m
WIDTH	
travelling	1.46 m
firing	2.65 m
HEIGHT	
travelling	1.22 m
elevation 45°	1.67 m
AXIS OF BORE (at 0°)	0.73 m
GROUND CLEARANCE	0.18 m
TRACK	1.3 m
TYRES	6.00 × 16
ELEVATION/DEPRESSION	+45°/-15°
TRAVERSE (total)	50°
RATE OF FIRE	25 rds/min
RANGE	
max	8750 m
normal	8600 m
CREW	6
TOWING VEHICLE	Zastava (4 × 4) AR-51 (the Italian FIAT Campagnola built under licence in Yugoslavia). It has now been replaced in production by a licensed version of the more recent FIAT 1107 (4 × 4) vehicle

Status: Production complete. In service with Burma, Guatemala, India, Indonesia, Romania, Sri Lanka and Yugoslavia.

Contractor: Federal Directorate of Supply and Procurement (SDPR)
PO Box 308
9 Nemanjina Street
Belgrade
Yugoslavia
Telephone: 621 522
Telex: 11360

Multiple rocket launchers

ARGENTINA

Argentinian Multiple Rocket Systems

The Instituto de Investigaciones Cientficas y Técnicas de la Fuerzas Armadas (CITEFA), the Research and Development Institute of the Argentine Armed Forces, has developed three multiple rocket systems for the Argentine Armed Forces.

SLAM-Pampero 105 mm Multiple Rocket System

The 105 mm calibre rocket has a solid propellant motor and is stabilised in flight by four wrap-round fins at its base. A 16-tube launcher is used with two rows each of eight launcher tubes, and a full 16-round salvo will fall within a 300 × 200 m area with maximum round dispersion CEP of less than 1 per cent. The launcher is mounted on the rear of a Unimog 416 (4 × 4) vehicle. To provide a stable firing platform two hydraulic stabilisers are lowered either side at the rear of the vehicle. The 16 rockets can be fired within 7.5 seconds out to a range of 11 000 m. The elevation/depression of the launcher assembly is +52°/0° whilst its traverse is 90° to either side of the vehicle's central axis. The single stage PHE-2 type solid propellent folding fin stabilised rocket uses instruction, exercise, HE-fragmentation, HE-Incendiary and HE-anti-personnel type warheads. The warshots have either contact or delayed action fuzes. If required, the launcher can be mounted on other vehicle types with differing numbers of tubes.

SPECIFICATIONS

CALIBRE	105 mm
WEIGHT	
rocket	28.5 kg
warhead	10.5 kg
LENGTH	1.45 m
MAX RANGE	11 000 m
NUMBER OF TUBES	16
LAUNCHER	Unimog 416 4 × 4 light truck

SAPBA-1 127 mm Multiple Rocket System

The SAPBA (Sistema de Proyectiles Balsticos Autopropulsados) was developed by CETMA as an area saturation MRS based on a 36-tube 127 mm calibre launcher system with a maximum range of 16 km. The fin-stabilised rocket is powered by a PHE-2 type solid propellant rocket motor developing a maximum thrust of 3500 kg. It carries an 18 kg warhead which can either be of the instruction, exercise, HE-incendiary or HE-anti-personnel type that scatters 4500 steel balls over an area of 6000 m². Warshot fuzes are either contact or delay action type.

The 36-tube launcher is subdivided into four three by three rifled barrel pods which are replaced after firing with new modules by the crane aboard the launch platform, a locally assembled FIAT 697N (6 × 6) truck. All 36 rockets can be ripple fired at 0.5 second intervals or singly. The launch vehicle is stabilised by jacks with the launcher aimed hydraulically to 90° right or left of the centreline at elevations between 0° and +60°. Manual traversing and elevation controls are also fitted. A crew of five is carried, a commander, aimer, two loaders and a crewman who deals with levelling the platform and connecting the 10 kg firing control box and its 50 m cable to the vehicle. The time taken for emplacement is two minutes and to come out of battery one minute. Reloading takes five minutes for all four pods.

CP 30 127 mm Multiple Rocket Launcher Systems

The CP 30 multiple rocket launcher system is an area saturation weapon which fires 127 mm calibre rockets of over 30 000 m maximum range. The 78.5 kg rocket uses a composite solid propellant fuel and can be fitted with a number of warhead types: instruction, exercise, HE-Fragmentation, HE-Incendiary or cargo (with either an anti-tank or anti-personnel submunition payload). The fuzing for the conventional rounds is either instant contact or delayed impact. The CEP value is below 1.5 per cent.

The launch platform can vary according to the number of rocket tubes in the launcher assembly. Typically it will be 36 for a FIAT 697N (6 × 6) truck and 27 for an ACMAT VLRA (8 × 8) truck.

Status: Pampero - in production and service with the Argentine Marine Corps. SAPBA-1 - in production and in service with Argentine Army. CP 30 - under advanced development.

Manufacturer: Instituto de Investigaciones Cientficas y Técnicas de la Fuerzas Armadas (CITEFA)
Zufrategut y Varela
1603 Villa Martelli
Provincia de Buenos Aires
Argentina

TAMSE VCLC Series Armoured Rocket Launchers

Development

During the early eighties the Argentinian firm of TAMSE (Tanque Argentino Mediano Sociedad del Estado) which had been formed to manufacture the German-designed TAM medium tank began to investigate the possibilities of a joint venture with Israel Military Industries to fit that company's modular rocket systems onto the TAM chassis.

In 1986 the first prototype vehicle, called the Vehiculo de Combate Lanza Cohetes (VCLC) underwent trials with the 160 mm LARS 160 rocket system. In early 1988 a second version was trialled with the 350 mm calibre MAR 350 system.

Description

The major differences between the VCLC and the TAM medium tank chassis (described under Germany in the *MBTs and medium tanks* section) include the modification of the chassis to incorporate additional hydraulic suspension and movement suppressor systems to absorb the recoil forces during rocket launching and the fitting of a new two-man turret.

On the turret either side at the rear are fitted the support arms for the rocket system modules which are traversed and elevated using the turret's electro-hydraulic power unit. No modifications are needed if a change between the rocket types is required. For self-protection and limited air-defence a pintle-mounted 7.62 mm MAG or 0.30 in Browning MG is mounted above the front of the turret.

When the LARS 160 rocket system is carried the vehicle is designated the VCLC-CAL. This has two 18-round 160 CAL pod container-launchers which, once fired, are unloaded and replaced in a secure area by a supporting vehicle. In Argentinian Army trials the vehicle used was an M809 Series (6 × 6) 5-ton flat-bed truck fitted with a 12 000 kg capacity hydraulically operated extendable crane and carrying two of the 160 CAL pods. The whole reloading operation takes around 10 minutes.

If fitted with the MAR 350 system then the vehicle is designated the VCLC-CAM and has two pairs of side-by-side cylindrical container-launchers for CAM 350 rockets. The resupply operation is similar to that used for the VCLC-CAL vehicle.

In combat the basic six-launcher VCLC battery will have a truck-mounted fire control radar system whose job is to monitor the first ranging

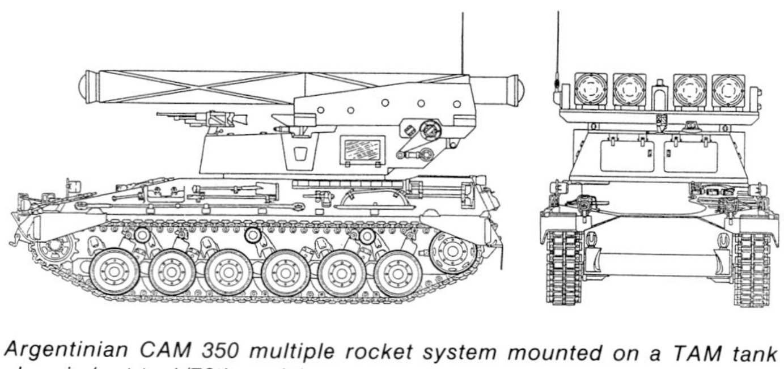

Argentinian CAM 350 multiple rocket system mounted on a TAM tank chassis (not to 1/76th scale)

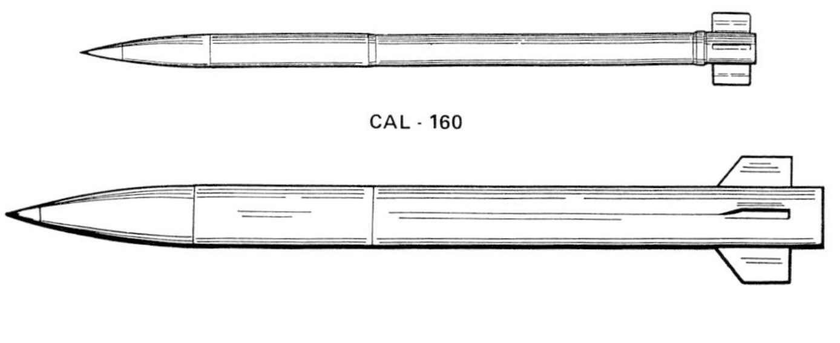

General arrangement drawings of CAL 160 (top) and CAM 350 (bottom) rockets (not to 1/76th scale)

shot for between two-thirds to three-quarters of its flight. The round is deliberately aimed offset from the target so as not to alert it. The tracking data obtained is then fed into the vehicle's fire control computer and used to calculate the launch information which is then transmitted to the battery launchers.

Status: Both the VCLC-CAL and VCLC-CAM are at the prototype stage. A total of five vehicles are believed to be in the trials programme, two in Argentina and the remainder in Israel.

Manufacturer: TAMSE
Av Rolón 1445-Boulogne Sur Mer
CP 1609
Buenos Aires
Argentina
Telephone: 763-2213/4937/1266/0714/2464
Telex: 26208

VCLC-CAL multiple rocket launcher based on a modified TAM tank chassis

SPECIFICATIONS		GROUND PRESSURE	0.84 kg/cm^2	ENGINE	super-charged MTU MB 833 Ka 500 6-cylinder diesel developing 720 hp at 2200 rpm (750 hp at 2400 rpm)
NUMBER OF ROCKETS	2 18-round 160 CAL or 4 single-round 350 CAM	LENGTH (chassis)	6.75 m		
		HEIGHT (with launcher system)	3.05 m		
CREW	3	WIDTH (chassis)	3.12 m		
COMBAT WEIGHT		GROUND CLEARANCE	0.45 m		
(with rocket system)	32 582 kg	TRACK	2.62 m	ARMAMENT	1 × 7.62 mm or 0.3 in MG
POWER-TO-WEIGHT RATIO	22.1 hp/tonne	FUEL CAPACITY	680 l plus 400 l in external auxiliary tanks		

AUSTRIA

130 mm M51 (32-round) Multiple Rocket System on Steyr 680 M3 (6 × 6) chassis

Development/Description
In the 1960s Austria procured a number of 130 mm M51 (32-round) MRSs from Czechoslovakia mounted on the rear of Praga V3S (6 × 6) truck chassis. Some years ago Austria removed the launchers from the Praga V3S (6 × 6) chassis and mounted them on the rear of Austrian Steyr 680 M3 (6 × 6) truck chassis.

No armour is provided for the cab so the launcher is traversed left or right before launching the rockets. Replacement rockets are carried in stowage boxes on either side of the hull under the launcher. The rockets are spin stabilised and it is believed that a rocket with an increased range has been introduced by Czechoslovakia in recent years, but it is not known if Austria has bought any. Romania uses the M51 launcher mounted on the rear of a ZIL-151/ZIL-157 (6 × 6) truck chassis (qv).

Status: Production complete. In service with Austrian Army with three batteries each with a total of six launchers.

Manufacturers: Rocket launcher, Czechoslovak state factories.
Truck, Steyr-Daimler-Puch AG
Werke Steyr
A-4400 Steyr
Austria

130 mm M51 (32-round) multiple rocket system mounted on rear of Steyr 680 M3 (6 × 6) truck chassis (Austrian Army)

SPECIFICATIONS		LENGTH (travelling)	6.73 m	ENGINE	Steyr WD 610.71 6-cylinder super-charged diesel developing 165 hp at 2800 rpm
CALIBRE	130 mm	WIDTH (travelling)	2.4 m		
NUMBER OF BARRELS	32	HEIGHT (travelling)	2.9 m		
LENGTH OF ROCKET	0.8 m	GROUND CLEARANCE	0.3 m		
WEIGHT OF ROCKET	24.2 kg	TRACK			
MAX RANGE	8200 m	front	1.81 m	GEARBOX	manual with 5 forward and 1 reverse gears
MAX VELOCITY	410 m/s	rear	1.72 m		
RELOAD TIME	2 mins	WHEELBASE	2.76 m + 1.2 m	TRANSFER BOX	2-speed
ELEVATION/DEPRESSION	+50°/0°	MAX SPEED	79.7 km/h	TYRES	9.00 × 20
TRAVERSE	240°	FUEL CAPACITY	180 l		
CREW	6	RANGE	500 km		
WEIGHT OF SYSTEM		FORDING	0.5 m		
(loaded)	8400 kg	GRADIENT	74%		

BELGIUM

70 mm (40-round) LAU97 Multiple Rocket Launcher System

Development/Description

Developed for export by Forges de Zeebrugge SA, the LAU97 is a light artillery rocket launcher system designed to provide saturation and holding fire at the regimental level. It consists of a launcher box with five horizontal rows of eight tubes and a 360° traversable base that allows elevation arcs of 0 to +55°. The system can be fitted on any military vehicle or trailer capable of carrying 1200 kg payloads. The launcher's azimuth and elevation laying is performed by either electric motors or hand wheels. A BR2 mortar sight or artillery panoramic sight is fitted for aiming the system with an electrically operated remote-control unit for firing. The rockets can be launched either singly or in ripple sequences up to the full 40-round salvo which takes less than six seconds. The rockets using the standard FZ-68 rocket motor and a 4.3 kg warhead have a maximum range of 7900 m at a launcher elevation of +40°; if the elevation is set at +15° then the range is reduced to 6000 m. If the Mk 40 rocket motor is used the maximum range at +40° elevation is reduced to 7500 m. A full complement of rockets can saturate an area measuring 200 × 300 m at maximum range. The FZ-100 6.2 kg cargo warhead with a payload of nine 0.48 kg anti-personnel/anti-tank bomblets is also available. The bomblets have a lethal radius of 10.5 m and are capable of penetrating up to 105 mm of armour plate. The submunitions are released by a delay-adjustable time fuze over the target area. The other warheads available include the 2.9 kg FZ-32 practice, the 3 kg FZ-49 anti-armour capable of piercing more than 400 mm of armour plate, the 4.25 kg FZ-58 penetration for use against buildings, bunkers and shelters, the 4.15 kg FZ-63 white phosphorous target marker/smoke screen type, the 4.3 kg FZ-71 anti-personnel which forms more than 8000 fragments on detonation and is lethal out to between 18 and 21 m radius, the 4.3 kg FZ-85 prefragmented anti-personnel anti-light armour, the 3.935 kg chaff which corresponds to a radar area of about 600 m^2 and the 4.9 kg M257 illuminating type with a 120 second duration one million candle lighting capacity. The launcher is operated by a crew of three, a layer, a commander and a gunner, who also reload the system by hand.

ACMAT VLRA Type TPK 4.15 LRM (4 × 4) truck fitted with LAU97 launcher

LAU97 multiple rocket launcher mounted on FZ-developed Stagricol trailer

Variants

This rocket launcher can also be mounted on a variety of tracked and wheeled chassis and in June 1984 was demonstrated on the rear of an Alvis Streaker high mobility load carrier in palletised form. The 70 mm (40-round) LAU97 multiple rocket system has been installed on a ACMAT VLRA Type TPK 4.15 LRM (4 × 4) truck, the Alvis Saladin armoured car (turret removed) and the Pinzgauer (6 × 6) vehicle. There is also a towed version.

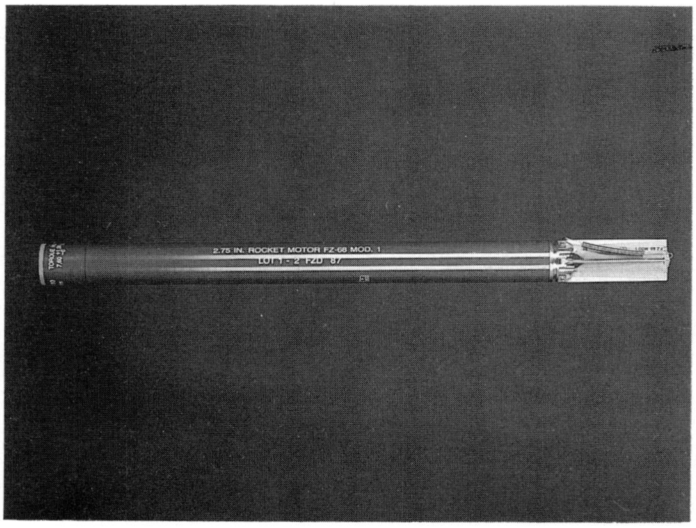

FZ-68 Mod 1 rocket without warhead fitted and showing folded fins (right)

SPECIFICATIONS (VLRA 4 × 4 ACMAT chassis)

CALIBRE	70 mm	CREW	3	RANGE	900 km plus		
NUMBER OF BARRELS	40	WEIGHT (loaded)	5350 kg	FORDING	0.9 m		
NUMBER OF ROCKETS CARRIED	160	LENGTH	5.325 m	GRADIENT	65%		
MAX RANGE	8000 m	WIDTH	1.936 m	ENGINE	Perkins 6.354.4 6-cylinder diesel 120 hp at 2800 rpm		
ELEVATION/DEPRESSION	+55°/0°	HEIGHT (travelling)	2.538 m				
		GROUND CLEARANCE	0.29 m				
		MAX SPEED	100 km/h				

Status: In production. In service with an unnamed Near East country on ACMAT chassis. From 1985 IPTN (Nusantara Aircraft Industries Ltd) of Indonesia started producing the 70 mm rocket system FFAR under licence from Forges de Zeebrugge SA.

Manufacturer: Forges de Zeebrugge Fz
63 rue En Bois,
B-4040 HERSTAL
Belgium
Telephone: (041) 48 20 77
Telex: 41312
Fax: (041) 48 2510

BRAZIL

Brazilian Multiple Rocket Systems

Development
Since the early 1960s Brazil has been developing military rockets. In the early stages there was not much inter-service co-operation and each of the three armed forces carried out its own research and development programme. However, both the Navy and the Air Force abandoned their programmes in favour of a Ministry of Aeronautics managed programme that resulted in the SONDA family of space-research rockets. The Army continued its own programme under the auspices of the Comissao Central de Mïsseis (Central Missile Commission), a branch of the Departmento de Estudos e Pesquisas Technologicas (Department of Studies and Technological Research). Since then the Army programme has been made the responsibility of the Instituto de Pesquisas e Desenvolvimento (Research and Development Institute) with its headquarters in Rio de Janeiro.

AVIBRAS 108-R

This rocket, which is manufactured by AVIBRAS, has the Brazilian Army designation FGT108-RA1 and is fitted with a single-stage solid propellant motor with a thrust of 1250 kg. It is spin stabilised in flight by one central and six canted rocket nozzles. The motor burns for half a second to attain a maximum velocity of 409 m/s at burn-out. The rocket is launched from the X2A1 trailer launcher which has 16 1.1 m long tubes. The launcher, also manufactured by AVIBRAS, weighs 530 kg empty, 802 kg loaded and is 2.95 m long, 1.6 m wide and 1.37 m high. The wheelbase is 1.4 m and the tyre size 5.00 × 16. The launcher is normally towed by a ¾-ton 4 × 4 vehicle. It can also be airlifted by helicopters. The maximum rate of fire is 2 rds/s. The elevation is -1° to +50° and the traverse is ±12°. A 50 m cable is attached to the launcher for firing purposes. The maximum height of the rocket's trajectory is 4000 m. The crew consists of four and reloads are carried in 16-round fibreglass containers. A battery will normally have four launchers with each launcher having at least 64 rockets available on its towing vehicle. Brief specifications follow:

Designation	**108-R**
CALIBRE	108 mm
WEIGHT	
rocket	16.8 kg
warhead	7.8 kg
explosive filler	2.5 kg
LENGTH	0.931 m
MAX SPEED	440 m/s
MAX RANGE	9100 m
NUMBER OF RAILS/TUBES	16 tubes
LAUNCHER	X2A1 trailer

AVIBRAS SBAT-70

This rocket is essentially a standard AVIBRAS 70 mm SBAT-70 folding-fin aircraft rocket. The 36-tube launcher has a crew of four. The maximum range is about 7500 m. The following warhead types can be fitted: AVC-70/AC high explosive anti-tank, AVC-70/AP high explosive fragmentation (anti-personnel), AVC-70/AC/AP high explosive anti-tank/anti-personnel, AVC-70/F flechette, AVC-70/FB white phosphorus smoke, AVC-70/EF practice smoke and AVC-70/E inert practice. The launcher weighs 700 kg when empty and approximately 1000 kg when loaded. The length is 3.17 m, width 1.67 m, height 1.19 m and wheelbase 1.47 m. The elevation is 0 to +50° and the traverse ±12°. A 50 m cable is attached to the launcher for firing purposes. The system is not operational with the Brazilian Army but is offered for export. Brief specifications follow:

Designation	**SBAT-70**
CALIBRE	70 mm
WEIGHT	
rocket	8.5 kg
warhead	3.2 kg
MAX RANGE	8500 m
NUMBER OF RAILS/TUBES	36 tubes
LAUNCHER	LM-70/36 trailer

AVIBRAS SBAT-127

This rocket, formerly manufactured by AVIBRAS, is an adaptation of the 127 mm SBAT-127 HVAR aircraft rocket. The 12-rail launcher was designed to be mounted either on a trailer or a vehicle. Two types of high explosive warhead of different weights are used, the heavier one produces a flight time of 70 seconds and the lighter one a flight time of 68 seconds for the rockets. Powered flight distances are 350 m and 275 m respectively. Maximum ranges are achieved at an elevation of +47°. The system was developed for export. Brief specifications follow:

Designation	**SBAT-127**
CALIBRE	127 mm
WEIGHT	
rocket	45/58 kg
warhead	19/32 kg
MAX RANGE	15 400/13 700 m
MAX SPEED	690/680 m/s
NUMBER OF RAILS/TUBES	12 rails
LAUNCHER	trailer or vehicle

AVIBRAS X-20

This rocket was the result of a collaboration programme between the Army (Instituto de Pesquisa e Desenvolvimento and Campo de Provas da Marambaia), the Air Force (Cento Técnico Aeroespacial) and private industry (AVIBRAS and Usimeca). The FGT X-20 (its Brazilian Army designation) is a single-stage armament research rocket with a 2100 kgf thrust solid propellant motor that burns for 4.2 seconds to give a final velocity of 1160 m/s at burn-out. The launcher trailer is 3.4 m long, 2.15 m wide and 1.85 m high. It can be towed by a 2½-ton 4 × 4 or 6 × 6 vehicle. The wheelbase is 1.85 m and tyre size 9.00 × 20. A three-rail system may be mounted on the chassis of the X1A1/X1A2 series light tank. Brief specifications follow:

Designation	**X-20**
CALIBRE	180 mm
WEIGHT	
rocket	127 kg
warhead	45 kg
explosive filler	20 kg
LENGTH	2.91 m
MAX RANGE	30 000 m
MAX SPEED	1070 m/s
NUMBER OF RAILS/TUBES	3 rails
LAUNCHER	XLF-40

AVIBRAS X-40

This was the largest of the Brazilian rockets. The programme started at the Army's Instituto de Engenharia (Military Engineering Institute) and participants include the same agencies as for the X-20. Manufactured by AVIBRAS, the X-40 is a single-stage armament research rocket fitted

X2A1 trailer launcher for 108-R rocket (Ronaldo S Olive)

SBAT-70 70 mm rocket being launched from its AVIBRAS LM-70/36 towed launcher

X-40 rockets on XLF-40 launcher elevated ready for launching

TECTRAN (6 × 6) AV-RMD resupply vehicle, right, loading new launcher pods onto TECTRAN (6 × 6) AV-LMU launcher, left

AVIBRAS ASTROS AV-VCC battalion command and control vehicle

ASTROS SS-30 multiple rocket system on TECTRAN (6 × 6) AV-LMU truck with shutters in position over cab windows and launcher elevated

AVIBRAS AV-LMU firing SS-60 rocket in Arabian Gulf

with a 7200 kg thrust solid propellant motor that burns for 7.8 seconds to give a terminal velocity of 1275 m/s at burn-out. A three-rail launcher system is mounted on the chassis of the X1A1/X1A2 series light tank, details of which are given in the *Light tanks* section. When travelling the launcher is traversed to the front of the vehicle. On arrival at the firing position a hydraulically operated stabiliser blade fitted to the front of the vehicle is lowered to the ground together with two stabiliser jacks at the rear of the hull. The launcher is mechanically controlled in azimuth and hydraulically in elevation. The combat weight of the whole system is 17 070 kg and there is a crew of three. Brief specifications follow:

Designation	**X-40**
CALIBRE	300 mm
WEIGHT	
rocket	511 kg
warhead	150 kg
explosive filler	87 kg
LENGTH	4.85 m
MAX RANGE	66 000 m
MAX SPEED	1310 m/s
NUMBER OF RAILS/TUBES	3 rails
LAUNCHER	XLF-40

AVIBRAS ASTROS II

AVIBRAS has developed the ASTROS (Artillery SaTuration ROcket System) for the Brazilian Army and the export market, with the first production units being completed in late 1983. Known users include several countries in the Arabian Gulf.

The ASTROS II system consists of the launching vehicles (Universal Multiple Launcher: AV-LMU), ammunition resupply vehicles (AV-RMD with two complete re-load sets for an AV-LMU) with onboard hydraulic cranes, a command and control vehicle (AV-VCC), a vehicle with radar for fire control (AV-UCF-optional) and several mobile workshop vehicles for electronic and mechanical field maintenance. All the vehicles are based on the same 6 × 6 10 000 kg cross-country truck (AV-VBA) designed and built by the AVIBRAS subsidiary company TECTRAN.

The launcher can fire three different types of rocket, the AVIBRAS SS-30 (32 rounds per launcher), SS-40 (16 rounds per launcher) and SS-60 (four rounds per launcher). The larger SS-40 and SS-60 rounds have cluster munition warheads available. The SS-40 warhead has 20 and the SS-60 warhead 65 dual-effect anti-armour anti-personnel bomblets with the area of effectiveness varying according to the electronic fuze setting which governs the height of release. The weight of the SS-60 warhead is 212 kg.

At the 1991 Paris Air Show AVIBRAS revealed that it has developed to the production stage three further submunition warheads for the SS-40/60 series rockets. These are:

(1) a high explosive incendiary type utilising white phosphorous for anti-personnel, rapid smoke deployment and anti-material incendiary employment

(2) a mine-deployment type capable of carrying three different mine types: anti-personnel with an effective radius of 30 m; anti-material with the ability to penetrate more than 100 mm of armour on its main axis and up to 20 mm of mild steel at distances of more than 100 m; anti-armour with the ability to penetrate more than 120 mm of armour plate

AVIBRAS ASTROS II rockets from top to bottom - SS-60, SS-40 and SS-30

(3) an airfield denial type with a number of heavy TNT filled delay action fuzed submunitions capable of penetrating more than 400 mm of reinforced concrete. Typical dimension for a crater in concrete would be: depth of explosion centre 340-650 mm, depth of crater formed (from the actual concrete surface) 150-300 mm and radius of crater (from the centre of explosion to where crater breaches original concrete surface) 550-860 mm.

The range of the rockets vary from 10-70 km according to type with all of them having wrap-around fins for improved accuracy. The command and control vehicle is equipped with military computers for fire control calculations and a radio set for data transfer and voice communication with the launchers, fire-control radar vehicle, ammunition resupply vehicles, forward observers and higher command echelons.

The ASTROS systems of the Iraqi army saw extensive combat service in the First and Second Gulf Wars. Saudi Arabian systems saw extensive service in the Second Gulf War against Iraq in 1991.

Designation	**SS-30**	**SS-40**	**SS-60**
CALIBRE	127 mm	180 mm	300 mm
WEIGHT OF ROCKET	69 kg	147 kg	576 kg
LENGTH	2.9 m	4.3 m	5.4 m
RANGE			
min	10 km	17 km	23 km
max	33 km	39 km	70 km
NUMBER OF TUBES	32	16	4
LAUNCHER	AVIBRAS/TECTRAN Universal Multiple Launcher		

AVIBRAS SS-300

Avibras has completed preliminary engine tests on a new medium-range tactical rocket, the SS-300, which will have a range of 300 km and a warhead of 1000 kg.

Status: The ASTROS II and SBAT-70 are in production for the Brazilian Army and export.

The ASTROS, 108-R, X-20 and X-40 are in Brazilian Army service. The SBAT-70 is in service with several unidentified countries and the ASTROS II is known to be in Iraqi (SS-30, SS-40 and SS-60, which are known locally under the names of Sijeel 30, Sijeel 40 and Sijeel 60) and Saudi Arabian service (including the SS-30).

Manufacturer: Enquiries to Avibrás Indústria Aerospacial S/A
International Sales Division
Antiga Estrada de Paraibuna
KM118 (CP 229)
12.200
Sao José dos Campos
Sao Paulo State
Brazil
Telephone: 123 21 7433
Telex: 123 3793
Fax: 123 22 1689

CANADA

Bristol Aerospace 70 mm Canadian Rocket Vehicle (CRV) 7 Advanced Rocket System

Development/Description

The 70 mm (2.75 in) CRV 7 rocket system was developed by the Canadian Department of National Defence and is produced by Bristol Aerospace Limited.

Although designed primarily for the air-to-ground role, it is also capable of being used in both the ground-to-ground and ground-to-air roles.

The primary launcher units are the 19-round disposable LAU-5003 which weighs 240.5 kg when loaded, or the reusable 19-round BAL19. Mounting options can include a towed system with six 19-round launchers on a converted 105 mm howitzer carriage bed or two 19-round launchers mounted either side of a turret unit on the rear of a High Mobility Multi-purpose Wheeled Vehicle (HMMWV).

The rocket motor used in the RLU-5002/B (or C15) which has a case-bonded high energy solid-propellant fuel that uses a hydroxyl terminated polybutadiene (HTPB) binder with an 88% solids content of ammonium perchlorate and ferric oxide. (Note: no aluminium powder is used in order to reduce the smoke emission characteristic).

The motor delivers 998 kg of thrust for 2.2 seconds and has a maximum range of 7.5 to 14 km in the ground-to-ground role, depending upon the warhead type carried and launch angle. The round is spin-stabilised in flight by a fluted exhaust nozzle and three spring-deployed aluminium wrap-round fins at the tail.

A HERO safe version of the C15 motor has been developed and tested. Designated the C16, this motor features a nozzle which is compatible with the HYDRA-70 series of launchers used by US forces.

The available warhead types include the 4.3 kg M151 HE-blast fragmentation variant which can be used against vehicles, light armour, soft ground targets, personnel, light naval craft and helicopters; the 7.7 kg M229 HE-blast fragmentation and 4.3 kg M247 HEDP versions for use against similar target types and ships; the 4.4 kg M156 White Phosphorus/incendiary variant for target marking; the 4.9 kg M257 flare version for target marking and night illumination; and the 2.7 kg Mk1 type for long-range HE bombardment against area targets.

Status: In production. In service with Canada, Norway, United Kingdom and at least six other countries as an air-to-gound weapon. On offer for export as a ground-to-ground weapon.

Bristol Aerospace towed 70 mm CRV7 system with six 19 round launchers and rocket being loaded

Manufacturer: Bristol Aerospace Limited
PO Box 874
Winnipeg
Manitoba
Canada
R3C 2S4
Telephone: (204) 775 8331
Telex: (07) 57774/57804
Fax: (204) 885 3195

CHILE

FAMAE/Royal Ordnance 160 mm Rayo Multiple Rocket System

Development/Description

The 160 mm Rayo project can be traced back to the mid-1980s when the Chilean Army issued a requirement for an MRS and a number of companies were invited to tender for the study and demonstration phase. Late in 1987 Royal Ordnance (RO) plc demonstrated a FFR-4 rocket in Chile, from a former RARDE technology demonstration programme. In January 1989 RO was awarded the contract for the design and development of the Rayo in conjunction with Chile's FAMAE (Fabricas y Naestranzas del Ejercito). In 1991 a company, known as FAMAE Ordnance, was established in Chile to market the Rayo MRS. Phase I of the project was the final development of the rocket itself, which met all technical objectives following firing trials in the UK and then Chile. Phase II involves the completion of the development of the warhead types, full development of the launcher and fire control system and the integration of a complete Rayo system onto a chassis. This is expected to be completed by 1992. Construction of a rocket manufacturing plant started in 1990 at FAMAE's Talagante facilities near Santiago. This industrialisation phase is also due to be finished in 1992.

RO Rocket Motors Division is taking the lead in the project whilst the Ammunition Division is supplying the HE warhead, which is based on its 155 mm RO22 projectile.

The Rayo system is of modular design so that it can be fitted to a wide range of tracked and wheeled chassis types. Typically a tracked chassis such as the Krauss-Maffei Puma armoured combat vehicle would have a traversable launcher on its hull rear fitted with two 16-round rocket pods, each in four layers of four. A light towed version is also under study which may have a single pod of 16 rockets.

The pods are protected against small arms fire and blast damage with each rocket carried and fired from its own environmentally sealed integral launch tube.

A conventional fire control system is fitted which is used in conjunction with a Global Positioning System (GPS) to give high first round accuracy without the need for firing a sighting rocket, which could easily give away the launcher's position.

Key information used by the system to calculate the firing solution includes target position data, weather, and vehicle altitude and angular displacement. The system will also allow the mixing of warhead types as well as a range of firing sequences from single to ripple launches.

Redundancy is built-in so that the launcher can be traversed and fired either automatically or by hand. Also if the vehicle's own crane is damaged then the crane on the resupply vehicle can be used. The system is deigned to be loaded, deployed and fired by a single person if required.

The Rayo rocket is 3.212 m long, has a warhead diameter of 160 mm and a nominal sea level range of 35 000 m. The rocket motor section case is of proven laminate aluminium design and uses aluminised HTPB solid propellant as the filling. Four fins at the rear unfold as the rockets leave the launcher. These have a 0.75° angle to maintain the stabilising spin produced inside the launcher tube by a helical spin rail.

Initially HE and bomblet warhead designs are being developed, but other warhead options could be produced to meet different market requirements.

Status: Development.

Manufacturer: FAMAE Ordnance, Chile.

Enquiries to: Royal Ordnance plc
Rocket Motors Division
Summerfield
Kidderminster
Worcestershire DY11 7RZ
UK
Telephone: (0562) 824061
Fax: (0562) 825063

CHINA, PEOPLE'S REPUBLIC

NORINCO 425 mm Type 762 (two-round) Mine Clearing Rocket System

Development/Description

This is currently the largest of the Chinese heavy rocket systems and is mounted on a cradle assembly that is carried on the rear of a tracked vehicle chassis which is derived from that of the Type 83 155 mm self-propelled gun/howitzer. The engine is mounted at the front and the crew compartment is located at the rear. The warhead section weighs 600 kg and probably contains a sectionalised payload of fuel-air explosive containers, in order to create the blast over-pressure required to clear a minefield. The range is from 800-1000 m using a 150 kg, fin-equipped, solid propellant rocket motor section, to attain a maximum flight velocity of 100 m/s. Each rocket can clear a passage some 130 m long and between 12 and 22 m wide. The vehicle is equipped with a 7.62 mm machine gun for self-defence purposes which can be dismounted for use on a special anti-aircraft tripod. Firing ports for rifles are located on either side of the crew compartment as well as at the rear where a door is provided for access. The launcher system is hydraulically controlled. Other systems available for the crew include a Type 803 intercom system, a Type 889 radio set, a panoramic sight Type WG002, a 0.5 m stereo-optical rangefinder system and a Type TYG-1 night vision set for the driver.

Status: In production. In service with the Chinese Army.

425 mm Type 762 mine clearing rocket system in action

Manufacturer: Chinese state arsenals
China North Industries Corporation (NORINCO)
7A, Yuetan Nanjie
Beijing
People's Republic of China
Telephone: (86) 6898; (86) 3461; (86) 3741; (86) 7570
Telex: 22339

SPECIFICATIONS

CALIBRE	425 mm	CREW	4	MAX CROSS-COUNTRY SPEED	30-32 km/h
NUMBER OF LAUNCH RAILS	2	COMBAT WEIGHT	27 000 kg	MAX ROAD RANGE	450 km
LENGTH OF ROCKET	4.7 m	POWER-TO-WEIGHT RATIO	19.3 bhp/tonne	FORDING	1.3 m
WEIGHT OF ROCKET AT LAUNCH	760 ± 5 kg	LENGTH (travelling)	7.162 m	GRADIENT	31°
WEIGHT OF WARHEAD	600 kg	WIDTH (travelling)	3.236 m	SIDE SLOPE	25°
MAX RANGE	800-1000 m	HEIGHT (travelling)	3.2 m	VERTICAL OBSTACLE	0.7 m
MAX VELOCITY	100 m/s	GROUND PRESSURE	0.62 kg/cm²	ARMAMENT	1 × 7.62 mm light machine gun
ELEVATION/DEPRESSION	+5°/+45°	TRACK	2.62 m	AMMUNITION	500 rounds
TRAVERSE	±90°	LENGTH OF TRACK ON GROUND	4.601 m		
		MAX ROAD SPEED	40-42 km/h		

CPMIEC 350 mm (2-round) M-1B Artillery Rocket System

Development/Description

China Precision Machinery Import and Export Corporation (CPMIEC) have developed a long range non-guided artillery rocket system called the M-1B.

According to available information the rocket is of 350 mm diameter, carries a 100 or 150 kg warhead and is capable of ranges up to 80-100 km.

The launch platform is a truck which is very similar in appearance to the TECTRAN (6 × 6) 10 000 kg cross-country truck used by the Brazilian AVIBRAS ASTROS II MRS system.

It carries two rectangular container-launcher boxes in a side-by-side configuration with powered elevation and limited traverse controls. It is possible that the rocket technology associated with the M-1B and the related WS-1(M-1) has come from Israel as there are certain similarities with the MAR-290 and MAR-350 systems (qv).

Status: Prototype trials. Believed to be on order for the Chinese Army. Offered for export.

Manufacturer: China Precision Machinery Import and Export Corporation (CPMIEC)
PO Box 845
Beijing
People's Republic of China
Telephone: (89) 5012, (83) 11804
Telex: 22484 CPMC CN
Fax: 8311657

CPMIEC 320 mm (4-round) WS-1 Artillery Rocket System

Development/Description

China Precision Machinery Import and Export Corporation (CPMIEC) has developed a long range non-guided artillery rocket weapon system called the WS-1 (alternative designation believed be M-1).

According to CPMIEC, the WS-1 system has been designed to bridge the gap between conventional towed and self-propelled artillery systems and tactical missile systems.

Mounted on the rear of a 6 × 6 cross country truck are four cylindrical launcher tubes for the WS-1 rocket. Each rocket is 4.52 m long, has a diameter of 320 mm and weighs 520 kg at launch, of which 150 kg is the warhead.

The WS-1 rocket has four fixed fins and reaches a maximum flight altitude of 30 000 m with maximum speed of Mach 3.6. Maximum range is quoted as 80 000 m with a minimum range of 20 to 30 000 m with probable deviation of 1 to 1.25 per cent of range. No details of the warhead type have been revealed although to be effective at maximum range it is logical that some type of cluster munition warhead would be carried.

The four round launcher is mounted on a turntable at the rear of the vehicle and has powered traverse and elevation. Hydraulically operated stabilisers have to be lowered at the rear of the vehicle to provide a more stable firing platform. Once the four rockets have been launched, new rockets are reloaded manually into the launcher tubes from a truck backed up to the rear of the firing platform.

Status: Production. In service with the Chinese Army. Offered for export.

Manufacturer: China Precision Machinery Import and Export Corporation (CPMIEC)
PO Box 845
Beijing
People's Republic of China
Telephone: 895012, 8311804
Telex: 22484 CPMC CN
Fax: 8311657

320 mm (4-round) WS-1 artillery rocket system

320 mm (4-round) WS-1 artillery rocket system (right) with rockets in foreground

NORINCO 284 mm (10-round) Type 74 Multiple Rocket-minelayer System

Development/Description

The Chinese have several heavy rocket systems in service, with one mounted on the rear platform of a CA-30A (6 × 6) truck. This launcher comprises a 4 m long five-rail I-shaped frame system for 10 rockets with a team of four required to load each rocket manually. The fin-stabilised Type 74 rockets each carry 10 plastic case anti-tank mines in a bulbous warhead. The Type 74 launchers are normally deployed in batteries of four which can lay a single 400 × 400 m minefield with one salvo. The maximum range of the rocket is 1500 m and all 10 can be fired within 15 seconds. The usual crew totals six, four of whom sit behind the driver's cab on an open bench seat. For firing the launcher has two stabiliser legs that are lowered at the rear of the chassis and steel shutters which are used to protect the cab windows.

A later rocket type used is the 284 mm Type 85 mine-laying rocket shell. This has a maximum range of 3000 m and contains 10 anti-tank mines. A complete salvo has a laying zone of 700 × 500 m.

SPECIFICATIONS

CALIBRE	284 mm
NUMBER OF RAILS	10
LENGTH OF RAIL	4 m
LENGTH OF ROCKET	2.47 m
WEIGHT OF ROCKET	127 kg
MAX RANGE	1500 m
ELEVATION/DEPRESSION	+48°/-7°
TRAVERSE	90° left, 45° right
CREW	6
WEIGHT OF SYSTEM	
LOADED	8780 kg
LENGTH	6.44 m
WIDTH	2.27 m
HEIGHT	2.93 m
MAX SPEED	40 km/h

Status: In production.In service with the Chinese Army.

Manufacturer: Chinese state factories
China North Industries Corporation (NORINCO)
7A, Yuetan Nanjie
Beijing
People's Republic of China
Telephone: (86) 6898; (86) 3461; (86) 3741; (86) 7570
Telex: 22339

284 mm Type 74 (10-round) rocket-minelayer on Chinese (6 × 6) CA-30A truck chassis

NORINCO 273 mm (4-round) Type 83 Multiple Rocket System

Development/Description

The system is based on the unarmoured Type 60-1 full-tracked artillery tractor with a five-man, fully enclosed cab at the front and a turntable with four 273 mm rockets at the very rear. Before launching the launch platform's two stabilisers are lowered at the rear of the chassis to provide a more secure firing platform.

Each rocket has a high explosive warhead, although other types of warhead could easily be accommodated in a rocket of this calibre. The rocket has a maximum velocity of 811 m/s and all four rockets can be salvo-fired in 7.5 seconds with time into action being one minute.

The launcher is traversed and elevated manually, with the controls mounted on its left side. Fire control equipment consists of one panoramic sight Type 85 and one indicator Type 85.

273 mm Type 83 MRS in firing position from rear with stabilisers lowered

Variants

273 mm (8-round) Multiple Rocket System

This system is a longer range version of the basic Type 83 mounted on the rear of an 8 × 8 cross-country truck chassis with a maximum speed of 68 km/h. An eight-round launcher assembly (in two banks of four 5.056 m long cylindrical container-launcher tubes) is fitted on a hydraulically raised rear deck turntable platform with elevation limits of 0° to +66° and traverse limits of 96° left or right.

The rocket is 4.586 m long, weighs 519 kg at launch and is fitted with a 150 kg HE warhead (compared to the 4.753 m and 484 kg of the earlier Type 83 rocket). The major difference is that the solid propellent fuel has changed to a case bonded central channel composite polybutadiene type. This fuel increases the maximum range capability to 80 000 m (compared to the range limits of 23 000 minimum and 40 000 m maximum of the Type 83). The firing interval is one round every 2 seconds.

A 150 kg submunition warhead is being developed, as is a mid-course correction ballistic guidance control system. The latter will enable real-time flight corrections to be made to enhance terminal accuracy. After such corrections are made the maximum range is expected to be not less than 75 000 m.

The long range rocket can also be used with the original 4-round Type 83 launch vehicle. The tracked Type 321 APC chassis can also be fitted with the 8-round launcher assembly.

273 mm (8-round) MRS on an 8 × 8 truck in the travelling configuration

SPECIFICATIONS (273 mm (4-round) Type 83)

CALIBRE	273 mm
NUMBER OF LAUNCH TUBES	4
WEIGHT OF ROCKET	484 kg
LENGTH OF ROCKET	4.753 m
MAX RANGE	40 000 m
MIN RANGE	23 000 m
MAX VELOCITY	811 m/s
ELEVATION/DEPRESSION	+56°/+5° 30'
TRAVERSE	10° left and right
CREW	5
WEIGHT OF SYSTEM	
loaded	17 542 kg
empty	15 134 kg
LENGTH TRAVELLING	6.19 m
WIDTH TRAVELLING	2.60 m
HEIGHT TRAVELLING	3.18 m (over launcher)
	2.81 m (top of cab)
GROUND CLEARANCE	0.40 m
TRACK	2.15 m
LENGTH OF TRACK ON GROUND	3.248 m
MAX ROAD SPEED	45 km/h
RANGE	400 km
MIN TURNING RADIUS	2.15 m
FORDING	1 m
GRADIENT	60%
SIDE SLOPE	30%
ENGINE	12150L-1 V-12 water-cooled 4-cycle direct injection diesel developing 300 hp at 1600 rpm

Status: In production. In service with the Chinese Army and being offered for export.

Manufacturer: Chinese state factories
China North Industries Corporation (NORINCO)
7A, Yuetan Nanjie
Beijing
People's Republic of China
Telephone: (86) 6898; (86) 3461; (86) 3741; (86) 7570
Telex: 22339

NORINCO 130 mm (30-round) Type 82 and Type 85 Multiple Rocket Systems

Development/Description

This multiple rocket system consists of a 6 × 6 cross-country truck with a four-door, fully-enclosed forward control cab. To the rear of this are 30 reserve rockets for immediate use and at the very rear of the chassis is a 130 mm rocket launcher with three layers of 10 tubes each. Two salvoes each of 30 rockets can be launched in five minutes.

The 130 mm rocket types available are:

DESIGNATION	Type 63 HE	Type 82 HE-fragmentation	Type 82 HE-I	Extended Range HE
DIAMETER	130.43 mm	130.43 mm	130.43 mm	130.43 mm
LENGTH	1.051 m	1.063 m	1.043 m	1.056 m
WEIGHT (complete)	33 kg	32 kg	32.8 kg	32.5 kg
MAX RANGE	10 100 m	10 217 m	10 100 m	15 000 m
WARHEAD TYPE	HE producing 1916 fragments lethal out to 24 m radius	HE-fragmentation with 2600 6.35 mm diameter steel spheres. It produces 3850 fragments lethal out to 42.1 m radius. Minimum rocket range is 3000 m	HE with 5000 incendiary pellets and burning radius of 32.8 m	HE-fragmentation with lethal radius of 32 m
FUZE TYPE AND SETTINGS	proximity or MJ-1 impact with instantaneous and delay actions			

There is also a 1.05 m long single manportable tripod-mounted launcher tube for 130 mm rocket which is designed for use by commando, guerrilla or special forces units. This can be broken down into small loads for carriage-launch tube 10.5 kg, tripod 11.9 kg, sight 1.2 kg and remote firing device 1.25 kg. Its elevation range is +5° to +40° and its traverse range 7° to the left or right. Time to bring the weapon into action is about 1.45 minutes and to disengage between 1½ to 2 minutes.

The launcher is also mounted on the top of the Type 85 tracked APC to improve its cross-country capability.

130 mm (30-round) Type 82 multiple rocket system in travelling configuration

SPECIFICATIONS

Model	Type 82	Type 85
CALIBRE	130.63 mm	130.63 mm
NUMBER OF BARRELS	30	30
LENGTH OF BARREL	1.050 m	1.050 m
NUMBER OF ROCKETS	60	60
RATE OF FIRE	30 rds in 14.5 -17.4 s	30 rds in 14.5 -17.4 s
CARRIAGE	6 × 6 truck	tracked Type 85 APC
WEIGHT IN FIRING POSITION	7500 kg	14 500 kg
LENGTH TRAVELLING	6.438 m	6.125 m
WIDTH TRAVELLING	2.250 m	3.060 m
HEIGHT TRAVELLING	2.260 m	3.014 m
ELEVATION	0-50°	0-50°
TRAVERSE	±75°	±15°
CREW	8	6

Status: In production. In service with the Chinese Army (Type 82 and Type 85) and Thailand (Type 85).

Manufacturer: Chinese state factories
China North Industries Corporation (NORINCO)
7A, Yuetan Nanjie
Beijing
People's Republic of China
Telephone: (86) 6898; (86) 3461; (86) 3741; (86) 7570
Telex: 22339

NORINCO 130 mm (19-round) Type 63 and Type 70 Multiple Rocket Systems

Development/Description

An indigenously designed and manufactured 19-tube 130 mm calibre Type 63 rocket launcher entered service in the Chinese Army mounted on the rear platform of the NJ-230 (4 × 4) truck. There are two variants of the Type 63 mounted on this truck, the major difference being that the second variant, Type 63-1 has a covered crew cabin. In both cases the firing units are grouped into batteries of six vehicles. The Type 70 version is mounted on the Type YW 531 APC and is assigned to rocket launcher units attached to the Armoured and Mechanised Infantry Divisions of the Chinese Army. The launch tubes are arranged in two rows, the top one of 10 over another of nine. The system is the replacement for the obsolete BM-13-16 systems used by the Chinese.

The 130 mm rocket types available are:

There is also a 1.05 m long single manportable tripod-mounted launcher tube for 130 mm rockets which is designed for use by commando, guerrilla or special forces units. This can be broken down into small loads for carriage-launch tube 10.5 kg, tripod 11.9 kg, sight 1.2 kg and remote firing device 1.25 kg. Its elevation range is +5° to +45° and its traverse range 7° to the left or right. Time to bring the weapon into action is about 1¾ minutes and to disengage between 1½ to 2 minutes.

DESIGNATION	**Type 63 HE**	**Type 82 HE-fragmentation**	**Type 82 HE-I**	**Extended Range-HE**
DIAMETER	130.43 mm	130.43 mm	130.43 mm	130.43 mm
LENGTH	1.051 m	1.063 m	1.043 m	1.056 m
WEIGHT (complete)	33 kg	32 kg	32.8 kg	32.5 kg
MAX RANGE	10 100 m	10 217 m	10 100 m	15 000 m
WARHEAD TYPE	HE producing 1916 fragments lethal out to 24 m radius	HE-fragmentation with 2600 6.35 mm diameter steel spheres. It produces 3850 fragments lethal out to 42.1 m radius. Minimum rocket range is 3000 m.	HE with 5000 incendiary pellets and burning radius of 32.8 m	HE-fragmentation with lethal radius of 32 m
FUZE TYPE AND SETTINGS	proximity or MJ-1 impact with instantaneous and delay actions			

130 mm Type 63-1 MRS with covered crew cab on NJ-230 (4 × 4) truck chassis (Andrew Li)

130 mm Type 70 MRS on Type YW 531 APC

SPECIFICATIONS

Model	**Type 63-1**	**Type 70**
CALIBRE	130.43 mm	130.43 mm
NUMBER OF BARRELS	19	19
LENGTH OF BARREL	1.050 m	1.050 m
NUMBER OF ROCKETS	38	39
RATE OF FIRE	19 rds in 9.5-11.5 s	19 rds in 9.5-11.5 s
CARRIAGE	4 × 4 cross country truck	tracked YW531 (Type 63) APC
WEIGHT IN FIRING POSITION	6887 kg	14 000 kg
LENGTH TRAVELLING	6.659 m	5.746 m
WIDTH TRAVELLING	2.140 m	2.978 m
HEIGHT TRAVELLING	2.453 m	2.625 m
ELEVATION	0-50°	0-50°
TRAVERSE	180°	180°
CREW	7	6

Status: In production. In service with China, North Korea and Vietnam.

Manufacturer: Chinese state factories
China North Industries Corporation (NORINCO)
7A, Yuetan Nanjie
Beijing
People's Republic of China
Telephone: (86) 6898; (86) 3461; (86) 3741; (86) 7570
Telex: 22339

NORINCO 122 mm Type 83 (40-round) SP MRS

122 mm Type 83 self-propelled multiple rocket system in travelling configuration

Development/Description

NORINCO has developed a self-propelled 122 mm rocket launcher on the same chassis as the Type 83 152 mm self-propelled gun/howitzer. This has been designed to accompany mechanised forces and unlike many other self-propelled multiple rocket systems has a fully armoured chassis.

The engine compartment is at the front right with the driver on the left-hand side. He is provided with a single piece hatch cover that opens to the rear, a heated day periscope WG501 and a TYG-1 night driving periscope.

The other four crew members are seated in the fully enclosed armoured compartment at the back with bullet-proof observation windows on either side and an entry door at the rear. It has torsion bar suspension and each side has six dual rubber tyred road wheels, a drive sprocket at the front, an idler at the rear and three track return rollers. The first and last road wheel stations have a hydraulic shock absorber.

The vehicle commander has a Type WG501 heated periscope while the gunner has a Type 59 tank periscope, Type 84 panoramic telescope and a Type 66-1 direct fire sight.

The 122 mm (40-round) launcher is mounted on a turntable over the rear of the vehicle and is almost identical to the Soviet 122 mm BM-21 (40-round)-system introduced in the early 1960s. The launcher can be traversed and elevated under full hydraulic power and manual controls are provided for emergency use. The rockets can be fired one by one or in a salvo with all 40 being fired in 20 seconds.

Mounted over the forward part of the vehicle is a reload pack with a further 40 122 mm rockets which can be reloaded into the launcher automatically. This reload feature is similar to that used on the Czechoslovakian 122 mm RM-70 (40-round) multiple rocket system which made its first appearance in 1972.

The 122 mm rocket types available are:

Status: Development complete, ready for production.

Manufacturer: Chinese state factories
China North Industries Corporation (NORINCO)
7A, Yuetan Nanjie
Beijing
People's Republic of China
Telephone: (86) 6898; (86) 3461; (86) 3741; (86) 7570
Telex: 22339

Type	HE	HE-fragmentation	HE-incendiary	HE-cargo	HE-fragmentation incendiary	minelaying	Type 84 minelaying
DIAMETER	122 mm	122 mm	122 mm	122 mm	122 mm	122 mm	122 mm
LENGTH	2.87 m	2.87 m	2.87 m	3.03 m	2.87 m	2.95 m	2.83 m
WEIGHT (complete)	66.83 ± 0.66 kg	66.88 ± 0.66 kg	66.83 ± 0.66 kg	67 kg	66.8 ± 0.66 kg	58.2 kg	63 kg
MAX RANGE	20 580 m	20 580 m	20 580 m	20 580 m	20 580 m	15 000 m	7000 m
WARHEAD WEIGHT	18.3 kg	18.3 kg	18.3 kg	19.25 kg	18.3 kg	25.7 kg	33 kg
WARHEAD TYPE	HE	6 kg HE with 4100 steel balls	HE with 6000 incendiary pellets and burning radius of 30 m	39 bomblets with 110 mm armour penetration and 7 m lethal radius	HE with 3400 steel balls and burning radius of 30 m	6 AT mines or 96 anti-personnel mines (900 × 800 m full salvo laying zone)	8 AT mines or 128 anti-personnel mines (650 × 650 m full salvo laying zone)
FUZE TYPE AND SETTINGS	impact with instantaneous, short or long delay	proximity or impact with instantaneous, short or long delay	impact with instantaneous, short or long delay	proximity	impact with instantaneous, short or long delay	proximity	proximity

Note: Extended Range (30 000 m) versions of the HE (lethal radius 28.2 m), HE cargo, HE-fragmentation, HE-incendiary and HE-fragmentation incendiary (lethal radius 44.2 m and burning radius of about 30 m) rocket rounds are available.

SPECIFICATIONS

CALIBRE	122.4 mm
NUMBER OF BARRELS	40
UNIT OF FIRE	40 + 40 rockets
LENGTH OF COMPLETE ROCKET	2.87 m
WEIGHT OF COMPLETE ROCKET	66.88 kg
WEIGHT OF PROPELLANT	20.6 kg
MAX RANGE	20 580 m
RATE OF FIRE	40 rockets in 18-20 s
ELEVATION/DEPRESSION	+55°/0°
TRAVERSE	66° (right) 102° (left)
CREW	5
COMBAT WEIGHT	29 900 kg
POWER-TO-WEIGHT RATIO	17.39 hp/tonne
GROUND PRESSURE	0.68 kg/cm²
LENGTH OVERALL	7.18 m
LENGTH OF CHASSIS	6.85 m
WIDTH	3.145 m
WIDTH OVER TRACKS	2.62 m
HEIGHT OVERALL	3.18 m
HEIGHT HULL TOP	1.47 m
GROUND CLEARANCE	0.45 m
LENGTH OF TRACK ON GROUND	4.601 m
TRACK	2.62 m
MAX ROAD SPEED	55 km/h
AVERAGE SPEED ROAD	40-42 km/h
AVERAGE SPEED CROSS-COUNTRY	30-32 km/h
MAX ROAD RANGE	450 km
FORDING	1.3 m
GRADIENT	60%
SIDE SLOPE	30%
VERTICAL OBSTACLE	0.7 m
TRENCH	2.6 m

NORINCO 122 mm (40-round) Type 90 Multiple Rocket System

Development/Description

The Type 90 40-round MRS is an indigenously designed and built system that is equipped with an automatic operating and laying system, an electric firing system and an automatically reloadable pack of 40 reload rockets. The chassis used is the Tiema SC2030 6 × 6 truck model with four rows of 10 tubes mounted above each other on a rotating cradle assembly. Elevation and traverse are performed by electrical systems.

A hydraulically operated collapsible awning is mounted over the rear deck for use as protection and camouflage. Conversion to the combat mode takes between 1½ and 2 minutes following arrival at the launch site.

The automatic operating and laying system consists of five main subsystems: an HJ-1 minicomputer; a launcher direction finder (which takes less than 3 seconds to display the launcher's magnetic azimuth angle); longitudinal and transversal slope sensors; launcher assembly elevation and traverse position sensors; and a special vehicle power supply interface powerpack unit. Total weight of the system is less than 30 kg.

The launcher is controlled by the computer through man/machine interaction. An interface is provided for connection to a battalion central command system. Instructions from which can either be input directly via the interface or manually input by push buttons. There are three modes of operating and laying the system: manual, electrical or automatic.

Hydraulic jacks at the rear and sides of the vehicle are used to minimise the effect of uneven ground or vibration during firing. The computer automatically corrects the firing solution for any deviation of the firing angle caused by ground slope.

The electrical firing system fires the rockets either singly or in a salvo with 0.5 seconds between rounds. The system comprises an electric igniter, remote firing device and ignition contacts and cables. Firing may be conducted both from the driver's cab or remotely outside the vehicle.

The fin stabilised rockets used are of the 122 mm calibre extended range composite solid propellant type and HE, Cargo (mine), steel ball HE, HE-incendiary and steel ball HE-incendiary warhead options.

The automatic loading system consists of elevator, rack and feeder. Loading of the rockets can be performed either automatically by push button operation in the driver's cab or manually by external operations. The loading mechanism can be rotated 90°. Automatic reloading of the launcher takes less than 3 minutes. A rocket-in-position and after-fire chamber display system in the driver's cab monitors the weapon status after loading or firing.

Type 90 122 mm (40) round multiple rocket system with cover retracted to rear of cab

SPECIFICATIONS		ELEVATION/DEPRESSION	+55°/0°	GROUND CLEARANCE	0.412 m
CALIBRE	122 mm	RELOAD TIME (automatic)	3 mins	MAX ROAD SPEED	85 km/h
NUMBER OF TUBES	40	TRAVERSE	±102°	MAX ROAD RANGE	600 km
LENGTH OF TUBE	3.0 m	COMBAT WEIGHT	approx 20 000 kg	FORDING	0.9 m
MAX RANGE	30 000 m	LENGTH (travelling)	9.84 m	MIN TURNING RADIUS	approx 12 m
UNIT OF FIRE	40 + 40 rockets	WIDTH (travelling)	2.5 m	ENGINE	air-cooled diesel developing 300 hp
RATE OF FIRE	40 rds/18-20 s	HEIGHT (travelling)	3.245 m		

Status: Production. In service with the Chinese Armed Forces.

Manufacturer: Chinese state factories
China North Industries Corporation (NORINCO)
7A, Yuetan Nanjie
Beijing
People's Republic of China
Telephone: (86) 6898; (86) 3461; (86) 3741; (86) 7570
Telex: 22339

NORINCO 122 mm (40-round) Type 81 Multiple Rocket System

Development/Description

The Chinese are known to have in service a locally built modified version of the Soviet BM-21 122 mm system. The 40-tube launcher is carried on a CQ261 Hongyan (a locally built variant of the French Berliet GB4) 6 × 6 truck chassis. The launcher tube is 3 m long.

In addition to these there is also a short 122 mm HE rocket which is used in conjunction with a 2.5 m long single-tube animal or manportable tripod-mounted 122 mm rocket launcher. This is designed for use by guerrilla, commando or special forces units.

The rocket weighs 46.3 kg complete and is 1.927 m long. Its maximum range is 13 400 m with the launcher having an elevation range of -3 to +50° and a traverse of +15°. A contact fuze with preselected instantanous, short or long delay actions is fitted.

Total weight of the launcher when assembled for firing is 59 kg (which is made of the 25 kg tripod, the 30 kg launch tube and 4 kg for the sight and remote firing device with its 25-35 m long connector cable).

The launcher can also be fixed on a special mount for use on a light cross-country vehicle. In this application the launcher can either be fired from the vehicle or dismounted for ground use. A total of 10-15 reload rounds would normally be carried.

The Type 81 multiple rocket launcher system fires the same family of rockets as the 122 mm Type 83 already described.

122 mm (40-round) Type 81 multiple rocket system

SPECIFICATIONS		MAX ROCKET RANGE	30 000 m	LENGTH (travelling)	7.12 m
CALIBRE	122 mm	ELEVATION	0° to +55°	WIDTH (travelling)	2.5 m
NUMBER OF ROCKET TUBES	40	TRAVERSE	102° left, 70° right	HEIGHT (travelling)	3.082 m
LENGTH OF ROCKET TUBE	3.0 m	RATE OF FIRE	40 rds/18-20 s	MAX SPEED	n/av
NUMBER OF ROCKETS	40	RELOAD TIME	8-10 mins	ROAD RANGE	n/av
		CREW	7		
		COMBAT WEIGHT	15 160 kg		

Status: The Type 81 is in production and in service.

Manufacturer: Chinese state factories
China North Industries Corporation (NORINCO)
7A, Yuetan Nanjie
Beijing
People's Republic of China.
Telephone: (86) 6898; (86) 3461; (86) 3741; (86) 7570
Telex: 22339

NORINCO 122 mm (24-round) Type 83 Multiple Rocket System and Type 84 Minelaying Rocket System

Development/Description

This consists of a 24-round version of the 122 mm (40-round) Type 81 multiple rocket system mounted on the rear of a 6 × 6 cross-country chassis. The engine is at the front, the four-door, fully enclosed cab in the centre and the rocket launcher at the rear. Reload rockets are carried by a dedicated resupply vehicle. A special 122 mm minelaying cargo round is available which carries eight small anti-tank mines or 128 anti-personnel mines and has a maximum range of 7000 m. In addition to these there is also a short 122 mm HE rocket which is used in conjunction with a 2.5 m long single-tube animal or portable tripod-mounted 122 mm rocket launcher. This is designed for use by guerrilla, commando or special forces units.

The rocket weighs 46.3 kg complete and is 1.927 m long. Its maximum range is 13 400 m with the launcher having an elevation range of -3° to +50° and a traverse of +15°. A contact fuze with preselected instantaneous, short or long delay actions is fitted.

Total weight of the launcher when assembled for firing is 59 kg (which is made of the 25 kg tripod, the 30 kg launch tube and 4 kg for the sight and remote firing device with its 25-35 m long connector cable).

The launcher can also be fixed on a special mount for use on a light cross-country vehicle. In this application the launcher can either be fired from the vehicle or dismounted for ground use. A total of 10-15 reload rounds would normally be carried. The elevation/depression of the launcher is +55°/0° and the traverse ±90° of the central axis. A full salvo takes between 15-20 seconds to fire.

When used in the minelaying role the system designation changes to the Type 84 minelaying rocket launcher.

122 mm (24-round) Type 83 multiple rocket system

The multiple rocket launcher system fires the same family of rockets as the 122 mm Type 83 system already described.

Status: Production. In service with Chinese Army. On order for Thailand who is to develop its own minelaying cargo round.

Manufacturer: Chinese state factories
China North Industries Corporation (NORINCO)
7A, Yuetan Nanjie
Beijing
People's Republic of China
Telephone: (86) 6898; (86) 3461; (86) 3741; (86) 7570
Telex: 22339

SPECIFICATIONS

CALIBRE	122 mm	MAX RANGE	HE rocket 20 580 m; minelaying rocket 7000 m	LENGTH TRAVELLING	6.51 m
NUMBER OF BARRELS	24			WIDTH TRAVELLING	2.2 m
LENGTH OF ROCKET TUBE	3 m			HEIGHT TRAVELLING	2.7 m
		CREW	6	GROUND CLEARANCE	0.285 m
		WEIGHT OF SYSTEM LOADED	6700 kg	MAX ROAD SPEED	60 km/h
				RANGE	550 km
				FORDING	0.85 m

NORINCO 107 mm (12-round) Type 63 and Type 81 Multiple Rocket Systems

Development/Description

The 107 mm (12-round) Type 63 MRS was developed in the late 1950s as the replacement for the 102 mm (6-round) Type 50-5 MRS which has now been withdrawn from service. The Type 63 is issued on the scale of 18 per Chinese infantry division.

The launcher has three banks of four 107 mm barrels and is mounted on a rubber-tyred split pole type carriage. In the firing position the wheels are removed and the launcher is supported by the two trails at the rear which are fitted with spades, which are carried on the top of the trails when travelling, and two short legs in the front of the carriage. The rockets are spin stabilised. Chinese supplied Type 63 MRS have found their way to the guerrilla forces fighting the Soviets and Afghan Government and are being used with considerable success.

The launcher is also mounted on a 4 × 4 truck, designated Type 81, with an enlarged cab to accommodate the launcher crew of four and 12 reload rounds. In transit the launcher is covered by a tarpaulin. There is a seat for one crew member outside the cab and to the rear. When firing the vehicle suspension is locked and the entire system can be removed from the vehicle if required and placed on a towing carriage. The launcher can be fired from inside the cab or from a remote position. Maximum road speed is 100 km/h. A pack model developed for use by airborne and mountain units weighs 281 kg in the firing position and can be dismantled into manpack loads. A lighter version of the Type 63 developed recently has been given the designation in the West of the Type 63-I. It is some 136 kg lighter than the basic model and is distinguishable from it by its smaller spoked roadwheels and its four banks of three rockets.

107 mm (12-round) Type 81 multiple rocket launcher on rear of 4 × 4 truck

China is known to use the Soviet 132 mm (16-round) BM-13-16 and the 140 mm (16-round) BM-14-16 MRSs, both of which are mounted on the rear of a ZIL-151 (6 × 6) truck chassis. Each Chinese Artillery Division has one rocket launcher regiment with 32 MRSs.

The 107 mm rocket types available are:

DESIGNATION	Type 63-II HE	Type 63-III HE	Type 63-I HE-incendiary	HE-fragmentation	Type 63 HE-Incendiary	Jamming Rocket
CALIBRE	107 mm	107 mm	107 mm	107 mm	107 mm	107 mm
LENGTH	0.841 m	0.897 m (with mechanical fuze) 0.947 m (with electronic fuze)	0.915 m	n/av	0.840 m	0.920 m
WEIGHT (complete)	18.84 kg	n/av	18.74 kg	18.9 kg	18.8 kg	18.3 kg
MAX RANGE	8500 m	10 000 m	8000 m	7800 m	8500 m	7800 m
WARHEAD TYPE	HE producing 1214 fragments with lethal radius of 18.5 m	HE with 12.5 m lethal radius	HE-incendiary with 40 s burning time	HE with some 1600 6.35 mm diameter steel spheres	HE with some 1600 incendiary pellets and a burning radius of over 21 m	parachute deployed jammer unit with duration of up to 15 mins
FUZE TYPE AND SETTINGS	radio frequency proximity or MJ-1 contact with instantaneous and delay actions					

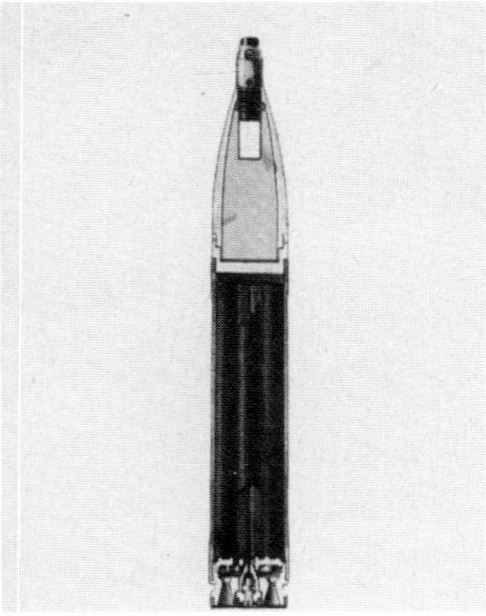

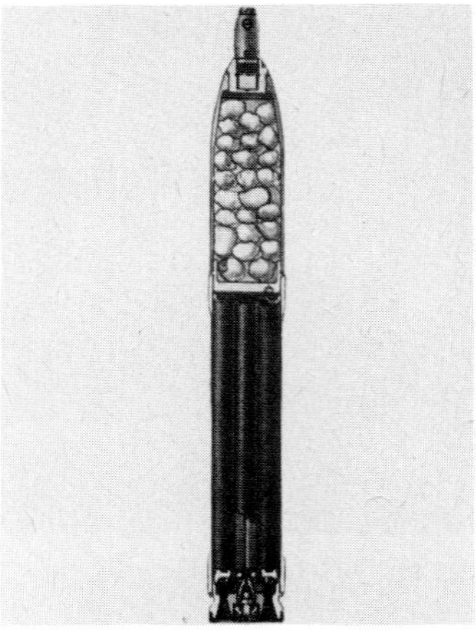

Rockets fired by Type 63 multiple rocket launcher, left: 107 mm HE; right: 107 mm incendiary

107 mm (12-round) Type 63 multiple rocket system captured in Lebanon in 1982 (Israeli Ministry of Defence)

There is also a 0.9 m long Type 85 single manportable tripod mounted launcher tube for 107 mm rockets with is designed for use by commando, guerrilla or special forces units. This weighs 22.5 kg in total and can be broken down into small loads for carriage. Its elevation range is +5° to +45° and its traverse range 7° to the left or right. Time to bring the weapon into action and to disengage is about ½ minute. Typical rate of fire is 3 rds/min. A remote firing device is used to ignite the rocket.

The 107 mm rockets can also be fired by laying on a suitable piece of sloped or raised terrain and pointing in the direction of the target. In this application a set of small batteries with an electrical cable can be used as the firing mechanism. The accuracy is not very good against point targets unless the launch occurs very close to them but is more than sufficient for area targets such as population centres or large military installations.

In Afghan Mojahedin service the Type 85 single-tube rocket launcher is referred to as the BM-1.

Variants

During the 1982 invasion of Lebanon a number of locally modified Type 63 MRSs were captured from the Palestinian guerrillas by the Israelis. These consisted of the basic 12-round launcher fitted to a crudely built pedestal and mounted on the rear of a light (4 × 4) truck chassis to increase mobility.

SPECIFICATIONS

Model	**Type 63**	**Type 81**
CALIBRE	107 mm	107 mm
NUMBER OF BARRELS	12	12
NUMBER OF ROCKETS	12	24
RATE OF FIRE	12 rds in 7-9 s	vehicle 12 rds in 10-12 s ground 12 rds in 7-9 s
RELOAD TIME	3 min	3 min
CARRIAGE	split trail	4 × 4 light truck
WEIGHT IN FIRING POSITION	611 kg	3370 kg
LENGTH TRAVELLING	2.6 m	4.6 m
WIDTH TRAVELLING	1.4 m	1.964 m
HEIGHT TRAVELLING	1.2 m	2.185 m
ELEVATION	0-60°	0-60°
TRAVERSE	±15°	±60°
CREW	5	4

Status: In production. In service with Albania, Burkina Faso, Cambodia, Chad, China, Iran (including licence built version), North Korea, (licence built versions including self-propelled), Syria, UNITA, Vietnam, Zaïre and Afghanistani Mojahedin Guerrillas (referred to by them as the BM-1).

Manufacturer: Chinese state factories
China North Industries Corporation (NORINCO)
7A, Yuetan Nanjie
Beijing
People's Republic of China
Telephone: (86) 6898; (86) 3461; (86) 3741; (86) 7570
Telex: 22339

Other Chinese MRL Systems

The Chinese have a number of other multiple rocket systems in service on which only limited information is available at present. These are:

305 mm (8-round) Type 79 Minelaying Rocket System

Mounted on the same CA-30A (6 × 6) truck chassis as the Type 74 minelaying system this rocket launcher uses welded tubular steel frames instead of rails.

The rockets are larger overall and have ring stabilisers around the tail fins.

The rocket types used are the following:

	Type 79 minelaying rocket shell	**Fuel Air Explosive anti-personnel/mine-sweeping rocket shell**
CALIBRE	305 mm	305 mm
RANGE	2500 m	2700 m
MINE CLEARANCE ZONE	n/app	48 m diameter circle
ANTI-PERSONNEL LETHAL ZONE	n/app	24-60 m diameter circle
PAYLOAD	10 plastic AT mines	n/app

SPECIFICATIONS (Type 79)

CALIBRE	305 mm
NUMBER OF LAUNCH RAILS	8
RATE OF FIRE	1 rd/1.75 s
SALVO MINELAYING ZONE	600 × 350 m
ELEVATION	+4° to +50°
TRAVERSE	95° left, 45° right
MAX SPEED	65 km/h

253 mm (10-round) Type 81 Mine Clearing Rocket System

This is carried on the rear of a (6 × 6) truck chassis with a combat weight of 6200 kg.

The 253 mm Type 81 mine-sweeping rocket system uses the following rocket types:

	Type 81 anti-personnel/ mine-sweeping rocket shell	**Type B1-II anti-personnel/mine-clearing rocket shell**
CALIBRE	253 mm	253 mm
MAX RANGE	1400 m	200 m (with jet efflux cone) 300 m (without jet efflux cone)
MINE CLEARING ZONE	n/av	35 m diameter circle

SPECIFICATIONS (Type 81)

CALIBRE	253 mm
NUMBER OF ROCKET RAILS	10
RATE OF FIRE	10 rds/12-14 s
ELEVATION	90° left, 75° right
MAX SPEED	35 km/h

180 mm (10-round) Type 71 Multiple Rocket Launcher System

This is a truck-mounted system and has 10 cage-type launch rails for 180 mm calibre rockets. These are 2.737 m long, weigh 134.8 kg at launch and carry a 45 kg HE warhead. Moveable tail fins are fitted for stabilisation. Maximum velocity is around 620 m/s which gives a maximum range of 20 000 m.

20-round Minelaying Rocket System
This is believed to be truck-mounted. No other details are available.

24-round 253 mm Type 81-II SP Mine Clearing Rocket System
This uses a rectangular turret mounting 24 launcher tubes to replace the normal turret of a Type 62 light tank. The Type 81-II rockets carried have long stand-off nose-mounted spikes that detonate the HE-warheads above the ground so as to create an over-pressure to clear the mines over a wide area.

10-round Mine Clearing Rocket System
This is of unknown calibre and is carried on the rear of a (4 × 4) truck chassis. No other details are available.

Status: All systems are believed to be in production and in service with the Chinese Armed Forces.

Manufacturer: Chinese State factories.

CZECHOSLOVAKIA

130 mm M51 (32-round) Multiple Rocket System on Praga V3S (6 × 6) chassis

Development/Description
The 130 mm M51 (32-round) MRS was developed in the 1950s and is a standard Praga V3S (6 × 6) truck chassis with the rocket launcher mounted in the rear. No armour protection is provided for the cab so the launcher is traversed left or right before launching the rockets. Spare rockets are carried in stowage boxes either side of the hull under the launcher. The rockets are spin stabilised and it is believed that a rocket with an increased range has been introduced in recent years.

Austria uses the M51 launcher mounted on the rear of a Steyr 680 M3 (6 × 6) truck (qv) and Romania uses it on the rear of a Soviet ZIL-151/ZIL-157 truck (qv).

The 130 mm M51 (32-round) is sometimes called the RM-130 and is issued on the scale of one battalion of 18 launchers in each Motorised Rifle Division and Tank Division.

Status: Production complete. In service with Austria (qv) (on Steyr 680 M3 chassis), Bulgaria, Cuba, Czechoslovakia (replaced by M1972), Egypt, Romania (qv) (on ZIL-151/ZIL-157 chassis) and Yemen.

Manufacturer: Czechoslovak state factories.

130 mm M51 (32-round) multiple rocket system on Praga V3S (6 × 6) chassis (Christopher F Foss)

SPECIFICATIONS		WEIGHT OF SYSTEM (loaded)	8900 kg	FUEL CAPACITY	120 l
CALIBRE	130 mm	LENGTH (travelling)	6.91 m	FORDING	0.8 m
NUMBER OF BARRELS	32	WIDTH (travelling)	2.31 m	GRADIENT	60%
LENGTH OF ROCKET	0.8 m	HEIGHT (travelling)	2.92 m	VERTICAL OBSTACLE	0.46 m
WEIGHT OF ROCKET	24.2 kg	GROUND CLEARANCE	0.4 m	ENGINE	Tatra T-912 6-cylinder in-line air-cooled diesel developing 98 hp at 2100 rpm
MAX RANGE	8200 m	TRACK			
MAX VELOCITY	415 m/s	front	1.87 m		
UNIT OF FIRE	96 rockets	rear	1.755 m	GEARBOX	manual with 4 forward and 2 reverse gears
RELOAD TIME	2 mins	WHEELBASE	3.58 m + 1.12 m		
ELEVATION/DEPRESSION	+50°/0°	MAX SPEED	62 km/h	TRANSFER BOX	2-speed
TRAVERSE	120°	RANGE	443 km	TYRES	8.25 × 20
CREW	6				

122 mm RM-70 (40-round) Multiple Rocket System

Development/Description
The 122 mm RM-70 (40-round) MRS, which has the NATO designation M1972, was first seen in public during a parade in Czechoslovakia following the SHIELD-72 manoeuvres. The system is essentially an armoured version of the Czechoslovak Tatra 813 (8 × 8) truck fitted with the same launcher as the Soviet BM-21 MRS at the rear of the hull, with an additional pack of 40 rockets to the rear of the cab for rapid loading. Main advantages of the RM-70 over the BM-21 are its rapid reloading capability, complete armour protection for the crew from small arms fire and shell splinters and much improved cross-country mobility. The vehicle is fitted with a central tyre pressure regulation system which enables the driver to adjust the tyre pressure to suit the type of ground being crossed. Some vehicles have been fitted with a hydraulically operated BZ-T dozer blade at the front of the hull for preparing fire positions and clearing obstacles. It is also fitted with a winch with a capacity of 22 000 kg.

In 1989 the then East German Army RM-70 systems were seen mounted on the Tatra T815 VVN (8 × 8) 12 000 kg truck chassis. The most notable change from the T813 is the unarmoured cab. A 12 000 kg capacity winch is fitted on the front of the vehicle with 85 m of cable. The vehicle can

RM-70 system mounted on rear of Tatra T815 VVN (8 × 8) cross-country truck

also be fitted with the SSP 1000 arrow-type snow plough or the BZ-T815 dozer blade if required. It is probable that the Czechoslovakian Army is also using this model.

RM-70 122 mm (40-round) Multiple Rocket System travelling, preparing to fire and firing

The spin and fin-stabilised 122 mm rocket types available are:

SOVIET DESIGNATIONS	**M-21-OF, DB-1B**	**M-21-OF**	**DKZ-B**
SOVIET INDUSTRIAL INDEX NUMBER	9M22U	9M22M	9M28
DIAMETER	122 mm	122 mm	122 mm
LENGTH	3.226 m	2.870 m	1.905 m
WEIGHT (complete)	77.5 kg	66 kg	45.8 kg
RANGE			
max	20 380 m	20 000 m	10 800 m
min indirect	1500 m	1500 m	2500 m
min direct	500 m	500 m	400 m
WARHEAD TYPES	HE-fragmentation, smoke, incendiary, HCN,GB and VX chemical		
WARHEAD WEIGHTS			
HE-fragmentation	19.4 kg	18.4 kg	19.4 kg
Chemical	19.3 kg	n/av	19.3 kg
FUZE SETTINGS	instant/single time delay/safe		
FUZE ARMING DISTANCE	150-400 m	150-400 m	150-400 m
BURN-OUT VELOCITY	690 m/s	700 m/s	450 m/s
LAUNCHERS	BM-21a/b RM-70 BM-21/SR-114 122 mm/DAC 665T	BM-21a/b BM-21V RM-70 BM-21/SR-114 122 mm/DAC 665T	BM-21a/b BM-21V RM-70 9P132 BM-21SR-114 122 mm/A.R.O 122 mm/Trailer 122 mm/DAC 665T

SPECIFICATIONS		LENGTH (travelling)	8.8 m	TRENCH	1.6 m
CALIBRE	122.4 mm	WIDTH (travelling)	2.5 m	ENGINE	Tatra T-930-3 V-12 air-cooled diesel developing 270 hp at 2700 rpm
NUMBER OF BARRELS	40	HEIGHT (travelling)	2.9 m		
HE rocket fuze	0.9 kg	GROUND CLEARANCE	0.4 m		
HE filler	6.6 kg	TRACK	2.03 m		
TYPE OF HE FILLER	86% TNT, 14% Al	WHEELBASE	1.65 m + 2.2 m + 1.45 m	GEARBOX	dual range 5-speed plus overdrive provides 20 forward and 4 reverse gears
UNIT OF FIRE	80 rockets				
RELOAD TIME	2-3 mins	MAX SPEED	75 km/h		
ELEVATION/DEPRESSION	+50°/0°	RANGE	600 km		
TRAVERSE	172°	FUEL CAPACITY	400 l	TYRES	15.00 × 20
CREW	6	FORDING	1.4 m		
WEIGHT OF SYSTEM (loaded)	33 700 kg	GRADIENT	60%		
		VERTICAL OBSTACLE	0.6 m		

Status: In service with Czechoslovakia, Germany, Libya, Poland and the Soviet Union.

Manufacturers: Chassis conversion, Czechoslovak state factories; rocket launcher, Soviet state factories.

EGYPT

325 mm SAKR-80 MEDIUM Artillery Rocket System

Development

The SAKR-80 was developed by the SAKR Factory of the Arab Organisation for Industrialisation (AOI) in conjunction with the French firm Societe Nationale des Poudres (SNPE) to meet an Egyptian Army requirement for a FROG-7 replacement.

SNPE is responsible for the propulsion propellant and Sakr for the various warheads, launcher and fire control system. Both companies are collaborating closely on the design, development and manufacturer of the propulsion subsystems. Mass production could start within a year of the procurement contract being signed.

Depending upon warhead type the SAKR-80 is designed to neutralise targets such as battlefield command control, and communications (C^3) centres, air defence batteries, troop and armour concentrations and second echelon units out to 80 km range by multiple rocket attack.

The tactic used is a 2.5 minute 'shoot and scoot' firing sequence to engage the target from a position some 10-15 km from the Forward-Line-of-Troops (FLOT), which is about half the distance possible with FROG-7 system.

Description

The SAKR-80 is an unguided single-stage fin-stabilised ballistic rocket. The maximum flight velocity is 1200 m/s with a flight time of 160 seconds to its maximum 80 000 m range. The highest altitude it can reach, which depends upon the launch angle, is 29 000 m.

Propulsion is by a two-part composite solid propellant rocket motor unit. The main or first stage uses 310 kg of the composite propellant and burns for six seconds producing 11 000 kg of thrust.

The spin rotation rocket motor is mounted to the rear of the main motor and is connected to it by an ignition train. This causes the motor to ignite 0.6 seconds after the main motor fires. The motor then burns for 1.5 seconds with the energy produced being used to impart the stabilising spin to the rocket's airframe.

A 200 kg warhead is fitted with a nose-mounted electronic fuzing system. The warhead type is chosen according to the target to be engaged. The available types are:

a) HE-fragmentation for airburst use against light armour, communications aerials, radar antenna or concealed targets. The warhead has an outer layer of steel balls around the HE charge which produce over 7500 fragments on detonation that are capable of penetrating up to 14 mm of steel plate at 30 m distance.

b) HE-hollow charge for use against protected targets.

c) Anti-personnel/anti-vehicle cluster munition type for use against widely dispersed personnel and armour concentrations. This carries 950 ribbon stabilised 36 mm diameter cylindrical shaped-charge bomblets each weighing 0.20 kg. Using an airburst detonation it can saturate a zone six hectares in area with the bomblets.

d) An anti-armour cluster munition type for use against armoured vehicles. This carries 65 anti-armour mines which are each 112 mm in diameter, 2.5 kg in weight and use an electronic sensor unit to detect the target. The shaped-charge is capable of defeating 120 mm of armour plate at 0.5 m range with the mine casing having a secondary shrapnel effect against soft-top vehicles and trucks. Airburst detonation of the warhead allows a six hectare area to be seeded with the mines.

For maximum effect of the system a real-time target acquisition system such as an RPV unit must be deployed with it. To exploit this, both automatic-fire control and automatic survey systems are being developed for use with the launcher.

These are a conversion of the existing FROG-7 BAZ-135L4 (8 × 8) launcher with a three/four launch rail assembly and a turretless T-54/T-55 MBT chassis with a four round elevatable container launcher unit on a turntable.

At a launch site the BAZ-135L4 must first be stabilised with hydraulic jacks (located between the first and second road wheels and at the truck rear) before a SAKR-80 can be fired.

The SAKR-80 rounds are carried in a rocket pod which is elevated by hydraulic arms before firing.

SPECIFICATIONS

Rocket

CALIBRE	325 mm
LENGTH	6.5 m
LAUNCH WEIGHT	660-700 kg
WARHEAD TYPES	HE-fragmentation, ICM, HE-shaped charge
WARHEAD WEIGHT	200 kg
MAX RANGE	80 000 m

Launcher

CHASSIS	BAZ-135L4 (8 × 8) wheeled
COMBAT WEIGHT	20 000 kg
LENGTH	10.25 m
WIDTH	2.8 m
HEIGHT	3.25 m
GROUND CLEARANCE	0.6 m
MAX SPEED	
CROSS-COUNTRY	60 km/h
FORDING	1.4 m
NUMBER OF ROCKETS	3

Status: Pre-production trials. On order for the Egyptian Army.

Manufacturer: Sakr Factory for Developed Industries
PO Box 33
Heliopolis
Cairo
Arab Republic of Egypt
Telephone: 660250/660230/660452/660479
Cable: CERVA HELIOPOLIS
Telex: 92175/92770 CERVA UN
Fax: 202-2901978/202-660 180

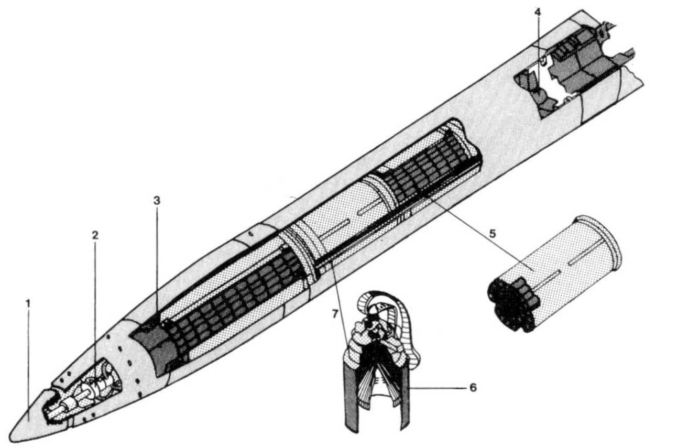

Dual purpose warhead for SAKR 80 showing **(1)** *nose-mounted electronic time fuze,* **(2)** *electronic pyrotechnic distribution unit,* **(3)** *annular cut off cable,* **(4)** *main gas generator,* **(5)** *cluster bomb unit,* **(6)** *dual purpose anti-armour/anti-personnel sub-munition and* **(7)** *longitudinal new cut-off cables*

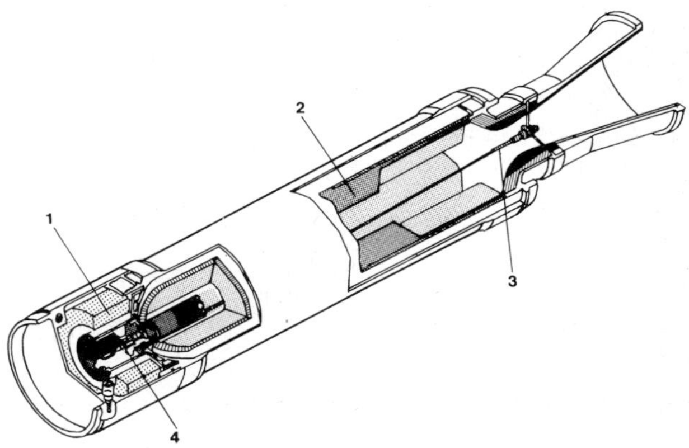

Rocket propulsion systems of SAKR 80 showing **(1)** *spin rotation motor,* **(2)** *main motor,* **(3)** *and* **(4)** *pyro-technic link between the two motors*

122 mm SAKR Series Multiple Rocket Launchers

SAKR-36 System

Following development work in the late seventies the SAKR Factory for Developed Industries successfully test-fired the SAKR-36 (or Hawk-36) 122 mm calibre multiple rocket system in May 1981. Based on the Soviet 122 mm Grad rocket the weapon is fitted with interface systems to allow it to be fired from a number of rocket launcher types including the Helwan Machine Tools Company's 40-round launcher on a ZIL truck chassis, the 30-round RL-21 launcher on a 3500 kg Soviet ZIL-131 6 × 6 or Japanese Isuzu truck chassis, the 21-round launcher on a 4000 kg modified Romanian 4 × 4 Bucegi SR-114 truck chassis, the lightweight 4/8-round light vehicle systems and the Soviet BM-21 system.

The major change is a considerable increase in range which has been achieved by the use of a lightweight rocket motor and case that is filled with 23 kg of composite bonded star grain propellant which burns for 3.2 seconds.

A number of rocket warheads are available and these can be fitted either with the standard electro-mechanical fuze or a more sophisticated programmable time fuze for the submunition-carrying warheads with several possible settings that allow dispensing between 800-1000 m altitude over the target.

The basic HE round is 2.58 m in length and weighs a total of 56.5 kg complete with its 17.5 kg HE-fragmentation warhead. Its range at 50° launcher elevation is 36 000 m.

The 3.1 m long anti-personnel bomblet rocket weights 61.5 kg complete and has a 28 kg warhead which contains 98 bomblets each with a lethal radius of 15 m. The remaining rocket type is of the mine dispensing type and is 3.16 m in length. The 28 kg warhead contains five anti-tank mines. Range of the submunition rockets at 50° launcher elevation is 33 000 m.

SAKR-18 System

To complement the SAKR-36 the Egyptians have also fielded the 122 mm calibre SAKR-18 system which uses a 2-second burn rocket motor with 20.45 kg of double base propellant grain as used in the Egyptian copy of the Soviet 122 mm Grad rocket to give a maximum range of 18 000-20 000 m at +48° launcher elevation. Rocket length is 3.25 m and weight complete is 67 kg. Two 23 kg warhead versions are available, both of the submunition dispensing category. The anti-personnel warhead contains 77 bomblets of the same type as carried in the SAKR-36 anti-personnel rocket. The second warhead type contains four anti-tank mines. The fuze fitted is an electro-mechanical type one which allows for dispensing altitudes between 1800 and 2000 m.

Lightweight Manportable and Vehicle-Mounted Systems

In 1987 the Helwan Machine Tools Company Factory No 999, which manufactures the 30-round 122 mm rocket launcher system, exhibited two types of manportable tripod-mounted 122 mm launchers for use by special forces and other mobile units. The single-round launcher is known as the PRL-111 whilst the three-round systems has the designation PRL-113. Both have an optical mortar-type sight on their left sides which can be used both for direct and indirect fire roles. They are capable of being traversed 7° left and right and elevated from -10° to +40°. The rocket type used is the 122 mm SAKR-10 which gives the system a minimum range of 3000 m and a maximum range of 10 800 m.

Early in 1988 the 122 mm four- and eight-round lightweight mobile launchers were revealed by the SAKR factory. For use with fast moving or airmobile units the launchers can be mounted on any type of 4 × 4 light truck capable of carrying the weight involved, such as the 1135 kg AM 720 Jeep.

The four-round launcher can also be used as a ground mount (total weight 266 kg, tube length 3 m, elevation 17-55°, traverse ±15°) with the unit breaking down into a number of loads for air dropping or man-portable purposes (11 pieces weighing between 15 and 36 kg). Crew requirement in this configuraiton is four. Time into action is around 10 minutes.

The mobile launchers require a crew of three - the driver, gunner and commander - and can be traversed 120° left or right and elevated from 0° to +55°. All the 122 mm family of rocket types can be fired including the SAKR-10, SAKR-18, SAKR-36 and the Grad A and Grad B copies to allow maximum engagement ranges from 10.8 to 36 km according to the operation being mounted. The sight used is a simple mortar-type.

Egyptian-built 30-round RL-21 122 mm multiple rocket system on rear of ZIL-131 (6 × 6) truck chassis (Christopher F Foss)

SPECIFICATIONS (30-round SAKR rocket launcher on 6 × 6 ZIL-131 chassis)

CALIBRE	122.4 mm
NUMBER OF BARRELS	30
LENGTH OF TUBE	3 m
MAX RANGE	18 000 to 36 000 m (according to rocket type)
ELEVATION/DEPRESSION	+55°/0°
TRAVERSE	72° right, 102° left
COMBAT WEIGHT	11 250 kg
LENGTH (travelling)	7.25 m
WIDTH (travelling)	2.6 m
HEIGHT (travelling)	2.9 m
MAX SPEED (fully loaded)	40 km/h

Status: SAKR-36, in production. In service with Egypt (with 21-round SAKR, 30-round SAKR and 40-round SAKR and BM-21 heavy weight rocket launchers and SAKR 4- and 8-round light weight launchers), Iraq and Sudan.

SAKR-18, in production. In service with Egypt (with 21-round SAKR, 30-round SAKR and 40-round SAKR and BM-21 heavy weight rocket launchers and SAKR 4- and 8-round light weight launchers).

SAKR-10, in production. In service with Egypt (with single-round PRL-111, 3-round PRL-113, SAKR 4-round and SAKR 8-round lightweight launchers). The SAKR 4-round launcher is also in service with the Afghan Mojahedin guerrillas.

Manufacturers: SAKR rocket systems and 4/8-round lightweight launchers: SAKR Factory for Developed Industries
PO Box 33
Heliopolis
Cairo
Arab Republic of Egypt
Telephone: 660250/660230/660452/660479
Cable: CERVA HELIOPOLIS
Telex: 92175/92770 CERVA UN
Fax: 202-2901978/202-660180

PRL-111/113, 21-, 30- and 40-round launchers: Helwan Machine Tools Company
Ein Helwan
PO Box 11726
Cairo
Arab Republic of Egypt
Telephone: 782588; 782657
Telex: 92737

SAKR-36 rocket being launched during trials

Enquiries to: Helwan Machine Tools Company
General Marketing Department
23 Talaat Harb Street
PO Box 1582
Cairo
Arab Republic of Egypt
Telephone: 744158; 741738

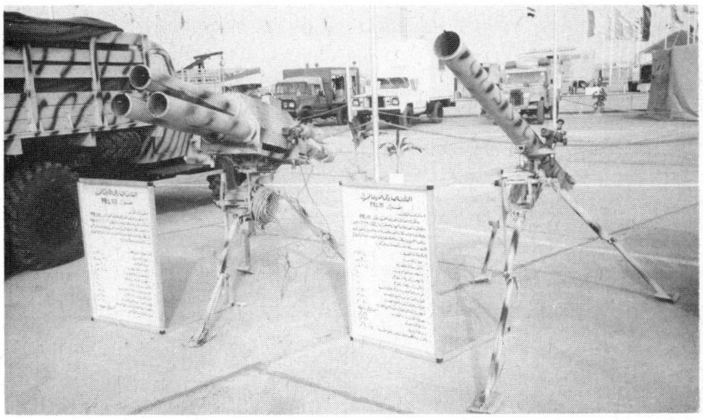

PRL-113 (3-round) and PRL-111 (single-round) rocket launchers which have a maximum range of 10 800 m (Christopher F Foss)

SAKR 80 mm VAP (12-round) Multiple Rocket Launcher

Development/Description

The SAKR Factory for Developed Industries developed the 12-round 80 mm VAP vehicle-mounted multiple rocket launcher for mobile units in the late seventies and began production in the early 1980s. Vehicles which can be fitted with the system include 4 × 4 jeeps and the Fahd 4 × 4 APC.

The rectangular-shaped launcher comprises three layers of four tubes and is mounted on a pedestal which is fitted in the rear of the vehicle. Both elevation and traverse controls are provided on the pedestal together with a mortar-type optical sight located midway along the launcher tube assembly to allow both direct and indirect fire missions. The rockets are fired by a remote-control off-vehicle system.

The 1.5 m long rocket weighs 12 kg complete and can be fitted with either HE-fragmentation or illumination warhead types. The former uses steel balls or cubes to produce the fragmentation effect. A contact fuze is fitted which is armed after 250 m of flight. Four folding fins are fitted aft which deploy upon launch to provide stabilisation. Maximum range with the 3 kg of one second burn double base grain solid propellant fuel in the rocket motor is 8000 m.

Status: In production. In service with Egypt and other unspecified countries.

VAP 80 mm light vehicle-mounted system with launcher elevated (Christopher F Foss)

Manufacturer: SAKR Factory for Developed Industries
PO Box 33
Heliopolis
Cairo
Arab Republic of Egypt
Telephone: 660250/660230/660452/660479
Telex: 92175/92770 CERVA UN
Fax: 202-2901978/202-660 180

Other Egyptian Multiple Rocket Systems

Development/Description

The Egyptian Army uses a number of rocket systems provided by Czechoslovakia and the Soviet Union, including 30-40 122 mm (40-round) BM-21s, 50 130 mm (32-round) M51s, 20 132 mm (16-round) BM-13-16s, 60 140 mm (16-round) BM-14-16s and 15-20 240 mm (12-round) BM-24. Egypt can no longer obtain rockets for these systems from Czechoslovakia or the Soviet Union so the Arab Organisation for Industrialisation is now making the 132 mm rocket for the BM-13-16 and the 122 mm Grad-A and Grad-B rockets for the BM-21 by reverse engineering.

The 122 mm rockets (including the SAKR types) are produced on a co-production basis with a number of European firms. The rocket body is supplied by Vickers, the composite SAKR rocket propellant by Bayern Chemie of Germany, the rocket motor by SNPE of France and the SAKR proximity fuze by Simpa of France (with its explosive content from E Lacroix). The warheads are provided by the Abu Zaabal Company for Speciality Chemicals Factory No 18 which is also responsible for manufacturing the explosive and filling the warhead. A copy of the Soviet MRV-U impact and short- or long-delay action fuze for 122 mm rockets is also produced at the El Maasara Company for Engineering Industries Factory No 45.

The Egyptian Army has also deployed locally made rocket launchers on tanks, two quadruple box-like launchers either side of a T-62's turret, and on wheeled APCs, a rectangular system of three superimposed quadruple rail launchers in the rear of a Walid (4 × 4) APC. The purpose of these systems is to lay smokescreens which last up to 15 minutes. The Walid-mounted launcher described above uses the 80 mm D-3000 rocket with a mechanical impact fuze and the complete 12-round salvo can form a smokescreen 1000 m wide. The rocket itself weighs 26.5 kg, is 1.51 m long and has a maximum range of 2500 m. The T-62 system also uses this rocket. A six-round launcher tube assembly on the Walid can use the 122 mm D-6000 model. This is a 122 mm rocket which is 2.25 m long, weighs 56 kg and has a maximum range of 6000 m. The amount of white smoke composition carried by the rockets is 7.8 kg and 15 kg respectively. The latter is sufficient to create a 400 m long smokescreen using the six rockets carried.

Locally designed D-6000 6-round rocket system of Egyptian Army mounted on rear of Walid (4 × 4) APC (Egyptian Army)

Locally designed 12-round D-3000 rocket system of Egyptian Army mounted on rear of Walid (4 × 4) APC (Egyptian Army)

Status: Egyptian manufactured rockets for the BM-13-16 and BM-21 systems are in service with the Egyptian Army. The 122 mm (40-round) has also been supplied in quantity to a number of other countries including Iraq, Morocco, Somalia and Sudan.

Manufacturer: SAKR Factory for Developed Industries
PO Box 33
Heliopolis
Cairo
Arab Republic of Egypt
Telephone: 660250/660230/660452/660479
Telex: 92175/92770 CERVA UN
Fax: 202-2901978/202-660 180

GERMANY

110 mm Light Artillery Rocket System

Development

The 110 mm Light Artillery Rocket System (LARS) was developed in the 1960s and accepted for service with the Federal German Army in 1969 with first production systems being delivered the following year. Prime contractor for the launcher was Wegmann and Company of Kassel with GUF responsible for the rockets. The system is called the Artillerie Raketenwerfer 110 SF2 by the Federal German Army and is issued on the scale of one battery of eight launchers per division, with each battery having two Fieldguard fire control systems (on 4 × 4 truck chassis) and a resupply vehicle which carries 144 rockets. The Federal German Army has 209 LARS in service.

The system has undergone upgrading from LARS-1 to the LARS-2 over the past few years. The programme has included a new fire control system, more rocket types and mobility improvement by fitting the launcher to a new MAN (6 × 6) truck chassis. An air-transportable, towed, two-wheel, 15-round launcher was also developed by Dynamit Nobel

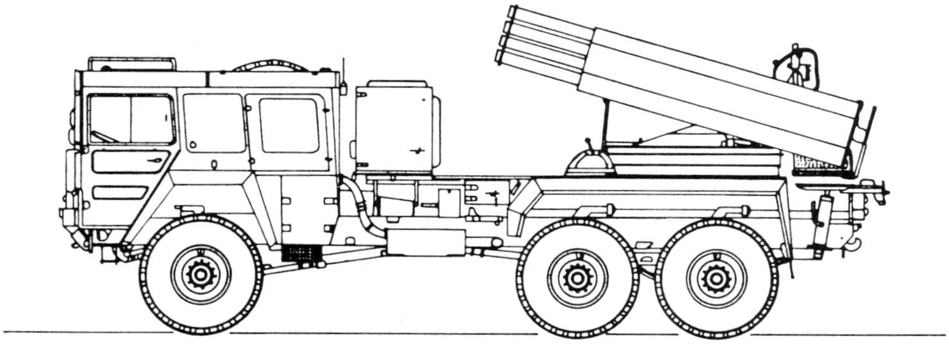

110 mm Light Artillery Rocket System 2 mounted on rear of MAN (6 × 6) 7000 kg truck

Contraves developed and produced Fieldguard fire control system on MAN 5000 kg (4 × 4) cross-country truck chassis. Radar operates in J-band with 300 to 20 000 m range (Michel C Klaver)

110 mm Light Artillery Rocket System on MAN (6 × 6) chassis with stabilisers lowered at rear

but was not adopted by the Federal German Army. The same launcher was also tested on the rear of a Swedish Army Bv 206 tracked vehicle.

With the introduction of the MLRS into the German Army, the LARS will be phased out of service and it is expected that 150 LARS will be passed onto Greece, Portugal and Turkey.

Description

The system essentially consists of a MAN (6 × 6) truck chassis fitted with a cab and a multiple rocket launcher at the rear. Mounted on the roof of the cab is a 7.62 mm MG3 machine gun with an elevation of +9° to +50°, a depression of 0° and a traverse of 360°. The crew of three, consisting of the commander, aimer and driver, all sit in the cab.

The launcher at the rear consists of two banks each with 18 barrels, and is balanced by a sprung equilibrator and elevated by a mechanical drive system. A drive is also provided for azimuth changes. The panoramic sight Model 59 is mounted between and above the two banks of 18 barrels and the aimer is provided with a seat which folds forwards for travelling. Before the rockets are fired from inside the cab two mechanical stabilisers are lowered at the rear of the vehicle and the covers at the end of each bank of launchers are folded forwards through 270° so that they rest on top of the launcher.

The fin-stabilised rockets can be fired singly, part rippled or in a complete ripple. When ripple firing there is a half-second gap between each rocket being launched and the complete load of 36 rockets can be launched in 17.5 seconds. The standard rocket has a solid propellant motor which burns for 2.2 seconds and has a minimum range of 6000 m and a maximum range of 14 000 m. More recently two higher-performance rockets have been built using the same rocket motor casing but with improved propellants. These have maximum ranges of 19 000 and 25 000 m respectively. Total length for all three rockets with their warheads attached is 2.263 m and their total weight is 35 kg. The interchangeable warheads used are all 0.95 m long, 0.109 m in diameter and 17.3 kg in weight. There is a wide variety available which includes the following LARS-1 types: the Diehl training warhead, the Diehl practice warhead, the Diehl DM-15 smoke warhead (containing 6 kg of a smoke generating compound for a 15-minute duration smokescreen), the Dynamit Nobel DM-11 fragmentation warhead with an impact fuze, the Dynamit Nobel DM-21 fragmentation warhead with a 20 m, altitude-actuated, ground-proximity fuze and the Diehl DM-701 dispensing warhead containing eight AT1 anti-tank mines. For the LARS-2, Dynamit Nobel has produced the DM-28 practice warhead (converted from the older DM-15 smoke warhead), the DM-39 radar target warhead (fitted with a radar reflector to act as the pilot shot for the Fieldguard radar system to measure actual trajectory paths for salvo firing), the DM-711 dispensing warhead containing five parachute-retarded 2.2 kg AT2 anti-tank mines and a new smoke warhead (containing 8.4 kg of a smoke generating compound for a 15-minute duration smokescreen). In addition the company has also developed for the German MoD a bomblet warhead (containing 65 M42 or M77 0.21 to 0.23 kg multi-purpose shaped charge fragmentation submunitions containing 0.03 kg of explosive each with an effective fragmentation radius of 3 m and an armour penetration capability of 65 to 100 mm) that has yet to be ordered for the Federal German Army. Once the 36 rockets have been launched the vehicle normally moves off to be reloaded to avoid retaliatory fire. The launcher takes 15 minutes to be reloaded.

As the Magirus-Deutz Jupiter chassis of the 110 mm LARS-1 is now obsolete within the Federal German Army the launchers have been removed from their chassis and installed on the rear of a MAN (6 × 6) 7-ton truck which has improved cross-country mobility. The modification started in 1980 and finished in 1983. Full details of this chassis are given in *Jane's Military Vehicles and Logistics 1991-92*, page 345.

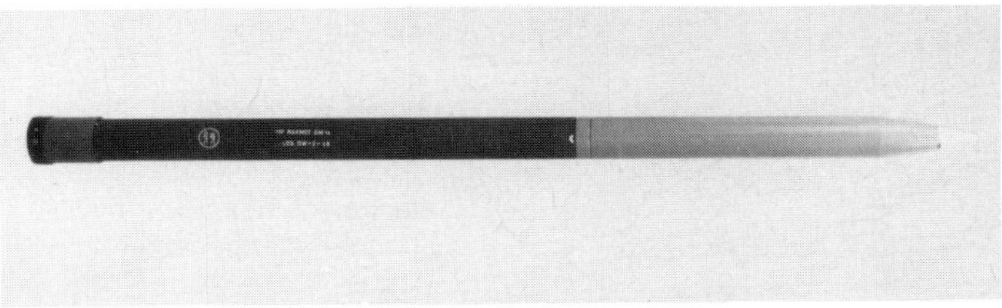

Rocket used in 110 mm LARS consists of two parts, rocket motor (left) and warhead (right)

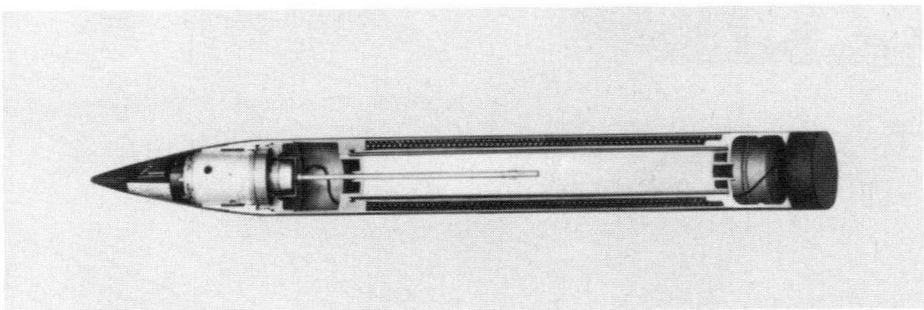

Fragmentation warhead containing about 5800 fragments and fitted with ground-proximity fuze

SPECIFICATIONS

CALIBRE	110 mm
NUMBER OF LAUNCHER TUBES	36
ELEVATION/DEPRESSION	+55°/0°
TRAVERSE	360°
WEIGHT OF SYSTEM (loaded)	17 480 kg
LENGTH (travelling)	8.28 m
WIDTH (travelling)	2.5 m
HEIGHT (travelling)	2.99 m

Status: Production complete. In service with Federal German Army. To be phased out of service with some 150 systems expected to be passed onto Greece, Portugal and Turkey free of charge.

Manufacturer: Launcher: Wegmann and Co GmbH
D-3500 Kassel
August-Bode-Strasse
Federal Republic of Germany
Telephone: 0561 1050
Telex: 99859
Fax: 0561 105 2208

Rockets: Dynamit Nobel Aktiengesellschaft
Postfach 12 61 D-5210 Troisdorf
Federal Republic of Germany

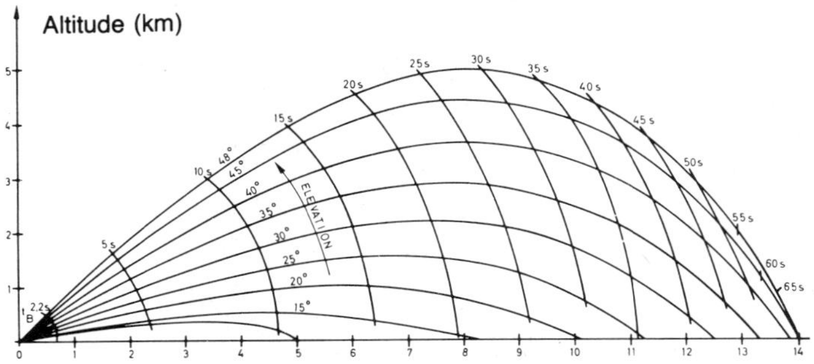

Range table for Standard 110 mm Light Artillery Rocket System

INDIA

122 mm (40-round) Multiple Rocket Launcher System

Development/Description

India is building an indigenously developed 40-round rocket launcher based on the Soviet 122 mm BM-21 design. Mounted on the locally produced Shaktiman (6 × 6) 5000 kg truck chassis, the launcher fires a new longer range 122 mm rocket, the LRAR, developed by the Armament Research and Development Establishment in Pune. The warhead contains 18 kg of TNT explosive and the rocket is fitted with a double-base solid propellant motor. The LRAR can also be fired from the BM-21. A full 40-round salvo can be fired within 20 seconds and reloading is thought to be manual.

Status: In production. In service with the Indian Army.

INTERNATIONAL

LTV 227 mm Multiple Launch Rocket System

Development

Early in 1976 the United States Missile Command at Redstone Arsenal Alabama, initiated feasibility and concept formulation of a General Support Rocket System, which would use a low cost rocket that would be handled like a round of conventional ammunition and have a high rate of fire. This system would be used against troops and light equipment, air defence systems and command centres.

In March 1976 five companies, Boeing, Emerson Electric, Martin Marietta, Northrop and Vought, were awarded contracts for the concept definition of the General Support Rocket System (GSRS).

After evaluating all five concept definition studies, in September 1977 Boeing Aerospace and the Vought Corporation (now known as LTV, Aerospace and Defense Company (LTVAD)) were each awarded contracts for the 29-month validation phase of the GSRS competition.

Under the terms of the contract each company delivered three prototype Self-Propelled Launcher Loaders (SPLLs) and between 140 and 160 rockets for competitive testing at White Sands Missile Range in New Mexico.

Early in 1978 the United States Army Missile Research and Development Command redirected the GSRS towards a standard NATO weapon that could be developed and produced in both the USA and Europe, and renamed the programme Multiple Launch Rocket System or MLRS.

In 1980, the LTVAD of Dallas, Texas, was selected as prime US contractor for MLRS and was awarded a full-scale development contract. Teamed with LTV were the Atlantic Research Corporation for the solid propellant rocket motor, Bendix Guidance Systems Division for the stabilisation reference package/position determining system, the Brunswick Corporation for launch tubes, Norden Systems for the fire control system and Vickers for the launcher drive system.

European participation was introduced by an MoU between the governments of the United States, Germany, the United Kingdom, France and Italy allowing for the joint development, production and subsequent deployment of a multiple-launch, free flight rocket system.

In 1983 LTV's first MLRS units entered service with the US Army and in the same year MLRS-Europaische Produktions Gesellschaft GmbH (MLRS-EPG), was established in Munich to undertake European production. Diehl of Germany, Hunting Engineering of the United Kingdom, Aerospatiale of France and SNIA BPD of Italy were invited to form this Prime Contract Consortium and sharing of manufacture was established in accordance with national offtake ratios.

The main European sub-contractors to EPG were Aerospatiale for launcher integretion, BPD for the rocket motor, Diehl for rocket integration and Thyssen Henschel for the carrier vehicle.

The first European-built MLRS systems were handed over late in 1989 and production will peak at 9 or 10 launchers per month. It is expected that the European MLRS programme will involve the manufacture of:

284 launchers
180 000 M77 rockets and practice rockets
12 500 AT-2 rockets.

In accordance with the MLRS MoU, the MLRS International Corporation (MIC) was established to market and sell the MLRS to non-participant nations worldwide. MIC is jointly owned by LTV Aerospace and Defense Company and MLRS-EPG. Sales to other countries will be allocated in accordance with the US/European offtake ratio.

Description

The development of the MLRS was planned in three distinct phases. Phase I covered the vehicle, launcher, and an M26 anti-materiel/anti-personnel warhead containing M77 submunitions. Phase II addresses the AT2

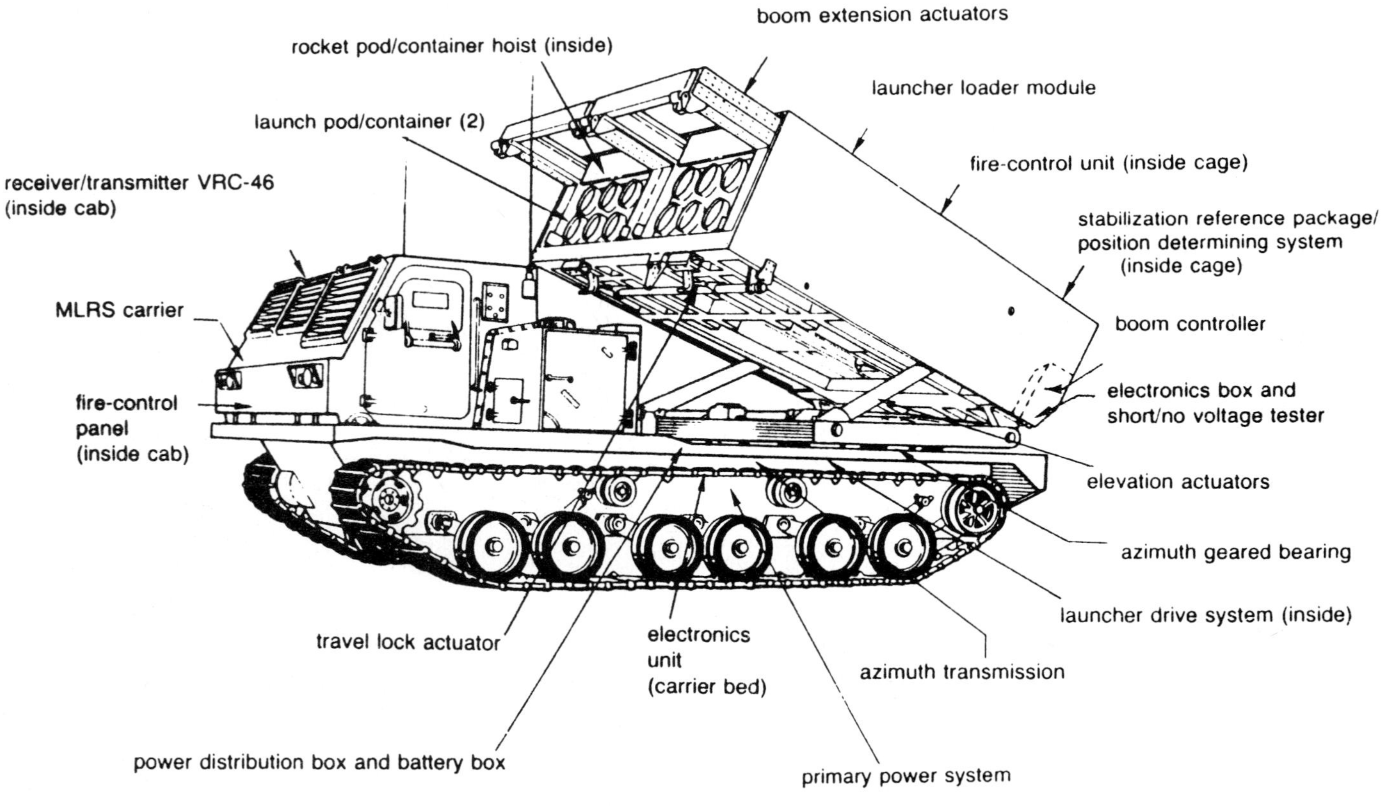

M270 Armored Vehicle-Mounted Rocket Launcher (AVMRL) (not to 1/76th scale)

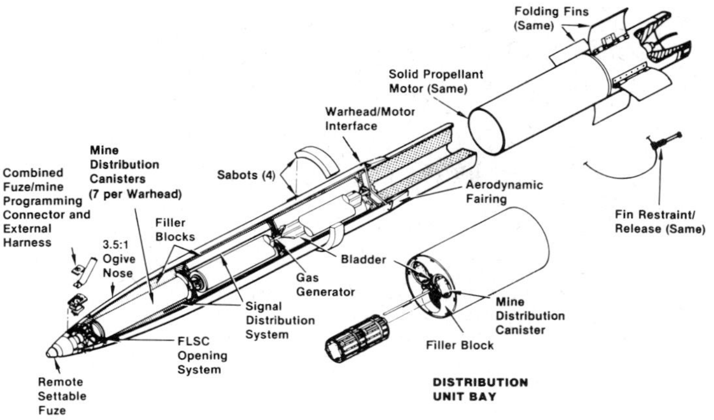

Major components of AT2 rocket (NB: remote settable fuze is electronic time type) (not to 1/76th scale)

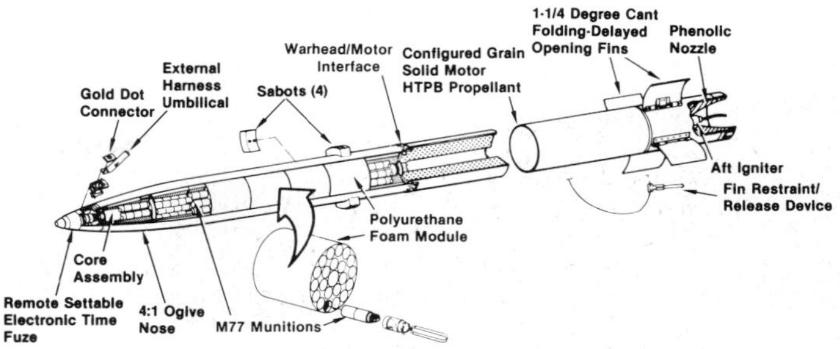

Major components of rocket with M26 warhead containing 644 M77 dual purpose shaped-charge blast fragmentation bomblets (not to 1/76th scale)

scatterable mine warhead, due to enter service in 1992. Phase III is the development of a terminally guided anti-armour warhead (TGW) which is scheduled to enter service in the mid-1990s.

Phase I-M270 Launcher

The lightly armoured M270 self-propelled launcher is a stretched version of the M2 Bradley infantry fighting vehicle. The Launcher Loader Module, mounted on the rear of the vehicle hull consists of a base, turret and cage, and contains the computerised fire control system, a stabilisation reference package/position determining system, a launcher drive system and a twin boom crane unit for self-loading and unloading.

The three man crew consists of the driver, gunner and section chief all of whom are seated in the cab at the front of the vehicle, which is protected against small arms fire and shell fragments by aluminium armour and louvred windows.

An over-pressure ventilation system prevents rocket fumes from entering the cab and the vehicle is also equipped with a filtration unit to protect the crew from chemical, biological and radiological particles.

Equipment capabilities permit a reduced crew, or even one person, to accomplish a complete fire mission including the loading and unloading operations.

MLRS Rocket Pod

The standard rocket pod, which is also the transportation and storage container, weighs 2308 kg when loaded and is comprised of six glass fibre tubes rigidly held in an aluminium case structure. Blow-off covers seal each rocket in its tube, so the rocket requires no maintenance, inspection or crew servicing from the factory to the launch site.

The rocket pods, each 4.219 m long, 1.051 m wide and 0.839 m high, are loaded by the integral twin boom loader system of the launcher.

M26 Warhead

The Phase I rocket has a minimum 10-year shelf-life, weighs 307 kg, is 3.937 m long and 0.227 m in diameter. The rocket is centred in the launch tube by a set of four forward sabots, two aft riders and two spin lugs that ride along helical rails. When the rocket is discharged, the sabots are discarded and four folding fins at the rear of the rocket motor are released by a cable-release initiated pyrotechnic delay device attached to the launch tube. The M26 warhead weighs 156 kg, has an aluminium skin, and carries 644 M77 Dual Purpose Improved Conventional Munition (DPICM) shaped-charge blast fragmentation bomblets weighing 213 g each, nestled with eight polyurethane foam supports. Armour penetration is 76 to 102 mm. An M445 remotely set, electronic time fuze in the warhead initiates dispersal of the ribbon-stabilised bomblets in mid-air over the target. The range of this version is approximately 32 km. The Phase I rocket was used operationally with great success in the 1991 Second Gulf War.

Phase II-AT-2 warhead

The AT-2 anti-tank mine was developed by Germany and has been in service in the ground scatterable version for some years. The mine warhead is 236.6 mm in diameter, weighs 107 kg and contains 28 AT2 DM1399 anti-tank mines. Due to its lighter 258.5 kg weight, the AT2 rocket has a range of 40 km, somewhat longer than the M77 rocket with the M26 warhead. The warhead is activated by a DM42 adjustable timer fuze and is opened in flight by a small explosive charge. The seven mine dispensers, each with four mines, are pneumatically ejected and then retarded to release the mines which are parachute stabilised in flight.

The 3.927 m long rocket usually dispenses the mines when overflying the target area at about 1200 m. The 128 m high AT2 anti-tank mine has a diameter of 103.5 mm, a deployed height of 165 mm with its ground orientation springs open and a weight of 2.2 kg. Detonation of the full-width attack mine occurs either when an armoured vehicle passes over it or when it reaches 97 per cent of its useful lifetime, determined before launch from one of six available settings. Armour penetration is in excess of 140 mm. An anti-handling/removal device is incorporated to disrupt mineclearing operations. The rocket has been ordered by Germany and the UK.

Phase III-Terminal Guidance Warhead

The TGW will have dimensions similar to the Phase II warhead but will carry three terminally guided submunitions which are scattered in mid-air by a time fuze to perform unguided dives until they establish their operational altitude. They then pull up, using onboard reaction controls and gyro platforms to perform a horizontal guide.

During this phase of flight, each submunition will use its active radar seeker to search for its target. The onboard target discriminator and selector logic systems ensure that only a tank will be attacked. Once a valid target is acquired, the submunition dives on it, using a contact-fuzed shaped tandem charge warhead to kill it.

The MLRS Phase III is being developed under an MoU signed by the United States, United Kingdom, France and Germany which allows for

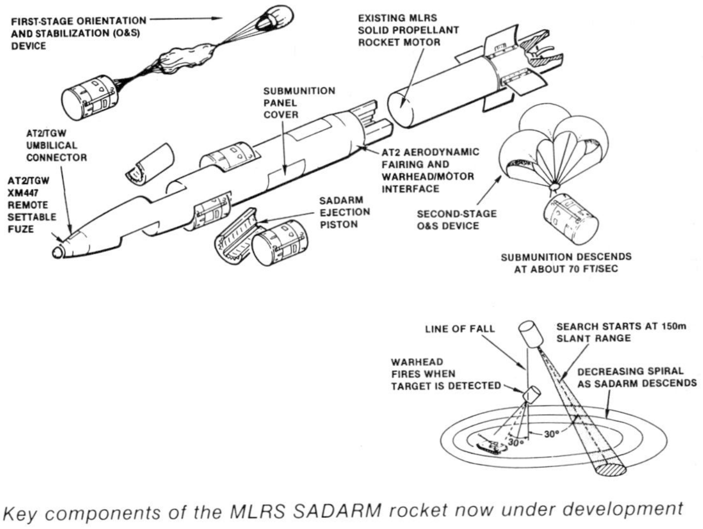

Key components of the MLRS SADARM rocket now under development

M270 AVMRL of the US Army in travelling configuration (US Army)

Royal Netherlands Army M270 AVMRL with launcher elevated

US Army M270 AVMRL being reloaded with six-round pack of rockets

a total transfer of technology. Although each partner will have the right of production, workshare is allocated among the partners of the MDTT consortium.

Following bids submitted by six multi-national contractors, the winning consortium was the one headed by Diehl of Germany, Thomson-CSF of France, THORN EMI of the UK and Martin Marietta of the USA.

Martin Marietta is responsible for the target detection algorithms, integration assembly and system testing. Diehl is responsible for the control section, wing design and fabrication and will also build the signal processor and handle power conditioning and distribution.

Thomson-CSF will make the radio frequency receiver and fabrication components of the warhead, design the warhead structure and dispenser, and carry out assembly.

THORN EMI will design and build the submunition radome antenna and gyroscope and will also develop the tracker algorithm and fabricate the fuze sensor and electronics together with instrumentation and telemetry. The System Demonstration Phase of the TGW is due to be completed in October 1992 but it is unlikely that the US Army will place production contracts for this version as it is also funding SADARM and BAT (qv) versions of the MLRS.

Variants

Sense and Destroy Armor (SADARM)

The MLRS-SADARM rocket will have a range of over 35 km and have similar aerodynamic and ballistic characteristics as the AT2 minelaying rocket, but will have six 175 mm diameter SADARM anti-armour submunitions as the payload.

The submunitions will be packaged in individual compartments, each containing a gas generator and piston for submunition expulsion. They will be ejected in opposing pairs to form a circular pattern so that no two search the same area and attack the same target. Upon ejection each SADARM deploys a two-stage orientation and stabilisation subsystem. The first stage decelerates the SADARM and the second establishes a controlled descent path for the spinning submunition. As it continues to fall at approximately 21.3 m/s, infra-red and millimetre-wave sensors are activated at about 150 m slant range to the ground to search a circular area with a decreasing spiral pattern. The multiple sensors then examine potential targets for signature date, and at a valid target confirmation, fires the warhead which propels the mass of an

M270 AVMRL fire control display panel

M77 DPICM submunition, 644 of which are carried by each rocket

explosively formed penetrator slug at hypersonic velocity. This destroys the target by top attack through the impact kinetic energy of the penetrator slug.

Aerojet Electro Systems is funded for full-scale development of the MLRS variant and the 155 mm howitzer variant. Both variants are scheduled to be in service in 1996 with the MLRS-SADARM production contract to be awarded in 1994.

US Army Tactical Missile System (Army TACMS)

Late in March 1986, the US Army announced that LTV had been selected to design and develop the Army Tactical Missile System (Army TACMS) and was subsequently awarded a $180.3 million contract.

Army TACMS is fired from the MLRS M270 launcher using off-axis guidance techniques to prevent enemy radars from plotting the trajectory and initiating counter-battery fire against its launch point.

Each missile is approximately 3.962 m long, 0.61 m in diameter and is powered by an advanced solid propellant fuel to ranges in excess of 100 km. Each warhead will contain 1000 M74 anti-personnel/anti-materiel bomblets which are dispersed over the target area to detonate upon impact.

Teamed with LTV are Atlantic Research Corporation for the solid rocket fuel propulsion, Honeywell Incorporated for the onboard inertial guidance system and Martin Marietta Corporation for the anti-armour warhead.

A follow-on warhead, currently under development with Martin Marietta Aerospace, would include precision-guidance sub-missiles for use against moving targets.

The first Army TACMS missile was fired in 1988 and the same year LTV was awarded an initial contract for 66 production missiles worth $44 million. Early in 1990 the company was awarded a $59.6 million fixed price contract for low-rate initial production of 104 missiles to be delivered by late in 1991.

The US Army expects to purchase 1000 Army TACMS (Block I) weapons and a similar number of Block II missiles. First missiles were deployed to Germany, in October 1990 as the replacement for the conventional Lance.

The US Army has awarded LTV an $83 million contract to integrate Army TACMS into the existing M270 AVMRL. The changes are principally in the software of the system's fire control computer and will allow the system to fire standard MLRS rockets or Army TACMS missiles.

Army TACMS will allow battlefield commanders to strike well behind enemy lines, beyond the range of existing artillery, rockets or missiles, so raising the nuclear threshold.

Around 30 TACMS rounds were successfully used by the US Army in the 1991 Second Gulf War.

US Army TSSAM-BAT Brilliant Submunition

Since 1984 the Northrop Corporation has been working on a classified US Army R & D programme to develop the BAT dual sensor (acoustic and infra-red) brilliant submunition for use in several deep attack

M270 AVMRL with full scale mock-up of Army TACMS missile and missile pod

M270 AVMRL firing

weapons, including the MGM-137 ground-launched version of the Tri-Service Stand-off Attack Missile (TSSAM).

The BAT programme was revealed in 1991 when the US Army awarded a $344 million contract to the Northrop Corporation Electronic Systems Division to complete its development. The contract will run for 42 months and provide a final design for production in the mid-1990s. Total programme cost is estimated to be around $2.2 billion.

Raytheon is the principal subcontractor to the programme. Two of the TSSAM weapons will be carried on an M270 launcher vehicle. The BAT is also likely to become a payload of the ATACMS.

The BAT programme has replaced the TACMS-IRTGSM programme in the US Army's funding of such systems and has caused US Army interest in the MLRS-TGW programme to be downgraded to the technologies involved rather than in developing that system.

The unpowered 20 kg BAT submunition is 914.4 mm long, 139.7 mm in diameter and is fitted with four foldable wings that each have a protruding acoustic sensor spike attached. A nose-mounted infra-red sensor is also fitted together with a shaped-charge warhead. Four flip-out wraparound tail fins are used to stabilise the weapon in flight.

Resupply

The resupply vehicles used by the US Army are a Heavy Expanded Mobility Tactical Truck (HEMTT) (8 × 8), which carries four fully loaded rocket pods, and its Heavy Expanded Mobility Ammunition Trailer (HEMAT), which carries an extra four fully loaded rocket pods. The truck is air-transportable by C-130 or C-141 aircraft. The US is currently developing a DROPS type system.

The European armies will choose their own national re-supply systems. The United Kingdom, for example, will use its Demountable Rack Off-loading and Pick-up System (DROPS). This is also a truck-trailer system consisting of the Leyland Medium Mobility Load Carrier (MMLC) and a trailer, both of which carry four rocket pods. Germany will utilise its indigenous MULTI system which is very similar to DROPS.

High Mobility Artillery Rocket System (HIMARS)

This is a US Army proposal for a lighter tracked or wheeled version of the MLRS system using a single six pack rocket launcher.

M28 Practice Rocket

The practice warhead is used as a training vehicle for the troops. This warhead uses steel pipe in place of the tactical munitions. Additionally, three smoke canisters are located in bay number 4. These canisters are made from plexiglass and contain 100 mg of titanium tetrachloride ($TiCl_4$).

At the warhead event, the canisters are broken by the central core burster and the $TiCl_4$ reacts with moisture in the air to produce smoke. This cloud must be visible for a minimum of 10 seconds from 6000 m.

SPECIFICATIONS		TRACK WIDTH	533 mm	ENGINE	Cummins VTA-903
CREW	3	LENGTH OF TRACK ON			turbo-charged
WEIGHT OF SYSTEM		GROUND	4.33 m		8-cylinder diesel
loaded	25 191 kg	MAX SPEED	64 km/h		developing 500 hp at
empty	20 189 kg	RANGE	483 km		2400 rpm
LENGTH (travelling)	6.972 m	FUEL CAPACITY	617 l	TRANSMISSION	General Electric
WIDTH (travelling)	2.972 m	FORDING	1.1 m		HMPT-500 hydro-
HEIGHT		GRADIENT	60%		mechanical
travelling	2.617 m	SIDE SLOPE	40%	STEERING	hydrostatic
elevated	5.925 m	VERTICAL OBSTACLE	1 m	SUSPENSION	torsion-bar-in-tube
GROUND CLEARANCE	0.43 m	TRENCH	2.29 m	ELECTRICAL SYSTEM	24 V

Status: In production and service. By early 1991 orders had been placed by the five participant nations for more than 1000 launchers and 750 000 rockets. The first European built systems were handed over late in 1989.
Bahrain: ordered 9 systems for delivery in 1992.
France: 1 from US with remainder being built in Europe.
Germany: 6 from US with remainder being built in Europe.
Italy: 1 from US with remainder being built in Europe.
Japan: is subject of a co-production agreement.
Netherlands: 22 purchased direct from the United States and already delivered.
Turkey: 12 purchased direct from US, of which 6 have been delivered with an additional 120 launchers and 36 000 rockets to be manufactured in Turkey by a new company jointly formed by Turkish Industry and LTVAD.
United Kingdom: 4 from US with remainder being built in Europe.
United States: 804 launchers and 556 110 rockets are under contract.

Note: The total of French, German, Italian and UK launchers ordered is 284, of which 55 are for France, 150 for Germany, 24 for Italy and 59 for the UK.

Manufacturer & Marketing: Prime contractor in USA: LTV Missiles and Electronics Group
Missiles Division
PO Box 650003
Dallas
Texas 75265-0003
USA

Prime contractor in Europe: MLRS-Europaische Produktions Gesellschaft mbH
Postfach 1163
Mehlbeerenstrasse 4
D8028 Taufkirchen
Federal Republic of Germany

Marketing enquiries to: MLRS International Corporation
79 Knightsbridge
London SWIX 7RB
UK
Telephone: 071-235-2122
Fax: 071-245-6325
Telex: 8952636 MLRSIC G

IRAN

Iranian Artillery Rocket Launcher Systems

From the mid-1980s onwards Iran has produced a number of unguided artillery rockets.

In 1985 members of the Iranian Government concerned with weapons procurement visited the People's Republic of China and North Korea to sign agreements for the transfer of missile and rocket technology.

One of the first systems to be produced by the Iranian Defence Industry Organisation (DIO) following the visit was the Oghab (Eagle) rocket. This was originally thought to be a straight copy of the Chinese 273 mm Type 83 rocket but is in fact an indigenous design using transferred technology. It is 4.82 m long, 230 mm in diameter and weighs 360 kg (of which 70 kg is a HE-fragmentation warhead). The maximum range is given as 80 000 m with the solid propellant rocket motor developing a maximum thrust of 7648 kN.

It is launched from an elevatable triple-rail launcher assembly mounted on the rear of a Mercedes-Benz LA911B (4 × 4) 4500 kg truck chassis. The truck is also known as the LA1113 and before 1963 as the LA 328. Before firing two pairs of stabiliser jacks have to be lowered.

First combat use of the Oghab occured in December 1986 when several rounds were fired at the Iraqi city of Basra. From then to the Gulf War ceasefire approximately 330 rounds were fired by rocket units of both the regular army and the ground forces of the Islamic Revolution Guards Corps (IRGC).

The Security and Army (SECARM) exhibition at Libreville, Gabon in early 1989 revealed three more unguided rocket designs.

The largest is the Nazeat which is 5.9 m long, 355 mm in diameter and weighs 950 kg with a 150 kg warhead. Maximum range is 90 000 m with the solid propellant rocket motor delivering 11 217 kN of thrust.

The Shahin 2 is smaller at 3.87 m long, 333 mm in diameter and 580 kg in weight but has a larger 180 kg HE warhead. Two versions apparently exist each having a maximum range of 20 000 m. The solid propellant rocket motor is said to generate 22 434 kN of thrust.

The third design remains unidentified but is approximately 180-200 mm in diameter and uses four radially folding fins.

There may be a larger unguided artillery rocket with a 130 km range which was used in the latter stages of the Gulf War by missile units of the regular air force and tentatively identified as the IRAN-130. Five of these weapons were fired in March/April 1988 at the Iraqi city of al-Amareh during the "War of the Cities".

The Iranian DIO are also producing a copy of the Chinese 107 mm Type 63 MRS and the rockets used by it. This is produced in towed and self-propelled versions. The latter including a variant with the launcher mounted on Toyota (4 × 4) light vehicles (and used in Lebanon by Hezbollah and the IRGC). Aid in building these probably came from both the People's Republic of China and North Korea. A copy of the 30-round 122 mm calibre rocket launcher produced by North Korea is also in production and is mounted on the Mercedes Benz LA911B (4 × 4) 4500 kg truck chassis.

Status: Oghab-in production. In service with the Iranian Army and IRGC.
Nazeat-in production. In service with the Iranian Armed Forces.
Shahin 2-in production. In service with the Iranian Armed Forces.
Type 63 copy-in production. In service with the Hezbollah (Shia Lebanese Militia Group), Iranian Army and IRGC.
122 mm 30-round copy in production. In service with the Iranian Army and IRGC.

Manufacturer: Enquiries to: Defence Industries Organisation
Ministry of Defence
Tehran
Islamic Republic of Iran

Trailer-mounted Iranian Shahin 333 rocket system

Iranian 230 mm diameter Oghab rocket

Iranian 355 mm diameter Nazeat rocket

Iranian 122 mm (30-round) MRS mounted on Mercedes Benz LA911B (4 × 4) 4500 kg truck chassis

IRAQ

Iraqi Multiple Rocket Launcher Systems

Apart from using large numbers of Soviet and Egyptian supplied multiple rocket launcher systems the Iraqi Army also uses a number of other systems. Available details are given below.

550 mm Laith 90 Artillery Rocket System

This is a locally built cluster munition warhead version of the Soviet Luna-M (Soviet designation R-75, NATO designation FROG-7a) artillery rocket system. Maximum range has been increased from 70 000 to 90 000 m.

400 mm Ababeel 100 Artillery Rocket System

This is a 400 mm calibre 4-round truck-mounted rocket launcher with a maximum range of 100 000 m. A submunition dispensing warhead is fitted which carries either 300 anti-tank/fragmentation bomblets or 25 anti-tank minelets. The latter weigh 800 kg each and have a 2 m radius magnetic influence fuzing system that can fire the warhead even if they are laying on their sides.

262 mm Ababeel 50 Multiple Rocket Launcher System

This is a co-produced version of the Yugoslavian M-87 (12-round) MRS developed with Iraqi assistance. A detailed description is given in the Yugoslavian part of this section (qv). This has a maximum range of 50 000 m. A few pre-production series vehicles were delivered before the Gulf War.

Babeel 100 (4-round) launcher in ground emplaced configuration

Barq rocket is a SA-3 SAM used in the ground-ground role

107 mm (12-round) multiple rocket system (Christopher F Foss)

127 mm Sajeel 30, 180 mm Sajeel 40 and 300 mm Sajeel 60 Multiple Rocket Launcher Systems

These are respectively locally license-built versions of the Brazilian ASTROS II SS-30, SS-40 and SS-60 launchers and rockets. Details differ in some ways from those given by Astros (qv Brazilian part of this section) and may reflect local variations.

DESIGNATION	**Sajeel 30**	**Sajeel 40**	**Sajeel 60**
CALIBRE	127 mm	180 mm	300 mm
WEIGHT			
launch	68.1 kg	152 kg	587.65 kg
warhead	21 kg	n/av	n/av
LENGTH			
complete	2.950 m	3.978 m	5.281 m
warhead	0.685 m	1.795 m	1.092 m
rocket motor	2.265 m	2.183 m	4.189 m
THRUST	1650-2000 kg	2070-2510 kg	4600-6400 kg
BURN TIME	2.25-2.75 s	4.1-5.1 s	6.2-8.8 s
MAX RANGE	30 000 m	35 000 m	60 000 m
MIN RANGE	9000 m	15 000 m	20 000 m
WARHEAD TYPE	HE	ICM	ICM

The Sajeel 40 warhead contains 20 anti-armour/anti-personnel bomblets whilst the Sajeel 60 warhead contains 65 bomblets of the same type. Both are fitted with a variable time delay which can be programmed before launch from 5.2 seconds onwards in order that the warhead detonates over the target area and not over friendly forces.

107 mm Portable Rocket Launcher System

The Iraqis have designed and built a portable 12-round launcher system for use with an indigenous 107 mm calibre rocket design. Aid may have come from the People's Republic of China in developing this system. Available details are:

CALIBRE	107 mm
NUMBER OF BARRELS	12
MAX RANGE	8000 m
WEIGHT	
system	385 kg
rocket complete	19 kg
LENGTH (firing)	3 m
WIDTH (firing)	1.7 m
HEIGHT (firing)	1.1 m
ELEVATION	0° to + 60°
TRAVERSE	30° (15° left, 15° right)
RATE OF FIRE	12 rockets/min

122 mm Multiple Rocket Launcher Systems

The Iraqis have taken locally built versions of the Egyptian Sakr-36 30-round 122 mm MRS launcher assembly and mounted them on the rear of the Soviet MT-LB tracked APC and Swedish SAAB-Scania SBA 111 series (4 × 4) truck.

30 mm al-Tawbiya Multiple Rocket launcher System

This weapon was used in quantity by the Iraqi Army during the latter stages of the Gulf War as an anti-personnel weapon to stem massed Iranian Infantry attacks. No other details are available at present.

Launcher 81

GAZ-66 (4 × 4) 2000 kg truck fitted with 35 round (five rows each of seven rockets) 81 mm rocket launcher at rear.

Launcher 81 is a GAZ-66 with an 81 mm multiple launcher mounted on rear

Kaser Rocket
SA-6 Gainful SAM used in ground-to-ground role.

Barq Rocket
SA-3 Goa SAM used in ground-to-ground role.

Fahd Rockets
Two versions, Fahd 250 and Fahd 500.

Kassir Rocket
Maximum range 150 km.

Al Rasheed Rocket
Maximum range 25 km.

Nissan 28 Rocket
Maximum range 110 km.

Nissir Rocket
Maximum range 800 to 10 000 m, this may be the Launcher 81 mentioned above.

Fahd 500 rocket

ISRAEL

Israel Military Industries 350 mm Artillery Rocket System MAR 350

Development/Description
Israel Military Industries has developed a 350 mm calibre artillery rocket system, the MAR 350, for use on the same range of chassis as the LAR 160. The initial firing trials were held in late 1986 and in early 1988 it was revealed that Argentina was offering the same system on its VCLC version of the TAM tank chassis (qv).

Two types of warhead are available: HE-COFRAM with impact or proximity fuze. Cluster munition containing 770 CL-3022S bomblets and a remotely set time fuze. A low cost guidance system is available for rockets to be fired to long ranges.

Fuzed rockets are factory sealed in their launch pod containers, two per container, which are loaded in the field for firing.

The basic mounting is the LAR-160 hydraulically operated cradle system with two cylindrical rocket container-launchers in a side-by-side arrangement. The VCLC carries two such systems along the lines of the 290 mm rocket mounted on Israeli Army Centurion tank chassis. Weight of the 2-round pod container is 2700 kg, length 6.2 m, width 0.97 m and height 0.526 m. All four rockets can be fired within 30 seconds.

SPECIFICATIONS

OVERALL LENGTH	5.80 m
ROCKET DIAMETER	0.35 m
OVERALL WEIGHT	850 kg
WARHEAD WEIGHT	300 kg
EFFECTIVE OPERATIONAL	
max range	80 km
min range	29.5 km
LPC WEIGHT (loaded)	2700 kg

Status: Development complete. Troop trials with the Israeli Army are believed to be in progress.

Manufacturer: Israel Military Industries Ltd
POB 1044
Ramat Hasharon 47100
Israel
Telephone: (03) 5485222
Telex: 33719
Fax: (03) 5406908

Israel Military Industries 290 mm (4-round) Artillery Rocket System MAR 290

Development/Description
In 1960 Israel Military Industries established a new department, the Rocket Systems Division, which started work in 1965 on a 290 mm artillery rocket with a 25 000 m range, the MAR 290. This entered Israeli Defence Force (IDF) service mounted in frame racks, holding four rockets, carried on Sherman tank chassis.

At one stage the 290 mm rocket was developed into a new version known as the MK4. This had a range of 40 000 m but was used as a research and development vehicle only and is still retained for this role.

The 290 mm MAR 290 rocket system is still in IDF use but is now mounted on a new launching system using Centurion tank hulls. The new system retains the same 290 mm rockets but they are now launched from four tubes.

The basic 290 mm rocket is 5.45 m long, weighs 600 kg at launch, has a warhead weight of 320 kg and a range of 25 000 m. The tube launcher can fire four rockets in 10 seconds. On the turretless Centurion chassis the four tubes are carried in a horizontal layer on a new superstructure on the original turret ring. The tubes can be traversed through a full 360° and elevation is from 0 to +60°. Each launching tube has a diameter of 700 mm to accommodate the tail span of 570 mm.

The Centurion carrier has a crew of four: commander, driver, and two fire control operators. The rockets are fired when all four crew members are under cover in the vehicle.

Launcher tube laying and reloading are hydraulically powered. The reloading sequence is as follows:

Once the launcher has fired all four rockets it moves to a pre-selected location for reloading. The fresh rockets are taken to the location by a truck which lowers them to the ground for the packaging to be removed. For this the truck uses its own handling crane and the same crane also lifts the prepared rockets onto a special reloading frame. When the launcher vehicle approaches it uses a system of two lugs and a bracket on the front hull to engage with corresponding items on the loading frame. The frame is then connected to the vehicle's hydraulic system

Production version of 290 mm MRS on Centurion chassis with rocket being launched

and a ram on the frame raises the entire frame off the ground into a raised horizonal position until the rockets are in line with the launching tubes which have been reversed to point over the rear of the vehicle. Using hydraulics again, the rockets are then slid into the tubes. The entire sequence is carried out by one crew member using a remote control unit and it takes about 10 minutes. After loading, the loading frame is lowered to the ground, disconnected mechanically and hydraulically and is then ready for more rockets to be loaded by the supply truck.

The height of the Centurion with the launching tubes lowered is 3.6 m and the weight is 50.8 tonnes. Each launching tube is 6 m long. The only armament now carried by the Centurion, apart from the launching tubes, is a single pintle-mounted 0.30 Browning machine gun in front of the driver's hatch.

Status: In production. In service with the Israeli Army.

Manufacturer: Israel Military Industries Ltd
POB 1044
Ramat Hasharon 47100
Israel
Telephone: (03) 5485222
Telex: 33719
Fax: (03) 5406908

Rockets being loaded on Centurion production version of 290 mm MRS

Israel Military Industries 160 mm LAR-160 Multiple Rocket Launcher System

Development/Description

After the successful development and production of 240 mm and 290 mm artillery rocket systems for the Israel Defense Forces in the 1970s, Israel Military Industries Ltd developed a new generation artillery rocket system, the LAR-160. Combat experience with the 240 mm and 290 mm systems was used in the development of the LAR-160 system.

The Light Artillery Rocket System consists of a tracked or wheeled all-terrain vehicle or trailer on which is mounted a multiple rocket launcher. Two factory sealed, expendable launch pod containers, containing fuzed, 160 mm diameter rockets in their launch tubes, are loaded on the launcher. The pods, which also serve as storage and transportation containers for the rockets, can be rapidly replaced in the field after the rockets have been fired. When the rockets are fired they burst frangible covers over the forward and aft ends of the launch tubes.

The Mark I system was deployed on AMX-13 light tank chassis by the Venezuelan Army. The Mark II system is currently in production for another country, and the next generation upgrades are in development for the Israel Defense Forces.

The rockets are 160 mm in diameter and 3.314 m long, employing composite solid propellants. Wraparound stabilising fins deploy when the rocket exits the launcher. The Mark I version of the LAR-160 rocket, weighing 100 kg, carries a 40 kg HE-COFRAM warhead activated by a super-quick impact fuze or proximity fuze. Maximum effective range is 30 km. Its launch pod containers hold 18 rockets each, with two pods on the launcher. The current production Mark II rocket version weighs 110 kg. Its 46 kg warhead can be an HE-COFRAM type or a cluster warhead containing 104 CL-3022S AP/AM submunitions. A remotely set electronic time fuze opens the canister at the appropriate height to give area coverage of about 31 400 m^2 for each cluster warhead. Thirteen rockets are loaded in each launch pod container. Maximum range is 34 km. The minimum range of 12 km and intermediate ranges are obtained through the use of spoilers, which produce a high trajectory that optimises bomblet distribution and allows the LAR-160 to fire over hills and obstacles.

Launchers are being mounted on heavy truck chassis in the current production lot. Suitable chassis for mounting the modular launcher are light tanks (such as AMX-13), medium tanks (such as M47 and M48), heavy trucks (such as M809, MAN, Steyr and Mercedes), self-propelled howitzer (M109), and a trailer towed by a heavy truck. Each launcher will accommodate two 13-rocket launch pod containers.

The LAR-160 can be operated with traditional fire control methods of tube-artillery or with radar fire control. Using the latter system, one to three pilot rockets are fired in the general direction of the target while being tracked by fire control radar. Pilot rockets can be destroyed in-flight along their trajectory or fired at an offset so as not to alert the target. Radar data is used by a computer to automatically calculate launcher aiming parameters.

Elevation (to +57°) and traverse of the launchers are performed by an electro-hydraulic system, backed up by a manual system. The fire control system is based on micro-processor technology. Optional fire control system components include inertial navigation and direction finding equipment in each launcher, digital data communications and display in each launcher, and a digital firing computer in each launcher, enabling each launcher to operate independently of the battery framework. There are also optional interfaces between the firing system and the elevation and traverse system so that aiming can be fully automated. Operators in each launcher are able to set the size of each salvo and ammunition type, remotely program warhead fuzes and rocket spoiler deployment, and fire rockets while under cover when launchers are mounted on armoured chassis. All 26 rockets can be fired in less than 60 seconds. Launchers are reloaded in less than five minutes by the truck driver and launcher crew from conventional army trucks equipped with a 15 ton-meter hydraulic crane.

Variants

The LAR-160 launcher is not restricted to firing 160 mm rockets. Any launch pod container with a suitable interface can be mounted on the launcher. Israel Military Industries is now completing development of an 80 km range, 350 mm diameter artillery rocket. Two of these rockets will be loaded in a launch pod. These pods are fully interchangeable with pods containing 160 mm rockets, so a commander has a wide range of options without having to field completely separate systems. As new types of submunitions and mines are developed they are being incorporated into submunition warheads.

Rocket being fired from LAR 160 mm system on AMX-13 light tank chassis

Pod of 18 rockets being loaded onto launcher mounted on AMX-13 chassis

SPECIFICATIONS

	Mark I	Mark II
ROCKET CALIBRE	160 mm	160 mm
ROCKET LENGTH	3.314 m	3.314 m
ROCKET WEIGHT	100 kg	110 kg
WARHEAD WEIGHT	40 kg	46 kg
RANGE		
min	12 km	12 km
max	30 km	34 km

Status: Mark I in service with Venezuela. Mark II in production for undisclosed country. Next generation upgrades in development for the Israel Defense Forces and other customers.

Manufacturer: Israel Military Industries Ltd
POB 1044
Ramat Hasharon 47100
Israel
Telephone: (03) 5485222
Telex: 33719
Fax: (03) 5406908

Trailer mounted LAR-160 system with 13 rockets to a pod deployed for firing

ITALY

BPD DIFESA E SPAZIO SpA Multiple Rocket Systems

BPD DIFESA E SPAZIO SpA has developed, as a private venture, three multiple rocket systems called the FIROS 6, FIROS 25/30 and FIROS 70. The primary role is area saturation.

FIROS 6 Multiple Rocket System

Development/Description
The FIROS 6 is a 48-round MRS mounted on the rear of light 4 × 4 vehicles such as the Land Rover, FIAT Campagnola, FIAT 40 PM10 or the FIAT-OTO 6614 APC. It is understood that a small number of FIROS 6 systems has been ordered by the Italian Army. The launcher is moved and aimed in traverse and elevation by hand wheels and partial or full ripples can be fired at the rate of 10 rockets a second. The crew can fire the rockets standing alongside the vehicle or from a remote position up to 30 m from the launcher via a cable. The launcher can be reloaded by hand within five minutes. The rockets used in this system are the same solid propellant, with four folding fins, BPD DIFESA E SPAZIO SpA two-inch rockets that are widely used in the air-to-ground role. When being used in the ground-to-ground role the rocket has a maximum range of about 6500 m. Technical details of the rocket are as follows:

BPD DIFESA E SPAZIO SpA 51 mm rocket

CALIBRE	51 mm
LENGTH	
without warhead	744 mm
with warhead	1.05 m
WEIGHT	
without warhead	2.6 kg
with warhead	4.8 kg
MAX RANGE	6550 m
TIME OF FLIGHT TO MAX RANGE	39 s
PROPELLANT TYPE	extruded double base
PROPELLANT WEIGHT	1.06 kg
TOTAL IMPULSE	210 kg/s
MAX CHAMBER PRESSURE	100 kg/cm^2
AVERAGE THRUST	200 kg
ACCELERATION	60 g
SPEED AT BURNOUT	515 m/s
MOTOR CASE	steel, cold extruded
NOZZLE FIN	with thrust deflectors and folding free rotating fins
WARHEAD TYPES	anti-tank/anti-personnel (AT/AP), high explosive incendiary (HEI), illuminating (III) (under development), pre-formed fragmentation (PFF), smoke (SM), target practice (TP), target practice flash (TP-FL), target practice smoke (TP-SM), target practice spotting (TP-SP)

The anti-tank/anti-personnel rocket can penetrate over 250 mm RHA steel in addition to the secondary anti-personnel effect.

FIROS 6 MRS on rear of LWB Land Rover (4 × 4) light vehicle with launcher being elevated

FIROS 6 launcher module data

NUMBER OF TUBES	48
LENGTH OF MODULE	2.08 m
WIDTH OF MODULE	0.67 m
HEIGHT OF MODULE	0.93 m
WEIGHT	
empty	250 kg
loaded	480 kg
ELEVATION/TRAVERSE	manual
ELEVATION LIMITS	−5° to +45°
TRAVERSE	360°
POWER SUPPLY	24 V DC
RATE OF FIRE	10 rockets/s

FIROS 25/30 Multiple Rocket Systems

Development/Description
The FIROS 25/30 Field Rocket System has been developed as a private venture by BPD DIFESA E SPAZIO SpA. It is an area saturation weapon system and is based on the 122 mm BPD DIFESA E SPAZIO SpA unguided rockets. The system is intended for use either in battery or isolated missions, and can cover large areas in the 8 to 34 km zone.

A typical battery would consist of a command post equipped with a fire direction system, six firing units and six to twelve escort units. The firing unit is a modular rocket launcher installed on a heavy (6 × 6) truck.

It is available in two different versions: a standard version, with a motorised/manual movement system simple to operate; an automatic version, equipped with an inertial navigator, a servo-control system and a ballistic computer; this firing unit configuration enables the crew to perform the topographic and ballistic calculation and the aiming procedures through an Automatic Control Unit installed in the vehicle cab.

The launcher consists of two removable modules with 20 launching tubes each. The escort unit can carry four launching modules. The loading operations are carried out by removing the empty module with a jib crane fitted on the support vehicle and installing the filled one ready to fire.

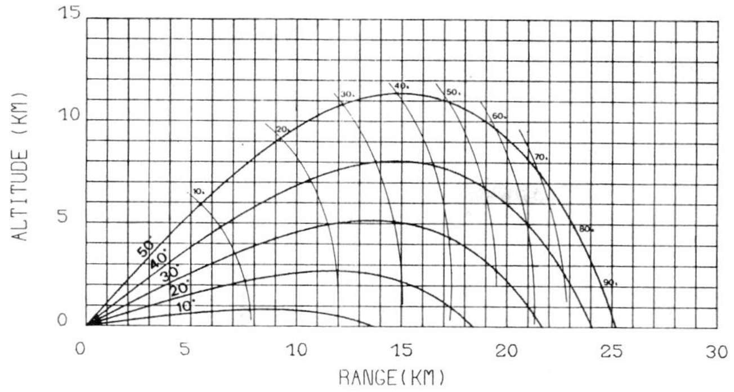

Trajectory diagram for FIROS 25 122 mm rocket with 17kg conventional warhead

FIROS 25 MRS on rear of Mercedes Benz (6 × 6) truck during launching preparation. This version is used by the United Arab Emirates

The firing unit has a fire control system and a link with the command post to acquire the necessary firing data in real time. The fire control system has been designed to enable selection of the number (from 1 to 40) and type of rockets to be fired.

The FIROS 25 and FIROS 30 rockets differ in propellant types and maximum range and can carry a variety of different warheads, from the conventional type (HE, PFF, WP) to the sub-ammunition types dispensing anti-tank or anti-personnel mines and bomblets.

Production versions of the FIROS 30 system for the Italian Army are mounted on an IVECO (6 × 6) chassis with a special armoured forward control cab. The United Arab Emirates use the FIROS 25 system on Mercedes Benz (6 × 6) truck chassis.

The 122 mm rockets launched by the 122 mm FIROS 25 and FIROS 30 systems can also be launched by the widely used Soviet 122 mm BM-21 systems.

BPD 122 mm rocket	**F25**	**F30**
CALIBRE	122 mm	122 mm
LENGTH (without warhead)	2.098 m	2.075 m
WEIGHT (without warhead)	39.3 kg	39.5 kg
MAX RANGE		
conventional warhead	25 000 m	34 000 m
sub-ammunition warhead	22 000 m	30 000 m
PROPELLANT WEIGHT	22.2 kg	24 kg
TOTAL IMPULSE	5000 kg/s	6000 kg/s
CHAMBER PRESSURE	190 kg/cm^2	210 kg/cm^2
AVERAGE THRUST	5000 kg	3000 kg
COMBUSTION TIME (at 50% max pressure)	0.96 s	2.05 s
PROPELLANT TYPE	double base	composite

Conventional warheads	**F25**	**F30**
WEIGHT (total)	18 kg	26 kg
LENGTH	658 mm	740 mm
TRAJECTORY SAFETY	150 m	150 m
TYPES OF WARHEAD	preformed fragmentation (PFF), white phosphorus (WP), high explosive (HE)	

Sub-ammunition warheads	
WEIGHT (total)	31 kg
LENGTH	1.246 m
TYPES OF WARHEAD	anti-tank mine (ATM) containing 6 anti-tank mines each with a magnetic sensor to detect the tank and a millimetric sensor to trigger firing when the vehicle is closest to the mine, anti-personnel mine (APM) containing 66 anti-personnel mines; anti-personnel anti-materiel bomblet (APAMB) containing 77 dual purpose bomblets

Battery command post of FIROS 30 multiple rocket system on IVECO (6 × 6) chassis

FIROS 30 multiple rocket launcher on IVECO chassis and fitted with fully armoured forward control cab

Characteristics	
TIME TO BRING THE SYSTEM INTO ACTION	5 mins (1 min using automatic firing unit)
RIPPLE TIME	16 s (40 rockets)
DISENGAGEMENT TIME	1 min (30 s using automatic firing unit)

Launcher Unit	
NUMBER OF TUBES	40 (20 per module)
ELEVATION	0° to +55°
AZIMUTH	±105°
CREW	3 (2 using automatic firing unit)

Loading Module	
LENGTH	3.7 m
WIDTH	0.82 m
HEIGHT	0.69 m
EMPTY WEIGHT	400 kg
LOADED WEIGHTS	F25: 1566 kg (conventional warhead), 1806 kg (submunition warhead) F30: 1710 kg (conventional warhead), 1810 kg (submunition warhead)

FIROS 70 Artillery Rocket System

This is at present at the feasibility phase and consists of a cross-country truck with a forward control cab and to the rear of this is a double twin-rail 315 mm launcher which will fire rockets to a maximum range of 70 000 m.

Status: FIROS 6 is in production. FIROS 25 is in production and service with the United Arab Emirates and two other undisclosed countries in the Middle East, FIROS 30 is in production for the Italian Army (20 launchers out of 200 systems required). FIROS 70 is still at the feasibility study phase.

Manufacturer: BPD DIFESA E SPAZIO SpA
via Sicilia
162-00187 Rome
Italy
Telephone: (0039-6) 9789 2918
Telex: 611434
Fax: (0039-6) 9789 2205

JAPAN

337 mm (Twin-round) Type 67 Rocket Launcher

Development/Description
The Type 67 Rocket Launcher was developed in 1965 by the Aerospace Division of the Nissan Motor Company Limited and entered service in 1968. It is used only by the Japanese Ground Self-Defence Force. Production of the Type 68 rocket is now complete.

Type 67 rocket launcher in firing position

The system consists of a Hino (6 × 6) 4000 kg truck chassis fitted with two launcher rails for the Type 68 rocket. The rocket is 4.5 m long, has a body diameter of 337 mm, weighs 573 kg at launch and has a range of 28 000 m. Before a rocket is launched three stabilisers are lowered to the ground, one either side of the vehicle and one at the rear. Another Hino (6 × 6) 4000 kg truck carries six additional rockets and is fitted with a hydraulic crane for transferring the rockets to the launcher. The Hino has a maximum road speed of 78 km/h and is powered by a six-cylinder diesel which develops 160 hp at 2400 rpm.

Status: Production complete. In service only with the Japanese Ground Self-Defence Force. The system is due for replacement by the coastal defence missile system Type 88 in the near future.

Manufacturer: Aerospace Division
Nissan Motor Company Limited
5-1 3-chome
Momoi
Suginami-ku
Tokyo 167
Japan
Telephone: 03 (3301) 6610
Fax: 03 (3301) 6714

130 mm (30-round) Type 75 Multiple Rocket System

Development
Development of a 130 mm self-propelled multiple rocket system began in 1965 by the Japanese Self-Defence Agency Technical Research and Development Institute. The first two prototypes were built in 1973 with Komatsu responsible for the chassis and the Aerospace Division of the Nissan Motor Company for the rocket. After operational tests the system was accepted for service with the Japanese Ground Self-Defence Force in 1975. By 1985 the Japanese Ground Self-Defence Force had 66 Type 75 MRS in service. The engine, suspension and tracks are the same as in the Type 73 APC.

Description
The crew compartment is at the front of the all-welded aluminium alloy hull with the driver sitting on the left and the commander on the right. The driver has three periscopes to his front and a single piece hatch cover that opens to the rear-right and contains a night vision infra-red periscope that can be traversed through a full 360°. The commander has a single piece hatch cover that opens to the left rear which contains six vision blocks. A 12.7 mm (0.50) M2 HB machine gun is provided at the front right commander's position for anti-aircraft defence. The operator is seated behind the commander and has a single piece hatch cover that opens forward and to the right.

The engine compartment is on the left side of the hull behind the driver with the air-inlet and air-outlet louvres in the roof and the exhaust pipe running along the left side of the hull. The exhaust pipe is fitted with a silencer that is covered by a box-like net structure on the left middle of the hull top. The launcher is mounted on the top of the hull at the back and in the rear of the hull is a large door that opens to the right. The torsion bar suspension consists of five dual rubber-tyred road wheels with the drive sprocket at the front and the idler at the rear. There are no track return rollers. The Type 75 MRS has no amphibious capability but is fitted with infra-red driving lights.

The rockets have a single-stage solid propellant motor and are fin stabilised. The rockets can be fired singly or ripple fired with the complete rounds being launched in 12 seconds. The launcher is electrically operated but can, in emergencies, be operated manually. Reload rockets are normally carried in accompanying trucks.

130 mm (30-round) Type 75 MRS from rear (Keiichi Nogi)

The Type 75 is accompanied by a Type 75 self-propelled ground-wind measuring unit, which is a derivative of the Type 73 APC, is fitted with a wind-measuring mast which can be extended to 12.5 m and can measure wind velocities of up to 30 m/s in two directions. Another variant is the Type 76 artillery location radar vehicle.

Status: Production complete. In service with the Japanese Ground Self-Defence Force. In the future Japan is expected to order the MLRS. At least 36 MLRS are expected to be procured with the first units being imported and the rest under licence.

Manufacturer: Aerospace Division
Nissan Motor Company Limited
5-1 3-chome
Momoi
Suginami-ku
Tokyo 167
Japan
Telephone: 03 (3301) 6610
Fax: 03 (3301) 6714

SPECIFICATIONS

CALIBRE	131.5 mm
NUMBER OF LAUNCHER FRAMES	30
LENGTH OF ROCKET	1.856 m
WEIGHT	
complete rocket	43 kg
warhead	15 kg
MAX RANGE	15 000 m
ELEVATION/DEPRESSION	+50°/0°
TRAVERSE	100° (50° left and right)
WEIGHT OF LAUNCHER	
empty	2000 kg
loaded	3200 kg
CREW	3
WEIGHT (system loaded)	16 500 kg
LENGTH (travelling)	5.78 m
WIDTH (travelling)	2.8 m
HEIGHT (travelling)	2.67 m
GROUND CLEARANCE	0.4 m
TRACK	2.37 m
TRACK WIDTH	400 mm
LENGTH OF TRACK ON GROUND	3.3 m
MAX ROAD SPEED	53 km/h
RANGE	300 km
GROUND PRESSURE	0.63 kg/cm^2
GRADIENT	60%
VERTICAL OBSTACLE	0.7 m
TRENCH	2.1 m
ENGINE	Mitsubishi 4ZF 2-stroke V4 air-cooled diesel developing 300 hp at 2200 rpm
SUSPENSION	torsion bar
ELECTRICAL SYSTEM	24 V
ARMAMENT	1 × 12.7 mm MG

KOREA, NORTH

North Korean Multiple Rocket Launcher Systems

North Korean BM-11 version of Soviet 122 mm MRS with two banks of 15 launcher tubes instead of single 40-round bank, mounted on rear of Japanese Isuzu (6 × 6) 2500 kg truck chassis (Israeli Ministry of Defence)

Development/Description

Initial production of multiple rocket launchers began in North Korea during the early 1960s with a locally built version of the 107 mm Chinese Type 63 system. Well over 2700 have been produced to date for the home and export market.

With the Korean People's Army (KPA) the towed 107 mm system is found in the Infantry Division and Brigade multiple rocket artillery battalions and batteries. In the Mechanised and Motorised Division MRL battalions and in the specialised Combined Arms Brigades battery the towed version is replaced by either a tracked version or a truck-mounted variant. The former has the launcher mounted on the licence built M1973 (YW531) APC.

Considerable numbers have been exported and these have seen combat service with the Palestinian Liberation Organisation (PLO) and Syria in Lebanon and with Iran in the Gulf War.

Following delivery during the late sixties of numbers of Soviet 122 mm BM-21 MRL systems the North Koreans began production of their own derivative in the mid 1970s. Known as the BM-11 this has two side-by-side 15-round banks of launcher tubes and is usually mounted on a locally produced version of the Soviet Ural-375D (6 × 6) 4000 kg truck, although for export purposes it can be mounted on other chassis. Examples of this are the PLO's Isuzu (6 × 6) 2500 kg systems that were captured by Israel during the 1982 war in Lebanon.

In KPA service the BM-11 is used at Army level within MRL Divisions (144 systems) and at Corps level within MRL Brigades and Regiments (up to 60 systems). One of the roles assigned to these units is to provide chemical warfare delivery capability at the strategic level using chemical warheads that have been indigenously designed, manufactured and filled with agents that include nerve gases. The BM-11 may also be found in the MRL battery of some of the Combined Arms Brigades replacing the six 107 mm systems found in that unit.

A second 122 mm MRL system, designated the M1977 by the US Army, is also in production but no other details are available at present.

For heavy firepower a 12-round 220-240 mm design believed to be similar in appearance to the Soviet BM-24 is known to be in production. This will presumably be replacing some of the older Soviet 140 mm and 200 mm systems that are still available.

The older systems are known to include varying numbers of truck-mounted 132 mm BM-13-32, 140 mm BM-14-16, 200 mm BMD-20 and 240 mm BM-24 weapons plus small numbers of the towed 140 mm RPU-14 for airborne force use. In addition China is known to have provided a number of its 130 mm Type 63 systems.

Status: 107 mm system in production. In service with Iran, Israel, Libya, North Korea, PLO and Syria.

122 mm BM-11 in production. In service with Algeria, Egypt, Iran, Israel, North Korea, Libya, Nicaragua, PLO, Polasario Front, Syria, Uganda and Zimbabwe.

122 mm M1977 in production. In service with North Korea.

220-240 mm system in production. In service with North Korea.

Manufacturer: North Korean state factories.

SPECIFICATIONS (BM-11 provisional)					
CALIBRE	122 mm	ROCKET WEIGHT (complete)	66 kg	WARHEAD TYPES	HE-fragmentation, smoke and chemical
NUMBER OF TUBES	30	WEIGHT OF HE FILLER	6.4 kg	CREW	5
RATE OF FIRE	30 rounds in 15 s	MAX RANGE	20 500 m	LAUNCHER	locally built version Ural-375D (6 × 6) truck chassis
RELOAD TIME	8-9 mins	MIN RANGE	2500 m	RELOAD TIME	8-9 min

KOREA, SOUTH

130 mm (36-round) Kooryong Multiple Rocket System

130 mm (36-round) Kooryong MRS in travelling configuration

Development/Description

The 130 mm 36-round Kooryong multiple rocket launcher has been designed and built in South Korea for Corps and Divisional level MRL units. Although fitted on the rear of the five-ton (6 × 6) KM809A1 truck chassis in South Korean Army service it can be fitted to any wheeled or tracked vehicle with a five-ton class load capacity.

The launcher uses a hydraulic motor to drive the azimuth, elevation and stabilising jack systems. If required all can be operated manually as well. The fire control system consists of three solid-state components, the firing control box, the ignition distribution box and a circuit tester unit. A resupply KM813A1 6 × 6 five-ton truck is used to carry 72 reload rounds and to act as the loading platform. Reloading takes about 10 minutes.

There are two kinds of rocket, the basic and the improved model. The basic round is 2.4 m long, weighs 54 kg in total and uses a 1.79 m long rocket motor section that is 33.3 kg in weight.

The explosive content of the 21 kg warhead is 6.5 kg. The warhead is of the prefragmented HE type and contains some 16 000 small steel balls embedded in its structure. The lethal area of the fragments is approximately 3600 m^2 when the warhead is detonated above ground by a proximity fuze. The range varies from 10 to 23 km with the lower values being attained by the use of either 85 mm or 102 mm diameter brake rings attached to the forward part of the warhead. These drag rings ensure both a reduction in the range dispersion factor and an enhancement in warhead effectiveness for short range fire missions.

The improved 64 kg round uses a 1.93 m long rocket motor section that is 43 kg in weight to give ranges of between 10 and 32 km. The warhead remains the same as does the use of the brake rings and the cruciform flat folding flight stabiliser fins that pop open on launch.

Single-round, partial or full ripple salvo firings can be undertaken using the fire control box from either the driver's seat or off-vehicle at a site several metres from the vehicle. A firing table or the fire direction calculator is used to provide the ballistic data. The latter provides a faster and more accurate method of calculation than the tables.

Surface wind measurements are made using a 10.25 m high mast installation mounted on the back of a 1¼-ton 4 × 4 KM450 light truck and connected to a microprocessor. This provides both digital average and instant wind speed data for the fire control calculations.

SPECIFICATIONS (Improved version - KM809A1 truck chassis)

CALIBRE	130 mm	WEIGHT OF WARHEAD	21 kg	LENGTH (travelling)	7.7 m
NUMBER OF BARRELS	36	RANGE	10-32 000 m	WIDTH (travelling)	2.5 m
LENGTH OF BARRELS	3.4 m	RATE OF FIRE	2 rounds/s	HEIGHT (travelling)	2.9 m
LENGTH OF ROCKET	2.54 m	ELEVATION/DEPRESSION	+55°/0°	MAX ROAD SPEED	80 km/h
WEIGHT OF ROCKET		TRAVERSE	102°	GRADIENT	55%
(complete)	64 kg	CREW	4	POWER SUPPLY	24 V DC
		COMBAT WEIGHT (loaded)	17 000 kg		

Status: In production. In service and type standardised with the South Korean Army.

Manufacturers: Daewoo Heavy Industries Ltd
Daewoo Centre Building
5-541 Namdaemun-Ro
Jung-Gu
Seoul
South Korea
Telex: 23301

Samsung Co Ltd
20th Floor
Samsung Main Building
CPO Box 1144
Seoul
South Korea
Telex: 23657; 23169; 23302

PAKISTAN

Pakistan 122 mm (30-round) Multiple Rocket System

In early 1989 Pakistan successfully test fired an indigenously made 122 mm lorry-mounted 30-round multiple rocket launcher system. No other details are known other than it has a range of 25 000 m. The system is mounted on the rear of a US supplied M35 series (6 × 6) truck chassis with the rockets being made by Pakistan Ordnance Factories.

Status: Production. In service with the Pakistani Army.

Manufacturer: Enquiries to: Pakistan Ordnance Factories
Wah Cantonnement
Pakistan
Telex: 5840 POFAC.PK

Pakistani 122 mm (30-round) MRS based on an M35 series (6 × 6) track

POLAND

140 mm (8-round) WP-8 Multiple Rocket System

Development/Description
The 140 mm (8-round) WP-8 MRS has been designed specifically for use by the Polish 6th Pomerarian Air Assault Division in its composite artillery battalion. The scale of issue is six WP-8 launchers with 12 D-30 122 mm Sonyushka howitzers. The launcher fires the same spin stabilised rockets as the Soviet 140 mm (16-round) RPU-14 MRS, the Soviet 140 mm (16-round) BM-14-16 and the 140 mm (17-round) BM-14-17 truck-mounted MRSs and the single-round 140 mm launcher used by a number of guerrilla forces.

The rocket types used are:

SOVIET DESIGNATIONS	**HE M-14-OF** **Smoke M-14D**
DIAMETER	139.7 mm
LENGTH	1.092 m
WEIGHT (complete rocket)	39.62 kg
MAX VELOCITY	402 m/s
RANGE	
max	9810 m
min indirect	3800 m
min direct	400 m
WARHEAD TYPES	HE-fragmentation smoke GB chemical
WARHEAD WEIGHTS	
HE-fragmentation	18.8 kg
GB chemical	18.3 kg
WEIGHT OF ROCKET FUZE	0.63 kg
FUZE SETTINGS	instant, short and long delay times, and safe
FUZE ARMING DISTANCE	150-400 m

SPECIFICATIONS

CALIBRE	140 mm	LENGTH (travelling)	3.294 m	CREW	5
NUMBER OF BARRELS	8	WIDTH (travelling)	1.634 m	TOWING VEHICLE	UAZ-469B (4 × 4) light vehicle
RELOAD TIME	2 mins	HEIGHT (travelling)	1.2 m		
CARRIAGE	split trial	GROUND CLEARANCE (travelling)	0.254 m		
WEIGHT		TRACK	1.4 m		
system loaded	687.6 kg	ELEVATION/DEPRESSION	+47°/-12°		
system unloaded	370 kg	TRAVERSE (total)	28°		

Status: Production complete. In service with Poland.

Manufacturer: Polish state factories.

140 mm (8-round) WP-8 multiple rocket system

ROMANIA

122 mm (40-round) Multiple Rocket Launcher System

Development/Description

At the first Baghdad International Exhibition for Military Production held in April 1989 Romania revealed that it was producing an indigeneous designed and built 122 mm 40-round multiple rocket launcher which was carried on the rear decking of a DAC 665T (6 × 6) 5000 kg truck chassis.

The launcher consists of two removable modules each with 20 launching tubes that are banked in four rows of five. A resupply vehicle based on the same chassis, which weighs 15 000 kg fully loaded, carries two reload modules and a rear mounted MH 90 hydraulically operated crane with a lifting capacity of 6000 kg at the minimum outreach of the arm and 2000 kg at its maximum outreach. Maximum lifting height is 6 m.

Replacing the expended modules takes about 10 minutes. The resupply vehicle can also tow a simple four wheeled two-axle trailer which carries four 20-round racks of 122 mm rockets that can be used to reload the modules by hand. To refill two 20-round launcher packs takes approximately 15 minutes. Total weight of a fully loaded trailer is 12 500 kg.

The rocket type normally used is the standard Soviet 122 mm M-21-OF although the other Soviet 122 mm types and the equivalent People's Republic of China and Yugoslavian models can also be fired. At maximum range the area covered by the 40-round salvo is 210 000 m^2, with small braking rings fitted to the rockets the area coverage at 15 000 m range is 100 000 m^2 whilst with large braking rings fitted the area coverage at 12 000 m is 75 000 m^2.

122 mm (40-round) MRS based on DAC 665T (6 × 6) truck chassis

Status: Production. In service with Iraq and the Romanian Army. Believed to be in service with Liberia.

Manufacturer: Romanian state factories.

SPECIFICATIONS

CALIBRE	122 mm	CREW	1-5	SIDE SLOPE	30%
NUMBER OF BARRELS	40	WEIGHT (system loaded)	17 250 kg	MAX SPEED	
LENGTH OF BARREL	3 m	LENGTH (travelling)	7.39 m	road	80 km/h
MAX RANGE	20 380 m	WIDTH (travelling)	2.51 m	cross country	25 km/h
ELEVATION	0 to +55°	HEIGHT (travelling)	3.25 m	ENGINE	6-cylinder, 10.39 l diesel developing 215 hp at 2200 rpm
TRAVERSE	180° (100° left, 80° right)	FORDING	1.2 m		
		GRADIENT	60%		

122 mm (18-round) Multiple Rocket Launcher Trailer System

Development/Description

Romania revealed that it was producing an indigenous designed and built 122 mm 18-round multiple rocket launcher trailer for use with the short version Soviet DKZ-B and Chinese 122 mm rockets at the first Baghdad International Exhibition for Military Production held in April 1989.

The launcher consists of three banks of six launcher tubes placed one on top of each other within a restraining metal framework. The assembly is then placed on top of a four-wheel two-axle trailer that has a towing bar at the front. Launcher traverse is a full 360° by use of a central pivoting section for the assembly. A braking unit is believed to be fitted to prevent launcher movement during firing.

The towing vehicle can be of various types including the 122 mm (12-round) A.S.O. (4 × 4) multiple rocket launcher vehicle.

SPECIFICATIONS

CALIBRE	122 mm	TRAVERSE	360°	LENGTH	
NUMBER OF BARRELS	18	CARRIAGE	two-axle four wheel trailer with tow bar	travelling (with tow bar extended)	3.412 m
LENGTH OF BARRELS	2.17 m			firing (with tow bar up)	2.278 m
MAX RANGE	14 000 m			WIDTH (travelling)	1.710 m
MIN RANGE	1500 m	WEIGHT		HEIGHT (travelling)	2.050 m
ELEVATION	0 to +52°	system loaded	2890 kg		
		system unloaded	2083 kg		

Status: Production. In service with the Romanian Army.

Manufacturer: Romanian state factories.

122 mm (18-round) trailer-mounted MRS

122 mm (12-round) Multiple Rocket Launcher System

122 mm (12-round) MRS mounted on rear of A.S.O (4 × 4) light chassis

Development/Description

The first Baghdad International Exhibition for Military production in April 1989 saw Romania reveal that it was producing an indigenous designed and built 122 mm 12-round multiple rocket launcher which was carried on the rear decking of an A.S.O. (4 × 4) light truck chassis.

The launcher is similar in appearance to that developed for the 18-round trailer system (qv) but only has two banks of six launcher tubes within the restraining metal framework.

The 122 mm rocket types normally used are the short version Soviet DKZ-B and equivalent Chinese round.

Status: Production. In service with the Romanian Army.

Manufacturer: Romanian state factories.

SPECIFICATIONS

CALIBRE	122 mm
NUMBER OF BARRELS	12
LENGTH OF BARRELS	2.17 m
MAX RANGE	14 000 m
MIN RANGE	1500 m
ELEVATION	0 to +55°
TRAVERSE	200° (90° left, 110° right)
LENGTH (travelling)	4.830 m
WIDTH (travelling)	1.775 m
HEIGHT	
travelling (launcher at 35°)	2.815 m
travelling (launcher at 0°)	1.850 m
MAX SPEED	
road	60 km/h
cross-country	15-20 km/h

122 mm (single round) Rocket Launcher

122 mm (single round) Rocket Launcher in firing position

Development/Description

In addition to the previous multiple rocket launchers there is also a short 122 mm HE rocket which is used in conjunction with a 2.5 m long single-tube man or animal portable tripod-mounted 122 mm rocket launcher. This is designed for use by guerrilla, commando and special forces units.

The rocket weighs 46.3 kg complete and is 1.927 m long. Its maximum range is 13 400 m with the launcher having an elevation range of -3° to +50° and a traverse of ±15°. A contact fuze with preselected instantaneous, short or long delay actions is fitted.

Total weight of the launcher when assembled for firing is 59 kg (which is made of the 25 kg tripod, the 30 kg launch tube and 4 kg for the sight and remote firing device with its 25-35 m long connector cable).

The launcher can also be fixed on a special mount for use on a light cross-country vehicle. In this application the launcher can either be fired from the vehicle or dismounted for ground use. A total of 10-15 reload rounds would normally be carried.

Status: Production. In service with the Romanian Army.

Manufacturer: Romanian state factories.

Other Romanian Multiple Rocket Systems

Development/Description

The Romanian Army used the 130 mm Czechoslovakian M51 (32-round) MRS mounted on the rear of a Soviet ZIL-151 (6 × 6) truck chassis rather than on the original Praga V3S (6 × 6) truck chassis. The systems have now been mounted on the rear of the more recent Soviet ZIL-157 truck chassis.

No armour protection is provided for the cab so the launcher is traversed left or right before launching the rockets. Spare rockets are carried in stowage boxes on either side of the hull. The rockets are spin stabilised and it is believed that a rocket with an increased range has been introduced. Romania also uses a 122 mm (21-round) BM-21 MRS launcher variant mounted on the rear of a Bucegi SR-114 (4 × 4) truck.

130 mm M51 (32-round) multiple rocket system on ZIL-151 (6 × 6) chassis

The spin and fin-stabilised 122 mm rocket types available for use with the various 122 mm Romanian rockets are identical to those fired by the Soviet 122 mm BM-21 system fully described under the USSR.

Status: Production complete, with about 600 built. In service with Romania.

Manufacturers: Chassis, Likhachev Motor Vehicle Plant, Moscow, USSR; rocket launcher, Czechoslovak state arsenals; conversion work, Romanian state arsenals.

Reloading 122 mm (21-round) BM-21 MRS launcher variant mounted on rear of modified Bucegi SR-114 (4 × 4) truck

SPECIFICATIONS (ZIL-157 chassis)

CALIBRE	130 mm	GROUND CLEARANCE	0.31 m	GEARBOX	manual with 5 forward and 1 reverse gears
NUMBER OF BARRELS	32	TRACK			
LENGTH OF ROCKET	0.8 m	front	1.755 m	TRANSFER BOX	2-speed
WEIGHT OF ROCKET	24.2 kg	rear	1.75 m	SUSPENSION	
MAX RANGE	8200 m	WHEELBASE	3.665 m + 1.12 m	front	longitudinal semi-elliptical springs with hydraulic double-acting shock absorbers
MAX VELOCITY	410 m/s				
UNIT OF FIRE	96 rockets	MAX SPEED	65 km/h		
RELOAD TIME	2 mins	RANGE	430 km		
ELEVATION/DEPRESSION	+50°/0°	FUEL CAPACITY	215 l	rear	bogie with semi-elliptical springs
TRAVERSE	240°	FORDING	0.85 m		
CREW	6	GRADIENT	55%	TYRES	12.00 × 18
WEIGHT OF SYSTEM (loaded)	8630 kg	VERTICAL OBSTACLE	0.6 m	ELECTRICAL SYSTEM	12 V
LENGTH (travelling)	6.684 m	ENGINE	ZIL-157 6-cylinder in-line water-cooled petrol developing 109 hp at 2800 rpm		
WIDTH (travelling)	2.315 m				
HEIGHT (travelling)	2.78 m				

SOUTH AFRICA

ARMSCOR 127 mm (40-round) Valkiri Mk II Multiple Artillery Rocket System

Development/Description

The 127 mm (40-round) Valkiri Mk II MRS is a second generation version of the Valkiri Multiple Artillery Rocket System mounted on an all-terrain Samil 100 (6 × 6) truck chassis fitted with a 5-man armoured (against small arms and shrapnel) and mine-protected crew cabin.

Known by the name Bataleur to the South African Defence Forces the first battery of eight launchers was delivered in March 1990 as the replacement for the 24-round Valkiri Mk I systems used by the mechanised brigades.

The larger Samil 100 chassis has sufficient space for all the launcher system equipment and rations, stores and water required for a 14 day operational cycle. The modular launcher pack unit consists of five rows of eight rocket launcher tubes fitted with electro-hydraulically operated elevation and traverse gear. A manual back-up system is also fitted. The Samil 100 chassis and suspension have been strengthened with the front axle being upgraded from 6500 to 7500 kg. Improvements to the engine and transmission give the vehicle 45 per cent better traction and 90 per cent better gradient ability than a standard Samil 100. Additional fuel tanks extend the range to 1000 km at a speed of 80 km/h.

The launcher unit can be mounted on other (6 × 6) 10 000 kg vehicle types if required. Sophisticated electronic navigation and north-seeking equipment is currently under development for the weapon system.

Firing is usually carried out with hydraulically operated stabiliser jacks lowered and is performed from within the armoured cab or via an optional remote hand-held firing console. Provision is also made for computation of corrected firing solution data on a computerised fire control system using inputs about the local meteorological conditions, wind velocity and target distance.

127 mm (40-round) Valkiri Mk II MRS deployed in the field with rocket being launched

The rockets may be fired either singularly or in ripples of 20 rockets at a time with a separation interval of one second. Total salvo time for all 40 rockets is 46 seconds.

The 2.8 m long, 127 mm rockets weight 59 kg overall and use a double base solid propellant rocket motor. They can be fitted with a number of warhead types including HE prefragmented anti-personnel, anti-armour submunition or various cargo heads of different types and payloads. The submunition and cargo warheads use a height adjustable altimeter fuze whilst the HE pre-fragmented warhead uses a 1.35 kg dual-purpose fuze with selector switch for either Direct Action or Proximity Action (factory preset for an 8 m burst height) but variable between 3 to 12 m) and with a super quick point detonating back-up function.

Salvoes can consist of a combination of rocket warhead types and fuze settings. Nominal range is 22 000 m which can be reduced to 8000 m by the use of aerodynamic brakes. An extended range rocket motor can also be used.

Reloading in the field is undertaken by using an ergonometrically designed loading platform.

Status: In production. In service with the South African Army.

Manufacturer: ARMSCOR
Private Bag X337
Pretoria 0001
South Africa
Telephone: (012) 428 1911
Telex: 32017

SPECIFICATIONS (Valkiri Mk II)		CREW	5	TYRES	16.00 × R20 radial ply
CALIBRE	127 mm	COMBAT WEIGHT	21 500 kg	ELECTRICAL SYSTEM	24 V DC
NUMBER OF TUBES	40	LENGTH (travelling)	9.3 m	ENGINE	4-stroke, V-10 air-cooled diesel developing 315 hp at 2500 rpm
LENGTH OF TUBE	3.0 m	WIDTH (travelling)	2.35 m		
LENGTH OF ROCKET	2.8 m	HEIGHT (travelling)	3.4 m		
MAX RANGE	22 000 m	MAX SPEED	93 km/h		
MIN RANGE	8000 m	FUEL CAPACITY	400 l	TRANSMISSION	manual 9-speed constant mesh gearbox with differential lock high/low range transfer box
RATE OF FIRE	1 rd/s	MAX RANGE	1000 km		
LAUNCH TIME	40 in 46 s	GROUND CLEARANCE	0.35 m		
RELOAD TIME	10 mins	FORDING DEPTH	1.2 m		
ELEVATION/DEPRESSION	+50°/0°	GRADIENT	35°		
TRAVERSE	110°	VERTICAL OBSTACLE	0.5 m		

ARMSCOR 127 mm (24-round) Valkiri Mk I 22 Multiple Rocket Launcher System

Development

Development of the Valkiri Mk I 22 started in November 1977 as a counter to the Soviet BM-21 122 mm MRS in service with neighbouring African countries. It was completed in March 1981 with the system entering series production in April and first deliveries beginning later in 1982. The Valkiri Mk I 22 is designed to be deployed either on its own in batteries of two troops with four launchers each or in conjunction with more conventional artillery against area targets such as guerrilla camps, troop concentrations and soft skinned vehicle convoys. Due to its high mobility it is well suited to the shoot and scoot technique whereby it would fire all its rockets at a target and then retire before counter battery fire is initiated. The Valkiri Mk I 22 has seen extensive combat service in southern Angola since it entered the South African Army inventory. The system was given the name 'Chindungu: the red pepper that bites' by UNITA.

Description

The launcher unit consists of a 24-round pack of three rows of eight launcher tubes mounted on the rear hull of a Unimog (4 × 4) truck chassis. Overhead canopy railings are fitted to camouflage the vehicle as a normal truck. The vehicle carries hydraulically operated elevation and traversing gear for the launcher and is stabilised during firing by a pair of hydraulic supports mounted at the vehicle rear. A conventional artillery sight is mounted on the side of the launcher and is stored in a special container on the vehicle when not in use. Firing is either by a firing panel within the cab or via a 50 m plug-in cable connected to a remote firing unit. A test circuit is incorporated to check the rocket status and circuitry prior to firing. The rockets are launched either singly or in ripples of 2 to 24. The firing rate is one rocket per second.

The 127 mm rocket has a double base propellant motor and a prefragmented anti-personnel warhead with a thin walled cast epoxy resin cylinder containing some 8500 steel balls to give a lethal area of 1500 m². A contact or proximity fuze is screwed to the front of the warhead. The range varies from 8000 to 22 000 m depending on whether spoiler rings are used on the rockets to slow them.

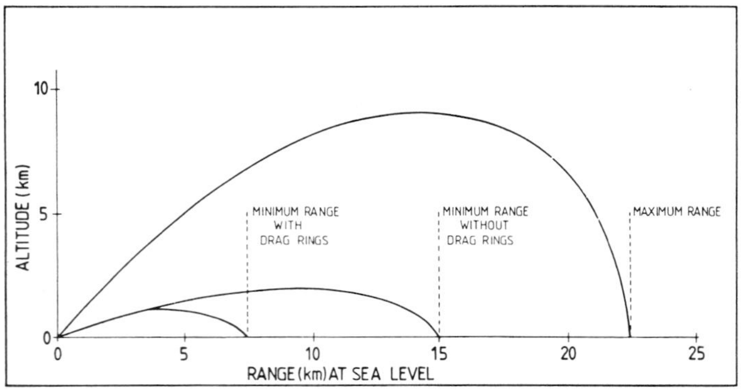

Range table for Valkiri Mk I 22 multiple rocket launcher

127 mm (24-round) Valkiri Mk I 22 MRLS in firing position

Valkiri Mk I 5 multiple rocket system firing a rocket during trials

127 mm (24-round) Valkiri Mk I 22 MRLS firing

A standard five-ton military truck fitted with a collapsible cargo assembly and carrying 48 reload rounds with their fuzes is assigned to each Valkiri.

The reloading is manual and takes about 10 minutes. Upper atmosphere and surface meteorological data required for firing in order to achieve a high first hit probability is obtained by means of weather balloons, a vehicle-mounted telescopic mast with wind vane and anemometer close to the launchers and a wind gun that fires small calibre projectiles to estimate the prevailing ground wind conditions. The time taken for the launcher to deploy in a presurveyed site is five minutes; to come out of action takes only two minutes. A radio is fitted within the vehicle cab as standard with an extension to the remote fire unit.

Variant

In March 1988 Armscor revealed that it had developed a short-range towed multiple rocket launcher based on its Valkiri Mk I 22 system known as the Valkiri Mk I 5 rocket system.

The launcher has 12 tubes, weighes 1360 kg and is towed by a light military truck. The tube pack has a built-in levelling bubble system and is elevated and traversed by handwheels. Launcher length is 3.0 m, wheelbase 1.9 m and height 1.8 m. Towing speed can be up to 120 km/h.

Firing is by means of a remote control unit attached to the launcher by cable. It is crank-powered and can launch the rockets either singly or in ripple mode at the rate of one per second. A direct-fire capability is available.

The rocket is 1.4 m long and weighs 30 kg at launch including the 18 kg prefragmented HE warhead. This contains 8500 steel balls cast in resin and covers an area of 1000 m^2 in a kidney-shaped pattern with lethal shrapnel following detonation at the preset height of about 8 m by proximity fuzes. Motor burn time is 0.8 seconds which gives a maximum velocity of 250 m/s. Maximum range is 5500 m although this can be shortened by the use of drag rings down to 3000 m. The system can be prepared for firing in three and a half minutes by its launch crew of two to four.

If the 32 kg rocket variant is used then the range coverage is increased to between 5000 and 7000 m. The trailer-mounted Valkiri 5 system is still at the prototype stage.

Status: Production complete. In service with the South African Army.

Manufacturer: ARMSCOR
Private Bag X337
Pretoria 0001
South Africa
Telephone: (012) 428 1911
Telex: 32017

SPECIFICATIONS (Valkiri Mk I 22)					
CALIBRE	127 mm	ELEVATION/DEPRESSION	+50°/0°	FORDING	0.6 m
NUMBER OF TUBES	24	TRAVERSE	110°	GRADIENT	30
LENGTH OF TUBE	3.0 m	CREW	2	VERTICAL OBSTACLE	0.35 m
LENGTH OF ROCKET	2.68 m	COMBAT WEIGHT	6400 kg	TYRES	12.50 × R20
MAX RANGE	22 000 m	LENGTH (travelling)	5.35 m	ELECTRICAL SYSTEM	24 V DC
MIN RANGE	8000 m	WIDTH (travelling)	2.3 m	ENGINE	4-stroke, 6-cylinder water-cooled diesel developing 99 hp at 2800 rpm
UNIT OF FIRE	96 rockets	HEIGHT (travelling)	2.32 m		
RATE OF FIRE	1 rd/s	MAX SPEED	90 km/h		
LAUNCH TIME	24 in 24 s	FUEL CAPACITY	90 l		
RELOAD TIME	10 mins	MAX RANGE	400 km	TRANSMISSION	2 × 4 gear shift with differential lock
		GROUND CLEARANCE	0.3 m		

SPAIN

SANTA BARBARA 140 mm (40-round) Teruel Multiple Rocket Launcher System

Development/Description

The 140 mm Teruel-MRS was developed by the Council for Rocket Research and Development of the Spanish MoD to replace the older MRS in Spanish Army service. The launcher has two 20-round packs arranged in five rows of four launcher tubes each. Mounted on the rear hull of an IASA-Pegaso 6 × 6 truck chassis the vehicle is fitted with an armoured six-man crew cab that can mount a light 7.62 mm machine gun for local defence and anti-aircraft purposes. The launcher is deployed in battalions (groups) of three batteries, each with six firing units. A longer rocket has also been developed and is fitted with a double grain solid fuel motor that, when fitted with aerodynamic airbrakes, allows three distinct trajectories to be flown using the same firing elevation. The rockets also have pop out fins for stabilisation in flight. The warhead can be either a 20 kg HE-fragmentation or, in the case of the long rocket a 21 kg submunition carrying warhead which comes in four versions carrying 42 anti-personnel grenades each with either 950 3.2 mm diameter steel balls, 60 g of composition B explosive and a PO model percussion fuze or 28 hollow charge grenades with contact fuze and capable of penetrating 110 mm of armour or six pressure activated anti-tank mines fitted with anti-disturbance features or 14 smoke grenades which each provide four minutes of smoke. The fuzes for the rockets can either be the PD M34A2 contact type, proximity or timed. All six rockets can be mixed within the launcher and can be selected according to the target to be engaged. A IASA-Pegaso 6 × 6 resupply vehicle with four blocks of 20 long rocket reload rounds or six blocks of 20 short rocket reload rounds is assigned to each vehicle as part of the battery. Reloading is manual and takes five minutes while the time taken to emplace a launcher is two minutes, to fire a complete salvo 45 seconds and to come out of action another two minutes. The short HE rocket is 2.044 m long, weighs 56.3 kg and has a maximum range of 18 000 m. The burn time of the motor is 1.6 seconds. The long rocket is 3.23 m long, weighs 76 kg and has a burn time of 2.7 seconds to give a maximum range of 28 000 m. Minimum range for both rockets is 6000 m.

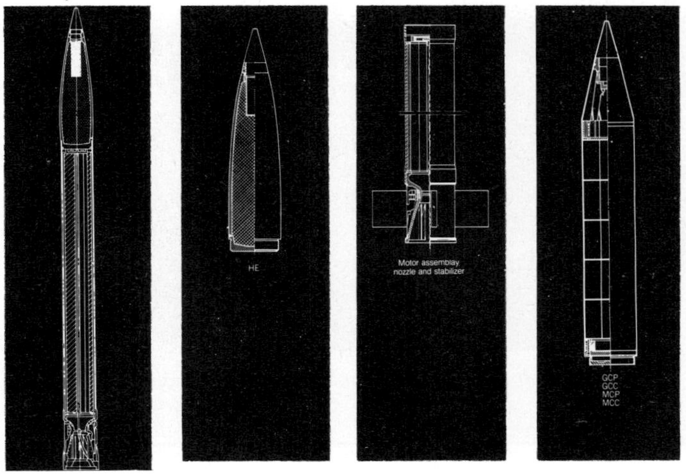

Left to right: complete rocket, HE warhead, motor assembly, nozzle and stabiliser and special warhead (anti-personnel grenades, HEAT anti-tank grenades and anti-tank mines)

140 mm (40-round) Teruel multiple rocket launcher in firing position with launcher elevated, stabiliser jacks lowered and cab closed up

Status: Teruel is in limited production. In service with Gabon and the Spanish Army (12 launchers have been delivered, 12 more on order and 36 are required to be in service by 1994 and 72 by the year 2000).

Manufacturer: SANTA BARBARA
Julian Camarillo 3228037
Spain
Telephone: (91) 58 50 100
Telex: (91) 44466 ENSAB E
Fax: (91) 58 50 268

Production version of Teruel 140 mm (40-round) multiple rocket system mounted on rear of Pegaso 3055 (6 × 6) truck

SPECIFICATIONS		WEIGHT OF ROCKET		MAX VELOCITY	687 m/s
CALIBRE	140.5 mm	WARHEAD		UNIT OF FIRE	80 rockets
NUMBER OF BARRELS	40	high explosive	18.6 kg	RELOAD TIME	5 mins
LENGTH OF ROCKET	3.23 m	submunition	21 kg	ELEVATION/DEPRESSION	+55°/0°
WEIGHT OF ROCKET		smoke	21 kg	TRAVERSE	240°
(high explosive)	76 kg	MAX RANGE	28 000 m	CREW	5 (normal)

TAIWAN

126 mm (40-round) Kung Feng III/IV Multiple Rocket Systems

Development/Description

The Kung Feng 126 mm MRSs were designed in Taiwan by Sun Yat-sen Scientific Research Institute in the early 1970s. The basic Kung Feng (Worker Bee) consists of two banks of 20 126 mm calibre rocket launcher tubes arranged in five rows of four. The tubes are over 1.6 m in length, and fire spin-stabilised rockets. The rocket length is about 0.8 m and the weight approximately 25 kg. The maximum range is estimated to be 9000 m, with the 40 electrically fired rockets being ripple fired in 16 seconds.

The launcher can be mounted on a variety of platforms. In Taiwan the following are known:

A towed version for use by the infantry. The launcher is mounted on a two-wheeled carriage with a tubular frame and towed by a ¾-ton truck. This system is believed to be designated Kung Feng III.

M113A1 rocket launcher

A self-propelled version on the M113A1 APC for mechanised and armoured units. The launcher is mounted just behind the troop compartment hatch pointing forward. The elevation and depression are hydraulically operated but there is no facility for traversing. A blast shield is fitted over the rear hull top to protect the dome ventilator. In order to fire the crew must disembark. Use of the troop compartment is limited by the launcher's angle of elevation. This system is known as Kung Feng IV.

LVTP5 rocket launcher

A self-propelled version on the LVTP5 amphibious assault personnel carrier for the Marine Corps. The launcher is fitted just to the rear of the commander's hatch and can be fired while the vehicle is afloat. This is also known as the Kung Feng IV.

An ECM chaff dispensing system known as Kung Feng Sea is in use on Taiwanese naval destroyers.

Status: In production. In service with the Taiwanese Armed Forces.

Kung Feng IV multiple rocket launcher on M113A1 APC chassis

Towed version of Kung Feng IV multiple rocket launcher on parade in October 1975

117 mm (45-round) Kung Feng VI Multiple Rocket Launcher System

Development/Description

The Kung Feng VI was designed by the Sun Yat-sen Scientific Research Institute in the late 1970s. Intended only as a self-propelled MRS the present model is designated Kung Feng VIA and is a 45-barrel launcher with five layers of nine tubes mounted on a traversable pedestal with hydraulic elevation equipment. The carrier vehicle normally used is the M52A1 (6 × 6) five-ton truck chassis with the launcher mounted on the rear. The cab is unarmoured so that the traverse is not normally beyond 90° when firing. A support vehicle carrying reload rounds is required. Reloading is manual and takes up to 15 minutes for all 45 barrels.

The rocket is about 1.8 m in length, weighs around 60 kg and is spin stabilised in flight by four fold-out fins at the tail. Maximum range is estimated to be 15 000 m, with a full salvo having a lethal area of some 30 000 m². The rockets are electrically fired with a 0.5 second delay between each to give a total ripple firing time of 22 seconds. A computer based fire control capability is under development.

Variant

An experimental version has been identified. A simple non-traversable launcher with hydraulic elevation controls was mounted in place of the twin 40 mm turret on an M42 tracked self-propelled gun chassis. The current status is not known.

117 mm Kung Feng VI 45-tube multiple rocket launcher mounted on rear of (6 × 6) cross-country truck (DTM)

Status: In production. In service with the Taiwanese Army.

THAILAND

Thai Multiple Rocket Launcher Systems

Development/Description

The Royal Thai Air Force's Directorate of Air Armament has developed a range of rocket launchers known as the Hau Fah (or Sky Cobra) series for the ground-to-ground role which utilises aircraft air-to-ground rocket types already in service with Thai aircraft and gunship helicopters.

The first system is the Hau Fah 1 which is tripod-mounted and has simple elevation controls. It fires a single rocket of the 70 mm (2.75 in) FFAR class and is designed for local defence and mobile operations.

The Hau Fah 3 launcher is much larger and is intended for use with the Thai Air Force developed 70 mm (2.75 in) FFAR unguided solid propellant powered rocket. It consists of three banks of launch tubes placed side-by-side in a single rectangular box unit that is mounted on a turn-table carried on the rear of a cross-country vehicle such as a light truck. Each rocket bank comprises 25 launcher tubes in three rows of eight, nine and eight tubes respectively. Aiming in elevation and azimuth is done manually by the gunner who is seated in a semi-open compartment next to the launcher assembly. Firing can either be done singly, in pairs or as complete bank salvoes.

The Hau Fah 4 is a fixed frame-mounted launcher with simple elevation controls for use as a perimeter defence weapon. Five 127 mm (5 in) calibre rockets are used with the system.

In conjunction with the Thai Air Force Directorate the Thai Army Weapon Production Centre (AWPC) has also developed a rocket launcher, the Hau Fah 2 or Marshal Thansom system with a 20-round capacity.

In production for the Thai Army the launcher has four rows of 105 mm calibre launch tubes laid on top of each other in a rectangular box shape. The whole system is then mounted on a two-wheeled carriage with simple elevation controls and towed by a light truck.

The AWPC passed the design over to Attawut Industries for manufacture following modifications to the launcher and the adoption of a more modern composite rocket motor propellant to increase the range.

Status: Hau Fah 1, 3 and 4 in limited production. In service with Royal Thai Air Force and Army.

Hau Fah 2 production. In service with Royal Thai Army.

TURKEY

70 mm (40-round) MAKSAM RA-7040 Multiple Rocket Launcher

Development/Description

As part of the Turkish Armed Forces weapons diversification programme MKEK have developed a 70 mm (2.75 in) multiple rocket launcher known as the RA-7040.

The system is of a modular type and is mounted on a two-wheeled trailer with four hydraulically operated jacks for stabilisation purposes during firing. Any vehicle with a 600 kg towing capacity can tow the launcher. It can be adapted to fit on both land force and marine corps vehicles if required.

The trailer-launcher weighs 1320 kg when empty and can fire all 40 rockets in either single or ripple launch modes within 10-40 seconds. All types of 70 mm rockets can be fired with a 40-round salvo typically covering an area of 200 × 300 m at a range of 7400 m.

The launcher has a full 360° traverse capability and is controlled by a remote-control device with 25 m of cable.

Status: Production. To date, 24 have been produced for the Turkish Army.

Manufacturer: Makina ve Kimya Endüstri Kurumu (MKEK)
General Directorate
06330 Tandogan
Ankara
Turkey
Telephone: (90-4) 213 4171/223 2011
Telex: 42223 MKGA TR/44363 MKEK TR
Cable: MAKKIM-ANKARA
Fax: (90-4) 223 01 40

70 mm (40-round) MAKSAM RA-7040 MRS deployed in firing position

SPECIFICATIONS		WEIGHT		WIDTH (travelling)	2.060 m
CALIBRE	70 mm	in firing position	1690 kg	HEIGHT (travelling)	2.135 m
NUMBER OF TUBES	40	empty	1320 kg	ELEVATION	0 to +55°
CARRIAGE	two wheel trailer	LENGTH (travelling)	3.470 m	TRAVERSE	360°
				POWER SUPPLY	24 V DC

UNION OF SOVIET SOCIALIST REPUBLICS

Free Rocket Over Ground (FROG) Artillery Rocket System

Development

The first Luna (Soviet name for moon, NATO designation FROG) unguided ballistic rocket models were tested in 1956 and 1957. Designed primarily to provide battlefield nuclear support fire at the Divisional level the Soviets accepted both the single rail Luna (NATO designation FROG-1) carried on a modified IS-1/2 heavy tank chassis and the shorter range single rail Luna-1 (NATO designation FROG-2) carried on a non-amphibious variant of the PT-76 light amphibious tank chassis in 1957 for operational deployment in the following year.

The missiles were grouped together in special missile-launcher battalions of two FROG vehicles and attached to Tank and Mechanised (later Motorised Rifle) Divisions.

By 1960 the cumbersome IS-1/2 Luna system was being replaced with the similar performance Luna-2 (NATO designations FROG-3, FROG-4 and FROG-5) based on a further modified version of the Luna-1 non-amphibious PT-76 chassis. The major differences between the Luna-2 variants which warranted the individual NATO designations was in the shape of the warheads.

In 1965 the single-rail Luna-M (Soviet designation R-75, Soviet Industrial Index number 9K21, NATO designation FROG-7a) launcher was deployed using a BAZ-135L4 (8 × 8) cross-country vehicle as the Tractor-Erector-Launcher (TEL). Coincidental with this, the Divisional artillery rocket battalion was doubled in size to a battalion of two firing batteries with two TELs each.

The FROG-7a was joined in service during 1968 by the FROG-7b variant which has a longer warhead section.

The FROG-1 and FROG-2 systems have been considered to be obsolete for a number of years but the FROG-3, FROG-4 and FROG-5 weapons can still be found in a number of armies as front-line weapons.

A few FROG-5 launchers can still be found with some Soviet units but these are now used only for training duties. The NATO designated FROG-6 weapon is a non-operational dummy training round.

Early FROG types were exported to the Warsaw Pact, Algeria, Cuba, Egypt, North Korea and Syria whilst the FROG-7 has been supplied to the Warsaw Pact, Algeria, Cuba, Egypt, Iraq, North Korea, Kuwait, Libya, Syria and Yemen. In late 1988/early 1989 Iraq shipped, via the port of Jounieh, two launchers and a number of FROG-7a rockets to the Lebanese Forces Christian militia besieged in Beirut for use against attacking Syrian forces and to threaten the Syrian capital of Damascus.

Iraq is also manufacturing its own variant of the FROG-7 called the Laith 90. This has a cluster munition warhead and an increased maximum range of 90 000 m. Iraq and North Korea are also known to be producing their own chemical warheads for FROG rockets. Iraq subsequently used its FROG-7 systems in the 1991 Second Gulf War against northern Saudi Arabia and against rebel positions in the following Civil War.

In 1985 the Soviet Army started using its FROG-7b systems in Afghanistan. Fitted with HE and submunition warheads the rockets were used against village targets as part of the Soviets effort to deny basic agricultural food supplies to the guerrilla force.

Two Soviet FROG-7 rockets moving forwards and showing steering on the first and last axles

In 1989 it was revealed that the Afghanistan Army had taken delivery of a number of FROG-7 launchers. These used life expired FROG-7a/b rounds from Soviet stocks and took the place of the 500 odd Scud-B rounds delivered previously and fired against various guerrilla targets.

The long-term replacement for the FROG-7 is the inertially guided US/NATO designated SS-21 Scarab (Soviet name Tochka, meaning 'point') missile system. First deployed operationally in 1976, the weapon had, by 1989, replaced the majority of the FROG-7 systems in Eastern Europe and had been exported to Czechoslovakia, the then East Germany and Syria.

Description (FROG-7)

A FROG-7 battalion consists of a HQ-battery and two firing batteries each with two FROG-7 TELs and a trailer-mounted D-band RMS-1 (NATO designation End Tray) long range meteorological radar. The TELs have no NBC protection for their crews.

Amongst its vehicles the HQ battery contains four BAZ-135L4 TZM (Transportna-zaryazhyushcha Machina: transporter-loader vehicle) which carry three reload FROG-7 rounds each. These are placed on the TEL by that vehicle's own onboard hydraulically operated goose-neck crane which is positioned on the right side of the launcher rail just behind and above the wheel arch of the centre pair of wheels. The reloading operation takes only 20 minutes or so in total.

Preparation time for firing can take 15-30 minutes depending upon what operations are needed to be performed and whether the unit has only just occupied the position. The Luna-M itself is aimed by raising the TEL's rail to the desired elevation and adjusting the weapon's speed brakes so as to operate during the flight to give the correct range. If needed a limited traverse angle can also be set on the launch rail assembly

Rear view of FROG-7 rocket clearly showing the small exhaust nozzles around the large exhaust venturi

to compensate for any last minute off-set required. For optimum usage the associated RMS-1 radar has to be used to establish the meteorological parameters that might effect the rocket's flight trajectory.

Offensive operations with FROG-7s require them to be positioned well forward, around 8-18 km from the line of contact in order to engage tactical targets in the enemy's rear. In defensive operations these distances may be up to 25-50% further back.

The FROG-7 round is of conventional single-stage design with a cylindrical warhead of the same diameter as the rocket body. The combined booster-sustainer rocket motor uses solid propellant fuel and has a large exhaust venturi surrounded by a number of small exhaust nozzles that are used on launch to impart a slow stabilising spin. Four tail-mounted fins further act to stabilise the rocket in flight.

The warheads available include nuclear, conventional HE, chemical and, in the case of the FROG-7b cluster munition.

The only official Soviet information available for a chemical warhead is that for a 540 mm calibre type weighing 436 kg and carrying a 216 kg payload of VX persistent nerve gas agent. This is dispersed as a dense aerosol/droplet cloud by a VT fuzed airburst at around 1500-2000 m altitude in order to disable unprotected personnel and contaminate material, terrain and/or engineering structures over approximately 6 km^2. Other bulk chemical warhead types are likely to be available for the FROG-7 including, according to French military sources, one filled with the blood gas agent hydrogen cyanide (HCN, agent designation AC).

During the First Gulf War the Iraqis apparently used the vesicant (blister) agent known as distilled mustard gas (sulphur mustard, agent designation HD) in FROG-7 warheads on several occasions against Iranian rear area troop concentrations. Although not as lethal as a nerve agent, mustard gas can penetrate normal clothing and rubber and is potentially very long lived in a variety of environmental conditions. It can persist on the ground, on foliage, equipment and buildings, thus it can be used to seal-off whole areas of a battlefield or provide flank protection during an offensive operation.

The mustard gas affects exposed skin to such an extent that the casualty can typically be away from duty for a minimum of two-three months, but usually much longer. If it is inhaled or swallowed than the effects are often fatal but if the casualty survives this internal exposure then he will almost certainly succumb to a secondary respiratory infection which can result in delayed death or medical problems which can last for years.

Hence mustard gas is in military terms an ideal chemical agent as it not only removes the man from active service but also causes him to occupy the services of other people involved in much needed transport and medical duties both at the front and in rear area facilities. The presence of large numbers of gassed victims can also cause a detrimental moral effect on the rear area non-combat personnel.

In order to survive in a heavily contaminated mustard gas area, a person must wear full NBC protective equipment and continually decontaminate his weapons and equipment.

The conventional HE warhead weighs 550 kg but coupled with the FROG-7s CEP of 500-700 m is only really suited for bombardment tasks against area targets.

SPECIFICATIONS (FROG-7)

LAUNCH WEIGHT	2300-2500 kg (depending upon warhead type)
LENGTH	FROG-7a 9.1 m FROG-7b 9.5 m
BODY DIAMETER	0.55 m
WARHEAD TYPES	1, 5 or 25 kt nuclear, HE, chemical (and FROG-7b cluster)
WARHEAD DIAMETER	HE 0.55 m, chemical 0.54 m
WARHEAD WEIGHT	HE 500 kg, nuclear 450 kg, chemical 436 kg
MAX RANGE	70 000 m
MIN RANGE	11 000 m (14 000 m with chemical warhead)
CEP	500-700 m
CREW	4
CARRIAGE	single rail launcher on BAZ-135L4(8 × 8) cross-country vehicle
COMBAT WEIGHT	23 000 kg
LENGTH (travelling)	10.75 m
WIDTH (travelling)	2.8 m
HEIGHT (travelling)	3.66 m
MAX SPEED	65 km/h
TRENCH	2.63 m
VERTICAL OBSTACLE	0.685 m
FORDING	0.58 m

Status: Production as required. In service with the following.

Country	Version	Number of Launchers
Afghanistan	FROG-7a/b	n/av
Algeria	FROG-4 FROG-7	18
Bulgaria	FROG-3 FROG-7	36
Cuba	FROG-4 FROG-7	65
Czechoslovakia	FROG-7	36[1]
Egypt	FROG-7	12 (reserve)[2]
Hungary	FROG-7	18
Iraq	FROG-7	n/av
Korea, North	FROG-5	9[5]
	FROG-7	18[5]
Libya	FROG-7	48
Poland	FROG-3 FROG-7	52[1]
Romania	FROG-3	32
Soviet Union	FROG-3	n/av (reserve)
	FROG-4	
	FROG-5	n/av (training)
	FROG-7	n/av[1]
Syria	FROG-7	18
Yemen	FROG-7	21
Yugoslavia	FROG-7	8

NOTES
[1] FROGs are being transferred to reserve as SS-21 Scarab enters service.
[2] Egypt is building the SAKR-80 to replace the FROG-7 and will modify the BAZ-135L4 launcher vehicle to carry it (qv).
[3] Iraq produced its own cluster munition variant of FROG-7 the Laith 90 with 90 000 m maximum range.
[4] Iraq produced its own chemical agent warheads for the FROG-7.
[5] North Korea produces its own chemical agent, and possibly biological warfare agent, warheads for its FROG-5 and FROG-7 rockets.

Manufacturer: Soviet State factories.

300 mm BM-30 Multiple Rocket System

Development

The 300 mm BM-30 is the largest calibre multiple rocket launcher used by the Soviet Army and was originally designated the M1983 by NATO until its correct Soviet designation became known. The Soviet industrial designation for the system is the 9A52 and it is commonly referred to as the Smerch (Tornado).

It is understood that the system is deployed at Front/Army level artillery brigades and divisions.

Under the Conventional Forces Europe (CFE) Treaty signed in 1990, the Soviet Union admitted to having the following quantities of the 300 mm BM-30 weapon deployed in the West:

Region	IV.1	IV.2	IV.3	IV.4	V.1
Quantity	154	103	94	nil	51

The above figures were the situation as of November 1990. For the Conventional Forces Europe Treaty, Region IV is defined as Central Europe while Region V covers the flanks of the Atlantic to the Urals (ATTU).

300 mm BM-30 MRS deployed in the firing position

300 mm BM-30 MRS resupply vehicle

Description
The system is based on an improved MAZ-543 (8 × 8) cross-country truck designated the MAZ-7310 which is also used for the resupply vehicle.

The launcher has a fully enclosed two door cab that is on the left side of the vehicle whereas the rocket resupply vehicle has a four door fully enclosed cab that is in two parts, one either side of the front mounted engine compartment. The launcher vehicle also has a radiator grill that extends to the right side of the chassis whereas the resupply vehicle has a smaller radiator that is in the centre of the vehicle.

On the launcher vehicle there is an additional fully enclosed compartment to the immediate rear of the driver's compartment at a slightly raised level.

Mounted on the rear of the chassis is the 300 mm 16-round multiple rocket launcher, this has one group of four rocket tubes either side and another group of four rocket tubes horizontally above.

Before the rockets are launched two stabilisers are lowered to the ground, these are positioned one either side between the rear two road wheels.

The unguided rocket has a maximum range of 70 000 m and is fitted with a cluster warhead that contains 72 HE fragmentation bomblets.

The BM-30 resupply vehicle has a similar chassis and carries a total of 14 rockets which are reloaded using an onboard crane. This vehicle has a three man crew and weighs 41.5 tonnes fully loaded. In addition there is also special drill and training equipment.

SPECIFICATIONS

CALIBRE	300 mm
NUMBER OF LAUNCHER TUBES	12
WEIGHT OF ROCKET	800 kg
WEIGHT OF WARHEAD	300 kg
MAX RANGE	70 000 m
MIN RANGE	12 000 m
CREW	4
WEIGHT OF SYSTEM LOADED	43 100 kg
MAX ROAD SPEED	65 km/h
RANGE	650 km

Status: In production. In service with the Soviet Army.

Manufacturer: Soviet state factories.

250 mm (6-round) BM-25 Multiple Rocket System

As far as it is known this system is no longer in front line service with any country apart from perhaps Yemen. Full details were given in *Jane's Armour and Artillery 1987-88*, pages 783-784.

220 mm BM-22 (16-round) Multiple Rocket System

Development/Description
The 220 mm BM-22 Uragan (Hurricane, 16-round) MRS was introduced into Soviet Army service in 1977 and is sometimes referred to in the West as the M1977. The system is found in the four BM-22 Uragan multiple rocket launcher battalions (each of one HQ, one support and three firing batteries, each with six BM-22s and 12 resupply vehicles) of the Rocket Launcher Brigade attached to the Front Level Artillery Division. It can also be found in the single MRL battalion of the Army Corps Artillery Brigade. It has been combat tested in Afghanistan where the Mojahedin refer to it as BM-40.

Under the Conventional Forces Europe (CFE) Treaty signed in 1990, the Soviet Union admitted to having the following quantities of 220 mm BM-22 weapons deployed in the West:

Region	**IV.1**	**IV.2**	**IV.3**	**IV.4**	**V.1**
Quantity	531	378	362	72	153

The Soviet industrial designation for the BM-22 is the 9P140 and the system is mounted on a modified ZIL-135 (8 × 8) cross-country truck chassis designated the BAZ-135L4 which is fitted with a central tyre pressure regulation system and has steering on the front and rear axles. Before the rockets are launched two stabilisers are lowered at the rear of the vehicle, one either side.

The BM-22 provides chemical, high explosive and submunition (including fragmentation bomblet, incendiary bomblet and minelet) supporting fire to manoeuvre units. The rocket weighs about 360 kg and has a range of 5 to 40 km. The launcher has a traverse of 240° and consists of one layer of four tubes and two layers of six tubes. A rapid reload capability is provided for by another ZIL-135 that carries 16 rockets arranged in two stacks positioned either side of the vehicle bed with a single round capacity reloading arm between them. For reloading the launcher is lowered to the horizontal and swung to one side, the new rockets being loaded with the aid of a rammer. Reloading time is estimated to take up to 20 minutes. Maximum rate of fire is one round every 0.5 seconds.

Status: In production. In service with Afghanistan, Syria and the Soviet Army.

Manufacturer: Soviet state factories.

220 mm BM-22 (16-round) MRL in travelling configuration

220 mm BM-22 rocket resupply vehicle which has a similar chassis to the MRL

SPECIFICATIONS (provisional)		MIN RANGE	5000 m	WIDTH (travelling)	2.8 m		
CALIBRE	220 mm	RELOAD TIME	15-20 mins	HEIGHT (travelling)	3.2 m		
NUMBER OF LAUNCHER TUBES	16	ELEVATION	0° to +55°	MAX SPEED	65 km/h		
LENGTH OF ROCKET	4.8 m	TRAVERSE	240°	RANGE	500 km		
WEIGHT OF ROCKET	360 kg	CREW	6	FUEL CAPACITY	770 l		
MAX RANGE	40 000 m	WEIGHT OF SYSTEM	22 750 kg				
		LENGTH (travelling)	9.3 m				

240 mm (12-round) BM-24 Multiple Rocket System

Full details of this truck-mounted system, which may still be in service with Algeria, Angola, Cuba, Egypt, Israel, Mozambique, North Korea, Somalia and Syria, were given in *Jane's Armour and Artillery 1990-91*, pages 738/739. Israel Military Industries manufactured rockets for this system which is used by the IDF, but production of these rockets has now ceased. Details were given in *Jane's Armour and Artillery 1989-90*, pages 693/694.

200 mm (4-round) BMD-20 Multiple Rocket System

Full details of this truck-mounted system, which may still be in service with Cuba, Ethiopia, Iraq and North Korea, were given in *Jane's Armour and Artillery 1987-88*, page 785.

140 mm (17-round) BM-14-17 Multiple Rocket System

Development/Description

The 140 mm (17-round) BM-14-17 MRS was first seen in public during the 1959 November parade in Moscow. It has a similar launcher to the 140 mm (16-round) BM-14-16 MRS (mounted on a 6 × 6 ZIL-151 truck chassis) but has an upper layer of nine barrels and a lower layer of eight, and is mounted on the rear of a GAZ-63A (4 × 4) truck chassis. The scale of issue of the MRS is uncertain, but it is probably the same as for the 140 mm (16-round) BM-14-16 MRS. As far as it is known this multiple rocket launcher system is no longer in front line service with the Soviet Army.

Before the rockets are launched shutters are released to cover the cab windows. Two of the crew of six are seated in the cab and the other four are seated two each side of the cab rear facing outwards.

140 mm (17-round) BM-14-17 multiple rocket system in travelling order

The 140 mm rockets, are spin stabilised and are the same type as used by the Polish 140 mm (8-round) WP-8 and Soviet 140 mm (16-round) RPU-14 towed MRSs and the 140 mm (16-round) BM-14-16 MRS.

The rocket types used are:

SOVIET DESIGNATIONS	**HE M-14-OF** **Smoke M-14D**
DIAMETER	139.7 mm
LENGTH	1.092 m
WEIGHT (complete rocket)	39.62 kg
MAX VELOCITY	402 m/s
RANGE	
max	9810 m
min indirect	3800 m
min direct	400 m
WARHEAD TYPES	HE-fragmentation smoke GB chemical
WARHEAD	
HE-fragmentation	18.8 kg
GB chemical	18.3 kg
WEIGHT OF ROCKET FUZE	0.63 kg
FUZE SETTINGS	instant, short and long delay times, and safe
FUZE ARMING DISTANCE	150-400 m

Status: Production complete. In service with Afghanistan, Afghan Mojahedin Guerrillas, Cuba, Poland, Somalia, and the USSR.

Manufacturer: Soviet state factories.

SPECIFICATIONS		WIDTH (travelling)	1.985 m	GRADIENT	50%
CALIBRE	140.4 mm	HEIGHT (travelling)	2.245 m	VERTICAL OBSTACLE	0.46 m
NUMBER OF BARRELS	17	GROUND CLEARANCE	0.27 m	ENGINE	GAZ-51A 6-cylinder water-cooled petrol developing 70 hp at 2800 rpm
RELOAD TIME	1.5-2 mins	TRACK			
ELEVATION/DEPRESSION	+52°/0°	front	1.588 m		
TRAVERSE	210° (145° left, 65° right)	rear	1.6 m		
		WHEELBASE	3.3 m	GEARBOX	manual with 4 forward and 1 reverse gears
CREW	6	MAX SPEED	65 km/h		
WEIGHT OF SYSTEM (loaded)	4570 kg	RANGE	650 km	TRANSFER BOX	2-speed
		FUEL CAPACITY	195 l	TYRES	10.00 × 18
LENGTH (travelling)	5.41 m	FORDING	0.8 m		

140 mm (16-round) BM-14-16 Multiple Rocket System

Development/Description

The 140 mm (16-round) BM-14-16 MRS was first shown in public during a parade in Red Square, Moscow, in 1953. When originally introduced it was mounted on the rear of a ZIL-151 (6 × 6) truck chassis but in recent years some have been observed mounted on the rear of the more modern ZIL-131 (6 × 6) truck chassis both in the Soviet Union and Angola. The BM-14-16 MRS is issued on the scale of one battalion of 18 launchers in each Motorised Rifle Division and Tank Division, but it has been replaced in most front line units by the 122 mm M1972 (40-round) and 122 mm BM-21 (40-round) systems. An additional six MRSs are attached to each Motorised Rifle regiment. Some Soviet Asian Military District units were seen in 1980 with the BM-14/ZIL-131 combination. In the Chinese Army each Artillery Division has one rocket launcher regiment

140 mm (16-round) BM-14-16 multiple rocket system mounted on ZIL-151 (6 × 6) truck chassis

with 32 BM-13-16 (16-round) or 32 BM-14-16 (16-round), or 32 Chinese BM-21 variants or 32 of the later Chinese developed 130 mm MRSs (qv China).

Before the rockets are launched two stabilising jacks are lowered at the rear of the vehicle and armoured shutters are released to cover the cab windows. Three of the crew of seven sit in the cab and the other four are seated two either side of the cab rear facing outwards. The BM-14's accuracy, at two-thirds maximum range, for projectile error in range is 115 m, in deflection 36 m and a CEP of 132 m. At maximum range the corresponding figures are 30 m, 85 m and 100 m.

The 140 mm rockets, are spin stabilised and are the same type as used by the Polish 140 mm (8-round) WP-8 and Soviet 140 mm (16-round) RPU-14 towed MRSs and the 140 mm (17-round) BM-14-17 MRS mounted on a GAZ-63A (4 × 4) truck chassis. One chemical round is known to have an 18.3 kg warhead filled with 2.2 kg of Sarin (agent GB) nerve gas. An impact fuze is fitted to detonate the warhead so that the light yellow Sarin liquid is dispersed as a vapour and fine aerosol. The nominal time to cause death in unprotected personnel is between 2 to 15 minutes.

The rocket types used are:

SOVIET DESIGNATIONS	**HE M-14-OF**
	Smoke M-14D
DIAMETER	139.7 mm
LENGTH	1.092 m
WEIGHT (complete rocket)	39.62 kg
MAX VELOCITY	402 m/s
RANGE	
max	9810 m
min indirect	3800 m
min direct	400 m
WARHEAD TYPES	HE-fragmentation
	smoke
	GB chemical
WARHEAD WEIGHTS	
HE-fragmentation	18.8 kg
GB chemical	18.3 kg
WEIGHT OF ROCKET FUZE	0.63 kg
FUZE SETTINGS	instant, short and long term delay times, and safe
FUZE ARMING DISTANCE	150-400 m

SPECIFICATIONS

CALIBRE	140.4 mm	HEIGHT (travelling)	3.17 m	VERTICAL OBSTACLE	0.64 m
NUMBER OF BARRELS	16	GROUND CLEARANCE	0.265 m	TRENCH	0.69 m
RELOAD TIME	3-4 mins	TRACK		ENGINE	ZIL-121 6-cylinder in-line water-cooled petrol developing 92 hp
ELEVATION/DEPRESSION	+52°/0°	front	1.59 m		
TRAVERSE	200° (140° left, 60° right)	rear	1.72 m		
CREW	7	WHEELBASE	3.665 m + 1.22 m	GEARBOX	manual with 5 forward and 1 reverse gears
WEIGHT OF SYSTEM (loaded)	8350 kg	MAX SPEED	60 km/h		
LENGTH (travelling)	6.92 m	RANGE	600 km		
WIDTH (travelling)	2.3 m	FUEL CAPACITY	300 l		
		FORDING	0.8 m	TRANSFER BOX	2-speed
		GRADIENT	50%	TYRES	8.25 × 20

Status: Production complete. In service with Algeria, Angola, Cambodia, China, Egypt, North Korea, Poland, Somalia, Syria, USSR and Vietnam. As far as it is known, this system is no longer in front line service with the Soviet Union, none were declared as being in the West under the CFE Treaty.

Manufacturer: Soviet state factories.

140 mm (16-round) BM-14-16 Multiple Rocket System

Development/Description

The 140 mm (16-round) RPU-14 MRS has been designed specifically for the use of airborne units and is issued on the scale of one battalion of 18 in the artillery regiment of each Soviet Airborne Assault Division. A replacement GAZ-66 truck-mounted 122 mm MRS (NATO designation M1975) is now in service with a number of these Divisions. The launcher fires the same spin-stabilised rockets, as the Polish 140 mm (8-round) WP-8 towed MRS, the Soviet 140 mm (16-round) BM-14-16 and 140 mm (17-round) BM-14-17 truck-mounted MRSs and the single-round 140 mm launcher used by a number of guerrilla forces.

140 mm (16-round) RPU-14 multiple rocket system in travelling order

As far as it is known, this system is no longer in front line service with the Soviet Union, none were declared as being in the West under the CFE Treaty. The rockets fired are the same as those used by the BM-14-16 covered in the previous entry.

SPECIFICATIONS

CALIBRE	140.4 mm
NUMBER OF BARRELS	16
RELOAD TIME	4 mins
CARRIAGE	split trail
WEIGHT (firing position)	1835 kg
LENGTH (travelling)	4.036 m
WIDTH (travelling)	1.8 m
HEIGHT (travelling)	1.6 m
GROUND CLEARANCE (travelling)	0.315 m
TRACK	1.45 m
TYRES	6.50 × 20
ELEVATION/DEPRESSION	+48°/0°
TRAVERSE (total)	30°
CREW	5
TOWING VEHICLE	GAZ-66 (4 × 4) truck

Status: Production complete. In service with Egypt, Iraq, North Korea, USSR and Yemen.

Manufacturer: Soviet state factories.

122 mm BM-21 Series Multiple Rocket System

Development/Description

The 122 mm BM-21a (Soviet system name Grad (Hail)), NATO identifier (40-round) MRS entered operational service in 1964 and was first seen in public during a parade in Moscow in November of that year. Since then it has been exported to many countries in Africa, the Middle East and the Far East. It is issued on the scale of one independent battalion of 18 in each of the Motorised Rifle Divisions and Tank Divisions. In the Soviet Army the battalion has now been integrated into the Divisional artillery regiment. In some Divisions six BM-21s or a later system are attached to each Motorised Rifle regiment to form a regimental battery. The BM-21a consists of a standard Ural-375D (6 × 6) truck chassis fitted with a 40-round launcher at the rear of the hull. Blast shields are not fitted to protect the cab windows when the rockets are being launched as on other Soviet MRS as these are strong enough to withstand the overpressure and other effects associated with firing.

It has been supplemented since 1972 by the 122 mm RM-70 (40-round) MRS which is essentially an armoured version of the Czechoslovak Tatra 813 (8 × 8) truck fitted with the same launcher as the BM-21a at the rear of the hull but with an additional pack of 40 rockets to the rear of the cab for rapid loading (qv).

In 1976 the Soviets fielded the slightly lower profile BM-21b (Soviet designation Grad-1, NATO designators M1976 and BM-21 Modified) which is believed to be issued to the independent MRL batteries of those Motor Rifle Regiments which operate in areas of rough terrain.

The primary difference from the BM-21a is that two centre tubes in each of the lower banks of the 122 mm launcher tubes have been removed to provide a total of 36. The launcher assembly is also mounted on the 3500 kg ZIL-131 (6 × 6) truck chassis which has rear corrugated sheet metal work fenders that rotate with the unit as it is aimed for firing. The crew numbers five with a manual reload time for the system of seven to eight minutes. A full 36-round salvo takes 18 seconds to fire.

Under the Conventional Forces Europe (CFE) Treaty signed in 1990, the Soviet Union admitted to having the following quantities of 122 mm BM-21 Grad weapons deployed in the West:

Region	**IV.1**	**IV.2**	**IV.3**	**IV.4**	**V.1**
Quantity	1571	1152	1008	393	419

The above figures were the situation as of November 1990. For the Conventional Forces Europe Treaty, Region IV is defined as Central Europe while Region V covers the flanks of the Atlantic to the Urals (ATTU).

For use in the composite artillery battalion allocated to each Soviet Airborne Assault Brigade and Airborne Division the 12-round BM-21V (V standing for vozdushnodesantii, or airborne) system was developed. This has the NATO designator M1975 and has two tiers of six tubes mounted on the lower chassis of the reduced weight airborne forces GAZ-66B (4 × 4) 2000 kg truck.

This has a collapsible canvas cab, removable doors and windscreen, telescopic steering wheel and tie down points for aiding parachute dropping and air landing. The launcher is traversed forward towards the cab for travelling and for firing uses two stabilising jacks at the rear of the vehicle.

The rockets can either be fired singularly or in a salvo of six seconds duration. Reloading is done manually by the two man crew and takes approximately five minutes. The launcher can traverse a full 360°.

The composite artillery battalion has one battery of six BM-21V vehicles alongside two six-gun batteries of 122 mm D-30 Sonyushka howitzers.

For Spesnatz and other specialist users there is also a single-round manportable tripod launcher system known by the Soviets as the Grad-P (Soviet industrial index number 9P132). This has a total weight of 49.44 kg (21.82 kg the tube 27.72 kg the tripod) and a launch tube length of 2.4435 m. The elevation limits are believed to be from 0 to +35° and the traverse ±15°. Emplacing the system takes three minutes and displacing it a 1/2 minute. The maximum rate of fire is 2 rds/min, with the manual reloading time being a 1/2 minute. Maximum range achieved in Vietnam by the DKZ-B rocket was 10 973 m.

The effective fragmentation radii of the 122 mm rocket against APC, IFV and SP guns are respectively 1, 0.5 and 0.75 m. It takes the battalion 23/32 minutes to occupy fire positions (day/night) and 11/14 minutes to leave the fire position. The battalion will normally be found some 5000 m behind the FEBA during offensive and defensive operations. Rocket types include smoke, HE fragmentation and chemical. The latter includes a version containing the non-persistent and highly volatile liquid blood agent hydrogen cyanide (chemical formula HCN, agent designation AC). The nominal time to cause death to unprotected personnel is from 30 seconds to 15 minutes depending on agent concentration and exposure time. Hydrogen cyanide also degrades the effectiveness of respirator filters thus opening up the possibility of a rapid follow up attack using more HCN or another non-persistent agent, which could cause heavy casualties even among fully protected personnel.

Other chemical warheads known to be in use include a 19.3 kg type with 2.9 kg of VX nerve gas as the payload and a 19.3 kg type with 3.1 kg of Sarin (agent GB) nerve gas as the payload. In the VX type a proximity fuze is fitted to detonate the warhead so that the chemical

Captured truck park with two vehicles in foreground showing nine-round and six-round locally built MRSs using parts of BM-21 system (Israeli Ministry of Defence)

122 mm BM-21 (40-round) MRS on Ural-375D (6 × 6) truck chassis with launcher covered up

Polish soldiers loading rockets onto 122 mm (40-round) BM-21 multiple rocket system

Loading 12-round BM-21V (NATO designation M1975) MRSs based on GAZ-66 (4 × 4) truck chassis

agent is dispersed as a dense aerosol and droplets in the case of the dark brown liquid filling. In the Sarin type an impact fuze is fitted to detonate the warhead so that the light yellow liquid filling is dispersed as a vapour and finely dispersed aerosol. The nominal times for causing death in unprotected personnel are between 4 and 10 minutes for VX and 2 to 15 minutes for Sarin.

Sarin is a non-persistant agent which in sunny conditions and a light breeze at around 20°C will be effective from 15 minutes up to 4 hours. In wet and windy conditions and at 10°C the time is from 15 minutes up to one hour and in calm, sunny conditions at around -10°C with snow lying around it can be as much as one to two days.

VX is a persistant agent which in the same conditions as quoted for Sarin is effective for respectively 3 to 21 days, 1 to 12 hours and from 1 to 16 weeks.

The primary use for Sarin-filled munitions is to disable personnel through their respiratory organs whilst for the VX munition it is to disable personnel through their unprotected areas of the skin and to contaminate engineering structures, terrain and material/vehicles.

The screw-in fuze has two settings, impact and delayed action, this has to be set manually before the rockets are launched.

The launcher can be traversed through 180° and the bank of 40 launcher tubes elevated from 0 to +55° electrically or manually. The rockets can be launched singly at any desired time interval by manual means or in partial or complete salvo at a fixed 0.5 second time interval from within the cab or up to 60 m away from the cab with the aid of a remote-control unit connected to the vehicle by a cable. The rockets must be fired with the launcher parked obliquely to the target to prevent blast damage to the unshielded cab.

The sequence of firing is as follows. When the sequence switch is closed current pulses of about one amp are applied to the projectile contact sector. This causes the powder igniters to fire and flame enters the central channel of the rocket and ignites the propellant charge. The gases generated tear off the contact cover and begin to flow through the seven nozzle apertures. When a thrust of 600 to 800 kg is attained, the projectile guide pin leaves the tube stop device and the rocket accelerates up the grooved tube which imparts a slow rotary motion. Once the rocket leaves the 3 m long tube the four stabiliser blades deploy, these are mounted at an angle of 1° to the longitudinal axis of the rocket. The actual fuze is not armed until the rocket is between 150 and 400 m from the launcher vehicle. The combination of spin and fin stabilisation ensures closely clustered salvoes at ranges up to 16 000 m.

The spin and fin-stabilised 122 mm rocket types available are:

122 mm (40-round) BM-21 multiple rocket launcher of the Soviet Army in travelling configuration

Variants

Romania uses a number of locally designed and built 122 mm rocket launcher systems (qv).

China is building a number of locally designed and built BM-21 and derivative rocket launcher systems (qv).

During the 1982 invasion of Lebanon the Israelis came across several other variants using BM-21 components. Two, a six-round and a nine-round version, were locally built and fitted with simple elevation and azimuth controls. Both were mounted on the rear of light truck chassis. The third, however, was a North Korean design. The designation is BM-11 (qv North Korean entry in this section) and it comprises two banks of three rows of five launcher tubes with all the necessary controls, mounted on the rear of a Japanese Isuzu (6 × 6) 2500 kg truck chassis. A significant number of these combinations were captured or destroyed during the campaign. Pakistan also has the BM-11/Isuzu combination in service.

Egypt (qv) now has considerable rocket launcher technology based on 122 mm calibre equipment originally supplied by the Soviet Union.

SOVIET DESIGNATIONS	**M-21-OF, DB-1B**	**M-21-OF**	**DKZ-B**
SOVIET INDUSTRIAL INDEX NUMBER	9M22U	9M22M	9M28
DIAMETER	122 mm	122 mm	122 mm
LENGTH	3.226 m	2.870 m	1.905 m
WEIGHT (complete)	77.5 kg	66 kg	45.8 kg
RANGE			
max	20 380 m	20 000 m	10 800 m
min indirect	1500 m	1500 m	2500 m
min direct	500 m	500 m	400 m
WARHEAD TYPES	HE-fragmentation, smoke, incendiary, HCN, GB and VX chemical		
WARHEAD WEIGHTS			
HE-fragmentation	19.4 kg	18.4 kg	19.4 kg
chemical	19.3 kg	n/available	19.3 kg
FUZE SETTINGS	instant/single time delay/safe		
FUZE ARMING DISTANCE	150-400 m	150-400 m	150-400 m
BURN-OUT VELOCITY	690 m/s	700 m/s	450 m/s
LAUNCHERS	BM-21a/b RM-70 BM-21/SR-114 122 mm/DAC665T	BM-21a/b BM-21V RM-70 BM-21/SR-114 122 mm/DAC665T	BM-21a/b BM-21V RM-70 9P132 BM-21/SR-114 122 mm/A.S.O 122 mm/Trailer 122 mm/DAC 665T

SPECIFICATIONS (BM-21a)

CALIBRE	122.4 mm
NUMBER OF BARRELS	40
WEIGHT OF HE ROCKET FUZE	0.9 kg
WEIGHT OF HE FILLER	6.4 kg
TYPE OF HE FILLER	86% TNT, 14% Al
UNIT OF FIRE	120 rockets
RELOAD TIME	8 mins
ELEVATION/DEPRESSION	+55°/0°
TRAVERSE	180° (120° left, 60° right)
EMPLACEMENT TIME	2.5 min
DISPLACEMENT TIME	0.5 min
CREW	5
WEIGHT OF SYSTEM (loaded)	13 300 kg
LENGTH (travelling)	7.35 m
WIDTH (travelling)	2.69 m
HEIGHT (travelling)	2.85 m
GROUND CLEARANCE	0.41 m
TRACK	2 m
WHEELBASE	3.5 m + 1.4 m
MAX SPEED	75 km/h
RANGE	405 km
FUEL CAPACITY	360 l
FORDING	1.5 m
GRADIENT	60%
VERTICAL OBSTACLE	0.65 m
TRENCH	0.875 m
TURNING RADIUS	10.5 m
ENGINE	ZIL-375 V-8 water-cooled petrol developing 180 hp at 3200 rpm
GEARBOX	manual with 5 forward and 1 reverse gears
TRANSFER BOX	2-speed
SUSPENSION	
front	longitudinal semi-elliptical springs with hydraulic shock absorbers
rear	bogie with longitudinal semi-elliptical springs
TYRES	14.00 × 20
BRAKES	air/hydraulic
ELECTRICAL SYSTEM	12 V

SPECIFICATIONS (BM-21b and BM-21V)

TYPE	**BM-21b**	**BM-21V**
CALIBRE	122.4 mm	122.4 mm
NUMBER OF BARRELS	36	12
UNIT OF FIRE	120 rockets	120 rockets
RELOAD TIME	7-8 min	5 min
CREW	5	2
VEHICLE	ZIL-131	GAZ-66B
COMBAT WEIGHT	10 500 kg	6000 kg
LENGTH (travelling)	6.9 m	5.65 m
WIDTH (travelling)	2.5 m	2.32 m
HEIGHT (travelling)	2.48 m	2.44 m
MAX SPEED	80 km/h	85 km/h
FUEL CAPACITY	340 l	210 l
RANGE	525 km	875 km
EMPLACEMENT TIME	2.5 min	3.5 min
DISPLACEMENT TIME	0.5 min	0.5-1.0 min

Status: In production. In service with Afghanistan, Algeria, Angola, Bulgaria, Chad, China (qv), Cuba, Egypt (qv), Ethiopia, Germany, Hungary, India, Iran, Iraq (qv and BM-21V), Israel, North Korea (qv), Libya, Morocco, Mozambique, Nicaragua, Pakistan, Peru, Poland, Somalia, Sudan, Syria, Tanzania, USSR (including naval infantry), Vietnam, Yemen and Zambia.

Manufacturer: Soviet state factories.

UNITED STATES OF AMERICA

70 mm Rapid Deployment Multiple Rocket Weapon System

Development

The Rapid Deployment Multiple Rocket Weapon System (RD-MRWS) was developed by the US Army and BEI Defense Systems Company, Inc of Fort Worth, Texas as an economical, ground-launch adaptation of the US Standard 70 mm HYDRA 70 aerial rocket system to provide heavy firepower in a lightweight system for land forces.

RD-MRWS adapts the aerial rocket system and warhead family for ground launch in fully air-transportable configurations for motorised, towed, and man or animal ground mobility. Micro-processor fire control and remote-set fuzing provide accurate fire-and-forget attack of aircraft, armour, material and personnel in Air Defence, Fire Support, Close Combat, Combat Support and Combat Service Support roles and special roles such as strikes, raids and lodgements; riverine/maritime operations; rear area security/protection; special operations; and urban combat. Fixed/semi-fixed systems with remote or at-site fire control in single/multiple installations can be configured for defence zones, shorelines, borders, corridors or approach routes.

Description

Up to six 19-tube replaceable launchers can be mounted on a towed M91 chemical rocket launcher chassis or a 105 mm howitzer bed. Other configurations include six 19-tube launchers on a 2½-ton truck, three 19-tube launchers on a High Mobility Multi-purpose Wheeled Vehicle, a 25-rocket square array universal mount, various armoured vehicle turret mounts, two 7- or 19-tube launchers on a light utility vehicle, two 7-tube launchers on a Fast Attack vehicle, four 19-tube launchers on an M42 carriage or quad 12.7 mm machine gun mount, and a manportable 4-tube launcher on a tripod mount. Launchers are reusable 7-tube M260 or 19-tube M261 lightweight aircraft pods or are specially configured for the tripod and universal mounts.

Up to eight fire units can be controlled by cable or radio from one location. The operator can select single, pair or four-round bursts from each launcher and he can select the total quantity to be fired and the firing rate for a salvo. Range is 700 m to over 15 km.

The crew for the larger systems is four with a typical six-pod launcher system taking 10 minutes to reload. A single 19-tube launcher can be fired in about 0.5 second in the quick fire mode and within 1.5 seconds in the slow fire mode. A full six-pod launcher takes less than seven seconds to empty in the slow mode.

Warheads for RD-MRWS include the M261 multi-purpose submunition type with nine M73 controlled-fragmentation, shaped-charge anti-armour, anti-helicopter, anti-materiel, anti-personnel bomblets; the 4.5 kg M151 High Explosive type with the remote set multi-function M433 fuze for superquick or variable penetration function; the M247 shaped-charge anti-armour type; the M257 illumination type for one million candlepower for about two minutes; the M264 Screening type for up to five minutes of smoke generation; the M259E1 anti-material, anti-personnel, and anti-aircraft flechette type; bomblet and unitary training types; and heavier flechette and chaff types.

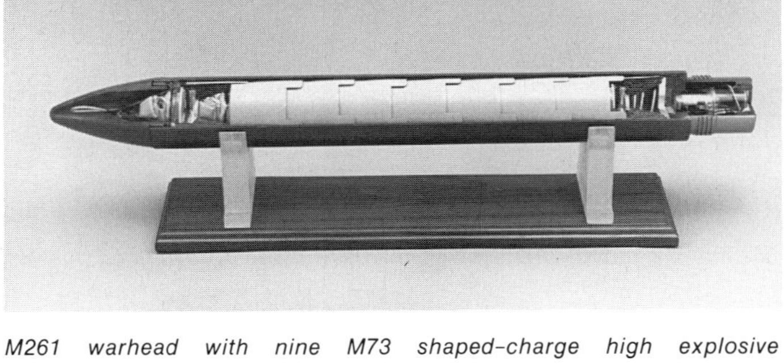

M261 warhead with nine M73 shaped-charge high explosive fragmentation submunitions

Status: All components type classified, ready for production.

Manufacturer: BEI Defense Systems Company
11312 S Pipeline Road
Fort Worth
Texas 76155
USA
Telephone: (817) 267 8191
Telex: 901 893 5022
Fax: (817) 540 3961

Towed RD-MLRS rocket system with six pods carrying 114 rockets

Host System Augmentation RD-MRWS on FMC Light Armored Vehicle/Air Defence with rocket being launched

YUGOSLAVIA

400 mm (8-round) Multiple Rocket System

Development/Description

The Yugoslavian defence industry has developed to the prototype stage a 400 mm calibre multiple rocket system to meet the requirements of the Yugoslavian Army. Based on technology associated with the 262 mm M87 Orkan system few details have been released to date.

All functions of the fire control system are fully automated using computer technology with high system accuracy achieved by using a sophisticated radar monitoring unit that corrects the launcher laying parameters by tracking the launch trajectory of the first rocket fired.

The eight tube launcher assembly is mounted on a vehicle with five drive axles. The 400 mm rocket carries a cluster munition warhead. System range is said to be 120 km.

Status: Development.

Manufacturer: Enquiries to: Federal Directorate of Supply and Procurement (SDPR)
PO Box 308
9 Nemanjina Street
Belgrade
Yugoslavia
Telephone: 621 522
Telex: 11360

262 mm (12-round) LRSV M-87 Orkan Multiple Rocket Launcher System

Development/Description

During late 1988 at a military exhibition in Iraq the Iraqi army displayed a 262 mm 12-round multiple rocket launcher system known as the Ababeel 50. Examination of available evidence indicated that the system was Yugoslavian in origin and mounted on the rear of the Yugoslavian FAP 2832 BS/AV (8 × 8) cross-country truck. Since then further information gained at the Iraqi's First Baghdad International Exhibition for Military Production in 1989 has shown that it is in fact a joint Iraqi-Yugoslavian project with Iraq contributing to the development and producing some of the parts locally.

A typical battery of M87 launchers consists of:

four 8 × 8 launchers
four 8 × 8 resupply vehicles (each with 24 reload rockets)
one 8 × 8 command post vehicle
two 4 × 4 topographic survey light vehicles
two 4 × 4 observation post light vehicles
one 4 × 4 meteorological survey vehicle.

The command post vehicle is fitted with a German Teldix land navigation system. Both the resupply and command post vehicles are on the same 8 × 8 chassis as the launcher vehicle and have a similar tarpaulin system to make identification of the launcher difficult.

Loading the launcher is performed by a semi-automatic system with the preparations to fire taking only two minutes. Firing the rockets can be done either singularly or in a ripple mode by means of an electronic trigger in the vehicle cab. If needed it can also be done off-vehicle by means of a remote-control box with attached cable. All the traverse and elevation functions are automatically performed via control units with manual systems available as a back up.

In Yugoslavian Army service the launcher is fitted with a 12.7 mm heavy AA machine gun on the cab roof and four 82 mm smoke dischargers in the front bumper. The five man crew has two 7.62 mm fully automatic rifles two 7.62 mm semi-automatic rifles and a 90 mm RBR-90 hand-held anti-tank rocket launcher for self defence.

The rocket, known as the M-87, is 4.2 m long, weighs 385 kg and uses a two-stage solid propellant rocket motor. The booster stage uses 10 kg of fuel and burns for 200 ms to generate 8000 kg of thrust to lift the rocket clear of the launcher tube. Once this happens the main sustainer motor cuts in with its 130 kg fuel load to give 18 000 kg of thrust during its five second burn time. This gives the rocket a maximum velocity of 1200 m/s. Flight time to the maximum range of 50 000 m is 110 seconds with the highest possible altitude reached, depending upon launch angle, being 22 000 m.

The device mounted on top of the left side of the launcher is a TV camera made by Bosch of Germany. This is used in a similar manner to that fitted to the German 110 mm LARS in that a single rocket is launched and at some point in its flight is detonated so as not to alert the target area. The camera monitors this and the data collected is used to calculate the correct firing solution so that the remainder of the battery or vehicle salvo can be fired.

The 95 kg warhead is of the submunition dispensing type with the following payloads available:

(a) cluster munition containing 300 dual-purpose bomblets that have an armour-piercing hollow-charge warhead which is surrounded by 420 steel spheres to enhance the fragmentation effect. Each bomblet has a lethal fragmentation radius of 10 m
(b) anti-armour with a total of 30 small cylindrical hollow-charge mines each fitted with four flip-out curved vanes to stabilise their flight to the ground and ensure correct orientation there. This payload is designed both to create an obstruction in front of aggressor forces to slow him down and to channel him into pre-selected 'killing grounds'.

A longer range 262 mm rocket is being developed. Designed to reach 70 000 m, the warhead size of the original is being reduced and the amount of solid propellant carried is being increased in order to achieve this.

The vehicle is fitted with a central tyre pressure regulation system and to provide a more stable firing platform four stabilisers are lowered to the ground, one either side to the rear of the second roadwheel and two at the very rear. When travelling the launcher is traversed to the front and the whole launcher is covered by a tarpaulin cover with integral bows.

Status: Production. In service with Iraq (2-3 systems delivered before Gulf War, as the Ababeel 50) and Yugoslavia.

Manufacturers: Enquiries to: Federal Directorate of Supply and Procurement (SDPR)
PO Box 308
9 Nemanjina Street
Belgrade
Yugoslavia
Telephone: 621 522
Telex: 11360

262 mm (12-round) LRSV M-87 Orkan MRLS in travelling configuration with bows and tarpaulin cover stowed to immediate rear of cab

262 mm (12-round) LRSV M-87 Orkan MRLS deployed in firing position

SPECIFICATIONS

CALIBRE	262 mm	MAX RANGE	50 000 m	HEIGHT (travelling)	3.84 m		
NUMBER OF BARRELS	12	ELEVATION/DEPRESSION	+56°/-5°	MAX SPEED	80 km/h		
LENGTH OF ROCKET	4.2 m	TRAVERSE	n/av	ARMAMENT	1 × 12.7 mm HMG, 4 × 82 mm smoke launchers, crew small arms		
WEIGHT		CREW	5				
complete rocket	385 kg	WEIGHT (system)	32 000 kg				
warhead	95 kg	LENGTH (travelling)	9 m				
		WIDTH (travelling)	2.64 m				

128 mm (32-round) M-77 Oganj Multiple Rocket System

Development/Description

The M-77 128 mm was developed in the early 1970s to meet the requirements of the Yugoslavian Army. It is normally found in batteries of six launchers. It performs a similar role to the Czechoslovakian 122 mm RM-70 (40-round) MRS but does not have an armoured cab nor the excellent cross-country performance of the Czechoslovakian vehicle which is based on the Tatra 813 (8 × 8) truck chassis.

The M-77 128 mm MRS basically consists of FAP series truck with a 32-barrel 128 mm rocket launcher mounted at the rear of the hull, with an additional pack of 32 rockets to the rear of the cab for rapid loading. The launcher fires a 128 mm calibre 2.6 m long rocket to a maximum range of 20 600 m. Weight of the rocket is 67 kg, with the 19.5 kg warhead containing 2562 6.35 mm diameter and 930 10.32 mm diameter steel spheres that are lethal out to a radius of over 30 m. The explosive content is 3.8 kg and a fuze with preselected superquick, inertia and delay action options is fitted as standard. The rockets are either launched singly or in a ripple salvo; the latter takes a total of 25.6 seconds. The rockets can be launched either by manual, semi-automatic or fully automatic control with an optional 25 metre off-vehicle remote-control box-cable system. Once the rockets have been launched the launcher is traversed to the rear and depressed into the horizontal, the pack of 32 rockets to the rear of the cab is raised until it lines up with the launcher and the rockets are pushed into it. The reloading system then goes back onto the floor of the truck and the launcher is traversed forwards ready for firing again. The complete reloading sequence takes two minutes with the second salvo being capable of launch within five minutes of the first. The reloading can also be performed manually. When travelling the launcher is lowered and the launcher and reloading pack are covered by a tarpaulin which is folded up under the rear in the firing position. The time required to occupy or leave a firing position is less than 20 seconds, with the time taken for occupying a position, firing both rocket loads and leaving being some four to five minutes in total.

It is believed that the original FAP-2220 BDS (6 × 4) chassis was used only for early production vehicles as it is not suitable for travelling across very rough country because of its low ground clearance. Recently a number of the FAP 2020 BS (6 × 6) and FAP-2026 BDS/AV (6 × 6) trucks have been observed fitted with the system. These vehicles have large tyres with single rather than dual rear wheels and are fitted with a central tyre-pressure regulation system that allows the driver to adjust the ground pressure to suit the type of ground being crossed.

First major production version of M-77 Oganj 128 mm MRL on rear of FAP 2026 BS (6 × 6) truck chassis with tarpaulin showing bank of 32 rockets lowered behind cab and rocket launcher at the rear. The stabilisers are mounted at the very rear of the vehicle and to the immediate rear of the front road wheels

128 mm Single Tube Launcher

To complement the M-77 Yugoslavia also produces a 128 mm light-weight single-tube rocket launcher for use by commando, special forces and territorial defence units.

Available details of the 5-man (two for the launcher, three for the rocket) bipod-mounted weapon are:

CALIBRE	128 mm
WEIGHT (in combat position)	50 kg
TRAVERSE	±10°
ELEVATION	+7° to +48°
MAX RANGE	20 600 m
SIGHT UNIT	NSB-4A

Firing of the launcher is performed by remote hand-controller and cable from a distance of 25 m.

Status: In production (latest version uses FAP-2026 BDS/AV (6 × 6) truck chassis). In service with Iraqi and Yugoslavian Armies.

Manufacturer: Enquiries to: Federal Directorate of Supply and Procurement (SDPR)
PO Box 308
9 Nemanjina Street
Belgrade
Yugoslavia
Telephone: 621 522
Telex: 11360

128 mm light-weight single-tube rocket launcher in firing position and showing bipod and baseplate

SPECIFICATIONS (with FAP-2026 BDS/AV 6 × 6 chassis)

CALIBRE	128 mm	CREW	5	ENGINE	V-8 direction injection water cooled diesel developing 256 hp
NUMBER OF BARRELS	32	LENGTH (travelling)	8.4 m		
MAX RANGE	20 600 m	WIDTH (travelling)	2.49 m		
ELEVATION/DEPRESSION	+50°/0°	HEIGHT (travelling)	3.1 m	TYRES	15.00 × 21
TRAVERSE	185° left, 185° right	TRACK	2.02 m		
RELOAD TIME (automatic)	2 mins	WHEELBASE	3.4 m + 1.4 m		
WEIGHT OF SYSTEM (loaded)	22 400 kg	MAX ROAD SPEED	80 km/h		
		RANGE	600 km		

128 mm (32-round) M-85 Multiple Rocket Launcher System

Development/Description
The first Baghdad International Exhibition for Military Production in Iraq revealed that the Yugoslavian Army has a 128 mm rocket launcher, the M-85, based on the TAM truck chassis in service.

The 128 mm rocket used weighs 25.5 kg and can be fired either singularly or in a ripple salvo with set intervals of 0.2, 0.4 or 0.6 seconds. All other available details are given below.

SPECIFICATIONS

CALIBRE	128 mm
NUMBER OF BARRELS	32
WEIGHT OF ROCKET	25.5 kg
MAX RANGE	12 500 m
COMBAT WEIGHT	9800 kg
ELEVATION/DEPRESSION	+48°/0°
TRAVERSE	360°

Status: Production. In service with Yugoslavia.

Manufacturer: Enquiries to: Federal Directorate of Supply and Procurement (SDPR)
PO Box 308
9 Nemanjina Street
Belgrade
Yugoslavia
Telephone: 621 522
Telex: 11360

128 mm (32-round) M-63 Plamen Multiple Rocket System

Development/Description
The 128 mm (32-round) M-63 MRS has been developed in Yugoslavia to meet the requirements of the Yugoslavian Army. It is normally found in MRL battalions of three batteries, each with four launchers.

The M-63 is mounted on a split-trail carriage with elevation, traverse, launch controls and sight on the left side of the launcher. The spin-stabilised M-63 rockets fired by the M-63 are 0.814 m long, weigh 23.1 kg, have a maximum velocity of 444 m/s and are fitted with a 7.55 kg warhead that is filled with 2.3 kg of high explosive. A UTI M63(OV) point detonating fuze with superquick-inertia action is used. A practice round of similar dimensions and weight for loading and unloading training is also used. The rockets are either fired singly or in a ripple salvo with set intervals of 0.2, 0.4 or 0.6 seconds.

An improved version of the M-63 with a wider carriage (2.212 m in travelling position) is in production and both 8- and 16-round launchers are available for export. If required these can be mounted on a self-propelled system.

All versions of the M-63 launcher can be adapted for towing by animals.

128 mm Single Tube Launcher

Yugoslavia also produces a 128 mm M-71 light-weight single-tube rocket launcher for use by commando, special forces and territorial defence units.

Available details of the 3-man crewed bipod-mounted weapon are:

CALIBRE	128 mm
LENGTH	
in travelling configuration	1.133 m
in combat configuration	1.13 m
WEIGHT	
in travelling configuration	21.5 kg
in combat configuration	
unloaded	18.5 kg
loaded	45 kg
ELEVATION	+5 to +45°
MAX RANGE	8564 m
MIN RANGE	800 m
SIGHT UNIT	NBS-4A

Time required for emplacement is five minutes and for disengagement (in fired state) three minutes. The rocket type used is the M-63 type previously described.

128 mm Improved Rocket

In 1989 Yugoslavia revealed the improved M-87 HE rocket for use with both the M-63 and M-71 launchers. This weighs 25.5 kg in total, is 0.96 long and uses an M-84 point detonating fuze with superquick inertia action. The warhead contains 3.3 kg of HE filler.

Maximum velocity is increased to 740 m/s with the 7.8 kg of solid propellent fuel carried giving a maximum range of 13 000 m.

128 mm (32-round) M-63 rocket launcher in firing position with trails open

Status: Production as required. In service with Yugoslavia.

Manufacturer: Enquiries to: Federal Directorate of Supply and Procurement (SDPR)
PO Box 308
9 Nemanjina Street
Belgrade
Yugoslavia
Telephone: 621 522
Telex: 11360

Close-up of rear of 128 mm (32-round) M-63 rocket launcher

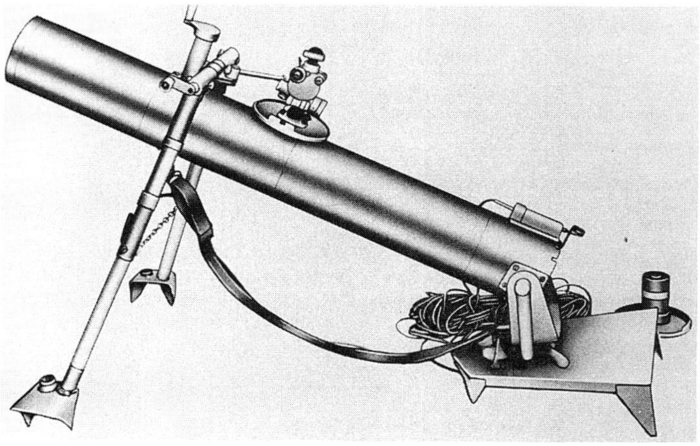

128 mm M-71 light-weight single-tube rocket launcher deployed in firing position and showing bipod and baseplate

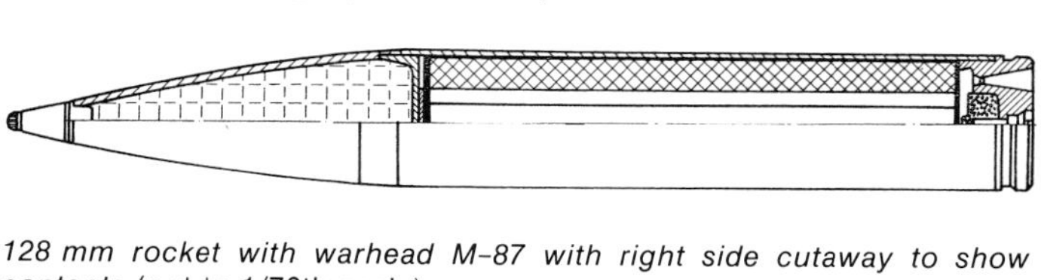

128 mm rocket with warhead M-87 with right side cutaway to show contents (not to 1/76th scale)

SPECIFICATIONS (M-63 rocket used)

CALIBRE	128 mm
NUMBER OF BARRELS	32
LENGTH OF BARREL	1.03 m
LENGTH OF ROCKET	0.814 m
WEIGHT OF ROCKET	23.1 kg
MAX RANGE	8545 m
MAX VELOCITY	444 m/s
RELOAD TIME	5 mins
CREW	7
CARRIAGE	split trail
WEIGHT IN FIRING POSITION	
unloaded	1395 kg
loaded	2134 kg
LENGTH	
travelling	3.682 m
firing	3.65 m
WIDTH (travelling)	2.212 m
HEIGHT (travelling)	1.278 m
GROUND CLEARANCE	0.268 m
TRACK	1.892 m
TYRES	8.25 × 20
ELEVATION/DEPRESSION	+48°/0°
TRAVERSE (total)	30° (15° left, 15° right)
RELOAD TIME	3 mins
TOWING VEHICLE	TAM 1500 (4 × 4) truck

122 mm M-88 Rocket System

Development/Description

Although the Yugoslavian Army has standardised on a 128 mm calibre rocket system for its own use the Federal Directorate of Supply and Procurement has commenced production of a 122 mm calibre HE-fragmentation rocket, the M-88, for the export market.

Available details of the rocket are given in the specifications table. The HE filler of the warhead is layered with 2780 6.35 mm diameter steel spheres to increase the lethality of the fragmentation pattern.

SPECIFICATIONS

CALIBRE	122 mm
LENGTH	2.87 m
LAUNCH WEIGHT	66 kg
WEIGHT OF HE FILLER	6.4 kg
MAX VELOCITY	690 m/s
MAX RANGE	20 400 m

Status: Production. In service with undisclosed countries.

Manufacturer: Enquiries to: Federal Directorate of Supply and Procurement (SDPR)
PO Box 308
9 Nemanjina Street
Belgrade
Yugoslavia
Telephone: 621 522
Telex: 11360

Coastal Artillery Guns and Missiles
(including Fixed Fortification Weapons)

ARGENTINA

Argentine Coastal Defence

Until fairly recently the Argentine Navy still manned some 10 coastal defence batteries, which comprised 12 M-1898 87 mm guns, 16 US supplied 155 mm M3 guns and 12 280 mm turret mounted guns. It is believed that these are no longer fully operational. During the 1982 Falklands Conflict Argentine units in the Falkland Islands used a field expedient trailer-mounted twin round MM38 Exocet surface-to-surface missile launcher against Royal Navy warships.

AUSTRIA

Austrian Static Defences

For many years the Austrian Army has been using turrets from obsolete tanks in the static defence role. These include M47 tanks (90 mm), T-34/85 tanks (85 mm), Charioteer tank destroyers (84 mm) and most recently Centurion tanks (105 mm). In the first three cases the turrets were taken from vehicles that had been withdrawn from service. But in the case of the Centurion the turrets were purchased from the German company of Krauss-Maffei who took Canadian Centurion tanks in part exchange for Leopard 1 MBTs. Early in 1984 Austria purchased 120 Centurion tanks from the Netherlands at $3800 per vehicle. Their turrets were subsequently removed for use in the static defence role.

*Centurion 105 mm tank turret fitted with training equipment used for static defence by the Austrian Army (*Austrian MoD*)*

BRAZIL

Brazilian Coastal Defence

Coastal artillery comes under the control of the Brazilian Army with 57 mm guns, 6-inch (152 mm) Vickers Mk 5 turret-mounted guns and 12 inch (304.8 mm) turret-mounted guns, but it is believed that these are no longer maintained in a fully operational condition. There is some interest in acquiring coastal defence missiles and this requirement could be met by the Brazilian designed SM-70 Barracuda system covered in the following entry.

AVIBRAS SM-70 Barracuda Coastal Defence Missile

Development/Description

The SM-70 (Superficie-Mar 70 kilometres) Barracuda has been developed by AVIBRAS Industria Aerospacial S/A since 1986 to meet an initial requirement for a coastal defence missile. Further versions may be procured for shipborne and airborne launch roles.

In general appearance the SM-70 looks like the Exocet with an overall length of 5.2 m, wing span of 1.2 m and a body diameter of 0.30 m. The range is said to be 70 000 m. A 150-200 kg HE warhead is likely to be used.

The basic operational configuration is to be a fire unit mounted on five TECTRAN (6 × 6) 10 000 kg cross-country trucks: three being launch vehicles with four container-launchers, a target search and acquisition vehicle with a retractable radar antenna and a command vehicle (with the FILA Command system based on the AVIBRAS C-2001 computer and a data link to the launcher vehicles). The missile will have an AVIBRAS SIS strapdown inertial navigation system and cruise at Mach 0.9. Each fire unit will also have 12 replacement launcher-containers.

Status: Development. First prototype round launch due 1991.

Manufacturer: AVIBRAS Industria Aerospacial S/A
International Sales Division
KM118 (Caixa Postal 229)
12-200
Sao Jose dos Campos
Sao Paulo State
Brazil
Telephone: 123 21 7433
Telex: 123 3793
Fax: 123 22 1689

BULGARIA

Bulgarian Coastal Defence

Coastal defence is the responsibility of the Navy and includes two artillery regiments with about 20 batteries with 150 guns, including 100 mm, 130 mm SM-4-1 and 152 mm types. There are also two battalions with a total of 36 launchers for the SSC-2b Samlet coastal defence missile.

CHILE

Chilean Coastal Artillery

Coastal artillery comes under the operational control of the Chilean Navy and consists of 16 155 mm GPFM-3 weapons which have been in service since well before the Second World War.

CHINA, PEOPLE'S REPUBLIC

Chinese Coastal Defence

Coastal defence is the responsiblity of the Navy and is believed to consist of between 35 and 40 independent artillery and surface-to-surface missile regiments deployed in 25 coastal defence regions. Artillery weapons include 85 mm, 100 mm and 130 mm guns, all of which are believed to be semi-mobile, plus surface-to-surface missiles.

The latter include the HY-1 (Sea Eagle-1), HY-2 (Sea Eagle-2), the HY-4 (Sea Eagle-4), the Ying Ji-1 (Hawk Attack-1) and the Ying Ji-2 (Hawk Attack-2).

Hai Ying-1 Coastal Defence Missile

Development

In 1959 the technology required for the license manufacture of the Soviet P-15 (US/NATO designation SS-N 2a Styx) ship-to-ship missile was transferred to the People's Republic of China. The licence built copy was given the domestic designation Shui Ying-1 (SY-1, Water Eagle-1). Production and final assembly was undertaken by the Nanching Aircraft Factory with the autopilot, terminal guidance radar, fuzing system, warhead, solid propellant rocket booster and the liquid fuel sustainer rocket engine components produced under licensed subcontracts by other factories.

Preparation for the licence production began in 1960 with component manufacturing beginning in October 1963. The initial ground firing trials were successfully conducted between November and December 1964 and the follow-on shipboard trials from August to November 1965. Fully instrumented flights were then conducted in mid-1966 with the missile certification trials being carried out in November 1966. Its design certification was given approval shortly afterwards together with authorisation for full scale production and operational use.

In 1974 design work began on modifying the SY-1 to improve its performance. The modifications concentrated on two key areas:

1) The replacement of the original aneroid air pressure altimeter and vertical speed sensor unit with a radar altimeter system. Although the missile had a high hit rate the original altimeter device always had a large altitude control error associated with it that invariably caused a loss in altitude from the preset height during the weapon's flight.

2) Replacement of the original conical scan I-band radar with its poor anti-sea clutter/counter-ECM performance by a centimetric monopulse transistorised set.

Trial firings of the modified weapon started in 1975 with three series of design certification test firings occurring between 1977 and 1980. All of these failed and in February 1981 a design review was conducted which led to a fourth and successful certification firing series in late 1982. Approval for production of the SY-1A was given in December 1983.

At the same time as the SY-1A modification was underway a study was being conducted into introducing a lower flight altitude capability of 30 m and below into the terminal homing phase of the missile's trajectory after the seeker had acquired the target. This feature was successfully flight tested in 1980 and, although not introduced operationally into the SY-1A missile was incorporated into later systems.

Development of the coastal defence version of the SY-1, the Hai Ying-1 (Sea Eagle-1), began in 1964. The design was chosen for production in April 1965 with the main contractor again being the Nanching Aircraft Factory and the same subcontractors producing suitably modified variants of the terminal radar seeker, autopilot, solid propellant rocket booster and the liquid fuel sustainer rocket engine.

In 1966 a small number of pre-production rounds were built for a series of three trial firings carried out late that year and in early 1967. All failed due to self-guidance radar problems which were initiated upon ignition of the booster. Technical modifications were then made to shorten the launcher's guide rail so as to reduce the initial shock and vibration interference levels from the booster and increase the rigidity of the missile body. In addition the maximum search range of the radar seeker was increased. In May 1972 the modified coastal defence weapon successfully completed its flight test programme and in 1974 its design and production certification was granted. Full-scale batch production began in 1975.

In 1967 the decision was taken to modify the HY-1 weapon into a shipboard system to equip the Type 051 (Luda class) missile destroyers. This was undertaken by the Nanching Aircraft Factory and involved the reworking of all the electrical circuitry (to protect it against the hostile marine environment), modification of several of the radar components and the installation of an additional unit in the autopilot.

Firing tests of this shipborne weapon took place between 1968 and 1975 with full design and production certification occurring in January 1976.

To reduce the number of HY-1 production variants and improve maintenance requirements the decision was taken in the mid-seventies to manufacture a common missile for both shipborne and coastal defence use. This entailed further modifications to the radar seeker, autopilot, electrical circuitry and missile body with the common missile entering production during 1978.

In 1983 the common HY-1 version underwent a further redesign to incorporate a centimetric monopulse J-band radar seeker with enhanced

Chinese Hai Ying-1 coastal defence misile on modified Type 63 light amphibious tank chassis

counter-ECM and anti-sea clutter features, an improved radar altimeter to reduce the terminal attack height to around 8 m or so and propulsion system improvements to increase the operational engagement range. Trial firings were conducted during 1985 with technical and production acceptance certification following in early 1986. Full scale production of this HY-1A variant started in October that year with operational deployment occurring in 1987.

A technology transfer of SY-1 and HY-1 production techniques was made to North Korea (qv) from 1969 onwards.

Description

The general configuration of the HY-1 variants is similar to that of a small aircraft, with fixed small aspect trapezoidal wings and a triple tail surface configuration. All the horizontal surface components of the weapon have trailing edge elevator surfaces with the vertical surface component having a movable rudder. A large jettisonable solid fuel booster motor is located in the ventral position and is used to boost the weapon in the initial launch and acceleration phases, after which the liquid fuel sustainer rocket motor cuts in to provide a cruising speed of Mach 0.85. Maximum length is around 6.6 m, body diameter about 0.75 m and launch weight 2300 kg. A 513 kg HE warhead is fitted.

The effective range limits for the HY-1 are 5560-29 560 m and for the HY-1A 5560-46 325 m.

Guidance for the mid-course flight phase is by autopilot with the terminal phase (on original variants) using one of six available I-band radar seeker head frequencies. The conical scan seeker is housed within a fibreglass reinforced plastic honeycomb sandwich composite radome and is switched on around some 15 000 m or so from the target's predicted position (a 50 per cent increase over the original SY-1 radar performance) with a preset wide or narrow gate facility so that the missile can either attack any ship in a formation or one in particular. On the later HY-1A variant a modern J-band active radar seeker is used.

The flight cruise altitude can also be preset in a multiple of 50 between 100 and 300 m using the onboard altimeter with, in the case of the earlier weapons, the terminal attack phase being essentially a shallow dive.

The launch sites can either be fixed or mobile. The launch rail being mounted on a semi-trailer or a specially modified Type 63 amphibious tank chassis. The later configuration has also been seen using SY-1 missiles in a coastal defence role.

Status: SY-1 production complete. In service with the Chinese Armed Forces.

SY-1A production complete. In service with the Chinese Armed Forces.

HY-1 production complete. In service with the Chinese Armed Forces. Also used by North Korea but believed replaced by HY-2.

HY-1A in production. In service with the Chinese Armed Forces.

Manufacturers: Chinese and North Korean state factories.

CPMIEC Hai Ying-2 Coastal Defence Missile

Development/Description

The HY-2 (Sea Eagle-2) is a Chinese developed product-improved version of the Soviet P-15 (US/NATO designation SS-N-2 Styx) anti-ship missile which is assigned the western designation Silkworm.

In general configuration it is similar to the HY-1 but is 6.25 m long, has a body diameter of 0.78 m, a wing span of 2.75 m and weighs some 2998 kg at launch. The warhead weighs 513 kg and contains conventional HE.

Guidance in the mid-course phase is by gyro autopilot with the terminal phase depending upon which model HY-2 is used. Maximum effective range is between the limits of 20 000 to 95 000 m.

The missile operates in conjunction with a 6 × 6 truck-mounted I/J-band target acquisition radar which has a peak power output of 200 kW, a pulse width of 0.2 μs and a range resolution of 60 m. The set can satisfactorily track a 7500 m² radar cross section target at 111 km distance.

Apart from its use with the Chinese Armed Forces the HY-2 has been exported to North Korea (qv, where it is also licence built). It has also been exported to both sides in the Iran/Iraq War, with the latter country using it and achieving mediocre results. Iraq used it during the 1991 Gulf War against Coalition Naval units (qv Iraq).

Hai Ying-2 coastal defence system on its launcher in travelling configuration

Hai Ying-2 on missile transloader being loaded onto launcher

Variants

Basic Active Radar Guided HY-2.

This uses an active I-band radar seeker for the terminal phase and cruises at an altitude of 200 m.

Infra-red Guided HY-2A

This uses a non-imaging infra-red seeker with a cassegrain lens arrangement for the terminal homing phase. Cruise altitude is reduced to 30 m.

Advanced Active Radar Guided HY-2G

This uses an advanced centimetric wavelength monopulse active radar seeker (probably operating in the J-band) for the terminal homing phase.

A radio frequency altimeter is also fitted which allows the missile to descend during the attack phase from its cruise height of 20 m to approximately 8 m in order to present a more difficult target to defending air defence systems.

Status: In production. In service with China, Iran, Iraq, North Korea and, reportedly, Zaïre.

Manufacturers: Chinese and North Korean state factories.

Enquiries to: China Precision Machinery Import and Export Corporation (CPMIEC)
PO Box 845
Beijing
People's Republic of China
Telephone: 895012; 8311804
Telex: 22484 CPMC CN
Fax: 8311657

Hai Ying-2 missile on ground trolley

CPMIEC C-101 and Hai-Ying 3 Coastal Defence Missiles

Development/Description

The HY-3 (Sea Eagle-3, alternative designation C-301) is derived from the C-101 ramjet-powered missile that was developed in the late seventies from the basic HY-1 family of airframes.

The first showing of the C-101 (in model form) occurred in 1986. It has two side-mounted ramjet engines at the rear, two jettisonable solid propellant boosters mounted beneath them, a V-shaped tail arrangement and a vertical fin with a control surface on its upper part. Length is 5.8 m, body diameter 0.56 m and a range in the order of 50 000 m. Guidance is by a mid-course inertial system with an active radar terminal homing seeker. The warhead carried weighs 400 kg and is of the semi-armour piercing HE type. Speed is said to be supersonic with the weapon cruising at around 300 m altitude. The launch weight is assessed at approximately 2200 kg.

The HY-3 is a heavier and longer range weapon and is similar in shape to the C-101, but has four solid propellant booster motors with stabiliser fins attached in the fashion of the BAe Bloodhound SAM mounted above and below the two ramjet sustainers. The overall length of the HY-3 with the four boosters attached is 9.85 m which reduces to the 9.46 m of the main missile body after they are jettisoned. Missile body diameter is 0.76 m. Speed is Mach 2 with a cruising height settable between 100 and 300 m. The warhead weighs approximately 512 kg and is probably similar to the standard conventional HE configuration fitted to the other HY family members. Launch weight is estimated at around 3400 kg and the weapon range is said to be 130 000 m.

Status: C-101 ready for production. Believed solely for export market and awaiting first order.

HY-3 final development phase. Expected to enter service with the Chinese Armed Forces in the mid-1990s.

CPMIEC Hai Ying-3 coastal defence missile on its launcher

Manufacturer: Chinese state factories.

Enquiries to: China Precision Machinery Import and Export Corporation (CPMIEC)
PO Box 845
Beijing
People's Republic of China
Telephone: 895012; 8311804
Telex: 22484 CPMC CN
Fax: 8311657

CPMIEC Hai Ying-4 Coastal Defence Missile

Development/Description

The HY-4 (Sea Eagle-4) is a further development of the HY-2 series to enhance the performance characteristics.

The general configuration remains essentially the same but the fuselage is lengthened and the liquid fuel sustainer rocket motor is replaced by a small liquid fuel turbojet propulsion unit. The air for which is drawn in through a ventral scoop intake located towards the rear of the under fuselage.

The initial launch and acceleration phases are still carried out by using a jettisonable solid fuel rocket motor attached immediately behind the scoop.

The guidance system is believed to be similar to that of the HY-2G with an autopilot for the mid-course phase and an active monopulse radar seeker (probably operating in the J-band) for the terminal attack.

HY-4 coastal defence missile on transport trolley

Cruise altitude can be varied between 70 and 200 m using the onboard radio frequency altimeter. Launch weight is 2000 kg with the cruising speed being in the order of Mach 0.8.

Effective range limits are from 35 000 to 135 000 m. Missile length is 7.36 m, body diameter 0.76 m and wing span 2.75 m. The warhead weighs 500 kg and contains conventional HE.

Status: In production. In service with the Chinese Armed Forces.

Manufacturer: Chinese state factories.

Enquiries to: China Precision Machinery Import and Export Corporation (CPMIEC)
PO Box 845
Beijing
People's Republic of China
Telephone: 895012; 8311804
Telex: 22484 CPMC CN
Fax: 8311657

HY-4 coastal defence missile being loaded onto launcher from missile transloader

---

CPMIEC Ying Ji-1 and Ying Ji-2 Coastal Defence Missiles

Development/Description

The YJ-1 (Hawk Attack-1, alternative designation C-801) missile is reminiscent of the MM38 Exocet in appearance and is fired from a box shaped container/launcher.

It uses a solid fuel sustainer rocket motor with a jettisonable solid fuel booster motor section for the propulsion system. Flight control is by four movable cruciform tail fins and four foldable mid-body wings.

Total missile length is 5.814 m, body diameter 0.36 m and maximum wing span 1.18 m. The launch weight is 815 kg of which 165 kg is the semi-armour piercing conventional HE warhead.

Mid-course guidance is by autopilot with the terminal phase being controlled by a monopulse active radar seeker (probably operating in the J-band). Cruise speed is Mach 0.9 with the cruise altitude being controlled by a radio altimeter between 20-30 m. For the terminal attack the missile is automatically commanded to descend down to around 5-7 m height. Maximum effective range is 50 000 m, minimum effective range is 8000 m.

In February 1989 the YJ-2 (alternative designation C-802) version was revealed. The only exernal difference to the YJ-1 is that an air inlet scoop has been added beneath the main missile body to serve the small Wopen air-breathing turbojet that has replaced the solid fuel sustainer rocket motor. The jettisonable solid fuel booster motor section is retained.

The Ying Ji-2 is 6.392 m long, 0.36 m in body diameter and weighs 715 kg at launch. Minimum effective range is 15 000 m and maximum effective range 120 000 m. The cruise speed is between Mach 0.8 and 0.9 with the cruising and terminal attack altitudes the same as those for the Ying Ji-1. In its coastal defence role the YJ-2 is launched from a large truck (of about 10 000 kg) which carries two to four container-launchers on an elevating cradle fitted to its rear decking.

Ying Ji-1 coastal defence missile out of its launcher with container/launcher in background

Status: HJ-1 in production. In service with the Chinese Armed Forces and several Middle Eastern countries including Iran.
HJ-2 in production. In service with the Chinese Armed Forces.

Manufacturer: Chinese state factories.

Enquiries to: China Precision Machinery Import and Export Corporation (CPMIEC)
PO Box 845
Beijing
People's Republic of China
Telephone: 895012; 8311804
Telex: 22484 CPMC CN
Fax: 8311657

Ying Ji-1 coastal defence missile out of its launcher

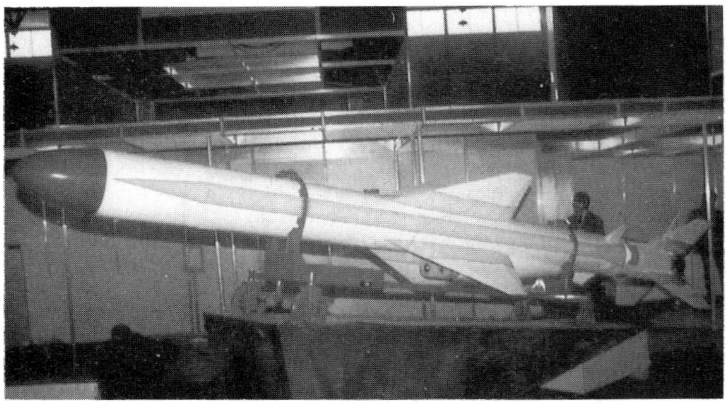

Ying Ji-2 coastal defence missile with fins and wings unfolded

NORINCO 155 mm Coastal Defence System

Development/Description
The NORINCO coastal defence system consists of a battery of six NORINCO 155 mm gun-howitzers Type WA 021, fully described in the *Towed guns & howitzers* section, fire control computer (FCC), advanced long range laser rangefinder (LRF) and an advanced phased array radar.

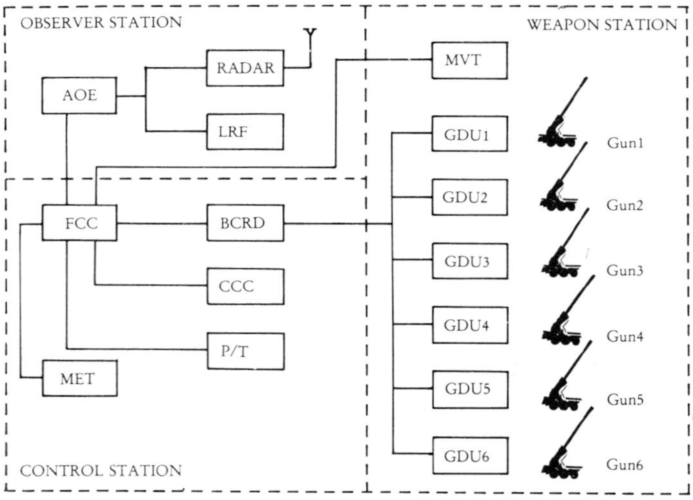

Block diagram of the NORINCO 155 mm Coastal Defence System

According to NORINCO, the system is capable of searching, intercepting, engaging and tracking surface targets under all weather conditions.

The two key parts of the NORINCO coastal defence system are the battery of six 155 mm gun-howitzers Type WA 021, each of which is provided with a gun display unit (GDU) and the fire control system.

The fire control system utilises an advanced military 16-bit microcomputer in the FCC and an advanced communications system.

It has two operating modes, interactive and automatic to input the various data in to the fire control computer.

Each fire control computer can support six guns of its own battery and can also take over the control of a further 12 guns if their fire control computers malfunction.

Information is fed into the central fire control computer from the observer's station (radar and laser rangefinder) and the meteorological station.

The WA 021 gun-howitzer has a maximum range of 30 000 m using ERFB projectiles and 39 000 m using ERFB-BB projectiles, traverse is through a total of 70° (30° left and 40° right) and a rate of fire of 5 rds/min can be achieved.

Status: Development complete. Ready for production on receipt of orders.

Manufacturer: NORINCO
China North Industries Corporation
7A Yuetan Nanjie
PO Box 2137
Beijing
People's Republic of China
Telephone: (86) 6898, (86) 3461, (86) 3471, (86) 7570
Telex: 22339 CNIN CN

DENMARK

Danish Coastal Defences

Coastal artillery is the responsibility of the Royal Danish Navy and retains a number of coastal defence artillery equipments, mainly for covering minefields and similar tasks.

Three types of gun are currently in service. The first is an ex-German Naval Weapon originally developed for use as the main sescondary armament for the 'Bismarck' and 'Scharnhorst' class capital ships. A number of spare weapons produced, designated 15 cm (5.9 inch) SKC/28 mounted in both twin (Zwillingslafette) barbettes and single Kusten Marine Pivot Lafette (KstMPL) mountings were used for coastal defence. In Danish service the SKC/28 gun is installed as two twin-gun installations (taken from the battlecruiser 'Gneisenau').

The second weapon used is also an ex-German naval design, the 15 cm (5.9 inch) SKL/45, dating from the First World War and used during the Second on a variety of second line warships and as a coastal defence weapon. A total of six single mounts are in service.

There are also a number of ex-American 3-inch/50 naval guns used in coastal defence mountings.

More recently it was stated that Harpoon surface-to-surface missiles removed from decommissioned Danish light frigates would be used for the coastal defence role from the early 1990s.

SPECIFICATIONS

	15 cm SKC/28	**15 cm SKL/45**
CALIBRE	149.1 mm	149.1 mm
WEIGHT (including breech mechanism)	9026-9080 kg	5730-6102 kg
LENGTH (overall)	8.2 m	6.71 m
PROPELLANT CHARGE	14.1 kg	14.35 kg
MUZZLE VELOCITY	875 m/s	835 m/s
PROJECTILE TYPES	HE base fuzed shell 44.8 kg HE noze fuzed shell 45.5 kg APC shell 45.3 kg	
MAX RANGE	23 500 m	

EGYPT

Egyptian Coastal Defence

Coastal defence is under the operational control of the Egyptian Navy although it is manned by the Egyptian Army. Equipment in service includes Soviet supplied 130 mm SM-4-1 guns, OTOMAT surface-to-surface missiles, the older Soviet supplied missiles are understood to have now been phased out of service.

FINLAND

Finnish Coastal Defence

Coastal defence is the responsibility of the Finnish Army and includes two coastal artillery regiments and three independent coastal artillery battalions, one of which is mobile. The country has recently completed a programme to modernise not only the guns but also the command and communications systems to co-ordinate accurate fire.

Weapons include 100 mm armed T-54/T-55 turrets set in concrete, a new Tampella 130 mm turret-mounted gun with a range of 25 000 m, and older 152 mm and 254 mm weapons, some of which are also turret-mounted. More recently Finland has taken delivery of a quantity of the Swedish RBS 15CD coastal defence missile system carried and launched from a Finnish SISU cross-country truck chassis.

VAMMAS 130 mm Coastal Defence Gun

Development/Description

This 130 mm coastal defence gun was developed to meet the requirements of the Finnish Army by VAMMAS, previously known as Tampella, and entered service in the mid-1980s.

The turret is of all steel construction and is normally located on top of a hardened emplacement with fire control being accomplished by a computerised fire control system connected to a surveillance radar system.

A roof-mounted sight is provided and if required the crew can leave the turret via doors located in the rear. The 52 calibre ordnance is fitted with a pepperpot type muzzle brake.

Variants

Although all production systems to date have been armed with a 130 mm ordnance, the turret system can also be fitted with 122 mm, 152 mm or 155 mm barrels.

SPECIFICATIONS

CALIBRE	130 mm
LENGTH OF BARREL	6818 mm
WEIGHT OF TURRET	16 000 kg
ELEVATION/DEPRESSION	+50°/-5°
TURRET TRAVERSE	360°
MAX RANGE	24 000 m (standard ammunition)
MAX RATE OF FIRE	6 rpm

Status: Production as required. In service in Finland.

Manufacturer: VAMMAS
POB 18
SF-38201 Vammala
Finland
Telephone: 358 32 1971
Telex: 22283 VAMAS SF
Fax: 358 32 41148

FRANCE

Aerospatiale MM40 Exocet Coastal Defence Missile System

Development/Description

The MM40 Exocet coastal defence system is derived from the MM40 shipborne anti-ship missile and is in operational service with Qatar which has bought four batteries equipped with Berliet 6 × 6 cross-country vehicles.

A battery typically comprises a mobile radar command post (RCP-equipped with a 4 × 4 truck with a two-three man command and control room shelter with one or two radar consoles and a weapons control system and a two-axle trailer carrying the extending mast surveillance and target acquisition radar antenna, and a 4 × 4 truck mounting the operations room shelter with a two-axle trailer carrying the RCP's generator unit); three 6 × 6 truck firing units each carrying four MM40 container-launcher canisters and towing a small two-axle mobile generator unit; and a resupply 6 × 6 truck with six reload container-launchers. If required existing radars can also be used.

The missile is 5.8 m long, has a body diameter of 0.35 m and a maximum wing span of 1.135 m. Launch weight is 870 kg with a 165 kg SERAT HE Shaped charge warhead and delay action contact fuze. An overflight proximity function is fitted as backup. If the sea state is high and the target small then use of the proximity function can be programmed into the warhead. This is activated when the missile to target distance, very accurately measured by the seeker, effectively gives a reading of zero. The warhead is then exploded above the target's superstructure with the blast and fragmentation effect causing severe damage to the exposed antenna arrays and weaponry thus depriving it of some or all of its operational capabilities.

The missile has a slender cylindrical shape with a sharply pointed nose. Flight control is by small tail control surfaces which are in-line with foldable swept cruciform long chord cropped wings at the midbody area.

Propulsion is by a 2.4 second burn solid propellant booster rocket motor coupled to a 200-220 second burn solid fuel sustainer rocket motor. Maximum speed reached is in excess of 300 m/s. Maximum effective range is around 70 000 m with the minimum range being 4000 m.

The missile's inertial guidance package is first pre-programmed for the mid-course phase of the flight via the interface unit. The missile can be fired at offset angles of better than 90° from the launcher axis.

The general launch, flight and guidance details of the MM40 are similar to those described for the MM38 missile in the UK Excalibur system entry of this section.

In 1988 Aerospatiale revealed an MM40 Block 2 missile for use in the coastal defence role. This has improved decision making support facilities, an increased offset firing angle of 105°, improved target selectability, multi-target engagement capabilities and an obstacle overflying capacity to allow inland firing positions screened by hills and so on to be used.

In the direct engagement role the MM40 Block 2 typically rises to around 30 m altitude at launch and then descends to 15 m for its mid-course cruise phase. At approximately 9000 m from the target the missile descends to the lowest permittable sea skimming height allowable by the sea-state it finds. At approximately 300 m to go before impact the missile adopts its final terminal descent to ensure a hit on the target's hull.

Coastal defence version of MM40 Exocet being launched

Status: In production. In service with four countries including the Qatari Armed Forces (four batteries on Berliet 6 × 6 cross-country trucks) and one NATO navy.

Manufacturer: Aerospatiale
Divisions Engins Tactiques
2 Rue Beranger
F-92322 Chatillon Cedex
France
Fax: 4746 2777

INTERNATIONAL

Matra/OTO Melara OTOMAT Coastal Defence System

Development/Description

The OTOMAT Coastal Defence System (OCDS) uses the OTOMAT antiship missile developed by Engins Matra of France and OTO-Melara of Italy. It has been sold to Egypt (which has three batteries in service and three of its naval Sea King Mk47 helicopters fitted for target updating of the system) and Saudi Arabia. Both countries use the Berliet GBD (6 × 6) truck as the carrying platform.

A typical OCDS comprises one or more OTOMAT Coastal Defence Units (OCDUs) which each have a command group and from two-four firing groups.

The command group consists of a single 6 × 6 vehicle with a control cabin that houses the tactical display console, missile control panel, various communications and telecommand transmitter; a radar operations truck with a trailer-mounted Thomson CSF TRS 3410 I-band 40 km range surveillance and target acquisition radar with track-while-scan facilities and an optional hoistable 15 m antenna mast; a command car and an operational maintenance vehicle. A VHF radio link of 20 km range is provided between all the vehicles of an OCDU allowing wide tactical dispersion.

A firing group consists of one launcher vehicle which is fitted with a launch cab containing a control panel for remote command group control or local firing, a north-seeking gyro assembly, an optical sighting station (with 20 km range for the local firing option) and two missiles in their ready to fire container launchers; and a resupply vehicle with crane and two reload OTOMAT missile container-launchers.

The missile itself has a cylindrical body with a blunt nose and belled exhaust nozzle at the tail. There are delta shaped folding wings at midbody with small horizontal truncated triangular shaped control fins at the rear.

Missile length is 4.46 m, body diameter 0.46 m and maximum wing span 1.35 m. Launch weight (complete) is 770 kg of which 620 kg is the missile with a 210 kg HE semi-armour piercing warhead with contact delay and proximity fuzing circuits.

Propulsion for the boost and acceleration phases is provided by two jettisonable Hotchkiss-Brandt/SNPE 3500 kg thrust 4-second burn composite solid propellant rocket booster motors mounted either side of the missile body aft. The sustainer rocket engine is a 400 kg thrust Turbomeca TR281 Arbizon III turbojet. Cruise speed is Mach 0.9.

Guidance is by a pre-programmed inertial autopilot and radio altimeter for the mid-course phase. Updating of the target position can be performed by either the command group using a data link and target information derived from external sources such as a helicopter. In the latter case a 100 km range 14 kg weight PRT 405 UHF tracking and data link to the command vehicle is mounted on the helicopter. Terminal homing is by a two-axis Thomson-CSF 'Col-Vert' I/J-band active radar seeker.

OTOMAT coastal defence missile being launched during trials from its (6 × 6) launcher vehicle

At launch the weapon climbs to an initial height of some 250 m at a distance of 600 m from the launcher. The autopilot/altimeter system then commands it to descend to its cruising altitude of 80 m for the mid-course phase. At a pre-determined point (some 12 000 m) from the target the radar seeker is activated in its elevation and 20° left/right azimuth scan search mode. When the target is found and the seeker lock-on for continuous tracking and the missile is commanded to climb at a distance-to-target of 4000 m to a height of 175 m whereupon when it reaches a distance-to-target of 2000 m it enters a 7° dive attitude to impact the target.

If electronic countermeasures are detected as being used by the seeker during its terminal attack phase then the missile climbs to the 175 m altitude at 5000 m to go and enters a 5° dive.

Italian supplied missiles use a single-axis homing seeker and conduct a sea-skimming attack profile in the terminal flight phase. Both the OCDS customers, however, are thought to use the French supplied weapon.

In the fire-and-forget role with only target designation at the launch phase the effective range of the missile is around 100 km. With the use of the fire-and-update facility the effective range is increased to 166 km. Salvo firing achieves a similar improvement to the effective range. Minimum engagement range is 6000 m. The initial missile course can be up to 200° away from the launcher platform heading with the missile adopting an angled flight path if a helicopter is used for a mid-course guidance update.

Status: Production as required. In service with Egypt, Kenya and Saudi Arabia.

Manufacturers: OTO-Melara SpA
Via Vladilocchi 15
I-19100 La Spezia
Italy
Telephone: (187) 53 01 11
Telex: 27368/211101 OTO I

Matra SA
4 Rue de Presbourg
F-75116
Paris
France
Telephone:(1) 4723 54 04
Telex: 620777 F

IRAN

Iranian Coastal Defences

These include Chinese supplied HY-2 (Sea Eagle 2) and the North Korean supplied version of the Chinese HY-2 which were delivered from 1986.

The first combat firing of a coastal defence HY-2 by Tehran in the Iran/Iraq War was on 3 September 1987 (one round, Kuwaiti mainland target - missed). Further firings occurred on 4 September (one round, Kuwaiti mainland target - missed), 5 September (one round, Kuwaiti mainland target - missed), 15 October (one round, 279 932 ton tanker 'Sungari' in Kuwaiti waters - hit), 16 October (one round, target tanker 'Sea Island City' in Kuwaiti waters - hit), 22 October (one round, target Kuwaiti off-shore 'Sea Island' oil loading terminal - hit), 6 December (one round, target 'Mina al-Ahmadi' oil terminal - missed because of a barge-mounted radar reflector decoy system), 20 April 1988 (one round, target on Kuwaiti City waterfront - missed).

A number of HY-2s were also apparently launched against US Navy ships during the battles that took place during Operation 'Praying Mantis' on 18 April 1988, but all were successfully seduced by a variety of ECM ship and airborne based defensive techniques.

IRAQ

Iraqi Coastal Defences

These include the Chinese HY-2 Silkworm and the Faw-70, Faw-150 and Faw-200 which are claimed to be locally built versions of the Soviet SSC-3 Styx coastal defence missile. During the Second Gulf War two Silkworms were fired against Coalition ships on 25 February. One suffered mechanical failure and crashed into the sea whilst the other was destroyed by two BAe Sea Dart SAMs fired from HMS *Gloucester* during its cruise flight.

Faw Coastal Defence Missile Systems

Development/Description

At the First Baghdad International Exhibition for Military Production Iraq showed a series of coastal defence missiles under the name Faw.

The smallest missile was designated Faw-70 and appears to be a locally built version of the Soviet P-20/P-21 (US/NATO designation SS-N-2c Styx) weapon with a maximum range of 70 000 m and a minimum of around 5400 m.

The general configuration is like that of a small aircraft, with delta planform foldable wings and a triple tail surface arrangement. A jettisonable solid fuel booster motor is used for the launch and acceleration phases, after which an internal liquid fuel sustainer rocket motor cuts in to provide a cruising speed of about Mach 0.9. Maximum length is 6.5 m, the body diameter 0.78 m and launch weight (complete) about 2500 kg.

Guidance for the mid-course phase is by autopilot with the terminal phase being either active I-band radar or passive infra-red depending upon the seeker head type fitted. The seeker is switched on at around 10 000 m from the predicted target position with alternative wide and narrow gate facilities so that it can either attack any ship in a formation or one in particular.

The flight altitude can be preset in a multiple of 50 between 50 and 300 m with the terminal trajectory being effectively a shallow dive. The warhead weighs 500 kg and contains conventional HE.

The Faw-150 is thought to be an improved version of the Faw-70 with increased length 7.4 m and enlarged fuel tanks for a maximum range of 150 km whilst the Faw-200 is a still longer, 8.0 m, maximum 200 km range variant.

To achieve maximum effectiveness with all three missile types an over-the-horizon targetting platform such as a ship, plane or helicopter equipped with radar and data link to the battery launch position is required.

Faw-70 coastal defence missile (Christopher F Foss)

Faw-150 coastal defence missile (Christopher F Foss)

Missile weights are believed to be in the order of 2500 kg, 2650 kg and 2740 kg for the weapons.

Not readily apparent is where the technological aid has come from in order for Iraq to construct this class of weapon. Based on the evidence available at present the most likely source would be Yugoslavia with some help from Egypt, China and the Soviet Union.

Status: In service with the Iraqi Armed Forces. Following the 1991 Middle East conflict it is not certain as to whether production facilities for the Faw-70, Faw-150 and Faw-200 systems still exist.

Manufacturer: Military Production Authority
P O Box 5757
Baghdad
Iraq
Telex: 213670 SAFAN IK; 212461 SOTI IK
Fax: 5384634; 7723436

*Faw-200 coastal defence missile (*Christopher F Foss*)*

JAPAN

Mitsubishi Type 88 Coastal Defence Missile

Development/Description

Development of the Type 88 (SSM-1) ground-launched coastal defence missile began in 1981 with the issue of the first proposal and has been the joint responsibility of the Japanese Defence Agency's Technical Research and Development Institute and Mitsubishi Heavy Industries.

Based on the Type 80 ASM-1 air-launched anti-ship missile, the first prototype weapons were built in 1983 with the first trial launches occuring in 1985. The test programme was completed in 1987 with first production round deliveries being made in 1988.

The Japanese Ground Self Defence Force is acquiring three fire units equipped with the Type 88. Each of which consists of six surveillance/targeting Type 73 (4 × 4) light vehicles carrying Type 85/JTPS-P11 radars, a command and control vehicle, four Type 73 (6 × 6) 3500 kg trucks fitted with fire control system shelters, 12 Type 73 data relay light trucks, 16 Type 74 special large trucks used as the missile tractor-erector-launcher (TEL) vehicles and 16 crane equipped reload carrying vehicles.

The first six TELs and associated fire controls were delivered in mid-1988 for use by the Japanese Self Defence Force's Fuji Instruction Brigade at the Fuji School, the Signal School and the Ordnance School.

The TEL has six hydraulically elevated cylindrical container-launcher canisters mounted in two layers of three. The six canisters are reloaded as one unit direct from the reloader vehicle.

Length of the missile is 5.08 m with its jettisonable solid propellant booster motor and 4 m without. The sustainer motor is an MHI TJM-2 microturbojet which gives a cruising speed of Mach 0.9. The body diameter is 0.35 m and the maximum wing span 1.16 m. Launch weight (complete) is 661 kg with the missile itself weighing 500 kg. The HE warhead weight is 225 kg.

The Type 88 will be launched from up to 50 km from the coast at targets up to 100 km out to sea. Once fired it adopts a terrain avoiding flight profile using an inertial guidance package until it clears the coastline. It then descends to a low altitude cruising height until the active radar seeker acquires a designated target. The missile finally adopts a sea-skimming flight profile for the terminal attack phase.

Mitsubishi Type 88 coastal defence launcher in travelling configuration

Status: In production. In service with the Japanese Ground Self Defence Force. (A planned total of 56 launchers is envisaged).

Manufacturer: Mitsubishi Heavy Industries
5-1 Marunouchi 2-chrome
Choyoda-Ku
Tokyo
Japan
Telephone: (03) 212 3111
Telex: 22282/22443

KENYA

Kenyan Coastal Defences

This is believed to comprise a number of truck-mounted OTOMAT coastal defence missiles (qv).

KOREA, NORTH

North Korean Coastal Defence Missile Systems

Development/Description

In 1968-69 North Korea took delivery of sufficient Soviet S-2 Sopka (US/NATO designation SS-C-2b Samlet) coastal defence missiles to equip five launch batteries on the east coast to protect major naval bases and ports.

Irregular deliveries of S-2s continued throughout the early seventies so that further batteries could be raised for both the east and west coast defensive missile belts. By this time the Soviets were supplying the missiles in 'knock-down' kit form and had trained the North Koreans both to assemble and test them.

However, due to political differences the Soviets declined to provide more modern missiles so the more ideological, sympathetic People's Republic of China was approached.

Part of the subsequent aid received was Chinese reverse engineered copies of Soviet missile systems and the transfer of indigenous Chinese missile research and development technology.

In the area of coastal defence missiles this resulted, during 1969-70, in the reorganisation and expansion of the local S-2 maintenance and assembly programme and the acquisition and deployment of the Hai Ying-1 (Chinese for Sea Eagle, US designation CSS-N-2) reverse engineered copy of the Soviet P-15 (US/NATO designation SS-N-2. Styx) anti-ship missile to supplement the S-2.

The initial HY-1 rounds were drawn direct from Chinese Navy stocks with some rounds being provided in 'knocked-down' component form for local assembly. This soon changed to the HY-1 being built in limited numbers at the SY-1 indigenous production facility that was set up in 1972, albeit with many major components still coming from Chinese factories.

In 1974-75 the SY-1/HY-1 production facility began to be converted to build China's Hai Ying-2 (HY-2, NATO codename Silkworm) anti-ship missile. Full production of the HY-2 is believed to have begun in 1976.

Operational HY-1 batteries are believed to have served on to around 1980 by which time they are all thought to have been converted to HY-2 systems.

By the early eighties North Korea was producing and assembling all the HY-2 sub-systems except the sustainer rocket motor and guidance system which still came from China.

Production of the HY-2 has continued through to 1989, but is expected to be phased out in lieu of a more modern longer ranged design such as the Chinese HY-3 or HY-4.

Currently North Korea deploys an estimated six coastal defence missile battalions with a total of 15 or so S-2 and HY-2 launch batteries.

Each battery has around 85 men and is organised with an HQ unit, a fire control element with radar and two or three launch platoons. The latter are further subdivided into two launcher sections each with one launcher. The battery tends to occupy a semi-fixed launch site and has access to reload rounds.

In 1987-88 following an agreement signed in 1986 North Korea began supplying Iran with batches of HY-2 Silkworm missiles.

These, together with several HY-2 batteries supplied direct from China in mid-1986, were used operationally in the northern part of the Gulf on various occasions, against Kuwaiti ships and off-shore oil production facilities, and along the straits of Hormuz during the 18 April 1988 aircraft carrier USS Enterprise and her Carrier Air Wing 11 rearrangement of the Iranian Navy by the US Navy as Operation 'Praying Mantis'.

Status: HY-2 variant-in production. In service with Iran and North Korea.

Manufacturer: North Korean state factories.

KOREA, SOUTH

South Korean Coastal Defences

The South Korean Marines have a number of truck-mounted Harpoon missiles used in the coastal defence role.

LIBYA

Libyan Coastal Defences

It is believed that Libya has a quantity of Soviet supplied SSC-3 Styx coastal defence missiles in service.

MEXICO

Mexican Coastal Defences

Coastal defence is the responsibility of the Mexican Army and includes 75 mm M1902/M1906 guns and 120 mm L/27 guns, but it is believed that these are probably now non-operational.

NORWAY

Norwegian Coast Defence

Coastal defence is the responsibility of the Norwegian Navy and includes 40 coastal defence forts: 25 in southern Norway and 15 in northern Norway (of which 19 are active with the remaining 21 in a care and maintenance state ready for war mobilisation) whose weapons include Bofors 75 mm SK75 turret-mounted guns, 105 mm, 120 mm ERSTA, 127 mm and 150 mm guns in a variety of mountings. In addition there are a number of cable mine and torpedo batteries.

In 1981 it was announced that the Norwegian Coastal Artillery had signed a contract with AB Bofors of Sweden initially worth NOK500 million for an undisclosed number (originally understood to be 12) 120 mm ERSTA guns.

The first unit was delivered in 1985 and test fired during 1987. These guns are replacing older German designed and built weapons that date back to the Second World War.

Each of the 120 mm ERSTA installations costs around NOK600 million so by 1989 the number of guns involved had been reduced to eight, despite an offer from AB Bofors to offset contracts in Norway to the value of the original contract.

Three of the guns are installed at Meloyvaer in Trondheim fjord, a further three will be installed to protect the harbour at Narvik and the remaining two will be used to protect the US Marine Corps depot in the province of Trondelag.

The Norwegian company of NFT has developed and placed in production a 20 mm gun turret for coast artillery, this is now in service with Norway. Details of this turret are given later in this section. There is also an ongoing competition for a new 155 mm coastal defence gun but as of mid-1991 no system had been selected.

Bofors 120 mm ERSTA coastal gun with ordnance elevated

NFT Penguin Coastal Defence Missile System

Development/Description

For a number of years the Norwegian Ministry of Defence has been studying coastal defence missile systems based on the use of a derivative of the Penguin Mk 3 air-to-surface missile. To date no production order has been placed.

The alternative system options available are believed to be:

a) a 6 × 6 truck-mounted self-contained mobile coast defence unit
b) a fixed site coast defence battery or
c) a light helicopter or truck transportable unit.

A battery would typically consist of one six-round or two four-round missile positions with a containerised fire control system and surveillance/target acquisition system. The missiles would be launched using data from either the radar or external sensors such as coast defence radars, maritime reconnaissance aircraft or naval vessels.

The launch site need not be on the coast as the Penguin has the capability to over-fly elevated terrain, thus making use of natural features to reduce the launch detection probability and the discovery of the battery launch site itself.

The missile is 2.96 m long, has a body diameter of 0.28 m and a maximum wing span of 1.42 m. It has a slender cylindrical body with a rounded nose and slight boattail. Flight control is by cruciform swept long chord tapered canard fins on the nose which align with similar shape, but larger, wings aft of the mid-body. Launch weight (complete) is 340 kg with a semi-armour piercing warhead weighing 120 kg.

Propulsion is by a two-stage dual thrust solid propellant booster and sustainer rocket motor with reduced smoke characteristics. Cruise speed is Mach 0.8 with the missile having an effective range to over 30 000 m.

Guidance to the predicted target position is by pre-programmed autopilot on a direct attack route or via any arbitrary way point. The latter allows an attack to be conducted around an island, to come from a misleading direction or for several missiles to attack simultaneously from different directions to confuse the defences.

Flight altitude of the weapon can be a designated altitude from around 60-100 m down to sea-skimming and is maintained by a radio altimeter.

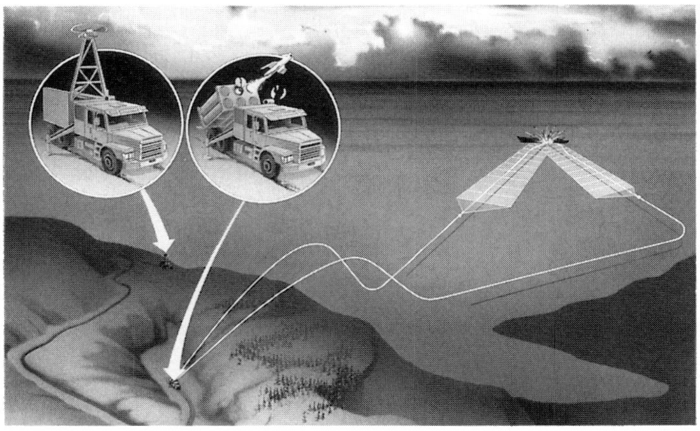

Artist's impression of the NFT Penguin coastal defence missile system in operation

When in range of the target the sea-skimming altitude is automatically adopted for the terminal attack phase and the passive infra-red homing NFT developed seeker is activated to conduct its pre-established search pattern. The seeker has the capability to distinguish between authentic targets and decoys.

Status: Concept studies.

Manufacturer: Norsk Forsvarsteknologi AS (NFT)
PO Box 1003
N-3601 Kongsberg
Norway
Telephone: +47 3 73 82 00
Telex: 71491 Vaapnn
Fax: +47 3 73 93 30

NFT 20 mm and 25 mm Coastal Defence Stationary Gun Mounts

Development/Description

The NFT-developed turrets used for these applications are identical except that one uses a 20 mm RH 202 automatic cannon and the other a 25 mm Mauser MK 25 Model E automatic cannon.

Installed over a single storey underground installation the turret can be used as a self-contained fort unit which can operate for a pre-determined period without being resupplied or be used as a transportable fort-unit for semi-permanent installation at prepared sites.

The turret armament can be used against ground, sea or air targets. In the latter role the 20 mm version is used by the Norwegian coastal artillery force to provide an anti-aircraft defence of its coastal defence installations.

The turret is armoured but can accept additional armouring and camouflage coatings as required. It can also be fitted with ventilation and air-conditioning systems for use in extreme environmental and NBC conditions. Optical sights for both ground and air targets are fitted.

NFT 20 mm coastal defence stationary gun mount

Artist's impression of NFT 20 mm coastal defence stationary gun mount and underground accommodation

SPECIFICATIONS (turret)

CREW	one (turret gunner), other team members are within the underground installation
WEIGHT	approx 1000 kg
TURRET BEARING (outer diameter)	1200 mm
ARMAMENT	1 × 20 mm RH 202 automatic cannon or 1 × 25 mm Mauser MK 25 Model E automatic cannon
RATE OF FIRE	900 rds/min
AMMUNITION	as required in installation
MUZZLE VELOCITY	20 mm Rh 202 1050 m/s; 25 mm MK 25 1100 m/s
OPTICS	
ground/sea target	
magnification	× 8
field-of-view	150 mins
air target	
optical system	
(a combined day and night sight is an option)	
ELEVATION/DEPRESSION	−10°/+50°
TRAVERSE	360°
DRIVE	electrical-servo with variable traversing speeds from a few mrads/s to a max 90°/s or manual

Status: 20 mm turret - production as required. In service with Norwegian Coastal Artillery Force (order placed January 1986 by Norwegian Naval Material Command).

25 mm turret - ready for production.

Manufacturer: Norsk Forsvarsteknologi AS (NFT)
Manufacturing Division
PO Box 1003
N-3601 Kongsberg
Norway
Telephone: +47 3 73 82 50
Telex: 71491 vaapn n
Fax: +47 3 73 85 86

PARAGUAY

Paraguayan Coastal Defences

These consisted of 76 mm M1911 guns and 152 mm (6-inch) Vickers Mk 5 guns, but they are now believed to be no longer operational.

POLAND

Polish Coastal Defences

In Poland coastal defence is the responsbility of the Polish Navy and includes six artillery batteries with 130 mm SM-4-1 and/or 152 mm gun-howitzer M1937 (ML-20) and three battalions with a total of 54 SSC-2b Samlet coastal defence missile launchers.

PORTUGAL

Portuguese Coastal Defences

Coastal defence is the responsibility of the Portuguese Army and includes a coastal artillery regiment among whose equipment includes a total of 27 coastal artillery weapons with calibres of 150 mm, 152 mm and 234 mm. The latter being British supplied 9.2-inch coastal guns. A number of British supplied twin six-pounder coastal guns are also in service. There are coastal artillery commands in the Azores and Madiera.

QATAR

Coastal Defence

The Navy mans four batteries each with three Aerospatiale MM40 Exocet coastal defence missile systems based on a Berliet 6 × 6 chassis. Each of these has four missiles.

ROMANIA

Romanian Coastal Defences

Coastal defence is the responsibility of the Romanian Navy and includes some 10 coastal artillery batteries with up to 100 guns of 130 mm SM-4-1, 150 mm and 152 mm types.

SAUDI ARABIA

Saudia Arabian Coastal Defences

Saudia Arabia has a number of OTOMAT surface-to-surface missile batteries for the coastal defence role.

SPAIN

Spanish Coastal Defences

In Spain, coastal defence is the responsibility of the Spanish Army Coast Artillery Command which consists of six mixed artillery regiments and a coast artillery group. The following weapons are known to be in service:
381 mm (15-inch) L/45 Vickers Model 1926, recent reports have indicated that only seven of these are now operational.
305 mm (12-inch) L/50 Vickers, originally a naval model, some may have come from obsolete warships but some were manufactured especially for coastal defence by Fabrica de San Carlos. It is believed that 14 of these weapons are still operational.
203 mm, 24 of these are still believed to be operational.
152 mm (6-inch) L/50 Vickers Model 1923 (Model W) in both single and twin mounts with a maximum range of 21 600 m with as many as 113 still operational.
Harpoon SSM - It is believed that Spain has recently taken delivery of a number of McDonnell Douglas Harpoon surface-to-surface missiles in the coastal defence role.

Many of the coastal artillery weapons defend key Spanish ports and the Straits of Gibraltar. There is also a coastal artillery group stationed in Cueta and Melilla, the Balearic Islands have one coastal artillery regiment while the Canary Islands have two coastal artillery regiments.

SANTA BARBARA has proposed that the ordnance of their new towed SB 155/39 could be used to replace the existing ordnance of the licenced manufactured Vickers-Armstrong 152 mm L/50 coastal guns. These weapons have been modernised over the years but conversion to 155 mm would enable them to fire standard NATO projectiles to ranges of between 22 500 and 24 000 m with enhanced range projectiles being fired to a range of 30 000 m. This conversion would allow the retention of the existing recoil system, cradle and other components.

The heavy guns are mounted singularly in armoured gun installations with the fire control duties being handled by fixed Inisel license built Swedish 9KA410 systems using a 9KR400 I/J-band radar. The guns are divided into groups of four installations (with two batteries per group).

SPECIFICATIONS

Model	**305 mm (12 in) Vickers L/50 gun**	**381 mm (15 in) Vickers L/45 Mk B gun**
MUZZLE VELOCITY	1015.5 m/s	762 m/s
WEIGHT OF SHELLS	AP 386.4 kg	AP 885 kg HE 801.8 kg
WEIGHT OF CHARGE	139.5 kg	200 kg
RANGE	19 835 m at 15° elevation	35 100 m at 40° elevation
BARREL LENGTH (including breech mechanism)	15.69 m	17.671 m
WEIGHT (including breech mechanism)	67 913-68 931 kg	88 300 kg

SWEDEN

Swedish Coastal Defence

The coastal defence of Sweden is the responsibility of the Royal Swedish Coastal Artillery which comes under naval control. This is organised into five brigades:
KA1 Vaxholms Kustartilleriregemente based at Vaxholm
KA2 Karlskronas Kustartilleriregemente based at Karlskrona
KA3 Gotlands Kustartilleriregemente based at Förosund
KA4 Älvsborns Kustartilleriregemente based at Goteborg
KA5 Härnösands Kustartilleriregemente based at Härnösand.

These brigades control 12 mobile and 53 static batteries and missile installations, infantry battalions, coastal patrol craft, minelayers and landing craft.

In recent years the Royal Swedish Coastal Artillery has been receiving many new weapons including the Bofors 120 mm ERSTA fixed system, Bofors 120 mm KARIN/CD 80 mobile gun and two missile systems, the Saab RBS 15KA and the manportable RBS 17 Hellfire. Sweden has ordered a total of 90 launcher sets and 700 RBS 17 Hellfire missiles for delivery by 1990.

In addition to the new weapons, new command and control systems as well as fire control systems have been ordered and are now fully operational.

The following weapons are now operational with the Royal Swedish Coastal Artillery:
Rb 08A coastal defence missile (to be phased out of service)
RBS 15KA coastal defence missile (operational from 1988)
RBS 17 Hellfire coastal defence missile (operational from 1988)
Rb 52 coastal defence missile (being phased out of service)
75 mm turret-mounted Bofors L/60 gun
75 mm mobile Bofors gun
105 mm gun
120 mm Bofors ERSTA turret-mounted gun
120 mm Bofors KARIN/CD 80 mobile coastal gun
152 mm static gun.

Swedish Army static defences

Sweden has installed a number of redundant tank turrets that are used to provide static defence points at key installations such as airports. The turrets are either permanently emplaced in concrete or carried on to trailers to prepared sites where they are then mounted on pre-prepared concrete bases. Tank turrets known to be in use include Centurion turrets armed with 20-pounder (83.4 mm guns) and Strv.74 light tank turrets armed with 75 mm guns.

Bofors 120 mm Mobile Coast Defence Gun KARIN/CD 80

Development

During the early 1970s AB Bofors was asked to develop a new mobile coast defence gun and as the FH-77A was under development at the same time it was decided to develop the two weapons using the same split trail carriage. Thus the FH-77A (see separate entry in this section) and KARIN both use the same basic carriage although KARIN uses a different ammunition handling system, an improved laying system which is fully remote-controlled and enables moving naval targets to be engaged with high precision. The first complete KARIN was delivered to the Royal Swedish Coast Artillery in 1978 (to Coast Artillery Regiment (KA2) at Karlskrona). In 1981 a revised model to operate with radar fire-control was delivered and during 1985 the first complete training using the production version was carried out. KARIN is now often referred to as CD 80 as it is employed in special battalions known as KA-bataljon m/80. These special mobile battalions are used to cover open stretches of coastline or reinforce fixed defences under attack. Each battalion has a battalion commander and his staff, a fire control and communications battery and two coastal artillery batteries. Each of the two coastal artillery batteries has four KARIN guns under the control of the battery commander and his headquarters platoon. The rest of the battery is made up of an ammunition supply and transport platoon, a fire control platoon, a self-defence infantry and traffic control platoon and an air defence platoon equipped with two 40 mm L/70 anti-aircraft guns. A support platoon has repair, catering and emergency medical facilities.

120 mm Mobile Coast Defence Gun KARIN/CD 80 in travelling configuration

Description

KARIN uses a 120 mm gun barrel that is 55 calibres long and can fire at a rate of up to 15 rds/min. The ammunition feed system, sometimes referred to as a conveyor, is on the right trail leg of the carriage. The aiming system is known as MARIA and is operated by one man seated to the left of the gun. The system receives digital fire-control data via wire or radio from a fire control station housed in a container-type shelter. Supplied by Philips Elektronikindustrier AB (PEAB) this system combines a search and fire control radar, computers, indicators and an optronic director. On the gun the layer has a panoramic sight, a gun computer, a layer's panel and a computer panel. As most of the fire control data is fed to the gun automatically all the layer normally has to do is make slight corrections and monitor the process.

For long dinstance moves KARIN is towed by a 6 × 6 tractor at speeds of up to 70 km/h. The auxiliary engine is used for local deployments at speeds of up to 7 km/h and the engine is also used to provide hydraulic power to move the trail legs, operate the ammunition feed system and lay the barrel.

KARIN fires the same projectiles as ERSTA, a special naval target projectile with a delay fuze to ensure that the charge detonates only after the projectile has passed through the target armour, and an HE. Both are fired at a maximum muzzle velocity of 830 m/s. Terminally guided projectiles are under development. When firing an enhanced range projectile known as HCER at a muzzle velocity of 900 m/s a range of more than 30 000 m is achieved; the normal range of KARIN using the naval target ammunition is 21 000 m. A complete round weighs 43 kg and the projectile 24 kg. An HE projectile is also available.

120 mm Mobile Coast Defence Gun KARIN/CD 80 in firing position showing power rammer on right

120 mm Mobile Coast Defence Gun KARIN/CD 80 deployed in firing position

SPECIFICATIONS

CALIBRE	120 mm
BARREL LENGTH	6.6 m
MUZZLE BRAKE	multi-baffle
BREECH MECHANISM	vertical sliding block
CARRIAGE	split trail with APU
SHIELD	no
WEIGHT	12 500 kg
LENGTH	
travelling	13.1 m
firing	12.15 m
WIDTH	
travelling	2.65 m
firing	9.05 m
HEIGHT (travelling)	2.96 m
AXIS OF BORE	1.6 m
GROUND CLEARANCE	0.42 m
TRACK	
driving wheels	2.2 m
castor wheels	2.2 m
FORDING	0.75 m
GRADIENT	50%
ELEVATION/DEPRESSION	+50°/−3°
TRAVERSE (total)	60°
RATE OF FIRE	up to 15 rpm
MAX RANGE	30 000 m
TOWING VEHICLE	6 × 6

Status: Deliveries to the Royal Swedish Coastal Artillery are complete.

Manufacturer: Swedish Ordnance (previously AB Bofors)
S-691 80 Bofors
Sweden
Telephone: (46) 586 810 00
Telex: 73210 bofors s
Fax: (46) 586 581 45

Bofors 120 mm static Coast Defence Gun ERSTA

Development

The 120 mm static coast defence gun ERSTA was developed by AB Bofors in the early 1970s to meet the Swedish defence force long term requirements for high firing endurance, high precision firing at long range, high protection levels for the gun and crew and high level effects on target, all relating to coastal defence. ERSTA is thus an automatic turret gun with an automatic ammunition transport and loading system.

The installation of an ERSTA gun is a considerable civil engineering project and is undertaken by the Swedish ABV concern. Installation involves blasting positions in solid rock and the three storied underground facilities include, quite apart from those of the gun and its associated mechanisms, a fire-control centre, kitchens, crew accommodation, washrooms, power plant (with two diesel generators - one acting as a spare - and several months fuel supply for continuous running), ventilation compartment and desalination equipment to provide fresh water for drinking and other purposes. The fire control centre associated with ERSTA has armoured observation cupolas and armoured covers for folding radar antennas.

Description

The gun used with ERSTA is 62 calibres long and capable of rates of fire of up to 25 rds/min. This rate of fire can be extended to the point where 200 rounds in 20 minutes can be fired. The gun barrel is water-cooled using a new system. The gun is protected by an armoured cupola which stands approximately 1.8 m above ground level and gives protection against both nuclear and conventional weapons. Extra protection is provided by lowering the cupola and the entire traversing portion of the gun by hydraulic power. This takes the cupola and the traverse portion off its normal spring-loader roller bearing and locks them onto the foundation. The barrel can be lowered into a concrete fork and the muzzle can be automatically sealed by a cover. Further protection is provided by special camouflage and possible the use of dummy guns and installations nearby.

The cupola, which is covered in a neutron-absorbing material, is emplaced on a long underground 'tube' which contains the ammunition transport system. The entire installation is set in concrete with the bulk of the installation underground, including a chamber for the ammunition magazine and handling system. Inside this magazine, which is located to one side of the tube and under about five metres of overhead protection, the fixed rounds are stored in racks that can be moved by overhead handling equipment. In use these racks are moved from their storage positions around the magazine walls to the start of the transport system. The rounds are placed horizontally onto a handling table from where they are moved into a cross-shaped device that turns to point the round nose-upwards for transport up a tube that runs through the centre of the cupola well. Half-way to the gun the round enters a chamber where it is loaded onto a turntable device ready for further lifting to a position under the gun where it is turned to the appropriate angle of elevation for automatic loading. After firing the spent case is ejected to a chamber at the bottom of the well. Under this chamber is a water reservoir for the barrel cooling system.

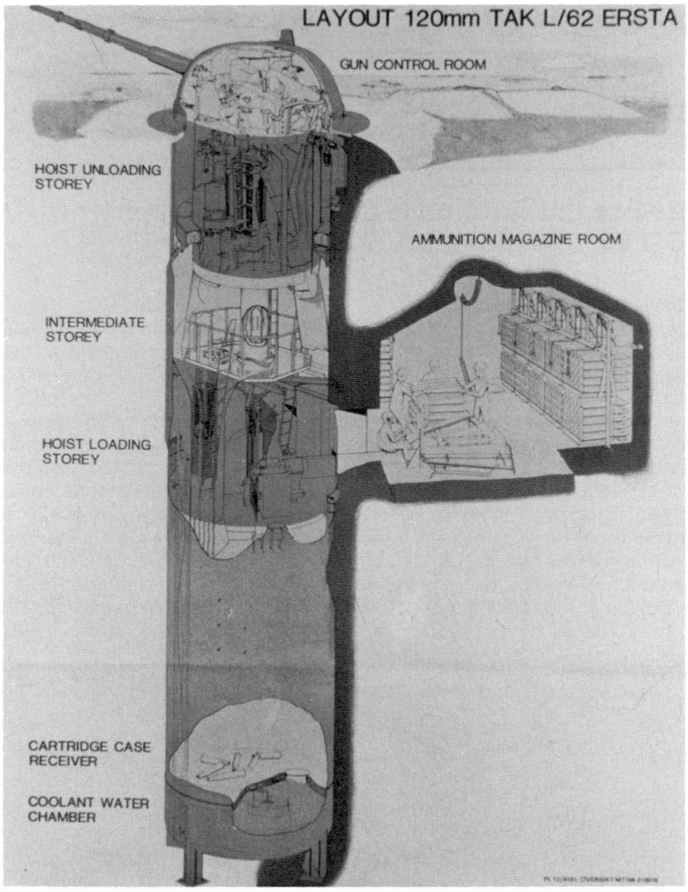

Layout of ERSTA 120 mm installation

ERSTA has a crew of eleven although for short periods the entire system can be operated by only three. Of these two are in the cupola operating zero-setting instruments to lay the gun using data provided by a modern fire-control system using a PEAB 9 KM 400 (KONDOR) surveillance radar.

Camouflaged ERSTA 120 mm coast defence gun

This has the same protection level as the gun and can be rapidly covered by armour plates. Other sensors used include a TV tracker and laser rangefinder.

The ammunition fired by ERSTA uses similar projectiles to those fired by the KARIN/CD 80 but allied to bigger cartridge cases to contain the larger propellant charges. The standard projectile is known as a naval target shell which is fuzed to burst only when it has penetrated a target and a standard HE projectile. An improved projectile using a base bleed unit and known as HCER (high capacity extended range - formerly known as PROBOS) is under development along with other range-enhancing projectiles. The HCER projectile weighs 26 kg and has a maximum range of 36 000 m. Details of the existing projectiles are as follows:

Type	Naval target	HE
WEIGHT OF SHELL	24.5 kg	24.6 kg
WEIGHT OF CHARGE	2.6 kg	3.9 kg
MUZZLE VELOCITY	880 m/s	880 m/s
MAX RANGE	27 000 m	27 000 m

Status: Production. In service with Norway (8) and Sweden.

Manufacturer: Swedish Ordnance (previously AB Bofors)
S-691 80 Bofors
Sweden
Telephone: (46) 586 810 00
Telex: 73210 bofors s
Fax: (46) 586 581 45

SPECIFICATIONS

CALIBRE	120 mm
LENGTH OF GUN	7.44 m
CREW	11
ELEVATION/DEPRESSION	+50°/-3°
TRAVERSE	±270°
LAYING SPEED	
TRAVERSE	10°/s slewing 0-3°/s tracking
ELEVATION	5°/s slewing manual tracking
RATE OF FIRE	25 rpm (200 rounds in 20 mins)
TOTAL WEIGHT	70.5 tons plus 103.5 tons concrete
CUPOLA ARMOUR THICKNESS	70 mm
EMPLACEMENT DIAMETER	4.04 m
POWER REQUIREMENTS	
AC	3 × 380 V, 50 Hz, 6 kW
DC	24V, 0.1 kW

Bofors 75 mm TK75 Coastal Defence Gun System

Description

The Bofors 75 mm TK75 semi-automatic turret gun was developed during the 1960s by AB Bofors to meet the requirements of the Royal Swedish Coastal Artillery. The gun is 60 calibres long and is mounted in a dome-like turret set into concrete with an underground ammunition department. The turret armour is 70 mm thick and provides protection against nuclear attack and is usually camouflaged. Ammunition is fed to the gun by an electrical hoist system that can have four or six steps according to the installation.

The gun has a crew of five to seven: traverse and elevation layers, a loader (in the turret) and two ammunition handlers (in the ammunition compartment). The ammunition handlers place the one-piece ammunition onto a five-compartment loading table from which the rounds are moved under power to a hoist drum. Using a worm and bevel gear drive the round is fed up the hoist in steps under microswitch control to a pendulum where it is moved to a loading tube. The loading tube is supported on the pendulum which moves around the elevation trunnion centre. The raising of the pendulum and loading tube with the round is effected by a torsion spring placed in the trunnion centre. The torsion spring is cocked by the recoil. The loading tube delivers the round to a spring-operated rammer for loading into the breech which uses a vertical sliding block mechanism. The gun is fired by the traverse layer using a foot pedal. After firing the spent case is ejected into a separate compartment. Using this semi-automatic ammunition hoist system the gun can fire at a rate of up to 25 rds/min. The hoist system uses three-phase 380 V AC supplied and a mains-derived 24 V DC supply. In the event of an electrical failure the hoist system can be operated manually.

The gun is hand-layed in traverse and elevation by two hand cranks in the turret. For indirect laying follow-the-pointer dials are used. For direct aiming a periscopic sight with dual eyepieces, one for each layer, is used. The fire control unit normally consists of one radar and 1-2 optical/laser rangefinders which have the same level of protection as the gun installation.

The gun fires four types of ammunition. Details are as follows:

Ammunition type	Naval Target Shell	Naval Target Shell ER	HE Shell	Practice Shell
WEIGHT OF PROJECTILE (with fuze)	5.73 kg	6.35 kg	5.52 kg	6.55 kg
WEIGHT OF CHARGE	0.62 kg	0.81 kg	0.73 kg	n/app
WEIGHT OF COMPLETE ROUND	11.5 kg	approx 12 kg	11.5 kg	12.2 kg
MUZZLE VELOCITY	850 m/s	830 m/s	870 m/s	850 m/s
TIME OF FLIGHT				
to 3000 m	4.6 s	4.3 s	4.1 s	4.3 s
to 6000 m	12.6 s	10.4 s	11.2 s	11.1 s
MAX RANGE (20° elevation)	11 400 m	13 300 m	12 000 m	12 200 m

Status: Production as required (over 100 built to date 1 July 1990). In service with Norway, Sweden and several other undisclosed countries.

Manufacturer: Swedish Ordnance (previously AB Bofors)
S-691 80 Bofors
Sweden
Telephone: (46) 586 810 00
Telex: 73210 bofors s
Fax: (46) 586 581 45

Bofors 75 mm TK 75 coast defence gun prior to the adding of camouflage

Bofors 75 mm TK 75 coast defence gun

SPECIFICATIONS

CALIBRE	75 mm
BARREL LENGTH	4.245 m
BREECH MECHANISM	vertical sliding block
SHIELD	70 mm thick turret
CUPOLA DIAMETER	2.3 m
HEIGHT OF CUPOLA (above ground level)	0.9 m
TOTAL WEIGHT	17.5 tons
WEIGHT	
4-step hoist	16 750 kg
6-step hoist	17 500 kg
ELEVATION/DEPRESSION	+20°/-5°
TRAVERSE	±270°
RATE OF FIRE	
electrical hoist	25 rpm
manual operation	15 rpm
MAX RANGE	11 400 to 13 300 m (depending upon ammunition type)
CREW	5-7

Bofors 75 mm Mobile Coastal Gun

Development/Description
This 75 mm mobile coastal gun has been developed by Bofors from an anti-aircraft gun to meet the requirements of the Swedish Navy which is responsible for coastal defence.

Status: Production complete. In service with Sweden.

Manufacturer: Swedish Ordnance (previously AB Bofors)
S-691 80 Bofors
Sweden
Telephone: (46) 586 810 00
Telex: 73210 bofors s
Fax: (46) 586 581 45

Bofors 75 mm mobile coastal gun in firing position

SAAB Rb 08A Coastal Defence Missile

Development/Description
Development of the Rb 08A began as a Royal Swedish Navy study in 1959. It was taken over as a project by the Swedish company Saab-Scania (now SAAB Missile AB) and the French company Nord-Aviation (later Aerospatiale) using the Nord CT20 RPV as its basis. Initial production began in 1966 for shipboard use on the destroyers Halland and Småland. A coastal defence version was then fielded with production of the 98 total (all versions) combat rounds finishing in 1970.

In comparison to the CT20 the Rb 08A is longer, heavier and has folding wings. The launch and acceleration phase is carried out by two jettisonable solid propellant rocket booster motors carried in a cradle attached to the missile body beneath the wings. The 400 kg thrust liquid fuel Turbomeca Marbore IID turbojet then cuts it for the cruise phase at Mach 0.7-0.8.

Guidance is by launcher bearing, autopilot and height lock followed by activation of the active radar seeker near to the target's projected position. This enables the missile's 200 km maximum range to be fully exploited.

Missile length is 5.72 m, body diameter 0.671 m and maximum wing span 3.018 m. Launch weight (complete) is 1227 kg including 315 kg for the jettisonable booster unit and 250 kg for the blast-fragmentation warhead.

Swedish Rb 08A coastal defence missile being launched

Status: Production completed 1970. In service in declining numbers with the Swedish Coast Defence Force, being replaced by RBS 15KA missile.

Manufacturer: SAAB Missiles AB
S-58188 Linkoping
Sweden
Telephone: 46 13 2860
Telex: 500 01 smablg s

SAAB RBS 15K Anti-ship Missile System

Development/Description
The first development contract for the coastal defence version of the RBS 15 anti-ship missile, designated the RBS 15K for the Swedish Coastal Artillery Force was signed in 1984.

Users may have their own system configuration. The basic tactical unit for Swedish use is the missile battery. This consists of two missile platoons each with one missile fire control section (with one fire control vehicle-MFCV-carrying a mission fire control panel, alphanumeric keyboard, a target data panel and a communication panel in a shelter) and two launch sections (with one launch vehicle carrying four container-launchers weighing 1530 kg each), one service maintenance platoon and a Guard platoon for local area security.

Target data is provided either from external sensors such as coastal defence radars, naval vessels and aircraft or, optionally, from a radar integrated into the launch platoon or battery.

Each launch vehicle can be deployed up to 1000 m from the MFCV. Target information is supplied over a data link. Launch orders for a pre-programmed target can be issued within a few seconds. The vehicles will normally only spend several minutes at a launch site and can be reloaded in the field very quickly.

Two or three of the batteries form a battalion which has its own target sensor unit, maintenance unit and command centre.

The coastal defence missile is identical to the ship-launched version and is 4.35 m long, 0.50 m in diameter and has a maximum unfolded wingspan of 1.40 m (folded span is 0.85 m). Launch weight (complete) is 770 kg with the two 85 kg weight jettisonable three second burn solid fuel booster motors. The sustainer motor used is a 370 kg thrust liquid fuel Microturbo TRI 60-1-077 turbojet which provides a high sub-sonic cruising speed. The missile itself weighs 600 kg with a 200 kg plus HE blast-fragmentation Swedish Ordnance warhead.

Initial guidance is by pre-programmed autopilot with a FM-CW radar altimeter for height control. Once the missile has flown a certain distance, defined by the distance actually flown and the projected target location, the Bofors Electronics AB J-band broad frequency-agile ECM resistant radar seeker is activated in its search mode. The search pattern set depends upon the target data available. The final lock-on mode can also be pre-set either as an active or active/passive operation.

Finnish RBS 15K missile being launched from a Sisu truck chassis

Flight control is by cruciform wings with ailerons for combined pitch and roll plus canard fins at the nose of the weapon for yaw.

Maximum effective range is 70+ km with a variable flight trajectory. Near to the target it adopts a sea skimming terminal flight trajectory. Maximum range is 150 km.

Status: Production. The RBS 15K is operational with the Swedish and Finnish Coastal Artillery forces.

Manufacturer: SAAB Missiles AB
S-581 88 Linkoping
Sweden
Telephone: 46 13 2860
Telex: 500 01 smablg s

RBS 17 Hellfire Shore Defence Missile System

Development/Description

In late 1984 Rockwell International was awarded an $8 million contract by the Swedish Defence Ministry to adapt its AGM-114 Hellfire anti-tank missile to the coastal defence role to replace the obsolete SS-11 (Rb 52) wire-guided weapons being used by the Swedish Coastal Defence Force.

Following the two year development programme a $65 million series production contract was awarded in 1987 to Rockwell for a total of 90 or so sets of two single rail launcher systems and 700 missiles. First deliveries were made in 1989 with final deliveries in 1990. Final assembly of the Hellfire missile is conducted by Bofors in Sweden under the direction of Rockwell Tactical Systems Division. A total of 25 battalions are due to form on the RBS 17 missile over the period 1989-95.

The launcher system consists of the modified Hellfire missile, launch control equipment, launcher, a Laser Designator unit and missile container.

Each RBS 17 set for a site has two single-rail launchers, a launch control box, two cable reels, two safety and arming boxes and four power supply batteries.

The launcher is in two parts, the base and three folding legs weighing 21.8 kg and the trunnion and launch rail weighing 26.2 kg. The total weight is 48.0 kg with dimensions of 1.9 m long, 2.15 m wide and 1.05 m high. It takes less than five minutes to assemble on terrain with a 10° slope. The assembly is stabilised with stakes or weights at the foot pads. Alignment and levelling is accomplished by levels on the launcher itself.

Once set up a two man crew can load the launcher in less than 60 seconds. Launching is possible at elevations up to +10° for direct fire or +20° for indirect fire, with a traverse capability of 60° without disturbing the tactical set-up.

The missile itself has a long cylindrical body with a blunted rounded ogive and short limited span fixed cruciform fins forward. Aft are the long short-span cruciform tail control surfaces.

Propulsion is by a single-stage solid propellant smokeless rocket motor that gives the missile a speed of around Mach 1.17. The effective range limits are from 300 to 10 000 m.

Missile length is 1.625 m, body diameter 0.1778 m and maximum wing span 0.362 m. Launch weight is 48.25 kg with a 9 kg AB Bofors HE blast fragmentation warhead and electro-mechnical fuze with pyrotechnic delay action.

Guidance is by a semi-active terminal homing seeker that tracks the reflected laser energy from the target being illuminated by the laser designator. The CEP value is said to be less than 1.5 m at extreme range. The RBS-17 can operate with a variety of laser designators and night sights. Typical characteristics of such systems are: designator - weight 11 kg, length 0.445 m, width 0.270 m, height 0.205 m; night sight - weight 7.3 kg, length 0.489 m, width 0.235 m, height 0.205 m.

The missile can be fired in one of two modes: Lock-on Before Launch (LOBL) for the +10° launch elevation or Lock-on After Launch (LOAL) for the +20° angle. The trajectory can be direct, low or high.

A full range of training equipment is available including a special training missile (STM), dummy missile and both classroom and field configuration training aids.

Status: Production. In service with the Swedish Coastal Artillery Force.

Manufacturer: Rockwell International
Tactical Systems Division
Defence Electronics Operations
1800 Satellite Boulevard
Duluth
Georgia 30136
USA

RBS 17 Hellfire shore defence missile with launcher

RBS 17 Hellfire shore defence missile being launched during trials

SWITZERLAND

Fortress Artillery

The Swiss Army includes in its war establishment three fortress and three redoubt brigades and 20 fortress guard companies. These are equipped with a wide range of artillery weapons and mortars. In 1983 and 1986 the Swiss Ministry of Defence announced that additional 120 mm Model 59/83 fortification mortars were to be purchased, the prime contractor being the Federal Weapons Factory at Berne. The ammunition for these 120 mm fortification mortars is provided by the Federal Ammunition Factory at Altdorf. The first 120 mm mortars were installed in the early 1960s and types of bombs fired include high explosive (Type 61), smoke and incendiary (Type 61) and illumination starshell (Type 76).

Further fortress armament has recently been introduced by using redundant 155 mm M126 howitzers and mountings produced when the Swiss Army's M109 self-propelled howitzers were converted to M109A2 standard. These short-barrelled howitzers have been installed in concealed concrete barbettes each with three compartments, one for the howitzer, the second for ammunition and the third for personnel. As far as can be determined the original M127 mounting elevation and traverse limits have been retained, allowing the howitzer to fire standard M107 HE 155 mm projectiles to a maximum range of 14 600 m.

12 cm Festungsminenwefer 59/83 (120 mm fortress mortar Model 59/83) dual mounting inside armoured cupola (Swiss Army)

155 mm Fortress Gun Bison

Development/Description

The Swiss Federal Armament Works (SFAW) has developed a new generation 155 mm fortress gun system called the Bison.

This has been designed to have a range of more than 30 000 m and at the same time, for cost and logistical reasons, be able to use the existing family of ammunition used with the 155 mm M109 series of self-propelled howitzer used by the Swiss Army in considerable numbers (qv *Self-propelled guns and howitzers* section).

The new 155 mm/52 calibre gun has a chromium plated barrel to reduce barrel erosion.

The prototype of the Bison carried out its first ballistic firing in 1987 and volume production is expected to commence in 1994.

The Bison will be fitted with a flick rammer loading system which is of a modular design and is suitable for both automatic and semi-automatic loading on fortress guns, field howitzers or self-propelled artillery systems.

Pending the availability of adequate modular charges, the Swiss Federal Construction Works is proposing the semi-automatic version for the 155 mm fortress gun Bison, for example projectile ramming but with the charges being loaded manually. It also has fully automated primer charge feeding.

With the installation of a flick rammer, the rate of fire can be significantly increased and five rounds can be fired within 25 seconds in the burst fire mode.

SPECIFICATIONS

CALIBRE	155 mm
BARREL LENGTH	8.1 m
MAX RANGE	30 000 m

Status: Prototype system. Not yet in production or service.

Manufacturer: Swiss Federal Armament Works
CH-3602 Thun
Allmendstrasse 86
Switzerland
Telephone: 033 28 11 11
Telex: 921 256
Fax: 033 28 20 47

Prototype of the SFAW 155 mm Bison fortress gun system on a trials mount

Close-up of the SFAW 155 mm Bison fortress gun system from rear with flick rammer loading system with charge on left and projectile on right

SYRIA

Syrian Coastal Defence

Coastal defence is the responsibility of the Syrian Army and is believed to include 130 mm SM-4-1 guns, one brigade of SSC-1b Sepal coastal defence missiles and one brigade with 18 launchers of SSC-3 Styx coastal defence missiles.

TAIWAN

Taiwanese Coastal Defences

The only known Taiwanese coastal defences are a number of trailer-mounted Hsiung Feng I (Proud Wind 1) surface-to-surface missiles which are manned by the army, although it is likely that a wide range of field artillery is available for use in the coastal defence role, especially on the islands of Matsu and Quemoy. Under development is the new Hsiung Feng II (Proud Wind II) surface-to-surface missile.

Hsiung Feng I Coastal Defence Missile

Development/Description
The Hsiung Feng I (Proud Wind, formerly Male Bee I) is a copy of the Israeli Gabriel MkII anti-ship missile fitted with a semi-active radar seeker head developed by Sun Yat Sen Institute.

The missile is part of the Honeywell H930 Mod 3 mobile coast defence system which entered service in the early eighties and is carried on semi-trailers. The system includes the Hsiung Feng triple reinforced glass-fibre container-launcher unit, a locally produced CS/SPG-24 fire control radar (similar in many respects to the American RCA R-76 40 km range I-band monopulse naval fire control radar system) for target tracking and illumination, a locally developed CS/UPS-60(X)C I/J-band surface search and target indication radar and a Command and Control centre (with three operators using two minicomputers and 11 microprocessors).

The missile itself has a cylindrical body with a pointed nose. Flight control is by cruciform tail fins which are in-line with larger rectangular midbody wings. A three second burn solid propellant booster and 200 second burn sustainer rocket motor is fitted. The semi-armour piercing warhead weighs 180 kg and contains some 75 kg of conventional HE. It is detonated by a delay action contact fuze.

Before launch the missile's guidance system is programmed with the target data obtained from the unit's own radars. The missile is then fired and under control of its twin gyro system assumes an initial cruise height of about 100 m. At a range of 7500 m from the launcher the onboard autopilot commands the missile to descend to 20 m altitude using a radar altimeter to maintain the height. At a predetermined distance of between 1000 to 2000 m from the target the seeker is switched on, the target is acquired and locked-on and the missile descends to one of its three possible pre-set attack altitudes for the final run in. The actual set altitude varies between 1 and 3 m and depends upon the sea state encountered at the time.

If the primary semi-active terminal homing capability is being affected by countermeasures then either the backup manual/Honeywell optical tracking (with a locally produced passive infra-red night vision device) guidance or the command guidance with radar tracking option can be selected.

*Hsiung Feng I coastal defence missile system during a parade in Taiwan with launcher covers open and radar vehicle in right background (*L J Lamb*)*

Effective missile range is about 44 000 m with a cruise speed of Mach 0.65. Missile length is 3.42 m, body diameter 0.34 m and maximum wingspan 1.34 m. Launch weight (complete) is 522 kg.

Status: In production. In service with Taiwan. (Five Honeywell H930 Mod3 systems were delivered from 1980 onwards).

Manufacturer: Taiwanese state factories.

Hsiung Feng II Coastal Defence Missile

Development/Description
The Hsiung Feng II (Proud Wind II) anti-ship missile was revealed in March 1988 and is a longer range and faster missile than the Hsiung Feng I. Few details are available at present other than it is in the Harpoon category. It is being produced in a coastal defence version.

Propulsion is by a Turbomeca Arbizon IV turbojet sustainer with the launch and acceleration phases being carried out by a jettisonable solid propellant booster rocket motor. Length is 4.6 m, body diameter 0.34 m and launch weight 685 kg.

Cruise speed is Mach 0.9 with the maximum engagement range being in excess of 160 000 m. The practical guidance range using the inertial reference system with its mid-course correction capability and the terminal attack phase active J-band radar seeker with its passive infra-red Automatic Target Recognition Image Tracking Module Mk 2 (mounted in a fairing above the nose-mounted seeker radome) is around 92 000 m.

Status: Entering production phase. On order for Taiwanese Armed Forces.

Manufacturer: Taiwanese state factories.

TURKEY

Turkish Coastal Artillery

The Turkish Army order of battle includes five Fortress Defence Regiments (Coastal) whose equipment includes 9.2-inch (234 mm) coastal guns, some of which were supplied by Canada in the 1950s with the US paying for their installation.

UNION OF SOVIET SOCIALIST REPUBLICS

USSR Coastal Defences

The Coastal Artillery and Rocket Troops number up to 8000 officers and men and come under the Soviet Navy. They are organised into a coastal artillery division with detachments to protect key Soviet Naval bases. Weapons included 130 mm SM-4-1 towed coastal guns, 152 mm and 305 mm turret-mounted guns. Missiles include the SSC-1b Sepal (with up to 100 launchers) and the SSC-3 Styx.

On the borders of China and the USSR the latter has deployed a number of tank turrets in the static defence role, these include 100 mm T-54/T-55 and 122 mm T-10, some of which are believed to be used in conjunction with radar systems for improved target acquisition when being used in the indirect fire support role.

Soviet 305 mm and 152 mm Coastal Defence Guns

Available details of the Second World War/First World War 305 mm gun designs still used as coastal defence guns are as follows:

CALIBRE	305 mm L/56	305 mm L/52
WEIGHT	n/av	51 580 kg
MAX RANGE	APC 43 900 m	APC 24 620 m
	HE 46 000 m plus	HE n/av
SHELL WEIGHTS	APC 479.9 kg	APC 405 kg
	HE 380 kg	HE 324 kg
MUZZLE VELOCITY	APC est 850 m/s	APC 762 m/s
	HE est 940 m/s	HE 925 m/s

Available details of the Pre-Second World War 152 mm B-38 design used as a coast defence gun are as follows:

CALIBRE	152 mm L/57
MAX RANGE	HE 29 000 m
SHELL WEIGHT	HE 55 kg
MUZZLE VELOCITY	HE 915 m/s

Status: Production complete. All three guns in operational service with the Coastal Artillery and Rocket Troops in limited numbers.

130 mm Coastal Gun SM-4-1

Development/Description
The 130 mm Coastal Gun SM-4-1 was developed in the post-Second World War period and is essentially a 130 mm naval gun on a large mobile carriage. In concept the SM-4-1 is similar to the German SKC/28M mobile coastal gun introduced into the German forces in 1940 but the Soviet weapon fires a lighter projectile much further.

The carriage of the SM-4-1 has removable front and rear axles each of which has four rubber-tyred road wheels. In action the axles are removed and the four stabilisers extended and staked to the ground. The stabilisers are mounted two either side of the carriage and when travelling the stakes are mounted on the front of the shield, two either side. The shield has slightly curved sides and a removable centre section which allows the weapon to obtain a maximum elevation of +45°. The long-stepped barrel is 58 calibres long and has an unusual muzzle brake and a travelling lock. Ammunition used with the 130 mm SM-4-1 is not interchangeable with the Soviet 130 mm Field Gun M-46 or the 130 mm anti-aircraft gun KS-30.

It fires an APHE projectile weighing 33.6 kg with a muzzle velocity of 1050 m/s which will penetrate 250 mm of armour at an incidence of 0° at a range of 1000 m and an HE projectile weighing 33.4 kg with a muzzle velocity of 1050 m/s.

The SM-4-1 is normally used in four-gun batteries in conjunction with a ZALP-B (Top Bow) radar fire control system which gives it the capability to engage targets in all weathers. Quantities of these weapons were captured from Egypt when Israeli forces captured Sharm el Sheikh in the 1967 Middle East War.

130 mm coastal gun SM-4-1 in travelling order

Status: Production complete. In service with Bulgaria, Egypt, North Korea, Poland, Romania, Syria, USSR, Yemen and Yugoslavia.

Manufacturer: Soviet state factories.

SPECIFICATIONS

CALIBRE	130 mm	LENGTH (travelling)	12.8 m	ELEVATION/DEPRESSION	+45°/-5°
BARREL LENGTH (overall)	7.6 m	WIDTH (travelling)	2.85 m	TRAVERSE	360°
WEIGHT		HEIGHT (travelling)	3.05 m	RATE OF FIRE	5 rpm
travelling	19 000 kg	AXIS OF BORE (at 0°)	2.03 m	RANGE	29 500 m
firing position	16 000 kg	GROUND CLEARANCE	0.3 m	TOWING VEHICLE	AT-T Heavy Tracked Artillery Tractor
		TYRES	10.50 × 20		

SSC-1b Sepal Coastal Defence Missile

Development/Description
Development of the S-35 (US/NATO designations SSC-1b Sepal) coastal defence missile was begun in the late fifties by the Chelomai design bureau as a land-based variant of the P-5/P-5D/P-6/P-7/P-35 (respectively US/NATO designations SS-N-3c Shaddock/SS-N-3c Shaddock/SS-N-3a Shaddock/SS-N-3a Shaddock/SS-N-3b Sepal) family of anti-ship missiles.

The original US/NATO designation for the system was the SSC-1a Shaddock which was assigned in 1961 when the weapon was first spotted by western intelligence in use as a naval coastal defence and an army tactical nuclear delivery system. Initial operational capability is believed to been achieved the previous year.

In the mid-sixties the definitive coastal defence SSC-1b variant entered service using a new version of the ZIL-135/BAZ-135 (8 × 8) tractor-erector-launcher (TEL) vehicle. This has a larger cab than the original vehicle.

By the end of the seventies the Soviet Coastal Missile Artillery Forces had about 100 launchers in independent coastal defence missile batteries of three launchers and coastal defence missile battalions with five-six batteries totalling 15-18 launchers. The latter are grouped at Fleet level into brigades of three battalions with supporting units. In 1988 the battalion dispositions were: Northern Fleet - three; Baltic Fleet - six; Black Sea Fleet - five; Pacific Fleet - five.

Initial target surveillance is by a target acquisition battery at brigade level which hands over the contact to the battalion HQ target acquisition section. This monitors the contact by using its own target tracking radar and interrogates it by a Square Head IFF system. If confirmed as hostile it is passed over to the battery's missile guidance platoon whereby a modified truck-mounted NATO designation Front Door D-band guidance radar accepts the target and a pair of missiles are launched from one of the battery's launcher sections.

The missile is carried in a sealed cylindrical container-launcher on the rear platform of the vehicle with a hydraulically operated elevation mechanism. For firing the launcher is elevated to +15-20°, the containers hemispherical front and rear ends hinged upwards and the S-35 fired. This then uses a pair of jettisonable solid fuel rocket booster motors, mounted either side of the sustainer, to eject itself clear for the launch and acceleration phases.

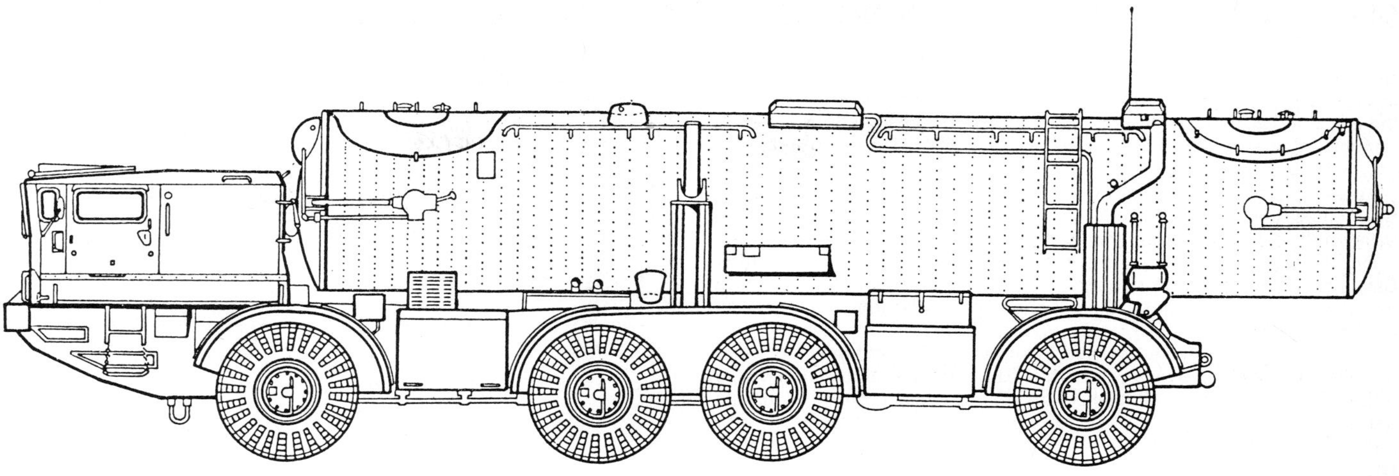

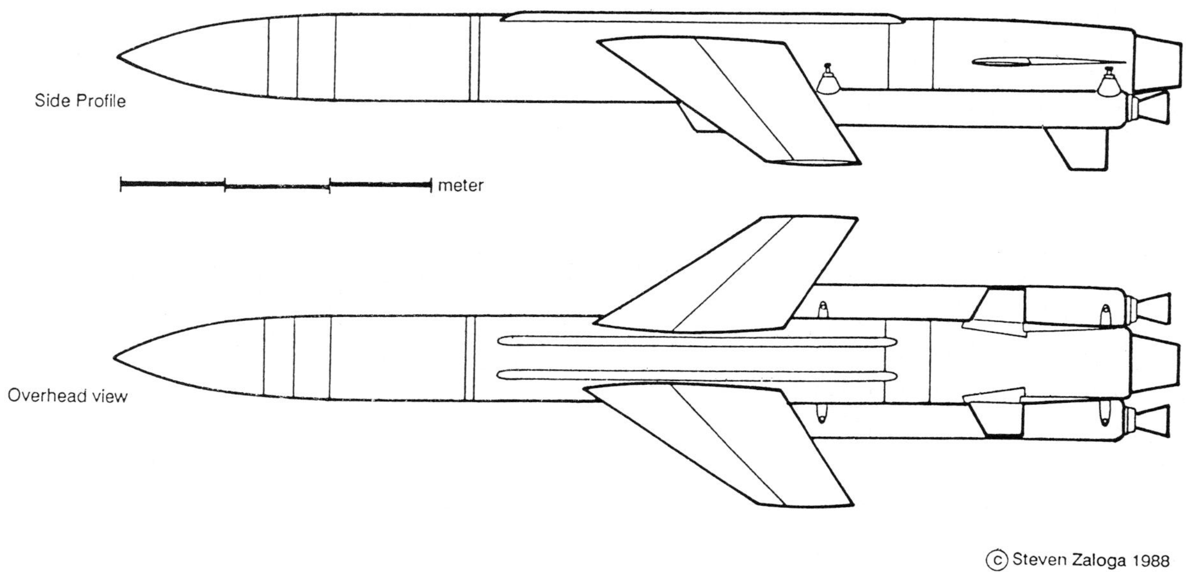

Latest version of SSC-1b Sepal (S-35) coastal defence missile launcher (top) with missile below (not to 1/76th scale) (Steven Zaloga)

Early version of SSC-1b Sepal (S-35) coastal defence missile launcher which has a higher cab roof

After booster burn-out the cruise propulsion is taken over by the liquid fuel turbojet sustainer with its air intake located, scoop fashion, beneath the belly of the weapon.

Guidance for the initial stages of the flight is inertial, with mid-course corrections sent by data link. If the target is located at the S-35's maximum effective range of 275 km then the I-band seeker is activated in its search mode at a range-to-target of about 165 km. The missile's cruising altitude is around 4500 m. The data derived from the seeker is sent back via the video data link to the command centre where it can be assessed and corrective guidance signals or target selection instructions re-transmitted to the missile as appropriate. A change to the target acquisition mode followed rapidly by the seeker lock-on is normally performed for this target range at a range-to-target distance of 92-93 km. The missile then adopts a 20° or so dive angle to reach the target's vicinity for the final terminal-phase pursuit course run-in at a shallow angle.

For maximum effect in an over-the-horizon engagement an airborne platform such as a Tupolev Tu-95 Bear D, Kamov Ka-25 Hormone-B or Tupolev Tu-16 Badger D fitted as a data relay platform to the command centre has to be used otherwise the missile's range is effectively limited to that of the radar horizon.

The warhead weighs 1000 kg and can either be of the conventional HE type or nuclear with a yield in the region of 350 kt. Missile length is 10.2 m, body diameter 0.85 m and maximum wing span 2.1 m whilst the launch weight is 5400 kg. Maximum possible range is 460 km with a CEP of 500-800 m.

Variants

There is a reconnaissance drone derivative of the family. Known as the DR-3, this is basically the cruise missile body fitted with a new camera package and guidance section. Launched from the same vehicle as used in the SSC-1b system the DR-3 is used by the Soviet Army and Syria (at least one having been shot down by the Israelis).

There is also evidence to suggest that a target drone conversion is also in service with the Soviet Armed Forces.

Status: Production as required. In service with Syria (one battalion of 18 launchers with conventional warhead equipped S-35), the USSR and Yugoslavia (several batteries).

Manufacturer: Soviet state factories.

SSC-2b Samlet Coastal Defence Missile

Development

Initial development of the S-2 Sopka (Soviet word for hill, with the Soviet industrial index number 4K87 and the US/NATO designations SSC-2b Samlet) coastal defence missile was begun in 1947 by the Mikoyan-Gurevich design bureau.

The radar beam-rider S-2 and a related bomber launched missile (the Soviet designation KS-1, US/NATO designation AS-1 Kennel) were developed under the programme codename Komet.

Operational deployment of the initial S-2 variant occurred in 1958 at Balalava and Kildin Island with the S-2 battalions (division) having three batteries each of 2-3 semi-fixed launchers in heavily protected concrete bunkers.

In 1960 a mobile Rocket Assisted Take-off (RATO) catapult launched a S-2 version was deployed by the BRAV (Baregovy Raketno-Artilleriiskie Voiska: Coastal Missile-Artillery Forces) in coastal defence missile battalions known as BROMs (Beregoviy oborony raketniy batalon). These battalions were increased in size to three batteries each of six mobile launchers. Guidance for a battery was provided by an I-band NATO designation Sheet Bend surveillance and fire control radar located near to the launch site.

The first mobile S-2 battalion entered service at Poti on the Black Sea. In 1962 the Cuban Armed Forces began receiving the S-2 and eventually took delivery of three battalions (54 launchers). Egypt followed in 1963 with at least four batteries (with 24 launchers). Other countries known to have received the weapon include Bulgaria (two battalions with 36 launchers), the former East Germany (three batteries with 18 launchers), North Korea (qv entry this section), the People's Republic of China and Poland (three battalions with 54 launchers).

Of these, it is believed that only Bulgaria, Poland and North Korea still retain the weapon in front-line service whilst Cuba and the Soviet Union may have some left in storage as a war reserve weapon.

The only recorded combat use of the S-2 is by Egypt which fired four Samlets on the night of 21/22 October during the 1973 Yom Kippur War against Israeli naval targets off Alexandria; all the missiles apparently missed.

Description

The S-2 has an airframe which is derived from the MiG-15 fighter plane. The main body is cigar shaped with delta swept wings at midbody that have two fences on either side. The aircraft-like tail assembly has a large tail plane well up the fin. There is also a large radome above the nose-mounted turbojet inlet. Missile length is 7.9 m, the body diameter 1.2 m and the maximum wing span 4.9 m.

The liquid fuel turbojet is a used RD-500K from a MiG-15 that only has around five hours of propulsion life left. It is mounted axially on the S-2 with straight through ducting. Initial launch and acceleration is achieved by the jettisonable solid propellant JATO motor pack fixed immediately beneath the tail jet pipe. After the JATO pack is jettisoned the turbojet is air started and the weapon climbs to its cruising altitude of 3-400 m and speed of Mach 0.9. Effective maximum range being approximately 90 000 m.

The launch platform is a four wheeled two-axle trailer with a short launch rail assembly with elevation and limited traverse controls. The towing vehicle used is the Zil-157V (6 × 6) 2500 kg tractor truck for semi-trailers.

Close-up of SSC-2b Samlet (S-2 Sopka) on its missile transloader vehicle

Stored missile life is said to be five years with annual maintenance checks. After that it has to be completely refurbished.

Guidance in the initial flight stages is by radar beam riding with radio commands to provide course updates. The terminal phase is carried out by a semi-active homing seeker which guides the weapon on a pursuit course trajectory until the S-2 impacts. Launch weight is 3100 kg with the conventional HE warhead weighing 600 kg.

Status: Production complete. In service with Bulgaria, North Korea and Poland.

Manufacturer: Soviet state factories.

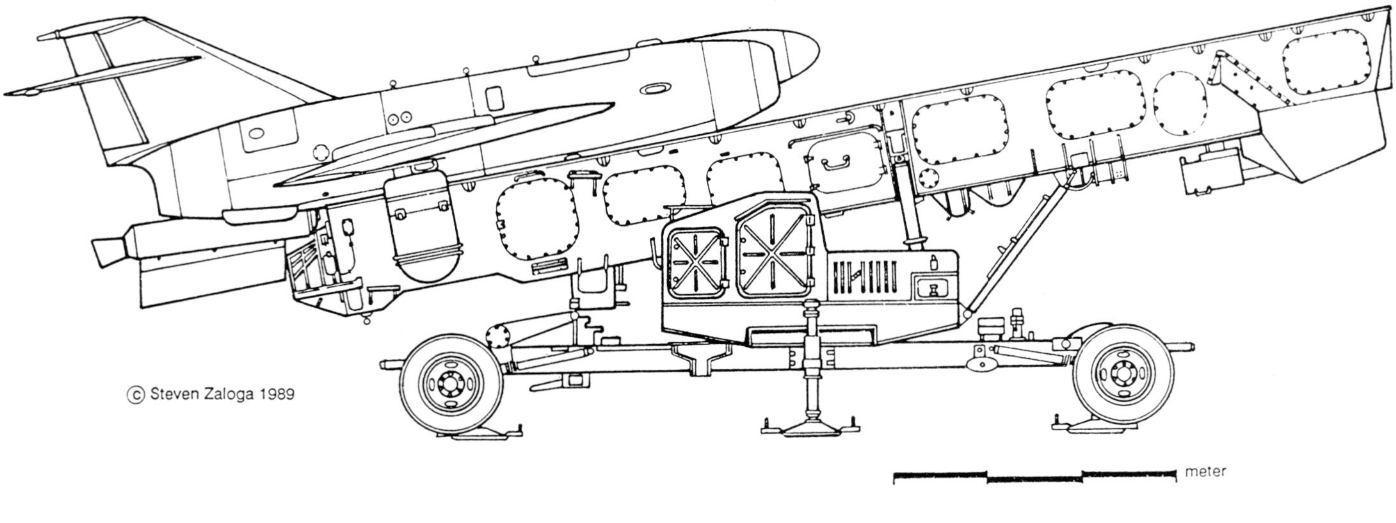

SSC-2b Samlet (S-2 Sopka) coastal defence missile on its launcher (not to 1/76th scale) (Steven Zaloga)

SSC-3 Styx Coastal Defence Missile

Development/Description

In the late seventies the Soviets began development of a coastal defence version of the P-20/P-21 (US/NATO designation SS-N-2c Styx) anti-ship missile. Operational trials were conducted in the early eighties with the first operational deployment occurring in 1984 when units were assigned to the Pacific Fleet. The first exports being made in the mid-eighties to the then East Germany (three batteries of six launchers each) and Yugoslavia. Since then other countries which have accepted SSC-2 shipments have included Syria (one battalion with 18 launchers) and Libya.

In Soviet service the SSC-3 is grouped at Fleet level into brigades of three battalions. Each battalion has three batteries with four tractor-erected-launcher (TEL) vehicles. The SSC-3 supplements rather than replaces the existing SSC-1b Sepal.

The TEL is based on the MAZ-7910 (8 × 8) heavy wheeled transporter and has a large three man engagement control compartment added behind the driver's station. Adjacent to this compartment on the starboard side of the vehicle is an elevatable I-band NATO designation Plank Shave search, track and missile guidance radar assembly. In the travelling configuration the radar antenna is telescoped down on to the roof of the driver's cab.

To the rear of the control compartment is a tandem cylindrical container-launcher assembly for two P-20 or P-21 missiles. For a missile launch they have to be rotated through 90° and elevated to about +20° The front and rear canister covers are then folded open and the weapon fired.

The general configuration of the P-20/P-21 is like that of a small aircraft with a cigar shaped body, a rounded nose and midbody truncated delta planform wings which are foldable. The tail assembly is of a triple surface type and has the liquid fuel turbojet sustained located beneath it. A jettisonable solid fuel rocket booster motor for the launch and acceleration phases is attached beneath the missile body just under the end of the wing roots.

Cruising speed is about Mach 0.9 with missile length at 6.5 m, body diameter 0.78 m, wing span 2.8 m and launch weight (complete) around 2500 kg. The warhead weighs 500 kg and contains conventional HE.

Guidance for the mid-course phase is by programmed autopilot with precision accelerometers. The terminal phase is handled, depending upon whether the missile is the P-20 or P-21, either by an active I-band radar or passive infra-red seeker head.

The flight altitude can be present in a multiple of 50 between 50 and 300 m.

The Plank Shave radar, when elevated to its maximum height, allows engagements out to a range of 35 000 m with the TEL's own facilities.

Two Yugoslav SSC-3 coastal defence missile units during a ceremonial parade. Front housing contains radar gear and operations cabin; rear housing is on a turntable and contains two Styx anti-ship missiles on separate launch ramps. Note radar scanner in lowered position for travel

The maximum effective range of the missile, however, is about 80 000 m which means that a forward observer system such as an airborne platform would have to be used to fully exploit the range capabilities.

For short range engagements the missile climbs at an initial angle of +45° and then settles to its preset cruise level altitude (between 50 to 150 m). For longer ranges it climbs to either 200, 250 or 300 m altitude. The terminal attack phase in all cases is effectively a shallow dive ending at approximately a 2-5 m sea-skimming height.

It is possible that the US/NATO designation SS-N-2d Styx variant (Soviet designation P-22) is also used with the SSC-3 system. This weapon has a maximum engagement range of over 100 000 m.

In all cases the system operator manually sets a range gate around the designated target and feeds the derived target bearing and range data to the missile's autopilot/guidance system before launch.

Status: In production. In service with Libya, Syria, the USSR and Yugoslavia.

Manufacturer: Soviet state factories.

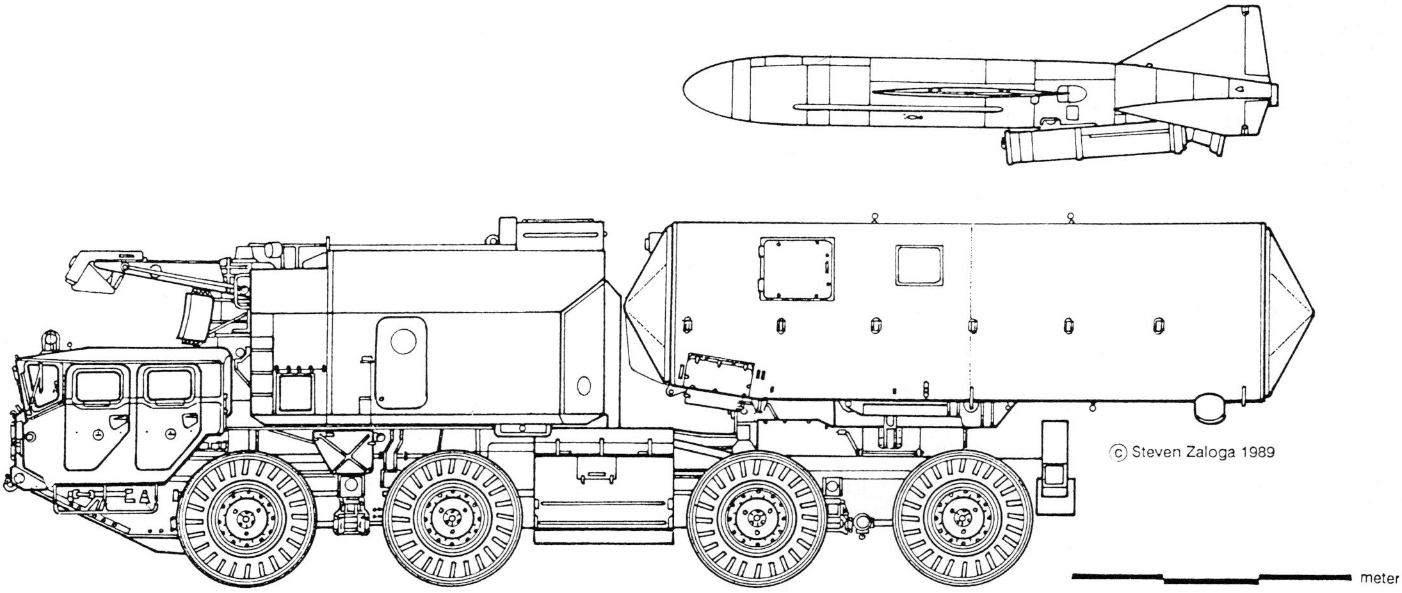

SSC-3 Styx (P-20/P-21 P-22) coastal defence missile system with its TEL based on a MAZ-7910 (8 × 8) chassis (not to 1/76th scale) (Steven Zaloga)

UNITED KINGDOM

Coastal Defences

Coastal defences in the UK and overseas were operated and maintained by the Royal Artillery but with the exception of a small number of 9.2 in coastal guns in Gibraltar, all of these weapons were phased out from 1957 without replacement. In the early 1980s, following the use of MM38 Exocet SSMs launched by Argentinian forces from the Falkland Islands against Royal Naval warships in the 1982 Falklands Conflict, the Excalibur system was developed. Details of this system are given in the following entry. Excalibur is a mobile system currently deployed to cover the Straits of Gibraltar.

Vosper Thornycroft Excalibur Coast Defence Missile System

Development/Description

The Excalibur system was developed by Vosper Thornycroft (UK) Ltd to meet a Royal Navy requirement that was perceived after the experiences of the 1982 Falklands Conflict for a mobile and air-transportable coastal defence missile system based on existing MM38 Exocet missiles. The system can be readily adapted to other missile types if required.

The trailer-mounted system for the Royal Navy comprises a control cabin which contains the missile fire control system, communications equipment, power supply converters and associated radar, gyro and system interface displays; two radar/launcher assemblies which carry two missiles in their container-launchers, a hydraulically extended Kelvin-Hughes Type 1007 I-band fire control radar antenna and a hydraulically operated ramp for accurate missile elevation and levelling purposes and a stores and accommodation cabin to support the launch crew and system in the field.

The Aerospatiale MM38 itself is 5.21 m long, has a body diameter of 0.348 m and a maximum wingspan of 1.04 m. Launch weight is 735 kg with a 165 kg SERAT HE shaped charge warhead and delay action contact fuze that is effective up to a 70° incidence angle. An overflight proximity radar fuze is fitted as a backup.

The missile has a slender cylindrical shape with a sharply pointed nose. Flight control is by small tail control surfaces which are in-line with swept cruciform long chord cropped wings at the midbody area.

Propulsion is by a 2.4 second burn SNPE Erpevier booster solid propellent rocket motor coupled to an SNPE Eole V 110 second burn solid propellant sustainer rocket motor which cuts out about 35 000 m from the launch platforms position. Maximum speed reached is 300 m/s which reduces to Mach 0.6 at the extreme effective range of around 40 000 m. Maximum range is 42-45 000 m. Minimum range is 5000 m.

The missile's inertial guidance package for the mid-course phase of the flight is first pre-programmed by the missile fire control system via the ITS. The missile can be fired at offset angles of up to 30° from the launcher axis.

If fired at around 12° elevation the missile climbs to a maximum height of about 30 m (highest altitude attainable at this stage is 75 m) where it flies level for a short time. The TRT AHV-7 radio altimeter then commands it to descend at a distance from the launch platform of 4 km to its cruising height of between 9 and 15 m. At a pre-set distance into the flight, and some 12-15 km from the projected target position, the ADAC I/J-band active radar seeker is activated to operate in its search mode. Once the target is detected and lock-on achieved the seeker switches over to automatic target tracking and causes the missile to follow a proportional navigation flight profile with progressive altitude drops to its final pre-set sea skimming altitude or about 2.5 m or less until impact. Time from seeker activation to lock-on is typically two seconds.

A 'super ADAC' seeker is being offered by Aerospatiale for backfit to the MM38.

Status: Excalibur system-production as required. In service with the Royal Navy (at least one battery).

Manufacturer: Vosper Thornycroft (UK) Ltd
Woolston Works
Woolston
Southampton
SO9 5GR
UK
Telephone: (0703) 445144
Telex: 47682 VT WOOL G
Fax: (0703) 421539

Excalibur coast defence missile system operationally deployed with control cabin in centre and fire control radar on left two round launcher

Excalibur coastal defence missile system on trial at RAE Aberporth (UK MoD/RAE Aberporth)

British Aerospace (Dynamics) Sea Skua Coastal Defence Missile

Development/Description

In mid-1988 British Aerospace announced the development of a coastal defence version of the Sea Skua anti-ship missile. It is basically a four-round coastal defence system which is mounted on a four-ton truck and uses elements of the ship-launched version.

The crew comprises the vehicle driver and a missile operator with 8-10 trucks comprising a coastal defence battery. Alternatively the system can be pallet-mounted. The weapons would also require an I-band target tracking and illumination radar.

The Sea Skua is 2.50 m long, has a body diameter of 0.25 m and a maximum wing span of 0.72 m. Launch weight (complete) is 147 kg with the semi-armour piercing warhead weighing 20 kg. A delay action impact fuze is fitted so the warhead detonates within the target. The effective range limits are from 1800 to 18 000 m. If the missile has not made contact after 70 seconds of flight it automatically self-destructs.

The missile body is cylindrical with a rounded nose and a taper to a smaller diameter aft. Electrically activated triangular cruciform control surfaces are located midbody with in-line truncated triangular stabilising fins aft.

Propulsion is by separate Imperial Metal Industries solid propellant booster and sustainer rocket motors which ignite simultaneously. The sustainer has two angled exhaust nozzles. Cruise speed is about Mach 0.85-0.90.

There is a two second delay between depressing the firing button and actual launch in order that the gyros can spin up, the thermal batteries activate and the guidance system tunes.

Guidance is by autopilot during the boost phase which stabilises it in roll and maintains the missile on a steady bearing towards the target at a steady height. At boost burn-out the lateral guidance loop is closed and the I-band Marconi semi-active radar seeker head is activated to provide a proportional navigation course in the horizontal plane to the target. Shortly after this the height loop is closed and the missile descends to the pre-selected sea-skimming height under radio altimeter control. Four sea-skimming flight altitudes are available with the selection being made according to the sea state extant at the moment of launch. Since the guidance systems are effectively independent then several missiles can be ripple fired at a target. Target engagement can be up to 80° off the launcher axis.

Side drawing of Sea Skua coastal defence missile system mounted on rear of 6 × 6 truck with search radar mounted on top of elevated arm (not to 1/76th scale)

Sea Skua anti-ship missile in flight

Status: Production of Sea Skua missile has been under way for many years and has been used operationally in the Falklands and more recently in the Middle East, in both cases it was launched from Westland Lynx helicopters of the Royal Navy. Coastal defence version has been proposed to a number of countries.

Manufacturer: British Aerospace (Dynamics) Limited
Six Hills Way
Stevenage
Hertfordshire
SG1 2DA
UK
Telephone: (0438) 312422
Telex: 825125/825126

UNITED STATES OF AMERICA

McDonnell Douglas Harpoon Coastal Defense System

Development

The latest variant of the Harpoon missile family is the Harpoon Coastal Defense System. The Harpoon Coastal Defense System is a ground mobile anti-ship weapon system that provides area defence of high value assets and control of shipping lanes from prepared and unprepared launch sites. The basic Harpoon Coastal Defense System comprises a single battery of one Missile Control Unit and one Launcher Transporter Unit. The battery can be configured with additional Launcher Transporter Units and/or an optional Launcher Replenishment Unit. The system is designed to operate over primary and secondary road systems with limited off-road capability, thus providing a highly survivable weapon system. The Harpoon Coastal Defense System is capable of transitioning from the over-the-road configuration to launch ready status in less than 30 minutes.

Description

The individual system components comprise:

(1) Missile Control Unit - The Missile Control Unit is the only manned vehicle in the system during tactical operations. It provides for all command, control and communications functions for the Harpoon Coastal Defense System and is comprised of the following:

(a) The Shelter which houses the launch control and communications equipment and the Environmental Control Subsystem (ECS) which provides a controlled environment for both equipment and personnel within the shelter.

(b) The Harpoon Ship Command Launch Control Set (HSCLCS) AN/SWG-1A which performs engagement planning, flight plan computations and initiation and launch of the Harpoon missile.

(c) Electrical Power Subsystem (EPS) which provides generation of electrical power, monitoring of power quality, conditioning and distribution of power, and circuit overload protection.

Artist's impression of McDonnell Douglas Harpoon Coastal Defense System deployed

McDonnell Douglas Harpoon missile in flight

(d) The Communications Subsystem (CS) which provides voice and optional data links for the transfer of target information, tactical unit co-ordination, and command and control direction for the battery.

(2) Launcher Transporter Unit - The Launcher Transporter Unit provides the support structure and performs the transportation and launching functions for the canistered Block 1C Harpoon missiles. The Launcher Transporter Unit is comprised of the following:

(a) The Launcher Erector Unit (LEU) which supports the four cannisterised missiles during transportation and raises the missiles to the required launch angle at the launch site.

(b) The Electrical Power Subsystem (EPS) which provides generation of electrical power, monitoring of power quality, conditioning and distribution of power, and circuit overload protection.

(c) The Cable Reel Unit (CRU) which performs storage, deployment and retrieval functions for the cables that interconnect the battery vehicles.

(d) The Launcher Parameters Set which determines the launcher heading and attitude and interfaces the data to the HSCLCS for engagement planning and flight plan computations.

(3) Launcher Replenishment Unit - The Launcher Replenishment Unit is used to resupply the Launcher Transporter Unit in the field. Typically, the Launcher Replenishment Unit has a four canister assembly which is transferred between the units by a hydraulically operated crane mounted on the Launcher Replenishment Unit transporter.

The Harpoon missiles used on the Harpoon Coastal Defense System are standard production RGM-84D Block 1C rounds in shock resistant canisters. First deliveries of this configuration occurred in 1984.

The missile itself has a cylindrical body of uniform diameter and a plastic coated nose radome. Flight control is by tail control fins which are in-line with midbody cruciform delta shaped stabilising fins. A 2.9 second burn 5600 kg thrust tandem-mounted solid propellant booster rocket motor is attached with in-line tail fins for the boost and acceleration phases.

The sustainer motor is a 300 kg thrust Teledyne CAE J402-CA-400 liquid fuel turbojet with a flush ventral air intake. Cruising speed is Mach 0.83.

The weapon is 4.628 m in length, 0.343 m in body diameter and has a maximum wing span of 0.914 m. Launch weight (complete) is 722 kg of which 221.6 kg is the Naval Weapons Center penetration HE-blast fragmentation warhead fitted with a contact time delay fuzing system.

The missile's onboard data processor is programmed before launch. A Northrop or Smiths attitude reference assembly and an IBM PSIP-OA digital computer form the Mid-course Guidance Unit (MGU) which acts as the autopilot controlling the mid-course phase of the flight. Altitude control is maintained by a short pulse Honeywell or Kollsman AN/APN-194 radar altimeter.

The final active flight phase is controlled by a Texas Instruments J-band frequency-agile PR-53/DSQ-28 two-axis solid state high resolution active radar seeker.

The missile is usually launched in this pre-set Range and Bearing Launch (RBL) mode, turning on the radar seeker only at the last moment to detect, evaluate, lock-on and maintain tracking to impact on the target. After lock-on the missile descends to its adaptive sea-skimming altitude. The radar can be set for small to large acquisition windows that determine the range-to-target at which the seeker is activated. The smaller the window the more precise the initial target data must be.

The alternative launch technique is the Bearing Only Launch (BOL) in which the missile is fired along the target's bearing and the radar is activated early in the trajectory to scan a 45° search sector either side of the missile's course heading.

In either launch mode once the target is detected and the seeker is locked-on in its tracking phase the Block 1C Harpoon can use a preselected terminal trajectory. This may be either a low apogee pop-up attack manoeuvre or a sea-skimming profile. The radar seeker and MGU continue to generate proportional navigation commands to the flight controls in order to ensure tracking and impact on even a high speed manoeuvring surface target.

The Block 1C Harpoon can also be programmed to fly single or multiple 'dog-legs' to the target (known as 'waypointing') to conceal the position of the battery and if several Harpoons are being fired together, to allow them the option of arriving simultaneously from different directions. The original missile launch angle can be up to 180° from the target bearing.

Maximum effective range of the Harpoon Coastal Defense System is said to be over 120 000 m.

In 1989 McDonnell Douglas Missile Systems Company was awarded a contract to further upgrade Harpoon. This Block 1D Harpoon will carry software and guidance system modifications - also available in a retrofit kit - that will permit the missile to re-attack a target by flying cloverleaf-shaped search patterns. The Block 1D Harpoon range will be almost double that of earlier units, be lengthened by approximately 0.61 m and weigh some 68 kg more.

Status: Production. Believed to be in service with South Korea and Spain. Denmark has contracted for the Harpoon Coastal Defense System and will use the missiles and launch components taken from decommissioned light frigates.

Manufacturer: Enquiries to: Harpoon Coastal Defense System
McDonnell Douglas Missile Systems Company
PO Box 516
St Louis
MO 63166
USA

YEMEN

Yemeni Coastal Defence

It is believed that Yemen has a small number of 130 mm SM-4-1 coastal defence guns still in service.

YUGOSLAVIA

Yugoslav Coastal Defence

Coastal defence is the responsibility of the Yugoslav Army and includes some 25 coastal artillery installations, both gun and missile. The former includes 45 D-44 guns, 55 German 88 mm guns (although these may well have been phased out of service), 75 122 mm M-37 guns, 85 130 mm guns and 85 152 mm D-20 gun howitzers. Missiles in service include the Soviet supplied SSC-3 Styx coastal defence missile system known locally as the BROM and SSC-1b Sepal coastal defence weapons.

Armour and artillery in service

Listings by country of equipment in service, in reserve and due to enter service by 1991 (*denotes a later in-service date). This data is based on the latest available information at the time of going to press and updates that in the main text. This section also excludes fixed fortifications and coastal defence systems which have their own section.

AF: Air Force	IS: Internal Security	N: Navy	P: Police
BG: Border Guard	Ma: Marines	NG: National Guard	R: Reserve
G: Gendarmerie	Mi: Militia	NI: Naval Infantry	T: Training

Europe and North America (including USA, USSR and Canada)

	Tanks	Reconnaissance vehicles	APCs	Tank Destroyers self-propelled guns and mortars	Towed and Rocket Artillery Systems	Anti-air systems
ALBANIA	T-34/85 T-54 Type 59 Type 63	BRDM-1	BTR-40 BTR-50 BTR-152 Type 531	SU-76 SPG	45 mm M1942 ATG 57 mm M1943 ATG 76 mm M1942 FG 85 mm D44 FG 85 mm Type 56 FG 122 mm M1931/7 gun 122 mm M1938 How 122 mm Type 60 How 130 mm Type 59-1 FG 152 mm Type 66 GH 152 mm M1937 GH 152 mm D1 How 107 mm Type 63 MRS	37 mm M1939 LAAG 57 mm S-60 AAG 85 mm KS-12 AAG 100 mm KS-19 AAG
AUSTRIA	Jagdpanzer SK 105 M60A1/M60A3		Saurer 4K 7FA Saurer 4K 4FA G-1 Saurer 4K 4FA G-2	155 mm M109/M109A2 SPH 81 mm GrW1 SPM	85 mm M52 FG 105 mm M18/40 FH 105 mm M101 How 155 mm M59 gun (static) 155 mm M114 How 155 mm GH N-45 130 mm M51/Steyr 680 M3 MRS	20 mm GAI-B01 LAAG (and AF) 35 mm GDF-002 LAAG (and AF) 40 mm M42 SPAAG 40 mm Bofors L/70 AAG
BELGIUM	Leopard 1	FN 4RM/62F AB(G) Scorpion Scimitar	AIFV BDX (G & AF) M113A1-B Spartan	Striker/Swingfire TD 90 mm Jagdpanzer Kanone TD M113A1-B MILAN TD 155 mm M109/M109A2 SPH 203 mm M110A2 SPH	105 mm M101 How 105 mm Model 56 P How 155 mm M114 How	12.7 mm M55 LAAG 20 mm twin Rh 202 LAAG 20 mm M167 LAAG 35 mm Gepard SPAAG
BULGARIA	T-34/85 (R and T) T-54 T-55 T-62 T-72	BRDM-1 BRDM-2 PSzH-IV (BG)	BTR-40 (R) BTR-50 BTR-60 BTR-152 (IS) MT-LB BMP-1 BMP-23 BMP-23D BMP-30	BRDM-2 Sagger TD SU-100 SPG 122 mm SO-122 SPH 152 mm SO-152 SPH 120 mm MT-LB SPM 120 mm 2S12 SPM	57 mm M1943 ATG (Mi) 76 mm M1942 FG 85 mm D44 FG 85 mm SD44 FG 100 mm M1944 FG 122 mm D30 How 122 mm M1931/7 gun 122 mm M1938 How 130 mm M46 FG 152 mm D20 GH 152 mm M1937 GH 122 mm BM-21 MRS 130 mm M51/Praga V3S MRS 132 mm BM-13 MRS FROG-3 ARS FROG-7 ARS	14.5 mm ZPU-2 LAAG 23 mm ZU-23 LAAG 37 mm M1939 LAAG 57 mm S-60 AAG 85 mm KS-12 AAG 100 mm KS-19 AAG 23 mm ZSU-23-4 SPAAG 57 mm ZSU-57-2 SPAAG (R) SA-4 Ganef SAM SA-6 Gainful SAM SA-9 Gaskin SAM SA-13 Gopher SAM
CANADA	Leopard C1	Lynx M113 C & R	Cougar (WFSV) Grizzly (WAPC) M113 Bison	155 mm M109A1/M109A2 SPH M113A2/TOW TD	105 mm C5 P How 105 mm C1 How 155 mm M114 How	35 mm GDF-005 LAAG ADATS

Country	Tanks	Reconnaissance vehicles	APCs	Tank Destroyers self-propelled guns and mortars	Towed and Rocket Artillery Systems	Anti-air systems
CZECHOSLOVAKIA	T-34/85 (R and T) T-54 T-55 T-72B/T-72G PT-76	BRDM-1 OT-65A BRM-1 VT-90	BVP-1 (and BG) BVP-2 OT-62 (and BG) OT-64 (and BG) OT-90 OT-810 MT-LB	BRDM-2 Sagger TD 122 mm SO-122 SPH 152 mm DANA SPG 203 mm SO-203 SPG 240 mm SM-240 SPM	57 mm M1943 ATG (R) 76 mm M1942 FG (R) 85 mm M52 FG 100 mm M53 FG 122 mm D30 How 122 mm M1931/7 gun 122 mm M1938 How 130 mm M46 FG 152 mm D20 GH 152 mm M1937 GH 122 mm RM70 MRS 130 mm M51 MRS FROG-7 ARS	14.5 mm ZPU-4 LAAG 23 mm ZU-23 LAAG 30 mm M53 LAAG 57 mm S-60 AAG 23 mm ZSU-23-4 SPAAG 30 mm M53/59 SPAAG SA-4 Ganef SAM SA-6 Gainful SAM SA-8 Gecko SAM SA-9 Gaskin SAM SA-13 Gopher SAM
DENMARK	Centurion Mk 5 and Mk 5/2 Leopard 1A3 M41		M113/M113A1	81 mm M125/M125A1 SPM 120 mm M106 SPM 155 mm M109A3 M113 TOW TD	105 mm M101 How 155 mm M59 gun 155 mm M114/39 203 mm M115 How	12.7 mm M55 LAAG 40 mm Bofors L/60 LAAG 40 mm Bofors L/70 (and AF) LAAG
FINLAND	T-54 T-55 T-55M T-72 PT-76		BTR-50 BTR-60 BMP-1 BMP-2 Sandringham PASI-83(XA-180) MT-LBV		76 mm m/02 FG 76 mm m/36 FG 105 mm m/41 FG 105 mm m37/10 FH 105 mm m61/37 FH 122 mm D30 How 122 mm m/38 FH 122 mm M-60 FG 130 mm m/54 FG 150 mm m/40 How 152 mm m/38 How 155 mm K83 GH	20 mm ITK Madsen LAAG 23 mm ZU-23 LAAG 30 mm HS831 LAAG 35 mm GDF-002 LAAG 40 mm Bofors L/60 LAAG 40 mm Bofors L/70 LAAG 57 mm S-60 AAG 57 mm ZSU-57-2 SPAAG Crotale NG SAM 35 mm Marksman SPAAG
FRANCE	AMX-30 AMX-30 B2 Leclerc	AML-60 and -90 (G) AMX-10RC ERC 90 F4 VBC-90(G) VBL	AMX-10P AMX-VCI (and G) VAB (and G) VXB-170(G)	105 mm Mk 61 SPH 155 mm Mk F3 SPG 155 mm GCT SPG 81 mm VTT/PM SPM 120 mm VTT/PM SPM VAB HOT TD	105 mm M101 How 105 mm Model 56 P How (R) 155 mm Model 50 How 155 mm TR gun 227 mm MLRS	20 mm 53T1 LAAG 20 mm Tarasque LAAG 20 mm Cerbere LAAG(AF) 30 mm LAAG 40 mm Bofors L/70 LAAG 30 mm AMX-13 DCA SPAAG Crotale SAM Roland SAM Mistral SAM
GERMANY	Leopard 2 Leopard 1A1A1 Leopard 1A2 Leopard 1A3 Leopard 1A4 Leopard 1A6 M48A2G2	Spahpanzer Luchs Wiesel	Marder Transportpanzer 1 MOWAG SW1 and SW2 (BG and P) UR-416 (P) M113A1 TM170 (IS) BMP1A1 OST	155 mm M109G SPH 203 mm M110A2 SPG Jagdpanzer Kanone TD Jagdpanzer Jaguar 1 HOT TD Jagdpanzer Jaguar 2 TOW TD 120 mm M113 SPM	105 mm Model 56 P How 105 mm M101 How 155 mm M114 How 155 mm FH-70 110 mm LARS MRS 227 mm MLRS	20 mm twin Rh 202 LAAG 20 mm FK 20-2 LAAG 35 mm Gepard SPAAG 40 mm Bofors L/60 (R) 40 mm Bofors L/70 LAAG Roland SAM (and AF, N)
GREECE	AMX-30 Leopard 1A4 M24 Chaffee M47 (and NG) M48A1/M48A3/ M48A5 M60A1 M26 (NG)	M8 and M20 (and R)	AMX-10P M2 and M3 Half-track (NG) M59 M113/M113A1/M113A2 Leonidas MOWAG Roland (G) UR-416 (G)	105 mm M52A1 SPH 105 mm M108 SPH 155 mm M44 SPH 155 mm M109/M109A1B/ M109A2 SPH 175 mm M107 SPG 203 mm M110 SPH M113A2 TOW TD M901 TOW TD 81 mm M125A1 SPM 107 mm M106A1 SPM	75 mm M116 P How (and R) 25-pounder FG (R) 105 mm M101 How (and CAR) 105 mm M102 How 105 mm Model 56 P How 155 mm M114 How 155 mm M198 How 203 mm M115 How 5.5-inch Gun	20 mm twin Rh 202 LAAG 20 mm M693 LAAG 30 mm twin Artemis LAAG (and N, AF) 35 mm GDF-002 LAAG (AF) 40 mm M1 LAAG (NG) 40 mm Bofors L/70 LAAG 40 mm Breda L/70 LAAG 75 mm M51 AAG 90 mm M118 AAG 90 mm M117 AAG 40 mm M42 SPAAG
HUNGARY	T-34/85 (R and T) T-54 T-55 T-72	FUG D-442 BRDM-2 BRM-1	BMP-1 MT-LB BTR-70 BTR-80 FUG D-944	122 mm SO-122 SPH 152 mm SO-152 SPH BRDM-2 Sagger TD	57 mm M1943 ATG (R) 76 mm M1942 FG (R) 85 mm D44 FG 100 mm T-12 ATG 122 mm D30 How 122 mm M1931/7 gun (R) 122 mm M1938 How 152 mm D-20 GH 122 mm BM-21 MRS 240 mm BM-24 MRS FROG-7 ARS	14.5 mm ZPU-2 LAAG 57 mm S-60 AAG 23 mm ZSU-23-4 SPAAG 57 mm ZSU-57-2 SPAAG SA-4 Ganef SAM SA-6 Gainful SAM SA-8 Gecko SAM SA-9 Gaskin SAM
IRELAND		AML-60 and -90 Scorpion	Panhard M3 Timoney Mk 3 Unimog		25-pounder FG 105 mm Light Gun	40 mm Bofors L/60 LAAG 40 mm Bofors L/70 LAAG
ITALY	Leopard 1 M47 (and IS) M60A1 C1*	FIAT 6616 (IS) M8 (IS) M6 (IS)	AMX-VCI LVTP7 (Ma) M113 (and IS) VCC-1 (and Ma) FIAT 6614 (AF, Army and P)	155 mm M109G SPH 155 mm M109L SPH 175 mm M107 SPG 203 mm M110 SPH 107 mm M106 SPM B1 (8 × 8) M113 TOW TD	105 mm Model 56 P How 105 mm M101 How 155 mm FH-70 155 mm M114 How 203 mm M115 How 122 mm FIROS 30 MRS 227 mm MLRS	12.7 mm M55 LAAG 20 mm Rh 202 LAAG 40 mm M1 LAAG 40 mm Bofors L/60 LAAG 40 mm Breda L/70 LAAG 40 mm Bofors L/70 LAAG 4 × 25 mm SIDAM SPAAG
MALTA			Humber FV1611			14.5 mm ZPU-4 LAAG 20 mm Oerlikon LAAG 40 mm Bofors L/70 AAG

Tanks	Reconnaissance vehicles	APCs	Tank Destroyers self-propelled guns and mortars	Towed and Rocket Artillery Systems	Anti-air systems
NETHERLANDS					
Leopard 1 Leopard 2	M113 C & R	M113 YP-408 (P) YPR 765 (AIFV) UR-416 (P)	155 mm M109A1/M109A2 SPH 203 mm M110A2 SPH 107 mm M106 SPM	105 mm M101 How 155 mm M114/39 227 mm MLRS	40 mm L/70 LAAG (and AF) 35 mm CA 1 SPAAG
NORWAY					
Leopard 1 M48A2/M48A5 NM116 (modified M24)		M113 NM135 (Mod M113)	155 mm M109G SPH	105 mm M101 How 155 mm M114 How 155 mm M114/39	12.7 mm M55 LAAG 20 mm FK 20-2 LAAG 40 mm M1 LAAG (R) 40 mm Bofors L/70 LAAG (and AF)
POLAND					
T-34/85 (R) T-54 T-55 T-72B/T-72G PT-76	BRDM-2 FUG	OT-62 TOPAS OT-62 TOPAS-2A OT-62 TOPAS-2AP (and Ma) OT-64 SKOT-1A OT-64 SKOT-2A OT-64 SKOT-2AP BTR-60 BTR-152 (R) BWP MT-LB	BRDM-2 Sagger TD 122 mm SO-122 SPH 122 mm DANA SPG 203 mm SO-203 SPG 120 mm 2S12 SPM	57 mm M1943 ATG (Mi) 85 mm SD44 ATG 76 mm M1942 FG (Mi) 85 mm D44 FG 100 mm M1944 FG 100 mm T-12 ATG 122 mm D30 How 122 mm M1931/7 gun 122 mm M1938 How 130 mm M46 FG 152 mm D1 How 152 mm M1937 GH 122 mm BM-21 MRS 122 mm RM70 MRS 140 mm BM-14-16 MRS 140 mm BM-14-17 MRS 140 mm WP-8 MRS 240 mm BM-24 MRS FROG-7 ARS	14.5 mm ZPU-2 LAAG 14.5 mm ZPU-4 LAAG 23 mm ZU-23 LAAG 57 mm S-60 AAG 23 mm ZSU-23-4 SPAAG 57 mm ZSU-57-2 SPAAG (R) SA-6 Gainful SAM SA-8 Gecko SAM SA-9 Gaskin SAM SA-13 Gopher SAM
PORTUGAL					
M48A5	AML-60 and -90 Saladin Ferret Mk 4 EBR-75 VBL	Chaimite (and Ma) Commando Mk 3 EBR-ETT Panhard M3 M3 Half-track M113 Shorland (IS) Condor	155 mm M109A2 SPH 81 mm M125A1 SPM 107 mm M106A1/A2 SPM M113A2 TOW TD	25-pounder FG 105 mm M101 How 5.5-inch Medium Gun (R) 155 mm M114 How	12.7 mm M55 LAAG 20 mm twin Rh 202 LAAG 20 mm M163A1 SPAAG 40 mm M1 LAAG 40 mm Bofors L/60 LAAG M730A2 Chaparral SAM
ROMANIA					
T-34/85 (R) T-54 T-55 T-72 TR-77 TR-125	BRDM-2 TAB-C	BTR-50 BTR-60 BTR-152 TAB-71 TAB-72 TAB-77 MLI-84 MLVN ABAL BMP-1	SU-76 SPG (R) SU-100 SPG ISU-152 SPG 122 mm SO-122 SPH 122 mm M1989 SPH 82 mm TAB-71AR SPM 82 mm TABC-79AR SPM 120 mm MLVM-AR SPM BRDM-2 Sagger TD	57 mm M1943 ATG 76 mm M1942 FG 76 mm M48 Mountain Gun 76.2 mm Mountain Gun 85 mm D48 ATG 100 mm Mountain Howitzer 100 mm Skoda (various) 105 mm Schneider 100 mm M ATG 122 mm M1931/7 gun 122 mm M1938 How 130 mm M1982 gun 150 mm CEH M1937 gun 152 mm M1947 gun 152 mm M1981 gun 152 mm M1981 GH 152 mm M1938 How 122 mm APR-21/Bucegi MRS 122 mm APR-40/Bucegi MRS 122 mm APR-40/Ural-375 MRS FROG-3 ARS FROG-7 ARS	14.5 mm ZPU-2 LAAG 2 × 30 mm LAAG 30 mm M53 LAAG 37 mm M1939 LAAG 57 mm S-60 AAG 85 mm KS-12 AAG 100 mm KS-19 AAG 57 mm ZSU-57-2 SPAAG SA-6 Gainful SAM
SPAIN					
AMX-30 M41 M47/M47E M48/M48E/ M48A5	AML-60 and -90 (R) VEC Scorpion (Ma)	BMR-600 Panhard M3 (R) M113 UR-416 (P) BLR	105 mm M52A1 SPH (Ma) 105 mm M108 SPH 155 mm M109/M109A1/ M109A1B SPH 155 mm M44 SPH (R) 175 mm M107 SPG 203 mm M110 SPH 81 mm M125A1/A2 SPM 120 mm M125A1 SPM	105 mm Model 56 P How (and Ma) 105 mm m/26 FH 155 mm M114 How 203 mm M115 How 140 mm Teruel MRS 216 mm L21/E2/E3 MRS 300 mm L-10/D3 MRS	12.7 mm M55 LAAG 20 mm M693 LAAG 20 mm KAA-1 LAAG 20 mm twin Rh 202 LAAG 20 mm GAI-B01 LAAG 35 mm GDF-002 LAAG 40 mm Bofors L/70 LAAG 90 mm M117 AAG Roland SAM
(marines also operate M48E, AML-60, M109A1B, BLR and Panhard M3)					
SWEDEN					
Centurion Strv 101/102/104 Stridsvagn 103B/103C		Pbv 302 SKPF m/42 SUSU XA-180 (UN) CV90*	155 mm Bandkanon 1A SPG Infanterikanon 91 Pvrbv 551 TOW TD	105 mm Type 4140 FH 150 mm m/39 FH 155 mm FH-77 155 mm FH-77B 155 mm Model 50 How	20 mm M/40-70 LAAG 20 mm HS 804 LAAG 40 mm m/36 LAAG 40 mm L/70 LAAG Lvrbv 701/RBS-70 SAM
SWITZERLAND					
Pz 61 Pz 68 Mk 1 to 4 Pz 87		M113	155 mm M109/M109A1 SPH 120 mm Mw Pz 64 SPM MOWAG Piranha TOW TD	105 mm Model 46 FG 155 mm Model 50 How	20 mm GAI-B01 LAAG (and AF) 20 mm KAA-01 (single and twin) LAAG 20 mm HS404 (triple) LAAG 35 mm GDF-002 LAAG (and AF) 40 mm Bofors L/70 LAAG

Tanks	Reconnaissance vehicles	APCs	Tank Destroyers self-propelled guns and mortars	Towed and Rocket Artillery Systems	Anti-air systems
TURKEY					
Leopard 1A3 M41 M47 (and R) M48/M48A1/M48A2C M48A5/M48T1 M48A5T1/M48A5T2	M8 (R)	Commando V-150 (G) M59 (R) M113/M113A1 UR-416 (G) Condor Shorland (G) AIFV Fox NBC	105 mm M52/M52A1 SPH (and R) 105 mm M7 SPH (and R) 105 mm M108 SPH 155 mm M44A1 SPH (and R) 155 mm M44T SPG 155 mm M109 SPH 175 mm M107 SPG 203 mm M110 SPH 203 mm M55 SPH (R) 107 mm M106/M106A1 SPM M113 TOW TD	75 mm M116 P How 105 mm M101 How (and R) 105 mm M102 How 155 mm M59 gun 155 mm M114 How 203 mm M115 How 227 mm MLRS	12.7 mm M55 LAAG 20 mm twin Rh 202 LAAG 20 mm GAI-D01 LAAG 35 mm GDF-001 LAAG 40 mm Bofors L/60 LAAG 40 mm Bofors L/70 LAAG (and AF) 40 mm M1 LAAG 75 mm M51 AAG 90 mm M117 AAG 40 mm M42 SPAAG 90 mm M118 AAG
UNION OF SOVIET SOCIALIST REPUBLICS					
T-34/85 (R and IS) T44M (R and IS) T-54 (and NI) T-55A/T-55B (and NI) T-62/T-62A T-64/T-64A/T-64B T-72M/T-72M1 T-72M2 (and NI) T-80 TO-55 (and NI) TO-62 PT-76 (and NI)	BRM/BRM-1/BRM-2 BRDM-1 (and NI) BRDM-2 (and NI) BTR-72 BTR-74	BTR-40 (and R, IS) BTR-50 P/PA/PK BTR-60 PA/PB (and NI, IS) BTR-70 BTR-80 (and NI) BTR-152 (and R, IS) BMP-1/BMP-1M BMP-2 BMP-3 AT-P MT-LB/MT-LBV (and NI) MT-L (and IS) BMD-1/BMD-1M BMD-2 GT-S (and IS) DT-20 (IS)	ASU-85 TD BRDM-1 Sagger TD BRDM-1 Swatter TD BRDM-2 Sagger TD (and NI) BRDM-2 Swatter TD BRDM-3 Spandrel TD (and NI) SU-100M SPG (R) 120 mm SO-120 SPM/H 120 mm 2S23 SPM/H 122 mm SO-122 SPH (and NI) 152 mm SO-152 SPH 152 mm SP-152 SPG 203 mm SO-203 SPG 82 mm BTR-60 SPM 82 mm MT-LB SPM 120 mm MT-LB SPM 240 mm SM-240 SPM	45 mm M1942 ATG (Mi) 57 mm M1943 ATG (Mi) 57 mm Ch-26 ATG 76 mm M1938 Mountain Gun 76 mm M1966 Mountain Gun 76 mm M1942 FG (Mi) 85 mm D48 ATG 85 mm D44 FG (R and Mi) 100 mm T-12 ATG (and NI) 120 mm 2B16 122 mm D-30 How 122 mm M1938 How 125 mm Rapira 2 ATG 130 mm M46 FG 152 mm D1 How 152 mm D20 GH 152 mm 2A36 152 mm 2A65 152 mm 2S19 SPG 180 mm S-23 gun 203 mm B-4/B-4M How (R) 122 mm BM-21 MRS (and NI) 122 mm M1975 MRS 122 mm M1976 MRS 140 mm RPU-14 MRS 140 mm BM-14-17 MRS (and R) 220 mm BM-22 MRS 240 mm BM-24 MRS 250 mm BM-25 MRS 300 mm BM-30 MRS FROG-3 ARS (R) FROG-5 ARS (R and T) FROG-7 ARS (and R)	14.5 mm ZGU-1 LAAG 14.5 mm ZPU-1 LAAG (R and Mi) 14.5 mm ZPU-2 LAAG (R and Mi) 14.5 mm ZPU-4 LAAG (R and Mi) 23 mm ZU-23 LAAG 37 mm M1939 (R and Mi) 57 mm S-60 AAG 85 mm KS-12 AAG (R) 100 mm KS-19 AAG (R) 130 mm KS-30 AAG (R) ZSU-23-4 SPAAG (and NI) ZSU-57-2 SPAAG SA-4 Ganef SAM SA-6a/b Gainful SAM SA-8a/b Gecko SAM (and NI) SA-9 Gaskin SAM (and R, NI) SA-10 Grumble SAM SA-11 Gadfly SAM SA-12a Gladiator SAM* SA-12b Giant SAM* SA-13 Gopher SAM (and NI) 30 mm 2S6/SA-19
UNITED KINGDOM					
Chieftain Challenger 1 Challenger 2*	Ferret Fox Scorpion (and AF) Scimitar (and AF) Shorland (IS)	FV432 Saracen Spartan (and AF) Humber FV1611 (IS) Saxon Warrior Piglet Fox NBC	Spartan/MILAN TD Striker/Swingfire TD 105 mm Abbot SPG 5.5-inch Gun (T) SPH 155 mm AS-90 203 mm M110A2 SPH 81 mm FV432 SPM	25-pounder FG (T) 105 mm Light Gun 105 mm PH (T) 155 mm FH-70 155 mm M109A1/M109A2 227 mm MLRS	35 mm GDF-002 LAAG (AF) Tracked Rapier SAM Stormer/HVM*
UNITED STATES OF AMERICA					
M60/M60A1/M60A3 M1/M1A1/M1A2* Abrams	M3 Bradley Fox NBC	Arrowpointe Dragoon M113/M113A1/M113A2/M113A3 Commando M706 Commando Ranger (AF and N) LVTP7/LVTP7A1 (Ma) M2 Bradley LAV-25 (Ma)	155 mm M106 SPH 155 mm M109A1/M109A2 and M109A3 SPH 203 mm M110A1/M110A2 SPH M113A1/M113A2 TOW TD LAV TOW TD (Ma) M901 TOW TD 81 mm M125/M125A1 SPM 107 mm M106/M106A1 SPM 81 mm LAV SPM (Ma)	105 mm M101 How 105 mm M102 How 105 mm M119 Light Gun 155 mm M114 How 155 mm M198 How 227 mm MLRS	20 mm M167 LAAG 20 mm M163 SPAAG 40 mm M42 SPAAG M730 Chaparral SAM ADATS*

(Marines also use M60A1, M109A1/M109A2/M109A3, M110A2, 155 mm M198)

Tanks	Reconnaissance vehicles	APCs	Tank Destroyers self-propelled guns and mortars	Towed and Rocket Artillery Systems	Anti-air systems
YUGOSLAVIA					
M47 (R) M4 Sherman (R) Sherman Firefly (R) M-84 T-55 T-54 T-72G PT-76	BRDM-2 M3A1 M8	BTR-40 BTR-50 BTR-60 BTR-152 M60 M80 BVP M80A OT-810 MT-LB TAB-71 (Mi) TAB-72 (Mi) BOV-M (and Mi)	BOV-1 Sagger TD M36B2 TD (R) M10 TD (R) M18 Hellcat TD (R) BRDM-2 Sagger TD SU-76 SPG SU-100 SPG 105 mm M7 SPH 122 mm 2S1 SPH	57 mm M1943 ATG 75 mm Pak 40 ATG 76 mm M48 Mountain Gun 76 mm M1942 FG 85 mm D44 FG 100 mm T-12 ATG 105 mm M56 How 105 mm M101 How 105 mm M18 FH 105 mm M18(M) FH 105 mm M18/40 FH 122 mm D30 How 122 mm M1931/37 gun 122 mm M1938 How 130 mm M46 FG 152 mm M1937 GH 152 mm D-20 GH 152 mm M84 How 155 mm M65 How 155 mm M59 gun 155 mm M114 How 128 mm M-63 MRS 128 mm M-77 MRS 128 mm M-85 MRS 262 mm M-87 MRS FROG-7 ARS	12.7 mm M55 LAAG 20 mm M55 LAAG 20 mm M30/M38/M38V LAAG 20 mm M75 LAAG 30 mm M53 LAAG 37 mm M1939 LAAG 40 mm Bofors L/70 LAAG 40 mm M1 LAAG 40 mm Mk 1 LAAG 57 mm S-60 AAG 85 mm KS-12 AAG 90 mm M117 AAG 20 mm BOV-3 SPAAG 23 mm ZSU-23-4 SPAAG 30 mm BOV-3 SPAAG 30 mm M53/59 SPAAG 57 mm ZSU-57-2 SPAAG SA-6 Gainful SAM SA-8 Gecko SAM SA-9 Gaskin SAM

Africa

For North Africa see page 774.

Country	Tanks	Reconnaissance vehicles	APCs	Tank Destroyers self-propelled guns and mortars	Towed and Rocket Artillery Systems	Anti-air systems
ANGOLA	T-34/85 T-54 T-55 T-62 PT-76	BRDM-1 BRDM-2	BTR-50 BTR-60 BTR-152 BMP-1	100 mm SU-100 SPG 122 mm SO-122 SPH	76 mm M1942 FG 85 mm D44 FG 122 mm D30 How 130 mm M46 FG 122 mm BM-21 MRS 140 mm BM-14-16 MRS	14.5 mm ZPU-1 LAAG 14.5 mm ZGU-1 LAAG 14.5 mm ZPU-2 LAAG 14.5 mm ZPU-4 LAAG 20 mm M55 LAAG 23 mm ZU-23 LAAG 37 mm M1939 LAAG 57 mm S-60 AAG 23 mm ZSU-23-4 SPAAG 57 mm ZSU-57-2 SPAAG SA-6 Gainful SAM SA-8 Gecko SAM SA-9 Gaskin SAM SA-13 Gopher SAM
BENIN	T-54 T-55 PT-76	M8 and M20 BRDM-2 Panhard VBL	M3 Half-track		105 mm M2 How 122 mm D30 How 130 mm M46 FG	14.5 mm ZPU LAAG 37 mm M1939 LAAG 57 mm S-60 AAG SA-9 Gaskin SAM
BOTSWANA		Shorland (IS) BRDM-2	Commando V-150 BTR-60		105 mm Light Gun 105 mm M56 P How	
BURKINA FASO		AML-60 and -90 Ferret M8 and M20 EE-9 Cascavel	Panhard M3		105 mm M101 How 105 mm Model 56 P How 107 mm Type 63 MRS	14.5 mm ZPU LAAG 20 mm LAAG
BURUNDI		AML-60 and -90 Shorland (IS)	Walid Panhard M3 BTR-40		75 mm M116 P How 76 mm Type 54 FG	14.5 mm ZPU-4 LAAG
CAMEROON		Ferret Mk 1/Mk 2 M8 and M20 VBL	Commando V-150 (and G) M3 Half-track	81 mm Commando V-150 SPM (and G)	75 mm M116 P How 85 mm Type 56 FG 130 mm Type 59 FG	14.5 mm Type 58 LAAG 35 mm GDF-002 LAAG 37 mm Type 63 LAAG 40 mm Bofors L/60 LAAG
CENTRAL AFRICAN REPUBLIC	T-55	Ferret BRDM-2	BTR-152 VAB			
CHAD	T-55	AML-60 and -90 BRDM-2 ERC 90 Lynx ERC 90 Sagaie EE-9 Cascavel	M3 Half-track Commando V-150S BTR-60		76 mm M1942 FG 105 mm M2 How 122 mm D-30 How 107 mm Type 63 MRS 122 mm BM-21 MRS 132 mm BM-13 MRS	14.5 mm ZPU-1 LAAG 14.5 mm ZGU-1 LAAG 14.5 mm ZPU-2 LAAG 14.5 mm ZPU-4 LAAG 20 mm LAAG 23 mm ZU-23 LAAG 30 mm LAAG SA-6 Gainful SAM SA-13 Gopher SAM
CONGO	T-34/85 (R) T-54 T-55 Type 59 Type 62 PT-76	BRDM-1 BRDM-2	BTR-50 BTR-60 BTR-152 OT-62		57 mm M1943 ATG 75 mm M116 P How 76 mm M1942 FG 85 mm Type 56 FG 85 mm D48 ATG 100 mm M1944 FG 122 mm M1938 How 122 mm Type 54 How 122 mm D30 How 130 mm M46 FG 122 mm BM-21 MRS	14.5 mm ZPU-4 LAAG 37 mm M1939 LAAG
DJIBOUTI		AML-60 and -90 BRDM-2 Panhard VBL	BTR-60	VCR/TH HOT TD	105 mm M56 P How	
ETHIOPIA	T-34/85 T-54 T-55 T-62 T-72 M47	BRDM-1 BRDM-2	BTR-40 BTR-60 BTR-152 BMP-1 M113A1 UR-416 (IS)	122 mm SO-122 SPH	75 mm M116 P How 85 mm D48 ATG 105 mm M101 How 122 mm D30 How 122 mm M1938 How 130 mm M46 FG 152 mm D1 How 155 mm M114 How 122 mm BM-21 MRS 200 mm BMD-20 MRS	14.5 mm ZPU-4 LAAG 23 mm ZU-23 LAAG 37 mm M1939 LAAG 23 mm ZSU-23-4 SPAAG 57 mm ZSU-57-2 SPAAG SA-6 Gainful SAM
EQUATORIAL GUINEA		BRDM-2	BTR-40 BTR-152		76 mm M1942 FG 105 mm m/26 FH 122 mm M1938 How 122 mm Type 54 How	
GABON		Ferret Mk 1/Mk 2 EE-9 Cascavel AML-60 and -90 M8 EE-3 Jararaca ERČ 90 Sagaie Panhard VBL	VXB 170 Commando V-150 Panhard M3 ACMAT VBLA EE-11 Urutu		76 mm Type 54 FG 76 mm M1942 FG 105 mm M2 How 140 mm Teruel MRS	23 mm ZU-23 LAAG 37 mm Type 55 LAAG 40 mm Bofors LAAG 20 mm ERC 2 Sagaie SPAAG

Tanks	Reconnaissance vehicles	APCs	Tank Destroyers self-propelled guns and mortars	Towed and Rocket Artillery Systems	Anti-air systems
GHANA					
	Saladin	MOWAG Piranha 4 × 4 and 6 × 6		76 mm M1942 FG 25-pounder FG 105 mm Light Gun	
GUINEA					
T-34/85 T-54 PT-76	BRDM-2	BTR-40 BTR-50 BTR-60 BTR-152		57 mm M1943 ATG 76 mm M1942 FG 85 mm D44 FG 122 mm M1938 How 122 mm Type 54 How 122 mm M1931/7 gun 130 mm M46 FG	14.5 mm ZPU-4 LAAG 30 mm M53 LAAG 37 mm Type 55 LAAG 37 mm M1939 LAAG 57 mm Type 59 AAG 57 mm S-60 AAG 100 mm KS-19 AAG SA-6 Gainful SAM SA-9 Gaskin SAM
GUINEA-BISSAU					
T-34/85 T-54 T-55 PT-76	BRDM-2	BTR-40 BTR-50 BTR-60 BTR-152 Type 56		85 mm D44 FG 122 mm M1938 How 122 mm D30 How 130 mm M46 FG	14.5 mm ZPU-1 LAAG 14.5 mm ZPU-2 LAAG 14.5 mm ZPU-4 LAAG 23 mm ZU-23 LAAG 37 mm M1939 LAAG 57 mm S-60 AAG SA-6 Gainful SAM SA-9 Gaskin SAM
IVORY COAST					
AMX-13	AML-60 and -90 ERC 90 Sagaie	Panhard M3 VAB (G) ACMAT VBLA (G)		105 mm M2 How	20 mm Bofors LAAG 40 mm Bofors L/60 LAAG 20 mm M3 VDA SPAAG
KENYA					
Vickers Mk 3	AML-60 and -90 Saladin Mk 2 Ferret Shorland (IS)	Panhard M3 UR-416		25-pounder FG 105 mm Light Gun 105 mm M56 P How	12.7 mm M55 LAAG 20 mm TCM-20 LAAG
LIBERIA					
		M3A1 Half-track MOWAG (4 × 4)		75 mm M116 P How 105 mm M101 How	
MADAGASCAR					
PT-76	Ferret M8 BDRM-2	M3A1 Half-track UR-416		76 mm M1942 FG 105 mm M2 How 105 mm M3 How 122 mm Type 60 FG 122 mm D30 How 130 mm Type 59-1 FG 130 mm M46 FG	14.5 mm ZPU-4 LAAG 23 mm ZU-23 LAAG 37 mm Type 55 LAAG SA-9 Gaskin SAM
MALAWI					
	Fox BRDM-2			25-pounder FG 105 mm Light Gun	14.5 mm ZPU LAAG
MALI					
T-34/85 T-54 T-55 Type 62	BRDM-2 M8	BTR-60 BTR-152		85 mm D44 FG 100 mm M1944 FG 122 mm D30 How 130 mm M46 FG 122 mm BM-21 MRS 132 mm BM-13 MRS	14.5 mm ZPU LAAG 37 mm M1939 LAAG 57 mm S-60 AAG SA-9 Gaskin SAM
MAURITANIA					
	AML-60 and -90 EBR-75 M3A1	Panhard M3 M3 Half-track			14.5 mm ZPU-1 LAAG 14.5 mm ZPU-2 LAAG 23 mm ZU-23 LAAG 37 mm M1939 LAAG SA-9 Gaskin SAM
MOZAMBIQUE					
T-34/85 T-54 T-55	BRDM-1 BRDM-2	BTR-40 BTR-60 BTR-152		76 mm M1942 FG 85 mm D44 FG 85 mm D48 ATG 85 mm Type 56 FG 100 mm M1944 FG 122 mm M1938 How 122 mm D30 How 130 mm M46 FG 152 mm D1 How 122 mm BM-21 MRS 240 mm BM-24 MRS	14.5 mm ZPU LAAG 20 mm M55 LAAG 23 mm ZU-23 LAAG 37 mm M1939 LAAG 37 mm Type 55 LAAG 37 mm Type 68 LAAG 57 mm S-60 AAG 23 mm ZSU-23-4 SPAAG 57 mm ZSU-57-2 SA-6 Gainful SAM SA-8 Gecko SAM SA-9 Gaskin SAM
NIGER					
	M8 ERC 60-20 AML-60 and -90 Panhard VBL	Panhard M3 M3 Half-track UR-416			20 mm M693 LAAG 37 mm LAAG
NIGERIA					
T-55 Vickers Mk 3	AML-60 and -90 Fox Saladin Mk 2 Scorpion EE-9 Cascavel Scimitar	Saracen Steyr 4K 7FA MOWAG Piranha (6 × 6) Panhard M3 Saxon	155 mm Palmaria SPG	76 mm M1942 FG 25-pounder FG 105 mm Light Gun 105 mm M56 How 105 mm Model 56 P How 122 mm D74 FG 130 mm M46 FG 155 mm FH-77B 122 mm BM-21 MRS	14.5 mm ZPU LAAG 20 mm LAAG 23 mm ZU-23 LAAG 37 mm M1939 LAAG 23 mm ZSU-23-4 SPAAG Roland SAM
RWANDA					
	AML-60 and -90 Panhard VBL	Panhard M3		105 mm M101 How	

Tanks	Reconnaissance vehicles	APCs	Tank Destroyers self-propelled guns and mortars	Towed and Rocket Artillery Systems	Anti-air systems
SENEGAL					
	AML-60 and -90 M8 and M20	Panhard M3 VXB-170 (G) M3 Half-track		75 mm M116 P How 105 mm M101 How 155 mm Model 50 How 155 mm M114 How	20 mm M693 LAAG 40 mm Bofors L/60 LAAG
SIERRA LEONE					
	Ferret Saladin Mk 2	MOWAG Piranha 4 × 4		25-pounder FG	
SOMALIA					
T-34/85 T-54 T-55 M41 M47	BRDM-2 FIAT 6616 AML-60 and -90 Ferret	BTR-40 BTR-50 BTR-60 BTR-152 Commando V-150 M113 FIAT 6614 Panhard M3	M113A2 TOW TD	76 mm M1942 FG 85 mm D44 FG 85 mm D48 ATG 100 mm M1944 FG 105 mm M56 PH 122 mm Type 60 FG 122 mm D30 How 122 mm M1938 How 122 mm BM-21 MRS 130 mm Type 59-1 FG 130 mm M46 FG 132 mm BM-13 MRS 140 mm BM-14-16 MRS 140 mm BM-14-17 MRS 152 mm D1 How 155 mm M114A1 How 155 mm M198 How 180 mm S-23 Gun 240 mm BM-24 MRS	14.5 mm ZPU-1 LAAG 14.5 mm ZPU-2 LAAG 14.5 mm ZPU-4 LAAG 20 mm Cerbere LAAG 20 mm M167 LAAG 23 mm ZU-23 LAAG 37 mm M1939 LAAG 37 mm Type 63 LAAG 37 mm Type P698 LAAG 57 mm S-60 AAG 100 mm KS-19 AAG 23 mm ZSU-23-4 SPAAG
SOUTH AFRICA					
Olifant (Centurion)	Eland-60 and -90 Ferret Swerwer (IS) Ribbok (IS) Rooikat	Ratel 20 and 90 Saracen Hippo (IS) Rhino (and IS) Buffel (and IS) Bulldog (and IS) Casspir (IS) Wolf (IS) Ingwe (IS)	81 mm Ratel SPM 155 mm G-6 SPGH Ratel ZT5 TD	25-pounder G-1 FG 140 mm (5.5-inch) G-2 gun 155 mm G-4 (Soltam M-71) Gun/How 155 mm G-5 GH 127 mm Valkiri MRS 127 mm Bataleur MRS	20 mm GAI-B01 LAAG 20 mm GAI-C01 LAAG 23 mm ZU-23 LAAG 35 mm GDF-002 LAAG 40 mm Bofors L/60 LAAG (R) 20 mm Ystervark SPAAG Crotale SAM (Cactus)
TANZANIA					
Type 59 Type 62 T-54	Scorpion BRDM-2	BTR-40 BTR-152 Type 56 Type 531		76 mm M1942 FG 76 mm Type 54 FG 85 mm Type 56 122 mm M1931/7 gun 122 mm D30 How 122 mm Type 54 How 122 mm Type 60 FG 130 mm Type 59-1 130 mm M46 FG 122 mm BM-21 MRS	14.5 mm ZPU-2 LAAG 14.5 mm ZPU-4 LAAG 23 mm ZU-23 LAAG 37 mm Type 55 LAAG 37 mm M1939 LAAG SA-6 Gainful SAM SA-9 Gaskin SAM
TOGO					
T-54 T-55	AML-60 and -90 M8 and M20 EE-9 Cascavel Scorpion 90 mm Panhard VBL	Panhard M3 UR-416 M3A1 Half-track		57 mm M1943 ATG 105 mm M2 How	14.5 mm ZPU-4 LAAG 37 mm M1939 LAAG
UGANDA					
M4 Sherman T-54 T-55 T-34/85	BRDM-2 Saladin Ferret	BTR-40 BTR-152 OT-64		76 mm M1942 FG 122 mm M1938 How 122 mm BM-11 MRS	23 mm ZU-23 LAAG 30 mm Oerlikon LAAG
ZAÏRE					
Type 62	AML-60 and -90 Ferret M3A1	Panhard M3 M3 Half-track M113 Type 531 BTR-40 BTR-50 BTR-60 BTR-152 Fahd (IS)	90 mm M3 Half-track SPTD	57 mm M1943 ATG 75 mm M116 P How 85 mm Type 56 FG 122 mm M1938 How 122 mm D30 How 122 mm Type 60 FG 130 mm Type 59-1 FG 152 mm Type 66 GH 107 mm Type 63 MRS	14.5 mm ZPU LAAG 23 mm ZU-23 LAAG 37 mm M1939 LAAG 37 mm Type 63 LAAG 40 mm Bofors L/60 LAAG 57 mm S-60 AAG
ZAMBIA					
T-34/85 T-54 T-55 Type 59 PT-76	BRDM-1 BRDM-2	BTR-60		76 mm M1942 FG 25-pounder FG 105 mm M101 How 105 mm Model 56 P How 122 mm D30 How 130 mm M46 FG 122 mm BM-21 MRS	14.5 mm ZPU LAAG 20 mm M55 LAAG 37 mm Type 55 LAAG 37 mm M1939 LAAG 57 mm S-60 AAG 85 mm KS-12 AAG
ZIMBABWE					
T-34/85 T-54 T-55 Type 63 Type 59	Eland -60 and -90 Ferret BRDM-2 EE-9 Cascavel	UR-416 Type 531 Hyena Leopard BTR-152 Buffel Hippo Crocodile		76 mm M1942 FG 25-pounder FG 105 mm Model 56 P How 105 mm Light Gun 122 mm Type 60 FG 5.5-inch gun 122 mm BM-11 MRS	14.5 mm ZPU-1 LAAG 14.5 mm ZPU-2 LAAG 14.5 mm ZPU-4 LAAG 20 mm M55 LAAG 23 mm ZU-23 LAAG 37 mm M1939 LAAG

Middle East & North Africa

Country	Tanks	Reconnaissance vehicles	APCs	Tank Destroyers self-propelled guns and mortars	Towed and Rocket Artillery Systems	Anti-air systems
ALGERIA	T-34/85 T-54 T-55 T-62 T-72	AML-60(G) BRDM-1 BRDM-2	BTR-50 BTR-60 BTR-152 Walid BMP-1 BMP-2 Panhard M3 (G)	122 mm SO-122 SPH 152 mm SO-152 SPH BRDM-2 Sagger TD	57 mm M1943 ATG 76 mm M1942 FG 85 mm D44 FG 122 mm M1931/7 How 122 mm M1938 How 122 mm D30 How 122 mm D74 FG 130 mm M46 FG 152 mm M1931/7 GH 152 mm D20 GH 122 mm BM-11 MRS 122 mm BM-21 MRS 140 mm BM-14-16 MRS 240 mm BM-24 MRS FROG-4 ARS FROG-7 ARS	14.5 mm ZPU LAAG 23 mm ZU-23 LAAG 23 mm TA-23 SPAAG 37 mm Twin LAAG 37 mm M1939 LAAG 57 mm S-60 AAG 85 mm KS-12 AAG (AF) 100 mm KS-19 AAG (AF) 130 mm KS-30 AAG (AF) 23 mm ZSU-23-4 57 mm ZSU-57-2 SA-4 Ganef SAM SA-6 Gainful SAM SA-8 Gecko SAM SA-9 Gaskin SAM
BAHRAIN	M60A3	Ferret Saladin AML-90	Saxon Panhard M3 Shorland		105 mm Light Gun 155 mm M198 How 227 mm MLRS*	
CYPRUS	T-34/85 AMX-30 B2	EE-9 Cascavel EE-3 Jararaca Shorland (IS)	BTR-50 BTR-152 VAB EE-11 Urutu Leonidas AMX VCA	EE-3 MILAN TD VAB HOT TD	75 mm M116 P How 76 mm M1942 FG 25-pounder FG 100 mm M1944 FG 105 mm M56 How 105 mm M101 How 128 mm M-63 MRS 155 mm Mk F3 SPH	20 mm M55 LAAG 35 mm GDF LAAG 40 mm Mk 1 LAAG
EGYPT	M1A1 T-54 T-55 T-62 M60A3 M60A1	BRDM-1 BRDM-2 Commando Scout	BTR-40 BTR-50 BTR-60 BTR-152 BMP-1 OT-62 OT-64 M113A1/A2 Walid Mk 1/Mk 2 (NG) BMR-600 Fahd Hotspur Hussar (IS) Hotspur 4984 (IS)	155 mm M109A2 SPH 120 mm M106A2 SPM 82 mm M125A2 SPM M901 TOW TD M113 TOW TD BRDM-2 Sagger TD	57 mm M1943 ATG 76 mm M1942 ATG 85 mm D44 FG 100 mm M1944 FG 122 mm M1931/7 gun 122 mm D74 FG 122 mm D30 How 122 mm M1938 How 122 mm Type 60 FG 130 mm SM-4-1 gun (coastal) 130 mm M46 FG 130 mm Type 59-1M FG 152 mm D1 How 152 mm M1937 GH 152 mm D20 GH 180 mm S-23 gun 240 mm M240 M 80 mm VAP-80 MRS 122 mm BM-21 MRS 122 mm Sakr-18 MRS 122 mm Sakr-36 MRS 130 mm M51/Praga V3S MRS 132 mm BM-13-16 MRS 140 mm BM-14-16 MRS 240 mm BM-24 MRS FROG-7 ARS SAKR-80 ARS	12.7 mm M53 LAAG 14.5 mm ZPU-2 LAAG 14.5 mm ZPU-4 LAAG 20 mm LAAG 23 mm ZU-23 LAAG 30 mm LAAG 35 mm GDF-002 LAAG 37 mm M1939 LAAG 37 mm Twin LAAG 57 mm S-60 AAG 85 mm KS-12 AAG 100 mm KS-19 AAG 23 mm ZSU-23-4 SPAAG 57 mm ZSU-57-2 SPAAG SA-6 Gainful SAM Crotale SAM M730A2 Chaparral
IRAN	Chieftain Mk 3/3 (P) Chieftain Mk 5/5 (P) M47/M47M M48/M48A5 M60A1 T-54 T-55 T-62 T-72 Type 59 PT-76 Type 69-2	Ferret Scorpion BRDM-2 EE-9 Cascavel	BTR-40 BTR-50 BTR-60 BTR-152 BMP-1 M113A1	155 mm M109/M109A1 SPH 175 mm M107 SPG 175 mm M1978 SPG 203 mm M110 SPH	75 mm M116 P How 85 mm D44 FG 105 mm M101 How 105 mm M56 How 122 mm Type 60 FG 122 mm Type 54 How 122 mm M1938 How 130 mm Type 59 FG 130 mm M46 FG 155 mm M114 How 155 mm GH N-45 GH 203 mm M115 How 107 mm Type 63 MRS 122 mm Type 81 MRS 122 mm BM-21 MRS Shatin-2 ARS Oghab ARS Nazeat ARS	14.5 mm ZPU-1 LAAG 14.5 mm ZPU-2 LAAG 14.5 mm ZPU-4 LAAG 23 mm ZU-23 LAAG 35 mm GDF LAAG 37 mm M1939 LAAG 40 mm M1 LAAG 40 mm Bofors L/70 LAAG 57 mm S-60 AAG 75 mm M51 AAG 85 mm KS-12 AAG 23 mm ZSU-23-4 SPAAG 57 mm ZSU-57-2 SPAAG

Tanks	Reconnaissance vehicles	APCs	Tank Destroyers self-propelled guns and mortars	Towed and Rocket Artillery Systems	Anti-air systems
IRAQ (status of some equipment is now questionable)					
T-34/85 (R) T-54 T-55 T-62 T-72 T-72B/T-72G Asad Babl Type 59 Mod Type 69-2 Chieftain PT-76 T-55 Mod	AML-60 and -90 BRDM-1 BRDM-2 EE-9 Cascavel EE-3 Jararaca ERC-90 PShZ-IV	BTR-50 BTR-60 BTR-152 BMP-1 BMP-2 EE-11 Urutu Panhard M3 OT-62 OT-64 Walid MOWAG Roland Type 531 MT-LB M-60P BMD-1	SU-100 SPG ISU-122 SPG 122 mm SO-122 SPH 152 mm SO-152 SPH 155 mm Mk F3 SPG 155 mm M109A1 SPH 155 mm GCT SPG 120 mm MT-LB SPM* 240 mm 2S4 SPM 160 mm T-54 SPM VCR/TR HOT TD BRDM-2 Sagger TD BRDM-2 Spandrel TD	75 mm M116 P How 85 mm D44 FG 105 mm Model 56 P How 105 mm M101 How 105 mm M56 How 130 mm Type 59 FG 122 mm D30 How 122 mm Saddam How 122 mm M1931/7 gun 122 mm M1938 How 122 mm D-74 FG 130 mm Type 59-1 FG 130 mm M46 FG 152 mm D-1 How 152 mm 2A36 G 152 mm M1937 GH 152 mm Type 66 GH 155 mm M46 Mod FG 155 mm M114 How 155 mm GH N-45 GH 155 mm G-5 GH 155 mm Type WA 021 GH 180 mm S-23 gun 107 mm MRS 108 mm FGT 108R MRS 122 mm BM-21 MRS 122 mm M1975 MRS 122 mm BM-11 MRS 127 mm Sajeel 30 MRS 122 mm Scania MRS 122 mm MT-LB MRS 132 mm BM-13 MRS 140 mm MRS 180 mm Sajeel 40 MRS 300 mm Sajeel 60 MRS 40 mm Ababeel 100 ARS FROG-7 ARS 262 mm Ababeel 50 MRS Layth 90 ARS	14.5 mm ZPU-1 LAAG 14.5 mm ZPU-2 LAAG 14.5 mm ZPU-4 LAAG 23 mm ZU-23 LAAG 37 mm M1939 LAAG 37 mm Type 63 LAAG 57 mm S-60 AAG 85 mm KS-12 AAG 100 mm KS-19 AAG 130 mm KS-30 AAG 23 mm ZSU-23-4 SPAAG 30 mm M53/59 SPAAG 30 mm M53/70 SPAAG 57 mm ZSU-57-2 SPAAG SA-6 Gainful SAM SA-8 Gecko SAM SA-9 Gaskin SAM SA-13 Gopher SAM Roland SAM
ISRAEL					
Centurion (upgraded) Merkava Mk 1/ Mk 2/Mk 3 M48A5 M60/M60A1/ M60A3 Sherman (R) (various) T-54 T-55 T-62 PT-76 (R)	BRDM-2 RBY-1	BTR-40 (BG) BTR-50 (R) BTR-60 (R) BTR-152 (BG) OT-62 (R) M2 and M3 Half- track (and R) M113A1/M113A2 Walid Mk 1 (BG) Centurion (conversion) Fox NBC	155 mm M109/M109A1 SPH 155 mm L33 Soltam GH 155 mm Sherman SPH 175 mm M107 SPG 203 mm M110 SPH 81 mm M125A2 SPM 120 mm M106A2 SPM 160 mm Sherman SPM M113 TOW TD RBY Mk 1 TOW TD RBY Mk 1 MAPATS TD BRDM-2 Sagger TD	105 mm M101 How 105 mm M102 How 122 mm D30 How 122 mm M1938 How (training) 130 mm M46 FG 155 mm M114 How 155 mm M-71 GH 122 mm BM-11 MRS 122 mm BM-21 MRS 160 mm LAR MRS 240 mm BM-24 MRS 290 mm MAR ARS 350 mm MAR ARS	20 mm M167 LAAG 20 mm TCM-20 LAAG 23 mm ZU-23 LAAG 40 mm M1 LAAG 40 mm Bofors L/70 LAAG 20 mm M163 SPAAG 20 mm TCM-20 SPAAG (and AF) 23 mm TCM-Mk 3 SPAAG 23 mm ZSU-23-4 SPAAG (AF) M730 Chaparral SAM
JORDAN					
Tariq (Centurion) M48 M60A1/ M60A3 Khalid	Saladin (P)	Saracen M113A1/M113A2 EE-11 Urutu BMP-2	155 mm M109/ M109A2 SPH 155 mm M44 SPH 203 mm M110A1/ M110A2 SPH 81 mm M125 SPM 107 mm M106 SPM M901 TOW TD	105 mm M101 How 105 mm M102 How 155 mm M59 gun 155 mm M114 How 155 mm GH N-45 GH 203 mm M115 How	12.7 mm M55 LAAG 20 mm M167 LAAG 40 mm Bofors LAAG 20 mm M163 SPAAG 23 mm ZSU-23-4 SPAAG 40 mm M42 SPAAG SA-8 Gecko SAM SA-13 Gopher SAM
KUWAIT (status of some equipment is now questionable)					
Chieftain Mk 5 (K) Vickers Mk 1 M-84	Ferret Saladin Scorpion	Saracen M113A1/M113A2 AMX-VCI Commando V-150 Commando V-300 Fahd Valkyr (IS) BMP-2	81 mm M125A2 SPM 155 mm Mk F3 SPG 155 mm M109A1B SPH M901 TOW TD M113 TOW TD	25-pounder FG 155 mm M114 How FROG-7 ARS	SA-8 Gecko SAM
LEBANON					
AMX-13 FL-10 M48A5 (and R) M41 T-54	Saladin (and R) AML-90 Ferret	Chaimite (IS) Panhard M3 M59 M113A1/M113A2 (and R) Saracen VAB AMX-VCI Shorland (IS)	81 mm M125A2 SPM 240 mm 2S4 SPM	105 mm M102 How 122 mm M1938 How 130 mm M46 FG 155 mm Model 50 How 155 mm M114 How 155 mm M198 How FROG-7 ARS	20 mm KAB-001 LAAG 20 mm M55 LAAG 23 mm ZU-23 LAAG 30 mm Oerlikon LAAG 40 mm M42 SPAAG
LIBYA					
T-34/85 (T) T-54 T-55 T-62 T-72	EE-9 Cascavel Saladin Mk 2 Shorland (IS) BRDM-2 FIAT 6616	BTR-50 BTR-60 BMP-1 OT-62 OT-64 Chaimite Saracen M113A1 FIAT 6614 EE-11 Urutu	ISU-122 TD ISU-152 TD BRDM-2 Sagger TD 122 mm SO-122 SPH 152 mm SO-152 SPH 152 mm DANA SPG 155 mm Palmaria SPG 155 mm M109 SPH	105 mm M101 How 130 mm M46 FG 122 mm D30 How 122 mm D74 FG 152 mm D20 GH 155 mm M114 How 155 mm GH N-45 GH 107 mm Type 63 MRS 122 mm BM-11 MRS 122 mm BM-21 MRS 122 mm RM70 MRS 130 mm M51 MRS Astros MRS FROG-7 ARS	14.5 mm ZPU-4 LAAG 20 mm M55 LAAG 23 mm ZU-23 LAAG 40 mm Bofors L/70 LAAG 57 mm S-60 AAG 100 mm KS-19 AAG 23 mm ZSU-23-4 SPAAG 57 mm ZSU-57-2 SPAAG 30 mm M53/59 SPAAG SA-6 Gainful SAM SA-8 Gecko SAM SA-9 Gaskin SAM SA-13 Gopher SAM Crotale SAM

	Tanks	Reconnaissance vehicles	APCs	Tank Destroyers self-propelled guns and mortars	Towed and Rocket Artillery Systems	Anti-air systems
MOROCCO						
	AMX-13 M48A3/M48A5 T-54 T-55 Jagdpanzer SK 105	AML-20, -60 and -90 EBR-75 FL-10 and FL-11 AMX-10RC RAM V-1 Eland-60 and -90 BRDM-2	Panhard M3 M3 Half-track M113A1 OT-62 OT-64 UR-416 VAB Ratel 20 and 90 EE-11 Urutu	BRDM-2 Sagger TD UR-416 Cobra TD SU-100 SPG 105 mm Mk 61 SPH 155 mm M109A1B SPH 155 mm Mk F3 SPG 81 mm M125A1 SPM	76 mm M1942 FG 85 mm D44 FG 105 mm P How 105 mm M101 How 105 mm Light Gun 122 mm D30 How 130 mm M46 FG 152 mm M1937 GH 155 mm M114 How 122 mm BM-21 MRS	14.5 mm ZPU-2 LAAG 14.5 mm ZPU-4 LAAG 20 mm M167 LAAG 23 mm ZU-23 LAAG 37 mm M1939 LAAG 57 mm S-60 AAG 100 mm KS-19 AAG 20 mm M163 SPAAG M730 Chaparral SAM
OMAN						
	Chieftain Mk 7/2C/Mk 15 M60A1	Scorpion VBC 90 Ferret OTO R 2.5	Saxon (and IS) VAB (G) Fahd	VAB MILAN TD	75 mm M116 P How 25-pounder FG 105 mm Light Gun 130 mm M46 FG 130 mm Type 59-1M FG	23 mm ZU-23 LAAG 20 mm VAB SPAAG (G)
QATAR						
	AMX-30S	Ferret	AMX-10P Saracen AMX-VCI VAB UR-416 (IS) Commando Mk 3	155 mm Mk F3 SPG VCAC HOT TD 81 mm VAB SPM	25-pounder FG 155 mm G-5	Roland SAM
SAUDI ARABIA						
	AMX-30S 105 mm M56 P How M60A3 M1A2*	AML-60 and -90 35 mm GDF-002 LAAG OTO R 2.5	AMX-10P Panhard M3 M113 Commando V-150 (NG) UR-416 BMR 600 (N) Glover Transac (IS)	155 mm M109A1B M109A2 SPH 155 mm GCT SPG 81 mm M125A1/ M125A2 SPM 81 mm Commando V-150 SPM (NG) 107 mm M106A1/M106A2 SPM VCC-1 TOW TD Commando V-150 TOW (NG) AMX-10 HOT TD	105 mm M101 How 105 mm M102 How (NG) 155 mm M114 How 155 mm FH-70 FH 155 mm M198 How Astros II MRS 227 mm MLRS*	20 mm M163 SPAAG 20 mm Commando-Vulcan SPAAG (NG) 30 mm AMX-30 SA SPAAG Shahine SAM Crotale SAM (AF)
SUDAN						
	M60A1/M60A3 T-54 T-55 Type 59 Type 62 Type 63 M47	Ferret Saladin Mk 2 BRDM-2 AML-90	BTR-40 BTR-50 BTR-152 Commando V-100/ V-150/V-150S OT-62 OT-64 Saracen M113 Type 531 Walid Mk 1/Mk 2 Panhard M3 AMX-VTT Fahd	155 mm Mk F3 SPG 81 mm M125A1/M125A2 SPM	25-pounder FG 76 mm Type 54 ATG 85 mm D44 FG 85 mm D48 ATG 100 mm M1944 FG 105 mm M56 PH 105 mm M101 How 122 mm M1938 How 122 mm Type 54 How 122 mm D30 How 122 mm Type 60 FG 130 mm Type 59-1 FG 130 mm M46 FG 155 mm M114 How 122 mm Sakr MRS 122 mm BM-21 MRS	14.5 mm ZPU LAAG 20 mm M167 LAAG 23 mm ZU-23 LAAG 37 mm M1939 LAAG 37 mm Type 63 LAAG 40 mm Bofors L/60 LAAG 85 mm KS-12 AAG 100 mm KS-19 AAG
SYRIA						
	T-34/85 (R and T) T-54 T-55 T-62 T-72 T-74M T-80 PT-76	BRDM-1 BRDM-2 Shorland (IS)	BTR-40 BTR-50 BTR-60 BTR-152 BMP-1 OT-64	BRDM-2 Sagger TD SU-100 SPG 100 mm M1944 T-34 SPG ISU-122 SPG ISU-152 SPG 122 mm SO-122 SPH 122 mm D-30 T-34 SPH 152 mm SO-152 SPH	85 mm D44 FG 100 mm M1944 FG 122 mm M1938 How 122 mm M1931/7 gun 122 mm D30 How 130 mm M46 FG 152 mm D1 How 152 mm M1937 GH 152 mm 2A36 G 180 mm S-23 gun 240 mm M-240 mortar 107 mm Type 63 MRS 122 mm BM-11 MRS 122 mm BM-21 MRS 140 mm BM-14-16 MRS 220 mm BM-22 MRS 240 mm BM-24 MRS FROG-7 ARS	14.5 mm ZGU-1 LAAG 14.5 mm ZPU-1 LAAG 20 mm LAAG 14.5 mm ZPU-2 LAAG 14.5 mm ZPU-4 LAAG 23 mm ZU-23 LAAG 37 mm M1939 LAAG 57 mm S-60 AAG 85 mm KS-12 AAG 100 mm KS-19 AAG 23 mm ZSU-23-4 SPAAG 57 mm ZSU-57-2 SPAAG SA-6a/b Gainful SAM SA-8 Gecko SAM SA-11 Gadfly SAM SA-13 Gopher SAM
TUNISIA						
	M48A3 M60A1/ M60A3 AMX-13 FL-10 Jagdpanzer SK 105 M41A1	AML-60 Saladin EE-3 Jararaca EE-9 Cascavel	M113A1 VXB-170(G) Commando V-150 FIAT 6614 (IS) EE-11 Urutu (IS)	105 mm M108 SPH 155 mm M109A2 SPH 81 mm M125A1 SPM M113A2 TOW TD 107 mm M106A2 SPM	105 mm M101 How 155 mm M114 How 155 mm Model 50 How	20 mm M55 LAAG 37 mm Type 55 LAAG 37 mm M1939 LAAG 37 mm Type 63 LAAG 40 mm M42 SPAAG Chaparral SAM 20 mm TA-20 SPAAG
UNITED ARAB EMIRATES						
	AMX-30 OF-40 Mk 2	AML-60 and -90 Ferret (R) Saladin (R) Scorpion Shorland (IS) EE-9 Cascavel	AMX-VCI Saracen (R) AMX-10P VAB Panhard VCR/TT Panhard M3 VTS EE-11 Urutu	155 mm Mk F3 SPG 155 mm G-6 SPH EE-11 TOW TD	25-pounder FG 105 mm Light Gun 105 mm Model 56 P How	35 mm GDF-002 LAAG 20 mm M3 VDA SPAAG Crotale SAM

	Tanks	Reconnaissance vehicles	APCs	Tank Destroyers self-propelled guns and mortars	Towed and Rocket Artillery Systems	Anti-air systems
YEMEN	T-34/85 T-54 T-55 T-62 M60A1	Ferret Saladin BRDM-2	BTR-40 BTR-60 BTR-152 M113/M113A1 Walid BMP-1	SU-100 SPG 107 mm M106 SPM	76 mm M1942 FG 85 mm D44 FG 105 mm M101 How 105 mm M102 How 122 mm M1938 How 130 mm M46 FG 122 mm M1931/7 gun 155 mm M114 How 122 mm BM-21 MRS 250 mm BMD-25 MRS FROG-7 ARS	14.5 mm ZPU-2 AAG 20 mm M167 LAAG 23 mm ZU-23 LAAG 37 mm M1939 LAAG 57 mm S-60 AAG 20 mm M163 SPAAG 23 mm ZSU-23-4 SPAAG 57 mm S-60 AAG 85 mm KS-12 AAG SA-6 Gainful SAM SA-9 Gaskin SAM

Far East & Australasia

	Tanks	Reconnaissance vehicles	APCs	Tank Destroyers self-propelled guns and mortars	Towed and Rocket Artillery Systems	Anti-air systems
AFGHANISTAN	T-34/85 T-54 T-55 T-62 PT-76	BRDM-2	BTR-40 BTR-50 BTR-60 BTR-70 BTR-80 BTR-152 BMP-1 BMP-2	BRDM-2 Sagger TD 120 mm SO-120 SPM/H	76 mm M1938 Mountain Gun 76 mm M1966 Mountain Gun 76 mm M1942 FG 85 mm D48 ATG 100 mm M1944 FG 122 mm D30 How 122 mm M1938 How 152 mm D1 How 152 mm D20 GH 122 mm BM-21 MRS 132 mm BM-13-16 MRS 140 mm BM-14-17 MRS 220 mm BM-22 MRS	14.5 mm ZPU-1 LAAG 14.5 mm ZPU-2 LAAG 14.5 mm ZPU-4 LAAG 23 mm ZU-23 LAAG 37 mm M1939 LAAG 57 mm S-60 LAAG 85 mm KS-12 AAG 100 mm KS-19 AAG 23 mm ZSU-23-4 SPAAG
AUSTRALIA	Leopard AS1		M113A1 LAV-25 (trials) Bison (trials)	81 mm M125A1 SPM	105 mm Model 56 P How 105 mm Hamel Light Gun 105 mm M101 How 155 mm M198 How	
BANGLADESH	T-54 T-55 Type 59 Type 62				6-pounder ATG 76 mm Type 54 FG 25-pounder FG 105 mm Model 56 P How 105 mm M101 How 122 mm Type 54 How	20 mm M55 LAAG
BRUNEI		Scorpion Shorland	VAB (4 × 4)(IS)		105 mm Light Gun	
BURMA	Comet	Ferret Humber Daimler	Bren carriers		6-pounder ATG 76 mm M48 Mountain Gun 17-pounder ATG 25-pounder FG 105 mm M101 How 5.5-inch gun	40 mm M1 LAAG
CAMBODIA	T-54 T-55 Type 59 Type 63 PT-76		BTR-40 BTR-60 BTR-152 M113 Mod		76 mm M1942 FG 122 mm D30 How 122 mm Type 60 FG 122 mm Type 54 How 122 mm Model 1931/7 gun 130 mm Type 59-1 FG 107 mm Type 63 MRS 122 mm BM-21 MRS 132 mm BM-13 MRS 140 mm BM-14 MRS	14.5 mm ZPU-1 LAAG 14.5 mm ZPU-2 LAAG 14.5 mm ZPU-4 LAAG 37 mm M1939 LAAG 57 mm S-60 AAG

	Tanks	Reconnaissance vehicles	APCs	Tank Destroyers self-propelled guns and mortars	Towed and Rocket Artillery Systems	Anti-air systems
CHINA, PEOPLE'S REPUBLIC	IS-2 T-34/85 Type 59 Type 59-II Type 59 Mod Type 60 Type 62 Type 63 Type 69-I/ Type 69-II Type 79 Type 80/ Type 80-II Type 85-II/ Type 85-IIA		BTR-60 Type 56 Type 63/ Type 63-I/Type 63-II Type 55 Type 77-1/77-2 Wz 501/WZ-501A Type 85 YW 531 B/C YW 534 YW 307 YW 309 WZ-551 WZ-501	122 mm Type 54-1 SPH 122 mm Type 83 SPH ISU-122 SPG ISU-152 SPG SU-100 SPG 152 mm Type 83 SPH YW 531 HHJ-8 TD WZ 501 HJ-73B TD WZ 551 HJ-8 TD 155 mm SPH 122 mm Type 85 SPH 122 mm YW 531C SPH 122 mm WZ 551 SPH 82 mm YW 304 SPM 120 mm YW 381 SPM 82 mm Type 85 SPM 120 mm Type 85 SPM	57 mm Type 55 ATG 76 mm Type 54 ATG 85 mm Type 56 FG 100 mm Type 73 ATG 100 mm Type 86 ATG 100 mm Type 59 FG 122 mm Type 60 FG 122 mm Type 54 How 122 mm Type 83 How 122 mm M1931 gun 130 mm Type 59 FG 130 mm Type 59-1 FG 152 mm Type 66 GH 152 mm Type 54 How 152 mm Type 83 FG 155 mm Type WAC-21 GH 107 mm Type 63 MRS 107 mm Type 63-1 MRS 107 mm Type 81 MRS 122 mm Type 81 MRS 122 mm Type 83 MRS 122 mm Type 84 MRS 122 mm Type 90 MRS 130 mm Type 85 MRS 130 mm Type 63/Type 63-I MRS 130 mm Type 70 MRS 132 mm BM-13-16 MRS 140 mm BM-14-16 MRS 180 mm Type 71 MRS 252 mm Type 85 MRS 253 mm Type 81 253 mm Type 81-II 273 mm Type 83 MRS 284 mm Type 74 MRS 305 mm Type 79 ARS 320 mm WS-1 ARS 350 mm M-1B ARS 425 mm Type 762 ARS	12.7 mm Type 77 LAAG 14.5 mm Type 58 LAAG 14.5 mm Type 56 LAAG 14.5 mm Type 75 /75-2 LAAG 14.5 mm Type 80 LAAG 23 mm LAAG 25 mm Type 85 LAAG 37 mm Type 63 LAAG 37 mm Type P793 LAAG 37 mm Type 74 LAAG 37 mm Type 55 LAA 57 mm Type 59 AAG 57 mm Type 66 AAG 85 mm Type 56 AAG 100 mm Type 59 AAG 37 mm Type 80 SPAAG* 57 mm ZSU-57-2 SPAAG 57 mm Type 80 SPAAG Type 551 HN-5C SAM*
HONG KONG			Saxon (IS)			
INDIA	T-54 T-55 T-72M1 Vijayanta PT-76	BRDM-2 Ferret (IS)	BTR-50 BTR-60 OT-62 OT-64 BMP-1 BMP-2 FV432	Sexton 25-pounder SPG 105 mm Abbot SPG 130 mm Vijayanta SPG	75 mm Mountain How 76 mm M48 Mountain Gun 85 mm D48 ATG 3.7-inch Mountain How 25-pounder FG 100 mm M1944 FG 105 mm FG 130 mm M46 FG 5.5-inch gun 105 mm Model 56 P How 152 mm D20 GH 155 mm FH-77B 180 mm S-23 gun 122 mm BM-21 MRS 122 mm LRAR MRS	20 mm LAAG 40 mm M1 LAAG 40 mm Mk 1 LAAG 40 mm Bofors L/70 LAAG 23 mm ZSU-23-4 SPAAG SA-6 Gainful SAM SA-8 Gecko SAM SA-9 Gaskin SAM SA-11 Gadfly SAM
INDONESIA	AMX-13 FL-10 PT-76 (and Ma)	Ferret Saladin AMX-10 PAC 90 (Ma) Commando Scout	AMX-10P (Ma) AMX-VCI BTR-40 BTR-50 (and Ma) BTR-152 Commando V-150 Saracen ACMAT VBLA Yari (P) Commando Ranger Condor		76 mm M48 Mountain Gun 76 mm M1942 FG 105 mm Light Gun 105 mm Model 56 P How 105 mm Mk 61 SPG 105 mm M101 How 122 mm M1938 How (and Ma)	20 mm twin Rh 202 LAAG 20 mm Oerlikon LAAG 40 mm M1 LAAG 40 mm Bofors L/70 LAAG (and Ma) 57 mm S-60 AAG
JAPAN	Type 61 Type 74 Type 90	Type 82 Type 87	Type 60 Type 73 Type 89	105 mm Type 74 SPH 155 mm Type 75 SPH 203 mm M110A2 SPH 81 mm Type SV 60 SPM 107 mm Type SX 60 SPM 106 mm Type 60 SPRR	105 mm M101 How 155 mm FH-70 130 mm Type 75 MRS 300 mm Type 67 MRS	12.7 mm M55 LAAG (AF) 20 mm Rh-202 LAAG 20 mm M167 LAAG 35 mm GDF-001 LAAG

	Tanks	Reconnaissance vehicles	APCs	Tank Destroyers self-propelled guns and mortars	Towed and Rocket Artillery Systems	Anti-air systems
KOREA, NORTH	T-34/85 T-54 T-55 T-62 Type 59 Type 62 PT-76 Type 63 M1985	BA-64 (IS) BRDM-2	BTR-40 BTR-50 BTR-60 BTR-152 BMP-1 Type 531 M1973	SU-76 SPG SU-85 SPG SU-100 SPG ASU-57 TD 122 mm M1981 SPG 122 mm M1985 SPG 130 mm M1975 SPG 152 mm M1974 SPH 152 mm M1977 SPH 175 mm M1978 SPG	45 mm M1942 ATG (Mi) 57 mm M1943 ATG 76 mm M1942 FG 85 mm D44 FG 85 mm Type 56 FG 85 mm D48 ATG 100 mm M1944 FG 122 mm M1931/7 gun 122 mm M1938 How 122 mm D74 FG 122 mm Type 60 FG 122 mm Type 54 How 122 mm D30 How 130 mm M46 FG 130 mm Type 59 FG 152 mm M1937 GH 152 mm M1938 How 152 mm Type 66 GH 107 mm Type 63 MRS 122 mm BM-11 MRS 122 mm BM-21 MRS 130 mm Type 63 MRS 140 mm RPU-14 MRS 140 mm BM-14-16 MRS 200 mm BMD-20 MRS 240 mm BM-24 MRS FROG-5 ARS FROG-7 ARS	14.5 mm ZGU-1 LAAG 14.5 mm ZPU-1 LAAG 14.5 mm ZPU-2 LAAG 14.5 mm ZPU-4 LAAG 14.5 mm SPAAG 23 mm ZU-23 LAAG 37 mm Type 55 LAAG 37 mm M1939 LAAG 37 mm SPAAG 57 mm S-60 AAG 57 mm Type 59 AAG 85 mm KS-12 AAG 100 mm KS-19 AAG 23 mm ZSU-23-4 SPAAG 37 mm (twin) SPAAG 57 mm ZSU-57-2 SPAAG
KOREA, SOUTH	M47 M48A2/M48A3/ M48A5 M60A1 Type 88		FIAT 6614 LVTP7 (marines) LVTP7A1 (marines) M113 M3 Half-track AIFV	M36 TD 105 mm M52 SPH 155 mm M109A2 SPH 175 mm M107 SPG 203 mm M110 SPH	105 mm M101 How 105 mm M102 How 105 mm KH-178 How 155 mm KH-179 How 155 mm M114 How 155 mm M59 gun 203 mm M115 How 130 mm KM809A1 MRS	12.7 mm M55 LAAG 20 mm M167A1 LAAG 35 mm GDF-002 LAAG 40 mm Bofors L/60 40 mm Bofors L/70 40 mm M1 LAAG 12.7 mm M55 SPAAG
LAOS	T-34/85 PT-76 T-54/T-55		BTR-40 BTR-152 M113		75 mm M116 P How 105 mm M101 How 105 mm M102 How 122 mm D30 How 155 mm M114 How	14.5 mm ZPU-1 LAAG 14.5 mm ZPU-2 LAAG 14.5 mm ZPU-4 LAAG 14.5 mm Type 58 LAAG 23 mm ZU-23 LAAG 37 mm M1939 LAAG 57 mm S-60 AAG 23 mm ZSU-23-4 SPAAG 57 mm ZSU-57-2 SPAAG
MALAYSIA		AML-60 and -90 Shorland (IS) Scorpion 90 Ferret	Saxon (IS) Commando V-100 and V-150 Panhard M3 SB 301 (IS) Stormer Condor SIBMAS		105 mm Model 56 P How 155 mm FH-70	40 mm Bofors L/70 LAAG 40 mm M1 LAAG
MONGOLIA	T-34/85 T-54 T-55 T-62	BRDM-2	BTR-60 BTR-152 BMP-1	SU-100 SPG BRDM-2 Sagger TD	76 mm M1942 FG 100 mm M1944 FG 122 mm D30 How 122 mm M1938 How 130 mm M46 FG 152 mm D1 How 122 mm BM-21 MRS	14.5 mm ZPU LAAG 37 mm M1939 LAAG 57 mm S-60 AAG
NEPAL	AMX-13	Ferret Mk 2/3			3.7-inch Mountain How	40 mm Bofors L/60 LAAG 37 mm LAAG
NEW ZEALAND		Ferret Scorpion	M113		105 mm Model 56 P How 5.5-inch gun 105 mm Hamel Light Gun	
PAKISTAN	M24 Chaffee M47M M48A5 T-54 T-55 Type 59 Type 60 Type 69	Ferret (IS)	M113A1 UR 416 (IS) Type 531 Glover Transac Shorland (IS)	105 mm M7 SPH 155 mm M109/M109A2 SPH 203 mm M110A2 SPH M36 TD M901 TOW TD	75 mm M116 P How 25-pounder FG 85 mm Type 56 FG 100 mm Type 59 FG 105 mm M101 How 105 mm Model 56 P How 122 mm Type 54 How 130 mm Type 59 FG 130 mm Type M46 FG 5.5-inch gun 152 mm M1937 GH 155 mm M59 gun 155 mm M114 How 155 mm M198 How 122 mm MRS	12.7 mm M55 LAAG 14.5 mm ZPU-1 LAAG 23 mm ZU-23 LAAG 37 mm M1939 LAAG 37 mm Type 55 LAAG 35 mm GDF-002 LAAG 40 mm M1 LAAG 40 mm Mk 1 LAAG 57 mm S-60 AAG 57 mm Type 59 AAG Crotale SAM M113/RBS 70 SAM M113/ANZI/14.5 mm SAM/ SPAAG
PHILIPPINES		Scorpion	AIFV Chaimite M113A1 LVTP5 (Ma) LVTH6 (Ma) LVTP7A1 (Ma) Commando V-150		105 mm Model 56 P How 105 mm M101 How (and Ma) 155 mm M114 How 155 mm M68 GH	20 mm M167 LAAG

	Tanks	Reconnaissance vehicles	APCs	Tank Destroyers self-propelled guns and mortars	Towed and Rocket Artillery Systems	Anti-air systems
SINGAPORE	AMX-13 SM-1 Centurion	AMX-10RC	Commando V-150 Commando V-200 (AF) M113A2	81 mm M125A1/A2 SPM 120 mm Commando V-200 SPM (AF)	105 mm LG1 FG 155 mm M68 GH(R) 155 mm M71 GH(R) 155 mm FH-88 GH	20 mm Oerlikon LAAG 35 mm GDF-002 LAAG (AF) 40 mm Bofors L/70 LAAG Commando V-200 RBS-70 SAM (AF)
SRI LANKA		Ferret Mk 1 Saladin Daimler Dingo Daimler Mk II	BTR-152 Sandringham Buffel		76 mm M48 Mountain Gun 85 mm Type 56 FG 25-pounder FG	40 mm Bofors L/60 LAAG
TAIWAN	M24 (Mod) M41A1/A3 M48A2/M48A3/ M48A5/M48H Type 64 M60A3 M24 flamethrower	M8	M3 Half-track M113A1/M113A2 LVT3C (Ma) LVTP5A1 (Ma) LVTH6 (Ma) CM-21 AIFV Commando V-150 MOWAG Piranha (unconfirmed)	M18 Hellcat TD 105 mm M108 SPH 155 mm M44 SPH 155 mm XT-69 SPH 155 mm XT-69 SPG 155 mm M109/ M109A2 SPH (and Ma) 203 mm M110A2 SPH (and Ma) 81 mm M125A2 SPM 107 mm M106A2 SPM 81 mm Commando V-150 SPM 107 mm Commando V-150 SPM Commando V-150 TOW TD M42A2 TOW TD M132 flamethrower	75 mm M116 P How 105 mm M101 How (and Ma) 105 mm T64 How 155 mm T65 How 155 mm M59 gun 155 mm M114 How (and Ma) 203 mm M115 How 240 mm How 117 mm Kung Feng VI MRS 126 mm Kung Feng III MRS 126 mm Kung Feng IV MRS (and Ma)	12.7 mm M55 LAAG 35 mm GDF AAG 40 mm M1 LAAG 40 mm M42/M42A1/ M42A2/M42A3/M42A4 SPAAG (and Ma) 90 mm M117 AAG M730 Chaparral SAM (and Ma)
THAILAND	M41/M41A1/ M41A2/M41A3 M48A5 Type 59 Type 69-2 Stingray	Scorpion Shorland Mk 3 EE-9 Cascavel	Commando V-150 LVTP7 (Ma) M113A1/A2 Saracen M3 Half-track Type 85	107 mm M106A1 SPM 155 mm M109A5 SPH*	105 mm Model 56 P How 105 mm M101 How 130 mm Type 59-1 FG 155 mm M71 GH 155 mm M114 How 155 mm M68 GH (Ma) 155 mm GC 45 GH (Ma) 155 mm M198 How 155 mm GH N-45 GH (Ma) 70 mm MRS (AF) 105 mm MRS 122 mm Type 83 MRS 130 mm Type 85 MRS	12.7 mm M55 LAAG 20 mm M167 LAAG 30 mm Arrow LAAG(AF) 40 mm M1 LAAG 20 mm M163A1 SPAAG 37 mm LAAG (AF) 40 mm M42A1 SPAAG 40 mm Bofors L/70
VIETNAM Status of American equipment is in doubt.	M48 T-34/85 T-54 T-55 T-62 Type 63 Type 59 PT-76	BRDM-2	Commando V-100 BTR-50 BTR-60 BTR-152 Type 531 Type 56 M113 Mod	SU-76 SPG SU-85 SPG SU-100 SPG 155 mm M109 SPH 203 mm M110 SPH	76 mm M1942 FG 85 mm D44 FG 85 mm D48 ATG 100 mm M1944 FG 105 mm M101 How 105 mm M102 How 122 mm D30 How 122 mm D74 FG 122 mm M1931/7 gun 122 mm M1938 How 122 mm Type 60 FG 130 mm M46 FG 130 mm Type 59 FG 130 mm Type 59-1 FG 152 mm Type 66 GH 152 mm D20 GH 152 mm M1937 How 152 mm D1 How 155 mm M114 How 107 mm Type 63 MRS 122 mm BM-21 MRS 130 mm Type 63 MRS 140 mm BM-14-16 MRS	12.7 mm M53 LAAG 14.5 mm ZPU-2 LAAG 14.5 mm ZPU-4 LAAG 23 mm ZU-23 LAAG 30 mm M53 LAAG 37 mm M1939 LAAG 40 mm M1 LAAG 57 mm S-60 AAG 85 mm KS-12 AAG 100 mm KS-19 AAG 23 mm ZSU-23-4 SPAAG 37 mm Type 63 SPAAG 40 mm M42 SPAAG 57 mm ZSU-57-2 SPAAG SA-6 Gainful SAM SA-9 Gaskin SAM

Latin America & Caribbean

Country	Tanks	Reconnaissance vehicles	APCs	Tank Destroyers self-propelled guns and mortars	Towed and Rocket Artillery Systems	Anti-air systems
ARGENTINA	AMX-13 FL-12 Jagdpanzer SK 105 M4 Sherman M4 Sherman Firefly M41 TAM	Shorland (P and G) ERC 90 Lynx (Ma) AML-90	AMX VCI BDX LVTP7 (Ma) M3 Half-track M113A1 MOWAG Roland (Ma, P and G) VCTP VCR-TT (Ma)	105 mm M7 SPH 155 mm Mk F3 SPG	105 mm Model 56 P How (and Ma) 105 mm M101 How (and Ma) 155 mm M59 gun 155 mm M114 How 155 mm CITEFA Mod 77 How 155 mm CITEFA Mod 81 How 105 mm SLAM-Pampero MRS 127 mm SAPBA-1 MRS	20 mm GAI-BO1 LAAG 20 mm twin Rh 202 LAAG (and Ma) 20 mm TCM 20 (AF) 30 mm Hispano-Suiza HS 816 LAAG (Ma) 35 mm GDF-001 LAAG (Ma and AF) 40 mm Bofors L/60 LAAG 40 mm Bofors L/70 LAAG 40 mm M1 LAAG 90 mm M117 AAG
BOLIVIA	Jagdpanzer SK 105	EE-9 Cascavel Mk 2	Commando V-100 M113/M113A1 MOWAG Roland EE-11 Urutu Steyr 4K 7FA		75 mm Bofors M1935 FG 75 mm M116 P How 105 mm M101 How 105 mm FH-18 F How	20 mm single Oerlikon 20 mm twin Oerlikon (and AF) 37 mm M2A1 LAAG
BRAZIL	M41A3/M41B/ M41C (and Ma) X1	EE-3 Jararaca EE-9 Cascavel (and Ma)	EE-11 Urutu (and Ma) M59 M113 LVTP7A1 (Ma)	105 mm M108 SPH 105 mm M7 SPH	105 mm M101 How (and Ma) 105 mm M102 How 105 mm Model 56 P How 105 mm Light Gun (Ma)* 155 mm M114 How 105 mm FH 18 F How (R) 57 mm gun (AT) 108 mm FGT 108R MRS (and Ma) ASTROS II MRS	12.7 mm M55 mod LAAG 20 mm twin HS 404 mod LAAG 35 mm GDF-001 LAAG 40 mm Bofors L/60 LAAG 40 mm Bofors L/70 LAAG 40 mm M1 LAAG 90 mm M117 AAG (R) 90 mm M118 AAG (R) Roland SAM
CHILE	AMX-13 AMX-30 M24 (60 mm) M3 Stuart M4 Mod Sherman M51 Sherman M41A3	EE-9 Cascavel	MOWAG Piranha (6 × 6) M113A1 EE-11 Urutu MOWAG Roland (Ma and P) LVTP5 (Ma) VTP-1 Orca VTP-2 Escarabajo (AF)	155 mm Mk F3 SPG 120 mm MOWAG Piranha (6 × 6) SPM	105 mm FH 18 F How 105 mm Model 56 P How 105 mm M101 How 155 mm M114 How (Ma)	12.7 mm M55 LAAG (AF) 20 mm twin LAAG (AF) 20 mm GAI-C01 LAAG 20 mm twin Rh 202 LAAG (and AF) 35 mm Oerlikon GDF-001 LAAG (and AF) 37 mm M1A1 LAAG (Ma and AF) 40 mm Bofors L/70 LAAG 40 mm M1 LAAG Crotale SAM (AF)
COLOMBIA	M3A1 Stuart	M8 and M20 Mods EE-9 Cascavel EE-3 Jararaca	M3A2 Half-track EE-11 Urutu M113A1 Panhard Buffalo	M8 Mod TOW TD	57 mm M18 ATG (R) 105 mm M101 How	35 mm GDF-002 LAAG 40 mm M1A1 LAAG
CUBA	T-34/35 T-54 T-55 T-62 T-72 PT-76 IS-2 (R)	BRDM-1 BRDM-2	BTR-40 BTR-60 BTR-152 BMP-1	SU-100 SPG BRDM-2 Sagger TD	57 mm M1943 ATG 57 mm Ch-26 ATG 76 mm M1942 FG 85 mm D44 FG 85 mm SD44 FG 85 mm D48 ATG 122 mm D74 FG 122 mm M1931/7 gun 122 mm D30 How 122 mm M1938 How 130 mm M46 FG 152 mm M1937 GH 152 mm D1 How 152 mm D20 GH FROG-5 ARS FROG-7 ARS	12.7 mm M53 LAAG 14.5 mm ZGU-1 LAAG 14.5 mm ZPU-1 LAAG 14.5 mm ZPU-2 LAAG 14.5 mm ZPU-4 LAAG 23 mm ZU-23 LAAG 30 mm M53 LAAG 37 mm M1939 LAAG 57 mm S60 AAG 85 mm KS-12 AAG 100 mm KS-19 AAG 14.5 mm BTR-152 SPAAG 23 mm ZSU-23-4 SPAAG 30 mm twin M-53 BTR-60 SPAAG 57 mm ZSU-57-2 SPAAG SA-6 Gainful SAM SA-9 Gaskin SAM SA-13 Gopher SAM
DOMINICAN REPUBLIC	AMX-13 FL-10 M41	AML-60 and -90 Landsverk Lynx	M3A1 Half-track Commando V-150		75 mm Bofors M12/33 How 105 mm M101 How 105 mm Bofors L/22 How	20 mm LAAG 40 mm Bofors L/60 LAAG
ECUADOR	AMX-13 FL-12 M3A1 Stuart (R) M41	AML-60 and -90 EE-9 Cascavel	AMX-VCI M113 UR-416 Condor VAB	155 mm Mk F3 SPG	105 mm M101 How 105 mm Model 56 P How 155 mm M198 How	20 mm Oerlikon LAAG (AF) 35 mm GDF-002 (AF) 40 mm M1 LAAG 40 mm Bofors L/70 LAAG (and AF) 40 mm M1 LAAG (and AF)
EL SALVADOR	AMX-13	AML-90	UR-416 M113 M114 Mod		105 mm M56 How 105 mm M101 How 105 mm M102 How 155 mm M114 How	20 mm Breda M35 LAAG 20 mm M55 LAAG 40 mm Bofors L/60 LAAG 20 mm M55/M114 SPAAG
GUATEMALA	M3A1 Stuart M41A3	M8 RBY-1	Commando V-150 M113 Armadillo		75 mm M56 75 mm M116 P How 105 mm M101 How 105 mm M102 How 105 mm M56 How	20 mm M55 LAAG 20 mm GAI-DO1 LAAG 40 mm M1 LAAG 40 mm M42 SPAAG

	Tanks	Reconnaissance vehicles	APCs	Tank Destroyers self-propelled guns and mortars	Towed and Rocket Artillery Systems	Anti-air systems
GUYANA		Shorland (IS)	EE-11 Urutu		130 mm M46 FG	
HAITI	M3A1 Stuart M5A1 Stuart		Commando V-150		37 mm M3A1 ATG 57 mm M1 ATG 75 mm M116 P How 105 mm M101 How	20 mm TCM-20 LAAG 40 mm M1 LAAG
HONDURAS		Saladin Scimitar Scorpion T17E1 Staghound RBY-1 M3A1			75 mm M116 P How 105 mm M101 How 105 mm M102 How 155 mm M198 How	20 mm LAAG
JAMAICA		Ferret	Commando V-150			
MEXICO	M3A1 Stuart M5A1 Stuart	M8 ERC-90 Lynx MAC-1 DN-3 Caballo Panhard VBL	HWK-11 M3 Half-track Panhard M3 VTS Panhard VCR Sedena 100	75 mm M8 SPH 105 mm M7 SPH Panhard VBL MILAN TD	37 mm M3A1 ATG 75 mm M116 P How 105 mm M101 How	12.7 mm M55 LAAG
NICARAGUA	T-54 T-55 PT-76	T17E1 Staghound BRDM-2	BTR-40 BTR-60 BTR-152 BTR-50	BRDM-2 Sagger TD	76 mm M1942 FG 105 mm M101 How 122 mm M1938 How 122 mm D30 How 152 mm D20 GH 122 mm BM-11 MRS 122 mm BM-21 MRS	14.5 mm ZGU-1 LAAG 14.5 mm ZPU-1 LAAG 14.5 mm ZPU-2 LAAG 14.5 mm ZPU-4 LAAG 20 mm LAAG 23 mm ZU-23 LAAG 37 mm M1939 LAAG 40 mm M1 LAAG 57 mm S-60 AAG 85 mm KS-12 AAG 100 mm KS-19 AAG SA-9 Gaskin SAM
PANAMA (status of equipment is uncertain)			Commando V-150 Commando V-300			
PARAGUAY	M3A1 (mod) M4A3 Sherman M4 Sherman Firefly	M8 (mod)	M2 (mod) Half-track		75 mm Bofors M1935 FG 75 mm Schneider M1927 Mountain Gun 105 mm Schneider M1927 Mountain Gun	20 mm LAAG 40 mm M1 LAAG
PERU	AMX-13 FL-10/ FL-12 T-54 T-55	M8 BRDM-2 (and AF) FIAT 6616	Chaimite (Ma) MOWAG Roland (and G) M113A1 UR-416 FIAT 6614	81 mm FIAT 6614 SPM 107 mm M106A1 SPM 122 mm 2S1 SPH	105 mm M101 How 105 mm Model 56 P How 122 mm D74 FG 122 mm D30 How 130 mm M46 FG 155 mm M114 How 122 mm BM-21 MRS	twin 20 mm LAAG (and Ma) 40 mm Bofors L/60 LAAG 40 mm Bofors L/70 LAAG 76 mm M3 AAG 23 mm ZSU-23-4 SPAAG
URUGUAY	M3A1 Stuart M24 Chaffee M41A1 (90 mm)	EE-3 Jararaca EE-9 Cascaval	M113 M3 Half-track Condor EE-11 Urutu		75 mm Bofors M1935 FG 105 mm M101 How 105 mm M102 How 155 mm M114 How	40 mm M1 LAAG 40 mm Bofors L/60 LAAG
VENEZUELA	AMX-30 AMX-13 FL-10 M48A1 (Ma) Scorpion 90	AML-60 and -90 M8 EE-3 Jararaca (Ma) EE-9 Cascavel (Ma)	AMX-VCI UR-416 LVTP7 (Ma) EE-11 Urutu (and Ma) inc AFSV Commando V-150 Transportpanzer 1 (Ma) Shorland (IS) Commando V-100 Dragoon	155 mm Mk F3 SPG M18 Hellcat TD	75 mm M116 P How 105 mm Model 56 P How (and Ma) 105 mm M101 How 160 mm LARS AMX-13	40 mm Bofors L/60 LAAG 40 mm twin Breda L/70 LAAG 40 mm Bofors L/70 LAAG 20 mm AML S530 SPAAG 40 mm M42A1 SPAAG (Ma) Roland SAM

ADDENDA

MBTs and medium tanks

SOUTH AFRICA

Olifant MBT

Development

South Africa is believed to have about 300 Centurion MBTs which have been modernised to the Olifant (Elephant) Mk 1A standard.

Modifications to the Mk 1A over the original Centurion include the installation of a 105 mm L7 type rifled tank gun, bank of six 81 mm electrically operated smoke dischargers either side of the turret and installation of a new powerpack that consists of a V-12 diesel engine developing 750 hp coupled to an automatic transmission. Passive night vision equipment has been installed as has a new commander's cupola.

The Olifant Mk 1A does not have a computerised fire control system with the commander being provided with an AFV No 4 sight and the gunner with an AFV No 18 sight with magnifications of ×1 and ×6, range drum, elevation indicator and a traverse indicator. The tank commander also has a hand-held laser rangefinder and this is used to provide accurate target information at longer ranges.

The driver can replace his standard day periscope with a locally designed passive night vision device.

The installation of the new powerpack greatly increases the operational range of the Olifant 1A compared to the original Centurion as the diesel engine is much more fuel efficient, in addition, fuel tanks with a significant increase in capacity have been fitted.

Development work on the Olifant Mk 1B commenced in 1985 with the first prototype being completed in 1987. Following trials with prototype vehicles by the South African Armoured Corps, the first production Olifant Mk 1B vehicles were completed in early 1991.

Description

The overall layout of the Olifant Mk 1A and Mk 1B is virtually identical to that of the basic Centurion MBT described under the UK in the *MBTs and medium tanks* section.

Whereas the Olifant Mk 1A is based on the original Centurion MBT, the Mk 1B is a complete rebuild and incorporates many features not previously fitted to the Centurion. These can be summarised as:

(1) Replacement of bogie type suspension by torsion bars which greatly improves cross-country mobility, total wheel travel is now 435 mm. The first and last two road wheel stations either side have a hydraulic damper with hydraulic bump stops at each wheel station.
(2) The hull now has a double floor with the torsion bars between the two floor plates, the double floor also provides increased protection against mines.
(3) New powerpack consisting of a V-12 diesel developing 940 hp coupled to a new AMTRA III automatic transmission with four forward and two reverse gears which also incorporates a double differential steering system with a two speed mechanical steering drive and hydraulic retarder.
(4) Fitting of new final drives with a ratio of 6.167/1.
(5) Installation of additional passive armour to glacis plate, nose, forward part of turret roof and turret front and sides. Turret front and side armour is placed well away from the actual turret.
(6) Fitting of modular side skirts that provide some protection against HEAT attack as well as helping to keep dust down.
(7) Redesign of the driver's compartment with new single piece lift and swing type hatch cover and three new periscopes, one of which can be replaced by a passive vision device. A hull escape hatch is fitted. The driver has a new instrument panel and tillers have been replaced by yoke type system.
(8) A turret bustle is fitted for increased stowage.
(9) Two new unity powered roof-mounted periscopes are fitted for the loader who also has a new one piece hatch cover.
(10) The 105 mm gun is now fitted with a thermal sleeve and a fume extractor.
(11) New solid state gun control equipment.
(12) Gunner's sight now has day/night capability and incorporates a laser rangefinder for improved first round hit capability. Commander has new day/night sight.
(13) The headlamps are now armoured with batteries in floor of vehicle.
(14) Mounted over main armament is an infra-red/white light searchlight that moves in elevation with main armament.
(15) Crew compartment is fitted with fire detection/suppression system.

The Olifant Mk 1B has a combat weight of 58 tonnes and a maximum speed of 58 km/h with a total of 68 rounds of 105 mm ammunition being carried.

Variants

Mine clearing vehicle

The Olifant Mk 1A and Mk 1B can both be fitted with plough or roller type mine clearing systems.

Close-up of hull and turret side of Olifant Mk 1B (Christopher F Foss)

Olifant Mk 1A MBTs of the South African Armoured Corps

Olifant Mk 1B of the South African Armoured Corps

Armoured recovery vehicle

An armoured recovery vehicle based on the chassis of the Olifant Mk 1A has been completed and is now in service with the South African Armoured Corps. Details are given in *Jane's Military Vehicles and Logistics 1991-92*, page 27.

Status: Conversions to Olifant Mk 1A standards have been completed, conversion to Olifant Mk 1B standard is underway.

Prime contractor: ARMSCOR
Private Bag X337
Pretoria 0001
South Africa
Telephone: (012) 428 1911
Telex: 320217

SPECIFICATIONS (Olifant Mk 1A)

CREW	4	FUEL CAPACITY	1240 l	ARMAMENT	
COMBAT WEIGHT	56 000 kg	MAX RANGE		main	1 × 105 mm gun
POWER-TO-WEIGHT		road	500 km	coaxial	1 × 7.62 mm MG
RATIO	13.39 hp/tonne	cross-country	240 km	anti-aircraft	1 × 7.62 mm MG
LENGTH	8.61 m	GRADIENT	60%	smoke dischargers	2 × 4 81 mm
WIDTH	3.39 m	TRENCH	3.45 m	AMMUNITION	
HEIGHT	2.94 m	ENGINE	V-12 diesel developing	main	72
MAX ROAD SPEED	45 km/h		750 hp at 2300 rpm	7.62 mm	5600

Light tanks

UNITED KINGDOM

Alvis Storm 105 Light Tank

Development

Late in August 1991, Alvis disclosed that they were developing a new light tank called the Storm 105 to meet the requirements of the US Army for an Armored Gun System.

The Storm 105 is based on similar automotive components (including engine, transmission and suspension) of the Alvis Stormer II vehicle that is already in production for the British Army as the platform for the Shorts Starstreak High Velocity Missile (HVM) system.

The vehicle will have a combat weight of 14 968 kg and be fully airdroppable in a Lockheed C-130 Hercules transport aircraft.

Description

The hull will be of all welded aluminium construction with the driver front left, powerpack to his right and two man turret mounted towards the rear. Over the frontal arc the hull will provide protection against attack from 30 mm rounds.

The driver has a single piece hatch cover that opens to the rear and in the immediate front of this is a single wide angle day periscope that can be replaced by a passive periscope for driving at night.

Suspension of the Storm is of the hybrid type with hydropneumatic suspension units on the first, second and sixth road wheel stations and torsion bar suspension on the third, fourth and fifth road wheel stations. Drive sprocket is at the front, idler at rear and top return rollers are also provided.

The powerpack consists of a Perkins six-cylinder diesel developing 300 hp coupled to a modified Self-Changing Gears T300 series seven speed transmission with automatic control.

As originally envisaged the Storm 105 would be fitted with a two man turret developed by Cadillac Gage Textron of the US for the US Marine Corps Light Armored Vehicle - Assault Gun (LAV - AG) which is fully described in the *Armoured personnel carriers* section under the United States.

In August 1991 it was reported that there was some doubt as to whether the LAV - AG would move into production owing to a shortage of funding by the US Marine Corps.

The two man power operated turret is armed with a 105 mm EX-35 rifled tank gun fitted with a muzzle brake and fed by a Fairey Hydraulics bustle-mounted automatic loader. A 7.62 mm machine gun is mounted coaxial with the main armament.

The 105 mm gun will fire all types of 105 mm ammunition including Armour Piercing Fin Stabilised Discarding Sabot, with a total of 30 rounds being carried.

Model of the Alvis Storm 105 which is being proposed to meet the requirements of the US Army for an airdroppable Assault Gun Vehicle armed with a 105 mm EX-35 rifled tank gun

Alvis are now moving towards building the first pre-production model of the Storm 105 chassis which incorporates technology from the companies LPXV vehicle announced in 1988 and the Stormer flatbed which they designed and built in 14 weeks for Operation Desert Storm.

Status: Development.

Manufacturer: Alvis Limited
Vehicles and Turrets Division
The Triangle
Walsgrave
Coventry
West Midlands CV2 2SP
UK
Telephone: (0203) 535455
Telex: 31459
Fax: (0203) 539280

UNITED STATES OF AMERICA

Armored Gun System (AGS)

Development

Late in August 1991, the United States Tank Automotive Command finally issued its request for Proposals for the Armored Gun System (AGS) with some 300 vehicles expected to be required to replace M551 Sheridan light tanks used by the 82nd Airborne Division and to form new battalions to operate with the light divisions.

The Sheridans are used by a single battalion of the 82nd Airborne Division and were deployed to Grenada and more recently to Saudi Arabia. They were the first tanks to arrive in Saudi Arabia until the arrival of the heavier M1 and M60 series MBTs.

A total of 1700 M551s were built between 1966 and 1970, all for the US Army and apart from the recent addition of thermal night vision devices they have never been upgraded.

The Request for Proposals are due back in mid-November and one contractor is expected to be selected next Spring to build a total of six prototype vehicles plus additional hulls and turrets for ballistic tests.

The first prototype of the AGS must be completed within 18 months of contract award with a low rate production decision expected in the Autumn of 1994 and full rate production three years later.

The AGS will be armed with a 105 mm EX-35 rifled tank gun which will be government furnished equipment (GFE) with a 7.62 mm machine gun being mounted coaxial with the main armament.

The AGS is required to be airportable in the new C-17 transport aircraft and will have a full range of night vision equipment and a NBC system. The vehicles used by the light divisions may well have increased armour protection over those vehicles used by the 82nd Airborne Division.

The AGS is a Non-Developmental Item and possible tracked contenders for this requirement include:

(1) Alvis of the UK with the new Storm 105, brief details of which are given in the previous entry.

(2) Hägglund Vehicles of Sweden with a vehicle based on the chassis of the Combat Vehicle 90 which was selected by the Swedish Army in 1991.

(3) FMC, Ground Systems Division with a development of their private venture Close Combat Vehicle - Light. The first prototype of this was completed in 1985 and in 1991 the company stated that the US Army had added a requirement for Low Velocity Air Drop (LVAD) and that this would further reduce the weight to 16 136 kg. Design modifications to the CCV L to meet this objective are underway. Modifications also include a new lighter weight track and replacement of the costly commander's independent thermal viewer with the same gunner's sight as installed on the M1 Abrams MBT with an extension for the gunner.

(4) Cadillac Gage Textron with the Stingray light tank which is already in service with Thailand.

(5) Creusot-Loire of France with the MARS-15 light tank. The prototype of which was completed in 1990 and fitted with a GIAT Industries TS-90 turret armed with a 90 mm gun.

(6) General Dynamics, Land Systems Division teamed with Teledyne Continental Motors who will offer a system that builds on the latters Direct Fire Support Vehicle, the first prototype of which was completed some years ago and armed with an externally mounted 105 mm gun fed by an automatic loader. General Dynamics, Land Systems Division, would be responsible for the chassis and systems integration with Teledyne Continental Motors being responsible for the turret.

(7) Thyssen Henschel of Germany with a new design.

While all of the above possible contenders are tracked, there are also a number of wheeled vehicles that could meet the requirement, these include:

(1) GIAT Industries AMX-10RC (6 × 6) which is in service with three countries.

(2) Cadillac Gage Textron V-600 (6 × 6) which is still at the prototype stage.

(3) Light Armored Vehicle - Assault Gun, under development by Cadillac Gage Textron for the US Marine Corps and consisting of a Diesel Division, General Motors of Canada 8 × 8 chassis fitted with a new Cadillac Gage Textron two man power operated turret armed with a 105 mm EX-35 gun fed by an automatic loader. In late August, however, there was some doubt as to whether the LAV - AG would move into the production phase due to a shortage of funding from the US Marine Corps.

(4) IVECO FIAT Centauro B1 (8 × 8) vehicle which is currently in production for the Italian Army.

Status: Not yet in production or service.

Armoured personnel carriers

IRELAND

Timoney 6 × 6 Armoured Personnel Carrier

Development

Following the production of 12 Timoney 4 × 4 armoured personnel carriers for the Irish Army, Timoney Technology built two prototypes of a 4 × 4 armoured reconnaissance vehicle.

The Mark 1 armoured reconnaissance vehicle was powered by a Perkins T6.3544 six-cylinder diesel developing 155 hp coupled to an Allison Transmission Division of General Motors MT-653 five-speed fully automatic transmission.

From the transmission, power was taken through a Timoney single stage transfer box to the front and rear differential assemblies. The latter incorporated oil-cooled disc brakes which were basically the same as those in the 4 × 4 armoured personnel carriers.

The vehicle had independent suspension consisting of upper and lower wishbones and coil springs which, like the hub reduction gears, was also the same as in the 4 × 4 armoured personnel carrier. It was also fitted with Trelleborg tyres with a Trelleborg run-flat insert system.

The Mark 1 was fitted with a two man turret of the Alvis Scorpion CVR (T) armed with a 76 mm gun.

The Mark 1 was followed by the Mark 2 which was powered by a Detroit Diesel Model 4-53T developing 194 hp and was fitted with two Dowty-Messier water-jet propulsion systems.

The company then developed a number of 4 × 4, 6 × 6 and 8 × 8 airport firefighting vehicles which have been produced in production quantities.

More recently Timoney Technology has built prototypes of a 6 × 6 armoured vehicle which incorporate improvements in the design of the independent suspension and axle modules and total wheel travel has increased from 300 mm to 400 mm. A tyre pressure control system has also been installed which allows the driver to adjust the tyre pressure to suit the terrain being crossed.

By early 1991 three prototypes of the Mark 8 6 × 6 vehicle had been built, with the last two having a different hull front.

Description

The hull is of all welded high hardness homogeneous steel armour plate of Brinell Hardness 500 BHN. The plate joints are formed using high hardness steel extrusions to give a smooth contour. Over the frontal arc, the hull provides protection against 14.5 mm AP attack.

According to Timoney Technology, this innovative construction method provides significant benefits both in construction and performance. In construction, plate preparation is simplified since a suitable weld preparation is provided by the simple flame cut edge and the extrusion. No machining preparation is required on the plate.

The armour extrusions are so designed as to provide extra plate thickness in the heat affected weld zone to compensate for any softening of the steel in this region. The extrusion overlap provides a double thickness in this region. This construction method eliminates the need for spall backing plates behind the weld so resulting in simpler manufacture and a cleaner finish.

The smooth exterior surface reduces the risk of accidental crew injury, radar signature is reduced and decontamination procedures are greatly simplified.

The powerpack consists of the engine, automatic transmission, transfer gearbox and the complete cooling system.

The engine, a Perkins Phaser developing 300 bhp or a Cummins 6 CTA 8.3 developing 350 bhp, is directly coupled to a ZF transfer gearbox incorporating a torque distributing locking differential which proportions the torque between the front and rear wheels.

The cooling system for the engine and transmission, including the hydraulically driven high efficiency axial fan, is mounted directly on the powerpack on anti-vibration mountings. The complete powerpack and cooling system can be removed as one assembly.

Latest prototype of the Timoney (6 × 6) armoured personnel carrier without armament fitted

The powerpack is carried on four anti-vibration mountings and is retained on these mountings by quick release over centre clamps. The powerpack is positioned by guideways for quick and easy installation and removal.

The three axles are composed of identical modules. In the front axle, the second output (through drive) location is blanked off and in the rear axle the water propulsion system drive is installed in this location. The commonality of the three axles reduces the logistic requirements of the vehicle.

The axle boxes are suspended from the armour floor of the vehicle by rubber mounting bushes to reduce the transmission of noise and vibration of the hull itself. To protect the axle housings from hostile action or from terrain damage, a removable armoured steel cover is located under the axles, resulting in a smooth and unbroken underbody surface from front to rear.

The brakes are of the air over hydraulic, ventilated disc type mounted directly on the axle housing. This inboard location of the brakes reduces the unsprung weight of the suspension and enables the brakes to be mounted within the protection of the body. Vulnerable hydraulic connections, which would normally extend to the wheels are eliminated. Service of the brakes is also greatly simplified and brake pads can be changed without having to remove the wheels.

The rear axle is fitted with an extension gearbox which drives the water propulsion propellers. This gearbox is engaged by a mechanical dog clutch and when engaged the transmission is locked into 1st and 2nd gear to ensure a correct relationship between wheel speed and propeller speed for optimum water borne performance and for maximum efficiency in entering and leaving the water.

The steerable propulsion propellers are of the open type to give maximum propulsive efficiency. When not in use, the propellers are rotated around the vertical axis into a park position which reduces vulnerability to damage from tree branches and other obstacles. Risk of injury to personnel is also reduced.

Each wheel is independently sprung with the suspension system consisting of an upper and lower wishbone arm with twin helical springs acting on the lower wishbone. A telescopic damper is located coaxially with one of the springs. The low mounting of the springs and extremely compact design of the suspension assembly eliminates any intrusion of the suspension into the available crew space, even though the vertical suspension movement is 400 mm.

The wheel assembly consists of a planetary drive which facilitates the installation of a central tyre inflation system which allows the driver to adjust the tyre pressure to suit the nature of the ground being crossed. A control system at the driver's station warns if the allowable speed at a particular pressure is being exceeded.

The troop compartment is at the rear of the hull with the 10 troops being seated five down each side facing each other. Between this area and the driver's position there is space for up to three more people or one person and communications equipment.

If required a turret mounting weapons up to 105 mm in calibre can be mounted on top of the hull, this also being armed with a 7.62 mm coaxial machine gun and a 7.62 mm anti-aircraft machine gun. The vehicle is also provided with firing ports and vision blocks for the crew.

Access to the rear is via a hydraulically operated ramp that can be fitted with an additional access door if required. In addition there is an access door in the side of the hull with the driver being provided with a single hatch that opens to the right. The driver also has a day periscope with a wiper, fully adjustable seat with 260 mm raise/lower facility and a fully adjustable steering column.

If required the vehicle can be fitted with a full collective filtration system using the cabin overpressure principle and an air-conditioning system can also be provided. In the engine compartment, a fire detection and suppression system can be installed, this will operate either automatically or manually as required.

Other vehicle options include heated driver's periscope, driver's night vision device and power operated winch.

Status: Prototype.

Manufacturer: Timoney Technology Limited
Trim
Co Meath
Ireland
Telephone: 046 31622
Telex: 43588 TIL Ei

SPECIFICATIONS

CREW	2 + 10
CONFIGURATION	6 × 6
COMBAT WEIGHT	15 000 kg
EMPTY WEIGHT	12 500 kg
PAYLOAD	2500 kg
LENGTH	6.45 m
WIDTH	2.70 m
HEIGHT	2.04 m
GROUND CLEARANCE	420 mm
TRACK	2.33 m
WHEELBASE	2.4 m + 1.4 m
ANGLE OF APPROACH/ DEPARTURE	45°/45°
MAX ROAD SPEED	100 km/h
MAX WATER SPEED	8 km/h
CRUISING RANGE	730 km
FORDING	amphibious
GRADIENT	70%
SIDE SLOPE	45%
VERTICAL OBSTACLE	0.6 m
TURNING RADIUS	8 m
ENGINE	Cummins 6 CTA 8.3 6-cylinder diesel developing 350 hp at 2500 rpm OR Perkins Phaser 300 Ti 6-cylinder diesel developing 300 hp at 2600 rpm
TRANSMISSION	ZF 6 HP 500 automatic with 6 forward and 1 reverse gears
TRANSFER BOX	ZF A10002/4D single speed
STEERING	powered, first axle
TYRES	14.00 × R20, run-flat cores, central tyre inflation system
SUSPENSION	all round independent, double wishbone twin coil spring, 400 mm wheel travel, shock absorbers on each wheel station
BRAKES	dual circuit, air over hydraulic, disc brakes

SOUTH AFRICA

TFM Mamba Light Armoured Vehicle

Development

The Mamba family of light armoured vehicles was originally conceived by MECHAM, part of the Vehicles and Weapons Industries Group of ARMSCOR.

At present there are four basic vehicles in the range:

Mamba 7 - a seven seater personnel carrier for both road and off-road use.
Mamba 8 - a heavier duty version of Mamba 7 with lower gear ratios, larger wheels and a turbocharged diesel engine.
Mamba 11 - an 11 seater with the same attributes as the Mamba 8.
Mamba weapons platform - a four seater based on the Mamba 8 but with a lower profile and platform at the rear, this is also referred to as the Mamba Sabre.

Description

Like all members of the Mamba family, the Mamba 11 has a hull of all welded steel that provides a high degree of protection against attack from mines with the upper part providing protection against penetration from 7.62 mm rifle fire, shrapnel and grenade fragments. The heavy duty commercial transmission is protected by a sheathing of armoured steel.

To reduce life cycle costs, proven production components have been used wherever possible in the design of the Mamba family of light armoured vehicles.

Mamba 11 (4 × 2) armoured personnel carrier with roof hatches open

The engine is at the front of the vehicle with the driver and vehicle commander to the immediate rear, five troops are seated down either side of the vehicle with the 11th man being provided with a fold-up seat to the immediate rear of the commander and driver.

There is a large entry door in the rear and there is an additional hatch in the lower part of the hull either side of the commander and driver.

Bullet proof windows are provided all round, some of which have firing ports and there are a total of nine roof hatches, one each above the commander and driver, a circular one for the 11th man and the remaining six over the rear troop compartment, three down each side.

If required a 7.62 mm machine gun can be mounted on the circular roof hatch above the 11th man's position.

A wide range of optional equipment for the Mamba family includes weapon mounts, firing ports, tyre inflation system, sliding roof, ambulance fittings, command vehicle fittings, full radio suppression, hydraulically controlled bomb disposal equipment and a teargas dispenser.

The Mamba weapons platform can be fitted with a 106 mm recoilless rifle at the rear with ammunition storage bins being provided in the mine protected hull under the rear of the vehicle with access to this via a hatch at the rear that opens downwards. A total of 21 rounds of 106 mm ammunition are carried.

The version used by the Bophuthatswana Defence Force is armed with a light machine gun and also carries a mortar for dismounted use. Trials are already being carried out with an 81 mm mortar.

Other weapons fits are however possible including 107 mm rocket system and a 20 mm cannon to name but a few.

Status: Production as required. Known to be in service with Bophuthatswana Defence Force.

Manufacturer: TFM (Pty) Ltd
PO Box 343
Olifantsfontein 1665
South Africa
Telephone: (012) 61 2106
Telex: 3-21041
Fax: (012) 61 1462

SPECIFICATIONS (Mamba II mine protected APC)		ANGLE OF APPROACH/ DEPARTURE	40°/35°	CLUTCH	single dry plate		
CREW	2 + 9	MAX SPEED	106 km/h	SUSPENSION	semi-elliptic leaf springs and double action telescopic hydraulic shock absorbers		
CONFIGURATION	4 × 2	FUEL CAPACITY	150 l	STEERING	ball and nut power		
COMBAT WEIGHT	6120 kg	RANGE		TURNING RADIUS	8.25 m		
UNLOADED WEIGHT	4980 kg	road	800 km	BRAKES			
PAYLOAD	1140 kg	cross-country	600 km	main	vacuum/hydraulic		
DISTRIBUTION		GRADIENT	40%	parking	mechanical on propshaft		
front load	2200 kg	ENGINE	4-cylinder in-line direct injection turbocharged water-cooled diesel developing 112 hp at 2800 rpm	TYRES	12.50 × 20		
rear load	3920 kg	TRANSMISSION	manual, 5 forward and 1 reverse gears	NUMBER OF TYRES	4 + 2 spare		
LENGTH	5.88 m			ELECTRICAL SYSTEM	12 V		
WIDTH	2.30 m			BATTERY	1 × 12 V		
HEIGHT	2.27 m						
FLOOR HEIGHT	0.30 m						
GROUND CLEARANCE	0.30 m						
WHEELBASE	3.38 m						
TRACK	1.8 m						

Self-propelled guns and howitzers

CZECHOSLOVAKIA

PRAM-S 120 mm Self-propelled Mortar System

Development

In 1990 Czechoslovakia completed the prototype of a 120 mm self-propelled mortar system called the PRAM-S. As this was not declared under the Conventional Forces Europe (CFE) Treaty it is probable that the system is still at the prototype stage.

The BMP-1 and the more recent BMP-2 have been built under licence in Czechoslovakia and additional details of variants of the BMP-1 are given in the *Armoured personnel carriers* section on pages 306 to 307.

Description

The PRAM-S 120 mm mortar system based on a modified BMP-1 chassis which has had its turret removed and roof modified to accept a new turret armed with a 120 mm mortar system that can be aimed, loaded and fired under complete armour protection.

The chassis of the BMP-1 has been lengthened and it now has a total of seven road wheels compared to the original BMP-1 that has six.

The driver is seated at the front left with the engine compartment to his right with the area to the rear being raised to provide sufficient space for the mortar, ammunition and its crew.

Mounted in the forward part of the roof is the 120 mm breech loaded mortar which has a traverse of 15° left and right and can be elevated from +40° to +80° with maximum range being 8000 m. When travelling the mortar is at the minimum elevation angle and when required for operation vertical armoured covers unfold either side of the mortar to allow for greater elevation.

The commander is seated on the right and has a hatch in the roof, three periscopes and an infra-red searchlight with the layer being seated on the left side and provided with a roof-mounted sighting system.

The vehicle commander also operates the R-123 UHF/VHF radio with the antenna for this being located forward and to the left of the driver's hatch.

There is also a hatch cover in the roof of the vehicle on which an externally mounted 12.7 mm NVS machine gun is fitted for anti-aircraft defence.

The loader is seated towards the rear of the mortar and is responsible for preparing the mortar bombs and filling the automatic loader from the ammunition containers to the left and right rear of the combat compartment.

In the rear of the vehicle is a door for ammunition resupply and it is estimated that about 80 rounds of 120 mm mortar bomb ammunition are carried.

The PRAM-S is fully amphibious being propelled in the water by its tracks and is fitted with a NBC system and a full range of night vision equipment.

SPECIFICATIONS

CREW	4
WEIGHT	16 000 kg
LENGTH	7.47 m
WIDTH	2.94 m
HEIGHT	2.25 m
RANGE	500 km

Status: Prototype. Not yet in production or service.

Manufacturer: ZTS Dubnica nad Vahom
Zavody Tazeho
Strojarstava
NP Kosice
Czechoslovakia

EQUIPMENT INDEX

A

C

D

E

N

O

T

MANUFACTURERS INDEX